MOLECULAR BIOLOGY OF
THE CELL

MOLECULAR BIOLOGY OF THE CELL

Bruce Alberts • Dennis Bray
Julian Lewis • Martin Raff • Keith Roberts
James D. Watson

Garland Publishing, Inc.
New York & London

"Long ago it became evident that the key to every biological problem must finally be sought in the cell, for every living organism is, or at sometime has been, a cell."

Edmund B. Wilson
The Cell in Development and Heredity
3rd edition, 1925, Macmillan, Inc.

Bruce Alberts received his Ph.D. from Harvard University and is currently a Professor in the Department of Biochemistry and Biophysics at the University of California Medical School in San Francisco. *Dennis Bray* received his Ph.D. from the Massachusetts Institute of Technology and is currently a Senior Scientist in the Medical Research Council Cell Biophysics Unit at King's College London. *Julian Lewis* received his D.Phil. from Oxford University and is currently a Lecturer in the Anatomy Department at King's College London. *Martin Raff* received his M.D. degree from McGill University and is currently a Professor in the Zoology Department at University College London. *Keith Roberts* received his Ph.D. from Cambridge University and is currently Head of the Department of Cell Biology at the John Innes Institute, Norwich. *James D. Watson* received his Ph.D. from the University of Indiana and is currently Director of the Cold Spring Harbor Laboratory. He is the author of *Molecular Biology of the Gene* and, with Francis Crick and Maurice Wilkins, won the Nobel Prize in Medicine and Physiology in 1962.

Library of Congress Cataloging in Publication Data
Main entry under title:

Molecular biology of the cell.

Includes bibliographies and index.
1. Cytology. 2. Molecular biology. I. Alberts, Bruce, 1938– [DNLM: 1. Cells. 2. Molecular biology. QH 581.2 M718]
QH581.2.M64 1983 574.87 82-15692
ISBN 0-8240-7282-0

Published by Garland Publishing, Inc.
136 Madison Avenue, New York, NY 10016

Printed in the United States of America

15 14 13 12 11 10 9 8 7 6 5

Preface

There is a paradox in the growth of scientific knowledge. As information accumulates in ever more intimidating quantities, disconnected facts and impenetrable mysteries give way to rational explanations, and simplicity emerges from chaos. Gradually the essential principles of a subject come into focus. This is true of cell biology today. New techniques of analysis at the molecular level are revealing an astonishing elegance and economy in the living cell and a gratifying unity in the principles by which cells function. This book is concerned with those principles. It is not an encyclopedia but a guide to understanding. Admittedly, there are still large areas of ignorance in cell biology and many facts that cannot yet be explained. But these unsolved problems provide much of the excitement, and we have tried to point them out in a way that will stimulate readers to join in the enterprise of discovery. Thus, rather than simply present disjointed facts in areas that are poorly understood, we have often ventured hypotheses for the reader to consider and, we hope, to criticize.

Molecular Biology of the Cell is chiefly concerned with eucaryotic cells, as opposed to bacteria, and its title reflects the prime importance of the insights that have come from the molecular approach. Part I and Part II of the book analyze cells from this perspective and cover the traditional material of cell biology courses. But molecular biology by itself is not enough. The eucaryotic cells that form multicellular animals and plants are social organisms to an extreme degree: they live by cooperation and specialization. To understand how they function, one must study the ways of cells in multicellular communities, as well as the internal workings of cells in isolation. These are two very different levels of investigation, but each depends on the other for focus and direction. We have therefore devoted Part III of the book to the behavior of cells in multicellular animals and plants. Thus developmental biology, histology, immunobiology, and neurobiology are discussed at much greater length than in other cell biology textbooks. While this material may be omitted from a basic cell biology course, serving as optional or supplementary reading, it represents an essential part of our knowledge about cells and should be especially useful to those who decide to continue with biological or medical studies. The broad coverage expresses our conviction that cell biology should be at the center of a modern biological education.

This book is principally for students taking a first course in cell biology, be they undergraduates, graduate students, or medical students. Although we assume that most readers have had at least an introductory biology course, we have attempted to write the book so that even a stranger to biology could follow it by starting at the beginning. On the other hand, we hope that it will also be useful to working scientists in search of a guide to help them pick their way through a vast field of knowledge. For this reason, we have provided a much more thorough list of references than the average undergraduate is likely to require, at the same time making an effort to select mainly those that should be available in most libraries.

This is a large book, and it has been a long time in gestation—three times longer than an elephant, five times longer than a whale. Many people have had a hand in it. Each chapter has been passed back and forth between the author who wrote the first draft and the other authors for criticism and revision, so that each chapter represents a joint composition. In addition, a small number of outside experts contributed written material, which the authors reworked to fit with the rest of the book, and all the chapters were read by experts, whose comments and corrections were invaluable. A full list of acknowledgments to these contributors and readers for their help with specific chapters is appended. Paul R. Burton (University of Kansas), Douglas Chandler (Arizona State University), Ursula Goodenough (Washington University), Robert E. Pollack (Columbia University), Robert E. Savage (Swarthmore College), and Charles F. Yocum (University of Michigan) read through all or some of the manuscript and made many helpful suggestions. The manuscript was also read by undergraduate students, who helped to identify passages that were obscure or difficult.

Most of the advice obtained from students and outside experts was collated and digested by Miranda Robertson. By insisting that every page be lucid and coherent, and by rewriting many of those that were not, she has played a major part in the creation of a textbook that undergraduates will read with ease. Lydia Malim drew many of the figures for Chapters 15 and 16, and a large number of scientists very generously provided us with photographs: their names are given in the figure credits. To our families, colleagues, and students we offer thanks for forbearance and apologies for several years of imposition and neglect. Finally, we owe a special debt of gratitude to our editors and publisher. Tony Adams played a large part in improving the clarity of the exposition, and Ruth Adams, with a degree of good-humored efficiency that put the authors to shame, organized the entire production of the book. Gavin Borden undertook to publish it, and his generosity and hospitality throughout have made the enterprise of writing a pleasure as well as an education for us.

We welcome readers' suggestions and corrections, which should be sent to us c/o Gavin Borden, Garland Publishing, Inc., 136 Madison Avenue, New York, NY 10016.

Contents in Brief

List of Topics

Introduction to the Cell

PART I

CHAPTER 1 The Evolution of the Cell

CHAPTER 2 Small Molecules, Energy, and Biosynthesis

CHAPTER 3

Macromolecules: Structure, Shape, and Information

The Molecular Organization of Cells

PART

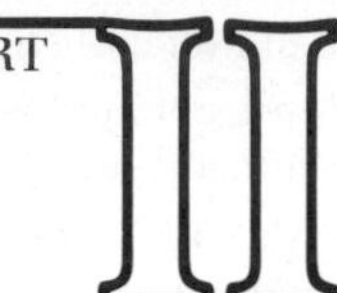

CHAPTER 5

Basic Genetic Mechanisms

CHAPTER 6

The Plasma Membrane

CHAPTER 9

Energy Conversion: Mitochondria and Chloroplasts

CHAPTER 11

Cell Growth and Division

CHAPTER 12

Cell-Cell Adhesion and the Extracellular Matrix

CHAPTER 13

Chemical Signaling Between Cells

PART 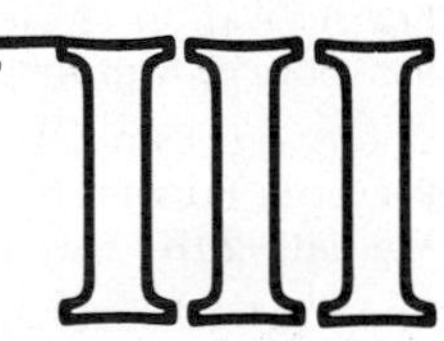

From Cells to Multicellular Organisms

CHAPTER 14

Germ Cells and Fertilization

CHAPTER 15

Cellular Mechanisms of Development

CHAPTER 16

Differentiated Cells and the Maintenance of Tissues

CHAPTER 17

The Immune System

CHAPTER 18

The Nervous System

CHAPTER 19

Special Features of Plant Cells

Acknowledgments

Chapter 1
Read in whole or in part by: Tom Cavalier-Smith (King's College London), Monroe Strickberger (University of Missouri, St. Louis).

Chapter 2
Read in whole or in part by: Charles Gilvarg (Princeton University), Phillips Robbins (Massachusetts Institute of Technology).

Chapter 3
Read in whole or in part by: Robert Fletterick (University of California, San Francisco), Phillips Robbins (Massachusetts Institute of Technology).

Chapter 4
Read in whole or in part by: David Sabatini (New York University School of Medicine), John Sedat (University of California, San Francisco), Patrick O'Farrell (University of California, San Francisco), Fred Richards (Yale University).

Chapter 5
Read in whole or in part by: Glenn Herrick (University of Utah), Patrick O'Farrell (University of California, San Francisco), Brian McCarthy (University of California, Irvine).

Chapter 6
Read in whole or in part by: Mark Bretscher (MRC Laboratory of Molecular Biology, Cambridge, England), Bastien Gomperts (University College Hospital Medical School, London), Juan Korenbrot (University of California, San Francisco), Dale Oxender (University of Michigan).

Chapter 7
Major contribution: James Rothman (Stanford University).

Read in whole or in part by: George Palade (Yale University), Daniel Friend (University of California, San Francisco), Gary Firestone (University of California, Berkeley).

Chapter 8
Read in whole or in part by: Michael Ashburner (Cambridge University), Pierre Chambon (University of Strasbourg), Patrick O'Farrell (University of California, San Francisco), Joseph Gall (Yale University), Larry Gerace (Johns Hopkins University), Glenn Herrick (University of Utah), E. G. Jordan (Queen Elizabeth College, London), Brian McCarthy (University of California, Irvine), Robert Perry (Institute of Cancer Research, Philadelphia), Abraham Worcel (University of Rochester).

Chapter 9
Major contribution: Peter Garland (University of Dundee, Scotland), Piet Borst (Jan Swammerdam Institute, University of Amsterdam).

Read in whole or in part by: Martin Brand (Cambridge University), Roderick Capaldi (University of Oregon), Richard McCarty (Cornell University),

Allison Smith (John Innes Institute, Norwich, England), Charles Yocum (University of Michigan).

Chapter 10
Additional material: Marc Kirschner (University of California, San Francisco).

Read in whole or in part by: Roger Cooke (University of California, San Francisco), Graham Dunn (MRC Cell Biophysics Unit, London), John Kendrick-Jones (MRC Laboratory of Molecular Biology, Cambridge, England), Marc Kirschner (University of California, San Francisco), Klaus Weber (Max Planck Institute for Biophysical Chemistry, Göttingen).

Chapter 11
Additional material: J. Michael Bishop (University of California, San Francisco), Jeremy Pickett-Heaps (University of Colorado).

Read in whole or in part by: Leland Hartwell (University of Washington), J. Murdoch Mitchison (University of Edinburgh), Zacheus Cande (University of California, Berkeley), John Wyke (Imperial Cancer Research Fund, London).

Chapter 12
Additional material: Günther Gerisch (Max Planck Institute for Biochemistry, Martinsried), Robert Trelstad (Rutgers Medical School).

Read in whole or in part by: Stephen Burden (Harvard Medical School), Max Burger (University of Basel), Bernie Gilula (Baylor University), Zach Hall (University of California, San Francisco), Richard Hynes (Massachusetts Institute of Technology), Colin Manoil (Harvard Medical School), Darwin Prockop (Rutgers Medical School), David Rees (National Institute for Medical Research, Mill Hill, London), John Scott (University of Manchester), Malcolm Steinberg (Princeton University).

Chapter 13
Additional material: Michael Wilcox (MRC Laboratory of Molecular Biology, Cambridge, England).

Read in whole or in part by: Peter Baker (King's College London), Philip Cohen (University of Dundee), Daniel Koshland (University of California, Berkeley), Anne Mudge (University College London), Jesse Roth (National Institutes of Health, Bethesda), Charles Stevens (Yale University).

Chapter 14
Major contribution: David Epel (Stanford University)

Read in whole or in part by: Adelaide Carpenter (University of California, San Diego), Jeffrey Hall (Brandeis University), Anne McLaren (University College London), Montrose Moses (Duke University), Duncan O'Dell (University College London), David Phillips (Rockefeller University), Lewis Tilney (University of Pennsylvania).

Chapter 15
Major contribution: Cheryll Tickle (Middlesex Hospital Medical School, London).

Additional material: Judith Kimble (University of Wisconsin).

Read in whole or in part by: John Gerhart (University of California, Berkeley), Peter Lawrence (MRC Laboratory of Molecular Biology, Cambridge, England), Anne McLaren (University College London), Norman Wessells (Stanford University), Lewis Wolpert (Middlesex Hospital Medical School, London).

Chapter 16
Additional material: Vernon Thornton (King's College London).

Read in whole or in part by: Barry Brown (King's College London), Judah

Folkman (Harvard Medical School), Peter Gould (Middlesex Hospital Medical School, London), Jay Lash (University of Pennsylvania), John Owen (University of Birmingham, England), Philippe Sengel (University of Grenoble), Cheryll Tickle (Middlesex Hospital Medical School, London), Rosalind Zalin (University College London).

Chapter 17
Read in whole or in part by: Leroy Hood (California Institute of Technology), Peter Lachmann (MRC Center, Cambridge, England), Avrion Mitchison (University College London), Alan Munro (Cambridge University), Hans Müller-Eberhard (Scripps Clinic and Research Institute), William Paul (National Institutes of Health, Bethesda), Robert Schreiber (Scripps Clinic and Research Institute), Martin Weigert (Institute of Cancer Research, Philadelphia).

Chapter 18
Major contribution: Charles Stevens (Yale University).

Additional material: Regis Kelly (University of California, San Francisco).

Read in whole or in part by: Jonathan Ashmore (University of Sussex, England), Peter Baker (King's College London), Darwin Berg (University of California, San Diego), Anne Mudge (University College London).

Chapter 19
Major contribution: Brian Gunning (Australian National University, Canberra).

Read in whole or in part by: Jim Dunwell (John Innes Institute, Norwich, England), Ray Evert (University of Wisconsin, Madison), Larry Fowke (University of Saskatchewan, Saskatoon), John Hall (University of Southampton, England), David Hanke (University of Cambridge, England), Andy Johnston (John Innes Institute, Norwich, England), Virginia Walbot (Stanford University), Trevor Wang (John Innes Institute, Norwich, England), John Watts (John Innes Institute, Norwich, England).

Cover photograph kindly provided by Michael Verderame and Robert Pollack of Columbia University. The fluorescein-phalloidin used to stain the actin cables was the generous gift of Drs. Theodore Wieland and A. Deboben of the Max Planck Institute, West Germany. The photograph is of a mouse fibroblast that had been transformed to anchorage-independent growth by the virus Simian Virus 40 (SV40) and subsequently selected for anchorage-dependent growth. This particular cell was stained for SV40 large T antigen (*red*) and fluorescein-phalloidin (*green*), which specifically stains F actin.

Prologue

It is all too easy now to underestimate cells. We have known about them for such large fractions of our lives that, for the most part, we cease being aware of how remarkable they really are. Almost as soon as we learn our first rudiments of science, we are told that all living beings are formed from cells, that cells come into existence from the growth and division of preexisting ones, and that they may exist either singly as unicellular organisms or as parts of immensely complicated organisms that may contain billions of interacting, highly specialized units. It has been their potential for great diversity of size, shape, and function that has allowed evolution to proceed in such strikingly different directions.

The mere cataloging of the different names and unique properties of cells has very limited intellectual appeal. But the textbook dry facts take on new meaning when we first use our simple school microscopes to look at the tiny one-celled creatures like the amoebae or paramecia that inhabit drops of pond water. Then the cell as an amazing moving body comes alive, and it is natural to wonder what exact molecules it is made of and how it can so regularly grow and divide to provide more of its kind. Until the 1950s, however, this objective seemed far beyond our capabilities as scientists. Up to then we had little choice but to focus upon the descriptive morphological approach, using better and better microscopes to reveal more and more cellular structures. To these we frequently gave fancy names, like the ergastoplasm or chondriosomes, without understanding why they were there or how they functioned. Not surprisingly, many found this approach unsatisfying and moved on from cells per se to explore the underlying chemical reactions that were becoming increasingly amenable to logical analysis.

These "biologists turned biochemists" soon discovered how cells use the energy in food molecules to build up new biological molecules, thereby discovering how cells can grow and divide without disobeying the thermodynamic dictum that all chemical reactions must move in the direction that maximizes the production of heat and disorder. This momentous achievement greatly encouraged the increasing number of scientists who thought the essence of cells lay entirely in their molecular organization and the enzymatically mediated pathways by which their molecules are either broken down or built up. There was, however, still great uncertainty about where genes fitted into the chemical picture, and particularly whether they had a direct role in correctly linking together the hundreds of amino acids that make a typical protein molecule. How this might occur was still conceptually quite unclear as late as the 1940s, and no one expected the incredibly rapid pace at which the nature and transmission of genetic information was worked out between 1953 and 1966. Then with the dominant role of DNA so clearly established, it was very tempting to say that by understanding the nucleic acids we had understood the essence of the living state and that the greatest challenges of biology had been surmounted.

This is a view that we do not share. The interconnecting pathways between ATP and DNA, marvelous as they are, do not give us the living cell. Even the simplest cells are far more complex objects than generally perceived and vastly more ingenious than any computerized intelligent machine yet designed. That this is so is strongly hinted at by simple observation. We do not have to see through the outer surfaces of cells to appreciate that the biological organization that permits them to act in such rational ways must indeed be incredibly subtle and versatile.

Consider for example the extraordinarily complex changes in cellular shape that accompany the movements of fibroblasts. These connective tissue cells are the principal makers of the extracellular matrix that helps glue together the tissues of multicellular organisms. Within animals, fibroblasts must constantly be prepared to move into areas of newly forming tissue. Removed from the animal and grown in culture, they are accessible to microscopic analysis, and the morphological changes that accompany their passage from one point to another have been extensively documented through cinematographic analysis.

The isolated well-fed single fibroblast is thus revealed to be a restless, apparently unsatisfied creature unable to stay quiet but instead internally programmed to move. It and all its progeny will continue moving and growing until the flat surface of the plastic dish is covered with a single layer of closely packed cells. Moving fibroblasts bear a resemblance to amoebae with their extending pseudopodia (false feet) that we first saw in our early school science classes, but the details of their respective motions are not the same.

Fibroblast movement is initiated by the rapid, virtually frenetic, throwing out of filamentous extensions (microspikes) and sheetlike projections (lamellipodia). Each of these projections can make firm attachments to the underlying surfaces ahead. Such attachments lead to a forward flow of the cell's cytoplasm and its enclosed nucleus. Many many more such locomotor protuberances are pushed out than ever make firm attachments, and those that fail to attach are swept up as "ruffles" in the backward flow of the upper cell surface that eventually sends them to the rear of the cell. Large numbers of potential adhesion points can thus be sampled, with firm union made only to the most favorable sites.

The capacity of fibroblasts for long persistent solo excursions is not a property of all cells. When, for example, single epithelial cells of the sort that line our intestines or skin are placed in culture, they show no tendency to move about. The locomotor behavior of a given cell type thus appears to be highly foreordained and, like virtually all other cellular events of consequence, is hardly ever left to chance. As a result, the exact final position of a given cell within a multicellular organism arises from a myriad of well-regulated biochemical steps that effectively give the cell no choice but to come together harmoniously with other cells in a particular configuration.

To understand how these logical steps unfold, we clearly must probe beneath the cell surface. Happily we now possess the highly sophisticated microscopic, biochemical, and genetic engineering procedures to let us tackle on virtually equal odds the cell's almost overwhelming complexity. We have already found that the apparently amorphous cytoplasmic mass contains interlaced patterns of specific fibrous protein aggregates. These filamentous structures, themselves built of smaller subunit protein molecules, are assembled into the elaborate scaffolds and molecular machines that give rise to directed cell movements.

As this book unfolds, we shall relate how various cell structural elements are built and maintained by specific interactions between complex molecules. And with less precision we shall outline how they enable a cell to grow and divide and how they generate the metamorphoses of the cell's architecture that we call cell movement and differentiation, which enable cells to participate in the construction of multicellular organisms. We hope that we shall also convey the sense of great mystery that surrounds the many problems that we do not yet know quite how to handle, the feeling of marvelous excitement that comes from the great achievements of today's cell biology, and, last but not least, the logical as well as the optical beauty of cells.

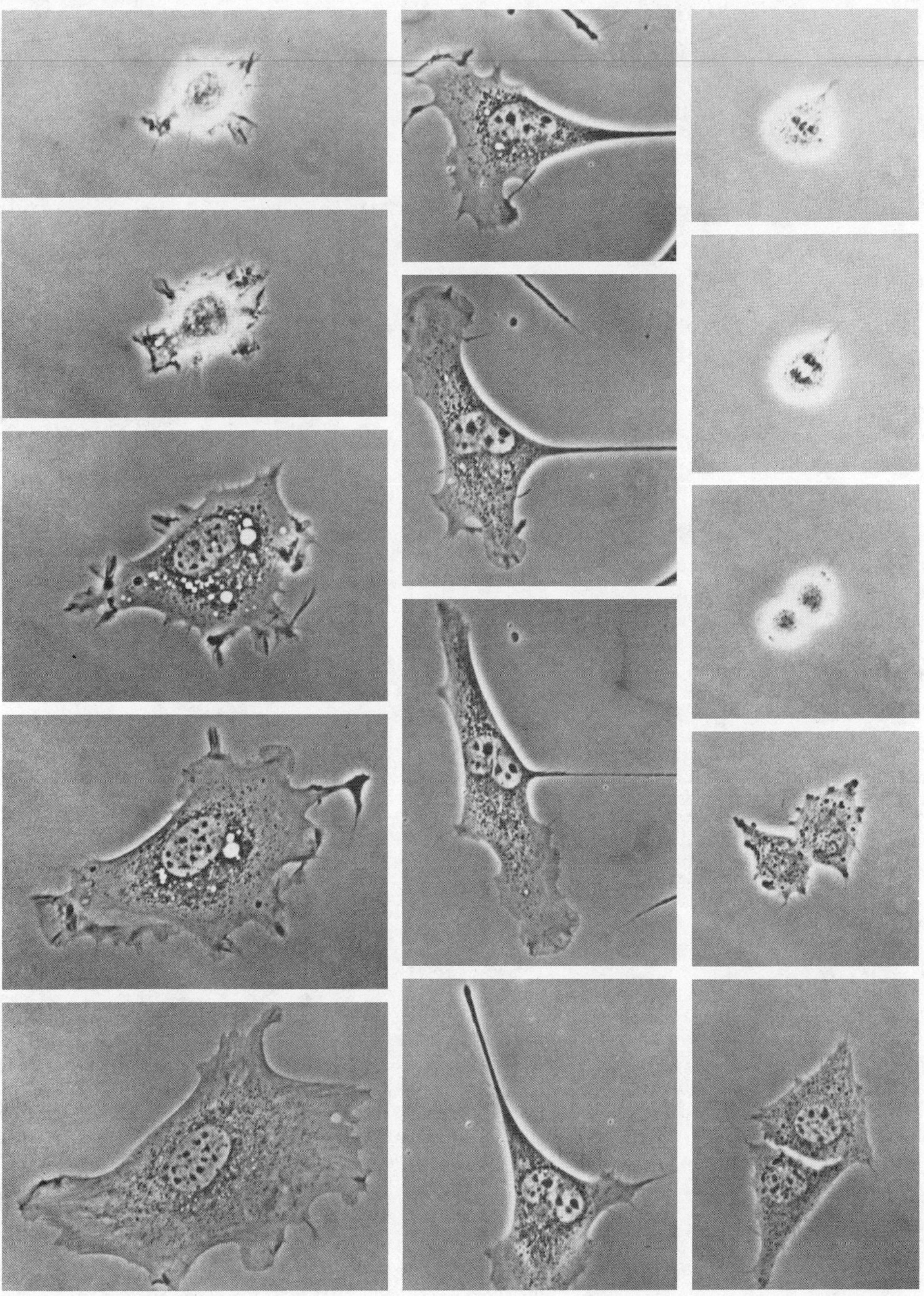

Mobile behavior of mouse fibroblasts (3T3 cells), as revealed by phase-contrast microscopy. (*Left panel*) As the cell flattens down upon a surface, needlelike microspikes and sheetlike lamellipodia are projected outward to seek suitable attachment sites. Intermittently, the lamellipodia fold back on themselves ("ruffle") before extending again. (*Center panel*) After flattening down, fibroblasts assume a polarized shape with leading lamellipodia and begin to crawl along the surface of the culture dish. In this series of four micrographs we observe an abrupt change in direction. (*Right panel*) In going through mitosis, a flattened cell rounds up prior to formation of the mitotic spindle; the two daughter cells reassume a flattened position following their separation. In each panel, successive micrographs taken at successive times are displayed from top to bottom. (Courtesy of Guenter Albrecht-Buehler.)

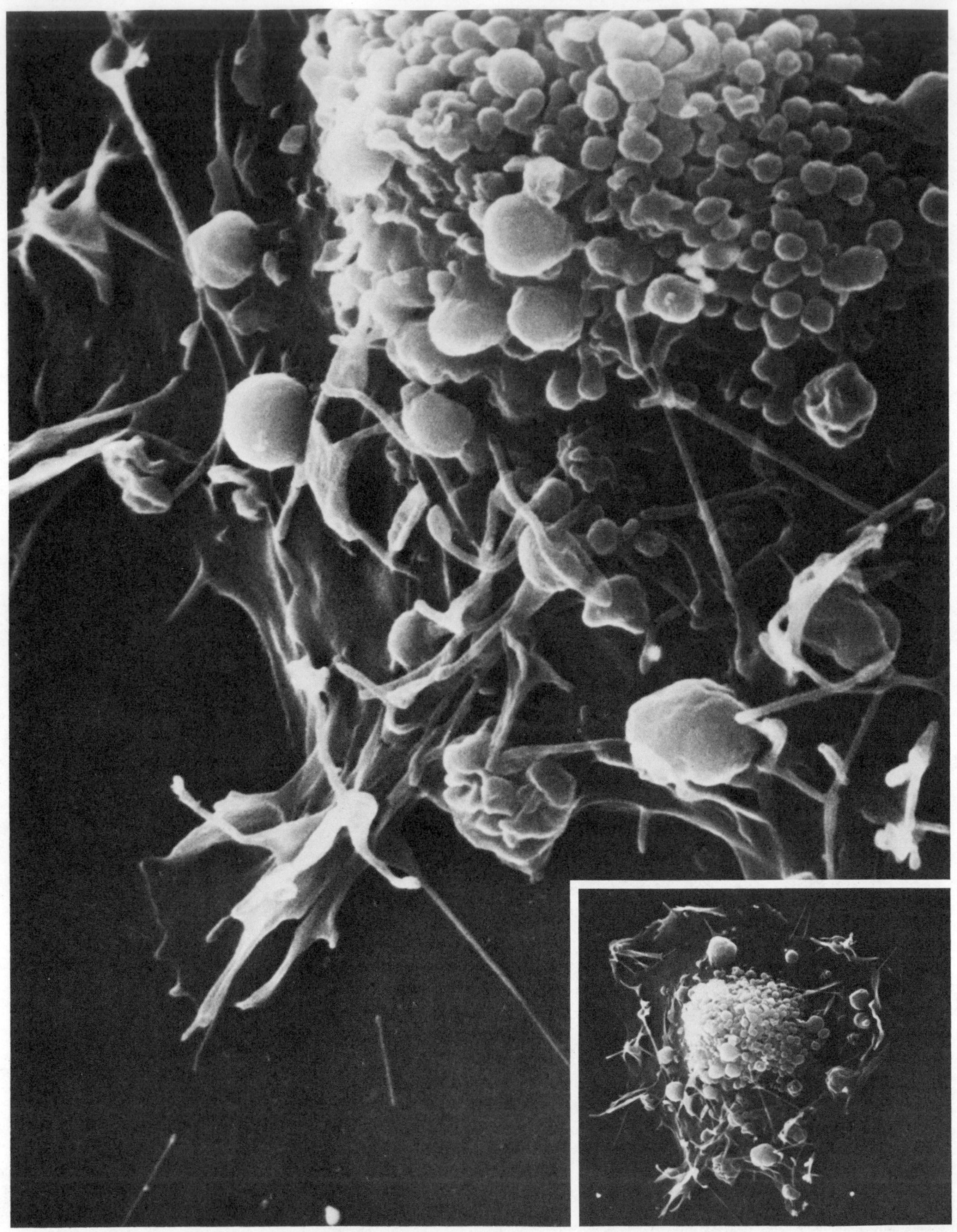

A scanning electron microscopic view of part of the surface of a mouse fibroblast in the process of flattening. Microspikes and lamellipodia project outward at the ruffling edges while large numbers of hemispherelike projections ("blebs") project from the region over the centrally located nucleus. The insert shows a lower-magnification view of the entire cell, whose diameter is about 25 μm. (Courtesy of Guenter Albrecht-Buehler.)

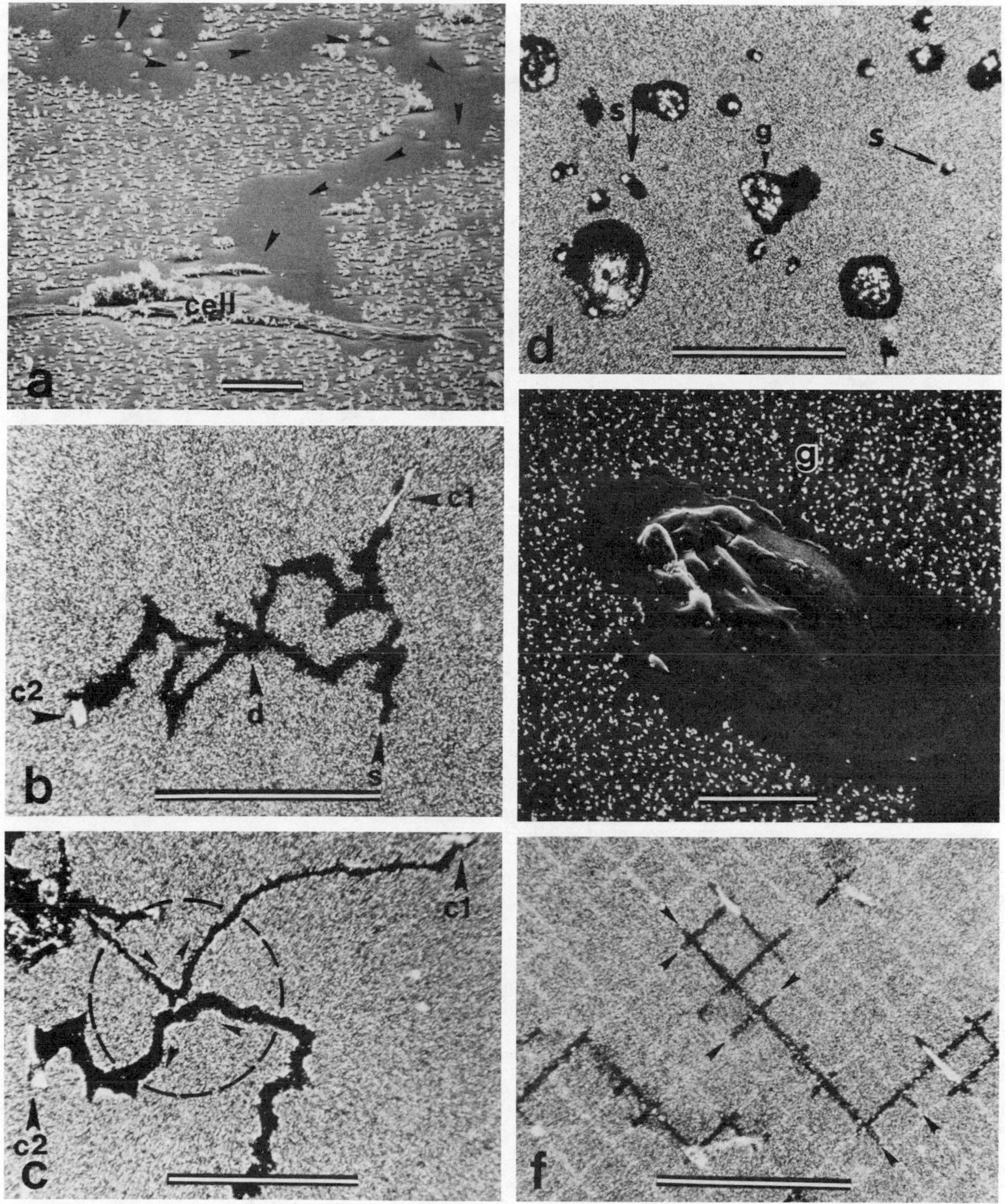

Paths of cell migration are revealed by cells moving on a surface coated with tiny gold particles. Extending microspikes and lamellipodia pick up the loose gold particles and help bring them back over the cell body where they are engulfed (phagocytosed) into the cell. Such events create clearings in regions through which cells have moved. In dark-field illumination, the tracks appear black while the gold-particle-filled cells glow brightly. (Courtesy of Guenter Albrecht-Buehler; reproduced from *J. Natl. Cancer Inst.* Monograph 60, 1982.)

(a) Scanning electron micrograph of a track, showing the many tiny gold particles on the substrate, the cell, and the particle-free track left behind by the cell (*arrows*). Bar indicates 20 μm. (b) Dark-field micrograph of the branching track of a mother 3T3 mouse cell that started at "s" and divided at "d" into the two similar tracks of the sister cells "c1" and "c2." Bar indicates 500 μm. (c) Collision between two 3T3 mouse cells (c1 and c2). Within the circled area the two cells have bounced off each other as if they were colliding billiard balls. Bar indicates 500 μm. (d, e) Dark-field light micrograph (d) and scanning electron micrograph (e) of migrating PTK 1 (rat kangaroo) cell groups (g). In (d), "s" points to single cells that migrated little. Bars indicate 500 μm (d) and 50 μm (e). (f) Tracks of guided 3T3 cells (*bright white structures*) on a checkerboard of guiding lines (*whitish lines*). The cells follow the lines but at the intersections probe into optional directions, as indicated by the small sideways "thorns" in the tracks at these points (arrowheads point to a few of these thorns). Bar indicates 500 μm.

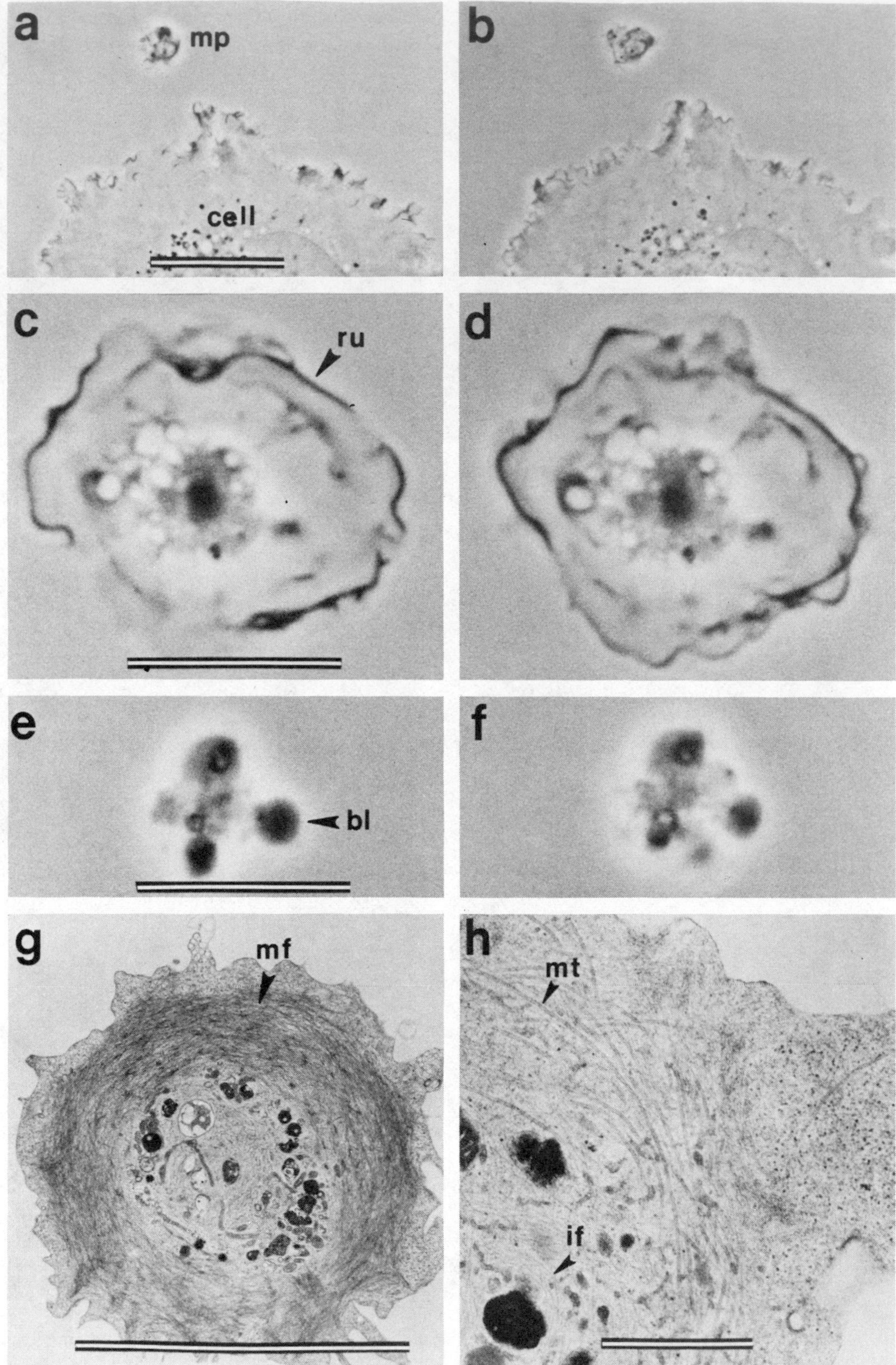

Very small autonomously moving cellular fragments (microplasts) are often generated following exposure of cells to the cytoskeleton-disrupting drug cytochalasin B. Though they lack a nucleus, microplasts can flatten, ruffle, and bleb, showing their possession of organized functional cytoskeletal elements. (Courtesy of Guenter Albrecht-Buehler; reproduced from *J. Natl. Cancer Inst.* Monograph 60, 1982.)

(a, b) A ruffling microplast (mp) near the edge of a flattening human cell (cell) for size comparison. Bar indicates 20 μm. Time lapse between the two pictures is 35 seconds, showing the movement. (c, d) Two photographs of a ruffling microplast taken 15 seconds apart. One of the many ruffles is indicated by "ru." Bar indicates 10 μm. (e, f) Two photographs of a blebbing microplast taken 10 seconds apart. One of the many blebs is indicated by "bl." Bar indicates 10 μm. (g, h) Electron micrographs of a typical microplast sectioned parallel to the flat surface on which it sits. Visible are peripheral actin-containing microfilaments (mf), microtubules (mt), and intermediate filaments (if). Bars indicate 10 μm (g) and 1 μm (h).

A Note to the Reader

Although the chapters of this book can be read independently of one another, they are arranged in a logical sequence of three parts. The first three chapters of **Part I,** which cover elementary principles and basic biochemistry, can serve either as an introduction for those who have not studied biochemistry or as a refresher course for those who have. Chapter 4, which concludes Part I, deals with the principles of the main experimental methods for investigating cells. It is not necessary to read this chapter in order to understand the later chapters, but a reader will find it a useful reference.

Part II represents the central core of cell biology and is concerned mainly with those properties that are common to most eucaryotic cells. It begins, in Chapter 5, with the fundamental molecular mechanisms of heredity—the replication and maintenance of DNA and the translation of its linear sequence of nucleotides into the linear sequences of amino acids in proteins. Details of chromosome structure and of topics related to the control of gene expression in eucaryotic cells are, however, reserved for discussion in Chapter 8 ("The Cell Nucleus"). Chapters 6 and 7 are devoted to the various membrane systems of the cell and the functions associated with them. Chapter 9 deals with the mitochondria and chloroplasts that provide for the cell's energy requirements, Chapter 10 with the cytoskeleton and cell movement, and Chapter 11 with the mechanisms of cell growth and division. Part II ends with two chapters on the ways in which cells interact with each other: Chapter 12 discusses cell-cell adhesion and the extracellular matrix, while Chapter 13 discusses chemical signaling between cells.

Part III follows the behavior of cells in the construction of multicellular organisms, from the formation of eggs and sperm through the processes of fertilization and embryonic development (Chapters 14 and 15) to the differentiated tissues of mature animals (Chapter 16). The cells of the immune system and of the nervous system are singled out for detailed discussion in Chapters 17 and 18, not only because they provide examples of the most complex and sophisticated tasks that assemblies of cells can perform, but also because many general insights into cell behavior derive from studies of these systems. Finally, Chapter 19 discusses the intriguing special features of plant cells.

The progress of cell biology has depended to a large extent on advances in experimental methods, and Chapter 4 includes several tables giving the dates of crucial developments along with the names of the scientists involved. Elsewhere in the book the policy has been to avoid naming individual scientists. The authors of major discoveries, however, can usually be identified by consulting the **lists of references** at the end of chapters. These references frequently include the original papers in which important discoveries were first reported. **Superscript numbers** that accompany many of the text headings refer to the numbered citations in the reference lists, providing a convenient means of following up specific topics.

Throughout the book, **boldface type** has been used to highlight key terms at the point in a chapter where the main discussion of them occurs. This may or may not coincide with the first appearance of the term in the text. *Italics* are used to set off important terms with a lesser degree of emphasis.

I

Introduction to the Cell

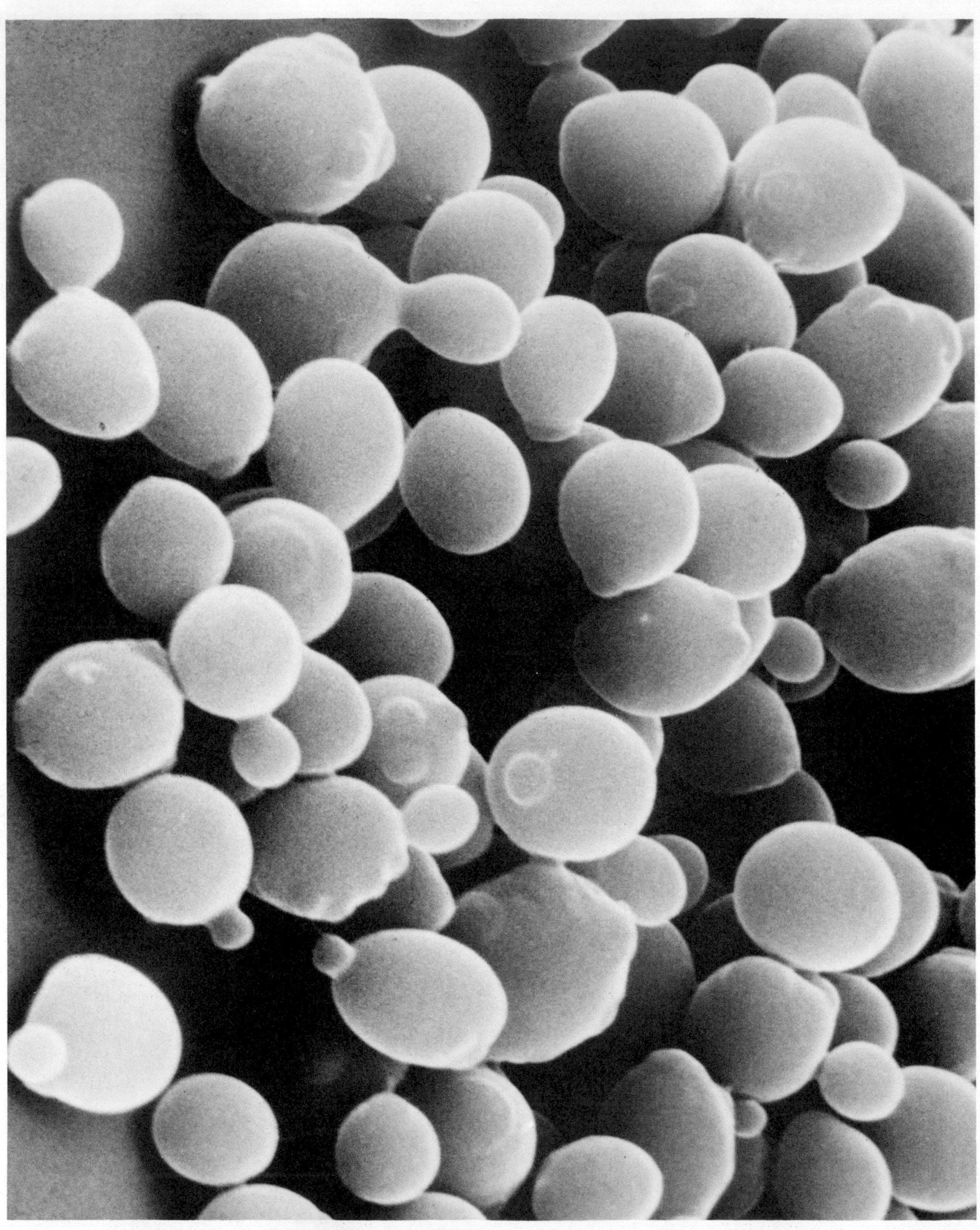

Scanning electron micrograph of growing yeast cells. These unicellular eucaryotes bud off small daughter cells as they multiply. (Courtesy of Ira Herskowitz and Eric Schabatach.)

1 The Evolution of the Cell

All living creatures are made of cells—small membrane-bounded compartments filled with a concentrated aqueous solution of chemicals. The simplest forms of life are solitary cells that propagate by dividing in two. Higher organisms, such as ourselves, are like cellular cities in which groups of cells perform specialized functions and are linked by intricate systems of communication. In a sense, cells are halfway between molecules and man. We study them to learn, on the one hand, how they are made from molecules and, on the other, how they cooperate to make an organism as complex as a human being.

All organisms, and all of the cells that constitute them, are believed to have descended from a common ancestor cell by *evolution.* Evolution involves two essential processes: (1) the occurrence of random *variation* in the genetic information passed from an individual to its descendants and (2) the *selection* of genetic information that helps its possessors to survive and propagate. Evolution is the central principle of biology, helping us to make sense of the bewildering variety in the living world.

This chapter, like the book as a whole, is concerned with the progression from molecules to multicellular organisms. It discusses the evolution of the cell, first as a living unit constructed from smaller parts, and then as a building block for larger structures. Through evolution, we introduce the cell components and activities that are to be treated in detail, in broadly similar sequence, in the chapters that follow. Beginning with the origins of the first cell on earth, we consider how the properties of certain types of large molecules allow hereditary information to be transmitted and expressed, and permit evolution to occur. Enclosed in a membrane, these molecules provide the essentials of a self-replicating cell. Following this, we describe the major transition that occurred in the course of evolution, from small bacteriumlike cells to much larger and more complex cells such as are found in present-day plants and animals. Lastly, we suggest ways in which single free-living cells might have given rise to large multicellular organisms, becoming specialized and cooperating in the formation of such intricate organs as the brain.

Clearly, there are dangers in an evolutionary approach: the large gaps in our knowledge can be filled only by speculations that are likely to be wrong in many details. But there is enough evidence from fossils and from comparative studies of present-day organisms and molecules to allow us to make intelligent guesses about the major stages in the evolution of life.

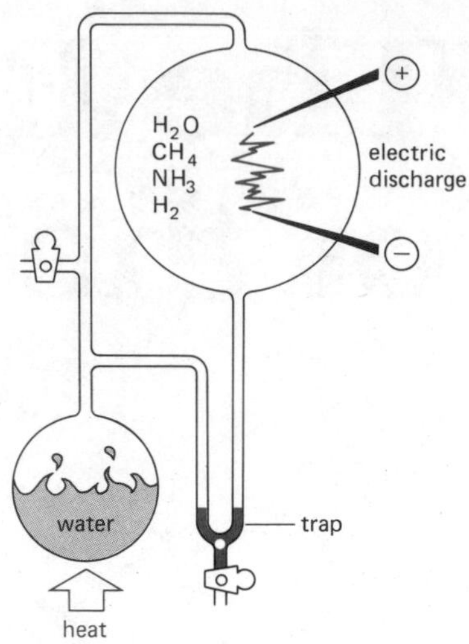

Figure 1–1 A typical experiment simulating conditions on the primitive earth. Water is heated in a closed apparatus containing CH_4, NH_3, and H_2, and an electric discharge is passed through the vaporized mixture. Organic compounds accumulate in the U-tube trap.

From Molecules to the First Cell[1]

Simple Biological Molecules Can Form Under Prebiotic Conditions

The conditions that existed on the earth in its first billion years are still a matter of dispute. Was the surface initially molten? Did the atmosphere contain ammonia, or methane? Everyone seems to agree, however, that the earth was a violent place with volcanic eruptions, lightning, and torrential rains. There was little if any free oxygen and no layer of ozone to absorb the harsh ultraviolet radiation from the sun.

Simple organic molecules (that is, molecules containing carbon) are likely to have been produced under such conditions. The best evidence for this comes from laboratory experiments. If mixtures of gases such as CO_2, CH_4, NH_3, and H_2 are heated with water and energized by electrical discharge or by ultraviolet radiation, they react to form small organic molecules—usually a rather small selection, each made in large amounts (Figure 1–1). Among these products are a number of compounds, such as hydrogen cyanide ($H{-}C{\equiv}N$) and formaldehyde $\left(\begin{smallmatrix}H & \\ & C{=}O \\ H & \end{smallmatrix}\right)$, that readily undergo further reactions in aqueous solution (Figure 1–2). Most important, the four major classes of small organic molecules found in cells—*amino acids, nucleotides, sugars,* and *fatty acids*—are generated.

While such experiments cannot reproduce the early conditions on the earth exactly, they make it plain that the formation of organic molecules is surprisingly easy. And the developing earth had immense advantages over any human experimenter; it was very large and could produce a wide spectrum of conditions. But above all, it had much more time—hundreds of millions of years. In such circumstances it seems very likely that, at some time and place, many of the simple organic molecules found in present-day cells accumulated in high concentrations.

$HCHO$	formaldehyde
CH_3COOH	acetic acid
NH_2CH_2COOH	glycine
$HCOOH$	formic acid
$CH_3CH(OH)COOH$	lactic acid
$NH_2CH(CH_3)COOH$	alanine
$CH_3NH{-}CH_2COOH$	sarcosine
$H{-}C{\equiv}N$	hydrogen cyanide
$NH_2{-}C({=}O){-}NH_2$	urea
$NH_2CH(CH_2COOH)COOH$	aspartic acid

Figure 1–2 A few of the compounds that might be formed in the experiment described in Figure 1–1. Compounds shown in color are important components of present-day living cells.

Polynucleotides Are Capable of Directing Their Own Synthesis

Simple organic molecules such as amino acids and nucleotides can associate to form large *polymers.* One amino acid can join with another by forming a peptide bond, while two nucleotides can join together by a phosphodiester bond. The repetition of these reactions leads to linear polymers known as **polypeptides** and **polynucleotides,** respectively. In present-day living organisms, polypeptides—known as *proteins*—and polynucleotides—in the form of both *ribonucleic acids* (*RNA*) and *deoxyribonucleic acids* (*DNA*)—are commonly viewed as the most important constituents. A restricted set of 20 amino acids constitute the universal building blocks of the proteins, while RNA and DNA molecules are constructed from four types of nucleotides each. One can only speculate as to why these particular sets of monomers should have been selected for biosynthesis in preference to others that are chemically similar.

The earliest polymers may have formed in several ways—for example, by the heating of dry organic compounds or by the catalytic activity of high concentrations of inorganic polyphosphates. The products of similar reactions in the test tube are polymers of variable length and random sequence in which the amino acid or nucleotide added at any point depends mainly on chance (Figure 1–3). However, once a polymer has formed, it is able to influence the formation of other polymers. Polynucleotides, in particular, have the ability to specify the sequence of nucleotides in new polynucleotides by acting as *templates* for the polymerization reactions. For example, a polymer composed of one nucleotide (polyuridylic acid, or poly U) can serve as a template for the synthesis of a second polymer composed of another type of nucleotide (polyadenylic acid, or poly A). Such templating depends on the fact that one polymer preferentially binds the other. By lining up the subunits required to make poly A along its surface, poly U promotes the formation of poly A (Figure 1–4).

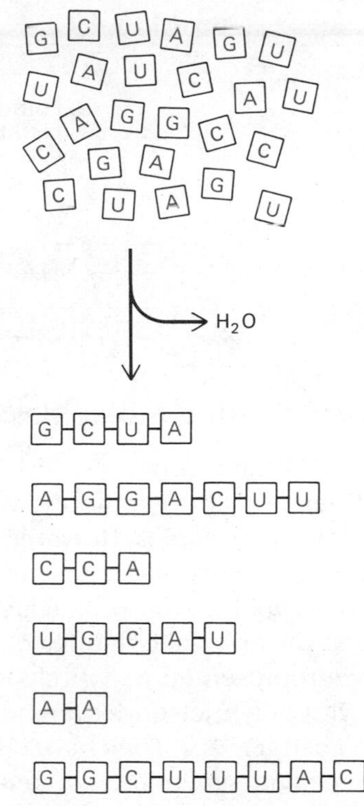

Figure 1–3 Nucleotides of four kinds (here represented by the single letters A, U, G, and C) can undergo spontaneous polymerization with the loss of water. The product is a mixture of polynucleotides that are random in length and sequence.

Specific pairing between complementary nucleotides probably played a crucial part in the origin of life. Consider, for example, a polynucleotide such as RNA, made of a string of four nucleotides, containing the bases uracil (U), adenine (A), cytosine (C), and guanine (G). Because of complementary pairing between the bases A and U and between the bases G and C, when RNA is added to a mixture of activated nucleotides under conditions that favor polymerization, new RNA molecules are produced in which nucleotides are joined in a sequence that is complementary to the first. That is, the new molecules are rather like a mold of the original, with each A in the original corresponding to a U in the copy, and so on. The sequence of nucleotides in the original RNA strand contains information that is, in essence, preserved in the newly formed complementary strands. A second round of copying, with the complementary strand as a template, restores the original sequence (Figure 1–5).

Such *complementary templating* mechanisms are elegantly simple, and they lie at the heart of information-transfer processes in biological systems. Genetic information contained in every cell is encoded in the sequences of nucleotides in its polynucleotide molecules, and this information is passed on (inherited) from generation to generation by means of complementary base-pairing interactions.

Rapid formation of polynucleotides in a test tube requires the presence of specific protein catalysts, or *enzymes*, which would not have been present in the "prebiotic soup." However, less efficient catalysts in the form of minerals or metal ions would have been present; and, in any case, catalysts only speed up reactions that would occur anyway given sufficient time. Since both time and a supply of chemically reactive nucleotide precursors were available in abundance, it is likely that slowly replicating systems of polynucleotides became established in the prebiotic conditions on earth.

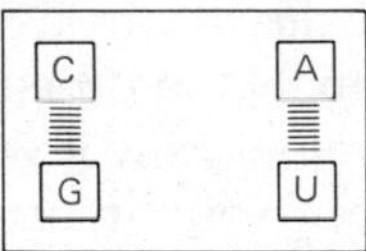

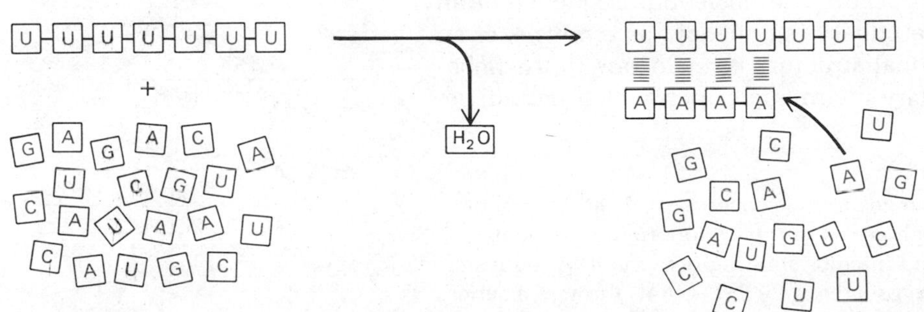

Figure 1–4 Preferential binding occurs between pairs of nucleotides (C with G and A with U) by relatively weak chemical bonds (*above*). This pairing enables one polynucleotide to act as a template for the synthesis of another (*left*).

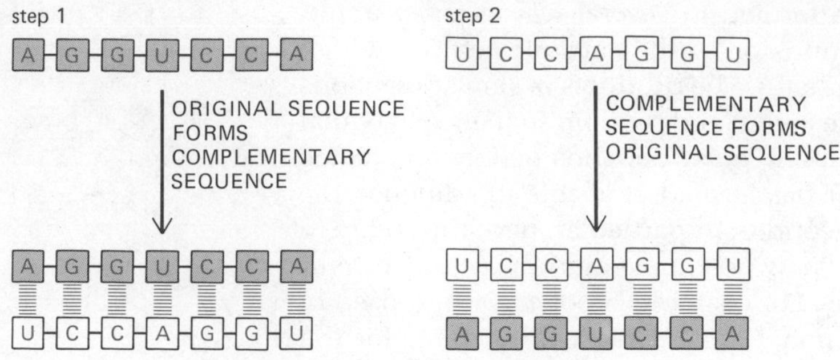

Figure 1–5 Replication of a polynucleotide sequence (here an RNA molecule). In step 1, the original RNA molecule acts as a template to form an RNA molecule of complementary sequence. In step 2, this complementary RNA molecule itself acts as a template, forming RNA molecules of the original sequence. Since each templating molecule can produce many copies of the complementary strand, these reactions can result in the "multiplication" of the original sequence.

Self-replicating Molecules Undergo Natural Selection

Under favorable conditions, a polynucleotide molecule in a rich soup of nucleotides is able to multiply, with each copy of the original serving as the parent for new copies. However, many errors will inevitably occur in the copying process, especially under primordial conditions. New and imperfect copies of the original will be propagated. In time, therefore, the sequence of nucleotides in the original polynucleotide molecule will change until the information it once represented is entirely lost (Figure 1–6).

But polynucleotides are not just strings of symbols that carry information in an abstract way. They have chemical personalities that affect their behavior. The specific sequence of nucleotides governs the properties of the whole molecule, especially how it folds up in solution. Just as the nucleotides in a polynucleotide can pair with free complementary nucleotides in their environment to form a new polymer, so they can pair with complementary nucleotide residues within the polymer itself. A sequence GGGG in one part of a polynucleotide chain can form a relatively strong association with a CCCC sequence in another region of the molecule. Such associations produce various three-dimensional folds, and the molecule as a whole takes on a unique shape that depends entirely on the sequence of its nucleotides (Figure 1–7).

The three-dimensional folded structure of a polynucleotide affects both its stability and its ability to replicate, so that not all polynucleotide shapes will be equally successful in a replicating mixture. Some will be too long or too tightly folded to act as efficient templates. Others might be unstable under the prevailing conditions. In fact, it has been demonstrated in laboratory studies that replicating systems of RNA molecules undergo a form of natural selection and that different favorable sequences will eventually predominate, depending on the exact conditions.

A polynucleotide such as an RNA molecule therefore has two important characteristics: it carries information encoded in its nucleotide sequence that it passes on by the process of replication, and it has a unique folded structure that determines how it will function and respond to external conditions. These two features—one informational, the other functional—are the two essential ingredients required for evolution to occur. The nucleotide sequence of an RNA molecule is analogous to the hereditary information, or *genotype,* of an organism. The folded three-dimensional structure is analogous to the *phenotype*—the expression of the hereditary information upon which natural selection operates.

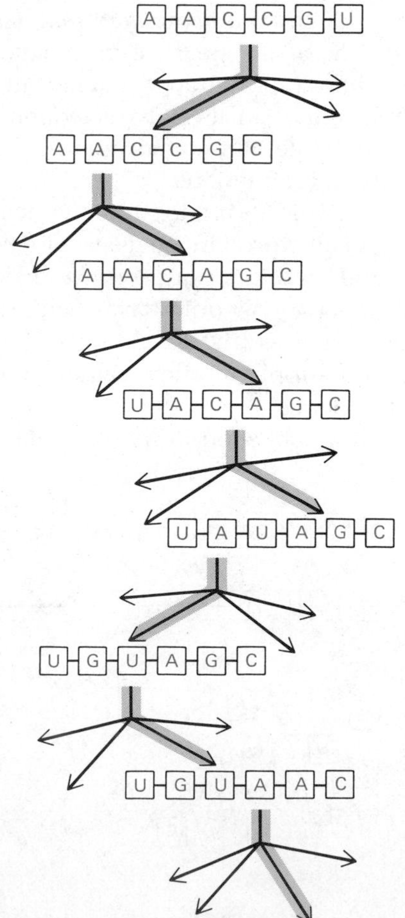

Figure 1–6 Changes in the sequence of an RNA molecule can occur through errors in replication. Here a particular "lineage" is traced in color showing how the RNA sequence AACCGU changes progressively to UGUAAC through a series of copying errors. Many other sequences will be generated at the same time, as indicated by the multiple arrows.

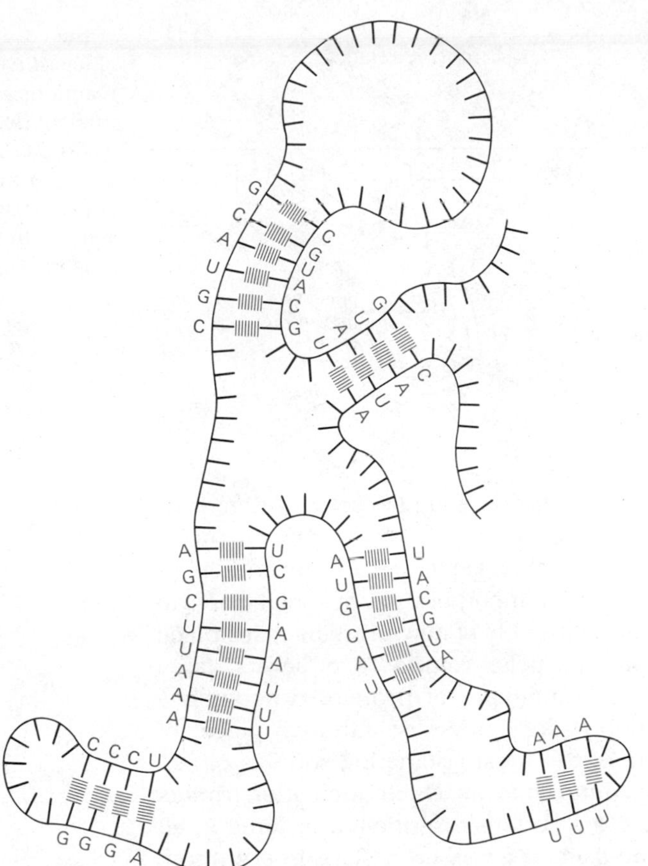

Figure 1–7 Nucleotide pairing between different regions of the same polynucleotide (RNA) chain causes the molecule to adopt a distinctive shape.

Information Flows from Polynucleotides to Polypeptides

The suggestion, therefore, is that between 3.5 and 4 billion years ago, somewhere on earth, self-replicating systems of polynucleotides began the process of evolution. Polymers with different nucleotide sequences competed for the available precursor materials to construct copies of themselves, just as organisms now compete; success depended on the accuracy and the speed with which the copies were made and on the stability of those copies.

However, while the structure of polynucleotides is well suited for information storage and replication, these molecules are not sufficiently versatile to provide all the structural and functional building blocks of a living cell. Polypeptides, on the other hand, are composed of many different amino acids, and, as will be discussed in Chapter 3, their diverse three-dimensional forms, which often bristle with reactive sites, make them ideally suited to a wide range of structural and chemical tasks. Even random polymers of amino acids produced by prebiotic synthetic mechanisms are likely to have had catalytic properties, some of which could have enhanced the replication of RNA molecules. Some classes of polypeptides would therefore have been extremely useful to a replicating system, especially if they could be tailor-made. Polynucleotides that helped guide the synthesis of specific polypeptides in their environment would have had a great advantage in the evolutionary struggle for survival (Figure 1–8).

Yet how could polynucleotides exert such control? How could the information encoded in their sequence specify the sequences of polymers of a different type? In present-day organisms, RNA directs the synthesis of polypeptides—that is, **protein synthesis**—but it is a process that requires remarkably elaborate biochemical machinery. One RNA molecule carries the

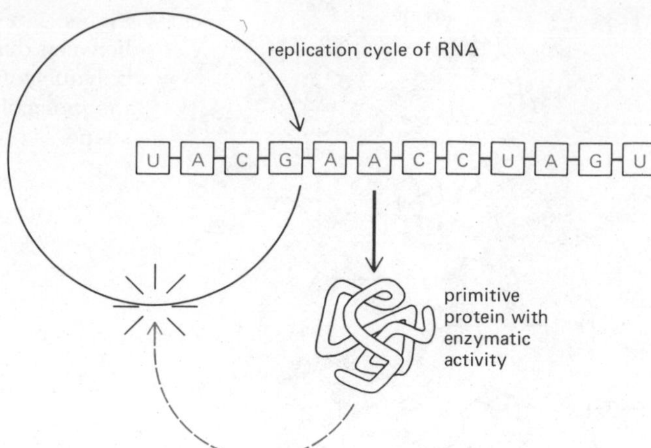

Figure 1–8 Proteins can act as efficient catalysts of chemical reactions such as the formation of nucleotides or their polymerization into RNA. Therefore, an RNA molecule that can direct the synthesis of an appropriate protein is able to accelerate its own replication, as illustrated schematically here.

genetic information for a particular polypeptide, while a set of other RNA molecules bind amino acids; the two types of RNA molecules form complementary base pairs with one another to enable sequences of nucleotides in the informational RNA molecule to direct the incorporation of specific amino acids into a growing polypeptide chain. Assembly of new proteins takes place on the surface of *ribosomes*—complex particles composed of several large RNA molecules and more than 50 different types of protein. How such a complex mechanism arose in evolution is still a mystery, although pieces of the puzzle are falling into place. One of the most fascinating sources of evidence is the genetic "dictionary," or *genetic code*, by which nucleotide triplets are translated into amino acids. Since the code is essentially the same in all living organisms, it must have become fixed at a very early stage in evolution, and it is likely to contain traces of the way that primordial translation was achieved.

Whatever the preliminary steps of evolution may have been, once RNA molecules were able to direct the synthesis of proteins, they had potentially at their disposal an enormous workshop of chemical tools. It was now possible in principle to synthesize enzymes that could catalyze a large range of chemical reactions, including the synthesis of more proteins and RNA molecules. Once the evolution of nucleic acids had thus advanced to the point of specifying enzymes to aid in their own manufacture, the proliferation of the replicating system would have been immensely speeded up. The potentially explosive nature of such an autocatalytic process can be seen today in the life cycle of some bacterial viruses: after they have entered a bacterium, such viruses direct the synthesis of proteins that catalyze selectively their own replication, so that within a short time they take over the entire cell (Figures 1–9 and 1–10).

Membranes Defined the First Cell

The appearance of protein synthesis controlled by nucleic acids was no doubt one of the crucial events leading to the formation of the first cell. Another must have been the development of an outer membrane. The proteins synthesized under the control of a certain species of RNA would not facilitate reproduction of that species of RNA unless they were retained in the neighborhood of the RNA; moreover, as long as these proteins were free to diffuse among the population of replicating RNA molecules, they could benefit equally any competing species of RNA that might be present. If a variant RNA arose that made a superior type of enzyme, the new enzyme could not contribute *selectively* to the survival of the variant RNA in its competition with its fellows. Selection of RNA molecules according to the quality of the proteins that they

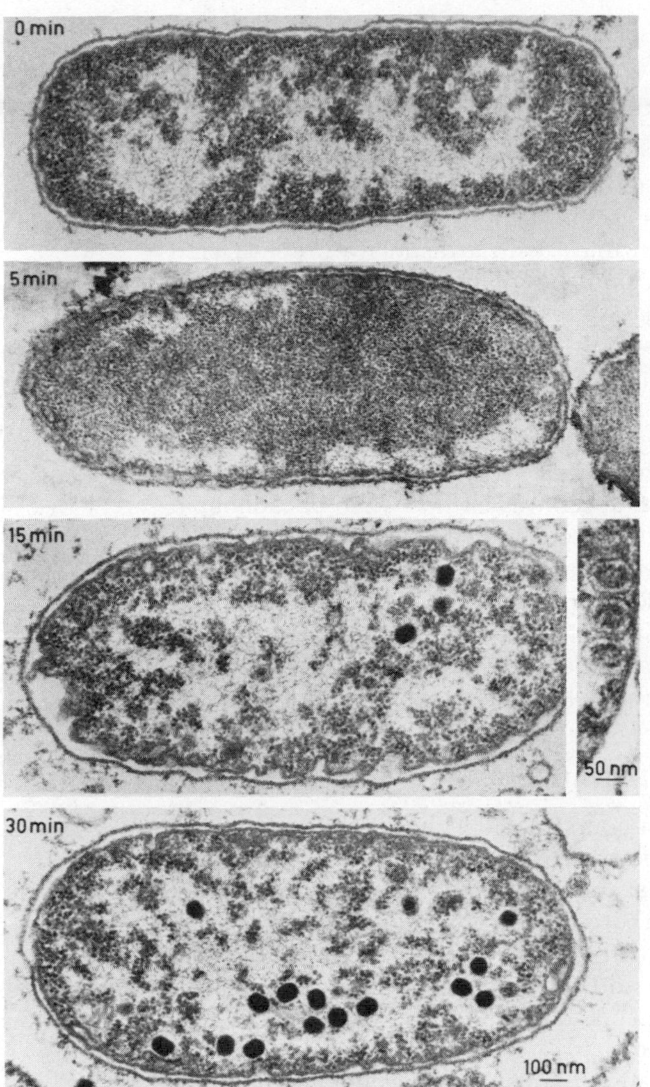

Figure 1–9 Sequence of electron micrographs showing the growth of a virus inside a bacterial cell. Infection begins when the virus attaches to the outside of the bacterium (see also Figure 1–10) and injects its DNA into the bacterial cell. Within 5 minutes, this DNA has directed the synthesis of a set of specific proteins, some of which degrade the DNA of the host bacterium, while others catalyze the replication of the viral DNA. The dense particles seen in the cell 15 minutes after infection are immature virus particles consisting of viral DNA packed into spherical shells of protein (the shells are first made separately, as shown in the inset). Virus particles continue to mature and accumulate in the cell, as seen in the 30-minute specimen. (Courtesy of E. Kellenberger.)

Figure 1–10 A higher magnification micrograph of a bacterial cell that has been infected with virus particles for more than an hour. The infectious cycle is almost complete, and the bacterial cell is about to burst open, releasing several hundred new infective virus particles to the surroundings. The virus shown in this micrograph and in the micrographs of Figure 1–9 is bacteriophage T4. (Courtesy of E. Kellenberger.)

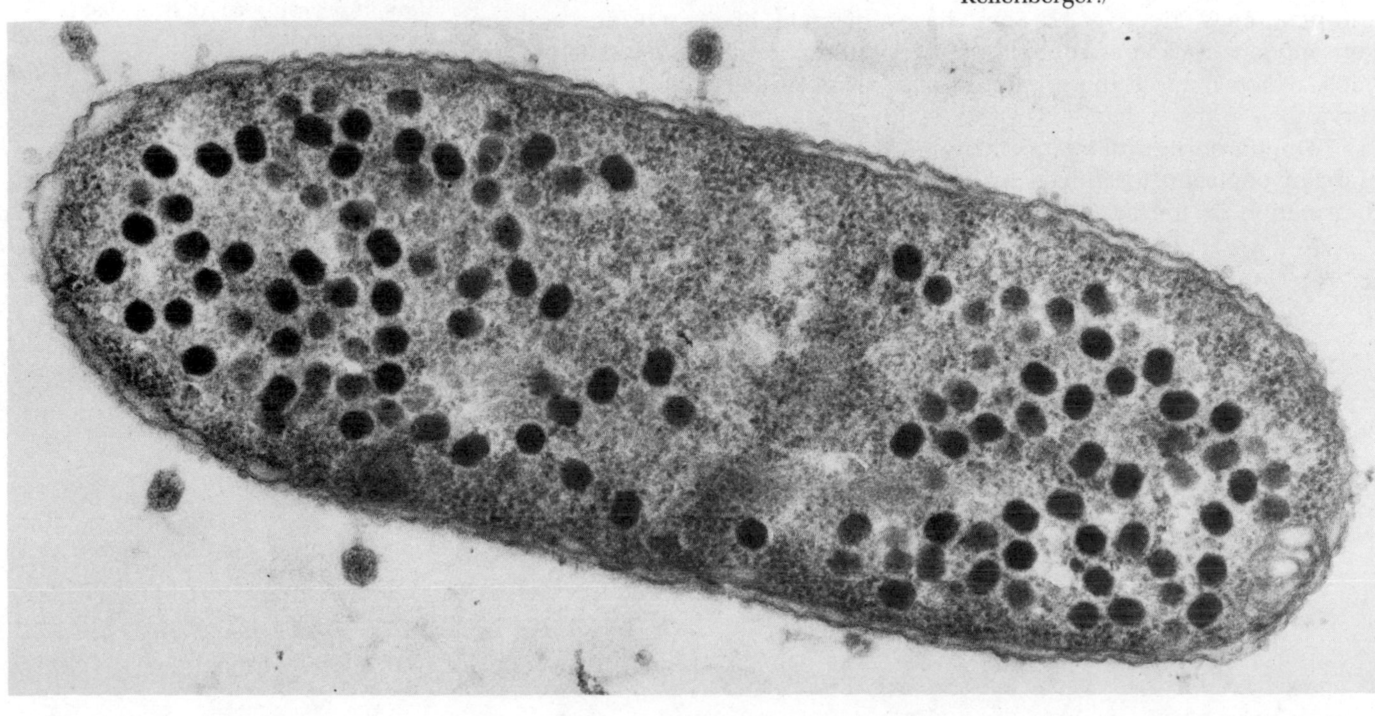

generated could not begin until some form of compartment evolved to contain the proteins made by an RNA molecule and thereby make these proteins primarily available for its own use (Figure 1–11).

All present-day cells are surrounded by a **plasma membrane,** composed of phospholipids and proteins. In the electron microscope such membranes appear as sheets about 7 nm thick, with a distinctive three-layered appearance due to the tail-to-tail packing of the phospholipid molecules. Artificial membranes with a very similar appearance can be made in the test tube simply by mixing phospholipids and water together. Under suitable conditions such artificial membranes round up into closed vesicles with diameters between 1 and 10 μm. Although these vesicles are inert, like soap bubbles, it is easy to imagine that by enclosing a distinct population of molecules they could form a spatially isolated functional unit.

It has been postulated that the first cell was formed when phospholipid molecules in the prebiotic soup spontaneously assembled into such membranous structures, enclosing a self-replicating mixture of RNA and protein molecules. Once sealed within a closed membrane, RNA molecules could begin to evolve, not merely on the basis of their own structure, but also according to the proteins they could make: the nucleotide sequences of the RNA molecules could now become expressed in the character of the cell as a whole.

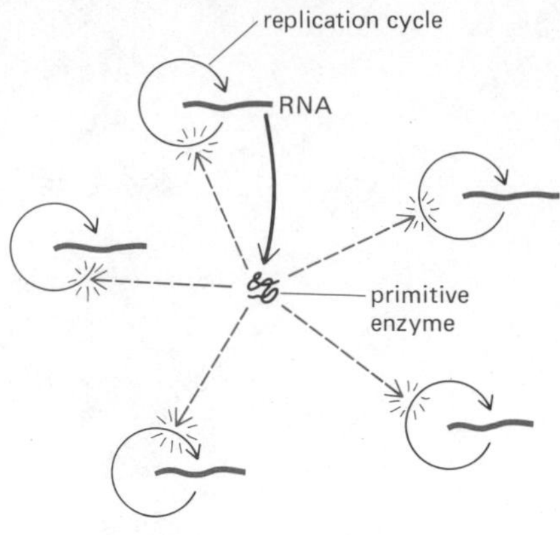

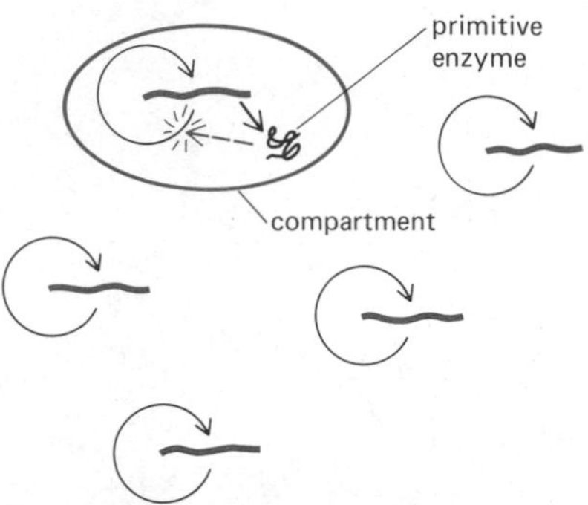

Figure 1–11 Schematic drawing showing the evolutionary advantage of cell-like compartments. In a mixed population of self-replicating RNA molecules capable of protein synthesis (as illustrated in Figure 1–8), any improved form of RNA that is able to produce a more useful protein must share this protein with all of its competitors. However, if the RNA is enclosed within a compartment, such as a lipid membrane, then any protein it makes is confined for its own use; the RNA can therefore be selected on the basis of its making a better protein.

Mycoplasmas Are the Simplest Living Cells

The picture we have presented is, of course, speculative: there are no fossil records that trace the origins of the first cell. Nevertheless, there is persuasive evidence from present-day organisms, and from experiments, that the broad features of this evolutionary story are correct. The prebiotic synthesis of small molecules, the self-replication of RNA molecules, the translation of RNA sequences into amino acid sequences, and the assembly of lipid molecules to form membrane-bounded compartments—all presumably occurred to generate the first cell 3.5 or 4 billion years ago.

It is useful to compare this putative first cell with the simplest present-day cells, the **mycoplasmas.** Mycoplasmas are small bacteriumlike organisms that normally lead a parasitic existence in close association with animal and plant cells (Figure 1–12). Some have a diameter of about 0.3 μm and contain enough nucleic acid to direct the synthesis of about 750 different proteins, which may be the minimum number of proteins that a cell needs to survive.

One important difference between the first cell as we have described it and a mycoplasma (or indeed any other present-day cell) is that the hereditary information in the latter is stored in DNA rather than RNA. Both types of

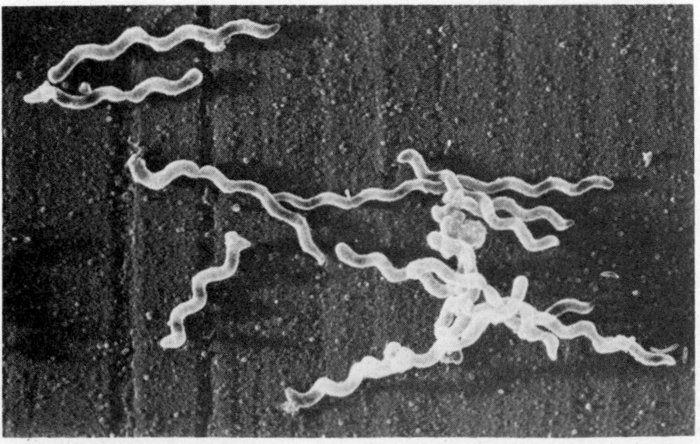

Figure 1–12 *Spiroplasma citrii,* a mycoplasma that grows in plant cells. (Courtesy of J. Burgess.)

polynucleotides are found in present-day cells, but they function in a collaborative manner, each having evolved to perform specialized tasks. Small chemical differences fit the two kinds of molecules for distinct functions. DNA serves as the permanent repository of genetic information. Unlike RNA, it exists principally in a double-stranded form composed of a pair of complementary polynucleotide molecules. Not only is genetic information that is stored in this way made more stable, but the double-stranded arrangement permits the operation of a repair mechanism: an intact strand serves as the template for the correction or repair of an associated damaged strand. DNA guides the synthesis of specific RNA molecules, again by the principle of complementary base-pairing, though now this pairing is between slightly different types of nucleotides. The resulting single-stranded RNA molecules then perform the two other primeval functions: they direct protein synthesis and in some situations they have a structural role not unlike that of proteins.

In addition to its various classes of polynucleotides, the mycoplasma cell contains many enzymes and structural proteins, some in its interior and some embedded in its membrane; these together synthesize essential small molecules that are not provided in the environment, redistribute the energy needed to drive biosynthetic reactions, and maintain appropriate chemical conditions inside the cell. The evolution of these latter metabolic functions will be discussed in the following section.

Summary

Living cells probably arose on earth by the spontaneous aggregation of molecules about 3.5 billion years ago. From our knowledge of present-day organisms and the molecules they contain, it seems that at least three steps must have occurred before the first cell emerged: (1) polymers of RNA capable of directing their own replication through complementary base-pairing interactions had to be formed; (2) mechanisms by which an RNA molecule could direct the synthesis of a protein had to be developed; and (3) a lipid membrane had to assemble to enclose the self-replicating mixture of RNA and protein molecules. At some later stage in the evolutionary process, DNA took the place of RNA as the hereditary material.

From Procaryotes to Eucaryotes[2]

It is thought that all organisms living now on earth derive from one single primordial cell born several billion years ago. This cell, outreproducing its competitors, took the lead in the process of cell division and evolution that would eventually cover the earth in green, change the composition of its atmosphere, and make it the home of intelligent life. The family resemblances between all organisms seem too strong to be explained in any other way. One important landmark along this evolutionary road occurred about 1.5 billion years ago, when there was a transition from small cells with a relatively simple internal structure—the so-called **procaryotes,** which include the various types of bacteria—to the larger and radically more complex *eucaryotic* cells such as are found in higher animals and plants.

Procaryotic Cells Are Structurally Simple But Biochemically Diverse

Bacteria are the simplest organisms found in most natural environments. They are spherical or rod-shaped cells, commonly several μm in linear dimension (Figure 1–13). They often possess a tough protective coat, called a *cell wall,* beneath which a plasma membrane encloses a single cytoplasmic compartment containing DNA, RNA, proteins, and small molecules. In the electron microscope this cell interior appears as a more or less uniform matrix (see top panel of Figure 1–9).

Bacteria are small and can replicate quickly by simply dividing in two by *binary fission.* When food is plentiful, "survival of the fittest" generally means survival of those that can divide the fastest. Under optimal conditions, a single procaryotic cell can divide every 20 minutes and thereby give rise to 4 billion cells (approximately equal to the present human population on earth) in less than 11 hours. The ability to divide quickly enables populations of bacteria to adapt rapidly to changes in their environment. Under laboratory conditions, for example, a population of bacteria maintained in a large vat will evolve within a few weeks by spontaneous mutation and natural selection to utilize new types of sugar molecules as a carbon source.

In nature, bacteria live in an enormous variety of ecological niches, and they show a corresponding richness in their underlying biochemical composition. Two distantly related groups can be recognized: the *eubacteria,* which are the commonly encountered forms that inhabit soil, water, and living organisms; and the *archaebacteria,* which are found in such incommodious environments as bogs, ocean depths, salt brines, and hot acid springs (Figure 1–14).

There exist species of bacteria that can utilize virtually any type of organic molecule as food, including sugars, amino acids, fats, hydrocarbons, polypeptides, and polysaccharides. Some are even able to obtain their carbon atoms from CO_2 and their nitrogen atoms from N_2. Despite their relative simplicity, bacteria have survived for longer than any other organisms and still constitute the most abundant type of cell on earth.

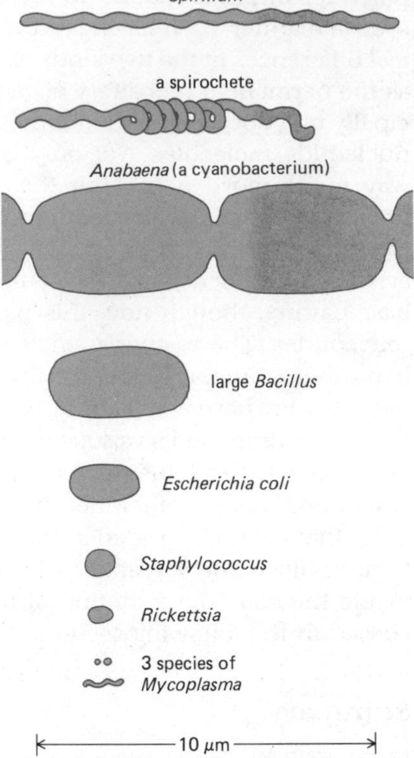

Figure 1–13 Some procaryotic cells drawn to scale.

Metabolic Reactions Evolve

A bacterium growing in a salt solution containing a single type of carbon source, such as glucose, must carry out a large number of chemical reactions. Not only must it derive from the glucose the chemical energy needed for many vital processes, it must also use the carbon atoms of glucose to synthesize every type of organic molecule that the cell requires. These reactions are catalyzed by hundreds of enzymes working in reaction "chains" so that the product of one reaction is the substrate for the next; such enzymatic chains, called *metabolic pathways,* will be discussed in the following chapter.

Originally, when life began on earth, there was probably little need for such metabolic reactions. Cells could survive and grow on the molecules in their surroundings—a legacy from the prebiotic soup. As these natural resources became exhausted, organisms that had developed enzymes to make

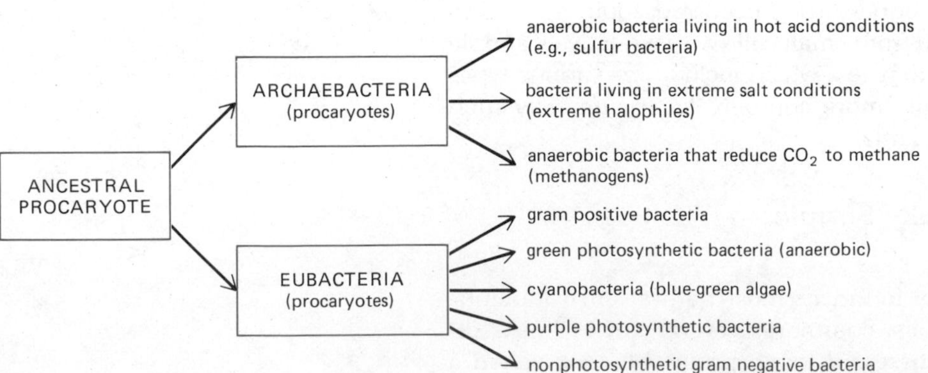

Figure 1–14 Family relationships between present-day bacteria (arrows indicate probable paths of evolution). The origin of eucaryotic cells is discussed later in the text.

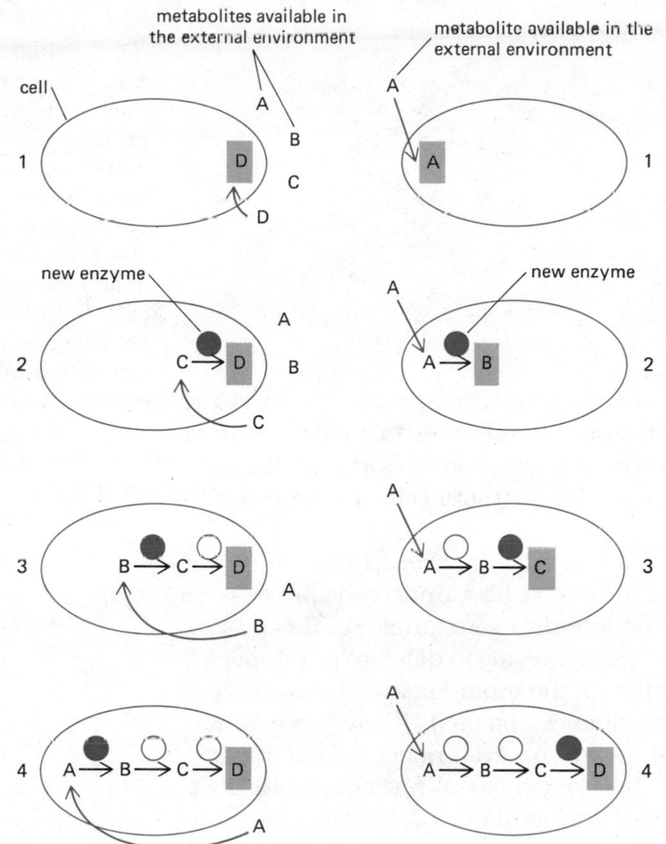

Figure 1–15 Schematic drawing showing two possible ways in which metabolic pathways might have evolved. The cell on the left is provided with a supply of related substances (A, B, C, D) produced by prebiotic synthesis. One of these, substance D, is metabolically useful. As the cell exhausts the available supply of D, a selective advantage is obtained by the evolution of a new enzyme that is able to produce D from the closely related substance C. By a series of similar steps, fundamentally important metabolic pathways may have evolved. On the right, a metabolically useful compound A is available in abundance. An enzyme appears in the course of evolution that, by chance, has the ability to convert substance A to substance B. Other changes then occur within the cell that enable it to make use of the new substance. The appearance of further enzymes can build up a long chain of reactions.

more organic molecules had a strong selective advantage. In this way, the complement of enzymes possessed by cells is thought to have gradually increased, generating the metabolic pathways of present organisms. Two plausible ways in which a metabolic pathway could arise in evolution are illustrated in Figure 1–15.

If metabolic reactions evolved by the sequential addition of new enzymatic reactions to existing ones, the most ancient reactions should, like the oldest rings in a tree trunk, be closest to the center of the "metabolic tree," where the most fundamental of the basic molecular building blocks are synthesized. This position in metabolism is firmly occupied by the transitions involving sugar phosphates, among which the most centrally placed of all is probably the sequence of reactions known as **glycolysis,** by which glucose can be degraded in the absence of oxygen (that is, *anaerobically*). The oldest metabolic pathways would have had to be anaerobic because there was no oxygen in the atmosphere of the primitive earth. Glycolysis occurs in virtually every living cell and drives the formation of the compound *adenosine triphosphate,* or *ATP,* which is used by all cells as a source of readily available chemical energy.

Connecting to the centrally placed reactions of sugar phosphates are hundreds of other chemical reactions. Some of these are responsible for the synthesis of small molecules, many of which in turn are utilized in further reactions to make the large polymers specific to the organism. Other reactions are used to degrade complex molecules, taken in as food, into simpler chemical units. One of the most striking features of these metabolic reactions is that they take place in all kinds of organisms. Certainly differences exist: the amino acid lysine is made in different ways in bacteria, in yeasts, and in green plants, and is not made at all in higher animals; and many specialized prod-

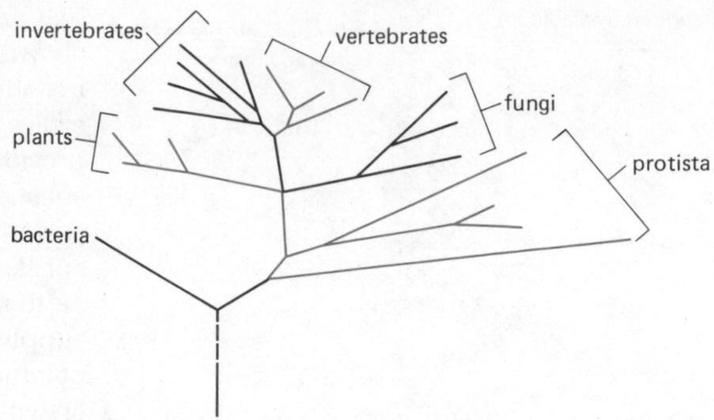

Figure 1–16 Evolutionary relationships of organisms deduced from the amino acid sequences of their cytochrome c (a protein involved in respiration). Each terminal branch of the tree represents a different species, and the total length of branches connecting any two species is proportional to the number of amino acids by which their cytochrome c differs. The evolutionary tree obtained in this way closely resembles that based on evidence from anatomical structures and the fossil record. (From M. O. Dayhoff and R. M. Schwartz. *Ann. N.Y. Acad. Sci.* 361:92–104, 1981.)

ucts of metabolism are restricted to certain genera or species. But in broad terms the majority of reactions and most of the enzymes that catalyze them are found in all living things, from bacteria to man; for this reason they are believed to have been present in the primitive ancestral cells that gave rise to all of these organisms.

The enzymes that catalyze the fundamental metabolic reactions, while continuing to serve the same essential functions, have undergone progressive modifications as organisms have evolved into divergent forms. For this reason, the amino acid sequence of the same type of enzyme in different living species provides an extremely valuable indication of the evolutionary relationship between these species (Figure 1–16). The evidence obtained closely parallels that from other sources, such as the fossil record. An even richer source of information is locked in the living cell in the sequences of nucleotides in DNA. Recently developed methods of analysis enable these *DNA sequences* to be determined in large numbers and compared between species; it is expected that they will enable the course of evolution to be followed with unprecedented accuracy.

Cyanobacteria Can Fix CO_2 and N_2

If the earliest metabolic steps evolved to fill the gaps in the supply of organic molecules from earlier prebiotic synthesis, what happened when such compounds were exhausted? A strong selective advantage would then belong to those organisms able to utilize carbon and nitrogen atoms (in the form of CO_2 and N_2) from the atmosphere. But while they are abundantly available, CO_2 and N_2 are also very stable. It therefore requires a large amount of energy as well as a number of complicated chemical reactions to convert them to a usable form—that is, into organic molecules, such as simple sugars.

In the case of CO_2, the mechanism that evolved to achieve this transformation was **photosynthesis,** in which radiant energy captured from the sun drives the conversion of CO_2 into organic compounds. The interaction of sunlight with a pigment molecule, *chlorophyll,* excites an electron to a more highly energized state. As the electron drops back to a lower energy level, the energy it gives up drives chemical reactions that are facilitated and directed by protein molecules.

One of the first sunlight-driven reactions was probably the phosphorylation of nucleotides to form the high-energy compound ATP. Another would have been the generation of "reducing power." The carbon and nitrogen atoms in atmospheric CO_2 and N_2 are in an oxidized and inert state. One way to make them more reactive, so that they participate in biosynthetic reactions, is to reduce them, that is, to give them a larger charge of electrons. In the process of reduction, electrons are removed from poor electron donors and transferred to a strong electron donor by chlorophyll in a reaction that requires light; the strong electron donor is then used to reduce CO_2 or N_2.

Comparison of the mechanisms of photosynthesis in various present-day bacteria suggests that one of the first sources of electrons was H_2S, from which the primary waste product would have been elemental sulfur. Much later the more difficult but ultimately more rewarding process of obtaining electrons from H_2O was accomplished, and O_2 began to accumulate in the earth's atmosphere as a waste product.

Cyanobacteria (also known as blue-green algae) are today a major route by which both carbon and nitrogen are converted into organic molecules and thus enter the *biosphere.* They include the most self-sufficient organisms that now exist. Able to "fix" both CO_2 and N_2 into organic molecules, they are, to a first approximation, able to live on water and air alone; the mechanisms by which they do this have probably remained essentially constant for over a billion years.

Bacteria Can Carry Out the Aerobic Oxidation of Food Molecules

Many people today are justly concerned about the environmental consequences of human activities. But in the past other organisms have caused revolutionary changes in the earth's environment (although very much more slowly). Nowhere is this more apparent than in the composition of the earth's atmosphere, which with the advent of photosynthesis has been transformed from a mixture containing practically no molecular oxygen to one in which oxygen represents 21% of the total.

Since oxygen is an extremely reactive chemical that can interact with most cytoplasmic constituents, it was probably toxic to many early organisms, just as it is to many present-day anaerobic bacteria. However, this reactivity also provides a source of chemical energy, and, not surprisingly, this has been exploited by organisms during the course of evolution. By using oxygen, organisms are able to oxidize more completely the molecules they ingest. For example, in the presence of oxygen glucose can be completely degraded to CO_2 and H_2O, while in the absence of oxygen it can be broken down only to lactic acid or ethanol, the end products of anaerobic glycolysis. In this way much more energy can be derived from each gram of glucose. The energy release in the aerobic oxidation of food molecules—usually called **respiration**—is used to drive the synthesis of ATP in much the same way that photosynthetic organisms produce ATP from the energy of sunlight. In both processes there is a series of electron transfer reactions that generates a H^+ gradient between the outside and inside of a tiny membrane-bounded compartment; the H^+ gradient then serves to drive the synthesis of the ATP. Today, respiration is used by the great majority of organisms, including most procaryotes.

Eucaryotic Cells Contain Several Distinctive Organelles

As molecular oxygen accumulated in the atmosphere, what happened to the remaining anaerobic organisms with which life had begun? In a world that was rich in oxygen, which they could not use, they were at a severe disadvantage. Some, no doubt, became extinct. Others either developed a capacity for respiration or found niches from which oxygen was largely absent, where they could continue an anaerobic way of life. It seems, however, that a third class discovered a strategy for survival more cunning, and vastly richer in implications for the future: they are believed to have formed an intimate association with an aerobic type of cell, living with it in *symbiosis.* This is the most plausible explanation for the origin of present-day cells of the **eucaryotic** type (Panel A), with which this book will be chiefly concerned.

ANIMAL CELL

thin section of a generalized animal cell

PLANT CELL

thin section of a generalized cell from a higher plant

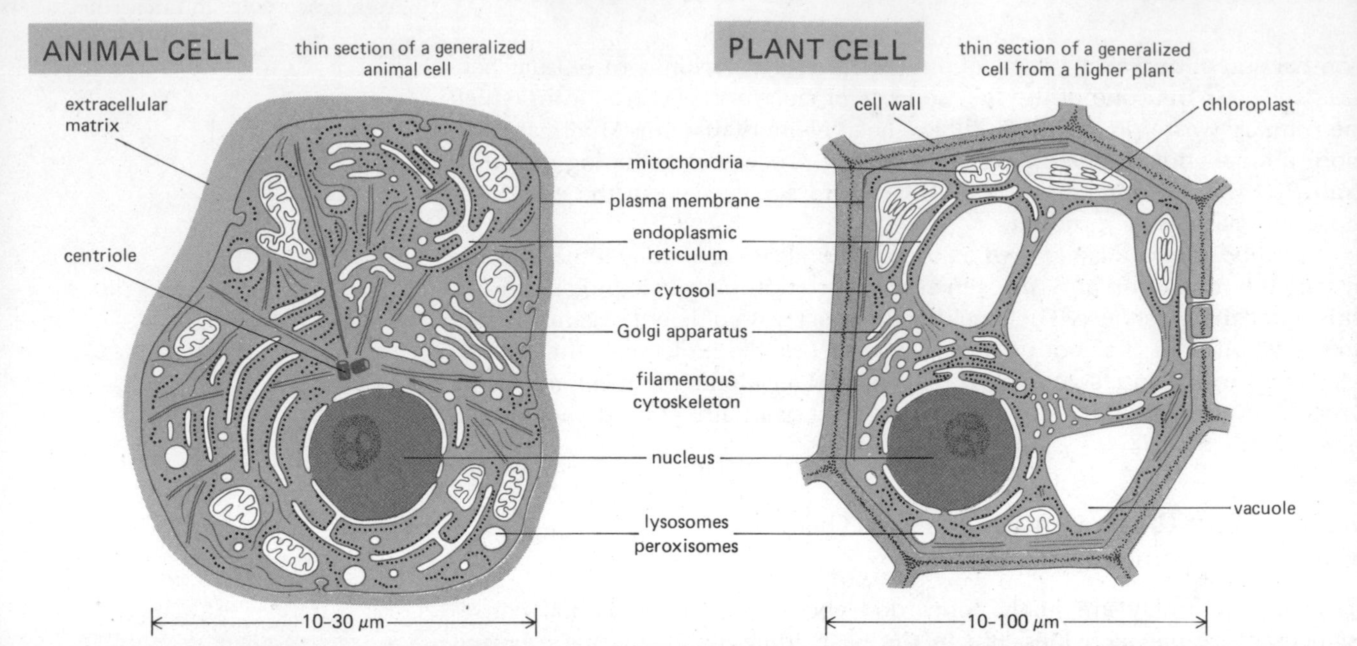

THE MEMBRANE SYSTEM OF THE CELL

PLASMA MEMBRANE

The outer boundary of the cell is the plasma membrane, a continuous sheet of lipid molecules about 4–5 nm thick in which various proteins are embedded.

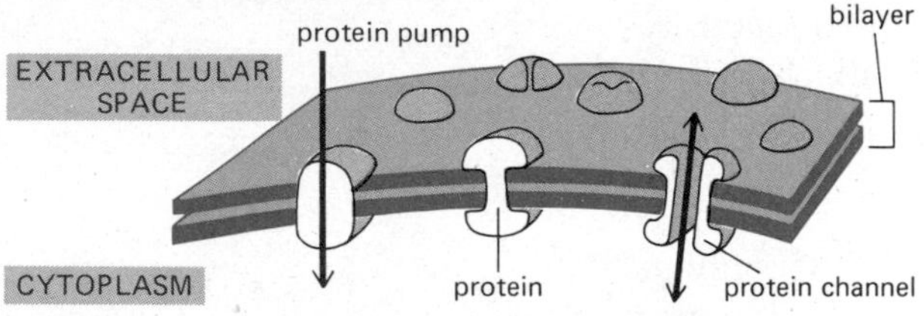

Some of these proteins serve as pumps and channels for transporting specific molecules into and out of the cell.

GOLGI APPARATUS

A system of stacked, membrane-bounded, flattened sacs involved in modifying, sorting, and packaging macromolecules for secretion or for delivery to other organelles.

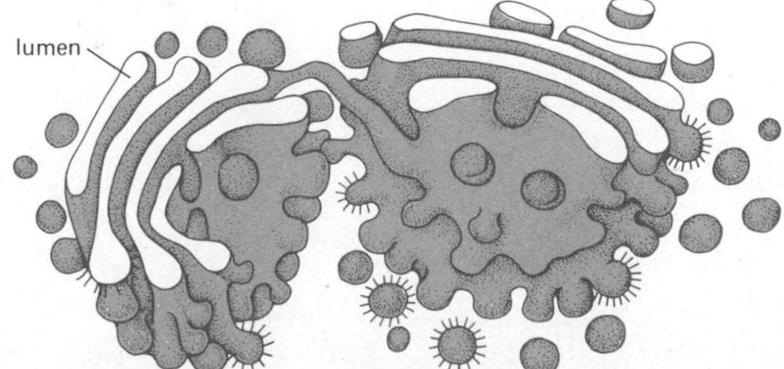

Around the Golgi apparatus are numerous small membrane-bounded vesicles (50 nm and larger). These are thought to carry material between the Golgi apparatus and different compartments of the cell.

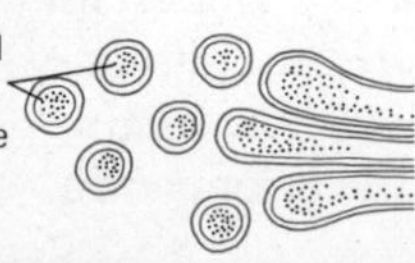

ENDOPLASMIC RETICULUM

Flattened sheets, sacs, and tubes of membrane extend throughout the cytoplasm of eucaryotic cells, enclosing a large intracellular space. The ER membrane is in structural continuity with the outer membrane of the nuclear envelope and it specializes in the synthesis and transport of lipids and membrane proteins.

The rough endoplasmic reticulum (rough ER) generally occurs as flattened sheets and is studded on its outer face with ribosomes engaged in protein synthesis.

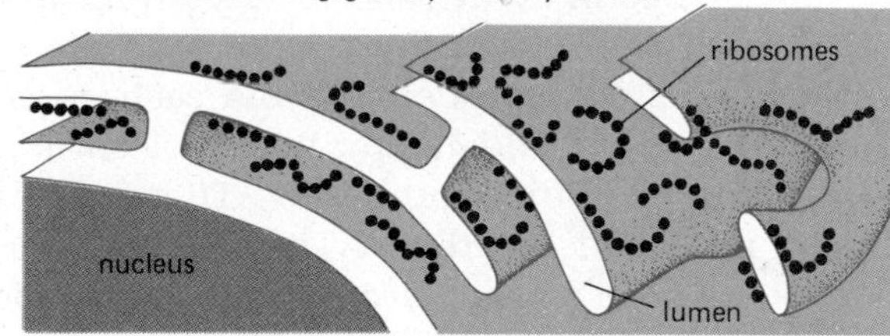

The smooth endoplasmic reticulum (smooth ER) is generally more tubular and lacks attached ribosomes. A major function is in lipid metabolism.

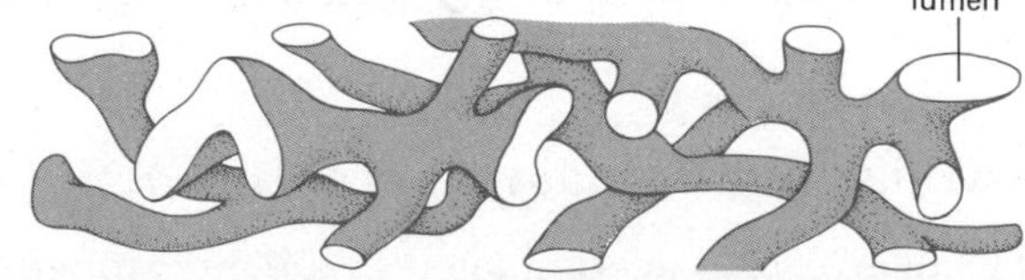

LYSOSOMES

membrane-bounded vesicles that contain hydrolytic enzymes involved in intracellular digestions

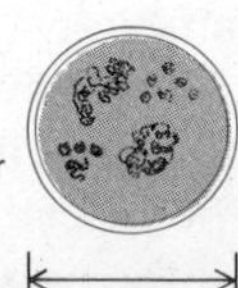

PEROXISOMES

membrane-bounded vesicles containing oxidative enzymes that generate and destroy hydrogen peroxide

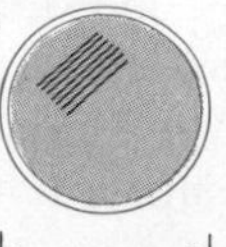

Panel A Eucaryotic cells: a survey of their principal organelles.

NUCLEUS

The nucleus is the most conspicuous organelle in the cell. It is separated from the cytoplasm by an envelope consisting of 2 membranes. All of the chromosomal DNA is held in the nucleus, packaged into chromatin fibers by its association with an equal mass of histone proteins. The nuclear contents communicate with the cytosol by means of openings in the nuclear envelope called nuclear pores.

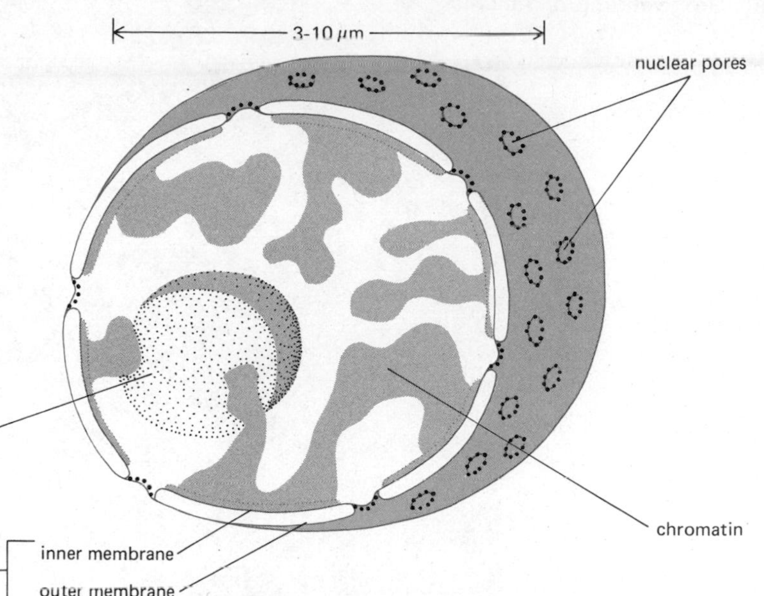

CYTOSKELETON

In the cytosol, arrays of protein filaments form networks that give the cell its shape and provide a basis for its movements. In animal cells the cytoskeleton is often organized from an area near the nucleus that contains the cell's pair of centrioles. 3 main kinds of cytoskeletal filaments are:

1. microtubules

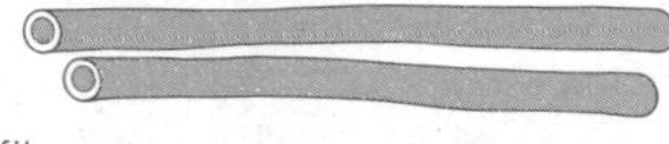

25-nm diameter

2. actin filaments

7-nm diameter

3. intermediate filaments

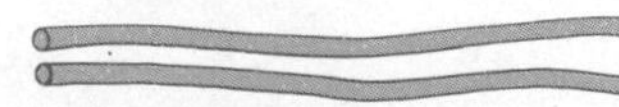

10-nm diameter

MITOCHONDRIA

About the size of bacteria, mitochondria are the power plants of all eucaryotic cells, harnessing energy obtained by combining oxygen with food molecules to make ATP.

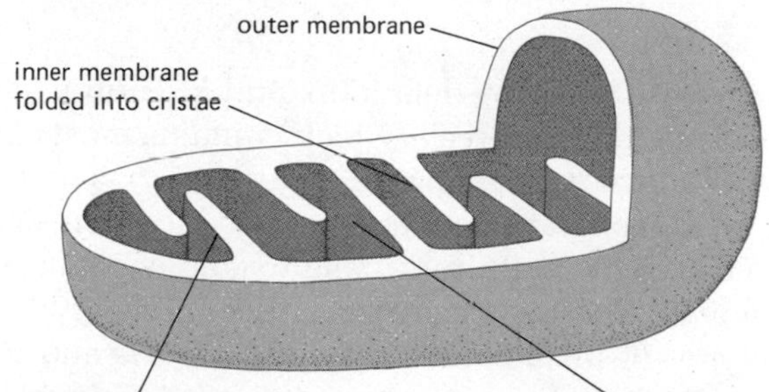

SPECIAL PLANT CELL ORGANELLES

chloroplasts—These chlorophyll-containing plastids are double-membrane bounded organelles found in all higher plants. An elaborate internal membrane system contains the photosynthetic apparatus.

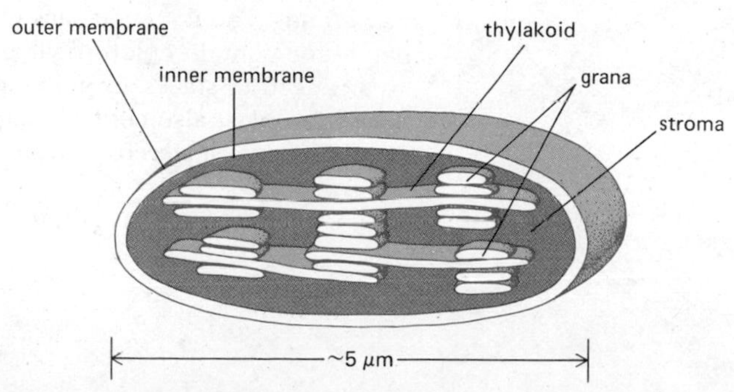

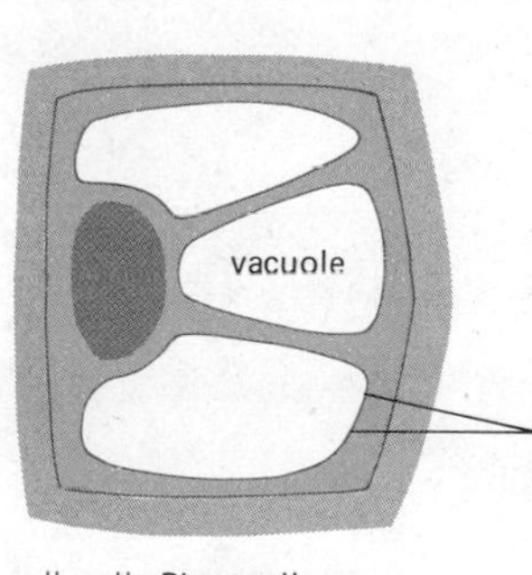

vacuole—A very large single-membrane-bounded vesicle occupying up to 90% of the cell volume, the vacuole functions in space-filling and also in intracellular digestion.

cell wall—Plant cells are surrounded by a rigid wall composed of tough fibrils of cellulose laid down in a matrix of other polysaccharides.

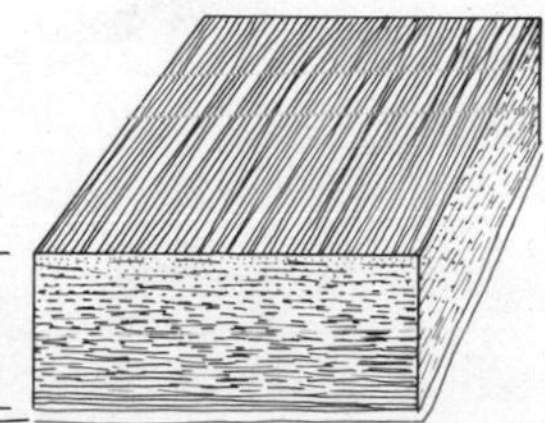

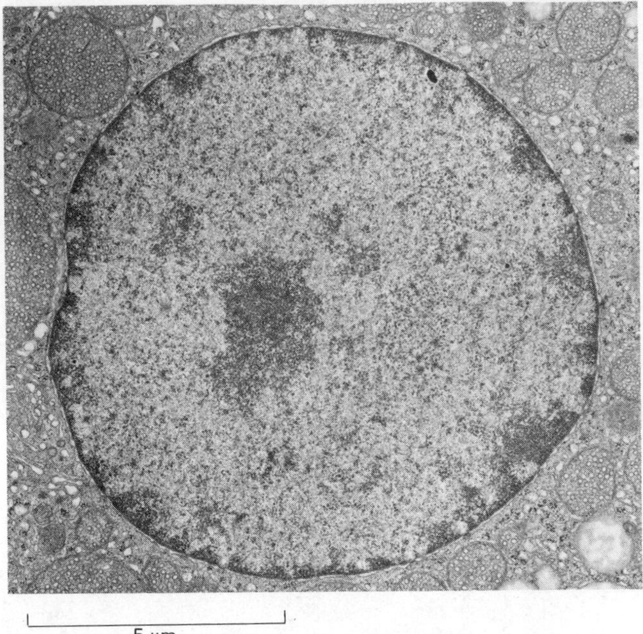

Figure 1–17 The nucleus contains most of the DNA of the eucaryotic cell. It is seen here in a thin section of a mammalian cell examined in the electron microscope. (Courtesy of Daniel S. Friend.)

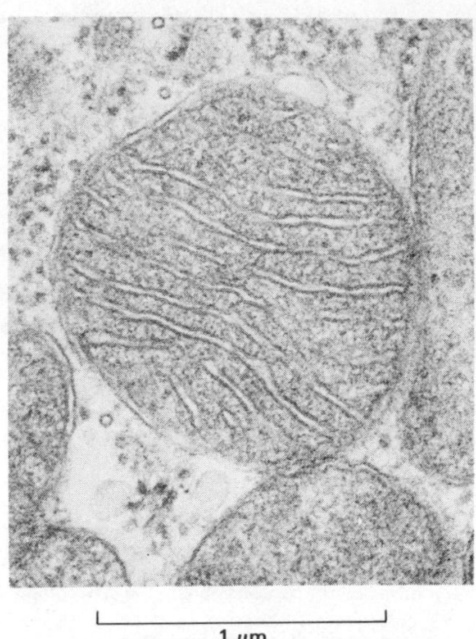

Figure 1–18 Mitochondria carry out the oxidative degradation of nutrient molecules in all eucaryotic cells. As seen in this electron micrograph, they possess a smooth outer membrane and a highly convoluted inner membrane. (Courtesy of Daniel S. Friend.)

Eucaryotic cells, by definition, and in contrast to procaryotic cells, have a *nucleus* ("caryon" in Greek), which contains most of the cell's DNA, enclosed by a double layer of membrane (Figure 1–17). The DNA is thereby kept in a compartment separate from the rest of the contents of the cell, the *cytoplasm*, where most of the cell's metabolic reactions occur. In the cytoplasm, moreover, many distinctive *organelles* can be recognized. Prominent among these are two types of small bodies, the *mitochondria* and *chloroplasts* (Figures 1–18 and 1–19). Each of these is enclosed in its own double layer of membrane that is chemically different from the membrane surrounding the nucleus. Mitochondria are an almost universal feature of eucaryotic cells, while chloroplasts are found only in those eucaryotic cells that are capable of photosynthesis—that is, in plants but not in animals or fungi. Both organelles are thought to have a symbiotic origin.

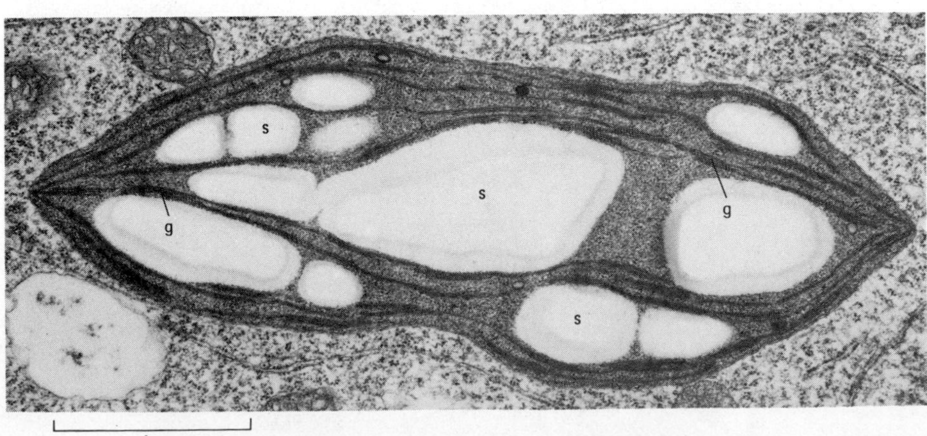

Figure 1–19 Electron micrograph of a chloroplast in a moss cell showing its extensive system of internal membranes. The flattened sacs of membrane contain chlorophyll and are arranged in stacks, or *grana* (g). This chloroplast also contains large accumulations of starch (s). (Courtesy of J. Burgess.)

Eucaryotic Cells Depend on Mitochondria for Their Oxidative Metabolism

Mitochondria show many similarities to free-living procaryotic organisms: for example, they often resemble bacteria in size and shape, they contain DNA, and they reproduce by dividing in two. By breaking up eucaryotic cells and separating their component parts, it is possible to show that mitochondria are responsible for respiration and that this process occurs nowhere else in the eucaryotic cell. Without mitochondria, the cells of animal and fungi would be anaerobic organisms, depending on the relatively inefficient and antique process of glycolysis for their energy. Many present-day bacteria can respire, and the mechanism by which they do so bears an unmistakable resemblance to that in mitochondria.

It seems probable, therefore, that eucaryotic cells are descendants of primitive anaerobic organisms, which survived in a world that had become rich in oxygen by engulfing aerobic bacteria. Rather than digest the bacteria, they nourished them and maintained them in symbiosis for the sake of their capacity to consume atmospheric oxygen and produce energy, just as we keep cows for their capacity to consume grass and produce milk. Of course, we cannot prove absolutely that this is what did happen, but certain present-day microorganisms provide strong evidence of the feasibility of such an evolutionary sequence: for example, an exceptional eucaryotic organism, the amoeba *Pelomyxa palustris,* lacks mitochondria and instead harbors aerobic bacteria in a permanent symbiotic relationship.

Chloroplasts May Be Descendants of Procaryotic Algae

Chloroplasts carry out photosynthesis in much the same way as procaryotic cyanobacteria, absorbing sunlight in the chlorophyll that is attached to their membranes. Some bear a close structural resemblance to the cyanobacteria, being similar in size and in the way that their chlorophyll-bearing membranes are stacked in layers (Figure 1–19). Moreover, chloroplasts reproduce by dividing and contain DNA. All this strongly suggests that chloroplasts have evolved from cyanobacteria that made their home inside eucaryotic cells, performing photosynthesis for their hosts in return for the sheltered and nourishing environment that their hosts provided for them. Symbiosis of photosynthetic cells with other cell types is, in fact, a common phenomenon, and a number of present-day eucaryotic cells can be observed to contain authentic cyanobacteria (Figure 1–20).

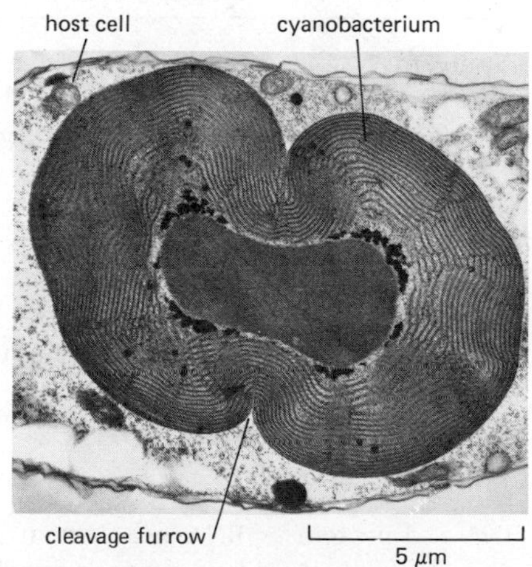

Figure 1–20 A close relative of present-day cyanobacteria that lives in a permanent symbiotic relationship inside another cell (the two organisms are known jointly as *Cyanophora paradoxa*). The "cyanobacterium" is undergoing cleavage. (Courtesy of Jeremy D. Pickett-Heaps.)

Figure 1–21 shows the evolutionary origins of the eucaryotes according to the symbiotic theory. It must be stressed, however, that mitochondria and chloroplasts show important differences from, as well as similarities to, present-day aerobic bacteria and cyanobacteria. Their quantity of DNA is very small, for example, and most of the molecules from which they are constructed are synthesized elsewhere in the eucaryotic cell and imported into the organelle. Assuming that they did originate as symbiotic bacteria, they have undergone large evolutionary changes and have become greatly dependent on their hosts.

Was the acquisition of mitochondria by some primitive anaerobic cell the crucial step in the genesis of the eucaryotes, bringing in its wake the evolution of their other special characteristics? We lack evidence to answer this question: existing eucaryotes have in common not only mitochondria but also a whole constellation of other features that distinguish them from procaryotes (Table 1–1). These function together to give eucaryotic cells a wealth of different capabilities, and it is impossible to say which of them evolved first.

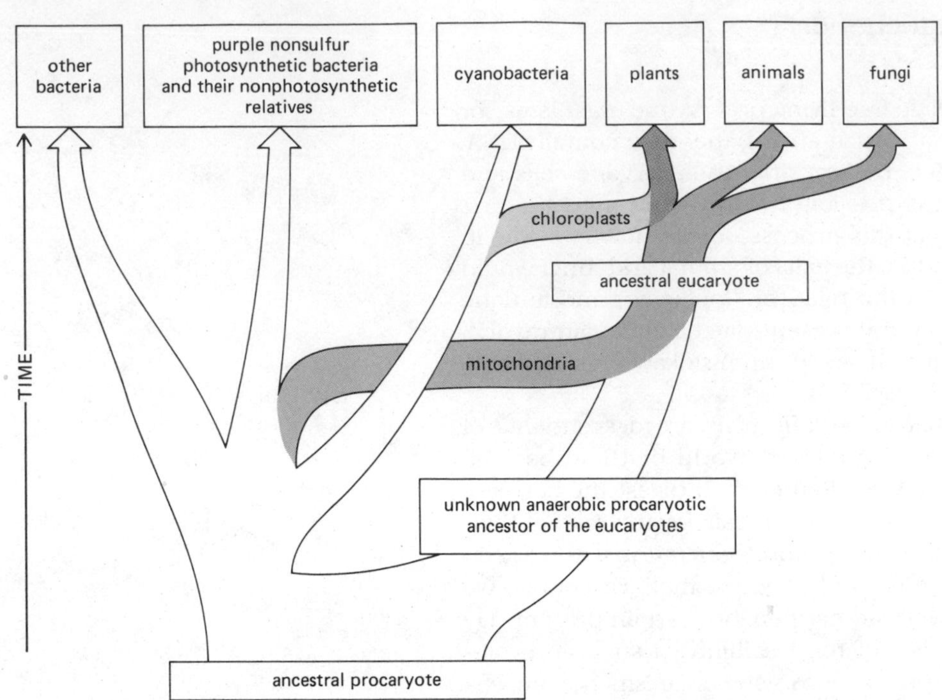

Figure 1–21 The postulated origin of the eucaryotes by symbiosis of aerobic with anaerobic procaryotes.

Table 1–1 Comparison of Procaryotic and Eucaryotic Organisms

	Procaryotes	**Eucaryotes**
Organisms	bacteria and cyanobacteria	protists, fungi, plants, and animals
Cell size	generally 1 to 10 μm in linear dimension	generally 10 to 100 μm in linear dimension
Metabolism	anaerobic or aerobic	aerobic
Organelles	few or none	nucleus, mitochondria, chloroplasts, endoplasmic reticulum, etc.
DNA	circular DNA in cytoplasm	very long DNA containing many noncoding regions; organized into chromosomes and bounded by nuclear envelope
RNA and protein	RNA and protein synthesized in same compartment	RNA synthesized and processed in nucleus; proteins synthesized in cytoplasm
Cytoplasm	no cytoskeleton, cytoplasmic streaming, endocytosis, or exocytosis	cytoskeleton composed of protein filaments; cytoplasmic streaming; endocytosis and exocytosis
Cell division	by binary fission	by mitosis (or meiosis)
Cellular organization	mainly unicellular	mainly multicellular, with differentiation of cells

Eucaryotic Cells Contain a Rich Array of Internal Membranes

Eucaryotic cells are usually much larger in volume than procaryotic cells, commonly by a factor of a thousand or more, and they carry a proportionately larger quantity of most cellular materials; for example, a human cell contains about 800 times as much DNA as a typical bacterium. This large size creates problems. Since all the raw materials for the biosynthetic reactions occurring in the interior of a cell must ultimately enter and leave by passing through the plasma membrane that covers its surface, and since the membrane is also the site of many important reactions, an increase of cell volume requires an increase of cell surface. But it is a fact of geometry that a simple scaling up of a structure increases the volume as the cube of the linear dimension, while the surface area is increased only as the square. Therefore, if the large eucaryotic cell is to keep as high a proportion of surface to volume as the procaryotic cell, it must supplement its surface area by means of convolutions, infoldings, and other elaborations of its membrane.

This probably explains in part the complex profusion of **internal membranes** that is a basic feature of all eucaryotic cells. Membranes surround the nucleus, the mitochondria, and (in plant cells) the chloroplasts. They form a labyrinthine compartment called the **endoplasmic reticulum** (Figure 1–22) where lipids and proteins of cell membranes as well as material destined for export from the cell are synthesized. They also form stacks of flattened sacs constituting the **Golgi apparatus** (Figure 1–23), which is likewise involved in the synthesis and transport of various organic molecules. Membranes surround **lysosomes,** in which stores of enzymes required for purposes of intracellular digestion are contained and so prevented from attacking the proteins and nucleic acids of the cell itself. In the same way, membranes surround **peroxisomes,** where dangerously reactive peroxides are generated and degraded. They also form small vesicles and, in plants, a large liquid-filled *vacuole.* All these membrane-bounded structures correspond to distinct internal compartments within the cytoplasm. In a typical animal cell, these compartments occupy nearly half of the total cell volume. The remaining compartment of the cytoplasm, which includes everything other than the membrane-bounded organelles, is usually referred to as the **cytosol.**

All of the membranous structures in our list lie in the interior of the cell. How, then, can they help to solve the problem we posed at the outset and provide the cell with a surface area that is adequate to its large volume? The answer depends on exchange between the internal membrane-bounded compartments and the outside of the cell. This is achieved by *endocytosis* and *exocytosis,* processes unique to eucaryotic cells. In endocytosis, portions of the external surface membrane invaginate and pinch off to form membrane-bounded cytoplasmic vesicles containing substances that were present in the external medium or were adsorbed onto the cell surface. Exocytosis is the reverse process, whereby membrane-bounded vesicles inside the cell fuse with the plasma membrane and release their contents into the external medium. In this way, membranes surrounding compartments deep inside the cell serve to increase the effective surface area of the cell for exchanges of matter with the external world.

As we shall see in later chapters, the various membranes and membrane-bounded compartments in eucaryotic cells have become highly specialized, some for secretion, some for absorption, some for specific biosynthetic processes, and so on.

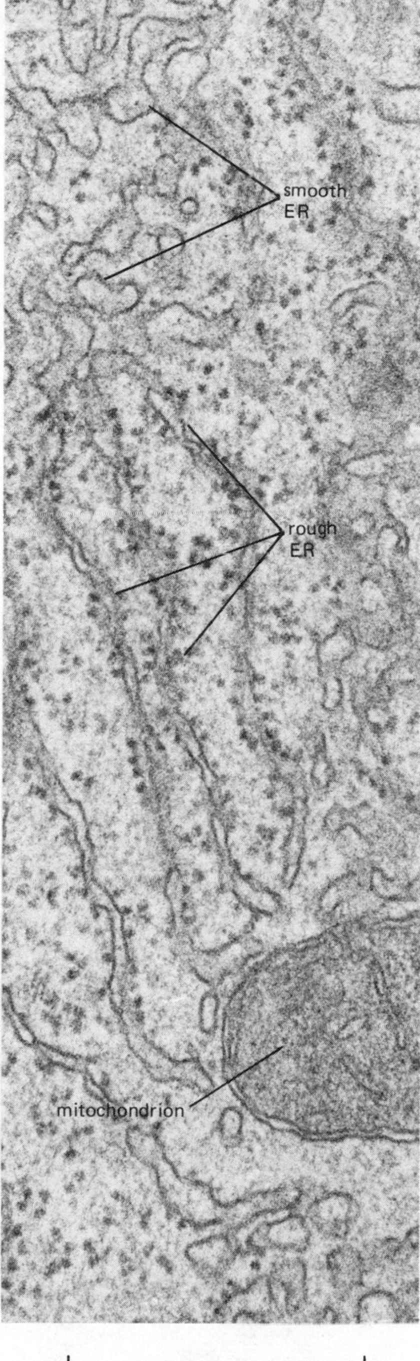

Figure 1–22 Electron micrograph of a thin section of a mammalian cell showing both smooth and rough regions of the endoplasmic reticulum (ER). (Courtesy of George Palade.)

Eucaryotic Cells Have a Cytoskeleton

The larger a cell is, and the more elaborate and specialized its internal structures, the greater its need to keep these structures in their proper places and

Figure 1–23 Electron micrograph of a thin section of a mammalian cell showing the Golgi apparatus, which is composed of flattened sacs of membrane arranged in multiple layers (see also Panel A, pages 16–17). The Golgi apparatus is involved in the synthesis and packaging of material destined to be secreted from the cell, as well as in the routing of newly synthesized proteins to the correct cellular compartment. (Courtesy of Daniel S. Friend.)

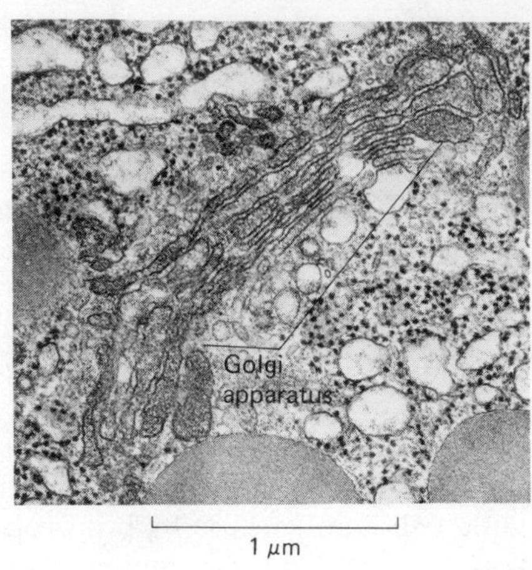

to control their movements. All eucaryotic cells have an internal skeleton, the **cytoskeleton,** that gives the cell its shape, its capacity to move, and its ability to arrange its organelles and transport them from one part of the cell to another. The cytoskeleton is composed of a network of protein filaments, two of the most important of which are *actin filaments* and *microtubules* (Figure 1–24). These two must date from a very early epoch in evolution since they are found almost unchanged in all eucaryotes. Both are involved in the generation of cellular movements; actin filaments, for example, participate in the contraction of muscle, while microtubules are the main structural and force-generating elements in *cilia* and *flagella*—the long projections on some cell surfaces, which beat like whips and serve as instruments of propulsion.

Actin filaments and microtubules are also essential for the internal movements that occur in the cytoplasm of all eucaryotic cells. Thus microtubules in the form of a *mitotic spindle* are a vital part of the usual machinery for partitioning DNA equally between the two daughter cells when a eucaryotic cell divides. Without microtubules, therefore, the eucaryotic cell could not reproduce. In this and other examples, movement by free diffusion would be either too slow or too haphazard to be useful. In fact, it has been suggested that most of the organelles in a eucaryotic cell are attached, directly or indirectly, to the cytoskeleton and that the only way they are able to move is along cytoskeletal tracks by an energy-requiring transport process.

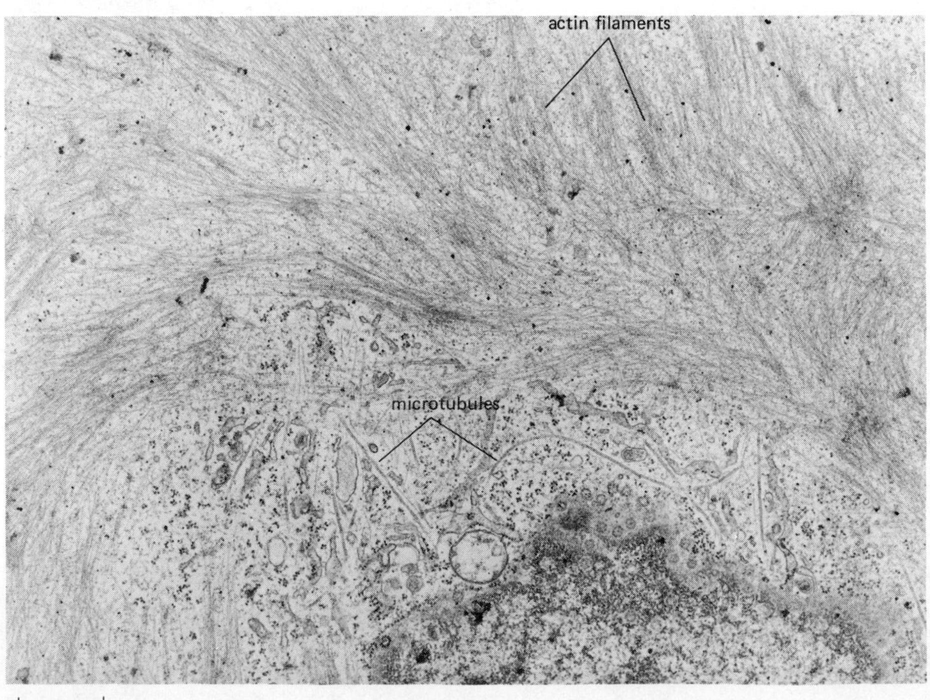

Figure 1–24 Microtubules and actin filaments, two prominent components of the cytoskeleton, are seen in this electron micrograph of an animal cell. (From B. S. Spooner, *Bioscience* 25:440–451, 1975. Copyright 1975 by the American Institute of Biological Sciences. Reprinted by permission of the copyright holder.)

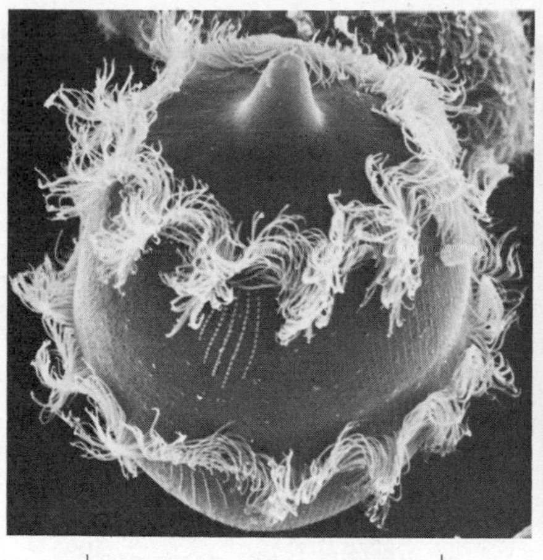

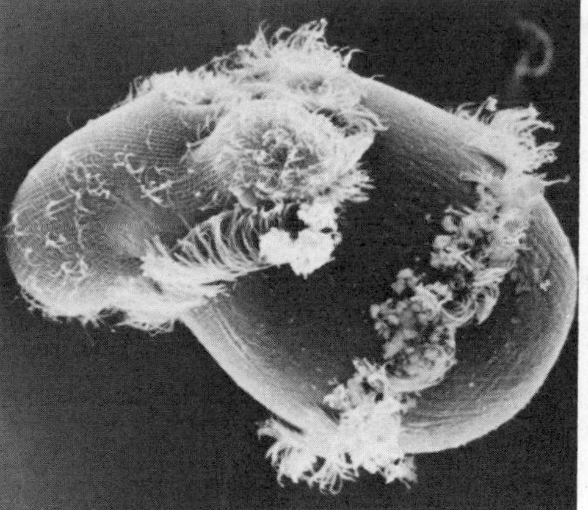

Figure 1–25 Scanning electron micrograph showing one protozoan eating another. Protozoans are single-cell animals that show an amazing diversity of form and behavior. *Didinium* (*left*), a ciliated protozoan, has two circumferential rings of motile cilia and a snoutlike protuberance at its leading end, with which it captures its prey. On the right, *Didinium* is shown engulfing another protozoan, *Paramecium*. (Courtesy of D. Barlow.)

Protozoa Include the Most Complex Cells Known

The complexity that can be achieved by a single eucaryotic cell is nowhere better illustrated than in *protists*. These are free-living, single-celled eucaryotes that exhibit a bewildering variety of different forms and behaviors: they can be photosynthetic or carnivorous, motile or sedentary. Their anatomy is often complex and includes such structures as sensory bristles, photoreceptors, flagella, leglike appendages, mouth parts, stinging darts, and musclelike contractile bundles. Although they are single cells, they can be as intricate and versatile as many multicellular organisms. This is particularly true of the group of protists known as **protozoa**—or "first animals."

Didinium is a carnivorous protozoan. It has a globular body, about 150 μm in diameter, encircled by two fringes of cilia; its front end is flattened except for a single protrusion rather like a snout (Figure 1–25). *Didinium* swims around at high speed in the water by means of the synchronous beating of its cilia. When it encounters a suitable prey, usually another type of protozoan, *Paramecium*, it releases numerous small paralyzing darts from its snout region. Then the *Didinium* attaches to and devours the *Paramecium*, inverting like a hollow ball to engulf the other cell, which is as large as itself. Most of this complex behavior—swimming, and paralyzing and capturing its prey—is generated by the cytoskeletal structures lying just beneath the plasma membrane. Included in this *cell cortex*, for example, are the parallel bundles of microtubules that form the core of each cilium and enable it to beat.

But the protozoa, for all their marvels, do not represent the peak of eucaryotic evolution. Greater things were achieved, not by concentrating every sort of complexity in a single cell, but by dividing the labor among different types of cells. *Multicellular organisms* evolved, in which cells closely related by ancestry became differentiated from one another, some developing one feature to a high degree, others another, so forming the specialized parts of one great cooperative enterprise.

Genes Can Be Switched On and Off

The various specialized cell types in a single higher plant or animal often appear radically different (Panel B). This seems paradoxical, since all of the cells in a multicellular organism are closely related, having recently descended from the same precursor cell—the fertilized egg. Common lineage implies similar genes; how then do the differences arise? In a few cases, cell specialization involves the loss of genetic material: an extreme example is the mam-

CELL TYPES

There are over 200 different types of cell in the human body. These are assembled into a variety of different types of tissue such as

- epithelia
- connective tissue
- muscle
- nervous tissue

Most tissues contain a mixture of cell types.

EPITHELIA

Epithelial cells form coherent cell sheets called epithelia, which line the inner and outer surfaces of the body. There are many specialized types of epithelia.

Absorptive cells have numerous hairlike microvilli projecting from their free surface to increase the area for absorption.

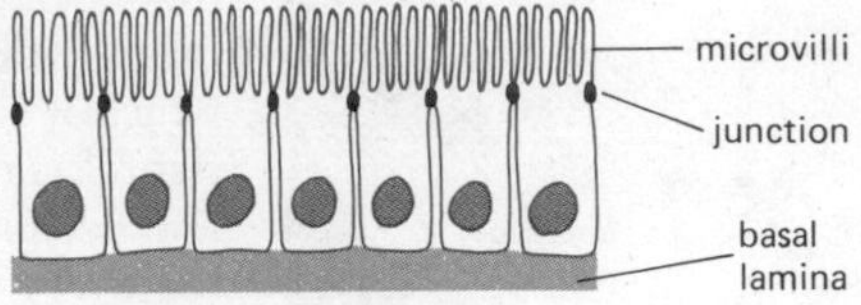

Adjacent epithelial cells are bound together by junctions that give the sheet mechanical strength and also make it impermeable to small molecules. The sheet rests on a basal lamina.

ciliated cells — cilia on their free surface beat in synchrony to move substances (such as mucus) over the epithelial sheet.

secretory cells — most epithelial layers have some cells that secrete substances onto the surface.

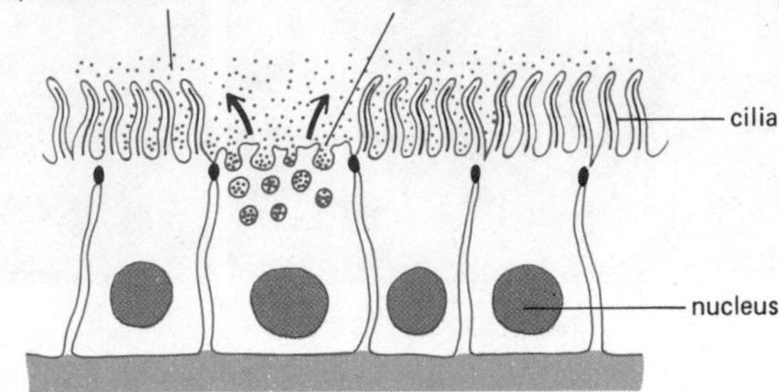

CONNECTIVE TISSUE

The spaces between organs and tissues in the body are filled with connective tissue made principally of a network of tough protein fibers embedded in a polysaccharide gel. This extracellular matrix is secreted mainly by fibroblasts.

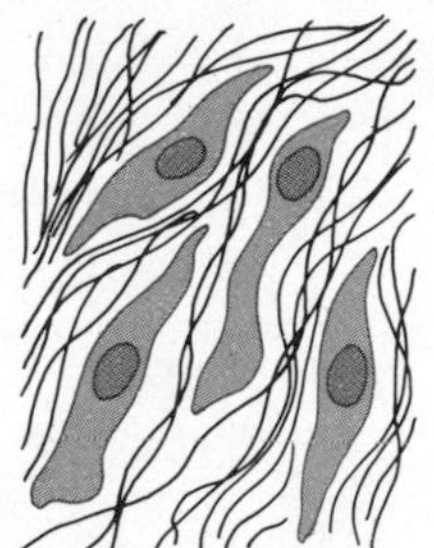

fibroblasts in loose connective tissue

2 main types of extracellular protein fiber are collagen and elastin.

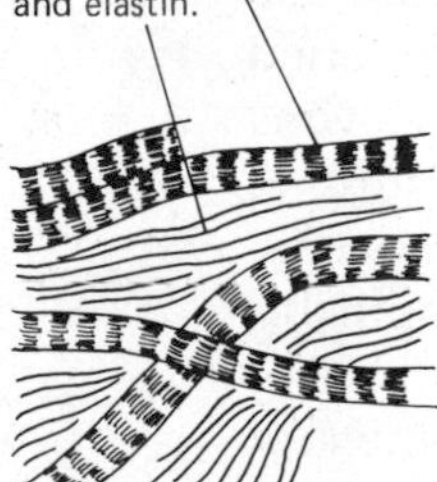

Bone is made by cells called osteoblasts. These secrete an extracellular matrix in which crystals of calcium phosphate are later deposited.

Calcium salts are deposited in the extracellular matrix.

osteoblasts linked together by cell processes

extracellular matrix

Adipose cells are among the largest cells in the body. These cells are responsible for the production and storage of fat. The nucleus and cytoplasm are squeezed to the cell periphery by a large lipid droplet.

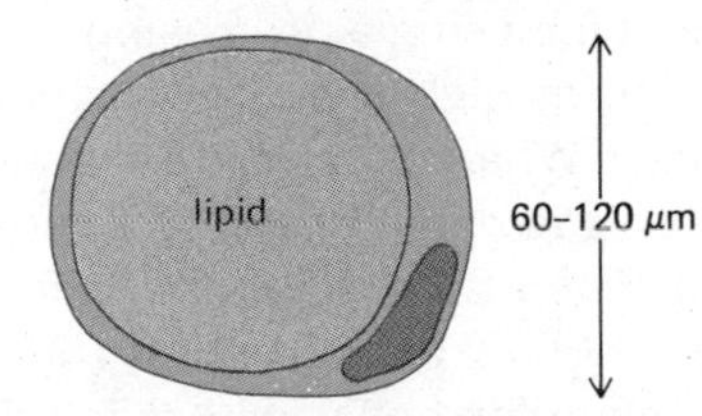

lipid

60–120 μm

NERVOUS TISSUE

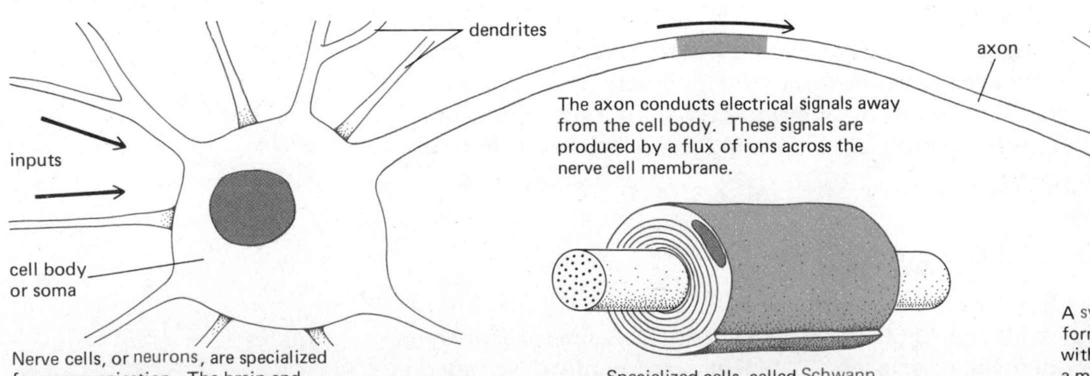

The axon conducts electrical signals away from the cell body. These signals are produced by a flux of ions across the nerve cell membrane.

Nerve cells, or neurons, are specialized for communication. The brain and spinal cord, for example, are composed of a network of neurons among supporting glial cells

Specialized cells, called Schwann cells, or oligodendrocytes, wrap around an axon to form a multilayered membrane sheath.

A synapse is where a neuron forms a specialized junction with another neuron (or with a muscle cell). At synapses, signals pass from one neuron to another (or from a neuron to a muscle cell).

Panel B Some of the different types of cells present in the vertebrate body.

Secretory epithelial cells are often collected together to form a gland that specializes in the secretion of a particular substance. As illustrated, exocrine glands secrete their products (such as tears, mucus, and gastric juices) into ducts. Endocrine glands secrete hormones into the blood.

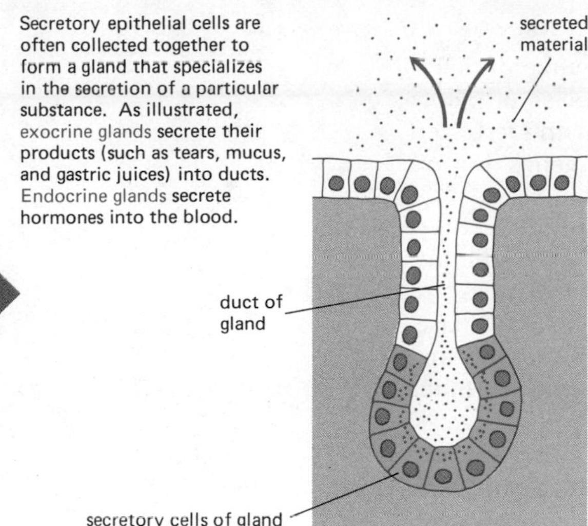

BLOOD

Erythrocytes (or red blood cells) are very small cells with no nucleus or internal membranes, and are stuffed full of the oxygen-binding protein hemoglobin.

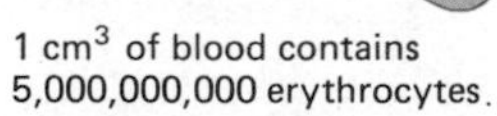

1 cm^3 of blood contains 5,000,000,000 erythrocytes.

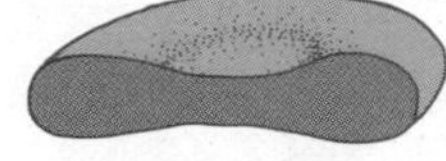

Their normal shape is a biconcave disc.

white blood cells (leucocytes) – there is about 1 leucocyte for every 1000 red blood cells. Although they travel in the circulation, they can pass through the walls of blood vessels to do their work in the surrounding tissues. There are several different kinds, including:

macrophages and neutrophils – these cells move to sites of infection where they ingest bacteria and debris.

wall of small blood vessel

bacterial infection in connective tissue

lymphocytes – responsible for immune responses such as the production of antibody and the rejection of tissue grafts.

GERM CELLS

A sperm from the male fuses with an egg from the female, which then forms a new organism by successive divisions. Both sperm and egg are haploid, i.e., they carry only 1 set of chromosomes.

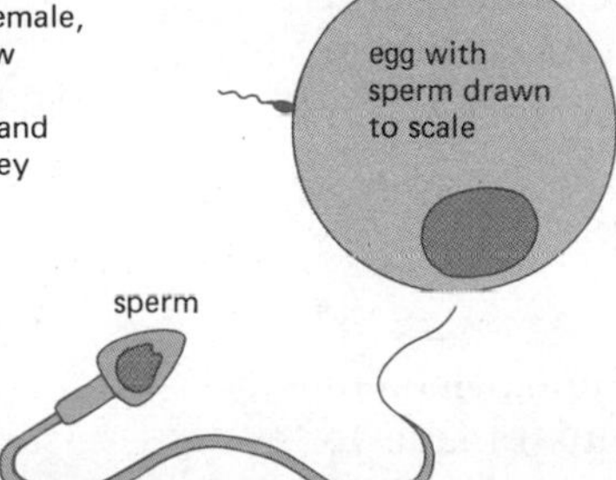

MUSCLE

Muscle cells produce mechanical force by their contraction. In vertebrates there are three main types:

skeletal muscle – this moves joints by its strong and rapid contraction. Each muscle is a bundle of muscle fibers, each of which is an enormous multinucleated cell.

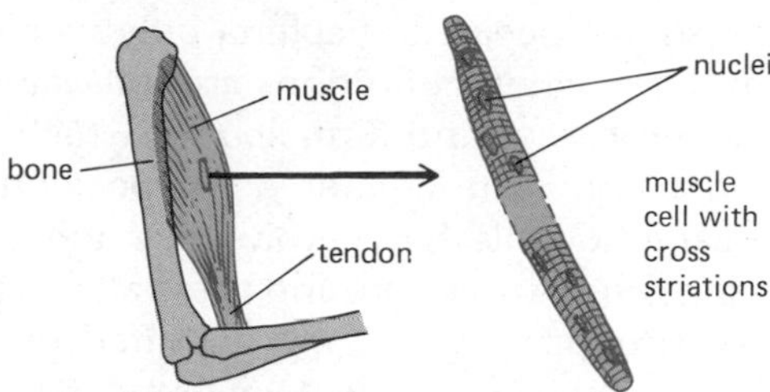

smooth muscle – present in digestive tract, bladder, arteries, and veins, it is composed of thin elongated cells (not striated) each with a single nucleus.

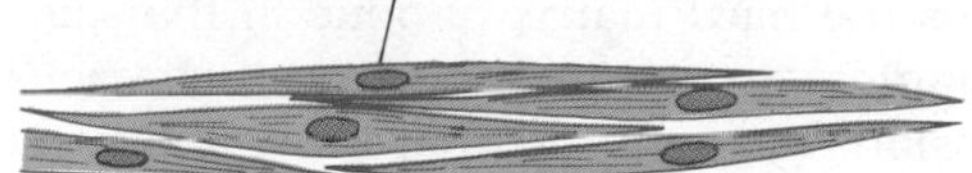

cardiac muscle – intermediate in character between skeletal and smooth muscle, it produces the heart beat. Adjacent cells are linked by electrically conducting junctions that cause the cells to contract in synchrony.

SENSORY CELLS

Among the most complex cells in the vertebrate body are those that detect external stimuli. Hair cells of the inner ear are primary detectors of sound. Modified epithelial cells, they carry special microvilli (stereocilia) on their surface. The movement of these in response to sound vibrations causes an electrical signal to pass to the brain.

stereocilia are very rigid because they are packed with actin filaments

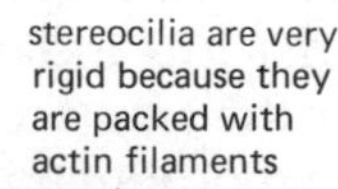

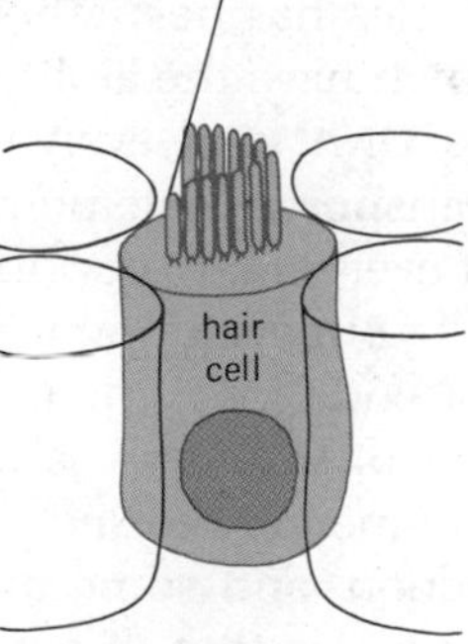

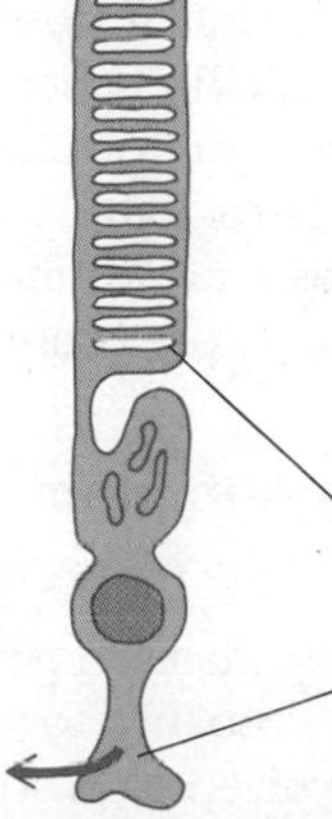

Rod cells in the retina of the eye are nerve cells specialized to respond to light. The photo-sensitive region contains many membranous discs in whose membranes the light-sensitive pigment rhodopsin is embedded. Light causes an electrical signal to pass to other nerve cells.

malian red blood cell, which loses its entire nucleus in the course of differentiation. But the overwhelming majority of cells in most species of plants and animals retain all of the genetic information contained in the fertilized egg. Specialization depends not on the loss or acquisition of genes, but on changes in *gene expression*.

Even bacteria do not make all of their types of protein all of the time, but are able to adjust the level of synthesis according to external conditions. Proteins required specifically for the metabolism of lactose, for example, are made by some species of bacteria only when this sugar is available for use. Other bacteria, when conditions are unfavorable, arrest most of their normal metabolic processes and form *spores*, which have tough, impermeable outer walls and a cytoplasm of altered composition.

Eucaryotic cells have evolved far more sophisticated mechanisms for controlling gene expression, and these affect entire systems of interacting gene products. Groups of genes are activated or repressed in response to both external and internal signals. Membrane composition, cytoskeleton, secretory products, even metabolism—all these and other features must change in a coordinated manner when cells become differentiated. Compare, for example, a skeletal muscle cell specialized for contraction, with an *osteoblast*, which secretes the hard matrix of bone in the same animal (Panel B, pages 24–25). Such radical transformations of cell character reflect stable changes in gene expression. The controls that bring about such changes have evolved in eucaryotes to a degree unmatched in procaryotes.

Eucaryotic Cells Have Vastly More DNA Than They Need for the Specification of Proteins

Eucaryotic cells contain a very large quantity of DNA; as we have said, in human cells there is almost a thousand times more DNA than in typical bacteria. Yet it seems that only a small fraction of this DNA—perhaps 1% in human cells—carries the specifications for proteins that are actually made. Why then is the remaining 99% of the DNA there? One hypothesis is that much of it acts merely to increase the physical bulk of the nucleus. Another is that it is in large part parasitic—a collection of DNA sequences that have over the ages accumulated in the cell, exploiting the cell's machinery for their own reproduction, and bringing no benefit in return. Indeed, the DNA of many species has been shown to contain sequences called *transposable elements*, which have the ability to "jump" occasionally from one location to another in the DNA, and even to insert additional copies of themselves at new sites. Transposable elements could thus proliferate like a slow infection, becoming an ever larger proportion of the genetic material.

But evolution is opportunistic. Whatever the origins of the DNA that does not code for protein, it is certain that it now has some important functions. Part of this DNA is structural, enabling portions of the genetic material to become condensed or "packaged" in specific ways, as described in the next section, and some of the DNA is regulatory and helps to switch on and off the genes that direct the synthesis of proteins, thus playing a crucial role in the sophisticated control of gene expression in eucaryotic cells.

In Eucaryotic Cells the Genetic Material Is Packaged in Complex Ways

The length of DNA in eucaryotic cells is so great that the risk of entanglement and breakage becomes severe. Probably for this reason proteins unique to eucaryotes, the *histones*, have evolved to bind to the DNA and wrap it up into

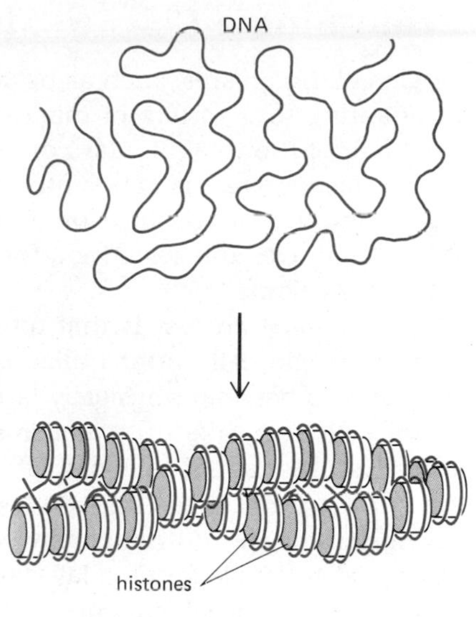

Figure 1–26 Schematic illustration of how the positively charged proteins called histones mediate the folding of DNA in chromosomes.

compact and manageable **chromosomes** (Figure 1–26). Tight packaging of the chromosomes is an essential part of the preparations for cell division in eucaryotes (Figure 1–27). All eucaryotes (with one minor exception) have histones bound to their DNA, and the importance of these proteins is reflected in the fact that they have been remarkably conserved in evolution: several of the histones of a pea plant are almost exactly the same, amino acid for amino acid, as those of a cow.

Many other proteins besides histones are bound to the DNA in eucaryotic cells. By altering the opportunities of the DNA to interact with other molecules, some of these DNA-binding proteins alter the patterns of gene expression from one type of specialized cell to another. For example, since genes contained in a tightly packed mass of DNA are not expressed, gene expression can be controlled through changes in the packaging of the DNA.

The membranes enclosing the nucleus in eucaryotic cells protect the delicate control machinery associated with DNA, sheltering it from the rapid movements and from many of the chemical changes that take place in the cytoplasm. They also allow the segregation of two crucial steps in gene expression: (1) the copying of DNA sequences into RNA sequences (*DNA transcription*) and (2) the use of these RNA sequences, in turn, to direct the synthesis of specific proteins (*RNA translation*). In procaryotic cells, there is no compartmentalization—the translation of RNA sequences into protein begins as soon as they are transcribed, even before their synthesis is completed. In eucaryotes, however (except in mitochondria and chloroplasts, which in this respect as in others are closer to bacteria), the two steps in the path from gene to protein are kept strictly separate: transcription occurs in the nucleus, translation in the cytoplasm. The RNA has to leave the nucleus before it can be used to guide protein synthesis. While it is in the nucleus, it undergoes elaborate processing, in which some parts of the RNA molecule are discarded and other parts are modified.

Because of these complexities, the genetic material of a eucaryotic cell offers many more opportunities for control than are present in bacteria.

Summary

Present-day living cells are classified as procaryotic (bacteria and their close relatives) or eucaryotic. Procaryotic cells are believed to resemble most closely the earliest ancestral cell. Although they have a relatively simple structure, they are biochemically diverse: for example, all of the major metabolic pathways can be found in bacteria, including the three principal energy-yielding processes of glycolysis, respiration, and photosynthesis. Eucaryotic cells are larger and more complex than procaryotic cells and contain more DNA, together with components that allow this DNA to be handled in elaborate ways. The DNA of the eucaryotic cell is enclosed in a membrane-bounded nucleus, while the cytoplasm contains many other membrane-bounded organelles. These include mitochondria, which carry out the terminal oxidation of food molecules, and, in plant cells, chloroplasts, which carry out photosynthesis. Various lines of evidence suggest that mitochondria and chloroplasts are the descendants of earlier procaryotic cells that established themselves as internal symbionts of a larger anaerobic cell. Eucaryotic cells are also unique in containing a cytoskeleton of protein filaments that help organize the cytoplasm and provide the machinery for movement.

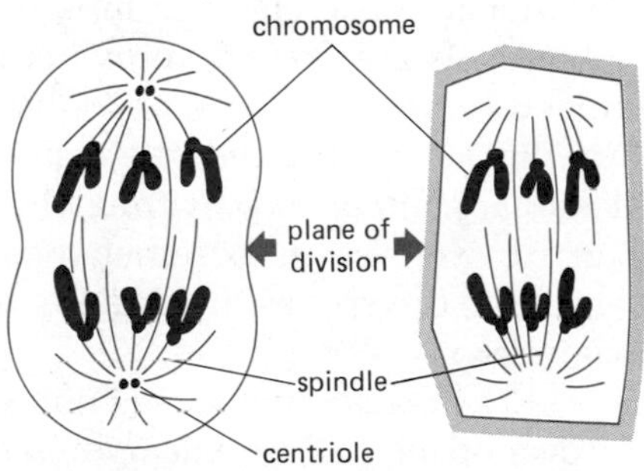

Figure 1–27 Schematic drawing of eucaryotic cells in mitosis. An animal cell is shown on the left and a plant cell on the right. The nuclear envelope has broken down, and the DNA, having replicated, has condensed into two complete sets of chromosomes. One set is distributed to each of the two newly forming cells by a mitotic spindle composed largely of microtubules.

From Single Cells to Multicellular Organisms[3]

Single-cell organisms, such as bacteria and protozoa, have been so successful in adapting to a variety of different environments that they comprise more than half of the total biomass on earth. Unlike higher animals, many of these unicellular organisms can synthesize all of the substances they need from a few simple nutrients, and some of them divide more than once every hour. What, then, was the selective advantage that led to the evolution of **multicellular organisms**?

The short answer is that multicellular organisms can exploit resources that no single cell could utilize so well. Multicellularity enables a tree, for example, to become physically large; to have roots in the ground, where one set of cells can take up water and nutrients; and to have leaves in the air, where another set of cells can efficiently capture the radiant energy from the sun. In the trunk of the tree are specialized cells that form channels for transporting water and nutrients between the roots and the leaves. Yet another set of specialized cells forms a layer of bark to prevent water loss and to provide a protected internal environment. The tree as a whole does not compete directly with unicellular organisms for its ecological niche; it has found a radically different way to survive and propagate.

As different animals and plants appeared, they changed the environment in which further evolution occurred. Survival in a jungle calls for different talents from those required for survival in the open sea. Innovations in movement, sensory detection, communication, social organization—all enabled eucaryotic organisms to compete, propagate, and survive in ever more complex ways.

Single Cells Can Associate to Form Colonies

It seems likely that an early step in the evolution of multicellular organisms was the association of unicellular organisms to form colonies. The simplest way of achieving this is for daughter cells to remain associated after each cell division. Even some procaryotic cells show such social behavior in a primitive form. Myxobacteria, for example, live in the soil and feed on insoluble organic molecules that they break down by secreting degradative enzymes. They stay together in loose colonies in which the digestive enzymes secreted by individual cells are pooled, thus increasing the efficiency of feeding. These cells indeed represent a peak of sophistication among procaryotes; for when food supplies are exhausted, the cells aggregate tightly together and form a multicellular *fruiting body*, within which the bacteria differentiate into spores that can survive even in extremely hostile conditions. When conditions are more favorable, the spores in a fruiting body germinate to produce a new swarm of bacteria.

Green algae (not to be confused with the procaryotic "blue-green algae" or cyanobacteria) are eucaryotes that exist as unicellular, colonial, or multicellular forms (Figure 1–28). Different species of green algae can be arranged in order of complexity, illustrating the kind of progression that probably occurred in the evolution of higher plants and animals. Unicellular green algae, such as *Chlamydomonas*, are similar to flagellated protozoa except that they possess chloroplasts, which enable them to carry out photosynthesis. In closely related genera, groups of flagellated cells live in colonies held together by a matrix of molecules secreted by the cells themselves. The simplest species (those of the genus *Gonium*) have the form of a concave disc made of 4, 8, 16, or 32 cells. Their flagella beat independently, but since they are all oriented in the same direction they are able to propel the colony through the water. Each cell is equivalent to every other, and each can divide to give rise to an entirely new colony. Larger colonies are found in other genera, the most spectacular being *Volvox,* some of whose species have as many as 50,000 or more cells linked together to form a hollow sphere. In *Volvox,* the individual cells forming a colony are connected by fine cytoplasmic bridges so that the beating of their flagella is coordinated to propel the entire colony along like a rolling ball (Figure 1–28). Within the *Volvox* colony there is some division of labor between cells, with a small number of cells being specialized for reproduction and serving as precursors of new colonies. The other cells are so dependent on each other that they cannot live independently, and the organism dies if the colony is disrupted.

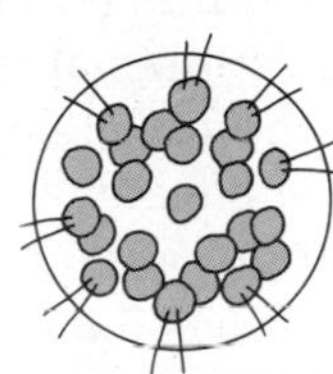

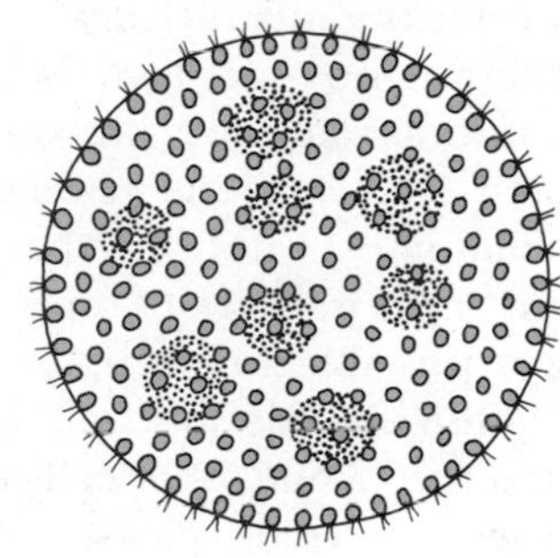

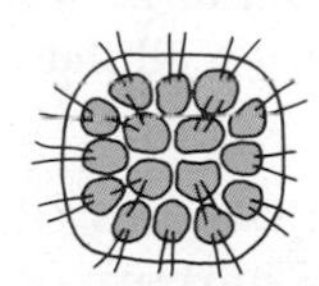

Figure 1–28 Four closely related genera of green algae, showing a progression from unicellular to colonial and multicellular organization.

The Cells of a Higher Organism Become Specialized and Cooperate

In some ways, *Volvox* is more like a multicellular organism than a simple colony. All of its flagella beat in synchrony as it spins through the water, and the colony is structurally and functionally polarized and can swim toward a distant source of light. The reproductive cells are usually confined to one end of the colony, where they divide to form new miniature colonies, which are initially sheltered inside the parent sphere. Thus in a primitive way *Volvox* displays the two essential features of all multicellular organisms: its cells become *specialized* and they *cooperate.* By specialization and cooperation, the cells combine to form a coordinated single organism with richer capabilities than any one of its component parts.

Organized patterns of cell differentiation occur even in some procaryotes. For example, many kinds of cyanobacteria remain together after cell division, forming filamentous chains that can be as much as a meter in length. At regular intervals along the filament, individual cells take on a distinctive character and become able to incorporate atmospheric nitrogen into organic molecules. These few specialized cells perform nitrogen fixation for their neighbors and share the products with them. But eucaryotic cells appear to be very much better at this sort of organized division of labor; they, and not procaryotes, are the living units from which all the more complex multicellular organisms are constructed.

Multicellular Organization Depends on Cohesion Between Cells

To form a multicellular organism, the cells must be somehow bound together, and eucaryotes have evolved a number of different devices that perform this function. In *Volvox,* as noted above, the cells do not separate entirely at cell division but remain connected by cytoplasmic bridges. In higher plants, the cells not only remain connected by cytoplasmic bridges (called *plasmodesmata*), but also are imprisoned in a rigid honeycomb of chambers walled with cellulose that the cells themselves have secreted (*cell walls*).

The cells of most animals do not have rigid walls, and cytoplasmic bridges are unusual. Instead, the cells are bound together by a relatively loose meshwork of large extracellular organic molecules (called the *extracellular matrix*) and by adhesions between their plasma membranes. In *sponges,* for example, which are commonly considered the most primitive of present-day animals, the body wall typically consists of a coherent sheet of cells comprising just five different specialized types; these form a system of channels and pores for the passage of water, from which food particles are filtered and ingested by the cells. Sponges grow indefinitely through cell proliferation, and their size and structure are not precisely fixed. They have no nervous system to coordinate the activities of their parts, and they have been described as "loose republics of cells"—to be contrasted with the more strictly disciplined cell communities that constitute higher animals. Nevertheless, a sponge is far from being a totally chaotic structure. If the sponge is forced through a fine sieve so that its individual cells are mechanically separated from one another, the cells will often spontaneously reassemble into an intact sponge, aggregating initially into a large mass and then eventually rearranging themselves into a coherent multicellular sheet. Such sheets of cells are called **epithelia.**

Epithelial Sheets of Cells Enclose a Sheltered Internal Environment

Of all the ways in which animal cells are woven together into multicellular tissues, the epithelial arrangement is perhaps the most fundamentally important. The epithelial sheet has much the same significance for the evolution of complex multicellular organisms that the cell membrane has for the evolution of complex single cells.

The importance of epithelial sheets is well illustrated in another lowly group of animals, the *coelenterates.* These stand a rung higher in the scale of evolution than do the sponges, for they have a nervous system of sorts; but among animals with nervous systems, they are probably the most primitive. The group includes sea anemones, jellyfish, and corals, as well as the small freshwater organism *Hydra.* Coelenterates are constructed from two layers of epithelium, the outer layer being the *ectoderm,* the inner being the *endoderm.* The endodermal layer surrounds a cavity, the *coelenteron,* in which food is digested (Figure 1–29). The cells are bound together in such a way that the epithelial sheets not only have mechanical strength, but also can serve as a barrier to the passage of molecules; they thus prevent food from escaping and make it possible to set up specialized chemical conditions for its digestion. Among the endodermal cells are some that secrete digestive enzymes into the coelenteron, while other cells absorb and further digest the nutrient molecules that these enzymes release. By forming a tightly coherent epithelial sheet that prevents all these molecules from being lost to the exterior, the endodermal cells create for themselves an environment in the coelenteron that is suited to their own digestive tasks. Meanwhile the ectodermal cells, facing the exterior, remain specialized for encounters with the outside world. In the ectoderm, for example, are cells that contain, coiled inside them, a poison dart, which can be unleashed to paralyze the small animals that *Hydra* feeds on.

Figure 1–29 A schematic view of the body plan of *Hydra.* The outer layer of cells (ectoderm) is primarily protective, while cells of the inner layer (endoderm) are engaged principally in digestion. Sandwiched between these two layers is a net of interconnected nerve cells. (Note that in the right-hand figure the nerve cells are shown disproportionately large.)

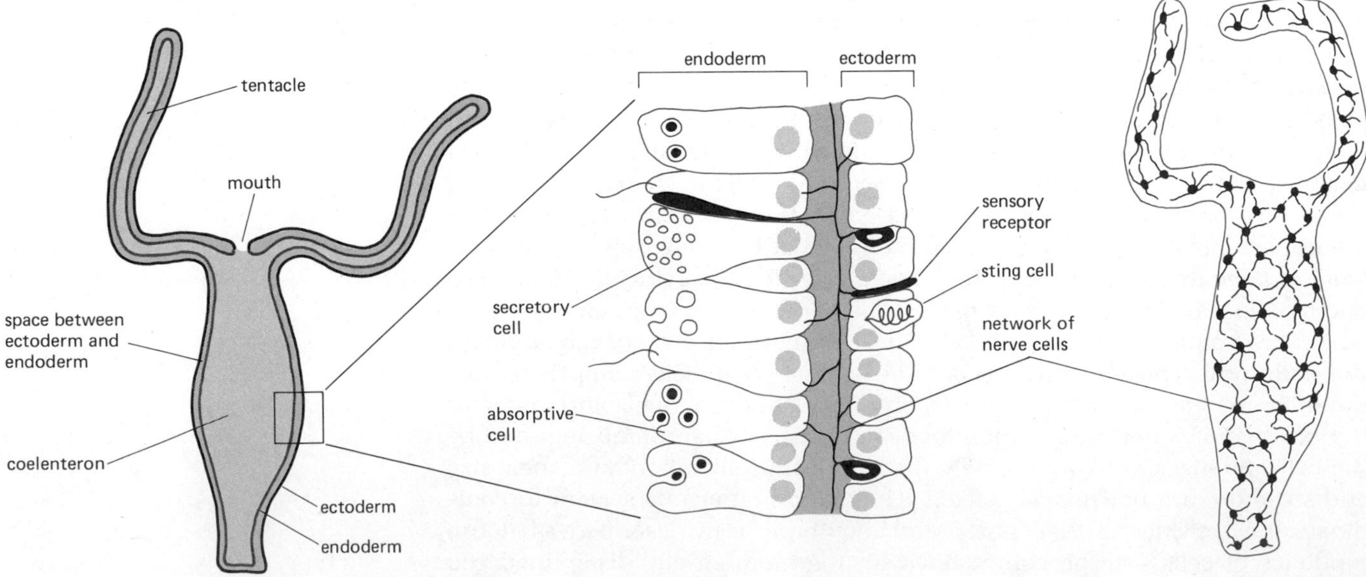

Sandwiched between the ectoderm and the endoderm is another compartment, separate both from the coelenteron and from the outside world. It is in this narrow space that the nerve cells chiefly lie, pressed close against the inner face of the ectoderm. The animal can change its shape and move by contractions of musclelike cells in the ectoderm and endoderm. The nerve cells convey electrical signals to control and coordinate these contractions (Figures 1–30 and 1–31). As we shall see later, the concentrations of simple inorganic ions in the medium surrounding a nerve cell are crucial for its function. Most nerve cells—our own included—are designed to operate when bathed in a solution with an ionic composition similar to that of sea water. This presumably reflects the conditions under which the first nerve cells evolved. Most coelenterates still live in the sea, but not all. *Hydra,* in particular, lives in fresh water. It has evidently been able to colonize this new habitat only because its nerve cells are contained in a space that is sealed and isolated from the exterior by sheets of epithelial cells that maintain the internal environment necessary for nerve cell function.

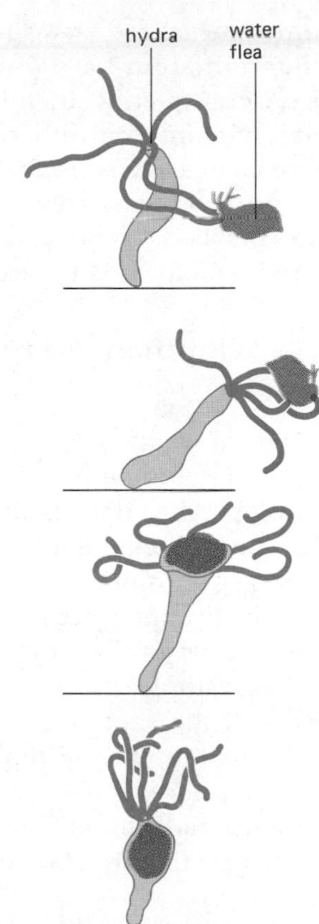

Figure 1–30 *Hydra* can perform a range of fairly complex activities. It is depicted here catching a small water flea in its tentacles and stuffing this prey into its coelenteron for digestion.

Cell-Cell Communication Controls the Spatial Pattern of Multicellular Organisms

The cells of *Hydra* are not only bound together mechanically and connected by junctions that seal off the interior from the exterior environment; they also communicate with one another along the length of the body. If one end of a *Hydra* is cut off, the remaining cells react to the absence of the amputated part by adjusting their characters and rearranging themselves so as to regenerate a complete animal. Evidently signals pass from one part of the organism to the other governing the development of its body pattern, with tentacles and a mouth at one end and a foot at the other. Moreover, these signals are independent of the nervous system. If a developing *Hydra* is treated with the drug colchicine, so that nerve cells are prevented from forming, the animal is unable to move, catch prey, or feed itself. However, its digestive system still functions normally, so it can be kept alive by anyone with the patience to stuff its normal prey into its mouth. In such force-fed animals the body pattern is maintained, and lost parts are regenerated just as well as in an animal that has an intact nervous system.

From humble ancestors resembling coelenterates the vastly more complex higher animals have evolved, and the latter owe their complexity to a more sophisticated exploitation of the same basic principles of cell cooperation that underlie the construction of *Hydra.* Epithelial sheets of cells line all external and internal surfaces in the body, creating sheltered compartments and controlled internal environments in which specialized functions are performed by differentiated cells. Specialized cells interact and communicate with one another, setting up signals to govern the character of each cell according to its place in the structure as a whole. To show how it is possible to generate multicellular organisms of such size, complexity, and precision as a human being, it is necessary, however, to consider more closely the sequence of events in **development.**

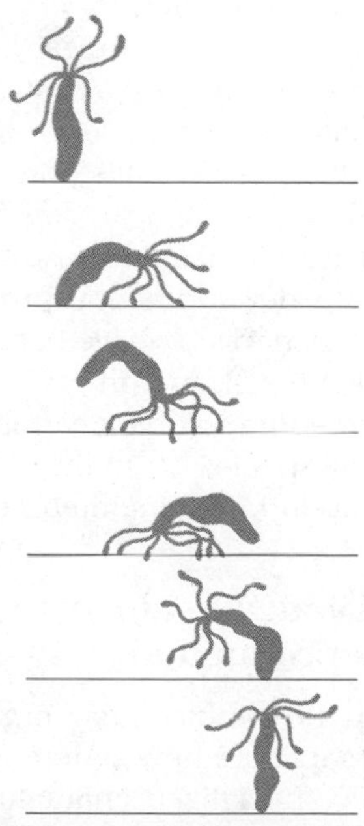

Figure 1–31 *Hydra* can swim, glide on its base, or, as shown here, travel by somersaulting.

Cell Memory Permits the Development of Complex Patterns

The cells of almost every multicellular organism are generated by repeated division from a single precursor cell; they constitute a *clone.* As proliferation continues and the clone grows, some of the cells, as we have seen, become differentiated from others, adopting a different structure, a different chemistry, and a different function, in response to cues from their neighbors. It is remarkable that eucaryotic cells and their progeny will usually persist in their differently specialized states even after the influences that originally directed their differentiation have disappeared—in other words, these cells have a *memory*. Consequently, their final character is not determined simply by their final environment, but rather by the entire sequence of influences to which they have been exposed in the course of development. Thus as the body grows and matures, progressively finer and finer details of the adult body pattern become specified, creating an organism of gradually increasing complexity whose ultimate form is the expression of a long developmental history.

Basic Developmental Programs Tend to Be Conserved in Evolution

The structure of an animal is also the outcome of its evolutionary history, which, like development, presents a chronicle of progress from the simple to the complex. What then is the connection between the two perspectives, of evolution on the one hand and development on the other?

During evolution, many of the developmental devices that evolved in the simplest multicellular organisms have been conserved as basic principles for the construction of their more complex descendants. We have already mentioned, for example, the organization of cells into epithelia. It is notable also that some of the same basic specialized cell types, such as nerve cells, are found throughout nearly the whole of the animal kingdom, from *Hydra* to man. Furthermore, the early developmental stages of animals whose adult forms appear radically different are often surprisingly similar; it takes an expert eye to distinguish, for example, a young chick embryo from a young human embryo.

Such observations are not difficult to understand. Consider the process by which a new anatomical feature—say an elongated beak—appears in the course of evolution. A random mutation occurs that changes the amino acid sequence of a protein and hence its biological activity. This altered protein may, by chance, affect the cells responsible for the formation of the beak in such a way that they make one which is longer. But the mutation must also be compatible with the development of the rest of the organism: only then will it be propagated by natural selection. There would be little selective advantage in forming a longer beak if, in the process, the tongue was lost or the ears failed to develop. A catastrophe of this type is far more likely if the mutation affects events occurring early in development than if it affects those at the end. The early cells of an embryo are like cards at the bottom of a house of cards—a great deal depends on them and even small changes in their properties are likely to result in disaster. Fundamental steps have been "frozen" into developmental processes just as the genetic code or protein synthetic mechanisms have become frozen into the basic biochemical organization of the cell. In contrast, cells produced at the end of development have more freedom to change. It is presumably for this reason that the embryos of different species so often resemble each other in their early stages and, as they develop, seem sometimes to replay the steps of evolution.

Eucaryotic Organisms Possess a Complex Machinery for Reproduction

Within the multicellular organism there must be some cells that serve as precursors for a new generation. In higher plants and animals these cells have a highly specialized character and are called *germ cells*. The propagation of the species depends on them, and there is a powerful selection pressure to adjust the structure of the organism as a whole to provide the germ cells with the best chance of survival. Other cells may die, but as long as germ cells survive, new organisms of the same sort will be produced. In this sense the most fundamental distinction to be drawn in a multicellular organism is the distinction between germ cells and the rest, that is, between germ cells and *somatic* cells.

Not all multicellular organisms reproduce by means of distinctive differentiated germ cells. Many simple animals, including sponges and coelenterates, can reproduce by budding off portions of their bodies, and many plants do likewise. Germ cells are, however, a necessity for **sexual reproduction.** This process is so familiar to us that we take it for granted, but it is by no means the obvious way to reproduce: it is far more complicated than asexual reproduction and requires a large diversion of resources. Two individuals of the same species but different sex produce germ cells of usually very different character—*eggs* from one, *sperm* from the other. An egg cell fuses with a sperm cell to form a *zygote*—the single precursor cell for the development of a new organism, whose genes represent a partly random reassortment of the genes of the two parents. While they may also reproduce in other ways, almost all eucaryotic species, unicellular as well as multicellular, are capable of reproducing sexually. Eucaryotic cells have evolved a complex machinery for sex; our lives revolve around it. Strong selective pressures must have operated to favor the evolution of sexual reproduction in preference to simpler strategies based on ordinary cell division. Although it is surprisingly difficult to say with certainty what those selection pressures were, it is at least plain that sexual reproduction brings new possibilities for manipulating and recombining the genes of a species. It may thus have played a crucial part in permitting the evolution of novel genes in novel combinations, and so in engendering the endless variety of forms and functions seen in plants and animals today.

The Cells of the Vertebrate Body Exhibit More Than 200 Different Modes of Specialization

The wealth of diverse specializations to be found among the cells of a higher animal is incomparably greater than any procaryote can show. In a vertebrate, more than 200 distinct **cell types** are plainly distinguishable, and many of these types of cells probably include, under a single name, a large number of more subtly different varieties. Panel B (pages 24–25) shows a small selection. In this profusion of specialized behaviors one can see displayed, in a single organism, the astonishing versatility of the eucaryotic cell. Each feature and each organelle of the prototype that we have outlined in Panel A (pages 16–17) is developed to an unusual degree or revealed with special clarity in one cell type or another. Much of current knowledge of the general properties of eucaryotic cells has depended on the study of such specialized types of cells, individually displaying to exceptionally good advantage particular features upon which all cells depend in some measure. To take one arbitrary example, consider the *neuromuscular junction,* where just three types of cells are involved: a muscle cell, a nerve cell, and a Schwann cell. Each has a very different role (Figure 1–32).

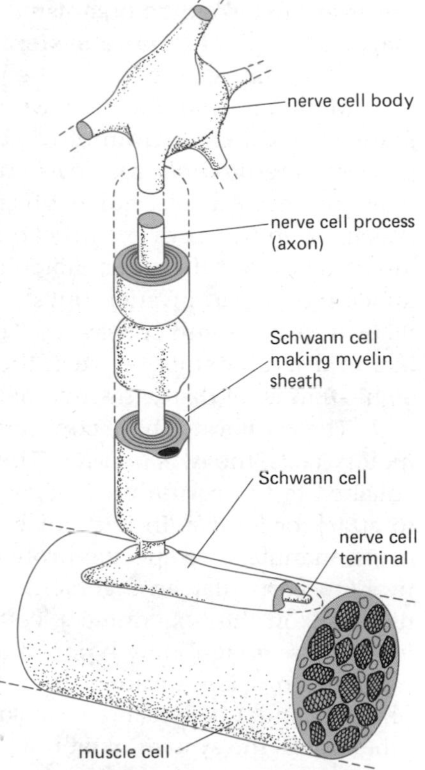

Figure 1–32 Schematic diagram showing a nerve cell, with its associated Schwann cells, contacting a muscle cell at a neuromuscular junction.

1. The muscle cell has made contraction its speciality. Its cytoplasm is packed with organized arrays of protein filaments, including vast numbers of actin filaments. There are also many mitochondria interspersed among the protein filaments, supplying ATP as fuel for the contractile apparatus.
2. The nerve cell stimulates the muscle to contract, conveying an excitatory signal to the muscle from the brain or spinal cord. The nerve cell therefore is extraordinarily elongated: its main body, containing the nucleus, may lie a meter or more from the junction with the muscle. The cytoskeleton is consequently well developed so as to maintain the unusual shape of the

cell and to transport materials efficiently from one end of the cell to the other. The most crucial specialization of the nerve cell, however, is its plasma membrane, which contains proteins that act as ion *pumps* and ion *channels,* causing a movement of ions that is equivalent to a flow of electricity. Whereas all cells contain such pumps and channels in their plasma membranes, the nerve cell has exploited them in such a way that a pulse of electricity can propagate in a fraction of a second from one end of the cell to the other, conveying a signal for action.

3. Lastly, Schwann cells are specialists in the mass production of plasma membrane, which they wrap around the elongated portion of the nerve cell, laying down layer upon layer of membrane like a roll of tape, to form a *myelin sheath* that serves as insulation.

Cells of the Immune System Are Specialized for the Task of Chemical Recognition

Among all the cell systems that have evolved in higher animals, there are two that stand out in different ways as pinnacles of complexity and sophistication: the *immune system* of the vertebrate is one, the *nervous system* the other. Each of them far surpasses the performance of any artificial device—the vertebrate immune system in its capacity for chemical discrimination, the nervous system in its capacities for perception and control. Each system comprises a large number of different cell types and depends on interactions between them.

The protected and well-nourished environment in the interior of a multicellular animal is as inviting to foreign organisms as it is congenial to the animal's own cells. Hence there is a need for such animals to defend themselves against invading organisms—particularly viruses and bacteria. The primary task of the **immune system** is to destroy any such foreign microorganisms that may gain entry to the body.

Many eucaryotic cells have an ability to engulf and digest particles of matter from their surroundings. When the particles are relatively large, the process is termed *phagocytosis.* Among the differentiated cells in higher animals, there are professional phagocytic cells, such as *macrophages,* which specialize in this activity and can swallow up and destroy bacteria and other foreign cells. But there is a difficulty: it is good that the phagocytic cell should attack the foreign invader, but it would be disastrous if it were to attack also its own relatives and colleagues. The immune system therefore faces the problem of discriminating between the animal's own cells and those that are foreign—that is, of distinguishing between self and nonself.

The vertebrates have consequently evolved a specialized class of discriminatory cells, the *lymphocytes.* These are not themselves phagocytic; but they collaborate to provide the phagocytic cells with cues that tell them whether to attack or let live. In particular, certain of the lymphocytes (the B lymphocytes) manufacture specific protein molecules, or *antibodies,* that bind selectively to particular arrangements of atoms on the surfaces of invading organisms or on the toxic molecules that they produce. To brand a new type of invader as foreign, new types of antibody must be produced; and since the variety of possible invaders is vast and essentially unpredictable, the B lymphocytes must be capable of making an endless variety of antibodies. On the other hand, the system must not produce antibodies that bind to the animal's own cells and molecules.

The vast diversity of antibodies is generated by random changes in the DNA coding for the specific binding sites of antibody molecules. In this way, through a kind of specialized mutation, millions of genetically different lymphocytes are created, each able to proliferate to form a clone whose members all produce the same distinctive antibody. Of these many potential clones, the ones that make antibodies that react with self molecules are destroyed or suppressed (by mechanisms still poorly understood), while those that make antibodies against foreign molecules are selected to survive and multiply. Thus the genesis of an individual animal's immune system, like the process of evolution, depends on a strategy of random variation followed by selection.

Nerve Cells Allow a Rapid Adaptation to a Changing World

The immune system of vertebrates places them in a class apart. Lower animals apparently do not have lymphocytes to help defend against invading microorganisms. A **nervous system,** on the other hand, is found in almost all multicellular animals and fulfills a still more fundamental need—the need for a quick adaptative response to external events.

Evolution acts over many generations to optimize the structure of an organism according to the environment in which it lives. In most ecological niches, though, there occur changes that are far too rapid for evolutionary adaptation to keep pace. The most successful organism, therefore, will be one that is capable of another sort of adaptation, requiring no genetic mutation yet producing optimal behavior when circumstances change. If the sequence of environmental changes is perfectly predictable, like the alternation of night and day or of summer and winter, the organism can be genetically programmed to change autonomously according to the appropriate timetable. Thus the photosynthetic protist *Gonyaulax* (belonging to the group of cells known as *dinoflagellates*) shows a 24-hour rhythm in its photosynthetic activities, which continues even if the cell is maintained for weeks on end in conditions of constant lighting. Such biological clocks exist in many other organisms, but their mechanism remains a profound mystery.

Most environmental changes, however, are not so predictable. Bacteria in the gut, for example, will experience irregular fluctuations in the nature and quantity of food that is available to them, and any bacterium that can adjust its metabolism to these changes will have an advantage over one that cannot. These organisms consequently have evolved the ability to sense the concentrations of nutrients in their environment and to react by adjusting the rates at which they synthesize their metabolic enzymes. Special intracellular control molecules (such as *cyclic AMP*) serve to couple the environmental stimulus to the appropriate response.

In a multicellular organism, the signal that couples a sensation to a response must generally pass between cells. Thus metabolic adjustments are often mediated by hormones that are released by one set of cells and travel through the tissues to produce a response in other sets of cells. But hormones take time to travel a long distance, and in doing so they diffuse widely. If a chemical signal is to be delivered fast, it must be released close to its target; and in that way it can also have a precisely localized action. But if the chemical signal is to be released close to its target, how can it be used to couple a sensation to a response in a remote part of the body? The nerve cell provides the answer. At one end, it is itself sensitive to a chemical or physical stimulus; at the other end, it can in turn release a chemical signal or *neurotransmitter* that acts on other cells. Stimulation at one end triggers an electrical excitation, which is propagated rapidly to the other end and, on arriving there, triggers release of the neurotransmitter. This rapid signaling device enables multicellular animals to make rapid responses to the changing world around them. It also enables them to coordinate precisely the activities of widely separated parts of the body.

Developing Nerve Cells Must Assemble to Form a Nervous System

A single nerve cell of a human being is not very different from a single nerve cell of a worm. The superiority of the human nervous system lies in the enormous number of its cells and, above all, in the way that they are connected together to transmit, combine, and interpret sensory inputs and to coordinate complex patterns of activity. Similarly, the capabilities of a computer depend not so much on the nature of the individual switches or memory elements as on how many there are and the way they are linked together into a system. In the case of the computer, an external agent—the manufacturer—assembles the components in the proper configuration. But for the nervous system, as for the rest of the body, there is no external manufacturer: the cells must assemble into a functional system themselves, following instructions carried in their DNA and adjusting the final product according to the external world. To understand the cellular basis for the evolution of the nervous system, one must therefore look to the mechanisms by which nerve cells develop their fantastically intricate shapes and form their precisely ordered patterns of connections.

Nerve cells begin their existence like any other sort of cell, relatively small and compact. Long processes are then sent out from the body of the nerve cell toward the targets with which it must connect (Figure 1–33). Each such process, known as an *axon* or a *dendrite*, according to whether it carries signals away from the cell body or toward it, is constructed by means of a *growth cone* (Figure 1–34). This organelle, like so many others, represents a

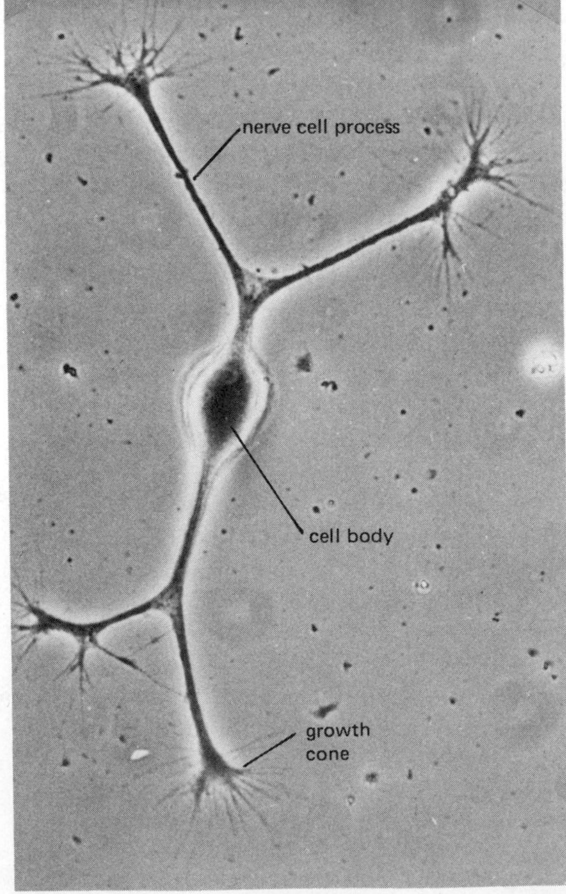

Figure 1–33 Light micrograph of a nerve cell that has been isolated from a chick embryo and put into a tissue-culture dish containing a nutrient solution. The cell is beginning to grow elongated processes. (Courtesy of Zoltan Gabor.)

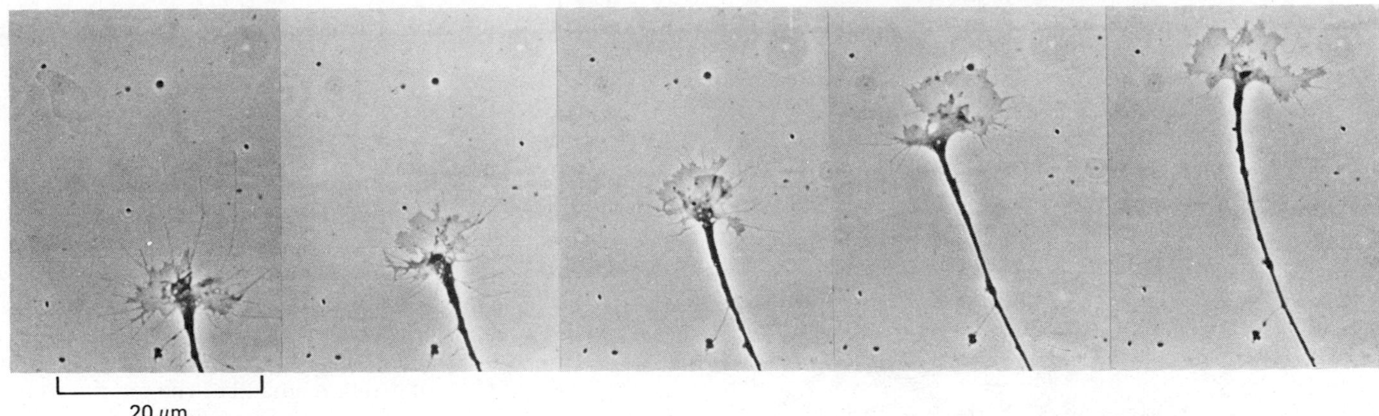

specialization of an apparatus that is common to eucaryotic cells as a device for locomotion. However, instead of pulling the entire nerve cell along, the growth cone leaves the cell body behind and pulls out an elongating axon or dendrite. The growth cone seems to crawl forward through the tissues like a dog on a leash, sniffing its way along the path that will lead to its quarry. In some cases, growth cones appear to be guided simply by physical means, advancing through preestablished channels or along specified tracks in the extracellular matrix. It seems, however, that the evolution of complex nervous systems has depended to a large extent on the development of chemical markers, whereby a particular nerve cell can recognize its proper target from among a mass of others that are inappropriate.

Figure 1–34 The tip of a nerve cell process extending along the surface of a tissue-culture dish as in the previous picture. Photographs taken at intervals of approximately 5 minutes show the elongation of the nerve cell process and the rapidly changing form of the growth cone. The remainder of the nerve cell lies to the bottom outside the micrograph; at this magnification the cell body would be 20 cm to 30 cm away. (Courtesy of Stephen Clark.)

Nerve Cell Connections Determine Patterns of Behavior

By mechanisms such as those just discussed, nervous systems of astonishing complexity are constructed. Look at the visual system of a fly, for example (Figures 1–35, 1–36, and 1–37); this entire structure is built to genetic specifications, and will develop even in the absence of light. Patterns of nerve connections, furthermore, constrain patterns of behavior. Without education, without need of experience, the male fly mates with the female, the spider

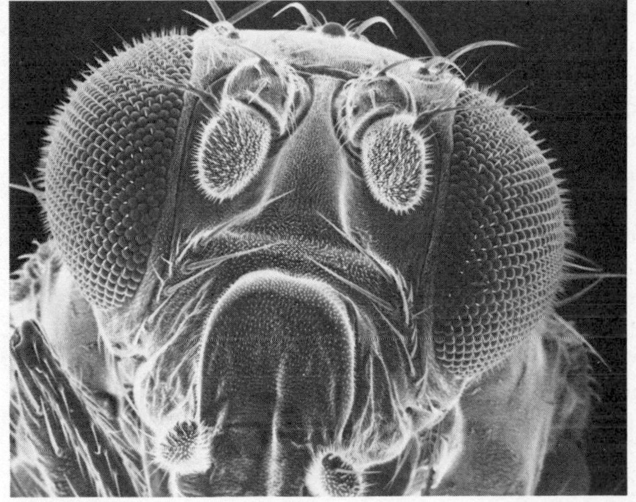

Figure 1–35 The head of a fruit fly (*Drosophila*) seen in a scanning electron microscope. Situated on either side of the head are two large compound eyes consisting of large numbers of units known as ommatidia. Each ommatidium has a separate lens that focuses the light onto a group of photosensitive receptor cells at its base (see Figure 1–36). (Courtesy of Rudi Turner and Anthony Mahowald.)

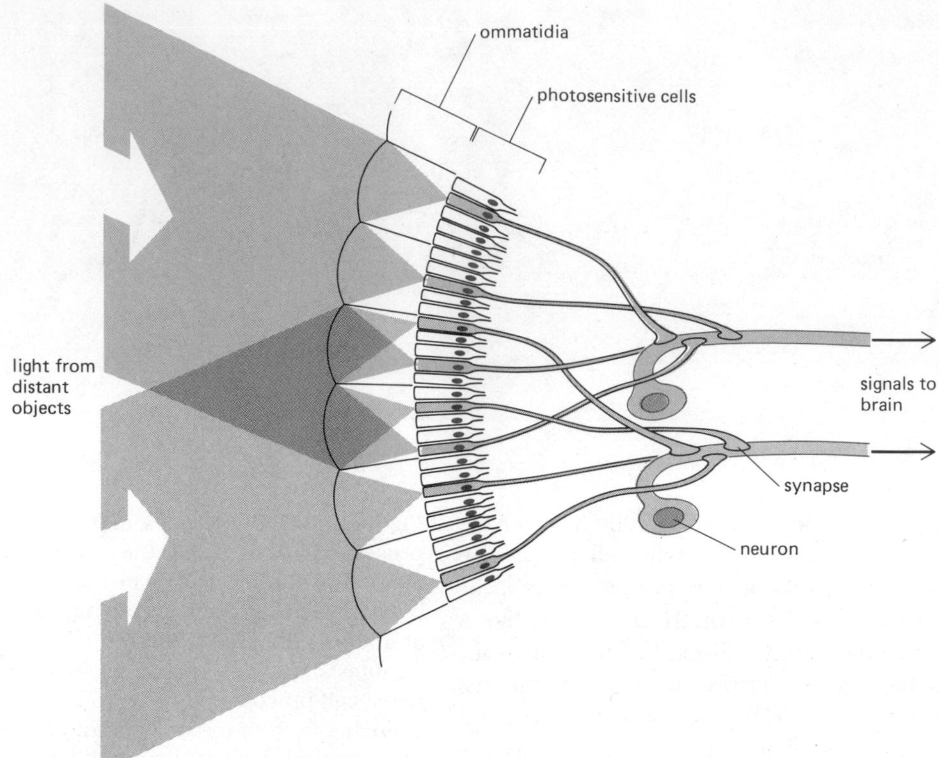

Figure 1–36 Schematic diagram of the neuronal connections in the outermost layer of the fly eye as seen in a vertical section. Light enters each of the ommatidia of the compound eye (see Figure 1–35) and is focused onto one of eight photosensitive receptor cells at its base (only five of which are shown here). Because of the curvature of the compound eye, light from a distant point source is focused onto a different photosensitive receptor cell in different ommatidia. The interweaving of the short axons of the photoreceptor cells, however, connects the photoreceptor cells that are "looking" at the same point to the same bundle of nerve axons that passes into the insect brain. More than a thousand such axon bundles are present in each fly eye, each of which is precisely wired in the course of development to the correct set of photoreceptor cells.

spins its web, the bird migrates to the south. All these activities are prescribed by the DNA of the species, acting through its control over the behavior of the individual cells, as they build the nervous system in the embryo and play their part in its functioning in the adult.

But not all behavior is genetically determined. The past experiences of an animal are important, as well as its DNA. Sensory deprivation during the development of a mammal can alter the microscopic structure of the brain, and mature animals of almost every species, from coelenterates to man, are to some degree capable of learning. Learning is by definition the outcome of experience, and therefore of electrical activity in nerve cells, and it must involve the production of lasting changes in neural connections. Beyond this, we understand very little of the mechanism. It is perhaps the central unsolved problem of neurobiology.

The brain connections that allow us to read and write and speak our native tongue are the outcome of education, and they represent an inheritance of a nongenetic kind. Learning and communication enable the human species to adapt itself over many generations, in a way that is possible for lower organisms only through genetic evolution. Yet even these sophisticated capacities, on which all our culture and society depend, can be seen to rest on the minutiae of cell behavior—on the rules by which nerve cells make lasting adjustments of their interconnections as a consequence of electrical activity.

Of course, we can no more understand the society or the multicellular organism by studying only single cells than we can understand the single cell by studying only isolated biological molecules. Yet if we do not understand the cell, we can never completely understand the organism. And if we do not understand the constituent molecules, we cannot properly understand the cell. Molecules, therefore, must be the starting point for our discussion of the living cell in the next chapter.

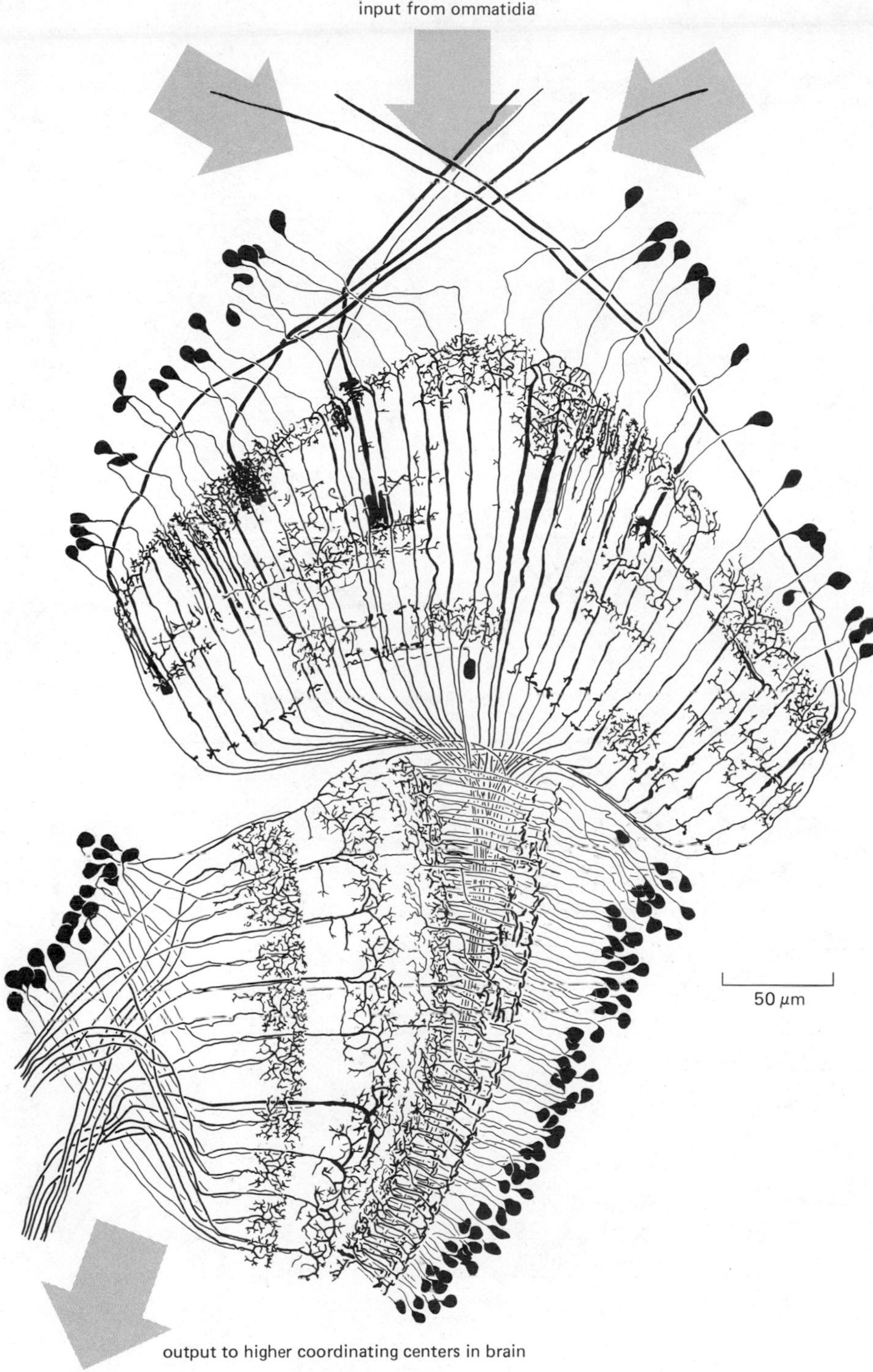

Figure 1–37 Portion of the brain of a fly, showing part of the network of nerve cells that process the input from the ommatidia (see Figure 1–36). (Redrawn from N. Strausfeld, Atlas of an Insect Brain. New York: Springer, 1976.)

Summary

The evolution of large multicellular organisms depended on the ability of eucaryotic cells to express their hereditary information in many different ways and to function cooperatively as a single organism. One of the earliest developments was probably that of epithelia, in which cells join together in sheets, separating the internal space of the animal from the exterior. In addition to

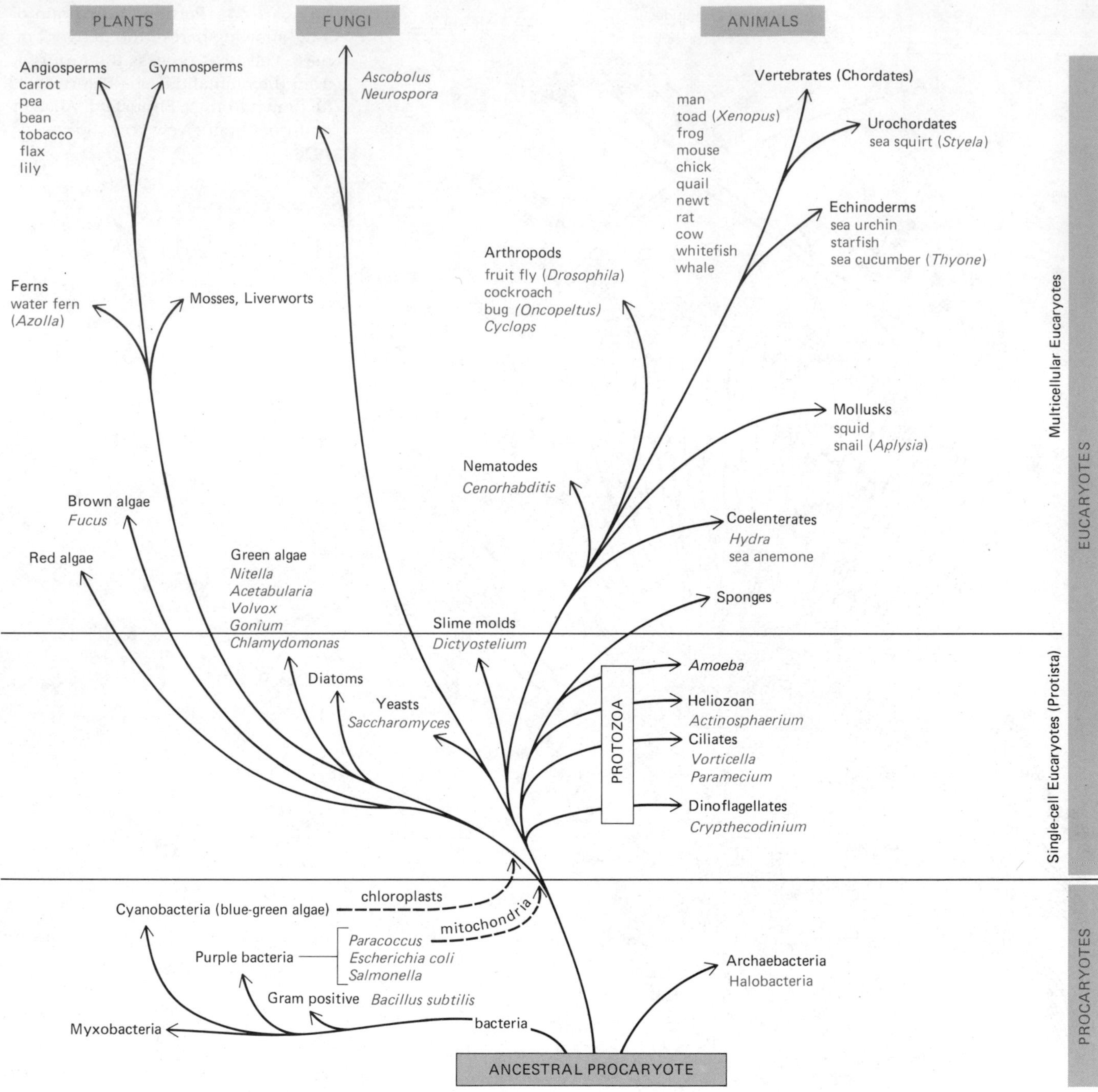

epithelial cells, primitive differentiated cell types would have included nerve cells, muscle cells, and connective tissue cells, all of which can be found in very simple present-day animals.

In the evolution of higher animals (Figure 1–38), the same fundamental developmental strategies were used to produce an increasing number of specialized cell types and more sophisticated methods of coordination between them. Two systems of cells in higher animals represent, in different ways, pinnacles of complexity in multicellular organization: one is the vertebrate immune system, the cells of which have the potential to produce millions of different protein antibodies; the other is the nervous system. In the lower animals

Figure 1–38 Evolutionary relationships among some of the organisms mentioned in this book. The branches of the tree show paths of common descent but (unlike the tree shown in Figure 1–16) do not indicate by their length the passage of time. (Note, similarly, that the vertical axis of the diagram shows major categories of organisms and not time.)

the pattern of connections between nerve cells is for the most part rigidly specified genetically, and behavioral patterns evolve by genetic mutation. In higher animals, up to and including humans, the performance and structure of the nervous system are increasingly subject to modification (learning) as a consequence of the capacity of nerve cells to alter their connections in response to electrical activity caused by environmental influences.

References

General

Attenborough, D. Life on Earth. Boston: Little Brown, 1979. (A beautifully illustrated evolutionary account of the living world.)

Curtis, H. Biology, 3rd ed. New York: Worth, 1979.

de Witt, W. Biology of the Cell: An Evolutionary Approach. Philadelphia: Saunders, 1977.

Evolution. *Sci. Am.* 239(3), 1978. (An entire issue devoted to the topic.)

Fawcett, D.W. The Cell, 2nd ed. Philadelphia: Saunders, 1981. (A picture book of the fine structure of eucaryotic cells.)

Keeton, W.T. Biological Science, 3rd ed. New York: Norton, 1980.

Luria, S.E.; Gould, S.J.; Singer, S. A View of Life. Menlo Park, Ca.: Benjamin-Cummings, 1981.

Maynard Smith, J. The Theory of Evolution, 3rd ed. New York: Penguin, 1975.

Raven, P.H.; Evert, R.F.; Curtis, H. Biology of Plants, 3rd ed. New York: Worth, 1981.

Thomas, L. The Lives of a Cell: Notes of a Biology Watcher. New York: Viking Press, 1974. (A collection of short, thought-provoking essays.)

Wilson, E.B. The Cell in Development and Heredity, 3rd ed. New York: Macmillan, 1928.

Wolfe, S.L. Biology of the Cell, 2nd ed. Belmont, Ca.: Wadsworth, 1981. (Includes a good final chapter on cell evolution.)

Cited

1. Eigen, M.; Gardiner, W.; Schuster, P.; Winkler-Oswatitsch, R. The origin of genetic information. *Sci. Am.* 244(4):88–118, 1981.

   Folsome, C.E., ed. Life: Origin and Evolution. (Readings from *Sci. Am.*) San Francisco: Freeman, 1979.

   Wong, J.T-F. Coevolution of genetic code and amino acid biosynthesis. *Trends Biochem. Sci.* 6:33–36, 1981.

2. Dickerson, R.E. Cytochrome c and the evolution of energy metabolism. *Sci. Am.* 242(3):136–153, 1980.

   Margulis, L. Origin of Eukaryotic Cells. New Haven: Yale University Press, 1970.

   Origins and Evolution of Eukaryotic Intracellular Organelles. *Ann. N.Y. Acad. Sci.* Vol. 361, 1981.

   Woese, C.R. Archaebacteria. *Sci. Am.* 244(6):98–122, 1981.

3. Buchsbaum, R. Animals Without Backbones, 2nd ed. Chicago: University of Chicago Press, 1976.

   Valentine, J.W. The evolution of multicellular plants and animals. *Sci. Am.* 239(3): 140–158, 1978.

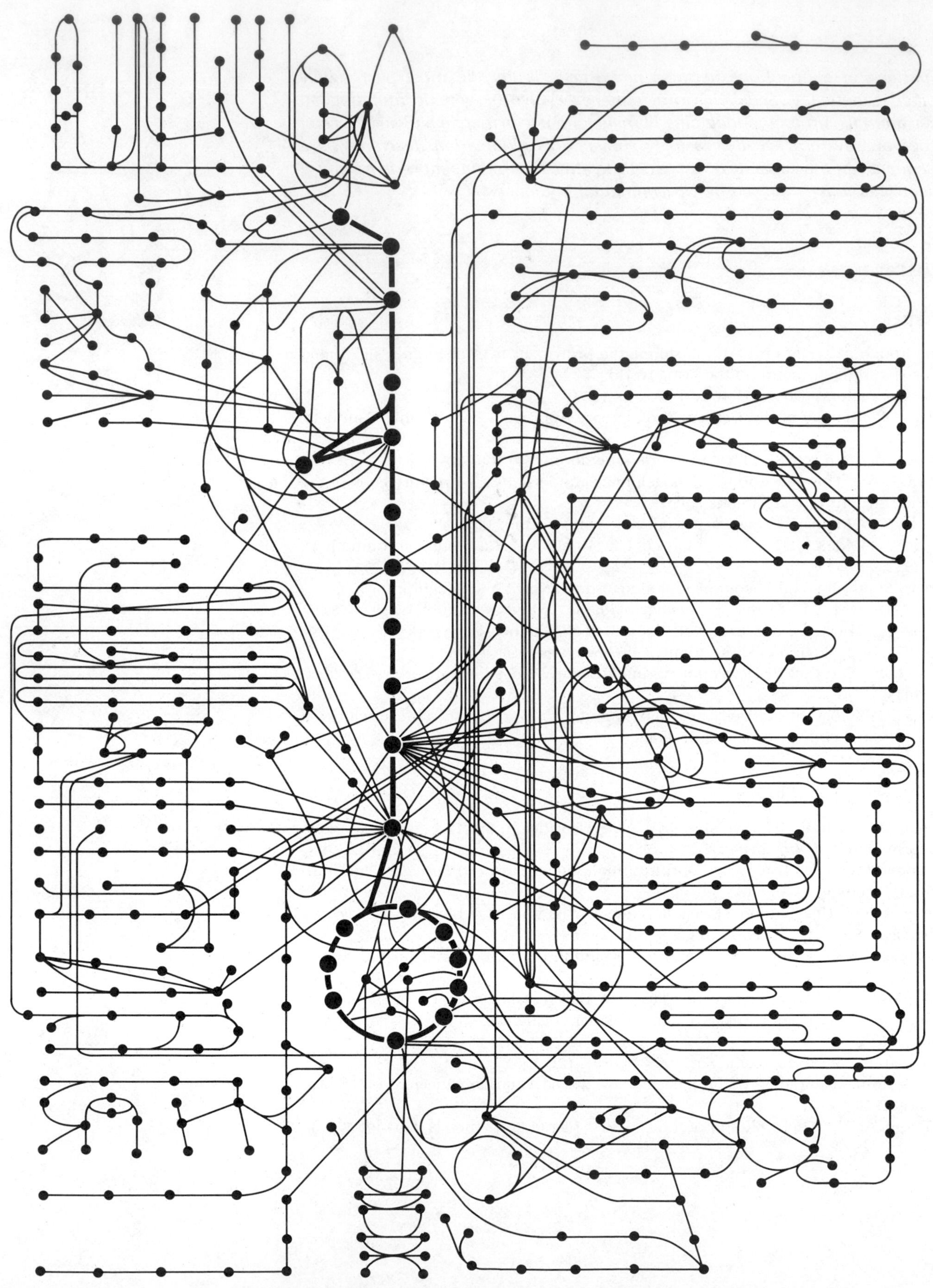

A maze illustrating the chemical reactions that interconvert small molecules in cells.

2

Small Molecules, Energy, and Biosynthesis

"I must tell you that I can prepare urea without requiring a kidney or an animal, either man or dog." This sentence, written 150 years ago by the young German chemist Wöhler, signaled an end to the belief in a special *vital force* that exists in living organisms and gives rise to their distinctive properties and products. But what was a revelation in Wöhler's time is commonplace today—living creatures are made of chemicals. There is no room in the contemporary view of life for vitalism—or for anything else outside the laws of chemistry and physics. This is not to say that no mysteries remain in biology: there are many areas of ignorance, as will become apparent in later chapters. But we should begin by emphasizing the truly enormous amount that is known.

We now have detailed information about the essential molecules of the cell—not just a small number of molecules but almost all of them. In many cases we know their precise chemical structures and exactly how they are made and broken down. We know in general terms how chemical energy drives the biosynthetic reactions of the cell, how thermodynamic principles operate in cells to create molecular order, and how the myriad of intracellular chemical changes occurring continuously within them are controlled and coordinated.

In this and the next chapter we give a brief survey of the chemistry of the living cell. Here we deal with the processes involving small molecules: those mechanisms by which the cell synthesizes its fundamental chemical ingredients and by which it obtains its energy. Chapter 3 describes the giant molecules of the cell, which are polymers of the small molecules and whose properties are responsible for the specificity of biological processes and the transfer of biological information.

The Chemical Components of a Cell

Cell Chemistry Is Based on Carbon Compounds[1]

A living cell is composed of a restricted set of elements, six of which (C, H, N, O, P, S) make up more than 99% of its weight. This composition differs markedly from that of the earth's crust and is evidence of a distinctive type of chemistry (Figure 2–1). What is this special chemistry, and how did it evolve?

The most abundant substance of the living cell is not special at all, since it covers two-thirds of the earth's surface. Water accounts for about 70% of the weight of cells, and most intracellular reactions occur in an aqueous environment. Life on this planet began in the ocean, and the conditions in that primeval environment put a permanent stamp on the chemistry of living things. All organisms have been designed around the unique properties of water, such as its polar character and hydrogen bonds, its high melting and boiling points, and its high surface tension (Panel C, pp. 46–47).

If we disregard water, all but a minor fraction of the molecules of a cell are carbon compounds, which are the subject matter of **organic chemistry.** Carbon is outstanding among all the elements on earth for its ability to form large molecules; only silicon comes anywhere close, and it is a poor second. The carbon atom, because of its small size and four outer-shell electrons, can form four strong covalent bonds with other atoms. Most important, it can join to other carbon atoms to form chains and rings and thereby generate large and complex molecules with no obvious upper limit to their size. Other abundant atoms in the cell are also small and able to make very strong covalent bonds (Panel D, pp. 48–49).

In principle, the simple rules of covalent bonding between carbon and other elements permit an astronomically large number of compounds. The number of different carbon compounds in a cell is very large, but it is only a tiny subset of what is theoretically possible. In some cases we can point to good reasons why this compound or that performs a given biological function; more often it seems that the actual "choice" was one among many reasonable

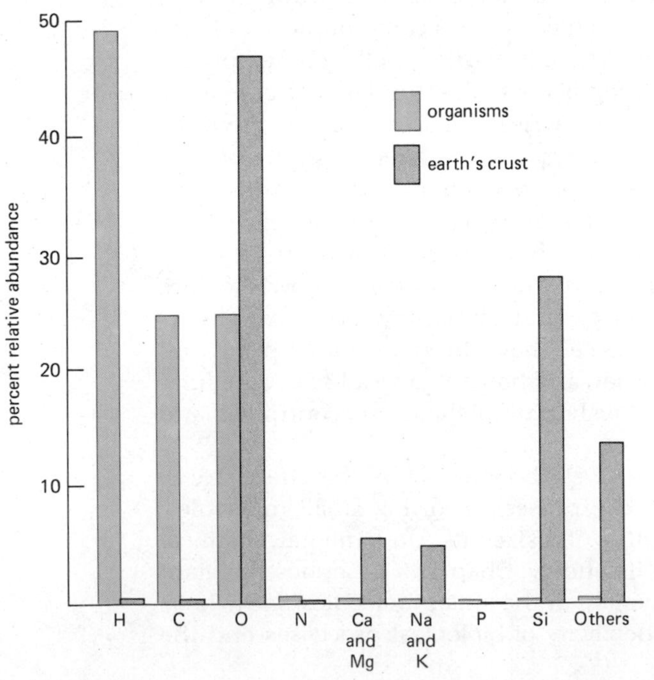

Figure 2–1 The relative abundance of chemical elements found in the earth's crust (the nonliving world) compared to that in the soft tissues of living organisms. The relative abundance is expressed as a percentage of the total number of atoms present.

alternatives, and something of an accident (Figure 2–2). Once established, certain chemical themes and patterns of reaction were preserved, with variations, during the course of evolution. Apparently the development of new classes of compounds was only rarely necessary or useful.

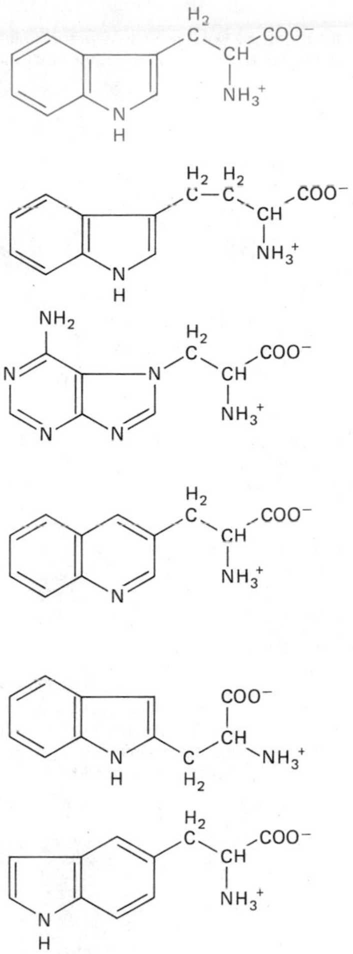

Cells Use Four Basic Types of Small Molecules

Certain simple combinations of atoms—such as the methyl (—CH_3), hydroxyl (—OH), carboxyl (—COOH), and amino (—NH_2) groups—recur repeatedly in biological molecules. Each such group has distinct chemical and physical properties that influence the behavior of whatever molecule the group occurs in. The main types of chemical groups and some of their salient properties are summarized in Panel D (pp. 48–49).

The so-called **small organic molecules** of the cell are carbon compounds with molecular weights in the 100 to 1000 range, containing up to 30 or so carbon atoms. Molecules of this kind are usually found free in solution in the cytoplasm, where they form a pool of intermediates from which large molecules, called *macromolecules*, are made. They are also essential intermediates in the chemical reactions that transform energy derived from food into usable forms (see below).

There are (at a rough estimate) nearly a thousand different kinds of small molecules in a cell, but many of them are related to each other in chemical structure. All biological molecules are synthesized from and broken down to the same simple compounds, synthesis and breakdown occurring through sequences of chemical changes that are limited in scope and follow definite rules. As a consequence, the compounds in a cell can be classified into a small number of distinct families. The large macromolecules in a cell, which form the subject of Chapter 3, are assembled from the small molecules and so belong to the same families.

Broadly speaking, cells contain four families of small organic molecules: the simple **sugars,** the **fatty acids,** the **amino acids,** and the **nucleotides.** Each of these families contains many different members with common chemical features. Although some cellular compounds do not fit into these categories, the four families together, including both the small molecules and the macromolecules made from them, account for a surprisingly large fraction of the cell mass (Table 2–1).

Figure 2–2 Living organisms synthesize only a small number of the organic molecules that they could in principle make. Of the six amino acids shown above, only the top one (tryptophan) is made by cells.

Table 2–1 The Approximate Chemical Composition of a Bacterial Cell

	Percent of Total Cell Weight	Number of Types of Each Molecule
Water	70	1
Inorganic ions	1	20
Sugars and precursors	3	200
Amino acids and precursors	0.4	100
Nucleotides and precursors	0.4	200
Lipids and precursors	2	50
Other small molecules	0.2	~200
Macromolecules (proteins, nucleic acids, and polysaccharides)	22	~5000

Adapted from S. E. Luria, S. J. Gould, and S. Singer, A View of Life. Menlo Park, Ca.: Benjamin-Cummings, 1981.

HYDROPHILIC AND HYDROPHOBIC MOLECULES

Because of the polar nature of water molecules, they will cluster around ions and other polar molecules.

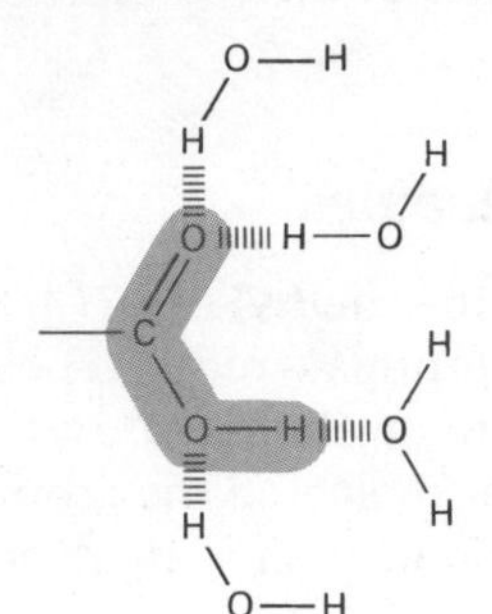

Substances that take part in water's hydrogen-bonded structures are therefore hydrophilic and relatively water-soluble.

Nonpolar molecules interrupt the H-bonded structure of water. They are therefore hydrophobic and quite insoluble in water.

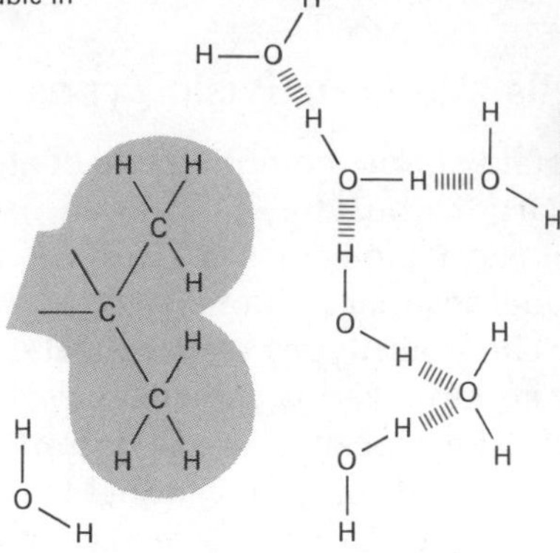

WATER

Although a water molecule has an overall neutral charge (having the same number of electrons and protons), the electrons are asymmetrically distributed, which makes the molecule polar

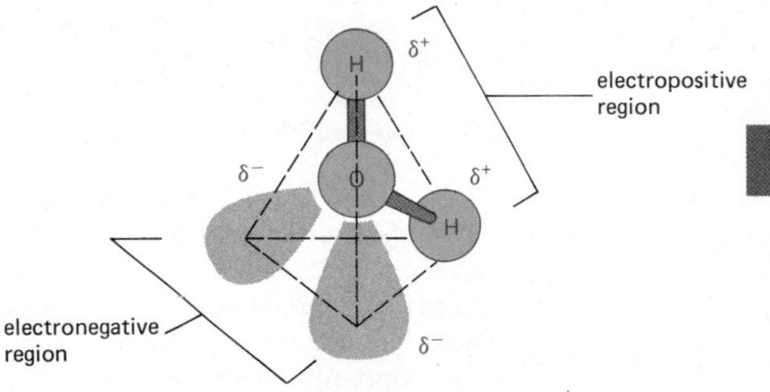

The oxygen nucleus partly draws electrons away from the hydrogen nuclei, leaving these nuclei with a small net positive charge. Weakly negative regions occur near the oxygen atom at the other 2 corners of an imaginary tetrahedron.

WATER STRUCTURE

Molecules of water join together transiently in a hydrogen-bonded lattice. Even at 37°C, 15% of the water molecules are joined to 4 others in a short-lived assembly known as a "flickering cluster."

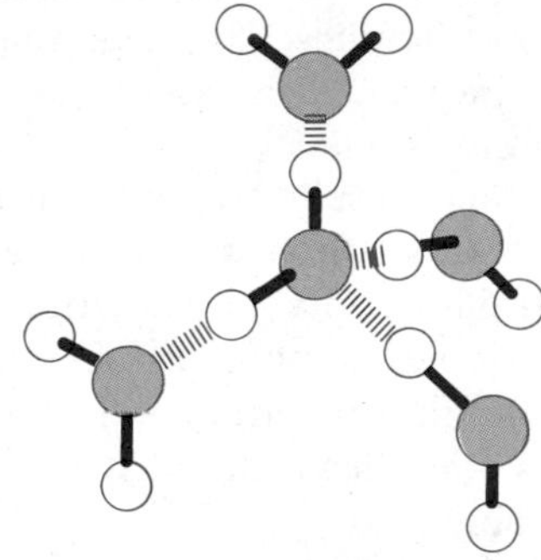

The cohesive nature of water is responsible for many of its unusual properties, such as high surface tension, specific heat, and heat of vaporization.

HYDROGEN BONDS

Because they are polarized, 2 adjacent H_2O molecules can form a linkage known as a hydrogen bond. Hydrogen bonds have only about 1/20 the strength of a covalent bond.

Hydrogen bonds are strongest when the 3 atoms lie in a straight line.

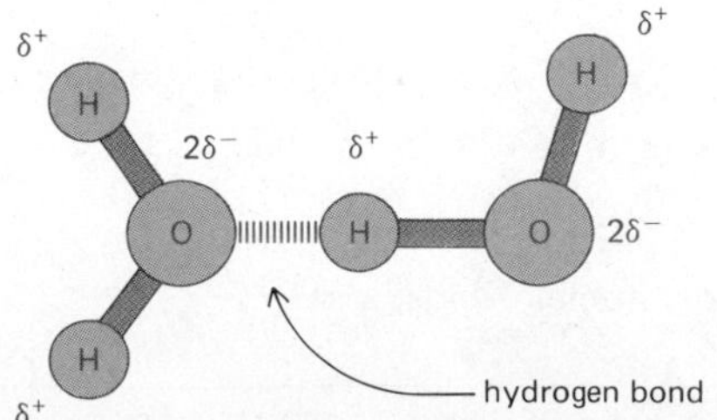

Panel C The chemical properties of water and their influence on the behavior of biological molecules.

HYDROPHOBIC INTERACTIONS CAN HOLD MOLECULES TOGETHER

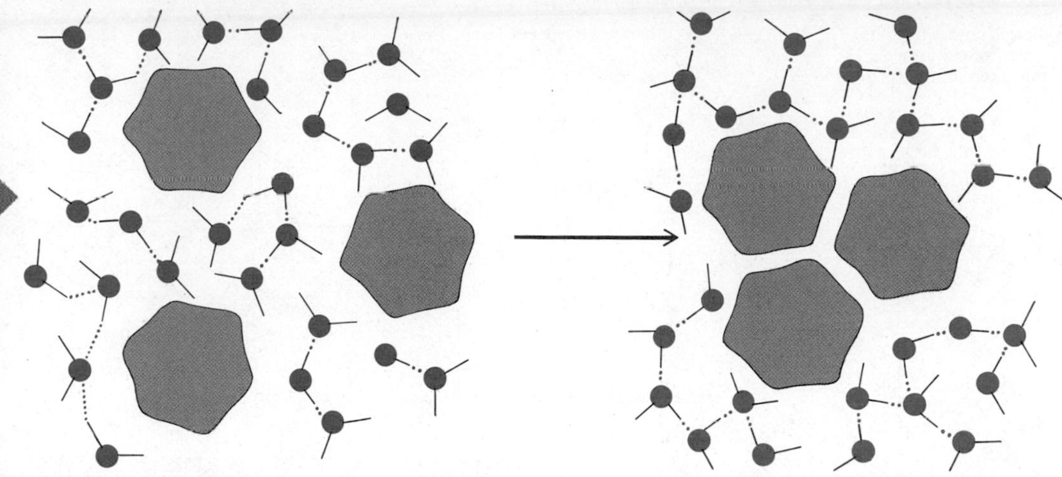

2 (or more) hydrophobic groups surrounded by water will tend to coalesce since they thereby cause less disruption to the hydrogen-bonded structure of water.

ACIDS AND BASES

An acid is a substance that releases an H^+ ion (proton) in solution.

e.g.,

$$\underset{\text{acid}}{CH_3COOH} \rightleftharpoons \underset{\text{base}}{CH_3COO^-} + \underset{\text{proton}}{H^+}$$

A base is a substance that accepts an H^+ ion (proton) in solution.

e.g.,

$$\underset{\text{base}}{CH_3{-}NH_2} + \underset{\text{proton}}{H^+} \rightleftharpoons \underset{\text{acid}}{CH_3{-}NH_3^+}$$

Water itself has a slight tendency to ionize, acting both as an acid and as a base.

$$H{-}O{-}H \cdots OH_2 \rightleftharpoons HO^{-} + H_3O^{+}$$

$$H_2O \rightleftharpoons H^+ + OH^-$$

pH

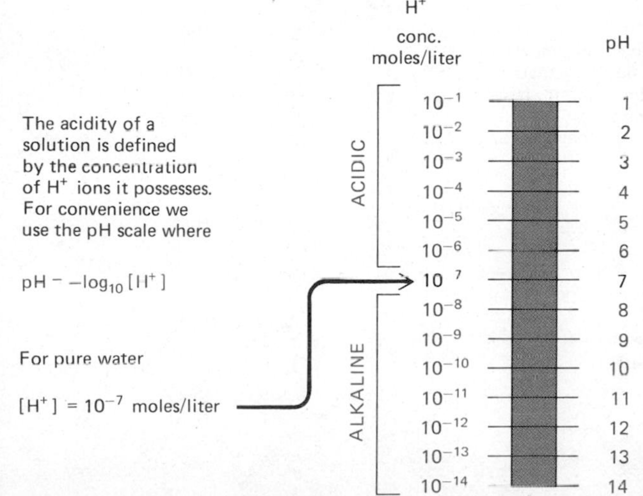

The acidity of a solution is defined by the concentration of H^+ ions it possesses. For convenience we use the pH scale where

$$pH = -\log_{10}[H^+]$$

For pure water

$[H^+] = 10^{-7}$ moles/liter

OSMOSIS

If 2 aqueous solutions are separated by a membrane that allows only water molecules to pass, water will move into the more concentrated solution by a process known as osmosis

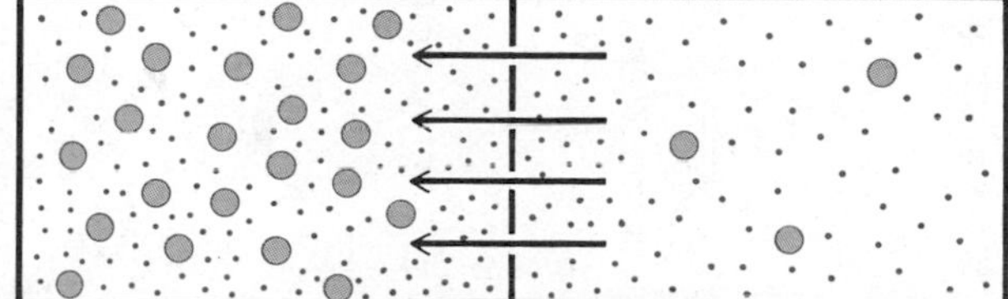

This movement of water from a hypotonic to a hypertonic solution causes an increase in hydrostatic pressure. Two solutions that are osmotically balanced are said to be isotonic.

CARBON SKELETONS

The unique role of carbon in the cell comes from its ability to form strong covalent bonds with other carbon atoms. Thus carbon atoms can join to form chains.

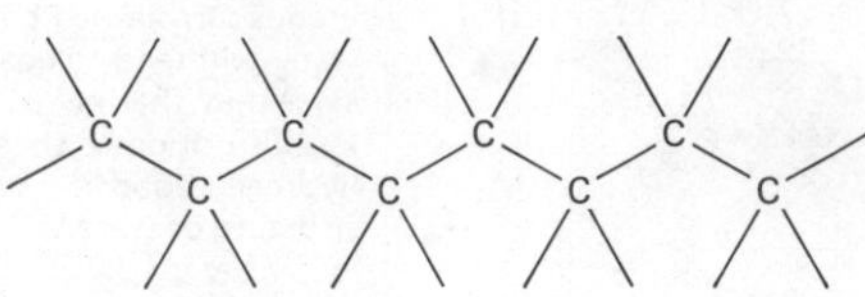

[also written as]

or branched trees

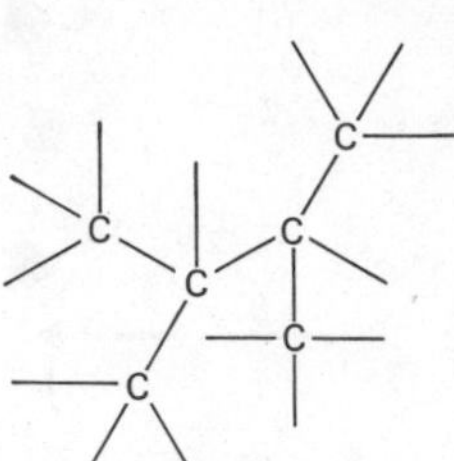

[also written as]

or rings

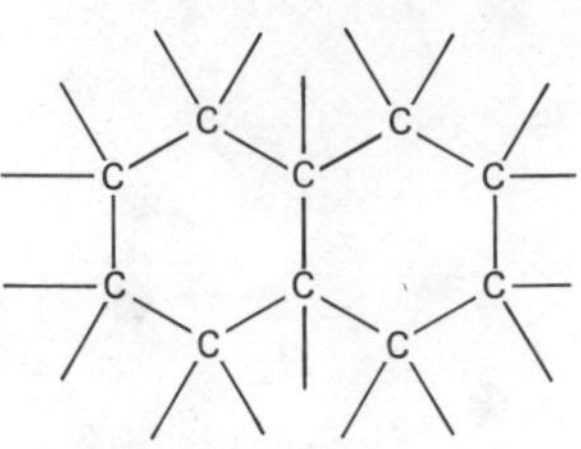

[also written as]

COVALENT BONDS

Atoms in biological molecules are usually joined by covalent bonds (formed by sharing pairs of electrons). Each atom can form a fixed number of such bonds in a definite spatial arrangement.

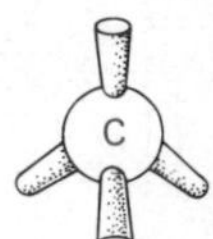

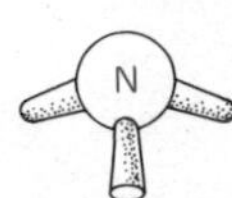

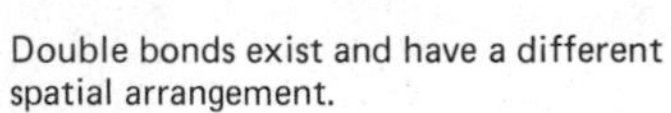

Double bonds exist and have a different spatial arrangement.

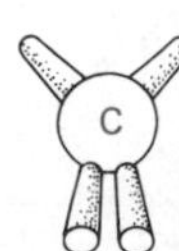

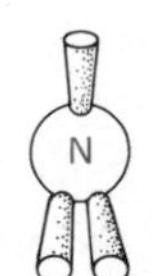

HYDROCARBONS

Carbon and hydrogen together make stable compounds called hydrocarbons. These are nonpolar, do not form hydrogen bonds, and are generally insoluble in water.

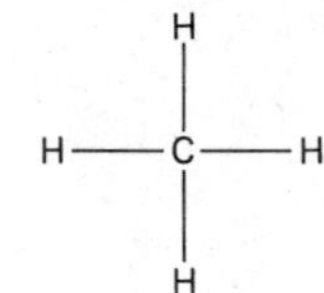

methane

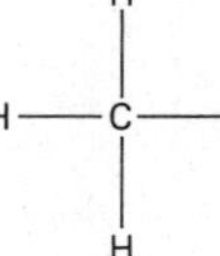

methyl group

Part of a fatty acid chain:

CH_3
H_2C
CH_2
H_2C
CH_2
H_2C
CH_2
H_2C
CH_2
H_2C
CH_2
H_2C
CH_2

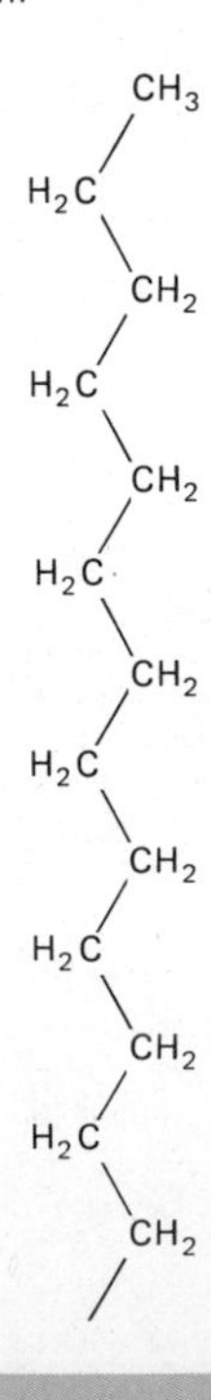

RESONANCE AND AROMATICITY

When resonance occurs throughout a ring compound, an aromatic ring is generated.

[often written as]

The carbon chain can include double bonds. If these are on alternate carbon atoms, the bonding electrons move within the molecule, stabilizing the structure by a phenomenon called resonance.

the truth is somewhere between these 2 structures

Panel D Chemical bonds and groups commonly encountered in biological molecules.

C–O COMPOUNDS

Many biological compounds contain a carbon bonded to an oxygen. For example:

alcohol

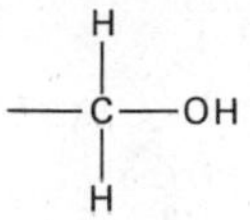

The –OH is called a hydroxyl group.

aldehyde

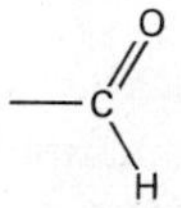

ketone

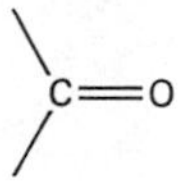

The C=O is called a carbonyl group.

carboxylic acid

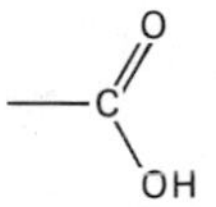

The –COOH is called a carboxyl group. In water this loses a H^+ ion to become $-COO^-$.

esters

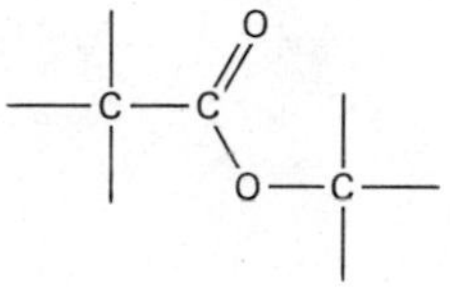

Esters are formed by combining an acid and an alcohol:

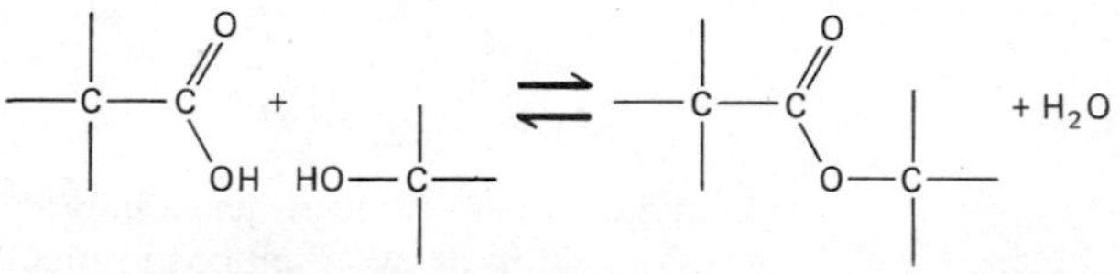

$+ H_2O$

C–N COMPOUNDS

Amines and amides are 2 important examples of compounds containing a carbon linked to a nitrogen.

Amines in water combine with a H^+ ion to become positively charged.

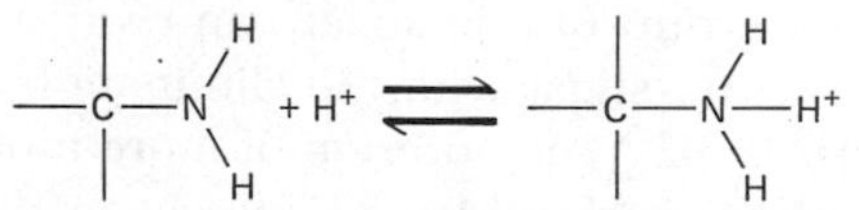

They are therefore basic.

Amides are formed by combining an acid and an amine. They are more stable than esters. Unlike amines, they are uncharged in water. An example is the peptide bond.

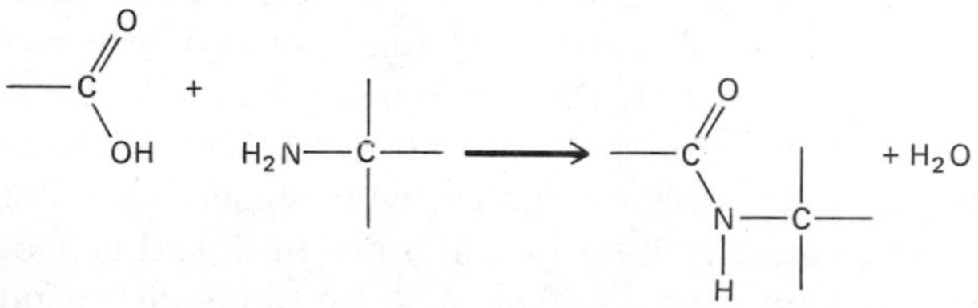

$+ H_2O$

Nitrogen also occurs in several ring compounds:

purines and pyrimidines

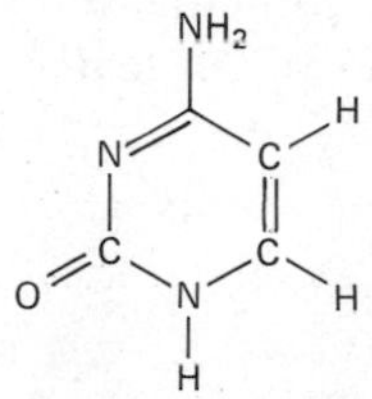

cytosine (a pyrimidine)

PHOSPHATES

Inorganic phosphate is a stable ion formed from phosphoric acid, H_3PO_4. It is often written as Ⓟ or P_i.

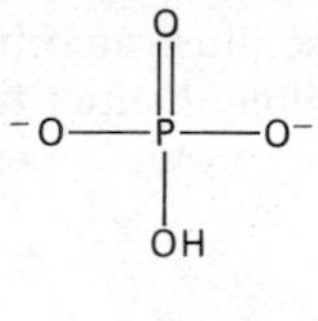

Phosphate esters can form between a phosphate and a free hydroxyl group.

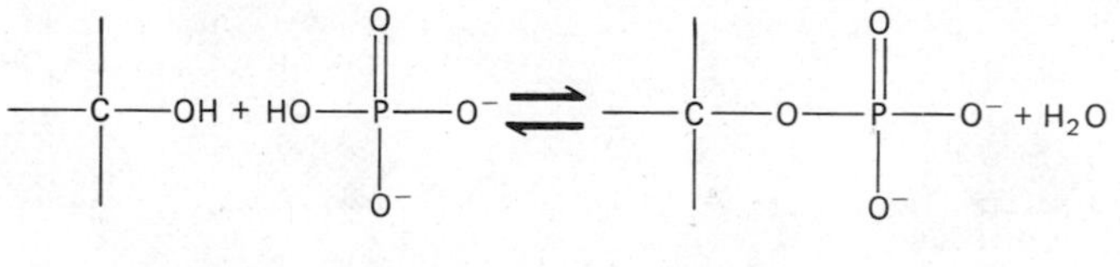

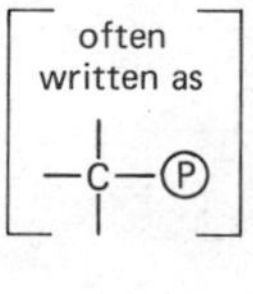

often written as

The combination of a phosphate and a carboxyl group, or 2 or more phosphate groups, gives an acid anhydride.

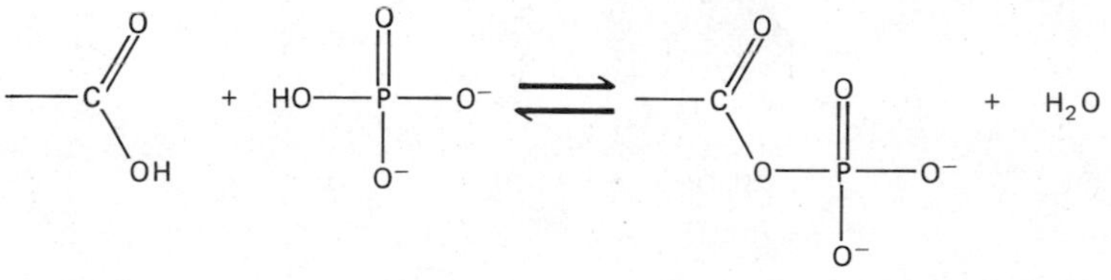

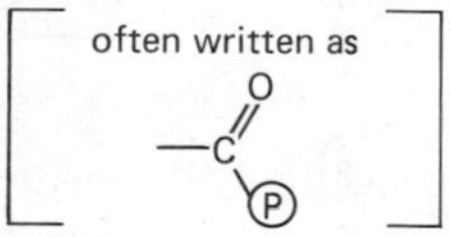

often written as

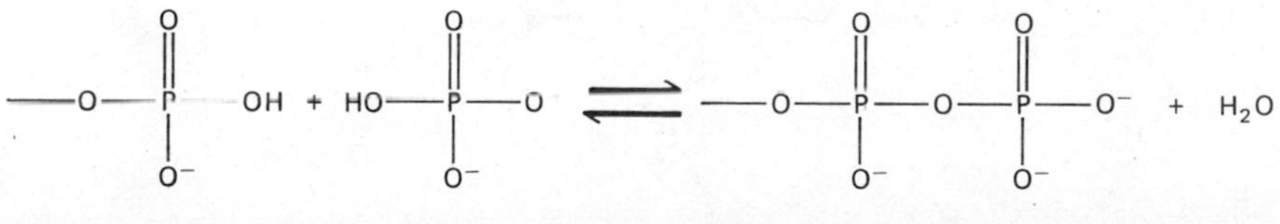

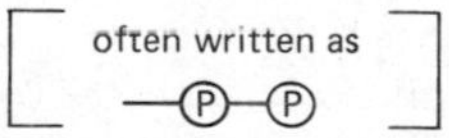

often written as

These reactions are very readily reversed, as the hydrolysis of acid anhydrides is highly favored.

Sugars Are Food Molecules of the Cell

The simplest type of sugars—the **monosaccharides**—are compounds with the general formula $(CH_2O)_n$, where *n* is an integer from three through seven. *Glucose*, for example, has the formula $C_6H_{12}O_6$ (Figure 2–3). All sugars contain hydroxyl groups and either an aldehyde ($\overset{}{_{H}}{>}C{=}O$) or a ketone ($>C{=}O$) group. The hydroxyl group of one sugar can combine with the aldehyde or ketone group of a second sugar with the elimination of water to form a disaccharide (Panel E, pp. 52–53). The addition of more monosaccharides in the same way results in **oligosaccharides** of increasing length (trisaccharides, tetrasaccharides, and so on) up to very large **polysaccharide** molecules with thousands of monosaccharide units (residues). Because each monosaccharide has several free hydroxyl groups that can form a link to another monosaccharide (or to some other compound), the number of possible polysaccharide structures is enormously large. Even a simple disaccharide consisting of two glucose residues can exist in 11 different varieties (Figure 2–4), while three different hexoses ($C_6H_{12}O_6$) can join together to make several thousand different trisaccharides. For this reason it is very difficult to determine the structure of any particular polysaccharide; with present methods it takes longer to determine the arrangement of half a dozen linked sugars (for example, those in a glycoprotein) than to determine the nucleotide sequence of a DNA molecule containing many thousands of nucleotides.

Glucose is the principal food compound of many cells. A series of oxidative reactions (see p. 70) leads from this hexose to various smaller sugar

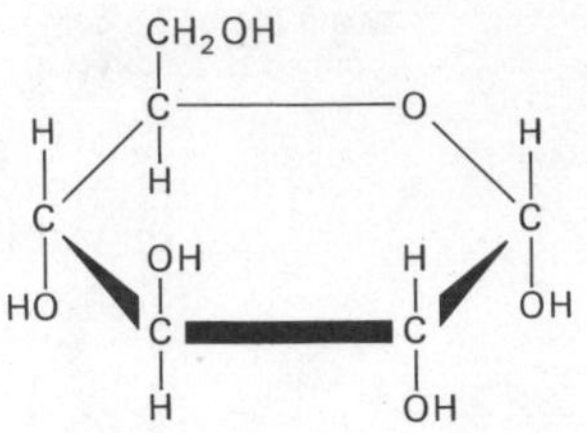

Figure 2–3 The structure of the monosaccharide glucose, a simple hexose sugar.

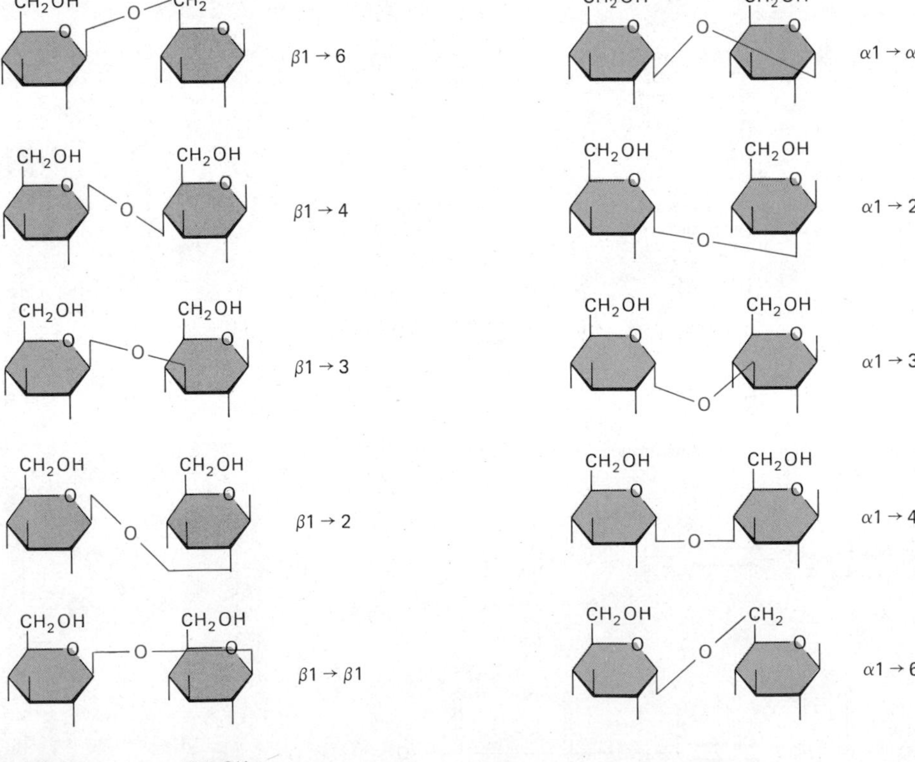

Figure 2–4 Eleven disaccharides consisting of two D-glucose units. Although these differ only in the type of linkage between the two glucose units, they are chemically distinct. Since the oligosaccharides associated with proteins and lipids may have six or more different kinds of sugar joined in both linear and branched arrangements through linkages such as those illustrated here, the number of possible distinct types of oligosaccharides is extremely large.

derivatives and eventually to CO_2 and H_2O. The net result can be written

$$C_6H_{12}O_6 + 6O_2 \rightarrow 6CO_2 + 6H_2O + \text{energy}$$

In the course of glucose breakdown, energy and "reducing power," both of which are essential in biosynthetic reactions, are salvaged and stored, mainly in the form of two crucial molecules, called **ATP** and **NADH,** respectively (see p. 68).

Simple polysaccharides, with a repeating structure composed of glucose—principally *glycogen* in animal cells and *starch* in plants—are used to store energy for future use. But sugars do not function exclusively in the production and storage of energy. Important extracellular structural materials (such as cellulose) are composed of simple polysaccharides, and smaller but more complex, nonrepeating sequences of sugar molecules are often covalently linked to proteins in *glycoproteins* and to lipids in *glycolipids.* The great variety possible in these short oligosaccharide chains of glycoproteins and glycolipids is thought to play a part in sophisticated recognition processes within and between cells.

Fatty Acids Are Components of Cell Membranes

A fatty acid molecule such as *palmitic acid* (Figure 2–5) has two distinct regions: a long hydrocarbon chain, which is hydrophobic (water insoluble) and not very reactive chemically, and a carboxylic acid group that is ionized in solution, extremely hydrophilic (water soluble), and readily forms esters and amides. In fact, almost all of the fatty acid molecules in a cell are covalently linked to other molecules by their carboxylic acid group. The many different fatty acids found in cells differ in such chemical features as the length of their hydrocarbon chains and the number and position of the carbon-carbon double bonds they contain (Panel F, pp. 54–55).

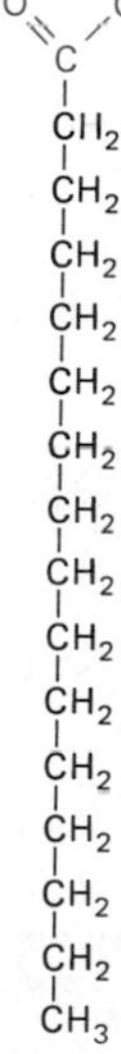

Figure 2–5 Palmitic acid. The carboxylic acid group (*color*) is shown in its ionized form.

Fatty acids are a valuable source of food since they can be broken down to produce more than twice as much ATP energy, weight for weight, as glucose. They are stored in the cytoplasm of many cells in the form of droplets of *triglyceride* molecules, which consist of three fatty acid chains, each joined to a glycerol molecule (Panel F, pp. 54–55). When required, the fatty acid chains can be released from triglycerides and broken down into two-carbon units. These two-carbon units, present as the acetyl group in a molecule called *acetyl CoA,* are then further degraded in various energy-yielding reactions, which will be described below.

But the most important function of fatty acids is in the construction of cellular membranes. These thin, impermeable sheets that enclose all cells and surround their internal organelles are composed largely of **phospholipids,** which are small molecules that resemble triglycerides in that they are constructed from fatty acids and glycerol. However, in phospholipids the glycerol is joined to two rather than three fatty acid chains. The remaining site on the glycerol is occupied by a phosphate group, which is in turn attached to another small hydrophilic compound such as *ethanolamine, choline,* or *serine.*

Each phospholipid molecule has a hydrophobic tail—composed of the two fatty acid chains—and a hydrophilic polar head group, where the phosphate is located. Thus phospholipid molecules are, in effect, detergents, and this is evident in their properties. A small amount of phospholipid will spread over the surface of water to form a *monolayer* of phospholipid molecules; in this thin film, the tail regions pack together very closely facing the air and the head groups are in contact with the water (Panel F, pp. 54–55). Two such films can combine tail to tail to make a phospholipid sandwich, or **lipid bilayer,** which is the structural basis of all cell membranes.

RING FORMATION

The aldehyde or ketone group of a sugar can react with a hydroxyl group.

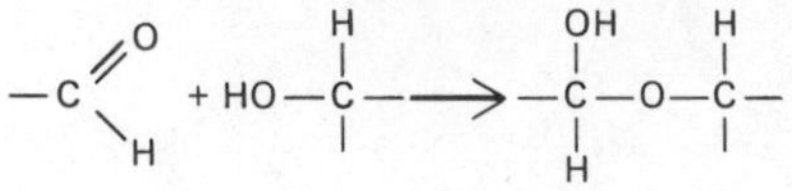

$$-C(=O)H + HO-C- \longrightarrow -C(OH)(H)-O-C-$$

For the larger sugars ($n > 4$) this can happen within the same molecule to form a 5- or 6-membered ring.

glucose

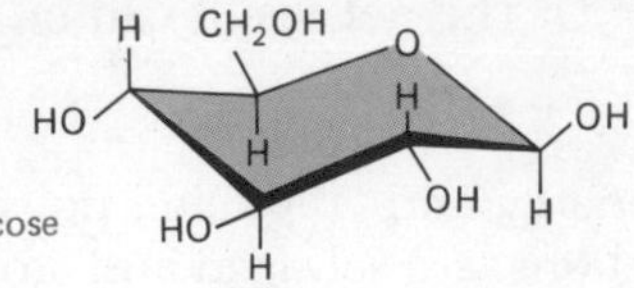

ribose

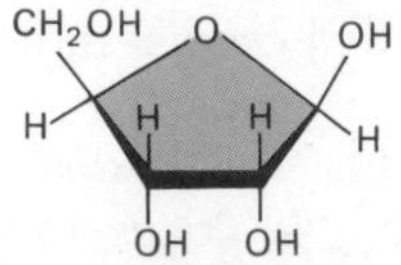

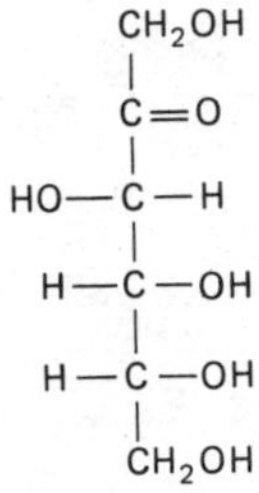

MONOSACCHARIDES

Monosaccharides are aldehydes or ketones

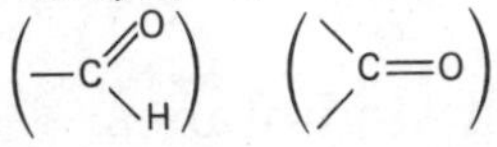

that also have 2 or more hydroxyl groups. Their general formula is $(CH_2O)_n$. The simplest are trioses ($n = 3$) such as

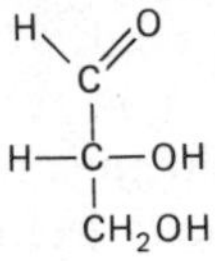

H–C(=O)–CH(OH)–CH_2OH

glyceraldehyde (an aldose)

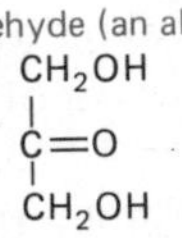

CH_2OH–C(=O)–CH_2OH

dihydroxyacetone (a ketose)

HEXOSES

$n = 6$ 2 common hexoses are

glucose: H–C(=O)–(H–C–OH)–(HO–C–H)–(H–C–OH)–(H–C–OH)–CH_2OH

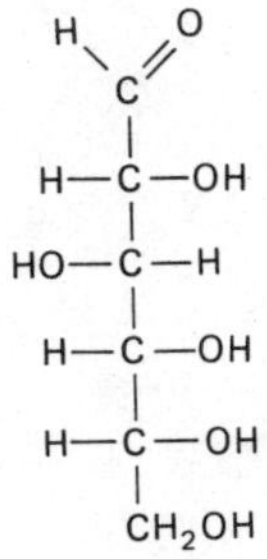

fructose: CH_2OH–(C=O)–(HO–C–H)–(H–C–OH)–(H–C–OH)–CH_2OH

PENTOSES

$n = 5$

a common pentose is

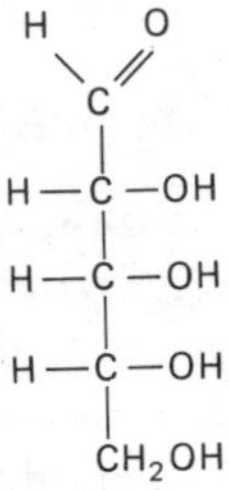

H–C(=O)–(H–C–OH)–(H–C–OH)–(H–C–OH)–CH_2OH

ribose

NUMBERING SYSTEM

The carbon atoms of a sugar are numbered from the end closest to the adehyde or ketone.

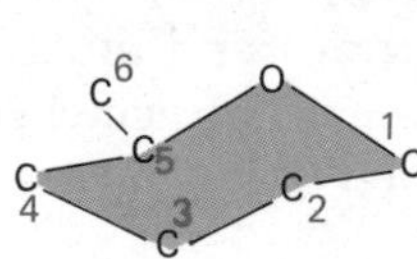

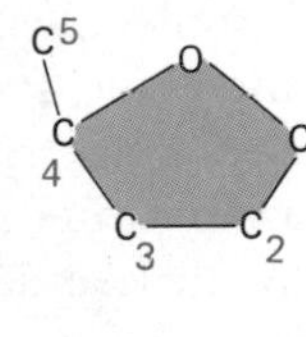

STEREOISOMERS

Monosaccharides have many isomers that differ only in the orientation of their hydroxyl groups – e.g., glucose, galactose, and mannose are isomers of each other.

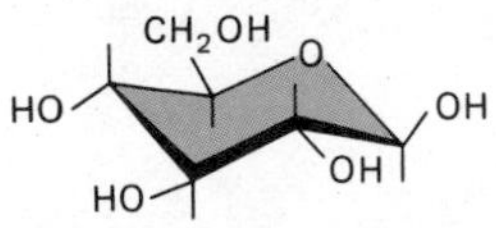

glucose

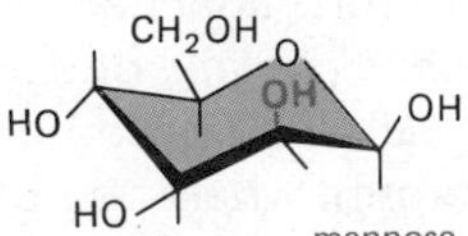

mannose

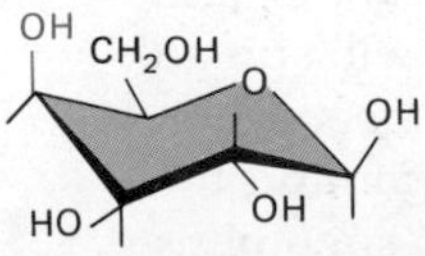

galactose

D and L FORMS

2 isomers that are mirror images of each other have the same chemistry and therefore are given the same name and distinguished by the prefix D or L.

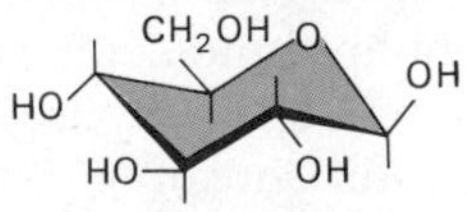

D-glucose

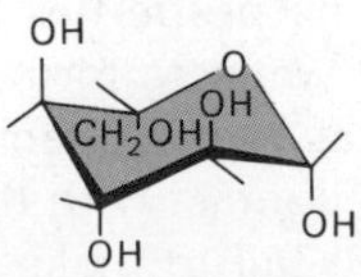

L-glucose

Panel E An outline of some of the types of sugars commonly found in cells.

α- AND β-LINKS

The hydroxyl group on the carbon that carries the aldehyde or ketone can rapidly change from one position to another. These 2 positions are called α- and β-.

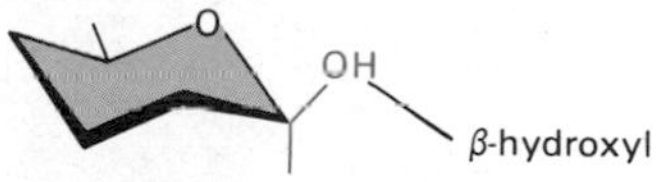

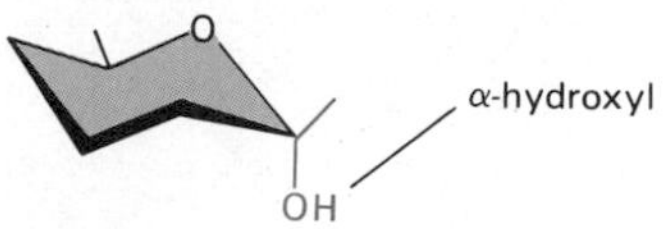

As soon as 1 sugar is linked to another, the α- or β-form is frozen.

SUGAR DERIVATIVES

The hydroxyl groups of a simple monosaccharide can be replaced by other groups. For example

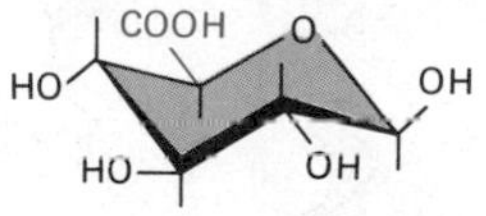

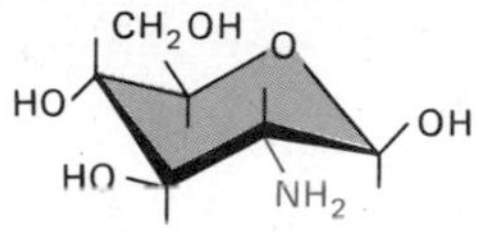

DISACCHARIDES

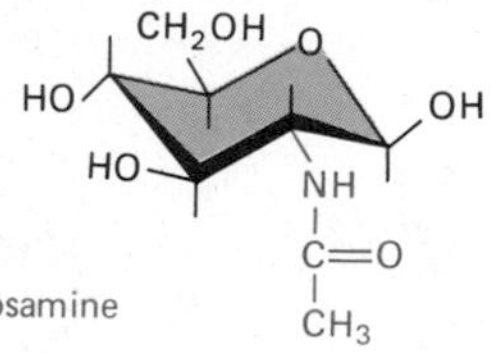

2 sugars can react with each other forming a glycosidic link. 3 common disaccharides are maltose (glucose α1,4 glucose), lactose (galactose β1,4 glucose), and sucrose (glucose α1,2 fructose).

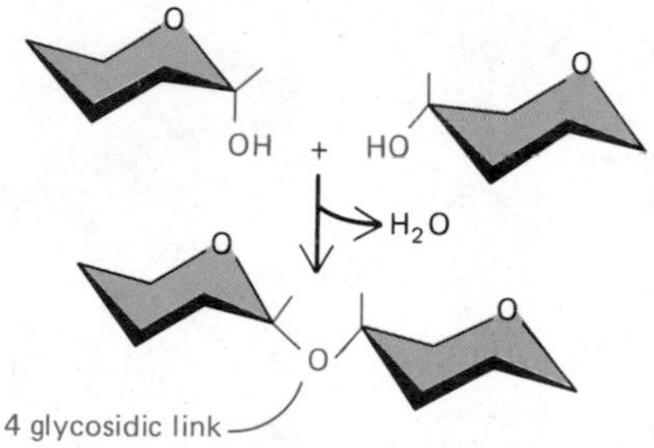

OLIGOSACCHARIDES AND POLYSACCHARIDES

Large linear and branched molecules can be made from simple repeating units. Short chains are called oligosaccharides, while long chains are called polysaccharides. Glycogen, for example, is a polysaccharide made entirely of glucose units joined together.

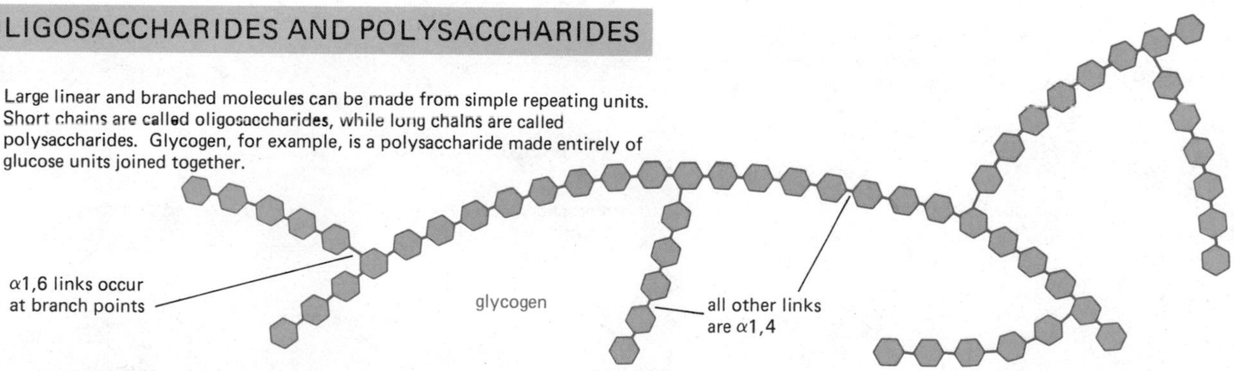

COMPLEX OLIGOSACCHARIDES

In many cases a sugar sequence is nonrepetitive. Very many different molecules are possible. Such complex oligosaccharides are usually linked to proteins or to lipids.

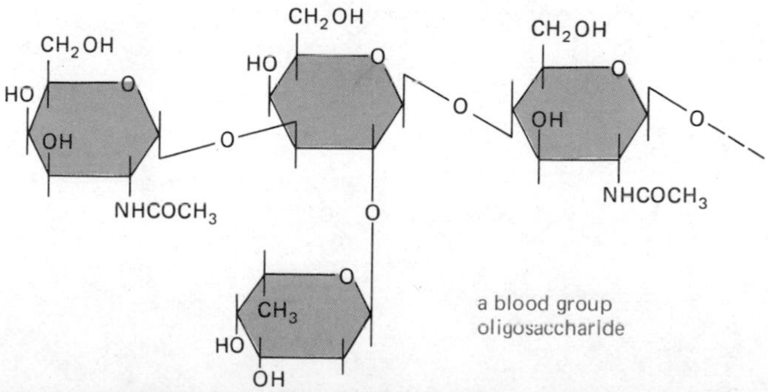

COMMON FATTY ACIDS

These are carboxylic acids with long hydrocarbon tails.

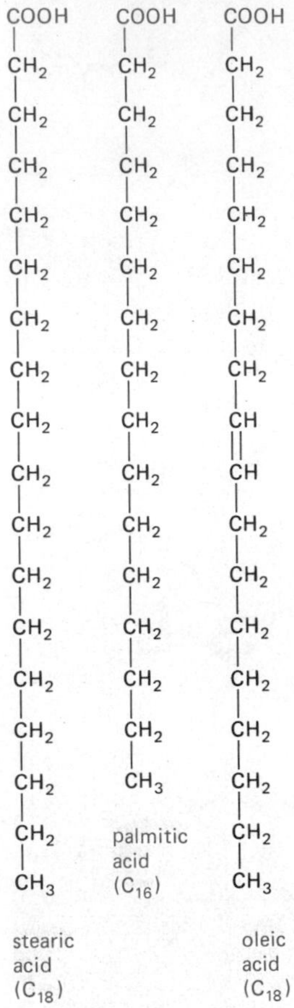

stearic acid (C_{18})	palmitic acid (C_{16})	oleic acid (C_{18})
COOH	COOH	COOH
CH_2	CH_2	CH_2
CH_2	CH_2	CH_2
CH_2	CH_2	CH_2
CH_2	CH_2	CH_2
CH_2	CH_2	CH_2
CH_2	CH_2	CH_2
CH_2	CH_2	CH_2
CH_2	CH_2	CH
CH_2	CH_2	‖ CH
CH_2	CH_2	CH_2
CH_2	CH_2	CH_2
CH_2	CH_2	CH_2
CH_2	CH_2	CH_2
CH_2	CH_2	CH_2
CH_2	CH_3	CH_2
CH_2		CH_2
CH_3		CH_3

Hundreds of different kinds of fatty acids exist. Some have 1 or more double bonds and are said to be unsaturated.

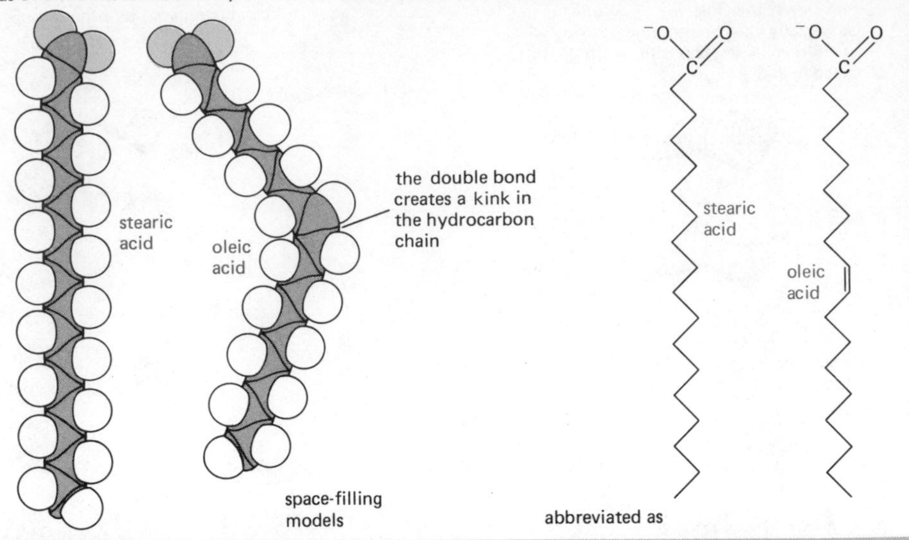

TRIGLYCERIDES

Fatty acids are stored as an energy reserve (fat) through linkage to glycerol to form triglycerides.

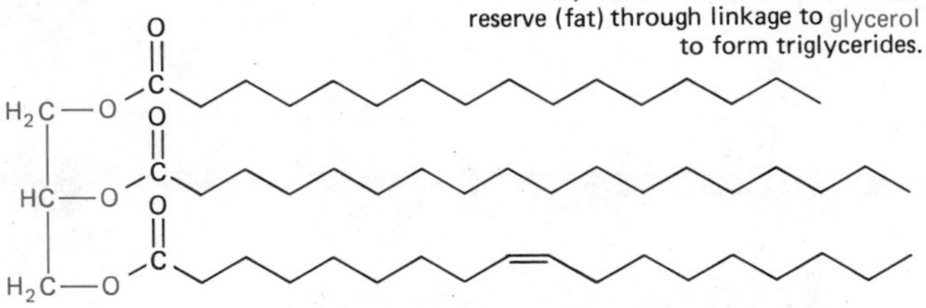

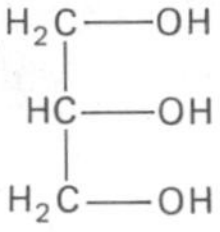

$H_2C—O—C(=O)—$
$HC—O—C(=O)—$
$H_2C—O—C(=O)—$

$H_2C—OH$
$HC—OH$
$H_2C—OH$
glycerol

CARBOXYL GROUP

If free, the carboxyl group of a fatty acid will be ionized.

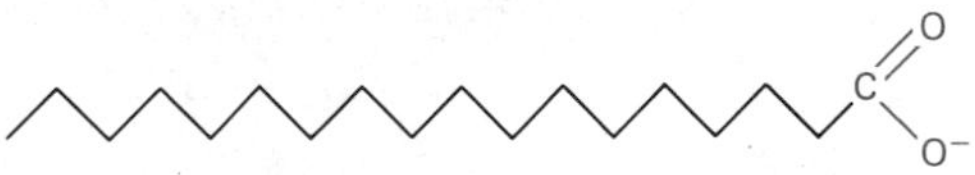

$—C(=O)O^-$

But more usually it is linked to other groups to form either esters

$—C(=O)—O—C—$

or amides

$—C(=O)—N(H)—$

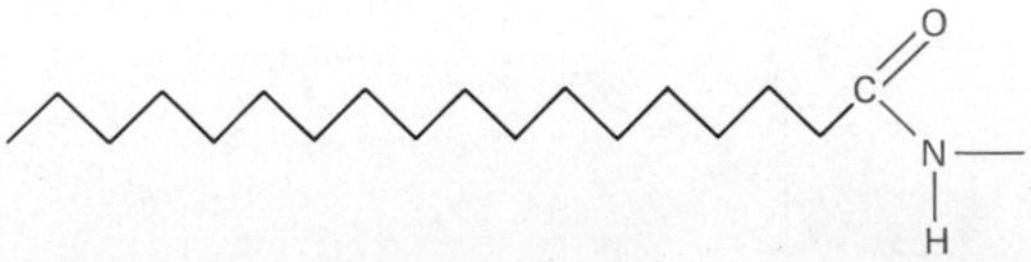

PHOSPHOLIPIDS

Phospholipids are the major constituent of cell membranes.

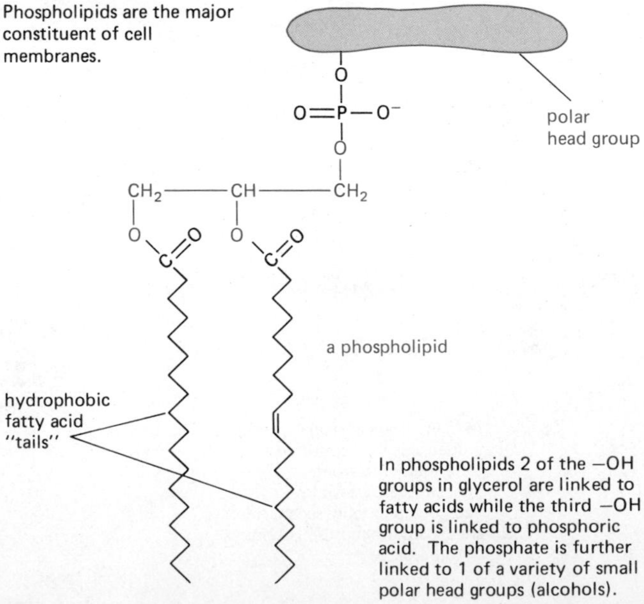

polar head group

O
$O=P—O^-$
O
$CH_2—CH—CH_2$
O O
C=O C=O

a phospholipid

hydrophobic fatty acid "tails"

In phospholipids 2 of the —OH groups in glycerol are linked to fatty acids while the third —OH group is linked to phosphoric acid. The phosphate is further linked to 1 of a variety of small polar head groups (alcohols).

Panel F An outline of some of the types of fatty acids commonly encountered in cells, and the structures that they form.

LIPID AGGREGATES

Fatty acids have a hydrophilic head and a hydrophobic tail.

In water they can form a surface film or form small micelles.

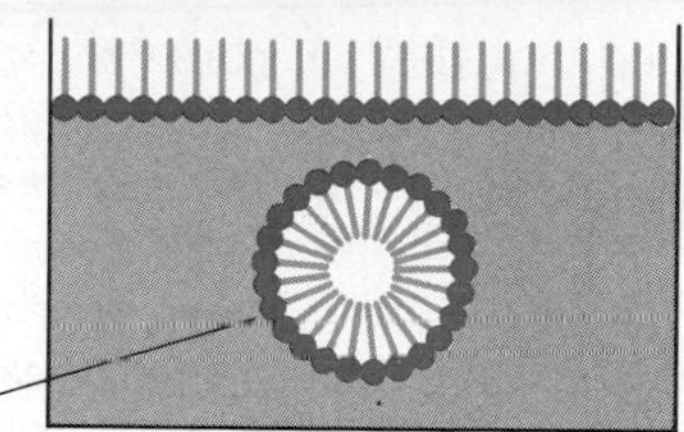

Their derivatives can form larger aggregates held together by hydrophobic forces:

Triglycerides form large spherical fat droplets in the cell cytoplasm.

1-200 nm or more

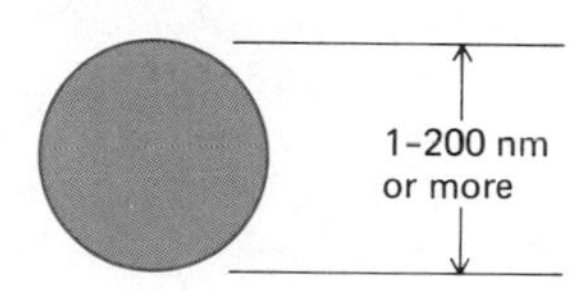

Phospholipids and glycolipids form self-sealing lipid bilayers that are the basis for all cellular membranes.

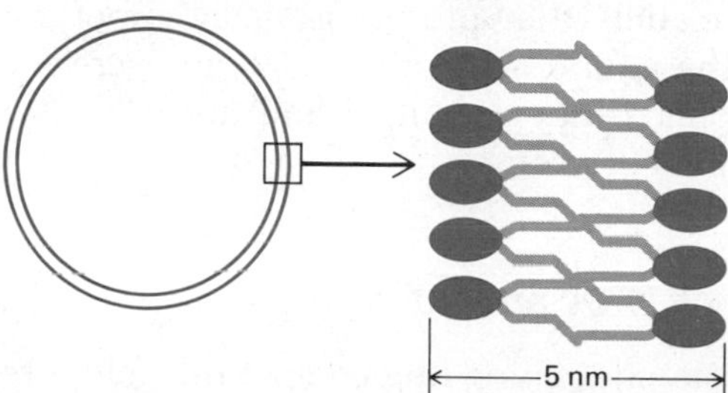

OTHER LIPIDS

Lipids are compounds that are soluble in organic solvents. 2 other common types of lipid are steroids and polyisoprenoids. Both are made from isoprene units.

CH_3—C(=CH_2)—CH=CH_2 isoprene

STEROIDS

Steroids have a common multiple-ring structure.

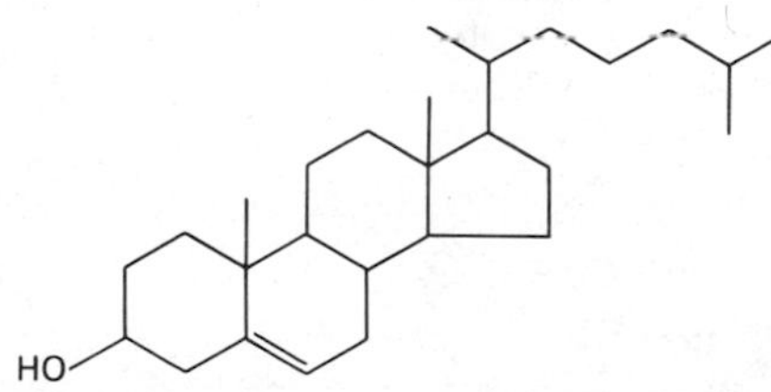

cholesterol — found in many membranes

OH

O

testosterone — male steroid hormone

GLYCOLIPIDS

Like phospholipids, these compounds are composed of a hydrophobic region, containing 2 long hydrocarbon tails, and a polar region, which now contains 1 or more sugar residues and no phosphate.

galactose

sugar residue

hydrophobic region

a simple glycolipid

POLYISOPRENOIDS

long chain polymers of isoprene

dolichol phosphate — used to carry activated sugars in the membrane-associated synthesis of glycoproteins and some polysaccharides.

Amino Acids Are the Subunits of Proteins

The common amino acids are chemically varied, but they all contain a carboxylic acid group and an amino group, both linked to a single carbon atom (Figure 2–6). They serve as subunits in the synthesis of **proteins,** which are long linear polymers of amino acids joined head to tail by a peptide bond between the carboxylic acid group of one amino acid and the amino group of the next (Figure 2–7). There are 20 common amino acids in proteins, each with a different *side chain* attached to the α-carbon atom (Figure 2–8 and Panel G, pp. 58–59). The same 20 amino acids occur over and over again in all proteins, including those made by bacteria, plants, and animals. While the choice of precisely these 20 amino acids is probably an example of an evolutionary accident, the chemical versatility they provide is vitally important. As we shall see, the properties of the amino acid side chains, in aggregate, determine the properties of the proteins they constitute and underlie all of the diverse and sophisticated functions of proteins.

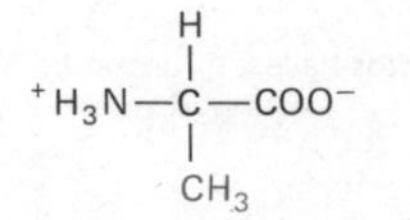

Figure 2–6 The amino acid alanine as it exists at pH 7 in its ionized form. When incorporated into a polypeptide chain, the charges on the amino and carboxyl groups of the free amino acid disappear.

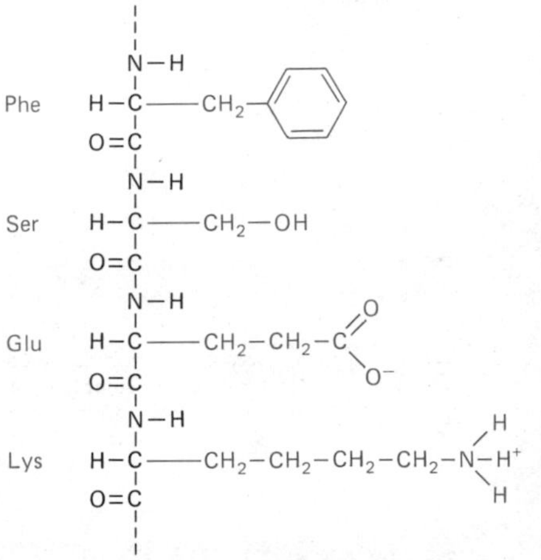

Figure 2–7 A small part of a protein molecule. The four amino acids shown are linked together by a type of covalent bond called a peptide bond. A protein is therefore also sometimes referred to as a *polypeptide.*

Nucleotides Are the Subunits of DNA and RNA

In nucleotides, one of several nitrogen-containing ring compounds (often referred to as *bases* because they can combine with H^+) is linked to a five-carbon sugar (either *ribose* or *deoxyribose*) that also carries a phosphate group. There is a strong family resemblance between the nitrogen-containing rings found in nucleotides. *Cytosine* (C), *thymine* (T), and *uracil* (U) are called **pyrimidine** compounds because they are all simple derivatives of a six-membered pyrimidine ring; *guanine* (G) and *adenine* (A) are **purine** compounds, with a second five-membered ring fused to the six-membered ring (Panel H, pp. 60–61).

Nucleotides can act as carriers of chemical energy. The triphosphate ester of adenine, **ATP** (Figure 2–9), above all others, participates in the transfer of energy between hundreds of individual cellular reactions. Its two reactive and easily hydrolyzable terminal phosphates are forced into their covalent linkages during the oxidation of foodstuffs, and the energy released by the hydrolysis of either or both of these "high-energy" phosphate groups can be used elsewhere to drive energetically unfavorable biosynthetic processes. Other nucleotide derivatives serve as carriers for the transfer of particular chemical groups, such as hydrogen atoms or sugar residues, from one molecule to another. And a cyclic-phosphate-containing adenine derivative, *cyclic AMP,* serves as a universal signaling molecule within cells and controls the speed of many different intracellular reactions.

As we have already mentioned in Chapter 1, the special significance of nucleotides is in the preservation of biological information. Nucleotides serve as building blocks for the construction of **nucleic acids,** long polymers in which nucleotide subunits are covalently linked by the formation of a phosphate ester between the 3′-hydroxyl group on the sugar residue of one nucleotide and the 5′-phosphate group on the next (Figure 2–10). There are two main types of nucleic acids, differing in the type of sugar in their polymeric backbone. Those based on the sugar *ribose* are known as **ribonucleic acids,** or **RNA,** and contain the bases A, U, G, and C. Those based on *deoxyribose* (in which the hydroxyl at the 2′ position of ribose is replaced by a hydrogen) are known as **deoxyribonucleic acids,** or **DNA;** these contain the bases A, T, G, and C. The sequence of bases in a DNA or RNA polymer represents the genetic information of the living cell. The ability of the bases from different nucleic acid molecules to recognize each other by noncovalent interactions (called **base-pairing**)—G with C, and A with either T or U—underlies all of heredity and evolution, as will be explained in the following chapter.

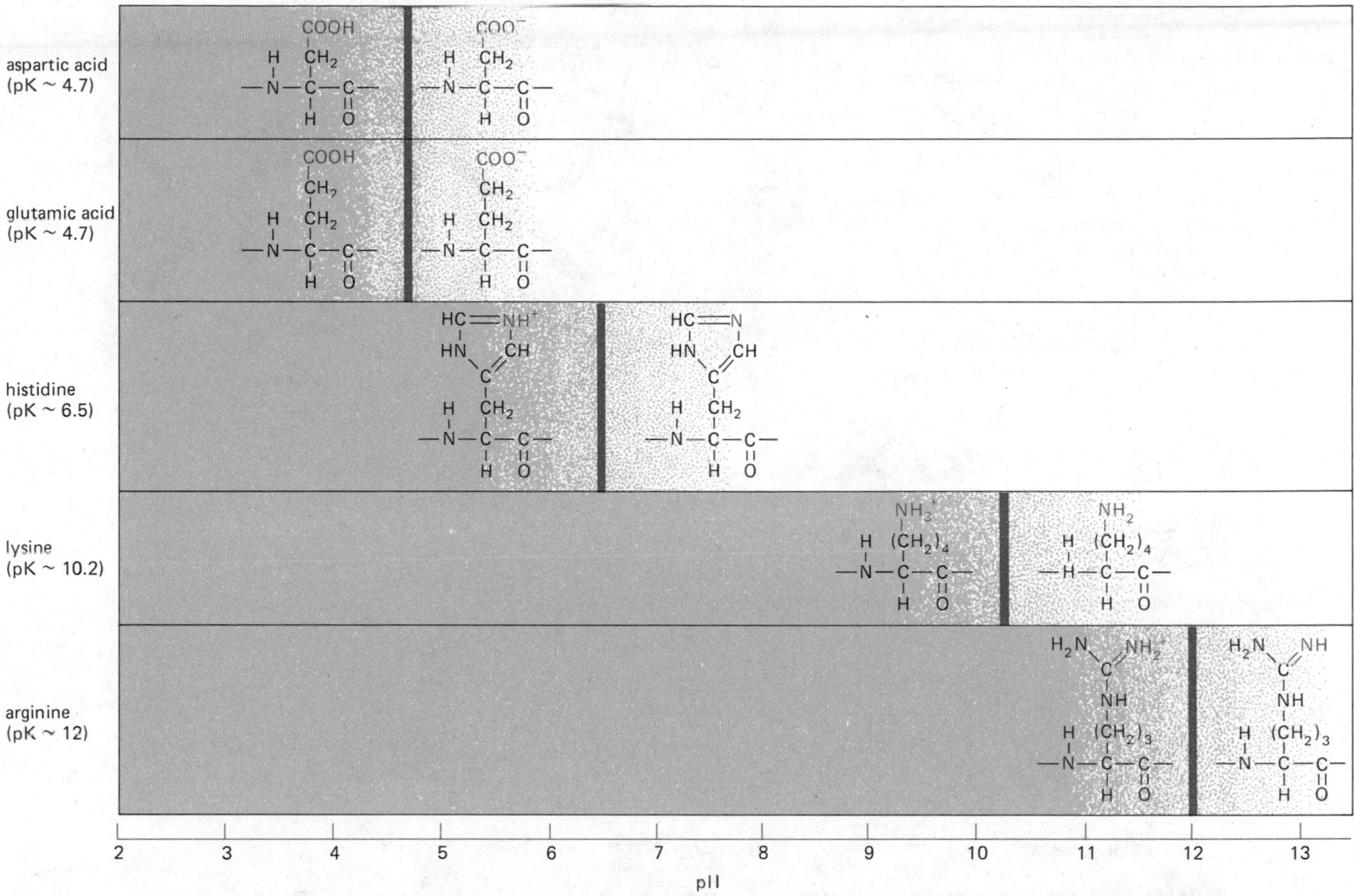

Figure 2–8 The charge on amino acid side chains depends on the pH. Carboxylic acids readily lose H^+ in aqueous solution to form a negatively charged ion that is denoted by the suffix "-ate," as in aspart*ate* or glutam*ate*. A comparable situation exists for amines, which in aqueous solution take up H^+ to form a positively charged ion (which does not have a special name). These reactions are rapidly reversible, and the amounts of the two forms, charged and uncharged, depend on the pH of the solution. At a high pH, carboxylic acids tend to be charged and amines uncharged; at a low pH, the opposite is true, the carboxylic acids are uncharged and amines charged. The pH at which exactly *half* of the carboxylic acid or amine residues are charged is known as the pK of that amino acid.

In the cell the pH is close to 7, and almost all carboxylic acids and amines are in their fully charged form.

triphosphate
ribose
adenine
adenosine
A-(P)-(P)-(P) *or* ATP

Figure 2–9 Chemical structure of adenosine triphosphate (ATP); two common abbreviations for this compound are also shown.

THE AMINO ACID

The general formula of an amino acid is

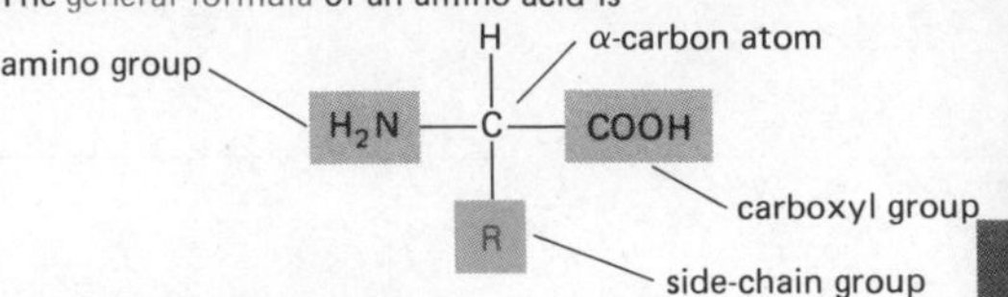

R is commonly one of 20 different side chains. At pH 7 both the amino and carboxyl groups are ionized.

$$^{+}H_3N—C(H)(R)—COO^{-}$$

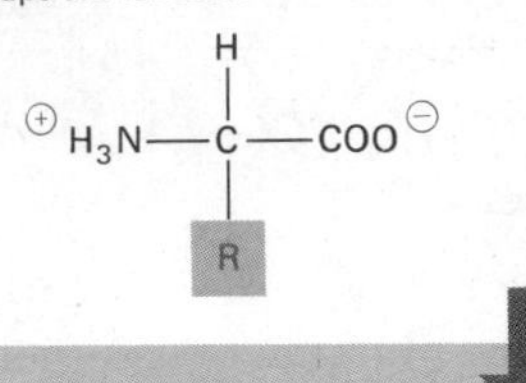

OPTICAL ISOMERS

The α-carbon atom is asymmetric, which allows for 2 mirror image (or stereo-) isomers, D and L.

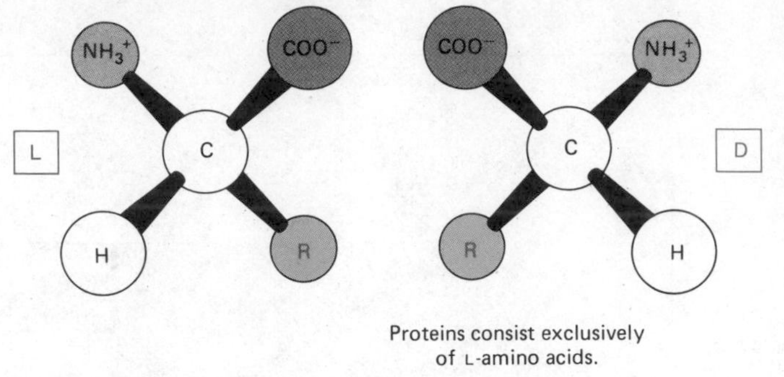

Proteins consist exclusively of L-amino acids.

PEPTIDE BONDS

Amino acids are commonly joined together by an amide linkage, called a peptide bond.

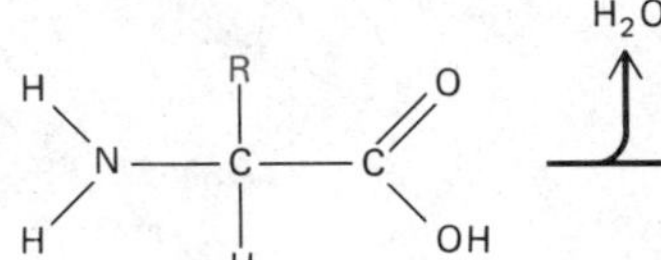
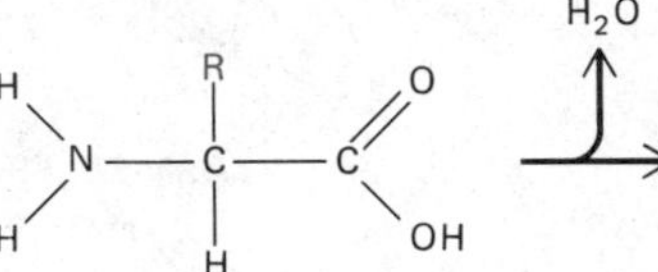

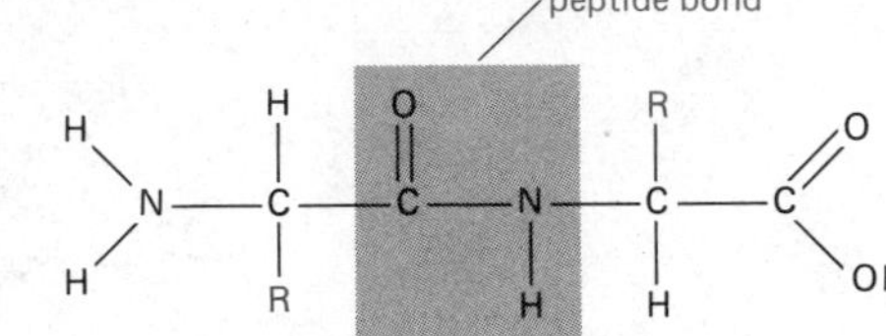

Proteins are long polymers of amino acids linked by peptide bonds, and they are always written with the *N*-terminus toward the left.

amino or *N*-terminus

$$^{+}H_3N—CH(CH_2\text{-imidazole})—CO—NH—CH(CH_2SH)—CO—NH—CH(CH(CH_3)_2)—COO^{-}$$

carboxyl or *C*-terminus

FAMILIES OF AMINO ACIDS

The common amino acids are grouped according to whether their side chains are

- acidic
- basic
- uncharged polar
- nonpolar

These 20 amino acids are given both 3-letter and 1-letter abbreviations.

Thus: alanine = Ala = A

BASIC SIDE CHAINS

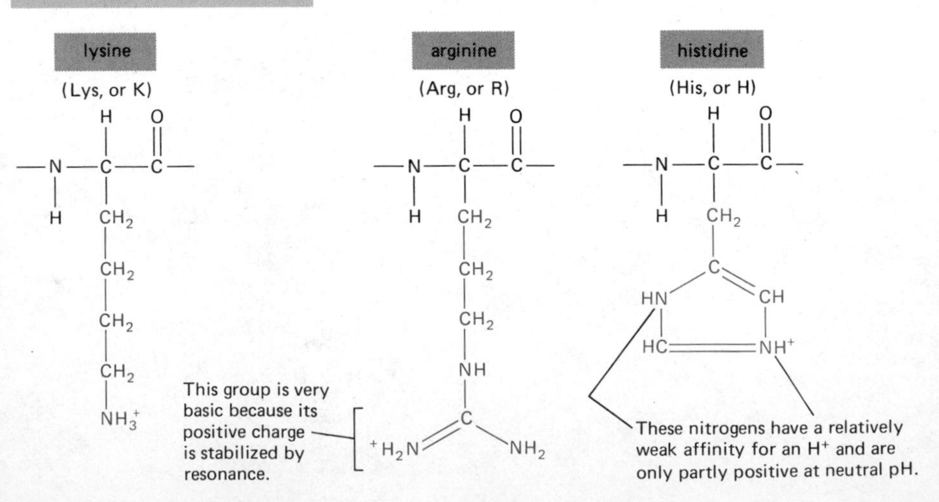

Panel G The 20 amino acids involved in the synthesis of proteins.

ACIDIC SIDE CHAINS

aspartic acid

(Asp, or D)

glutamic acid

(Glu, or E)

UNCHARGED POLAR SIDE CHAINS

glycine

(Gly, or G)

asparagine

(Asn, or N)

glutamine

(Gln, or Q)

Although the amide N is not charged at neutral pH, it is polar.

cysteine

(Cys, or C)

Paired cysteines allow disulfide bonds to form in proteins.

—CH_2—S—S—CH_2—

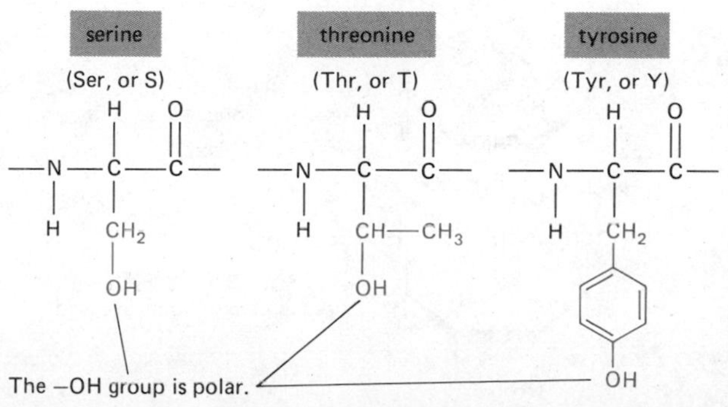

NONPOLAR SIDE CHAINS

alanine

(Ala, or A)

valine

(Val, or V)

leucine

(Leu, or L)

isoleucine

(Ileu, or I)

proline

(Pro, or P)

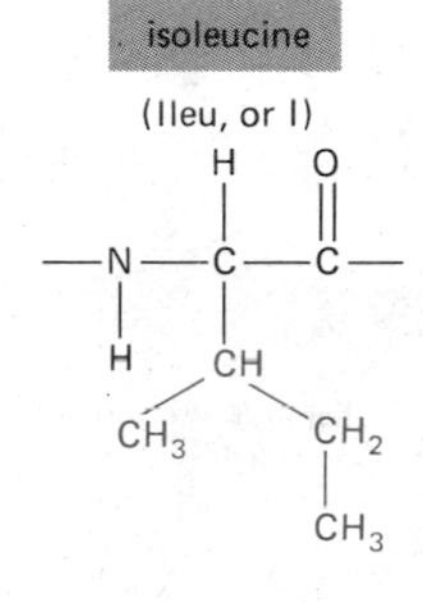

(actually an imino acid)

phenylalanine

(Phe, or F)

methionine

(Met, or M)

tryptophan

(Trp, or W)

Nonpolar side chains in proteins tend to cluster together on the inside of the molecule.

BASES

The bases are N-containing ring compounds, either purines or pyrimidines.

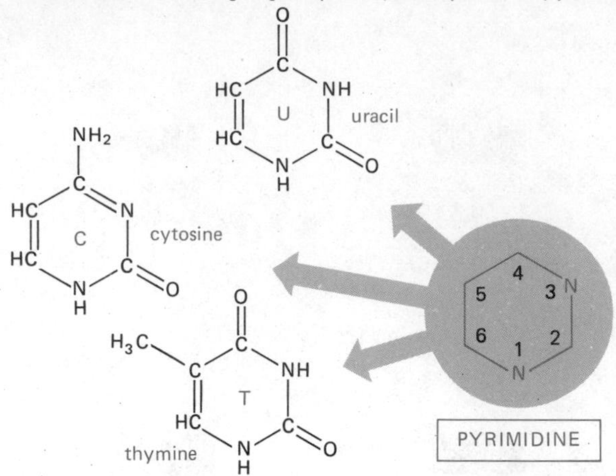

PHOSPHATES

The phosphates are normally joined to the C5 hydroxyl of the ribose or deoxyribose sugar. Mono-, di-, and triphosphates are common.

as in AMP

as in ADP

as in ATP

The phosphate makes a nucleotide negatively charged.

NUCLEOTIDES

A nucleotide consists of a nitrogen-containing base, a 5-carbon sugar, and 1 or more phosphate groups.

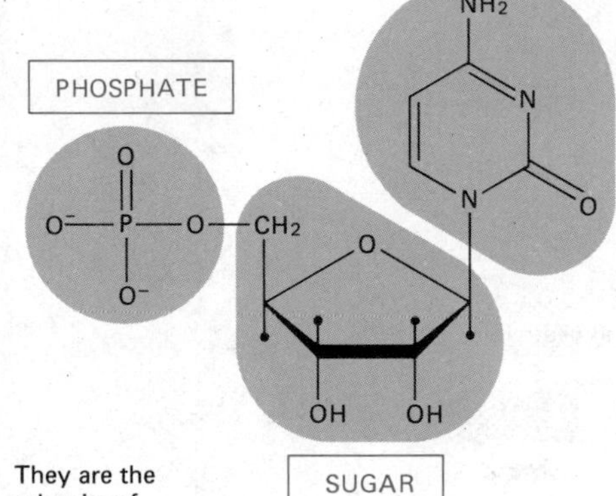

They are the subunits of the nucleic acids.

BASE-SUGAR LINKAGE

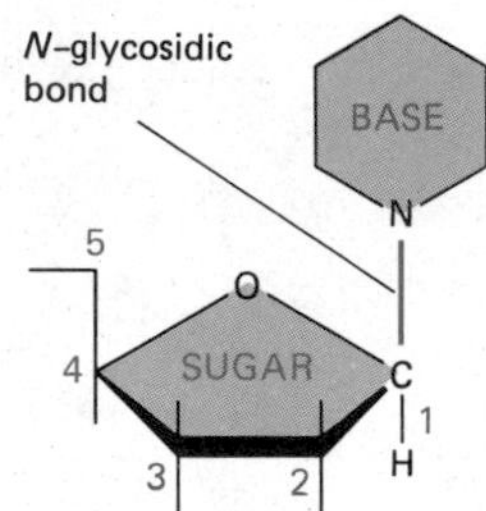

The base is linked to the same carbon (C1) used in sugar-sugar bonds.

SUGARS

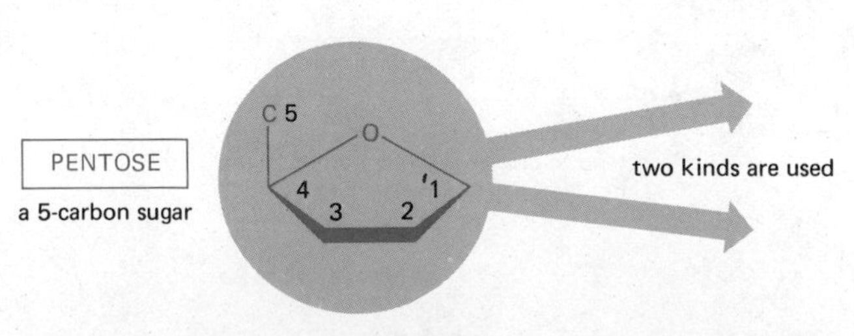

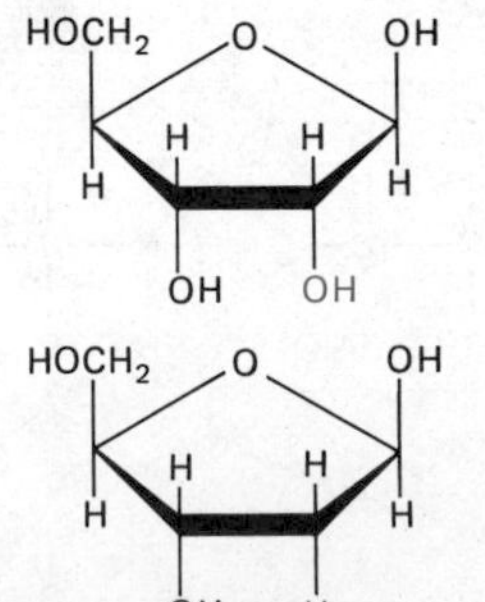

β-D-RIBOSE
used in ribonucleic acid

β-D-2-DEOXYRIBOSE
used in deoxyribonucleic acid

Panel H A survey of the major types of nucleotides and their derivatives encountered in cells.

NOMENCLATURE

The names can be confusing, but the abbreviations are clear.

BASE + SUGAR = NUCLEOSIDE

BASE + SUGAR + PHOSPHATE = NUCLEOTIDE

BASE	NUCLEOSIDE	ABBR.
adenine	adenosine	A
guanine	guanosine	G
cytosine	cytidine	C
uracil	uridine	U
thymine	thymidine	T

Nucleotides are abbreviated by 3 capital letters as follows:

AMP = adenosine monophosphate
dAMP = deoxyadenosine monophosphate
UDP = uridine diphosphate
ATP = adenosine triphosphate
etc.

NUCLEIC ACIDS

Nucleotides are joined together by a single type of phosphodiester linkage to form nucleic acids.

The linear sequence of nucleotides in a nucleic acid chain is commonly abbreviated by a 1-letter code, A–G–C–T–T–A–C–A, with the 5′ end of the chain written at the left.

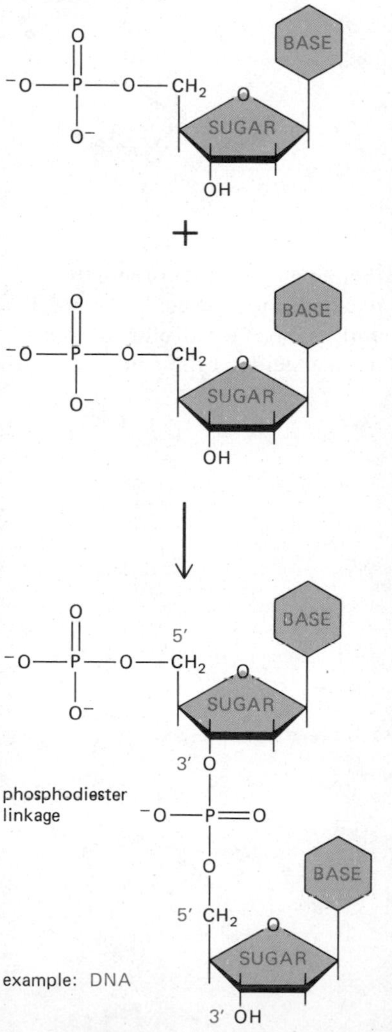

phosphodiester linkage

example: DNA

NUCLEOTIDES HAVE MANY OTHER FUNCTIONS

1. They carry chemical energy in their easily hydrolyzed acid-anhydride bonds.

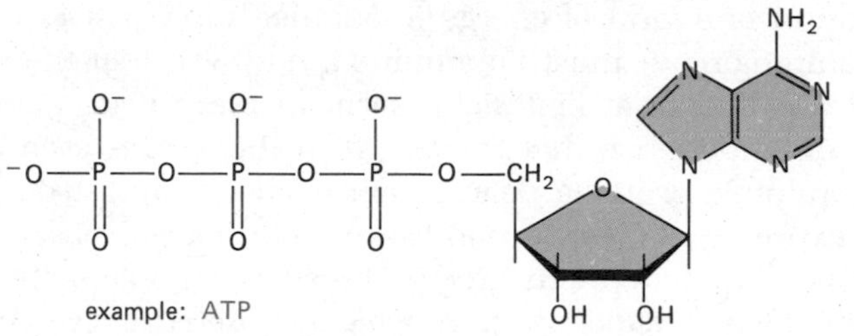

example: ATP

2. They combine with other groups to form coenzymes.

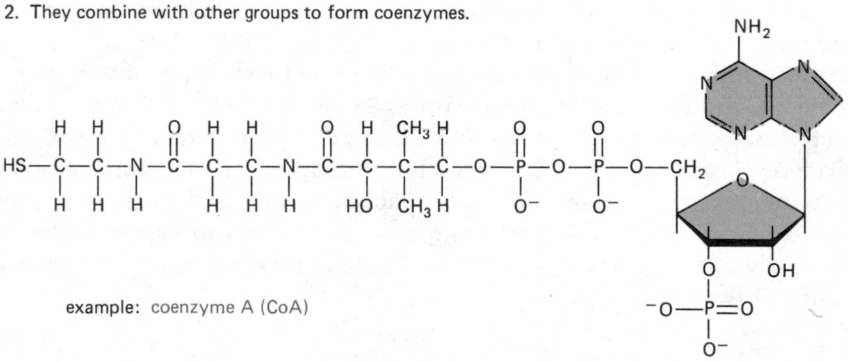

example: coenzyme A (CoA)

3. They are used as specific signaling molecules in the cell.

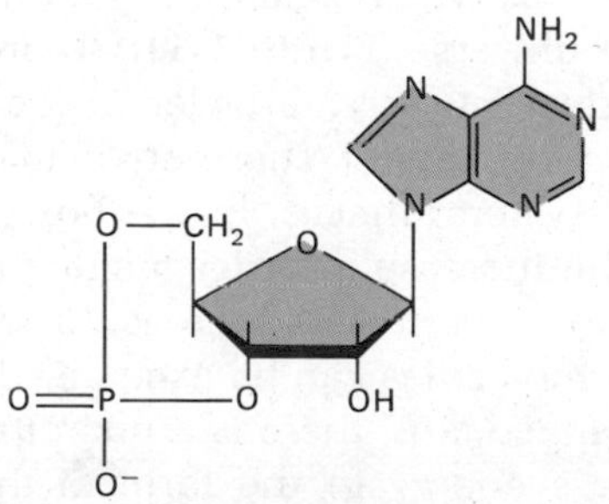

example: cyclic AMP

Summary

Living organisms are autonomous, self-propagating chemical systems. They are made from a distinctive but restricted set of carbon-based small molecules that are essentially the same for every living species. The main categories are sugars, fatty acids, amino acids, and nucleotides. Sugars are a primary source of chemical energy for cells and are incorporated into polysaccharides for energy storage. Fatty acids are also important for energy storage, but their most significant function is in the formation of cellular membranes. Polymers consisting of amino acids constitute the remarkably diverse and versatile macromolecules known as proteins. Nucleotides are involved in intracellular signaling and play a central part in energy transfer, but their unique role is as the subunits of the informational macromolecules, RNA and DNA.

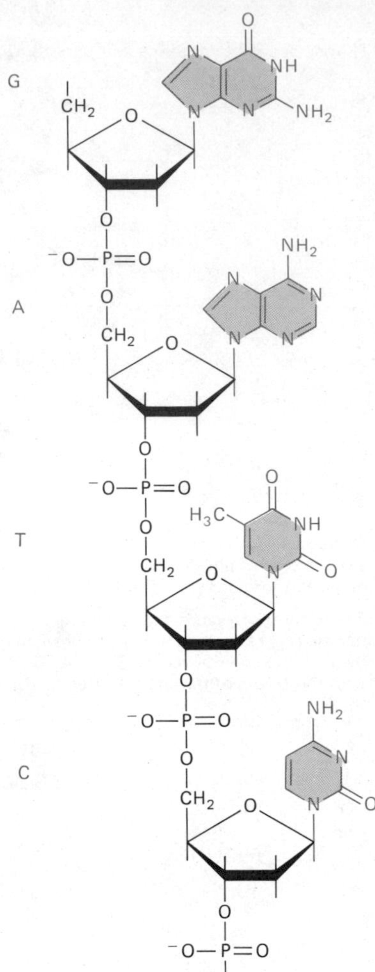

Figure 2–10 A short length of deoxyribonucleic acid, or DNA. DNA and its close relative RNA are the nucleic acids of the cell.

Biological Order and Energy

Cells must obey the laws of physics and chemistry. The rules of mechanics and of the conversion of one form of energy to another apply just as much to a cell as to a steam engine. It must be admitted, however, that there are puzzling features of a cell that, at first sight, seem to place it in a special category. It is common experience that things left to themselves eventually become disordered: buildings crumble, dead organisms decay, and so on. This general tendency is expressed in the *Second Law of Thermodynamics*, which states that in any isolated system, the degree of disorder can only increase.

The puzzle is that living organisms are very highly ordered, at every level. Order is strikingly apparent in large structures such as a butterfly wing or an octopus eye, in subcellular structures such as a mitochondrion or a cilium, and in the shape and arrangement of molecules from which these structures are built. The large number of atoms in each molecule of protein or nucleic acid have been captured, ultimately, from their highly disorganized state in the environment and locked together into a precise structure. Every time large molecules are made from smaller ones, as when a living cell grows, order is created out of chaos. Even a nondividing cell requires constant ordering or repair processes for survival, since all of its organized structures are subject to spontaneous accidents and side reactions. How is this possible thermodynamically? We shall see that the answer lies in the fact that the cell is constantly releasing heat to its environment, and therefore it is not an isolated system in the thermodynamic sense.

Biological Order Is Made Possible by the Release of Heat Energy from Cells[3]

For thermodynamic purposes, the cell and its immediate environment can be thought of as a sealed box, which is sitting in a uniform sea of matter representing the rest of the universe (Figure 2–11). In order to grow and maintain itself, the cell must constantly create order inside the box. As we have just seen, however, the Second Law of Thermodynamics states that the amount of order in the entire system (that is, in the box plus the sea) must always decrease. Therefore, the increase in order inside the box must be more than compensated for by an even greater increase in *dis*order in the rest of the universe. Although no molecules can be exchanged between the box and the sea, heat can be exchanged, and there is a quantitative relationship between heat and order. Heat is energy in the form of the random commotion of molecules, and it therefore represents energy in its most disordered form. If the cell releases heat to the sea, it increases the intensity of molecular motions in the sea—thereby increasing their randomness, or disorder.

The quantitative relationship between heat and order, first recognized in the late nineteenth century, enables us to calculate in principle exactly how much heat a cell must release (in kilocalories) in order to compensate for a given amount of ordering within it (such as the assembly of proteins from amino acids)—so that the net process increases the total disorder of the universe. This relationship can be derived by considering the changes of molecular motions that result when a given amount of heat energy is transferred from a hot body to a cold one. While the quantitative details need not concern us here, it is important to note that the chemical reactions that generate the heat must be intimately associated at the molecular level with the order-producing events themselves. Such associated reactions are said to be *coupled,* as will be explained subsequently.

Figure 2–11 illustrates in a highly schematic way how such coupled reactions release heat energy, which disorders the environment, thereby compensating for the increase in order they create in the cell. Because the heat release makes these reactions possible, it can be considered to *drive* the ordering processes.

Energy cannot be created or destroyed in chemical reactions. The continual heat loss from the cell that drives the production of biological order therefore requires a continual input of energy into the cell. This energy must be in a form other than heat. For plants, the energy is initially derived from the electromagnetic radiation of the sun, whereas for animals it is derived from the energy stored in the covalent bonds of the organic molecules they eat. However, since these organic nutrients are themselves produced by photosynthetic organisms, such as a green plant, the sun serves as the ultimate energy source for both types of organisms.

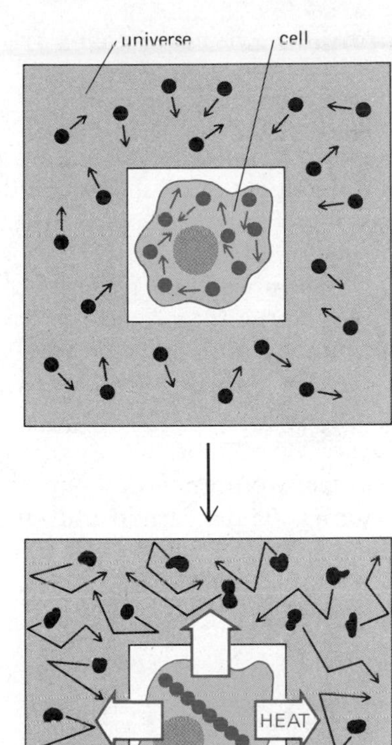

Figure 2–11 For a simple thermodynamic analysis of a living cell, it is useful to consider it and its immediate environment enclosed in a sealed box that allows heat but not molecules to be exchanged with the rest of the universe. In the upper diagram, the molecules of both the cell and the rest of the universe are depicted in a relatively disordered state. In the lower diagram, heat has been released from the cell by a reaction that orders the molecules it contains. The increase in random motion, including bond distortions, of the molecules in the rest of the universe creates a disorder that more than compensates for the increased order in the cell, as required by the laws of thermodynamics for spontaneous processes. In this way the release of heat by a cell to its surroundings allows it to become more highly ordered internally at the same time that the universe as a whole becomes more disordered.

Photosynthetic Organisms Use Sunlight to Synthesize Organic Compounds[4]

Solar energy enters the living world (the *biosphere*) by means of the **photosynthesis** carried out by photosynthetic organisms—either plants or bacteria. In photosynthesis, electromagnetic energy is converted into chemical bond energy. But, at the same time, part of the energy of sunlight is converted into heat energy, and it is the release of the heat to the environment that increases the disorder of the universe and thereby drives the process.

The reactions of photosynthesis are described in detail in Chapter 9; in broad terms, they occur in two distinct stages. In the first stage (the *light reactions*), the visible radiation excites an electron in a pigment molecule that, in returning to a lower energy state, provides the energy needed for the synthesis of the molecules ATP and NADPH. In the second stage (the *dark reactions*), the ATP and NADPH are used to drive a series of "carbon-fixation" reactions in which CO_2 from the air is used to form sugar molecules (Figure 2–12).

The *net* result of photosynthesis, so far as the green plant is concerned, can be summarized by the equation

$$\text{energy} + CO_2 + H_2O \rightarrow \text{sugar} + O_2$$

which is the reverse of the oxidative decomposition of a sugar. However, this simple equation hides the complex nature of the dark reactions, which involve many linked reaction steps. Furthermore, while the initial fixation of CO_2 results in sugars, subsequent metabolic reactions soon convert these sugars into the other small and large molecules essential to the plant cell.

Chemical Energy Passes from Plants to Animals

Animals and other nonphotosynthetic organisms cannot capture energy from sunlight directly and so have to survive on "second-hand" energy obtained by eating plants. The organic molecules made by plant cells provide both building blocks and fuel to the organisms that feed on them. All types of plant molecules can serve this purpose—sugars, proteins, polysaccharides, lipids, and many others.

The transactions between plants and animals are not all one-way. Plants, animals, and microorganisms have existed together on this planet for so long that many of them have become an essential part of the other's environment. The oxygen released by photosynthesis is consumed in the combustion of organic molecules by nearly all organisms, and some of the CO_2 molecules that are "fixed" today into larger organic molecules by photosynthesis in a green leaf were yesterday released into the atmosphere by the respiration of an animal. Thus, carbon utilization is a cyclic process that involves the biosphere as a whole and crosses boundaries between individual organisms (Figure 2–13). Similarly, atoms of nitrogen, phosphorus, and sulfur can, in principle, be traced from one biological molecule to another in a series of similar cycles.

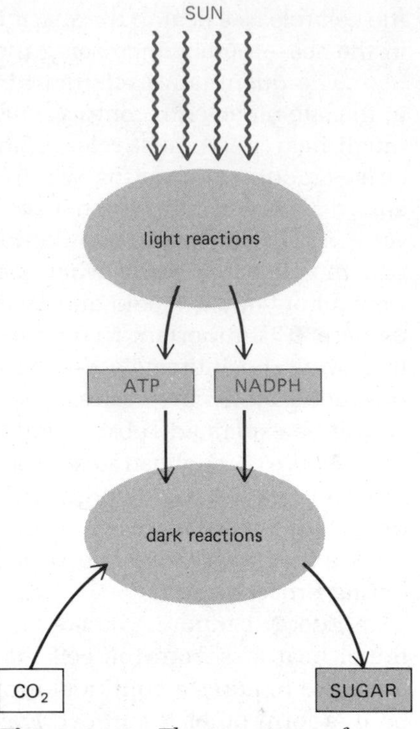

Figure 2–12 The two stages of photosynthesis in a green plant.

Cells Obtain Energy by the Oxidation of Biological Molecules

The carbon and hydrogen atoms in a cell are not in their most stable form. Because the earth's atmosphere contains a great deal of oxygen, the most energetically stable form of carbon is as CO_2 and of hydrogen is as H_2O. A cell is therefore able to obtain energy from glucose or protein molecules by allowing their carbon and hydrogen atoms to combine with oxygen to produce CO_2 and H_2O, respectively. However, the cell does not oxidize molecules in one

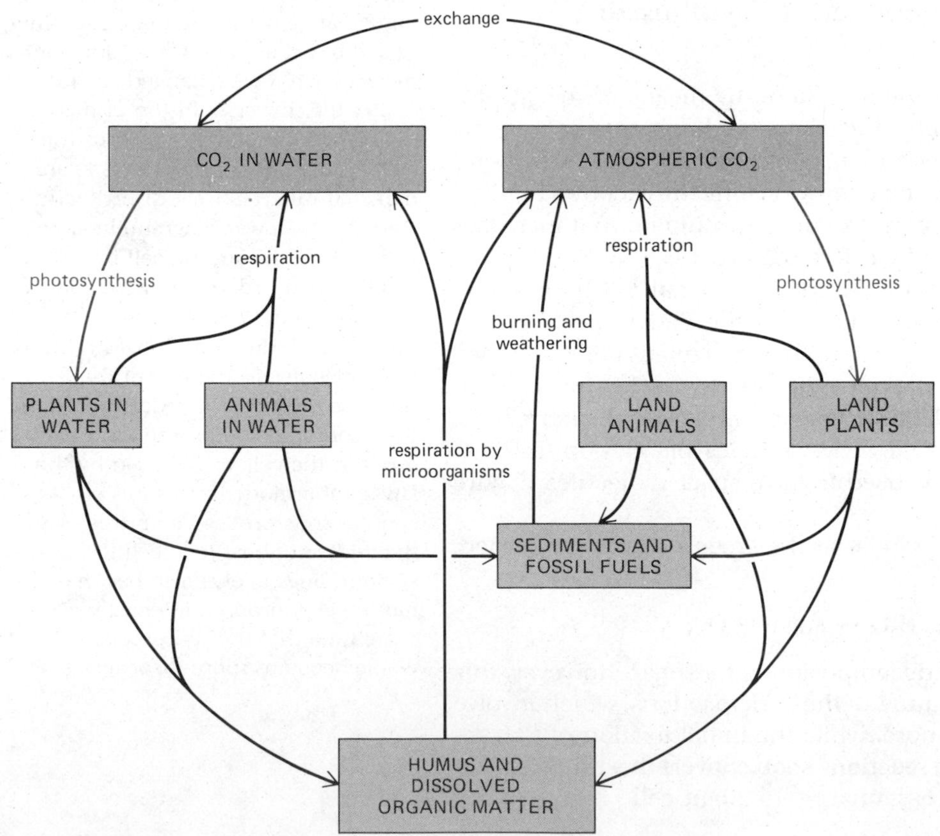

Figure 2–13 The carbon cycle. Individual carbon atoms are incorporated into organic molecules of the living world by the photosynthetic activity of plants, bacteria, and marine algae. They pass to animals, microorganisms, and organic material in soil and oceans in cyclic paths. CO_2 is restored to the atmosphere when organic molecules are oxidized by cells or burned by humans as fossil fuels.

step, as occurs in a fire. It takes them through a large number of reactions that only rarely involve the direct addition of oxygen. Before we can trace these reactions and understand the driving force behind them, we need to have a clear idea of the process of oxidation.

Oxidation, in the sense used above, does not mean only the addition of oxygen atoms; rather it applies more generally to any reaction in which electrons are transferred from one atom to another. **Oxidation** in this sense refers to the removal of electrons and **reduction**—the converse of oxidation—means the addition of electrons. Thus, Fe^{2+} is oxidized if it loses an electron to become Fe^{3+}, and a chlorine atom is reduced if it gains an electron to become Cl^-. The same terms are used when there is only a partial shift of electrons between atoms linked by a covalent bond. For example, when a carbon atom covalently binds an electronegative atom such as oxygen, chlorine, or sulfur, it gives up more than its equal share of electrons—acquiring a partial positive charge—and so is said to be oxidized. Conversely, a carbon atom in a C-H linkage has relatively more electrons and so it is said to be reduced (Figure 2–14).

The combustion of food materials in a cell converts the C and H atoms in organic molecules (where they are in a relatively electron-rich, or reduced, state) to CO_2 and H_2O, where they have given up electrons and are, therefore, highly oxidized. The shift of electrons from carbon and hydrogen to oxygen allows all of these atoms to achieve a more stable state and hence is energetically favorable. This shift of electrons occurs in sequential steps by a series of intermediate reactions that, in the majority of cases, involve the transfer of the components of a hydrogen atom between one molecule and another.

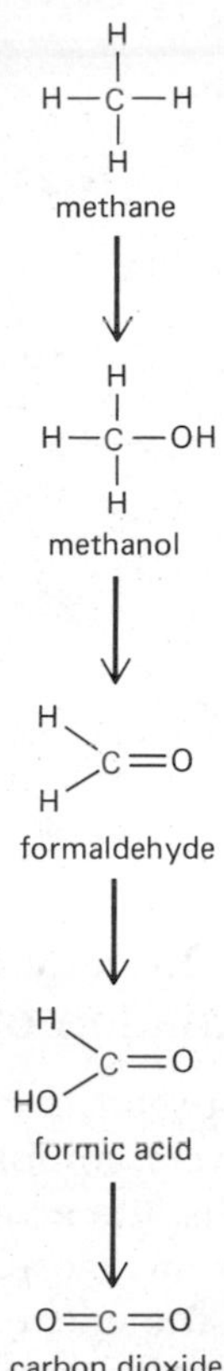

Figure 2–14 The carbon atom of methane may be converted to that of carbon dioxide by the successive removal of its hydrogen atoms. With each step, electrons are shifted away from the carbon atom as it passes to a more energetically stable state (that is, it becomes more highly oxidized).

The Breakdown of Organic Molecules Takes Place in Sequences of Enzyme-catalyzed Reactions

Although the most energetically favorable form of carbon is as CO_2 and that of hydrogen is as H_2O, a living organism does not disappear in a puff of smoke for the same reason that the book in your hands does not burst into flame: both exist in metastable energy troughs (Figure 2–15) and require *activation energy* before they can pass to more stable configurations. In the case of the book, the activation energy can be provided by a lighted match. For a living cell the same end result can be achieved in a less drastic and destructive fashion. Highly specific protein catalysts, or **enzymes,** combine with biological molecules in such a way that they reduce the activation energy of particular reactions that the bound molecules can undergo. By selectively lowering the activation energy of one pathway or another, enzymes determine which of several alternative reaction paths is followed (Figure 2–16). In this way, various cellular molecules are directed through specific reaction pathways.

The success of living forms is largely attributable to the cell's ability to make a large number of specific enzymes. Each enzyme is a protein with a unique three-dimensional conformation that creates an *active site* for binding a particular set of other molecules (substrates) to its surface. An enzyme binds its substrate in such a way as to speed up a particular one of the many chemical reactions that the substrate can undergo, typically by a factor of 10^6 or more. No other type of catalyst can match the specificity and efficiency of enzymes.

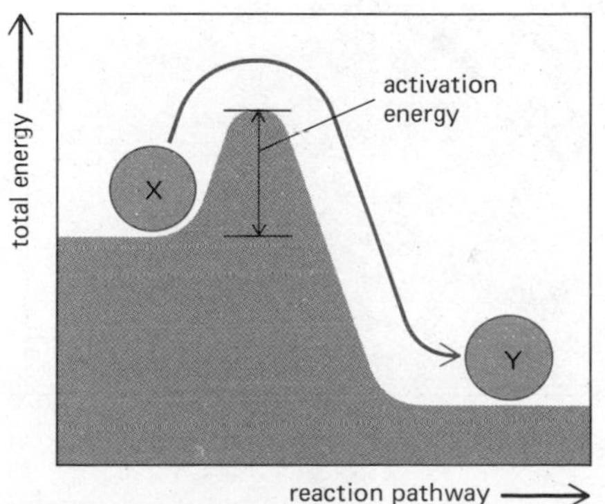

Figure 2–15 Schematic diagram illustrating the principle of activation energy. Compound X can achieve a lower, more favorable energy state by being converted to compound Y. However, this transition will not take place unless X can acquire enough activation energy to undergo the reaction.

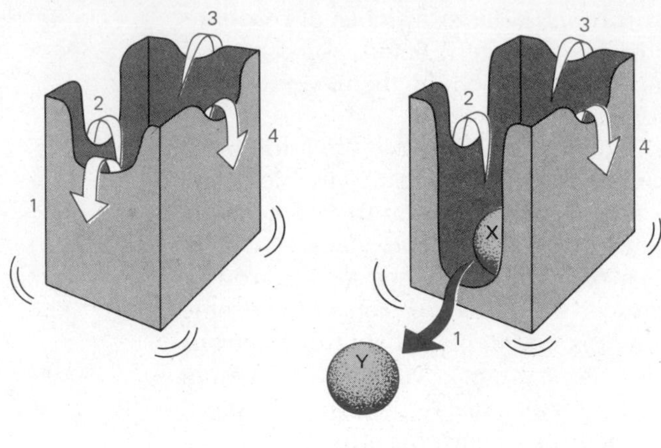

Figure 2–16 A "jiggling-box" model illustrating how enzymes direct molecules along desired reaction pathways. In this model the colored ball represents a potential enzyme substrate that is bouncing up and down due to the constant bombardment of colliding water molecules. The four walls of the box represent the activation energy barriers for four different chemical reactions that are energetically favorable. In the left-hand box, none of these reactions occur because the energy available from collisions is insufficient to surmount any of the energy barriers. In the right-hand box, enzyme catalysis lowers the activation energy for reaction number 1 only and thereby allows this reaction to proceed with available energies.

Part of the Energy Released in Oxidation Reactions Is Coupled to the Formation of ATP

Cells derive useful energy from the "burning" of glucose only because they burn it in a very complex and controlled way. By means of enzyme-directed reaction paths, the synthetic, or *anabolic,* chemical reactions that create biological order are closely coupled to the degradative, or *catabolic,* reactions that provide the energy. The crucial difference between a **coupled reaction** and an uncoupled catabolic reaction is illustrated by the mechanical analogy in Figure 2–17, where an energetically favorable chemical reaction is represented by rocks falling from a cliff. The kinetic energy of falling rocks would normally be entirely wasted in the form of heat generated when they hit the ground (section A). But, by careful design, part of the kinetic energy could be used to drive a paddle wheel that lifts a bucket of water (section B). Because the rocks can reach the ground in section B only by moving the paddle wheel, we say that the spontaneous reaction of rock falling has been directly coupled to the nonspontaneous reaction of lifting the bucket of water. Note that because part of the energy is now used to do work in section B, the rocks hit

Figure 2–17 Schematic diagram of a mechanical model illustrating the principle of coupled chemical reactions. The spontaneous reaction shown in (A) might serve as an analogy for the direct oxidation of glucose to CO_2 and H_2O, which produces heat only. In (B) the same reaction is coupled to a second reaction; the second reaction might serve as an analogy for the synthesis of ATP. The more versatile form of energy produced in (B) can be used to drive other cellular processes, as in (C).

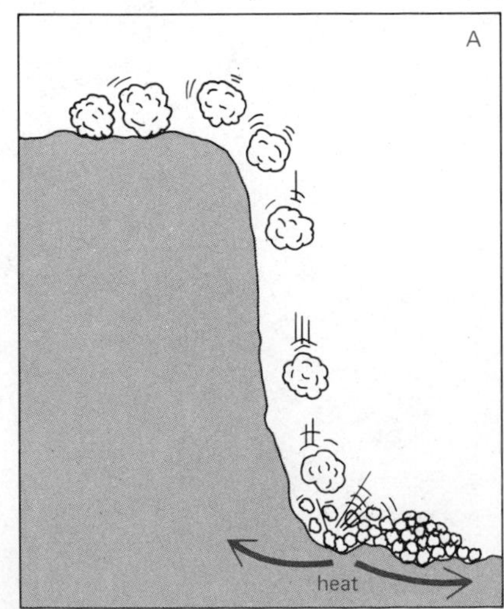

kinetic energy transformed into heat energy only

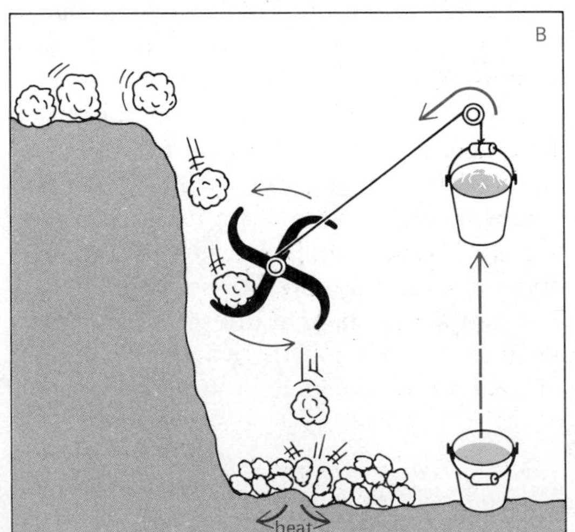

part of the kinetic energy is used to lift a bucket of water, and a correspondingly smaller amount is released as heat

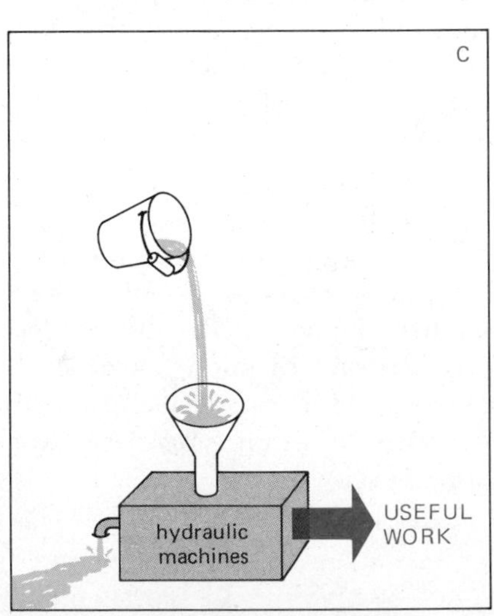

the potential kinetic energy stored in the elevated bucket of water can be used to drive a wide variety of different hydraulic machines the ground with less velocity than in section A, and therefore correspondingly less energy is wasted as heat.

In cells, enzymes play the role of paddle wheels in our analogy and couple the spontaneous burning of foodstuffs to reactions that generate the nucleoside triphosphate, ATP. Just as the energy stored in the elevated bucket of water in Figure 2–17 can be dispensed in small doses to drive a wide variety of different hydraulic machines (section C), ATP serves as a convenient and versatile store, or currency, of energy for driving many different chemical reactions that the cell needs.

The Hydrolysis of ATP Generates Order in Cells

ATP acts as a carrier of chemical energy because it is relatively unstable. Under the conditions existing in the cytoplasm, the breakdown of ATP by hydrolysis to release inorganic phosphate (P_i) occurs very readily and releases a great deal of usable energy (see p. 75). Consequently, the bonds broken in this hydrolysis reaction are sometimes described as *high-energy bonds,* although there is nothing special about the covalent bonds themselves. Other reactions can be driven by the energy released by ATP hydrolysis provided they can be somehow coupled to this process.

Among the many hundreds of reactions driven by the hydrolysis of ATP are those involved in the synthesis of biological molecules, in the active transport of molecules across cell membranes, and in the generation of force and movement. These three types of processes play a vital part in the establishment of biological order. The macromolecules formed in biosynthetic reactions carry information, catalyze specific reactions, and are assembled into highly ordered structures within cells and in the extracellular space. Membrane-bound pumps maintain the special internal composition of cells and permit signals to pass within and between cells. Finally, the production of force and movement enables the cytoplasmic contents of cells to become organized and the cells themselves to move about and assemble into organized tissues.

Summary

Living cells are highly ordered and must create order within themselves to grow and survive. This is thermodynamically possible only because of a continual input of energy, part of which is released from the cells to their environment as heat. The energy comes ultimately from the electromagnetic radiation of the sun, which drives the formation of organic molecules in photosynthetic organisms such as green plants. Animals obtain their energy by taking up these organic molecules and oxidizing them in a series of enzyme-catalyzed reactions that are coupled to the formation of ATP. ATP is a common currency of energy in all cells, and its hydrolysis is coupled to other reactions to drive a variety of energetically unfavorable processes that create order.

Food and the Derivation of Cellular Energy[2,5]

Food Molecules Are Broken Down in Three Stages to Give ATP

The proteins, lipids, and polysaccharides that make up the major part of the food we eat must be broken down into smaller molecules before our cells can use them. The enzymatic breakdown, or catabolism, of these molecules may be regarded as proceeding in three stages (Figure 2–18). We shall give, first, a short outline of these stages and then discuss two of them in more detail.

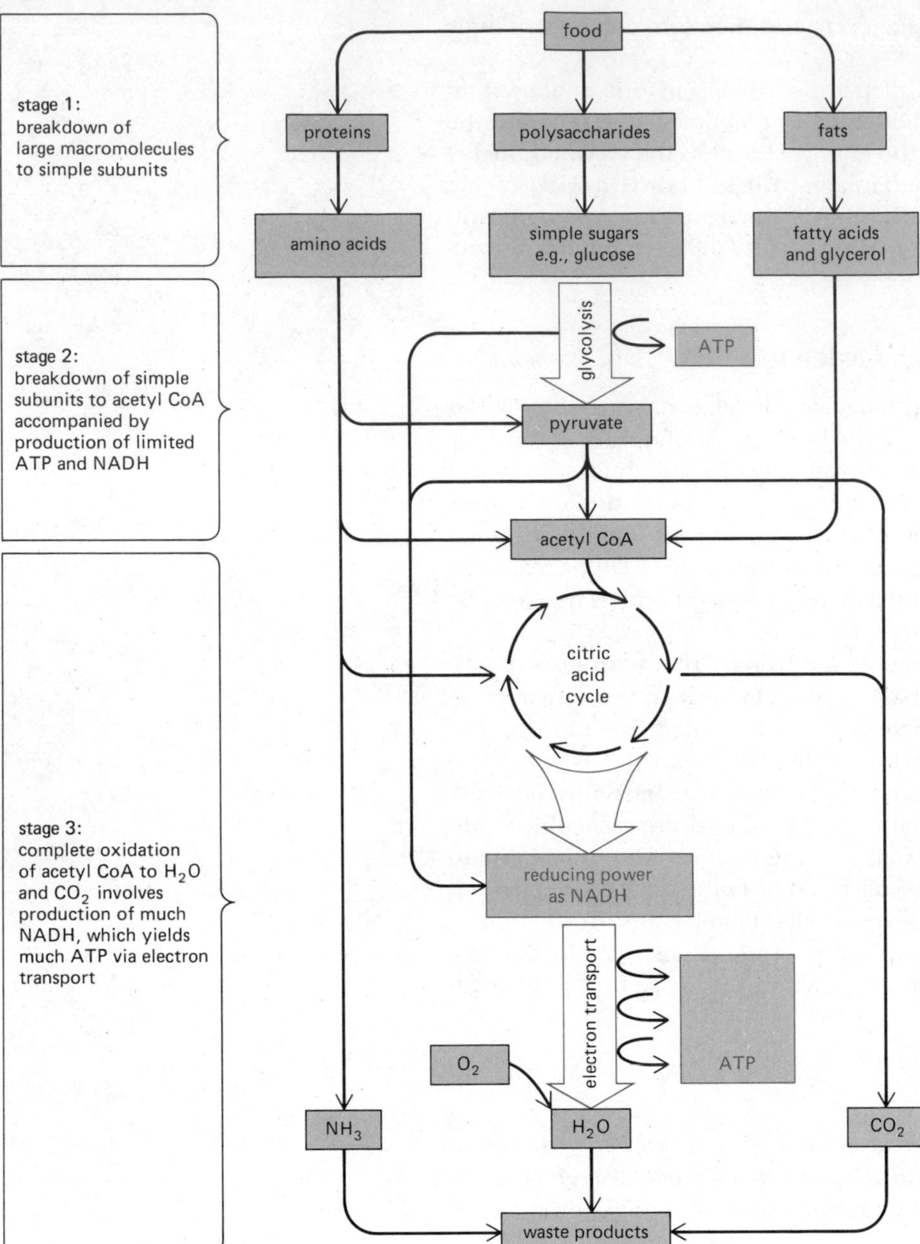

Figure 2–18 Simplified diagram of the three stages of catabolism that lead from food to waste products. This series of reactions produces ATP, which is then used to drive biosynthetic reactions and other energy-requiring processes in the cell.

In stage 1, large polymeric molecules are broken down into their monomeric subunits—proteins into amino acids, polysaccharides into sugars, and fats into fatty acids and glycerol. These preliminary processes, which we know as *digestion,* occur mainly outside cells through the action of secreted enzymes. In stage 2, the resultant small molecules enter cells and are further degraded in the cytoplasm. Most of the carbon and hydrogen atoms of sugars are converted into *pyruvate,* which then enters mitochondria, where it is converted to the acetyl groups of *acetyl coenzyme A* (*acetyl CoA,* Figure 2–19). Acetyl CoA, like ATP, is a chemically reactive compound that releases a great deal of energy when it is hydrolyzed. Major amounts of acetyl CoA are also produced by the oxidation of fatty acids.

The last stage of catabolism, stage 3, occurs when the acetyl group of acetyl CoA is completely degraded to CO_2 and H_2O. It is in this final stage that most of the ATP is generated. Through a series of coupled chemical reactions,

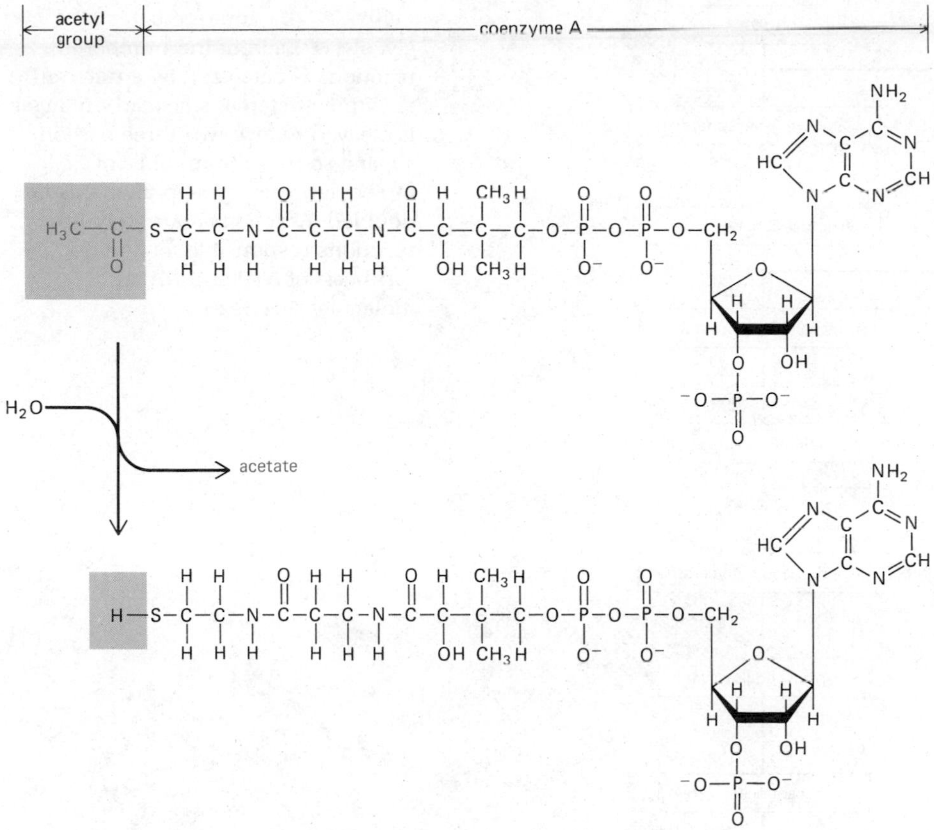

Figure 2–19 The structure of the crucial metabolic intermediate acetyl coenzyme A (acetyl CoA). Acetyl groups produced in stage 2 of catabolism (see Figure 2–18) are covalently linked to coenzyme A (CoA).

more than half of the energy theoretically derivable from the combustion of carbohydrates and fats to H_2O and CO_2 is channeled into driving the energetically unfavorable reaction P_i + ADP → ATP. Because the rest of the combustion energy is released by the cell as heat, the generation of ATP in this way creates net disorder in the universe, in conformity with the Second Law of Thermodynamics.

Through the generation of ATP, the energy originally derived from the combustion of carbohydrates and fats is redistributed as a conveniently packaged form of chemical energy that is easily released. Roughly 10^9 molecules of ATP are in solution throughout the intracellular space in a typical cell, where their energetically favorable hydrolysis back to ADP and phosphate provides the driving energy for a variety of energetically unfavorable reactions.

Glycolysis Can Produce ATP Even in the Absence of Oxygen

The most important part of stage 2 of catabolism is a sequence of reactions known as **glycolysis**—the lysis (splitting) of glucose. In glycolysis, a glucose molecule with six carbon atoms is converted into two molecules of pyruvate, each with three carbon atoms. This conversion requires a sequence of nine enzymatic reactions that involve a series of phosphate-containing intermediates (Figure 2–20). Logically, the sequence can be divided into three parts: (1) in reactions 1 to 4, glucose is converted to the three-carbon aldehyde *glyceraldehyde 3-phosphate*—a conversion that requires an investment of energy in the form of ATP; (2) in reactions 5 and 6, the aldehyde group of the glyceraldehyde 3-phosphate is oxidized to a carboxylic acid, and the energy from this reaction is coupled to the creation of a new high-energy phosphate linkage

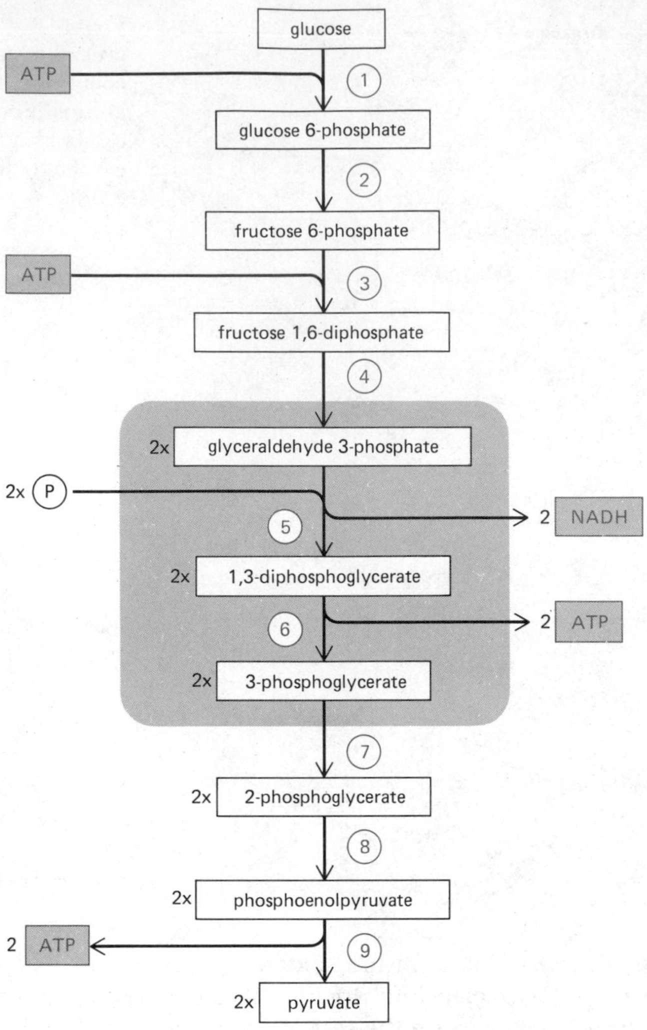

Figure 2–20 Intermediates of glycolysis. Each of the numbered reactions is catalyzed by a different enzyme. At step 4, a six-carbon sugar is cleaved to give two three-carbon sugars, so that the number of molecules at every step after this is doubled. Reactions 5 and 6 are the reactions responsible for the net synthesis of ATP and NADH molecules (see text).

in ATP; and (3) in reactions 7, 8, and 9, the original investment of ATP made in the first reaction sequence is repaid.

At the end of glycolysis, the ATP balance sheet shows a net profit of the two molecules of ATP (per glucose molecule) that were produced in reactions 5 and 6. These two steps, therefore, lie at the heart of glycolysis. As the only reactions in the sequence in which a high-energy phosphate linkage is created from inorganic phosphate, they provide an excellent illustration of the way in which reactions in the cell can be coupled together to harvest the energy released by oxidations (Figure 2–21). In reaction 5 of glycolysis, hydrogen (as a hydride ion: a proton plus two electrons) is removed from the aldehyde group in glyceraldehyde 3-phosphate and transferred to the carrier molecule NAD^+ (Figure 2–22). At the same time, as shown in Figure 2–21, a phosphate ion from solution becomes linked by a highly reactive bond to the new carbonyl group produced. In reaction 6 of glycolysis, this reactive phosphate group is transferred to ADP to form ATP, leaving a free carboxylic acid group. The overall result of these two reactions is, therefore, that a sugar aldehyde is oxidized to a carboxylic acid, an inorganic phosphate group is transferred to a high-energy linkage on ATP, and a molecule of NAD^+ is reduced to NADH. This elegant pair of coupled reactions was probably among the earliest metabolic steps to appear in the evolving cell. In addition to forming the central part of glucose metabolism, they are driven in reverse during photosynthesis

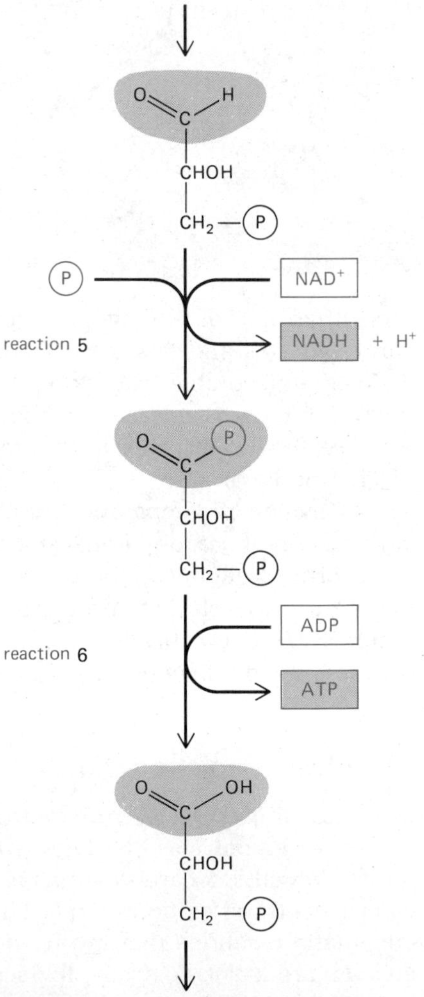

Figure 2–21 Reactions 5 and 6 of glycolysis: the oxidation of an aldehyde to a carboxylic acid is coupled to the formation of ATP and NADH (see Figure 2–20).

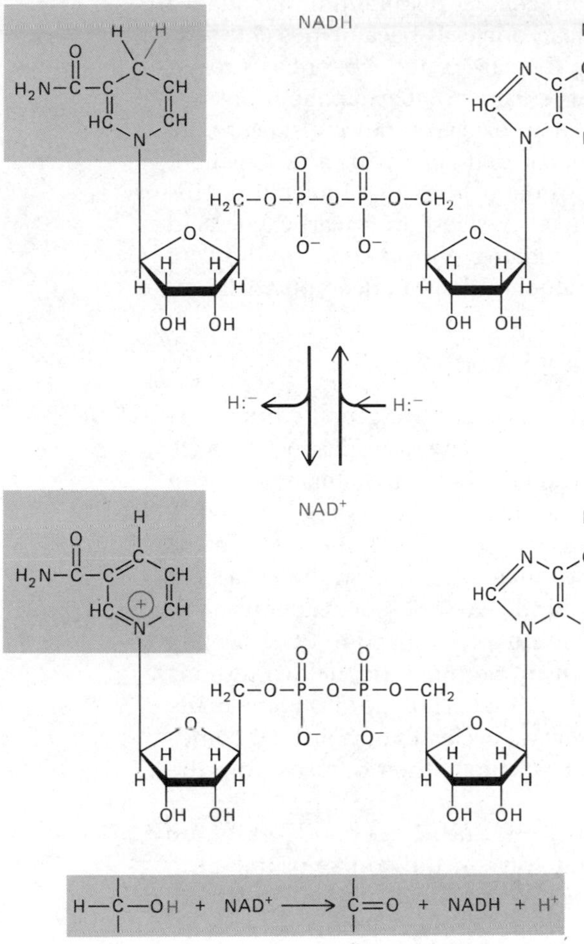

NADH

$H^{:-}$ $H^{:-}$

NAD^+

$$\mathrm{H{-}\overset{|}{\underset{|}{C}}{-}OH + NAD^+ \longrightarrow \overset{|}{\underset{|}{C}}{=}O + NADH + H^+}$$

Figure 2–22 The structure of NADH and NAD^+, the most important carriers of hydrogen in catabolic reactions. The part of the NAD^+ molecule known as the nicotinamide ring (in the colored box) is able to accept a hydrogen atom together with an additional electron (a hydride ion, H^-), forming NADH. In this reduced form, the nicotinamide ring has a reduced stability because it is no longer stabilized by resonance. As a result, the added hydride ion is easily transferred to other molecules.

In the biological oxidation of a substrate molecule such as an alcohol (see lower part of figure), two hydrogen atoms are lost from the substrate. One of these is added as a hydride ion to NAD^+, producing NADH, while the other is released into solution as a proton (H^+). (See also Figure 9–16, p. 494.)

by the NADPH and ATP produced by the light-activated reactions, and they thereby play a pivotal role in the photosynthetic carbon-fixation process (p. 515).

For most animal cells, glycolysis is only a prelude to stage 3 of catabolism, since the pyruvic acid that is formed quickly enters the mitochondria to be completely oxidized to CO_2 and H_2O. However, in the case of anaerobic organisms (those that do not utilize molecular oxygen) and for tissues, such as skeletal muscle, that can function under anaerobic conditions, glycolysis can become a major source of the cell's ATP. Here, instead of being degraded in mitochondria, the pyruvate molecules stay in the cytosol and, depending on the organism, can be converted into ethanol plus CO_2 (as in yeast) or into lactate (as in muscle), which are then excreted. In such anaerobic energy-yielding reactions, called **fermentations,** the further reaction of pyruvate is required in order to use up the reducing power produced in reaction 5 of glycolysis, thereby regenerating the NAD^+ required for glycolysis to continue (see p. 523).

Oxidative Catabolism Yields a Much Greater Amount of Usable Energy

The anaerobic generation of ATP from glucose through the reactions of glycolysis is relatively inefficient. The end products of anaerobic glycolysis still contain a great deal of chemical energy that can be released by further oxidation. The evolution of *oxidative catabolism* (cellular respiration) in oxidative microorganisms and in mitochondria of eucaryotic cells became possible only after molecular oxygen had accumulated in the earth's atmosphere as a result of photosynthesis by the cyanobacteria. Earlier, anaerobic catabolic processes had dominated life on earth. The addition of an oxygen-requiring stage to the catabolic process (stage 3 in Figure 2–18) provided cells with a much more powerful and efficient method for extracting energy from food molecules. This third stage begins with the *citric acid cycle* (also called the tricarboxylic acid cycle, or the Krebs cycle) and ends with *oxidative phosphorylation,* both of which occur in aerobic bacteria and the mitochondria of eucaryotic cells.

Metabolism Is Dominated by the Citric Acid Cycle[6]

The primary function of the **citric acid cycle** is to oxidize acetyl groups that enter the cycle in the form of acetyl CoA molecules. The reactions form a cycle because the acetyl group is not oxidized directly, but only after it has been covalently added to a larger molecule, *oxaloacetate,* which is regenerated at the end of one turn of the cycle. As illustrated in Figure 2–23, the cycle begins with the reaction between acetyl CoA and oxaloacetate to form the tricarboxylic acid molecule called *citric acid* (or *citrate*). A series of reactions then occurs in which two of the six carbons of citrate are oxidized to CO_2, forming another molecule of oxaloacetate to repeat the cycle. (Because the two carbons that are newly added in each cycle enter a different part of the citrate molecule from that oxidized to CO_2, it is only after several cycles that their turn comes to be oxidized.) The CO_2 produced in these reactions then diffuses from the mitochondrion and leaves the cell.

The energy made available when the C—H and C—C bonds in citrate are oxidized is captured in several different ways in the course of the citric acid cycle. At one step in the cycle (succinyl CoA to succinate) a high-energy phosphate linkage is created by a mechanism resembling that described for glycolysis above. (Although this reaction produces GTP rather than ATP, all nucleoside triphosphates are equivalent energetically because of exchange reactions such as ADP + GTP $\rightleftarrows$ ATP + GDP.) All of the remaining energy of oxidation that is captured is channeled into the conversion of hydrogen-carrier molecules to their reduced forms; for each turn of the cycle, three molecules of NAD^+ are converted to NADH and one *flavin adenine nucleotide* (FAD) is converted to $FADH_2$.

The additional oxygen atoms required to make CO_2 from the acetyl groups entering the citric acid cycle are supplied not by molecular oxygen but by water. Three molecules of water are split in each cycle, and their oxygen atoms are used to make CO_2. Some of their hydrogen atoms enter substrate molecules and are raised to a higher energy and removed (together with the hydrogen atoms of the acetyl groups) to carrier molecules such as NADH. At another site in the mitochondrion, the energy carried by these activated hydrogen atoms is harnessed in reactions generating ATP; it is the latter reactions of *oxidative phosphorylation* (to be considered in more detail on the following page) that require molecular oxygen from the atmosphere.

The mitochondrion is, therefore, both the power house of the cell and the place where the carbon and hydrogen atoms of food molecules are finally oxidized. It is the center toward which all catabolic processes lead, whether they begin with sugars, fats, or proteins. For, in addition to pyruvate, fatty acids and some amino acids also pass from the cytosol into mitochondria, where they are converted into acetyl CoA or one of the other intermediates of the citric acid cycle.

In addition to producing the ATP required for many biosyntheses, the mitochondrion functions as the starting point for biosynthetic reactions by

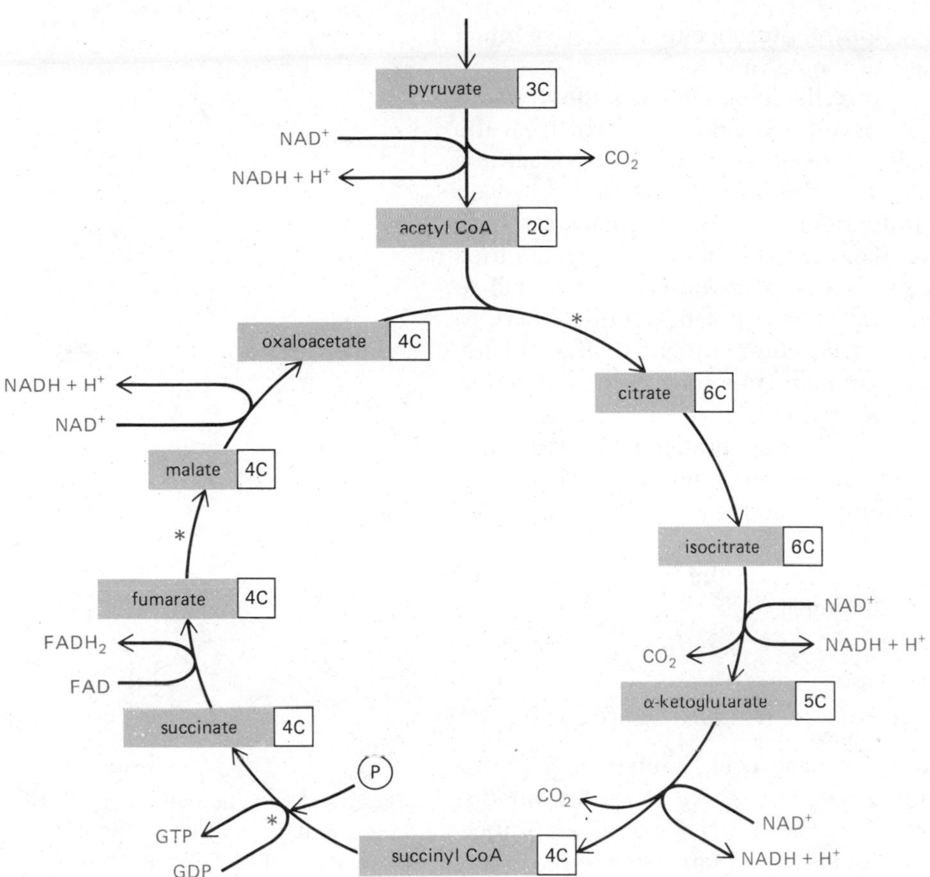

Figure 2–23 The citric acid cycle. In mitochondria and in aerobic bacteria, the acetyl groups produced from pyruvate are further oxidized. The carbon atoms of the acetyl groups are converted to CO_2, while the hydrogen atoms are transferred to the carrier molecules NAD^+ and FAD. Additional oxygen and hydrogen atoms enter the cycle in the form of water at the steps marked with an asterisk (*). For details, see Figure 9–12, p. 491.

producing vital carbon-containing intermediates, such as *oxaloacetate* and *α-ketoglutarate.* These substances are transferred back from the mitochondrion to the cytosol, where they serve as precursors for the synthesis of essential molecules, such as amino acids.

The Transfer of Electrons to Oxygen Drives ATP Formation[7]

Oxidative phosphorylation is the last step in catabolism, and the point at which the major portion of metabolic energy is released. In this process, molecules of NADH and $FADH_2$ transfer the electrons that they have gained from the oxidation of food molecules to molecular oxygen, O_2. The reaction, which is formally equivalent to the burning of hydrogen in air to form water, releases a great deal of chemical energy. Part of this energy is used to make ATP; the rest is liberated as heat.

Although the overall chemistry of NADH and $FADH_2$ oxidation involves a transfer of hydrogen to oxygen, complete hydrogen atoms are not transferred directly. It is the *electrons* from the hydrogen atoms that are important. This is because a hydrogen atom can be readily dissociated into its constituent electron and proton (H^+). The electron can then be transferred separately to a molecule that accepts only electrons, while the proton remains in aqueous solution (Figure 2–24). By the same reasoning, if an electron alone is donated to a molecule with a strong affinity for hydrogen, then a hydrogen atom will be automatically reconstituted by the withdrawal of a proton from solution. In the course of oxidative phosphorylation, electrons from NADH and $FADH_2$

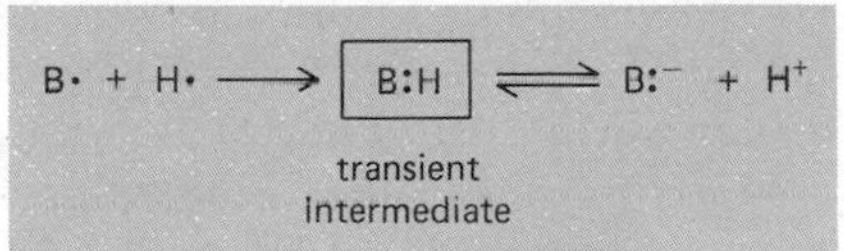

Figure 2–24 The transfer of an electron between one molecule and another can be equivalent to the transfer of a hydrogen atom. In this example, molecule B is reduced by the addition of a hydrogen atom, but if B-H is an acidic molecule, the hydrogen atom readily dissociates to give a proton (or H^+ ion) in solution. The essential part of the reduction is the transfer of the electron, which is retained by B.

pass down a chain of carrier molecules, but the presence or absence of intact hydrogen atoms depends on the nature of the carrier.

This series of electron transfers along the **electron-transport chain** takes place on the inner membrane of the mitochondrion, in which all the carrier molecules are embedded. At each step of the transfer, the electrons fall to a lower energy state, until at the end they are transferred to oxygen molecules that have diffused into the mitochondrion. Oxygen molecules have the highest affinity of all for electrons, and electrons bound to oxygen are thus in their lowest energy state. The energy released as these electrons fall to lower energy states is harnessed, in a way that is not fully understood, to pump protons from the inner mitochondrial compartment to the outside (Figure 2–25). An *electrochemical proton gradient* is thereby generated across the inner mitochondrial membrane. This gradient, in turn, drives a flux of protons back through an enzyme complex in the membrane that adds a phosphate group to ADP, generating ATP inside the mitochondrion. The newly made ATP is transferred from the mitochondrion to the rest of the cell, where it, in turn, drives a variety of metabolic reactions.

The nature of the electron-transport chain and the mechanism of ATP synthesis will be described in detail in Chapter 9.

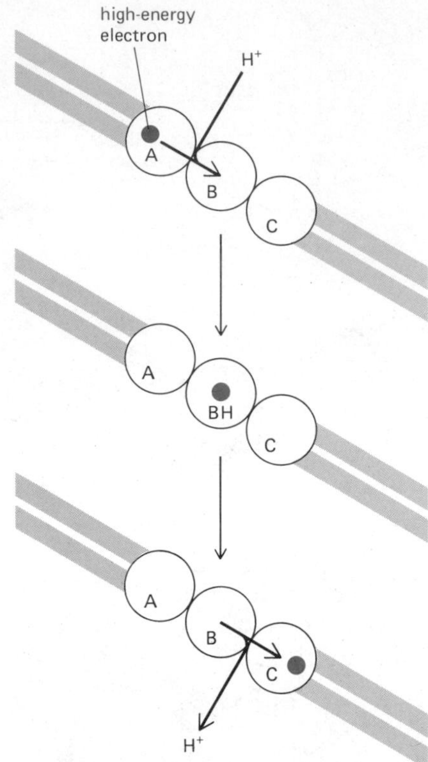

Figure 2–25 Schematic diagram of the generation of a H^+ gradient across a membrane by electron-transport reactions. A high-energy electron (derived, for example, from the oxidation of a metabolite) is passed sequentially by carriers A, B, and C to a lower energy state. In this diagram, carrier B is arranged in the membrane in such a way that it takes up H^+ from one side and releases it to the other as the electron passes. The resulting H^+ gradient represents a form of stored energy that is harnessed by other membrane proteins in the mitochondrion to drive the formation of ATP (see also Figure 9–34, p. 508).

Amino Acids and Nucleotides Are Part of the Nitrogen Cycle

The metabolic changes described thus far have been limited to four elements—carbon, hydrogen, oxygen, and phosphorus. We have not yet discussed the metabolism of nitrogen or sulfur. These two elements are important constituents of proteins and nucleic acids, which are the two most vital macromolecules of the cell and make up approximately two-thirds of its dry weight. Atoms of nitrogen and sulfur pass from compound to compound and between organisms and their environment in a series of reversible cycles.

Although molecular nitrogen is abundant in the earth's atmosphere, it is chemically unreactive. Only a few living species are able to incorporate it into organic molecules, a process called **nitrogen fixation.** Nitrogen fixation occurs in certain microorganisms and by some geophysical processes, such as lightning discharge. It is essential to the biosphere as a whole, for without it life on this planet would not exist. Only a small fraction of the nitrogenous compounds in today's organisms, however, represents fresh products of nitrogen fixation. Most organic nitrogen has been in circulation for some time, passing from one living organism to another. Thus nitrogen-fixing reactions can be said to perform a "topping-up" function for the total nitrogen supply. Vertebrates, for example, receive virtually all of their nitrogen in their dietary intake of proteins and nucleic acids. In the body, these macromolecules are broken down to component amino acids and nucleotides, which are then utilized to make other molecules.

Amino acids that are not utilized in biosynthesis can be oxidized to generate metabolic energy. Most of their carbon and hydrogen atoms eventually form CO_2 or H_2O, while their nitrogen atoms are shuttled through various forms and eventually appear as urea, which is excreted. Each amino acid is processed differently, and a whole constellation of enzyme reactions exists for their catabolism. Conversely, a different series of reactions allows intermediates of the citric acid cycle to be used to synthesize a number of amino acids. About half of the 20 amino acids found in proteins can be made by vertebrates; the others must be supplied in the diet. For this reason, the latter are called *essential amino acids* (Figure 2–26). They are made in other organisms, usually by long and energetically expensive pathways that have been lost in the course of vertebrate evolution.

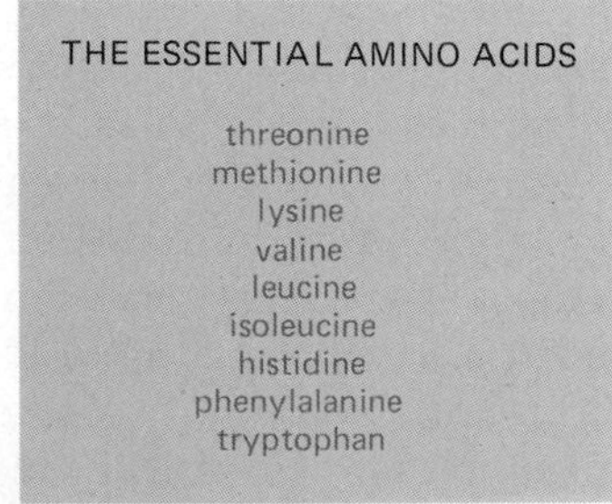
THE ESSENTIAL AMINO ACIDS

threonine
methionine
lysine
valine
leucine
isoleucine
histidine
phenylalanine
tryptophan

Figure 2–26 The nine essential amino acids, which cannot be synthesized by human cells and so must be supplied in the diet.

Summary

Animal cells can be considered to derive energy from food in three stages. In stage 1, proteins, polysaccharides, and fats are broken down by extracellular reactions to small molecules. In stage 2, these small molecules are degraded within cells to produce acetyl CoA and a limited amount of ATP and NADH. These are the only reactions that can yield energy in the absence of oxygen. In stage 3, the acetyl CoA molecules are degraded in mitochondria to give CO_2 and hydrogen atoms that are linked to carrier molecules such as NADH. Electrons from the hydrogen atoms are passed through a complex chain of carriers that leads eventually to the reduction of molecular oxygen to form water. Driven by the energy released in these electron-transfer steps, hydrogen ions (H^+) are transported out of the mitochondrion. The resulting electrochemical proton gradient across the inner mitochondrial membrane is harnessed to drive the synthesis of most of the cell's ATP.

Biosynthesis and the Creation of Order

Thousands of different chemical reactions are occurring in a cell at any instant of time. The reactions are linked together in chains in which the product of one reaction becomes the substrate of the next. It is possible in principle to go from any one compound to any other. However, like motor traffic along the routes of a major city, much of the metabolic traffic in a cell tends to be either "inward" or "outward." The inward traffic consists of catabolic reactions that convert food molecules into sugars and sugar phosphates, as described earlier. The outward traffic consists of biosynthetic reactions that begin with the intermediate products of glycolysis and the citric acid cycle (and their related compounds) and generate the larger and more complex molecules of the cell.

The Energy for Biosynthesis Comes from the Hydrolysis of ATP

Throughout the cell, reactions that would otherwise be unfavorable, including all biosynthetic reactions, are driven directly or indirectly by the hydrolysis of ATP molecules. The large amount of energy released in this reaction derives from several sources, including the high stability of phosphate (abbreviated as P_i or Ⓟ) in its free form and the release of unfavorable charge repulsion between two adjacent phosphates in an ATP molecule (Figure 2–27). An alternative pathway for ATP hydrolysis, in which two Ⓟ-Ⓟ bonds are hydrolyzed, releases about twice as much energy; in this case, ATP is hydrolyzed to AMP (*adenosine monophosphate*) and Ⓟ-Ⓟ (*pyrophosphate*), with subsequent hydrolysis of the released Ⓟ-Ⓟ to free phosphate (Figure 2–28).

We have thus far used the term "energy" quite loosely: what determines whether a reaction will occur is actually the change in **free energy.** As we have said earlier, the release of energy as heat, by increasing the violence of molecular motions and distorting molecules, creates disorder; and according to the Second Law of Thermodynamics, it is only those reactions that result in a net increase in disorder in the universe that can take place spontaneously. The change in free energy that occurs during a reaction, denoted $\boldsymbol{\Delta G}$, is defined in such a way that it provides a direct measure of the amount of disorder created in the universe when that reaction takes place. Reactions that release a large quantity of free energy are those that have a very large *negative* ΔG and create much disorder. Such reactions will have a strong tendency to

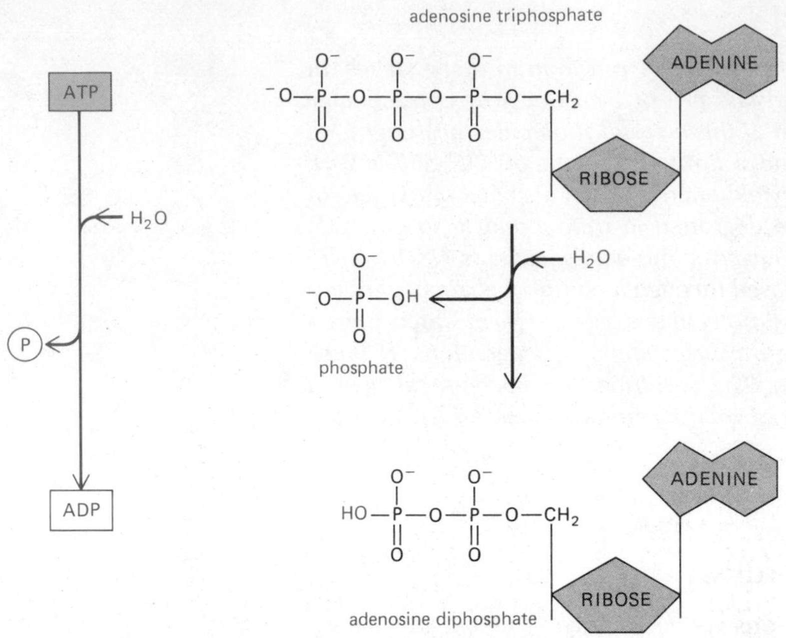

adenosine triphosphate

ATP

ADENINE

RIBOSE

H_2O

phosphate

ADP

ADENINE

RIBOSE

adenosine diphosphate

Figure 2–27 In the hydrolysis of ATP, the terminal phosphate can be cleaved to yield between 11 and 13 kilocalories per mole of usable energy, depending on intracellular conditions.

occur, although the rate at which they do so will depend on other factors, such as the availability of specific enzymes (see below). Conversely, reactions with a *positive* value of ΔG create net order in the universe and cannot occur spontaneously. Such energetically unfavorable reactions will happen only if they are coupled to a second reaction with a negative ΔG so large that the ΔG of the entire process is negative.

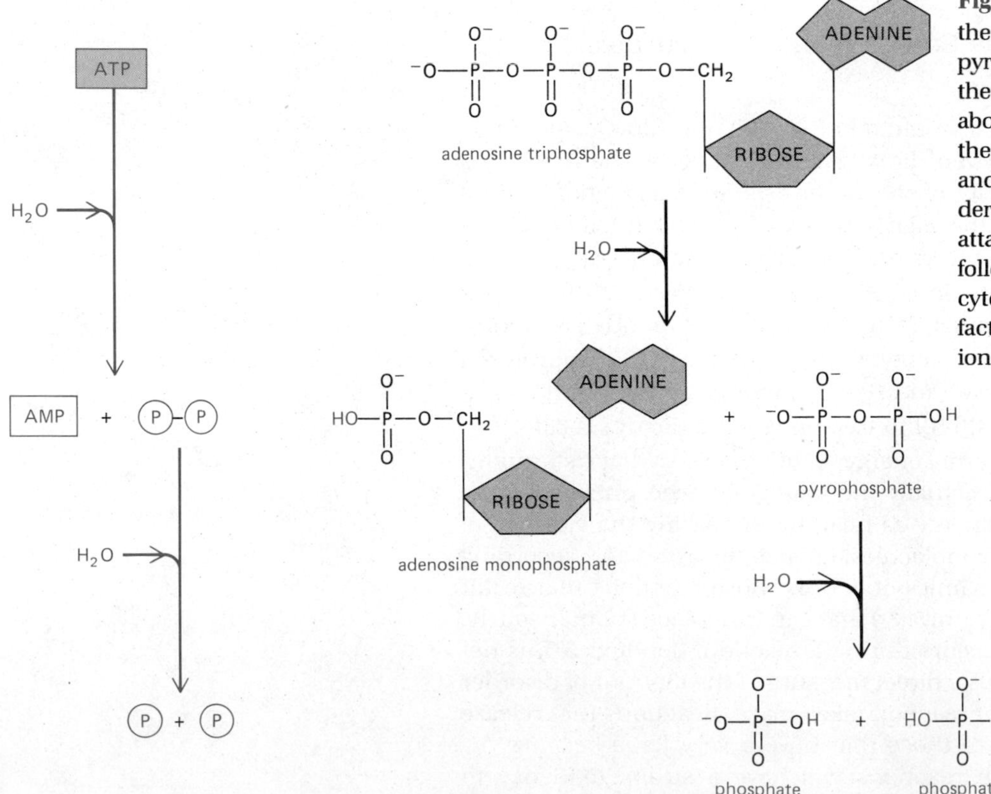

Figure 2–28 An alternative route for the hydrolysis of ATP, in which pyrophosphate is first formed and then hydrolyzed. This route yields about twice as much usable energy as the reaction in Figure 2–27. In this and the previous figure, the H atoms derived from water are shown attached to the phosphate groups following hydrolysis. At the pH of the cytoplasm, however, most of these in fact dissociate to form free hydrogen ions, H^+.

Biosynthetic Reactions Are Often Directly Coupled to ATP Hydrolysis

While enzymes speed up energetically favorable reactions, they cannot force energetically unfavorable reactions to occur. In terms of a water analogy, enzymes by themselves cannot make water run uphill. But in order to grow and divide, cells must do just that: they must build large and complex molecules from small and simple ones. We have seen that, in a general way, this is done through enzymes that couple the release of chemical energy (derived originally from the sun) to the completion of energetically unfavorable reactions. Let us examine in greater detail how such coupling is achieved.

Imagine a typical biosynthetic reaction in which two monomers, A and B, are to be joined in a *dehydration* (also called *condensation*) reaction, in which water is released:

$$\text{A-H} + \text{B-OH} \rightarrow \text{A-B} + H_2O$$

(For convenience, we shall refer to A-H and B-OH as Ⓐ and Ⓑ.) Almost invariably the reverse reaction (called *hydrolysis*), in which water breaks the covalently linked compound A-B, will be the energetically favorable one. This is the case, for example, in the hydrolysis of proteins, nucleic acids, and polysaccharides into their subunits.

The general strategy that allows the cell to make A-B from Ⓐ and Ⓑ is the same as that which allows it to make ATP from burning glucose: a multiple-step pathway couples the energetically unfavorable synthesis of the desired compound to an even more energetically favorable reaction (see Figure 2–17). In many cases the energetically favorable reaction exploited is the hydrolysis of an ATP molecule.

In the coupled pathway from Ⓐ and Ⓑ to A-B, energy from ATP hydrolysis first converts Ⓑ to a higher-energy intermediate compound, which then reacts directly with Ⓐ to give A-B. The simplest mechanism involves the transfer of a phosphate from ATP to Ⓑ to make Ⓑ-OPO_3^{2-} (or Ⓑ-Ⓟ), in which case the reaction pathway would contain only two steps:

1. Ⓑ + ATP → Ⓑ-Ⓟ + ADP
2. Ⓐ + Ⓑ-Ⓟ → A-B + Ⓟ

Since the intermediate Ⓑ-Ⓟ is formed and broken down again (possibly very quickly, while still bound to the surface of an enzyme molecule), the overall reactions that occur are

Ⓐ + Ⓑ → A-B and ATP → ADP + Ⓟ

Note that the first reaction has been forced to occur by being directly coupled to the second reaction.

The Outcome of Coupled Reactions Depends on the Total Change in Free Energy

The course of most reactions can be predicted quantitatively. A large body of thermodynamic data has been collected that makes it possible to calculate the change in free energy, ΔG, for most of the important metabolic reactions of the cell. The overall free-energy change for a pathway is then simply the sum of the energy changes in each of its component steps.

If we imagine two reactions

$$A \rightarrow B \quad \text{and} \quad C \rightarrow D$$

where the ΔG values are $+1$ and -9 kilocalories per mole, respectively, then if these two reactions can be coupled together, the ΔG for the coupled reaction will be −8 kilocalories per mole. (Recall that a mole is 6×10^{23} molecules of a substance.) This tells us that even a reaction with a positive ΔG, which will not occur spontaneously, can be driven by a second reaction. However, this requires that the latter reaction have a sufficiently large negative ΔG and that a mechanism exist by which the two reactions can be coupled together.

The ΔG for the hydrolysis of ATP to ADP and inorganic phosphate depends on the concentrations of all of the reactants (see p. 499), but under the usual conditions in a cell it is between −11 and −13 kilocalories per mole. In principle, this hydrolysis reaction can be used to drive an unfavorable reaction with a ΔG of, perhaps, +10 kilocalories per mole, provided that a suitable reaction path is available. For many biosynthetic reactions, however, even −13 kilocalories per mole is not enough, as in the synthesis of nucleic acids and in the activation of amino acids preparatory to protein synthesis. In these and other cases, the path of ATP hydrolysis is altered so that it initially produces AMP and Ⓟ-Ⓟ (pyrophosphate) (see Figure 2–28). Pyrophosphate is then itself hydrolyzed in a second step that makes an additional −13 kilocalories per mole available. Many biosynthetic pathways are effectively irreversible only because the pyrophosphate required for the reverse reaction is rapidly removed (Figure 2–29).

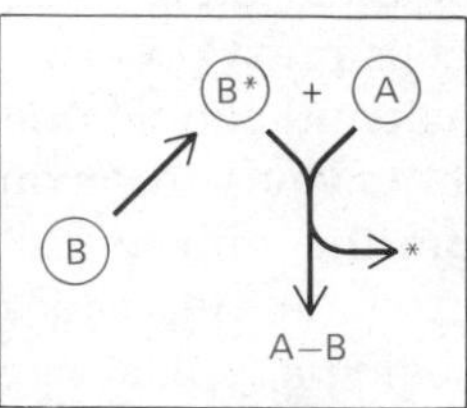

Figure 2–29 Examples of dehydration reactions of the type Ⓐ + Ⓑ → A−B. A schematic outline that applies in all cases is shown above; in general, nucleotide hydrolysis activates compound Ⓑ to Ⓑ* in order to drive an otherwise unfavorable reaction. In one of the two examples shown (below), the synthesis of the amino acid glutamine from glutamic acid and ammonia, a single phosphate bond is hydrolyzed. In the other example, two phosphate bonds are hydrolyzed in order to add each nucleotide to DNA or RNA (polynucleotide synthesis).

1. GLUTAMINE SYNTHESIS

2. POLYNUCLEOTIDE SYNTHESIS

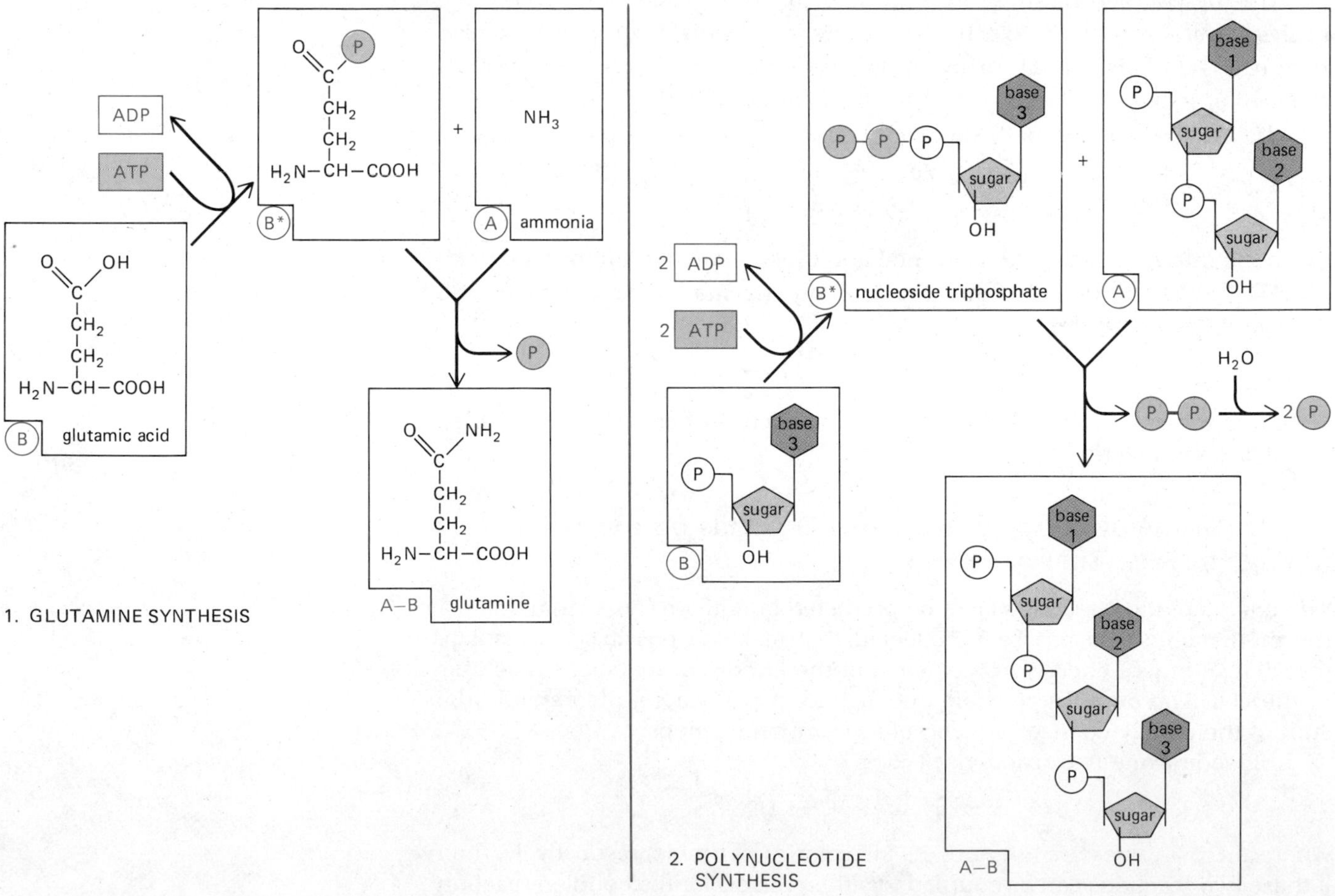

Table 2–2 Some Coenzymes Involved in Group-Transfer Reactions

Coenzyme	Group Transferred
ATP	phosphate
NADH, NADPH	hydrogen & electron (hydride ion)
Coenzyme A	acetyl
Biotin	carboxyl
S-Adenosylmethionine	methyl
UDP-glucose	glucose

Coenzymes are small molecules that are associated with some enzymes and are essential for their activity. Each one listed is a carrier molecule for a small chemical group, and it participates in various reactions in which that group is transferred to another molecule. Some coenzymes are covalently linked to their enzyme; others are less tightly bound.

Coenzymes Are Involved in the Transfer of Specific Chemical Groups

ATP usually drives biosynthetic reactions by reacting with a second molecule to form a highly reactive phosphorylated intermediate, as we have just seen. Because the new phosphate linkage is easily cleaved with release of free energy, the second molecule can readily be joined to other molecules. This general principle is not confined to ATP-mediated reactions: a wide variety of other chemically labile linkages also work in this way. For example, specific carrier molecules are involved in the transfer of chemical groups such as acetyl groups or methyl groups (Table 2–2). The same carrier molecule will often participate in many different biosynthetic reactions in which its group is needed.

Acetyl coenzyme A (acetyl CoA), which is produced in the breakdown of glucose, is an example of such a carrier molecule. It carries an acetyl group

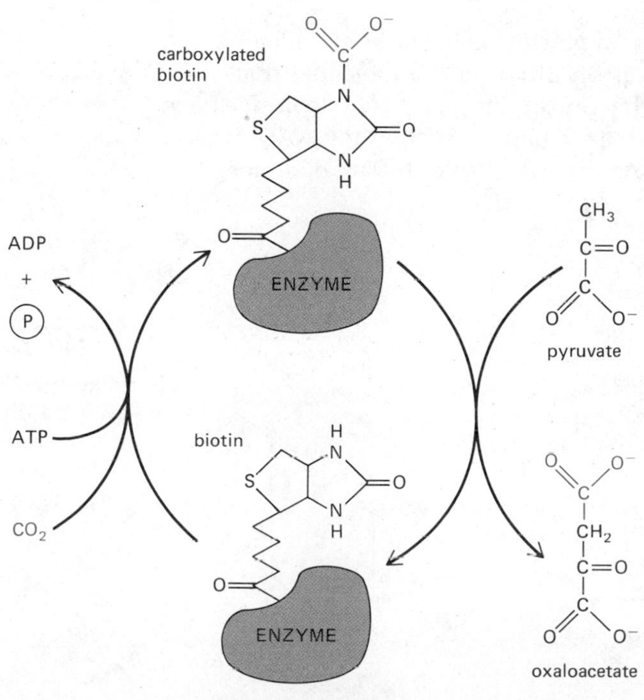

Figure 2–30 Transfer of a carboxyl group by the coenzyme biotin. Biotin acts as a carrier molecule for the carboxyl group (—COO$^-$). In the sequence of reactions shown, biotin is covalently bound to the enzyme pyruvate carboxylase. An activated carboxyl group derived from a bicarbonate ion (HCO_3^-) is coupled to biotin in a reaction that requires an input of energy from the hydrolysis of an ATP molecule. Subsequently, this carboxyl group is transferred to the methyl group of pyruvate to form oxaloacetate.

linked to CoA through a reactive thioester bond (see Figure 2–19). This acetyl group is readily transferred to another molecule, such as a growing fatty acid molecule. Another important example is biotin, which carries a carboxyl group in many biosynthetic reactions (Figure 2–30). Molecules such as acetyl CoA, biotin, and ATP are known as **coenzymes** because they are bound tightly to various enzyme surfaces and are essential for the activity of the enzyme. Many of the small molecules known as *vitamins,* which are required in trace amounts in the diet, are converted to coenzymes in the body.

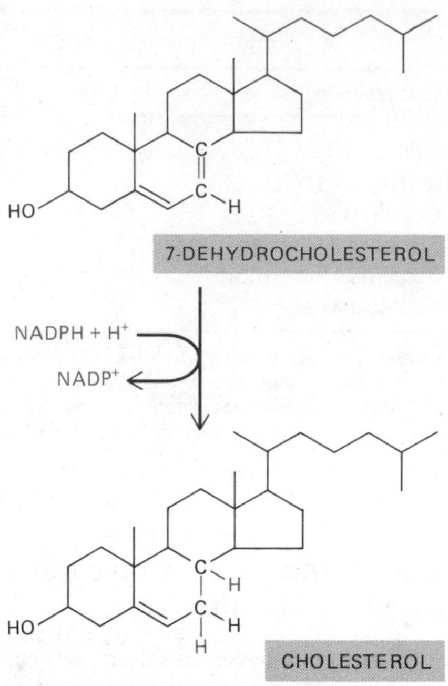

Biosynthesis Requires Reducing Power

We have seen that oxidation and reduction reactions occur continuously in cells. The chemical energy in food molecules is released by oxidative processes that are a form of combustion, while, in order to make biological molecules, the cell needs—among other things—to carry out a series of reduction reactions that require an input of chemical energy. By using the same principle of coupled reactions that operates in the synthesis of ATP, chemical energy is channeled into the synthesis of the high-energy bond between hydrogen and the nicotinamide ring in NADH. This high-energy bond then provides the energy for otherwise unfavorable enzyme reactions that transfer hydrogen (as a hydride ion) to another molecule. NADH, and the NADPH to which it can be readily converted, are therefore said to carry "reducing power."

To see how this works in practice, consider just one biosynthetic step: the last reaction in the synthesis of the lipid molecule *cholesterol.* In this reaction two hydrogen atoms are added to the polycyclic steroid ring in order to reduce a carbon-carbon double bond (Figure 2–31). As in most biosynthetic reactions, the constituents of the two hydrogen atoms required in this reaction are supplied as a hydride ion from NADPH and a proton (H^+) from the solution ($H:^- + H^+ = 2H\cdot$). As in NADH, the hydride ion to be transferred from NADPH is part of a nicotinamide ring and is easily lost because the ring can achieve a more stable aromatic state without it (see Figure 2–22). Therefore, NADH and NADPH both hold this hydride ion in a high-energy linkage from which it can be transferred to another molecule when a suitable enzyme is available to catalyze the transfer.

The difference between NADH and NADPH is trivial in chemical terms: NADPH has an extra phosphate group on a part of the molecule that is far from the active region (Figure 2–32). This phosphate group is of no importance to the reaction as such, but it serves as a handle for binding NADPH as a coenzyme to appropriate enzymes. As a general rule, NADH operates with

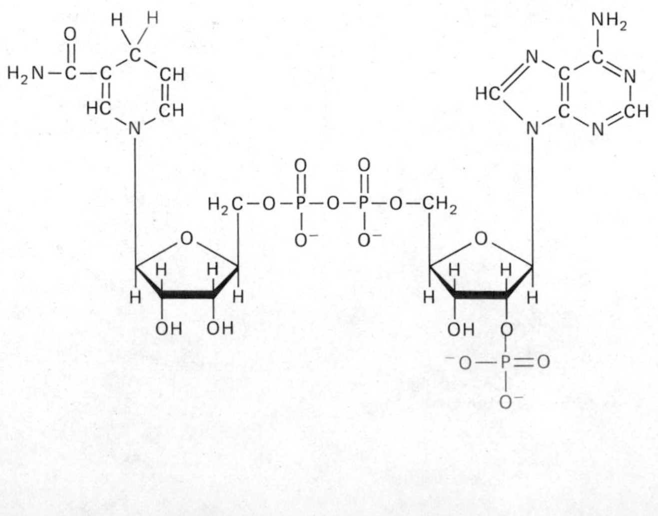

Figure 2-31 The final stage in one of the biosynthetic routes leading to cholesterol. The reduction of the C=C bond is achieved by the transfer of a hydride ion from the carrier molecule NADPH, plus a proton (H^+) from the solution.

Figure 2–32 The structure of NADPH, which differs from NADH (Figure 2–22) only in the presence of an extra phosphate group that allows it to be selectively recognized by certain enzymes (usually those involved in biosynthesis).

enzymes catalyzing catabolic reactions, while NADPH operates with enzymes that catalyze biosynthetic reactions. This means that catabolic and biosynthetic pathways can be regulated separately by alterations in the levels of NADH and NADPH, respectively.

Biological Polymers Are Synthesized by Repetition of Elementary Dehydration Reactions

The principal macromolecules synthesized by cells are polynucleotides (DNA and RNA), polysaccharides, and proteins. They are enormously diverse in structure and include the most complex molecules known. Despite this, they are synthesized from a relatively small number of small molecules (referred to as either *monomers* or *subunits*) by a restricted repertoire of chemical reactions.

The addition of monomers to proteins, polynucleotides, and polysaccharides is shown in Figure 2–33. Although the synthetic reactions for each polymer involve a different kind of covalent bond and different enzymes and cofactors, there are strong underlying similarities. The addition of subunits in

Figure 2–33 Schematic diagram of the polymerization reactions by which three kinds of biological polymer are synthesized. Although each reaction involves a number of different enzymes and other intermediates, there are underlying similarities. Synthesis in every case involves the loss of water (dehydration), the consumption of high-energy nucleoside triphosphates, and the production of inorganic pyrophosphate. The reverse reaction—the breakdown of all three types of polymer—occurs by the simple addition of water (hydrolysis).

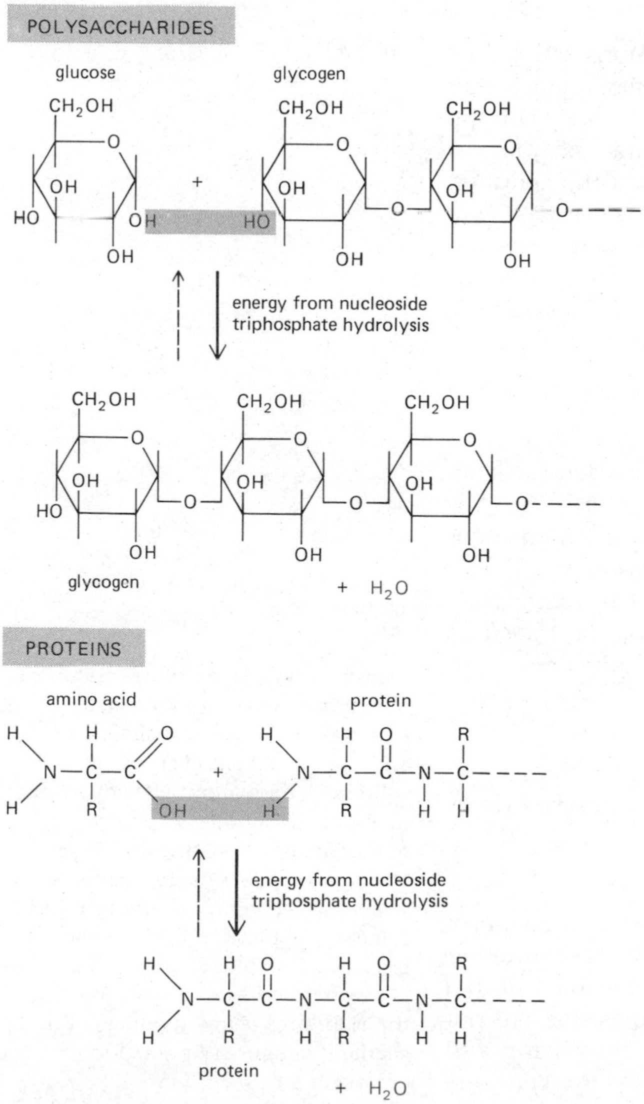

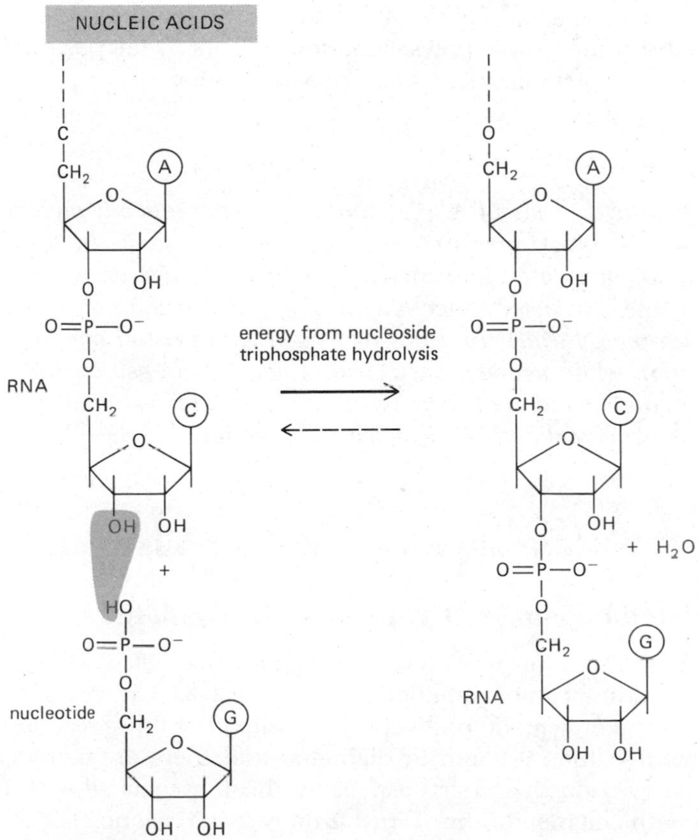

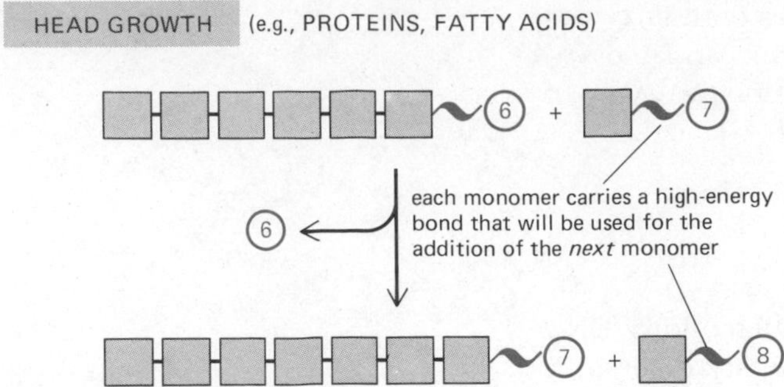

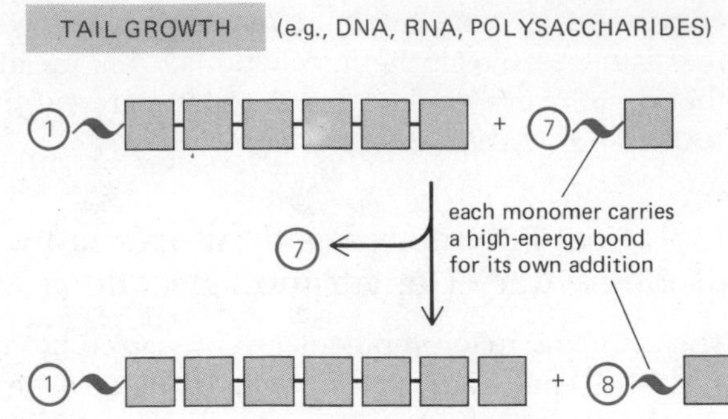

Figure 2–34 Head growth compared to the tail growth of polymers.

each case occurs by a dehydration reaction, involving the removal of a molecule of water from the reactants.

As in the more general case discussed previously (p. 77), the formation of these polymers requires the input of chemical energy, which is ultimately achieved by the standard strategy of coupling the biosynthetic reaction to the energetically favorable hydrolysis of a nucleoside triphosphate. In every case, at least one of the nucleoside triphosphates involved is cleaved to produce pyrophosphate, which is subsequently hydrolyzed to add extra driving force to the reaction (Figure 2–28).

The activated intermediates in the polymerization reactions can be oriented in one of two ways, giving rise to either head polymerization or tail polymerization. In *head polymerization,* the activated linkage is carried on the end of the growing polymer and must therefore be regenerated each time a monomer is added. In this case, each monomer brings with it the activated group that will be used to react with the next monomer in the series (Figure 2–34). In *tail polymerization,* the activated linkage carried by each monomer is used instead for its own addition. While the synthesis of polynucleotides and some simple polysaccharides occurs by tail polymerization, the synthesis of proteins occurs by head polymerization.

Summary

The hydrolysis of ATP is coupled to energetically unfavorable reactions, such as the biosynthesis of macromolecules, usually by the formation of reactive phosphorylated intermediates. Other reactive carrier molecules, called coenzymes, transfer other chemical groups in the course of biosynthesis: for example, NADPH transfers hydrogen as a proton plus two electrons (a hydride ion), while acetyl CoA transfers acetyl groups. Polymeric molecules such as proteins and nucleic acids are assembled from small activated precursor molecules by repetitive dehydration reactions.

The Coordination of Catabolism and Biosynthesis[8]

Metabolism Is Organized and Regulated

Some idea of how cleverly designed the cell is when viewed as a chemical machine can be obtained from Figure 2–35, which is a chart of a large number of the enzymatic pathways in a cell. All of these reactions occur in a cell that is less than 0.1 mm in diameter, and there are many enzymes that are not shown on this chart (especially those associated with the cytoskeleton and with cell membranes). Furthermore, each reaction requires a different enzyme

Figure 2–35 Some of the chemical ▶ reactions occurring in a cell. (A) Radiating from the glycolytic pathway and the citric acid cycle (shown in solid color) are about 500 common metabolic reactions. A typical mammalian cell synthesizes over 10,000 proteins, a major proportion of which are enzymes. In the arbitrarily selected segment of this metabolic maze that is color shaded, cholesterol is synthesized from acetyl CoA. To the right and below the maze, this segment is shown in detail in an enlargement (B).

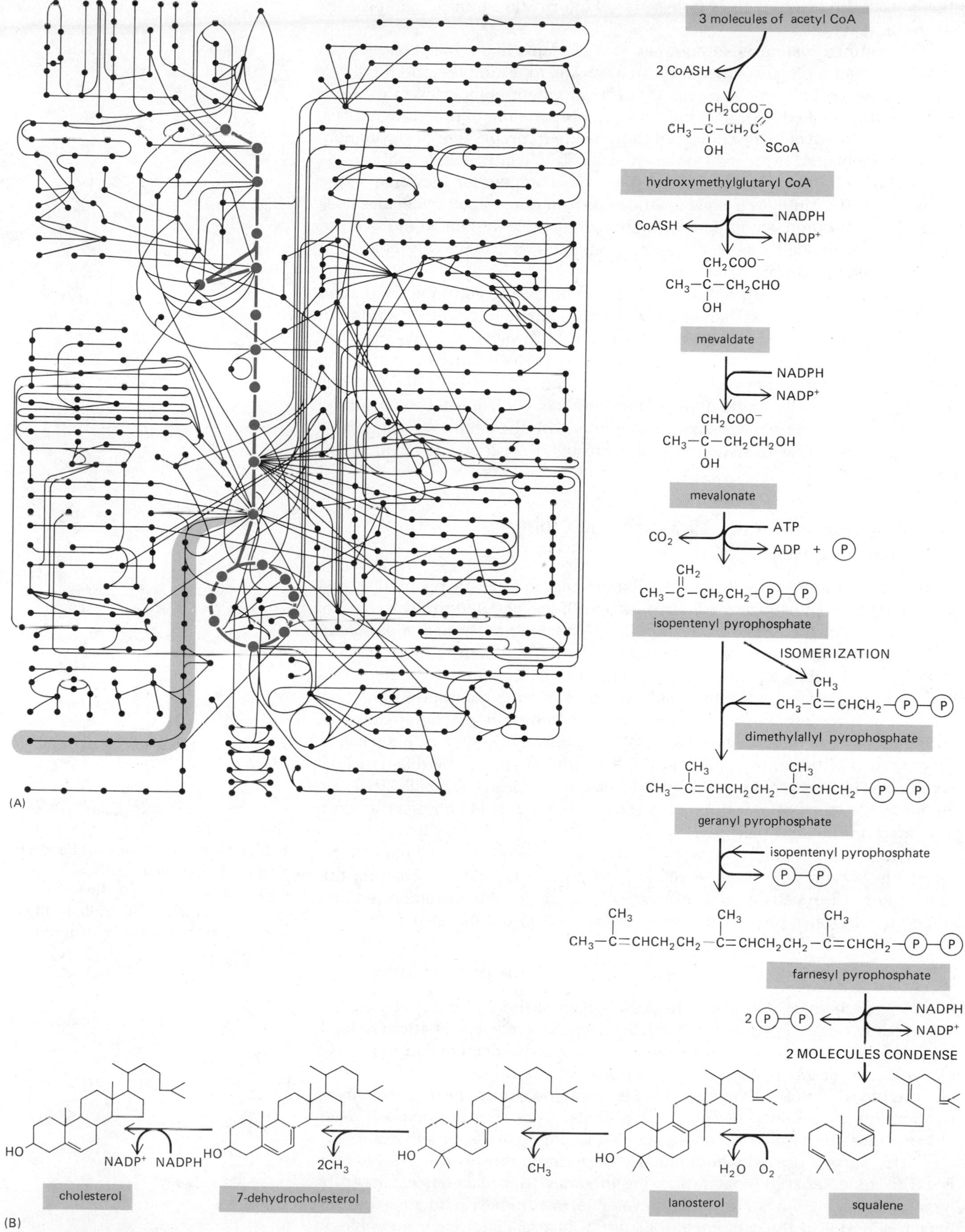

3 molecules of acetyl CoA
2 CoASH
hydroxymethylglutaryl CoA
CoASH
NADPH
NADP+
mevaldate
NADPH
NADP+
mevalonate
ATP
CO2
ADP + P
isopentenyl pyrophosphate
ISOMERIZATION
dimethylallyl pyrophosphate
geranyl pyrophosphate
isopentenyl pyrophosphate
farnesyl pyrophosphate
NADPH
NADP+
2 MOLECULES CONDENSE
NADP+ NADPH
2CH3
CH3
H2O O2
cholesterol
7-dehydrocholesterol
lanosterol
squalene
(A)
(B)

that is itself the product of a whole series of information-transfer and protein-synthesis reactions.

The entire system is so complex that it seems like a metabolic jungle. Take any small molecule—the amino acid *serine,* for example—and there will be half a dozen or more enzymes that can modify it chemically in different ways: it can be linked to AMP (adenylated) in preparation for protein synthesis, or degraded to glycine, or converted to pyruvate in preparation for oxidation; it can be acetylated by acetyl CoA or transferred to a fatty acid to make phosphatidyl serine. All of these different pathways compete for the same serine molecule, and a similar struggle for thousands of other small molecules goes on at the same time. One might think that the whole system would need to be so finely balanced that any minor upset, such as a temporary change in dietary intake, would be disastrous.

In fact, the cell is amazingly stable. It can adapt and continue to function in a coherent way during starvation or disease. Mutations of many kinds can lead to the elimination of particular reaction pathways, and yet—provided that certain minimum requirements are met—the cell survives. It does so because an elaborate network of control mechanisms regulates the chemical reactions within cells. Some of the higher levels of control will be considered in later chapters. Here we are concerned only with the simplest mechanisms that regulate the flow of small molecules through the various metabolic pathways in a cell.

Metabolic Pathways Are Regulated by Changes in Enzyme Activity

The concentrations of the various small molecules in a cell are buffered against major changes by a process known as **feedback regulation.** This type of regulatory mechanism fine-tunes the flux of metabolites through a particular pathway by temporarily increasing or decreasing the activity of crucial enzymes. For example, the first enzyme of a series of reactions is usually inhibited by the final product of that pathway: thus, if large quantities of the final product accumulate, further entry of precursors into the reaction pathway is automatically inhibited (Figure 2–36). Where pathways branch or intersect, as they often do, there are usually multiple points of control by different final products. The complexity of such feedback control processes is illustrated in Figure 2–37, which shows the pattern of enzyme regulation observed in a set of related amino acid pathways.

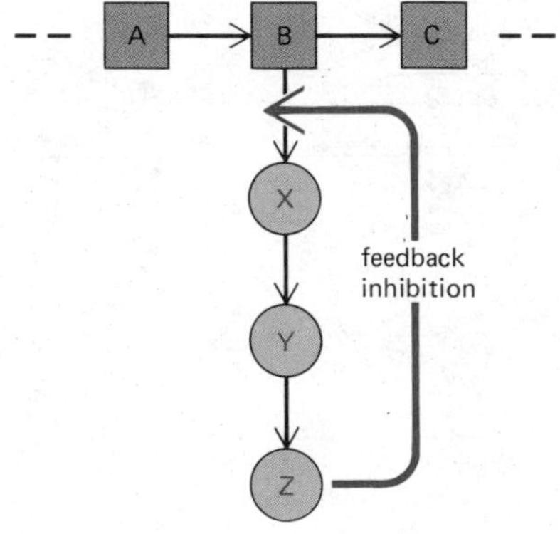

Figure 2–36 Feedback inhibition of a single biosynthetic pathway. The end product Z inhibits the first enzyme that is unique to its synthesis and thereby regulates its own level in the cell.

Feedback regulation can work almost instantaneously, and it may involve reversible enzyme activators as well as inhibitors. The molecular basis for this type of control in cells is well understood, but since an explanation requires some knowledge of protein structure, it will be deferred until Chapter 3.

Catabolic Reactions Can Be Reversed by an Input of Energy[9]

Large-scale changes that affect the metabolism of the entire cell can also be achieved by regulating a few enzymes. For example, a special pattern of feedback regulation enables a cell to switch from glucose degradation to glucose biosynthesis, or *gluconeogenesis.* The need for this reverse pathway is especially acute in periods of violent exercise, when the glucose needed for muscle contraction is generated by liver cells, and also in periods of starvation when glucose must be formed from fatty acids and amino acids for survival.

The normal breakdown of glucose to pyruvate during glycolysis is catalyzed by nine separate enzymes acting in series. The reactions catalyzed by most of these enzymes are readily reversible, but three reaction steps (numbers 1, 3, and 9 in the sequence of Figure 2–20) are effectively irreversible. In

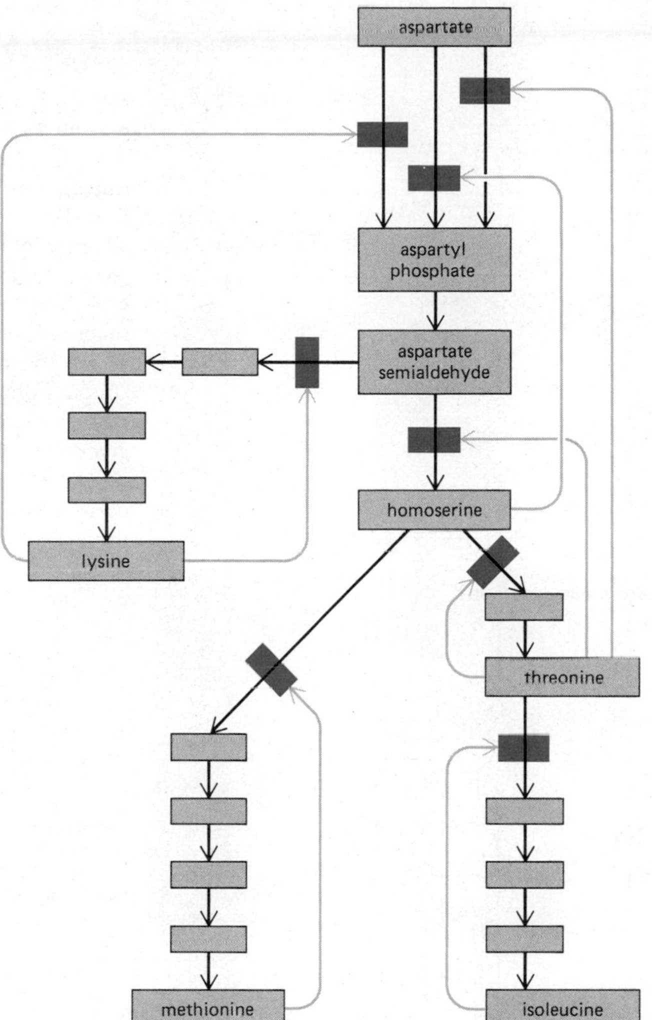

Figure 2–37 Feedback inhibition in the synthesis of the amino acids lysine, methionine, threonine, and isoleucine in bacteria. The colored arrows indicate positions at which products "feed back" to inhibit enzymes. Note that three different enzymes (called *isozymes*) catalyze the initial reaction, each inhibited by a different product.

fact, it is the large negative free-energy change that occurs in these reactions that normally drives the breakdown of glucose. For the reactions to proceed in the opposite direction and make glucose from pyruvate, each of these three reactions must be bypassed. This is achieved by substituting three alternate enzyme-catalyzed bypass reactions that are driven in the uphill direction by an input of chemical energy (Figure 2–38). Thus, while two ATP molecules are generated as each molecule of glucose is degraded to two molecules of pyruvate, the reverse reaction during gluconeogenesis requires the hydrolysis of four ATP and two GTP molecules. This is equivalent, in total, to the hydrolysis of six molecules of ATP for every molecule of glucose synthesized.

The bypass reactions in Figure 2–38 must be closely controlled so that glucose is broken down only when energy is needed and is synthesized only when the cell is nutritionally replete. If both forward and reverse reactions were allowed to proceed without restraint, they would shuttle metabolites backward and forward in futile cycles that consumed large amounts of ATP to no purpose.

The elegance of these control mechanisms can be illustrated by a single example. Step 3 of glycolysis is one of the reactions that must be bypassed during glucose formation. Normally the step involves the addition of a phosphate group to fructose 6-phosphate from ATP and is catalyzed by the enzyme *phosphofructokinase*. This particular enzyme is activated by AMP and ADP

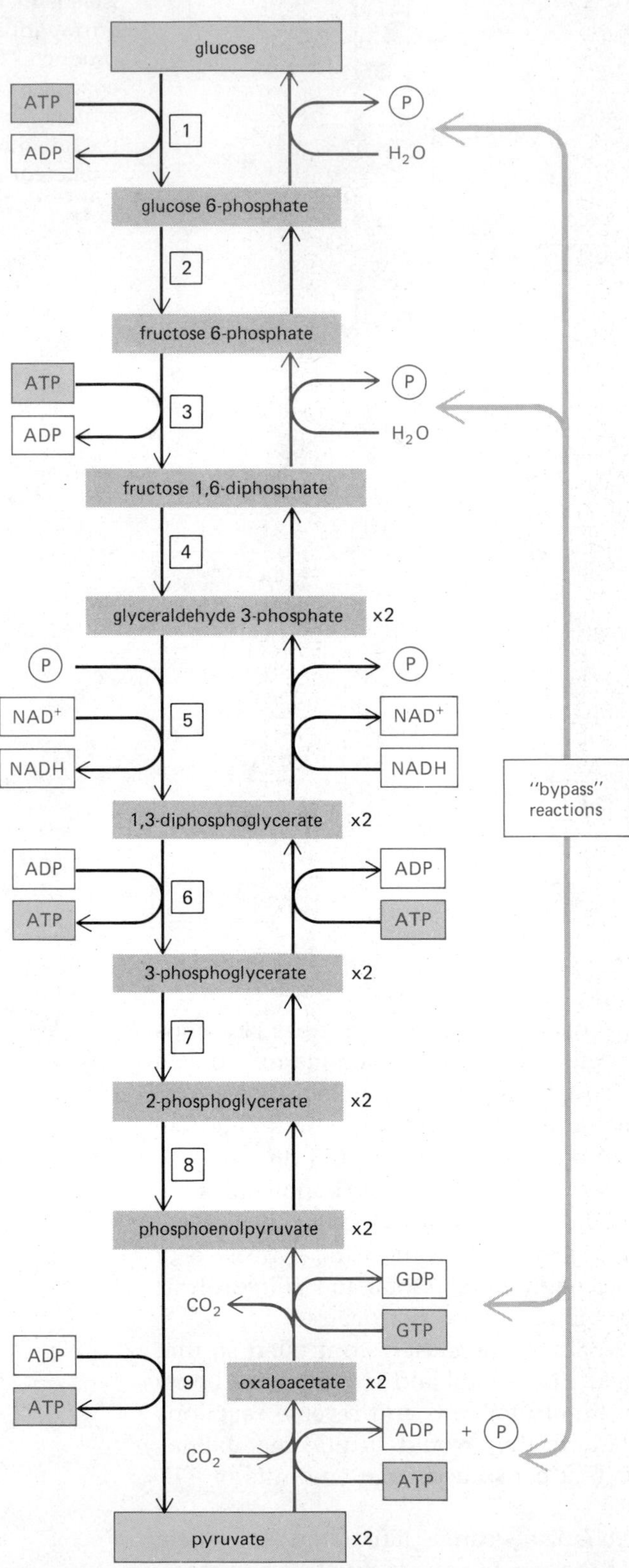

Figure 2–38 Comparison of the reactions that produce glucose during gluconeogenesis with those that degrade glucose. The degradative (glycolytic) reactions are energetically favorable (the free-energy change is less than zero), while the synthetic reactions require an input of energy. To synthesize glucose, different "bypass enzymes" are needed that bypass reactions 1, 3, and 9 of glycolysis. The overall flux of reactants is determined by feedback control mechanisms that operate at these crucial steps.

and inhibited by ATP, citrate, and fatty acids. In other words, the enzyme is activated when energy supplies are low and AMP and ADP accumulate, and it is inactivated when energy (in the form of ATP) or food supplies such as fatty acids or citrate (derived from amino acids) are abundant. The enzyme that catalyzes the reverse (bypass) reaction that leads to the formation of glucose is *fructose diphosphatase.* This enzyme is regulated in the opposite way by the same feedback control molecules, so that it works when the phosphofructokinase does not.

Note that phosphofructokinase is activated by ADP, which is a product of the reaction it catalyzes (ATP + fructose 6-P → ADP + fructose 1,6-diphosphate), and is inhibited by ATP, which is one of its substrates. As a result, this enzyme is subject to a complex form of positive feedback control. Under certain circumstances such feedback control gives rise to striking oscillations in the activity of the enzyme, causing corresponding oscillations in the concentrations of various glycolytic intermediates (Figure 2–39). While the physiological significance of these particular oscillations is not known, they illustrate how a biological oscillator can be produced by a few enzymes. In principle, such oscillations could provide an internal clock, enabling a cell to "measure time" and, for example, to perform certain functions at fixed intervals.

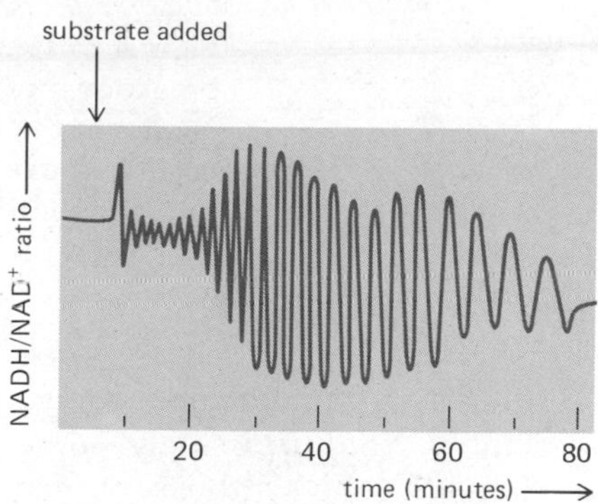

Figure 2–39 The abrupt addition of glucose to an extract containing the enzymes and cofactors required for glycolysis can produce large cyclic fluctuations in the levels of intermediates such as NADH. These metabolic oscillations arise, in part, from the positive feedback control of the glycolytic enzyme phosphofructokinase.

Enzymes Can Be Switched On and Off by Covalent Modification[10]

The types of feedback control just described permit the rates of reaction sequences to be continuously and automatically regulated in response to second-by-second fluctuations in metabolism. Cells have different devices for regulating enzymes when longer lasting changes in activity, occurring over minutes or hours, are required. These involve reversible covalent modification of enzymes, which is often, but not always, accomplished by the addition of a phosphate group to a specific serine, threonine, or tyrosine residue in the enzyme. The phosphate comes from ATP and its transfer is catalyzed by enzymes known as *protein kinases.*

We shall describe in the following chapter how phosphorylation alters the shape of an enzyme in such a way as to increase or inhibit its activity. The subsequent removal of the phosphate group, which reverses the effect of the phosphorylation, is achieved by a second enzyme, called a *phosphoprotein phosphatase.* Covalent modification of enzymes adds another dimension to metabolic control, because it allows specific reaction pathways to be regulated by signals (such as hormones) that are unrelated to the metabolic intermediates themselves.

Reactions Are Compartmentalized Both Within Cells and Within Organisms[11]

Not all of a cell's metabolic reactions occur within the same subcellular compartment. Because different enzymes are found in different parts of the cell, the flow of chemical components is physically as well as chemically channeled.

The simplest form of such spatial segregation occurs when two enzymes that catalyze sequential reactions form an enzyme complex, and the product of the first enzyme does not have to diffuse through the cytoplasm to encounter the second enzyme. As soon as the first reaction is over, the second begins. Some large enzyme aggregates carry out whole series of reactions without losing contact with the substrate. For example, the conversion of pyruvate to acetyl CoA proceeds in three chemical steps, all of which take place on the same large enzyme complex (Figure 2–40), and in fatty acid

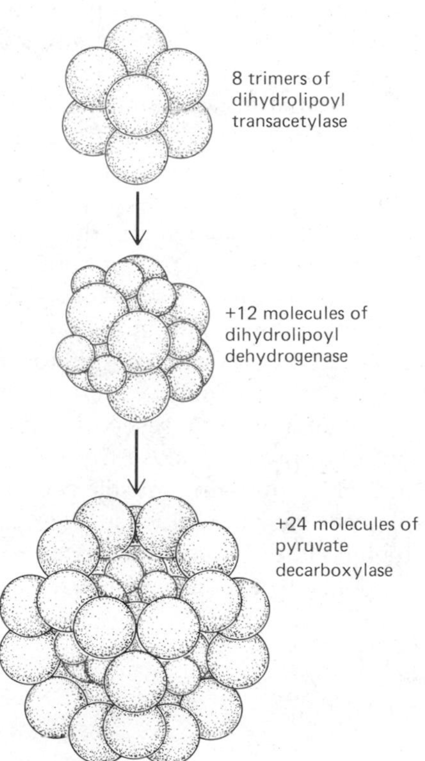

Figure 2–40 The structure of pyruvate dehydrogenase—an example of a large multienzyme complex in which reaction intermediates are passed directly from one enzyme to another. This enzyme complex catalyzes the conversion of pyruvate to acetyl CoA.

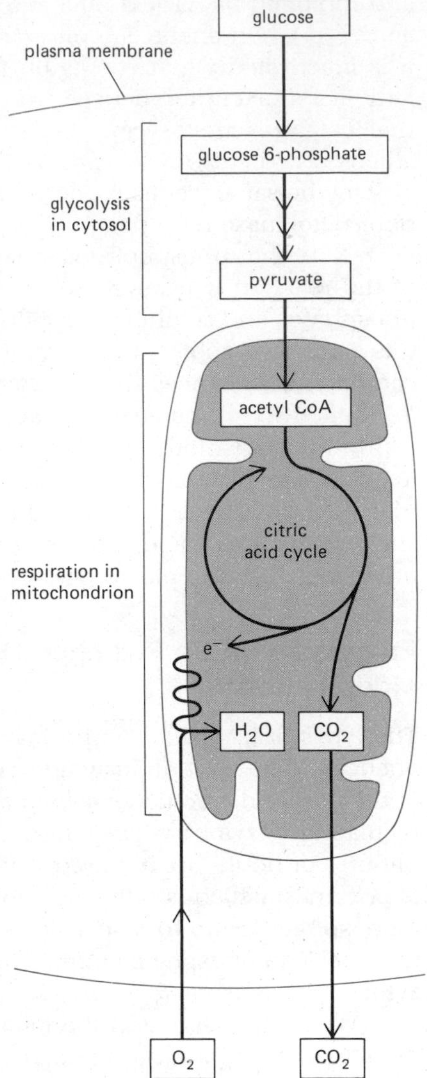

Figure 2–41 Segregation of the various steps in the breakdown of glucose in the eucaryotic cell. Glycolysis occurs in the cytosol, whereas the reactions of the citric acid cycle and oxidative phosphorylation take place only in mitochondria.

synthesis an even longer sequence of reactions is catalyzed by a single enzyme assembly. Not surprisingly, some of the largest enzyme complexes are concerned with the synthesis of macromolecules such as proteins and DNA.

The next level of spatial segregation in cells involves the confinement of functionally related enzymes within the same membrane or within the aqueous compartments of organelles that are bounded by membranes. The oxidative metabolism of glucose is a good example (Figure 2–41). After glycolysis, pyruvate is actively taken up from the cytosol into the inner compartment of the mitochondrion, which contains all of the enzymes and metabolites involved in the citric acid cycle. Moreover, the inner mitochondrial membrane itself contains all of the enzymes that catalyze the subsequent reactions of oxidative phosphorylation, including those involved in the transfer of electrons from NADH to O_2 and in the synthesis of ATP. The entire mitochondrion can therefore be regarded as a small ATP-producing factory. In the same way, other cellular organelles, such as the nucleus, the Golgi apparatus, and the lysosomes, can be viewed as specialized compartments where functionally related enzymes are confined to perform a specific task. In a sense, the living cell is like a modern city, with many specialized services concentrated in different areas that are extensively interconnected by various paths of communication.

Spatial organization in multicellular organisms extends beyond the individual cell. The different tissues of the body have different sets of enzymes and contribute in distinct ways to the survival of the organism as a whole. In addition to differences in specialized products such as hormones or antibodies, there are significant differences in the "common" metabolic pathways between various types of cells in the same organism. Although virtually all cells contain the enzymes of glycolysis, the citric acid cycle, lipid synthesis and breakdown, and amino acid metabolism, the levels of these processes in different tissues are subject to fine-tuning in response to the needs of the organism. Nerve cells, which are probably the most fastidious cells in the body, maintain almost no reserves of glycogen or fatty acids and rely almost entirely on a supply of glucose from the bloodstream. Liver cells supply glucose to actively contracting muscle cells and recycle the lactic acid produced by muscle cells back into glucose (Figure 2–42). All types of cells have their distinctive metabolic traits and cooperate extensively in the normal state as well as in response to exercise, stress, and starvation.

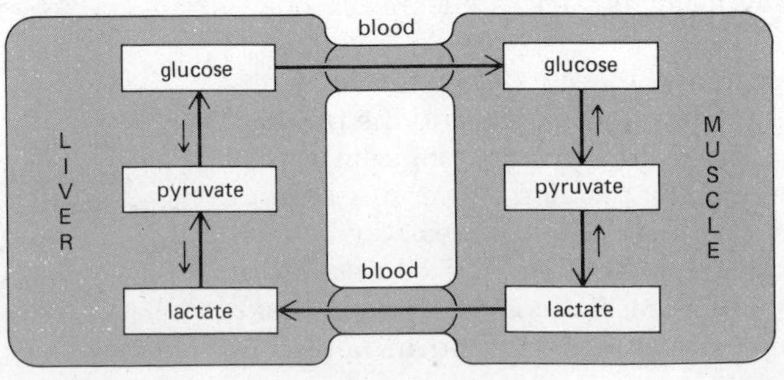

Figure 2–42 Schematic view of the metabolic cooperation between liver and muscle cells. The principal fuel of actively contracting muscle cells is glucose, much of which is supplied by liver cells. Lactic acid, the end product of anaerobic glucose breakdown in muscle, is converted back to glucose in the liver.

Summary

The many thousands of distinct chemical reactions carried out simultaneously by a cell are closely coordinated. A variety of control mechanisms regulate the activities of key enzymes in response to the changing conditions in the cell. One very common form of regulation is a rapidly reversible feedback inhibition exerted on the first enzyme of a pathway by the final product of that pathway. A longer-lasting form of regulation involves the chemical modification of one enzyme by another, often by phosphorylation. Combinations of regulatory mechanisms can produce major and long-lasting changes in the metabolism of the cell. Not all cellular reactions occur within the same intracellular compartment, and spatial segregation by internal membranes permits organelles to specialize in their biochemical tasks.

References

General

Lehninger, A.L. Principles of Biochemistry. New York: Worth, 1982.

Stryer, L. Biochemistry, 2nd ed. San Francisco: Freeman, 1981.

Wood, W.B.; Wilson, J.H.; Benbow, R.M.; Hood, L.E. Biochemistry: A Problems Approach, 2nd ed. Menlo Park, Ca.: Benjamin-Cummings, 1981.

Cited

1. Henderson, L.J. The Fitness of the Environment. Boston: Beacon, 1927; reprinted 1958. (A classical and readable analysis.)

   Masterton, W.L.; Slowinski, E.J. Chemical Principles, 4th ed. New York: Holt, Rinehart and Winston, 1977.

2. Lehninger, A.L. Bioenergetics: The Molecular Basis of Biological Energy Transformations, 2nd ed. Menlo Park, Ca.: Benjamin-Cummings, 1971. (An elegant short treatment.)

3. Klotz, I.M. Energy Changes in Biochemical Reactions. New York: Academic Press, 1967. (Basic thermodynamics.)

   Schrödinger, E. What is Life? Mind and Matter. Cambridge, Eng.: Cambridge University Press, 1969. (A physicist's view of order and disorder in biological systems, first published in 1944.)

4. Raven, P.H.; Evert, R.F.; Curtis, H. Biology of Plants, 3rd ed. New York: Worth, 1981. (Chapter 6 on photosynthesis.)

5. McGilvery, R.W. Biochemistry: A Functional Approach, 2nd ed. Philadelphia: Saunders, 1979. (Particularly good for understanding metabolic pathways: Chapters 23 and 26 cover glycolysis and the citric acid cycle.)

   Racker, E. A New Look at Mechanisms in Bioenergetics. New York: Academic Press, 1976. (A personal and historical account by a pioneer in the field.)

6. Krebs, H.A. The history of the tricarboxylic acid cycle. *Perspect. Biol. Med.* 14:154–170, 1970.

7. Racker, E. From Pasteur to Mitchell: a hundred years of bioenergetics. *Fed. Proc.* 39:210–215, 1980.

   Hinkle, P.C.; McCarty, R.E. How cells make ATP. *Sci. Am.* 238(3):104–123, 1978.

8. Newsholme, E.A.; Start, C. Regulation in Metabolism. New York: Wiley, 1973.

9. Hess, B. Oscillating reactions. *Trends Biochem. Sci.* 2:193–195, 1977.

10. Cohen, P. Control of Enzyme Activity. London: Chapman and Hall, 1976.

11. Banks, P.; Bartley, W.; Birt, L.M. The Biochemistry of the Tissues, 2nd ed. New York: Wiley, 1976.

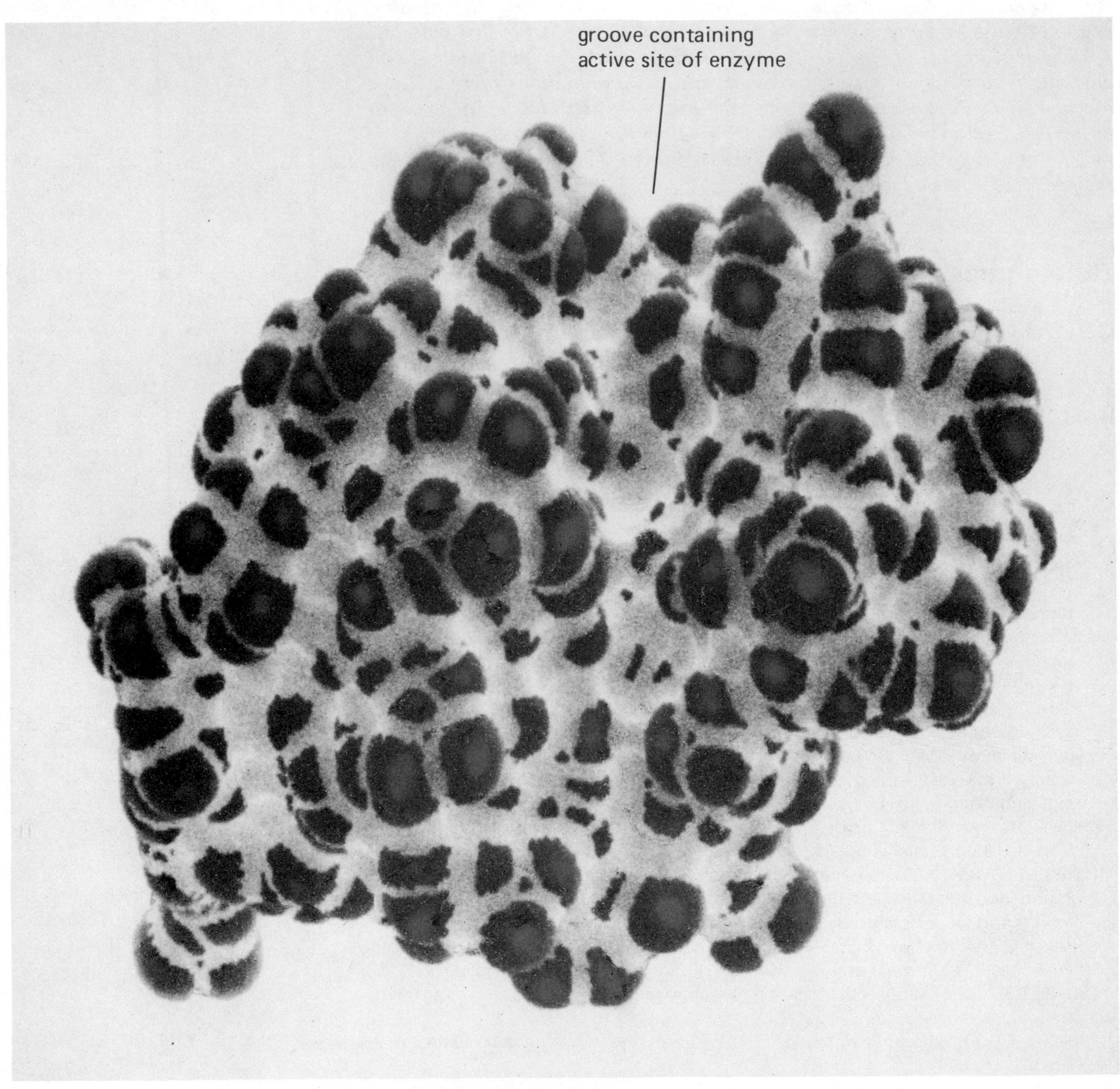

Computer-drawn model of the surface of egg-white lysozyme, a small protein of 129 amino acids. (Courtesy of Michael Connolly.)

Macromolecules: Structure, Shape, and Information

3

The principal macromolecules of the cell—proteins, nucleic acids, and polysaccharides—are synthesized from the small molecules described in the previous chapter as polymers of amino acids, nucleotides, and sugars, respectively. Their complex and precisely defined structures give them unique properties that enable them to carry out all of the most distinctive functions of the cell. Macromolecules are responsible for the assembly of cellular components, the catalysis of chemical transformations, the production of movement, and—most fundamental of all—for heredity. These vital functions depend on the fact that biological macromolecules carry information. Written into the structure of a protein or DNA molecule is a biological message that can be "read" in its interactions with other macromolecules. The principles by which this information is stored, transferred, and used are the subject of this chapter.

Molecular Recognition Processes[1]

Apart from water, macromolecules constitute the major mass of a cell (Table 3–1). They typically have molecular weights between about 10,000 and 1 million and are intermediate in size between the organic molecules of the cell discussed in Chapter 2 and larger units of biological activity, such as ribosomes and viruses (Figure 3–1).

As described in Chapter 2, a macromolecule is assembled from low-molecular-weight subunits that are added one after the other to form a long, chainlike polymer (see Figure 2–33). Usually only one family of subunits goes into the construction of each chain: amino acids are linked to other amino acids to form proteins; nucleotides are linked to other nucleotides to form nucleic acids; and sugars are linked to other sugars to form polysaccharides. Because the precise sequence of subunits is crucial to the function of a macromolecule, its biosynthesis requires mechanisms that determine exactly which subunit goes into the polymer at each position in the chain.

sugars, amino acids, and nucleotides ~0.5–1 nm

globular proteins ~2–10 nm

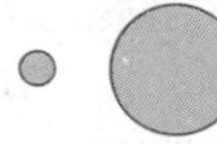

ribosome ~30 nm

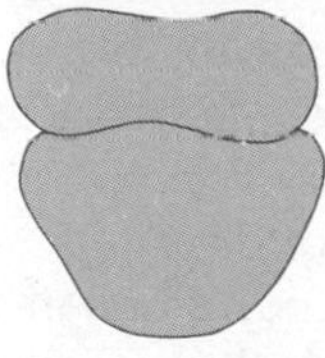

bacterial cell ~1000 nm

Figure 3–1 The size of protein molecules compared to other cell components.

Table 3–1 Approximate Chemical Compositions of a Typical Bacterium and a Typical Mammalian Cell

	Percent of Total Cell Weight	
Component	E. Coli *Bacterium*	*Mammalian Cell*
H_2O	70	70
Inorganic ions (Na^+, K^+, Mg^{2+}, Ca^{2+}, Cl^-, etc.)	1	1
Miscellaneous small metabolites	3	3
Proteins	15	18
RNA	6	1.1
DNA	1	0.25
Phospholipids	2	3
Other lipids	—	2
Polysaccharides	2	2
Total cell volume:	2×10^{-12} cm^3	4×10^{-9} cm^3
Relative cell volume:	1	2000

Proteins, polysaccharides, DNA, and RNA are macromolecules. Lipids are not generally classed as macromolecules even though they share some of their features; for example, most are synthesized as linear polymers of a smaller molecule (the acetyl group on acetyl CoA) and self-assemble into larger structures (membranes).

The Information Carried by a Macromolecule Is Expressed by Means of Weak Noncovalent Bonds

Macromolecular chains are held together by *covalent* bonds, which are strong enough to preserve the sequence of subunits in the macromolecule for long periods of time. But the information carried by this sequence is expressed through much weaker, *noncovalent* bonds. These weak bonds form between different parts of the same macromolecule and between different macromolecules. They therefore determine both the three-dimensional structure of macromolecular chains and how these structures interact with one another.

The noncovalent bonds encountered in biological molecules are usually classified into three types—**ionic bonds, hydrogen bonds,** and **van der Waals attractions.** Another important weak force is created by the three-dimensional structure of water, which tends to force hydrophobic groups together in order to minimize their disruptive effect on the hydrogen-bonded network of water molecules. This expulsion from the aqueous solution generates what is sometimes thought of as a fourth kind of weak noncovalent bond, commonly called a **hydrophobic bond.** These four types of weak bonds are individually discussed and explained in Panel I, pp. 94–95.

In an aqueous environment, noncovalent bonds are about 100 times weaker than covalent bonds (Table 3–2) and only slightly stronger than the average energy of thermal collisions at 37°C. A single noncovalent bond—unlike a single covalent bond—is therefore too weak to withstand the thermal motions that tend to pull molecules apart, and large numbers of noncovalent bonds are needed to hold two molecular surfaces together. Because large numbers of these bonds can form only when the two surfaces are precisely matched to each other (Figure 3–2), they account for the specificity of biological recognition, such as the kind that occurs between an enzyme and its substrates.

Similarly, weak noncovalent forces determine how different regions of the *same* molecule fit together. In principle, a long flexible chain such as a protein can fold in an enormous number of different ways, in each of which

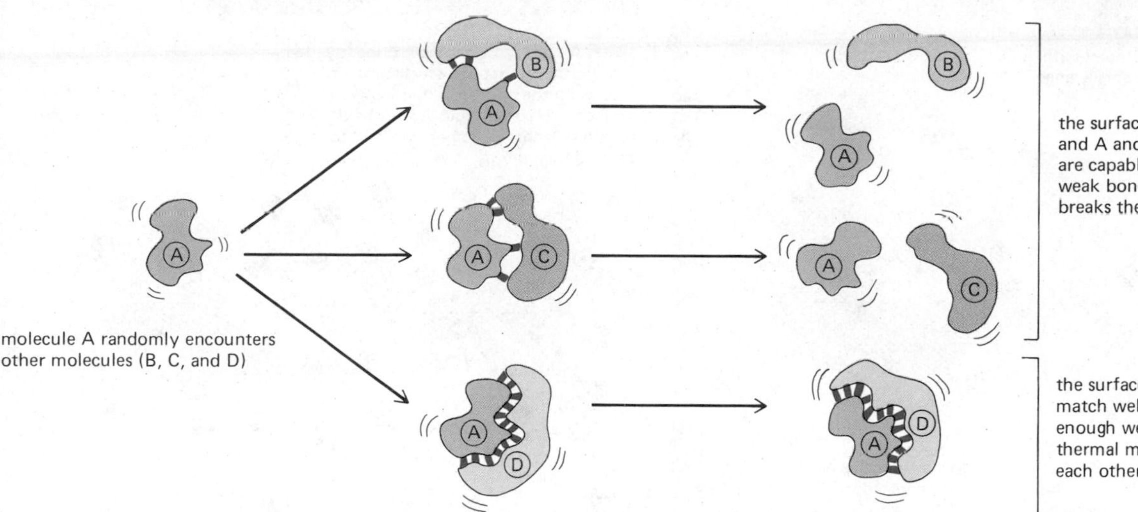

Figure 3–2 Schematic diagram illustrating how weak bonds mediate recognition processes between macromolecules.

Table 3–2 Covalent and Noncovalent Chemical Bonds

		Strength (kcal/mole)	
Bond Type	**Length (nm)**	*In Vacuum*	*In Water*
Covalent	0.15	90	90
Ionic	0.25	80	1
Hydrogen	0.30	4	1
Van der Waals attraction	0.20	1	1

The strength of a bond can be measured by the energy required to break it, here given in kilocalories per mole (kcal/mole). (One *kcal*, or *kilocalorie*, is the quantity of energy needed to raise the temperature of 1000 g of water by 1°C. An alternative unit in wide use is the kilojoule, equal to 0.24 kcal.) Individual bonds vary a great deal in strength, depending on the atoms involved and their precise environment, so that the above values are only a rough guide. Note that the aqueous environment in a cell will greatly weaken both the ionic and hydrogen bonds between nonwater molecules (Panel I, pp. 94–95).

the chain will establish a different set of weak interactions. In practice, however, most proteins in a cell stably fold in only one way: the sequence of amino acid subunits has been selected in the course of evolution, so that one three-dimensional arrangement of atoms (or *conformation*) is able to form many more weak interactions than any other.

Diffusion Is the First Step to Molecular Recognition

Before two molecules recognize each other, and hence express the information on their surfaces, they must come into close contact. This is achieved by the thermal motions that cause molecules to wander, or *diffuse*, from their starting positions. As the molecules in a liquid rapidly collide and bounce off each other, an individual molecule moves first one way and then another, its path making a "random walk" (Figure 3–3). The average distance that such a molecule moves is proportional to the square root of the time: that is, if it takes a molecule 1 second on average to move 1 μm, it will take 4 seconds to move 2 μm, 9 seconds to move 3 μm, and so on. Diffusion is therefore an efficient way for molecules to move for limited distances.

Figure 3–3 A random walk. Molecules in solution move in a random fashion due to the continual buffeting that they receive in collisions with other molecules. This movement causes molecules to diffuse over intracellular distances in a surprisingly short time.

WEAK CHEMICAL BONDS

Organic molecules can interact with other molecules through short-range noncovalent forces.

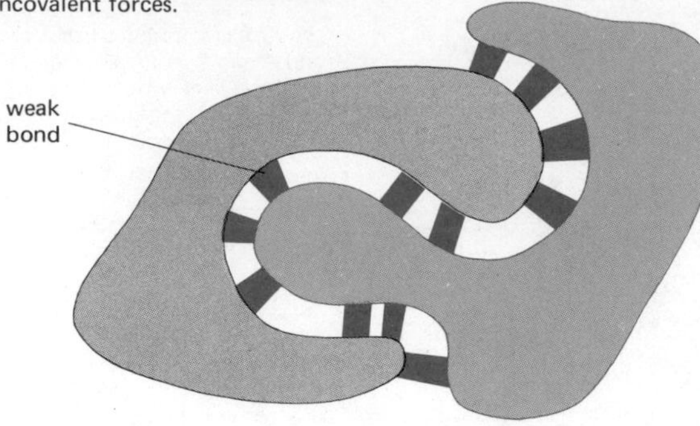

Weak chemical bonds typically have less than 1/20 the strength of a covalent bond. They are strong enough to provide tight binding only when many of them are simultaneously formed.

HYDROPHOBIC INTERACTIONS

Water forces hydrophobic groups together in order to minimize their disruptive effects on the hydrogen-bonded water network. Hydrophobic groups held together in this way are often said to be held together by "hydrophobic bonds"

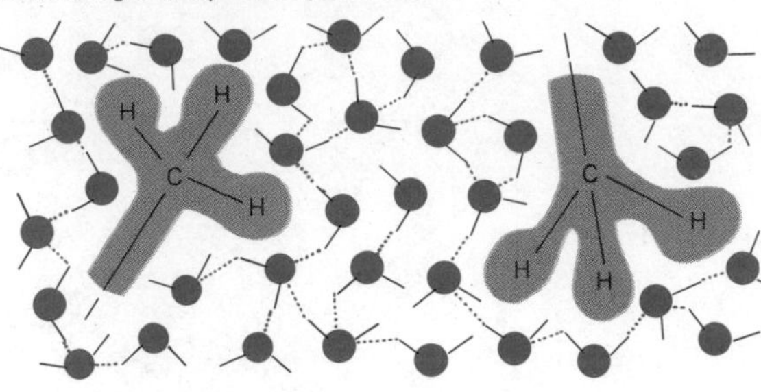

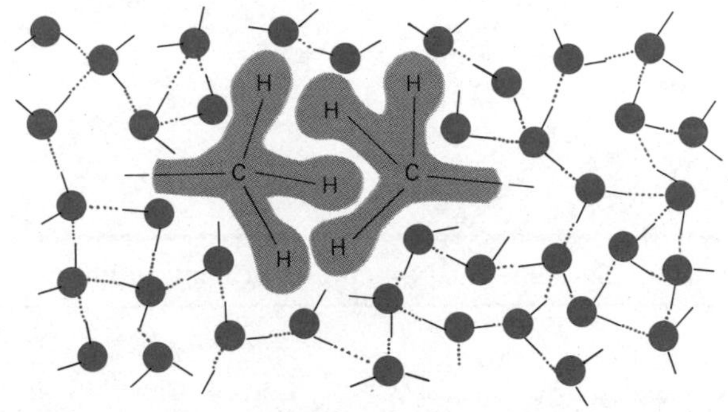

HYDROGEN BONDS

A hydrogen atom is shared between 2 other atoms (both electronegative, such as O and N) to give a hydrogen bond

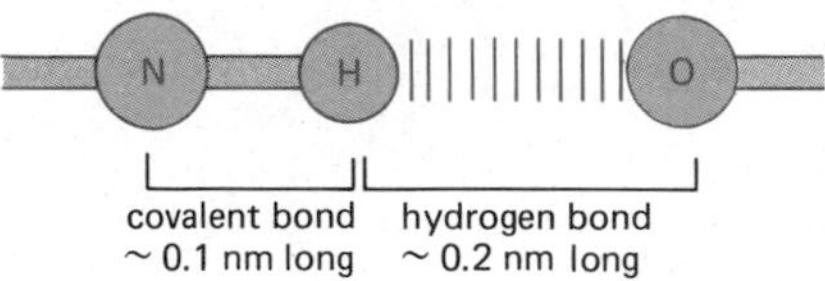

Hydrogen bonds are strongest when the three atoms are in a straight line:

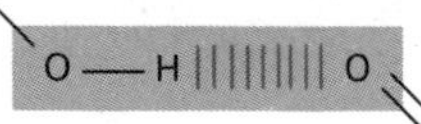

Examples in macromolecules:

2 polypeptide chains H-bonded together

2 bases, G and C, H-bonded in DNA or RNA

HYDROGEN BONDS IN WATER

Since the forming of any new hydrogen bonds between 2 dissolved molecules results in the breaking up of preexisting hydrogen bonds to molecules of water, the hydrogen bonds are relatively weak.

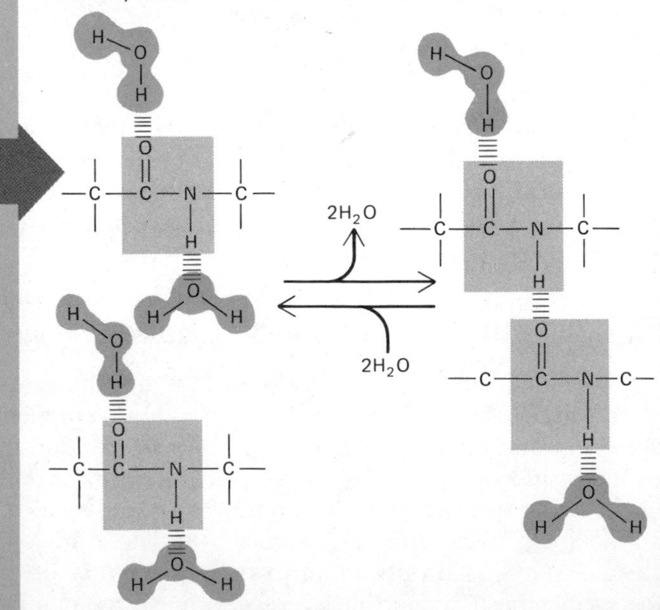

Panel I A summary of the principal types of weak noncovalent bonds that hold macromolecules together.

IONIC BONDS

Ionic interactions occur between either fully charged groups (ionic bond) or between partially charged groups.

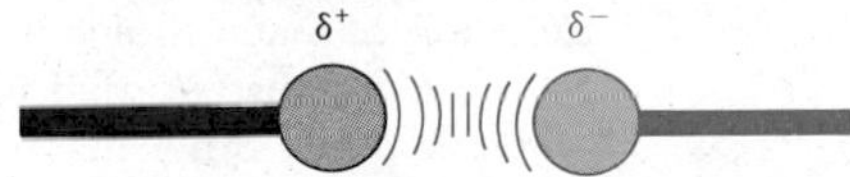

The force of attraction between the 2 charges δ^+ and δ^- is

$$\text{force} = \frac{\delta^+ \delta^-}{r^2 D} \text{ (Coulomb's law)}$$

where D = dielectric constant (1 for vacuum; 80 for water)

r = distance of separation

In the absence of water, ionic forces are very strong. They are responsible for the strength of minerals such as marble and agate.

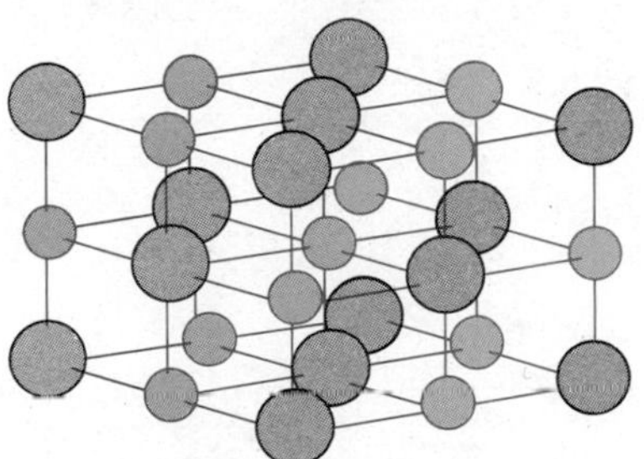

NaCl crystal

IONIC BONDS IN WATER

Charged groups are shielded by their interactions with water molecules. Ionic bonds are therefore quite weak in aqueous solution (about the same strength as a hydrogen bond).

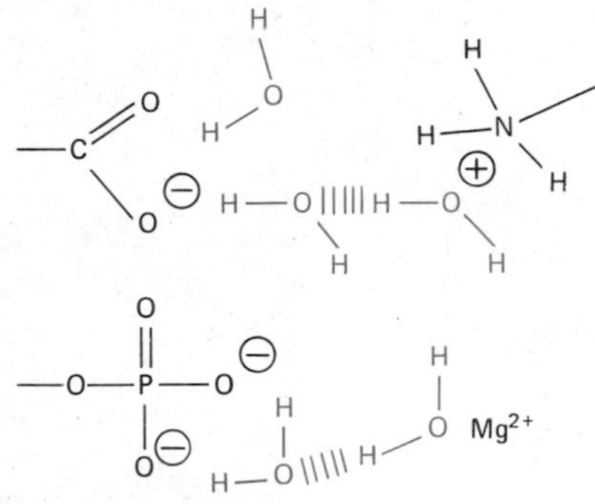

Nevertheless, ionic bonds are very important in biological systems; an enzyme that binds a positively charged substrate will often have a negatively charged amino acid side chain at the appropriate place.

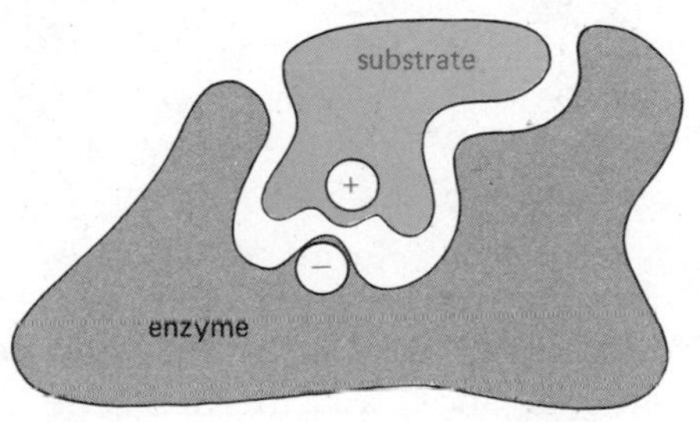

VAN DER WAALS FORCES

At very close distances, any 2 atoms show a weak bonding interaction due to their fluctuating electrical charges. This force is known as a van der Waals attraction.

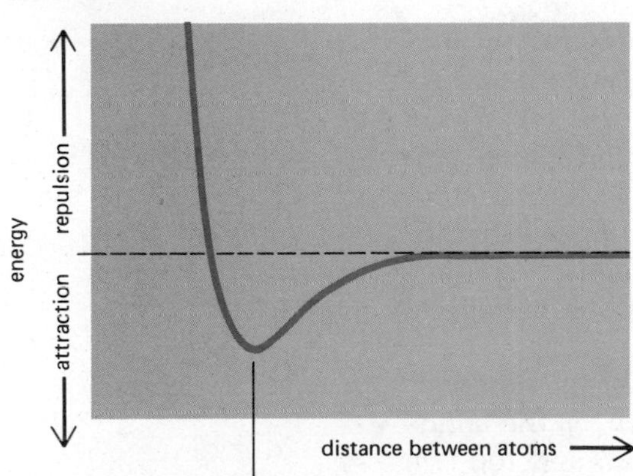

Each type of atom has a radius at which van der Waals forces are optimal.

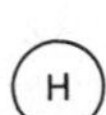

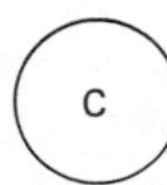

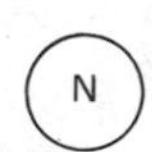

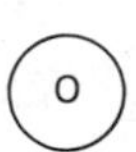

H	C	N	O
1.2 Å (0.12 nm)	2.0 Å (0.2 nm)	1.5 Å (0.15 nm)	1.4 Å (0.14 nm)

Although they are individually very weak, van der Waals attractions can become important when 2 macromolecular surfaces fit very closely together.

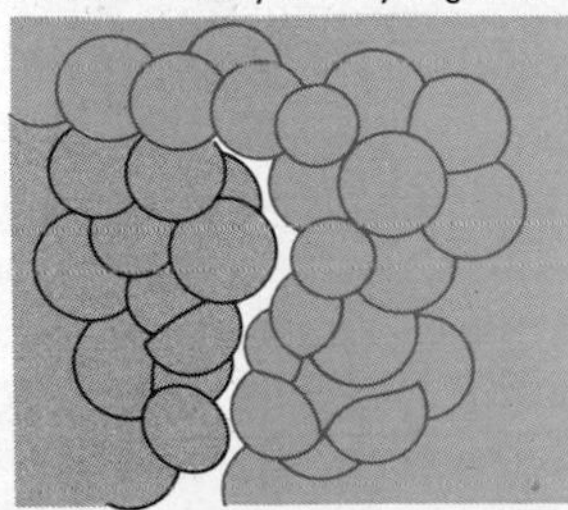

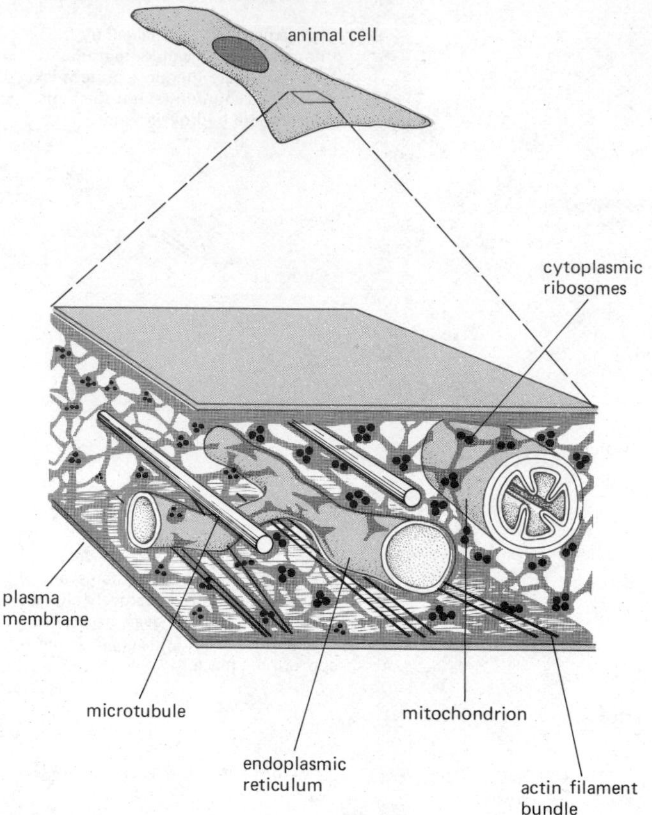

Figure 3–4 Schematic diagram of a small section of an animal cell illustrating the high density of protein filaments and membrane-bounded organelles that it contains. The diffusion of macromolecules is retarded because of their interaction with the many other macromolecules in the cytoplasm, whereas small molecules diffuse nearly as rapidly as they do in water. The cell shown has been flattened by its attachment to a surface.

Experiments performed by injecting fluorescent dyes and other labeled molecules into cells show that the diffusion of small molecules through the cytoplasm is nearly as rapid as it is in water. A molecule the size of ATP will require only about 0.2 second to diffuse an average distance of 10 μm—the diameter of a small animal cell. Macromolecules, however, move much more slowly. Not only do they have an intrinsically slower diffusion rate, but their movement is retarded by frequent collisions with many other macromolecules that are constrained in position in the cytoplasm (Figure 3–4).

Thermal Motions Not Only Bring Molecules Together But Also Pull Them Apart

Encounters between two macromolecules or between a macromolecule and a small molecule occur randomly through simple diffusion. An encounter may lead immediately to the formation of a complex (in which case the rate of formation is said to be *diffusion-limited*), or the rate of formation of the complex may be slower, requiring that there be some adjustment of the structure of one or both molecules before the interacting surfaces can fit together. In either case, once two interacting molecules have come sufficiently close, they form multiple weak bonds with each other that persist until random thermal motion causes the molecules to dissociate again.

In general, the stronger the binding of the molecules in the complex, the slower their rate of dissociation. At one extreme, the energy of the bonds formed are negligible compared with that of thermal motion, and the two molecules dissociate as rapidly as they came together. At the other extreme, the bonds are of such high energy that dissociation rarely occurs. The precise strength of the bonding between two molecules is a useful index of the specificity of the recognition process.

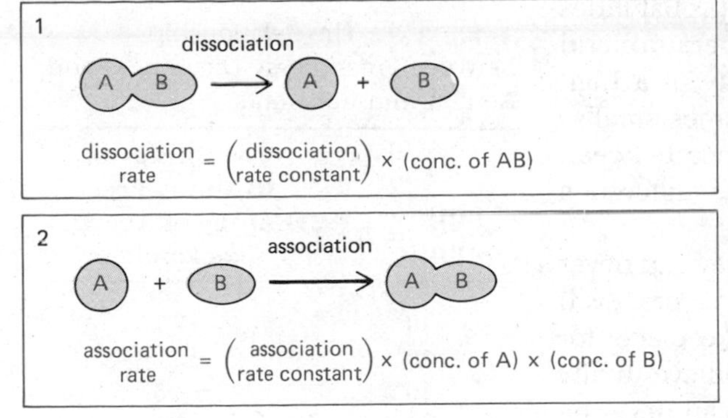

AT EQUILIBRIUM: association rate = dissociation rate

$$\left(\text{association rate constant}\right) \times (\text{conc. of A}) \times (\text{conc. of B}) = \left(\text{dissociation rate constant}\right) \times (\text{conc. of AB})$$

$$\frac{\text{(conc. of AB)}}{\text{(conc. of A)} \times \text{(conc. of B)}} = \frac{\text{association rate constant}}{\text{dissociation rate constant}} = \text{equilibrium constant } K$$

$$\frac{[AB]}{[A][B]} = K$$

Figure 3–5 The principle of equilibrium. The equilibrium between molecules A and B and the dimer AB is maintained by a balance between the two opposing reactions shown in (1) and (2). The ratio of the rate constants for the association and the dissociation reactions is equal to the equilibrium constant for the reaction, *K*. Because molecules A and B must collide in order to react, in reaction (2) the synthesis rate is proportional to the product of their individual concentrations. As a result, the product [A] × [B] appears in the final expression for *K*.

In the form shown, *K* will be referred to as the affinity constant. Its value is larger the stronger the binding between A and B. This form of the equilibrium constant is also referred to as the *association constant* (units of liters per mole). The equilibrium constant can alternatively be defined as the inverse of the constant shown here, which is referred to as the *dissociation constant* (units of moles per liter).

To see how the binding strength is measured, consider a reaction in which molecule A binds to molecule B. This reaction will proceed until it reaches an *equilibrium point,* at which the amounts of complex AB dissociating and forming are equal. The concentrations of A, B, and of the complex AB at this point can be used to determine an **equilibrium constant (*K*)** for the reaction, as explained in Figure 3–5. This constant is sometimes termed the **affinity constant** and is commonly employed as a measure of the strength of binding between two molecules; the *stronger* the binding, the *larger* the value of the affinity constant.

The equilibrium constants of reactions in which two molecules bind to each other are directly related to the standard free energy change for the binding ($\Delta G°$) by the equation described in Table 3–3, which has also been used there to calculate the $\Delta G°$ values corresponding to a range of *K* values. Affinity constants for simple binding interactions in biological systems often range between 10^3 and 10^{10} liters per mole, representing binding energies in the range of 4 to 14 kilocalories per mole.

As an example of a more complex interaction, consider mitosis, where each chromosome may need to stay associated continuously with its spindle fibers from metaphase to telophase, a period of perhaps 30 minutes. If a successful partitioning of one chromosome into the correct daughter cell occurs with less than one failure in every 1000 mitoses, this chromosome must be able to stay associated with its spindle for at least 30 minutes × 1000 mitoses = 3×10^4 minutes. Since failure is a random process, for all of the 46 chromosomes in a human cell to segregate without a mistake in 1000 mitoses, their association with the spindle must last 46 times as long, or 1.4×10^6 minutes. This gives a maximum permissible value for the rate of chromosome dissociation. Moreover, we can determine the maximum rate of association from the diffusion rate of a chromosome, and therefore the *minimum* value of the affinity constant *K* may be calculated (as described in Figure 3–5). The value obtained—at least 10^{16} liters per mole—represents a free-energy change of at least −23 kilocalories per mole, which could arise from 23 average hydrogen bonds.

Molecular Recognition Processes Can Never Be Perfect

The rapid movements of molecules as they diffuse in a solution is not the only form of energy they possess. At normal temperatures all atoms, including those that make up a cell, also possess energy in the form of vibrations, rotations, and electron distributions. This energy is randomly distributed to different atoms, mainly through molecular collisions, so that, while most atoms will have energy levels close to the average, a small proportion will possess very low or very high energy. It is possible to calculate the relative probabilities of a molecule existing in one of two energy states, given the temperature and the difference of energy between the two states. The probability of a high energy state, like that of a chromosome falling off a spindle, becomes smaller and smaller relative to a low energy state as the energy difference between the two increases. However, it becomes zero only when this energy difference becomes infinite (Table 3–3).

Because of the random element in molecular interactions, they can never be absolutely reliable. As a consequence, a cell continually makes errors. Even reactions that are energetically unfavorable will occasionally take place; for example, two atoms joined to each other by a covalent bond will eventually fall apart. Similarly, the specificity of an enzyme for its substrate can never be absolute because the recognition of one molecule as distinct from another can never be perfect. Mistakes could be avoided completely only if the cell evolved mechanisms with infinite energy differences between alternatives. Since cells have only a finite amount of energy at their disposal, they are forced to tolerate a certain level of failure and they use special repair reactions to correct many of their errors (p. 214 and p. 334).

On the other hand, errors are essential. If it were not for occasional mistakes in making DNA, as will be described shortly, evolution would probably not occur.

Table 3–3 The Relationship Between Free-Energy Differences and Equilibrium Constants

$\frac{[AB]}{[A][B]} = K$	**Free Energy of AB Minus Free Energy of A + B (kcal/mole)**
10^5	−7.1
10^4	−5.7
10^3	−4.3
10^2	−2.8
10	−1.4
1	0
10^{-1}	1.4
10^{-2}	2.8
10^{-3}	4.3
10^{-4}	5.7
10^{-5}	7.1

If the reaction $A + B \leftrightharpoons AB$ is allowed to come to equilibrium, the relative amounts of A, B, and AB will depend on the free-energy difference, ΔG°, between them. The above values are given for 37°C and are calculated from the equation

$$\Delta G^\circ = -RT \ln \frac{[AB]}{[A][B]}$$

or

$$\frac{[AB]}{[A][B]} = e^{-\Delta G^\circ/RT} = e^{-\Delta G^\circ(1.623)}$$

where ΔG° is in kcal/mole and represents the free-energy difference under standard conditions (where all components are present at a concentration of 1.0 mole per liter).

Summary

A macromolecule contains information in its sequence of subunits that determine the three-dimensional contours of its surface. These contours in turn govern the recognition between one molecule and another, or between different parts of the same molecule, by means of weak noncovalent bonds. Molecules recognize each other by a process in which they first meet by random diffusion and then bind with a strength that can be expressed as an equilibrium constant. Since the only way to make recognition infallible is to make the energy of binding infinitely large, living cells constantly make errors; these are corrected, where necessary, by special repair processes.

Nucleic Acids[2]

Genes Are Made of DNA

It has been obvious for as long as humans have sown crops or raised animals that each seed or fertilized egg must contain a hidden plan, or design, for the development of the organism. In modern times, the science of genetics grew up around the premise of invisible information-containing elements, called **genes,** that are distributed into the two progeny cells produced by each cell division. Before any cell divides, it has to make a copy of its genes in order to give a complete set to each of its daughter cells. The genes in the sperm and egg cells carry the hereditary information from one generation to the next.

Although the inheritance of biological characteristics seems mysterious, logically it must involve patterns of atoms that follow the laws of physics and chemistry: in other words, genes must consist of molecules. At first, the nature of these molecules was hard to imagine. What kind of molecule could be stored in a cell, direct the activities of a developing organism, and also be capable of accurate and almost unlimited replication?

By the end of the nineteenth century, biologists had recognized that the *chromosomes* that become visible in the nucleus as a cell begins to divide were the carriers of inherited information. But the evidence that the deoxyribonucleic acid (DNA) in these chromosomes is the substance of which genes are made came only much later, from studies on bacteria. In 1944, it was shown that the addition of the purified DNA of one strain of bacterium conferred heritable properties characteristic of that strain to a second, slightly different bacterial strain. This discovery came as a surprise; indeed, it was not generally accepted until the early 1950s because it had been commonly believed that only proteins had enough conformational complexity to carry the information stored in genes. Today the idea that DNA carries genetic information—stored in its long chain of nucleotides—is so fundamental to contemporary biological thought that it is sometimes difficult to realize the enormous intellectual gap that it filled.

DNA Molecules Consist of Two Long Complementary Chains Held Together by Base Pairs[3]

The difficulty that geneticists had in accepting DNA as the substance of genes is understandable, considering the simplicity of its chemistry. A DNA chain is a long unbranched polymer composed of only four different subunits. These are the deoxyribonucleotides containing the bases adenine (A), cytosine (C), guanine (G), and thymine (T). Nucleotides are linked together by covalent phosphodiester bonds that join the 5′ carbon of one deoxyribose group to the 3′ carbon of the next (Panel H, pp. 60–61). The four bases are attached to this repetitive sugar-phosphate chain almost like four different kinds of beads hung on a necklace (Figure 3–6).

How can a long chain of nucleotides encode the instructions for an organism or even a cell? And how can these messages be copied from one generation of cells to the next? The answers lie in the three-dimensional structure of the DNA molecule.

Early in the 1950s, x-ray diffraction analyses of specimens of DNA pulled into fibers suggested that the DNA molecule was a helical polymer composed of two strands. The helical structure of DNA was not surprising since a helix will often form if each of the neighboring subunits in a polymer is regularly oriented. But the fact that DNA was two-stranded was crucial. It provided the basic clue that led, in 1953, to the construction of a model that fitted the observed x-ray diffraction pattern and solved the structure of DNA.

A vital feature of the model was that all of the bases of the DNA molecule were on the *inside* of the double helix, with the sugar phosphates on the outside (Figures 3–6 and 3–7). From this it followed that the bases on one strand must come extremely close to those on the other. In fact, the fit proposed was so close that it required that each base on one chain specifically pair through noncovalent bonds with a complementary base on the other chain. In order for the two chains to fit together neatly, the model predicted that such **complementary base-pairing** would have to involve a large purine base (A or G, which each have double rings) pairing with a smaller pyrimidine base (T or C, which each have single rings). Model building also revealed that the numbers of effective hydrogen bonds that could be formed between G and C or between A and T were greater than for any other combinations. Complementary base-pairing between A and T and between G and C in the DNA double helix provided an explanation for earlier biochemical analyses of DNA preparations from different species. These analyses had shown that, although the nucleotide composition of DNA varies a great deal (for example, from 13% A residues to 36% A residues in the DNA of different types of bacteria), there is a general rule that, quantitatively, G = C and A = T.

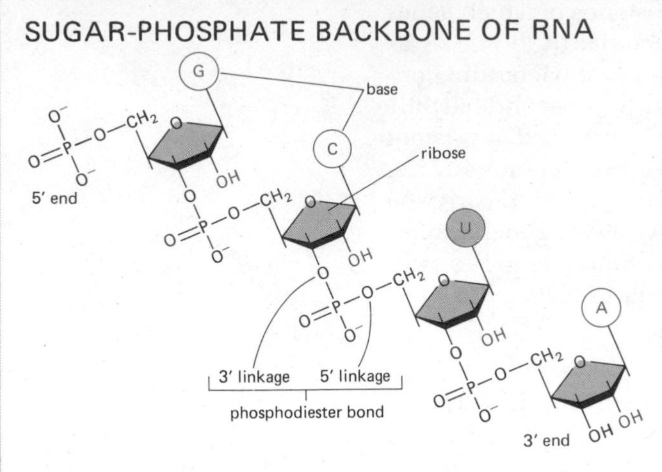

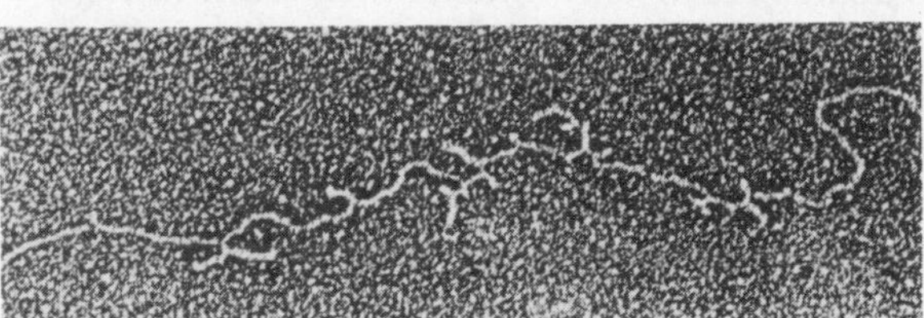

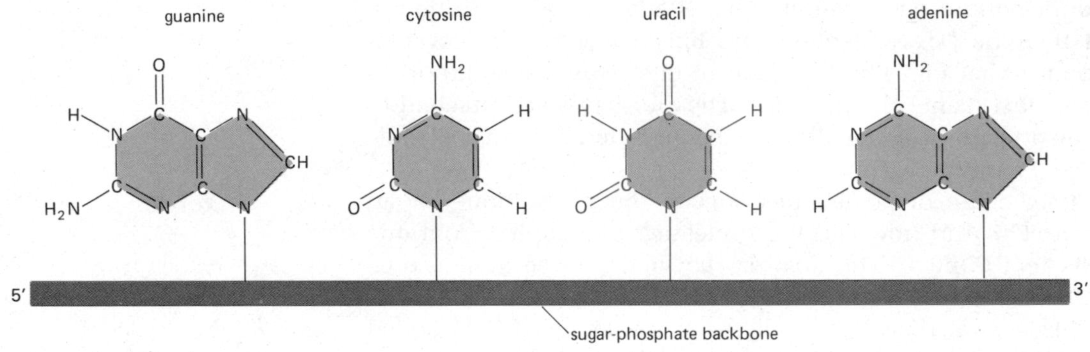

RNA SINGLE STRAND

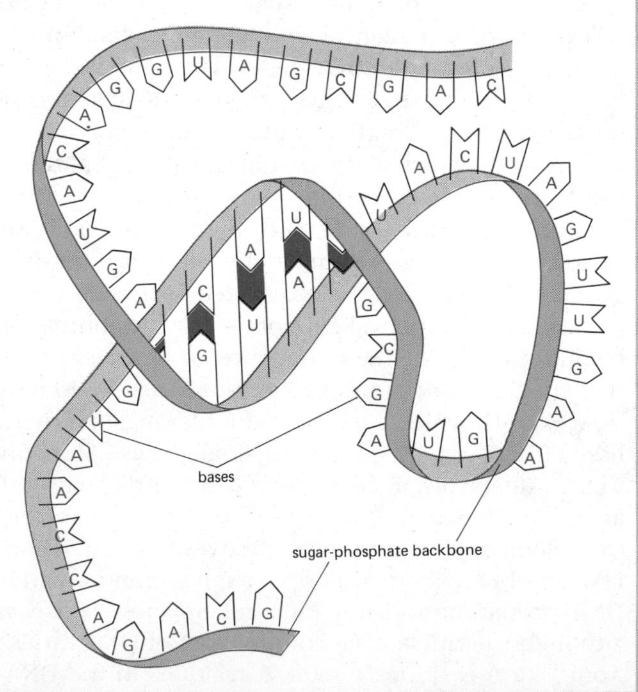

Figure 3–6 The structure of DNA and RNA. Both DNA and RNA are linear polymers of nucleotides (see Panel H, pp. 60–61). DNA differs from RNA in two ways: (1) the sugar-phosphate backbone contains deoxyribose rather than ribose, and (2) it contains the base thymine (T) instead of uracil (U). Specific hydrogen bonding between G and C and between A and T (A and U in RNA) generates complementary base-pairing. In a DNA molecule two antiparallel strands that are complementary in their nucleotide sequence are paired in a right-handed double helix, forming about 10 base pairs per helical turn. RNA is single-stranded, but it contains local regions of short complementary base-pairing that can form from a random matching process. The conformation of these double-helical regions in RNA is shown in Figure 3–7. (Micrographs courtesy of Mei Lie Wong [DNA] and Peter Wellauer [RNA]).

SUGAR-PHOSPHATE BACKBONE OF DNA

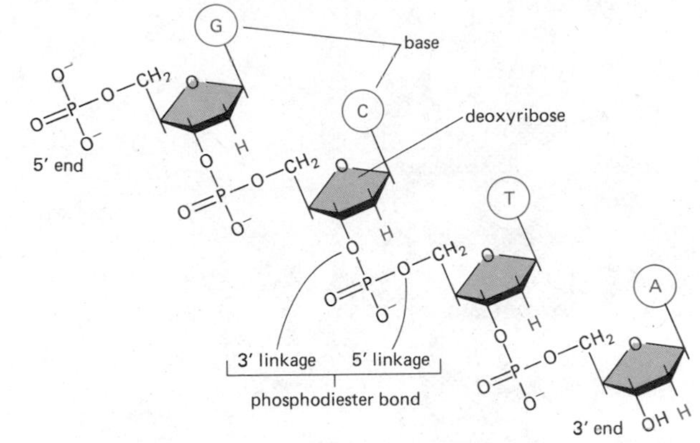

FOUR BASES AS BASE PAIRS OF DNA

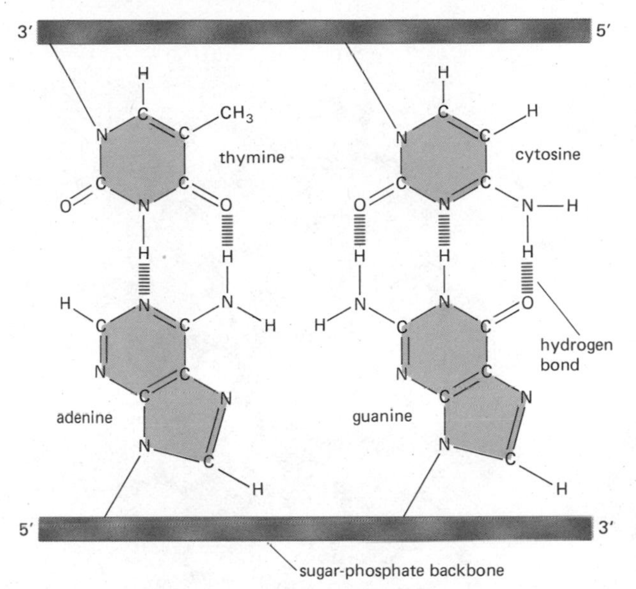

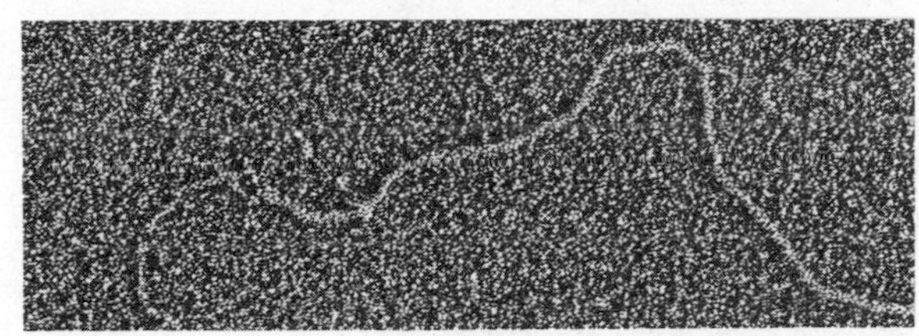

ELECTRON MICROGRAPH OF DNA

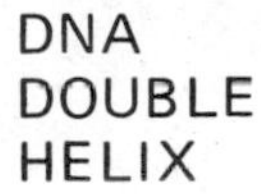

DNA DOUBLE HELIX

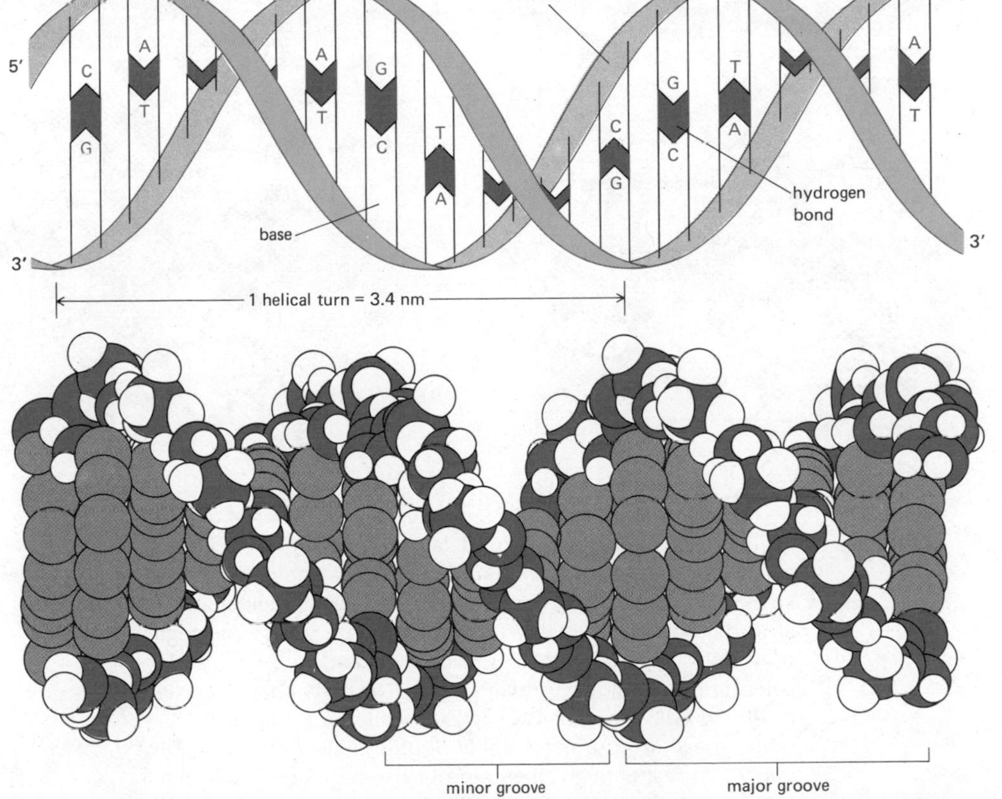

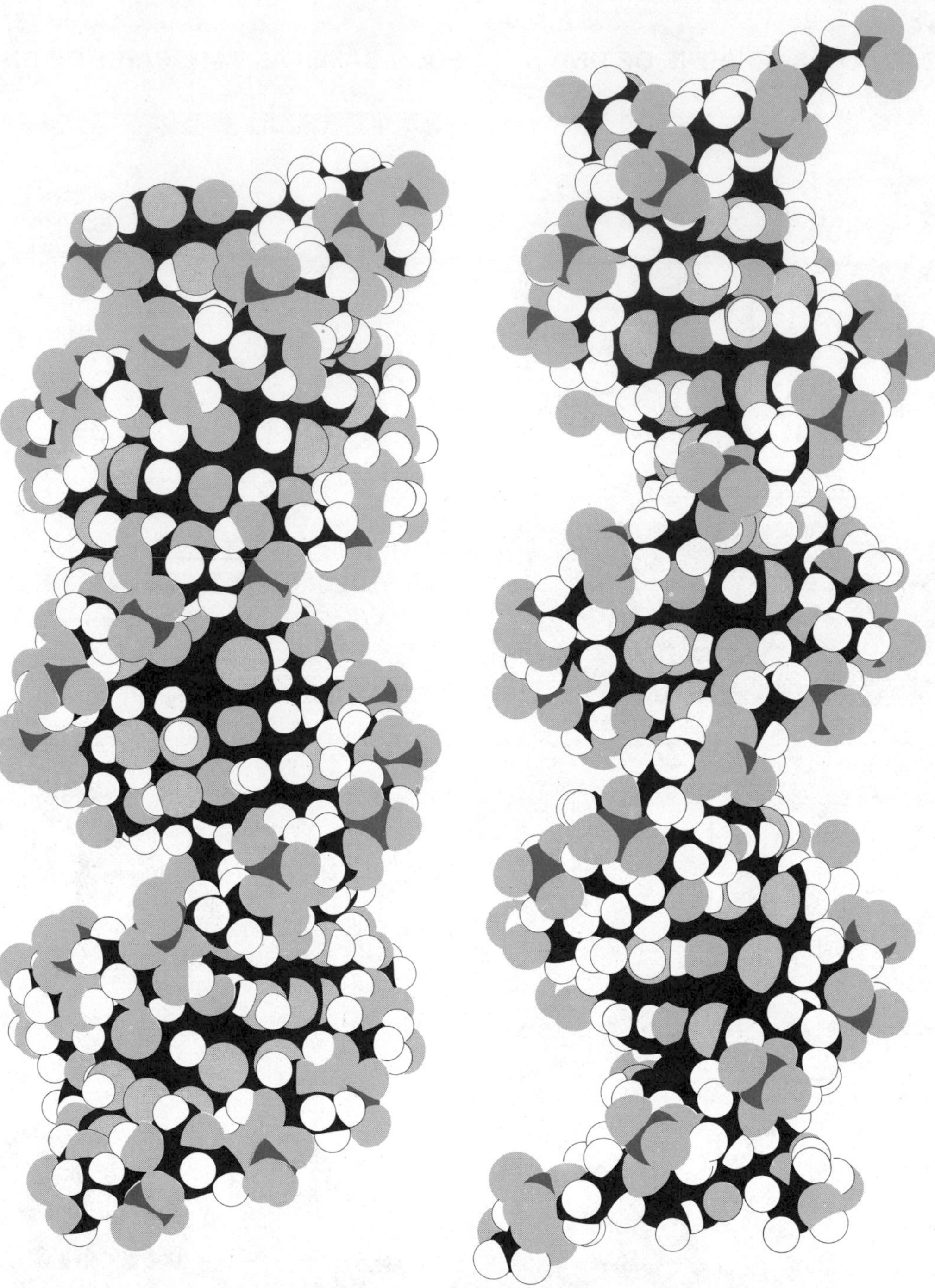

Figure 3–7 Space-filling models of the DNA double helix *(right)* and a region of double-stranded RNA helix *(left)*. Both structures are formed from two antiparallel nucleic acid strands and held together by complementary base pairing; they differ slightly in their helical pitch and in the orientation of their base pairs. The DNA helix shown is called a *B-form helix* and the RNA helix is called an *A-form helix*; while these are the most stable double-helical forms for DNA and RNA, respectively, they are not the only ones possible (see p. 395). (Redrawn from photograph courtesy of Sung-Hou Kim.)

The Structure of DNA Provides an Explanation for Heredity[4]

A gene carries biological information in a form that must be precisely copied and transmitted to all of its progeny cells. The implications of the discovery of the double-stranded structure of DNA were enormous because the structure immediately suggested the general type of mechanism by which this crucial information-transfer process is accomplished. Since each strand contains a nucleotide sequence that is exactly *complementary* to the nucleotide sequence of its partner strand, both strands actually carry the same genetic information. If we designate the two strands as A and A′, strand A can serve as a mold or *template* for making a new strand A′, while strand A′ can serve in the same way to make a new strand A. Thus, genetic information can be copied by a process in which strand A separates from strand A′ so as to allow each of these strands to serve as a template for the production of a new complementary partner strand.

As a direct consequence of the base-pairing mechanism, it becomes evident that DNA carries information by means of the linear sequence of its nucleotides. Each nucleotide—A, C, T, or G—can be considered as a letter in a simple four-letter alphabet that is used to write out biological messages in a linear "ticker-tape" form. Animals of different species differ because the respective DNA molecules in their cells carry different nucleotide sequences and thereby different biological messages.

Since the number of different possible sequences in a DNA chain that is n nucleotides long is 4^n, the amount of biological variety that can be generated using even a modest length of DNA is enormous, and a typical animal cell contains a meter of DNA (3×10^9 nucleotides). Written in a linear script of four letters, the hereditary material of a tiny bacterial virus occupies a full page of text (Figure 3–8), while in this form the genetic information carried in an animal cell would fill a book of more than 500,000 pages!

Although the principle underlying gene replication is both elegant and simple, the actual machinery by which this copying is carried out in the cell is complicated and involves many different proteins. The fundamental reaction is that shown in Figure 3–9, in which an enzyme called *DNA polymerase* catalyzes the addition of a deoxyribonucleotide to the 3′ end of a DNA chain. Each nucleotide added to the chain is in fact a *deoxyribonucleoside triphosphate;* the release of pyrophosphate from this activated nucleotide and its subsequent hydrolysis provide the energy for the **DNA replication** reaction and make it effectively irreversible (pp. 77–78).

The replication of the DNA helix begins with the local separation of its two complementary DNA strands. Each strand then acts as a template for the formation of a new DNA molecule by the sequential addition of deoxyribonucleoside triphosphates (Figure 3–10). The nucleotide added at each step is selected by a process that requires it to form a complementary base pair with the next nucleotide in the parental template strand, thereby generating a DNA daughter strand that is complementary in sequence to the template strand. The genetic information is duplicated in its entirety—that is, two complete DNA double helices are eventually formed, each identical in nucleotide sequence to the parental DNA helix that served as the template. Since each of the original strands ends up in a different daughter molecule at the end of the process, the mechanism of DNA replication is said to be "semiconservative" (Figure 3–11).

Errors in DNA Replication Cause Mutations[5]

The most impressive feature of DNA replication is its accuracy, which is achieved by a complex of proteins that forms a "replication machine." This machinery serves three purposes: (1) it selects the correct nucleotide by means of base-

GAGTTTTATCGCTTCCATGACGCAGAAGTTAACACTTTCGGATATTTCTGATGAGTCGAA
AAATTATCTTGATAAAGCAGGAATTACTACTGCTTGTTTACGAATTAAATCGAAGTGGAC
TGCTGGCGGAAAATGAGAAAATTCGACCTATCCTTGCGCAGCTCGAGAAGCTCTTACTTT
GCGACCTTTCGCCATCAACTAACGATTCTGTCAAAAACTGACGCGTTGGATGAGGAGAAG
TGGCTTAATATGCTTGGCACGTTCGTCAAGGACTGGTTTAGATATGAGTCACATTTTGTT
CATGGTAGAGATTCTCTTGTTGACATTTTAAAAGAGCGTGGATTACTATCTGAGTCCGAT
GCTGTTCAACCACTAATAGGTAAGAAATCATGAGTCAAGTTACTGAACAATCCGTACGTT
TCCAGACCGCTTTGGCCTCTATTAAGCTCATTCAGGCTTCTGCCGTTTTGGATTTAACCG
AAGATGATTTCGATTTTCTGACGAGTAACAAAGTTTGGATTGCTACTGACCGCTCTCGTG
CTCGTCGCTGCGTTGAGGCTTGCGTTTATGGTACGCTGGACTTTGTGGGATACCCTCGCT
TTCCTGCTCCTGTTGAGTTTATTGCTGCCGTCATTGCTTATTATGTTCATCCCGTCAACA
TTCAAACGGCCTGTCTCATCATGGAAGGCGCTGAATTTACGGAAAACATTATTAATGGCG
TCGAGCGTCCGGTTAAAGCCGCTGAATTGTTCGCGTTTACCTTGCGTGTACGCGCAGGAA
ACACTGACGTTCTTACTGACGCAGAAGAAAACGTGCGTCAAAAATTACGTGCGGAAGGAG
TGATGTAATGTCTAAAGGTAAAAAACGTTCTGGCGCTCGCCCTGGTCGTCCGCAGCCGTT
GCGAGGTACTAAAGGCAAGCGTAAAGGCGCTCGTCTTTGGTATGTAGGTGGTCAACAATT
TTAATTGCAGGGGCTTCGGCCCCTTACTTGAGGATAAATTATGTCTAATATTCAAACTGG
CGCCGAGCGTATGCCGCATGACCTTTCCCATCTTGGCTTCCTTGCTGGTCAGATTGGTCG
TCTTATTACCATTTCAACTACTCCGGTTATCGCTGGCGACTCCTTCGAGATGGACGCCGT
TGGCGCTCTCCGTCTTTCTCCATTGCGTCGTGGCCTTGCTATTGACTCTACTGTAGACAT
TTTTACTTTTTATGTCCCTCATCGTCACGTTTATGGTGAACAGTGGATTAAGTTCATGAA
GGATGGTGTTAATGCCACTCCTCTCCCGACTGTTAACACTACTGGTTATATTGACCATGC
CGCTTTTCTTGGCACGATTAACCCTGATACCAATAAAATCCCTAAGCATTTGTTTCAGGG
TTATTTGAATATCTATAACAACTATTTTAAAGCGCCGTGGATGCCTGACCGTACCGAGGC
TAACCCTAATGAGCTTAATCAAGATGATGCTCGTTATGGTTTCCGTTGCTGCCATCTCAA
AAACATTTGGACTGCTCCGCTTCCTCCTGAGACTGAGCTTTCTCGCCAAATGACGACTTC
TACCACATCTATTGACATTATGGGTCTGCAAGCTGCTTATGCTAATTTGCATACTGACCA
AGAACGTGATTACTTCATGCAGCGTTACCATGATGTTATTTCTTCATTTGGAGGTAAAAC
CTCTTATGACGCTGACAACCGTCCTTTACTTGTCATGCGCTCTAATCTCTGGGCATCTGG
CTATGATGTTGATGGAACTGACCAAACGTCGTTAGGCCAGTTTTCTGGTCGTGTTCAACA
GACCTATAAACATTCTGTGCCGCGTTTCTTTGTTCCTGAGCATGGCACTATGTTTACTCT
TGCGCTTGTTCGTTTTCCGCCTACTGCGACTAAAGAGATTCAGTACCTTAACGCTAAAGG
TGCTTTGACTTATACCGATATTGCTGGCGACCCTGTTTTGTATGGCAACTTGCCGCCGCG
TGAAATTTCTATGAAGGATGTTTTCCGTTCTGGTGATTCGTCTAAGAAGTTTAAGATTGC
TGAGGGTCAGTGGTATCGTTATGCGCCTTCGTATGTTTCTCCTGCTTATCACCTTCTTGA
AGGCTTCCCATTCATTCAGGAACCGCCTTCTGGTGATTTGCAAGAACGCGTACTTATTCG
CCACCATGATTATGACCAGTGTTTCCAGTCCGTTCAGTTGTTGCAGTGGAATAGTCAGGT
TAAATTTAATGTGACCGTTTATCGCAATCTGCCGACCACTCGCGATTCAATCATGACTTC
GTGATAAAAGATTGAGTGTGAGGTTATAACGCCGAAGCGGTAAAAATTTTAATTTTTGCC
GCTGAGGGGTTGACCAAGCGAAGCGCGGTAGGTTTTCTGCTTAGGAGTTTAATCATGTTT
CAGACTTTTATTTCTCGCCATAATTCAAACTTTTTTTCTGATAAGCTGGTTCTCACTTCT
GTTACTCCAGCTTCTTCGGCACCTGTTTTACAGACACCTAAAGCTACATCGTCAACGTTA
TATTTTGATAGTTTGACGGTTAATGCTGGTAATGGTGGTTTTCTTCATTGCATTCAGATG
GATACATCTGTCAACGCCGCTAATCAGGTTGTTTCTGTTGGTGCTGATATTGCTTTTGAT
GCCGACCCTAAATTTTTTGCCTGTTTGGTTCGCTTTGAGTCTTCTTCGGTTCCGACTACC
CTCCCGACTGCCTATGATGTTTATCCTTTGAATGGTCGCCATGATGGTGGTTATTATACC
GTCAAGGACTGTGTGACTATTGACGTCCTTCCCCGTACGCCGGGCAATAACGTTTATGTT
GGTTTCATGGTTTGGTCTAACTTTACCGCTACTAAATGCCGCGGATTGGTTTCGCTGAAT
CAGGTTATTAAAGAGATTATTTGTCTCCAGCCACTTAAGTGAGGTGATTTATGTTTGGTG
CTATTGCTGGCGGTATTGCTTCTGCTCTTGCTGGTGGCGCCATGTCTAAATTGTTTGGAG
GCGGTCAAAAAGCCGCCTCCGGTGGCATTCAAGGTGATGTGCTTGCTACCGATAACAATA
CTGTAGGCATGGGTGATGCTGGTATTAAATCTGCCATTCAAGGCTCTAATGTTCCTAACC
CTGATGAGGCCGCCCCTAGTTTTGTTTCTGGTGCTATGGCTAAAGCTGGTAAAGGACTTC
TTGAAGGTACGTTGCAGGCTGGCACTTCTGCCGTTTCTGATAAGTTGCTTGATTTGGTTG
GACTTGGTGGCAAGTCTGCCGCTGATAAAGGAAAGGATACTCGTGATTATCTTGCTGCTG
CATTTCCTGAGCTTAATGCTTGGGAGCGTGCTGGTGCTGATGCTTCCTCTGCTGGTATGG
TTGACGCCGGATTTGAGAATCAAAAAGAGCTTACTAAAATGCAACTGGACAATCAGAAAG
AGATTGCCGAGATGCAAAATGAGACTCAAAAAGAGATTGCTGGCATTCAGTCGGCGACTT
CACGCCAGAATACGAAAGACCAGGTATATGCACAAAATGAGATGCTTGCTTATCAACAGA
AGGAGTCTACTGCTCGCGTTGCGTCTATTATGGAAAACACCAATCTTTCCAAGCAACAGC
AGGTTTCCGAGATTATGCGCCAAATGCTTACTCAAGCTCAAACGGCTGGTCAGTATTTTA
CCAATGACCAAATCAAAGAAATGACTCGCAAGGTTAGTGCTGAGGTTGACTTAGTTCATC
AGCAAACGCAGAATCAGCGGTATGGCTCTTCTCATATTGGCGCTACTGCAAAGGATATTT
CTAATGTCGTCACTGATGCTGCTTCTGGTGTGGTTGATATTTTTCATGGTATTGATAAAG
CTGTTGCCGATACTTGGAACAATTTCTGGAAAGACGGTAAAGCTGATGGTATTGGCTCTA
ATTTGTCTAGGAAATAACCGTCAGGATTGACACCCTCCCAATTGTATGTTTTCATGCCTC
CAAATCTTGGAGGCTTTTTTATGGTTCGTTCTTATTACCCTTCTGAATGTCACGCTGATT
ATTTTGACTTTGAGCGTATCGAGGCTCTTAAACCTGCTATTGAGGCTTGTGGCATTTCTA
CTCTTTCTCAATCCCCAATGCTTGGCTTCCATAAGCAGATGGATAACCGCATCAAGCTCT
TGGAAGAGATTCTGTCTTTTCGTATGCAGGGCGTTGAGTTCGATAATGGTGATATGTATG
TTGACGGCCATAAGGCTGCTTCTGACGTTCGTGATGAGTTTGTATCTGTTACTGAGAAGT
TAATGGATGAATTGGCACAATGCTACAATGTGCTCCCCCAACTTGATATTAATAACACTA
TAGACCACCGCCCCGAAGGGGACGAAAAATGGTTTTTAGAGAACGAGAAGACGGTTACGC
AGTTTTGCCGCAAGCTGGCTGCTGAACGCCCTCTTAAGGATATTCGCGATGAGTATAATT
ACCCCAAAAAGAAAGGTATTAAGGATGAGTGTTCAAGATTGCTGGAGGCCTCCACTATGA
AATCGCGTAGAGGCTTTGCTATTCAGCGTTTGATGAATGCAATGCGACAGGCTCATGCTG
ATGGTTGGTTTATCGTTTTTGACACTCTCACGTTGGCTGACGACCGATTAGAGGCGTTTT
ATGATAATCCCAATGCTTTGCGTGACTATTTTCGTGATATTGGTCGTATGGTTCTTGCTG
CCGAGGGTCGCAAGGCTAATGATTCACACGCCGACTGCTATCAGTATTTTTGTGTGCCTG
AGTATGGTACAGCTAATGGCCGTCTTCATTTCCATGCGGTGCACTTTATGCGGACACTTC
CTACAGGTAGCGTTGACCCTAATTTTGGTCGTCGGGTACGCAATCGCCGCCAGTTAAATA
GCTTGCAAAATACGTGGCCTTATGGTTACAGTATGCCCATCGCAGTTCGCTACACGCAGG
ACGCTTTTTCACGTTCTGGTTGGTTGTGGCCTGTTGATGCTAAAGGTGAGCCGCTTAAAG
CTACCAGTTATATGGCTGTTGGTTTCTATGTGGCTAAATACGTTAACAAAAAGTCAGATA
TGGACCTTGCTGCTAAAGGTCTAGGAGCTAAAGAATGGAACAACTCACTAAAAACCAAGC
TGTCGCTACTTCCCAAGAAGCTGTTCAGAATCAGAATGAGCCGCAACTTCGGGATGAAAA
TGCTCACAATGACAAATCTGTCCACGGAGTGCTTAATCCAACTTACCAAGCTGGGTTACG
ACGCGACGCCGTTCAACCAGATATTGAAGCAGAACGCAAAAAGAGAGATGAGATTGAGGC
TGGGAAAAGTTACTGTAGCCGACGTTTTGGCGGCGCAACCTGTGACGACAAATCTGCTCA
AATTTATGCGCGCTTCGATAAAAATGATTGGCGTATCCAACCTGCA

Figure 3–8 The entire hereditary information of the simple bacterial virus ϕX174 is contained in the single strand of DNA whose nucleotide sequence is given here. (The sequence should be read from left to right in successive lines down the page, as if it was normal text.)

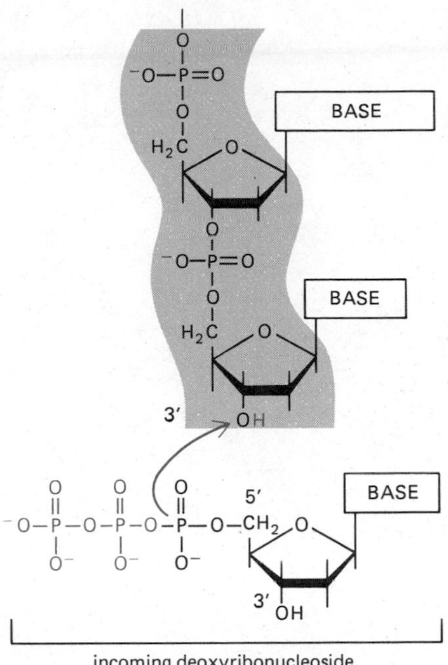

Figure 3–9 The addition of a deoxyribonucleotide to the 3′ end of a DNA chain is the fundamental reaction in which new DNA is synthesized.

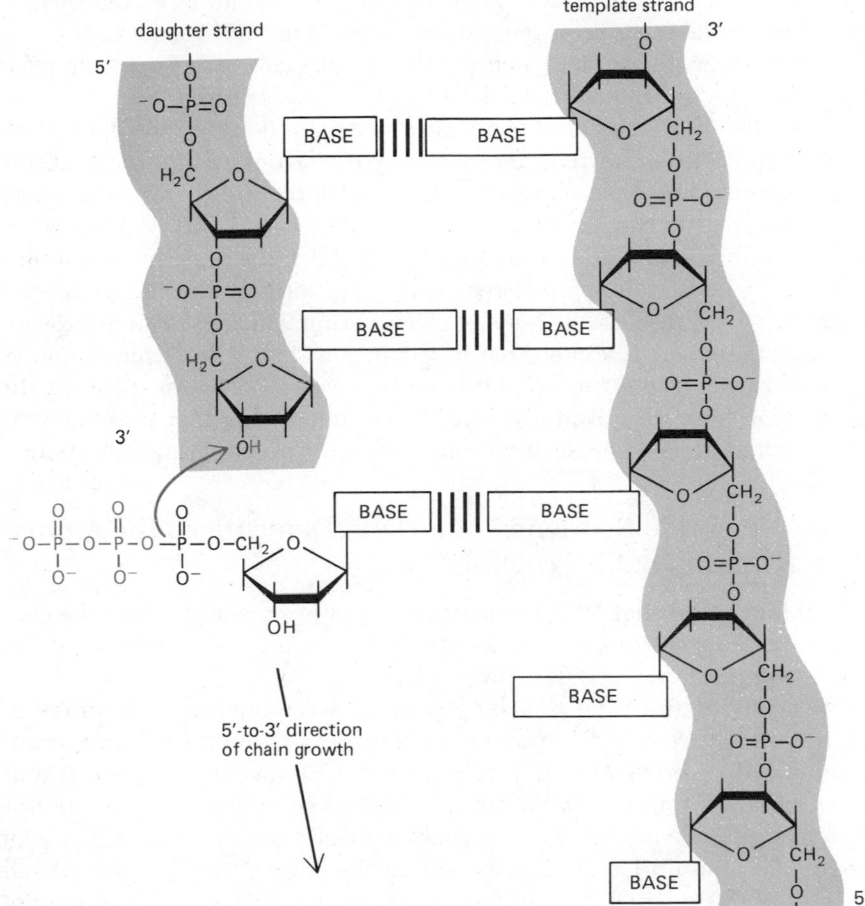

Figure 3–10 Base-pairing between an incoming deoxyribonucleotide and an existing strand of DNA (the *template* strand) guides the formation of a *daughter* strand of DNA with a complementary base sequence.

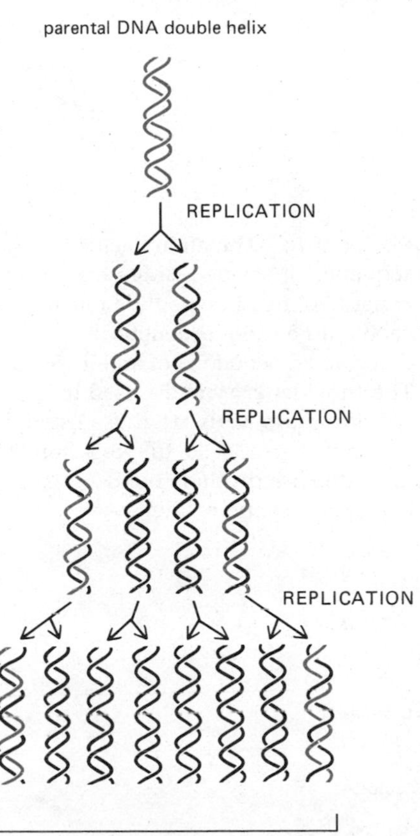

Figure 3–11 The semiconservative replication of DNA. In each round of replication, each of the two strands of DNA is used as a template for the formation of a newly synthesized complementary DNA strand. The individual strands therefore retain their integrity through many cell generations.

pairing with the template strand; (2) it links each new nucleotide to the end of the growing chain by catalyzing the formation of a covalent bond; and (3) it provides a correction mechanism that eliminates imperfectly positioned nucleotides. The precision of the replication machinery is such that the sequence of bases in a DNA molecule is copied with less than one mistake in 10^9 nucleotides added.

Very rarely, however, the replication machinery skips or adds a few bases, or it puts a T where it should have put a C, or an A instead of a G. Any change of this kind in the DNA sequence constitutes a genetic mistake, called a **mutation.** The consequences of such errors are twofold. First, the mistake will be copied in all future cell generations, since "bad" DNA sequences are copied as faithfully as good ones. Second, the mistake can have important effects on the cell, depending on where the mutation has occurred.

As was demonstrated conclusively by geneticists in the early 1940s, the units of heredity known as genes specify the structure of individual proteins. A mutation in a gene, caused by an alteration in its DNA sequence, may therefore lead to the inactivation of a crucial protein and result in cell death. In that case, the altered DNA sequence is lost. On the other hand, the mutation may occur in a nonessential region and be without effect, a so-called *silent mutation.* Very rarely, a mistake in base-pairing will lead to an improved gene. It may change a protein so that it becomes a "better" enzyme or contributes to a more effective structure. In these rare cases, organisms carrying the mutation will have an advantage, and the mutated gene may eventually replace the original gene in most of the population through natural selection.

The Nucleotide Sequence of a Gene Determines the Amino Acid Sequence of a Protein[6]

The information that DNA contains does not immediately affect the cell. Only when it is used to direct the synthesis of proteins is it able to determine a cell's chemical and physical properties.

At about the time that biophysicists were analyzing the three-dimensional structure of DNA by x-ray diffraction, biochemists were intensively studying the chemical structure of proteins. It was already known that they were chains of amino acids joined together by sequential peptide linkages, but it was not yet certain that each type of protein consisted of a unique sequence of amino acids. It was only in the early 1950s that this was discovered, when the small protein *insulin* was sequenced and shown to contain a precisely defined string of amino acids (Figure 3–12).

Just as solving the structure of DNA was seminal in understanding the molecular basis of genetics and heredity, so sequencing insulin provided a key to understanding the structure and function of proteins. If insulin had a definite, genetically determined sequence, then so presumably did every other protein. It seemed reasonable to suppose, moreover, that the properties of a protein would depend on the precise order in which its constitutent amino acids were arranged.

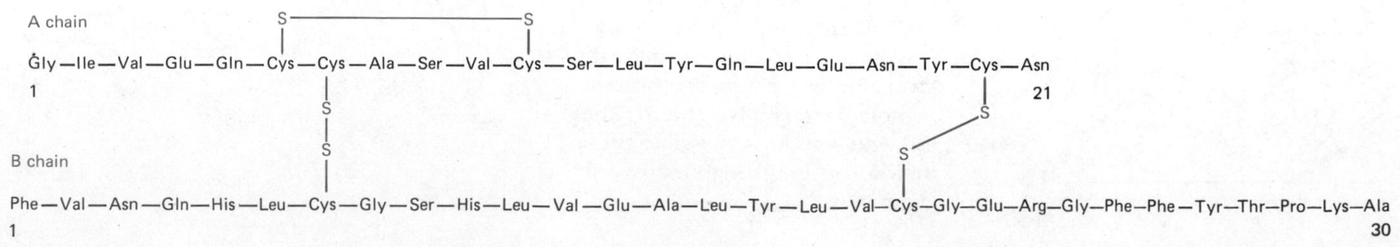

Figure 3-12 The amino acid sequence of bovine insulin. Insulin consists of two polypeptide chains, each with a unique, genetically determined sequence of amino acids. The three-letter symbols used to specify amino acids are those listed in Panel G, pp. 58–59; the -S-S- bonds indicated are disulfide bonds between cysteine residues.

Both DNA and protein are composed of a linear sequence of subunits, and eventually the biochemical analysis of the proteins made by mutant genes demonstrated that the two sequences were colinear—that is, the nucleotides in DNA were arranged in an order corresponding to the order of the amino acids in the protein that they specify. The central question in molecular biology then became how a cell performed such a biochemically sophisticated transformation, translating a nucleotide sequence in DNA into an amino acid sequence in a protein.

Portions of the DNA Sequence Are Copied into RNA to Make Protein[7]

The synthesis of a protein involves the copying of specific regions of DNA (called *coding regions,* or *genes*) into a chemically and functionally different type of polynucleotide known as ribonucleic acid, or RNA. Like DNA, RNA is composed of a linear sequence of nucleotides, but it has two small chemical differences: (1) the sugar-phosphate backbone of RNA contains ribose instead of a deoxyribose sugar, and (2) the base thymine (T) is replaced by the very closely related base uracil (U) (see Figure 3–6).

RNA retains all of the information of the DNA sequence from which it was copied, as well as the base-pairing properties of DNA—since U pairs with A in the same manner as T does (Figure 3–6). RNA molecules are synthesized by a process known as **DNA transcription,** which is similar to DNA replication in many respects. One of the two strands of DNA acts as a template on which the base-pairing abilities of incoming ribonucleotides are tested. When a good match is achieved with the DNA template, a ribonucleotide is incorporated as a covalently bonded unit in the growing RNA chain.

DNA transcription differs from DNA replication in a number of important ways. For one, the RNA product does not remain as a strand annealed to DNA. As soon as the RNA copy is completed, the original DNA helix re-forms and the RNA molecule is released. Thus RNA molecules are single-stranded. Moreover, RNA molecules are relatively short compared to DNA molecules since they are copied from a limited region of the DNA—enough to make one or a few proteins. The amount of RNA made from a particular region of DNA can be controlled, and some genes are used to make RNA in very large quantities while others are not transcribed at all. To a large extent, the control of gene expression is made possible by *gene regulatory proteins* that determine which segments of DNA are copied (p. 438).

Thousands of RNA transcripts can be made from the same DNA segment during each cell generation. In a eucaryotic cell, many of these RNA molecules undergo major chemical changes before they leave the nucleus to serve as the **messenger RNA** (or **mRNA**) molecules that direct the synthesis of proteins in the cytoplasm. Because each mRNA molecule can be translated into many thousands of copies of a polypeptide chain, the information contained in a small region of DNA can direct the synthesis of large quantities of a specific protein. For example, consider the protein *fibroin,* which is the major component of silk: in each silk gland cell, a single fibroin gene makes 10^4 copies of its mRNA, each of which directs the synthesis of 10^5 molecules of fibroin—producing a total of 10^9 molecules of fibroin per cell in a period of four days.

Sequences of Nucleotides in mRNA Are "Read" in Sets of Three and Translated into Amino Acids[8]

The rules by which a polynucleotide sequence of a gene is translated into the amino acid sequence of a protein, the so-called **genetic code,** were deciphered in the early 1960s. The sequence of nucleotides in the mRNA molecule that acts as an intermediate proved to be read in serial order in groups of three. Each triplet of nucleotides, called a **codon,** specifies one amino acid, and, in principle, each RNA sequence can be translated in any one of three different *reading frames* depending on exactly where on the molecule the decoding process begins (Figure 3–13). In almost every case, only one of these reading frames will produce a functional protein. Since there are no punctuation signals except at the beginning and end of the RNA message, the reading frame is set at the initiation of the translation process and is maintained thereafter.

Since RNA is a linear polymer of four different nucleotides, there are $4^3 = 64$ possible codon triplets (remember that it is the *sequence* of nucleotides in the triplet that is important). Only 20 different amino acids are commonly found in proteins, so that most amino acids must be specified by several codons; that is, the genetic code is *degenerate.* The code is shown in Figure 3–14: it has been highly conserved and is the same in organisms as diverse as bacteria, plants, and man.

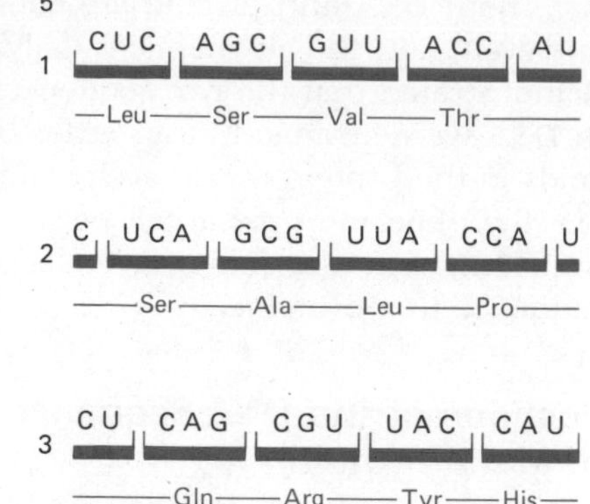

Figure 3–13 The three possible reading frames in protein synthesis. A sequence of nucleotides in RNA is read from the 5′ to the 3′ end in sequential sets of three nucleotides and thereby translated into amino acids. The same RNA sequence can therefore specify three completely different amino acid sequences, depending on the "reading frame."

tRNA Molecules Match Amino Acids to Groups of Nucleotides[9]

The mRNA codons do not directly recognize the amino acids that they specify in the way that an enzyme recognizes a substrate. Translation uses "adaptor" molecules that recognize both an amino acid and a group of nucleotide bases. These adaptors consist of a set of small RNA molecules known as **transfer RNAs** (or **tRNAs**), each of which is only 70 to 90 nucleotides in length.

Each tRNA molecule has a folded three-dimensional conformation held together by the same noncovalent interactions that hold together the two strands of the DNA helix. As noted earlier, a DNA helix is formed by the combined strength of many complementary base pairs. In a single-stranded polynucleotide such as tRNA, the same type of complementary base-pairing interactions occur between some of the nucleotide residues in the *same* chain.

1st position (5′ end) ↓	2nd position U	2nd position C	2nd position A	2nd position G	3rd position (3′ end) ↓
U	Phe	Ser	Tyr	Cys	U
	Phe	Ser	Tyr	Cys	C
	Leu	Ser	STOP	STOP	A
	Leu	Ser	STOP	Trp	G
C	Leu	Pro	His	Arg	U
	Leu	Pro	His	Arg	C
	Leu	Pro	Gln	Arg	A
	Leu	Pro	Gln	Arg	G
A	Ile	Thr	Asn	Ser	U
	Ile	Thr	Asn	Ser	C
	Ile	Thr	Lys	Arg	A
	Met	Thr	Lys	Arg	G
G	Val	Ala	Asp	Gly	U
	Val	Ala	Asp	Gly	C
	Val	Ala	Glu	Gly	A
	Val	Ala	Glu	Gly	G

Figure 3–14 The genetic code. Sets of three nucleotides in RNA (codons) are translated into amino acids in the course of protein synthesis according to the rules shown. For example, the codons GUG and GAG are translated into valine and glutamic acid, respectively. Note that those codons with U or C as the second nucleotide tend to specify the more hydrophobic amino acids (compare with Panel G, pp. 58–59).

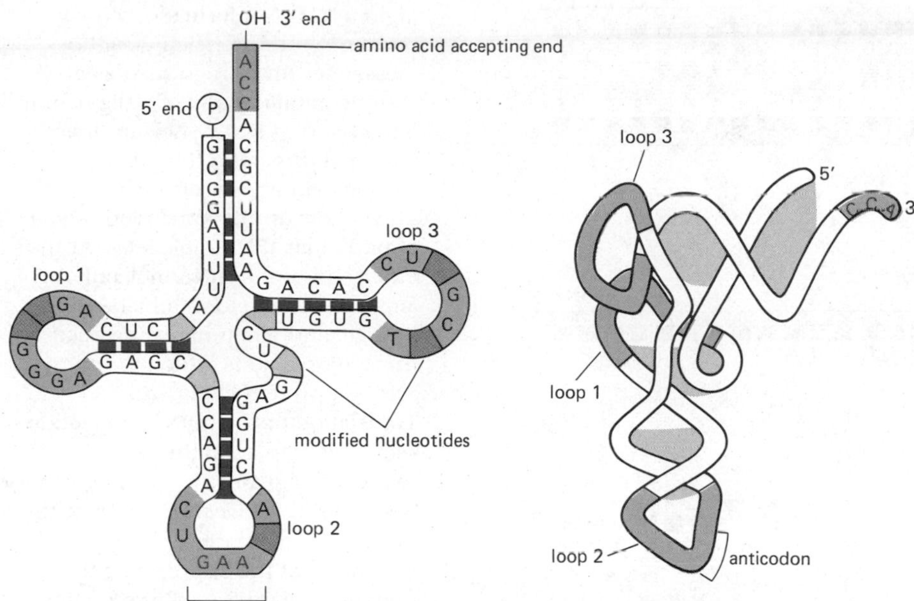

Figure 3–15 Phenylalanine tRNA of yeast. In the drawing on the left, the nucleotides are arranged to show the base-pairing that forms internal helical regions in the tRNA molecule. A schematic drawing of the *actual* shape of the molecule, based on x-ray-diffraction analysis, is depicted on the right. Here the regions where base pairs hold the strands together are lightly shaded in gray.

This causes the tRNA molecule to fold up in a unique way that is important for its function as an adaptor. Four short segments of the molecule contain a double-helical structure like that in Figure 3–7. But of special importance are three unpaired nucleotide residues at either end of the molecule: one such triplet of varying sequence forms the *anticodon* that can base-pair to a complementary triplet in an mRNA molecule, while the triplet at the free 3′ end of the molecule (the *CCA sequence*) is attached covalently to a specific amino acid (Figure 3–15).

Although they are much larger and not as compactly folded, there is experimental evidence that other types of RNA molecules, including mRNA and the **ribosomal RNA (rRNA)** molecules that form the structural core of a *ribosome* (see below), also contain regions with distinct three-dimensional conformations.

The RNA Message Is Read from One End to the Other by a Ribosome[10]

The codon recognition process that transfers genetic information from mRNA to protein depends on the same base-pair interactions that mediate the transfer of genetic information from DNA to DNA and from DNA to RNA (Figure 3–16). But the mechanics of ordering the tRNA molecules on the mRNA are complicated and require the **ribosome,** a complex of almost a hundred different proteins associated with several structural RNA molecules (rRNAs). Each ribosome serves as a large biochemical machine on which tRNA molecules position themselves so as to read the genetic message encoded in the mRNA. The ribosome first binds at a specific site on the mRNA molecule to set the reading frame and determine the amino-terminal end of the protein. Then, as the ribosome moves along an mRNA molecule, it translates one codon at a time, using tRNA molecules to add amino acids to the growing end of the polypeptide chain (Figure 3–17). When a ribosome reaches the end of the message, both it and the freshly made carboxyl end of the protein are released from the 3′ end of the mRNA molecule into the cell cytoplasm. Further details

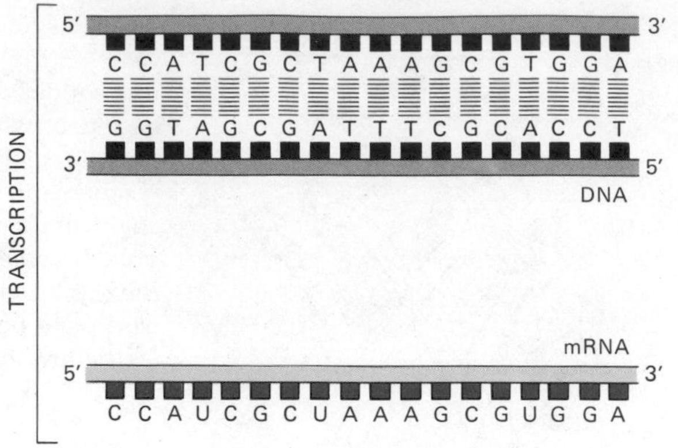

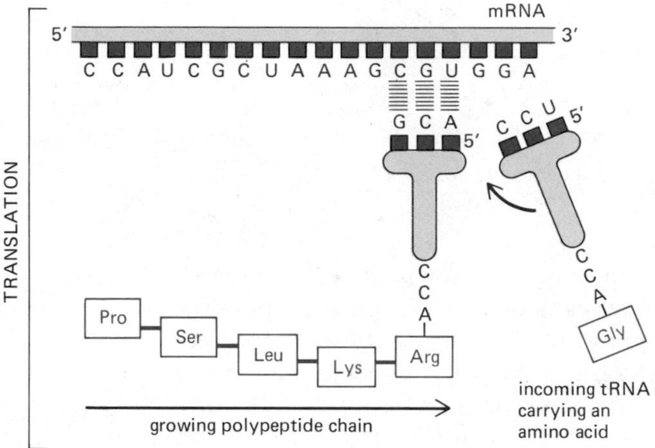

Figure 3–16 Information flow in protein synthesis. The nucleotides in messenger RNA are formed as a complementary copy of a segment of one strand of DNA. They are then matched three at a time to a complementary set of three nucleotides in the anticodon region of particular tRNA molecules. At the other end of the tRNA molecule, an amino acid is held in a high-energy linkage, and when matching occurs, this amino acid is added to the growing protein chain end. Translation of the mRNA nucleotide sequence into an amino acid sequence depends on complementary base-pairing between a codon of the mRNA and the corresponding anticodon of the appropriate tRNA. The molecular basis of information transfer in translation is therefore closely similar to that in DNA replication and transcription.

of ribosome structure and the mechanism of protein synthesis are given in Chapter 5.

Ribosomes operate with remarkable efficiency: in one second a single bacterial ribosome adds 20 amino acids to a growing polypeptide chain. And in one second the human body makes about 5×10^{14} copies of hemoglobin, a protein containing a unique sequence of 574 amino acids.

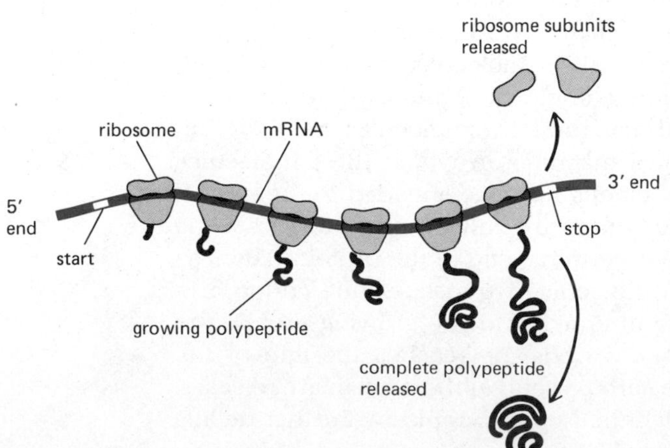

Figure 3–17 Schematic diagram of the synthesis of proteins on ribosomes. Ribosomes become attached to a start signal near the 5′ end of an mRNA molecule and then move toward the 3′ end, synthesizing protein as they go. A single mRNA will often have a number of ribosomes traveling along it at the same time; the entire assembly is known as a *polyribosome.*

Summary

Genetic information is carried in the linear sequence of nucleotides in DNA. Each molecule of DNA contains two complementary strands of nucleotides that are matched by the hydrogen bonds of the G-C and A-T base pairs. DNA replication, which duplicates the genetic information, occurs by the polymerization of a new complementary strand onto each of the old strands.

The expression of genetic information involves the translation of the linear sequence of nucleotides in DNA into a colinear sequence of amino acids in proteins. A limited segment of DNA is first copied into a complementary strand of mRNA, which is then translated into protein in a reaction catalyzed by a large complex known as a ribosome. The amino acids used for protein synthesis are attached to a family of small tRNA molecules, each of which recognizes, by complementary base-pairing interactions, a set of three nucleotides in mRNA. The sequence of nucleotides in mRNA is read from one end to the other in sets of three, according to a universal genetic code.

Protein Structure[11]

To a large extent cells are made of proteins, which constitute more than half of the dry weight of the cell (Table 3–1). Proteins determine the shape and structure of the cell and also serve as instruments of molecular recognition and catalysis. Although it is true that DNA stores the information required to make a cell, it does so in an "academic" fashion, the DNA itself having little direct influence on cellular processes. The gene for hemoglobin, for example, cannot carry oxygen: that is a property of the protein specified by it. In computer terminology, the nucleic acids represent the "software"—instructions that a cell receives from its parent. Proteins make up the "hardware"—the physical apparatus that executes the program stored in the memory.

These different functions of nucleic acids and proteins are reflected in the chemical nature of the subunits from which they are made. DNA and RNA consist of a series of nucleotides, all being chemically very similar to one another, assembled into giant molecules having much the same chemistry whatever their sequence. In contrast, proteins are made from an assortment of 20 very different amino acids, each with a distinct chemical personality (see Panel G, pp. 58–59). This variety allows for enormous versatility in the chemical properties of different proteins.

The Shape of a Protein Molecule Is Determined by Its Amino Acid Sequence[12]

Many of the bonds in a long polypeptide chain allow free rotation of the atoms they join, giving the protein backbone great flexibility. In principle then, any protein molecule can adopt an enormous number of different shapes, or **conformations.** However, under biological conditions, most polypeptide chains fold into only one of these conformations. This is because the side chains of the different amino acids associate with one another and with water to form various weak noncovalent bonds (see Panel I, pp. 94–95). Depending on which side chains are present and on their position in the chain, a large force is developed that gives one particular conformation of a protein unusual stability.

Most proteins fold spontaneously into their correct shape. For example, a protein can be unfolded, or *denatured,* to give a flexible polypeptide chain that has lost its original conformation. However, if the denaturing treatment is gentle enough, it can usually be reversed, and the unfolded polypeptide chains spontaneously refold into their original conformations. This behavior confirms that all of the information determining the conformation must be contained in the amino acid sequence itself.

One of the most important factors governing the folding of a polypeptide is the distribution of its polar and nonpolar side chains. As a protein is being synthesized, its many hydrophobic side chains tend to be pushed together in the interior of the molecule, which enables them to avoid contact with the aqueous environment (just as oil droplets coalesce after being mechanically dispersed in water). At the same time, all of the polar side chains tend to arrange themselves near the outside of the protein molecule, where they can interact with water and other polar groups (Figure 3–18). Since peptide bonds are themselves quite polar, they tend to interact both with one another and with polar side chains to form hydrogen bonds (Figure 3–19); nearly all polar residues that are buried within the protein are paired in this way. Hydrogen bonds thus play a major part in holding together different regions of polypeptide chain in a folded protein molecule, and they are crucially important for many of the binding interactions observed on protein surfaces.

Once they are outside the cytoplasm, secreted or cell-surface proteins often form additional covalent intrachain bonds. For example, the formation of **disulfide bonds** (also called S-S bridges) between neighboring cysteine-SH groups in a folded polypeptide chain (Figure 3–20) frequently serves to stabilize the three-dimensional structure of extracellular proteins, although they are not required for the specific folding of the molecule. Such S-S bridges are rarely formed in protein molecules that are still in the cell cytosol because the high intracellular concentration of the -SH reducing agent *glutathione* breaks most such bonds.

The net result of all the individual amino acid interactions is that most protein molecules fold up spontaneously into a unique conformation, usually compact and globular but sometimes long and fibrous. The inner core is composed of clustered hydrophobic side chains—packed into a tight, nearly

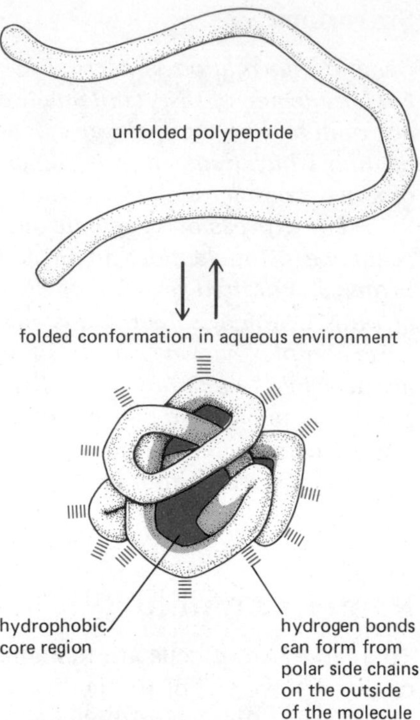

Figure 3–18 Schematic illustration of the folding of a protein into a globular conformation. The polar amino acid side chains tend to be exposed on the outside of the protein; the nonpolar amino acid side chains are buried on the inside to form a hydrophobic core "hidden" from the water.

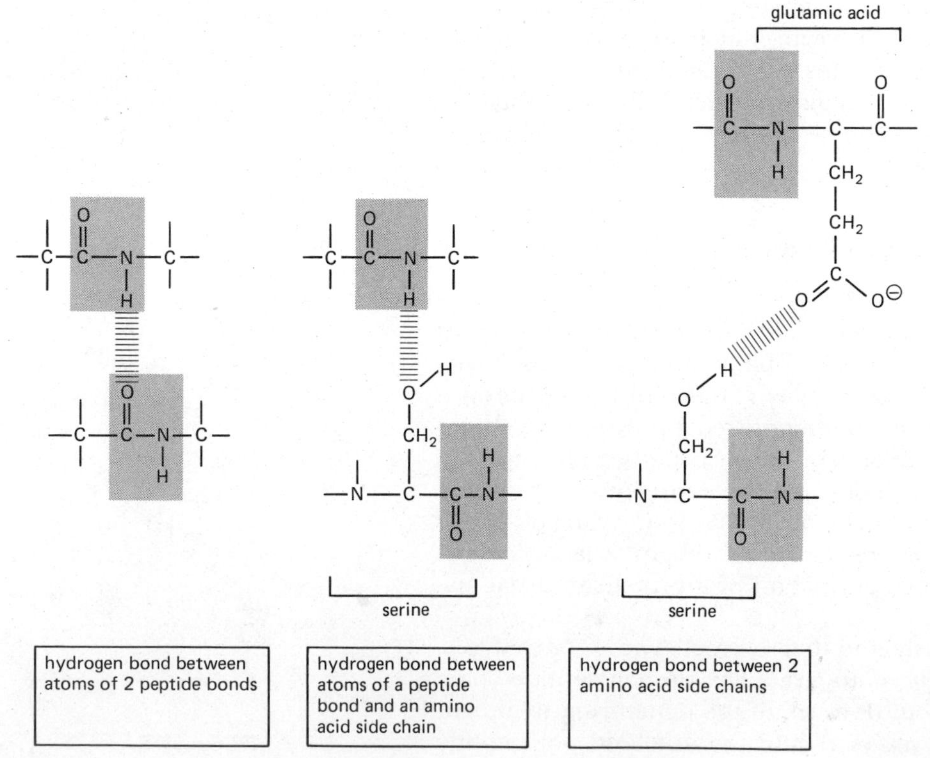

Figure 3–19 Some of the hydrogen bonds (indicated by thick broken lines) that form between the amino acids in a protein.

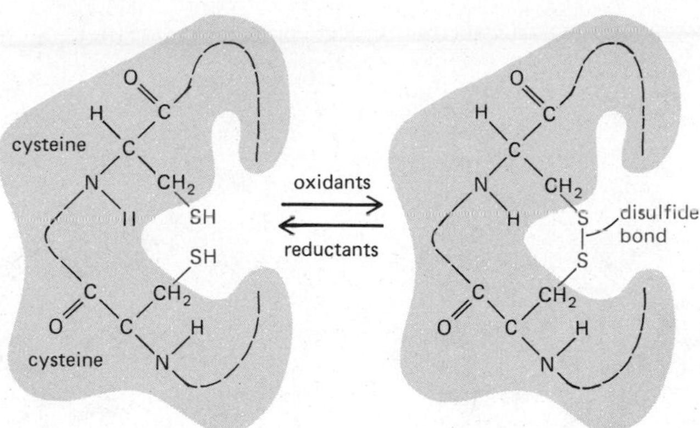

Figure 3–20 The formation of a covalent disulfide bond between neighboring cysteine side chains in a protein.

crystalline arrangement—while a very complex and irregular exterior surface is formed by the more polar side chains. The positioning and chemistry of the different atoms on this intricate surface make each protein specific for the binding of both other macromolecular surfaces and small molecules (see below). From a chemical standpoint, proteins are the most sophisticated molecules known.

Common Folding Patterns Recur in Different Protein Chains[13]

Although all of the information required for the folding of a protein chain is contained in its amino acid sequence, we have not yet learned how to "read" this information so as to predict the detailed three-dimensional structure of a protein whose sequence is known. Consequently, the folded conformation can be determined only by an elaborate *x-ray diffraction analysis* performed on crystals of the protein. So far, more than 200 proteins have been completely analyzed by this technique. Each of these proteins has a specific conformation so irregular that it would take a chapter to describe in full three-dimensional detail.

The complete structure of a protein may be represented by either a space-filling or a wire-type model, as illustrated and explained in Figure 3–21A and B for the enzyme lysozyme. Because the mass of detail seen in both types of model will often obscure the path of the polypeptide backbone, protein conformations are often better represented by more schematic models, such as those in Figure 3–21C and D. Here all of the side chains and actual atoms have been omitted, making it easier to follow the general course of the main chain.

When the three-dimensional structures of different protein molecules are compared, it becomes clear that, although the overall conformation of each protein is unique, several folding patterns recur repeatedly in parts of these macromolecules. Two patterns are particularly common because they result from regular hydrogen-bonding interactions between the peptide bonds themselves instead of depending on a unique pattern of side-chain interactions. Both patterns were correctly predicted in 1951 from model-building studies based on the different x-ray diffraction patterns of silk and hair. The two regular folding patterns discovered are now known as the *β-sheet*, which is adopted by the protein fibroin, found in silk; and the *α-helix*, which occurs in the protein α-keratin, found in skin and its appendages, such as hair, nails, and feathers.

The structure of the **β-sheet** makes up extensive regions of the core of most (though not all) globular proteins. In the example illustrated in Figure 3–22, which shows part of an antibody molecule, an *antiparallel β-sheet* is

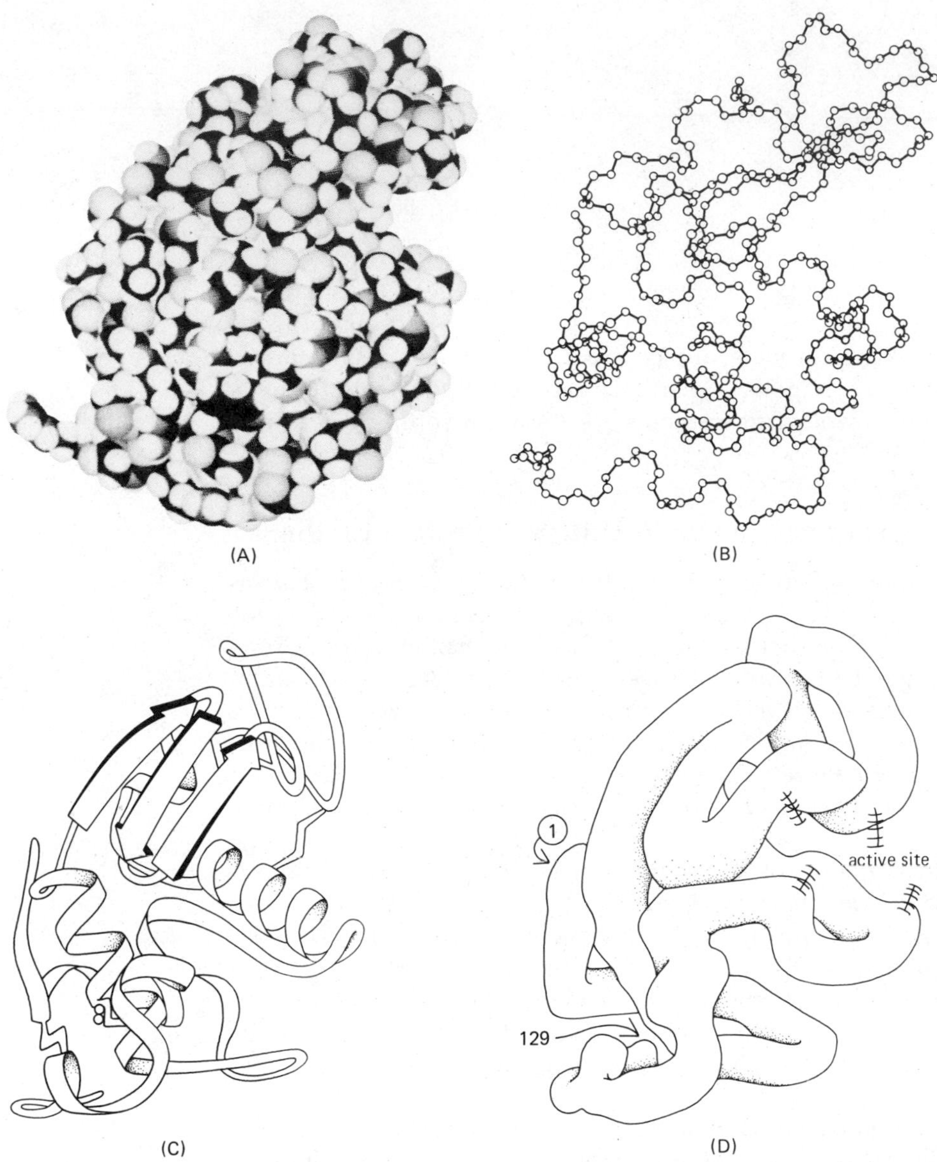

Figure 3–21 The three-dimensional conformation of the protein egg-white lysozyme as seen in four commonly used representations. (A) Space-filling model showing the radii of all atoms. (B) Backbone wire model composed of lines that connect atoms along the polypeptide backbone. (C) "Ribbon model," which represents all regions of regular hydrogen-bonded interactions present as either helices (α-helices) or sets of arrows (β-sheets). (D) "Sausage model," which shows the course of the polypeptide chain but omits all detail. Note that the core of all globular proteins is densely packed with atoms. Thus the impression of an open structure produced by models (B), (C), and (D) is incorrect. (A and B, courtesy of Richard J. Feldmann; C, courtesy of Jane Richardson.)

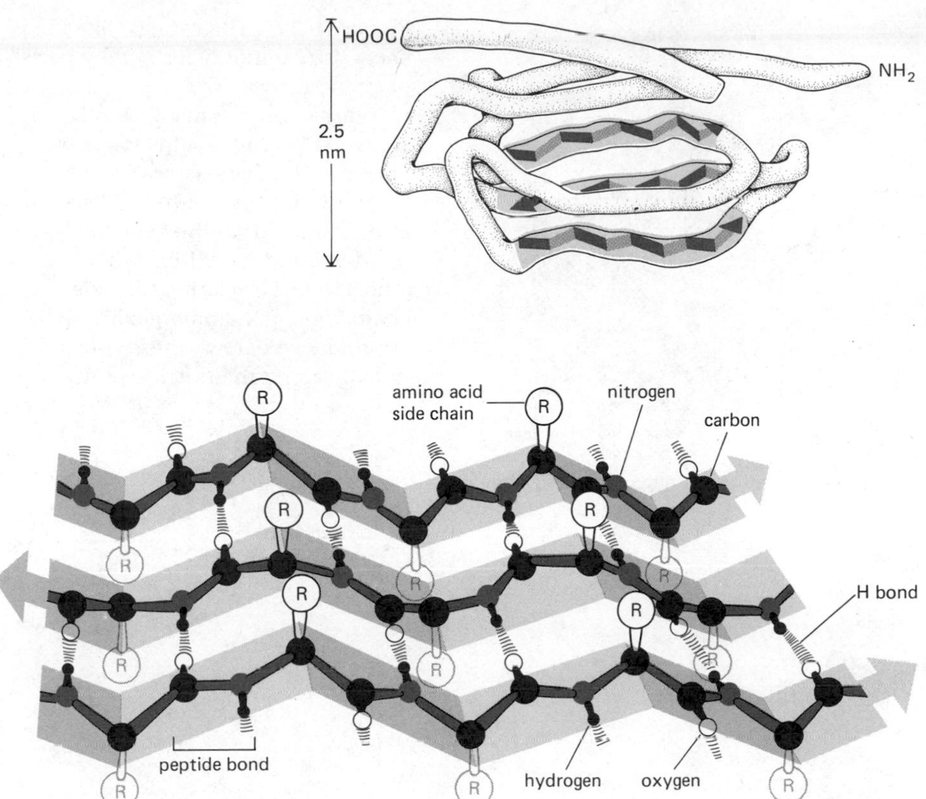

Figure 3–22 The β-sheet is a common structure formed by parts of globular proteins. At the top, a domain of 115 amino acids from an immunoglobulin molecule is shown, with one region of sheet structure outlined in color. At the bottom, a perfect antiparallel β-sheet is shown in detail. Note that every peptide bond is hydrogen-bonded to a neighbor. The actual sheet structures in globular proteins are usually somewhat less regular than the β-sheet shown here, and many sheets are slightly twisted.

formed when an extended polypeptide chain folds back and forth upon itself, with each section of the chain running in the direction opposite to that of its immediate neighbors. This gives a very rigid structure that is held together by hydrogen bonds between the peptide bonds in neighboring chains, and antiparallel β-sheets often form the framework on which a globular protein is constructed. The closely related *parallel β-sheets* are often covered on both sides by α-helices to form a layered structure, and this unit likewise serves as a framework around which many globular proteins are formed.

An **α-helix** is generated when a single polypeptide chain turns regularly about itself to make a rigid cylinder in which each peptide bond is regularly hydrogen-bonded to other peptide bonds elsewhere in the chain. Many globular proteins contain short regions of such α-helices (Figure 3–23), while long, rodlike sections of α-helix are found in many structural proteins, such as the intracellular α-keratin fibers that reinforce skin and its appendages. Space-filling models of an α-helix and a β-sheet are shown with and without their side chains in Figure 3–24.

Proteins Are Enormously Versatile in Structure

Because of the variety of their amino acid side chains, proteins are remarkably versatile with respect to the type of structures they can form. One extreme case is the family of **collagen** molecules found in the extracellular space. In collagen three separate polypeptide chains, each rich in the amino acid proline and containing the amino acid glycine at every third residue, are wound around each other to generate a regular triple helix. Further regular packing of these collagen molecules generates connective tissues such as tendons, in

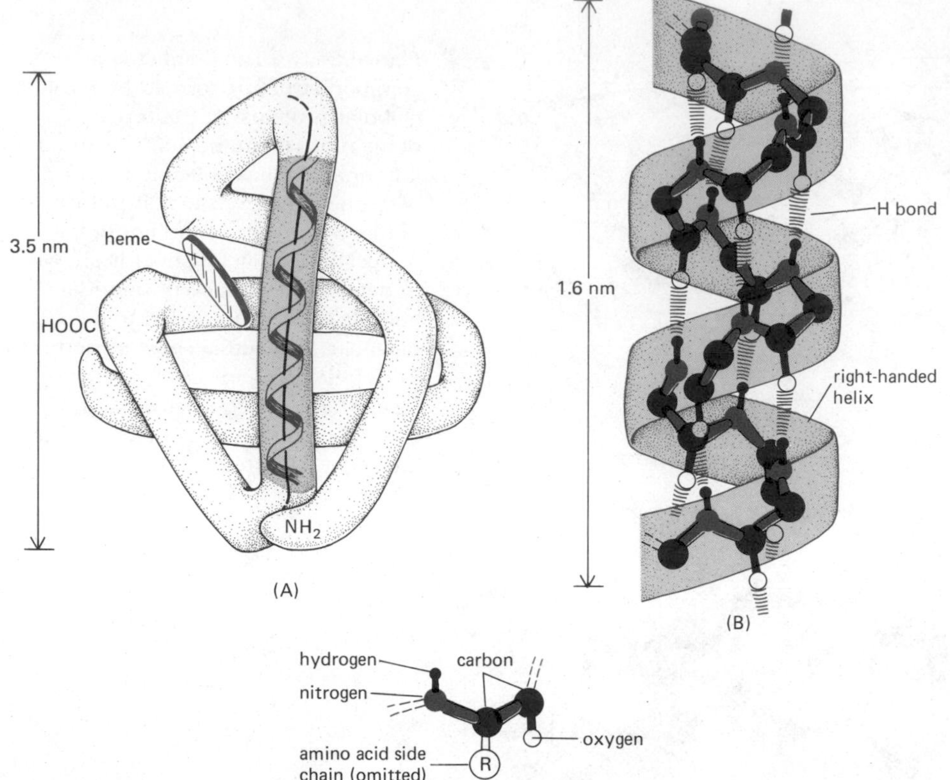

Figure 3–23 The α-helix, like the β-sheet, is commonly formed by parts of globular proteins. (A) The oxygen-carrying molecule myoglobin is shown (153 amino acids long), with one region of α-helix outlined in color. (B) A perfect α-helix is shown in detail. As in the β-sheet, every peptide bond is hydrogen-bonded to a neighbor. For clarity, the side chains on each amino acid have been omitted (see below); these protrude radially along the outside of the helix.

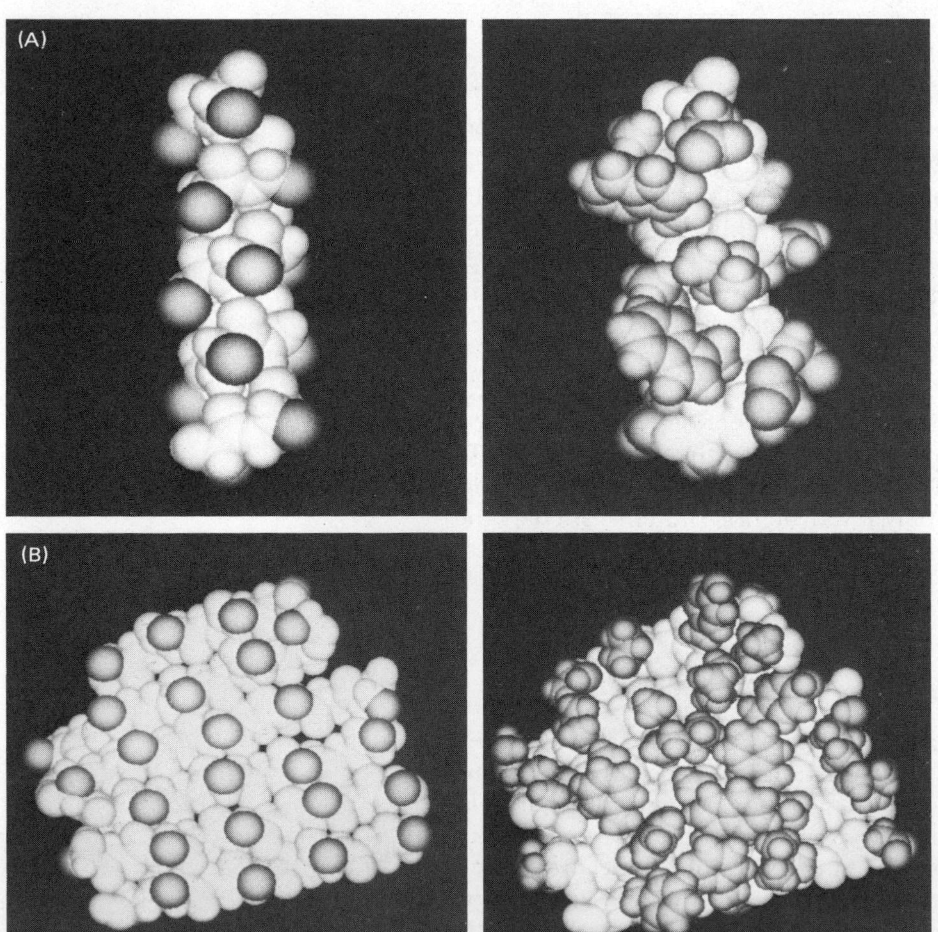

Figure 3–24 Space-filling models of an α-helix and a β-sheet with *(right)* and without *(left)* their amino acid side chains. (A) An α-helix (part of the structure of myoglobin). (B) A region of β-sheet (part of the structure of concanavalin A). (Courtesy of Richard J. Feldmann.)

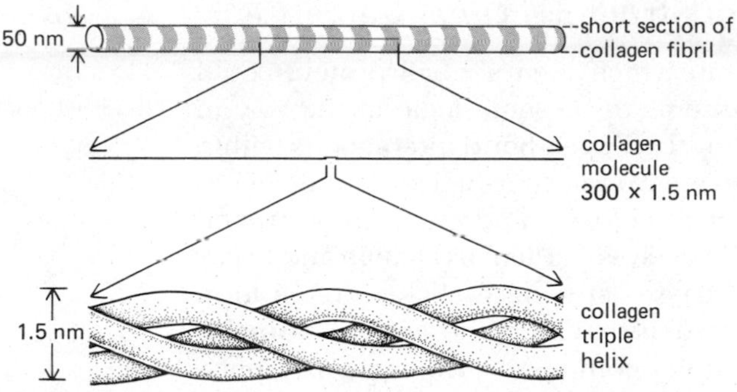

Figure 3–25 Collagen is a triple helix formed by three extended protein chains that are arranged in parallel. Many rodlike collagen molecules are cross-linked together to form tough and inextensible collagen fibrils and fibers.

which adjacent collagen molecules are tied together by covalent cross-links between neighboring lysine residues to generate fibers with enormous tensile strength (Figure 3–25).

At the other extreme is the extracellular protein **elastin,** in which relatively unstructured polypeptide chains are covalently cross-linked to generate an elastic material much like rubber. As illustrated in Figure 3–26, the elasticity is due to the ability of individual protein molecules to uncoil reversibly whenever a stretching force is applied. Like collagen, elastin is secreted into the extracellular space, where it enables tissues like arteries and lungs to deform and stretch without damage (see Chapter 12).

It is amazing that the same basic chemical structure—a chain of amino acids—can form so many different structures: an efficient rubberlike material (elastin), a steel-like cable (collagen), or the wide variety of catalytic surfaces on the globular proteins that function as enzymes. Figure 3–27 illustrates and compares the range of different shapes that could, in theory, be adopted by a polypeptide chain 300 amino acids long. As we have already emphasized, the conformation actually adopted depends entirely on the amino acid sequence.

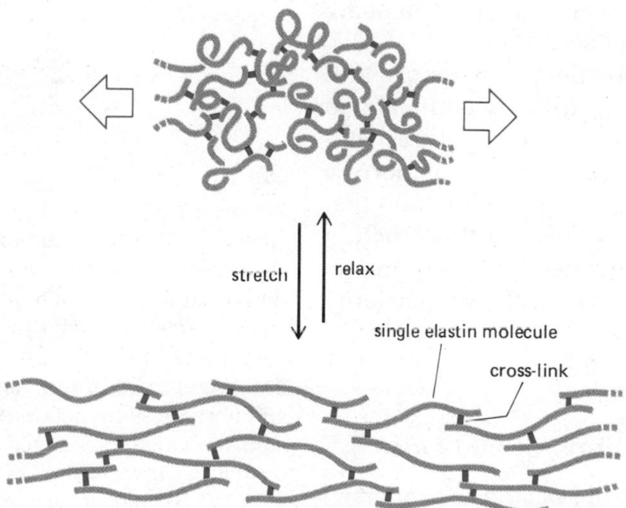

Figure 3–26 Elastin is composed of flexible polypeptides that are cross-linked together to create a rubbery matrix, with each molecule uncoiling into a more extended conformation when the matrix is stretched. The striking contrast between the physical properties of elastin and collagen is due to their very different amino acid sequences.

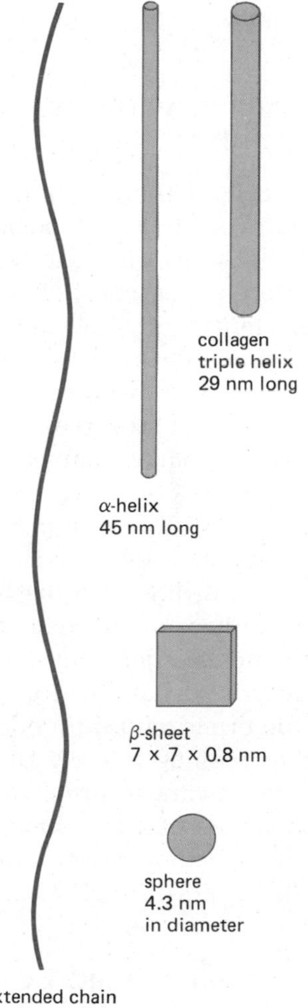

Figure 3–27 Possible sizes and shapes of a protein molecule that contains 300 amino acids. The structure formed is determined by the amino acid sequence. (Adapted from D. E. Metzler, Biochemistry. New York: Academic Press, 1977.)

Proteins Show Different Levels of Structural Organization

Though the number of different ways in which even a small protein could fold up is astronomically large, there seems to be some logic in the way in which it occurs. At the first level of folding, hydrogen-bond interactions within contiguous stretches of polypeptide chain give rise to α-helices and β-sheets, which comprise the protein's *secondary structure.* In addition, some combinations of α-helices and β-sheets are themselves particularly stable and occur frequently in many different proteins. These "structural clichés" often form the basis for the next higher level of structure, called the protein **domain.** Domains are relatively small globular units composed of a section of polypeptide chain containing 150 amino acids or less, and they seem to be the modular units from which globular proteins are constructed (see below). A number of different domains are usually strung together by relatively open lengths of polypeptide chain to make a globular protein. Finally, individual globular proteins themselves often assemble into larger *protein aggregates.*

The structure of a large protein can thus be resolved into several different levels, each of which is constructed from the one below it in a hierarchical fashion (Figure 3–28). These levels of organization may correspond to the steps by which a newly synthesized protein folds into its native structure.

sequence of amino acids

—Glu—Asp—Val—Ser—Lys—Gly—Pro—

1. secondary structure

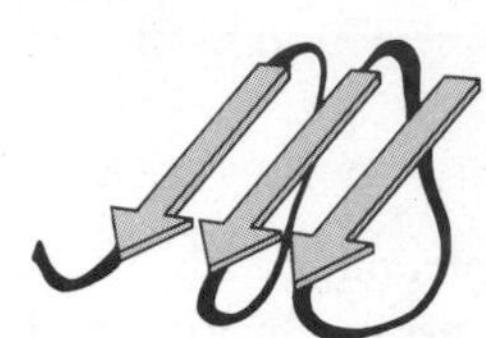

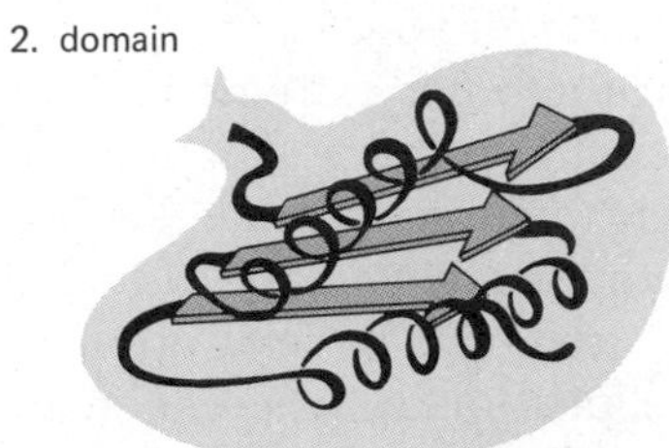

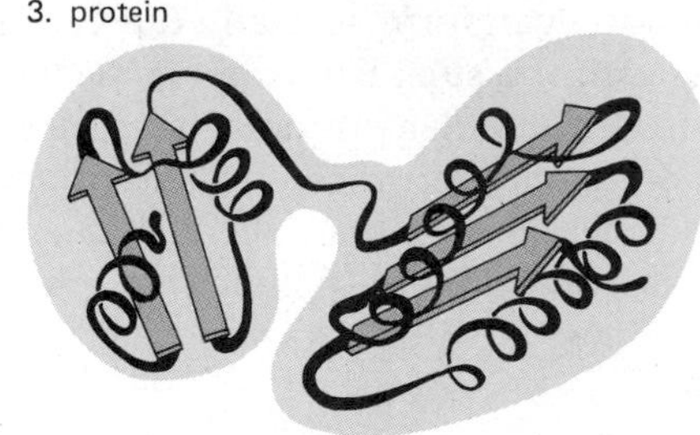

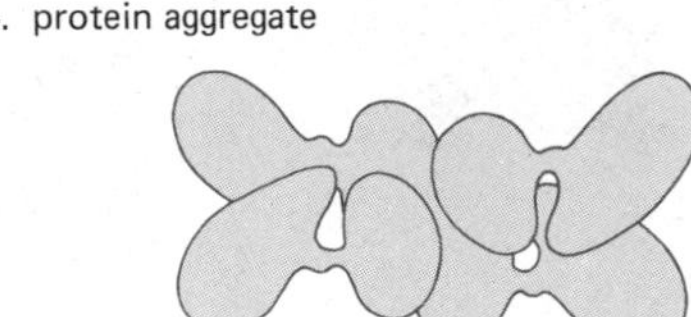

Figure 3–28 Levels of protein structure. The three-dimensional structure of a protein can be described in terms of four different levels of folding, each of which is constructed from the preceding one in hierarchical fashion. According to an alternate nomenclature, the amino acid sequence itself is termed the *primary structure,* the first folding level the *secondary structure,* the combination of the second and third folding levels the *tertiary structure,* and the fourth level the *quaternary structure* of a protein.

Relatively Few of the Many Possible Polypeptide Chains Would Be Useful

Since each of the 20 amino acids is chemically distinct and each can, in principle, occur at any position in a protein chain, there are $20 \times 20 \times 20 \times 20 = 160{,}000$ different possible polypeptide chains 4 amino acids long, or 20^n different possible polypeptide chains n amino acids long. For a typical protein length of about 300 amino acids, more than 10^{390} different proteins can be made.

We know, however, that only a very small fraction of the possible proteins will adopt a useful three-dimensional conformation. The vast majority would be expected to have many different conformations of about equal energy, each with different chemical properties. Since such instability is not compatible with the high degree of order needed to maintain a cell, such proteins have been eliminated by natural selection in the course of evolution.

Present-day proteins have an amazingly sophisticated structure and chemistry because of their unique folding properties. Not only is the amino acid sequence such that one conformation is extremely stable, but this conformation also has the correct shape and chemical properties to perform a catalytic or structural function in the cell. Proteins are so precisely built that the change of even a few atoms in one amino acid can disrupt the structure and cause a catastrophic change. But one should remember that they are the rare survivors of a very long evolutionary process in which the vast majority of proteins had more random conformations, were less useful, and were therefore discarded through natural selection.

New Proteins Often Evolve by Minor Alterations of Old Ones[14]

Cells have genetic mechanisms that allow genes to be duplicated and modified in the course of evolution (see p. 470). Consequently, once a protein sequence with a unique three-dimensional conformation and useful surface properties has evolved, its basic structure can be incorporated in many other proteins. Different proteins of related function often have similar amino acid sequences in present-day organisms. It is believed that such families of proteins evolved from a single ancestral gene that in the course of evolution duplicated to give rise to other genes in which mutations gradually accumulated to produce proteins with new functions.

A well-known example is the family of protein-cleaving (proteolytic) enzymes called **serine proteases.** This family includes the digestive enzymes chymotrypsin, trypsin, and elastase, and many of the proteases, such as thrombin, that control the blood-clotting process. If any two of these enzymes are compared, about 40% of the positions in their amino acid sequences are occupied by the same amino acid. The similarity of their three-dimensional conformations as determined by x-ray crystallography is even more striking: most of the detailed twists and turns in these polypeptide chains, which are several hundred amino acids long, are identical (Figure 3–29).

Nonetheless, the various serine proteases have quite distinct functions. Some of the amino acid changes that make these enzymes different were presumably selected in the course of evolution because they resulted in changes in substrate specificity and regulatory properties, giving them the different functional properties they have today. Other amino acid changes may be "neutral," in that they survived only because they did not alter the basic structure and function of the enzyme. Since mutation is a random process, there must also have been deleterious changes that altered the three-dimensional structure of these enzymes sufficiently to inactivate their function. Such altered proteins would have been lost because the individual organisms making them would have been at a disadvantage and would have been eliminated by natural selection.

It is not at all surprising, then, that cells contain whole sets of structurally related polypeptide chains that have common ancestors but different functions.

Figure 3–29 Comparison of the three-dimensional conformations of elastase (A) and chymotrypsin (B). Only the amino acids corresponding to the residues of the polypeptide chain shaded in color are the same in these two evolutionarily related proteases. Nevertheless, their conformations are extremely similar. The active sites are circled; both contain an activated serine residue (see Figure 3–46).

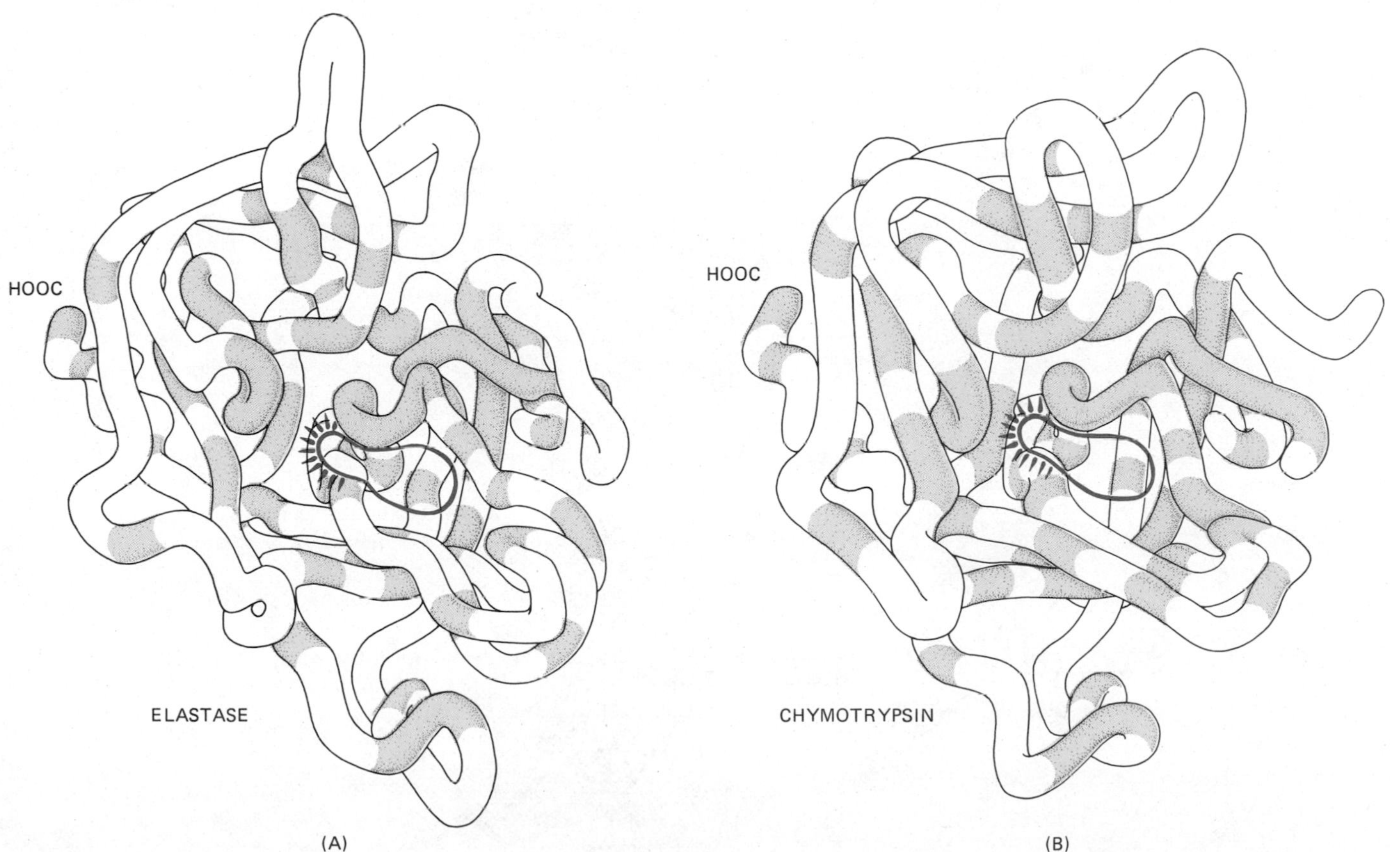

New Proteins Often Evolve Through the Combination of Different Polypeptide Domains[15]

Once a number of stable protein surfaces have been made in a cell, new surfaces with different binding properties can be generated by the joining of two or more individual proteins together by noncovalent interactions between them. In the schematic example shown in Figure 3–30, three different binding sites have been generated by combining different parts of one protein with three other proteins. This complexing together of globular proteins to make larger, functional protein aggregates is common in cells: although a typical polypeptide chain has a molecular weight of 40,000 to 50,000 (about 300 to 400 amino acids), and very few polypeptide chains are more than three times this size, many protein aggregates have molecular weights of 1 million or more.

A related but distinct way of making a new protein from existing chains is to join the corresponding DNA sequences so that a single larger polypeptide chain results (p. 471). Proteins that are believed to have evolved in this way can be recognized by the fact that different parts of their polypeptide chain fold independently into separate globular domains. Many proteins have such "multidomain" structures, and, as might be expected from the evolutionary considerations discussed above, an important binding site for another molecule frequently lies at the site where the separate domains are juxtaposed

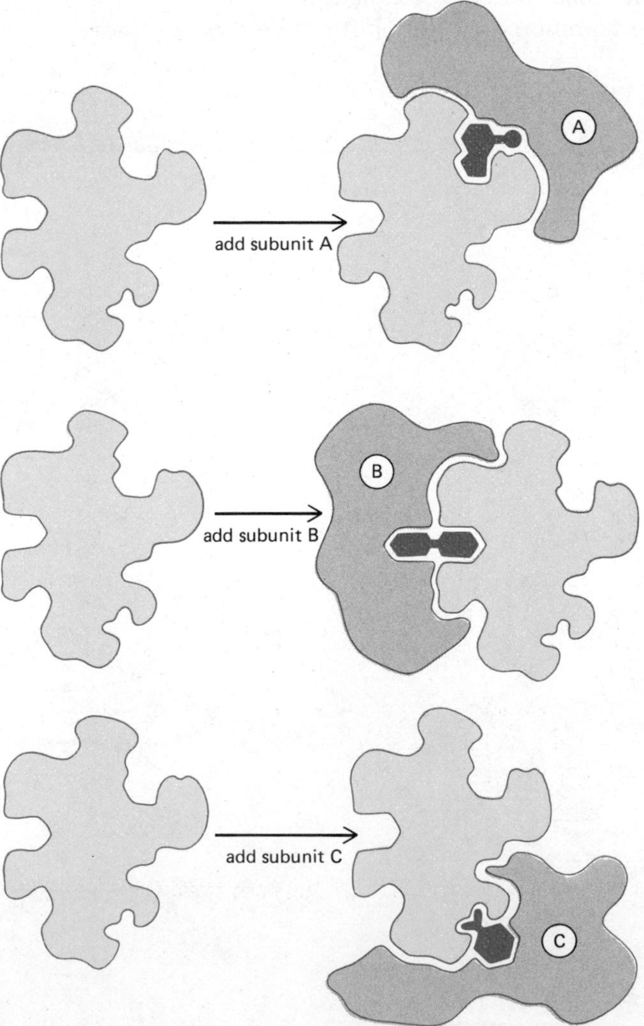

Figure 3–30 Combinations of protein domains. A schematic representation of the general principle by which the juxtaposition of separate protein surfaces, in the course of evolution, could give rise to new binding sites for small molecules called *ligands*.

(Figure 3–31). The structure of one particular multidomain protein is shown in Figure 3–32.

Putting together amino acid sequences by joining preexisting ones (Figures 3–30 and 3–31) is clearly a much more efficient evolutionary strategy for a cell than the alternative of deriving new protein sequences from scratch by random DNA mutation. Most polypeptides created by random mutations are bound to lack a unique conformation and will therefore be of little use. By contrast, the joining together of established protein domains or fragments should at least produce a new combination of unique surfaces. In fact, eucaryotic genomes seem to be constructed in a way that facilitates the occasional rearrangement of DNA sequences to create a new gene that codes for protein domains in new combinations (see Chapter 8).

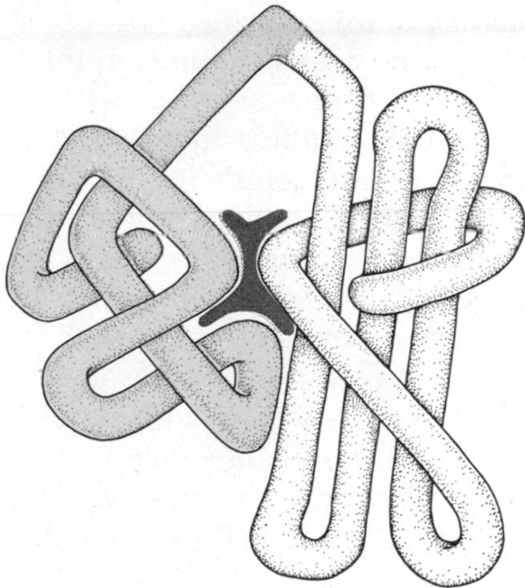

Figure 3–31 Schematic view of a single protein containing two separately folded domains. As indicated here, the binding sites for ligands often lie between such domains.

Protein Subunits Can Self-assemble into Large Structures in the Cell[16]

The same principles that enable several protein domains to associate to form binding sites operate to generate much larger structures in the cell. Supramolecular structures such as enzyme complexes, ribosomes, protein filaments, viruses, and membranes are not made as single, giant, covalently linked molecules; instead they are noncovalently assembled as aggregates of preformed macromolecular subunits.

There are several advantages to the use of smaller subunits to build larger structures: (1) building a large structure from one or a few repeating smaller subunits reduces the amount of genetic information required; (2) since the subunits associate through multiple bonds of relatively low energy, both assembly and disassembly can be readily controlled; and (3) subunit assembly minimizes errors in the synthesis of the structure, since correction mechanisms can operate in the course of assembly to exclude malformed subunits.

A Single Type of Protein Subunit Can Interact with Itself to Form Geometrically Regular Aggregates[17]

If a protein has a binding site that is complementary to a region of its own surface, it will assemble spontaneously into an aggregate structure. In the simplest possible case, a binding site recognizes itself and forms a symmetrical

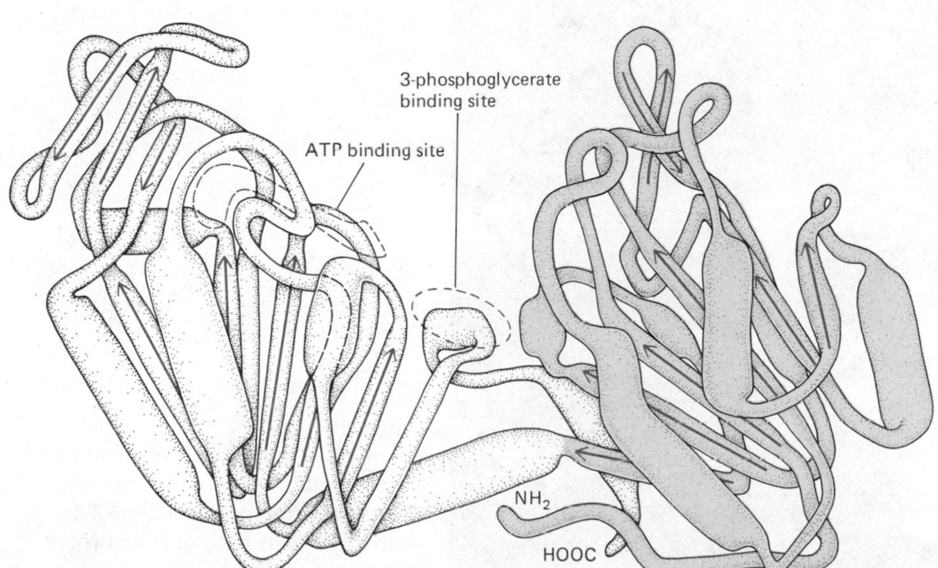

Figure 3–32 The structure of a protein composed of two domains. In this sausage model, regions of α-helix are represented by widened cylinders, while regions of β-sheet are indicated by the colored arrows. The enzyme shown is phosphoglycerate kinase (416 amino acids long), and the proposed binding sites for its two substrates are indicated. The two domains move toward each other when both substrates bind. This movement forms the active site. (After R. D. Banks, et al., *Nature* 279:773–777, 1979. Modified by H. C. Watson.)

dimer. Many enzymes and other proteins form dimers of this kind, which frequently act as subunits in the formation of larger aggregate structures (Figures 3–33 and 3–34).

If the binding site of a protein is complementary to a region of its surface that does not include the binding site itself, a chain of subunits will be formed. For some orientations of the two binding sites, the chain will soon run into itself and terminate, forming a closed ring of two, three, four, or more subunits (Figure 3–35). More commonly, an indefinitely extended polymer of subunits will result, and, provided that each subunit is bound to its neighbor in an identical way, the subunits in the polymer will be arranged in a **helix** (Figure 3–35). A helix is, therefore, a very simple structure to make.

An example of a helical structure commonly encountered in cells is the **actin filament,** which has two helical strands wound around each other and is formed from a single globular protein subunit called *actin*. Having two

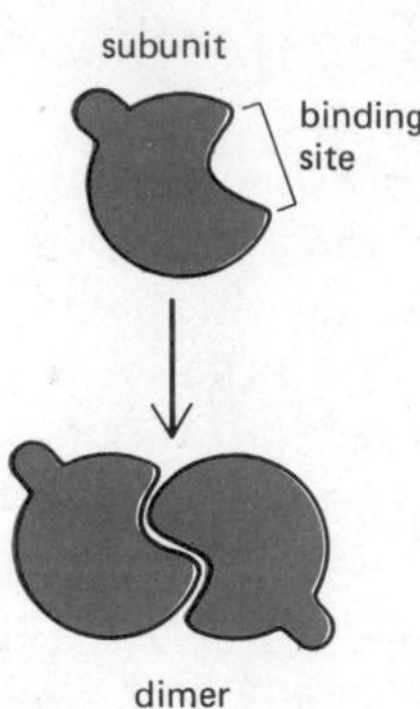

Figure 3–33 Schematic illustration of the formation of a dimer from a single type of protein subunit. A binding site that recognizes itself will produce symmetrical dimers.

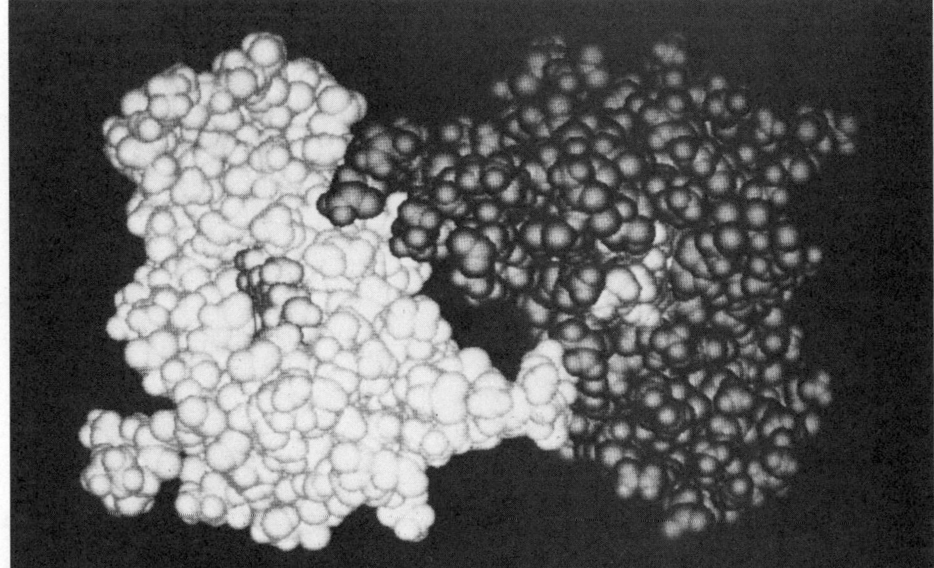

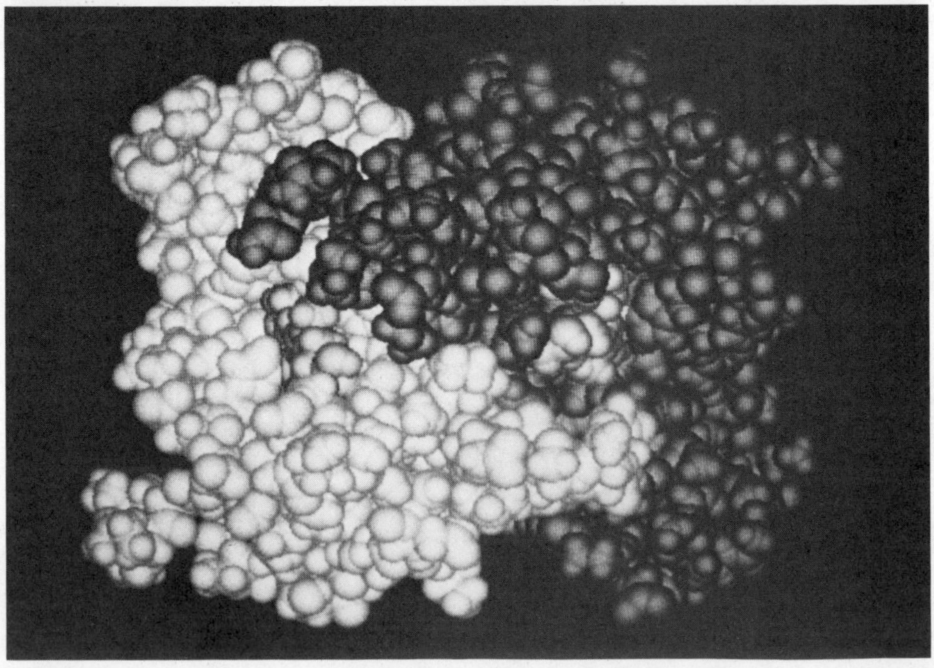

Figure 3–34 Space-filling model of the formation of a dimer from two identical protein monomers. The protein shown here is cytochrome c′. (Courtesy of Richard J. Feldmann.)

Figure 3–35 Rings or helices can form from a single type of protein subunit that interacts with itself repeatedly in the manner shown.

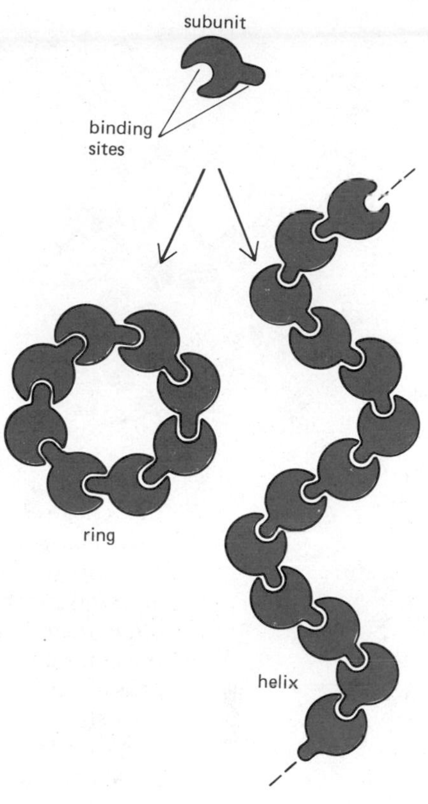

strands adds to the stability and strength of the assembly, since each subunit can interact with subunits in the opposite strand, as well as with its neighbors in the same strand. Molecules in which mechanical strength is especially important are usually made from fibrous rather than globular subunits, since these permit even more extensive regions of protein-protein contact (Figure 3–36).

Protein subunits may also assemble into flat sheets in which the subunits are arranged in hexagonal arrays. Specialized membrane transport proteins are sometimes arranged in this way in lipid bilayers, as, for example, in gap junctions between vertebrate cells. With a slight change in the geometry of the individual subunits, a hexagonal sheet can be converted into a tube (Figure 3–37); such cylindrical tubes are involved in the formation of the protein coats of some elongated viruses.

The formation of closed structures, such as rings or tubes or spheres, provides additional stability because it increases the total number of bonds that can form between the protein subunits. This principle is dramatically illustrated in the protein shells of many simple viruses that take the form of a hollow sphere. These coats are often made of hundreds of identical protein subunits that enclose and protect the viral nucleic acid inside (Figures 3–38 and 3–39). The protein in such coats must have a particularly adaptable structure, since it must make many different kinds of contacts and also change its arrangement to let the nucleic acid out and initiate viral multiplication.

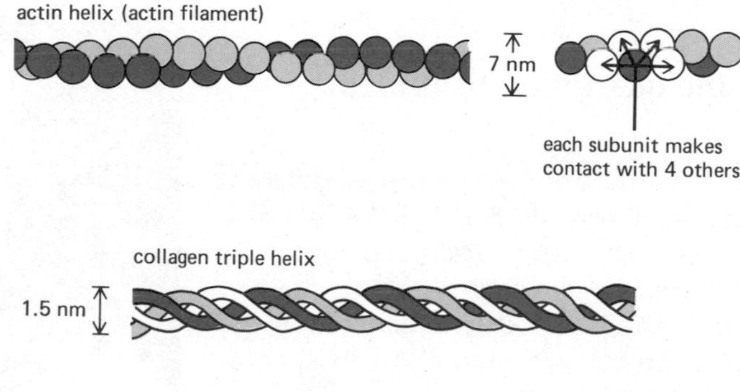

Figure 3–36 Two simple protein helices formed by cells. The actin helix contains a double strand of globular protein subunits in which each subunit makes contact with four other subunits. A collagen helix is composed of three extended protein chains that intertwine over an extensive distance to give a very strong rodlike structure. (Note the difference in scale between the two diagrams.)

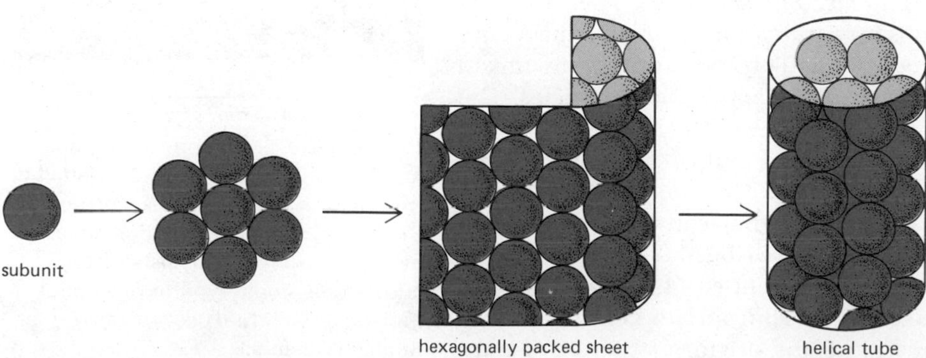

Figure 3–37 Hexagonally packed subunits can form a flat sheet or a tube.

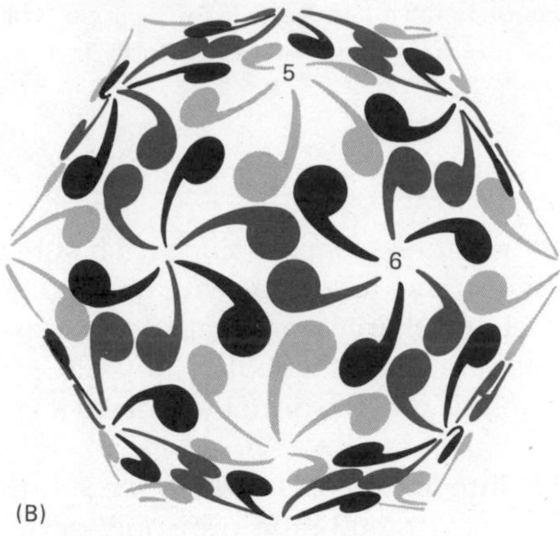

Figure 3–38 In some viruses, large numbers of identical protein subunits pack together to create a spherical shell that encloses the molecule of nucleic acid, either RNA or DNA. For geometrical reasons no more than 60 identical subunits can pack together in a precisely symmetrical way—an arrangement shown diagrammatically in (A)—and this limits the quantity of nucleic acid that is carried. However, if the subunits are not in precisely equivalent positions, then greater numbers may be incorporated into a larger shell. In (B) a shell is shown that is built from 180 identical subunits, each of which lies in one of three different "quasi-equivalent" packing arrangements. As may be seen in this figure, one color of subunit is arranged in a pentamer, while the other two colors alternate in a hexamer. In order to make such a shell, the protein subunit must be sufficiently flexible to fit into the three slightly different packing arrangements, shown here as three different colors of subunits. (Drawn from photographs supplied by Arthur J. Olson.)

Self-assembling Aggregates Can Include Different Protein Subunits and Nucleic Acids[18]

Many large cellular structures are constructed from a mixture of different proteins. In some cases, such as viruses and ribosomes, RNA or DNA may also form part of the structure. As for the simpler structures discussed above, the information for the assembly of many of these complex aggregates is contained in the macromolecular subunits themselves. This can be demonstrated by the fact that, under appropriate conditions, the isolated subunits can spontaneously assemble in a test tube into the final structure.

The first large macromolecular aggregate shown to be capable of self-assembly from its component parts was *tobacco mosaic virus*. This virus is a long rod in which a cylinder of protein is arranged around a helical RNA core (Figures 3–40 and 3–41). If the dissociated RNA and protein subunits are mixed together in solution, they recombine to form fully active virus particles. The assembly process is unexpectedly complex and involves the formation of double rings of protein that serve as *intermediates* that add to the growing virus coat.

Another complex macromolecular aggregate that can reassemble from its component parts is the bacterial ribosome. These ribosomes are composed of about 55 different protein molecules and 3 different rRNA molecules. If the 58 individual components are incubated under appropriate conditions in a test tube, they spontaneously re-form to give the original structure (Figure

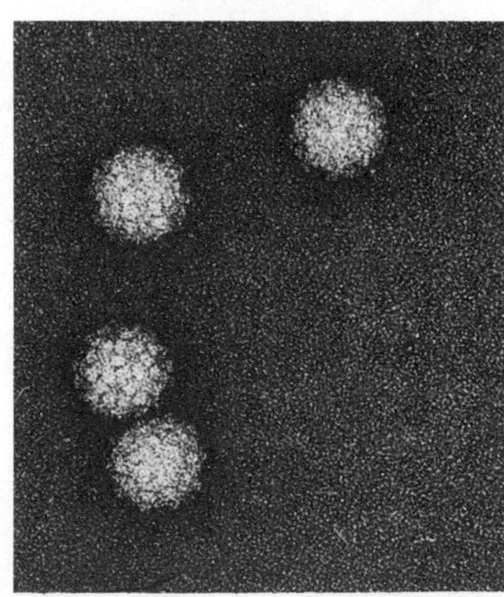

Figure 3–39 Electron micrograph of tomato bushy stunt virus. The shell of this virus is constructed according to the quasi-equivalent packing scheme in Figure 3–38B. It consists of 180 copies of a single type of protein of molecular weight 41,000. (Courtesy of Robley Williams.)

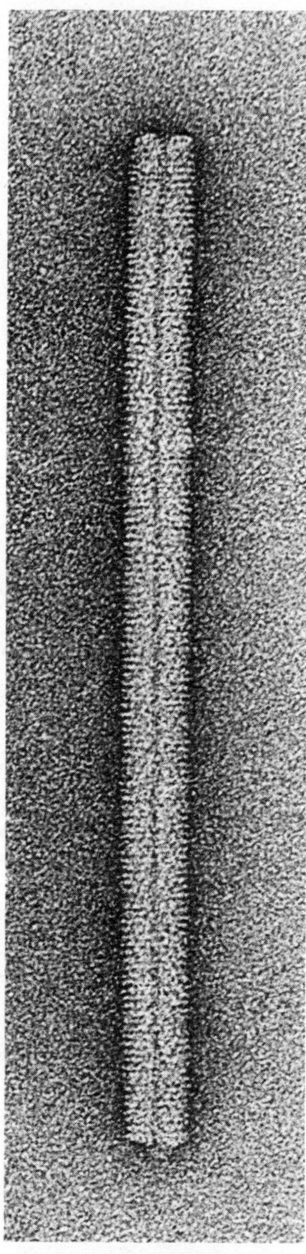

Figure 3–40 Electron micrograph of tobacco mosaic virus (TMV). This virus consists of a single long RNA molecule enclosed in a cylindrical protein coat that is composed of a tight helical array of identical protein subunits. Fully infective virus particles of normal appearance may be made in the test tube simply by mixing purified RNA and coat protein and allowing them to self-assemble. (Courtesy of Robley Williams.)

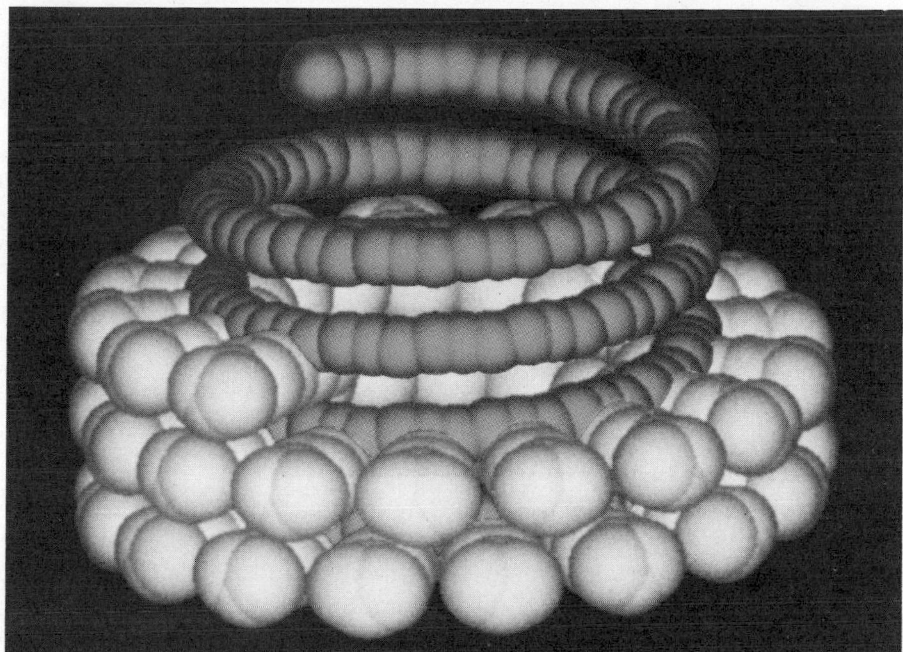

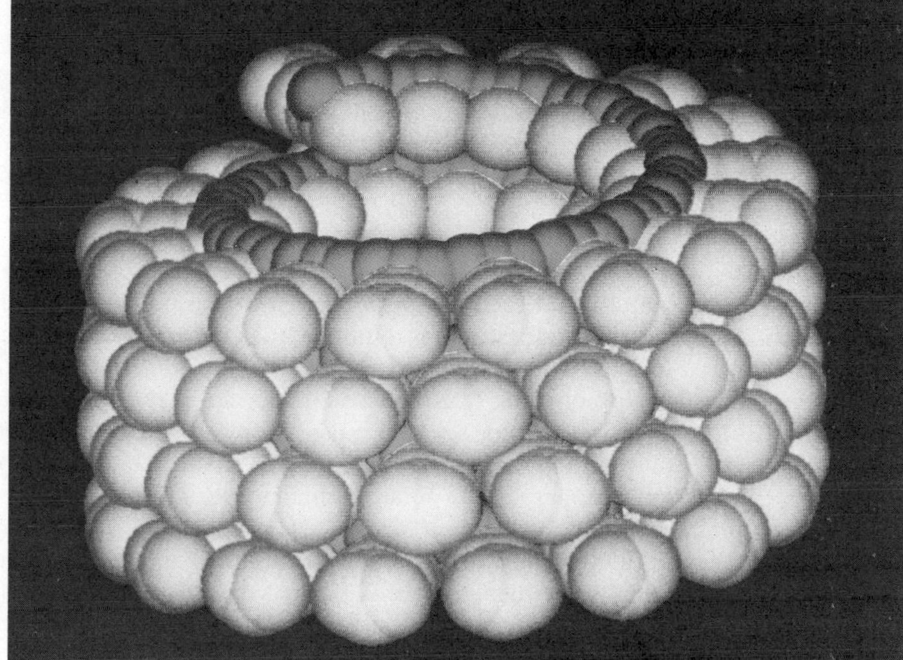

Figure 3–41 A model of tobacco mosaic virus showing two stages in its self-assembly. The coat protein molecules bind to the single-stranded RNA molecule *(color)*, forming a helical complex. (Courtesy of Richard J. Feldmann.)

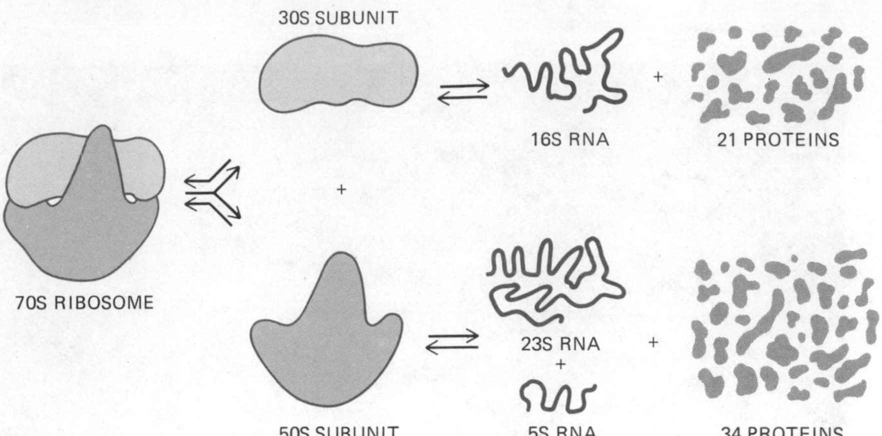

Figure 3–42 Self-assembly of bacterial ribosomes. If the 55 different proteins and 3 different RNA molecules are incubated under appropriate conditions, they reassemble into a functional ribosome.

3–42). Most important, such reconstituted ribosomes are able to carry out protein synthesis. As might be expected, the reassembly of ribosomes follows a specific pathway: certain proteins first bind to the RNA; this complex is then recognized by other proteins and so on until the final structure is made.

It is still not clear how some of the more elaborate self-assembly processes are regulated. For example, many structures in the cell appear to have a precisely defined length that is many times greater than that of their component macromolecules. How such length determination is achieved is a mystery, but one hypothetical mechanism, based on the vernier principle, is illustrated in Figure 3–43. Here, two rodlike molecules form a staggered complex. Because the molecules have slightly different lengths, there comes a point at which their ends precisely match. At this point further growth stops.

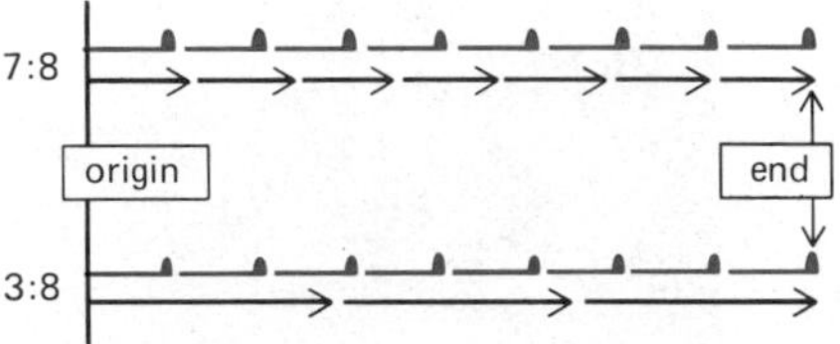

Figure 3–43 How the lengths of protein aggregates could, in theory, be determined by a vernier principle. Here black and colored rodlike molecules form a staggered complex that grows until their ends exactly match. In the top example this point is reached after exactly seven black and eight colored molecules interact; in the bottom example, after three black and eight colored molecules interact.

There Are Limits to Self-assembly[19]

It is logical to ask whether all cellular structures that are held together by noncovalent bonds are capable of self-assembly. Can a mitochondrion, a cilium, a myofibril, or even a whole cell form spontaneously out of a solution of their component macromolecules? The answer is no, because for large and complex structures part of the information for assembly is provided by special enzymes and by other cellular proteins that perform the function of jigs or templates that do not appear in the final assembled structure. Even some small structures lack some of the ingredients necessary for their own assembly. For example, in the formation of one small bacterial virus, the head structure, which is composed of a single protein subunit, is assembled on a scaffold composed of a second protein. Since this second protein does not appear in the final virus particle, the head structure cannot spontaneously reassemble once it is taken apart. Other examples are known in which proteolytic cleavage is an essential and irreversible step in the process of assembly. This is true for the coats of some bacterial viruses and even for some simple protein aggregates, including the structural protein collagen and the hormone insulin (Figure 3–44). From these relatively simple examples, it seems very likely that the assembly of a structure as complex as a mitochondrion or a cilium will involve both temporal and spatial ordering imparted by other cellular components, as well as irreversible processing steps catalyzed by degradative enzymes.

In the case of some complex organelles, it is probable that the information stored in the structure itself is required to construct a new copy. Thus the macromolecular components of an organelle—such as a mitochondrion or a Golgi complex—are normally assembled into the corresponding preexisting structures as the cell grows, by a process that involves the specific recognition of the organelle membrane (Chapter 7). If a cell were dissociated into its component macromolecules, essential information would probably be irretrievably lost, since the components of these organelles are unlikely to be capable of reassembling on their own. In this sense, therefore, it may not be accurate to say that *all* of the information needed to make a cell is contained in its DNA.

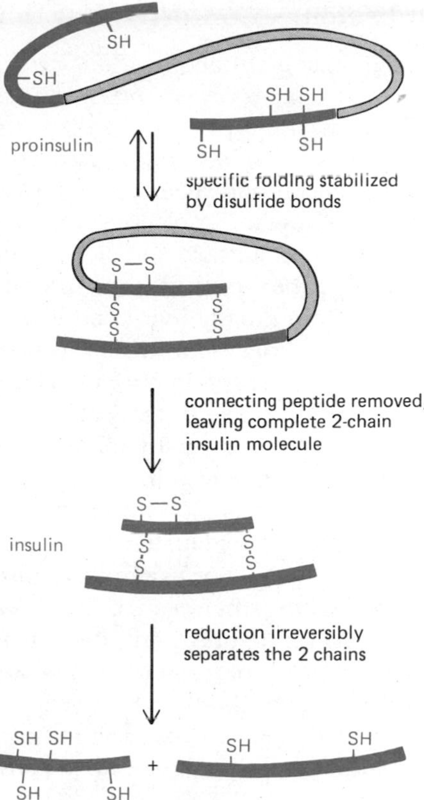

Figure 3–44 Because it is synthesized as a larger protein (*proinsulin*) that is cleaved by a proteolytic enzyme after it has folded into a specific shape, the polypeptide hormone insulin cannot spontaneously re-form its original conformation following the reduction of its disulfide bonds. Excision of part of the polypeptide chain of proinsulin causes an irretrievable loss of information needed for spontaneous reassembly of the molecule.

Summary

The three-dimensional conformation of a protein molecule is determined by its amino acid sequence. Particular folded structures are stabilized by noncovalent interactions between different parts of the polypeptide chain. The amino acids with hydrophobic side chains tend to cluster in the interior of the molecule, and local hydrogen-bond interactions between peptide backbones give rise to α-helices and β-sheets. Small globular regions known as domains are the modular units from which many proteins are constructed; they are linked together through short lengths of polypeptide chain to make a globular protein.

Proteins are brought together into aggregate structures by the same forces that determine protein folding. Proteins with binding sites for their own surface can assemble into dimers or larger oligomers, closed rings, spherical shells, or helical polymers. Mixtures of many different proteins, which can also include structural nucleic acids, can spontaneously assemble into large complex structures in the test tube. However, in many assembly processes irreversible steps occur so that not all structures in the cell are capable of spontaneous reassembly if they are dissociated into their component parts.

Protein Function[20]

The chemical properties of a protein molecule depend almost entirely on its exposed surface residues, which are able to form different types of weak noncovalent bonds with other molecules. An effective interaction of a protein molecule with another molecule (referred to as a **ligand**) requires that a number of weak bonds be formed simultaneously between them. Therefore, the only ligands that can bind tightly to a protein are those that fit precisely onto its surface.

The region of a protein that associates with a ligand, known as its **binding site,** usually takes the form of a cavity formed by a specific arrangement of amino acids on the protein surface. These amino acids often belong to widely separated parts of the polypeptide chain (Figure 3–45), and they represent only a minor fraction of the total amino acids present. The rest of the protein molecule is necessary to maintain the polypeptide chain in the correct position and to provide additional binding sites for regulatory purposes; the interior of the protein is often important only insofar as it gives the surface of the molecule the appropriate shape and rigidity.

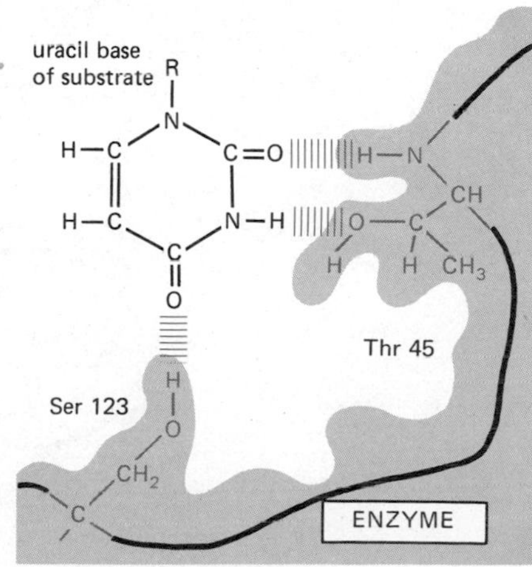

Figure 3–45 Hydrogen bonding between a protein and its ligand. The interaction shown here illustrates how the enzyme ribonuclease holds on to part of its RNA substrate, as determined by x-ray crystallographic analysis of an enzyme-substrate complex.

A Protein's Conformation Determines Its Chemistry

Neighboring surface residues on a protein often interact in a way that alters the chemical reactivity of selected amino acid side chains. These interactions are of several types.

First, neighboring parts of the polypeptide chain may interact in a way that restricts the access of water molecules to other parts of the protein surface. Since water molecules tend to form hydrogen bonds, they will compete with ligands for selected side chains on the protein surface. The tightness of hydrogen bonds (and ionic interactions) between proteins and their ligands is therefore greatly increased if water molecules are excluded. At first sight it is hard to imagine a mechanism that would exclude a molecule as small as water from a protein surface without affecting the access of the ligand itself. However, because of their strong hydrogen-bonding tendencies, water molecules exist in a large hydrogen-bonded network (Panel I, pp. 94–95), and it is often energetically unfavorable for individual molecules to break away from this network to reach into a crevice on the protein surface.

Second, the clustering of neighboring polar amino acid side chains alters their reactivity. For example, a number of negatively charged side groups can be forced together against their mutual repulsion by the way that the protein folds. When that happens, the affinity of each side chain for a positively charged ion is greatly increased. Selected amino acid side chains also interact through hydrogen bonds, thereby making normally unreactive side groups (such as the —CH_2OH on the serine shown in Figure 3–46) highly reactive so that they are able to enter into reactions that make or break selected covalent bonds.

The surface of each protein molecule, therefore, has a unique chemical reactivity that depends not only on which amino acid side chains are exposed but also on their exact orientation relative to each other. Because of this, even two slightly different conformations of the same protein molecule may differ drastically in their chemistry.

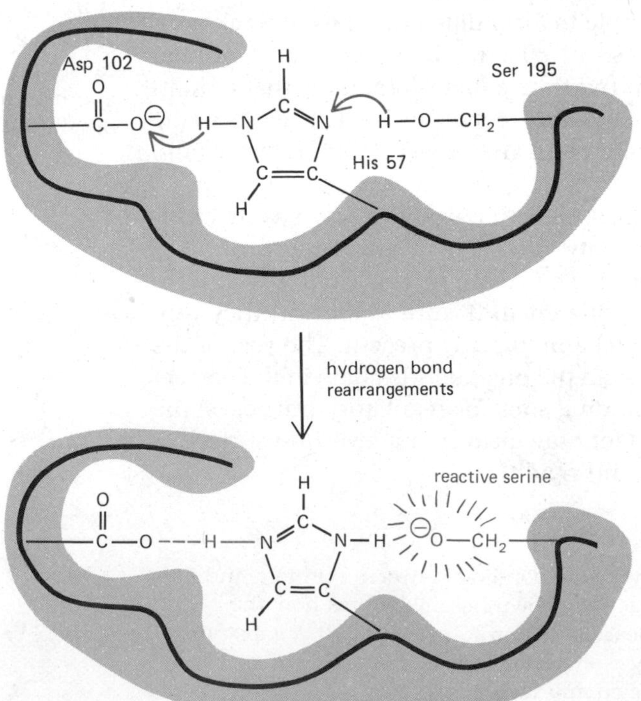

Figure 3–46 An unusually reactive amino acid at the active site of an enzyme. The example shown here is the "charge relay system" found in enzymes such as chymotrypsin and elastase that activates a serine at their active sites when a substrate is bound (see Figure 3–29). This serine then forms a transient covalent linkage with the enzyme substrate to catalyze hydrolysis of a peptide bond, as illustrated in Figure 3–49.

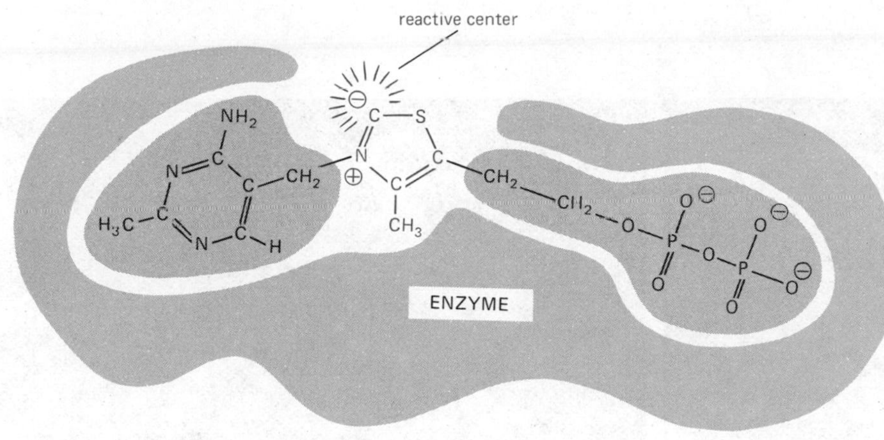

Figure 3–47 Coenzymes, such as thiamine pyrophosphate (TPP) depicted here, are small molecules that bind to an enzyme's surface and enable it to catalyze specific reactions. The reactivity of TPP centers on its "acidic" carbon atom, which readily exchanges its hydrogen atom for a carbon atom of a substrate molecule. Other regions of the TPP molecule probably act as "handles" by which the enzyme holds the coenzyme in the correct position.

Where side-chain reactivities are insufficient for the task at hand, proteins often enlist the help of selected nonpolypeptide molecules that the proteins bind to their surface. These ligands often serve as **coenzymes** in enzyme-catalyzed reactions, and they may be so tightly bound to the protein that they are effectively part of the protein itself. Examples are the iron-containing *hemes* in hemoglobin and cytochromes, *thiamine pyrophosphate* in enzymes involved in aldehyde-group transfers, and *biotin* in enzymes involved in carboxyl-group transfers (see p. 79). Each coenzyme has been selected for the unique chemical reactivity it confers when bound to a protein surface. Besides its reactive center, a coenzyme has other residues that bind it to its host protein (Figure 3–47). Coenzymes are often very complex organic molecules, and their exact chemistry when protein-bound is not always understood. Space-filling models of two enzymes bound to their coenzymes are shown in Figure 3–48A and B.

Substrate Binding Is the First Step in Enzyme Catalysis[21]

One of the most important functions of proteins is to act as enzymes that catalyze specific chemical reactions. The ligand in this case is the substrate molecule, and the binding of the substrate to the enzyme is an essential prelude to the chemical reaction (Figure 3–48C and D). Extremely high rates of chemical reactions are achieved by enzymes—far higher than for any man-made catalysts. This efficiency is attributable to several factors. The enzyme serves, first, to increase the local concentration of the substrate molecules at the catalytic site and to hold the appropriate atoms in the correct orientation for the reaction that is to follow. But, most important, some of the binding energy contributes directly to the catalysis: substrate molecules pass through a series of intermediate forms of altered geometry and electron distribution before forming the ultimate products of the reaction. The free energies of these intermediate forms, and especially of those in the most unstable *transition states,* are greatly reduced when the molecules are bound to the enzyme surface. Enzymes usually have a much greater affinity for the unstable transition states of substrates than for their stable forms. By using the energy available in this binding interaction, enzymes help their substrates attain a particular transition state, and thus greatly accelerate one particular reaction.

Some enzymes interact covalently with one or more of their substrates so that the substrate is linked to an amino acid (such as serine, cysteine, histidine, or lysine) or to a coenzyme molecule (such as pyridoxal phosphate). Such enzyme reactions often proceed in stages in which one substrate enters

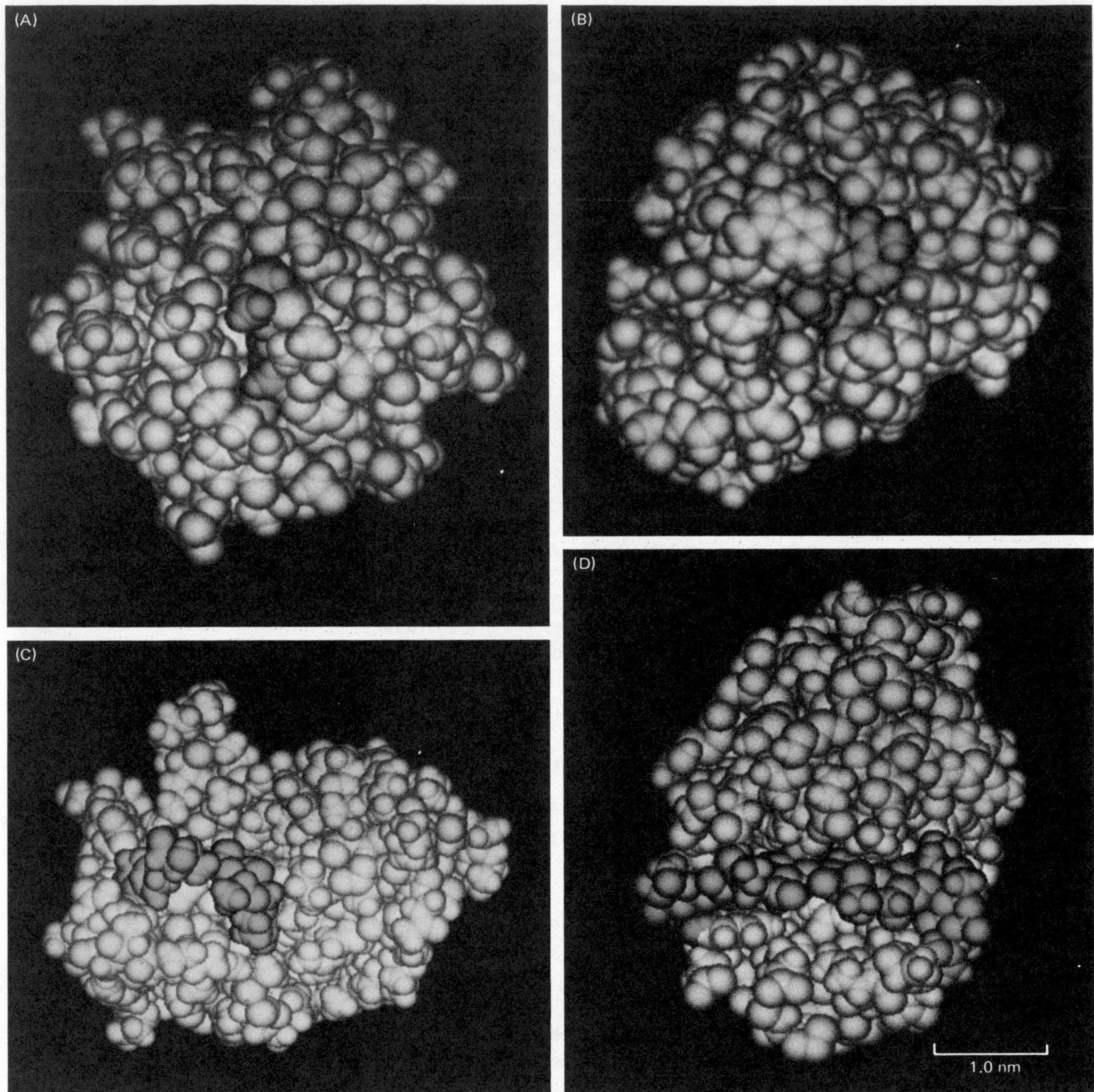

Figure 3–48 Computer-generated space-filling models of four different enzymes bound to a coenzyme or to a substrate. In each case, the bound ligand is shown in color: (A) cytochrome c with its bound heme, (B) flavodoxin with its bound flavin mononucleotide, (C) ribonuclease with a bound dinucleotide, and (D) egg-white lysozyme with a bound oligosaccharide. (Courtesy of Richard J. Feldmann.)

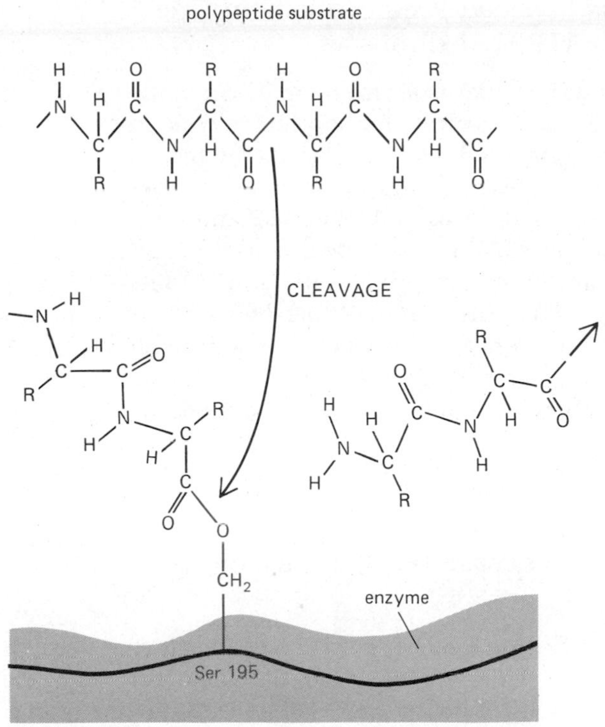

Figure 3–49 Enzymes sometimes form transient covalent bonds with their substrates. In the example shown here, a carboxyl group in a polypeptide chain forms a covalent bond at an activated serine residue of a protease as the backbone chain is broken (see Figure 3–46). When the unbound portion of the polypeptide chain has diffused away, a second step occurs (not shown) in which a water molecule hydrolyzes the newly formed covalent bond, thereby releasing the portion of the polypeptide bound to the enzyme surface and restoring a hydrogen atom to the serine at position 195.

the binding site, becomes covalently bound, and then reacts with a second substrate on the enzyme surface (Figure 3–49).

Because of the way that enzymes work, there is a limit to the amount of a substrate that a single enzyme can process in a given time. If the concentration of substrate is increased, the rate at which product is formed also increases up to a maximum value (Figure 3–50). At that point the enzyme molecule is saturated with substrate, and the rate of reaction depends only on how rapidly the substrate molecule can be processed. This rate is expressed as a *turnover number,* which for many enzymes is in the range of 1000 substrate molecules per second.

The other kinetic parameter frequently used to characterize an enzyme is its $\boldsymbol{K_M}$, which is the substrate concentration that allows one-half of the maximum rate of reaction (Figure 3–50). A *low* K_M value means that the enzyme reaches its maximum catalytic rate at a *low concentration* of substrate and generally indicates that the enzyme binds its substrate very tightly.

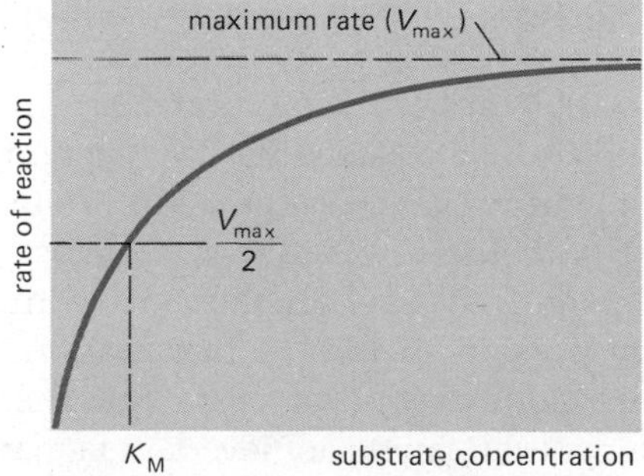

Figure 3–50 The rate of an enzyme reaction (V) varies with the substrate concentration, increasing with an increased concentration until a maximum value (V_{max}) is reached. At this point, all enzyme molecules are fully occupied and the rate of reaction is limited by the rate of the catalytic process on the enzyme surface. For most enzymes, the concentration of substrate at which the reaction rate is half maximal (K_M) is a measure of how tightly the substrates are bound, with a large value of K_M representing a weak binding, and vice versa.

Enzymes Accelerate Reaction Rates Without Shifting Equilibria

No matter how sophisticated an enzyme becomes, it cannot make the chemical reaction that it catalyzes either more or less energetically favorable. Like the simple binding interactions already discussed, any given chemical reaction has an *equilibrium point,* at which the backward and forward reaction fluxes are equal; therefore, no further net change occurs (see Figure 3–5). If an enzyme speeds up the rate of the forward reaction, A + B → AB, by a factor of 10^8, it must speed up the rate of the backward reaction, AB → A + B, by 10^8-fold as well. The *ratio* of the forward to the backward rate of reaction depends only on the concentrations of A, B, and AB. No combination of strong enzyme binding of the substrate molecules A and B and weak binding of the product molecule AB can alter this elementary chemical fact. The equilibrium point remains precisely the same whether or not the reaction is catalyzed by an enzyme.

Many Enzymes Make Reactions Proceed Preferentially in One Direction by Coupling Them to ATP Hydrolysis[22]

The living cell represents a chemical system that is far from equilibrium: the product of each enzyme usually serves as a substrate for another enzyme in the metabolic pathway and is rapidly consumed. More important, by means of enzyme-catalyzed pathways, previously described in Chapter 2, many reactions are driven by being *coupled* to the hydrolysis of ATP to ADP and inorganic phosphate (p. 66). To make this possible, the ATP pool is itself maintained at a level far from its equilibrium point, with a high ratio of ATP to its hydrolysis products (see Chapter 9, p. 498). This ATP pool thereby serves as a "storage battery" that keeps energy and atoms continually passing through the cell, directed along pathways that are determined by the enzymes present. For a living system, the approach to chemical equilibrium represents decay and death.

Multienzyme Complexes Help to Increase the Rate of Cell Metabolism[23]

Both the efficiency of enzymes in accelerating metabolic reactions and their organization in the cell are crucial to the maintenance of life. Cells in effect must race against the unavoidable processes of decay toward chemical equilibrium. If the forward rates of important reactions were not greater than the reverse rates, a cell would soon die. Some idea of the rate at which cellular metabolism proceeds can be obtained from the fact that a typical mammalian cell turns over (that is, completely degrades and replaces) its entire ATP pool once every one or two minutes. For each cell, this turnover represents the utilization of roughly 10^7 molecules of ATP per second (or, for the human body, about a gram of ATP every minute).

The rates of cellular reactions are so rapid because of the effectiveness of enzyme catalysts. Many important enzymes have become so efficient that any further increase in their efficiency is without effect because the rate-limiting steps in the reactions they catalyze are the steps in which the enzymes collide with their substrates; that is, the reaction rates are *diffusion-limited.*

If a reaction is diffusion-limited, its rate will depend on the concentration of both the enzyme and its substrate. For a sequence of reactions to occur very rapidly, each metabolic intermediate and enzyme involved must therefore be present in high concentration. Given the enormous number of different reactions carried out by a cell, there are limits to the concentrations of reactants that can be achieved. In fact, most metabolites are present in micromolar (10^{-6} M) concentrations, and most enzyme concentrations are much less than one micromolar. How is it possible, therefore, to maintain very fast metabolic rates?

The answer lies in the organization of cell components. Reaction rates can be increased without raising substrate concentrations by bringing the various enzymes involved in a reaction sequence together to form a large **multienzyme complex.** In this way, the product of enzyme A is passed directly to enzyme B and so on to the final product, and diffusion rates need not be limiting even when the concentration of substrate in a cell is very low. Such enzyme complexes are very common; the structure of one, pyruvate dehydrogenase, was shown in Figure 2–40. Large multienzyme complexes are also involved in RNA, DNA, and protein synthesis, although their structures are less well characterized.

Two additional mechanisms that have evolved to overcome kinetic problems in cells depend on intracellular membranes.

Intracellular Membranes Increase the Rates of Diffusion-limited Reactions[24]

The extensive intracellular membranes of eucaryotic cells act in at least two distinct ways to increase the rates of reactions that would otherwise be limited by the speed of diffusion. First, membranes can segregate certain substrates and all the enzymes that act on them into the same membrane-limited compartment, as in the mitochondrion or the cell nucleus. Assuming that each of these compartments occupies a total of 10% of the volume of a typical cell, the concentration of reactants in these organelles can be 10 times greater than in an identical cell with no compartmentalization (Figure 3–51). And as the total volume of each type of organelle compartment decreases relative to the total cell volume, the acceleration in the rate of its diffusion-limited reactions can increase accordingly.

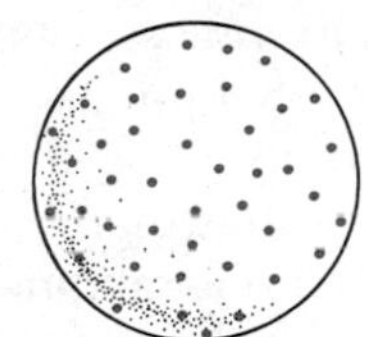
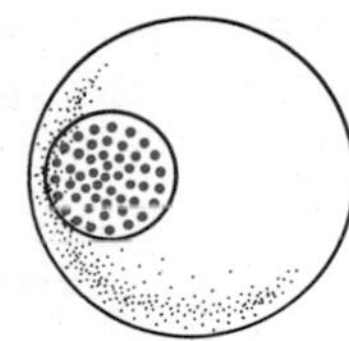

Figure 3–51 Membrane-limited compartments. A large increase in the concentration of interacting molecules is caused by confining them to one subcellular compartment in a eucaryotic cell.

Second, membranes can restrict the diffusion of reactants to the two dimensions of the membrane itself so that enzymes and their substrates are much more likely to collide with each other than if they were diffusing in three dimensions. Although the rate of diffusion of molecules in a membrane is slower by a factor of 100 than in aqueous solution, two reactants in a typical eucaryotic cell will come together much more rapidly if they are first bound to an internal membrane, such as the endoplasmic reticulum (Figure 3–52). The indicated membrane capture process seems certain to operate in the case of the enzymes and substrates involved in the synthesis of lipid molecules, where the substrates dissolve directly in the lipid bilayer; it probably also operates to accelerate many other reactions that utilize membrane-bound enzymes as well.

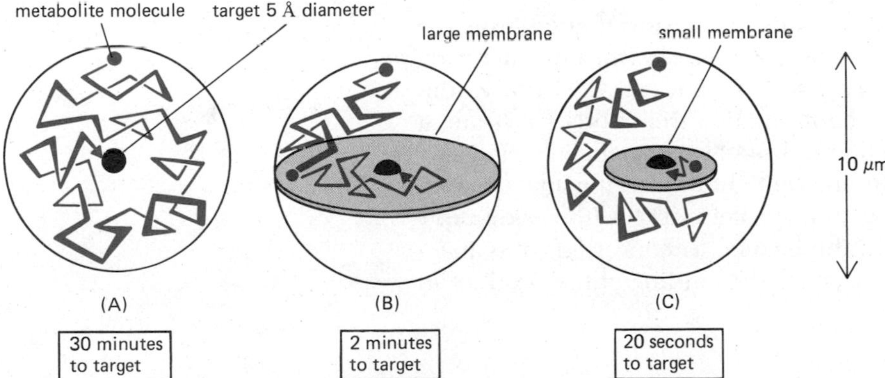

Figure 3–52 Membrane-accelerated collisions. Reaction rates increase when membranes convert diffusion in three dimensions to diffusion in two dimensions. The results of a series of theoretical calculations are shown here. (A) Diffusion without a membrane requires 30 minutes for an average molecule to find any single "target" inside a sphere of 10 μm. (B) The diffusion time is greatly reduced when the target is fixed in a membrane. Here it takes about 1 second for an average molecule to hit a large internal membrane, followed by a mean time of 2 minutes of diffusion in the membrane to find the target. (C) With an internal membrane that has a tenfold reduced surface area, an average molecule will require 10 seconds to find the membrane but will now hit the target by diffusion in the membrane about 10 times more quickly than in (B). Thus, the efficiency of the collision process is greatest in this smaller membrane, being increased by nearly 100-fold compared to the situation in (A).

Protein Molecules Can Reversibly Change Their Shape[25]

Because the mechanics of cellular life generally demand stable molecular structures, evolutionary pressures have worked against the ability of polypeptides to change randomly from one conformation to another. However, instead of eliminating this ability altogether, selective pressures have limited it in a highly specific way so that many (if not most) protein molecules are able to shift reversibly between several different but related stable conformations. Proteins with this property are known as **allosteric proteins.** Such a protein, for example, may be able to form within itself several alternative sets of hydrogen bonds of about equal energy, each alternative set requiring a change in the spatial relationships between different folds of the polypeptide chain. Only certain distinct conformations of the molecule are energetically favorable, and any intermediate conformations are unstable. The reason for distinct conformations is schematically illustrated in Figure 3–53; a conformation halfway between those denoted as A and B is unlikely since it would be unable to form either of the alternative sets of favorable hydrogen bonds.

Each distinct conformation of an allosteric protein has a somewhat different surface and thus a different ability to interact with other molecules. Often only one of two conformations has a high affinity for a particular ligand; in such a case the presence or absence of the ligand determines the conformation that the protein adopts (Figure 3–54). When there are two distinct ligands, each specific to a different surface of the same protein, the concentration of one molecule commonly changes the affinity of the protein for the other. Such allosteric changes are fundamental to the regulation of many biological processes.

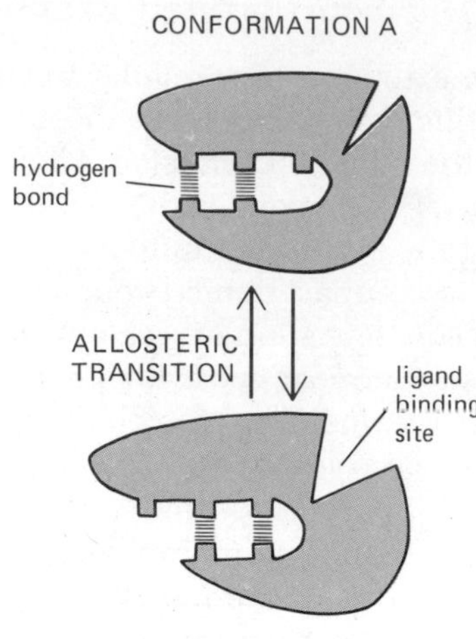

Figure 3–53 Alternative conformations for an allosteric protein. Conformation A and conformation B each contain two hydrogen bonds; therefore, they are more stable than any intermediate conformation.

Allosteric Proteins Are Involved in Metabolic Regulation

Allosteric proteins are essential to the **feedback regulation** that controls the flux through a metabolic pathway (p. 84). For example, enzymes that act early in a pathway are almost always allosteric proteins that can exist in two different conformations. One is the active conformation that binds substrate at its **active site** and catalyzes its conversion to the next substance in the pathway. The other is the inactive conformation that tightly binds the final product of the same pathway at a different place on the protein surface (the **regulatory site**). As the final product accumulates, the enzyme is converted to its inactivated conformation since this is stabilized by the binding of the product to

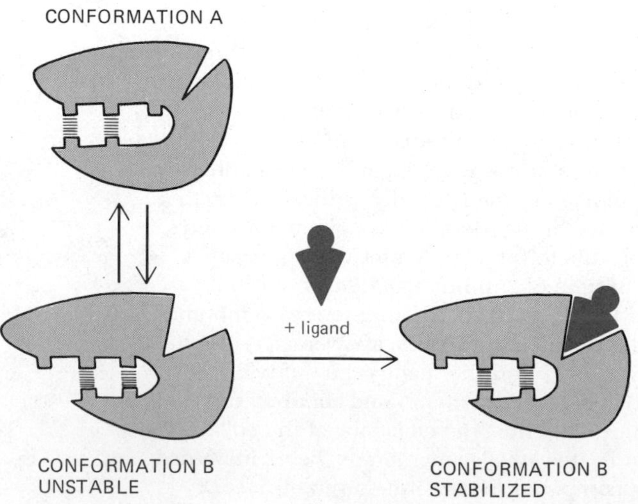

Figure 3–54 Stabilization of a protein conformation by the binding of a ligand. The tight binding of a ligand to only one conformation of an allosteric protein will shift it into the conformation that best binds the ligand.

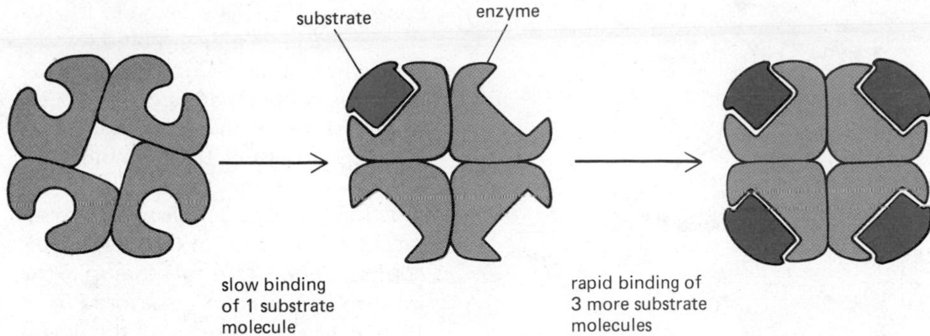

Figure 3–55 Cooperative binding of a substrate molecule to an enzyme. In an aggregate of allosteric subunits, the conformation of one subunit often influences that of its neighbors. In this way, the binding of a single substrate molecule to an enzyme may affect the affinity with which other substrate molecules bind.

the regulatory site (see Figure 3–54). In other cases, an enzyme involved in a metabolic pathway is *activated* by an allosteric transition that occurs when it binds a ligand that accumulates when the cell is deficient in a product of the pathway. In these ways, a cell makes a given product only when that product is needed, and a relatively constant concentration of all metabolites is automatically maintained.

Allosteric Proteins Are Vital for Cell Signaling[25]

As we have noted, allosteric proteins such as those involved in feedback regulation have at least two binding sites, one for the enzyme substrate and one or more for regulatory ligands. These sites occupy different regions of the protein surface and recognize totally different small molecules or ions. But, as we have just seen, the binding of a ligand to one site can affect another by changing the protein's conformation. This fact introduces a very powerful general principle of cell organization, for it means that any enzyme reaction or metabolic process can be regulated by any other in the cell, regardless of its chemical nature. For example, the production and breakdown of glycogen in muscle cells is linked to the concentration of Ca^{2+} by means of allosteric enzymes that alter their activity when the concentration of Ca^{2+} changes.

Allosteric proteins are especially effective signaling devices when, as is often the case, they exist as aggregates of identical subunits. The conformation of one subunit can then influence that of neighboring subunits, producing an effect similar to amplification. For example, a substrate molecule (or other ligand) bound to a site on one protein subunit can stabilize the conformation of other subunits and make it easier for other substrate molecules to bind to the enzyme (Figure 3–55). This behavior will produce a distinctive sigmoidal increase in enzyme activity as the substrate concentration is raised (Figure 3–56). Allosteric proteins of this kind act rather like a switch—flipping from one state to another with minimal time between the two (see Figure 13–38, p. 752).

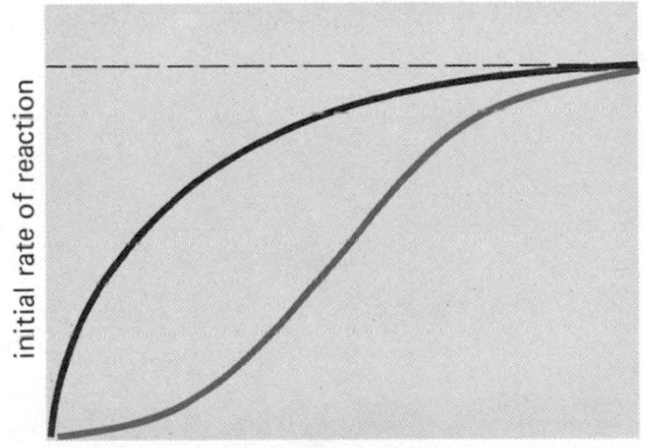

Figure 3–56 The activity of the allosteric enzyme illustrated in Figure 3–55 changes with substrate concentration as shown here. The "cooperative" binding of substrate molecules produces a distinctive sigmoidal response curve *(color)*, in contrast to the simple saturation kinetics of a nonallosteric enzyme *(black)*. (See also Figure 3–50.)

Proteins Can Be Pushed or Pulled into Different Shapes[25]

All of the directed movements in cells depend on forces generated by proteins. But how can a protein molecule be made to move in a controlled fashion? Before we answer this question, we must discuss some of the ways in which cells regulate the conformation of allosteric proteins. Consider an allosteric protein that can adopt two alternative conformations—a low-energy form C (inactive) and a high-energy form C* (active)—that differ in energy by about 4.3 kilocalories per mole (the energy available from forming about four hydrogen bonds on a protein surface). Given this energy difference, conformation C will be favored by about 1000 to 1 (see Table 3–3), and the protein will almost

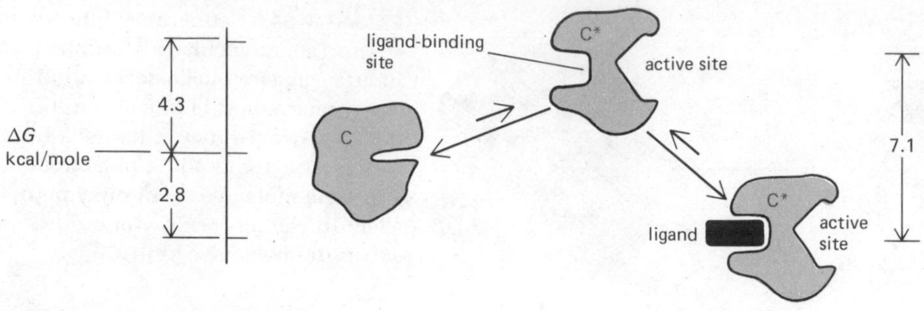

Figure 3–57 This schematic diagram depicts a ligand binding to the active conformation of an allosteric protein with a ΔG of -7.1 kilocalories per mole (equivalent to an affinity constant of 10^5 liters mole^{-1}). The ligand binding thereby "pulls" the enzyme from its normally favored inactive conformation C to its active conformation C*. In this example, the inactive conformation is favored by 1000 to 1 in the absence of the ligand, while the active conformation is favored by 100 to 1 in the presence of the ligand (see Table 3–3).

always be in its inactive conformation. However, there are two distinct ways in which the protein can be forced to adopt the active conformation.

The molecule can, in a sense, be "pulled" into the active C* conformation by binding to a low-molecular-weight ligand. Provided this ligand binds only to C*, the energy of this conformation will be selectively reduced without affecting that of C (Figure 3–57). Since the ligand binds relatively weakly to the protein (most of its binding energy having been used up in pulling the shape of the protein to fit the ligand), it can readily dissociate, and so this conformational change in the protein is perfectly reversible.

Alternatively, an input of chemical energy can be used to "push" conformation C to the active form C* in a much less reversible manner. A common mechanism involves the transfer of a phosphate group from ATP to a serine, threonine, or tyrosine residue in the protein, forming a covalent linkage. Suppose that this creates a charge repulsion unfavorable to conformation C. If this repulsion is reduced in the active C* form, then the change from C to C* will tend to occur (Figure 3–58). Controlled protein phosphorylation is commonly observed to activate or inhibit the function of specific proteins in eucaryotic cells (p. 743); in fact, about one-tenth of all the different proteins made in a mammalian cell contain covalently bound phosphate.

It is sometimes observed that ATP can be made *in vitro* by adding ADP to such a phosphorylated protein. When that occurs, it demonstrates directly that a great deal of the energy released by ATP hydrolysis has been stored in

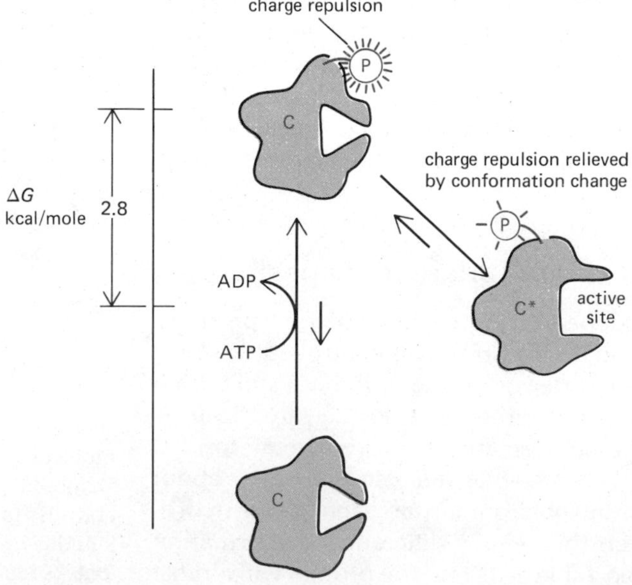

Figure 3–58 Many proteins are phosphorylated by ATP. In this diagram, the phosphorylation of the inactive form of an allosteric protein is imagined to produce an unfavorable charge repulsion, part of which is relieved by a shift to the active conformation C*. Phosphorylation has thereby "pushed" the enzyme into an active conformation. Note, however, that this is not the only way in which phosphorylation can act. In the best-studied case, that of glycogen phosphorylase, the phosphorylation instead creates a charge *attraction* that causes the allosteric change by bringing two separated parts of the protein together.

straining a protein's conformation during the initial phosphorylation event. But how do such energy-driven changes in protein conformation produce movement and, thereby, do useful work in a cell?

Energy-driven Changes in Protein Conformations Can Do Useful Work[26]

Suppose a protein is required to "walk" along a narrow thread, such as a microtubule or a DNA molecule. How an allosteric protein might do this by adopting different conformations is shown in Figure 3–59. With nothing to drive these conformational changes in an orderly way, the shape changes will be perfectly reversible and the protein will wander randomly back and forth along the thread or filament in an aimless fashion.

Since directional protein movement does net work, thermodynamic laws demand that it must deplete free energy from some other source (otherwise it could be used to make a perpetual motion machine). Therefore, no matter what modifications we make to the model in Figure 3–59, such as adding ligands that favor certain conformations, the protein molecule shown cannot go anywhere without an added source of energy.

What is needed is some means of making the series of protein conformation changes unidirectional. For example, the entire cycle would proceed in one direction if any one of the steps could be made irreversible. One way to do this is through the mechanism just discussed for driving allosteric changes in a protein molecule by a phosphorylation-dephosphorylation cycle. However, allosteric changes in proteins may also be driven by ATP hydrolysis without involving a phosphorylation of the protein. For example, in the modified cyclical walking scheme shown in Figure 3–60, ATP binding "pulls" the protein from conformation 1 to conformation 2; the bound ATP is then hydrolyzed to produce bound ADP and P_i, pushing the change from conformation 2 to conformation 3; and finally, the release of the bound ADP and P_i drives the protein back to conformation 1 again.

Because the transitions $1 \rightarrow 2 \rightarrow 3 \rightarrow 1$ are driven by the energy of ATP hydrolysis, this series will be effectively irreversible under physiological conditions (that is, the probability that ADP will recombine with P_i to form ATP by the route $1 \rightarrow 3 \rightarrow 2 \rightarrow 1$ is extremely low). This means that the entire cycle will go in only one direction and that the protein molecule will move continuously to the right in this schematic example. Examples of proteins that generate directional movement in this way include the important muscle protein *myosin* and *DNA helicase* (a protein that plays an essential part in DNA replication).

Similar mechanisms are employed by a large number of different protein machines to create coherent movement. All of these proteins have the ability to go through cyclic changes in shape during which they hydrolyze ATP; some are transiently phosphorylated in the process while others are not.

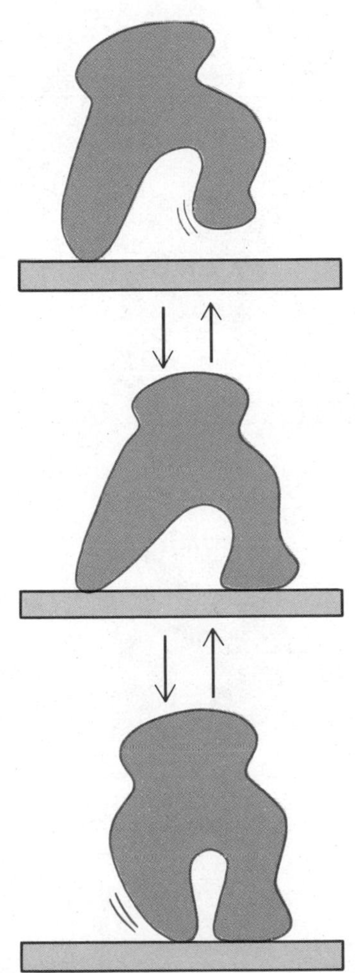

Figure 3–59 Schematic illustration of an allosteric "walking" protein. Although its three different conformations allow it to wander randomly back and forth while remaining bound to a filament, the protein cannot move uniformly in a single direction.

ATP-driven Membrane-bound Allosteric Proteins Can Act as Pumps[27]

Besides generating mechanical force, allosteric proteins can use the energy of ATP hydrolysis to do other forms of work, such as pumping specific ions into or out of the cell. For example, an allosteric protein, known as **Na^+-K^+ ATPase,** which is found in the plasma membrane of all animal cells, pumps $3Na^+$ out of the cell and $2K^+$ in during each cycle of conformational change driven by ATP-mediated phosphorylation. The importance of this ATP-driven pump is reflected in the fact that it consumes more than 30% of the total energy requirement of most cells. This frantic pumping of Na^+ and K^+ creates a cell

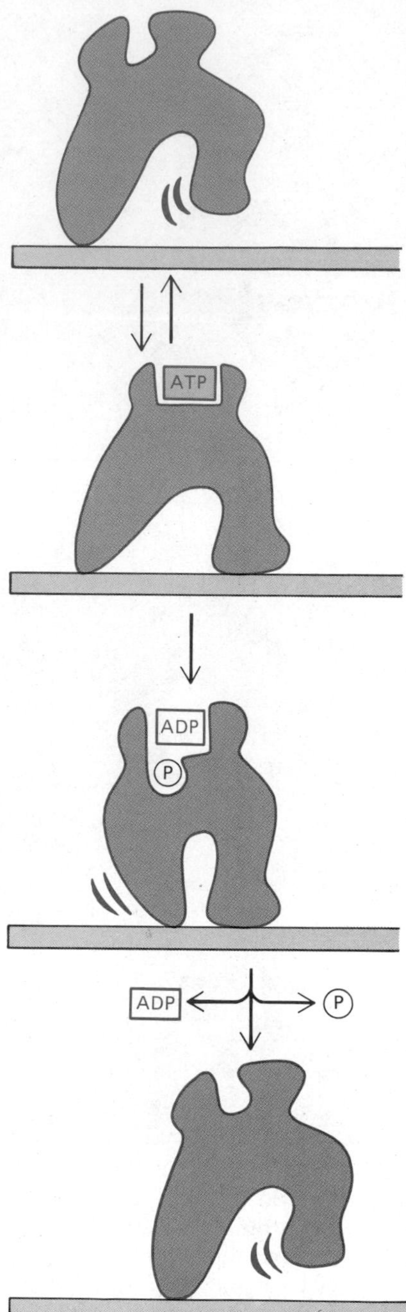

Figure 3–60 A "walking" protein in which an orderly transition between three different conformations is driven by the hydrolysis of a bound ATP molecule. One of these transitions is coupled to the hydrolysis of ATP, which makes the cycle essentially irreversible. In consequence, the protein moves consistently to the right along the filament.

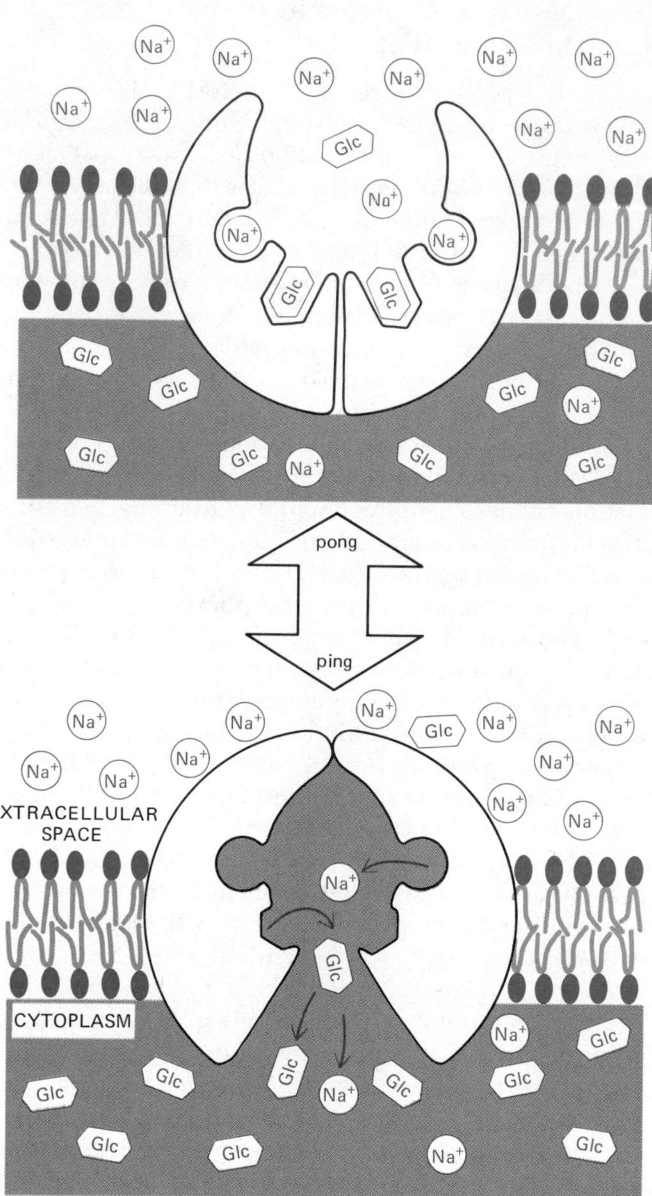

Figure 3–61 Schematic illustration of how a glucose pump could, in principle, be driven by a Na^+ gradient. The pump oscillates randomly between two alternate states, "pong" and "ping." In the "pong" state the protein is open to the extracellular space; in the "ping" state it is open to the cytoplasm. While Na^+ binds equally well to the protein in either state, the binding of Na^+ induces an allosteric transition in the protein that greatly increases its affinity for glucose. Since the Na^+ concentration is higher in the extracellular space than in the cytoplasm, glucose is much more likely to bind to the pump in the "pong" state; therefore, both Na^+ and glucose enter the cell (via a pong → ping transition) much more often than they leave it (via a ping → pong transition). As a result, the system carries both glucose and Na^+ into the cell.

interior in which the Na^+ concentration is low and the K^+ concentration is high with respect to the cell exterior, thereby generating two ion gradients across the plasma membrane, each the reverse of the other. As we shall now discuss, the energy stored in these and other ion gradients is, in turn, harnessed to drive conformational changes in a variety of other membrane-bound allosteric proteins, enabling them to do useful work for the cell.

Protein Molecules Can Harness Ion Gradients to Do Useful Work[27]

Although they are of enormous importance, ATP and the other nucleoside triphosphates are not the only sources of readily available energy for useful work. An important alternative exists in the ion gradients across various cell membranes, which can store and release energy in a fashion analogous to differences of water pressure on either side of a dam. For example, the large Na^+ gradient across the plasma membrane generated by the Na^+-K^+ ATPase is used to drive other plasma-membrane-bound protein pumps that transport glucose or amino acids into the cell.

The details of how such ion-driven protein pumps work are not yet known (and they seem certain to be complex). But the simple drawing in Figure 3–61 illustrates one possible way in which conformational changes in a plasma-membrane-bound protein might be driven by a Na^+ gradient, enabling the protein to pump glucose into a cell.

Membrane-bound allosteric pumps that are driven by the hydrolysis of ATP can also work in reverse and employ the energy in the ion gradient to synthesize ATP. In fact, we shall see in Chapter 9 that the energy available in the H^+ (proton) gradient across the inner mitochondrial membrane is used to synthesize most of the ATP used in the animal world by just such a mechanism.

Summary

The biological function of a protein depends on the detailed chemical properties of its surface. Binding sites are formed as surface cavities in which precisely positioned amino acid chains are brought together by protein folding. Enzymes catalyze chemical changes in bound substrate molecules, often employing small, tightly bound coenzyme molecules to extend the range of their reactions. Rates of enzyme reactions are often limited by diffusion and are increased when both enzyme and substrate are confined to the same small compartment of the cell.

Allosteric proteins reversibly change their shape when ligands bind to their surface. The changes produced by one ligand can affect the binding of a second ligand, thereby providing a mechanism for regulating various cell processes. Such changes in protein shape are often driven in an unidirectional manner by the expenditure of chemical energy. For example, by coupling allosteric changes to ATP hydrolysis, proteins can do useful work—such as generating a mechanical force or pumping ions across a membrane.

References

General

Judson, H.F. The Eighth Day of Creation: Makers of the Revolution in Biology. New York: Simon & Schuster, 1979. (A scholarly account of the history of molecular biology.)

Lehninger, A.L. Principles of Biochemistry. New York: Worth, 1982.

Schulz, G.E.; Schirmer, R.H. Principles of Protein Structure. New York: Springer, 1979.

Stryer, L. Biochemistry, 2nd ed. San Francisco: Freeman, 1981.

Watson, J.D. Molecular Biology of the Gene, 3rd ed. Menlo Park, Ca.: Benjamin-Cummings, 1976.

Cited

1. Cantor, C.R.; Schimmel, P.R. Biophysical Chemistry, Part I and Part III. San Francisco: Freeman, 1980.

   Eisenberg, D.; Crothers, D. Physical Chemistry with Applications to the Life Sciences. Menlo Park, Ca.: Benjamin-Cummings, 1979.

   Pauling, L. The Nature of the Chemical Bond, 3rd ed. Ithaca, N.Y.: Cornell University Press, 1960. (A classical account of the principles of interatomic bonding in organic molecules.)

2. Stent, G.S. Molecular Genetics: An Introductory Narrative. San Francisco: Freeman, 1971.

   Olby, R. The Path to the Double Helix. Seattle: University of Washington Press, 1974.

3. Watson, J.D.; Crick, F.H.C. Molecular structure of nucleic acids. A structure for deoxyribose nucleic acid. *Nature* 171:737–738, 1953.

4. Watson, J.D.; Crick, F.H.C. Genetical implications of the structure of deoxyribonucleic acid. *Nature* 171:964–967, 1953.

   Meselson, M.; Stahl, F.W. The replication of DNA in *E. coli. Proc. Natl. Acad. Sci. USA* 44:671–682, 1958.

   Sanger, F.; et al. Nucleotide sequence of bacteriophage ϕX174 DNA. *Nature* 265:687–695, 1977.

5. Drake, J.W. The Molecular Basis of Mutation. San Francisco: Holden-Day, 1970.

6. Thompson, E.O.P. The insulin molecule. *Sci. Am.* 192(5):36–41, 1955. (Review of the first sequence determination.)

   Yanofsky, C. Gene structure and protein structure. *Sci. Am.* 216(5):80–94, 1967. (The evidence for colinearity.)

7. Brenner, S.; Jacob, F.; Meselson, M. An unstable intermediate carrying information from genes to ribosomes for protein synthesis. *Nature* 190:576–581, 1961.

8. Crick, F.H.C. The genetic code: III. *Sci. Am.* 215(4):55–62, 1966.

9. Rich, A.; Kim, S.H. The three-dimensional structure of transfer RNA. *Sci. Am.* 238(1):52–62, 1978.

10. Lake, J.A. The ribosome. *Sci. Am.* 245(2):84–97, 1981.

11. Schulz, G.E.; Schirmer, R.H. Principles of Protein Structure. New York: Springer, 1979. (Chapters 1 through 5.)

    Cantor, C.R.; Schimmel, P.R. Biophysical Chemistry. Part I. The Conformation of Biological Macromolecules. San Francisco: Freeman, 1980. (Chapters 2 and 5.)

    Dickerson, R.E.; Geis, I. The Structure and Action of Proteins. New York: Harper & Row, 1969.

12. Anfinsen, C.B. Principles that govern the folding of protein chains. *Science* 181:223–230, 1973.

13. Richardson, J.S. The anatomy and taxonomy of protein structure. *Adv. Protein Chem.* 34:167–339, 1981.

    Pauling, L.; Corey, R.B.; Branson, H.R. The structure of proteins: two hydrogen-bonded helical configurations of the polypeptide chain. *Proc. Natl. Acad. Sci. USA* 37:205–211, 1951.

    Pauling, L.; Corey, R.B. Configurations of polypeptide chains with favored orientations around single bonds: two new pleated sheets. *Proc. Natl. Acad. Sci. USA* 37:729–740, 1951.

14. Hartley, B.S. Homologies in serine proteases. *Philos. Trans. R. Soc. Lond. (Biol.)* 257:77–87, 1970.

    Smith, E.L. Evolution of enzymes. In The Enzymes, 3rd ed., Vol. 1 (P.D. Boyer, ed.), pp. 267–339. New York: Academic Press, 1970.

    Doolittle, R.F. Protein evolution. In The Proteins, 3rd ed., Vol. 4 (H. Neurath, R.L. Hill, eds.), p. 1–118. New York: Academic Press, 1979.

15. Rossmann, M.G.; Argos, P. Protein folding. *Annu. Rev. Biochem.* 50:497–532, 1981.
Blake, C.C.F. Do genes-in-pieces imply proteins-in-pieces? *Nature* 273:267, 1978.
Banks, R.D.; et al. Sequence, structure and activity of phosphoglycerate kinase: a possible hinge-bending enzyme. *Nature* 279:773–777, 1979.

16. Metzler, D.E. Biochemistry. New York: Academic Press, 1977. (Chapter 4 describes how macromolecules pack together into large assemblies.)

17. Caspar, D.L.D.; Klug, A. Physical principles in the construction of regular viruses. *Cold Spring Harbor Symp. Quant. Biol.* 27:1–24, 1962.
Harrison, S.C. Structure of simple viruses: specificity and flexibility in protein assemblies. *Trends Biochem. Sci.* 3:3–7, 1978.
Harrison, S.C. Virus crystallography comes of age. *Nature* 286:558–559, 1980.

18. Fraenkel-Conrat, H.; Williams, R.C. Reconstitution of active tobacco mosaic virus from its inactive protein and nucleic acid components. *Proc. Natl. Acad. Sci. USA* 41:690–698, 1955.
Nomura, M. Assembly of bacterial ribosomes. *Science* 179:864–873, 1973.
Butler, P.J.G.; Klug, A. The assembly of a virus. *Sci. Am.* 239(5):62–69, 1978.
King, J. Regulation of structural protein interactions as revealed in phage morphogenesis. In Biological Regulation and Development, Vol. 2 (R.F. Goldberger, ed.), pp. 101–132. New York: Plenum, 1980.

19. Steiner, D.F.; Kemmler, W.; Tager, H.S.; Peterson, J.D. Proteolytic processing in the biosynthesis of insulin and other proteins. *Fed. Proc.* 33:2105–2115, 1974.

20. Metzler, D. Biochemistry: The Chemical Reactions of a Living Cell. New York: Academic Press, 1977. (Chapter 6 describes the general features of enzyme catalysis and of allosteric proteins.)
Fersht, A. Enzyme Structure and Mechanism. San Francisco: Freeman, 1977.

21. Wolfenden, R. Analog approaches to the structure of the transition state in enzyme reactions. *Accounts Chem. Res.* 5:10–18, 1972.

22. Wood, W.B.; Wilson, J.H.; Benbow, R.M.; Hood, L.E. Biochemistry, A Problems Approach, 2nd ed. Menlo Park, Ca.: Benjamin-Cummings, 1981. (Chapters 9 and 15 and associated problems.)

23. Reed, L.J.; Cox, D.J. Multienzyme complexes. In The Enzymes, 3rd ed., Vol. 1 (P.D. Boyer, ed.), pp. 213–240. New York: Academic Press, 1970.
Reed, L.J. Multienzyme complexes. *Accounts Chem. Res.* 7:40–46, 1974.
Srere, P.A.; Mosbach, K. Metabolic compartmentation: symbiotic, organellar, multienzymic, and microenvironmental. *Annu. Rev. Microbiol.* 28:61–83, 1974.

24. Adam, G.; Delbrück, M. Reduction of dimensionality in biological diffusion processes. In Structural Chemistry and Molecular Biology (A. Rich and N. Davidson, eds.), pp. 198–215. San Francisco: Freeman, 1968.

25. Monod, J.; Changeux, J.-P.; Jacob, F. Allosteric proteins and cellular control systems. *J. Mol. Biol.* 6:306–329, 1963.
Edelstein, S.J. Introductory Biochemistry. San Francisco: Holden-Day, 1973. (Chapter 10 on protein aggregates and allosteric interactions.)
Cantor, C.R.; Schimmel, P.R. Biophysical Chemistry. Part III: The Behavior of Biological Macromolecules. San Francisco: Freeman, 1980. (Chapters 15 and 17.)

26. Hill, T.L. A proposed common allosteric mechanism for active transport, muscle contraction, and ribosomal translocation. *Proc. Natl. Acad. Sci. USA* 64:267–274, 1969.
Hill, T.L. Biochemical cycles and free energy transduction. *Trends Biochem. Sci.* 2:204–207, 1977.

27. Kyte, J. Molecular considerations relevant to the mechanism of active transport. *Nature* 292:201–204, 1981.
Hokin, L.E. The molecular machine for driving the coupled transports of Na^+ and K^+ is an (Na^+ + K^+)-activated ATPase. *Trends Biochem. Sci.* 1:233–237, 1976.
Hopfer, U.; Groseclose, R. The mechanism of Na^+-dependent D-glucose transport. *J. Biol. Chem.* 255:4453–4462, 1980.

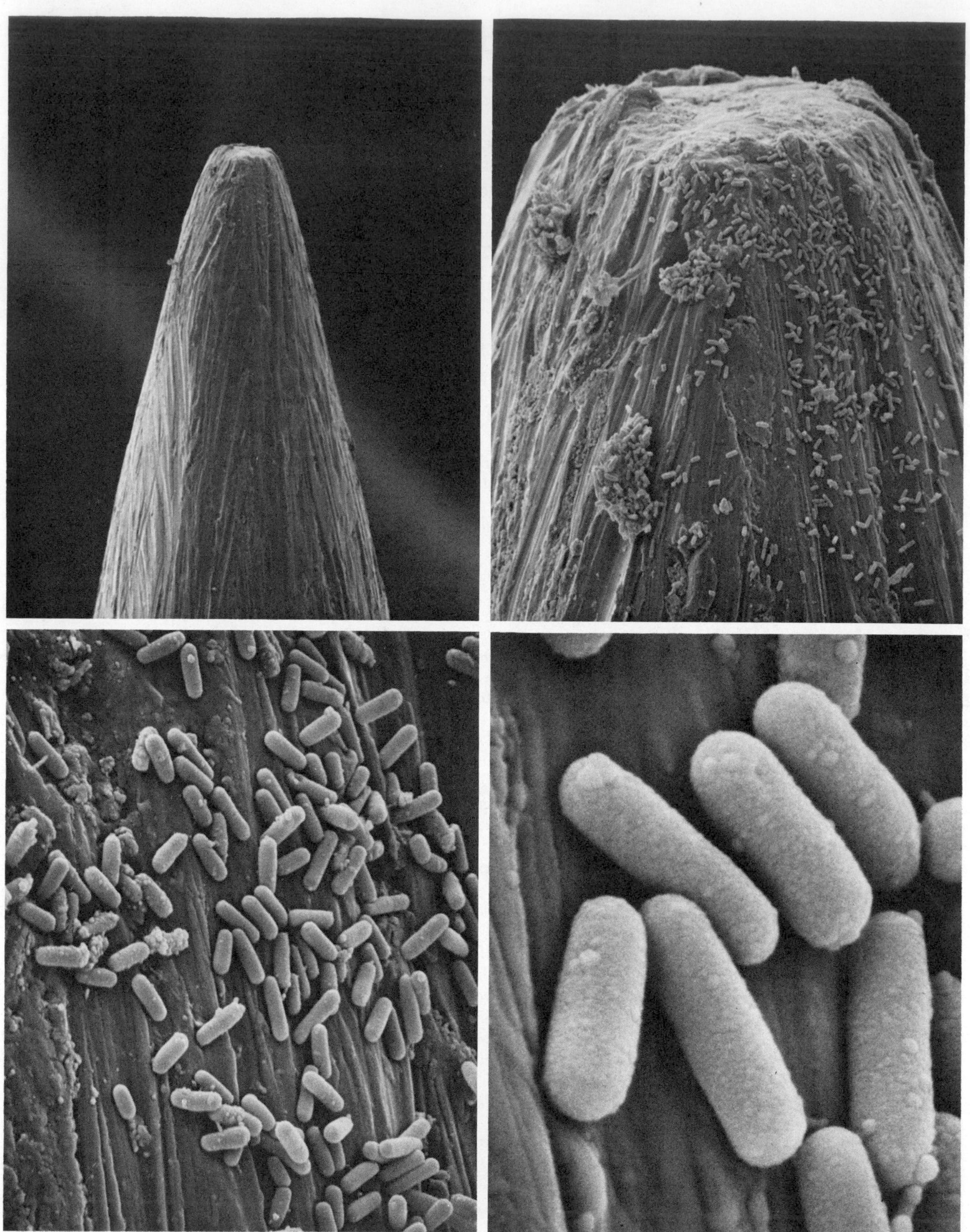

A sense of scale. These scanning electron micrographs, taken at progressively higher magnifications, show bacterial cells on the point of an ordinary domestic pin. (Courtesy of Tony Brain and the Science Photo Library.)

How Cells Are Studied

4

Cells are small and complex: it is hard to see their structure, hard to discover their molecular composition, and harder still to find out how their various components function. An enormous variety of experimental techniques have been developed to study cells, and the strengths and limitations of these techniques have largely determined our present conception of the cell. Most advances in cell biology—including the most exciting ones of recent years—have sprung from the introduction of new methods. To understand cell biology, therefore, one must understand something of its experimental techniques.

In this chapter, we shall briefly review the most important methods of contemporary cell biology and explain the principles and reasoning that underlie them. Many of these methods will be referred to in later chapters, where in some cases they will be discussed in greater detail.

Microscopy[1]

A typical animal cell is 10 to 20 μm in diameter, or about five times smaller than the smallest visible particle. It was, therefore, not until good light microscopes became available, in the early part of the nineteenth century, that plant and animal tissues were discovered to be aggregates of individual cells. This discovery, proposed as the **cell doctrine** by Schleiden and Schwann in 1835, marks the formal birth of cell biology.

Animal cells are not only tiny, they are colorless and translucent; consequently, the discovery of their fine structural details and their larger internal organelles was possible only with the development, in the latter part of the nineteenth century, of a variety of stains that provided sufficient contrast to render them visible. Eventually, with the development of the electron microscope in the early 1940s and the associated techniques for the preservation and staining of cells, the full extent of the internal complexity of the cell began to emerge.

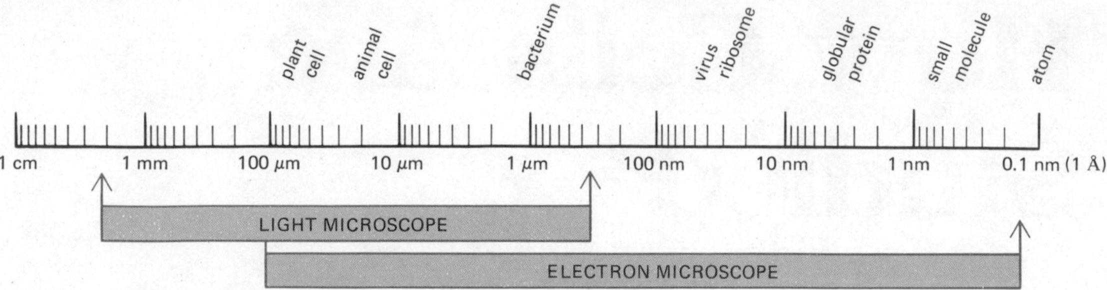

Figure 4–1 Sizes of cells and their components drawn on a logarithmic scale, together with the usable range of the light and electron microscopes. The following units of length are commonly employed in microscopy:

μm (micrometer) = 10^{-6} m
nm (nanometer) = 10^{-9} m
Å (Ångström unit) = 10^{-10} m

Figure 4–1 compares the range of resolution of modern light and electron microscopes. The appearance of cells and their contents in such microscopes depends on the behavior of light and electrons and on the response of cells to different staining and preparative procedures. These topics are discussed in the following sections.

The Light Microscope Can Resolve Details 0.2 μm Apart[2]

In general, radiation of a given wavelength cannot be used to probe structural details much smaller than its own wavelength: this represents a fundamental limitation of all microscopes. The ultimate limit to the resolution of a light microscope is, therefore, set by the wavelength of light, which ranges from about 0.4 μm to 0.7 μm, depending on its color. In practical terms, bacteria and mitochondria, which are about 500 nm (0.5 μm) wide, are the smallest objects that can be clearly seen in the light microscope; details smaller than this are obscured by optical diffraction effects. To understand why, we must follow what happens to a beam of light waves as it passes through the lenses of a microscope.

Optical diffraction is due to the interference between waves that have traveled by slightly different paths through an optical system. If two trains of waves are precisely in phase, with crest matching crest and trough matching trough, they will reinforce each other so as to increase brightness. On the other hand, if the trains of waves are out of phase, they will interfere with each other in such a way as to cancel each other (Figure 4–2). The interaction of light with an object will change the phase relationships of the light waves and produce such interference effects. For example, the shadow of a straight edge illuminated with light of uniform wavelength appears at high magnification as a set of parallel lines, while that of a circular spot appears as a set of concentric rings (Figure 4–3). For the same reason, a single point seen through a microscope appears as a blurred disc, and two point objects close together give overlapping images and merge into one. No amount of refinement of the lenses can overcome this limitation imposed by the wavelike nature of light. Thus, for visible light, the limiting separation at which two objects can still be seen as distinct—the so-called **limit of resolution**—is 0.2 μm. This limit was achieved by microscope makers at the end of the nineteenth century and is only rarely matched in contemporary, factory-produced microscopes. Although it is possible to *enlarge* an image as much as one

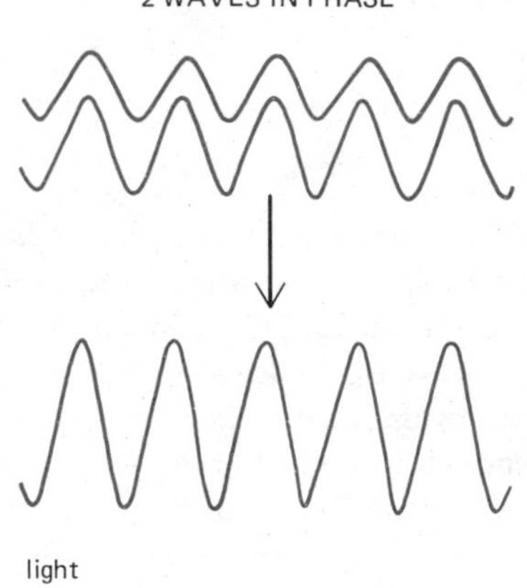

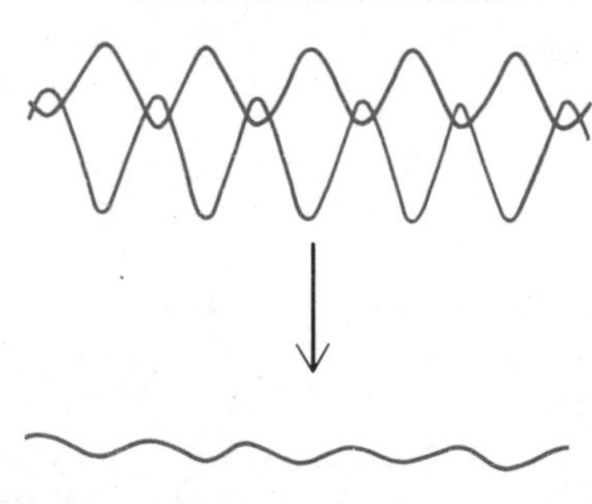

Figure 4–2 Interference between light waves. When two light waves combine *in phase,* the amplitude of the resultant wave is larger and the brightness is increased. Two light waves *out of phase* cancel each other out and produce a wave whose amplitude, and therefore brightness, is decreased.

wants—for example, by projecting it onto a screen—it is never possible to see objects in the light microscope with finer detail than about 0.2 μm.

Diffraction is not always a barrier to the study of cells: on the contrary, we shall see later how interference due to diffraction can be exploited to study living cells in the microscope and to reveal regular periodic features in biological structures. But first we shall discuss the use of chemical stains to enhance the visibility of cells.

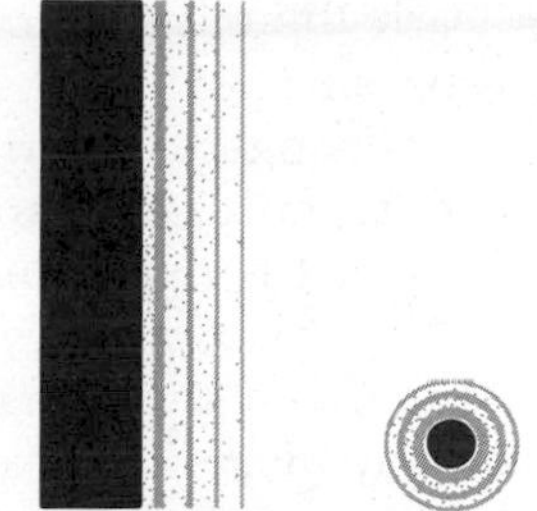

Figure 4–3 The interference effects observed at high magnification when light passes the edges of a solid object placed between the light source and observer.

Different Components of the Cell Can Be Selectively Stained[1,3]

There is little in the contents of most cells (which are 70% water by weight) to impede the passage of light rays. Thus most untreated cells are almost invisible in an ordinary light microscope. One way to make them visible is to stain them with selective organic dyes.

In the early nineteenth century, the demand for dyes to stain textiles led to a fertile period for organic chemistry. Some of the dyes were found to stain biological tissues and, unexpectedly, often showed a preference for particular parts of the cell—the nucleus or membranes, for example—allowing the internal structure of the cell to be seen with greater clarity. Today a rich variety of organic dyes are available, with such colorful names as *Malachite green, Sudan black,* and *Coomassie blue,* each of which has some specific affinity for particular subcellular components. The dye *hematoxylin,* for example, has an affinity for negatively charged molecules and therefore reveals the distribution of DNA and RNA in a cell. The chemical basis for the specificity of many dyes, however, is not known.

More recently, staining reagents have become available that are much more specific than organic dyes. For example, some enzymes can be located in cells by exposing them to substrate molecules that produce a localized visible product on reaction with the specific enzyme. Antibodies against specific macromolecules can be covalently coupled to such an enzyme, or to a fluorescent dye, and then used to mark the distribution of the macromolecule in the cell.

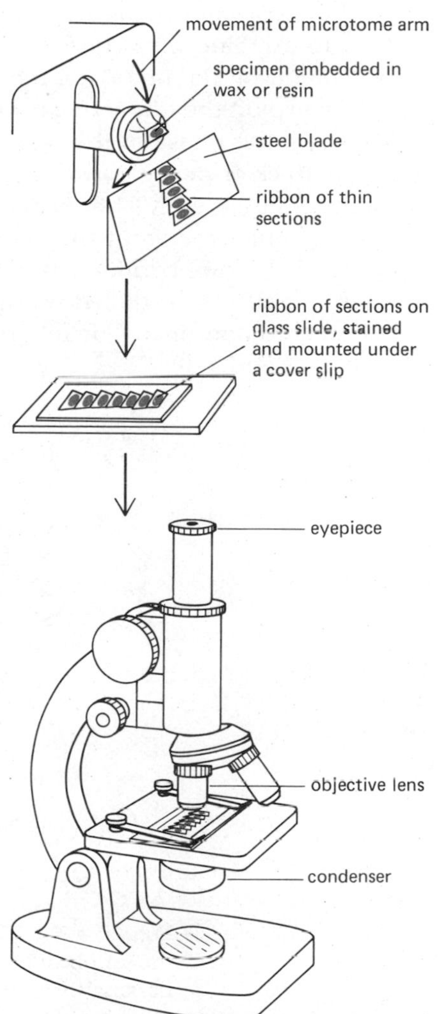

Figure 4–4 Schematic diagram showing how an embedded tissue is sectioned with a microtome in preparation for examination in the light microscope.

Tissues Are Usually Fixed and Sectioned for Microscopy

Before they can be stained, most biological tissues must be fixed. **Fixation** makes cells permeable to the stains and cross-links their macromolecules so that they are stabilized and locked in position. Some of the earliest fixation procedures involved a brief immersion in acids or organic solvents such as methanol, while current procedures usually include exposure of the cell to reactive aldehydes, particularly formaldehyde and glutaraldehyde. The latter compounds form covalent bonds with the free amino groups of proteins and thereby link adjacent molecules together.

Fixation and staining are not the only preparative steps used by microscopists. Most tissues are too thick to allow their cells to be examined directly with high resolution. The tissues are, therefore, first cut into thin slices (**sections**), each of which is then laid flat on the surface of a glass microscope slide. Sections are cut with a *microtome,* a machine that operates rather like a meat slicer. The sections for viewing in a light microscope are typically 1 to 10 μm thick and cut with a sharp metal blade (Figure 4–4).

Tissues are generally too soft to be cut into thin sections directly. Therefore, after fixation they are usually **embedded** in a liquid wax or plastic resin that permeates and surrounds the entire tissue. This embedding medium is then hardened (by cooling or by polymerization) to a solid block, which is readily sectioned by the microtome.

Clearly there is a serious danger that any treatment used for fixation may distort the structure of the cell. An alternative method of preparation that lessens this danger is rapid freezing, which obviates either fixation or embedding. The frozen tissue can be cut directly with a special microtome (called a cryostat) that is maintained in a cold chamber. Although **frozen sections** produced in this way have the advantage that they represent a more native form of the tissue, they are difficult to prepare and the presence of ice crystals causes many morphological details to be lost.

Living Cells Can Be Seen in a Phase-Contrast or Differential-Interference-Contrast Microscope[2]

The possibility that some components of the cell are lost or distorted during fixation and staining has never ceased to worry microscopists. Some reassurance can be gained by examining cells with a variety of techniques: if all of them produce the same image, this image is likely to represent a real structure and not an artifact.

The most convincing way to establish that a structure exists in living cells is to examine them under a microscope while they are alive, without prior fixation or staining. This requires special kinds of optical systems designed to exploit the diffracting properties of cells. When light passes through a living cell, the phase of the light wave is changed: light passing through a relatively thick or dense part of the cell, such as the nucleus, is retarded and its phase consequently shifted relative to light that has passed through an adjacent thinner region of the cytoplasm. Both the **phase-contrast microscope** and the **differential-interference-contrast microscope** exploit the interference effects produced when these two sets of waves recombine. With either microscope, many details of a living cell become clearly delineated (Figures 4–5 and 4–6).

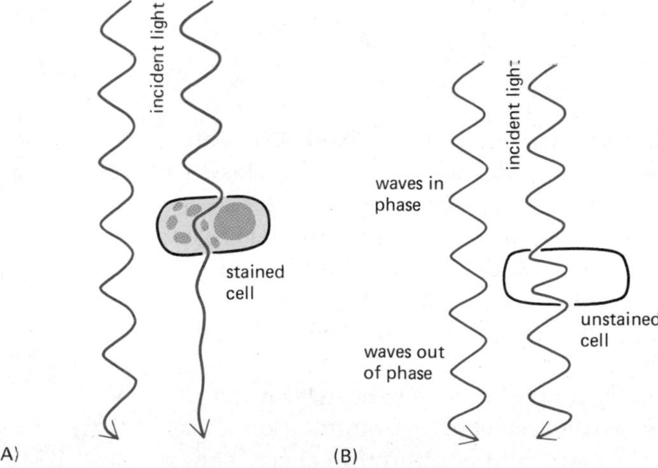

Figure 4–5 The stained portions of the cell in (A) reduce the amplitude of light waves of particular wavelengths passing through them. A colored image of the cell is thereby obtained that is visible by direct observation. Light passing through the unstained, living cell (B) does not undergo a major change in amplitude, and many details cannot, therefore, be seen directly; however, changes occur in the phase of this light that are exploited in phase-contrast and differential-interference-contrast microscopy to produce a high-contrast image.

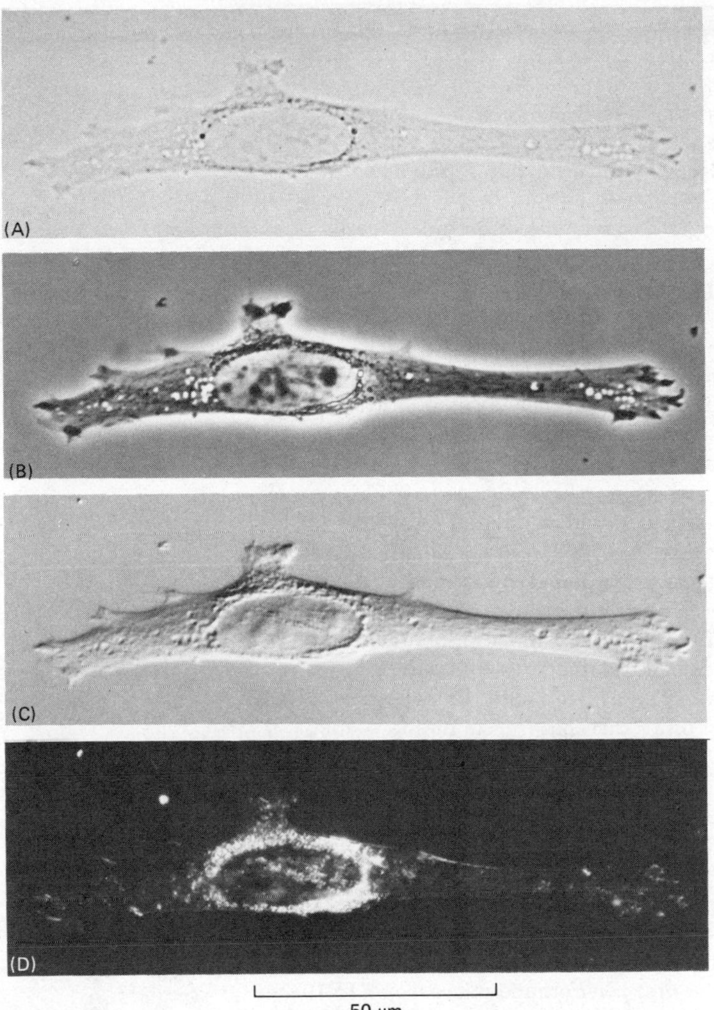

Figure 4–6 A fibroblast in tissue culture seen by four different types of light microscopy. The image in (A) was obtained by the direct transmission of light through the cell, a technique known as *bright-field microscopy*. The other images were obtained by techniques discussed in the text: (B) phase-contrast microscopy, (C) Nomarski differential-interference-contrast microscopy, and (D) dark-field microscopy. Most modern microscopes allow all four types of image to be obtained simply by interchanging optical components.

A simpler way to see the details of an unstained cell is to use the light that is scattered by the various cell components. In the **dark-field microscope,** the illuminating rays of light are directed from the side, so that only scattered light enters the microscope lenses. Consequently, the cell appears as an illuminated object against a black background. A similar arrangement can be used to detect fluorescence in a specimen, as exploited in the technique of fluorescence microscopy to be described later. Images of the same cell obtained by four different kinds of light microscopy are shown in Figure 4–6.

One of the great advantages of these types of microscopy is that they make it possible to see cells in action and to study the movements involved in such processes as mitosis and cell migration. Since many cellular motions are too slow to be seen in real time, it is often helpful to take time-lapse motion pictures (*microcinematography*) or video recordings. Here, successive frames separated by a short time delay are recorded, so that when the resulting film or video tape is projected or played at normal speed, events appear to be greatly speeded up. In this way it is possible to determine precise patterns and rates of movements of cells and their organelles.

Some of the landmarks in the development of light microscopy are outlined in Table 4–1.

Table 4–1 Some Important Discoveries in the History of Light Microscopy

1611	**Kepler** suggested a way of making a compound microscope.
1655	**Hooke** used a compound microscope to describe small pores in sections of cork that he called "cells."
1674	**Leeuwenhoek** reported his discovery of protozoa. He saw bacteria for the first time 9 years later.
1833	**Brown** published his microscopic observations of orchids, clearly describing the cell nucleus.
1835	**Schleiden and Schwann** proposed the cell theory, stating that the nucleated cell is the unit of structure and function in plants and animals.
1857	**Kolliker** described mitochondria in muscle cells.
1876	**Abbé** analyzed the effects of diffraction on image formation in the microscope and showed how to optimize microscope design.
1879	**Flemming** described with great clarity chromosome behavior during mitosis in animal cells.
1881	**Retzius** described many animal tissues with a detail that has not been surpassed by any other light microscopist. In the next two decades, he, **Cajal,** and other histologists developed staining methods and laid the foundations of microscopic anatomy.
1882	**Koch** used aniline dyes to stain microorganisms and identified the bacteria that cause tuberculosis and cholera. In the following two decades, other bacteriologists, such as **Klebs** and **Pasteur,** identified the causative agents of many other diseases by examining stained preparations under the microscope.
1886	**Zeiss** made a series of lenses, to the design of **Abbé,** that enabled microscopists to resolve structures at the theoretical limits of visible light.
1898	**Golgi** first saw and described the Golgi apparatus by staining cells with silver nitrate.
1924	**Lacassagne** and collaborators developed the first autoradiographic method to localize radioactive polonium in biological specimens.
1930	**Lebedeff** designed and built the first interference microscope. In 1932, **Zernicke** invented the phase-contrast microscope. These two developments allowed unstained living cells to be seen in detail for the first time.
1941	**Coons** used antibodies coupled to fluorescent dyes to detect cellular antigens.
1952	**Nomarski** devised and patented the system of differential interference contrast for the light microscope that still bears his name.

The Electron Microscope Resolves the Fine Structure of the Cell[1,4]

The limit of resolution imposed by the wavelength of visible light can be reduced by using electrons instead of light, since electrons have a much shorter wavelength. The wavelength of an electron decreases the faster the electron moves: in an electron microscope with an accelerating voltage of 100,000 volts, the wavelength of an electron is 0.004 nm. In theory, the resolution of such a microscope should be about 0.002 nm; however, because the aberrations of electron lenses are considerably greater than the aberrations of glass lenses, the practical resolving power of most modern electron microscopes is, at best, 0.1 nm (1 Å) (Figure 4–7). Furthermore, problems of specimen preparation, contrast, and radiation damage effectively limit the resolution of biological objects to about 2 nm (20 Å). However, this is still 100 times better than the resolution of the light microscope.

Figure 4–7 Electron micrograph of a thin layer of gold showing the individual atoms as bright spots. The distance between adjacent gold atoms is about 0.2 nm (2 Å). (Courtesy of Graham Hills.)

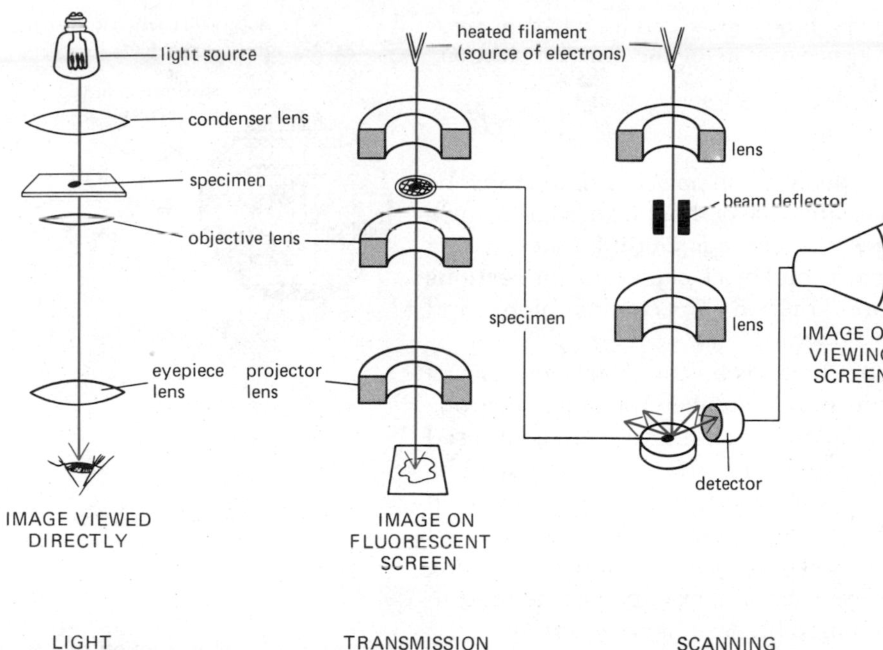

Figure 4–8 Schematic diagram showing the principal features of a light microscope, a transmission electron microscope, and a scanning electron microscope, drawn to emphasize the similarities of overall design. The two types of electron microscopes require the specimen to be placed in a vacuum.

In its overall design, the **transmission electron microscope (TEM)** is not unlike a light microscope—only larger and upside down (Figure 4–8). The source of illumination is a filament or cathode that emits electrons at the top of a cylindrical column about two meters high. If a linear electron beam is to be formed, air must first be pumped out of the column, creating a vacuum. The electrons are then accelerated from the filament by a nearby anode and passed through a tiny hole to form an electron beam that passes down the column. Magnetic coils placed at intervals along the column focus the electron beam, just as glass lenses focus the light in a light microscope. The specimen is put into the vacuum through an air lock and is then exposed to the focused beam of electrons. Some electrons passing through the specimen are scattered, according to the local density of the material, and the remainder are focused to form an image—in a manner analogous to that by which an image is formed in a light microscope—either on a photographic plate or on a phosphorescent screen. Because the scattered electrons are lost from the image, the dense regions of the specimen show up as areas of reduced electron flux.

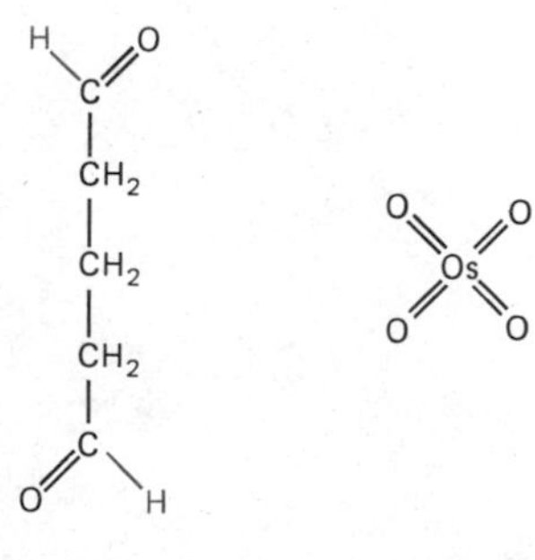

Figure 4–9 Glutaraldehyde and osmium tetroxide are common fixatives used for electron microscopy. The two reactive aldehyde groups of glutaraldehyde enable it to cross-link various types of molecules together, forming covalent bonds that displace the hydrogen atoms shown in color. Osmium tetroxide is reduced by many organic compounds with which it forms cross-linked complexes. It is especially useful for cell membranes since it reacts with the C═C double bonds present in many fatty acids.

Biological Specimens Require Special Preparation for the Electron Microscope[4,5]

In the early days of its application to biological materials, the electron microscope revealed many previously unimagined structures in cells. But in order to study biological specimens, electron microscopists had to develop new procedures for embedding, cutting, and staining tissues.

The preparation of specimens presents various problems. Since the tissue is exposed to a very high vacuum, it must be dried at some stage in its preparation. To avoid distortion by drying, the wet tissue is usually fixed with a mixture of the bifunctional cross-linking agent *glutaraldehyde,* which covalently couples protein molecules to their neighbors, and *osmium tetroxide,* which binds and stabilizes lipid bilayers as well as tissue proteins (Figure 4–9).

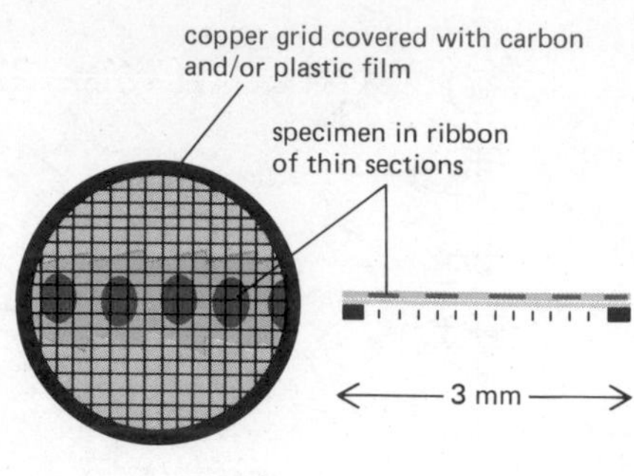

Figure 4–10 Diagram of the copper grid used to support the thin sections of a specimen in the transmission electron microscope.

Because electrons have very limited penetrating power, a tissue must be sliced into sections 50 to 100 nm thick (about 1/200 of the thickness of a single cell). To be able to cut such *thin sections,* the tissue is first infiltrated with a monomeric resin that polymerizes to form a solid block of plastic. The sections are then cut with a fine glass or diamond knife on a special microtome and supported on a small circular metal grid (Figure 4–10).

Contrast in the electron microscope depends on the atomic number of the atoms in the specimen: the higher the atomic number, the more electrons are scattered and the greater the contrast. Biological molecules are composed of atoms of very low atomic number (mainly carbon, oxygen, and hydrogen). To make them visible, therefore, thin sections of biological materials are stained by exposure to the salts of heavy metals, such as uranium or lead. Different cellular constituents are revealed with various degrees of contrast according to their degree of impregnation with these salts. For example, lipids tend to stain darkly, revealing the location of cell membranes (Figure 4–11).

In some cases, particular macromolecules can be specifically labeled in thin sections of cells. For example, some enzymes can be localized by incubation of sections with a substrate whose reaction leads to the local deposition of an electron-dense precipitate (Figure 4–12). Or antibodies can be coupled to such enzymes or to an electron-dense molecule (such as ferritin, which contains iron) and then used to locate the macromolecules they recognize.

Figure 4–11 Thin section of a root tip cell from a grass. Easily seen are the cell wall, nucleus, vacuoles, mitochondria, endoplasmic reticulum, Golgi apparatus, and ribosomes. (Courtesy of Brian Gunning.)

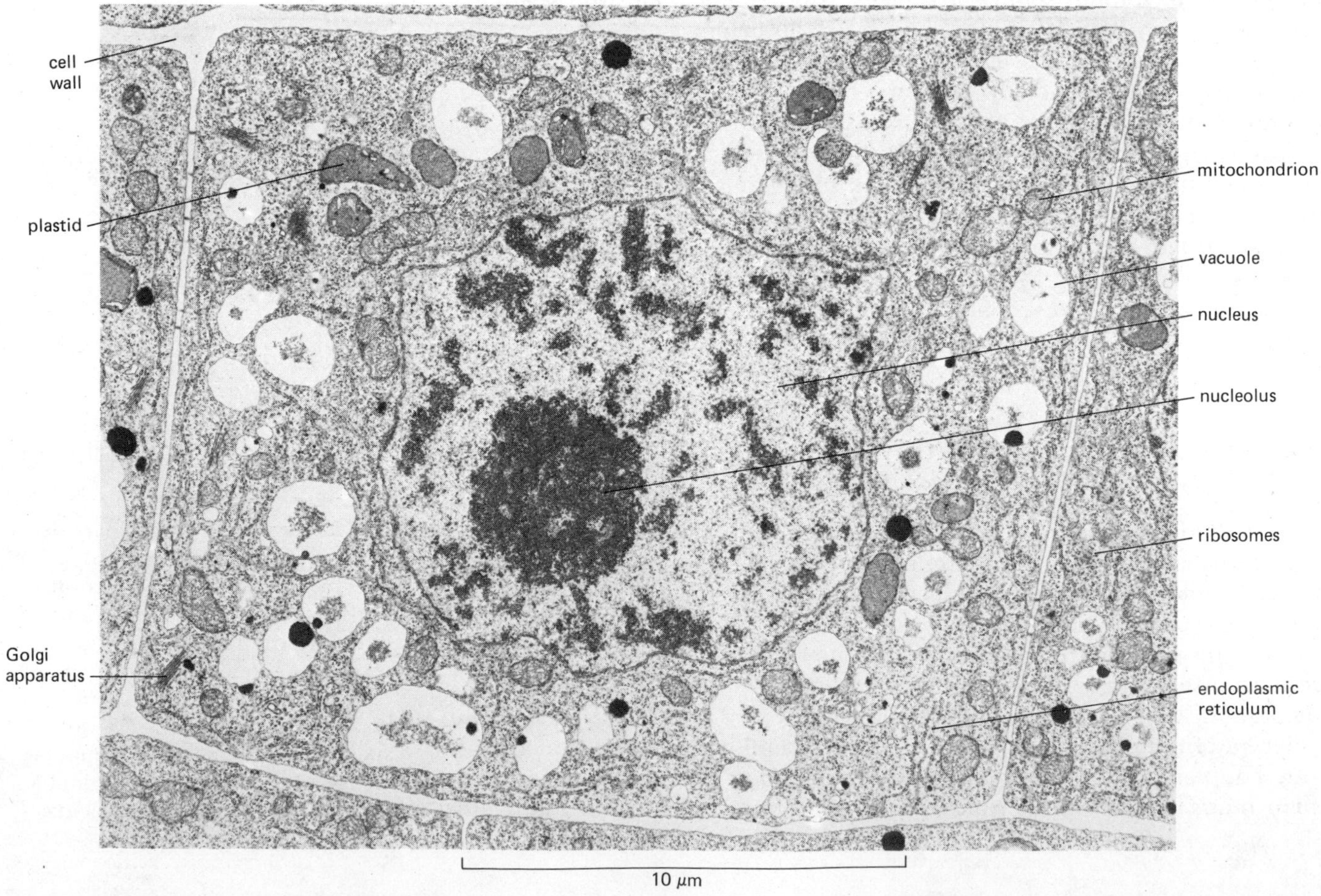

cell wall
plastid
Golgi apparatus
mitochondrion
vacuole
nucleus
nucleolus
ribosomes
endoplasmic reticulum
10 μm

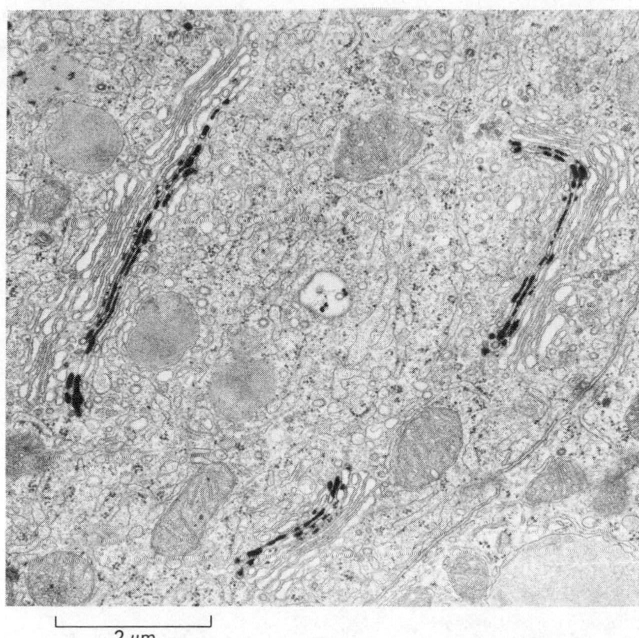

Figure 4–12 Electron micrograph of a cell showing the location of a particular enzyme (thiamine pyrophosphatase) in the Golgi apparatus. A thin section of the cell was incubated with a substrate that formed an electron-dense precipitate on reaction with the enzyme. (Courtesy of Daniel Friend.)

Three-dimensional Images Can Be Obtained by Electron Microscopy[6]

Thin sections are effectively two-dimensional slices of tissue and fail to convey the three-dimensional arrangement of cellular components. Although the third dimension can, in principle, be reconstructed from hundreds of serial sections (Figure 4–13), this is a tedious and lengthy process.

Fortunately, there are more direct means to obtain a three-dimensional image. One is to examine a specimen in a **scanning electron microscope (SEM).** Here the electron beam is reflected from the surface of the specimen rather than passing through it as happens in the transmission electron microscope. The specimen to be examined is fixed and dried and then coated with a thin layer of heavy metal evaporated onto it in a vacuum—a process known as *shadowing.* In the SEM (which is usually a smaller and simpler device than a transmission electron microscope) the specimen is scanned with a focused beam of electrons: as the beam hits the specimen, "secondary" electrons are produced from the metallic surface, and these are detected and converted into an image on a television screen. Since the amount of electron scattering depends on the relative angle of the beam to the surface, the image has bright high points and dark shadows that give it a three-dimensional appearance (Figure 4–14). In most forms of SEM, the resolution attainable is not very high (about 10 nm, with an effective magnification of up to 20,000 times); as a result, the technique finds its main use in the size range between intact single cells and small organisms.

To a limited degree, a three-dimensional image can be obtained from conventional thin sections by tilting the specimen in the electron beam and photographing it from two different angles. When the resulting pair of micrographs is examined through stereo glasses, it creates a three-dimensional image. The depth of specimen that can be examined in this way depends on the penetrating power of the electrons and, hence, on their energy. For this reason, **high-voltage electron microscopes** have been built that accelerate the electron beam through 1,000,000, rather than 100,000, volts. These giant machines enable sections as thick as 1 μm to be examined.

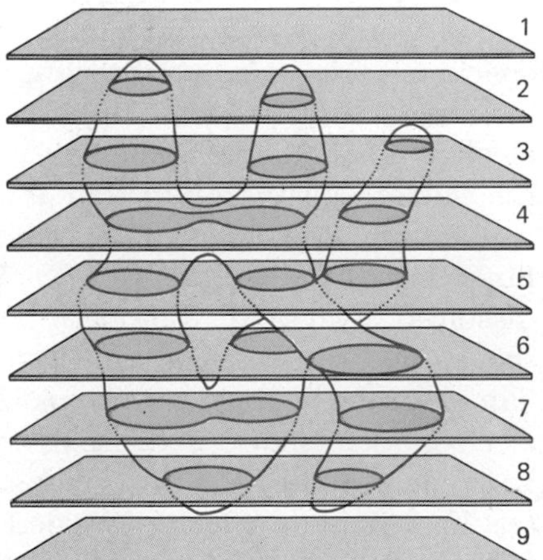

Figure 4–13 Schematic diagram illustrating how single thin sections sometimes give misleading impressions. In this example, most sections through a cell containing a branched mitochondrion will appear to contain two or three separate mitochondria. Sections 4 and 7, moreover, might be interpreted as showing a mitochondrion in the process of dividing. However, the true three-dimensional shape can be reconstructed from serial sections.

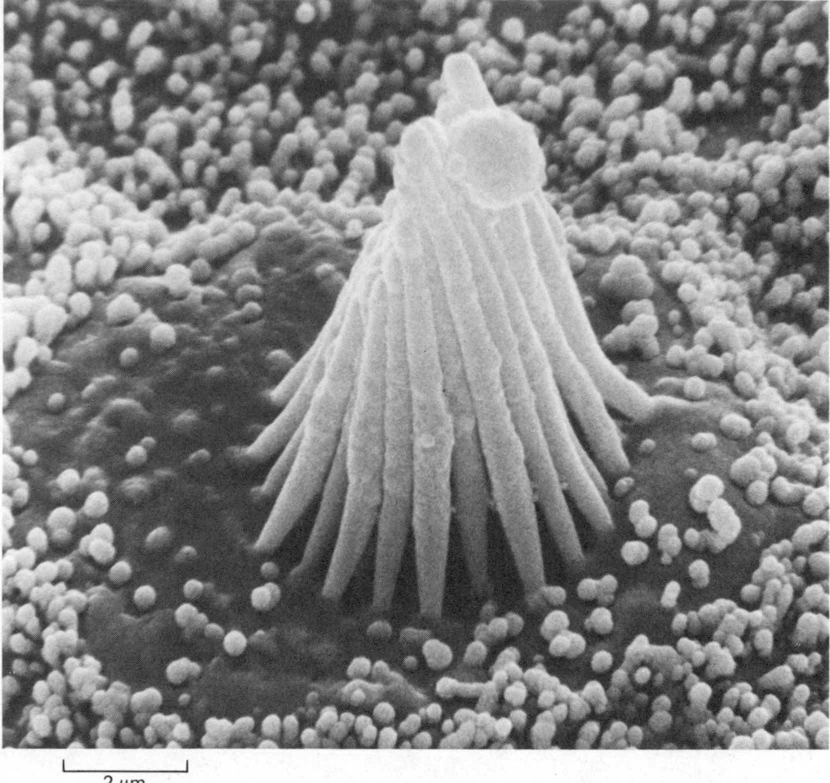

Figure 4–14 Scanning electron micrograph of the organ-pipe-like arrangement of stereocilia projecting from the surface of hair cells in the inner ear. (Courtesy of R. Jacobs and A. J. Hudspeth.)

Freeze-Fracture and Freeze-Etch Electron Microscopy Provide Novel Views of the Cell[7]

The evaporation of a thin film of a heavy metal such as platinum onto a dried specimen, known as **shadowing**, is used not only for the scanning electron microscope but also for certain specimens to be examined in the transmission electron microscope. Some specimens are thin enough or small enough for the electron beam to penetrate them directly after they have been shadowed; this is the case for individual molecules, viruses, and cell walls (Figure 4–15). But other specimens are too thick; therefore, the organic material must be dissolved away after they have been shadowed so that only the thin metal *replica* of the surface of the specimen is left. The replica is reinforced with a film of carbon so that it can be picked up on a grid and examined in a standard electron microscope. Because the metal is evaporated onto the specimen from an angle, the metal is thicker in some places than others (Figure 4–16). As a result, a shadow effect is created that gives the image a three-dimensional appearance.

Two other replica methods have been particularly useful. One of these, **freeze-fracture** electron microscopy, provides the only way of visualizing the interior of cell membranes. Cells are frozen at the temperature of liquid nitrogen (−196°C) in the presence of a *cryoprotectant* (antifreeze) to prevent distortion from ice crystal formation, and then the frozen block is cracked with a knife blade. The fracture plane often passes through the hydrophobic middle of lipid bilayers, thereby exposing the interior of cell membranes (Figure 4–17). The resulting fracture faces are shadowed with platinum, the organic material is dissolved away, and the replicas are floated off and viewed in the electron microscope. Such replicas are found to be studded with small particles (called *intramembrane particles*) thought to be composed mainly of large proteins that traverse the lipid bilayer (Figure 4–18). This technique pro-

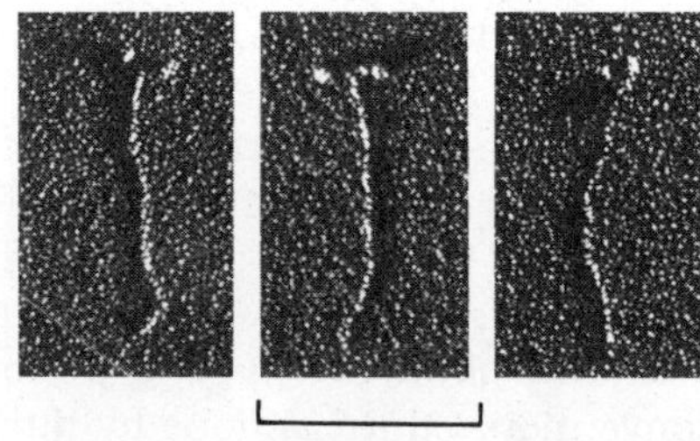

Figure 4–15 Electron micrographs of individual myosin protein molecules that have been shadowed with platinum. Myosin is a major component of the contractile apparatus of muscle. (Courtesy of Arthur Elliot.)

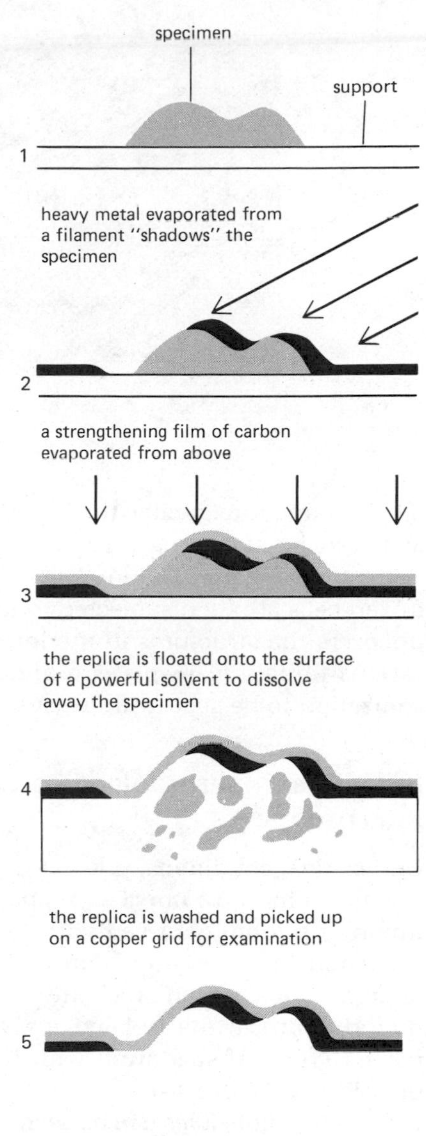

Figure 4–16 Schematic drawing showing how a metal-shadowed replica of the surface of a specimen is prepared. Note that the thickness of the metal reflects the surface contours of the original specimen.

vided the first demonstration that many of the *intercellular junctions,* where cells make contact with each other, are composed of organized arrays of specialized membrane proteins.

The closely related technique of **freeze-etch** electron microscopy makes it possible to examine the true exterior surface of cells and their membranes (rather than the interior of their membranes). In this technique, the cells are again frozen at the temperature of liquid nitrogen and the frozen block cracked with a knife blade. But now the ice level is lowered around the cells (and to a lesser extent within the cells) by the sublimation of water in a vacuum as the temperature is raised (a process called *freeze-drying*) (Figure 4–17). The parts of the cell exposed by the *etching* process are then converted into a platinum replica as in conventional freeze-fracture electron microscopy.

Cryoprotectants cannot be used in freeze-etching since they are nonvolatile and are left behind in the specimen as the water sublimes. Consequently, ice crystal formation has, in the past, severely limited the resolution obtainable with this procedure. A recent modification of the method prevents the formation of large ice crystals by freezing the specimen extremely rapidly

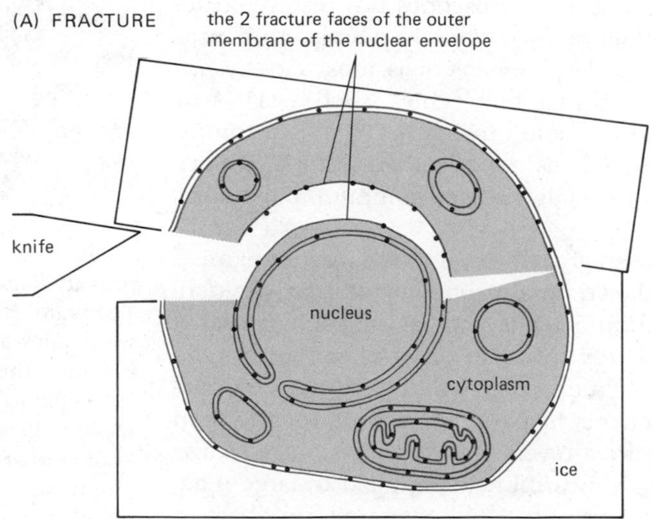

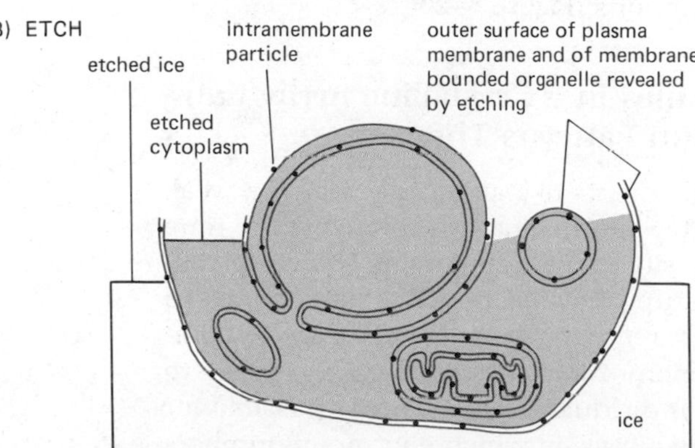

Figure 4–17 In freeze-fracture electron microscopy (A), the frozen specimen is fractured with a knife. The fracture plane tends to pass through the middle of lipid bilayers as shown. In freeze-etching (B), the specimen is first fractured as in (A) and then the ice level is lowered by the sublimation of water in a vacuum, thus exposing part of the external surface of the cell. Following these steps, a replica of the frozen surface is prepared (as described in Figure 4–16) and examined in the transmission electron microscope.

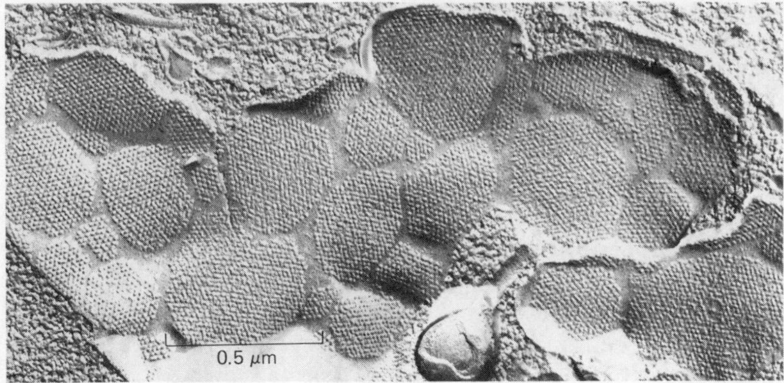

Figure 4–18 Freeze-fracture electron micrograph of the plasma membrane of an epithelial cell of rat bladder. The plane of fracture has passed through the middle of the lipid bilayer, thereby exposing a regular array of intramembrane particles, thought to be large membrane proteins. Although intramembrane particles are seen in all plasma membranes, they are rarely as densely packed as in these cells. (Courtesy of N. J. Severs.)

(at a cooling rate greater than 20°C per millisecond). Such **rapid freezing** is achieved by slamming a sample of cells against a copper block cooled to −269°C with liquid helium. Particularly impressive effects are achieved if the frozen cells are then subjected to extensive freeze drying. Such **deep etching** uncovers the structures in the interior of the cell. For example, it brings the various protein filaments of a muscle cell into stark relief, enabling their organization to be seen with exceptional clarity (Figure 4–19).

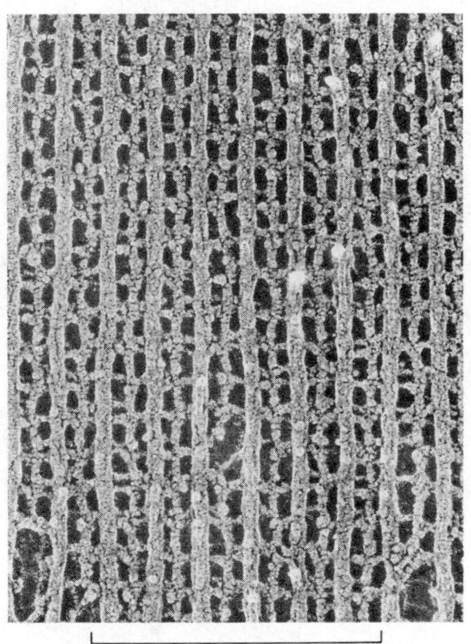

Figure 4–19 Regular array of protein filaments in an insect muscle. To obtain this image, the muscle cells were quick-frozen in liquid helium, fractured through the cytoplasm, and subjected to deep etching, and a metal replica was then prepared and examined at high magnification. (Courtesy of Roger Cooke and John Heuser.)

Individual Macromolecules Can Be Resolved in the Electron Microscope

Under ideal conditions, present-day electron microscopes can resolve structures 0.1 nm to 0.2 nm (1–2 Å) apart (Figure 4–7). While in practice this resolution is seldom achieved with biological specimens, it is most closely approached in the examination of isolated macromolecules, such as DNA or large proteins, which is a routine and valuable procedure. To create the required contrast, the isolated macromolecules are adsorbed onto a carbon-coated grid and shadowed with heavy metals such as platinum, palladium, or gold (see Figure 4–15).

Macromolecules can be seen in even greater detail by the use of **negative staining.** Here macromolecules are placed on a carbon-coated grid, drained, and washed with a concentrated solution of a heavy metal salt such as uranyl acetate. After the sample has dried, a very thin film of metal salt covers the grid everywhere except where it has been excluded by a macromolecule. Because the macromolecule allows electrons to pass much more readily than does the surrounding heavy metal stain, a reversed or negative image of it is created. Negative staining is especially powerful when applied to large macromolecular aggregates, such as viruses or ribosomes, or to uniform fibers or two-dimensional sheets of repeating subunits (Figure 4–20).

The Detailed Structure of Molecules in a Crystalline Array Can Be Calculated from the Diffraction Patterns They Create

A single molecule, even with these refinements of staining, gives only a weak and ill-defined image. Efforts to get better information by prolonging the time of inspection or by increasing the intensity of the illuminating beam are self-defeating because they damage and disrupt the object under examination. To discover the details of molecular structure, therefore, it is necessary to combine the information obtained from many molecules in such a way as to average out the random errors in the individual images. This is possible for substances that form regular crystalline arrays, in which enormous numbers

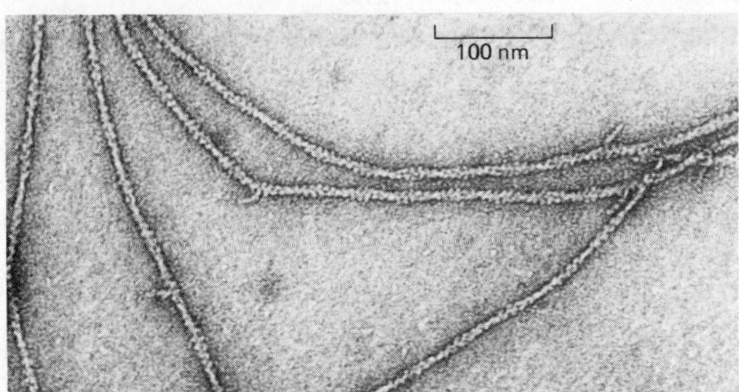

Figure 4–20 Electron micrograph of negatively stained actin filaments. Each filament is about 6 nm in diameter and is seen, on close inspection, to be composed of two helical chains of globular actin molecules. (Courtesy of Roger Craig.)

of molecules are held in identical orientation and in regularly spaced positions. The standard technique for extracting the desired information depends on diffraction.

Consider first a single object (such as a single molecule) placed in a beam of radiation of any sort whose wavelength is small compared to the dimensions of the object. The object will scatter some of the radiation. The scattered radiation can be thought of as consisting of a family of overlapping waves, each emanating from a different part of the object. As the waves overlap, they undergo interference, producing a distribution of radiation known as a **diffraction pattern.** The diffraction pattern can be recorded on a screen placed at some distance from the object and can be described in terms of the amounts of scattered radiation sent out by the object in different directions (Figure 4–21). The structure of the object determines the form of the diffraction pattern. Conversely, given a full description of the diffraction pattern, it is possible, in theory, to calculate the structure of the object that produced it. In practice, the diffraction pattern due to a single molecule would be far too faint and erratic for such a purpose.

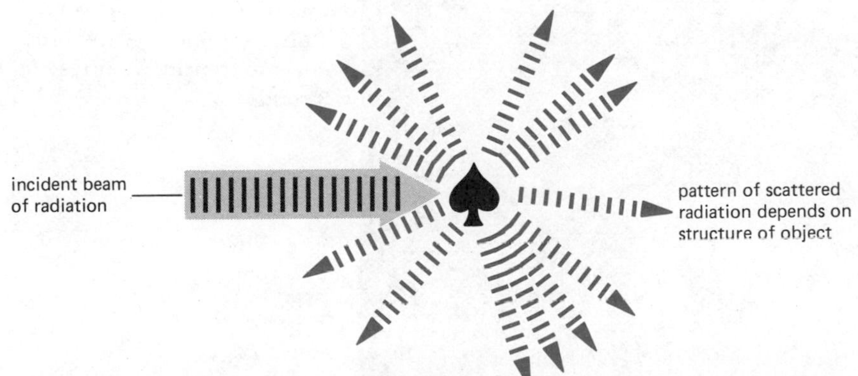

Figure 4–21 The scattering of radiation by a single object whose dimensions are comparable with the wavelength of the radiation. Radiation falling on the object is scattered with different intensities in different directions. The intensity of the scattered beam in a given direction depends on the way in which radiation scattered from one part of the object interferes with that scattered from another part. In the diagram, the resultant intensity of scattering in the various possible directions is indicated by the density of colored arrows radiating from the object.

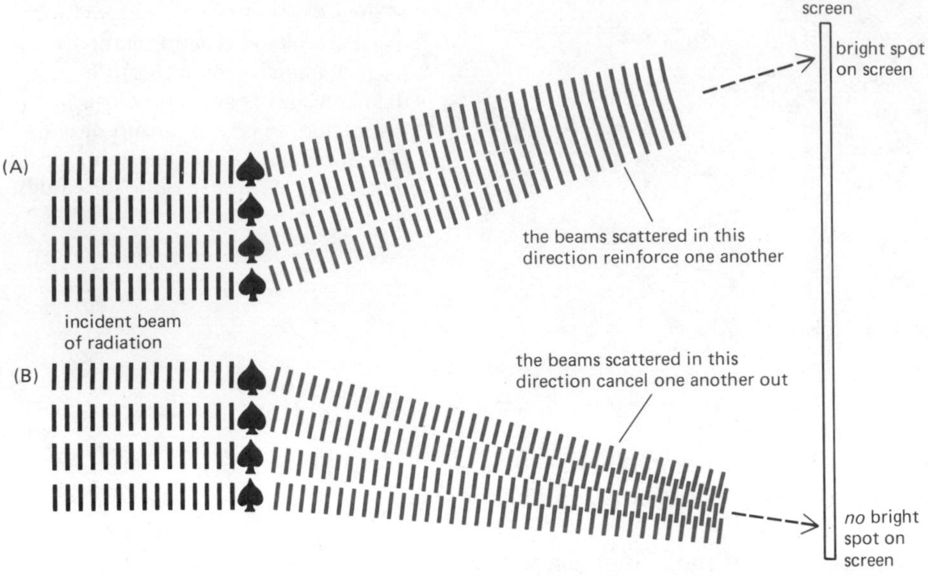

Figure 4–22 Scattering of radiation by a crystal. When many identical objects are arranged in a crystalline array, the radiation scattered by each object interferes with that scattered by the others. Only in certain directions (dependent on the spacing of the objects in the array) do the individual scattered beams reinforce, producing bright spots in the diffraction pattern of the array. The intensity of a given bright spot depends on the intensity with which each object in the array would scatter radiation in that direction if the object were examined in isolation, as in Figure 4–21.

Suppose now that many identical objects are arranged in a crystalline array and again illuminated with a beam of radiation (Figure 4–22). The radiation scattered by each object will interfere with that scattered by the others. Only in certain directions, depending on the spacing of the objects in the array, will the individual scattered beams reinforce, producing a bright spot in the diffraction pattern. The complete diffraction pattern of the crystalline array will thus consist of many such discrete bright spots of differing intensities (Figure 4–23).

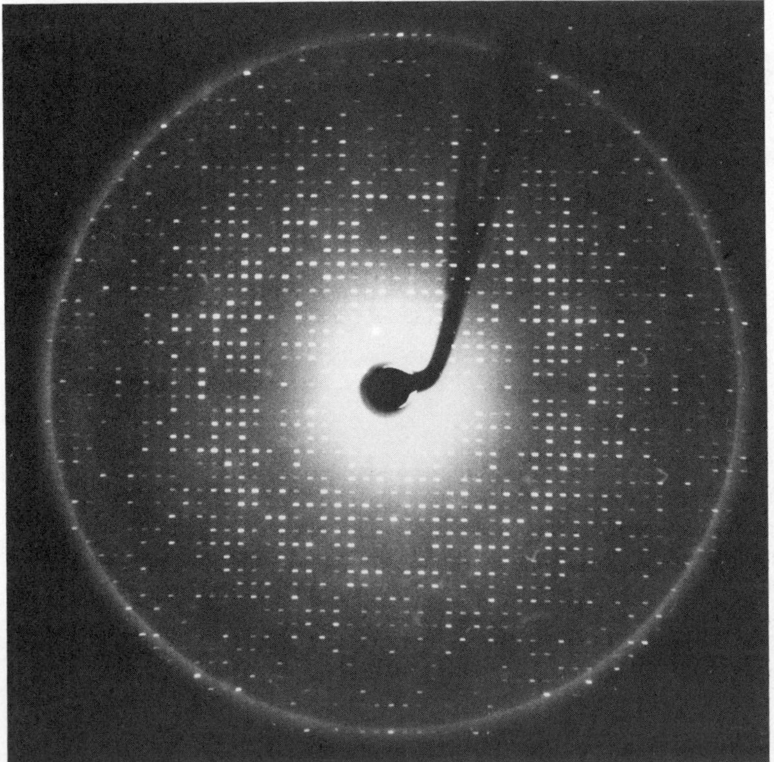

Figure 4–23 Part of the x-ray diffraction pattern obtained from a protein crystal. This particular crystal was used to help determine the atomic structure of the proteolytic enzyme trypsin. (Courtesy of Robert Stroud.)

The relative intensities of the various spots in the diffraction pattern depend on the scattering properties of the individual objects in the array. In fact, the intensity of a given spot is proportional to the average intensity of the radiation that would be scattered in the same direction from a representative single object standing alone. Thus, while the *positions* of the spots in the diffraction pattern depend on the arrangement of objects in the array, the *intensities* of the spots give information as to the internal structure of a representative single object. This information, moreover, is precise and plentiful because it is obtained by combining contributions from a very large number of equivalent sources. In fact, from a full description of the diffraction pattern of the crystalline array, it is often possible to calculate the structure of the individual objects from which it is built, as we shall see.

Image Reconstruction Techniques Based on Diffraction Can Be Used to Extract Additional Information from Electron Micrographs[8]

In principle, the procedure for reconstructing an image from a diffraction pattern can be carried out with any kind of radiation whose wavelength is appropriate to the object under examination. Figure 4–24 shows how the method is used to extract relatively detailed information about molecular structure from an apparently fuzzy electron micrograph of molecules in a crystalline array. The micrograph (not the array of molecules itself) is placed in the path of a beam of laser light, and an image of a typical individual molecule in the array is reconstructed from the *optical diffraction pattern*. Alternatively, and more conveniently, one can dispense with the beam of light and instead use a computer to *calculate* the diffraction pattern that would be produced by the micrograph. The image is then reconstructed from the computed diffraction pattern.

Such image reconstruction techniques average the information from all of the molecules in the array and thereby reveal details not apparent in the original micrograph, where they are obscured by the random "noise" in the images of the individual molecules. The method is particularly valuable for molecules that form crystalline arrays in two dimensions, such as some membrane proteins. In favorable cases, image reconstruction based on electron microscopy gives the shape of an individual protein molecule to a resolution of about 0.7 nm (7 Å). But this still falls short of what is required for a full description of molecular structure, for the atoms in a molecule are separated by distances of only 0.1 or 0.2 nm.

Some of the landmarks in the development of electron microscopy are outlined in Table 4–2.

Figure 4–24 The use of diffraction for averaging the electron microscope images of molecules in a two-dimensional array. (A) Original micrograph of a highly ordered array of glycoprotein molecules present in the cell wall of a small alga, with the dark regions representing the stain and the white regions representing the glycoprotein molecules. The calculated diffraction pattern of this image is shown in (B), and the reconstructed image of a small portion of this image—determined from the diffraction pattern—is shown in (C). Details of the shape of the molecules are clearly seen in the reconstruction (in which the dark regions now represent the glycoprotein molecules) but are hard to discern in the original micrograph. (Photographs courtesy of Graham Hills and Peter Shaw.)

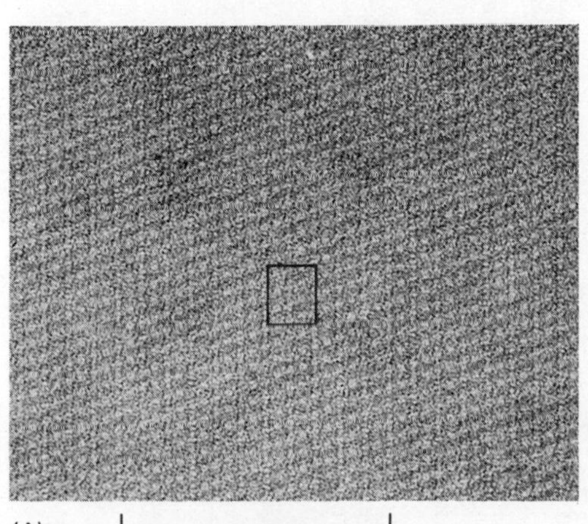

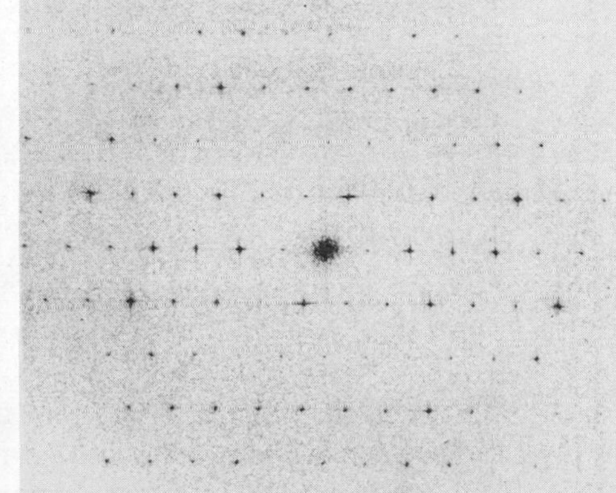

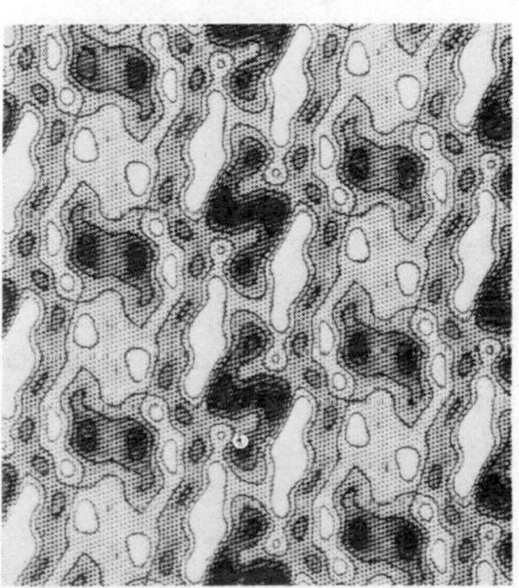

Table 4–2 Major Events in the Development of the Electron Microscope and Its Application to Cell Biology

1897	**J. J. Thomson** announced the existence of negatively charged particles, later termed *electrons.*
1924	**de Broglie** proposed that a moving electron has wavelike properties.
1926	**Busch** proved that it was possible to focus a beam of electrons with a cylindrical magnetic lens, laying the foundations of electron optics.
1931	**Ruska** and colleagues built the first transmission electron microscope.
1935	**Knoll** demonstrated the feasibility of the scanning electron microscope; three years later a prototype instrument was built by **Von Ardenne.**
1939	**Siemens** produced the first commercial transmission electron microscope.
1944	**Williams and Wyckoff** introduced the metal shadowing technique.
1945	**Porter, Claude,** and **Fullam** used the electron microscope to examine cells in tissue culture after fixing and staining them with OsO_4.
1948	**Pease and Baker** reliably prepared thin sections (0.1 to 0.2 μm thick) of biological material.
1952	**Palade, Porter,** and **Sjöstrand** developed methods of fixation and thin-sectioning that enabled many intracellular structures to be seen for the first time. In one of the first applications of these techniques, **H. E. Huxley** showed that skeletal muscle contains overlapping arrays of protein filaments, supporting the "sliding filament" hypothesis of muscle contraction.
1953	**Porter and Blum** developed the first widely accepted ultramicrotome, incorporating many features introduced by **Claude** and **Sjöstrand** previously.
1956	**Glauert** and associates showed that the epoxy resin *Araldite* was a highly effective embedding agent for electron microscopy. **Luft** introduced another embedding resin, *Epon,* five years later.
1957	**Robertson** described the trilaminar structure of the cell membrane, seen for the first time in the electron microscope.
1957	Freeze-fracture techniques, initially developed by **Steere**, were perfected by **Moore and Mühlethaler**. Later (1966), **Branton** demonstrated that freeze-fracture allows the interior of the membrane to be visualized.
1959	**Brenner and Horne** developed the negative staining technique, invented four years previously by **Hall,** into a generally useful technique for visualizing viruses, bacteria, and protein filaments.
1963	**Sabatini, Bensch, and Barrnett** introduced glutaraldehyde (usually followed by OsO_4) as a fixative for electron microscopy.
1965	**Cambridge Instruments** produced the first commercial scanning electron microscope.
1968	**de Rosier and Klug** described techniques for the reconstruction of three-dimensional structures from electron micrographs.
1979	**Heuser, Reese,** and colleagues developed a high-resolution, deep-etching technique based upon very rapid freezing.

X-ray Diffraction Reveals the Three-dimensional Arrangement of the Atoms in a Molecule[9]

The most important applications of diffraction to the analysis of molecular structure have employed x-rays. X-ray diffraction provides better resolution than even the most sophisticated techniques of electron microscopy. Using x-rays with a wavelength of about 0.1 nm to generate a diffraction pattern from a crystal, it is possible to deduce the arrangement of individual atoms in the molecules from which the crystal is built. Unlike visible light or beams of

Figure 4–25 Crystals of the enzyme glycogen phosphorylase viewed in a light microscope. (Courtesy of Robert Fletterick.)

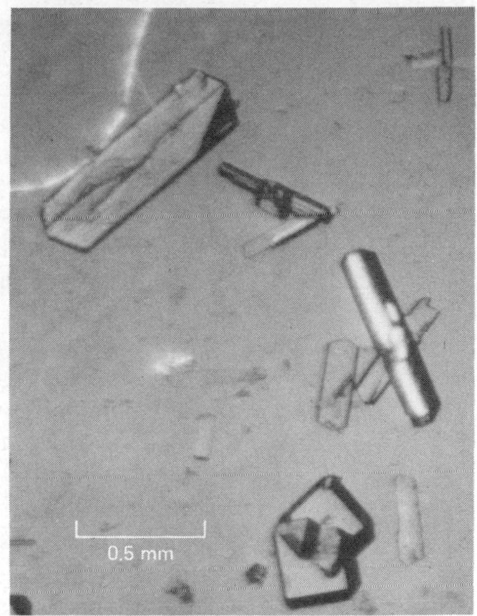

electrons, x-rays cannot be focused to form an image of the usual sort after passing through a specimen, but the diffraction pattern is easy to record on photographic film. Moreover, x-rays have greater penetrating power than electrons, and so much thicker specimens can be used. Because hydrated biological material can be examined by x-ray diffraction, the distortions induced by the preparative procedures necessary for electron microscopy are avoided.

For high-resolution studies, large, highly ordered crystals are required. The x-rays that penetrate the crystal are scattered chiefly by the electrons present in the sample. Therefore, large atoms with many electrons scatter x-rays more than do small atoms, so that the atoms of C, N, O, and P are more readily detected than atoms of H. For the same reason, any metal atoms, such as the iron atom in a hemoglobin molecule, create an intense beacon of scattered x-rays. It is a complex task to work back from the diffraction pattern to the three-dimensional arrangement of atoms in a representative molecule. The solution of x-ray diffraction patterns produced by molecules as large and irregular as proteins was not achieved until 1960 (Table 4–3). Even today it is a major undertaking that involves months of automated data collection and many hours on large computers. Often, an even slower step is the production of suitable, large crystals of the proteins (Figure 4–25). In order to obtain enough diffraction information, several sets of crystals are required, some of which have to be modified by inserting heavy metal atoms at specific sites. The growing of protein crystals is an empirical art, and success with any given molecule can take years.

Despite these difficulties, x-ray diffraction is a widely used technique because it is the only way to determine the detailed arrangement of atoms in most molecules. For example, with good crystals the structure of a protein can be calculated to a resolution of 0.3 nm to 0.4 nm, revealing the main course of the polypeptide chain but few other details. With high-quality crystals and very hard work, a resolution of 0.15 nm is obtainable, revealing the position of almost all of the nonhydrogen atoms in the protein.

Summary

A wide range of microscopic techniques is available for studying cells. In the light microscope, living cells can be seen by using phase-contrast, interference, or dark-field optics, while fixed dead cells can be stained with various dyes or more specific reagents that bind to particular components in the cell. The transmission electron microscope allows cells to be examined at much higher resolution and shows the arrangement of their organelles, membranes, and protein filaments. Electron-dense labeling reagents can be used to locate specific macromolecules in cells or on cell surfaces. The interior of cell membranes can be visualized by freeze-fracture electron microscopy, while the scanning electron microscope reveals the contours of the cell surface in three dimensions.

The shape of individual macromolecules that have been shadowed with a heavy metal or outlined by negative staining can be visualized by transmission electron microscopy. But the precise location of each atom in a molecule can be determined only if the molecule will assemble into large crystals. In this case, a beam of x-rays is passed through a crystal, and the three-dimensional arrangement of the atoms in the constituent molecules is calculated from the pattern of diffracted x-rays.

Table 4–3 Landmarks in the Development of X-ray Crystallography and Its Application to Biological Molecules

1864	**Hoppe-Seyler** crystallized, and named, the protein hemoglobin.
1895	**Röntgen** observed that a new form of penetrating radiation, which he named x-rays, was produced when cathode rays (electrons) hit a metal target.
1912	**Von Laue** obtained the first x-ray diffraction patterns by passing x-rays through a crystal of zinc sulfide.
	W. L. Bragg and W. H. Bragg proposed a simple relationship between an x-ray diffraction pattern and the arrangement of atoms in a crystal that produces the pattern.
1926	**Sumner** obtained crystals of the enzyme urease from extracts of jack beans and demonstrated that proteins possess catalytic activity.
1931	**Pauling** published his first essays on "The Nature of the Chemical Bond," detailing the rules of covalent bonding.
1934	**Bernal and Crowfoot** presented the first detailed x-ray diffraction patterns of a protein obtained from crystals of the enzyme pepsin.
1935	**Patterson** developed an analytical method for determining interatomic spacings from x-ray data.
1941	**Astbury** obtained the first x-ray diffraction pattern of DNA.
1951	**Pauling and Corey** proposed the structure of a helical conformation of a chain of L-amino acids—the α-helix—and the structure of the β-sheet, both of which were later found in many proteins.
1953	**Watson and Crick** proposed the double-helix model of DNA, based on x-ray diffraction patterns obtained by **Franklin and Wilkins.**
1954	**Perutz** and colleagues developed heavy-atom methods to solve the phase problem in protein crystallography.
1960	**Kendrew** described the first detailed structure of a protein (sperm whale myoglobin) to a resolution of 0.2 nm, and **Perutz** proposed a lower-resolution structure of the larger protein hemoglobin.
1966	**Phillips** described the structure of lysozyme, the first enzyme to be analyzed in detail.
1976	**Kim and Rich** and **Klug** and colleagues described the detailed three-dimensional structure of tRNA determined by x-ray diffraction.
1977–1978	**Holmes** and **Klug** determined the structure of tobacco mosaic virus (TMV), and **Harrison** and **Rossman** determined the structure of two small spherical viruses.

Cell Culture[10]

Given the appropriate conditions, most kinds of plant and animal cells will survive, multiply, and even express differentiated properties in a tissue-culture dish. Consequently, it is possible to determine the effects on cell behavior of adding or removing specific molecules such as hormones or growth factors and to obtain homogeneous populations of cells for biochemical analysis or for studying the interactions between one cell type and another. Experiments on cultured cells are often said to be carried out *in vitro* (literally, "in glass"); the term is also used in a different sense by biochemists to refer to biochemical reactions occurring outside of living cells. While it is often a great advantage to be able to study the complex behavior of cells in the strictly defined conditions of a culture dish, the observations must sooner or later be checked against the behavior of cells in their natural environment *in vivo.*

Cells Can Be Grown in a Culture Dish[11]

Tissue culture began in 1907 with an experiment designed to settle a controversy in neurobiology. The hypothesis under examination was known as the *neuronal doctrine,* which states that each nerve fiber is the outgrowth of a single nerve cell and not the product of the fusion of many cells. To test this contention, small pieces of spinal cord were placed on clotted plasma in a warm moist chamber and observed at regular intervals under the microscope. After a day or so, individual nerve cells could be seen extending long, thin processes into the plasma clot. Thus the neuronal doctrine was validated, and the foundations for the cell-culture revolution were laid.

The original experiments in 1907 involved the culture of small tissue fragments, or **explants.** Today, cultures are more commonly made from suspensions of cells dissociated from tissues. Such **dissociated cell cultures** (Figure 4–26) have the great advantage of enabling the experimenter to purify individual cell types from the mixture of cell types always present in a tissue and thus to examine them in isolation.

Unlike bacteria, most tissue cells are not adapted to growth in suspension and require a solid surface on which to grow and divide. This mechanical support was originally provided by the plasma clot but is now usually replaced by the surface of a plastic tissue-culture dish. Cells vary, however, in their requirements, and some will not grow or differentiate unless the culture dish is coated with extracellular matrix components, such as collagen.

Cultures prepared directly from the tissues of an organism are called **primary cultures.** In most cases, cells in primary cultures can be removed from the culture dish and used to form a large number of **secondary cultures;** they may be repeatedly subcultured in this way for weeks or months. Such cells often display the differentiated properties of the tissue from which they were obtained: fibroblasts continue to secrete collagen; cells derived from embryonic skeletal muscle fuse to form giant muscle fibers, which spontaneously contract in the culture dish; nerve cells extend axons that are electrically excitable and make synapses with other nerve cells; and epithelial cells form extensive sheets with many of the properties of an intact epithelium. Since these phenomena occur in culture, they are accessible to study in ways that are not possible in intact tissues.

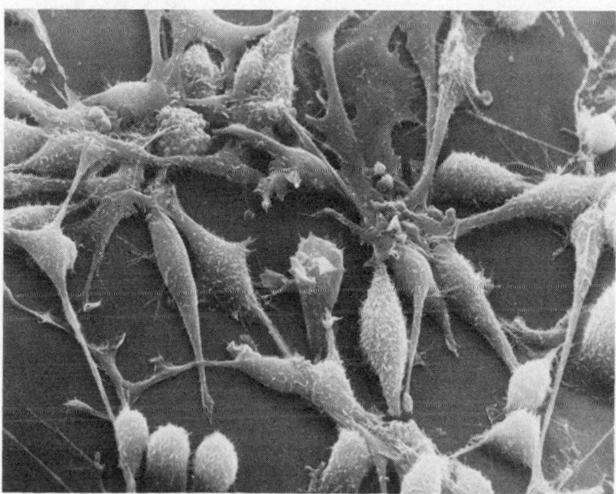

Figure 4–26 Scanning electron micrograph of rat fibroblasts growing in tissue culture. (Courtesy of Gunter Albrecht-Buehler.)

Chemically Defined Media Permit Identification of Specific Growth Factors[12]

Until the early 1970s, tissue culture was something of a blend of science and witchcraft. Although plasma clots were replaced by plastic culture dishes and liquid media containing a well-defined mixture of salts, amino acids, and vitamins, most media also contained a proportion of some poorly defined biological material, such as horse serum or fetal calf serum or a crude extract made from chick embryos. While such media are still used today for most routine tissue culture (Table 4–4), they are inappropriate for determining the specific requirements for the growth and differentiation of a particular type of cell.

This has led to the development of various chemically defined media that support the growth of different types of cells. These media often contain one or more of the various protein **growth factors** that many cells require in order to survive and proliferate in culture: for example, certain nerve cells need trace amounts of *nerve growth factor (NGF)* to differentiate and survive in culture, as well as in the intact animal. Other factors of this kind that play a vital role in the development and maintenance of specific cell types have also been discovered, and the search for new ones has been made very much easier by the availability of chemically defined media.

Eucaryotic Cell Lines Are a Convenient Source of Homogeneous Cells

Most vertebrate cells die after a finite number of divisions in culture: for example, human skin cells divide only 50 to 100 times before they die. This limited life span is believed to reflect the limited life span of the animal from which they are derived. For example, cells from humans suffering from a

Table 4–4 Composition of a Typical Medium Suitable for the Cultivation of Mammalian Cells

Amino Acids	Vitamins	Salts	Miscellaneous
Arginine	Biotin	NaCl	Glucose
Cystine	Choline	KCl	Penicillin
Glutamine	Folate	NaH_2PO_4	Streptomycin
Histidine	Nicotinamide	$NaHCO_3$	Phenol red
Isoleucine	Pantothenate	$CaCl_2$	Whole serum
Leucine	Pyridoxal	$MgCl_2$	
Lysine	Thiamine		
Methionine	Riboflavin		
Phenylalanine			
Threonine			
Tryptophan			
Tyrosine			
Valine			

Glucose is used at a concentration of 5 to 10 mM. The amino acids are all in the L form and, with one or two exceptions, are used at concentrations of 0.1 or 0.2 mM; vitamins are used at a 100-fold lower concentration, that is, about 1 μM. Serum, which is usually from horse or calf, is added to make up 10% of the total volume. Penicillin and streptomycin are antibiotics added to suppress the growth of bacteria. Phenol red is a pH indicator dye whose color is monitored to assure a pH of about 7.4.

Cultures are usually grown in a plastic or glass container with a suitably prepared surface that allows the attachment of cells. The containers are kept in an incubator at 37°C in an atmosphere of 5% CO_2, 95% air.

disease that causes premature aging and death have a more limited life span in culture. Occasionally, however, variant cells that are effectively immortal arise in culture. Such cells can be propagated indefinitely as a **cell line** (Table 4–5). They usually grow best when attached to solid surfaces and typically cease growing when they have formed a continuous, or *confluent*, layer over the surface of the tissue-culture dish.

Variant cells that divide indefinitely in culture are usually distinct, however, from cancer cells, which divide indefinitely *in vivo* as well as *in vitro*. Unlike other cell lines, cancer cells often grow without the need to attach to a surface, and they proliferate to a very much higher density than normal cells in a tissue-culture dish. Similar properties can be experimentally induced in normal cells by *transformation* with a tumor-inducing virus or chemical. The resulting **neoplastically transformed cell lines** are capable of causing tumors if injected into an animal. Both transformed and untransformed cell lines are extremely useful in cell research as sources of very large numbers of cells of a uniform type, especially since they can be stored at −70°C for an indefinite period and still produce viable cells when thawed.

The uniformity of cell lines can be further increased by **cloning.** A *clone* is a population of cells derived from a single ancestor cell. One of the most important uses of cell cloning has been the isolation of mutant cell lines with defects in specific genes: studying cells that are defective in a specific protein often reveals a good deal about the function of that protein in normal cells.

Table 4–5 Some Commonly Used Cell Lines

Cell Line	Cell Type and Origin
3T3	fibroblast (mouse)
BHK 21	fibroblast (Syrian hamster)
HeLa	epithelial cell (human)
PTK 1	epithelial cell (rat kangaroo)
L 6	myoblast (rat)
PC 12	chromaffin cell (rat)
SP 2	plasma cell (mouse)

Many of these cell lines were derived from tumors. All of them are capable of indefinite replication in tissue culture and express at least some of the differentiated properties of their cell of origin. BHK 21 cells, HeLa cells, and SP 2 cells are capable of growth in suspension; the other cell lines require a solid culture substratum in order to multiply.

Cells Can Be Fused Together to Form Hybrid Cells[13]

It is possible to fuse one cell with another to form a combined cell with two separate nuclei called a **heterocaryon.** The fusion is usually done by treating a suspension of cells with certain inactivated viruses or polyethylene glycol, either of which alters the plasma membranes of cells in such a way that they tend to fuse with each other. Heterocaryons provide a way of mixing the components of two separate cells—the plasma membranes, cytoplasms, and nuclei—in order to study their interactions. For example, the inert nucleus of a chicken red blood cell is reactivated to make RNA, and eventually to replicate its DNA, when it is exposed to the cytoplasm of a growing tissue-culture cell by fusion. And the first direct evidence that membrane proteins are able to move in the plane of the plasma membrane came from an experiment in which mouse cells and human cells were fused: while the mouse and human cell-surface proteins were initially confined to their own halves of the heterocaryon plasma membrane, they quickly diffused and mixed over the entire surface of the cell.

Eventually the heterocaryon proceeds to mitosis and produces a **hybrid cell** in which the separate nuclear envelopes break up and all the chromosomes come together in a single large nucleus (Figure 4–27). Such hybrid cells can be cloned to produce hybrid cell lines, and because they tend to be unstable and lose chromosomes, they have been extremely useful for mapping the genes on human chromosomes. The usual mapping procedure involves fusing mouse and human cells. For unknown reasons, these hybrids randomly lose human chromosomes, giving rise to a variety of different mouse-human hybrid cell lines, each of which contains only one or a few human chromosomes. Analysis of many of these lines has made it possible to assign particular biochemical functions to particular human chromosomes. For example, only hybrid cells containing human chromosome 1 synthesize the human version of the enzyme *uridine monophosphate kinase,* indicating that the gene encoding this enzyme is located on chromosome 1.

Some important discoveries in the development of tissue culture are outlined in Table 4–6.

Table 4–6 Some Landmarks in the Development of Tissue Culture

Year	Landmark
1885	**Roux** showed that embryonic chick cells could be maintained alive in a saline solution outside the animal body.
1907	**Harrison** cultivated amphibian spinal cord in a lymph clot, thereby demonstrating that axons are produced as extensions of single nerve cells.
1910	**Rous** induced a tumor by using a filtered extract of chicken tumor cells, later shown to contain an RNA virus (Rous sarcoma virus).
1913	**Carrel** showed that cells could grow for long periods in culture provided they were fed regularly under aseptic conditions.
1948	**Earle** and colleagues isolated single cells of the L cell line and showed that they formed clones of cells in tissue culture.
1952	**Gey** and colleagues established a continuous line of cells derived from human cervical carcinoma, which later became the well-known *HeLa* cell line.
1954	**Levi-Montalcini** and associates showed that nerve growth factor (NGF) stimulated the growth of axons in tissue culture.
1955	**Eagle** made the first systematic investigation of the essential nutritional requirements of cells in tissue culture and found that animal cells could propagate in a defined mixture of small molecules supplemented with a small proportion of serum proteins.
1956	**Puck** and associates selected mutants with altered growth requirements from cultures of HeLa cells.
1958	**Temin** and **Rubin** developed a quantitative assay for the infection of chick cells in culture by purified *Rous sarcoma virus.* In the following decade, the characteristics of this and other types of viral transformation were established by **Stoker, Dulbecco, Green,** and other virologists.
1961	**Hayflick and Moorhead** showed that human fibroblasts die after a finite number of divisions in culture.
1964	**Littlefield** introduced HAT medium for the selective growth of somatic cell hybrids. Together with the technique of cell fusion, this made somatic-cell genetics accessible.
	Kato and Takeuchi obtained a complete carrot plant from a single carrot root cell in tissue culture.
1965	**Ham** introduced a defined, serum-free medium able to support the clonal growth of certain mammalian cells.
	Harris and Watkins produced the first heterocaryons of mammalian cells by the virus-induced fusion of human and mouse cells.
1968	**Augusti-Tocco and Sato** adapted a mouse nerve cell tumor (neuroblastoma) to tissue culture and isolated clones that were electrically excitable and that extended nerve processes. A number of other differentiated cell lines were isolated at about this time, including skeletal-muscle and liver cell lines.
1975	**Köhler and Milstein** produced the first monoclonal-antibody-secreting hybridoma cell lines.
1976	**Sato** and associates published the first of a series of papers showing that different cell lines require different mixtures of hormones and growth factors to grow in serum-free medium.

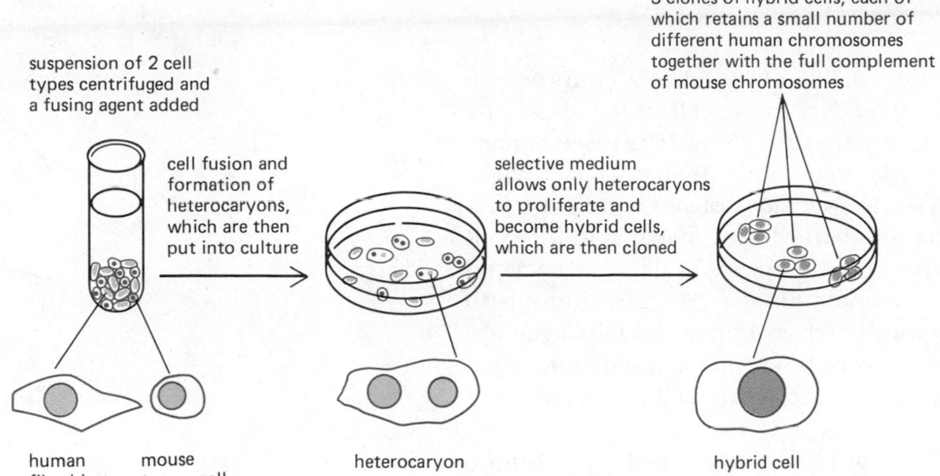

Figure 4–27 Schematic drawing showing how human cells and mouse cells are fused to produce heterocaryons (each with two or more nuclei) that lead eventually to hybrid cells (each with one fused nucleus). These particular hybrid cells are useful for mapping human genes on specific human chromosomes because most of the human chromosomes are quickly lost in a random manner, leaving clones that retain only one or a few. The hybrid cells produced by fusing other types of cells often retain most of their chromosomes.

Summary

Many animal and plant cells will survive and often proliferate in a culture dish if they are provided with a suitable nutrient medium; different cell types require different nutrients, including one or more specific protein growth factors. Although most animal cells die after a finite number of divisions, rare variant cells arise spontaneously in culture that can be maintained indefinitely as cell lines. Clones derived from a single ancestor cell can be obtained from cell lines, making it possible to isolate mutant cells with defects in a single protein. Two different types of cell can be fused to produce heterocaryons (cells with two nuclei), which eventually form hybrid cells (with one fused nucleus). Such cells are useful for studying the interactions between the components of two different cells and provide a convenient method for assigning genes to specific chromosomes.

The Fractionation of Cells and Their Contents[14]

While one can determine the arrangement of organelles and large macromolecular aggregates in cells and tissues by microscopy, and even locate specific molecules using specific staining procedures, a detailed molecular understanding of a cell requires biochemical analysis. Such analyses usually entail the disruption of the cell and the consequent obliteration of its delicate anatomy. To retain as much information as possible about the original location of the molecule under analysis, biologists have developed techniques for disrupting tissues and cells in a controlled fashion, so that different cells and different components of cells can be separated before biochemical analysis.

Cells Can Be Isolated from a Tissue and Separated into Different Types[15]

Many types of differentiated cells are not readily obtained as a cultured cell line. In any case, it is usually cheaper and quicker to use cells isolated directly from an animal or plant for large-scale biochemical analyses. The disadvantage is that all tissues in a higher animal or plant contain a mixture of cell types, which must be separated before analysis. Suspensions of single cells are first prepared from the tissue by disrupting the extracellular matrix and intercellular junctions that hold the cells together. The best yields of viable dissociated cells are usually obtained from fetal or neonatal tissues. The procedure is to treat the tissues with proteolytic enzymes (such as trypsin and collagenase) and agents that bind, or *chelate,* Ca^{2+} (such as ethylenediaminetetraacetic acid, or EDTA) and then to dissociate them into single cells by gentle mechanical disruption.

Several approaches are used to separate the different cell types from a mixed cell suspension. One is to exploit the differences in the cells' physical properties. For example, large cells can be separated from small cells and dense cells from light cells by sedimentation or centrifugation; these techniques will be described when we discuss the separation of organelles and macromolecules, for which they were originally developed. Another approach is based on the fact that some cells adhere strongly to glass or plastic and therefore can be separated from cells that adhere less strongly.

An important refinement of this last technique depends on the specific binding properties of antibodies. Antibodies that bind specifically to the surface of only one cell type in a tissue can be coupled to various matrices—such as collagen, polysaccharide beads, or plastic—to form an "affinity surface" to which only cells recognized by the antibodies will adhere. The bound cells are then recovered by gentle shaking or, in the case of a digestible matrix (such as collagen), by degrading the matrix with enzymes (such as collagenase).

The most sophisticated cell-separation technique involves labeling specific cells with antibodies coupled to a fluorescent dye and then separating the labeled cells from the unlabeled ones in an electronic **fluorescence-activated cell sorter.** Here, individual cells traveling in single file in a fine stream are assessed for their fluorescence by passing them through a laser beam. Slightly further downstream, tiny droplets, most containing either one or no cells, are formed by a vibrating nozzle. The droplets containing a single cell are automatically given a positive or a negative charge at the moment of formation, depending on whether they contain a fluorescent cell; they are then deflected by a strong electric field into an appropriate container. Occasional clumps of cells, detected by their increased light scattering, are left uncharged and are discarded into a waste container (Figure 4–28). Such machines can select 1 cell in 1000 and sort about 5000 cells each second.

Organelles and Macromolecules Can Be Separated by Ultracentrifugation[16]

The cells in a purified population can be disrupted in various ways: by osmotic shock, by ultrasonic vibration, by forcing the cells through a small orifice, or by grinding them up. These procedures break many of the membranes of the cell (including the plasma membrane and membranes of the endoplasmic reticulum and Golgi apparatus) into fragments that immediately reseal to form small, closed vesicles. But, if carefully applied, the disruption procedures leave organelles such as nuclei, mitochondria, lysosomes, and peroxisomes intact.

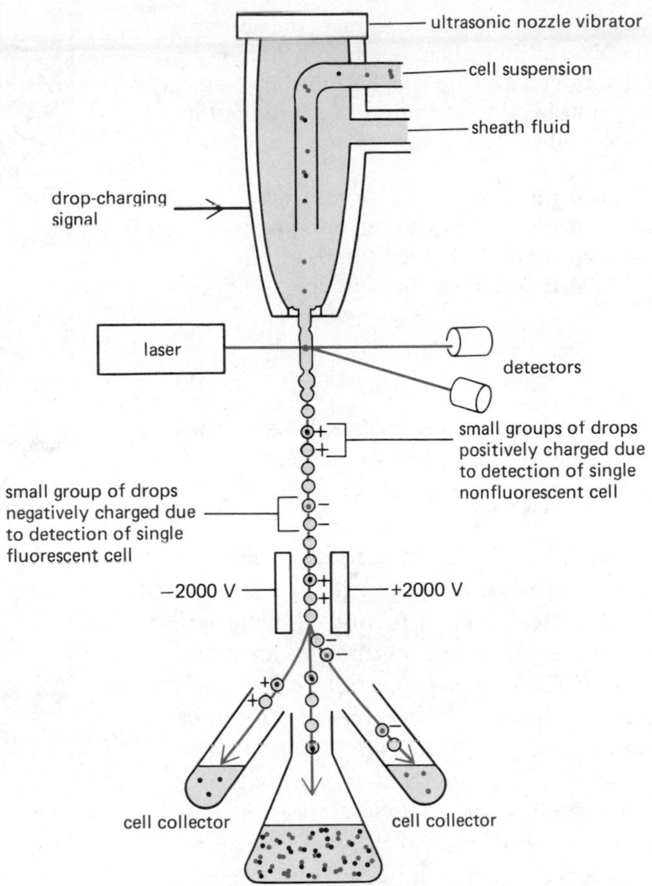

Figure 4–28 Schematic diagram of a fluorescence-activated cell sorter. When a cell passes through the laser beam, it is monitored for fluorescence. Droplets containing single cells are given a negative or positive charge, depending on whether the cell is fluorescent or not. Droplets are then deflected by an electric field into collection tubes according to their charge. Note that the cell concentration must be adjusted so that most droplets contain no cell; most of these flow to a waste container together with any cell clumps.

The population of cells is thereby reduced to a soluble extract containing a thick suspension of membrane-bounded particles, each with a distinctive size, charge, and density. Provided that the homogenization medium has been carefully chosen (this requires extensive trial and error for each organelle), the various particles retain most of the biochemical properties of the original organelles in the intact cell.

Separating the various components in this mixture became possible only after the commercial development in the early 1940s of an instrument known as the **preparative ultracentrifuge,** in which extracts of broken cells are rotated at high speeds (Figure 4–29). At a relatively low speed, large components, such as nuclei and unbroken cells, sediment rapidly and form a pellet at the bottom of the centrifuge tube; at a slightly higher speed, a pellet of mitochrondria is deposited; and at even higher speeds and longer periods of

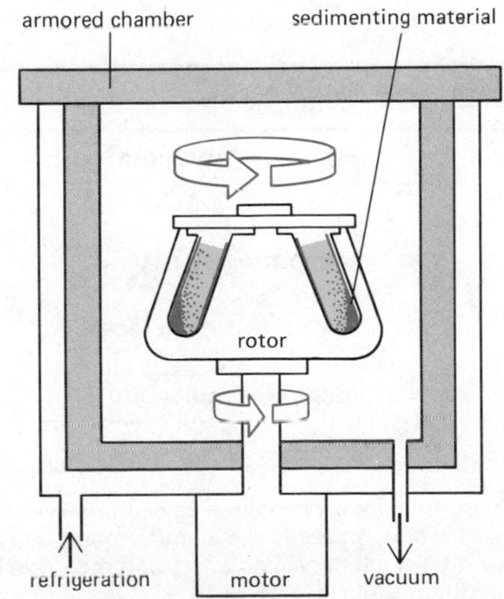

Figure 4–29 Diagram of a preparative ultracentrifuge. The sample is contained in tubes that are inserted into a ring of cylindrical holes in a metal rotor. The rapid rotation of the rotor generates enormous centrifugal forces that cause particles in the sample to sediment. The vacuum reduces friction; it thereby prevents heating of the rotor and allows the refrigeration system to maintain the sample at 4°C.

Figure 4–30 Schematic drawing showing how extracts of cells are repeatedly centrifuged at progressively higher speeds in order to fractionate their components. In general, the smaller the subcellular component, the greater the centrifugal force required to sediment it. Typical values for the various centrifugation steps referred to in the figure are

low speed:	1,000 times gravity for 10 minutes
medium speed:	20,000 times gravity for 20 minutes
high speed:	80,000 times gravity for 1 hour
very high speed:	150,000 times gravity for 3 hours

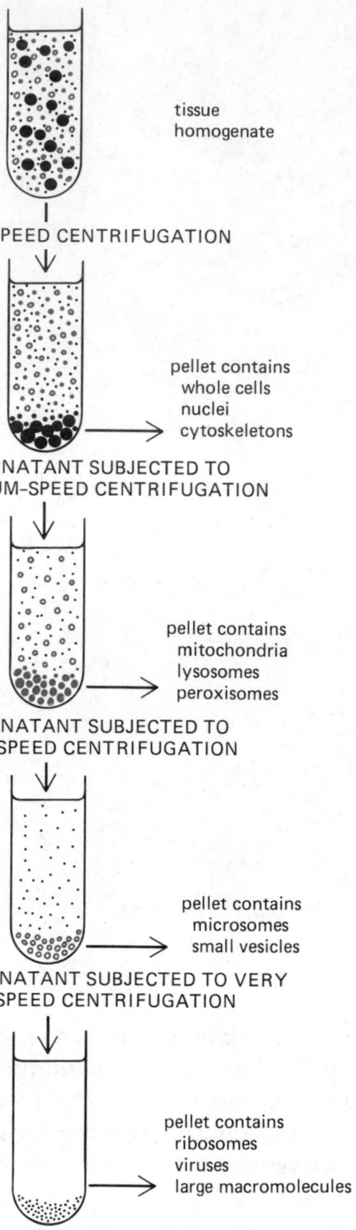

centrifugation, first the small, closed vesicles and then the ribosomes can be collected (Figure 4–30). All of these fractions are impure, but resuspending the pellet and repeating the centrifugation procedure several times removes many of their contaminants.

A finer degree of separation can be achieved by layering the cell homogenate as a narrow band on top of a salt solution in a centrifuge tube. To stabilize the sedimenting component against convective mixing, the salt solution beneath the band contains an increasingly dense solution of an inert, highly soluble material such as sucrose (a *density gradient*). Under these conditions, the different fractions sediment at different rates, forming distinct bands that can be individually collected (Figure 4–31). The rate at which each component sediments depends on its size and shape and is normally expressed as its *sedimentation coefficient* or *s value* (see Table 4–7). Present-day ultracentrifuges rotate at speeds up to 80,000 rpm and produce forces up to 500,000 times gravity. At these enormous forces, even relatively small macromolecules, such as tRNA molecules and simple enzymes, separate from one another on the basis of their size.

The ultracentrifuge is also used to separate cellular components on the basis of their *buoyant density* rather than their size. In this case, the sample is sedimented through a steep gradient that contains a very high concentration of sucrose or cesium chloride. The cellular components move down the gradient until they reach a position that is equal to their own density, and at this point they float and can move no further. This method can be so sensitive that it is capable of resolving macromolecules that have incorporated heavy

Table 4–7 Some Typical Sedimentation Coefficients

Particle or Molecule	Sedimentation Coefficient
Lysosome	9400S
Tobacco mosaic virus	198S
Ribosome	80S
Ribosomal RNA molecule	28S
tRNA molecule	4S
Hemoglobin molecule	4.5S

Sedimentation coefficients (*s*), in units of seconds, are given by $\frac{dx/dt}{\omega^2 x}$, where x is the distance from the center of rotation in centimeters, dx/dt is the speed of sedimentation in centimeters per second, and ω is the angular rotation of the centrifuge rotor in radians per second. Because such coefficients are extremely small numbers, they are normally expressed in Svedberg units (S), where $1S = 1 \times 10^{-13}$ sec.

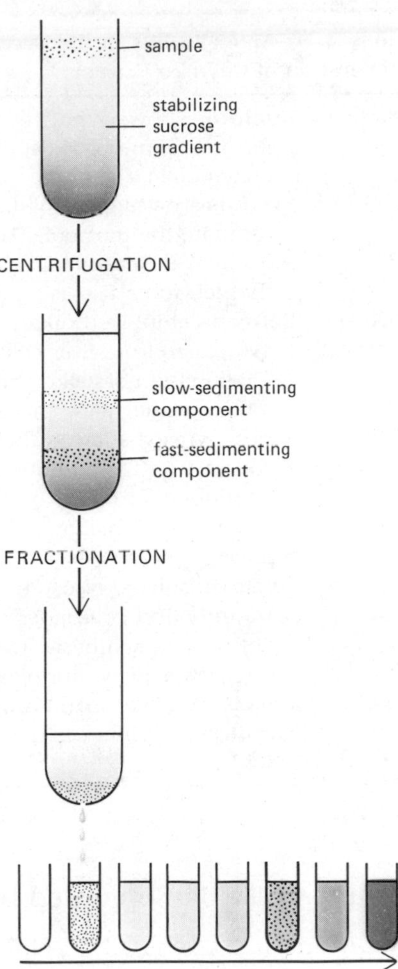

Figure 4–31 Subcellular components sediment at different speeds according to their size when layered over a sucrose-containing solution. In order to stabilize the sedimenting bands against convective mixing, the tube contains a continuous gradient of sucrose that increases in concentration toward the bottom of the tube (typical values might run from 5% to 20% sucrose). Following centrifugation, the different components may be collected individually, most simply by puncturing the plastic centrifuge tube and collecting drops from the bottom, as illustrated here.

isotopes, such as ^{13}C or ^{15}N, from the normal unlabeled species. In fact, the cesium chloride method was developed in 1957 to separate the labeled and unlabeled DNA produced after exposing a growing population of bacteria to nucleotide precursors containing ^{15}N; this classic experiment provided direct evidence for the semiconservative replication of DNA (p. 103).

The Molecular Details of Complex Cellular Processes Can Be Determined Only by Studies of Fractionated Cell Components[17]

As one of the principal means of isolating organelles and other large subcellular components, the ultracentrifuge has played a very important part in determining the *functions* of different components of the cell. For example, it was the availability of purified fractions derived from mitochondria and chloroplasts that led to an understanding of the central function of these organelles in energy interconversions. Similarly, resealed vesicles formed from the endoplasmic reticulum and the Golgi apparatus can be separated from each other and used as miniature versions of the intact organelle—allowing the many functions of each organelle to be analyzed in detail.

Fractionated cell extracts (also called **cell-free systems**) are widely used to study the processes that occur in cells. Only in this way can a biological process be isolated and studied free from all of the complex side reactions that occur in a cell and the detailed molecular mechanisms thereby be established. An early triumph of this approach was the elucidation of the mechanism of protein synthesis. The starting point was a crude cell extract that could translate RNA molecules into protein. Fractionation of this extract, step by step, produced in turn the ribosomes, tRNA, and various enzymes that together constitute the protein synthetic machinery. Once individual, pure components were available, they could be separately added or withheld, allowing their exact role in the process to be defined. The same system was later used to decipher the genetic code by using synthetic polyribonucleotides of known sequence as the "messenger RNA" (mRNA) to be translated. Today a variety of *in vitro* translation systems are used to determine the proteins encoded by purified preparations of mRNA in association with various procedures used for DNA cloning (see p. 188 and Table 4–13). Some landmarks in the development of methods used in the preparation of fractionated cell extracts are outlined in Table 4–8.

A major goal of cell biological research is to develop cell-free systems that carry out even very complex processes, such as the segregation of chromosomes on the mitotic spindle. But ultimately the analysis of cell-free systems depends on the complete separation of each of the individual protein components of the system. How this is achieved is discussed in the following sections.

Table 4–8 Major Events in the Development of the Ultracentrifuge and the Preparation of Cell-free Extracts

1897	**Buchner** showed that cell-free extracts of yeast can ferment sugars to form carbon dioxide and ethanol, laying the foundations of enzymology.
1926	**Svedberg** developed the first analytical ultracentrifuge and used it to estimate the molecular weight of hemoglobin as 68,000.
1935	**Pickels and Beams** introduced several new features of centrifuge design that led to its use as a preparative instrument.
1938	**Behrens** employed differential centrifugation to separate nuclei and cytoplasm from liver cells, a technique further developed for the fractionation of cell organelles by **Claude, Brachet, Hogeboom,** and others in the 1940s and early 1950s.
1949	**Szent-Györgyi** showed that isolated myofibrils from skeletal muscle cells contract on the addition of ATP. In 1955, a similar cell-free system was developed for ciliary beating by **Hofmann-Berling.**
1951	**Brakke** used density gradient centrifugation in sucrose solutions to purify a plant virus.
1953	**de Duve** isolated lysosomes and, later, peroxisomes by centrifugation.
1954	**Zamecnik** and colleagues developed the first cell-free system to carry out protein synthesis. This was followed by a decade of intense research activity, during which the *genetic code* was elucidated.
1957	**Meselson, Stahl, and Vinograd** developed density gradient centrifugation in cesium chloride solutions for separating nucleic acids.

Proteins Can Be Separated by Chromatography

One of the most generally useful methods of protein fractionation involves **chromatography,** a technique that was originally developed for the fractionation of low molecular weight components, such as sugars and amino acids. The most common type of chromatography used to separate small molecules is known as **partition chromatography.** In the most convenient form of this technique, a drop of the sample is applied as a spot to a sheet of absorbent paper (in *paper chromatography*) or to a sheet of plastic or glass that has been covered with a thin layer of inert absorbent material such as cellulose or silica gel (*thin-layer chromatography*). Then a mixture of solvents, such as water and an alcohol, is allowed to permeate the sheet from one edge. As the solvents move across the sheet, they pick up those molecules in the sample that are soluble in them. The solvents are selected so that one of them adsorbs more strongly to the absorbent material than the other. Consequently, molecules that are most soluble in the adsorbed solvent are relatively retarded, while those that are most soluble in the other solvent move more quickly. To detect the location of the different molecules, the chromatograph is dried and then stained (Figure 4–32). Various forms of paper and thin-layer chromatography are still widely used to analyze many different small molecules.

Proteins are most often fractionated by *column chromatography,* in which a mixture of proteins in solution is passed through a column containing a porous solid matrix and the different proteins are retarded to different extents by their interaction with the matrix. In this way, different proteins can be separately collected as they flow out of the bottom of the column (Figure 4–33). Many types of matrices have been developed that allow proteins to be separated without altering their native structure. These matrices discriminate between different proteins by charge, size, or their ability to bind to particular chemical groups on the matrix.

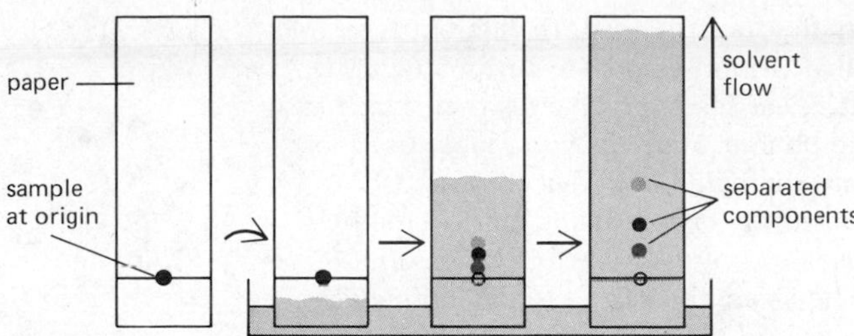

Figure 4–32 The separation of small molecules by paper chromatography. After the sample has been applied to the origin and dried, a solution containing a mixture of two solvents is allowed to flow slowly through the paper by capillary action. Different components in the sample move at different rates in the paper according to their relative solubility in the solvent that is preferentially adsorbed by the paper.

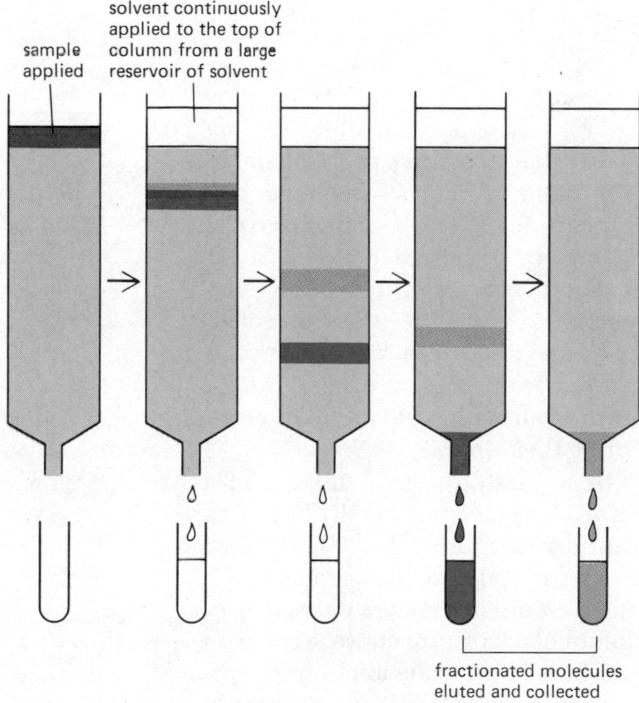

Figure 4–33 The separation of molecules by column chromatography. The sample is applied to the top of a cylindrical column made of glass or plastic containing a permeable solid matrix immersed in solvent. Then a large amount of solvent is pumped slowly through the column and is collected in separate tubes as it emerges from the bottom. Various components of the sample travel at different rates through the column and are thereby fractionated.

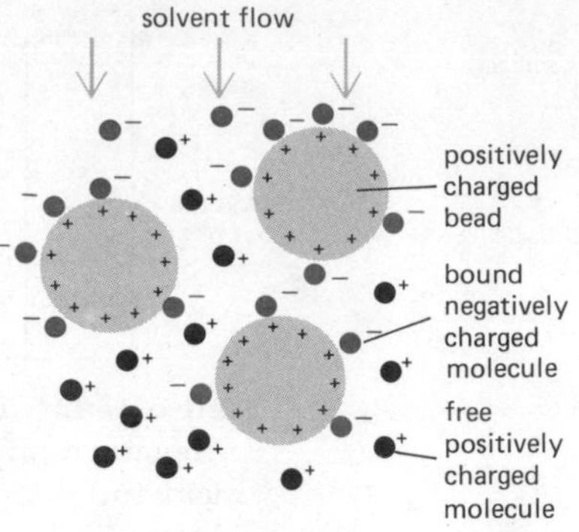

(A) ION-EXCHANGE CHROMATOGRAPHY

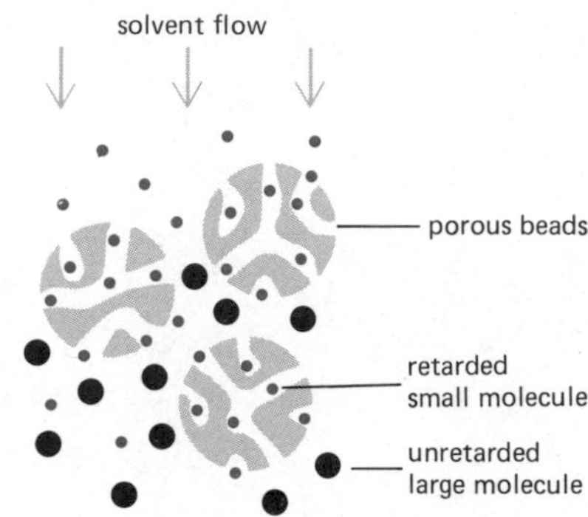

(B) GEL-FILTRATION CHROMATOGRAPHY

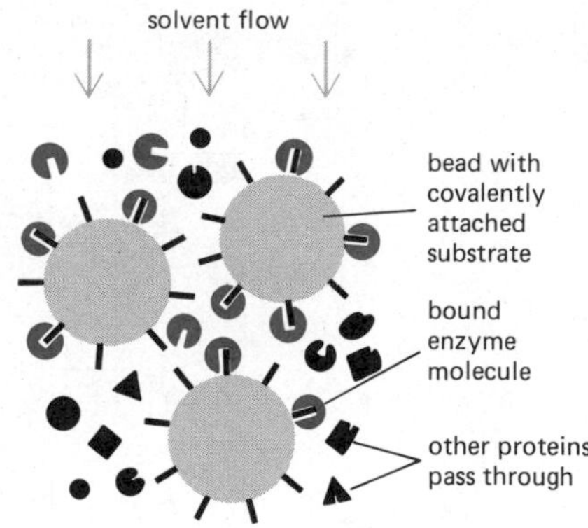

(C) AFFINITY CHROMATOGRAPHY

Figure 4–34 Schematic drawing of some different types of chromatographic matrix. In ion-exchange chromatography (A), the insoluble matrix carries ionic charges that retard molecules of opposite charge. Matrices commonly used for separating proteins are diethylaminoethylcellulose (DEAE-cellulose), which is positively charged; and carboxymethylcellulose (CM-cellulose) and phosphocellulose, which are negatively charged. The strength of the association between the dissolved molecules and the ion-exchange matrix depends on both the ionic strength and the pH of the eluting solution, which may therefore be varied in a systematic fashion (as in Figure 4–35) to achieve an effective separation. In gel-filtration chromatography (B), the matrix is inert but porous. Molecules that are small enough to penetrate into the matrix have a larger volume of solvent available to them and therefore travel more slowly through the column. Beads of cross-linked polysaccharide (dextran or agarose) are available commercially in a wide range of pore sizes, making them suitable for the fractionation of molecules of various molecular weights, from less than 500 to over 5×10^6. Affinity chromatography (C) utilizes an insoluble matrix that is covalently linked to a specific ligand, such as an antibody molecule or an enzyme substrate, that will bind a specific protein. Enormous purifications are often achieved in a single pass through such an affinity column.

Many different types of matrices are commercially available for this purpose (Figure 4–34). *Ion-exchange columns* are packed with small beads that carry either a positive or negative charge, so that proteins are fractionated according to the arrangement of charges on their surface. *Hydrophobic columns* are packed with beads from which hydrophobic side chains protrude, so that proteins with exposed hydrophobic regions are retarded. *Gel-filtration columns* are packed with tiny porous beads and separate proteins according to their size: molecules that are small enough to enter the pores percolate inside successive beads as they travel, while large molecules remain between the beads and therefore flow more rapidly through the column, emerging first. In addition to providing a means of separating molecules, gel-filtration chromatography is a convenient way to determine their size.

A particular protein cannot usually be separated from a mixture in a single step by column chromatography because each step generally increases the proportion of the protein in the mixture by no more than 20-fold. Since most proteins represent less than 1/1000 of the total cellular protein, it is usually necessary to use several different types of column in succession in order to purify them (Figure 4–35). A far more efficient procedure, known as **affinity chromatography,** takes advantage of the biologically important binding interactions that occur on protein surfaces. For example, if an enzyme substrate is covalently coupled to an inert matrix, such as a polysaccharide bead, the enzyme can often be retained by the matrix along with very few other proteins. In a similar way, specific antibodies can be coupled to a matrix, which can then be used to purify molecules recognized by the antibodies. Because of the great specificity of such *affinity columns,* 1000- to 10,000-fold purifications can sometimes be achieved in a single pass through the column.

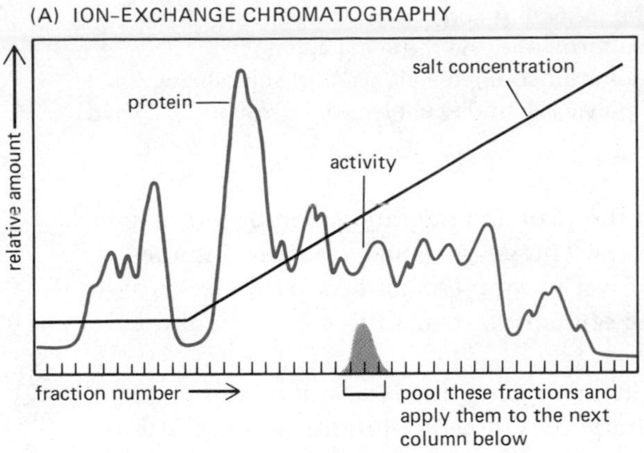

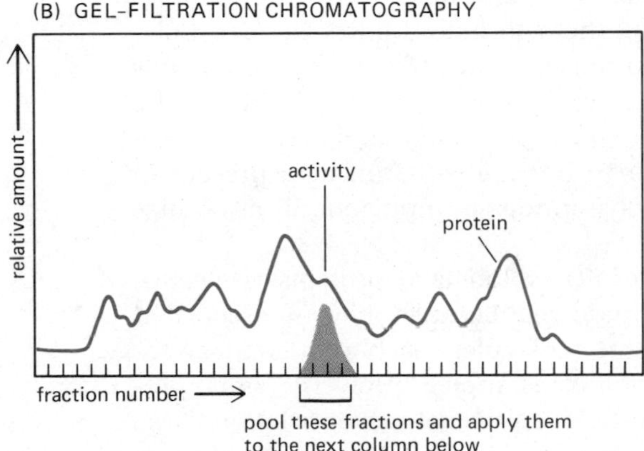

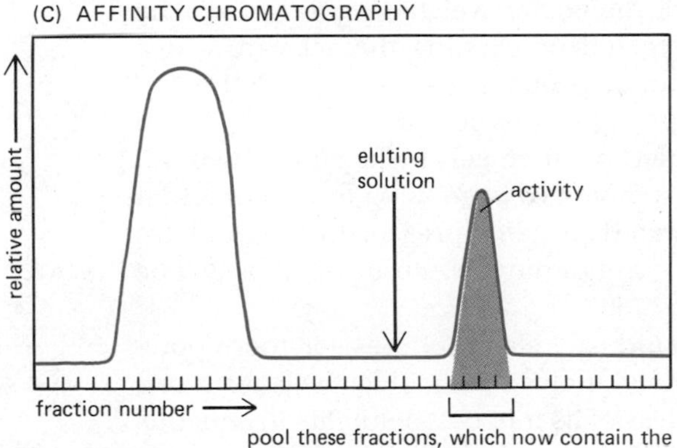

Figure 4–35 Typical results obtained when three different chromatographic steps are used in succession to purify a protein. In this example, a whole cell extract was first fractionated by allowing it to percolate through an ion-exchange resin packed into a column (A). The column was washed and the bound proteins were then eluted by passing a solution containing a gradually increasing concentration of salt onto the top of the column. Proteins with the lowest affinity for the ion-exchange resin passed directly through the column and were collected in the earliest wash fractions eluted from the bottom of the column. The remaining proteins were eluted in sequence according to their affinity for the resin—those proteins binding the tightest to the resin requiring the highest concentration of salt to remove them. The protein of interest eluted in a narrow peak and was detected by its enzymatic activity. The fractions with activity were pooled and then applied to a second, gel-filtration column (B). The elution position of the still impure protein was again determined by its enzymatic activity and the active fractions pooled and purified to homogeneity on an affinity column (C) that contained an immobilized substrate of the enzyme.

The Size and Subunit Composition of a Protein Can Be Determined by SDS Polyacrylamide-Gel Electrophoresis[18]

Proteins usually have a net positive or negative charge due to the negatively and positively charged groups of the amino acids on their surface. If an electric field is applied to a solution containing a protein molecule, the protein will migrate at a rate that depends on its net charge and on its size and shape. This technique, known as **electrophoresis,** is used to separate mixtures of proteins either in free aqueous solution or in solutions held in a solid porous matrix such as starch.

In the mid-1960s a modified version of this method—known as **SDS polyacrylamide-gel electrophoresis** (or SDS-PAGE)—was developed. In this procedure, which has revolutionized the way that proteins are routinely ana-

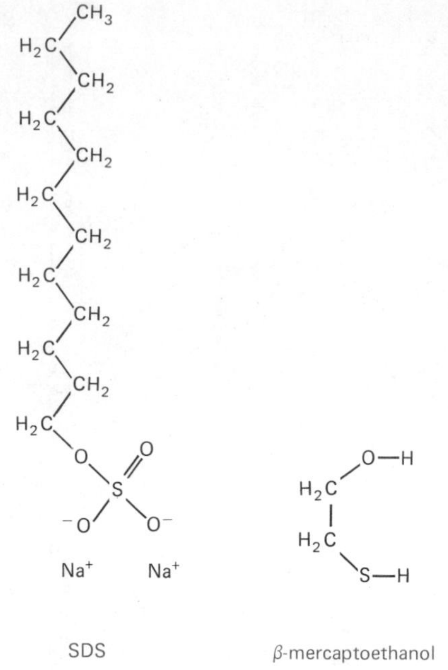

Figure 4–36 The detergent sodium dodecyl sulfate (SDS), here in its ionized form, and the reducing agent β-mercaptoethanol are two chemicals used in solubilizing proteins for SDS polyacrylamide-gel electrophoresis.

SDS

β-mercaptoethanol

lyzed, the inert matrix through which the proteins migrate is composed of a highly cross-linked gel of polyacrylamide. The gel is usually prepared immediately before use by polymerization from monomers, and its pore size can be adjusted so that the pores are small enough to retard the migration of the protein molecules of interest.

The proteins themselves are not in a simple aqueous solution but in one that includes a powerful, negatively charged detergent, **sodium dodecyl sulfate,** or **SDS** (Figure 4–36). Because this detergent binds to hydrophobic regions of the protein molecules, causing them to unfold into extended polypeptide chains, the individual protein molecules are freed from their associations with other proteins or lipid molecules and rendered freely soluble in the detergent solution. In addition, a reducing agent such as *mercaptoethanol* (Figure 4–36) is usually added in order to break any S-S linkages present in the proteins so that the constituent polypeptides in multisubunit molecules can be analyzed.

What happens when a mixture of SDS-solubilized proteins is electrophoresed through a slab of polyacrylamide gel? Because they are associated with many negatively charged detergent molecules, each protein migrates toward the positive electrode when a voltage is applied. However, small proteins move much more readily through holes in the gel meshwork than large ones, and, as a result, a series of discrete protein bands is produced from a complex mixture, arranged in order of molecular weight (Figure 4–37). The major proteins present are readily detected by staining the gel with a dye such as Coomassie blue, while even minor proteins are seen in gels treated with a silver stain (where as little as 10 ng of protein can be detected in a band). A specific protein can be identified on such gels by labeling it with an antibody that has been coupled to a radioactive isotope or to a fluorescent dye; this is often done after the separated proteins present in the gel have been transferred (by "blotting") onto a sheet of nitrocellulose paper, as will be described later for nucleic acids (see p. 190).

The procedure of SDS polyacrylamide-gel electrophoresis is more powerful than any previous method of protein analysis principally because it can be used to separate any protein, regardless of its inherent solubility in aqueous solution. Membrane proteins, protein components of the cytoskeleton, and proteins that are part of large macromolecular aggregates can be resolved as separate species. Since the method separates polypeptides strictly according to size, it also provides information about the molecular weight and the subunit composition of any protein complex (Figure 4–37).

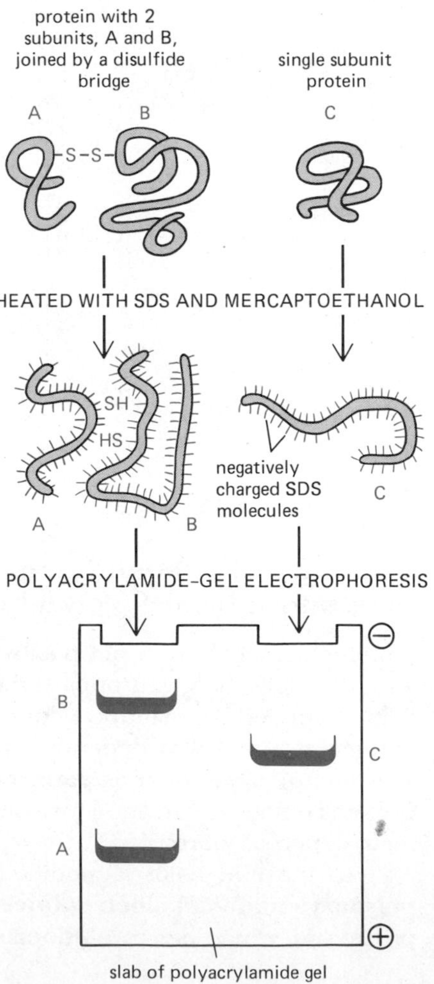

Figure 4–37 Schematic illustration of SDS polyacrylamide-gel electrophoresis, a powerful method for protein fractionation. Individual polypeptide chains form a complex with negatively charged molecules of sodium dodecyl sulfate (SDS) and therefore migrate as a negatively charged SDS-protein complex through a porous gel of polyacrylamide. Since the speed of migration under these conditions is greater the smaller the size of the polypeptide, this technique can be used to determine the approximate molecular weight of a polypeptide chain as well as the subunit composition of a protein.

More Than 1000 Proteins Can Be Resolved on a Single Gel by Two-dimensional Polyacrylamide-Gel Electrophoresis[19]

Closely spaced protein bands or peaks tend to overlap, so that any one-dimensional separation method, such as SDS polyacrylamide-gel electrophoresis or chromatography, can resolve only a relatively small number of proteins (generally fewer than 50). A new method for analyzing proteins, called **two-dimensional gel electrophoresis,** was developed in 1975. Combining the features of two different separation procedures, it can resolve more than 1000 different proteins in the form of a two-dimensional protein map.

In the first step, the sample is dissolved in a small volume of a solution containing a nonionic detergent, the denaturing reagent urea, and mercaptoethanol. This solution solubilizes, denatures, and dissociates all of the polypeptide chains without changing their intrinsic charge. The polypeptide chains are then fractionated by a procedure called **isoelectric focusing,** which depends on the fact that the net charge on a protein molecule varies with the pH of the surrounding solution. At a pH that is characteristic for each protein, there exists an *isoelectric point* at which the protein has no net charge and therefore will not migrate in an electric field. In isoelectric focusing, proteins are electrophoresed in a narrow tube of polyacrylamide gel in which a gradient of pH is established by mixtures of special buffers. Each protein moves to the position in the gradient that corresponds to its isoelectric point and stays there (Figure 4–38). This is the first dimension of two-dimensional gel electrophoresis.

In the second step, the narrow gel containing the separated proteins is soaked in SDS and the proteins are further fractionated according to size by electrophoresis through a slab of SDS polyacrylamide gel. Each polypeptide chain now migrates to a discrete spot on the gel according to its molecular weight. This is the second dimension of two-dimensional gel electrophoresis. The only proteins left unresolved will be those that have both an identical size and an identical isoelectric point, a relatively rare situation. Trace amounts of each polypeptide chain present can be detected by various staining procedures or by autoradiography if the protein sample was labeled with a radioisotope (see p. 180). Up to 2000 individual polypeptide chains can be resolved on a single two-dimensional gel—enough to account for most of the proteins in a bacterium (Figure 4–39). The resolving power is so great that two nearly identical proteins, differing in a single charged amino acid, can be distinguished.

Some landmarks in the development of chromatography and electrophoresis are outlined in Table 4–9.

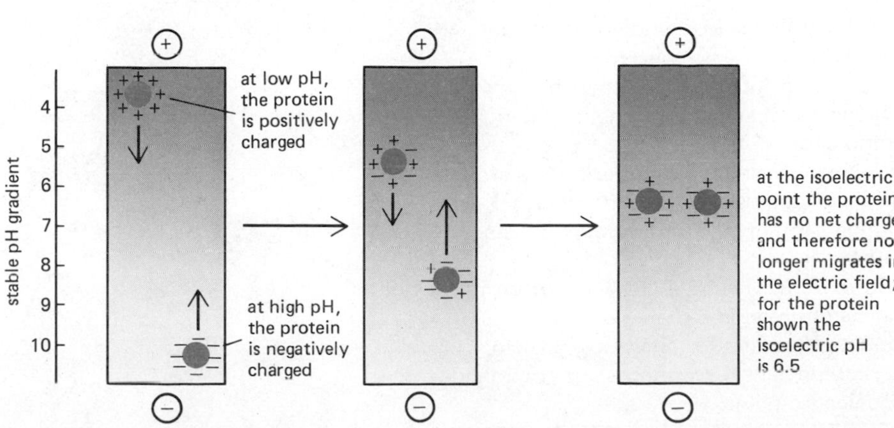

Figure 4–38 The separation of protein molecules by isoelectric focusing. At low pH (high H^+ concentration), the carboxylic acid groups of proteins tend to be uncharged (—COOH) and their nitrogen-containing basic groups fully charged (for example, $—NH_3^+$), giving most proteins a net positive charge. At high pH, the carboxylic acid groups are negatively charged ($—COO^-$) and the basic groups tend to be uncharged (for example, $—NH_2$), giving most proteins a net negative charge (see Figure 2–8, p. 57). At its *isoelectric pH,* a protein has no net charge. Thus, when a tube containing a fixed pH gradient is subjected to a strong electric field, each protein species present will migrate until it forms a sharp band at its isoelectric pH, as shown.

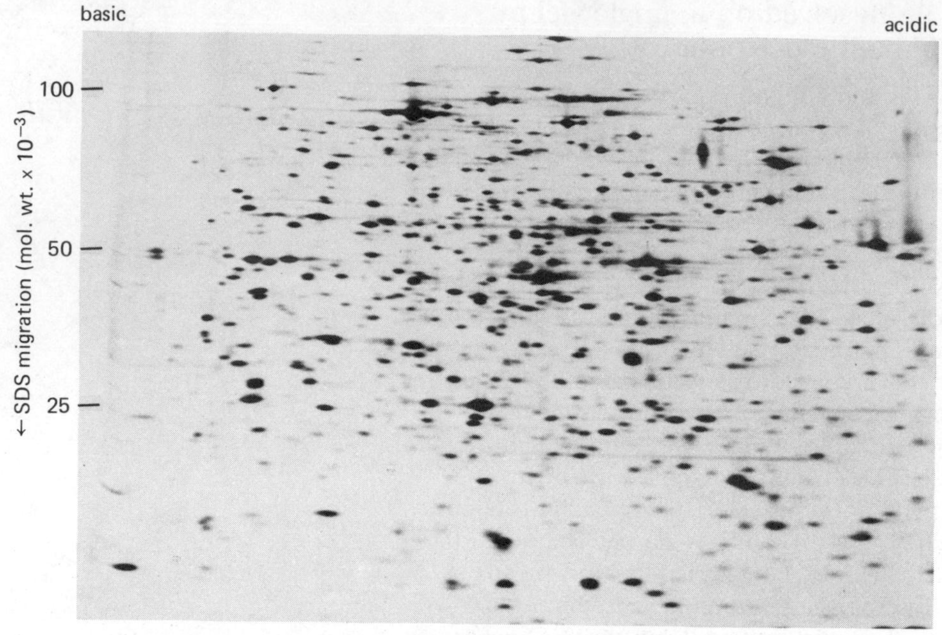

Figure 4–39 The fractionation of all of the proteins present in an *E. coli* bacterial cell by two-dimensional polyacrylamide gel electrophoresis. Each spot corresponds to a different polypeptide chain. The proteins were first separated according to their isoelectric points by isoelectric focusing from left to right. These proteins were then further fractionated according to their polypeptide molecular weights by electrophoresis from top to bottom in the presence of SDS. Note that different proteins are present in very different amounts. (Courtesy of Patrick O'Farrell)

Table 4–9 Landmarks in the Development of Chromatography and Electrophoresis and Their Application to Biological Molecules

1833	**Faraday** described the fundamental laws concerning the passage of electricity through ionic solutions.
1850	**Runge** separated inorganic chemicals by their differential adsorption to paper, a forerunner of later chromatographic separations.
1906	**Tswett** invented column chromatography, passing petroleum extracts of plant leaves through columns of powdered chalk.
1933	**Tiselius** introduced electrophoresis for separating proteins in solution.
1942	**Martin and Synge** developed partition chromatography, leading to paper chromatography two years later.
1946	**Stein and Moore** determined for the first time the amino acid composition of a protein, initially using column chromatography on starch and later developing chromatography on ion-exchange resins.
1955	**Smithies** used gels made of starch to separate serum proteins by electrophoresis.
	Sanger completed the analysis of the amino acid sequence of bovine insulin, the first protein to be sequenced.
1956	**Ingram** produced the first protein fingerprints, showing that the difference between sickle-cell hemoglobin and normal hemoglobin is due to a change in a single amino acid.
1959	**Raymond** introduced polyacrylamide gels, which are superior to starch gels for separating proteins by electrophoresis; improved buffer systems allowing high-resolution separations were developed in the next few years by **Ornstein** and **Davis.**
1966	**Maizel** introduced the use of sodium dodecyl sulfate (SDS) for improving polyacrylamide-gel electrophoresis of proteins.
1975	**O'Farrell** devised a two-dimensional gel system for analyzing protein mixtures in which SDS polyacrylamide-gel electrophoresis is combined with separation according to isoelectric point.

Table 4–10 Some Reagents Commonly Used to Cleave Peptide Bonds in Proteins

	Amino Acid 1	Amino Acid 2
Enzyme		
Trypsin	Lys or Arg	any
Chymotrypsin	Phe, Trp, or Tyr	any
V8 protease	Glu	any
Chemical		
Cyanogen bromide	Met	any
2-Nitro-5-thiocyanobenzoate	any	Cys

The specificity for the amino acids on either side of the cleaved bond is indicated. The carboxyl group of amino acid 1 is released by the cleavage; this amino acid is to the left of the peptide bond as normally written (see Panel G, pp. 58–59).

Selective Cleavage of a Protein Generates a Distinctive Set of Peptide Fragments[20]

While the molecular weight and isoelectric point are distinctive features of a protein, unambiguous identification ultimately depends on determining its amino acid sequence. The first stage of this process, which involves cleaving the protein into smaller fragments, can itself provide a great deal of useful information about a protein. Proteolytic enzymes and chemical reagents are available that will cleave proteins between specific amino acid residues (Table 4–10). The enzyme *trypsin,* for instance, cuts on the carboxyl side of lysine or arginine residues, whereas the chemical *cyanogen bromide* cuts peptide bonds next to methionine residues. Since these enzymes and chemicals cleave at relatively few sites in a protein, they tend to produce rather large peptides. If such a mixture of peptides is then separated by chromatographic or electrophoretic procedures, the resulting pattern, or **peptide map,** is diagnostic of the protein from which the peptides were generated and is sometimes referred to as the protein's "fingerprint" (Figure 4–40).

Protein fingerprinting was developed in 1956 as a way of comparing normal hemoglobin with the mutant form of the protein found in patients suffering from *sickle-cell anemia.* A single peptide difference was found and eventually traced to a single amino acid change, thus providing the first demonstration that a mutation can change a single amino acid in a protein.

A modified form of protein fingerprinting is commonly used in conjunction with SDS gel electrophoresis. Single protein species from such a gel are excised and mixed with a proteolytic enzyme such as papain. If the quantity of proteolytic enzyme and the duration of the digestion are carefully controlled, a set of large, incompletely fragmented polypeptides is produced. When analyzed on a second SDS gel, this mixture gives a simple but distinctive set of bands that is often sufficient to determine, for example, whether two proteins that have been isolated in different ways are identical.

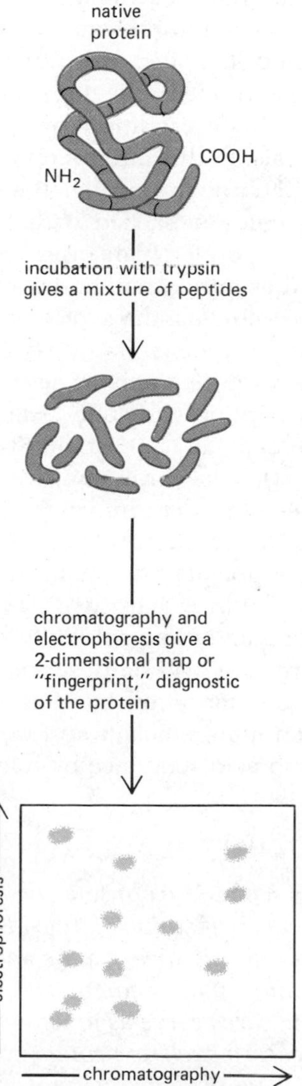

Figure 4–40 Production of a peptide map, or fingerprint, of a protein. In this case the protein is digested with trypsin to generate a mixture of many small polypeptide fragments, which is then fractionated in two dimensions by electrophoresis and partition chromatography. The pattern of spots obtained is diagnostic of the protein analyzed.

Short Amino Acid Sequences Can Be Analyzed by Automated Machines[21]

Once a protein has been cleaved into smaller peptides, the next logical step in the analysis is to determine the amino acid sequence of each isolated peptide fragment. This is accomplished by a repeated series of chemical reactions originally devised in 1967. First the peptide is exposed to a chemical that reacts only with the free amino group at the amino terminus of the peptide. This chemical is then further activated by exposure to a weak acid so that it specifically cleaves the peptide bond attaching the amino-terminal amino acid to the peptide chain; the released amino acid is then identified by chromatographic methods. The remaining peptide, which is shorter by one amino acid, is then submitted to the same sequence of reactions, and so on, until every amino acid in the peptide has been determined.

The reiterative nature of these reactions lends itself to automation, and machines, called **amino acid sequenators,** are commercially available for automatic determination of the amino acid sequence of peptide fragments. The final step is to arrange the sequences of the various peptide fragments in the order in which they occur in the intact polypeptide chain. This is achieved by comparing the sequence of different sets of overlapping peptide fragments obtained by cleaving the same protein with different proteolytic enzymes.

Recent improvements in protein sequencing technology have greatly increased its speed and sensitivity. This is in part the result of miniaturizing the components of the sequenator and in part due to improvements in a rapid, high-resolution form of column chromatography used for amino acid analysis. In this form of chromatography, known as **HPLC** (high-performance liquid chromatography), the solution to be analyzed is passed through a long, thin column packed with tiny beads. By forcing the solution through the column under high pressure, chromatographic separations are achieved in minutes rather than hours and the resolution is enhanced. As a result of these improvements, the sensitivity of amino acid analysis has been increased by a factor of thousands over that of the original method. It is now possible to obtain the sequence of several dozen amino acids at the amino-terminal end of a peptide with only a few micrograms of protein—the amount available from a single band on an SDS polyacrylamide gel.

Despite these improvements and the fact that the process is semiautomated, the determination of the entire amino acid sequence of a protein remains a major undertaking, the difficulty of which increases with the length of the polypeptide chain. A protein of 100 residues can often be sequenced in a month of hard work, but the difficulty of occasionally handling peptide fragments of low solubility prevents the process from being routine. Consequently, as will be described later, the sequences of most proteins are now being determined by cloning the gene for the protein and then using the much more efficient and rapid methods of DNA sequencing to decipher the amino acid sequence by reference to the genetic code.

Summary

Cells isolated from fetal or neonatal tissue usually serve as starting material for the purification of individual cell types. Such purified cells, or homogeneous cultured cell lines, can be analyzed biochemically by disrupting them and fractionating their contents by ultracentrifugation. Fractionated cell extracts in many cases serve as functional cell-free systems for analyzing complex cellular processes such as protein synthesis and DNA replication.

The proteins present in relatively large amounts in soluble cell extracts can readily be purified by column chromatography. The development of different types of column matrices allows biologically active proteins to be separated according to their molecular weight, their charge characteristics, or their affinity for other molecules. In a typical purification, the sample is passed through several such columns in turn, the enriched fractions obtained from one column being applied to the next. Once a protein has been purified to homogeneity, its amino acid sequence can be determined. The protein is first cleaved into a set of smaller peptides whose amino acid sequences are then determined by sensitive automated procedures.

The molecular weight and subunit composition of even very small amounts of protein can be determined by SDS polyacrylamide-gel electrophoresis. In two-dimensional gel electrophoresis, proteins are resolved as separate spots by isoelectric focusing in one dimension followed by SDS polyacrylamide-gel electrophoresis in a second dimension. These electrophoretic separations can be applied even to proteins that are normally insoluble.

Tracing Cellular Molecules with Radioactive Isotopes and Antibodies

Almost any property of a molecule—physical, chemical, or biological—can in principle be used as a means of detecting it. In cell biological studies, molecules are often detected either by their optical properties—whether as a pure substance or in combination with a stain or dye—or by their biochemical activity. In this section we consider two detection methods that have been particularly useful: those involving *radioisotopes* and those utilizing *antibodies*. Each of these methods is capable of detecting specific molecules in a complex mixture. Both methods are potentially very sensitive, and under optimal conditions they can detect fewer than 1000 molecules in a sample.

Radioactive Atoms Can Be Detected with Great Sensitivity[22]

Most naturally occurring elements are a mixture of slightly different *isotopes*, which differ from each other in the weight of their atomic nuclei but, because they have the same external set of electrons, have the same chemical properties. The nuclei of radioactive isotopes, or **radioisotopes,** are unstable and undergo random disintegration to produce different atoms. In the course of these disintegrations, energetic particles such as electrons or radiation such as γ-rays are given off.

Although naturally occurring radioisotopes are rare (because of their instability), radioactive atoms can be produced in large amounts in nuclear reactors in which other atoms are bombarded with high-energy particles (Table 4–11), and many common biological elements are readily available in radioisotopically labeled form. The radiation they emit is detected in various ways. Electrons (β particles) can be detected in a *Geiger counter* by the ionization they produce in a gas or in a *scintillation counter* by the small flashes of light they induce in a scintillation fluid. These methods make it possible to measure the quantity of a particular isotope present in a biological specimen. It is also possible to localize the isotope through its action on the grains of silver in a photographic emulsion, which is subsequently developed into visible spots. These methods of detection are capable of extreme sensitivity; in favorable circumstances, nearly every disintegration—and, therefore, every radioactive atom that decays—can be detected.

Table 4–11 Some Radioisotopes in Common Use in Biological Research

Isotope	Half-Life
^{32}P	14 days
^{131}I	8.1 days
^{35}S	87 days
^{14}C	5570 years
^{45}Ca	164 days
^{3}H	12.3 years

The isotopes are arranged in decreasing order of the energy of the β radiation (electrons) they emit. ^{131}I also emits γ radiation. The *half-life* is the time required for 50% of the atoms of an isotope to disintegrate.

Radioisotopes Are Used to Trace Molecules in Cells and Organisms[22,23]

One of the earliest uses of radioactivity in biology was to trace the chemical pathway of carbon during photosynthesis. Unicellular green algae were maintained in an atmosphere containing radioactively labeled CO_2 ($^{14}CO_2$), and at various times after they had been exposed to sunlight their soluble contents were separated by paper chromatography. Small molecules containing ^{14}C derived from CO_2 were detected by placing a sheet of photographic film over the dried paper chromatogram. In this way, most of the principal components in the photosynthetic pathway from CO_2 to sugar were identified.

Radioactive molecules can be used to follow the course of almost any process in cells. In a typical experiment, a precursor in radioactive form is added to cells so that the radioactive molecules mix with the existing unlabeled ones; both are treated identically by the cell since they differ only in the weight of their atomic nuclei. Changes in the location or chemical form of the radioactive molecules can be followed as a function of time. The resolution of such experiments is often sharpened by using a **pulse-chase** labeling protocol, in which the radioactive material (the *pulse*) is added for only a very brief period and then washed away and replaced by nonradioactive molecules (the *chase*). Samples are taken after various periods of time and the chemical form or location of the radioactivity identified at each time point (Figure 4–41).

By these methods, even the fate of stable and unreactive molecules, such as the proteins of cartilage and bone, can be followed. The use of radioactive tracers, in fact, showed that almost all of the molecules in a living cell are continually being degraded and replaced. Without radioisotopes such slow turnover processes would be almost impossible to detect.

Today nearly all common small molecules are available in radioactive form from commercial sources, and virtually any biological molecule, no matter how complicated, can be labeled with radioactivity. Compounds are often made with radioactive atoms incorporated at particular positions in their structure, enabling the separate fates of different parts of the same molecule to be followed (Figure 4–42).

One of the important uses of radioactivity in cell biology is in the localization of radioactive compounds in sections of whole cells or tissues by **autoradiography.** In this procedure, living cells are briefly exposed to a "pulse" of a specific radioactive compound and incubated for a variable period before being fixed and processed for light or electron microscopy. Each preparation is then overlaid with a thin film of photographic emulsion. After remaining in the dark for a number of days—during which time the radioisotope decays—the emulsion is developed. The position of the radioactivity in each cell can be determined by the position of dark silver grains. For example, by incubating cells with a radioactive DNA precursor ([^3H]*thymidine*), DNA is seen to be made in the nucleus and to remain there. By contrast, labeling cells with a radioactive RNA precursor ([^3H]*uridine*) reveals that RNA is initially made in the nucleus but then rapidly accumulates in the cell cytoplasm.

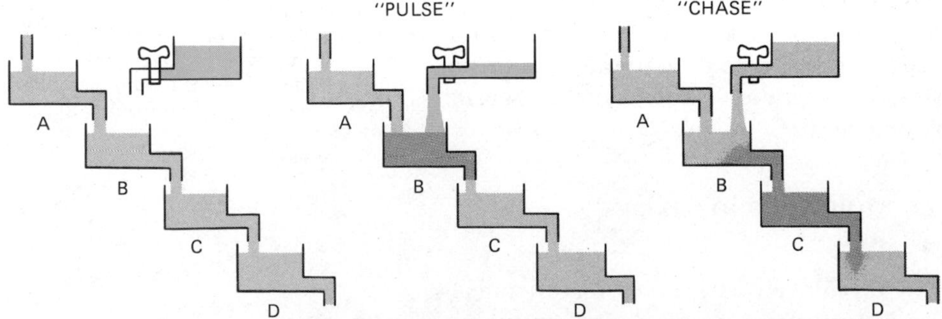

Figure 4–41 Schematic diagram outlining the logic of a typical pulse-chase experiment using radioisotopes. The chambers labeled A, B, C, and D represent either different compartments in the cell (detected by autoradiography or by cell fractionation experiments) or different chemical compounds (detected by chromatography or other chemical methods).

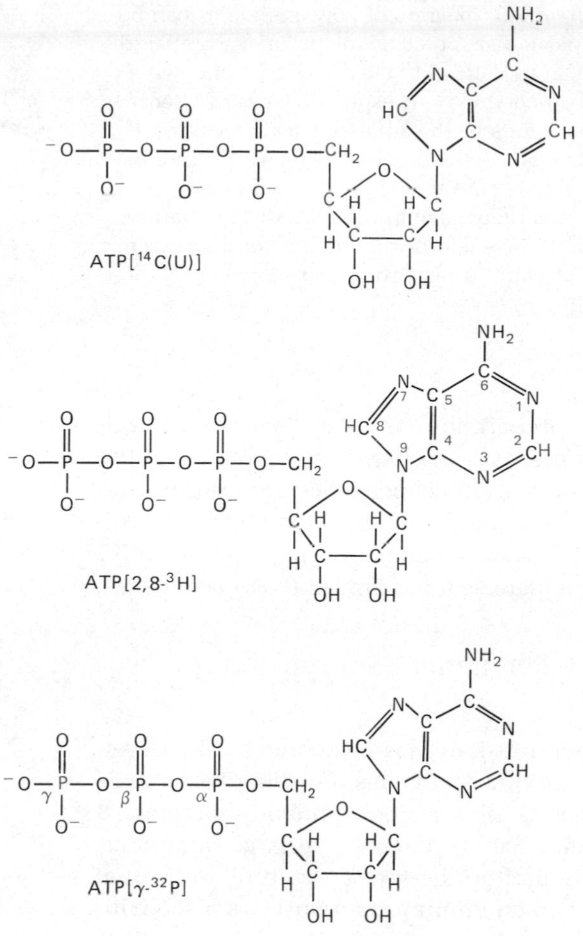

ATP[^{14}C(U)]

ATP[2,8-^{3}H]

ATP[γ-^{32}P]

Figure 4–42 Three commercially available radioactive forms of ATP, with the radioactive atoms shown in color. The nomenclature used to identify the position and type of the radioactive atoms is also shown.

Antibodies Can Be Used to Detect and Isolate Specific Molecules[24]

Antibodies are proteins produced by vertebrates as a defense against infection. They are unique among proteins because they are made in millions of different forms, each with a different binding site that specifically recognizes the molecule (called an *antigen*) that induced its production. The precise antigen specificity of antibodies makes them powerful tools for the cell biologist (Figure 4–43). Labeled with fluorescent dyes, they are invaluable for locating specific molecules in cells by fluorescence microscopy (Figure 4–44); coupled to electron-dense molecules such as *ferritin,* they are used to locate cellular antigens at high resolution in the electron microscope. As biochemical tools, they are used to identify specific proteins after they have been fractionated by electrophoresis in polyacrylamide gels. Moreover, antibodies can be coupled to an inert matrix to produce an affinity column that then is used either to purify specific macromolecules from a crude cell extract or, if the antigen is on the cell surface, to pick out specific types of living cells from a heterogeneous population.

Conventionally, antibodies are made by injecting a sample of the antigen several times into an animal, such as a rabbit or goat, and then collecting the antibody-rich serum. This *antiserum* contains a heterogeneous mixture of antibodies, each produced by a different antibody-secreting cell (a B lymphocyte). The different antibodies recognize various parts of the antigen molecule,

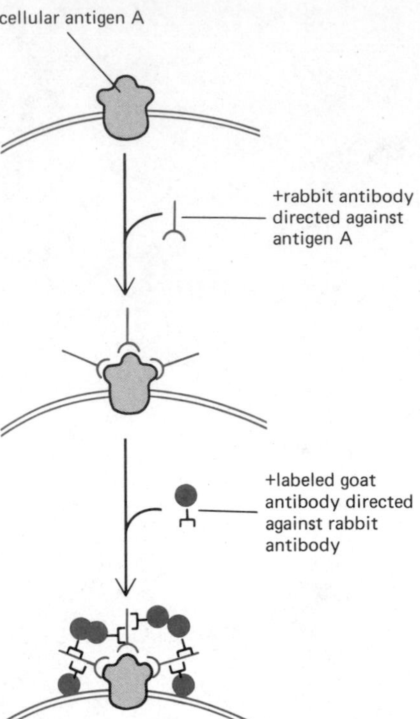

Figure 4–43 Schematic diagram of how antibodies can be used to locate a particular molecule in specimens prepared for microscopy. The modification illustrated here (known as *indirect immunocytochemistry*) is especially sensitive because each antibody that binds to the antigen is itself recognized by many antibodies of a second type that recognize the first. The second antibody is covalently linked to a marker molecule so that it is detectable in the microscope. Marker molecules include fluorescein or rhodamine (for fluorescence microscopy) and ferritin (for electron microscopy).

as well as impurities in the antigen preparation. The specificity of an antiserum for a particular antigen sometimes can be sharpened by removing the unwanted antibody molecules that bind to other molecules; for example, an antiserum produced against protein X can be passed through an affinity column of antigens Y and Z to remove any contaminating anti-Y and anti-Z antibodies. Still, the heterogeneity of such antisera has limited their usefulness.

Hybridoma Cell Lines Provide a Permanent Source of Monoclonal Antibodies[25]

In 1976 the problem of antiserum heterogeneity was overcome by the development of a new technique that revolutionized the use of antibodies as tools in cell biology. The technique involves cloning a single antibody-secreting B lymphocyte so that uniform antibodies can be obtained in large quantities. Since B lymphocytes normally have a limited life-span in culture, individual antibody-producing B lymphocytes from an immunized mouse are fused with cells derived from an "immortal" B lymphocyte tumor. From the resulting heterogeneous mixture of hybrid cells, those hybrids that have both the ability to make a particular antibody and the ability to multiply indefinitely in tissue

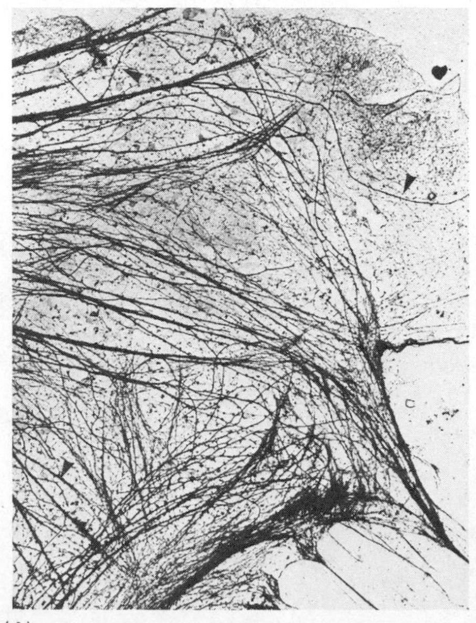

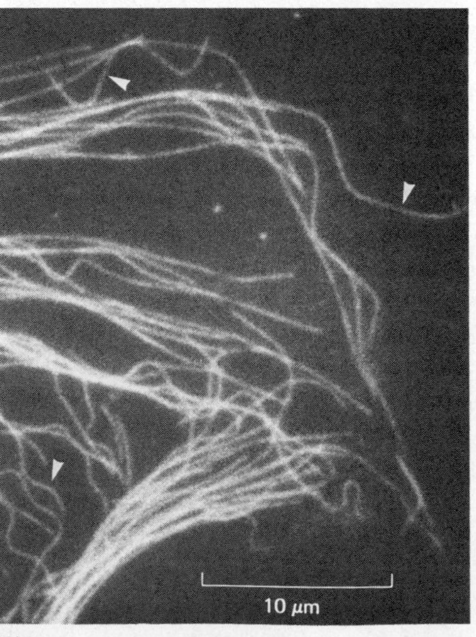

Figure 4–44 (A) An electron micrograph of the periphery of a cultured epithelial cell showing the distribution of microtubules and other filaments. (B) The same area stained with a fluorescent antibody to tubulin, the subunit protein of microtubules, using the technique of indirect immunocytochemistry (see Figure 4–43). Arrows indicate individual microtubules that are readily recognizable in the two figures. (From M. Osborn, R. Webster, and K. Weber, *J. Cell Biol.* 77:R27–R34, 1978. Reproduced by copyright permission of the Rockefeller University Press.)

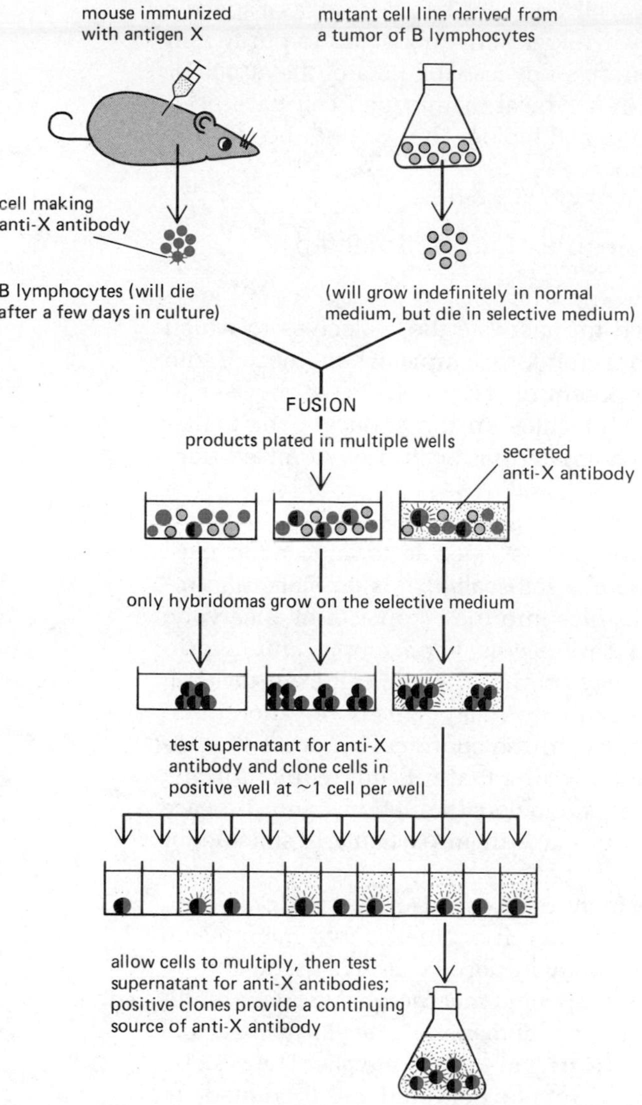

Figure 4–45 Schematic diagram showing how hybrid cells, or hybridomas, that secrete homogeneous monoclonal antibodies against a particular antigen (X) are prepared. The selective growth medium used contains an inhibitor (aminopterin) that blocks the normal biosynthetic pathways by which nucleotides are made. The cells must therefore use a bypass pathway in order to synthesize their nucleic acids, and this pathway is defective in the mutant cell line to which the normal B lymphocytes are fused. Because neither cell type used for the initial fusion can grow on its own, only the cell hybrids survive.

culture are selected. These **hybridomas** are propagated as individual clones, each of which provides a permanent and stable source of a single **monoclonal antibody** (Figure 4–45).

Since they are the product of a single B lymphocyte, monoclonal antibodies consist of a population of identical antibody molecules, each with an identical antigen-binding site. This site will recognize, for example, a defined conformation of a particular group of five or six amino acid side chains on a protein or a similar number of sugar residues in a polysaccharide. Because of their uniform specificity, monoclonal antibodies have an enormous advantage over conventional antisera, which usually contain antibodies recognizing a variety of different antigenic sites on even a small macromolecule.

But the most important advantage of the hybridoma technique is that monoclonal antibodies can be made against unpurified molecules that constitute only a minor component of a complex mixture. This advantage derives from the fact that individual hybridoma clones producing a particular antibody can be selected from a large mixture of different hybrid cells producing a variety of different antibodies. In principle, monoclonal antibodies therefore can be made against any protein in the cell, and each can be used as a specific probe—both to localize the protein by cytological methods and to purify it in order to study its structure and function. Since less than 5% of the 10,000 or so proteins estimated to be present in a typical mammalian cell have been isolated, it is hard to think of an area of cell biology that will not be affected by the monoclonal antibody revolution.

Antibodies and Other Macromolecules Can Be Injected into Living Cells[26]

Antibody molecules can often be used to inactivate the molecules to which they bind. Antibodies to nerve growth factor, for example, when injected into a newborn mouse, prevent the development of certain classes of nerve cells. Similarly, antibodies that react with molecules on the surface of particular types of certain cells can be used to eliminate those cells from a mixed population.

Proteins inside living cells cannot be reached by antibodies added externally because the plasma membrane is impermeable to large molecules. However, because the plasma membrane is self-sealing, it is possible to introduce antibodies and other macromolecules into the cytoplasm of eucaryotic cells by injecting them through a fine glass needle. For example, anti-myosin antibodies injected into a sea urchin egg prevent the egg cell from dividing in two even though nuclear division occurs normally. This observation demonstrates that myosin plays a crucial part in the contractile process that divides the cytoplasm during mitosis but suggests that it is not required in the mitotic spindle. The high specificity of monoclonal antibodies and the ease of producing them in concentrated form make them particularly suitable for this type of application.

Microinjection is being increasingly used to introduce not only antibodies but also other macromolecules and even organelles into mammalian cells; a skilled experimenter can inject several hundred cells in a tissue-culture dish in less than an hour. Alternatively, specific macromolecules can be introduced simultaneously into large numbers of mammalian cells by means of lipid vesicles or the empty shells (or "ghosts") of red blood cells. The vesicles or ghosts are first filled with the substance to be delivered and then made to fuse with the plasma membrane of the living cells so that their contents are delivered to the cytoplasm of the target cells. Under favorable conditions, over 50% of the cells in a population can be made to take up molecules supplied in this form.

Summary

Any molecule in the cell may be "labeled" by the incorporation of one or more radioactive atoms. These unstable atoms disintegrate, emitting radiation that allows the molecule to be traced. Among the many applications of radioisotopes to cell biology have been the analysis of metabolic pathways and the location of individual molecules in a cell by autoradiography.

Antibodies are also versatile and sensitive tools for detecting and localizing specific biological molecules. Vertebrates make millions of different antibody molecules, each with a binding site that recognizes a specific region of a molecule. The hybridoma technique allows monoclonal antibodies of a single specificity to be obtained in virtually unlimited amounts. In principle, monoclonal antibodies can be made against any cell macromolecule and can be used to localize and purify the molecule and, in some cases at least, to analyze its function.

Recombinant DNA Technology[27]

In the early 1970s DNA was the most difficult cellular compound for the biochemist to analyze. Enormously long and chemically monotonous, the nucleotide sequence of the hereditary material could be approached only by indirect means—such as through protein or RNA sequencing or by genetic analysis. Today the situation has entirely changed. From being the hardest macromolecule of the cell to analyze, DNA has become the easiest. It is now possible to excise specific regions of DNA, to obtain them in essentially unlimited quantities, and to determine the sequence of their nucleotides at a rate of several hundred nucleotides a day.

The new **recombinant DNA technology** has provided powerful and novel approaches to understanding the complex mechanisms by which eucaryotic gene expression is regulated, and it has largely superceded conventional methods for determining the amino acid sequence of a protein. Elaborations of the same methods offer great commercial promise for the large-scale economical production of protein hormones and vaccines, available at present only with great labor and cost.

Recombinant DNA technology comprises a mixture of techniques, some new and some borrowed from other fields such as microbial genetics (Table 4–12). The most important ones are (1) specific cleavage of DNA by *restriction nucleases,* (2) *nucleic acid hybridization,* which makes it possible to identify specific sequences of DNA or RNA with great accuracy and sensitivity by their ability to bind a complementary nucleic acid sequence, (3) *DNA cloning,* whereby a specific DNA fragment is integrated into a rapidly replicating genetic element (plasmid or virus) so that it can be amplified in bacteria or yeast cells, and (4) *DNA sequencing* of the nucleotides in a cloned DNA fragment.

Restriction Nucleases Hydrolyze DNA Molecules at Specific Nucleotide Sequences[28]

Many bacteria make enzymes called **restriction nucleases,** which protect them by degrading any invading foreign DNA molecules. Each enzyme recognizes a specific sequence of four to six nucleotides in DNA. The correspond-

Table 4–12 Major Steps in the Development of Recombinant DNA Technology

1869	**Miescher** isolated DNA for the first time.
1944	**Avery** provided evidence that DNA, rather than protein, carries the genetic information during bacterial transformation.
1953	**Watson and Crick** proposed the double-helix model for DNA structure based on x-ray results of **Franklin and Wilkins.**
1961	**Marmur and Doty** discovered DNA renaturation, establishing the specificity and feasibility of nucleic acid hybridization reactions.
1962	**Arber** provided the first evidence for the existence of DNA restriction enzymes, leading to their later purification and use in DNA sequence characterization by **Nathans and H. Smith.**
1966	**Nirenberg, Ochoa,** and **Khorana** elucidated the genetic code.
1967	**Gellert** discovered DNA ligase, the enzyme used to join DNA fragments together.
1972–1973	DNA cloning techniques were developed by the laboratories of **Boyer, Cohen, Berg,** and their colleagues at Stanford University and the University of California at San Francisco.
1975–1977	**Sanger and Barrell** and **Maxam and Gilbert** developed rapid DNA-sequencing methods.

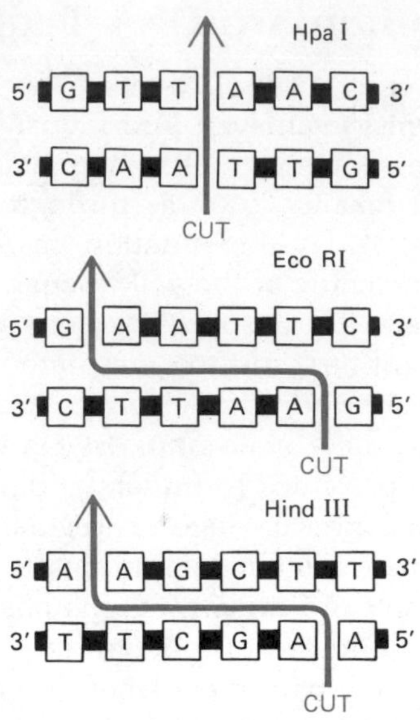

Figure 4–46 The DNA nucleotide sequences recognized by three widely used restriction nucleases. Such sequences are often, as in these examples, six base pairs long and "palindromic"—that is, the nucleotide sequences of the two strands are the same in the recognized region. The two strands of DNA are cut at or near the recognition sequence, often with a staggered cleavage that creates a cohesive end—as for Eco RI and Hind III. Restriction nucleases are obtained from various species of bacteria: Hpa I is from *Hemophilus parainfluenzae*; Eco RI, *Escherichia coli;* and Hind III, *Hemophilus influenzae.*

ing sequences in the genome of the bacterium itself are "camouflaged" by methylation at an A or a C residue, but any foreign DNA molecule that enters the cell is immediately recognized by the nuclease, and both strands of its DNA helix are cut (Figure 4–46). Many restriction nucleases have been purified from different species of bacteria, and more than 100, most of which recognize different nucleotide sequences, are now commercially available.

A particular restriction nuclease will cut any long length of DNA double helix into a series of fragments known as **restriction fragments.** By comparing the sizes of the DNA fragments produced from a particular genetic region after treatment with a combination of different restriction nucleases, a **restriction map** can be constructed that shows the location of each cutting (restriction) site in relation to its neighbors. Since such maps reflect the arrangement of selected nucleotide sequences in the region, a comparison of such maps for two or more related genes will give a rough estimate of the homology between them. For example, the restriction maps, and therefore presumably the nucleotide sequences, of the entire chromosomal regions coding for hemoglobin chains in man, orangutan, and chimpanzee have remained largely unchanged during the 5 to 10 million years since these species first diverged (Figure 4–47).

Figure 4–47 Restriction maps of human and various primate DNAs in a cluster of genes coding for hemoglobin. The two squares in each map indicate the positions of the DNA corresponding to the α-globin genes. Each letter stands for a site cut by a different restriction nuclease. The location of each cut was determined by comparing the sizes of the DNA fragments generated by treating the DNAs with the various restriction nucleases, individually and in combinations. (Courtesy of Elizabeth Zimmer and Alan Wilson.)

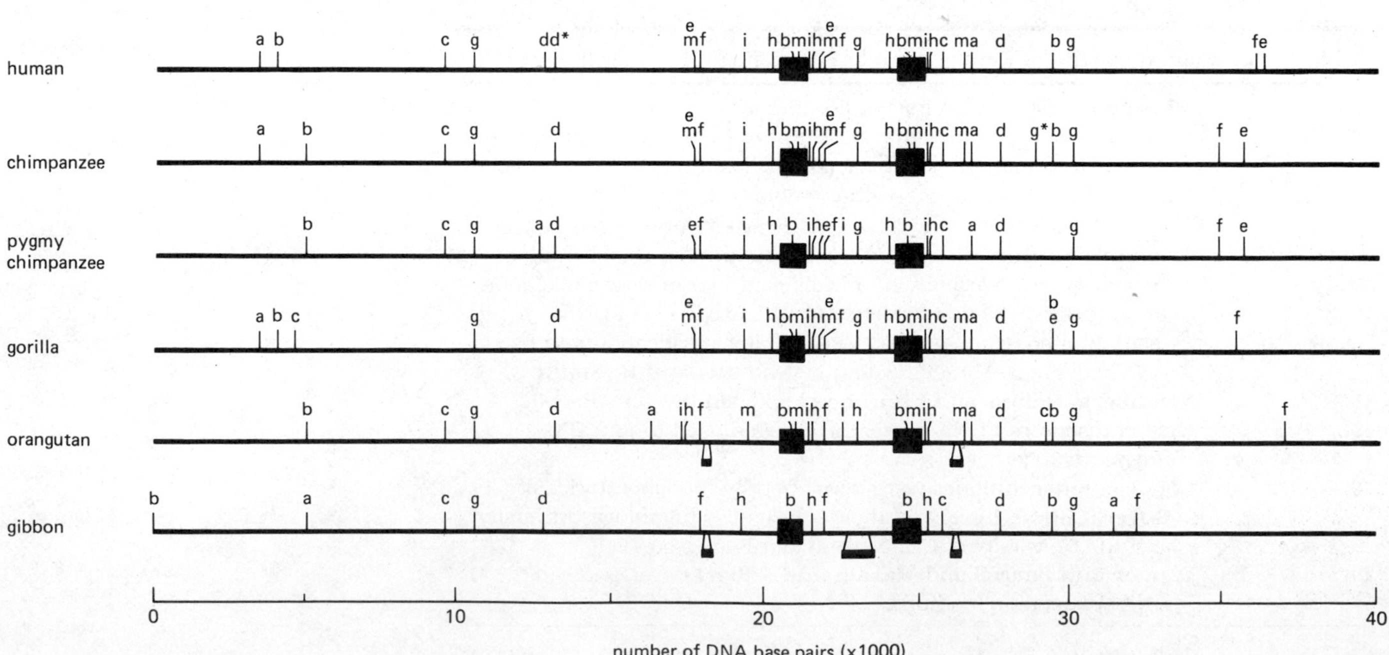

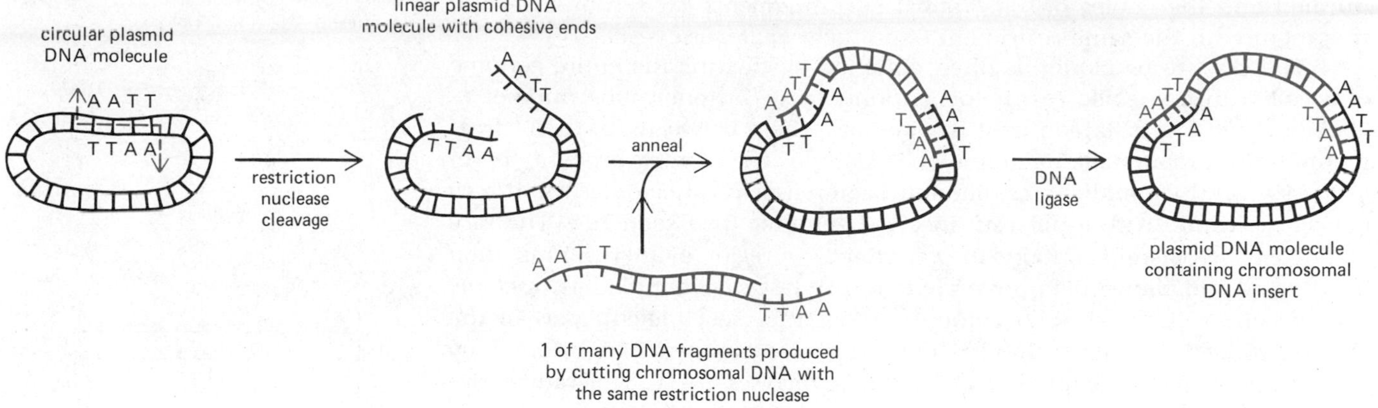

Figure 4–48 The cohesive ends produced by many kinds of restriction nucleases (see Figure 4–46) allow two DNA fragments to be joined by complementary nucleotide base-pair interactions. DNA fragments that are joined in this way can be covalently linked in a reaction catalyzed by the enzyme DNA ligase. In this example, a hybrid plasmid DNA molecule that contains a chromosomal DNA insert is formed.

Many restriction nucleases produce staggered cuts, which leave short, single-stranded ends on both fragments. These are known as *cohesive ends* since they can form complementary base pairs with any other end produced by the same enzyme. A circular DNA molecule that is cut at a single site by this type of restriction nuclease will therefore tend to re-form a circle by the annealing (base-pairing) of its cohesive ends. The cohesive ends generated by restriction enzymes have been very important in recombinant DNA technology because they enable any two DNA fragments to be joined, provided that they were generated with the same restriction nuclease, and thus have complementary cohesive ends. Once the two ends have joined by complementary base-pairing, they can be sealed by an enzyme known as a *DNA ligase*, which forms covalent phosphodiester bonds between the opposing ends of each strand of DNA (Figure 4–48). The combined use of restriction enzymes and DNA ligase has made it possible to graft fragments of any DNA into self-replicating elements.

Selected DNA Sequences Are Produced in Large Amounts by Cloning[29]

Fragments of DNA from any source can be amplified more than a millionfold by inserting them into a *plasmid* or a bacterial virus (*bacteriophage*) and then growing these in bacterial (or yeast) cells—a process called **DNA cloning.** Plasmids are small circular molecules of double-stranded DNA that occur naturally in both bacteria and yeast, where they replicate as independent units as the host cell proliferates. Although they generally account for only a small fraction of the total host cell DNA, they often carry vital genes, such as those that confer resistance to antibiotics. These genes, and the relatively small size of the plasmid DNA, are exploited in recombinant DNA technology.

Because it is so much smaller, plasmid DNA can easily be separated from the DNA of the host cell and purified. For use as *cloning vectors,* such purified plasmid DNA molecules are cut once with a restriction nuclease and then annealed to the DNA fragment that is to be cloned. The hybrid plasmid DNA molecules produced are then reintroduced into bacteria that have been made transiently permeable to macromolecules. Only some of the treated cells will take up a plasmid. They can be selected by the antibiotic resistance conferred on them by the plasmid since they alone will grow in the presence of antibiotic. As these bacteria divide, the plasmid also replicates to produce an enormous number of copies of the original DNA fragment (Figure 4–49). At the end of the period of proliferation, the hybrid plasmid DNA molecules are purified and the copies of the original DNA fragments excised by a second treatment with the same restriction endonuclease (Figure 4–50).

The DNA to be cloned is often obtained by cleaving the entire genome of a cell with a specific restriction endonuclease. An enormous number of DNA fragments is obtained in this way—anywhere between 10^5 to 10^7 fragments from a mammalian genome, for example. The cloning process, therefore, may produce millions of different bacterial or yeast colonies, each harboring a plasmid with a different inserted genomic DNA sequence. The rare colony whose plasmid contains the genomic DNA region of interest must then be selected and allowed to proliferate to form a large cell population, or *clone*. The selection of the desired colony is often the most difficult part of the cloning procedure. The technique normally used for identifying the colony containing a specific cloned DNA fragment involves the use of radioactive nucleic acid probes complementary to the cloned DNA. We shall now discuss how such probes are commonly made.

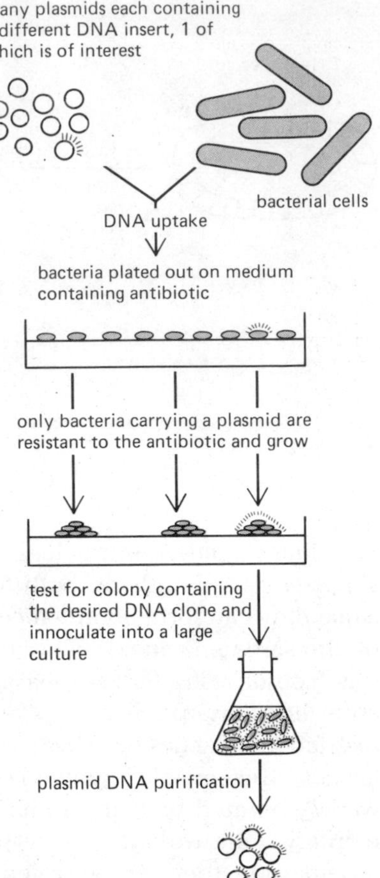

Figure 4–49 Purification and amplification of a specific DNA sequence by DNA cloning in a bacterium. DNA fragments are cloned in yeast cells by a similar procedure.

Copies of Specific mRNA Molecules Can Be Cloned[30,31]

The cloning procedure just described is sometimes called a "shotgun" approach because the entire genomic DNA is cut into an enormous number of fragments that are randomly placed with respect to genes. As a result, some will contain parts of genes and many will contain only *noncoding* DNA and thus no genes at all. An alternative strategy is to begin the cloning process by selecting only those DNA sequences that are transcribed into RNA. This is done by extracting the mRNA (or a purified subfraction of the mRNA) from cells and then making a *DNA copy* (called a **cDNA molecule**) of each mRNA molecule present. This is made possible by an enzyme known as *reverse transcriptase* because, instead of catalyzing the transcription of DNA into RNA, it catalyzes the reverse process of synthesizing a complementary DNA chain on an RNA template. The single-stranded cDNA molecules synthesized by reverse transcriptase can be converted into double-stranded cDNA molecules (by using the enzyme *DNA polymerase*), inserted into plasmids, and cloned (Figure 4–51).

It is possible to construct plasmids in a way that allows the cloned cDNA to direct the synthesis within a cell of large amounts of the particular protein that the cDNA specifies. By means of such "genetic engineering," bacteria or yeast can be induced to make useful proteins, such as human insulin, growth hormone, and interferon, in enormous quantities.

Alternatively, cDNA can be used to identify the rare gene-containing clones produced by the shotgun approach. Here the procedure is to make single-stranded cDNA using radioactive nucleotide precursors. The resulting radiolabeled DNA can be hybridized to the complementary genomic clone in a way that will be described below. Because the cDNA is made from mRNA, it will

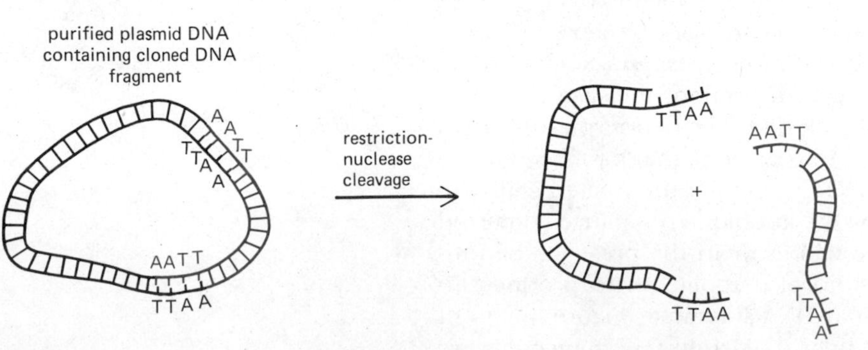

Figure 4–50 Recovery of a cloned DNA fragment from a plasmid containing a recombinant DNA molecule. The fragment is cut out of the plasmid by the same restriction nuclease that created the DNA fragments for the initial cloning (see Figure 4–48).

correspond to genomic DNA that codes for a protein, and its hybridization to the DNA in a clone marks that clone as one containing part of a gene encoding the mRNA molecule.

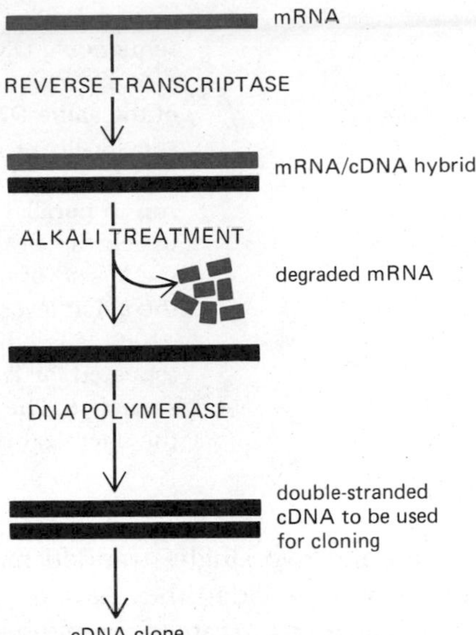

Figure 4–51 A DNA copy (cDNA) of an mRNA molecule is produced by the enzyme reverse transcriptase, a viral enzyme that uses an RNA strand as a template for the synthesis of a complementary DNA strand, thereby forming a DNA/RNA hybrid helix. Treatment of the DNA/RNA hybrid with alkali selectively degrades the RNA strand into nucleotides. The remaining single-stranded cDNA is then copied into double-stranded cDNA by the enzyme DNA polymerase.

Cloned DNA Fragments Can Be Rapidly Sequenced[32]

It has recently become possible to determine the nucleotide sequence of cloned DNA fragments simply and quickly. The principle underlying one of these methods is illustrated in Figures 4–52 and 4–53. As a result of this new technology, the complete DNA sequences of more than 100 mammalian genes have already been determined, including those coding for insulin, hemoglobin, interferon, and cytochrome c. At present, the easiest and most accurate way to sequence the amino acids in a protein is by sequencing its gene and then using the genetic code as a dictionary to convert the nucleotide sequence back to a protein sequence. Although there are, in principle, six different reading frames in which any DNA sequence can be read into protein (three on each strand), the correct one is usually recognized as the only one lacking frequent stop codons (see p. 108). The volume of DNA sequence information is already so large (>10^6 nucleotides) that computers must be used to store and analyze it.

Nucleic Acid Hybridization Reactions Provide a Sensitive Way of Detecting Specific Nucleotide Sequences[33]

When an aqueous solution of DNA is heated at 100°C or exposed to a very high pH (pH ≥ 13), the complementary base pairs that normally hold the two strands of the double helix together are disrupted and the double helix rapidly dissociates into two single strands. This process, called *DNA denaturation,* was for many years thought to be irreversible. However, in 1961 it was discovered that complementary single strands of DNA will readily re-form double helices (a process called **DNA renaturation** or **hybridization**) if they are kept for a prolonged period at 65°C. Similar hybridization reactions will occur be-

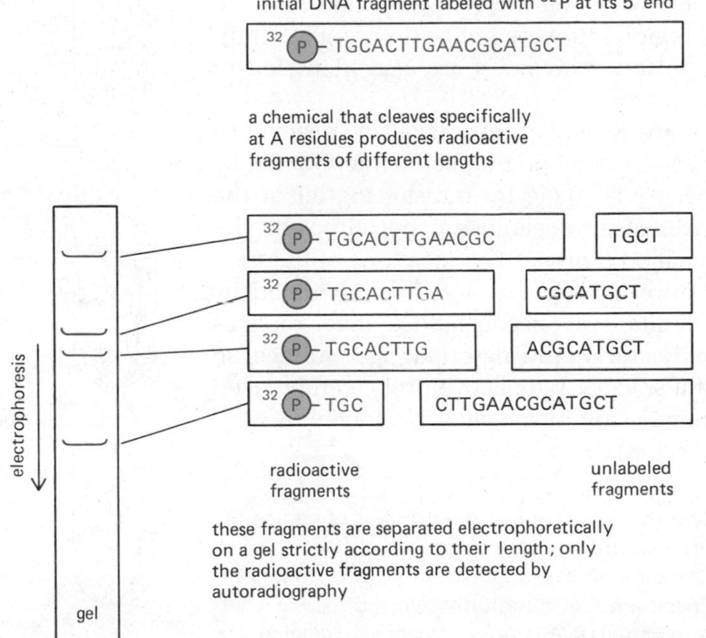

Figure 4–52 The generation of a family of DNA fragments by random cleavage of a DNA chain at a particular type of nucleotide. Each cleavage is produced by a mild chemical treatment that eliminates one nucleotide from the chain while leaving intact most of the nucleotides of the type eliminated. Only the left-hand fragments, possessing a 5′ terminal [^{32}P]phosphate group, are radioactive.

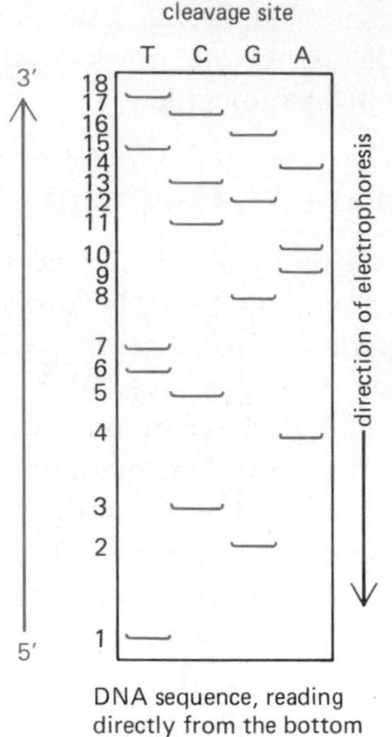

Figure 4–53 Schematic diagram showing one method for sequencing DNA. The type of procedure described in Figure 4–52 is carried out simultaneously on four separate samples of the same DNA using chemicals that cleave DNA specifically at T for the first sample, C for the second, G for the third, and A for the fourth. The resulting fragments are run in parallel lanes of the same gel, giving a pattern from which the DNA sequence is read. The nucleotide closest to the 5′ end of the sequence is determined by looking across the gel at level 1 (at the bottom of the gel) and seeing in which lane a band appears (T). The same procedure is repeated for level 2, then 3, and so on, to obtain the sequence. The method has been idealized here; in actuality the chemical treatments are less specific than shown.

tween any two single-stranded nucleic acid chains (DNA:DNA, RNA:RNA, or RNA:DNA), provided they have a complementary nucleotide sequence.

Because the rate of double-helix formation is limited by the rate at which two complementary nucleic acid chains happen to collide, the concentration of DNA molecules carrying a particular nucleotide sequence can be measured by the rate at which the DNA preparation of interest hybridizes to a radiolabeled cloned DNA probe of complementary sequence. This is such a stringent test that even complementary sequences present in a concentration of one molecule per cell can be detected (Figure 4–54). From such measurements it can be determined how many copies of the DNA sequence contained in the cloned probe are present in the DNA of a cell. While most sequences turn out to be present in only one or a few copies per haploid genome, others are present in hundreds of thousands of copies—the so-called *repeated DNA sequences.*

Alternatively, hybridization studies can be carried out with RNA isolated from cells to determine whether the DNA sequence that has been cloned is one of those transcribed into RNA and, if so, how many copies of the RNA are made per cell and in which types of cells and tissues. Somewhat more elaborate procedures identify the exact region of the cloned probe that hybridizes with cellular RNA molecules and thereby define the start and stop sites for RNA transcription (Figure 4–55); the regions that are cut out of the RNA transcripts during *RNA processing* (the intron sequences) are also identified in this way.

Radioactive cloned DNA probes are widely used to localize specific nucleic acid sequences in mixtures of DNA restriction fragments fractionated by gel electrophoresis. A replica of the gel is made by transferring all of the fractionated DNA fragments to a sheet of nitrocellulose paper either by diffusion or electrophoresis, a process called *blotting.* The locations of the fragments that hybridize to the radioactive DNA probe are then identified by autoradiography (Figure 4–56). In a similar way, nitrocellulose paper replicas can be made of crowded colonies of bacteria growing on an agar surface, so that hybridization of the paper with a specific radioactive probe can be used to identify the few cells carrying a newly cloned specific DNA fragment.

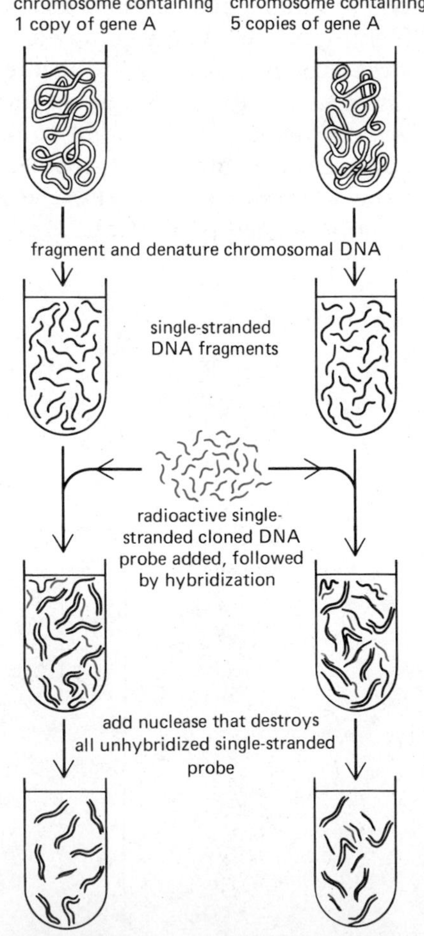

Figure 4–54 The measurement of the number of copies of a specific gene in a sample of DNA by means of DNA hybridization. The radioactive single-stranded DNA fragment used in such experiments is commonly referred to as a *DNA probe;* the chromosomal DNA is not radioactively labeled here.

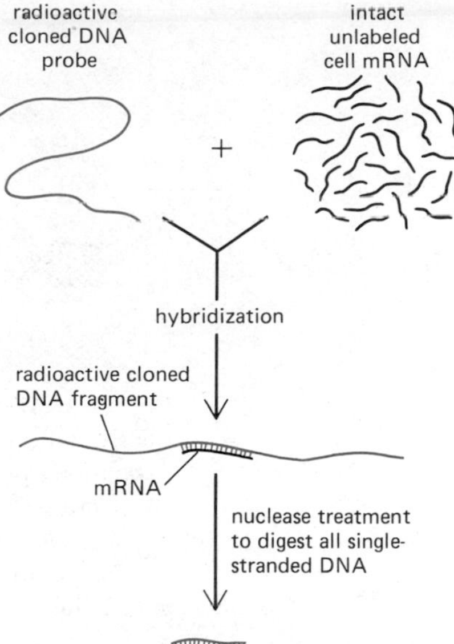

Figure 4–55 The use of nucleic acid hybridization to determine the region of a cloned DNA fragment that is transcribed into mRNA. The existence of intervening sequences (introns) in eucaryotic genes was discovered by this type of procedure.

In Situ Hybridization Techniques Are Used to Localize Specific Nucleic Acid Sequences in Chromosomes and Cells[34]

Nucleic acids, no less than other macromolecules, occupy precise positions within cells and tissues, and a great deal of potential information is lost when these molecules are extracted from cells by homogenization. For this reason, techniques have been developed in which nucleic acid probes are used in much the same way as labeled antibodies to localize specific nucleic acid sequences *in situ,* either in chromosomes or particular types of cells. In the original *in situ* hybridizations, highly radioactive nucleic acid probes were hybridized to squashed, fixed chromosomes that had been exposed briefly to a very high pH in order to disrupt their DNA base pairs. After extensive washing, the chromosomal regions that bound the radioactive probe were visualized by autoradiography (Figure 4–57). Recently, the spatial resolution of this technique has been improved by the development of special methods for labeling the nucleic acid probes with fluorescent dyes.

Similar *in situ* hybridization methods have been useful for detecting the presence of particular growing RNA transcripts on unusually large "lampbrush chromosomes"; here the chromosomes are not exposed to a high pH, so the chromosomal DNA itself remains double-stranded and thus cannot bind the probe. Comparable methods can be used on fixed tissue sections to determine which cells in a complex tissue contain cytoplasmic RNA molecules complementary to a particular DNA probe.

Recombinant DNA Techniques Allow Even the Minor Proteins of a Cell to Be Studied[31,35]

Until very recently, the only proteins that could be readily studied were relatively abundant components of the cell. Starting with several hundred grams of cells, a major protein—one that constitutes 1% or more of the total cellular protein—can be purified by a series of simple chromatographic and electro-

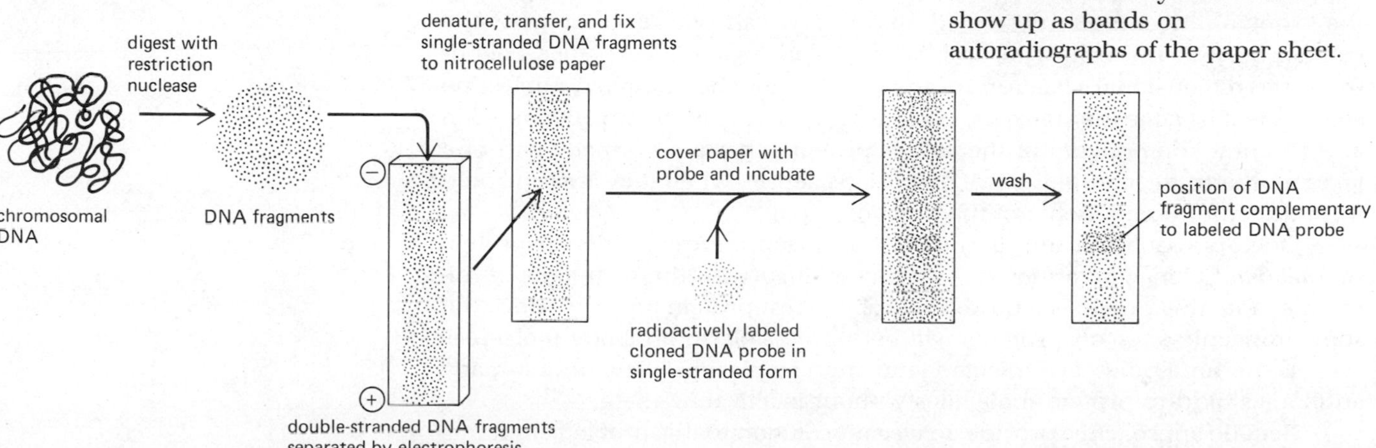

Figure 4–56 After a DNA sample has been cleaved by a restriction nuclease and then separated by electrophoresis, the many different DNA fragments present are transferred to nitrocellulose paper by blotting and then are exposed to a radioactive DNA probe for a prolonged period under annealing conditions. The sheet is washed extensively so that only those DNA fragments that hybridize to the probe remain radioactively labeled and show up as bands on autoradiographs of the paper sheet.

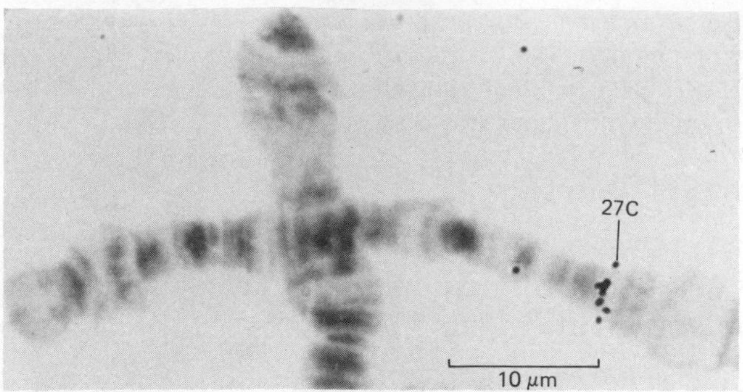

Figure 4–57 Localization of a *Drosophila* gene by *in situ* hybridization of a radioactive cloned DNA probe with *Drosophila* polytene chromosomes. The cluster of darkened silver grains detected in this autoradiograph is located at chromosome map position 27C, as indicated. Parts of two of the four giant chromosomes present in each larval salivary gland cell are shown. (Courtesy of Steven Henikoff.)

phoretic procedures to yield perhaps 0.1 g (100 mg) of pure protein. This quantity of protein is sufficient for conventional amino acid sequencing, detailed analysis of its biological or enzymatic activity (if any), and the production of antibodies, which can then be used to localize the protein in the cell. Moreover, if suitable crystals can be grown, the three-dimensional structure of the protein can be analyzed by x-ray diffraction crystallography. In this way, the structure and function of many abundant proteins have been determined, including hemoglobin, trypsin, immunoglobulin, and lysozyme.

The vast majority of the thousands of different proteins in a eucaryotic cell, including many of the most interesting ones, are present in only very small amounts. For most of them it is extremely difficult, if not impossible, to obtain more than a few micrograms of pure material. In principle, however, recombinant DNA technology has now made essentially any protein in the cell, including the minor ones, accessible to the same structural and functional studies that were previously possible only for a rare few. A summary of the steps that make this possible is given in Table 4–13.

Mutant Genes Can Now Be Made to Order[36]

Suppose one isolates a new protein from a cell extract and clones its gene by the "shotgun" technique described above. How can one discover what the protein does in the cell? The problem is surprisingly difficult since neither the three-dimensional structure of the protein nor the complete nucleotide sequence of its gene identifies the protein's function. And many proteins, such as structural components of the cell or proteins that are normally part of a large multienzyme complex, have no obvious activity when they are separated from the other components of the functional unit.

One approach already discussed is to inactivate the particular protein by means of a specific antibody. When combined with the technique of microinjection, this provides a powerful probe to test protein function. However, some antigenic sites on proteins will be inaccessible to antibody molecules even if the antibodies are injected into the cytoplasm. Furthermore, many antibodies bind to protein molecules without inactivating them.

Genetic approaches provide an elegant solution to this problem. Mutants that lack a particular protein or, more usefully, synthesize a temperature-

Table 4–13 Steps in the Purification of Large Amounts of a Minor Protein of the Cell Using Recombinant DNA Technology

1. Fractionate the cell extract by a series of conventional chromatographic procedures until the protein of interest is sufficiently enriched that a microgram can be obtained in pure form by cutting it out of a gel following high-resolution gel electrophoresis.
2. Analyze the denatured protein on a microsequenator to determine the sequence of the first 30 amino acids at its amino terminus.
3. Use the genetic code to predict the nucleotide sequences in mRNA corresponding to the above amino acid sequence. Using rapid chemical methods, synthesize a set of short DNA fragments, 15 to 20 nucleotides long, 1 of which will form complementary base pairs with part of the mRNA sequence. (There will be some ambiguity here since several different codons code for the same amino acid—p. 108).
4. Hybridize these short DNA fragments to total cellular mRNA and use them to direct reverse transcriptase to the mRNA molecules with complementary sequences. The reverse transcriptase then copies these complementary mRNA molecules to produce long cDNA molecules (Figure 4–51).
5. Produce large amounts of DNA containing the sequence of each of these cDNA molecules by cloning (Figures 4–49 and 4–50).
6. Hybridize DNA prepared from each cDNA clone to total cellular mRNA and thereby select and purify mRNA molecules that are complementary in sequence to each cloned cDNA sequence.
7. Translate each mRNA preparation obtained into protein by cell-free protein synthesis in order to determine which one codes for the desired protein.
8. Sequence the appropriate cDNA (Figure 4–53) and use the genetic code to determine the protein's complete amino acid sequence and where the coding sequence for the protein begins and ends.
9. Insert the cloned cDNA sequence into a specially engineered plasmid DNA vector containing inserted transcription and translation start signals. Use bacterial or yeast cells containing this new plasmid clone as the starting material for the isolation of large amounts (100 mg or more) of the purified protein.

sensitive version of the protein that is inactivated by a small increase (or decrease) in temperature may quickly reveal the function of the normal molecule. While this approach has been immensely useful, for example, in elucidating the principal metabolic pathways of bacteria, it has been mainly restricted to very rapidly replicating organisms—such as bacteria, yeast, nematode worms, and fruit flies—where very large numbers of mutants can be quickly isolated and then screened for a particular defect of interest.

In principle, the genetic approach can now be made more generally applicable by creating specific "mutations" outside the cell. With recently developed methods, a copy of an isolated cloned gene can be altered slightly by biochemical means and then put back into a cell, which now synthesizes an altered protein. In bacterial and yeast cells, this mutant gene will recombine with the normal gene often enough to make it possible to select for cells in which the mutant gene has replaced the single copy of the normal gene. In this way, cells carrying a specific protein in mutant form are made to order and the phenotype of a cell that lacks the normal gene thereby determined. Similar methods are not yet available for inserting a cloned mutant gene back into mammalian cells in place of the normal gene, but with the extraordinarily rapid rate of progress in recombinant DNA technology, it would not be surprising if this soon becomes possible.

Summary

Recombinant DNA technology has revolutionized the study of the cell. Any region of the cell's DNA can now be excised with restriction nucleases and produced in virtually unlimited quantities by DNA cloning and then sequenced at rates of hundreds of nucleotides a day. As a result, many genes and noncoding regions of the eucaryotic genome have already been sequenced.

By using nucleic acid hybridization methods, mRNA molecules corresponding to cloned DNA molecules can be detected, isolated, and translated into protein in cell-free systems. Furthermore, it is possible, in principle, to work backward from a protein to the gene that encodes it: by using a short stretch of amino acid sequence from the protein, specific DNA probes can be synthesized that will hybridize with the mRNA and DNA encoding the protein.

The practical consequences of recombinant DNA technology are far-reaching. Bacteria or yeast can be engineered to make a mammalian protein in virtually unlimited quantities, making it possible to analyze the structure and function of the protein or to use the protein as a vaccine or drug for medical purposes.

References

General

Cantor C.R.; Schimmel P.R. Biophysical Chemistry (3 vols.). San Francisco: Freeman, 1980. (A comprehensive account of the physical principles underlying many biochemical and biophysical techniques.)

Freifelder, D. Physical Biochemistry. San Francisco: Freeman, 1976.

Prescott, D., ed. Methods in Cell Biology. New York: Academic Press. (A multivolume series containing reviews of current techniques.)

Work, T.S.; Work, E.; Burden, R.H., eds. Laboratory Techniques in Biochemistry and Molecular Biology. Amsterdam: Elsevier/North-Holland Biomedical Press. (A multivolume series of practical guides to specialized biochemical procedures. Recent volumes include Sequencing of Proteins and Peptides, 1981; Gel Filtration Chromatography, 1980; and An Introduction to Affinity Chromatography, 1979.)

Cited

1. Bradbury, S. The Evolution of the Microscope. Elmsford, N.Y.: Pergamon, 1967.

   Bloom, W.; Fawcett, D.W. A Textbook of Histology, 10th ed. Philadelphia: Saunders, 1975. (A beautifully illustrated description of the anatomy of cells, as seen by light microscopy and transmission electron microscopy. Chapter 1 introduces the principal methods employed.)

2. Spencer, M. Fundamentals of Light Microscopy. Cambridge, Eng.: Cambridge University Press, 1982.

3. Nairn, R.C. Fluorescent Protein Tracing, 4th ed. New York: Churchill Livingstone, 1976.

   Lillie, R.D. Biological Stains, 8th ed. Baltimore: Williams & Wilkins, 1969.

4. Wischnitzer, S. Introduction to Electron Microscopy, 3rd ed. Elmsford, N.Y.: Pergamon, 1981.

   Weakley, B.S. A Beginner's Handbook in Biological Transmission Electron Microscopy, 2nd ed. New York: Churchill Livingstone, 1981.

5. Pease, D.C.; Porter, K.R. Electron microscopy and ultramicrotomy. *J. Cell Biol.* 91:287s–292s, 1981. (A short historical account.)

6. Everhart, T.E.; Hayes, T.L. The scanning electron microscope. *Sci. Am.* 226(1):54–69, 1972.

   Hayat, M.A. Introduction to Biological Scanning Electron Microscopy. Baltimore: University Park Press, 1978.

   Kessel, R.G. Tissues and Organs. San Francisco: Freeman, 1979. (An atlas of vertebrate tissues seen by scanning electron microscopy.)

7. Pinto da Silva, P.; Branton, D. Membrane splitting in freeze-etching. *J. Cell Biol.* 45:598–605, 1970.

   Heuser, J. Quick-freeze, deep-etch preparation of samples for 3-D electron microscopy. *Trends Biochem. Sci.* 6:64–68, 1981.

8. Unwin, P.N.T.; Henderson, R. Molecular structure determination by electron microscopy of unstained crystalline specimens. *J. Mol. Biol.* 94:425–440, 1975.

9. Glusker, J.P.; Trueblood, K.N. Crystal Structure Analysis: A Primer. Oxford, Eng.: Oxford University Press, 1972.

   Kendrew, J.C. The three-dimensional structure of a protein molecule. *Sci. Am.* 205(6):96–111, 1961.

   Perutz, M.F. The hemoglobin molecule. *Sci. Am.* 211(5):64–76, 1964.

10. Paul, J. Cell and Tissue Culture, 5th ed. New York: Churchill Livingstone, 1975.

11. Harrison, R.G. The outgrowth of the nerve fiber as a mode of protoplasmic movement. *J. Exp. Zool.* 9:787–848, 1910. (Possibly the first use of tissue culture.)

12. Ham, R.G. Clonal growth of mammalian cells in a chemically defined, synthetic medium. *Proc. Natl. Acad. Sci. USA* 53:288–293, 1965.

    Hayashi, I.; Larner, J.; Sato, G. Hormonal growth control of cells in culture. *In Vitro* 14:23–30, 1978.

13. Harris, H.; Watkins, J.F. Hybrid cells derived from mouse and man: artificial heterokaryons of mammalian cells from different species. *Nature* 205:640–646, 1965.

    Ruddle, F.H.; Creagan, R.P. Parasexual approaches to the genetics of man. *Annu. Rev. Genet.* 9:407–486, 1975.

14. Colowick, S.P.; Kaplin, N.O., eds. Methods in Enzymology, Vols. 1–84. New York: Academic Press, 1955–1982. (A multivolume series containing general and specific articles on most procedures commonly employed in the biochemical analysis of cells.)

    Cooper, T.G. The Tools of Biochemistry. New York: Wiley, 1977.

    de Duve, C.; Beaufay, H. A short history of tissue fractionation. *J. Cell Biol.* 91:293s–299s, 1981.

15. Jovin, T.M.; Arndt-Jovin, D.J. Cell separation. *Trends Biochem. Sci.* 5:214–219, 1980.

    Herzenberg, L.A.; Sweet, R.G.; Herzenberg, L.A. Fluorescence-activated cell sorting. *Sci. Am.* 234(3):108–117, 1976.

16. de Duve, C. Exploring cells with a centrifuge. *Science* 189:186–194, 1975.

    Palade, G. Intracellular aspects of the process of protein synthesis. *Science* 189:347–358, 1975.

    Claude, A. The coming of age of the cell. *Science* 189:433–435, 1975.

    (A. Claude, C. de Duve, and G. Palade shared a Nobel Prize in 1974 for their work on tissue fractionation.)

    Meselson, M.; Stahl, F.W. The replication of DNA in *Escherichia coli. Proc. Natl. Acad. Sci. USA* 44:671–682, 1958. (Density gradient centrifugation was used to show the semiconservative replication of DNA.)

    Scheeler, P. Centrifugation in Biology and Medical Science. New York: Wiley, 1981.

17. Nirenberg, N.W.; Matthaei, J.H. The dependence of cell-free protein synthesis in *E. coli* upon naturally occurring or synthetic polyribonucleotides. *Proc. Natl. Acad. Sci. USA* 47:1588–1602, 1961.

    Zamecnik, P.C. An historical account of protein synthesis, with current overtones—a personalized view. *Cold Spring Harbor Symp. Quant. Biol.* 34:1–16, 1969.

    Racker, E. A New Look at Mechanisms in Bioenergetics. New York: Academic Press, 1976. (Cell-free systems in the working out of energy metabolism.)

18. Andrews, A.T. Electrophoresis. New York: Oxford University Press, 1981. (A comprehensive guide to the theory, techniques, and biochemical applications of electrophoresis.)

19. O'Farrell, P.H. High resolution, two-dimensional electrophoresis of proteins. *J. Biol. Chem.* 250:4007–4021, 1975.

20. Ingram, V.M. A specific chemical difference between the globins of normal human and sickle-cell anaemia haemoglobin. *Nature* 178:792–794, 1956. (The original description of protein fingerprinting.)

    Cleveland, D.W.; Fischer, S.G.; Kirschner, M.W.; Laemmli, U.K. Peptide mapping by limited proteolysis in sodium dodecyl sulfate and analysis by gel electrophoresis. *J. Biol. Chem.* 252:1102–1106, 1977.

21. Walsh, K.A.; Ericsson, L.H.; Parmelee, D.C.; Titani, K. Advances in protein sequencing. *Annu. Rev. Biochem.* 50:261–284, 1981.
22. Rogers, A.W. Techniques of Autoradiography, 3rd ed. New York: Elsevier/North-Holland, 1979.
23. Calvin, M. The path of carbon in photosynthesis. *Science* 135:879–889, 1962. (A pioneer's account of one of the earliest uses of radioisotopes in biology.)
24. Coons, A.H. Histochemistry with labeled antibody. *Int. Rev. Cytol.* 5:1–23, 1956.
    Hudson, L.; Hay, F.C. Practical Immunology, 2nd ed. Oxford, Eng.: Blackwell, 1980.
    Eisen, H.N. Immunology, 3rd ed. New York: Harper & Row, 1981.
    Anderton, B.H.; Thorpe, R.C. New methods of analyzing for antigens and glycoproteins in complex mixtures. *Immunol. Today* 2:122–127, 1980.
25. Milstein, C. Monoclonal antibodies. *Sci. Am.* 243(4):66–74, 1980.
    Yelton, D.E.; Scharff, M.D. Monoclonal antibodies: a powerful new tool in biology and medicine. *Annu. Rev. Biochem.* 50:657–680, 1981.
26. Mueller, C.; Graessmann, A.; Graessmann, M. Microinjection: turning living cells into test tubes. *Trends Biochem. Sci.* 5:60–62, 1980.
    Furusawa, M. Cellular microinjection by cell fusion: technique and applications in biology and medicine. *Int. Rev. Cytol.* 62:29–67, 1980.
27. Glover, D.M. Genetic Engineering: Cloning DNA. New York: Chapman and Hall, 1980. (A brief, 80-page summary of methodology.)
    Williamson, R., ed. Genetic Engineering, Vols. 1–3. New York: Academic Press, 1979, 1981, 1982.
    Watson, J.D.; Tooze, J. The DNA Story: A Documentary History of Gene Cloning. San Francisco: Freeman, 1981.
28. Smith, H.O. Nucleotide sequence specificity of restriction endonucleases. *Science* 205:455–462, 1979.
29. Novick, R.P. Plasmids. *Sci. Am.* 243(6):102–107, 1980.
    Cohen, S.N. The manipulation of genes. *Sci. Am.* 233(1):24–33, 1975.
    Maniatis, T.; et al. The isolation of structural genes from libraries of eucaryotic DNA. *Cell* 15:687–701, 1978.
30. Maniatis, T.; Kee, S.G.; Efstratiadis, A.; Kafatos, F.C. Amplification and characterization of a β-globin gene synthesized *in vitro*. *Cell* 8:163–182, 1976.
31. Abelson, J.; Butz, E., eds. Recombinant DNA. *Science* 209:1317–1438, 1980. (A collection of articles by leaders in the field.)
32. Sanger, F. Determination of nucleotide sequences in DNA. *Science* 214:1205–1210, 1981.
    Gilbert, W. DNA sequencing and gene structure. *Science* 214:1305–1312, 1981.
33. Hood, L.E.; Wilson, J.H.; Wood, W.B. Molecular Biology of Eucaryotic Cells: A Problems Approach, pp. 56–61, 192–201. Menlo Park, Ca.: Benjamin-Cummings, 1975. (Hybridization analyses clearly explained.)
    Southern, E.M. Detection of specific sequences among DNA fragments separated by gel electrophoresis. *J. Mol. Biol.* 98:503–517, 1975.
34. Pardue, M.L.; Gall, J.G. Molecular hybridization of radioactive DNA to the DNA of cytological preparations. *Proc. Natl. Acad. Sci. USA* 64:600–604, 1969.
    Hennig, W. *In situ* hybridization of nucleic acids. *Trends Biochem. Sci.* 1:285–287, 1976.
35. Gilbert, W.; Villa-Komaroff, L. Useful proteins from recombinant bacteria. *Sci. Am.* 242(4):74–94, 1980.
    Itakura, K. Synthesis of genes. *Trends Biochem. Sci.* 5:114–116, 1980.
36. Shortle, D.; Nathans, D. Local mutagenesis: a method for generating viral mutants with base substitutions in preselected regions of the viral genome. *Proc. Natl. Acad. Sci. USA* 75:2170–2174, 1978.
    Hinnen, A.; Hicks, J.B.; Fink, G.R. Transformation of yeast. *Proc. Natl. Acad. Sci. USA* 75:1929–1933, 1978.
    Berg, P. Dissections and reconstructions of genes and chromosomes. *Science* 213:296–303, 1981.
    Anderson, W.F.; Diacumakos, E.G. Genetic engineering in mammalian cells. *Sci. Am.* 245(1):106–121, 1981.

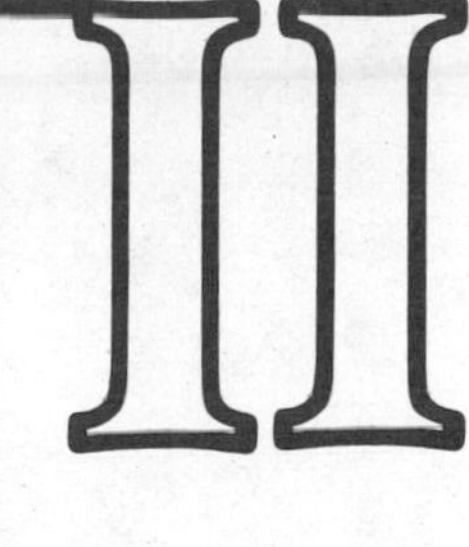

The Molecular Organization of Cells

Electron micrograph of DNA spilling out of a single disrupted human chromosome. Only about half of the DNA in one chromosome is shown. (Courtesy of James Paulson and Ulrich Laemmli.)

Basic Genetic Mechanisms

5

The capacity of cells to maintain a high degree of order in a chaotic universe stems from the genetic information that is expressed, maintained, replicated, and occasionally improved by four basic genetic processes—*protein synthesis, DNA repair, DNA replication,* and *genetic recombination.* These processes, which produce and maintain the proteins and the nucleic acids of a cell, are one-dimensional: in each of them, the information in a linear sequence of nucleotides is used to produce or alter either another linear chain of nucleotides (a DNA or RNA molecule) or a linear chain of amino acids (a protein molecule). Genetic events are therefore conceptually simple compared with most other cellular processes, which involve the expression of information contained in the three-dimensional surfaces of protein molecules. Perhaps that is why we understand genetic mechanisms in far greater detail than most other cellular events.

We shall begin with a detailed discussion of some of the mechanisms of protein synthesis that were introduced in Chapter 3. Then the mechanisms involved in the maintenance of DNA sequence integrity by DNA repair and its propagation by DNA replication are described. This leads to a discussion of viruses as self-replicating genetic entities that parasitize the genetic machinery of the cell, in which either DNA or RNA molecules serve as the primary repository of genetic information. The chapter concludes with a discussion of the mechanisms of genetic recombination in which DNA sequences are reassorted, a process of great importance to species adaptation in a changing environment.

Protein Synthesis

The Decoding of DNA into Protein[1]

Proteins generally constitute somewhat more than half of the total dry mass of a cell, and their synthesis is central to cell maintenance, growth, and development. Although protein synthesis is usually considered to start with the copying of DNA into *messenger RNA (mRNA)*, a number of other preparatory steps are also required: a specific *transfer RNA (tRNA)* molecule must be attached to each of the 20 common amino acids, and ribosomal subunits must be preloaded with auxiliary molecules. In the process of protein synthesis, all of these components are brought together in the cell cytoplasm in a ribosome complex. Here a single mRNA molecule is moved stepwise through a *ribosome* so that its sequence of nucleotides can be translated into a corresponding sequence of amino acids to create a distinctive protein chain.

RNA Polymerase Copies DNA into RNA[2]

The synthesis of an RNA copy of the nucleotide sequence in a limited region of DNA is catalyzed by the enzyme **RNA polymerase.** This process, called **DNA transcription,** is crucial for the transfer of information from DNA to protein. The RNA polymerase has been most thoroughly studied in procaryotes, where a single species of the enzyme mediates all RNA synthesis.

RNA polymerase initiates the transcription process after binding to a specific DNA sequence, called the **promoter,** that signals where RNA synthesis should begin. After binding to the promoter, the RNA polymerase unwinds about one turn of the DNA helix to expose a short stretch of single-stranded DNA that will act as a template for complementary base-pairing with incoming ribonucleotides. It then joins two of the incoming ribonucleoside triphosphate monomers together to begin an RNA chain. The RNA polymerase molecule then moves along the DNA template strand, extending the growing RNA chain in the 5′-to-3′ direction by one nucleotide at a time (Figure 5–1). The enzyme

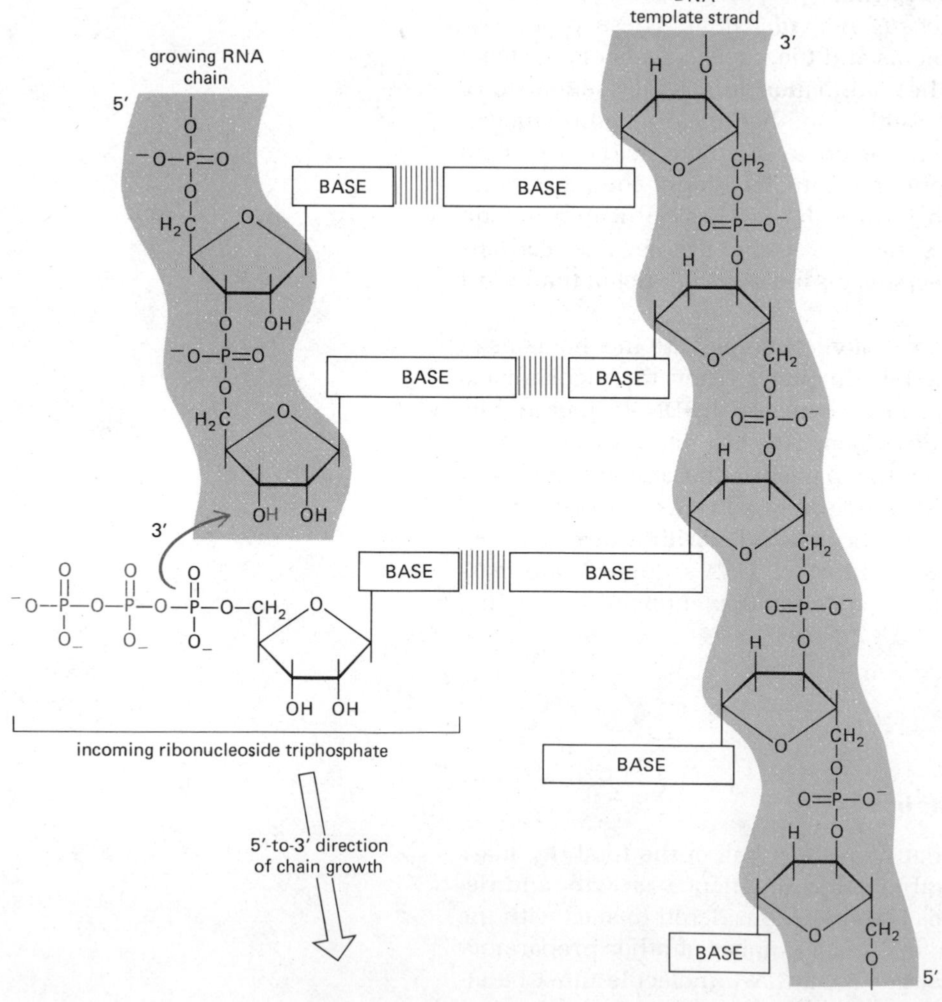

Figure 5–1 The reaction catalyzed by an RNA polymerase enzyme. In each step, an incoming ribonucleoside triphosphate is selected for its ability to base-pair with the exposed DNA template strand; a ribonucleoside monophosphate is then added to the growing 3′-OH end of the RNA chain (*colored arrow*). The new RNA chain therefore grows in the 5′-to-3′ direction and is complementary in sequence to the DNA template strand.

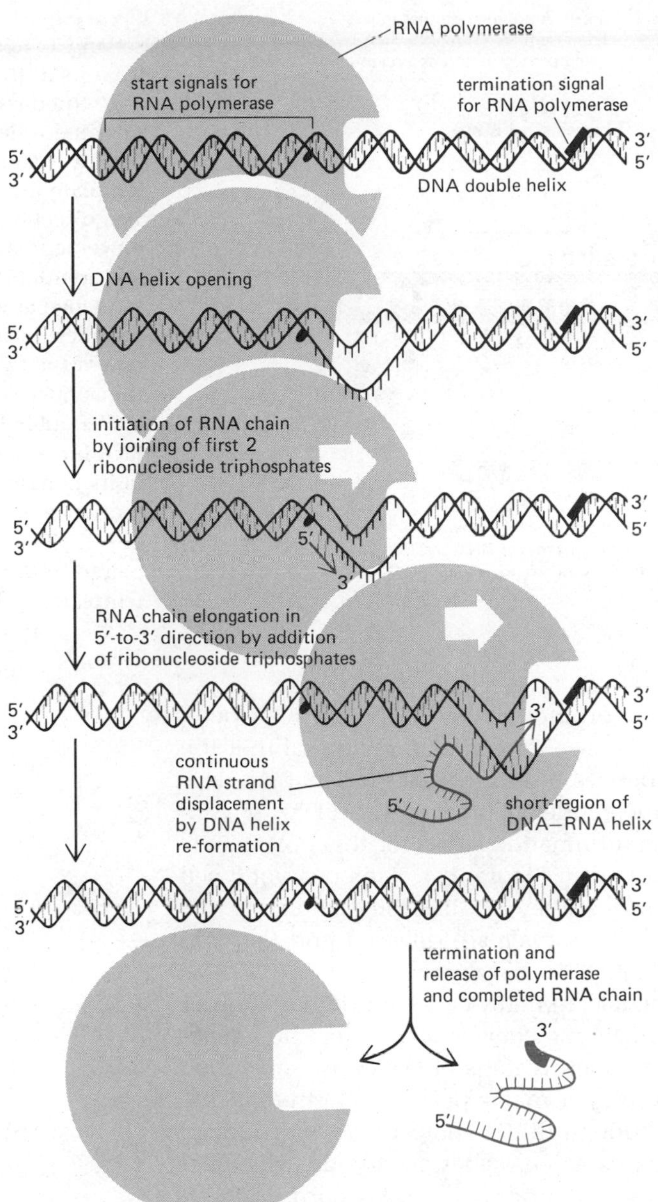

Figure 5–2 A diagram illustrating how RNA polymerase begins its synthesis at a special start signal on the DNA called a promoter and completes its synthesis at a special termination signal, whereupon both the polymerase and its completed RNA chain are released. Polymerization rates average about 30 nucleotides per second at 37°C. Therefore, an RNA chain of 5000 nucleotides takes about three minutes to complete. The RNA polymerase and the DNA helix are drawn approximately to scale.

continues to add nucleotides until it encounters a second special sequence in the DNA, the **termination signal,** at which point the polymerase releases both the DNA template and the newly made RNA chain. As the enzyme moves, a short RNA-DNA double helix is formed, but this is less stable than the DNA-DNA helix, which soon reestablishes itself, displacing the RNA. As a result, each completed RNA chain is released from the DNA template as a free, single-stranded RNA molecule (Figure 5–2).

In principle, any region of DNA could be copied into two different mRNA molecules—one from each of the two DNA strands. In fact, only one DNA strand is copied, although which it is can vary between neighboring genes. The "signpost" that indicates the strand to be copied is the promoter DNA sequence (the start signal). The promoter is oriented in such a way that it sets the RNA polymerase off in a particular direction across a given genetic region, and this automatically determines which of the two strands will be read (Figure 5–3).

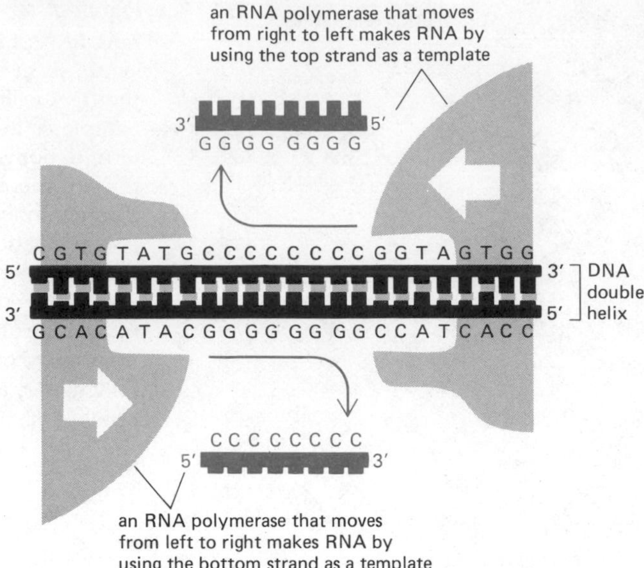

Figure 5–3 Because the DNA strand serving as template must be traversed from its 3′ to its 5′ end (see Figure 5–1), the direction of RNA polymerase movement determines which of the two DNA strands will serve as a template for the synthesis of RNA. The direction of polymerase movement is, in turn, determined by the orientation of the promoter sequence at which the RNA polymerase starts. In general, only one of the two DNA strands will be transcribed in any one region of the DNA double helix. Note that each RNA molecule made will be identical in its polarity and nucleotide sequence (except for the substitution of U for T) to the DNA strand that is paired to the strand used as template.

The DNA sequences that serve as promoters for the *E. coli* RNA polymerase have been well characterized. Their common features reveal that the polymerase recognizes two DNA sequences 6 nucleotides long that are separated from each other by about 25 nucleotides of unrecognized DNA (Figure 5–4). The DNA sequences that create a termination signal for this polymerase also share common features, as described in Figure 5–4. The eucaryotic cell contains three different RNA polymerases; they are basically similar to the bacterial enzyme, but their start and stop signals are different and they are less well characterized (see Chapter 8, p. 407).

The above outline omits many details: in many cases, a number of other complex steps must occur before an mRNA molecule is produced. Thus, *gene-regulatory proteins* help to determine which regions of DNA are transcribed by the RNA polymerase and thereby play a major part in determining the proteins made by a cell. Moreover, although mRNA molecules are produced directly by DNA transcription in procaryotes, in higher eucaryotic cells most RNA transcripts are altered extensively—by a process called *RNA processing*—before they leave the cell nucleus and enter the cytoplasm to become mRNA. All of these specialized aspects of mRNA production will be discussed in detail in Chapter 8, when we consider the cell nucleus. For the present, let us assume that functional mRNA molecules have been produced by a cell and proceed to examine how they direct protein synthesis.

Protein Synthesis Is Inherently Very Complex[3]

The molecular processes that underlie protein synthesis are very complex. Although we can describe many of them, they do not make conceptual sense in the way that DNA transcription, DNA repair, and DNA replication do. For example, we now know that not one but three main classes of RNA molecules (mRNA, tRNA, and rRNA) are involved in protein synthesis, but we do not fully understand why this must be so. Thus, the details of protein synthesis must largely be learned as fact without an obvious conceptual framework.

A true understanding of the mechanism of protein synthesis should help to illuminate the early events by which life itself came into existence. Here

(A) START SIGNAL

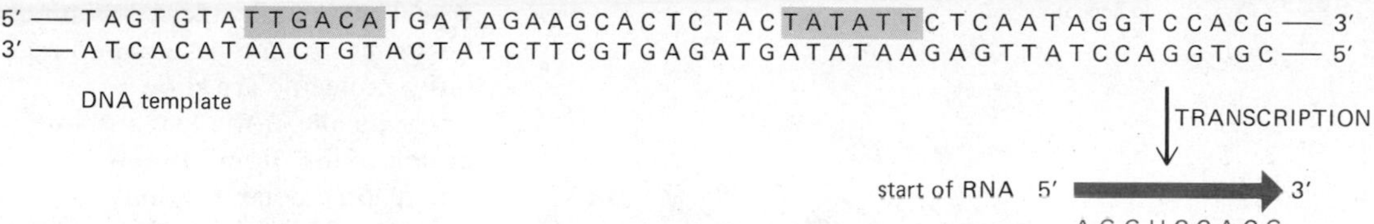

(B) STOP SIGNAL

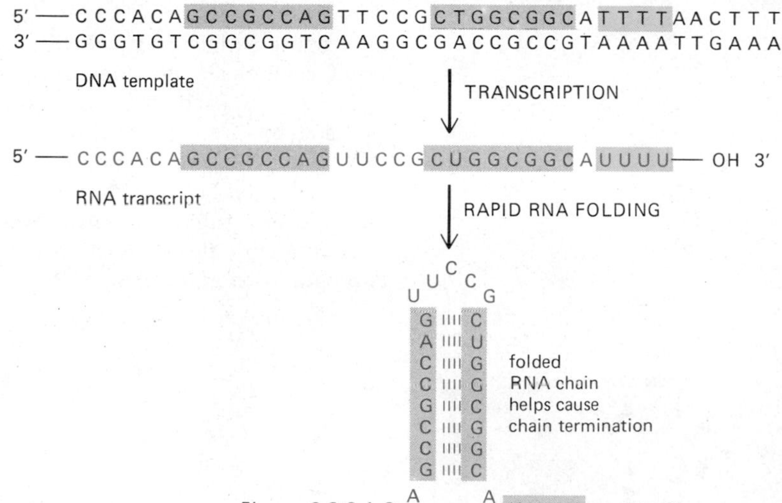

Figure 5–4 The start and stop signals for RNA synthesis by bacterial RNA polymerase of *E. coli*.

(A) The polymerase starts at a promoter sequence. The two short sequences about −35 and −10 nucleotides from the start of the RNA chain (*color*) are thought to define the promoter. Only limited modification of each of the two 6-nucleotide sequences shown can occur without eliminating promoter activity. The rest of the DNA sequence shown is not thought to be important.

(B) The polymerase stops when it synthesizes a run of U residues on the template that is located immediately following a self-complementary nucleotide sequence. The appropriate combination of DNA sequences therefore constitutes a termination signal. As indicated, the self-complementary sequence (*gray*) will rapidly form a hairpin helix in the newly synthesized RNA chain. Many different self-complementary sequences have been shown to function in this way; therefore, the hairpin helix rather than the nucleotide sequence is what seems to be required to create this part of the termination signal.

the key relationship must be that between RNA chain chemistry and polypeptide synthesis. How did it come about that the ordering of amino acids in polypeptides became determined by the sequences of nucleotides in RNA chains? So far we know of no chemical features of RNA that could lead to its preferential association with amino acids, not only providing the energy to make peptide bonds, but also dictating which amino acids are linked together. Until we do, we must attempt to clarify still further the details of protein synthesis as it occurs today and hope that the primordial moments of life may thus somehow be revealed.

The central agents in protein synthesis are the tRNA (transfer RNA) molecules to which amino acids are attached prior to their polymerization into polypeptides (Figure 5–5). By becoming attached at their carboxyl ends, amino acids are activated to high-energy forms from which peptide bonds form spontaneously to yield polypeptides. Such activation processes are obligatory for protein synthesis since free amino acids cannot be added directly to a growing polypeptide chain. (Only the reverse process in which a peptide bond is hydrolyzed by the addition of water occurs spontaneously.)

Only the tRNA molecule, and not its attached amino acid, determines where the amino acid is added during protein synthesis. This was established by an ingenious experiment in which an amino acid attached to a specific tRNA was chemically converted into a different amino acid (cysteine → alanine). If such hybrid tRNA molecules are allowed to instruct protein synthesis in a cell-free system, the wrong amino acid is inserted at every point in the protein chain where that tRNA is used (Figure 5–6). Thus the success of de-

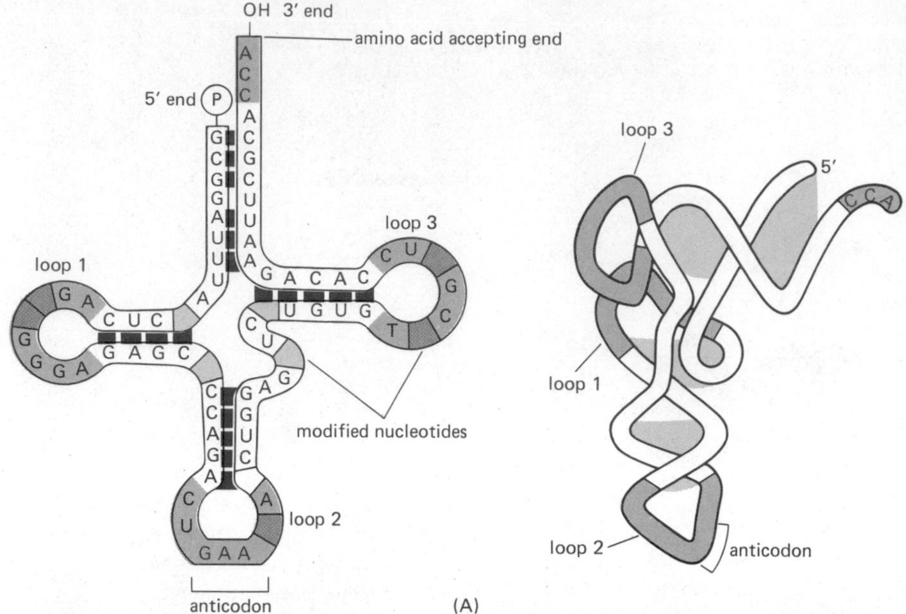

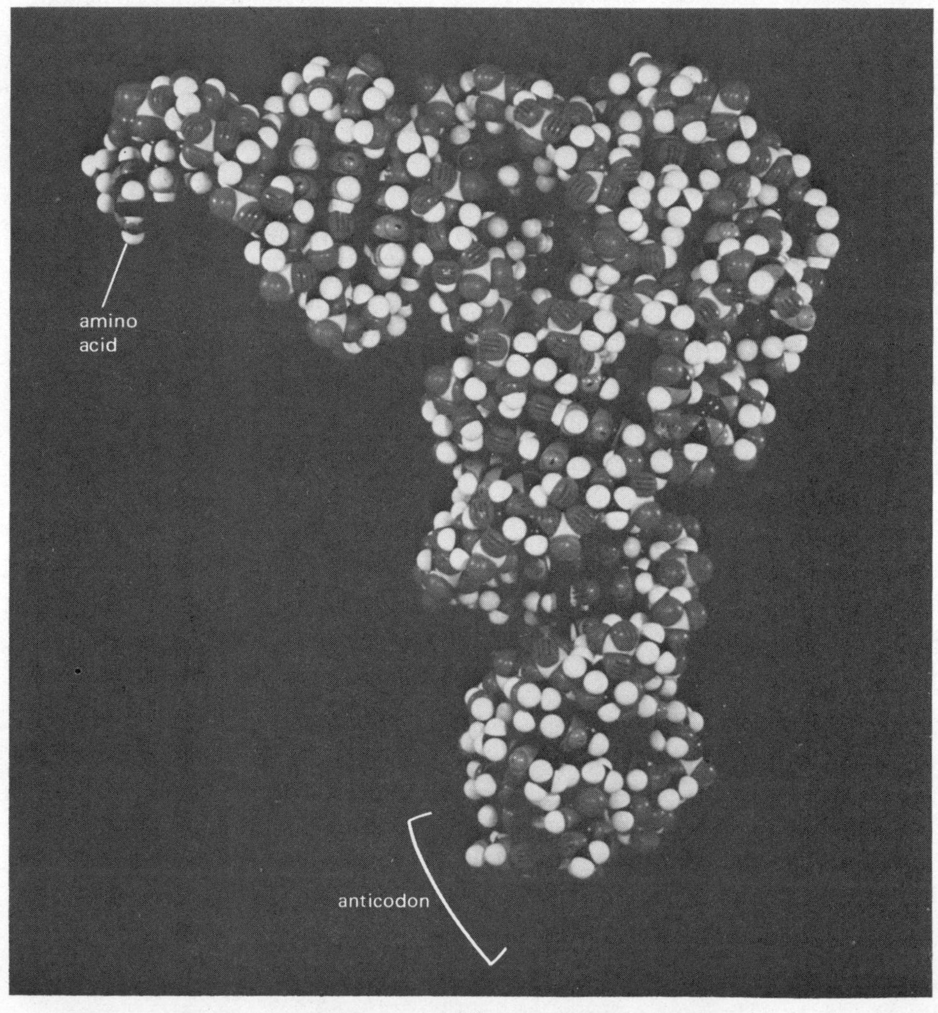

(B)

Figure 5–5 The structure of a typical tRNA molecule. (A) Outlines of the structure. The base-paired regions in the molecule are shown schematically at the left, and an outline of the overall three-dimensional conformation determined by x-ray diffraction is shown at the right. (B) A space-filling model of the structure outlined in (A). There are more than 20 different tRNA molecules, including at least 1 for each different amino acid. Although they differ in nucleotide sequence, they are all folded in a similar way. In each case the amino acid is attached to the A residue of a CCA sequence at the 3′ end of the tRNA molecule. The particular tRNA molecule shown binds phenylalanine and is therefore denoted as $tRNA^{Phe}$. (Photograph courtesy of Sung-Hou Kim.)

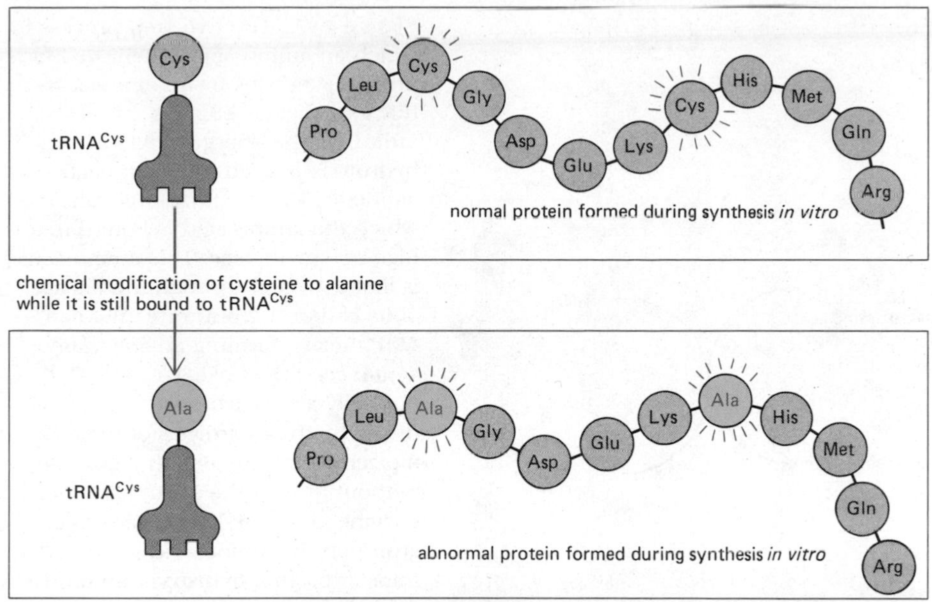

Figure 5–6 An experiment demonstrating that only tRNA, and not its attached amino acid, "recognizes" the site at which each amino acid is to be added during protein synthesis.

coding depends in part on the accuracy of the mechanism that normally links each activated amino acid to its corresponding tRNA molecule.

Equally important to successful mRNA decoding is the accuracy of the base-pairing between the *codons* in mRNA and the *anticodons* in each tRNA molecule (see Figure 5–9). These base pairs form effectively only when the mRNA and tRNA molecules are attached to appropriate binding slots on ribosomes. Unfortunately, we may wait many years before we deeply understand these ribosome-mediated binding interactions. As shown on page 209, ribosomes are very complex molecular aggregates, and the exact determination of their structure at the molecular level is not likely in the near future.

Specific Enzymes Couple Each Amino Acid to Its Appropriate tRNA Molecule[4]

How does each tRNA molecule recognize its corresponding amino acid? The answer is that a special set of enzymes, called **aminoacyl-tRNA synthetases,** couple each amino acid to its appropriate tRNA molecule. There is a different synthetase enzyme for every amino acid: one attaches glycine to tRNA^{Gly}, another attaches alanine to tRNA^{Ala}, and so on. The coupling reaction occurs in two steps, as illustrated in Figure 5–7, to create an **aminoacyl-tRNA** molecule (Figure 5–8).

The tRNA molecules serve as the final "adaptors" that convert nucleic-acid-sequence information into protein-sequence information. However, the aminoacyl-tRNA synthetase enzymes comprise a second set of specific adaptors of equal importance to the decoding process. Thus, the genetic code is translated by means of two linked sets of adaptors, each matching one molecular surface to another with great specificity; their combined action serves to identify each amino acid with a particular sequence of three nucleotides in the mRNA molecule—a codon for that amino acid (Figure 5–9).

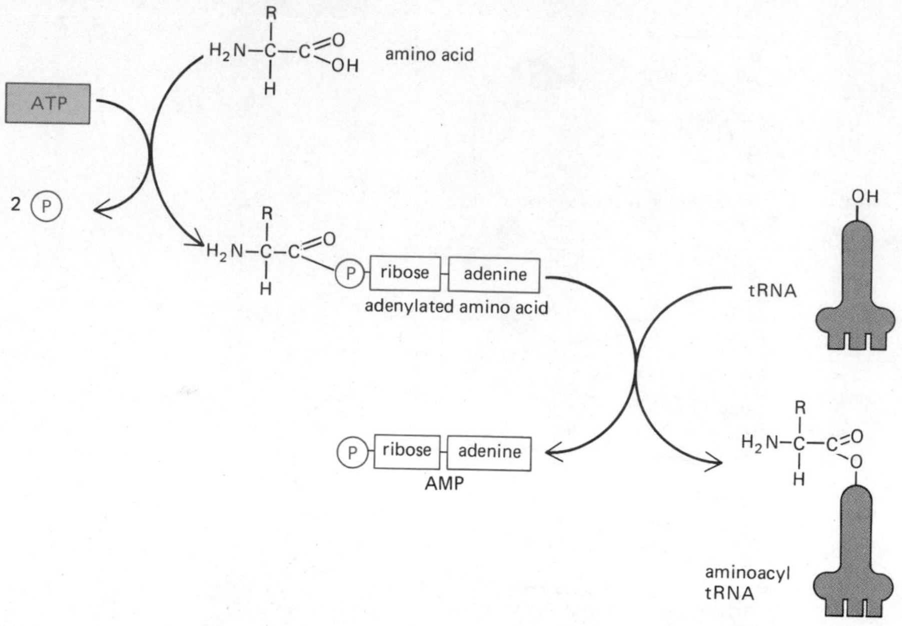

Figure 5–7 The two-step process in which an amino acid is activated for protein synthesis by an aminoacyl-tRNA synthetase enzyme. As indicated, the energy of ATP hydrolysis is used to attach each amino acid to its tRNA molecule, to which the amino acid is bound in a high-energy linkage. This amino acid is first activated through the linkage of its carboxyl group directly to an AMP moiety, forming an *adenylated amino acid;* the linkage of the AMP, normally an unfavorable reaction, is driven by the hydrolysis of the ATP molecule that donates the AMP. Without leaving the synthetase enzyme, the AMP-linked carboxyl group on the amino acid is then transferred to a hydroxyl group on the sugar at the 3′ end of the tRNA molecule. This transfer joins the amino acid by an activated ester linkage to the tRNA and forms the final aminoacyl-tRNA molecule.

Amino Acids Are Added to the Carboxyl-Terminal End of a Growing Polypeptide Chain

The fundamental reaction of protein synthesis is the formation of a peptide bond between the carboxyl group at the end of a growing polypeptide chain and a free amino group on an amino acid. A protein chain is consequently synthesized stepwise from its amino-terminal end to its carboxyl-terminal end. Throughout the entire process the growing carboxyl end of the polypeptide chain remains activated by the covalent attachment of a tRNA molecule (a *peptidyl-tRNA* molecule). The covalent linkage is disrupted in each cycle but is immediately replaced by an identical linkage carried in by the new amino acid (Figure 5–10). Thus, in protein synthesis each amino acid added carries with it the activation energy for the addition of the *next* amino acid rather than for its own addition. (This is an example of "head growth," described in Chapter 2 [Figure 2–34, p. 82].)

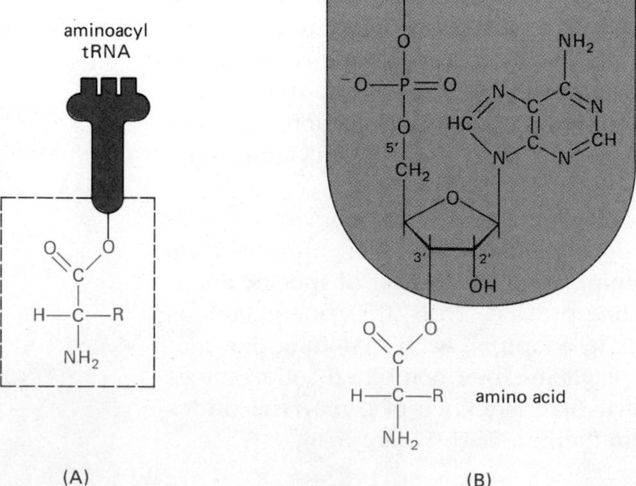

Figure 5–8 The structure of the aminoacyl-tRNA linkage. The carboxyl end of the amino acid has formed an ester bond to the ribose (at either the 2′ or the 3′ oxygen of the ribose). (A) Schematic of the structure. (B) Actual structure corresponding to boxed region in (A).

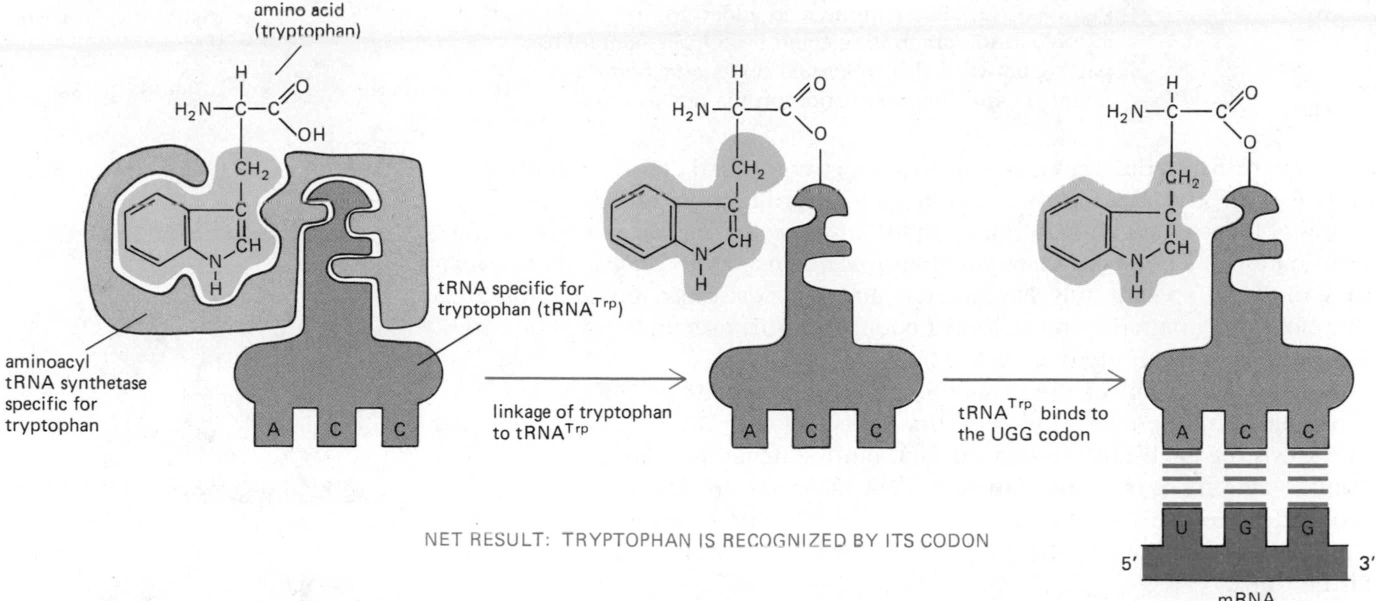

Figure 5–9 Schematic diagram illustrating how the genetic code is translated by means of two linked "adaptors": the aminoacyl-tRNA synthetase enzyme, which couples a particular amino acid to its corresponding tRNA, and the tRNA molecule, which then binds to the appropriate nucleotide sequence on the mRNA.

Each Amino Acid Added Is Selected by a Complementary Base-pairing Interaction Between Its Linked tRNA Molecule and an mRNA Chain[5]

In the course of translation, the mRNA sequence is read three nucleotides at a time as the translation machinery moves in the 5′-to-3′ direction along the mRNA molecule. Each amino acid is specified by the triplet of nucleotides (**codon**) in the mRNA molecule that pairs with a sequence of three complementary nucleotides at the anticodon tip of a tRNA molecule. Because only one of the many different types of tRNA molecules in a cell can base-pair with each codon, the codon determines the specific amino acid residue to be added to the growing polypeptide chain end (Figure 5–11).

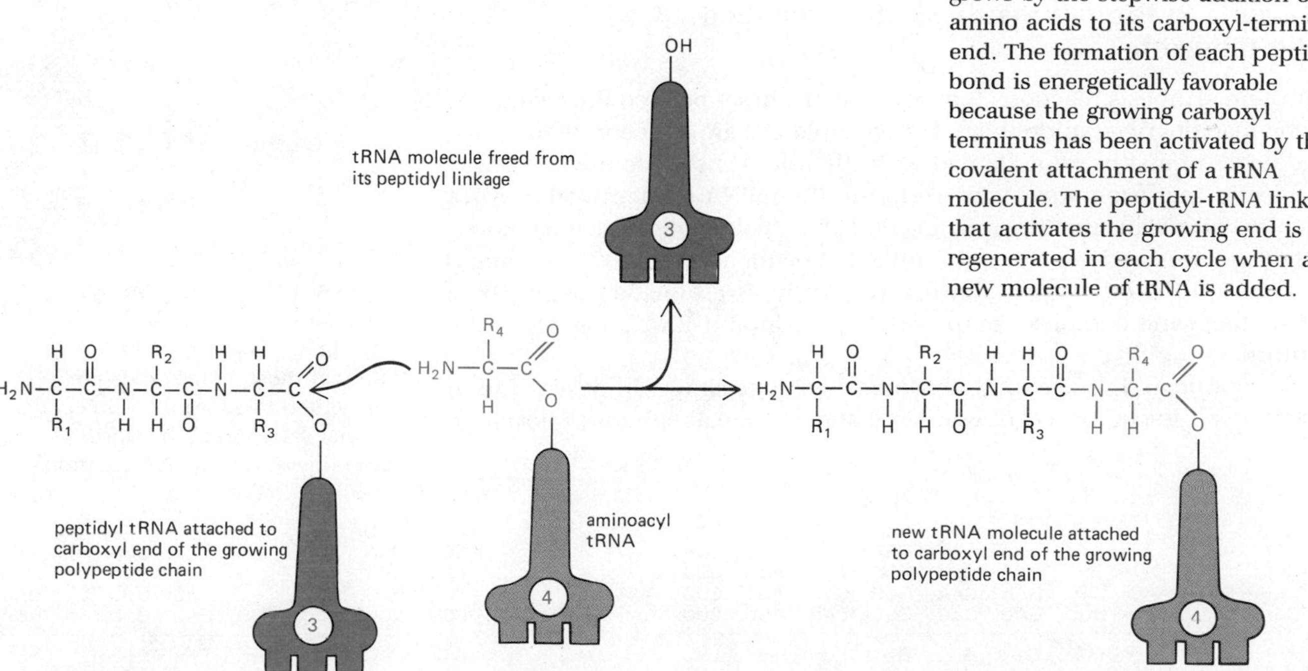

Figure 5–10 A polypeptide chain grows by the stepwise addition of amino acids to its carboxyl-terminal end. The formation of each peptide bond is energetically favorable because the growing carboxyl terminus has been activated by the covalent attachment of a tRNA molecule. The peptidyl-tRNA linkage that activates the growing end is regenerated in each cycle when a new molecule of tRNA is added.

Figure 5–11 Each amino acid added to the growing end of a polypeptide chain is selected by complementary base-pairing between the anticodon on its attached tRNA molecule and the next codon on the mRNA chain.

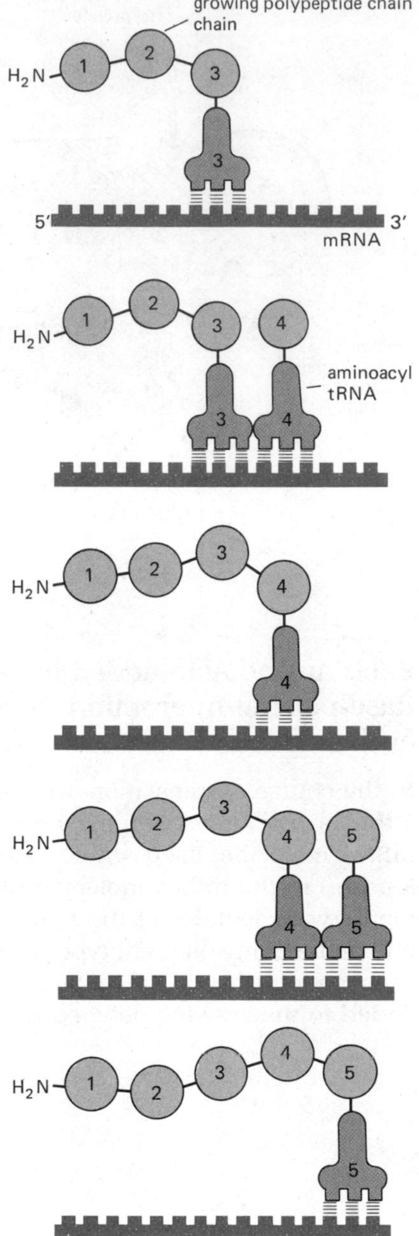

There are 64 different possible sequences composed of three nucleotides (4 × 4 × 4), and most of them occur somewhere in most mRNA molecules. Three of these 64 codons do not code for amino acids but instead specify the termination of a polypeptide chain; they are known as *stop codons*. That leaves 61 codons to specify only 20 different amino acids: most amino acids are therefore represented by more than 1 codon. For this reason, the genetic code is said to be *degenerate* (Figure 5–12).

The degeneracy of the genetic code means that either (1) a single tRNA molecule can base-pair with more than one codon or (2) there is more than one tRNA for each amino acid. In fact, both statements are true. For some amino acids there is more than one tRNA molecule. In addition, some tRNA molecules are constructed so that they require accurate base-pairing only at the first two positions of the codon and can tolerate a mismatch (or *wobble*) at the third.

The nucleotides in a completed nucleic acid chain (like the amino acids in proteins, p. 333) can be covalently modified in order to modulate the biological activity of the nucleic acid. In particular, many different modified nucleotides exist in a population of tRNA molecules, each produced by covalent modification after the tRNA has been synthesized: a few examples are illustrated in Figure 5–13. Some of the unusual nucleotides affect the conformation and base-pairing of the anticodon and thereby facilitate the recognition of the appropriate mRNA codon (or codons) by the tRNA molecule. Another type of posttranscriptional modification is also common to tRNA molecules: tRNAs are initially made as part of a somewhat longer RNA transcript, from which they are first cut out and then completed by an enzyme that adds the three nucleotides CCA to their 3′ end. Both types of tRNA processing reactions occur in both procaryotes and eucaryotes; in contrast, an extensive processing of mRNA precursors occurs only in eucaryotes (see Chapter 8).

The Events in Protein Synthesis Are Catalyzed on the Ribosome[6]

The protein synthesis reactions that we have just described require a complex catalytic machinery to guide them. For example, the growing end of the polypeptide chain must be kept aligned with the mRNA molecule in such a way as to ensure that each successive codon in the mRNA engages with a tRNA molecule. This means that the growing end of the polypeptide must be moved exactly three nucleotides along the mRNA after the addition of each amino acid. This and the other events in protein synthesis are made possible by a large multienzyme complex composed of protein and RNA molecules—the **ribosome.**

Eucaryotic and procaryotic ribosomes are quite similar in design and in function. Each is composed of one large and one small subunit. Eucaryotic

Figure 5–12 The genetic code. Codons are written with the 5′-terminal nucleotide on the left. Note that most amino acids are represented by more than one codon and that variation at the third nucleotide in a codon is common.

Ala	Arg	Asp	Asn	Cys	Glu	Gln	Gly	His	Ileu	Leu	Lys	Met	Phe	Pro	Ser	Thr	Trp	Tyr	Val	stop
	AGA									UUA					AGC					
	AGG									UUG					AGU					
GCA	CGA						GGA			CUA				CCA	UCA	ACA			GUA	
GCC	CGC						GGC		AUA	CUC				CCC	UCC	ACC			GUC	UAA
GCG	CGG	GAC	AAC	UGC	GAA	CAA	GGG	CAC	AUC	CUG	AAA		UUC	CCG	UCG	ACG		UAC	GUG	UAG
GCU	CGU	GAU	AAU	UGU	GAG	CAG	GGU	CAU	AUU	CUU	AAG	AUG	UUU	CCU	UCU	ACU	UGG	UAU	GUU	UGA

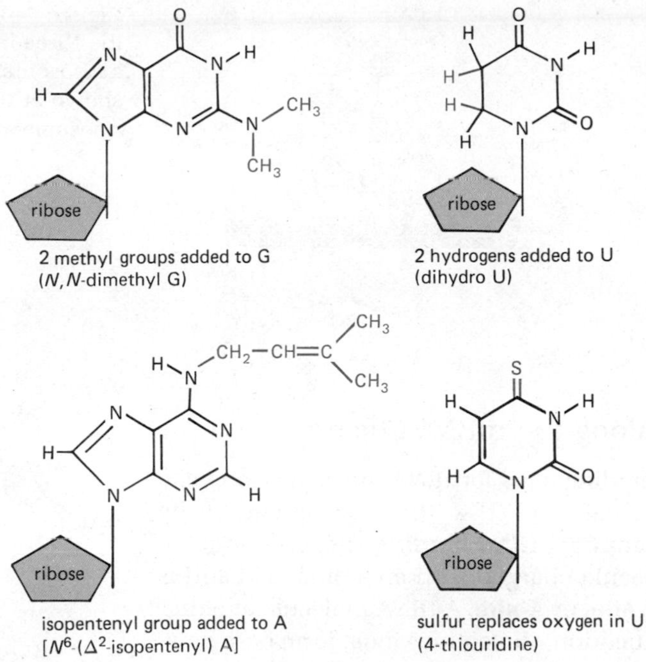

2 methyl groups added to G
(N,N-dimethyl G)

2 hydrogens added to U
(dihydro U)

isopentenyl group added to A
[N^6-(Δ^2-isopentenyl) A]

sulfur replaces oxygen in U
(4-thiouridine)

Figure 5–13 A few of the unusual nucleotides found in tRNA molecules produced by covalent modification of a normal nucleotide after it has been incorporated into a polynucleotide chain. In most tRNA molecules, about 10% of the nucleotides are modified in this way (see Figure 5–5A).

ribosomes are roughly half RNA by weight: the small subunit consists of 1 ribosomal RNA (rRNA) molecule bound to about 33 different ribosomal proteins, while the large subunit consists of 3 different rRNA molecules bound to more than 40 different ribosomal proteins. Procaryotic ribosomes are slightly smaller and contain fewer components (Figure 5–14). Both types of ribosomes have a groove that accommodates a growing polypeptide chain and a groove that accommodates an mRNA molecule. These grooves cover a stretch of about 30 amino acids and about 35 RNA nucleotides, respectively.

Figure 5–14 A comparison of the structures of procaryotic and eucaryotic ribosomes. Despite the differences in structure, they function in very similar ways.

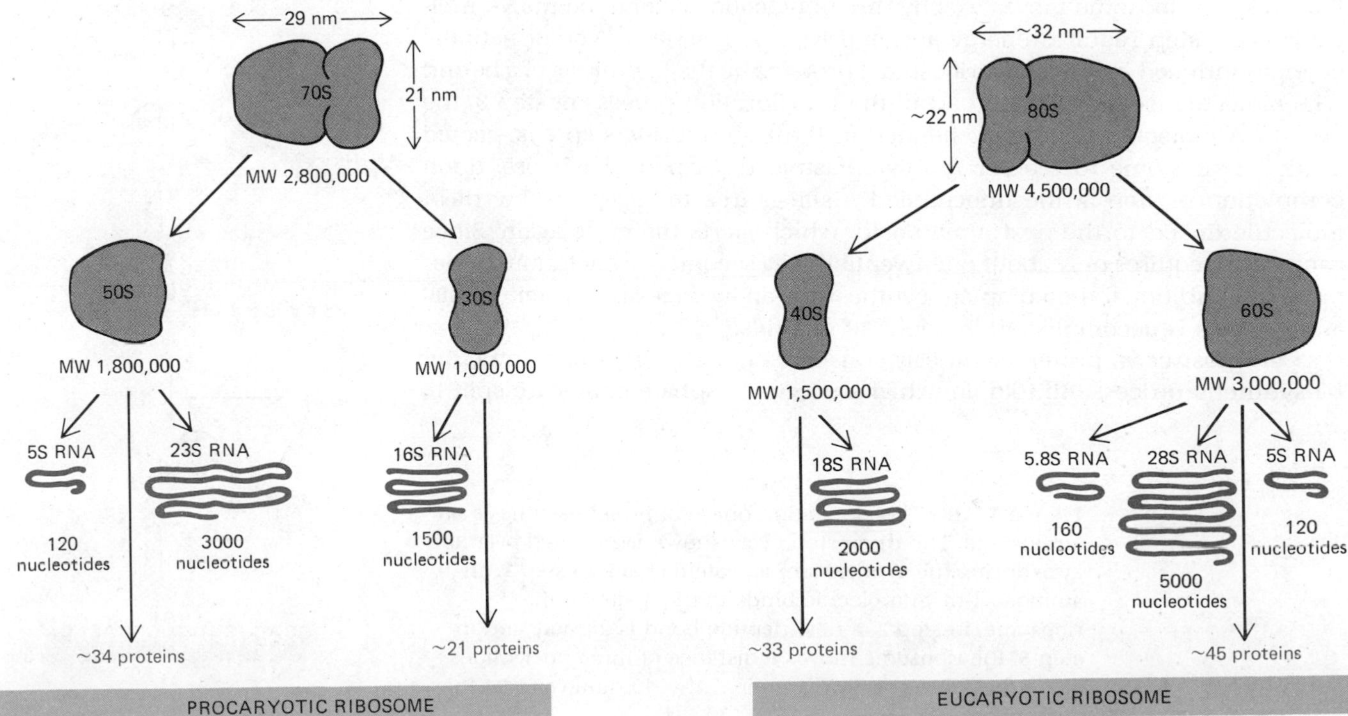

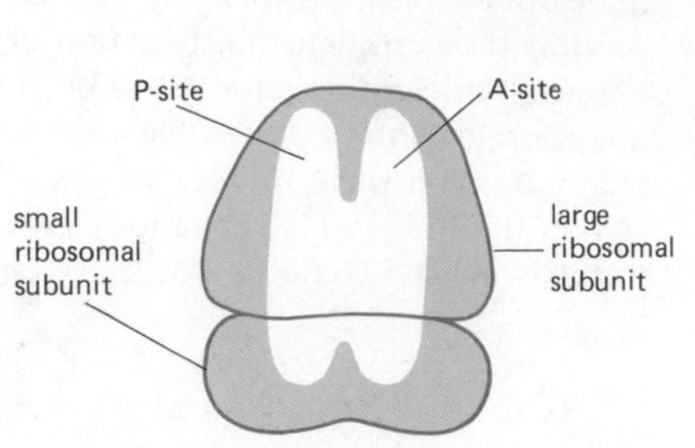

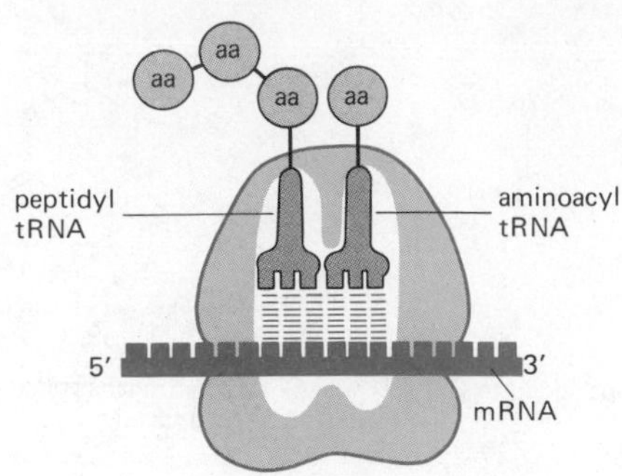

Figure 5–15 Schematic diagram of the three major RNA binding sites on a ribosome. An empty ribosome is shown at the left and a loaded ribosome at the right.

A Ribosome Moves Stepwise Along the mRNA Chain[6,7]

A ribosome contains two different binding sites for tRNA molecules: one site holds the tRNA molecule that is normally linked to the growing end of the polypeptide chain and is called the **peptidyl-tRNA binding site,** or **P-site;** the other holds the incoming tRNA molecule charged with an amino acid and is called the **aminoacyl-tRNA binding site,** or **A-site.** A tRNA molecule at either site is held in such a way that the anticodon nucleotides must form base pairs with a complementary mRNA codon in order to fit into the site. The A- and P-sites are so close together that the tRNA molecules held in each site form base pairs with adjacent codons in the mRNA molecule (Figure 5–15).

The process of polypeptide chain elongation on a ribosome can be considered as a cycle with three discrete steps (Figure 5–16). In step 1, an aminoacyl-tRNA molecule becomes bound to a vacant ribosomal A-site adjacent to an occupied P-site by forming base pairs with the three mRNA nucleotides exposed at the A-site. In step 2, the carboxyl end of the polypeptide chain is uncoupled from the tRNA molecule in the P-site and joined by a peptide bond to the amino acid linked to the tRNA molecule in the A-site. This reaction is catalyzed by **peptidyl transferase,** an enzyme that is tightly bound to the ribosome. In step 3, the new peptidyl-tRNA in the A-site is translocated to the P-site as the ribosome moves exactly three nucleotides along the mRNA molecule. This step requires energy and is driven by a series of conformational changes induced in one of the ribosomal proteins by the hydrolysis of a bound GTP molecule (see p. 137). As part of the translocation process of step 3, the free tRNA molecule that was generated in the P-site during step 2 is ejected from the ribosome to reenter the cytoplasmic tRNA pool. Therefore, upon completion of step 3, the unoccupied A-site is free to accept a new tRNA molecule linked to the next amino acid, which starts the cycle again. Since each cycle requires only about one-twentieth of a second in a bacterium under optimal conditions, the complete synthesis of an average-size protein of 400 amino acids is accomplished in about 20 seconds.

In most cells, protein synthesis consumes more energy than any other biosynthetic process. All told, four high-energy phosphate bonds are split to

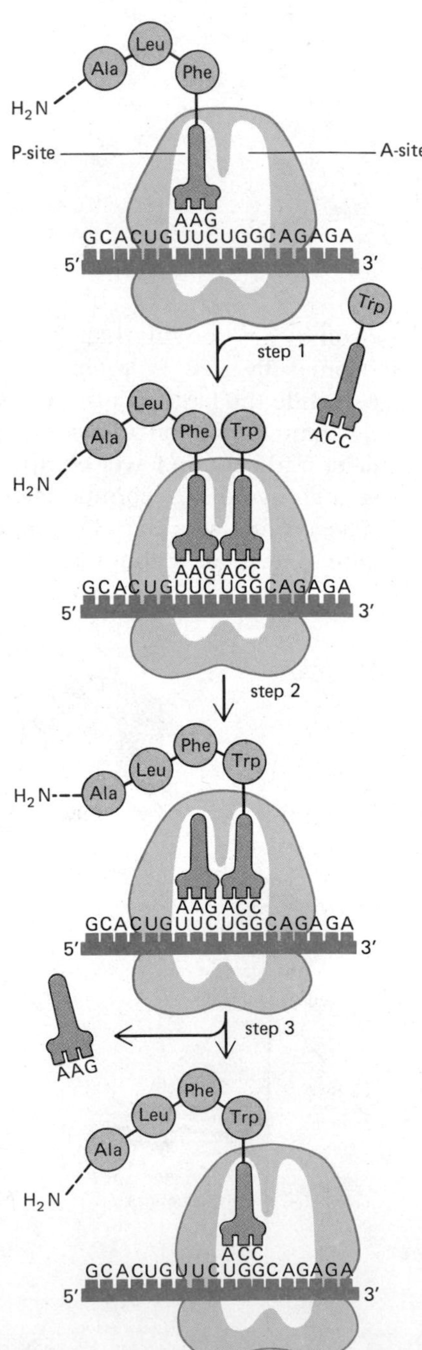

Figure 5–16 The elongation phase of protein synthesis on a ribosome. The three-step cycle shown is repeated over and over during the synthesis of a protein chain. In step 1, an aminoacyl-tRNA molecule binds to the A-site on the ribosome; in step 2, a new peptide bond is formed; and in step 3, the ribosome moves a distance of three nucleotides along the mRNA chain, "resetting" the ribosome so that the cycle can be repeated.

make each new peptide bond. Two of these are required to charge each tRNA molecule with an amino acid (Figure 5–7). And two more drive two of the cyclic reactions occurring on the ribosome during synthesis itself: one for the aminoacyl-tRNA binding in step 1, and one for the ribosome translocation in step 3.

A Protein Chain Is Released from the Ribosome Whenever One of Three Different Termination Codons Is Reached[6,8]

As already noted, three of the codons in an mRNA molecule are **stop codons,** which terminate the translation process. A protein called *release factor* binds directly to any stop codon (UAA, UAG, or UGA) that reaches the A-site on the ribosome. This binding disturbs the activity of the nearby peptidyl transferase enzyme, causing it to catalyze the addition of a water molecule instead of the free amino group of an amino acid to the peptidyl-tRNA. As a result, the carboxyl end of the growing polypeptide chain is freed from its attachment to a tRNA molecule. Since it is only this attachment that normally holds the growing polypeptide to the ribosome, the completed protein chain is released into the cell cytoplasm (Figure 5–17).

The Initiation Process Sets the Reading Frame for Protein Synthesis[6,9]

In principle an RNA sequence can be decoded in any one of three different *reading frames,* each of which will specify a completely different polypeptide chain (see Figure 3–13, p. 108). Which of the three frames is actually read is determined when a ribosome engages with an mRNA molecule to form an *initiation complex.* This complex is assembled at the exact spot on the mRNA where the polypeptide chain is to begin.

The initiation process is complicated, involving a number of steps catalyzed by proteins called **initiation factors,** many of which are themselves composed of several polypeptide chains. Because of their complexity, many of the details of initiation are still uncertain. However, it is clear that each ribosome is assembled onto an mRNA chain as two separate subunits, the small ribosomal subunit being added first. Before the mRNA is bound, a special **initiator tRNA** molecule, recognizing the codon AUG and carrying methionine, is loaded onto each small subunit. This loading reaction is catalyzed by one of the initiation factors, called *initiation factor 2,* or *IF-2.* In some eucaryotic cells the overall rate of protein synthesis is controlled by this factor (see the following page).

The small ribosomal subunit binds to the region of the mRNA molecule where protein synthesis is to begin by pairing its bound initiator tRNA molecule with a particular AUG start codon (Figure 5–18). An mRNA molecule usually contains many AUG sequences, each of which codes for methionine. But the vast majority of these will not serve as start codons. As explained elsewhere, which AUG is recognized as a start codon depends on other parts of the mRNA nucleotide sequence (p. 332).

At the completion of the initiation process, all of the initiation factors associated with the small ribosomal subunit up to this point are discharged to make way for the binding of a large ribosomal subunit to the small one. In this way, a complete functional ribosome is formed. The initiator tRNA molecule ends up bound to the P-site of the ribosome, so that the synthesis of a protein chain can begin directly with the binding of a second aminoacyl-tRNA molecule to the A-site of the ribosome (Figure 5–18). Further steps in the elongation phase of protein synthesis then proceed as described previously (see step 2 of Figure 5–16).

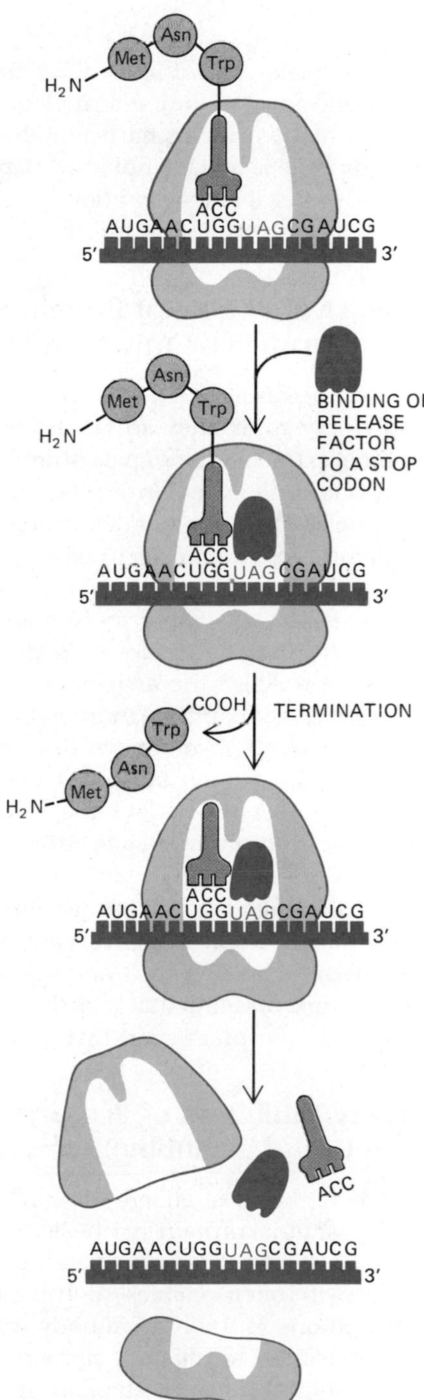

Figure 5–17 The final phase of protein synthesis. The binding of release factor to a stop codon terminates translation: the completed polypeptide is released, and the ribosome dissociates into its two separate subunits.

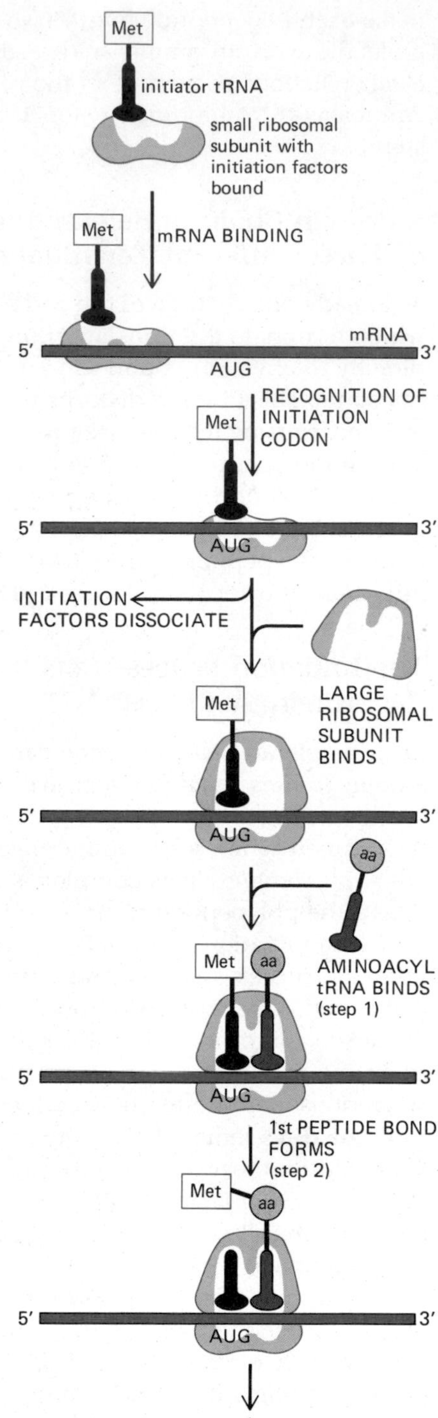

Figure 5–18 The initiation phase of protein synthesis. Events are illustrated as they occur in eucaryotes, but a very similar process occurs in bacteria. Step 1 and step 2 refer to steps in the elongation reaction shown in Figure 5–16.

Because an initiator tRNA molecule always carries the amino acid methionine (or the amino-formyl derivative of methionine in procaryotes), all newly made proteins have a methionine as their amino-terminal residue. This methionine is usually not important for protein function and is often removed shortly after its incorporation by the limited proteolysis of a few residues that occurs at the amino terminus of many proteins.

The Overall Rate of Protein Synthesis in Eucaryotes Is Controlled by Initiation Factors[10]

As will be discussed in Chapter 11, the cells in a multicellular organism multiply only when they are in an appropriate environment. The mechanism by which extracellular signals stimulate cells to grow and divide is not known, but one of their major effects must be to increase the overall rate of protein synthesis (p. 617). What determines this rate? Direct studies in tissues are very difficult, but when cells in tissue culture are starved of essential amino acids, glucose, or serum, there is a marked inhibition of their rate of polypeptide-chain *initiation,* which can be shown to result from inactivation of the protein synthesis initiation factor IF-2. Moreover, in at least one type of cell (immature red blood cells), the activity of IF-2 is known to be reduced in a controlled way by the phosphorylation of one of its three protein subunits. This suggests that eucaryotic protein synthesis rates are controlled in part by specific protein kinases, which in their active form inhibit the initiation of protein synthesis. One can speculate that unknown growth signals in tissues might cause cells to multiply by inactivating such protein kinases when cells are in the proper environment.

The initiation factors required for protein synthesis are more numerous and more complex in eucaryotes than in procaryotes, even though they perform the same basic functions. Many of the extra components could be regulatory proteins that coordinate cell growth in multicellular eucaryotes by controlling protein synthesis.

Many Inhibitors of Procaryotic Protein Synthesis Are Useful as Antibiotics[11]

Many of the most effective antibiotics used in modern medicine act by inhibiting bacterial protein synthesis. A number of these drugs exploit the structural and functional differences between procaryotic and eucaryotic ribosomes. Their selectivity often enables such compounds to be used at relatively high concentrations in the human body without undue toxicity. Because different antibiotics bind to different protein subcomplexes on bacterial ribosomes, they often inhibit different steps in the synthetic process. Some of the more common compounds in this group are listed in Table 5–1, along with their specific effects. Also listed in this table are several other commonly used inhibitors of protein synthesis, some of which act on eucaryotic cells. Because they can be used to block specific steps in the processes that lead from DNA to protein, the compounds listed are widely used for a variety of biochemical and cell biological studies.

Table 5–1 Inhibitors of Protein or RNA Synthesis

Inhibitor	Specific Effect
*Acting Only on Procaryotes**	
Tetracycline	blocks binding of aminoacyl-tRNA to A-site of ribosome
Streptomycin	prevents the transition from initiation complex to chain-elongating ribosome
Chloramphenicol	blocks the peptidyl transferase reaction on ribosomes (step 2 in Figure 5–16)
Erythromycin	blocks the translocation reaction on ribosomes (step 3 in Figure 5–16)
Rifamycin	blocks initiation of RNA chains by binding to RNA polymerase (prevents RNA synthesis)
Acting on Procaryotes and Eucaryotes	
Puromycin	causes the premature release of nascent polypeptide chains by its addition to growing chain end
Actinomycin D	binds to DNA and blocks the movement of RNA polymerase (prevents RNA synthesis)
Acting Only on Eucaryotes	
Cycloheximide	blocks the peptidyl transferase reaction on ribosomes (step 2 in Figure 5–16)
α-Amanitin	blocks mRNA synthesis by binding preferentially to RNA polymerase II

*The ribosomes of eucaryotic mitochondria (and chloroplasts) often resemble those of procaryotes in their sensitivity to inhibitors.

Summary

Before the synthesis of a particular protein can begin, the corresponding mRNA molecule must be produced by DNA transcription processes and exported to the cell cytoplasm. An extensive machinery is then called into play. The process begins with the binding of a small ribosomal subunit to an mRNA molecule. A unique initiator tRNA molecule positions the small ribosomal subunit over a special start codon on the mRNA. A large ribosomal subunit is added to complete the ribosome, and the elongation phase of protein synthesis ensues. Each amino acid is added to the carboxyl-terminal end of the growing polypeptide by means of a cycle of three sequential steps: aminoacyl-tRNA binding, followed by peptide bond formation, followed by ribosome translocation. The ribosome progresses from codon to codon in the 5′-to-3′ direction along the mRNA molecule until one of three stop codons is reached. A release factor then binds to the stop codon, terminating translation and releasing the completed polypeptide from the ribosome.

An aminoacyl-tRNA molecule serves as a decoding device that allows a particular sequence of three ribonucleotides in the mRNA to be translated as a unique amino acid in the newly synthesized protein. Each of the 20 amino acids is fitted to a particular codon by means of a two-step recognition process: in the first step the amino acid is recognized by a unique aminoacyl-tRNA synthetase enzyme that links it to a specific tRNA molecule; and in the second step a particular sequence of three nucleotides in the mRNA chain is recognized by the anticodon of the tRNA molecule. There is at least one specific aminoacyl-tRNA synthetase and at least one specific tRNA for each amino acid.

DNA Repair Mechanisms[12]

While long-term survival of a species may be enhanced by changes in its genetic inheritance, its short-term survival absolutely demands that the genetic record be accurately maintained. Adequate maintenance of the genetic material requires not only an extremely accurate mechanism for copying DNA sequences once in every cell generation but also a mechanism for repairing the many accidental lesions that occur spontaneously in DNA. Before examining these mechanisms of replication and repair, we shall discuss briefly just how faithfully DNA sequences are maintained from one generation to the next.

DNA Sequences Are Maintained with a Very High Fidelity[13]

The rate at which DNA sequences change (the mutation rate) can be estimated only indirectly. One way is to compare the amino acid sequence of the same protein in several different species: the fraction of the amino acids that are different is compared with the estimated number of years since each pair of species diverged from a common ancestor as determined by fossil records. In this way one can calculate the average number of years required to generate a stable change in 1% of the protein's amino acids. Because each such change will, in general, reflect a single alteration in the DNA sequence of the gene coding for that protein, this value can be used to estimate the average number of years required to generate a single, stable mutation in the gene.

Such estimates will always be underestimates of the actual mutation rate because some mutations will compromise the function of the protein and vanish from the population under selective pressure. But one family of proteins that has been studied is nearly free from this disadvantage. These proteins are the **fibrinopeptides:** 20-residue-long fragments that are discarded from the protein *fibrinogen* when it is activated to form *fibrin* during blood clotting. Since the fibrinopeptides have no direct function, they can tolerate almost any amino acid change. Analysis of the fibrinopeptides (Figure 5–19) indicates that an average size protein 400 amino acids long would be randomly altered by an amino acid change roughly once every 200,000 years. More recently, DNA-sequencing technology (see p. 189) has made it possible to determine the similarity of DNA sequences in homologous noncoding regions of the genome for several different mammalian species. The resulting estimates of mutation rate are in excellent agreement with those obtained from the fibrinopeptide studies.

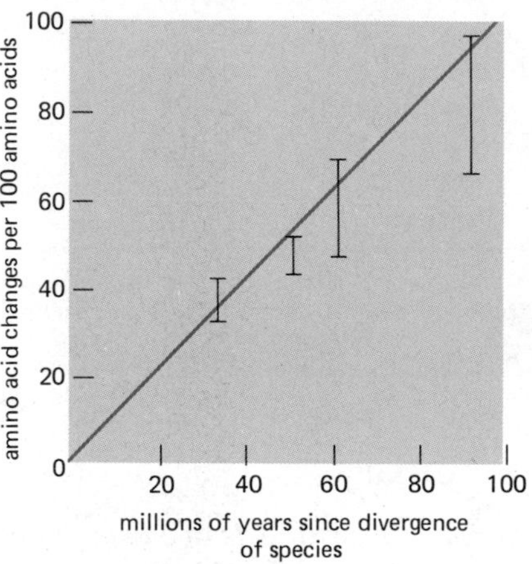

Figure 5–19 The rate of amino acid change during the evolution of the fibrinopeptides.

Directly Detected Spontaneous-Mutation Rates Are Consistent with Evolutionary Estimates[14]

The frequency of error in the replication of DNA sequences can be estimated more directly by observing the rate at which spontaneous changes arise in the genome of growing cells. This can be done either by estimating the frequency with which new mutants arise in very large animal populations (for example, fruit fly or mouse colonies) or by screening for specific enzyme changes in cells growing in tissue culture. Although they are only approximate, the numbers obtained in both cases are consistent with an error frequency of one base-pair change in roughly 10^9 base-pair replications. In mammals this figure is equivalent to about three changes per haploid genome for each cell generation. Consequently, a gene encoding an average-size protein containing about 10^3 coding base pairs would require about 10^6 cell generations to accumulate a mutation. This number is in accord with the evolutionary mutation rate estimate of one mutation in an average gene every 200,000 years, if we assume about five cell generations per year in an average germ line (from parental egg to daughter egg).

Most Mutations in Proteins Are Deleterious and Are Eliminated by Natural Selection[13]

When the number of amino acid differences in a particular protein in two different organisms is plotted against the time since the organisms diverged, the result is a reasonably straight line. That is, the longer the period since divergence, the larger the number of differences. For convenience, the slope of this line can be expressed in terms of the "unit evolutionary period" for that protein (the average time required for one amino acid change to appear in a sequence of 100 amino acid residues). When this is done for different proteins, each shows a different but characteristic rate of evolution (Figure 5–20). Since all base pairs must be subject to the same rate of random mutation, it is argued from such data that changes in the amino acid sequence affect the function of some proteins much more than they affect that of others. From Table 5–2 we can estimate that about 4 of every 5 random amino acid changes are harmful in hemoglobin; about 17 of every 18 amino acid changes are harmful in cytochrome c; and virtually all amino acid changes are harmful in histone H4. It can be assumed that the individuals who carried such harmful mutations were eliminated from the population by natural selection.

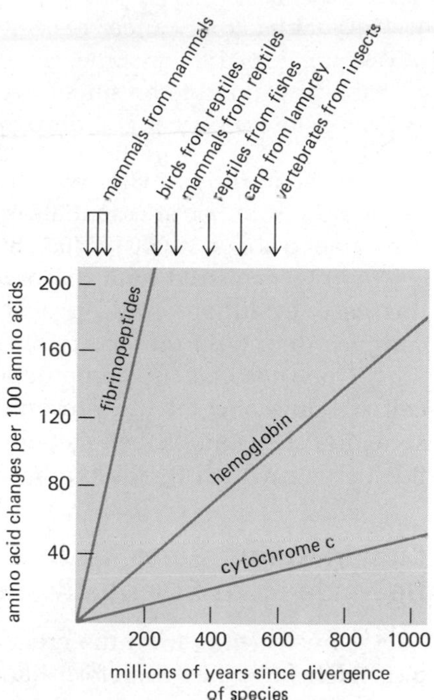

Figure 5–20 A comparison of the rates of amino acid change found in hemoglobin and cytochrome c with the rate found in the fibrinopeptides. Hemoglobin and cytochrome c have changed much more slowly during evolution than the fibrinopeptides.

Table 5–2 Observed Rates of Change of the Amino Acid Sequences in Various Proteins over Evolutionary Time

Protein	Unit Evolutionary Time (in millions of years)
Fibrinopeptide	1.2
Hemoglobin	6.1
Cytochrome c	21
Histone H4	600

The "unit evolutionary time" is defined as the average time required for one *acceptable* amino acid change to appear in the indicated protein for every 100 amino acids that it contains.

The Observed Low Mutation Rates Are Necessary for Life as We Know It[15]

Implicit in the evidence for elimination of mutant proteins is the principle that no species can afford to allow mutations to accumulate at a high rate in its germ cells. But somatic cells too must be protected from genetic change. Nucleotide changes in somatic cells can lead, for example, to the uncontrolled cell growth known as cancer, which causes about 20% of the premature deaths in the Western Hemisphere. The evidence that these deaths are due largely to the accumulation of changes in the DNA sequences of somatic cells includes the following:

1. Most cancers can be shown to arise as clones of cells descended from a single abnormal precursor cell.

2. Most carcinogens of man (or their metabolic breakdown products) raise the frequency of mutation events in bacteria.

3. A class of human genetic diseases called *xeroderma pigmentosum* is associated with defects in a particular type of DNA repair enzyme (see p. 219). Individuals with these diseases have abnormally high rates of skin cancer, caused by the damaging effects of the sun's ultraviolet rays on DNA.

4. The incidence of cancer rises steeply in older individuals. Statistical analysis suggests that for some cancers three to six independent random events (each presumed to be some sort of mutation) must accumulate in a line of cells before a cancer cell develops (Figure 5–21).

We shall discuss later why the observed mutation frequency is thought to limit the number of essential proteins that any organism can encode in its germ line to about 30,000 (p. 405). By the same argument, a mutation frequency tenfold higher would limit the organism to about 3000 essential proteins. In this case, evolution would probably have had to stop at an organism less complex than the fruit fly.

Thus, both for the perpetuation of a species with 30,000 proteins (germ-cell stability) and for the prevention of cancer resulting from mutations in somatic cells (somatic-cell stability), eucaryotes require the remarkably high fidelity with which DNA sequences are known to be maintained.

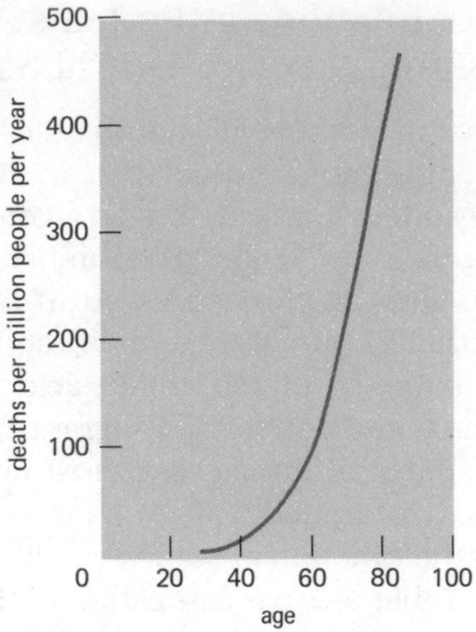

Figure 5–21 Annual United States death rates from cancer of the large intestine in relation to age. The exponential rise suggests that several independent heritable events are required to give rise to this type of cancer cell.

Low Mutation Rates Mean That Related Organisms Must Be Made from Essentially the Same Proteins[13]

As a genus distinct from the great apes, humans are only a few million years old. Each gene has therefore had the chance to accumulate relatively few nucleotide changes since our inception, and most of these will have been eliminated by natural selection. A great deal of our genetic heritage must have been formed long before *Homo sapiens* appeared, during the evolution of mammals (starting about 3×10^8 years ago) and even earlier. It is not very surprising, therefore, that the proteins of mammals as different as whales and humans are very similar. The evolutionary changes that have produced the striking morphological differences among mammals have had to do so with surprisingly few changes in the materials from which we are made.

If Left Uncorrected, Spontaneous DNA Damage Would Change DNA Sequences Rapidly[16]

The physicist Erwin Schroedinger pointed out in 1945 that, whatever its chemical nature (at that time unknown), a gene must be extremely small and composed of few atoms. Otherwise, the very large number of genes thought to be necessary to generate an organism would not fit in the cell nucleus. On the other hand, because it was so small, a gene could be expected to undergo significant changes due to spontaneous reactions induced by random thermal collisions with solvent molecules. This posed a serious dilemma, since the genetic data imply that genes are remarkably stable substances in which spontaneous changes (mutations) occur only extremely rarely.

Schroedinger's dilemma is real. The molecules in the gene do undergo major changes due to thermal fluctuations. We now know, for example, that about 5000 purine bases (adenine and guanine) are lost per day from the DNA of each human cell because of the thermal disruption of their *N*-glycosyl linkages to deoxyribose (*depurination*). Similarly, spontaneous deaminations of cytosine to uracil in DNA are estimated to occur at a rate of 100 per genome per day (Figure 5–22). DNA bases are also subject to change by reactive metabolites that alter their base-pairing abilities and by ultraviolet light from the sun, which can promote a covalent linkage of two adjacent thymine bases in DNA (forming *thymine dimers*). These are only a few of many changes that occur spontaneously in our DNA. Most of them would be expected to lead either to deletion of one or more base pairs in the daughter DNA chain after DNA replication or to a base-pair substitution (for example, each $C \rightarrow U$ deam-

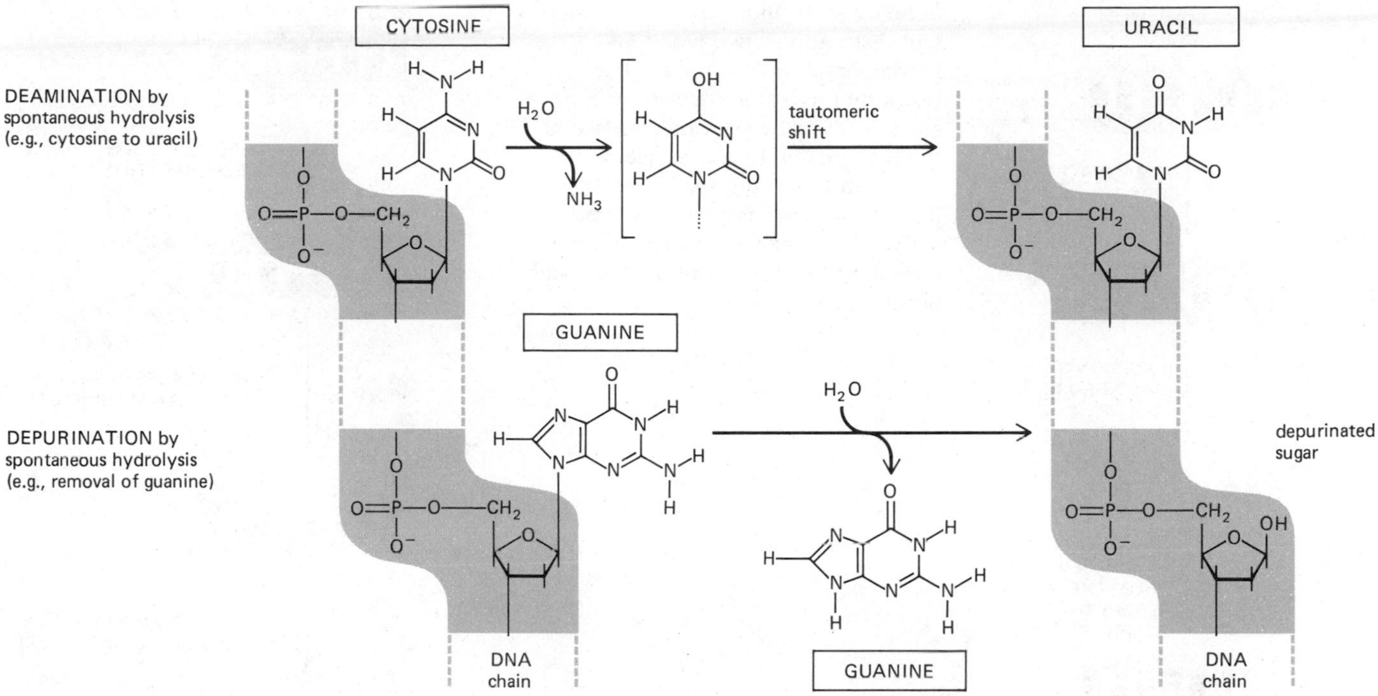

Figure 5–22 Two frequent spontaneous chemical reactions known to create serious DNA damage in cells are deamination and depurination. One specific example of each type of reaction is shown.

ination would eventually change a C-G base pair to a T-A base pair, since U closely resembles T and forms a complementary base pair with A; see p. 100). As we have seen, these changes would have disastrous consequences for living organisms.

The Stability of Genes Is Due to DNA Repair[17]

Despite the thousands of random changes caused by heat energy in the DNA of a human cell every day, at most only a few stable changes accumulate in the DNA sequence of each cell in a year. The explanation is that lesions are eliminated with remarkable efficiency by the process of **DNA repair.** The various repair mechanisms all depend on the existence of two copies of the genetic information, one on each strand of the DNA double helix. As illustrated schematically in Figure 5–23, the altered portion of a damaged strand is recognized and removed by one set of enzymes and then replaced in its original form by another enzyme, *DNA polymerase* (p. 103), which copies the information stored in the "good" strand by means of complementary base-pairing. Finally, an enzyme called *DNA ligase* seals a nick that remains in the DNA helix to complete the restoration of an intact DNA strand.

By far the most frequent lesion that occurs in DNA is **depurination.** This type of damage is very efficiently repaired according to the scheme outlined in Figure 5–24. First, the presence of the missing base is recognized by a repair nuclease that cuts the DNA phosphodiester backbone at the altered site. After the neighboring nucleotides (including the damaged one) have been removed by further cuts around the initial site of incision, an undamaged DNA sequence is restored, as illustrated in Figure 5–23.

Another very important repair pathway involves a battery of different enzymes called **DNA glycosylases,** each of which recognizes a single type of altered base in DNA and catalyzes its hydrolytic removal from the deoxyribose sugar (Figure 5–25). At least 20 different such enzymes are thought to exist, including those for removing deaminated C's, deaminated A's, different types

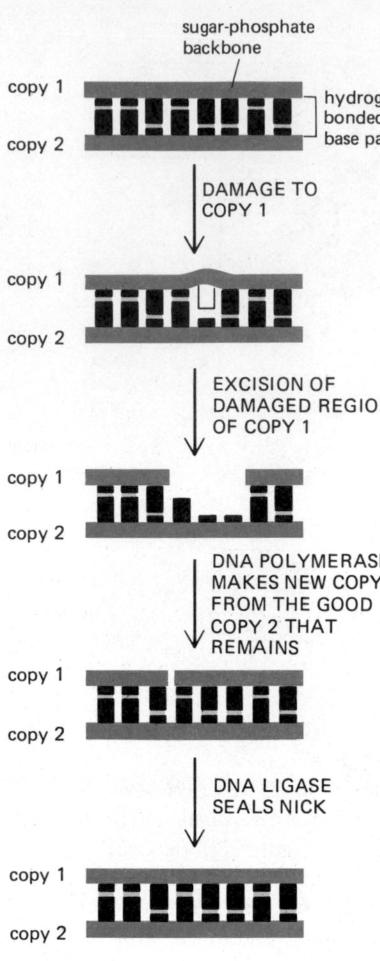

Figure 5–23 Illustration of excision and restoration—two reactions fundamental to DNA repair. The restoration reaction involves two steps: a filling by DNA polymerase of the gap created by the excision events and then the sealing by DNA ligase of a nick left in the repaired strand. Nick sealing consists of the re-formation of a broken phosphodiester bond.

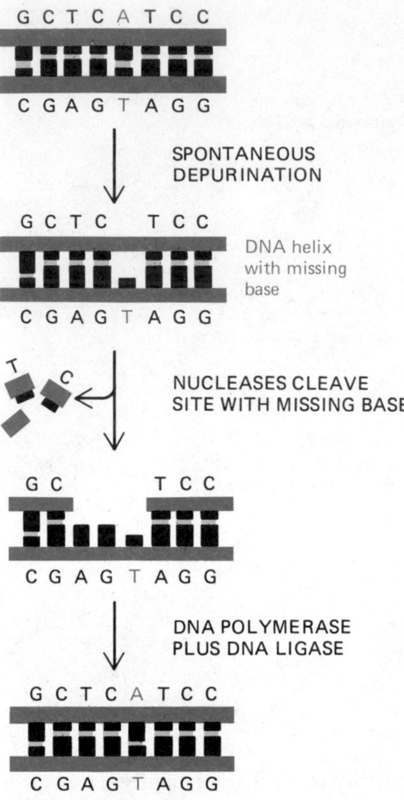

Figure 5–24 The repair of a depurinated site, the most frequent type of spontaneous DNA lesion. The excision step involves the recognition of a site with a missing base: after a phosphodiester bond is cleaved there, repair nucleases remove a few of the nearby nucleotides, as indicated. Subsequent steps proceed as in Figure 5–23.

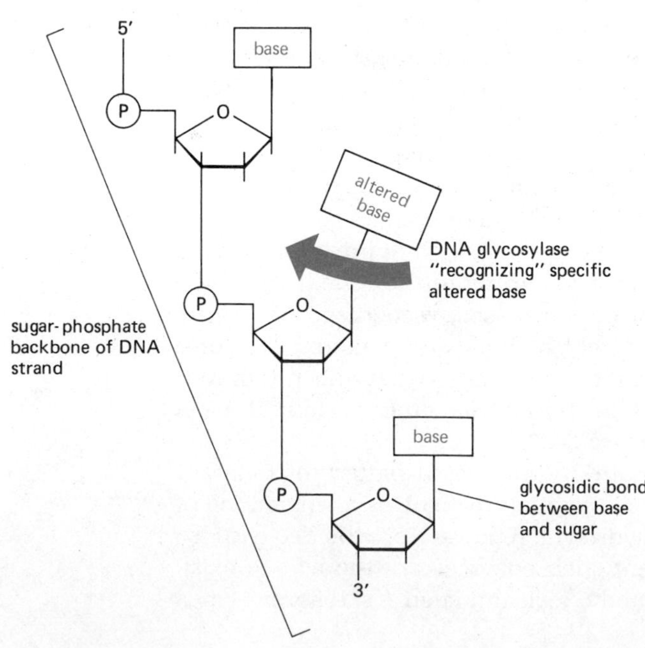

Figure 5–25 The reaction catalyzed by all DNA glycosylase enzymes. There are many different DNA glycosylases, each recognizing a different altered base.

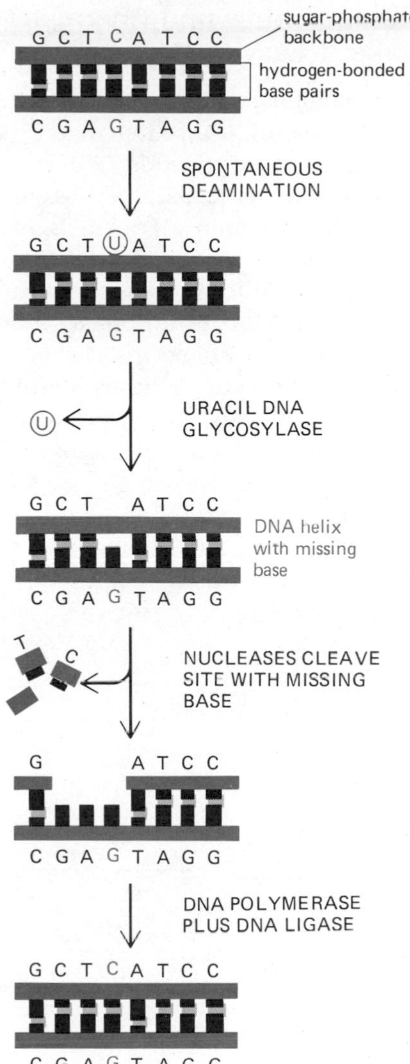

Figure 5–26 The repair pathway involving the enzyme uracil DNA glycosylase, which leads to the restoration of an accidentally deaminated cytosine in DNA.

of alkylated bases, bases with opened rings, and bases in which a carbon-carbon double bond has been accidentally converted to a carbon-carbon single bond.

As an example of the general mechanism that operates in all cases, the removal of a deaminated C is shown in Figure 5–26. First, the enzyme uracil DNA glycosylase removes the defective base. Then the sugar with the missing base is recognized by a second enzyme (the same enzyme that recognizes the depurinated DNA sites shown in Figure 5–24) that cuts the phosphodiester backbone at that point on the DNA. After the damaged residue and its immediate neighbors on the same strand have been cut out by yet another enzyme, a DNA polymerase finishes the job by catalyzing repair DNA synthesis. As a result, the U that was created by accidental deamination is restored to a C (Figure 5–26).

The importance of removing accidentally deaminated DNA bases has been directly demonstrated in two different ways. First, in mutant bacteria lacking the enzyme uracil DNA glycosylase, the normally low spontaneous rate of change of a C-G to a T-A base pair is elevated about twentyfold. The second example involves the fact that some of the C bases are normally methylated to produce 5-methylcytosine at specific points in the DNA sequence. (This occurs both in bacteria and in higher cells.) In studies of a particular bacterial gene, such methylations were found to increase the probability of a mutation at the methylated site. This result is attributed to the fact that the spontaneous deamination of 5-methyl C yields a T residue, which is a normal component of DNA, rather than a U residue, which is not. As a result, these particular C deaminations are not recognized by the cell's uracil DNA glycosylase, and they are consequently not repaired (Figure 5–27).

Since different DNA glycosylases recognize a wide variety of different types of base damage in DNA, most abnormal bases are quickly removed to create a site with a missing base. Each of these sites is then repaired by the same efficient repair system that restores depurinated bases.

Cells have a separate multiple-step excision pathway capable of removing almost any type of DNA damage that creates a very large lesion. Such "bulky lesions" include those created by the covalent reaction of DNA bases with large hydrocarbons, such as benzpyrene, that have carcinogenic potential and the thymine dimers caused by sunlight. In these cases, a large multienzyme complex recognizes the large distortion in the DNA double helix that results, rather than a specific base change. The first step is cleavage of the phosphodiester backbone next to the distortion. The second step is excision of the lesion and resynthesis of DNA in its place (see Figure 5–23).

The importance of these repair processes to life is reflected in the large investment that cells make in DNA repair enzymes. For example, a genetic analysis of human patients with xeroderma pigmentosum suggests that at least five different enzymes are required for the excision of "bulky lesions" alone. A much more comprehensive genetic analysis of a yeast suggests that there are more than 50 different types of DNA repair enzymes, each monitoring the DNA of each cell at all times.

cytosine —DEAMINATION→ uracil

5-methylcytosine —DEAMINATION→ thymine

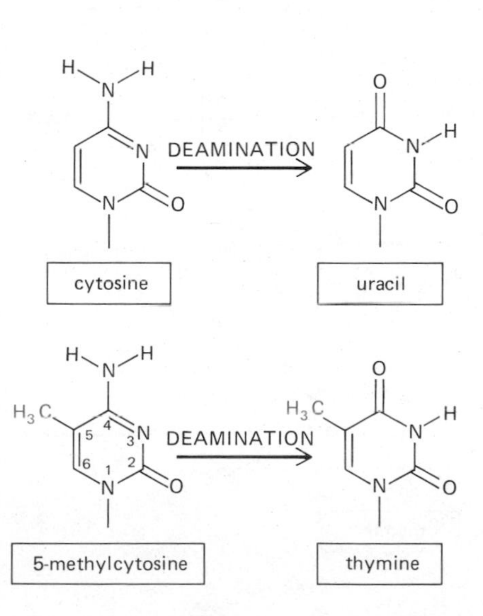

Figure 5–27 The deamination of a methylated cytosine residue in DNA produces thymine instead of uracil, which cannot be recognized and removed by uracil DNA glycosylase.

The Structure and Chemistry of the DNA Double Helix Make It Easy to Repair

The mechanism we have described for removing deaminated C's (Figure 5–26) could not work if U rather than T (which is 5-methyl U) were the fourth DNA nucleotide. Spontaneous C deamination yields a U, and therefore a repair enzyme that recognized and excised such accidents could also remove the normal U's in such a DNA (Figure 5–28). Similarly, the simplest purine base capable of pairing specifically with C is hypoxanthine, with which two hydrogen bonds are formed. But hypoxanthine is the direct deamination product of A (Figure 5–28). By adding a second amino group to hypoxanthine to create G, evolution produced the G-C base pair with its three hydrogen bonds. As a result, the two purine bases A and G are not interconvertible by spontaneous deamination. Thus, every possible deamination event in DNA yields an unnatural base, which can therefore be directly recognized and removed by a special DNA glycosylase (Figure 5–28).

The chemistry of the bases thus ensures that deamination will be detected. But accurate repair—and the fundamental answer to Schroedinger's dilemma—depends on the existence of separate copies of the genetic infor-

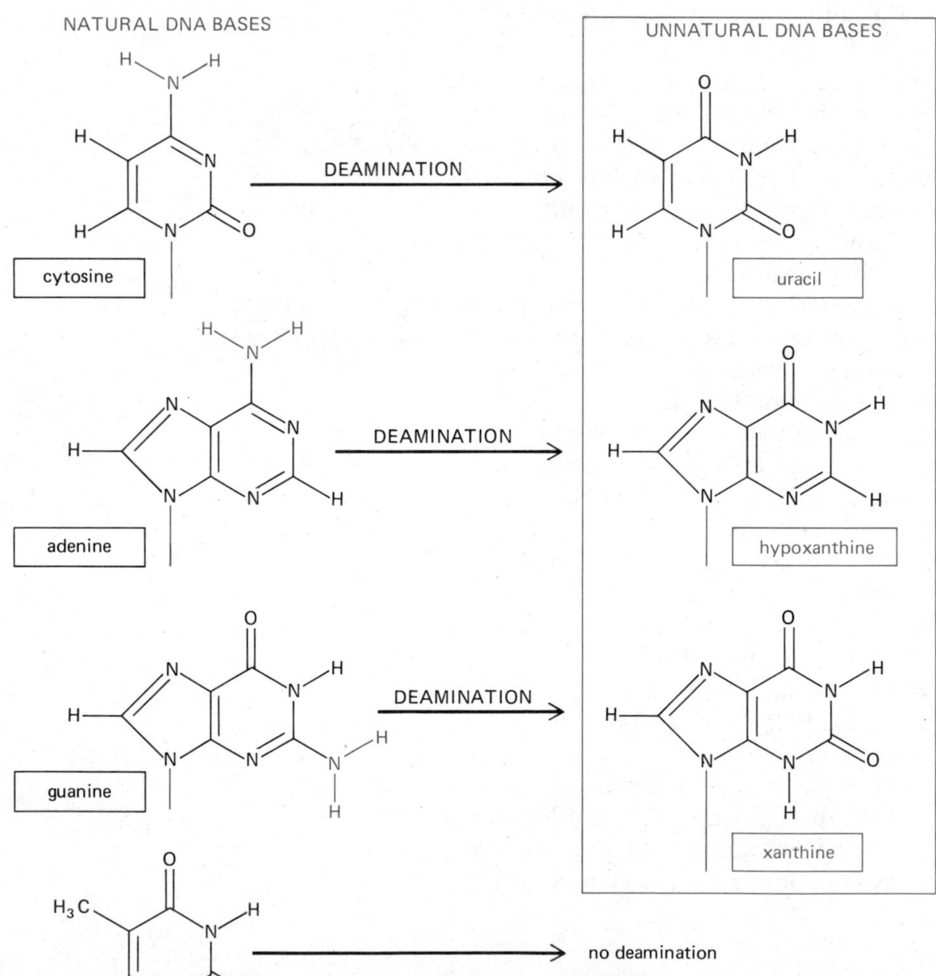

Figure 5–28 Illustration showing that all of the spontaneous deamination products of the DNA bases are recognizable as unnatural when they occur in DNA.

mation in the two strands of the double helix. Only in the very unlikely event that both copies are damaged simultaneously at the same base pair is the cell ever left without one good copy to serve as a template for DNA repair.

Genetic information can also be stored in single-stranded DNA, and some very small viruses have single-stranded genomes. But DNA repair processes cannot operate on such DNA, and the mutation rates of these viruses are high. Only organisms with very tiny genomes can afford to encode their genetic information in a structure other than a DNA double helix.

Summary

The fidelity with which DNA sequences are maintained in higher eucaryotes can be estimated from the rates of change of nonessential protein and DNA sequences over evolutionary time. This fidelity is so high that in an average year the mammalian genome of 3×10^9 base pairs is subjected to only about 15 base-pair changes affecting the germ line. Yet, inevitable chemical decay processes damage thousands of DNA nucleotides every day in a genome of this size. Genetic information can be stably stored in DNA sequences only because a large variety of different DNA repair enzymes are continuously scanning the DNA and removing damaged nucleotides.

The process of DNA repair depends on the fact that a separate copy of the genetic information is stored in each strand of the DNA double helix. An accidental lesion on one strand can therefore be cut out by a repair enzyme and a good strand resynthesized from the information in the undamaged strand.

DNA Replication Mechanisms[18]

Besides maintaining the integrity of DNA sequences by DNA repair, living organisms must duplicate their DNA accurately before every cell division. **DNA replication** involves polymerization rates of about 500 nucleotides per second in bacteria and about 50 nucleotides per second in mammals. Clearly, replication enzymes must be both accurate and fast. Speed and accuracy are achieved by means of a multienzyme complex of several different proteins that guides the process and constitutes an elaborate "replication machine."

Base-pairing Underlies DNA Replication as well as DNA Repair[19]

DNA templating is defined as a process in which the nucleotide sequence of DNA (or selected portions of DNA) is copied by complementary base-pairing (A with T or U, and G with C) into a complementary nucleic acid sequence (either DNA or RNA). The process entails the recognition of each nucleotide in DNA by an unpolymerized complementary nucleotide and requires that the two strands of the DNA helix be separated, at least transiently, so that the hydrogen-bond donor and acceptor groups on each base become exposed for base-pairing. The appropriate incoming single nucleotides (A, G, C, or T) are thus aligned for their enzyme-catalyzed polymerization into a new nucleic acid chain. In 1956 the first such nucleotide polymerizing enzyme was discovered and designated a **DNA polymerase.** The substrates for this enzyme were shown to be deoxyribonucleoside triphosphates, which are polymerized on a single-stranded DNA template according to the stepwise mechanism illustrated in Figure 5–29. We have already seen how this enzyme functions in DNA repair (Figure 5–23).

During DNA replication each old DNA strand functions as a template for formation of a new strand. The enormously long DNA sequence is therefore

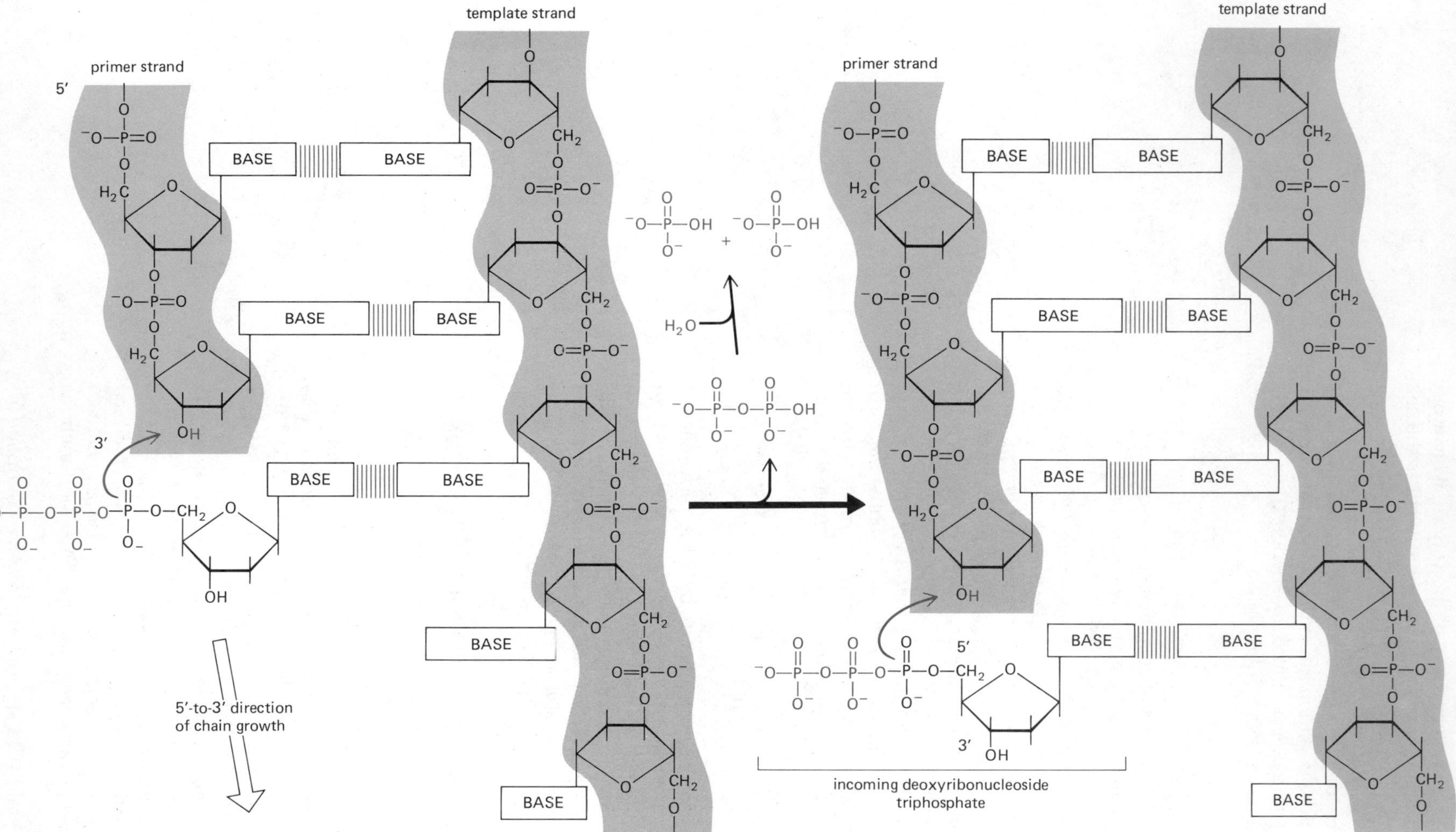

Figure 5–29 Two consecutive steps in the reaction catalyzed by DNA polymerase. The reaction is driven by the hydrolysis of the pyrophosphate released when each nucleotide is added (see p. 77). As indicated, nucleotides are selected one at a time by base-pairing to an exposed template strand and then are covalently joined in the enzymatically catalyzed reaction indicated by the colored arrow. As a result, the new DNA strand grows in the 5′-to-3′ direction. In addition to the template strand, the reaction requires a preexisting polynucleotide chain to whose 3′-OH end new nucleotides are added; this chain is referred to as a *primer* for the DNA polymerase.

said to be replicated "semiconservatively," and the two daughters of a dividing cell inherit a new DNA double helix containing one old and one new strand (see Figure 3–11, p. 105).

Autoradiographic analyses carried out in the early 1960s on whole replicating chromosomes labeled with a short pulse of $[^3H]$thymidine revealed a localized region of replication that moves along the parental DNA helix. Because of its Y-shaped structure, this active region is called a DNA **replication fork.** At each such region, the DNA of both new daughter helices is being synthesized.

The DNA Replication Fork Is Asymmetrical[20]

In the 1960s, the simplest mechanism of DNA replication appeared to be continuous growth of both new strands, nucleotide by nucleotide, at the replication fork as it moves from one end of a DNA molecule to the other. But because of the antiparallel orientation of the two DNA strands in the DNA helix (see p. 100), this mechanism would require one daughter strand to grow in the 5′-to-3′ direction and the other in the 3′-to-5′ direction (Figure 5–30). Such a replication fork would require two different DNA polymerase enzymes. One would polymerize in the 5′-to-3′ direction, as indicated in Figure 5–29, where each incoming deoxyribonucleoside triphosphate monomer carries the triphosphate activation needed for its own addition. The other DNA polymerase, moving in the 3′-to-5′ direction, would have to work by so-called "head growth," in which the end of the growing DNA chain carries the triphosphate activation required for the addition of each subsequent nucleotide. In fact no such 3′-to-5′ DNA polymerase exists, despite the fact that several "head growth" polymerization processes are found elsewhere in biochemistry (see Figure 2–34, p. 82).

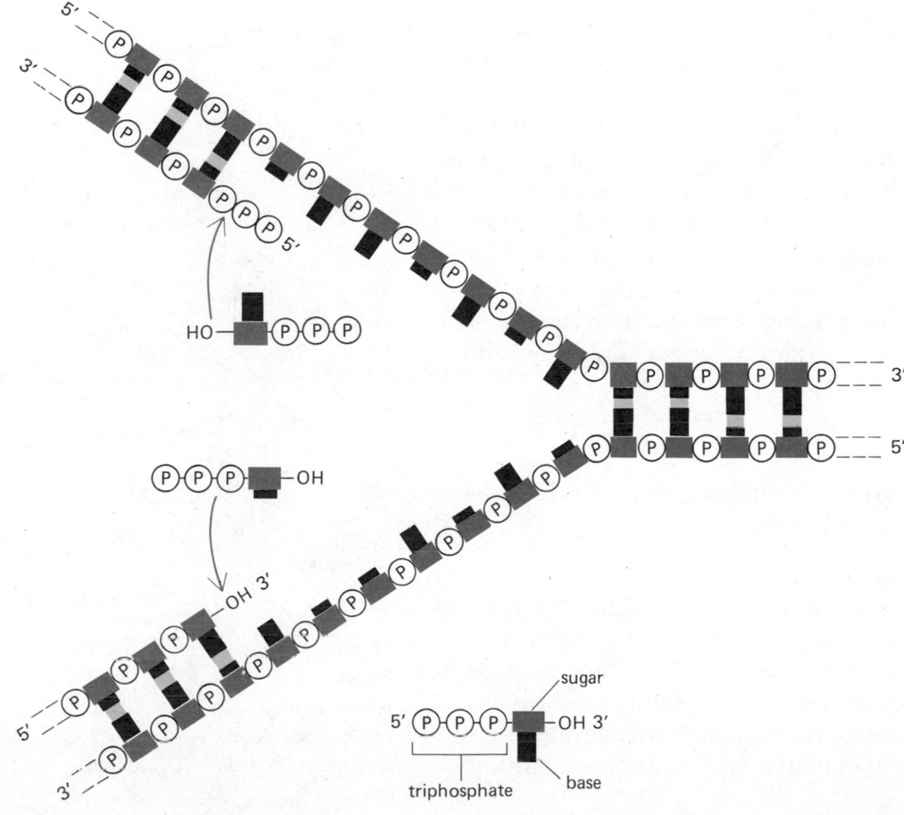

Figure 5–30 Intuitively, the simplest mechanism for DNA replication would be the (incorrect) scheme shown here. Both daughter DNA strands would grow continuously, requiring both 5′-to-3′ and 3′-to-5′ nucleotide polymerization, as indicated.

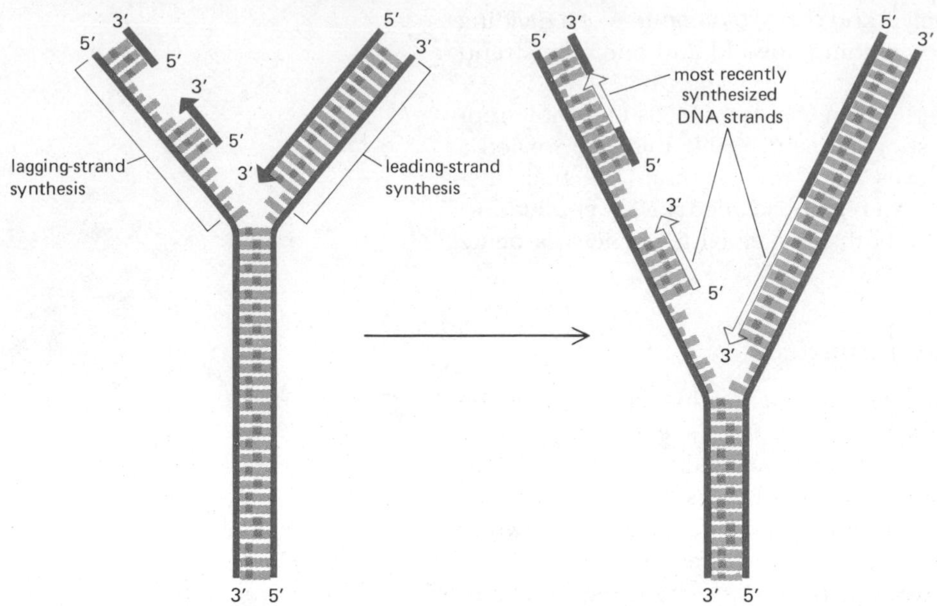

Figure 5–31 The structure of an actual replication fork, in which both daughter DNA strands are synthesized in the 5′-to-3′ direction, thereby requiring that the DNA synthesized on the lagging strand be made as a series of short pieces.

How, then, is 3′-to-5′ DNA synthesis achieved? The answer was first suggested in the late 1960s by experiments with radiolabeled DNA precursors. If highly radioactive [^{3}H]thymidine is added to dividing cells for a few seconds, only the most recently replicated DNA, just behind the replication fork, becomes labeled. This selective labeling method revealed the transient existence of pieces of DNA 1000 to 2000 nucleotides long at the bacterial growing fork (for historical reasons, such pieces are commonly known as *Okazaki fragments;* they are only 100 to 200 nucleotides long in eucaryotes). Shortly thereafter, these DNA pieces were shown to be synthesized in the 5′-to-3′ chain direction only and to be joined together after their synthesis to create long DNA chains by the same *DNA ligase* enzyme that seals nicks in the DNA helix during DNA repair (Figure 5–23).

The replication fork is now known to have the asymmetric structure shown in Figure 5–31. The DNA daughter strand that is synthesized continuously is known as the **leading strand,** and its synthesis precedes the synthesis of the daughter strand that is synthesized discontinuously, which is known as the **lagging strand.** The synthesis of the lagging strand is delayed because, while it is made *overall* in the 3′-to-5′ direction, each piece is made in the 5′-to-3′ direction. Due to the discontinuous, "backstitching" mechanism by which the DNA on the lagging side of the fork is made, only the 5′-to-3′ type of DNA polymerase is required at the fork.

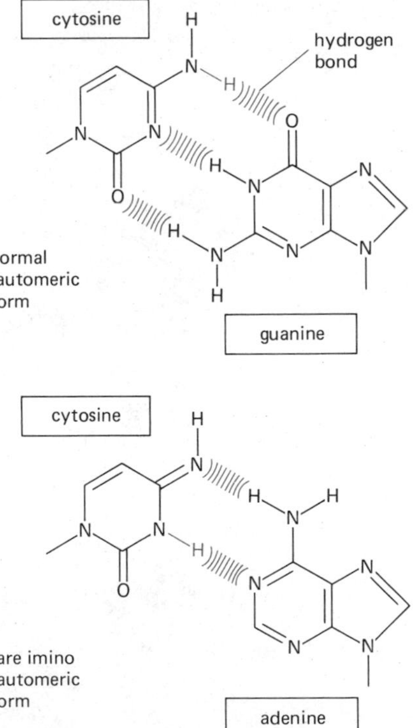

Figure 5–32 An example of an expected rare incorrect base pair: when cytosine is in its unfavored tautomeric form, it can form effective hydrogen bonds with adenine.

The High Fidelity of DNA Replication Requires a "Proofreading" Mechanism[21]

The fidelity of copying during DNA replication is such that only about one error is made in every 10^9 base-pair replications, as required to maintain the mammalian genome of 3×10^9 DNA base pairs (p. 214). Yet the standard complementary base pairs are not the only ones possible. Rare tautomeric forms of the four DNA bases are expected to occur transiently in normal DNA in ratios of one part to 10^4 or 10^5. These forms mispair. For example, the rare tautomeric form of C pairs with A instead of with G and thus causes a mutation

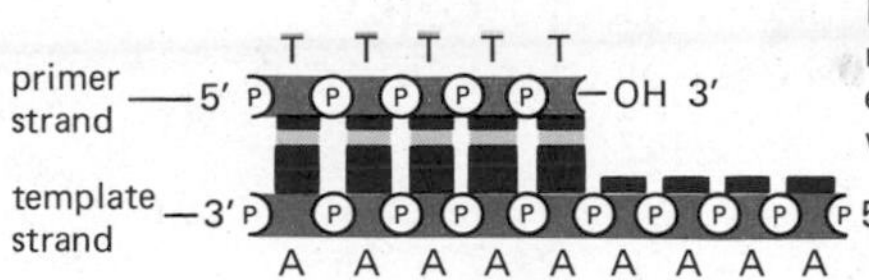

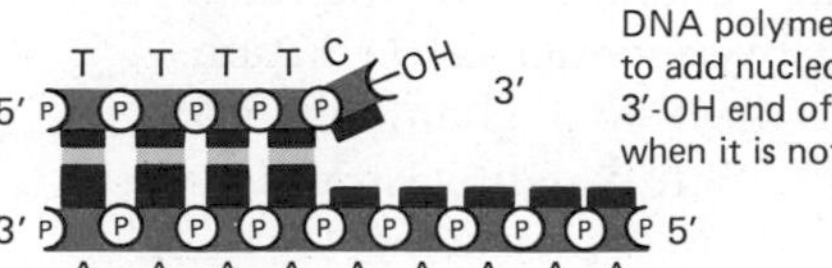

Figure 5–33 Examples of two synthetic DNA molecules that have been tested as primer templates for DNA polymerase. In such tests, the 3′-OH end of a primer strand can be extended only when it is base-paired.

(Figure 5–32). Consequently, high fidelity in DNA replication depends on "proofreading" mechanisms.

The proofreading process depends to a large extent on special properties of the DNA polymerase enzyme. Unlike RNA polymerases, DNA polymerases do not begin a new polynucleotide chain by linking two nucleoside triphosphates together. They absolutely require the 3′-OH end of a base-paired polynucleotide strand on which to add further nucleotides (see Figure 5–29). This preexisting strand to which nucleotides are added is called a *primer*. When a variety of synthetic DNA molecules are constructed and tested as templates for DNA polymerases, only those that have a primer strand with a base-paired 3′-OH end associated with them are directly copied (Figure 5–33). DNA molecules with a mismatched (not base-paired) 3′-OH end on their primer strand are not effective as templates. When confronted with such mismatched DNAs, known replicative DNA polymerases make use of a built-in 3′-to-5′ exonuclease activity to clip off any unpaired residues at the primer terminus by hydrolysis. Clipping continues until enough nucleotides have been removed from the 3′ end to regenerate a base-paired terminus and create an active template primer.

In this way, DNA polymerase functions as a "self-correcting" enzyme, removing its own polymerization errors as it moves along the DNA. This type of proofreading is thought to allow the removal of rare tautomer base pairings, as schematically illustrated in Figure 5–34.

It is now clear that there is an important advantage in a DNA polymerase that will not join two deoxyribonucleoside triphosphates together to start a new polynucleotide chain in the absence of a primer. The requirement for a perfectly base-paired terminus is what provides the DNA polymerase with its self-correcting properties. For such an enzyme to start synthesis in the complete absence of a primer without losing any of its discrimination between base-paired and unpaired growing 3′-OH termini is apparently not possible. By contrast, the RNA polymerase enzymes involved in gene transcription (p. 200) need not be self-correcting, since relatively high error rates can be tolerated in making RNA transcripts. RNA polymerases are able to start new poly-

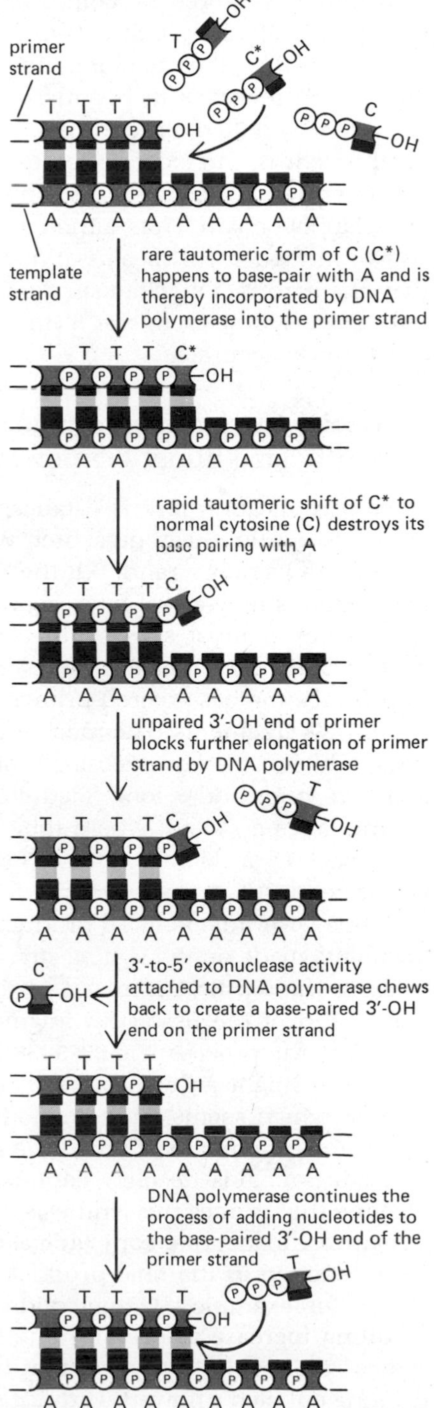

Figure 5–34 Illustration of the proofreading process known to remove errors during DNA synthesis catalyzed by procaryotic DNA polymerases. As yet, it has not been possible to demonstrate a mechanism of this type for most DNA polymerases isolated from higher eucaryotes.

nucleotide chains without a primer, and an error frequency of about 1 in 10^4 is found, both in RNA synthesis and in the separate process of translating mRNA sequences into protein sequences.

DNA Replication in the 5′-to-3′ Direction Is Intrinsically More Accurate

The preceding line of reasoning also suggests an explanation for the failure to find a second DNA polymerase that adds deoxyribonucleoside triphosphates in such a way as to cause chains to grow in the 3′-to-5′ chain direction, as would be required for continuous synthesis on both daughter strands (as in Figure 5–30). For such a 3′-to-5′ polymerase (where the growing 5′-chain end rather than the incoming mononucleotide carries the activating triphosphate), the mistakes in polymerization cannot be simply hydrolyzed away, since the bare 5′-chain end thus created would immediately terminate DNA synthesis. It is, therefore, much easier to correct a mismatched base that has just been added to the 3′ end than one that has just been added to the 5′ end of a DNA chain. Thus, although the type of mechanism for DNA replication shown in Figure 5–31 seems at first sight much more complex and unwieldy than the incorrect mechanism depicted in Figure 5–30, the fact that it involves only DNA synthesis in the 5′-to-3′ direction makes the actual mechanism much more accurate.

A Special Nucleotide Polymerizing Enzyme Is Needed to Synthesize Short Primer Molecules on the Lagging Strand

Once a replication fork is established, the DNA polymerase on the leading strand is continuously presented with a base-paired chain end on which to synthesize its new strand. But the DNA polymerase on the lagging side of the fork requires only about four seconds to complete each short DNA fragment, after which it must start synthesizing a completely new fragment at a site further along the template strand (Figure 5–31). A special mechanism is needed to produce the base-paired primers required to start the synthesis of each of these DNA fragments. The mechanism involves a polymerase called *RNA primase,* which uses ribonucleoside triphosphates to synthesize short **RNA primers** about 10 nucleotides long (Figure 5–35). Such primers are made at intervals on the lagging strand, where they are elongated by the DNA polymerase to begin each Okazaki fragment (see Figure 5–31). This DNA polymerase molecule continues until it runs into the RNA primer attached to the 5′ end of the previous DNA fragment. To produce a continuous DNA chain from the many fragments made on the lagging strand, a special DNA repair system acts quickly to erase the old RNA primer and replace it with DNA. DNA ligase then joins the 3′ end of the new DNA fragment to the 5′ end of the previous one to complete the process (Figure 5–36).

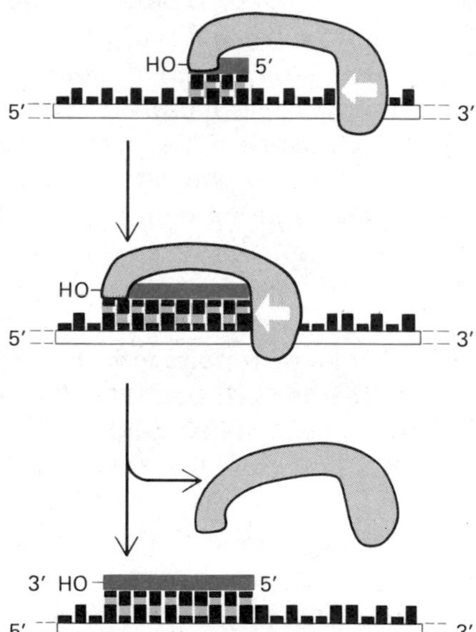

Figure 5–35 A schematic view of the reaction catalyzed by RNA primase, the enzyme that synthesizes the short RNA primers made on the lagging strand. Unlike DNA polymerase, this enzyme can start a new polynucleotide chain by joining two nucleoside triphosphates together. As indicated, the RNA primase stops after a short polynucleotide has been synthesized and makes the 3′ end of this primer available for DNA polymerase addition.

Why might an erasable RNA primer be preferred to an unerased DNA primer, which seems more economical? The argument that a self-correcting polymerase cannot start chains *de novo* also implies its converse: an enzyme that starts chains *de novo* cannot be efficient at self-correcting. Thus, any enzyme that primes the synthesis of Okazaki fragments will of necessity make a relatively inaccurate copy (at least 1 error in 10^5). Even if the amount of this copy retained in the final product constitutes as little as 5% of the total genome (for example, 10 nucleotides per 200-nucleotide DNA fragment), the resulting increase in overall mutation rate would be enormous. It therefore seems reasonable to suggest that the evolution of RNA rather than DNA for priming entailed a powerful advantage, since the ribonucleotides in the primer automatically mark these sequences as "bad copy" to be removed.

Special Proteins Help Open Up the DNA Double Helix in Front of the Replication Fork[22]

The DNA double helix must be opened up rapidly ahead of the replication fork so that the incoming deoxyribonucleoside triphosphates can base-pair with the parent template strand. However, the DNA double helix is very stable under normal conditions, the base pairs being so strongly locked in place that temperatures approaching that of boiling water (90°C) are required to separate the two strands in a test tube. For this reason, most DNA polymerases can copy only a DNA molecule in which the template strand has already been made single-stranded. Additional proteins are needed to help open the double helix and thus provide the appropriate exposed DNA template strand for the DNA polymerase to copy. They are of two types:

1. *Helix-destabilizing proteins*—also called *single-strand DNA-binding (SSB) proteins*. These are proteins that bind in a cooperative manner (Figure 8–7, p. 390) to DNA single strands. They thereby line up in long rows on the single-stranded DNA, extending its backbone and leaving the DNA bases exposed for base-pairing. Helix-destabilizing proteins do not bind directly to DNA in its double-helical form, but they strongly destabilize the DNA double helix by binding to DNA single strands, thereby favoring any helix-opening process. Moreover, their presence disrupts the weak, hairpin helices that would otherwise form in regions of DNA already made single-stranded. The actual template for the DNA polymerase molecule on the lagging strand is a segment of single-stranded DNA that has been straightened by protein binding (Figure 5–37). Helix-destabilizing proteins are thought to cover most regions of single-stranded DNA in the cell in this way and thereby also play an important part in DNA repair and genetic recombination (see p. 243).

2. *Replication proteins that are ATP-driven, DNA-walking machines.* The hydrolysis of ATP can change the shape of a protein so that it can perform mechanical work, as in muscle. A diagram illustrating how a hypothetical DNA-binding protein might utilize ATP to "walk" along DNA was shown on page 138. Such proteins, called *DNA helicases,* have in fact been shown to participate in the unwinding of DNA helices at the DNA replication fork. Figure 5–38 illustrates schematically how a helix-destabilizing protein and an ATP-driven DNA helicase are thought to cooperate in DNA-helix unwinding during DNA replication.

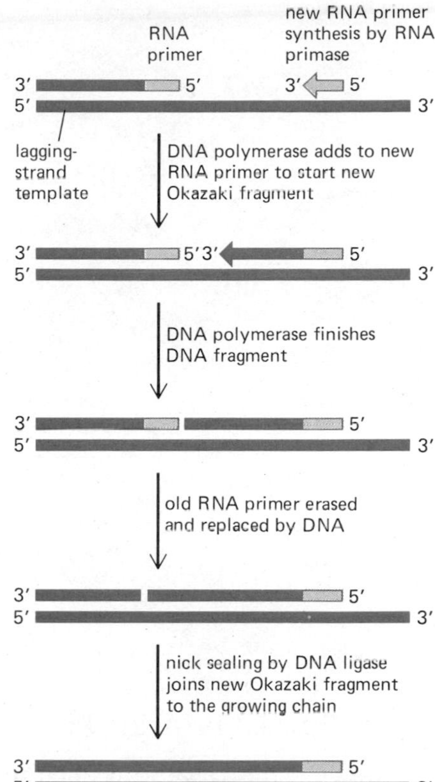

Figure 5–36 The steps involved in the synthesis of each DNA fragment on the lagging strand. In eucaryotes the RNA primers are made at intervals spaced by about 200 nucleotides on the lagging strand, and each RNA primer is 10 nucleotides long. The start signals for the RNA primase have not yet been characterized, but if a specific template nucleotide sequence is involved, it must be a very short one.

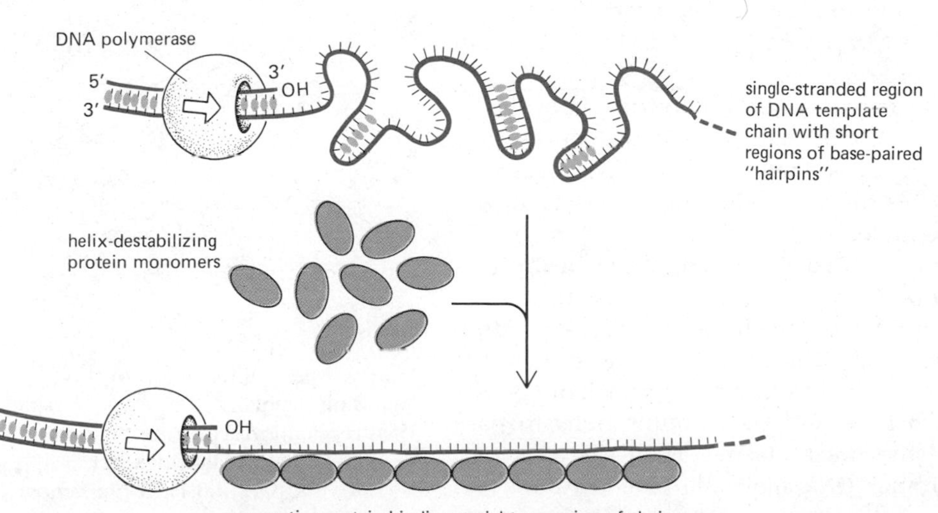

Figure 5–37 Illustration of the effect of helix-destabilizing proteins on the structure of single-stranded DNA. By the type of cooperative DNA binding shown, these proteins form long clusters that straighten out DNA template strands and facilitate the DNA polymerization process. The "hairpins" shown in the bare single-stranded DNA form from a chance matching of short regions of complementary nucleotide sequence.

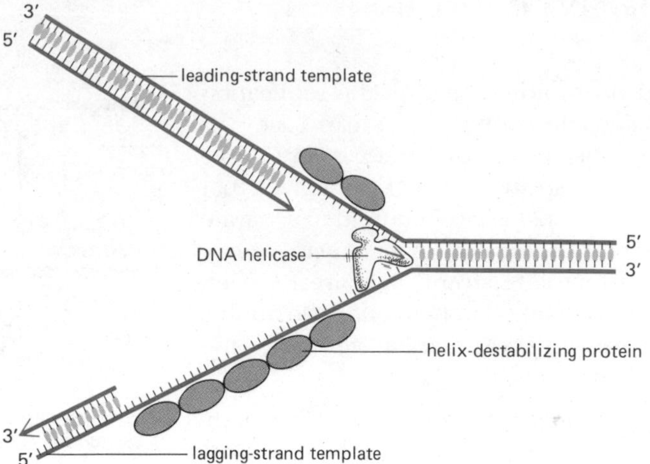

Figure 5–38 The DNA helix ahead of a replication fork is thought to be opened at a rapid rate by the combined action of a DNA helicase enzyme and helix-destabilizing proteins.

Other Special Proteins Prevent DNA Tangling[23]

By drawing the DNA helix incorrectly as a flat "ladder," we have thus far ignored the "winding problem." In fact, for every 10 base pairs replicated at the fork, the parental double helix must make one complete turn about its axis (Figure 5–39). Therefore, for a replication fork to move, the entire chromosome ahead of the fork would normally have to rotate rapidly, a process that would require the input of large amounts of energy in long chromosomes. It is thought that by forming a "swivel" in the helix, members of a class of proteins known as **DNA topoisomerases** solve this problem as well as the analogous problem that arises during DNA transcription.

A DNA topoisomerase can be viewed as a "reversible nuclease" that first breaks the DNA chain and then adds itself covalently to the broken end. The transient "nick" created by the topoisomerase allows the DNA helix on either side of the nick to swivel about the phosphodiester bond opposite the nick and relieve any accumulated winding strain (Figure 5–40). Because the covalent protein-DNA linkage is of relatively high energy, retaining the energy of the broken phosphodiester bond, the nicking reaction is reversible and the nick is closed as soon as the protein leaves. Resealing is therefore rapid and does not require additional energy input.

This mechanism is very different from that of the enzyme DNA ligase, which reseals preexisting nicks during DNA repair processes (Figures 5–23 and 5–36). Unlike a topoisomerase, DNA ligase must use an ATP molecule to form a high-energy linkage that activates the exposed 5′ end at a nick, before the two DNA ends at the nick can be joined.

Various types of DNA topoisomerases exist. Those that act in the way described above are designated *type I DNA topoisomerases* because they make a reversible single-strand break in the DNA double helix (Figure 5–40). More recently, *type II DNA topoisomerases* have been identified. These enzymes bind covalently to both strands of the DNA so as to make a transient *double-strand break* in the DNA helix, cause a second DNA double helix to pass through this break, and then reseal the break cleanly. They are thus very efficient at unknotting a knotted circular DNA molecule or separating two interlocked DNA circles (Figure 5–41).

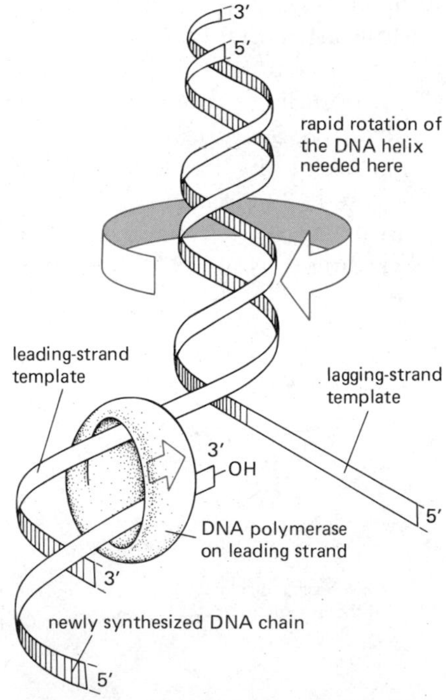

Figure 5–39 Illustration of the "winding problem" that arises during DNA replication. For a replication fork moving at 500 nucleotides per second, the parental DNA helix must rotate at 50 revolutions per second.

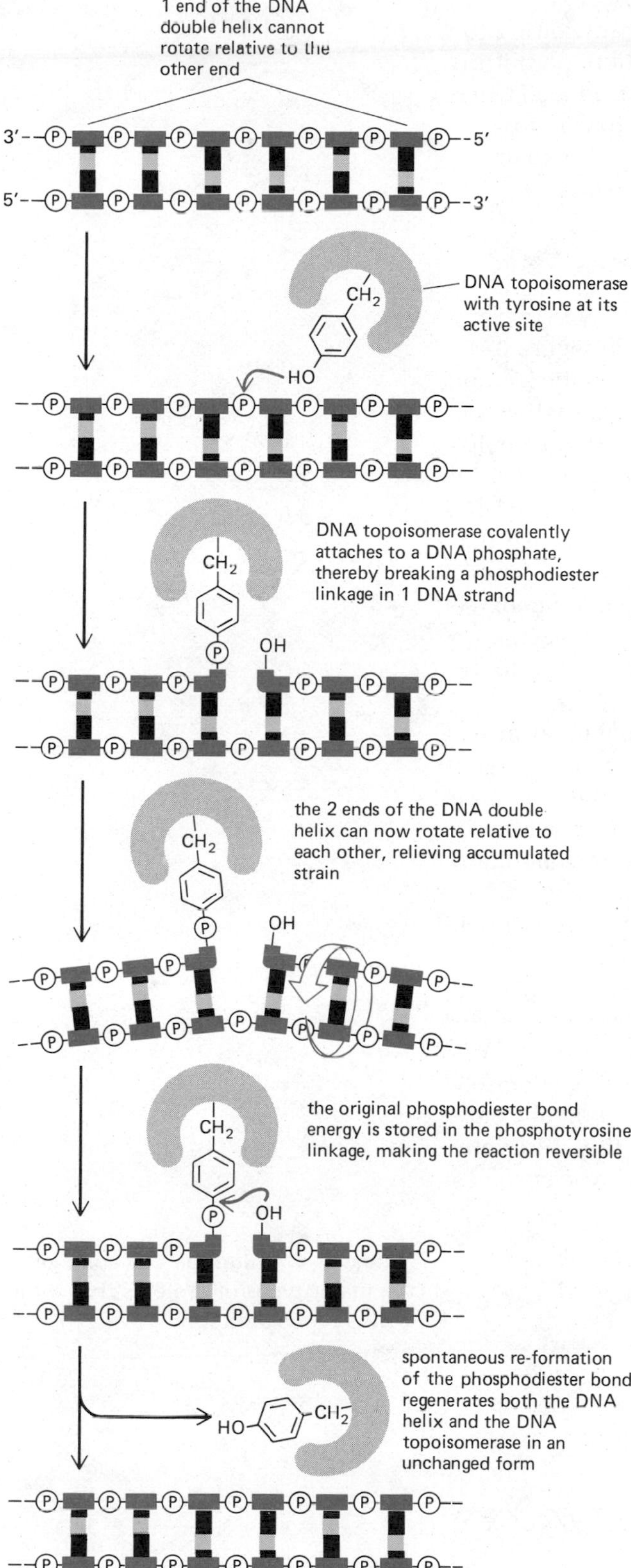

Figure 5–40 Schematic view of the reversible nicking reaction catalyzed by a DNA topoisomerase enzyme. As indicated, these enzymes form a transient covalent bond with DNA that generates a phosphotyrosine residue. The enzyme shown is a type I DNA topoisomerase.

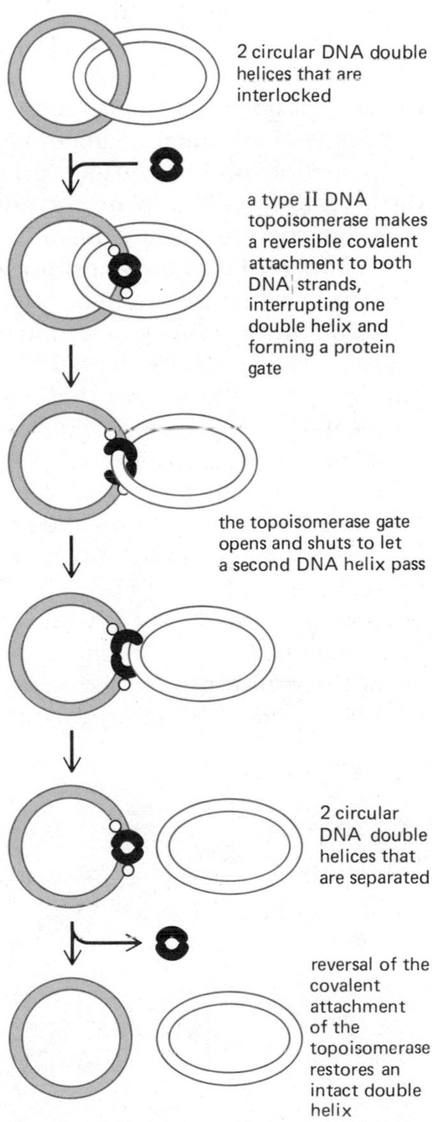

Figure 5–41 An example of the type of DNA-helix-passing reaction catalyzed by a type II DNA topoisomerase. These enzymes are found in both procaryotes and eucaryotes and probably have many functions.

Type II DNA topoisomerases probably play a crucial part in solving the various DNA tangling problems that arise during DNA replication, permitting the complete separation of the two long DNA daughter helices produced when each chromosome is replicated. The usefulness of such an enzyme for this purpose is readily appreciated by anyone who has tried to remove a bad tangle from a fishing line without the aid of scissors.

The Replication of DNA in Eucaryotes and Procaryotes Is Basically Similar[24]

Some of the major steps in DNA replication are summarized in Figure 5–42. Two identical DNA polymerase molecules work at the fork, one on the leading strand and one on the lagging strand. The DNA helix is opened when the DNA polymerase molecule on the leading strand acts in concert with helix-destabilizing proteins and a DNA helicase enzyme. While the DNA polymerase molecule on the leading strand can proceed in a continuous fashion, the DNA polymerase molecule on the lagging strand must restart at intervals, so that the DNA product on this strand is made as a series of short pieces (Okazaki fragments). Each such fragment begins when DNA polymerase adds deoxyribonucleotides to a short RNA primer made by an RNA primase enzyme on the lagging-strand template and ends when DNA ligase seals the 3′ end to the 5′ end of the previous fragment.

Most of the information presented in Figure 5–42 was obtained in the late 1970s, when purified bacterial and bacteriophage multienzyme systems capable of *in vitro* DNA replication became available. The development of these systems was greatly facilitated by the availability of mutants in a variety of different replication genes that could be exploited for the identification and purification of the corresponding enzymes (Figure 5–43).

Largely because the necessary mutants have been much more difficult to obtain, the detailed enzymology of DNA replication in eucaryotes has not yet been worked out. Nevertheless, the basic replication scheme, including the fork geometry and the use of an RNA primer, seems to be identical for procaryotes and eucaryotes. The major difference is that eucaryotic DNA is replicated not as bare DNA but as *chromatin,* in which the DNA is closely associated with tightly bound proteins called *histones.* As described in Chapter 8, these histones form disclike complexes around which the eucaryotic DNA is wound, creating a regular structure called a *nucleosome.* Nucleosomes

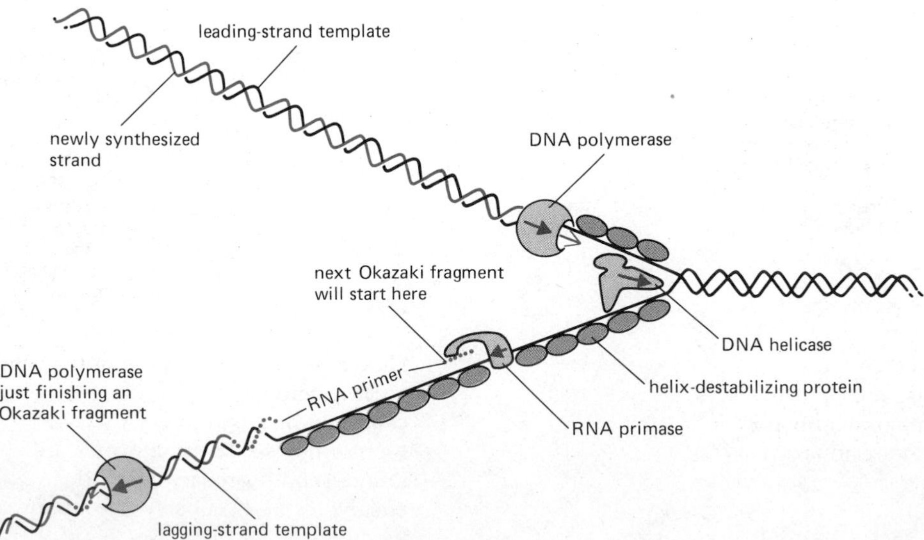

Figure 5–42 Summary of some of the major types of proteins that act at a DNA replication fork.

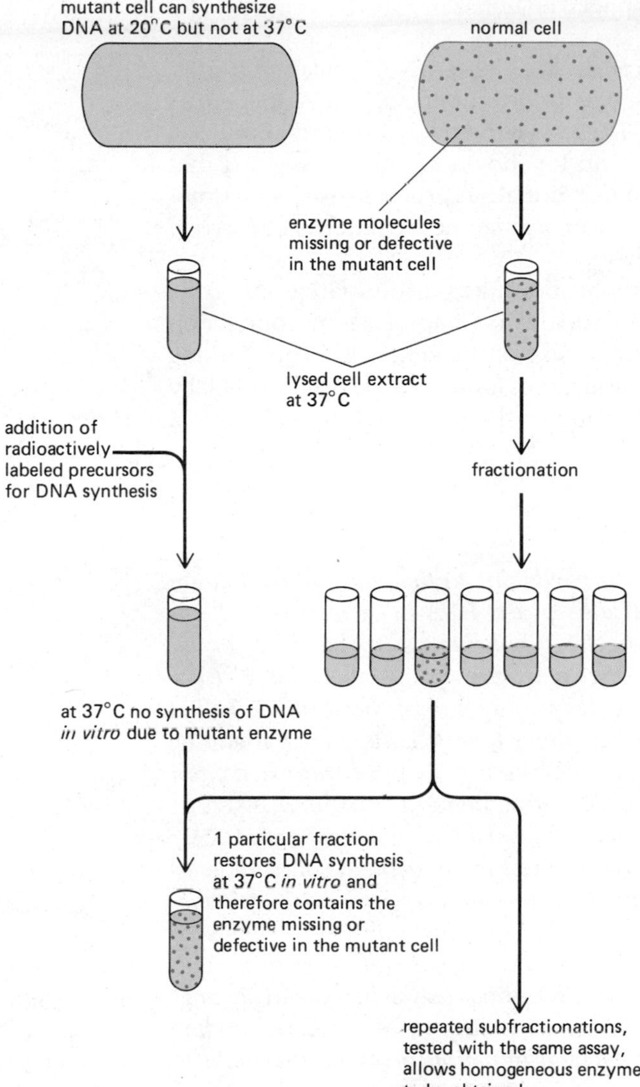

Figure 5–43 The availability of bacterial and bacteriophage mutants defective in DNA replication made it possible to detect and purify enzymes of unknown function essential for procaryotic DNA replication. In a modified form, such "*in vitro* complementation assays" have also been useful for the biochemical investigation of many other processes for which mutants can be isolated.

are spaced at 200-base-pair intervals along the DNA, which may account for the fact that the new DNA pieces on the lagging strand start at intervals about 10 times shorter in eucaryotes (nucleotide spacings of 100 to 200) than they do in bacteria (nucleotide spacings of 1000 to 2000). Moreover, if the nucleosomes act as barriers that temporarily stop the polymerase, the presence of chromatin rather than bare DNA could explain why eucaryotic replication forks move only about one-tenth as fast as bacterial forks.

Replication Forks Are Created at Replication Origins[25]

In organisms as diverse as bacteria and mammals, replication forks seem to originate at a structure called a *replication bubble,* a local region where the two strands of the parental DNA helix have separated from each other and served as templates for the synthesis of short regions of two complementary daughter strands (Figure 5–44, bottom). In principle, a fork can be started in this way if the base pairs in a local region of DNA helix are disrupted sufficiently by fork-initiation proteins to expose a DNA single strand on which an RNA primer can be synthesized by an RNA primase enzyme (Figure 5–44). However, this fork-initiation process has been exceptionally difficult to reproduce in cell extracts, and neither the special enzymes involved nor the precise manner by which it is accomplished are known.

For bacteria, as well as for several viruses that grow in eucaryotic cells, the replication bubbles have been shown to form at special DNA sequences that are called **replication origins.** These origin sequences are about 300 nucleotides long and in several cases can be shown to contain regions that bind specific proteins required for the fork-initiation process. Analogous replication origins with protein binding sites are thought to exist in the eucaryotic genome, although definite proof is lacking.

The general manner by which replication forks are initiated and grow in eucaryotic cells will be discussed in detail in Chapter 11 in connection with a description of the DNA-synthesis phase of the cell cycle. The working out of the exact mechanism by which replication forks are formed is especially important because it may be closely related to the process by which animal cell growth is controlled.

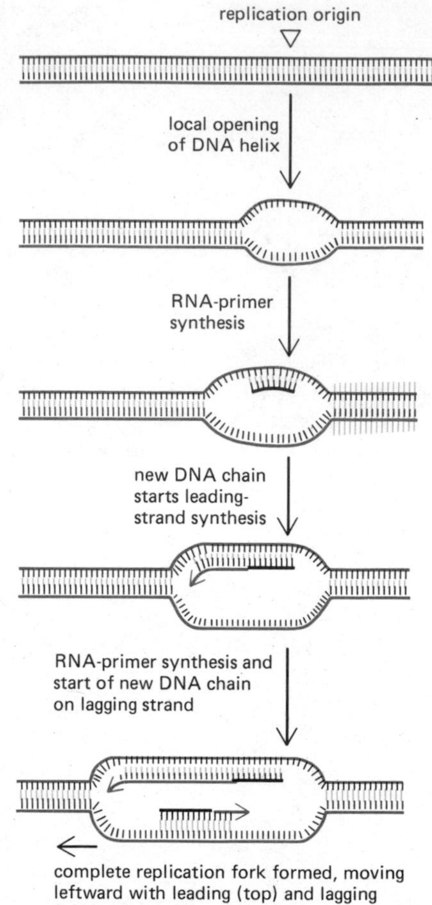

Figure 5–44 An outline of the type of process thought to be involved in the initiation of replication forks at replication origins. The details of the initiation process are not known.

Summary

The amazing fidelity of DNA replication is partly due to the fact that nucleotide polymerization takes place on both strands of the DNA helix in a 5′-to-3′ direction, which enables a self-correcting DNA polymerase to be used for all synthesis. Since the two strands of a DNA helix are antiparallel, this 5′-to-3′ DNA synthesis can take place continuously on only one of the strands (called the leading strand). On the lagging strand, short DNA fragments are made by a "backstitching" process. Because the self-correcting DNA polymerase cannot start a new chain, these lagging-strand DNA fragments are primed by short RNA primer molecules that are subsequently erased and replaced with DNA.

DNA replication requires the cooperation of many proteins, including (1) DNA polymerase and RNA primase enzymes to catalyze nucleoside triphosphate polymerizations, (2) DNA helicases and helix-destabilizing proteins to help open up the DNA helix to be copied, (3) DNA ligase and an enzyme that degrades RNA primers to seal together the discontinuously synthesized lagging-strand DNA fragments, (4) DNA topoisomerases to help relieve helical winding problems, and (5) as yet poorly characterized initiation proteins that help to generate a new replication fork at a replication origin.

Viruses

Thus far in this chapter we have discussed the fundamental processes of DNA transcription and protein synthesis, and have explained how genetic information is conserved from one generation to the next by DNA repair and DNA replication. We shall now describe how viruses exploit the universal genetic mechanisms of cells, emphasizing some of their unique variations on the mechanism of DNA replication.

Viruses Are Mobile Genes[26]

Viruses were first described as disease-causing agents that multiply only in cells and that by virtue of their tiny size pass through ultrafine filters that hold back even the smallest of bacteria. Before the advent of the electron microscope, their nature was obscure, though it was suspected that they might be naked genes that had acquired the ability to move from one cell to another. The use of ultracentrifuges in the 1930s made it possible to separate viruses from host-cell components, and by the early 1940s the generalization was beginning to emerge that all viruses contain nucleic acids. This heightened speculation that viruses and genetic material carried out similar functions. These ideas were confirmed by studies on the viruses of bacteria (**bacteriophages**). In 1952, it was shown that only the bacteriophage DNA, and not the bacteriophage protein, entered the bacterial host cell and initiated the replication events that led to the production of several hundred progeny viruses within every infected cell.

Viruses should thus be regarded as genetic elements, enclosed by a protective coat, that can move from one cell to another. Virus multiplication *per se* is often lethal to the cells in which it occurs. With many viruses the infected cell breaks open (*lyses*) and thereby allows the progeny viruses access to nearby cells. Many of the clinical manifestations of viral infection reflect these cytolytic properties of the virus.

The structure of the viral coat, the type of nucleic acid it contains, the mode of entry of the virus into the cell, and the mechanism of its replication once inside the cell all vary considerably, depending on the virus. Some of the structural differences between viruses are illustrated in Figures 5–45 and 5–46.

RNA VIRUSES

single-stranded RNA, e.g., TMV and poliovirus

double-stranded RNA, e.g., reovirus

DNA VIRUSES

single-stranded DNA, e.g., parvovirus

single-stranded circular DNA, e.g., bacteriophages ϕX174 and M13

double-stranded DNA, e.g., bacteriophage T4 and herpesvirus

double-stranded circular DNA, e.g., SV40 and polyoma virus

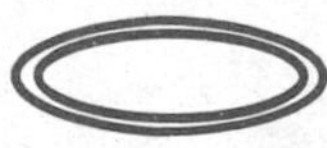

double-stranded DNA with covalently linked terminal protein, e.g., adenovirus

double-stranded DNA with each end covalently sealed, e.g., poxvirus

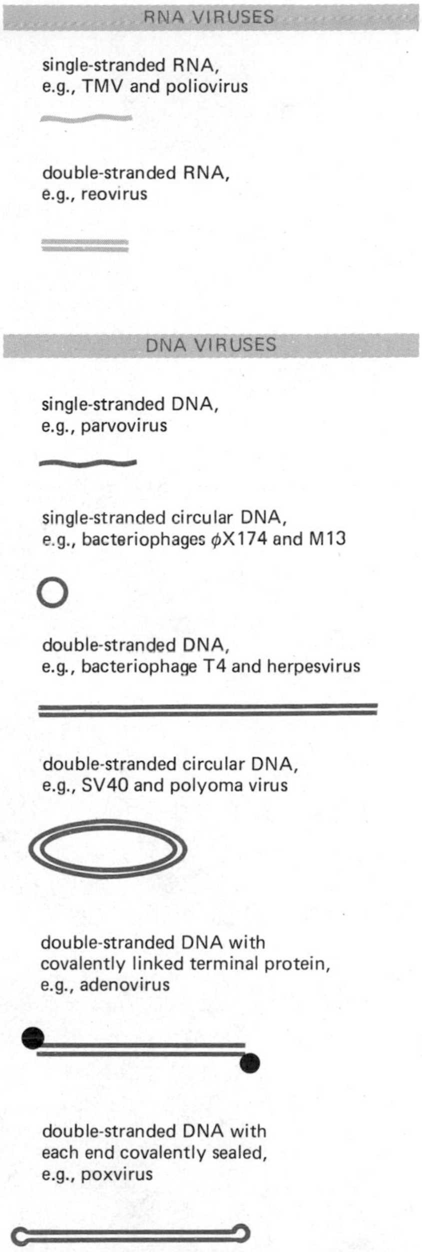

Figure 5–45 Schematic drawings (not to scale) of several different types of viral genomes. The largest viruses contain hundreds of genes and have a double-stranded DNA genome. The peculiar ends on some of these DNA molecules (as well as the circular forms) probably evolved because of the difficulty in replicating the last few nucleotides at the end of a DNA chain.

The Genetic Component of a Virus Is Either DNA or RNA[27]

The viral DNA encodes the virus-specific protein molecules found on the outside of all DNA viruses. At first it was thought likely that the viral DNA molecule was in effect a single gene coding for a single protein, but it is now clear that nucleic acid molecules from even the smallest known viruses carry the information for several proteins. The larger viral genomes, such as those of bacteriophage T4, code for several hundred proteins. Thus, the nucleic acid components of viruses should be regarded as tiny chromosomes.

The significance of the RNA found in some viruses, such as the tobacco mosaic virus (TMV), was at first much less clear. When RNA was present, DNA was not, so the best guess was that the RNA must be the viral genetic component. In the early 1950s, however, this conjecture flew in the face of the belief that the sole role of RNA was the transfer of genetic information from DNA to protein. But by 1956 it became possible to obtain purified RNA preparations from TMV that were infectious in the absence of any TMV protein. Soon afterward, many other viruses—such as polio, influenza, and measles—were found to contain RNA, and today the potential of RNA for carrying genetic information is indisputable.

The Outer Coat of a Virus May Be a Protein Capsid or a Membrane Envelope

The outer coat of viruses was initially thought to be constructed solely from protein molecules, perhaps only of one major type. Viral infections were believed to start with the dissociation of the viral chromosome from its protein coat, followed by self-replication of the chromosome to form many identical copies and by the synthesis of virus-specific coat proteins using virally coded RNA chains as messengers. Formation of the progeny virus particles would then occur by spontaneous assembly of the viral coat proteins around the progeny viral chromosomes (Figure 5–47).

We now realize that these original ideas, though correct in general, were a vast oversimplification of the extraordinarily diverse ways in which viruses can complete their life cycles. In the first place, the protein coat (or **capsid**) of most viruses contains more than one type of polypeptide chain, often arranged in several layers. In the second place, in many viruses the protein capsid is further enclosed in a membrane that contains lipid as well as protein and is similar in construction to the plasma membrane that surrounds all

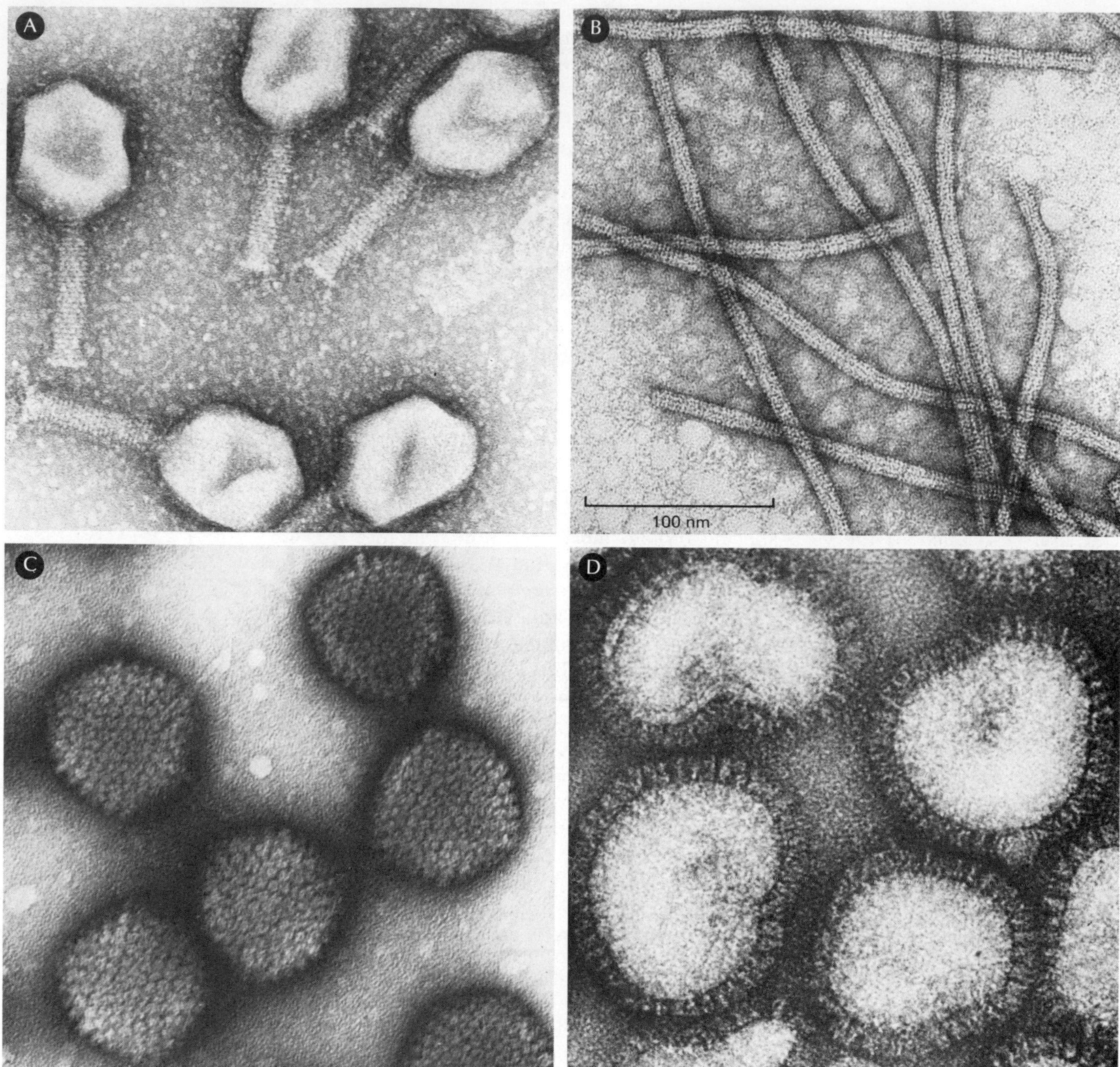

Figure 5–46 Electron micrographs (all at the same magnification; see scale indicated in [B]) of negatively stained virus particles. (A) *Bacteriophage T4,* a large DNA-containing virus that infects the *E. coli* bacterium. The DNA is stored in the bacteriophage head and injected into the bacterium through the cylindrical tail. (Courtesy of James Paulson.) (B) *Potato virus X,* a filamentous plant virus that contains an RNA genome. (Courtesy of Graham Hills.) (C) *Adenovirus,* a DNA-containing virus that can infect human cells. The protein capsid forms the outer surface of this virus. (Courtesy of Mei Lie Wong.) (D) *Influenza virus,* a large DNA-containing animal virus whose protein capsid is further enclosed in a membranous envelope. (Courtesy of R. C. Williams and H. W. Fisher.)

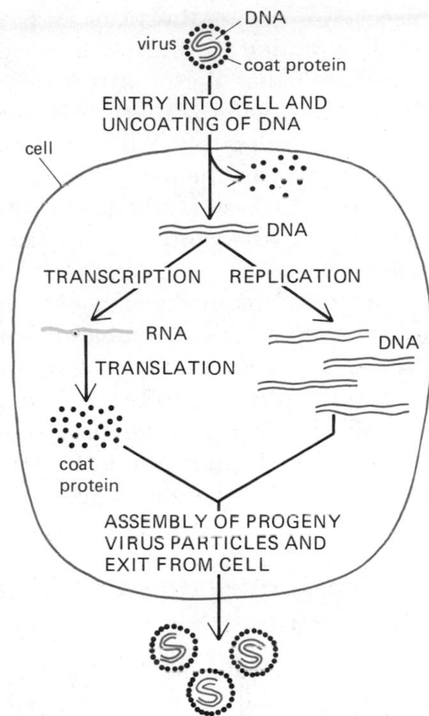

Figure 5–47 The simplest of all viral life cycles. The hypothetical virus shown consists of a small double-stranded DNA molecule that codes for a single viral capsid protein. No known viruses are as simple as this.

cells. While the lipid components of these *enveloped viruses* are identical to those found in the plasma membrane of the host cell, the proteins are virus-specific. Many enveloped viruses assemble their outermost coat in the plasma membrane of the host cell, the progeny particles pinching off as buds (Figure 5–48). This budding process allows virus particles to leave the cell without disrupting the plasma membrane and, therefore, without killing the cell; it is discussed in more detail in Chapter 7.

Viral Chromosomes Usually Code for One or More Enzymes Involved in the Replication of the Viral Nucleic Acid[28]

Viral chromosomes usually code not only for the viral coat proteins but also for the enzymes needed to replicate the viral nucleic acid. For example, the DNA of the relatively large bacteriophage T4 codes for at least 30 different enzymes that ensure the selective and rapid replication of the T4 chromosome in preference to the DNA of its *E. coli* host (Figure 5–49). These proteins mediate continuous rounds of T4 DNA replication as well as the selective incorporation of 5-hydroxymethylcytosine, which replaces cytosine in T4 DNA. Other proteins encoded by the T4 genome are nucleases that selectively degrade *E. coli* DNA. Still others alter host bacterial RNA polymerase molecules so that they transcribe different sets of bacteriophage genes at different stages of infection.

Smaller DNA viruses, such as the monkey virus SV40 and the tiny bacteriophage ϕX174, carry much less genetic information and rely much more on host cell enzymes to carry out their protein and DNA synthesis. Even these viruses, however, code for enzymes that function to initiate their own DNA synthesis selectively, since for a virus to be successful it must override the cellular control signals that would otherwise prevent the viral DNA from doubling more than once in each cell cycle.

Viral Genomes Come in a Variety of Forms[29]

When the DNA double helix was discovered, it seemed logical to believe that DNA should exist only in this form: if one polynucleotide chain is accidentally damaged, its complement can hold the broken pieces together until the damage is repaired. This concern, however, need not bother viral chromosomes—particularly if they are relatively short in length and contain only several thousand nucleotides—for the chance of accidental damage is very small compared with the risk to a cell genome containing millions of nucleotides. Moreover, each viral life cycle produces hundreds of progeny DNA molecules, so damage to a few would be of little real consequence. Thus, the

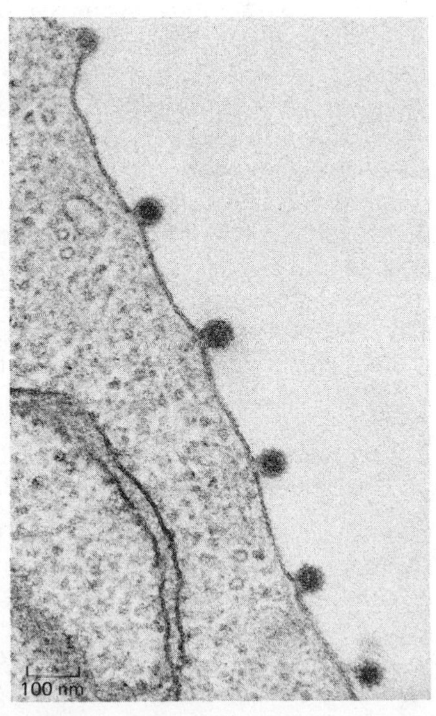

Figure 5–48 Electron micrograph of a thin section of an animal cell from which several copies of an enveloped virus (Semliki forest virus) are budding. (Courtesy of M. Olsen and G. Griffiths.)

discoveries during the 1950s and 1960s that many small DNA viruses contain single-stranded DNA are, in retrospect, not very surprising. For a time it was speculated that these viruses might have a fundamentally different way of replicating their DNA, but it is now known that they begin by producing a double-stranded replicating intermediate, and it is believed that all DNA replication involves the formation of complementary strands.

The first well-studied viral chromosomes were simple linear DNA double helices, but viral genomes may also be circular DNA double helices (SV40 and polyoma viruses), circular single-stranded DNA chains (M13 and ϕX174 bacteriophages), or linear single-stranded DNA chains (parvoviruses). Recently, more complex linear double helices have been discovered. Certain bacteriophages and the adenoviral group of animal viruses have protein molecules covalently attached to the 5′ ends of their DNA strands, and the DNA from the very large poxviruses have the opposite strands at each end covalently linked through phosphodiester linkages (Figure 5–45). Each of these forms of DNA requires unique enzymatic tricks for self-replication.

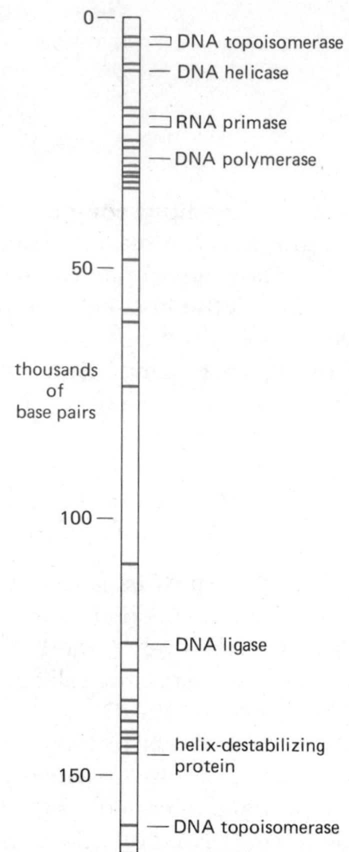

Figure 5–49 The T4 bacteriophage chromosome, showing the positions of the more than 30 genes involved in T4 DNA replication. The genome of bacteriophage T4 consists of more than 160,000 base pairs, encoding more than 150 different proteins, including those involved in DNA replication. The remaining proteins include many that are involved in the bacteriophage head and tail assembly (see Figure 5–46A).

Viral Chromosomes Can Integrate Themselves into Host Chromosomes[30]

The desired end result of the entry of a viral chromosome into a cell was initially always thought to be its immediate multiplication as a prelude to the formation of large numbers of progeny viruses. Many viruses can, however, exist in a *latent* state in which their genomes are present but inactive in the cell, and no progeny are produced. The molecular basis of viral latency remained obscure until the early 1950s, when it was discovered that many apparently uninfected bacteria could be induced to produce progeny bacteriophages after exposure to ultraviolet light. Subsequent experiments showed that these *lysogenic bacteria* carry within their chromosomes a complete viral chromosome. Such integrated viral chromosomes are called *proviruses.*

Bacteriophages that can integrate themselves into bacterial chromosomes are known as **lysogenic bacteriophages.** The best-known example is bacteriophage *lambda.* When lambda infects a suitable *E. coli* host cell, it normally multiplies to produce several hundred progeny lambda particles that are released when the bacterial cell lyses; this is called a *lytic infection.* Much more rarely, the linear infecting DNA molecules circularize and become integrated into the circular host *E. coli* chromosome by a site-specific recombination event (see p. 247). After successful integration, the resulting lysogenic bacterium, carrying the proviral lambda chromosome, multiples normally until it is subjected to an environmental insult, such as an exposure to ultraviolet light or ionizing radiation. Such insults signal the integrated provirus to leave the host chromosome and begin a normal cycle of viral replication. In this way the integrated provirus need not expire in its damaged host, but instead it has a chance to escape to a nearby undamaged *E. coli* cell (Figure 5–50).

RNA Viruses Also Replicate Through the Formation of Complementary Chains[26,31]

Multiplication of RNA viruses also involves the formation of complementary strands. For many of the best-studied RNA viruses, such as R17 bacteriophage or polio virus, the viral chromosome is single-stranded—although double helical RNA viral chromosomes exist, for example in the reoviruses that infect organisms as diverse as yeasts and mammals. In all cases, RNA replication is mediated by specific RNA-dependent RNA polymerase enzymes (*replicases*)

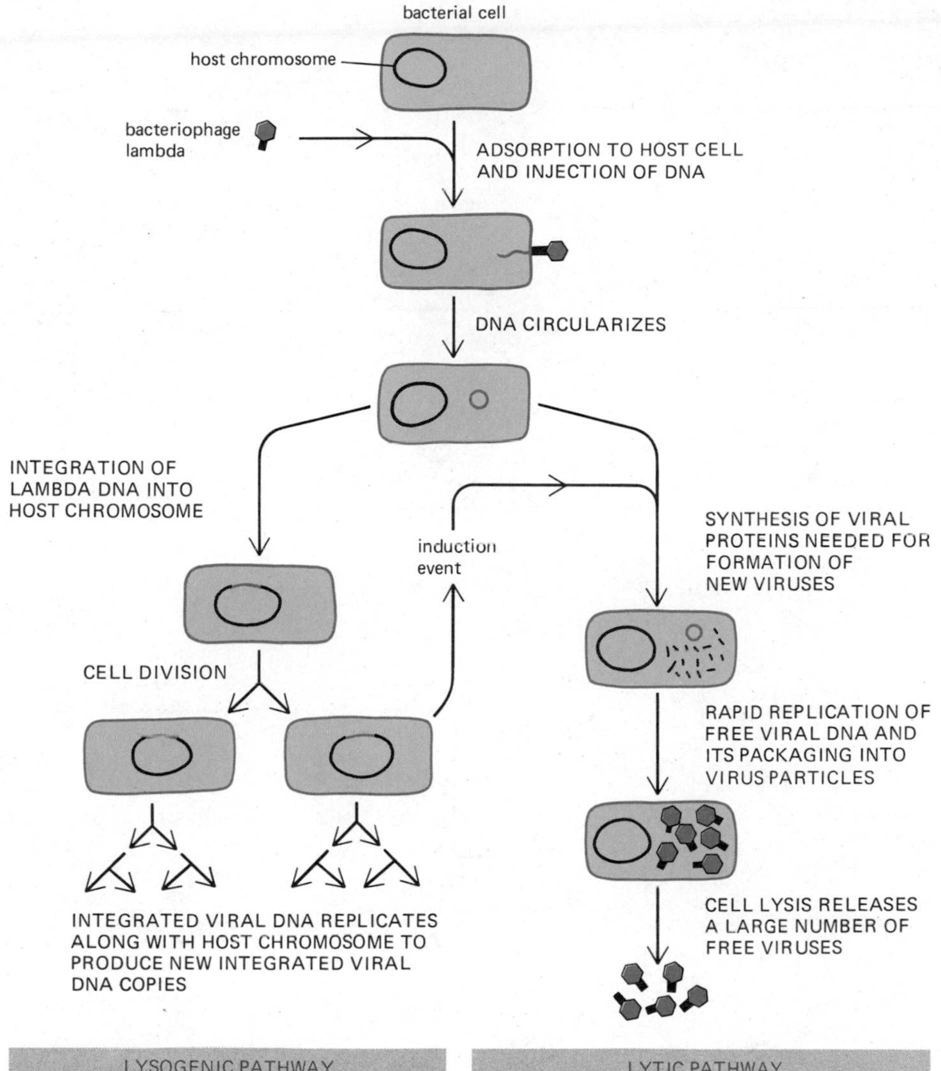

Figure 5–50 The life cycle of bacteriophage lambda. The lambda genome contains about 50,000 base pairs and encodes about 50 proteins. Its double-stranded DNA can exist in both linear and circular forms. As shown, the bacteriophage can multiply by either a lytic or a lysogenic pathway. When growing in the lysogenic state, damage to the cell's DNA causes the integrated viral DNA (provirus) to exit from the host chromosome and shift to lytic growth. The entrance and exit of the DNA from the chromosome are site-specific genetic recombination events catalyzed by the lambda *integrase* protein (see p. 247).

encoded by the viral RNA chromosome. These enzymes often become incorporated into the progeny virus particles, and upon viral infection they immediately become available to begin replicating the viral RNA. Packaging of replicases into progeny viral particles always occurs for the so-called *negative-strand viruses,* such as influenza or vesicular stomatitis virus. With these viruses, the infecting strand does not code for any proteins. Only its complementary strand carries useful genetic information, including the sequences coding for its viral replicase and its coat proteins. Thus, the infecting strand remains impotent without the required replicase.

The synthesis of viral RNA always begins at the 3′ end of the RNA template (the 5′ end of the new viral RNA molecule) and progresses in the 5′-to-3′ direction until the 5′ end of the template is reached. There are no error-correcting mechanisms for viral RNA synthesis, but this is not a serious deficiency as long as the RNA chromosome is relatively short. The genomes of all RNA viruses are small relative to those of the large DNA viruses, which is an obligatory consequence of their more primitive replication mechanism.

Some of the various ways in which viruses are known to replicate their genomes are outlined in Figure 5–51.

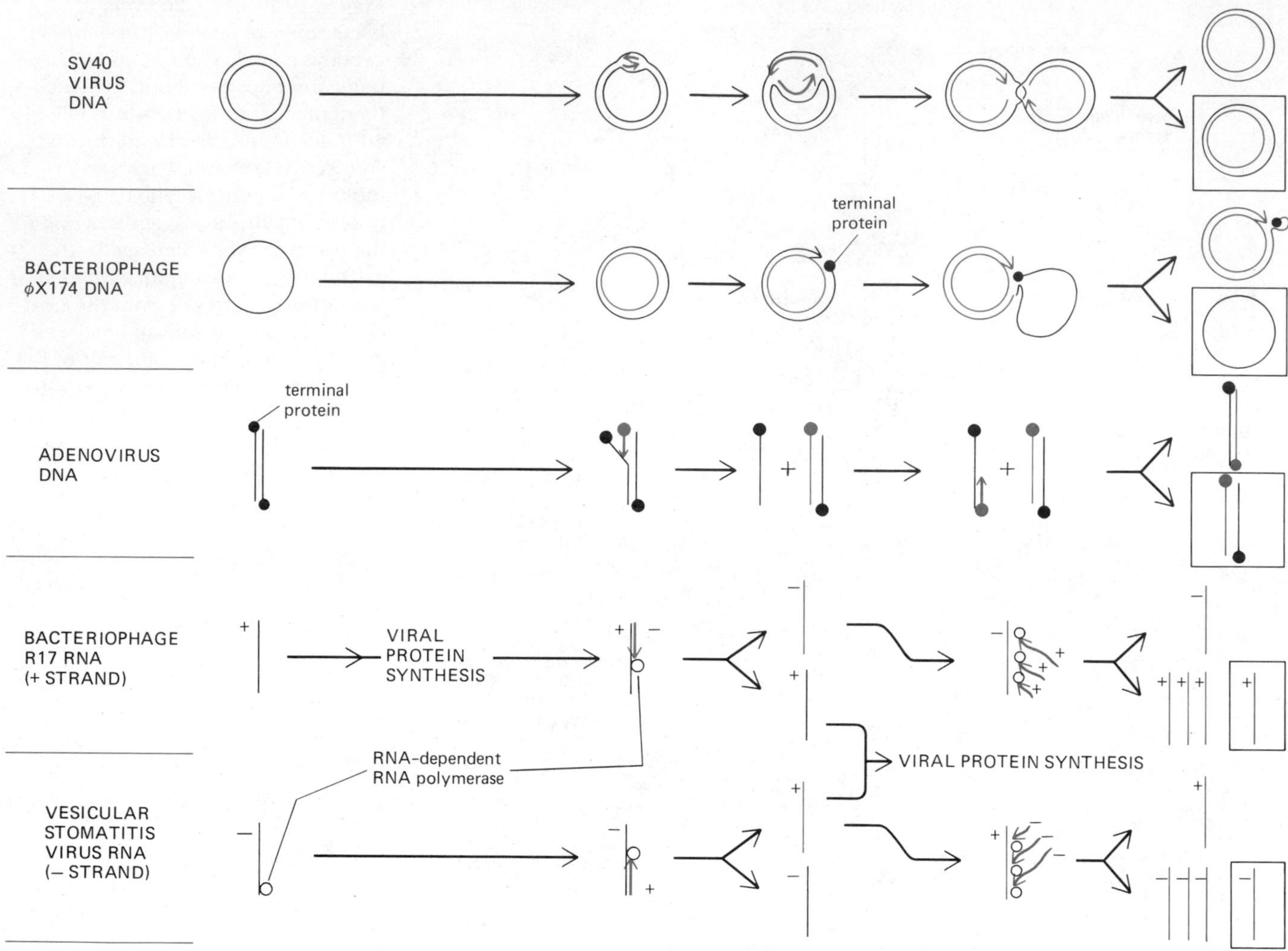

Figure 5–51 Some of the diverse strategies used by different viruses for replication of their genomes. Where indicated, *terminal proteins* are covalently attached to the ends of DNA chains; these proteins play an important role in the respective replication process. Note that the major difference between the life cycles of *positive-strand* and *negative-strand* RNA viruses is that the latter must synthesize a positive RNA strand before making viral proteins. For this purpose, a negative-strand virus must carry within its capsid one or more molecules of the viral polymerase, since its host cells contain no comparable enzyme.

Viral Genetic Elements Can Make Cells Cancerous[32]

Animal cells in which DNA viruses can multiply lytically are called *permissive* cells. In other cells, known as *nonpermissive* cells, viral replication aborts. In a small percentage of cases, the DNA virus instead transforms the nonpermissive cells into their cancerous equivalents. In such *neoplastic transformation,* one or more of the viral genes becomes integrated into the genome of the host cell. There it introduces the synthesis of proteins that are normally made only during viral multiplication. We still do not know exactly why the viral products make cells cancerous, but protein kinases are involved in some cases; this will be explained in more detail in Chapter 11 when the genetics of tumor viruses is discussed.

More puzzling initially were the so-called RNA tumor viruses, where viral infection leads simultaneously to a nonlethal release of progeny viral products from the cell surface and a permanent change in the infected cell that makes it cancerous. How RNA virus infection could lead to a permanent genetic alteration was unclear until it was observed that both viral multiplication and the cancerous transformation can be blocked by inhibitors of DNA synthesis. This hinted that DNA synthesis is a necessary feature of RNA tumor virus multiplication, a hypothesis proved correct in 1970 with the discovery of the

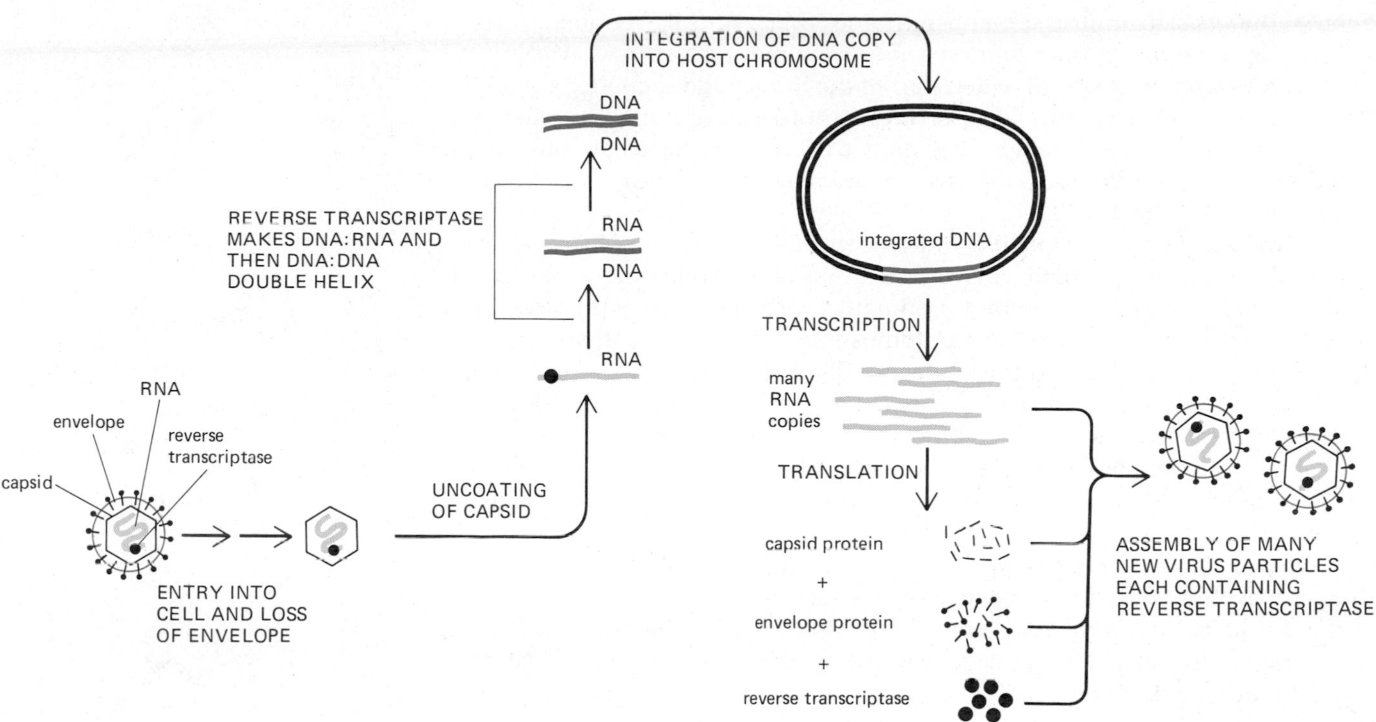

Figure 5–52 The life cycle of a retrovirus. The enzyme *reverse transcriptase* is a DNA polymerase that first makes an initial DNA copy of the viral RNA molecule and then a second DNA strand, generating a double-stranded DNA copy of the RNA genome. The integration of this DNA double helix into the host chromosome, catalyzed by unknown host enzymes, is required for the synthesis of new viral RNA molecules.

enzyme *reverse transcriptase,* through which the infecting RNA chains are transcribed into their DNA complements. For this reason, RNA tumor viruses, which include the first well-known tumor virus, the Rous sarcoma virus, are known as *retroviruses* (see p. 622). The life cycle of a retrovirus is outlined in Figure 5–52.

How Did Viruses Evolve?

Even the largest viruses depend heavily on their host cells for biosynthesis; for example, no known virus makes its own ribosomes or generates the ATP it requires. Clearly, therefore, cells must have evolved before viruses. The precursors of the first viruses were probably small nucleic acid fragments that developed the ability to multiply independently of the chromosomes of their host cells. Perhaps the precursors of RNA viruses resembled the *viroids* found in some plant cells; these small RNA circles—only 300 to 400 nucleotides long—are replicated entirely by plant cell enzymes since they do not code for any protein. Lacking a capsid, viroids exist only as naked RNA molecules and pass from plant to plant only when both donor and recipient cells are damaged. (It is the capsid that, in other viruses, makes entry into intact cells possible.) DNA viruses may have evolved from small circular fragments accidentally created from a piece of a cell chromosome. Under the pressure of natural selection, such independently replicating elements—whether RNA or DNA—could be expected to acquire nucleotide sequences from the host cell that would facilitate their own multiplication, including some sequences that code for proteins. In this way, RNA or DNA molecules resembling the small DNA *plasmids* described in Chapter 4 (see p. 187) would have resulted. Plasmids are analogous to viroids in that they lack a capsid and therefore cannot move from cell to cell in the way that viruses do.

The first virus probably appeared when a plasmid acquired genes coding for capsid proteins. But a capsid can enclose only a limited amount of nucleic acid, so that its size limits the number of genes that a virus can contain. Forced to make optimal use of their limited genomes, some small viruses (like ϕX174) evolved *overlapping genes,* in which part of the nucleotide sequence encoding one protein is used (in the same or a different reading frame) to encode a second protein. Other viruses evolved larger capsids and consequently could acquire new useful genes, partly by recombination with other viruses and partly from their host cells.

Having the unique capability of transferring DNA across species barriers, viruses have almost certainly played an important part in the evolution of other organisms. Most recombine frequently with their host chromosomes, picking up random pieces of host chromosomes and carrying them to different cells or organisms. Moreover, integrated copies of viral DNA (proviruses) have become a permanent part of the genome of most organisms. Although this DNA often becomes altered so that it is unable to produce a complete virus, it may still encode proteins, some of which are useful to the cell. Viruses thereby allow evolution to occur by the mixing of gene pools of different organisms.

Viruses have also been important in cell biology in quite a different way. Because of their relative simplicity, studies of their reproduction have progressed unusually rapidly and have illuminated many of the basic molecular mechanisms in cells. This is especially true for the mechanism of genetic recombination, which we shall consider next.

Summary

Viruses are infectious particles that consist of a DNA or RNA molecule (the viral genome) packaged in a protein capsid. A virus can multiply only inside a host cell, whose genetic mechanisms it subverts for its own reproduction. Both the structure of the viral genome and its mode of replication differ widely between viruses. The usual outcome of a viral infection is the lysis of the infected cell, with the release of viral particles. Some viruses, however, can instead become integrated into the host chromosome, where their genes are replicated along with those of the host cell. Because viruses sometimes carry host DNA sequences from one species to another, they make possible an occasional mixing of different gene pools during evolution.

Genetic Recombination Mechanisms[33]

In the preceding sections we have discussed the mechanisms by which the DNA sequences in cells are maintained from generation to generation with very little change and have described how viruses use variations of these mechanisms to replicate their own genomes. We shall now describe in detail the mechanisms that can result in the rearrangement of nucleotide sequences. They are collectively called **genetic recombination** and constitute a group of related but different genetic exchange reactions occurring between two separate segments of DNA helix.

Genetic recombination events can be divided into two broad classes—*general recombination* and *site-specific recombination.* In general recombination, genetic exchange takes place between homologous DNA sequences, most commonly between two copies of the same chromosome. One of the best known examples is the exchange of sections of homologous chromosomes (homologs) in the course of *meiosis,* an event that occurs between tightly synapsed chromosomes early in the development of eggs and sperm (p. 777). General recombination has the effect of increasing the diversity of gene combinations on the chromosomes of the mating population, thereby increasing the chances of survival in a changing environment (p. 771). While meiosis is limited to eucaryotes, the advantage of this type of gene mixing is so great that mating and reassortment of genes by general recombination has evolved also in procaryotic organisms.

Site-specific recombination differs from general recombination in that DNA homology is not required. Instead, exchange occurs at short, specific nucleotide sequences on one or both of the two participating DNA helices. Site-specific recombination therefore alters the arrangement of the nucleotide sequences in genomes. Sometimes these changes are scheduled and organized, as when an integrated bacterial virus exits from a bacterial chromosome (p. 236). But, in other instances they can be quite haphazard, as when a transposable element enters a genome (p. 248).

It has not yet been possible to unravel the biochemistry of genetic recombination in higher eucaryotes. As with DNA replication, most of what we know about the molecular details of these processes comes from studies on more accessible biological systems, such as the bacterium *E. coli* and its viruses.

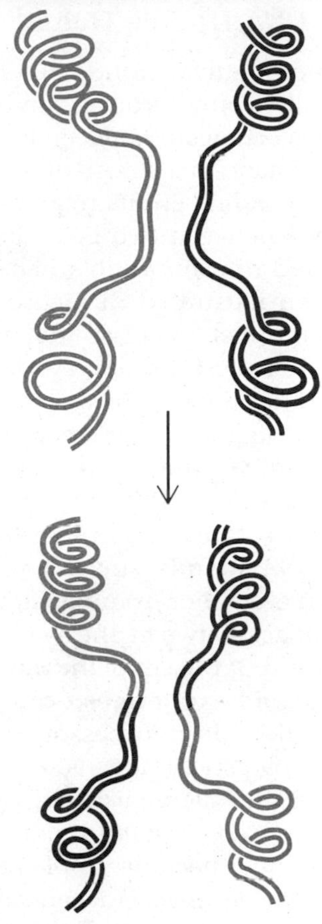

Figure 5–53 The breakage and reunion of two homologous DNA double helices creates two chromosomes that have crossed over.

General Recombination Processes Are Guided by Base-pairing Interactions Between Complementary Strands of Homologous DNA Helices[34]

General recombination involves DNA strand-exchange intermediates that require some effort to comprehend. Moreover, the exact pathway followed is likely to be different in different organisms. However, detailed genetic analyses of mating bacteria, viruses, and fungi suggest that the major outcome of general recombination is always the same:

1. The final result of the process is that two homologous DNA double helices are broken and the two broken ends are rejoined to their opposite partners to re-form two intact double helices, each of which contains parts of both of the two initial DNA molecules (Figure 5–53).
2. The site of exchange (that is, where a colored helix is joined to a black helix in Figure 5–53) can occur anywhere in the homologous nucleotide sequences of the participating chromosomes.
3. At the site of exchange, a strand of one helix has become base-paired to a strand of the second helix to create a *staggered joint* between the two different DNA helices (Figure 5–54). These staggered joints can be thousands of base pairs long; how they come about will be explained later.
4. No nucleotide sequences are altered at the place of exchange; the breaking and reunion process is so precise that not a single nucleotide is lost, gained, or changed.

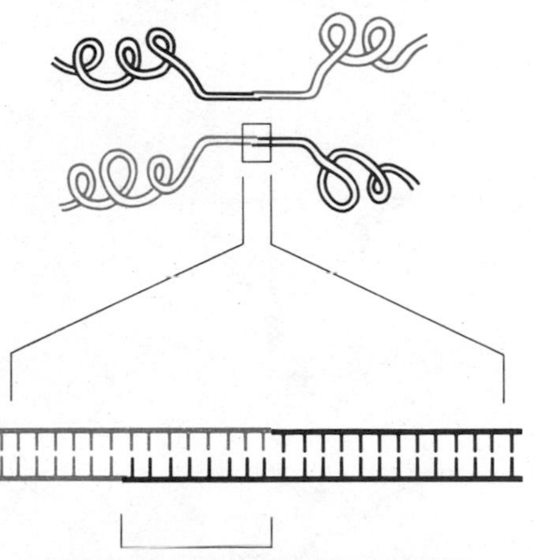

Figure 5–54 A staggered joint unites two chromosomes where they have crossed over. Such a joint is often thousands of nucleotides long.

The general recombination pathway is designed to ensure that only two regions of DNA helix with extensive DNA-sequence homology undergo an exchange reaction. The staggered joint formed at the place of exchange should ensure this because it requires that a long base-paired region form between strands from the two original helices. But how does this staggered joint arise, and how do the two homologous DNA helices that will pair recognize their sequence homology? As we shall see, the homologous regions first recognize each other directly by a base-pairing interaction. The formation of base pairs between complementary strands from the two DNA helices then guides the general recombination process so that it occurs only between long regions of matching DNA sequence and without alterations in chromosome organization.

General Recombination Is Initiated at a Nick in One Strand of a DNA Double Helix[34,35]

Each of the two strands in a DNA double helix is not only base-paired but is helically wound around the other. As a result, no significant base-pair interactions can occur between two homologous DNA double helices unless a nick is first made in a strand of one of them, freeing that strand for the unwinding and rewinding events required. For the same reason, any *mutual* exchange of strands between two DNA double helices requires at least two nicks, one in a strand of each double helix. Finally, to produce the type of breakage and reunion illustrated in Figure 5–53, it is clear that each of the four strands present must be cut so that each can be rejoined to a different partner. How are all of these nicking events accomplished and coordinated so as to guarantee that they occur only when two DNA helices share an extensive region of matching DNA sequence?

A clue comes from the observation, made in many different organisms, that a single nick in only one strand of a DNA helix is sufficient to initiate general recombination events. Thus, agents known to introduce such nicks into DNA strands, such as γ- or x-irradiation, can trigger a genetic recombination event. For example, in vertebrate cells such DNA-nicking agents induce exchanges between the two sister chromatids in each chromosome during the S and G_2 phases of the mitotic cell cycle (Figure 5–55). Because such **sister-chromatid exchanges** occur between identical copies of the same chromosome, they do not reassort different genomes; therefore, they cannot be detected by genetic means. However, a unique type of staining procedure can be used to distinguish the two sister chromatids from each other, and exchanges can then be readily detected by the striking "harlequin chromosomes" that become visible at metaphase (Figure 5–56).

During meiosis in eucaryotes and the analogous recombination events in bacteria, it is probable that an enzyme creates a nick at the scattered (but nonrandom) locations where pairing begins. The general manner in which such a nick is thought to induce an initial base-pairing interaction between two complementary stretches of DNA double helix is shown in Figure 5–57. Although the details are still unclear, evidence for a step of this type in the general recombination pathway comes from a study of some of the proteins involved, as will now be described.

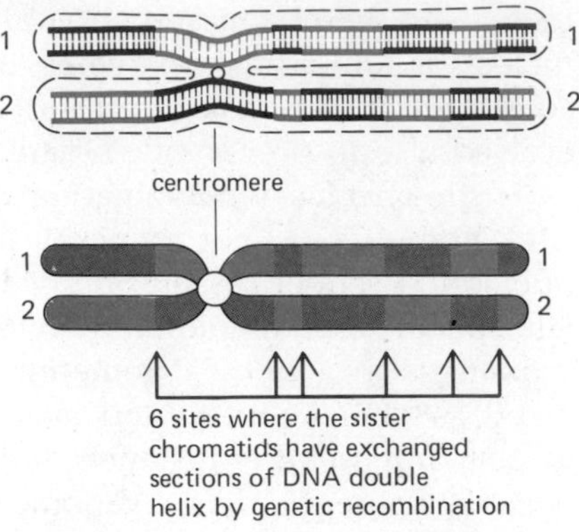

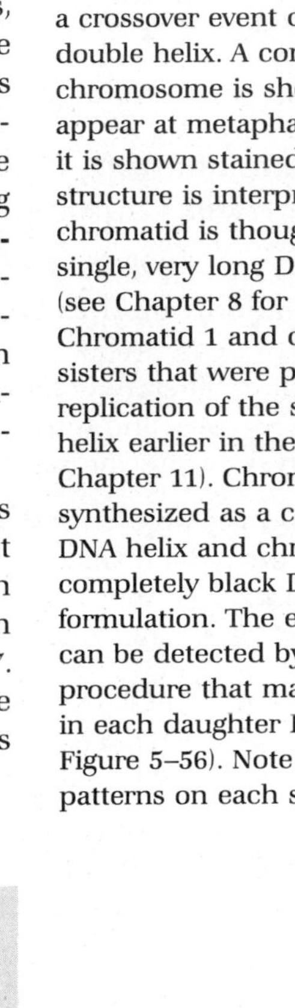

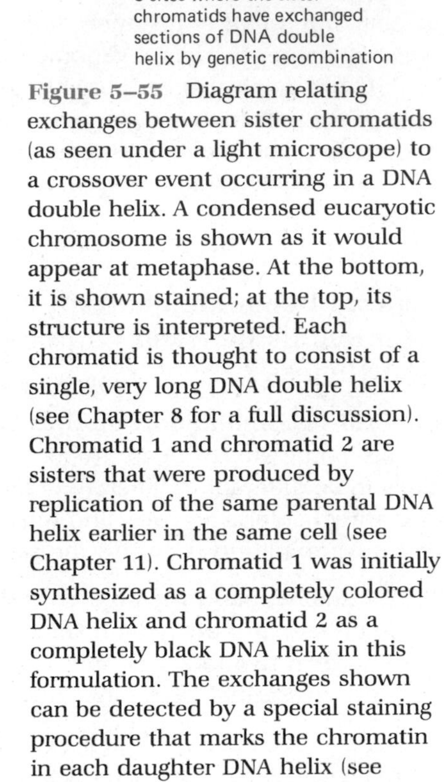

Figure 5–55 Diagram relating exchanges between sister chromatids (as seen under a light microscope) to a crossover event occurring in a DNA double helix. A condensed eucaryotic chromosome is shown as it would appear at metaphase. At the bottom, it is shown stained; at the top, its structure is interpreted. Each chromatid is thought to consist of a single, very long DNA double helix (see Chapter 8 for a full discussion). Chromatid 1 and chromatid 2 are sisters that were produced by replication of the same parental DNA helix earlier in the same cell (see Chapter 11). Chromatid 1 was initially synthesized as a completely colored DNA helix and chromatid 2 as a completely black DNA helix in this formulation. The exchanges shown can be detected by a special staining procedure that marks the chromatin in each daughter DNA helix (see Figure 5–56). Note the reciprocal patterns on each sister chromatid.

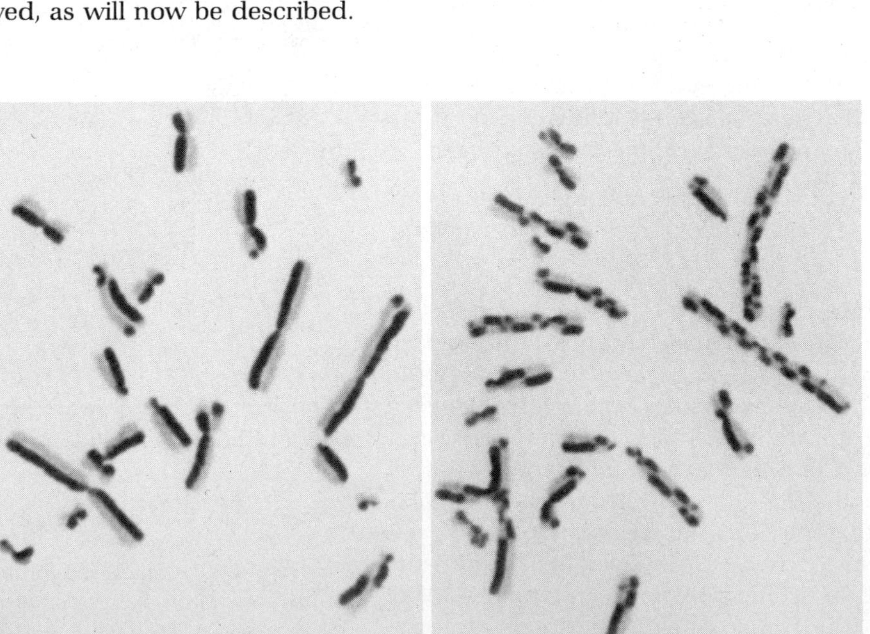

5 μm

Figure 5–56 Sister-chromatid exchanges visualized in metaphase chromosomes. The chromosomes on the left are from control cells and have experienced only a few chromatid exchanges. The chromosomes on the right are from cells treated to induce nicks in the DNA, and each chromatid has experienced many chromatid exchanges. Note the reciprocal staining patterns on each sister chromatid. (See also Figure 5–55.) (Courtesy of Judy Bodycote and Sheldon Wolff.)

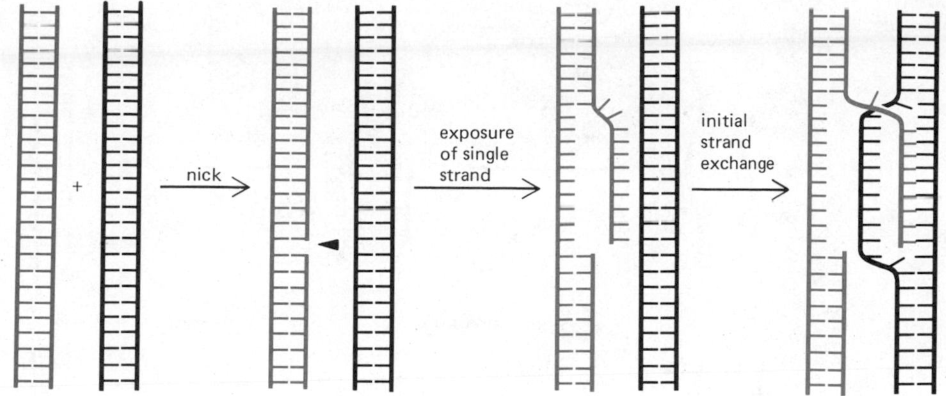

Figure 5–57 Schematic illustration of the initial strand exchange between two homologous DNA double helices undergoing general recombination. A nick in a single DNA strand frees the strand, which then invades the second helix to form a short pairing region. Only two DNA molecules that are complementary in nucleotide sequence can base-pair in this way and thereby initiate a general recombination event.

Special Proteins Enable DNA Single Strands to Pair with a Homologous Region of DNA Double Helix[36]

Most of what we know about the enzymology of the genetic recombination process is derived from studies of *E. coli* mutants defective in the general recombination pathway. Studies of these bacterial mutants have identified some of the special proteins required for these events. One of the proteins is the *E. coli SSB protein,* the same helix-destabilizing protein that is also essential for *E. coli* DNA replication (p. 227). By binding tightly in cooperatively formed clusters (p. 389) to the sugar-phosphate backbone of all of the DNA single strands in a cell, this small protein (19,000 daltons) expands the otherwise collapsed conformation of the strands and holds their bases in an exposed position (Figure 5–58). In the expanded conformation, the DNA single strands readily form complementary base pairs with either a nucleoside triphosphate (in DNA replication) or a complementary section of another DNA single strand, as occurs in general recombination.

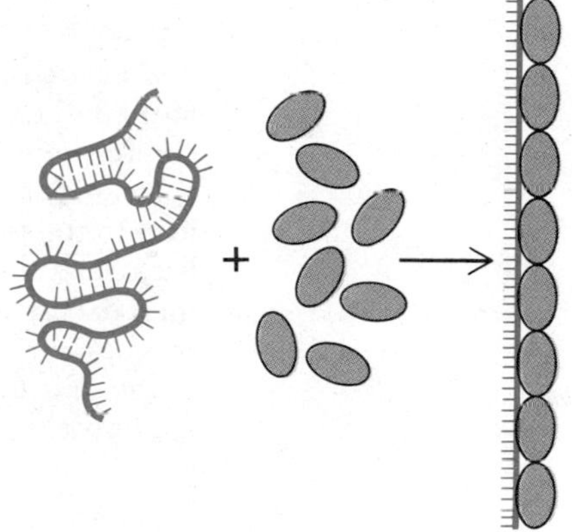

Figure 5–58 The binding of a helix-destabilizing protein prepares a DNA single strand for base-pairing interactions. The protein monomers help their neighbors to bind and thereby line up in long clusters on the DNA strand. The same helix-destabilizing protein that functions in DNA replication (see Figure 5–37) seems to function in genetic recombination also.

The base-pairing interaction that is central to general recombination can be mimicked in a test tube by the renaturation of DNA that has been separated into single strands. Such **DNA renaturation** occurs when a random collision juxtaposes the complementary nucleotide sequences on two matching DNA single strands, nucleating the formation of a short stretch of double helix between them. This relatively slow *helix nucleation* step is followed by a very rapid "zippering" step as the double helix grows to maximize base-pairing interactions (Figure 5–59). When the *E. coli* helix-destabilizing protein (SSB protein) binds to such DNA strands and expands their conformation, it speeds up the rate of helix nucleation and thereby the overall rate of strand rejoining by a factor of more than 1000 under physiological conditions. But this catalyzed pairing reaction takes place between two DNA single strands and is not in itself sufficient to account for the initial pairing events in genetic recombination, which involve double helices. At least one other protein is known to be involved in *E. coli:* the **RecA protein.**

The RecA protein is named after a gene that was identified in 1965 as having a central role in chromosome pairing. Long sought by biochemists, this important but elusive gene product was finally purified to homogeneity in 1976 and shown to be a protein of 38,000 daltons that has several interesting enzymatic activities. Like the helix-destabilizing protein, the RecA protein binds tightly to DNA single strands, and it can help pull a nicked strand out of a DNA double helix (as required in Figure 5–57). It also can bind simultaneously to a DNA double helix and hold the single strand and the double helix together (Figure 5–60A). Moreover, the RecA protein is a DNA-dependent ATPase, having an additional site for binding and hydrolyzing ATP. It can thus drive a

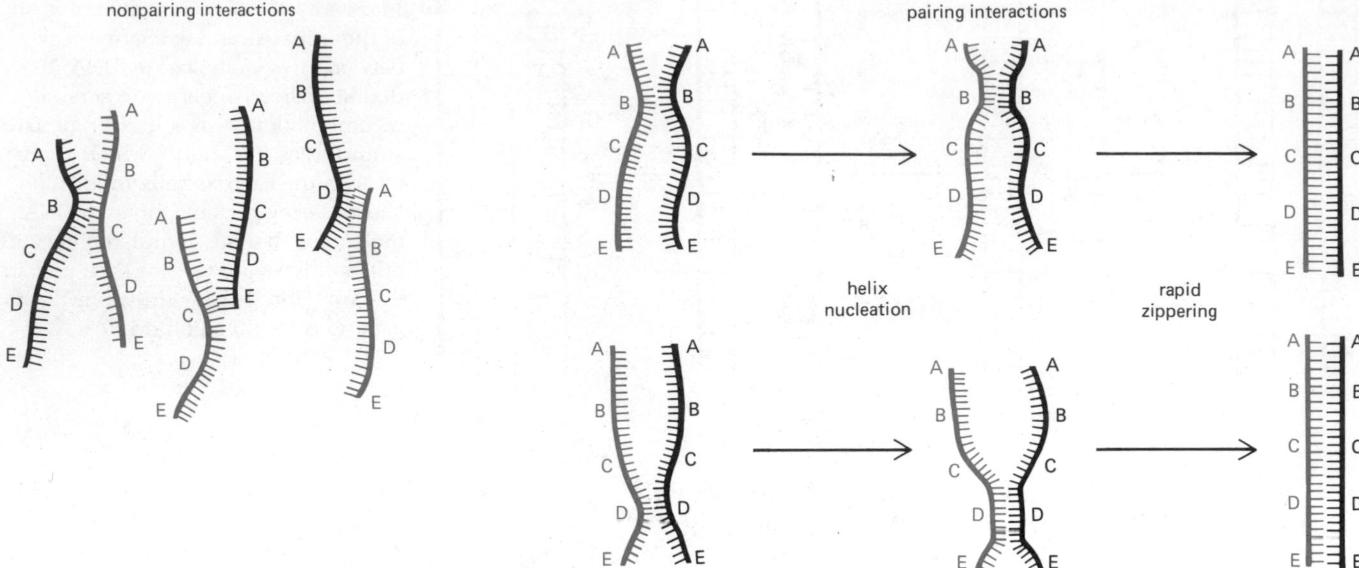

Figure 5–59 During DNA renaturation *in vitro,* DNA double helices are re-formed from their separated strands. Helix re-formation depends on the random collision of two complementary strands (see p. 189). While most such collisions are not productive, as shown at the left, a few will lead to the formation of a short region where complementary base pairs have formed (helix nucleation). A rapid zippering then occurs to complete each helix. By this trial-and-error process, one DNA strand can find its complementary partner in the midst of millions of nonmatching DNA strands. Trial-and-error recognition of a complementary partner appears to initiate all general recombination events.

series of conformational changes in the protein that convert the type of triple-stranded complex shown in Figure 5–60A to that shown in Figure 5–60B. This last reaction is the crucial one because it generates the direct base-pairing interaction between the single strand and the double helix that is thought to begin the central events in recombination. Although the details of this crucial reaction are not yet clear, according to one model, the energy of ATP hydrolysis is used by the RecA protein to help move a single strand along a double helix, promoting a search for a region of nucleotide sequence homology that is required for base-pairing to begin.

Studies *in vitro* have demonstrated that the *E. coli* helix-destabilizing protein (the SSB protein) cooperates with the RecA protein to facilitate pairing reactions. This may be the reason why genetic recombination is greatly reduced in the *E. coli* cell when either of these proteins is defective.

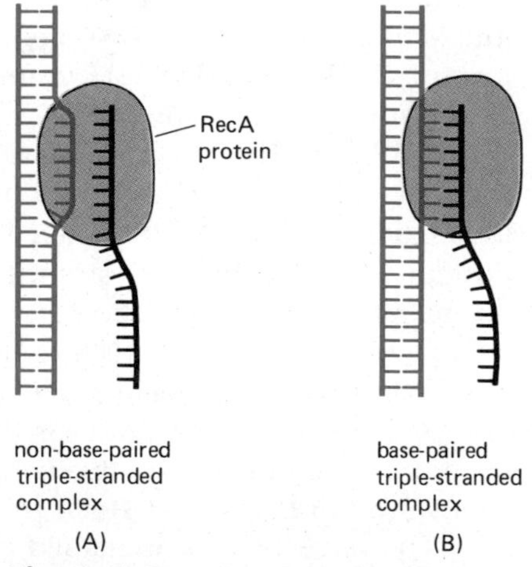

Figure 5–60 The two types of complexes formed between a DNA single strand and a double helix by the RecA protein. The formation of a non-base-paired complex precedes the formation of a base-paired complex when DNA molecules are incubated with the RecA protein *in vitro.*

Genetic Recombination Usually Involves a Cross-Strand Exchange[37]

The slow and difficult step in a genetic recombination event is thought to be the initial establishment of a single-strand exchange between two double helices (Figure 5–57). Because the homologous DNA sequences of the two interacting helices will be in close register after this initial exchange, extension of

Figure 5–61 The development of a cross-strand exchange. There are many different possible pathways that lead from the structure in Figure 5–57 to a cross-strand exchange, but only one is shown.

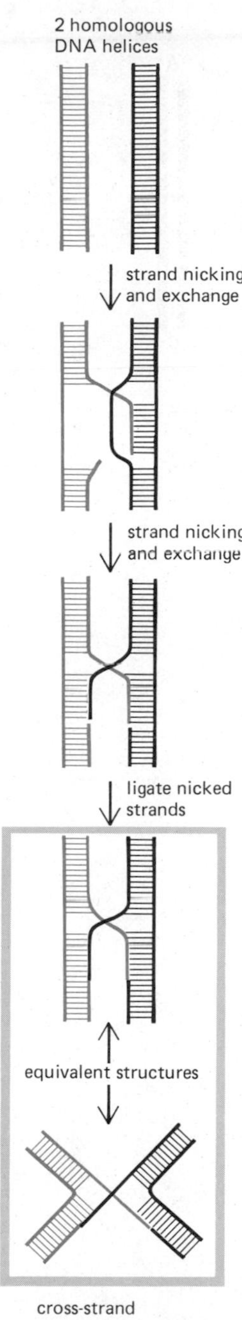

the region of pairing and the establishment of further strand exchanges between the two helices should occur rapidly. During these events, a limited amount of nucleotide excision and local DNA resynthesis often occurs, resembling some of the events in DNA repair. But because of the large number of possibilities, different detailed pathways are likely to be followed at this stage in different organisms, or even in the same cells during different individual events. In most of the pathways, an intermediate structure, the **cross-strand exchange,** is formed by the participating DNA helices. One of the simplest ways in which this structure can be formed is shown in Figure 5–61.

In the cross-strand exchange, two homologous DNA helices that initially paired are now held together by mutual exchange of two of the four strands present, one originating from each of the helices. No disruption of base-pairing is necessary to maintain this structure, which has several interesting and important properties: (1) The point of exchange between the two homologous DNA helices, which is located where the two strands cross between them, can migrate rapidly back and forth along the helices by a spontaneous process known as **branch migration** (Figure 5–62). Such branch migration can greatly extend the region of base-pairing between the interacting strands that initially belonged to different DNA helices, and it probably accounts for the considerable length of the pairing between sections that have recombined (see Figure 5–54). (2) The cross-strand exchange contains two crossing and two noncrossing strands. The structure can *isomerize* by undergoing the series of rotational movements shown in Figure 5–63. This isomerization alters the positions of the two pairs of strands so that the two original noncrossing strands become crossing strands, and vice versa.

In order to regenerate two separate DNA helices, and thus terminate the pairing process, the two crossing strands must be cut. If these two crossing strands are cut *prior* to an isomerization of the cross-strand exchange, the two original DNA helices will separate from each other nearly unaltered, with only a very short piece of single-stranded DNA being exchanged (Figure 5–63). However, if the two crossing strands are cut *after* the isomerization process, one section of each original DNA helix will become linked by a staggered joint to a section of the other DNA helix: in other words, the two DNA helices will have crossed over (Figure 5–63).

Figure 5–64 illustrates this process realistically, as it might occur between two sister chromatids in mitotic cells or between two homologous nonsister chromatids in meiotic cells (p. 778). While the crucial cross-strand exchange isomerization shown here and in Figure 5–63 should occur spontaneously at some rate, it may also be enzymatically driven or otherwise regulated by cells. Some kind of regulation seems most likely to operate during meiosis, when the two DNA double helices that pair are constrained in an elaborate structure called the *synaptonemal complex* (p. 780).

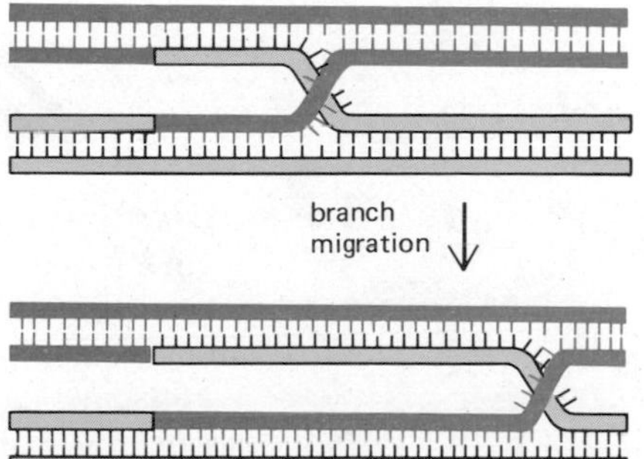

Figure 5–62 Branch migration occurs rapidly at a cross-strand exchange, thereby increasing the amount of base-pairing between strands from two different DNA double helices.

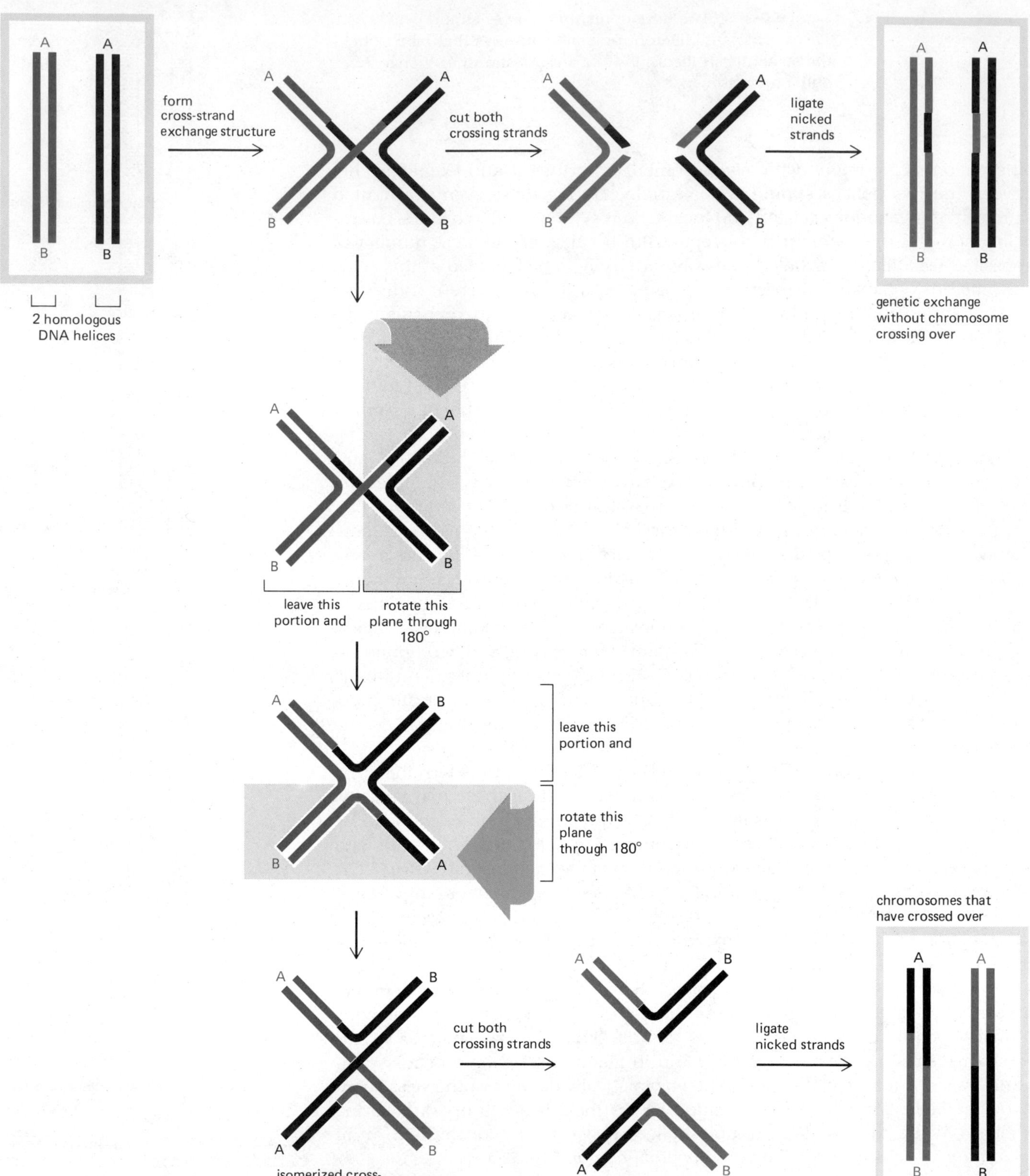

Figure 5–63 The isomerization of a cross-strand exchange. Without isomerization, the cutting of the two crossing strands terminates the exchange without crossing over (*top*). With isomerization, the cutting of the two crossing strands creates two chromosomes that have crossed over (*bottom*).

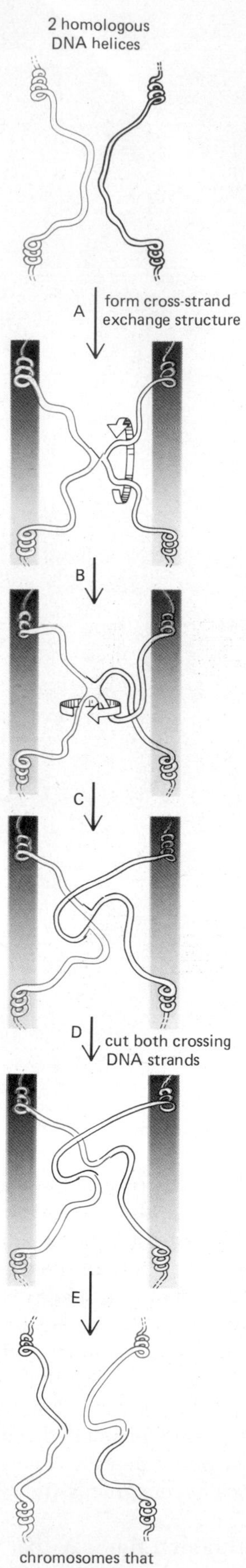

Figure 5–64 Illustration of a genetic recombination event between two homologous chromosomes. The isomerization of the cross-strand exchange occurs as described in Figure 5–63.

General Recombination Aids DNA Repair Processes[38]

Some forms of irradiation or reactive chemicals have a substantial probability of damaging both DNA strands simultaneously at the same site in the DNA helix. When this happens there may be no chance of repairing the lesion by any of the DNA repair mechanisms mentioned earlier. There is good evidence, though, from studies carried out with bacteria that some severe types of DNA damage that cannot otherwise be repaired are removed by special DNA repair processes that occur during genetic recombination events. While the repair pathway is not known, four copies of the genetic information (four DNA strands) are brought together transiently during the central pairing process (Figure 5–63). In principle, this makes repair possible between sister chromatids even if *both* DNA strands are damaged near the same base pair in one of the two DNA helices involved. Thus, the sister-chromatid exchanges discussed previously (Figures 5–55 and 5–56) may serve a function, despite the fact that they occur in somatic cells and reassort only identical genes.

Site-specific Recombination Enzymes Move Special Sequences of DNA in and out of Genomes[39]

Another type of genetic recombination is much more restricted in its range of action than is the general recombination system just described. Here the entire recombination process—including the initial recognition step that brings two DNA helices together—is guided by a recombination enzyme, and base-pairing between the two recombining molecules need not be involved (even in those systems where base-pairing does occur, a staggered joint that is only a few base pairs long is formed). This form of recombination is called **site-specific recombination** because the recombination enzyme recognizes specific nucleotide sequences present on one or both of the recombining DNA molecules, and it is at these sequences that recombination occurs.

The first such system to be discovered was the one that enables a bacterial virus, bacteriophage *lambda,* to move itself in and out of the *E. coli* chromosome. The enzyme responsible for this reaction, called *lambda integrase,* is encoded by one of the viral genes. The purified enzyme carries out the reaction shown in Figure 5–65. The integrase protein recognizes two different nonhomologous DNA sequences, one located on the circular bacteriophage DNA double helix and one located on the bacterial chromosome. In the recombination reaction, both strands of the bacteriophage DNA double helix are transiently broken and rejoined to the ends of similarly broken strands in the bacterial DNA double helix. The enzyme brings the specific sites on the bacterial and the bacteriophage chromosomes close together and then initiates the required DNA-cutting and -resealing reactions. The various steps in this process occur so quickly that it is normally not possible to detect any of the suspected intermediate DNA forms.

As indicated in Figure 5–66, the same type of site-specific recombination mechanism can also be carried out in *reverse* by the lambda bacteriophage, enabling it to exit from its integration site in the *E. coli* chromosome in order to multiply rapidly within the bacterial cell. A similar type of reaction is thought to occur in developing B lymphocytes, in which the deletion of specific DNA sequences rearranges the immunoglobulin genes (p. 981). It is also believed to

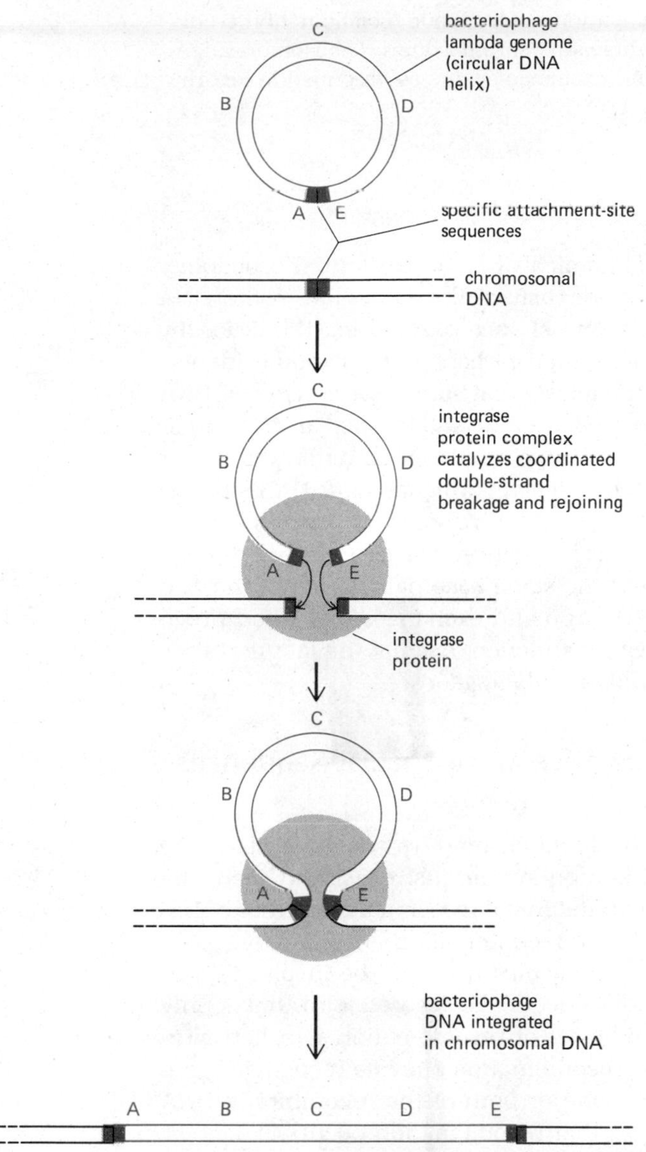

Figure 5–65 The site-specific recombination event that inserts bacteriophage lambda DNA into the *E. coli* host chromosome. The specific sites recognized by the integrase are the DNA sequences shown as colored squares (see also Figure 5–50, p. 237).

cause the inversion of a coat-controlling gene of the *Salmonella* bacterium (p. 462).

A large group of *mobile genetic elements* have been discovered in both procaryotic and eucaryotic chromosomes that have the ability to move in and out of their host genomes in a manner similar to the lambda bacteriophage. Like lambda, some of these are viruses, with the ability to move from one cell to another as a particle in which the DNA (or RNA) genome of the virus has been protected by a coat that enables the virus to exit from and reenter cells. But many mobile genetic elements do not form viral particles and cannot leave the cell. As a result, these DNA sequences are constrained to move from place to place within the chromosomes of a single cell and its progeny, and they are therefore called **transposable elements.** Like many viruses, transposable elements move by means of a site-specific recombination system that specifically recognizes their DNA, even when it is integrated in the host genome.

Although the way that viruses and transposable elements insert themselves into genomes is still incompletely understood, it is clear that different

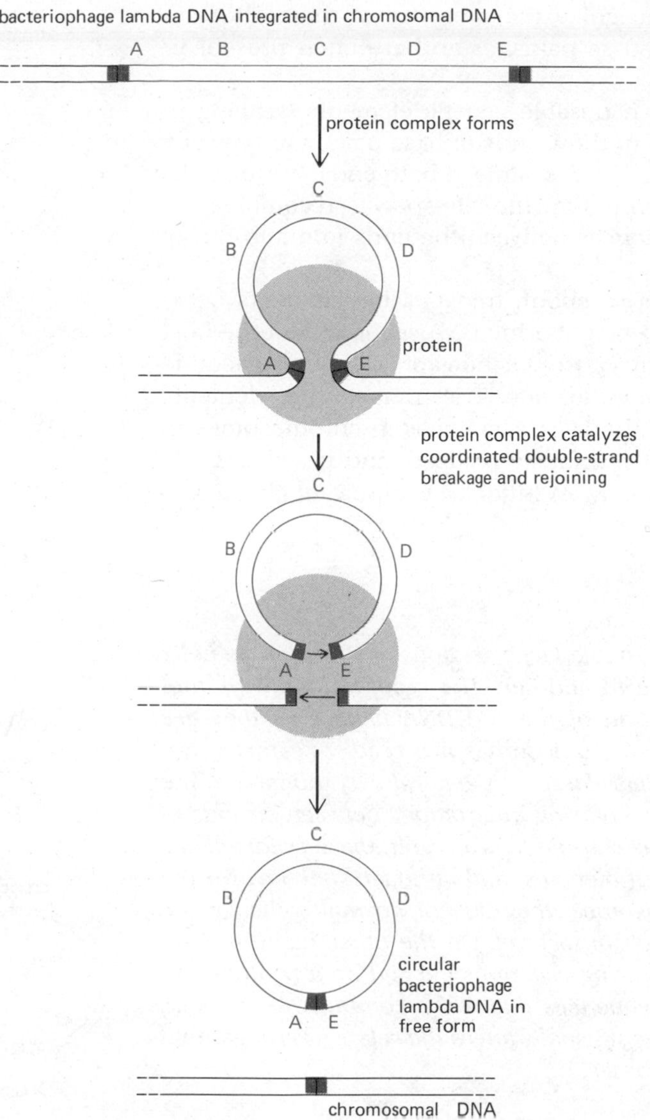

Figure 5–66 Bacteriophage lambda DNA exits from the host chromosome by a reaction that is the reverse of that shown in Figure 5–65. This reaction is catalyzed by a complex of the integrase with a second bacteriophage protein. The activity of this second protein is regulated to control the entrance and exit from the chromosome (see Figure 5–50, p. 237).

mechanisms are used for different mobile elements. In contrast to the specific chromosomal "attachment site" recognized by bacteriophage lambda, many elements integrate at a more or less randomly selected site on the chromosome. Although the integration of some mobile elements probably involves an integrase similar to that of bacteriophage lambda, it is known that some of the transposable elements studied in bacteria move by a somewhat more complicated site-specific recombination process. For these elements, the integration event is accompanied by the replication of the transposable element DNA sequence, which allows this DNA sequence to become inserted at a new chromosomal site without leaving the old one (Figure 5–67).

Both integrated forms of viruses and transposable genetic elements are thought to be scattered throughout the genomes of all cells, although transposable elements have thus far been well characterized only in bacteria, yeast, maize, and fruit flies. Many different types of transposable elements exist, and each is occasionally activated to move in or out of its genome site by unknown signals, in a process catalyzed by a variety of as yet poorly characterized site-specific enzyme complexes. Since such movement may occur only very rarely (for example, once in every 10^5 cell generations for many bacterial transposable elements), it is often difficult to recognize the presence of these DNA sequences. They are sometimes viewed as parasites and are often present in multiple copies per cell.

The DNA sequences of the transposable genetic elements (ranging in length from a few hundreds to tens of thousands of base pairs) are typically flanked by nucleotide sequences that are the same at both ends. Presumably, it is by recognizing both of these ends that the site-specific recombination enzymes act to move the DNA sequences between the ends into and out of chromosomes.

In addition to moving themselves about, transposable elements occasionally act to move or rearrange the neighboring DNA sequences of the host genome. For example, they frequently cause deletions of adjacent nucleotide sequences or move with them to another site. The transposable elements therefore cause the arrangement of the DNA sequences in chromosomes to be much less stable than had been previously realized, and it seems likely that they have been responsible for many evolutionarily important changes in genomes (see p. 473).

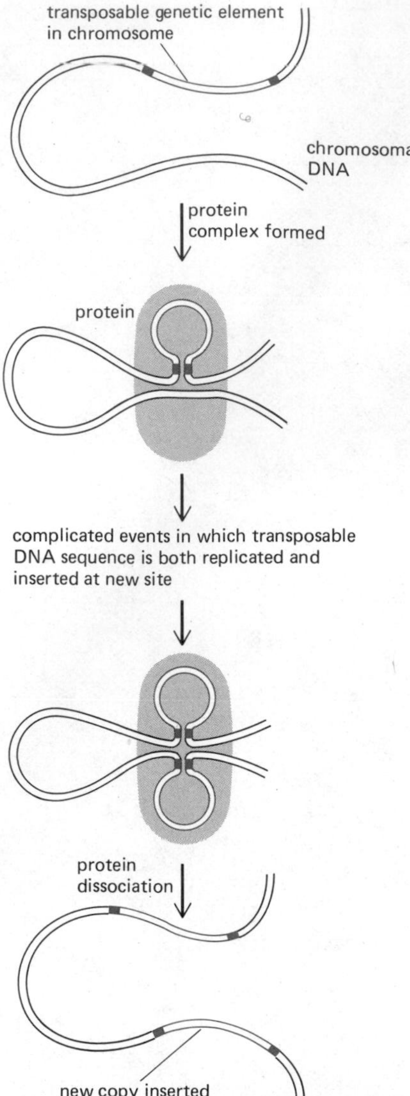

Figure 5–67 The movement of a transposable genetic element within a chromosome. For the type of element shown here, movement occurs without displacement of the DNA sequence from its original site. The two identical DNA sequences that commonly flank the two ends of such transposable elements are shown as colored squares.

Summary

Genetic recombination mechanisms enable large sections of DNA double helix to be moved from one chromosome to another. The reaction pathways that have evolved cause minimal disruption of the two DNA helices as they are broken and rejoined, and two intact chromosomes are readily restored. Recombination events fall into two broad classes. In general recombination, the initial reactions rely on extensive base-pairing interactions between strands of the two DNA double helices that will recombine. As a result, these events occur only between two homologous DNA molecules; and although they involve the exchange of genes between chromosomes, they do not normally change the overall sequence of the genes in a chromosome. On the other hand, in site-specific recombination events, the pairing reactions depend on a protein-mediated recognition of the two DNA sequences that will recombine. As a result, this second class of recombination event usually alters the relative arrangement of nucleotide sequences in chromosomes.

References

General

Stent, G.S. Molecular Genetics: An Introductory Narrative. San Francisco: Freeman, 1971.

Watson, J.D. Molecular Biology of the Gene, 3rd ed. Menlo Park, Ca.: Benjamin-Cummings, 1976.

Cited

1. Stryer, L. Biochemistry, 2nd ed. San Francisco: Freeman, 1981. (Chapters 25, 26, and 27).

   Hershey, J.W.B. The translational machinery. In Cell Biology: A Comprehensive Treatise, Vol. 4 (D.M. Prescott and L. Goldstein, eds.). New York: Academic Press, 1980.

2. Chamberlin, M.J. RNA polymerase: an overview. In RNA Polymerase (R. Losick and M. Chamberlin, eds.), pp. 17–67. Cold Spring Harbor, N.Y.: Cold Spring Harbor Laboratory, 1976.

   Pribnow, D. Genetic control signals in DNA. In Biological Regulation and Development. Vol. 1 (R. Goldberger, ed.), pp. 219–277. New York: Plenum, 1979.

Rosenberg, M.; Court, D. Regulatory sequences involved in the promotion and termination of RNA transcription. *Annu. Rev. Genet.* 13:319–353, 1979.

3. Rich, A.; Kim, S.H. The three-dimensional structure of transfer RNA. *Sci. Am.* 238(1):52–62, 1978.
4. Schimmel, P.R.; Söll, D. Aminoacyl tRNA synthetases: general features and recognition of transfer RNAs. *Annu. Rev. Biochem.* 48:601–648, 1979.
5. Crick, F.H.C. The genetic code: III. *Sci. Am.* 215(4):55–62, 1966.
   Schimmel, P.R.; Söll, D.; Abelson, J., eds. Transfer RNA. Cold Spring Harbor, N.Y.: Cold Spring Harbor Laboratory, 1980.
6. Nomura, M.; Tissières, A.; Lengyel P., eds. Ribosomes. Cold Spring Harbor, N.Y.: Cold Spring Harbor Laboratory, 1974.
   Lake, J.A. The ribosome. *Sci. Am.* 245(2):84–97, 1981.
   Chambliss, G.; et al., eds. Ribosomes: Structure, Function and Genetics. Baltimore: University Park, 1980.
7. Clark, B. The elongation step of protein biosynthesis. *Trends Biochem. Sci.* 5:207–210, 1980.
8. Caskey, C.T. Peptide chain termination. *Trends Biochem. Sci.* 5:234–237, 1980.
9. Hunt, T. The initiation of protein synthesis. *Trends Biochem. Sci.* 5:178–181, 1980.
   Steitz, J.A. Genetic signals and nucleotide sequences in messenger RNA. In Biological Regulation and Development, Vol. 1 (R. Goldberger, ed.), pp. 349–399. New York: Plenum, 1979.
10. Ochoa, S; deHaro, C. Regulation of protein synthesis in eucaryotes. *Annu. Rev. Biochem.* 48:549–580, 1979.
    Ranu, R.S. Regulation of protein synthesis in eucaryotes by the protein kinases that phosphorylate initiation factor eIF-2. *FEBS Lett.* 112:211–215, 1980.
11. Jiménez, A. Inhibitors of translation. *Trends Biochem. Sci.* 1:28–30, 1976.
12. Lindahl, T. DNA repair enzymes. *Annu. Rev. Biochem.* 51:61–88, 1982.
13. Dickerson, R.; Geis, I. The Structure and Action of Proteins, pp. 59–66. New York: Harper & Row, 1969.
    Wilson, A.C.; Carlson, S.S.; White, T.J. Biochemical evolution. *Annu. Rev. Biochem.* 46:573–639, 1977.
    Jukes, T.H. Silent nucleotide substitutions and the molecular evolutionary clock. *Science* 210:973–978, 1980.
14. Drake, J.W. Comparative rates of spontaneous mutation. *Nature* 221:1132, 1969.
15. Cairns, J. Cancer: Science and Society. San Francisco: Freeman, 1978. (Chapter 7 discusses the role of mutation in the development of cancer.)
    Ohta, T.; Kimura, M. Functional organization of genetic material as a product of molecular evolution. *Nature* 233:118–119, 1971. (The mutation rate limits the maximum number of genes in an organism.)
16. Schrödinger, E. What is Life? Cambridge, Eng.: Cambridge University Press, 1945.
17. Lindahl, T. New class of enzymes acting on damaged DNA. *Nature* 259:64–66, 1976.
    Kornberg, A. DNA Replication. San Francisco: Freeman, 1980. (Chapter 16 covers the enzymology of DNA repair.)
    Coulondre, C.; Miller J.H.; Farabaugh, P.J.; Gilbert, W. Molecular basis of base substitution hotspots in *E. coli*. *Nature* 274:775–780, 1978.
    Arlett, C.F.; Lehmann, A.R. Human disorders showing increased sensitivity to the induction of genetic damage. *Annu. Rev. Genet.* 12:95–115, 1978.
18. Kornberg, A. DNA Replication. San Francisco: Freeman, 1980.
    Alberts, B.M.; Sternglanz, R. Recent excitement in the DNA replication problem. *Nature* 269:655–661, 1977.
19. Meselson, M.; Stahl, F.W. The replication of DNA in *E. coli*. *Proc. Natl. Acad. Sci. USA* 44:671–682, 1958.
    Cairns, J. The bacterial chromosome and its manner of replication as seen by autoradiography. *J. Mol. Biol.* 6:208–213, 1963.
20. Inman, R.B.; Schnös, M. Structure of branch points in replicating DNA: presence of single-stranded connections in lambda DNA branch parts. *J. Mol. Biol.* 56:319–325, 1971. (Electron microscopy demonstrates that the replication fork is asymmetric.)
    Ogawa, T.; Okazaki, T. Discontinuous DNA replication. *Annu. Rev. Biochem.* 49:421–457, 1980.
21. Topal, M.; Fresco, J.R. Complementary base pairing and the origin of substitution mutations. *Nature* 263:285–289, 1976.

Brutlag, D.; Kornberg, A. Enzymatic synthesis of deoxyribonucleic acid: a proofreading function for the 3′ to 5′ exonuclease activity in DNA polymerases. *J. Biol. Chem.* 247:241–248, 1972.

Fersht, A.R. Enzymatic editing mechanisms in protein synthesis and DNA replication. *Trends Biochem. Sci.* 5:262–265, 1980.

22. Alberts, B.M.; Frey, L. T4 bacteriophage gene 32: a structural protein in the replication and recombination of DNA. *Nature* 227:1313–1318, 1970.

Coleman, J.E.; Oakley, J.L. Physical chemical studies of the structure and function of DNA binding (helix-destabilizing) proteins. *Crit. Rev. Biochem.* 7:247–290, 1979.

Abdel-Monem, M.; Hoffmann-Berling, H. DNA unwinding enzymes. *Trends Biochem. Sci.* 5:128–130, 1980.

23. Wang, J.C. Interaction between DNA and an *E. coli* protein ω. *J. Mol. Biol.* 55:523–533, 1971. (Discovery of the first DNA topoisomerase.)

Wang, J.C. DNA topoisomerases. *Sci. Am.* 247(1):94–109, 1982.

Gellert, M. DNA topoisomerases. *Annu. Rev. Biochem.* 50:879–910, 1981.

24. Edenberg, H.J.; Huberman, J.A. Eukaryotic chromosome replication. *Annu. Rev. Genet.* 9:245–284, 1975.

25. Kriegstein, H.J.; Hogness, D.S. Mechanism of DNA replication in *Drosophila* chromosomes: structure of replication forks and evidence for bidirectionality. *Proc. Natl. Acad. Sci. USA* 71:135–139, 1974.

Harland, R. Initiation of DNA replication in eucaryotic chromosomes. *Trends Biochem. Sci.* 6:71–74, 1981.

26. Luria, S.E.; Darnell, J.E.; Baltimore, D.; Campbell, A. General Virology, 3rd ed. New York: Wiley, 1978. (The introductory chapter provides a historical overview of the development of virology.)

27. Gierer, A.; Schramm, G. Infectivity of ribonucleic acid from tobacco mosaic virus. *Nature* 177:702–703, 1956.

28. Cohen, S.S. Virus-Induced Enzymes. New York: Columbia University Press, 1968.

29. Kornberg, A. DNA Replication. San Francisco: Freeman, 1980. (Chapters 14 and 15 provide up-to-date, beautifully illustrated descriptions of the varieties of ways phage and animal viral DNA's replicate.)

30. Lwoff, A. Lysogeny. *Bacteriol. Rev.* 17:269–337, 1953.

31. Zinder, N.D. RNA Phages. Cold Spring Harbor, N.Y.: Cold Spring Harbor Laboratory, 1975.

Fenner, F.; McAuslan, B.R.; Mims, C.A.; Sambrook, J.; White, D.O. The Biology of Animal Viruses, 2nd ed. New York: Academic Press, 1974.

32. Temin, H.M. The participation of DNA in Rous sarcoma virus production. *Virology* 23:486–494, 1964.

Temin, H.M.; Mizutani S. RNA-dependent DNA polymerase in virions of Rous sarcoma virus. *Nature* 226:1211–1213, 1970.

Baltimore, D. Viral RNA-dependent DNA polymerase in virions of RNA tumor viruses. *Nature* 226:1209–1211, 1970.

33. Stahl, F.W. Genetic Recombination. San Francisco: Freeman, 1979.

Fincham, J.R.S.; Day, P.R.; Radford, A. Fungal Genetics, 4th ed. Berkeley: University of California Press, 1979. (Chapter 11.)

Alberts, B.M., ed. Mechanistic Studies of DNA Replication and Genetic Recombination, ICN-UCLA Symposia of Molecular and Cellular Biology, Vol. 19. New York: Academic Press, 1980.

Dressler, D.; Potter, H. Molecular mechanisms of genetic recombination. *Annu. Rev. Biochem.* 51:727–762, 1982.

34. Holliday, R. A mechanism for gene conversion in fungi. *Genet. Res.* 5:282–304, 1964.

Meselson, M.S.; Radding, C.M. A general model for genetic recombination. *Proc. Natl. Acad. Sci. USA* 72:358–361, 1975.

35. Wolff, S. Sister chromatid exchange. *Annu. Rev. Genet.* 11:183–201, 1977.

36. Radding, C.M. Genetic recombination: strand transfer and mismatch repair. *Annu. Rev. Biochem.* 47:847–880, 1978.

Radding, C.M. Recombination activities of *E. coli* RecA protein. *Cell* 25:3–4, 1981.

37. Sigal, N.; Alberts, B. Genetic recombination: the nature of a crossed strand-exchange between two homologous DNA molecules. *J. Mol. Biol.* 71:789–793, 1972.

Sobell, H.M. Concerning the stereochemistry of strand equivalence in genetic recombination. In Molecular Mechanisms in Genetic Recombination (R.F. Grell, ed.), pp. 433–438. New York: Plenum, 1975.

38. Howard-Flanders, P. Inducible repair of DNA. *Sci. Am.* 245(5):72–80, 1981.

39. Campbell, A. Some general questions about movable elements and their implications. *Cold Spring Harbor Symp. Quant. Biol.* 45:1–9, 1981.

Landy, A.; Ross, W. Viral integration and excision: structure of the lambda *att* sites. *Science* 197:1147–1160, 1977.

Nash, H.A. Integration and excision of bacteriophage lambda: the mechanism of conservative site specific recombination. *Annu. Rev. Genet.* 15:143–167, 1981.

Cohen, S.N.; Shapiro, J.A. Transposable genetic elements. *Sci. Am.* 242(2):40–49, 1980.

Calos, M.P.; Miller, J.H. Transposable elements. *Cell* 20:579–595, 1980.

Bukhari, A.I. Models of DNA transposition. *Trends Biochem. Sci.* 6:56–59, 1981.

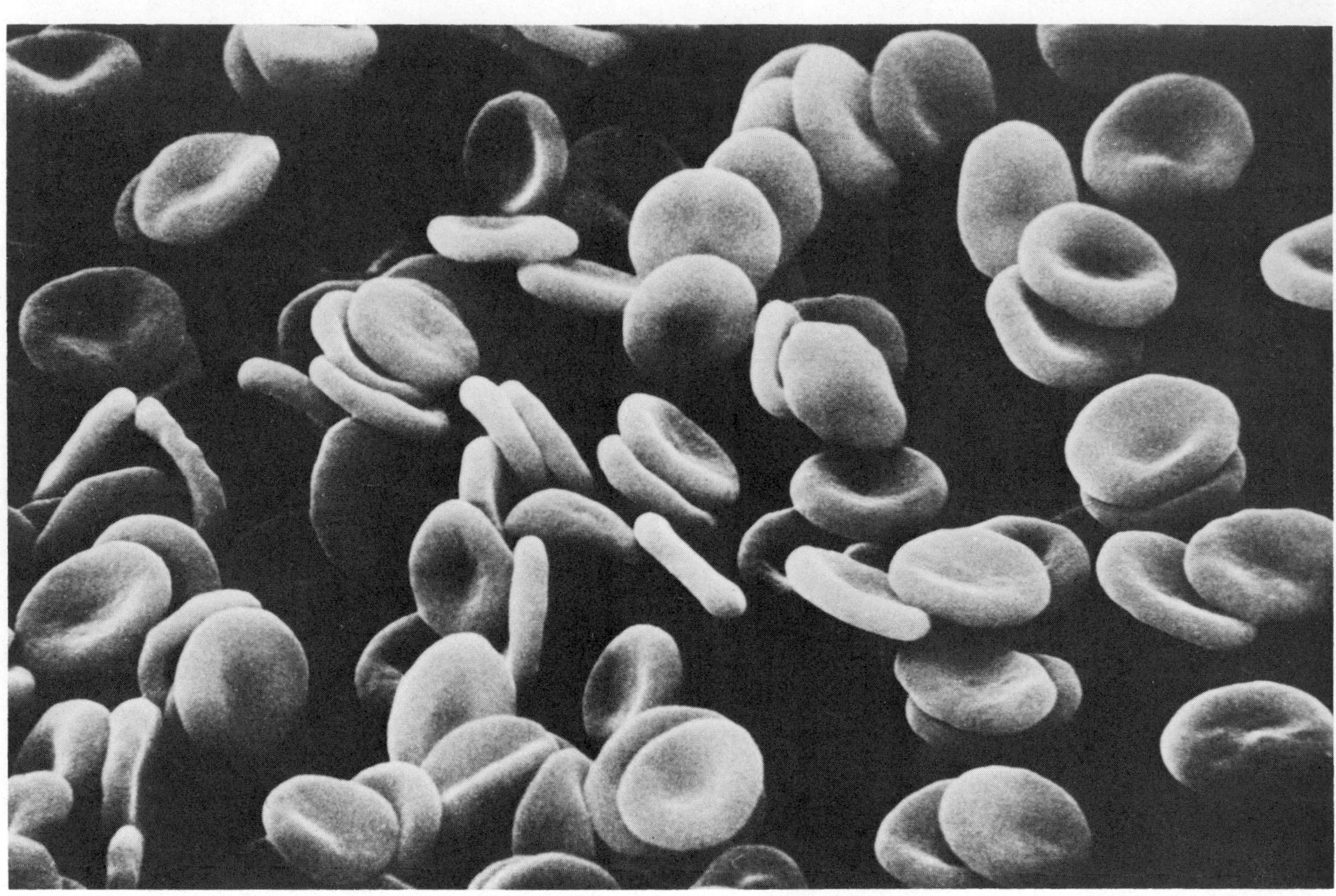

Scanning electron micrograph of human red blood cells. Since the plasma membrane is the only membrane in these cells, it is relatively easy to study, and so more is known about its structure than about any other eucaryotic cell membrane. (Courtesy of Bernadette Chailley.)

The Plasma Membrane

6

The development of the plasma membrane was a crucial step in the generation of the earliest forms of life—without it, cellular life is impossible. The **plasma membrane** that encloses every cell defines the cell's extent and maintains the essential differences between its contents and the environment. But this membrane is more than a passive barrier. It is also a highly selective filter that maintains the unequal concentration of ions on either side and allows nutrients to enter and waste products to leave the cell.

All biological membranes, including the plasma membrane and the internal membranes of eucaryotic cells, have a common overall structure: they are assemblies of lipid and protein molecules held together by noncovalent interactions. As shown in Figure 6–1, the lipid molecules are arranged as a continuous double layer 4 to 5 nm thick. This **lipid bilayer** provides the basic structure of the membrane and serves as a relatively impermeable barrier to the flow of most water-soluble molecules. The protein molecules are "dissolved" in the lipid bilayer and mediate the various functions of the mem-

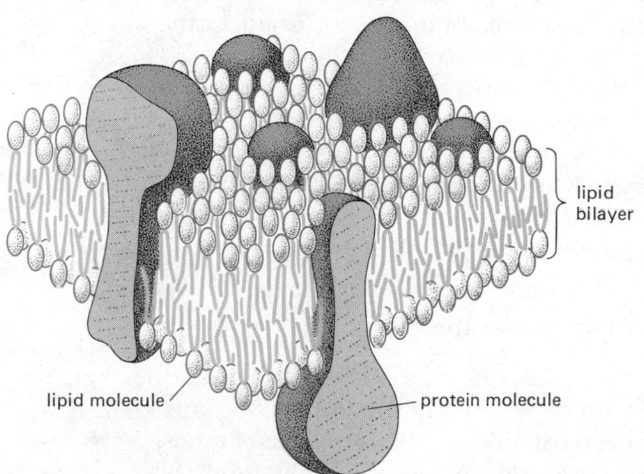

Figure 6–1 Schematic three-dimensional view of a small section of a cell membrane, about 10 nm square.

brane: some serve to transport specific molecules into or out of the cell; others are enzymes that catalyze membrane-associated reactions; and still others serve as structural links between the cell's cytoskeleton and the extracellular matrix, or as receptors for receiving and transducing chemical signals from the cell's environment.

All cell membranes are dynamic, fluid structures: most of their lipid and protein molecules are able to move about rapidly in the plane of the membrane. Membranes are also asymmetrical structures: the lipid and protein compositions of the two faces differ from one another in ways that reflect the different functions performed at the two surfaces.

Although the specific lipid and protein components vary greatly from one type of membrane to another, most of the basic structural and functional concepts discussed in this chapter are applicable to intracellular membranes as well as to plasma membranes. After considering the structure and organization of the main constituents of biological membranes—the lipids, proteins, and carbohydrates—we will discuss the mechanisms cells employ to transport small molecules across their plasma membranes and the very different mechanisms they use to transfer macromolecules and larger particles across this membrane.

The Lipid Bilayer[1]

The first indication that the lipid molecules in biological membranes are organized in a bilayer came from an experiment performed in 1925. Lipids from red blood cell membranes were extracted with acetone and floated on the surface of water. The area they occupied was then decreased by means of a movable barrier until a monomolecular film (a monolayer) was formed. This monolayer occupied a final area about twice the surface area of the original red blood cells. Because the only membrane in a red blood cell is the plasma membrane, the experimenters concluded that the lipid molecules in this membrane must be arranged as a continuous bilayer. The conclusion was right, but it turned out to be based on two wrong assumptions that fortuitously compensated for each other. On the one hand, the acetone did not extract all of the lipid. On the other, the surface area calculated for the red blood cells was based on dried preparations and was substantially less than the true value seen in wet preparations. Nonetheless, the conclusions drawn from this experiment had a profound influence on cell biology; as a result, the lipid bilayer became an accepted part of most models of membrane structure, long before its existence was actually established.

X-ray diffraction studies of highly ordered biological membranes show that the bulk of the lipid molecules in these membranes are present in the form of a bilayer. Evidence that all biological membranes are lipid bilayers is the fact that they all can be mechanically split down the middle (between the two lipid monolayers) when they are frozen—as in freeze-fracture electron microscopy (see p. 272).

Why are all cell membranes constructed as double layers of lipid molecules? One reason is that the special properties of these lipid molecules cause them to self-assemble into bilayers even outside the cell.

Membrane Lipids Are Amphipathic Molecules That Spontaneously Form Bilayers[2]

Lipid molecules are insoluble in water but dissolve readily in organic solvents. They constitute about 50% of the mass of most animal cell plasma membranes. There are approximately 5×10^6 lipid molecules in a section of lipid bilayer

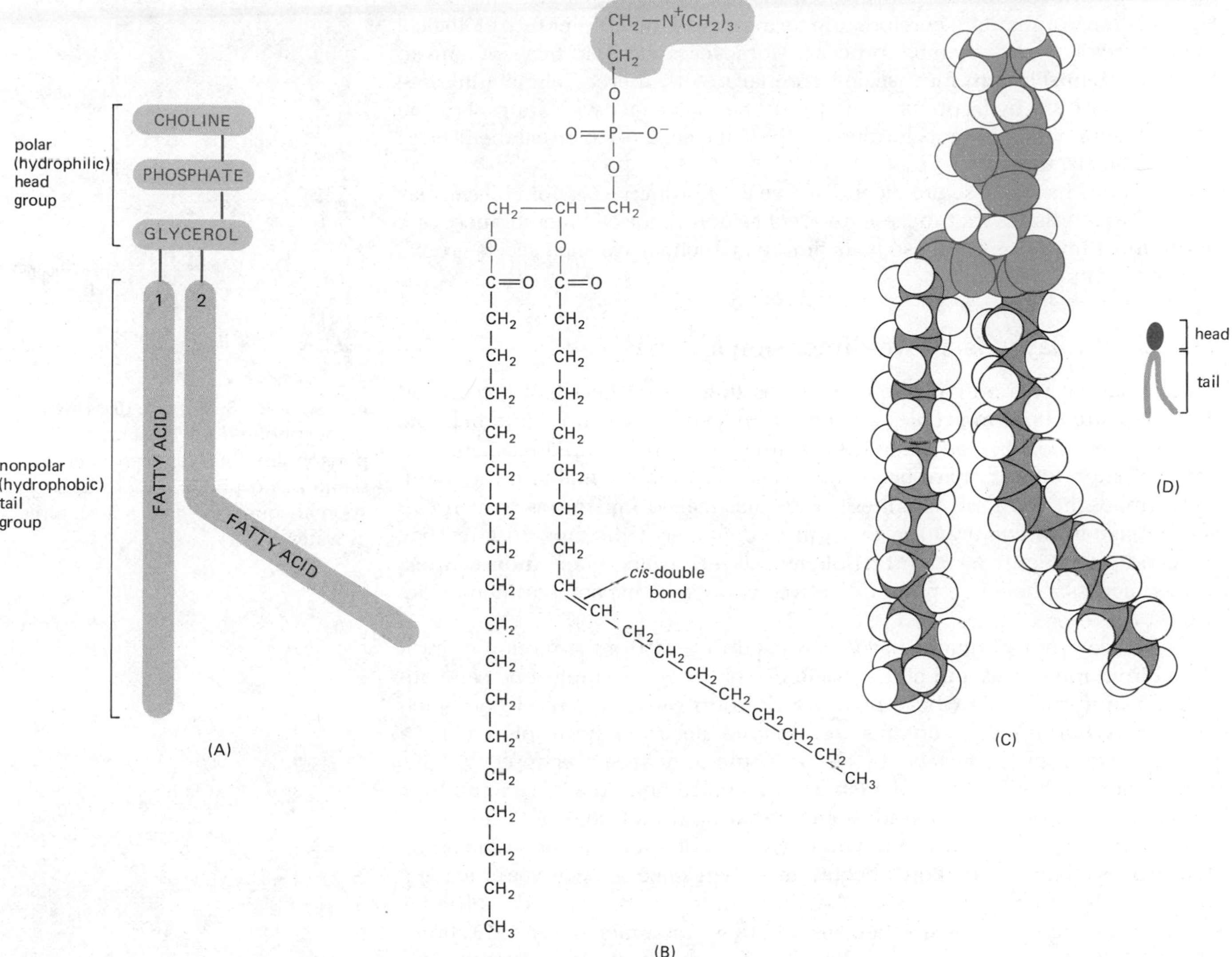

Figure 6–2 The parts of the phospholipid molecule phosphatidylcholine represented schematically (A), in formula (B), as a space-filling model (C), and as a symbol (D).

In order to distinguish the unsaturated fatty acid chain from the saturated one, in this drawing (and those that follow) the unsaturated fatty acid chain is drawn with a distinct bend in it, rather than with a small kink. In actuality, only the double bond is rigid in the unsaturated fatty acid. Because all of the single carbon-carbon bonds in the rest of the chain are free to rotate, both the saturated and unsaturated fatty acid chains will tend to pack in parallel arrays in each phospholipid monolayer.

1 μm × 1 μm, or about 10^9 lipid molecules in the plasma membrane of a small animal cell. The three major types of lipids in cell membranes are *phospholipids* (the most abundant), *cholesterol,* and *glycolipids.* All three are *amphipathic*—that is, they have a hydrophilic ("water-loving," or polar) end and a hydrophobic ("water-hating," or nonpolar) end. For example, the typical **phospholipid** molecule illustrated in Figure 6–2 has a *polar head group* and two *hydrophobic hydrocarbon tails.* The tails vary in length (normally from 14 to 24 carbon atoms), and one usually contains one or more *cis*-double bonds (that is, it is *unsaturated*), while the other does not (that is, it is *saturated*). As indicated in Figure 6–2, each double bond creates a bend in the tail. Such differences in tail length and saturation are important because they influence the fluidity of the membrane (see below).

When amphipathic molecules are surrounded on all sides by an aqueous environment, they tend to aggregate so as to bury their hydrophobic tails and leave their hydrophilic heads exposed to water. They do this in one of two ways: they can form spherical *micelles,* with the tails inward, or they can form bimolecular sheets, or *bilayers,* with the hydrophobic tails sandwiched between the hydrophilic head groups. These two possibilities are illustrated in Figure 6–3.

Most phospholipids and glycolipids spontaneously form bilayers in aqueous environments. Therefore, the formation of the lipid part of biological membranes is a self-assembly process. Moreover, such lipid bilayers tend to close on themselves to form sealed compartments, thereby eliminating free edges, where the hydrophobic tails would be in contact with water. For the same reason, compartments formed by lipid bilayers tend to reseal themselves when they are torn.

Besides its self-assembling and self-sealing properties, a lipid bilayer has other characteristics that make it an ideal structure for cell membranes. One of the most important of these is its fluidity, which, as we shall see, is crucial to many of its functions.

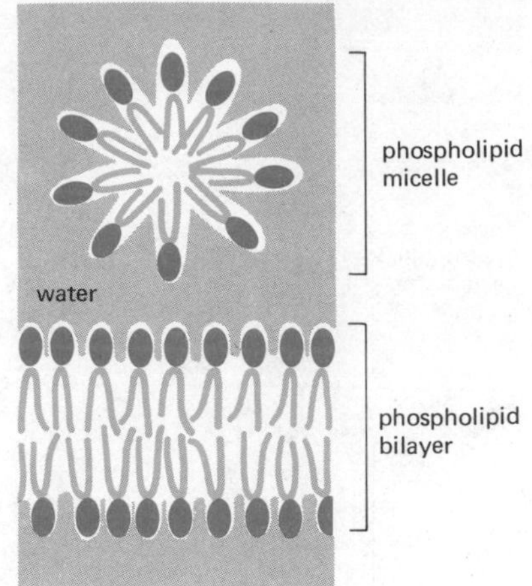

Figure 6–3 Schematic drawing of a phospholipid micelle and a phospholipid bilayer seen in cross-section. Phospholipid molecules spontaneously form such structures in water.

The Lipid Bilayer Is a Two-dimensional Fluid[3]

Surprisingly, it was only in the early 1970s that researchers first recognized that individual lipid molecules are able to diffuse freely within lipid bilayers. The initial demonstration came from studies of synthetic lipid bilayers. Two types of such bilayers have been very useful in experimental studies: (1) bilayers made in the form of spherical vesicles, called **liposomes,** which can vary in size from about 25 nm to 1 μm in diameter depending on how they are produced (Figure 6–4), and (2) planar bilayers, called **black membranes,** formed across a hole in a partition between two aqueous compartments (Figure 6–5).

A variety of techniques have been used to measure the motion of individual lipid molecules and of their different parts. For example, one can construct a lipid molecule whose polar head group carries a "spin-label," such as a nitroxyl group; this contains an unpaired electron whose spin creates a paramagnetic signal detectable by electron spin-resonance spectroscopy (ESR). The motion and orientation of such a spin-labeled lipid in a bilayer can thus be readily measured. Such studies show that lipid molecules in artificial bilayers very rarely migrate from the monolayer on one side to that on the other; this process, called "flip-flop," occurs less than once in two weeks for any individual lipid molecule (Figure 6–6). On the other hand, lipid molecules readily exchange places with their neighbors *within* a monolayer ($\sim 10^7$ times a second). This gives rise to a rapid lateral diffusion, with a diffusion coefficient (D) of about 10^{-8} cm^2 $second^{-1}$, which means that an average lipid molecule diffuses the length of a large bacterial cell (~ 2 μm) in about one second. In addition, these studies indicate that individual lipid molecules rotate very rapidly about their long axes and that their hydrocarbon chains are flexible, the greatest degree of flexion occurring near the center of the bilayer and the smallest adjacent to the polar head group (Figure 6–6).

Similar studies of labeled lipid molecules in isolated biological membranes, and in relatively simple whole cells such as mycoplasma, bacteria, and nonnucleated red blood cells (erythrocytes), have demonstrated that the be-

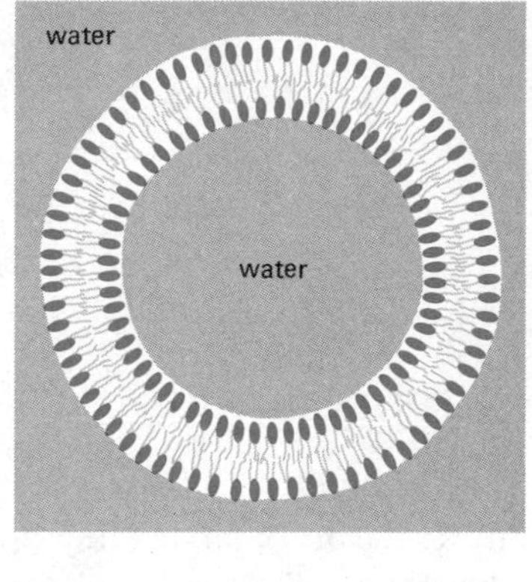

Figure 6–4 A schematic cross-sectional view of a small, spherical liposome. Liposomes are commonly used as model membranes in experimental studies.

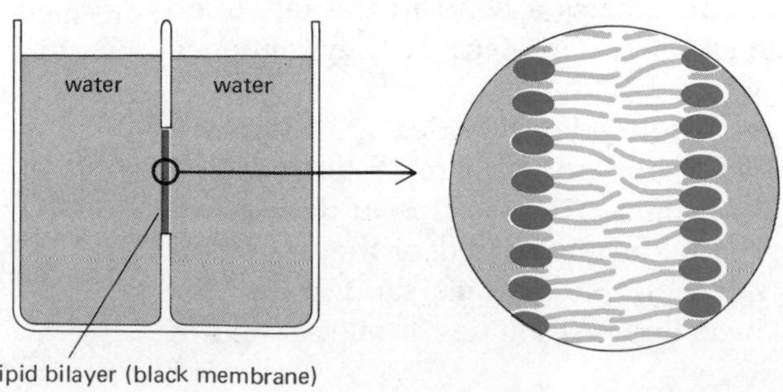

Figure 6–5 A schematic cross-sectional view of an artificial lipid bilayer, called a black membrane, formed across a small hole in a partition that separates two aqueous compartments. Black membranes are used to measure the permeability properties of artificial membranes.

havior of lipid molecules in cell membranes is generally the same as in synthetic bilayers: the lipid component of a biological membrane is a two-dimensional liquid in which the constituent molecules move rapidly but are largely confined to their own monolayer.

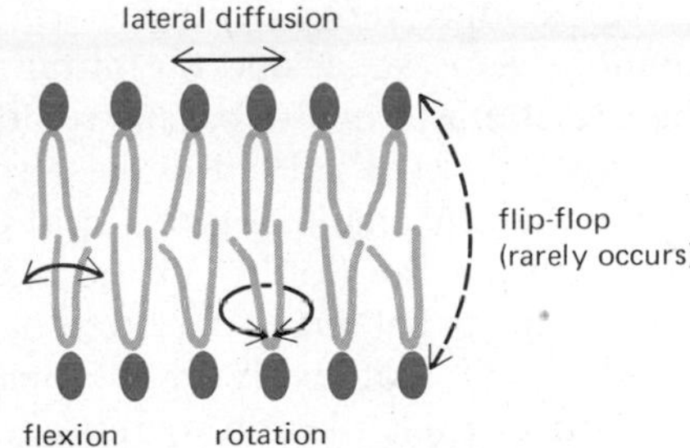

Figure 6–6 Different types of movement possible for phospholipid molecules in a lipid bilayer.

The Fluidity of a Lipid Bilayer Depends on Its Composition[4]

An artificial lipid bilayer made from a single type of phospholipid changes from a liquid state to a rigid crystalline (or gel) state at a sharp and characteristic freezing point. This change of state is called a *phase-transition,* and the temperature at which it occurs is lower (that is, the membrane becomes more difficult to freeze) if the hydrocarbon chains are short or have double bonds. A shorter chain length reduces the tendency of the hydrocarbon tails to interact with each other, and *cis*-double kinks produce kinks in the hydrocarbon chains that make it more difficult for them to pack together (Figure 6–7).

In artificial bilayers containing a mixture of phospholipids with varying degrees of saturation (and therefore different phase-transition points), *phase separations* can occur, with individual phospholipid molecules of the same type forming clusters of gel when their freezing points are reached. Since in biological membranes both saturated and unsaturated fatty acid chains are usually found in the same lipid molecule (that is, one chain is unsaturated while the other is not), they cannot separate in this way.

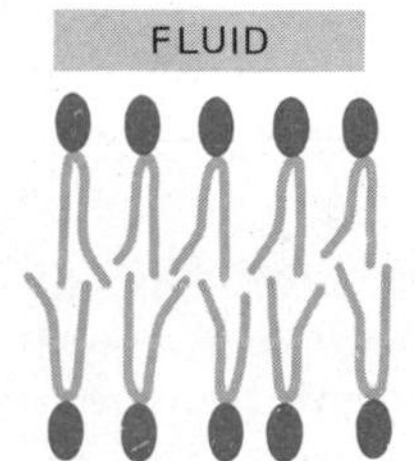

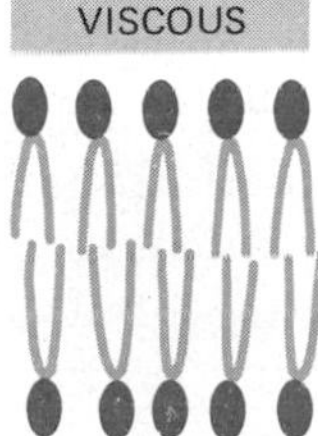

Figure 6–7 Double bonds in unsaturated hydrocarbon chains increase phospholipid bilayer fluidity by making it more difficult to pack the chains together.

Another determinant of membrane fluidity is **cholesterol.** Eucaryotic plasma membranes contain relatively large amounts of cholesterol, as much as one molecule for every phospholipid molecule. In addition to regulating fluidity, cholesterol is thought to enhance the mechanical stability of the bilayer. Cholesterol molecules orient themselves in the bilayer with their hydroxyl groups close to the polar head groups of the phospholipid molecules; their platelike steroid rings interact with—and partly immobilize—those regions of the hydrocarbon chains closest to the polar head groups, leaving the rest of the chain flexible (Figure 6–8). At the concentrations found in most eucaryotic plasma membranes, cholesterol also has the effect of preventing the hydrocarbon chains from coming together and crystallizing. In this way, cholesterol inhibits temperature-induced phase transitions and so prevents the drastic decrease in fluidity that would otherwise occur at low temperatures.

Figure 6–8 Cholesterol represented by a formula (A) and schematic drawing (B), and depicted interacting with two phospholipid molecules in a monolayer (C).

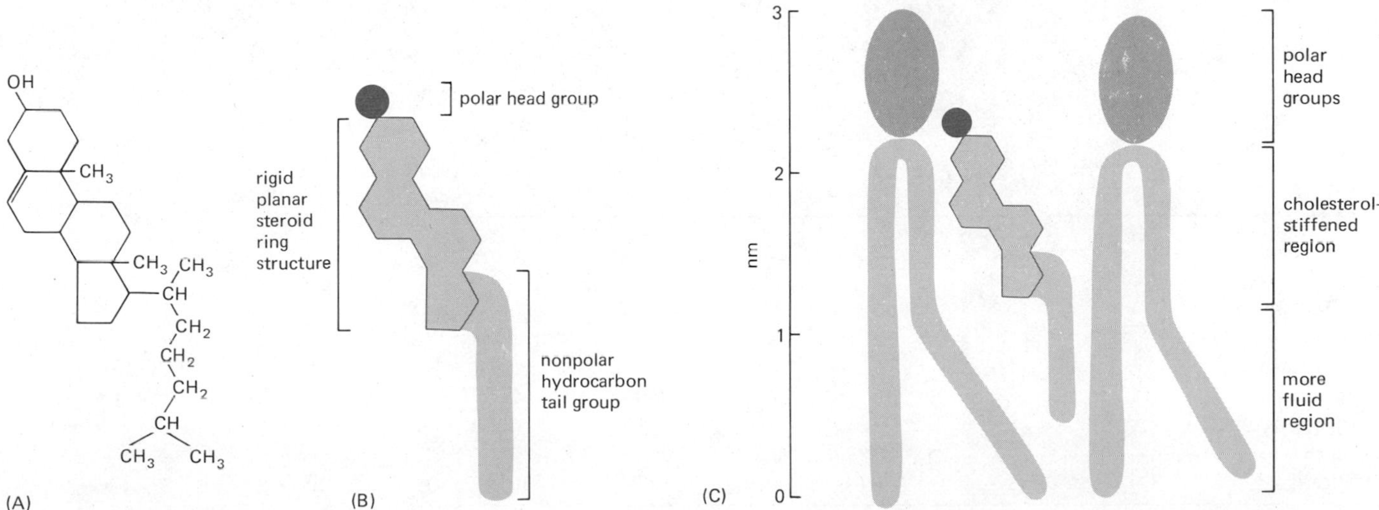

The importance of cholesterol in maintaining the mechanical stability of membranes is suggested by mutant animal cell lines that are unable to synthesize cholesterol. Such cells rapidly lyse unless cholesterol is added to the culture fluid. The added cholesterol is incorporated into the plasma membrane, thereby stabilizing the lipid bilayer and enabling the cells to survive.

The precise fluidity of the plasma membrane must be biologically important, since bacteria, yeast, and other poikilothermic organisms, whose temperatures fluctuate with that of their environment, change the fatty acid composition of their plasma membranes so as to maintain a relatively constant fluidity. It remains to be established which essential membrane processes depend most critically on the fluid state of the bilayer, although certain transport processes and membrane-enzyme activities are known to cease when the bilayer viscosity increases beyond a threshold level.

The Lipid Bilayer Serves as a Solvent for Membrane Proteins[5]

The lipid compositions of several different biological membranes are compared in Table 6–1. Unlike bacterial plasma membranes, which are often composed of one main type of phospholipid and contain no cholesterol, the plasma membranes of most eucaryotic cells contain not only large amounts of cholesterol but also a variety of phospholipids. For example, the plasma membrane of the human erythrocyte contains four major phospholipids—*phosphatidylcholine, sphingomyelin, phosphatidylserine,* and *phosphatidylethanolamine.* The structures of these molecules are shown in Figure 6–9; note that only phosphatidylserine carries a net negative charge, while the other three are electrically neutral at physiological pH.

One may wonder why the eucaryotic plasma membrane contains such a variety of phospholipids, with head groups that differ in size, shape, and charge. In seeking an answer, it may be helpful to think of the membrane lipids as constituting a two-dimensional solvent for the proteins in the membrane, just as water constitutes a three-dimensional solvent for proteins in an aqueous solution. It may be that some membrane proteins can function only in the presence of specific phospholipid head groups, just as many enzymes

Table 6–1 Approximate Lipid Compositions of Different Cell Membranes

	Percentage of Total Lipid by Weight					
Lipid	*Liver Plasma Membrane*	*Erythrocyte Plasma Membrane*	*Myelin*	*Mitochondrion (inner and outer membranes)*	*Endoplasmic Reticulum*	*E. coli*
Cholesterol	17	23	22	3	6	0
Phosphatidyl-ethanolamine	7	18	15	35	17	70
Phosphatidylserine	4	7	9	2	5	trace
Phosphatidyl-choline	24	17	10	39	40	0
Sphingomyelin	19	18	8	0	5	0
Glycolipids	7	3	28	trace	trace	0
Others	22	13	8	21	27	30

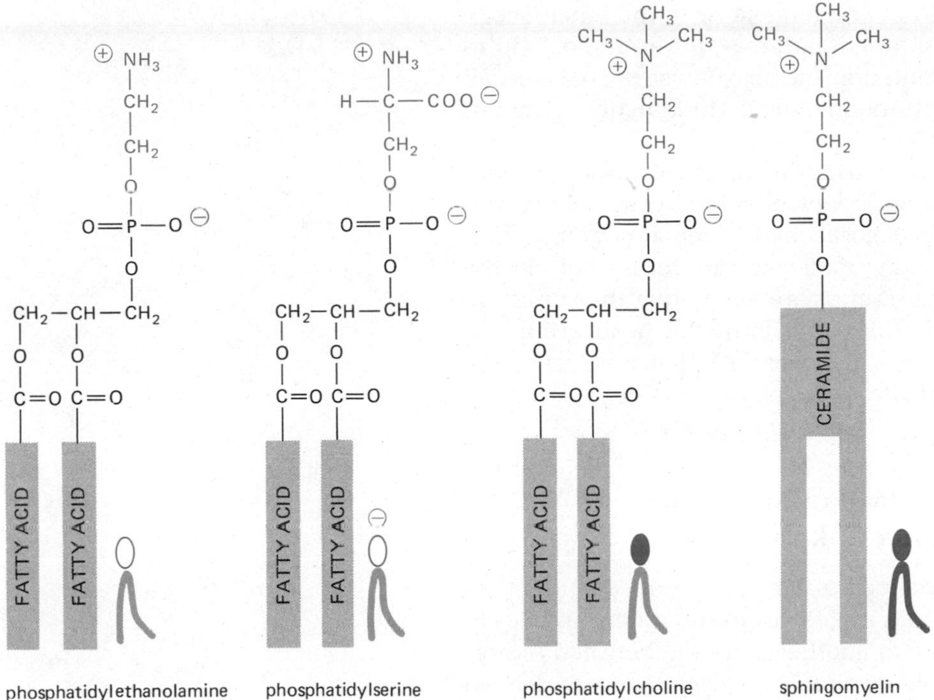

Figure 6–9 Formulas and symbols for the major phospholipids in human red blood cell membranes. Note that different head groups are represented by different symbols. The structure of ceramide is shown in Figure 6–11.

in aqueous solution require a particular ion for activity. Consistent with this view is the finding that in some systems where functional membrane proteins are inserted into artificial lipid bilayers, the proteins seem to require certain specific phospholipids to function optimally.

The Lipid Bilayer of the Plasma Membrane Is Asymmetrical[6]

The lipid compositions of the two halves of the lipid bilayer in those plasma membranes that have been analyzed are strikingly different. In the human red blood cell membrane, most of the lipid molecules that terminate in choline—$(CH_3)_3N^+CH_2CH_2OH$—are in the outer half of the lipid bilayer, while most of the phospholipids that contain a terminal primary amino group are in the inner half (Figure 6–10). Since the former are more saturated than the latter, the asymmetry in the distribution of the head groups is accompanied by an asymmetry in the distribution of hydrocarbon tails as well. In addition, the negatively charged phosphatidylserine is located in the inner half. Consequently, there is a significant difference in charge between the two monolayers.

Since lipid molecules do not spontaneously flip-flop between the monolayers at an appreciable rate, it is probable that the asymmetry is generated during biosynthesis of the bilayer in the endoplasmic reticulum, by enzymes

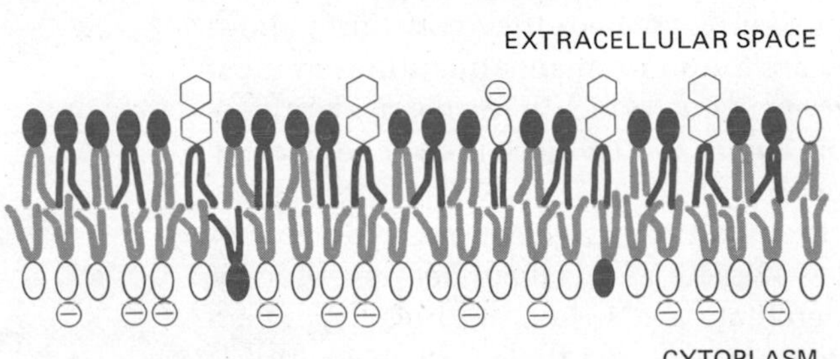

Figure 6–10 Schematic illustration depicting the asymmetrical distribution of phospholipids and glycolipids (see next page) in the lipid bilayer of human red blood cells. The symbols used for the phospholipids are those shown in Figure 6–9. Glycolipids are shown with hexagonal polar head groups. Cholesterol is not shown.

that transfer specific lipid molecules from one monolayer to the other (see p. 351). The lipid asymmetry must have a function. It may, for example, help to keep membrane proteins properly oriented in the bilayer; as we shall see, all membrane proteins are associated with the membrane in a highly asymmetrical fashion, a feature crucial to their function.

The lipid molecules that show the most striking and consistent asymmetry in distribution in the plasma membranes of animal cells belong to a class of lipids not yet considered—the oligosaccharide-containing lipid molecules called **glycolipids.** These intriguing molecules are found only in the outer half of the bilayer, and their sugar groups are exposed at the surface of the cell (Figure 6–10). The oligosaccharides that form the head groups of glycolipid molecules can be complex; this, together with their strategic position on the cell surface, suggests that they may play a part in intercellular communication, but this function has yet to be demonstrated.

Glycolipids Are Found on the Surface of All Plasma Membranes, But Their Function Is Unknown[7]

Glycolipids probably occur in all animal cell plasma membranes where they generally constitute about 5% of the lipid molecules in the outer monolayer. They vary remarkably from one species to another and even between tissues in the same species. In bacteria and plants almost all glycolipids are derived from *glycerol,* whereas in animal cells they are almost always derived from *sphingosine,* a long amino alcohol (Figure 6–11). The latter, called *glycosphingolipids,* have a general structure that is similar to that of the glycerol-based phospholipids, having a polar head group and two hydrophobic hydrocarbon chains (although one of the chains is derived from sphingosine and the other is a fatty acid chain). However, all glycolipid molecules are distinguished by the fact that the polar head group consists of one or more sugar residues.

Among the most widely distributed glycolipids in the plasma membranes of both eucaryotic and procaryotic cells are the **neutral glycolipids,** whose polar head groups consist of anywhere from 1 to 15 or more neutral sugars. Some neutral glycolipids are found only in certain mammals and usually only in certain tissues of those species. One notable example is *galactocerebroside,* one of the simplest glycolipids, which has only galactose as its polar head group (Figure 6–11). It is the main glycolipid in *myelin,* the multilayered membrane sheath that insulates nerves. Myelin is nothing more than many concentric layers of plasma membrane wound around a nerve fiber by a specialized myelinating cell (see p. 1032). A distinguishing feature of these myelinating cells is the large amount of galactocerebroside in their plasma membrane, where it constitutes almost 40% of the outer monolayer. Since it is not present in significant amounts in other membranes, galactocerebroside may play an important part in the membrane wrapping process that is unique to myelination.

The most complex of the glycolipids, the **gangliosides,** contain one or more sialic acid residues (also known as *N*-acetylneuraminic acid, or NANA), which gives them a net negative charge (Figure 6–12). Gangliosides are most abundant in the plasma membrane of neurons, where they constitute about 6% of the total lipid mass, though they are found in smaller quantities in most cell types. So far, more than 30 different gangliosides have been identified. Some common examples are shown in Figure 6–13, where the nomenclature used to describe them is also introduced.

There are only hints as to what the functions of glycolipids may be. For example, the ganglioside G_{M1} (Figure 6–13) acts as a cell-surface receptor for the bacterial toxin that causes the debilitating diarrhea of cholera; cholera toxin binds to and enters only those cells with G_{M1} on their surface, including sphingosine → ceramide → galactocerebroside

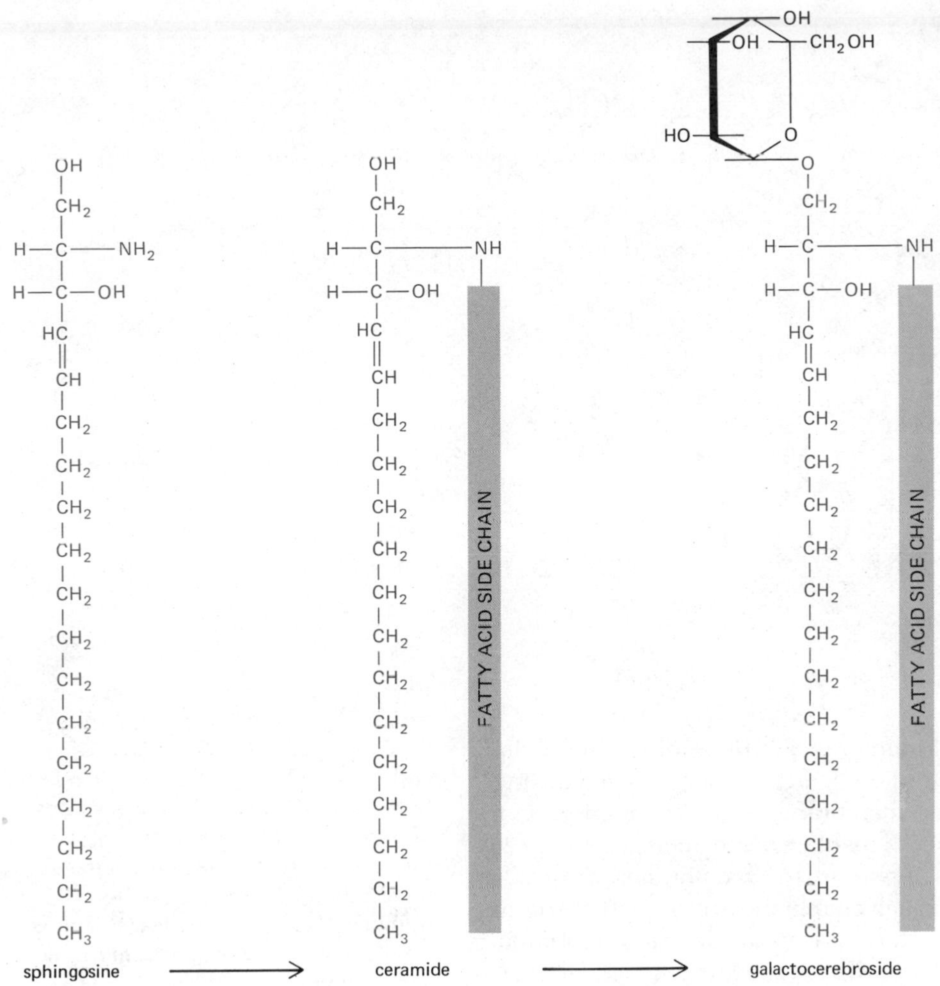

FATTY ACID SIDE CHAIN

Figure 6–11 Final steps in the synthesis of the simple glycosphingolipid, galactocerebroside.

intestinal epithelial cells (see p. 739). Although binding bacterial toxins cannot be the *normal* function of gangliosides, such observations suggest that they may also serve as receptors for normal signaling between cells. Since an increasing number of glycolipids can now be distinguished by specific antibodies, it should be possible to use such antibodies to examine the functions of these mysterious molecules; in principle, one can bind antibodies to one type of glycolipid molecule on the surface of a cell and test for inhibition or stimulation of specific cellular functions.

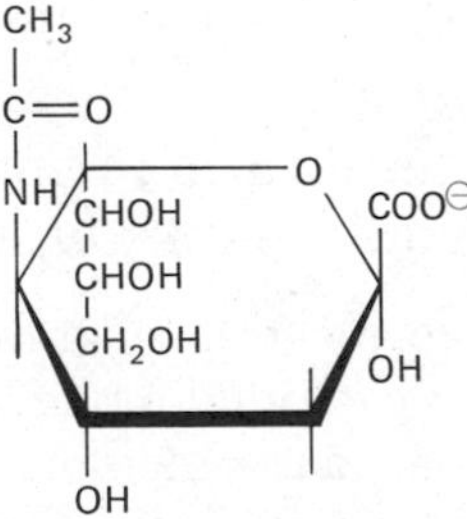

Figure 6–12 The structure of sialic acid (*N*-acetylneuraminic acid, or NANA).

Summary

All biological membranes consist of a continuous double layer of lipid molecules in which various membrane proteins are embedded. This lipid bilayer is fluid, with individual lipid molecules able to diffuse rapidly within their own monolayer. However, lipid molecules very rarely transfer spontaneously from one monolayer to the other. Membrane lipid molecules are amphipathic, and most of them spontaneously form bilayers when placed in water. For this reason, cellular lipid bilayers form by self-assembly and reseal if torn. There are three major classes of lipid molecules in the plasma membrane bilayer—phospholipids, cholesterol, and glycolipids—and the lipid compositions of the inner and outer monolayers are different. In addition, the different membranes of a single eucaryotic cell have distinct lipid compositions.

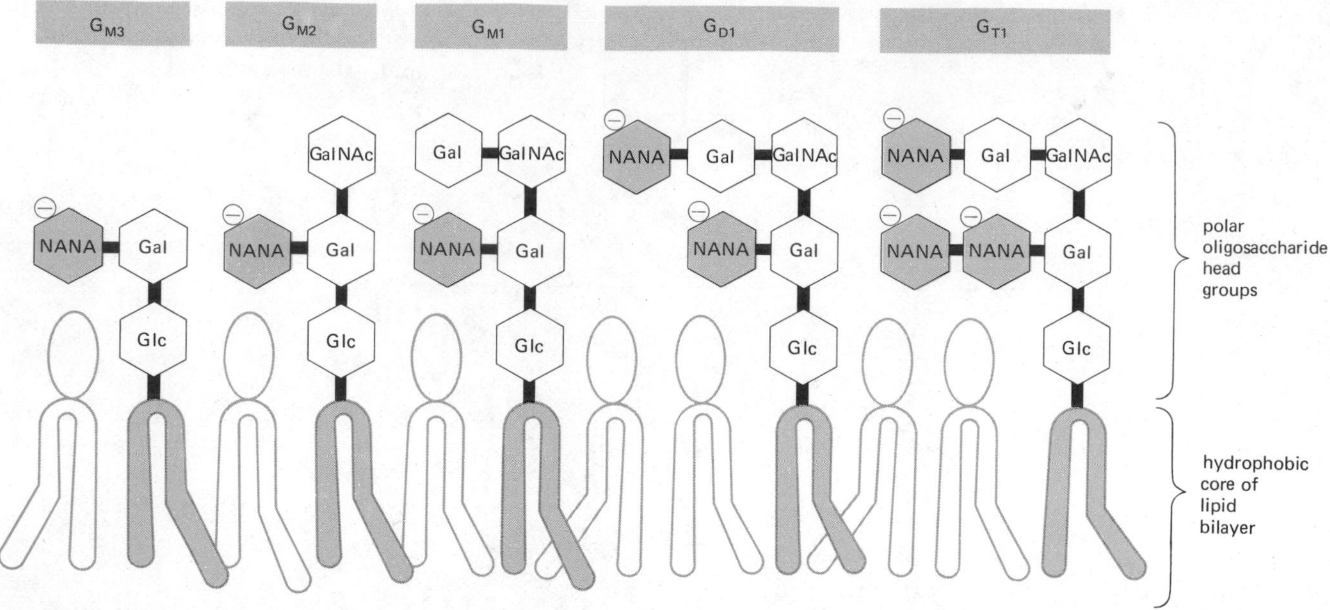

Figure 6–13 Some representative gangliosides with their standard designations. In G_{M1}, G_{M2}, G_{M3}, G_{D1}, and G_{T1}, the letters M, D, and T refer to the number of sialic acid residues (mono, di, and tri, respectively), while the number that follows the letter is determined by subtracting the number of uncharged sugar residues from 5. NANA = *N*-acetylneuraminic (sialic) acid; Gal = galactose; Glc = glucose; GalNAc = *N*-acetylgalactosamine. Gal, Glc, and GalNAc are all uncharged, while NANA carries a negative charge (see Figure 6–12).

Membrane Proteins

While the basic structure of biological membranes is determined by the lipid bilayer, their specific functions are carried out largely by proteins. Accordingly, the amounts and types of proteins in a membrane reflect its function: in the myelin membrane, which serves mainly to insulate nerve fibers, less than 25% of the membrane mass is protein, whereas in the membranes involved in energy transduction (such as the internal membranes of mitochondria and chloroplasts) approximately 75% is protein; the usual plasma membrane is somewhere in between, with about 50% of the mass being protein. Of course, the size of a lipid molecule is very small compared with that of a protein molecule, and so there are always many more lipid than protein molecules in membranes. For example, there is approximately one protein molecule for every 50 lipid molecules in a membrane that is 50% protein by mass.

Many Membrane Proteins Are Held in the Bilayer by Hydrophobic Interactions with Lipid Molecules[8]

Until quite recently scientists mistakenly believed that membrane proteins formed an extended monolayer on both surfaces of the lipid bilayer. It is now known that many membrane proteins are inserted directly into the bilayer itself (Figure 6–14). Like their lipid neighbors, these membrane proteins are usually amphipathic: they have hydrophobic regions that interact with the hydrophobic tails of the lipid molecules in the interior of the bilayer and hydrophilic regions that are exposed to water on one or, more usually, both sides of the membrane. The hydrophobicity of some membrane proteins is increased by the covalent attachment of one or more fatty acid chains that help anchor these proteins in the bilayer.

The ease with which membrane proteins can be removed from the membrane varies widely, some being released by gentle procedures (such as extraction by a salt solution) and others only after total disruption of the bilayer with detergents or organic solvents. These two extreme types are often referred to as *peripheral* and *integral* proteins, respectively. However, this operational classification should not be mistaken for a molecular description of how these proteins are associated with the bilayer, which usually is not known.

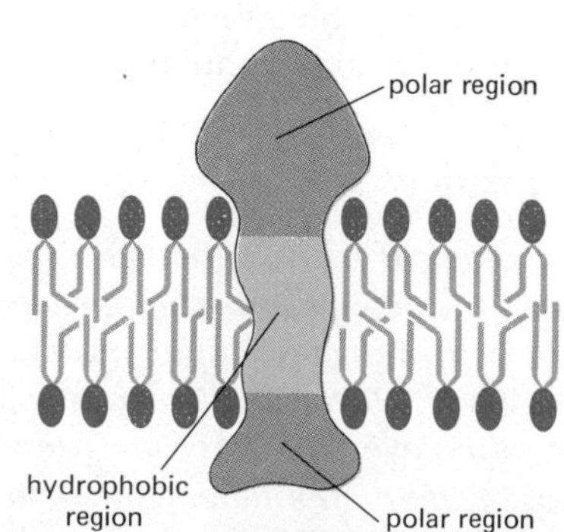

Figure 6–14 Schematic diagram of an amphipathic membrane protein held in the lipid bilayer by hydrophobic interactions with the hydrocarbon chains of the lipid molecules.

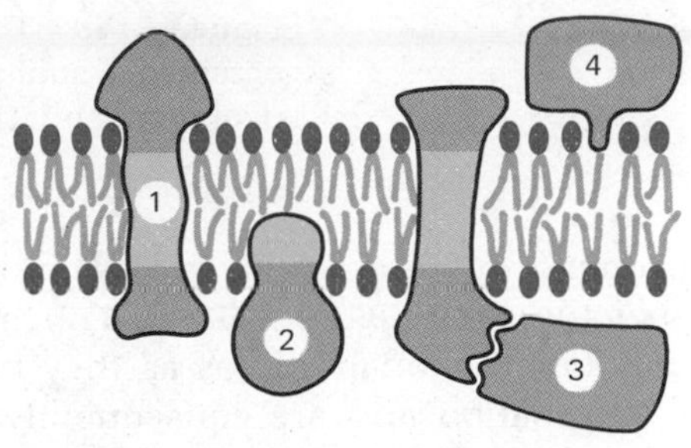

Figure 6–15 Schematic drawing of four ways in which membrane proteins may be associated with the lipid bilayer. While some proteins are known to extend across the bilayer (1) and some to be held by noncovalent interactions with other membrane proteins (3), it has not been definitively demonstrated that any have their polypeptide chain extending partway across the bilayer (2). Recently it has been shown that some membrane proteins have one or more covalently attached fatty acid chains that help anchor the protein in one monolayer or the other; while most of these are transmembrane proteins, others may not be (4).

Many proteins extend across the bilayer and are exposed to an aqueous environment on both sides of the membrane; these are called **transmembrane proteins.** Other proteins are exposed to water only on one side of the bilayer; some of these are anchored to the membrane by noncovalent interactions with transmembrane proteins, while others may be anchored by means of covalently attached fatty acid chains that extend into one monolayer or the other (Figure 6–15).

Because water is largely excluded from lipid bilayers, polypeptide chains in membranes are thought to be arranged mainly as α-helices or β-sheets. The exclusion of water means that the polar groups of a polypeptide chain within a membrane, which would otherwise form hydrogen bonds to water, have an enhanced tendency to form hydrogen bonds with each other instead. The hydrogen bonding between the peptide bonds is maximized if the polypeptide chain forms an α-helix or β-sheet as it passes through the bilayer. For the same reason, a polypeptide chain that enters the bilayer is likely to pass entirely through it before changing direction, since chain bending requires a loss of regular hydrogen-bonding interactions. As a result, very few membrane proteins are thought to extend only part way across the lipid bilayer.

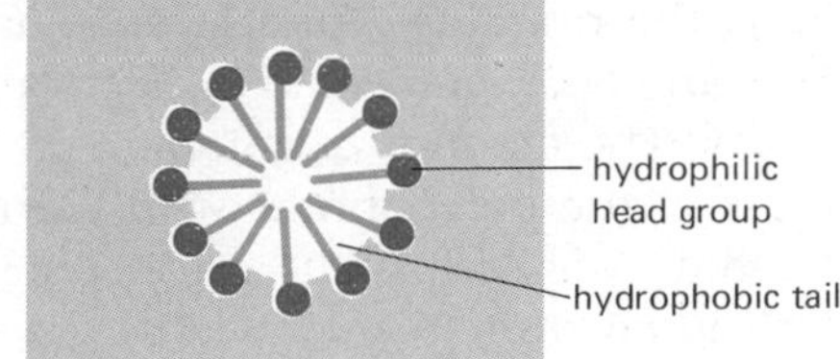

Figure 6–16 A detergent micelle in water, shown in cross-section. Detergent molecules are said to be amphipathic, since they have both polar and nonpolar ends.

In general, transmembrane proteins (and some other tightly bound membrane proteins) can be solubilized only by agents that disrupt hydrophobic associations and destroy the bilayer. The most useful among these for the membrane biochemist are **detergents,** small amphipathic molecules that tend to form micelles in water (Figure 6–16). When mixed with membranes, the hydrophobic ends of detergents bind to the hydrophobic regions on the exterior of membrane proteins, thereby displacing the lipid molecules. Since the other end of the detergent molecule is polar, this binding tends to bring the membrane proteins into solution as detergent-protein complexes (although some tightly bound lipid molecules also remain bound; see Figure 6–17). The polar ends of detergents can either be charged (ionic), as in the case of *sodium dodecyl sulfate (SDS),* or uncharged (nonionic), as in the case of the *Triton* detergents. The structures of these two commonly used detergents are illustrated in Figure 6–18.

When the detergent is removed, solubilized membrane proteins usually become highly insoluble and precipitate as a mixture of many different species in aqueous medium (Figure 6–19). Consequently, membrane proteins are much more difficult to purify and study than are water-soluble proteins. A further complication is that strong detergents (particularly ionic ones) often unfold

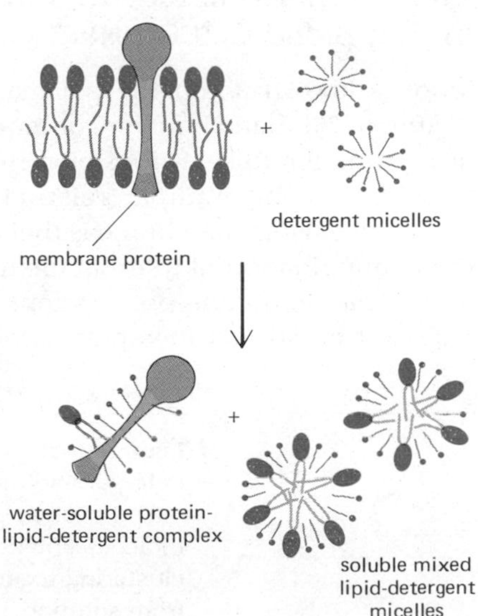

Figure 6–17 Solubilizing membrane proteins with detergent. The detergent disrupts the lipid bilayer and brings the proteins into solution as protein-lipid-detergent complexes. The phospholipids in the membrane are also solubilized by the detergent.

Figure 6–18 The structure of two commonly used detergents: sodium dodecyl sulfate (SDS), an anionic detergent, and Triton X-100, a nonionic detergent.

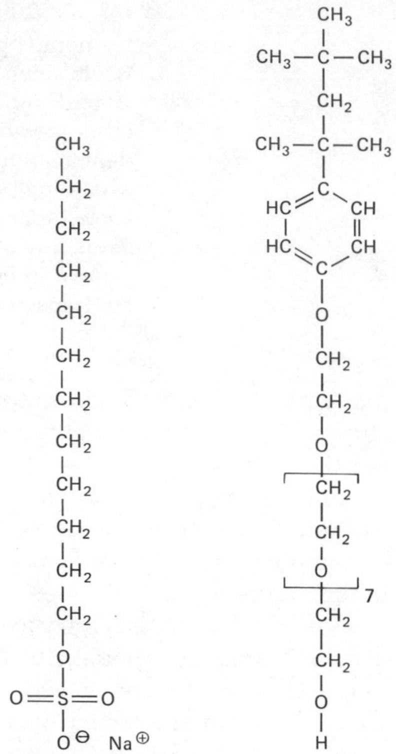

sodium dodecyl sulfate (SDS)

Triton X-100

proteins by complexing to the interior "hydrophobic core" of their native structure (see Figure 3–18, p. 112); such solubilized proteins are often of no use for functional studies as they have lost their normal three-dimensional conformation and are consequently inactive. Proteins unfolded in this way are said to be *denatured.* In the most favorable cases, however, membrane proteins can be kept soluble with a low concentration of a mild detergent and studied in an active, if not entirely normal, form.

The Use of SDS Polyacrylamide-Gel Electrophoresis Has Revolutionized the Study of Membrane Proteins

Detergent solubilization has been used with great success to study the major proteins in various membranes. Typically, the membranes are first heated to 100°C in a 1% solution of the strong detergent SDS (which disrupts all noncovalent protein-protein and protein-lipid interactions and completely unfolds the proteins), then layered on top of a polyacrylamide gel containing SDS, and lastly subjected to electrophoresis in a strong electric field for several hours. The amount of anionic SDS bound to protein is roughly equal to the mass of the polypeptide, with the result that the negative charge of the detergent overwhelms the intrinsic charge of the protein. Consequently, each individual protein migrates in the electric field at a rate determined by its molecular weight: the larger the protein, the more it is retarded by the complex meshwork of polyacrylamide molecules that constitute the gel and, therefore, the more slowly it moves. When stained with a dye that complexes with protein, such as Coomassie blue, the major proteins present can be seen as individual bands on the gel (see Figure 4–37, p. 174).

More recently, two-dimensional gel techniques (see p. 175) have been applied to studying membrane proteins; thus far, more than 50 different proteins from a single type of plasma membrane have been resolved by this technique. Since species present in small amounts are difficult to detect, there are undoubtedly many other proteins as well in a typical plasma membrane.

The Cytoplasmic Side of Membrane Proteins Can Be Studied in Red Blood Cell Ghosts[9]

More is known about the plasma membrane of the human red blood cell (Figure 6–20) than about any other eucaryotic membrane. There are a number of reasons for this: (1) Red blood cells are available in large numbers (from blood banks, for example) relatively uncontaminated by other cell types. (2) Since the plasma membrane is their only membrane, it can be isolated without the contamination of internal membranes, which presents a serious problem in plasma membrane preparations from other cell types. (3) It is easy to prepare red blood cell membranes, or "ghosts," by exposing the cells to a hy-

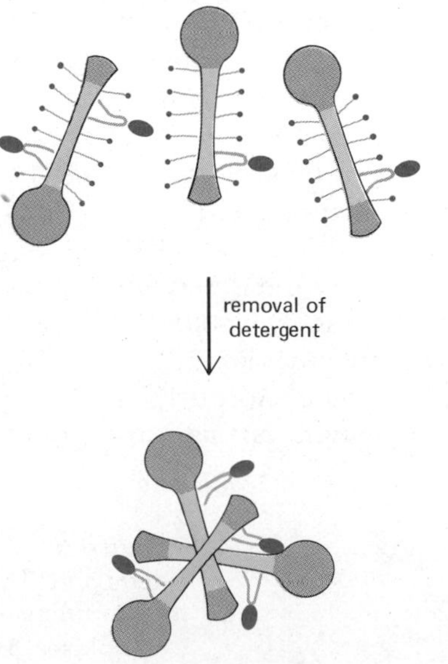

Figure 6–19 When the detergent is removed from detergent-solubilized membrane proteins, the relatively naked protein molecules (with only a few attached lipid molecules) tend to bury their hydrophobic regions by clustering together, forming large aggregates that precipitate from solution.

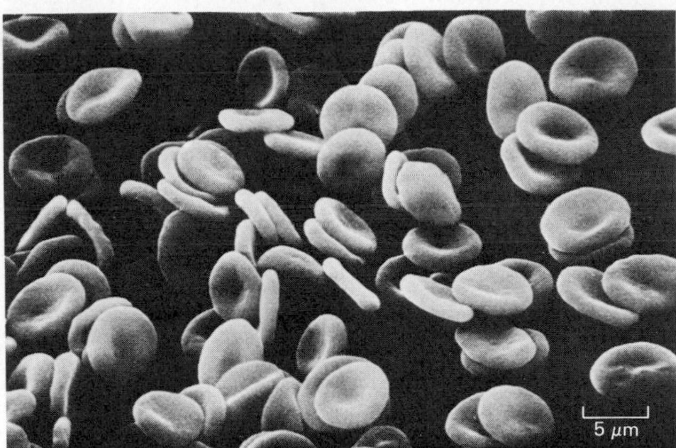

Figure 6–20 A scanning electron micrograph of human red blood cells. The cells have a biconcave shape and lack nuclei. (Courtesy of Bernadette Chailley.)

potonic salt solution. Because the solution has a lower salt concentration than the cell interior, water flows into the red cells, causing them to swell and burst (lyse) and release their hemoglobin (the only major nonmembrane protein). (4) Membrane ghosts either can be studied while they are still leaky (in which case reagents can interact with molecules on both faces of the membrane) or can be allowed to reseal themselves so that reagents can interact only with the external face. Moreover, since sealed inside-out vesicles can also be prepared from red cell ghosts (Figure 6–21), the external side and internal (cytoplasmic) side of the membrane can be studied separately. It was the use of sealed and unsealed red cell ghosts that first made it possible to demonstrate that some membrane proteins extend all the way through the lipid bilayer and that the lipid bilayer is asymmetrical.

The "sidedness" of a membrane protein can be determined in two ways. One way is to use a covalent labeling reagent (for example, one that is radioactive or fluorescent) that cannot penetrate membranes and therefore attaches covalently to specific groups only on the exposed side of the membrane (see below). The membranes are then solubilized, the proteins separated by polyacrylamide-gel electrophoresis, and the labeled proteins detected either by their radioactivity in autoradiographs of the gels or by their fluorescence in ultraviolet light. By such *vectorial labeling* it is possible to determine how a particular protein (a band on a gel) is oriented in the membrane: if it is labeled from both the external side (when intact cells or sealed ghosts are labeled) and the internal (cytoplasmic) side (when sealed inside-out vesicles are labeled), then it must be a transmembrane protein. The alternative approach is to expose either the external or cytoplasmic surface to membrane-impermeable proteolytic enzymes: if a protein is partially digested from both surfaces, it must be a transmembrane protein.

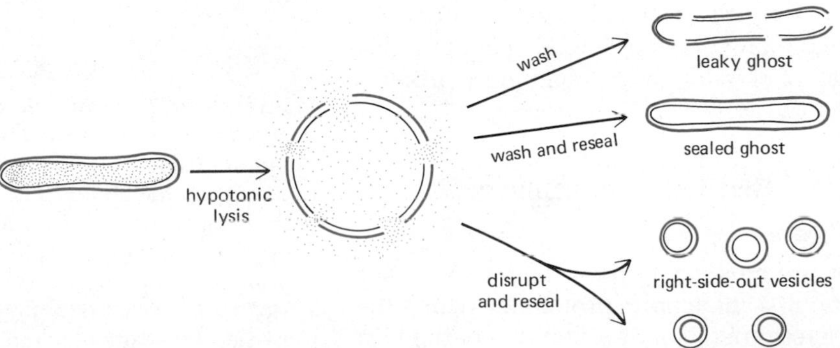

Figure 6–21 The preparation of sealed and unsealed red blood cell ghosts and of right-side-out and inside-out vesicles. The orientation of the membrane in the vesicles is determined by the ionic conditions used during the disruption procedure.

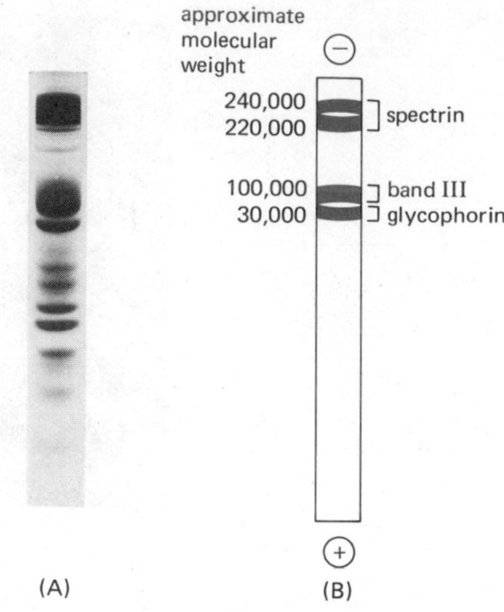

Figure 6–22 SDS polyacrylamide-gel electrophoresis pattern of the proteins in the human red blood cell membrane stained with Coomassie blue (A). The position of the three major proteins in the gel is indicated in the drawing in (B); the other bands in the gel, the less abundant membrane proteins, are omitted from the drawing. The large amount of carbohydrate in glycophorin molecules slows their migration so that they run almost as slowly as the much larger band III molecules. (Courtesy of Vincent Marchesi.)

When the plasma membrane proteins of the human red blood cell are studied by one-dimensional SDS polyacrylamide-gel electrophoresis, approximately 15 major protein bands are detected, varying in molecular weight from 15,000 to 250,000. Three of these proteins—*spectrin, glycophorin,* and *band III*—account for more than 60% (by weight) of the total membrane protein (Figure 6–22). Each of these three proteins is arranged in the membrane in a different manner. We shall, therefore, take them as examples of three principal known ways that proteins are associated with membranes.

Spectrin Is Loosely Associated with the Cytoplasmic Side of the Red Blood Cell Membrane[10]

Most of the protein molecules in the human red blood cell membrane are exposed to an aqueous environment only on the inner (cytoplasmic) side of the membrane, and most of these can be released from the membrane in very low ionic strength solutions. The most abundant such protein is **spectrin,** a long fibrous molecule that constitutes about 30% of the membrane protein mass, there being about 3×10^5 copies per cell (Figure 6–23). Spectrin is composed of a complex of two very large polypeptide chains (about 220,000 and 240,000 daltons each), arranged in a filamentous meshwork on the cytoplasmic surface of the membrane (Figure 6–24). There is evidence that actin and at least one other protein also participate in the formation of the meshwork, which is thought to be involved in maintaining the biconcave shape of red blood cells, while at the same time allowing them to deform as necessary to pass through narrow capillaries. Spectrin interacts indirectly, via a protein called *ankyrin,* with the cytoplasmic end of band III molecules (see below); this greatly reduces the rate of diffusion of these band III molecules in the plane of the membrane.

Thus, spectrin might be considered a component of the red blood cell cytoskeleton rather than of the plasma membrane; and the current interest in the protein stems largely from the hope that spectrin may tell us something about how other filamentous proteins present in nucleated cells interact with plasma membrane proteins. (Although spectrin itself appears to be confined to red blood cells, related proteins have recently been discovered in other cell types.) As we shall see later, intracellular filamentous proteins seem to be able to control the distribution and mobility of certain plasma membrane proteins in nucleated cells.

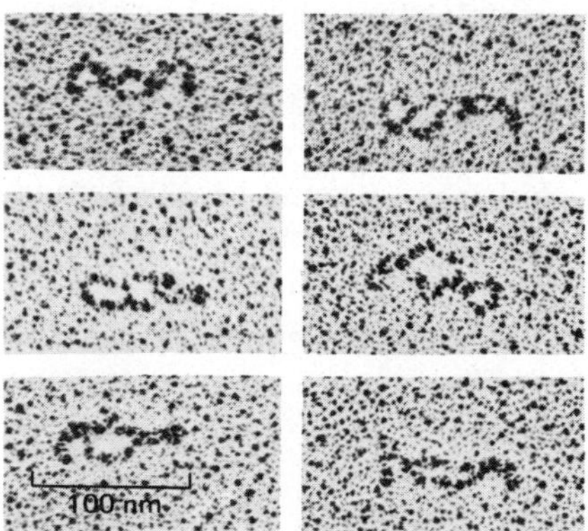

Figure 6–23 Electron micrographs of spectrin molecules shadowed with platinum. Each molecule consists of two polypeptide chains attached at both ends. (Courtesy of David M. Shotton. With permission from D. M. Shotton, B. E. Burke, and D. Branton, *J. Mol. Biol.* 131:303–329, 1979. Copyright by Academic Press Inc. [London] Ltd.)

Glycophorin Extends Through the Red Cell Lipid Bilayer as a Single α-Helix[11]

Glycophorin is one of the two major proteins exposed on the outer surface of the human red blood cell. It was the first membrane protein for which the complete amino acid sequence was determined, and it is still one of the best characterized eucaryotic plasma membrane proteins. Glycophorin is a trans-

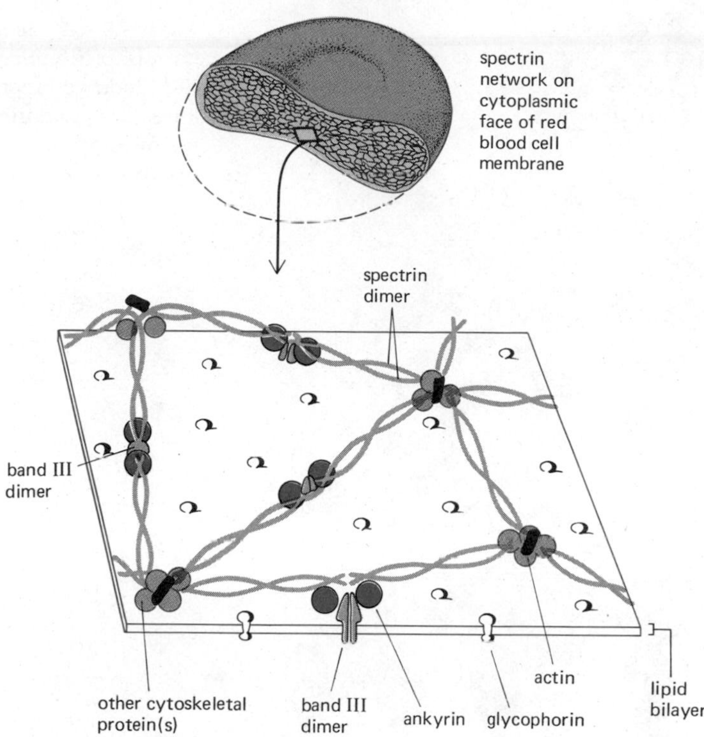

Figure 6–24 Schematic drawing of the probable organization of spectrin molecules on the internal (cytoplasmic) surface of the human red blood cell membrane. The two spectrin polypeptide chains are aligned in parallel and are variably coiled around each other to form flexible dimers (see Figure 6–23). The spectrin dimers join head to head to form tetramers that are linked together into a network by short actin filaments and other proteins on the cytoplasmic surface of the membrane. Spectrin molecules also interact indirectly with band III proteins via another protein called ankyrin.

membrane glycoprotein, 131 amino acids long, with most of its mass on the external surface of the membrane, where its hydrophilic amino-terminal end is located. This part of the protein carries all of the carbohydrate—about 100 sugar residues on 16 separate oligosaccharide side chains—and accounts for 60% of the molecule's mass. In fact, the great majority of the total surface carbohydrate (including more than 90% of the sialic acid and, therefore, most of the negative charge of the cell surface) is carried by glycophorin molecules. Glycophorin's hydrophilic carboxyl-terminal tail is exposed to the cytoplasm, while a hydrophobic α-helical segment about 20 amino acids long spans the nonpolar bilayer (Figure 6–25).

Like spectrin, glycophorin has been found only in red blood cells. Despite there being more than 6×10^5 glycophorin molecules per cell, its function remains unknown. Indeed, individuals whose red cells lack this protein appear to be perfectly healthy.

Band III of the Human Red Blood Cell Membrane Is a Transport Protein[12]

Unlike glycophorin, band III is known to play an important part in the specialized function of the cell. It is called **band III** because of its position relative to the other membrane proteins after electrophoresis in SDS polyacrylamide gels. Like glycophorin, band III is a transmembrane protein, but it traverses the membrane in a more folded or globular conformation, with the polypeptide chain probably extending across the bilayer several times (Figure 6–26). It has a molecular weight of ~100,000 (about 800 amino acids) and the little carbohydrate that it contains is exposed on the external surface. It appears to be a dimer, there being about 5×10^5 dimers per cell.

The main function of red blood cells is to carry O_2 from the lungs to the tissues and CO_2 from the tissues to the lungs. There is good evidence that band III is instrumental in this exchange. Red cells dispose of CO_2 by exchanging HCO_3^- for Cl^- as the cells move through the lungs. There is a special

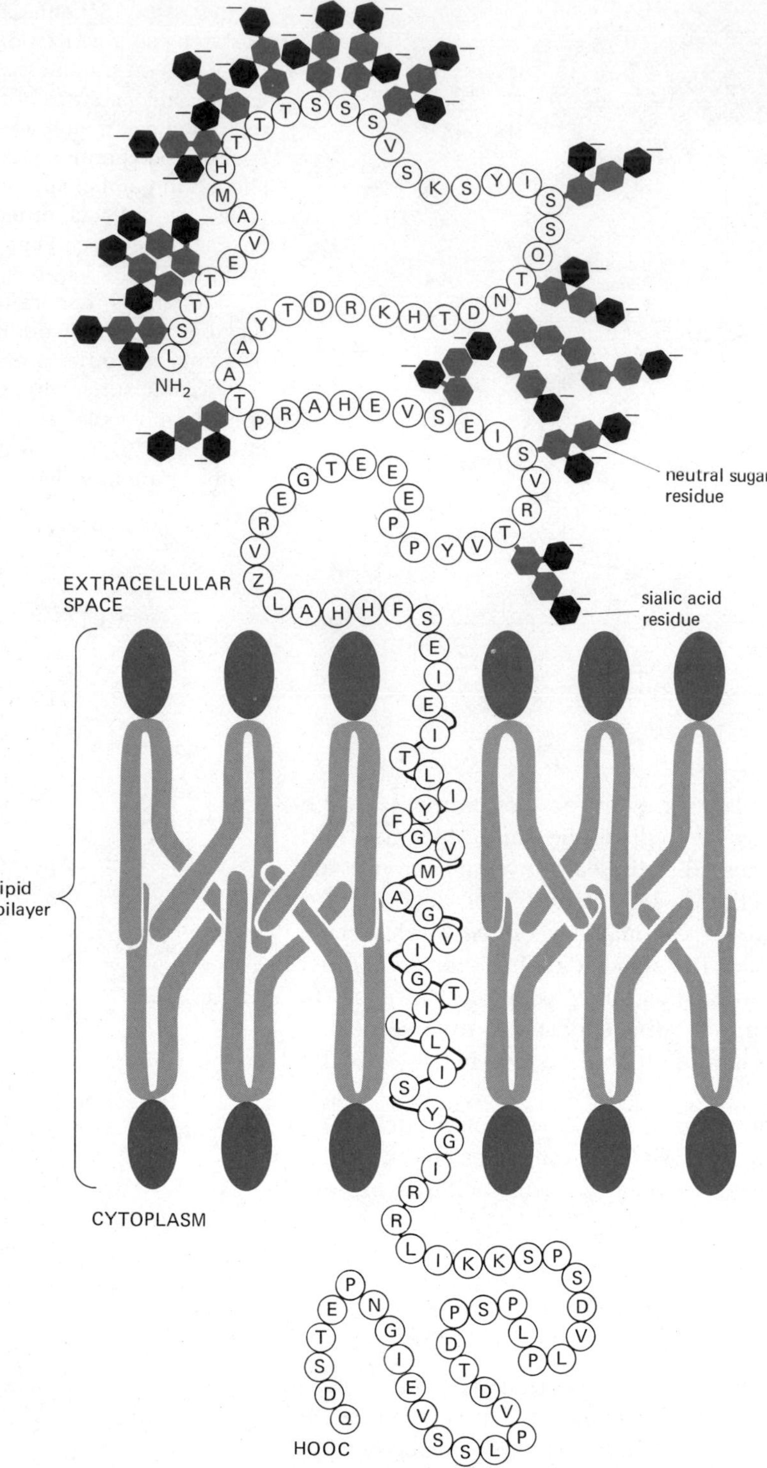

Figure 6–25 Schematic drawing of a glycophorin molecule in the human red blood cell membrane. The single-letter designations for amino acid residues are those given in Panel G, pp. 58–59.

anion channel in the membrane through which the exchange takes place, and it can be blocked by a specific inhibitor that binds to the protein that forms the channel. By modifying the inhibitor so that it radioactively labels the protein to which it binds, it has been possible to identify the protein of the anion channel in the red blood cell membrane as band III.

In a general way, it is not difficult to imagine how a transmembrane protein such as band III, with much of its mass in the lipid bilayer, could

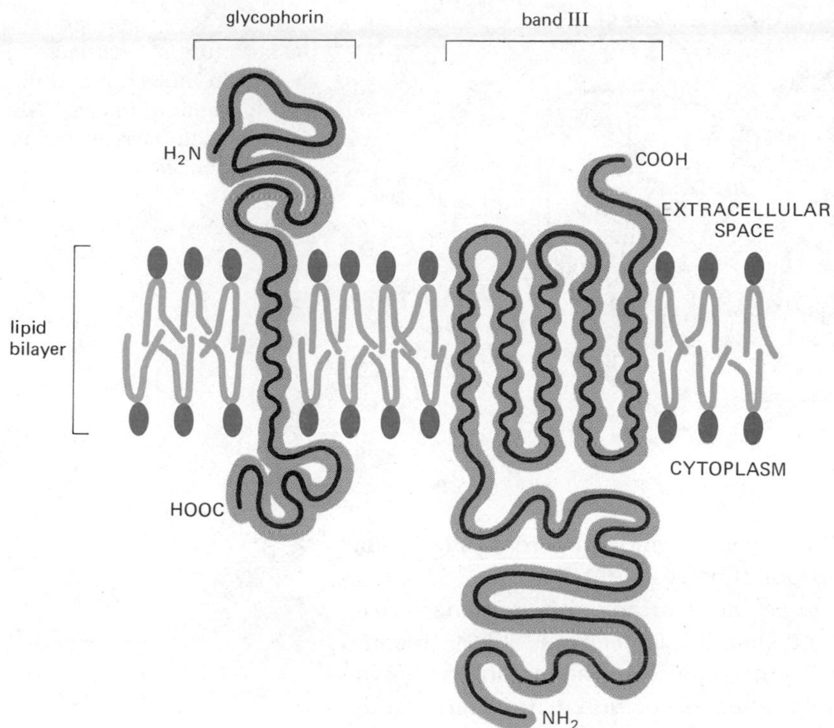

Figure 6–26 Schematic drawing of a possible arrangement of band III in the human red blood cell membrane; glycophorin is shown for comparison. While it is unknown exactly how band III is arranged in the membrane, there is evidence that it crosses the bilayer several times. Note that, while the amino terminus of glycophorin is outside the cell, the amino terminus of band III is thought to be inside.

mediate the passive transport of polar molecules across the nonpolar bilayer. For example, band III (or its dimer) could create a hydrophilic channel of the right size and charge to allow Cl^- and HCO_3^- to pass down their respective concentration gradients across the bilayer. On the other hand, it is difficult to see how a molecule such as glycophorin, which spans the bilayer as a simple α-helix, could mediate this type of transport on its own. Accordingly, there are good reasons for believing that proteins directly involved in the active or passive transport of polar molecules across membranes are more like band III than glycophorin in the way that they are associated with the lipid bilayer. An understanding of how such molecules work requires precise information about their three-dimensional structure in the bilayer. This information is not yet available for band III. The only membrane transport protein for which such detail is known is a protein called *bacteriorhodopsin,* which is found in the plasma membrane of certain bacteria where it serves as a light-activated proton (H^+) pump. Because of its likely relevance to membrane transport proteins of higher organisms, we will make a brief digression here to discuss what is known about the structure and function of this bacterial membrane protein.

Bacteriorhodopsin Is a Proton Pump That Traverses the Bilayer as Seven α-Helices[13]

The "purple membrane" of the bacterium *Halobacterium halobium* is a specialized patch in the plasma membrane (Figure 6–27). It contains a single species of protein molecule, **bacteriorhodopsin,** composed of 248 amino acids. Each molecule contains a single light-absorbing prosthetic group, or chromophore (called *retinal*), that is identical to the chromophore found in rhodopsin of the vertebrate retinal rod cell (see p. 1063). Retinal is covalently linked to a lysine side chain of the protein; when activated by a single photon of light, the excited chromophore causes a conformational change in the protein that results in the transfer of two H^+ from the inside to the outside of

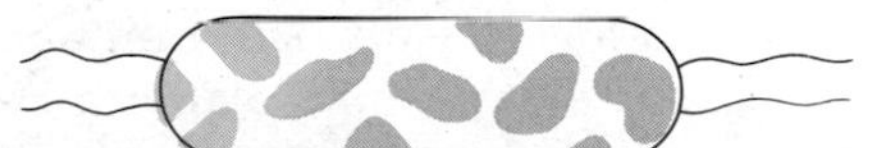

Figure 6–27 Schematic drawing of the bacterium *Halobacterium halobium* showing the patches of purple membrane that contain bacteriorhodopsin molecules.

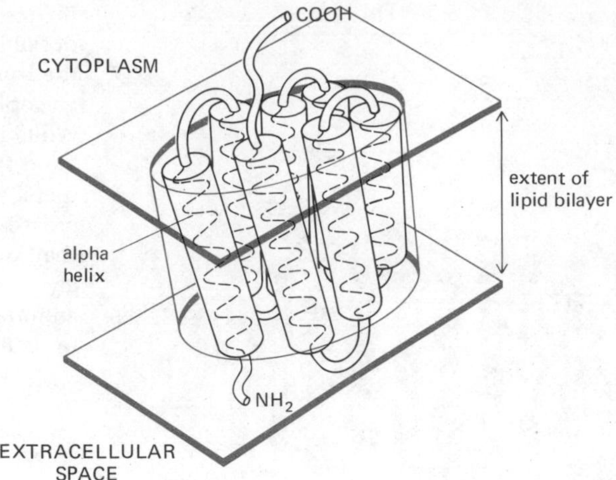

Figure 6–28 Schematic drawing of the structure of one bacteriorhodopsin molecule and its relationship to the lipid bilayer. The polypeptide chain crosses the bilayer as seven α-helices.

the cell. This transfer establishes a H^+ and voltage gradient across the plasma membrane, which in turn drives the production of ATP.

Because the bacteriorhodopsin molecule is arranged in a planar crystalline lattice (like a two-dimensional crystal), it has been possible to reconstruct its three-dimensional structure and orientation in the membrane down to a resolution of 0.7 nm by the combination of low-intensity electron microscopy and sophisticated low-angle electron diffraction analysis. The latter procedure is analogous to the study of three-dimensional crystals of soluble proteins by x-ray diffraction analysis (although one viral membrane glycoprotein has been studied by x-ray diffraction [see Figure 7–30, p. 346], so far, suitable three-dimensional crystals of cell membrane proteins have not been obtainable). As illustrated in Figure 6–28, these studies have shown that each bacteriorhodopsin molecule is folded into seven closely packed α-helices (each containing about 30 amino acids) that pass roughly at right angles through the lipid bilayer. Since recent studies suggest that these α-helices do not enclose a large aqueous channel, it seems likely that the protons are passed by the chromophore along a relay system set up by the side chains of the α-helices, although the molecular details are unknown. Since only protons (and electrons) could be transported in this way, bacteriorhodopsin is unlikely to provide a general model for the transport of other ions across membranes.

Membrane Transport Proteins Can Be Visualized by Freeze-Fracture Electron Microscopy[14]

Freeze-fracture electron microscopy is the only method available for looking at the hydrophobic interior of membranes. In this procedure, cells are frozen in a block of ice at the temperature of liquid nitrogen (−196°C) and the ice is then fractured. The fracture plane tends to pass through the hydrophobic middle of any lipid bilayer structure, including all biological membranes, thereby separating the bilayer into its two monolayers. The exposed *fracture faces* are then shadowed with platinum and carbon, the organic material is digested away, and the resulting platinum replica is examined with an electron microscope. As illustrated in Figure 6–29, two different fracture faces are exposed and replicated in this technique—the face representing the hydrophobic interior of the cytoplasmic (or protoplasmic) half of the bilayer (called the **P face**) and the face representing the hydrophobic interior of the external half of the bilayer (called the **E face**).

Human red blood cell membranes prepared in this way are studded with small bumps, called **intramembrane particles,** on both the P and E

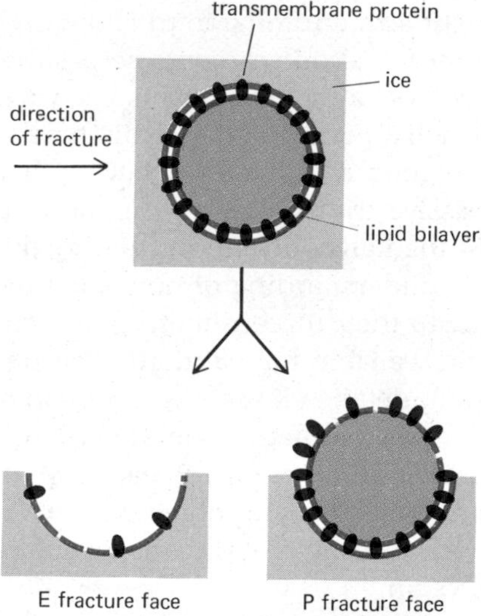

Figure 6–29 How freeze-fracture electron microscopy provides images of the hydrophobic interior of the cytoplasmic (or protoplasmic) half of the bilayer (called the P face) and the external half of the bilayer (called the E face). After the fracturing process shown here, the exposed fracture faces are shadowed with platinum and carbon, the organic material is digested away, and the resulting platinum replica is examined in the electron microscope.

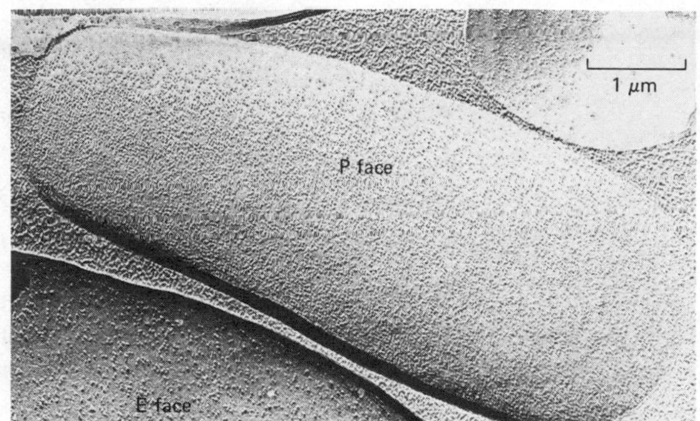

Figure 6–30 Freeze-fracture electron micrograph of human red blood cells. Note that the density of intramembrane particles on the protoplasmic (P) face is higher than on the external (E) face. (Courtesy of L. Engstrom and D. Branton.)

fracture faces. The particles are relatively homogeneous in size (7.5 nm in diameter), randomly distributed, and more concentrated on the P face than on the E face (Figure 6–30). Since protein-free artificial lipid bilayers generally lack intramembrane particles and appear as smooth surfaces when fractured, it is likely that these particles in biological membranes consist primarily of protein. In order to appear as a particle in freeze-fracture electron micrographs, a polypeptide, either by itself or as part of a larger aggregate, must traverse the membrane and have sufficient bulk in the interior of the bilayer to "cast a shadow." Although it is still uncertain which proteins constitute the intramembrane particles in human red cell membranes, it is likely that they are principally band III: the number of band III dimers (~500,000/cell) is approximately the same as the number of particles seen; and when lipid bilayers are artificially reconstituted with purified band III protein molecules, typical 7.5 nm intramembrane particles are observed when the bilayers are fractured.

What happens to transmembrane proteins during freeze-fracture? As the bilayer splits into two monolayers, either the inside or outside half of such a protein must pull out of the frozen monolayer with which it is associated. Alternately, the polypeptide chain could split (if covalent bonds are broken) where it crosses the bilayer (in which case the protein would not appear as an intramembrane particle). Since band III molecules are seen as intramembrane particles, it seems that they usually do not split during freeze-fracturing. Presumably they remain intact because a significant proportion of their mass lies in the bilayer. Moreover, since band III molecules have much more of their polypeptide chain extending from the cytoplasmic surface than from the outside surface of the bilayer, they would be expected to remain more often with the inner half of the frozen bilayer during fracturing than with the outer half, which would explain the higher frequency of intramembrane particles associated with the inner (P) half of the red cell membrane. On the other hand, a protein such as glycophorin could either break or pull its small carboxyl-terminal tail out of the frozen inner monolayer during the fracture process. In either case, it probably would lack sufficient mass to appear as an intramembrane particle (Figure 6–31).

Recently it has become possible to reconstruct functioning membrane transport systems in synthetic bilayers with a number of isolated transport proteins, including band III. So far, every such transport protein studied can be visualized as a particle in freeze-fracture electron microscopy. Since it is mainly such transport proteins that have sufficient mass in the bilayer to be resolved from the phospholipid background of the bilayer, it is likely that most intramembrane particles represent proteins that have some transport function.

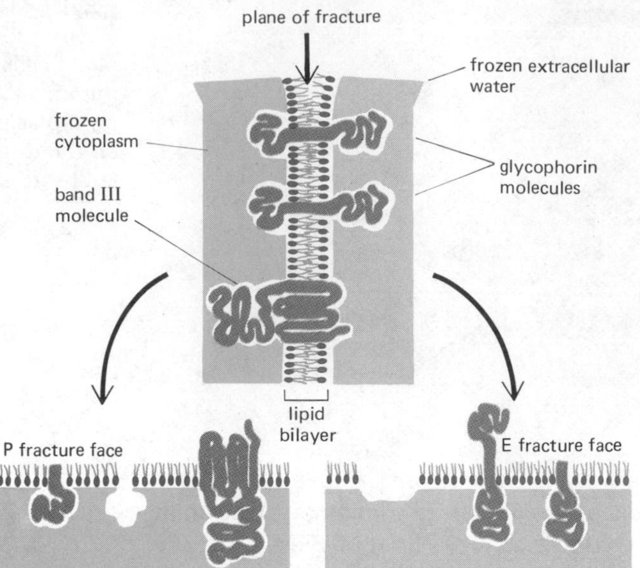

Figure 6–31 Schematic diagram showing the possible fates of glycophorin and band III molecules in the human red cell membrane during freeze-fracture. Band III molecules usually remain with the inner (P) fracture face and have sufficient mass above the fracture plane to be seen as intramembrane particles. Glycophorin molecules probably either break or remain with the outer (E) fracture face and, in either case, are probably not seen as intramembrane particles.

Vectorial Labeling Reagents Can Be Used to Study Some Plasma Membrane Proteins of Nucleated Cells[15]

Unlike the human red blood cell, nucleated cells have a variety of internal membranes, including those of the nucleus, endoplasmic reticulum, Golgi apparatus, mitochondria, lysosomes, and various vesicles. The plasma membrane often contains less than 10% of the total membrane area of these cells. For this reason, such plasma membranes can be studied directly only after they have been separated from the internal membranes. But separation is very difficult to achieve, and nearly all plasma membrane preparations are impure. The alternative is to use the membrane-impermeable **vectorial labeling** reagents discussed above, which label, and thereby identify, only those proteins exposed at the cell surface. While many plasma membrane proteins are labeled by this procedure, proteins dissolved in the cytoplasm or associated with internal membranes are not. A reagent commonly used in vectorial labeling is the enzyme *lactoperoxidase,* which, in the presence of hydrogen peroxide (H_2O_2) and radioactive iodide ($^{125}I^-$), oxidizes the iodide to a more reactive, enzyme-bound form. This activated species iodinates tyrosine and histidine residues of other proteins. However, since the active form of iodine is not released from the lactoperoxidase, and since the large enzyme molecule itself cannot penetrate the membrane, only proteins exposed on the cell surface become iodinated (and thereby labeled) when intact cells are treated (Figure 6–32). A disadvantage of the procedure is that it identifies only a subclass of the total plasma membrane proteins—those with tyrosine or histidine residues exposed on the external side of the bilayer. It is very much more difficult to label selectively the cytoplasmic surface of the plasma membrane of nucleated cells. A current approach to this problem involves methods for obtaining nonleaky, inside-out vesicles from isolated plasma membranes. The impurity of such preparations, however, remains a problem.

It is clear from vectorial labeling studies that the plasma membrane of a typical nucleated cell generally contains many more different proteins than the human red blood cell membrane. For example, more than 50 unique species can be detected by lactoperoxidase-catalyzed iodination of the outside of typical nucleated cells in culture. Thus far, little is known about the dis-

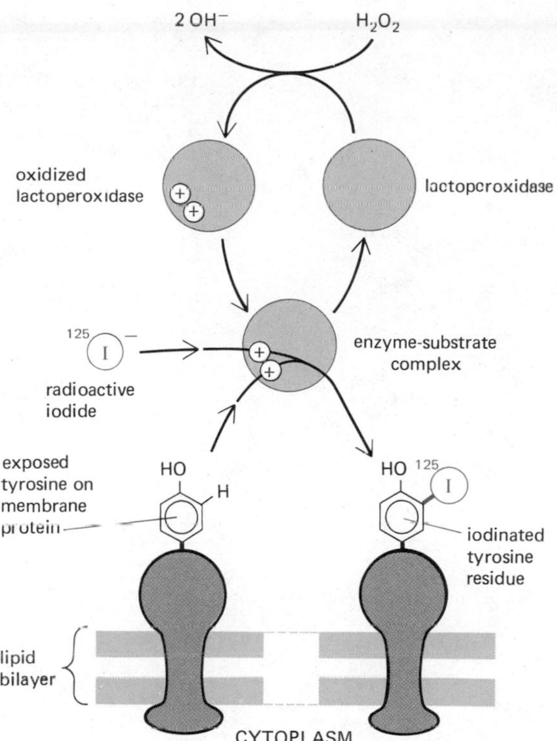

Figure 6–32 Selective labeling of proteins on the outside of the plasma membrane by lactoperoxidase-catalyzed iodination. As indicated, all reactions involving iodine are confined to the enzyme surface. The enzyme contains an iron atom at its active site that generates a free radical (in the tyrosine to be iodinated) as an intermediate in the reaction.

position of most of these proteins in the membrane or about their function. However, as we shall now discuss, it has recently become clear that many of them are mobile in the plane of the membrane.

When Two Cells Are Fused Together, Their Plasma Membrane Proteins Rapidly Mix[16]

Direct evidence that some plasma membrane proteins are mobile in the plane of the membrane was provided in 1970 by an elegant experiment using hybrid cells (*heterocaryons,* see p. 163) artificially produced by fusing mouse cells and human cells together. The fact that such cells are viable in culture shows that the structural and functional integrity of the fused plasma membranes remains intact. One can therefore ask whether the two different plasma membrane regions remain segregated or whether the two sets of molecules (in particular, the membrane proteins) mix. The experiment that dramatically answered this question made use of two sets of antibodies, one that bound to proteins on human cells and another that bound to proteins on mouse cells. The human proteins were visualized by antibodies coupled to rhodamine, a red fluorescent dye, while the mouse proteins were visualized by antibodies coupled to fluorescein, a green fluorescent dye. In most heterocaryons, the human and mouse proteins were initially confined to their own halves, forming fluorescent "harlequins." But within one hour, the two sets of proteins had spread and mixed over the entire surface of most cells, indicating that they were able to diffuse rapidly in the plane of the mixed membrane (Figure 6–33).

The fluid structure of the lipid bilayer enables membrane proteins to diffuse rapidly and to interact with one another, and it provides a simple means of distributing membrane constituents from sites where they are in-

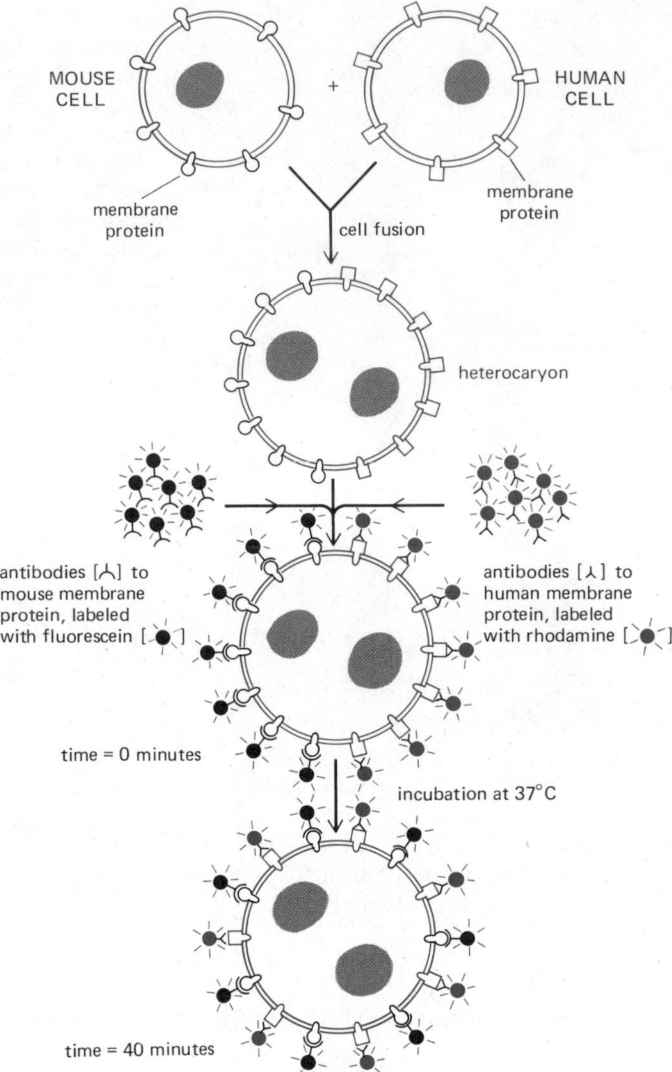

Figure 6–33 Schematic drawing of the experiment demonstrating the mixing of plasma membrane proteins on mouse-human hybrid cells. The mouse and human proteins are initially confined to their own halves of the newly formed heterocaryon plasma membrane, but intermix with time.

serted into the bilayer after their synthesis to other regions of the cell. It allows membranes to fuse with one another without loss of permeability control and ensures that membrane molecules are divided evenly between daughter cells at the time of cell division. It is hard to imagine how a cell could live, grow, and reproduce if its membranes were not fluid.

Membrane Proteins Cluster into Patches When They Are Cross-linked by Antibodies[17]

When fluorescent antibodies bind to specific plasma membrane proteins exposed on the surface of cells such as lymphocytes (the special class of white blood cells responsible for immune reactions, see Chapter 17), they induce these proteins to aggregate in patches. Since antibody molecules are *bivalent*—that is, they have two binding sites for *antigen* (the protein or other substance to which they bind)—they are able to cross-link soluble antigens and form large antibody-antigen complexes that precipitate out of solution. The clustering of antigens on the surface of cells, induced by the binding of antibody, is analogous to such a precipitation reaction—but in this case it occurs in two dimensions, within the plane of the membrane. However, in order for antibodies to cross-link membrane proteins into large complexes, these proteins must be free to move laterally in the lipid bilayer, which means that the membrane must be fluid.

Before it was realized that membranes are fluid, labeled antibodies had been widely used to visualize proteins on the lymphocyte surface, and the clustered distributions observed were erroneously interpreted as the usual distributions of these membrane proteins. That the clusters are, in fact, a direct consequence of the binding of antibodies can be shown by the use of *monovalent* antibody fragments. These are produced by digesting antibodies with the proteolytic enzyme papain to produce antibody fragments that have only one antigen-binding site each and therefore cannot cross-link neighboring membrane proteins (see p. 965). When such fragments are coupled to a fluorescent dye, it is found that although they still bind to the protein molecules on the lymphocyte surface, their distribution is diffuse; they fail to form the clusters induced by bivalent antibody molecules (Figure 6–34).

The phenomenon of antibody-induced clustering of membrane molecules has been demonstrated for many different proteins (and glycolipids) in all nucleated cells studied to date. Such **patching** is induced not only by antibodies, but also by proteins of a completely different class called **lectins.** These proteins, usually isolated from plants, have two or more sites that bind to specific carbohydrate residues on cell-surface glycoproteins and glycolipids; therefore, like antibodies, they can cross-link molecules that have the appropriate carbohydrate residues.

Protein distributions in membranes can be visualized at the molecular level by using antibodies or lectins (such binding molecules are collectively referred to as *ligands*) that have been coupled to the iron-containing protein *ferritin*. Because ferritin is electron dense, individual molecules can be visualized with an electron microscope. In most cases, when ferritin-coupled ligands are used to map membrane macromolecules under conditions where ligand-induced clustering is prevented (by using monovalent ligands or by chemically "fixing" the cell membrane so that its proteins are immobilized),

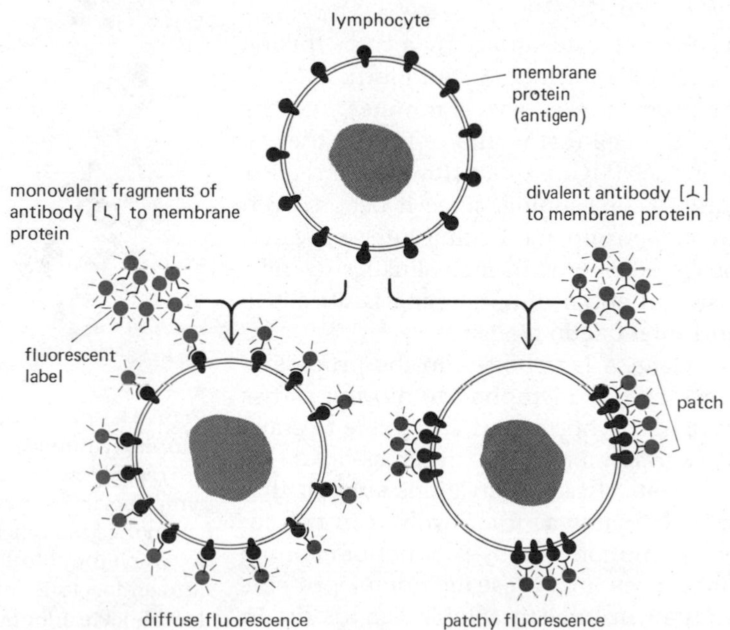

Figure 6–34 Schematic drawing of antibody-induced patching (clustering) of antigens on the surface of a lymphocyte. Whereas monovalent antibodies remain diffusely distributed on the lymphocyte surface, bivalent antibodies cross-link the antigens and cluster them into patches.

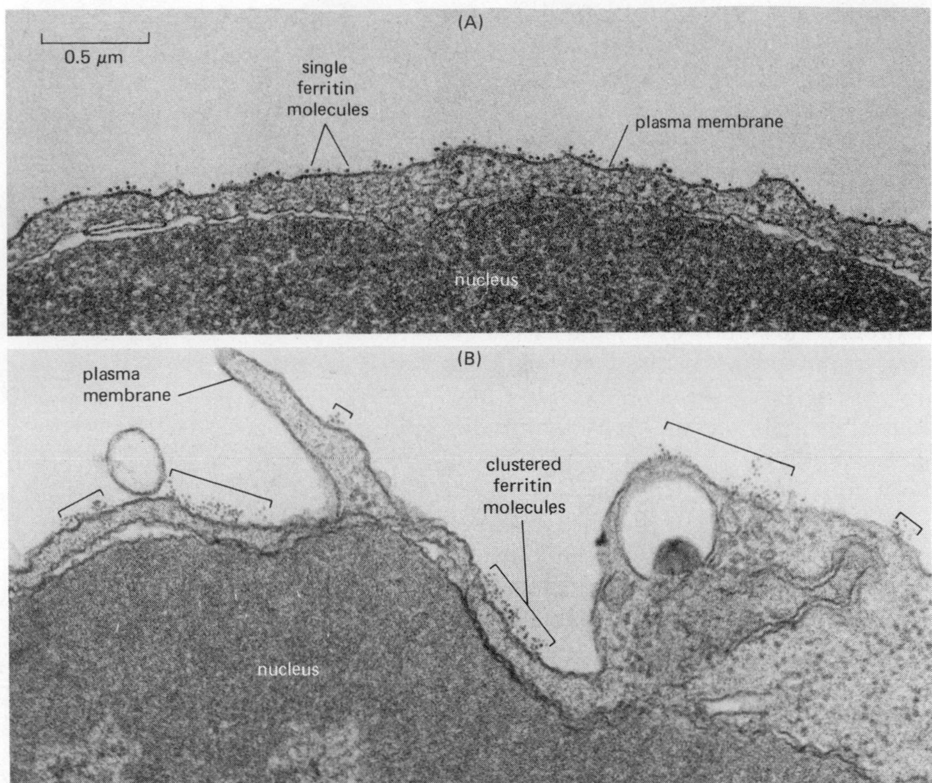

Figure 6–35 Electron micrographs showing diffuse distribution of ferritin-coupled monovalent antibodies (A) and clustering of ferritin-coupled bivalent antibodies (B) on the surface of a lymphocyte. Each ferritin molecule appears as a black dot. (Courtesy of Durward Lawson.)

the macromolecules in the undisturbed plasma membrane of single cells suspended in fluid are found to be diffusely distributed (Figure 6–35A). On the other hand, when ferritin-coupled cross-linking ligands are used on unfixed membranes, the labeled molecules redistribute into clusters (Figure 6–35B).

Cross-linked Membrane Proteins Are Actively Swept to One Pole of the Cell in the Process of "Capping"[18]

After patches have been induced on a lymphocyte surface by a cross-linking ligand, they rapidly collect over one pole of the lymphocyte to form a "cap" (Figures 6–36 and 6–37). This **capping** process takes several minutes, and the cap almost always forms at that end of the cell that would represent the tail of a motile lymphocyte moving over a surface (Figure 6–36). However, it is clear that such cell locomotion is not required for capping, since it occurs even when lymphocytes are maintained in suspension and cannot interact with a surface. Like patching, capping is observed only with cross-linking ligands; but unlike patching, which is a passive phenomenon, capping is an active process that requires energy (ATP) and intact actin filaments.

The mechanism of capping is unclear. A lymphocyte in the process of capping frequently takes on the appearance of a lymphocyte moving across a surface. Moreover, while most types of eucaryotic cells that are able to crawl along a surface form caps when their surface molecules are cross-linked by multivalent ligands, nonmotile cells do not. These observations suggest that the mechanism involved in capping is the same as that involved in cell locomotion. It is even possible to view locomotion as a by-product of capping if one considers that an adherent surface cross-links those membrane proteins that interact with it in the same way that multivalent soluble ligands do. In this view, as the cell attempts to move the adherent surface to its tail, the cell

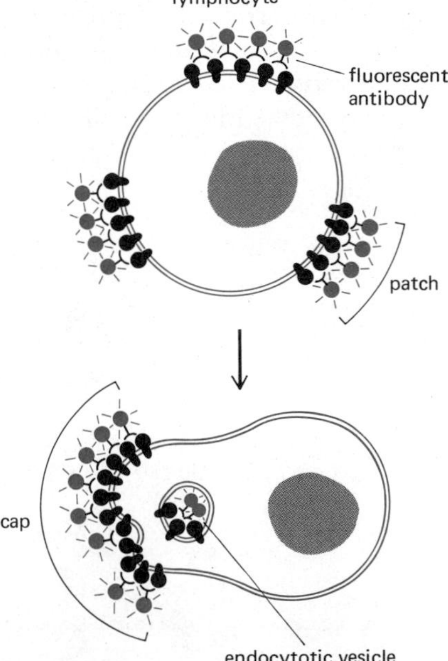

Figure 6–36 Schematic drawing of capping. Antibody-induced patches are rapidly swept to the tail of the lymphocyte, where they form a "cap." Often the antibodies (and the proteins to which they bind) are then ingested into endocytotic vesicles and the lymphocyte adopts a shape that is characteristic of locomotion.

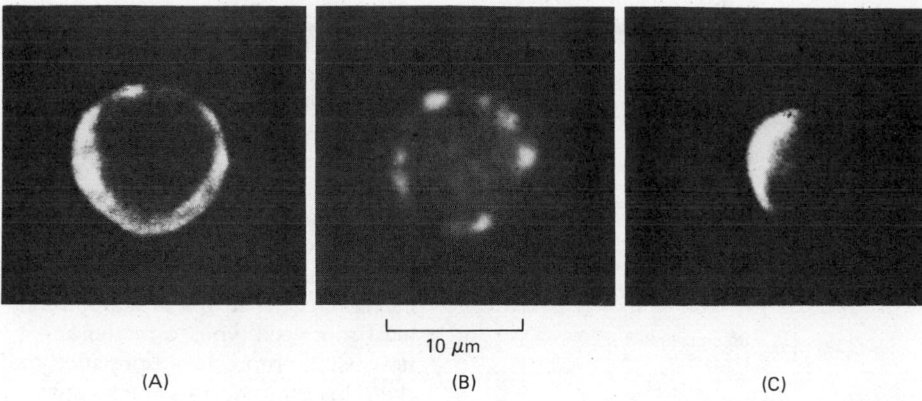

Figure 6–37 Immunofluorescence micrographs showing diffuse, patchy, and capped distributions of fluorescent antibodies on lymphocytes. Monovalent antibodies show diffuse staining (A), while bivalent antibodies initially cluster into patches (B) and then cap (C). (Courtesy of Stefanello dePetris.)

moves forward. From another perspective, one might instead view capping in terms of locomotion. When a soluble cross-linking ligand binds to the membrane of a cell in suspension, the cell responds as if it were interacting with a solid surface: it activates its machinery for locomotion, and capping ensues. Looked at from this perspective, capping provides a convenient means of studying cell locomotion.

Membrane molecules that have been cross-linked by antibodies or other ligands are often ingested by the cell by a process called **ligand-induced endocytosis**. Although in lymphocytes most of this endocytosis usually occurs only after capping has been completed (Figure 6–36), it depends very much on the specific cell-surface molecule to which the ligand binds. In some cases the ligand and the molecule to which it is bound are shed from the cell surface rather than endocytosed. Ligand-induced endocytosis (also called *receptor-mediated or adsorptive endocytosis)* is discussed in more detail on page 309.

Conflicting Views on How Cells Cap: Flow Versus Pull Hypotheses[19]

Two types of hypotheses have been proposed to explain capping. One type suggests that patches are swept along to the back of the cell by a continuous flow of membrane generated by the cell's ingesting bits of plasma membrane at the tail (by endocytosis), transporting them as vesicles across the cytoplasm and reinserting them at the front (by exocytosis, see p. 303). According to this hypothesis, individual protein molecules in the plasma membrane diffuse rapidly enough to maintain a nearly random distribution despite the membrane flow. However, when the membrane proteins are cross-linked to form large aggregates, they diffuse only slowly, and so the membrane flow sweeps them to one pole of the cell (Figure 6–38). Actin filaments and ATP are required for the directed transport of vesicles, which is hypothesized to drive the membrane flow. A similar hypothesis has been proposed to explain the locomotion of fibroblasts.

The other hypothesis proposed to explain capping suggests that membrane proteins clustered by cross-linking ligands interact with contractile systems of actin filaments in the cytoplasm and are actively pulled to the tail of the cell by these filaments (Figure 6–39). One difficulty with this model is that virtually any protein (and even glycolipids) can be capped by an appropriate multivalent ligand. The model would seem to require, then, that the submembranous actin filaments somehow recognize clustered membrane molecules and bind to them, whatever their type. Such recognition is difficult to imagine, but it could arise if a special class of membrane proteins interacts with the

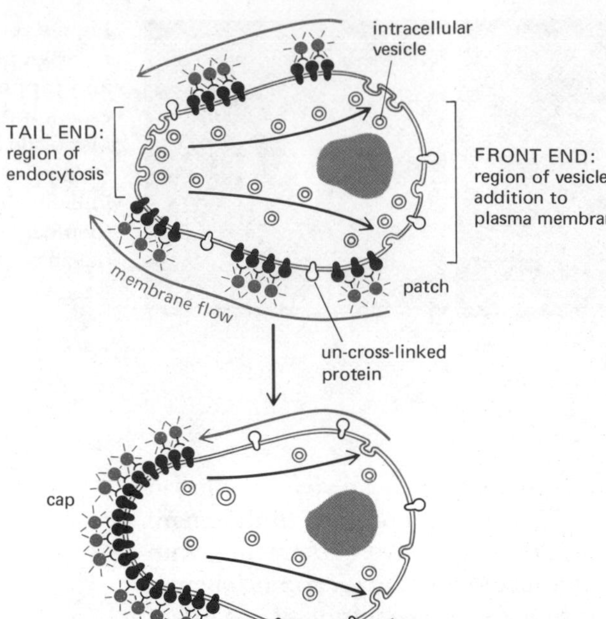

Figure 6–38 The membrane-flow hypothesis of capping. Membrane vesicles are ingested at the tail of a lymphocyte, transported across the cytoplasm, and reinserted at the front. This creates a flow of membrane on the cell surface that sweeps large patches of cross-linked protein to the tail of the cell. Even if endocytosis occurs over the entire cell surface, as seems to happen in at least some cell types, a membrane flow will be reproduced provided that the reinsertion occurs in a localized region.

actin filaments and if other membrane proteins (or glycolipids), when they are clustered, bind to or trap these special proteins.

There are still no definitive experiments that permit a choice between the "flow" and the "pull" hypotheses, although the demonstration of actin and myosin accumulations in the region of patches and caps has been used to support the pull models. Whatever the answer turns out to be, interest in capping has helped to focus attention on important interactions that occur between certain plasma membrane proteins and parts of the cytoskeleton. A variety of other experiments have shown that the plasma membrane is not an independent organelle; it is structurally and functionally connected to the complex cytoskeleton of the cell (see Chapter 10).

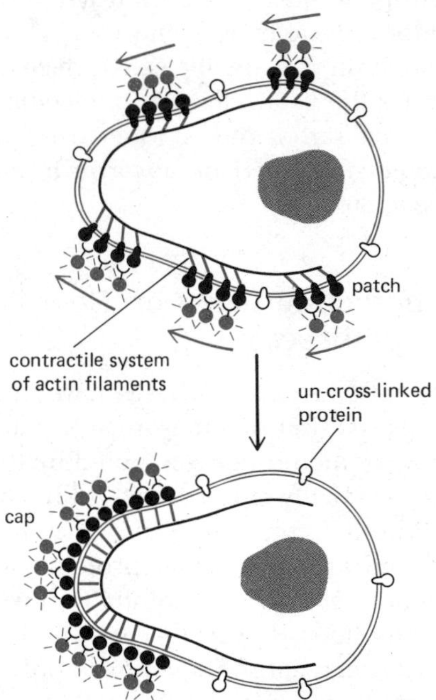

Figure 6–39 The "pull" hypothesis of capping. After cell-surface proteins become cross-linked into patches, they interact directly or indirectly with a contractile system of intracellular actin filaments that pulls the patches to the tail of the lymphocyte.

Antibody-induced Redistribution Can Be Used to Determine Whether Two Polypeptides Are Associated with Each Other in the Plasma Membrane

An important goal in membrane biochemistry is to determine the precise relationships between the different molecules in a particular membrane. For example, one may ask whether two different membrane polypeptides, A and B, are associated with each other. One approach to answering this question has been to use labeled antibodies directed against the two different polypeptides: antibodies against A can be used to patch and cap A molecules, and then monovalent antibody fragments against B can be used to see what has happened to B molecules (Figure 6–40). If B moves with A, the two molecules are assumed to be associated in the membrane, whereas if B is left behind, the two polypeptides are clearly not associated. The two antibodies can be visualized simultaneously by coupling them to different fluorescent dyes—such as fluorescein and rhodamine. The great majority of proteins that have been studied in this way (mainly those in the lymphocyte plasma membrane) cap independently, which implies the absence of stable, long-range ordering of most membrane proteins.

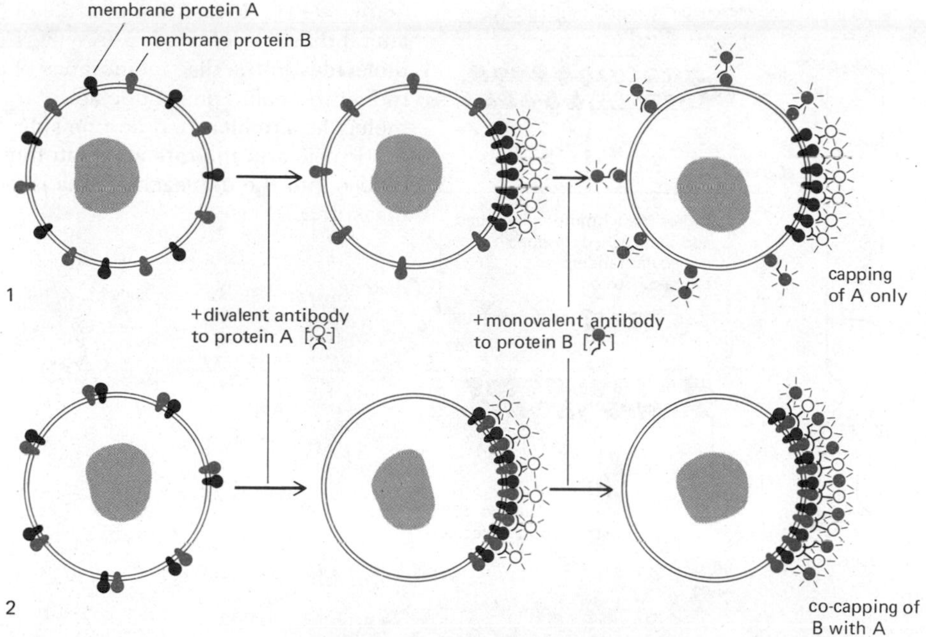

Figure 6–40 A method used to determine whether two plasma membrane proteins, A and B, are physically associated. If they are not associated, when A is capped with anti-A antibodies, B remains diffusely distributed (1). If A and B are associated, B is "co-capped" with A (2). Note that the distribution of B is detected with monovalent antibody fragments to avoid direct antibody-induced clustering of B.

Lateral Diffusion Rates of Membrane Proteins Can Be Quantified[20]

Like membrane lipids, membrane proteins are able to rotate about an axis perpendicular to the plane of the bilayer *(rotational diffusion)* and to move laterally in the membrane *(lateral diffusion)*, but they do not tumble *(flip-flop)* across the bilayer. Measurements of lateral diffusion rates for various protein molecules in a variety of different membranes have given a remarkably wide range of values, with diffusion coefficients (D) varying between 5×10^{-9} cm^2 sec^{-1} and 10^{-12} cm^2 sec^{-1}. (For comparison, recall that the diffusion coefficient for a phospholipid molecule in a membrane is about 10^{-8} cm^2 sec^{-1}.) Perhaps the most accurate measurement has been made on individual rhodopsin molecules in the disc membranes of vertebrate rod cells. The method used involves quickly bleaching the chromophore (retinal) in rhodopsin molecules on one side of a rod cell with an intense, highly focused beam of light and then measuring the time it takes for the bleached molecules to diffuse to the unbleached side (Figure 6–41). The diffusion coefficient obtained by this technique is about 5×10^{-9} cm^2 sec^{-1}. Recently a similar approach has been used to study the diffusion rates of specific protein molecules in the plasma membranes of living cells. Here fluorescent antibodies bound to cell-surface molecules are bleached in a small area by a laser beam, and the time taken for adjacent unbleached molecules to diffuse into the bleached area is measured. The diffusion rates of various plasma membrane proteins when measured in this way have been found to be 10 to 1000 times slower than those of rhodopsin molecules. One possible interpretation of these findings is that the

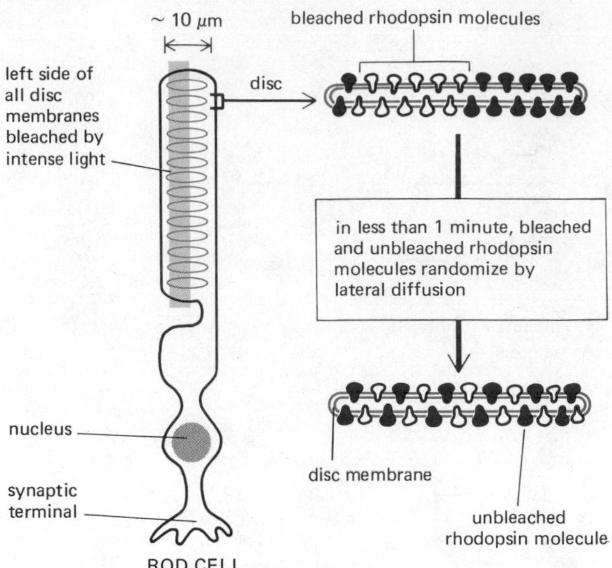

Figure 6–41 Measuring the rate of lateral diffusion of rhodopsin molecules in the disc membranes of a retinal rod cell. The rhodopsin molecules are bleached on one side of the cell, and the rate at which they diffuse into the unbleached area is measured.

diffusion rates of certain plasma membrane proteins are decreased because the proteins interact with other macromolecules, either in the membrane or adjacent to it. On the other hand, it is possible that the high-intensity laser beam damages the membrane and thereby causes abnormally low diffusion rates.

Cells Have Ways of Restricting the Lateral Mobility of Certain of Their Membrane Proteins

While the recognition that biological membranes are two-dimensional fluids represents a major advance in understanding membrane structure and function, the picture of a biological membrane as a lipid sea in which proteins float freely is oversimplified. Many membrane proteins are known to be restricted in their lateral mobility. For example, in epithelial cells, which line various cavities in the body, certain plasma membrane enzymes and transport proteins are confined to the apical surface of the cells, while others are confined to the basal and lateral surfaces by the barriers set up by a specialized type of cell junction (called *tight junctions,* see Figure 6–42 and p. 684). However, most of these protein molecules are presumably free to diffuse within the respective membrane areas to which they are confined. A more drastic restriction of movement occurs at cell junctions themselves, for the specific

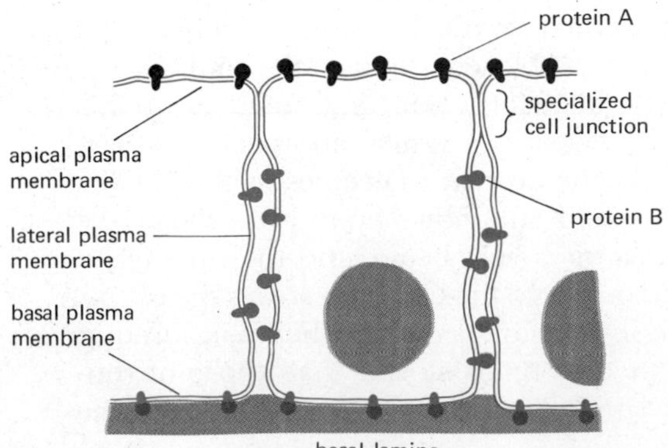

Figure 6–42 Schematic diagram of an epithelial cell showing how different plasma membrane proteins may be restricted to particular regions of the membrane. Proteins A (in the apical membrane) and B (in the basal and lateral membranes) can diffuse laterally in their own areas of membrane but are prevented from entering the other area by a specialized cell junction called a tight junction.

membrane proteins used in their construction generally do not diffuse laterally in the interacting membranes.

In principle, there are several ways that the lateral mobility of membrane proteins might be controlled. Three are shown in Figure 6–43. One way is exemplified by the purple membrane of *Halobacterium,* where the bacteriorhodopsin molecules assemble into large two-dimensional crystals in which the individual protein molecules are relatively fixed in relationship to one another. Large aggregates of this kind diffuse so slowly that they are effectively immobile (Figure 6–43A). In other cases the mobility of plasma membrane proteins is restricted by interactions with structures outside the cell, as when specific proteins in the plasma membranes of two interacting cells assemble to form a specialized cell junction (Figure 6–43B). Finally, as previously discussed, the movement and distribution of membrane protein molecules may be controlled by interactions with protein assemblies in the cytoplasm, such as actin filaments (Figure 6–43C), in the same way that spectrin tethers certain proteins in the plasma membrane of red blood cells (Figure 6–24). The molecular details of these putative interactions between plasma membrane proteins and various cytoplasmic elements in nucleated cells are unknown, but they are the focus of very active research.

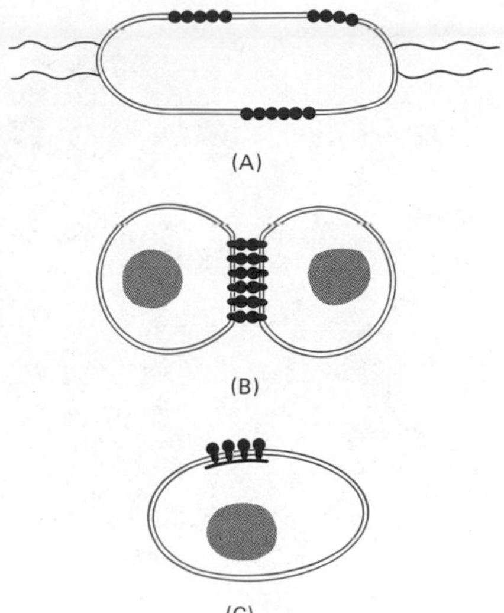

Figure 6–43 Three ways in which plasma membrane proteins can be restricted in their lateral mobility. They can assemble into large aggregates (such as bacteriorhodopsin in the purple membrane of *Halobacterium*) (A), or they can be tethered by interactions with proteins outside (B) or inside (C) the cell.

Summary

Whereas the lipid bilayer determines the basic structure of biological membranes, proteins are responsible for most membrane functions, serving as specific receptors, enzymes, or transporters. Many membrane proteins have hydrophobic surfaces that make them insoluble in aqueous solution. They associate with the lipid bilayer in various ways: many extend across the bilayer, either as a single α-helix or in a more folded conformation in which the polypeptide crosses the bilayer several times. Most of the latter are probably transport proteins and have sufficient bulk in the plane of the lipid bilayer to be visualized as intramembrane particles in freeze-fracture electron micrographs. Other membrane proteins do not span the bilayer and are exposed to an aqueous environment on one or the other side of the membrane. These are probably bound to the membrane by covalently attached fatty acid chains or by noncovalent interactions with transmembrane proteins.

Like the lipid molecules in the bilayer, many membrane proteins are able to diffuse in the plane of the membrane. One manifestation of this mobility is the clusters formed when membrane proteins are cross-linked by multivalent ligands such as antibodies or lectins; in some cells these clusters are actively swept to one pole of the cell to form a "cap." However, cells can restrict the lateral mobility of specific plasma membrane proteins by tethering them to other macromolecules on either the external or the cytoplasmic surface of the membrane.

Membrane Carbohydrate

All eucaryotic cells have carbohydrate on their surfaces, mostly as oligosaccharide side chains covalently bound to membrane proteins (glycoproteins) and to a lesser extent bound to lipids (glycolipids). While the great majority of the plasma membrane protein molecules exposed at the cell surface are thought to carry sugar residues, fewer than one in ten lipid molecules in the outer lipid monolayer of most plasma membranes carries carbohydrate. Moreover, while a single glycoprotein can have many oligosaccharide side chains, each glycolipid molecule has only one. In total, the proportion of carbohydrate in plasma membranes varies between 2% and 10% by weight.

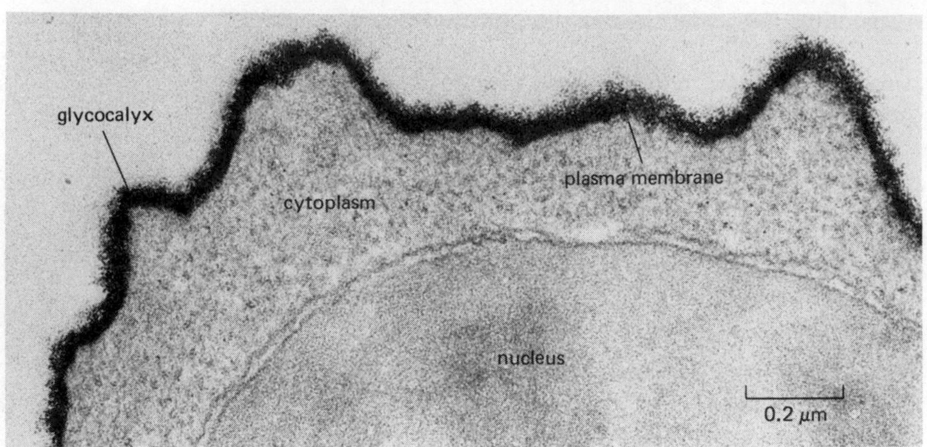

Figure 6–44 Electron micrograph of the surface of a lymphocyte stained with ruthenium red to show the cell coat (glycocalyx). (Courtesy of A. M. Glauert and G. M. W. Cook.)

The Carbohydrate in Biological Membranes Is Confined to the Noncytoplasmic Surface[21]

As we have seen, biological membranes are strikingly asymmetrical: the lipids of the outer and inner lipid monolayers (of the plasma membrane, at least) are different, and the exposed polypeptides on the two surfaces are always very different. The distribution of carbohydrate is even more asymmetrical since the oligosaccharide side chains of the glycolipids and glycoproteins of both internal and plasma membranes are located exclusively on the noncytoplasmic surface: in plasma membranes, the sugar residues are all exposed on the outside of the cell, while in internal membranes they are all facing inward toward the lumen of the membrane-bounded compartment.

The terms **cell coat** or **glycocalyx** are often used to describe the carbohydrate-rich peripheral zone at the surface of most eucaryotic cells. This

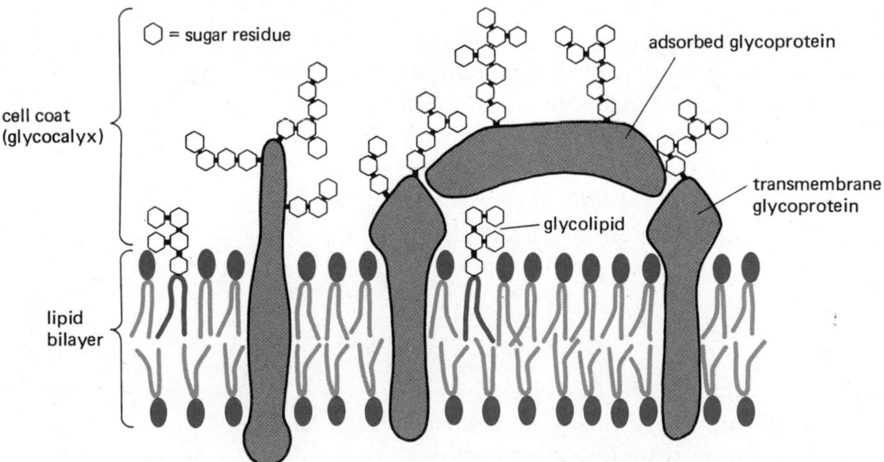

Figure 6–45 Schematic diagram of the cell coat (glycocalyx), which is made up of the oligosaccharide side chains of intrinsic membrane glycolipids and glycoproteins, as well as of adsorbed glycoproteins and proteoglycans. The adsorbed proteoglycans are not shown. Note that all of the carbohydrate is on the outside of the membrane.

zone can be visualized by a variety of stains, such as ruthenium red (Figure 6–44). The carbohydrate consists of the oligosaccharide side chains of plasma-membrane-bound glycoproteins and glycolipids, although it often includes, in addition, both glycoproteins and proteoglycans that have been secreted and then adsorbed on the cell surface (Figure 6–45). (Proteoglycans consist of many polysaccharide chains linked to a protein core, see p. 703.) Some of these adsorbed macromolecules are components of the extracellular matrix so that it is largely a matter of semantics as to where the plasma membrane ends and the extracellular matrix begins. Although the high concentration of cell-surface carbohydrate must have an important influence on many functions of the plasma membrane, the nature of these influences is not yet known.

Cell-Surface Carbohydrate Is Suspected to Be Important in Cell-Cell Interactions, But This Has Been Difficult to Demonstrate[22]

Of the more than 100 different monosaccharides found in nature, only 9 occur in membrane glycoproteins and glycolipids. The principal ones are galactose, mannose, fucose, galactosamine, glucosamine, glucose, and sialic acid. The sialic acid residues are usually found at the ends of the carbohydrate side chains, and they are mainly responsible for the net negative surface charge that characterizes all eucaryotic cells. The oligosaccharide side chains of glycoproteins and glycolipids can be complex. Although they usually contain fewer than 15 sugar residues, they are often branched, with the sugars bonded together by a variety of different linkages. In principle, even 3 sugar residues can be put together to form more than 1000 different trisaccharides. It is still technically very difficult to determine the sequence of complex oligosaccharide side chains on membrane proteins and lipids; thus the details of their structures are in many cases unknown.

The function of the oligosaccharide side chains in membrane glycolipids and glycoproteins is unclear. It is possible that those in certain transmembrane glycoproteins help to anchor and orient the proteins in the membrane by preventing them from slipping into the cytosol or from tumbling across the bilayer. The carbohydrate also may play a role in stabilizing the folded structure of a glycoprotein. In addition, carbohydrate may play a role in guiding a membrane glycoprotein to its appropriate destination in or on the cell, just as the special carbohydrate chains on lysosomal enzymes direct these soluble glycoproteins to lysosomes (see p. 372). However, these cannot be the only functions of carbohydrate in membrane glycoproteins, for if cells are treated with an antibiotic that inhibits the glycosylation of many membrane glycoproteins (*tunicamycin*, see p. 375), some of these sugar-free proteins are inserted normally into the plasma membrane and seem to be normally oriented and stable. Moreover, functions such as orienting, anchoring, stabilizing, and targeting cannot account for the carbohydrate in glycolipid molecules nor for the complexity of some of the carbohydrate chains in glycoproteins.

The complexity of some of the oligosaccharides on plasma membrane glycoproteins and glycolipids, taken together with their exposed position on the cell surface, suggests that they may play an important part in sophisticated cell-to-cell recognition processes. Some cells have surface proteins that bind specific oligosaccharides; theoretically such lectins could recognize oligosaccharides on the surface of other cells and thereby play a role in guiding cell-to-cell interactions. While there is evidence for this kind of interaction in plants, it has been difficult to prove that cell-surface carbohydrate functions in the same way in animals, despite increasing indirect evidence that this is the case.

Summary

In the plasma membrane of all eucaryotic cells, most of the proteins exposed on the cell surface and some of the lipid molecules in the outer lipid monolayer have oligosaccharide chains covalently attached to them. These carbohydrate chains can be complex, and in many cases their precise sequences are unknown. Although the function of this carbohydrate remains to be discovered, it seems likely that at least some of the oligosaccharide chains play a part in cell-to-cell recognition processes.

Membrane Transport of Small Molecules

Because of its hydrophobic interior, the lipid bilayer serves as a highly impermeable barrier to most polar molecules, thereby preventing most of the water-soluble contents of the cell from escaping. But for this very reason cells have had to evolve special ways of transferring polar molecules across their membranes. For example, cells must ingest essential nutrients and must excrete metabolic waste products. They must also regulate intracellular ion concentrations, which means transporting specific ions into or out of the cell. Transport of small molecules across the lipid bilayer is achieved by specialized transmembrane proteins, each of which is responsible for the transfer of a specific molecule or group of closely related molecules. Cells have also evolved the means to transport macromolecules (such as proteins), and even large particles, across their plasma membranes. But the mechanisms involved are very different from those used for transferring small molecules, and they will be discussed in a later section.

In this section we shall see that the selective permeability of biological membranes to simple ions creates large differences in the ionic composition of the cell interior compared to the extracellular fluid (Table 6–2). This enables cell membranes to store potential energy in the form of ion gradients. Transmembrane ion gradients are used to make ATP, to drive various transport processes, and to convey electrical signals. Before discussing specific transport proteins and the ion gradients that some of them generate, it is important to know something of the permeability properties of protein-free lipid bilayers.

Table 6–2 Comparison of Ion Concentrations Inside and Outside a Typical Mammalian Cell

Component	Intracellular Concentration (mM)	Extracellular Concentration (mM)
Cations		
Na^+	5–15	145
K^+	140	5
Mg^{2+}	30	1–2
Ca^{2+}	1–2 ($\leq 10^{-7}$ M is free)	2.5–5
H^+	4×10^{-5} ($10^{-7.4}$ M or pH 7.4)	4×10^{-5} ($10^{-7.4}$ M or pH 7.4)
Anions*		
Cl^-	4	110

*Because the cell must contain equal + and − charge (that is, be electrically neutral), the large deficit in intracellular anions reflects the fact that most cellular constitutients are negatively charged (HCO_3^-, PO_4^{3-}, proteins, nucleic acids, metabolites carrying phosphate and carboxyl groups, etc.).

Protein-free Lipid Bilayers Are Impermeable to Ions But Freely Permeable to Water[23]

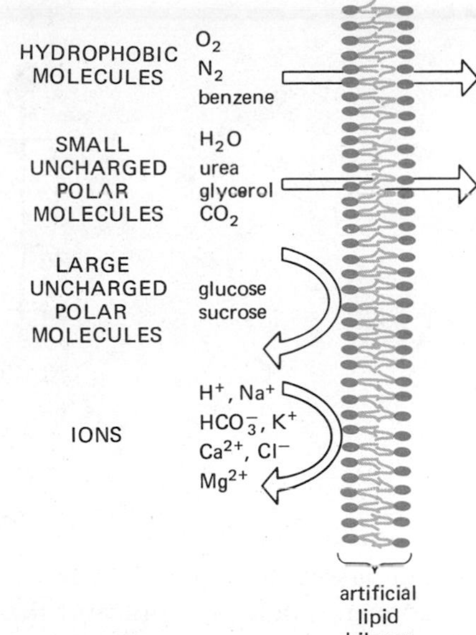

Figure 6–46 The relative permeability of an artificial lipid bilayer to different classes of molecules.

Given enough time, essentially any molecule will diffuse across a protein-free lipid bilayer down its concentration gradient. The rate at which a molecule diffuses across such a lipid bilayer, however, varies enormously, depending largely on the size of the molecule and its relative solubility in oil. In general, the smaller the molecule and the more soluble it is in oil (that is, the more hydrophobic or nonpolar it is), the more rapidly it will diffuse across a bilayer. *Small nonpolar* molecules readily dissolve in lipid bilayers and therefore rapidly diffuse across them. *Uncharged polar* molecules also diffuse rapidly across a bilayer if they are small enough. For example, CO_2 (44 daltons), ethanol (46 daltons), and urea (60 daltons) cross rapidly; glycerol (92 daltons) less rapidly; and glucose (180 daltons) hardly at all (Figure 6–46). Importantly, water (18 daltons) diffuses very rapidly across lipid bilayers even though water molecules are relatively insoluble in oil. This rapid rate of diffusion results in part from the fact that water molecules are small and uncharged; in addition, it is thought that the dipolar structure of the water molecule allows it to cross the regions of the bilayer containing the lipid head groups unusually rapidly.

In contrast, lipid bilayers are highly impermeable to all *charged* molecules (ions), no matter how small: the charge and high degree of hydration of such molecules prevents them from entering the hydrocarbon phase of the bilayer. Consequently, artificial bilayers are 10^9 times more permeable to water than to even such small ions as Na^+ or K^+ (Figure 6–47).

Membrane-bound Transport Proteins Transfer Specific Small Molecules Across Cell Membranes[24]

Like artificial lipid bilayers, cell membranes allow water and nonpolar molecules to permeate by simple physical diffusion. However, cell membranes are also permeable to various polar molecules, such as ions, sugars, amino acids, nucleotides, and many cell metabolites that pass across artificial lipid bilayers only very slowly. It is now known that specific membrane proteins are responsible for transferring such solutes across cell membranes. These proteins, referred to as **membrane transport proteins**, occur in many forms and in all types of biological membranes. Each different protein is designed to transport a different class of chemical compound (such as ions, sugars, or amino acids) and often only a specific molecular species of the class. The specificity of transport proteins was first indicated by studies in which single gene mutations were found to abolish the ability of bacteria to transport specific sugars across their plasma membranes. Similar mutations have now been discovered in humans suffering from a variety of inherited diseases affecting the transport of a specific solute in the kidney or intestine.

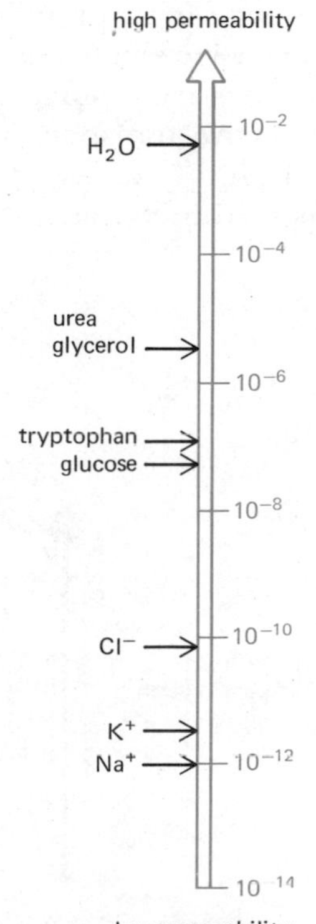

Figure 6–47 Permeability coefficients (cm/sec) for the passage of various molecules through artificial lipid bilayers.

Some transport proteins simply transport one solute from one side of the membrane to the other; they are called **uniports**. Others function as **cotransport** systems, in which the transfer of one solute depends on the simultaneous or sequential transfer of a second solute, either in the same direction **(symport)** or in the opposite direction **(antiport)** (Figure 6–48). For example, the transport of sugars into many bacteria occurs by the inward symport of H^+ along with the sugar molecules, while the Na^+-K^+ pump in eucaryotic plasma membranes (see below) operates as an antiport, pumping Na^+ out of the cell and K^+ in.

Many membrane transport proteins allow specific solutes to move across the lipid bilayer by a process called **passive transport.** If the transported molecule is uncharged, then only the difference in its concentration on the two sides of the membrane (its *concentration gradient*) determines the direc-

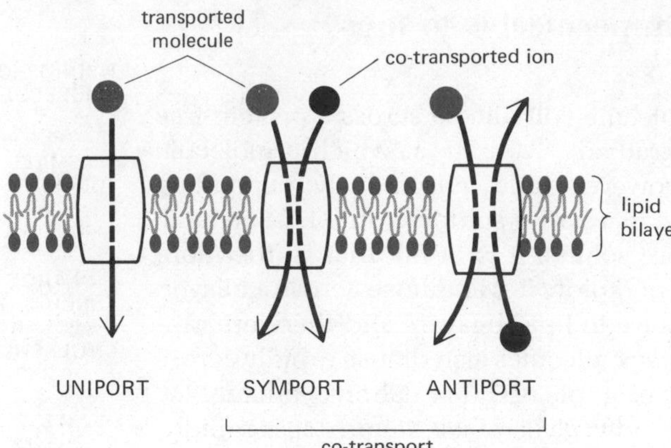

Figure 6–48 Schematic diagram of transport proteins functioning as uniport, symport, and antiport systems.

tion of passive transport. However, if the solute carries a net charge, then both its concentration gradient and the total electrical gradient across the membrane (the *membrane potential*) influence its transport. Both gradients together constitute the **electrochemical gradient**. In fact, all plasma membranes have electrical potentials (voltage gradients) across them, with the inside negative compared to the outside. This potential facilitates the entry of positively charged ions into cells but opposes the entry of negatively charged ions.

Some of the transport proteins that mediate passive transport form *aqueous channels* that permit solutes of appropriate size and charge to cross the bilayer by simple diffusion; these are called **channel proteins.** Others, called **carrier proteins** (or carriers or transporters) bind the specific molecule to be transported and transfer it across the membrane, a process called *facilitated diffusion.* Some carrier proteins function as pumps that actively drive the movement of specific solutes uphill against their electrochemical gradients, by so-called **active transport.** Unlike passive transport that can occur spontaneously, active transport must be tightly coupled to a source of metabolic energy (Figure 6–49). Most often this involves ATP hydrolysis by the carrier proteins or the co-transport of Na^+ or H^+ down their electrochemical gradients.

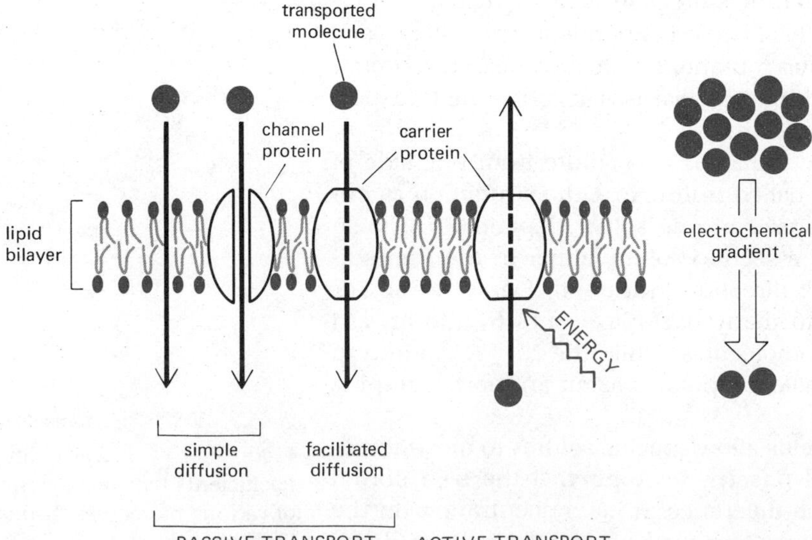

Figure 6–49 Schematic diagram of passive transport down an electrochemical gradient and active transport against an electrochemical gradient. Whereas passive transport, either by simple diffusion or by facilitated diffusion, occurs spontaneously, active transport requires an input of metabolic energy. Although small uncharged molecules can directly cross the lipid bilayer, the transfer of charged molecules (including all small ions) occurs at significant rates only through protein channels or carriers.

Transport Proteins Form a Continuous Protein Pathway Across the Lipid Bilayer[25]

All membrane transport proteins that have been studied in sufficient detail to establish their orientation in the membrane have been found to traverse the lipid bilayer, or at least to be part of a larger structure that traverses the bilayer. Moreover, they have sufficient mass in the lipid bilayer to be readily visualized as intramembrane particles in freeze-fracture electron micrographs. Since transport proteins form a continuous protein pathway across the membrane, the specific solutes they transport do not have to come into direct contact with the hydrophobic interior of the lipid bilayer.

Progress in understanding the detailed structure and function of transport proteins has been slow, largely because it is so difficult to isolate them in soluble form. Whereas the complete amino acid sequences of many hundreds of soluble proteins are known, only recently has the first transmembrane transport protein (bacteriorhodopsin) been completely sequenced. Moreover, none of these proteins has been crystallized in a form suitable for x-ray diffraction analysis of their structure. Nonetheless, it has been possible to purify some of them and, in favorable cases, to reinsert them into artificial lipid bilayers and reconstitute their transport function. Bacteriorhodopsin, discussed earlier, is a special case because its two-dimensional crystalline arrangement in the purple membrane of *Halobacterium* (where it serves as a light-activated H^+ pump) has made it possible to determine many aspects of its three-dimensional structure *in situ,* without the need to isolate or crystallize the protein (see p. 272).

Carrier Proteins Behave Like Membrane-bound Enzymes[24]

A carrier protein specifically binds and transfers a solute molecule across the lipid bilayer. This process resembles an enzyme-substrate reaction, and the carriers involved behave like specialized membrane-bound enzymes. Each type of carrier protein has a specific binding site for the solute (substrate). When the carrier is saturated (that is, when all such binding sites are occupied), the rate of transport is maximal. This rate, referred to as V_{max}, is characteristic of the specific carrier. In addition, each carrier protein has a characteristic binding constant for its solute, K_M, equal to the concentration of solute when the transport rate is half its maximal value (Figure 6–50). The solute binding can be blocked specifically by competitive inhibitors (which compete for the same binding site and may or may not be transported by the carrier) or by noncompetitive inhibitors (which bind elsewhere and specifically alter the structure of the carrier). The analogy with an enzyme-substrate reaction is limited, however, since the transported solute is usually not covalently modified by the carrier protein.

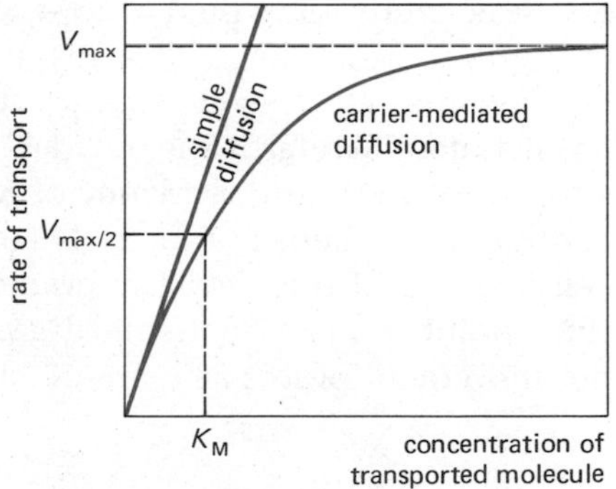

Figure 6–50 Kinetics of simple diffusion compared to carrier-protein-mediated diffusion. Whereas the rate of the former is always proportional to the solute concentration, the rate of the latter reaches a maximum (V_{max}) when the carrier protein is saturated. The concentration when transport is at half its maximal value is assumed to be equal to the solute's binding constant (K_M) for the carrier. While transport through channel proteins is usually referred to as occurring by simple diffusion, at very high solute concentrations the rate of transport often decreases.

Although the molecular details of how carrier proteins function are unknown, it is highly unlikely that the process involves the protein tumbling or shuttling back and forth across the lipid bilayer. It is more likely that they are transmembrane proteins that undergo a reversible conformational change in transferring the solute across the bilayer. A schematic model of how such a transfer might take place in facilitated diffusion is shown in Figure 6–51.

It requires only a relatively small modification in this type of model to link the carrier protein to a source of energy, such as ATP hydrolysis (see Figure 6–54) or an ion gradient, so that it can function to pump a solute uphill against its electrochemical gradient. An important example of a carrier protein that uses the energy of ATP hydrolysis to pump ions is the *Na^+-K^+ pump,* which plays a crucial part in generating the membrane potential across the plasma membranes of animal cells.

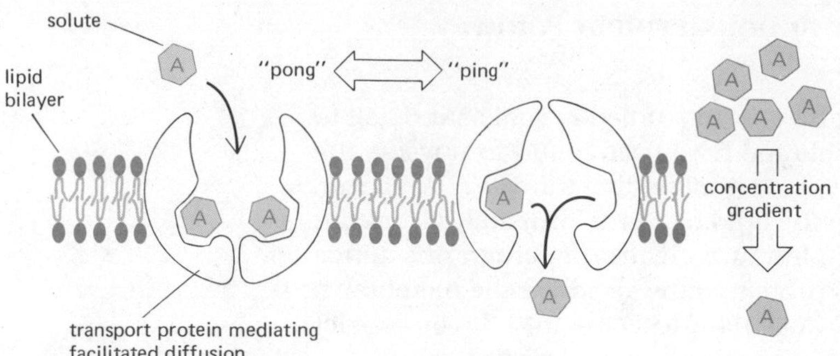

Figure 6–51 Highly schematic drawing of how a conformational change in a carrier protein could mediate the facilitated diffusion of an uncharged solute A. The protein can exist in two different conformational states: in state "pong," the binding sites for A are exposed on the outside of the bilayer; in state "ping," the same sites are exposed on the other side of the bilayer. Although this process is completely reversible, if the concentration of A is higher on the outside of the bilayer, more A will bind to the carrier protein in the pong conformation and there will be a net transport of A down its concentration gradient.

The Membrane Potential That Exists Across the Plasma Membrane Is Maintained by a Na^+-K^+ Pump[26]

The ionic basis of the membrane potential was first determined by studies on the squid giant axon, a nerve process so large (~1 mm in diameter) that a canula (small tube) is easily inserted into it and the cytoplasm replaced with a defined salt solution. The ion concentrations on both sides of the plasma membrane can thus be manipulated to show that (1) the membrane potential is determined largely by a K^+ concentration gradient and (2) the resting plasma membrane is more permeable to K^+ than it is to Na^+ or anions. The K^+ gradient and the differences in ion permeability are, in turn, determined by the properties of the specific transport proteins in the plasma membrane itself.

Two transport proteins are crucial for generating and maintaining the membrane potential (Figure 6–52). The first is the **Na^+-K^+ pump,** which actively pumps Na^+ out of the cell and K^+ in, so that the concentration of K^+ is higher inside than outside while the reverse is true for Na^+. The other is a **K^+ leak channel,** which allows K^+ to leak out of the cell down its steep concentration gradient. As a result of the net efflux of K^+, the inside of the cell becomes electrically negative relative to the outside. The resulting membrane potential retards the outward movement of K^+ through the leak channel and, at -75 mV, the tendency of K^+ to move out of the cell due to its concentration gradient is exactly balanced by the tendency of K^+ to enter the cell because of the electrical gradient (membrane potential). At this equilibrium point the electrochemical gradient for K^+ disappears, and there is no net movement of K^+: as many K^+ ions enter the axon as leave.

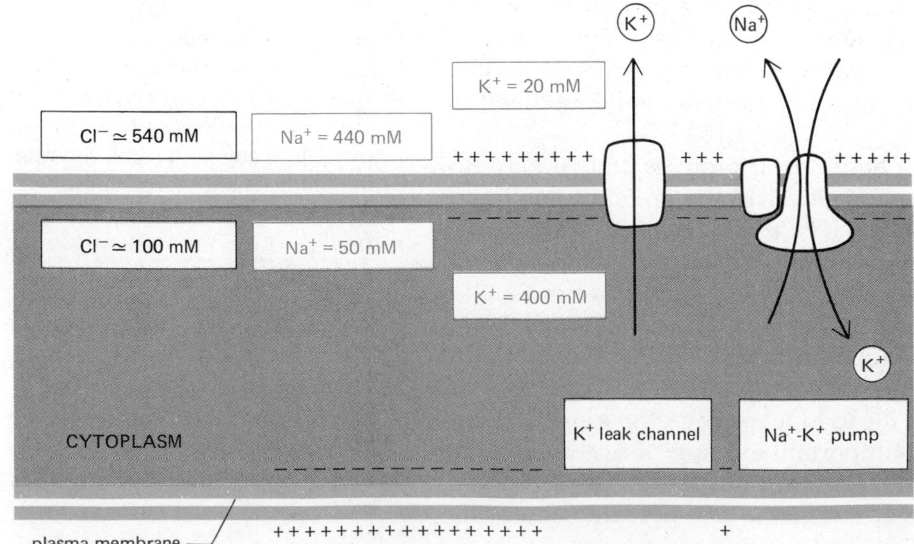

Figure 6–52 Origin of the membrane potential in the squid axon. The Na^+-K^+ pump generates K^+ and Na^+ concentration gradients in opposite directions. K^+ leak channels make the membrane much more permeable to K^+ than to Na^+ or anions, and K^+ leaks out of the cell down its concentration gradient. This results in the inside (cytoplasm) becoming negative with respect to the outside (extracellular space). Other cations and anions besides Na^+, K^+, and Cl^- are present in quantities such that the total concentration of positive ions is exactly balanced by the total concentration of negative ions everywhere except in the surface layer adjacent to the plasma membrane.

Because the K^+ leak channel is slightly permeable to Na^+, some Na^+ leaks slowly into the cell down its electrochemical gradient. This net influx of Na^+ makes the inside of the cell more positively charged and, hence, decreases the membrane potential, allowing still more K^+ to leave the cell. This would lead in the end to the equalization of the Na^+ and K^+ concentrations across the membrane and to the disappearance of the membrane potential were it not for the Na^+-K^+ pump, which maintains the Na^+ and K^+ concentration gradients across the membrane by pumping Na^+ out and K^+ in.

In cells other than the squid neuron, similar principles operate to generate the membrane potential (which varies between −20 mV and −200 mV, depending on the species and cell type), although the plasma membrane permeabilities to certain ions are different. For example, the plasma membranes of many cells are permeable to Cl^-, which therefore makes an important contribution to the membrane potential.

The Ubiquitous Plasma Membrane Na^+-K^+ Pump Is an ATPase[27]

The plasma membranes of virtually all animal cells contain a Na^+-K^+ pump that operates as an antiport, actively pumping Na^+ out of the cell and K^+ in against their concentration gradients (and in the case of Na^+, against an electrical gradient as well). The Na^+ and K^+ gradients maintained by the Na^+-K^+ pump are responsible not only for the cell's membrane potential, but also for controlling cell volume and for driving the active transport of sugars and amino acids (see below). It is, therefore, not surprising that more than one-third of an animal cell's energy requirement is consumed in fueling this pump. In electrically active nerve cells, which must continually reestablish their membrane potentials following depolarization, this figure approaches 70% of the cell's total energy requirement.

A major advance in understanding the Na^+-K^+ pump came with the discovery in 1957 that an enzyme that hydrolyzes ATP to ADP and phosphate requires Na^+ and K^+ for optimal activity. An important clue linking this **Na^+-K^+ ATPase** with the Na^+-K^+ pump was the observation that a known inhibitor of the pump, *ouabain*, also inhibited the ATPase. But the crucial evidence that ATP hydrolysis somehow provided the energy for driving the pump came from studies of resealed red blood cell ghosts, in which the concentrations of ions, ATP, and drugs on either side of the membrane could be varied and the effects on ion transport and ATP hydrolysis observed. It was found that (1) the transport of Na^+ and K^+ is tightly coupled to ATP hydrolysis, so that one cannot occur without the other; (2) ion transport and ATP hydrolysis can occur only when Na^+ and ATP are present inside the ghosts and K^+ is present on the outside; (3) ouabain is inhibitory only when present outside the ghosts, where it competes for the K^+ binding site; and (4) for every molecule of ATP hydrolyzed (100 ATP molecules can be hydrolyzed by each ATPase molecule each second), $3Na^+$ are pumped out and $2K^+$ are pumped in (Figure 6–53).

Although these experiments provided compelling evidence that ATP supplies the energy for pumping Na^+ and K^+ ions across the plasma membrane, they did not explain how ATP hydrolysis is coupled to ion transport. A partial explanation was provided by the finding that the terminal phosphate group of the ATP is transferred to an aspartyl residue of the ATPase in the presence of Na^+. This phosphate group is subsequently hydrolyzed in the presence of K^+, and it is this last step that is inhibited by ouabain. The Na^+-dependent phosphorylation presumably changes the conformation of the ATPase, somehow resulting in the transport of Na^+ out of the cell, while the K^+-dependent dephosphorylation results in the transport of K^+ into the cell and a return of

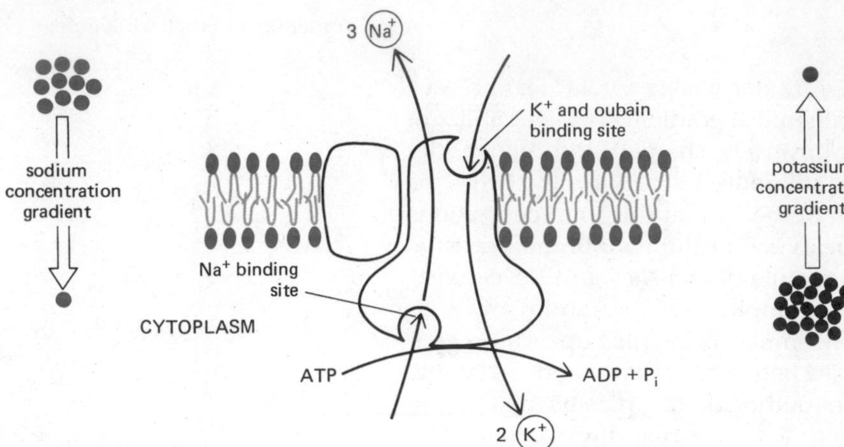

Figure 6–53 Schematic diagram of the Na^+-K^+ ATPase actively pumping Na^+ out and K^+ into a cell against their concentration gradients. For every molecule of ATP hydrolyzed, $3Na^+$ are pumped out and $2K^+$ are pumped in. Note that K^+ and ouabain compete for the same site on the external side of the ATPase. Although the Na^+-K^+ ATPase is shown as a dimer, it probably exists in the membrane as at least a tetramer of two small and two large subunits.

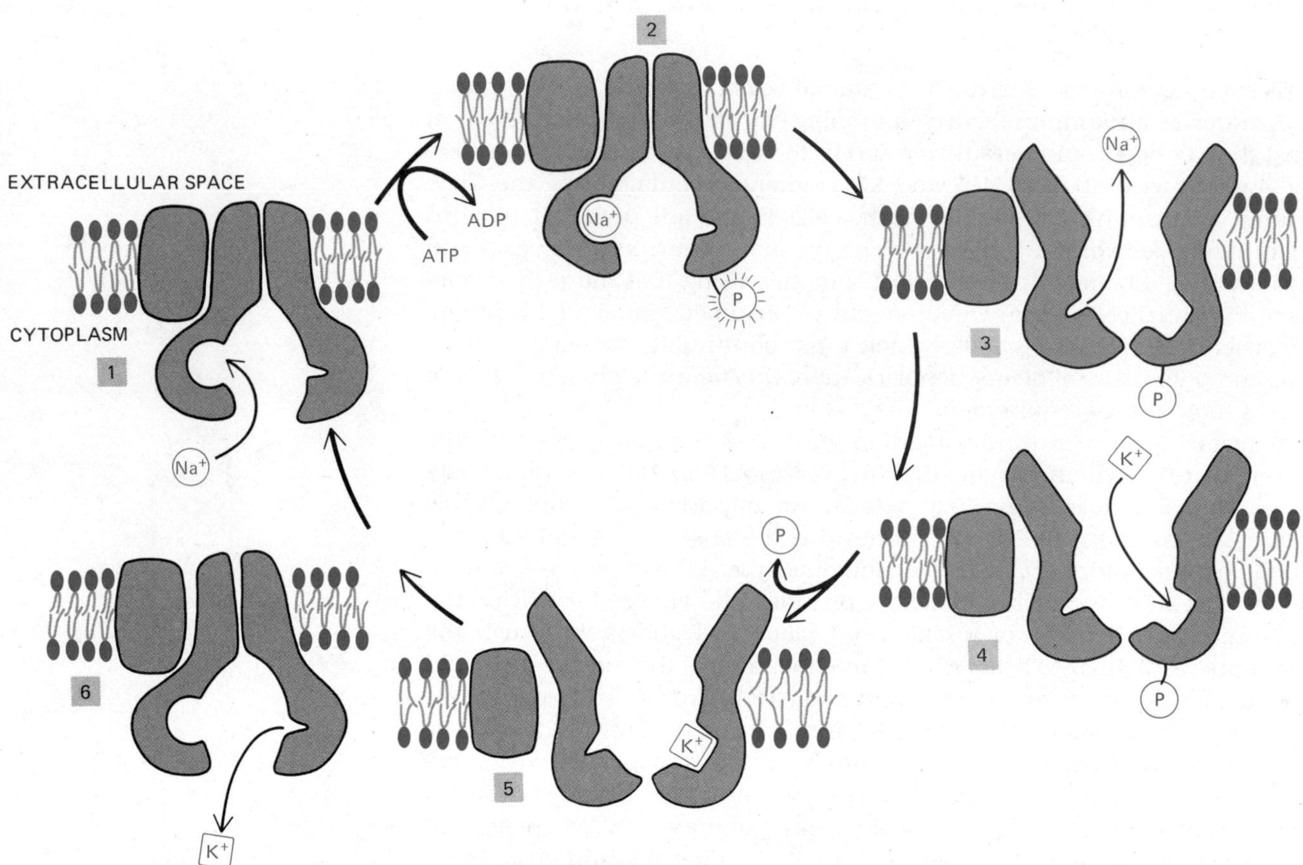

Figure 6–54 A schematic model of the Na^+-K^+ ATPase. The binding of Na^+ (1) and the subsequent phosphorylation (2) on the cytoplasmic face of the ATPase induce the protein to undergo a conformational change that transfers the Na^+ across the membrane and releases it on the outside (3). Then the binding of K^+ on the external surface (4) and the subsequent dephosphorylation (5) return the protein to its original conformation, which transfers the K^+ across the membrane and releases it into the cytoplasm (6). These changes in conformation are analogous to the ping ⇄ pong transitions shown in Figure 6–51, except that here the phosphorylation and dephosphorylation of the protein induce these transitions in an orderly manner so as to cause the protein to do useful work.

the ATPase to its original conformation (Figure 6–54). In fact, the Na^+-K^+ pump in red blood cell ghosts can be reversed: when the Na^+ and K^+ gradients are increased to such an extent that the pump can no longer cope, these ions move down their concentration gradients and ATP is synthesized from ADP and phosphate within the ghost. This suggests that the phosphorylated form of the ATPase is in a highly strained conformation and can relax and give up its energy (and phosphate) either to pump K^+ into the ghost or (when the Na^+ and K^+ gradients are extreme) to synthesize ATP (see p. 136).

The Na^+-K^+ ATPase has now been purified and found to consist of a transmembrane catalytic subunit (~100,000 daltons) and an associated glycoprotein (~45,000 daltons). The former has binding sites for Na^+ and ATP on its cytoplasmic surface and for K^+ and ouabain on its external surface and is reversibly phosphorylated and dephosphorylated. The function of the glycoprotein is unknown. A functional Na^+-K^+ pump can be reconstituted from the purified ATPase in the following way: the ATPase is solubilized in detergent, purified, and mixed with appropriate phospholipids; when the detergent is removed by dialysis, membrane vesicles are formed that pump Na^+ and K^+ in opposite directions in the presence of ATP and Mg^{2+} (Figure 6–55).

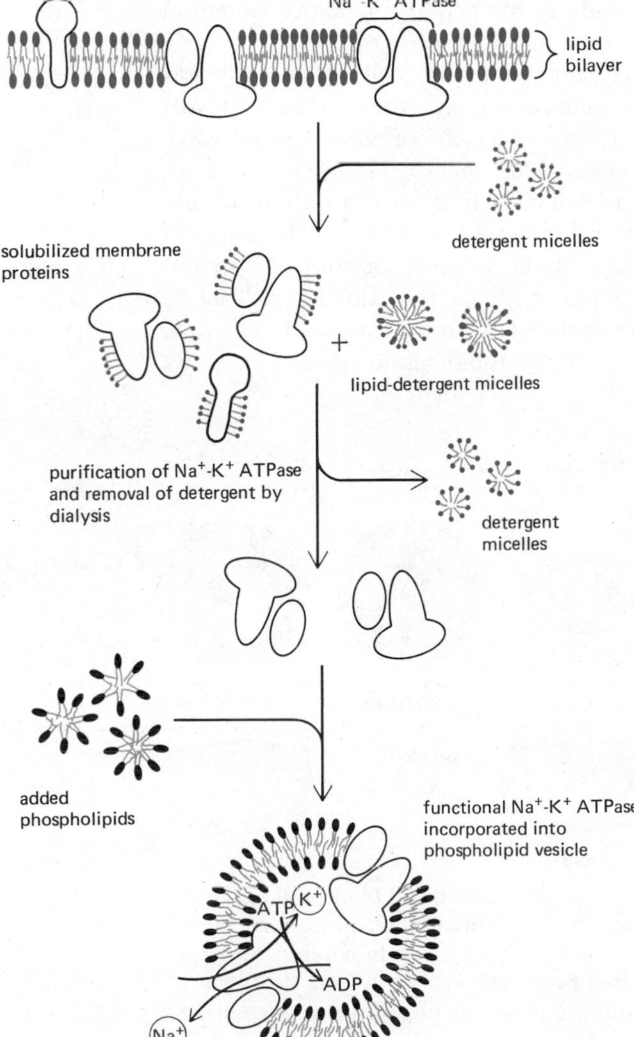

Figure 6–55 Solubilizing, purifying, and reconstituting functional Na^+-K^+ ATPase molecules into phospholipid vesicles. The reconstitution is carried out in the presence of high Na^+ and ATP so that these molecules end up in sufficiently high concentration inside the vesicles to allow the ATPase to function as a pump.

The Na^+-K^+ ATPase Helps Control Cell Volume by Controlling the Solute Concentration Inside Cells[28]

Since the Na^+-K^+ ATPase pumps 3 Na^+ out of the cell for every 2 K^+ it pumps in, it is "electrogenic"; that is, it tends to generate an electrical potential across the plasma membrane, with the inside negative relative to the outside. In this way, the Na^+-K^+ ATPase makes a small (~20%), direct contribution to the membrane potential. However, as we have seen, most of the membrane potential (~80%) is due to the K^+ and Na^+ gradients maintained by the Na^+-K^+ ATPase, combined with the greater permeability of the plasma membrane to K^+ than to Na^+ or anions.

The Na^+-K^+ ATPase also helps to regulate cell volume, since it controls the solute concentrations inside the cell and thereby the osmotic forces that would tend to make the cell swell or shrink. The macromolecules confined inside the cell exert an osmotic pressure on the plasma membrane; furthermore, since these macromolecules are mostly charged, they are necessarily accompanied by counterions, such as K^+, which add to the effect. This pressure on the inner face of the plasma membrane is counterbalanced by the osmotic pressure due to the molecules in the extracellular fluid—chiefly Na^+ and Cl^-. The Na^+ and the Cl^-, however, tend to leak into the cell down their concentration gradients, upsetting the balance and causing the cell to swell. The Na^+-K^+ ATPase solves this problem neatly: it directly pumps out of the cell the Na^+ ions that leak in, and at the same time it helps to prevent Cl^- ions from leaking in by keeping the inside of the cell at a negative potential that counteracts the effect of the Cl^- concentration gradient.

The importance of the Na^+-K^+ ATPase in controlling cell volume is shown by the fact that animal cells swell and sometimes burst if they are treated with ouabain, which inhibits the Na^+-K^+ ATPase. There are, of course, other ways for a cell to cope with its osmotic problems: plant cells and many bacteria are prevented from bursting by the semirigid cell walls that surround their plasma membranes; in amoebae the excess water that flows in osmotically is collected in contractile vacuoles, which discharge their contents periodically to the exterior. But for most cells in multicellular animals, the Na^+-K^+ ATPase is crucial. Figure 6–56 illustrates the way that cells with and without cell walls respond when exposed to hypotonic and hypertonic solutions.

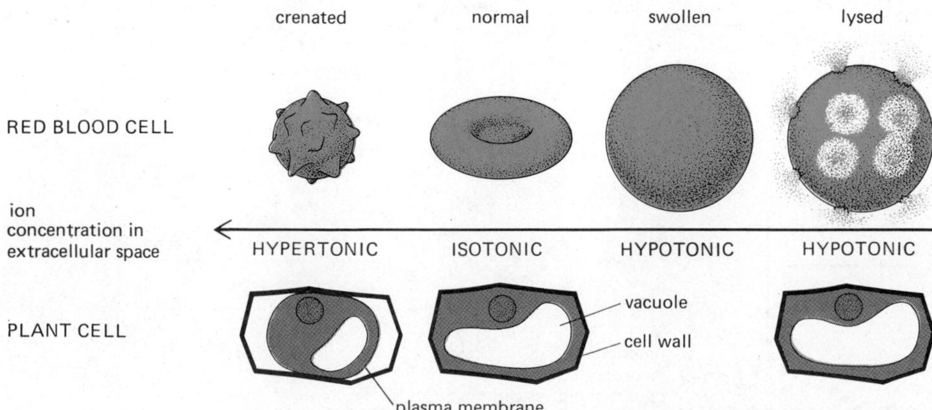

Figure 6–56 Response of a human red blood cell and a plant cell to changes in osmolarity of the extracellular fluid. Whereas the red cell swells and lyses in hypotonic fluid, the plant cell is protected by its cell wall. Both cells shrink in hypertonic fluid, and the plant cell membrane retracts from the cell wall.

Some Ca^{2+} Pumps Are Also Membrane-bound ATPases[29]

Eucaryotic cells maintain very low concentrations of Ca^{2+} in their cytosol ($\leq 10^{-7}$ M) in the face of very much higher extracellular Ca^{2+} concentrations ($\sim 10^{-3}$ M). The flow of Ca^{2+} ions down this enormous gradient in response to extracellular signals is an important means of transmitting such signals across the plasma membrane to the cell's interior. The gradient is in part maintained by the Ca^{2+} pumps in the plasma membrane that actively transport Ca^{2+} out of the cell. Although some of these pumps are now known to be ATPases, they have not been isolated in pure form, and relatively little is known about their structure or function.

More is known about a similar membrane-bound Ca^{2+} pump in the *sarcoplasmic reticulum* of muscle cells. The sarcoplasmic reticulum forms a network of fine channels in muscle cells and serves as an intracellular store of Ca^{2+}. The Ca^{2+} pump is responsible for pumping Ca^{2+} from the cytosol into the sarcoplasmic reticulum. (When a nerve impulse depolarizes the muscle cell membrane, Ca^{2+} is released from the sarcoplasmic reticulum into the cytosol, stimulating the muscle to contract.) Like the Na^+-K^+ pump, the Ca^{2+} pump is an ATPase that is phosphorylated and dephosphorylated during its pumping cycle. Each molecule can hydrolyze up to 10 ATP molecules per second on its cytoplasmic surface and pumps 2 Ca^{2+} ions into the sarcoplasmic reticulum for every ATP hydrolyzed. Since this Ca^{2+} ATPase is the only major protein in the membrane of the sarcoplasmic reticulum (accounting for about 90% of its total protein), it has been relatively easy to purify. It is a single polypeptide chain containing about 1000 amino acid residues; when incorporated into phospholipid vesicles (as described above for the Na^+-K^+ ATPase), this ATPase pumps Ca^{2+} as it hydrolyzes ATP.

Membrane-bound Enzymes That Synthesize ATP Are Transport ATPases Working in Reverse[30]

Enzymes very similar to the two transport ATPases discussed above are found in the plasma membranes of aerobic bacteria and in the inner mitochondrial and chloroplast membranes of eucaryotic cells, where they normally work in reverse: instead of ATP hydrolysis driving ion transport (as is the case for the Na^+-K^+ ATPase and Ca^{2+} ATPase), H^+ gradients across these membranes drive the synthesis of ATP from ADP and phosphate; for this reason they are called *ATP synthetases.* The H^+ gradients are generated during the electron transport steps of oxidative phosphorylation (in bacteria and mitochondria) or photosynthesis (in chloroplasts) or by the light-activated H^+ pump (bacteriorhodopsin) in *Halobacterium.* In all of these cases, the enzymes can work in either direction, depending on the conditions: they can hydrolyze ATP and pump H^+ across the membrane, or they can synthesize ATP when H^+ flows through the enzymes in the reverse direction. These enzymes are discussed in more detail in Chapter 9.

Active Transport Can Be Driven by Ion Gradients[24]

Many active transport systems are driven by the energy stored in ion gradients rather than directly by ATP hydrolysis. All of these function as co-transport systems—some as symports, others as antiports. In animal cells the co-transported ion is usually Na^+. For example, the active transport of some sugars and amino acids into animal cells is powered by the Na^+ gradient across the plasma membrane. Glucose uptake into intestinal and kidney cells is achieved by a symport system in which glucose and Na^+ bind to different sites on the glucose carrier protein; Na^+ tends to move into the cell down its electro-

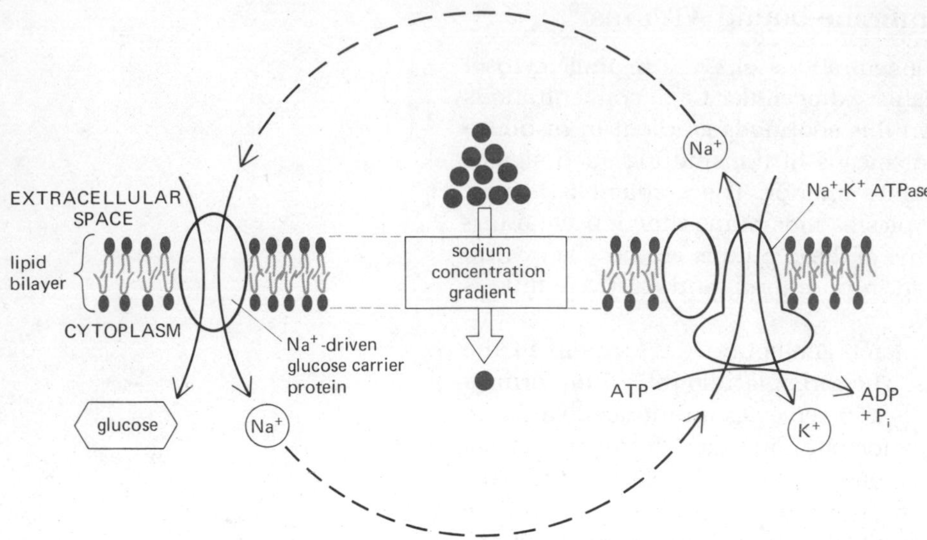

Figure 6–57 Schematic drawing of glucose symport, showing how the active transport of glucose is driven by a Na^+ gradient that is generated and maintained by the Na^+-K^+ ATPase.

chemical gradient and, in a sense, "drags" the glucose in with it. The greater the Na^+ gradient, the greater the rate of glucose entry. Conversely, if the Na^+ concentration in the extracellular fluid is markedly reduced, glucose transport stops. The Na^+ that enters the cell with glucose is pumped out by the Na^+-K^+ ATPase which, by maintaining the Na^+ gradient, indirectly drives the transport of glucose (Figure 6–57). A hypothetical (and oversimplified) model of how such a symport system could work is shown in Figure 3–61, page 138.

There are at least five different carrier proteins for amino acids in the plasma membranes of many animal cells (each specific for a group of closely related amino acids) that operate as symport systems with Na^+. The Na^+ gradient can also drive antiport systems. For example, when a sea urchin egg is fertilized, a large increase in intracellular pH activates protein and DNA synthesis in the egg; this change in pH is caused by an efflux of H^+ that is driven by an influx of Na^+.

In many cells, Na^+-linked carrier proteins serve to bring metabolites (like sugars and amino acids) into the cell. However, in some epithelial cells, such as those involved in absorbing nutrients from the gut, transport proteins are distributed asymmetrically in the plasma membrane and thereby contribute to the **transcellular transport** of absorbed solutes. As shown in Figure 6–58, Na^+-linked symports, located in the plasma membrane on the apical (absorptive) surface of the epithelial cell, transport nutrients into the cell, while Na^+-independent transport proteins in the basal and lateral membrane of the cell allow nutrients to leave the cell down their concentration gradients. The Na^+ gradient across the plasma membrane of these cells is maintained by a Na^+-K^+ ATPase located in the basal and lateral membranes. Related mechanisms are used by kidney and intestinal epithelial cells to pump water from one extracellular space to another.

In many of these epithelial cells, the plasma membrane area is greatly increased by the formation of thousands of **microvilli,** which extend as thin, fingerlike projections from the apical surface (Figure 6–58). Such microvilli can increase the total absorptive area of a cell by as much as 25-fold, thereby greatly increasing its transport capabilities.

In bacteria, most active transport systems driven by ion gradients use H^+ rather than Na^+ as the co-transported ion. For example, the active transport of most sugars and amino acids into bacterial cells is driven by the H^+

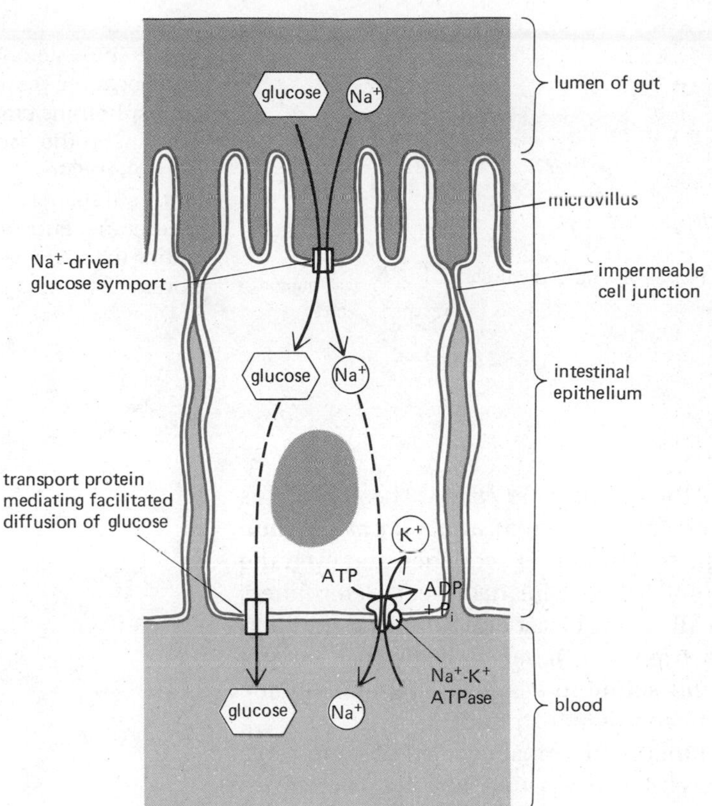

Figure 6–58 Asymmetrical distribution of transport proteins in an intestinal epithelial cell results in the transcellular transport of glucose from the gut lumen to the blood. Glucose is pumped into the cell through the cell's apical membranc by a Na^+-powered glucose symport, and glucose passes out of the cell (down its concentration gradient) by facilitated diffusion mediated by a different glucose carrier protein in the basal and lateral membrane. The Na^+ gradient driving the glucose symport is maintained by the Na^+-K^+ ATPase in the basal and lateral plasma membrane, which keeps the internal concentration of Na^+ low.

gradient across the plasma membrane. The *lactose carrier protein* (permease, or M protein) is the most extensively studied example. It is a single transmembrane protein (30,000 daltons) that functions as an H^+ symport: one proton is co-transferred for every lactose molecule transported into the cell.

Active Transport in Bacteria Can Occur by "Group Translocation" and Can Involve Water-soluble Binding Proteins[24,31]

Thus far we have seen that active transport can be driven by light (as in bacteriorhodopsin), by ATP hydrolysis, or by ion gradients. In principle, a fourth strategy would be to "trap" a molecule that has entered the cell passively by modifying the molecule in such a way that it cannot escape through the same channel. In fact, some bacteria use such a mechanism, called **group translocation.** For example, the active transport of sugars into some bacteria involves the active phosphorylation of the sugars during their transfer across the plasma membrane. Because they are ionized and cannot leak out, the resulting sugar phosphates accumulate in the cell. Moreover, by phosphorylating the transported sugars, the concentration of unphosphorylated sugars inside the cell is kept very low so that a concentration gradient of sugar continues to push these molecules into the cell. The most extensively studied system occurs in bacteria, where the phosphorylation mechanism is complex, involving at least four separate membrane proteins and phosphoenolpyruvate (rather than ATP) as the high-energy-phosphate donor (Figure 6–59).

Many active sugar and amino acid transport systems in bacteria involve a special class of water-soluble proteins located in the space between the

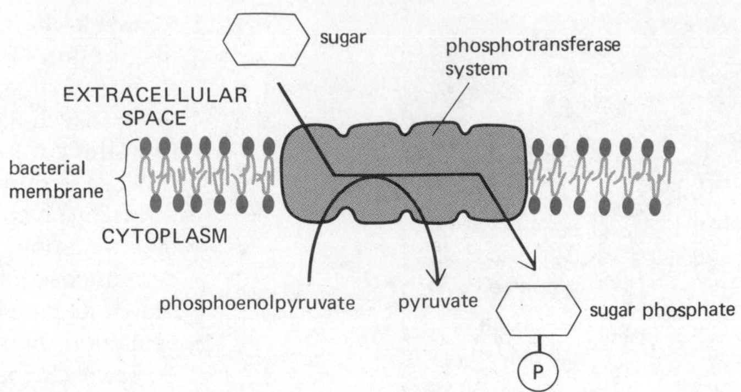

Figure 6–59 Active transport of sugars into bacteria by group translocation. A special "phosphotransferase system" of proteins in the bacterial membrane phosphorylates the sugar during the sugar's transport through the membrane. Phosphoenolpyruvate rather than ATP is the phosphate donor.

plasma membrane and the cell wall—the *periplasmic space.* These proteins have specific binding sites for the solute and are therefore called **periplasmic binding proteins.** Although the complete amino acid sequence and even the three-dimensional structure of some of these proteins have been determined, it is still unclear how they function. In all cases at least one protein in addition to the binding protein is required for transport. It seems likely that each of these binding proteins passes its bound solute to a specific transmembrane transport protein, but this has not yet been demonstrated.

The same binding proteins that function in transport also seem to serve as receptors in *chemotaxis,* an important process that enables bacteria to swim towards an increasing concentration of a specific nutrient. In chemotaxis, many different binding proteins converge on a smaller number of membrane proteins that transmit signals to the cell interior (see p. 759). It would make sense for the many binding proteins to show the same economy in their transport function by sharing a small number of transmembrane transport proteins.

Some Transmembrane Protein Channels Are "Gated" and Open Only Transiently[32]

We have seen that two classes of membrane transport proteins function in passive transport: (1) carrier proteins that have the characteristics of membrane-bound enzymes and (2) proteins that form transmembrane hydrophilic channels through which solutes of appropriate size and charge can pass by relatively simple diffusion. Not surprisingly, transport through channel proteins can occur at a very much faster rate than transport mediated by carrier proteins.

While some channels formed by transport proteins are continuously open, others open only transiently. The latter channels are said to be "gated." Some gated channels open in response to an extracellular ligand binding to a specific cell-surface receptor and are called **ligand-gated channels;** others open in response to a change in the membrane potential and are called **voltage-gated channels** (Figure 6–60). Still others open in response to changes in the intracellular concentration of specific ions; for example, some K^+ channels open when the concentration of free Ca^{2+} in the cytosol increases.

In many cases, gated channels have self-closing mechanisms that rapidly reclose the channels even if the stimulus that originally opened them is still operating. Consider the neuromuscular junction, where an impulse traveling down a nerve stimulates a muscle to contract. This apparently simple response involves the sequential opening and closing of at least four different sets of gated channels—all within less than one second (Figure 6–61):

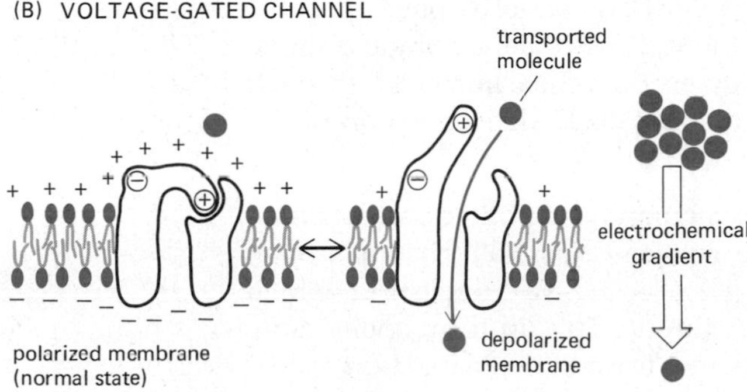

Figure 6–60 Schematic diagram of two types of gated ion channels that allow ions to pass down their electrochemical gradients only when the proteins are in the "open" configuration. The channel in (A) is ligand-gated and opens when an extracellular ligand binds to it (or to an associated membrane protein), while the channel in (B) is voltage-gated and opens when the membrane is depolarized.

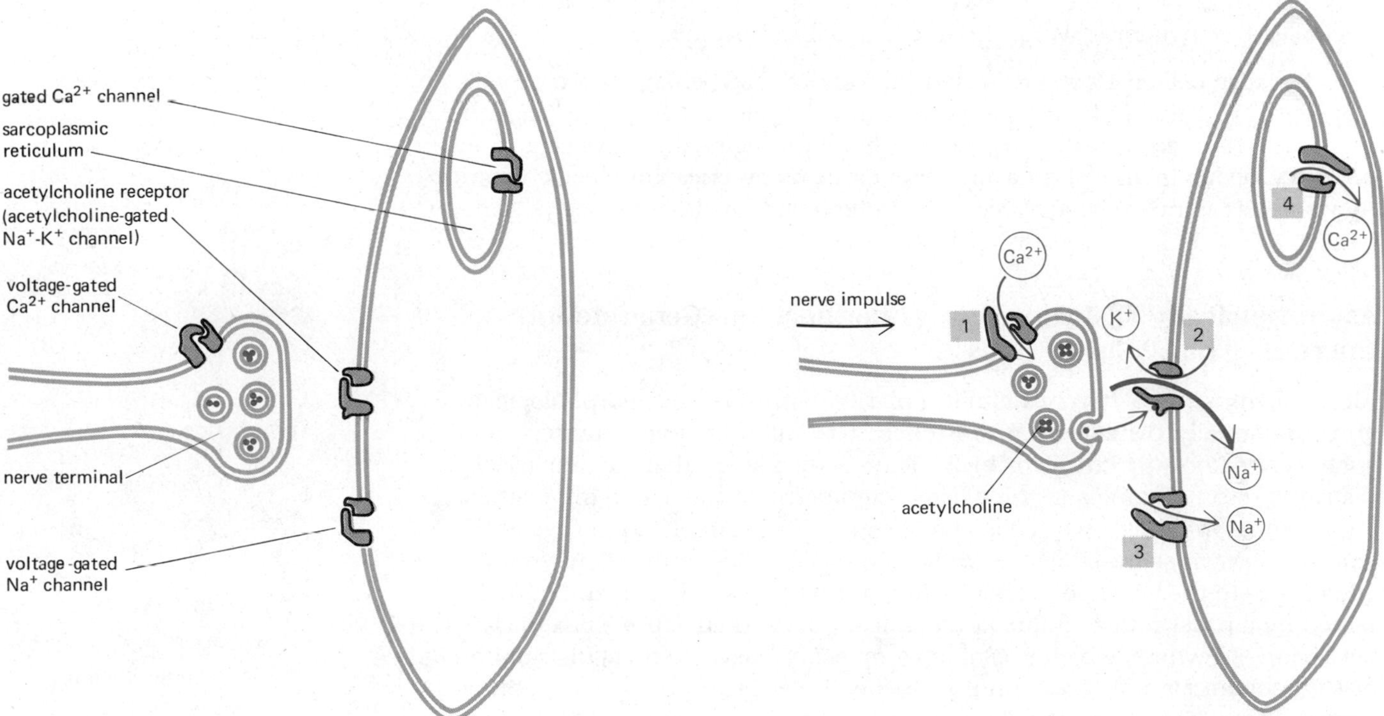

Figure 6–61 Schematic diagram of a neuromuscular junction illustrating some of the gated channels involved in the stimulation of muscle contraction by a nerve impulse. The various gated channels are numbered in the sequence in which they open.

1. The process is initiated when the nerve impulse reaches the nerve terminal; the decrease in the membrane potential *(depolarization)* transiently opens voltage-gated Ca^{2+} channels in the plasma membrane of this terminal. Since the Ca^{2+} concentration outside cells is always more than 1000 times greater than the free concentration inside, Ca^{2+} rushes into the nerve terminal, stimulating the terminal to secrete the neurotransmitter *acetylcholine.*
2. The released acetylcholine binds to acetylcholine receptor proteins on the opposing muscle cell plasma membrane. These receptors are ligand-gated cation channels, permeable to Na^+ and K^+, that are transiently opened by the binding of acetylcholine. Since the concentration of Na^+ is higher outside than inside, while the reverse is true for K^+, there is a net flow of Na^+ into and K^+ out of the muscle cell while the channel is open (about 1 msec). Because the electrochemical gradient across the membrane is much steeper for Na^+, the influx of Na^+ greatly exceeds the efflux of K^+ through the acetylcholine-activated channel and a localized depolarization of the muscle membrane occurs.
3. Contained within the same muscle cell membrane are other protein channels permeable mainly to Na^+. These voltage-gated Na^+ channels open briefly when the membrane depolarizes, allowing Na^+ to rush in, which further depolarizes the muscle cell membrane. This, in turn, opens more voltage-gated Na^+ channels and results in a wave of depolarization (called the *action potential)* spreading to involve the entire muscle membrane.
4. As a result of this action potential, Ca^{2+} channels in the sarcoplasmic reticulum membrane transiently open, allowing Ca^{2+} to escape into the cytosol. It is the sudden increase in intracellular Ca^{2+} concentration that causes the myofibrils within the muscle cell to contract.

Because cells with voltage-gated Na^+ and/or Ca^{2+} channels are capable of generating action potentials, they are said to be "electrically active." Ligand-gated ion channels, however, are not confined to electrically active cells: they probably occur in the plasma membranes of many eucaryotic cells, where they serve to convey signals from outside the cell to its interior (see p. 742).

Asymmetrically Distributed Ion Channels Can Generate Ion Currents That Polarize Cells[33]

Although most cells have a definite polarity, with one end morphologically and functionally distinct from the other, it is still a complete mystery how such cells "know" front from back. There is evidence that asymmetrically distributed ion channels in the plasma membrane can generate intracellular ion currents that help to establish cell polarity. For example, in the egg of the common brown seaweed, *Fucus*, there is a measurable current flow from the apical pole to the basal pole. This current seems to be generated and/or maintained by a passive Ca^{2+} influx at the apical pole and an active efflux of Ca^{2+} occurring elsewhere, which presumably reflects an asymmetrical distribution of Ca^{2+} channels and Ca^{2+} pumps, respectively, in the plasma membrane (Figure 6–62). This current flow across the egg could move highly charged intracellular molecules by electrophoresis, and there is evidence that it orients the cytoplasm in such a way as to affect subsequent embryonic development. Although a typical somatic cell is very much smaller than an egg, it would not be surprising if localized intracellular ion flows, generated by an asymmetric distribution of membrane transport proteins, were involved in the polarization of some somatic cells as well.

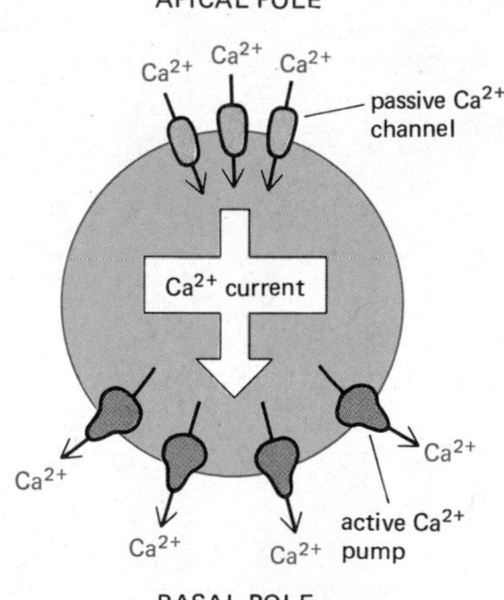

Figure 6–62 Schematic diagram showing how the asymmetrical distribution of transport proteins in the plasma membrane of the *Fucus* egg cell may generate an intracellular current. The current is thought to help polarize the egg and thereby influence subsequent embryonic development.

Ionophores Increase the Ion Permeability of Synthetic and Biological Membranes[23,34]

Ionophores are small hydrophobic molecules that dissolve in lipid bilayers and increase the ion permeability of the bilayer. Most are synthesized by microorganisms (presumably as biological weapons to weaken their competitors), and some have been used as antibiotics. They have been widely employed to increase membrane permeability to specific ions in studies on synthetic bilayers, cell organelles, and more recently, on intact cells. There are two classes of ionophores—**mobile ion carriers** and **channel formers** (Figure 6–63). Both types operate by shielding the charge of the transported ion so that it can penetrate the hydrophobic interior of the lipid bilayer. Since they are not coupled to energy sources, they only permit net movement of ions down their electrochemical gradients.

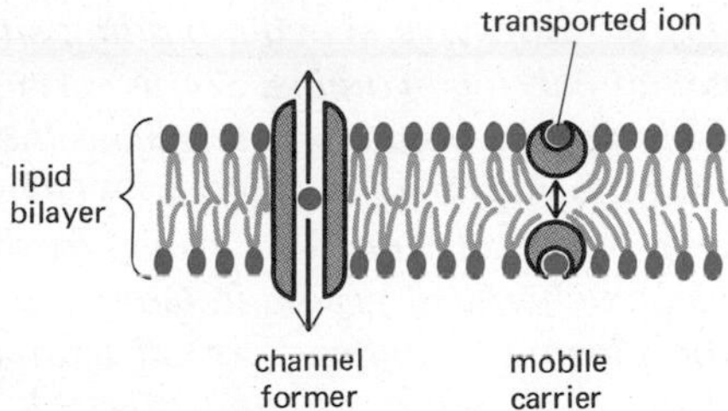

Figure 6–63 Schematic diagram of a mobile ion carrier and a channel-forming ionophore. In both cases, net ion flow occurs only down an electrochemical gradient.

Valinomycin is an example of a mobile ion carrier. It is a ring-shaped polymer that increases the permeability of a membrane to K^+. The ring has a hydrophobic exterior (made up of valine side chains), which contacts the hydrocarbon core of the lipid bilayer, and a polar interior where a single K^+ can precisely fit (Figure 6–64). Valinomycin transports K^+ down its electrochemical gradient by picking up a K^+ on one side of the membrane, diffusing across the bilayer, and releasing K^+ on the other side. Since the shuttle operates in both directions, net transport occurs only if there are more potassium ions that bind to the carrier when it is moving in one direction than in the other.

The ionophore *A23187* is another example of a mobile ion carrier, but it transports divalent cations such as Ca^{2+} and Mg^{2+}. At the low concentrations generally involved, this ionophore acts as an ion-exchange shuttle, carrying two H^+ out of the cell for every divalent cation it carries in. Therefore, unlike valinomycin, it does not depolarize the membrane. When living cells are exposed to A23187, Ca^{2+} rushes into the cytosol down a steep electrochemical gradient. Accordingly, this ionophore is widely used in cell biology to increase the concentration of free Ca^{2+} in the cytosol.

If the temperature of the membrane is lowered to below its freezing point, mobile carriers can no longer diffuse across the lipid bilayer and transport stops. This temperature dependence serves to identify an ionophore as a mobile carrier, since, by contrast, channel-forming ionophores continue to transport normally when the bilayer is frozen.

Gramicidin A is an example of a channel-forming ionophore. It is a linear peptide of 15 amino acids, all with hydrophobic side chains. Two such molecules are thought to come together in the bilayer to form a transmembrane channel that selectively allows monovalent cations to flow down their electrochemical gradients. These dimers are unstable and are constantly forming and dissociating, so that the average open time for a channel is about one second. With a large electrochemical gradient, Gramicidin A can transport about 2×10^7 cations per open channel in one second, which is a thousand times more than can be transported by a single mobile carrier molecule in the same time.

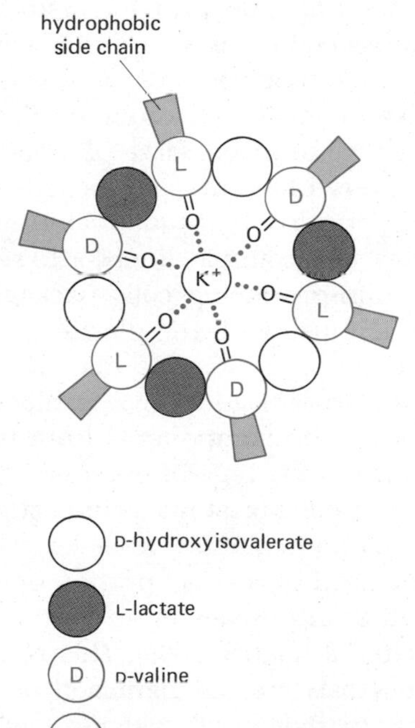

Figure 6–64 A valinomycin molecule with a K^+ ion bound by six oxygen atoms at the center of the ring structure.

Summary

Lipid bilayers are highly impermeable to most polar molecules. In order to transport such molecules into or out of cells, plasma membranes contain many specific transport proteins, each of which is responsible for transferring a particular solute across the membrane. These transport proteins form continuous protein pathways across the bilayer. Many transport proteins act like enzymes in that they have specific binding sites for the transported molecule; these are called carrier proteins. Some carrier proteins only catalyze the facilitated diffusion of the bound solute across the bilayer. Others are driven through a series of conformational changes by ATP hydrolysis or by ion binding, and are thereby able to act as pumps to transport the bound solute actively "uphill" against its electrochemical gradient. One very important pump is the Na^+*-*K^+ *ATPase in the plasma membrane of all animal cells, which uses the energy of ATP hydrolysis to pump* Na^+ *out of and* K^+ *into the cell.*

A different type of transport protein forms an open channel across the bilayer through which small molecules can move down their electrochemical gradients by simple diffusion. These passive channels permit much larger fluxes across the membrane than do channel proteins. Some of these channels are continuously open, while others are gated, opening and closing transiently in response to extracellular signaling ligands or to changes in membrane potential or intracellular ion concentrations.

Membrane Transport of Macromolecules and Particles: Exocytosis and Endocytosis

While transport proteins mediate the passage of many small polar molecules across cell membranes, they cannot transport macromolecules, such as proteins, polynucleotides, or polysaccharides. Yet most cells are able to eject and take in specific macromolecules across their plasma membranes; some even manage to ingest large particles. The mechanisms by which the cells do this are very different from those that mediate small solute and ion transport, for they involve the sequential formation and fusion of membrane-bounded vesicles. For example, in order to secrete insulin across their plasma membranes, insulin-producing cells package insulin molecules in intracellular vesicles, which fuse with the plasma membrane and open to the extracellular space, thereby releasing the insulin to the exterior. This fusion process is called **exocytosis** (Figure 6–65). In all eucaryotic cells, other vesicles carry new plasma membrane components from the Golgi apparatus to the plasma membrane by the same type of process.

Cells ingest macromolecules and particles by a similar mechanism, only with the sequence reversed. The substance to be ingested is progressively enclosed by a small portion of the plasma membrane, which first invaginates and then pinches off to form an intracellular vesicle containing the ingested material (Figure 6–65). This process is called **endocytosis.** Two types of endocytosis are distinguished on the basis of the size of the vesicles formed: **pinocytosis** ("cell drinking"), which involves the ingestion of fluid and/or solutes via small vesicles, and **phagocytosis** ("cell eating"), which involves the ingestion of large particles, such as microorganisms or cell debris, via large vesicles (often called *vacuoles*). While most kinds of cells are continually ingesting fluid and solutes by pinocytosis, large particles are ingested mainly by specialized phagocytic cells. For this reason, the terms pinocytosis and endocytosis tend to be applied to most cells interchangeably.

An important feature of both exocytosis and endocytosis is that the secreted or ingested macromolecules are sequestered in vesicles and do not generally mix with other macromolecules or organelles in the cell. By unknown mechanisms, each vesicle fuses only with specific membrane structures, ensuring an orderly transfer of macromolecules between the outside and inside of the cell. A very similar process operates in the transfer of newly synthesized macromolecules between various compartments within cells; here vesicles bud (Figure 6–65) from the endoplasmic reticulum and Golgi complex, migrate, and fuse with another membrane inside the cell. Although it is clear

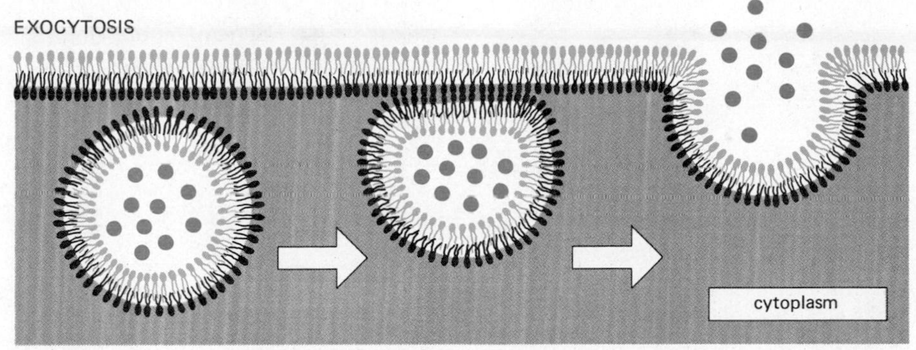

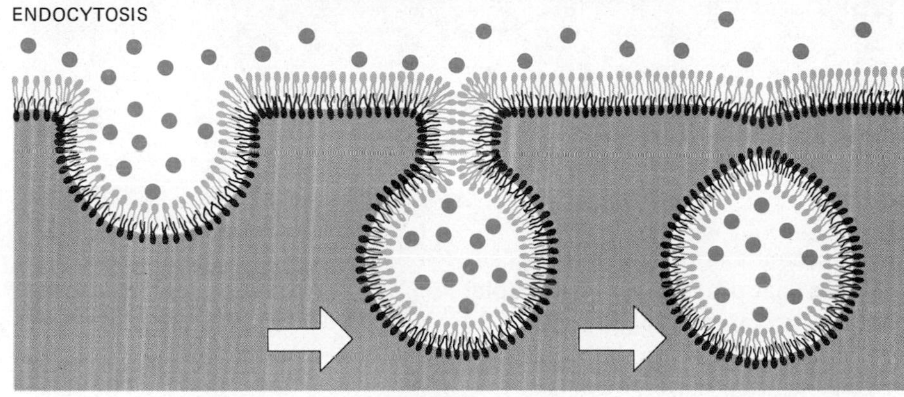

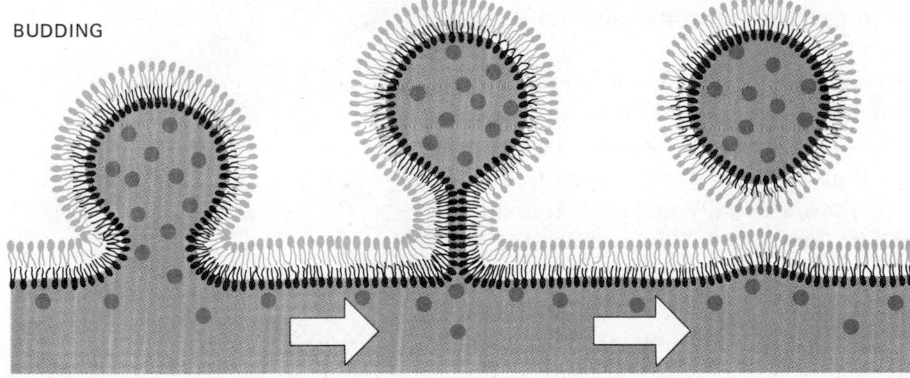

Figure 6–65 Schematic drawing of bilayer adherence and bilayer joining in three membrane fusion processes—exocytosis, endocytosis, and budding. Exocytosis and endocytosis are discussed in this chapter; budding, which is involved in the formation of intracellular transport vesicles, is discussed in Chapter 7. Note that because of the bilayer adherence step, endocytosis and exocytosis are not simply the reverse of each other: in exocytosis two cytoplasmic-side monolayers of the plasma membrane adhere, while in endocytosis two noncytoplasmic-side monolayers of the plasma membrane adhere. This difference presumably allows endocytosis and exocytosis to be regulated separately.

that the rapid, large-scale formation and fusion of vesicles is a fundamental feature of all eucaryotic cells, much remains to be learned about the molecular mechanisms involved in driving and guiding this traffic along specific paths.

Exocytosis Occurs by the Fusion of Intracellular Vesicles with the Plasma Membrane[35]

Most cells secrete macromolecules (and often smaller molecules) to the exterior. In eucaryotic cells, secretion almost always occurs by exocytosis. Some of the secreted molecules adhere to the cell surface and become part of the cell coat, others are incorporated into the extracellular matrix, while still others diffuse into the interstitial fluid and/or blood to nourish or signal other cells.

As described in Chapter 7, proteins to be secreted are synthesized on ribosomes attached to the rough endoplasmic reticulum (ER). They pass into the lumen of the ER and are transported to the Golgi complex by ER-derived

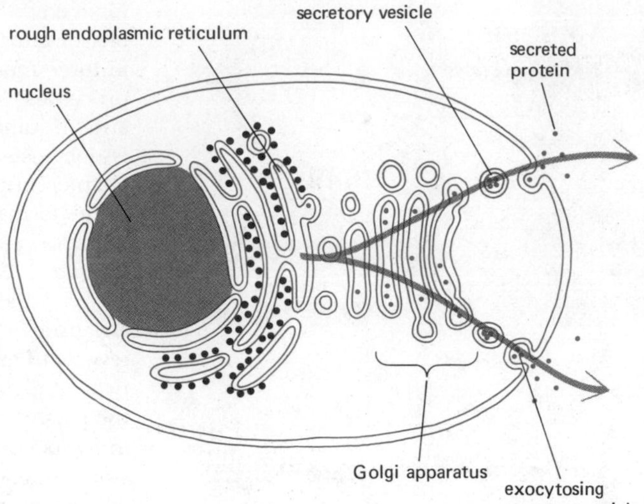

Figure 6–66 Schematic drawing showing the pathway followed by a secreted protein.

transport vesicles. In the Golgi complex, the proteins are modified, concentrated, sorted, and finally packaged into vesicles that pinch off from the Golgi apparatus and eventually fuse with the plasma membrane (Figure 6–66). By contrast, small molecules to be secreted are actively transported into secretory vesicles that have already formed. This process is often driven by an ion gradient. The small molecules are often complexed to specific macromolecules in the vesicles so that they can be stored at high concentration without generating an excessive osmotic gradient.

Some substances are continuously secreted by the cells that make them, while others are stored in **secretory vesicles,** to be released only when the cell is triggered by an extracellular signal. The signal to secrete is often a chemical messenger, such as a hormone, that binds to receptors on the cell surface. The resulting activation of the receptors usually causes a transient increase in the concentration of free Ca^{2+} in the cytosol. The increased Ca^{2+} in turn initiates the process of exocytosis, causing the secretory vesicles to fuse with the plasma membrane, thereby releasing their contents to the extracellular space. The secretory vesicle membrane becomes incorporated into the plasma membrane (Figure 6–65) and is later specifically retrieved by endocytosis, probably to be reincorporated in new secretory vesicles. The amount of vesicle membrane that is temporarily added to the plasma membrane can be enormous: in a pancreatic acinar cell, about 900 μm^2 of vesicle membrane is inserted into the apical plasma membrane (whose area is only 30 μm^2) when the cell is stimulated to secrete.

Triggered Exocytosis Is a Localized Response of the Plasma Membrane and Its Underlying Cytoplasm[36]

It is likely that Ca^{2+} acts *regionally* in secretory cells to initiate a localized response in a part of the cytoplasm and its overlying plasma membrane. This conclusion is suggested by experiments with mast cells, which secrete histamine when triggered by specific ligands that bind to receptors on their surface. It is the histamine secreted by mast cells that is responsible for many of the unpleasant symptoms, such as itching or sneezing, that accompany allergic reactions. When mast cells are incubated in a medium containing a soluble stimulant, exocytosis occurs all over the cell surface (Figure 6–67). However, if the stimulating ligand is attached to a solid bead so that it can interact only with a localized region of the mast cell surface, exocytosis is

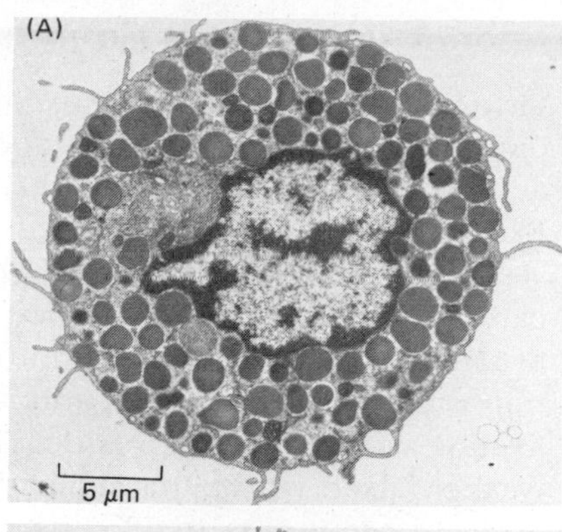

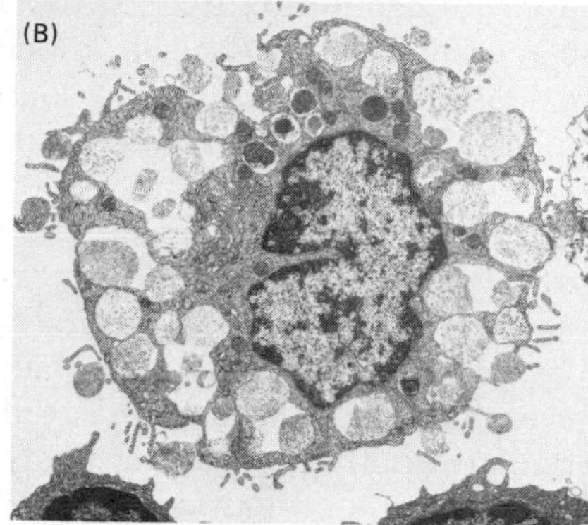

Figure 6–67 Electron micrographs of exocytosis in rat mast cells. The cell in (A) has not been stimulated. The cell in (B) has been activated to secrete its stored histamine by a soluble extracellular ligand. Histamine-containing vesicles are dark; those that have released their histamine are light. The material remaining in the spent vesicles consists of a macromolecular network to which the stored histamine is normally bound.

Once a secretory vesicle has fused with the plasma membrane, it can then fuse with a limited number of other vesicles within the cell. In this way, many vesicles in mast cells are able to open to the extracellular space indirectly through other open vesicles. As a consequence, the cell in (B) contains several large cavities lined by the fused membranes of many spent vesicles, which are now in continuity with the plasma membrane. All of these cavities are open to the extracellular space; however, this is not always apparent in one plane of section through the cell. (From D. Lawson, C. Fewtrell, B. Gomperts, and M. Raff, *J. Exp. Med.* 142:391–402, 1975. By copyright permission of the Rockefeller University Press.)

restricted to the region where the cell actually contacts the bead (Figure 6–68). Clearly, the mast cell does not respond as a whole when it is triggered; the activation of receptors, the resulting Ca^{2+} influx, and the subsequent exocytosis must all be localized in the particular region of the cell that has been excited. This demonstrates an important property of plasma membranes: individual segments can function independently of the rest of the membrane.

How does a localized increase of Ca^{2+} concentration in the cytosol cause secretory vesicles in the vicinity to fuse with the plasma membrane? Because liposomes constructed entirely from phospholipids with negatively charged polar head groups can be induced to fuse in the presence of very high concentrations of Ca^{2+} ($\geq 10^{-4}$ M), it has been argued that the Ca^{2+} can act on its own to induce membrane fusion within cells. However, in all cases studied, ATP is required for exocytosis in addition to the influx of Ca^{2+}, and it is unlikely that Ca^{2+} levels within cells ever become high enough to induce fusion directly. Whatever the mechanism, it seems probable that a Ca^{2+}-binding protein, such as *calmodulin* (which undergoes a large conformational change when it binds Ca^{2+}) is somehow involved (see p. 748).

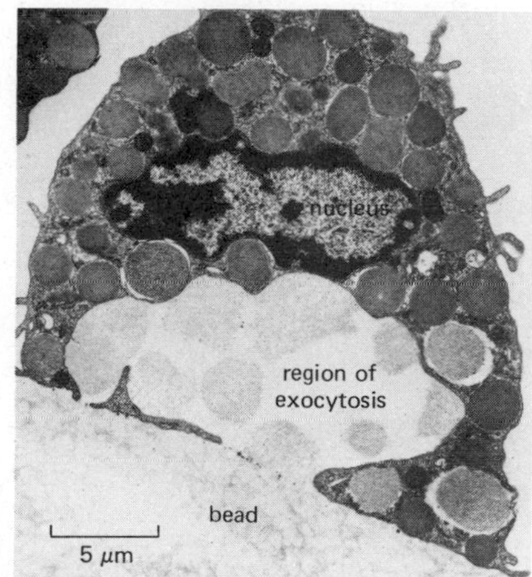

Figure 6–68 Electron micrograph of a mast cell that has been activated to secrete histamine by a stimulant coupled to a solid bead. Exocytosis has occurred only in the region of the cell that is in contact with the bead. (From D. Lawson, C. Fewtrell, and M. Raff, *J. Cell Biol.* 79:394–400, 1978. By copyright permission of The Rockefeller University Press.)

Membrane Fusion Involves Bilayer Adherence Followed by Bilayer Joining[35]

Because of their bilayer structure, cell membranes seen in cross-section in electron micrographs look like a sandwich, with two parallel dense lines separated by a clear space. When two membranes fuse, they first come into close apposition *(bilayer adherence)* producing a 5-layered image in which the central dense line represents the adherent monolayers of the two membranes (Figure 6–69). Such 5-layered images are only occasionally seen in electron micrographs of cells in the process of exocytosis, presumably because they are transient. In the fusion process, the adherent bilayers rapidly reorganize and join *(bilayer joining)* to form one continuous membrane, thereby opening the secretory vesicle to the extracellular space (see Figure 6–65). Bilayer adherence and bilayer joining (which together constitute **membrane fusion**) are fundamental cell-membrane phenomena that occur during cell division and cell fusion, as well as in exocytosis and endocytosis (Figure 6–70).

Endocytosis Occurs Continually in Most Cells[37]

Virtually all eucaryotic cells are continually ingesting bits of their plasma membranes in the form of small endocytotic (pinocytotic) vesicles. Processes such as this, that are continual (rather than induced), are said to be *constitutive*. Extracellular fluid (and everything dissolved in it) becomes trapped in the vesicles and is ingested as well. This process of bulk **fluid-phase endocytosis** can be visualized and quantified if a tracer, such as the enzyme peroxidase (isolated from horseradish), is introduced into the extracellular fluid. The reaction product of peroxidase and its substrate can be made electron-dense by fixation with osmium tetroxide. Therefore, if cells incubated in peroxidase are washed, fixed with glutaraldehyde, exposed to the substrate, and then fixed again in osmium tetroxide, the reaction product in the endocytotic vesicles can easily be seen in electron micrographs (Figure 6–71). The amount of peroxidase internalized increases linearly with its concentration in the extracellular fluid and increases continuously with time.

The rate of constitutive endocytosis varies from cell type to cell type, but it is usually surprisingly large. For example, a macrophage ingests 25% of its own volume each hour. This means that it ingests 3% of its plasma membrane each minute, or 100% in about half an hour. While fibroblasts endocytose at less than one-third this rate, some amoebae ingest their plasma membrane at even higher rates than a macrophage. Since a cell's surface area and volume remain unchanged during this process, it is clear that membrane is being added to the cell surface by exocytosis as fast as it is being removed by endocytosis. In view of the rapid rates involved and the lack of evidence for comparable rates of plasma membrane degradation, the lipids and proteins of the endocytosed membrane must be recycled and returned to the plasma membrane rather than degraded. How the membrane is recycled is unclear; nor is it known exactly which of the plasma membrane constituents are internalized so as to become involved in this massive recycling process. Large-scale recycling of the plasma membrane is a central feature of the "membrane-flow" hypotheses that has been proposed to explain capping and cell locomotion (see p. 279).

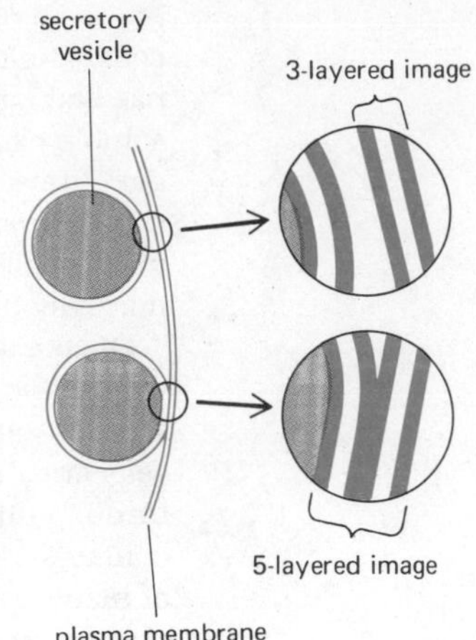

Figure 6–69 Schematic drawing of the 5-layered image seen with an electron microscope when the bilayer of a secretory vesicle adheres to the plasma membrane, just prior to bilayer joining.

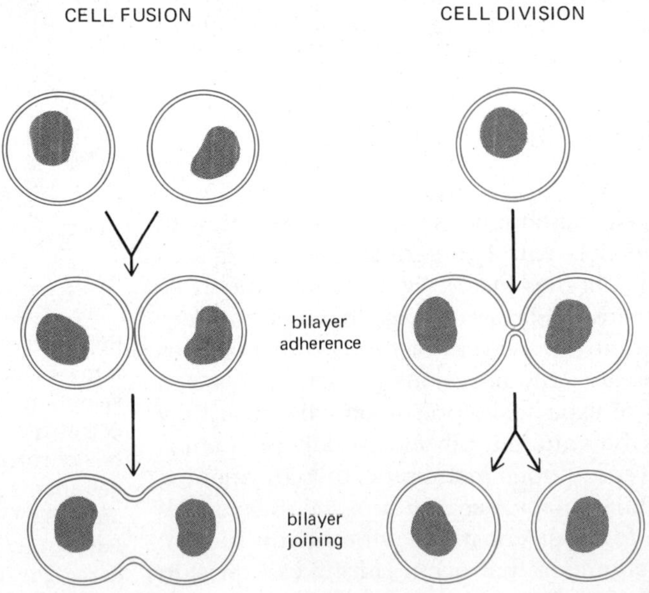

Figure 6–70 Schematic drawing of membrane fusion (consisting of bilayer adherence and bilayer joining) in the processes of cell division and cell fusion. Natural cell fusions occur between sperm and egg during fertilization and between myoblasts in the formation of multinucleated skeletal muscle cells.

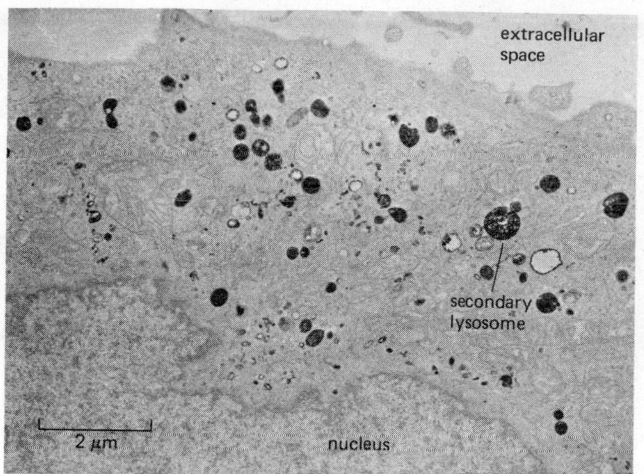

Figure 6–71 Electron micrograph of endocytotic vesicles in a cultured fibroblast that was incubated in a solution containing peroxidase for 60 minutes. The peroxidase is identified in endocytotic vesicles by the electron-dense reaction product generated after the cells were fixed and exposed to the substrate with which peroxidase reacts. (Courtesy of Ralph M. Steinman.)

Most Endocytotic Vesicles Ultimately Fuse with Lysosomes[37,38]

Endocytotic vesicles are formed in a variety of sizes and shapes (diameters generally between 50 and 400 nm) and usually enlarge by fusing with each other and/or with other intracellular vesicles. In most cells, the great majority of endocytotic vesicles (visualized by the peroxidase method described above) ultimately fuse with small vesicles called **primary lysosomes** to form **secondary lysosomes,** which are specialized sites of intracellular digestion (Figure 6–72). Because lysosomes contain a wide variety of degradative (hydrolytic) enzymes, most of the macromolecular contents of vesicles that fuse with them are rapidly broken down; the small breakdown products, such as amino acids, sugars, and nucleotides, are transported across the lysosomal membrane into the cytosol, where they can be used by the cell. On the other hand, most of the constituents of the endocytotic vesicle membrane escape degradation; they are retrieved by some unknown mechanism (probably by budding) and returned (directly, or indirectly via the Golgi apparatus) to the plasma membrane (Figure 6–72).

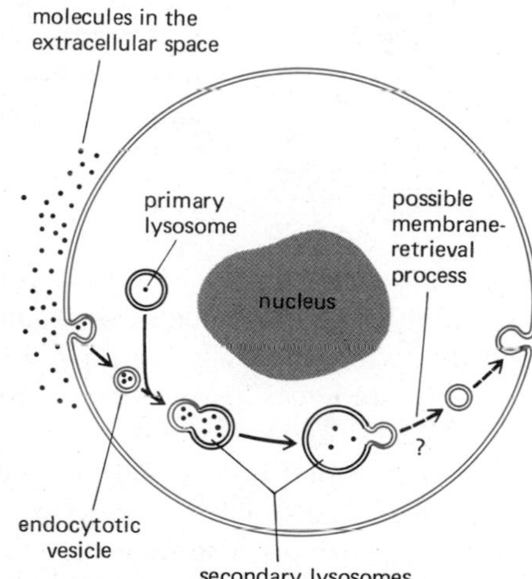

Figure 6–72 Schematic diagram of a cell showing the fusion of endocytotic vesicles with primary lysosomes to form secondary lysosomes. Although not shown, most endocytotic vesicles are thought to fuse with other vesicles in the cytoplasm before fusing with lysosomes. The right-hand side of the figure, showing the selective retrieval of the endocytotic membrane from the secondary lysosome and its direct return to the plasma membrane, is only one possible pathway of retrieval (see Figure 6–77 for an alternate pathway).

Not all endocytotic vesicles, however, fuse with lysosomes. In some cells, specialized endocytotic vesicles traverse the cytoplasm and release their contents by exocytosis at another surface. In bypassing the lysosomes, these vesicles shuttle material from one surface of the cell to another. This mechanism of bulk transcellular transport is one of the means by which the endothelial cells that line small blood vessels transfer substances out of the blood stream into the surrounding extracellular fluid. In vertebrate endothelial cells, the process is mediated by small vesicles (70 nm in diameter) that sometimes fuse simultaneously with each other and with both surfaces of the cell to form a transcellular, membrane-lined channel.

Many Endocytotic Vesicles Are Coated[39]

Most eucaryotic cells contain a specialized class of vesicles that vary in diameter from 50 to 250 nm and appear in conventional electron micrographs to be coated with bristlelike structures on their cytoplasmic surface (Figure 6–73). While some of these **coated vesicles** are thought to be involved in intracellular vesicular transport between the cell's organelles, others are continually generated by the invagination and pinching off of coated regions of the plasma membrane: these regions, called **coated pits** (Figures 6–73 and 6–74), constitute about 2% of the cell surface of a cultured fibroblast. Within

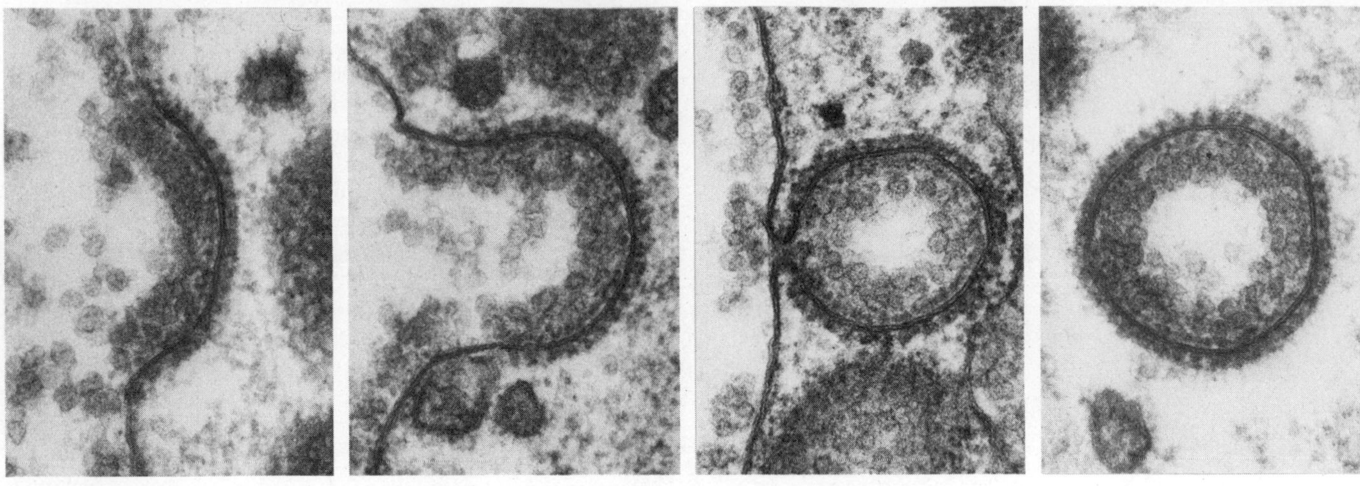

Figure 6–73 Electron micrographs illustrating the sequence of events thought to occur in the formation of a coated vesicle from a coated pit. The coated pits and vesicles shown are involved in ingesting lipoprotein particles into a very large hen oocyte to form yolk. They are very much larger than those seen in normal-size cells. (Courtesy of M. M. Perry and A. B. Gilbert, from *J. Cell Sci.* 39:257–272, 1979.)

seconds of being formed, endocytotic coated vesicles lose their coats and then fuse with other intracellular vesicles. Ultimately, most of them deliver their contents to primary lysosomes.

Coated vesicles have been isolated and their membranes found to contain several major proteins. The best characterized is **clathrin,** a highly conserved, fibrous protein (180,000 daltons), which, together with a smaller polypeptide (~35,000 daltons), forms a characteristic polyhedral coat on the surface of coated vesicles. The basic assembly unit of the coat is a three-legged protein complex (a *triskelion*) consisting of three clathrin polypeptides and three of the smaller polypeptides; the triskelions are arranged as a basketlike network of hexagons and pentagons on the surface of the coated vesicle (Figure 6–75). Under appropriate conditions, isolated triskelions spontaneously reassemble into typical polyhedral baskets (even in the absence of vesicles). It is thought that other proteins associated with the coated vesicle membrane are responsible for binding the clathrin coat to the vesicle and for trapping plasma membrane receptors in coated pits and vesicles (see next section).

In some cells, but not all, coated pits preferentially overlie cytoplasmic actin filament bundles (so-called *stress fibers*). This relationship indicates that coated regions of the plasma membrane can interact with the cytoskeleton of the cell and is reminiscent of the interactions postulated in the model for capping illustrated previously in Figure 6–39.

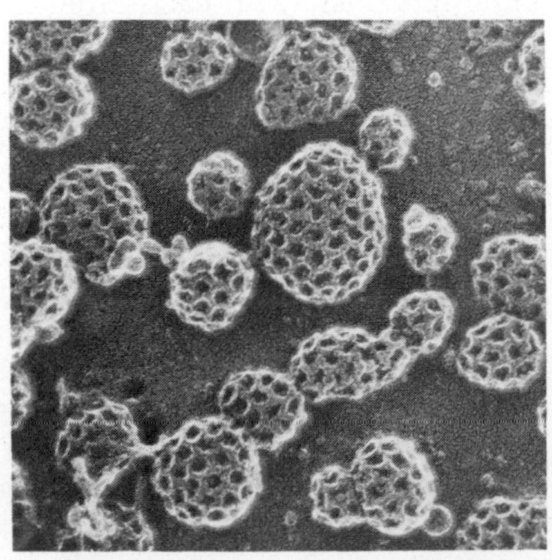

Figure 6–74 Electron micrograph of numerous coated pits and vesicles on the inner surface of the plasma membrane of cultured fibroblasts. The cells were rapidly frozen in liquid helium, fractured, and deep-etched to expose the cytoplasmic surface of the plasma membrane. (Reproduced from J. Heuser, *J. Cell Biol.* 84:560–583, 1980. By copyright permission of the Rockefeller University Press.)

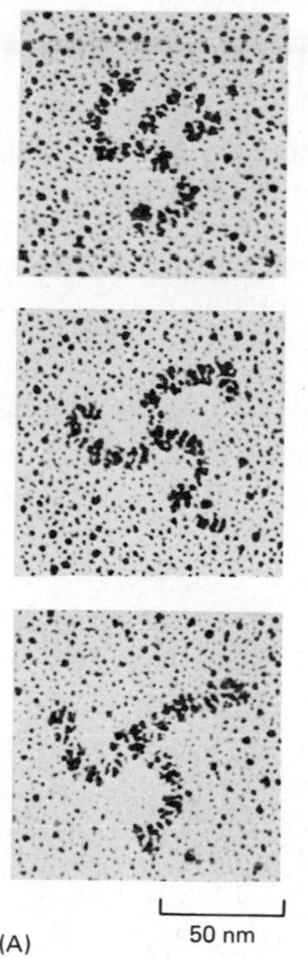

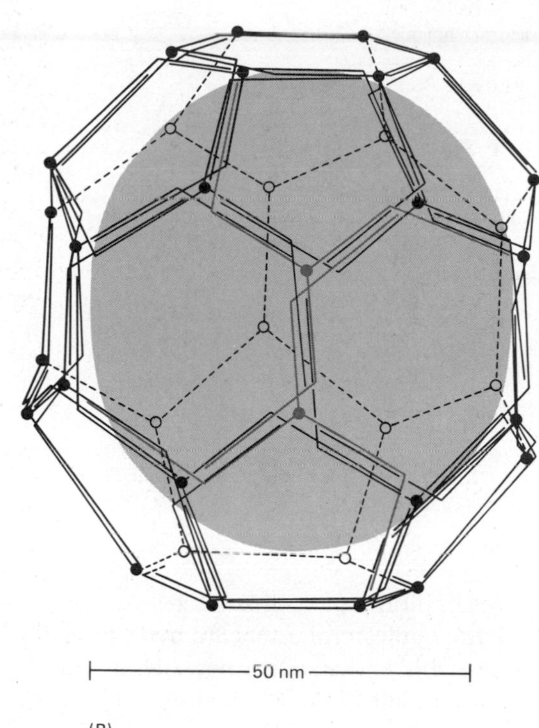

Figure 6–75 The structure of the clathrin coat of a coated vesicle. (A) Electron micrographs of clathrin triskelions shadowed with platinum. Each triskelion is thought to be composed of 3 clathrin polypeptide chains and 3 smaller polypeptide chains. (B) A schematic drawing of the probable arrangement of the triskelions on the cytoplasmic surface of a coated vesicle. Thirty-six triskelions are organized in a network of 12 pentagons and 8 hexagons. Two triskelions are shown in color. Other sizes and shapes of coated vesicles are thought to be similarly constructed from 12 pentagons plus a variable number of hexagons. The overlapping arrangement of the flexible triskelion arms provides both mechanical strength and flexibility. (A, from E. Ungewickell and D. Branton, *Nature* 289:420–422, 1981, © 1981 MacMillan Journals Ltd.; B, based on R. A. Crowther and B. M. F. Pearse, *J. Cell Biol.* 91:790–797, 1981.)

Coated Pits and Vesicles Provide a Specialized Pathway for Receptor-mediated Endocytosis of Specific Macromolecules[39, 40]

In most animal cells, coated pits and vesicles provide a specialized pathway for taking up specific macromolecules from the extracellular fluid, a process called **receptor-mediated endocytosis** (or *adsorptive endocytosis*). The macromolecules that bind to specific cell-surface receptors are internalized via coated pits at a much greater rate than the rate at which substances dissolved in the extracellular fluid enter cells by fluid-phase endocytosis. Receptor-mediated endocytosis is therefore a selective concentrating mechanism enabling cells to ingest large amounts of specific ligands without taking in a correspondingly large volume of extracellular fluid.

An important process known to occur through receptor-mediated endocytosis is the uptake of cholesterol by animal cells. Cholesterol uptake provides cells with most of the cholesterol they require for membrane synthesis. If this uptake is blocked, cholesterol accumulates in the blood and can contribute to the formation of atherosclerotic plaques in blood vessel walls. Most cholesterol is transported in the blood bound to protein in the form of complexes known as **low-density lipoproteins,** or **LDL.** These large spherical particles (22 nm in diameter) each contain a core of about 1500 cholesterol molecules esterified to long-chain fatty acids; the cholesterol ester core is surrounded by a lipid bilayer containing a single species of protein (Figure 6–76).

When an animal cell needs cholesterol for membrane synthesis, it makes receptor proteins for LDL and inserts them into its plasma membrane. Most of these receptor proteins associate spontaneously with coated pits and those that do not are induced to migrate to coated pits by the binding of LDL. Since coated pits are constantly pinching off to form coated vesicles, all of the LDL particles that bind to LDL receptors are rapidly internalized. The coated vesicles rapidly lose their coats and fuse with other vesicles to form larger vesicles called *endosomes;* these in turn fuse with primary lysosomes to form second-

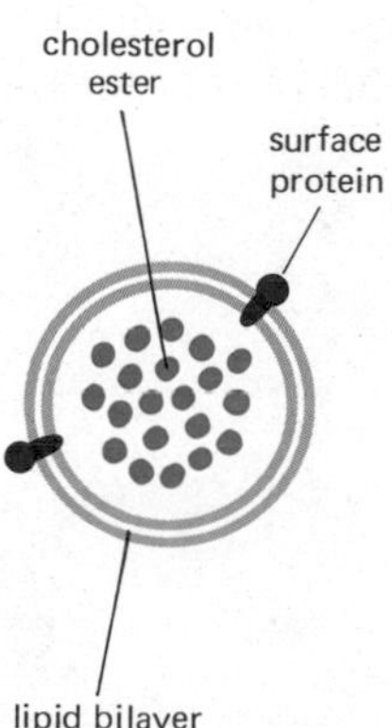

Figure 6–76 Schematic drawing of a low-density lipoprotein (LDL) particle seen in cross-section. Each particle contains two copies of a 250,000-dalton protein, which is responsible for the binding of LDL to cell-surface receptors.

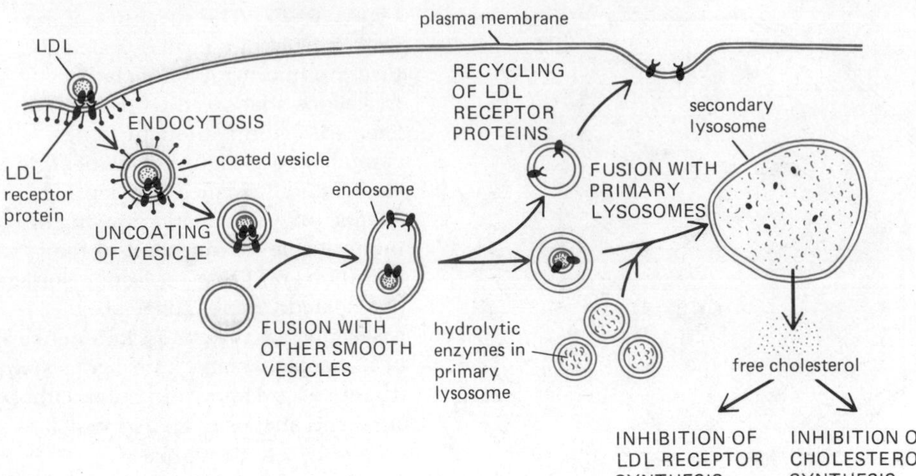

Figure 6–77 Receptor-mediated endocytosis of LDL via coated pits and vesicles regulates cholesterol levels in the cell. Note that while the LDL is broken down in secondary lysosomes, the LDL receptor proteins are retrieved from the endosomes and recycled to the plasma membrane.

ary lysosomes. Thus, within 10-15 minutes of binding to cell-surface receptors the LDL is delivered to lysosomes, where the cholesterol esters are hydrolyzed to free cholesterol and thereby become available to the cell for new membrane synthesis. If too much free cholesterol accumulates in a cell, it shuts off both the cell's own cholesterol synthesis and its synthesis of LDL receptor proteins, so that less cholesterol is made and less is taken up by the cell (Figure 6–77).

This pathway is disrupted in certain individuals who inherit defective genes for making LDL receptor proteins; consequently, their cells cannot take up LDL from the blood. The resulting high levels of blood cholesterol predispose these individuals to premature atherosclerosis and most die at an early age from coronary artery disease. The abnormality can involve either a loss of the binding site of the receptor for LDL or a loss of the LDL receptor's binding site for coated pits. In the latter case, normal numbers of LDL-binding receptor proteins are present, but they are not localized in the coated regions of the plasma membrane (Figure 6–78). Although LDL binds to the surface of these mutant cells, it is not internalized. This directly demonstrates the importance of coated pits in the receptor-mediated endocytosis of cholesterol.

Like other recycled components of the endocytotic vesicle membrane, most of the LDL receptor proteins that enter the cell are not destroyed in lysosomes; they rapidly return to the plasma membrane, reassociate with coated pits, and enter the cell again (Figure 6–77).

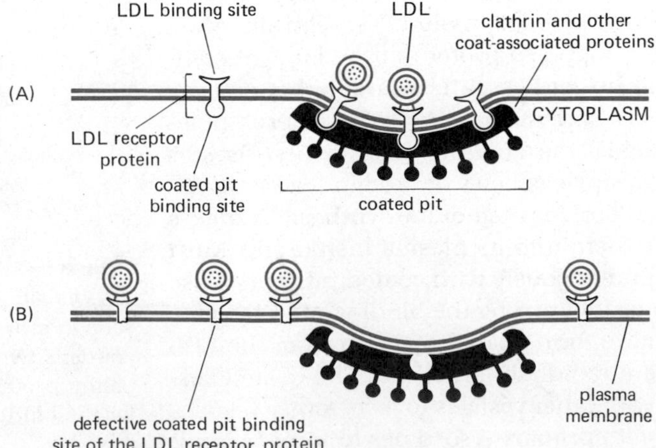

Figure 6–78 Schematic drawing of LDL receptor proteins binding to a coated pit region of the plasma membrane of a normal cell (A). Note that some LDL receptors localize in the coated pit even in the absence of LDL binding. A mutant cell is shown in (B); the LDL receptor proteins are abnormal and lack the sites that enable them to bind to coated pits. Such cells bind LDL but cannot ingest it.

Many Cell-Surface Receptors Associate with Coated Pits Only After Ligand Binding[40]

Many cell-surface receptors that enter cells via coated pits do not spontaneously associate with coated pits before a ligand binds to them. For example, the receptor protein for the hormone insulin is distributed diffusely on the surface of fibroblasts. After insulin binds to the receptors the insulin-receptor complexes associate with coated pits and become internalized. It seems that insulin binding alters the conformation of the insulin receptor protein so that it is now recognized by a protein component of the coated pit. In this case, many of the internalized insulin receptors are degraded in lysosomes along with the insulin, so that the concentration of insulin receptors on the cell surface is decreased. By this process, called *receptor down regulation,* the concentration of hormone in the extracellular fluid can regulate the concentration of its receptor on the target cell surface (see p. 744).

Not all endocytosis involves coated pits: endocytotic vesicles also form from uncoated regions of the plasma membrane. However, it is still uncertain what proportion of fluid-phase endocytosis is mediated by smooth vesicles and what proportion is mediated by coated vesicles.

Some Macromolecules Can Penetrate Cell Membranes Directly[41]

Although most macromolecules are transported across cell membranes by endocytosis or exocytosis, there are special cases in which large molecules pass directly through cell membranes. In bacteria, specific channels enable DNA molecules to cross both the cell wall and the plasma membrane in order to mediate genetic exchanges: for example, in the process of *genetic transformation,* the genetic constitution of some types of bacteria can be altered by exposing them to purified DNA molecules. In eucaryotic cells, secreted proteins are directly transferred across the membrane of the rough endoplasmic reticulum as they are being synthesized (see p. 343). In addition, some protein bacterial toxins are able to penetrate animal cell membranes and exert their effects within the cytosol. In all of these examples, the macromolecules seem to cross the membrane without the formation of vesicles, but in no case is the mechanism understood.

Specialized Phagocytic Cells Ingest Particles That Bind to Specific Receptors on Their Surface[37,42]

In protozoa, phagocytosis is a form of feeding: large particles are ingested in phagocytic vacuoles that fuse with lysosomes. The products of the subsequent digestive processes pass into the cytosol to be utilized as food. Most cells in multicellular organisms are unable to ingest large particles efficiently and leave this task to "professional" phagocytes. In mammals, there are two classes of white blood cells that act as phagocytes: **macrophages** (which are widely distributed in tissues as well as in the blood) and **polymorphonuclear leucocytes** (also called polymorphs or neutrophils). These two types of cell defend us against infection by ingesting invading microorganisms (Figure 6–79). Macrophages also play an important part in scavenging senescent and damaged cells and cellular debris. In quantitative terms, the latter function is far more important: in each of us, macrophages phagocytose more than 10^{11} senescent red blood cells every day. As in protozoa, the phagocytic vacuoles in phagocytes fuse with primary lysosomes to form secondary lysosomes, where the ingested material is degraded. Indigestible substances remain in secondary lysosomes, forming *residual bodies.*

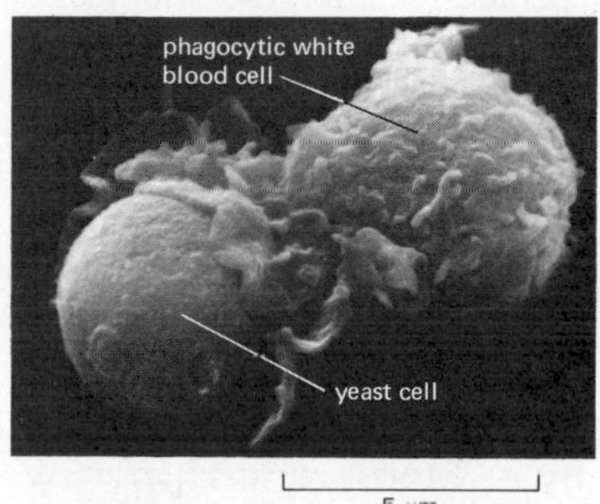

Figure 6–79 Scanning electron micrograph of a polymorphonuclear leucocyte phagocytosing a yeast cell. (From J. Boyles and D. F. Bainton, *Cell* 24:905–914, 1981. © M.I.T. Press.)

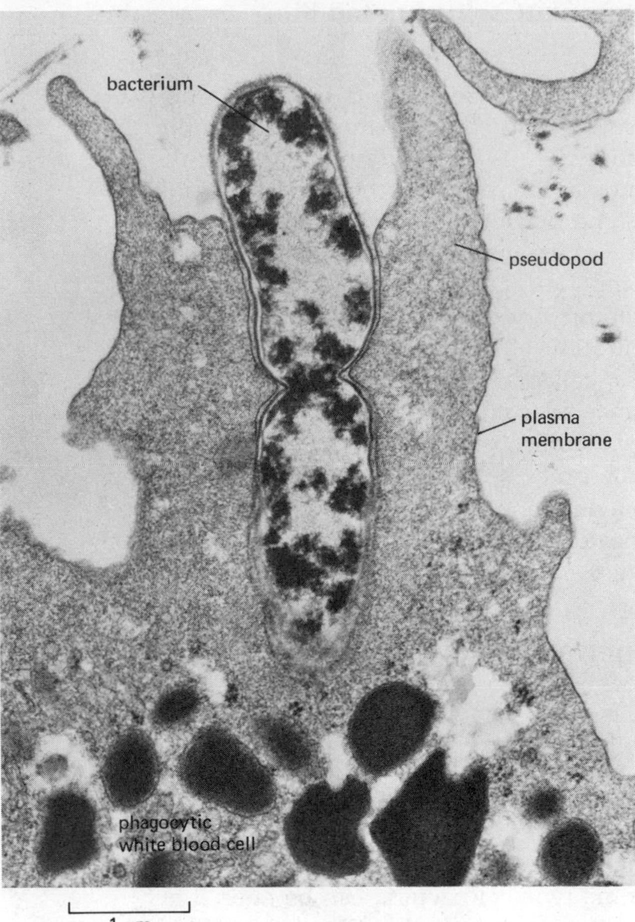

Figure 6–80 Electron micrograph of a polymorphonuclear leucocyte phagocytosing a bacterium, which is in the process of dividing. (Courtesy of Dorothy F. Bainton.)

In order to be phagocytosed, particles must first bind to the surface of the phagocyte. However, not all particles that bind are ingested. Phagocytes have a variety of specialized surface receptors that are functionally linked to the phagocytic machinery of the cell. The best characterized of these receptors are those that recognize antibody molecules. One of the principal ways in which antibodies protect us against infectious microorganisms is by binding to the microorganism's surface to form a coat that is then recognized by specific receptors on the surface of macrophages and polymorphonuclear leucocytes. The binding of antibody-coated particles to these receptors induces the cell to extend pseudopods, which engulf the particle and fuse to form a phagocytic vacuole (Figure 6–80).

Very little is known about other types of receptors that mediate phagocytosis. While it is clear that phagocytes can recognize and ingest senescent or damaged cells and certain microorganisms, what they recognize when antibody is not involved remains to be discovered.

Phagocytosis Is a Localized Response That Proceeds by a "Membrane-zippering" Mechanism[43]

It is possible to treat red blood cells in such a way that they will bind to the surface of macrophages without being phagocytosed. If macrophages that have bound such red cells are then allowed to phagocytose antibody-coated bacteria, only the bacteria are ingested; even red cells bound immediately next to a site of active phagocytosis are not ingested. This indicates that phago-

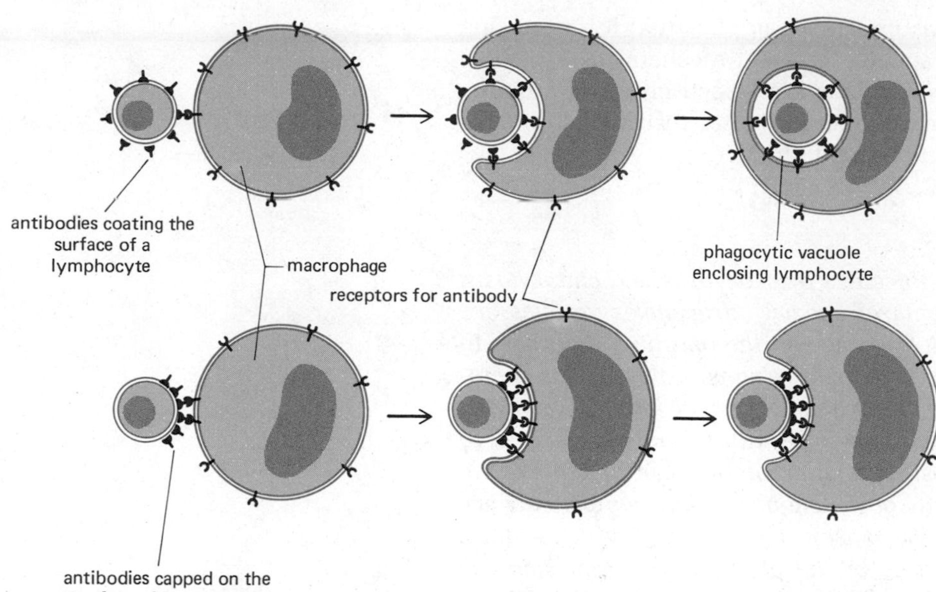

Figure 6–81 Schematic drawing of an experiment suggesting that phagocytosis proceeds by a "membrane-zippering" mechanism (see text).

cytosis, like exocytosis, is a localized response of a region of the plasma membrane and its underlying cytoplasmic structures.

If a macrophage is exposed to lymphocytes that have been evenly coated with antibody, it ingests the coated cells. However, if the antibody molecules have been confined to one pole of the lymphocyte by capping (see p. 278), the plasma membrane of the macrophage spreads over the surface of the lymphocyte only in the region of the cap and the cell is not ingested (Figure 6–81). It is clear, therefore, that the initial interaction of antibody-coated target cells with the receptors on the macrophage surface is insufficient to trigger ingestion. Instead, it initiates a spreading process of membrane apposition that requires the continuous contact of receptors with antibody if the target cell is to be fully enclosed by a phagocytic vacuole. This suggests that phagocytosis proceeds by a continuous *membrane-zippering* mechanism.

Membrane Vesicular Traffic: How Is It Powered, Guided, and Regulated?

Vesicles rapidly and continually shuttle between the various membranous organelles of a eucaryotic cell. Exocytosis and endocytosis occurring across the plasma membrane are only the most accessible examples of this intricate transport process. Despite its importance, remarkably little is known about the molecular mechanisms involved in driving, guiding, and regulating this vesicular traffic. It is clear that these processes require metabolic energy, probably in the form of ATP hydrolysis, but it is not known how the energy is used. There is evidence that actin and myosin (and some of their accessory proteins) are involved in the formation and movement of pseudopods during phagocytosis, but it is uncertain whether these proteins also play a part in fluid-phase and receptor-mediated endocytosis, or in exocytosis, or in moving vesicles from one membrane to another within the cell.

How do vesicles know where to go? There must be molecules on their cytoplasmic surfaces to guide the vesicles to the appropriate membrane, but nothing is known about their nature or diversity. The continual membrane fusions that occur in vesicular transport might be expected to lead to large-scale membrane intermixing, yet the various membranes within cells maintain their characteristic compositions. This constancy implies that if intermixing occurs, it must be accompanied by specific retrieval mechanisms. But how are specific membrane components recognized and selectively retrieved from, for example, the endosomal or plasma membranes after vesicle fusion? This problem will be discussed in detail in the next chapter.

Summary

Most cells secrete and ingest macromolecules by exocytosis and endocytosis, respectively. In exocytosis, the contents of special intracellular vesicles are released to the outside when the vesicles fuse with the plasma membrane. In endocytosis, the sequence is reversed: localized regions of the plasma membrane invaginate and pinch off to form small (pinocytotic) or large (phagocytic) vesicles. Most endocytotic vesicles eventually fuse with primary lysosomes to form secondary lysosomes, where most of the macromolecular contents of the vesicles are digested, while the majority of vesicle membrane components are somehow retrieved and returned to the plasma membrane. Both exocytosis and endocytosis are localized responses of the plasma membrane and the underlying cytoplasm.

Most animal cells are continuously endocytosing bits of their plasma membranes to form endocytotic vesicles and in this way can ingest extracellular fluids and solutes—a process called fluid-phase endocytosis. Some of the vesicles form from smooth regions of the plasma membrane while others form from coated regions of the membrane, called coated pits. Coated pits and vesicles provide a special pathway for concentrating and ingesting those extracellular macromolecules that bind to specific receptor proteins that localize in coated pits—a process called receptor-mediated endocytosis. Large extracellular particles, such as cellular debris or infecting organisms, are endocytosed mainly by "professional" phagocytic cells, which have different types of specialized receptors on their surface that mediate the phagocytic process.

References

General

Bretscher, M.S.; Raff, M.C. Mammalian plasma membranes. *Nature* 258:43–49, 1975.

Finean, J.B.; Coleman, R.; Michell, R.H. Membranes and Their Cellular Functions, 2nd ed., pp. 42–67. London: Blackwell, 1978.

Harrison, R.; Lunt, G.G. Biological Membranes. Glasgow: Blackie, 1980.

Quinn, P.J. The Molecular Biology of Cell Membranes. Baltimore: University Park Press, 1976.

Singer, S.J.; Nicolson, G.L. The fluid mosaic model of the structure of cell membranes. *Science* 175:720–731, 1972.

Weissmann, G.; Claiborne, R., eds. Cell Membranes: Biochemistry, Cell Biology and Pathology. New York: Hospital Practice, 1975.

Cited

1. Gorter, E.; Grendel, F. On bimolecular layers of lipoids on the chromocytes of the blood. *J. Exp. Med.* 41:439–443, 1925.
2. Ansell, G.B.; Hawthorne, J.N.; Dawson, R.M.C. Form and Function of Phospholipids, 2nd ed. Amsterdam: Elsevier, 1973.
3. Bangham, A.D. Models of cell membranes. In Cell Membranes: Biochemistry, Cell Biology and Pathology (G. Weissmann, R. Claiborne, eds.), pp. 24–34. New York: Hospital Practice, 1975.

Kornberg, R.D.; McConnell, H.M. Lateral diffusion of phospholipids in a vesicle membrane. *Proc. Natl. Acad. Sci. USA* 68:2564–2568, 1971.

Quinn, A.J., ed. The Molecular Biology of Cell Membranes, pp. 47–75. Baltimore: University Park Press, 1976.

Cantor, C.R.; Schimmel, P.R. Biophysical Chemistry, Part III, pp. 1348–1377. San Francisco: Freeman, 1980.

4. Chapman, D. Lipid dynamics in cell membranes. In Cell Membranes: Biochemistry, Cell Biology and Pathology (G. Weissmann, R. Claiborne, eds.), pp. 13–22. New York: Hospital Practice, 1975.

Quinn, A.J.; Chapman, D. The dynamics of membrane structure. *CRC Crit. Rev. Biochem.* 8:1–117, 1980.

Kimelberg, H.K. The influence of membrane fluidity on the activity of membrane-bound enzymes. In Dynamic Aspects of Cell Surface Organization. Cell Surface Reviews (G. Poste, G.L. Nicolson, eds.), Vol. 3, pp. 205–293. Amsterdam: Elsevier, 1977.

5. Stryer, L. Biochemistry, 2nd ed., pp. 206–215. San Francisco: Freeman, 1981.

6. Bretscher, M. Membrane structure: some general principles. *Science* 181:622–629, 1973.

Rothman, J.; Lenard, J. Membrane asymmetry. *Science* 195:743–753, 1977.

7. Fishman, P.H.; Brady, R.O. Biosynthesis and function of gangliosides. *Science* 194:906–915, 1976.

Hakomori, S. Glycosphingolipids in cellular interaction, differentiation, and oncogenesis. *Annu. Rev. Biochem.* 50:733–764, 1981.

Wiegandt, H. The gangliosides. *Adv. Neurochem.* 4:149–223, 1982.

8. Singer, S.J.; Nicolson, G.L. The fluid mosaic model of the structure of cell membranes. *Science* 175:720–731, 1972.

Bretscher, M.S.; Raff, M.C. Mammalian plasma membranes. *Nature* 258:43–49, 1975.

Helenius, A.; Simons, K. Solubilization of membranes by detergents. *Biochim. Biophys. Acta* 415:29–79, 1975.

9. Bretscher, M. Membrane structure: some general principles. *Science* 181:622–629, 1973.

Steck, T.L. The organization of proteins in the human red blood cell membrane. *J. Cell Biol.* 62:1–19, 1974.

10. Marchesi, V.T. Spectrin: present status of a putative cyto-skeletal protein of the red cell membrane. *J. Membr. Biol.* 51:101–131, 1979.

Branton, D.; Cohen, C.M.; Tyler, J. Interaction of cytoskeletal proteins on the human erythrocyte membrane. *Cell* 24:24–32, 1981.

11. Marchesi, V.T.; Furthmayr, H.; Tomita, M. The red cell membrane. *Annu. Rev. Biochem.* 45:667–698, 1976.

12. Steck, T.L. The band 3 protein of the human red cell membrane: a review. *J. Supramol. Struct.* 8:311–324, 1978.

Cabantchik, Z.I.; Knauf, P.A.; Rothstein, A. The anion transport system of the red blood cell. The role of membrane protein evaluated by the use of 'probes.' *Biochim. Biophys. Acta* 515:239–302, 1978.

13. Henderson, R.; Unwin, P.N.T. Three-dimensional model of purple membrane obtained by electron microscopy. *Nature* 257:28–32, 1975.

14. Branton, D. Fracture faces of frozen membranes. *Proc. Natl. Acad. Sci. USA* 55:1048–1056, 1966.

McNutt, N.S. Freeze-fracture techniques and applications to the structural analysis of the mammalian plasma membrane. In Dynamic Aspects of Cell Surface Organization. Cell Surface Reviews (G. Poste, G.L. Nicolson, eds.), Vol. 3, pp. 95–126. Amsterdam: Elsevier, 1977.

15. Juliano, R.L. Techniques for the analysis of membrane glycoproteins. *Curr. Top. Memb. Transp.* 11:107–144, 1978.

16. Frye, L.D.; Edidin, M. The rapid intermixing of cell surface antigens after formation of mouse-human heterokaryons. *J. Cell Sci.* 7:319–335, 1970.

17. de Petris, S.; Raff, M.C. Normal distribution, patching and capping of lymphocyte surface immunoglobulin studied by electron microscopy. *Nature New Biol.* 241:257–259, 1973.

Raff, M.C. Cell-surface immunology. *Sci. Am.* 234(5):30–39, 1976.

18. Taylor, R.B.; Duffus, W.P.H.; Raff, M.C.; de Petris, S. Redistribution and pinocytosis of lymphocyte surface immunoglobulin molecules induced by anti-immunoglobulin antibody. *Nature New Biol.* 233:225–229, 1971.

Schreiner, G.F.; Unanue, E.R. Membrane and cytoplasmic changes in B lymphocytes induced by ligand-surface immunoglobulin interaction. *Adv. Immunol.* 24:38–165, 1976.

de Petris, S. Distribution and mobility of plasma membrane components on lymphocytes. In Dynamic Aspects of Cell Surface Organization. Cell Surface Reviews (G. Poste, G.L. Nicolson, eds.), Vol. 3, pp. 643–728. Amsterdam: Elsevier, 1977.

Abercrombie, M.; Heaysman, J.E.M.; Pegrum, S.M. The locomotion of fibroblasts in culture. III. Movements of particles on the dorsal surface of the leading lamella. *Exp. Cell Res.* 62:389–398, 1970.

19. Bretscher, M.S. Directed lipid flow in cell membranes. *Nature* 260:21–23, 1976.

de Petris, S. Distribution and mobility of plasma membrane components on lymphocytes. In Dynamic Aspects of Cell Surface Organization. Cell Surface Reviews (G. Poste, G.L. Nicolson, eds.), Vol. 3, pp. 643–728. Amsterdam: Elsevier, 1977.

20. Poo, M.; Cone, R.A. Lateral diffusion of rhodopsin in the photoreceptor membrane. *Nature* 247:438–441, 1974.

Jacobson, K.; Elson, E.; Koppel, D.; Webb, W. Fluorescence photobleaching in cell biology. *Nature* 295:283–284, 1982.

21. Hirano, H.; Parkhouse, B.; Nicolson, G.L.; Lennox, E.S.; Singer, S.J. Distribution of saccharide residues on membrane fragments from a myeloma-cell homogenate: its implications for membrane biogenesis. *Proc. Natl. Acad. Sci. USA* 69:2945–2949, 1972.

Luft, J.H. The structure and properties of the cell surface coat. *Int. Rev. Cytol.* 45:291–382, 1976.

22. Roseman, S. Sugars of the cell membrane. In Cell Membranes: Biochemistry, Cell Biology and Pathology (G. Weissmann, R. Claiborne, eds.), pp. 55–64. New York: Hospital Practice, 1975.

Kornfeld, R.; Kornfeld, S. Structure of glycoproteins and their oligosaccharide units. In The Biochemistry of Glycoproteins and Proteoglycans (W.J. Lennarz, ed.), pp. 1–84. New York: Plenum, 1980.

Neufeld, E.F.; Ashwell, G. Carbohydrate recognition systems for receptor-mediated pinocytosis. In The Biochemistry of Glycoproteins and Proteoglycans (W.J. Lennarz, ed.), pp. 241–266. New York: Plenum, 1980.

23. Gomperts, B.D. The Plasma Membrane: Models for Its Structure and Function. New York: Academic Press, 1976.

24. Finean, J.B.; Coleman, R.; Michell, R.H. Membranes and Their Cellular Functions, 2nd ed., pp. 42–67. London: Blackwell, 1978.

Christensen, H.N. Biological Transport, 2nd ed. Reading, Pa.: Benjamin, 1975.

Wilson, D.B. Cellular transport mechanisms. *Annu. Rev. Biochem.* 47:933–965, 1978.

25. Singer, S.J. Thermodynamics, the structure of integral membrane proteins and transport. *J. Supramol. Struct.* 6:313–323, 1977.

Hobbs, A.S.; Albers, R.W. The structure of proteins involved in active membrane transport. *Annu. Rev. Biophys. Bioeng.* 9:259–291, 1980.

26. Kuffler, S.W.; Nicholls, J.G. From Neuron to Brain: A Cellular Approach to the Function of the Nervous System, pp. 88–98. Sunderland, Ma.: Sinauer, 1977.

27. Skou, J.C.; Norby, J.G., eds. Na^+-K^+ ATPase: Structure and Kinetics. New York: Academic Press, 1979.

Racker, E. Reconstitution and mechanism of action of ion pumps. In A New Look at Mechanisms in Bioenergetics, pp. 127–152. New York: Academic Press, 1976.

28. Sweadner, K.J.; Goldin, S.M. Active transport of sodium and potassium ions: mechanism, function and regulation. *N. Engl. J. Med.* 302:777–783, 1980.

29. MacLennan, D.H.; Campbell, K.P. Structure, function and biosynthesis of sarcoplasmic reticulum proteins. *Trends Biochem. Sci.* 4:148–151, 1979.

de Meis, L.; Vianna, A.L. Energy interconversion by the Ca^{2+}-dependent ATPase of the sarcoplasmic reticulum. *Annu. Rev. Biochem.* 48:275–292, 1979.

30. Racker, E. A New Look at Mechanisms in Bioenergetics. New York: Academic Press, 1976.

31. Postma, P.W.; Roseman, S. The bacterial phosphoenolpyruvate: sugar phosphotransferase system. *Biochim. Biophys. Acta* 457:213–257, 1976.

Oxender, D.; Quay, S. Binding proteins and membrane transport. *Ann. N.Y. Acad. Sci.* 264:358–372, 1975.

32. Stevens, C.F. The neuron. *Sci. Am.* 241(3):54–65, 1979.
33. Jaffe, L.F. Control of development by steady ionic currents. *Fed. Proc.* 40:125–127, 1981.
34. Pressman, B.C. Biological applications of ionophores. *Annu. Rev. Biochem.* 45:501–530, 1976.
35. Palade, G. Intracellular aspects of the process of protein synthesis. *Science* 189:347–358, 1975.

    Rubin, R.P. The role of calcium in the release of neurotransmitter substances and hormones. *Pharmacol. Rev.* 22:389–428, 1970.
36. Lawson, D.; Fewtrell, C.; Raff, M. Localized mast cell degranulation induced by concanavalin A-sepharose beads: implications for the Ca^{2+} hypothesis of stimulus-secretion coupling. *J. Cell Biol.* 79:394–400, 1978.
37. Silverstein, S.C.; Steinman, R.M.; Cohn, Z.A. Endocytosis. *Annu. Rev. Biochem.* 46:669–722, 1977.
38. Simionescu, N.; Simionescu, M.; Palade, G.E. Permeability of muscle capillaries to small heme-peptides: evidence for the existence of patent transendothelial channels. *J. Cell Biol.* 64:586–607, 1975.
39. Roth, T.F.; Porter, K.R. Yolk protein uptake in the oocyte of the mosquito, *Aedas aegypti. J. Cell Biol.* 20:313–332, 1964.

    Goldstein, J.L.; Anderson, R.G.W.; Brown, M.S. Coated pits, coated vesicles, and receptor-mediated endocytosis. *Nature* 279:679–685, 1979.

    Pearse, B.M.F.; Bretscher, M.S. Membrane recycling by coated vesicles. *Annu. Rev. Biochem.* 50:85–101, 1981.

    Ungewickell, E.; Branton, D. Assembly units of clathrin coats. *Nature* 289:420–422, 1981.
40. Pastan, I.H.; Willingham, M.C. Receptor-mediated endocytosis of hormones in cultured cells. *Annu. Rev. Physiol.* 43:239–250, 1981.
41. Smith, H.O.; Danner, D.B.; Deich, R.A. Genetic transformation. *Annu. Rev. Biochem.* 50:41–68, 1981.
42. Stossel, T.P. Phagocytosis. *N. Engl. J. Med.* 290:717–723, 774–780, 833–839, 1974 (in three parts).
43. Griffin, F.M., Jr.; Silverstein, S.C. Segmental response of the macrophage plasma membrane to a phagocytic stimulus. *J. Exp. Med.* 139:323–336, 1974.

    Griffin, F.M., Jr.; Griffin, J.A.; Silverstein, S.C. Studies on the mechanism of phagocytosis. II. The interaction of macrophages with anti-immunoglobulin IgG-coated bone marrow-derived lymphocytes. *J. Exp. Med.* 144:788–809, 1976.

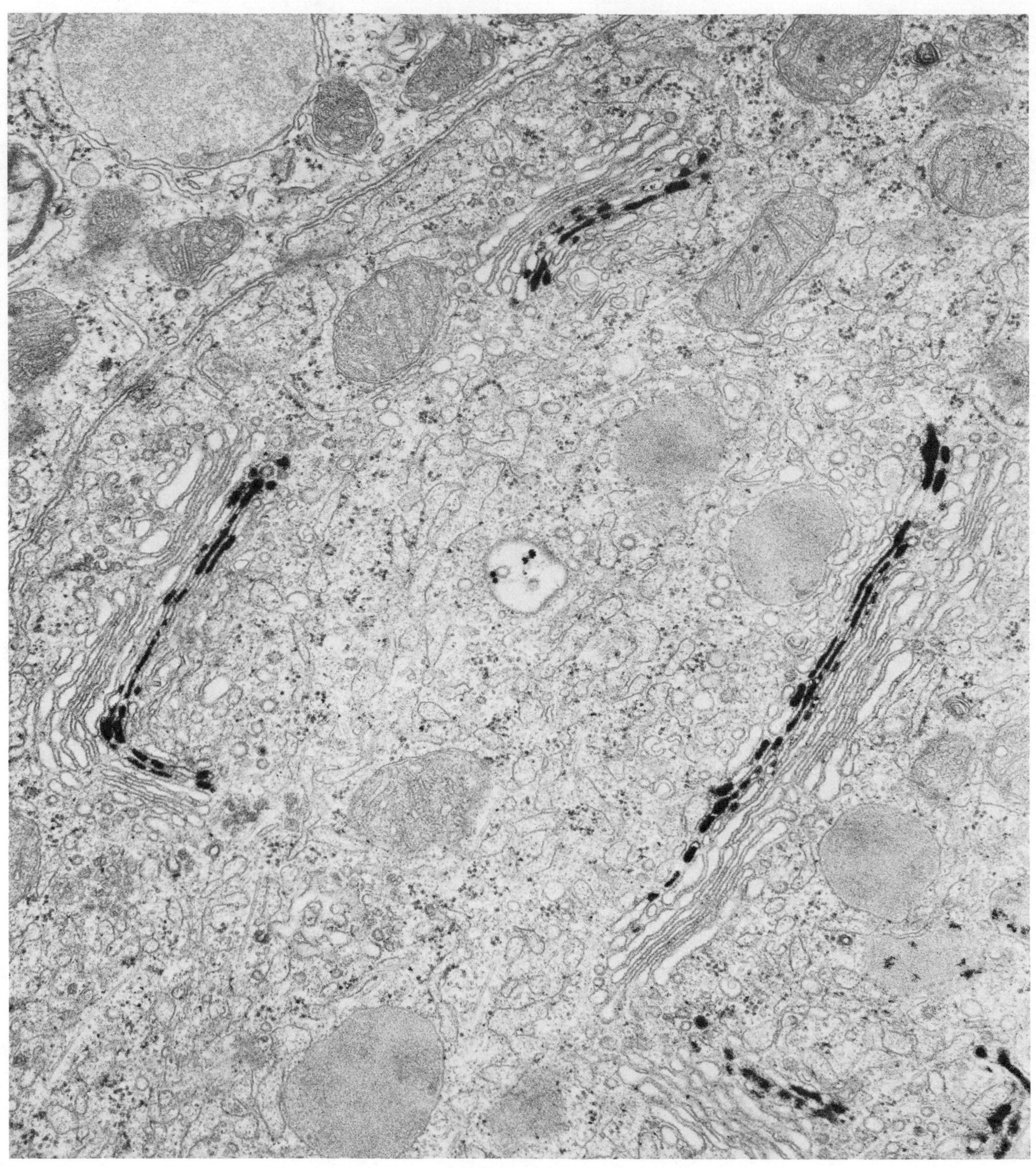

Electron micrograph of part of the cytoplasm of a mammalian cell. The cisternae on the *trans* side of each stack of Golgi cisternae have been darkened by a stain for a specific enzyme. (Courtesy of Daniel S. Friend.)

7 Internal Membranes and the Synthesis of Macromolecules

The cytoplasm accounts for most of a cell's mass. It is an enormously complex mixture of large and small molecules, being about 70% water and 15% to 20% protein by weight, and in a typical animal cell it contains about 10 billion (10^{10}) protein molecules of perhaps 10,000 different sorts. When a living cell is viewed at high magnification in the light microscope, its cytoplasm looks like a rather amorphous, gel-like substance in which scattered particles dart about. But the cytoplasm is much more highly organized than this light microscope image suggests. When viewed by electron microscopy, it is seen to contain many different *organelles,* each of which contains a specific set of proteins and is specialized for a particular function. The cytoplasm outside of these organelles is called the cell *cytosol,* and it is permeated by a dense network of protein filaments that collectively constitute the cell *cytoskeleton.* Many proteins are directly or indirectly bound to this cytoskeleton. In addition, nearly all proteins have selective affinities for certain other proteins, so that groups of protein molecules bind together to form large clusters that continuously appear and disappear. Because much of the organization of the cytoplasm is destroyed when attempts are made to study it, the details of its complex structure are poorly understood.

Yet one must understand the cytoplasm in order to understand the cell. Where, then, should one begin? One approach is to take the cytoplasm apart protein by protein, determining the detailed structure and chemistry of each. A different approach is to dissect the cytoplasm into its various internal compartments so as to study the functions and chemistry of each compartment in isolation. It is predominantly the latter approach that will guide us in this chapter, as we consider in turn the functions of the cytosol, the endoplasmic reticulum, the Golgi apparatus, the peroxisome, and the lysosome—leaving the two energy-converting organelles, the mitochondrion and the chloroplast, for detailed consideration in Chapter 9 and the cytoskeleton for discussion in Chapter 10.

The Compartmentalization of Higher Cells[1]

Large Eucaryotic Cells Need Internal Membranes

Many vital biochemical processes take place in or on membrane surfaces. For example, oxidative phosphorylation and photosynthesis both require a semipermeable membrane in order to couple the transport of protons to the synthesis of ATP. Moreover, membranes themselves provide the framework for the synthesis of new membrane components. In a procaryote, such as a bacterium, all of these functions (and many others) are carried out by the plasma membrane, which is usually the only membrane in the cell.

The most obvious distinguishing feature of eucaryotic cells besides the nucleus is their diverse array of intracellular membrane-bounded organelles and vesicles. Internal membrane systems are thought to be required in cells of large size (p. 133). A typical eucaryotic cell is 10 times larger in linear dimensions than a typical bacterium like *E. coli.* This means that, relative to their volume, eucaryotic cells have a much smaller area of membrane exposed at their external surface. Apparently, the plasma membrane alone cannot provide enough surface area or house enough membrane-bound enzyme molecules to support the vital functions that membranes sustain. Large quantities of additional membrane must, therefore, be contained within the cell itself.

Internal Membranes Divide the Cell into Specialized Compartments

In principle, the surface area of cellular membranes in eucaryotic cells could have been increased simply by repeated invaginations (or evaginations) of the plasma membrane. However, the internal membranes of eucaryotes are physically and chemically distinct from the plasma membrane as well as from each other. They partition the cell into functionally distinct compartments, each with boundaries established by sealed, selectively permeable membranes. Each compartment is a separate subcellular reaction vessel endowed with specialized functions that are carried out by the unique set of enzymes concentrated and held within it by its limiting membrane.

Even Complex Eucaryotic Cells Have Only a Few Major Intracellular Compartments[2]

There are seven major cellular compartments common to most eucaryotic cells (Figure 7–1). The nucleus is the site of DNA and RNA synthesis. The surrounding cytoplasm consists of the **cytosol** and the cytoplasmic organelles suspended in it. Most of the intermediary metabolism of the cell, including protein synthesis on free ribosomes, takes place in the cytosol. The **endoplasmic reticulum (ER),** with its membrane-bound ribosomes, synthesizes new integral membrane proteins and proteins destined for export, as well as new membrane lipid for the rest of the cell. The **Golgi apparatus** acts like a policeman, directing the macromolecular traffic that passes through it to the correct intracellular destinations, often with concurrent covalent modification (such as selective protein cleavage and glycosylation). **Mitochondria** and (in plants) **chloroplasts** generate most of the ATP utilized to drive the energetically unfavorable biosynthetic reactions. **Lysosomes** contain a variety of digestive enzymes that primarily degrade macromolecules and particles taken up from outside the cell. Finally, **peroxisomes** (also known as microbodies) are small vesicular compartments containing enzymes involved in a variety of oxidative reactions. In general, each membrane-bounded compartment (organelle) has a set of properties common to all cell types as well as added

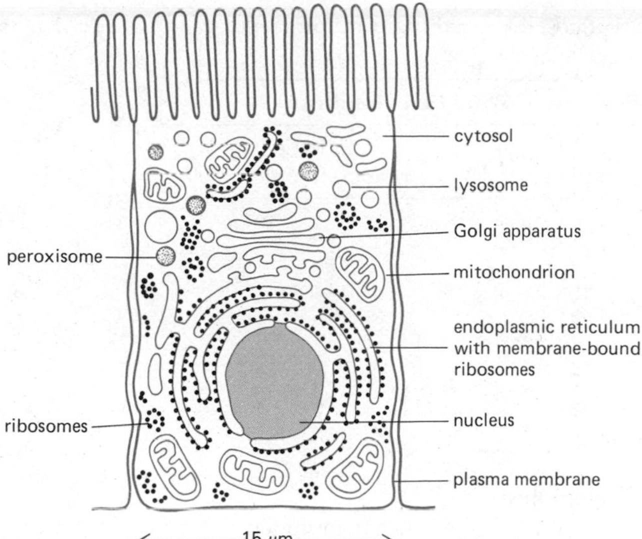

Figure 7–1 A schematic view of a typical animal cell emphasizing the major intracellular compartments. The cytosol, endoplasmic reticulum, Golgi apparatus, nucleus, mitochondrion, lysosome, and peroxisome represent distinct compartments isolated from the rest of the cell by at least one selectively permeable membrane.

activities that vary from cell type to cell type. Together, these organelles occupy a major fraction of the total cell volume (Table 7–1). Other kinds of organelles with more specialized functions (such as vesicles storing enzymes or hormones for triggered release) are present only in certain highly specialized cells.

The combined amount of intracellular membrane in all of these organelles is considerable. For example, in the two different mammalian cells analyzed in Table 7–2, the endoplasmic reticulum has a total membrane surface that is, respectively, 25 times and 12 times that of the plasma membrane. In terms of both area and mass, the plasma membrane is only a minor membrane in the great majority of eucaryotic cells (Figure 7–2).

Table 7–1 The Relative Volumes Occupied by the Major Intracellular Compartments in a Typical Liver Cell (Hepatocyte)

Intracellular Compartment	Percent of Total Cell Volume	Approximate Number per Cell*
Cytosol	54	1
Mitochondria	22	1700
Rough endoplasmic reticulum cisternae	9	1
Smooth endoplasmic reticulum cisternae plus Golgi cisternae	6	
Nucleus	6	1
Peroxisomes	1	400
Lysosomes	1	300

*All of the cisternae of the rough and smooth endoplasmic reticulum are thought to be joined to form a single large compartment. In contrast, the Golgi apparatus is organized into a number of discrete sets of stacked cisternae in each cell, and the extent of interconnection between these sets has not been clearly established.

Table 7–2 Most of the Membrane in a Eucaryotic Cell Is Used to Form Intracellular Organelles

	Percent of Total Cell Membrane	
Membrane Type	*Liver Hepatocyte*	*Pancreatic Exocrine Cell*
Plasma membrane	2	5
Rough ER membrane	35	60
Smooth ER membrane	16	<1
Golgi apparatus membrane	7	10
Mitochondria		
Outer membrane	7	4
Inner membrane	32	17
Nucleus		
Inner membrane	0.2	0.7
Secretory vesicle membrane	not determined	3
Lysosome membrane	0.4	not determined
Peroxisome membrane	0.4	not determined

These two cells are of very different sizes, since the average hepatocyte has a volume of about 5000 μm^3 compared to about 1000 μm^3 for the exocrine cell. Total cell membrane areas are estimated at about 110,000 μm^2 and 13,000 μm^2, respectively.

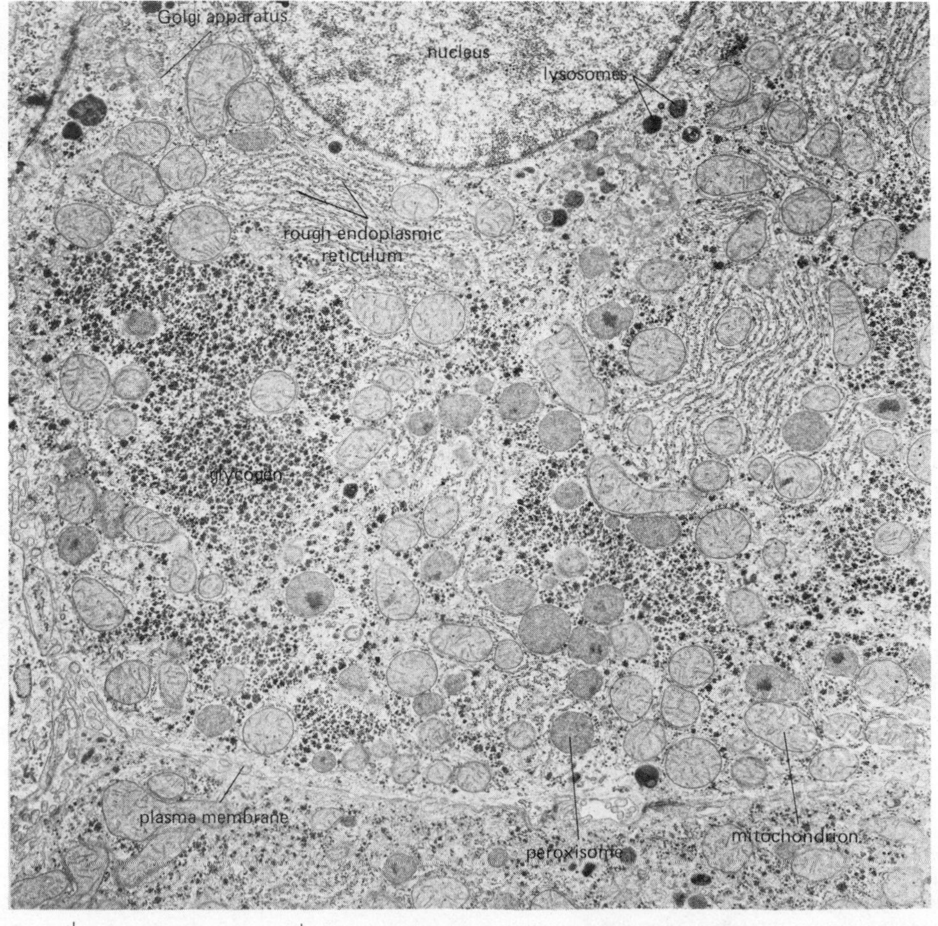

Figure 7–2 Electron micrograph of part of a liver cell seen in cross-section. Examples of each of the major intracellular compartments are indicated. (Courtesy of Daniel S. Friend.)

Intracellular Compartments Permit the Cell to Carry Out Many Incompatible Chemical Reactions Simultaneously

One function of the intracellular compartments is to prevent competing enzymatic reactions from interfering with each other. For example, essential proteins are synthesized in the cytosol, but unwanted proteins are hydrolyzed in lysosomes. Fatty acids are synthesized in the cytosol for use in membrane biosynthesis and as an energy store but are broken down in the mitochondria as an energy source. As will be described in detail in this chapter, various sugars are added to proteins in the endoplasmic reticulum and removed or modified later in the Golgi apparatus.

Peroxisomes provide an especially good example of the value of compartmentalization. They are small (0.1 to 1 μm in diameter) vesicles that house a number of important oxidative enzymes (p. 373). One product of these oxidative reactions is the potentially lethal oxidant hydrogen peroxide (H_2O_2), generated by reactions of the type

$$RH_2 + O_2 \rightarrow R + H_2O_2$$

where R is a small organic molecule undergoing oxidation. However, peroxisomes also contain the enzyme catalase, which splits H_2O_2 to yield H_2O and O_2. Thus, peroxide is destroyed at its site of synthesis and never has a chance to wreak havoc upon the cytosol.

Viruses Reveal the Existence of Highly Organized Pathways Between Host Cell Compartments[3]

Many animal viruses have a very limited amount of nucleic acid in their genome, sufficient for no more than four or five genes. Most of these genes code for structural proteins of the mature viral particle (or virion). The extremely limited genetic capacity of such viruses demands that they parasitize pathways of the host cell for most of the steps in their replication. In the course of its life cycle, the virus usually follows a distinct sequential route through the cellular compartments, thereby revealing how essential synthetic reactions are compartmentalized inside the host cell itself.

Enveloped animal viruses, defined as those whose genome is enclosed in a lipid bilayer membrane, have exploited the compartmentalization of the cell to an especially fine degree. To follow the life cycle of an enveloped virus is to take a tour through the cell. A well-studied example, *Semliki forest virus,* consists of an RNA-containing nucleocapsid and a surrounding membrane that, like any biological membrane, is composed of a lipid bilayer and proteins (Figure 7–3). A regularly arranged icosahedral (20-faced) shell of a protein (called C protein) forms the **capsid** inside which the genomic RNA is packaged (p. 233). This capsid is surrounded by a closely apposed lipid bilayer that contains only three proteins (called E1, E2, and E3). These so-called **envelope proteins** are glycoproteins that span the lipid bilayer and interact with the C protein of the nucleocapsid underneath to link the membrane and nucleocapsid together. The glycosylated portions of the envelope proteins are always on the outside of the lipid bilayer, and complexes of these proteins form the "spikes" that can be seen in electron micrographs to project outward from the surface of the virus (Figure 7–3A).

As diagrammed in Figure 7–4, infection is initiated when the virus binds to protein receptors on the host cell plasma membrane. As a result of this binding, the virions become selectively incorporated into "coated pits" on the surface of the plasma membrane, which invaginate to form coated vesicles inside the cell. The vesicles bearing endocytosed virions rapidly fuse with smooth-surfaced intermediary vesicles called *endosomes,* which in turn fuse with lysosomes.

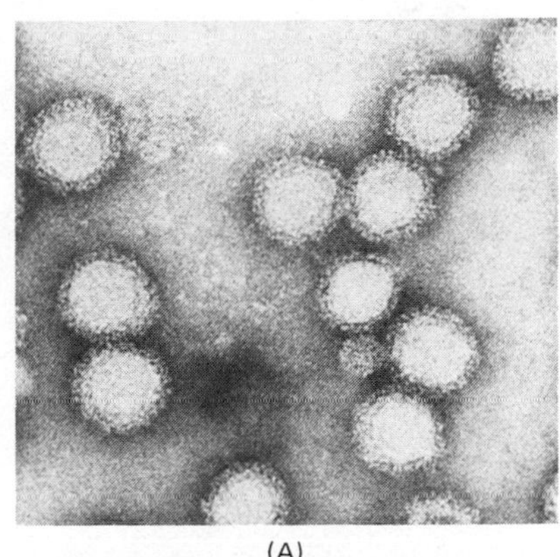

(A)

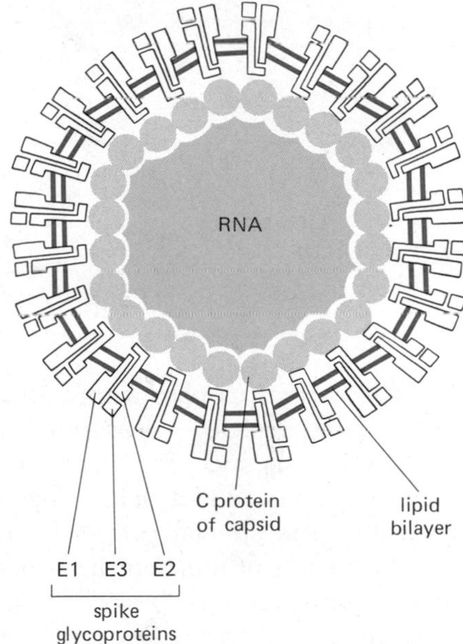

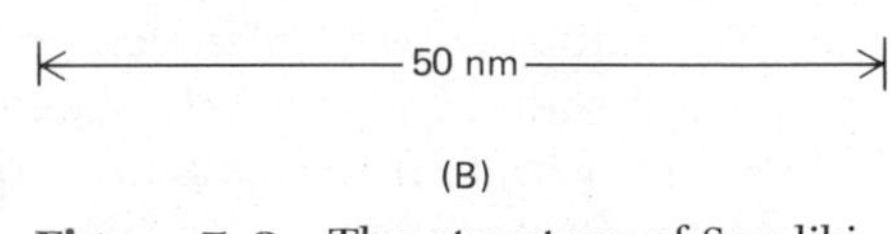

(B)

Figure 7–3 The structure of Semliki forest virus seen in a negatively stained electron micrograph of several whole virus particles (A) and in a schematic drawing of a cross-section of the virus (B). There are about 240 copies of the capsid C protein and 240 copies of the spike glycoprotein complex in each virus particle. The outer envelope of the virus consists of the spike glycoproteins embedded in a lipid bilayer membrane. (Micrograph courtesy of Kai Simons.)

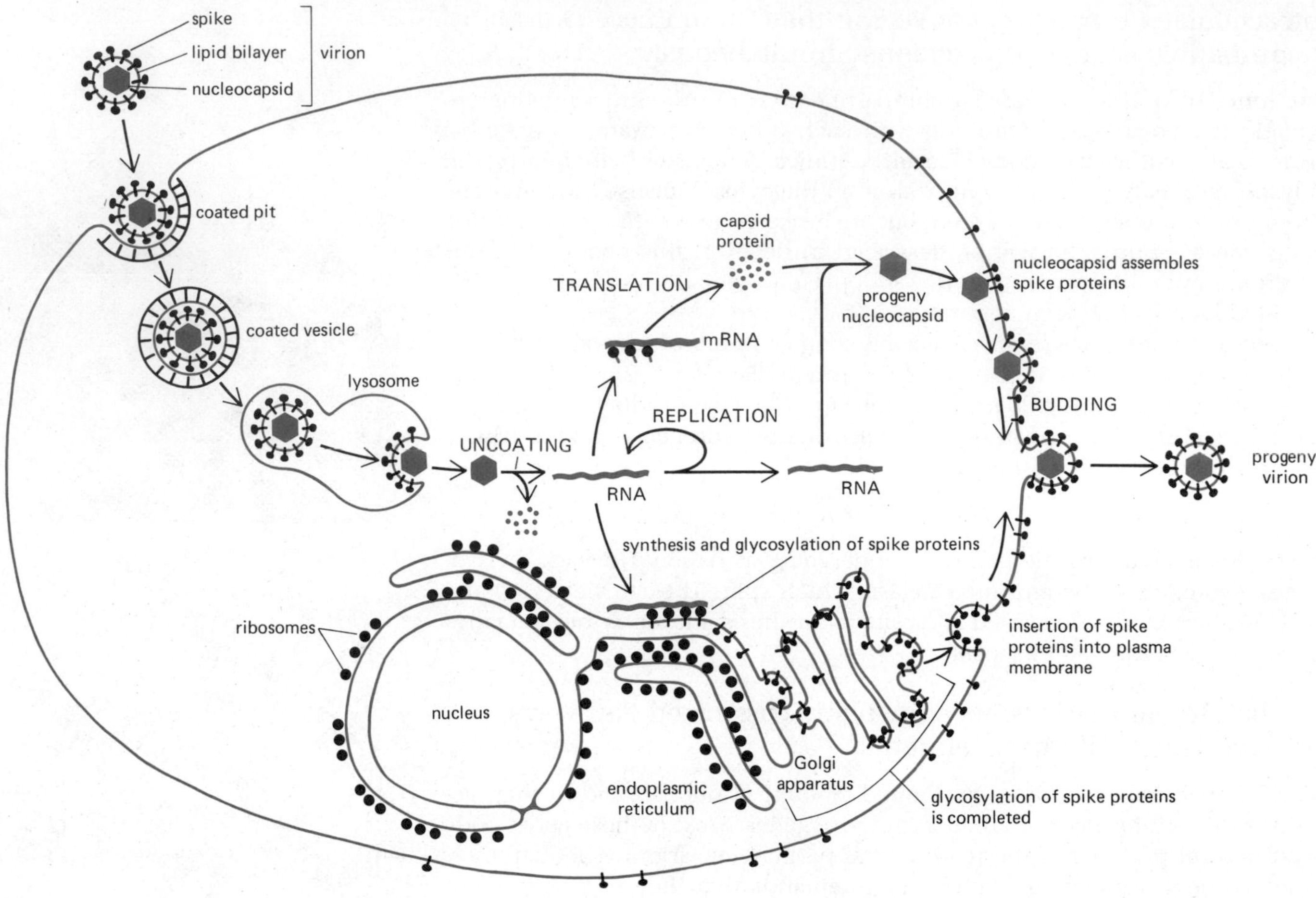

Figure 7–4 Schematic diagram of the life cycle of Semliki forest virus. The virus parasitizes the host cell for all of its biosyntheses.

The virus thus uses a specialized pathway to enter the cell—the coated pit → coated vesicle → lysosome pathway (p. 309)—through which the cell normally ingests specific proteins and small particles that bind to its external surface. But instead of being digested in the lysosomes, the virus escapes because of the special properties of its envelope proteins. These proteins promote fusion of adjacent lipid bilayers at the acidic pH inside lysosomes. Consequently, as soon as the virus is delivered into a lysosome, the viral envelope fuses with the lysosome membrane, allowing the bare nucleocapsid to escape into the cell cytosol (Figure 7–4).

The nucleocapsid is "uncoated" in the cytosol, and the released genomic RNA is translated by host cell ribosomes to produce a viral-coded RNA polymerase. This in turn makes many copies of the RNA, some of which serve as messenger RNA molecules to direct the synthesis of the four structural proteins of the virus, namely, the capsid C protein and the three envelope proteins E1, E2, and E3.

The capsid and envelope proteins are synthesized in different compartments, and they follow separate pathways through the cytoplasm. The envelope proteins, like normal cellular membrane glycoproteins, are synthesized by membrane-bound ribosomes attached to the rough endoplasmic reticulum, while the capsid protein, like a normal cytosolic protein, is synthesized by free ribosomes in the cytosol.

The newly synthesized capsid protein is located in the cell cytosol, where it binds to the recently replicated genomic RNA to form new nucleocapsids. In contrast, the envelope proteins are inserted into the membrane of the endoplasmic reticulum, where they become glycosylated. Remaining membrane-bound at all times, the envelope proteins are transported first to the Golgi apparatus (where they are further glycosylated) and then to the plasma membrane (Figure 7–4).

It is at the plasma membrane that the viral nucleocapsids and the envelope proteins finally meet. As the result of a specific interaction with the envelope proteins, the nucleocapsid becomes wrapped in a portion of the plasma membrane that is highly enriched in the virus-coded envelope proteins embedded in host cell lipids. Then, in a poorly understood process, the nucleocapsid buds out of the cell and pinches off in such a way as to acquire a complete envelope (Figure 7–4).

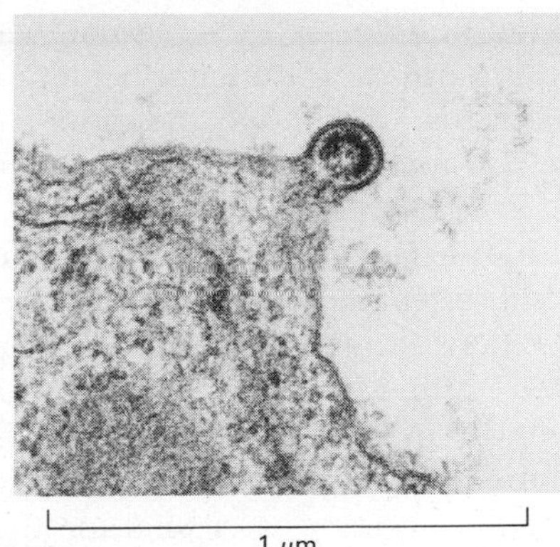

Figure 7–5 An enveloped virus caught in the act of budding from the plasma membrane. (Courtesy of Daniel S. Friend.)

Different Viruses Follow Different Pathways Through the Cell[4]

An enveloped virus such as Semliki forest virus has evolved to exploit the cell's own compartmental design. There are numerous variations on this theme: depending on the specific virus, some compartments are used and others are not. For example, most DNA-containing viruses replicate and undergo transcription to produce RNA in the nucleus rather than in the cytosol. Other viruses appear to enter the cell by fusing directly with the plasma membrane rather than by entering indirectly via lysosomes. In specialized cells that maintain distinct apical and basolateral cell membranes (see p. 282), some viruses bud exclusively from the apical plasma membrane, whereas others bud only from the basolateral plasma membrane (Figures 7–5 and 7–6). Other viruses

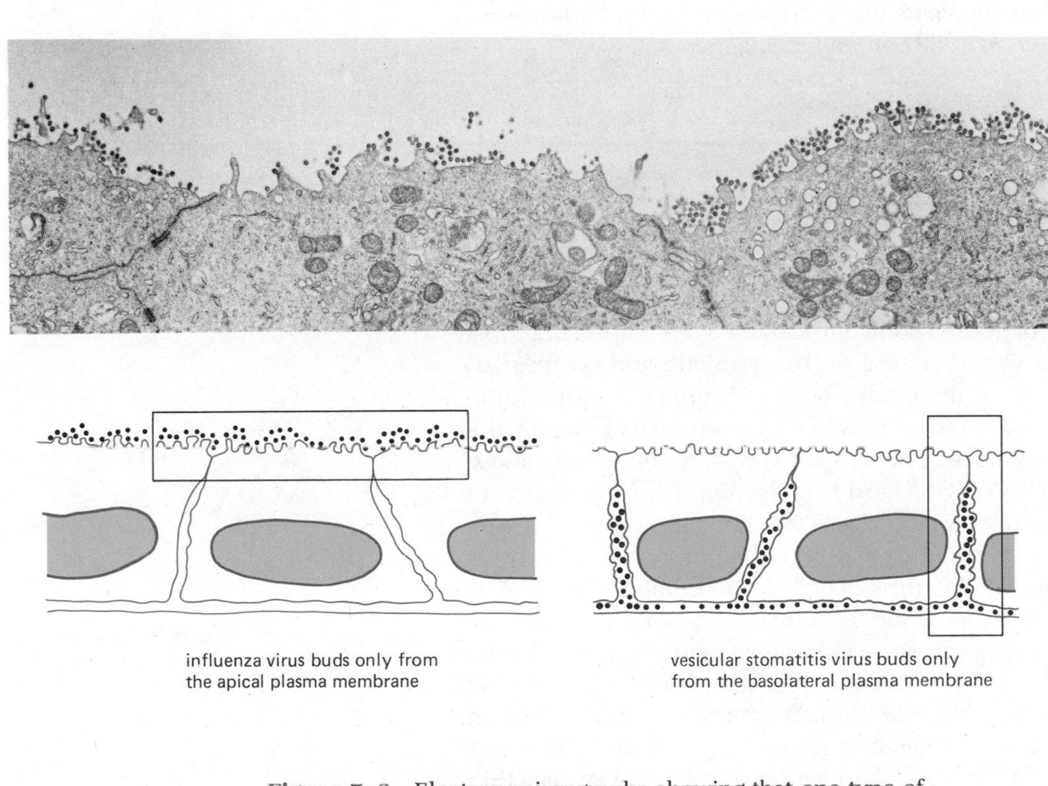

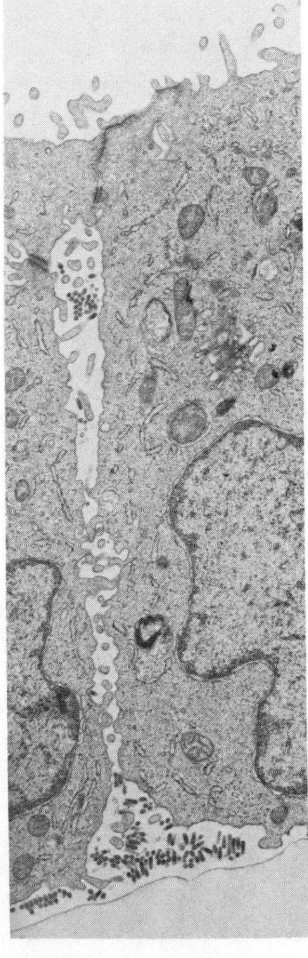

Figure 7–6 Electron micrographs showing that one type of enveloped virus buds from the apical plasma membrane, while another type buds from the basolateral plasma membrane, of the same epithelial cell line grown in tissue culture. These cells grow with their basal surface attached to the culture dish. (Courtesy of E. Rodriquez-Boulan and D. D. Sabatini.)

do not bud through the plasma membrane at all; instead they may bud through the endoplasmic reticulum membrane, through the membrane of the Golgi apparatus, or through the inner nuclear membrane. In these cases, exit from the cell involves either the exocytotic pathway used for secretion (p. 303) or the induction of cell lysis.

The fact that enveloped viruses bud specifically through different cell membranes demonstrates directly that these membranes have distinct properties and that there can be multiple distinct domains even within the same membrane. Moreover, it shows that membrane-fusion events can be specifically directed to one membrane or another depending on the particular envelope proteins that a virus contains. We shall see later that the same type of specific targeting is crucial for the generation of the different cellular compartments themselves. Thus the study of enveloped viruses helps to reveal the mechanisms by which the biosynthetic reactions of the host cell are directed and organized.

Summary

Eucaryotic cells contain extensive sets of intracellular membranes that enclose nearly half their total volume in separate intracellular compartments, known as cell organelles. Six main types of organelles are present in nearly all cells: endoplasmic reticulum, Golgi apparatus, nucleus, mitochondria, lysosomes, and peroxisomes. Plant cells contain the chloroplast as an additional major organelle. The remaining intracellular compartment is the cytosol. When viruses infect a eucaryotic cell, their biosyntheses follow a distinct sequential route through the cellular compartments. Virus life cycles thereby reveal how essential biosynthetic reactions are normally compartmentalized in cells.

The Cytosol[5]

Most Intermediary Metabolism Takes Place in the Cytosol

The term "intermediary metabolism" refers to the large set of chemical reactions through which the cell degrades some small molecules (capturing most of their chemical energy, mainly as ATP or a proton gradient) and synthesizes others as precursors for the macromolecules needed for the structure, function, and growth of the cell. All intermediary metabolism takes place in the cytoplasm, most of it in the **cytosol.** The cytosol generally represents about 55% of the total cell volume (Table 7–1), and it is teeming with thousands of enzymes that catalyze the reactions of glycolysis and gluconeogenesis, as well as the biosynthesis of sugars, fatty acids, nucleotides, and amino acids. In addition, it contains a variety of different cytoskeletal proteins that impart shape to a cell, cause coherent cytoplasmic movements, and provide a general framework that is used to help organize the many enzymatic reactions in the cytosol (see Chapter 10). About 20% of the weight of the cytosol is protein, so that it is more like a highly organized gelatinous mass than a simple solution of enzyme molecules, despite its rather amorphous appearance in electron micrographs. Thus, for example, we know from certain cytological studies that the cytosol immediately surrounding the Golgi apparatus is not identical to the cytosol that closely encircles the cell nucleus. However, because this type of organization is very difficult to preserve once cells are broken open, we know almost nothing about the mechanisms by which it is achieved.

The metabolic pathways that convert several important compounds to storage forms generate products that are readily visible in the microscope.

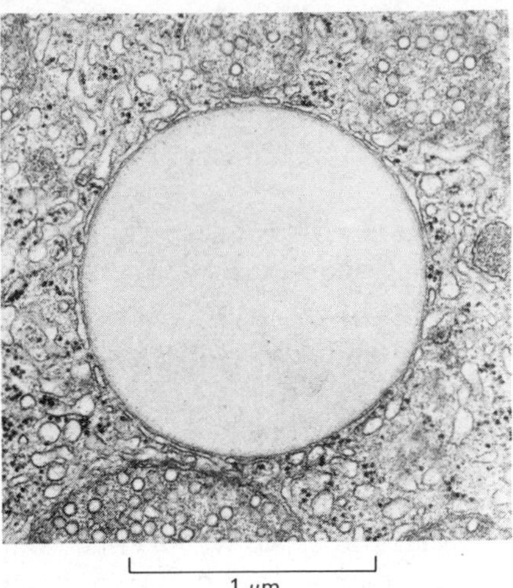

Figure 7–7 Electron micrograph of a lipid droplet containing triglyceride, the main storage form of fat. These droplets in the cell cytoplasm serve as an energy store. (Courtesy of Daniel S. Friend.)

The most visually obvious, as well as the most important, example is fat (triglycerides), the storage form of fatty acids. As shown in Figure 7–7, in many cells these insoluble triglycerides coalesce in the cytosol to form large, anhydrous droplets (from 0.2 μm to 5 μm in diameter). In adipocytes, the cells specialized for fat storage, these droplets can be as large as 80 μm, occupying virtually the entire cytosol.

Another major visible stored molecule is glycogen, a polymer of glucose. Glycogen is the major storage form of carbohydrate, and in many cells large individual molecules (some 10 to 40 nm in diameter) appear as granules in electron micrographs. The enzymes needed to carry out the synthesis and degradation of glycogen are bound to the surface of these glycogen granules (Figure 7–8).

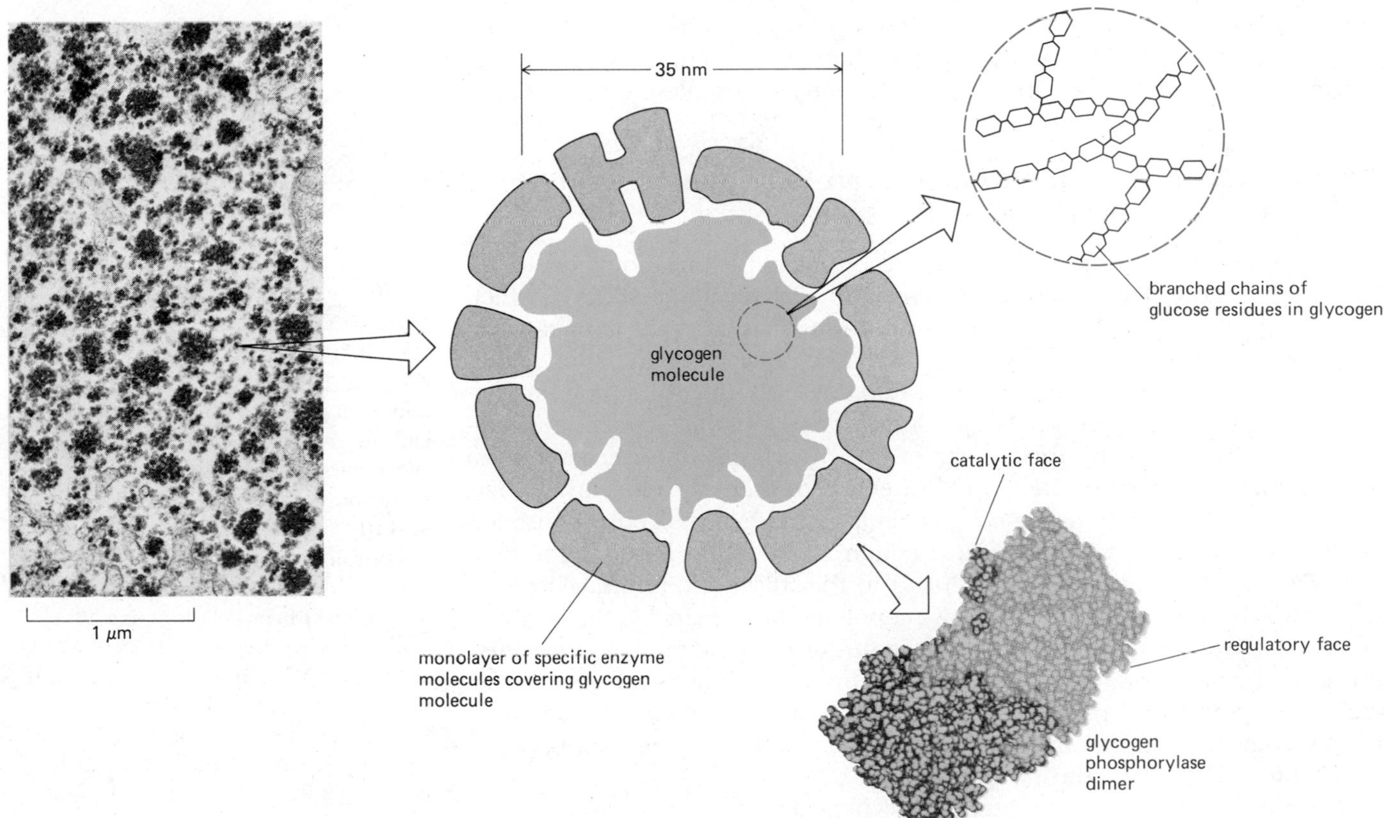

Figure 7–8 Electron micrograph and schematic drawing of a glycogen granule, the major storage form of carbohydrate in vertebrates. The synthesis and degradation of glycogen are catalyzed by enzymes bound to the granule surface, including the synthetic enzyme glycogen synthase and the degradative enzyme glycogen phosphorylase. The three-dimensional structure of the latter enzyme is shown. (Courtesy of Robert Fletterick and Daniel S. Friend.)

Many Proteins Are Synthesized by Ribosomes in the Cytosol[6]

The enzymes catalyzing the chemical reactions of intermediary metabolism are among the many proteins synthesized in the cytosol. Each protein is synthesized by a process in which *messenger RNA (mRNA)* molecules emerge from the cell nucleus and become attached to large ribonucleoprotein particles called **ribosomes.** Here the mRNA sequence is translated into a corresponding sequence of amino acids. Since the detailed mechanism of protein synthesis has been described in Chapter 5, only those features most relevant to the organization of the eucaryotic cell cytoplasm will be mentioned here.

Each ribosome is composed of a large and a small subunit that dissociate reversibly after each round of protein synthesis. Figure 7–9 illustrates the general size and shape of a bacterial ribosome. Eucaryotic ribosomes are somewhat larger, but they seem to be constructed similarly and to function in much the same way as bacterial ribosomes (see Figure 5–14, p. 209).

Ribosomes are complex structures about 30 nm in their largest dimension, containing somewhat more RNA than protein by weight. The *ribosomal RNA (rRNA)* molecules form a framework on which dozens of species of proteins, each present at one copy per ribosome, spontaneously assemble. In addition to playing an important part in assembling the ribosomal proteins, the rRNA may form specific complementary base pairs with selected regions of *transfer RNA (tRNA)* and mRNA as part of the crucial ribosome-mRNA-tRNA recognition process (p. 210). The ribosome is properly viewed as a large protein-synthesizing machine that not only brings together the reagents needed for protein synthesis in the proper spatial relationships, but also provides the enzymes that catalyze the entire process of polypeptide-bond formation.

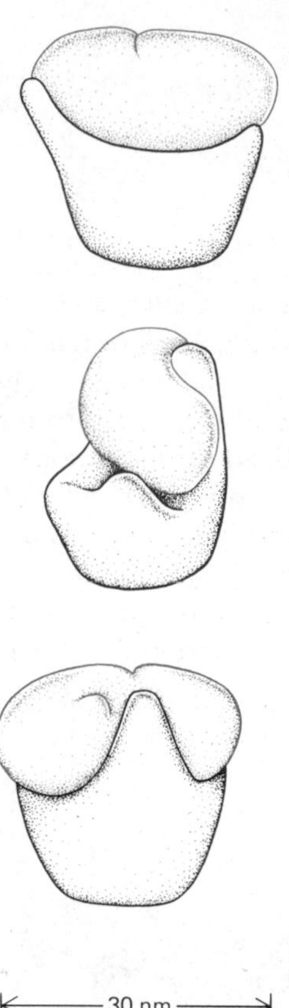

Figure 7–9 Three views of a three-dimensional model of the bacterial ribosome showing the small (*colored*) subunit and the large (*black*) subunit paired. In eucaryotes the large subunit contains one large and two small rRNA molecules plus about 45 proteins; the small subunit contains a different large rRNA molecule plus about 33 proteins. Based on their sedimentation rates in a centrifuge, these large and small subunits are commonly referred to as the "60S" and the "40S" subunits, respectively (a complete ribosome sediments at 80S).

A typical liver cell contains 10^7 ribosomes, each with a mass of 4.5×10^6 daltons (about one-half rRNA and one-half protein). Ribosomes, therefore, constitute about 5% of the total dry weight of such a cell. Cells less active in protein synthesis contain proportionately fewer ribosomes.

The Binding of Many Ribosomes to an Individual Messenger RNA Molecule Generates Polysomes[7]

Protein synthesis has three distinct stages: initiation, elongation, and termination (Chapter 5). Two different signals are needed for initiation. One, which helps to bind the mRNA molecule to the ribosome, is different in eucaryotes and procaryotes (as is discussed on p. 332). The other is the initiation codon AUG, which is recognized by a special initiator tRNA (incorporating methionine) that must pair with the AUG codon in order to initiate each new polypeptide chain. Driven by the energy of GTP hydrolysis, the ribosome then moves along the mRNA in the 5′-to-3′ direction to begin the elongation stage. At each step of elongation, the specific aminoacyl tRNA molecule whose anticodon is complementary to the next codon in the mRNA binds to the ribosome (see p. 210); the amino acid linked to that tRNA is then transferred to the carboxyl-terminal end of the growing polypeptide chain. Synthesis therefore always proceeds from the amino terminus of a protein toward its carboxyl terminus. Termination occurs when one of three stop codons is encountered, releasing the finished product from the ribosome. Completed proteins are usually active as soon as they are released, since most of the folding of the polypeptide takes place during synthesis.

Amino acids are incorporated in polypeptides at a rate as fast as 20 per second, so that the complete synthesis of an average-sized protein takes 20 to 60 seconds. Even during this very short period multiple initiations take place, with a new ribosome hopping onto the 5′ end of a molecule of mRNA almost as soon as the preceding ribosome has translated enough of the amino acid sequence to get out of the way. Under physiological conditions, therefore, actively translated mRNA is found in **polyribosomes,** or **polysomes,** formed by several ribosomes spaced as close as 80 nucleotides apart along a single messenger molecule (Figures 7–10 and 7–11). Polyribosomes are a common

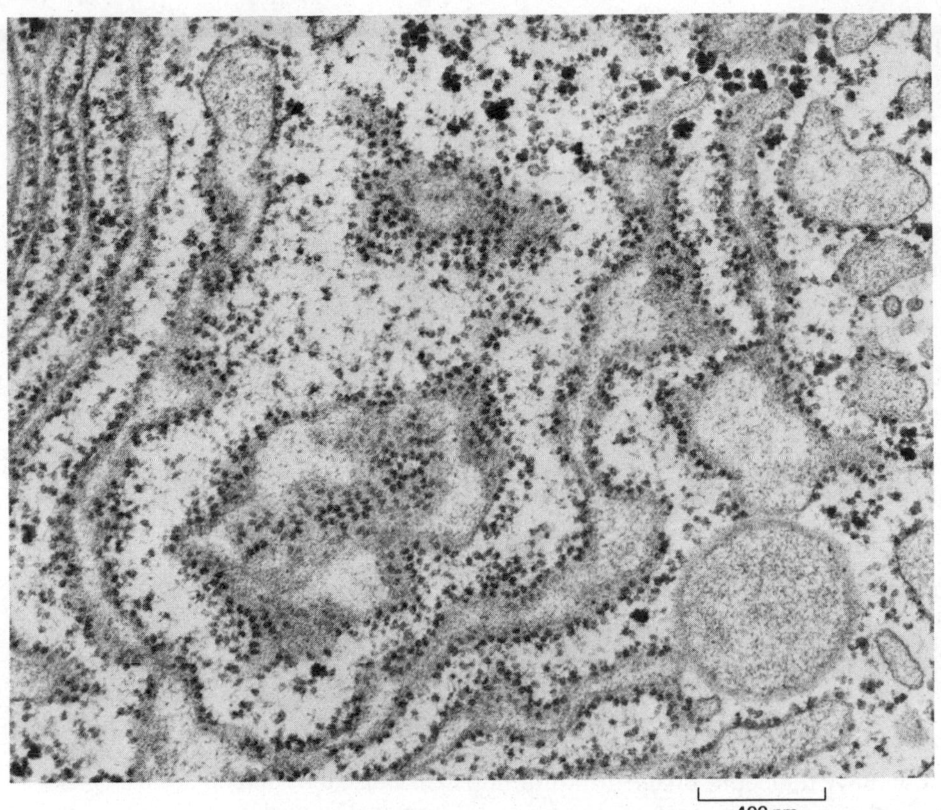

Figure 7–10 Electron micrograph of typical polyribosomes in action during protein synthesis. The cell cytoplasm is generally crowded with such polyribosomes, some free in the cytosol and some membrane-bound. (Courtesy of George Palade.)

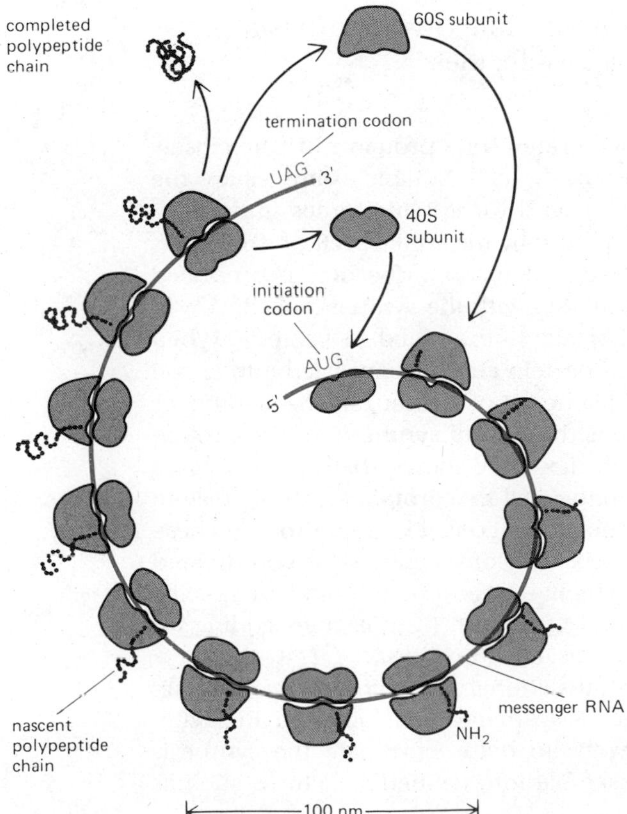

Figure 7–11 Schematic drawing of a polyribosome. Protein synthesis begins with the binding of the small subunit to the appropriate site on the messenger RNA molecule. When the polypeptide chain is completed, both subunits dissociate from the mRNA (see Chapter 5).

feature of nearly all cells. After cell lysis, the polysomes can be isolated and separated from single ribosomes in the cytosol by ultracentrifugation (Figure 7–12). The identification of a particular species of mRNA molecule in such isolated polysomes provides the best evidence that the corresponding coding sequence is being actively translated in the cells from which the polysomes were prepared. Such identifications can be accomplished by decoding the mixture of mRNAs isolated from polysomes by using the mixture to direct cell-free protein synthesis. Alternatively, nucleic acid hybridization reactions can be used to test these mRNAs for sequence homology with a particular cloned DNA probe (p. 189).

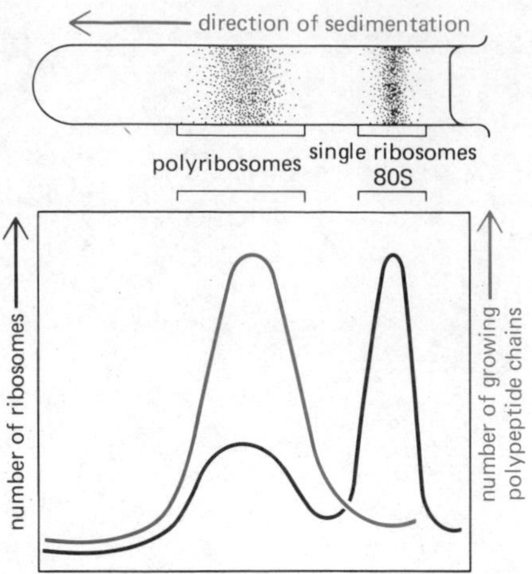

Figure 7–12 Diagram illustrating how polyribosomes are separated from single ribosomes and their subunits by sedimentation in a centrifuge. This method is based on the fact that large molecular aggregates move faster than small ones in a strong gravitational field. Generally the sedimentation is carried out through a gradient of sucrose to stabilize the solution against convective mixing.

Protein Synthesis Is Blocked by Specific Inhibitors

There are many drugs and toxins that specifically affect each of the major steps of protein synthesis, and these have been very useful for working out cellular mechanisms. Among the most commonly used drugs in experimental studies are *chloramphenicol, cycloheximide,* and *puromycin.* The first two drugs reversibly inhibit *peptidyl transferase,* the ribosomal enzyme that forms peptide bonds. However, in a eucaryotic cell, chloramphenicol inhibits protein synthesis only on the ribosomes in the mitochondria (and in chloroplasts of plants), while cycloheximide affects only cytosolic ribosomes. The differences in sensitivity to these two drugs reflect differences in the structure of the two classes of ribosomes, and it provides a powerful way to determine whether a particular protein is translated in mitochondria (or chloroplasts, see p. 528). Puromycin is a structural analogue of a tRNA molecule linked to an amino acid; the ribosome mistakes it for an authentic amino acid and covalently incorporates it at the carboxyl terminus of the growing polypeptide chain, causing the premature termination and release of this polypeptide. (See Table 5–1, p. 213.)

Some Proteins Regulate the Rate of Their Own Synthesis by Binding to the Messenger RNA Molecules on Which They Are Made[8]

In certain cases the intracellular concentration of a protein must be closely regulated because an excess would harm the cell. Notable examples are the ribosomal proteins, many of which will bind to RNA molecules indiscriminately if more are present than can be bound by the available rRNA. How does the cell ensure that each of the 80 or so proteins in a ribosome is produced in approximately equal amounts and in step with the synthesis of rRNA?

This problem has been most thoroughly examined in bacteria. When multiple copies of one of the ribosomal protein genes are inserted into *E. coli* (using a plasmid vector, see p. 187), the amount of messenger RNA coding for that protein increases dramatically, but the rate of synthesis of the protein itself increases only slightly. Further studies have shown that in many cases the addition of a free ribosomal protein to a cell-free protein synthesis system specifically inhibits the translation of the mRNA coding for that protein. These striking observations suggest that bacterial ribosome synthesis is coordinated by a feedback regulation system in which excess proteins bind to specific sequences or structures near the 5′ end of the mRNA molecules coding for them and thereby block their own synthesis (Figure 7–13).

Although very few examples of translational regulation of this sort are known, it would be surprising if the basic mechanism outlined in Figure 7–13 did not also operate in eucaryotes to help coordinate the synthesis of sets of interacting proteins that assemble into defined structures such as ribosomes.

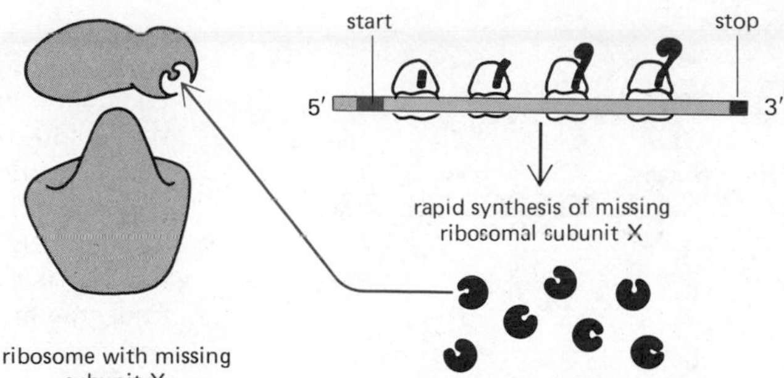

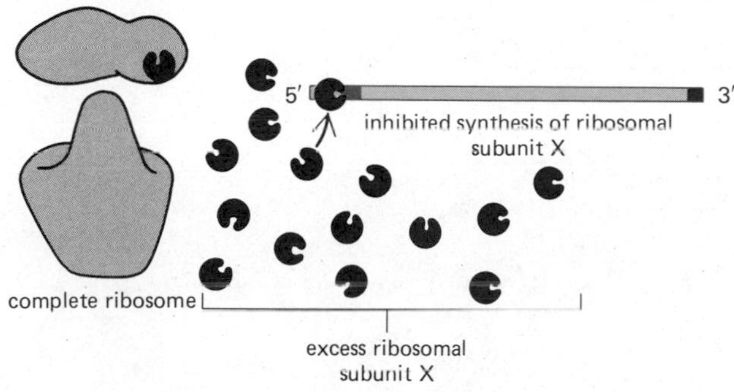

Figure 7–13 Highly schematic view of a mechanism for coordinating the synthesis of ribosomal proteins in bacteria. This is one of the few well-established examples of this type of specific translational regulation.

In Eucaryotes Only One Species of Polypeptide Chain Can Be Synthesized on Each Messenger RNA Molecule[9]

Eucaryotic and procaryotic messenger RNAs differ somewhat in both structure and function. Eucaryotiç mRNAs (except those synthesized in mitochondria and/or chloroplasts) are modified in the nucleus immediately after their transcription (see p. 412). As a result, they usually carry both a unique "cap" structure, composed of a 7-methylguanosine residue linked to a triphosphate, at their 5′ end (Figure 7–14) and a run of about 200 adenylic residues ("poly

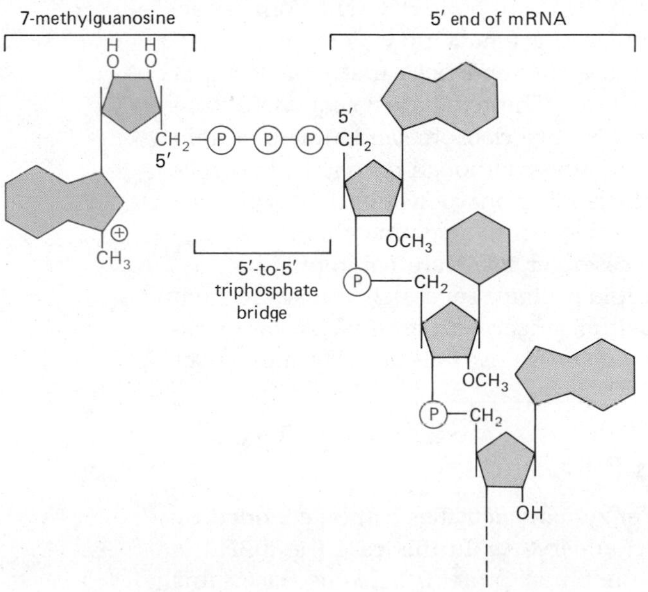

Figure 7–14 The structure of the 5′ cap found at the 5′ end of eucaryotic messenger RNA molecules. Note the unusual linkage to the positively charged 7-methylguanosine and the methylation of the 2′ hydroxyl group on the ribose sugar. (The second sugar is not always methylated.)

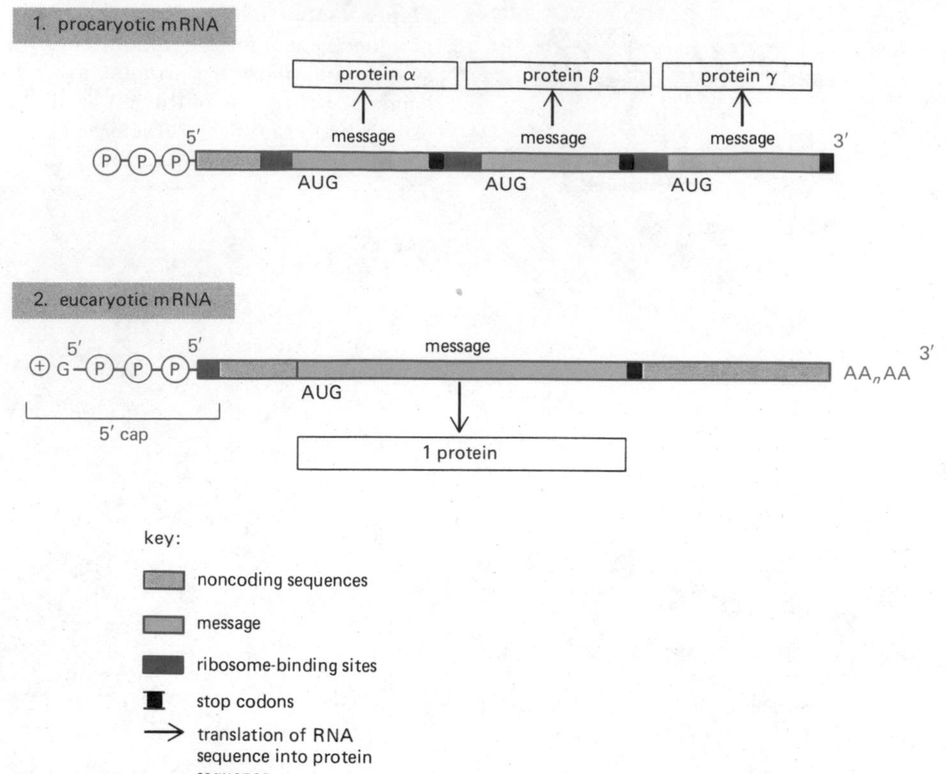

Figure 7–15 A comparison of the structure of procaryotic and eucaryotic messenger RNA molecules. While both mRNAs are synthesized with a triphosphate group at their 5′ end, the eucaryotic RNA molecule immediately acquires a 5′ cap. Since this cap is required for initiating ribosome binding, ribosomes bind and initiate synthesis near the 5′ end of eucaryotic mRNAs. By contrast, in procaryotes there can be multiple ribosome-binding sites in the interior of an mRNA chain.

A") at their 3′ end. What part, if any, the poly A plays in the translation process is uncertain. But the 5′ cap structure is crucial, being specifically recognized by ribosomes as an initiation signal for protein synthesis. Elegant genetic experiments in yeast have shown that the synthesis of a eucaryotic protein will generally begin at whatever AUG codon is nearest to the 5′ cap and proceed by decoding the messenger RNA in phase until a stop codon is encountered. None of the many other AUG codons in the chain can serve as initiation sites. Because no messenger RNA sequences beyond (on the 3′ side of) the first stop codon can be translated, only a single species of polypeptide chain can be synthesized from each mRNA molecule.

In all these respects, procaryotic mRNAs are quite different from eucaryotic mRNAs (Figure 7–15). Bacterial mRNAs have no 3′ poly A or 5′ cap structure. Instead they have special initiation-site sequences that can occur at multiple points in the same mRNA molecule. These initiation sequences form base pairs with a specific region of the rRNA in a ribosome and initiate protein synthesis at an adjacent AUG codon. Moreover, although bacterial ribosomes recognize stop codons as signaling the end of one polypeptide chain, they are able to slide past them to initiate a second polypeptide chain at a subsequent AUG. As a result, bacterial messenger RNAs are commonly *polycistronic,* meaning that they encode multiple proteins separately translated from the same mRNA molecule. In contrast, all eucaryotic mRNAs are *monocistronic,* only one species of polypeptide chain being translated per messenger molecule.

Polyproteins Are Often Made in Eucaryotes[10]

Despite the above constraint, related enzymatic activities can be encoded by the same messenger RNA molecule in eucaryotes. In this case the mRNA is translated into a single, large multifunctional protein. In some cases this *polyprotein* is cleaved by specific proteases to yield distinct enzymes. In other cases, however, it remains intact as a single multifunctional polypeptide.

As illustrated in Figure 7–16, a series of related enzymatic activities can be linked together in a polyprotein in eucaryotes, while in procaryotes the same enzymatic activities often involve the noncovalent complexing of separate enzyme subunits. In part, this difference may reflect the greater difficulty of bringing together subunits in the eucaryotic cell, with its approximately 1000-fold increased volume. However, it could also be due to the differences in mRNA translation mechanisms illustrated in Figure 7–15.

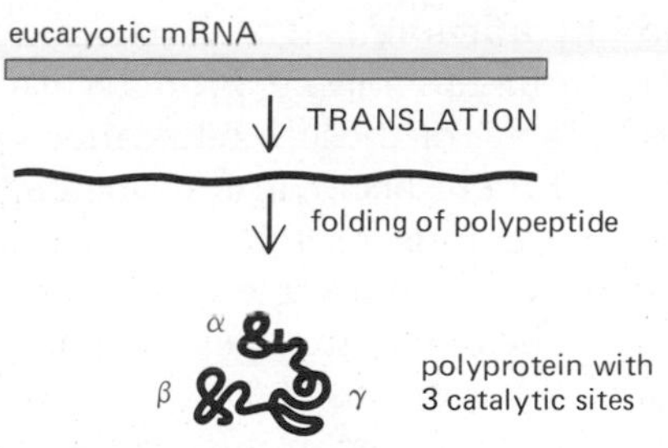

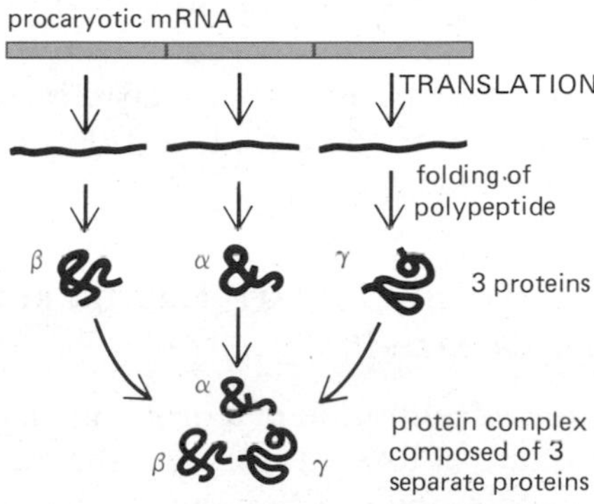

Figure 7–16 Proteins with multiple enzymatic activities (polyproteins) are thought to be relatively common in eucaryotes. In bacteria similar enzymes are often assembled from separate protein chains, which may be translated from different regions of a single, long mRNA molecule.

Many Proteins Undergo Covalent Modification After Their Synthesis[11]

Many, if not most, proteins undergo some type of covalent modification after being released from the ribosome. More than 100 different such modifications of the amino acid side chains are known. These are often reversible modifications that regulate the biological activity of the protein (see p. 747). A very common example is the phosphorylation of a selected free OH group of the amino acid side chain of a serine, a threonine, or a tyrosine residue in a protein (Figure 7–17). Depending on the protein, this phosphorylation can either increase or decrease its functional activity. Enzymes called **protein kinases** utilize ATP to phosphorylate proteins in this way. The activity of such kinases is closely regulated by key intracellular signals, such as the concentration of cyclic AMP or of Ca^{2+}. Protein phosphorylation is reversible because of the abundant phosphatases present in the cytosol. When protein kinases are turned off, these phosphatases remove the phosphate from the phosphorylated enzyme, and the enzyme activity returns to its former level. The adenylation of tyrosine side chains (in which an AMP moiety from ATP is reversibly added to the tyrosine OH group) provides another important regulatory system, as does the reversible methylation of glutamic acid side chains.

Other covalent modifications of proteins that occur in the cytosol are permanent and required for activity. For example, many enzymes have a covalently attached coenzyme (such as biotin, lipoic acid, or pyridoxal phosphate) that plays an essential part in their reaction mechanism (p. 129). There are numerous other irreversible modifications whose exact functions are less clear. For example, the amino terminus of many proteins in the cytosol is modified ("blocked") by acetylation, and the amino group on a few selected lysines can be methylated.

Many other permanent covalent modifications occur only to those proteins that leave the cytosol for other intracellular or extracellular destinations. For example, in the endoplasmic reticulum almost half of the proline residues in the secreted protein collagen are hydroxylated to hydroxyproline, a reaction that increases the strength of the collagen triple helix (p. 696). Another important example of this kind is the addition of a second carboxyl group to glutamic acid residues in proteins to form γ-carboxyglutamate, a dicarboxylic acid with a high affinity for Ca^{2+}. This device is utilized to create the Ca^{2+}-binding sites used by a number of Ca^{2+}-binding proteins involved in blood-clotting reactions. In addition, most secreted proteins are glycoproteins with

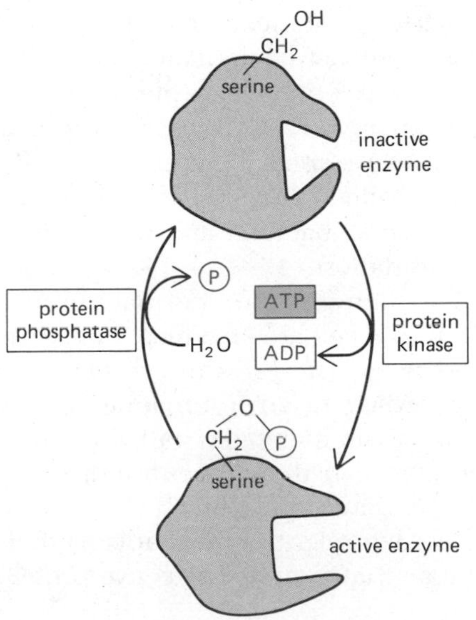

Figure 7–17 Schematic illustration of the opposing actions of a protein kinase and a protein phosphatase in phosphorylating and dephosphorylating a hydroxyl group on a protein side chain. As indicated, the phosphorylation generally causes an allosteric change in the protein's conformation that either increases or decreases its activity.

carbohydrates linked to asparagine or, more rarely, serine or threonine residues, and membrane proteins can have fatty acids covalently attached to them. Some of these modifications occur in the endoplasmic reticulum, and some of them occur in the Golgi apparatus, as will be described later.

Numerous proteins are secreted by specialized secretory cells: for example, some endocrine cells secrete polypeptide hormones, and pancreatic acinar cells secrete digestive enzymes. Many of these proteins are initially made as larger proteins, which are cleaved by specialized proteases into the active protein. Depending on the protein, this cleavage can occur in the Golgi apparatus, in developing secretory vesicles, or after secretion. Some of these covalent modifications can be considered to be regulatory. In particular, many secreted enzymes (including digestive proteases) that would be harmful if synthesized in an active form inside a cell are synthesized and secreted as inactive precursors called **zymogens.** These enzymes become active only after their modification outside the cell by one or more specific proteolytic cleavages.

Some Proteins Are Degraded Soon After They Have Been Synthesized[12]

The concentration of a protein in a cell is determined by the balance between the rate of its synthesis and the rate of its degradation. As a result, there is a constant **protein turnover.** The rate of synthesis of a protein is usually controlled by regulating the amount of its mRNA available for translation. In addition, a cell can control the concentration of a protein by regulating the rate at which the protein is destroyed.

Protein degradation seems to be a random process, since an old protein molecule is no more likely to be degraded than a new molecule during a particular time interval. Nevertheless, since the different species of proteins in a single cell have a different inherent susceptibility to the degradation process, they turn over at remarkably different rates. A "typical" protein molecule is degraded, on average, about two days after it has been synthesized, but the average degradation times for individual proteins range from several minutes to months or even years.

Most proteins are compact structures, inherently somewhat resistant to proteolytic attack and, therefore, to degradation. One important function of protein degradation is to rid the cell of defective proteins. These can either be coded for by defective genes, result from biosynthetic errors, or be created by the spontaneous denaturation of previously functional polypeptides. The less compact conformations of these "abnormal" proteins probably make them more susceptible to proteolytic attack. As part of this finely controlled process, the energy of ATP hydrolysis appears to be utilized to form a covalent linkage between a special polypeptide and the protein to be degraded. This polypeptide, called *ubiquitin* because of its widespread occurrence in organisms as diverse as bacteria and man, evidently marks the attached protein for rapid degradation.

Many of the "normal" proteins that are subject to rapid degradation within a cell are enzymes that catalyze a rate-determining step in a metabolic pathway. The rates of synthesis of these key proteins are usually regulated according to environmental conditions so as to promote the most efficient use of the metabolic pathway. Only if they are continuously and rapidly degraded can the concentrations of these enzymes adjust rapidly to the new levels determined by a change in their rate of synthesis (p. 750). It is not yet clear whether the mechanisms that degrade these proteins are different from those that degrade defective proteins.

Not All Proteins Synthesized in the Cytosol Remain There

As will be discussed in the next section, some ribosomes involved in translation bind to the endoplasmic reticulum to form membrane-bound ribosomes. Many of the proteins destined for a variety of intracellular organelles and their membranes, as well as all known proteins destined for secretion, are synthesized by such ribosomes.

But some of the proteins translated by free ribosomes in the cytosol also leave the cytosol. For example, histones, as well as many other chromatin proteins, rapidly diffuse through large nuclear pores and bind to structures in the nucleus following their synthesis in the cytoplasm (p. 432). Most of the proteins in chloroplasts and mitochondria, and at least some of those in peroxisomes, are translated by free ribosomes in the cytosol and then are transported across the membranes surrounding their respective organelles.

Summary

The cytosol consists of all of the space outside the cellular organelles and generally represents 50% to 60% of the total cell volume in eucaryotes. Most intermediary metabolism and the protein synthesis required for cell growth and maintenance occur in the cytosol. The amount of any particular protein in a cell depends on the balance between the rate at which it is synthesized (which is largely controlled by the rate at which its mRNA is synthesized in the nucleus) and the rate at which it is degraded. Some proteins are degraded rapidly and continuously in cells, presumably to allow rapid changes in their concentrations in response to regulatory signals. The reversible covalent modification of proteins provides an important mechanism for regulating the activity of specific proteins in cells. For example, the activities of many cellular proteins are controlled by cycles of phosphorylation and dephosphorylation.

The Endoplasmic Reticulum[13]

All cells contain an **endoplasmic reticulum (ER),** whose membrane typically constitutes more than half of the total membrane in a cell (Table 7–2). Although it is highly convoluted, this membrane is thought to form a single continuous sheet enclosing a single closed sac. This internal space, called the **ER lumen** or the ER cisternal space, often occupies more than 10% of the total cell volume (Table 7–1). The lumen of the ER is separated from the cytosol by a single membrane (the ER membrane), which mediates the communication between these two compartments. Because the ER membrane is continuous with the outer nuclear membrane, the ER lumen and the interior of the nucleus are also separated by only a single membrane—in this case, the inner nuclear membrane. By contrast, the lumens of the ER and Golgi compartments are separated from one another by two membranes, and the extensive macromolecular traffic between them is thought to require transport vesicles (Figure 7–18).

The ER provides the cell with a mechanism for separating those newly synthesized molecules that belong in the cytosol from those that do not. In addition, the ER plays a central part in the biosynthesis of the macromolecules used to construct other cellular organelles. Lipids, proteins, and complex carbohydrates destined for transportation to the Golgi apparatus, to the plasma membrane, to lysosomes, or to the cell exterior are all synthesized in association with the ER. For this reason, an understanding of the biochemistry of

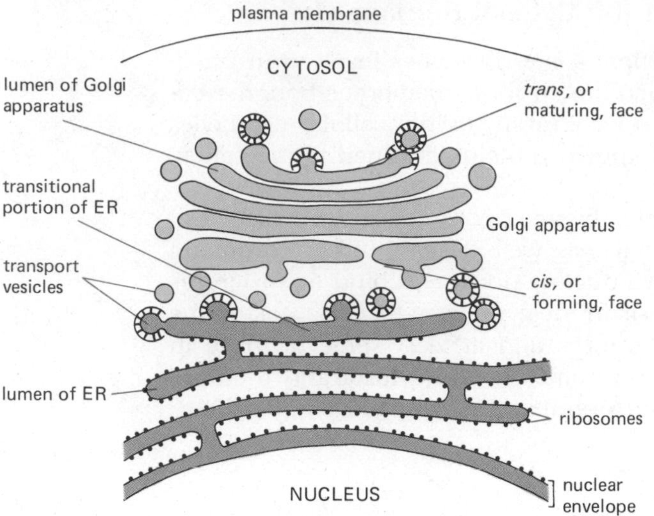

Figure 7–18 Schematic diagram showing the separation of the ER lumen from three other intracellular compartments with which the ER interacts. Note that the ER lumen is separated from both the cytosol and the nucleus by only a single membrane, whereas two membranes separate the lumen of the ER and the lumen of the Golgi apparatus. The relative orientation of the Golgi apparatus shown is that found in secretory cells; in many other cells this orientation is reversed, and the *trans* side of the Golgi apparatus faces the cell nucleus.

the ER is crucial to understanding the origins of many other cellular structures. Recent progress in elucidating the biosynthetic pathways utilized by the ER has been made possible by the development of cell-free systems in which the relevant biochemical steps can be studied *in vitro.*

Attached Ribosomes Define "Rough" Regions of ER

Two functionally distinct regions of ER can readily be distinguished in electron micrographs of some cells: the **rough ER,** which is studded with ribosomes on the cytoplasmic side of the membrane, and the **smooth ER,** which is physically a portion of the same membrane but lacks any attached ribosomes. These two regions also differ considerably in shape: whereas rough ER is organized in stacks of flattened sacs, called *cisternae,* smooth ER consists of a meshwork of fine tubules (Figures 7–19 and 7–20). The outer membrane of the nuclear envelope is always studded with ribosomes and is continuous with rough ER membrane. The fact that ribosomes are never found on the luminal side of the ER was the first indication that its biosynthetic functions (in this case, protein synthesis) are asymmetrically distributed across the ER membrane.

Although it is present in all nucleated cells except sperm, rough ER is especially abundant in cells specialized for protein secretion (such as the

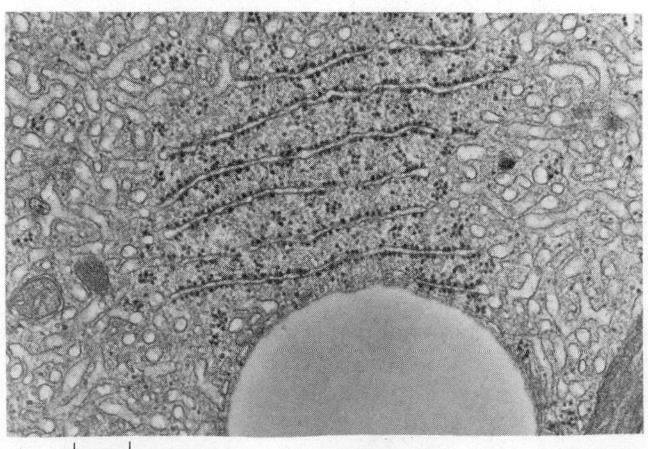

Figure 7–19 Electron micrograph showing the very different morphologies of the rough and smooth ER. This is a Leidig cell, which produces steroids in the testes and has an extensive smooth ER. Part of a large spherical lipid droplet is also seen. (Courtesy of Daniel S. Friend.)

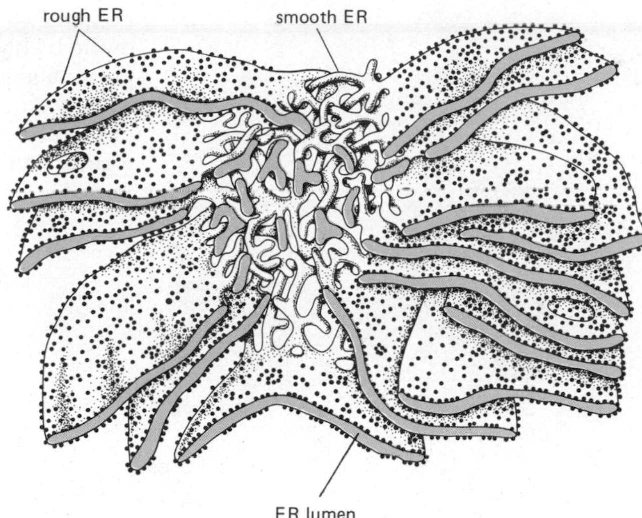

Figure 7–20 Three-dimensional reconstruction of a region of smooth and rough ER in a liver cell. The rough ER forms oriented stacks of flattened cisternae, each having a luminal space 20 nm to 30 nm wide. Connected to these cisternae are the smooth ER membranes, which form a fine network of tubules 30 nm to 60 nm in diameter. It is thought that the ER membrane is continuous and encloses a single lumen. (After R. V. Krstić, Ultrastructure of the Mammalian Cell. New York: Springer-Verlag, 1979.)

pancreatic acinar cells and antibody-secreting plasma cells) or for extensive membrane synthesis (such as the immature egg cell or the retinal rod cell). In such cases, as many as half of the total ribosomes in the cell are bound to the ER; and in the case of protein-secreting cells the ER lumen is often grossly dilated by the protein secreted into it from these ribosomes (Figures 7–21 and 7–22).

Smooth ER is not involved in protein synthesis. Although it is plentiful in some specialized cells (see next section), in the great majority of cells (including most secretory cells) smooth ER is really nothing more than a small

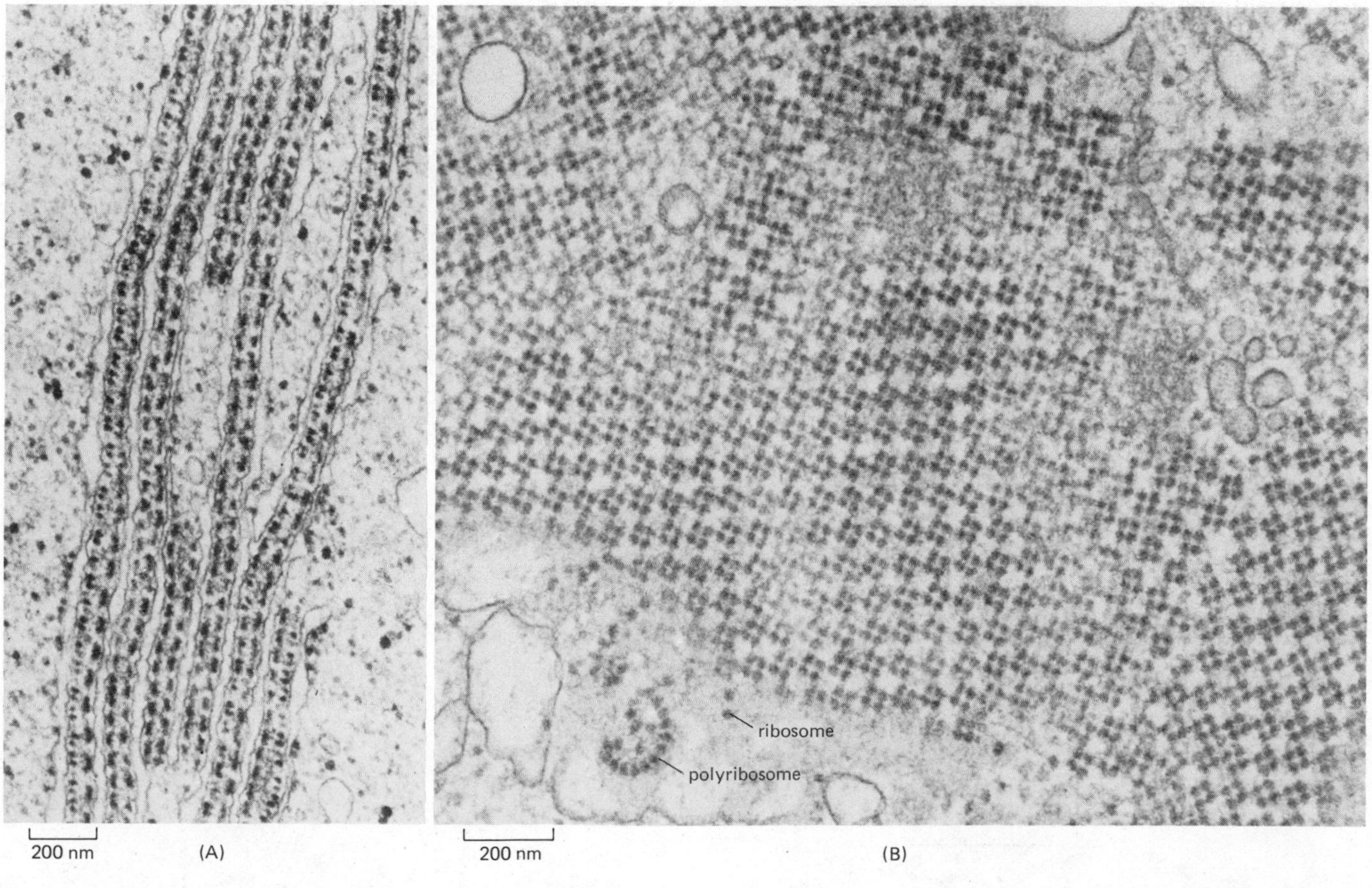

Figure 7–21 Electron micrographs of the unusually highly ordered array of ribosomes that cover the extensive rough ER in a lizard oocyte (immature egg cell). (A) View of a section cut perpendicularly to the stacked rough ER cisternae. (B) Section showing surface view. (Courtesy of Nigel Unwin.)

ribosome
polyribosome
200 nm (A)
200 nm (B)

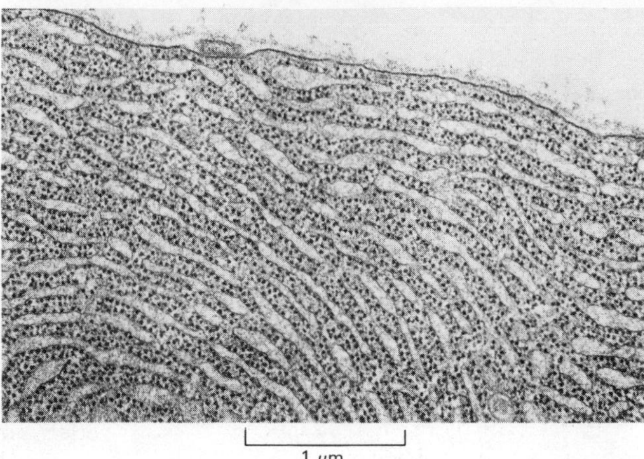

Figure 7–22 Electron micrograph of the highly active rough ER in a pancreatic acinar cell; the ER lumen is dilated in this cell specialized for protein secretion (compare with Figure 7–19). (Courtesy of Daniel S. Friend.)

ribosome-free region of the rough ER (Figure 7–23). Such regions are usually called **transitional ER** (rather than smooth ER), and they represent the specialized region of ER from which the vesicles carrying newly synthesized proteins and lipids bud off for intracellular transport (described below, p. 352).

Smooth ER Is Abundant in Certain Specialized Cells[14]

Smooth ER is sometimes a predominant organelle in cells specializing in lipid metabolism. For example, the main type of cell in the liver, the *hepatocyte*, is the principal site of production of lipoprotein particles for export. The enzymes that synthesize the lipid components of lipoproteins are located in the membranes of the smooth ER. The smooth ER of the hepatocyte also contains enzymes that catalyze a series of reactions to detoxify both drugs and harmful compounds produced by metabolism. The most extensively studied of the *detoxification reactions* are catalyzed by an enzyme called *cytochrome P450.* This protein uses high-energy electrons received from NADPH—transferred to it by a special reductase enzyme in the ER—to add hydroxyl groups to any one of a variety of potentially harmful water-insoluble hydrocarbons dissolved in the bilayer. Other enzymes located in the ER membrane then add negatively charged, water-soluble molecules (such as sulfate or glucuronic acid) to these

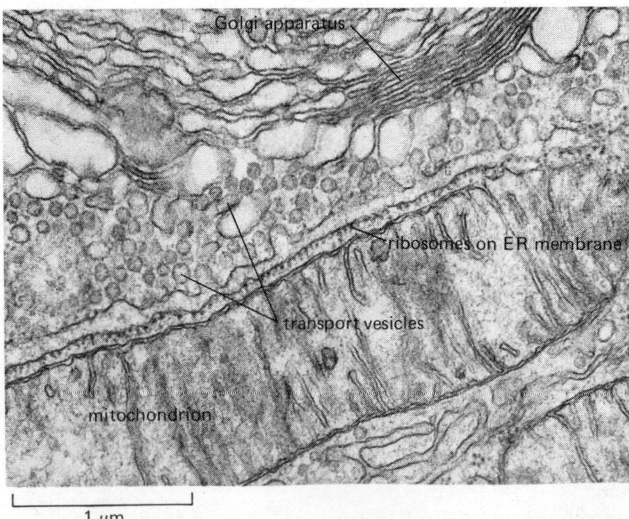

Figure 7–23 A region of transitional ER from which transport vesicles are seen to bud in a mouse epithelial cell. See also Figure 7–18. (Courtesy of Daniel S. Friend.)

hydroxyl groups. After a series of such reactions, a water-insoluble drug or metabolite that might otherwise be stuck permanently in cell membranes is rendered sufficiently water-soluble that it leaves the cell and is excreted in the urine. Because the rough ER alone cannot house enough of these and other necessary enzymes, a major portion of the membrane in a hepatocyte normally consists of smooth ER (Table 7–2).

When large quantities of certain compounds, such as the drug phenobarbital, enter the circulation, detoxification enzymes are synthesized in unusually large amounts in the liver, and the smooth ER doubles in surface area within a few days. After drug removal, the excess smooth ER membranes appear to be removed specifically by a lysosomal process (involving structures called autophagic vacuoles), and the smooth ER returns to normal in five days. How these dramatic changes are regulated is not known.

Other types of cells also have extensive smooth ER. For example, steroid hormones are synthesized from cholesterol, and the cells that synthesize these hormones (such as those in the testis) have an expanded smooth ER compartment to accommodate the enzymes needed to make cholesterol and to modify it to form steroid hormones (an example was shown in Figure 7–19). Muscle cells have a specialized and elaborate smooth ER, called the *sarcoplasmic reticulum,* that sequesters Ca^{2+} from the cytosol. The Ca^{2+}-ATPase that pumps in Ca^{2+} (see p. 295) is the major membrane protein present in this smooth ER. The removal of Ca^{2+} from the cytosol permits the relaxation of the myofibrils following each round of muscle contraction.

Rough and Smooth Regions of ER Can Be Physically Separated[15]

In order to study the functions and biochemistry of the ER, it is necessary to separate the ER membranes from other components of the cell. Intuitively, this would seem a staggering task since the ER is interleaved extensively with other components of the cell cytoplasm. Fortunately, when tissues or cells are disrupted by homogenization, the ER is fragmented into many smaller (~ 100 nm diameter) closed vesicles called **microsomes,** which are relatively easy to purify.

Microsomes derived from rough ER are studded with ribosomes and are called *rough microsomes.* Ribosomes are always found on the *outside* surface of such microsomes, demonstrating that their interior is biochemically equivalent to the luminal space of the ER (Figure 7–24). Many vesicles of a size similar to that of rough microsomes, but lacking attached ribosomes, are also found in these homogenates. Such *smooth microsomes* are derived in part

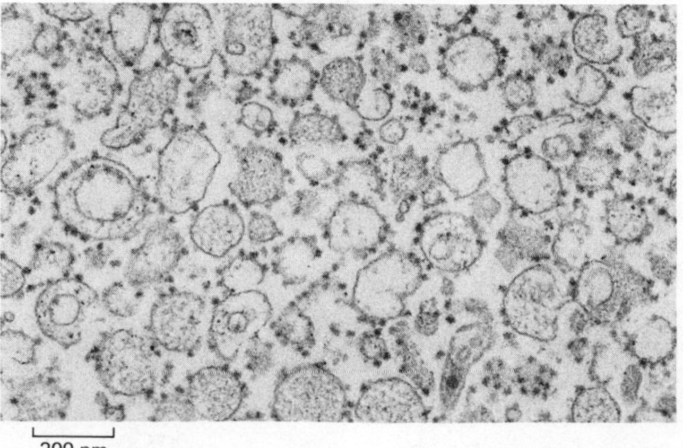

Figure 7–24 Electron micrograph of a thin section of a purified fraction of rough microsomes. These are spherical vesicles with ribosomes attached to their outside surfaces. (Courtesy of George Palade.)

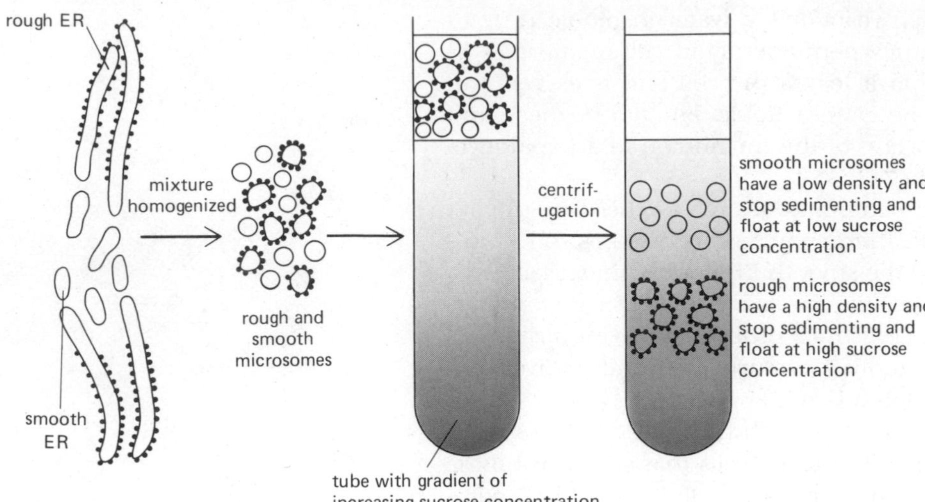

Figure 7–25 Schematic illustration of the isolation procedure used to purify rough and smooth microsomes from the ER.

from smooth portions of the ER and in part from vesiculated fragments of plasma membrane, Golgi apparatus, and mitochondria (the ratio depending on the tissue). Thus, while rough microsomes can be equated with rough portions of ER, the origins of smooth microsomes cannot be so easily assigned. An outstanding exception is the liver. Because of the exceedingly large quantities of smooth ER in the hepatocyte, most of the smooth microsomes in liver homogenates are derived from smooth ER.

Ribosomes, which contain large amounts of RNA, make rough microsomes more dense than smooth microsomes. As a result, the rough and smooth microsomes can be separated from each other by sedimenting the mixture to equilibrium in sucrose density gradients (Figure 7–25). When the separated rough and smooth microsomes of a tissue such as liver are compared with respect to such properties as enzyme activity or polypeptide composition, they are remarkably similar, although not identical. It, therefore, seems that most of the components of the ER membrane can diffuse freely between rough and smooth regions of the ER membrane, as would be expected for a fluid, continuous membrane system.

Because they can be readily purified in functional form, rough microsomes represent a specially useful preparation for studying the many different processes carried out by the ER. The vesicles are topologically sealed in the same manner as the rough ER, leaving their cytoplasmic surface easily accessible to components that can be added *in vitro.* To the biochemist, rough microsomes represent small authentic versions of the endoplasmic reticulum, still capable of protein synthesis, glycosylation, and membrane synthesis.

Rough Regions of ER Contain Specific Proteins Responsible for the Binding of Ribosomes[16]

Because the ER membrane, like all membranes, is a two-dimensional fluid, most proteins and lipids will equilibrate freely between rough and smooth regions in the absence of special restraints. The fact that there are, nonetheless, some components present in rough ER that are not present in smooth ER when isolated liver microsomes are compared indicates that some special restraining mechanisms must exist. In particular, it seems that the rough ER contains special nonequilibrating proteins in its membranes that both bind ribosomes to it and help give it its flattened appearance (see Figure 7–20).

The ribosomes of the rough ER are held on the membrane in part by their growing polypeptide chains, which are threaded across the ER membrane as they are synthesized (see below). However, when the synthesis of polypeptide chains is terminated with a drug like puromycin, these ribosomes still remain tightly bound to the membrane of rough microsomes. Such ribosomes can be dislodged by suspending the microsomes in a solution of high salt concentration, which breaks the interactions between the ribosomes and special membrane proteins of the rough microsomes. Futhermore, ribosomes can reassociate with rough microsomal membranes from which ribosomes have been removed. They bind rapidly and with a high affinity to a limited number of sites on the membrane, so that the "stripped" membrane will regain the same number of ribosomes it had when originally isolated.

The binding site on the ribosome is located on the large subunit and seems to attach to two specific glycoproteins in the rough ER membrane. These two proteins, called *ribophorins*, span the ER membrane, and they are found exclusively in the rough regions of ER. It is not clear how the ribophorins are prevented from diffusing into the membrane of the smooth ER. They could be segregated by forming large two-dimensional crystals in the bilayer, or they could be tethered in place by fibrous proteins on either side of the bilayer (see p. 283). Whatever the mechanism, it probably accounts for the flattened appearance of the rough ER in comparison with that of the smooth ER.

Membrane-bound Ribosomes Synthesize Proteins That Pass Through the Membrane During Their Translation[17]

Many proteins at some stage must cross the permeability barrier presented by a cell membrane. Proteins secreted from the cell (secretory proteins) are the most obvious of numerous examples. In this case, the binding of the ribosomes to the membrane of the ER helps to solve the problem of getting these water-soluble proteins across a hydrophobic membrane: the proteins thread their way across the ER membrane as they are synthesized before they have had a chance to fold into their final hydrophilic conformations. Once inside the lumen of the rough ER, the proteins to be secreted are transported through the Golgi apparatus and then to the outside of the cell. Thus, secretory proteins become permanently segregated from the cytosol as soon as they have been synthesized.

A similar but more subtle problem is presented by membrane proteins destined to span the lipid bilayer. Often, large hydrophilic portions of these polypeptide chains end up being exposed on the luminal side of intracellular membranes or on the topologically equivalent extracellular side of the plasma membrane (Figure 7–26). Unlike secreted proteins, membrane proteins are not made in large amounts, so their mode of synthesis is more difficult to study.

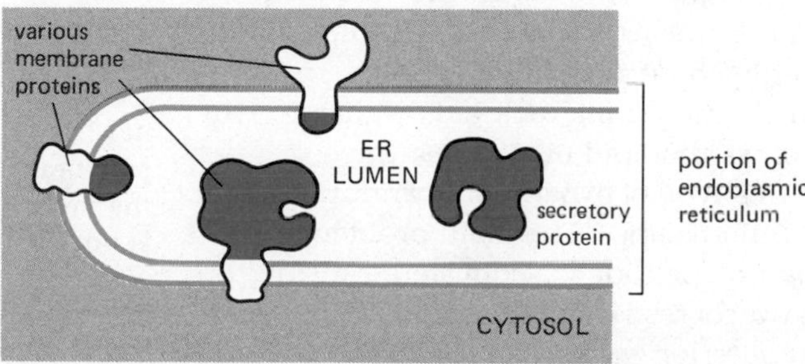

Figure 7–26 Schematic diagram of four different proteins, emphasizing the portion of each protein that has passed through the lipid bilayer of the ER during protein synthesis. Three different membrane proteins and a secretory protein are shown. The colored region of each has passed through the bilayer.

However, one membrane protein produced in large amounts by cells is the *G protein* of the enveloped *vesicular stomatitis virus* (VSV). When this virus infects a cell, only one membrane protein, the G protein, is made by the cell, instead of the hundreds or thousands of different kinds that are synthesized in uninfected cells. This makes it relatively easy to follow its synthesis. It has been shown that most of the G-protein polypeptide chain is transferred across the ER membrane as it is synthesized, just as in the case of a secreted protein. However, instead of being released into the lumen of the ER, the protein remains anchored in the bilayer by a short hydrophobic segment of the polypeptide chain. The G protein thereby remains membrane-bound; the 500 amino acids that have crossed the ER membrane will form an external "spike," which protudes from the lipid bilayer coating the final virus particle.

Membrane-bound ribosomes whose translation products cross the ER membrane during synthesis are said to be engaged in **vectorial discharge.** This term is useful because it emphasizes two important features of the process. First, the ribosome is found only on the cytosolic side of the ER membrane, so that the transfer of the polypeptide chain across the membrane must be "vectorial," in the direction from cytosol to lumen. (Transfer is also vectorial in the sense that the polypeptide is passed with its amino-terminal end first, since this is the direction of protein synthesis [p. 206].) Second, the term "discharge" emphasizes that this must be an active, energy-requiring process. The mechanism that permits the growing polypeptide to cross the membrane is still unknown, as is the source of the required energy, but these issues can now be approached biochemically using cell-free systems. It is possible that a novel sort of transport protein, such as an energy-requiring polypeptide pump, is involved in vectorial discharge.

Direct Evidence Favoring Vectorial Discharge Has Come from Experiments in Bacteria[18]

The concept of vectorial discharge derives from experiments first performed in 1966 showing that many of the nascent polypeptides released from microsomes by puromycin remain sequestered in the microsome lumen. But the most convincing experiments were not done until 1977, and, as is so often the case, they involved bacterial cells. Gram-positive bacteria, such as *Bacillus subtilis,* secrete large quantities of enzymes. There being no ER in bacteria, these enzymes are synthesized by ribosomes bound to the plasma membrane. In the experiment illustrated in Figure 7–27, it was found that one end of these secreted polypeptide chains could be selectively hydrolyzed by proteolytic enzymes added to intact bacterial cells whose cell walls had been removed, while the other end, which was still being synthesized by ribosomes inside the cell, could not be hydrolyzed. This showed that one end of the growing polypeptide chain must be exposed on the outside of the cell. Moreover, almost all of the completed portion of a nascent polypeptide was attacked by the protease, leaving only a small undigested segment 50 amino acid residues long. Since a resistant segment of about 30 amino acid residues is left within the ribosome when free ribosomes are subjected to mild proteolysis, these observations suggest that only about 20 residues of each polypeptide are within the membrane at any one time during vectorial discharge. This length of polypeptide is sufficient to cross the lipid bilayer just once.

Although it has been attempted, this type of experiment cannot be done as convincingly with eucaryotic rough microsomes. The main problem is that the topology is backward; the growing polypeptide is sequestered inside the microsome, and there is at present no way of restricting the action of an added protease exclusively to the microsome interior.

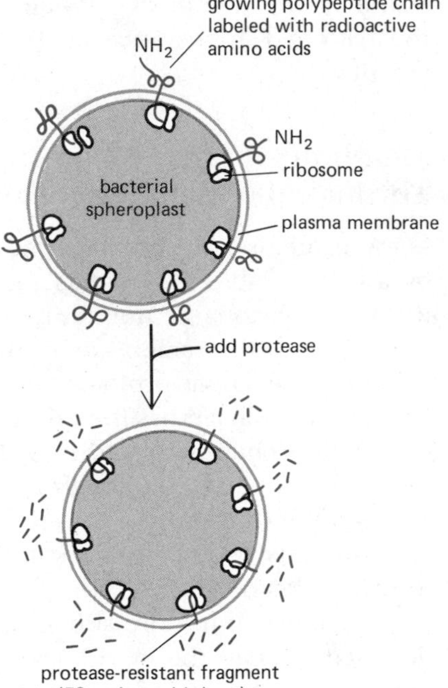

Figure 7–27 Schematic illustration of vectorial transport in bacteria. After the cell walls have been removed (to form spheroplasts), the cells are briefly exposed to radioactive amino acids to label the growing polypeptide chains synthesized by plasma-membrane-bound ribosomes. Part of the growing polypeptide chain has been transferred across the membrane and can be digested by the protease treatment. The drawing is not to scale; in reality the diameter of the ribosome is only about one-fortieth of the diameter of the bacterial cell.

Membrane-bound Ribosomes Are Derived from Free Ribosomes That Are Directed to the ER Membrane by Special Signal Sequences[19]

Although the membrane-bound ribosomes of the rough ER synthesize a selected class of proteins destined for vectorial discharge, there is no evidence that the ribosomes themselves differ from those free in the cytoplasm. In the current view, membrane-bound ribosomes are simply ribosomes that are specifically directed to the ER membrane because of the polypeptide chain they happen to be translating.

An important insight into how ribosomes are directed to the ER came from detailed studies of antibody (immunoglobulin) synthesis. The light and heavy chains of immunoglobulin constitute the principal secretory products of the plasma cells of the immune system (Chapter 17). When mRNA encoding the light chain of an immunoglobulin is translated *in vitro* by free ribosomes, the light chain is synthesized in the form of a precursor containing an extra 20 amino acid residues not found in the secreted product. The extra "leader" peptide is located at the extreme amino terminus of the precursor immunoglobulin. But chains synthesized *in vitro* by ribosomes attached to rough microsomes lack this extra amino-terminal peptide. These discoveries provided evidence for an earlier proposal that the *leader sequence* peptide acts as a "signal" to direct the ribosome to the rough regions of the ER membrane (Figure 7–28). Some receptor that recognizes proteins containing this *signal peptide* must therefore be present exclusively in the rough regions of the ER membrane.

This viewpoint has been greatly fortified by the subsequent discovery that the major secretory proteins of the pancreas, when synthesized *in vitro* on free ribosomes, are translated as precursor polypeptides ("presecretory" proteins) containing a similar amino-terminal leader sequence of 16 residues not found *in vivo*. Since then, precursors to dozens of secretory and integral membrane proteins in procaryotes and eucaryotes have been described. In all of these cases, the amino-terminal **signal sequence** contains large numbers of hydrophobic amino acid residues, even though the actual sequences

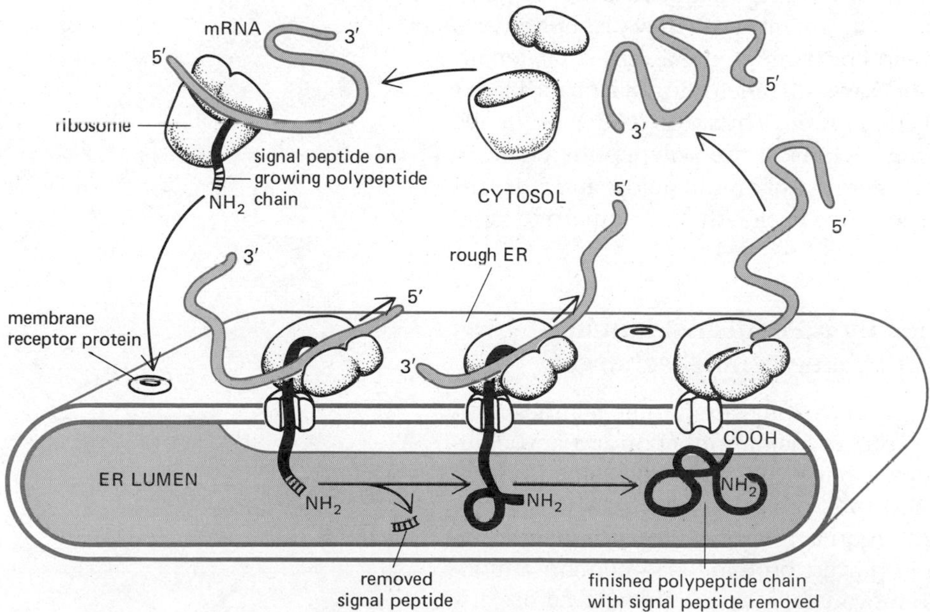

Figure 7–28 A schematic view of the signal peptide hypothesis. For simplicity, only a single ribosome, rather than a polyribosome, is shown. The indicated binding of a cytosolic ribosome to the rough ER is now thought to require two separate interactions: an affinity of the ribosome itself for special membrane proteins in the ER (presumably the ribophorins, see p. 341) and the interaction of an amino-terminal signal peptide on the growing polypeptide chain with a membrane receptor protein. However, once protein transfer across the membrane begins, the growing polypeptide chain itself helps to anchor the ribosome, replacing the second interaction.

are quite variable. It has now been shown that these signal sequences are, in fact, removed on the luminal side of the rough ER, even before the synthesis of the polypeptide is completed, by a specific protease found only in rough microsomes. This type of amino-terminal sequence that is cleaved off after protein synthesis has never been found on a protein that remains in the cytosol.

Ribosomes in the cytosol that begin synthesizing a protein destined for the ER must be brought to the rough ER to begin the vectorial discharge process. But how are these ribosomes specifically recognized by the ER membrane? Recent evidence has shown that a multisubunit protein present in the cytosol binds to these ribosomes shortly after they have synthesized the appropriate signal peptide. This protein, called *signal recognition protein,* halts further protein synthesis by the ribosome until it has become linked to the rough ER membrane. Presumably, this protein recognizes both the amino terminus of the newly synthesized protein and some receptor on the rough ER membrane, thereby binding the ribosome to the ER. Protein synthesis by the ribosome then resumes, as vectorial discharge begins. While the nature of the receptor in the ER membrane is unknown, the ribophorins are thought to be involved in stabilizing the ribosome-ER complex.

There Is Genetic Evidence for the Signal Hypothesis[20]

Genetic analysis of procaryotes has been used in two different ways to test the idea that leader sequences act as the proposed signals for the vectorial discharge of proteins. (1) Many mutants of an *E. coli* membrane protein have been isolated in which the protein is retained in the cytoplasm instead of being inserted in the membrane. Most of these mutations have been found to alter the extreme amino-terminal leader sequence region of the protein. (2) In specially designed strains of *E. coli,* the DNA segments coding for polypeptide chains of integral membrane proteins and of cytosolic proteins have been fused to create new hybrid proteins. When a substantial length of the amino-terminal portion of a membrane protein of *E. coli* is fused with most of the carboxyl-terminal portion of a cytosolic protein, the hybrid protein is found in the membrane. Thus, only the amino-terminal portion seems to be required for passage into a membrane.

Experiments in *E. coli* using hybrid proteins have demonstrated another important fact about signal sequences: although these sequences are required for directing the growing polypeptide to the membrane, they are not always sufficient in themselves. A hybrid protein containing only a very short length of the amino terminus of a membrane protein—including all of its leader sequence—was not inserted into the membrane. Thus, the leader sequence may normally interact with neighboring regions of the polypeptide chain to create a unique three-dimensional arrangement of amino acids, and it is this arrangement—rather than the leader sequence itself—that is required for specific membrane insertion.

Some Proteins Cross Membranes by a Posttranslational Import Mechanism Rather Than by Vectorial Discharge[21]

Some proteins can cross certain membranes long after their synthesis on ribosomes. For example, most of the proteins inside mitochondria and chloroplasts are coded for by nuclear genes (Chapter 9). These proteins are synthesized and released in the cytosol and then are pumped across the appropriate organelle membrane in an energy-requiring process (**posttranslational import**). Interestingly, most of these imported proteins also contain amino-terminal leader sequences that are removed following transport across the mitochondrial or chloroplast membranes. These leader sequences have been

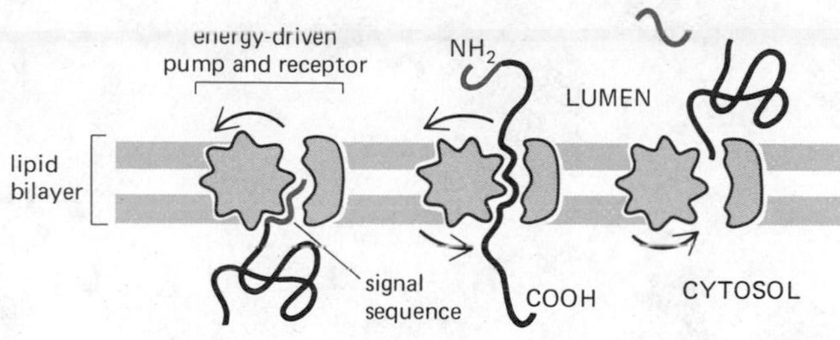

Figure 7–29 Schematic view of an energy-driven protein pump. After a receptor recognizes some special feature of an amino-terminal peptide, the pump is activated to force the entire protein through the membrane. A mechanism of this type is apparently needed to account for the observed transfer of already synthesized proteins from the cytosol into mitochondria and chloroplasts.

shown to be required to provide the "signal" that is recognized to direct the protein to the correct site. The membranes of both of these cellular organelles must contain a specific receptor for the appropriate amino-terminal signal. These receptors would seem to be functionally linked to an energy-driven pump that forces the recognized protein across the membrane, possibly unfolding it transiently in the process (Figure 7–29).

Some membrane proteins may enter and partially cross the ER membrane by a posttranslational import mechanism. This is probably the case for proteins that have a large amount of their mass in the lipid bilayer and only small portions exposed on the noncytoplasmic side of the membrane. It is easy to imagine how such a hydrophobic membrane protein could simply dissolve in the lipid bilayer after being released into the cytosol or even during its synthesis. The energy required to transfer a relatively small hydrophilic portion of the protein across the lipid bilayer could come from the gain in free energy associated with the transfer of the hydrophobic portion from an aqueous environment to the hydrophobic interior of the bilayer. Some of the proteins located in the outer membrane of mitochondria appear to enter their membrane in this way following their synthesis on ribosomes in the cytosol (see p. 539).

Even if we consider only the few examples just discussed, it is obvious that many different types of signal sequences must exist on proteins. Some signals will direct a protein to the ER, others to a mitochondrion, and yet others to a chloroplast (in a plant cell). The recognition systems involved must have a high degree of selectivity, since proteins seem to be directed quite reliably to their correct intracellular address.

Most Proteins Synthesized in the Rough ER Are Glycosylated[22]

Glycosylation is one of the major biosynthetic functions of the ER. It is a striking fact that most proteins sequestered in the lumen of the ER before being secreted from the cell or transported to other intracellular destinations (such as the Golgi apparatus, lysosomes, or plasma membrane) are **glycoproteins** (Figure 7–30). In contrast, the soluble proteins of the cytosol are not glycosylated.

An important advance in understanding the process of glycosylation was the discovery that mainly one species of oligosaccharide (composed of *N*-acetylglucosamine, mannose, and glucose) is transferred to proteins in the ER, and that this oligosaccharide is always linked to the NH_2 group on the side chain of an asparagine residue of the protein (Figure 7–31). All of the diversity of the asparagine-linked oligosaccharide structures on mature glycoproteins results from extensive modifications of this single precursor structure, most of which occur during subsequent transit through the Golgi apparatus (p. 357). The asparagine-linked oligosaccharides (***N*-linked oligosaccharides**) are by far the most common ones found in glycoproteins. Often, but less frequently, oligosaccharides are linked to the OH group on the side

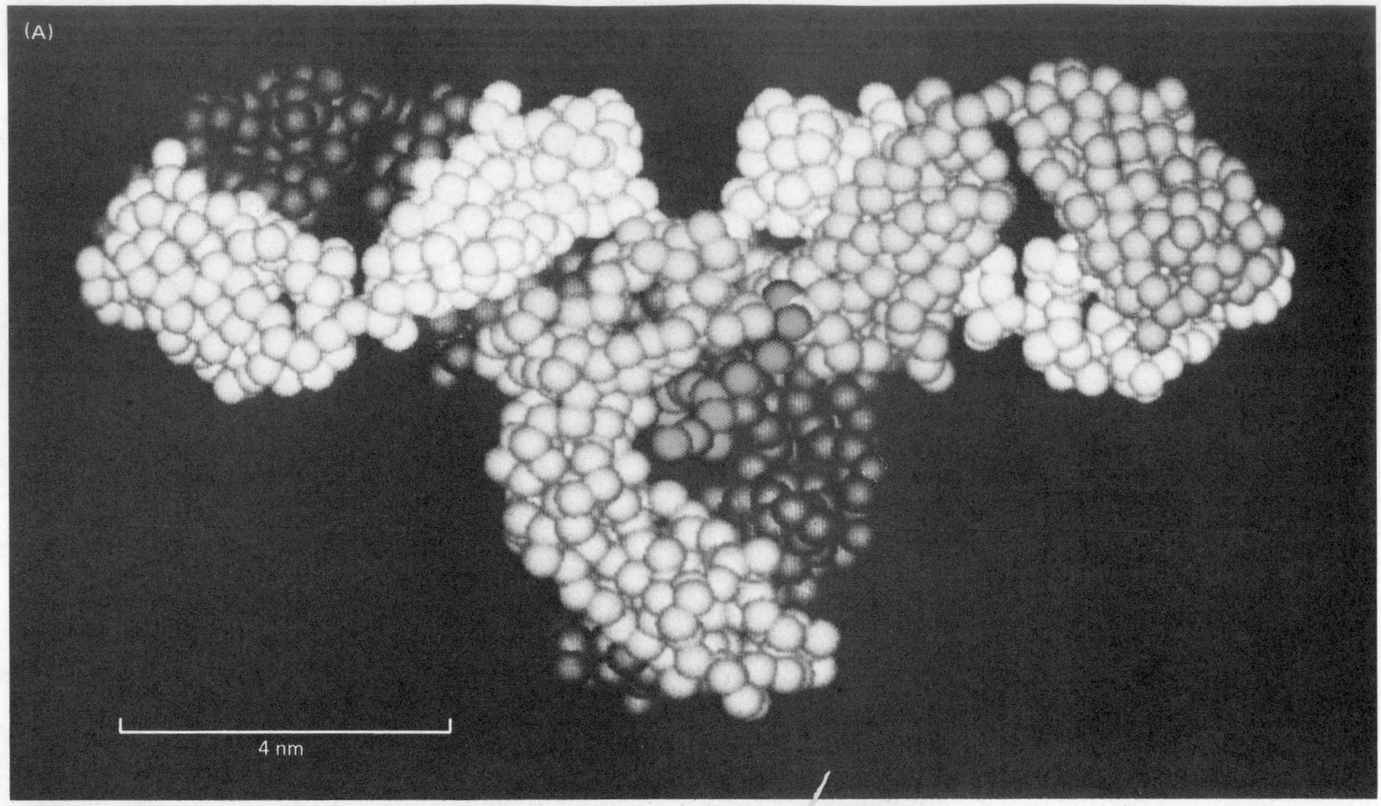

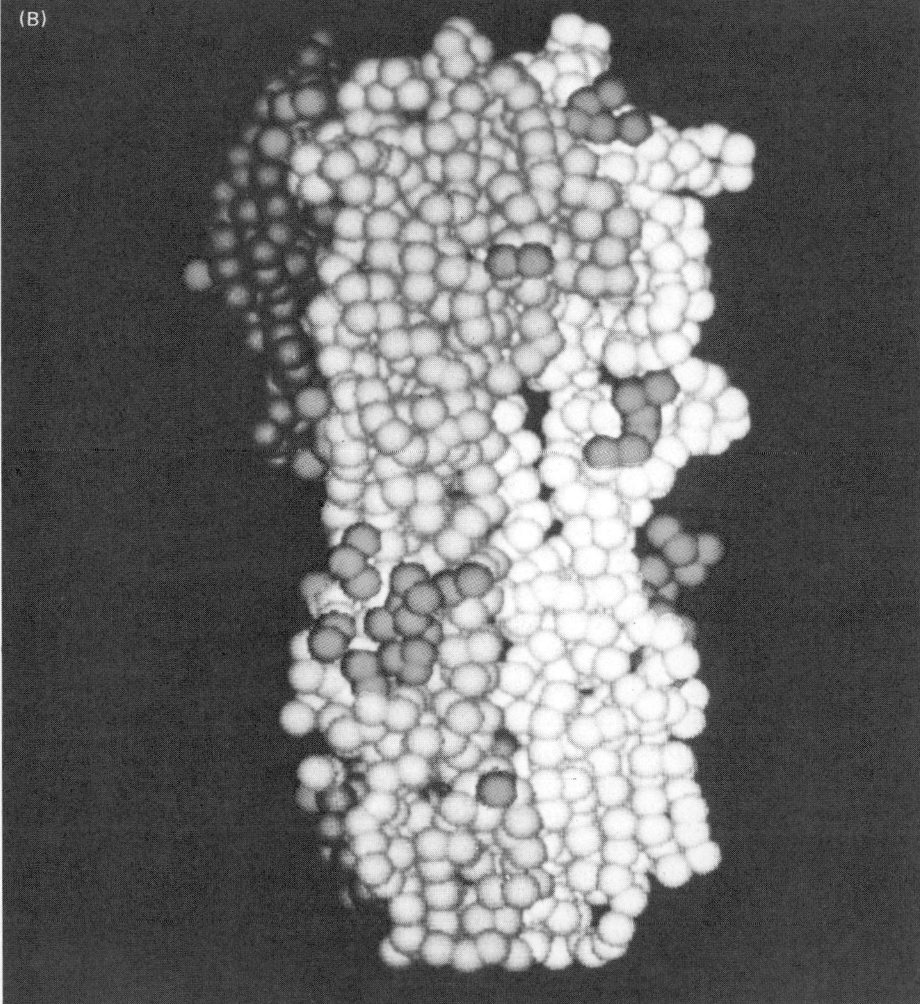

Figure 7–30 The three-dimensional structure of two glycoproteins, showing the structure and location of their covalently attached oligosaccharide chains (*colored atoms*). (A) An immunoglobulin (antibody) molecule. (B) A membrane glycoprotein: the influenza virus hemagglutinin. Glycoproteins are either inserted into cell membranes or secreted from cells, and they can contain from as little as 1% to as much as 85% carbohydrate by weight. Some carbohydrate-rich glycoproteins contain tens or even hundreds of attached oligosaccharide chains per molecule. (Photographs courtesy of Richard J. Feldmann.)

chain of a serine, threonine, or hydroxylysine residue (***O*-linked oligosaccharides**); it is not yet clear whether this glycosylation begins in the ER or occurs only in the Golgi apparatus, and the details are as yet poorly understood.

Intracellular sites of glycosylation can be identified by autoradiography. For example, when slices of thyroid are briefly incubated with [^{3}H]mannose, most of the ^{3}H is incorporated into an oligosaccharide attached to thyroglobulin, a major glycoprotein synthesized by thyroid cells. To determine where in the cell this reaction takes place, cells are labeled briefly with [^{3}H]mannose and then processed for electron microscopy by the method illustrated in Figure 7–32. Thin sections of these cells are coated with a thin layer of photographic emulsion so that sites of ^{3}H-disintegration in the section are recorded as silver grains in the overlying film (Figure 7–33). When the section is examined in the electron microscope, the location of ^{3}H relative to known cellular structures can be determined. The result of this type of experiment is clear-cut: [^{3}H]mannose is incorporated only in the ER (Figure 7–34).

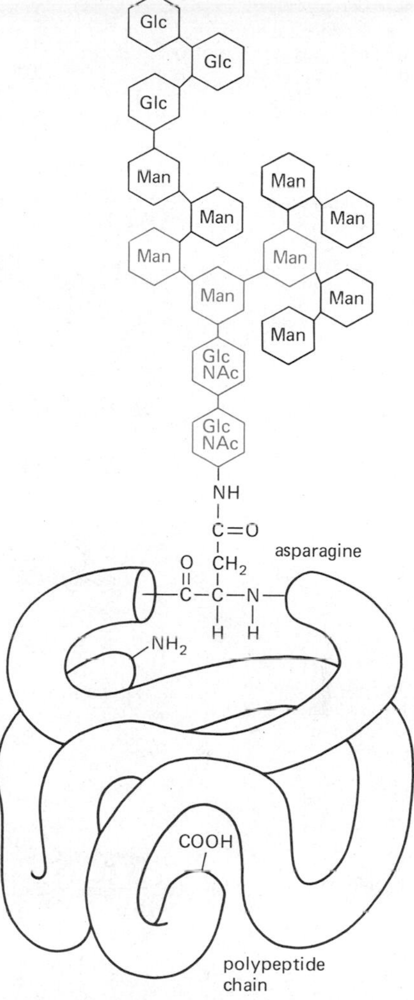

Figure 7–31 The structure of the asparagine-linked oligosaccharide that is added to most proteins on the luminal side of the ER membrane. The sugars shown in color form the "core region" of this oligosaccharide. For many glycoproteins, only the core sugars survive the extensive oligosaccharide trimming process in the Golgi apparatus (see Figure 7–50).

The Oligosaccharide Is Added to the Growing Polypeptide Chain on the Luminal Side of the ER

The actual transfer of the oligosaccharide to the asparagine is believed to take place on the luminal side of the ER membrane, and the enzyme catalyzing this event is a membrane-bound protein with its active site exposed on the luminal surface. This fact explains why cytosolic proteins, which never encounter the luminal side of the ER, are not glycosylated. As illustrated in Figure 7–35, the oligosaccharide is preformed in its entirety and is transferred to the target asparagine residue on the protein in a single enzymatic step almost as soon as that residue emerges on the luminal side of the ER membrane. This scheme ensures maximum access to the target asparagine (Asn) residues, which are those in the sequences Asn-X-Ser or Asn-X-Thr (where X is any amino acid). These two sequence combinations occur much less frequently in glycoproteins than in nonglycosylated cytoplasmic proteins. Evidently there has been selective pressure against these sequences during the evolution of glycoproteins, no doubt because glycosylation at many sites would interfere with protein folding.

If glycosylation occurred on the cytoplasmic surface, it might pose serious difficulties for the vectorial discharge mechanism, which would now have to accommodate bulky oligosaccharide chains. A system in which the oligosaccharide and the polypeptide are separately transported across the membrane by distinct mechanisms before being joined together on the luminal side would seem to simplify matters.

The Oligosaccharide Is Donated to the Polypeptide by an Activated Lipid and Then Almost Immediately Modified[23]

The sequence of events shown in Figure 7–35 requires that the oligosaccharide to be transferred to an asparagine residue must be present on the luminal side of the ER in an "activated" form. Activation is achieved by linking the oligosaccharide to a donor molecule via a high-energy bond. A major advance came in the early 1970s with the discovery that the activated donor is a special lipid molecule, **dolichol,** to which the oligosaccharide is linked via a pyrophosphate bridge.

The oligosaccharide is built up sugar by sugar on this membrane-bound lipid molecule. Sugars are first activated in the cytosol by the formation of *nucleotide-sugar intermediates,* which then donate their sugar (directly or indirectly) to the lipid in an orderly sequence (Figure 7–36). Dolichol is very

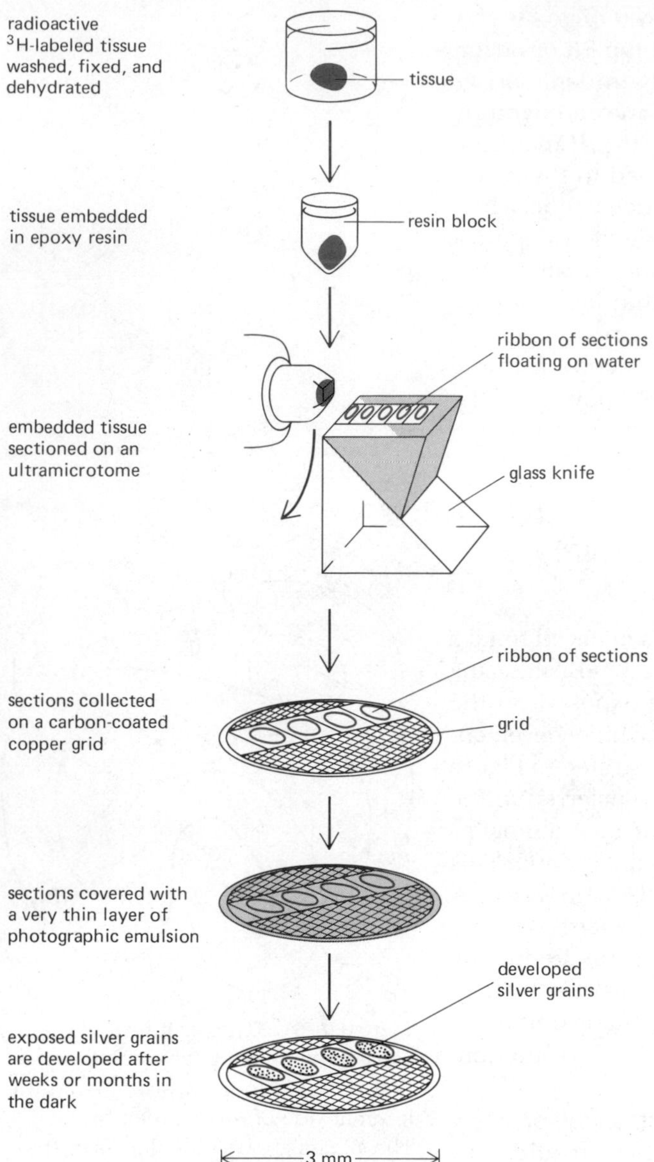

Figure 7–32 Diagram illustrating how thin sections of tissues are processed in order to localize radioactivity by electron microscopic autoradiography. It is important that all radioactivity not incorporated into the product be removed by the fixation and washing steps prior to embedding, while the product itself must remain in the tissue.

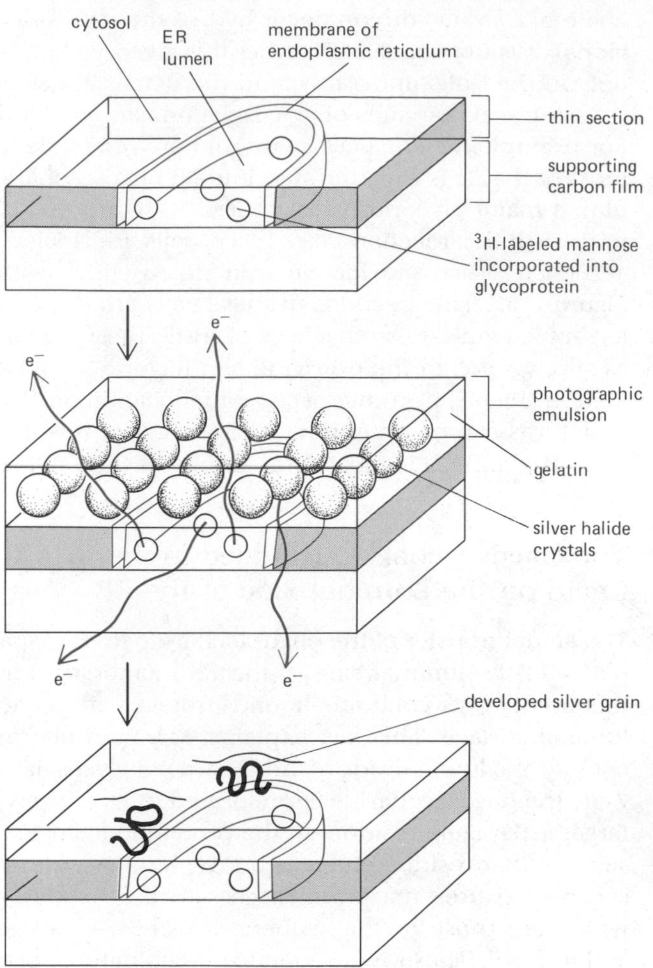

Figure 7–33 Illustration of the principles involved in autoradiography. Here the location of silver grains relative to the image produced by the electron microscope indicates the location of the [^{3}H]mannose incorporated into glycoprotein. Because the electrons emitted by ^{3}H-decay can generate developed silver grains slightly displaced from the point of emission, the localization of a radioactive molecule is imprecise compared to the resolution of the electron microscope.

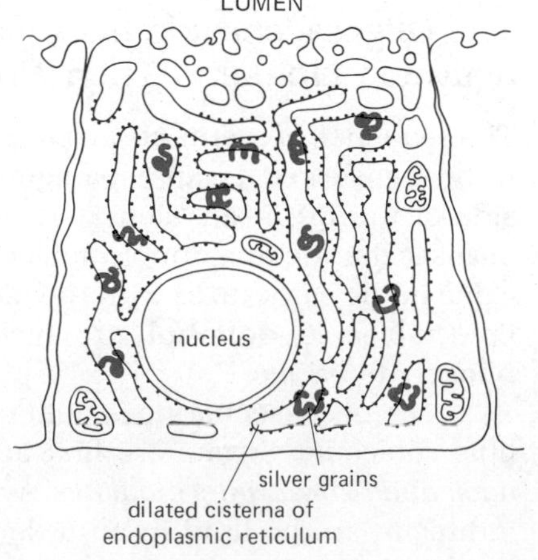

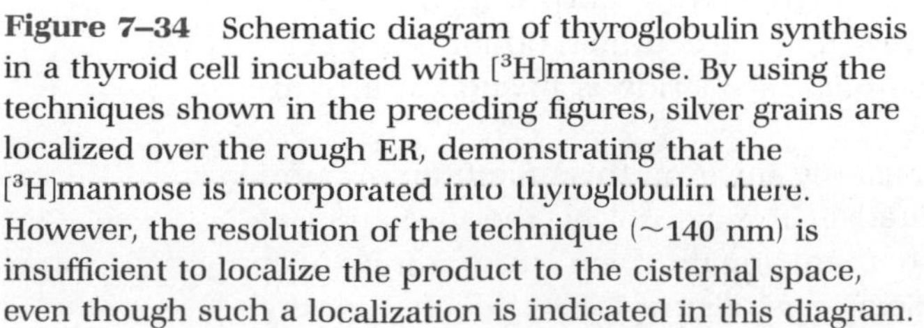

Figure 7–34 Schematic diagram of thyroglobulin synthesis in a thyroid cell incubated with [^{3}H]mannose. By using the techniques shown in the preceding figures, silver grains are localized over the rough ER, demonstrating that the [^{3}H]mannose is incorporated into thyroglobulin there. However, the resolution of the technique (~140 nm) is insufficient to localize the product to the cisternal space, even though such a localization is indicated in this diagram.

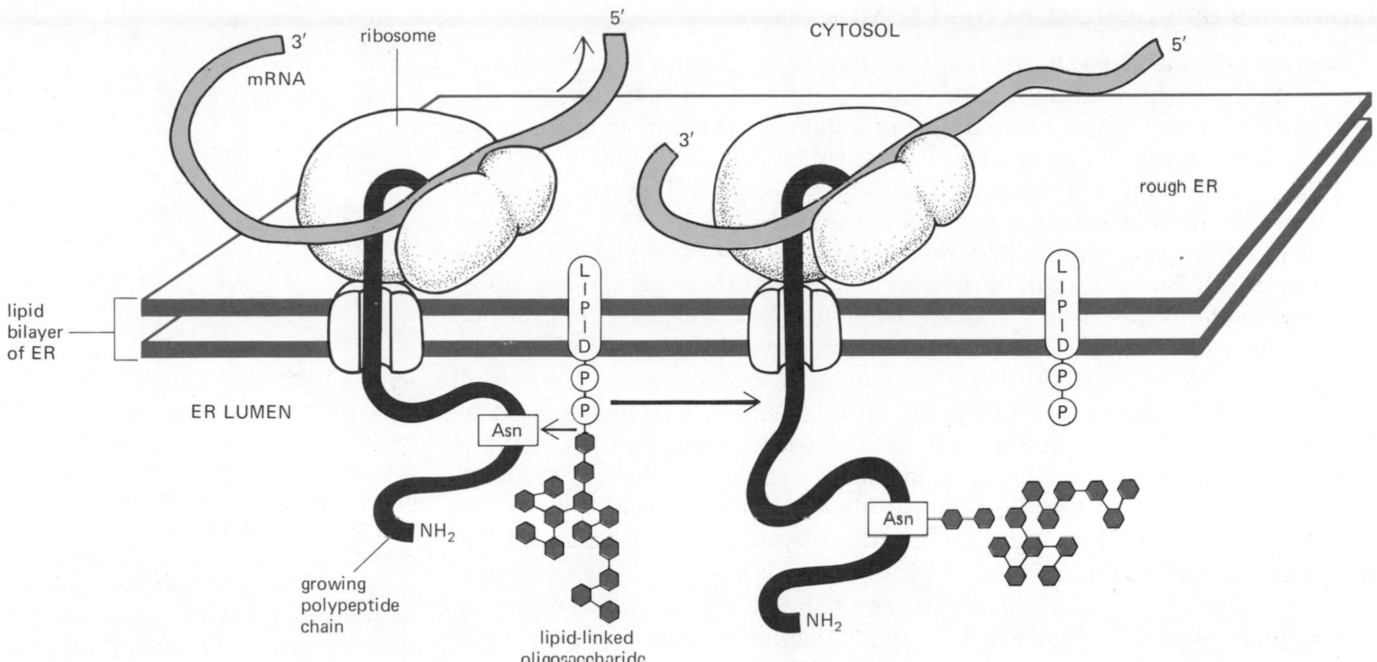

hydrophobic and quite long; with its 22 five-carbon units, it can span the thickness of a lipid bilayer more than three times. Since it is firmly embedded in the membrane, the attached oligosaccharide is firmly anchored to the luminal surface of the ER, where glycosylation must take place.

The initial glycosylation of a protein in the ER is followed by the first steps in what will ultimately be an extensive process of remodeling the oligosaccharide. The three glucose residues are quickly removed from the oligosaccharides of most, but probably not all, glycoproteins. Because this oligosaccharide "trimming" or "processing" continues in the Golgi apparatus, we shall defer our discussion of it to the next section.

Figure 7–35 Schematic illustration of protein glycosylation in the ER. As soon as a polypeptide chain enters the ER lumen, it is glycosylated on its susceptible asparagine residues. As indicated, the oligosaccharide shown in Figure 7–31 is transferred to the asparagine as an intact unit in a reaction catalyzed by a membrane-bound *glycosyl transferase* enzyme.

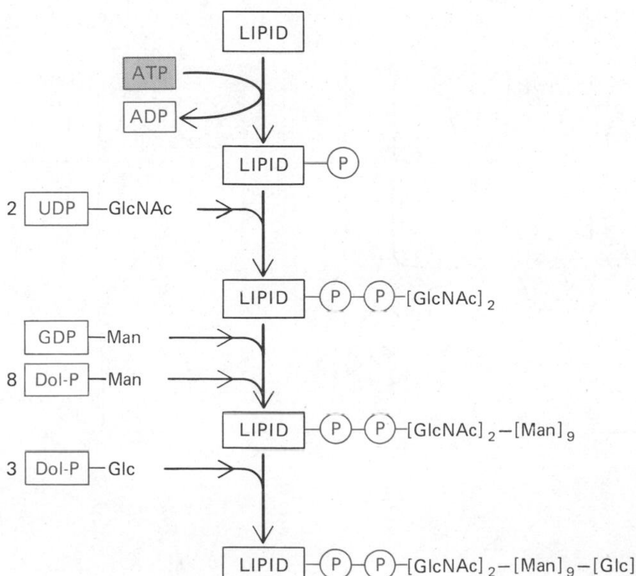

Figure 7–36 The synthesis of the lipid-linked oligosaccharide in the ER membrane occurs while the intermediates shown are embedded in the membrane. The lipid shown here is dolichol. The pyrophosphate linkage to the lipid serves to activate the oligosaccharide for subsequent transfer to proteins.

Lipids Are Synthesized in the ER Membrane

Phospholipids and cholesterol, the principal building blocks of all lipid bilayers, are both synthesized on the membranes of the ER. In fact, except for the fatty acids and two mitochondrial phospholipids (see below), all of the lipids made in the cell are synthesized on these membranes. Thus, one of the chief functions of the ER is to produce nearly all of the lipids and proteins required for the elaboration of new cellular membranes.

All of the principles underlying the assembly of lipid bilayers can be illustrated by a consideration of phospholipid biosynthesis. The major phospholipid made in the ER is phosphatidylcholine (also called lecithin), which can be formed in three steps from two fatty acids plus glycerol phosphate and choline (Figure 7–37). Each step is catalyzed by an enzyme that is itself dissolved in the lipid bilayer of the ER with its active site facing the cytosol. In the first step, acyl transferase condenses fatty acids and glycerol phosphate from the cytosol to produce phosphatidic acid, a compound sufficiently water-insoluble to remain in the lipid bilayer after it has been synthesized. This phosphatidic acid in the bilayer acts as the substrate for the next enzyme in the pathway, and so on.

From the viewpoint of membrane assembly, the most important part of the pathway shown in Figure 7–37 is the actual incorporation of new lipids into the bilayer, which occurs during step 1. Later steps modify the head group of these lipids, and therefore the chemical nature of the bilayer, but do not result in net membrane growth.

Figure 7–37 The synthesis of phosphatidylcholine in the ER membrane. The reactions are catalyzed on the cytosolic side of the bilayer by membrane-bound enzymes, and the lipid intermediates diffuse rapidly in the bilayer from enzyme to enzyme. The lipid bilayer is expanded only in step 1, in which relatively soluble precursors react to produce a hydrophobic lipid molecule. The compound CDP-choline is abbreviated here as C-Ⓟ-Ⓟ-choline.

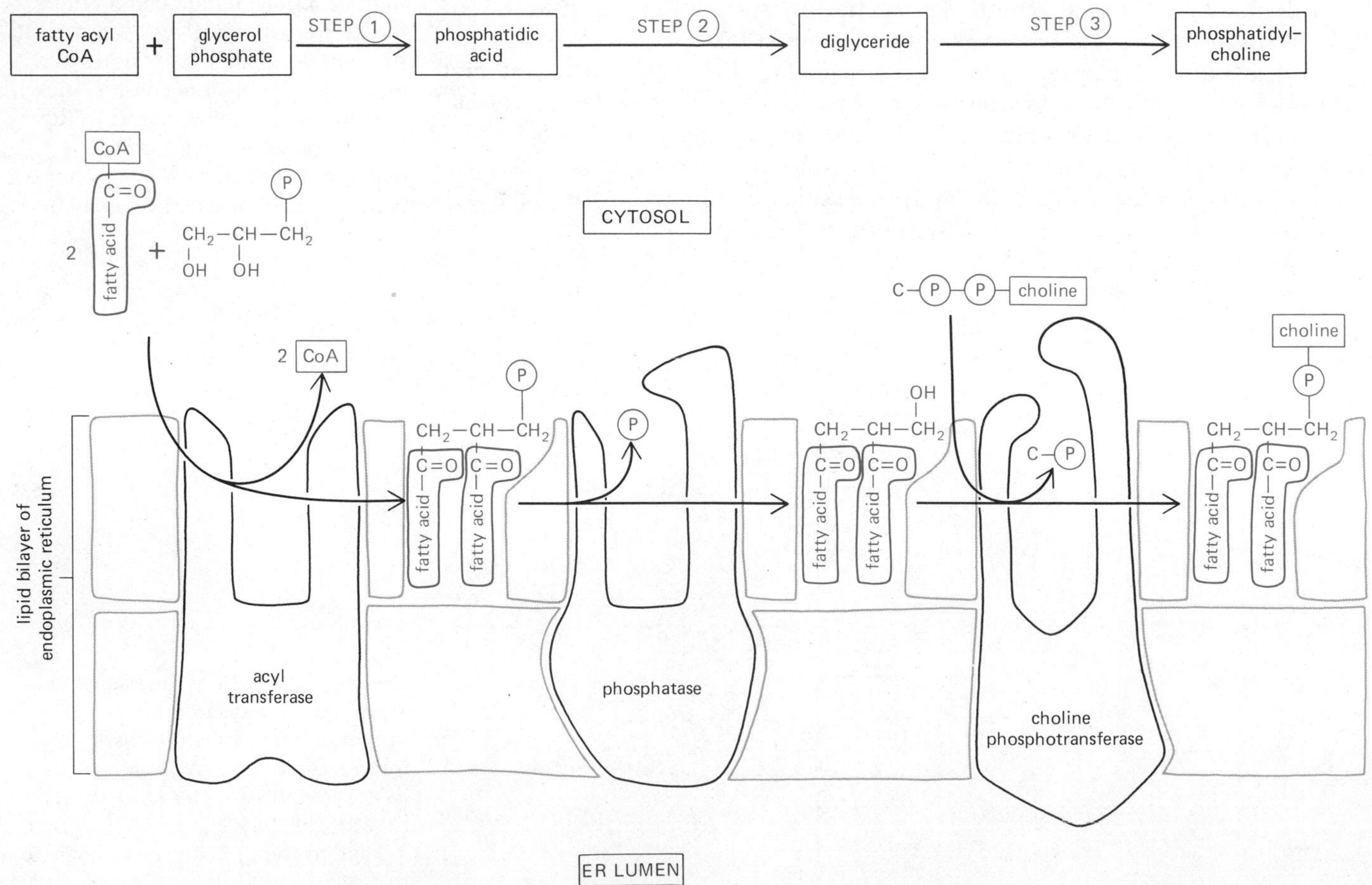

Lipid Biosynthesis Is Asymmetric[24]

The actual relationship of lipid biosynthesis to the growth of the lipid bilayer was first clarified by an experiment in bacteria, where phospholipid synthesis takes place in the plasma membrane since there are no internal membranes. It was discovered that newly synthesized phospholipid molecules are initially present only in the cytoplasmic half (monolayer) of the lipid bilayer of the plasma membrane.

This finding led to a simple picture of how lipid bilayers grow both in bacteria and in eucaryotic cells. Since almost every step in phospholipid synthesis utilizes metabolites found only in the cytosol (that is, fatty acyl CoA, glycerol phosphate, CDP-choline), the active sites of the enzymes that catalyze these steps must also face the cytosol. Thus, even after the initial "fixation" of an early product into the bilayer (as occurs in step 1 of Figure 7–37), the subsequent modifications of the newly inserted lipid to form other lipids all take place on the same side of the bilayer.

Such a mechanism would tend to turn any lipid bilayer into a monolayer since the insertion of new lipids, and therefore membrane expansion, is restricted to the cytosolic half of the bilayer. Clearly, there must be a mechanism whereby phospholipids rapidly equilibrate across the ER bilayer so that the opposing monolayer will expand in parallel with the cytosolic monolayer. In synthetic lipid bilayers, lipids do not "flip-flop" in this way (p. 258). But phospholipids do equilibrate across ER and bacterial membranes within a matter of a few minutes, which is almost 100,000 times faster than can be accounted for by spontaneous "flip-flop." Movement of phospholipids from one monolayer to another on this time scale has so far been observed only in those membranes capable of lipid biosynthesis; it does not occur in the plasma membranes of animal cells, for example. Therefore, some mechanism that causes this movement of lipid molecules must exist in the ER (and in the plasma membrane of bacteria). An attractive but unproven idea is that the equilibration is catalyzed by a specific transfer protein found only in these membranes (Figure 7–38).

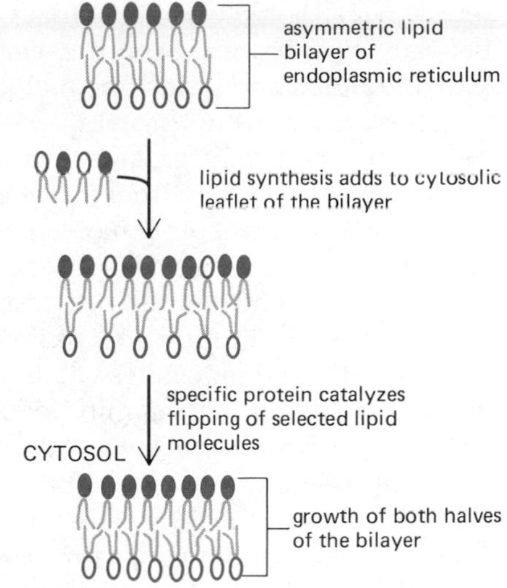

Figure 7–38 Schematic drawing of the "flipping" of lipid molecules from one monolayer ("leaflet") of the ER bilayer to the other. Since new lipid molecules are added only to the cytoplasmic leaflet, some special mechanism (presumably involving membrane-bound transfer proteins) must transfer some of these molecules to the luminal leaflet so that the membrane grows as a bilayer. The two leaflets of at least most bilayers contain a different mixture of lipid species. Such an asymmetrical bilayer would be generated in the ER if the transfer proteins preferentially recognize certain species of lipid.

Special Proteins Transfer Phospholipids from the ER to Mitochondria[25]

Mitochondria have a special problem with phospholipids since they possess only a limited ability to synthesize them. Most membranes, such as those of the Golgi apparatus, lysosomes, and plasma membrane, receive both their proteins and their lipids (synthesized in the ER) together in the form of small vesicles (see next section). Mitochondrial membranes, however, do not grow by accepting quanta of membranes synthesized elsewhere. Instead, those lipids and proteins made outside of the mitochondria are incorporated separately into mitochondrial membranes after being delivered in "soluble" rather than in membrane-bound form. While some proteins are derived by synthesis on the special ribosomes inside the mitochondria themselves (p. 532), most mitochondrial proteins are acquired by import of water-soluble polypeptide precursors synthesized on free ribosomes in the cytosol (see p. 538). Similarly, two negatively charged lipids found in mitochondria are synthesized in mitochondrial membranes: phosphatidylglycerol and cardiolipin. Other lipids, such as phosphatidylcholine and phosphatidylethanolamine, must be imported from the ER (where they are synthesized) or from other cellular membranes that have received these lipids from the ER. How are the lipid molecules transferred to the mitochondrial membranes?

The problem seems to be solved by special *phospholipid transfer pro-*

teins—water-soluble proteins that transfer individual phospholipid molecules between membranes. Each transfer protein recognizes only specific types of phospholipids and binds one molecule. Transfer between membrane bilayers is achieved when the protein "extracts" a molecule of phospholipid from a membrane and diffuses away with this lipid buried within its binding site. When it encounters another membrane with suitable properties, it discharges the bound phospholipid molecule into the new lipid bilayer. Under many conditions *in vitro*, transfer proteins act to distribute phospholipids at random among any and all membranes present. But even a random exchange process can result in a net transport of lipids from a lipid-rich to a lipid-poor membrane, since lipid molecules will tend to be preferentially deposited in a membrane that has a high affinity for lipids while being preferentially removed from membranes with lower affinities. For example, by this process alone, phosphatidylcholine and phosphatidylethanolamine molecules could be transferred from the ER, where they are synthesized, to a mitochondrial membrane. Whether the transfer proteins can target phospholipids in a more specific manner is still unknown.

The Luminal Side of an Internal Organelle Is Topologically Equivalent to the Outside of the Cell

Macromolecules transported from the ER to other sites are packaged in small **transport vesicles** that pinch off from the transitional portion of the ER (Figure 7–39). These vesicles are enclosed by a lipid bilayer containing lipids and proteins from the ER membrane and contain soluble proteins from the lumen of the ER. When such vesicles fuse with a specific target membrane (either of an internal organelle or the plasma membrane), the constituents of the vesicle membrane become part of the target membrane (Figure 7–40). The soluble proteins inside are simultaneously delivered to the lumen of the target organelle (or secreted outside the cell if the target is the plasma membrane). Although poorly understood in biochemical terms, repeated cycles of vesicle budding and fusion seem to play a major part in transporting macromolecules from one cell organelle to another. It is likely that the transport vesicles in-

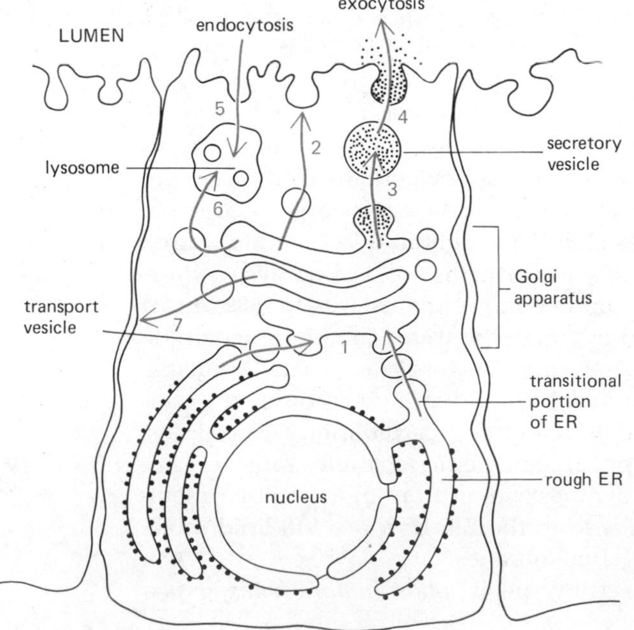

Figure 7–39 Schematic view of vesicle-mediated transport of membrane-bound and soluble materials from the ER to various intracellular sites. Transport vesicles bud from the transitional portion of the ER and fuse with cisternae of the Golgi apparatus (1). From the Golgi apparatus, further cycles of vesicle budding and fusion transport these materials throughout the cell. A total of seven different directed pathways for vesicle transport are illustrated. Pathways 2 and 7 supply different sets of membrane proteins to the apical and the basolateral plasma membranes, respectively. In addition, materials made in the ER that are essential for ER function are selectively retained there.

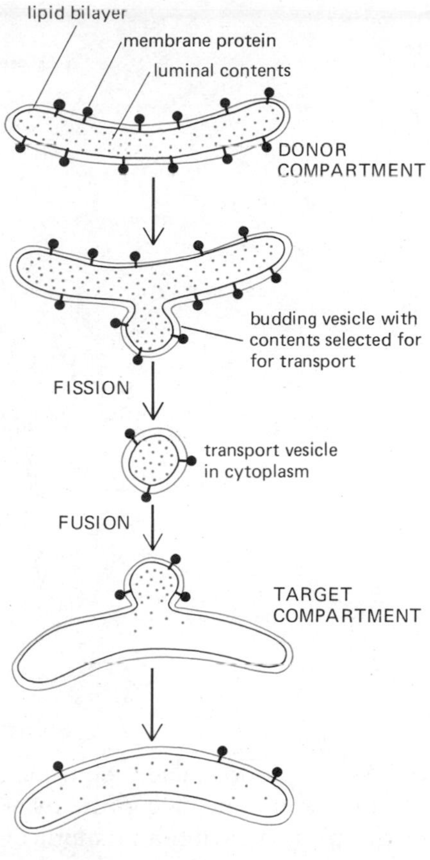

Figure 7–40 Schematic diagram showing how the sidedness of membranes is preserved during vesicle-mediated transport. Note that the original orientation of both proteins and lipids in the donor-compartment membrane is preserved in the target-compartment membrane and that soluble materials are transferred from lumen to lumen.

volved in such intracellular transport are among the *coated vesicles* described earlier (p. 307).

As shown in Figure 7–40, the processes of vesicle budding and fusion necessarily preserve the sidedness of membranes. This means that any transmembrane asymmetry introduced in the ER is conserved during intracellular transport and imparted to target membranes. This in turn leads to a fundamental simplification of the cell biologist's view of the interrelationships among the various organelles. Since repeated cycles of budding and fusion can allow a particle on the luminal side of any one organelle to reach either the luminal side of any other organelle or the outside of the cell, the luminal spaces of each of the cytoplasmic organelles are topologically equivalent to the outside of the cell (Figure 7–41). This important fact is also helpful in understanding the variety of endocytotic and exocytotic processes that occur in cells (see p. 302).

Membrane Growth by Continuous Expansion of the ER Ensures the Propagation of Transmembrane Asymmetry

The ER is truly a membrane factory, synthesizing membrane proteins and lipids for export to other parts of the cell. The topography of biosynthesis of glycoproteins in this factory provides a straightforward explanation for the origin of their assymetric orientation in membranes. For example, carbohydrates are incorporated into proteins on the luminal side of the ER only. Since

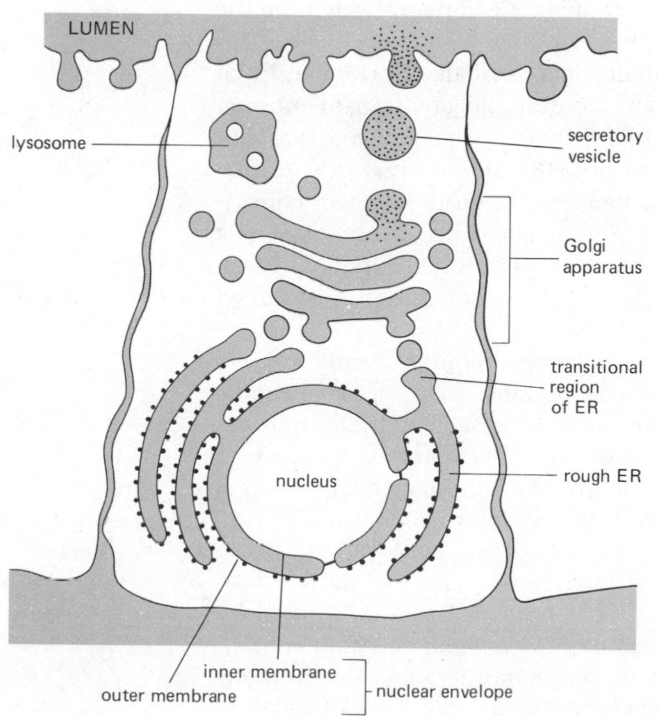

Figure 7–41 Schematic view of a cell with topologically equivalent spaces shown in color. In principle, cycles of vesicle budding and fusion permit any lumen to communicate with any other and with the cell exterior.

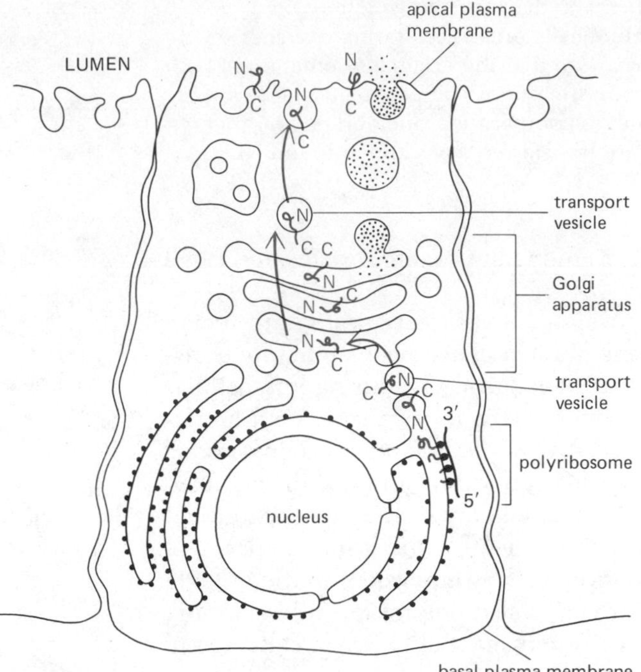

Figure 7–42 Schematic view of how the oriented insertion of a transmembrane protein into the ER membrane is preserved when that protein is transported to other membranes.

this side is topologically equivalent to the outside of the cell and since all plasma membrane glycoproteins are glycosylated in the ER, the carbohydrate of the plasma membrane should be found only on the extracellular side of the plasma membrane, as is the case (p. 284). No carbohydrate should be found on the cytoplasmic surface of membranes.

It is also easy to see why the polypeptide chains of membrane proteins are asymmetrically arranged. Proteins are always inserted into the ER membrane from the cytosolic side, where the ribosomes synthesizing these proteins are found. Since membrane proteins once inserted cannot reorient across the bilayer, each copy of a given polypeptide species will necessarily have the same asymmetric orientation as every other copy.

The principle of topological equivalence lets us relate the orientation of a protein in any one membrane to another. For example, any transmembrane protein inserted into the ER by the vectorial discharge mechanism discussed earlier (Figure 7–28) will have its amino terminus on the luminal side and its carboxyl terminus on the cytosolic side. Because of the topological considerations illustrated in Figure 7–42, such a protein, after intracellular transport to the plasma membrane, would have its amino terminus outside and its carboxyl terminus inside the cell. In fact, this is the most commonly observed orientation for transmembrane proteins.

As new membrane macromolecules are incorporated molecule by molecule into a preexisting, asymmetrical ER membrane, they rapidly become indistinguishable from the preexisting components because of their rapid lateral diffusion. The ER membrane thus grows by continuous expansion of a closed vesicle with only one accessible side (that facing the cytosol), creating a membrane asymmetry that is propagated throughout the cell.

Summary

The ER serves as a factory for the production of the protein and lipid components of most of the cell's organelles. Its extensive membranes contain many different biosynthetic enzymes, including those responsible for almost all of the cell's lipid synthesis and for the addition of a unique asparagine-linked oligosaccharide to most ER proteins. Newly synthesized proteins destined for secretion, as well as those to be delivered to several other intracellular organelles, must first be delivered to the ER. The selection process is usually guided by the presence of a "signal sequence" at the amino terminus of the growing polypeptide chain. A cytosolic ribosome making a polypeptide carrying such a signal sequence will bind to receptor proteins in the ER membrane. All or part of the polypeptide is then actively transported through the ER membrane into the ER lumen. Proteins that are destined for function in the ER lumen, for secretion, or for transfer to the lumen of other cell organelles pass completely into the ER lumen. Many proteins are not released into the ER lumen but retain part of their polypeptide chain on the cytosolic side of the ER membrane. Some of these transmembrane proteins remain in the ER membrane, while others are thought to be selectively exported to construct the Golgi, lysosome, peroxisome, plasma, and inner nuclear membranes. The asymmetry of the protein insertion and glycosylation processes in the ER establishes the polarity of the membrane proteins in these other organelles as well.

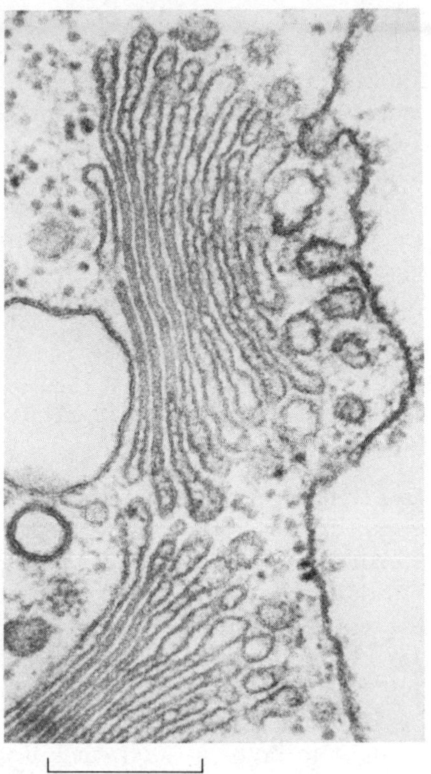

Figure 7–43 Electron micrograph of a Golgi apparatus in a plant cell seen in cross-section (the green alga *Chlamydomonas*). In plant cells the Golgi apparatus is generally more distinct and clearly separated from other intracellular membranes than in animal cells. (Courtesy of George Palade.)

The Golgi Apparatus[26]

The Golgi Apparatus Consists of Stacks of Disc-shaped Cisternae with Associated Small Vesicles

The **Golgi apparatus** is usually located near the cell nucleus, and in animal cells it is frequently disposed about the centriole pair that defines the cell center. It is normally composed of numerous sets of membrane-bounded, smooth-surfaced cisternae. Each set of flattened, disc-shaped cisternae forms a structure that resembles a stack of plates, called a **Golgi stack,** or **dictyosome,** about 1 μm in diameter (Figures 7–43 and 7–44). A stack typically contains about 6 cisternae, although the number can be as high as 30 or more in lower eucaryotes.

The number of Golgi stacks per cell varies enormously, depending on the cell type—from as few as one to the hundreds. The Golgi apparatus can even account for a large fraction of the cell volume in some specialized cells. One example is the goblet cell of the intestinal epithelium, which secretes mucus into the gut; the glycoproteins in mucus are glycosylated principally in the Golgi apparatus.

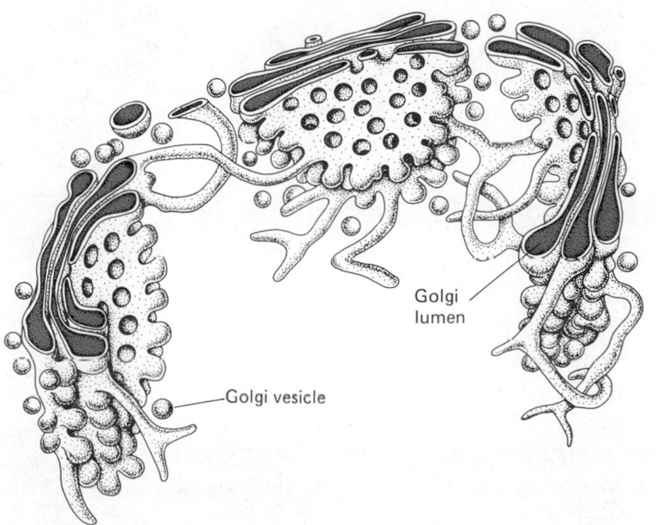

Figure 7–44 Three-dimensional representation of a Golgi apparatus as derived from electron micrographs of an animal cell. Note that the stacks of flattened cisternae have dilated edges from which vesicles appear to be budding. (After R. V. Krstić, Ultrastructure of the Mammalian Cell. New York: Springer-Verlag, 1979.)

Swarms of small (~50 nm in diameter) membrane-bound vesicles are always associated with the Golgi stacks, being clustered on the side abutting the ER, as well as along the circumference of a stack near the dilated rims of each cisterna (Figure 7–44). Some of these "Golgi vesicles" are "coated" by a polyhedral lattice (Figure 7–45; see also p. 308). Such *coated vesicles* are often seen budding from the Golgi cisternae. Many specialized cells that elaborate large quantities of a secretory product have, in addition to the small Golgi vesicles, numerous large (~1000 nm in diameter) *secretory vesicles,* also called secretory granules or vacuoles (Figure 7–46). These much larger vesicles are located on the side of the Golgi apparatus closest to the plasma membrane, and they contain the concentrated product that the cell secretes.

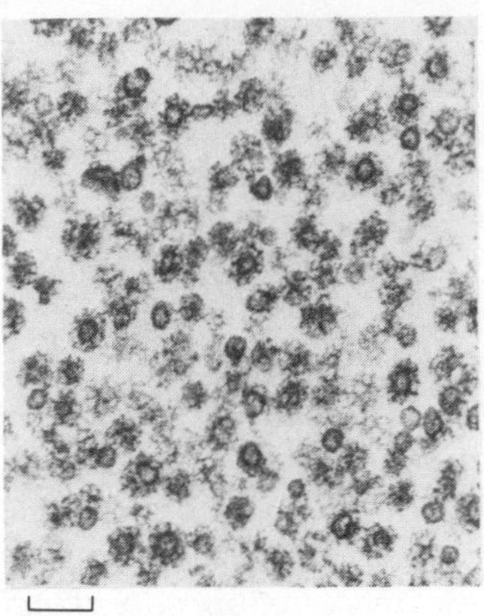

Figure 7–45 An electron micrograph of a preparation of purified coated transport vesicles, isolated from bovine brain. These vesicles are thought to shuttle both to and from the Golgi cisternae in all cells. (From S. R. Pfeffer and R. B. Kelly, *J. Cell Biol.* 91:385–395, 1981. By copyright permission of the Rockefeller University Press.)

The Golgi Apparatus Is Structurally and Biochemically Polarized[27]

The Golgi apparatus has two distinct faces: a ***cis,*** or **forming, face** and a ***trans,*** or **maturing, face.** The *cis* face is closely associated with a smooth transitional portion (p. 338) of the rough ER. In secretory cells, the *trans* face is the face closest to the plasma membrane; here, the large secretory vesicles are found exclusively in association with the *trans* face of a Golgi stack, and the membrane of a forming secretory vesicle is often continuous with that of the *trans* face of the last ("*trans*-most") cisterna. In contrast, the small Golgi vesicles are localized more evenly along the stack. Proteins are commonly thought to enter a Golgi stack from the ER on the *cis* side and to exit for multiple destinations on the *trans* side; however, neither their exact path through the Golgi apparatus nor how they travel from cisterna to cisterna along each stack are known.

The two faces of the Golgi apparatus are biochemically distinct. For example, a variation in the thickness of the Golgi membranes can be detected across the stack in certain cases, with those at the *cis* side being thinner (ER-like) and those at the *trans* side being thicker (plasma-membrane-like). More striking are the results obtained when certain histochemical tests are used in conjunction with electron microscopy to localize particular proteins within the Golgi apparatus. Some of these tests reveal membrane-bound enzyme activities that show a distinct polarity in their localization within the Golgi stack (Figure 7–47). A particularly intriguing biochemical finding was the discovery that lysosomal enzymes, such as acid phosphatase, are concentrated within the *trans*-most cisterna of the Golgi stack and within some of the coated vesicles nearby. This suggests that specific vesicles leaving for lysosomes are assembled in this region (see p. 370).

Secretory proteins are found by histochemical methods in all of the stacked cisternae, even though the large secretory vesicles in which these products are concentrated are associated only with the *trans*-most Golgi cisterna.

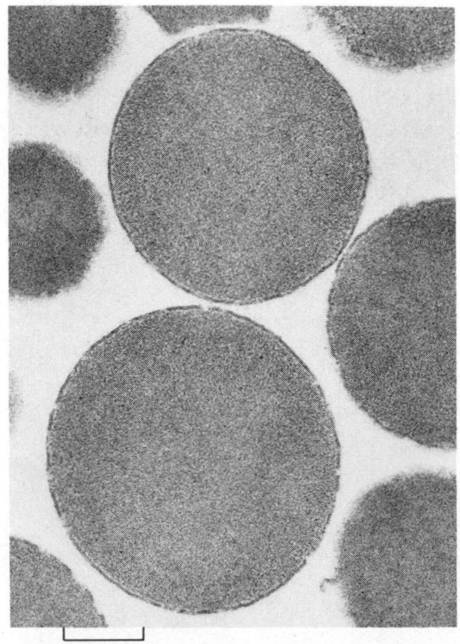

Figure 7–46 Electron micrograph of a purified preparation of large secretory vesicles. These vesicles are found only in cells specialized for secretion. (Courtesy of Daniel S. Friend.)

The Golgi Apparatus Is Not Yet Understood in Biochemical Terms

The Golgi apparatus is probably the principal director of macromolecular traffic in the cell. Many types of molecules pass through some portion of the Golgi structure at some stage in their maturation, usually shortly after their synthesis in the ER. These include secreted proteins, glycoproteins, and proteoglycans (see p. 702); glycolipids; plasma membrane glycoproteins; proteins of lysosomes; and cell-wall material in plants. It is also clear that the Golgi apparatus covalently modifies macromolecules during their passage, most commonly by altering the asparagine-linked oligosaccharide attached to proteins in the ER (see below) and also by specific proteolysis, by glycosylation of selected serines and threonines on proteins, by sulfation, and by fatty acid addition. In some cases the biochemical steps involved in these modifications are known and some of the enzymes have even been purified. The precise relationship between the biochemical events, the patterns of molecular traffic (by which macromolecules are sorted into different cellular compartments), and the characteristic polarized morphology of the Golgi apparatus, however, still remain a mystery that poses a formidable challenge to today's biochemists.

Carbohydrate Structures Are Modified in the Golgi Apparatus[28,29]

While different proteins can be very differently glycosylated, the copies of a single type of polypeptide chain usually contain the same oligosaccharides at each stage in their maturation. In fact, the exact pattern of glycosylation on a particular newly synthesized protein can be used to trace the pathway of a protein through the intracellular compartments. As described on page 345, the first step in glycosylation takes place in the ER, where mainly one species of oligosaccharide is attached to proteins. Most of the differences in the oligosaccharide structures found attached to different mature proteins are generated by subsequent modifications made during their passage through the Golgi apparatus.

Two broad classes of asparagine-linked oligosaccharides, the **complex oligosaccharides** and **high-mannose oligosaccharides,** are found in mature glycoproteins (Figure 7–48). Sometimes both types are attached (in different places) to the same polypeptide chain. High-mannose oligosaccharides contain only mannose and *N*-acetylglucosamine. Complex oligosaccharides have, in addition, a variable number of galactose and sialic acid residues, and they may contain fucose. Sialic acid is of special interest as it is the only sugar residue of glycoproteins that bears a net negative electrical charge (see p. 263).

It is useful to distinguish between the "core" and "terminal" regions of complex oligosaccharides because they are composed of sugars added in the ER and the Golgi apparatus, respectively. The inner *core region,* linked to asparagine, contains two *N*-acetylglucosamine and three mannose residues. While fewer in number than those originally added in the ER, all of these core sugars were present in the original lipid-linked oligosaccharide. The *terminal region* of complex oligosaccharides consists of a variable number of *N*-acetylglucosamine–galactose–sialic acid trisaccharide units linked to the core mannose residues. Frequently the terminal region is truncated, containing only a disaccharide (*N*-acetylglucosamine–galactose) or even just *N*-acetylglucosamine. In addition, a fucose residue may or may not be added, usually to the core *N*-acetylglucosamine residue attached to the asparagine. None of these sugars is present in the lipid-linked oligosaccharide in the ER; all are added in the Golgi.

In contrast, the high-mannose class of mature oligosaccharides contains just two *N*-acetylglucosamines and many mannose residues, often approaching the number originally present in the lipid-linked oligosaccharide precursor. All of the sugar residues found in the mature high-mannose oligosaccharides are those added as part of the lipid-linked precursor in the ER. The missing mannose residues, both here and in the complex oligosaccharides, are removed by specific mannosidases present in the Golgi apparatus.

Clearly, the modifications giving rise to the final form of a glycoprotein require a series of specific mannosidase and glycosyl transferase enzymes. The Golgi membranes can be isolated as vesicles of unusually low density after a differential centrifugation of homogenates of such tissues as liver and

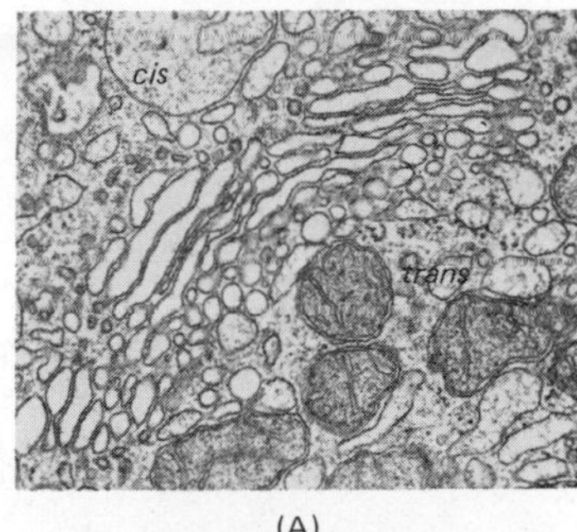

(A)

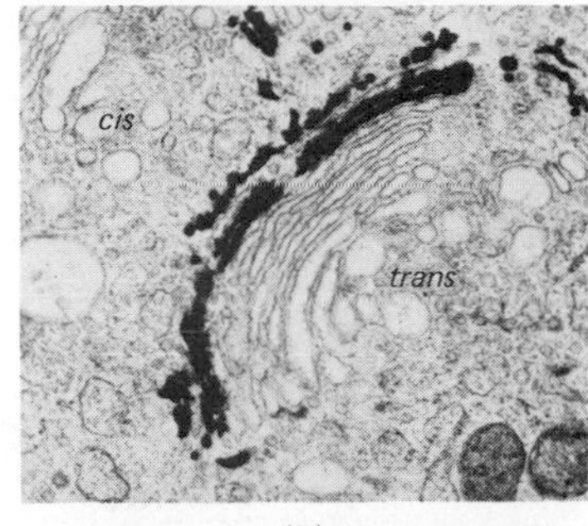

(B)

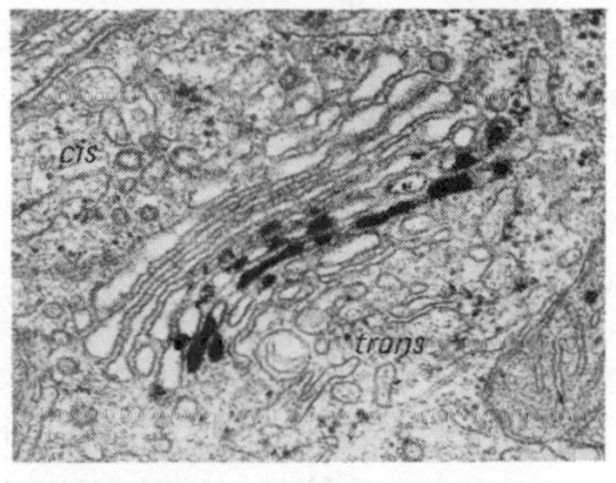

(C)

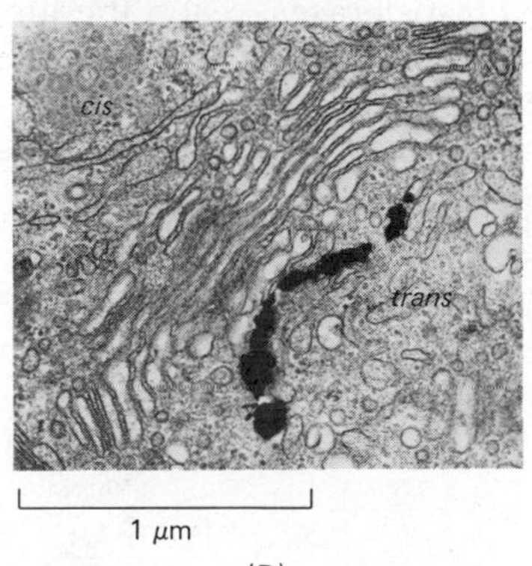

(D)

Figure 7–47 Histochemical stains demonstrate that the Golgi apparatus is biochemically polarized. (A) Unstained. (B) Osmium is preferentially reduced by the cisternae on the *cis* side. (C) Localization of the enzyme thiamine pyrophosphatase. (D) Localization of the enzyme acid phosphatase in the *trans*-most cisterna. For the type of method used to produce these stains, see Figure 7–57. (Courtesy of Daniel S. Friend.)

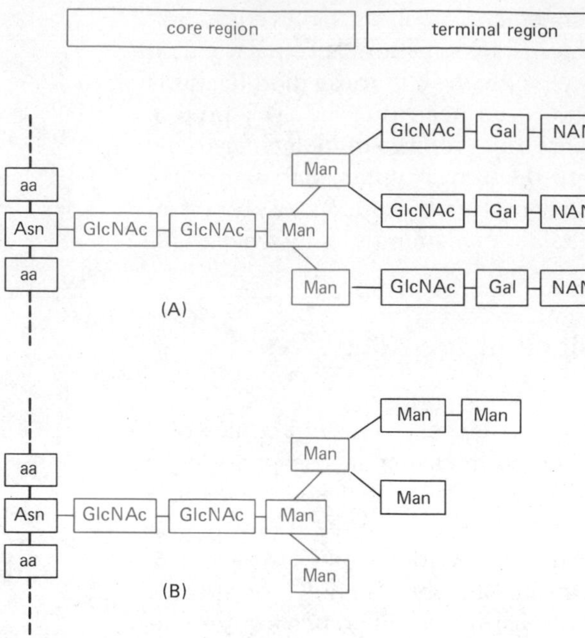

Figure 7–48 The two main classes of asparagine-linked oligosaccharides: a "complex" oligosaccharide (A) and a "high-mannose" oligosaccharide (B). The following abbreviations are used: asparagine, Asn; mannose, Man; *N*-acetylglucosamine, GlcNAc; *N*-acetylneuraminic acid (sialic acid), NANA; galactose, Gal; amino acid, aa.

kidney, and they are found to contain these enzymes. In fact, the marker generally used to identify these Golgi-derived vesicles is the presence of the enzyme *galactosyl transferase*.

Three types of glycosyl transferases acting in a rigidly determined sequence add the terminal three sugars (in the sequence *N*-acetylglucosamine–galactose–sialic acid) to generate mature complex oligosaccharides (Figure 7–49). All of these reactions take place on the luminal side of the Golgi apparatus. How the sugar nucleotides, which serve as substrates for the glycosyl transferases, reach the luminal side of the Golgi is not yet known.

This multiglycosyl transferase system illustrates the most general principle of oligosaccharide synthesis: wherever oligosaccharides are made, sugars are joined together in a specific sequence by virtue of the fact that the product of glycosylation at each step is recognized as the exclusive substrate for the next enzyme in the series. By contrast, the sequences found in the other principal macromolecules (DNA, RNA, and protein) are copied from a template in a repeated series of steps using the same enzyme(s). For complex carbo-

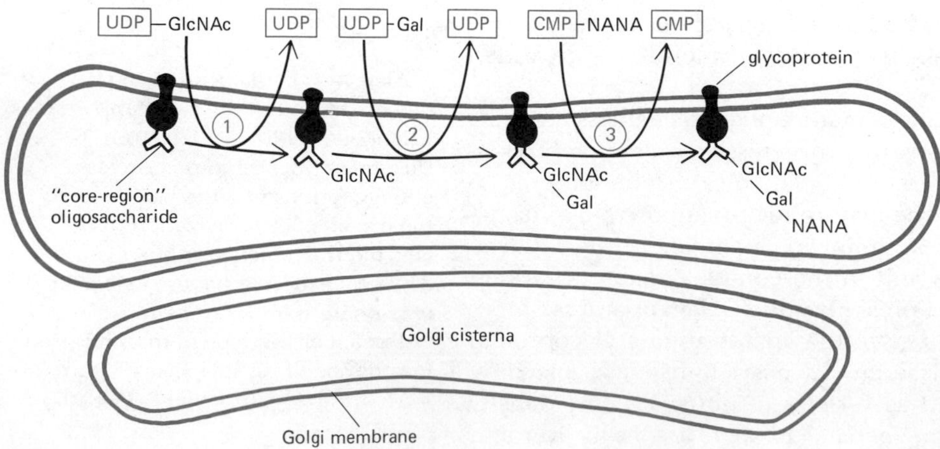

Figure 7–49 Schematic illustration of the stepwise addition of sugar residues that occurs in the Golgi apparatus to generate complex oligosaccharides. Three different glycosyl transferase enzymes act sequentially, using sugars activated by linkage to a nucleotide as substrates. It is not known how the sugar nucleotides are transported from the cytosol to the lumen of the Golgi apparatus. For abbreviations used, see Figure 7–48.

hydrates, a different enzyme is required at each step to provide the appropriate degree of specificity, as well as to catalyze the formation of the variety of different covalent linkages found in small oligosaccharides of complex sequence. While the same general strategy is followed in the synthesis of large polymers of simple repeating sugar sequences, only a small number of different specific enzymes is needed in these cases (as in the synthesis of glycogen or plant cell-wall polysaccharides).

Oligosaccharide-trimming Pathways Are Elaborate and Precisely Programmed[29]

The mature form of the glycoprotein (G protein) of vesicular stomatitis virus (p. 342) contains a complex oligosaccharide with three mannose residues in its core (Figure 7–48A). Therefore, six of its original nine mannose residues have been removed in the Golgi apparatus. However, when this virus infects certain mutant cells whose Golgi apparatus lacks the enzyme that adds the first *N*-acetylglucosamine to the "core" region (step 3 in Figure 7–50), the oligosaccharide on its G protein now contains a core of five mannose residues instead of three. This indicates that the removal of the last two mannose residues is contingent upon the prior addition of the first *N*-acetylglucosamine to the core. Subsequent biochemical studies have shown that oligosaccharide processing occurs by the highly ordered sequential pathway shown in Figure 7–50; the mannosidases involved will only remove the two extra mannoses if an *N*-acetylglucosamine is first transferred to the particular mannose that is colored in this figure.

These important findings tell us that the trimming pathway follows the same general principle utilized for the synthesis of specific sequences of sugars and indicate how elaborate these pathways can be. In a closely programmed series of cleavages, the product of one step is the exclusive substrate of the next enzyme in the system (Figure 7–50).

The Correct Program of Trimming Is Set by an Irreversible "Switch" Thrown Early in Oligosaccharide Processing

An examination of the variety of mature oligosaccharide structures produced reveals that while all of these arrangements are derived by alteration of the original precursor lipid-linked oligosaccharide, the trimming pathways required to achieve them are often mutually exclusive. No one pathway of trimming can generate all the known structures. What sets the "switch" that determines which processing program any particular glycoprotein will follow? The information for this decision must ultimately be contained in the attached polypeptide since different proteins, all starting with the same oligosaccharides, end up with different ones. The details of the decision-making process should soon be known since the trimming enzymes are now being purified and intensively studied.

Carbohydrate Modification in the Golgi Apparatus Can Be Detected by Autoradiography[30]

Electron microscopic autoradiography of cells after a brief incubation with [^{3}H]mannose shows that the ER is the sole site of incorporation of this sugar into glycoproteins (p. 348). Similar experiments have also been carried out with other sugars. For example, a brief incubation with [^{3}H]*N*-acetylglucosamine results in the simultaneous labeling of both ER and Golgi, as one might expect from its presence in both "core" and "terminal" regions of complex oligosaccharides (Figure 7–48). But for [^{3}H]galactose and [^{3}H]sialic acid, ex-

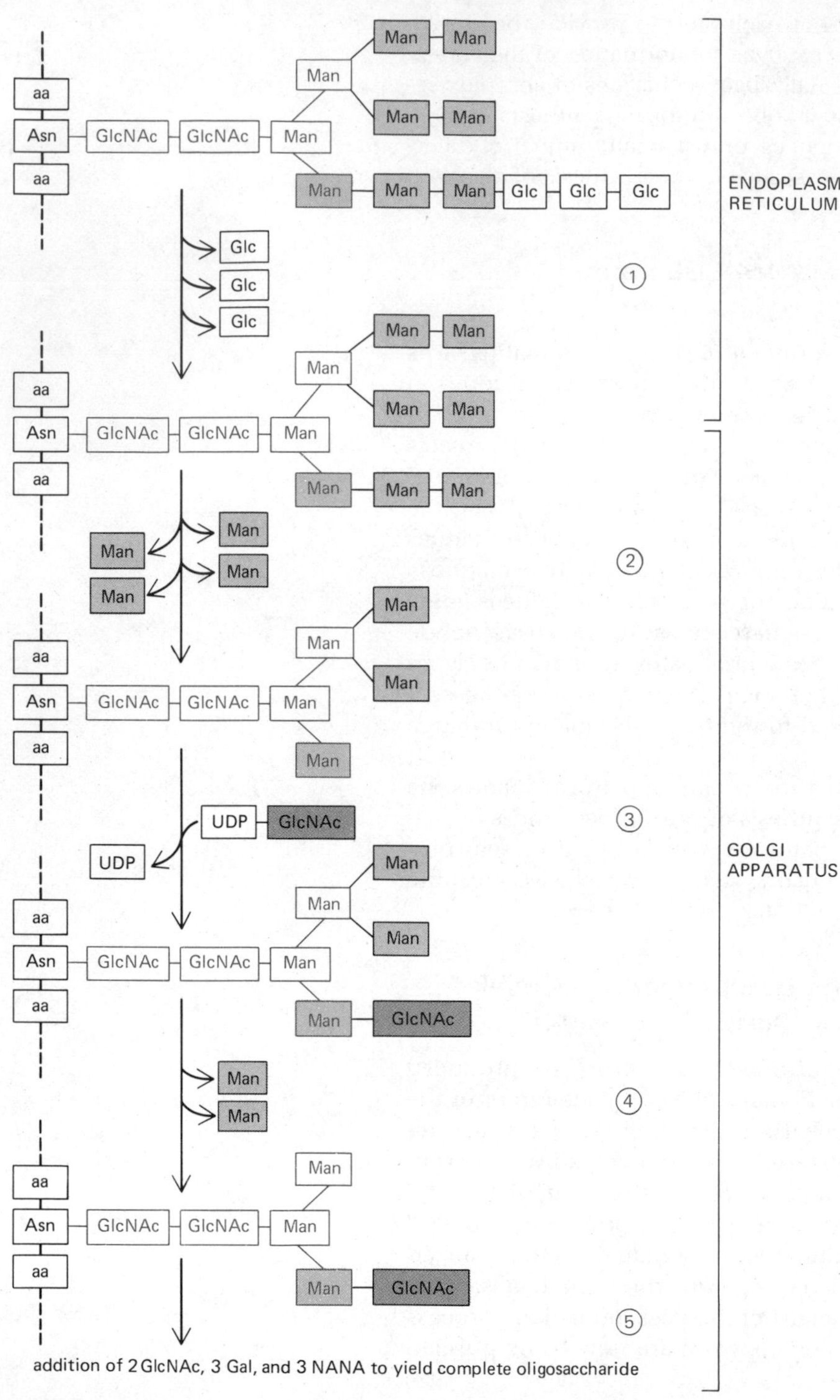

Figure 7–50 Diagram of the oligosaccharide processing that takes place in the ER and the Golgi apparatus. The process is highly ordered so that each step shown is dependent on the previous reaction in the series. Thus, in the mutant cells described in the text, steps 4 and 5 are blocked because the enzyme that catalyzes step 3 is missing. Step 5 here is actually the *series* of sequential reactions shown in the previous figure. For abbreviations used, see Figure 7–48.

posed silver grains are initially found only over the Golgi apparatus. Indeed, this was the first (and still the best) evidence that these sugars are added to oligosaccharides only in the Golgi.

When a brief incubation ("pulse") with [^{3}H]mannose is followed by a period of time ("chase") in nonradioactive medium, ^{3}H-labeled glycoproteins are observed to move rapidly from the ER to the Golgi region and then to many other destinations—including all portions of the plasma membrane, lysosomes, and the cell exterior—revealing the complexity of the molecular traffic directed by the Golgi apparatus. Transit times are such that a typical newly synthesized protein requires about 10 minutes to pass from the ER to the Golgi and 30 to 60 minutes to travel through the Golgi to its final destination.

Proteins Destined for Secretion Are Packaged into Golgi-associated Secretory Vesicles That Then Fuse with the Plasma Membrane[31]

All cells have a Golgi apparatus. This important organelle is required for the final stages in the synthesis of many proteins, during which the initial polypeptide chain made in the ER is "matured" by covalent modification. As we shall see, the Golgi apparatus is also required for the segregation process that makes possible the subsequent intracellular transport of these proteins to specific sites in the cell. In addition to these functions, highly specialized secretory cells use their Golgi apparatus to concentrate and store large amounts of one or a few products in Golgi-associated *secretory vesicles.* These vesicles are designed so that their contents can be quickly released to the cell exterior whenever the cell is stimulated by a specific signal. Although many cells lack such secretory vesicles, a great deal has been learned in general about the function of the Golgi apparatus from studies of specialized secretory cells.

Much of our knowledge of the process of secretion comes from autoradiographic studies of the acinar cell of the pancreas. This exocrine cell is specialized for the secretion of a variety of digestive enzymes and zymogens (inactive enzymes that are later activated by specific proteolytic cleavage); the proteins secreted include trypsinogen, chymotrypsinogen, amylase, lipase, deoxyribonuclease, and ribonuclease. The intracellular pathway for processing and packaging these proteins has provided a model for understanding secretion and related processes in a wide variety of biological contexts.

Because the bulk of the protein synthesized in the pancreatic acinar cell is destined for secretion, it is possible to trace the intracellular pathway of secreted proteins from their site of synthesis to their discharge from the cell using autoradiography in conjunction with electron microscopy. (In fact, this combination of techniques was developed for this purpose.) By pulse-labeling with [^{3}H]amino acids, followed by a chase period in nonradioactive medium for varying times, the newly synthesized, radioactive proteins have been found to move from the rough ER through the smooth transitional portion of the rough ER and then to the Golgi apparatus (Figure 7–51). This part of the intracellular transport pathway is energy dependent; secretory proteins do

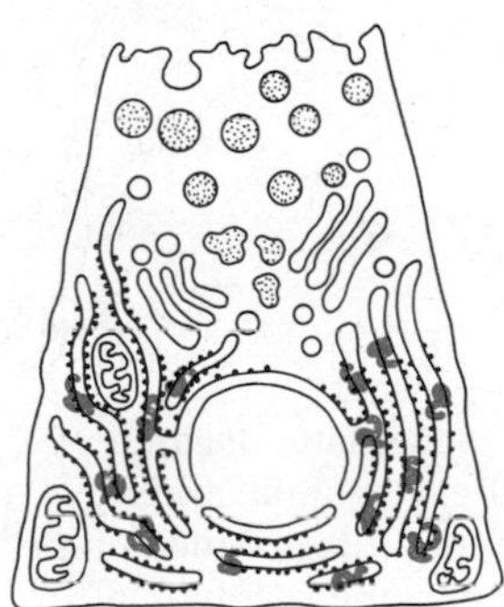

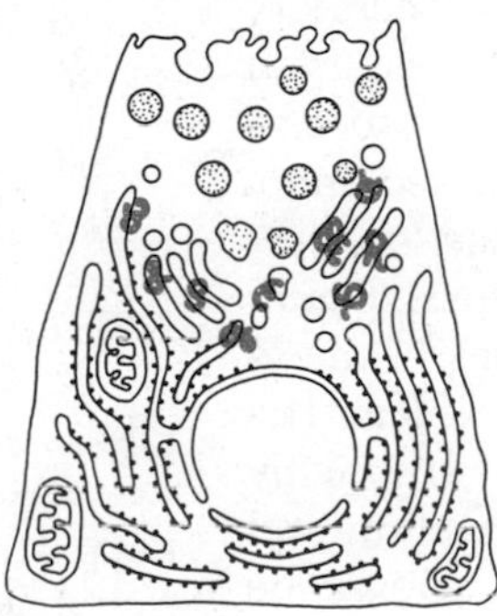

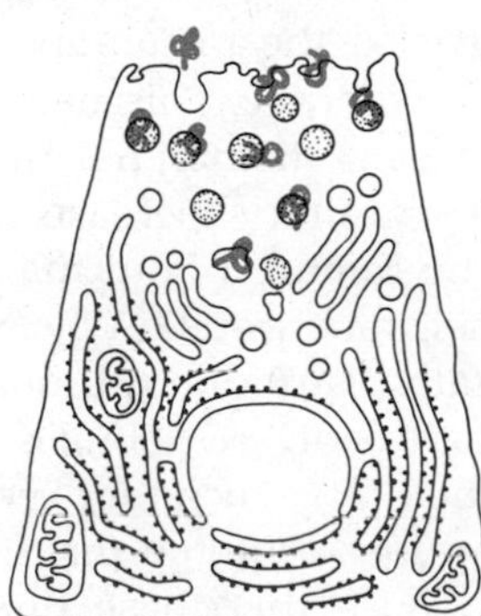

Figure 7–51 Schematic illustration of the results obtained when electron microscopic autoradiography (Figure 7–32) is used to examine a pancreatic acinar cell that has been briefly pulse-labeled with [^{3}H]amino acids and then incubated in unlabeled media ("chased") for various times. The colored spots represent developed silver grains that, with time, are found ever closer to the cell exterior, indicating the path followed by newly synthesized protein molecules. This cell is unusual in that most of the protein it synthesizes is secreted from the cell.

not move from the ER to the Golgi apparatus when ATP synthesis is blocked with drugs. This is not surprising, since any process that follows a vectorial pathway must be energy dependent at one or more steps (see p. 137).

The [^{3}H]proteins are next found in **condensing vacuoles,** which are large immature secretory vesicles associated with the *trans* face of the Golgi stack (Figure 7–52). In some types of secretory cells the condensing vacuole is distinct from the Golgi stacks; in most others the vacuole is actually continuous with the dilated rims of the *trans*-most Golgi stack and believed to form from it. The secretory proteins reach the condensing vacuoles as a dilute solution. With time they become concentrated (hence the name "condensing" vacuoles), resulting in mature **secretory vesicles** (called zymogen granules in the case of the pancreas), which are easily identified in electron micrographs by the high density of their contents (Figures 7–46 and 7–52). The process whereby the secretory proteins are concentrated in condensing vacuoles does not require energy. In the pancreatic acinar cell, concentration appears to result from the interaction of the positively charged pancreatic secretory proteins with a negatively charged proteoglycan, a large protein-linked polysaccharide that is sulfated in the Golgi apparatus. This spontaneous electrostatic interaction leads to the formation of osmotically inactive precipitates, resulting in the passive efflux of water from the condensing vacuole. In other cells, the same type of concentration process is driven by other means. For example, in preparation for insulin secretion, the osmotic activity of insulin is reduced by the formation of a metal complex that causes the protein to crystallize.

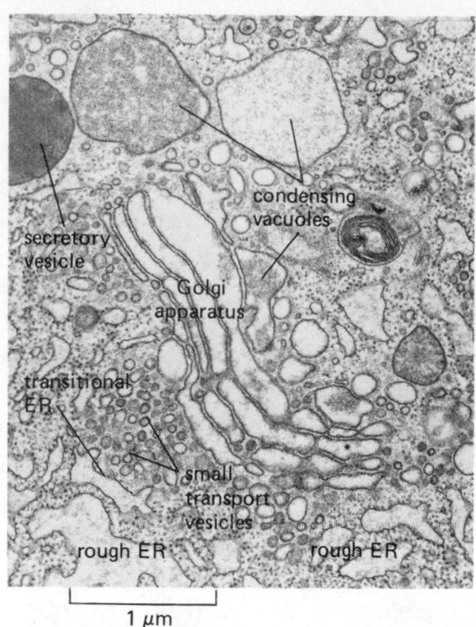

Figure 7–52 Electron micrograph of the Golgi apparatus in a pancreatic acinar cell, showing condensing vacuoles in various stages of maturation. (Courtesy of George Palade.)

The concentrated secretory proteins are stored in secretory vesicles that reside in the apical region of the cell between the Golgi apparatus and the lumen of a secretory duct. The secretory proteins are discharged from the cell by exocytosis, a process in which the secretory vesicles fuse with the plasma membrane to release their contents to the outside (p. 304). Exocytosis is a highly specific and closely regulated process. Secretory vesicles fuse only with the apical portions of the plasma membrane, thereby avoiding fruitless and dangerous discharges into the spaces between cells or into other organelles in the cell. Furthermore, exocytosis occurs only in response to an appropriate signal. The stimulus for exocytosis in the pancreatic acinar cell is either acetylcholine (a neurotransmitter) or cholecystokinin (a hormone). These signaling molecules are released by nerves or by intestinal cells, respectively, when pancreatic enzymes are needed for digestion.

Membrane Components Are Recycled

When a secretory vesicle fuses with the plasma membrane, not only are its contents discharged from the cell, but also the membrane of the secretory vesicle becomes a part of the plasma membrane. This necessarily results in an increase in the surface area of the plasma membrane. And yet the surface area of a secreting cell shows only a transitory increase during secretion. Clearly membrane components must be removed from the surface, or *recycled,* at a rate nearly equal to that of their incorporation by exocytosis (Figure 7–53). Besides maintaining a steady-state distribution of quantities of membrane among the various cellular compartments, recycling is economical. The membranes of all transport vesicles (including secretory vesicles) are highly specialized, possessing unique polypeptides required for the formation and fusion of the vesicles. It makes good sense that these proteins should be recycled and used in several rounds of transport rather then being utilized only once and then discarded.

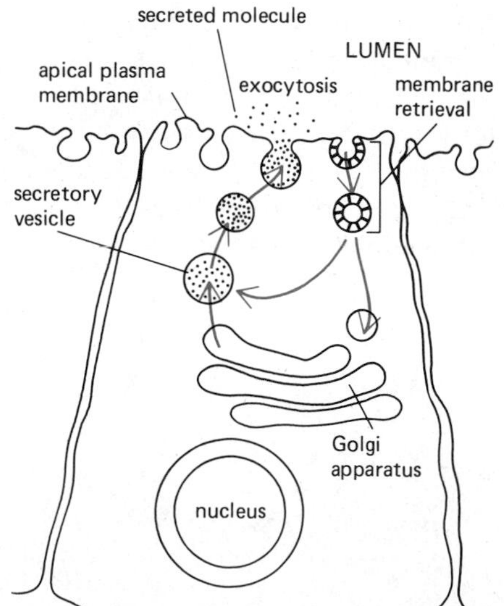

Figure 7–53 Schematic diagram of a secretory epithelial cell illustrating the postulated recycling of membrane components following secretion at the apical surface. In some cases membrane retrieval seems to be mediated by coated vesicles, although the precise mechanism is unclear.

Recycling can be expected in every route of intracellular transport. It is best to think of all transport vesicles as reusable two-way shuttle systems that

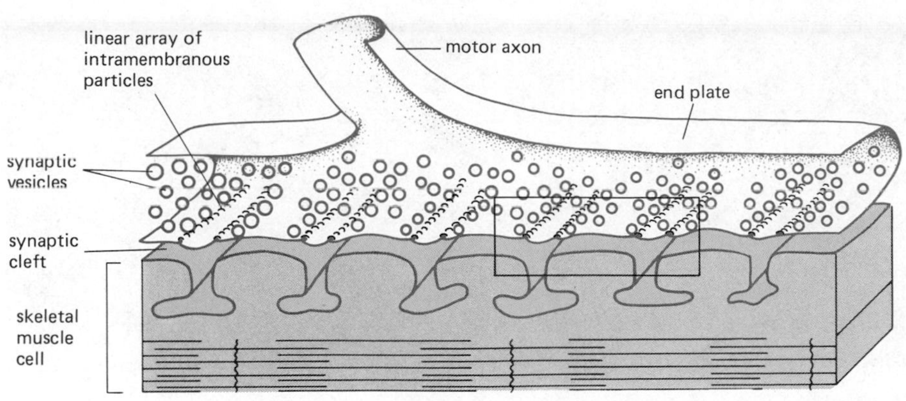

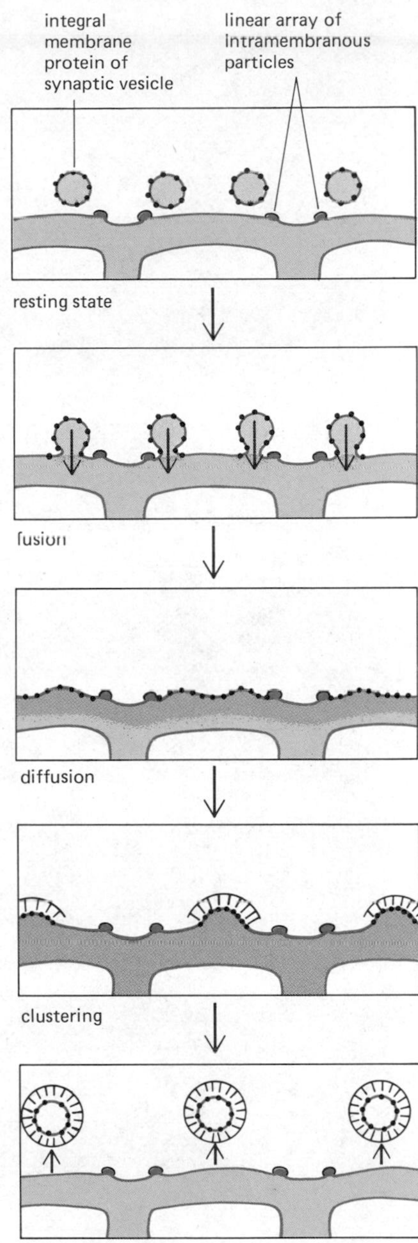

Figure 7–54 Three-dimensional drawing of a neuromuscular junction, a specialized synapse between a nerve and a skeletal muscle cell. The boxed area is enlarged and shown in cross-section in the above panels, where the events that follow the triggering of the nerve are schematically illustrated. (See also Figure 7–55.)

deliver their "cargo" by fusion to their target in the first half of a cycle and then bud off again (minus their cargo) to re-fuse with the organelle from which they originated (see Figure 7–56). This raises the fundamental and unanswered question of how the cargo is irreversibly delivered to its destination rather than simply equilibrated between the two organelles connected by the shuttle system. The answer is crucial to an understanding of the mechanism by which specific proteins are transported within cells.

Membrane Fusion and Recycling Can Be Demonstrated by Electron Microscopy[32]

In the neuromuscular junction (Figure 7–54), one highly synchronized cycle of a transport vesicle shuttle system can be followed by electron microscopy. The synaptic region of the nerve axon (that is, the nerve terminal) at the neuromuscular junction is crammed with hundreds of uniform (~50 nm in diameter) vesicles, each containing the neurotransmitter acetylcholine. Stimulation of the nerve induces a synchronized burst of exocytosis in which the vesicles fuse with the plasma membrane of the nerve terminal, thereby discharging acetylcholine from the terminal and initiating contraction of the adjoining muscle cell (p. 1036). These fusion events are triggered by a local increase in Ca^{2+} concentraton within the nerve terminal induced by a change in membrane potential. The nerve terminal is designed so that each triggering event increases the rate of exocytosis more than 10,000-fold, yet does so for only a few milliseconds.

The cycle of events at the synapse, initiated synchronously by a single nerve impulse, has been captured by very rapidly freezing the tissue in liquid helium before preparation for electron microscopy. Some of the results are shown in Figure 7–55. Within five milliseconds after stimulation, distinct openings appear in the nerve terminal plasma membrane, each representing the point of fusion of one neurotransmitter-containing vesicle. Fusion is limited to specific regions of the plasma membrane (called "active zones") that are adjacent to two linear arrays of intramembranous particles. Within another two milliseconds fusion is complete and the exocytotic openings have disappeared; the protein components of the vesicle membrane can be seen to have been inserted into the plasma membrane, for the intramembrane particles in synaptic vesicles are distinctly larger than those initially found in the plasma membrane. These large particles move further and further away from the initial points of fusion in the subsequent few hundred milliseconds. The diffusion coefficient that can be estimated by electron microscopy for these larger particles (2×10^{-10} cm^2/sec) suggests that the vesicle proteins move randomly in the bilayer, mixing freely with plasma membrane components.

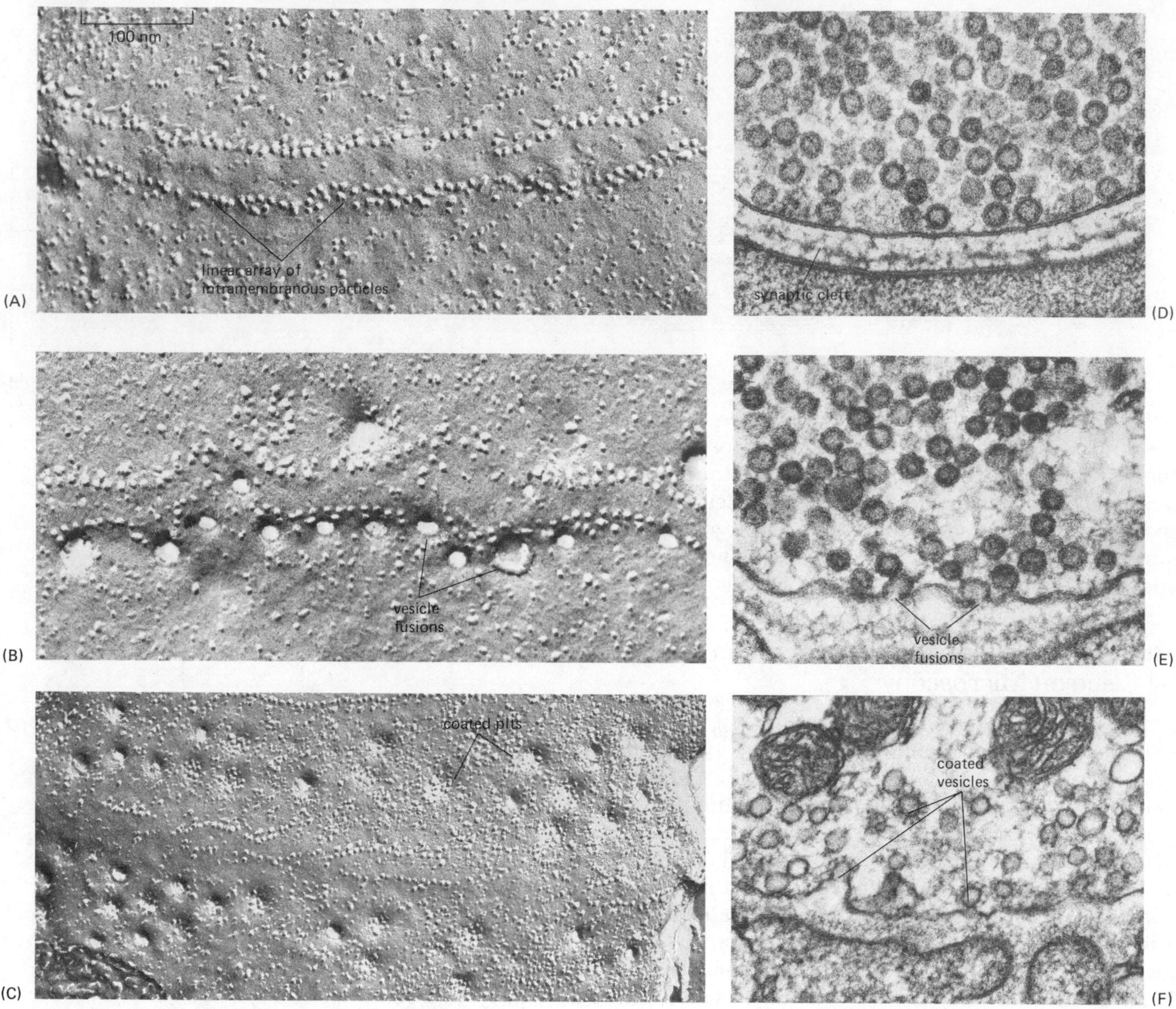

Figure 7–55 The cycle of membrane events at a stimulated neuromuscular junction, viewed by electron microscopy. Freeze-fracture electron micrographs of the cytoplasmic half of the presynaptic membrane ("P face") are shown on the left, while thin-section micrographs are shown on the right. (A, D) Resting state. (B, E) Fusion of synaptic vesicles near the linear array of intramembranous particles. (C, F) Clustering and recapture of synaptic vesicle components via coated pits and coated vesicles. Compare to Figure 7–54. (Courtesy of John Heuser.)

The stage is now set for retrieval from the plasma membrane of the specific protein components of the synaptic vesicles (such as the large intramembrane particles). Coated regions (or "coated pits") become numerous on the plasma membrane of the nerve terminal within 10 seconds after secretion, and the large intramembrane particles collect into them. Within 20 seconds after stimulation, the coated pits pinch off by endocytosis to form coated vesicles (see Figure 7–45). These vesicles contain the original protein components of the synaptic vesicle, even though several seconds earlier those proteins were thoroughly intermixed with other plasma membrane components. The cycle ends when the coat dissociates from the coated vesicle, which refills with acetylcholine to form a smooth-surfaced, regenerated synaptic vesicle. This scheme probably accounts for the strikingly uniform size of the synaptic vesicles, a size defined by the dimensions of the latticelike coat of clathrin (see p. 307).

Further evidence for the above retrieval scheme can be obtained by stimulating the nerve in the presence of electron-dense extracellular markers such as ferritin. These markers quickly appear within coated vesicles and eventually show up in synaptic vesicles.

Coated Vesicles Are Thought to Play a Major Part in the Intracellular Sorting of Proteins[33]

In addition to mediating a variety of endocytotic processes (p. 309), coated vesicles appear to be responsible for the transport of membrane proteins between the ER and Golgi apparatus and between the Golgi apparatus and the plasma membrane. Different sorts of coated vesicles would seem to be necessary to account for their diverse array of functions.

As previously described in detail (p. 307), coated vesicles are remarkable because of their almost perfectly polyhedral "basket" or "coat." The coat itself is composed primarily of the structural protein *clathrin* and a smaller polypeptide, which are associated with a variety of accessory proteins in the transport-vesicle membrane. The accessory proteins must serve at least two important functions in intracellular transport events. First, they must provide binding sites on the noncytosolic (luminal) surface of the coated pit that are responsible for trapping specific "cargo" molecules, thereby ensuring that these molecules are packaged into coated vesicles. Second, the fact that there is accurate delivery of the contents of each transport vesicle to the correct intracellular membrane "address" suggests that there must be distinct subpopulations of transport vesicles, each with unique accessory proteins ("docking markers") on their surface that are recognized by complementary "acceptors" on the target membranes. One possible view of this process is illustrated in Figure 7–56.

Coated vesicles may be viewed as miniature sorting machines. Their latticelike coat appears to be designed to allow them to bud off from a membrane, carrying a load of the appropriate cargo. Their selectivity must be due to other accessory proteins in the coated vesicle membrane that trap specific cargo molecules and, at the same time, exclude other membrane components. In fact, coated-vesicle-mediated transport could be the principal means by which cells execute their molecular sorting decisions.

Why Have a Golgi Apparatus?[34]

At first glance it would appear that the Golgi apparatus should be dispensable. Could not all of the cell's sorting and packaging problems be solved by the ER, where proteins originate, before being transported to the Golgi apparatus? Such sorting can certainly take place at the level of the ER, since many of its membrane-bound proteins (for example, ribophorins and enzymes involved in lipid and oligosaccharide synthesis) are selectively retained rather than exported.

A general answer may be suggested by considering the magnitude of the problem of sorting new macromolecules. The maintenance of the functional (for example, enzymatic) distinctions among different cellular compartments requires that the sorting of proteins be done with a very low error rate, which may not be achievable with only a single step of sorting in the ER. It has thus been suggested that a Golgi apparatus interposed between the ER and the ultimate designations could serve as an editor, allowing several successive sorting steps to take place, thereby reducing the overall error rate.

It is also worth noting the antagonistic nature of the known enzymatic reactions carried out in the ER and Golgi. Oligosaccharides are *synthesized* in

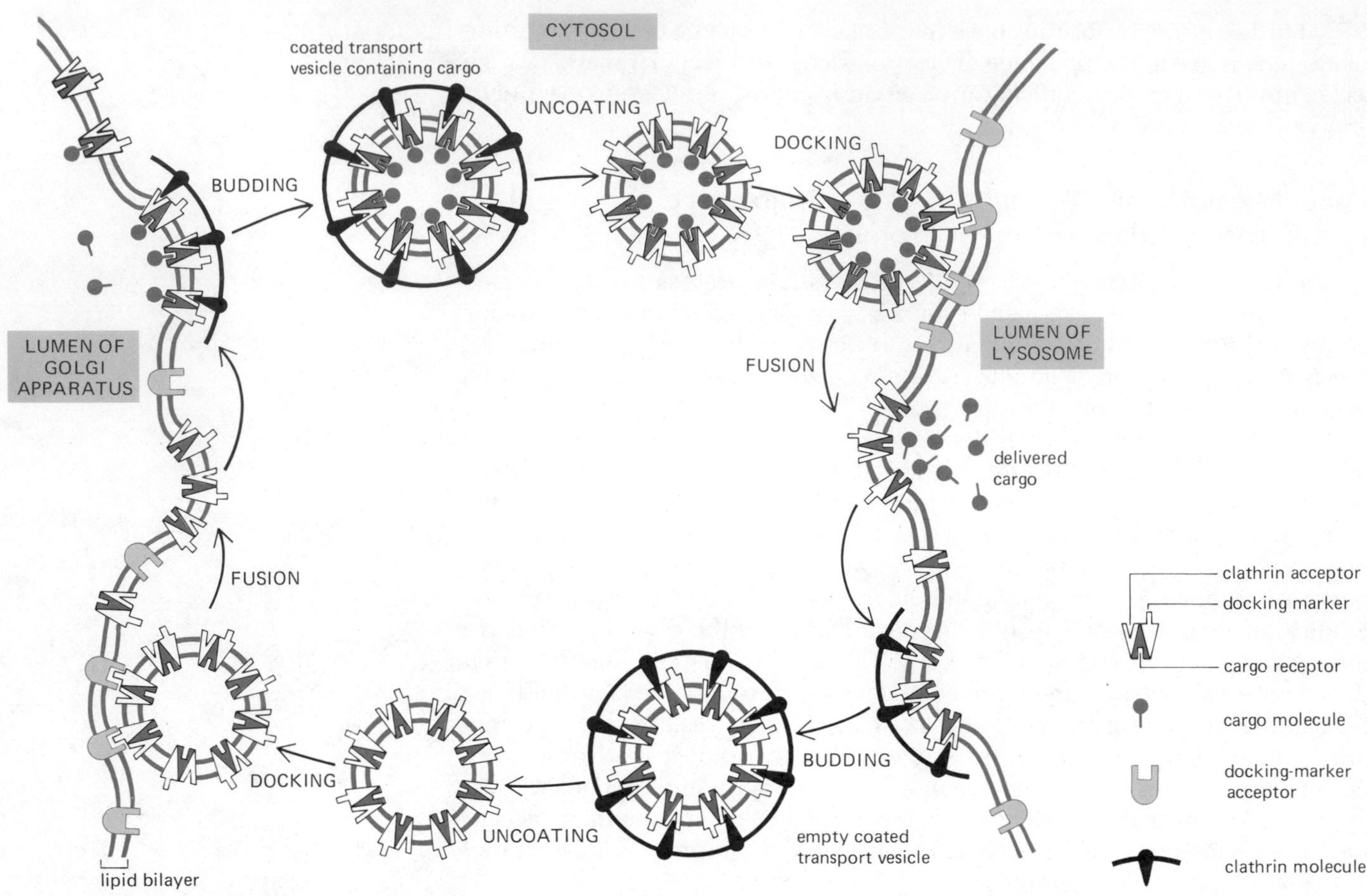

Figure 7–56 Highly schematic view of a possible mechanism allowing the targeting of coated transport vesicles to a specific intracellular membrane (see text for details). In this example, the net transport of "cargo" from the lumen of the Golgi apparatus to the lumen of the lysosome is shown.

the ER, while extensive portions of the same structures are *degraded* in the Golgi. As we shall see, these competing sets of reactions are crucial, since they can play an important part in determining the intracellular compartment to which a newly synthesized macromolecule is sent.

Summary

The Golgi apparatus is a polarized structure consisting of an oriented stack of disc-shaped cisternae surrounded by a swarm of small vesicles. Near the cis face of a Golgi stack, proteins from the lumen and membranes of the ER are transferred to the lumen and membranes of the Golgi cisternae, probably by means of coated vesicles. Once exposed to the Golgi lumen, these proteins are covalently modified in a variety of ways to produce their final mature forms. In particular, the asparagine-linked oligosaccharide chains, previously coupled to proteins in the ER, are extensively modified by removal and addition of selected sugar residues. These various covalent modifications follow one of a branching series of highly ordered enzymatic pathways, the choice of pathway being determined by an unknown feature of each individual polypeptide chain.

Proteins are usually exported from the Golgi apparatus near the trans-most face of a stack to a variety of intracellular and extracellular destinations. The accurate sorting of proteins for selective export is one of the major functions of the Golgi apparatus. This sorting is thought to be mediated largely by the coated vesicles that surround the cisternae and to depend on sets of receptors in the Golgi membrane that recognize specific markers on the proteins to be transported.

Lysosomes and Peroxisomes[35]

Lysosomes Are the Principal Sites of Intracellular Digestion

In the course of enzymatic studies of carbohydrate metabolism in liver homogenates in 1949, certain irregularities in the assay of acid phosphatase (an enzyme that was being used as a control) were noticed. For example, hydrolase activity was higher in extracts prepared with distilled water than with an osmotically balanced sucrose solution. Enzyme activity was higher in aged than in fresh preparations and was no longer associated with sedimentable particles. Similar findings were soon reported for several other hydrolytic enzymes. These observations led to the discovery of a new organelle, named the **lysosome,** a membranous bag of hydrolytic enzymes used for the controlled intracellular digestion of macromolecules. Damage to the membrane, induced by osmotic lysis or aging, explains the release of the enzymes in a nonsedimentable form in cell extracts. But the permeability barrier represented by the lysosomal membrane is sufficient to prevent the enzymes from escaping into the cell.

Some 40 enzymes are now known to be contained in lysosomes. They are all hydrolytic enzymes, including proteases, nucleases, glycosidases, lipases, phospholipases, phosphatases, and sulfatases. And all are **acid hydrolases,** optimally active near the pH of 5 maintained within this organelle. Although the membrane of the lysosome is normally impermeable to these enzymes, the fact that the enzymes require an acid pH for optimal activity protects the cytoplasm against damage should leakage occur.

Like all other intracellular organelles, the lysosome not only contains a unique collection of enzymes but also has a unique surrounding membrane. As expected, this membrane permits the final products of the digestion of macromolecules to escape so that they can be either excreted or reutilized by the cell. In addition, the membrane is thought to contain a special transport protein that utilizes the energy of ATP hydrolysis to pump H^+ into the lumen of the lysosome, thereby maintaining the pH of the lumen at 5. It must also contain "docking-marker acceptor" proteins that mark a lysosome as a target for fusion with specific transport vesicles in the cell (Figure 7–56).

Some small molecules penetrate the lysosome membrane in an uncharged form and then become charged by picking up a proton in the acidic environment of the lysosome lumen. Once charged, such molecules are more hydrophilic and therefore less able to cross the lipid bilayer of the lysosome. Therefore, these molecules enter the lysosome more rapidly than they leave and become highly concentrated inside. An important example of such a substance is the antimalarial drug *chloroquine,* which increases the pH inside the lysosome when added to intact cells and is therefore commonly used in the laboratory as an inhibitor of lysosome functions.

Histochemical Staining Demonstrates That Lysosomes Are Heterogeneous Organelles[36]

Lysosomes were identified by electron microscopic cytochemistry about a decade after their first biochemical description. Because of their extraordinary diversity of shapes and sizes, it is often necessary to use the electron-dense precipitate formed by the reaction product of a hydrolase (such as a phosphatase or sulfatase) to identify a particular organelle as a lysosome in electron micrographs (Figure 7–57). By this criterion, lysosomes are found in all eucaryotic cells. Liver lysosomes have been the most extensively studied. They are spherical vesicles about 0.5 μm in diameter with an electron-dense core

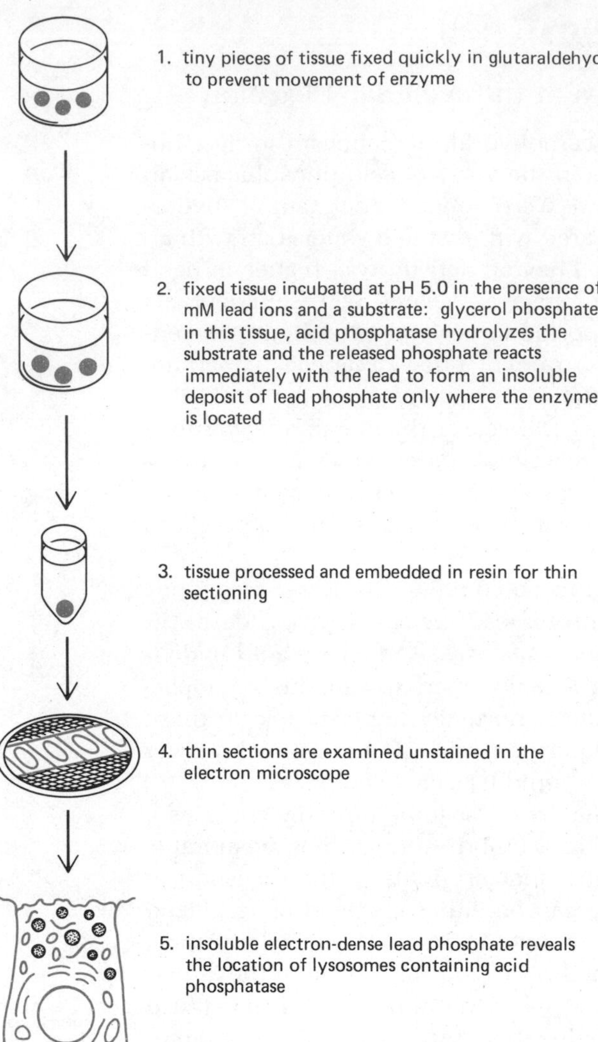

Figure 7–57 An outline of the cytochemical method commonly used to demonstrate the location of acid phosphatase, a marker enzyme for lysosomes, in electron micrographs.

that is packed with hydrolases (the latter constituting over 60% of the core by weight).

The heterogeneity of lysosomal morphology contrasts with the relatively uniform ultrastructure of all other cellular organelles. This diversity reflects the wide array of different digestive functions mediated by acid hydrolases, including the digestion and turnover of intra- and extracellular constituents, programmed cell death in embryogenesis, digestion of phagocytosed microorganisms, and even cell nutrition (since lysosomes are the principal site of cholesterol assimilation from endocytosed serum lipoprotein, see p. 309). For this reason, lysosomes are best viewed as a collection of distinct organelles whose common feature is a high content of hydrolytic enzymes. Two general classes of lysosomes are usefully distinguished: **primary lysosomes,** which are newly formed and therefore have not yet encountered substrate for digestion, and **secondary lysosomes,** which are membranous sacs of diverse morphology that contain substrates and hydrolytic enzymes (Figure 7–58). Secondary lysosomes result from the repeated fusion of primary lysosomes with a variety of membrane-bounded substrates (Figure 7–59). The morphology of secondary lysosomes will therefore vary in as many ways as there are for internalizing and packaging different substrates—large ones will result from phagocytosis, small ones from endocytosis, and so on.

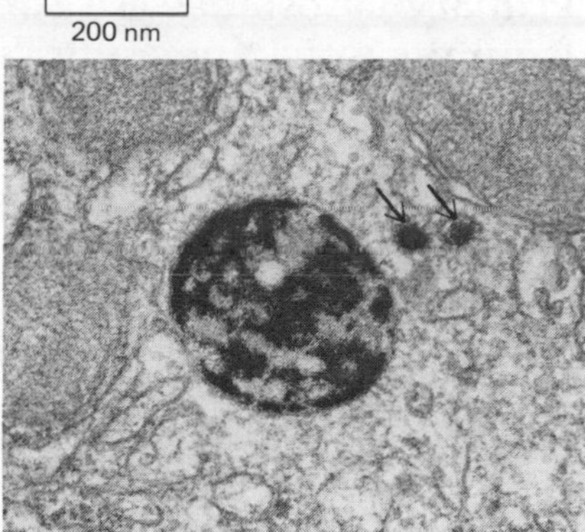

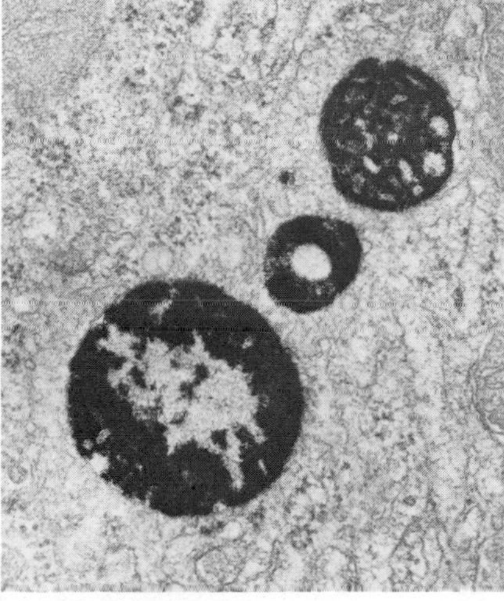

Figure 7–58 Electron micrographs of two sections of a cell stained to reveal the location of acid phosphatase. The larger membrane-bounded organelles containing dense precipitate are secondary lysosomes, whose diverse morphology reflects variations in the amount and nature of the digestive material they contain. Two primary lysosomes are indicated by arrows in the top panel. (Courtesy of Daniel S. Friend.)

Because of the diverse morphology of secondary lysosomes, they are often given special names: (1) large *digestive vacuoles* result from phagocytosis of large particles (like bacteria); (2) *multivesicular bodies* are membranous sacs containing numerous vesicles ~50 nm in diameter; and (3) *autophagic vacuoles* are lysosomal structures containing (and presumably digesting) intracellular membranes or organelles, such as mitochondria or secretory vesicles. And there are numerous others (Figure 7–59).

To avoid confusion, it is best to concentrate on primary lysosomes. How are they formed, and what directs their specific fusion with packaged substrates?

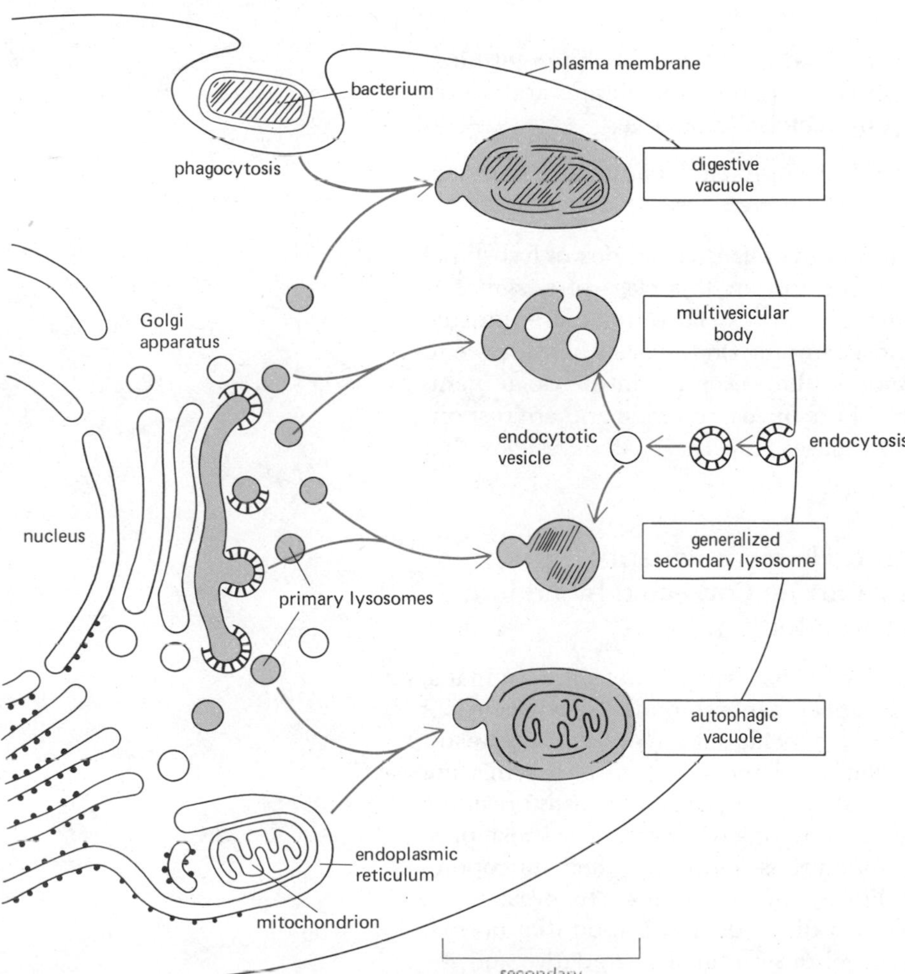

Figure 7–59 Schematic view of some of the ways in which secondary lysosomes are thought to be formed by the fusion of target vesicles with a series of small primary lysosomes containing newly synthesized hydrolytic enzymes. The precise pathways followed, and how they differ, are not known.

Primary Lysosomes are Formed by Budding from the Golgi Apparatus[37]

There is cytological evidence that primary lysosomes are formed by budding from the *trans*-most cisternae of the Golgi apparatus. Numerous Golgi-associated vesicles (coated and smooth), about 50 nm in diameter, contain acid hydrolase activity in this region. When "caught in the act" of budding from Golgi sacs, these vesicles always appear to be coated with clathrin. In the simplest view, primary lysosomes are derived from coated vesicles that have pinched off from the Golgi apparatus; the clathrin coat is thought to dissociate from the vesicle after budding. If this view is correct, the problem of lysosome biogenesis is reduced to the question of how the acid hydrolases become specifically clustered and packaged in coated vesicles in one region of the Golgi. However, even in the *trans*-most region of the Golgi, only a fraction of coated and other small vesicles have demonstrable hydrolase activity. This reflects the fact that coated vesicles are involved in the packaging and transport of macromolecules along many different intracellular routes. Thus, lysosomal enzymes must somehow find the correct population of coated vesicles out of the many other subpopulations that form from the Golgi apparatus.

Three lines of indirect evidence suggest that lysosomal hydrolases are synthesized on the ER and transferred to its lumen before being transported to the Golgi apparatus:

1. Cytochemical stains demonstrate some acid hydrolase activity within both the ER and elements of the Golgi apparatus.
2. Lysosomal enzymes are initially synthesized with the same amino-terminal leader peptides (signal sequences) used by plasma membrane and secretory proteins to enter the rough ER by vectorial discharge.
3. Virtually all lysosomal hydrolases are glycoproteins, containing oligosaccharides that must have been acquired in the ER.

Further studies of the composition of the oligosaccharides of lysosomal enzymes have provided an important clue to how this class of enzymes is packaged within the cell. The lysosomal enzymes so far purified are unusual in containing a phosphorylated mannose residue. On the basis of evidence to be discussed below, it has been concluded that receptors in the Golgi membrane recognize these unique portions of the oligosaccharide and are responsible for the packaging of specific hydrolases into the coated vesicles that become primary lysosomes.

Lysosomal Function in Cultured Cells from Patients Having a Lysosomal Enzyme Deficiency Can Be Corrected by Adding the Missing Enzyme to the Culture Medium[38]

There are numerous inherited disorders of lysosomal metabolism in humans, many of them accompanied by severe clinical manifestations. These disorders are characterized by lysosomes that lack a specific acid hydrolase. The result is the massive intralysosomal accumulation of the substrate of the missing enzyme, either as an intact macromolecule or as a partially digested residue. Depending on which enzyme is missing, any one of a variety of substances may accumulate, including glycosaminoglycans (formerly called mucopolysaccharides), glycoproteins, glycogen, lipids, and glycolipids. The presence of abnormal accumulations in lysosomes is a direct demonstration that normal cellular function entails a substantial level of lysosomal degradative activity. The engorged lysosomes full of abnormally accumulated substances interfere with normal cell functions and thereby cause the clinical manifestations of the specific disease (Figure 7–60). The biochemical study of these genetic diseases has provided an unusually good example of how an investigation of a medical problem at the molecular level can provide important insights into fundamental aspects of cell biology.

Much of our current thinking about lysosome biogenesis is based on an elegant series of experiments whose initial purpose was to elucidate the molecular defects in a type of lysosomal disorder called **mucopolysaccharidosis.** This disorder is characterized by the massive accumulation of glycosaminoglycans (mucopolysaccharides, see p. 702) in lysosomes. One example of mucopolysaccharidosis is *Hurler's disease,* in which characteristic bone deformities develop. Electron microscopy has demonstrated that most cells of patients with this disease contain large digestive vacuoles filled with glycosaminoglycans, normally components of the extracellular matrix. Fibroblasts cultured from the skin of patients with Hurler's disease can be grown in tissue culture and shown to have the same lysosomal deficiency as they have in the body. When normal fibroblasts are grown together with those from a Hurler patient, however, the abnormal cells no longer accumulate glycosaminoglycans. The "corrective factor" for the defective cell can be separately purified from tissue-culture medium in which normal cells are grown and thereby identified as the enzyme α-L-iduronidase, the particular hydrolase missing from the lysosomes of the defective cells.

Since a whole spectrum of different lysosomal enzyme deficiencies can be corrected by different enzymes in this way, it is clear that most lysosomal hydrolases are released in small amounts into the medium by normal cells and that they can be taken up again by the same or other cells. It is not yet known how or why the release of these hydrolases occurs. But it has been shown that the uptake of hydrolases from the medium depends on specific cell-surface proteins and occurs by receptor-mediated endocytosis (p. 309). Presumably, the enzyme-receptor complexes are internalized in coated vesicles, which ultimately fuse with the deficient lysosomes to supply the needed hydrolase. Thus, Hurler fibroblasts can internalize almost half of the α-L-iduronidase added, while only a trace of the bulk fluid of the medium is taken up concurrently.

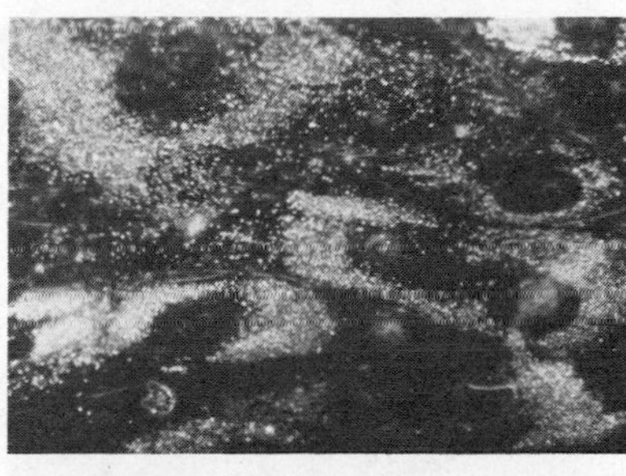

(A)

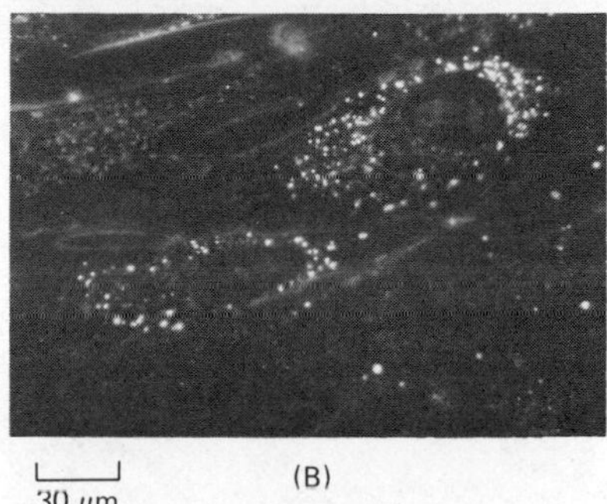

30 μm (B)

Figure 7–60 Appearance of living, cultured fibroblast cells viewed by dark-field microscopy. Dense cytoplasmic inclusions in lysosomes show up as white specks that surround the dark cell nucleus. (A) Cells from a patient with a lysosomal storage disease. (B) Normal cells. (Courtesy of George H. Thomas, from H. A. Taylor et al., *Clin. Genet.* 4:388–397, 1973.)

Only Lysosomal Hydrolases with a Mannose-Phosphate-containing Oligosaccharide Are Internalized[39]

There must be a common feature of all lysosomal hydrolases that is responsible for their high-affinity binding to the specific cell-surface receptors and, therefore, for their efficient internalization. The concept of such a specific "recognition marker" was greatly strengthened by parallel studies of another disorder of lysosomes, *I-cell disease.* Cultured fibroblasts from patients homozygous for this single gene mutation have almost no hydrolases in their lysosomes; however, these enzymes are found in the extracellular fluids in very high concentrations. A crucial finding is that none of the hydrolases released by I cells into the medium is recognized by the receptors on normal cells; consequently, these hydrolases cannot be internalized. For example, α-L-iduronidase prepared from the culture medium of I-cell-disease fibroblasts does not correct cultured Hurler cells; nor is it internalized, even though it is fully active as a hydrolase. By contrast, I-cell-disease fibroblasts bind, endocytose, and retain hydrolases released by normal cells. The simple explanation is that a single gene defect in I-cell disease affects an enzyme that adds the recognition marker to many different hydrolases. As a result, the hydrolases re-

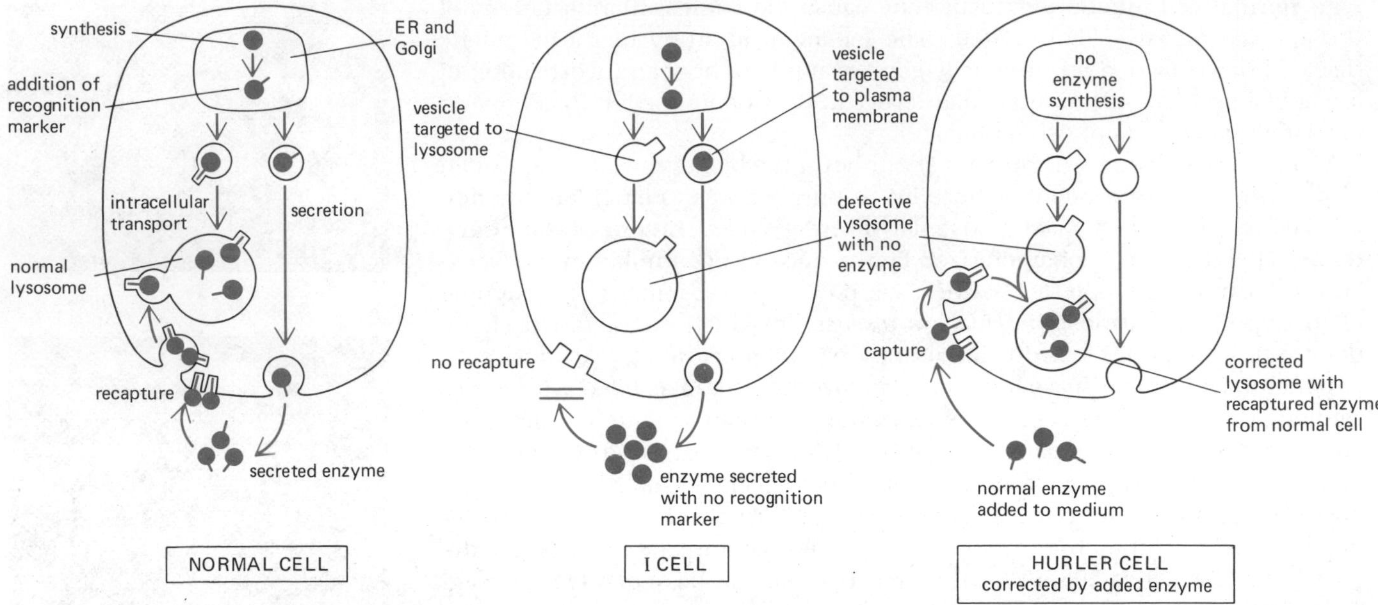

Figure 7–61 Schematic diagram illustrating the secretion and recapture of lysosomal hydrolytic enzymes by mutant and normal cells grown in culture. Specific receptor proteins located on the plasma membranes of all three of these cells allow them to take up hydrolytic enzymes into their lysosomes when these enzymes are added to the culture medium. Hurler cells fail to make α-L-iduronidase, and the defect can be corrected by adding this enzyme (which is released by normal cells) to the medium. All of the hydrolytic enzymes are made by I cells, but they are all secreted because a recognition marker that is required both for their extracellular uptake and for their intracellular transport to lysosomes is missing.

leased from I cells are not recognized for uptake either by I cells or by normal cells (Figure 7–61).

Comparison of the hydrolases released by normal cells and by I cells led to the discovery that the marker recognized by the cell-surface receptor is an uncommon oligosaccharide containing *mannose 6-phosphate.* Treatment of normal hydrolases with an enzyme that removes terminal phosphates converts them to so-called "low-uptake" forms indistinguishable from those released by I cells. Moreover, when mannose 6-phosphate itself is added to the tissue-culture medium, it binds to the cell-surface receptors and thereby acts as a potent competitive inhibitor of the internalization of a variety of different normal lysosomal hydrolases.

Intracellular Hydrolases Are Probably Directed to Primary Lysosomes by the Mannose-Phosphate Marker[40]

When normal fibroblasts are grown for many cell generations in medium containing enough mannose 6-phosphate to block the internalization of hydrolases from the cell exterior, their lysosomes function normally and contain all the usual hydrolases. This rules out the possibility that newly synthesized lysosomal enzymes normally reach new lysosomes by an extracellular "secretion-recapture" route. The absence of hydrolases in I-cell lysosomes would, therefore, imply that the mannose-phosphate-containing oligosaccharide is recognized as a marker by a specific receptor inside the cell. Indeed, the Golgi and ER membranes are now known to contain the same mannose-phosphate receptor identified in the plasma membrane. It is this *intracellular* receptor that must direct the specific packaging of hydrolases into primary lysosomes. When, as in I-cell disease, the correct recognition marker is missing, the hydrolases, instead of being packaged, pass through the Golgi apparatus and are secreted continuously to the cell exterior.

Interestingly, experiments have shown that once hydrolases reach the lysosome they lose their mannose-phosphate marker. This might help to fix these enzymes in the lysosome, preventing them from being retrieved by the receptors that recognize the marker during the membrane recycling process (see Figure 7–56).

The Membrane of the Peroxisome Is Formed by Budding from the Smooth ER[41]

Peroxisomes are organelles present in almost all eucaryotic cells. Normally they cosediment with lysosomes in density gradient centrifugation. Their existence as a separate intracellular organelle became generally recognized only in the early 1960s. By that time a combination of biochemistry and electron microscopy had identified a different organelle bounded by a single membrane, with a diameter of about 0.5 μm, as a concentrated source of at least three oxidative enzymes in liver cells: *D-amino acid oxidase, urate oxidase,* and *catalase.* The later development of a histochemical stain for catalase, an enzyme that constitutes up to 40% of the total peroxisomal protein, revealed that small (0.15 to 0.25 μm) peroxisomes ("microperoxisomes") are ubiquitous in mammalian cells. The larger peroxisomes are confined to relatively few cell types. They sometimes stand out in electron micrographs due to the presence of a "crystalloid" core, a tubular structure containing highly concentrated peroxisomal enzymes (Figure 7–62), but often such a core is absent.

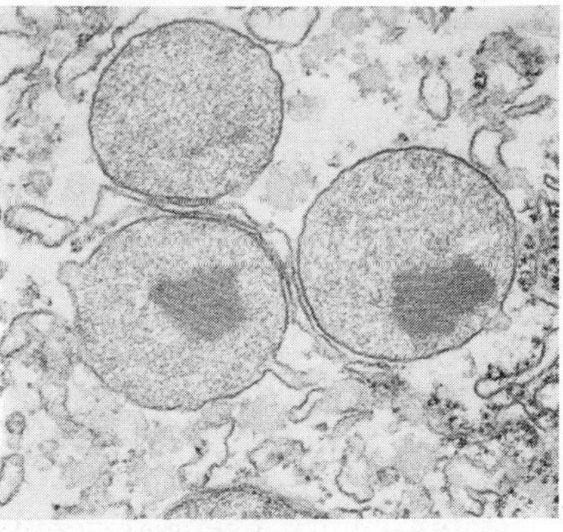

Figure 7–62 Electron micrograph of three peroxisomes in a liver cell. Both a crystalline electron-dense inclusion and a close association with the ER membranes are frequently seen. (Courtesy of Daniel S. Friend.)

Like the mitochondrion, the peroxisome is a major site of oxygen (O_2) utilization. In fact, the peroxisome is thought by some to represent the vestige of an ancient organelle that carried out all of the oxygen metabolism of primitive preeucaryotic cells when oxygen entered the atmosphere. According to this view, the later development of mitochondria rendered the peroxisome largely obsolete since many of the same reactions, which had been carried out in peroxisomes without producing energy, were now coupled to ATP formation by means of oxidative phosphorylation. The oxidative reactions still carried out by peroxisomes would therefore be those that remained useful to the cell despite the presence of mitochondria.

Consistent with this hypothesis is the relatively primitive mechanism by which new peroxisomes appear to be generated. It is thought that certain integral membrane proteins unique to the peroxisome are synthesized in the ER and cluster in the ER membrane to form a preorganelle; this preorganelle then forms a bud from a region of smooth ER. Several major peroxisomal enzymes, including catalase and urate oxidase, are synthesized in the cytosol and transported into the peroxisome as it is forming, by a process that involves a specific recognition of the peroxisomal membrane. Because mature peroxisomes are often seen to be joined by a thin sleevelike projection to the smooth ER (Figure 7–62), it has been suggested that many of them remain attached rather than budding off to become free organelles.

Peroxisomes Use Oxygen to Carry Out Catabolic Reactions[42]

Peroxisomes contain most of a cell's **catalase.** In addition, they contain one or more enzymes that use molecular oxygen to remove hydrogen atoms from specific substrates, carrying out the oxidative reaction

$$RH_2 + O_2 \rightarrow R + H_2O_2$$

Catalase utilizes the H_2O_2 generated by these other enzymes to oxidize a variety of substrates—including phenols, formic acid, formaldehyde, and alcohol—by the "peroxidative" reaction: $H_2O_2 + R'H_2 \rightarrow R' + 2H_2O$. In addition, at low concentrations of $R'H_2$, catalase will convert H_2O_2 to H_2O:

$$2H_2O_2 \rightarrow 2H_2O + O_2$$

This latter reaction is sometimes considered a safety device that prevents a dangerous accumulation of the strong oxidizing agent H_2O_2 in the absence of a sufficient supply of hydrogen donors ($R'H_2$). Important to peroxisome function is the fact that its membrane is unusually permeable, permitting inorganic ions and all low-molecular-weight substrates up to the size of sucrose to pass with ease.

The large peroxisomes in liver and kidney cells are thought to be important in detoxifying various molecules. For example, almost half of the ethanol we drink is oxidized to acetaldehyde in this way. Peroxisomes normally account for at least 10% of the total oxygen uptake of the liver.

In addition to detoxification reactions, most peroxisomes catalyze the breakdown of fatty acids to acetyl CoA using a special H_2O_2-producing enzyme. The acetyl CoA produced can be transported via the cytosol to mitochondria to feed the citric acid cycle, or it can be used for biosynthetic reactions elsewhere. It is estimated that between one-fourth and one-half of fatty acid breakdown occurs in peroxisomes, the remainder occurring in mitochondria (see p. 488).

The Enzyme Content of Peroxisomes Varies with Cell Type[42]

It is difficult to make many general statements about peroxisome metabolism because the composition of this organelle varies so widely. Thus, the peroxisomes in different cells of the same organism often contain very different sets of enzymes. Moreover, in some cells the number and content of peroxisomes vary, depending on the conditions. For example, yeast cells grown on sugar contain very few peroxisomes; those grown in the presence of methanol develop large numbers of special peroxisomes that oxidize methanol; and those grown on fatty acids contain numerous peroxisomes that cause the breakdown of fatty acids to acetyl CoA.

Two very different types of peroxisomes have been extensively studied in plants. One type is present in leaves, where it catalyzes the oxidation of a side product of the reaction that fixes CO_2 in carbohydrate; this process is called *photorespiration* because it uses up O_2 and liberates CO_2 (see p. 516). A very different type of peroxisome is present in germinating seeds, where it serves to convert the fatty acids stored in seed lipids into the sugars needed for the production of the materials of the young plant. Because this is accomplished by a series of reactions known as the **glyoxylate cycle,** these peroxisomes are also called **glyoxysomes.** In the glyoxylate cycle, two molecules of acetyl CoA produced by fatty acid breakdown in the peroxisome are used to make succinic acid, which leaves the peroxisome and is converted into glucose. The glyoxylate cycle does not occur in animal cells, which therefore cannot convert fats into carbohydrates.

A Sorting Mechanism Similar to That Used in the Lysosomal Pathway May Operate for Other Intracellular Compartments[43]

The striking experimental observations on the sorting of acid hydrolases into lysosomes suggest that the intracellular sorting of many proteins may proceed in the same general way. The following type of model might apply, with the use of either a common feature of protein structure or an oligosaccharide as an "address marker" to specify the packaging of proteins in the correct class of coated vesicles (Figure 7–63):

1. As newly synthesized proteins (and other macromolecules) move through the ER, the oligosaccharides on some of them are selectively trimmed. Proteins containing certain types of oligosaccharides or other markers are retained by receptors in the ER membrane. All other proteins are automatically exported to the Golgi apparatus.

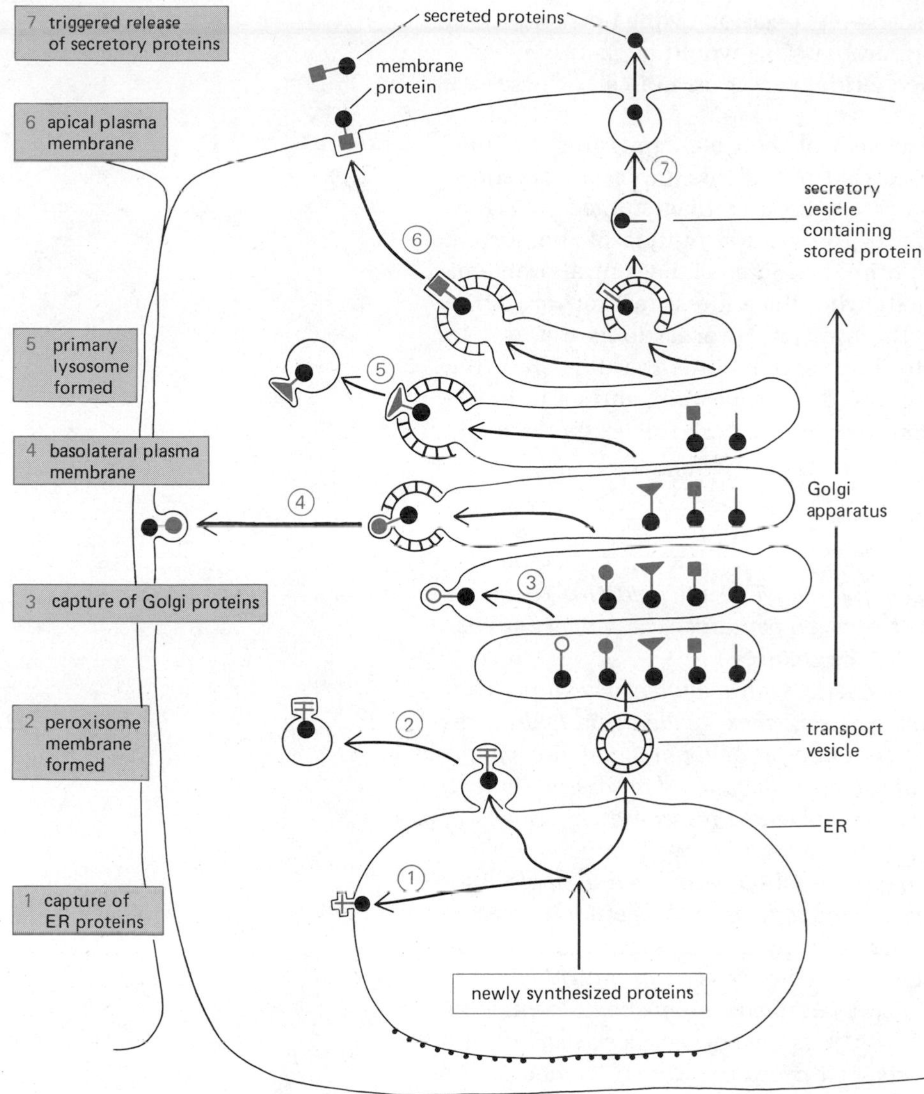

Figure 7–63 An outline of the intracellular sorting of proteins that are made in the ER. Proteins are targeted to different locations within the cell on the basis of the "address markers" they contain. Except for the lysosome, the markers recognized are unknown, and experiments with *tunicamycin*, an inhibitor of *N*-linked glycosylation, demonstrate that *N*-linked oligosaccharides are not required in many cases. Note that the peroxisome membrane is thought to form by direct budding from the ER (see Figure 7–62). Once formed, this membrane provides an import system that pumps selected soluble proteins made in the cytosol into the peroxisome; thus, peroxisomes obtain many of their proteins in much the same way as do mitochondria.

2. Many of the proteins transported to the Golgi apparatus are recognized by an array of oligosaccharide-modifying enzymes that further alter their carbohydrate side chains in specific ways. Some of these modified oligosaccharide side chains bind to specific receptor proteins located in the Golgi membrane. Since related sets of proteins have the same oligosaccharide marker, they are captured by the same type of receptor. Similarly, other types of Golgi receptors recognize and thereby capture proteins marked in other ways.

3. Some types of Golgi receptor molecules retain their captured proteins in the Golgi apparatus. Others segregate in the Golgi membrane and are packaged in specific classes of coated vesicles that transport the receptors and their bound cargo to a specified intracellular compartment.

4. All macromolecules in the Golgi apparatus that fail to bind to a specific receptor are transported to the exterior of the cell.

In this model, any macromolecule not captured by a specific receptor in the ER or Golgi, whether by accident or not, will be transported through the ER and Golgi and eventually to the plasma membrane. Thus the secretion of a soluble protein requires only a "hands-off" decision by the normal sorting apparatus in the ER and the Golgi apparatus. This would explain why lysosomal hydrolases that lack an oligosaccharide marker (as in I-cell disease) are secreted.

More than 80 years after the discovery of the Golgi apparatus, the molecular details of the sorting process carried out in this important organelle are now beginning to emerge. Although the process diagrammed in Figure 7–63 is speculative, this diagram reflects the general outline of the sorting mechanism emerging from current experiments. Some of the central problems that remain are (1) to understand exactly what the address markers are; (2) to discover how they are altered during the transport process to fix the protein at the correct address (for example, to fix an enzyme in either the ER or the Golgi membranes), or to route it to a more distant location; and (3) to determine how this address code is deciphered by the coated vesicles that mediate transport.

Summary

Peroxisomes are organelles specialized for carrying out oxidative reactions using molecular oxygen. They generate hydrogen peroxide, which they both use and destroy by means of the catalase they contain.

Lysosomes are organelles specialized for intracellular digestion. They contain a wide variety of different hydrolytic enzymes—called acid hydrolases because they characteristically operate best at a pH of about 5, the internal pH of the lysosome. The hydrolases are synthesized in the ER and processed through the Golgi apparatus, whence they are transported to lysosomes by means of special coated vesicles.

Serious genetic diseases result from the absence of even a single lysosomal hydrolase, demonstrating the importance of lysosomal digestion processes for normal cell function. One disease—the I-cell disease—results in the absence of all of the hydrolases from lysosomes. Studies of cells from individuals with this disease led to the discovery that a special mannose-phosphate-containing oligosaccharide is normally attached to the lysosomal hydrolases as an address marker, allowing them to be sorted out by a receptor protein located in the Golgi membranes. It appears that different oligosaccharides and other features of the protein surface are used as address markers for the intracellular sorting of proteins.

Organelles with Double Membranes: The Nucleus, Mitochondria, and Chloroplasts

The discussion of intracellular transport mechanisms has thus far ignored the special problem associated with transferring newly synthesized proteins into the nucleus, mitochondria, and (in plants) chloroplasts. All of these intracellular organelles are enclosed by two separate membranes, which are said to constitute the *envelope* of the organelle. The two membranes contain lipid bilayers, but they have very different protein compositions reflecting their different functions. In addition, the presence of two concentric membranes creates two separate spaces inside the organelle: an *intermembrane space* between the outer and inner membranes, and a larger compartment surrounded by the inner membrane. In the discussion that follows, we shall refer to this innermost compartment as the organelle *lumen;* the lumen is commonly called the *matrix space* in mitochondria and the *stroma* in chloroplasts. The intracellular sorting mechanisms must not only deliver the appropriate newly synthesized proteins to each double-membrane organelle, but must

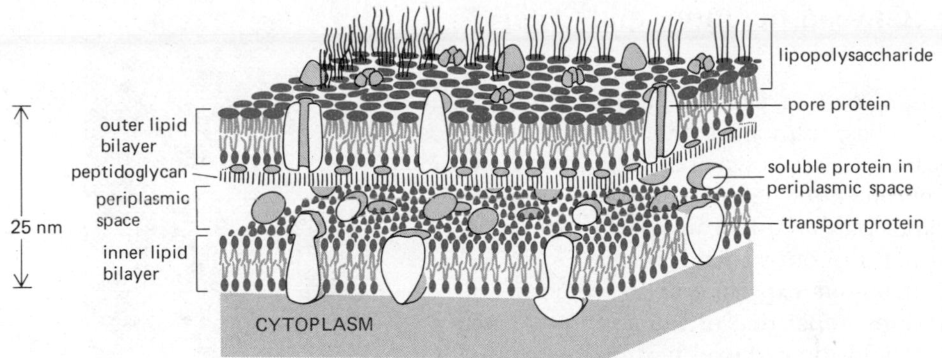

Figure 7–64 Schematic view of a small section of the envelope of an *E. coli* bacterium. Note that two concentric lipid bilayer membranes separate the cell cytoplasm from the outside world. The rigid peptidoglycan layer is highly porous, and it is covalently linked to a small protein that ties it to the outer membrane. The black threads represent oligosaccharide chains attached to the special lipopolysaccharide molecules that form the outer monolayer of the outer membrane. For clarity, only a few of these oligosaccharide chains are shown.

also direct them to one of the four different suborganelle compartments: the outer membrane, the inner membrane, the intermembrane space, or the lumen.

Many of the proteins in these double-membrane organelles are made outside the organelle and then transported across both outer and inner membranes to gain access to the lumen, or across the outer membrane to reach their final location in the inner membrane or in the intermembrane space. Specialized *adhesion sites* between the outer and inner membranes are thought to facilitate the transport of these molecules.

Bacteria with Inner and Outer Cell Membranes May Be Viewed as Models of the Double-Membrane Organelles[44]

Because of the power of genetic analysis in bacteria and the fact that bacteria are free-growing organisms, more is known about the double-membrane system of *E. coli* and *Salmonella* bacteria than about the double membranes of eucaryotic organelles. As illustrated in Figures 7–64 and 7–65, the *E. coli* envelope is a complex structure that contains a proteoglycan layer (which imparts rigidity to the cell) sandwiched between an inner and an outer lipid-bilayer-based membrane. The inner and outer membranes contain distinct sets of membrane proteins and have very different functions. The two membranes are brought together at discrete regions called **adhesion sites,** which can be seen in electron micrographs of all *E. coli* cells, but are most readily revealed as the sites where various bacterial viruses attach to the bacterial envelope (Figure 7–66).

If we ignore the rigid middle layer, the bacterial double membrane provides a good model for analyzing the transport across the two membranes of eucaryotic double-membrane organelles. However, the directions of protein transport are opposite: in bacteria some proteins made in the lumen (the cytoplasm) move outward, whereas in eucaryotic cells many proteins made on the outside (in the cytosol) move in toward the lumen of the organelle (Figure 7–67).

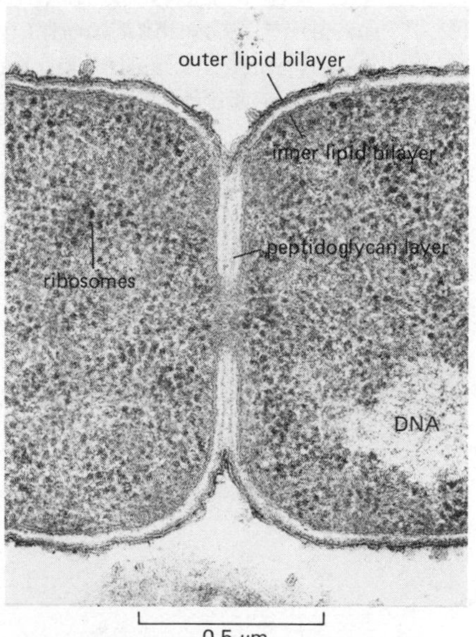

Figure 7–65 Electron micrograph of a dividing *E. coli* bacterium seen in cross-section. The cell is in the process of cleaving, and the arrangement of the double membrane and peptidoglycan layer is clearly visible. (From I. D. J. Burdett and R. G. E. Murray, *J. Bacteriol.* 119:303–324, 1974.)

Membrane Proteins May Move Between the Inner and Outer Bilayers at Adhesion Sites[20, 45]

It is possible to induce the synthesis of certain bacterial outer-membrane proteins and then, using specific antibodies, to monitor their appearance in the outer membrane. Studies of this type have shown that at least some of the proteins first enter the outer membrane at or near the special membrane adhesion sites, suggesting that these sites play an important part in the transfer of macromolecules between the inner and outer bilayers.

We have already discussed the ingenious experiments in which *E. coli* are "engineered" to make hybrid proteins consisting of the amino-terminal portion of an *E. coli* membrane protein and the carboxyl-terminal portion of a cytosolic protein (see p. 344). When hybrid molecules are synthesized that contain the amino terminus of an *E. coli outer*-membrane protein, some are found to be inserted mainly into the inner membrane rather than the outer membrane. This suggests that, in *E. coli*, proteins destined for the outer membrane are first inserted into the inner membrane and that their subsequent transfer to the outer membrane requires a second signal that is defective in some of the hybrid proteins.

These observations raise the possibility that membrane proteins are transported across eucaryotic double membranes in two separate steps. In this view, newly synthesized organelle membrane proteins carry a signal sequence that enables them to bind and then insert into the outer membrane of the appropriate organelle. By diffusing in the plane of the bilayer, all proteins in the outer membrane would eventually encounter adhesion sites. Receptor proteins located at adhesion sites may then selectively transfer those membrane proteins carrying a specific marker to the inner membrane (Figure 7–68). This transfer process would presumably be energy dependent and unidirectional. The only proteins remaining in the outer membrane of eucaryotic organelles would, therefore, be those lacking the appropriate marker for transfer. At present this type of model for selective intermembrane transfer is only a working hypothesis, as the postulated receptors at adhesion sites have not so far been identified.

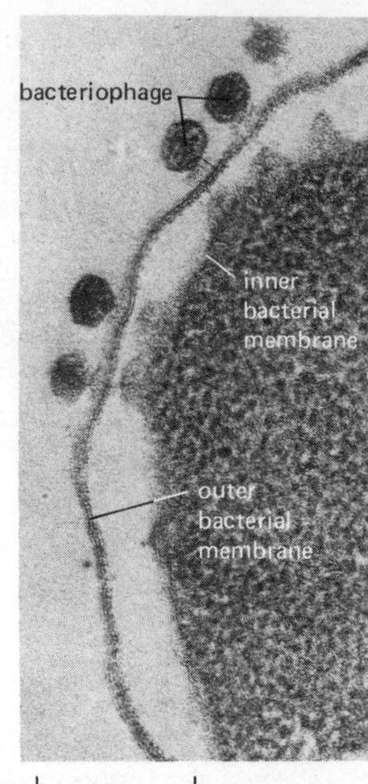

Figure 7–66 Electron micrograph of bacterial viruses that have attached to adhesion sites in the *E. coli* double membrane. The virus appears to inject its DNA into the cell interior at these sites. (From M. Bayer, H. Thurow, and M. H. Bayer, *Virology* 94:95–118, 1979.)

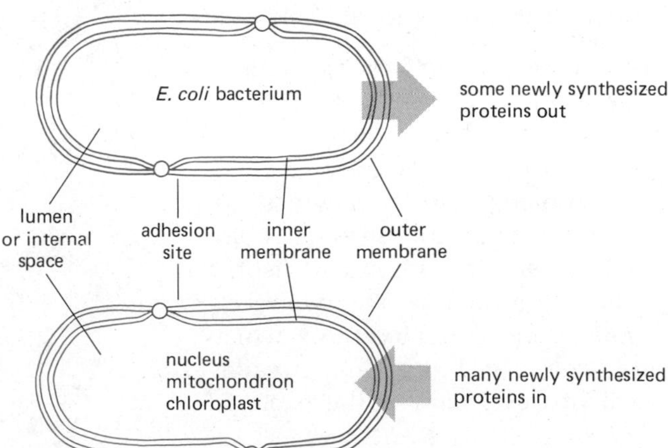

Figure 7–67 An analogy between the *E. coli* bacterium and some eucaryotic organelles. The internal space (here referred to as the lumen) is enclosed by an envelope composed of two concentric membranes, each being a lipid bilayer in which membrane proteins are embedded. The transfer of proteins into (and across) the two membranes may occur by similar mechanisms in bacteria and eucaryotes.

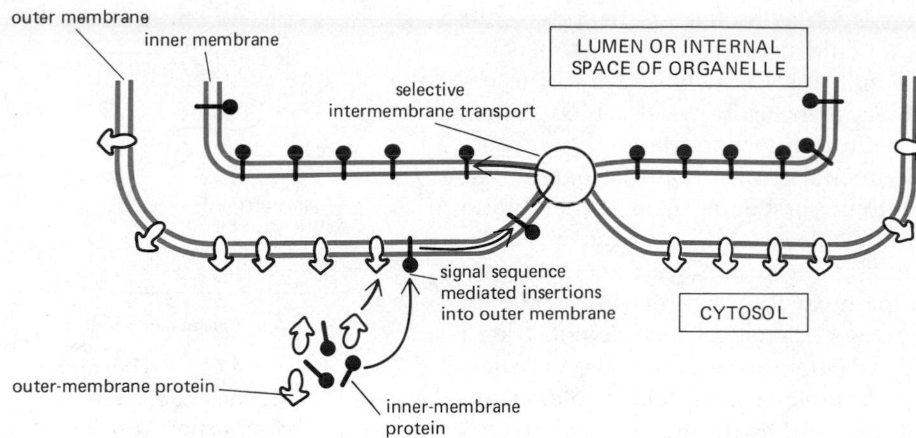

Figure 7–68 A drawing illustrating how adhesion sites in double-membrane organelles might selectively transfer lipids and membrane proteins between the outer and inner bilayers.

Double-membrane organelles must transport soluble proteins, as well as membrane-bound ones, from the cytosol into their lumens. We shall now consider how proteins can pass through two membranes in succession.

Adhesion Sites May Also Play a Part in Transporting Water-soluble Molecules into the Lumen of Double-Membrane Organelles[46]

The transport pathway suggested in Figure 7–68 for membrane proteins could also be followed by lipids and other small hydrophobic molecules that are soluble in the lipid bilayer. However, water-soluble molecules cannot enter organelles in this way. Small water-soluble molecules are normally transported across single bilayer membranes by specific transport proteins (p. 287), and protein pores are present in the outer membrane of double-membrane organelles to allow small soluble molecules access to the intermembrane space. A second set of transport proteins must be present in the inner membrane to permit molecules in the intermembrane space to enter the lumen (Figure 7–69). Because these two sets of transport proteins are different, the molecules in the lumen and intermembrane space are also very different (see Chapter 9).

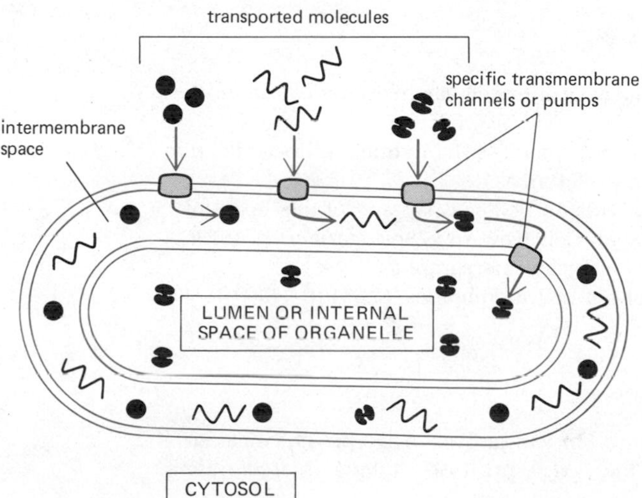

Figure 7–69 Diagram showing one way in which water-soluble molecules are transported into double-membrane organelles. These organelles can accumulate soluble molecules in their intermembrane space and selectively transfer some of them to the lumen. Any molecule for which there is a transport protein can be transferred in this way.

But how are *large* soluble molecules such as proteins transported into these organelles? While in principle this could be achieved by cytoplasmic transport vesicles, this would leave the molecules transported into the organelle lumen still within vesicles (pathway 1 in Figure 7–70), which is presumably why transport vesicles are not sent to mitochondria or chloroplasts from the Golgi apparatus. Moreover, experiments in mitochondria suggest that proteins destined for the lumen (matrix) do not first enter the intermembrane space (see p. 538). In principle, large molecules could enter the organelle lumen from the cytosol via a vesicle budding and fusion cycle occurring entirely within the intermembrane space (pathway 2 in Figure 7–70); however, this type of vesicle-mediated transport process has not yet been demonstrated to occur between the double membranes of either bacteria or cell organelles.

Perhaps the most likely mechanism for transporting soluble proteins into the lumen of double-membrane organelles would be through special protein channels extending across both membranes at adhesion sites. Unique structures joining the inner and outer membranes of the cell nucleus, called *nuclear pore complexes,* are known to transport proteins and nucleic acids into and out of the cell nucleus (see p. 432). More relevant to the situation in mitochondria and chloroplasts, in *E. coli* many bacterial viruses bind specifically to adhesion sites (Figure 7–66) and inject their DNA directly into the cell through these sites. While adhesion sites have been seen in mitochondria and chloroplasts, there is no direct evidence so far that they serve as selective pores, and they have been very difficult to isolate for biochemical studies. Perhaps, as is thought to be the case in *E. coli,* the adhesion sites in mitochondria and chloroplasts are labile, continually breaking up and re-forming at different places.

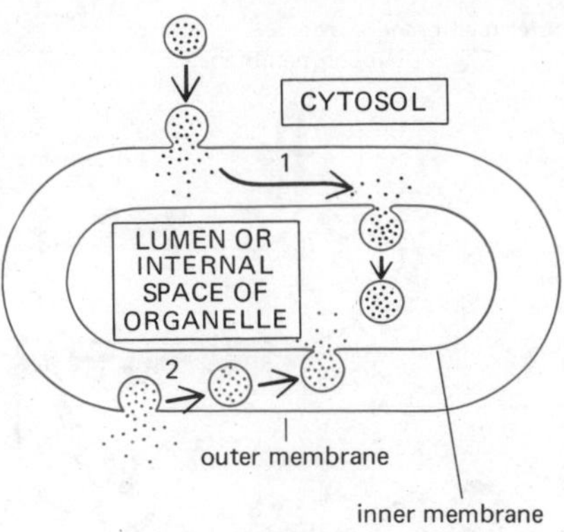

Figure 7–70 Schematic illustration of the fact that transport vesicles from the cytosol cannot be used to transfer water-soluble molecules into the lumen of double-membrane organelles. Although molecules can be transferred to the intermembrane space via a transport vesicle and could then enter the lumen inside of another vesicle (pathway 1), they would not, as a result, be released into the lumen. However, in principle, water-soluble molecules could be directly transferred from the cytosol to the lumen (and vice versa) by cycles of vesicle budding and fusion that occur entirely within the intermembrane space (pathway 2). However, this type of vesicle transport is not known to occur.

Summary

Unlike the ER, Golgi apparatus, and lysosome, the nucleus, mitochondria, and chloroplasts are bounded by two separate bilayer membranes. Although molecules transported into the lumen could pass through each membrane independently, special adhesion sites joining the two bilayers seem likely to be involved in such transport. For mitochondria and chloroplasts, this transport process would seem to be more specific than that operating in the nuclear envelope, where large nuclear pore complexes permit the rapid transfer of macromolecules into and out of the nucleus.

References

General

DePierre, J.W.; Ernster, L. Enzyme topology of intracellular membranes. *Annu. Rev. Biochem.* 46:201–262, 1977.

Fawcett, D. The Cell, 2nd ed. Philadelphia: Saunders, 1981. (An outstanding collection of micrographs, with limited text and references.)

Karp, G. Cell Biology. New York: McGraw-Hill, 1979. (See Chapter 7 of this textbook.)

Krstić, R.V. Ultrastructure of the Mammalian Cell. New York: Springer-Verlag, 1979. (A superb atlas of drawings made from electron micrographs.)

Lodish, H.F.; Rothman, J.E. The assembly of cell membranes. *Sci. Am.* 240(1):48–63, 1979.

Cited

1. Palade, G.E.; Farquhar, M.G. Cell biology. In Pathophysiology: The Biological Principles of Disease (L.H. Smith, S.O. Thier, eds.), pp. 1–56. Philadelphia: Saunders, 1981.

2. Weibel, E.R.; Stäubli, W.; Gnägi, H.R.; Hess, F.A. Correlated morphometric and biochemical studies on the liver cell. *J. Cell Biol.* 42:68–91, 1969.

Bolender, R.P. Stereological analysis of the guinea pig pancreas. *J. Cell Biol.* 61:269–287, 1974.

3. Helenius, A.; Kartenbeck, J.; Simons, K.; Fries, E. On the entry of Semliki forest virus into BHK-21 cells. *J. Cell Biol.* 84:404–420, 1980.

Simons, K.; Garoff, H.; Helenius, A. How an animal virus gets into and out of its host cell. *Sci. Am.* 246(2):58–66, 1982.

Miller, D.; Lenard, J. Antihistiminics, local anesthetics, and other amines as antiviral agents. *Proc. Natl. Acad. Sci. USA* 78:3605–3609, 1981.

4. Lenard, J.; Compans, R.W. The membrane structure of lipid-containing viruses. *Biochim. Biophys. Acta* 344:51–94, 1974. (A review.)

Rodriguez Boulan, E.; Sabatini, D.D. Asymmetric budding of viruses in epithelial monolayers: a model system for study of epithelial polarity. *Proc. Natl. Acad. Sci. USA* 75:5071–5075, 1978.

Bergmann, J.E.; Tokuyasu, K.T.; Singer, S.J. Passage of an integral membrane protein, the vesicular stomatitis virus glycoprotein, through the Golgi apparatus in route to the plasma membrane. *Proc. Natl. Acad. Sci. USA* 78:1746–1750, 1981.

5. Stryer, L. Biochemistry, 2nd ed. San Francisco: Freeman, 1981. (Chapters 27 and 29.)

Fulton, A.B. How crowded is the cytoplasm? *Cell* 30:345–347, 1982.

6. Lake, J.A. The ribosome. *Sci. Am.* 245(2):84–97, 1981.

Wool, I.G. The structure and function of eukaryotic ribosomes. *Annu. Rev. Biochem.* 48:719–754, 1979.

7. Rich, A. Polyribosomes. *Sci. Am.* 209(6):44–53, 1963.

8. Dean, D.; Nomura, M. Feedback regulation of ribosomal protein gene expression in *E. coli*. *Proc. Natl. Acad. Sci. USA* 77:3590–3594, 1980.

9. Kozak, M. How do eucaryotic ribosomes select initiation regions in mRNA? *Cell* 15:1109–1123, 1978.

Sherman, F.; Stewart, J.W.; Schweingruber, A.M. Mutants of yeast initiating translation of iso-1-cytochrome c within a region spanning 37 nucleotides. *Cell* 20:215–222, 1980.

Lewin, B. Gene Expression, Vol. 2, Eucaryotic Chromosomes, 2nd ed. New York: Wiley, 1980. (Chapter 23 describes the structure of eucaryotic messenger RNA.)

10. Steiner, D.F.; Quinn, P.S.; Chan, S.J.; Marsh, J.; Tager, H.S. Processing mechanisms in the biosynthesis of proteins. *Ann. N.Y. Acad. Sci.* 343:1–16, 1980.

Stark, G.R. Multifunctional proteins: one gene—more than one enzyme. *Trends Biochem. Sci.* 2:64–66, 1977.

11. Wold, F. *In vivo* chemical modification of proteins (post-translational modification). *Annu. Rev. Biochem.* 50:783–814, 1981.

Schlesinger, M.J. Proteolipids. *Annu. Rev. Biochem.* 50:193–206, 1981.

Neurath, H.; Walsh, K.A. Role of proteolytic enzymes in biological regulation. *Proc. Natl. Acad. Sci. USA* 73:3825–3832, 1976. (A review.)

Chock, P.B.; Rhee, S.G.; Stadtman, E.R. Interconvertible enzyme cascades in cellular regulation. *Annu. Rev. Biochem.* 49:813–843, 1980.

12. Hershko, A.; Ciechanover, A. Mechanisms of intracellular protein breakdown. *Annu. Rev. Biochem.* 51:335–364, 1982.

13. DePierre, J.W.; Dallner, G. Structural aspects of the membrane of the endoplasmic reticulum. *Biochim. Biophys. Acta* 415:411–472, 1975.

Fawcett, D. The Cell, 2nd ed. Philadelphia: Saunders, 1981. (See pages 303–352.)

14. Jones, A.L.; Fawcett, D.W. Hypertrophy of the agranular endoplasmic reticulum in hamster liver induced by phenobarbital. *J. Histochem. Cytochem.* 14:215–232, 1966. (Includes a review of the functions of the smooth ER in liver.)

Mori, H.; Christensen, A.K. Morphometric analysis of Leydig cells in the normal rat testis. *J. Cell Biol.* 84:340–354, 1980. (The smooth ER contributes 60% of the total cell membrane of this steroid-hormone-producing cell.)

15. de Duve, C. Tissue fractionation past and present. *J. Cell Biol.* 50:20d–55d, 1971.

Dallner, G. Isolation of rough and smooth microsomes—general. *Methods Enzymol.* 31:191–201, 1974.

16. Adelman, M.; Sabatini, D.; Blobel, G. Ribosome-membrane interaction. *J. Cell Biol.* 56:206–229, 1973.

Kreibich, G.; Ulrich, B.; Sabatini, D. Proteins of rough microsomal membranes related to ribosome binding. *J. Cell Biol.* 77:464–487, 1978.

17. Sabatini, D.; Blobel, G. Controlled proteolysis of nascent polypeptides in rat liver cell fractions: location of the polypeptides in rough microsomes. *J. Cell Biol.* 45:146–157, 1970.

Sabatini, D.; Kreibich, G.; Morimoto, T.; Adesnick, M. Mechanisms for the incorporation of proteins in membranes and organelles. *J. Cell Biol.* 92:1–22, 1982. (A review.)

18. Redman, C.; Sabatini, D. Vectorial discharge of peptides released by puromycin from attached ribosomes. *Proc. Natl. Acad. Sci. USA* 56:608–615, 1966.

Smith, W.P.; Tai, P.C.; Davis, B.D. Interaction of secreted nascent chains with surrounding membrane in *Bacillus subtilis. Proc. Natl. Acad. Sci. USA* 75:5922–5925, 1978.

19. Milstein, C.; Brownlee, G.; Harrison, T.; Mathews, M.B. A possible precursor of immunoglobulin light chains. *Nature New Biol.* 239:117–120, 1972.

Devillers-Thiery, A.; Kindt, T.; Scheele, G.; Blobel, G. Homology in amino-terminal sequence of precursors to pancreatic secretory proteins. *Proc. Natl. Acad. Sci. USA* 72:5016–5020, 1975.

Blobel, G.; Dobberstein, B. Transfer of proteins across membranes, II. Reconstitution of functional rough microsomes from heterologous components. *J. Cell Biol.* 67:852–862, 1975.

Walter, P.; Blobel, G. Translocation of proteins across the endoplasmic reticulum, III. Signal recognition protein (SRP) causes signal sequence-dependent and site-specific arrest of chain elongation that is released by microsomal membranes. *J. Cell Biol.* 91:557–561, 1981.

20. Emr, S.D.; Hall, M.N.; Silhavy, T.J. A mechanism of protein localization: the signal hypothesis and bacteria. *J. Cell Biol.* 86:701–711, 1980.

Davis, B.D.; Tai, P.C. The mechanism of protein secretion across membranes. *Nature* 283:433–438, 1980.

21. Wickner, W. Assembly of proteins into membranes. *Science* 210:861–868, 1980.

Kreil, G. Transfer of proteins across membranes. *Annu. Rev. Biochem.* 50:317–348, 1981.

22. Kornfeld, R.; Kornfeld, S. Comparative aspects of glycoprotein structure. *Annu. Rev. Biochem.* 45:217–238, 1976.

Gibson, R., Kornfeld, S.; Schlesinger, S. A role for oligosaccharides in glycoprotein biosynthesis. *Trends Biochem. Sci.* 5:290–294, 1980.

Wagh, P.V.; Bahl, O.P. Sugar residues on proteins. *CRC Crit. Rev. Biochem.* 10:307–377, 1981.

23. Struck, D.; Lennarz, W. The function of saccharide lipids in synthesis of glycoproteins. In The Biochemistry of Glycoproteins and Proteoglycans (W.J. Lennarz, ed.), Chapter 2. New York: Plenum, 1980.

24. Rothman, J.E.; Kennedy, E.P. Rapid transmembrane movement of newly synthesized phospholipids during membrane assembly. *Proc. Natl. Acad. Sci. USA* 74:1821–1825, 1977.

Coleman, R.; Bell, R.M. Evidence that biosynthesis of phosphatidylethanolamine, phosphatidylcholine, and triacylglycerol occurs on the cytoplasmic side of microsomal vesicles. *J. Cell Biol.* 76:245–253, 1978.

25. Wirtz, K.W.A. Transfer of phospholipids between membranes. *Biochim. Biophys. Acta* 344:95–117, 1974. (A review.)

26. Whaley, W.G. The Golgi Apparatus. New York: Springer-Verlag, 1975.

Farquhar, M.; Palade, G. The Golgi apparatus (complex)—(1954-1981) from artifact to center stage. *J. Cell Biol.* 91:77s–103s, 1981.

27. Novikoff, A. The endoplasmic reticulum: a cytochemist's view (a review). *Proc. Natl. Acad. Sci. USA* 73:2781–2787, 1976.

28. Schachter, H.; Roseman, S. Mammalian glycosyltransferases: their role in the synthesis and function of complex carbohydrates and glycolipids. In The Biochemistry of Glycoproteins and Proteoglycans (W. J. Lennarz, ed.), Chapter 3. New York: Plenum, 1980.

29. Hubbard, S.C.; Ivatt, R.J. Synthesis and processing of asparagine-linked oligosaccharides. *Annu. Rev. Biochem.* 50:555–583, 1981.

Tabas, I.; Kornfeld, S. The synthesis of complex-type oligosaccharides, III. *J. Biol. Chem* 253:7779–7786, 1978.

30. Leblond, C.; Bennett, G. In International Cell Biology (B. Brinkley, K. Porter, eds.), pp. 326–336. New York: Rockefeller University Press, 1977.
31. Palade, G. Intracellular aspects of the process of protein synthesis. *Science* 189:347–358, 1975.
32. Heuser, J.E.; et al. Synaptic vesicle exocytosis captured by quick freezing and correlated with quantal transmitter release. *J. Cell Biol.* 81:275–300, 1979.

    Heuser, J.; Reese, T. Structural changes after transmitter release at the frog neuromuscular junction. *J. Cell Biol.* 88:564–580, 1981.
33. Pearse, B.M.F.; Bretscher, M.S. Membrane recycling by coated vesicles. *Annu. Rev. Biochem.* 50:85–101, 1981.

    Anderson, R.G.W.; Brown, M.S.; Goldstein, J.L. Role of the coated endocytic vesicle in the uptake of receptor-bound low density lipoprotein in human fibroblasts. *Cell* 10:351–364, 1977.

    Rothman, J.; Fine, R. Coated vesicles transport newly synthesized membrane glycoproteins from endoplasmic reticulum to plasma membrane in two successive stages. *Proc. Natl. Acad. Sci. USA* 77:780–784, 1980.
34. Rothman, J. The Golgi apparatus: two organelles in tandem. *Science* 213:1212–1219, 1981.
35. de Duve, C. Exploring cells with a centrifuge. *Science* 189:186–194, 1975.

    Bainton, D. The discovery of lysosomes. *J. Cell Biol.* 91:66s–76s, 1981.
36. Holtzman, E. Lysosomes: A Survey. New York: Springer-Verlag, 1976.
37. Hasilik, A. Biosynthesis of lysosomal enzymes. *Trends Biochem. Sci.* 5:237–240, 1980.

    Erickson, A.H.; Conner, G.E.; Blobel, G. Biosynthesis of a lysosomal enzyme. *J. Biol. Chem.* 256:11224–11231, 1981.

    Waheed, A.; Pohlmann, R.; Hasilik, A.; von Figura, K. Subcellular location of two enzymes involved in the synthesis of phosphorylated recognition markers in lysosomal enzymes. *J. Biol. Chem.* 256:4150–4152, 1981.
38. Neufeld, E.F.; Lim, T.W.; Shapiro, L.J. Inherited disorders of lysosomal metabolism. *Annu. Rev. Biochem.* 44:357–376, 1975.
39. Kaplan, A.; Achord, D.T.; Sly, W.S. Phosphohexosyl components of a lysosomal enzyme are recognized by pinocytosis receptors on human fibroblasts. *Proc. Natl. Acad. Sci. USA* 74:2026–2030, 1977.

    Neufeld, E.F.; Ashwell, G. Carbohydrate recognition systems for receptor-mediated pinocytosis. In The Biochemistry of Glycoproteins and Proteoglycans (W.J. Lennarz, ed.), Chapter 6. New York: Plenum, 1980.

    Reitman, M.; Varki, A.; Kornfeld, S. Fibroblasts from patients with I-cell disease and pseudo-Hurler polydystrophy are deficient in uridine 5′-diphosphate-*N*-acetylglucosamine: glycoprotein *N*-acetylglucosaminylphosphotransferase activity. *J. Clin. Invest.* 67:1574–1579, 1981.
40. von Figura, K.; Weber, E. An alternative hypothesis of cellular transport of lysosomal enzymes in fibroblasts. *Biochem. J.* 176:943–950, 1978.

    Fischer, H.D.; Gonzalez-Noriega, A.; Sly, W.S.; Morré, D.J. Phosphomannosyl-enzyme receptors in rat liver. *J. Biol. Chem.* 255:9608–9615, 1980.

    Varki, A.; Kornfeld, S. Structural studies of phosphorylated high mannose-type oligosaccharides. *J. Biol. Chem.* 255:10847–10858, 1980.

    Reitman, M.L.; Kornfeld, S. Lysosomal enzyme targeting; *N*-acetylglucosaminylphosphotransferase selectively phosphorylates native lysosomal enzymes. *J. Biol. Chem.* 256:11977–11980, 1981.
41. Novikoff, A.B.; Novikoff, P.M. Microperoxisomes. *J. Histochem. Cytochem.* 21:963–966, 1973

    de Duve, C. Biochemical studies on the occurrence, biogenesis and life history of mammalian peroxisomes. *J. Histochem. Cytochem.* 21:941–948, 1973.

    de Duve, C. Evolution of the peroxisome. *Ann. N.Y. Acad. Sci.* 168:369–381, 1969.

    Goldman, B.M.; Blobel, G. Biogenesis of peroxisomes: intracellular site of synthesis of catalase and uricase. *Proc. Natl. Acad. Sci. USA* 75:5066–5070, 1978.
42. de Duve, C.; Baudhuin, P. Peroxisomes (microbodies and related particles). *Physiol. Rev.* 46:323–357, 1966.

Tolbert, N.E.; Essner, E. Microbodies: peroxisomes and glyoxysomes. *J. Cell Biol.* 91:271s–283s, 1981.

43. Elbein, A.D. The tunicamycins—useful tools for studies on glycoproteins. *Trends Biochem. Sci.* 6:219–221, 1981.

Green, R.F.; Meiss, H.K.; Rodriguez Boulan, E. Glycosylation does not determine segregation of viral envelope proteins in the plasma membrane of epithelial cells. *J. Cell Biol.* 89:230–239, 1981.

44. Inouye, M., ed. Bacterial Outer Membranes: Biogenesis and Functions. New York: Wiley, 1980.

Osborn, M.J.; Wu, H.C.P. Proteins of the outer membrane of gram-negative bacteria. *Annu. Rev. Microbiol.* 34:369–422, 1980.

45. Smit, J.; Nikaido, H. Outer membrane of gram-negative bacteria; electron microscopic studies on porin insertion sites and growth of cell surface of *Salmonella typhimurium*. *J. Bacteriol.* 135:687–702, 1978.

46. Bayer, M.E.; Thurow, H.; Bayer, M.H. Penetration of the polysaccharide capsule of *E. coli* by bacteriophage K29. *Virology* 94:95–118, 1979.

Neupert, W.; Schatz, G. How proteins are transported into mitochondria. *Trends Biochem. Sci.* 6:1–4, 1981.

The Cell Nucleus

8

The cell's complement of DNA is sequestered in the nucleus, which is delimited by a double membrane called the *nuclear envelope*. This envelope is the defining feature of eucaryotic cells, distinguishing them from procaryotes. The nuclear envelope isolates the central genetic processes of DNA replication and RNA synthesis from the cytoplasmic ribosomes, where the genetic message is translated into protein. This separation of RNA and protein synthesis appears to be one of its most important functions. Otherwise, the evolution of nuclear membranes is difficult to account for. There are other major differences between procaryotes and eucaryotes, but they appear to be secondary. For example, the procaryotic genome is usually much smaller, and eucaryotic DNA is tightly complexed with specialized proteins—*histones*—that "package" it and help to regulate its activity. But neither a large genome nor the presence of histones is invariably associated with the possession of a nucleus. The largest DNA complements of bacteria are as large as the smallest DNA complements of eucaryotes; and one class of eucaryotes, the unicellular dinoflagellates, has a nuclear envelope but no histones.

It seems then that the separation of nuclear and cytoplasmic compartments has advantages even for small cells with relatively small amounts of DNA. Why should this be, if the essential tenets of the *central dogma* of molecular genetics—DNA→RNA→protein—apply to all living cells? The best answer we have at present is that in eucaryotic cells it is important to separate RNA synthesis (*transcription*) both temporally and spatially from protein synthesis (*translation*). In procaryotic cells, transcription and protein synthesis occur concurrently—ribosomes translate the 5′ end of an RNA molecule while it is still being synthesized at the 3′ end. Consequently, there is little opportunity for altering the RNA transcripts before they are translated into protein. In eucaryotes, by contrast, RNA transcripts can be extensively altered in the nucleus before being translated into protein. This alteration, called *RNA processing*, represents an important intermediate step in the transfer of genetic information in eucaryotes, which is followed by RNA transport out of the nucleus to the cytoplasmic ribosomes (Figure 8–1). We shall see that at

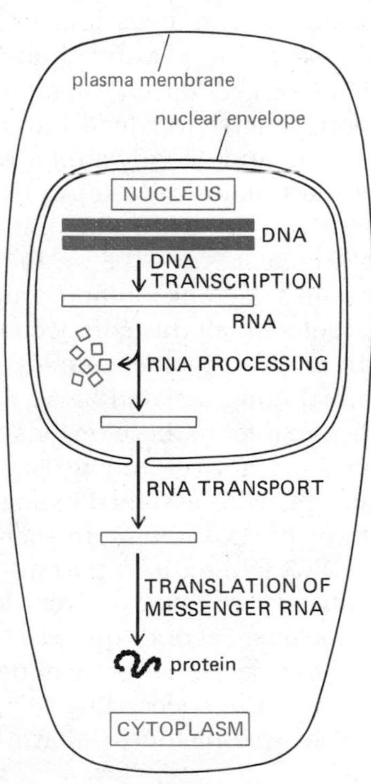

Figure 8–1 A schematic view of the expanded process of protein synthesis ("DNA → RNA → protein") in eucaryotes. Because of the nuclear envelope, RNA processing and RNA transport steps are interposed between transcription and the translation of messenger RNA into protein.

least one type of RNA processing—called *RNA splicing*—has important advantages for cell function. Moreover, RNA splicing has allowed eucaryotic cells to evolve genes with interrupted structures, which has advantages for the evolution of new proteins. It is therefore not difficult to provide reasons for the fact that eucaryotic cells contain a nucleus.

A great deal of information is presented in this chapter. In separate sections, we describe how the DNA of eucaryotes is packaged as chromosomes and outline the basic mechanisms involved in RNA synthesis and RNA processing. The third section describes the structure of the nuclear envelope and is followed by an extensive discussion of the mechanisms that control eucaryotic gene expression. Finally, the chapter ends with a section on the nature of the evolutionary processes that occur in eucaryotic genomes.

The Organization of DNA into Chromosomes[1]

For the first 40 years of this century, biologists tended to dismiss the possibility that DNA could carry the genetic information in chromosomes, partly because nucleic acids were erroneously believed to contain only a simple repeating tetranucleotide sequence (such as *AGCTAGCTAGCT*. . .) We now know, however, that DNA is an enormously long, unbranched, linear polymer that can contain many millions of nucleotides arranged in an irregular but nonrandom sequence, the genetic information being contained in the exact order of the nucleotides. The linear four-letter genetic code written in words of three nucleotides (p. 107) is a straightforward solution to the problem of storing a large amount of genetic information in a small amount of space. In DNA, every million "letters" (nucleotides) take up a linear distance of only 3.4×10^5 nm (0.034 cm) and occupy a total volume of about 10^6 nm^3 (10^{-15} cm^3). Thus, for a typical human cell, 20 μm in diameter—with a (haploid) DNA sequence of 3×10^9 nucleotides—all of the genetic information could in theory be packed into a cube 1.5 μm (1.5×10^{-4} cm) on each side. By comparison, 3×10^9 letters in this book would occupy about a million pages.

Yet with all this compactness, it now seems that only a small proportion of the DNA in higher animals and plants actually codes for proteins or for essential nontranslated RNAs, such as transfer and ribosomal RNAs. It is customary to refer to these regions as *coding DNA*. The functions of the noncoding regions are not yet fully understood, but some of them undoubtedly contain sequences with essential biological functions, including the determination of patterns of DNA folding in each chromosome.

DNA folding is important in eucaryotic cells for two reasons. First, it is essential for packing the very long DNA molecules in an orderly way in the cell nucleus. Second, the exact manner in which a region of the genome is folded in a particular cell can determine the activity of the genes in that region (p. 458). In the succeeding sections, we introduce the specialized eucaryotic proteins—the histones—that play a central part in the folding of eucaryotic DNAs.

Histones Are Among the Most Highly Conserved Proteins Known[2]

The DNA of all organisms is closely associated with a wide variety of different DNA-binding proteins. These proteins act as enzymes to mediate such important functions as RNA synthesis, and act as structural proteins to organize the DNA in the cell nucleus. It is traditional to divide the DNA-binding proteins in eucaryotes into two general classes: the **histones** and the **nonhistone chromosomal proteins.** The latter designation serves as a catchall for hundreds or thousands of different proteins with many different functions. In contrast, the histones constitute a well-defined class of structural proteins. Compared to other chromosomal proteins, histones occur in enormous quantity (about 60 million copies of each type per cell). In fact, their total mass is about equal to that of the cell's DNA. For historical reasons, DNA that is associated with histones (which is thought to be all of the nuclear DNA in eucaryotic cells) is known as **chromatin.**

Histones are relatively small proteins with a very high proportion of positively charged amino acids (lysine and arginine); the positive charge helps the histones to bind tightly to DNA, regardless of its nucleotide sequence. Histones probably remain bound to the DNA at all times, and so are likely to play an important part in many of the reactions that involve the genome.

The five types of histones in eucaryotic cells fall into two main groups. The first group comprises the *nucleosomal histones,* which are small proteins responsible for folding the DNA into *nucleosomes* (as will be discussed in detail in the next section). They are designated **histones H2A, H2B, H3, and H4.** These four histones are among the most highly conserved of all known proteins: for example, there are only two differences in the amino acid sequences of histone H4 in peas and cows (Figure 8–2). Such conservation sug-

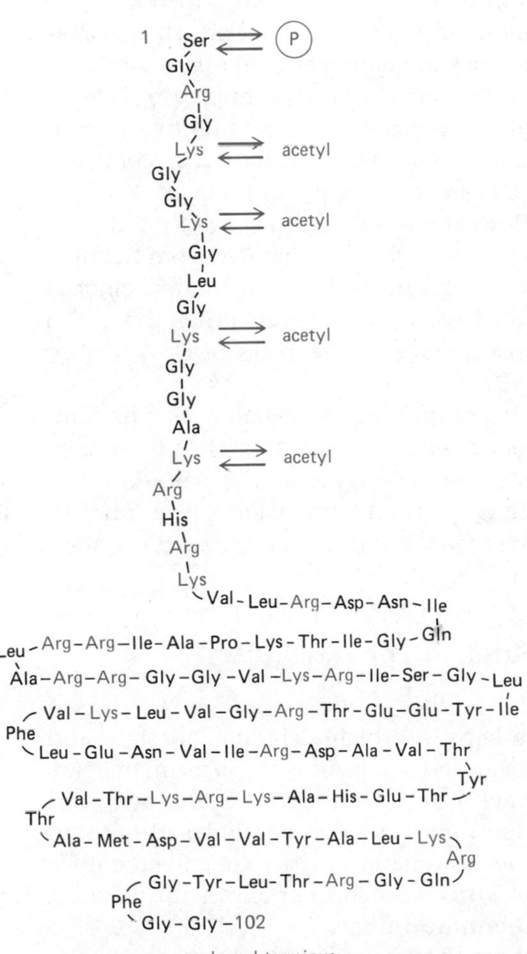

Figure 8–2 The amino acid sequence of histone H4, one of the four nucleosomal histones. The positively charged amino acids are colored for emphasis. Also indicated is the fact that, as for the other three nucleosomal histones, an elongated amino-terminal "tail" is reversibly modified within the cell by the acetylation of selected lysines and the phosphorylation of serine (see Figures 8–9 and 8–10). The bovine sequence is shown; in peas the sequence is the same except that one valine is changed to an isoleucine, and one lysine is changed to an arginine.

gests that these histones have crucial functions involving nearly all of their amino acids (see p. 215).

The second group consists of the *H1 histones,* of which there are several different but closely related varieties in each cell. Except for a central core region, the amino acid sequences of the H1 histones have been much less conserved during evolution than those of the four nucleosomal histones.

The Association of Histones with DNA Leads to the Formation of Nucleosomes, the Unit Particles of Chromatin[3]

If stretched out, the DNA double helix in each human chromosome would be on average about 5 cm long. The histones are responsible for packing this long DNA molecule into a nucleus only a few μm in diameter. There are several orders of packing, and a major advance in our present understanding of chromatin structure came in 1974 with the discovery of the fundamental packing unit known as the **nucleosome.** The nucleosome gives chromatin its "beads-on-a-string" appearance in electron micrographs taken after treatments that unfold the higher-order packing (Figure 8–3B).

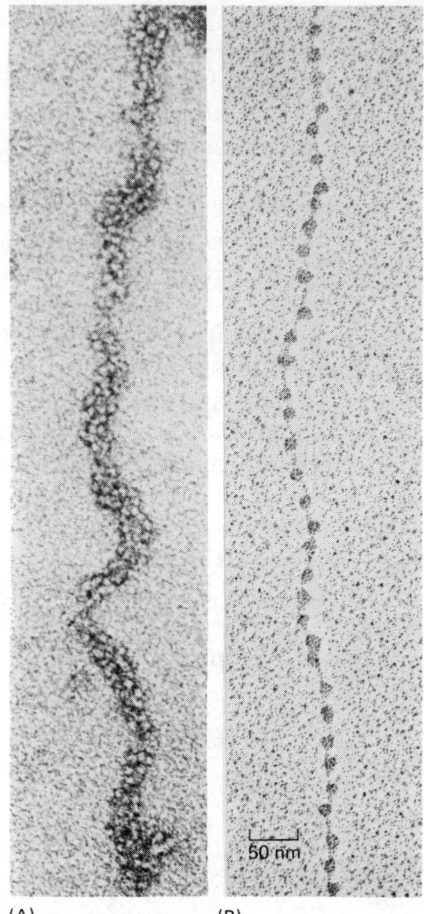

Figure 8–3 Electron micrographs of chromatin strands before and after treatments that decondense the native structure to produce the beads-on-a-string form. (A) shows the native structure, known as the 30-nm fiber. In (B), the decondensed beads-on-a-string form of chromatin is shown at the same magnification. For schematic drawings of both of these chromatin forms, see Figure 8–24. These electron micrographs were taken by modifications of the procedure outlined in Figure 8–35. (A, courtesy of Barbara Hamkalo; B, courtesy of Victoria Foe.)

The nucleosome "bead" can be removed from the long DNA "string" by digestion with enzymes that degrade DNA. (Enzymes that degrade both DNA and RNA are called *nucleases;* enzymes that degrade only DNA are *deoxyribonucleases,* or *DNases.*) The nuclease that is used to detach individual nucleosomes is a bacterial enzyme, micrococcal nuclease. After digestion for a short period with this enzyme, only the DNA between the nucleosome beads is degraded. The rest is protected from digestion by its bound histones and remains as double-stranded DNA fragments 146 base pairs long. These DNA-histone complexes, the nucleosome beads, appear in electron micrographs as disc-shaped particles with a diameter of about 11 nm. Each nucleosome bead contains a set of 8 histone molecules—2 copies of each of the 4 highly conserved nucleosomal histones, H2A, H2B, H3, and H4. These form a protein core around which the double-stranded DNA fragment is wound (Figure 8–4).

In undigested chromatin, the DNA extends as a continuous thread from nucleosome to nucleosome. Each nucleosome bead is separated from the next by a region of *linker DNA,* which varies in length but is approximately 60 base pairs long. This linker DNA plus the nucleosome bead constitutes the entire nucleosome, which therefore contains about 200 base pairs of DNA (Figure 8–4).

An average-size protein molecule contains about 400 amino acids (molecular weight about 50,000) and therefore requires about 1200 DNA base pairs to code for it. Since there is 1 nucleosome for every 200 base pairs in chromatin, these 1200 base pairs of coding DNA will be associated with 6 nucleosomes and the human haploid genome of 3×10^9 DNA base pairs will contain 1.5×10^7 nucleosomes.

Nucleosomes Are Packed Together to Form Regular Higher-Order Structures[3,4]

In the living cell, chromatin must be kept in a highly compacted state and therefore probably rarely adopts the extended beads-on-a-string form. Instead, the nucleosomes are packed upon each other to generate regular arrays in which DNA is highly condensed. When nuclei are very gently lysed onto an electron microscope grid, most of the chromatin is seen as a fiber with a diameter of about 30 nm (Figure 8–3A); this basic form of chromatin packing is commonly known as the **30-nm chromatin fiber.**

Two different models proposed for the packing of nucleosomes in the basic 30-nm chromatin fiber are illustrated in Figure 8–5. In both models, the

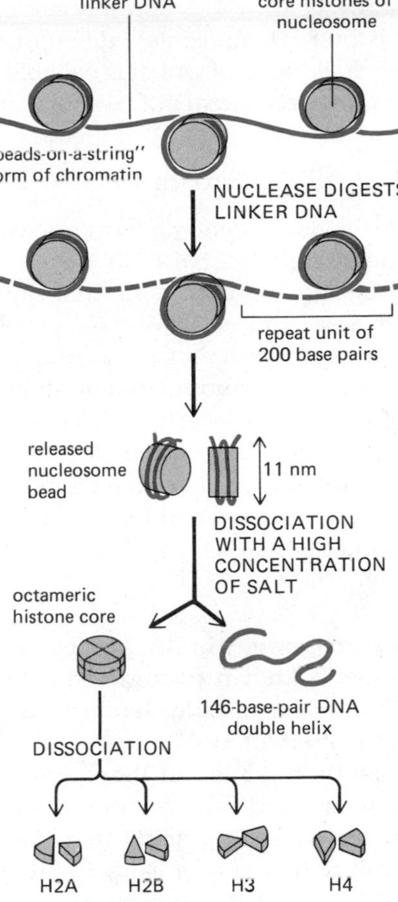

Figure 8–4 Schematic diagram illustrating the structure of the nucleosome bead. Beads are released from chromatin by limited digestion of the linker DNA with micrococcal nuclease. In each nucleosome bead, 146 base pairs of DNA double helix are wound around an octameric histone core. This protein core is composed of two each of histones H2A, H2B, H3, and H4. These histones contain 102 to 135 amino acids each. In the beads-on-a-string form of chromatin, each nucleosome bead is connected to its neighbor beads by an exposed "string" of linker DNA.

chromatin of a human chromosome containing 5 cm of DNA would be about 1.2 mm long if extended, whereas it would be about 2 cm long in the beads-on-a-string conformation.

Histone H1 Proteins Help Pack Nucleosomes Together[5]

Histone H1 molecules appear to be responsible for packing nucleosomes into the 30-nm fiber. Each H1 molecule has a globular central region linked to extended amino-terminal and carboxyl-terminal "arms" (Figure 8–6). The globular portion binds to a unique site on each nucleosome, and the arms are thought to extend to cover the linker DNA near the point where it joins the nucleosome bead as well as to contact the histone cores of adjacent nucleosomes. In this way, histone H1 proteins pull the nucleosomes together into a regular, repeating array (Figure 8–6).

An important feature of histone H1 binding not shown in Figure 8–6 is that it tends to bind to DNA in clusters of eight or more histone H1 molecules. This type of interaction, called **cooperative binding,** results from a strong tendency on the part of any individual histone H1 molecule to bind next to another that is bound, rather than in isolation. Many other protein molecules that bind to the genome also bind in a cooperative manner; all such proteins will form clusters of many molecules whenever they bind. This general feature of cooperative binding is illustrated in Figure 8–7.

We shall see that chromatin seems to be packaged in large units, each of which can change structure by decondensing as the genes it contains are

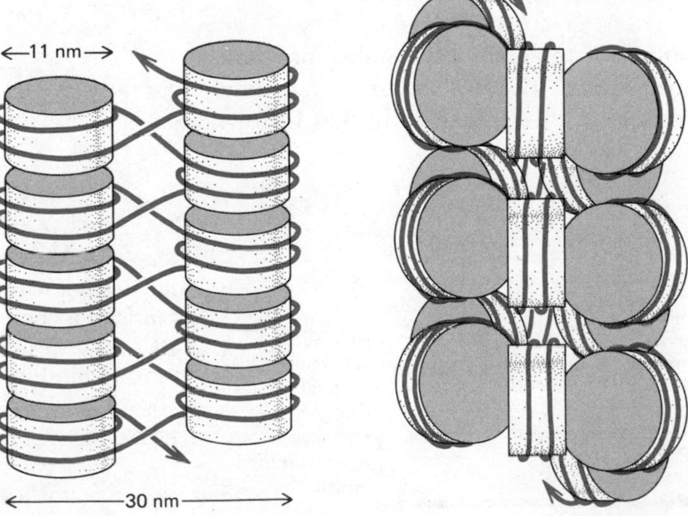

Figure 8–5 Two different models have been suggested for the packing of the beads-on-a-string form of nucleosomes into the basic 30-nm fiber. The actual structure of this form of chromatin, seen in the electron micrograph in Figure 8–3A, is not known. It is even possible that different types of packing are used in different regions of the chromosome.

activated (p. 450). If chromatin is organized by cooperative interactions between histone H1 molecules, different domains of chromatin would resemble tiny crystals, each of which is capable of expanding suddenly with the alteration or removal of a cluster of H1 molecules during gene activation (Figure 8–8).

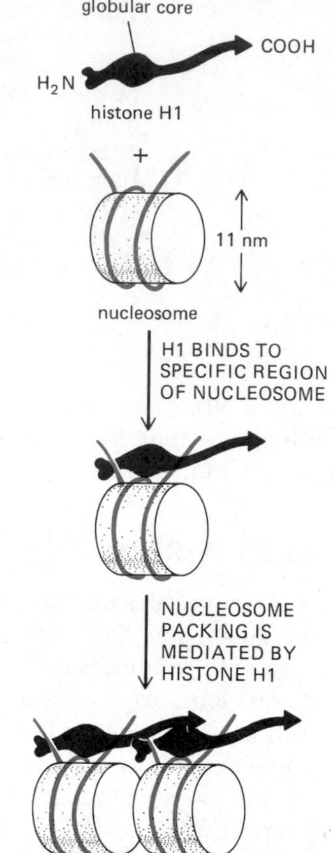

Figure 8–6 Schematic illustration of the manner in which histone H1 is thought to help pack adjacent nucleosomes together. Neither the exact structure of histone H1 nor the precise sites of interaction of its amino-terminal and carboxyl-terminal arms with the nucleosome are known. However, the globular core of H1 appears to be bound to each nucleosome as indicated. Each molecule of H1 histone contains about 220 amino acids.

Not All Nucleosomes Are Packed in Precisely the Same Way[2,6]

Purified histones will form nucleosomes in a test tube with any DNA molecule, including bacterial DNA that is not normally associated with histones. It is therefore not surprising that nucleosomes are characteristic of all regions of DNA seen in electron micrographs of chromatin. However, nucleosomes are not as uniform as they appear in these micrographs. In fact, important DNA functions are modulated by subtle alterations of the nucleosomal organization of chromatin in selected regions of the genome. Although much remains to be learned about the molecular basis for these subtle but biologically important differences in organization, a few of the causes of heterogeneity in chromatin structure can be identified. Some of these alterations appear to occur primarily in regions of the chromatin that contain active genes (see also p. 449).

Heterogeneity in the Nucleosomal Histones. Each of the nucleosomal histones is known to undergo a series of covalent modifications. For example, selected lysine side chains are acetylated (Figure 8–9) and selected serines are phosphorylated (Figure 8–10). Only a fraction of each type of histone is covalently modified in these ways (see Figure 8–2), so that some nucleosomes contain modified histones while others do not. Because these modifications occur on amino acids that have been especially highly conserved during evolution, it is very likely that they have an important function in the cell.

Several of the histone modifications are rapidly reversed after they are formed. Most notably, the acetyl groups on lysines are constantly being added by the enzyme *histone acetylase* and then removed by *histone deacetylase.* As a result, an acetyl group on a histone lasts, on average, for only about 10 minutes. This means that a cell is able to change the pattern of acetylation of its nucleosomes rapidly in response to appropriate signals (see p. 750). *Histone acetylation* is thought to be enhanced in the chromatin of active genes.

Another reversible histone modification is responsible for the covalent linkage of a highly conserved 74-amino-acid-long protein, called *ubiquitin,* to about one-tenth of the histone H2A molecules in the cell (Figure 8–11). These special uH2A molecules seem to be confined to the interphase chromatin of active genes, and they disappear temporarily during each mitosis as the chromosomes condense.

Not all modifications of nucleosomal histones are reversible. For example, many cells synthesize one or more minor "histone variants" that differ slightly in amino acid sequence from most of the molecules of that histone. The significance of these histone variants is unknown.

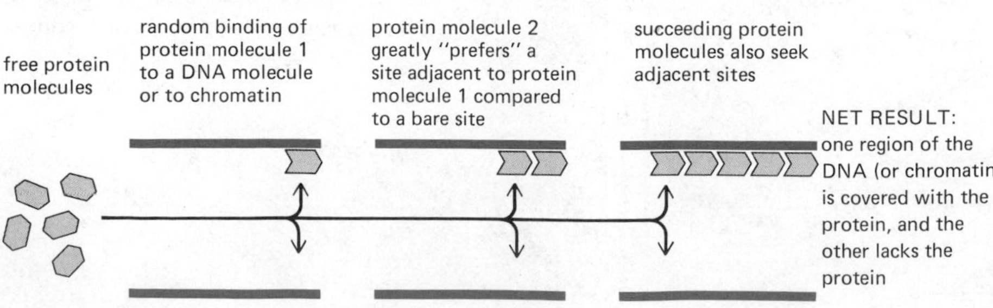

Figure 8–7 General scheme illustrating why any protein that binds in a cooperative manner to DNA or to proteins in chromatin will tend to bind in large clusters, showing "all-or-none" binding to a given region.

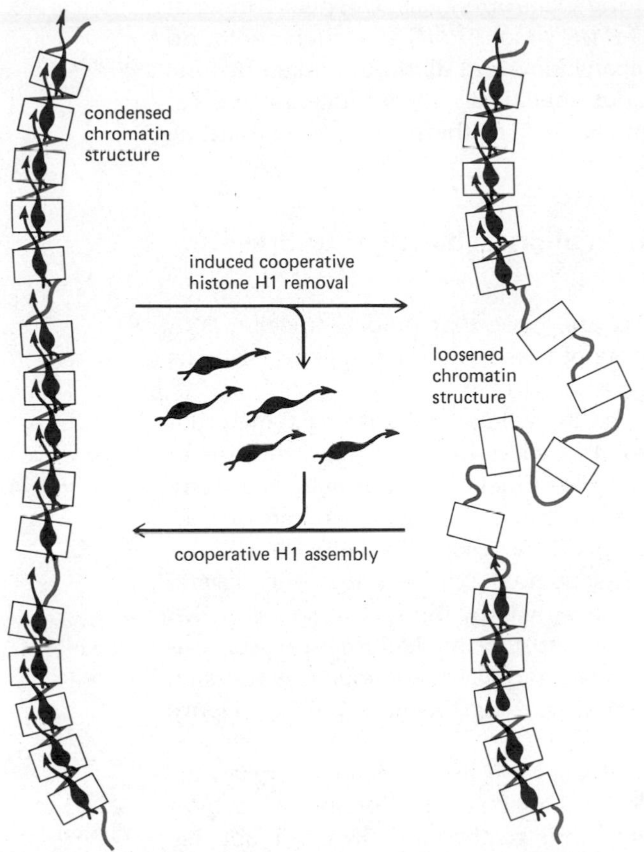

Figure 8–8 A local region of chromatin should act as a structural unit because of the cooperative packing of its nucleosomes caused by histone H1. This diagram depicts a unit being decondensed suddenly upon receiving some external regulatory signal. A chromatin decondensation of this type is thought to underlie the process of gene activation (see Figure 8–93).

Heterogeneity in Histone H1. Although there is about one H1 molecule per nucleosome, not all of the H1 molecules are identical. In some cells there are at least five different subtypes, all of which can be phosphorylated on specific amino acid side chains. The association of these different types of histone H1 with nucleosomes probably produces functional differences in the chromatin.

Heterogeneity Due to the Association of Abundant Nonhistone Proteins with Selected Nucleosomes. Most of the nonhistone chromosomal proteins are present in very small amounts. For example, there are only about 10,000 molecules of some known *gene regulatory proteins* per cell or about one copy per 3000 nucleosomes. Such rare proteins are very difficult to isolate in sufficient amounts for study. For this reason, only the most abundant nonhistone proteins have thus far been investigated in detail. These are present in quantities as great as 1 molecule for every 10 nucleosomes (over a million molecules per cell), so it is unlikely that most recognize specific DNA sequences. Instead, they probably influence more general structural features of chromatin, such as those required for gene transcription.

The abundant nonhistone proteins include the "high-mobility group," or **HMG proteins,** so called because they are relatively small and highly charged,

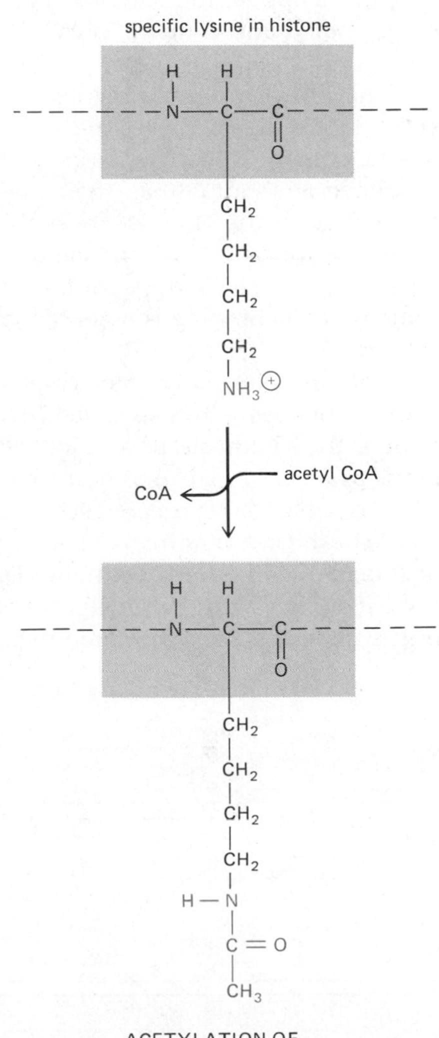

Figure 8–9 The reversible acetylation of selected lysines on histones removes a positive charge. As indicated previously for histone H4 (Figure 8–2), the lysines that undergo this type of modification are located within the first 30 amino acids of the amino terminus of each of the four nucleosomal histones.

and they, therefore, move quickly during electrophoresis. Of special interest are two closely related proteins called *HMG 14* and *HMG 17*, which are found in all mammalian cells. Like histone acetylation and ubiquitin-linked histone H2A molecules, they appear to seek out specifically those nucleosomes associated with active genes, with two molecules of either HMG protein binding directly to each nucleosome bead.

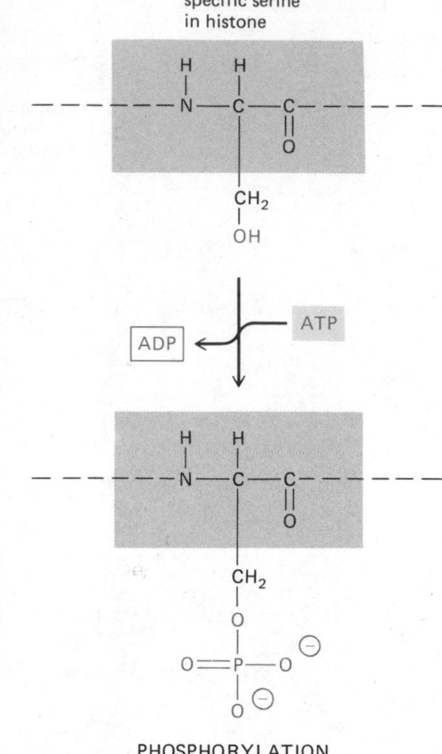

Figure 8–10 The reversible phosphorylation of selected serines on histones adds negative charge. Those nucleosomes containing either a phosphorylated or an acetylated amino acid will be different from the rest.

The DNA Helix Is Punctuated by Proteins That Bind to Specific DNA Sequences[7]

Among the minor nonhistone proteins are those that bind to specific DNA sequences along the chromosome. Some of these sequence-specific proteins control the way in which the DNA sequence is read.

Such control was first demonstrated in bacteria when it was shown that the *lactose repressor protein* binds to DNA so as to inhibit the synthesis of RNA by a cluster of genes involved in lactose metabolism. It acts as a **gene regulatory protein** and is removed from the DNA only when the protein products of these genes are actually needed (see p. 439). The lactose repressor protein binds tightly to a specific DNA sequence 21 base pairs long. Several other bacterial proteins that control gene activity by binding to specific DNA sequences have been extensively characterized. Like the lactose repressor protein, these appear to recognize a particular short DNA sequence from outside of the helix, and bind to DNA without disturbing the base pairing (Figure 8–12).

Recently, two gene regulatory proteins have been isolated from eucaryotic cells. One of these, the so-called *T antigen*, is a protein coded by a monkey virus, SV40. This large protein binds to the viral DNA at a specific sequence of about 36 base pairs, which is serially repeated three times near the points on the DNA where both viral DNA replication and RNA synthesis begin (Figure 8–13). Like the lactose repressor protein, the T antigen binds to DNA so as to block transcription of specific viral genes (see p. 439); at the same time, its binding is essential for the initiation of viral DNA replication at a nearby site.

The other well-characterized eucaryotic protein that binds to specific DNA sequences is the so-called *5S transcription factor*, which is required to initiate the synthesis of a small ribosomal RNA—the 5S RNA. This protein binds as an oligomer to a sequence of about 35 base pairs near the middle of the very small 5S RNA gene (see Figure 8–34).

These DNA-binding proteins are thought to be prototypes for hundreds or thousands of others, yet to be discovered, that recognize different specific DNA sequences and help determine which RNAs are made. In a multicellular organism, a specific set of such regulatory proteins is thought to play a major

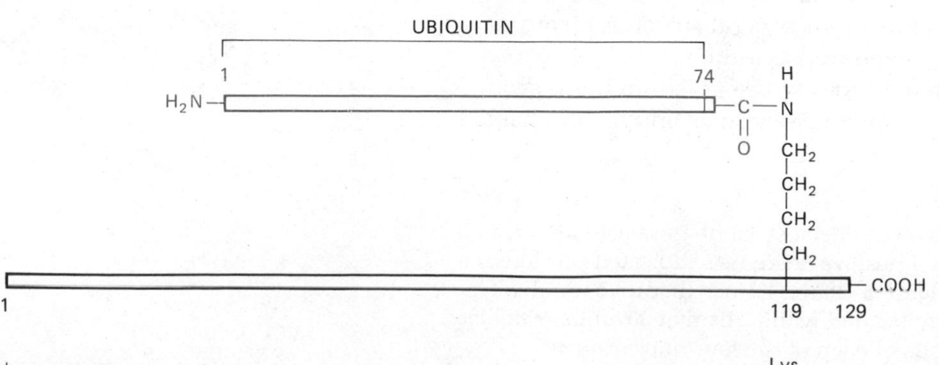

Figure 8–11 Some of the histone H2A molecules in nucleosomes are covalently modified by addition of the carboxyl-terminal end of the protein ubiquitin to a lysine side chain. The nucleosomes containing this so-called uH2A histone are different from the rest.

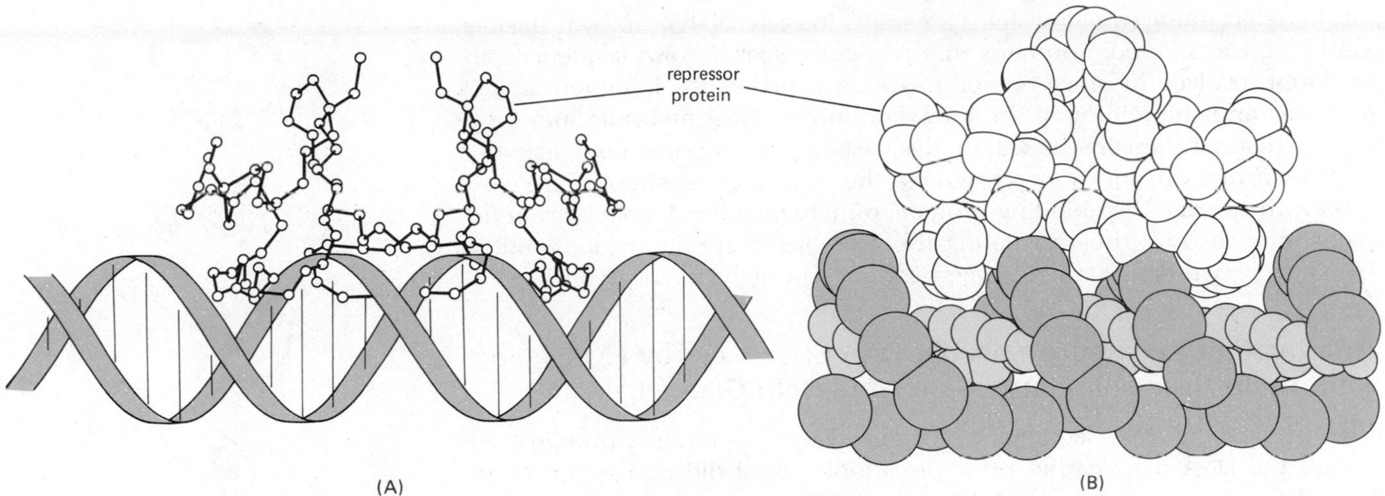

Figure 8–12 The structure of a bacteriophage gene regulatory protein bound to DNA. The protein molecule shown is a dimer of the bacteriophage lambda cro protein, whose three-dimensional structure has been determined by x-ray diffraction. The DNA binding site of the cro protein has been inferred from model-building studies. (A) Wire model bound to a schematic DNA helix. (B) Space-filling model of the DNA-protein complex shown in (A). In this model, each amino acid in the protein has been represented by a sphere, and the colored balls represent the DNA backbone. (Courtesy of Brian W. Matthews, redrawn from W. F. Anderson, D. H. Ohlendorf, Y. Takeda, and B. W. Matthews, *Nature* 290:754–758, 1981. © 1981 MacMillan Journals Ltd.)

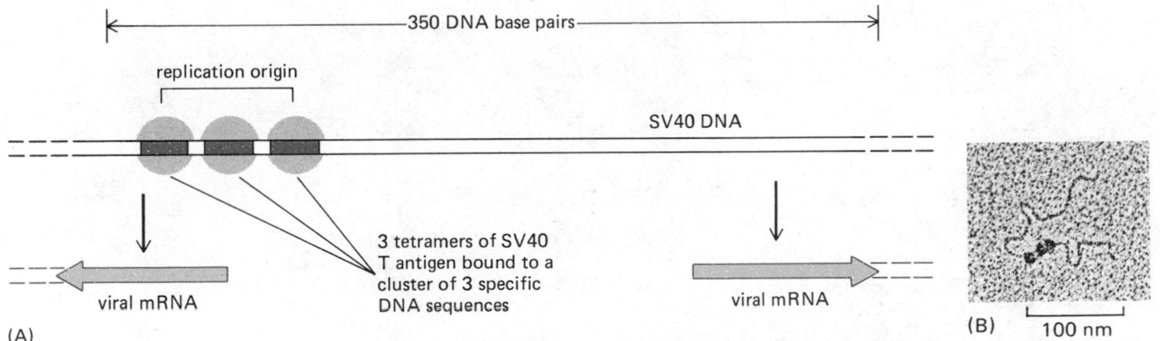

Figure 8–13 The binding of the T-antigen protein of the SV40 virus to a specific region of viral DNA. (A) A recognized DNA sequence about 36 nucleotides long is repeated three times near the point where both viral DNA replication and viral RNA synthesis begin, and the repeated sequences bind three tetrameric molecules of the T antigen. (B) Electron micrograph of three molecules of T antigen bound to purified SV40 DNA. Although the viral DNA is initially circular, it has been cut with a restriction nuclease to produce the small linear DNA fragment shown. Each protein sphere is a tetramer of 400,000 daltons. (Courtesy of Richard Myers, Robert Tjian, and Robley Williams.)

part in determining which proteins are made in each cell type (p. 444). Besides regulating transcription, proteins that recognize specific DNA sequences are no doubt involved in other genetic processes, such as the initiation of DNA synthesis and the folding of the long, continuous DNA molecule into functionally distinct domains, as will be discussed below. A section of eucaryotic DNA freed of its histones may resemble the schematic illustration in Figure 8–14, with specific DNA-binding proteins punctuating the double helix at frequent intervals and thus regulating the activities of specific regions. But at this stage very little is known concerning such proteins in eucaryotes.

Figure 8–14 A schematic view of a section of DNA helix punctuated by the binding of different proteins to specific DNA sequences.

Proteins Can Recognize Specific DNA Sequences by Hydrogen Bonding to Base Pairs and by Sensing Helix Geometry

How are specific DNA sequences recognized by DNA-binding proteins? Although the DNA helix can in principle adopt several different forms, by far the most stable is the *B-form DNA* helix, whose structure is shown in Figure 8–15A. Portions of each base pair are exposed in two separate "grooves," called the *major* (wide) and the *minor* (narrow) grooves (Figure 8–15B). In these grooves, each of the four possible base-pair arrangements (A-T, T-A, G-C, or C-G) can be uniquely recognized by the different arrangement of the atoms that protrude. Thus the amino acid side chains on each sequence-specific DNA-binding protein are thought to be arranged so as to maximize the protein's hydrogen bond interactions with a particular base-pair sequence. One

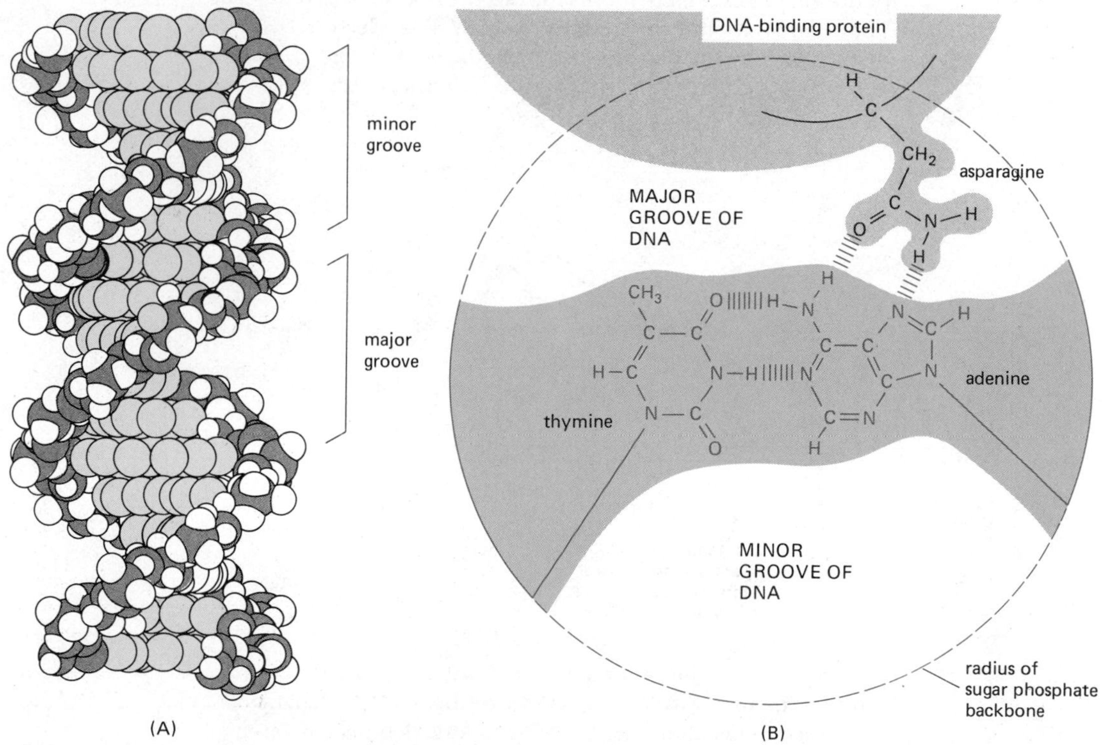

Figure 8–15 The recognition of DNA base pairs by DNA-binding proteins. (A) The DNA double helix (B-form) with its major and minor grooves colored. The edges of the base pairs protrude into these grooves, enabling DNA-binding proteins to recognize different DNA sequences from outside the helix by hydrogen bond interactions. A proposed interaction in the major groove between an amino acid and an A-T base pair is illustrated in (B), viewed along the helix axis. The B-form DNA helix is right-handed.

possible interaction between an amino acid and a base pair is illustrated in Figure 8–15B.

But this is probably not the only criterion by which a specific DNA sequence can be recognized. X-ray diffraction studies of small synthetic DNAs show that the exact geometry of the DNA helix can vary depending on the nucleotide sequence. While most DNA is in the B-form shown in Figure 8–15A, local regions of special sequence can form a slightly different right-handed helix known as *A-form DNA* (see Figure 3–7, p. 102). Moreover, sequences composed of alternating purines and pyrimidines, such as GCGCGCGC, can form a very different left-handed double helix known as *Z-form DNA*. Regions containing such altered DNA helix conformations are expected to be rare in chromosomes, but they could be specifically recognized by proteins and thereby have important biological roles.

More generally relevant is the fact that in B-form DNA both the exact tilt of the bases and the helical twist angle between base pairs have been found to depend on which nucleotides are adjacent to each other in the sequence. These small variations in the helix will cause atoms (including DNA phosphates) to be displaced from their idealized positions by perhaps ±0.1 nm (Figure 8–16). All DNA-binding proteins that recognize specific sequences would be expected to sense these displacements.

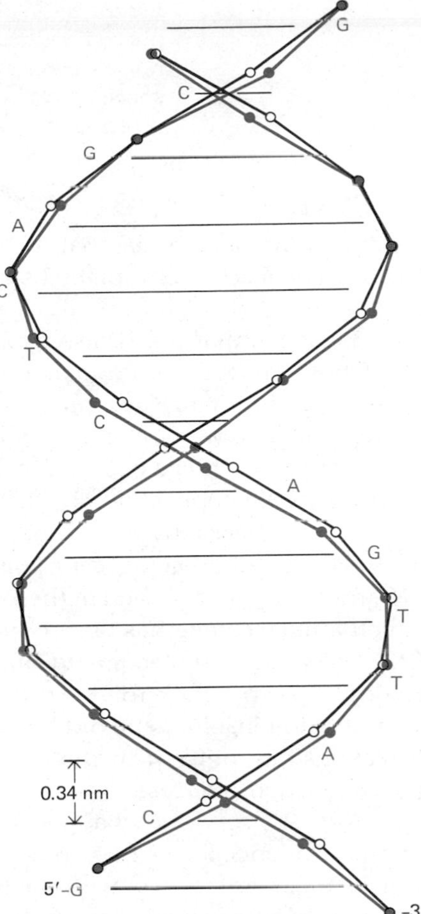

Figure 8–16 Drawing of the phosphodiester backbone of a B-form DNA helix showing how the conformation of the helix varies due to variations in the helical twist angle between adjacent base pairs. The actual path of phosphates is shown schematically (*colored line*) for the nucleotide sequence 5′-GCATTGACTCAGCG-3′. For comparison, the path of phosphates for an idealized regular B-form helix is shown (*black line*). The twist angles for different combinations of adjacent base pairs (dinucleotides along the nucleotide sequence) average about 34°, but range between 28° and 40° for different neighbors. (Courtesy of Edward Trifonov, data from W. Kabsch, C. Sander, and E. Trifonov, *Nucleic Acids Res.* 10:1097–1104, 1982.)

Histones Restrict the Accessibility of DNA to Other DNA-binding Proteins and Can Thereby Affect Gene Regulation[8]

Although all of the DNA in nucleosomes is freely available for collision with small molecules, its availability to larger molecules is necessarily restricted where it contacts the nucleosome bead. This means that nucleosomes can limit the access of regulatory proteins to the specific DNA sequences to which they bind. In the case of some regulatory proteins, binding may be possible only if the local region of DNA is quite free of nucleosomes. But other gene regulatory proteins can probably bind to their specific DNA sequences provided that those sequences occur in the relatively accessible linker regions between beads.

Because the location of nucleosome beads in chromatin could affect gene expression and other processes by interfering with the DNA binding of other proteins, some regions of chromatin might be expected to show a nonrandom positioning of nucleosomes, leaving relatively exposed those DNA sequences that must be recognized by other proteins in the cell. Because the spacing between nucleosomes is more or less regular, the exact placement of one nucleosome on the DNA will also affect the position of its neighbors. Thus, the whole repeating nucleosomal structure in a region may have to be positioned uniquely so as to leave a particular DNA sequence either free of nucleosomes or in a region of linker DNA. The sequence-specific placing of nucleosomes is known as *nucleosome phasing*, and such nucleosomes are said to be *phased* (Figure 8–17).

In fact, as we shall see, in at least some cases nucleosomes are nonrandomly placed on DNA, and occasional "gaps" (missing nucleosomes) as well as nucleosome phasing are observed.

Nucleosome Beads Can Be Nonrandomly Positioned in Chromatin[9]

Regions of DNA that are free of nucleosomes can be detected by treating cell nuclei with trace amounts of a deoxyribonuclease (DNase I) that at low concentrations will digest free DNA but not the linker DNA associated with nucleosomes. Chromatin treated in this way is preferentially cut at short regions spaced thousands of base pairs apart in the DNA sequence. These *DNase-*

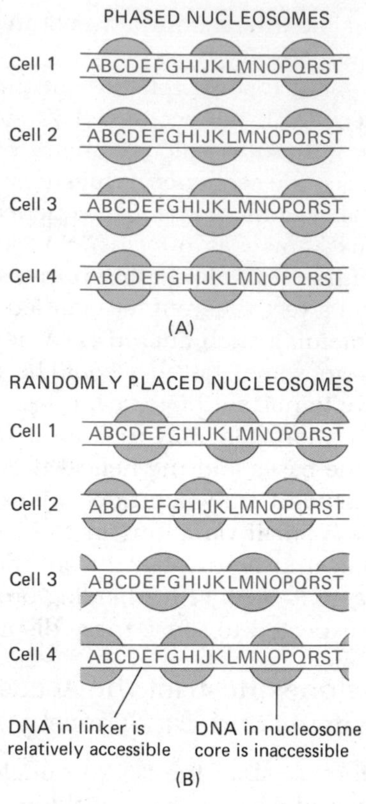

Figure 8–17 Diagram illustrating the difference between phased and randomly placed nucleosomes. (A) When nucleosomes are phased in a region of chromatin, the same specific DNA sequences become accessible for recognition by proteins in all cells of the same type. (B) In another region of chromatin, the pattern of nucleosome placement might vary.

hypersensitive sites reveal that the basic 30-nm chromatin fiber is really more like a chain of large "chromatin beads" than a continuous ropelike fiber (Figure 8–18).

Evidence that the DNase-hypersensitive sites may be important in gene regulation has come from experiments with SV40 virus, whose DNA is organized in a chromatin structure containing nucleosomes. SV40 DNA contains a single nucleosome-free region about 300 base pairs long that is rapidly digested by DNase I. The exposed region is located very near the points at which both viral DNA replication and RNA synthesis begin (Figure 8–13). Similarly, many of the DNase-hypersensitive regions in cell chromatin are located just to the 5′ side of an active gene, within 200 base pairs of the point where the RNA transcript starts. Each of these sites is believed to represent a region from which a nucleosome has been removed to allow regulatory molecules and/or RNA polymerase molecules to bind and initiate the synthesis of new RNA chains. In accord with this view, a particular DNase-hypersensitive site at the 5′ side of the insulin gene has been found in the chromatin of a cell type that makes insulin, but not in the chromatin of several other cell types that do not. How such nuclease-hypersensitive sites are formed is not known.

Nor is it yet clear what part is played by nucleosome phasing, although there is evidence for its occurrence. The evidence comes from experiments in which nuclei are treated with micrococcal nuclease or with a higher dose of DNase I, which preferentially cuts the linker DNA between adjacent nucleosome beads. Such experiments have shown that nucleosomes are precisely placed with regard to specific DNA sequences in at least some regions of chromatin. It is not yet known what causes nucleosomes to become so precisely positioned or phased, nor exactly what phasing means for chromatin function. But it is intriguing to speculate that in different cell types nucleosomes could be arranged in different phases on the same gene and in this way determine whether the gene is expressed or not.

Each Chromosome Probably Contains One Very Long DNA Molecule Organized in a Series of Looped Domains[10,11]

The uncoiled DNA of an entire eucaryotic chromosome is thought to consist of a single linear DNA molecule. Assuming that this is true of all 23 chromosomes in the human haploid genome, then each is a DNA molecule with an average length of $(3 \times 10^9) \div 23 = 1.3 \times 10^8$ nucleotides. This represents an average of nearly 5 cm of uncoiled DNA and thus poses a formidable packaging problem in a nucleus that is typically about 5 μm (5×10^{-4} cm) in diameter. Since packaging the DNA into 30-nm chromatin fibers only reduces the 5 cm to just over 1 mm, there must be several higher orders of folding. The probable

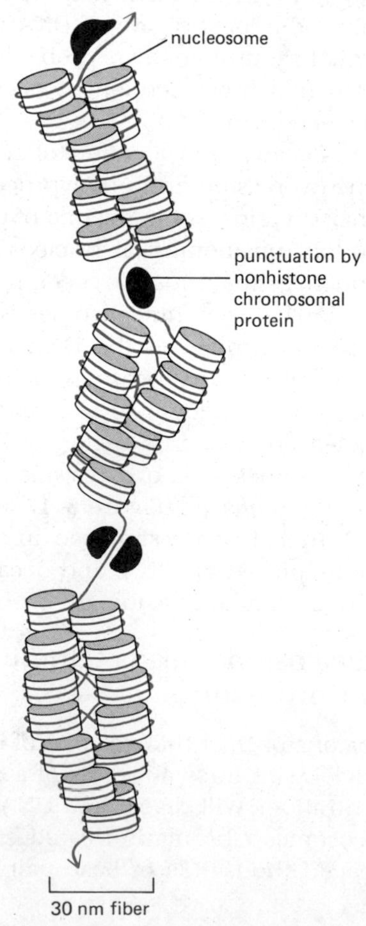

Figure 8–18 A schematic section of chromatin illustrating the interruption (punctuation) of its regular nucleosomal structure by short regions in which DNA is unusually vulnerable to digestion by DNase I. These regions could be freed of nucleosomes by the binding of special nonhistone chromosomal proteins to the DNA.

nature of the next level of folding was originally suggested by the appearance of certain special chromosomes—the *lampbrush chromosomes* in many oocytes and the *polytene chromosomes* of certain insect cells. As we shall discuss later, these two types of chromosomes appear to be organized as a series of **looped domains.** In 1972, the *E. coli* bacterial chromosome—a circular DNA molecule about 1 mm in length and lacking histones—was also shown to be folded into a similar series of loops. It now appears that this form of chromosome organization may be general.

It has been hypothesized that the looped domains in chromatin are established and maintained by DNA-binding proteins that clamp two regions of the 30-nm fiber together, presumably by recognizing specific DNA sequences that will form the neck of each loop (Figure 8–19). Organisms as diverse as *Drosophila* and man seem, from indirect estimates, to have looped domains of similar average size: although different loops vary widely in size, a length of DNA in the loop of between 20,000 and 80,000 base pairs is typical. This means that a typical human chromosome might contain $(1.3 \times 10^8) \div (5 \times 10^4) = 2600$ looped domains, in which each loop of chromatin is formed from an average length of about 400 nm (0.4 μm) of the 30-nm fiber (Figure 8–20).

But the 5-cm-long DNA molecule in each human chromosome would still be about 100 μm long in the looped domain form illustrated in Figure 8–20. Therefore, it must be further coiled still to fit into the cell nucleus.

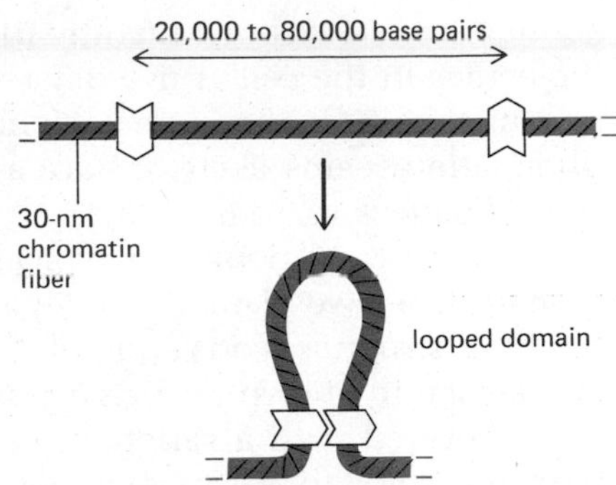

Figure 8–19 The folding of a region of 30-nm chromatin fiber into a looped domain.

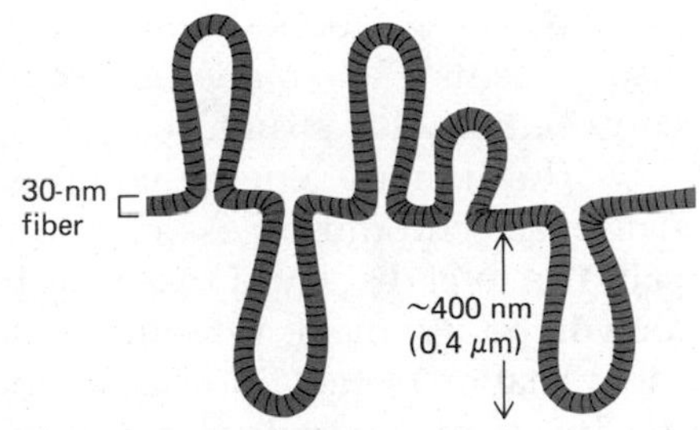

Figure 8–20 Schematic view of a section of a chromosome folded into a series of looped domains, each containing perhaps 20,000 to 80,000 base pairs of double-helical DNA condensed in a 30-nm chromatin fiber.

Bands on Mitotic Chromosomes Reveal an Even Higher Level of Organization[11,12]

Lampbrush chromosomes and polytene chromosomes are interphase chromosomes that are visible because of special properties that will be explained later. Most chromosomes are too extended and thin during interphase for their loops to be detectable and are readily amenable to structural investigation only during mitosis, when they coil up to form much more condensed structures (Figure 8–21). This coiling, which reduces a 5-cm length of DNA to

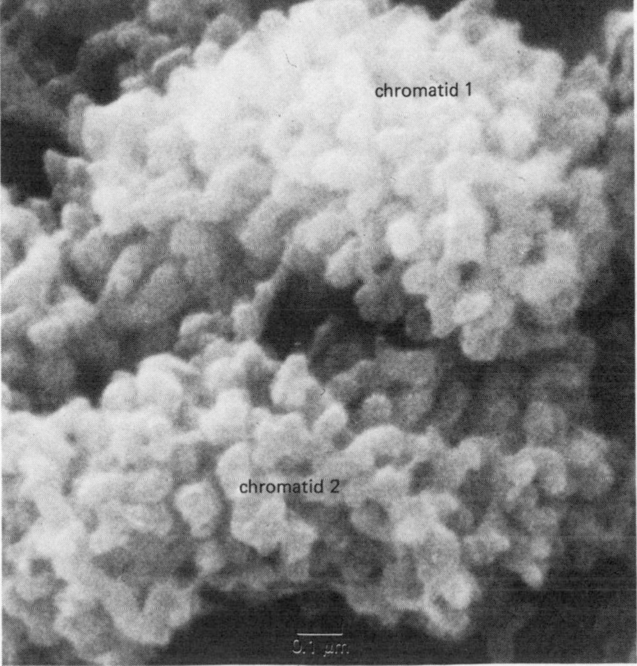

Figure 8–21 Scanning electron micrograph of a portion of a typical highly condensed chromosome that forms during mitosis. Before mitosis each chromosome had duplicated, and it therefore consists of two identical paired chromatids. (From M. P. Marsden and U. K. Laemmli, *Cell* 17:849–858, 1979. © M.I.T. Press.)

about 5 μm, is accompanied by the phosphorylation of all of the histone H1 molecules in the cell at five different specific serine residues. Because of the part histone H1 plays in packing nucleosomes together (Figure 8–6), its phosphorylation seems likely to have a causal role in chromosome condensation during mitosis (p. 644).

Figure 8–22 depicts a typical chromosome at the metaphase stage of mitosis. The two daughter DNA molecules are separately folded to give a bipartite structure consisting of the two daughter *chromatids* held together at a structure known as a *centromere.* These mitotic chromosomes are normally covered with a variety of molecules, including large amounts of ribonucleoproteins (p. 427). Once this covering is stripped away, each chromatid can be seen in electron micrographs to be organized into loops of chromatin emanating from a central axis (Figures 8–21 and 8–23). Such micrographs have supplied strong supporting evidence for the idea that all chromosomes are constructed from a series of looped domains. The arrangement of the DNA in each chromatid is still controversial, but to illustrate how the organized folding of chromatin may be achieved, we have drawn each chromatid as a closely packed series of looped domains wound in a tight helix in Figure 8–24, which also presents a schematic view of all the different folding processes that would contribute to this structure.

The human genome comprises 23 pairs of chromosomes (including 1 pair of sex chromosomes), making a total of 46 chromosomes in every diploid cell. The total display of these 46 chromosomes at mitosis is called the human **karyotype.** Cytological methods developed since 1970 permit unambiguous identification of each individual human chromosome. Some of these methods involve staining mitotic chromosomes with dyes that fluoresce only when they bind to certain types of DNA sequences. Although these dyes have very low specificity and appear to distinguish mainly DNA rich in A-T base pairs from

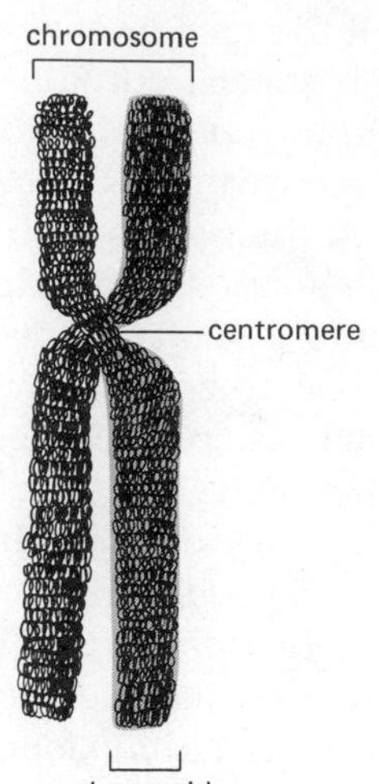

Figure 8–22 Schematic drawing of a typical chromosome at the metaphase stage of mitosis. Each chromatid contains one of two identical daughter DNA molecules (one of which is colored here) created earlier in the cell cycle by DNA replication.

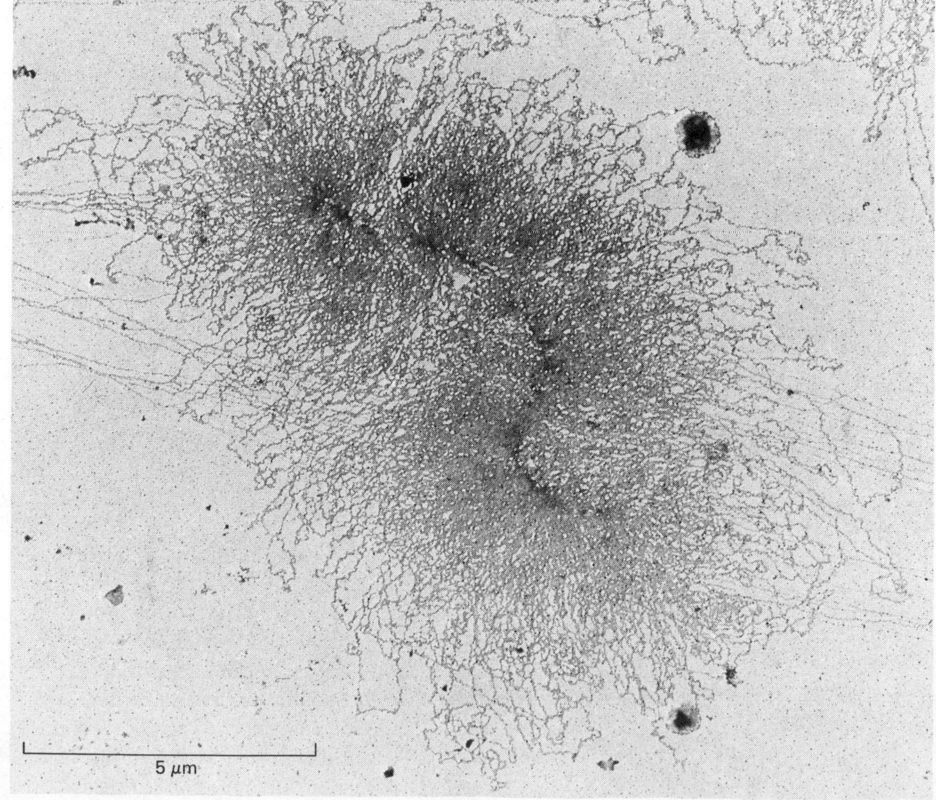

Figure 8–23 Electron micrograph of a single chromatid of a mitotic chromosome from an insect (*Oncopeltus*) treated to reveal loops of chromatin fibers that emanate from a central axis of the chromatid. (Courtesy of Victoria Foe.)

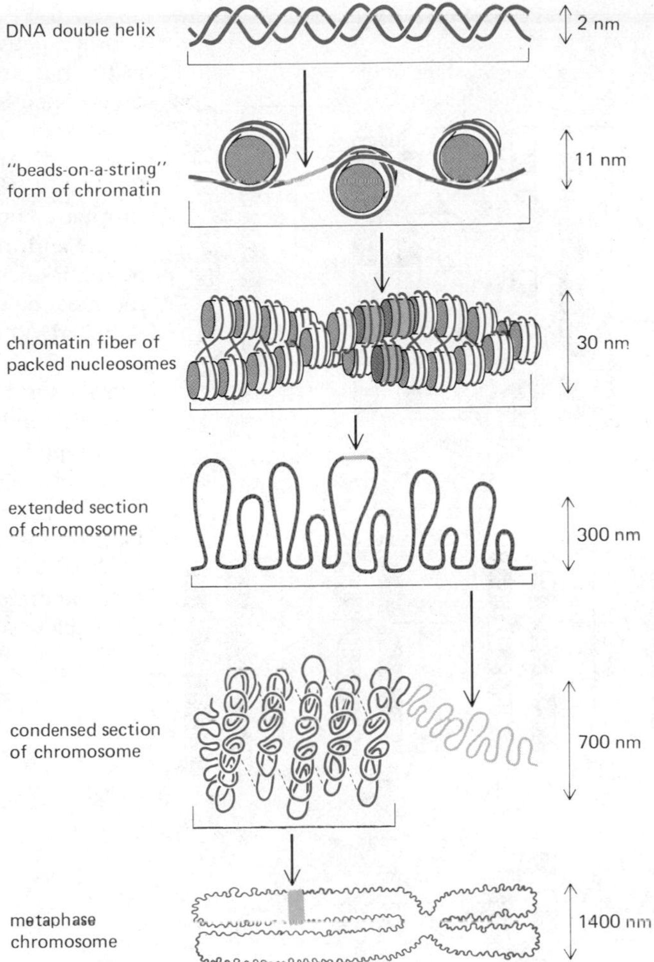

Figure 8–24 Schematic illustration of the many different orders of chromatin packing postulated to give rise to the highly condensed metaphase chromosome.

DNA rich in G-C base pairs, they produce a striking and reproducible pattern of fine bands on each mitotic chromosome (Figure 8–25). Because there is a unique pattern of bands on each type of chromosome, each chromosome can be identified and numbered, as illustrated in Figure 8–26.

By examining human chromosomes very early in mitosis, when they are less condensed than at metaphase, it has been possible to estimate that the total haploid complement contains at least 2000 distinct bands of A-T-base-pair-rich DNA. These bands progressively coalesce as condensation proceeds during mitosis to produce fewer and thicker bands (Figure 8–26).

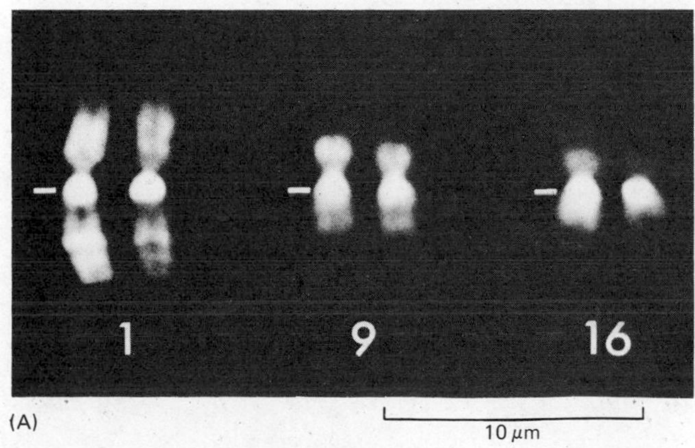

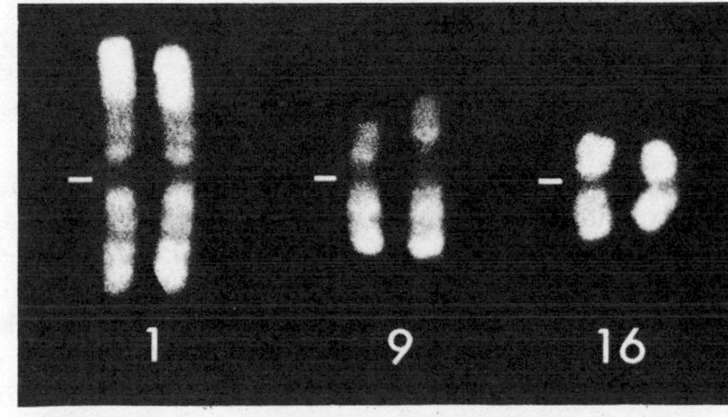

Figure 8–25 Fluorescense micrographs of three pairs of human mitotic chromosomes stained with (A) the A-T-base-pair-specific dye Hoescht 33258 and (B) the G-C-base-pair-specific dye olivomycin. The bars indicate the position of the centromere. Note that the banding patterns complement each other, since the bands that are bright in (A) are dark in (B), and vice versa. (From K. F. Jorgenson, J. H. van de Sande, and C. C. Lin, *Chromosoma* 68:287–302, 1978.)

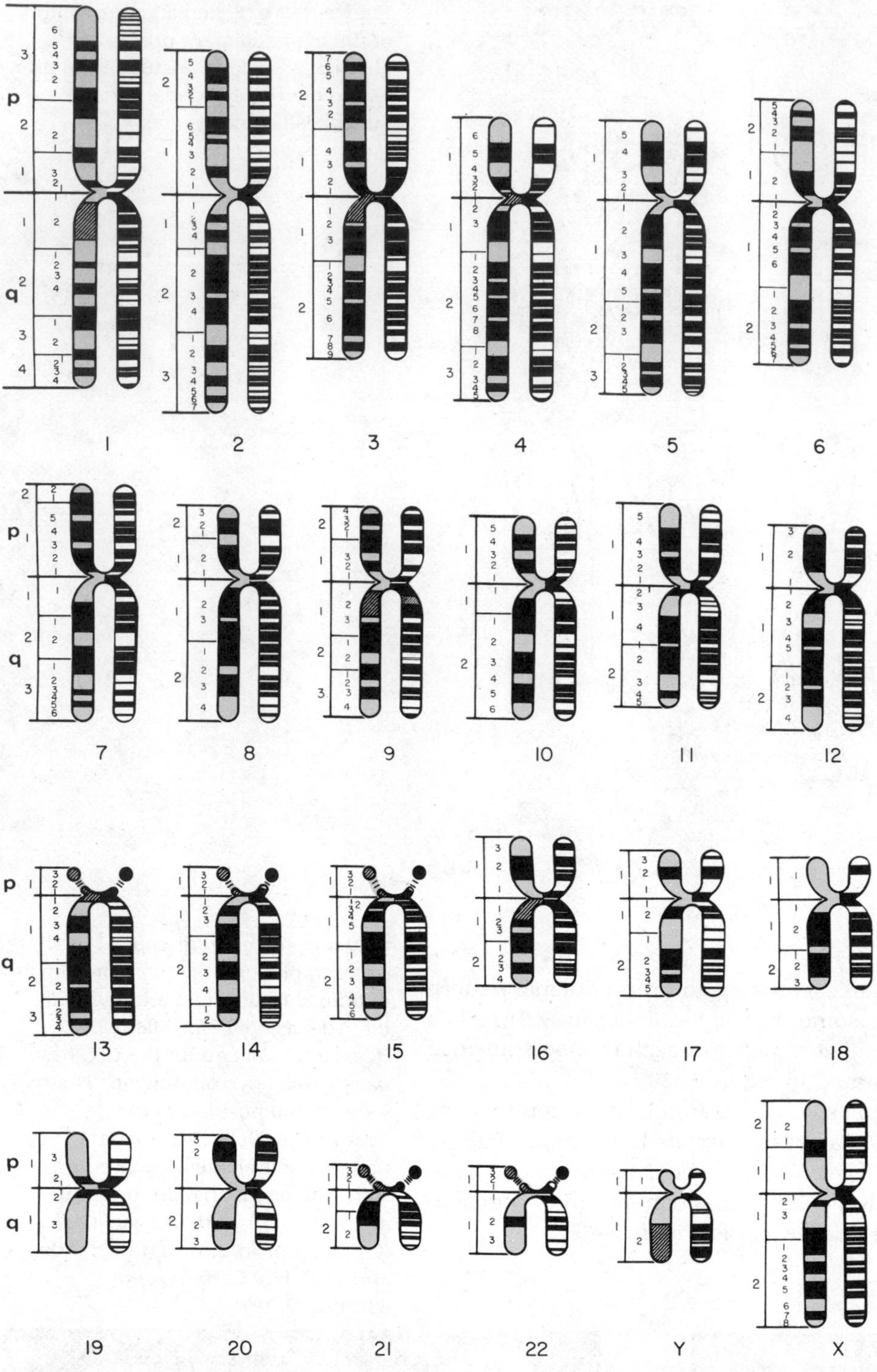

Figure 8–26 A standard map of the banding pattern of each chromosome in the human *karyotype*, as determined both at the metaphase stage (*color-tinted chromosomes*) and at the early prophase stage of mitosis (*untinted chromosomes*). The early prophase chromosomes are much longer and thinner than the metaphase chromosomes, and many more bands can therefore be detected. All of the bands shown here stain with reagents that appear to be specific for A-T-rich DNA sequences. Note that only one chromosome of each type is shown, whereas two chromosomes of each type are actually present during mitosis. (From J. J. Yunis, *Science* 191:1268–1270, 1976. © 1976 by the American Association for the Advancement of Science.)

Why such bands exist at all is a mystery. Even the thinnest of the bands diagrammed in Figure 8–26 should contain 30 or more looped domains, and the average base-pair composition should, presumably, be random over such long stretches of DNA sequence (more than a million base pairs, or nearly the size of a typical bacterial genome). One suspects that eucaryotic genomes may function in such packets. For example, a clustering of neighboring looped domains of similar chromatin structure could give rise to the bands (Figure

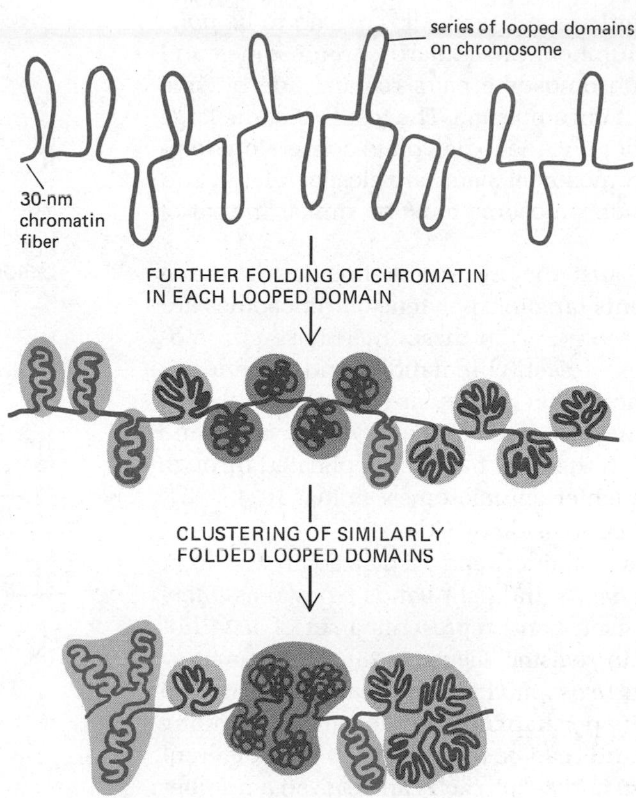

Figure 8–27 Schematic illustration of the way in which a clustering of neighboring looped domains that are similarly packaged in chromatin could produce the bands seen on mitotic chromosomes. As indicated, these bonds are postulated to represent an important organizational unit throughout interphase, even though they can only be clearly seen when the chromosomes begin to condense in preparation for mitosis.

8–27) and might be necessary to generate a single functional unit of chromatin during interphase (p. 637).

The presence of numerous chromosome bands is a general feature of mitotic chromosomes from species as diverse as man and *Drosophila.* Moreover, the exact pattern of bands in a chromosome has remained unchanged over long periods of evolutionary time. For example, every human chromosome has a clearly recognizable counterpart with a nearly identical banding pattern in the chromosomes of the chimpanzee, gorilla, and orangutan (although there has been a single chromosome fusion that gives humans 46 chromosomes instead of the apes' 48). This strongly suggests that the long-range spatial organization of chromosomes is important to proper gene expression and that the existence of bands reflects an important type of organizational unit.

Interphase Genes Can Be Seen in Polytene Chromosomes[13]

The banding patterns of mitotic chromosomes are suggestive and interesting but do not provide any detailed insight into the organization or function of individual genes, since even the narrowest discernable band is probably 10 to 100 genes long. To be able to detect changes in chromatin structure at the level of individual genes, it would be necessary to examine the stretched-out interphase chromosome. In ordinary cells this is not possible because the interphase chromatin strands are too fine and too tangled. But in some insect cells, because of a specialization known as *polyteny,* a banding pattern on a scale that may correspond to single genes is clearly visible. **Polytene chromosomes** are characteristic of some of the secretory cells of insect larvae; these cells grow to an enormous size through multiple cycles of DNA synthesis without cell division. The resulting giant cells contain as much as several thousand times the normal DNA complement. Such cells are said to be *polyploid* when they contain increased numbers of standard chromosomes and *polytene* when all the homologous chromosome pairs remain side by side with each other to create a single giant chromosome. The fact that some large insect cells eventually undergo a direct polytene to polyploid conversion demonstrates that these two different chromosomal states are closely related and that the basic structure of a polytene chromosome must be similar to that of a normal chromosome.

Because they are so large and because the precise side-to-side adherence of individual chromatin strands prevents tangling, polytene chromosomes are easy to see in the light microscope. Moreover, in these interphase chromosomes, the individual looped domains in each chromatid strand are linearly arranged rather than coiled up on each other as they are normally. Polyteny has been most studied in the four chromosomes of the larval salivary gland cells of the fruit fly *Drosophila*, in which the DNA has been replicated through 10 cycles without separation of the daughter chromosomes, so that 1024 ($=2^{10}$) identical strands of chromatin are lined up side by side.

When polytene chromosomes are stained and visualized in the light microscope, distinct alternating dark *bands* and light bands (known as *interbands*) are visible. It is believed that each band represents a set of 1024 homologous looped domains arranged in register (Figure 8–28). Approximately 5000 bands are detectable in the total *Drosophila* genome, each estimated to contain 3000 to 300,000 DNA base pairs per haploid chromosome (depending on the size of the band). Since these bands can be recognized by their different thicknesses and spacings (Figures 8–29 and 8–30), each can be given a number to generate a polytene chromosome "map." The structure and significance of the relatively small amounts of DNA in the interbands is not known.

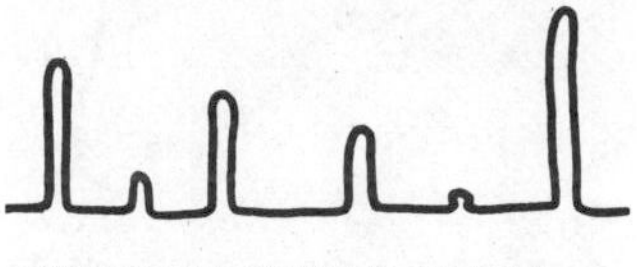

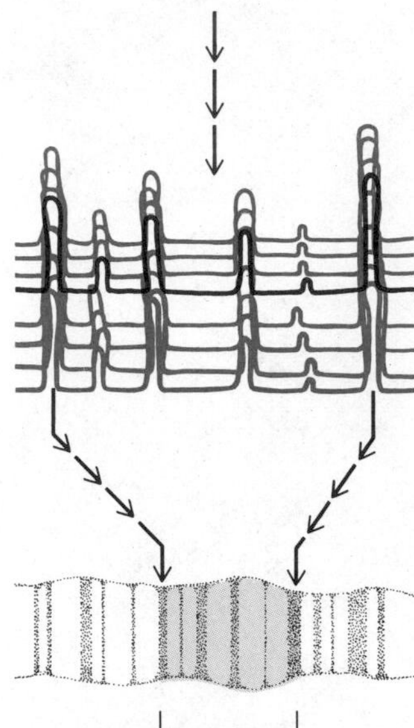

Figure 8–28 Schematic diagram indicating how the bands seen on polytene chromosomes are thought to be generated by side-to-side packing of homologous looped domains. Within each band, the loops of 30-nm chromatin fiber appear to pack together tightly to form much more condensed structures than those indicated here.

Each Looped Domain of Chromatin May Correspond to a Separate Unit of Function[14]

Through analysis of mutant flies that have a recognizably altered chromosome, it has been possible to locate individual genes on the *Drosophila* polytene chromosomes and to show that each band represents a unique genetic region. Other genetic studies have suggested that *Drosophila* makes only about 5000 essential proteins, a number that corresponds to the number of chromosome bands. For example, the region on the *Drosophila melanogaster* chromosome between two well-studied genes—*zeste* and *white*—contains 14 identifiable bands. After an intensive effort to isolate as many mutants as possible in this region, hundreds of mutants were mapped, but they were found to define a total of only 18 genetically independent groups ("complementation groups") or genes. Similarly, about 50 genes have been mapped on another chromosomal region with about 50 visible bands. It is not possible with these techniques to determine whether a particular gene lies in a band or an interband region. However, there is at least a rough correspondence between a gene and a looped domain of chromatin in *Drosophila*, suggesting that the average looped domain contains the DNA coding sequences for only one essential protein.

The evidence that each band in the *Drosophila* polytene chromosome represents a separate unit of function is supported by the observation that individual bands appear to unfold as a gene starts to synthesize RNA (p. 450). Such unfolded bands, called *chromosome puffs*, appear and disappear in a characteristic sequence reflecting the ordered pattern of gene activation occuring at different stages of *Drosophila* development (see Figure 8–97).

However, if we divide the total *Drosophila* genome into 5000 parts, we find that there are 20,000 DNA base pairs in an average gene, which is much more than the 1200 base pairs required to specify an average-size protein of right arm of chromosome 3

X chromosome

normal mitotic chromosomes at the same scale

chromosome 4

region where the 2 homologous chromosomes are separated

chromocenter

right arm of chromosome 2

left arm of chromosome 3

left arm of chromosome 2

20 μm

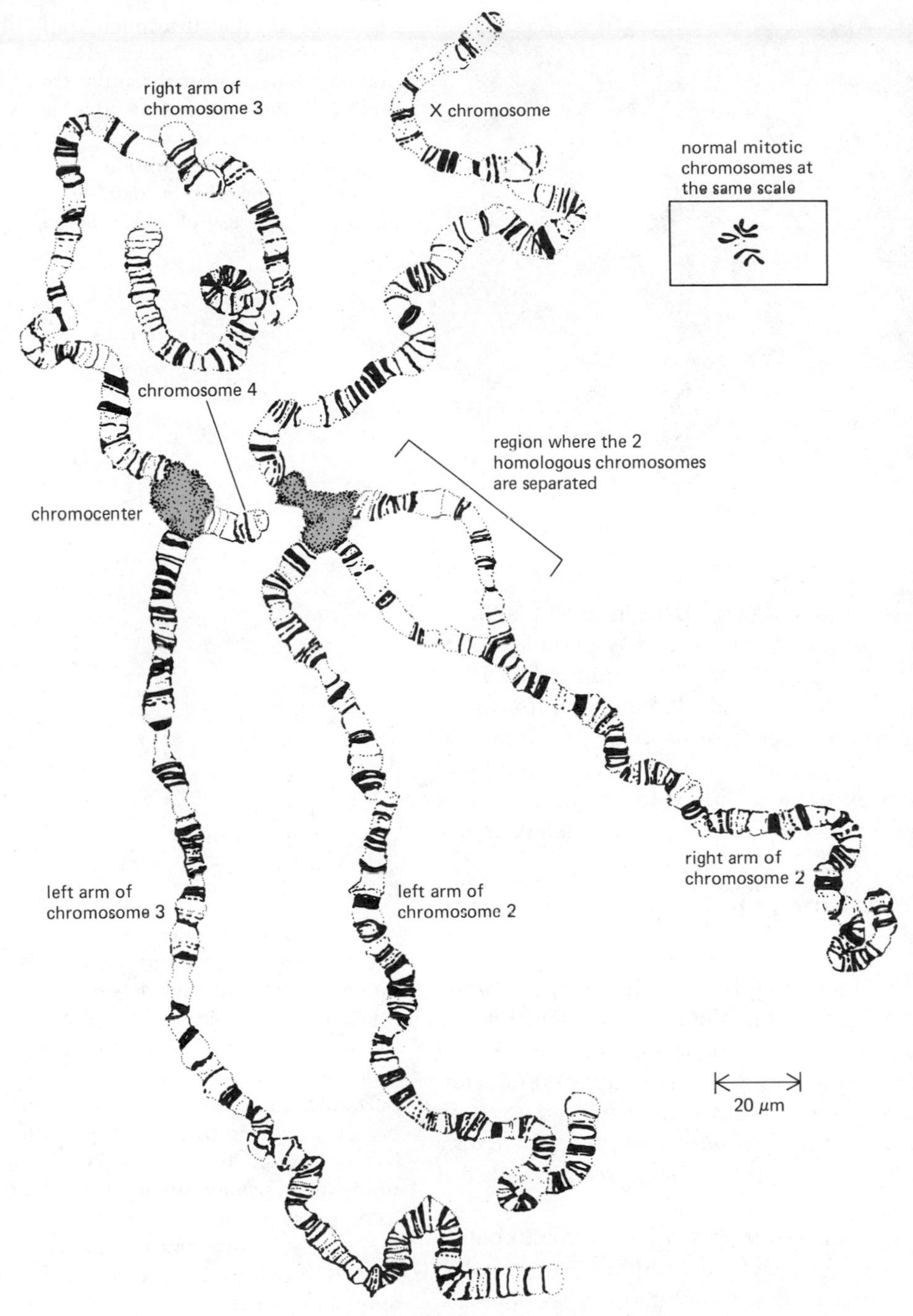

Figure 8–29 A detailed sketch of the entire set of polytene chromosomes in one *Drosophila* salivary cell. These chromosomes have been spread out for viewing by squashing them against a microscope slide. Note that there are four different chromosome pairs present. Each chromosome is tightly paired with its homolog, and the four chromosome pairs are linked together by regions near their centromeres that have aggregated to create a single large "chromocenter." Here, this chromocenter has been split into two halves by the squashing procedure used. (Modified from T. S. Painter, *J. Hered.* 25:465–476, 1934.)

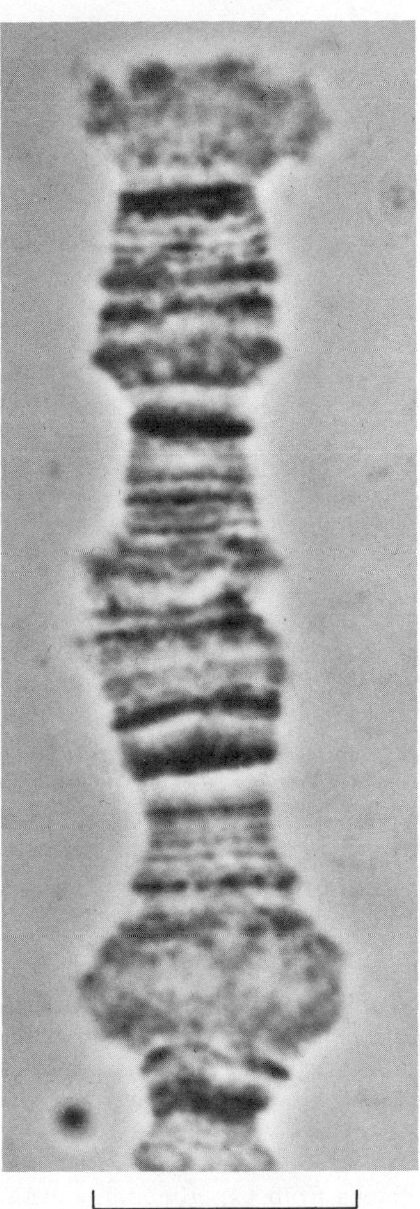

Figure 8–30 Light micrograph of a portion of a polytene chromosome from *Drosophila* salivary glands showing the distinct patterns recognizable in different chromosome bands. These bands occur in interphase chromosomes and are a special property of the giant polytene chromosome. They should not be confused with the much coarser bands diagrammed in Figure 8–26, which are revealed by special staining techniques on normal mitotic chromosomes. (Courtesy of Joseph G. Gall.)

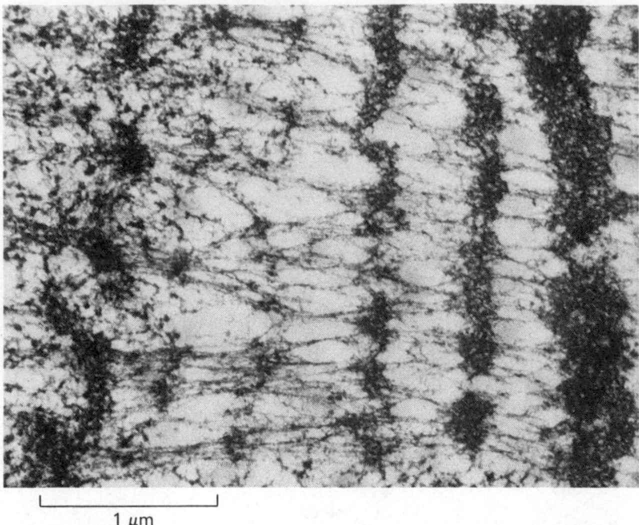

Figure 8–31 Electron micrograph of a small section of a *Drosophila* polytene chromosome, seen in thin section. Different chromosome bands have a different appearance, suggesting that they contain chromatin packed in a distinct manner. (Courtesy of Viekko Sorsa.)

400 amino acids. As we shall shortly see, there is more DNA in most genomes than is needed to code for essential proteins. Some of it may code for nonessential proteins and some of it is noncoding. Some of the noncoding DNA may play a structural part in the organization of each domain as a loop.

By analogy with the polytene chromosomes in *Drosophila,* it seems reasonable to suggest that looped domains of chromatin also serve as functional units in higher eucaryotic cells, such as those of man. In these organisms also, each looped domain probably contains only one or perhaps a few genes.

Different Domains of Chromatin Appear to Contain Differently Organized Nucleosomes[15]

The individual bands on polytene chromosomes have a characteristic morphology in both the light and electron microscopes (Figures 8–30 and 8–31). This suggests that the different bands—corresponding to different looped domains—have somewhat different chromatin structures. A similar conclusion is reached from studies in which fluorescent antibodies are used to stain *Drosophila* polytene chromosomes: while antihistone antibodies stain the bands uniformly, antibodies against some nonhistone proteins stain some bands but not others (Figure 8–32).

Independent evidence for structural differences between chromatin bands comes from experiments in which isolated polytene chromosomes are exposed to elevated concentrations of inorganic ions, which tend to decondense chromatin. Some bands, however, seem to be particularly sensitive to high concentrations of K^+, whereas others are particularly sensitive to high concentrations of Mg^{2+}, and so on.

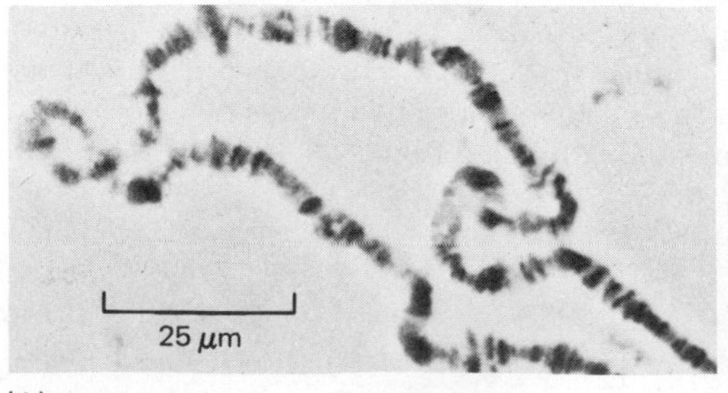

(A)

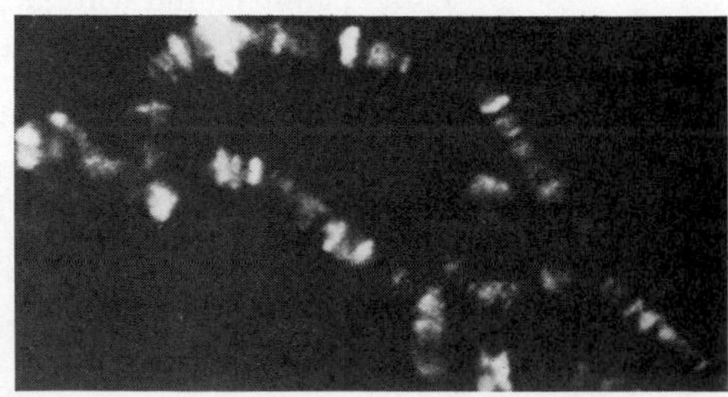

(B)

Figure 8–32 The same region of a polytene chromosome viewed by phase-contrast microscopy (A) and by the fluorescence of a labeled monoclonal antibody that detects a particular nonhistone chromosomal protein (B). Note that a subset of the bands stain for the presence of the nonhistone protein, which is believed to be present only in regions of potentially "active" chromatin. (From G. C. Howard, S. M. Abmayr, L. A. Shinefeld, V. L. Sato, and S. C. R. Elgin, *J. Cell Biol.* 88:219–225, 1981. © 1981 Rockefeller University Press.)

These types of differences are unlikely to be confined to insects, but only insects have polytene chromosomes that are easy to see with a microscope. At a much coarser level of resolution, differences between the chromatin structures of different mammalian genes can be detected by the altered susceptibility of their DNA to digestion with certain nucleases (p. 448). As we shall see, the particular chromatin structure of a mammalian gene appears to affect the ability of that gene to be expressed in a cell.

Most Chromosomal DNA Does Not Code for Essential Proteins[16]

There are several reasons for believing that most of the DNA in the genome of higher organisms does not code for essential proteins. Long before recombinant DNA technology made it possible to examine the nucleotide sequence of chromosomal DNA directly, several paradoxes were evident. For example, the relative amount of DNA in the haploid genome of different organisms bears no systematic relationship to the complexity of the organism. Human cells contain 800 times more DNA than the bacterium *E. coli,* but then some amphibian and plant cells contain 30 times as much DNA as do human cells (Figure 8–33). This paradox cannot be dismissed by postulating multiple copies of each gene in some organisms, for it is clear from DNA renaturation studies (p. 189) on numerous organisms that there is only one copy of most DNA sequences in each haploid DNA complement. The failure of DNA content to correlate with genetic complexity is dramatically illustrated by the vast differences in DNA content sometimes seen between closely related species. The most striking example is in amphibians, where the amount of DNA can vary 100-fold between different species.

Population biologists have estimated just how much of the DNA of higher organisms actually codes for essential proteins (or is involved in the regulation of genes coding for such proteins). In outline the argument runs as follows: Mutation is an accidental process in which randomly selected nucleotides in the DNA sequence are altered at a low but finite rate. Since most such mutations will be deleterious to the organism when they occur in an essential DNA sequence, there is a limit to the number of essential genes that can be

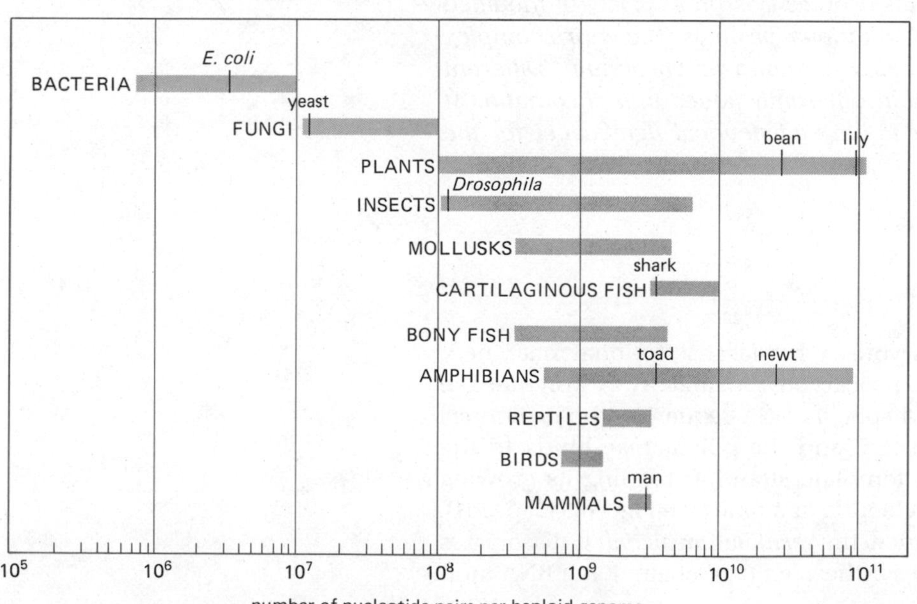

Figure 8–33 The amount of DNA in a haploid genome varies over a 100,000-fold range from the smallest procaryotic cell, the mycoplasma, to the large cells of some plants and amphibia. Note that man has a genome size (3×10^9 nucleotide pairs) much smaller than that of some other organisms.

stably maintained. It has been estimated from the observed mutation rate that no more than about 1% of the mammalian genome can be involved in regulating or coding for essential proteins.

The most important implication of these considerations is that, despite the fact that the mammalian genome is large enough, in principle, to code for nearly 3 million average-size proteins (3×10^9 nucleotides), no organism is likely to be able to maintain a repertoire of more than about 30,000 essential proteins. From this point of view, man is unlikely to be more than about six times more complex than the fruit fly *Drosophila*, with its 5000 essential genes.

What does the rest of the DNA do in higher eucaryotic genomes? We have already suggested that some of it may have a purely structural function in the organization of chromatin. In later sections of this chapter we shall discuss more recent evidence on the nature of the noncoding DNA and some other hypotheses about its function. Whatever the answer(s), the data shown in Figure 8–33 make it clear that it is not a great handicap for a higher cell to carry along a great deal of extra DNA, which suggests that there has been little pressure to minimize DNA content to include only the essential regions.

Summary

The mammalian genome contains about 3×10^9 DNA nucleotides, but only a small fraction of this DNA is thought to code for essential proteins. Although the function of the remainder is unknown, some of it is likely to be involved in forming the higher-order structure of eucaryotic chromosomes. To a first approximation, a chromosome may be considered a single, enormously long DNA molecule that is organized into a series of looped domains. Despite the fact that it contains tens of thousands of base pairs, each looped domain may contain only one (or a few) protein-coding sequences.

Almost all of the DNA in eucaryotes is tightly bound to an equal mass of histones, forming a string of small disclike DNA-protein particles called nucleosomes. The nucleosome structure appears to be interrupted at intervals by short stretches of "free" DNA that may be binding sites for regulatory proteins. Some of these proteins may help determine the placing, or phasing, of nucleosomes so that specific DNA sequences are left accessible between adjacent particles.

Normally, nucleosomes are packed together to form more condensed structures. In these structures, the nucleosomes exist in a variety of modified forms and are associated with different additional proteins. The entire complex of DNA and all of its associated proteins is known as chromatin. Different regions of the genome can be somewhat differently folded into chromatin. At least some of these differences appear to have a functional significance for the expression of genes.

RNA Synthesis and RNA Processing[17]

RNA synthesis appeared first in procaryotes, whose transcriptional machinery is simpler than that of eucaryotes. In procaryotes, a single **RNA polymerase** enzyme catalyzes all RNA synthesis. A specific DNA sequence (the *promoter*) signals where RNA synthesis is to begin, and the polymerase binds to the promoter and moves along the DNA template strand, extending its growing chain in the 5′-to-3′ direction one nucleotide at a time (see Figure 5–2, p. 201). When it reaches a second DNA sequence, the *termination signal*, it dissociates from the DNA and releases the newly synthesized RNA chain. Each RNA molecule thus represents a single-strand copy of the nucleotide sequence of one DNA strand in a relatively short region of the genome.

Although the general principle of eucaryotic DNA transcription is the same as that of procaryotes, the machinery is considerably more complex. First, there are three different RNA polymerases, each of which transcribes different sets of genes. Each of the eucaryotic RNA polymerases is more complex than its bacterial analogue: the procaryotic enzyme is composed of 5 polypeptide chains, while the eucaryotic enzymes contain 9 to 11. Moreover, whereas the purified bacterial RNA polymerase works well in the test tube, requiring only the addition of bare DNA to produce faithful transcription, eucaryotic RNA polymerases have additional requirements. Partly for this reason, it was not until 1979 that systems became available in which eucaryotic transcriptional mechanisms could be analyzed *in vitro*.

Three Different RNA Polymerases Make RNA in Eucaryotes[18]

Of the three eucaryotic RNA polymerases, only *RNA polymerase II* transcribes the genes that will be translated into proteins. The other two synthesize RNAs that form part of the protein synthetic machinery: *polymerase I* makes the large ribosomal RNAs and *polymerase III* makes a variety of very small, stable RNAs—including the tRNAs and the small 5S RNA of the ribosome.

The three RNA polymerases can be distinguished by their sensitivity to α-amanitin, a poison isolated from mushrooms. RNA polymerase I is unaffected by α-amanitin; RNA polymerase II is very sensitive to this poison; and RNA polymerase III is moderately sensitive to it. All three RNA polymerases, like their bacterial counterpart, have a molecular weight of about half a million.

As might be expected, the three enzymes recognize different start signals on the DNA. Both polymerase I and polymerase II recognize DNA sequences upstream from the start sites for transcription. The polymerase II recognition signal, for example, includes two short DNA sequences, one beginning about 25 nucleotides upstream from the initial point of RNA synthesis and another about 50 nucleotides further upstream. This organization of recognition sites is very much like that found for the bacterial RNA polymerase (see Figure 5–4, p. 203). By contrast, RNA polymerase III seems to recognize specific gene regulatory proteins, including the 5S transcription factor (p. 392); the binding of an oligomeric form of this protein to a sequence in the *middle* of the 5S RNA gene is both necessary and sufficient for RNA polymerase III to begin RNA synthesis about 45 nucleotides away (Figure 8–34). It would seem from this figure that the RNA polymerase III molecule should collide with the regulatory protein as the 5S RNA is synthesized, but, in fact, it seems that the protein remains bound without interfering with transcription as the polymerase passes.

The cells of "higher" eucaryotes contain about 40,000 molecules of RNA polymerase II, about the same number of RNA polymerase I molecules, and about 20,000 molecules of RNA polymerase III. But the precise concentration of all of the RNA polymerases varies, depending on the rate of cell growth. Much remains to be learned about these three eucaryotic RNA polymerases, which so far can be made to initiate RNA synthesis *in vitro* only in the presence of other poorly characterized proteins. It remains to be discovered why all eucaryotes, including organisms as different as yeast and man, should require three different RNA polymerases for transcription.

Because RNA polymerase II makes all of the mRNA precursors and thus determines which proteins a cell will make, we shall confine most of our discussion to the synthesis and fate of the RNA transcripts synthesized by this polymerase.

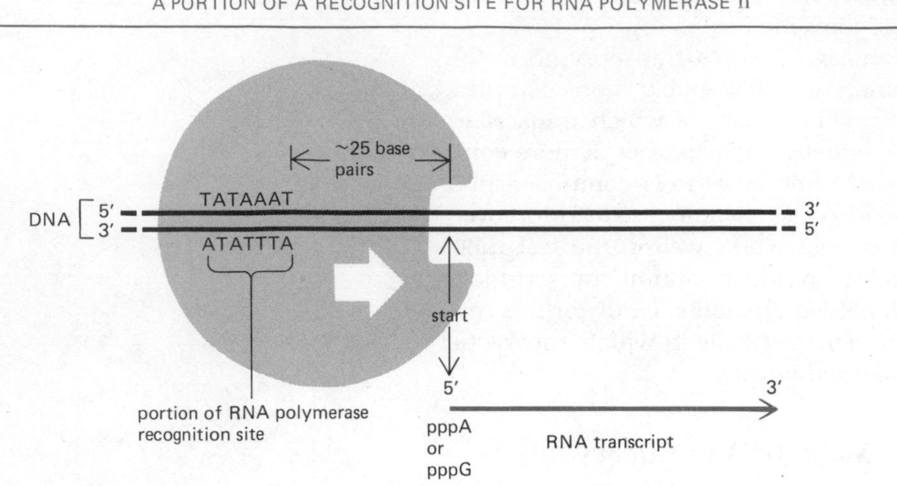

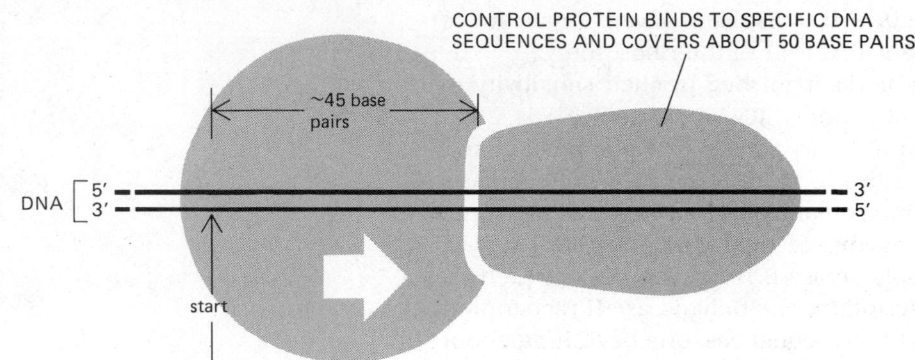

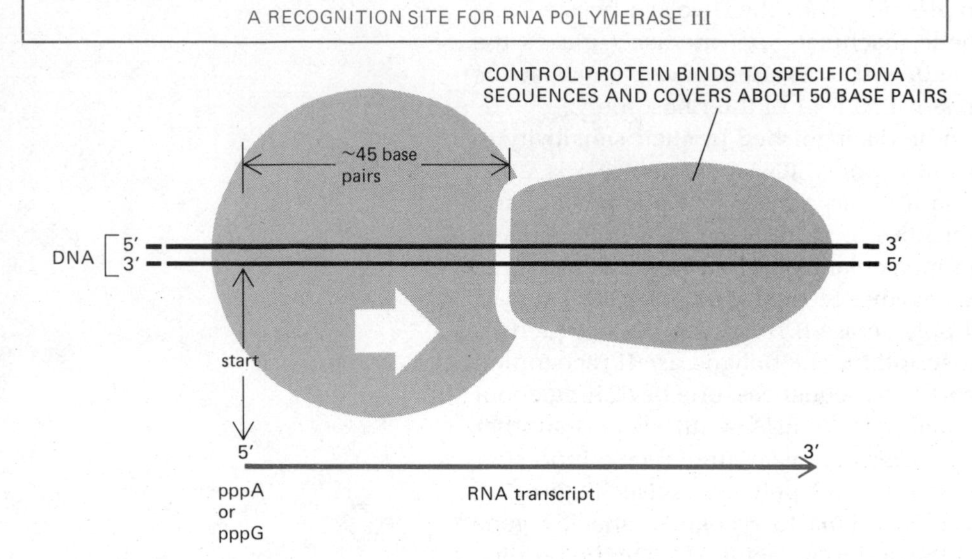

Figure 8–34 The different types of start sites recognized by eucaryotic RNA polymerases II and III. Like bacterial RNA polymerase and RNA polymerase I, RNA polymerase II appears to recognize specific DNA sequences located upstream from the point at which RNA synthesis begins. However, it appears that parts of the sequence recognized can extend as far as 150 nucleotides from the start of the RNA chain (only a small portion of the RNA polymerase II promoter sequence is illustrated here). In contrast, RNA polymerase III is unique in recognizing a gene regulatory protein (here the 5S transcription factor) that binds to a sequence downstream from the point where the polymerase begins its RNA synthesis.

RNA Polymerase II Transcribes Some DNA Sequences Much More Often Than Others[19,20]

Experiments with purified polymerases *in vitro* are essential for establishing the biochemical requirements for transcription, but they do not give a clear picture of the structure and dynamics of the transcriptional apparatus as it operates in the cell. Electron micrographs of cell chromatin, on the other hand, can catch large stretches of DNA with its bound RNA polymerases "in the act" of transcription.

Ordinary thin-section electron micrographs simply show granular clumps of chromatin, demonstrating that there is a good deal of substructure within the nucleus (see Figure 8–61), but they reveal very little about how genes are transcribed. A much more detailed picture emerges if the nucleus is ruptured and its contents spilled out onto an electron microscope grid (Figure 8–35). At the furthest point from the center of the lysed nucleus, the chromatin is diluted sufficiently to make individual chromatin strands visible in the expanded, beads-on-a-string form shown previously in Figure 8–3 (Figure 8–36).

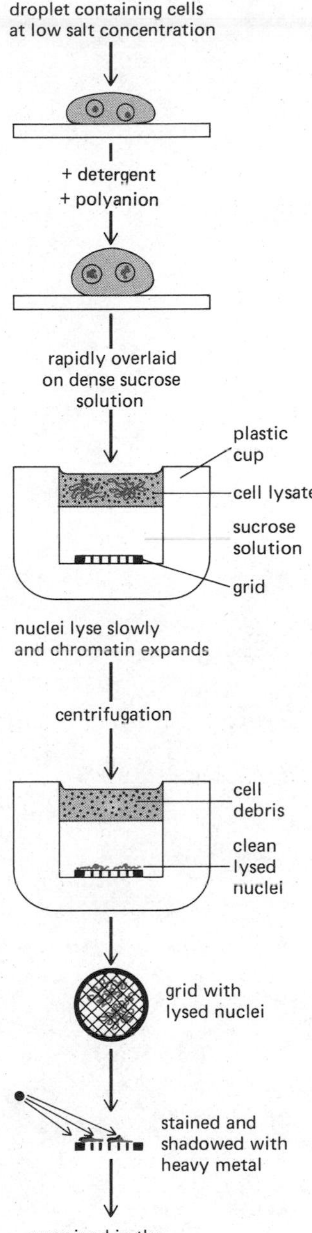

Figure 8–35 A method for examining the chromatin of a cell nucleus by electron microscopy after it has been gently spread out and freed from cellular debris.

RNA polymerase molecules actively engaged in transcription appear as large, globular particles with a single RNA molecule trailing behind (Figure 8–37). Particles representing active RNA polymerase II molecules are usually seen as single units, without any nearby neighbors. This indicates that most genes are transcribed into mRNA precursors only infrequently, so that one polymerase finishes transcription before another one begins. Occasionally, however, many polymerase particles (and RNA transcripts) are seen clustered together. These clusters occur on the relatively few genes that are transcribed at high frequency (Figure 8–38). The length of the attached RNA molecules in such a cluster increases in the direction of transcription, producing a characteristic "Christmas-tree" appearance. Each cluster defines a unique RNA polymerase II start site and stop site for a specific **transcription unit** (Figure 8–39).

Extensive biochemical studies have confirmed and extended the results obtained by electron microscopy, leading to three major conclusions:

1. Eucaryotic RNA polymerase molecules, like those in procaryotes, begin and end transcription at specific sites on the chromosome.
2. The average length of the finished RNA molecule produced by RNA polymerase II in a transcription unit is about 8000 nucleotides, and RNA molecules 10,000 to 20,000 nucleotides long are quite common. These lengths, much longer than the 1200 nucleotides of RNA needed to code for an average protein of 400 amino acids, reflect the peculiar structure of eucaryotic genes, which will be examined in detail later. For the moment, however, it is important to note that a transcription unit need not be precisely equivalent to a gene: some RNA transcripts may have no function, while others will give rise to products of more than one gene (see p. 418).
3. Different RNA polymerase II start sites function with very different efficiencies, and thus some genes are transcribed at much higher rates than others. The pattern of transcription observed in electron micrographs agrees well with the results of biochemical studies showing that while many different messenger RNA molecules accumulate in a cell, most of them are present at relatively low frequency (Table 8–1).

Table 8–1 The Population of mRNA Molecules in a Typical Mammalian Cell

	Copies per Cell of Each mRNA Sequence		Number of Different mRNA Sequences in Each Class		Total Number of mRNA Molecules in Each Class
Abundant class	12,000	×	4	=	48,000
Intermediate class	300	×	500	=	150,000
Scarce class	15	×	11,000	=	165,000

This division of mRNAs into just three discrete classes is somewhat arbitrary. In some cells a more continuous spread in abundances is seen. However, a total of 10,000 to 20,000 different mRNA species is normally observed in each cell, most species being present at a low level (5 to 15 molecules per cell). Usually 95% to 97% of the total cytoplasmic RNA is rRNA and only 3% to 5% is mRNA, a ratio consistent with the presence of about 10 ribosomes per mRNA molecule. This particular cell type contains a total of about 360,000 mRNA molecules in its cytoplasm.

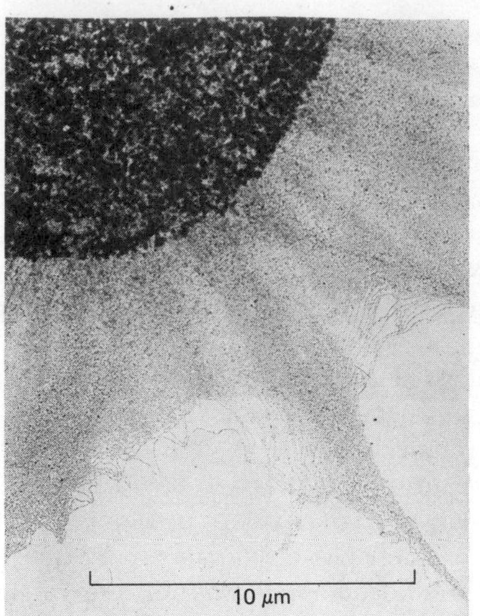

Figure 8–36 A typical cell nucleus visualized by electron microscopy by the procedure shown in Figure 8–35. An enormous tangle of chromatin can be seen spilling out of the lysed nucleus; only the chromatin at the outermost edge of this tangle will be sufficiently dilute for meaningful examination at higher power. (Courtesy of Victoria Foe.)

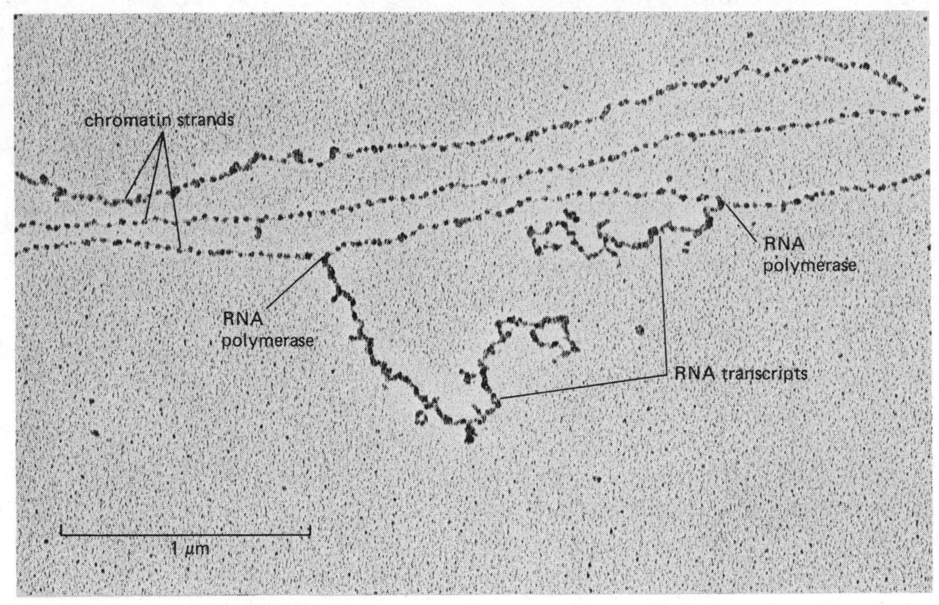

Figure 8–37 A region of chromatin in its beads-on-a-string form, showing two RNA polymerase II molecules engaged in the process of transcription. Most of the chromatin in a higher eucaryotic nucleus lacks such RNA transcripts; in fact, with about 40,000 RNA polymerase II molecules per cell, there is only about 1 RNA polymerase molecule for every 750 nucleosomes (150,000 DNA base pairs). (Courtesy of Victoria Foe.)

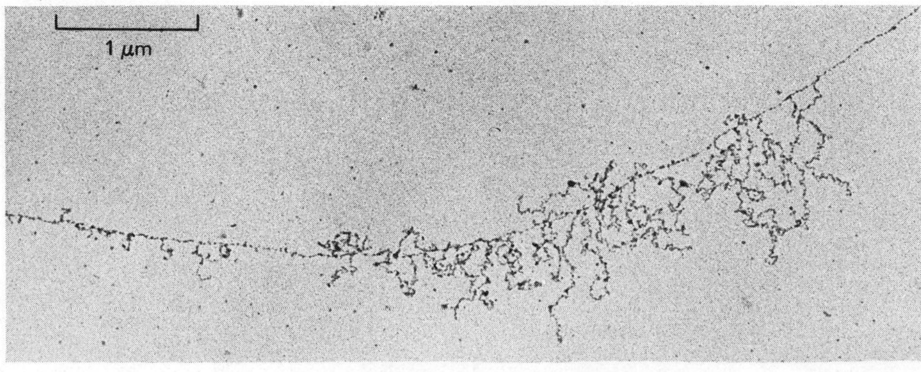

Figure 8–38 An unusual region of chromatin containing a gene that is being transcribed at high frequency, so that many RNA polymerase II molecules with their growing RNA transcripts are visible at the same time. (From V. E. Foe, L. E. Wilkinson, and C. D. Laird, *Cell* 9:131–146, 1976. © 1976 M.I.T. Press.)

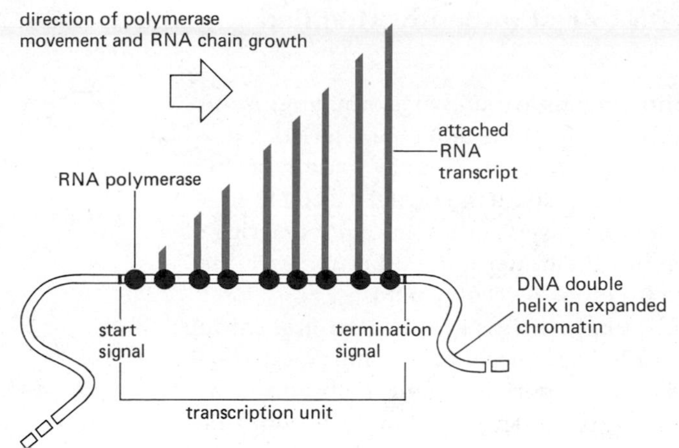

Figure 8–39 An idealized transcription unit, showing how the electron microscope appearance is used to determine the direction of RNA chain growth.

Transcription Occurs on DNA Bound Up in Nucleosomes[20]

The complex formed by the four nucleosomal histones (H2A, H2B, H3, and H4) is so firmly associated with DNA that each histone octamer (Figure 8–4) rarely, if ever, leaves the particular region of DNA to which it is bound. It is now clear that the DNA remains bound up in nucleosomes even during DNA transcription and DNA replication. Therefore, it is not surprising that electron micrographs of spread chromatin normally reveal a regular pattern of nucleosome beads throughout both the transcribed and untranscribed regions of the chromatin (see Figures 8–37 and 8–38).

The fact that RNA in eucaryotes must be transcribed from a DNA-histone complex, rather than from bare DNA as in procaryotes, seems to have important implications for the way in which eucaryotic genes are regulated. Even in the case of the expanded beads-on-a-string form of chromatin, it is difficult to imagine RNA polymerase transcribing the DNA bound up in a nucleosome without some temporary change in the conformation of the nucleosome itself. It is inconceivable that DNA in a more condensed form of chromatin, such as the 30-nm fiber, could be transcribed by the RNA polymerase without a major disruption of the nucleosome packing (Figure 8–40).

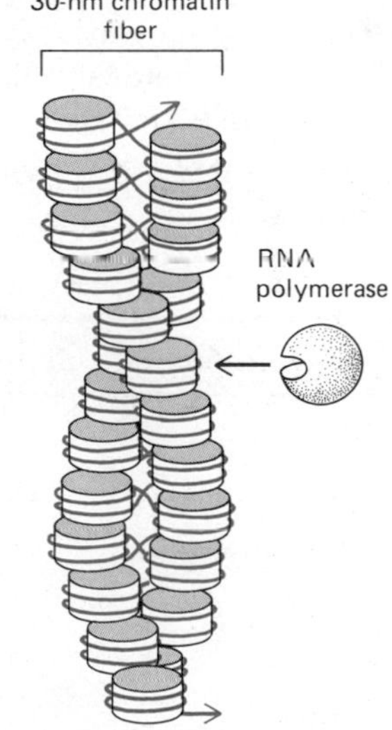

Figure 8–40 Schematic illustration of an RNA polymerase molecule approaching a 30-nm chromatin fiber, drawn roughly to scale. The number of nucleosomes shown is approximately that present in an average transcription unit. Somehow the polymerase must gain access to the DNA without displacing the histone octamers from the chromatin. A major unfolding of the chromatin is required to make this possible.

New RNA Is Packaged in Ribonucleoprotein Particles[21]

Newly made RNA in eucaryotes, unlike that in bacteria, has a distinct beaded appearance in the electron microscope (see Figure 8–37). These beads represent RNA-protein complexes—not unlike the DNA-protein complexes of nucleosomes—that form along the RNA chain, serving to condense and package each RNA transcript. Among the many different kinds of such *ribonucleoprotein* packages, one predominates. The predominant particle has been characterized as a short region of RNA transcript covered by a few proteins of 30,000 to 40,000 molecular weight. The proteins in this particle are among the most abundant proteins in the cell nucleus.

The spacing and size of the particles seen in RNA transcripts varies, but the pattern tends to be similar in adjacent RNA molecules from the same transcription unit. This suggests that there is considerable specificity in the RNA packaging process. Just as the packaging of DNA into chromatin is important for its function, the as yet poorly understood packaging of RNA by nuclear proteins is thought to guide the primary RNA transcripts through the subsequent RNA processing and transport events, which we shall describe next.

The Precursors of Messenger RNA Are Covalently Modified at Both Ends[22]

All RNA polymerase II transcripts in the nucleus are known as **heterogeneous nuclear RNA (hnRNA)** molecules because one of the first characteristics used to distinguish them from other RNAs in the nucleus was the variety of their sizes. Many of these transcripts are destined to leave the nucleus as messenger RNA (mRNA) molecules. Before they leave, however, they undergo a series of covalent modifications that are related to their later function and clearly distinguish them from transcripts made by other RNA polymerases.

First, the 5′ end of the RNA molecule (which is synthesized first during transcription) is *capped* by the specialized structure that will later mediate its binding to a ribosome (p. 331). Capping occurs immediately, before the transcription of the rest of the molecule is complete. The RNA molecule continues to grow in the 5′-to-3′ direction at a rate of about 30 nucleotides per second until it reaches the termination signal in the chromatin, halting transcription. Second, for most transcripts destined to become mRNA molecules, a separate *poly-A polymerase* enzyme adds 100 to 200 residues of adenylic acid (as *poly A*) to the 3′ end of the RNA chain to complete the **primary RNA transcript** (Figure 8–41). The site of polyadenylation (poly-A addition) is created either by a cleavage of the growing chain or by chain termination by the RNA polymerase. The function of the poly-A tail is not known for certain, but it seems to help mediate subsequent RNA processing and the export of the mature mRNA from the nucleus.

Because only RNA polymerase II transcripts have 5′ caps and 3′ poly-A tails, it has been suggested that the capping or poly-A addition reactions (or both) are mediated by enzymes that interact with polymerase II and not with polymerases I or III. The importance of marking the ends of the mRNA precursors in a distinct way may explain why they are synthesized by a separate type of RNA polymerase molecule.

Even though polymerase II transcripts comprise more than half of the RNA synthesized by a cell (Table 8–2), we shall see below that these transcripts

Table 8–2 Selected Data on Amounts of RNA in a Typical Mammalian Cell

	Steady-State Amount (percent of total cell RNA)	Percent of Total RNA Synthesis
Nuclear rRNA precursors ↓	4	39
Cytoplasmic rRNA	71	—
Nuclear hnRNA ↓	7	58
Cytoplasmic mRNA	3	—
Small stable RNAs (mostly tRNAs)	15	3

The figures shown here were derived from the analysis of mouse fibroblast tissue culture cells (L cells), each of which contained 26 pg of RNA (5×10^{10} nucleotides of RNA), of which about 14% was located in the cell nucleus. (The cell nucleus thus contains about twice as much DNA as RNA.) An average of about 200×10^6 nucleotides is polymerized into RNA every minute during interphase. This is about 20 times the average rate at which DNA is synthesized during the S phase. Note that although most of the RNA synthesized is hnRNA, only a small fraction of this RNA escapes degradation. As a result, the mRNA produced from the hnRNA is only a minor fraction of the total RNA in the cell. (Modified from B. P. Brandhorst and E. H. McConkey, *J. Mol. Biol.* 85:451–563, 1974.)

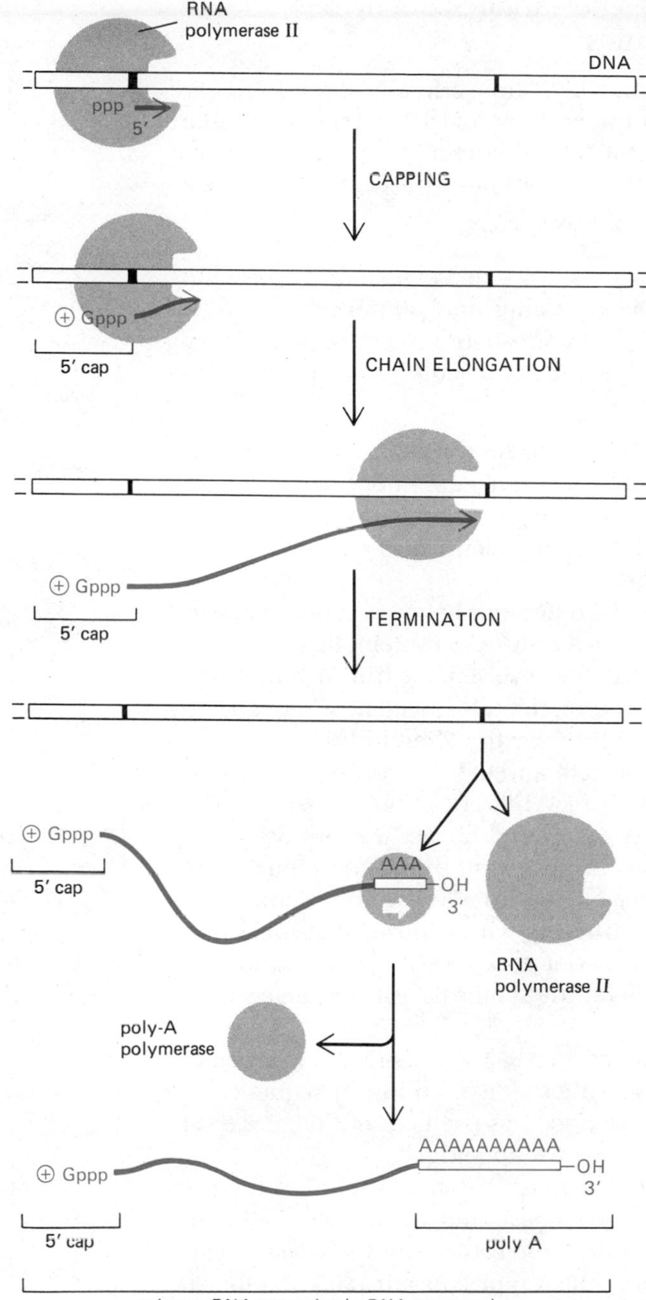

Figure 8–41 RNA transcripts produced by RNA polymerase II are covalently modified by capping at the 5′ end and by the addition of poly A at the 3′ end. The (+) charge on the cap is produced by methylation of the nitrogen at the 7-position of the guanine base, which occurs shortly after the guanyl-transfer step (not shown). The detailed cap structure was illustrated in Figure 7–14, page 331.

are unstable and short-lived. Consequently, hnRNA in the cell nucleus and the cytoplasmic mRNA derived from it constitute only a minor fraction of the total RNA present in a cell (Table 8–2). Fortunately for biochemists, the long stretch of poly A characteristic of the 3′ end of these RNA molecules provides a convenient way of purifying them despite their relative scarcity. When the total cellular RNA is passed through a column containing poly dT linked to a solid support, the complementary base-pairing between T and A residues selectively binds the molecules with poly-A tails to the column; the bound molecules can then be released for further analysis (Figure 8–42). This procedure is widely used for separating the hnRNA and mRNA molecules from the ribosomal and transfer RNA molecules that predominate in cells.

RNA Processing Removes Long Nucleotide Sequences from the Middle of RNA Molecules[22]

The first evidence that the RNA polymerase II transcripts in the nucleus are unstable came from studies in which the addition of [^{3}H]uridine over a short period was used to introduce radioactivity into these hnRNA molecules, which could then be followed over a longer period of time. These and later experiments resulted in two remarkable discoveries:

1. The length of the newly made RNA molecules decreases rapidly, reaching the size of cytoplasmic mRNA molecules after only 30 minutes or so. On average, the primary RNA transcripts contain about 6000 nucleotides, while mature mRNA molecules contain about 1500 nucleotides in the tissue culture cells studied.
2. After about 30 minutes, radioactively labeled RNA molecules begin to leave the nucleus as mRNA molecules. However, only about 5% of the mass of the labeled RNA ever reaches the cell cytoplasm. The remainder is degraded over a period of an hour or so into small fragments in the cell nucleus.

These observations point to a crucial difference between procaryotes and higher eucaryotes in the pathway from DNA to RNA to protein. In procaryotes, in which the basic genetic mechanisms first evolved, each RNA molecule is composed of a continuous string of the nucleotides needed to encode the amino acids of a protein—the nucleotides near the 5′ end of the RNA chain specifying the amino terminus of the protein and those near the 3′ end specifying its carboxyl terminus (Figure 8–43, top). While eucaryotic transcripts have the same polarity (nucleotides toward the 5′ end of the RNA always specify amino acids at the amino terminus of the protein), they often contain long insertions of noncoding RNA sequences copied from regions of a gene known either as **intervening sequences** or as **introns.** These intron sequences must be cut out of each RNA transcript in order to convert the transcript to a messenger RNA molecule that can code for the synthesis of a complete protein (Figure 8–43, bottom).

The discovery of interrupted genes in 1977 was entirely unexpected. Because the coding RNA sequences on either side of an intron sequence are joined to each other after the intron sequence has been cut out (Figure 8–44), the RNA processing reaction is known as **RNA splicing.**

RNA splicing, which appears to occur in the cell nucleus before the RNA is exported to the cytoplasm, may be extensive enough to account for the conversion of the very long nuclear hnRNA molecules (up to 50,000 nucleotides) to the much shorter cytoplasmic mRNA molecules (usually 500 to 3000 nucleotides). Thus the nuclear envelope in eucaryotes seems to serve to keep newly made RNA molecules away from the cytoplasmic ribosomes until RNA processing reactions have sorted out the sections that are to be translated into proteins.

Analyses based on nucleic acid hybridization (p. 189) show that most of the rapidly degraded nuclear RNA sequences are different from RNA sequences that survive. The removal of introns helps to account for the fact that while 7% to 10% of the total DNA sequences in a typical mammalian cell are transcribed into RNA in the nucleus, the cytoplasmic mRNA represents only about 1% or 2% of the DNA sequences. Much of the difference is thought to be due to the discarded intron sequences.

Before there can be a discussion of how and why such a complicated genetic system may have come about, it is necessary to explain what is known about intron sequences and the mechanics of their removal.

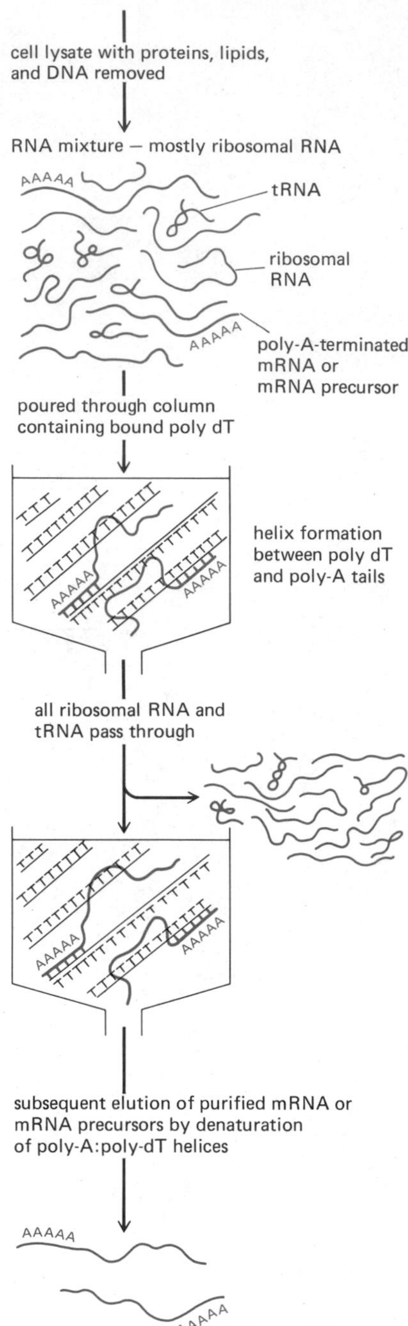

Figure 8–42 The purification of mRNA molecules and their precursors by the selective annealing of the poly A at their 3′ ends to a column containing a covalently bound synthetic polynucleotide composed of T or U residues. In the example shown, polydeoxythymidylic acid (poly dT) is linked to the column.

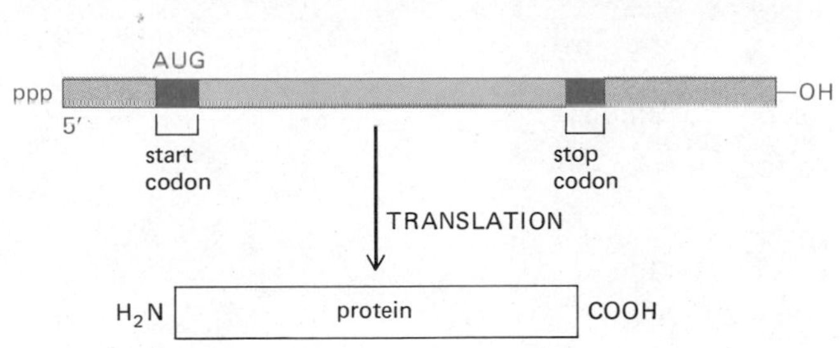

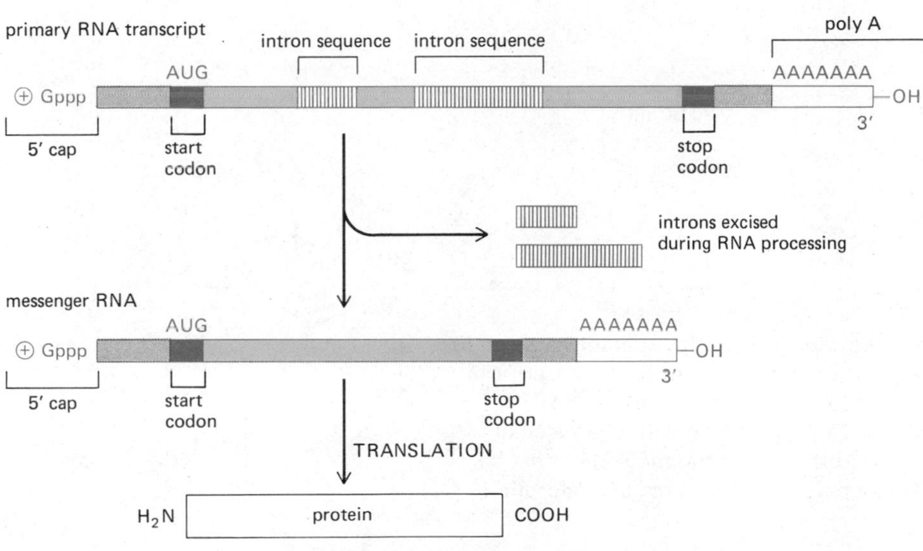

Figure 8–43 Comparison of the nucleotide sequence of procaryotic and eucaryotic RNA transcripts. Only the eucaryotic transcripts contain intervening (intron) sequences, which must be cut out before an mRNA molecule is translated into a protein.

Small Nuclear Ribonucleoprotein Particles May Help Guide RNA Processing[22,23]

Introns range in size from fewer than 100 nucleotides to 10,000 nucleotides or more. They differ dramatically from coding sequences in that most of the nucleotide sequence of an intron can often be experimentally altered without greatly affecting gene function. Moreover, introns seem to have accumulated mutations rapidly during evolution, leading to the suggestion that they have no function at all and are largely genetic "junk"; this proposition will be examined later. There are, however, a few nucleotides at each end of an intron that cannot be altered without disrupting gene function, which are nearly the same in all introns whose sequences are known. These boundary sequences are believed to be the signals for RNA splicing. At present, we have only one important clue about how they work: the conserved boundary sequences have proved to be complementary to a region of the RNA molecule contained in an abundant **small nuclear ribonucleoprotein particle (snRNP).** This particle is a minor component of the ribonucleoprotein that forms on RNA transcripts and may participate in RNA splicing by forming an RNA-RNA double

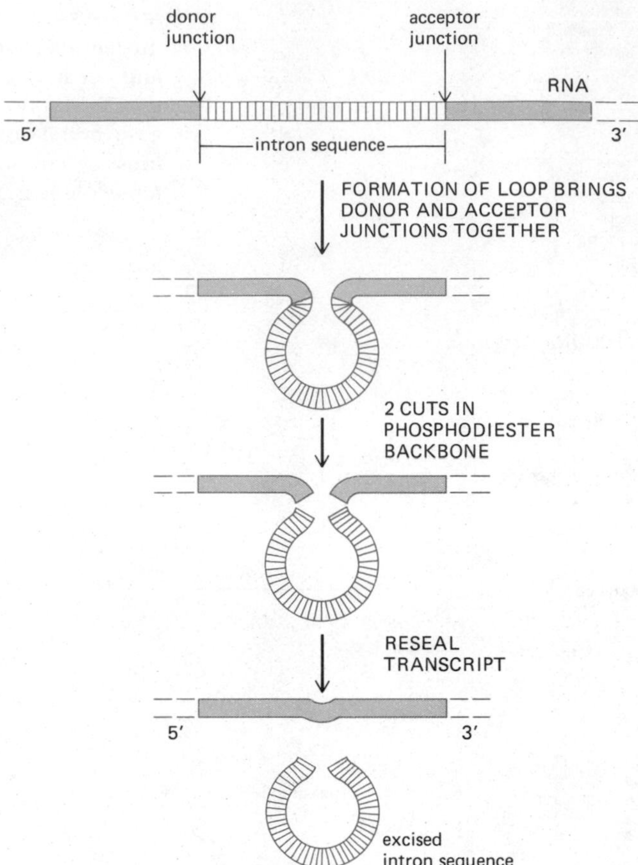

Figure 8–44 Schematic view of the RNA splicing reaction in which an intron sequence is cut out of an RNA transcript to produce a continuous coding sequence in an mRNA molecule. During this RNA processing reaction, a piece of RNA must be excised and the rest of the molecule must be resealed at paired sequences called "donor" and "acceptor" splice junctions.

helix with the ends of each intron sequence so as to bring them together (Figure 8–45). The formation of this structure is thought to be accompanied by enzymatic cleavage and resealing reactions that remove the intron sequence, but leave the coding sequences intact (Figure 8–44). These reactions must be carried out precisely because an error of even one nucleotide would shift the reading frame in the resulting mRNA molecule and make nonsense of its message.

Besides the snRNP particle just discussed, there are many other classes of small ribonucleoprotein particles. Like the particles thought to function in RNA splicing (present in 10^6 copies per cell), these other particles are abundant, and many of them are located in the cell nucleus. It is likely that some of these ribonucleoprotein particles play a part in guiding specific RNAs out of the cell nucleus (p. 422), while others help direct other types of RNA processing events, including those associated with assembly of new ribosomes in the nucleolus (p. 424).

In many respects, all of these ribonucleoprotein particles resemble ribosomes. Although they are much smaller (about 250,000 daltons each, compared with 4.5 million daltons for a ribosome), they too contain multiple poly-

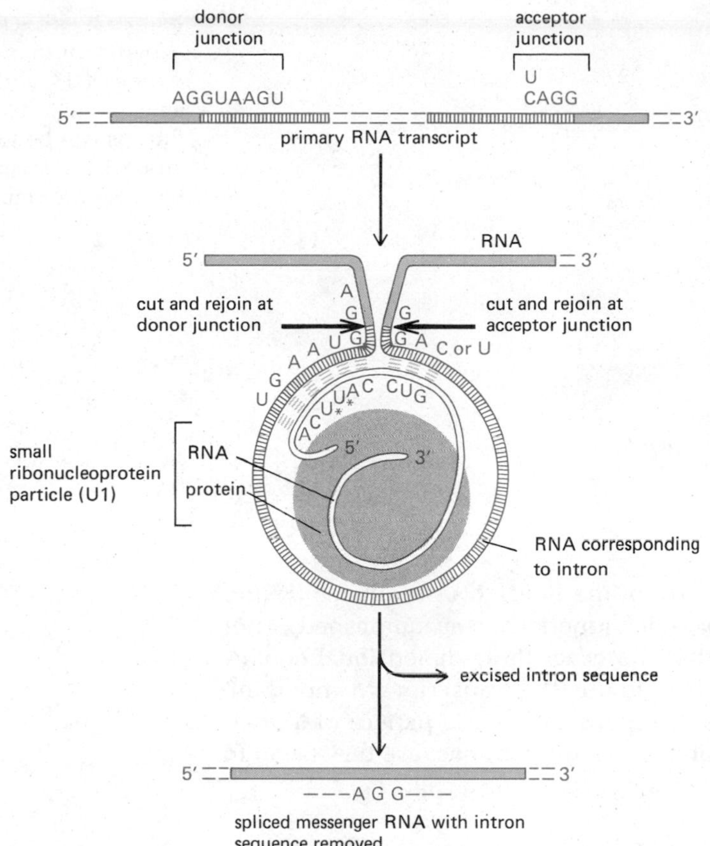

Figure 8–45 Schematic view of the postulated participation of a specific small nuclear ribonucleoprotein particle, known as *U1*, in the RNA splicing reaction. This particle contains a small bound RNA molecule (165 nucleotides long), part of whose sequence is complementary to nucleotide sequences at the intron boundaries on hnRNA molecules. By complementary base-pairing, it has been suggested that U1 RNA could hold the donor and acceptor splice junctions together as a preliminary to the splicing reactions shown in Figure 8–44.

peptide chains complexed to a stable RNA molecule. Like ribosomes, small ribonucleoprotein particles are believed to recognize specific nucleic acid sequences through RNA-RNA base-pair complementarity, thereby helping to position protein subunits that catalyze various complex reactions.

Multiple Intron Sequences Can Be Removed from a Single RNA Transcript[22,24]

If intron sequences are removed from RNA in the way shown in Figure 8–45, the sequence at the 5′ end of any one intron (called for convenience the *donor junction*) should be able to join with that at the 3′ end of any other intron (called for convenience the *acceptor junction*) in the splicing process. This prediction has been confirmed by experiments in which the donor and acceptor halves of two different introns are combined. The resulting hybrid intron sequence is recognized by the RNA-splicing enzymes and removed (Figure 8–46).

In view of the apparent equivalence of all donor and of all acceptor junctions of introns in RNA splicing, it is surprising that many of the vertebrate genes thus far cloned and analyzed contain large numbers of introns (7 in the case of the egg-white protein, ovalbumin, and more than 50 in the case of a procollagen α-chain, to give an extreme example). If donor and acceptor junctions were randomly paired for splicing, functional mRNA sequences between introns would be lost, with disastrous consequences. In order to make a particular protein, the RNA processing machinery must guarantee that each donor junction usually pairs only with the acceptor junction that is closest to

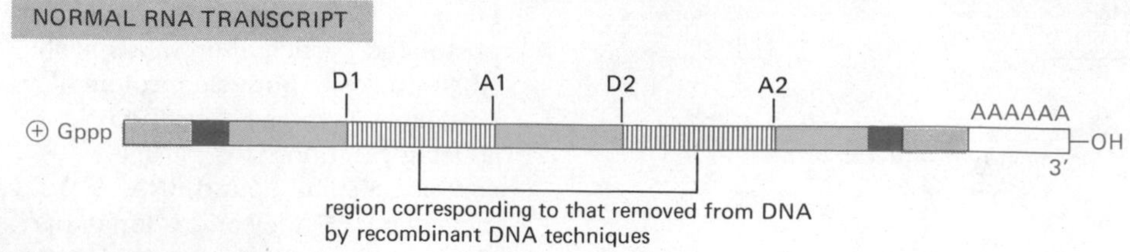

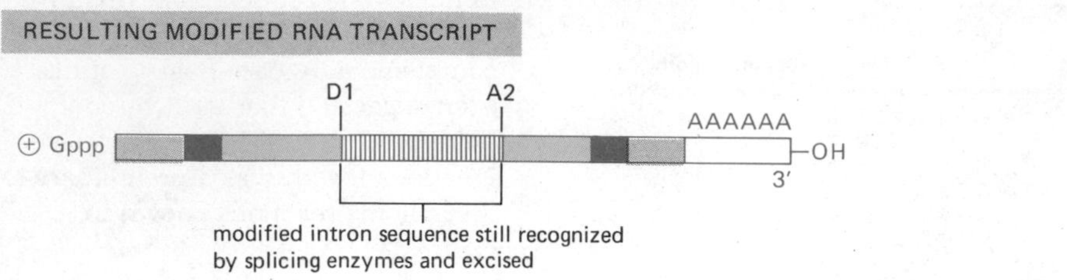

Figure 8–46 Outline of an experiment demonstrating that a donor ("D1") and an acceptor ("A2") splice junction from two different introns can be joined to each other in an RNA transcript if their normal partners are removed.

it in the downstream (5′-to-3′) direction of the linear RNA sequence (Figure 8–47). How this sequential pairing of splice junctions is accomplished is not yet known, although it is thought that the exact three-dimensional conformations adopted by the intron sequences in the RNA transcript are important. However, we shall now see that in some cases the simple pattern of 5′-to-3′ splicing does not hold. These exceptions are important because they begin to show what advantages interrupted genes may have for a cell.

The Same RNA Transcript Can Be Processed in Different Ways to Produce mRNAs Coding for Several Proteins[25]

Although most intron sequences themselves appear to have no specific function, there is increasing evidence that the existence of the splicing apparatus associated with them confers extra genetic flexibility on the cell. This flexibility

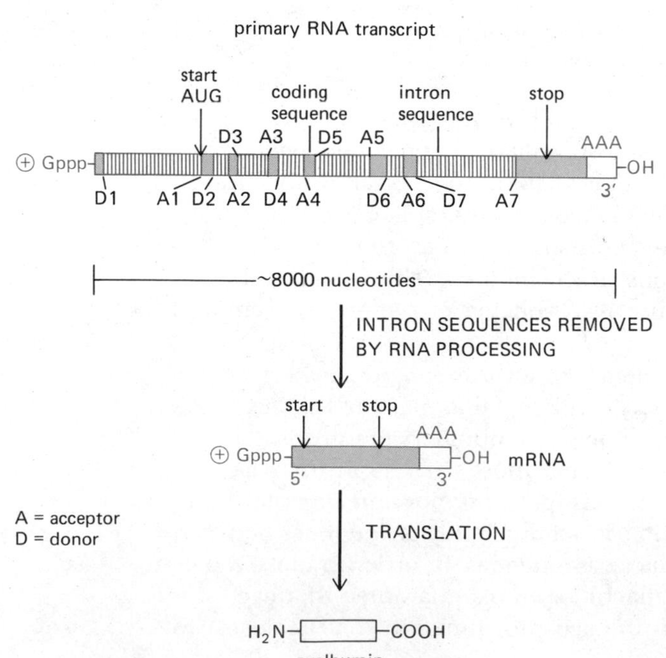

Figure 8–47 The primary RNA transcript for the chicken ovalbumin gene, showing the organized removal of seven introns required to obtain a functional mRNA molecule. As in Figure 8–46, donor splice junctions are denoted by "D," and acceptor splice junctions by "A."

is possible because different patterns of RNA splicing can generate several different proteins from the same RNA transcript. Changes in the pattern of splicing of some transcripts are known to occur in the course of cell differentiation, so that the same DNA coding sequences are put to different uses as the cell develops.

The versatility conferred by RNA splicing was first discovered in adenovirus, a large virus that infects animal cells. The adenovirus genome directs the synthesis of several very long RNA transcripts, each of which contains the coding sequences for a number of different essential proteins. Alternative RNA processing pathways then generate different mRNA molecules from identical primary transcripts, allowing the same 5′-cap sequence to be used for initiating the synthesis of several different proteins. Thus the rule of sequential pairing of splicing signals is plainly violated.

Recall that the single capped nucleotide at the 5′ end of each RNA transcript made by an RNA polymerase II molecule later serves in a mRNA molecule as a signal for the initiation of protein synthesis (p. 331). Usually, even if mRNA molecules contain multiple start and stop codons, nothing beyond the first stop codon downstream from the 5′ cap will be translated because the stop codon reached automatically detaches the mRNA from the ribosome. For this reason, an individual RNA transcript can code only for one protein once it has been processed into mRNA. On the other hand, the same nuclear transcript can produce several different messenger RNAs, if the RNA processing machinery treats some of the coding sequences as introns and removes them. Now the 5′ cap can be spliced to the beginning of a variety of coding sequences and serve as the initiation signal for the synthesis of several different proteins (Figure 8–48). How the variable processing of identical RNA transcripts is controlled is not known.

Possibly the advantage to a virus of adopting this alternative RNA processing strategy is economy: viruses need to minimize the size of their genome in order to package it in virus particles.

Different Proteins May Be Made from a Single DNA Coding Sequence at Different Stages of Cell Development[26]

At least one type of cellular coding sequence is now also known to produce an RNA transcript that is processed in more than one way. Instead of producing an entirely different protein, however, a programmed change in the primary RNA transcript during the development of the cell causes a change in the pattern of splicing, which produces a new variant of the same protein.

The change takes place in the production of antibody molecules by lymphocytes (see p. 985). Early in the life history of the lymphocyte, the antibody is anchored in the plasma membrane. Later the same antibody is secreted. The secreted form is identical to the membrane-bound form except at the extreme carboxyl terminus, where the membrane-bound form has a long string of hydrophobic amino acids that anchors it in the membrane, while the secreted form has a much shorter string of water-soluble amino acids. The switch from membrane-bound to secreted antibody therefore requires a different nucleotide sequence at the 3′ end of the mRNA, which codes for the carboxyl terminus of the protein.

The membrane-bound form of the protein is generated by the transcription of all of the coding sequences into nuclear RNA to make a long primary transcript. The nucleotides coding for the hydrophobic carboxyl terminus of the membrane-bound protein are located near the end of this long transcript, just beyond an acceptor splice junction (Figure 8–49). The normal RNA splicing processes will join these nucleotides to the nucleotides coding for the rest of the protein. During this joining event, the nucleotides coding for the water-

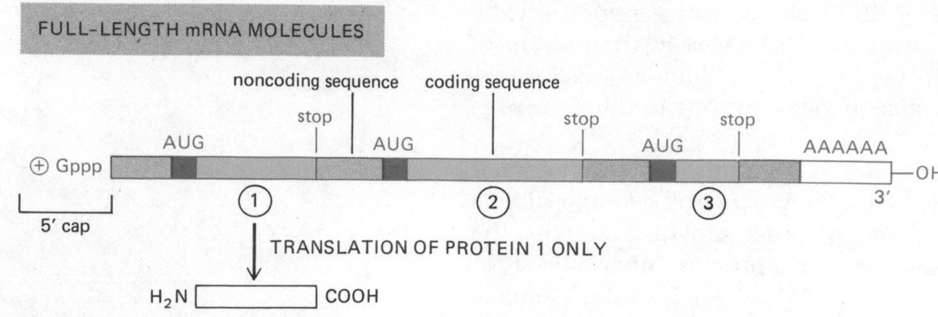

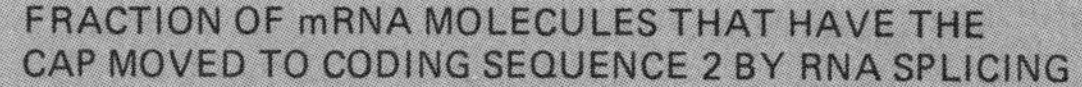

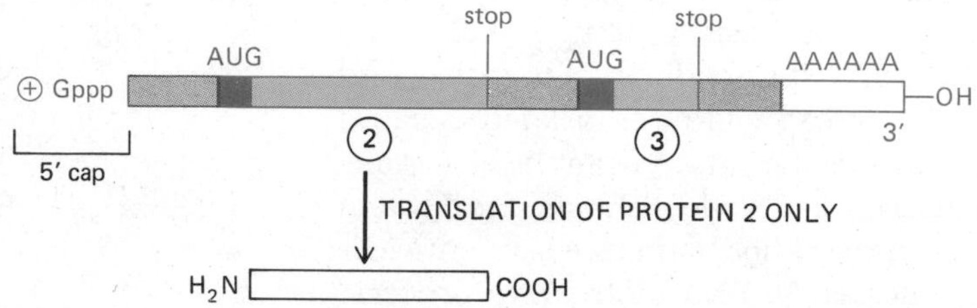

FRACTION OF mRNA MOLECULES THAT HAVE THE
CAP MOVED TO CODING SEQUENCE 3 BY RNA SPLICING

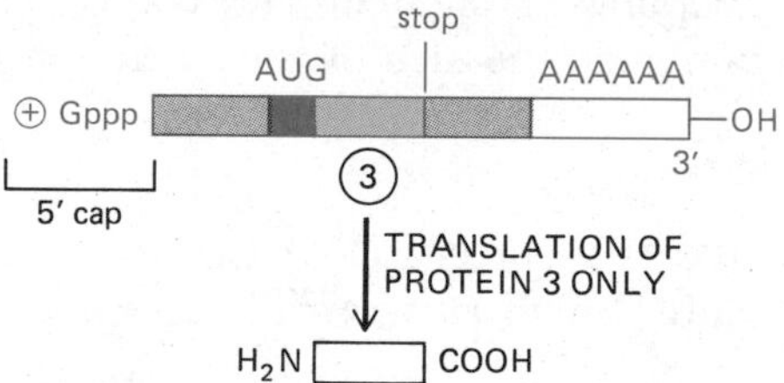

Figure 8–48 For some viruses the same primary RNA transcript is processed in several different ways, being exported to the cell cytoplasm to produce three (or more) different mRNA molecules, each coding for a different protein. In each case, only the coding sequence closest to the 5′ cap is translated from the mRNA molecule.

soluble tail of the secreted molecule are removed: the last donor splicing signal in the transcript falls just before the nucleotides coding for these water-soluble amino acids, and these nucleotides are therefore treated as part of the intron that is spliced out (Figure 8–49).

The secreted form of the molecule, on the other hand, is generated from a shorter primary transcript that ends before the last acceptor splice junction in the long transcript, thereby eliminating the nucleotides coding for the hydrophobic carboxyl terminus of the membrane form, which lie beyond. Since no acceptor site now remains to combine with the donor site adjacent to the nucleotides coding for the water-soluble amino acids, they remain in the final mRNA molecule and are decoded (Figure 8–49).

There are two ways in which the shorter transcript for the secreted molecule could be generated: by the premature arrest of transcription or by the cleavage of the long primary transcript followed by the addition of poly A to the new 3′ end created. Experiments with viruses have shown that different 3′ ends of a primary transcript can be generated by cleavage and the addition of poly A at different points. But exactly how the transcript length of RNAs coding for antibodies is controlled to produce the programmed kind of switch just described is not known.

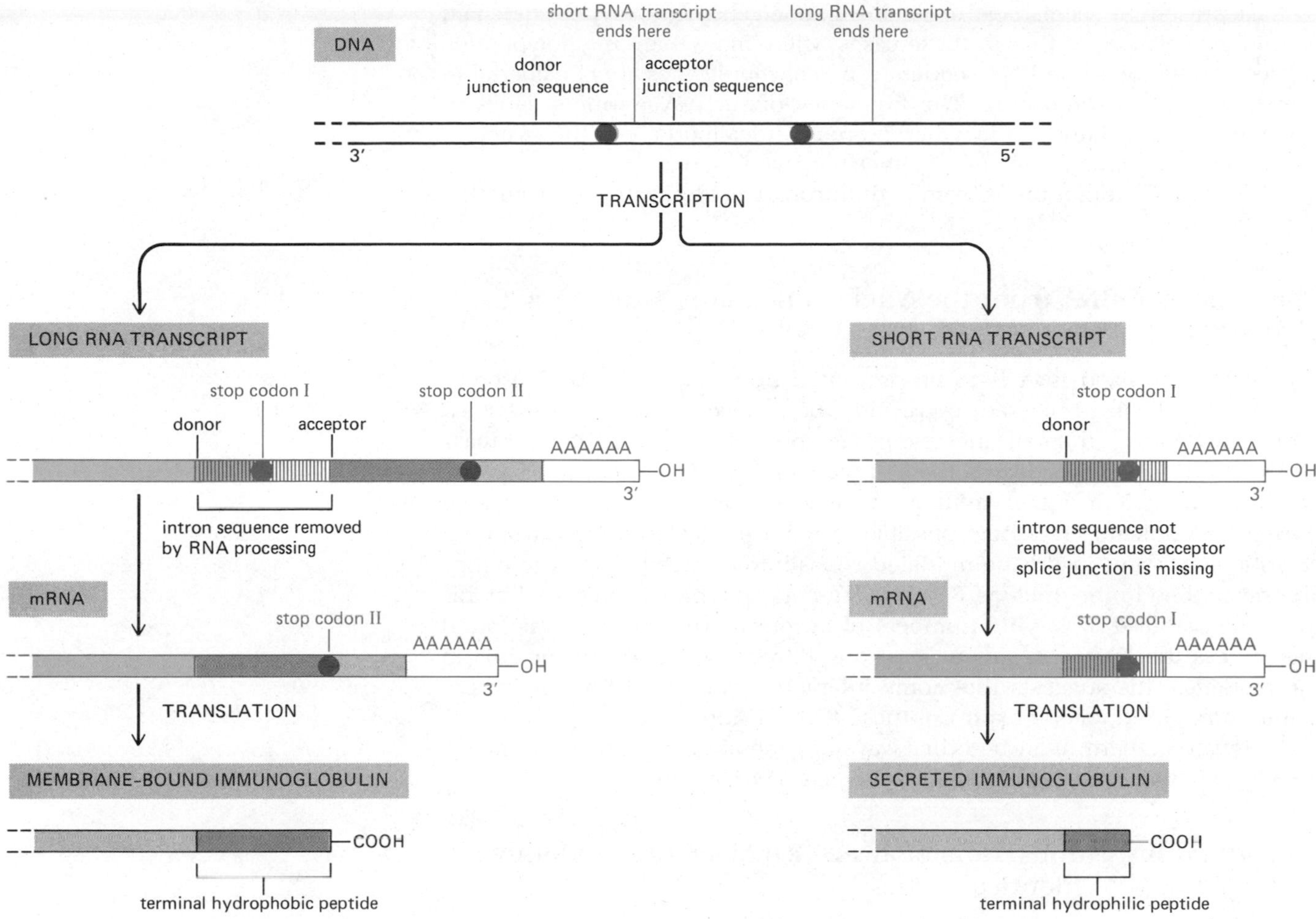

Figure 8–49 The synthesis of two closely related proteins from the same DNA sequence. The intron sequence near the 3′ end of the long primary RNA transcript made from an antibody gene is removed by RNA splicing. It thereby gives rise to an mRNA molecule that codes for a membrane-bound antibody molecule (immunoglobulin). In contrast, a shortened primary RNA transcript is produced, after antigen stimulation, that lacks the indicated acceptor splice junction. The adjacent intron sequence that was removed from the long transcript is therefore not removed from the short transcript. As a result, a hydrophilic carboxyl-terminal portion of the antibody molecule is synthesized by the translation of mRNA sequences that were previously located in the excised intron sequence. A secreted rather than a membrane-bound antibody molecule is the result.

What Is a Gene?[27]

The discovery that eucaryotic genes contain introns and that their coding sequences can be put together in more than one way has raised new questions about the precise definition of a gene. The gene was first clearly defined in molecular terms in the early 1940s from work carried out on the biochemical genetics of the fungus *Neurospora.* Much earlier, a **gene** had been defined operationally as a region of the genome that segregates as a single unit during meiosis and gives rise to a definable phenotypic trait such as a red or a white eye in *Drosophila* or a round or wrinkled seed in peas. After the work on *Neurospora,* it became clear that most genes correspond to a region of the genome that directs the synthesis of a single enzyme: this led to the central principle of *"one gene, one polypeptide chain."* This hypothesis proved to be enormously fruitful for subsequent research. Thus, as more was learned about the mechanism of gene expression in the 1960s, a gene became identified as that stretch of DNA that was transcribed into the RNA coding for a single polypeptide chain (or a single structural RNA such as a tRNA or rRNA molecule). The discovery of split genes in the late 1970s could be readily accommodated by the original definition of a gene, provided that a single polypeptide chain was specified by the RNA transcribed from any one DNA sequence. But it is now known that some DNA sequences, because of variable RNA splicing, participate in the production of at least two different mRNA molecules and therefore of at least two different proteins with distinct biological roles. How then is a gene to be defined?

At present, it seems best to retain the one-gene–one-polypeptide-chain definition. This means that in those cases where more than one polypeptide is specified by the same DNA sequence, two or more genes are considered to overlap on the chromosome. The frequency of such overlapping genes in eucaryotes will be known only when more is understood about the structure and function of large numbers of higher eucaryotic genes.

In the meantime, the existence of introns has implications for another important process.

The Export of RNA from the Nucleus Requires Molecular Signals[28]

If newly synthesized RNA is to be prevented from encountering ribosomes before it has been processed, there must be a selective mechanism for exporting RNAs from the cell nucleus. In fact, only about 5% of the total mass of RNA transcribed ever leaves the cell nucleus. Specific export signals have not been identified, but in some cases at least they seem to be generated during RNA splicing. It is now possible to remove selected introns from eucaryotic genes and insert the modified genes into cultured cells, where they are transcribed in the nucleus. Several different genes have been tested in this way after removal of varying numbers of introns. Surprisingly, it was found that unless the DNA contains at least one intron, the transcript remains in the nucleus. This suggests that some interaction with the RNA-splicing enzymes is required for the export of these RNAs (Figure 8–50).

However, there must be other ways of generating export signals for an mRNA molecule, since not all eucaryotic genes contain introns.

Ribosomal RNAs and Transfer RNAs Are Made on Tandemly Arranged Sets of Identical Genes[29]

Thus far we have discussed the fundamental mechanisms involved in gene transcription and the RNA processing events that take place afterward. We shall now direct our attention to how some special gene products required in especially large amounts are synthesized in cells. It turns out that many of the most abundant proteins of a differentiated cell, including the hemoglobin

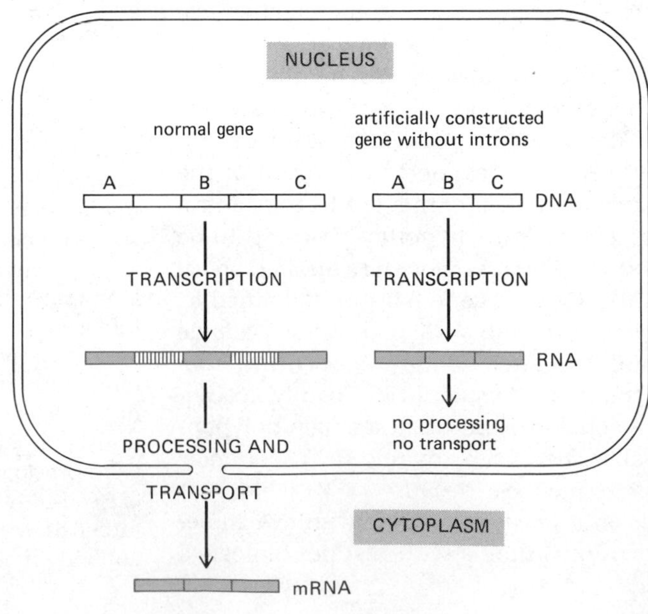

Figure 8–50 Schematic outline of an experiment demonstrating that the actual excision of an intron sequence—and not just its absence—is required for the transport of some mRNA molecules to the cell cytoplasm.

in the red blood cell and the myoglobin in a muscle cell, are synthesized from genes that are present in only a single gene per haploid genome. The abundance of these proteins is due to the fact that each of the many mRNA molecules transcribed from the gene can be translated into as many as 10 protein molecules per minute. This will normally produce more than 10,000 protein molecules per mRNA molecule each cell generation. However, a growing cell must synthesize 10 million copies of each type of ribosomal RNA molecule in each cell generation in order to construct its 10 million ribosomes, and there is no translation step at which production can be amplified, since RNA is the final gene product. Adequate quantities of ribosomal RNAs can, in fact, be produced only because the cell contains multiple copies of the genes coding for ribosomal RNAs (**rRNA genes**).

Human cells contain about 200 rRNA gene copies per haploid genome, spread out in small clusters on five different chromosomes (see p. 428), while cells of the frog *Xenopus* contain about 600 rRNA gene copies per haploid genome in a single cluster on one chromosome. But the general pattern of rRNA gene organization and rRNA synthesis is identical in all eucaryotes. The multiple copies of the highly conserved rRNA genes on a given chromosome are located in a tandemly arranged series in which each gene is separated from the next by a nontranscribed region known as *spacer DNA,* which varies in length and sequence. Because of this repeating arrangement, and because they are transcribed at a very high rate, the tandem arrays of rRNA genes can easily be seen in spread chromatin preparations (Figure 8–51). The RNA

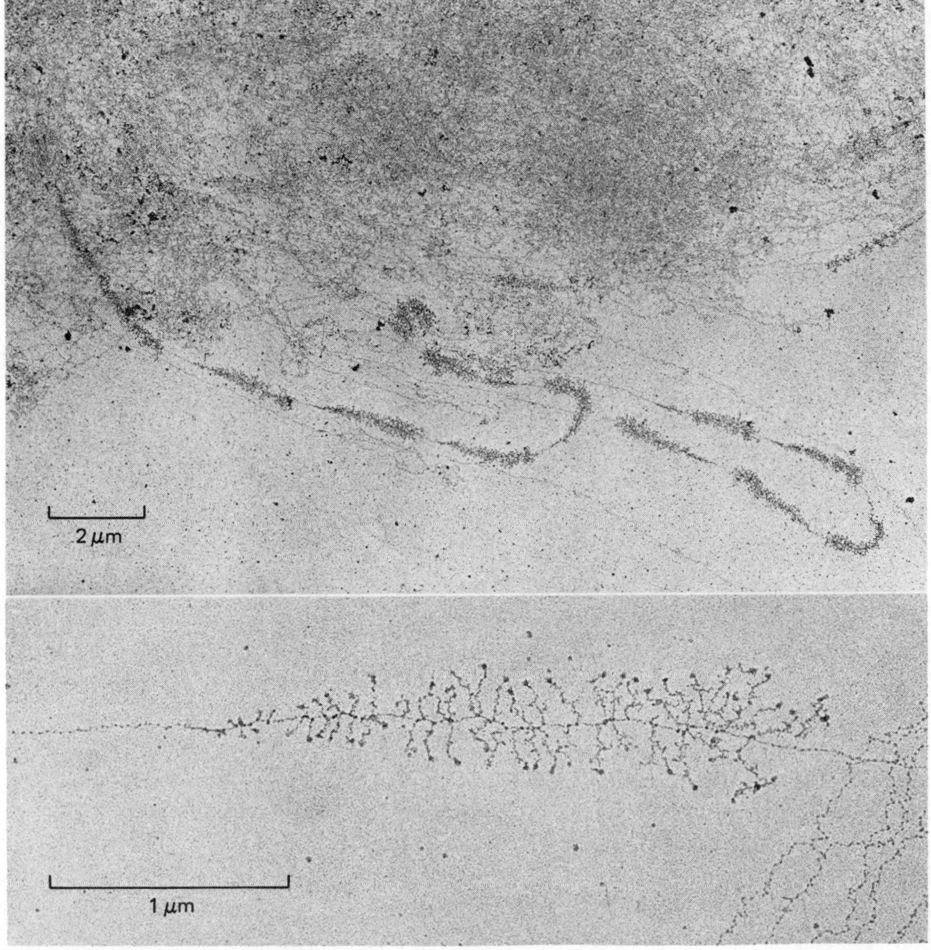

Figure 8–51 The transcription of the 45S RNA ribosomal RNA precursor on tandemly arranged rRNA genes, as visualized in the electron microscope. Note the pattern of alternating transcribed gene and nontranscribed spacer in the lower magnification view in the upper panel. (From V. E. Foe, *Cold Spring Harbor Symp. Quant. Biol.* 42:723–740, 1978.)

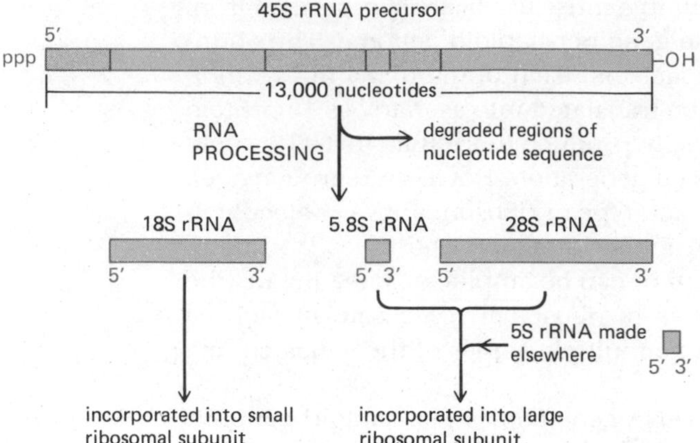

Figure 8–52 The pattern of processing of a 45S rRNA precursor molecule into three separate ribosomal RNAs. Note that nearly half of the nucleotide sequences in this precursor are degraded in the nucleus.

polymerase molecules and their associated transcripts are so densely packed (typically about 100 per gene) that they fan out perpendicularly from the DNA to give each transcription unit a "Christmas-tree" appearance. As noted earlier (Figure 8–39), the tip of each of these "trees" represents the point on the DNA at which transcription begins, while the end of the rRNA gene is sharply demarcated by the sudden disappearance of RNA polymerase molecules and their transcripts.

The rRNA genes are transcribed by RNA polymerase I, and each gene produces the same primary RNA transcript. In humans this RNA transcript, known as *45S RNA,* is about 13,000 nucleotides long. Before it leaves the nucleus in assembled ribosomal particles, the 45S RNA is cleaved to give one copy each of the 28S RNA (about 5000 nucleotides), the 18S RNA (about 2000 nucleotides), and the 5.8S RNA (about 160 nucleotides) of the final ribosome (see p. 209). The derivation of these three different rRNAs from the same primary transcript assures that they will be made in equal quantities. The remaining part of each primary transcript (about 6000 nucleotides) is degraded in the nucleus (Figure 8–52). It is thought that these extra RNA sequences play a transient part in ribosome assembly, which begins as soon as the 45S RNA is synthesized.

Other tandemly arranged genes with similar nontranscribed spacers include those that code for the 5S RNA of the large ribosomal subunit (the only rRNA that is transcribed separately) and those that code for the many different tRNAs. Both classes of genes are transcribed by RNA polymerase III. The histone genes are among the few examples of tandemly arranged genes that code for proteins, and they are transcribed by RNA polymerase II.

The Nucleolus Is a Ribosome-producing Machine[30]

Continuous transcription of duplicated genes ensures an adequate supply of rRNA. The newly synthesized rRNA is immediately packaged with ribosomal proteins to generate the ribosomes. This packaging takes place in the nucleus, in a large, diffuse structure called the **nucleolus.** The nucleolus contains large loops of DNA whose rRNA genes are transcribed at a furious rate by RNA polymerase I. Such a loop of DNA is known as a **nucleolar organizer** region. The very beginning of the rRNA packaging process can be seen in electron micrographs: the 5′ tail of each ribosomal RNA transcript in each "Christmas tree" is encased by a protein-rich granule (Figure 8–51). These granules, which do not appear on other types of RNA transcripts, presumably reflect the first of the protein-RNA interactions that take place in the nucleolus.

The biosynthetic functions of the nucleolus can be traced by means of a brief radioactive labeling of newly made RNA. After various intervals of further incubation, a cell fractionation procedure is employed to isolate the radioactive nucleoli. Such experiments show that the intact 45S transcript is first packaged into a large complex containing many different proteins imported from the cytoplasm, where all proteins are synthesized (Figure 8–53). These proteins include most of the 70 different polypeptide chains that will make up the ribosome, as well as other proteins and small RNA molecules that are believed to catalyze the construction of ribosomes and to remain in the nucleolus when the ribosomal subunits are exported to the cell cytoplasm in finished form.

As the 45S rRNA molecule is processed, its large ribonucleoprotein particle gradually loses some of its RNA and protein and then splits to form separate precursors of the large and small ribosomal subunits. Within 30 minutes of radioactive pulse labeling, the first mature small ribosomal subunits, containing their 18S rRNA, emerge from the nucleolus and appear in the cell cytoplasm. The assembly of the mature large ribosomal subunit, with its 28S, 5.8S, and 5S rRNAs, takes longer to complete (about an hour), and the nucleolus therefore contains many more incomplete large ribosomal subunits than small ones.

The very last steps in ribosome maturation occur only as these subunits are transferred to the cell cytoplasm. This delay prevents functional ribosomes from gaining access to the incompletely processed hnRNA molecules in the nucleus.

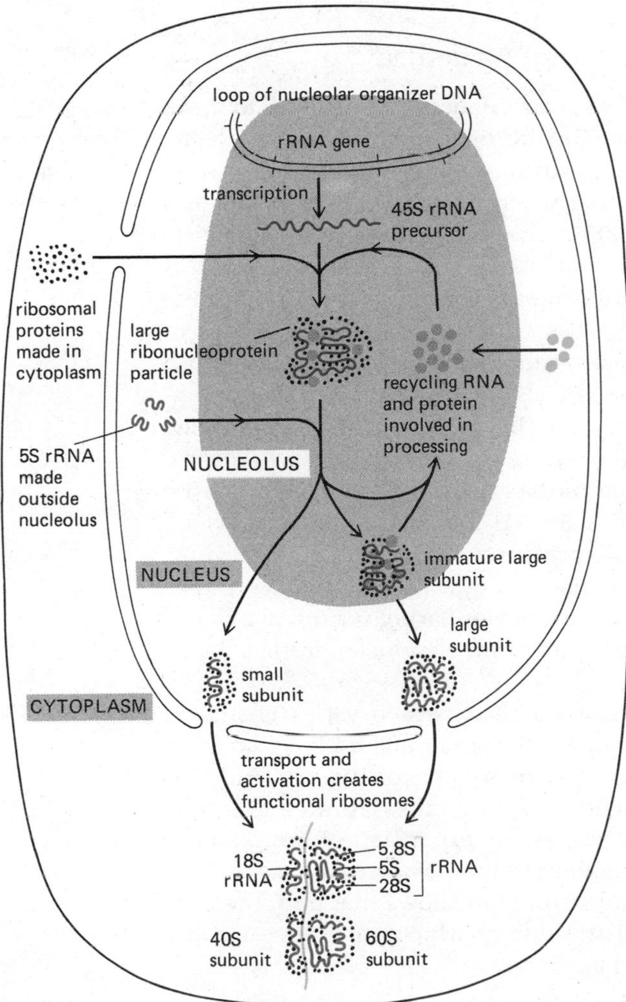

Figure 8–53 Schematic view of the function of the nucleolus in ribosome synthesis. The large 45S rRNA transcript is packaged in a large ribonucleoprotein particle containing many ribosomal proteins imported from the cell cytoplasm. While this particle remains in the nucleolus, selected pieces are discarded as it is processed into immature large and small ribosomal subunits. These two subunits are thought to attain their final functional form only as they are individually transported through the nuclear pores into the cell cytoplasm.

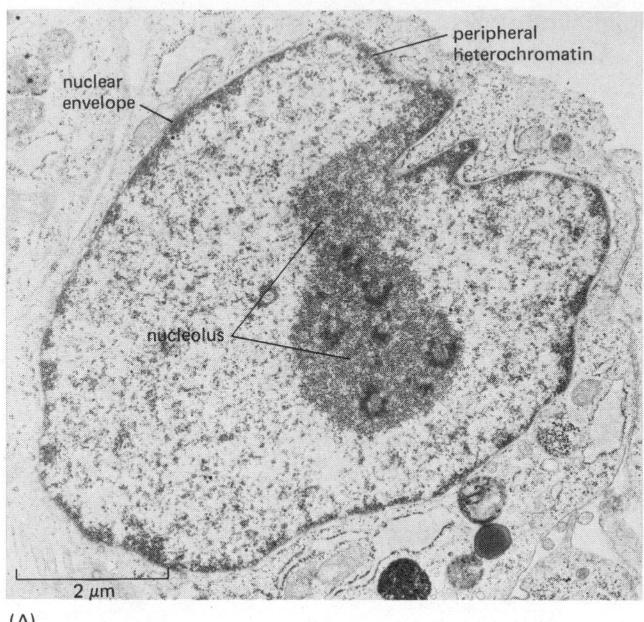

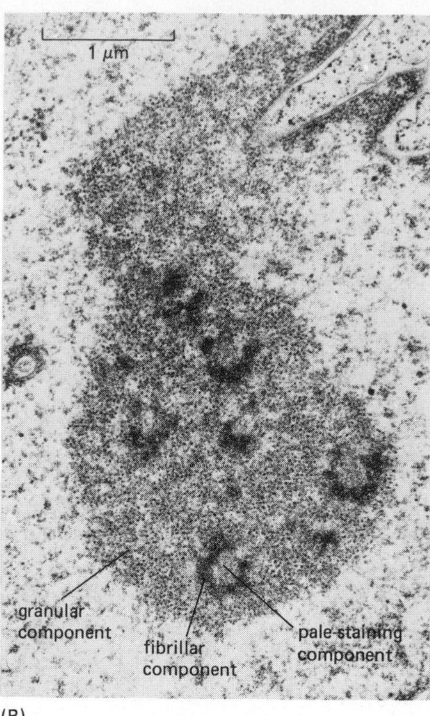

Figure 8–54 Electron micrograph of a thin section of a nucleolus in a human fibroblast cell, showing its three distinct zones. (A) View of entire nucleus. (B) Detail. (Courtesy of E. G. Jordan and J. McGovern.)

The Nucleolus Has a Highly Organized Structure[31]

In view of what is now known about nucleolar function, it is possible to make sense of some of its structure. As seen in the light microscope, the large, spheroidal nucleolus is the most obvious structure in the nucleus of a nonmitotic cell. Consequently, it was so closely scrutinized by early cytologists that an 1898 review could list some 700 references. By the 1940s, cytologists had demonstrated that the nucleolus contains high concentrations of RNA and proteins; but its major function in ribosomal RNA synthesis and ribosome assembly was not discovered until the 1960s.

Some of the details of nucleolar organization can be seen in the electron microscope. Unlike the cytoplasmic organelles, the nucleolus has no membrane to keep it together; instead it seems to be constructed by the specific binding of unfinished ribosome precursors to each other by unknown means. In a typical electron micrograph, three partially segregated regions can be distinguished in the nucleolus (Figure 8–54): (1) a *pale-staining component,* which contains DNA from the nucleolar organizer region of a chromosome; (2) a *granular component,* which contains 15-nm-diameter particles representing the most mature of the ribosomal precursor particles; and (3) a dense *fibrillar component* composed of many fine, 5-nm ribonucleoprotein fibers, representing RNA transcripts.

The size of the nucleolus reflects its activity, which varies greatly in different cells and can change in a single cell. For example, it is very small in some dormant plant cells but can occupy up to 25% of the total nuclear volume in cells that are making unusually large amounts of protein. The differences in size are due largely to contraction or expansion of the granular component, which is probably controlled at the level of ribosomal gene transcription: electron microscopy of spread chromatin shows that both the fraction of activated ribosomal genes and the rate at which each gene is transcribed can vary according to circumstances.

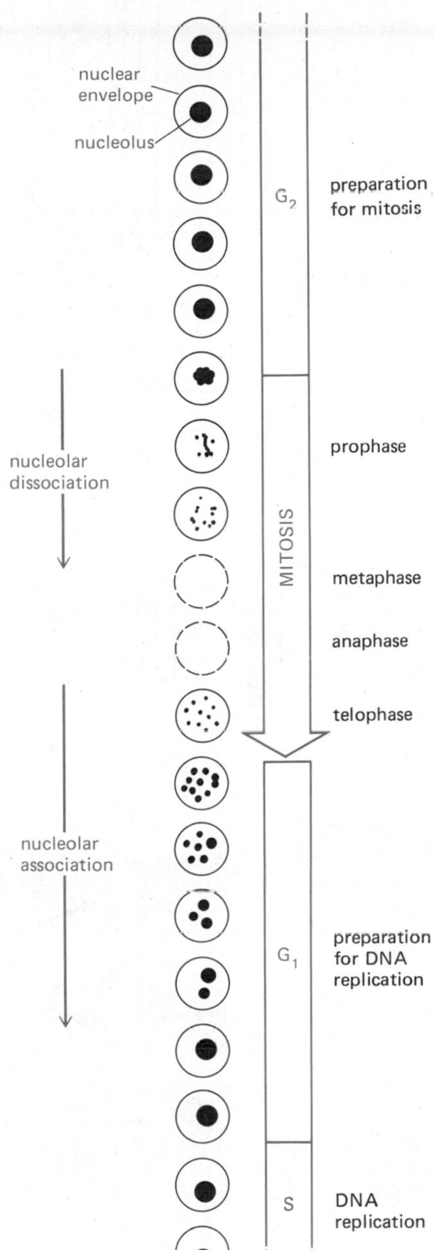

Figure 8–55 The appearance of the nucleolus in a human cell changes during the cell cycle. Only the cell nucleus is represented in this diagram.

The Nucleolus Is Reassembled on Specific Chromosomes After Each Mitosis[32]

The appearance of the nucleolus changes dramatically during the different phases of the cell cycle (Figure 8–55). As the cell approaches mitosis, the nucleolus first decreases in size and then disappears as the chromosomes condense and all RNA synthesis stops. In general, there is no nucleolus in a metaphase cell. When ribosomal RNA synthesis restarts at the end of mitosis (in telophase), tiny nucleoli reappear at the chromosomal locations of the ribosomal RNA genes.

In man the ribosomal RNA genes are located near the tips of each of the 5 different chromosomes shown in Figure 8–56 (that is, on 10 of the 46 chromosomes). Correspondingly, 10 small nucleoli form after mitosis in a human cell, though they are rarely seen because they quickly grow and fuse to form the single large nucleolus typical of an interphase cell (Figure 8–57). Thus, in a diploid human cell prior to DNA synthesis, a single nucleolus contains 10 separate loops of DNA, each contributed by 1 of 10 separate chromosomes. When nucleoli are isolated from cells, these DNA loops are sheared off to free them from the chromosomes (Figure 8–58).

What happens to the RNA and protein components of the disassembled nucleolus during mitosis? It seems that at least some of them become distributed over the surface of all of the metaphase chromosomes and are carried as cargo to each of the two daughter cell nuclei. As the chromosomes decondense at telophase, these "old" nucleolar components help reestablish the newly emerging nucleoli.

Summary

Much of the cell's mRNA is produced by a complex process beginning with the synthesis and processing of heterogeneous nuclear RNA (hnRNA) in the nucleus, followed by selective RNA transport to the cytoplasm. The primary hnRNA transcript is made by an RNA polymerase II molecule; it is capped by the addition of a special nucleotide to its 5′ end and is usually polyadenylated at its 3′ end. These events are often followed by one or more RNA splicing events, in which regions called intervening sequences (or intron sequences) are removed from the middle of the hnRNA. In this process, most of the mass of the primary RNA transcript is removed and degraded in the nucleus. As a result, although the rate of production of hnRNA typically accounts for about half of a cell's RNA synthesis, the mRNA it produces represents only about 3% of the steady-state level of RNA in a cell.

Unlike most genes coding for proteins, the genes coding for the structural RNAs synthesized by RNA polymerase I and III are usually repeated many times in the genome and are often clustered in tandem arrays. RNA polymerase III molecules make tRNAs and the small 5S RNA of the ribosome. The synthesis of the large rRNA precursor molecule containing the major rRNAs is catalyzed by RNA polymerase I molecules in a distinct intranuclear organelle called the nucleolus, which contains the rRNA genes (rDNA) from the nucleolar organizer regions of several separate chromosomes. The nucleolus is the site of assembly of all the cell's ribosomes.

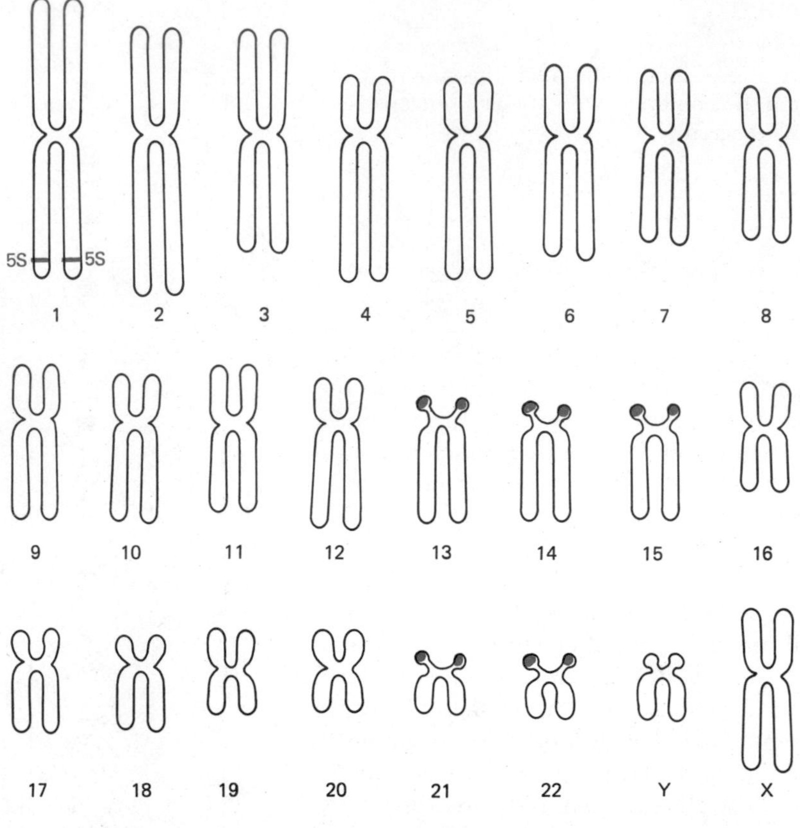

Figure 8–56 The complement of human chromosomes as seen at mitosis, showing the location of the ribosomal RNA genes on chromosomes 13, 14, 15, 21, and 22 (the nucleolar organizer regions are shown as colored dots) and the 5S RNA genes on chromosome 1. These locations have been determined by *in situ* hybridization of mitotic chromosomes to the appropriate radioactively labeled RNA molecules (p. 191). Note that only one of each type of chromosome is shown, each containing two paired chromatids. In actuality, there will be two copies of each numbered chromosome and either two X chromosomes (female) or one X and one Y chromosome (male) in all diploid cells at mitosis, each composed of a pair of identical chromatids (see Figure 8–22).

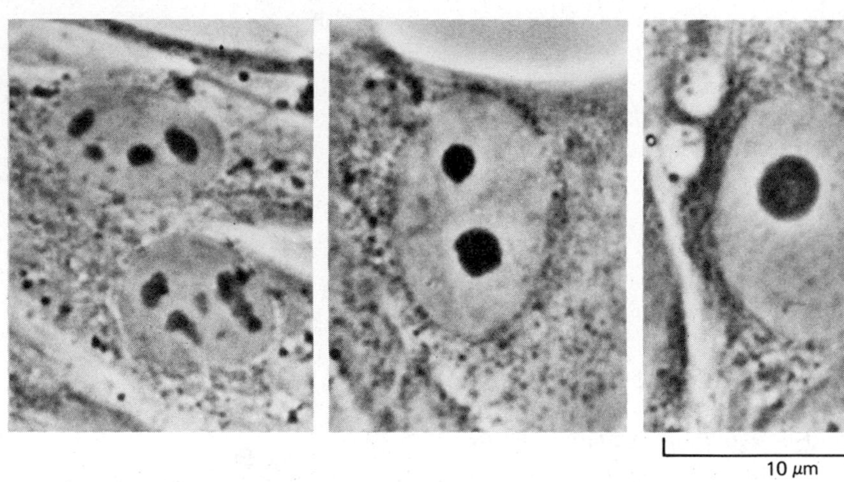

Figure 8–57 Light micrographs of human fibroblast cells grown in culture, showing various stages of nucleolar fusion. (Courtesy of E. G. Jordan and J. McGovern.)

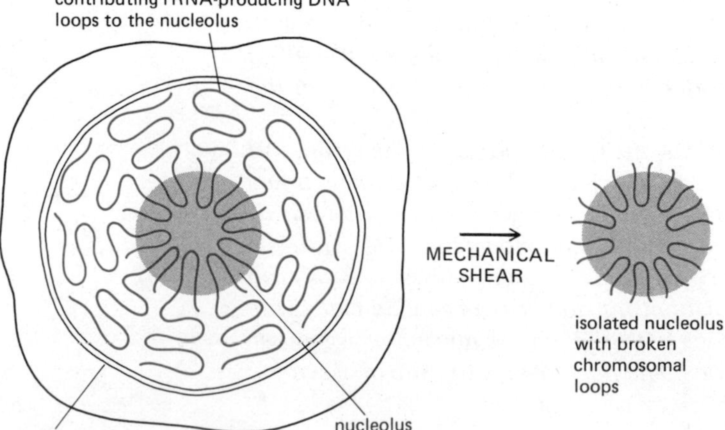

Figure 8–58 Highly schematic view of a human cell, showing the contributions to a single large nucleolus of loops of chromatin containing rRNA genes from 10 separate chromosomes. Purified nucleoli are very useful for biochemical studies of nucleolar function; to obtain such nucleoli, the loops of chromatin are mechanically sheared from their chromosomes, as shown.

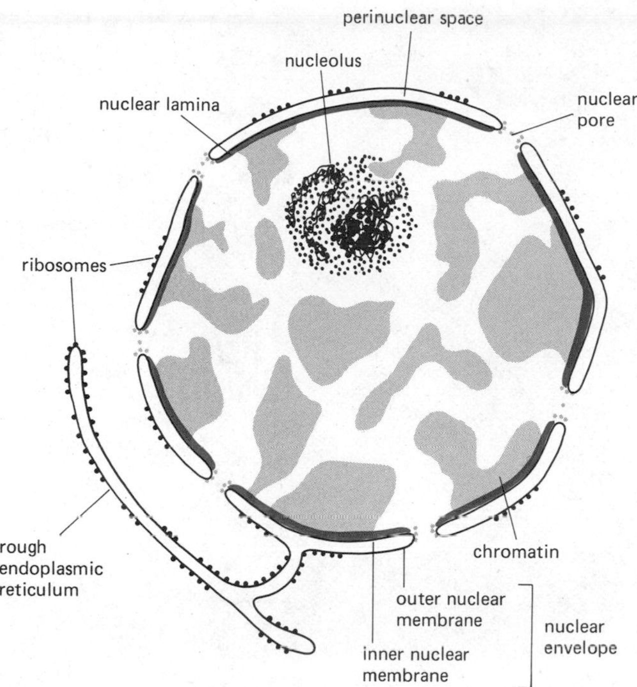

Figure 8–59 Schematic diagram of a cross-section of a typical cell nucleus. The nuclear envelope is a double membrane whose outer membrane is continuous with the endoplasmic reticulum membrane. The lipid bilayers of the inner and outer nuclear membranes are fused at the nuclear pores.

The Nuclear Envelope[33]

The Nucleus Is Enclosed by a Double Membrane[34]

The nuclear contents (the **nucleoplasm**) are separated from the cytoplasm by the membranes of the **nuclear envelope.** The nuclear envelope is a double membrane composed of two lipid bilayers separated by a gap of 20 to 40 nm known as the **perinuclear space** (Figures 8–59 and 8–60). In thin-section electron micrographs, the **outer nuclear membrane** can be seen to be continuous with the endoplasmic reticulum (ER) membrane, and, like the membrane of the rough endoplasmic reticulum (RER), the outer surface of the outer nuclear membrane is often studded with ribosomes engaged in protein synthesis (Figure 8–61). Since the lumen of the ER is continuous with the perinuclear space (Figure 8–60), both the perinuclear space and the outer nuclear membrane can be regarded as small (see Table 7–2, p. 322) and specialized regions of the endoplasmic reticulum.

At localized regions called *nuclear pores* the outer membrane is connected to the **inner nuclear membrane.** Thus, the entire membrane surface is physically continuous (Figure 8–62)—although, as we shall see, the inner and outer layers are very different functionally and biochemically. Because the nuclear envelope is continuous with the ER, the nuclear membrane grows, and can expand and contract rapidly, by the direct exchange of material with the ER membrane (Figure 8–62). This may facilitate the rapid changes in nuclear membrane area that accompany the breakdown and re-formation of the nuclear membrane during mitosis (p. 659). The nuclear membrane area also expands dramatically when a formerly quiescent nucleus begins to synthesize RNA or DNA rapidly.

One advantage of the double-membrane system is that the inner and outer nuclear membranes are specialized for interacting with components in the nucleoplasm or cytoplasm, respectively. For example, specific proteins in the inner membrane are thought to interact with a set of proteins that form the underlying *nuclear lamina.*

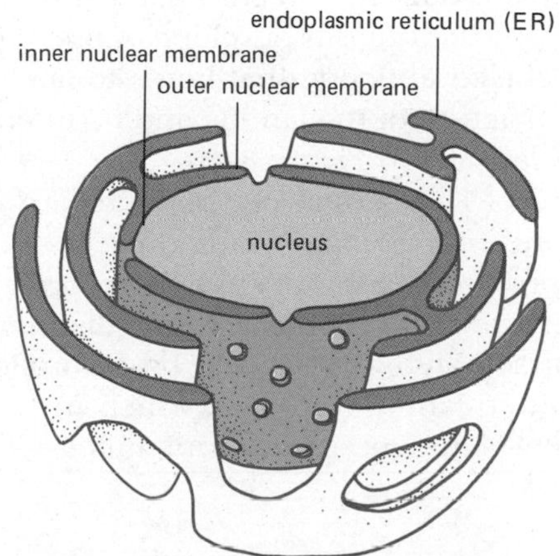

Figure 8–60 A three-dimensional sketch illustrating the relationship between the endoplasmic reticulum and the nuclear membranes.

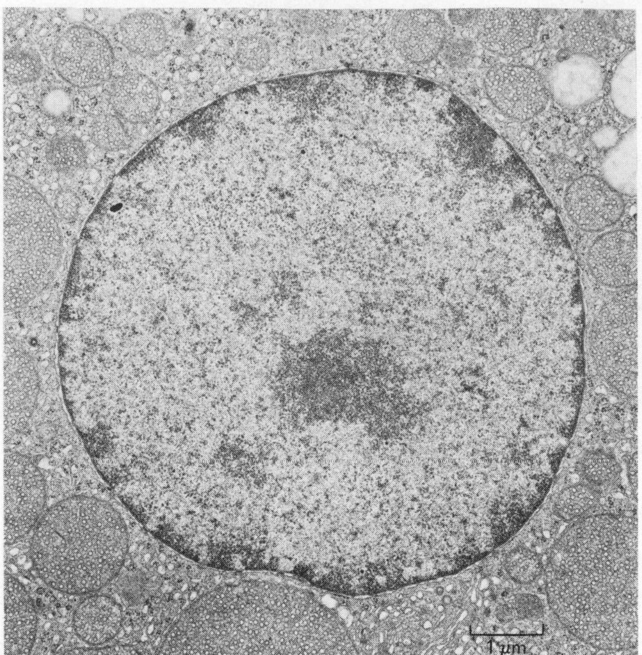

Figure 8–61 Thin-section electron micrograph of an adrenal cortex cell nucleus showing some of its fine structure. Since most of the chromatin is in its extended interphase form, the individual chromosomes cannot be discerned. (Courtesy of Daniel S. Friend.)

The Nuclear Lamina Helps to Determine Nuclear Shape[35]

The nuclear envelope has an electron-dense layer lying on the nucleoplasmic side of its inner membrane (Figure 8–63). This **fibrous lamina** (or the **nuclear lamina**) varies in thickness in different cells; in some it cannot be detected at all by microscopic techniques. Nevertheless, there is biochemical evidence that a nuclear lamina is present in almost all eucaryotic cells, and it is thought to play a crucial part in organizing both the nuclear envelope and the underlying chromatin.

The nuclear lamina is composed of a fibrous meshwork that can be isolated as a membrane-free sheet containing specialized proteins that form pores in the nuclear membrane (Figure 8–64). In vertebrates, the lamina is formed largely by spontaneous assembly of three major polypeptides that are thought to bind to specific proteins embedded in the lipid bilayer of the inner membrane. Other components associated with the lamina are believed to bind to specific sites on chromatin and thereby guide the interactions of chromatin with the nuclear envelope. A micrograph of cells stained with a fluorescently labeled antibody that binds to lamina proteins is shown in Figure 8–65. The structure of the lamina and its interaction with chromatin are outlined schematically in Figure 8–66.

The lamina polypeptides are probably instrumental in the dissolution and re-formation of the nuclear envelope that occur during each mitosis. At prometaphase, most of these proteins are released from the nuclear membrane and become diffusely distributed in the cytoplasm. This reversible disassembly is believed to be controlled by a transient phosphorylation of the three lamina proteins, which in turn causes the breakdown of the nuclear envelope observed during mitosis (see p. 659).

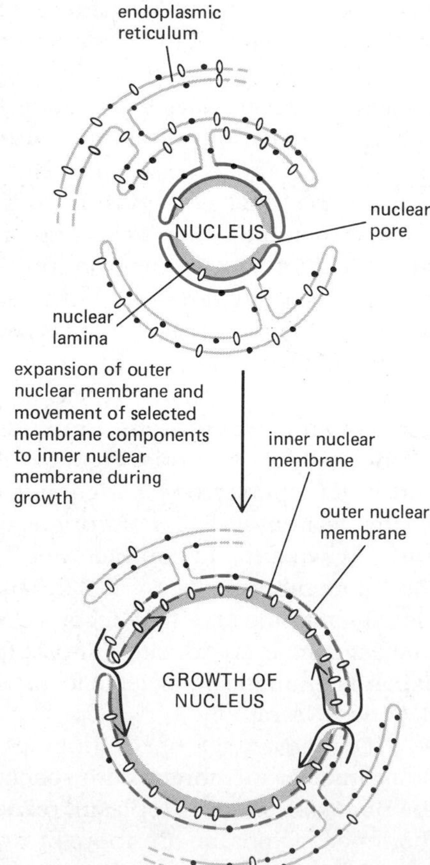

Figure 8–62 Schematic view of the rapid expansion of the nuclear envelope that occurs just after telophase in a normal cell cycle or when quiescent cells become active in RNA or DNA synthesis.

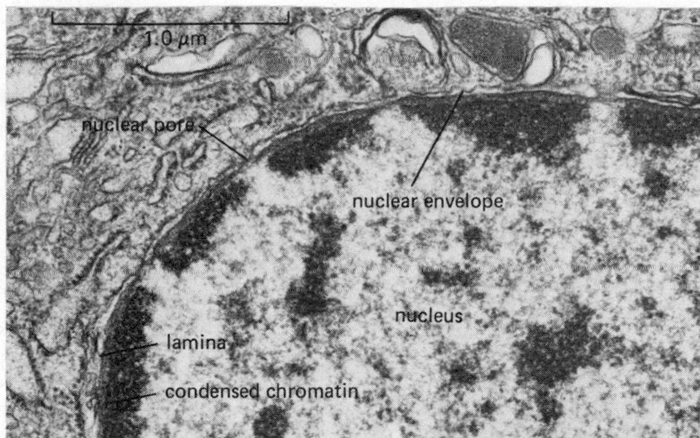

Figure 8–63 Electron micrograph of a mammalian cell nucleus that has a distinct nuclear lamina about 40 nm thick. In many cells the lamina is much thinner and cannot be seen in this type of electron micrograph. (Courtesy of Larry Gerace.)

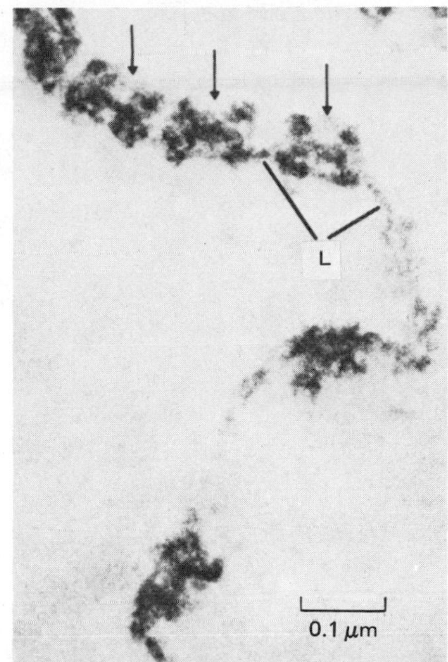

Figure 8–64 Electron micrograph of a piece of isolated nuclear lamina (L) with its attached nuclear pore complexes (arrows). (From N. Dwyer and G. Blobel, *J. Cell Biol.* 70:581–591, 1976. © 1976 Rockefeller University Press.)

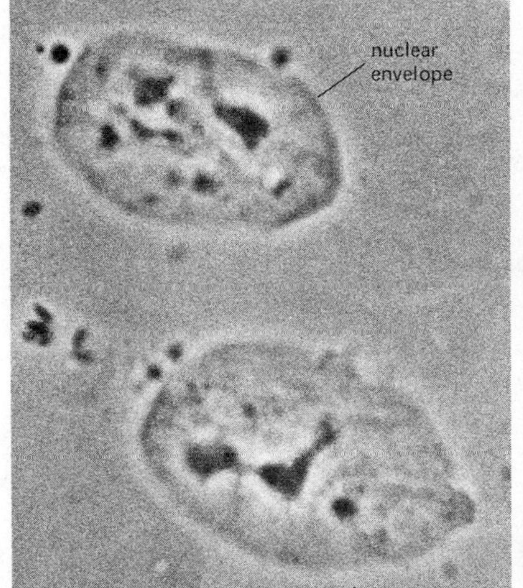

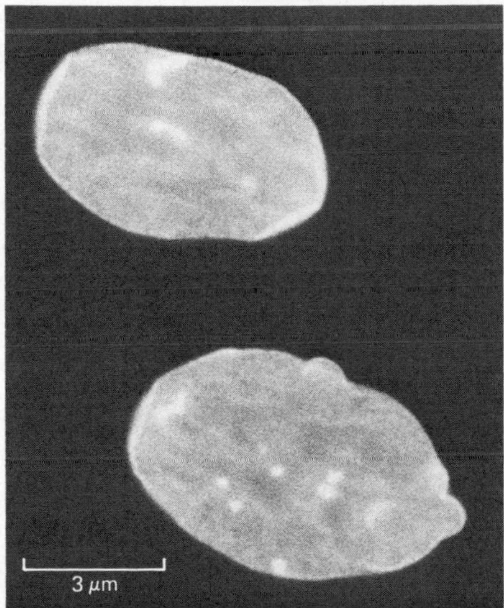

Figure 8–65 (*Left*) Light micrograph of two mammalian cells grown in tissue culture. (*Right*) Fluorescence micrograph of the same cells with their lamina proteins stained by immunofluorescent methods. Only the nuclei can be seen in these micrographs. (Courtesy of Frank McKeon and Marc Kirschner.)

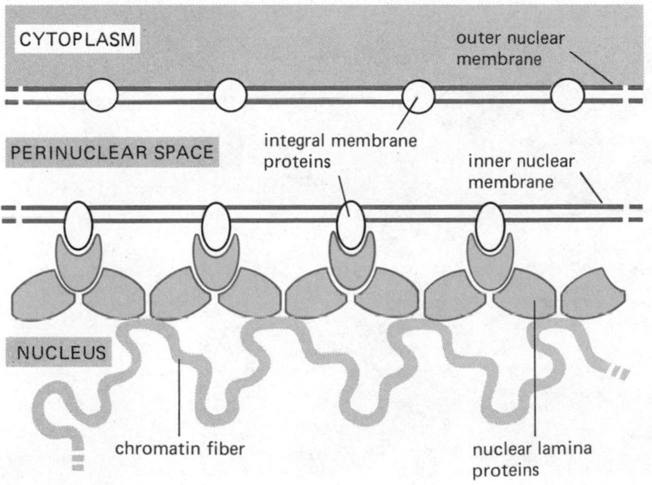

Figure 8–66 Highly schematic view of the nuclear lamina as a fibrous network of lamina protein subunits that both organizes the nuclear envelope and binds to specific sites on chromatin. The three major polypeptides of the lamina in vertebrates have molecular weights of 60,000 to 70,000.

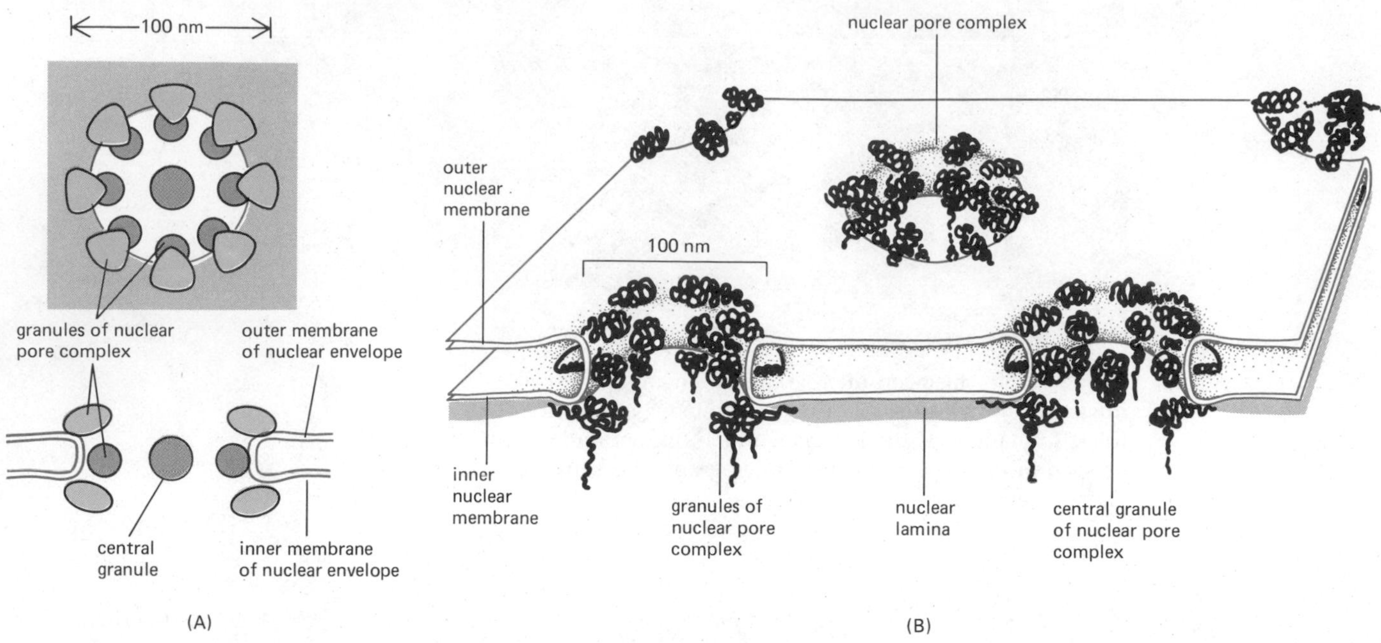

Nuclear Transport Occurs Through Nuclear Pores[36]

For reasons already outlined (p. 376), double-membrane systems such as the nuclear envelope pose special problems for molecular transport. To overcome these problems, the nuclear envelopes of all eucaryotes, from yeast to man, are perforated by **nuclear pores,** each of which is surrounded by a large disclike structure known as the **nuclear pore complex** (~80 nm inner diameter). On both surfaces of the envelope each complex is defined by eight large protein granules arranged in an octagonal pattern (Figure 8–67A and Figure 8–68). In sections perpendicular to the plane of the membranes, the edge of the pore complex is seen to cross the perinuclear space, bringing the two lipid bilayers of the inner and outer membrane together around the margins of each pore (Figure 8–67). These regions where the inner and outer nuclear membranes are continuous probably enable lipid-soluble materials dissolved in the membrane (lipids and membrane proteins) to flow from the ER membrane where they are synthesized into the inner nuclear membrane (Figure 8–62).

Figure 8–67 Schematic diagrams showing the arrangement of the nuclear pore complexes in the nuclear envelope. (A) A top view and a central vertical section. The "central granule" shown in brown is seen in some pores but not others; while these granules could be part of the pore, they are often considered to be large complexes caught in transit through it. (B) Three-dimensional view of a small region of the nuclear envelope.

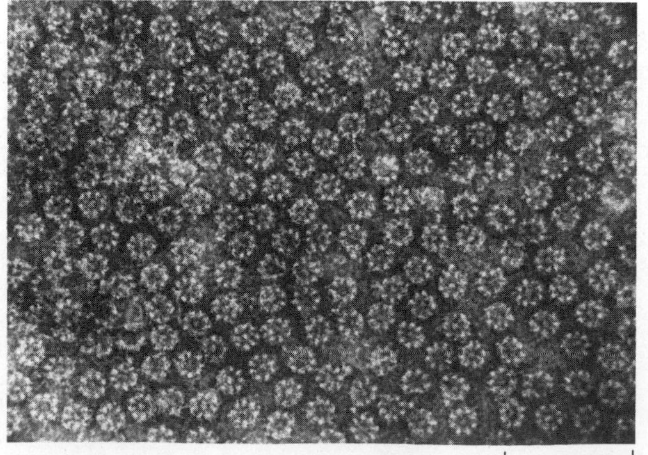

Figure 8–68 Electron micrograph of a negatively stained preparation of nuclear pore complexes. Note the ring of eight large granules which surrounds each pore; each of these granules is larger than a ribosome. (Courtesy of A. C. Fabergé.)

The hole seen in the center of each complex is thought to be an aqueous pore that provides the main channel by which water-soluble molecules shuttle between the nucleus and the cytoplasm. It is often seen to be plugged by a large central granule believed to represent newly made ribosomes and other particulate matter caught in transit (Figure 8–67). While direct proof of transport through pores is lacking, there is a great deal of indirect evidence supporting it. For example, small molecules (5000 daltons or less) injected into cells diffuse across the nuclear envelope so fast that the nuclear membrane can be considered to be freely permeable to them. A protein of 17,000 daltons equilibrates between cytoplasm and nucleus within 2 minutes, while a protein of 44,000 daltons takes 30 minutes to equilibrate. An injected spherically shaped protein larger than 60,000 daltons seems hardly able to enter the nucleus at all. A quantitative analysis of data from such experiments shows that the nuclear envelope behaves as would be expected if the nuclear pore complex contained a water-filled cylindrical channel 9 nm in diameter and 15 nm long (Figure 8–69), dimensions that are not inconsistent with the size of the small, irregular, open channel seen in some electron micrographs.

The nuclear envelope acts to shield the nucleoplasm from many of the particles, filaments, and large molecules that function in the cytoplasm. For example, mature cytoplasmic ribosomes are too large to pass through the putative 9-nm channels, thus guaranteeing that all protein synthesis is confined to the cytoplasm.

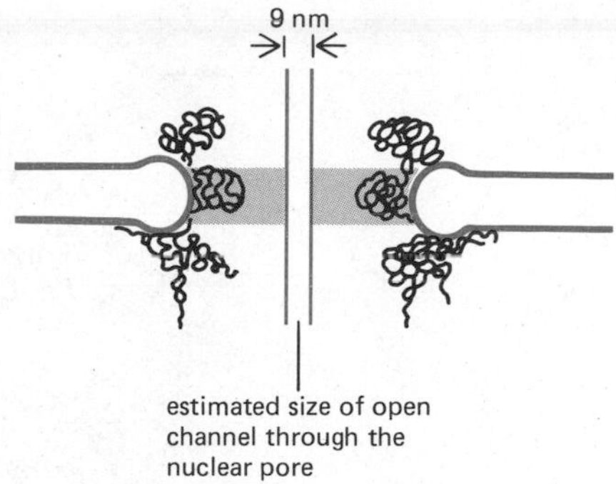

Figure 8–69 Schematic diagram of a cross-section of a nuclear pore complex. The idealized cylinder that has been superimposed on the pore represents the relative size of the open channel estimated from transport measurements. In some electron micrographs, poorly visible strands appear to fill most of the interior of the pore, and these are probably responsible for confining the effective opening to the 9-nm channel shown.

Nuclear Pores Probably Transport Large Particles Selectively[36]

Some idea of the amount of material transported through the nuclear pores can be obtained by considering that a typical mammalian cell contains 3000 to 4000 pore complexes in its nuclear envelope (about 11 pores per square μm of membrane area, compared with a maximum density of about 60 pores per square μm if pores were packed as closely as possible in the envelope). A cell synthesizing DNA needs to import about 10^6 histone molecules from the cytoplasm every 3 minutes in order to package newly made DNA into nucleosomes: this means transporting about 100 histone molecules per minute per pore. If it is growing rapidly, the same cell will transport about 3 newly assembled ribosomes per minute per pore to the cytoplasm. And that is only a very small part of the total traffic involved.

The nuclear export of new ribosomal subunits is particularly problematic. Since the diameter of these particles is about 15 nm, they are much too large to pass through the 9-nm channel. It is commonly assumed that they are specifically transported through the pores by an active transport system and that they are greatly distorted in the process. Electron micrographs showing particles apparently in progress through the pores support this view, as does the fact that the nuclear pore complexes stain heavily for ribonucleoprotein. Messenger RNA molecules, complexed with special proteins to form ribonucleoprotein particles, are also thought to be selectively exported in this way.

Similar problems arise in connection with large macromolecules, such as DNA and RNA polymerases, that have to be imported into the nucleus in spite of sizes of at least 100,000 to 200,000 daltons. If these proteins were to adopt a rodlike rather than a spherical conformation, they could diffuse through pores end first. Alternatively, they could pass through pores by interacting with some type of receptor protein on the pore margin, which transports them through the pore by enlarging the channel.

In some, but not all cells, complexes identical to nuclear pore complexes appear in the cell cytoplasm in separate membranous stacks called **annulate lamellae** (Figure 8–70). These structures are most often observed in germ cells

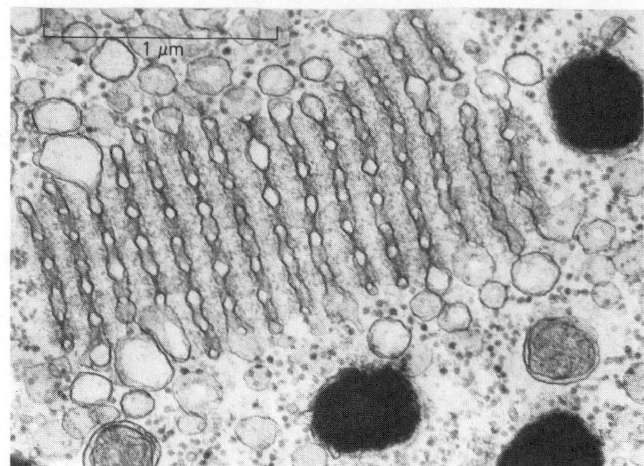

Figure 8–70 Stacks of annulate lamellae as seen in cross-section in the cytoplasm of a mature frog oocyte. (Courtesy of Richard G. Kessel, from *J. Ultrastruct. Res.*, Supplement 10, 1968.)

Figure 8–71 Freeze-fracture electron micrograph of the elongated nuclear envelope of a fern spore, illustrating the ordered arrangement of the nuclear pore complexes in parallel rows. (Courtesy of Don H. Northcote, from K. Roberts and D. H. Northcote, *Microsc. Acta* 71:102–120, 1971.)

and in cells with high protein-synthetic activity, including most tumor cells. While they may represent extra nuclear envelope held in reserve, their existence has led to the suggestion that the nuclear pore complex might have a second function distinct from transport, such as a direct role in some step of RNA or ribosome processing. Unfortunately, the nuclear pore complex is one of several important structures in the cell that has yet to yield to modern biochemistry.

The Inner Surface of the Nuclear Lamina Helps to Organize the Chromosomes[37]

It is probable that the outer surface of the nuclear lamina plays an important part both in holding the nuclear pores in place and in shaping the nuclear envelope, while the inner surface holds regions of the interphase chromosomes against the inner nuclear membrane. These two different organizing roles are clearly coordinated, inasmuch as the chromatin that lines the inner nuclear membrane (which is unusually condensed chromatin and therefore clearly visible in electron micrographs) can be seen to be specifically excluded from a considerable region beneath and around each nuclear pore (see Figure 8–63). This exclusion of chromatin clears a path between the cytoplasm and the nucleoplasm and must reflect a change in the binding properties of the lamina in the region around the pores.

Thus far the nuclear envelope has been discussed as if it were a highly disordered structure, with pores scattered about randomly on its surface and pieces of chromatin adhering to its inner surface in a haphazard way. But it is almost always a mistake to think of an important intracellular structure as being randomly ordered in space, and the nucleus is no exception. For example, in some cases the nuclear pores are found to be highly organized in the nuclear envelope. Figure 8–71 shows the parallel rows of nuclear pores observed in the elongated nucleus of a fern spore. In other cells, either concentrated clusters of nuclear pores or unusual areas free of nuclear pores have been detected in the nuclear envelope, and these are specifically oriented with respect to the cell itself (Figure 8–72). Such ordering presumably reflects some corresponding organization within the nuclear lamina.

Figure 8–72 Freeze-fracture electron micrograph of the clustered nuclear pores at one end of a sperm nucleus. (Courtesy of Daniel S. Friend.)

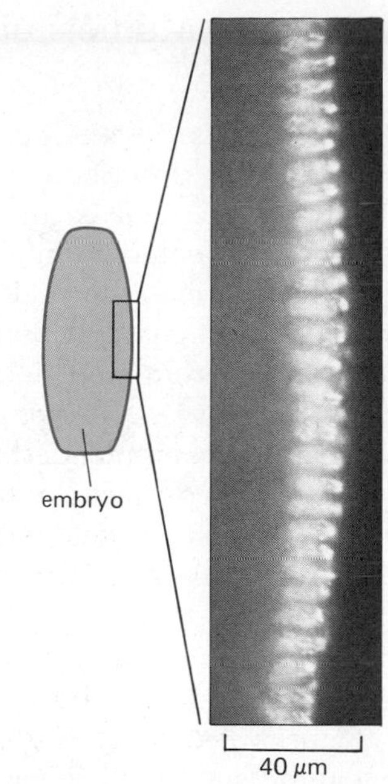

Figure 8–73 Low-magnification light micrograph of fluorescently stained chromosomes in the interphase cells of a *Drosophila* embryo at the cellular blastoderm stage. Note that the most brightly staining region (the chromocenter), which is known to contain specific regions of each of the four chromosomes (see Figure 8–29), is oriented toward the outer surface of the embryo in every cell. (Courtesy of John Sedat.)

In some cells, specific regions of particular chromosomes are readily recognized in interphase because of their unusual condensation or staining properties. For example, in *Drosophila,* regions of all four chromosomes aggregate to form a *chromocenter,* which stains unusually brightly with a DNA-specific fluorescent dye. If cells are examined in young *Drosophila* embryos, each cell is found to have its chromocenter pointed outward toward the exterior surface (Figure 8–73). Similarly, the giant interphase chromosomes present in the polytene cells of *Drosophila* larvae have been found to be similarly oriented and arranged in three dimensions in neighboring cells (Figure 8–74). These observations suggest that in at least some cell nuclei specific regions of chromosomes bind to special sites on the nuclear lamina. In this way, the inner surface of the nuclear envelope could serve to help organize the chromosomes inside.

Summary

The nucleus is enclosed by an envelope consisting of a concentric pair of membranes. The outer nuclear membrane is continuous with the ER membrane, and the perinuclear space between it and the inner nuclear membrane is continuous with the ER lumen. All of the cell's RNA molecules and ribosomes are made in the nucleus and are exported to the cytosol, while all of the proteins that function in the nucleus are synthesized in the cytosol and must be imported. The exchange of materials between nucleus and cytosol is thought to occur through the nuclear pores, which are channels formed through sites of connection between the inner and outer nuclear membranes. The nuclear pore complex at these sites is a large supramolecular structure, which appears to be organized and held in place by the nuclear lamina, a fibrous protein network that lines the inside of the inner nuclear membrane and helps to shape the nucleus and organize its chromosomes.

Basic Elements in the Control of Gene Expression[38]

Many of the individual cells of a higher organism are strikingly distinct, both in morphology and in function (p. 24). Because highly specialized cells often synthesize large amounts of one or a few proteins, cell differentiation might have involved the amplification of the particular DNA sequences that code for those proteins. Conversely, since cell differentiation is nearly always irreversible, some biologists suspected that it involved a selective loss of some DNA sequences (or genes). However, neither of these expectations has been borne out. With very few exceptions, which we shall describe in detail later, cell differentiation seems to depend on changes in the selection of DNA sequences for decoding into proteins, without any irreversible change in the DNA sequences themselves.

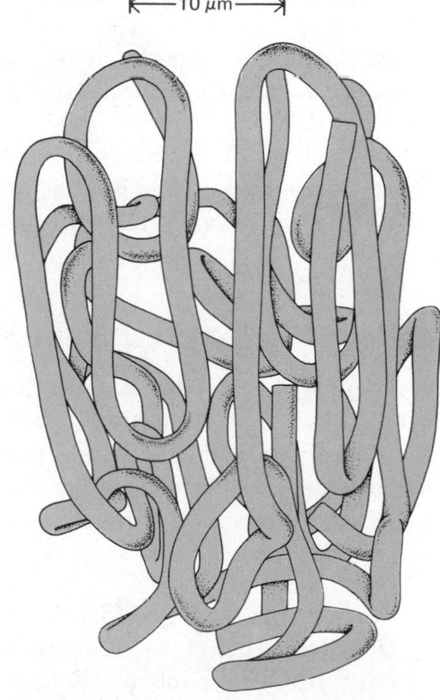

Figure 8–74 The large polytene chromosomes of *Drosophila* salivary gland cells are folded in a complex way in each cell nucleus. Discovery of the organization shown required a sophisticated analysis of a series of optical sections of the fluorescently stained chromosomes in a single cell. (Redrawn from photographs supplied by David Agard and John Sedat.)

The Cells of a Multicellular Organism All Contain the Same DNA[39]

The best evidence for the constancy of the animal genome comes from experiments to be described in Chapter 15, in which the nucleus of a fully differentiated frog cell is injected into a frog egg whose nucleus has been removed. The injected "donor" nucleus is capable of programming the recipient egg to produce a normal tadpole. The tadpole now contains a full range of differentiated cells that derived their DNA sequences from the nucleus of the original differentiated donor cell. Clearly, therefore, the donor DNA had lost no important DNA sequences in the course of *its* differentiation. A similar conclusion has been reached in experiments carried out with various plants. Here differentiated pieces of tissue are cultured in synthetic media and then dissociated into single cells. Often, one of these individual cells can regenerate an entire adult plant (Figure 8–75).

Further evidence that no large blocks of DNA are lost or rearranged during the developmental process in vertebrates comes from comparisons of the detailed banding patterns detectable in condensed chromosomes at mitosis (Figures 8–25 and 8–26) in different cell types. By this criterion, the chromosome sets of all differentiated cells in the human body appear to be identical.

Much more sensitive tests based on recombinant DNA technology (p. 185) have made it possible to extend this line of enquiry to exclude frequent DNA amplification as a mechanism of differentiation as well. The best evidence to date indicates that only a single copy of each gene codes for each of the abundant proteins characteristic of particular differentiated cells, even in cells devoted almost exclusively to the synthesis of these proteins. For example, there is only one functional gene for a hemoglobin subunit in chick red blood cells, for silk fibroin in silk gland cells of the moth, and for ovalbumin (egg white protein) in chicken oviduct gland cells. Moreover, in each case, these genes and the nucleotide sequences surrounding them appear to be identical in all of the animal's cells that have been tested, whether or not those cells make the proteins. Similar studies have indicated that most other DNA sequences coding for proteins are the same in all differentiated cells.

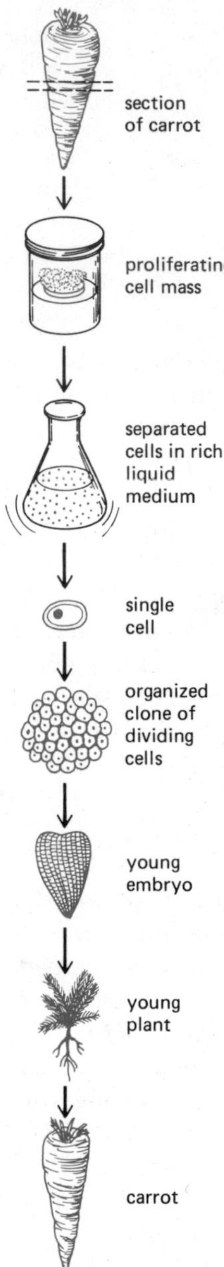

Figure 8–75 Schematic illustration of a type of experiment that can be carried out in many plants. Many differentiated plant cells retain the ability to "dedifferentiate," forming a clone of progeny cells that can later organize to reproduce the entire plant (see Chapter 19).

Different Cell Types Synthesize Different Sets of Proteins[40]

The different cell types in a multicellular organism become different from each other because they synthesize and accumulate different sets of proteins. But how extensive are the protein differences required to create two cells as different as a muscle cell and a fibroblast, for example? While we still do not know the answer to this fundamental question, certain general statements can be made:

1. Inasmuch as many functions are common to all cells, any two cells in a single organism have many proteins in common. These include some abundant proteins such as those of the cytoskeleton, the major histone and nonhistone chromosomal proteins, some of the proteins that are essential to the endoplasmic reticulum and Golgi membranes, the proteins of the nuclear lamina, RNA packaging proteins, ribosomal proteins, and so on. Abundant proteins are relatively easy to characterize, and many of them appear to be identical in different cell types. Many of the different enzymes involved in the central reactions of metabolism (p. 83) are also the same from one cell type to another.
2. Certain abundant proteins are highly specialized. Some of these proteins cannot be detected, even by sensitive tests, except in the specialized cells

in which they function. For example, hemoglobin can be detected only in red blood cells, although the DNA sequences that code for the various hemoglobin chains are present in all mammalian cells.

3. Although the unique morphology and behavior of each type of specialized cell might suggest the existence of many cell-type-specific proteins, only the 2000 or so most abundant proteins (those present in quantities of 50,000 or more copies per cell) can be compared between different cell types, even using two-dimensional polyacrylamide gel electrophoresis (p. 175). This technique produces similar results whether the comparison involves two different cell lines grown in tissue culture (such as muscle and nerve cells) or cells of two different tissues of young rodents (such as liver, kidney, or lung). The great majority of the proteins detected are synthesized in both cell types examined, and at rates that differ by less than a factor of five. Only a few percent of the proteins are synthesized at very different rates or exclusively in one cell type or the other.

Studies of the number of different mRNA sequences suggest that a typical higher eucaryotic cell synthesizes 10,000 to 20,000 different proteins (Table 8–1, p. 409), many of which are too rare to be detected by two-dimensional gel electrophoresis. If these minor cell proteins show the same pattern of similarities and differences as the more abundant proteins, a rather small number of protein differences must suffice to create very large differences in cell behavior. However, even in this case, it would be true that the different cell types in higher eucaryotes make a substantial number of different proteins, despite their apparently identical genomes.

Different Cell Types Transcribe Different Sets of Genes[41]

Given that the differences between different types of cells depend on the particular proteins that they make, it remains to be determined at what level the control of protein synthesis is exercised. Any or all of the steps in the pathway leading from DNA to protein could be involved (Figure 8–76). In principle, cell types could develop differences by (1) controlling how or when a given gene is transcribed **(transcriptional control),** (2) controlling how the initial RNA transcript is processed **(processing control),** (3) selecting which completed mRNAs in the cell nucleus are exported to the cytoplasm **(transport control),** (4) selecting which mRNAs in the cytoplasm are translated by ribosomes **(translational control),** or (5) selectively stabilizing certain mRNA molecules in the cytoplasm **(mRNA degradation control).** Although control may operate at any of these levels depending on the protein and the type of cell, we now know that a very important part of the control process occurs at the first step in the pathway—the synthesis of the primary RNA transcript.

The most convincing evidence for the importance of transcriptional control has come from an experiment in which RNA made in liver nuclei is compared with that made in brain nuclei. The experiment depends on the use of cloned cDNA "probes" made from liver mRNA (see p. 188). First, nuclei from both liver and brain cells are isolated and incubated with highly radioactive RNA precursors (ribonucleoside triphosphates; see p. 200), so that the RNA transcripts synthesized in these nuclei become radioactively labeled. These RNA molecules are then assayed as described in Figures 8–77 and 8–78.

The assay is designed to test whether hnRNA made in a brain cell nucleus contains nucleotide sequences corresponding to the mRNA molecules found in the liver cell cytoplasm, but not in the brain cell cytoplasm. The liver cell hnRNA serves as a positive control, since it must contain such sequences. Within the limited sensitivity of the assay, no brain cell hnRNA sequences could be found to correspond to any one of 11 different liver-specific mRNAs

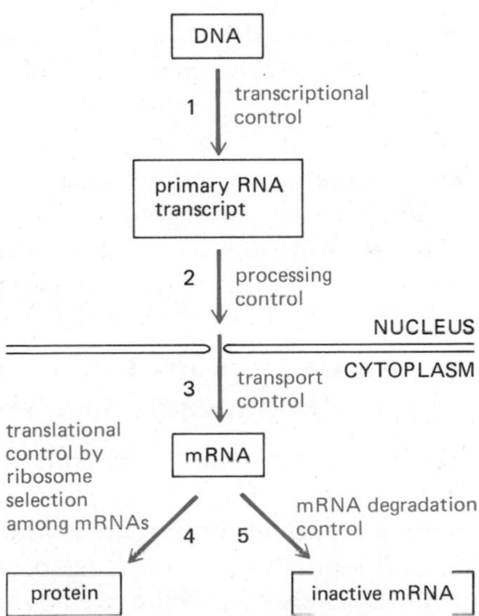

Figure 8–76 Diagram illustrating the five different levels at which gene expression might be controlled in eucaryotes.

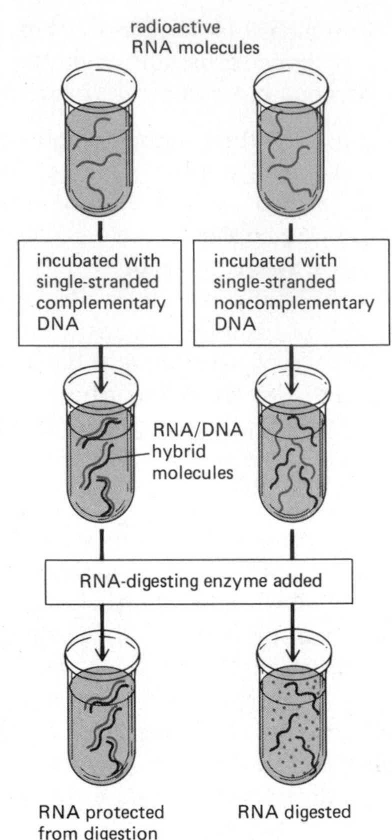

Figure 8–77 Schematic illustration of a general method used to detect radioactive RNA molecules of specific sequence. If the RNA molecules can form RNA/DNA-hybrid double helices with the specific single-stranded DNA segment used as a probe, they will be left undigested after the enzyme treatment and detected by their radioactivity. In the experiment described in the text, pure DNA segments corresponding to a variety of different genes were obtained by cloning the cDNA molecules produced by reverse transcription of a mixture of liver mRNAs (see p. 188), and each was used as a probe in a separate experiment (see Figure 8–78).

tested. The absence of these mRNAs, and the proteins for which they code, from the brain cell cytoplasm is therefore primarily due to a block in the transcription of the corresponding genes in the brain cell nuclei. In other words, for these 11 randomly selected liver-specific mRNAs, the differences between a brain cell and a liver cell are largely determined by transcriptional controls.

The sensitivity of the assay used in the above experiment was limited. Depending on the particular probe used, a 10-fold to 50-fold reduced level of transcription of a liver-specific gene in the brain cell nuclei would have escaped detection. In a few favorable cases, it has been possible to test with much greater sensitivity whether a DNA sequence that is plentifully transcribed in one cell type is transcribed at all in another cell type. Experiments of this kind have revealed rates of gene transcription between two cell types differing by as much as a factor of 1 million.

If such results hold generally, they would indicate that unexpressed eucaryotic genes show much less "leakage" of transcription than do unexpressed genes in bacteria (in which the largest known differences in transcription rates between expressed and unexpressed gene states are about 1000-fold). This suggests that at least some unexpressed genes in higher eucaryotes are turned off by mechanisms that are different and more effective than those used for turning off unneeded bacterial genes.

We begin our discussion of transcriptional control mechanisms by considering a general mode of gene control that applies to both bacteria and eucaryotes. In bacteria, gene regulatory proteins act by binding to specific DNA sequences to either turn off or turn on the transcription of a gene. Such *gene repressor* and *gene activator* proteins are well understood and seem likely to provide important insight into the more complex mechanisms used for the regulation of gene transcription in eucaryotes.

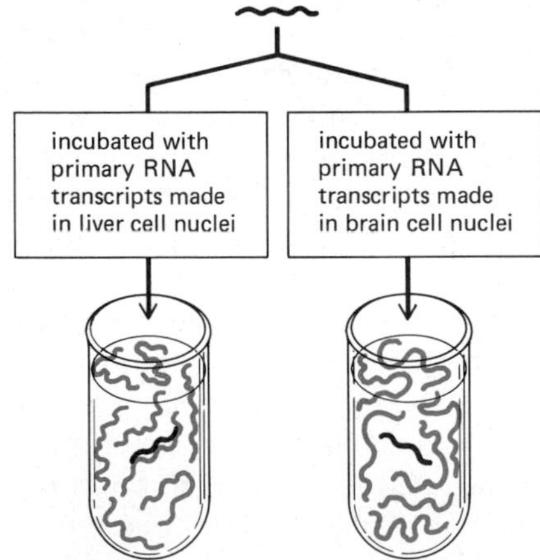

Synthesis of complementary RNA molecules detected only in liver nuclei. CONCLUSION: the DNA sequence corresponding to this liver-specific mRNA molecule is not transcribed in brain cell nuclei.

Figure 8–78 Schematic illustration of an experiment showing that gene expression in mammalian cells is controlled largely at the level of gene transcription (see also Figure 8–77).

Repressor Proteins Inhibit the Transcription of Specific Genes in Bacteria[42]

The chromosome of *E. coli* consists of a single circular DNA molecule of about 3×10^6 base pairs. This is enough DNA to code for about 2500 different proteins, though only a fraction of these proteins are made at any one time. *E. coli* regulates the expression of many of its genes according to the intracellular levels of specific metabolites.

Nearly all gene regulation in procaryotes is known to occur at the transcriptional level. Gene-control mechanisms were first worked out in the 1950s and 1960s in studies of the synthesis of the enzymes involved in the hydrolysis of the disaccharide lactose to the monosaccharides glucose and galactose in

E. coli. These experiments led to the eventual purification and characterization of the **lactose repressor protein,** whose binding to a specific DNA sequence has already been discussed (p. 392). This protein turns off specific gene transcription by binding to a specific **operator** DNA sequence of 21 base pairs that is present only once in the *E. coli* genome and overlaps an adjacent RNA polymerase binding site (the **promoter** site). When the repressor binds to the operator sequence, it blocks the access of the RNA polymerase to its binding site, thereby preventing transcription of the adjacent region of DNA (Figure 8–79).

The lactose repressor protein is itself regulated by a small sugar molecule called allolactose, which is formed when lactose is present in the cell. When allolactose reaches a high enough concentration, it induces an allosteric conformational change (p. 134) in the repressor protein, causing the latter to loosen its hold on the DNA and allow transcription to proceed: the gene is then said to be *derepressed.* As a result, an *E. coli* cell makes the enzymes it needs for the breakdown of lactose only when lactose is present (Figure 8–80).

Many other examples of this type of specific gene repression in bacteria are now known. In each case, a specific signaling ligand (such as allolactose) is recognized by a different **gene repressor protein** with an affinity for a specific operator sequence in the DNA. The binding of this ligand to the protein turns on a transcription unit adjacent to the operator by decreasing the repressor protein's affinity for its specific DNA sequence. Conversely, a specific signaling ligand can just as well be used to turn *off* an adjacent transcription unit. In this case, the allosteric change caused by ligand binding to the repressor protein *increases,* instead of decreases, the affinity with which the repressor protein binds to its specific DNA sequence. Note that, in the first case, an increase in the concentration of the signaling ligand activates

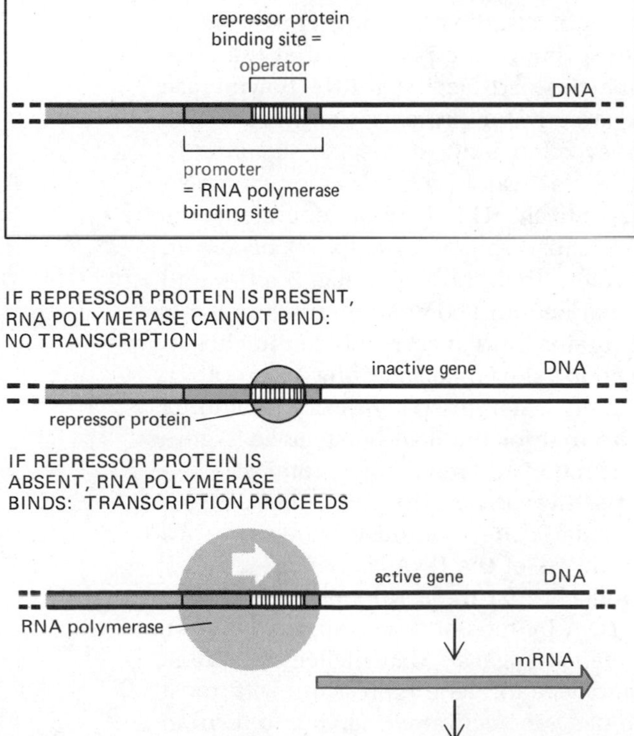

Figure 8–79 Schematic illustration of the mechanism by which specific gene repressor proteins control gene transcription in procaryotes. The repressor protein and the RNA polymerase enzyme bind tightly to different specific nucleotide sequences in the DNA, which are designated as operator and promoter sites, respectively.

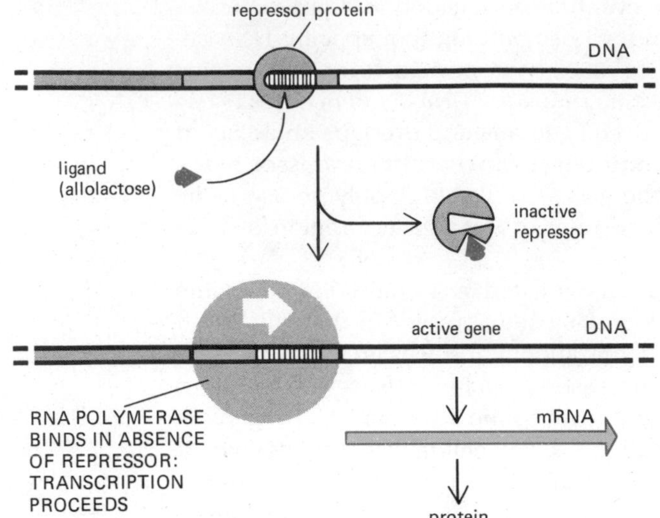

Figure 8–80 Schematic illustration of gene derepression in bacteria. A specific small molecule (here allolactose) acts as a ligand that binds to a repressor protein (here the lactose repressor protein) to remove the repressor protein from the DNA and thereby activate specific genes. The single RNA transcript synthesized in this case codes for three different proteins involved in lactose metabolism; because the three genes that encode these proteins are adjacent and coordinately controlled, each is said to be part of the same lactose *operon.*

transcription of a specific gene, whereas in the second case it suppresses specific gene transcription. Nevertheless, in principle the two mechanisms are identical because more transcription always occurs in the absence of the regulatory protein than in its presence. This type of gene control is therefore called **negative regulation.**

Gene Activator Proteins Probably Predominate in Higher Eucaryotic Cells[42]

An alternative to negative regulation is regulation by **gene activator proteins** (Figure 8–81). Some RNA transcription units in *E. coli* have relatively weak RNA polymerase binding sites that are activated by the binding of a second protein to an adjacent specific DNA sequence. Whereas repressor proteins bind to DNA sites in such a way that the repressor blocks the RNA polymerase binding site, the binding sites for gene activator proteins are placed just at the edge of the polymerase site, and the protein instead facilitates RNA polymerase binding. But in all other respects gene activator proteins closely resemble repressor proteins. Like repressors, they often bind to specific signaling ligands that either increase or decrease the activator protein's affinity for DNA and thereby turn genes on or off, respectively. This type of gene control is called **positive regulation,** however, because more transcription occurs in the presence of the gene regulatory protein than in its absence. Positive and negative modes of regulation are summarized and compared in Figure 8–82.

Some bacterial gene regulatory proteins bind at several different sites in the genome and repress transcription at one site while activating it at another. These different actions result from slightly different spacings of the binding site for the regulatory protein relative to that for the RNA polymerase (Figure 8–83). There is indirect evidence that some of the regulatory proteins in eucaryotes can also function as either positive or negative regulators of gene transcription, depending on the particular gene controlled. However, in a typical higher eucaryotic cell, only about 7% of the DNA sequences are ever transcribed into RNA. It seems very unlikely that transcription is specifically blocked on the remaining 93% of the DNA by tens of thousands of different, specific repressor proteins. Common sense suggests that higher cells must instead employ some general purpose device for gene repression, with most specific gene regulatory proteins acting as gene activators, serving to turn on particular genes for transcription.

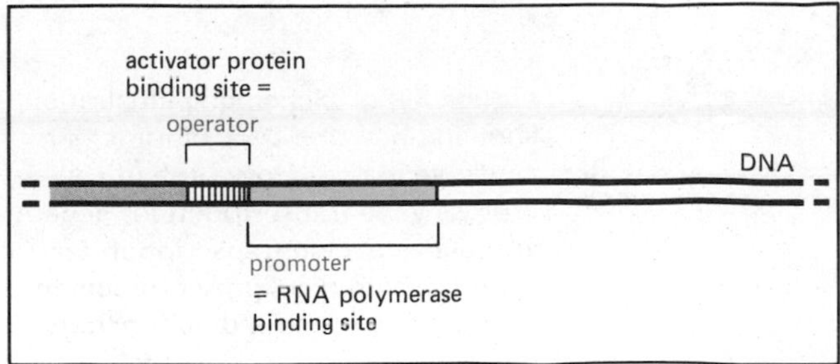

Figure 8–81 Schematic illustration of the mechanism by which specific gene activator proteins control gene transcription in procaryotes. The indicated promoter site DNA sequence binds the RNA polymerase only very weakly unless the gene activator protein is simultaneously bound to its nearby operator site. This mechanism should be compared with that of gene repression (Figure 8–79).

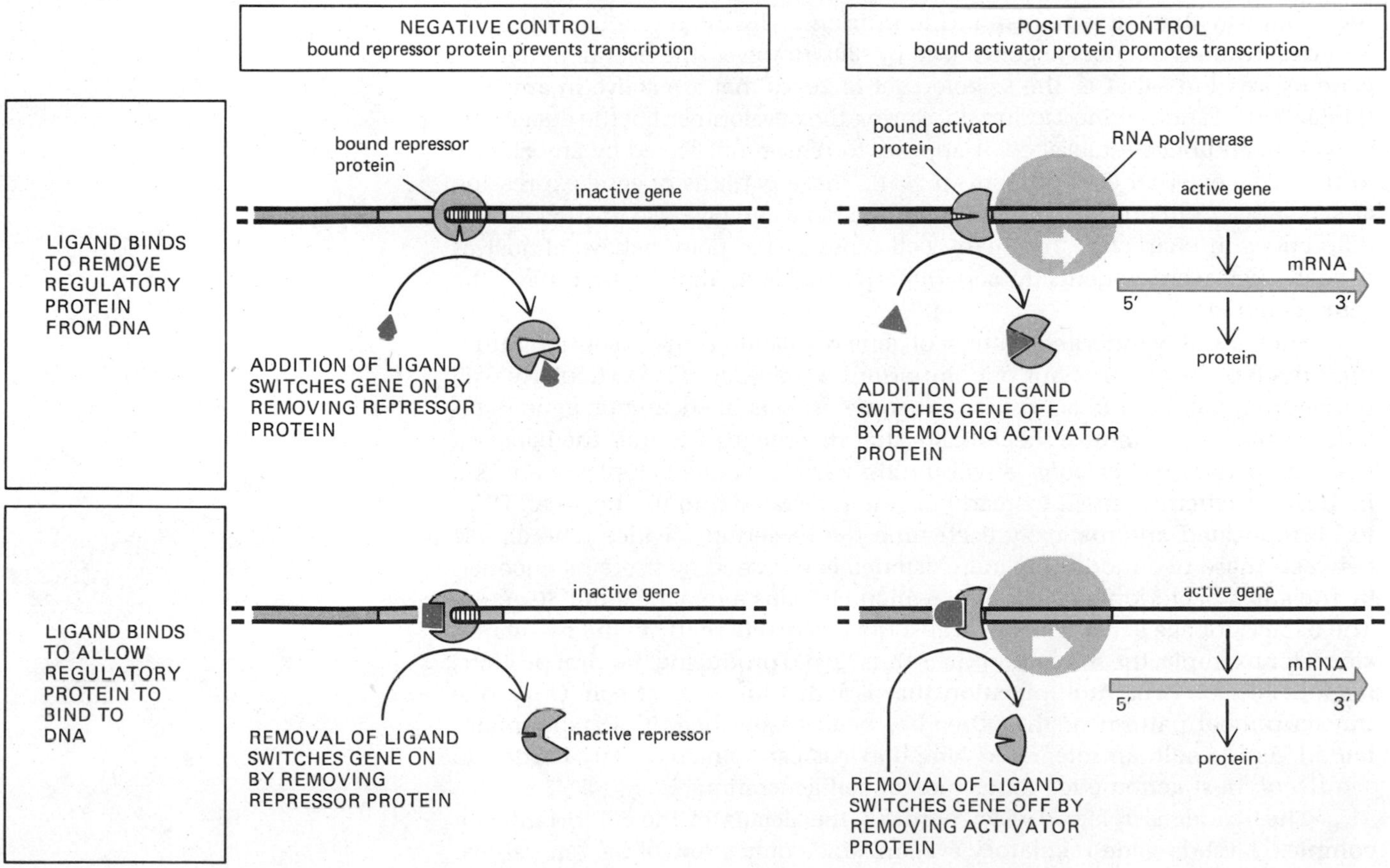

IF ACTIVATOR PROTEIN IS PRESENT, RNA POLYMERASE BINDS: TRANSCRIPTION PROCEEDS

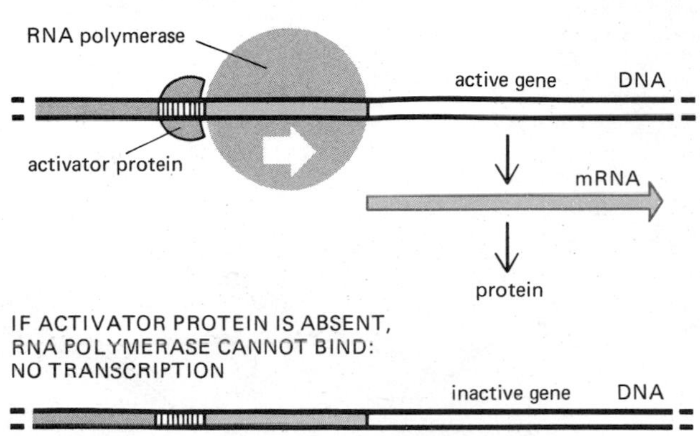

Figure 8–82 Summary of the different mechanisms by which specific gene regulatory proteins control gene transcription in procaryotes. Note that the addition of an inducing ligand can turn on a gene either by removing a gene repressor protein from the DNA (*upper left panel*) or by causing a gene activator protein to bind (*lower right panel*). Likewise, the addition of an inhibitory ligand can turn off a gene either by removing a gene activator protein from the DNA (*upper right panel*) or by causing a gene repressor protein to bind (*lower left panel*).

NEGATIVE CONTROL
bound repressor protein prevents transcription

POSITIVE CONTROL
bound activator protein promotes transcription

LIGAND BINDS TO REMOVE REGULATORY PROTEIN FROM DNA

bound repressor protein
inactive gene
ADDITION OF LIGAND SWITCHES GENE ON BY REMOVING REPRESSOR PROTEIN

bound activator protein
RNA polymerase
active gene
mRNA
5′
3′
protein
ADDITION OF LIGAND SWITCHES GENE OFF BY REMOVING ACTIVATOR PROTEIN

LIGAND BINDS TO ALLOW REGULATORY PROTEIN TO BIND TO DNA

inactive gene
inactive repressor
REMOVAL OF LIGAND SWITCHES GENE ON BY REMOVING REPRESSOR PROTEIN

active gene
mRNA
5′
3′
protein
REMOVAL OF LIGAND SWITCHES GENE OFF BY REMOVING ACTIVATOR PROTEIN

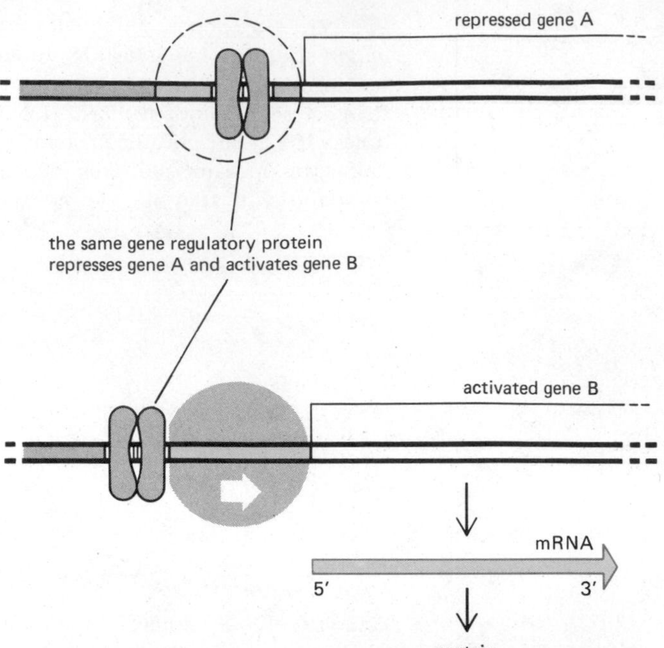

Figure 8–83 Schematic illustration showing how the DNA binding of a single gene regulatory protein can repress gene transcription for gene A and activate gene transcription for gene B. Some regulatory proteins in bacteria have such a dual function.

Two Interacting Gene Regulatory Proteins That Repress Each Other's Synthesis Produce a Stable "Memory" in a Bacterial Cell[43]

Eucaryotic cells respond to environmental signals such as hormones in much the same way that bacteria respond to nutrients—by the reversible activation and deactivation of specific genes (see p. 730). However, the overall pattern of gene expression—that is, the selected set of genes that are active in any particular cell—is determined gradually during the development of the eucaryotic organism, and once established, it appears to remain unaffected by any change in the cell's external environment (p. 834). These patterns of gene expression are so stable from one cell generation to the next, and give rise to such radical differences in character, that many cell types in the body behave almost as though they were genetically separate species, even though they have the same genome.

Such a stably inherited pattern of gene regulation is the exception rather than the rule in bacteria, but one important example that is particularly well understood will help to shed light on some aspects of eucaryotic gene regulation. This is a switchlike mechanism that determines whether the bacterial virus *lambda* (λ) *bacteriophage* will multiply in the *E. coli* cytoplasm and kill its host, or whether it will instead become integrated into the host cell DNA, to be replicated automatically each time the bacterium divides. The choice between these two modes of viral existence is governed by proteins encoded by the small bacteriophage genome, which contains a total of about 50 genes. The bacteriophage genes must be transcribed very differently in the two states, since, for example, the integrated virus must avoid producing the viral proteins responsible for virus multiplication that is lethal to the host cell. Once one transcriptional pattern or the other has been established, it is stably maintained. As a result, an integrated lambda virus can remain quietly hidden in the *E. coli* host genome for thousands of cell generations.

The reader is referred elsewhere for the details of the important, but complex, lambda gene regulatory system, since only a few of its general fea-

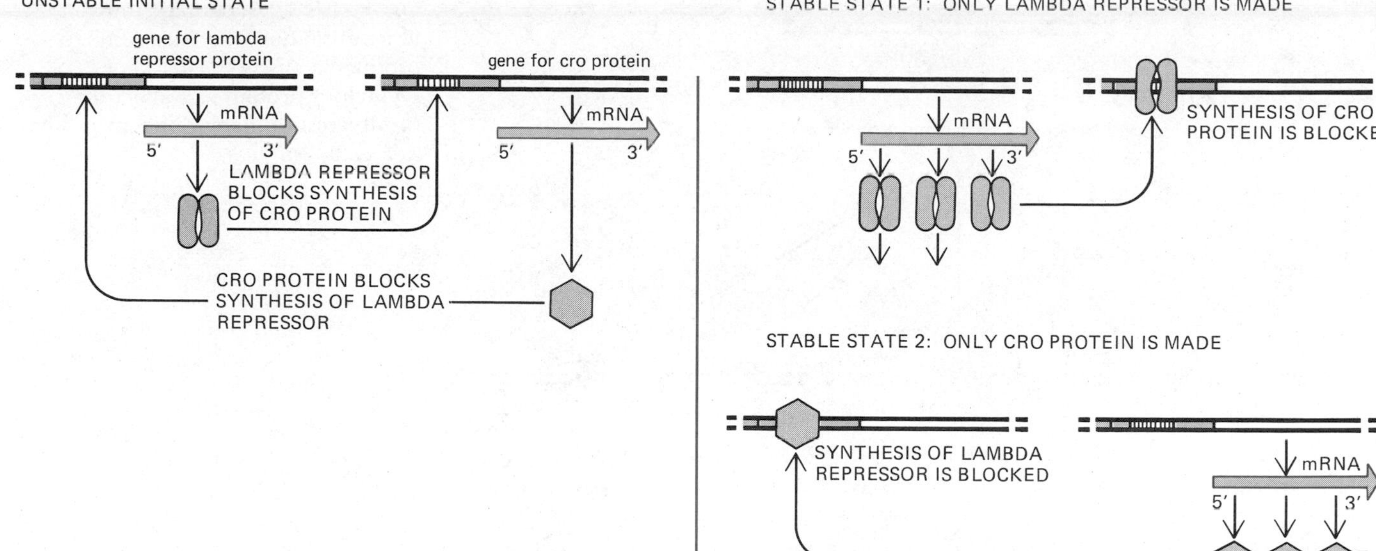

Figure 8–84 A simplified version of the switchlike regulatory system that determines the mode of growth of bacteriophage lambda in the *E. coli* host cell. In state 1 (the lysogenic state), large amounts of lambda repressor protein are synthesized. This gene regulatory protein turns off the synthesis of several bacteriophage proteins, including the cro protein. As a result, the viral DNA becomes integrated in the *E. coli* chromosome and is duplicated automatically as the bacterium grows. In state 2 (the lytic state), large amounts of cro protein are synthesized. This gene regulatory protein turns off the synthesis of the lambda repressor protein. As a result, many bacteriophage proteins are made and the viral DNA replicates freely in the *E. coli* cell, eventually producing many new bacteriophage particles and killing the cell.

tures need be stressed here. At the heart of this system are two regulatory proteins synthesized by the virus: the **lambda repressor protein** and the **cro protein,** each of which blocks the synthesis of the other when bound to the operator site of the other's gene. There are only two stable states. In state 1 (the *lysogenic state*), the lambda repressor dominates, and the lambda repressor but not the cro protein is synthesized. In state 2 (the *lytic state*), the cro protein dominates, and the cro protein but not the lambda repressor is synthesized (Figure 8–84). The predominance of one or the other of these proteins in turn controls a set of other genes, so that most of the DNA of the stably integrated bacteriophage is quiescent in state 1, while it is extensively transcribed, replicated, packaged into new bacteriophage, and released by host cell lysis in state 2.

The mode of growth adopted by an infecting lambda bacteriophage is not randomly determined. For example, when the host bacteria are growing well, the virus tends to adopt state 1, allowing the DNA of the virus to multiply quickly along with the host chromosome. On the other hand, when an integrated virus finds itself in a sickly cell, it converts from state 1 to state 2 in order to multiply in the cell cytoplasm and make a quick exit (see p. 236).

Gene Regulatory Proteins That Act on Many Genes Simultaneously Are Probably Used in Combinations to Generate Different Tissues in Eucaryotes

The important lesson to be learned from bacteriophage lambda is that an amazingly complex pattern of behavior can be achieved with only a few gene regulatory proteins that reciprocally affect each other's synthesis and activities. With the number of gene regulatory elements available to eucaryotes, the possibilities for gene regulation are staggering.

Virtually nothing is known for certain about the molecular mechanisms that switch large numbers of genes on and off in an orderly way during the development of complex multicellular eucaryotes. However, some idea of the nature of the controls involved can be deduced from certain gene mutations that have striking effects on embryogenesis. In particular, as discussed in detail in Chapter 15, a number of single gene mutations in the fruit fly *Dro-*

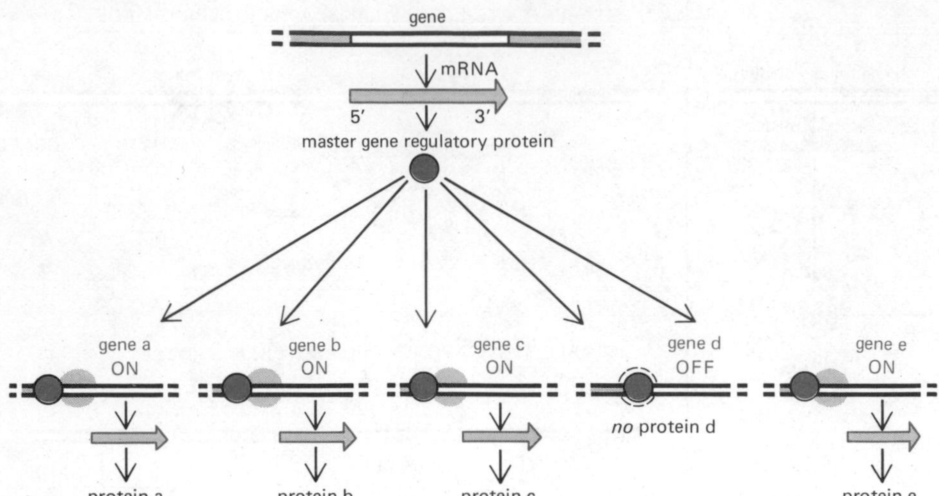

Figure 8–85 Schematic drawing illustrating how the "decision" to produce a single master gene regulatory protein can affect the production of many different proteins in a cell.

sophila convert one part of the body to another. For example, a mutation called *Antennapedia* causes the whole group of cells that would normally make an antenna to switch to making a leg instead, producing a fly with a leg growing out of its head (p. 840). The class of mutations that brings about such wholesale transformations are known as *homeotic mutations.* It is assumed that the proteins coded for by these mutant genes are master gene regulatory proteins, each controlling a large number of other genes in a way that may be similar to the control exerted by the lambda repressor and cro proteins but on a larger scale (Figure 8–85).

The remarkable feature of the homeotic mutations in *Drosophila* is that by interfering with the action of a single gene product they cause the formation of a complex alternative structure, rather than just an amorphous mass of tissue. This is one of several observations that suggest that a large number of differently organized tissues are produced in *Drosophila* by combining the activities of a relatively small number of master gene regulatory proteins (see p. 840). That this may also be the case in mammals, including man, is suggested by the observation that the absence of a single gene regulatory protein (the receptor protein for testosterone) causes an individual with a male (XY) genotype to develop as an almost perfect female (p. 732). As we shall now discuss, such combinatorial gene regulation may enable complex organisms to develop through the action of a relatively small number of different master gene regulatory proteins.

In Principle, Many Different Cell Types Can Be Efficiently Specified by Combinations of a Few Gene Regulatory Proteins[44]

The essence of **combinatorial gene regulation** is illustrated in Figure 8–86, in which each numbered element represents a regulatory protein made by a different controlling gene. In this purely hypothetical scheme, one initial cell type gives rise to two different types of cells, denoted as A and B, which differ only in respect to whether or not gene regulatory protein ① is made. This difference is analogous to the two states of bacteriophage lambda, which depend on whether or not lambda repressor protein is produced. The subsequent development of each of these cells leads to the additional production in some cells first of gene regulatory proteins ② and ③ and later of gene regulatory proteins ④ and ⑤. In the end, 8 different cell types (cell G through cell N) have been created with 5 different regulatory proteins.

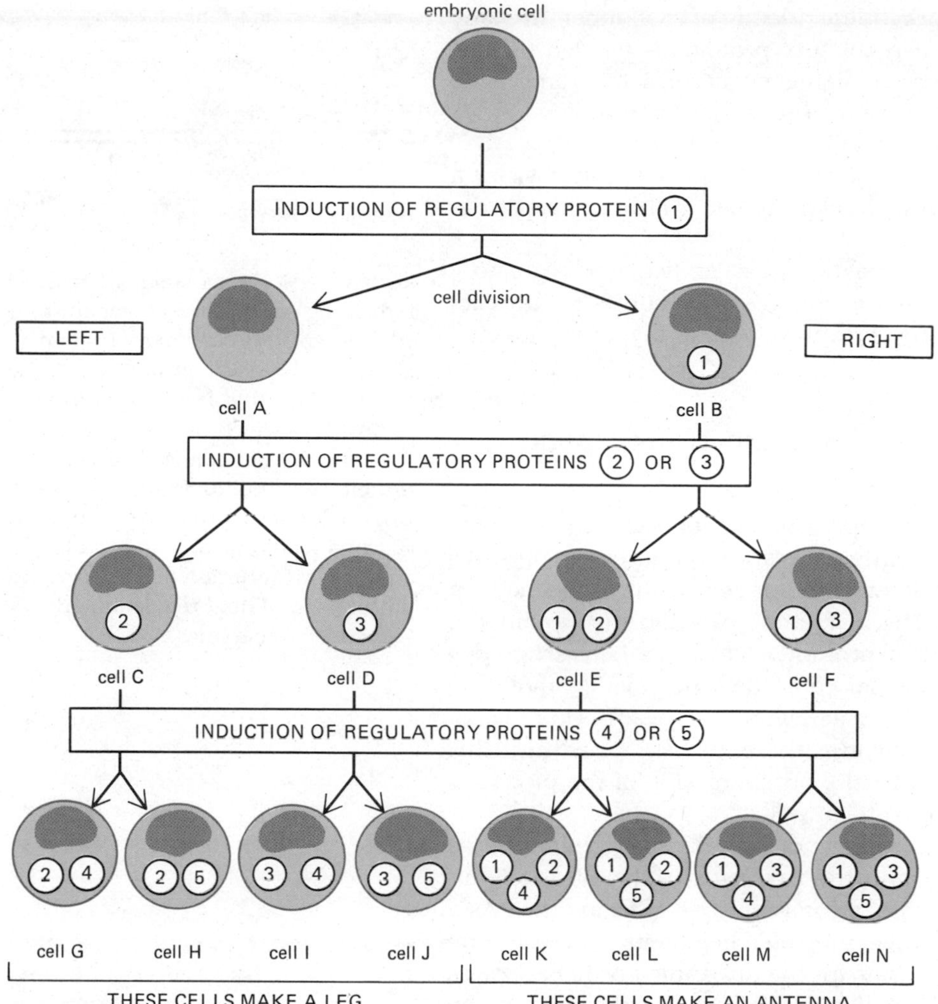

Figure 8–86 A highly schematic scheme for cell development illustrating how combinations of a few master gene regulatory proteins can generate many different cell types in embryos.

In this simple scheme, a "decision" to make one of a pair of different master gene regulatory proteins (shown as numbered circles) is made after each cell division. Sensing its relative position in an embryonic field, the daughter cell toward the left side of the embryo (the left side of the page) always induces the synthesis of the even-numbered protein of each pair, while the daughter cell toward the right side of the embryo induces the synthesis of the odd-numbered protein. This decision is similar to the decision of bacteriophage lambda to make either the lambda repressor protein or the cro protein (Figure 8–84). The production of each master gene regulatory protein is assumed, like the bacteriophage lambda decision, to be self-perpetuating. Therefore, the cells in the enlarging clone contain an increasing number of regulatory proteins, each of which is assumed to control a whole battery of genes (Figure 8–85).

With the addition of 2 more gene regulatory proteins to the system shown in Figure 8–86 (⑥ and ⑦), 16 cell types would be generated at the next step. And by the time 10 more such steps occurred, slightly more than 10,000 different cell types would have been specified, in principle, through the action of only 25 different regulatory proteins. Combinatorial regulation of this sort is thus very efficient. Moreover, it can explain the homeotic mutants of *Drosophila.* If a mutation eliminates one of the gene regulatory proteins that would normally appear early in development, a whole group of cells will be respecified. The mutant cells may now contain exactly the same set of gene regulatory proteins that are present in the cells of a different structure and accordingly produce that structure. For example, a *Drosophila* antenna would be converted into a leg with the elimination of protein ① in Figure 8–86.

Input from Several Different Gene Regulatory Proteins May Be Needed to Turn On a Single Gene[45]

At first glance, the scheme for combinatorial gene regulation just discussed would appear to predict that cell E differs from cell B by the presence or absence of exactly the same proteins that distinguish cell C from cell A, since in both pairs the difference between the two cells is the addition of regulatory protein ② to the daughter cell (see Figure 8–86). But even in bacterial systems, regulatory proteins do not always work independently of each other. In many cases, the interaction of two different regulatory proteins is needed to turn on a single bacterial gene. Moreover, in higher eucaryotes the same gene activator protein (a steroid hormone receptor protein) has been shown to turn on the synthesis of a different set of proteins in different types of cells (p. 731). Thus, the change in gene expression caused by inducing the synthesis of a particular gene regulatory protein in a developing cell will be different depending on the cell's past history.

Two of many possible modes of interaction between two gene regulatory proteins are diagrammatically illustrated in Figures 8–87 and 8–88. Interactions of this type impart an enormous versatility to gene regulatory mechanisms.

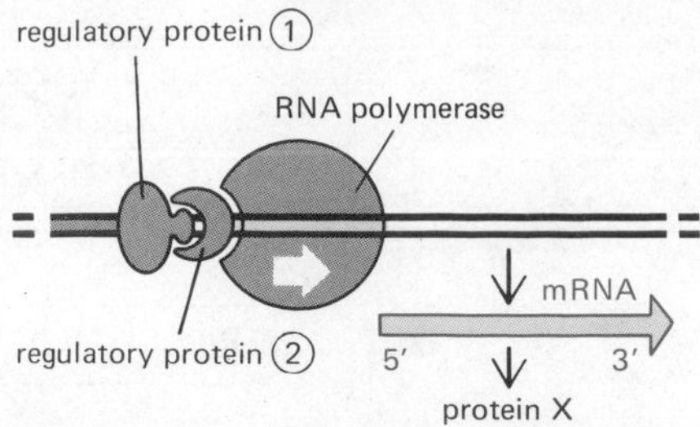

Figure 8–87 A schematic diagram indicating one possible mechanism in which two gene regulatory proteins are needed to turn on the transcription that leads to synthesis of protein X. Here, binding of protein ① to the DNA is required before protein ② can bind. With some other DNA sequences, protein ① or protein ② alone would be able to bind to activate transcription. Clear examples of this type of dual regulation are known in procaryotes.

In Bacteria, DNA Supercoiling Facilitates the Transcription of Genes[46]

Thus far our concern has been how gene repressor and gene activator proteins function in bacteria and how similar proteins might work in eucaryotes. We shall now consider another quite different type of gene control mechanism known to operate in bacterial cells. This regulatory mechanism exploits the fact that about 10 DNA base pairs at a promoter site must be disrupted—allowing RNA-DNA base pairs to form—before an RNA polymerase molecule can begin to synthesize a new RNA chain (see Figure 5–2, p. 201). One turn of the DNA helix must be unwound to expose these 10 base pairs, and this requires that the two segments of DNA helix on either side of the promoter rotate relative to each other by one turn (Figure 8–89A).

But the DNA helix in a bacterial cell is known to be folded into a series of looped DNA domains (see p. 397), and the two ends of the DNA helix in each loop are relatively fixed and unable to rotate relative to each other. As a result, each time that an RNA polymerase molecule begins its synthesis, one **DNA supercoil** should form to compensate for the unwinding of 10 base pairs in the DNA double helix (Figure 8–89B). Since DNA supercoiling bends the stiff DNA helix, the process shown in Figure 8–89B creates a tension in the DNA that makes it energetically unfavorable compared to the process shown in Figure 8–89A. Therefore, in the absence of compensating factors, RNA synthesis will occur less often on DNA constrained in a looped domain than on DNA that is free to rotate.

Bacteria, however, are able to exploit the absence of free rotation in looped domains to promote rather than hinder their RNA synthesis. A bacterial enzyme called **DNA gyrase** uses the energy of ATP hydrolysis to pump supercoils continuously into the DNA, thereby maintaining each looped domain under tension. These supercoils are *negative supercoils*, having the opposite handedness from the positive supercoils that form when RNA polymerase binds. As a result, the local unwinding of a region of DNA helix by RNA polymerase normally *removes* a negative supercoil from the DNA rather than adding a positive one. Because superhelical tension in the DNA is reduced during the initiation of RNA synthesis, this synthesis is energetically favored compared to RNA synthesis on DNA that is not supercoiled (Figure 8–90).

A single nick or gap in a region of DNA helix that forms a looped domain will create a "swivel" in the DNA that causes the superhelical tension in the entire loop to dissipate (see Figure 5–40, p. 229). Since many promoter sequences on the bacterial genome will function very poorly, if at all, without this superhelical tension, RNA synthesis rates are depressed throughout a damaged looped domain until a completely intact double helix is restored. This limitation is probably advantageous to the cell because it tends to prevent

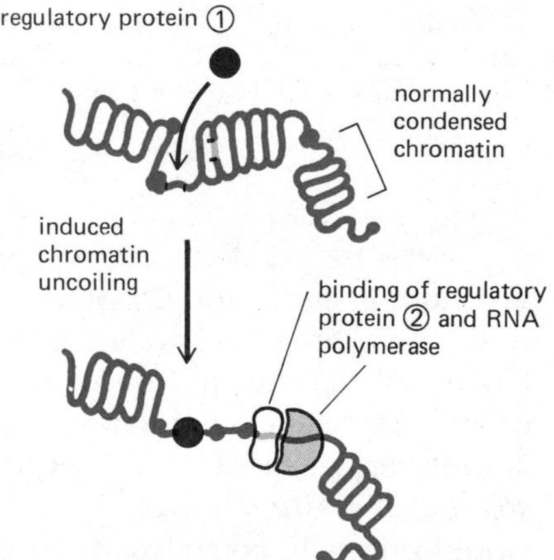

Figure 8–88 A schematic diagram indicating a different possible mechanism from that shown in Figure 8–87 in which two gene regulatory proteins are needed to turn on the transcription that leads to synthesis of protein X. Here the binding of protein ① is required to decondense (or otherwise alter) the normal chromatin structure before protein ② can bind. There is evidence that such alterations in chromatin structure can control gene expression in eucaryotes (see Figure 8–93).

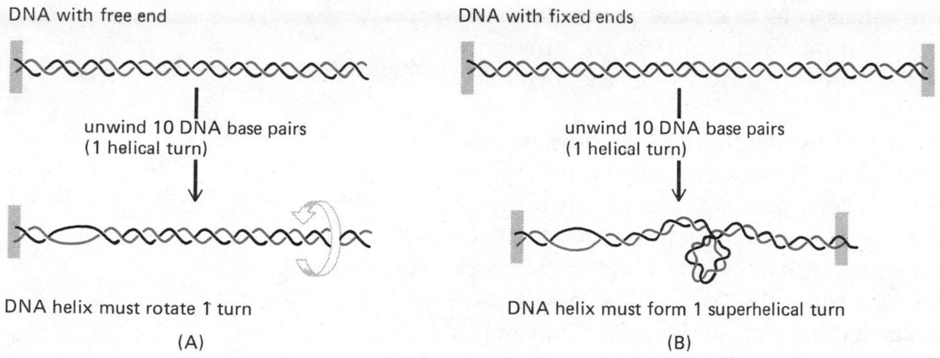

DNA with free end

unwind 10 DNA base pairs (1 helical turn)

DNA helix must rotate 1 turn

(A)

DNA with fixed ends

unwind 10 DNA base pairs (1 helical turn)

DNA helix must form 1 superhelical turn

(B)

Figure 8–89 Effects transmitted along the DNA helix when it opens to initiate transcription. (A) The DNA double helix rotates by one turn for every 10 base pairs opened. (B) If rotation is prevented, one DNA supercoil forms in the DNA double helix for every 10 base pairs opened. The supercoil formed here is defined as a "positive supercoil."

moving RNA polymerase molecules from interfering with other genetic processes that create nicks or gaps in DNA—such as DNA repair, DNA replication, and genetic recombination processes (see Chapter 5).

Although the majority of the DNA in a bacterial cell is normally kept under negative superhelical tension, apparently this is not the case for most of the DNA in eucaryotic cells. It remains possible, however, that superhelical tension (either negative or positive) is imparted to individual looped domains of chromatin under special circumstances—as when a eucaryotic gene is activated or when its chromatin structure is altered. As we shall see, in-depth studies of eucaryotic gene regulatory mechanisms have only just begun.

The Regulation of Gene Expression in Eucaryotic Cells Includes Types of Controls Not Found in Bacteria

In principle, all of the gene regulatory mechanisms discussed so far could operate equally well in procaryotes and eucaryotes. But eucaryotic genes and chromosomes, as we have seen, are not exactly like procaryotic genes and

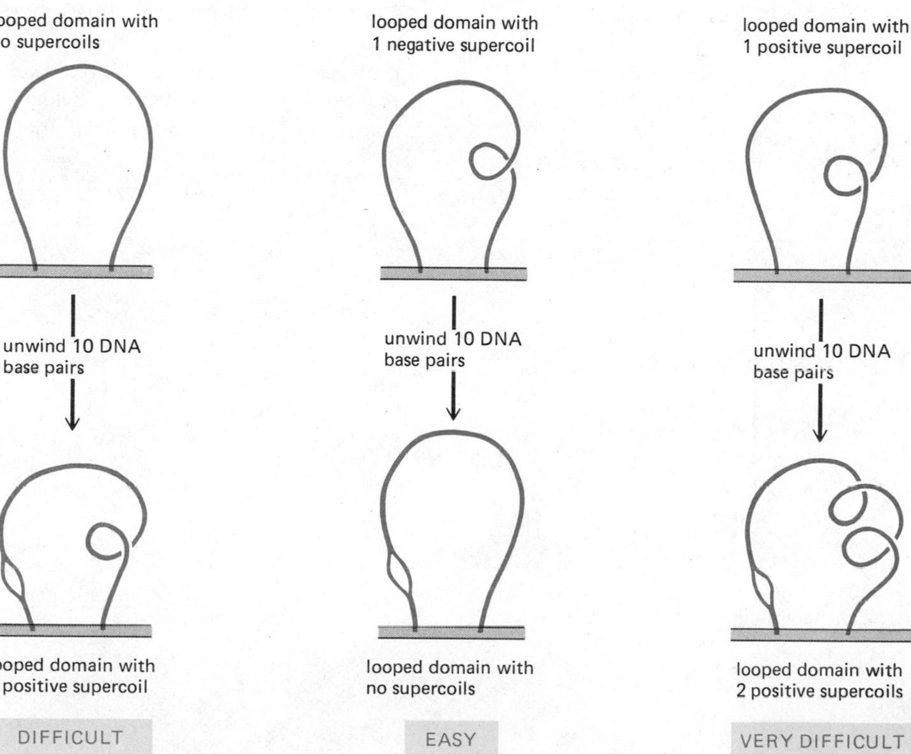

Figure 8–90 The effects of superhelical tension on helix opening in a looped domain. The looped domains in bacterial DNA are "negatively supercoiled," making the initiation of RNA synthesis easier than it would be on a nonsupercoiled chromosome (see text).

chromosomes. Two features unique to eucaryotes that have particularly important regulatory implications are the tight packaging of DNA with histones to form chromatin, and the extensive RNA splicing involved in the formation of mRNA molecules.

We shall first consider the part played by tightly bound chromosomal proteins in the regulation of eucaryotic gene expression. Eucaryotic DNA is packaged by histones that never leave the DNA molecules to which they are bound. This means that eucaryotic RNA polymerases must transcribe the DNA while it remains bound up in nucleosomes. However, the majority of chromatin fibers are so highly condensed that most of the DNA is inaccessible (see Figure 8–40). The available evidence suggests that a given domain of chromatin cannot be transcribed unless it is first decondensed. This requirement would seem to add to the gene activation process in eucaryotes an extra step that is not required in bacteria.

Eucaryotic Gene Activation Mechanisms Appear to Loosen Chromatin Structure All Along a Gene[47]

An important piece of evidence for the selective decondensation of the chromatin of active genes comes from experiments in which isolated nuclei are treated with pancreatic DNase I, an enzyme that at high enough concentrations can degrade DNA even in nucleosomes. At moderate concentrations of this enzyme, not all of the DNA is digested, and the sequences that are preferentially degraded belong to the 7% or so that are transcribed. Moreover, different DNA sequences are preferentially degraded in different cells of the same organism, in a pattern corresponding to the different pattern of RNAs that the cells make. Remarkably, even genes that are transcribed only a few times in every cell generation are sensitive to DNase I, indicating that the special state of the chromatin, rather than the process of RNA transcription itself, makes these regions unusually accessible to the enzyme (Figure 8–91).

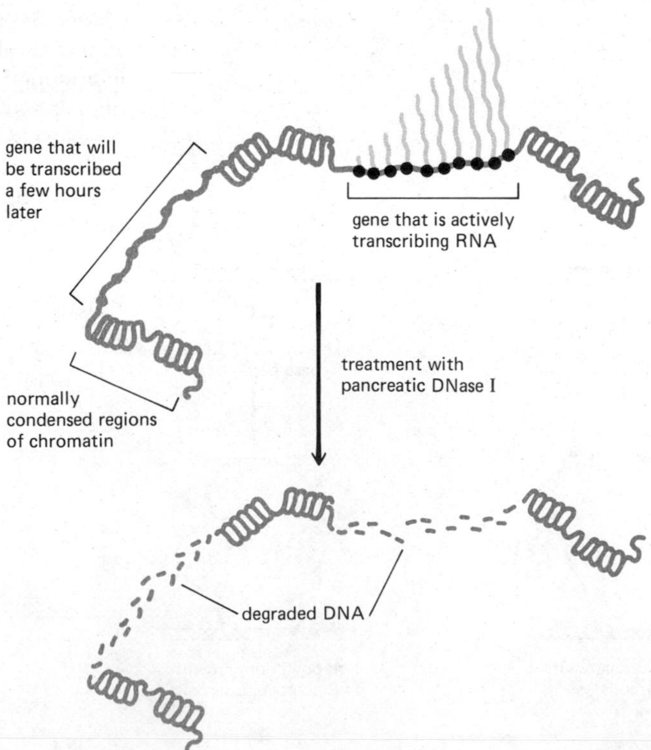

Figure 8–91 Schematic illustration of a mild digestion of chromatin with pancreatic DNase I, which selectively degrades the DNA of both actively transcribing genes and potentially active genes.

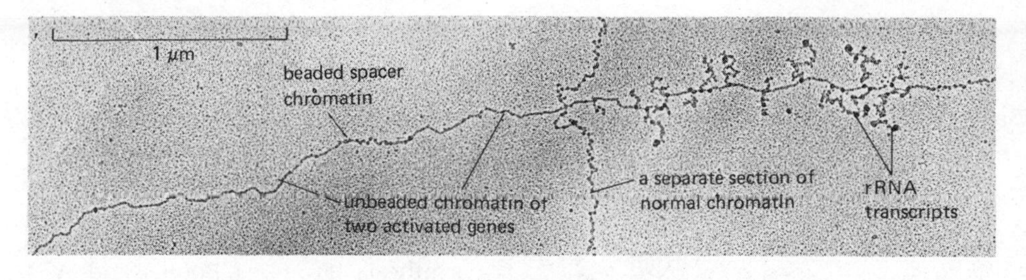

Figure 8–92 Electron micrograph of spread chromatin from an insect embryo, showing a visible change in the chromatin structure of two tandemly arranged rRNA genes. A change from a beaded to an unbeaded form of chromatin appears to precede the initiation of rRNA synthesis in these embryos. The unbeaded chromatin is thought to represent a region where the DNA has uncoiled from around the nucleosome beads, in preparation for transcription. (From V. E. Foe, *Cold Spring Harbor Symp. Quant. Biol.* 42:723–740, 1978.)

Usually the difference between the so-called transcriptionally active nucleosomes and inactive nucleosomes cannot be detected in the electron microscope. A striking exception is found in certain insect embryos that display conspicuous tandem arrays of apparently unbeaded chromatin in the region of activated rRNA genes (Figure 8–92). Because these genes remain covered by histonelike proteins, it has been suggested that they represent regions of the chromatin where each nucleosome has opened up into an extended conformation, uncoiling the two turns of the DNA helix that are normally wound around its core to facilitate transcription.

What is the biochemical basis for the structural difference between active and inactive chromatin? We have referred to this question in earlier sections on nucleosomes. Histone H1, which binds to nucleosomes to help pack them together, seems to be less tightly bound to at least some active chromatin, although the four nucleosomal histones are present in normal amounts. A second difference is that the histones in transcriptionally active nucleosomes appear to be unusually highly acetylated (see p. 390). An enhanced acetylation of neighboring nucleosomes has been shown to decrease their tendency to pack together *in vitro*. Thus, both the reduced affinity of active chromatin for histone H1 and its enhanced acetylation should help to convert the 30-nm chromatin fiber to a beads-on-a-string conformation. Finally, the nucleosomes of active genes selectively bind to two small, abundant, nonhistone proteins, HMG 14 and HMG 17 (p. 392) and seem to be highly enriched in the form of histone H2A that is covalently linked to ubiquitin (p. 390). All of these changes may play an important part in uncoiling the chromatin of active genes.

Eucaryotic Gene Activation Probably Occurs in Two Stages

All of the nucleosomes in a transcription unit seem to be converted to an active form in those cells in which the transcription unit is active. Although we know nothing about how this process is controlled, a special set of gene regulatory proteins, quite different from those in procaryotes, must be involved. These proteins must not only recognize a specific site on the chromatin (presumably a DNA sequence), but also cause a structural change in the neighboring chromatin that is propagated from the recognized site through an average of at least 40 adjacent nucleosomes (8000 DNA base pairs)—and perhaps throughout an entire looped domain of chromatin (see below). The better understood type of gene regulatory proteins that resemble the gene activator and repressor proteins of bacteria are thought to regulate the subsequent transcription of specific genes within regions of exposed active chromatin. Eucaryotic gene activation thus probably occurs in two stages: first, the chromatin structure is modified to decondense it; second, DNA transcription is activated in selected subregions of the decondensed chromatin (Figure 8–93).

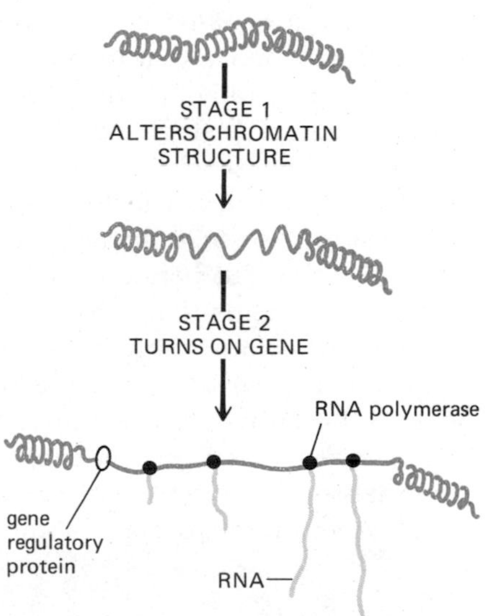

Figure 8–93 Schematic diagram illustrating the general nature of the two stages believed to be involved in eucaryotic gene activation. (A) In stage 1 the structure of a local region of chromatin is modified in preparation for transcription. (B) In stage 2, gene regulatory proteins bind to specific sites on the altered chromatin to induce RNA synthesis. According to this view, transcription in procaryotes provides a model only for stage 2 of the eucaryotic gene-control process.

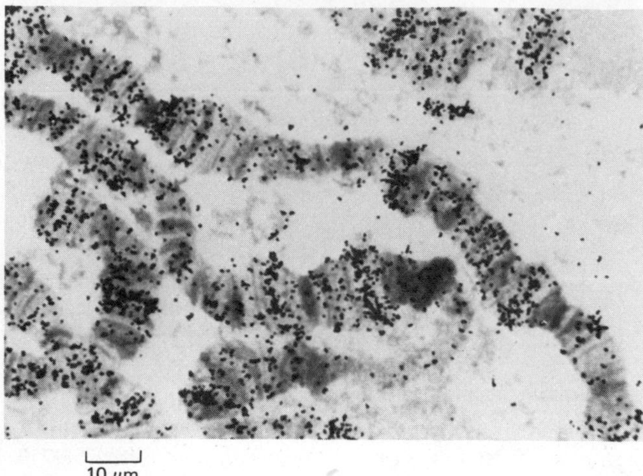

Figure 8–94 An autoradiogram that illustrates the location of [^{3}H]uridine incorporation sites on polytene chromosomes. The highest grain densities are found over expanded regions on the chromosomes (chromosome puffs), indicating that these are the most active sites of RNA synthesis. (From J. J. Bonner and M. L. Pardue, *Cell* 12:227–234, 1977. © M.I.T. Press.)

Looped Chromatin Domains Unfold as They Are Transcribed in Polytene Chromosomes[13,48]

Long before chromatin structure was understood, studies of the polytene chromosomes of the giant cells of *Drosophila* larval salivary glands revealed that a major change in DNA packing accompanies gene transcription. We have already described how these chromosomes consist of 1024 parallel chromatids generated by repeated chromosomal replication (p. 402). We have also discussed the evidence that the bands on the resulting giant chromosomes correspond structurally to looped domains of chromatin and functionally to individual genes. We shall now consider how the appearance and biochemistry of individual bands change as the genes are transcribed.

One of the main factors controlling the activity of genes in polytene chromosomes of *Drosophila* is the insect hormone *ecdysone*, the levels of which rise and fall periodically, inducing the transcription of various genes coding for proteins that the larval insect requires for each molt and for pupation. The regions being transcribed at any instant can be identified by labeling the cells briefly with the radioactive RNA precursor [^{3}H]uridine and locating the growing RNA transcripts by autoradiography (Figures 8–94 and 8–95). This analysis reveals that the most active chromosomal regions are decondensed, forming the distinctive **chromosome puffs** briefly mentioned earlier. Electron microscopy of thin sections of these puffs (Figure 8–96) shows

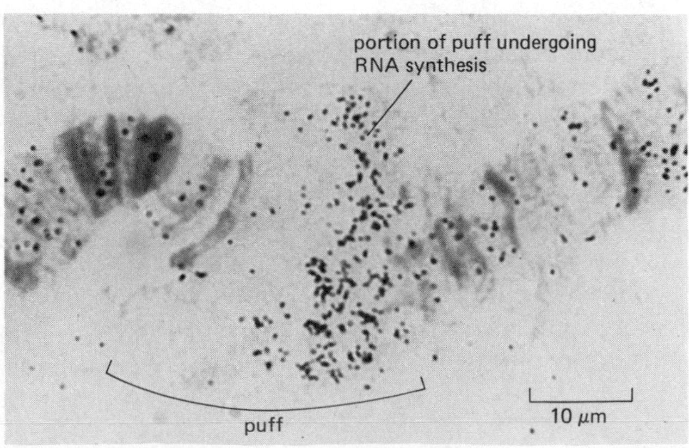

Figure 8–95 Autoradiograms of large puffs labeled as in Figure 8–94 reveal that RNA can in some cases be synthesized only in part of the puff, with the remainder of the puff containing decondensed chromatin that is apparently not transcribed. (Courtesy of Jose Bonner.)

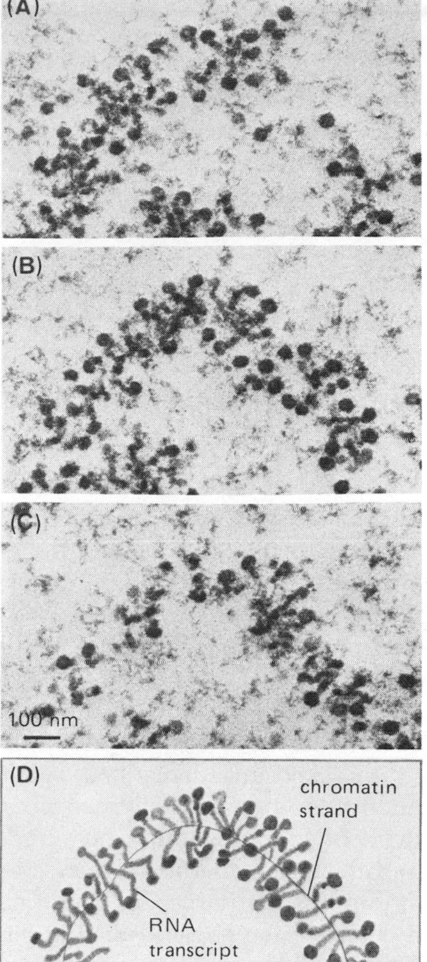

Figure 8–96 Electron micrographs of serial thin sections through a large chromosome puff, showing the conformation of a prominent, unusually long transcription unit. Bar in (C) indicates 100 nm. A three-dimensional reconstruction of part of the transcription unit is shown in (D), revealing RNA transcripts with knobbed ends attached to a single chromatin fiber. (From K. Andersson, B. Björkroth, and B. Daneholt, *Exp. Cell Res.* 130:313–326, 1980.)

that the DNA in the chromatin is much less condensed than it would be in the 30-nm chromatin fiber.

As the organism progresses through different developmental stages, new puffs arise and old puffs recede as transcription units are activated and deactivated, and different mRNAs and proteins are made (Figure 8–97). From inspection of each puff when it is relatively small and the banding pattern of the chromosome is still discernable, it seems that most puffs arise from the uncoiling of a single chromosome band (Figure 8–98). This suggests that the looped domains, which are thought to be folded to form each chromosome band (Figure 8–28), function as a unit during transcription.

Ecdysone binds to a specific receptor protein, which in turn binds to specific sites on chromatin to induce RNA transcription (see p. 730). Immunofluorescence staining of the chromosomes shows that most of the bands whose puffing can be induced by ecdysone have at least one specific nonhistone protein in common. This protein may help fold the chromatin into a conformation in which the DNA can be recognized by the ecdysone receptor protein, thus acting at stage 1 of the general scheme for eucaryotic gene activation shown previously in Figure 8–93—the ecdysone receptor protein itself acting at stage 2.

An intriguing observation that bears on the mechanism of local chromosome decondensation is that puffing can be "infectious." This infectivity

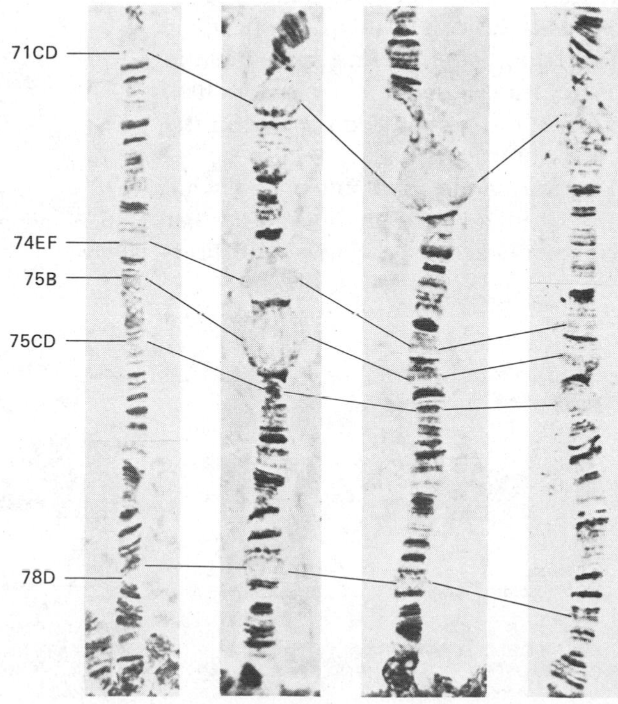

Figure 8–97 A temporal series of photographs illustrating how puffs arise and recede in the polytene chromosomes of *Drosophila melanogaster*. A region of the left arm of chromosome 3 is shown, which exhibits five very large puffs in salivary gland cells, each active for only a short developmental period (see lines). The series of changes shown occur over a period of 22 hours, appearing in a reproducible pattern as the organism develops. (Courtesy of Michael Ashburner.)

depends on the fact that the two homologous polytene chromosomes present in these diploid cells are normally tightly paired. Sometimes one of the two homologs contains a defective gene that fails to puff in response to ecdysone, while the gene on the other homolog is normal. Some defective genes can be induced to puff when brought into close apposition to the normal gene (Figure 8–99). To explain such results, it has been suggested that each band is constructed like a tiny crystal and that the chromatin packing of one band can affect the packing of its homologous partner band in the same way that a seed crystal will direct the growth of larger crystals. It is possible that a local change in chromatin structure induced by a regulatory protein spreads throughout an entire chromosome band by the same mechanism.

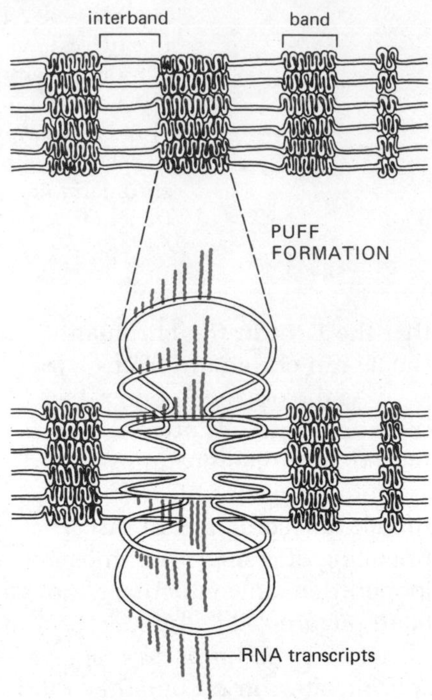

Figure 8–98 A highly schematic view of the process of puff formation in a polytene chromosome.

Lampbrush Chromosomes Display Their Transcription Units in Extended Looped Domains[49]

Immature eggs, or oocytes, are another source of highly visible, active chromosomes. At an early stage in oocyte differentiation, each chromosome replicates to begin meiosis, and the homologous replicated chromsomes pair to form a structure containing a total of four chromatids. This structure may persist for months or years as the oocyte builds up a supply of mRNA and other materials required for its ultimate development into a new individual (p. 784). During this period the meiotically paired chromosomes often expand and become very active in RNA synthesis. This activity is accompanied by the extension of long chromatin loops, which are covered with newly transcribed RNA packed into dense RNA-protein complexes. These so-called **lampbrush chromosomes** are visible in the light microscope, even though they are neither polytene nor unusually condensed (Figure 8–100). Since their discovery in the late nineteenth century, the unusually large lampbrush chromosomes of amphibians have been the most extensively studied. The organization of the lampbrush chromosome is schematically illustrated in Figure 8–101.

The pattern of transcriptional activity varies from one loop to another. Many loops are transcribed continuously from end to end and consequently increase gradually in thickness from one end to the other, with the thin end of the loop containing the shortest RNA transcripts. Other loops are discontinuously transcribed, so that their thickness varies irregularly along their length, as one transcription unit begins and another ends. And some include an extended section of chromatin that is not transcribed at all (Figure 8–102). The majority of the chromatin is not in loops and remains condensed in the chromomeres; in general, this chromatin is not transcribed.

Lampbrush chromosome loops, like the bands that puff on polytene chromosomes, seem to correspond to fixed units of chromatin folding that have decondensed and become transcriptionally active. Nucleic acid hybrid-

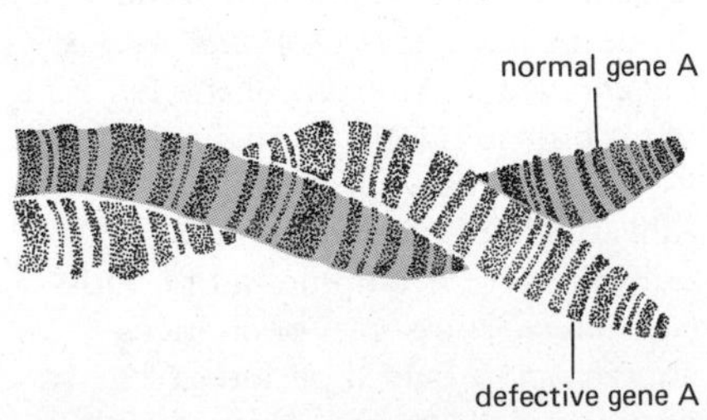

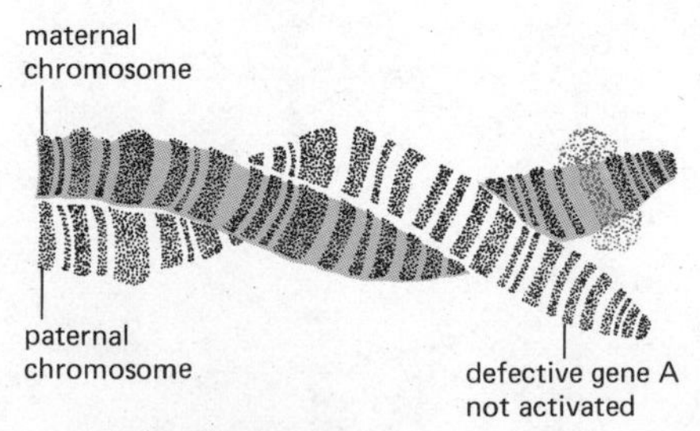

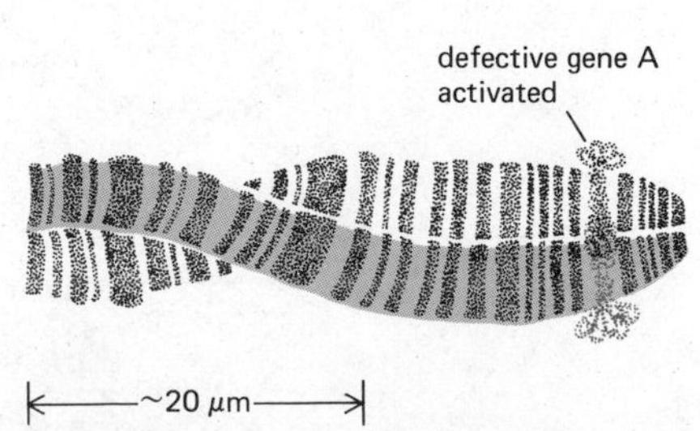

Figure 8–99 The "infectious puff" phenomenon, as detected in the polytene chromosomes of *Drosophila*.

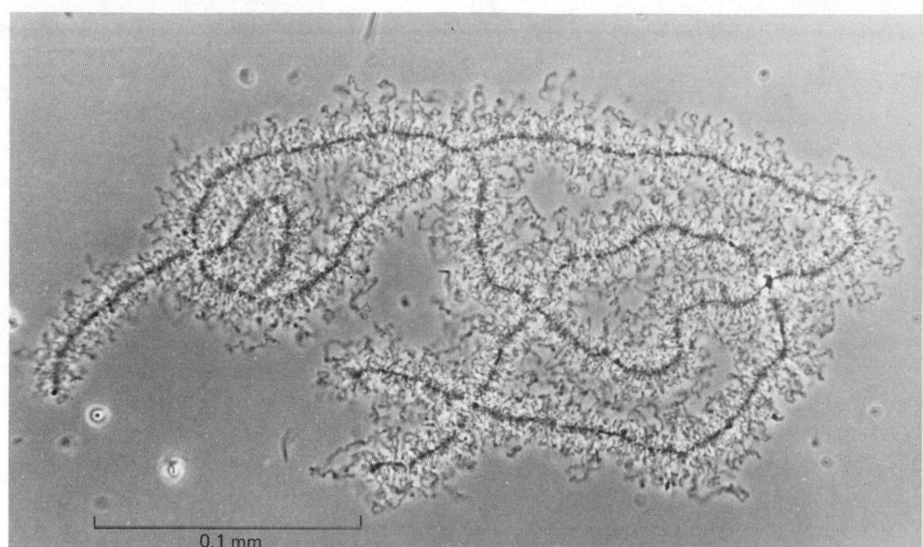

Figure 8–100 Light micrograph of the highly extended lampbrush chromosomes seen in an amphibian oocyte. (Courtesy of Joseph G. Gall.)

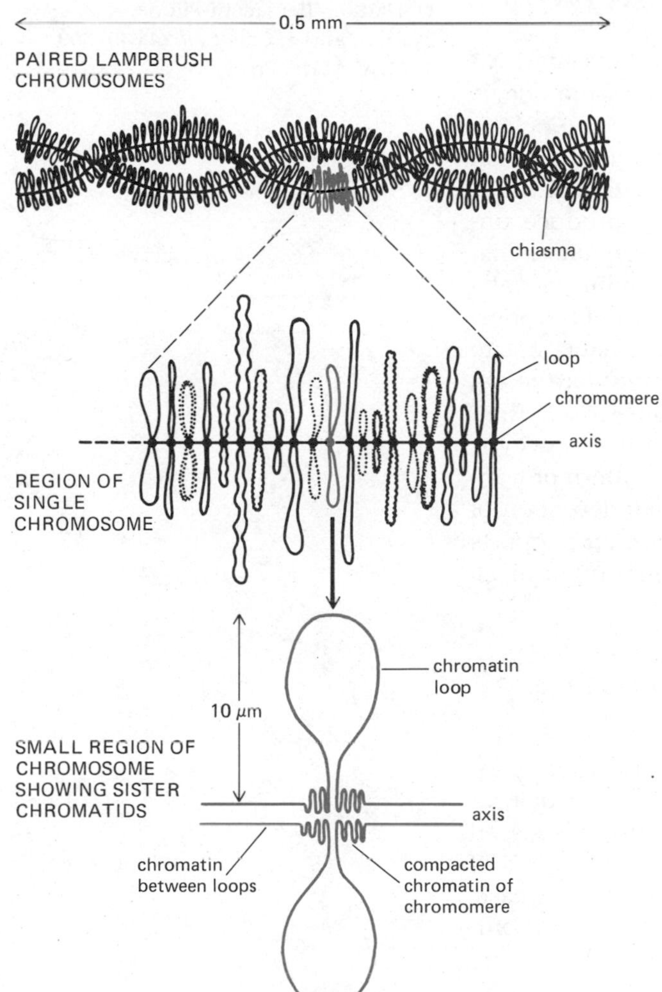

Figure 8–101 (*Left*) Drawings of lampbrush chromosome structure. It has been estimated that the set of lampbrush chromosomes contains a total of about 10,000 different chromatin loops in many amphibians, with the remainder of the DNA being highly condensed in the chromomeres. Note that each loop corresponds to a particular DNA sequence, and that four copies of each loop are present in each cell, since the structure shown at the top consists of two paired homologous chromosomes and each chromosome is composed of two closely apposed sister chromatids. This four-stranded structure is characteristic of this stage of development of the oocyte (the diplotene stage of meiosis—see p. 784).

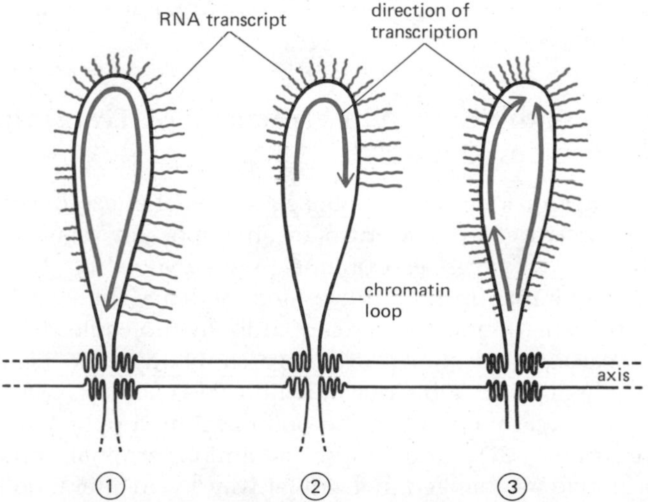

Figure 8–102 Drawings of several types of lampbrush loop morphologies, as detected by electron microscopy.

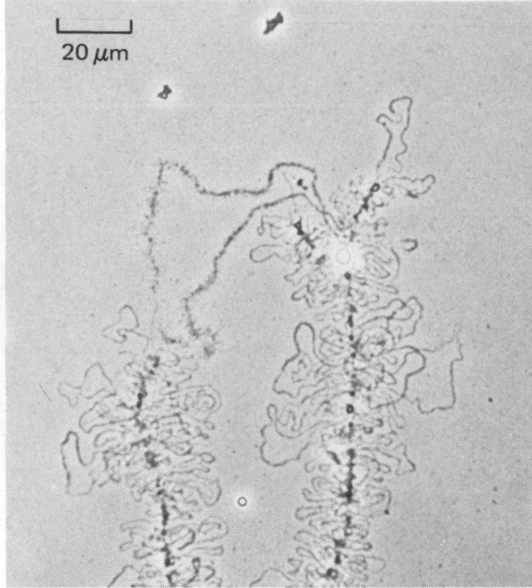

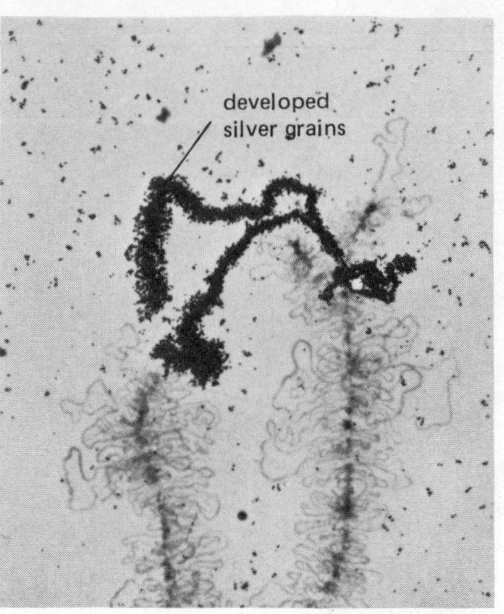

Figure 8–103 The use of nucleic acid hybridization to demonstrate the unusual transcription on amphibian lampbrush chromosomes. A single-stranded DNA probe corresponding to a particular highly repeated DNA sequence was prepared in radioactive form. The chromosomes in (A) were annealed with this probe, washed extensively, and then subjected to autoradiography (B). The extended loop that becomes radioactive here is synthesizing unusually long RNA transcripts that contain copies of several clustered histone genes. The fact that these long RNA transcripts hybridize with the DNA probe reveals that the repeated DNA sequence of the probe is copied into RNA, even though in other cells this sequence serves as a *nontranscribed* spacer between the histone genes. (From M. O. Diaz, G. Barsacchi-Pilone, K. A. Mahon, and J. Gall, *Cell* 24:649–659, 1981. © M.I.T. Press.)

ization experiments show that a given loop always contains the same DNA sequence and that it remains extended in the same manner as the oocyte grows.

In some respects, the pattern of transcriptional activity on lampbrush chromosomes does not seem to be typical of an ordinary somatic cell. First, all of the loops are unusually heavily transcribed, and many produce unusually long primary transcripts. Second, highly repeated DNA sequences that are thought not to be transcribed in normal cells are transcribed from some lampbrush loops (Figure 8–103). Finally, at least in *Xenopus,* most of the enormous transcriptional activity of the lampbrush chromosomes seems not to lead to mRNA production, since all of the stored mRNA molecules for the egg have already been made by the time the lampbrush chromosomes reach their largest and most active phase. Thus, the function of much of this activity is mysterious and may be unique to the oocyte. But even if the pattern of transcriptional activity is peculiar to the oocyte, the selective decondensation of individual loops that accompanies their transcriptional activation supports the hypothesis that looped chromatin domains are functional units in all chromosomes.

Altering the Ends of a Primary RNA Transcript Can Change the Protein Made[26,50]

Having considered some consequences of eucaryotic chromatin structure for gene activity, we now turn to an important regulatory implication of RNA splicing, which is also unique to eucaryotes. We have previously discussed how a change in RNA processing mediates the switch from the synthesis of membrane-bound to secreted antibody molecules (Figure 8–49). In that case, the carboxyl-terminal portion of an antibody molecule is replaced with a completely different sequence of amino acids because of a difference in RNA splicing caused by changing the point at which poly A is added to the 3′ end of the primary RNA transcript. The amino-terminal portion of a protein can in principle be changed in a similar way, by an alteration in the site of transcriptional initiation leading to a change in RNA splicing near the 5′ end of the transcript.

Regulation based on changes in RNA processing is unlikely to be confined to the control of antibody synthesis. In a number of other cases, two different cell types in the same organism have been shown to produce a different primary RNA transcript from the same gene. Even in the same cell, some of the transcripts from a given genetic region can differ in length from the rest, with primary RNA transcripts being formed with different 5′ and 3′ ends. In many of these cases, RNA splicing would be expected to produce altered mRNA molecules, which in turn will code for variant proteins.

These and other observations reveal that the gene control processes operating during development determine not only the rate at which a given DNA sequence is transcribed, but also the position of the 5′ cap and the 3′ poly-A sequence in the primary transcript. These termini in turn affect RNA processing and thus can alter the structure of the protein that is synthesized. The decision as to where DNA transcription will begin and end may consequently be at least as important to the function of a genetic region as is the rate of its transcription.

Are Important Controls Also Exerted by Alterations in the Specificity of RNA Processing and Export?

An average primary RNA transcript seems to be at most 10 times longer than the mature mRNA molecule generated from it by RNA splicing. Yet only about one-twentieth of the total mass of the hnRNA made ever leaves the cell nucleus (p. 414). Simple bookkeeping therefore suggests that a substantial fraction of the primary transcripts (perhaps half) may be completely degraded in the cell nucleus without ever generating an exported mRNA molecule. The discarded RNAs may be those whose sequence cannot be made into an mRNA molecule. On the other hand, some of them may represent a store of potential mRNA molecules that are appropriately processed into mRNA only in other cell types. In this case, direct RNA processing and export controls would be important for cell development. For example, different types of RNA-splicing enzymes might operate in different cell types, producing different mRNAs from the same primary RNA transcript. However, we have already described the negative result of testing for this kind of control in liver and brain cells (p. 437), and, so far, no clear examples have been discovered.

Summary

Specialized higher eucaryotic cells differ from each other by virtue of the proteins that they make. Most of the control of gene expression that underlies these differences seems to operate at the level of DNA transcription. In procaryotes, gene regulatory proteins bind to specific DNA sequences close to RNA polymerase start sites and thereby either repress or activate the transcription of adjacent genes. The control of gene expression in higher eucaryotes is thought to involve a relatively small set of master gene regulatory proteins that interact in a combinatorial fashion to define a large number of different cell and tissue types. A system of interacting gene regulatory proteins of the procaryotic type could, in principle, lead to self-sustaining patterns of selective gene activity and so explain the clonal inheritance of gene expression typical of eucaryotic cell development. However, the actual mechanisms by which most eucaryotic gene regulatory proteins act are not known, and some of them seem certain to be more complex than those in procaryotes. In particular, the chromatin of a eucaryotic gene must be decondensed before it can be transcribed, and gene regulatory proteins that alter chromatin structure are thought to be the basis for an additional important level of control of eucaryotic gene expression.

Gene Regulatory Mechanisms and Cell "Memory"

Once a pattern of gene regulation has been established in a higher eucaryotic cell, it tends to remain unchanged throughout subsequent cell generations. In other words, there is a strong propensity for genes that are off to remain off and genes that are on to remain on in all of the daughter cells of a clone. This cell "memory" cannot be maintained simply by a gene regulatory component permanently bound to a control site in the DNA, because such a mechanism cannot provide a memory that persists through many cell generations. If the differentiated state is to be permanently heritable, daughter cells must receive new copies of the regulatory component (or set of regulatory components) that was present in the parent cell. The regulatory component itself must, therefore, be made anew in each cell generation. The general rule must be, "If the gene regulatory component is already present, make another of the same kind; if it is not present, don't." In other words, the element that controls gene expression must also control its own production by means of a positive feedback loop. It is this positive feedback loop that endows the system with memory.

We have already dealt with one such cell-memory system in the bacteriophage lambda, involving two gene regulatory proteins that bind to operator sites on the DNA and mutually repress each other's synthesis (p. 442). In this section, we shall outline several other types of gene regulatory systems that are suspected to play a role in the generation of the stably inherited patterns of gene expression in higher eucaryotic cells. We begin by returning to the role that chromatin structure plays in controlling eucaryotic gene expression.

A Highly Condensed Fraction of Interphase Chromatin Contains Specially Inactivated Genes

We have seen that loops of chromatin may represent functional units whose transcriptional activity requires certain forms of histone and nonhistone proteins that help decondense the chromatin, making its DNA accessible to other gene regulatory proteins or to RNA polymerase itself. Microscopy can detect differences in the state of condensation of chromatin at the level of individual loops only in lampbrush and polytene chromosomes. Differences in the ordinary chromosomes of other kinds of cells can also be seen, but only over much more extensive regions of chromatin. On the basis of their appearance in the microscope, two general classes of chromatin have been distinguished in higher eucaryotic cells: condensed chromatin, or **heterochromatin,** and less condensed chromatin, or **euchromatin.**

Heterochromatin was originally identified by light microscopy because it remains condensed during interphase, apparently maintaining the type of structure adopted by the rest of the chromatin only during mitosis. It was later found that, like mitotic chromatin, heterochromatin is inactive in DNA transcription: autoradiography of cells that have been incubated with radioactive RNA precursors shows that the bulk of RNA is synthesized over regions of euchromatin, the label being absent from the heterochromatin (Figure 8–104).

The DNA in heterochromatin is replicated very late in the S phase of each cell cycle (see p. 632). This probably reflects its unusually condensed state, since a local decondensation of chromatin strands is required in the vicinity of any DNA replication fork. The biochemical basis for the difference between heterochromatin and euchromatin is unknown.

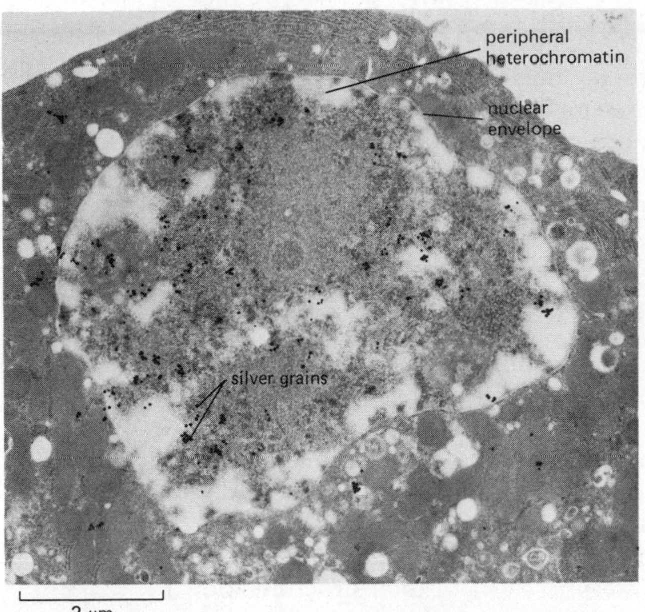

Figure 8–104 Autoradiograph of a thin section of a cell nucleus from a cell that has been pulse-labeled with [^{3}H]uridine in order to label sites of RNA synthesis (silver grains). The white areas are regions of heterochromatin, which tend to pack along the inside of the nuclear envelope. Because of the particular method used for sample preparation, areas containing heterochromatin have been bleached. Most of the RNA synthesis occurs in the euchromatin that borders these heterochromatic areas. (Courtesy of Stan Fakan.)

Two Subclasses of Heterochromatin Can Be Distinguished[51]

In most cells 90% or so of the chromatin is thought to be transcriptionally inactive (p. 414). While this chromatin has a more condensed conformation than the 7% to 10% of the chromatin in transcriptionally active regions, only a fraction of it is packed in the highly condensed conformation known as heterochromatin. Heterochromatin is therefore a special class of transcriptionally inactive chromatin, and it is believed to contain genes that are permanently turned off in a cell and all of its progeny. Some chromosomal regions are condensed into heterochromatin in all cells, and these are called the **constitutive heterochromatin** of the organism. Other regions are condensed into heterochromatin only in some cells and not in others; these regions are the **facultative heterochromatin** of the cell.

Constitutive heterochromatin, which is thought to contain DNA that is never transcribed in any cell, is the only form of heterochromatin distinguishable from normal chromatin even in highly condensed mitotic chromosomes. In human chromosomes it is localized around the centromere of each mitotic chromosome, where it can be detected as darkly staining bands by special staining procedures (Figure 8–105). In some other mammals constitutive heterochromatin is located in bands throughout the chromosomes, and in some cases makes up an entire arm of a chromosome. During interphase, regions of constitutive heterochromatin can aggregate to form *chromocenters.* In mammals, the number and arrangement of such chromocenters varies with cell type and developmental stage.

Most regions of constitutive heterochromatin contain relatively simple, serially repeated DNA sequences. For historical reasons, these highly repeated sequences are known as *satellite DNAs* (p. 468). While some satellite DNAs seem to have a role in chromosome pairing in meiosis, their function, and that of the condensed interphase chromatin structure that they form, is otherwise obscure.

The significance of facultative heterochromatin is more obvious. It almost certainly reflects some of the stable differences in genetic activity adopted by different cell types. The total amount of facultative heterochromatin is very different in different cells, embryonic cells seeming to have very little and

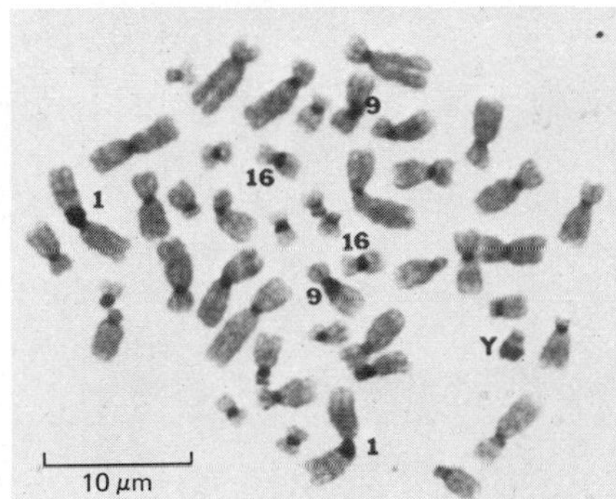

Figure 8–105 Human chromosomes at metaphase, stained by a special technique that darkens the constitutive heterochromatin. The numbers and letter mark specific chromosomes. (Courtesy of James German.)

some highly specialized cells a great deal. This suggests that, as cells develop, more and more genes are permanently inactivated through being packaged in a condensed form in which they are no longer accessible to gene activator proteins. Facultative heterochromatin is not known to contain large amounts of highly repeated DNA sequences, nor does it stain differentially in mitotic chromosomes. Most of what is known about facultative heterochromatin comes from studies of the inactivation of one of the two X chromosomes in female mammalian cells.

An Inactive X Chromosome Is Inherited[52]

All female mammalian cells contain two X chromosomes, while male cells contain one X and one Y chromosome. Presumably because a double dose of X chromosome products would be lethal, the female cells have evolved a mechanism for permanently inactivating one of the two X chromosomes in each cell. In mice this occurs between the third and the sixth day of development, when one or the other of the two X chromosomes in each cell is condensed at random into heterochromatin. This compact chromosome is seen in the light microscope during interphase as a distinct structure known as a *Barr body,* and is located near the nuclear membrane. It replicates in late S phase, and most of its DNA is not transcribed in any of the progeny cells. Because the inactive X chromosome is faithfully inherited, every female is a mosaic composed of clonal groups of cells in which only the paternally inherited X chromosome (X_p) is active and a roughly equal number of groups of cells in which only the maternally inherited X chromosome (X_m) is active. In general, the cells expressing X_p and those expressing X_m are distributed in clusters in the adult animal, reflecting the fact that sister cells tend to remain close neighbors during embryonic development and growth. For example, in the skin especially large patches can be detected that contain the progeny of single cells from the early embryo (Figure 8–106).

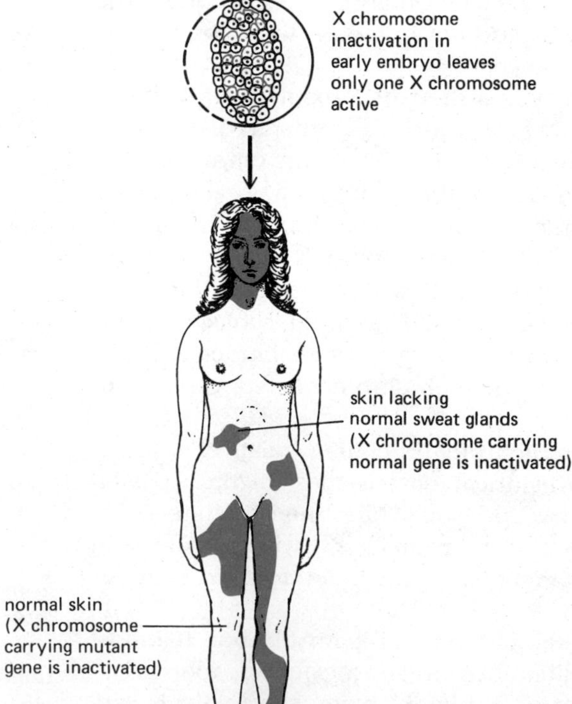

Figure 8–106 The pattern of gene expression in the skin of a woman carrying one normal X chromosome and one mutant X chromosome. In any one cell, only the normal or the mutant X chromosome is active. The mutant X chromosome carries a gene causing a skin disease, anhidrotic ectodermal displasia, representing a deficiency in the sweat glands. The defective skin is found in patches, which reveal the clonal origin of those cells in which the X chromosome carrying the normal allele of this gene has been inactivated. These clones are caused by the fact that, once inactivated, an inactive X chromosome is faithfully inherited in all daughter cells (see Figure 8–107).

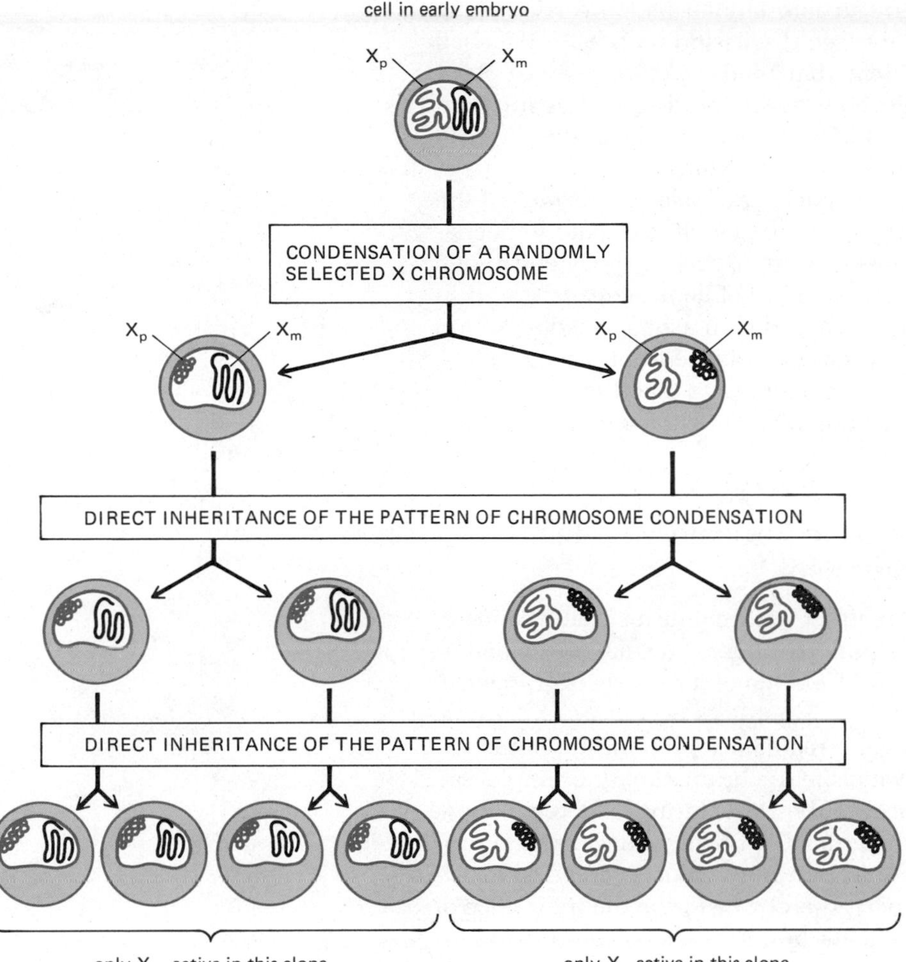

Figure 8–107 Schematic illustration of the clonal inheritance of a condensed inactive X chromosome that occurs in female mammals.

It is not known how the condensation of an X chromosome occurs, but it is known to have a tendency to spread from one region to another along a chromosome. This has been demonstrated by studies with mutant animals in which one of the X chromosomes has become joined to the end of an *autosome* (a nonsex chromosome). In the mutant chromosomes, regions of the autosome adjacent to an inactivated X chromosome are often condensed into heterochromatin and the genes they contain thereby inactivated in a heritable way. This suggests that the phenomenon of X chromosome inactivation is analogous to a chromatin crystallization process, which spreads from a particular nucleation site on the X chromosome. Once the especially tight chromatin structure is established randomly on one of the two X chromosomes, it is faithfully inherited during all subsequent replications of that chromosome (Figure 8–107).

Eucaryotic Genes Can Be Turned Off by a Novel Heritable Mechanism[52,53]

Conventional gene control mechanisms (such as the action of diffusible gene regulatory proteins like those that act in bacteria) cannot explain the heritability of X inactivation, since identical DNA sequences on the active and inactive X chromosomes are quite differently regulated in the same cell nucleus. One possibility is that the unusual chromatin structure of the inactivated X chromosome is passed directly from the parental to the daughter DNA helices during DNA replication by means of the structural information carried along with the inherited parental nucleosomes (see discussion on p. 635). We shall see below that any chromosomal protein that binds cooperatively to chromatin and never leaves the DNA could be inherited and serve this function equally well. We shall also explore several other possible explanations for such directly inherited states of gene expression in eucaryotes.

Regardless of its molecular basis, the packing of selected regions of the genome into heterochromatin is clearly a type of genetic regulatory mechanism not available to bacteria. The crucial feature of this uniquely eucaryotic form of gene regulation is that the stable memory of gene states is stored in a directly inherited chromatin structure rather than in a stable feedback loop of self-regulating gene activator proteins that can diffuse from place to place in the nucleus. Whether the mechanism that operates to inactivate these large regions of chromatin also operates at the level of an individual looped domain of chromatin is not yet known.

Cooperatively Bound Clusters of Gene Regulatory Proteins Can, in Principle, Be Directly Inherited

Several relatively simple types of gene regulatory mechanisms that have been proposed to account for an inherited pattern of gene regulation in higher eucaryotes are capable of explaining the inheritance of condensed inactive chromatin we have just discussed.

One such scheme is based on a gene regulatory protein that binds cooperatively in multiple copies to a specific site in chromatin (for example, see Figure 8–7). If the cluster remains bound to the DNA during DNA replication, part of it could be inherited by both daughter DNA helices. Because the protein binding is cooperative, each inherited part of the cluster could bind additional free regulatory protein monomers, thereby restoring the initial size of the cluster. In this way, the regulatory state of a gene would be directly inherited via its bound chromosomal proteins (Figure 8–108). In principle, such directly inherited protein clusters could act either to turn off or to turn on a single gene in a stable manner. There are as yet no demonstrated examples of this type of gene regulatory mechanism, but the appropriate experimental tests are difficult, and such schemes remain untested.

Directly Inherited Patterns of DNA Methylation Help to Control Some Mammalian Genes[54]

The only gene regulatory mechanism capable of explaining cell memory for which there is some direct evidence in mammalian cells involves the methylation of DNA. The DNA of many higher eucaryotes, including man, contains

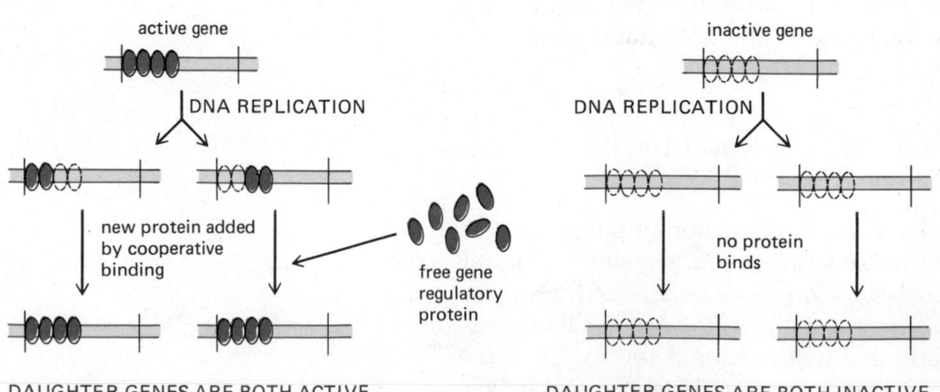

Figure 8–108 Diagram illustrating a general scheme that permits the direct inheritance of states of gene expression during DNA replication. In this hypothetical model, portions of a cooperatively bound cluster of gene regulatory proteins are transferred directly from parental to daughter DNA helices. The inherited cluster causes each of the daughter DNA helices to bind additional copies of the same gene regulatory protein. However, because the binding is cooperative (see Figure 8–7), DNA synthesized from an identical parental DNA helix that lacks this bound gene regulatory protein will remain free of the protein. If the bound gene regulatory protein turns off gene transcription, then the inactive gene state will be directly inherited, as in the case of X-chromosome inactivation discussed in the text. Another bound gene regulatory protein of this type could turn on gene transcription; in that case, the active gene state would be directly inherited as indicated here.

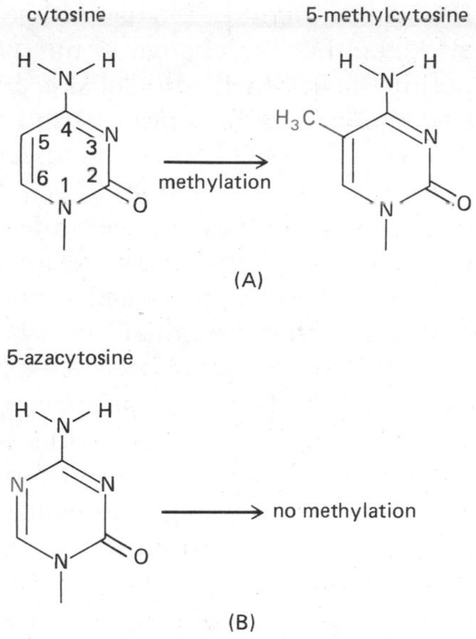

Figure 8–109 (A) The formation of 5-methylcytosine occurs by the methylation of a cytosine base in the DNA double helix. This event is confined to selected cytosine, or C, residues located in the sequence CpG. (B) A synthetic nucleotide (5-aza C) carrying the base 5-azacytosine cannot be methylated. Also, when small amounts of 5-aza C are incorporated into DNA, they inhibit the methylation of normal C residues.

a small proportion of 5-methylcytosine that is formed by the methylation of cytosine bases in the DNA double helix (Figure 8–109). This methylation is known to be inherited automatically because of the special characteristics of the enzyme that methylates DNA in mammalian cells. The enzyme seems to be restricted in two ways: first, it methylates C bases mainly in a CpG sequence; second, the CpG sequence must be base-paired with a CpG sequence that is already methylated. Thus, preexisting DNA methylation patterns are directly inherited following DNA replication (Figure 8–110), and a region of DNA important for gene activation can be stably maintained through repeated cell divisions in either a methylated or an unmethylated form.

In general, the DNA of inactive genes tends to be more heavily methylated than the DNA of active genes. Moreover, in several cases, an inactive gene that contains methylated DNA has been shown to become unmethylated after gene activation. However, these observations leave open the possibility that reduced

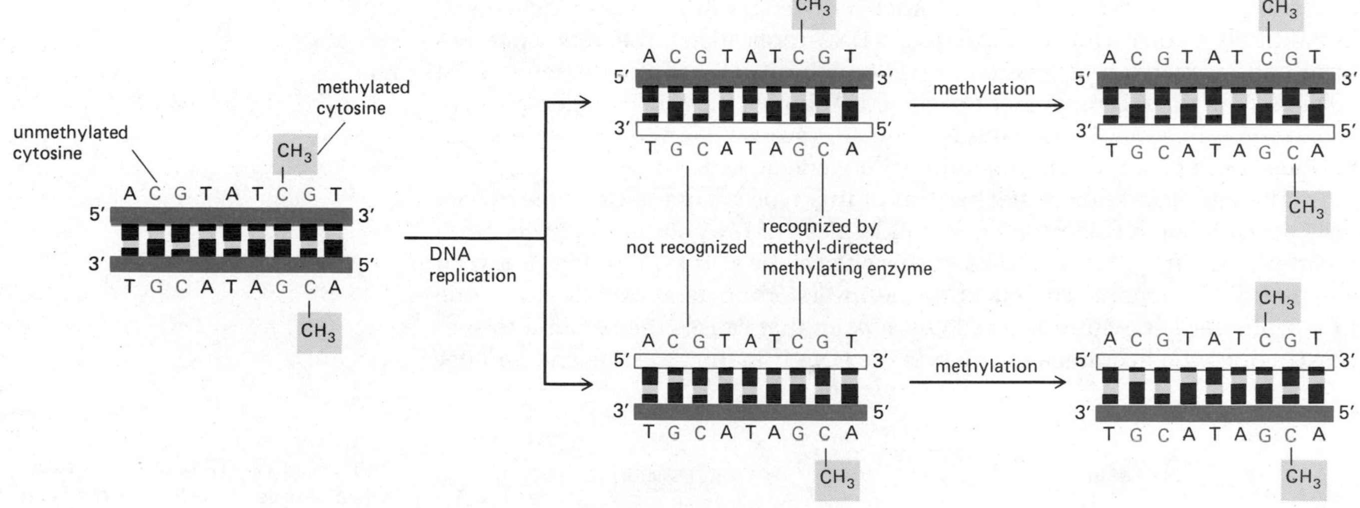

Figure 8–110 Suggested scheme by which DNA methylation patterns are thought to be faithfully inherited. In eucaryotic DNAs, a large fraction of the cytosine bases in the sequence CpG are methylated (see Figure 8–109). Because of the existence of methyl-directed methylating enzymes, once a pattern of DNA methylation is established, the sites of methylation tend to be inherited in the progeny DNA, as shown. This means that directed changes in DNA methylation patterns could be used to turn on or turn off specific genes in a clonally inherited manner. There is evidence that some genes in mammalian cells are controlled in this way, with the DNA of inactive genes being more highly methylated than the DNA of active genes.

DNA methylation is a cause rather than an effect of gene activation. The best evidence that the change in methylation plays a causal part comes from experiments in which a nucleoside containing the base analogue 5-aza C (Figure 8–109) is added for a brief period to tissue-culture cells. The 5-aza C, which cannot be methylated, is incorporated into DNA where it depresses general levels of DNA methylation. In cells treated this way, selected genes that were inactive can be shown to become actively transcribed and, at the same time, to acquire unmethylated C residues. Since the active gene state can usually be maintained for many cell generations in the absence of 5-aza C, it seems quite likely that the initial methylation of the gene caused its inactivity.

It is not yet clear how extensively DNA methylation controls genes. For example, whether it plays a part in the inactivation of the X chromosome in female mammals is unknown. In some well-studied cases, the expected changes in the methylation of a gene do not accompany a change in its expression. Moreover, *Drosophila,* a sophisticated multicellular eucaryote, seems to have no methylated C residues in its DNA at all. Therefore, it is likely that DNA methylation is one of several different mechanisms that are used to create inherited patterns of gene regulation in higher eucaryotic cells.

In Special Cases, Local DNA Sequences Are Reversibly Rearranged to Turn Genes On and Off[55]

In some single-cell organisms, a much more radical mechanism operates to create a stably inherited pattern of gene regulation. The DNA of some procaryotes and lower eucaryotes can be rearranged in such a way that specific genes are activated or inactivated. Since all changes in a DNA sequence will be faithfully copied during subsequent DNA replications, the gene state determined by such a rearrangement will be inherited by all the progeny of the cell in which the rearrangement occurred. Although these rearrangements are stable through several generations, they are reversible and over long enough periods produce an alternating pattern of gene activity.

The simplest known mechanism of this type occurs in *Salmonella* bacteria, in which a specific 1000-nucleotide piece of DNA occasionally becomes inverted (Figure 8–111). The DNA at this site can be said to "flip-flop" between a (+) and a (−) state. The effect of the inversion on gene expression is due to a promoter site within the 1000 base pairs that causes the bacteria to synthesize one type of surface coat when the DNA is in the (+) state and another

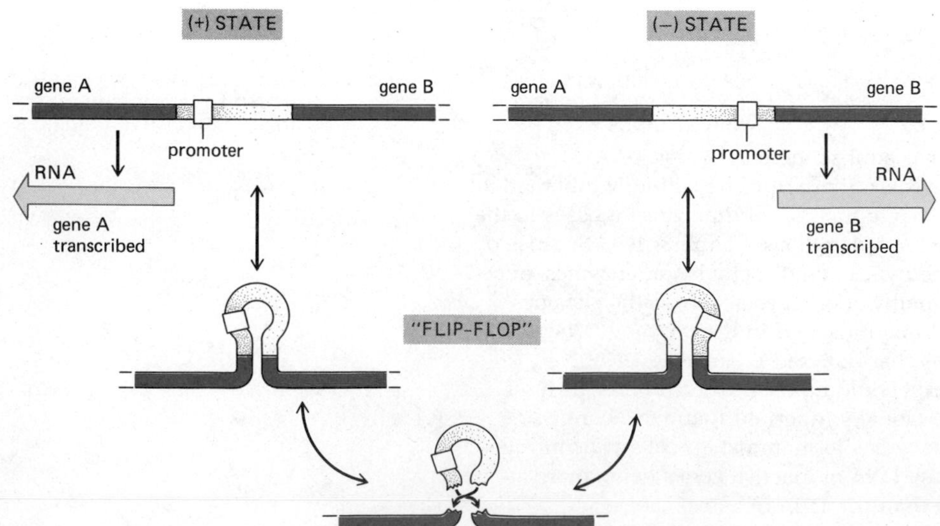

Figure 8–111 Diagram illustrating how a simple "flip-flop" process that inverts a small DNA segment causes alternating transcription of gene A or gene B in a cell. The flip-flop occurs by a site-specific recombination mechanism described elsewhere (p. 247), which is activated only rarely. Therefore, the (+) and (−) states shown tend to be faithfully inherited in each clone of cells. A control mechanism of this type operates in *Salmonella* bacteria, although the mechanism is somewhat more complex.

when it is in the (−) state. Since flip-flops occur only rarely, whole clones of bacteria will grow up with one type of surface coat or the other. This phenomenon, known as *phase variation,* almost certainly evolved because it protects the bacterial population against the immune response of its hosts. If the host makes antibodies against one type of coat, a few bacteria whose coats have been altered by gene inversion will still be able to survive and multiply.

A related but more complex, and in some ways more interesting, mechanism occurs in some yeast strains. Yeasts are single-cell eucaryotes that exist in either a haploid or a diploid state, diploidy being achieved by a process known as *mating,* in which two haploid cells fuse. In order for two haploid cells to mate, they must differ in *mating type.* In the common baker's yeast, *Saccharomyces cerevisiae,* there are two mating types, α and a, determined by which of two genes, the *α* gene or the *a* gene, occupies a particular site, called the *mating-type locus,* on the yeast chromosome. Because mating type is inherited, close relatives will not mate with each other. However, haploid yeast cells periodically switch their mating type. The change in mating type results from a change in the DNA sequence at the mating-type locus. But the change is not a simple inversion causing either the *α* or the *a* sequence to be expressed: instead each time an α-type cell changes to an a-type cell, the *α*-type gene in the mating locus is excised and replaced by a newly synthesized *a*-type gene copied from a silent "master" *a*-type gene elsewhere in the genome. Because the change involves the removal of one gene from an active "slot" and its replacement by another, this mechanism has become known as the *cassette mechanism.*

The change is reversible because—although the original *α*-type gene at the mating-type locus is discarded—there is also a silent master *α*-type gene elsewhere in the genome. At infrequent intervals a new copy of the *α*-type gene is generated and reinserted to restore the original state of the genome. In this way, copies made from the master *a*-type and the master *α*-type genes function as disposable DNA cassettes that are inserted at intervals into the mating-type locus (which acts as a "playing head"), causing the yeast cells to shift back and forth between the two different mating types (Figure 8–112).

It is believed that the difference between the silent sites, which contain the master mating-type genes, and the active locus that contains the expressed gene lies in the structure of the chromatin. Apparently the genes at the silent sites are not transcribed because their DNA lies in a chromatin structure inaccessible to activator proteins. However, when moved to the mating-type locus, the same DNA sequence adopts a more open chromatin conformation, as a result of effects that spread from a nearby chromosomal region. In principle, any one of a large number of different DNA sequences could be inserted at a single site by such a cassette mechanism, and many different directly inheritable states of a gene could thus be made available to the cell.

Recently, a related type of mechanism has been found to cause the many changes observed in the major protein that covers the surface of trypanosomes. These unicellular eucaryotes cause serious diseases, such as sleeping sickness, in both humans and livestock. As in the case of *Salmonella,* the surface changes enable the trypanosome colony to escape the host's immune defenses.

The experimentally demonstrated regeneration of a whole organism from the genetic information present in a single somatic cell (p. 436) suggests that the differentiation of higher eucaryotic cells need not be accompanied by any *irreversible* change in the DNA sequence. But observations of this type leave open the possibility that *reversible* DNA sequence changes—resembling those already observed in *Salmonella,* yeast, and trypanosomes—are responsible for some of the inherited changes in gene expression characteristic of developing somatic cells.

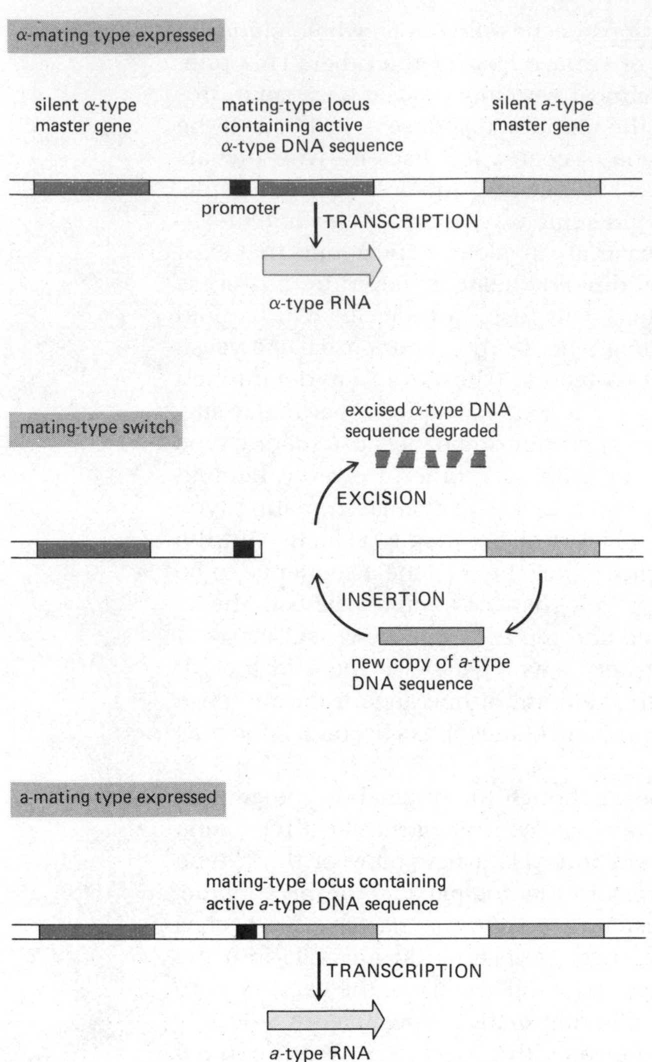

Figure 8–112 Schematic diagram of the "cassette mechanism" that enables yeast to switch their mating type between α-type and a-type. In principle, such a cassette mechanism could also insert gene copies from more than two silent sites, generating a whole range of gene-expression patterns. As in the mechanism in Figure 8–111, the fact that a DNA rearrangement is involved means that the switch in gene expression will be directly inherited by the progeny cells.

Selected Portions of Chromosomes Can Be Either Deleted or Amplified in Somatic Cells[56]

Although the somatic cells of most organisms seem to retain all their DNA intact in the course of development and differentiation, there is no a priori reason why they should. In principle, only the cells of the germ line (the eggs and sperm) must contain a complete genome. In fact, there are a few cases in which extensive changes in the DNA do occur.

Perhaps the most dramatic examples occur in some primitive eucaryotes, such as the worm *Ascaris* and the *copepods*—a class of minute crustaceans. Very early in their development, after the egg has cleaved no more than a few times, all the dividing cells but one eliminate specific heterochromatic sections of their chromosomes, which are consequently reduced to about half their original size. The eliminated fragments are quickly lost from the cells. In this way, about half the total DNA is eliminated from the somatic cell lineage, and the single pair of daughter cells that retains intact chromosomes will give rise to all of the germ cells for the next generation. Presumably, the DNA lost from the remaining cells contains sequences that are necessary for germ cells only.

At the molecular level, the elimination of specific pieces of chromosomes must be due to cutting of the DNA helix at specific points in the DNA sequence. In some organisms only the tips of chromosomes are lost in this way. But in one copepod, *Cyclops strenuus*, the heterochromatic pieces are interspersed all along the chromosomes, so that their removal requires many cuts, each followed by specific DNA rejoining to restore an intact chromosome (Figure 8–113).

Similar DNA rearrangements occur on a much smaller scale in a specialized vertebrate cell, the lymphocyte that produces antibody molecules. During lymphocyte development, the genes that will code for antibody molecules are rearranged in such a way as to generate a wide variety of different antibody proteins. As in the case of the copepod, the cuts come at specific DNA sequences, they are accompanied by a DNA rejoining process, and they are programmed to occur only at certain stages of cell development (see Chapter 17). However, only a tiny fraction of the total genome is lost in an antibody-producing cell. (This rearrangement, which occurs at the level of the DNA, should not be confused with the change in RNA processing events that later mediate the switch from membrane-bound antibody to secreted antibody production in these lymphocytes.)

Conversely, in some specialized cells specific genes are amplified. Here a small region of the genome is replicated many times during a single cell generation to produce multiple extra copies of a specific DNA sequence. This occurs in two types of cells that have to make an unusually large amount of one gene product relatively quickly. One is the developing egg (oocyte) of many animals, which must make and store huge numbers of ribosomes. For example, rRNA genes are amplified about 1000-fold in very young *Xenopus* oocytes, producing many free rDNA molecules that cluster together to form about 1000 extra nucleoli (Figure 8–114). In these nucleoli, large amounts of rRNA are synthesized and ribosomes are assembled, as the egg grows to its mature size (see Figure 14–29, p. 789).

A second example of specific gene amplification occurs in the follicle cells that synthesize and secrete the proteins that form the hard coat, or chorion, of insect eggs. In *Drosophila*, for example, just before the chorion proteins are needed, the DNA sequences coding for them are amplified about 30-fold (Figure 8–115).

According to our current understanding, although selected DNA sequences in somatic cells can be deleted, altered, or amplified, such irreversible changes in the genome take place only occasionally in the development of higher plants and animals.

Major Rearrangements of Chromosomes Are Usually Deleterious[57]

Not surprisingly, substantial changes in the number and arrangement of chromosomes often result in serious abnormalities. Humans must have two and only two copies of each autosomal chromosome in order to be normal. Embryos that accidentally inherit only a single copy of one of the autosomes are not viable, and those individuals who survive with three copies of one autosome (*trisomy*) are always strikingly abnormal: only individuals with trisomy of chromosome 21 (which produces mongolism, or Down's syndrome), chromosome 13, or chromosome 18 can survive at all. An exact number of chromosomes is clearly important for normal cell function and development.

Usually rearrangements within or between chromosomes will also produce embryos that are not viable and die long before birth. But some kinds of chromosome rearrangements that do not affect the total amount of genetic material have been found in humans who are otherwise entirely normal. Fusions may occur between the tips of any two of the five pairs of single-armed chromosomes (chromosomes 13, 14, 15, 21, and 22) to create one two-armed

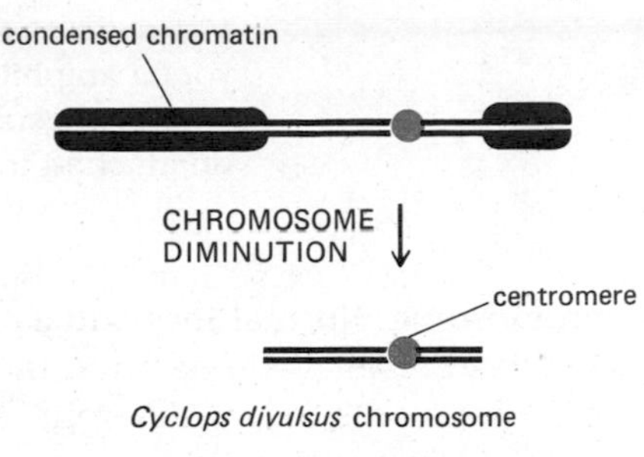

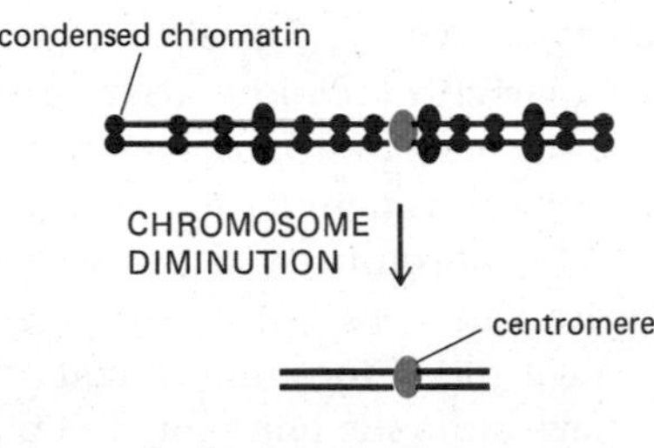

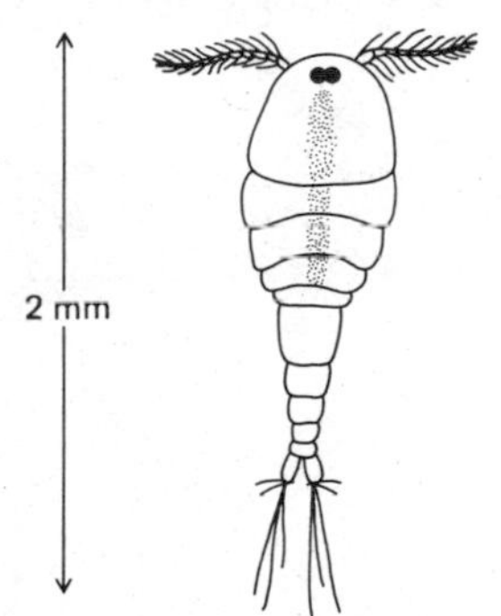

Figure 8–113 Comparison of the manner in which selected portions of chromosomes are eliminated in two different copepods. In both cases the eliminated chromatin becomes unusually condensed before its elimination, as schematically illustrated.

Figure 8–114 Illustration of how the rRNA genes in very young amphibian oocytes are thought to be amplified. The amplified genes are synthesized as DNA molecules unattached to the chromosomes.

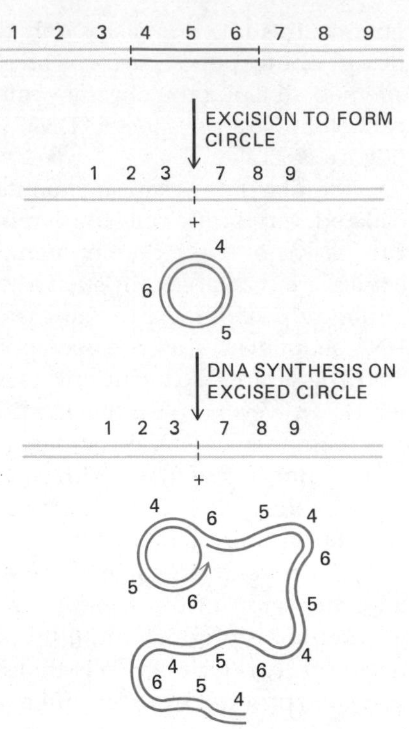

chromosome. Normal individuals in which this has happened therefore have 45 instead of 46 chromosomes. In addition, **chromosome translocations,** in which segments of two different chromosomes are exchanged without loss of material, can be found in normal individuals (Figure 8–116). A most remarkable illustration of how little effect such changes may have is found in the barking deer, where two closely related species have almost exactly the same genetic material distributed in either 46 small or 6 large chromosomes. However, the particular chromosome changes that have been observed in any animal are by no means a random selection of all of those that are possible; apparently only some arrangements can produce a normal organism.

The chromosomal rearrangements that have just been discussed are inherited and present in every cell. But identical types of rearrangements can also take place in somatic cells during the lifetime of an animal, and these can have a drastic effect on cell behavior. Virtually all human tumor cells have various chromosomal abnormalities, including translocations, deletions, and duplications. Most seem to be random, but in some cases a specific chromosomal change is found reproducibly in a particular type of tumor. For example, the cancer cells of most patients with a particular type of leukemia

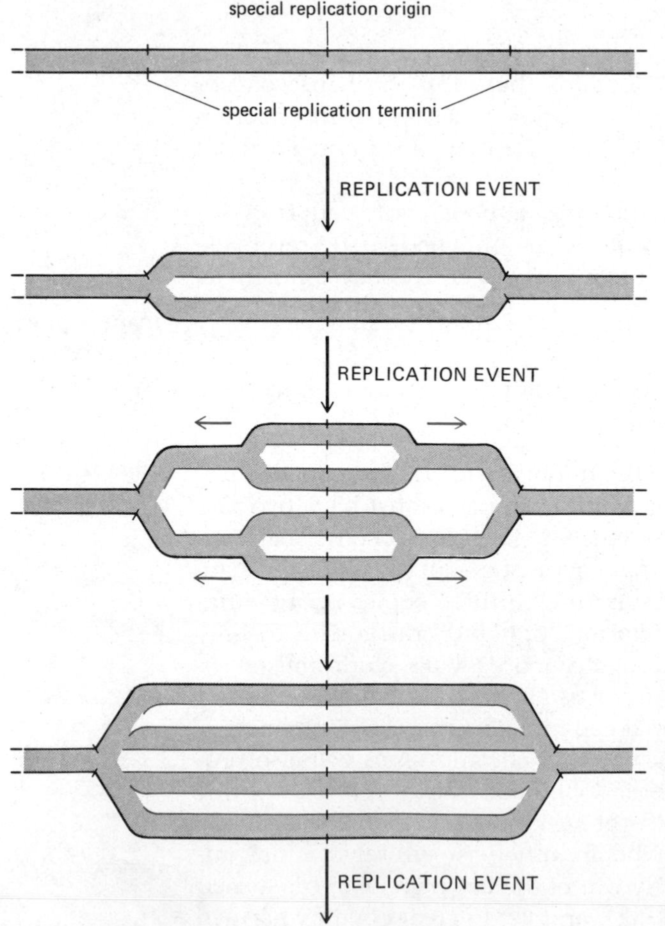

Figure 8–115 Illustration of the type of process thought to be involved in the specific amplification of chorion genes in the follicle cells of *Drosophila*. Here the amplified genes appear to remain attached to the chromosomes, as shown.

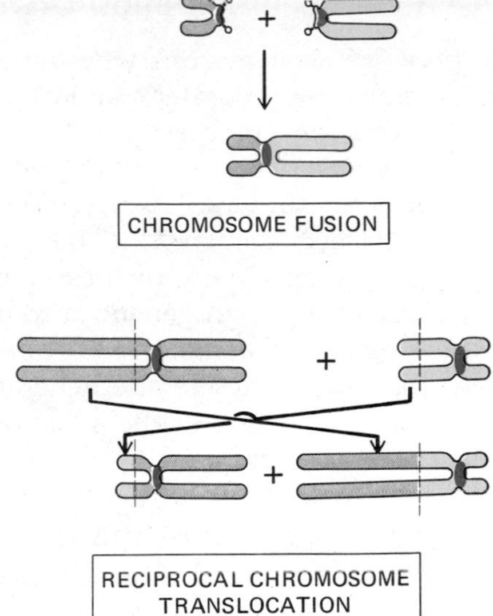

Figure 8–116 Illustration of two types of chromosomal rearrangements commonly found in humans. In a *chromosome fusion,* two chromosomes join to become one. In a *reciprocal chromosome translocation,* two chromosomes exchange parts to produce two rearranged chromosomes.

(chronic myelogenous leukemia) contain a translocation between the distal ends of chromosomes 22 and 9. Likewise, the cancer cells of most patients with a cancer known as Burkitt's lymphoma contain a specific translocation between chromosomes 8 and 14, and so on. It has been postulated that these reproducible chromosome changes move a gene coding for a regulatory protein next to a gene that is highly expressed and thereby cause the regulatory protein to be made in abnormally large amounts. This overproduction could be caused by changes in the chromatin structure of the translocated gene, and it is thought to play an important part in enabling the particular cell from which the tumor arises to escape its normal growth controls. The nature of those controls and the means of escape from them will be discussed in Chapters 11 and 16.

Chromosomal rearrangements are never stable unless they contain the tip of a normal chromosome at each chromosome end, and cells carrying a chromosome with a broken end eventually die. The tips of chromosomes, called **telomeres,** must be specially constructed in order to ensure that the end of the DNA double helix located there will finish each cycle of DNA replication completely. Recent experiments conducted with yeast cells suggest that the telomeres on the tips of each of the 17 yeast chromosomes contain a closely related DNA sequence that is about 3000 nucleotides long. Why the proper capping of a chromosome end should require such a long DNA sequence is not known.

Summary

Some of the mechanisms of cell "memory" in higher eucaryotes seem to act at the level of chromosome structure. In particular, the packing of genes into heterochromatin seems to be associated with their heritable inactivation, as dramatically illustrated by the inactivation of one of the two X chromosomes in the cells of female mammals. This type of inherited gene control cannot be caused by diffusible gene regulatory proteins of the procaryotic type, since the two identical copies of a gene in a single cell are differently affected. Likely mechanisms include the direct inheritance of cooperatively bound chromosomal proteins during DNA replication, the direct inheritance of patterns of DNA methylation, and the inheritance of locally rearranged DNA sequences.

The Organization and Evolution of DNA Sequences[58]

A great deal of evolutionary history is recorded in the genomes of present-day organisms. Some of that history can now be inferred from the numerous DNA sequences determined through recombinant DNA technology (p. 189). As many more genes are sequenced, it may be possible to see quite clearly how the DNA sequences coding for specific proteins have evolved over hundreds of millions of years. But for the present, we must deal in a much more general way with the relationship between existing genomes and the processes by which they have evolved. By studying the behavior of the DNA in living organisms, we can make enlightened guesses concerning evolutionary events.

Genetic Recombination Drives Evolutionary Changes

There are clear reasons why the DNA nucleotide sequences coding for proteins must be accurately replicated and conserved, as pointed out in connection with the processes of DNA replication and DNA repair (Chapter 5). Although most of the DNA sequences that are known in detail are extraordinarily stable (they undergo about 1 random base-pair change per 1000 base pairs every 200,000 years, p. 214), there are some that undergo rearrangements surprisingly often. Many of these are highly repeated sequences that become rearranged through genetic recombination. Although the full significance of these rearrangements is not completely understood, some of them contribute in important ways to the generation of diversity during evolution. In this section, we shall describe some of the repeated DNA sequences and the alterations they undergo.

A Large Fraction of the DNA of Most Eucaryotes Consists of Repeated Nucleotide Sequences[58]

The discovery of highly repeated nucleotide sequences in higher eucaryotic DNAs came as a great surprise. They were first revealed by hybridization techniques that measure the number of gene copies. The procedure involves breaking the genome into short pieces of DNA double helix about 1000 nucleotides long and then denaturing these pieces to produce DNA single strands. The speed with which the mixture of single-stranded fragments will reanneal under conditions in which the double-helical conformation is stable depends on how many complementary strands each fragments finds. For the most part, the reaction is very slow because about 6 million different 1000-nucleotide-long DNA fragments must be added together to reconstitute the haploid complement of a mammalian cell. Any gene present in only two copies in a diploid cell must randomly collide with 6 million noncomplementary strands for every matching partner strand that it happens to find.

When the DNA from a typical mammalian cell is analyzed in this way, about 70% of the DNA strands reanneal as slowly as one would expect for a large collection of nonrepeated DNA sequences, requiring days for complete annealing. But most of the remaining 30% of the DNA strands anneal much more quickly. These strands contain sequences that are repeated many times in the genome, and they thus collide with a complementary partner relatively rapidly. Such highly repeated DNA sequences, which are not found in most bacteria, are of two types—*satellite DNAs* and *interspersed repeated DNAs.*

Frequent Genetic-Recombination Events Expand and Contract Serially Repeated Satellite DNA Sequences[59]

The most rapidly annealing DNA strands in an experiment of the sort just described usually contain serial repetitions of a short nucleotide sequence (Figure 8–117). The repeat unit in a sequence of this type may be composed of only one or two nucleotides, but a repeat length of 170 to 250 nucleotides is the most common. These serially repeated DNA sequences are called **satellite DNAs** because the first DNAs discovered of this type had an unusual ratio of nucleotides that made possible their separation from the bulk of the cell's DNA as a minor component (or "satellite"). Satellite DNA sequences appear not to be transcribed and are most often located in the heterochromatin associated with the centromeric regions of chromosomes.

In some mammals, a single type of satellite DNA sequence constitutes 10% or more of the DNA and may even occupy a whole chromosome arm (millions of copies of the repeat length per cell). Satellite DNA sequences seem to have changed unusually rapidly, and even to have shifted their positions

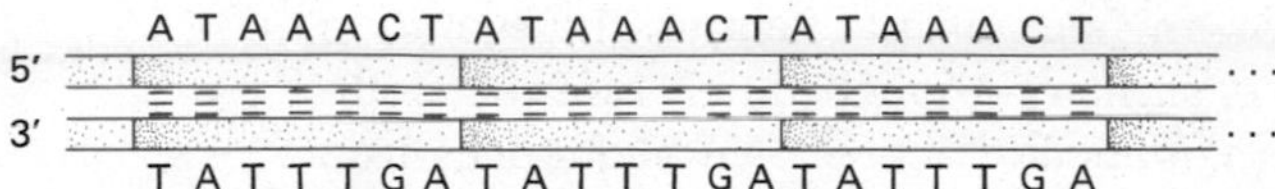

Figure 8–117 A simple satellite DNA sequence consisting of many serially arranged repetitions of a sequence seven nucleotides long. This particular DNA sequence is found in *Drosophila*.

on chromosomes in the course of evolution. For example, the human genome contains at least three predominant satellite DNA sequences, a different mixture of which is present at each centromere. When the two homologous mitotic chromosomes of any human are compared, some of the arrangements of the satellite DNA sequences inherited from the mother and father are usually found to be strikingly different from each other.

Moreover, there are usually marked differences in the satellite DNA sequences of any two closely related species. This contrasts with the high degree of conservation of DNA sequences elsewhere in the genome. One reason for the rapid evolution of satellite DNA is that its repetitive nature encourages duplication and deletion of large blocks of DNA during genetic recombination. Most of this recombination seems to be "incestuous," occurring between two newly replicated daughter DNA helices (Figure 8–118). It is not known whether

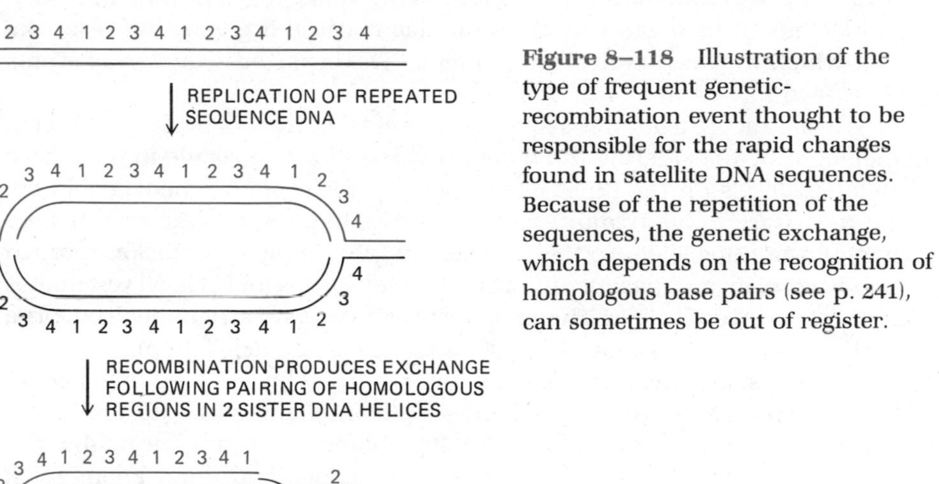

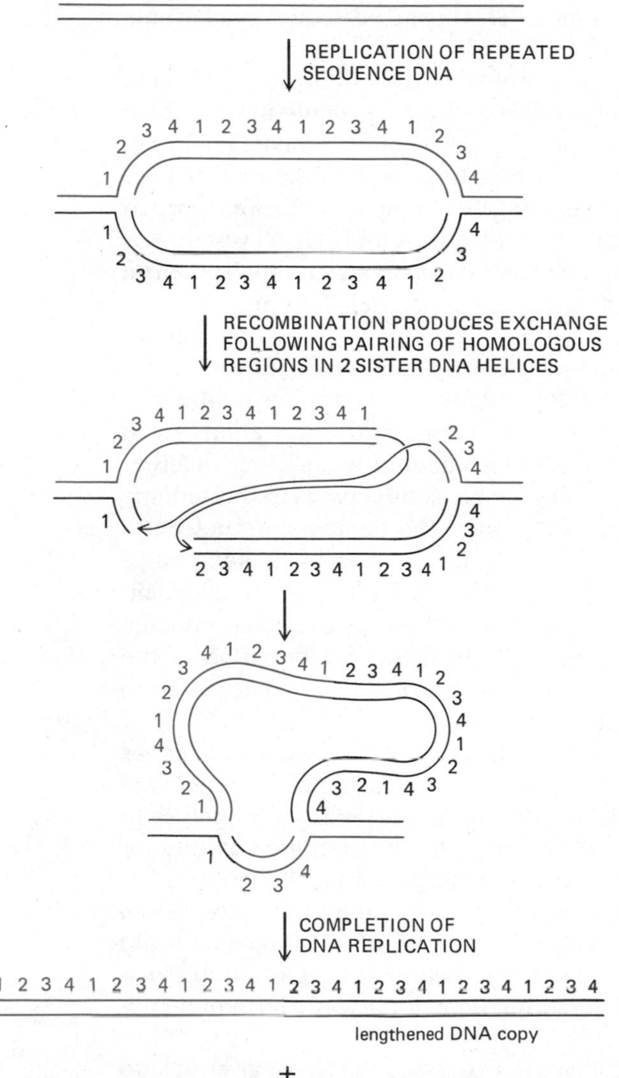

Figure 8–118 Illustration of the type of frequent genetic-recombination event thought to be responsible for the rapid changes found in satellite DNA sequences. Because of the repetition of the sequences, the genetic exchange, which depends on the recognition of homologous base pairs (see p. 241), can sometimes be out of register.

the genetic lability of satellite DNA has any significant effect on the evolution of organisms. But as we shall now see, similar—though much rarer—recombination events duplicate and delete genetically active DNA sequences in a way that can have important evolutionary consequences.

The Evolution of Globins Shows How Random DNA Duplications Contribute to the Evolution of Organisms[60]

Once an amino acid sequence has evolved to form a useful and uniquely folded protein domain, the DNA sequence that codes for it can be duplicated and the duplicate copy modified to make additional, somewhat different proteins. This picture derives from the unmistakable homologies in amino acid sequence and structure between various proteins with different functions (p. 119).

As we have already emphasized, most mutations are harmful (p. 215). In haploid organisms, an essential gene must be duplicated before it can undergo any major modification without jeopardizing the survival of the organism. In contrast, diploid organisms always carry a second "extra" copy of each gene that can undergo mutation without necessarily killing the organism. In theory, diploidy thus gives organisms an advantage because it greatly increases the rate at which they can evolve new proteins. The expected sequence of events is discussed in Chapter 14 (p. 775).

We can reconstruct this sequence of events for the red blood cell protein hemoglobin by considering the different forms of the molecule in organisms at different levels on the phylogenetic scale. A molecule like hemoglobin became necessary to allow multicellular animals to grow to large sizes, where they could no longer derive their oxygen supply simply by diffusion through the body coverings; consequently, a similar molecule is found in all vertebrates and in many invertebrates. The most primitive oxygen-carrying molecule is a globin composed of a single chain of about 150 amino acids. In many marine worms, insects, and primitive fish, oxygen is carried by this kind of globin. In higher vertebrates, however, two kinds of globin chains make up the hemoglobin molecule. It appears that about 500 million years ago, during the evolution of the higher fish, a series of gene mutations and duplications must have occurred. These events led to the establishment of two slightly different globin genes, coding for the α (alpha) and β (beta) globin chains in the genome of each individual. In modern higher vertebrates, the hemoglobin molecules are composed of a complex of four of these chains: two α-chains and two β-chains (Figure 8–119). This structure is much more efficient than single-chain globins because the four oxygen binding sites in the $\alpha_2\beta_2$ molecule interact, causing a cooperative allosteric change (p. 752) in the conformation of the molecule as it binds and releases oxygen. This enables it to deliver a much larger fraction of its bound oxygen to the tissues.

Still later, during the evolution of mammals, one of the two β-chain genes apparently underwent mutation and duplication once again giving rise to a γ (gamma) chain that is synthesized specifically in the embryo (or fetus) to produce an $\alpha_2\gamma_2$ hemoglobin. This *fetal hemoglobin* has a higher affinity for oxygen than adult hemoglobin and is thus advantageous for the fetus. A further duplication occurred still later, during primate evolution, to give rise to a δ (delta) globin gene and thus to a second minor form of hemoglobin ($\alpha_2\delta_2$) found only in adult primates (Figure 8–120). Sometime during evolution an ϵ (epsilon) globin gene also appeared, resulting in an embryonic form of hemoglobin ($\alpha_2\epsilon_2$).

The end result of the gene duplication processes that have given rise to the diversity of hemoglobin chains is clearly seen in the arrangement of the genes coding for the different functional polypeptide chains that arose from

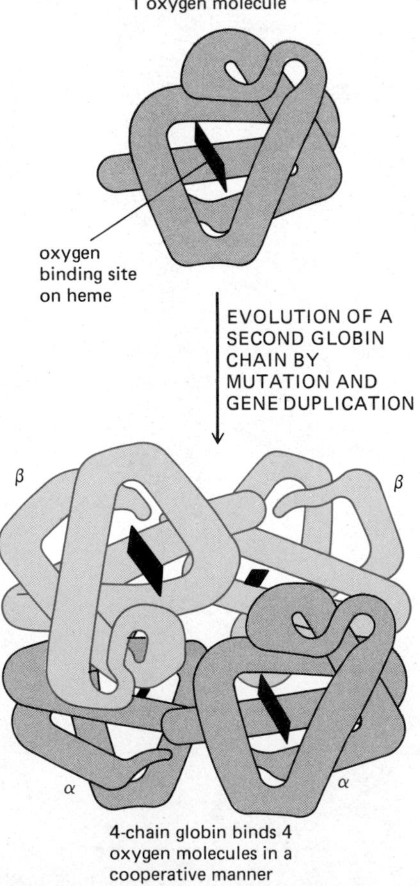

Figure 8–119 A comparison of the structure of 1-chain and 4-chain globins. The 4-chain globin shown is hemoglobin, which is a complex of two α and two β globin chains.

the original β-chain. Detailed analysis of these genes using recombinant DNA technology has shown that they are arranged as a series of homologous DNA sequences located within 50,000 base pairs of each other on one human chromosome (Figure 8–121). The same analysis has also revealed that some duplicated globin DNA sequences in this region do not correspond to genes. These sequences, known as *pseudo-genes,* have an unmistakable homology to functional genes but have been disabled by mutations that prevent them from being expressed. The existence of such pseudo-genes should not be surprising—not every DNA duplication will lead to a new functional gene.

A great deal of our evolutionary history may be discernable in our chromosomes once the DNA sequences of such genes have been carefully compared throughout the long lineage from worms to fish to mammals to man (see also Figure 4–47, p. 186).

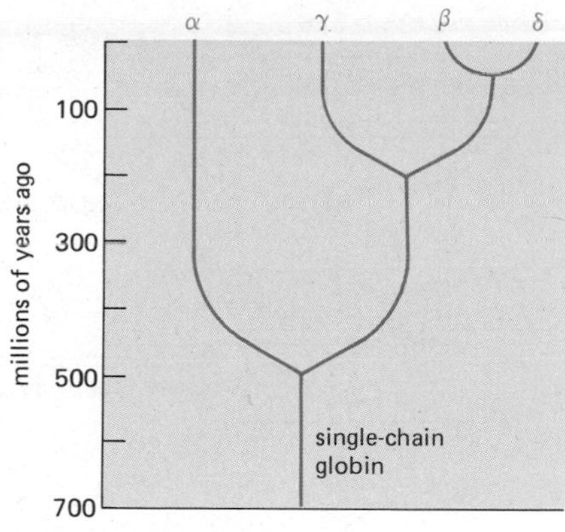

Figure 8–120 An evolutionary tree showing the suspected origins of some of the different human globin chains. This information can be derived from detailed comparisons of either the corresponding protein or DNA sequences in a variety of species.

Recombination Events Also Create New Types of Protein Chains by Joining Coding Domains[61]

We have just seen how a series of gene duplications gave rise to several different hemoglobin molecules composed of two interacting polypeptide chains. Similar DNA duplications have also led to the evolution of new single genes. The proteins encoded by such genes can be recognized because they contain repeating, similar protein domains that are covalently linked and serially arranged. For example, the immunoglobulins (Figure 8–122), albumins, and collagens have been generated in this way.

It is likely that the presence of intervening sequences (introns) in eucaryotic DNA has made it possible for the duplications necessary to form a single gene coding for proteins of this type to accumulate much faster than they could in their absence. Without introns there would be only a few sites in an initial DNA sequence at which a recombinational exchange between sister DNA molecules could duplicate a useful protein domain. On the other hand, long introns on either side of a domain provide many different potential recombination sites (Figure 8–123). For this reason, introns can greatly increase the probability of a favorable duplication event occurring.

In a similar way, introns should help to generate useful new proteins by increasing the probability that a chance recombination event will join two initially separated DNA sequences that code for distinct regions of two different proteins. In general, then, the introns present in eucaryotic coding sequences should increase the rate at which new proteins evolve.

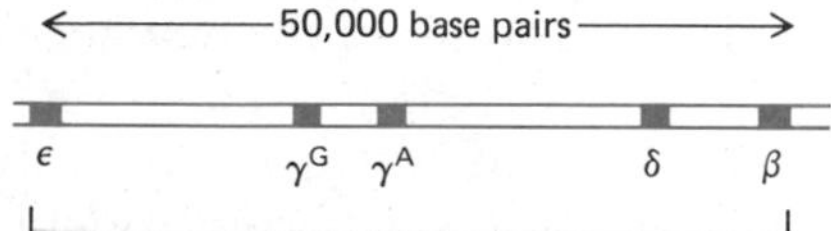

Figure 8–121 A summary of the clustered arrangement of the β-like globin genes in man. The DNA between the indicated genes is not thought to code for proteins. The α genes are located in a second cluster on a different chromosome, having presumably been separated from the β gene by some very early recombination event.

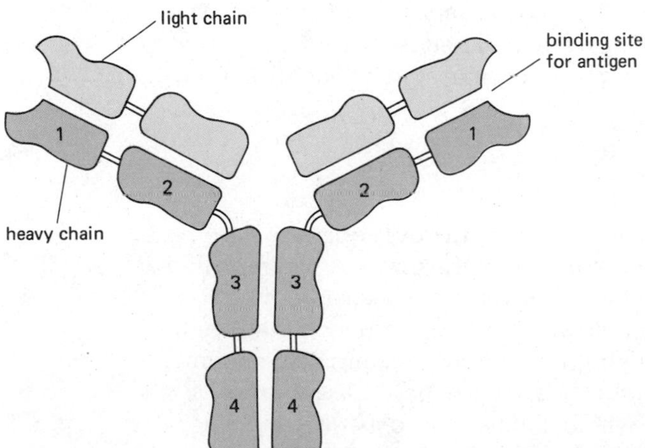

Figure 8–122 A schematic view of an immunoglobulin molecule. This molecule is a complex formed from four chains: the two heavy chains are each made of four similar covalently linked domains (numbered 1 through 4), while the two light chains each contain two such domains. All of these protein domains appear to have evolved from the serial duplication of a single short ancestral protein sequence.

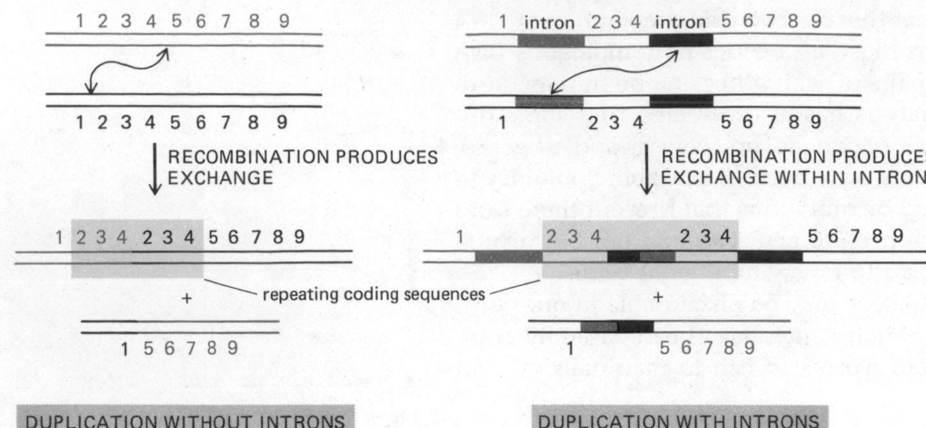

Figure 8–123 Schematic drawing showing how the presence of introns would facilitate the accidental creation of a new protein containing the duplicated protein domain that is designated here as "2 3 4." Note that the new intron that forms can function with many different internal sequences (Figure 8–46). This means that recombination at many sites can lead to a protein that contains the indicated duplication.

Short, Interspersed Repeated Sequences Are Common in Eucaryotic Genomes[62]

Satellite DNA sequences constitute only about one-third of the total repeated DNA in most mammals. The remaining repeated sequences (comprising about 20% of the total DNA in a cell) anneal together more slowly than satellite DNA, but more rapidly than single-copy sequences. A few essential genes are known to be repeated in the genome (the genes coding for rRNA, 5S RNA, the tRNAs, and the histones) and they account for some of this repeated DNA fraction. But these known sequences represent less than 1% of the genome. Like satellite DNA, the vast majority of this repeated DNA has no known function.

In both human and rodent cells, a single large family of **interspersed repeated DNA sequences** has been discovered to constitute a major fraction of the nonsatellite repeated DNA. This group of related DNA sequences differs from satellite sequences in several respects:

1. Rather than being serially repeated at a few chromosomal sites, these sequences are interspersed throughout the genome in hundreds of thousands of individual copies, each about 300 nucleotides long.

2. Many of the sequences are efficiently transcribed into RNA, both as part of long RNA transcripts and as short RNA molecules.

3. These sequences have changed in sequence and location much more slowly during mammalian evolution than have satellite DNA sequences.

Many different functions can be imagined for these interspersed repeated DNA sequences. For example, they could serve as origins for DNA replication, signals for RNA processing, regulatory sequences for RNA transcription, and/or structural elements of chromosomes. But as yet, their functions, if any, are unknown.

The Concept of "Selfish" DNA[63]

It is possible that most of the interspersed repeated DNA sequences have no function at all. In particular, the 10^5 to 10^6 copies of the interspersed repeated DNA sequence described are likely to have arisen long ago—before the divergence of humans and rodents—from a single ancestral sequence that multiplied and spread like a parasite throughout some early mammalian genome. Such an ancestral DNA sequence would thus have behaved like a virus or a *transposable element* (p. 248), both of which can be viewed as DNA sequences that parasitize their host cells. Both multiply by generating identical copies

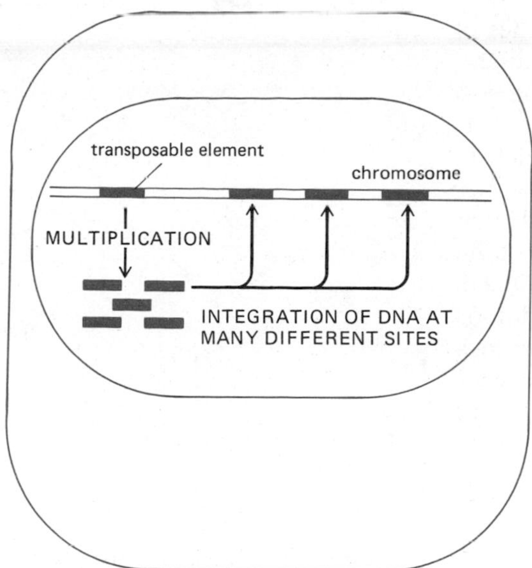

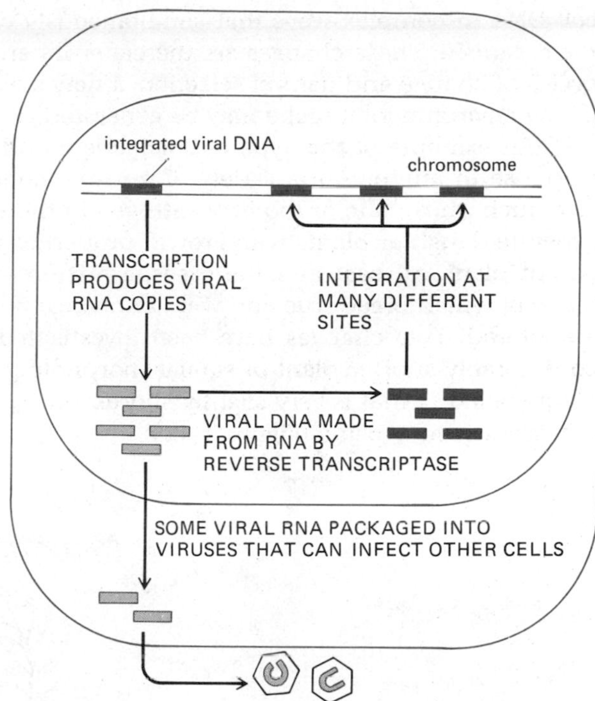

of themselves, which can then be inserted into the host's DNA and carried along passively by the replication of the host chromosome (Figure 8–124). It has therefore been suggested that many of the highly repeated DNA sequences found in the genome of today's organisms should be viewed as "selfish DNA" that has been selected during evolution only for its ability to spread and duplicate itself in the genome of higher organisms with minimal damage to the organism itself. Such DNA sequences are expected to spread especially rapidly in a sexual species, since maternally derived chromosomes can be "infected" by a transposable element carried in the paternally derived genome, and vice versa.

Such selfish DNA would then constitute part of the vast excess of noncoding DNA, whose sequences are not essential for the survival of the cell. But even if the primordial "function" of an interspersed repeated DNA sequence was its own survival, it seems likely that in the course of time its relationship with its host cell would have become symbiotic. For example, in order to minimize their deleterious effect on the host (and thereby maximize their own chance of survival), some transposable elements might have evolved terminal DNA sequences corresponding to RNA splice junctions; in this way, if they inserted into a coding DNA sequence they would leave the original mRNA sequence intact. We have already noted how the insertion of multiple introns into the coding sequences of higher eucaryotes could have created new possibilities for these organisms, with respect both to gene control mechanisms (p. 454) and to the evolution of new proteins (p. 471). A selfish DNA sequence that created new introns could therefore have become advantageous to its host.

Figure 8–124 Diagram illustrating the possible origin of repeated DNA sequences from the multiplication of either transposable elements or viruses in cells. Unlike the viral DNA, the DNA of transposable elements is not packaged into particles that leave the cell. Therefore, if a transposable element multiplies, it does so within a single cell.

Some repeated DNAs in genomes correspond in sequence to cellular mRNA molecules—lacking the introns found in the gene that codes for that mRNA and containing a poly-A "tail." It therefore appears that RNA sequences other than viral ones are occasionally reverse transcribed and can enter the genome by the viral pathway shown here (see also Figure 5–52, p. 239).

Cataclysmic Changes in Genomes Can Increase Biological Diversity[64]

There are probably many reasons for the colonization of eucaryotic genomes by a variety of apparently nonessential DNA segments. Especially interesting is evidence that suggests that some of these "guests" in the genome of yeast cells, insects, and certain plants can be activated at intervals to alter the host cell DNA in complex ways that simultaneously change several properties of the organism. These changes are thereafter inheritable, so that, by successive cycles of change and natural selection, a new species of organism adapted to a new environmental niche may be generated.

An example of this type of change is found in the flax plant, which in response to any one of a variety of unusual conditions (for example, either too much phosphate or too little nitrogen in the soil) can produce seeds that generate daughter plants with growth properties that differ from those of the parent plant. At least in some cases, a mixture of plant types is produced, many of which breed true and will not change further, even if soil conditions are altered. Two changes have been investigated in detail: one results in a considerably shorter plant of similar morphology (called S) and another in a plant (called L) that is very slightly shorter but has many more branches and a hairless seed capsule (Figure 8–125).

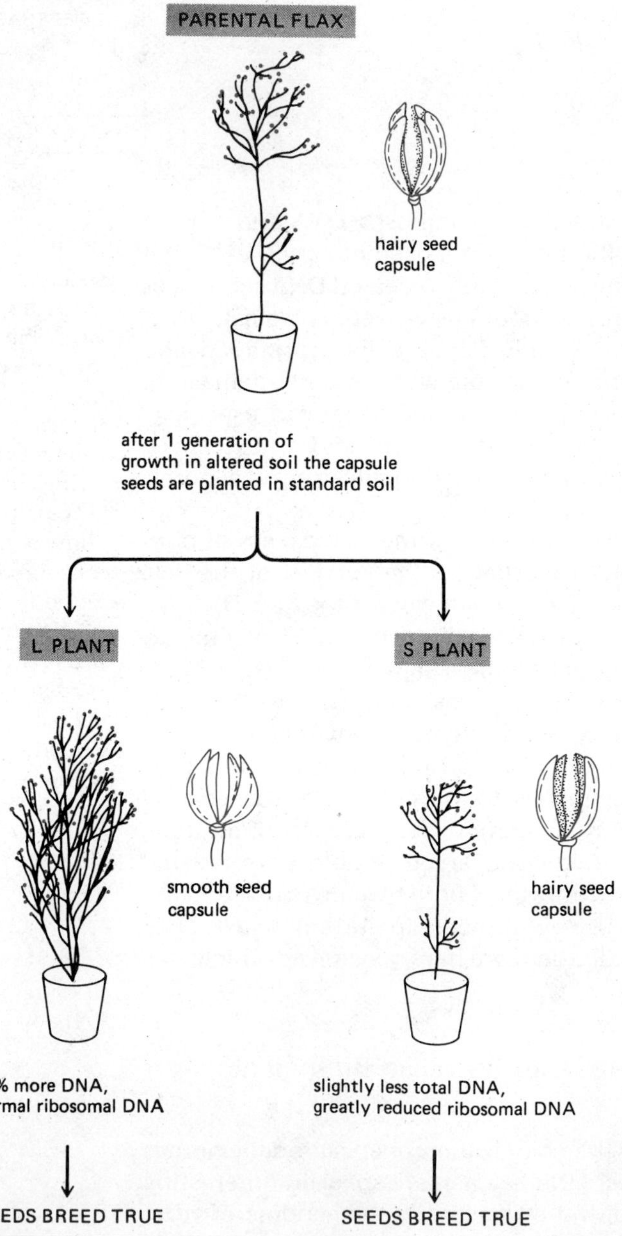

Figure 8–125 Diagram illustrating some dramatic changes that can occur in the flax plant, as described in the text.

Biochemical analysis shows that major changes in the DNA have occurred in the sudden transition from the parent plant to the S or L plant. Although no change in the morphology of the chromosomes is detectable, the L plant contains 10% more DNA in each cell than does the parent because of the differential amplification of many different DNA sequences. The S plant has about 6% less DNA per cell than the parent, with a selective deletion of certain DNA sequences. For example, the number of ribosomal RNA genes is halved.

It is not known which of the many different changes found in the DNA are responsible for the stable alterations in the morphology of the flax plant. Nor is it clear how these changes arise. However, an analogous type of induced hereditary change was observed more than 30 years ago in maize, in which an elegant genetic and cytological analysis revealed alterations in the activity and location of genetic units called *controlling elements.* These controlling elements are now thought to be analogous to the transposable elements that have been studied at the molecular level in bacteria, *Drosophila,* and yeast (p. 248). The situation in *Drosophila* is probably the most analogous to that in maize. In crosses between certain strains of flies, an extensive movement of a particular class of transposable element is induced, which gives rise to a high frequency of mutants in the offspring (a phenomenon known as *hybrid dysgenesis*). Moreover, as earlier observed in maize, the insertion of a transposable element into the *Drosophila* chromosome can lead to a striking change in the pattern of expression of a neighboring gene (Figure 8–126).

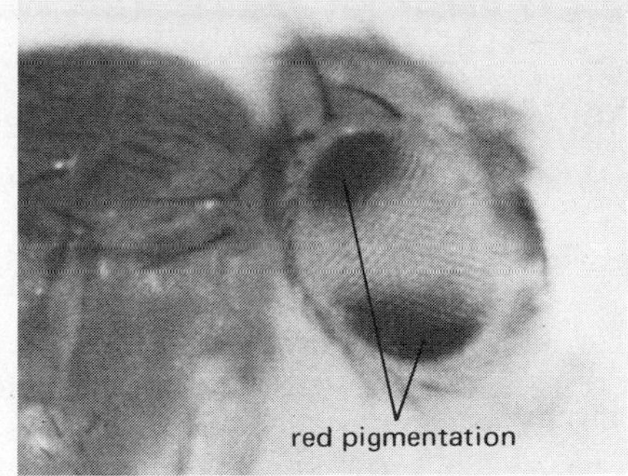

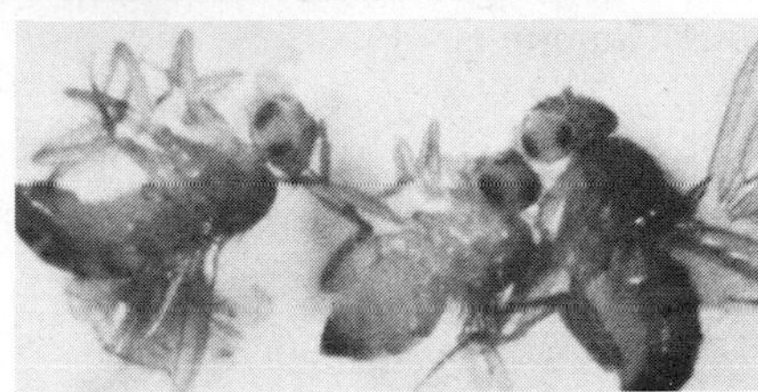

Figure 8–126 Photographs of the eye of an adult *Drosophila,* showing a change in the pattern of eye color caused by a transposable element. The element has inserted next to the *white* gene, which is required for the production of the fly's red eye pigmentation. Because of a change in the local DNA sequence caused by the element, the altered *white* gene becomes active only in those cells located near the dorsal and ventral margins of the eye, producing a pale yellow eye with two noncontiguous patches of dark red along its edges. If a similar spatially regulated change was induced in the activity of a gene that controls the proliferation of individual eye cells, it could drastically alter the shape of the eye and be of major adaptive value for the mutant organism. (Courtesy of Robert Levis and Gerald M. Rubin.)

The picture that emerges is one in which either chance or severe environmental stresses change the internal environment in an organism and activate certain unknown components that cause transposable-element-like DNA sequences to move in and out of chromosomes, sometimes amplifying, deleting, rearranging, or changing the pattern of expression of neighboring DNA sequences in the process. The exact changes caused by each such cataclysmic event will be determined at least partly by chance, and the level, timing, or pattern of expression of important genes in the progeny organisms will be stably altered in different ways.

It seems likely that the ability to undergo such occasional dramatic DNA rearrangements makes an important contribution to the evolution of organisms. If so, it may be in part for this reason that cells succeed so well in carrying the repeated DNA sequences that appear to be parasitizing their genomes.

Summary

It can be argued that the most sophisticated organisms arise from cells that have the most adaptable genomes. Surprisingly frequent DNA rearrangements occur in many eucaryotic genomes; some of these rearrangements probably helped to create a wider diversity of eucaryotic organisms. Repeated DNA sequences in eucaryotes play a part in these DNA rearrangements, which may explain why most eucaryotes have so many repeated sequences in their genomes.

Some of the repeated sequences are interspersed as single copies throughout the genome. Others are serially arranged as "satellite DNAs." The latter have evolved unusually rapidly, probably by the same type of out-of-register genetic recombination events that create occasional gene duplications elsewhere in the genome. Such random gene-duplication events are thought to be important in the evolution of new proteins.

References

General

DNA Structures. *Cold Spring Harbor Symp. Quant. Biol.*, Volume 47, 1983. (A collection of original articles.)

Hood, L.E.; Wilson, J.H.; Wood, W.B. Molecular Biology of Eucaryotic Cells. Menlo Park, Ca.: Benjamin-Cummings, 1975.

Lewin, B. Gene Expression, Vol. 2, Eucaryotic Chromosomes, 2nd ed. New York: Wiley, 1980. (Parts 2, 3, and 4 cover most of the material in this chapter at a more advanced level; also, an excellent source of literature references.)

Watson, J.D. The Molecular Biology of the Gene, 3rd ed. Menlo Park, Ca.: Benjamin-Cummings, 1976.

Cited

1. Gall, J.G. Chromosome structure and the C-value paradox. *J. Cell Biol.* 91:3s-14s, 1981. (A very readable review.)

   Hsu, T.C. Human and Mammalian Cytogenetics: A Historical Perspective. New York: Springer-Verlag, 1979.

   Felsenfeld, G. Chromatin. *Nature* 271:115–122, 1978. (A review.)

   Chambon, P. Summary: the molecular biology of the eukaryotic genome is coming of age. *Cold Spring Harbor Symp. Quant. Biol.* 42:1209–1234, 1978.

   Bostock, C.J.; Sumner, A.T. The Eucaryotic Chromosome. Amsterdam: North-Holland, 1978. (An advanced treatment.)

   Igo-Kemenes, T.; Hörz, W.; Zachau, H.G. Chromatin. *Annu. Rev. Biochem.* 51:89–122, 1982.

2. Isenberg, I. Histones. *Annu. Rev. Biochem.* 48:159–191, 1979.

3. Kornberg, R.D.; Klug A. The nucleosome. *Sci. Am.* 244(2):52–64, 1981.

   McGhee, J.D.; Felsenfeld, G. Nucleosome structure. *Annu. Rev. Biochem.* 49:1115–1156, 1980.

4. Worcel, A.; Strogatz, S.; Riley, D. Structure of chromatin and the linking number of DNA. *Proc. Natl. Acad. Sci. USA* 78:1461–1465, 1981.

   McGhee, J.D.; Rau, D.C.; Charney, E.; Felsenfeld, G. Orientation of the nucleosome within the higher order structure of chromatin. *Cell* 22:87–96, 1980.

5. Thoma, F.; Koller, T.; Klug, A. Involvement of histone H1 in the organization of the nucleosome and of the salt-dependent superstructures of chromatin. *J. Cell Biol.* 83:403–427, 1979.

   Simpson, R.T. Structure of the chromatosome, a chromatin particle containing 160 base pairs of DNA and all the histones. *Biochemistry* 17:5524–5531, 1978.

6. Allfrey, V.G. Post-synthetic modifications of histone structure. In Chromatin and Chromosome Structure (H.J. Li, R. Eckhardt, eds.), pp. 167–191. New York: Academic Press, 1977.

   Goodwin, G.H.; Walker, J.M.; Johns, E.W. The high mobility group (HMG) non-histone chromosomal proteins. In The Cell Nucleus, Vol. 6 (H. Busch, ed.), pp. 181–219. New York: Academic Press, 1978.

   Sandeen, G.; Wood, W.I.; Felsenfeld, G. The interaction of high mobility proteins HMG 14 and 17 with nucleosomes. *Nucleic Acids Res.* 8:3757–3778, 1980.

   Ajiro, K.; Borun, T.W.; Shulman, S.D.; McFadden, G.M.; Cohen, L.H. Comparison of the structures of human histones 1A and 1B and their intramolecular phosphorylation sites during the HeLa S-3 cell cycle. *Biochemistry* 20:1454–1464, 1981.

7. Ptashne, M.; Gilbert, W. Genetic repressors. *Sci. Am.* 222(6):36–44, 1970.

   Ohlendorf, D.H.; Anderson, W.F.; Fisher, R.G.; Takeda, Y.; Matthews, B.W. The molecular basis of DNA-protein recognition inferred from the structure of cro repressor. *Nature* 298:718–723, 1982.

   Tjian, R. T antigen binding and the control of SV40 gene expression. *Cell* 26:1–2, 1981.

   Engelke, D.R.; Ng, S.-Y.; Shastry, B.S.; Roeder, R.G. Specific interaction of a purified transcription factor with an internal control region of 5S RNA genes. *Cell* 19:717–728, 1980.

8. Zachau, H.G.; Igo-Kemenes, T. Face to phase with nucleosomes. *Cell* 24:597–598, 1981.

9. Elgin, S.C.R. DNase I—hypersensitive sites of chromatin. *Cell* 27:413–415, 1981.

Wu, C.; Gilbert, W. Tissue-specific exposure of chromatin structure at the 5′ terminus of the rat preproinsulin gene. *Proc. Natl. Acad. Sci. USA* 78:1577–1580, 1981.

McGhee, J.D.; Wood, W.I.; Dolan, M.; Engel, J.D.; Felsenfeld, G. A 200 base pair region at the 5′ end of the chicken adult β-globin gene is accessible to nuclease digestion. *Cell* 27:45–55, 1981.

Samal, B.; Worcel, A.; Louis, C.; Schedl, P. Chromatin structure of the histone genes of *D. melanogaster*. *Cell* 23:401–409, 1981.

10. Worcel, A.; Burgi, E. On the structure of the folded chromosome of *E. coli*. *J. Mol. Biol.* 71:127–148, 1972.

Benyajati, C.; Worcel, A. Isolation, characterization, and structure of the folded interphase genome of *Drosophila melanogaster*. *Cell* 9:393–408, 1976.

11. Marsden, M.; Laemmli, U.K. Metaphase chromosome structure: evidence for a radial loop model. *Cell* 17:849–858, 1979.

Georgiev, G.P.; Nedospasov, S.A.; Bakayev, V.V. Supranucleosomal levels of chromatin organization. In The Cell Nucleus, Vol. 6 (H. Busch, ed.), pp. 3–34. New York: Academic Press, 1978.

12. Lewin, B. Gene Expression, Vol. 2, Eucaryotic Chromosomes, 2nd ed., pp. 428–440. New York: Wiley, 1980.

Yunis, J.J. High resolution of human chromosomes. *Science* 191:1268–1270, 1976.

13. Beermann, W. Chromosomes and genes. In Developmental Studies on Giant Chromosomes (W. Beermann, ed.), pp. 1–33. New York: Springer-Verlag, 1972.

Bostock, C.J.; Sumner, A.T. The Eucaryotic Chromosome, pp. 233–265. Amsterdam: North-Holland, 1978.

14. Lewin, B. Gene Expression, Vol. 2, Eucaryotic Chromosomes, 2nd ed., pp. 479–502. New York: Wiley, 1980.

Judd, B.H.; Young, M.W. An examination of the one cistron: one chromomere concept. *Cold Spring Harbor Symp. Quant. Biol.* 38:573–579, 1974.

15. Levinger, L.; Varshavsky, A. Selective arrangement of ubiquitinated and D1 protein-containing nucleosomes within the *Drosophila* genome. *Cell* 28:375–385, 1982.

Howard, G.C., Abmayr, S.M.; Shinefeld, L.A., Sato, V.L.; Elgin, S.C.R. Monoclonal antibodies against a specific nonhistone chromosomal protein of *Drosophila* associated with active genes. *J. Cell Biol.* 88:219–225, 1981.

Lezzi, M.; Robert, M. Chromosomes isolated from unfixed salivary glands of Chironomus. In Developmental Studies on Giant Chromosomes (W. Beermann, ed.), pp. 35–56. New York: Springer-Verlag, 1972.

16. Ohta, T.; Kimura, M. Functional organization of genetic material as a product of molecular evolution. *Nature* 233:118–119, 1971.

Gall, J.G. Chromosome structure and the C-value paradox. *J. Cell Biol.* 91:3s–14s, 1981.

17. Darnell, J.E., Jr. Transcription units for mRNA production in eukaryotic cells and their DNA viruses. *Prog. Nucleic Acid Res. Mol. Biol.* 22:327–353, 1979.

Perry, R.P. RNA processing comes of age. *J. Cell Biol.* 91:28s–38s, 1981.

18. Chambon, P. Eucaryotic nuclear RNA polymerases. *Annu. Rev. Biochem.* 44:613–638, 1975.

Brown, D.D. Gene expression in eucaryotes. *Science* 211:667–674, 1981.

Grosveld, G.C.; de Boer, E.; Shewmaker, C.K.; Flavell, R.A. DNA sequences necessary for transcription of the rabbit β-globin gene *in vivo*. *Nature* 295:120–126, 1982.

McKnight, S.L.; Kingsbury, R. Transcriptional control signals of a eucaryotic protein-coding gene. *Science* 217:316–324, 1982.

19. Hastie, N.D.; Bishop, J.O. The expression of three abundance classes of mRNA in mouse tissues. *Cell* 9:761–774, 1976.

Lewin, B. Gene Expression, Vol. 2, Eucaryotic Chromosomes, 2nd ed., pp. 708–719. New York: Wiley, 1980.

20. Miller, O.L. The nucleolus, chromosomes, and visualization of genetic activity. *J. Cell Biol.* 91:15s–27s, 1981. (A review.)

Foe, V.E.; Wilkinson, L.E.; Laird, C.D. Comparative organization of active transcription in *Oncopeltus fasciatus*. *Cell* 9:131–146, 1976.

21. Beyer, A.L.; Christensen, M.E.; Walker, B.W.; LeStourgeon, W.M. Identification and characterization of the packaging proteins of core 40S hnRNP particles. *Cell* 11:127–138, 1977.

Beyer, A.L.; Miller, O.L.; McKnight, S.L. Ribonucleoprotein structure in nascent hnRNA is nonrandom and sequence-dependent. *Cell* 20:75–84, 1980.

22. Perry, R.P. RNA processing comes of age. *J. Cell Biol.* 91:28s–38s, 1981. (Includes a historical review.)

Chambon, P. Split genes. *Sci. Am.* 244(5):60–71, 1981.

Crick, F. Split genes and RNA splicing. *Science* 204:264–271, 1979.

Darnell, J.E., Jr. Variety in the level of gene control in eucaryotic cells. *Nature* 297:365–371, 1982.

23. Lerner, M.R.; Steitz, J.A. Snurps and scyrps. *Cell* 25:298–300, 1981.

24. Wozney, J.; Hanahan, D.; Tate, V.; Boedtker, H.; Doty, P. Structure of the pro α2 collagen gene. *Nature* 294:129–135, 1981.

25. Ziff, E.B. Transcription and RNA processing by the DNA tumor viruses. *Nature* 287:491–499, 1980. (A review.)

26. Early, P.; et al. Two mRNAs can be produced from a single immunoglobulin μ gene by alternative RNA processing pathways. *Cell* 20:313–319, 1980.

27. Beadle, G. Genes and the chemistry of the organism. *Am. Sci.* 34:31–53, 1946.

28. Gruss, P.; Lai, C.-J.; Dhar, R.; Khoury, G. Splicing as a requirement for biogenesis of functional 16S mRNA of simian virus 40. *Proc. Natl. Acad. Sci. USA* 76:4317–4321, 1979.

29. Long, E.O.; Dawid, I.B. Repeated genes in eucaryotes. *Annu. Rev. Biochem.* 49:727–764, 1980.

Miller, O.L. The nucleolus, chromosomes, and visualization of genetic activity. *J. Cell Biol.* 91:15s–27s, 1981.

30. Perry, R.P. Processing of RNA. *Annu. Rev. Biochem.* 45:605–629, 1976. (Emphasizes ribosomal RNA.)

Jordan, E.G. The Nucleolus, 2nd ed. Oxford, Eng.: Oxford University Press, 1978. (A clear brief review).

31. Ghosh, S. The nucleolar structure. *Int. Rev. Cytol.* 44:1–28, 1976.

Fawcett, D.W. The Cell, 2nd ed., pp. 243–265. Philadelphia: Saunders, 1981.

32. McClintock, B. The relation of a particular chromosomal element to the development of the nucleoli in Zea Mays. *Z. Zellforsch. Mikrosk. Anat.* 21:294–323, 1934.

Anastassova-Kristeva, M. The nucleolar cycle in man. *J. Cell Sci.* 25:103–110, 1977.

33. Franke, W.W.; Scheer, U.; Krohne, G.; Jarasch, E.-D. The nuclear envelope and the architecture of the nuclear periphery. *J. Cell Biol.* 91:39s–50s, 1981. (A review.)

Wolfe, S.L. Biology of the Cell, 2nd ed. Belmont, Ca.: Wadsworth, 1981. (Chapters 11 and 12.)

34. Fawcett, D.W. The Cell, 2nd ed., pp. 266–302. Philadelphia: Saunders, 1981.

Krstic, R.V. Ultrastructure of the Mammalian Cell, pp. 2–10. New York: Springer-Verlag, 1979.

35. Gerace, L.; Blobel, G. The nuclear envelope lamina is reversibly depolymerized during mitosis. *Cell* 19:277–287, 1980.

36. Bonner, W.M. Protein migration and accumulation in nuclei. In The Cell Nucleus, Vol. 6 (H. Busch, ed.), pp. 97–148. New York: Academic Press, 1978.

Paine, P.L.; Horowitz, S.B. The movement of material between nucleus and cytoplasm. In Cell Biology: A Comprehensive Treatise, Vol. 4 (D.M. Prescott, L. Goldstein, eds.), pp. 299–338. New York: Academic Press, 1980.

37. Bostock, C.J.; Sumner, A.T. The Eucaryotic Chromosome, pp. 196–208. Amsterdam: North-Holland, 1978.

38. Watson, J.D. Molecular Biology of the Gene, 3rd ed. Menlo Park, Ca.: Benjamin-Cummings, 1976. (Chapters 14 and 17).

Brown, D.D. Gene expression in eucaryotes. *Science* 211:667–674, 1981.

39. Gurdon, J.B. The developmental capacity of nuclei taken from intestinal epithelium cells of feeding tadpoles. *J. Embryol. Exp. Morphol.* 10:622–640, 1962.

Steward, F.C.; Mapes, M.O.; Mears, K. Growth and organized development of cultured cells. *Am. J. Bot.* 45:705–713, 1958.

40. Garrels, J.I. Changes in protein synthesis during myogenesis in a clonal cell line. *Dev. Biol.* 73:134–152, 1979.

Van Nest, G.; MacDonald, R.J.; Raman, R.K.; Rutter, W.J. Proteins synthesized and secreted during rat pancreatic development. *J. Cell Biol.* 86:784–794, 1980. (Two-dimensional gels of developing tissues.)

41. Derman, E.; et al. Transcriptional control in the production of liver-specific mRNAs. *Cell* 23:731–739, 1981.
42. Miller, J.H.; Reznikoff, W.S., eds. The Operon. New York: Cold Spring Harbor Laboratory, 1978. (Reviews of gene regulatory proteins in bacterial systems.)
43. Ptashne, M.; et al. How the lambda repressor and cro work. *Cell* 19:1–11, 1980.

Herskowitz, I.; Hagen, D. The lysis-lysogeny decision of phage lambda: explicit programming and responsiveness. *Annu. Rev. Genet.* 14:399–445, 1980.

44. Gierer, A. Molecular models and combinatorial principles in cell differentiation and morphogenesis. *Cold Spring Harbor Symp. Quant. Biol.* 38:951–961, 1974.
45. Ogden, S.; Haggerty, D.; Stoner, C.M.; Kolodrubetz, D.; Schleif, R. The *E. coli* L-arabinose operon: binding sites of the regulatory proteins and a mechanism of positive and negative regulation. *Proc. Natl. Acad. Sci. USA* 77:3346–3350, 1980.

Yamamoto, K.R.; Alberts, B. Steroid receptors: elements for modulation of eukaryotic transcription. *Annu. Rev. Biochem.* 45:721–746, 1976.

46. Smith, G.R. DNA supercoiling: another level for regulating gene expression. *Cell* 24:599–600, 1981.

Sinden, R.R.; Pettijohn, D.E. Chromosomes in living *E. coli* cells are segregated into domains of supercoiling. *Proc. Natl. Acad. Sci. USA* 78:224–228, 1981.

Wang, J.C. Superhelical DNA. *Trends Biochem. Sci.* 5:219–221, 1980.

47. Weintraub, H.; Groudine, M. Chromosomal subunits in active genes have an altered conformation. *Science* 193:848–856, 1976.

Garel, A.; Zolan, M.; Axel, R. Genes transcribed at diverse rates have a similar conformation in chromatin. *Proc. Natl. Acad. Sci. USA* 74:4867–4871, 1977.

Foe, V.E. Modulation of ribosomal RNA synthesis in *Oncopeltus fasciatus:* an electron microscopic study of the relationship between changes in chromatin structure and transcriptional activity. *Cold Spring Harbor Symp. Quant. Biol.* 42:723–740, 1978.

Weisbrod, S.; Groudine, M.; Weintraub, H. Interaction of HMG 14 and 17 with actively transcribed genes. *Cell* 19:289–301, 1980.

Mathis, D.; Oudet, P.; Chambon, P. Structure of transcribing chromatin. *Prog. Nucleic Acid Res. Mol. Biol.* 24:1–55, 1980.

Weisbrod, S. Active chromatin (a review). *Nature* 297:289–295, 1982.

48. Ashburner, M.; Chihara, C.; Meltzer, P.; Richards, G. Temporal control of puffing activity in polytene chromosomes. *Cold Spring Harbor Symp. Quant. Biol.* 38:655–662, 1974.

Lamb, M.M.; Daneholt, B. Characterization of active transcription units in Balbiani rings of *Chironomus tentaus. Cell* 17:835–848, 1979.

Korge, G. Direct correlation between a chromosome puff and the synthesis of a larval saliva protein in *Drosophila melanogaster. Chromosoma* 62:155–174, 1977. (The "infectious puff" phenomenon.)

49. Sommerville, J.; Malcolm, D.B.; Callan, H.G. The organization of transcription on lampbrush chromosomes. *Philos. Trans. R. Soc. Lond.* (*Biol.*) 283:359–366, 1978.

Bostock, C.J.; Sumner, A.T. The Eucaryotic Chromosome, pp. 347–374. Amsterdam: North-Holland, 1978.

50. Hagenbüchle, O.; Tosi, M.; Schibler, U.; Bovey, R.; Wellauer, P.K.; Young, R.A. Mouse liver and salivary gland α-amylase mRNAs differ only in 5′ non-translated sequences. *Nature* 289:643–646, 1981.

Marie, J.; Simon, M.-P.; Dreyfus, J.-C.; Kahn, A. One gene, but two messenger RNAs encode liver L and red cell L′ pyruvate kinase subunits. *Nature* 292:70–72, 1981.

Amara, S.G.; Jonas, V.; Rosenfeld, M.G.; Ong, E.S.; Evans, R.M. Alternative RNA processing in calcitonin gene expression generates mRNAs encoding different polypeptide products. *Nature* 298:240–244, 1982.

51. Brown, S.W. Heterochromatin. *Science* 151:417–425, 1966.

Hsu, T.C.; Cooper, J.E.K., Mace, M.L., Brinkley, B.R. Arrangement of centromeres in mouse cells. *Chromosoma* 34:73–87, 1971.

52. Lyon, M.F. X-chromosome inactivation and developmental patterns in mammals. *Biol. Rev.* 47:1–35, 1972.

Cattanach, B.M. Control of chromosome inactivation. *Annu. Rev. Genet.* 9:1–18, 1975.

Martin, G.R. X-chromosome inactivation in mammals. *Cell* 29:721–724, 1982.

53. Alberts, B.; Worcel, A.; Weintraub, H. On the biological implications of chromatin structure. In The Organization and Expression of the Eukaryotic Genome (E.M. Bradbury, K. Javaherian, eds.), pp. 165–191. New York: Academic Press, 1977.

54. Razin, A.; Riggs, A.D. DNA methylation and gene function. *Science* 210:604–610, 1980.

Naveh-Many, T.; Cedar, H. Active gene sequences are undermethylated. *Proc. Natl. Acad. Sci. USA* 78:4246–4250, 1981.

Groudine, M; Eisenman, R.; Weintraub, H. Chromatin structure of endogenous retroviral genes and activation by an inhibitor of DNA methylation. *Nature* 292:311–317, 1981.

55. Simon, M.; Zeig, J.; Silverman, M.; Mandel, G.; Doolittle, R. Phase variation: evolution of a controlling element. *Science* 209:1370–1374, 1980.

Kushner, P.J.; Blair, L.C.; Herskowitz, I. Control of yeast cell types by mobile genes: a test. *Proc. Natl. Acad. Sci. USA* 76:5264–5268, 1979.

Borst, P.; Cross, G.A.M. Molecular basis for trypanosome antigenic variation. *Cell* 29:291–303, 1982.

56. Beermann, S. The diminution of heterochromatic chromosomal segments in *Cyclops* (Crustacea, Copepoda). *Chromosoma* 60:297–344, 1977.

Spradling, A.C.; Mahowald, A.P. Amplification of genes for chorion proteins during oogenesis in *Drosophila melanogaster*. *Proc. Natl. Acad. Sci. USA* 77:1096–1100, 1980.

Gall, J.G. Chromosome structure and the C-value paradox. *J. Cell Biol.* 91:3s–14s, 1981. (Includes a review of gene amplifications.)

Sakano, H.; Hüppi, K.; Heinrich, G.; Tonegawa, S. Sequences at the somatic recombination sites of immunoglobulin light-chain genes. *Nature* 280:288–293, 1979.

57. Hsu, T.C. Human and Mammalian Cytogenetics: A Historical Perspective. New York: Springer-Verlag, 1979. (Chapters 22 and 23.)

Klein, G. The role of gene dosage and genetic transpositions in carcinogenesis. *Nature* 294:313–318, 1981. (A review.)

Cairns, J. The origin of human cancers. *Nature* 289:353–357, 1981.

Szostak, J.W.; Blackburn, E.H. Cloning yeast telomeres on linear plasmid vectors. *Cell* 29:245–255, 1982.

58. Lewin, B. Gene Expression, Vol. 2, Eucaryotic Chromosomes, 2nd ed., pp. 503–569, 861–930. New York: Wiley, 1980.

59. Hsu, T.C. Human and Mammalian Cytogenetics: A Historical Perspective. New York: Springer-Verlag, 1979.

Craig-Holmes, A.P.; Shaw, M.W. Polymorphism of human constitutive heterochromatin. *Science* 174:702–704, 1971.

John, B.; Miklos, G.L.G. Functional aspects of satellite DNA and heterochromatin. *Int. Rev. Cytol.* 58:1–114, 1979.

60. Ingram, V. Gene evolution and the hemoglobins. *Nature* 189:704–708, 1961.

Efstratiadis, A.; et al. The structure and evolution of the human β-globin gene family. *Cell* 21:653–668, 1980.

Proudfoot, N. Pseudogenes. *Nature* 286:840–841, 1980.

Schimke, R.T. Gene amplification and drug resistance. *Sci. Am.* 243(5):60–69, 1980.

61. Doolittle, R.F. Protein evolution. In The Proteins, Vol. 4, 3rd ed. (H. Neurath, R.L. Hill, eds.), pp. 1–118. New York: Academic Press, 1979.

Darnell, J.E. Implications of RNA-RNA splicing in evolution of eucaryotic cells. *Science* 202:1257–1260, 1978.

62. Jelinek, W.R.; Schmid, C.W. Repetitive sequences in eukaryotic DNA and their expression. *Annu. Rev. Biochem.* 51:813–844, 1982.

Jagadeeswaran, P.; Forget, B.G.; Weissman, S.M. Short interspersed repetitive DNA elements in eucaryotes: transposable DNA elements generated by reverse transcription of RNA Pol III transcripts? *Cell* 26:141–142, 1981.

63. Orgel, L.E.; Crick, F.H.C. Selfish DNA: the ultimate parasite. *Nature* 284:604–607, 1980.

Doolittle, W.F.; Sapienza, C. Selfish genes, the phenotype paradigm and genome evolution. *Nature* 284:601–603, 1980.

Cohen, S.N.; Shapiro, J.A. Transposable genetic elements, *Sci. Am.* 242(2):40–49, 1980.

64. McClintock, B. Controlling elements and the gene. *Cold Spring Harbor Symp. Quant. Biol.* 21:197–216, 1956.

Cullis, C.A. Molecular aspects of the environmental induction of heritable changes in flax. *Heredity* 38:129–154, 1977. (A review.)

Spradling, A.C.; Rubin, G.M. *Drosophila* genome organization: conserved and dynamic aspects. *Annu. Rev. Genet.* 15:219–264, 1981.

Roeder, G.S.; Farabaugh, P.J.; Chaleff, D.T.; Fink, G.R. The origins of gene instability in yeast. *Science* 209:1375–1380, 1980.

Bingham, P.M.; Kidwell, M.G.; Rubin, G.M. The molecular basis of P-M hybrid dysgenesis: the role of the P element, a P-strain-specific transposon family. *Cell* 29:995–1004, 1982.

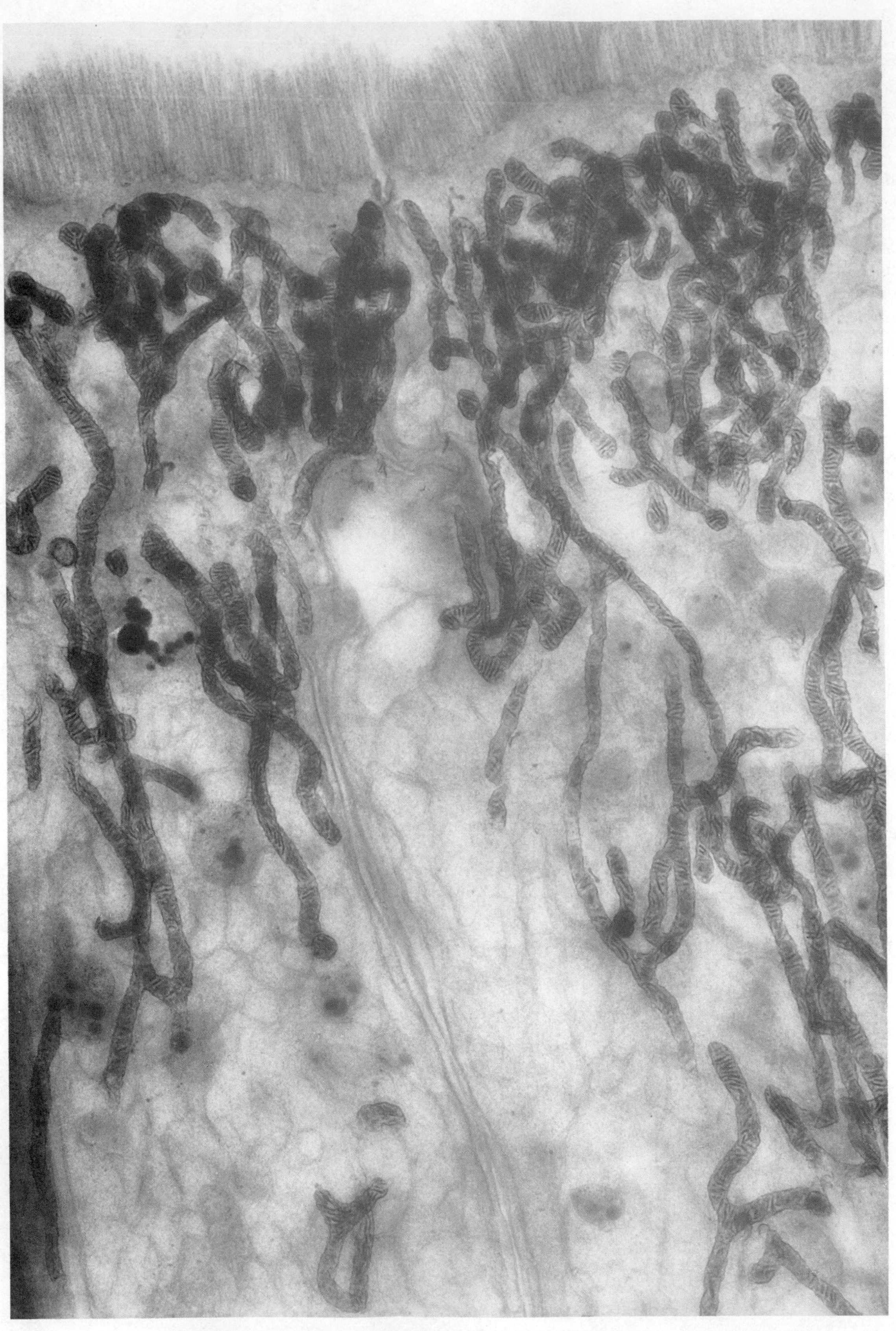

Snakelike mitochondria of a snail epithelial cell, as visualized in a high-voltage electron microscope. (Courtesy of Pierre Favard.)

Energy Conversion: Mitochondria and Chloroplasts

The nucleus is not the only organelle in eucaryotic cells that contains DNA: **mitochondria,** present in all eucaryotic cells, and **plastids** (most notably **chloroplasts**), which are confined to plants, contain their own DNA—which specifies part of the machinery essential to their function. Although these organelles are considerably smaller than the nucleus, they are numerous and thus often occupy a large volume in the cell. Both types of organelles convert energy to forms that can be used to drive vital cellular reactions. Their specialized function is reflected in the morphological feature that most strikingly distinguishes mitochondria and chloroplasts from the nucleus—the large amount of internal membrane they contain. This membrane plays two crucial roles in the function of these "energy organelles." First, it provides the framework for the electron-transport processes that convert the energy of oxidation reactions into more useful forms—in particular into ATP. Second, it creates a large internal compartment in each organelle, in which special enzymes that catalyze other important cellular reactions are confined.

Without mitochondria, animal cells would be dependent on anaerobic glycolysis for all their ATP. But glycolysis, which splits glucose to pyruvate (p. 69), releases only a small fraction of the total free energy available from the oxidation of sugars. In mitochondria, the metabolism of sugars (and of fatty acids) is completed with their oxidation by molecular oxygen (O_2) to CO_2 and H_2O. The energy available is harnessed so efficiently that about 36 molecules of ATP are produced for each molecule of glucose, while glycolysis alone would produce only 2 molecules of ATP.

Chloroplasts, like mitochondria, are very efficient ATP-producing machines, although their source of energy is quite different. Chloroplasts depend on sunlight rather than sugars or fatty acids and might therefore be expected to show fundamental differences in the way they are organized. But in fact, both mitochondria and chloroplasts are organized according to the same general principles, and both synthesize ATP in the same way.

This striking conclusion has emerged from painstaking studies—carried out over the past 20 years—that have now shown conclusively that the central pathway by which energy is harnessed for biological purposes is the same not only in mitochondria and chloroplasts, but in bacteria as well. This pathway operates by a process known as **chemiosmosis.** The pathway begins with high-energy electrons that have either been excited by sunlight or trapped in electron-rich foodstuffs. The electrons are passed along a chain of proteins known as an **electron-transport chain,** embedded in an ion-impermeable membrane. In the course of their transport along the chain, the electrons fall to successively lower energy levels. The energy released by the electrons in their descent down the chain is harnessed to pump protons from one side of the membrane to the other. The result is an *electrochemical proton gradient* across the membrane, and the energy stored in the gradient can in turn be harnessed by other proteins in this membrane (Figure 9–1). In mitochondria and chloroplasts, most of the energy is used for the chemical conversion of ADP and P_i to ATP, but some of it drives the transport of specific metabolites into and out of the organelles. In bacteria, the electrochemical gradient itself is as important a store of directly usable energy as is the ATP that it generates. For example, not only does this gradient drive many transport processes, but it also enables a bacterium to swim, inasmuch as a backflow of protons drives the rapid rotation of the bacterial flagellum (p. 758).

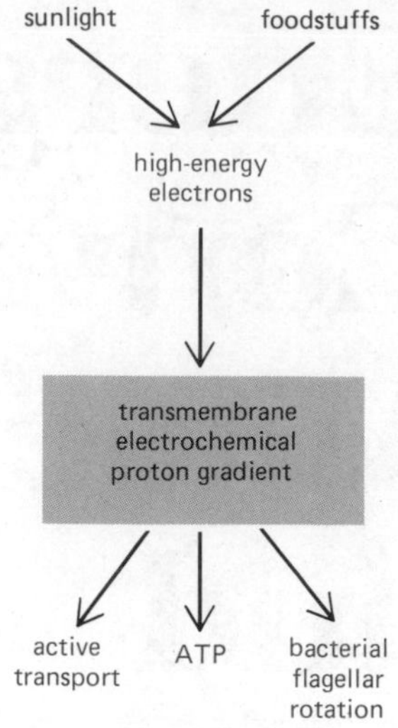

sunlight
foodstuffs
high-energy electrons
transmembrane electrochemical proton gradient
active transport
ATP
bacterial flagellar rotation

Figure 9–1 Chemiosmosis as a universal mechanism: energy is first harnessed to create an electrochemical proton gradient across a membrane. This gradient serves as a versatile energy store, being used in a variety of different ways in mitochondria, chloroplasts, and bacteria.

It is generally believed that the energy-converting organelles of eucaryotes evolved directly from procaryotes that were engulfed by primitive eucaryotic cells early in evolutionary history and developed a symbiotic relationship with them. This would explain why mitochondria and chloroplasts contain their own DNA. But in the billion or so years since the first eucaryotic cells appeared, mitochondria and chloroplasts have lost much of their own genome and have thereby become heavily dependent on proteins that are encoded by the nuclear genome, synthesized in the cell cytosol, and then imported into each organelle. Conversely, the host cells are now dependent on the organelles for a supply of ATP to carry out the constant work of biosynthesis, ion and solute pumping, and movement that keeps them alive.

The Mitochondrion[1]

Mitochondria occupy a substantial fraction of the cytoplasm of virtually all eucaryotic cells. Although they are large enough to be seen in the light microscope and were first identified in the nineteenth century, real progress in elucidating their function had to wait until 1948, when a procedure was developed for isolating intact mitochondria. For technical reasons, most biochemical studies have been carried out with purified mitochondria extracted from the liver, where each cell contains 1000 to 2000 of the organelles occupying roughly a fifth of the total cell volume.

Mitochondria are usually depicted as stiff, elongated cylinders with a diameter of 0.5 to 1 μm. However, time-lapse microcinematography of living cells reveals a mobility and plasticity that belies their static image in electron micrographs (Figure 9–2). As they move about in the cytoplasm, mitochondria often appear to be associated with the microtubules of the cytoskeleton (Figure 9–3). This association may determine the unique orientation and distribution of mitochondria in different types of cells. Thus, the mitochondria of some cells form long moving filaments or chains (Figure 9–3), while in other types of cells they are fixed in position near a site of unusually high ATP consumption: for example, they are packed between the adjacent myofibrils in a cardiac muscle cell and are tightly wrapped around the flagellum in a sperm (Figure 9–4).

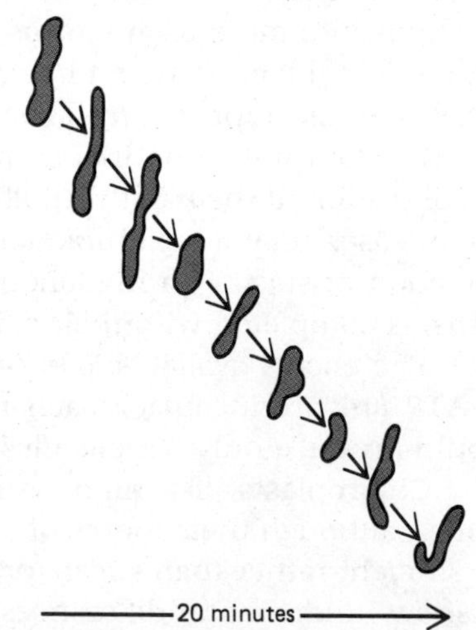

Figure 9–2 Illustration of the rapid changes of shape observed when mitochondria are visualized in living cells.

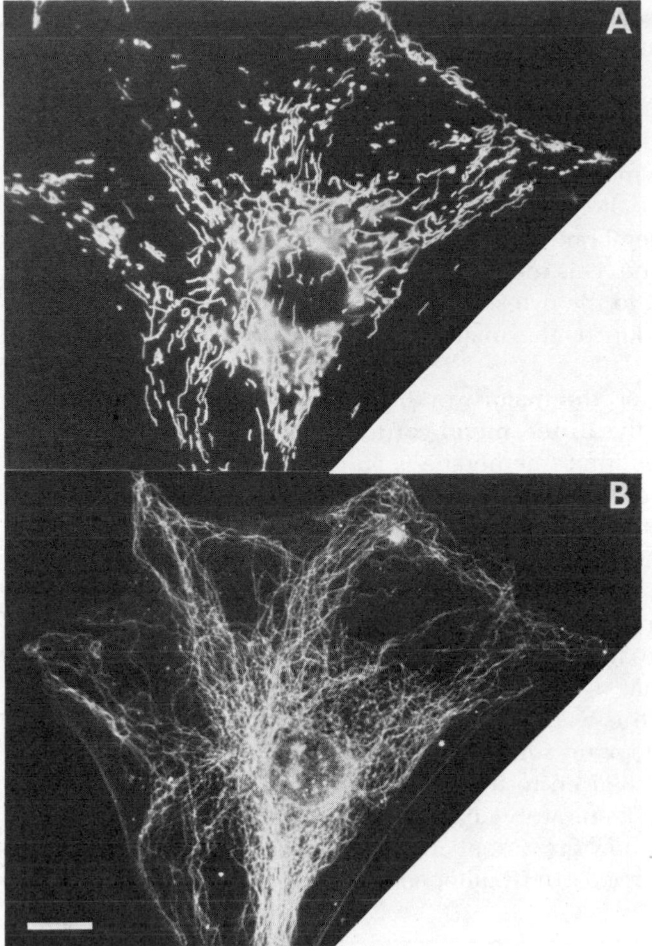

Figure 9–3 (A) Light micrograph of chains of elongated mitochondria seen in a large, living mammalian tissue-culture cell that has been stained with a vital fluorescent dye. The dye used (rhodamine 123) is specific for mitochondria. (B) Fluorescently stained microtubules in the same cell; note the alignment of mitochondria along microtubules. The bar indicates 25 μm. (Courtesy of Lan Bo Chen.)

The Mitochondrion Contains an Outer Membrane, an Inner Membrane, and Two Internal Compartments[2]

The mitochondrion is bounded by a pair of highly specialized membranes that play a crucial part in its activities. Each of the two lipid bilayers contains a unique collection of proteins, and together they enclose and define two separate mitochondrial compartments: the internal **matrix space** and a narrower **intermembrane space.** If purified mitochondria are gently disrupted

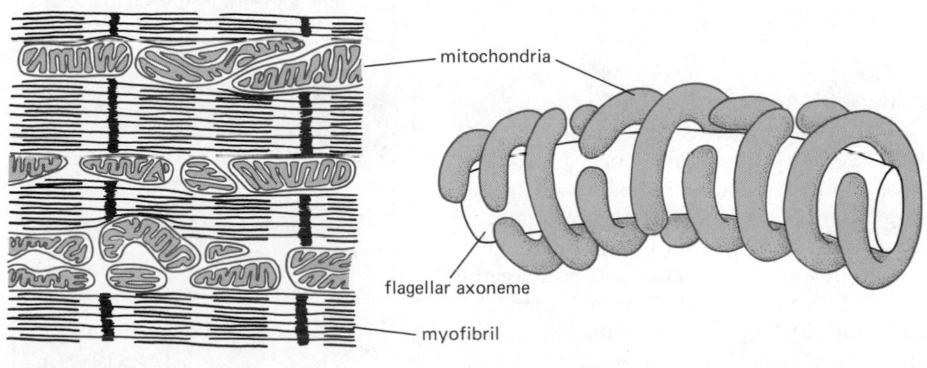

Figure 9–4 Two examples of the specific localization of mitochondria near sites of high ATP utilization. For a discussion of how the ATP produced is hydrolyzed to create movement in muscle (*left*) and in eucaryotic flagella (*right*), see Chapter 10. During the development of the flagellum of the sperm tail, helically wound microtubules temporarily encircle the axoneme and appear to help localize the mitochondria in the tail.

and then fractionated into separate components (Figure 9–5), the biochemical composition of each of the four different parts of a mitochondrion can be determined; some of the results are shown in Figure 9–6.

These and other studies have demonstrated that the **outer membrane** contains many copies of a transport protein that forms large aqueous channels through the lipid bilayer. The outer membrane thus resembles a net that is permeable to all molecules of 10,000 daltons or less, including small proteins. Such molecules can enter the intermembrane space, but most of them cannot pass the impermeable inner membrane. This means that, while the intermembrane space is chemically equivalent to the cytosol with respect to the small molecules it contains, the composition of the matrix space is much more highly specialized.

As we shall explain in detail later, the major working part of the mitochondrion is the matrix space and the **inner membrane** that encloses it. Among the specialized features of the inner membrane is an unusually high proportion of *cardiolipin,* a phospholipid that accounts for more than 10% of the membrane's lipid content and is thought to help make it unusually impermeable to ions. The membrane also contains a variety of transport proteins that make it selectively permeable to those small molecules that are metabolized by the many mitochondrial enzymes concentrated in the matrix space. The matrix enzymes include those that metabolize pyruvate and fatty acids to produce acetyl CoA and those that oxidize acetyl CoA in the citric acid cycle. The principal end products of this oxidation are CO_2, which is released from the cell, and NADH, which is the main source of electrons for transport along the *respiratory chain*—the name given to the electron-transport chain in mitochondria. The enzymes of the respiratory chain are embedded in the inner mitochondrial membrane, and they are essential to the whole process of *oxidative phosphorylation* that is carried out in mitochondria and that generates most of the animal cell's ATP.

The Inner Membrane Is Folded into Cristae[3]

The inner membrane is usually highly convoluted, forming a series of infoldings, known as *cristae,* in the matrix space. Because of these convolutions, the total area of inner membrane in liver cell mitochondria is nearly five times the area of outer membrane and constitutes about a third of the total membrane of the cell (see Table 7–2, p. 322). The number of cristae is threefold greater in the mitochondrion of a cardiac muscle cell than in the mitochondrion of a liver cell, presumably reflecting the greater demand for ATP in heart tissue. The cristae of mitochondria in different cell types also show striking morphological differences whose significance is unknown (Figure 9–7). Moreover, mitochondria often contain special enzymes that better enable them to

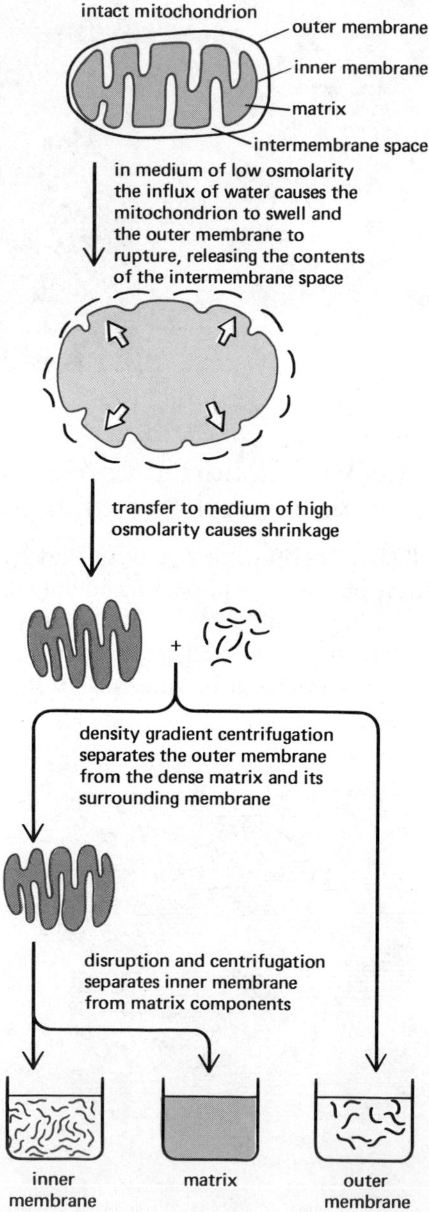

Figure 9–5 Techniques for fractionating purified mitochondria into separate components have made it possible to study the different proteins in each mitochondrial compartment. The method shown, which allows the processing of large numbers of mitochondria at the same time, takes advantage of the fact that in media of low ionic strength water flows into mitochondria and greatly expands the matrix space. While the cristae of the inner membrane allow it to unfold to accommodate the expansion, the outer membrane—which has no folds to begin with—breaks, releasing a structure composed of only the inner membrane and the matrix.

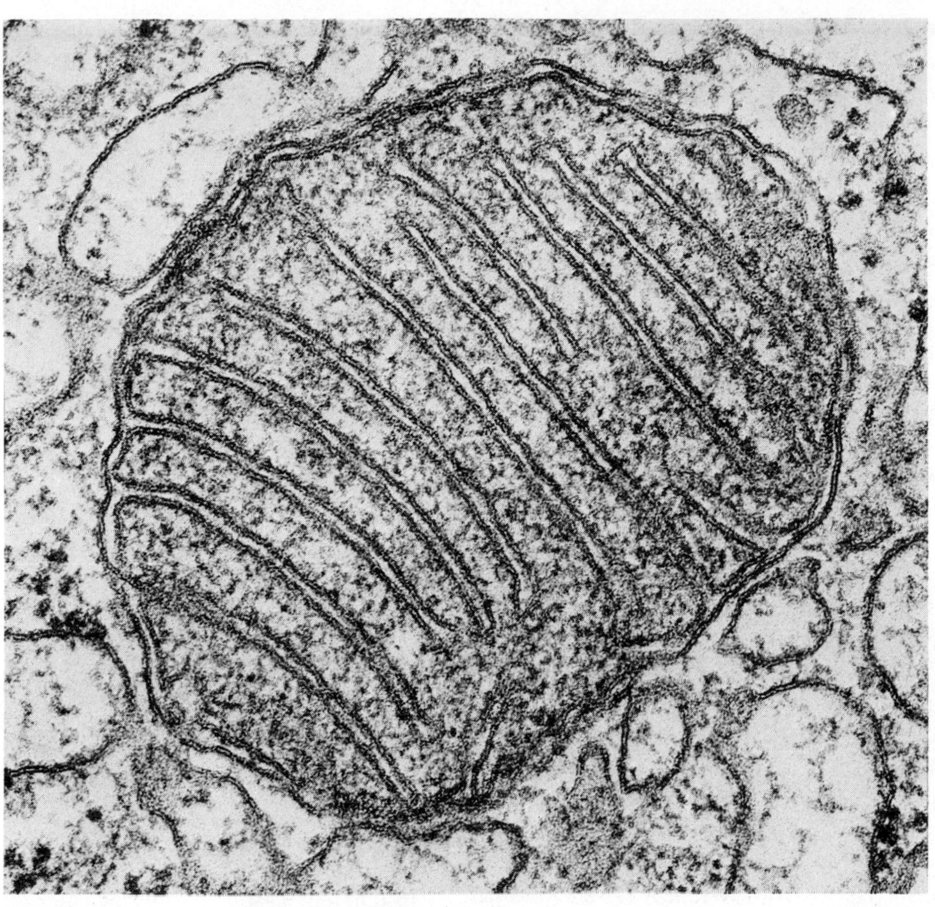

Figure 9–6 The general organization of a mitochondrion. In the liver, an estimated 67% of the total mitochondrial protein is located in the matrix, 21% in the inner membrane, 6% in the outer membrane, and 6% in the intermembrane space. As indicated below, each of these four different regions contains a separate set of proteins specifically suited to its functions. (Micrograph courtesy of Daniel S. Friend.)

Matrix. The matrix contains a highly concentrated mixture of hundreds of different enzymes, including those required for the oxidation of pyruvate and fatty acids and for the citric acid cycle. It also contains several identical copies of the mitochondrial DNA genome, special mitochondrial ribosomes, tRNAs, and various enzymes that are required for the expression of the mitochondrial genes.

Inner Membrane. The inner membrane is folded into numerous cristae, which increase its total surface area. It contains three major types of proteins: (1) those that carry out the oxidation reactions of the respiratory chain, (2) an enzyme complex called *ATP synthetase* that makes ATP in the matrix, and (3) specific transport proteins that regulate the passage of metabolites into and out of the matrix. Since an electrochemical gradient that drives the ATP synthetase is established across this membrane by the respiratory chain, it is important that the membrane be impermeable to most small ions.

Outer Membrane. Because it contains a large channel-forming protein, the outer membrane is permeable to all molecules of 10,000 daltons or less. Other proteins in this membrane include enzymes that convert lipid substrates into forms that are subsequently metabolized in the matrix.

Intermembrane Space. This space contains several enzymes that use the ATP that passes out of the matrix to phosphorylate other nucleotides.

serve the particular cell in which they reside. In this chapter, however, we shall ignore such differences and focus instead on the enzymes and properties common to all mitochondria. Before explaining in detail the electron-transport processes that are the unique feature of mitochondrial activity, we shall briefly trace the metabolic steps that create the substrates of the respiratory chain.

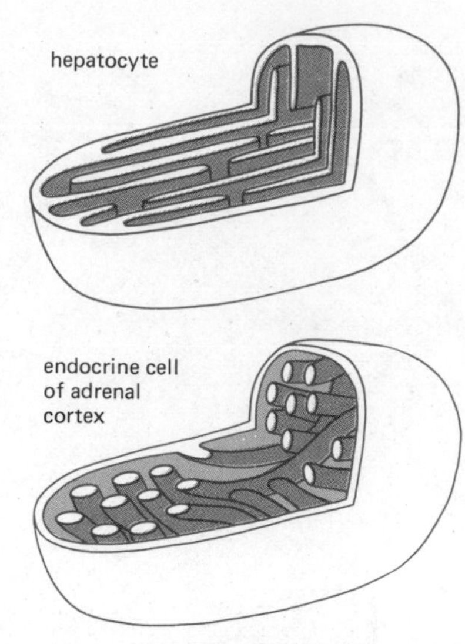

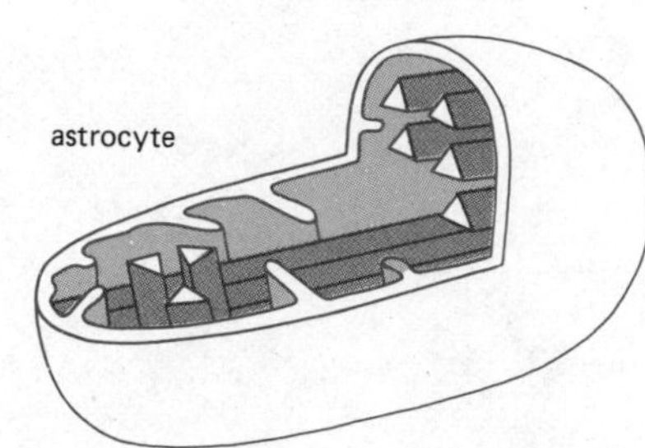

Figure 9–7 Some morphologically distinct mitochondrial cristae found in different tissues of the rat. The effect of these differences on mitochondrial function is unknown.

Mitochondrial Oxidation Begins When Large Amounts of Acetyl CoA Are Produced in the Matrix Space from Pyruvate and from Fatty Acids[4]

Oxidative metabolism in mitochondria, as we have mentioned, is largely fueled by fatty acids and by pyruvate that is produced by glycolysis in the cytosol. These compounds are selectively transported from the cytosol into the mitochondrial matrix, where they are broken down into the two-carbon acetyl group on acetyl CoA (Figure 9–8); the acetyl group is then fed into the citric acid cycle for further degradation, and the process ends with the passage of acetyl-derived high-energy electrons along the respiratory chain.

To ensure a continuous supply of fuel for oxidative metabolism, fatty acids and pyruvate exist in animal cells in storage forms: fats, which store fatty acids, and glycogen, which stores glucose that will be broken down to pyruvate. Fats are quantitatively by far the more important, in part because their oxidation releases more than six times as much energy as the oxidation of an equal mass of glycogen in its hydrated form. An average adult human has enough glycogen to last for only about a day of normal activities, but enough fat to last for nearly a month. If our main fuel reservoir was carried as glycogen instead of as fat, body weight would need to be increased by an average of about 60 pounds.

Most of our fat is stored in adipose tissue, from which it is released into the blood stream for use by other cells when needed. This need arises after a period of not eating; even a normal overnight fast results in the mobilization of fat, so that in the morning most of the acetyl CoA that enters the citric acid cycle is derived from fatty acids rather than from glucose. After a meal, however, most of the acetyl CoA entering the citric acid cycle comes from glucose derived from foodstuffs, and any excess glucose is used to replenish depleted glycogen stores or to synthesize fats. (Note that while sugars are readily converted to fats, fats cannot be converted to sugars in animal cells.)

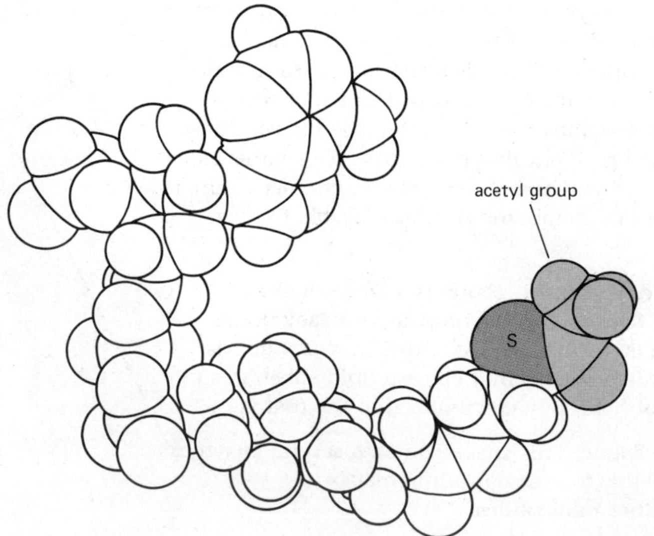

Figure 9–8 A space-filling model of acetyl CoA. The acetyl group is only a small part of the mass of this coenzyme. The atom labeled with an S is a sulfur atom. (See also Figure 2–19, p. 69.)

A fat molecule is composed of three molecules of fatty acid held in ester linkage to glycerol. Such *triglycerides* (*triacylglycerols*) have no charge and are virtually insoluble in water, coalescing into droplets in the cell cytosol. A single large droplet occupies nearly the entire cell volume of *adipocytes,* the large cells specialized for fat storage in adipose tissue. Much smaller fat droplets are common in cells such as cardiac muscle cells that rely on the breakdown of fatty acids for their energy supply; these droplets are often closely associated with mitochondria (Figure 9–9). In all cells, enzymes in the outer and inner mitochondrial membranes mediate the movement of fatty acids derived from fat molecules into the mitochondrial matrix. There, each fatty acid molecule is broken down completely by a cycle of reactions that trims two carbons at a time from its carboxyl end, generating one molecule of acetyl CoA in each turn of the cycle (Figure 9–10). The acetyl CoA is then fed into the citric acid cycle to be oxidized further.

The six-carbon sugar glucose is initially broken down in the cytosol to two molecules of the three-carbon pyruvate, which still retains most of the total energy that can be derived from the oxidation of glucose. This energy is harvested by the processes that begin when the pyruvate enters the mitochondrion. There it encounters an enzyme complex larger than a ribosome, *pyruvate dehydrogenase* (p. 87). This complex—containing multiple copies of three different catalytic enzymes, five different coenzymes, and two regulatory enzymes—rapidly converts pyruvate to acetyl CoA (Figure 9–11). This acetyl CoA joins the acetyl CoA produced from fatty acids to feed the citric acid cycle.

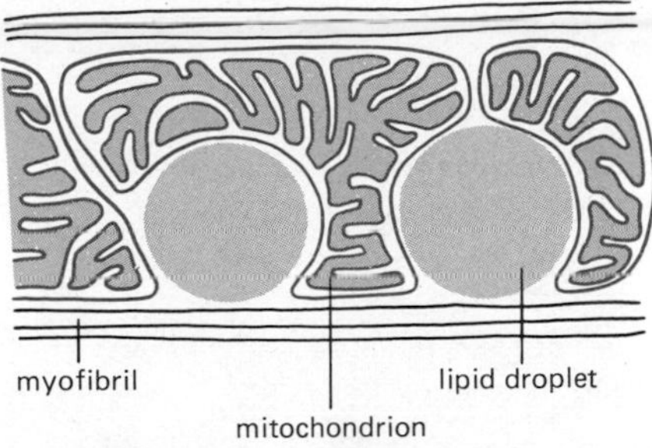

Figure 9–9 Fat droplets in a cardiac muscle cell are surrounded by mitochondria that oxidize the fatty acids derived from triglycerides.

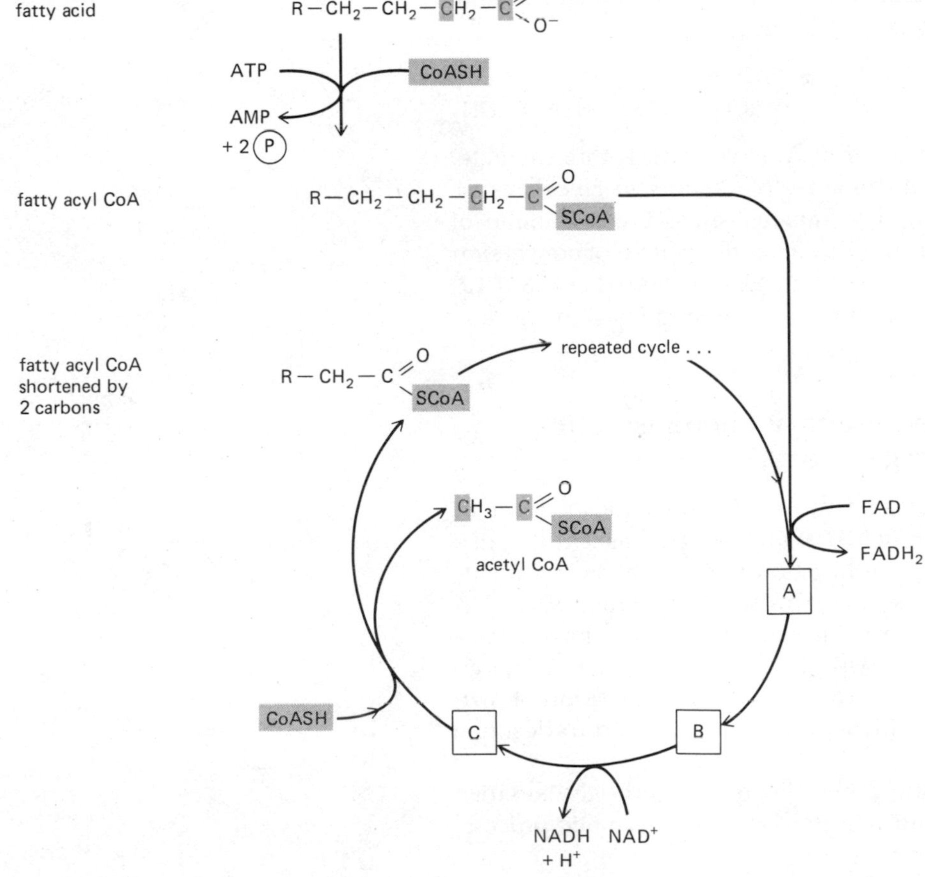

Figure 9–10 The fatty acid oxidation cycle that is catalyzed in the mitochondrial matrix. Each turn of the cycle shortens the fatty acid by two carbons, as indicated, and generates one molecule of acetyl CoA and one molecule each of NADH and $FADH_2$.

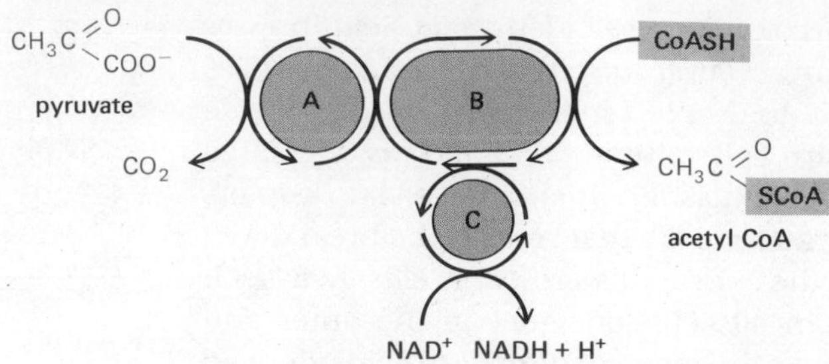

Figure 9–11 Schematic illustration of the coupled reactions that convert pyruvate to acetyl CoA in the mitochondrial matrix. The structure of the large enzyme complex that catalyzes these reactions (pyruvate dehydrogenase) was illustrated previously in Figure 2–40, page 87.

The Citric Acid Cycle Oxidizes the Acetyl Group on Acetyl CoA to Generate NADH and $FADH_2$ for the Respiratory Chain[5]

In the nineteenth century, biologists noticed that in the absence of air, cells produce lactic acid (or ethanol), while in its presence they produce CO_2 and H_2O. Efforts to define the pathways of aerobic metabolism eventually focused on the oxidation of pyruvate and led in 1937 to the discovery of the **citric acid cycle,** also known as the *tricarboxylic acid cycle,* or the *Krebs cycle.* The citric acid cycle is now known to account for about two-thirds of the total oxidation of carbon compounds in most cells. Its major end products are CO_2, NADH, and $FADH_2$. NADH and $FADH_2$ feed their electrons into the respiratory chain, at the end of which these electrons are used to reduce O_2 to H_2O.

The citric acid cycle begins with the condensation of acetyl CoA with the four-carbon compound *oxaloacetate* to produce the six-carbon *citric acid* for which the cycle is named. Then, as a result of seven sequential enzymatically mediated reactions, two carbon atoms are removed as CO_2, and oxaloacetate is regenerated. Each such turn of the cycle produces two CO_2 molecules from two carbon atoms that entered in *previous* cycles (Figure 9–12); the net result insofar as the acetate is concerned is

$$CH_3COOH \text{ (as acetyl CoA)} + 2H_2O + 3NAD^+ + FAD \rightarrow 2CO_2 + 3NADH + FADH_2$$

This reaction also produces one molecule of ATP, via a GTP intermediate, through a phosphorylation reaction of the kind that occurs in glycolysis (p. 70). But its most important contribution to metabolism is the extraction of high-energy electrons from acetyl CoA that occurs in the course of conversion of its two carbon atoms into CO_2. These electrons, held by NADH and $FADH_2$, are then passed to the respiratory chain in the inner mitochondrial membrane (Figure 9–13).

On the Inner Mitochondrial Membrane, a Chemiosmotic Process Converts Oxidation Energy into ATP[6]

Although the citric acid cycle constitutes part of aerobic metabolism, none of the reactions leading to the production of NADH and $FADH_2$ makes direct use of molecular oxygen. That is the exclusive function of the final series of catabolic reactions that takes place on the inner mitochondrial membrane. In these reactions, the electrons that have been removed from oxidized substrates by NAD^+ and FAD are combined with molecular oxygen. The energy made available by these reactions is harnessed to drive the conversion of ADP + P_i to ATP: hence the term **oxidative phosphorylation** is used to describe the entire process (Figure 9–14).

As we have already mentioned, the generation of ATP through oxidative phosphorylation via the respiratory chain depends on a chemiosmotic process.

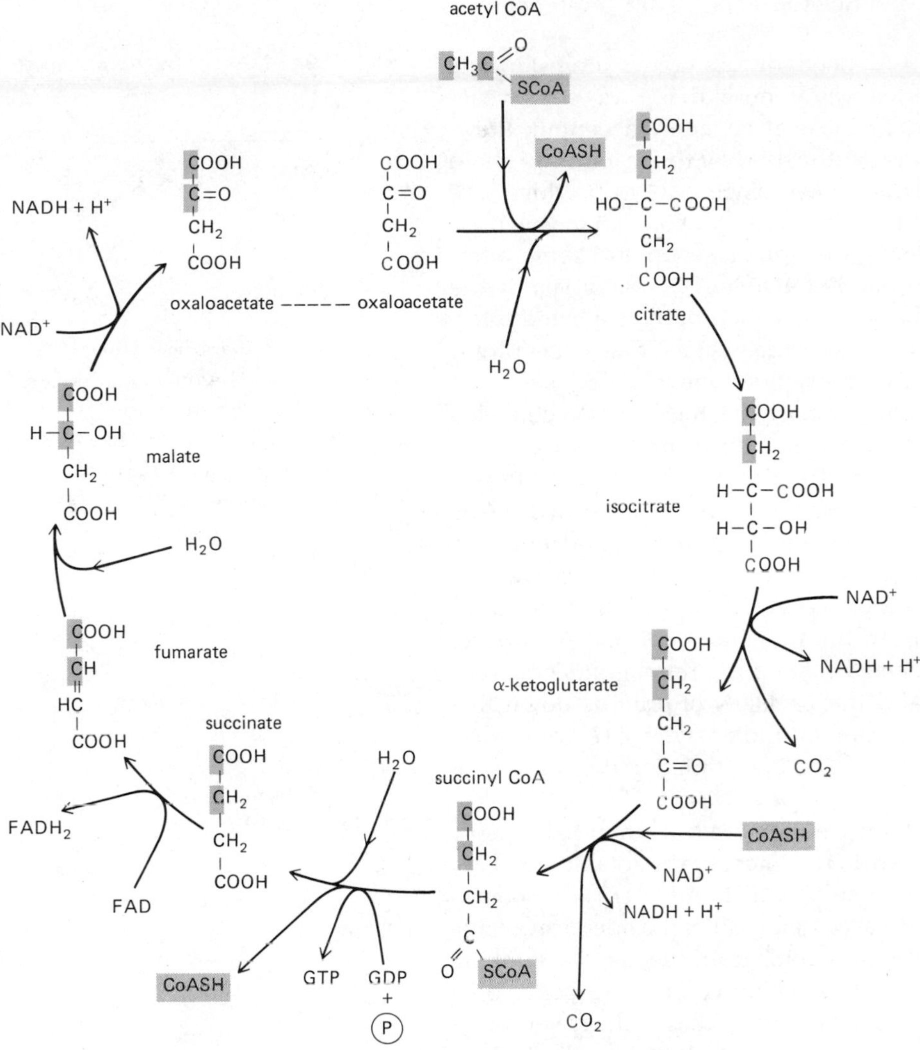

Figure 9–12 The citric acid cycle. The intermediates are shown as their free acids, although they are actually ionized. Each of the indicated steps is catalyzed by a different enzyme located in the mitochondrial matrix. The two carbons that enter the cycle from acetyl CoA will be converted to CO_2 in subsequent turns of the cycle.

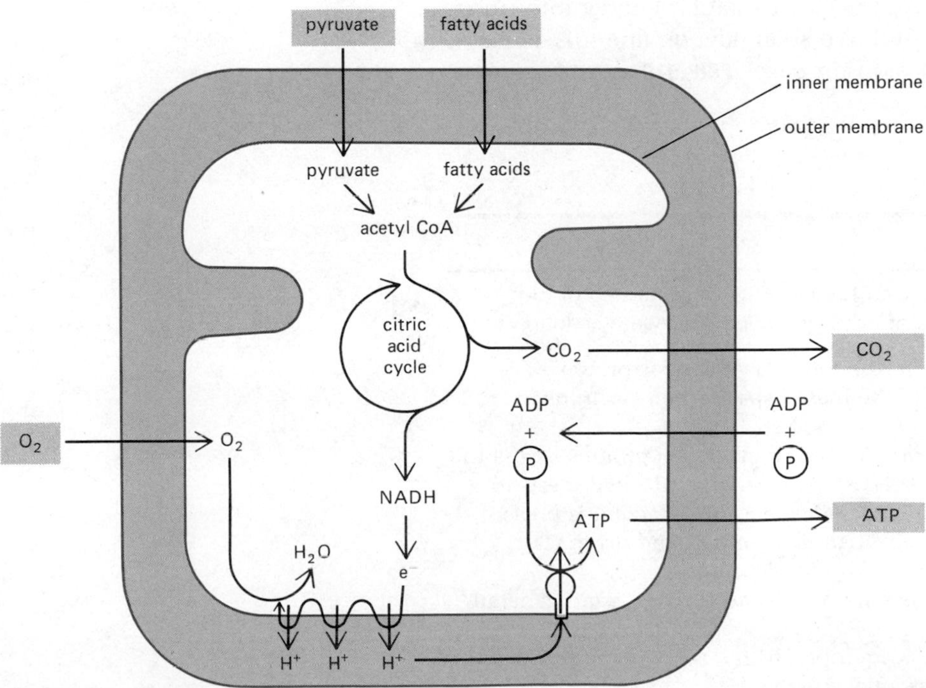

Figure 9–13 The flow of major reactants in and out of the mitochondrion. In addition to the exchanges shown, electrons from the NADH generated by glycolysis in the cytosol are fed to the electron-transport chain. This transfer is accomplished indirectly by means of one of several "shuttle" systems that transport a reduced compound into the matrix, oxidize it to produce NADH (or $FADH_2$), and then transport it out again to be reduced back to the original compound by NADH in the cytosol. The mitochondrial inner membrane is not permeable to NADH directly.

When it was first proposed in 1961, this mechanism solved a long-standing problem in cell biology; however, the idea was so novel that it was some years before enough evidence accumulated to make it generally accepted. Previously, it had been believed that ATP was synthesized by the respiratory chain by the same kind of mechanism already known since 1937 to produce ATP during glycolysis: that is, it was thought that the energy of oxidation was used to produce a high-energy bond between a phosphate group and some intermediate compound and that the conversion of ADP to ATP was driven by the energy released when this bond was broken. Detection of the expected intermediates in the pathway, however, proved extremely difficult; the difficulty is hardly surprising because, as it turns out, there are none.

According to the *chemiosmotic hypothesis,* the high-energy chemical intermediates are replaced by a link between chemical processes ("chemi") and transport processes ("osmosis"—from the Greek *osmos,* push): hence *chemiosmosis* (Table 9–1). As the high-energy electrons from the hydrogens on NADH and $FADH_2$ are transported down the electron-transport chain in the inner mitochondrial membrane (the respiratory chain), the energy released as they pass from one carrier molecule to the next is used to pump protons across the inner membrane from the mitochondrial matrix into the intermembrane space. This creates an *electrochemical proton gradient* across the inner mitochondrial membrane, and the backflow of protons down this gradient is in turn used to drive a membrane-bound enzyme, *ATP synthetase,* that catalyzes the conversion of ADP + P_i to ATP, completing the process of oxidative phosphorylation.

The electrons in NADH and $FADH_2$ molecules are held in a high-energy linkage, and only when they combine with O_2 is most of the total free energy of cellular oxidations released. Thus, the use of NADH and $FADH_2$ as electron carriers in the earlier stages of oxidation saves nearly all of the energy available from burning carbohydrates, fats, and amino acids until the last stage. Then, in one final crescendo of activity, the final oxidation is forced to occur by a pathway that makes use of electron-transport intermediates in the inner membrane that harvest the energy as it is released and use it ultimately to drive the reaction ADP + $P_i \rightarrow$ ATP. In this way, the inner membrane functions as an *energy-conversion machine,* changing NADH oxidation energy into phosphate-bond energy (as ATP) (Figure 9–14). We shall now outline this series of reactions, saving the details for discussion on pages 500–510.

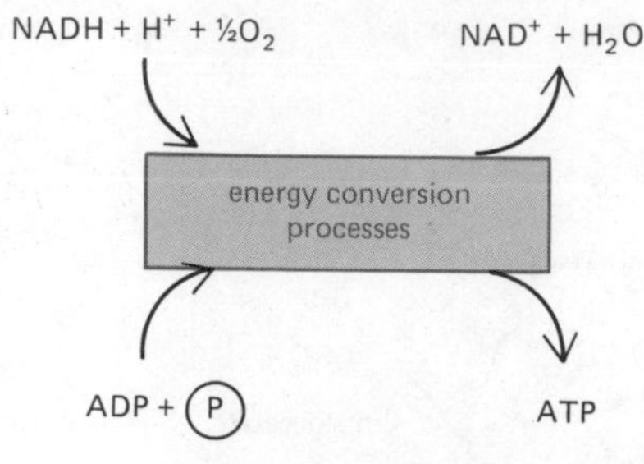

Figure 9–14 The major net energy conversion catalyzed by the mitochondrion. The process in which part of the energy of NADH oxidation becomes phosphate-bond energy in ATP is called oxidative phosphorylation.

Table 9–1 The Chemiosmotic Hypothesis

The chemiosmotic hypothesis, as proposed in the early 1960s, consisted of four independent postulates. In terms of mitochondrial function they were as follows:

1. The mitochondrial respiratory chain in the inner membrane is proton-translocating, and it pumps H^+ out of the matrix space when electrons are transported along the chain.
2. The mitochondrial ATP synthetase complex also translocates protons across the inner membrane: being reversible, it can use the energy of ATP hydrolysis to pump H^+ across the membrane, but if a large enough H^+ gradient is present, protons flow in the reverse direction through the complex and drive ATP synthesis.
3. The inner mitochondrial membrane is impermeable to H^+, OH^-, and generally to anions and cations.
4. The inner mitochondrial membrane is equipped with a set of carrier proteins that mediate the entry and exit of essential metabolites.

The Respiratory Chain Transfers Electrons from NADH to Oxygen[7]

The **respiratory chain** embedded in the mitochondrial inner membrane depends on the same general mechanisms for harvesting energy that underlie other catabolic reactions. The reaction $H_2 + \frac{1}{2}O_2 \rightarrow H_2O$ is made to occur in many small steps, so that most of its energy can be converted into a storage form instead of being released to the environment as heat. As in glycolysis or the citric acid cycle, this involves employing an indirect pathway for the reaction. But there is an important difference: the molecules to be oxidized are hydrogens, rather than carbohydrates, and the indirect pathway involves first separating hydrogen atoms into protons and electrons. The electrons are passed to the respiratory chain, while the protons escape into the aqueous surroundings and are returned only after the electrons reach the end of the respiratory chain, when they neutralize the negative charges created by the final addition of the electrons to the oxygen molecule (Figure 9–15).

We shall trace the oxidation process starting from NADH since most of the reactions begin with the hydrogen on NADH rather than $FADH_2$. Each hydrogen atom (which we shall denote as H·) consists of one electron (e^-) and one proton (H^+). Each molecule of NADH carries not a single hydrogen atom but rather a *hydride ion* (a hydrogen atom plus an extra electron, $H{:}^-$). However, because protons are freely available in aqueous solutions, this is effectively equivalent to carrying the two hydrogen atoms of a hydrogen molecule ($H{:}^- + H^+ \rightarrow H_2$). Correspondingly, in the reactions that produce NADH, two hydrogen atoms are simultaneously removed from a substrate molecule, creating one hydride ion and one proton, as illustrated in Figure 9–16.

Electron transport along the respiratory chain begins when a hydride ion ($H{:}^-$) is removed from NADH to regenerate NAD^+, and the hydride ion is converted into a proton and two electrons ($H{:}^- \rightarrow H^+ + 2e^-$). These two electrons are passed to the first of the many electron carriers embedded in the inner mitochondrial membrane. At this stage the electrons are at very high

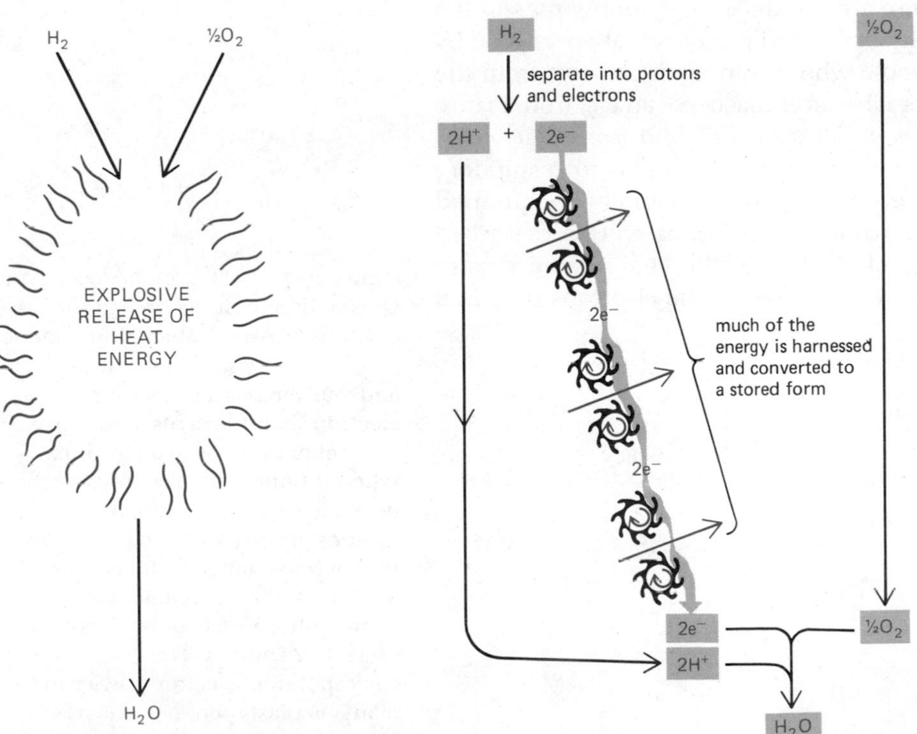

Figure 9–15 Schematic illustration of how most of the energy that would be released as heat if hydrogen were burned (*left panel*) is instead harnessed and stored in a form useful to the cell by means of the electron-transport chain in the mitochondrial inner membrane (*right panel*). The rest of the oxidation energy is released as heat by the mitochondrion.

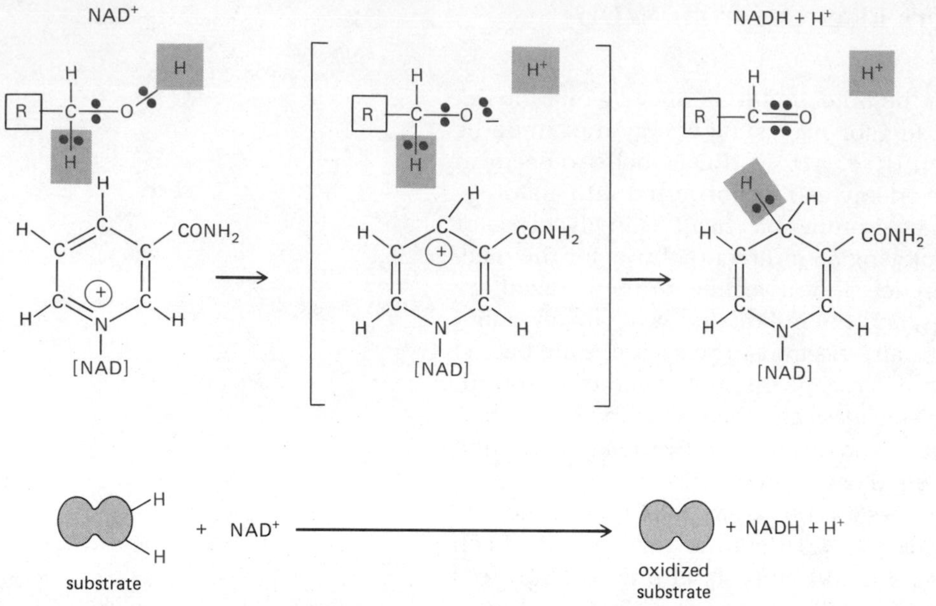

Figure 9–16 The biological oxidation of an alcohol to an aldehyde is thought to proceed in the manner shown. The components of two complete hydrogen atoms are lost from the alcohol: a hydride ion is transferred to NADH, and a proton escapes to the aqueous solution. The illustrated steps in the process occur on a protein surface, being catalyzed by specific chemical groups on the enzyme alcohol dehydrogenase (not shown). (Modified from P. F. Cook, N. J. Oppenheimer, and W. W. Cleland, *Biochemistry* 20:1817–1825, 1981.)

energy. Their energy is gradually reduced as they pass along the sequence of more than 15 different electron carrier molecules in the respiratory chain. For the most part, the electrons pass from one metal atom to another. Each of these metals is tightly bound in a different way to a protein surface, which alters the electron affinity of the bound atom. However, the simplest of the electron carriers is a small hydrophobic molecule known as **ubiquinone,** or *coenzyme Q.* Ubiquinone can pick up or donate either one or two electrons, and it temporarily picks up a proton from the medium along with each electron that it carries (Figure 9–17).

The best understood of the many proteins in the respiratory chain is *cytochrome c,* whose three-dimensional structure has been determined by x-ray crystallography (Figure 9–18). There are five different **cytochromes** in the respiratory chain, constituting a family of colored proteins that are related by the presence of a bound heme molecule whose iron atom changes from the ferric (Fe III) to the ferrous (Fe II) state whenever it accepts an electron (Figure 9–19). In addition to the cytochromes, at least six different *iron-sulfur complexes,* two copper atoms, and two flavins are tightly bound to respiratory chain proteins and carry electrons. Several neighboring carriers are grouped together to form large *respiratory enzyme complexes,* as will be described later (p. 504). Each successive group of carriers in the chain has a greater affinity for electrons than its predecessor. As a result, the electrons pass in a

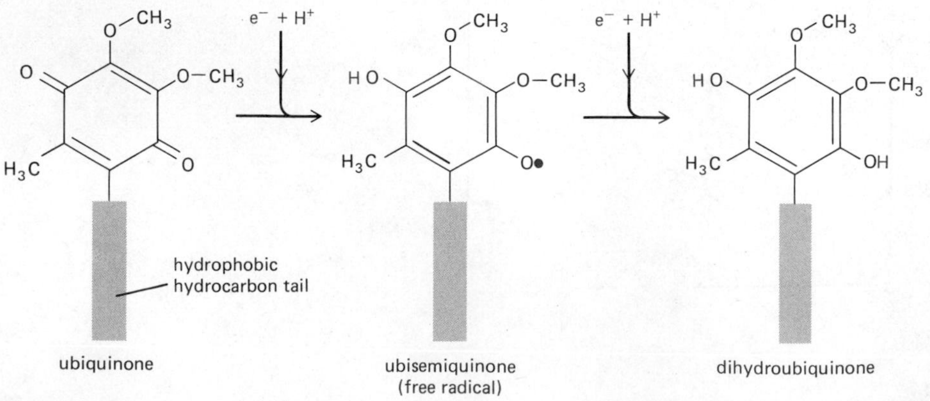

Figure 9–17 Ubiquinone (coenzyme Q) is an important carrier in the electron-transport chain. Ubiquinone picks up one proton from the aqueous environment for every electron that it accepts, and it can carry either one or two electrons. When it donates its electrons to the next carrier in the chain, these protons are released. The hydrophobic tail, which confines ubiquinone to the membrane, commonly consists of ten isoprene units in mammals. The corresponding electron carrier in plants is plastiquinone, which is almost identical to ubiquinone.

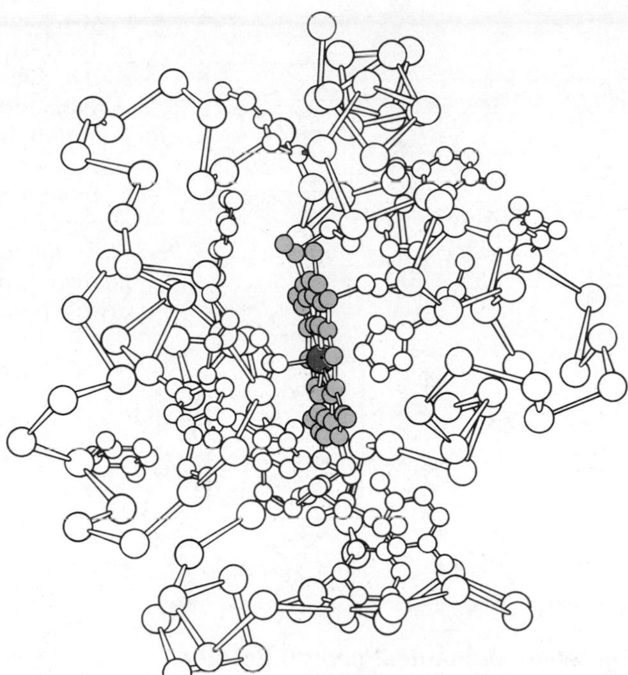

Figure 9–18 The structure of cytochrome c, an electron carrier in the electron-transport chain. This small protein contains just over 100 amino acids and is held rather loosely on the membrane (see Figure 9–31). The iron atom (*dark color*) on the bound heme (*light color*) can carry a single electron.

cascade from NADH to lower and lower energy levels, moving from one enzyme complex to another until they are finally transferred to oxygen, which has the greatest affinity of all for electrons.

The cytochromes, iron-sulfur complexes, and copper atoms can carry only one electron at a time. Yet each NADH donates two electrons, and each O_2 molecule must receive four electrons to produce water. There are several electron-collecting and electron-dispersing points along the electron-transport chain, at which these changes in electron number are accommodated.

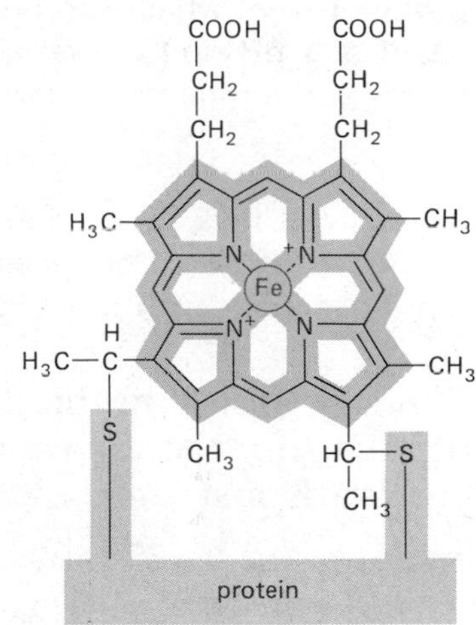

Figure 9–19 The structure of the heme group covalently attached to cytochrome c. Because hemes in different cytochromes have a slightly different structure and are held by the protein in different ways, each of the cytochromes has a different affinity for an electron.

Energy Released by the Passage of Electrons Along the Respiratory Chain Is Stored as an Electrochemical Proton Gradient Across the Inner Membrane[8]

As previously discussed, much of the energy released during electron transport is harnessed by components of the respiratory chain to move protons (H^+) from the matrix space to the intermembrane space and thus to the outside of the mitochondrion. This movement of protons has two major consequences:

1. It generates a pH gradient across the inner mitochondrial membrane, with the H^+ concentration in the matrix significantly lower than in the rest of the cell, where the pH is generally close to 7 (remember that small molecules equilibrate freely across the outer membrane of the mitochondrion so that the pH in the intermembrane space will be equivalent to that of the cytosol).
2. It generates a voltage gradient (membrane potential) across the inner mitochondrial membrane, with the inside negative and the outside positive (as a result of the net outflow of positive ions).

The pH gradient (ΔpH) drives H^+ back into the matrix and pushes OH^- out of the matrix. The membrane potential ($\Delta\psi$) acts to attract any positive ion into the matrix and push any negative ion out—thereby reinforcing the effect of the pH gradient on the movement of OH^- and H^+ ions. Together,

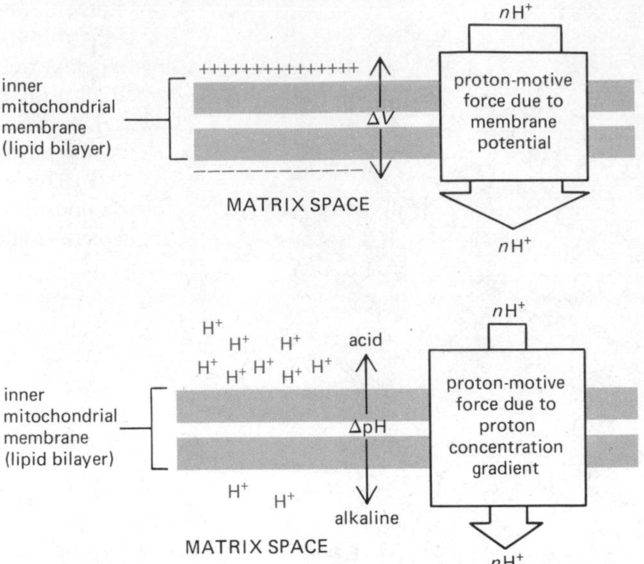

Figure 9–20 The two components of the electrochemical proton gradient across the inner mitochondrial membrane. The total proton-motive force consists of two components: a large force due to the membrane potential (designated either as ΔV or $\Delta\psi$) and a smaller force due to the proton concentration gradient (ΔpH). Both forces act to drive a proton into the matrix space.

these two forces are said to constitute an **electrochemical proton gradient** (Figure 9–20).

The electrochemical gradient exerts what is commonly known as a **proton-motive force,** which can be measured in units of millivolts (mV). Since each ΔpH of 1 pH unit has an effect equivalent to a membrane potential of about 60 mV, the total proton motive force is given by

$$\text{proton-motive force} = \Delta\psi - 60(\Delta\text{pH})$$

In a typical cell the proton-motive force in respiring mitochondria is about 220 mV and is made up of a membrane potential of about 160 mV and a pH gradient of about −1 pH unit.

The Energy Stored in the Electrochemical Proton Gradient Across the Inner Membrane Is Harnessed to Produce ATP and to Transport Metabolites into the Matrix Space[9,10]

The mitochondrial inner membrane contains an unusually high proportion of protein, being approximately 70% protein and 30% lipid by weight. Many of the proteins belong to the electron transport chain, which establishes the electrochemical proton gradient across the membrane. Another major component is the enzyme that catalyzes the synthesis of ATP. This enzyme, the **ATP synthetase,** is a large protein complex through which protons flow back down the electrochemical gradient into the matrix. Like a turbine, this protein complex converts one form of energy to another, synthesizing ATP from ADP and P_i in the mitochondrial matrix in a reaction that is coupled to the inward flow of protons (Figure 9–21).

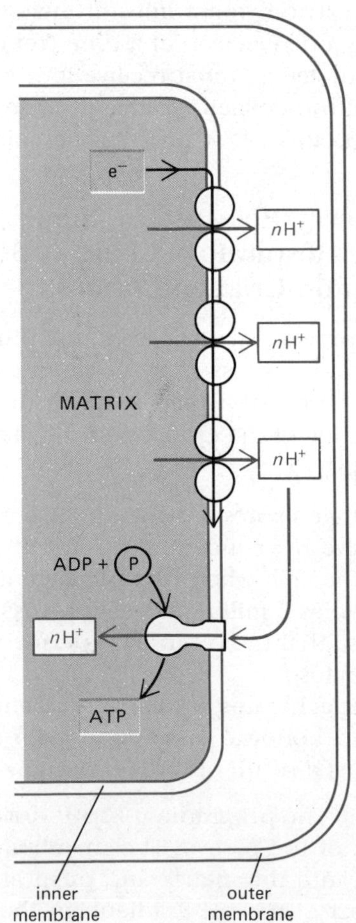

Figure 9–21 The general mechanism of oxidative phosphorylation. As a high-energy electron is passed along the electron-transport chain to lower energies, protons are pumped out of the matrix space at each of three energy-conserving sites. These protons create an electrochemical proton gradient across the inner membrane that drives protons back through the ATP synthetase, a transmembrane protein complex that uses the energy of the proton flow to synthesize ATP from ADP and P_i in the matrix.

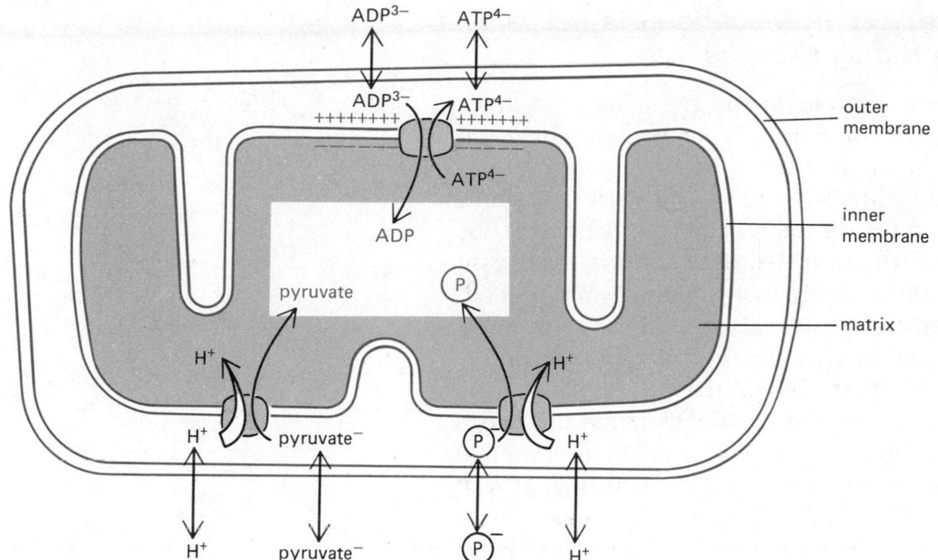

Figure 9–22 Some of the active transport processes that are driven by the electrochemical proton gradient across the mitochondrial inner membrane. The outer membrane is freely permeable to all of these compounds. For a discussion of symport and antiport transport mechanisms, see Chapter 6.

But ATP synthesis is not the only process driven by the electrochemical gradient across the inner mitochondrial membrane. The enzymes in the mitochondrial matrix, where the citric acid cycle and other metabolic reactions take place, must be supplied with high concentrations of substrates, and ATP synthetase must be supplied with ADP and phosphate. This means that many different charged substrates must pass through the generally impermeable inner membrane of the mitochondrion. The exchange with the cytosol is mediated by a variety of important transport proteins embedded in the inner membrane. In many cases, these proteins actively transport specific molecules against their electrochemical gradients, a process requiring an input of energy. For most metabolites this energy comes from *co-transporting* another molecule *down* its electrochemical gradient (see p. 287). For example, ADP and phosphate are both transported into the matrix space. ADP is carried by an ADP-ATP antiport system: for each ADP molecule moved in, an ATP molecule moves out down its electrochemical gradient. At the same time, a symport system couples the inward movement of phosphate to the inward flow of H^+; the H^+ ions enter the matrix down their electrochemical gradient, "dragging" the phosphate with them. Pyruvate, an important mitochondrial fuel, is also transported into the matrix in this way (Figure 9–22).

A further vital function of the electrochemical gradient across the inner mitochondrial membrane is to pump Ca^{2+} out of the cytosol. It is essential that the concentration of free Ca^{2+} in the cytosol be kept very low ($<10^{-7}$ M) because small changes in cytosolic Ca^{2+} concentration are signals that regulate various cellular processes (see p. 739). The inner membrane contains a transport protein that efficiently moves Ca^{2+} into the matrix, using the energy stored in the voltage gradient across the inner membrane (Figure 9–23).

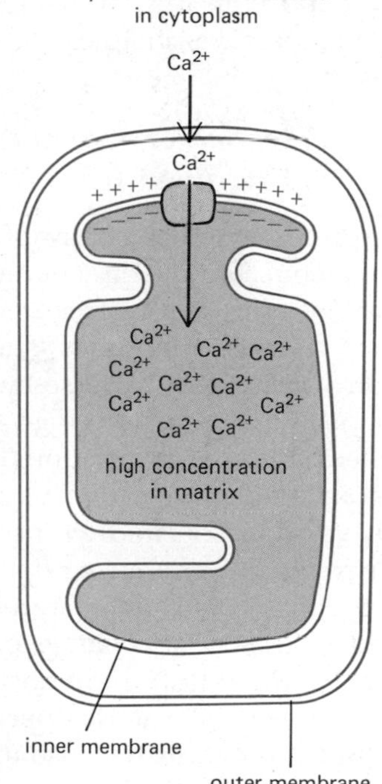

Figure 9–23 Calcium is actively moved into mitochondria by a transport process that is driven by the membrane potential. Because most of the calcium inside the mitochondrion is thought to be precipitated as calcium phosphate, it seems likely that only a relatively small amount of work is necessary to accumulate Ca^{2+}.

The more energy from the electrochemical gradient that is used to transport molecules and ions into the mitochondrion, the less is available to drive the ATP synthetase. If substantial concentrations of Ca^{2+} are added to respiring mitochondria, for example, the mitochondria cease ATP production completely, all the energy in their electrochemical gradient being diverted to pumping Ca^{2+} instead. Clearly, there must be mechanisms that channel the energy stored in the electrochemical gradient toward those processes for which it is most needed at any given time.

The Rapid Conversion of ADP to ATP in Mitochondria Keeps the ATP Pool in the Cytoplasm Highly Charged

We have seen that a special transport protein in the mitochondrial inner membrane functions as an antiport system, using the energy stored in the electrochemical gradient across the inner membrane to move ADP into the matrix space in exchange for ATP (see Figure 9–22). As a result, ADP molecules produced by ATP hydrolysis in the cytosol rapidly enter mitochondria for recharging, while the ATP molecules formed in the mitochondrial matrix by oxidative phosphorylation rapidly escape to the cytosol where they are needed. A typical ATP molecule in the human body shuttles in and out of a mitochondrion for recharging thousands of times per day, keeping the concentration of ATP in a cell 5 to 10 times higher than that of ADP.

We have also seen how biosynthetic enzymes in cells guide their substrates along specific reaction paths, often driving energetically unfavorable reactions by coupling them to the energetically favorable hydrolysis of ATP (see Figure 2–29, p. 78). The highly charged ATP pool is thereby used to drive cellular processes in much the same way that a battery can be used to drive electric engines: if the activity of the mitochondria is halted, the quantity of ATP decreases and the cell's battery runs down. Eventually, the ATP level falls to the point where energetically unfavorable reactions can no longer be driven by ATP hydrolysis.

It might at first sight seem that this state of affairs would not be reached until the concentration of ATP was zero. In fact, it is reached much sooner than that, at a finite concentration of ATP that depends on the concentrations of ADP and P_i in the cell. To explain why, we must turn to some elementary thermodynamic principles.

The Difference Between $\Delta G°$ and ΔG: A Large Negative Value of ΔG is Required for ATP Hydrolysis to Be Useful to the Cell[11]

According to the Second Law of Thermodynamics, chemical reactions proceed spontaneously in the direction that corresponds to an increase in the disorder of the *universe*. In Chapter 2 we noted that reactions that release energy as heat to their surroundings (such as the hydrolysis of ATP) tend to increase the disorder of the universe by increasing the violence of random molecular motions in their surroundings. Reactions can also affect the degree of disorder by altering the concentrations of reacting chemicals. The net change of disorder in the universe due to a reaction is reflected in the **change of free energy, ΔG,** associated with the reaction: reactions that bring about a large *decrease* in free energy (so that ΔG is very negative) create the most disorder in the universe and proceed most readily.

The free-energy change for hydrolyzing ATP to ADP plus inorganic phosphate (P_i) under the conditions that normally obtain inside the cell is between −11 and −13 kilocalories per mole, making the hydrolysis extremely favorable. But this is only because the concentration of ATP in the cell is kept very high compared with the concentration of ADP and P_i. In so-called "standard conditions," when ATP, ADP, and P_i are all present at the same concentration of 1 mole per liter, the ΔG for ATP hydrolysis—called the **standard free-energy change** or **$\Delta G°$** of the reaction—is only −7.3 kilocalories per mole. At still lower concentrations of ATP relative to ADP and P_i, ΔG will become equal to zero. At this point, the rate at which ADP and P_i will join to form ATP will just equal the rate at which ATP will hydrolyze to form ADP and P_i. In other words, when $\Delta G = 0$, the reaction is at *equilibrium* (Figure 9–24).

While the value of $\Delta G°$ is a constant that depends only on the nature of the reactants, the value of ΔG is a variable that depends on the concentration

1. ATP $\xrightarrow{\text{hydrolysis}}$ ADP + Ⓟ

$$\text{hydrolysis rate} = \left(\begin{matrix}\text{hydrolysis}\\ \text{rate constant}\end{matrix}\right) \times \left(\begin{matrix}\text{concentration}\\ \text{of ATP}\end{matrix}\right)$$

2. ADP + Ⓟ $\xrightarrow{\text{synthesis}}$ ATP

$$\text{synthesis rate} = \left(\begin{matrix}\text{synthesis}\\ \text{rate constant}\end{matrix}\right) \times \left(\begin{matrix}\text{conc. of}\\ \text{phosphate}\end{matrix}\right) \times \left(\begin{matrix}\text{conc. of}\\ \text{ADP}\end{matrix}\right)$$

3. AT EQUILIBRIUM: synthesis rate = hydrolysis rate

$$\therefore \left(\begin{matrix}\text{synthesis}\\ \text{rate constant}\end{matrix}\right) \times \left(\begin{matrix}\text{conc. of}\\ \text{phosphate}\end{matrix}\right) \times \left(\begin{matrix}\text{conc. of}\\ \text{ADP}\end{matrix}\right) = \left(\begin{matrix}\text{hydrolysis}\\ \text{rate constant}\end{matrix}\right) \times \left(\begin{matrix}\text{conc. of}\\ \text{ATP}\end{matrix}\right)$$

$$\frac{(\text{conc. of ATP})}{\left(\begin{matrix}\text{conc. of}\\ \text{ADP}\end{matrix}\right) \times \left(\begin{matrix}\text{conc. of}\\ \text{phosphate}\end{matrix}\right)} = \frac{\left(\begin{matrix}\text{synthesis}\\ \text{rate constant}\end{matrix}\right)}{\left(\begin{matrix}\text{hydrolysis}\\ \text{rate constant}\end{matrix}\right)} = \text{equilibrium constant } K$$

$$\frac{[\text{ATP}]}{[\text{ADP}][\text{Ⓟ}]} = K$$

4\.

For the reaction

ATP ⟶ ADP + Ⓟ

the following equation applies:

$$\Delta G = \Delta G^\circ - RT \ln \frac{[\text{ATP}]}{[\text{ADP}][\text{Ⓟ}]}$$

Where ΔG and ΔG° are in kilocalories per mole, R is the gas constant (2×10^{-3} kcal mole^{-1} $^\circ\text{K}^{-1}$), T is the absolute temperature (°K), and all the concentrations are in moles per liter.

When the concentrations of all reactants are at 1 M, $\Delta G = \Delta G^\circ$ (since $RT \ln 1 = 0$). ΔG° is thus a constant defined as the standard free-energy change for the reaction.

At equilibrium the reaction has no net effect on the disorder of the universe, so that $\Delta G = 0$. Therefore, at equilibrium

$$RT \ln \frac{[\text{ATP}]}{[\text{ADP}][\text{Ⓟ}]} = \Delta G^\circ$$

But the concentrations of reactants at equilibrium must satisfy the equilibrium equation:

$$\frac{[\text{ATP}]}{[\text{ADP}][\text{Ⓟ}]} = K$$

Therefore, at equilibrium $\Delta G^\circ = RT \ln K$

We therefore see that whereas ΔG° indicates the equilibrium point for a reaction, ΔG reveals how far away the reaction is from equilibrium.

Figure 9–24 The basic relationship between free-energy changes and equilibrium, as illustrated by the ATP hydrolysis reaction. The equilibrium constant shown here, K, is in units of liters per mole.

of the reactants. In fact, the value of ΔG indicates how far a reaction is from equilibrium. Therefore, it is ΔG, not ΔG°, that reveals whether a reaction can be used to drive other reactions. The high concentration of ATP in cells compared to that of ADP and P_i, maintained by the efficient conversion of ADP to ATP in mitochondria, causes ΔG to have a strongly negative value by keeping the ATP hydrolysis reaction in cells very far out of equilibrium. Without this disequilibrium, ATP hydrolysis could no longer be used to direct the reactions of the cell, and many biosynthetic reactions would run backward as well as forward.

Cellular Respiration Is Remarkably Efficient

By means of oxidative phosphorylation, each pair of electrons in NADH provides energy for the formation of about three molecules of ATP during its transfer to oxygen (the pair of electrons in $FADH_2$, being at a somewhat lower energy, generates only about 2 ATP molecules). In all, about 12 molecules of ATP are formed from each molecule of acetyl CoA that enters the cycle, which means that 24 ATP molecules are produced from 1 molecule of glucose and 96 ATP molecules from 1 molecule of palmitate, a 16-carbon fatty acid. If one includes the energy-yielding reactions that occur before acetyl CoA is formed, the complete oxidation of 1 molecule of glucose gives a net yield of about 36 ATPs, while about 129 ATPs are obtained from the complete oxidation of 1 molecule of palmitate. These numbers are approximate maximal values, since the actual amount of ATP made in the mitochondrion depends on what fraction of the electrochemical gradient energy is used for purposes other than ATP synthesis.

When the free-energy changes for burning fats and carbohydrates directly to CO_2 and H_2O are compared with the total amount of energy stored in the phosphate bonds of ATP during the corresponding biological oxidations, it turns out that the efficiency with which oxidation energy is converted into ATP bond energy is often greater than 50%. This is considerably better than the efficiency of most man-made energy conversion devices. If cells were to work at the efficiency of an electric motor or a gasoline engine (10 to 20%), an organism would have to eat far more voraciously to maintain itself. Moreover, since all wasted energy is liberated as heat, large organisms would need to develop more efficient mechanisms of giving up heat to the environment.

In learning about cellular respiration, students may wonder why chemical interconversions in cells often involve such complex pathways. The citric acid cycle and many of the steps in the respiratory chain could certainly have been bypassed, allowing the final oxidation of sugars to CO_2 plus H_2O to be accomplished more directly. While this would have made respiration much easier to learn, it would have been a disaster for the cell. Only through oxidation pathways involving many intermediates, with each compound differing from the preceding one only slightly in energy, can the huge amounts of free energy released by oxidations be parceled out into small packets, so that it can be converted efficiently to useful forms by means of coupled reactions (see Figure 2–17, p. 66).

Summary

The mitochondrion carries out most cellular oxidations and produces the bulk of the cell's ATP. The mitochondrial matrix space contains a large number of different enzymes, including those that oxidize pyruvate and fatty acids to acetyl CoA, and the citric acid cycle enzymes that use this acetyl CoA to produce large amounts of NADH (and $FADH_2$). The energy available from combining oxygen with the reactive electrons carried by these latter compounds is then harnessed by an electron-transport chain embedded in the mitochondrial inner membrane, which pumps protons out of the matrix to create a transmembrane electrochemical proton gradient. This gradient—including contributions from both a membrane potential and a pH difference—is in turn used by a protein complex in the inner membrane to synthesize ATP. The gradient also directly drives the active transport of selected metabolites into and out of the matrix, powering an ATP-ADP exchange that keeps the cell's ATP pool highly charged.

The Respiratory Chain[7,9]

Having considered in general terms how mitochondria function, let us now look in more detail at the respiratory chain—the electron-transport chain that is so crucial to all oxidative metabolism. Because the components of the chain are also intrinsic components of the inner mitochondrial membrane, their study has made major contributions to membrane biology, providing the clearest examples of the complex interactions that occur between the individual protein components of biological membranes.

Functional Inside-Out Particles Can Be Isolated from Mitochondria[12]

In order to study the respiratory chain, which is relatively inaccessible to experimental manipulation in intact mitochondria, it has been necessary to break up mitochondria. By disrupting mitochondria with ultrasound, it is possible to isolate *submitochondrial particles,* which consist of broken cristae that have resealed into small closed vesicles about 100 nm in diameter (Figure

9–25). Electron micrographs of negatively stained submitochondrial particles reveal that their outside surface is studded with spheres about 9 nm in diameter attached to the membrane by stalks (Figure 9–26). In intact mitochondria, these "lollipops"—ATP synthetase—are seen to coat the *inner* (matrix-facing) side of the inner membrane. Thus, the submitochondrial particles are inside-out vesicles of inner membrane, with what was previously their matrix-facing surface exposed to the surrounding medium. As a result, they are freely accessible to metabolites that would normally be provided in the matrix space. When NADH, ADP, and inorganic phosphate are added, such particles transport electrons from NADH to O_2, and couple this oxidation to ATP synthesis.

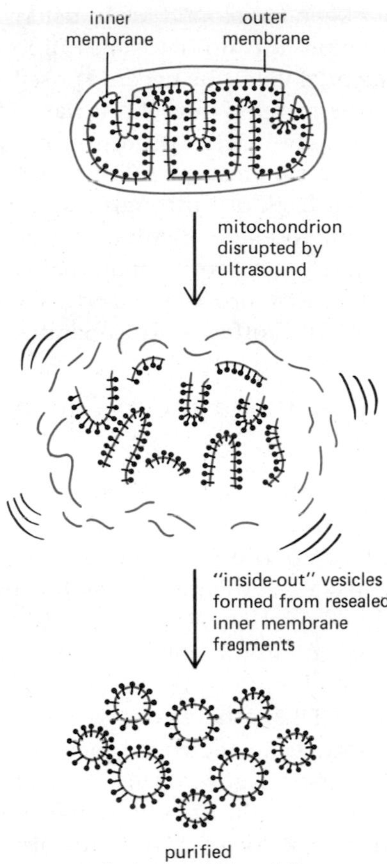

Figure 9–25 Outline of the procedure used to prepare submitochondrial particles from purified mitochondria. The particles are pieces of broken-off cristae that form closed vesicles.

ATP Synthetase Can Be Purified and Added Back to Membranes in an Active Form[12,13]

The first experiments showing that the different membrane proteins that catalyze oxidative phosphorylation could be separated from each other without destroying their activity were performed in 1960. The 9-nm protein spheres studding the surface of submitochondrial particles were stripped from the particles and purified in soluble form. The stripped particles could still oxidize NADH in the presence of oxygen, but they could no longer synthesize ATP. On the other hand, the purified spheres on their own acted as ATPases, hydrolyzing ATP to ADP and P_i. When purified spheres (referred to as *F_1ATPases*) were added back to stripped submitochondrial particles, their ATPase activity was reversed; that is, the reconstituted submitochondrial particles once again made ATP from ADP and P_i.

Subsequent work showed that the F_1ATPase is part of a larger transmembrane complex (about 500,000 daltons) containing nine different polypeptide chains, which is now known as ATP synthetase (also called *F_0F_1ATPase*). This complex comprises about 15% of the total inner membrane protein. Very similar ATP synthetases are present in both chloroplast and bacterial membranes. This ubiquitous protein complex contains a transmembrane proton channel, and it synthesizes ATP when protons flow through the channel down their electrochemical gradient.

One of the most convincing demonstrations of the function of ATP synthetase came from an experiment performed in 1974. By that time, methods

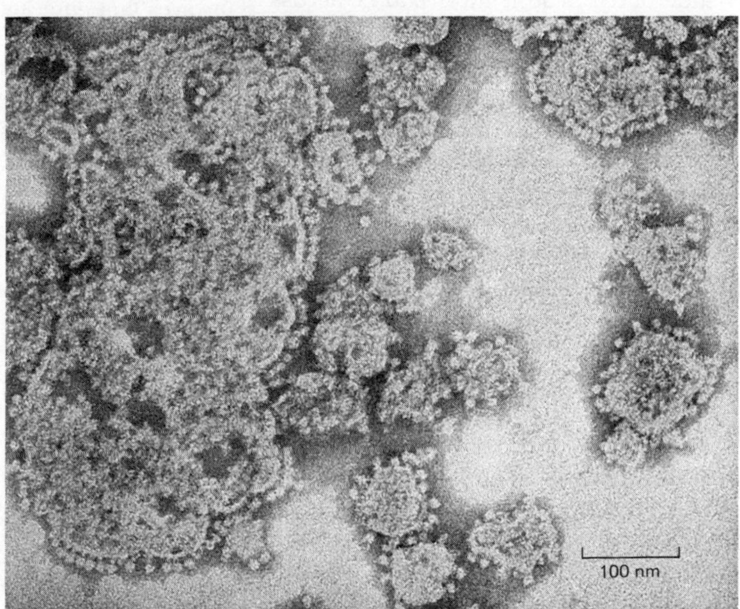

Figure 9–26 Electron micrograph of submitochondrial particles. (Courtesy of Ephraim Racker.)

had been developed for transferring detergent-solubilized integral membrane proteins into lipid vesicles (liposomes) formed from purified phospholipids (see p. 293). It thus became possible to form a hybrid membrane that contained both a purified mitochondrial ATP synthetase and bacteriorhodopsin—a bacterial light-driven, proton pump (p. 227). When these vesicles were exposed to light, the protons pumped into the vesicle lumen by the bacteriorhodopsin flowed back out through the ATP synthetase, causing ATP to be made in the solution outside (Figure 9–27). Because a direct interaction between a bacterial proton pump and a mammalian ATP synthetase seems highly unlikely, this experiment provides strong support for the view that proton translocation and ATP synthesis are separate events in oxidative phosphorylation.

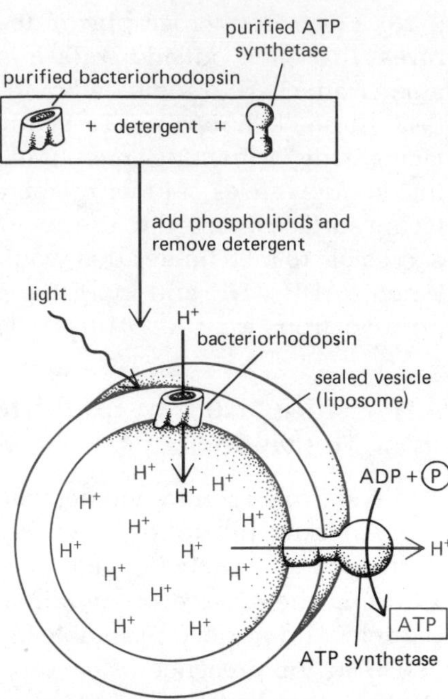

Figure 9–27 Outline of an important experiment demonstrating that the ATP synthetase can be driven by a simple proton flow. By combining a light-driven bacterial proton pump (bacteriorhodopsin), an ATP synthetase purified from ox heart mitochondria, and phospholipids, vesicles were produced that synthesized ATP in response to light. There is a controversy among experts as to whether the proton pathway shown is often short-circuited in mitochondria—with the protons that are pumped entering ATP synthetase by flowing along the membrane rather than by a pathway that involves first entering the bulk fluid phase. However, this experiment makes it unlikely that a direct transfer of protons from a respiratory enzyme complex to ATP synthetase is involved.

ATP Synthetase Can Function in Reverse to Hydrolyze ATP and Pump H^+ [12,14]

ATP synthetase is a reversible enzyme complex; it can either use the energy of ATP hydrolysis to pump protons across the inner mitochondrial membrane or it can harness the flow of protons down an electrochemical proton gradient to make ATP (Figure 9–28). It thereby acts as a *reversible coupling device* for interconverting electrochemical-proton-gradient and chemical-bond energies. Its direction of action depends on the balance between the steepness of the electrochemical proton gradient and the local ΔG for ATP hydrolysis.

ATP synthetase derives its name from the fact that it is normally driven by the large electrochemical proton gradient maintained by the respiratory chain (see Figure 9–20) to make most of the cell's ATP. Although the exact number is still disputed, one molecule of ATP is probably made by the ATP synthetase for every two to three protons driven through it. To facilitate the calculations to be described below, we shall assume that this number is three.

Whether the ATP synthetase works in its ATP-synthesizing or its ATP-hydrolyzing direction at any instant depends on the exact balance between the favorable free-energy change for moving the three protons across the membrane into the matrix space (ΔG_{3H^+}, which is less than zero) and the unfavorable free-energy change for ATP *synthesis* in the matrix ($\Delta G_{\text{ATP synthesis}}$, which is greater than zero). As previously discussed, the value of $\Delta G_{\text{ATP synthesis}}$ will depend on the exact concentrations of the three reactants ATP, ADP, and P_i in the mitochondrial matrix space (Figure 9–24). The value of ΔG_{3H^+}, on the other hand, will be proportional to the value of the proton-motive force across the inner mitochondrial membrane. The following example will help to explain how the balance between these two free-energy changes affects ATP synthetase.

As explained in the legend to Figure 9–28, a single proton moving into the matrix down an electrochemical gradient of 220 millivolts (mV) liberates 5.06 kilocalories per mole (kcal/mole) of free energy, while the movement of three such particles liberates three times this much free energy ($\Delta G_{3H^+} = -15.2$ kcal/mole). Thus, if the proton-motive force remains constant at 220 mV, the ATP synthetase will synthesize ATP until a ratio of ATP to ADP and P_i is reached where $\Delta G_{\text{ATP synthesis}}$ is just equal to $+15.2$ kcal/mole (at this point $\Delta G_{\text{ATP synthesis}} + \Delta G_{3H^+} = 0$). Now there will be no further net ATP synthesis or hydrolysis by the ATP synthetase.

Suppose that a great deal of ATP is suddenly hydrolyzed by energy-requiring reactions in the cell cytosol—causing the ATP:ADP ratio in the matrix to fall. Now the value of $\Delta G_{\text{ATP synthesis}}$ will decrease (see Figure 9–24), and ATP synthetase will begin to synthesize ATP again to restore the original ATP:ADP ratio. Alternatively, if the proton-motive force drops suddenly and is then maintained at a constant 200 mV, ΔG_{3H^+} will change to -13.8 kcal/mole. As a result, ATP synthetase will start hydrolyzing some of the ATP in the ma-

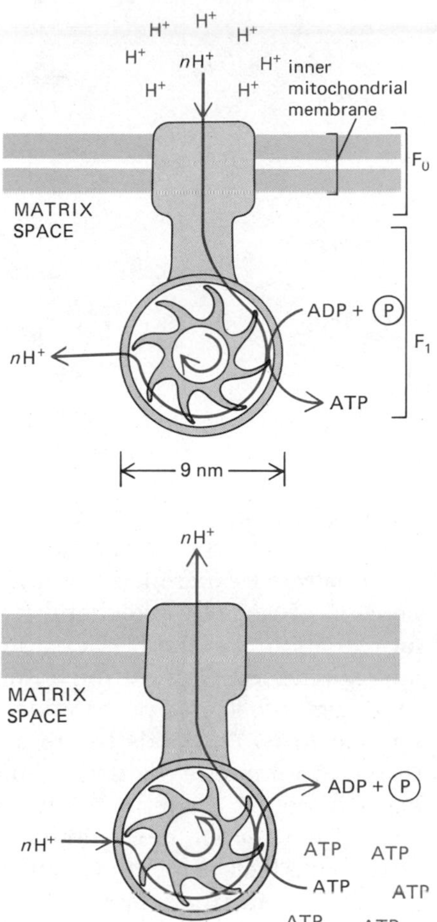

Figure 9–28 ATP synthetase is a large complex, with a molecular weight of about half a million, composed of at least nine different polypeptide chains. The entire complex is also known as the F_0F_1ATPase. Five of its polypeptide chains make up the spherical head of the complex, which can be isolated and is known as the F_1ATPase. As indicated, ATP synthetase serves as a reversible coupling device that interconverts electrochemical-proton-gradient and chemical-bond energies. As explained in the text, its direction of operation at any given instant depends on the net free-energy change for the coupled processes of proton translocation across the membrane and the synthesis of ATP from ADP and P_i.

We have previously shown how the free-energy change (ΔG) for ATP hydrolysis depends on the concentrations of the three reactants ATP, ADP, and P_i (Figure 9–24); the ΔG for ATP synthesis is the negative of this value. The ΔG for proton translocation across the membrane is the sum of (1) the ΔG for moving a mole of any ion through a difference in membrane potential $\Delta\psi$ and (2) the ΔG for moving a mole of a molecule between any two compartments in which its concentration differs. The equation for the proton-motive force on p. 496 combines these same two components, replacing the concentration difference by an equivalent increment in the membrane potential to produce an "electrochemical potential" for the proton. Thus, the ΔG for proton translocation and the proton-motive force measure the same potential, one in kilocalories and the other in millivolts. The conversion factor between them is the Faraday. Thus,

$$\Delta G_{H^+} = -0.023 \text{ (proton-motive force)}$$

where ΔG_{H^+} is in kilocalories per mole (kcal/mole) and the proton-motive force is in millivolts (mV). For an electrochemical proton gradient of 220 mV, $\Delta G_{H^+} = 5.06$ kcal/mole.

trix—until a new balance of ATP to ADP and P_i is reached (where $\Delta G_{\text{ATP synthesis}} = +13.8$ kcal/mole)—and so on.

In many bacteria, ATP synthetase is routinely reversed in a transition between aerobic and anaerobic metabolism, as we shall see later. The reversibility of the ATP synthetase is a property shared by other membrane proteins that couple ion movement to ATP synthesis or hydrolysis. For example, both the Na^+-K^+ pump and the Ca^{2+} pump, described in Chapter 6, hydrolyze ATP and use the energy released to pump specific ions across a membrane (p. 291). If either of these pumps is exposed to an abnormally steep gradient of the ions that it transports, it will act in reverse—synthesizing ATP from ADP and P_i instead of hydrolyzing it. Thus, like ATP synthetase, such pumps are able to convert the electrochemical energy stored in a transmembrane ion gradient directly into phosphate bond energy in ATP.

The Respiratory Chain Can Be Shown to Pump H^+ Across the Inner Mitochondrial Membrane

Whereas the ATP synthetase does not normally transport H^+ out of the matrix space across the inner mitochondrial membrane, the respiratory chain embedded in this membrane normally does, thereby generating the electro-

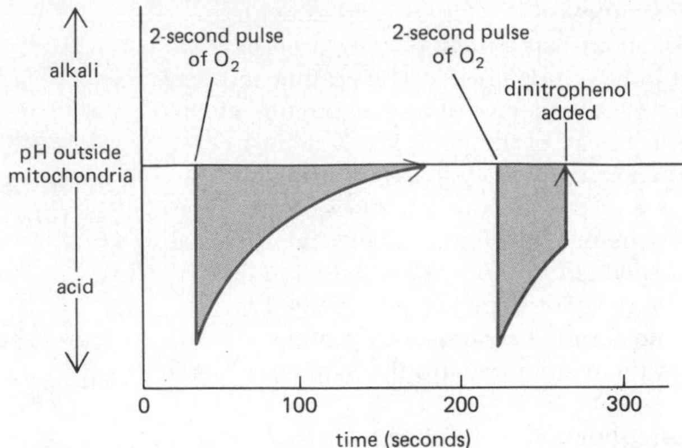

Figure 9–29 Experimental demonstration of respiratory-driven H^+-translocation in mitochondria. A small sealed chamber contains an anaerobic suspension of purified mitochondria, a source of electrons for respiration, and an ionophore that abolishes the membrane potential (valinomycin plus K^+; see p. 301). When a pulse of oxygen is injected through a syringe needle, a change in pH results, as shown. The subsequent addition of dinitrophenol abolishes the pH change by allowing the rapid flow of protons back through the mitochondrial inner membrane, as will be described later.

chemical proton gradient that drives ATP synthesis. The ability of the respiratory chain to translocate protons outward from the matrix space can be demonstrated experimentally under special conditions. For example, a suspension of isolated mitochondria can be provided with a suitable substrate for oxidation, and the H^+ flow through ATP synthetase can be blocked by an appropriate inhibitor. In the absence of air, the injection of a small amount of oxygen into such a preparation causes a brief burst of respiration that lasts for one to two seconds before all of the oxygen is consumed. During this respiratory burst, a sudden acidification of the medium resulting from the extrusion of H^+ from the matrix space can be measured with a sensitive pH electrode. Over the next minute or two, the pH returns to its original value, as H^+ leaks back across the membrane by various slow pathways (Figure 9–29).

A similar experiment can be carried out with a suspension of submitochondrial particles. In this case, the medium becomes more basic when oxygen is injected, since protons are pumped *into* each vesicle because of its inside-out orientation.

The Respiratory Chain Contains Large Enzyme Complexes Embedded in the Inner Membrane[15,16]

Purified preparations of inner mitochondrial membrane have been used as the starting material for purification of the various proteins that comprise the respiratory chain. Relatively mild ionic detergents were used to disrupt the membrane sufficiently to allow preferential solubilization of individual components (Figure 9–30), leading eventually to the identification of three major membrane-bound **respiratory enzyme complexes** in the pathway from NADH to oxygen:

1. The **NADH dehydrogenase complex,** consisting of at least 12 polypeptide chains, accepts electrons from NADH and passes them on to ubiquinone (uQ). Ubiquinone is a small lipid-soluble molecule (p. 494) that transfers its electrons to a second respiratory enzyme complex, the b-c_1 complex.
2. The **b-c_1 complex** contains 8 different polypeptide chains and is thought to exist as a dimer of about 500,000 daltons. This complex, which contains two cytochromes itself, accepts electrons from ubiquinone and passes them on to cytochrome c. Cytochrome c is a small, peripheral membrane protein (p. 495) that carries its electron to the cytochrome oxidase complex.
3. The **cytochrome oxidase complex** (cytochrome aa_3) is the best characterized of the three complexes. Consisting of 7 different polypeptide chains, it is isolated as a dimer of about 250,000 daltons. Each monomer has been shown to sit in the membrane in the manner illustrated in Figure 9–31.

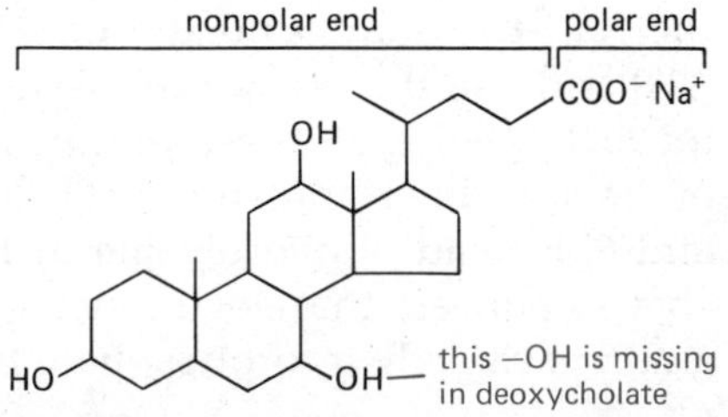

nonpolar end
polar end
$COO^- Na^+$
OH
HO
OH— this —OH is missing in deoxycholate

Figure 9–30 The structure of the relatively mild anionic detergents cholate and deoxycholate. While both are strong enough to solubilize membrane proteins, they can often be used without damaging enzyme activities.

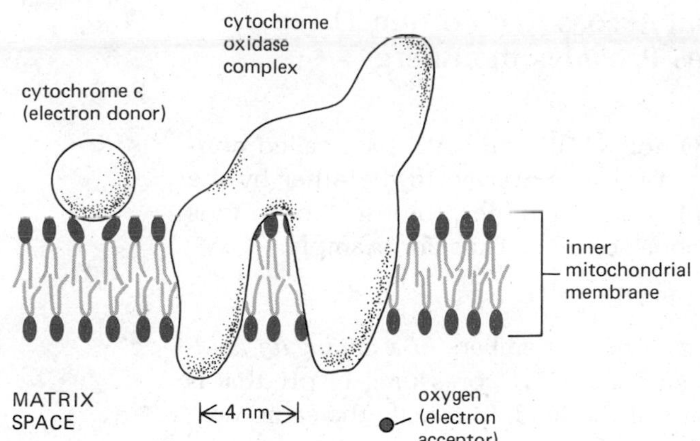

Figure 9–31 The relative sizes and shapes of the cytochrome oxidase complex and its electron donor and acceptor molecules. The low-resolution three-dimensional structure of the cytochrome oxidase complex shown was obtained from images of two-dimensional crystals (crystalline sheets) viewed in the electron microscope at various angles.

Cytochrome oxidase accepts four electrons from cytochrome-c molecules (one at a time) and ultimately transfers them to a single bound O_2 molecule, thereby forming two water molecules. As is also true for the other respiratory chain complexes, there are multiple electron carriers in each cytochrome oxidase unit. In cytochrome oxidase, two electrons enter the low potential redox centers (the heme of cytochrome a and a protein-linked copper atom, Cu_a) before electrons are passed to the oxygen binding site, which in turn involves another copper atom and the heme of cytochrome a_3. Exactly how each bound oxygen molecule reacts with four electrons and four protons to generate two molecules of water is not known.

The cytochrome oxidase reaction is estimated to account for 90% of the total oxygen uptake in most cells. Because cyanide, azide, and carbon monoxide all bind tightly to this complex, thereby blocking all electron transport, these compounds are highly toxic.

Electron Transfers Are Mediated by Random Collisions in the Bilayer Between Diffusing Donors and Acceptors[17]

The three major enzyme complexes of the respiratory chain appear to diffuse as independent entities in the plane of the inner membrane bilayer; in 5 msec they are estimated to diffuse about 40 nm on average, a distance that is roughly six times their own diameter. The two components that carry electrons between the complexes—ubiquinone and cytochrome c—are smaller and move about ten times faster than the enzyme complexes; they have diffusion coefficients similar to those of phospholipid molecules ($\sim10^{-8}$ $cm^2 sec^{-1}$). Calculations suggest that the expected random collisions between these carriers and the enzyme complexes are sufficient to permit the observed rates of electron transfer along the respiratory chain (each complex donates and receives an electron about once every 5 to 20 msec). Thus, there is no need to postulate a structurally ordered chain of electron-transfer proteins in the lipid bilayer.

This view is supported by the observation that the various components of the respiratory chain are present in quite different amounts: for each molecule of NADH dehydrogenase complex there are estimated to be 3 molecules of b-c_1 complex, 7 molecules of cytochrome oxidase complex, 9 molecules of cytochrome c, and 50 molecules of ubiquinone. In addition, it has been possible to dilute the inner membrane proteins with excess phospholipids, by fusing liposomes with submitochondrial particles. Studies of the membranes of such particles by freeze-fracture electron microscopy suggest that the respiratory complexes are much more widely separated than in normal submitochondrial particles, and no ordered arrays are seen.

A Large Drop in Redox Potential Across Each of the Three Respiratory Enzyme Complexes Provides the Energy Needed to Pump Protons[15, 18]

Pairs of compounds like H_2O and $\frac{1}{2}O_2$ (or NADH and NAD^+) are called **conjugate redox pairs,** since one compound is converted to the other by the addition of one or more electrons plus one or more protons—the protons being readily available from any aqueous solution. Thus, for example:

$$\tfrac{1}{2}O_2 + 2e^- + 2H^+ \rightarrow H_2O$$

It is well known that a 50:50 mixture of both members of a *conjugate acid-base* pair acts as a buffer, maintaining a defined "H^+ pressure," or pH, that is a measure of the dissociation constant of the acid. In exactly the same way, a 50 : 50 mixture of both members of a conjugate redox pair maintains a defined "electron pressure," or **redox** (oxidation-reduction) **potential,** ***E,*** that is a measure of the electron carrier's affinity for electrons.

By placing electrodes in contact with solutions containing the appropriate conjugate redox pairs, one can measure the redox potential of each of the various electron carriers that participate in biological oxidation-reduction reactions. Pairs of compounds having the most negative redox potentials have the weakest affinity for electrons, and therefore contain carriers with the strongest tendency to donate electrons and the least tendency to accept them. Thus, a 50 : 50 mixture of NADH and NAD^+ has a redox potential of −320 mV, indicating that NADH has a strong tendency to donate electrons, while a 50 : 50 mixture of H_2O and $\frac{1}{2}O_2$ has a redox potential of +820 mV, indicating that O_2 has a strong tendency to accept electrons.

The redox potentials measured along the respiratory chain are shown in Figure 9–32. The potentials drop in three large steps, one across each major enzyme complex. The change in redox potential between any two electron carriers is directly proportional to the free energy released by an electron transfer between them (Figure 9–32). Each complex acts as an energy conversion device to harness this free-energy change, pumping protons across the membrane to create an electrochemical gradient as electrons pass through. This conversion can be directly demonstrated by incorporating each purified complex separately into liposomes (see Figure 9–27). When any one of the

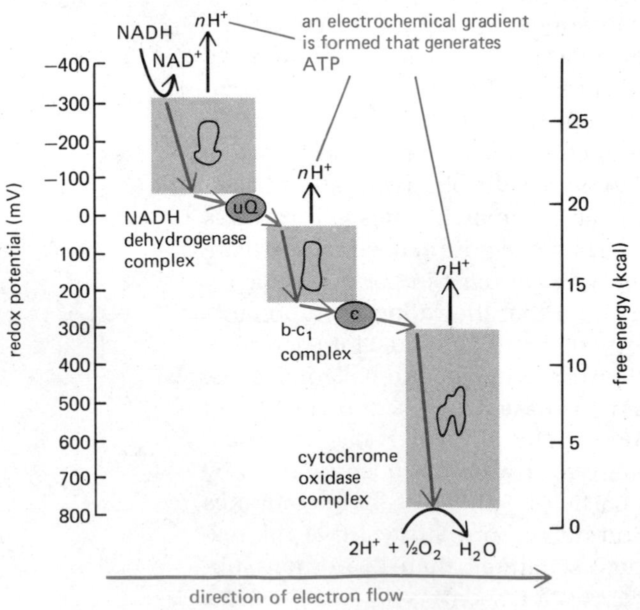

Figure 9–32 The decline in redox potential as electrons flow down the respiratory chain to oxygen. The standard symbol used for redox potential is either E_0' or E_h. The standard free-energy change for the transfer of each of the two electrons donated by an NADH molecule can be obtained from the right-hand ordinate. The two electrons transported from $FADH_2$ produce less useful energy than the two electrons transported from NADH. The reason is that the $FADH_2$ bound to proteins has a different redox potential than does NADH (in the range of 0 mV). Thus, its electrons are passed to ubiquinone by a process that does not conserve energy. The electron transport from $FADH_2$ to oxygen therefore causes protons to be pumped at only two sites rather than three (not shown).

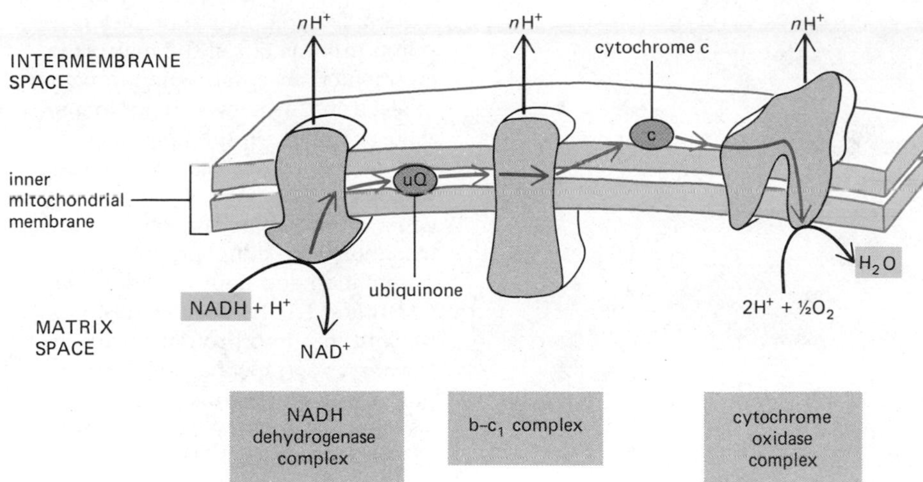

Figure 9–33 Schematic illustration of the flow of electrons through the three major respiratory enzyme complexes during the transfer of *two* electrons from NADH to oxygen. Ubiquinone and cytochrome c serve as carriers between the complexes.

three isolated complexes is presented with an appropriate electron donor and acceptor so that electrons can pass through it, protons are translocated across the liposome membrane.

The energy conversion mechanism underlying oxidative phosphorylation requires that each protein complex be inserted across the lipid bilayer membrane in a fixed orientation so that all protons are pumped in the same direction out of the matrix space (Figure 9–33). Such a *vectorial organization* of membrane proteins has been demonstrated by using membrane-impermeable probes to label each complex from only one side of the membrane or the other (see p. 274). A unique orientation in the bilayer is a common feature of all membrane proteins, and one that is essential to their function.

The Mechanisms of Respiratory Proton Pumping Are Not Well Understood[19]

In oxidative phosphorylation, a maximum of about three ATP molecules can be synthesized for each NADH molecule oxidized (that is, for every *two* elections that pass through all three of the enzyme complexes in the respiratory chain). If we assume that three protons flow back through the ATP synthetase for every ATP molecule that it synthesizes, we can conclude that about 1.5 protons are pumped by an average enzyme complex for each electron that it transports (that is, some enzyme complexes pump one proton per electron, while others pump two).

The molecular mechanism by which electron transport is coupled to proton pumping is likely to be different for different respiratory enzyme complexes. For example, protons seem to be pumped directly by ubiquinone when this carrier donates its electrons to the b-c_1 complex. This is possible because ubiquinone picks up a H^+ from the aqueous medium along with each electron that it carries, and it liberates this H^+ when it releases the electron (see Figure 9–17). Ubiquinone is freely soluble in the lipid bilayer, and it is thought to accept electrons near the inside surface of the membrane and to donate them to the b-c_1 complex near the outside surface, thereby transferring one proton across the bilayer for every electron that it transports (Figure 9–34). One proton per electron can be pumped in this way.

In contrast to the above type of mechanism, there is no obvious electron carrier that also carries a proton for the cytochrome oxidase complex, and in this case it seems that electron transport induces an orderly allosteric change in protein conformation, which causes part of the protein complex itself to pump protons. Exactly how this is accomplished is not known.

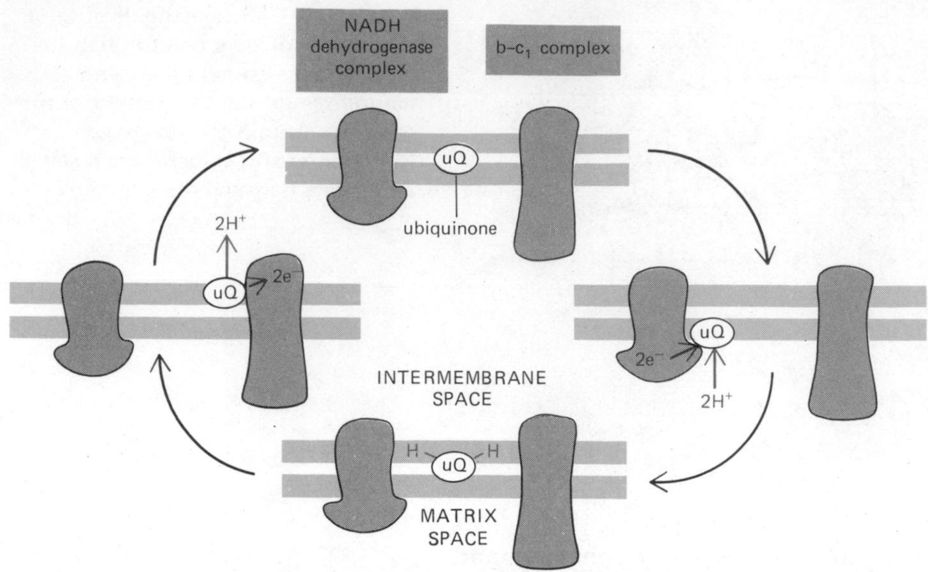

Figure 9–34 Ubiquinone (uQ) can pump protons out of the matrix by an oriented acceptance and donation of electrons. As shown, ubiquinone diffuses rapidly in the lipid bilayer and can carry two electrons and two protons at a time (see Figure 9–17). Other electron carriers whose oxidation and reduction involve protonation and deprotonation are NADH and $FADH_2$; they are also thought to pump protons in this way. However, such mechanisms cannot pump enough protons per electron to account for all of the proton fluxes observed (see text).

H^+ Ionophores Dissipate the H^+ Gradient and Thereby Uncouple Electron Transport from ATP Synthesis[20]

Since the 1940s, several different lipophilic weak acids have been known to act as *uncoupling agents.* The addition of these low-molecular-weight organic compounds to cells stops ATP synthesis by mitochondria without blocking their oxygen uptake. In the presence of an uncoupling agent, electron transport continues at a rapid rate, but no H^+ gradient is generated. The explanation for this effect is both simple and elegant: uncoupling agents act as H^+ carriers or ionophores (protonophores) and provide an alternative pathway to the ATP synthetase for the flow of H^+ across the inner mitochondrial membrane. As a result of this "short-circuiting," the proton-motive force is completely dissipated, and ATP can no longer be synthesized. The mechanism involved is illustrated for the well-studied uncoupler 2,4-dinitrophenol in Figure 9–35.

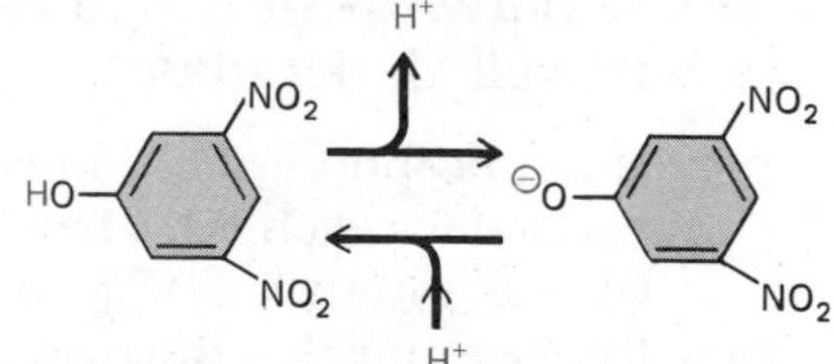

Respiratory Control Normally Restrains the Electron Flow Through the Chain[12]

When an uncoupler such as dinitrophenol is added to cells, mitochondria increase their rate of oxygen uptake substantially because of an increased rate of electron transport. This increase reflects the existence of **respiratory control.** The control is thought to act via a direct inhibitory influence of the electrochemical proton gradient on the rate of electron transport. When the electrochemical gradient is collapsed by an uncoupler, electron transport is free to run unchecked at the maximal rate allowed by the supply of substrates. However, when the gradient increases, electron transport becomes more difficult and the process slows. Moreover, if an artificially large electrochemical

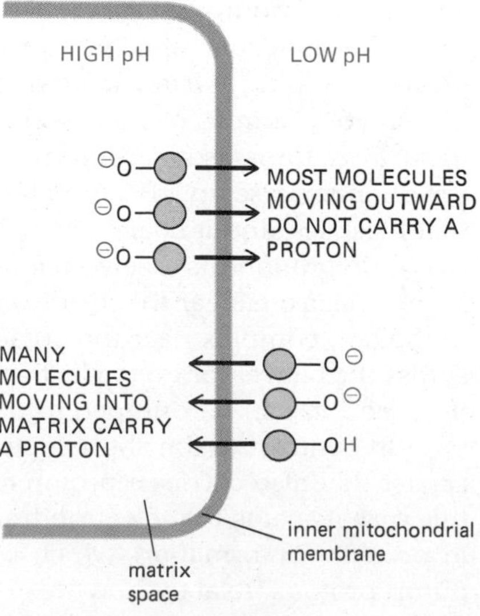

Figure 9–35 Proton conduction through the inner mitochondrial membrane by the uncoupling agent dinitrophenol (DNP). Both the uncharged (protonated) and the charged species of dinitrophenol can diffuse across the lipid bilayer without requiring transport proteins. Because of the electrochemical proton gradient, the dinitrophenol molecules will carry more protons in than out, until the proton-motive force has completely disappeared.

gradient is experimentally created across the inner membrane, normal electron transport stops completely and a *reverse electron flow* can be detected in some sections of the respiratory chain! This last observation suggests that respiratory control reflects a simple balance between the free-energy change for electron-transport-linked proton pumping and the free-energy change for electron transport—in other words, that the magnitude of the electrochemical proton gradient affects both the rate and the direction of electron transport in much the same way that it affects the directionality of ATP synthetase (see p. 502).

Respiratory control is just one part of an elaborate interlocking system of feedback controls that coordinate the rates of glycolysis, fatty acid breakdown, the citric acid cycle, and electron transport. The rates of all of these processes are adjusted to the ATP:ADP ratio, increasing whenever an increased utilization of ATP causes ADP levels to rise. To focus on just one part of this control system, consider the ATP synthetase in the inner mitochondrial membrane. This enzyme works faster as the concentration of its substrates ADP and P_i increase. As it speeds up, the enzyme lets more protons flow into the matrix and thereby dissipates the electrochemical gradient more rapidly. And the falling gradient, in turn, enhances the rate of electron transport.

Similar controls, including feedback inhibition of several key enzymes by ATP, act to adjust the rates of NADH production to the rate of NADH utilization by the respiratory chain, and so on. As a result of these many control mechanisms, the body oxidizes fats and sugars 5 to 10 times more rapidly during a period of strenuous exercise than during a period of rest.

Natural Uncouplers Convert the Mitochondria in Brown Fat into Heat-generating Machines[21]

In some specialized fat cells, mitochondrial respiration is naturally uncoupled from the ATP synthesis mechanism. In these cells, known as brown fat cells, the energy of oxidation is dissipated entirely as heat instead of being converted into ATP. The inner membranes of the large mitochondria present contain a special transport protein that allows protons to move freely down their electrochemical gradient, uncoupling electron transport from respiratory control. As a result, the cells oxidize their fat stores at a rapid rate and produce heat rather than ATP. These tissues thereby serve as "heating pads" that revive hibernating animals and protect sensitive areas of newborn human babies from the cold.

Chemiosmotic Mechanisms Represent a Common Theme Among the Diversity of Bacteria[22]

Bacteria use enormously diverse types of energy sources. Some bacteria are very much like animal cells, synthesizing ATP from sugars, which they oxidize to CO_2 and H_2O by glycolysis and the citric acid cycle. These bacteria have a respiratory chain in their plasma membrane very similar to that in the mitochondrial inner membrane. Other types of bacteria are strict anaerobes that derive their energy from glycolysis alone (fermentation) or from an electron-transport chain that employs a molecule other than oxygen as the final electron acceptor. These alternative electron acceptors can be nitrogen compounds (nitrate or nitrite), sulfur compounds (sulfate or sulfite), or organic compounds (fumarate or carbonate). The electrons are transferred to these acceptors by a series of electron carriers in the plasma membrane that are comparable to those present in mitochondrial respiratory chains.

Despite this diversity, all bacteria studied to date have embedded in their plasma membrane an ATP synthetase that is very similar to that found in mitochondria (and chloroplasts). In anaerobic bacteria that lack an electron-transport chain, this ATP synthetase works in reverse, using the ATP produced by glycolysis to establish a proton-motive force across the bacterial plasma membrane. In all other bacteria, an electron-transport chain establishes the proton-motive force that drives ATP synthetase to make ATP.

The fact that all bacteria, including the strictly fermentative types, maintain a proton-motive force across their plasma membrane reflects the essential role of the electrochemical proton gradient in transporting substances across this membrane against their concentration gradients. Thus Na^+ is pumped out of bacteria by a Na^+-H^+ antiport system that takes the place of the Na^+-K^+ ATPase of eucaryotic cells. Similarly, active transport systems that bring nutrients into bacteria tend to be H^+-symport systems in which desired metabolites are dragged into the cell along with one or more protons by means of a specific carrier protein. Most amino acids and many sugars are transported in this way (Figure 9–36). Other sources of energy—including ATP hydrolysis or an inward symport with Na^+—drive active transport in some bacterial carrier proteins, but they are relatively rare. In animal cells, by contrast, most inward transport across the plasma membrane is driven by the Na^+ gradient established by the Na^+-K^+ ATPase (see p. 295).

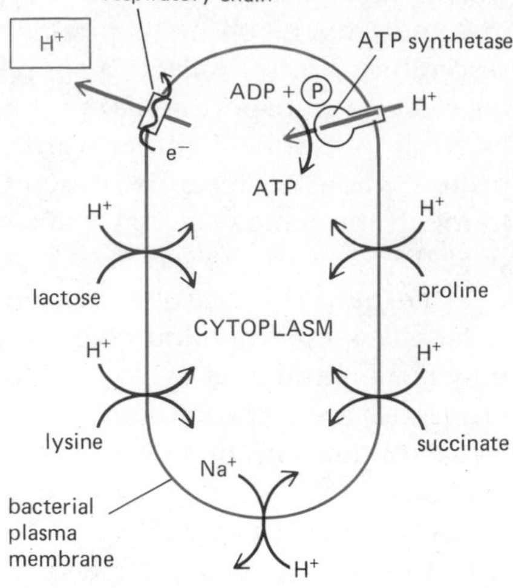

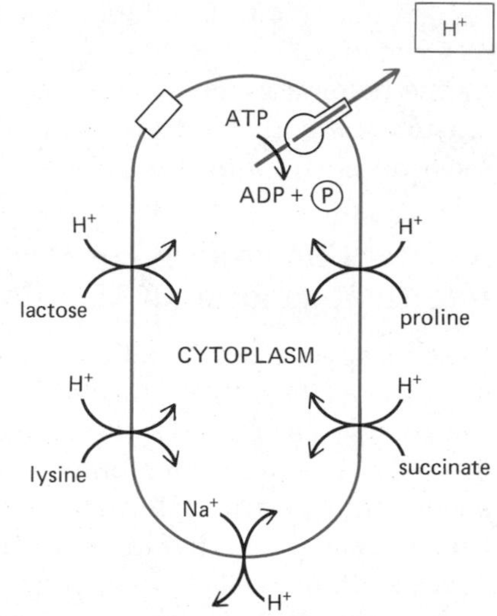

Figure 9–36 A proton-motive force generated across the plasma membrane of bacterial cells pumps nutrients into the cell and expels sodium. In (A) the electrochemical proton gradient is being generated in an aerobic bacterium by a respiratory chain and is then used by ATP synthetase to make ATP. In (B) the same bacterium growing under anaerobic conditions will derive its ATP from glycolysis. Part of this ATP is hydrolyzed by ATP synthetase to establish the transmembrane proton-motive force that drives transport processes. (As described in the text, other bacteria have electron-transport chains that will pump protons under anaerobic conditions, using a molecule other than oxygen as the final acceptor for electrons.)

Summary

In the pathway from NADH to O_2, the respiratory chain contains three major enzyme complexes that are embedded in the inner mitochondria membrane. Each of these complexes can be purified, inserted into synthetic lipid vesicles, and shown to pump protons when electrons are transported through it. In the native membrane, the mobile electron carriers ubiquinone and cytochrome c complete the electron transport chain. The path of electron flow is

NADH → NADH dehydrogenase complex → ubiquinone → b-c_1 complex → cytochrome c → cytochrome oxidase complex → molecular oxygen (O_2)

The respiratory enzyme complexes are inserted in the membrane with a defined orientation, so that all protons are pumped out of the matrix.

The electrochemical proton gradient produced by the respiratory chain is harnessed to make ATP by another transmembrane protein complex called ATP synthetase. ATP synthetase is a reversible coupling device that normally converts a proton flow into the matrix into ATP phosphate-bond energy, but it can also hydrolyze ATP to pump protons outward. Its universal appearance in mitochondria, chloroplasts, and bacteria testifies to the central importance of chemiosmotic mechanisms in all cells.

The Chloroplast[23]

All animals and most microorganisms rely on the continual uptake of large amounts of organic compounds from their environment. These compounds provide the carbon skeletons for biosynthesis and, through controlled oxidation, the metabolic energy that drives all cellular processes. It is believed that the first organisms on the primitive earth had access to an abundance of organic compounds produced by geochemical processes (p. 4). But most of these compounds were used up billions of years ago. Since that time, virtually all of the organic materials required by living cells have been produced by *photosynthetic organisms*—which include many different types of photosynthetic bacteria. The most advanced of the photosynthetic bacteria are the cyanobacteria, which have a minimal requirement for nutrients. They use electrons from water and the energy of sunlight to convert atmospheric CO_2

into organic compounds. Moreover, in the course of splitting water, in the reaction: $nH_2O + nCO_2 \xrightarrow[\text{light}]{} (CH_2O)_n + nO_2$, they liberate into the atmosphere the oxygen required for the reactions of oxidative phosphorylation. As we shall explain later, it was the evolution of cyanobacteria from more primitive photosynthetic bacteria that first made possible the development of aerobic life forms.

An important later development was the plants, in which photosynthesis is carried out in an intracellular organelle—the chloroplast. Chloroplasts specialize in energy metabolism during the daylight hours. During the night, they cease their production of energy-rich metabolites, and the plant cell relies on its mitochondria to generate ATP. Plant mitochondria closely resemble their counterparts in animal cells.

Largely on the basis of biochemical evidence, it is widely believed that chloroplasts are the descendants of cyanobacteria that were endocytosed to live in symbiosis with a primitive eucaryotic cell. In this view, the many differences between chloroplasts and mitochondria are due in part to their being derived from different bacterial ancestors, and in part to evolutionary changes that reflect their different roles in eucaryotic cells. Nevertheless, the fundamental mechanisms involved in light-driven ATP synthesis in chloroplasts and in respiration-driven ATP synthesis in mitochondria are very much the same.

Chloroplasts Have an Extra Compartment but Still Resemble Mitochondria[9,24]

Chloroplasts carry out their energy interconversions by chemiosmotic mechanisms in much the same way that mitochondria do and are organized on the same principles (Figures 9–37, 9–38, 9–39, and 9–40): they have a highly permeable outer membrane, a much less permeable inner membrane in which some special transport proteins are embedded, and a narrow intermembrane space between. The inner membrane surrounds a large central space called the **stroma,** which is analogous to the mitochondrial matrix and contains many different soluble enzymes.

There is an important difference, though, between chloroplasts and mitochondria in organization. The inner chloroplast membrane is not folded into cristae and does not contain an electron-transport chain. Instead, the

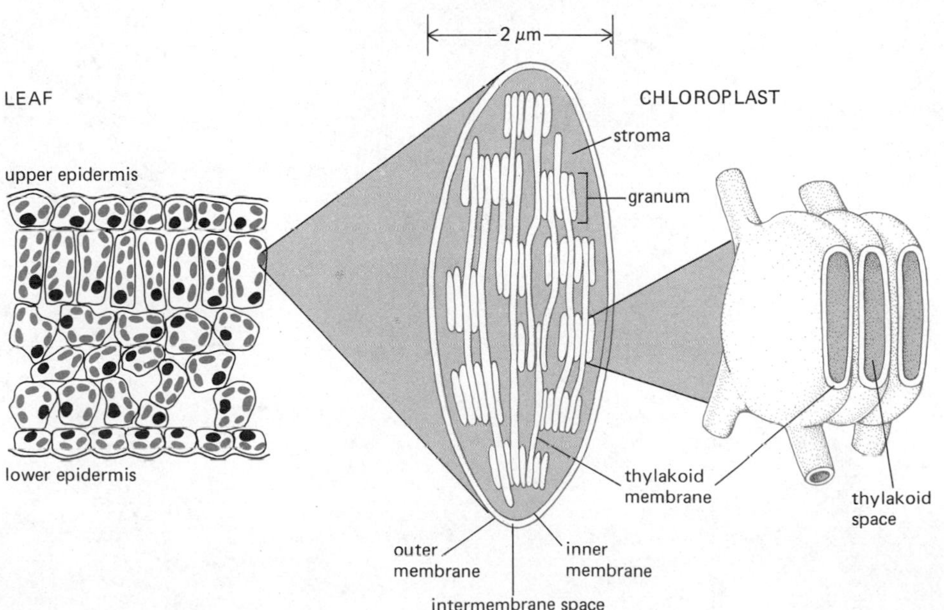

Figure 9–37 The chloroplast contains three distinct membranes (the outer membrane, inner membrane, and thylakoid membrane) that delimit three separate internal compartments—the intermembrane space, the stroma, and the thylakoid space. The thylakoid membrane contains all of the energy-generating systems of the chloroplast. As indicated, the individual thylakoids are interconnected, and they tend to stack to form aggregates called grana.

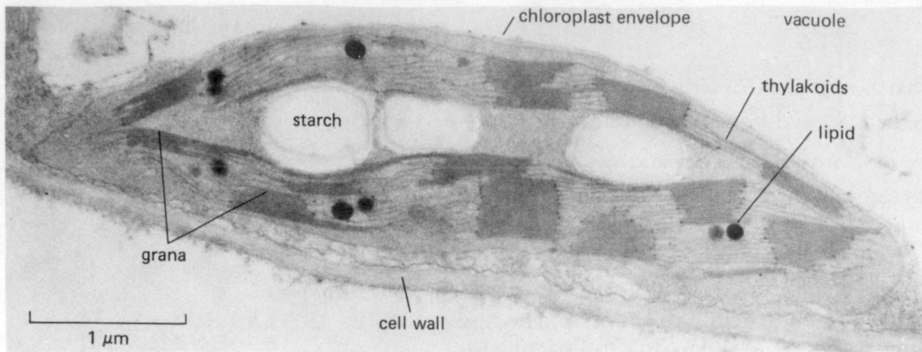

Figure 9–38 Electron micrograph of a thin section of a typical chloroplast, showing not only its various membranes, but also the stores of starch granules and lipid droplets that it has accumulated from its biosyntheses. (Courtesy of Kitty Plaskitt.)

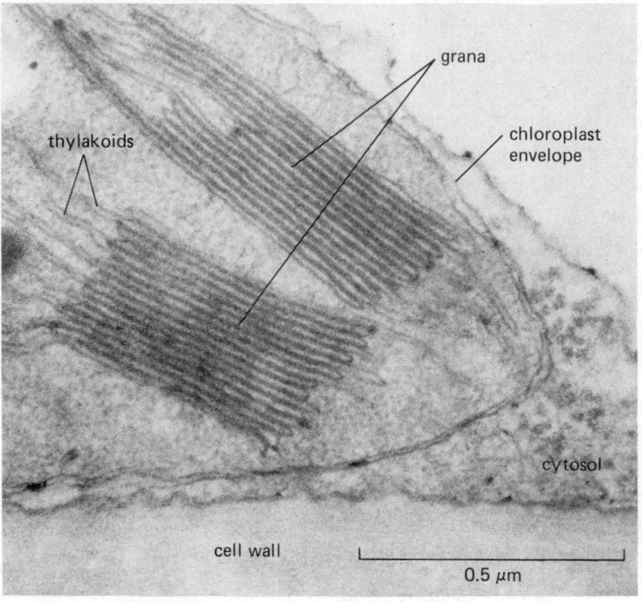

Figure 9–39 The internal membranes of a chloroplast at higher magnification than in Figure 9–38. (Courtesy of Kitty Plaskitt.)

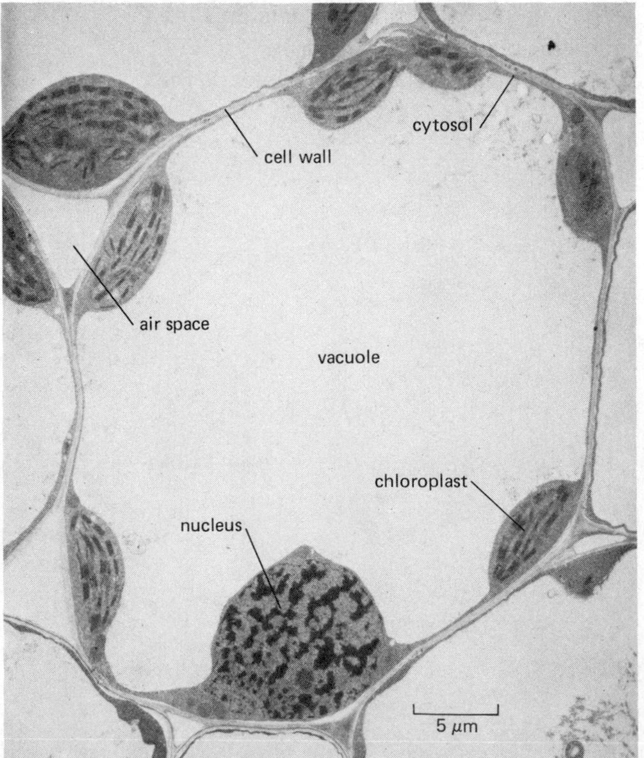

Figure 9–40 An electron micrograph of a wheat leaf cell. Note the ring of cytoplasm-containing chloroplasts that surrounds a large vacuole. (Courtesy of Kitty Plaskitt.)

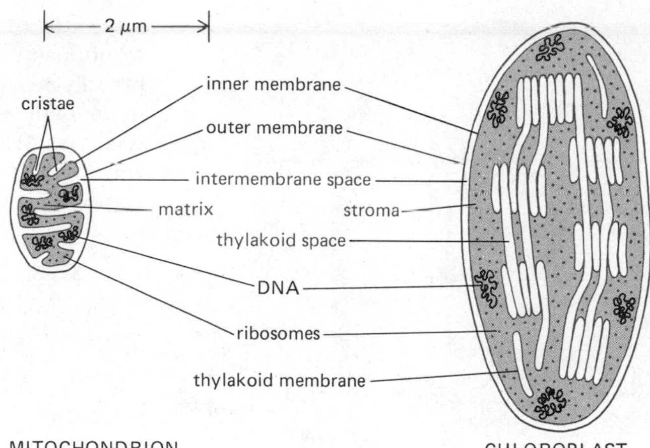

Figure 9–41 Comparison of a mitochondrion and a chloroplast. Note that the chloroplast is generally much larger and that it contains the thylakoid membrane and the thylakoid space as additional features. However, only the mitochondrion has an inner membrane that is folded into cristae.

photosynthetic light-absorbing system, the electron-transport chain, and an ATP synthetase are all contained in a third distinct membrane that forms a set of flattened disclike sacs, the **thylakoids** (Figure 9–37). The lumen of each thylakoid is connected with the lumen of a number of other thylakoids, thereby defining a third internal compartment called the *thylakoid space,* which is separated from the stroma by the ion-impermeable *thylakoid membrane.*

The structural similarities and differences between mitochondria and chloroplasts are illustrated in Figure 9–41. In a general way, one might imagine the conversion of a mitochondrion into a chloroplast by a process that greatly enlarges the mitochondrion and at the same time nips off its cristae and releases them into the matrix space as submitochondrial particles. It is important to note that the knobbed end of the chloroplast ATP synthetase, where ATP is made, protrudes from the thylakoid membrane into the stroma, just as it protrudes into the matrix from the membrane of each mitochondrial crista (see Figure 9–52, p. 521).

Two Unique Reactions Occur in Chloroplasts: The Light-driven Production of ATP and NADPH and the Conversion of CO_2 to Carbohydrate

The many different reactions that occur in photosynthesis can be grouped into two broad categories: (1) In the **light-dependent reactions,** energy derived from sunlight energizes an electron in chlorophyll, enabling it to move along an oxidation chain in the thylakoid membrane in much the same way that an electron moves along the respiratory chain in mitochondria. This electron-transport process is used to pump protons across the thylakoid membrane, and the resulting proton-motive force drives the synthesis of ATP. At the same time, the process generates high-energy electrons that convert $NADP^+$ to NADPH; in this process, water is oxidized to provide the electrons to NADPH, and O_2 is liberated. (2) In the **dark reactions,** the ATP and NADPH made in the light-dependent reactions serve as the source of energy and reducing power, respectively, and drive the conversion of CO_2 to carbohydrate (**carbon fixation**). These reactions begin in the chloroplast stroma and continue in the cell cytosol. They are called dark reactions because, although they require the products of the light-dependent reactions, they do not involve light directly.

Thus, the formation of oxygen (which requires light) and the conversion of carbon dioxide to carbohydrate are separate photosynthetic processes (Figure 9–42).

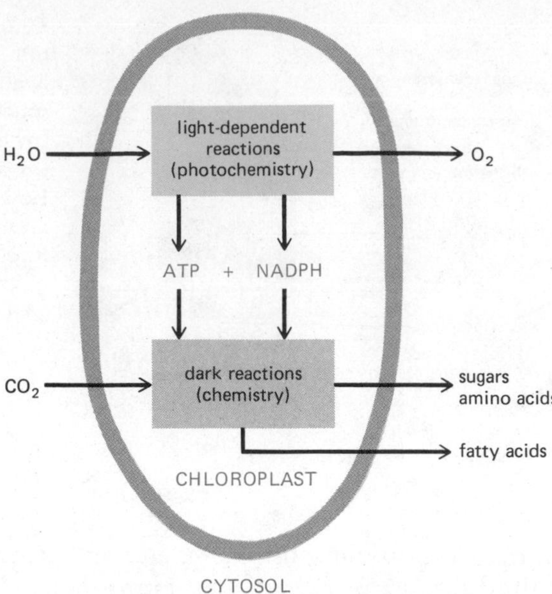

Figure 9–42 The photosynthetic reactions in a chloroplast can be broadly separated into light-dependent reactions and dark reactions. Water is oxidized in the light-dependent reactions, while carbon dioxide is fixed in the dark reactions.

Carbon Fixation Is Catalyzed by Ribulose Bisphosphate Carboxylase, the World's Most Abundant Enzyme[25]

We have seen earlier in this chapter how cells use the large amount of free energy released when carbohydrates are oxidized to CO_2 and H_2O to produce ATP. Clearly, therefore, the combining of CO_2 and H_2O to make carbohydrates must be a very unfavorable reaction that can be achieved only by coupling this synthesis to other very favorable reactions.

The central reaction in which an atom of inorganic carbon (as CO_2) is converted to organic carbon (as *3-phosphoglycerate*, an intermediate in glycolysis) is illustrated in Figure 9–43. Discovered in 1948, this carbon-fixing reaction is catalyzed in the chloroplast stroma by a large enzyme called *ribulose bisphosphate carboxylase* (~500,000 daltons); for each molecule of CO_2 that reacts with the five-carbon compound ribulose 1,5-bisphosphate, two molecules of the three-carbon compound 3-phosphoglycerate are formed. Since each molecule of the ribulose bisphosphate carboxylase works rather sluggishly (it processes about 3 molecules of substrate per second compared to 1000 molecules per second for a typical enzyme), many copies of it are needed in each chloroplast. As a result, the enzyme often represents more than 50% of the total chloroplast protein, and it is widely claimed to be the most abundant protein in the world!

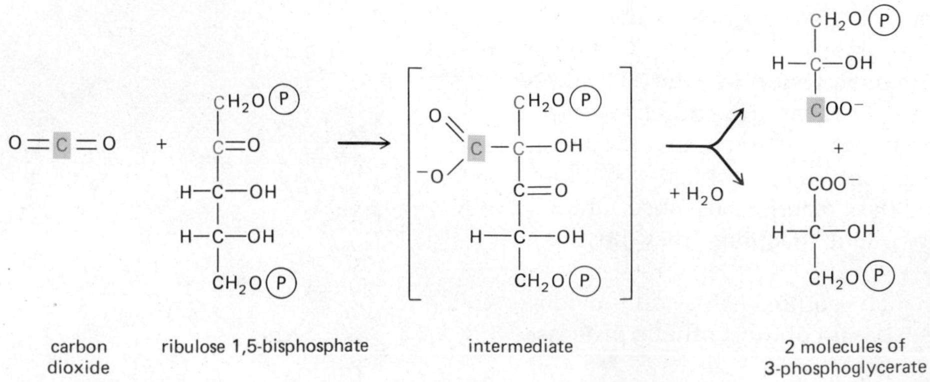

carbon dioxide + ribulose 1,5-bisphosphate → [intermediate] + H_2O → 2 molecules of 3-phosphoglycerate

Figure 9–43 The initial reaction in which carbon dioxide is converted into organic carbon. This reaction is catalyzed in the chloroplast stroma by the abundant enzyme ribulose bisphosphate carboxylase. (This enzyme is alternatively called ribulose diphosphate carboxylase.)

In the Carbon-Fixation Cycle, Three Molecules of ATP and Two Molecules of NADPH Are Consumed for Each CO_2 Molecule Fixed[26]

While the actual reaction in which CO_2 is fixed does not require an extra input of energy, it relies on a continuous supply of the energy-rich compound *ribulose 1,5-bisphosphate,* to which each molecule of CO_2 is added (Figure 9–43). The working out of the elaborate pathway by which this compound is regenerated was one of the most successful early uses of radioisotopes. As outlined in Figure 9–44, 3 molecules of CO_2 are fixed by ribulose bisphosphate carboxylase to produce 6 molecules of 3-phosphoglycerate (containing $6 \times 3 = 18$ carbon atoms in all: 3 from the CO_2 and 15 from ribulose 1,5-bisphosphate). The 18 carbon atoms then undergo a cycle of reactions that regenerate the 3 molecules of ribulose 1,5-bisphosphate (containing $3 \times 5 = 15$ carbon atoms in all) that were used up in the initial carbon-fixation step, leaving one molecule of *glyceraldehyde 3-phosphate* (3 carbon atoms) as the net gain. In this **carbon-fixation cycle** (or Calvin-Benson cycle), 3 molecules of ATP and 2 molecules of NADPH are used up for each CO_2 molecule that is converted into carbohydrate (Figure 9–44).

The glyceraldehyde 3-phosphate produced in chloroplasts by the carbon-fixation cycle is a three-carbon sugar that also serves as a central inter-

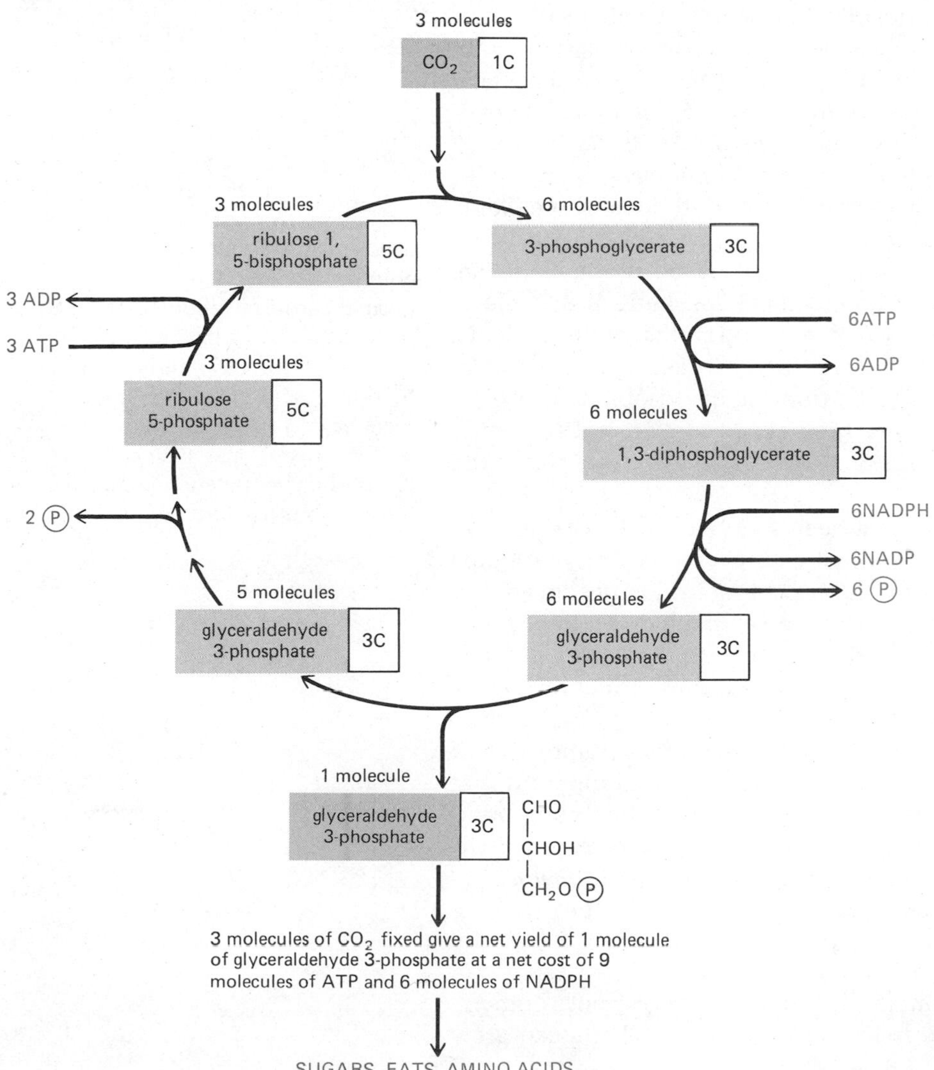

Figure 9–44 The carbon-fixation cycle. The many intermediates between glyceraldehyde 3-phosphate and ribulose 5-phosphate are omitted for clarity.

mediate in glycolysis (p. 69). Some of it is converted into fatty acids, amino acids, and starch by important biosynthetic reactions that occur in the chloroplast stroma. The rest is exported to the cell cytoplasm, where most of it is rapidly converted into fructose 6-phosphate and glucose 6-phosphate by reversal of several reactions in glycolysis (p. 86). These two sugar phosphates are then joined together, eventually producing the disaccharide **sucrose.** The function of glucose in animal cells is served in plants by sucrose, the major form in which sugar is transported between plant cells. It is exported from the leaves via vascular bundles (Figure 9–45) to provide the carbohydrate required by the rest of the plant, just as glucose is transported in the blood of animals.

In Tropical Plants, Carbon Fixation Is Compartmentalized to Facilitate Growth at Low CO_2 Concentrations[27]

An important specialization characterizes the leaves of many plants that are adapted to hot, dry environments, among them corn, sugar cane, and crab grass. The specialization is useful because a property of the first enzyme in the carbon-fixation pathway interferes with a mechanism evolved by the plants to avoid water loss. The difficulty with the enzyme, ribulose bisphosphate carboxylase, is that it can use O_2 as well as CO_2 as the substrate to be added to ribulose 1,5-bisphosphate. Although CO_2 is strongly preferred, O_2 is quite frequently added to ribulose 1,5-bisphosphate, creating one molecule of 3-phosphoglycerate (instead of two) and one molecule of the two-carbon compound glycolate. The glycolate is shuttled into plant peroxisomes, which begin the process of converting two molecules of glycolate into one molecule of 3-phosphoglycerate (three carbons) plus one molecule of CO_2. Because the entire process uses up O_2 and liberates CO_2, it is termed *photorespiration.*

Photorespiration becomes dominant in hot, dry conditions where plants are forced to close their stomata (the gas exchange pores in the leaves) in order to prevent excessive water loss, causing the CO_2 levels in the leaf to fall precipitously. Many plants have evolved an ingenious mechanism that allows them to grow efficiently even at such low CO_2 concentrations. In these plants, the carbon-fixation cycle occurs only in the chloroplasts of specialized *bundle-sheath cells,* which contain all of the plant's ribulose bisphosphate carboxylase. This enzyme is kept supplied with a high concentration of CO_2 by the CO_2-pumping activity of surrounding mesophyll cells, thereby preventing the wasteful reaction with O_2 (Figure 9–45).

The CO_2 "pump" involves a cycle that begins with a special CO_2-fixation step catalyzed in the cytosol of the mesophyll cells by an enzyme that binds carbon dioxide with high affinity. A four-carbon compound is produced and transported into the bundle-sheath cells, where it is broken down to produce one molecule of CO_2 and one molecule of a three-carbon compound. The latter is then returned to the mesophyll cells where it picks up another CO_2 molecule to restart the pumping cycle.

When a pulse of radioactive $^{14}CO_2$ is given to a plant that pumps CO_2, it encounters the mesophyll cells first, so that the first organic compound that is labeled contains four carbons, in contrast to the three-carbon compound that labels first in all other plants (see Figure 9–43). For this reason, CO_2-pumping plants are called *C_4 plants,* while all others are called *C_3 plants.*

As for any vectorial transport process, the cycle that causes the pumping of CO_2 into the bundle-sheath cells in C_4 plants costs energy. In hot, dry environments, this cost is often much less than the loss by photorespiration in C_3 plants, and C_4 plants predominate. But in a cool, moist environment, the energy devoted to CO_2 pumping is less useful, and C_3 plants can grow quite well under these conditions.

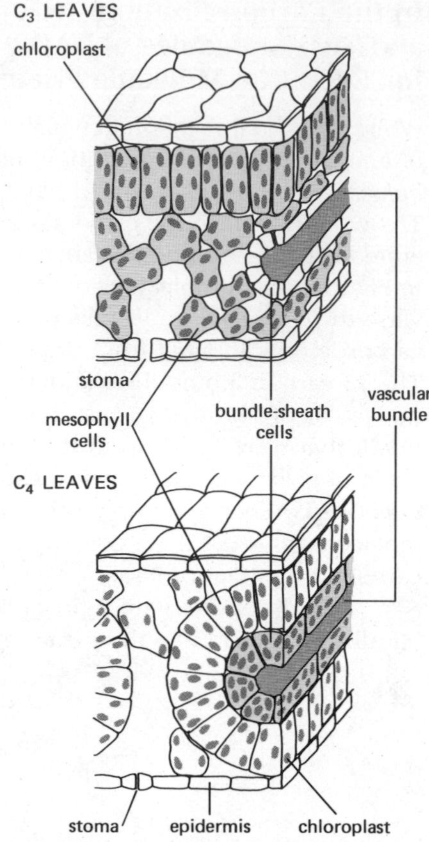

Figure 9–45 A comparison of the anatomy of the leaf in a C_3 plant and a C_4 plant. In both cases, the colored cells are those that contain chloroplasts able to carry out the normal carbon-fixation cycle. In C_4 plants, the mesophyll cells are specialized for CO_2 pumping rather than for carbon fixation, and they create a high $CO_2:O_2$ ratio in the bundle-sheath cells. The vascular bundles carry the sucrose made in the leaf to other tissues.

Light Energy Captured by Chlorophyll Is Used to Produce a Strong Electron Donor from a Weak One[28]

We now return to the question of how the very unfavorable reactions that produce carbohydrates from CO_2 and H_2O are driven (Figure 9–44). In the chloroplast, the required energy is derived from sunlight absorbed by **chlorophyll** molecules (Figure 9–46). For purposes of light harvesting, these light-absorbing molecules (pigments) are grouped into clusters of several hundred molecules—called an **antenna complex**—by means of special proteins that hold them tightly on the thylakoid membrane. Depending on the plant, varying amounts of accessory pigments called *carotenoids* are also located in each complex. The special proteins to which the clustered chlorophyll molecules are bound alter the chlorophyll molecules in such a way that the light energy absorbed by any one of them is funneled to a special chlorophyll molecule in the complex, called the *reaction-center chlorophyll.* Thus, each antenna complex serves as a "funnel" that collects light energy and directs it to a single reaction center (Figure 9–47).

The light energy absorbed by an electron in an isolated chlorophyll molecule in solution is released as light (fluorescence) and heat as the excited electron rapidly returns to its original energy level (Figure 9–48). But a reaction-center chlorophyll molecule is closely associated with an *electron acceptor* and an *electron donor,* and the three molecules together form the heart of a **photosystem.** In a reaction mediated by associated proteins, the excited electron in the reaction center is passed to the electron acceptor, leaving a positively charged hole with a very high affinity for electrons in the chlorophyll. The hole is rapidly filled by an electron that is pulled away from the nearby electron donor (Figure 9–48).

Such a photosystem enables light to generate a net electron transfer from a weak electron donor (a molecule with a strong electron affinity) to a molecule that becomes, as a result of the transfer, a strong electron donor. Thus, the excitation energy that would have normally been released as fluorescence and heat has gone instead to raise the energy of an electron, and it has thereby created a strong electron donor where none had been before.

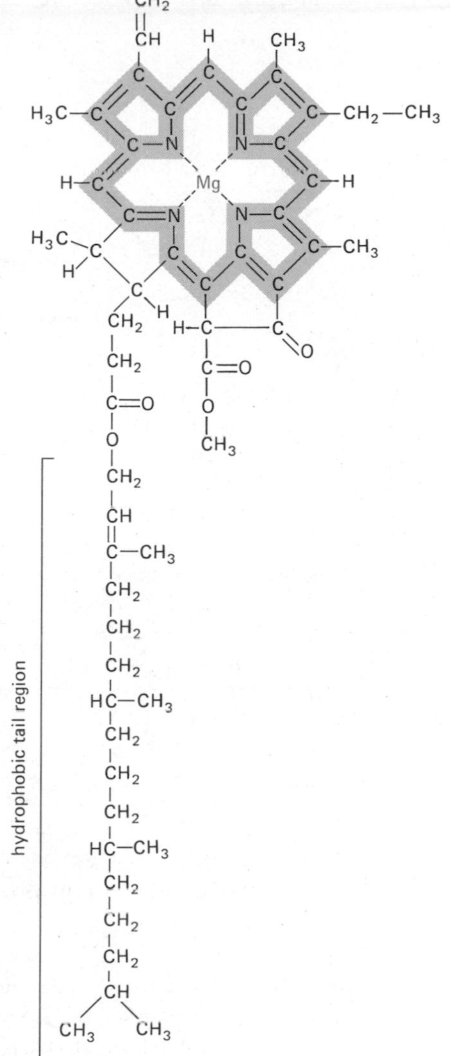

Figure 9–46 The structure of chlorophyll. Electrons are delocalized over the bonds shown in color.

Noncyclic Photophosphorylation Produces Both NADPH and ATP[9,29]

In the most complex form of photosynthesis, which produces both ATP and NADPH (**noncyclic photophosphorylation**), two photosystems in series energize each electron. In the first (called photosystem II for historical reasons), four electrons are removed from a water molecule by a poorly understood water-splitting enzyme to fill the holes created by light in reaction-center

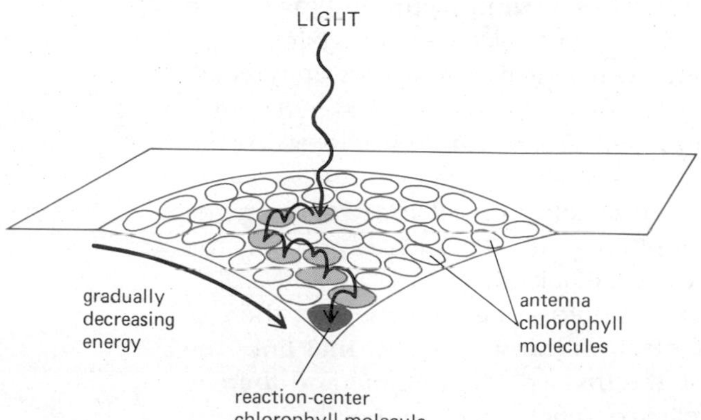

Figure 9–47 The antenna complex in the thylakoid membrane serves as a funnel for transferring an electron excited by light to the reaction center.

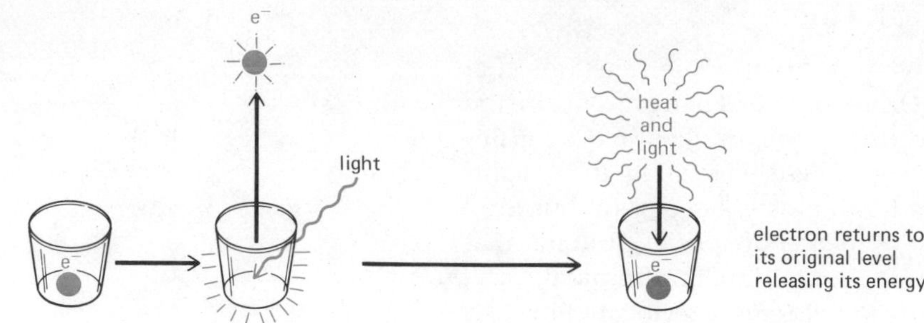

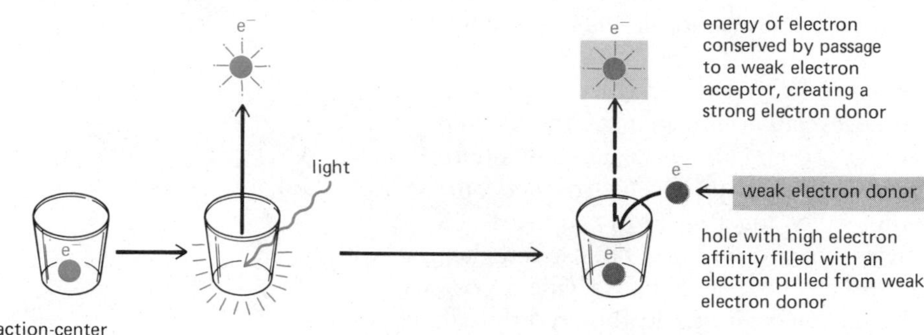

Figure 9–48 The light energy absorbed by an electron in an isolated chlorophyll molecule is released as light and heat (A). In contrast, a photosystem uses the light energy absorbed by chlorophyll to create a strong electron donor from a weak one (B).

chlorophyll molecules, thereby producing molecular oxygen by the reaction: $H_2O \rightarrow 4H^+ + 4e^- + O_2$. For each water molecule split, four strong electron donors are produced and four quanta of light are used. Each of the strong electron donors then passes its electron down an electron-transport chain that closely resembles part of the respiratory chain found in the inner mitochondrial membrane. Along the way, protons are pumped across the thylakoid membrane into the thylakoid space, creating an electrochemical gradient that drives the synthesis of ATP by an ATP synthetase complex (Figures 9–49 and 9–50). The final electron acceptor in this electron transport chain is the second photosystem in the scheme (photosystem I), which accepts the electron into the hole left by light excitation of its reaction-center chlorophyll molecule. In the end, the electron that leaves photosystem I has been boosted to such a high energy level by the two light quanta by which it has been sequentially activated that two of these electrons (plus a proton from the medium) can drive the reduction of $NADP^+$ to NADPH (Figure 9–49).

The zig-zag scheme for photosynthesis shown in Figure 9–49 is known as the "**Z-scheme.**" By means of its two separate electron-energizing steps, one catalyzed by each photosystem, an electron is passed from water, which normally holds onto its electrons very tightly (redox potential = +820 mV), to NADPH, which normally holds onto its electrons rather loosely (redox potential = −320 mV). The two steps are necessary in part because a single quantum of visible light cannot energize an electron all the way from the bottom of photosystem II to the top of photosystem I, which is probably the energy change required to pass an electron efficiently from water to $NADP^+$. More important, the use of two separate photosystems means that there is enough energy left over to enable the electron-transport chain that links the two photosystems to pump H^+ across the thylakoid membrane and thereby to harness some of the light-derived electron energy for producing ATP.

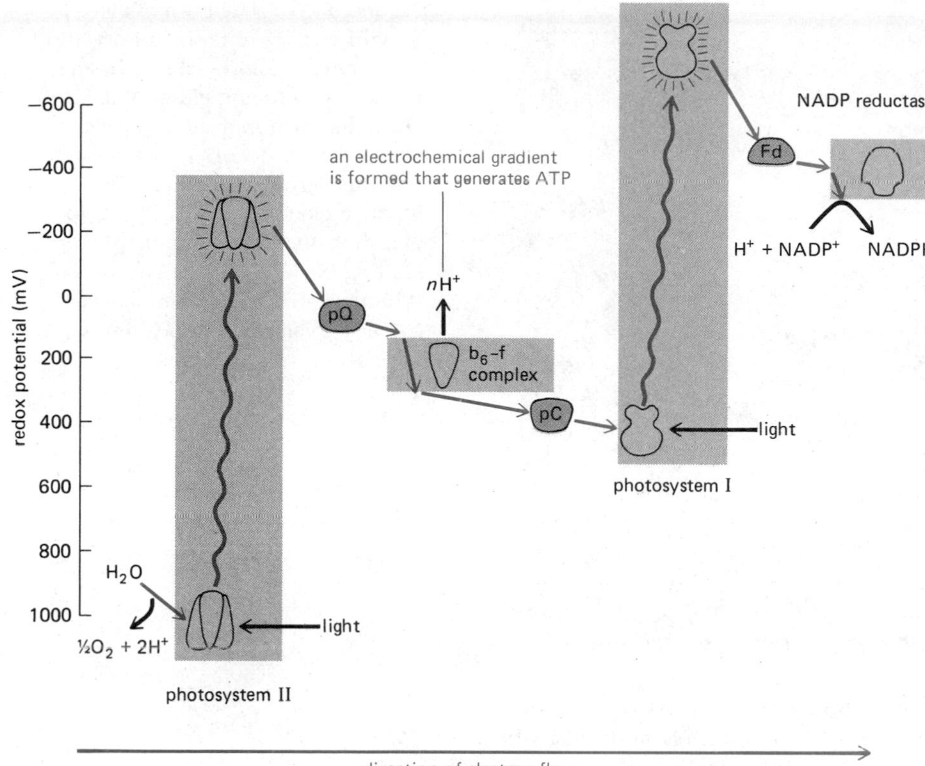

Figure 9–49 Redox potential changes for the passage of electrons through the Z-scheme for the photosynthetic production of NADPH and ATP. The net electron flow from water to $NADP^+$ produces NADPH. In addition, ATP is synthesized by an ATP synthetase that harnesses the electrochemical proton gradient produced by the electron-transport chain linking photosystem II and photosystem I. Although not shown, the H^+ from water oxidation also contributes to the generation of the electrochemical gradient that drives ATP synthesis. The Z-scheme for ATP production is called noncyclic photophosphorylation in order to distinguish it from the cyclic scheme shown in Figure 9–51. For further details, see Figure 9–50.

Chloroplasts Can Make ATP by Cyclic Photophosphorylation Without Making NADPH

In the noncyclic photophosphorylation scheme just discussed, slightly more than one molecule of ATP is synthesized for every pair of electrons that passes from H_2O to $NADP^+$ to produce a molecule of NADPH. But considerably more ATP than NADPH is needed for carbon fixation (see Figure 9–44). Chloroplasts can produce the extra ATP because they have a way of using light energy to make ATP without making NADPH, a process called **cyclic photophosphorylation.** This involves a cyclic electron flow in which high-energy electrons created by photosystem I are transferred back to an earlier step in the electron-transport chain rather than being passed on to $NADP^+$. As the electrons

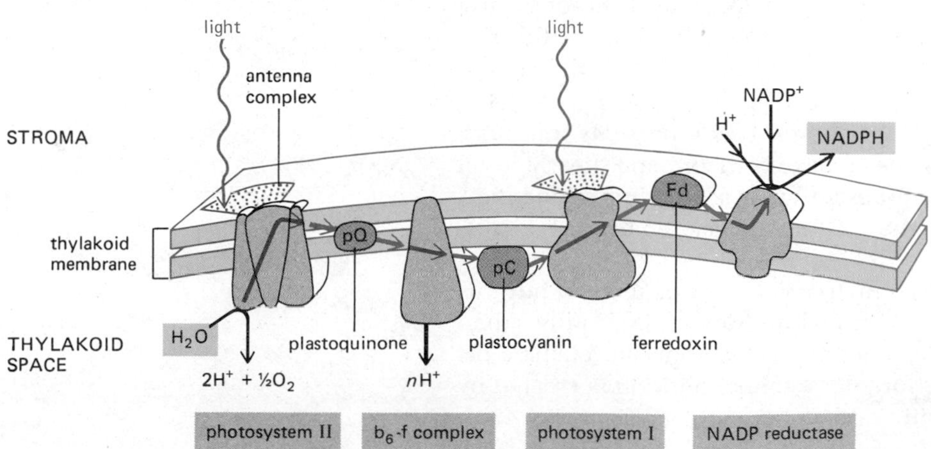

Figure 9–50 Electron flow in the Z-scheme for photosynthesis as it might occur in the thylakoid membrane. The highly mobile electron carriers in the chain are plastoquinone (which closely resembles the ubiquinone of mitochondria), plastocyanin (a small copper-containing protein), and ferredoxin (a small iron-containing protein). The *b_6-f complex* closely resembles the b-c_1 complex of mitochondria. Note that protons are pumped out of the stroma by the b_6-f complex, just as they are pumped out of the matrix by the b-c_1 complex (see Figure 9–52).

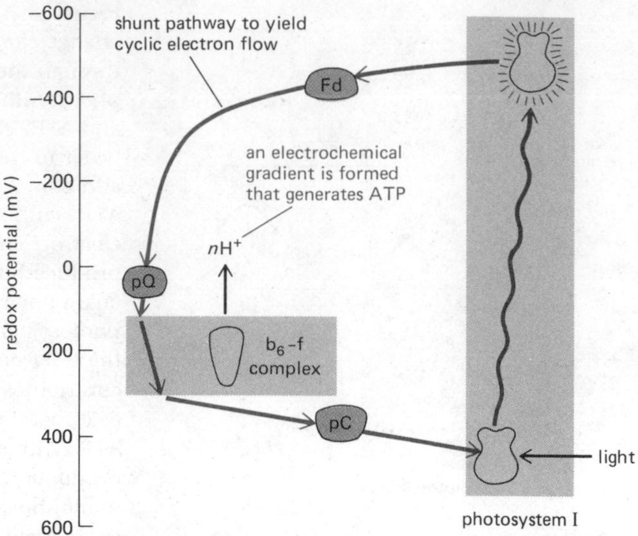

Figure 9–51 The path of electron flow in cyclic photophosphorylation. This scheme allows ATP to be made without producing either NADPH or O_2. Whether noncyclic or cyclic electron flow occurs depends on whether ferredoxin donates its reactive electron to $NADP^+$, as in Figure 9–49, or to components leading back to the b_6-f complex. Whenever NADPH accumulates, $NADP^+$ levels will be low, tending to favor the cyclic scheme.

flow through the electron-transport chain, protons are pumped across the thylakoid membrane, and the resulting electrochemical gradient drives the synthesis of ATP (Figure 9–51).

In summary, noncyclic photophosphorylation involves the photoreduction of $NADP^+$ by water, is mediated by the combined action of photosystem I and photosystem II, and produces NADPH, ATP, and O_2. In contrast, cyclic photophosphorylation involves only photosystem I and produces ATP without the formation of either NADPH or O_2. Thus, the relative activities of cyclic and noncyclic electron flows will determine how much light energy is converted into reducing power (NADPH) and how much into high-energy phosphate bonds (ATP). At least in part, the balance is automatically regulated according to the need for NADPH. Note that whether the electron flow is noncyclic or cyclic depends on whether ferredoxin donates its reactive electron to $NADP^+$ or to components that lead back to the *b_6-f complex* (compare Figures 9–49 and 9–51). At low concentrations of $NADP^+$, caused by an accumulation of NADPH, the cyclic scheme that produces only ATP will thus be favored.

The Geometry of Proton Translocation Is Similar in Mitochondria and Chloroplasts[9]

The presence of the thylakoid space separates a chloroplast into three rather than two internal compartments and makes it seem quite different from a mitochondrion. However, the geometry of proton translocation in the two organelles is surprisingly similar. As illustrated in Figure 9–52, in chloroplasts protons are pumped out of the stroma (pH 8) into the thylakoid space (pH of about 5), creating a gradient of 3 to 3.5 pH units. This represents a proton-motive force of about 200 mV across the thylakoid membrane (nearly all of which is contributed by the pH gradient rather than by a membrane potential), which then drives ATP synthesis by the ATP synthetase embedded in this membrane.

Like the stroma, the mitochondrial matrix has a pH of about 8, but this is created by pumping protons out of the mitochondrion entirely, rather than into an interior space in the organelle. Since the cell cytoplasm retains a pH of about 7, most of the proton-motive force in a mitochondrion is created by the resulting membrane potential rather than by a pH gradient. For both mitochondria and chloroplasts, however, the catalytic end of ATP synthetase is at a pH of about 8 and located in the large organelle compartment that is packed full of soluble enzymes (matrix and stroma, respectively). Consequently, it is here that all of the organelle's ATP is made (Figure 9–52).

While there are similarities between mitochondria and chloroplasts, chloroplasts are constructed in a way that makes their electron and proton-transport processes easier to study. Thus, by breaking both the inner and outer membranes of a chloroplast, isolated thylakoid discs can be obtained intact. These thylakoids resemble submitochondrial particles in having a membrane whose electron-transport chain has its $NADP^+$-, ADP-, and phosphate-utilization sites all freely accessible to the outside. But thylakoids represent an undisturbed native structure, and they are much more active than are the submitochondrial particles created artificially from mitochondria. As a result, several of the first experiments that demonstrated the central role of chemiosmotic mechanisms were carried out with chloroplasts rather than with mitochondria.

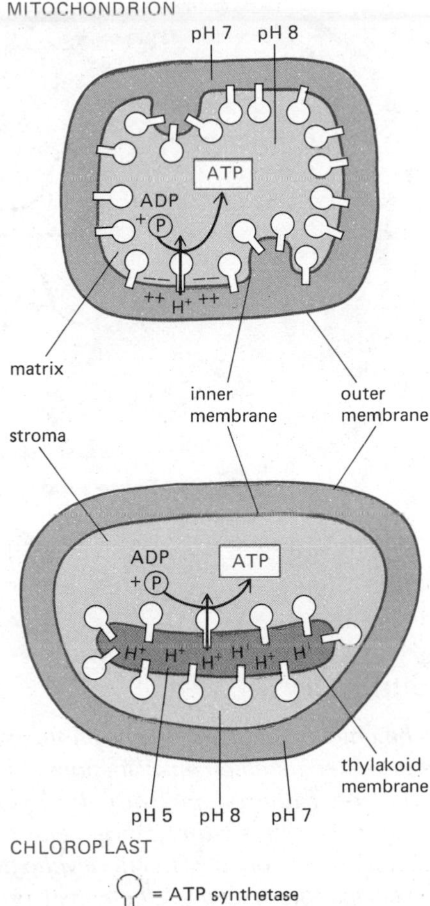

Figure 9–52 Comparison of proton flows and ATP synthetase orientations in mitochondria and chloroplasts. Those compartments with similar pH have been colored the same (see text).

Like the Inner Mitochondrial Membrane, the Inner Chloroplast Membrane Contains Carrier Proteins to Facilitate Metabolite Exchange with the Cytosol[30]

Although the light-driven reactions of photosynthesis are most readily studied in chloroplast preparations in which the inner and outer membranes have been broken or removed, such chloroplasts fail to carry out photosynthetic CO_2 fixation because of the absence of important substances that are normally present in the stroma. Chloroplasts can also be isolated in a way that leaves their inner membrane intact. In such chloroplasts, the inner membrane can be shown to have a selective permeability that reflects the fact that it contains specific carrier proteins. Most notably, much of the glyceraldehyde 3-phosphate produced by the CO_2 fixation that they catalyze is transported out of the chloroplasts by an efficient antiport system which exchanges three-carbon sugar-phosphates for inorganic phosphate.

Glyceraldehyde 3-phosphate normally provides the cytosol with an abundant source of carbohydrate that is used by the cell as the starting point for many other biosyntheses. But this is not all that this three-carbon compound provides. Once it reaches the cytosol, it is readily converted back to 3-phosphoglycerate by means of a reaction pathway that generates one molecule of ATP and one of NADPH (the reverse of the two-step reaction that forms glyceraldehyde 3-phosphate in the carbon-fixation cycle, see Figure 9–44). As a result, the export of glyceraldehyde 3-phosphate from the chloroplast also provides the main source of NADPH and ATP for the rest of cellular metabolism (Figure 9–53).

Chloroplasts Also Carry Out Other Biosyntheses[31]

In addition to photosynthesis, the chloroplast carries out many biosynthetic processes of importance to the plant cell. For example, all of the cell's fatty acids are made by enzymes located in the chloroplast stroma, using the ATP, NADPH, and carbohydrate readily available there. Moreover, the reducing power of light-activated electrons drives the reduction of nitrite (NO_2^-) to ammonia (NH_3) in the chloroplast; this ammonia provides the plant with nitrogen required for the synthesis of amino acids and nucleotides. The metabolic importance of the chloroplast therefore extends far beyond its unique role in photosynthesis, and it is reflected in the large fraction of the total cytoplasmic volume that these organelles typically occupy (Figure 9–40).

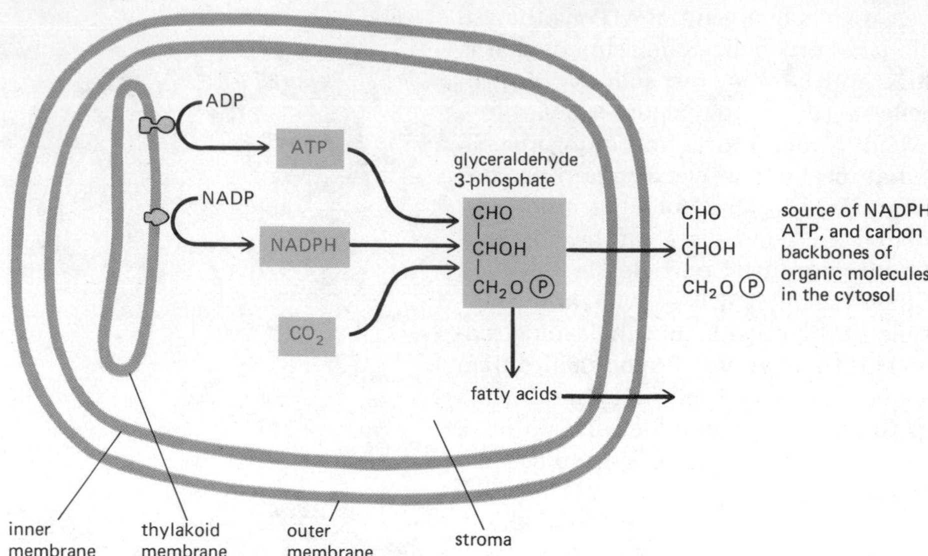

Figure 9–53 An outline sketch of the central pathways of chloroplast metabolism. Many other reactions that take place in chloroplasts and contribute to important cellular processes are not shown here. Note that the ATP and NADPH produced by the chloroplast are used directly for biosynthesis in the chloroplast stroma. Unlike the mitochondrion, the chloroplast supplies the cytosol with ATP only indirectly, as shown.

Summary

The chloroplast carries out photosynthesis, obtaining the high-energy electrons that enter its electron-transport chain by using sunlight to excite a chlorophyll electron rather than from the oxidation of foodstuffs. In consequence, it is primarily a biosynthetic organelle rather than a degradative one. Two types of electron flow occur: (1) a noncyclic flow is mediated by two photosystems linked in series that transfer electrons from water to NADP$^+$ to produce NADPH, with the concomitant production of ATP; and (2) a cyclic flow is mediated by a single photosystem through which electrons circulate in a closed loop, with the production only of ATP. Both of these electron-transport processes occur in the thylakoid membrane, and they cause protons to be pumped into the thylakoid space. A backflow of protons through ATP synthetase then produces all of the chloroplast's ATP in the stroma.

The ATP and NADPH made by photosynthesis drive many different biosynthetic reactions in the chloroplast stroma, including the all-important carbon-fixation cycle that creates carbohydrate from CO_2. This carbohydrate is exported to the cell cytosol as a three-carbon sugar-phosphate, along with numerous other products of chloroplast metabolism.

The Evolution of Electron-Transport Chains

Much of the structure, function, and evolution of cells and organisms can be related to their need for energy. We have seen that the fundamental mechanisms for harnessing energy from such disparate sources as light and the oxidation of glucose are the same. Apparently an effective method for synthesizing ATP arose early in evolution and has since been conserved with only small variations. How did the crucial individual components—ATP synthetase, redox-driven proton pumps, and photosystems—first arise? Hypotheses about events occurring on an evolutionary time scale are inevitably speculative and not readily tested. But clues abound, both in the many different primitive electron-transport chains that survive in some present-day bacteria and in geological evidence concerning the environment of the earth billions of years ago.

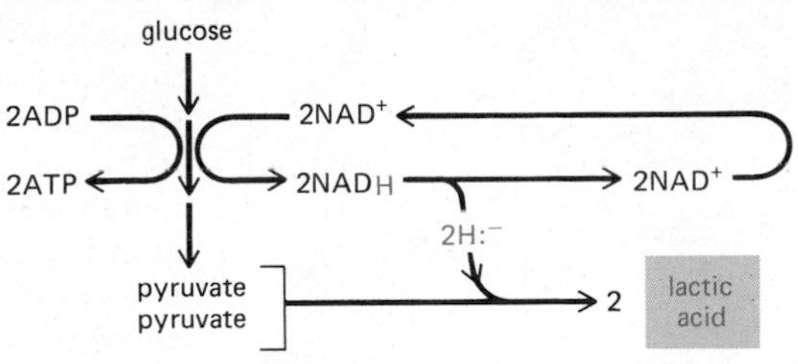

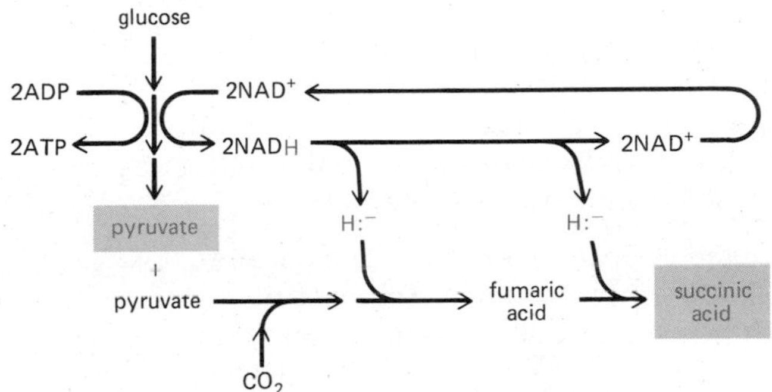

Figure 9–54 Outline of two different types of fermentation processes. (A) The NAD^+ used up in glycolysis is regenerated by the transfer of a hydride ion from NADH (plus a H^+ from the solution) to pyruvate to produce lactic acid (the reverse of the type of reaction shown previously in Figure 9–16). The lactic acid is excreted. (B) The NAD^+ used up in glycolysis is regenerated by two successive transfers of hydride ions from NADH (plus protons from solution) to compounds that produce succinic acid. The succinic acid is excreted and a molecule of pyruvate is saved for biosynthesis.

The Earliest Cells Produced ATP by Fermentation Processes[32]

As explained in Chapter 1, the first living cells are thought to have arisen in an environment that lacked oxygen but was rich in geochemically produced organic molecules. The earliest metabolic pathways producing ATP must therefore have resembled present-day forms of fermentation.

In the process of **fermentation,** ATP is made by a substrate-level phosphorylation event (p. 70) that harnesses the energy released by a reaction pathway in which a hydrogen-rich organic molecule, such as glucose, is partly oxidized. Without oxygen available to serve as a hydrogen acceptor, the hydrogens lost from the oxidized molecules must be transferred (via NADH or NADPH) to a different organic molecule (or to a different part of the same molecule), which thereby becomes more reduced. At the end of the fermentation process, one (or more) of the organic molecules produced is excreted into the medium as a metabolic waste product; others, such as pyruvate, are retained by the cell for biosynthesis.

In different organisms, the excreted end products are different, but they tend to be organic acids (carbon compounds carrying a COOH group). The most important such products in bacterial cells include lactic acid (which also accumulates in anaerobic mammalian glycolysis; see p. 71) and formic, acetic, propionic, butyric, and succinic acids. Two different fermentation pathways of present-day bacteria are illustrated in Figure 9–54.

The Evolution of Energy-conserving Electron-Transport Chains Enabled Anaerobic Bacteria to Use Nonfermentable Organic Compounds as a Source of Energy[33]

The early fermentation processes would have provided not only the ATP but also the reducing power (as NADH or NADPH) required for essential biosyntheses, and many of the major metabolic pathways probably evolved while

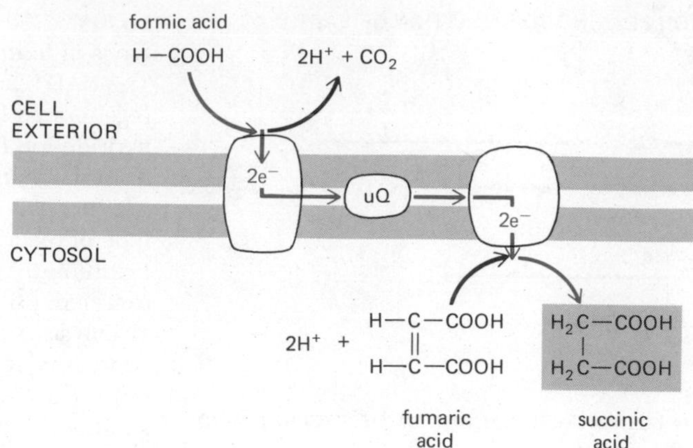

Figure 9–55 The oxidation of formic acid by fumaric acid as mediated by an energy-conserving electron-transport chain. Present-day bacteria, including *E. coli,* can pump protons across their plasma membranes using this membrane-bound electron-transport system. The redox potential of the formic-acid–CO_2 pair is −420 millivolts, while that of the fumaric-acid–succinic-acid pair is +30 millivolts.

fermentation was the only mode of energy production. In the course of time, however, the metabolic activities of these procaryotic organisms must have changed the local environment, leading to the evolution of new biochemical pathways. The following series of changes seems likely to have been the eventual consequence of the accumulation of fermentative waste products:

Stage 1. Because of the continuous excretion of acids, the pH of the environment was lowered and transmembrane proton pumps evolved to pump H^+ out of the cell to prevent death from intracellular acidification. One of these pumps may well have used the energy available from ATP hydrolysis and could have been the ancestor of the present-day ATP synthetase.

Stage 2. At the same time that the nonfermentable organic acids were accumulating and favoring the evolution of an ATP-consuming proton pump, the supply of ready-made fermentable nutrients that provided the energy both for the pumps and for all other vital processes was dwindling. The resulting selective pressures strongly favored bacteria that could excrete H^+ without hydrolyzing ATP, allowing the ATP to be conserved for other cellular activities. Such pressures probably led to the first membrane-bound proteins that could use electron transport between molecules of different redox potential as the energy source for transporting H^+ across the plasma membrane. Some of these proteins must have found their electron donors and electron acceptors among the nonfermentable organic acids that had accumulated. Such electron-transport proteins can in fact be found in present-day bacteria: for example, Figure 9–55 shows how some present-day bacteria pump protons by using the relatively small amount of redox energy derived from the transfer of electrons from formic acid to fumaric acid.

Stage 3. Eventually some bacteria evolved electron-transport systems that were efficient enough to harness more redox energy than needed just to maintain their internal pH. A large electrochemical gradient generated by excessive H^+ pumping enabled protons to leak back into the cell through the ATP-driven proton pumps, thereby running them in reverse so that they functioned as ATP synthetases and made ATP. Because such bacteria required much less of the increasingly scarce supply of fermentable nutrients, they proliferated at the expense of their neighbors.

These three hypothetical stages in the evolution of oxidative phosphorylation mechanisms are summarized in Figure 9–56.

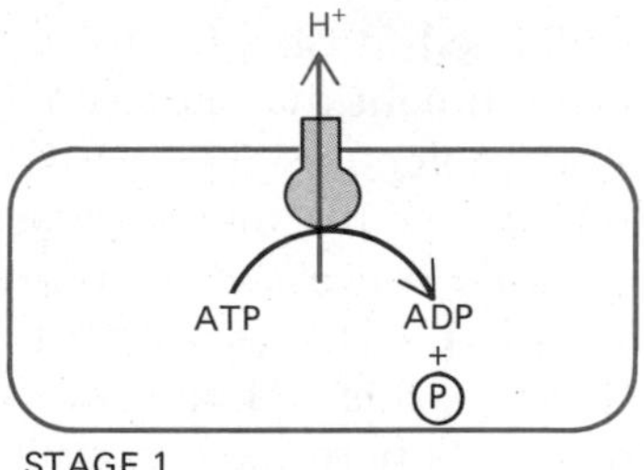

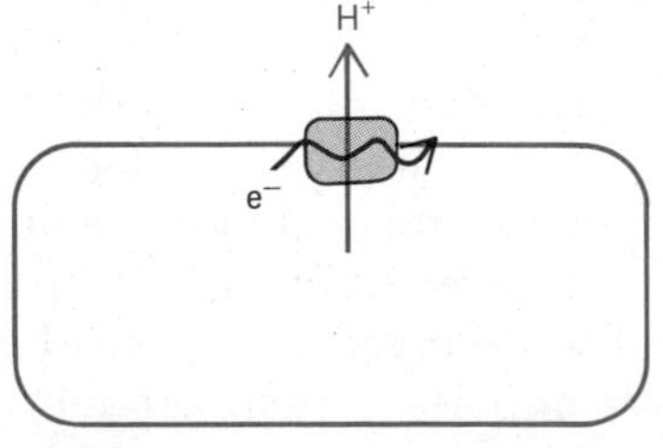

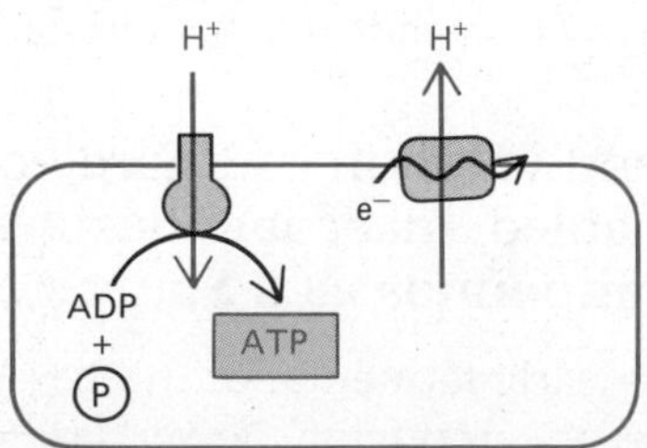

Figure 9–56 The three hypothetical stages for the evolution of oxidative phosphorylation mechanisms (see text).

By Providing an Inexhaustible Source of Reducing Power, Photosynthetic Bacteria Overcame a Major Crisis in the Evolution of Cells[34]

While the evolutionary steps just outlined would have solved the problem of maintaining both a neutral cellular pH and an abundant store of energy, they would have left unsolved another equally serious problem. The depletion of fermentable organic nutrients meant that some alternative source of carbon had to be found to make the sugars that served as the precursors of so many other cellular molecules. The carbon dioxide in the atmosphere provided an abundant potential carbon source. But to convert carbon dioxide into an organic molecule such as a carbohydrate requires that the fixed carbon dioxide be reduced by a strong hydrogen donor, such as NADH or NADPH, which can provide the two electrons needed to generate each (CH_2O) unit from CO_2 (see Figure 9–44). Early in cellular evolution, such strong reducing agents would have been plentiful as products of fermentation processes. But, as the supply of fermentable nutrients dwindled and a membrane-bound ATP synthetase began to produce most of the ATP, the plentiful supply of NADH would also have disappeared. It thus became imperative for cells to evolve a new way of generating a source of strong reducing power.

The main electron donors available in the environment after the disappearance of the fermentable molecules were the organic acids produced by the anaerobic metabolism of carbohydrates, hydrogen sulfide (H_2S), and water. But the reducing power of all of these molecules is far too weak to be useful for carbon dioxide fixation. It is thought that this problem was first solved by the ancestors of the present-day green sulfur bacteria, which made the necessary evolutionary breakthrough by developing a mechanism that made use of energy from light. Present-day green sulfur bacteria use light energy to transfer hydrogen atoms (as an electron plus a proton) from hydrogen sulfide to NADPH, thereby creating the strong reducing power required for carbon fixation (Figure 9–57). Because the electrons removed from H_2S are at a much more negative redox potential than those of water (-230 mV compared with $+820$ mV for water), one quantum of light absorbed by the single photosystem present in these bacteria is sufficient to achieve a high enough redox potential to generate NADPH, and a relatively simple photosynthetic electron-transport chain suffices.

The next step, which came with the development of the cyanobacteria some 2×10^9 years ago, was the evolution of organisms capable of using water

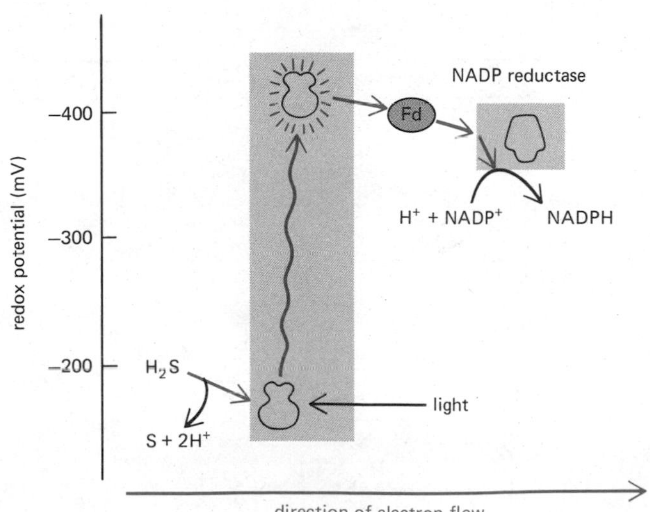

Figure 9–57 The general flow of electrons in a relatively primitive form of noncyclic photosynthesis observed in present-day green sulfur bacteria.

as the hydrogen source for carbon dioxide reduction. This entailed the addition of a second photosystem, acting in series with the first, to bridge the enormous gap in redox potential between H_2O and NADPH (see Figure 9–49). The biological consequences of this evolutionary step were extremely far-reaching. For the first time there were organisms that made only very minimal chemical demands on their environment; they could therefore spread and evolve in ways denied to the earlier photosynthetic bacteria, which needed H_2S or organic acids as a source of electrons. In consequence, large amounts of biologically synthesized reduced organic materials accumulated, and for the first time oxygen entered the atmosphere. This opened the way for the subsequent development of bacteria and the more complex life forms that relied on aerobic metabolism to make their ATP—harnessing the large amount of energy released by breaking down carbohydrates and other reduced organic molecules all the way to CO_2 and H_2O. An outline of the suspected evolutionary pathways involved is presented in Figure 9–58.

Evolution is always conservative, taking parts of the old and building upon them to create something new. Thus, parts of the electron-transport chain that were derived to service anaerobic bacteria more than 3 billion years ago are probably what we see, in an altered form, in the mitochondria and chloroplasts of higher eucaryotes. There is an especially striking homology between the function of the enzyme complex that pumps protons in the central segment of the mitochondrial respiratory chain (the b-c_1 complex) and corresponding segments of the electron-transport chains of both bacteria and chloroplasts today (Figure 9–59).

Figure 9–58 A phylogenetic tree of the probable evolution of mitochondria and chloroplasts and their bacterial ancestors. As indicated, oxygen respiration seems to have evolved independently from photosynthesis at least three times: in the green, purple, and blue-green (cyanobacterial) lines of photosynthetic bacteria. An aerobic purple bacterium gave rise to the mitochondrion while an aerobic blue-green bacterium (a cyanobacterium) gave rise to the chloroplast.

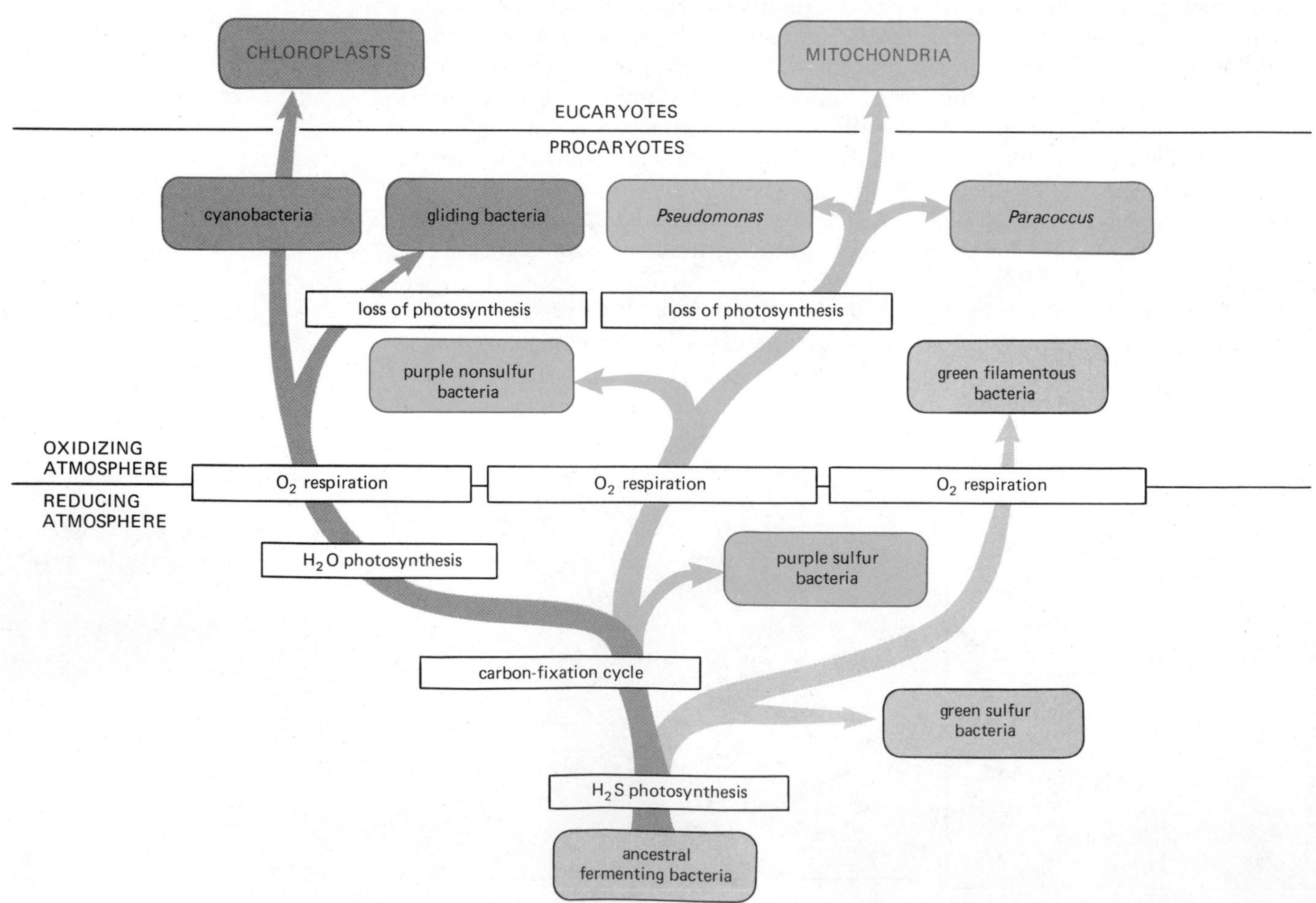

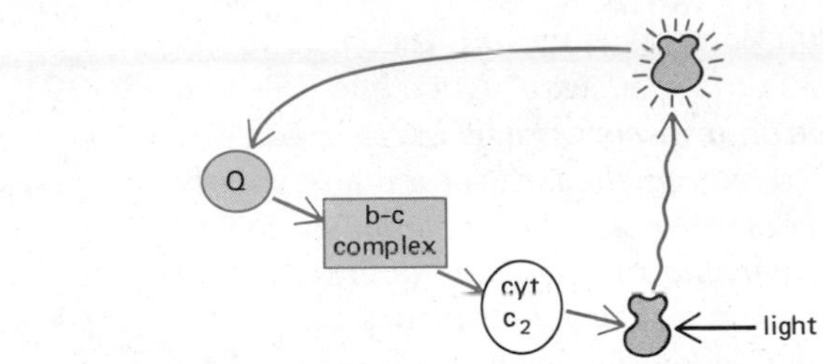

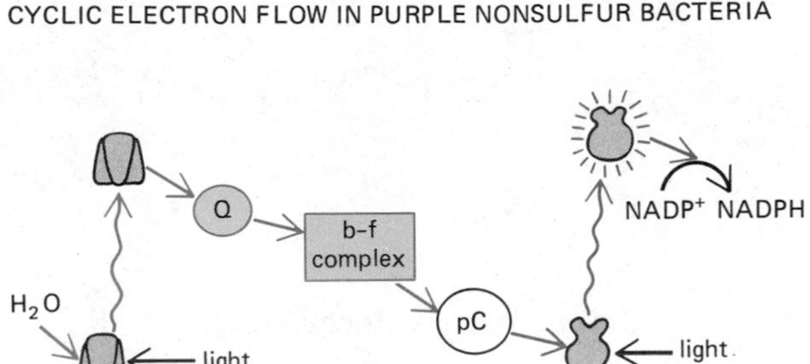

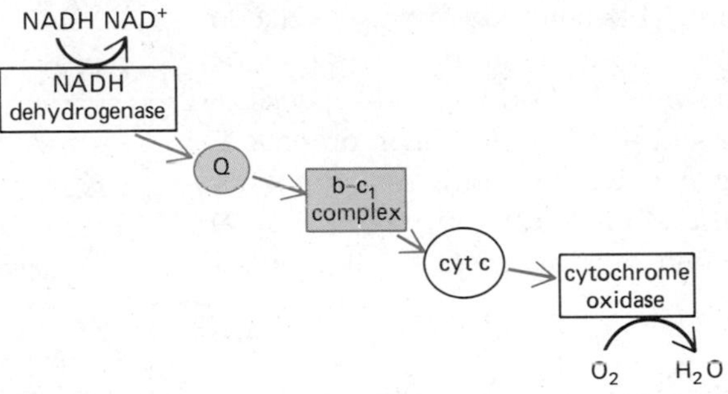

Figure 9–59 Bacteria, chloroplasts, and mitochondria all contain an electron-transport complex that closely resembles the b-c_1 complex of mitochondria. The complexes all accept electrons from a ubiquinonelike carrier (here designated as Q) and pump protons across their respective membranes. They are assumed to be evolutionarily related.

Summary

Early cells were bacteriumlike organisms living in an environment rich in highly reduced organic molecules that had been formed by natural processes over the course of hundreds of millions of years. They derived most of their ATP by converting these molecules to a variety of organic acids that were released as waste products. These fermentations caused an acidification of the medium that may have led to the evolution of the first membrane-bound H^+-pumps as a means of maintaining neutral pH in the cell interior. It is clear from the properties of present-day bacteria that the cooperation of an electron-transport-driven H^+-pump and a reversible ATP-driven H^+-pump first arose in this anaerobic environment. In this way, the relatively small amounts of energy released by redox reactions between accumulated nonfermentable compounds could be harnessed for generating ATP.

The proliferation of bacteria that used preformed organic molecules as the source of both carbon and reducing power could not proceed far, because these sources were replenished by geochemical processes only very slowly. The depletion of fermentable organic nutrients therefore led to the evolution of photosynthetic bacteria that could use carbon dioxide to make carbohydrates. By combining parts of the electron-transport chains that had developed earlier, light energy was harvested by primitive cells resembling present-day green sulfur bacteria to generate the reducing power required for carbon fixation. The subsequent appearance of the more complex photosynthetic electron-transport chains of the cyanobacteria caused large amounts of oxygen to appear in the atmosphere. Moreover, because life could proliferate over large areas of the earth, reduced organic molecules accumulated again. Once both organic molecules and oxygen were abundant, electron-transport chains became adapted for the transport of electrons from NADH to oxygen; the large amount of energy released was converted to useful forms, as efficient aerobic metabolism developed in many bacteria. Exactly the same aerobic mechanisms operate in the mitochondria of eucaryotic cells, and there is considerable evidence that these mitochondria have derived from early independent bacteria.

The Biogenesis of Mitochondria and Chloroplasts[35]

New cytoplasmic organelles must be generated to keep pace with cell growth and division as well as to compensate for the continuing degradation (turnover) of organelles in resting cells. Organelle biosynthesis requires the ordered synthesis of the requisite proteins and lipids and the delivery of each component to the correct cellular compartment. This complex process is accomplished in two basically different ways. One is to make an entirely new organelle; for example, lysosomes are made *de novo* by the budding off of specialized portions of the Golgi apparatus, as discussed in Chapter 7. The other is for an existing organelle to grow and divide in two, which is how new mitochondria and chloroplasts arise. Because of the similar ways in which they are produced, these energy organelles will often be discussed together in this section.

Mitochondria and Chloroplasts Contain Separate Genetic Systems That Are Required for Their Replication[36]

The biosynthesis of both mitochondria and chloroplasts involves the contribution of two separate genetic systems: relatively few proteins are encoded by organelle DNA and made on ribosomes within the organelle, while many are encoded by nuclear DNA, made on cytosolic ribosomes, and then imported into the organelle.

In the intact cell, the contributions of the two genetic systems to the construction of an organelle are closely coupled by poorly understood feedback mechanisms. Fortunately for the study of energy-organelle biosynthesis, this coupling is not absolute: isolated organelles will continue to make organelle DNA, RNA, and proteins for brief periods in a test tube. This has provided one means of determining which genes are present in organelle DNA and which proteins are made on organelle ribosomes. The other is to use specific inhibitors on intact cells. For example, the drug *cycloheximide* inhibits cytosolic protein synthesis but does not affect protein synthesis by chloroplast or mitochondrial ribosomes. Conversely, various antibiotics—such as chloramphenicol, tetracycline, and erythromycin—inhibit protein synthesis in energy organelles, but they have no major effect on cytosolic protein synthesis. Some of the inhibitors that differentially affect cytosolic and organelle protein synthesis are indicated in Figure 9–60.

The traffic of proteins between the cytosol and the energy organelles is all in one direction: specific proteins are imported into mitochondria and chloroplasts, but no protein is known to be exported from these organelles. Moreover, RNA and DNA molecules are not exchanged between the organelles and the rest of the cell; therefore, if a functional genetic system is lost from an organelle, the cell cannot replace it.

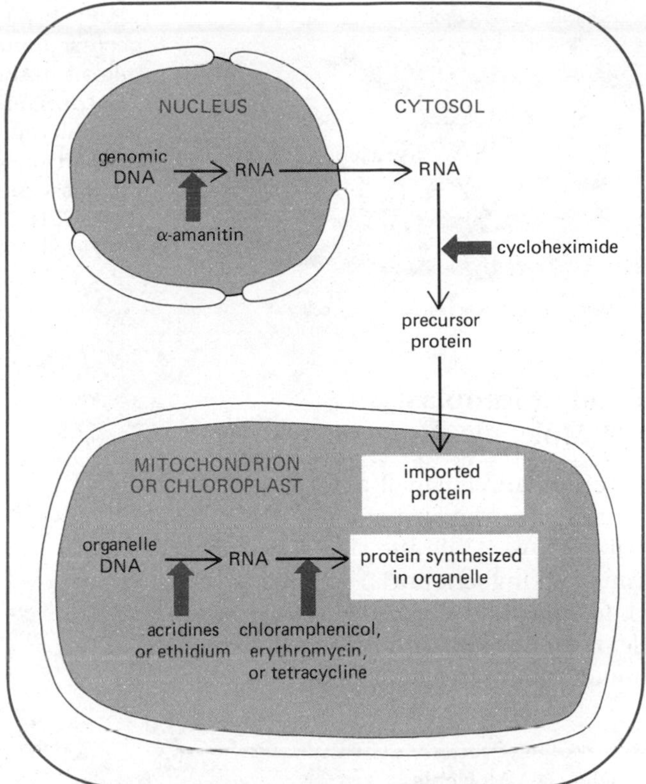

Figure 9–60 An overview of the biosynthesis of mitochondrial and chloroplast proteins. Each thick arrow indicates the site of action of an inhibitor that is specific for either mitochondrial or cytosolic protein synthesis. See text for other details.

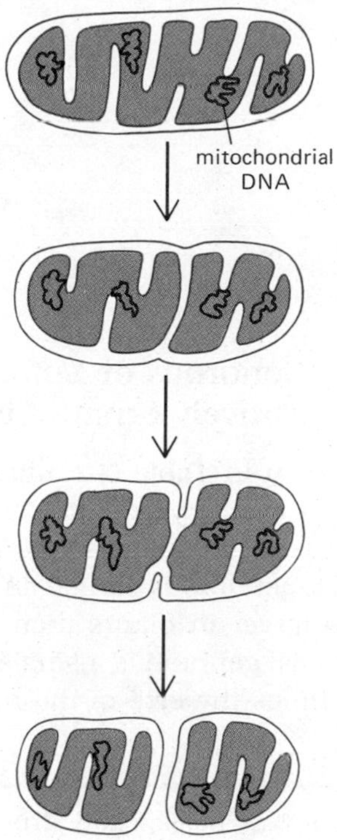

mitochondrial DNA

Figure 9–61 Diagram of a dividing mitochondrion. The pathway shown has been postulated from static views of dividing mitochondria like that in Figure 9–62.

As a Cell Grows, the Number of Mitochondria and Chloroplasts It Contains Is Increased by Organelle Division[37]

New mitochondria and chloroplasts are never made *de novo;* they always arise by the growth and division of existing mitochondria and chloroplasts. Observations of living cells indicate that mitochondria not only divide but occasionally also fuse with each other. On average, however, each organelle must double in mass and then divide in half once each cell generation. Electron micrographs suggest that organelle division begins by an inward furrowing of the inner membrane, as occurs in cell division in many bacteria (Figures 9–61 and 9–62); thus, it appears to be a controlled process rather than an event caused by an accidental pinching in two.

In most cells, individual energy organelles divide throughout interphase, out of phase with either the division of the cell or with each other. Similarly, the replication of organelle DNA is not limited to the nuclear DNA synthesis phase (S phase) but occurs throughout the cell cycle. Although individual DNA molecules seem to enter replication at random—so that in a given cell cycle some may replicate more than once and others not at all—the total number of organelle DNA molecules will double in every cell cycle, keeping the amount of organelle DNA per cell constant.

There are, however, exceptions to this general pattern: in some algae that contain only one or a few chloroplasts, the organelle divides just prior to cytokinesis in a plane that is identical to the future plane of cell division. Therefore, it is clear that organelle division can be precisely controlled by the cell in some circumstances (Figure 9–63).

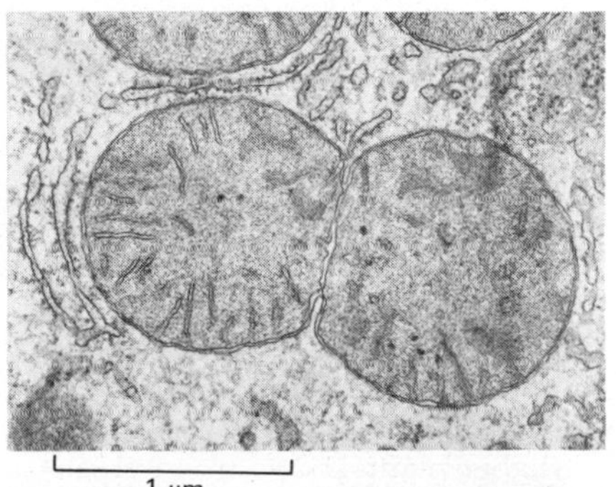

Figure 9–62 Electron micrograph of a dividing mitochondrion in a liver cell. (Courtesy of Daniel S. Friend.)

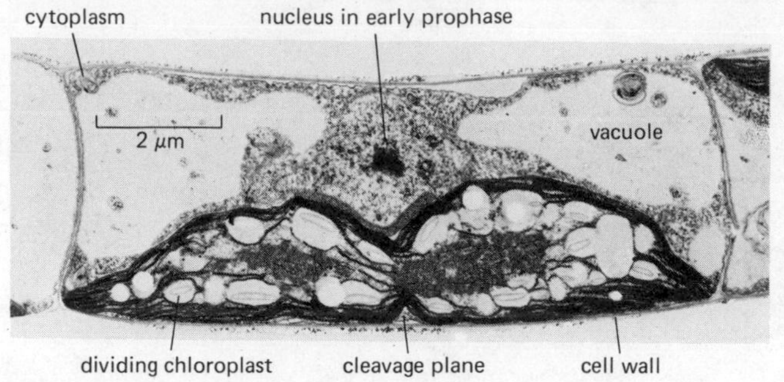

Figure 9–63 Chloroplast division occurs at a unique point early in mitosis in the simple filamentous alga *Klebsormidium*. This cell has only one chloroplast, and its plane of division coincides with the plane of the subsequent division of the cell. (From J. D. Pickett-Heaps, *Cytobios* 6:167–183, 1972.)

Most Genomes of Mitochondria and Chloroplasts Are Relatively Small, Circular DNA Molecules[38]

As shown in Table 9–2, organelle DNAs are relatively small and simple, and in higher eucaryotes they are circular molecules. The size of many organelle DNAs is in the same range as that of viral DNAs. While the size of the chloroplast genome is similar in all organisms, the mitochondrial genome is very much larger in plants than in animals. In mammals (Figure 9–64), the mitochondrial genome is a single DNA circle of about 11 million daltons (less than 10^{-5} times the size of the nuclear genome). However, plants contain a mito-

Table 9–2 The Size and Structure of Organelle DNA Molecules

Species	Structure	Mass ($\times 10^6$ daltons)*	Comments
Mitochondrial DNA			
Animals (from flatworm to man)	circular	9 to 12	in any one species, only a single size of molecule is found
Higher plants	circular	varies	in all plants studied there are several size classes of DNA circles, with a total genetic information content of 300 to 1000 $\times$ 10^6 daltons, depending on the species
Fungi			
Baker's yeast (*Saccharomyces*)	circular	50	
Kluyveromyces	circular	22	
Protozoa			
Plasmodium (Malaria)	circular	18	
Paramecium	linear	27	
Chloroplast DNA			
Algae			
Chlamydomonas	circular	120	
Euglena	circular	90	
Higher plants	circular	85 to 97	in any one species, only a single size of molecule is found

*There are approximately 1500 DNA base pairs per 10^6 daltons of DNA double helix.

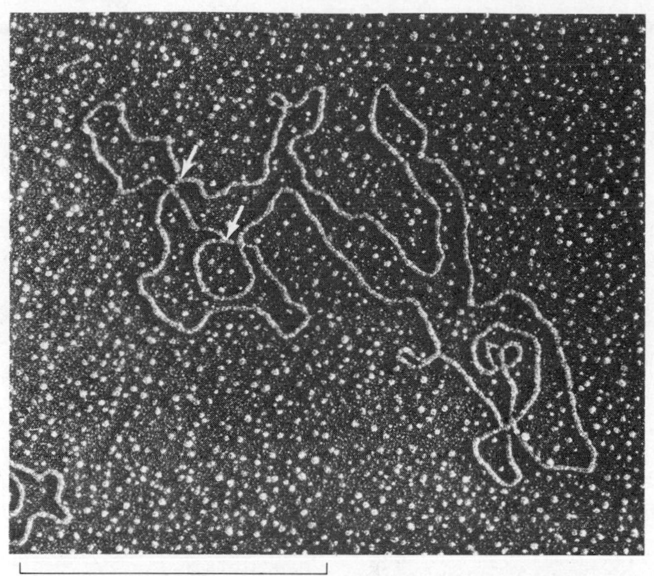

Figure 9–64 Electron micrograph of a mammalian mitochondrial DNA molecule caught during the process of DNA replication. The circular DNA genome has replicated only between the two points marked by arrows. (Courtesy of David Clayton.)

chondrial genome that is 30 to 100 times larger, composed of a number of different DNA circles. The significance of this difference between higher plants and animals is not known.

All mitochondria and chloroplasts contain multiple copies of the organelle DNA molecule, as shown by the examples in Table 9–3. This DNA is usually distributed in several separate clusters in the matrix of mitochondria and in the stroma of chloroplasts, where it it thought to be attached to the inner membrane. There are no histones in organelles, and although it is not known how the DNA is packaged, the genome structure is likely to resemble that in bacteria rather than eucaryotic chromatin.

In mammalian cells, mitochondrial DNA makes up less than 1% of the total cellular DNA. However, in other cells—such as the leaves of higher plants or the very large egg cells of amphibia—a much larger fraction of the cellular DNA may be in organelles (Table 9–3), and a corresponding larger fraction of cellular RNA and protein synthesis takes place in them.

Table 9–3 The Relative Amounts of Organelle DNA in Some Cells and Tissues

Organism	Tissue or Cell Type	DNA Molecules per Organelle	Organelles per Cell	Organelle DNA as Percent of Total Cellular DNA
Mitochondrial DNA				
Rat	liver	5–10	1000	1
Mouse	L-cell line	5–10	100	<1
Frog	egg	5–10	10^7	99
Yeast	vegetative diploid	2–50	2–50	15
Chloroplast DNA				
Chlamydomonas	vegetative diploid	80	2	7
Maize	leaves	20–40	20–40	15

The large variation in the number and size of mitochondria per cell in yeast is due to mitochondrial fusion and fragmentation.

Mitochondria and Chloroplasts Contain Complete Genetic Systems[39,40]

Despite the small number of proteins encoded in their genomes, energy organelles carry out their own DNA replication, DNA transcription, and protein synthesis. These processes take place in an internal compartment of the organelle: the matrix in mitochondria and the stroma in plastids. Although the proteins that mediate them are unique to the organelle, most of them are encoded not in the organelle DNA but in the nuclear genome (see p. 542). This fact is all the more surprising because the protein synthetic machinery of organelles resembles that of bacteria rather than that of eucaryotes. The resemblance is particularly close in the case of chloroplasts:

1. Chloroplast ribosomes are very similar to *E. coli* ribosomes, both in their sensitivity to various antibiotics (such as chloramphenicol, streptomycin, erythromycin, and tetracycline) and in their structure. For example, parts of the nucleotide sequences of the ribosomal RNAs of chloroplasts (from maize) and *E. coli* bacteria are strikingly similar.
2. Protein synthesis in chloroplasts starts with *N*-formylmethionine, as in bacteria, and not with methionine, as in the cytosol of eucaryotic cells.
3. Chloroplast ribosomes are able to use bacterial tRNAs in protein synthesis; it is even possible to make functional hybrid ribosomes by combining a chloroplast small subunit with an *E. coli* large subunit. In all these respects, chloroplast ribosomes differ from the ribosomes found in the cytosol of the same plant cell.
4. Chloroplast mRNAs are efficiently translated by a protein-synthesizing extract made from *E. coli* bacteria.

The mitochondrial translation system also shares a number of properties with bacterial systems: mitochondrial ribosomes are sensitive to antibacterial antibiotics, and protein synthesis starts with *N*-formylmethionine. However, there are also some fundamental differences. The most striking of these has emerged from a comparison of the nucleotide sequences of mitochondrial genes with the amino acid sequences of the proteins encoded by them. Thus, UGA, which is a stop codon in the "universal" genetic code used for nuclear-encoded proteins and in bacteria, is read as tryptophan in both mammalian and yeast mitochondria. Moreover, there are differences in the meaning of several other codons, and some codes of mammalian and yeast mitochondria are not the same (Table 9–4). These differences are a consequence of peculiarities in the mitochondrial tRNAs, which are encoded by the mitochondrial and not the nuclear genome, and which we shall discuss in more detail later. Why mitochondrial genetic codes should differ from that of bacteria and the eucaryotic nucleus is not known.

Table 9–4 Differences Between the "Universal" Genetic Code and Two Mitochondrial Genetic Codes*

Codon	Mammalian Mitochondrial Code	Yeast Mitochondrial Code	"Universal" Code
UGA	*Trp*	*Trp*	STOP
AUA	*Met*	*Met*	Ile
CUA	Leu	*Thr*	Leu
AGA, AGG	*STOP*	Arg	Arg

*Italic type indicates that the code differs from the "universal" code.

Yeasts Have Many Advantages for the Study of Mitochondrial Biogenesis, Including the Availability of Genetic Analysis[41,42]

Most experiments on the mechanism of mitochondrial biogenesis are now performed with *Saccharomyces carlsbergensis* (brewer's yeast) and *Saccharomyces cerevisiae* (baker's yeast), for several reasons. First, when grown on glucose, these yeasts have a unique ability to live by glycolysis alone and therefore without functional mitochondria. This makes it possible to study mutations in both mitochondrial and nuclear DNA that drastically interfere with mitochondrial biogenesis; such mutations are lethal in nearly all other organisms. Second, yeasts are simple unicellular eucaryotes, easy to grow and characterize biochemically. Finally, yeast cells can grow as either haploids or diploids, and they normally reproduce asexually by budding (asymmetrical mitosis). But they can also reproduce sexually: periodically two haploid cells fuse by sexual mating to form a diploid zygote, which can grow mitotically or divide by meiosis to produce new haploid cells. Experimental control of the alternation between asexual and sexual reproduction has made possible an extensive genetic analysis of the genes that affect mitochondrial function. It has also been possible by these means to determine whether such genes are located in the nucleus or in the mitochondria, because mutations in mitochondrial genes are not inherited according to the Mendelian rules that govern the inheritance of nuclear genes.

To understand the **non-Mendelian (cytoplasmic) inheritance** of mitochondrial genes, consider what happens to such genes when two haploid yeast cells mate to form a diploid zygote. In the example illustrated in Figure 9–65, one yeast carries a mutation that makes mitochondrial protein synthesis resistant to chloramphenicol, while the other is wild-type (chloramphenicol-sensitive). These genes can easily be identified in yeast growing on a substrate such as glycerol, which can be metabolized only by yeast with intact mitochondria; in the presence of chloramphenicol, only the cells carrying the mutant mitochondrial gene will grow. The diploid zygote starts out with a mixture of mutant and wild-type mitochondria, but when the zygote undergoes mitosis to produce a diploid daughter by budding, only a limited number of mitochondria enter the bud. With continuing mitotic replication, an occasional bud will receive all mutant or all wild-type mitochondria. Thereafter, all of the progeny from that bud will have mitochondria that are genetically identical. By this random process, diploid yeast progeny with only a single type of mitochondrial DNA are eventually produced—a process called *mitotic segregation.* When such a diploid cell containing a single type of mitochondrial DNA undergoes meiosis to form four haploid daughter cells, each daughter receives the same mitochondrial genes. This type of inheritance is called *non-Mendelian* or *cytoplasmic* to contrast it with the Mendelian inheritance of nuclear genes (Figure 9–65). When it occurs, it demonstrates that the gene being tested is located in the mitochondria.

In Mammals, Mitochondrial Genes Are Maternally Inherited[43]

The consequences of cytoplasmic inheritance are more profound for some animals, including ourselves, than they are for yeasts. In yeast the two haploid cells that mate are equal in size, and they contribute equal amounts of mitochondrial DNA to the zygote. Mitochondrial inheritance in yeast is therefore *biparental,* both parents contributing equally to the mitochondrial gene pool of the progeny (although, as we have just seen, after several generations of vegetative growth the *individual* progeny often contain mitochondria from one parent only). In higher animals, by contrast, the egg cell always contributes much more cytoplasm to the zygote than the sperm, and in some animals

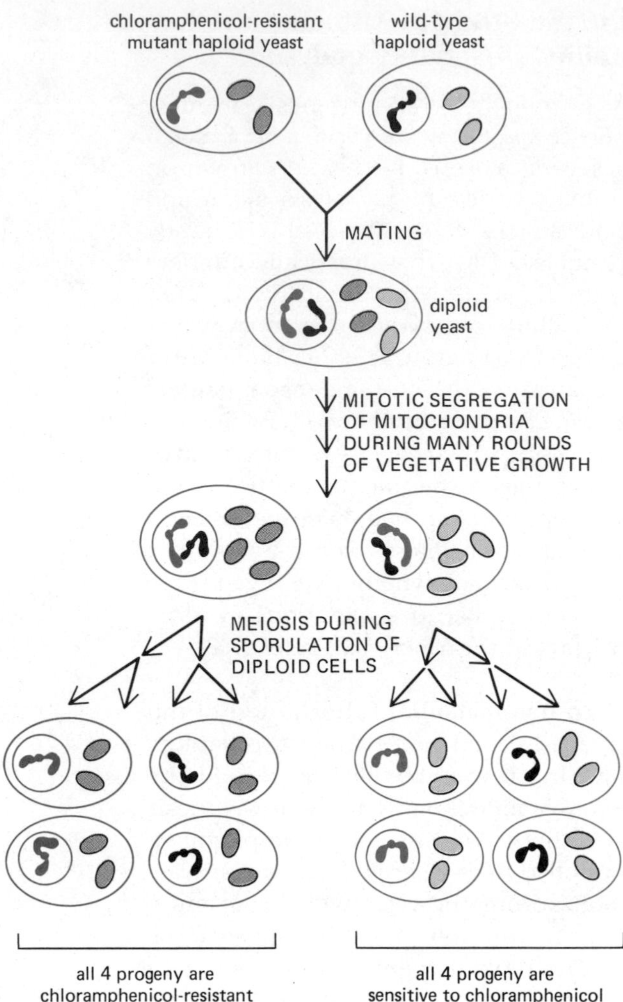

Figure 9–65 A schematic representation of the difference in pattern of inheritance between mitochondrial and nuclear genes of yeast. For each nuclear gene, two of the four cells that result from meiosis inherit the gene from one parent cell and the remaining two cells inherit the gene from the other parent cell (*Mendelian inheritance*). In contrast, because of the mitotic segregation of mitochondria during vegetative growth (see text), all four of the cells that result from meiosis can inherit their mitochondrial genes from only one of the two parent cells (*non-Mendelian* or *cytoplasmic inheritance*).

the sperm may contribute no cytoplasm at all. Therefore, one would expect mitochondrial inheritance in higher animals to be *uniparental* (or more precisely, *maternal*); and, indeed, maternal inheritance has been demonstrated. For example, different strains of laboratory rat were found that have two types of mitochondrial DNA (A and B) that differ slightly in nucleotide sequence. When rats carrying type A were crossed with rats carrying type B, the progeny contained only the maternal type of mitochondrial DNA.

Petite Mutants in Yeast Demonstrate the Overwhelming Importance of the Cell Nucleus in Mitochondrial Biogenesis[44]

One class of mitochondrial DNA mutants in yeast contains large deletions that abolish all mitochondrial protein synthesis, so that the yeast cannot make functional mitochondria. Because they form small colonies when grown in media with low glucose, they are called *cytoplasmic petite mutants.*

Although petite mutants cannot make any mitochondrial proteins and therefore cannot make functional mitochondria, they nevertheless contain *promitochondria,* which look rather like normal mitochondria, having a normal outer membrane and an inner membrane with poorly developed cristae (Figure 9–66). Promitochondria contain many of the imported enzymes that are specified by nuclear genes and made on cytosolic ribosomes, including DNA and RNA polymerases, all of the citric acid cycle enzymes, and many

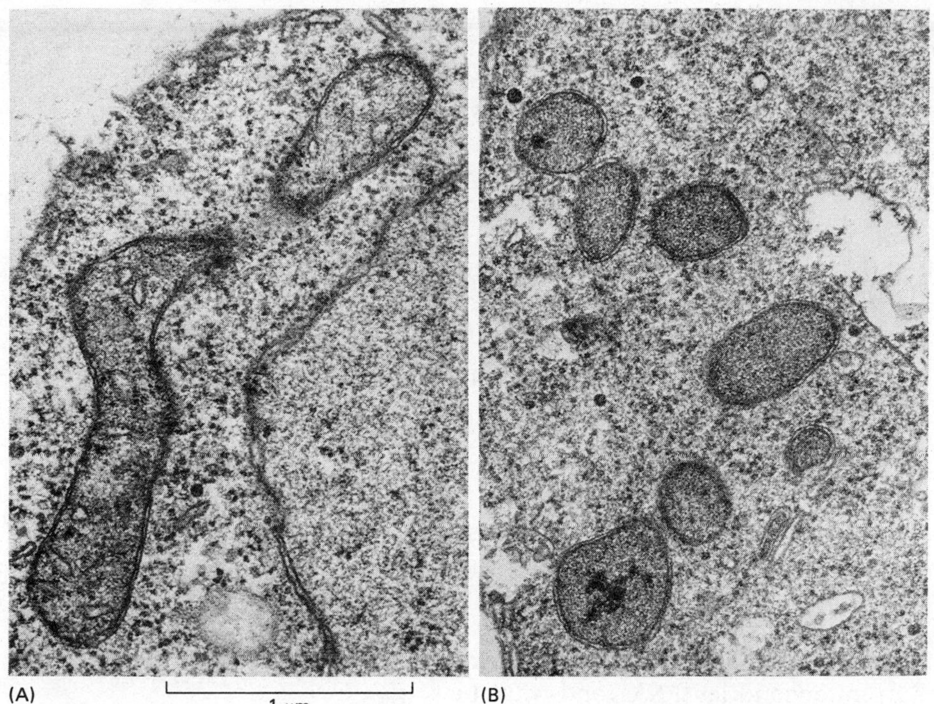

Figure 9–66 Electron micrographs of thin sections of yeast cells showing the structure of normal mitochondria (A) and the structure of mitochondria in a petite mutant in which all of the mitochondrion-encoded gene products are missing (B). In the latter case, the organelle is constructed entirely from nucleus-encoded proteins. (Courtesy of Barbara Stevens.)

inner membrane proteins. This observation dramatically emphasizes the overwhelming importance of the nucleus in mitochondrial biogenesis.

As a sidelight, it is interesting that even though the fraction of the mitochondrial DNA genome deleted in such petite mutants varies from 20% to more than 99.9%, the total amount of mitochondrial DNA per cell is always about the same as in wild-type yeast. This similarity is due to a poorly understood DNA-amplification process that results in the formation of "tandemly repeated" DNA molecules of about the same size as wild-type mitochondrial DNA: thus a petite mutant that contains 50% of the wild-type mitochondrial DNA sequence will have two copies of this segment per DNA molecule, while one that contains 0.1% of the wild-type mitochondrial DNA sequence will have 1000 copies of this segment per DNA molecule. Petite mutants consequently are useful for obtaining large amounts of specific segments of mitochondrial DNA that have, in effect, been cloned by nature.

A Large Fraction of the DNA in Yeast Mitochondria Is Noncoding[41]

Genetic methods combined with other approaches have identified 2 rRNA genes, about 30 tRNA genes, and 8 genes coding for proteins in the yeast mitochondrial genome. Of the 8 proteins, 7 are located in the inner mitochondrial membrane, while the other is a ribosomal protein. Each of these proteins is part of a larger enzyme complex made up largely of subunits imported from the cytosol.

All of the known gene products of yeast mitochondrial DNA account for less than 20% of the potential coding content of this DNA, yet an intensive search for additional genes has been unsuccessful. The reason for the discrepancy is the very large amount of noncoding DNA in the yeast genome, which falls into three categories: (1) Some yeast mitochondrial genes contain large intervening sequences (introns); for example, the genes for cytochrome b and for subunit I of cytochrome oxidase are nearly 10 times the size expected from the size of the protein product. There are striking differences in both the number and size of such introns in different yeast strains. (2) Some mitochondrial mRNAs have noncoding "leader" sequences that are much longer than the coding sequences. (3) About half of the yeast mitochondrial DNA consists of segments containing more than 95% A-T base pairs. Although the function of these segments is unknown, much of this DNA is unlikely to code for any protein or structural RNA.

Since so much yeast mitochondrial DNA appears to be noncoding, the smaller mitochondrial DNAs listed in Table 9–2 could contain most of the *same* genes with less noncoding DNA. The results have so far tended to confirm this supposition, although the comparison is incomplete.

The Complete Nucleotide Sequence of the Human Mitochondrial Genome Is Known, and It Has Some Surprising Features[45]

The type of genetic analysis that made the yeast mitochondrial genome so accessible to study is not possible with human cells. However, the relatively small size of human mitochondrial DNA makes it attractive material for the application of modern DNA sequencing technology (see p. 189), and in 1981 the complete sequence of its 16,569 nucleotides was published. By comparing this sequence with known sequences in mitochondrial tRNAs, and with the limited amino acid sequences available for proteins encoded by the mitochondrial DNA, it has been possible to locate most of the human mitochondrial genes on the circular DNA molecule (Figure 9–67).

The mapping of the 2 rRNA genes, the 22 tRNA genes, and 13 protein-coding sequences of the human mitochondrial genome has revealed several surprising features: (1) Unlike the yeast mitochondrial genome, nearly the entire nucleotide sequence of the human mitochondrial genome appears to code for protein and RNA. (2) Whereas at least 31 different tRNAs specify amino acids in the cytosol, 22 different tRNAs suffice for mitochondrial protein synthesis: the normal codon–anticodon pairing rules are relaxed in mitochondria, so that many tRNA molecules recognize any one of the four nucleotides in the third (wobble) position (p. 208). Such "2 out of 3" reading allows one tRNA to pair with any one of four different codons. This permits protein synthesis with a smaller number of different tRNA molecules, but it may be less accurate. (3) As we have mentioned before, the genetic code is altered, so that 4 of the 64 codons have different "meanings" than they have in cytosolic protein synthesis (Table 9–4).

Figure 9–67 The organization of the human mitochondrial genome, as determined by its complete nucleotide sequence. The DNAs of the bovine and mouse mitochondrial genomes have also been sequenced and have the same organization. Note that while there are 13 protein-coding regions, the functions of only 5 of the proteins produced are known (3 cytochrome oxidase subunits, 1 ATP synthetase subunit, and cytochrome b).

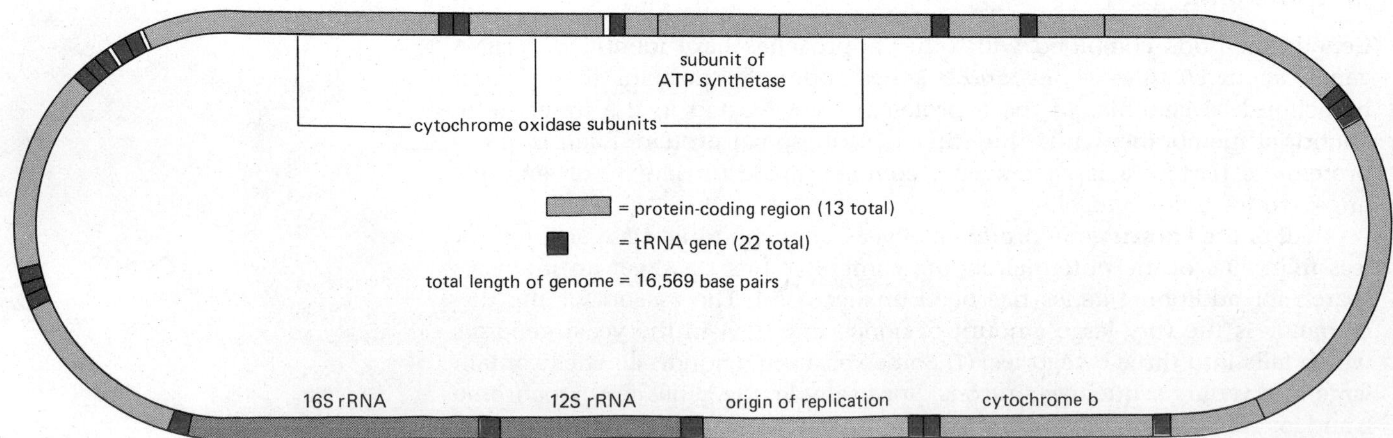

Chloroplasts Have a More Complex Genome Than Yeast and Animal Mitochondria[39,46]

Cytoplasmic inheritance was first discovered in studies of the segregation of mutant chloroplast genes in plants. Our knowledge of this process at the molecular level is still deficient for two reasons. First, plants contain two separate cytoplasmic genetic systems, and it is often difficult to determine whether a particular mutant cytoplasmic gene resides in mitochondria or in chloroplasts. Second, chloroplast DNA is larger and more complex than animal and yeast mitochondrial DNA (see Table 9–2), a difference that makes the analysis of genes in chloroplast DNA more difficult. However, there is one advantage: the close similarity of chloroplast and bacterial genetic systems allows chloroplast DNA and mRNAs to be accurately transcribed or translated (or both) in an efficient bacterial transcription-translation system. The *in vitro* decoding of chloroplast DNA in extracts of bacterial cells provides a useful method for identifying genes encoding specific proteins.

Available results show that many proteins are made in chloroplasts, but few have been identified with specific functions. Like mitochondrial DNA, the chloroplast genome contributes subunits to many multienzyme complexes that also contain subunits made on the ribosomes in the cytosol.

Whereas mitochondria exist only as mature forms, chloroplasts are present in an immature proplastid form in young plant embryos (p. 1120). If plants are grown in the dark, these proplastids fail to mature. Chloroplast development is also blocked at the proplastid stage when chloroplast DNA is defective or absent. While such chloroplast mutations are lethal in higher plants, mutant cells without functional chloroplasts are perfectly viable in unicellular algae—such as *Euglena*—if oxidizable substrates are provided. Such chloroplast mutants are analogous to the mitochondrial petite mutants of yeast.

In about two-thirds of higher plants the chloroplasts from the male parent (contained in pollen grains) do not enter the zygote, and chloroplast inheritance is therefore strictly maternal. In other plants, the pollen chloroplasts enter the zygote, making chloroplast inheritance biparental. Defective chloroplasts are a cause of *variegation* in plants: a mixture of normal and defective chloroplasts in a zygote may sort out by mitotic segregation (see p. 533) during growth and produce alternating green and white patches in leaves; the green patches contain normal chloroplasts, while the white patches contain defective chloroplasts.

The RNA Transcripts Made on Mitochondrial DNA Are Extensively Processed After Their Synthesis[47]

There is little that distinguishes RNA synthesis in chloroplasts from RNA synthesis in *E. coli*. Like bacterial mRNAs, chloroplast mRNAs do not contain a cap structure at their 5′ end or the long 3′-poly-A tails characteristic of most eucaryotic mRNA. Moreover, extensive RNA processing does not seem to be involved in the production of most mature mRNAs.

In contrast, the processing of precursor RNAs plays an important role in the two mitochondrial systems thus far studied in detail—human and yeast. In human cells both strands of the mitochondrial DNA are transcribed at the same rate from a single promoter on each strand, producing two giant RNA molecules, each containing a full-length copy of one DNA strand. Transcription is, therefore, completely symmetric. The transcripts made on one strand—called the *heavy strand (H strand)* because of its density in CsCl—are extensively processed by nuclease cleavage to yield the two rRNAs, most of the tRNAs, and about ten poly-A-containing RNAs (Figure 9–68). In contrast, the *light strand (L strand)* transcript is processed to produce only eight tRNAs and one small poly-A-containing RNA; the remaining 90% of this transcript seems to contain no useful information and is degraded. The poly-A-containing RNAs are thought to be the mitochondrial mRNAs; they lack a cap structure at their 5′ end and carry a poly-A tail of about 55 nucleotides at their 3′ end. The tail is added posttranscriptionally by a mitochondrial poly-A polymerase.

Unlike human mitochondrial genes, some yeast mitochondrial genes contain introns, which must be removed by RNA splicing. However, there is no evidence for symmetric transcription in yeast: all genes but one are transcribed from the same DNA strand, with separate promoters for several genes.

The discovery of introns in yeast mitochondrial genes (as well as in a few chloroplast genes) is surprising in view of the endosymbiont theory of mitochondrial origin, since introns have not been found in bacterial genes. Moreover, introns are often present in a mitochondrial gene in one strain of yeast and absent from the same gene in another yeast strain. For this and other reasons, it has been suggested that the introns found in the DNA of some energy organelles represent the vestiges of transposable elements (see p. 472). However, neither the origin of the split genes observed in organelle DNAs nor their effect on the organism is clear.

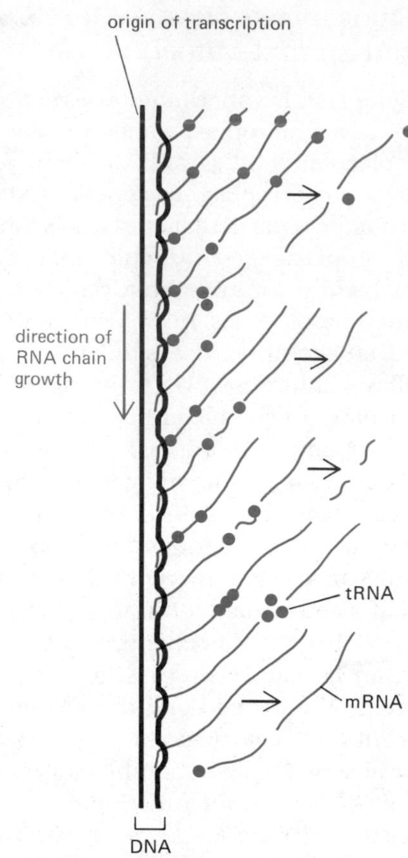

Figure 9–68 Schematic diagram of the processing of a single large transcript synthesized on the H strand of mitochondrial DNA into the 2 rRNAs, 14 tRNAs, and 10 poly-A-containing RNAs that it produces. As indicated, most of this RNA processing occurs while the RNA molecule is still being transcribed. Dots indicate short regions of RNA containing tRNA sequence. (Modified from D. Ojala, J. Montoya, and G. Attardi, *Nature* 290:470–474, 1981.)

Proteins Are Imported into Mitochondria and Chloroplasts by an Energy-requiring Process[48]

Most of the proteins found in chloroplasts and mitochondria are imported from the cell cytosol (p. 377). This raises two related questions: how does the cell direct proteins to the appropriate organelle, and how do they enter the organelle?

A partial answer has been provided by studies on the import of the small subunit (S) of the abundant enzyme *ribulose 1,5-bisphosphate carboxylase* into the chloroplast stroma. When mRNA isolated from the cytoplasm of the unicellular alga *Chlamydomonas,* or from pea leaves, is used to program a protein-synthesizing system *in vitro,* one of the many proteins produced reacts with a specific anti-S antibody. The S protein made *in vitro* is called pro-S, as it is larger than mature S by about 50 amino acids. When the completed pro-S protein is incubated with intact chloroplasts, it is taken up into the organelle and converted into mature S by an endopeptidase in the chloroplast. Mature S then associates with the large subunit of ribulose 1,5-bisphosphate carboxylase, which is made on chloroplast ribosomes, to form the active enzyme in the chloroplast stroma.

The mechanism of pro-S protein import is unknown. It is thought that pro-S binds to a receptor protein on the chloroplast outer membrane, possibly at an *adhesion site* between the outer and inner membranes (p. 378), and is then transferred into the stroma through a transmembrane pore by an energy-dependent process.

The import of proteins into the mitochondrion is generally similar. If purified yeast mitochondria are incubated with cell extracts containing newly synthesized yeast proteins in a radioactive form, the nuclear-encoded mitochondrial proteins are distinguished from the nonmitochondrial proteins in the cytosol and selectively incorporated into the mitochondria in a manner that faithfully mimics their selective uptake within the cell. Thus, outer membrane proteins, inner membrane proteins, matrix proteins, and proteins of the intermembrane space each find their way to their own special compartments within the mitochondrion.

Most of the newly synthesized proteins destined for the inner membrane, the matrix, and the intermembrane space contain an amino-terminal leader peptide that is cleaved during the import process by a specific protease located in the mitochondrial matrix. In addition, the proteins that are targeted to these three different mitochondrial compartments all require the electrochemical proton gradient across the inner membrane for their import. The mechanism for import of proteins destined for the outer mitochondrial membrane is different, since it neither requires energy nor results in a proteolytic cleavage of a longer precursor protein. These and other observations have suggested that the four different classes of mitochondrial proteins are imported according to the general scheme shown in Figures 9–69 and 9–70. It has been postulated that all the proteins except those destined for the outer membrane are inserted into the inner mitochondrial membrane by an energy-dependent process that occurs at a membrane adhesion site. The proteolytic cleavages that follow this initial insertion are thought to cause a conformational change in each protein, which either fixes it in the inner membrane, or thrusts it out of this membrane into the matrix space or into the intermembrane space, as appropriate (Figure 9–70).

The transport of proteins through the mitochondrial and chloroplast membranes is in principle analogous to the transport of proteins through the membrane of the endoplasmic reticulum described in Chapter 7. However, there are several important differences. First, transport into the matrix or the stroma involves passage through both outer and inner organelle membranes, whereas transport into the endoplasmic reticulum involves passage through only a single membrane (p. 355). In addition, whereas the import of proteins into the endoplasmic reticulum occurs by a *vectorial discharge* mechanism that begins while the incomplete protein is still being synthesized on the ribosome (*co-translational import,* p. 341), the import of proteins into both chloroplasts and mitochondria is thought to occur after the synthesis of the protein has been completed *(posttranslational import).*

Despite these differences, in each case the cell makes a precursor protein containing a "signal sequence" that directs the protein to the proper membrane. In many cases it appears that this signal sequence is cleaved from the precursor after the transport process is complete. However, some proteins are

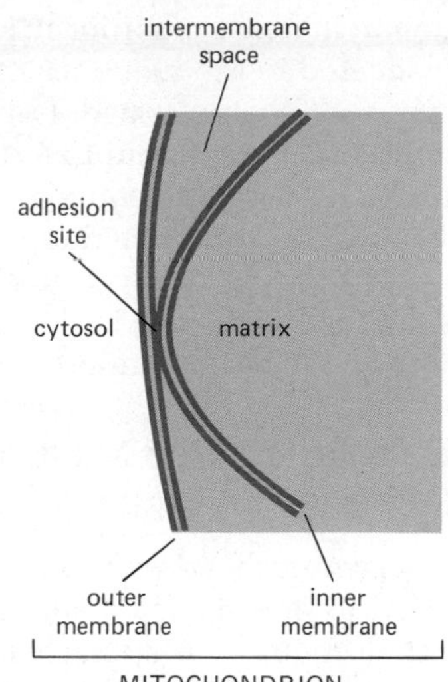

Figure 9–69 Schematic drawing of a small region of a mitochondrion that contains a membrane adhesion site thought to be involved in selective protein import into the mitochondrion. The proteins imported are those encoded by the cell nucleus and synthesized in the cytosol.

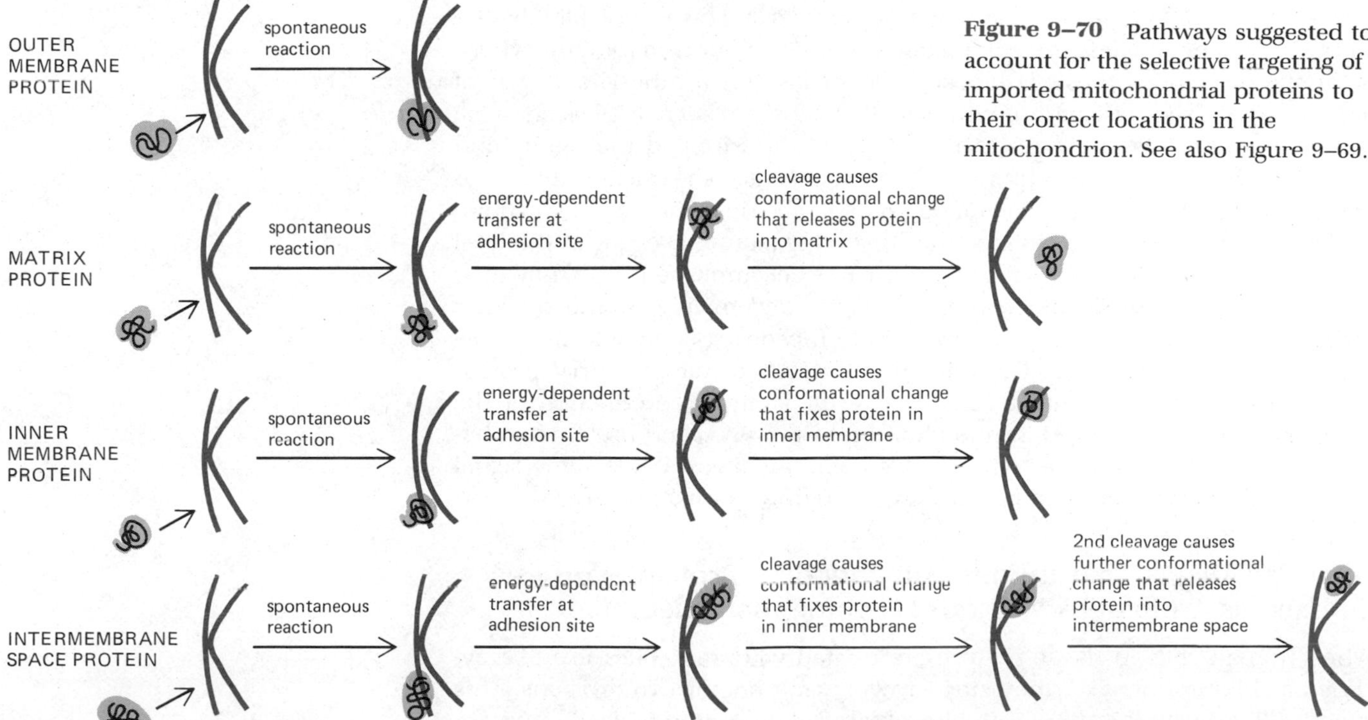

Figure 9–70 Pathways suggested to account for the selective targeting of imported mitochondrial proteins to their correct locations in the mitochondrion. See also Figure 9–69.

not made in a precursor form; in these cases the signal sequence is presumed to be located in the sequence of the mature protein. The signal sequences are not yet well characterized, but they must be of several different types, each type directing its protein to a different intracellular destination. Consider, for example, a plant cell: some of the proteins that begin their synthesis in the cytosol will be transported into mitochondria, others into chloroplasts, others into peroxisomes, and yet others into the endoplasmic reticulum. The complex processes involved in the correct intracellular routing of proteins are only beginning to be understood, and they are discussed in detail in Chapter 7.

Chloroplasts Make Most of Their Own Lipids, While Mitochondria Rely Mainly on Import[31]

The biosynthesis of new mitochondria and chloroplasts requires lipids in addition to nucleic acids and proteins. Chloroplasts tend to make the lipids that they require. In spinach leaves, for example, all cellular fatty acid synthesis takes place in the chloroplast, although desaturation of the fatty acids occurs outside. Even the major glycolipids of the chloroplast are synthesized locally.

Mitochondria, on the other hand, import most of their lipids. In animal cells, phospholipids synthesized in the endoplasmic reticulum are transferred to the outer membrane of mitochondria by phospholipid transfer proteins (see p. 351) and then move into the inner membrane, presumably at adhesion sites. The main reaction of lipid biosynthesis catalyzed by the mitochondria themselves is the conversion of phosphatidic acid to cardiolipin, a phospholipid found mainly in the mitochondrial inner membrane, which constitutes about 20 percent of its total lipid.

The Biosynthesis of Mitochondria and Chloroplasts Is Largely Controlled by the Nucleus

Little is known about how the nuclear and organelle genetic systems communicate in order to coordinate their contributions to energy-organelle synthesis. But overall control is clearly with the nucleus. This is indicated by the fact that "pro-organelles" are still made in normal amounts in mutants whose organelle protein synthesis is blocked. Organelle DNA synthesis and part of organelle RNA synthesis continue normally in these pro-organelles, showing that the synthesis and import of the enzymes involved and the timing and control of these processes requires only nuclear genes. The nucleus must also regulate the amount of proteins made on organelle ribosomes; for some chloroplast proteins, there is evidence that this regulation may occur at the level of chloroplast DNA transcription, but how it is accomplished is unknown.

Although organelle biosynthesis is controlled mainly by the nucleus, there is suggestive evidence for some kind of feedback control from the organelles, at least in the case of mitochondria. When mitochondrial protein synthesis is blocked in intact cells, the result is an overproduction of the imported enzymes involved in mitochondrial DNA, RNA, and protein synthesis, as though the cell were trying to overcome the block. While some signal from the mitochondria must be involved, its nature is unknown.

How Can Drugs That Inhibit Mitochondrial Protein Synthesis Be Used as Antibiotics Without Harming the Patient?[49]

When human cells in tissue culture are treated with antibiotics like tetracycline or chloramphenicol, they stop growing after one or two divisions. This effect is due to inhibition of mitochondrial protein synthesis, leading to abnormal mitochondria and an inadequate supply of ATP. How, then, can these drugs be used to treat bacterial infections without killing the patient? There are several answers:

1. Some antibiotics (like erythromycin) cannot pass the impermeable inner membrane of mammalian mitochondria.
2. Most cells in our body do not divide, or divide at a low rate, so that existing mitochondria are replaced only slowly (in most tissues half of the mitochondria are replaced every five days or more). Thus, a complete block of mitochondrial protein synthesis must be maintained for many days to deplete the mitochondria severely.
3. Local tissue conditions prevent some drugs from reaching the mitochondria of especially sensitive cells. For example, the high Ca^{2+} concentration in bone marrow results in the formation of Ca^{2+}-tetracycline complexes that cannot enter the rapidly dividing (and therefore vulnerable) blood cell precursors.

A combination of these factors allows some drugs that inhibit mitochondrial biosynthesis to be used as antibiotics in higher animals. There are only two instances in which side effects might be attributable to an effect on mitochondria: prolonged treatment of patients with high doses of chloramphenicol can lead to the inability of bone marrow to make either red or white blood cells, and prolonged tetracycline treatment may lead to defects in the gut epithelium. In both cases, it is still disputed whether these side effects are actually caused by the expected block in mitochondrial biogenesis.

Mitochondria and Chloroplasts Have Probably Evolved from Endosymbiotic Bacteria[50]

We have already stressed the "procaryotic" character of the organelle genetic systems, which is especially striking for chloroplasts, and we have suggested that it may be due to their origin as endosymbiotic bacteria, as discussed in Chapter 1. According to the **endosymbiont hypothesis,** eucaryotic cells started out in evolution as primitive organisms without mitochondria or chloroplasts and then established a stable endosymbiotic relation with a bacterium, whose oxidative phosphorylation system they subverted for their own use. Since animal and plant mitochondria are very similar, the endocytotic event that led to the development of mitochondria is presumed to have occurred very early in the evolution of the eucaryotic cell, before animals and plants separated. Later, chloroplasts were probably derived by a separate endocytotic event involving a cyanobacterium, creating the first plant cell (see Figure 1–21, p. 20).

Since most of the genes that encode present-day energy-organelle proteins are in the cell nucleus, it seems likely that an extensive gene transfer from organelle to nuclear DNA occurred early in eucaryote evolution. This would explain why some of the nuclear genes coding for mitochondrial proteins have retained a marked similarity to bacterial genes: for example, the amino acid sequence of the amino terminus of the mitochondrial enzyme *superoxide dismutase* in a chicken resembles the corresponding segment of the same bacterial enzyme much more than it resembles the amino terminus of the same type of enzyme found in the cytosol of eucaryotic cells.

What type of bacterium gave rise to the mitochondrion? Complete amino acid sequence and three-dimensional x-ray crystallographic analyses of cytochromes from many different types of bacteria have provided an important clue by showing that these proteins are all closely related to each other and to the cytochrome c of animal and plant mitochondrial respiratory chains. These data and other biochemical information have suggested the evolutionary tree shown previously in Figure 9–58. It appears that mitochondria are descendants of a particular type of purple photosynthetic bacterium, which had lost its ability to carry out photosynthesis and was left with only a respiratory chain (p. 526).

Why Do Mitochondria and Chloroplasts Have Their Own Genetic Systems?

Why do mitochondria and chloroplasts require their own separate genetic systems when other organelles, such as peroxisomes and lysosomes, do not? The question is not trivial, for the maintenance of a separate genetic system is costly in terms of the additional nuclear genes required: ribosomal proteins (at least 53), aminoacyl-tRNA synthetases, DNA and RNA polymerases, RNA processing and modifying enzymes, and so on—at least 90 proteins in all—must be encoded by nuclear genes for this purpose (Figure 9–71). The amino acid sequences of most of the proteins studied in mitochondria and chloroplasts differ from those of their counterparts elsewhere, and there is reason to think that these organelles have very few proteins in common with the rest of the cell. This means that the nucleus must provide at least 90 genes just

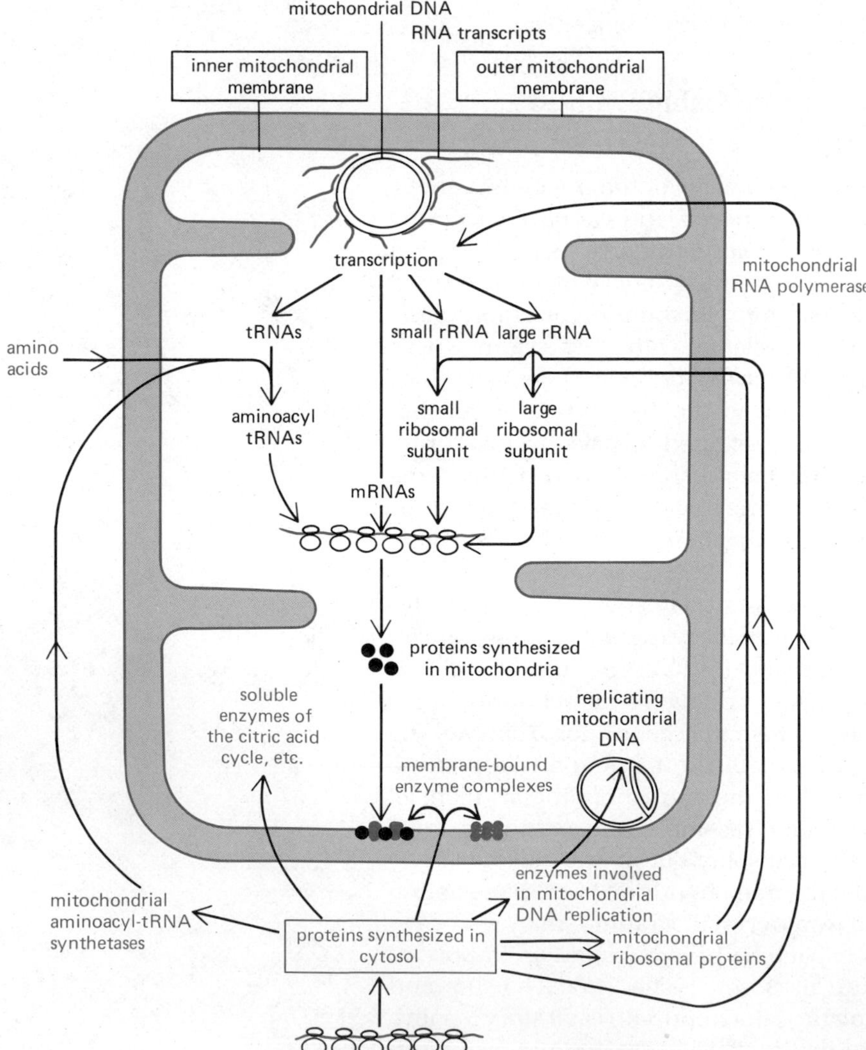

Figure 9–71 The proteins synthesized in the cytosol and then imported into the mitochondrion play a major part in running the genetic system of the mitochondrion in addition to contributing most of the organelle protein.

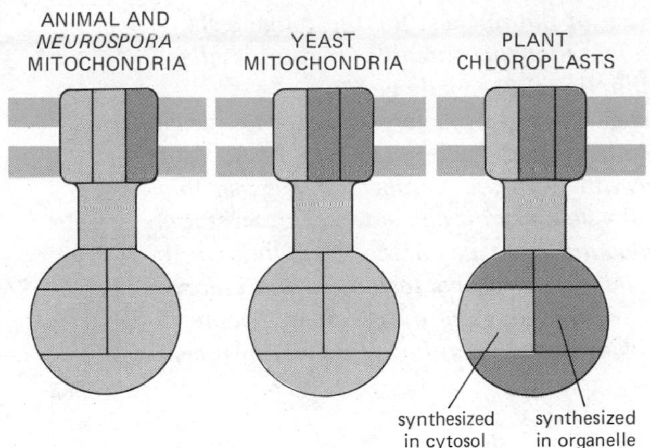

Figure 9–72 The site of synthesis of the subunits of ATP synthetase varies for different mitochondria and chloroplasts. The figure shows a schematic structure for the synthetase complex, disregarding the size and stoichiometry of subunits.

for the maintenance of the genetic system of each energy organelle. The reason for such a costly arrangement is not clear, and the hope that the nucleotide sequence of mitochondrial DNA would provide the answer has proved unfounded. We cannot think of compelling reasons why the proteins made in mitochondria should be made there rather than in the cytosol.

The usual explanation for the existence of genes in energy organelles is that some proteins have to be made inside the organelle because they are too hydrophobic to get to their site in the membrane from the outside. The results with the ATP synthetase complex, summarized in Figure 9–72, make this explanation implausible. Although the individual protein subunits in the complex are highly conserved in evolution, the site of synthesis of the subunits is not. In chloroplasts, several fairly hydrophilic proteins—including four out of five subunits of the F_1ATPase part of the complex—are made on chloroplast ribosomes. Conversely, in the fungus *Neurospora* and in animal cells, a very hydrophobic subunit (subunit 9) of the membrane part of the ATPase is made on cytosolic ribosomes and imported. The diversity in the location of the genes coding for the subunits of functionally equivalent proteins in different organisms (Figure 9–72) is difficult to explain by any hypothesis that postulates a specific evolutionary advantage of present-day mitochondrial or chloroplast genetic systems.

All of which leaves, by a process of elimination, the unsatisfactory possibility that these organelle genetic systems are an evolutionary dead end. In terms of the endosymbiont hypothesis, it would mean that the process whereby the endosymbionts transferred most of their genes to the nucleus stopped before it was complete.

Summary

Mitochondria and chloroplasts grow and divide in two, in a coordinated process that requires the contribution of two separate genetic systems: that of the organelle and that of the cell nucleus. Most of the proteins in these organelles are encoded by nuclear DNA, synthesized in the cytosol, and then individually imported into the organelle. However, a relatively small number of organelle proteins—along with all of their RNAs—are encoded by the organelle DNA and synthesized in the organelle itself. The determination of the complete sequence of the more than 16,000 nucleotides in the human mitochondrial genome has revealed that it encodes a total of 2 ribosomal RNAs, 22 transfer RNAs, and 13 different polypeptide chains. Chloroplast genomes are about 10 times larger than the human mitochondrial genome and are thought to contain many more genes. But the overwhelming importance of the nucleus for the biogenesis of both organelles is demonstrated by the fact that recognizable "pro-organelles" will form even in mutants that lack a functional organelle genome.

The ribosomes of chloroplasts closely resemble bacterial ribosomes, while mitochondrial ribosomes show both similarities and differences that make their origin more difficult to trace. However, other protein similarities suggest that both organelles originated when a primitive eucaryotic cell entered into a stable endosymbiotic relationship with a bacterium: a purple bacterium is thought to have given rise to mitochondria, and (later) a cyanobacterium to the chloroplast. This implies that, although many of the genes of these ancient bacteria still function to make organelle proteins, most of them have for unknown reasons become integrated into the nuclear genome.

References

General

Karp, G. Cell Biology, pp. 342–417. New York: McGraw-Hill, 1979.

Lehninger, A.L. Principles of Biochemistry. New York: Worth, 1982. (Chapters 16, 17, and 23.)

Nicholls, D.G. Bioenergetics: An Introduction to the Chemiosmotic Theory. New York: Academic Press, 1982.

Stryer, L. Biochemistry, 2nd ed. San Francisco: Freeman, 1981. (Chapters 13, 14, and 19.)

Cited

1. Ernster, L.; Schatz, G. Mitochondria: a historical review. *J. Cell Biol.* 91:227s–255s, 1981.

   Tzagoloff, A. Mitochondria. New York: Plenum, 1982.

   Fawcett, D.W. The Cell, 2nd ed., pp. 410–485. Philadelphia: Saunders, 1981.

   Whittaker, P.A.; Danks, S.M. Mitochondria: Structure, Function and Assembly. New York: Longman, 1979.

2. DePierre, J.W.; Ernster, L. Enzyme topology of intracellular membranes. *Annu. Rev. Biochem.* 46:201–262, 1977.

   Srere, P.A. The infrastructure of the mitochondrial matrix. *Trends Biochem. Sci.* 5:120–121, 1980.

3. Krstić, R.V. Ultrastructure of the Mammalian Cell, pp. 28–57. New York: Springer-Verlag, 1979.

   Pollak, J.K.; Sutton, R. The differentiation of animal mitochondria during development. *Trends Biochem. Sci.* 5:23–27, 1980.

4. McGilvery, R.W. Biochemistry: A Functional Approach, 2nd ed. Philadelphia: Saunders, 1979. (Chapter 28.)

5. Krebs, H.A. The history of the tricarboxylic acid cycle. *Perspect. Biol. Med.* 14:154–170, 1970.

6. Mitchell, P. Coupling of phosphorylation to electron and hydrogen transfer by a chemi-osmotic type of mechanism. *Nature* 191:144–148, 1961.

   Racker, E. From Pasteur to Mitchell: a hundred years of bioenergetics. *Fed. Proc.* 39:210–215, 1980.

7. Capaldi, R.A., ed. Membrane Proteins in Energy Transduction. New York: Dekker, 1979.

8. Wood, W.B.; Wilson, J.H.; Benbow, R.M.; Hood, L.E. Biochemistry: A Problems Approach, 2nd ed. Menlo Park, Ca.: Benjamin-Cummings, 1981. (See problems, Chapters 9, 12, and 14.)

   Nicholls, D.G. Bioenergetics: An Introduction to the Chemiosmotic Theory. New York: Academic Press, 1982. (Chapter 3.)

9. Hinkle, P.C.; McCarty, R.E. How cells make ATP. *Sci. Am.* 238(3):104–123, 1978.

10. Klingenberg, M. The ADP, ATP shuttle of the mitochondrion. *Trends Biochem. Sci.* 4:249–252, 1979.

Durand, R.; Briand, Y.; Touraille, S.; Alziari, S. Molecular approaches to phosphate transport in mitochondria. *Trends Biochem. Sci.* 6:211–214, 1981.

Nicholls, D. Some recent advances in mitochondrial calcium transport. *Trends Biochem. Sci.* 6:36–38, 1981.

LaNoue, K.F.; Schoolwerth, A.C. Metabolite transport in mitochondria. *Annu. Rev. Biochem.* 48:871–922, 1979.

11. Eisenberg, D.; Crothers, D. Physical Chemistry with Applications to the Life Sciences. Menlo Park, Ca.: Benjamin-Cummings, 1979. (Chapters 4 and 5.)

12. Racker, E. A New Look at Mechanisms in Bioenergetics. New York: Academic Press, 1976. (A personal account of the concepts and history.)

13. Racker, E.; Stoeckenius, W. Reconstitution of purple membrane vesicles catalyzing light-driven proton uptake and adenosine triphosphate formation. *J. Biol. Chem.* 249:662–663, 1974.

14. Racker, E. Structure and function of ATP-driven ion pumps. *Trends Biochem. Sci.* 1:244–247, 1976.

15. Fillingame, R.H. The proton-translocating pumps of oxidative phosphorylation. *Annu. Rev. Biochem.* 49:1079–1113, 1980.

16. Fuller, S.D.; Capaldi, R.A.; Henderson, R. Structure of cytochrome c oxidase in deoxycholate-derived two-dimensional crystals. *J. Mol. Biol.* 134:305–327, 1979.

17. Hackenbrock, C.R. Lateral diffusion and electron transfer in the mitochondrial inner membrane. *Trends Biochem. Sci.* 6:151–154, 1981.

18. White, A.; Handler, P.; Smith, E.L.; Hill, R.L.; Lehman, I.R. Principles of Biochemistry, 6th ed., pp. 271–282. New York: McGraw-Hill, 1978.

19. Wikström, M.; Krab, K.; Saraste, M. Proton-translocating cytochrome complexes. *Annu. Rev. Biochem.* 50:623–655, 1981.

20. Hanstein, W.G. Uncoupling of oxidative phosphorylation. *Trends Biochem. Sci.* 1:65–67, 1976.

21. Nicholls, D.G. Brown fat mitochondria. *Trends Biochem. Sci.* 1:128–130, 1976.

Lin, C.S.; Klingenberg, E.M. Isolation of the uncoupling protein from brown adipose tissue mitochondria. *FEBS Lett.* 113:299–303, 1980.

22. Konings, W.N.; Michels, P.A.M. Electron-transfer-driven solute translocation across bacterial membranes. In Diversity of Bacterial Respiratory Systems, Vol. 1 (C.J. Knowles, ed.), pp. 33–86. Boca Raton, Fl.: CRC Press, 1980.

23. Bogorad, L. Chloroplasts. *J. Cell Biol.* 91:256s–270s, 1981. (A historical review.)

Haliwell, B. Chloroplast Metabolism—The Structure and Function of Chloroplasts in Green Leaf Cells. Oxford, Eng.: Clarendon, 1981.

Hatch, M.D.; Boardman, N.K., eds. Photosynthesis. In The Biochemistry of Plants: A Comprehensive Treatise, Vol. 8. New York: Academic Press, 1981.

24. Miller, K.R. The photosynthetic membrane. *Sci. Am.* 241(4):102–113, 1979.

25. Ellis, R.J. The most abundant protein in the world. *Trends Biochem. Sci.* 4:241–244, 1979.

26. Bassham, J.A. The path of carbon in photosynthesis. *Sci. Am.* 206(6):88–100, 1962.

27. Björkman, O.; Berry J. High-efficiency photosynthesis. *Sci. Am.* 229(4):80–93, 1973. (C_4 plants.)

Chollet, R. The biochemistry of photorespiration. *Trends Biochem. Sci.* 2:155–159, 1977.

Heber, U.; Krause, G.H. What is the physiological role of photorespiration? *Trends Biochem. Sci.* 5:32–34, 1980.

28. Govindjee; Govindjee, R. The absorption of light in photosynthesis. *Sci. Am.* 231(6):68–82, 1974.

Bennett, J. The protein that harvests sunlight. *Trends Biochem. Sci.* 4:268–271, 1979.

Thornber, J.P.; Markwell, J.P. Photosynthetic pigment-protein complexes in plant and bacterial membranes. *Trends Biochem. Sci.* 6:122–125, 1981.

29. Levine, R.P. The mechanism of photosynthesis. *Sci. Am.* 221(6):58–70, 1969.

30. Heber, U.; Walker, D.A. The chloroplast envelope—barrier or bridge? *Trends Biochem. Sci.* 4:252–256, 1979.

Heber, U.; Heldt, H.W. The chloroplast envelope: structure, function, and role in leaf metabolism. *Annu. Rev. Plant Physiol.* 32:139–168, 1981.

31. Stumpf, P.K. Plants, fatty acids, compartments. *Trends Biochem. Sci.* 6:173–176, 1981.

Raven, J.A. Division of labor between chloroplast and cytoplasm. In The Intact Chloroplast (J. Barber, ed.), pp. 403–443. Amsterdam: Elsevier, 1976.

32. Gest, H. The evolution of biological energy-transducing systems. *FEMS Microbiol. Lett.* 7:73–77, 1980.

Gottschalk, G. Bacterial Metabolism. New York: Springer-Verlag, 1979. (Chapter 8 covers fermentations.)

33. Knowles, C.J., ed. Diversity of Bacterial Respiratory Systems, Vol. 1. Boca Raton, Fl.: CRC Press, 1980.

34. Gromet-Elhanan, Z. Electrochemical gradients and energy coupling in photosynthetic bacteria. *Trends Biochem. Sci.* 2:274–277, 1977.

Dickerson, R.E. Cytochrome c and the evolution of energy metabolism. *Sci. Am.* 242(3):136–153, 1980.

Blankenship, R.E.; Parson, W.W. The photochemical electron transfer reactions of photosynthetic bacteria and plants. *Annu. Rev. Biochem.* 47:635–653, 1978.

35. Gillham, N.W. Organelle Heredity. New York: Raven, 1978.

Kirk, J.T.O.; Tilney-Basset, R.A. The Plastids: Their Chemistry, Structure, Growth and Inheritance, 2nd ed. Amsterdam: Elsevier, 1979.

Lewin, B. Gene Expression, Vol. 2, Eucaryotic Chromosomes, 2nd ed. New York: Wiley, 1980. (Chapter 21 covers organelle genomes.)

Kroon, A.M.; Saccone, C., eds. The Organization and Expression of the Mitochondrial Genome. Amsterdam: Elsevier/North-Holland, 1980.

36. Tzagoloff, A.; Macino, G.; Sebald, W. Mitochondrial genes and translation products. *Annu. Rev. Biochem.* 48:419–441, 1979.

37. Posakony, J.W.; England, J.M.; Attardi, G. Mitochondrial growth and division during the cell cycle in HeLa cells. *J. Cell Biol.* 74:468–491, 1977.

38. Attardi, G.; Borst, P.; Slonimski, P.P. Mitochondrial Genes. Cold Spring Harbor, N.Y.: Cold Spring Harbor Laboratory, 1982.

Cummings, D.J.; Borst, P.; Dawid, I.B.; Weissman, S.M.; Fox, C.F., eds. Extrachromosomal DNA: ICN-UCLA Symposia on Molecular and Cellular Biology, Vol. 15, New York: Academic Press, 1979.

Clayton, D.A. Replication of animal mitochondrial DNA. *Cell* 28:693–705, 1982.

39. von Wettstein, D. Chloroplast and nucleus: concerted interplay between genomes of different cell organelles. In International Cell Biology, 1980–1981 (H.G. Schweiger, ed.), pp. 250–272. New York: Springer-Verlag, 1981.

40. Buetow, D.E.; Wood, W.M. The mitochondrial translation system. *Subcell. Biochem.* 5:1–85, 1978.

Edwards, K.; Kössel, H. The rRNA operon from Zea mays chloroplasts: nucleotide sequence of 23S rDNA and its homology with *E. coli* 23S rDNA. *Nucleic Acids Res.* 9:2853–2869, 1981.

41. Borst, P. The biogenesis of mitochondria in yeast and other primitive eukaryotes. In International Cell Biology, 1980–1981 (H.G. Schweiger, ed.), pp. 239–249. New York: Springer-Verlag, 1981.

42. Birky, C.W., Jr. Transmission genetics of mitochondria and chloroplasts. *Annu. Rev. Genet.* 12:471–512, 1978.

43. Giles, R.E.; Blanc, H.; Cann, H.M.; Wallace, D.C. Maternal inheritance of human mitochondrial DNA. *Proc. Natl. Acad. Sci. USA* 77:6715–6719, 1980.

44. Bernardi, G. The petite mutation in yeast. *Trends Biochem. Sci.* 4:197–201, 1979.

Locker, J.; Lewin, A.; Rabinowitz, M. The structure and organization of mitochondrial DNA from petite yeast. *Plasmid* 2:155–181, 1979.

Montisano, D.F.; James, T.W. Mitochondrial morphology in yeast with and without mitochondrial DNA. *J. Ultrastruct. Res.* 67:288–296, 1979.

45. Anderson, S.; et al. Sequence and organization of the human mitochondrial genome. *Nature* 290:457–465, 1981.

Bibb, M.J.; Van Etten, R.A.; Wright, C.T.; Walberg, M.W.; Clayton, D.A. Sequence and gene organization of mouse mitochondrial DNA. *Cell* 26:167–180, 1981.

46. Ciferri, O. The chloroplast DNA mystery. *Trends Biochem. Sci.* 3:256–258, 1978.

47. Attardi, G. Organization and expression of the mammalian mitochondrial genome: a lesson in economy. *Trends Biochem. Sci.* 6:86–89, 100–103, 1981.

Ojala, D.; Montoya, J.; Attardi, G. tRNA punctuation model of RNA processing in human mitochondria. *Nature* 290:470–474, 1981.

Levens, D.; Ticho, B.; Ackerman, E.; Rabinowitz, M. Transcriptional initiation and 5′ termini of yeast mitochondrial RNA. *J. Biol. Chem.* 256:5226–5232, 1981.

Borst, P.; Grivell, L.A. One gene's intron is another gene's exon. *Nature* 289:439–440, 1981. (Review of the introns found in organelle DNAs.)

48. Chua, N.-H.; Schmidt, G.W. Post-translational transport into intact chloroplasts of a precursor to the small subunit of ribulose-1,5-bisphosphate carboxylase. *Proc. Natl. Acad. Sci. USA* 75:6110–6114, 1978.

Schatz, G. How mitochondria import proteins form the cytoplasm. *FEBS Lett.* 103:203–211, 1979.

Neupert, W.; Schatz, G. How proteins are transported into mitochondria. *Trends Biochem. Sci.* 6:1–4, 1981.

Gasser, S.M.; et al. Imported mitochondrial proteins, cytochrome b_2 and cytochrome c_1 are processed in two steps. *Proc. Natl. Acad. Sci. USA* 79:267–271, 1982.

Ellis, R.J. Chloroplast proteins: synthesis, transport, and assembly. *Annu. Rev. Plant Physiol.* 32:111–138, 1981.

49. van den Bogert, C.; Kroon, A.M. Tissue distribution effects on mitochondrial protein synthesis of tetracyclines after prolonged continuous intravenous administration to rats. *Biochem. Pharmacol.* 30:1706–1709, 1981.

50. Whatley, J.M.; John, P.; Whatley, F.R. From extracellular to intracellular: the establishment of mitochondria and chloroplasts. *Proc. R. Soc. Lond. (Biol.)* 204:165–187, 1979.

Doolittle, W.F. Revolutionary concepts in evolutinary cell biology. *Trends Biochem. Sci.* 5:146–149, 1980.

Dickerson, R.E. Cytochrome c and the evolution of energy metabolism. *Sci. Am.* 242(3):136–153, 1980.

Schwartz, R.M.; Dayhoff, M.O. Origins of prokaryotes, eukaryotes, mitochondria, and chloroplasts. *Science* 199:395–403, 1978.

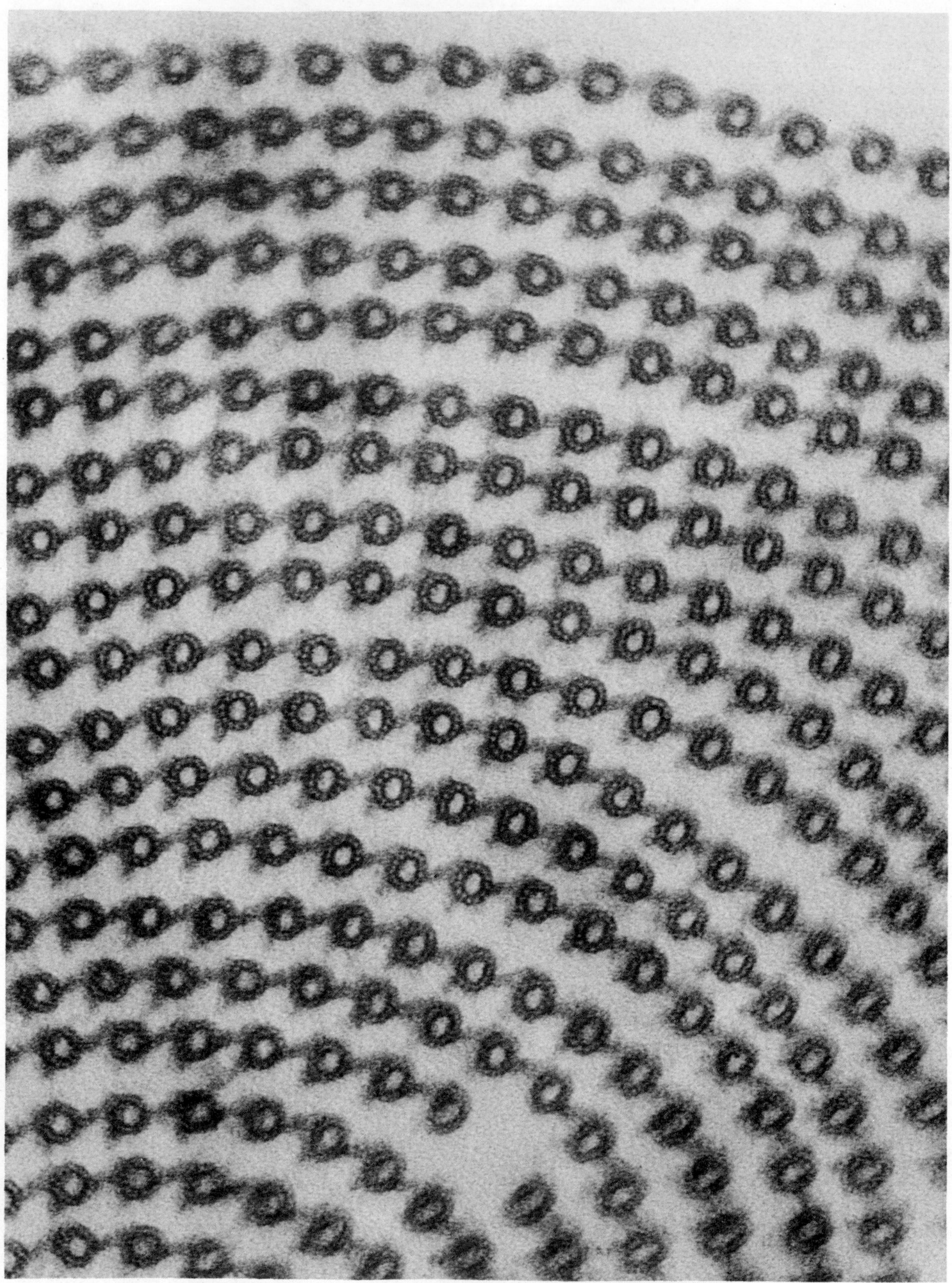

A bundle of microtubules cut in cross-section and examined by electron microscopy. The microtubules, about 25 nm in outer diameter, are the main structural components of the contractile rod, or *axostyle,* found in many flagellated protozoa. (From D. Woodrum and R. Linck, *J. Cell Biol.* 87:404–414. Reproduced by copyright permission of the Rockefeller University Press.)

The Cytoskeleton

10

Eucaryotic cells have distinct shapes and a high degree of internal organization. Moreover, they are capable of changing their shape, of repositioning their internal organelles, and, in many cases, of migrating from one place to another. These properties of shape, internal organization, and movement depend on complex networks of protein filaments in the cytoplasm that serve as the "bone and muscle" of the eucaryotic cell—the cell's **cytoskeleton.**

The two most important types of filaments of the cytoskeleton are *actin filaments* (sometimes referred to as microfilaments) and *microtubules.* Both are made of globular protein subunits that can assemble and disassemble rapidly in the cell. Delicate mechanisms exist for controlling their assembly from pools of unpolymerized subunits in the cytoplasm. A third class of protein filaments, intermediate in diameter between actin filaments and microtubules and therefore called *intermediate filaments,* is found in most animal cells; they are made of fibrous protein subunits and are much more stable than most actin filaments and microtubules.

In addition to the three major types of protein filaments, the cytoskeleton also contains many different accessory proteins that either link the filaments to one another or to other cell components such as the plasma membrane, or influence the rate and extent of the filament polymerization. Specific sets of accessory proteins interact with protein filaments to produce movements; the two best-understood examples are found in the the contraction of muscle, which depends on actin filaments, and the beating of cilia, which depends on microtubules. Although these movements involve different sets of proteins, they both depend on ATP hydrolysis and the sliding of one protein filament against another.

The filament arrays responsible for the movements of muscle and cilia are unusually stable and highly ordered, features that have enabled their components to be identified and their mechanism of action to be understood in far greater detail than for other cytoskeletal assemblies. This chapter, therefore, begins with descriptions of muscle and cilia, which serve to introduce the major components of the cytoskeleton. The next section examines actin filaments and microtubules in their less organized and less stable arrangements and discusses how their assembly and disassembly are controlled. Subsequent sections describe intermediate filaments and the interactions between the major filaments of the cytoskeleton, and the chapter ends with two of the most challenging questions in contemporary cell biology: how a cell maintains its spatial organization, and how it moves.

Muscle Contraction[1]

Of all the kinds of movement shown by living things, those based on muscle contraction are both the most familiar and the best understood. Running, walking, swimming, and flying all depend on the ability of the muscular tissue of the vertebrate body to contract rapidly on its scaffolding of bone. **Skeletal muscle** is an amazingly effective apparatus for the production of rapid movements under voluntary control. Closely related forms of muscle, *cardiac* and *smooth muscle,* have become specialized for other, involuntary movements, such as heart pumping and gut peristalsis.

All forms of muscle produce movement by means of active contraction. This contraction is achieved by a sophisticated and powerful intracellular protein apparatus that, in a more rudimentary form, generates movements in almost all cells. Most powerful of all is skeletal muscle (and the comparable invertebrate tissues, such as insect flight muscle), in which the contractile apparatus is so highly organized that its mechanism of action is clearly reflected in its ultrastructure. Consequently, we now have a uniquely detailed understanding of muscle contraction and a good model of how living cells use chemical energy to do mechanical work.

A Myofibril Is the Contractile Element of a Skeletal Muscle Cell

Skeletal muscles consist of long, thin **muscle fibers,** each of which is a single, unusually large cell formed by the fusion of many separate cells. But while a muscle fiber therefore contains many nuclei, about two-thirds of its dry mass is made up of **myofibrils,** long cylindrical elements 1 μm to 2 μm in diameter that extend the entire length of the cell (Figure 10–1). Since isolated myofibrils contract in the presence of ATP, it is clear that they are the contractile elements of muscle cells. As we shall see, each myofibril itself consists of a chain of tiny contractile units composed of repeating assemblies of cytoskeletal proteins.

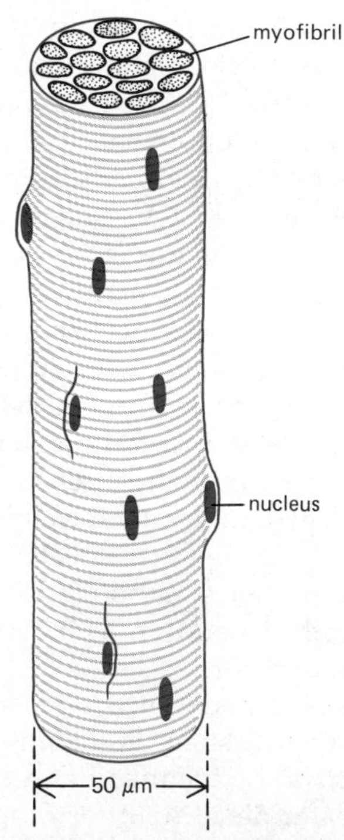

Figure 10–1 Schematic drawing of a short section of a skeletal muscle cell (or fiber).

Myofibrils Are Composed of Repeating Assemblies of Thick and Thin Filaments

The contractile units of the myofibril are readily visible in the light microscope, giving the myofibril a banded or striated appearance. Each of the regular repeating units, or **sarcomeres,** is about 2.5 μm long. Sarcomeres of adjacent myofibrils are aligned in register, so that the whole muscle cell has a similarly striated appearance (Figure 10–2). In a longitudinal section viewed at higher magnification, a series of light and dark stripes can be seen in each sarcomere. The darker stripes are called *A bands*; the lighter stripes, *I bands*; while the dense line in the center of the I band that separates one sarcomere from the next is known as the Z line or **Z disc** (Figure 10–3).

The molecular basis of the cross-striations, and a strong clue to their functional significance, was revealed in 1953 in one of the first applications of the electron microscope to thin sections of biological material. Each sarcomere was found to contain two sets of parallel and partly overlapping protein filaments: **thick filaments** (each about 1.6 μm long and 15 nm in diameter),

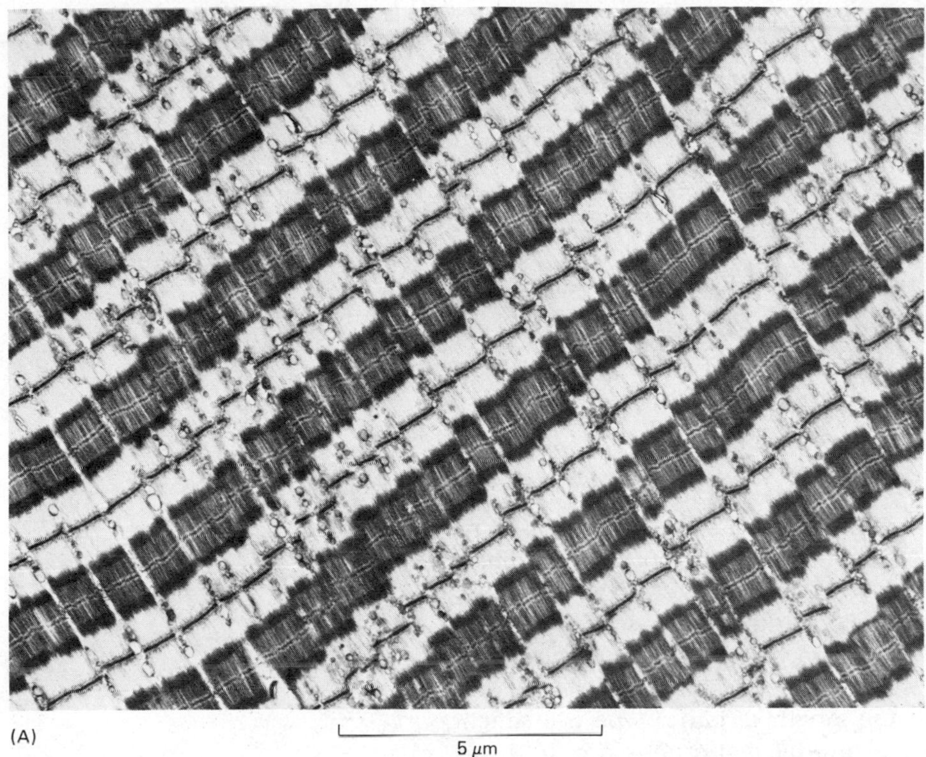

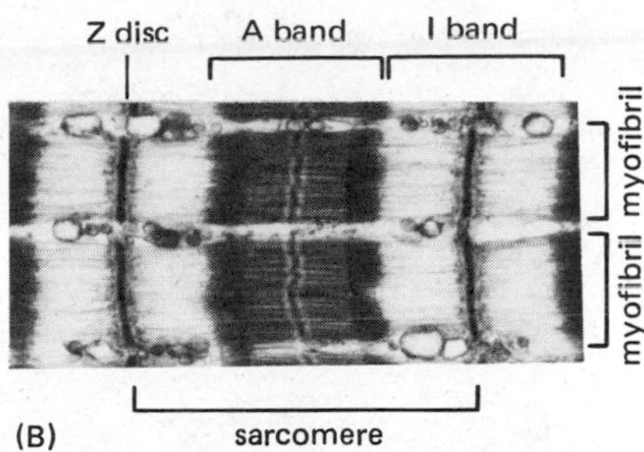

Figure 10–2 (A) Low-magnification view of a skeletal muscle fiber (from rabbit psoas muscle) showing the regular pattern of cross-striations. The muscle fiber and the many myofibrils it contains (see Figure 10–1) run diagonally from upper left to lower right. (B) Detail from the electron micrograph in (A) showing portions of two adjacent myofibrils and the definition of a sarcomere. (Courtesy of Roger Craig.)

extending from one end of the A band to the other; and **thin filaments** (each about 1.0 μm long and 8 nm in diameter), extending across the I bands and part way into the A band (Figure 10–3). A cross-section of muscle in the region of the A band where thick and thin filaments overlap shows that the thick filaments are arranged in a regular hexagonal lattice, with the thin filaments placed in a regular manner between them (Figure 10–4).

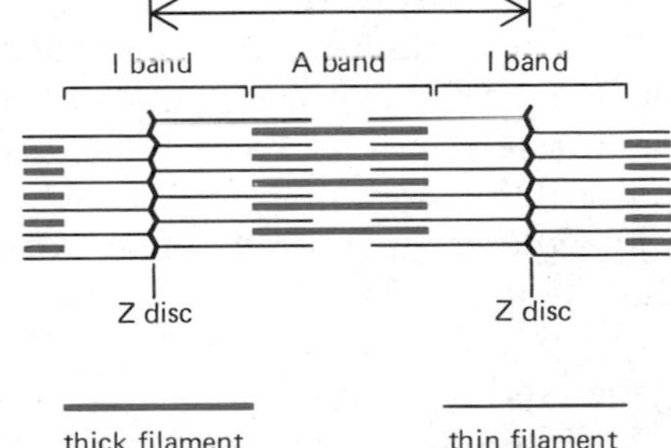

Figure 10–3 Schematic diagram of a single sarcomere showing the origin of the I and A bands seen in micrographs of a skeletal muscle cell (see Figure 10–2B).

Contraction Occurs as Filaments Slide Past Each Other[2]

If a source of monochromatic light, such as a laser, is directed through a living muscle cell, a series of interference fringes is produced that provides a sensitive measure of sarcomere spacing. Measurements reveal that each sarcomere shortens proportionately as the muscle contracts; thus, if a myofibril con-

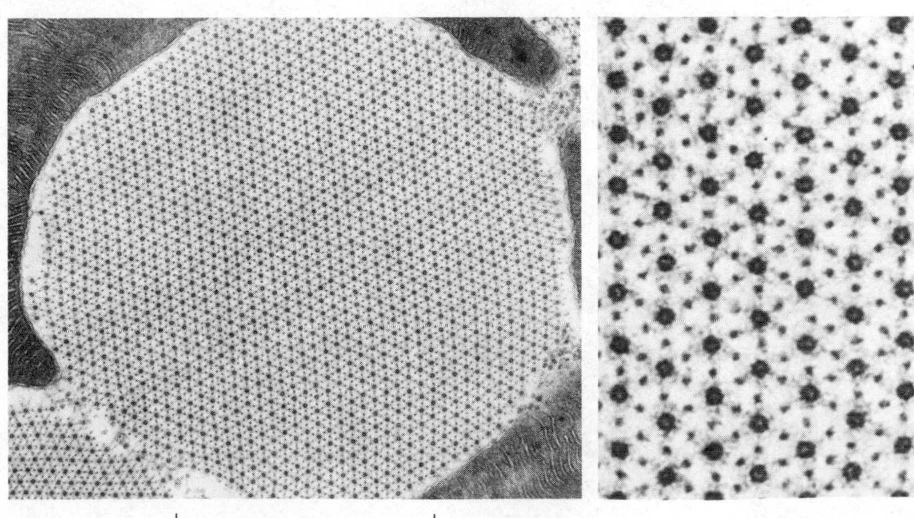

Figure 10–4 Electron micrographs of an insect flight muscle viewed in cross-section show how the thick and thin filaments are packed together with crystalline regularity. Unlike their vertebrate-muscle counterparts, the thick filaments have a hollow center, as seen in the enlargement on the right. (From J. Auber, *J. de Microsc.* 8:197–232, 1969.)

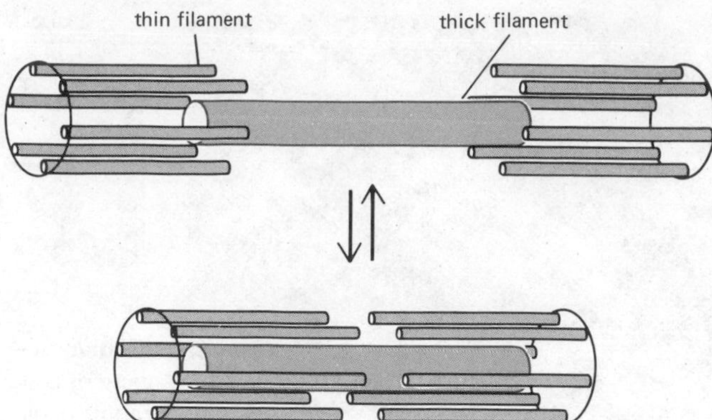

Figure 10–5 Schematic diagram of the sliding filament model of muscle contraction, in which the thin and thick filaments slide past one another without shortening.

taining a chain of 20,000 sarcomeres contracts from 5 cm to 4 cm (that is, by 20%), the length of each of the 20,000 sarcomeres decreases correspondingly from 2.5 μm to 2.0 μm.

When a sarcomere shortens, only the I band decreases in length: the dense A band remains unchanged. The altered pattern of bands can be explained very simply if the contraction is caused by thick filaments sliding past the thin filaments with no change in the length of either type of filament (Figure 10–5). This simple **sliding filament model,** first proposed in 1954, has been of pivotal importance in understanding the contractile mechanism. In particular, it directed attention to the molecular interactions between adjacent thick and thin filaments that cause them to slide.

The sliding filament model is supported by several different lines of evidence. The fact that the filaments themselves do not change in length as a muscle shortens has been demonstrated by electron microscopy, while x-ray diffraction analyses have shown that the internal packing of subunit molecules in each kind of filament remains unchanged. The mechanical tension that can be produced by a muscle varies according to the amount of overlap between the thick and thin filaments, as it should if the tension is generated by sliding interactions between them.

The ultrastructural basis for the force-generating interaction is visible at very high magnifications in electron micrographs, in which the thick filaments can be seen to possess numerous, tiny side arms or *cross-bridges* that extend across the gap of about 13 nm to make contact with adjacent thin filaments (Figure 10–6). It is now known that in muscle contraction the thick and thin filaments are pulled past each other by the cross-bridges acting cyclically, rather like banks of tiny oars. The interacting proteins of the thin and thick filaments have been identified, respectively, as *actin,* the most abundant of all the cytoskeletal proteins, and *myosin,* which is commonly found in association

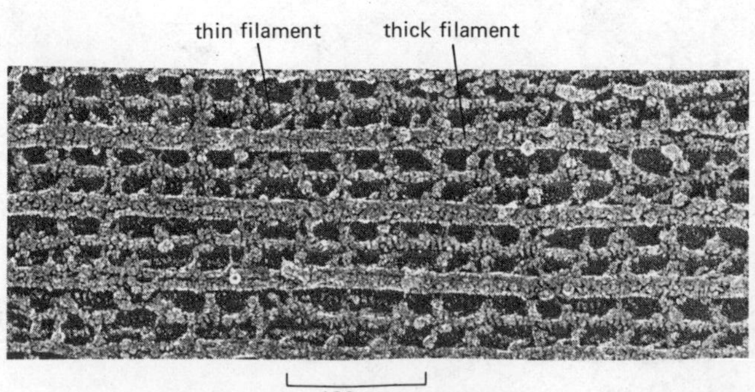

Figure 10–6 Electron micrograph of an insect flight muscle that has been subjected to rapid freezing, fracturing, and deep etching, showing the nearly crystalline array of thick myosin filaments and thin actin filaments. The cross-bridges spanning the two types of filaments are myosin heads. (Courtesy of John Heuser and Roger Cooke.)

with actin in cellular structures that produce movement. Most of what we know about the properties of these two important proteins, which are present in virtually all eucaryotic cells, has been learned from biochemical experiments on the actin and myosin extracted from muscle.

Thin Filaments Consist Principally of Actin

The **actin** from skeletal muscle is only one of at least six different types synthesized in vertebrate cells. All eucaryotic species, except the most primitive ones such as yeast, have multiple actin genes encoding slightly different proteins expressed in the cells of different tissues or at different stages of development. But actin genes are very highly conserved, and the differences between the proteins are so slight that, at least in tests *in vitro*, actin molecules from widely divergent sources are functionally interchangeable. The principal properties of actin molecules extracted from skeletal muscle are therefore common to actin molecules from all other sources.

Actin is usually isolated by treating dry powdered muscle with very dilute salt solutions. This treatment breaks down the actin filaments into their globular subunits, which are composed of a single polypeptide of molecular weight 41,800 sometimes known as globular actin or **G actin.** Each molecule of G actin is associated with one tightly bound Ca^{2+}, which stabilizes its globular conformation, and one molecule of noncovalently bound ATP. The terminal phosphate of the bound ATP is hydrolyzed when the G actin polymerizes to form actin filaments, also called filamentous actin or **F actin.** Polymerization can be induced by simply raising the salt concentration to a level closer to that found in cells: the actin solution, which in dilute salt is only slightly more viscous than water, then undergoes a rapid and impressive increase in viscosity as the actin molecules associate to form filaments.

Although ATP is hydrolyzed during the polymerization of actin, the process does not require energy, and actin will still polymerize in the presence of a nonhydrolyzable analogue of ATP. ATP hydrolysis does, however, greatly increase the rate of actin polymerization and has an important effect on its dynamics, which will be explained later when we discuss cellular activities that depend on the controlled polymerization of actin. The contractile activity of muscle does not depend on these properties but rather on the properties of stable actin filaments.

Actin filaments as seen in electron micrographs consist of two strands of globular molecules about 4 nm in diameter twisted into a helix with 13.5 molecules per turn (Figures 10–7 and 10–8). Such filaments have been identified as the principal components of thin filaments of skeletal muscle both by their x-ray diffraction pattern and by fluorescent anti-actin antibodies that stain the I-band region of the sarcomere. The thin filaments of muscle are composed of more than just actin, however; they contain several other proteins as described below (p. 559).

Actin filaments are polar structures; their two ends are therefore different. The polarity of the actin filament, which is essential to its function in cell motility, is detectable in the way it interacts with myosin. But before discussing this interaction, we must introduce some of the properties of the myosin molecule.

Thick Filaments Are Composed Principally of Myosin

Myosin can be extracted from almost every cell type in the vertebrate body and is always present where actin filaments form contractile bundles in cells. It has been much less highly conserved during evolution than actin and is found in several different forms. The special properties of myosin from non-

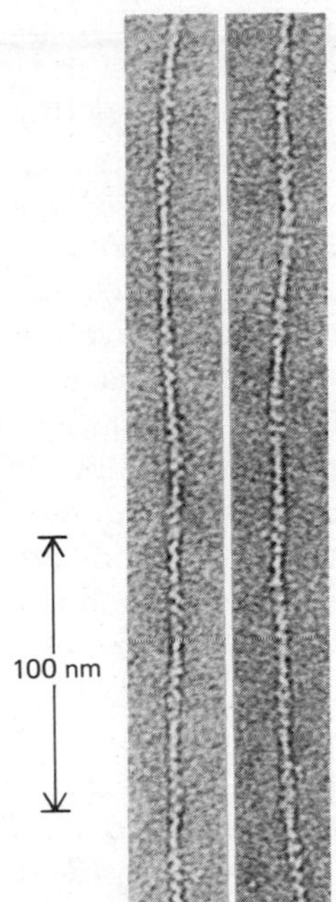

Figure 10–7 Electron micrographs of negatively stained actin filaments. (Courtesy of Roger Craig.)

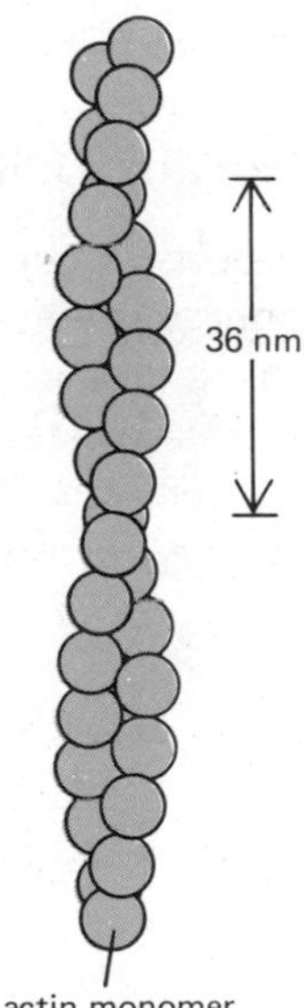

Figure 10–8 Schematic drawing of an actin filament showing the helical arrangement of the globular actin molecules. Although they are portrayed here as spheres, x-ray diffraction studies indicate that each actin molecule is, in fact, pear-shaped, about 6.5 nm long, and about 4 nm across at its widest part (see also Figure 10–7).

muscle cell types will be discussed in later sections. The principal distinguishing feature of skeletal muscle myosin is that it spontaneously polymerizes *in vitro* to form polymers that are much larger than those formed by other kinds of myosin.

Myosin can be extracted from skeletal muscle by treatment with concentrated salt solutions, in which myosin is soluble. Muscle treated in this way loses only its thick filaments, which fragment into their constituent myosin molecules. These have a molecular weight of about 500,000 and are seen in electron micrographs to be long rodlike molecules, each with two globular heads (Figure 10–9). Each molecule can be further broken down by treatment with high concentrations of urea or detergent into six polypeptide chains: two identical **heavy chains** of 200,000 daltons each and two pairs of **light chains** of about 20,000 daltons and 16,000 daltons each. The heavy chain consists of a long α-helical section attached to a globular head. In the intact myosin molecule, the long α-helices of two heavy chains coil around each other to form the rodlike tail from which two heads project; each of these heads is a complex of the globular head of one heavy chain with one molecule of each type of light chain (Figure 10–10).

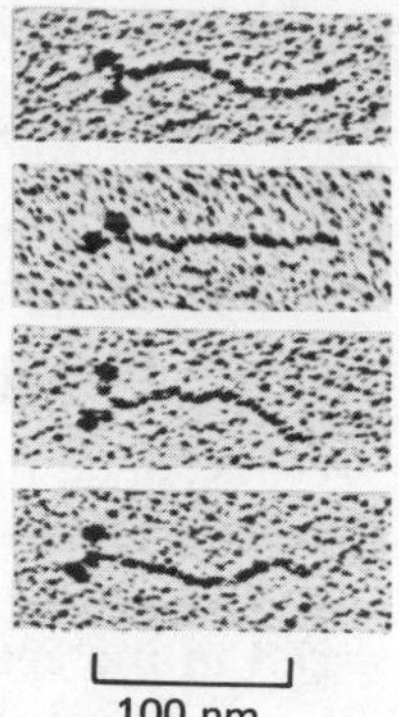

Figure 10–9 Electron micrographs of myosin molecules shadowed with platinum. Note that each molecule is composed of two globular heads attached to a single fibrous tail. (Courtesy of David Shotton.)

Even at elevated salt concentrations, isolated myosin molecules tend to form dimers, and at physiological ionic strengths, a myosin solution becomes turbid as the dimers aggregate to form large fibers. Under some conditions the latter closely resemble the natural thick filaments of muscle. Unlike actin filaments, myosin filaments are not formed by the end-to-end addition of subunits. Instead, the myosin molecules associate by their tails. The body of the natural thick filament is composed of some hundreds of myosin tails packed together in a regular staggered array from which the myosin heads project in a repeating pattern (Figure 10–11A and B). The structure is bipolar, with a bare central region where two oppositely oriented sets of myosin tails come together. The globular heads of the myosin molecules interact with actin and form the cross-bridges between the thick and thin filaments of muscle.

ATP Hydrolysis Drives Muscle Contraction[3]

The interactions of actin and myosin produce a coherent contractile force as chemical energy is converted into mechanical work. The efficiency with which skeletal muscle accomplishes this conversion is very high: only 30% to 50% of the energy is wasted as heat. An automobile engine typically wastes 80% to 90% of the energy available from gasoline.

The energy for muscle contraction comes from the hydrolysis of ATP to ADP and inorganic phosphate (P_i). Yet no major difference is detected in the ATP level in a resting muscle and an actively contracting one. The reason for this apparent paradox is that a very efficient backup system exists in the muscle cell cytoplasm for the regeneration of ATP. High levels of another active phosphate compound, *phosphocreatine* (Figure 10–12), are present, and the

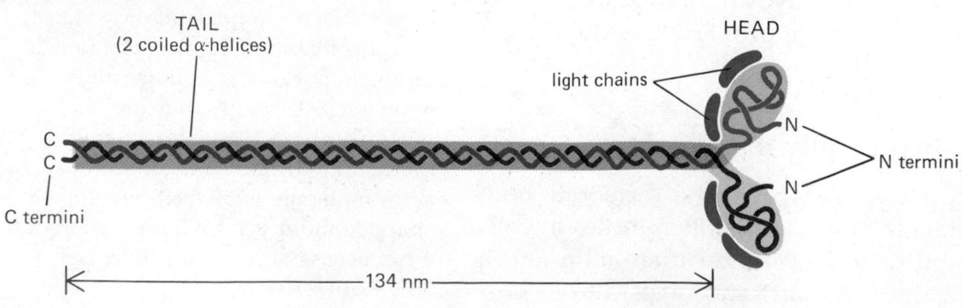

Figure 10–10 Schematic drawing of a myosin molecule showing the two heavy chains and four light chains of which it is composed. The light chains are of two different types, and one molecule of each is present on each of the two myosin heads.

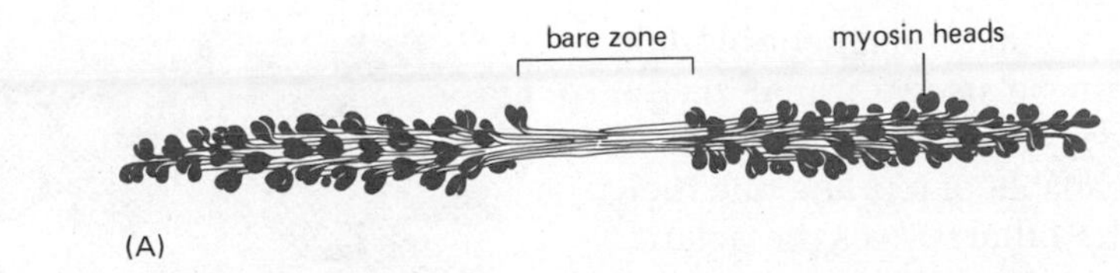

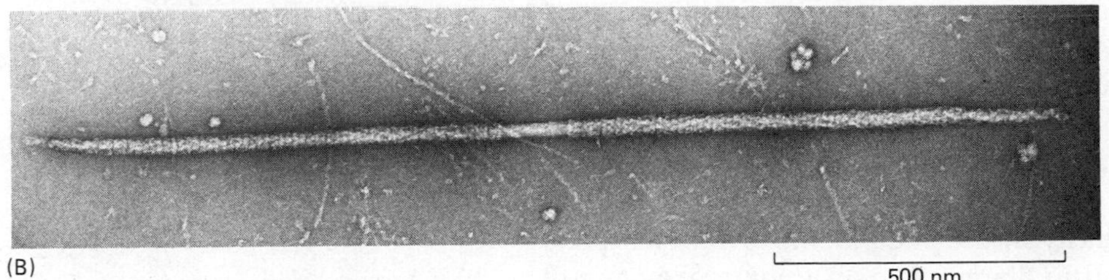

Figure 10–11 (A) Schematic diagram of a myosin thick filament from skeletal muscle, not drawn to scale. The myosin molecules aggregate together by means of their tail regions, with their heads projecting to the outside. Note the "bare zone" in the center of the filament consisting entirely of myosin tails. (B) Electron micrograph of a myosin thick filament isolated from scallop muscle. Note the central bare zone. (Courtesy of Roger Craig.)

enzyme *creatine kinase* catalyzes a reaction between this compound and ADP to form creatine and ATP. So, after a short burst of activity, it is the intracellular level of phosphocreatine that drops—even though the contractile machinery itself consumes ATP. The pool of phosphocreatine thus serves as a store of ATP energy much like a battery, being recharged (when the muscle is resting) from the new ATP that is generated by cellular oxidations.

$$^{-}O-P(=O)(O^{-})-NH-C(=NH)-N(CH_3)-CH_2-COO^{-}$$

phosphocreatine

Figure 10–12 Phosphocreatine acts as a storage form of high-energy phosphate groups in vertebrate muscle and other tissues. Its high-energy phosphate group (*color*) is transferred to ADP by the enzyme creatine kinase.

Myosin Is an Actin-activated ATPase[1,2]

ATP hydrolysis during muscle contraction is a direct consequence of the interaction of myosin with actin. Even on its own, myosin acts as an enzyme that can hydrolyze ATP (an ATPase). But purified myosin by itself is not a very active ATPase: each molecule takes about 30 seconds to split an ATP molecule. The rate-limiting step is not the initial binding of ATP to myosin nor the hydrolysis itself—both of which are extremely rapid—but the release of the products of ATP hydrolysis (ADP and inorganic phosphate, P_i), which remain tightly bound in a noncovalent complex with the myosin molecule and prevent further ATP binding and hydrolysis.

The rate of ATP hydrolysis by myosin is powerfully stimulated by actin. In the presence of actin filaments, each myosin molecule hydrolyzes 5 to 10 molecules of ATP every second, which is comparable to the rates measured in contracting muscle. The stimulation of myosin ATPase by actin filaments reflects a physical association between the two. This association does not affect the step in which ATP is hydrolyzed by myosin; instead it causes a more rapid release of ADP and P_i from the myosin molecule, which is thus freed to bind another molecule of ATP and start the reaction again, as described in the next two sections.

Myosin Heads Bind to Actin Filaments[1,2]

Both the binding to actin filaments and the hydrolysis of ATP occur at the globular head of the myosin molecule. If the myosin molecule is cleaved into large pieces by the limited action of proteolytic enzymes, the fragments can be separated and analyzed. The protease papain, for example, cuts the myosin subunit at the base of the head region releasing an almost complete tail—called the *myosin rod*—and two separate *myosin heads* (Figure 10–13). The myosin rod, like the intact myosin molecule, self-assembles at physiological salt concentrations into large, ordered aggregates, which, however, lack side extensions. The two separate heads produced by papain cleavage are each about 120,000 daltons and are called **myosin subfragment 1 (S1 fragments)**.

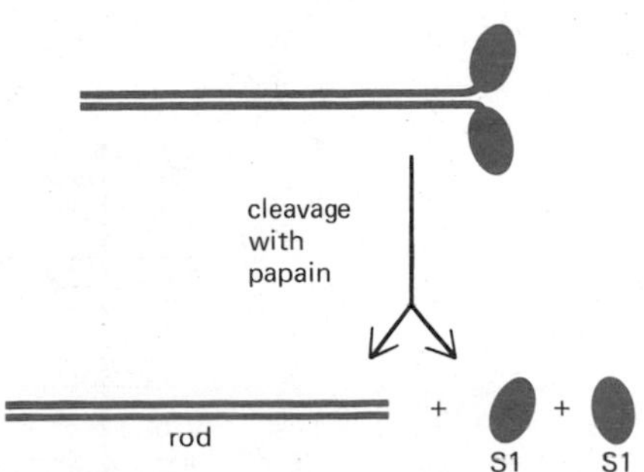

Figure 10–13 Limited digestion with the proteolytic enzyme papain cleaves the myosin molecule between its head and tail, generating a myosin rod and two S1 fragments.

These heads retain all of the ATPase activity and the actin-filament-binding properties of the intact myosin molecule and can be used to analyze the interaction between actin and myosin.

Each actin molecule in an actin filament is capable of binding one molecule of S1. It is the complex of actin filaments with S1 that reveals the polarity of the actin filament. In negatively stained preparations in the electron microscope (Figure 10–14), such complexes have a regular and distinctive form: each S1 fragment forms a lateral projection, and the superimposed image of many such projections gives the appearance of arrowheads along the actin filament. The arrows on these *decorated actin filaments* all point the same way on both strands of the filamentous actin double helix, demonstrating its inherent structural polarity.

The existence of a structural polarity in the actin filament makes it possible to predict the orientation of the thin filaments in a sarcomere. As noted above, the myosin head regions (which constitute the S1 fragment) have opposite orientations on either side of the bare region of the thick filament (that is, on either side of the sarcomere A band). Since these head regions must interact with the thin filaments in the region of overlap, it would be reasonable to assume that the thin filaments on either side of the sarcomere are of opposite polarity. This prediction can be tested by using S1 to decorate fragments of muscle consisting of isolated Z discs and their attached actin filaments. When this is done, all the S1 arrowheads are found to point away from the Z discs and toward the center of the sarcomere, as expected (Figure 10–15).

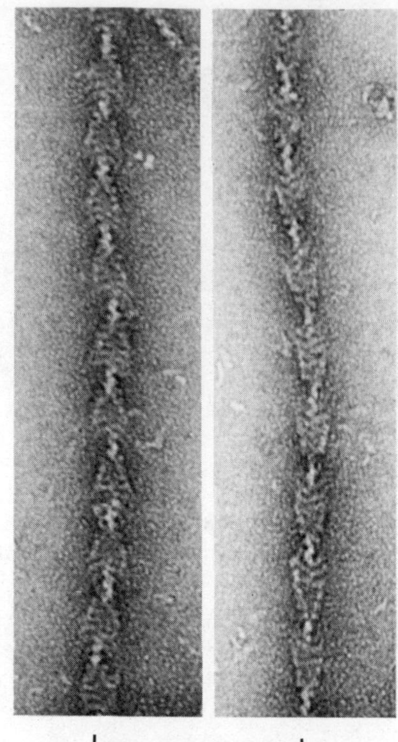

Figure 10–14 Electron micrograph of actin filaments decorated with S1 fragments of myosin. The helical arrangement of the bound S1 fragments, which are tilted in one direction, gives the appearance of arrowheads and indicates the polarity of each actin filament. (Courtesy of Roger Craig.)

Myosin Heads "Walk" Along Actin Filaments

What is the molecular mechanism by which the thin and thick filaments slide past each other to produce the mechanical force in muscle contraction? A great deal of evidence—from electron microscopy, x-ray diffraction, kinetic analysis, and other studies—is consistent with the following model: A myosin head carrying the products of a previous ATP hydrolysis (ADP and P_i) moves from its position on the thick filament into close proximity to a neighboring actin subunit. This movement is thought to occur by random diffusion and to require flexibility in the myosin molecule. As soon as the myosin binds to actin, the ADP and P_i are released from the myosin head, causing it to tilt and thereby pull on the rest of the thick filament. At the end of this *power stroke,* a fresh molecule of ATP binds to the head and detaches it once more from the actin filament. Hydrolysis of the bound ATP quickly relaxes the myosin head to its original conformation and prepares it for a second cycle (Figure 10–16). For a discussion of the general principles involved, see page 137.

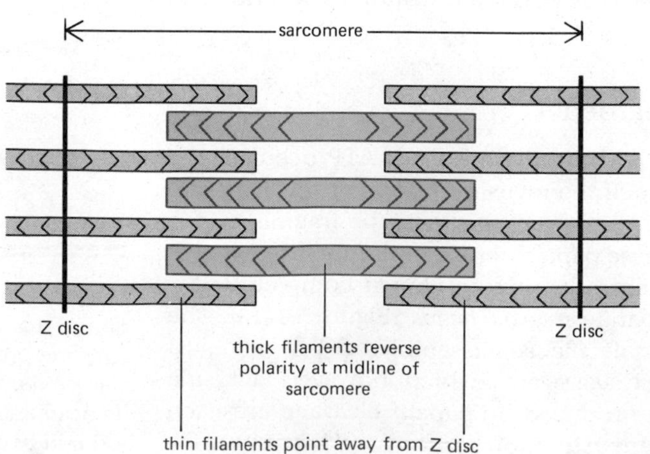

Figure 10–15 Schematic diagram of a sarcomere showing that the thick and thin filaments overlap with the same relative polarity on either side of the midline.

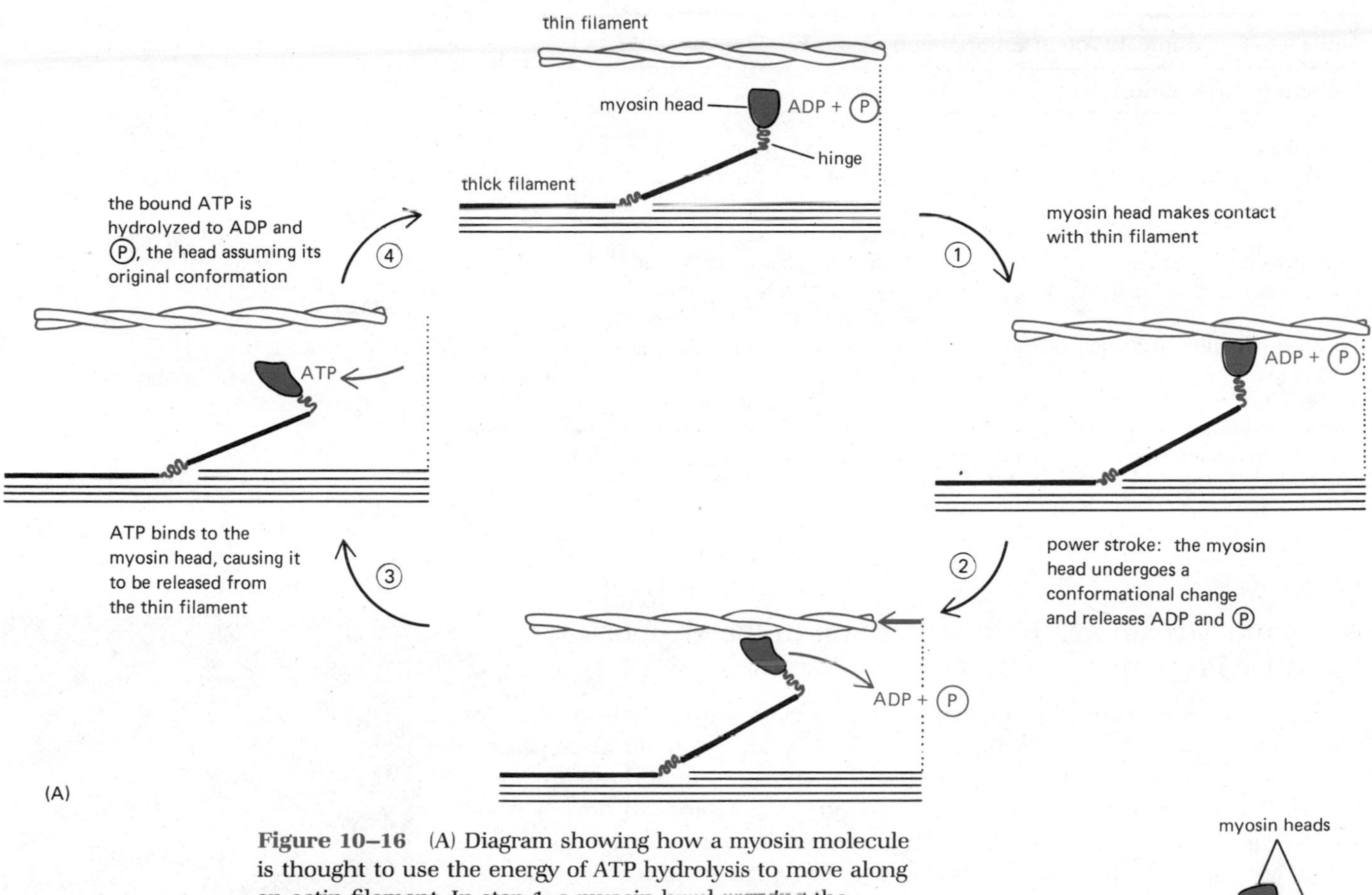

Figure 10–16 (A) Diagram showing how a myosin molecule is thought to use the energy of ATP hydrolysis to move along an actin filament. In step 1, a myosin head carrying the products of ATP hydrolysis (ADP + P_i) moves to an adjacent thin filament; this movement is thought to occur by diffusion and to depend on the flexible hinge regions of the myosin molecule (see part B). In the power stroke (step 2) the myosin head undergoes a change in shape accompanied by the release of ADP + P_i, causing the myosin head to pull against the thin filament. In step 3, an ATP molecule binds to the myosin head, causing it to be released from the thin filament. The cycle is completed in step 4, in which the bound ATP molecule is hydrolyzed and the myosin head returns to its original conformation. Each of the two heads on each myosin molecule is thought to cycle independently of the other. (B) Schematic representation of the flexible hinge regions in the myosin molecule that allow the heads to move freely.

Each individual myosin head is, therefore, thought to "walk" along an adjacent actin filament. As it undergoes a cyclical change in conformation, the myosin head pulls against the actin filament, causing it to slide against the thick filament. Once an individual myosin head has detached from the actin filament, it is carried along by the action of other myosin heads in the same thick filament, so that a snapshot picture of an entire thick filament in a contracting muscle would show some of the myosin heads attached to actin filaments while others were detached. A certain amount of springlike elasticity in the myosin molecule is essential to allow this to happen. Each thick filament carries about 500 myosin heads, and each of these cycles about five times per second in the course of a rapid contraction.

Table 10–1 Principal Protein Components of the Myofibril

Protein (polypeptide MW)	Function
Actin (42,000)	major component of thin filaments
Myosin (2 × 200,000 + 4 × 20,000)	major component of thick filaments and an ATPase that interacts with actin to produce contraction
α-Actinin (2 × 95,000)	anchors actin filaments at the Z disc
Tropomyosin (2 × 35,000)	binds along the length of actin filaments; involved in Ca^{2+} regulation
Troponin (18,000 + 21,000 + 37,000)	associated with thin filaments where it prevents the interaction of actin and myosin in the absence of Ca^{2+}

Other proteins associated with the myofibril of vertebrate skeletal muscle include *C protein,* which binds to the thick filaments, and *desmin,* an intermediate filament protein associated with the Z discs (see Table 10–6).

Actin and Myosin Are Held in Position in the Myofibril by Other Proteins

Although actin and myosin are the principal factors in muscle contraction, they are not the only proteins in the myofibril. A large number of "accessory" muscle proteins are also present (Table 10–1). While some of these are known to regulate contraction, as will be discussed, others seem to have a purely structural role. For example, **α-actinin** and **desmin** found in the region of the Z disc are believed to help determine the arrangement of filaments in each sarcomere and to tie adjacent sarcomeres together, respectively. At the end of each sarcomere, parallel actin filaments are embedded in one side of the Z disc. Since α-actinin binds to the ends of actin filaments *in vitro* and is specifically localized at the Z disc in muscle cells, it is thought to anchor the actin filaments in the Z disc. Desmin, an intermediate filament protein (p. 593), is found in high concentration in the regions between Z discs of adjacent myofibrils and is thought to help keep the myofibrils in register there (Figure 10–17).

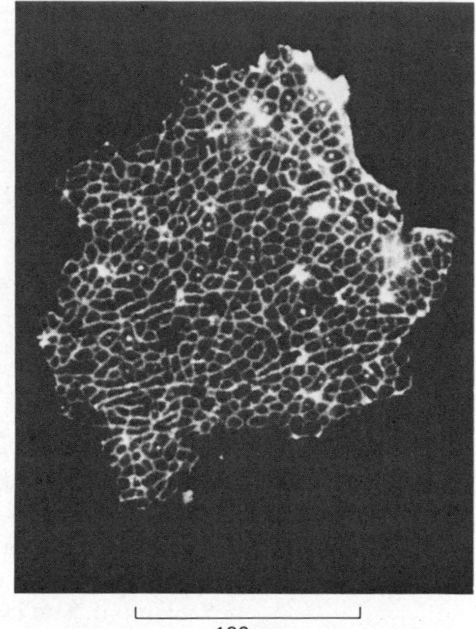

Figure 10–17 Immunofluorescence micrograph of a sheet of skeletal muscle Z discs stained with anti-desmin antibody. Because the Z discs of adjacent myofibrils are linked together, they can be isolated as a flattened sheet. The Z discs appear as unstained pores, while the regions between them are brightly stained, suggesting that desmin may play a role in linking Z discs together. (Courtesy of Bruce Granger and Elias Lazarides.)

Muscle Contraction Is Initiated by a Rise in Intracellular Ca^{2+} [4]

All skeletal muscles are under voluntary control and contract only when a signal passes to them from their motor nerve. The chain of events leading from the excitation of the nerve to the contraction of the muscle depends on important specializations of the muscle cell membranes through which the signal is relayed. The firing of the nerve triggers an action potential in the muscle-cell plasma membrane, and this electrical excitation spreads rapidly into a series of membranous folds, the *transverse tubules* or *T tubules,* that extend inward from the plasma membrane to surround each myofibril at its Z disc (Figure 10–18). Here the electrical signal is somehow transferred to the **sarcoplasmic reticulum,** an adjacent sheath of anastomosing, flattened vesicles derived from the endoplasmic reticulum, which surrounds each myofibril rather like a net stocking.

When activated by the electrical signal carried by the T tubules, the sarcoplasmic reticulum releases into the cytosol large amounts of the Ca^{2+} stored in its lumen: the resulting sudden rise in free Ca^{2+} concentration initiates myofibril contraction. Because the signal from the muscle-cell plasma membrane is passed (via the T tubules and sarcoplasmic reticulum) to every sarcomere in the cell within milliseconds, all of the myofibrils in the cell contract simultaneously.

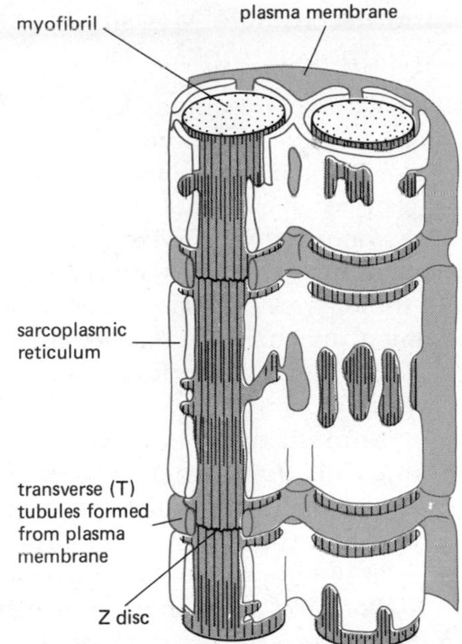

Figure 10–18 Schematic diagram showing the system of membrane channels involved in relaying the signal to contract from the muscle-cell plasma membrane to all of the myofibrils in that cell.

Troponin and Tropomyosin Mediate the Ca^{2+} Regulation of Muscle Contraction[5]

The Ca^{2+} dependence of muscle contraction, and hence its dependence on motor commands transmitted via nerves, is entirely due to a set of specialized accessory proteins closely associated with actin filaments. When myosin is mixed with pure actin filaments in a test tube, myosin ATPase is activated whether or not Ca^{2+} is present. However, in a normal myofibril, the activity of myosin ATPase depends on Ca^{2+}. This difference reflects the fact that the regulation of contraction by Ca^{2+} depends on a special set of accessory proteins, and these are absent from the artificially reconstituted filaments (Figure 10–19).

One of these accessory proteins is a rigid, rod-shaped protein called **tropomyosin,** which lies in the long pitched grooves on either side of the actin filament, thereby stiffening the filament.

The other major protein involved in Ca^{2+} regulation is **troponin,** a complex of three polypeptides (troponins T, I, and C). *Troponin T* has a binding site for tropomyosin and is thought to be responsible for positioning the complex (Figure 10–19). When *troponin I* is added to troponin T and tropomyosin, the complex inhibits the interaction of actin and myosin, even in the presence of Ca^{2+}. The further addition of *troponin C* completes the complex, and the interaction of actin and myosin remains inhibited in the absence of Ca^{2+}. However, troponin C binds up to four molecules of Ca^{2+} and, with Ca^{2+} bound, relieves this inhibition. The Ca^{2+}-induced reversal of the tropomyosin-troponin inhibition of myosin binding to actin underlies the induction of myofibril contraction by Ca^{2+}.

How do troponin and tropomyosin control the interaction of myosin with actin? There is only one Ca^{2+}-binding polypeptide (troponin C) for every seven actin molecules along the filament (Figure 10–19), while control is exerted along the whole length of the actin filament. Detailed structural studies reveal that in the *resting state,* when no contractile force is being generated, the rodlike tropomyosin molecules shield the position on the actin filaments that is occupied in an actively contracting muscle by the myosin heads; the tropomyosin may therefore sterically block the interaction of actin and myosin. When the level of Ca^{2+} is raised, the tropomyosin molecules shift their position slightly (presumably in response to a change in shape of the troponin molecules to which they are bound), and this shift is thought to allow the myosin heads to interact with actin filaments (Figure 10–20). It is as though muscle contraction is controlled by a tiny wedge that is driven between actin and myosin to prevent them from interacting—the wedge being removed only in response to a Ca^{2+} signal.

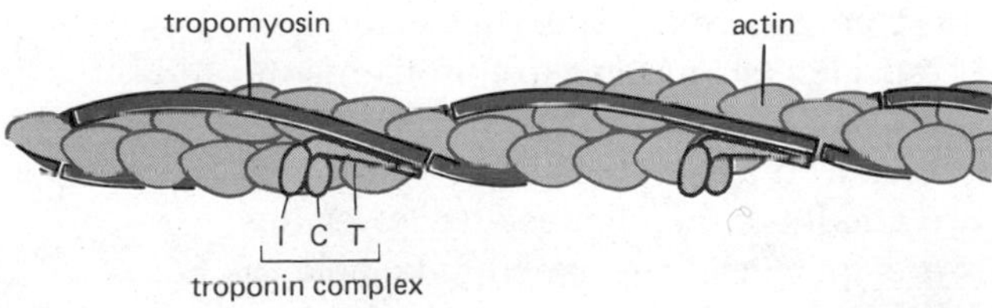

Figure 10–19 Schematic diagram of a muscle thin filament showing the position of tropomyosin and troponin along the actin filament. (Drawn from data supplied by Carolyn Cohen.)

Smooth Muscle Myosin Is Activated by Ca^{2+}-dependent Phosphorylation[6]

We have been describing in detail only one of the three major types of muscle present in the vertebrate body—skeletal muscle. The two others are *cardiac muscle,* which is responsible for the pumping action of the heart (which contracts perhaps 3 billion times in the course of an average human life span), and *smooth muscle,* which produces the slower and longer-lasting contractions of such tissues as the walls of the stomach, intestines, and blood vessels. All three types of muscle contain actin and myosin and contract by a sliding filament mechanism. Cardiac muscle has a striated appearance like that of skeletal muscle, reflecting a very similar organization of actin filaments and myosin filaments.

Smooth muscle, by contrast, has no striations. It consists of long, tapering cells that have a single nucleus and contain both thick and thin filaments aligned with the long axis of the cell. However, these filaments are not arranged in the strictly ordered pattern found in skeletal and cardiac muscle and do not appear to form myofibrils.

Both the actin and the myosin of smooth muscle are of a kind special to that tissue. The actin differs slightly in amino acid sequence from the actin of skeletal or cardiac muscle, although the differences have no known functional significance. But although smooth muscle myosin closely resembles that of skeletal muscle, it differs functionally in two very important ways: (1) The level of its ATPase activity, even under optimal conditions, is tenfold lower than that of myosin of skeletal muscle, and this activity is subject to a more direct Ca^{2+}-regulation. (2) The myosin of smooth muscle, in common with the myosins of nonmuscle cells, is able to interact with actin filaments and thereby cause contraction only when its light chains are phosphorylated; when the myosin light chains are dephosphorylated, the myosin cannot interact with actin and the muscle relaxes.

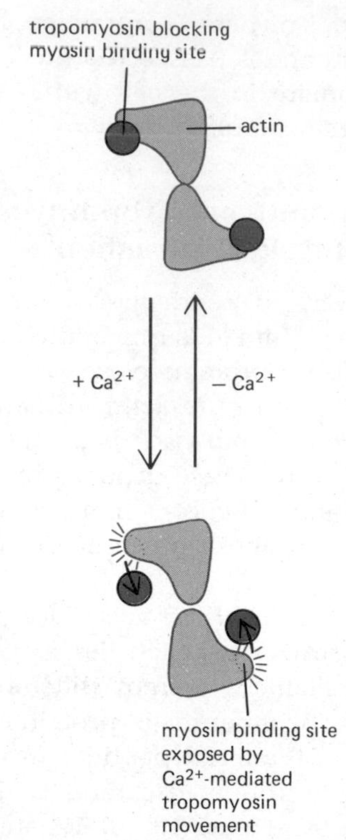

Figure 10–20 A cross-sectional view of a thin filament showing how, in the absence of Ca^{2+}, tropomyosin is thought to block the interaction of the myosin head with actin.

Both the phosphorylation and the dephosphorylation of smooth muscle myosin light chains are carried out by specific enzymes. Smooth muscle myosin ATPase is Ca^{2+}-dependent because the phosphorylating enzyme—**myosin light-chain kinase**—is regulated by Ca^{2+} levels. The effect of Ca^{2+} is mediated by a Ca^{2+}-binding protein, *calmodulin,* that is structurally and functionally very similar to troponin C (see p. 748). The complex of calmodulin with Ca^{2+} activates myosin light-chain kinase, which otherwise has little tendency to phosphorylate myosin (Figure 10–21). Several other factors also affect the activity of the myosin light-chain kinase—for example, the level of cyclic AMP in the cell, which is regulated in turn by hormones that influence smooth muscle contraction. Smooth muscle, therefore, is triggered to contract by an influx of Ca^{2+}, as is skeletal muscle. But the mechanism of triggering is different, and it is activated not by impulses from voluntary nerves but by nerves of the autonomic system or by hormones.

The enzymatic steps that follow Ca^{2+} binding to the soluble protein calmodulin act relatively slowly to activate smooth muscle contraction. Rapid activation is not required, since the much slower myosin cross-bridge cycle in smooth muscle cells only allows them to contract slowly. While calmodulin is an ubiquitous protein that mediates Ca^{2+}-dependent processes in many cells (see Chapter 13), troponin C is best considered a specialized form of calmodulin that has evolved in skeletal muscle cells to bind to thin filaments directly and thereby provide a more rapid Ca^{2+} regulation of contraction.

Smooth muscle cells should not be viewed as slow, poorly constructed versions of skeletal muscle cells that are adequate merely because fewer demands are made on them. In fact, they are specifically designed to be able to maintain tension for prolonged periods while hydrolyzing five- to tenfold less

ATP than would be required by a skeletal muscle cell that performed the same task. The slow myosin cross-bridge cycle in smooth muscle prevents it from contracting quickly, but it enables the muscle to maintain a constant tension with much greater efficiency.

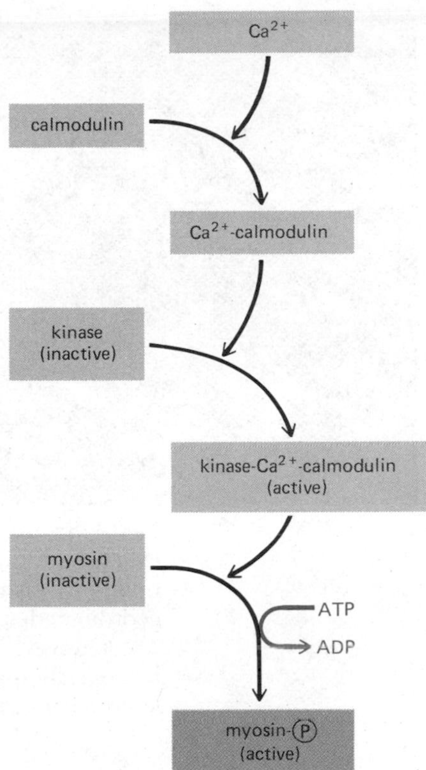

Figure 10–21 The cascade of reactions by which the contraction of smooth muscle is activated in the presence of Ca^{2+}. The kinase in this diagram is the enzyme that catalyzes the phosphorylation of a particular site on one of the two different types of myosin light chains. Nonmuscle myosin molecules are regulated by the same mechanism.

Summary

Muscle contraction is produced by the sliding of actin filaments against myosin filaments. The head regions of myosin molecules, which project from myosin filaments, undergo an ATP-driven cycle in which they attach to adjacent actin filaments, undergo a conformational change that pulls one filament against the next, and then detach. This cycle is facilitated by accessory muscle proteins that hold the actin and myosin filaments in parallel overlapping arrays with the correct orientation and spacing for sliding to occur. Two other accessory proteins—troponin and tropomyosin—allow the contraction of skeletal muscle to be regulated by Ca^{2+}. At low Ca^{2+} levels, these bind to the actin filaments in such a way that they block access of myosin heads. When Ca^{2+} levels rise in response to an electrical stimulus, the position of the troponin-tropomyosin complex changes slightly so as to allow myosin heads to bind to actin filaments, and contraction ensues.

Actin and myosin are also found in smooth muscle where they produce contraction in fundamentally the same way as in skeletal muscle. In such cells the two types of filaments are less highly ordered than in skeletal muscle, and their movement is dependent on a Ca^{2+}-regulated phosphorylation of myosin.

Ciliary Movement[7]

Next to muscle contraction, the best understood type of cellular movement is ciliary beating. **Cilia** are tiny hairlike appendages about 0.25 μm in diameter that contain a bundle of parallel microtubules at their core. They extend from the surface of many kinds of cells and are found in most animal species and some lower plants. Their primary function is to move fluid over the surface of a cell or to propel single cells through a fluid. Protozoa, for example, employ cilia both for the collection of food particles and for locomotion. In the human body, huge numbers of cilia (10^9 per cm^2 or more) on the epithelial cells lining the respiratory tract sweep layers of mucus, together with trapped particles of dust and dead cells, up toward the mouth, where they are swallowed and eliminated; cilia also help to sweep ova along the oviduct.

Like the actin filament-based contractile mechanism in muscle, the microtubule-based motile mechanism of the cilium is plainly reflected in its ultrastructure. Our knowledge of how cilia beat provides an important insight into how arrays of microtubules might produce other types of movement, such as intracellular transport and mitosis.

Cilia and Flagella Propagate Bending Movements

Fields of cilia bend in coordinated, unidirectional waves (Figure 10–22). Each cilium moves as a tiny whip: a forward active stroke, in which the cilium is fully extended and able to exert maximal force on the surrounding liquid, is followed by a recovery phase, in which the cilium returns to its original position by an unrolling movement that minimizes the viscous drag (Figure 10–23). Each cycle requires 0.1 second to 0.2 second. The cycles of adjacent cilia are almost but not quite in synchrony, and the small delay produces the striking patterns shown in Figure 10–22.

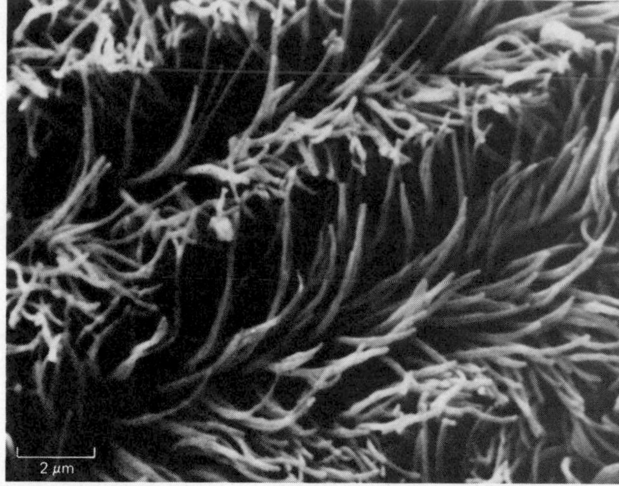

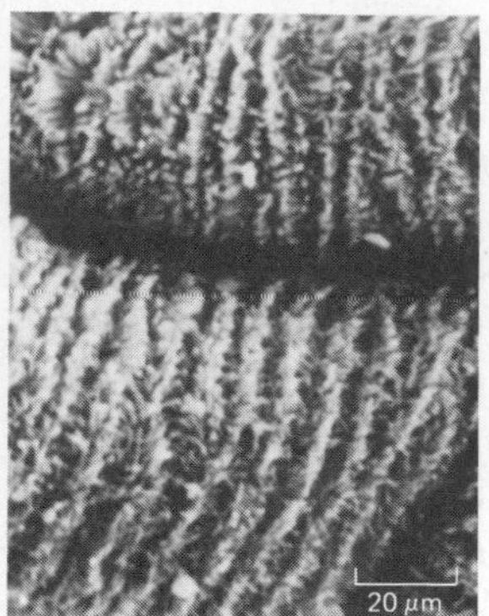

Figure 10–22 Scanning electron micrographs of a field of cilia from the gut of a marine worm. Although the cilia are uniformly distributed, the beat of adjacent cilia is coordinated, giving rise to a series of unidirectional waves. These waves can be seen most clearly at a lower magnification (*right*). (From J. S. Mellor and J. S. Hyams, *Micron* 9:91–94, 1978.)

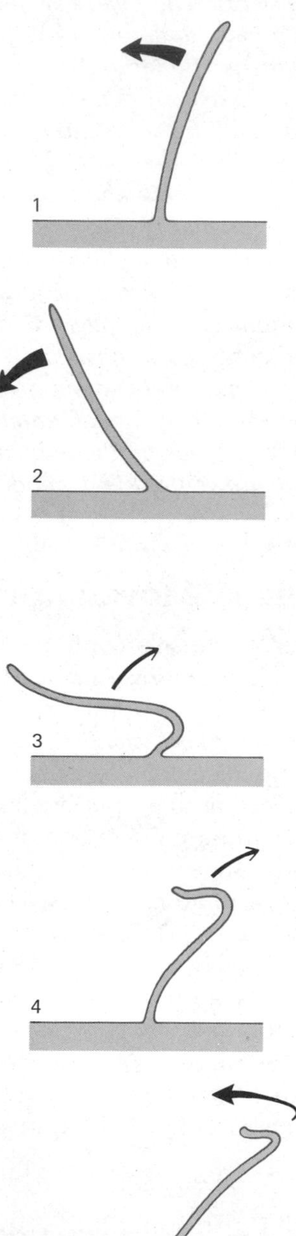

Figure 10–23 Schematic diagram of the movement of a cilium on the surface of a eucaryotic cell. The active stroke (stages 1 and 2), in which fluid is driven over the surface of the cell, is followed by a recovery stroke (stages 3, 4, and 5).

The simple **flagella** of sperm and of many protozoa are very much like large cilia in their internal structure, but they are usually very much longer and, instead of whiplike movements, they usually propagate quasi-sinusoidal waves. Nevertheless, the basis of this movement is almost certainly the same as that in cilia, and for most purposes the terms cilia and flagella can be used synonomously in eucaryotes. (It is, however, important not to confuse the flagella of eucaryotes with those of bacteria, which are quite different. Bacterial flagella are described on page 757.)

The Core of a Cilium Contains a Bundle of Parallel Microtubules in a "9 + 2" Arrangement

Ciliary movement is produced by the bending of the ciliary core, or **axoneme,** a complex structure composed entirely of microtubules and their associated proteins. The **microtubules** that form the principal structural component of the axoneme are hollow tubes of protein with an outer diameter of 25 nm. They are arranged in a pattern whose curious and distinctive appearance was one of the most striking revelations of early electron microscopy. The ciliary axoneme consists of a sheaf of nine doublet microtubules arranged in a ring around a pair of single microtubules (Figure 10–24). This "9 + 2" array is characteristic of almost all forms of cilia and eucaryotic flagella, from those of protozoa to those found in humans. The microtubules extend continuously for the length of the cilium, which is usually about 10 μm but can be as great as 200 μm in some cells.

While each member of the pair of singlet microtubules (the *central pair*) is a complete microtubule, each of the outer doublets is composed of one complete and one partial microtubule—known as the *A and B subfibers,* respectively—fused together so that they share a common tubule wall. In transverse sections, each complete microtubule is seen to be formed from a ring of 13 globular subunits, while the incomplete B subfiber of the outer doublet microtubules is formed from only 10.

Microtubules Are Hollow Tubes Formed from Tubulin[8]

Microtubules consist of molecules of **tubulin,** a globular polypeptide of 50,000 daltons. Because of the large numbers of microtubules that course along the axons of nerve cells, the most abundant source of tubulin for biochemical studies is vertebrate brain. Tubulin extracted from this source is a dimer of about 100,000 daltons, each dimer being composed of two polypeptides, *α-tubulin* and *β-tubulin,* which have closely related amino acid sequences. When tubulin molecules assemble into microtubules, they form *protofilaments* of tubulin polypeptides aligned in rows—with the β-tubulin of one dimer joined to the α-tubulin of the next. Usually 13 such protofilaments are arranged side by side around a central core that appears to be empty in electron micrographs (Figure 10–25). Optical and x-ray diffraction techniques show that the tubulin polypeptides in adjacent protofilaments are not aligned in register but are staggered, producing a regular lattice of tubulin molecules in the wall of each microtubule (Figure 10–26).

The assembly of tubulin molecules into microtubules resembles the assembly of actin into filaments in several important features. It occurs spontaneously *in vitro* and is normally accompanied by the hydrolysis of one molecule of bound nucleotide, though in the case of tubulin the nucleotide is GTP rather than ATP (Table 10–2). As with actin, the nucleotide hydrolysis has a crucial influence on the kinetics of polymerization. Microtubule assembly in cells is organized by various specialized structures that provide a base from which the microtubules can grow. These *organizing centers* nucleate a variety of transient arrays of microtubules that will be discussed later.

The microtubules of eucaryotic cilia and flagella are organized in the axoneme bundle by a specialized structure known as the *basal body.* This acts as a template for the "9 + 2" arrangement and gives rise to the doublet microtubules, which do not arise by spontaneous assembly *in vitro.* Axoneme microtubules are further stabilized by various accessory proteins, making them relatively permanent structures. As in the case of actin filaments, the ability of ciliary microtubules to produce movement depends on their polarity. To understand the importance of microtubule polarity to ciliary movement, we must now return to a more detailed examination of the ciliary axoneme.

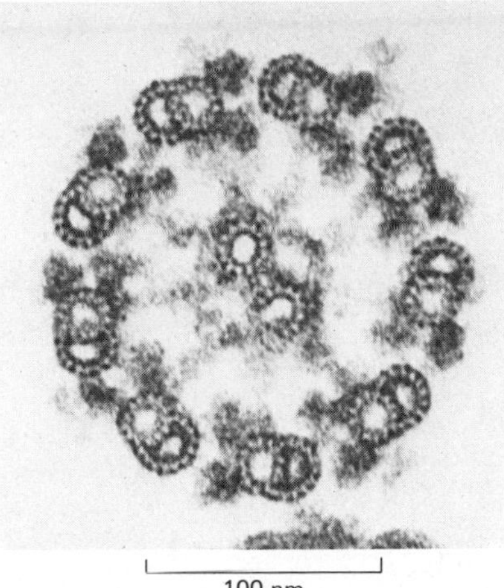

Figure 10–24 Electron micrograph of the flagellum of a green algal cell (*Chlamydomonas*) shown in cross-section. The distinctive "9 + 2" arrangement of microtubules occurs in almost all cilia and eucaryotic flagella. A schematic diagram of this structure seen in cross-section, illustrating the principal structural components, is shown in Figure 10–27. (Micrograph courtesy of Lewis Tilney.)

The Ciliary Axoneme Contains Links, Spokes, and Sidearms Made of Protein

Associated with the microtubules of the axoneme are many other protein structures whose interactions provide the power for the ciliary motor and harness it so as to generate wavelike movements. The most important of these structures are sets of arms that project from each doublet and extend toward an adjacent doublet in the outer ring. These arms occur in pairs (Figure 10–27) and are spaced along the microtubule at regular 24-nm intervals. The

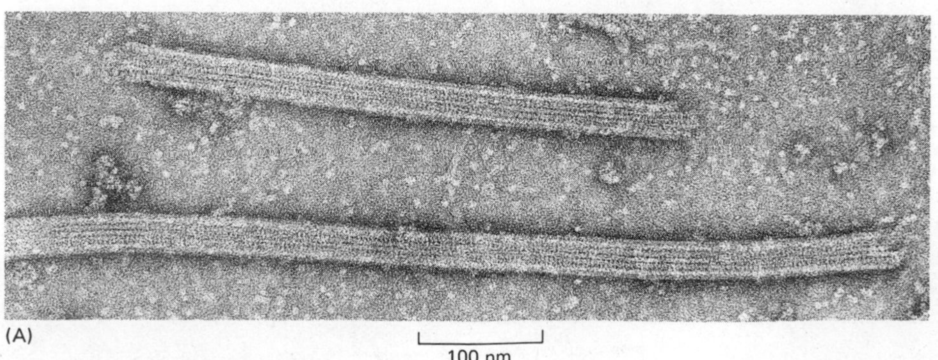

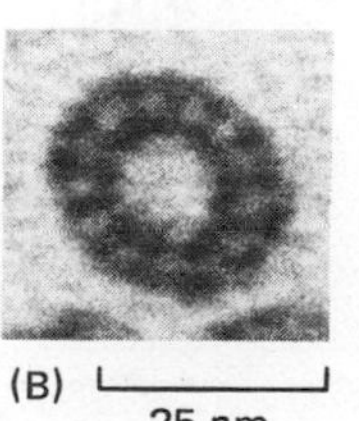

Figure 10–25 (A) Electron micrograph of negatively stained microtubules. (B) A microtubule, seen in cross-section, with its 13 distinct subunits, each of which corresponds to a separate tubulin polypeptide. (A, courtesy of Robley Williams; B, courtesy of Richard Linck.)

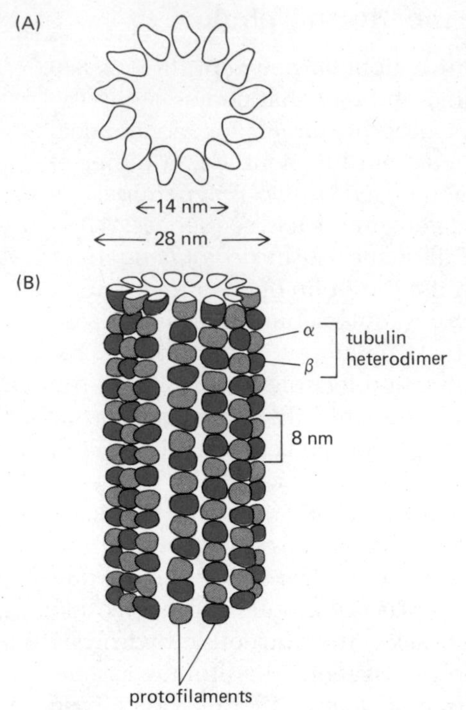

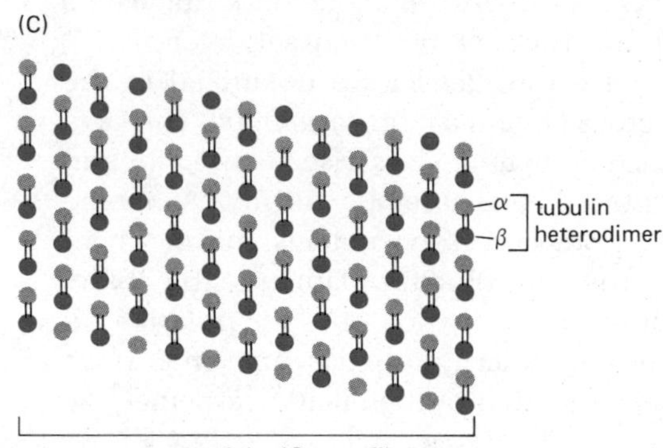

Figure 10–26 Schematic diagram of a microtubule showing how the tubulin polypeptides pack together to form the cylindrical wall. (A) shows the 13 polypeptides in a cross-section. (B) A side view of a short section of a microtubule, with the tubulin polypeptides aligned into rows or *protofilaments.* (C) A portion of the microtubule wall "unrolled" to show the packing of the two kinds of tubulin polypeptides; each of the 13 protofilaments is composed of a series of αβ heterodimers. (Drawn from data supplied by Linda Amos.)

Table 10–2 Comparison of Actin and Tubulin

	Actin	**Tubulin**
Polypeptide molecular weight	42,000	50,000 (α-tubulin) 50,000 (β-tubulin)
Unpolymerized form	globular, monomer	globular, dimer (1 α + 1 β)
Bound nucleotide (in unpolymerized state)	ATP (1 per monomer)	GTP (2 per dimer)
Factors needed for polymerization	Ca^{2+} or Mg^{2+} NaCl	Mg^{2+} chelator to remove Ca^{2+} NaCl
Form of polymer	2-stranded helix	hollow tube composed of 13 protofilaments
Diameter of filament	7 nm	25 nm

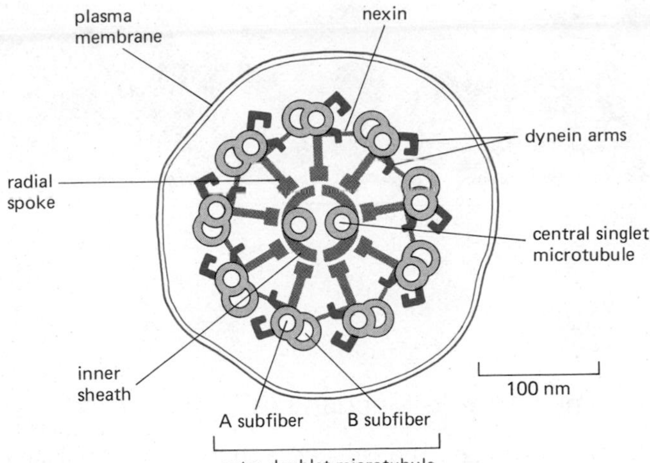

Figure 10–27 Schematic diagram of a cilium shown in cross-section (an electron micrograph of this type of structure is shown in Figure 10–24.) The various projections from the microtubules occur at regular intervals along the cilium as described in Table 10–3.

arms, which are composed of a protein called *dynein,* play an essential part in the movement of the cilium, as described below. At more widely spaced intervals, another protein, called *nexin,* forms links between the adjacent doublets; these are thought to be highly elastic, and they form "straps" around the entire axoneme, rather like hoops around a barrel.

Projecting inward from each doublet is a *radial spoke,* which ends in a globular portion very near the *inner sheath.* The inner sheath is composed of slender protein arms that project outward from the central pair of microtubules and curve around them (Figure 10–27). It apparently enables the central pair to help regulate the movement of the axoneme. Looked at from the side of the axoneme, each of these structures—the dynein side arms, the nexin links, the radial spokes, and the arms of the inner sheath—can be seen as a series of regular projections, each with its own periodicity (Table 10–3).

Cilia and flagella end inside the cell in the **basal body** (Figure 10–28). The basal body is composed of parallel microtubules that form a short cylinder with the same outer diameter and ninefold symmetry as the axoneme itself. Each of the outer microtubule doublets of the axoneme extends into the basal body, where it is joined by a third partial microtubule. As a result, the basal body contains a ring of nine fused *triplet* microtubules. The two central microtubules of the cilium terminate before they reach the basal body, which therefore has no microtubules in its core (see also Figures 10–49 and 10–51, pp. 579 and 581).

Table 10–3 Major Protein Structures of the Ciliary Axoneme

Axoneme Component (periodicity along axoneme)	**Function**
Tubulin dimers (8 nm)	principal component of microtubules
Dynein arms (24 nm)	project from microtubule doublets and interact with adjacent doublets to produce bending
Nexin links (86 nm)	hold adjacent microtubule doublets together
Radial spokes (29 nm)	extend from each of the 9 outer doublets inward to the central pair
Sheath projections (14 nm)	project as a series of side arms from the central pair of microtubules; together with the radial spokes these regulate the form of the ciliary beat

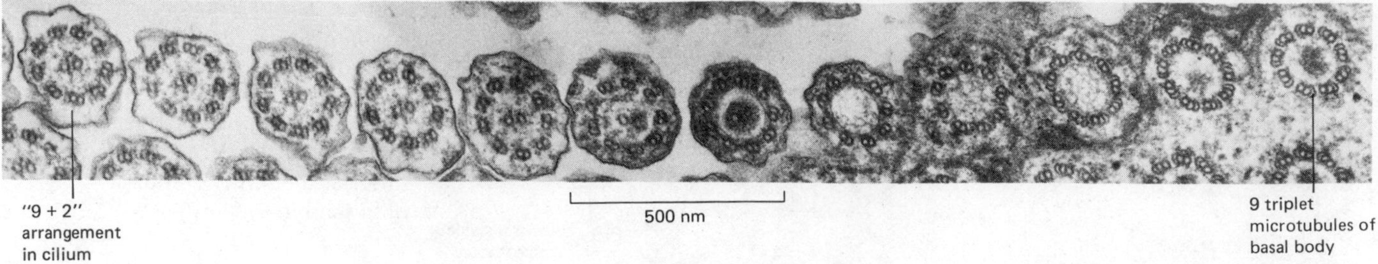

The Axoneme Moves by a Sliding Microtubule Mechanism[9]

If a flagellum is severed from a cell by a laser beam, the isolated structure continues to propagate bending movements in a normal way. It is therefore clear that the motile machinery is contained in the axoneme itself and that its movements do not depend on a motor at its base (as does the bacterial flagellum—see p. 758). Indeed, an isolated axoneme will still propagate bending movements even after removal of its plasma membrane, provided that it is in a salt solution containing ATP and either Mg^{2+} or Ca^{2+}.

Direct evidence that the bending force is produced by the sliding of microtubules has been obtained by exposing isolated axonemes to proteolytic enzymes. This treatment disrupts both the nexin links and the radial spokes, while the dynein arms and the microtubules themselves appear to remain intact. The partially digested structure is extremely sensitive to ATP, and in the presence of as little as 10 μM ATP it will disintegrate lengthwise into its constituent doublet microtubules. ATP causes the axoneme to elongate until it is up to nine times its original length—the component fibers in the axoneme telescoping out of the loosened structure (Figure 10–29). It seems that the adjacent outer doublets can actively slide against each other once they are freed of their lateral cross-links (such as those made of nexin). In the intact structure the sliding movement is converted to bending, as shown diagramatically in Figure 10–30.

In support of a **sliding microtubule mechanism** for ciliary bending, electron micrographs of cilia at various stages of bending show that the microtubules do not change appreciably in length. Moreover, the rows of radial spokes, which occur in a regular repeating pattern and therefore serve as markers for each outer doublet, show a relative displacement in the region of the bend that indicates that the microtubules slide relative to each other within the axoneme (Figure 10–31).

Figure 10–28 Electron micrograph of a cross-section through the cortex of a ciliated protozoan, *Tetrahymena.* The plane of section grazes the surface of the cell, showing in successive sections how the "9 + 2" arrangement of microtubules in the axoneme leads into the nine triplet microtubules of the basal body.

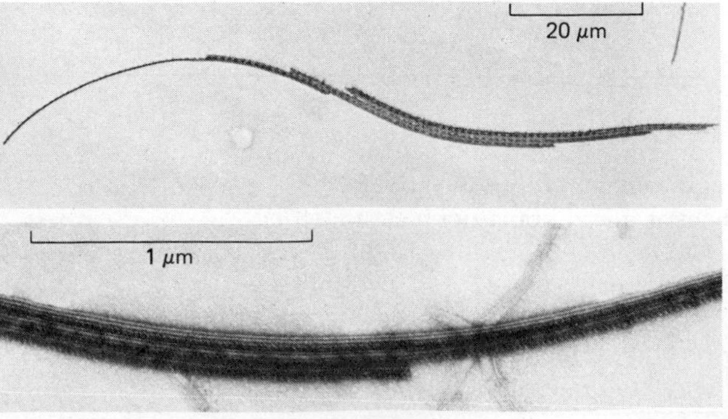

Figure 10–29 Electron micrographs of an isolated axoneme (from a cilium of *Tetrahymena*) that has been briefly exposed to the proteolytic enzyme trypsin to loosen the protein ties that normally hold it together. Following treatment with ATP, the individual microtubule doublets, which are seen in greater detail in the lower part of the figure, slide against each other, causing the original structure to increase in length by as much as nine times. (From F. D. Warner and D. R. Mitchell, *J. Cell Biol.* 89:35–44, 1981. Reproduced by permission of the Rockefeller University Press.)

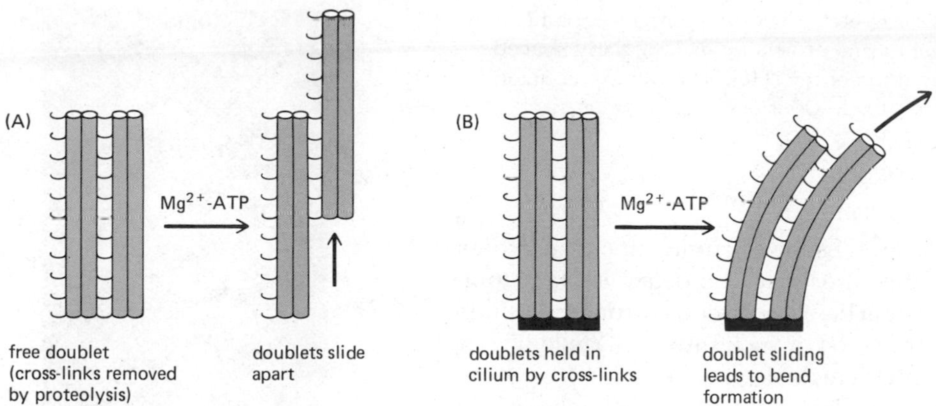

Figure 10–30 Schematic illustration of how the relative sliding of outer microtubule doublets against each other shown in (A) causes bending if the doublets are tied to each other at one end (B).

Dynein Is Responsible for the Sliding[10]

Since the adjacent outer doublets slide actively against each other, a force must be generated between them. One point of contact in the intact axoneme is that of nexin links, but since these are destroyed by the protease treatment described above and the sliding is unimpaired, they cannot be required for movement. This leaves the **dynein arms.** Normally, these arms do not quite extend all of the way from one doublet to the next, but if the cilium is allowed to use up all of its ATP, the dynein arms make contact with the adjacent doublet (Figure 10–32). In addition, there is a change in orientation of the dynein arms: they point downward toward the basal body in the presence of ATP and at right angles to the microtubule doublets in its absence.

There is good evidence that the dynein arms are required for ciliary movement. When the arms are selectively extracted from the naked axoneme under carefully controlled conditions, their progressive disappearance is accompanied by a parallel slowing of axonemal movement. When all of the arms are lost, the movement stops. Moreover, the procedure can be reversed: purified dynein added to the outer doublets of extracted axonemes reconstitutes the side arms and restores the ability of the microtubules to undergo ATP-dependent sliding.

But how do dynein arms generate the force to slide adjacent microtubules past each other? An important advance in answering this question was the discovery that dynein is itself an ATPase. Dynein is a very large protein composed of multiple polypeptides, the major ones having molecular weights of about 400,000. The ATPase activity of the isolated protein is low, but it increases when purified microtubules are added—an effect similar to, but not as large as, the stimulation of myosin ATPase by the addition of actin filaments. Each dynein arm generates about as much force as a myosin molecule in muscle, the force acting so as to move the adjacent microtubule doublet that it contacts toward the tip of the cilium.

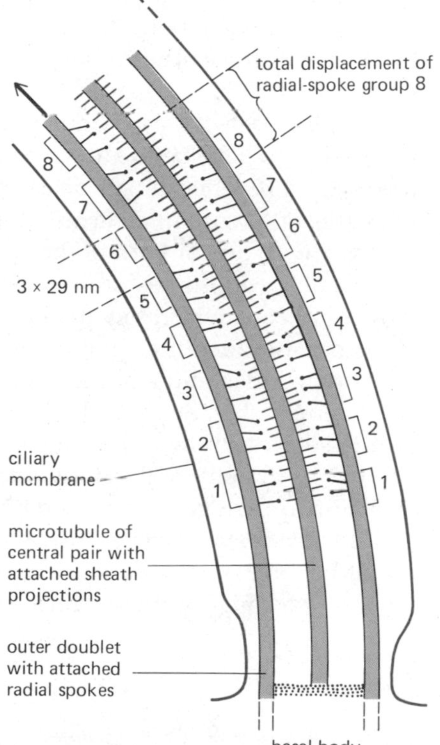

Figure 10–31 Schematic diagram of the base of a bending cilium showing how the radial spokes (numbered in groups of three) act as markers along the length of each doublet microtubule. The progressive displacement in the radial spokes on either side of the cilium indicates that the two microtubule doublets are sliding with respect to each other. This sliding is also revealed at the tip of the axoneme, where the doublets on the inside of the bending cilium protrude beyond those on the outside (see Figure 10–30B).

Cilia Can Be Dissected Genetically[11]

It is generally easy to select mutant organisms that have lost the power of movement. After the normal ones have been allowed to swim, crawl, fly, or slide away, those left are either dead or paralyzed. And if movement depends on only one type of motile element, such as skeletal muscle or cilia, the full force of genetic dissection can be brought to bear in analyzing its internal structure.

The favorite object of such studies is the unicellular alga *Chlamydomonas reinhardii,* which has two flagella that propel it through the water. Many nonmotile mutants have been isolated: in some, the mechanism of flagellar

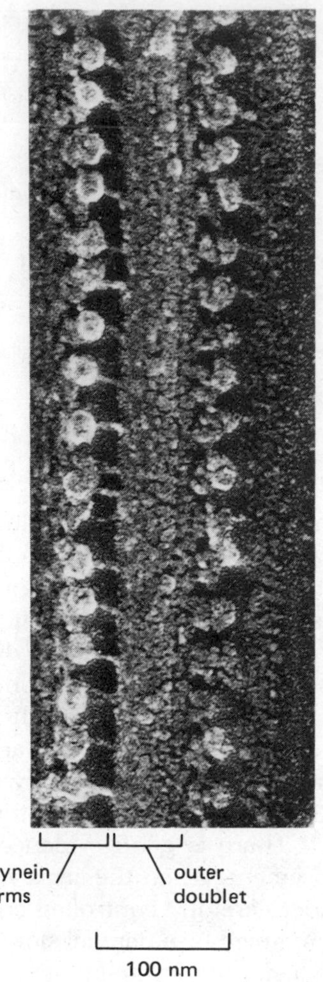

Figure 10–32 Freeze-etch electron micrograph of a cilium showing the dynein arms projecting like regularly spaced lollipops from the doublet microtubules to make contact with an adjacent doublet. (Courtesy of John Heuser.)

assembly is defective and the flagella are absent or rudimentary; in others, flagella are present but immobile. In the latter case, the defect is likely to be in a protein component of the motor mechanism. Various structural abnormalities are apparent in electron micrographs of such mutant flagella (Figure 10–33). In one class of mutants the only detectable change is the loss of the dynein arms. In a second class the mutants lack only the radial spokes, while in a third they lack both the central pair of microtubules and the inner sheath. In all three classes the isolated membrane-free axonemes fail to move in the presence of ATP.

Ciliary defects also occur in humans, where various hereditary forms of male sterility have been shown to be due to nonmotile sperm. Depending on the form of this hereditary disease, the sperm flagella lack both dynein arms, only the outer or inner arms, the radial spoke heads, or the inner sheath together with one or both of the central pair of microtubules. Exactly the same defects occur in the respiratory cilia of such individuals, who commonly have long histories of respiratory tract disease—with recurrent bronchitis and chronic sinusitis—because their immotile respiratory cilia are unable to clear mucus from their lungs and sinuses.

Remarkably, about half of the individuals with *immotile cilia syndrome* also have an extremely rare condition known as *situs inversus*, in which the normal asymmetry of the body—as seen in the position of the heart, intestine, liver, and appendix—is reversed (the whole complex of abnormalities is known as *Kartagener's syndrome*). It has therefore been suggested that the unidirectional beating of cilia during early human development could play a critical part in determining the normal left-right asymmetry of the body.

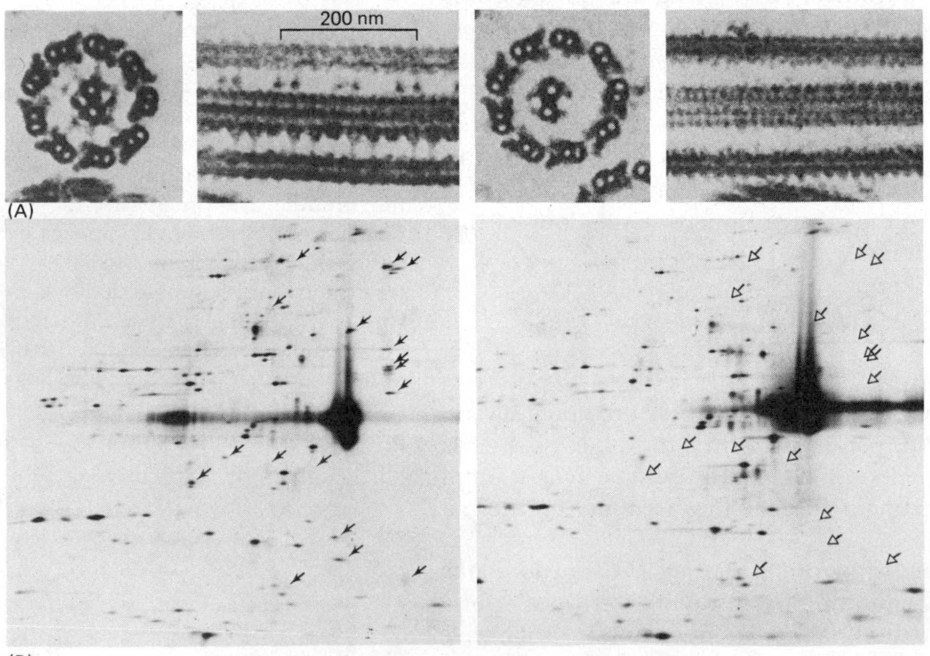

Figure 10–33 (A) Electron micrographs comparing a spokeless flagellum from a nonmotile *Chlamydomonas* mutant (*right*) to a normal flagellum from a wild-type organism (*left*) in transverse and longitudinal sections. (B) When the proteins of these flagella are analyzed by two-dimensional gel electrophoresis, a number of proteins are found to be missing in the mutant. The positions of 17 of these, all components of the radial spoke, are indicated by arrows. Only one of the 17 proteins is altered in the mutant, but because this protein is required for spoke assembly, the flagella that form lack all of the proteins. (From G. Piperno, B. Huang, Z. Ramanis, and D. Luck, *J. Cell Biol.* 88:73–79, 1981. Reprinted by permission of Rockefeller University Press.)

The Conversion of Microtubule Sliding to Ciliary Bending Depends on the Inner Sheath[12]

The radial spokes and the inner sheath appear to control the activities of the dynein arms so that they produce a coherent wave of movement in the cilium. If all the dynein arms were to become active at the same time—as the myosin heads do in contracting muscle—the axoneme would simply twist into a tight helix. In order to produce a local bend that propagates from the base of the cilium to the tip in a coherent wave, there must be controls that coordinate the action of the dynein arms. These controls cannot depend on fluxes of Ca^{2+} or other ions since, as mentioned above, the axoneme will beat without its plasma membrane. Presumably the activation of individual dynein arms must depend on the mechanical movements of other parts of the axoneme, relayed through protein-protein interactions.

The inner sheath of the cilium appears to be involved in this "slide-to-bend" conversion. In some cilia the direction of bending appears to be determined by the orientation of the central microtubule pair, the power stroke being in the plane that bisects the two central microtubules. Other axonemes naturally lack the central pair of microtubules (and therefore have a "9 + 0" arrangement). These either are immobile, as is the aberrant solitary cilium (the so-called *primary cilium*) that forms during interphase in many types of cells, or move very slowly as do the sperm tails of some insects. Further evidence with regard to the control of bending comes from *Chlamydomonas* mutants that lack either the inner sheath or the radial spokes. Although the membrane-free axonemes are paralyzed in these mutants, the doublet microtubules can still slide relative to each other if these axonemes are treated with proteolytic enzymes. This fact suggests that the activation of the dynein arms is normally inhibited by other proteins, and that signals relayed by the radial spokes and the inner sheath relieve this inhibition, causing an organized pattern of activation of the dynein arms throughout the flagellum.

Summary

Cilia have a cylindrical core containing a ring of nine doublet microtubules, which cause the cilium to bend by sliding against each other. Side arms composed of the protein dynein extend from each microtubule doublet. Contact between these dynein arms and a neighboring doublet is believed to activate ATP hydrolysis by the dynein and generate a sliding force between the microtubules. Accessory proteins, such as nexin, bundle the ring of microtubule doublets together and limit the extent of their sliding. Other accessory proteins act—in association with a pair of single microtubules at the center of the cilium—to regulate the sliding of the doublet microtubules at the periphery and thereby produce the cyclical ciliary bends that underlie ciliary beating.

General Features of Microtubules and Actin Filaments as Dynamic Assemblies

Myofibrils and cilia are relatively permanent structures specialized to produce repetitive movement. But most cellular movements depend on labile structures that appear at specific stages of the cell cycle or in response to external signals and then disappear again. The most familiar of these are the mitotic spindle and the contractile ring that form during cell division. These and other motile structures in cells are formed from microtubules and actin filaments, which are assembled from pools of soluble tubulin and actin in the cytoplasm and disassembled when they are no longer required.

This section details some of the movements that depend on such transient assemblies and considers the properties of tubulin and actin that make these movements possible.

Microtubules Are Highly Labile Structures That Are Sensitive to Specific Antimitotic Drugs[13]

Many of the microtubule arrays in cells are labile, and there are some whose function depends on this lability. During interphase, microtubules radiate from the *cell center* throughout the cytoplasm, but at the onset of mitosis these cytoplasmic microtubules disassemble as the microtubules of the mitotic spindle begin to form. At the end of mitosis the process is reversed. A more specialized case is found in the feeding tentacles of heliozoans—single-cell organisms closely related to *Amoeba.* Each tentacle contains several hundred microtubules packed together in a parallel bundle (Figure 10–34). In the course of feeding, the tentacles often retract rapidly to the cell body, drawing with them any attached prey; this retraction is caused by a rapid depolymerization of the microtubules. Depolymerization is also produced by exposure to low temperatures or to Ca^{2+}; when the temperature is raised or Ca^{2+} is removed, both the tentacles and the microtubules rapidly re-form.

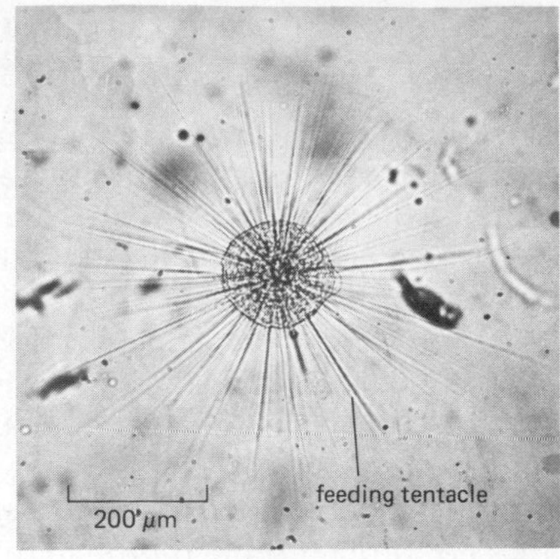

(A)

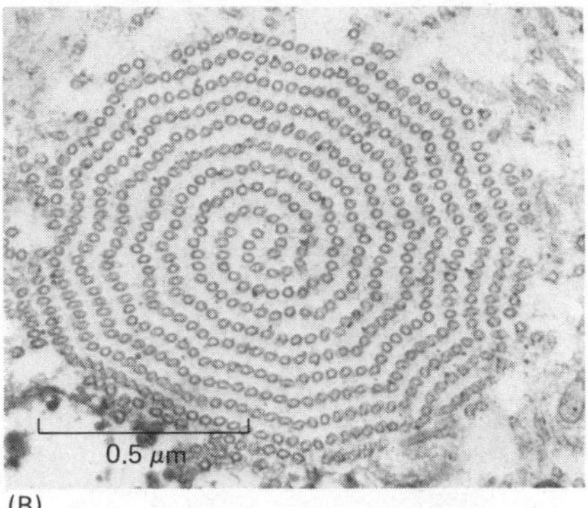

(B)

Figure 10–34 (A) Light micrograph of the heliozoan *Actinosphaerium.* (B) An electron micrograph showing one of the feeding tentacles of *Actinosphaerium* in cross-section. The core of each tentacle is formed by two concentric spirals of microtubules. (Courtesy of Lewis Tilney.)

Processes such as these depend on an exchange of tubulin molecules between the microtubules and a soluble cytoplasmic pool: it has been estimated that only about 50% of the tubulin in a typical cultured cell is polymerized under normal growth conditions. But the stability of the different microtubular structures varies considerably—the mitotic spindle being one of the most labile. This lability can be seen in the extreme sensitivity of the spindle to various drugs that bind to tubulin and prevent it from polymerizing. Of these, **colchicine,** an alkaloid extracted from the meadow saffron, has the longest history, having been used medicinally since ancient Egyptian times as a treatment for gout. One molecule of colchicine (Figure 10–35) will bind tightly to one tubulin dimer and thereby prevent its polymerization. Addition of colchicine (or a related compound, *colcemid*) to a dividing cell causes the disappearance of the mitotic spindle and blocks cells in mitosis (p. 647). For this reason, such compounds are known as *antimitotic drugs* (Table 10–4). Removal of the drug allows the spindle to re-form and mitosis to proceed. *Vinblastine* and *vincristine,* which also inhibit microtubule formation, have been widely used as anticancer drugs; they are useful because the disruption of mitotic spindle microtubules preferentially kills rapidly dividing cells.

Addition of a different type of drug, **taxol,** increases (rather than decreases) the polymerization of tubulin *in vitro:* when added to cells, it causes much of the free tubulin to assemble into microtubules. Similarly, D_2O (heavy water) causes a shift toward tubulin polymerization and thus increases the number of microtubules in a mitotic spindle, for example. Surprisingly, the additional stabilization of the microtubules in the cell caused by either taxol or D_2O is as deleterious to the function of microtubules as their depolymerization: a cell with microtubules stabilized by either taxol or D_2O is unable to progress through its growth cycle and it stops at mitosis. It therefore seems that at least some microtubules must be in a state of dynamic assembly and disassembly in order to function properly.

Factors that regulate the assembly of microtubules or affect their stability are discussed in a later section. We now turn to some cellular processes that depend on the assembly of actin filaments and to the dynamics of actin and tubulin polymerization.

H_3CO H_3CO H_3CO $NHCOCH_3$ O OCH_3

Figure 10–35 Chemical structure of colchicine.

Table 10–4 Some Drugs That Bind to Actin or Tubulin

	Mode of Action
Tubulin-binding drugs	
Colchicine, colcemid	inhibit the addition of tubulin molecules to microtubules, leading to microtubule depolymerization
Vinblastine, vincristine	induce the formation of paracrystalline aggregates of tubulin
Taxol	stabilizes microtubules
Actin-binding drugs	
Cytochalasins (B, D, E, etc.)	inhibit the addition of actin molecules to actin filaments, leading to filament depolymerization
Phalloidin	stabilizes actin filaments

Actin Filaments Are Continually Formed and Broken Down in Cells[14]

There are several clear cases in which actin is rapidly assembled into filaments in cells. One of the most impressive is the *acrosomal reaction* of invertebrate sperm. As the sperm of the sea cucumber, for example, approaches an egg, it suddenly sends out a thin *acrosomal process,* which punctures the coating of the egg like a harpoon and allows the sperm and egg membranes to fuse (Figure 10–36; see also p. 802). The mature acrosomal process contains about 25 long actin filaments produced by an almost explosive polymerization of actin molecules into filaments that extend at a rate of 10 μm per second. Another striking example occurs in blood platelets. Platelets are rounded cell fragments that lack a nucleus and ribosomes but are capable of rapidly changing shape and secreting local chemical mediators in response to an injury to a blood vessel (see Figure 16–35, p. 922). In the course of this response, platelets develop numerous thin projections that play a part in the formation and contraction of blood clots. These projections contain a large number of actin filaments that quickly form from a pool of unpolymerized actin. Their subsequent contraction is not brought about by disassembly but by interaction of the actin filaments with myosin.

Many types of cells extend and retract fingerlike *microvilli* on their surfaces (Figure 10–37). These structures contain a core of actin filaments, and their extension and retraction are caused by actin polymerization and depolymerization, respectively. For example, the surface of an unfertilized sea urchin egg is covered with some 130,000 short microvilli. When the egg is fertilized, the fusion of cortical granules with the plasma membrane increases the surface area of the egg more than twofold, and the extra membrane is accommodated by the elongation of the microvilli. Each microvillus increases its length from 0.3 μm to 1 μm in seconds. Biochemical experiments indicate that this lengthening is accompanied by a large-scale polymerization of actin.

In cells such as the unactivated sperm and the rounded blood platelet, a large proportion of the unpolymerized actin is bound to a protein known as **profilin.** A similar actin-binding protein has been found in virtually all vertebrate and invertebrate cells. This 16,000-dalton protein forms a rapidly reversible, one-to-one complex with actin molecules, thereby retarding their polymerization. It seems that by sequestering actin molecules, profilin acts as a buffer to keep the effective concentration of free actin at a constant level; moreover, regulatory mechanisms that weakened the profilin-actin interaction

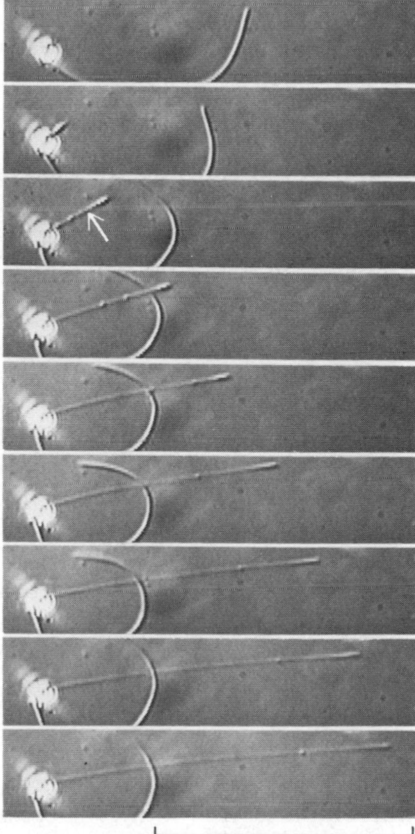

Figure 10–36 Light micrographs showing stages in the elongation of the acrosomal process of sea urchin sperm. The photographs were taken at intervals of 0.75 second, beginning two seconds after the sperm was artificially activated to undergo the acrosomal reaction. The arc to the right of the sperm head is a portion of its tail that has curved around out of the field of the micrograph. (From L. G. Tilney and S. Inoué, *J. Cell Biol.* 93:820–827, 1982. Reproduced by permission of the Rockefeller University Press.)

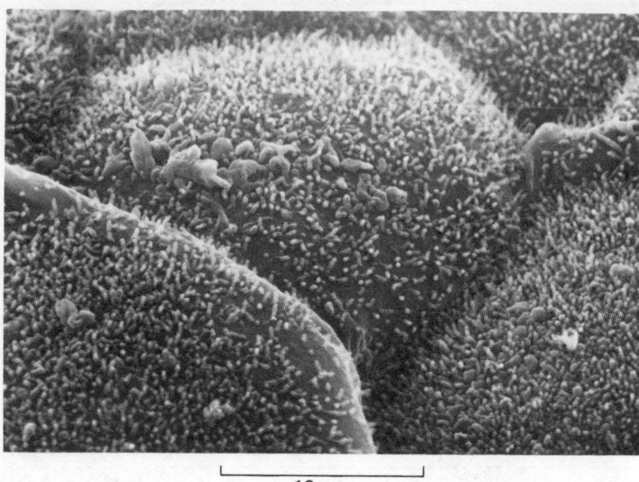

Figure 10–37 Scanning electron micrograph of short, actin-containing microvilli covering the surface of cancer cells in culture. (Courtesy of Günter Albrecht-Buehler.)

could, in principle, trigger the polymerization of actin molecules in cells. In fact cells contain many other actin-binding proteins, some of which collaborate with profilin in controlling the formation and state of aggregation of actin filaments in the cytoplasm (p. 592).

Specific Drugs Change the State of Actin Polymerization and Thereby Affect Cell Behavior[15]

It is probable that a dynamic equilibrium exists between actin molecules and actin filaments in all cells and that this plays a part in various cell movements such as locomotion and phagocytosis. But changes in actin polymerization in such situations are difficult to detect since—unlike the activation of sperm or platelets—they do not produce a long-lasting and easily identified effect.

Drugs that affect the state of actin polymerization can, however, be shown to disrupt many of these cell movements (Table 10–4). For example, the **cytochalasins** (Figure 10–38), a family of metabolites excreted by various species of molds, paralyze many different kinds of vertebrate cell movement—inhibiting cell locomotion, phagocytosis, cytokinesis, the production of thin cell extensions known as microspikes (p. 584), and the folding of epithelial sheets into tubes. They do not, however, inhibit mitosis, which principally involves microtubules, nor do they affect muscle contraction, which does not involve the assembly and disassembly of actin filaments. The cytochalasins act by binding specifically to one end of actin filaments (corresponding to the "barbed" end of an actin filament decorated with S1), thereby preventing the addition of actin molecules to that end.

Phalloidin is a highly poisonous alkaloid produced by the toadstool *Amanita phalloides.* By contrast with the cytochalasins, it stabilizes actin filaments and inhibits their depolymerization. The drug does not readily cross the plasma membrane, and it must be injected into a cell before it can act. When this is done, phalloidin is found to block the migration of both protozoan amoebae and various vertebrate cells in culture, suggesting that the dynamic assembly and disassembly of actin filaments is crucial for these movements. Because phalloidin stabilizes actin filaments by binding to them in a highly specific fashion along their lengths, fluorescent derivatives of the drug have been useful for staining actin filaments inside cells (see Figure 10–78).

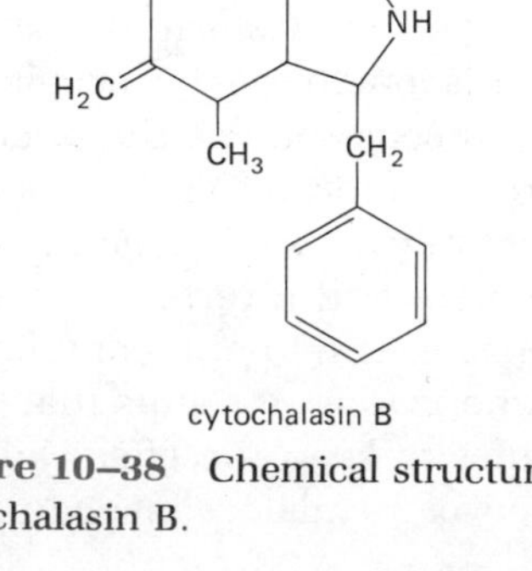

cytochalasin B

Figure 10–38 Chemical structure of cytochalasin B.

The Polymerization of Actin and Tubulin Can Be Studied *in Vitro*[16]

The properties of purified actin in a test tube provide important clues to how the polymerization of this protein is controlled in the living cell. As already noted, skeletal muscle actin is capable of spontaneously assembling into filaments *in vitro*, and actins from other tissues behave in a very similar way. Indeed, actins from sources as widely separated as slime molds and vertebrates will co-polymerize into mixed actin filaments. We shall now examine the kinetics of this spontaneous reaction.

Efficient polymerization of actin requires ATP. This nucleotide is bound to the unpolymerized actin molecule and is hydrolyzed each time an actin molecule is added to the polymer. Polymerization does not proceed at a uniform rate but begins with a *lag phase* in which there is no observable increase in viscosity (Figure 10–39). (Recall that actin polymerization is initiated by an increase in the ionic strength of the actin solution and can be monitored by measurements of viscosity.) The lag phase is thought to reflect the fact that the first step in filament formation, in which two or three actin molecules must come together in a specific geometric conformation, is particularly difficult. This step is known as *nucleation*. Once it has been achieved, the addition of further actin molecules to the end of the filament proceeds rapidly.

Eventually the concentration of actin monomers drops to a point at which molecules come on and off the filament at the same rate—a concentration of free actin known as the *critical concentration*. Below the critical concentration actin assembly cannot proceed and assembled filaments begin to depolymerize (Figure 10–40). Some of the proteins associated with actin in cells can be shown *in vitro* to nucleate actin polymerization and to influence actin filament formation (p. 591).

Assembly experiments similar to those described for actin can be carried out with purified tubulin, the soluble protein that forms microtubules. As noted earlier, tubulin normally exists as a dimeric molecule containing one α-tubulin and one β-tubulin polypeptide and is normally associated with GTP, which is hydrolyzed in the course of polymerization. But in addition to the exchangeable GTP that is hydrolyzed to GDP during polymerization, each tubulin dimer contains one firmly bound molecule of GTP that does not participate in the polymerization reaction and whose significance is unknown.

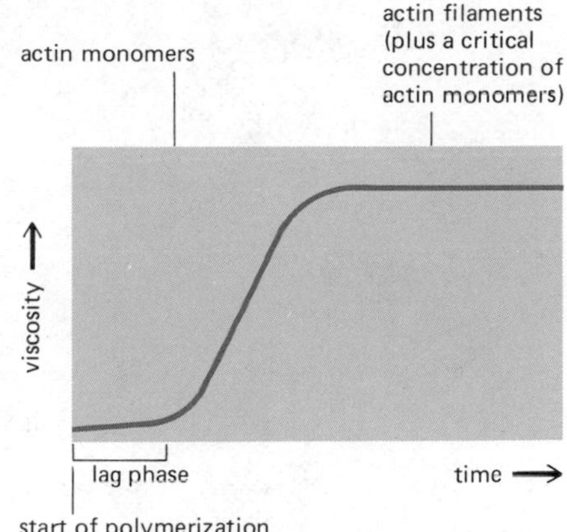

Figure 10–39 Time course of actin polymerization as measured by increasing viscosity. Polymerization is normally initiated by an increase in the ionic strength and, as indicated here, usually proceeds after an initial lag phase.

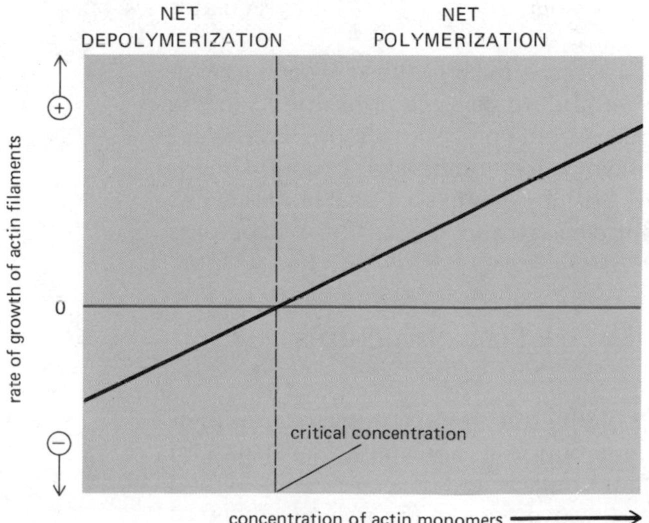

Figure 10–40 Rate of growth of actin filaments at different concentrations of free actin molecules. Below the critical concentration, no filaments are formed and any that are present undergo negative growth (i.e., they depolymerize). Above the critical concentration, filaments are formed and grow by the addition of actin molecules to their ends. As the free actin molecules are used up, their concentration falls until it reaches the critical concentration. At this point the net growth rate is zero, and free actin molecules and filaments are at equilibrium.

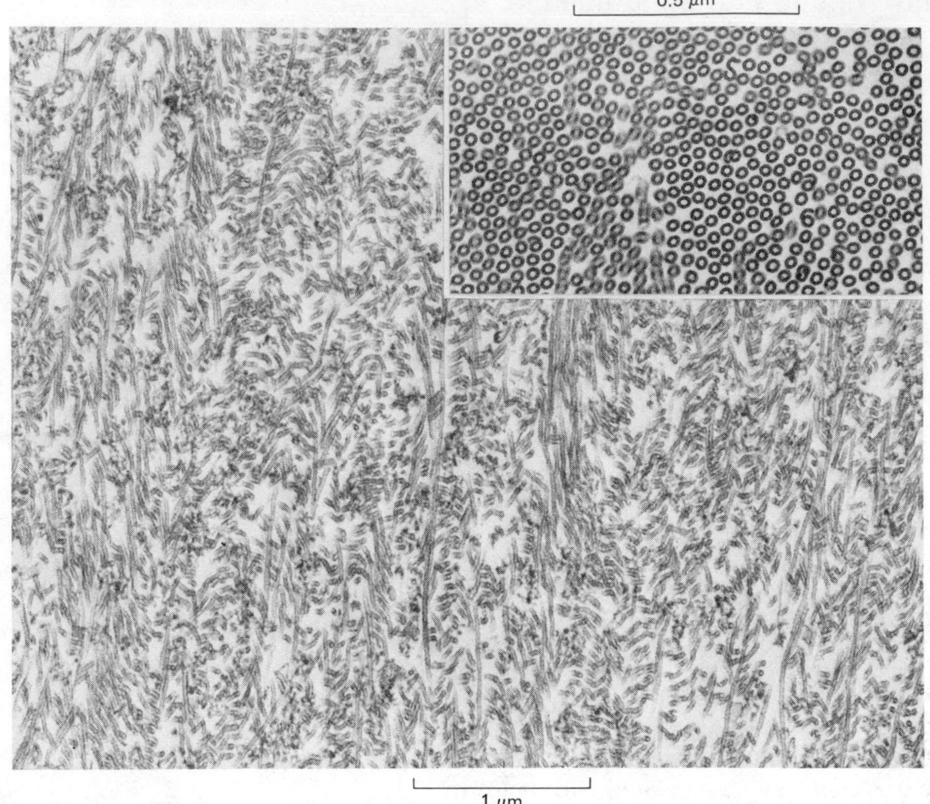

Figure 10–41 Electron micrograph of microtubules prepared by polymerization in the test tube. The microtubules were centrifuged into a pellet and sectioned before examination in the electron microscope. (Courtesy of Olav Behnke.) In the inset, a similar preparation of microtubules is seen at somewhat higher magnification and mainly in cross-section. (Courtesy of Richard Linck.)

The resulting microtubule consequently contains one molecule of GTP and one molecule of GDP for each tubulin dimer.

The microtubules produced by polymerization in the test tube are similar to those found in cells (Figure 10–41). They polymerize at high temperatures (37°C) and depolymerize at low temperatures (4°C). Drugs that depolymerize microtubules in the cell, such as colchicine or vinblastine, and drugs that stabilize microtubules in the cell, like taxol, have corresponding effects in the *in vitro* system. In addition, providing that microtubule accessory proteins (p. 578) and the calcium regulatory protein *calmodulin* (p. 748) are present, microtubule assembly *in vitro* is completely blocked or reversed by micromolar amounts of Ca^{2+}, which strongly suggests that Ca^{2+} also regulates microtubule assembly in cells.

The kinetics of tubulin assembly are very similar to those of actin assembly, shown previously in Figures 10–39 and 10–40. A lag phase due to the formation of an initial fragment of a microtubule precedes a more rapid elongation phase. Once again there is a critical concentration of free tubulin below which polymerization cannot proceed. Structures such as basal bodies act as nucleating centers for microtubules both *in vitro* and in cells (these structures are discussed later in this chapter).

Actin Filaments and Microtubules Are Polar Structures That Grow at Different Rates from Their Two Ends[17]

Polymers of actin and tubulin have a polarity due to the arrangement of their asymmetric subunits in a specific orientation in the polymer. The structural polarity of the polymer is essential in both muscle and cilia for the production of organized movement. It also makes the two ends of the polymer different in a way that is important for the control of polymer growth.

The binding of specific proteins to both actin and tubulin has been used effectively to determine their polarity. S1 fragments of myosin containing the head region of the myosin molecule decorate actin in the characteristic arrowhead pattern shown in Figure 10–14. While microtubules do not have as convenient a marker, it has recently been found that under certain unusual experimental conditions free tubulin molecules will add to the outside of existing microtubules to form curved protofilament sheets that in cross-section resemble hooks. Depending on the polarity of the microtubules, the hooks appear to be pointing either clockwise or counterclockwise. Microtubules have been shown to extend with uniform and identical polarity in nerve cell axons, in cilia, and from the centriole-containing region in mitotic and interphase cells.

In both actin filaments and microtubules, the two opposite ends grow and depolymerize at very different rates. The asymmetry of growth of microtubules can be observed by allowing tubulin molecules to polymerize for a short time on short fragments of ciliary axoneme and then examining them in the electron microscope (Figure 10–42). In this way, one end can be seen to elongate at three times the rate of the other end. The fast-growing end (defined as the *(+) end*) has been shown to be the end that points away from the basal body in a cilium and also the end to which tubulin molecules add to the growing cilium in a cell. For actin filaments, a similar type of experiment has been performed that demonstrates that the fast-growing (+) end corresponds to the end of the actin filament that is embedded in the Z disc in muscle. It seems likely that in all cells proteins are present that bind selectively to one or the other end of actin filaments and microtubules. Such *capping proteins* are expected to play a crucial role in determining the spatial arrangement of the cytoskeleton—allowing filaments to be attached by the bound end to other structures in the cell and controlling the addition or loss of subunits at either end independently. The spatial organization and stability of both actin filaments and microtubules may thus largely depend on the proteins attached to their two ends.

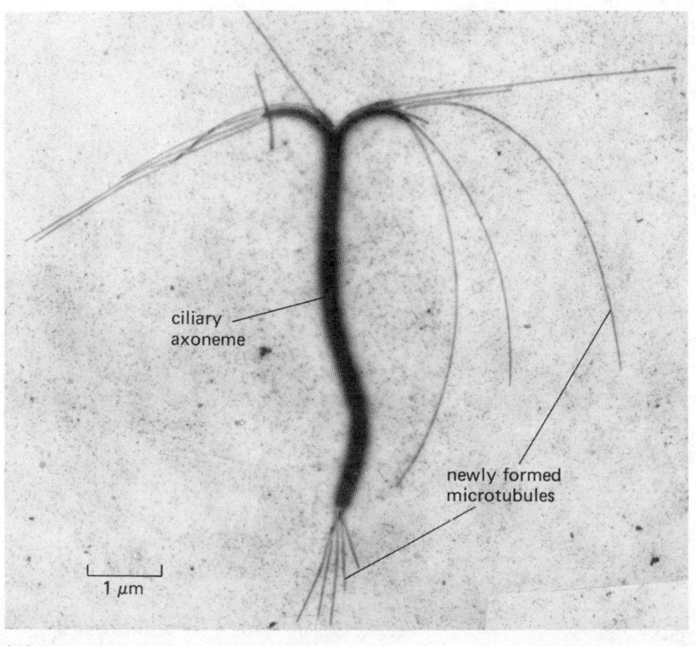

(A)

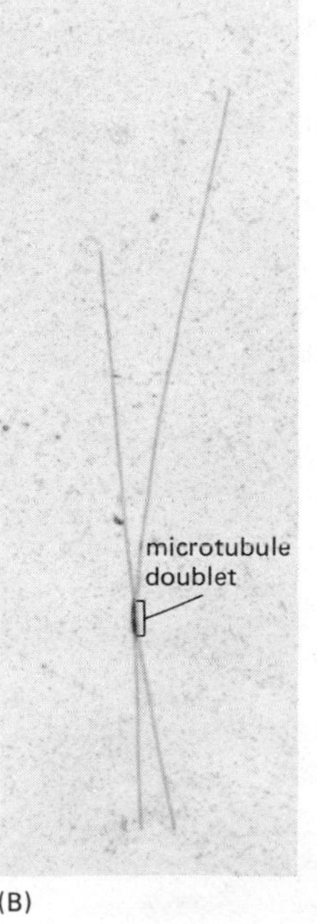

(B)

Figure 10–42 Electron micrographs showing the preferential growth of microtubules at one end. When a fragment of ciliary axoneme is incubated with brain tubulin under polymerizing conditions, the microtubules produced from one end of the axoneme (corresponding to the end farthest from the basal body in the intact cilium) grow faster than those from the other end (A). In (B), microtubules produced from short fragments of ciliary outer doublets also show asymmetry of growth. (Courtesy of Gary Borisy.)

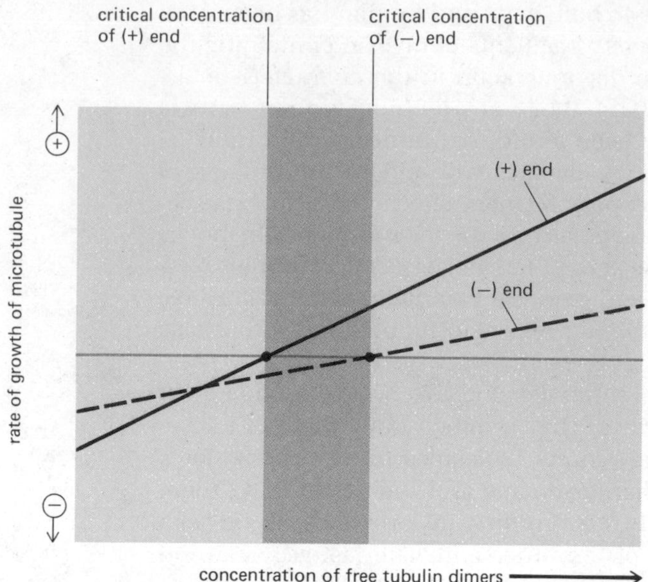

Figure 10–43 The rate of growth of microtubules at different concentrations of free tubulin dimers. This graph is similar to Figure 10–40, except that the two ends of the filament (in this case a microtubule) have different growth rates and different critical concentrations. As a consequence, there is a range of free tubulin concentrations (*brown*) in which one end (+) of the microtubule is polymerizing while the other end (−) is depolymerizing. When free tubulin is in equilibrium with microtubules *in vitro,* no net growth is occurring, since the rate at which tubulin molecules are coming off one end is equal to the rate at which they are going on the other. Under these conditions, microtubules do not change their length but undergo a treadmilling of tubulin molecules (see Figure 10–44).

Actin and Tubulin Polymers Can Undergo a "Treadmilling" of Subunits[18]

Actin and tubulin do not actually require energy for polymerization—assembly occurs even in the presence of nonhydrolyzable nucleoside triphosphate analogues. But nucleoside triphosphate hydrolysis during polymerization allows the shape of the actin and tubulin molecules to change as they enter the polymer, just as the conformation of the myosin head changes in response to its ATP hydrolysis. As a result, the affinity of free molecules for one end of the polymer becomes greater than for the other: in other words, the *critical concentration*—defined as the concentration at which the rate of addition balances the rate of dissociation (see p. 573)—will be lower at one end [the (+) end] of the filament than at the other [the (−) end]. This has the important consequence that, at intermediate concentrations of free actin or tubulin molecules, the filament may exist in a steady state in which molecules come off *predominantly* (but not exclusively) at one end and are put on predominantly at the other, a phenomenon known as **treadmilling** (Figure 10–43). In this steady state, the rate of addition to the (+) end equals the rate of loss from the (−) end and net polymer growth is zero. However, even though the net length of the polymer does not change, individual molecules are being translocated from one end of the polymer to the other in a continuous process (Figure 10–44). Treadmilling of isolated actin filaments or microtubules re-

Figure 10–44 Schematic drawing showing the treadmilling of tubulin molecules in a microtubule. At equilibrium, tubulin molecules (as αβ heterodimers) are continually adding to and being lost from both ends of the microtubule. In the presence of GTP, however, there is a net addition at one end (+) and a net loss from the other end (−). Consequently, there is a continual movement of tubulin (as indicated here by dark color) from one end of the microtubule to the other. A similar situation may exist for actin filaments in which treadmilling is driven by ATP hydrolysis.

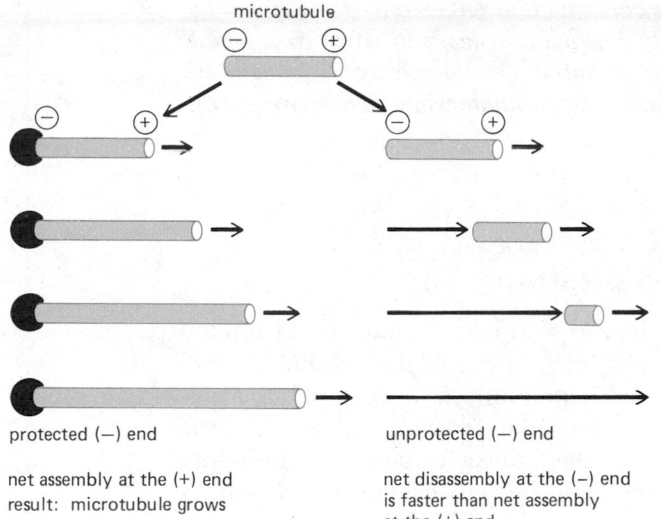

Figure 10–45 A model for the spatial control of microtubules in the cytoplasm. The concentration of free tubulin is assumed to be maintained at a value such that a microtubule with both ends free will depolymerize, whereas a microtubule with only its (+) end free will grow. In this case, microtubules could exist only if the cell contains anchorage sites to which the (−) end of the microtubules can bind, and free microtubules would not occur.

quires energy, which is provided by nucleoside triphosphate hydrolysis: otherwise it could represent a form of perpetual motion machine, violating the laws of thermodynamics. Although treadmilling has been demonstrated only under rather special conditions *in vitro,* it may well occur in the cell where, if provided with various links and harnesses, it could in principle be used to push or pull other materials.

The existence of (+) and (−) ends having different critical concentrations for polymerization also has important implications for the spatial organization of filaments in the cell. In the case of tubulin, there is evidence that the concentration of free molecules in the cytoplasm is so low that an unattached microtubule would lose molecules faster from its depolymerizing (−) end than it would add them to its (+) end and would, in consequence, eventually disappear. Thus, only a microtubule anchored at its (−) end would be expected to be stable in the cell (Figure 10–45). Microtubules are in fact found to grow from special regions, called organizing centers, that protect their (−) end from depolymerization. In the next section we examine some of these structures, as well as other components that regulate the activities of microtubules in cells.

Summary

Except where they are assembled into such stable structures as myofibrils or cilia, actin filaments and microtubules are labile structures. They can be formed rapidly from a pool of soluble subunits in the cell and rapidly disassembled when no longer needed. This balance between assembly and disassembly is disturbed by specific drugs that enhance or inhibit polymerization and lead to abnormal accumulations of either the protein filament or the unpolymerized protein molecule, respectively.

Actin filaments and microtubules form spontaneously from purified actin or tubulin molecules in the test tube. The polymerization processes are similar in that both show an initial lag phase due to the need to form small nucleating fragments, and both are accompanied by the hydrolysis of a nucleoside triphosphate—ATP for actin and GTP for tubulin. The assembled polymers are polar structures that grow at different rates at their two ends. There are indications that both the anchorage of filaments to other structures in the cell and their assembly or disassembly can be controlled at either end. The nucleotide hydrolysis that accompanies assembly can lead, at least in vitro, *to a steady state of "treadmilling" in which actin or tubulin molecules, on average, are adding to one end of an actin filament or microtubule while dissociating from the other.*

Microtubule Organizing Centers and Microtubule-associated Proteins

Microtubules alone, as flexible rods in the cytoplasm, would be of limited value to the cell. They must be linked to other parts of the cell before they can act as a structural framework or participate in cell movement. Most microtubules are anchored at one end: cilia terminate in basal bodies, while the cytoplasmic microtubules in interphase cells appear to terminate in a specialized region of the cell adjacent to the cell nucleus called the *cell center.* The basal body and the cell center act as nucleating centers from which microtubules regrow after they have been depolymerized by experimental treatments; for this reason they are referred to as **microtubule organizing centers.** In addition, many microtubules appear to be stabilized or linked to other structures by accessory proteins that are attached along their length.

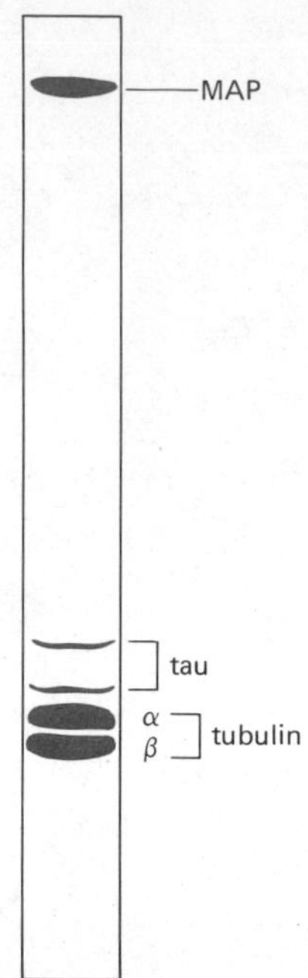

Figure 10–46 Drawing of an SDS polyacrylamide-gel electrophoretic pattern showing the polypeptides present in partly purified microtubules from brain.

Cytoplasmic Microtubules Are Associated with Other Proteins[19]

Highly purified tubulin assembles into microtubules spontaneously in the presence of GTP, but it does so much less efficiently than tubulin in less pure preparations. The impurities that enhance the polymerization of brain microtubules are accessory proteins that fall into two major classes: **MAPs** (microtubule-associated proteins), which have a polypeptide chain molecular weight of 200,000 to 300,000, and **tau proteins,** which have a molecular weight of 60,000 to 70,000 (Figure 10–46). When purified, both classes of accessory proteins induce polymerization of purified tubulin and become bound to the newly formed microtubules. The tau proteins appear to bind to several tubulin molecules simultaneously, thereby enhancing tubulin polymerization. MAPs act similarly, but they differ in being composed of two domains: while one binds to the microtubule, the other extends outward and may be involved in cross-linking the microtubule to other cell components (Figure 10–47). Antibodies to both classes of proteins bind along the entire length of cellular microtubules. MAPs and tau proteins are found in most vertebrate cells, and it is thought—though not proven—that they help to regulate microtubule assembly in these cells.

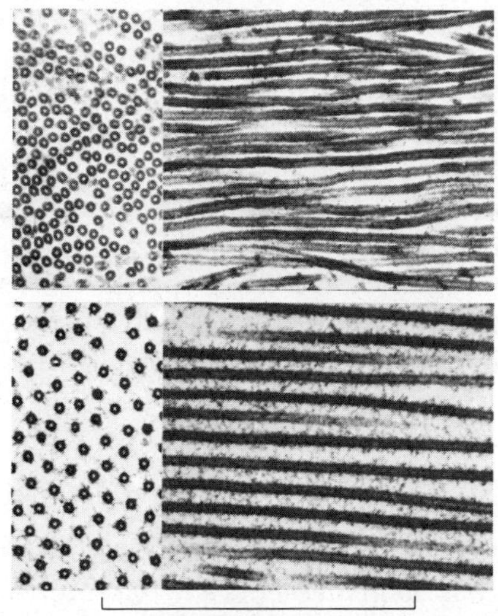

Figure 10–47 Electron micrographs showing microtubules in transverse and longitudinal section polymerized in the absence (*top*) or presence (*bottom*) of MAPs. Note that the MAPs bind along the length of the microtubules and increase the spacing between adjacent microtubules. (From H. Kim, L. I. Binder, and J. L. Rosenbaum, *J. Cell Biol.* 80:266–276, 1979. Reproduced by permission of Rockefeller University Press.)

Microtubules Grow from Discrete Organizing Centers[20]

Microtubules in the cytoplasm of an interphase cell in tissue culture can be made visible by labeling the cells with fluorescent anti-tubulin antibodies after the cells have been fixed. The microtubules are present in greatest density around the nucleus and radiate out to the cell periphery in fine lacelike threads. (see Figure 10–61). The origin of the microtubules is seen most clearly if they are first depolymerized with colchicine and then allowed to regrow (Figure 10–48). The regenerating microtubules first appear as one or two small starlike structures called *asters* and then elongate towards the cell periphery until the original distribution is reestablished. If the microtubules in cultured cells are decorated with microtubule hooks to determine their polarity, they are seen to all have their (+) ends facing away from the centers of the original asters—the microtubule organizing centers.

In the middle of the major microtubule organizing center of almost all animal cells, called the **cell center** or the **centrosome,** is a *centriole pair*. The structure of a **centriole** is shown in Figure 10–49; it is a cylinder 0.1 μm in diameter and 0.3 μm long, just barely detectable in the light microscope. The ninefold array of triplet microtubules in a centriole is the same as that in the basal body of cilia and flagella (Figure 10–49B). However, not all microtubule organizing centers contain centrioles. In mitotic cells of higher plants, for example, the microtubules terminate in poorly defined regions of electron density that are completely devoid of centrioles. Similarly, centrioles are not detected in the meiotic spindle of mouse oocytes, although they appear later in the early developing embryo.

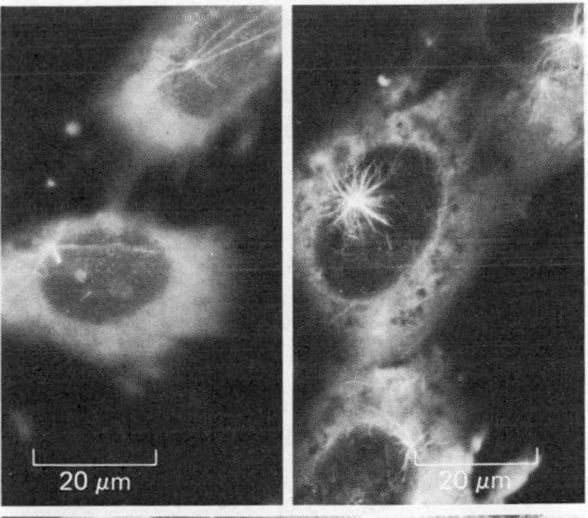

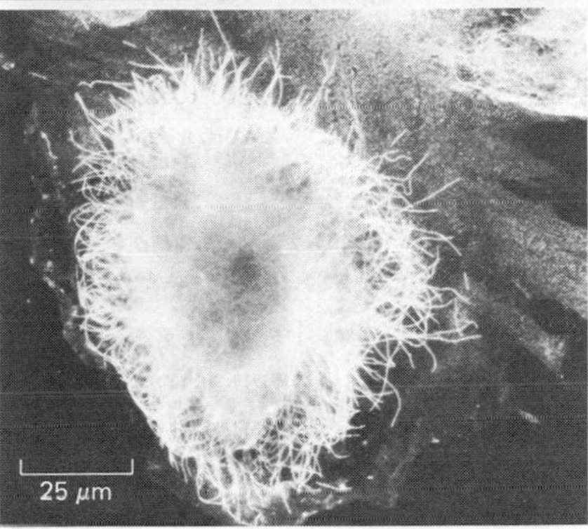

Figure 10–48 Immunofluorescence micrographs showing the recovery of microtubules in cultured cells following exposure to colcemid for one hour. After washing away the colcemid, the cells were stained at different times with anti-tubulin antibodies. Microtubules appear first in starlike asters and then elongate toward the periphery of the cell. (From M. Osborn and K. Weber, *Proc. Natl. Acad. Sci. USA* 73:867–871, 1976.)

Centrioles and Basal Bodies Are Structurally Identical and Functionally Interconvertible[21]

Centrioles and basal bodies have very similar, if not identical, structures and in many instances are interconvertible. For example, the single-cell green alga *Chlamydomonas* has two flagella, each with a basal body. At the onset of mitosis, the flagella are resorbed and the basal bodies migrate to a position near the nucleus, where they organize the mitotic spindle. After mitosis, the centrioles again become basal bodies and the flagella reappear. Vertebrate cells in tissue culture often have a **primary cilium** growing out of their cell center, with a "9 + 0" arrangement of microtubules. This cilium is immotile, and it is produced from only one of the pair of centrioles—disappearing before the cell commences mitosis.

In many organisms the egg does not seem to possess active centrioles, while the sperm contains basal bodies (or centrioles) that are donated to the egg during fertilization. These sperm centrioles act to organize the mitotic spindle for the first cleavage divisions. A comparable organizing effect is produced experimentally by the injection of purified basal bodies from *Chlamydomonas* or from the ciliated protozoan *Tetrahymena* into frog eggs. The fact that these induce the formation of mitotic asters in the cytoplasm demonstrates that a basal body can act as a centriole when present in the cytoplasm of a cell that is predisposed to divide.

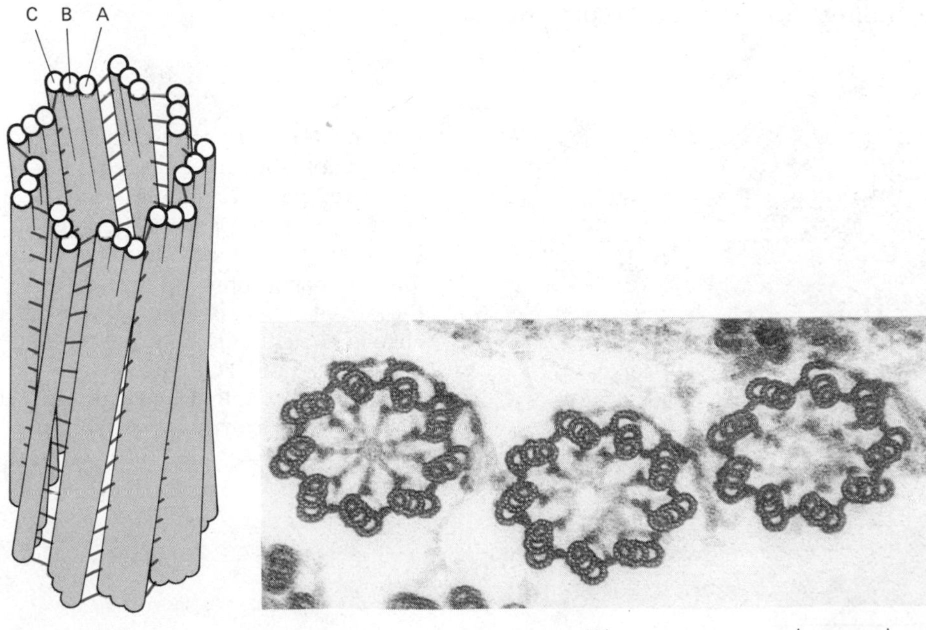

Figure 10–49 (A) Schematic drawing of a centriole. It is composed of nine sets of *triplet* microtubules, each triplet containing one complete microtubule (the A subfiber) fused to two incomplete microtubules (the B and C subfibers). Other proteins form links that hold the cylindrical array of microtubules together (*color*). Centriolar structure is essentially identical to that of a basal body of a cilium. (B) An electron micrograph of a section through three basal bodies in the cortex of the protozoan *Saccinobacculus ambloaxostylus.* (Micrograph courtesy of D. T. Woodrum and R. W. Linck.)

Centrioles Usually Arise by Duplication[22]

The otherwise continuous increase in cell mass throughout the animal cell cycle is punctuated by two discrete duplication events: the replication of DNA and the doubling of the centrioles—the latter process creating the spindle poles that allow the cell nucleus to divide in two at mitosis. In cultured fibroblasts, centriole doubling commences at around the time that DNA synthesis begins. The centrioles normally occur in pairs aligned at right angles to each other. A new pair of centrioles is formed in the vicinity of an existing centriole pair in a process that is the same in virtually all eucaryotic organisms. First, the two members of a pair separate; then, a daughter centriole is formed perpendicular to each original centriole at its basal end (the end that would be furthest from the cilium in a basal body) (Figure 10–50; also Figure 11–19, p. 630). An immature centriole contains a ninefold symmetric array of *single* microtubules; each microtubule presumably acts as a template for the assembly of the triplet microtubules of mature centrioles.

In ciliated vertebrate cells, which may contain hundreds of cilia, the centrioles of the precursor cells are intimately involved in forming the many basal bodies required. For example, during differentiation of the ciliated epithelial cells that line the oviduct and the trachea, the centriole pair migrates from its normal location near the nucleus to the apical region of the cell where the cilia will form. There, instead of forming a single daughter centriole in the typical manner, each centriole in the pair forms numerous electron-dense "satellites." Many basal bodies then arise from these satellites and migrate to the membrane to initiate the formation of cilia.

There are also clear cases where centrioles seem to arise *de novo*. For example, although unfertilized eggs of some species appear to lack functional centrioles and use the sperm centriole for the first mitotic division (see above), under certain conditions (such as extreme ionic imbalance or electrical stimulation), an unfertilized egg can produce a variable number of centrioles. Each of these centrioles nucleates the formation of a small aster, one of which can be used by the egg for division so that a haploid organism develops in a process called *parthenogenesis*. These studies suggest that some precursor to centrioles may exist in the cytoplasm of unfertilized eggs.

The unusual mode of duplication and the continuity of generations of centrioles has suggested that they may be self-replicating, autonomous organelles and may even carry their own genome. Although there is no direct evidence to support this view, there is indirect evidence for the presence of functional RNA in centrioles.

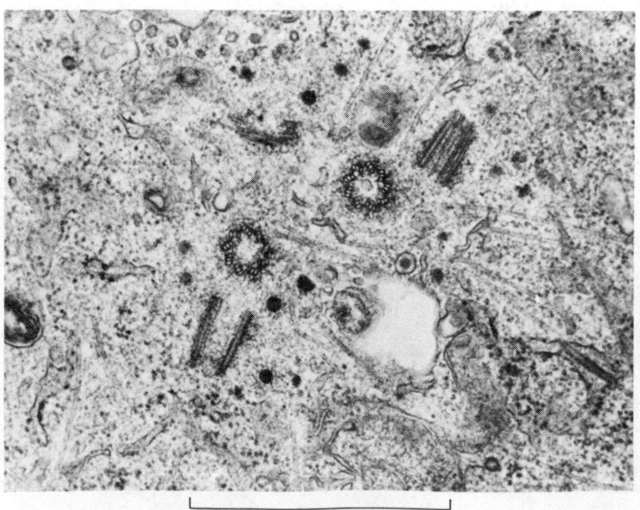

Figure 10–50 An electron micrograph showing a newly replicated pair of centrioles. One centriole of each pair has been cut in cross-section and the other in longitudinal section, indicating that the two members of each pair are aligned at right angles to each other. (From M. McGill, D. P. Highfield, T. M. Monahan, and B. R. Brinkley, *J. Ultrastruct. Res.* 57:43–53, 1976.)

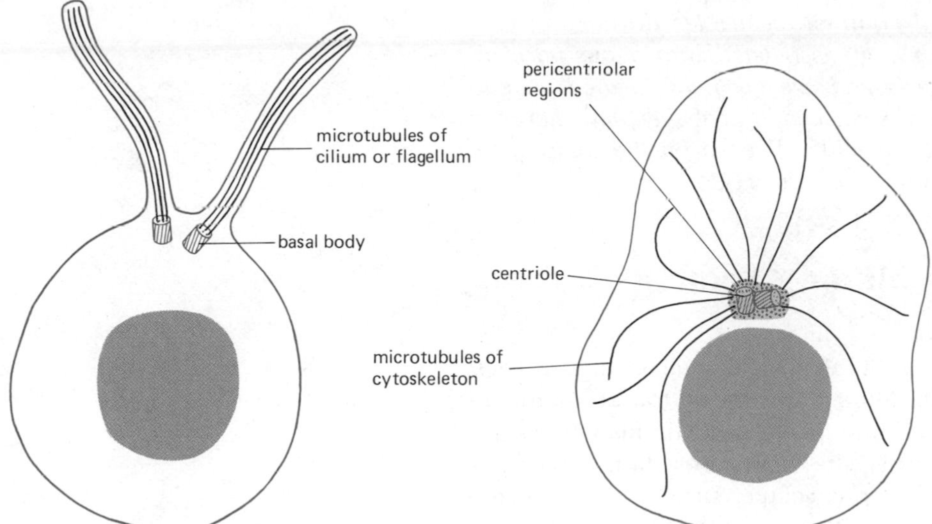

Figure 10–51 Schematic drawing showing that while centrioles and basal bodies are extremely similar, if not identical, in structure, they act as organizing centers for very different types of microtubule arrays.

What Is the Function of Centrioles?

It is relatively easy to envisage how the ninefold symmetry of a centriole or a basal body might nucleate the assembly of a cilium, with its ninefold symmetric array of doublet microtubules. But this rigid symmetry seems inappropriate for nucleating the hundreds of microtubules present in the approximately radial or spherical arrays in the interphase or mitotic cell (Figure 10–51). In fact, close examination reveals that most cytoplasmic microtubules do not arise directly from the centriole but emanate from a densely staining material, the **pericentriolar material,** that surrounds the centriole. In cells that do not have centrioles, microtubules also terminate in such densely staining material, which must represent the true organizing center for cytoplasmic microtubules.

What, then, is the role of the centriole? One extreme view is that the centriole is an organelle that serves only to organize cilia and flagella and that its association with cell centers and with the poles of the mitotic spindle merely reflects a mechanism designed to distribute the potential to form basal bodies equally to each daughter cell. However, mitotic spindles that contain centrioles at each pole are highly focused and have elaborate astral microtubules running in all directions, while those lacking centrioles, such as plant cells, have no asters and are less focused at the poles. Moreover, it seems from other evidence that the centrioles in animal cells act as master organizers that determine the precise position of the pericentriolar material and thereby the overall polarity of a cell (see p. 601).

Summary

Most microtubules are associated with accessory proteins, such as the MAPs and tau proteins, that are thought to regulate microtubule assembly and influence the interactions of microtubules with other components of the cell. Other microtubule accessory proteins must form part of the microtubule organizing centers, to which the microtubules in a cell are attached by one end. The most prominent of these organizing centers is the cell center, which surrounds the cell's centriole pair.

Basal bodies and centrioles are indistinguishable in structure and interchangeable in function but differ in the way they nucleate microtubules. Basal bodies lie at the base of the ciliary axoneme and their microtubules are con-

tinuous with those of the axoneme. In contrast, centrioles do not nucleate microtubule growth directly but only through associated amorphous pericentriolar material. Centrioles exist in pairs and are normally formed next to existing centrioles. The centriole pair at the cell center probably plays a crucial role in organizing the cytoskeleton through its influence on the many cytoplasmic microtubules that radiate outward from this region.

Actin Filaments and Actin-binding Proteins in Nonmuscle Cells[23]

Actin constitutes a substantial proportion of the protein in all eucaryotic cells. In fibroblasts, for example, it makes up almost 10% of the total protein, approximately half of which is polymerized into filaments. Actin filaments serve at least two functions in nonmuscle cells: they form cross-linked bundles, which provide mechanical support for various cellular structures and extensions, and together with myosin they form the various contractile systems thought to be responsible for many cellular movements.

In this section we shall first examine bundles of actin filaments that function as mechanical supports and then discuss examples of actin and myosin contractile systems in nonmuscle cells. Finally, we shall consider a number of actin-binding proteins, which organize actin filaments into three-dimensional networks in the cell.

Microvilli Contain Bundles of Actin Filaments[24]

Many kinds of cellular extensions have a core of cross-linked actin filaments. While some of these extensions are capable of movement, such as rapid extension and retraction, the major function of the actin core appears to be that of a mechanical stiffener.

Microvilli are the best known examples of such structures. They cover the exposed surfaces of many kinds of epithelial cells, especially where cellular function requires a maximum surface area for absorption, such as in the intestine or kidney. These fingerlike extensions are about 1 μm in length and 0.1 μm in diameter, and are often packed together like the bristles of a brush. A single epithelial cell in the human small intestine has several thousand microvilli in its **brush border** (Figure 10–52), making its absorptive surface area about 25 times greater than it would be without them.

The core of a microvillus contains about 40 actin filaments that run in a parallel bundle along its length. At the tip of the microvillus the actin filaments are embedded in an ill-defined cap of amorphous material, while at their base they extend into a perpendicular network, composed largely of actin filaments, called the *terminal web*. The terminal web also contains myosin, and part of its function may be to create the tension required to maintain the stiff microvilli in their upright position.

The actin filaments of the microvillus core have a uniform polarity as judged by their binding of myosin S1 fragments, with their "barbed" (+) ends anchored at the tip of the microvillus (Figure 10–53). This is the usual orientation of actin filaments attached to membranes, and it also corresponds to the polarity of actin filaments at the Z disc of skeletal muscle (Figure 10–53). Linkage to the microvillus membrane occurs not only at the tip but also along its length: a helical "staircase" of fine lateral links extends from the core of actin filaments to the microvillus membrane.

Cross-links also bind adjacent actin filaments together in the rigid microvillus core. An important component of these cross-links is the actin-bind-

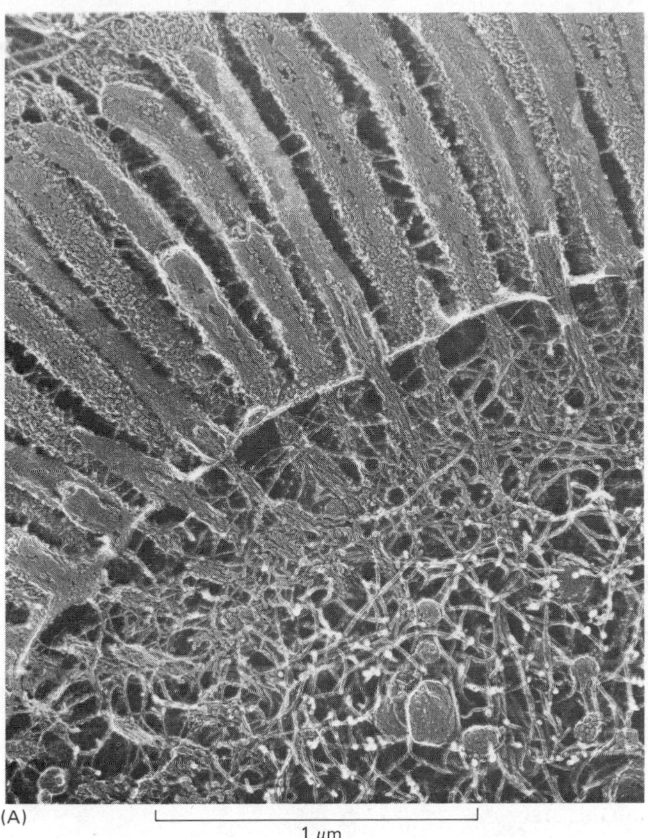

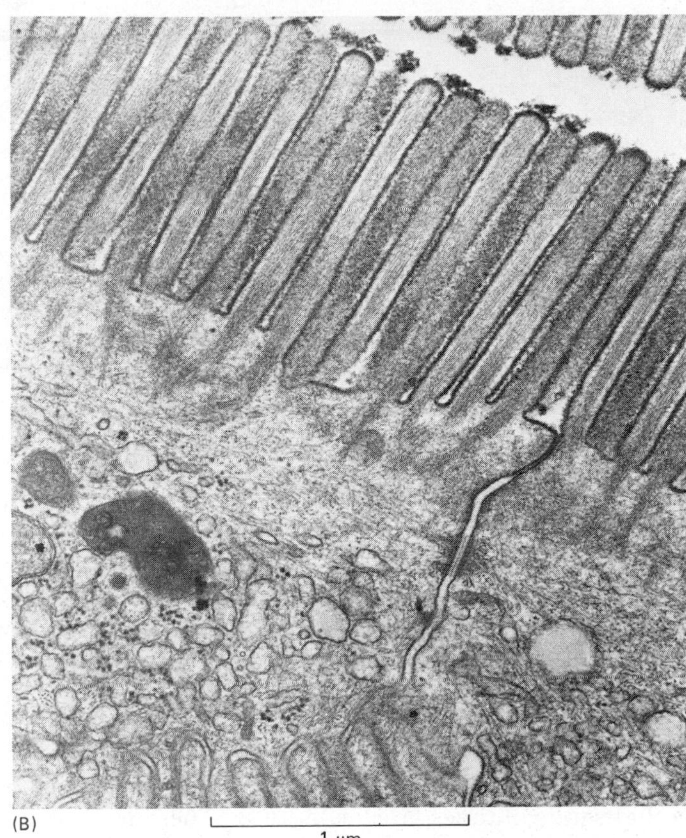

Figure 10–52 Two views of the intestinal brush border. (A) was obtained by rapid freezing and deep etching following detergent extraction. (B) was obtained by conventional thin-section electron microscopy. (A, courtesy of John Heuser; B, from P. T. Matsudaira and D. R. Burgess, *Cold Spring Harbor Symp. Quant. Biol.* 46:845–854, 1982.)

ing protein *fimbrin,* which is also found in other actin-containing extensions of the cell surface. There are various other actin-binding proteins in the microvillus, the most extensively studied of which is *villin,* but their function is unclear. These and other actin-binding proteins will be discussed in more detail later.

Rigid Arrays of Cross-linked Actin Filaments Play a Part in the Detection of Sound[25]

Rigid, actin-containing microvilli contribute to functions other than absorption. In the eye of the squid, for example, each retinal cell carries hundreds of thousands of tightly packed microvilli, each covered with a photosensitive membrane. The regular alignment of these microvilli, which lie at right angles in adjacent cells, allows the animal to detect the plane of polarization of the incoming light (Figure 10–54).

An even more remarkable specialization occurs on the *hair cells* found in the cochlea and vestibule of the inner ear. Hair cells are exquisitely sensitive to very small movements, whether caused by the vibrations of incoming sound

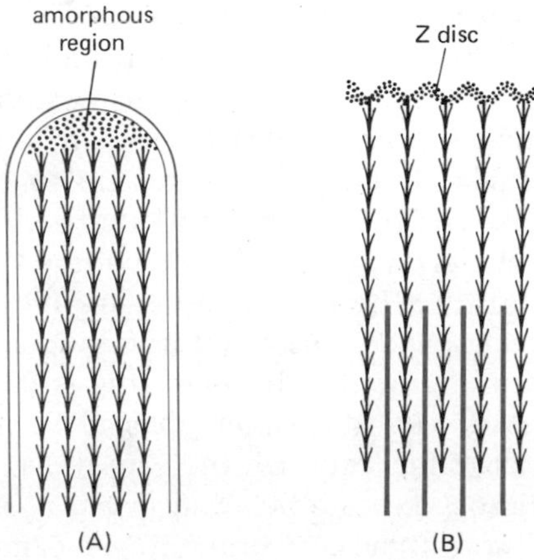

Figure 10–53 Schematic drawing showing the polarity of actin filaments in a microvillus (A) and in a muscle sarcomere (B). The arrowheads indicate the polarity of decoration with S1 fragments of myosin. The (+) ends of these filaments are at the tip of the microvillus and in the Z disc of the sarcomere.

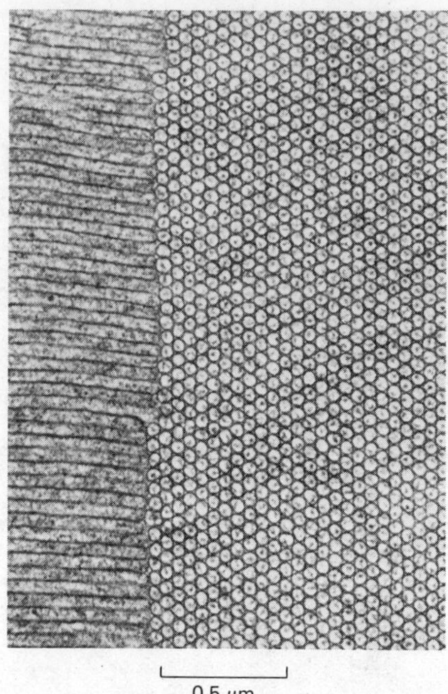

Figure 10–54 Electron micrograph of microvilli in the retina of the squid eye. The microvilli are covered with a photosensitive membrane and are packed together in parallel arrays. The arrays from two adjacent retinal cells are shown sectioned vertically with respect to the retina. The microvilli on one cell are aligned at right angles to those of its neighboring cells—an arrangement that allows the plane of polarization of incoming light to be detected. (Courtesy of Helen Saibil.)

or the displacement of fluid in the semicircular canals produced by a change in head position. Such movements are detected by means of very large specialized microvilli on the cell surface (Figure 10–55). They are called **stereocilia,** although this term is misleading since they contain a core of actin filaments like microvilli rather than a core of microtubules like true cilia. Rows of stereocilia of decreasing length are arranged together on the surface of each hair cell, giving the appearance of a pipe organ (Figure 10–56). Individual stereocilia are rigid structures that taper to a point where they meet the cell surface. In the detection of sound, tiny movements of the stereocilia are converted into an electrical signal in the hair cell, which transmits the signal to the brain (see p. 1061).

It is likely that the particular mechanical properties of stereocilia—their stiffness and ability to bend at their base—are essential to the delicately adjusted sensitivity of the hair cells, enabling them, for example, to respond to sound of a particular frequency against a background of random noise that is thousands of times greater (p. 1063). These mechanical properties may depend on the way the actin filaments are cross-linked in the core of the stereocilium. The hundreds of actin filaments in a lizard stereocilium, for example, are seen in cross-section to be irregularly spaced; yet *longitudinal* sections show that they are held in strict longitudinal alignment, so that the crossover points of the actin helices all lie in the same plane. As explained in Figure 10–57, this arrangement is produced by a cross-linking protein that binds to sterically precise positions on two adjacent actin helices, causing them to form regular actin bundles without requiring that they be evenly spaced in cross-section.

Dynamic Actin-Filament Structures Occur on the Surface of Many Cells[26]

Microvilli in the intestinal brush border and stereocilia in the ear are relatively permanent specializations of certain types of epithelial cells, but dynamic actin-filament extensions are extremely common in eucaryotic cells, as in the earlier example of the microvilli rapidly extended from the surface of sea urchin eggs following fertilization. Cells in tissue culture also commonly put out a large number of hairlike extensions called **microspikes,** each about 0.1 μm wide and 5 μm to 10 μm long, as they settle onto the tissue-culture surface, migrate, or round up to divide (Figure 10–58). Even longer microspikes—up to 50 μm in length and sometimes called *filopodia*—are formed at the growing tip (growth cone) of a developing nerve axon (see Figure 18–63, p. 1073). These extensions are motile structures that form and retract with great speed—possibly through local polymerization and depolymerization of actin filaments, although this is far from certain. Actin filaments are all oriented with the same polarity in microspikes as they are in the intestinal microvillus, although they are much less orderly in arrangement (Figure 10–59). Microspikes are believed to act as sensory devices, or feelers, by which cells explore their environment (see Chapter 18).

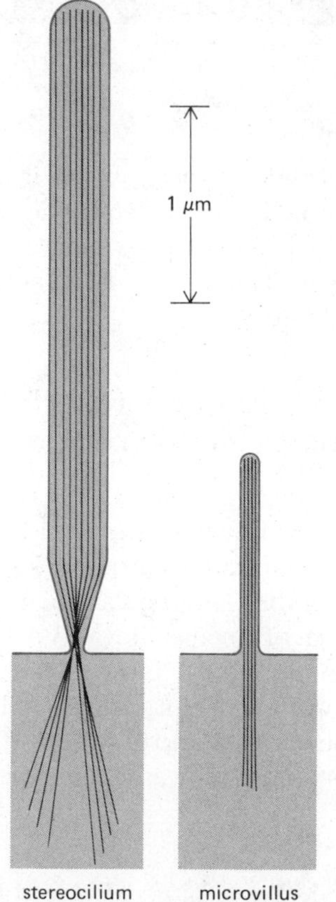

Figure 10–55 Schematic illustration comparing the sizes of a typical stereocilium and a typical microvillus, both of which are supported by bundles of actin filaments in their core.

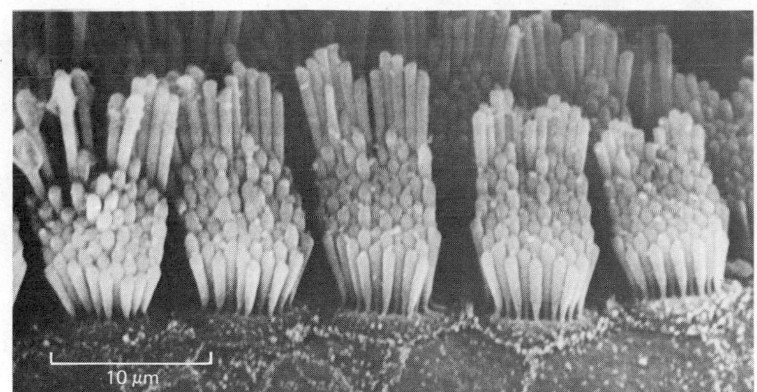

Figure 10–56 Scanning electron micrograph of the surface of hair cells showing the "pipe-organ" arrangement of stereocilia. (Courtesy of Lewis Tilney.)

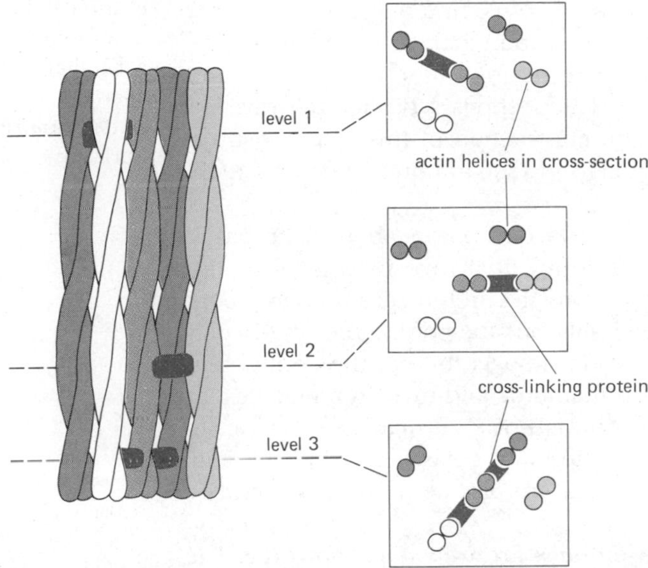

Figure 10–57 Schematic diagram of cross-linked actin filaments. The five actin filaments depicted on the left of the diagram are not only parallel but aligned longitudinally so that their helices are in register. (For simplicity, the individual actin molecules are not shown.) The right-hand part of the figure shows transverse sections made through a bundle of such aligned actin filaments. It can be seen that a cross-linking protein such as *fimbrin* (*solid color*) can join together adjacent actin helices at several levels even though the side-by-side packing of filaments is not regular.

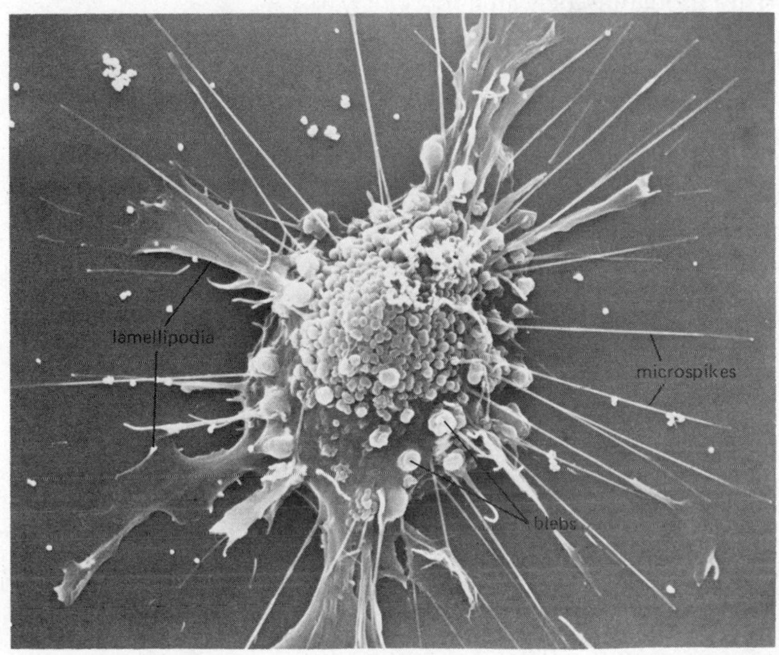

Figure 10–58 Scanning electron micrograph of a fibroblast settling onto a culture dish surface. Extending from the cell surface are three distinct types of protrusions: *blebs,* flattened sheets (or *lamellipodia*), and *microspikes* (see Figure 10–86). The latter appear to act as sensory devices enabling the cell to detect features of its surroundings. (Courtesy of Günter Albrecht-Buehler.)

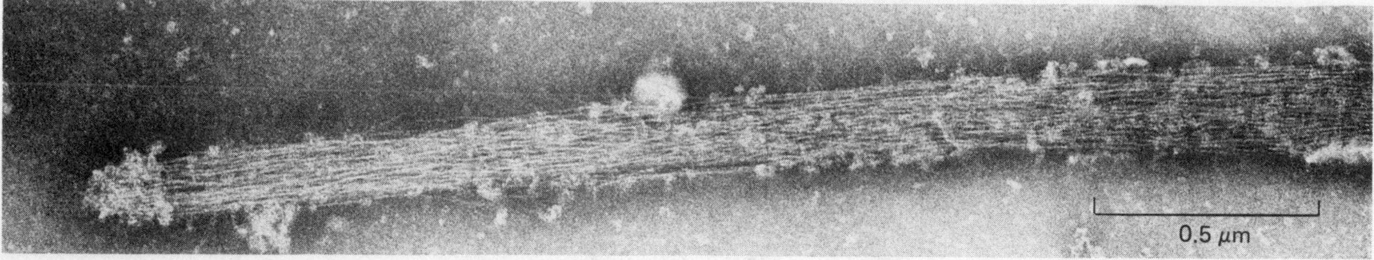

Figure 10–59 Electron micrograph showing a high-magnification view of a microspike extended by a human glial cell in culture. The cell membrane and the soluble proteins of the cell have been removed by treatment with detergent and the actin filaments revealed by negative staining. (Courtesy of Uno Lindberg, Anna-Stina Höglund, and Roger Karlsson.)

Nonmuscle Cells Contain Small "Musclelike" Assemblies[27]

Oriented arrays of actin filaments are not restricted to surface protrusions such as microvilli and microspikes. Almost all eucaryotic cells contain, in addition, bundles of actin filaments in their cytoplasm. Through their interaction with myosin, these bundles are capable of contracting and producing a mechanical force, although very much weaker than that generated in muscle. One of the clearest examples is the beltlike bundle of actin filaments and myosin known as the **contractile ring,** which appears during cell division just beneath the plasma membrane. The contraction of this ring constricts the middle of the cell and eventually leads to the separation of the two daughter cells (Figure 10–60; see also p. 662).

The contractile ring illustrates two important points about actin-based contractions in nonmuscle cells. First, it shows that myosin is involved, for when anti-myosin antibodies are injected into sea urchin eggs that are about to divide, nuclear division proceeds normally but the contractile ring is paralyzed, and the cytoplasm fails to divide. This suggests that the contraction is based on an interaction between actin filaments and myosin molecules, as it is in muscle. Second, the contractile ring is not a permanent structure; it must be assembled at the start of cell division and disassembled at the end. Many other actin-based contractile systems are also temporary structures, formed only when they are needed.

Not all nonmuscle contractile assemblies are transitory, however. The circular bundles of actin filaments associated with the structures known as *belt desmosomes* (p. 683), for example, are more lasting. They are found near the apical surface of epithelial cells and are thought to play an important part in changes in cell shape, particularly during embryogenesis. For example, the coordinated contraction of these actin bundles constricts the apical end of the cell, giving rise to the regular foldings of epithelial cell sheets that take place during early embryonic development (this topic is discussed in more detail in Chapter 15).

While the role of the contractile ring and belt desmosomes is reasonably clear, other bundles of actin filaments have a less obvious function. A good example is the organized bundles of actin filaments known as **stress fibers,** which are prominent components of the cytoskeleton of cells in culture (Figure 10–60). Stress fibers are bundles of actin filaments and associated proteins, typically 0.5 μm wide and about 5 μm long, that lie in the cytoplasm close to the lower surface of the cell, adjacent to the surface of the culture dish. These fibers can be separated from other components of the cell and will contract if exposed to ATP. They are revealed in a particularly striking fashion by immunofluorescence (Figure 10–61) and in this way have been shown to contain actin, myosin, α-actinin, and tropomyosin. Some of these proteins, including myosin, are distributed with a regular periodicity along the length of a stress fiber, but the details of their organization—as with all other nonmuscle contractile assemblies—is unknown. We do, however, know something of the special properties of nonmuscle myosin.

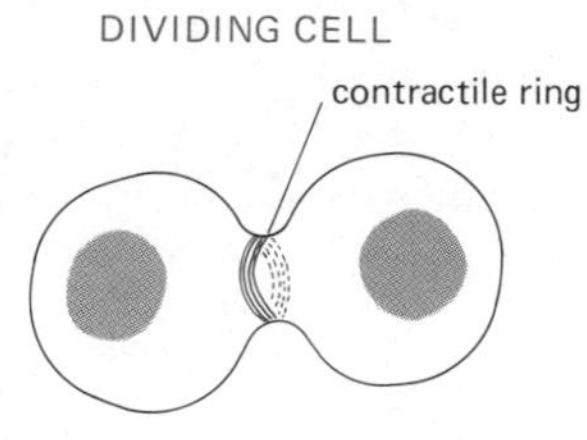

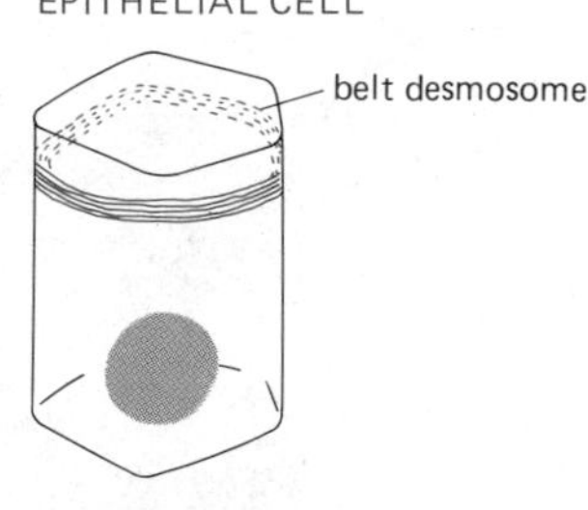

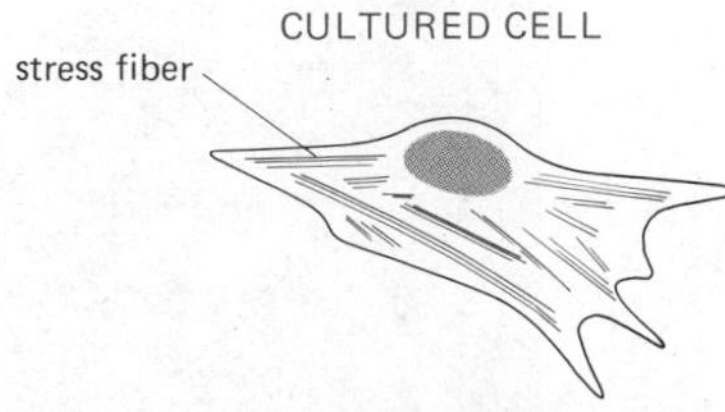

Figure 10–60 Schematic diagram showing three different examples of contractile bundles of actin filaments in nonmuscle cells. These bundles also contain myosin and other actin-binding proteins.

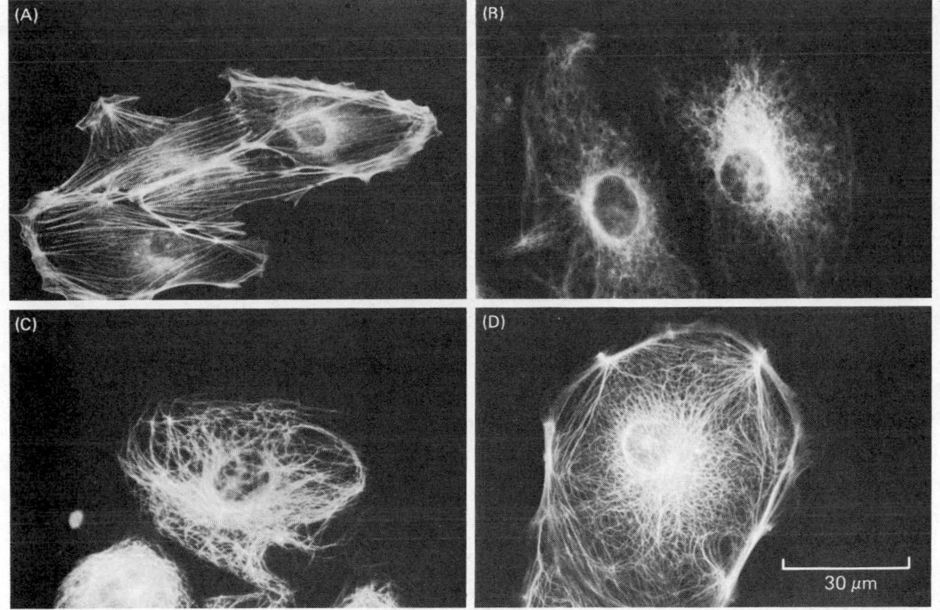

Figure 10–61 Immunofluorescence micrographs of rat kangeroo epithelial cells in interphase. These have been labeled with (A) anti-actin antibodies to show the actin filament bundles known as stress fibers; (B) antibodies to vimentin, a constituent of one type of intermediate filaments (see p. 594); (C) anti-tubulin antibodies to show cytoplasmic microtubules; and (D) anti-keratin antibodies to show a second type of intermediate filament in these cells. (Courtesy of Mary Osborn.)

The Assembly of Myosin in Nonmuscle Cells Is Ca^{2+}-dependent[28]

Myosin is found in various forms in almost all eucaryotic cells. Although all myosins bind actin with a resulting increase in their ATPase activity, it is only in skeletal and cardiac muscle that myosin thick filaments are seen in the electron microscope. In other cells, myosin is detectable by immunofluorescence, especially where actin filaments are organized into contractile bundles, but the detailed arrangement of myosin molecules cannot be seen in the electron microscope. Nonmuscle cells contain much less myosin, relative to actin, than muscle cells, and their myosin filaments are very much smaller and more labile than the thick filaments of skeletal muscle cells—not surprisingly, since the forces they must generate are very much weaker.

The relatively small size of the filaments formed by **nonmuscle myosin** is only one of three important characteristics that distinguish it from skeletal muscle myosin. The others are that the activation of nonmuscle myosin, as with smooth muscle myosin, depends on the phosphorylation of the myosin light chains (see p. 560) and that phosphorylation induces the assembly of nonmuscle myosin molecules into small bipolar aggregates in which 10 to 20 myosin molecules (compared with 500 or so myosin molecules in a skeletal muscle thick filament) are held together by the aggregation of their tail regions (Figures 10–62 and 10–63). As with smooth muscle myosin, phosphorylation

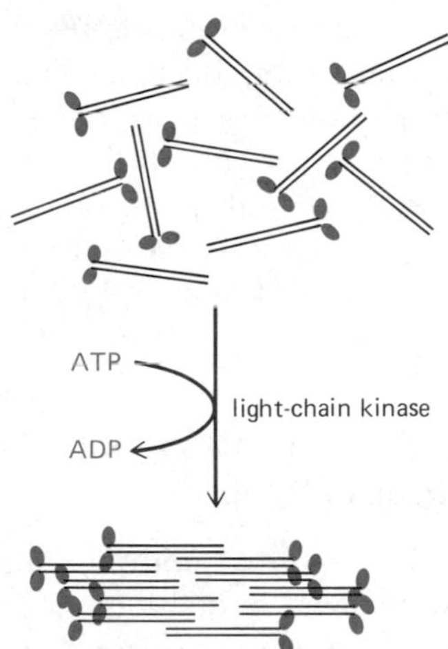

Figure 10–62 Phosphorylation of a light chain of nonmuscle myosin by a Ca^{2+}-dependent protein kinase not only activates the myosin (as described for smooth muscle myosin in Figure 10–21), but also causes the myosin molecules to assemble into short bipolar aggregates.

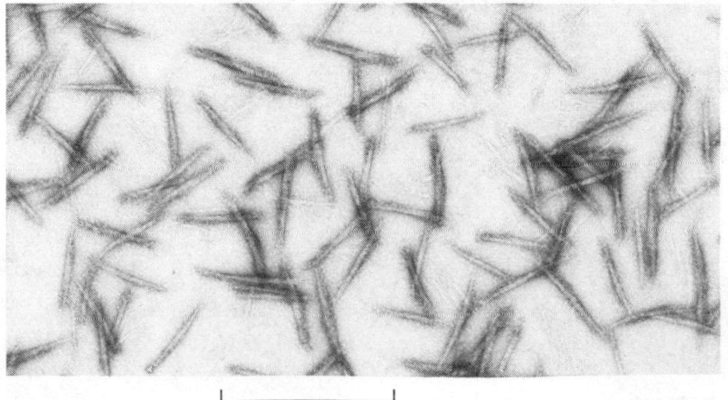

Figure 10–63 Electron micrograph of negatively stained short filaments of nonmuscle myosin that have been induced to assemble by the phosphorylation of their light chains. (Courtesy of John Kendrick-Jones.)

is catalyzed by a myosin kinase that is stimulated by Ca^{2+}, so that both the state of aggregation of nonmuscle myosin and its ability to interact with actin filaments are altered by small changes in the concentration of free Ca^{2+} in the cytosol. Such changes in intracellular Ca^{2+} commonly occur in response to extracellular signals (see p. 742).

Actin Filaments Are Often Anchored in Cell Membranes[29]

In order to exert a mechanical force, a contractile assembly must be anchored to other cellular components. Thus many, if not most, actin filaments are anchored at one end to cell membranes. This is true even for noncontractile actin assemblies (for example, recall that the actin filaments in the core of a microvillus are attached to the overlying plasma membrane both at their tips and along their lengths). Stress fibers in cultured cells terminate at specialized regions of the plasma membrane, known as *adhesion plaques* (Figure 10–64). The functional significance of such membrane attachments is clearest, however, in the actin bundles of the contractile ring and the belt desmosomes; these must be directly or indirectly connected to the plasma membrane since their contraction constricts the cell surface.

How actin filaments attach to membranes is not known. In some cases distinct wisps of material are seen to extend from the end of actin filaments to the plasma membrane (Figure 10–65). In others there is strong circumstantial evidence for the involvement of a specific linking protein known as **vinculin.** Vinculin has been identified by fluorescent antibodies both in the adhesion plaques at the ends of stress fibers and in belt desmosomes. Although this suggests that it may act as an anchoring protein linking actin filaments to the plasma membrane, vinculin is not itself a membrane protein, so it presumably serves as a link between actin and an integral component of the membrane.

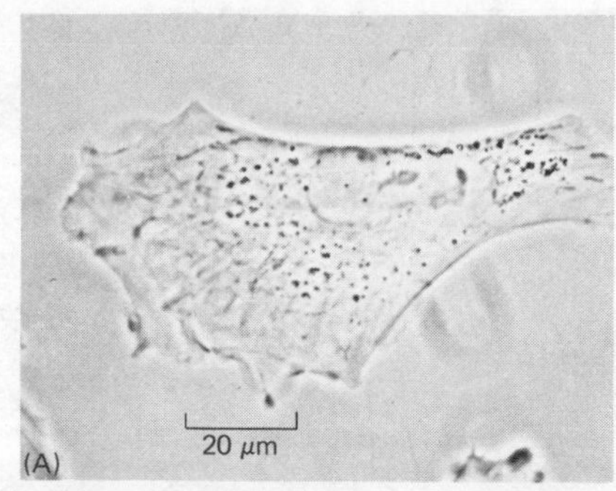

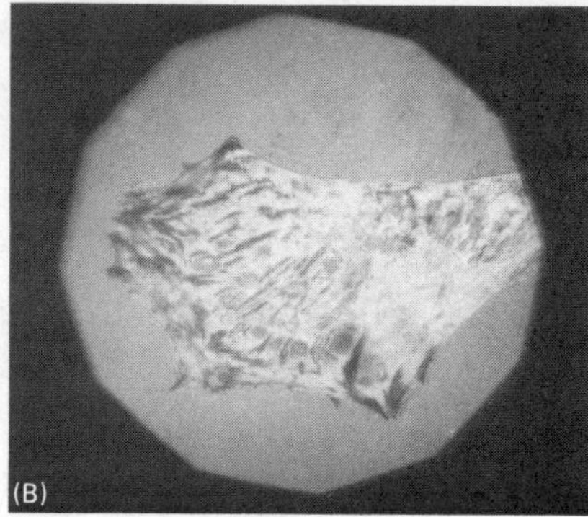

Figure 10–64 Phase-contrast (A) and reflection-interference (B) micrographs of cultured fibroblasts showing adhesion plaques. Adhesion plaques are regions on the surface of a cell where it comes into close contact with a surface. They are best seen by reflection-interference microscopy (B), in which light is reflected from the lower surface of a cell attached to a glass slide; the adhesion plaques appear as darkened patches. (Courtesy of Julian Heath.)

Amoeboid and Fibroblast Locomotion Depend on Actin Filaments[30]

Perhaps the most complex and dynamic assemblies of actin filaments are those involved in cell locomotion. As the final section of this chapter emphasizes, cell movements such as locomotion involve the coordinated action of many components of the cytoskeleton. However, they depend most immedi-

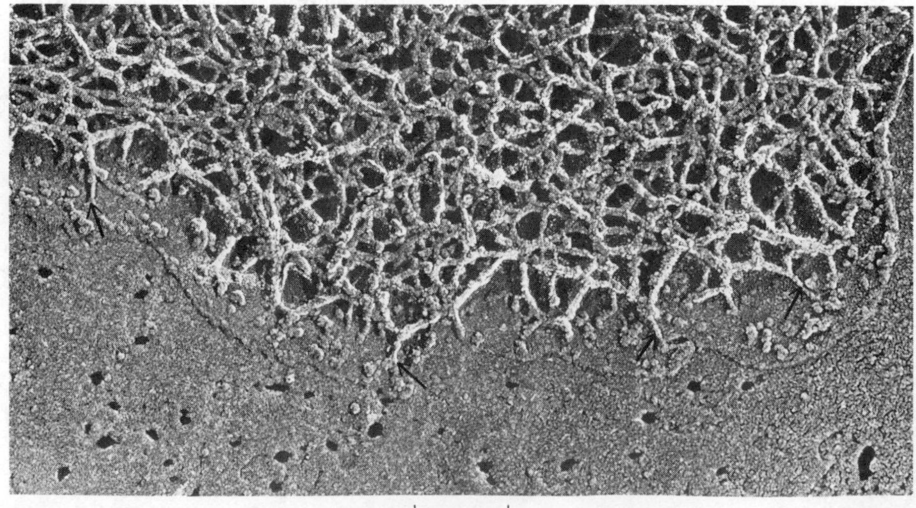

Figure 10–65 Freeze-etch electron micrograph showing the attachment of actin filaments to the plasma membrane of a hair cell of the chick ear, in a region distant from its stereocilia. The fracture plane of this deep-etched specimen passes through the plasma membrane into the underlying network of actin filaments. Thin wisps of protein (indicated by arrows) link the ends of the actin filaments to the plasma membrane. (Courtesy of N. Hirokawa and L. G. Tilney.)

ately on a thin layer of relatively disorganized actin filaments that lie just beneath the plasma membrane of all cells. The activity of this **cortical layer** can be studied most readily in certain primitive eucaryotes, such as *Amoeba* and the plasmodial slime molds. *Amoeba* moves by continually extending and retracting stubby processes, called **pseudopods,** that appear to support it like little legs (Figure 10–66). The pseudopods are so large (more than 100 μm in a giant amoeba) that their cortical layer and other internal contents can be seen in the light microscope, and their movements are so rapid that they can be followed by direct observation.

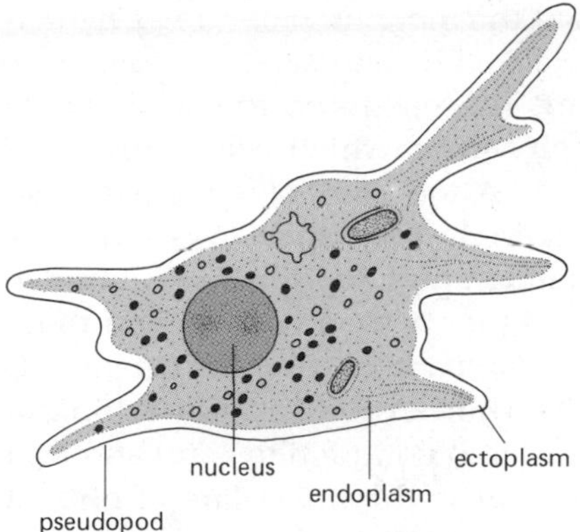

Figure 10–66 Diagram of a moving amoeba. Giant amoeba can be 0.5 mm in diameter and visible to the naked eye.

The locomotion of most vertebrate cells appears quite different from that of *Amoeba,* and it is very much slower. Vertebrate cell locomotion has been most extensively studied in *fibroblasts*—connective tissue cells that readily grow and migrate in culture. They are 10 times smaller in linear dimension than *Amoeba,* and their movement is so slow that it is most easily examined with the use of time-lapse microcinematography or video recording. As a fibroblast moves, it continuously forms thin sheetlike extensions, called **lamellipodia,** from its leading edge. Some of these form permanent attachments to the substratum, but many are carried back in a sweeping, wavelike motion (sometimes referred to as **ruffling**) on the upper surface of the cell. Microspikes may extend from a lamellipodium like the fingers of a hand, and these too are carried back on the surface of ruffles (Figure 10–67). Although ruffling probably occurs only when the cell has a free upper surface, as in tissue culture, it presumably reflects a flow of surface membrane and/or intracellular movements that are responsible for the translocation of the cell under other circumstances also.

Despite the apparent differences between amoeboid and fibroblast locomotion, it is possible that fundamentally similar mechanisms underlie them both. Indeed, some vertebrate cells, such as white blood cells, also move by extending pseudopodia, although these are much smaller than in *Amoeba* and their movement is much less rapid. Furthermore, both fibroblast and amoeboid locomotion are sensitive to the actin-binding drugs cytochalasin B and phalloidin, which prevent the polymerization and depolymerization of actin filaments, respectively. This suggests that both forms of locomotion depend on the assembly and disassembly of actin filaments.

To understand how actin assembly and disassembly may be involved in cell locomotion, we must examine the cytoplasmic movements that accompany amoeboid locomotion before turning to some interactions between actin and various actin-binding proteins *in vitro* that may underlie these movements.

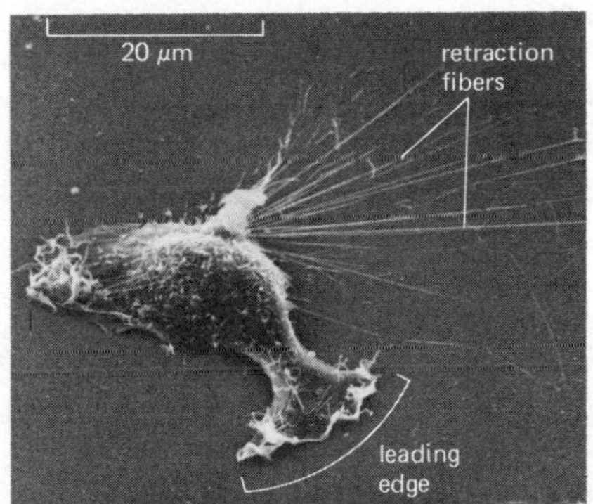

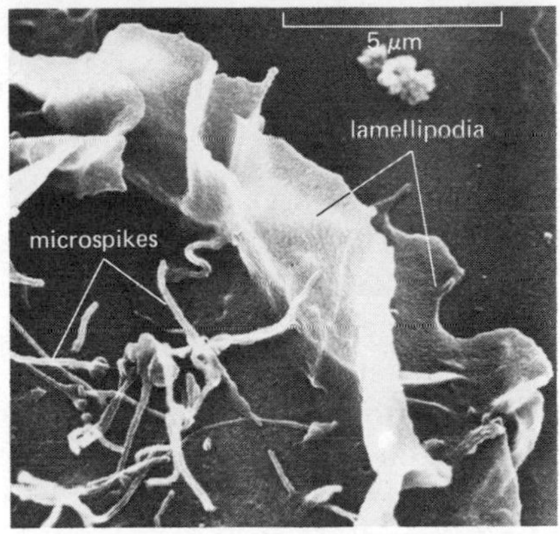

Figure 10–67 Scanning electron micrographs of human glial cells crawling on the surface of a tissue-culture dish. The lower panel is a close-up view of the leading edge of such a cell, showing the lamellipodia and microspikes. (Courtesy of Ulf Brunk and Peter Collins.)

Amoeboid Movement Is Accompanied by Gel-to-Sol Transitions in the Cytoplasm[31]

Two regions of cytoplasm in *Amoeba* can be distinguished by light microscopy: a central core of fluid cytoplasm, called the **endoplasm** (also called the **plasma sol**), which is full of refractile particles and organelles; and a surrounding transparent, gel-like **ectoplasm** (or **plasma gel**), which represents an unusually thick variant of a normal cell cortical layer. As a pseudopod extends from the cell, the inner endoplasm can be seen to stream in the direction of extension, appearing to congeal into more solid ectoplasm at its tip; elsewhere in the cell, ectoplasm appears to be changing into more fluid endoplasm and streaming towards the pseudopod. These transitions between viscous ectoplasm and more fluid endoplasm are known as **gel-to-sol** transitions. Although it is still controversial, the *Amoeba* is thought to move, in part, by contraction of the thick cortical layer of ectoplasm, which squeezes out the more fluid endoplasm, thereby creating a strong **cytoplasmic streaming** that causes each pseudopod to extend. Pronounced cytoplasmic streaming also occurs in the acellular slime mold *Physarum* and in the giant algal cell *Nitella,* which will be discussed in detail in Chapter 19 (p. 1131).

A good deal of circumstantial evidence suggests that cytoplasmic streaming in *Amoeba* and *Physarum* is an actin-based movement: (1) actin filaments are far and away the most abundant filamentous proteins in these organisms, and they are concentrated in the ectoplasm; (2) when the actin-binding drug *phalloidin* is injected into them, cytoplasmic streaming stops immediately; and (3) crude extracts of the cytoplasm of these cells are rich in actin and can undergo contractions and show some streaming movements.

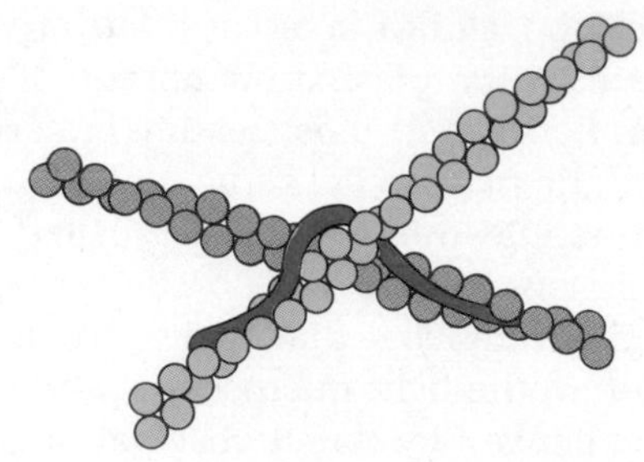

Figure 10–68 By forming a flexible link between two adjacent actin filaments, filamin produces a three-dimensional network of actin filaments with the physical properties of a gel.

It is this third line of evidence that has made it possible to study the molecular basis of cytoplasmic streaming. Actin-rich extracts of not only *Amoeba* but also of vertebrate cells such as macrophages and fibroblasts can gelate, solate, contract, and show vigorous streaming in response to small changes in the ionic composition of the surrounding medium. While they are biochemically complex, these extracts all contain actin as their principal filamentous component and have a number of distinct actin-binding proteins.

Cross-linking Proteins Can Produce an Actin Gel[23,32]

Small, compact cross-linking molecules, like **fimbrin,** can hold actin filaments rigidly together in parallel register to produce tight bundles of actin filaments (see Figure 10–57). But not all actin cross-linking proteins act in this way: some are long and flexible and can link any actin filaments that come into close proximity, whatever their orientation relative to each other, thus producing a random three-dimensional network (Figure 10–68). Two such proteins are **α-actinin** and **filamin,** originally isolated from muscle but now known to have close relatives in many other types of cells. Filamin, for example, is a long, flexible molecule with a molecular weight of about 250,000 and a tendency to form dimers in solution. Each monomer has a binding site for an actin filament, and so the dimer is ideally suited for the production of random three-dimensional networks; α-actinin has a similar effect, as illustrated in Figure 10–69. Even a relatively small number of filamin or α-actinin molecules added to a solution of actin filaments has a dramatic effect on its physical properties—changing it from a viscous fluid to a solid gel.

Fragmenting Proteins Can Produce a Ca^{2+}-dependent Liquefaction of Actin Gels[23,33]

Gel-to-sol transitions in crude cytoplasmic extracts of *Amoeba* and other cells are extremely sensitive to the levels of Ca^{2+}. An increase in the free Ca^{2+} concentration from 10^{-7} M to 10^{-5} M causes a change from gel to fluid and often produces vigorous streaming movements.

Reconstituted gels containing actin filaments and filamin exhibit no such Ca^{2+} sensitivity. This means that some other factor must be required to mediate the effect of Ca^{2+}. A number of protein candidates have been identified; when added to a gel of actin filaments and filamin, they cause it to change into a more fluid state in response to Ca^{2+}. It has been suggested that in the presence of Ca^{2+} these proteins bind so strongly to actin that they are able to insert themselves between the subunits in the actin filaments, causing them to disassemble (Figure 10–70), so that the cross-linked network of actin filaments produced by filamin is broken up. The addition of less than one molecule of such an actin-fragmenting protein for every 400 actin molecules can cause a precipitous drop in the viscosity of a gelled actin filament solution.

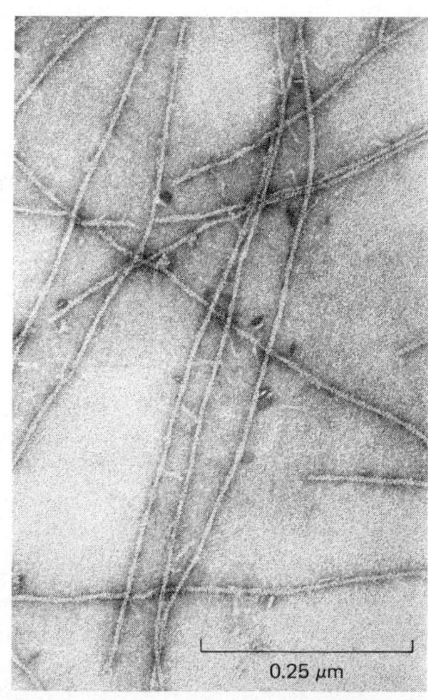

Figure 10–69 Electron micrograph of negatively stained actin filaments cross-linked by α-actinin. (From B. M. Jockusch and G. Isenberg, *Proc. Natl. Acad. Sci. USA* 78:3005–3009, 1981.)

Significantly, this binding occurs only at levels of Ca^{2+} that are above 10^{-6} M, concentrations that occur only transiently in the cytosol, usually when cells are responding to changes in their environment.

Ca^{2+}-dependent actin-fragmenting proteins have been identified in almost every type of vertebrate cell. One of the best known is **gelsolin,** originally isolated from macrophages; another is **villin,** one of the major proteins of the intestinal brush border microvilli. Paradoxically, these proteins can also hold a small number of actin molecules in such a way that they can act as potent nucleating agents for actin polymerization if added to a solution of free actin molecules. Whether the primary function of these proteins in the living cell is to shorten actin filaments or to initiate their assembly is at present unclear.

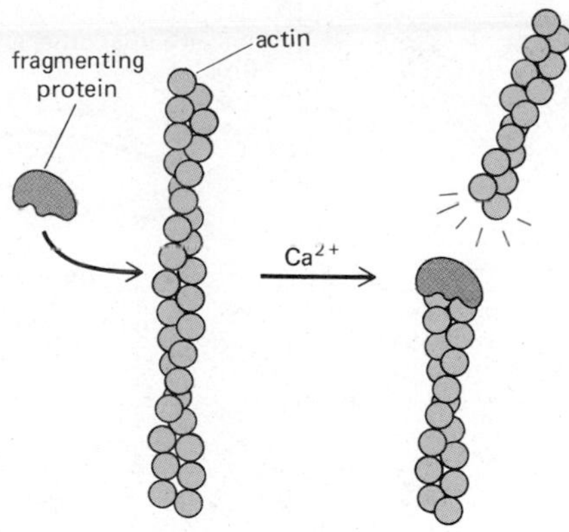

Figure 10–70 Fragmenting proteins reduce the length of actin filaments, possibly by binding so strongly to actin molecules that they sever actin filaments. Such proteins are generally Ca^{2+}-dependent in their action. Examples of proteins thought to act in this way are villin and gelsolin.

The Contraction of Cytoplasmic Gels Is Mediated by Myosin[34]

A mixture of actin filaments, filamin, and gelsolin is capable of undergoing Ca^{2+}-dependent gel-to-sol transitions. But, unlike the cruder actin-rich gels obtained from cells, such an artificial mixture will not contract or display streaming movements. The missing factor appears to be nonmuscle *myosin,* since if myosin is selectively removed from the crude actin-rich gels, contractions and streaming no longer occur, although the extracts are still able to undergo Ca^{2+}-dependent liquefaction. This suggests that the motive force underlying cytoplasmic streaming is a Ca^{2+}-activated interaction between actin and myosin, as is the case in muscle contraction.

How can actin and myosin produce coherent movements when the filaments are distributed in an apparently random three-dimensional network? At least at one level, the answer seems to be that an actin filament has a well-defined polarity and that myosin heads can bind and move along an actin filament only if they are oriented in the correct direction with regard to the filament's polarity. Thus the small bipolar aggregates of nonmuscle myosin molecules (see Figure 10–63) probably form miniature sarcomeres by pulling one set of actin filaments against another, even though the actin filaments and myosin aggregates are not part of a highly ordered array (Figure 10–71).

Although we are still far from a detailed understanding of the molecular basis of cytoplasmic streaming, the interactions of myosin, filamin, and gelsolin with actin filaments provide models for the essential elements of this phenomenon. A gel-like network of actin filaments can be produced by an interaction with a cross-linking protein such as filamin. A rise in Ca^{2+} will cause a precipitous drop in the viscosity of the actin-filamin network through the action of actin-fragmenting proteins such as gelsolin. The same rise in Ca^{2+} will activate myosin molecules to pull actin filaments against one another, thereby generating vigorous fluid streaming.

The properties of some of the actin-binding proteins of vertebrate nonmuscle cells are summarized in Figure 10–72 and Table 10–5.

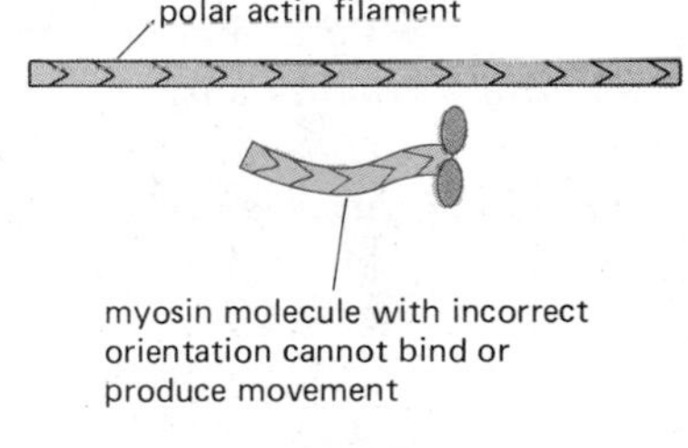

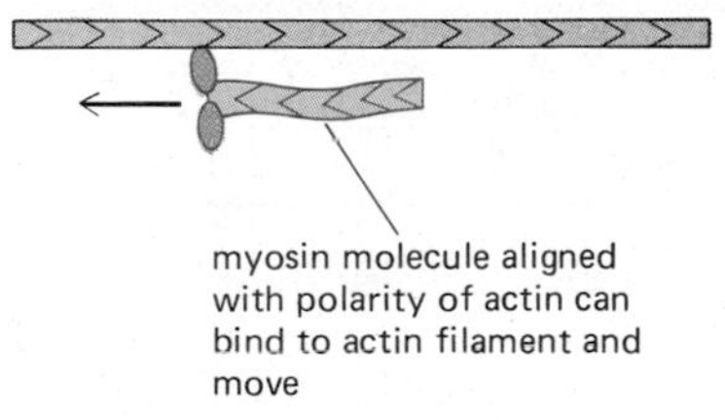

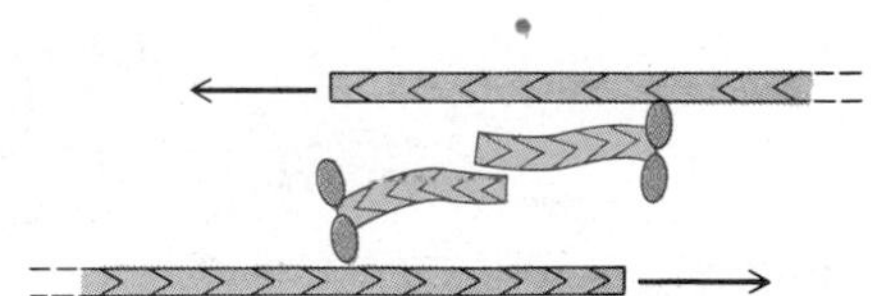

Figure 10–71 Schematic illustration of how a bipolar aggregate of nonmuscle myosin molecules can produce a sliding of two actin filaments of opposite polarity, as in muscle.

Summary

Actin is found in many different structures in the cell, and it associates with a large number of different actin-binding proteins. Rigid bundles of parallel actin filaments, cross-linked by proteins such as fimbrin, are present in the core of microvilli and stereocilia, where they perform a largely structural role. Bundles of actin filaments associated with short bipolar aggregates of nonmuscle myosin are found in specific regions of the cell where musclelike contractions are needed, such as the contractile ring of a dividing cell, the belt desmosomes at the apical region of an epithelial cell, and the stress fibers of a flattened tissue culture cell. In addition, less well-organized networks of actin filaments with

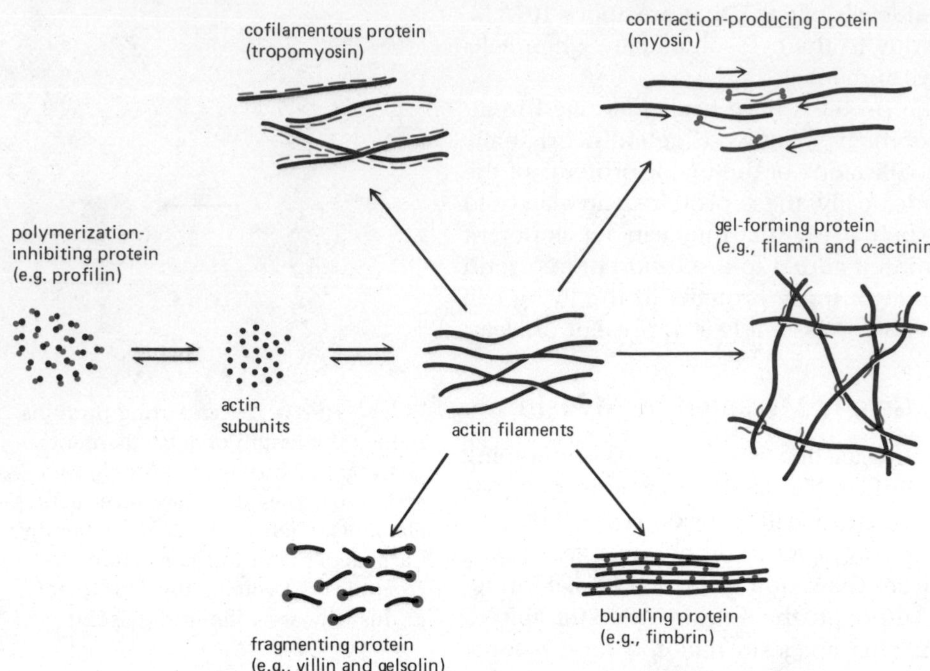

Figure 10–72 Highly schematic diagram summarizing the various changes in the state of actin aggregation produced by actin-binding proteins (Table 10–5). For simplicity, only the principal types of actin-binding protein are shown, and each is illustrated as having only a single mode of action on actin. Not shown here are anchoring proteins (such as vinculin) that tie actin filaments to other cell components, and various capping proteins that have been less well characterized.

the properties of a gel are found throughout the cytoplasm, concentrated in a cortical layer located just beneath the plasma membrane. These networks appear to be formed by flexible, actin-cross-linking proteins such as filamin, and they undergo a Ca^{2+}-induced decrease in viscosity that is mediated by actin-fragmenting proteins such as gelsolin. By interacting with nonmuscle myosin and with proteins that anchor them to the plasma membrane, these networks are thought to be responsible for a variety of cell-surface movements and to play a crucial part in the complex process of cell locomotion.

Table 10–5 Major Types of Actin-binding Proteins in Nonmuscle Cells

Protein	Action on Actin *in Vitro*
Nonmuscle myosin	an ATPase that interacts with actin filaments to produce movement as in muscle
Nonmuscle tropomyosin	a rodlike protein that binds along the length of actin filaments
Filamin, α-actinin	proteins that form flexible cross-links between actin filaments, producing a three-dimensional network
Fimbrin	a protein that joins adjacent actin filaments together to generate parallel bundles
Profilin	a protein that binds to actin monomers, thereby restricting their polymerization
Villin, gelsolin	Ca^{2+}-dependent actin-fragmenting proteins that also nucleate actin filament assembly
Capping proteins	proteins that bind to one or the other end of actin filaments and prevent the addition or loss of actin monomers

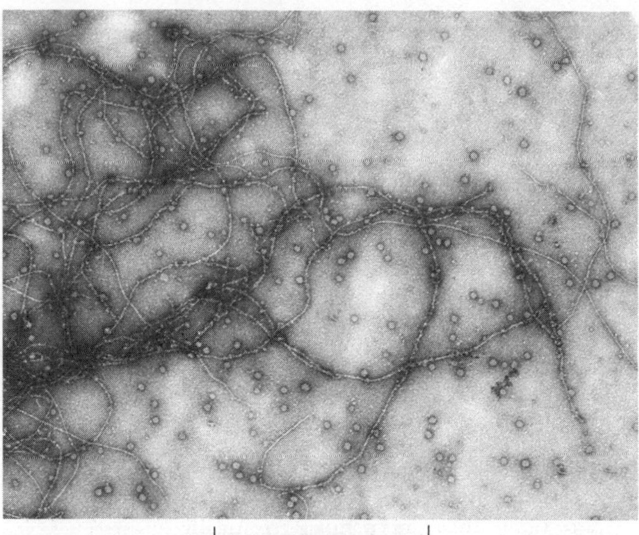

Figure 10–73 Electron micrograph of negatively stained neurofilaments present in a sample of axoplasm from the giant axon of a marine worm. The circular objects are nonfilamentous protein aggregates in the preparation. (Courtesy of David Gilbert.)

Intermediate Filaments

Intermediate filaments are tough and durable protein fibers that appear as straight or gently curving arrays in electron micrographs of most eucaryotic cells (Figure 10–61). Their diameter is characteristically 8 nm to 10 nm and, therefore, intermediate between those of actin filaments and microtubules (Figure 10–73). They are particularly prominent in those parts of a cell that are subject to mechanical stress, such as along the length of a nerve cell process, close to the spot desmosomes between adjacent epithelial cells, and throughout the cytoplasm of a smooth muscle cell. Their role in keeping the Z discs of the adjacent sarcomeres in skeletal muscle in register has already been discussed (p. 558). Intermediate filaments are the most stable components of the cytoskeleton and the least soluble constituents of the cell: when cells are extracted with solutions of high or low ionic strength or with nonionic detergents, intermediate filaments remain behind while most of the cytosol and other protein filaments are lost. In fact, the term "cytoskeleton" was originally coined to describe these unusually insoluble fibers.

Intermediate Filaments Consist of Fibrous Polypeptides That Vary Greatly in Size[35]

When shadowed with metal and examined in the electron microscope, the proteins of intermediate filaments appear as irregular threadlike molecules, in contrast to actin or tubulin, which appear as small globular structures. Within the assembled intermediate filament, these fibrous polypeptides are believed to associate side by side to form a ropelike structure similar to that of a collagen molecule. An arrangement of this kind can account for many of the distinctive properties of intermediate filaments. For example, it produces filaments of high tensile strength that can, if necessary, be further strengthened by the addition of covalent bonds between the subunits.

Moreover, if only a part of each fibrous protein is involved in the interactions that form intermediate filaments, the remaining part can vary considerably without affecting the overall filament structure. As discussed below, intermediate filaments in fact are composed of polypeptides of a surprisingly wide range of sizes (from about 40,000 to over 200,000 daltons) that vary both between different cell types and, in the same cell type, between different animal species. The intermediate filaments of mammalian axons (neurofilaments), for example, contain three different polypeptides, with molecular weights of 70,000, 140,000, and 210,000. Squid neurofilaments are composed of only two polypeptides of 60,000 and 200,000 daltons.

In striking contrast, actin and tubulin have not changed appreciably in size in billions of years. The most plausible explanation of this stability is that the assembly of a polymer from globular subunits imposes stringent limitations on the variability of the three-dimensional form of the subunits—even small changes in size or shape would interfere with the assembly process. Intermediate filament proteins are not similarly restricted.

It is not known how the individual polypeptides are arranged in an intermediate filament, but it has been suggested that they may be grouped in sets of three so that the subunit of the filament is a trimer. Biochemical studies indicate that each polypeptide has two distinct regions of α-helix, and it is thought that these regions of the three polypeptides in a trimer are wound around each other in a coiled-coil arrangement (Figure 10–74). Other regions of the polypeptide chains, less rich in α-helices, may serve both to stabilize the filament and to project from the filament surface where they can interact with other cytoplasmic constituents. It is these regions, which are not directly involved in filament formation, that are likely to show considerable variation from polypeptide to polypeptide within the family of intermediate filament proteins.

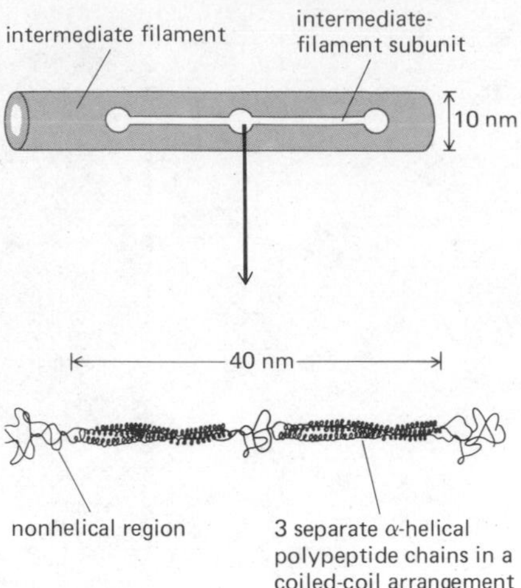

Figure 10–74 One model of intermediate-filament structure in which the basic subunit is composed of three fibrous polypeptides. (Adapted from P. Steinert, *J. Mol. Biol.* 123:49–70, 1978.)

Disassembly of Intermediate Filaments May Require Filament Destruction[36]

Under the conditions existing in the cytoplasm, the assembly of intermediate filaments is probably irreversible. There is no evidence for a pool of unpolymerized intermediate filament proteins in the cell, nor for a dynamic equilibrium between their soluble and polymerized forms, as there is for actin and tubulin. This stability raises important questions of control. Is the cell able to regulate the number or length of intermediate filaments it contains? If so, how is this achieved?

A partial answer to these questions has been provided by the discovery of proteolytic enzymes that are able to degrade specifically one class of intermediate filaments or another. Most of these enzymes are activated *in vitro* by Ca^{2+}, but the circumstances under which they work in the living cell are unknown. However, it is possible that the only way cells can depolymerize their intermediate filaments is by cleaving their polypeptides into smaller fragments.

Different Cell Types Contain Intermediate Filaments of Distinct Composition[37]

There are many different classes of intermediate filaments, each composed of a distinctive set of subunit proteins (Table 10–6). In general, any one cell type contains only one class of intermediate filament (Figure 10–75): (1) Neurons contain **neurofilaments,** which in vertebrates consist of three different intermediate filament polypeptides—the so-called neurofilament triplet. (2) Epithelial cells contain **keratin filaments** (also called *tonofilaments*), which are made up of a variable number of closely related keratin proteins. (3) Most other cells have intermediate filaments composed of a 55,000-dalton protein called **vimentin,** which may be co-polymerized with other, cell-type-specific subunits. For example, muscle cells have intermediate filaments composed of vimentin plus a closely related protein, **desmin,** while astrocytes (a special class of supporting, or glial, cell in the central nervous system) have large

Table 10–6 Intermediate Filaments of Vertebrate Cells

Intermediate Filament	Component Polypeptide (MW)	Cell Type
Keratin filaments	various keratin proteins (40,000–65,000)	epithelial cells
Neurofilaments	neurofilament triplet proteins (70,000; 140,000; 210,000)	neurons
Vimentin-containing filaments	vimentin (55,000)	fibroblasts and many other cell types
	vimentin + glial fibrillary acidic protein (50,000)	some glial cells
	vimentin + desmin (51,000)	muscle cells

bundles of intermediate filaments (referred to as *glial filaments*) composed of vimentin plus a 50,000-dalton protein called **glial fibrillary acidic protein.** Intermediate filaments in fibroblasts and in many other cell types seem to be constructed mainly of vimentin.

Why are there so many different intermediate filament proteins? The heterogeneity implies that these filaments mediate somewhat different interactions in different cell types. It is possible that much of the variability conferred on actin filaments and microtubules in different cell types by the addition of different accessory proteins is imparted to intermediate filaments by differences in their constituent polypeptides.

Keratin Filaments Strengthen Epithelial Cell Sheets[38]

In epithelial cell sheets, neighboring cells are held together mechanically by strong, rivetlike junctions called *spot desmosomes* (see p. 683). On their inner (cytoplasmic) face, these junctions serve as anchorage sites for keratin filaments that form an irregular network throughout each epithelial cell (Figure 10–61D). Since another type of filament extends from cell to cell across the extracellular space to connect the halves of a spot desmosome, the keratin filaments are part of a continuous network of protein filaments that extends through the entire epithelial cell sheet and serves to give it tensile strength.

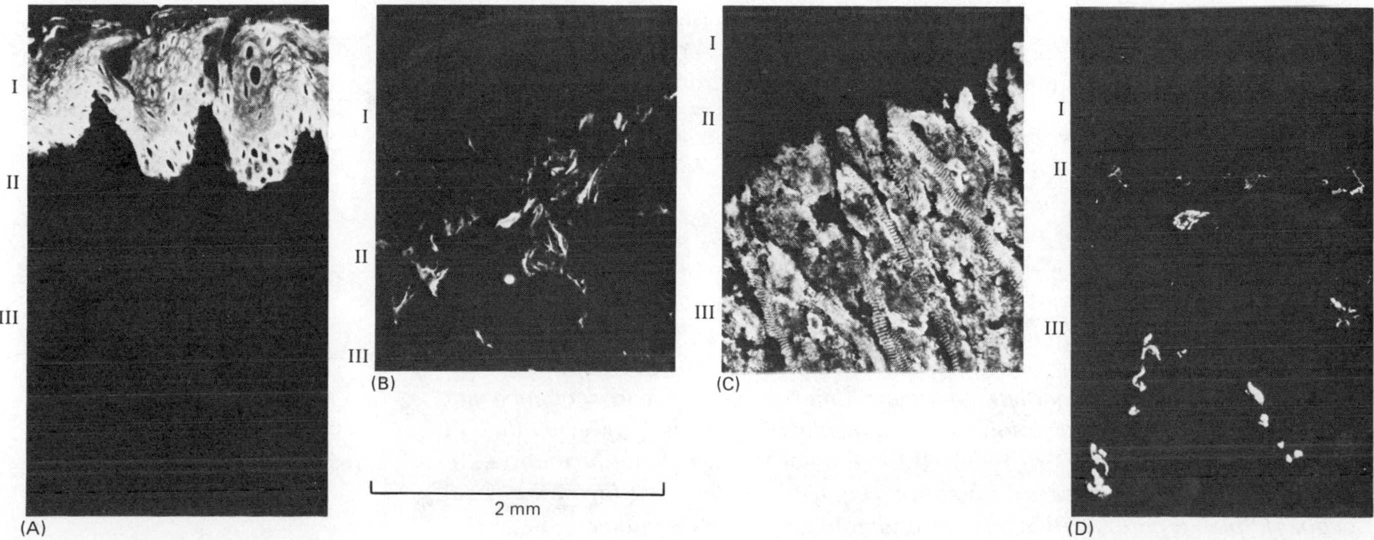

Figure 10–75 Immunofluorescence micrographs of frozen sections of rat tongue stained with antibodies against various intermediate proteins. (A) *Anti-keratin* antibodies stain epithelial cells of layer I; (B) *anti-vimentin* antibodies stain fibroblasts and blood vessels, mostly in layer II; (C) *anti-desmin* antibodies stain muscle cells in layer III; and (D) *anti-neurofilament* antibodies stain nerve processes in layers II and III. (From M. Osborn, N. Geisler, G. Shaw, G. Sharp, and K. Weber, *Cold Spring Harbor Symp. Quant. Biol.* 46:413–429, 1982.)

Keratin filaments are also found in the tough outer covering of higher animals. These filaments accumulate in the cytoplasm of mature epithelial cells and become progressively cross-linked to each other and to associated proteins, in part by means of disulfide bonds. The cells eventually die, but their skeletons persist to form a tough protective outer layer on the animal (p. 912). Hair and nails, as well as the outer layer of the skin, are formed this way.

Keratin filaments can be solubilized by disrupting their disulfide bonds and extracting the proteins in a denaturing solvent such as urea. The polypeptides thus obtained are remarkably diverse; even keratins from a single tissue such as bovine snout consist of six or more related proteins. Moreover, keratins from different parts of the same animal, such as hoof and snout, have a slightly different composition and are coded for by distinct families of related keratin genes. The diversity of the keratin proteins is perhaps not surprising in view of the important and varied roles that these proteins play: they provide the animal with its primary barrier against heat and water loss, as well as supplying it with camouflage, armament, and decoration.

Do Intermediate Filaments Have Nonstructural Functions?[39]

It is very likely that some of the other classes of intermediate filaments also have structural roles. For example, intermediate filaments in skeletal muscle appear to hold the adjacent myofibrils in register by "tying" the edges of the Z discs together (see Figure 10–17), and it is thought that neurofilaments serve to strengthen nerve cell axons, which may be more than a meter in length and normally persist as relatively fixed structures for the lifetime of an animal.

In other cells, however, the distribution of intermediate filaments gives little clue to their function. In cultured fibroblasts and white blood cells, for example, they are present as random networks (Figure 10–61). If intermediate filaments are simply tension-bearing elements, why are there so many different kinds of polypeptides? And what is the function of those regions of a polypeptide that do not appear to be necessary for the formation of the filament itself?

Until recently it has been very difficult to explore the functions of intermediate filaments because of the lack of reagents that can specifically disrupt them in living cells—as colchicine and cytochalasins specifically disrupt microtubules and actin filaments, respectively. It is now possible, however, to inject cultured cells with monoclonal antibodies that react with intermediate filaments. Such antibodies injected into a fibroblast cause the collapse of the vimentin-containing filaments into a tight cap close to the nucleus. Surprisingly, the antibodies have no effect on the growth, division, movement or shape of the cells. It seems, therefore, that the function of these intermediate filaments may be too subtle to be discerned by examining cultured cells in a microscope.

Summary

Intermediate filaments are ropelike polymers of fibrous polypeptides that play a structural or tension-bearing role in the cell. A variety of tissue-specific forms are known that differ in the type of polypeptide they contain: these include the keratin filaments of epithelial cells, the neurofilaments of nerve cells, and the vimentin filaments of fibroblasts and most other cells. The polypeptides of the different types of intermediate filaments differ in amino acid sequence and frequently show large variations in their molecular weight. However, they all contain homologous regions believed to be involved in filament formation. Regions of the polypeptides that are not required for filament formation itself probably play a role in other, as yet undefined, cellular functions.

Organization of the Cytoskeleton[40]

Up to this point, microtubules, actin filaments, and intermediate filaments have been treated as though they were independent components of the cell. But it is obvious that the different parts of the cytoskeleton must be linked together and their functions coordinated in order to mediate changes in cell shape and produce various types of cell movements. For example, when a fibroblast in culture rounds up to divide, the entire cytoskeleton is reorganized: stress fibers and cytoplasmic microtubules are disassembled, while a mitotic spindle and then a contractile ring are formed, all as part of a controlled sequence of events. Unfortunately, remarkably little is known about the interactions between the three major filament systems of the cytoskeleton and even less about the molecular mechanisms that coordinate the many changes that occur in these systems.

The Cytoskeleton Can Be Seen in Three Dimensions in the Electron Microscope[41]

One of the oldest controversies in cell biology concerns the structural organization of the cytoplasm. The earliest microscopic observations of living cells revealed that the cytoplasm is a viscous fluid that can change from a more fluid state to one resembling a deformable solid. In the period between 1870 and 1885, it was widely believed—because of the appearance of cells following fixation and staining—that the cytoplasm contained a three-dimensional network of protein fibers. This view was fiercely opposed by some histologists who contended that the network seen after fixation was an artifact and consisted of coagulated proteins that had been produced by the harsh treatment of the cell. Over 100 years later, similar arguments can still be heard, although they now concern the existence of fibers of much smaller diameter that can be seen only in electron micrographs.

Conventional thin-section electron microscopy is not a good way to examine the three-dimensional arrangement of protein filaments in the cytoplasm. The sections are too thin to give any indication of the overall geometry of the filaments, and it is very difficult to tell whether there are connections between the filaments, or even, in many cases, to distinguish between filament types. There are, however, several ways to examine thicker layers of cytoplasm in the electron microscope. Representative views obtained by these methods are shown (all approximately at the same magnification) in Figure 10–76.

The "cleanest" images are obtained when cells are extracted with a nonionic detergent, since the phospholipid and soluble proteins of the cell can be washed away. Cells treated in this way and then quickly frozen and deep-etched reveal a particularly striking view of the cytoskeleton (Figure 10–76A). Individual actin filaments and intermediate filaments remain in their original position and, under special conditions, the microtubules may also be preserved. The different types of protein filaments can be identified by their diameter and, in some cases, by the arrangement of their protein subunits. A similar view of the cytoskeleton can be obtained by staining cultured cells with heavy metals after the cells have been extracted with a detergent: the edge of the cell is so thin that it can often be viewed in the electron microscope without sectioning (Figure 10–76B). Thick sections of the cytoplasm can also be visualized in the high-voltage electron microscope (Figure 10–76C). In these images the three types of protein filaments appear to be largely independent and linked together, if at all, by only occasional cross-bridges.

The cytoskeletal network looks quite different, however, in cells that have not been treated with detergent. Such cells can be examined either by high-

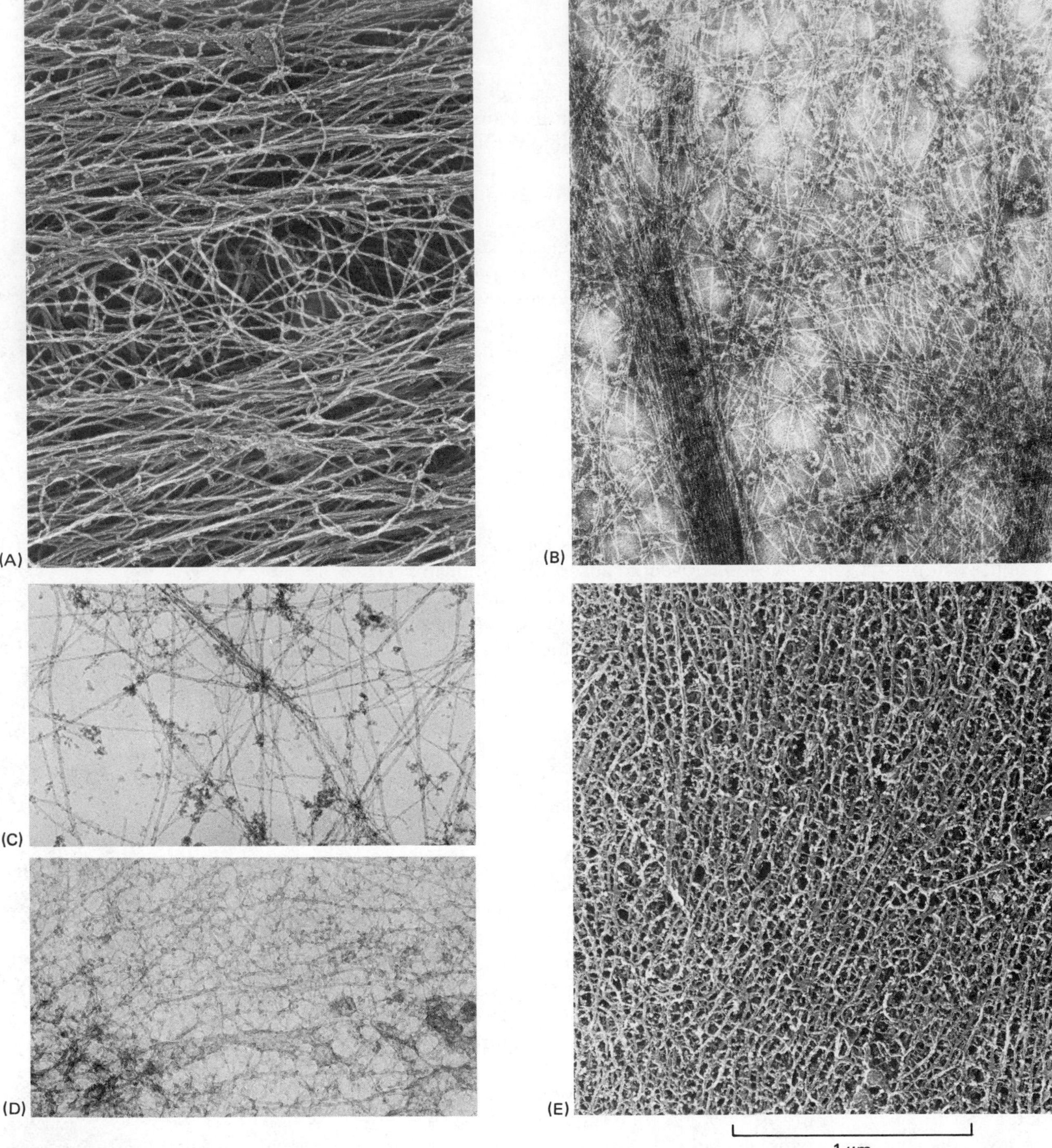

Figure 10–76 Views of the cytoplasm obtained by various preparative procedures (all at the same magnification).

(A) Deep-etch electron micrograph of the cytoplasm of a fibroblast extracted with a nonionic detergent. Most of the straighter filaments arranged in loose bundles running from left to right are actin filaments, while those that crisscross (center of the micrograph) are predominantly intermediate filaments. (Courtesy of John Heuser and Marc Kirschner.)

(B) Electron micrograph of negatively stained protein filaments in a thin region of a human glial cell in culture. Because the cell was fixed in the presence of a nonionic detergent, much of its soluble protein has been lost. (Courtesy of Uno Lindberg, Anna-Stina Höglund, and Roger Karlsson.)

(C) High-voltage electron micrograph of a portion of a detergent-extracted cultured fibroblast. The actin filaments are decorated with S1 fragments of myosin. (Courtesy of Manfred Schliwa.)

(D) High-voltage electron micrograph of a portion of a cell prepared as in Figure 10–76C but without extraction in detergent. The cytoplasm is filled by an extensive three-dimensional network of fine protein filaments—the so-called *microtrabecular network.* (Courtesy of Manfred Schliwa.)

(E) Deep-etch electron micrograph of a rat axon that has not been extracted with detergent. The protein filaments are extensively interconnected by cross-links, which in places appear to be periodically arranged. (Courtesy of S. Tsukita and H. Ishikawa.)

voltage electron microscopy (Figure 10–76D) or by rapid freezing and deep etching (Figure 10–76E). The principal filaments of the cytoskeleton now appear to be extensively interconnected by a three-dimensional network of fine threads, presumably composed of the proteins that are removed in detergent-treated cells. This network has been called the *microtrabecular network.* It is at present uncertain whether this fine network is part of the cytoskeleton in living cells or whether it is produced by the aggregation of soluble macromolecules in the course of fixation and dehydration. However, there is independent evidence that the various fiber systems of the cytoskeleton are interconnected and that even many components normally thought of as soluble and freely diffusing are, in fact, associated with the cytoskeleton in the living state.

Organelles and Soluble Proteins Can Be Associated with the Cytoskeleton[42]

The concentration of actin in a cell is extremely high—over 50 mg/ml in certain parts. For this reason alone, any protein that has a tendency to bind to actin, even if only weakly, will be found at least temporarily associated with actin or actin filaments or both.

Indeed there are many indications that other components of the living cell are associated with various parts of the cytoskeleton. Membrane-bounded organelles such as mitochondria and lysosomes often move individually in a highly characteristic fashion, known as **saltatory movement,** in which they travel in rapid bursts along straight but usually invisible tracks, stopping briefly now and then before setting off again, often back along the same track but sometimes in a different direction. Thin links can often be seen extending from these organelles to adjacent protein filaments in electron micrographs. Clusters of cytosolic ribosomes are frequently observed in association with protein filaments; and when cells are extracted with nonionic detergents, much of the protein-synthesizing machinery remains behind with the cytoskeleton. Even more surprising, soluble enzymes, such as some of those involved in glycolysis, appear to be bound to the actin filaments in myofibrils and to the stress fibers of fibroblasts, where they can be detected by immunofluorescence.

An association of soluble proteins with actin is also strongly suggested by studies of the transport of protein components in the nerve cell axon. When nerve cells are incubated briefly with radioactive amino acids, cytoskeletal proteins synthesized in the cell body become labelled and then transported along the axon to the nerve terminal at a slow but steady rate of between 1 mm and 5 mm per day—a process called *slow axonal transport.* Two principal rates of slow transport have been identified within this range: a very slow component composed almost entirely of tubulin and neurofilament proteins, and a slightly faster component, which includes actin together with a large collection of other proteins (Figure 10–77). Interestingly, among the proteins that travel with actin are some that are normally considered to be soluble enzymes, such as creatine kinase and enolase. Since these proteins move as a discrete radioactive band for weeks or even months, they obviously cannot diffuse freely in the cytoplasm. Moreover, since they move at the same rate as actin, it seems reasonable to suppose that they are in some way physically associated with actin filaments.

Taken together, these observations introduce a new level of order in our picture of the cell. Not only are the insoluble components such as protein filaments linked together through specific binding interactions, but so too are many of the components in the cytosol that were previously thought to be soluble and freely diffusing.

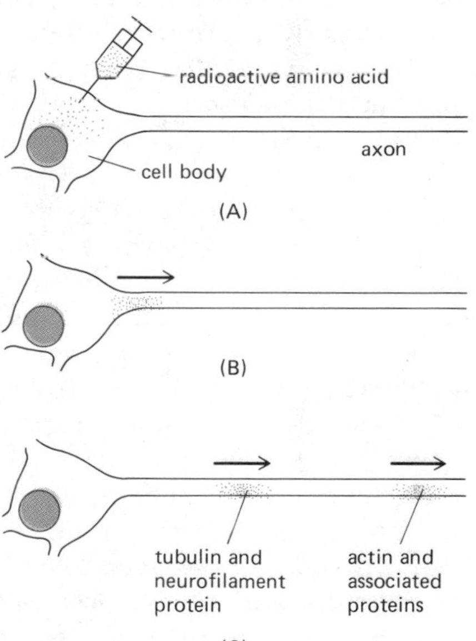

Figure 10–77 Schematic illustration of how the slow transport of proteins in a nerve axon can be measured. (A) The proteins in a nerve cell are labeled by a pulse of radioactive amino acid applied to the cell body. (B) Some of the radioactive proteins are transported along the axon at a steady rate of 1 mm to 5 mm per day. (C) After several days, two distinct peaks of radioactivity can be resolved that carry different cytoskeletal components: the faster moving of these includes actin together with a large number of other proteins, some of which are normally considered to be soluble components of the cytoplasm.

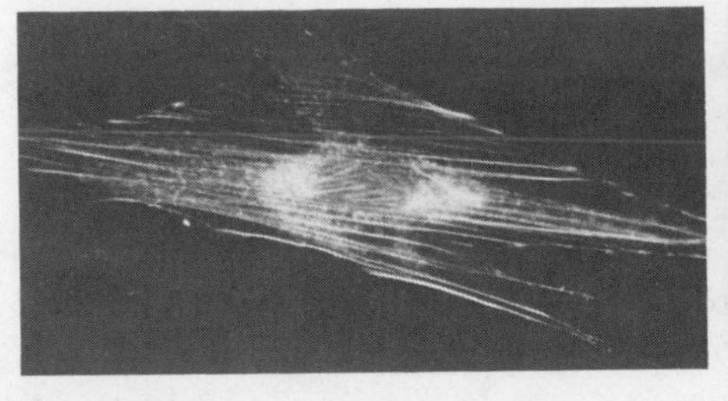
(A)

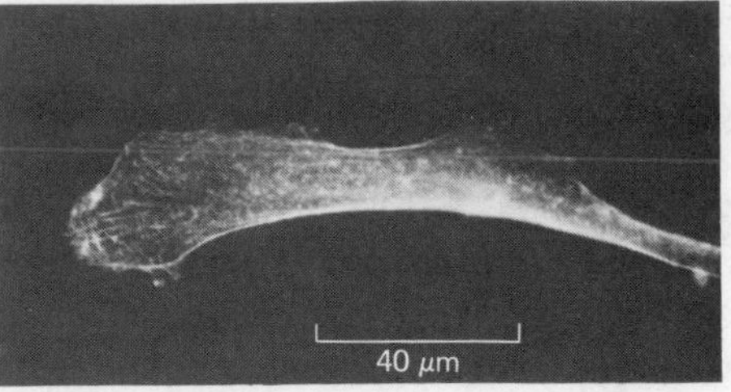

(B)

Figure 10–78 Immunofluorescence micrographs of a normal fibroblastic cell (A) and a tumor-virus-transformed cell (B), both stained with a fluorescent derivative of the actin-binding drug phalloidin. While the normal cell has many prominent stress fibers, the neoplastically transformed cell contains very few, if any. (From M. Verderame, D. Alcorta, M. Egnor, K. Smith, and R. Pollack, *Proc. Natl. Acad. Sci. USA* 77:6624–6628, 1980.)

Extensive Changes in the Cytoskeleton Are Produced by a Single Gene of a Transforming Virus[43]

The differences in growth potential between cancer cells and normal cells in tissue culture are often associated with conspicuous cytoskeletal changes. Cancer cells grow more readily in low levels of serum and are able to grow while suspended in an agar gel; moreover, they do not stop growing when they have covered the bottom of a tissue-culture dish, as normal cells do, but continue to pile up on each other until they reach a very high density (see p. 620). The reorganization of the cytoskeleton accompanying these alterations in behavior is reflected in two changes: cancer cells are usually more rounded, and stress fibers are reduced in number or even absent (Figure 10–78).

These and other changes, known collectively as **neoplastic transformation,** can be produced in normal cells by infection with a tumor virus such as Rous sarcoma virus. This simple virus, which causes cancer in chickens, has only four genes, one of which (known as the *src* gene) is the sole cause of the transformation: when this gene is active, cells are transformed and tumors are formed; when it is inactive, the cells appear to be normal (see p. 625).

Recently the product of the *src* gene has been shown to be a protein kinase with an unusual specificity. The kinase catalyzes the phosphorylation of tyrosine residues in a particular subset of cell proteins, the most interesting of which, from the point of view of the cytoskeleton, is *vinculin.* As discussed above, this protein is associated with adhesion plaques and is thought to play a part in anchoring actin filament bundles to the plasma membrane. Its modification by the *src* gene product may play a causative role in the changes in the cytoskeleton observed after cell transformation by Rous sarcoma virus. Most importantly, these observations suggest that cell growth and division may normally be regulated by signals received via an organized cell cytoskeleton.

Microtubules May Be the Overall Organizers of the Cytoskeleton[44]

All eucaryotic cells have a distinct spatial geometry that can be recognized by the position of their organelles and by features of their external surfaces. While all components of the cytoskeleton reflect this geometry, microtubules often seem to play a unique part in determining it. It is a common observation, for example, that microtubules are aligned with the long axis of cells, and in many cases their presence is essential for the maintenance of the asymmetrical cell shape. For example, exposure to colchicine arrests the development of long processes in growing nerve cells and the scheduled elongation of certain embryonic epithelial cells—apparently by retarding the polymerization of their microtubules. Colchicine has no effect on the shape of some mature elongated cells, such as muscle fibers or lens cells. But here it appears that the transient presence of microtubules at an earlier stage creates a shape that is later maintained by other elements.

Microtubules clearly influence the distribution of intermediate filaments in most cells in culture. In cultured fibroblasts, for example, intermediate filaments spread out in a radial pattern quite similar to that of cytoplasmic microtubules, extending from a region near the cell nucleus toward the periphery. If the cells are treated with colchicine, the microtubule network depolymerizes very rapidly (typically in less than ten minutes), and over the next several hours the intermediate filament network gradually collapses into a dense filamentous cap lying adjacent to the nucleus. If the drug is removed, the microtubules rapidly repolymerize and the intermediate filaments slowly return to their normal distribution.

There is some evidence that microtubules also influence the distribution of actin filaments in the cell. Perhaps the clearest example is the contractile ring, whose action completes the process of cell division. The contractile ring always forms in a plane perpendicular to the mitotic spindle equator, so that—as the two halves of the spindle draw the two sets of chromosomes apart—one set of chromosomes is enclosed in each daughter cell (see Chapter 11). In this case it is clearly vital that the activity of the actin-based contractile ring be coordinated with those of the microtubule-based spindle. In fact, if the position of the forming mitotic spindle is displaced mechanically, the position at which the contractile ring subsequently forms is correspondingly changed (Figure 10–79).

There is a great deal of evidence that microtubules frequently function as a temporary scaffold to organize the other components of the cytoplasm. This role is especially evident in plant cells, as described in Chapter 19. For example, a transient *preprophase band* of cortical microtubules determines the plane of cell division, while other microtubules specify the place and orientation of fibrous cell wall deposition (see pp. 1136 and 1128). In mammals, spiral rings of microtubules become wrapped about the axoneme of a developing sperm tail and then disappear—leaving a spiral ring of mitochondria, which supply the axoneme with ATP (see Figure 9–4, p. 485). Finally, a particularly striking example of such transient scaffolding by microtubules occurs during the early development of insect flight muscle, where microtubules form in regular arrays together with the developing thin and thick filaments (Figure 10–80). These microtubules later disappear, leaving the highly organized arrangement of filaments characteristic of this specialized muscle.

In cells in culture, various types of local surface activity that are actin-based, such as membrane ruffling, microspike formation, and phagocytosis, often appear to occur in regions of the cell close to the end of microtubules, and the site of these movements in the cell is markedly changed by treatment with colchicine. In the presence of colchicine, for example, the normally coherent movements of cells in culture become disorganized; membrane ruffling now occurs around the entire periphery of the cell, which wanders "like a ship without a rudder" rather than moving in a straight line.

It seems, then, that cytoplasmic microtubules determine the cell's polarity and coordinate the various parts of the cytoskeleton responsible for complex cell movements. But, as previously discussed (p. 579), many microtubules are in turn organized by the cell center, which therefore might be considered the "command post" of the cell. In support of this idea, it has been observed that the cell center of migrating cells in tissue culture is usually on the same side of the nucleus as the advancing ruffling membrane, and that one of the two centrioles of this center is usually aligned with the direction of migration while the other centriole is at right angles to the culture substratum.

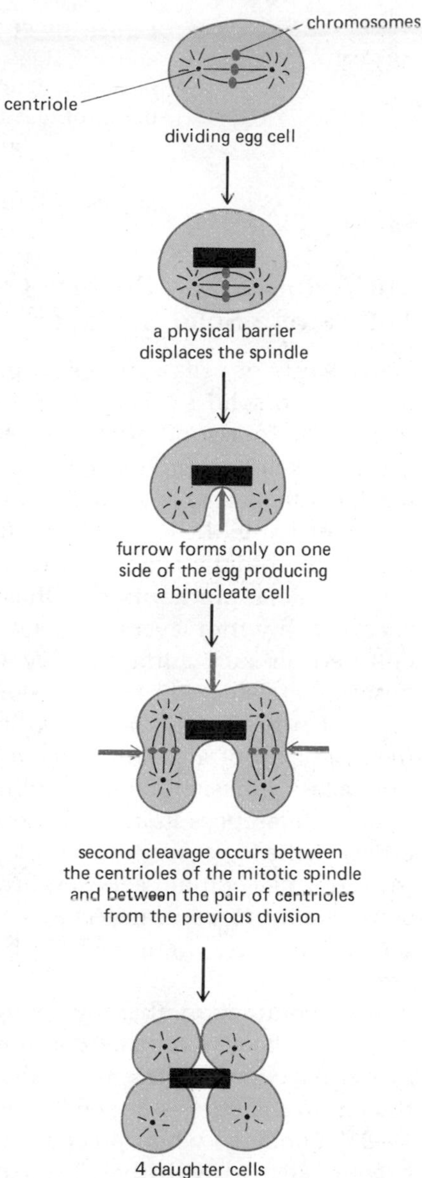

Figure 10–79 Schematic drawing of an experiment that shows the influence of the centrioles on the plane of cell division. If the spindle of a mitotic cell is pushed to one side, cleavage fails to occur on the opposite side of the cell. Subsequent cleavages occur not only between the two mitotic spindles but also between the two adjacent centrioles, which in this abnormal cell share the same cytoplasm but are not linked by a mitotic spindle. It seems that the contractile bundle of actin filaments that produces the cleavage furrow always forms in the region midway between two centrioles.

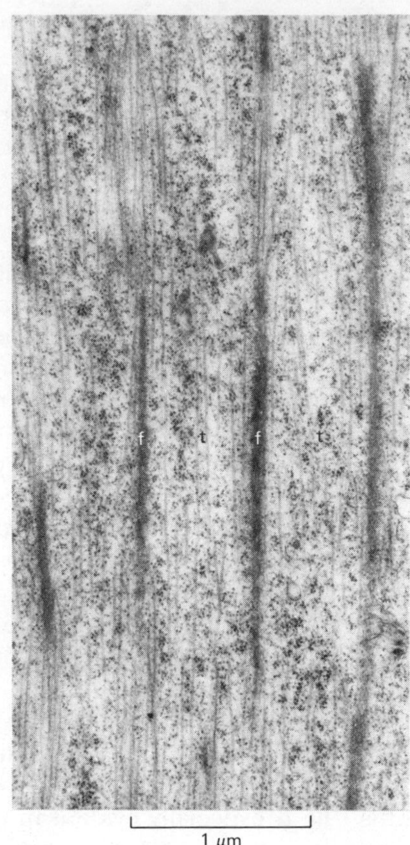

Figure 10–80 Electron micrograph of an insect muscle in an early stage of development. Aggregates of thick and thin filaments (f) are formed in parallel with cytoplasmic microtubules (t). As the aggregates coalesce into myofibrils, the microtubules are lost, suggesting that the microtubules act as temporary scaffolding to align the thick and thin filaments. (From J. Auber, *J. de Microsc.* 8:197–232, 1969.)

The Cytoskeleton Enables Cells to Respond to the Physical Nature of a Solid Surface[45]

Many kinds of cells synthesize proteins and proliferate only if they are attached to a solid surface. How this control is exerted is unknown, but it is reasonable to assume that the cytoskeleton is involved in some way (p. 619). The cytoskeleton and hence the shape of a cell are both radically altered when a cell contacts a solid surface. Two physical properties of a surface appear to be important in these responses: the stickiness or adhesiveness of the surface and its three-dimensional contours.

Artificial gradients of adhesiveness can conveniently be prepared by spraying very thin layers of metal on glass or plastic surfaces. When cells are cultured on such surfaces, they show a strong preference for the more adhesive, metal-coated parts. Fibroblasts have been shown to migrate *up* a gradient of adhesiveness produced in this way; and, in a similar type of experiment, the tip of a growing nerve cell axon, called a *growth cone,* will cross from a less adhesive to a more adhesive region on a culture dish. The response to local differences in adhesiveness can be explained by the contractile properties of the cytoskeleton. The long thin microspikes that protrude from a migrating cell contain a loosely arranged bundle of actin filaments and exhibit some ability to contract. They can, for example, pick up loose particles of debris and carry them back to the cell. It seems likely that these structures are able to test the adhesiveness of the surrounding environment by means of their contraction, thereby acting as "feelers" for the cell (Figure 10–81).

The effect of surface contours on cell movement can be seen in fibroblasts placed on an inverted V-shaped "rooftop" of glass. They will move over the top to the other side only if the rooftop angle is sufficiently obtuse (Figure 10–82). Similarly, when placed on the surface of a glass cylinder, fibroblasts become aligned with the long axis of the cylinder and move predominantly in this direction, but only when the diameter of the cylinder is less than about 200 μm. In both situations the cells migrate along the path of least curvature; one possible mechanism by which cells might do this depends on the organization of stress fibers, as schematically illustrated in Figure 10–82.

Cytoskeletal Organization Can Be Passed from a Parent Cell to Its Daughters[46]

In the short term, the organization and function of the cytoskeleton do not depend on the nucleus. A cell that has lost its nucleus will still attach to a substratum, change its shape, migrate, ingest particles, and so on. If small

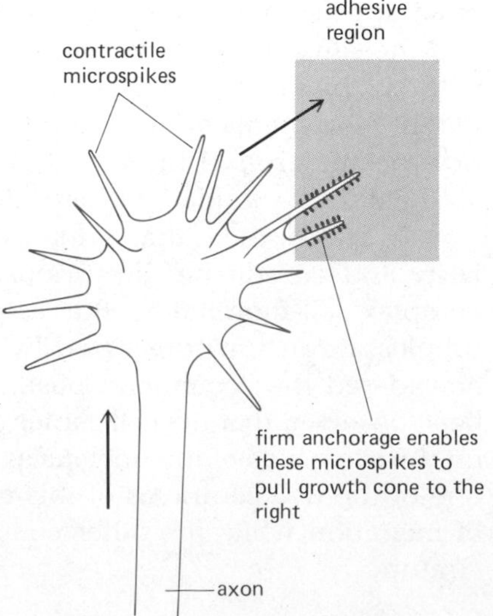

Figure 10–81 Schematic drawing of a growth cone at the tip of a growing axon, which has extended many actin-containing microspikes (filopodia). These extend at random onto the surrounding surface and then retract back into the growth cone. Microspikes that contact a more adhesive region, as shown, cannot retract so easily into the cell and, therefore, pull the growing axon in that direction.

fragments are broken off from a fibroblast, the plasma membrane reseals around each one, and the fragments (which may be less than 1% of the volume of the cell) continue to display various types of movement for hours. Some will only extend ruffling membranes, others will only protrude and retract microspikes, and still others will continuously generate blebs on their surface. They will not, however, translocate, suggesting that although actin-based motility will occur in local cytoskeletal subassemblies, overall movement requires some general coordination of the local movements, possibly provided by microtubules emanating from the cell center.

There is other evidence for a form of "memory" in the cytoskeleton. For example, when a cell divides in culture, a similar arrangement of stress fibers and cell extensions often appears in the two daughter cells; moreover, as they move apart, the two daughter cells may show similarities in their paths of migration. But the best-documented example of cytoskeletal inheritance comes from studies of *Paramecium*, a large single-celled organism whose surface is covered with rows of motile cilia. Normally, all of the rows are aligned in parallel, enabling a highly coordinated beating of the cilia. By experimental manipulations, it is possible to disturb this pattern and produce some inverted rows of cilia that beat in the opposite direction to that of their neighbors (Figure 10–83). Once established, such altered patterns are passed on to the cells of successive generations, apparently forever. This form of heredity has nothing to do with DNA: the modified cells inherit a particular pattern of ciliary rows via their cortical cytoskeleton.

The ability to inherit directly a particular cytoskeletal organization could have far-reaching consequences for the morphogenetic processes responsible for shaping the adult body plan of an organism. A clear example is seen in the development of the snail *Limulaea peregra*, whose coiled shell is genetically determined to be either a right-handed or a left-handed spiral. The handedness of the adult is observable as early as the eight-cell stage of development, in the asymmetric pattern of cell divisions known as *spiral cleavage* (Figure 10–84). This handedness of the embryo is determined by a slightly skewed position of the mitotic spindle during the early cleavages, which in turn is determined by *maternal inheritance:* the genotype of the mother snail (whether right- or left-handed) determines the form of the egg, independent of the form of her own shell. In the course of oogenesis, a factor is presumably produced by the mother snail that specifies the position of the mitotic spindle in the mature egg, and this factor determines the handedness of the embryo independent of the embryo's own genes.

Examples such as these show that the cytoskeleton of one cell can influence that of its daughter cells and affect even the morphology of an entire organism. How such organizational information is propagated from cell to cell is still a mystery.

Cytoskeletal Organization Can Be Transmitted Across Cell Membranes

The cytoskeleton of one cell can influence not only the cytoskeleton of its daughter cells but that of neighboring cells in the same tissue. One mechanism by which this influence might operate is through the formation of intercellular junctions that connect cells together and act as anchor points for protein filaments in the cytoplasm of the adjacent cells. In the epithelial cell sheet shown in Figure 10–85, for example, the intermediate filaments run in a continuous pattern across the sheet even though the sheet is composed of many distinct cells. Presumably this continuity in the pattern of intermediate filaments reveals that they are anchored at corresponding points in adjacent cell membranes.

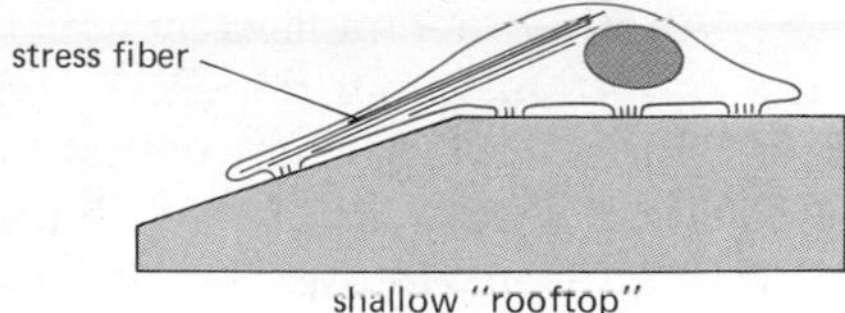

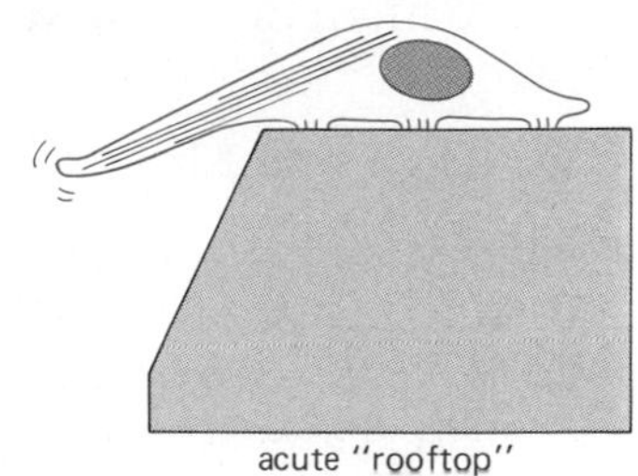

Figure 10–82 A fibroblast will not cross a "rooftop" of glass if the angle is too acute. This may be because it cannot then form a linear bundle of actin filaments (stress fiber) at its leading edge.

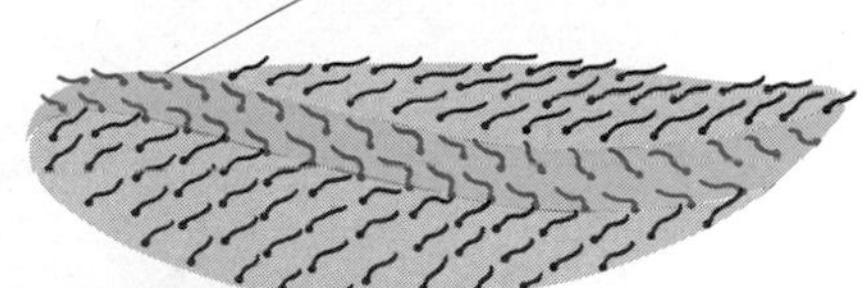

Figure 10–83 Schematic diagram of the rows of cilia on the surface of a normal paramecium and on a paramecium in which rows of cilia have been inverted so that they beat in the opposite direction. Such altered patterns are propagated indefinitely as the paramecium divides, even though the information in the DNA is unchanged.

Another mechanism by which the cytoskeleton of one cell can influence that of its neighbors depends on interactions between a cell's cytoskeleton and the extracellular matrix that the cell secretes. As described in Chapter 12, a cell with a polarized cytoskeleton tends to secrete an oriented extracellular matrix, and this, in turn, will influence the orientation of other cells in contact with the matrix (see Figure 12–71, p. 713). Because of the interactions between cells mediated by both cell junctions and the extracellular matrix, it is probable that the cytoskeletons of cells in many tissues are organized according to the pattern of the entire tissue rather than as independent units.

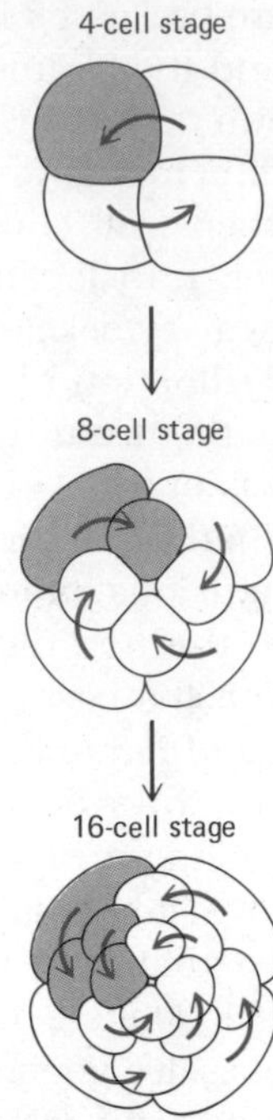

Figure 10–84 Schematic representation of the spiral cleavage of a fertilized snail egg. Each arrow connects a pair of cells formed at the previous division (the diagram should be imagined in three dimensions with the arrowheads pointing toward the observer). All the progeny of the colored cell at the 4-cell stage are shown in color at the 8-cell or 16-cell stages.

How Do Cells Move?[47]

Despite the great advances in understanding the cytoskeleton in recent years, the molecular mechanisms of such important movements as phagocytosis, mitosis, saltatory movements, and cell locomotion are still unknown. One of the main reasons for our ignorance is that the biochemical machinery responsible for these movements is not localized to a single structure, such as a cilium or myofibril, but is widely distributed throughout the cell. Furthermore, because the machinery is labile and easily disrupted as the cell is broken open, it is difficult to isolate in a functional form.

Consider, for example, a fibroblast crawling along the surface of a tissue-culture dish (Figure 10–86). It continually extends lamellipodia and microspikes from its leading edge, some of which adhere to the dish while others are carried back along the upper surface of the cell in the process of ruffling. Any particles adhering to the cell surface are carried back at the same rate as the ruffles. While ruffling is occurring at the front of the cell, portions at its rear remain stuck to the culture dish and get drawn out into long *retraction fibers* as the cell moves forward. These break suddenly and retract into the body of the cell, often leaving behind a fragment of adherent plasma membrane and cytoplasm. This cycle of movements is repeated over and over as the cell crawls steadily forward at a rate of about 40 μm per hour.

In order to travel across a surface, a fibroblast must carry out at least *three* processes: (1) its leading edge must extend over the surface, (2) this edge must make an attachment to the surface, and (3) the attached edge must pull the remainder of the cell forward (Figure 10–87). Of these three, the pulling action is the most easily understood, since the cell contains arrays of actin filaments which can undergo contraction through the action of myosin. However, the detailed arrangement of the molecules in these arrays and the way the filaments are linked, as they must be, to the tissue-culture dish and to the rest of the cytoskeleton are not known.

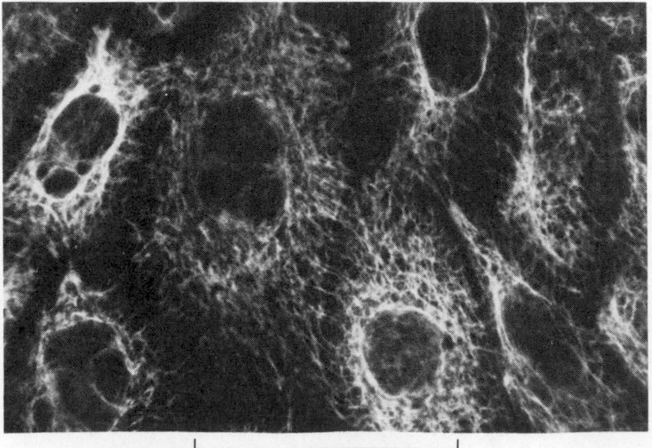

Figure 10–85 Immunofluorescence micrograph of a sheet of epithelial cells in culture stained with an antibody against intermediate filaments. Many of the filaments appear to run continuously from cell to cell, even though the cells are separated by their plasma membranes and the intervening extracellular matrix. (Courtesy of Michael Klymkowski.)

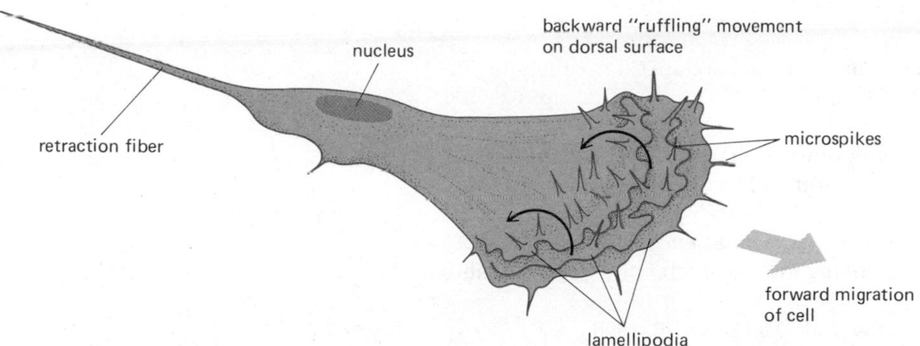

Figure 10–86 Schematic drawing of a fibroblast migrating along the surface of a tissue-culture dish. (See also Figure 10–67.)

How the leading edge extends forward is even more difficult to explain. Various mechanisms can be envisioned: for example, an active squeezing by actin filaments and myosin in the cortex could generate a forward stream of cytoplasm (as may be the case in *Amoeba*), or, alternatively, the controlled polymerization of actin filaments could extend the leading edge in a manner analogous to the formation of an acrosomal process in marine invertebrate sperm. The plasma membrane needed to surround the new extension might be carried forward by the protruding cytoskeleton, or it might be provided by the preferential addition of membrane (by exocytosis) at the leading edge. The possibility that plasma membrane is continually recycled due to both exocytosis at the leading edge of a cell and ingestion by endocytosis elsewhere—the *membrane flow* hypothesis—was discussed in Chapter 6 in relation to the mechanism of "capping" (see p. 279).

The most mysterious aspect of fibroblast locomotion, however, is related to its control. How are the various localized types of movement, such as ruffling and the extension and retraction of lamellipodia and microspikes, coordinated so that the entire cell moves forward in a coherent way? How is the cell able to alter these movements in response to changes in its environment? These remain among the most challenging questions in cell biology today.

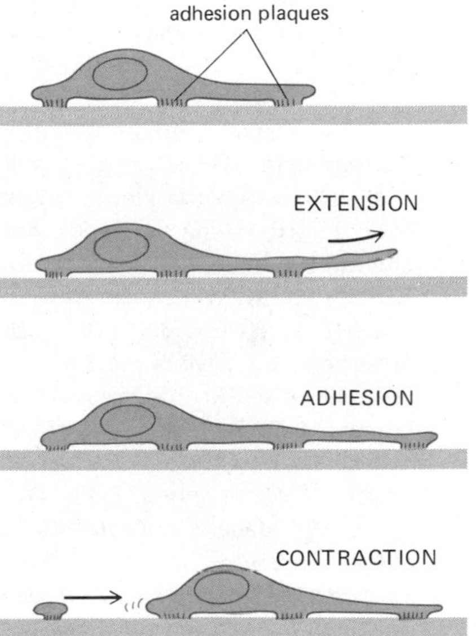

Figure 10–87 Highly schematic illustration of the principal events in the locomotion of a fibroblast on a surface. Extension, adhesion, and contraction follow each other in a cycle (see text).

Summary

Actin filaments, microtubules, intermediate filaments, and their associated proteins are regulated by unknown mechanisms to produce changes in cell shape and various cell movements. In addition, the cytoskeleton seems to organize the cytoplasm by binding various membrane-bounded organelles and soluble proteins. Microtubules emanating from the cell center determine the distribution of intermediate filaments and appear to be responsible for establishing and maintaining cell polarity. The organization of a cell's cytoskeleton can be influenced by that of its neighbors either through intercellular junctions or by the extracellular matrix, and it can be passed on to its daughters cells when the cell divides. Despite the many recent advances in dissecting the molecular composition of the various components of the cytoskeleton, the molecular mechanisms underlying most cell movements remain a mystery.

References

General

Goldman, R.; Pollard, T.; Rosenbaum, J., eds. Cell Motility. Cold Spring Harbor, N.Y.: Cold Spring Harbor Laboratory, 1976.

Inoué, S.; Stephens, R.E. Molecules and Cell Movement. New York: Raven, 1975.

Organization of the Cytoplasm. *Cold Spring Harbor Symp. Quant. Biol.* 46, 1982.

Stebbings, S.; Hyams, J.S. Cell Motility. Harlowe, Eng.: Longman, 1979.

Cited

1. Offer, G. The molecular basis of muscular contraction. In Companion to Biochemistry (A.T. Bull, J.R. Lagnado, J.O. Thomas, D.R. Tipton, eds.), pp. 623–671. New York: Longman, 1974.
   Squire, J. The Structural Basis of Muscle Contraction. New York: Plenum, 1981.
   Huxley, A.F. Reflections on Muscle. Princeton, N.J.: Princeton University Press, 1980.
2. Huxley, H.E. The mechanism of muscular contraction. *Science* 164:1356–1366, 1969.
3. Bessman, S.P.; Geiger, P.J. Transport of energy in muscle: the phosphorylcreatine shuttle. *Science* 211:448–452, 1981.
4. Katz, B. Nerve, Muscle and Synapse. New York: McGraw-Hill, 1966.
5. Murray, J.M.; Weber, A. The cooperative action of muscle proteins. *Sci. Am.* 230(2):59–71, 1974.
   Taylor, K.A.; Amos, L.A. A new model for the geometry of the binding of myosin crossbridges to muscle thin filaments. *J. Mol. Biol.* 147:297–324, 1981.
6. Adelstein, R.S.; Eisenberg, E. Regulation and kinetics of the actin-myosin-ATP interaction. *Annu. Rev. Biochem.* 49:921–956, 1980.
   Hartshorne, D.J. Phosphorylation of myosin and the regulation of smooth-muscle actomyosin. In Cell and Muscle Motility II (R.M. Dowben, J.W. Shay, eds.), pp. 188–220. New York: Plenum, 1982.
7. Satir, P. How cilia move. *Sci. Am.* 231(4):44–63, 1974.
8. Dustin, P. Microtubules. *Sci. Am.* 243(2):66–76, 1980.
   Dustin, P. Microtubules. New York: Springer-Verlag, 1978.
   Roberts, K.; Hyams, J.S., eds. Microtubules. New York: Academic Press, 1979.
9. Summers, K.E.; Gibbons, I.R. ATP-induced sliding of tubules in trypsin-treated flagella of sea-urchin sperm. *Proc. Natl. Acad. Sci. USA* 68:3092–3096, 1971.
   Warner, F.D.; Satir, P. The structural basis of ciliary bend formation. *J. Cell Biol.* 63:35–63, 1974.
10. Warner, F.D.; Mitchell, D.R. Dynein, the mechanochemical coupling adenosine triphosphatase of microtubule-based sliding filament mechanisms. *Int. Rev. Cytol.* 66:1–43, 1980.
   Satir, P.; Wais-Steider, J.; Lebduska, S.; Nasr, A.; Avolio, J. The mechanochemical cycle of the dynein arm. *Cell Motility* 1:303–327, 1981.
11. Afzelius, B.A.; Eliasson, R. Flagellar mutants in man: on the heterogeneity of the immotile-cilia syndrome. *J. Ultrastruct. Res.* 69:43–52, 1979.
   Afzelius, B.A. Genetic disorders of cilia. In International Cell Biology, 1980–1981 (H.G. Schweiger, ed.), pp. 440–447. New York: Springer-Verlag, 1981.
12. Brokaw, C.J.; Luck, D.J.L.; Huang, B. Analysis of the movement of *Chlamydomonas* flagella: the function of the radial-spoke system is revealed by comparison of wild-type and mutant flagella. *J. Cell Biol.* 92:722–732, 1982.
   Sugino, K.; Naitoh, Y. Simulated cross-bridge patterns corresponding to ciliary beating in *Paramecium. Nature* 295:609–611, 1982.
13. Wilson, L. Action of drugs on microtubules. *Life Sci.* 17:303–310, 1975.
   Schliwa, M.; Euteneuer, U.; Bulinski, J.C.; Izant, J.C. Calcium lability of cytoplasmic microtubules and its modulation by microtubule-associated proteins. *Proc. Natl. Acad. Sci. USA* 78:1037–1041, 1981.
   Schiff, P.B.; Horwitz, S.B. Taxol stabilizes microtubules in mouse fibroblast cells. *Proc. Natl. Acad. Sci. USA* 77:1561–1565, 1980.
   Inoué, S; Sato, H. Cell motility by labile association of molecules: the nature of mitotic spindle fibers and their role in chromosome movement. *J. Gen. Physiol.* 50:259–292, 1967.
14. Tilney, L.G.; Hatano, S.; Ishikawa, H.; Mooseker, M.S. The polymerization of actin: its role in the generation of the acrosomal process of certain echinoderm sperm. *J. Cell Biol.* 59:109–126, 1973.
   Carlsson, L.; Nyström, L.-E.; Sundkvist, I.; Markey, F.; Lindberg, U. Actin polymerizability is influenced by profilin, a low molecular weight protein in non-muscle cells. *J. Mol. Biol.* 115:465–483, 1977.
15. Lin, D.C.; Tobin, K.D.; Grumet, M.; Lin, S. Cytochalasins inhibit nuclei-induced actin polymerization by blocking filament elongation. *J. Cell Biol.* 84:455–460, 1980.
   Wieland, T. Modification of actins by phallotoxins. *Naturwissenschaften* 64:303–309, 1977.

Brenner, S.L.; Korn, E.D. Substoichiometric concentrations of cytochalasin D inhibit actin polymerization. *J. Biol. Chem.* 254:9982–9985, 1979.

16. Weisenberg, R.C. Microtubule formation *in vitro* in solutions containing low calcium concentrations. *Science* 177:1104–1105, 1972.

Timasheff, S.N. The *in vitro* assembly of microtubules from purified brain tubulin. *Trends Biochem. Sci.* 4:61–65, 1979.

Kirschner, M.W. Microtubule assembly and nucleation. *Int. Rev. Cytol.* 54:1–71, 1978.

Pollard, T.D.; Craig, S.W. Mechanism of actin polymerization. *Trends Biochem. Sci.* 7:55–58, 1982.

17. Woodrum, D.T.; Rich, S.A.; Pollard, T.D. Evidence for biased bidirectional polymerization of actin filaments using heavy meromyosin prepared by an improved method. *J. Cell Biol.* 67:231–237, 1975.

Bergen, L.G.; Borisy, G.G. Head to tail polymerization of microtubules *in vitro:* electron microscope analysis of seeded assembly. *J. Cell Biol.* 84:141–150, 1980.

18. Wegner, A. Head to tail polymerization of actin. *J. Mol. Biol.* 108:139–150, 1976.

Kirschner, M.W. Implications of treadmilling for the stability and polarity of actin and tubulin polymers *in vivo*. *J. Cell Biol.* 86:330–334, 1980.

Margolis, R.L.; Wilson, L. Microtubule treadmills—possible molecular machinery. *Nature* 293:705–711, 1981.

19. Kim, H.; Binder, L.I.; Rosenbaum, J.L. The periodic association of MAP-2 with brain microtubules *in vitro*. *J. Cell Biol.* 80:266–276, 1979.

Amos, L.A. Arrangement of high molecular weight associated proteins on purified mammalian brain microtubules. *J. Cell Biol.* 72:642–654, 1977.

Connolly, J.A.; Kalnins, V.I. The distribution of tau and HMW microtubule-associated proteins in different cell types. *Exp. Cell Res.* 127:341–350, 1980.

20. Weber, K.; Osborn, M. Intracellular display of microtubular structures revealed by indirect immunofluorescence microscopy. In Microtubules (K. Roberts, J.S. Hyams, eds.), pp. 279–313. New York: Academic Press, 1979.

21. Heidemann, S.R.; Kirschner, M.W. Aster formation in eggs of *Xenopus laevis:* induction by isolated basal bodies. *J. Cell Biol.* 67:105–117, 1975.

22. Kuriyama, R.; Borisy, G.G. Centriole cycle in Chinese hamster ovary cells as determined by whole-mount electron microscopy. *J. Cell Biol.* 91:814–821, 1982.

Sorokin, S.P. Reconstructions of centriole formation and ciliogenesis in mammalian lungs. *J. Cell Sci.* 3:207–230, 1968.

Wheatley, D.N. The Centriole: A Central Enigma of Cell Biology. New York: Elsevier, 1982.

23. Schliwa, M. Proteins associated with cytoplasmic actin. *Cell* 25:587–590, 1981.

Korn, E.D. Actin polymerization and its regulation by proteins from nonmuscle cells. *Physiol. Rev.* 62:672–737, 1982.

Weeds, A. Actin-binding proteins—regulators of cell architecture and motility. *Nature* 296:811–816, 1982.

24. Hirokawa, N.; Heuser, J.E. Quick-freeze, deep-etch visualization of the cytoskeleton beneath surface differentiations of intestinal epithelial cells. *J. Cell Biol.* 91:399–409, 1981.

Mooseker, M.S.; Tilney, L.G. Organization of an actin-filament-membrane complex: filament polarity and membrane attachment in the microvilli of intestinal epithelial cells. *J. Cell Biol.* 67:725–743, 1975.

Matsudaira, P.T.; Burgess, D.R. Organization of the cross-filaments in intestinal microvilli. *J. Cell Biol.* 92:657–664, 1982.

25. DeRosier, D.J.; Tilney, L.G.; Egelman, E. Actin in the inner ear: the remarkable structure of the stereocilium. *Nature* 287:291–296, 1980.

26. Albrecht-Buehler, G.; Goldman, R.D. Microspike-mediated particle transport towards the cell body during early spreading of 3T3 cells. *Exp. Cell Res.* 97:329–339, 1976.

Albrecht-Buehler, G. Filopodia of spreading 3T3 cells: do they have a substrate-exploring function? *J. Cell Biol.* 69:275–286, 1976.

Burgess, D.R.; Schroeder, T.E. Polarized bundles of actin filaments within microvilli of fertilized sea urchin eggs. *J. Cell Biol.* 74:1032–1037, 1977.

27. Lazarides, E.; Weber, K. Actin antibody: the specific visualization of actin filaments in non-muscle cells. *Proc. Natl. Acad. Sci. USA* 71:2268–2272, 1974.

Sanger, J.M.; Sanger, J.W. Banding and polarity of actin filaments in interphase and cleaving cells. *J. Cell Biol.* 86:568–575, 1980.

28. Scholey, J.M.; Taylor, K.A.; Kendrick-Jones, J. Regulation of non-muscle myosin assembly by calmodulin-dependent light chain kinase. *Nature* 287:233–235, 1980.

Adelstein, R.S. Calmodulin and the regulation of the actin-myosin interaction in smooth muscle and nonmuscle cells. *Cell* 30:349–350, 1982.

29. Heath, J.P.; Dunn, G.A. Cell to substratum contacts of chick fibroblasts and their relation to the microfilament system. A correlated interference-reflexion and high-voltage electron-microscope study. *J. Cell Sci.* 29:197–212, 1978.

Geiger, B.; Tokuyasu, K.T.; Dutton, A.H.; Singer, S.J. Vinculin, an intracellular protein localized at specialized sites where microfilament bundles terminate at cell membranes. *Proc. Natl. Acad. Sci. USA* 77:4127–4131, 1980.

30. Wehland, J.; Osborn, M.; Weber, K. Phalloidin-induced actin polymerization in the cytoplasm of cultured cells interferes with cell locomotion and growth. *Proc. Natl. Acad. Sci. USA* 74:5613–5617.

31. Taylor, D.L.; Condeelis, J.S. Cytoplasmic structure and contractility in amoeboid cells. *Int. Rev. Cytol.* 56:57–144, 1979.

32. Wang, K. Filamin, a new high-molecular weight protein found in smooth muscle and nonmuscle cells. Purification and properties of chicken gizzard filamin. *Biochemistry* 16:1857–1865, 1977.

33. Yin, H.L.; Stossel, T.P. Control of cytoplasmic actin gel-sol transformation by gelsolin—a calcium-dependent regulatory protein. *Nature* 281:583–586, 1979.

34. Stossel, T.P.; Hartwig, J.H. Interactions of actin, myosin, and a new actin-binding protein of rabbit pulmonary macrophages: role in cytoplasmic movement and phagocytosis. *J. Cell Biol.* 68:602–619, 1976.

35. Lazarides, E. Intermediate filaments as mechanical integrators of cellular space. *Nature* 283:249–256, 1980.

Steinert, P.M. Structure of the three-chain unit of the bovine epidermal keratin filament. *J. Mol. Biol.* 123:49–70, 1978.

Anderton, B.H. Intermediate filaments: a family of homologous structures. *J. Muscle Res. Cell Motility* 2:141–166, 1981.

36. Nelson, W.J.; Traub, P. Properties of a Ca^{2+}-activated protease specific for the intermediate-sized filament protein vimentin in Ehrlich Ascites tumour cells. *Eur. J. Biochem.* 116:51–57, 1981.

37. Lasek, R.J.; Krishnan, N.; Kaiserman-Abramof, I.R. Identification of the subunit proteins of 10nm neurofilaments isolated from axonplasm of squid and *Myxicola* giant axons. *J. Cell Biol.* 82:336–346, 1979.

Franke, W.W.; Schmid, E.; Osborn, M.; Weber, K. Different intermediate-sized filaments distinguished by immunofluorescence microscopy. *Proc. Natl. Acad. Sci. USA* 75:5034–5038, 1978.

Franke, W.W.; Schmid, E.; Winter, S.; Osborn, M.; Weber, K. Widespread occurrence of intermediate-sized filaments of the vimentin-type in cultured cells from diverse vertebrates. *Exp. Cell Res.* 123:25–46, 1979.

38. Sun, T.-T.; Shih, C.; Green, H. Keratin cytoskeletons in epithelial cells of internal organs. *Proc. Natl. Acad. Sci. USA* 76:2813–2817, 1979.

39. Klymkowsky, M.W. Intermediate filaments in 3T3 cells collapse after intracellular injection of a monoclonal anti-intermediate filament antibody. *Nature* 291:249–251, 1981.

40. Cohen, C. Cell architecture and morphogenesis. I. The cytoskeletal proteins. *Trends Biochem. Sci.* 4:73–77, 1979.

Cohen, C. Cell architecture and morphogenesis. II. Examples in embryology. *Trends Biochem. Sci.* 4:97–101, 1979.

Oliver, J.M.; Berlin, R.D. Mechanisms that regulate the structural and functional architecture of cell surfaces. *Int. Rev. Cytol.* 74:55–94, 1982.

41. Heuser, J.; Kirschner, M.W. Filament organization revealed in platinum replicas of freeze-dried cytoskeletons. *J. Cell Biol.* 86:212–234, 1980.

Wolosewick, J.J.; Porter, K.R. Microtrabecular lattice of the cytoplasmic ground substance: artifact or reality. *J. Cell Biol.* 82:114–139, 1979.

Small, J.V. Organization of actin in the leading edge of cultured cells: influence of osmium tetroxide and dehydration on the ultrastructure of actin meshworks. *J. Cell Biol.* 91:695–705, 1981.

Schliwa, M.; van Blerkom, J. Structural interaction of cytoskeletal components. *J. Cell Biol.* 90:222–235, 1981.

42. Lasek, R.J. The dynamic ordering of neuronal cytoskeletons. In Cytoskeletons and the Architecture of Nervous Systems. *Neurosci. Res. Program Bull.* 19:7–32, 1981.

Masters, C.J. Interactions between soluble enzymes and subcellular structure. *Trends Biochem. Sci.* 3:206–208, 1978.

Fulton, A.B. How crowded is the cytoplasm? *Cell* 30:345–347, 1982.

43. Hunter, T. Proteins phosphorylated by the RSV transforming function. *Cell* 22:647–648, 1980.

Bishop, J.M. Oncogenes. *Sci. Am.* 246(3):68–78, 1982.

44. Byers, B.; Porter, K.R. Oriented microtubules in elongating cells of the developing lens rudiment after induction. *Proc. Natl. Acad. Sci. USA* 52:1091–1099, 1964.

Vasiliev, J.M.; et al. Effect of colcemid on the locomotory behaviour of fibroblasts. *J. Embryol. Exp. Morphol.* 24:625–640, 1970.

Albrecht-Buehler, G.; Bushnell, A. The orientation of centrioles in migrating 3T3 cells. *Exp. Cell Res.* 120:111–118, 1979.

Gotlieb, A.I.; May, L.M.; Subrahmanyan, L.; Kalnins, V.I. Distribution of microtubule organizing centers in migrating sheets of endothelial cells. *J. Cell Biol.* 91:589–594, 1981.

45. Carter, S.B. Principles of cell motility: the direction of cell movement and cancer invasion. *Nature* 208:1183–1187, 1965.

Dunn, G.A.; Heath, J.P. A new hypothesis of contact guidance in tissue cells. *Exp. Cell Res.* 101:1–14, 1976.

Benecke, B.-J.; Ben-Ze'ev, A.; Penman, S. The control of mRNA production, translation and turnover in suspended and reattached anchorage-dependent fibroblasts. *Cell* 14:931–939, 1978.

46. Albrecht-Buehler, G. Daughter 3T3 cells. Are they mirror images of each other? *J. Cell Biol.* 72:595–603, 1977.

Aufderheide, K.J.; Frankel, J.; Williams, N.E. Formation and positioning of surface-related structures in protozoa. *Microbiol. Rev.* 44:252–302, 1977.

Freeman, G. The multiple roles which cell division can play in the localization of developmental potential. In Determinants of Spatial Organization (S. Subtelny, ed.), pp. 53–76. New York: Academic Press, 1979.

47. Bellairs, R.; Curtis, A.; Dunn, G., eds. Cell Behaviour. Cambridge, Eng.: Cambridge University Press, 1982.

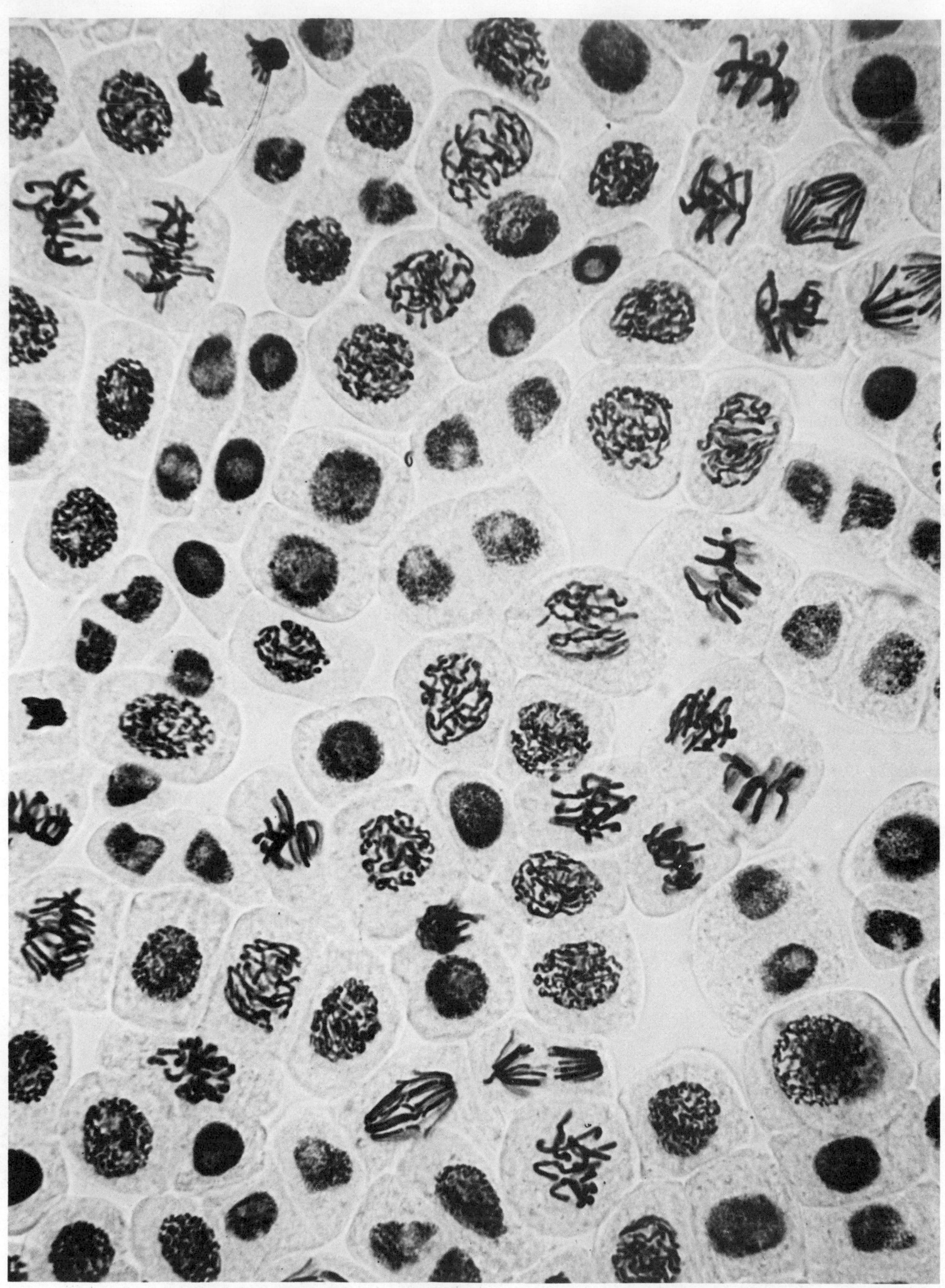

Cells from a root tip of a plant in various stages of cell division.
(Courtesy of John McLeish.)

Cell Growth and Division

11

As highly organized units in a universe favoring disorder, cells are subject to wear and tear as well as to accidents. Any individual cell is therefore bound to die. If an organism is to continue to live, it must create new cells at a rate as fast as that at which its cells die. For this reason, cell division is central to the life of all organisms. In an adult human, for example, millions of cells must divide every second simply to maintain the status quo.

The process of cell division itself is strikingly visible in the microscope; it consists of two sequential processes: nuclear division (called **mitosis**) and cytoplasmic division (called **cytokinesis**). But before a typical cell can divide, it must double its mass and duplicate all of its contents. Only in this way will the two new daughter cells contain all of the components that they need to begin their own cycle of cell growth followed by division. Most of the work involved in preparing for division goes on invisibly during the growth phase of the cell cycle, which is, quite misleadingly, denoted as **interphase.**

Although a cell spends most of its lifetime in interphase and only occasional periods in the cell-division phase, most early work on the cell cycle focused on the brief division events (mitosis and cytokinesis), largely because they could be studied by direct microscopic examination. More recently, through the use of more indirect and sophisticated techniques, we have learned a considerable amount about the interphase part of the cell cycle as well. In this chapter we shall describe some of the methods currently used to study the cell cycle, consider cell-cycle regulation, and discuss several of the main events occurring during each of its different phases. Although our knowledge of the molecular basis of the cell cycle is fragmentary, wherever possible we shall try to discuss the mechanisms that are likely to be involved.

The Control of Cell Division[1,2]

Most cell components are made continuously throughout the interphase period between cell divisions. It is, therefore, difficult to define distinct stages in the progression of the growing cell through interphase. One outstanding exception is DNA synthesis, since the DNA in the cell nucleus is replicated only

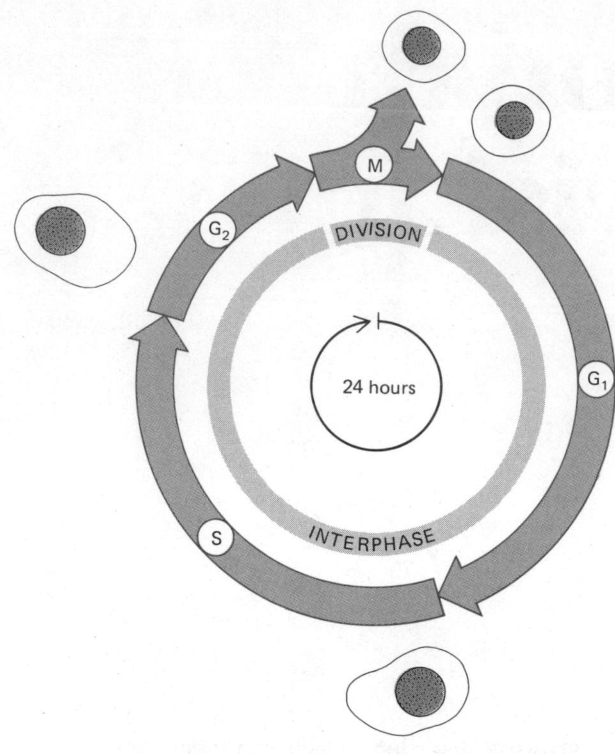

Figure 11–1 The four successive phases of the cell cycle. After the *M phase,* which consists of nuclear division (mitosis) and cytoplasmic division (cytokinesis), the daughter cells begin interphase of a new cycle. Interphase starts with the *G_1 phase,* in which the cells, whose biosynthetic activities have been greatly slowed during mitosis, resume a high rate of biosynthesis. The *S phase* begins when DNA synthesis starts, and ends when the DNA content of the nucleus has doubled and the chromosomes have replicated (each chromosome is now said to consist of two identical "sister chromatids"). The cell then enters the *G_2 phase,* which ends when mitosis starts. The M phase begins with mitosis (hence the "M") and ends with cytokinesis. During the early part of the M phase, the replicated chromosomes condense from their extended interphase condition and are easily seen in the light microscope. The nuclear envelope breaks down, and each chromosome undergoes precisely orchestrated movements that result in the separation of its pair of sister chromatids as the nuclear contents are divided. Two new nuclear envelopes then form, and the cytoplasm divides to generate two daughter cells, each with a single nucleus. This process of cytokinesis terminates the M phase and marks the beginning of the interphase of the next cell cycle. Although a typical 24-hour cycle is illustrated here, cell-cycle times in higher eucaryotic cells vary widely, with most of the variability being in the length of the G_1 phase (see text).

during a limited portion of interphase. This period is denoted as the **S phase** (S = synthesis) of the cell cycle. The other distinct stage of the cycle is, of course, the cell-division phase, which includes both nuclear division (mitosis) and the cytoplasmic division (cytokinesis) that follows. The entire cell-division phase is denoted as the **M phase** (M = mitotic). This leaves the period between the M phase and the start of DNA synthesis, which is called the **G_1 phase** (G = gap), and the period between the completion of DNA synthesis and the next M phase, which is called the **G_2 phase.** Interphase is thus composed of successive G_1, S, and G_2 phases, and it normally comprises 90% or more of the total cell cycle time. For example, in rapidly dividing cells of higher eucaryotes, the successive cell divisions (M phases) that interrupt interphase generally occur only once every 16 to 24 hours, and each M phase itself lasts only 1 to 2 hours. A typical cell cycle with its four successive phases is illustrated in Figure 11–1, and some of the major sequential events are outlined in the legend.

The Cells in a Multicellular Organism Divide at Very Different Rates[3]

In unicellular organisms, such as bacteria and protozoa, there is a strong selective pressure for each individual cell to grow and divide as rapidly as possible. For this reason the rate of cell division is generally limited only by the rate at which nutrients can be taken up from the medium and converted to cellular materials. The situation in multicellular animals is quite different. To varying degrees, different cell types have given up their potential for rapid division so that their numbers can be kept at a level that is optimal for the organism as a whole: it is the survival of the organism that is paramount, not the survival of any of its individual cells. As a result, the 10^{13} cells of the human body divide at very different rates. Some cells, such as neurons, skeletal-muscle cells, and red blood cells, do not divide at all once they are mature. Other cells, such as the epithelial cells that line the inside and outside surfaces of the body (for example, in the intestine, lung, and skin) divide continuously and rapidly throughout the life of the organism. Some of these cells go through their entire growth and division cycle in as little as 8 hours. However, the behavior of most animal cells falls somewhere between these two extremes: although they can divide, they do so only rarely. The observed cell-cycle times (also called *generation times*) range from 8 hours to 100 days or more.

These great differences in the division rates of cells in different tissues can be observed and the cell-cycle times measured by exploiting the technique of *autoradiography.* Autoradiography provides a way of specifically marking those cells that are synthesizing DNA in preparation for division. An animal is repeatedly injected with tritiated thymidine, a radioactive precursor of a compound every cell uses exclusively for the synthesis of DNA. At a certain time after the injections, tissues are removed from the animal, washed to free them from unincorporated thymidine, fixed for microscopy, and cut into thin slices about one cell thick. The tissue slices are then covered with a thin film of photographic emulsion and exposed for days or weeks, after which they are developed like ordinary photographic film. Those cells that have synthesized DNA at any time during the labeling period (and thus have been in S phase) can be identified by the developed silver grains that appear over their nuclei (Figure 11–2). From the fraction of cells labeled in this way following successively longer exposures of an animal to radioactive thymidine, it is possible to estimate the period between successive S phases. Experiments of this kind demonstrate that cell-cycle times differ strikingly, even among cells in the same tissue. For example, Figure 11–3 illustrates how these times differ among the epithelial cells of the intestinal lining, depending on the position of the cells in the epithelial cell sheet.

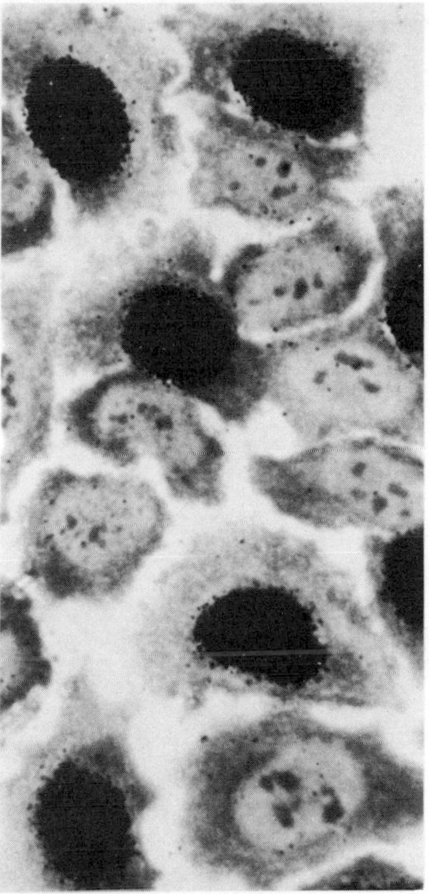

Figure 11–2 An autoradiograph obtained after a short period of incubation of cells grown in tissue culture with tritiated thymidine. The presence of exposed silver grains over a cell nucleus (blackened area) indicates that the cell incorporated radioactive thymidine into its DNA, and thus was in S phase, sometime during the labeling period. (Courtesy of James Clever.)

Differences in Cell-Cycle Times Are Due Mainly to Variations in the Length of G_1[1,4]

The principal difference between cells that divide rapidly and those that divide slowly is the length of time they spend in the G_1 phase of the cell cycle. Some cells divide very slowly, staying in G_1 for days or even years. By contrast, the time taken for a cell to progress from the beginning of S through mitosis is remarkably constant, irrespective of its rate of division.

Much more detailed measurements of the cell cycle can be made when cells are growing in culture, where their environment is readily controlled and manipulated. Cell division in culture can be slowed or stopped by limiting the supply of essential nutrients, by depriving the cells of essential protein growth factors, by adding low levels of protein-synthesis inhibitors, or by allowing the cells to become overcrowded. In every instance, the cell cycle is arrested in the G_1 phase. This finding implies that once a cell has passed out of G_1, it is committed to completing the S, G_2, and M phases. In fact, experiments have shown that the point of no return—known as the **restriction point** (**R point,** or **R**)—occurs late in G_1. After cells have passed this point, they will complete the rest of the cycle at their normal rate regardless of external conditions (Figure 11–4).

Determination of Cell-Cycle Times[5,6]

How does one determine the length of each phase of the cell cycle? First it is necessary to establish the length of the total cell cycle, which can easily be done in a homogeneous population of cultured cells by periodically counting the number of cells present under a microscope and recording the number

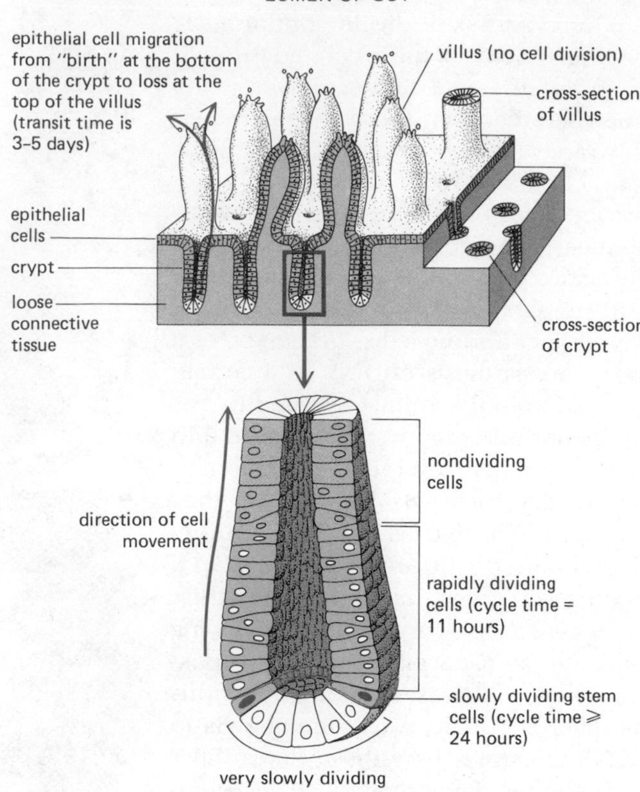

Figure 11–3 Schematic representation of cell division and migration in the epithelium lining the small intestine of the mouse. All cell division is confined to the bottom portion of the flask-shaped epithelial infoldings known as "crypts." Newly generated cells move upward to form the epithelium that covers the villi, which function in the digestion and absorption of foodstuffs from the lumen of the gut. Most epithelial cells have a very short lifetime, being shed from the tip of a villus within five days of emerging from the crypt. However, a ring of about 20 slowly dividing "immortal" cells (shown in dark color) remain anchored in a unique niche near the base of each crypt. These "stem cells" will usually divide to give rise to two daughter cells: one that remains in place as an undifferentiated stem cell and one that migrates upward to differentiate and join the villus epithelium. Occasionally, however, two differentiated daughter cells will be produced from a single stem-cell division. The resulting deficiency in stem cells is believed to be made up by the divisions of one or a few of the stem cells in the ring; such a cell is thought to divide even more slowly than its neighbors and to give rise to two stem cells when it divides. Experiments in which the normal organization of the intestinal epithelium is disturbed reveal that the large differences in cell-cycle times found there are determined in part by the relative position of an epithelial cell in the tissue and in part by a cell's developmental history. (Adapted from C. S. Potten, R. Schofield, and L. G. Lajtha, *Biochim. Biophys. Acta* 560:281–299, 1979.)

of hours required for the total cell number to double. (Alternatively, the total cell mass can be monitored.) Once this interval is known, the length of the S phase can be estimated by adding tritiated thymidine to the tissue culture medium for a brief period that is much shorter than the S phase itself (typically, 30 minutes or less). The cells are then prepared for autoradiography, and the fraction of the cells that have incorporated the radioisotope into their DNA is determined by counting the fraction of cells with exposed silver grains over their nucleus (see Figure 11–2). As explained in Figure 11–5, this fraction multiplied by the total cell-cycle time is roughly equal to the average length of the S phase in the population.

The length of the M phase can be determined in an analogous fashion by scanning the cell population by light microscopy and determining the fraction of cells containing condensed chromosomes at any one time (this fraction is known as the **mitotic index**). Different small correction factors must be used in these S-phase and M-phase calculations, to allow for the fact that there are always more "young" cells than "old" cells in a continuously growing population ("young" and "old" referring to the time elapsed since the last cell

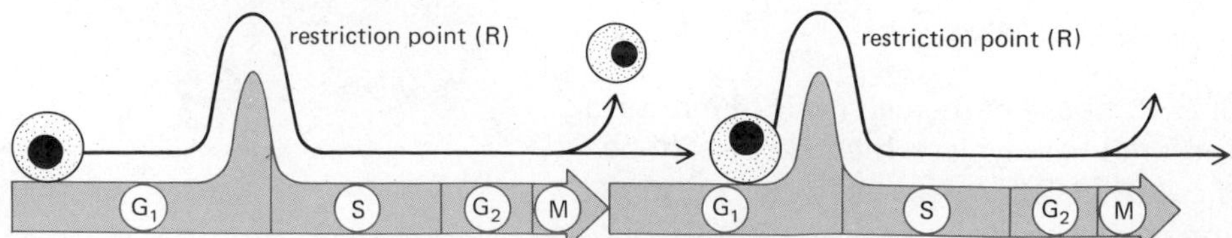

Figure 11–4 Schematic illustration showing that cells normally stop dividing when they reach a point late in G_1, called the restriction point (R), unless signaled to go through another whole cycle.

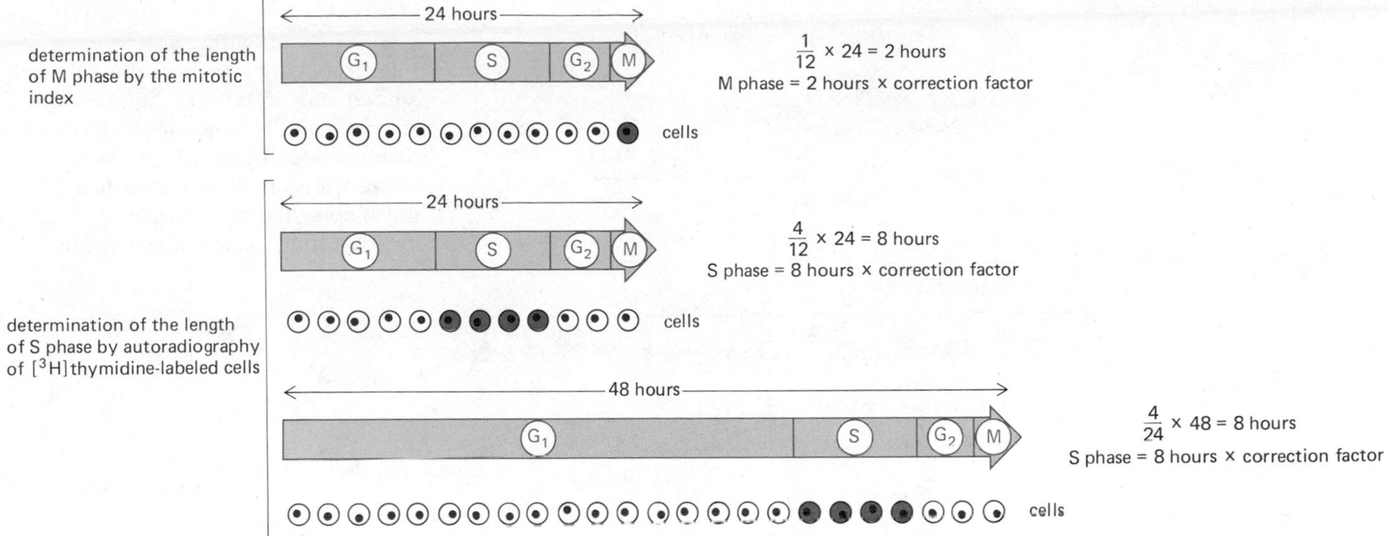

Figure 11–5 Illustration of the general principle that the length of each phase of the cell cycle is approximately equal to the fraction of the cells in that phase at any instant multiplied by the total cell-cycle time. This calculation is based on the assumption that all the cells in the population are growing at the same rate. The indicated "correction factors" range from 0.7 for early G_1 cells to 1.4 for mitotic cells, with an intermediate value for S-phase cells. A correction factor is needed because there are always more young cells than old cells in a continuously dividing population. The exact age distribution is given by the equation $y = 2^{(1-x)}$, where y (varying from 2 to 1) is the relative number of cells at cell-cycle age x (where x varies from 0 for the earliest G_1 cells to 1 for the latest M-phase cells).

division). For example, because a single old cell becomes two young cells as soon as it divides, there will be twice as many cells in early G_1 as in late M phase.

Measuring the lengths of the G_1 and G_2 phases of the cycle is more complicated because there is no way of specifically marking G_1- or G_2-phase cells. This problem is overcome by synchronizing the growth of cells in culture and measuring the time that elapses between phases that can be specifically marked. The simplest way of acquiring a synchronously growing population of mammalian cells is by taking advantage of the fact that cells in M phase undergo changes in their cytoskeleton that cause the cells to "round up." Consequently, when cells are grown on the surface of a tissue-culture dish, the M-phase cells adhere so weakly to the dish that they can be removed by gentle agitation (Figure 11–6).

Mitotic cells that have been freshly collected by removing them from the dish in this way constitute a *synchronous cell population* that will almost immediately enter the G_1 phase of their cycle. The time between the collection of the cells and the first substantial incorporation of tritiated thymidine into DNA is equal to the length of the G_1 phase. While in principle the length of the G_2 phase of the cycle could be estimated by continuing to follow this synchronized cell population, the cells tend to become asynchronous with time. A more accurate measure of the length of G_2 can be obtained indirectly by studying asynchronously growing cells using the method described in Figure 11–7. Alternatively, the length of the G_2 phase can be estimated by subtracting the G_1, S, and M intervals from the total cell-cycle time, once their values are known.

Cell-cycle analyses have been made much easier in recent years through the use of a product of modern electronics called the *fluorescence-activated cell analyzer*. In this complex machine, a cell suspension is forced through a fine nozzle at a rate of several thousand cells per second, and an optical measurement is made and recorded for each individual cell as it briefly passes a tiny window (see p. 166). An asynchronously growing cell population is analyzed by treating the cells with a fixative (to arrest cell division and make their membranes permeable) and then staining them with a dye that becomes fluorescent only when it binds to DNA. When a cell is treated in this way, the

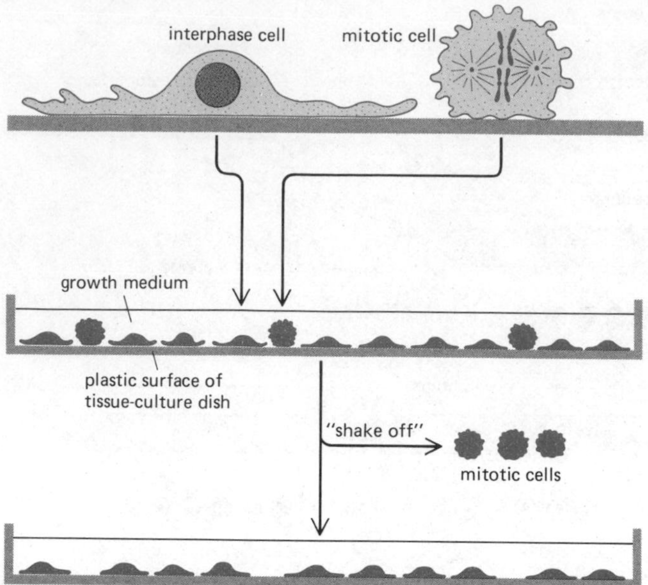

Figure 11–6 A commonly used method for obtaining a synchronously growing population of animal cells in culture. Mitotic cells are collected by shaking them off the dish on which they are growing. When transferred to a new dish, these collected cells continue through their cycles in synchrony.

intensity with which it fluoresces is directly proportional to the amount of DNA that it contains. By passing such cells through the fluorescence analyzer, one can rapidly determine the relative fluorescence of a large number of cells and, therefore, their relative amounts of DNA. Those cells with the least amount of DNA are in G_1, those with double this amount are either in G_2 or M, while cells in S have intermediate amounts (shaded area in Figure 11–8). The length of the G_1, G_2 plus M, and S phases of the cell cycle can be readily calculated from the fraction of cells in each of these three categories (see Figure 11–5).

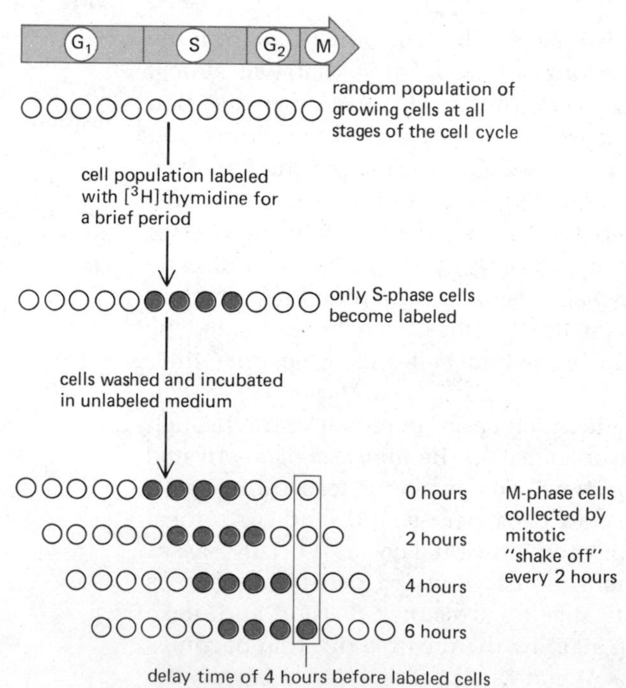

Figure 11–7 A method commonly used to measure the length of the G_2 phase. An asynchronously growing cell population is given a brief pulse of tritiated thymidine. After washing away the excess tritiated thymidine, mitotic cells are collected by the shake-off method at various time intervals, and the time at which radioactively labeled DNA is first found in these mitotic cells is determined. The "delay time" is equal to the length of the G_2 phase. Alternatively, autoradiography can be used after the same intervals to determine the fraction of cells with thymidine-labeled nuclei that also contain condensed mitotic chromosomes. Only the autoradiographic method is applicable to cells in tissues.

A Special "Trigger Protein" May Control Cell Division[1,7]

It is not known what determines whether a mammalian cell will go beyond the restriction point (R) in G_1 and start a new cycle. One plausible hypothesis is based on the observation that cells behave as if they need to accumulate a threshold amount of some unstable *trigger protein* (also called *U protein,* for unstable protein) in order to pass through R and thereby be triggered to make DNA and divide. Because of its instability, this hypothetical protein would reach a concentration great enough to initiate a cycle of cell division only when synthesized relatively rapidly. In addition, its concentration would drop precipitously during the M phase, when protein synthesis is greatly reduced, and would build up toward the threshold level in G_1.

Such a trigger-protein model for the control of cell growth may be an oversimplified view of the actual mechanism involved, but it has been useful for interpreting observed cell behavior. For example, according to this model, any condition that reduces the general rate of protein synthesis should delay the accumulation of threshold levels of the U protein, thus lengthening G_1 and reducing the rate of cell division. In fact, when cells are cultured in the presence of varying concentrations of protein-synthesis inhibitors, their cell-cycle times are greatly extended without a substantial change in the time required to pass through S, G_2, or M phases. The observed increases in the length of G_1 fit with the results predicted from the model, provided that each molecule of U protein remains active in the cell for only a few hours. This model also accounts for the growth-inhibiting effects of cell crowding or starvation, which are known to depress protein synthesis rates and arrest the cell at the most sensitive point in G_1, which is R.

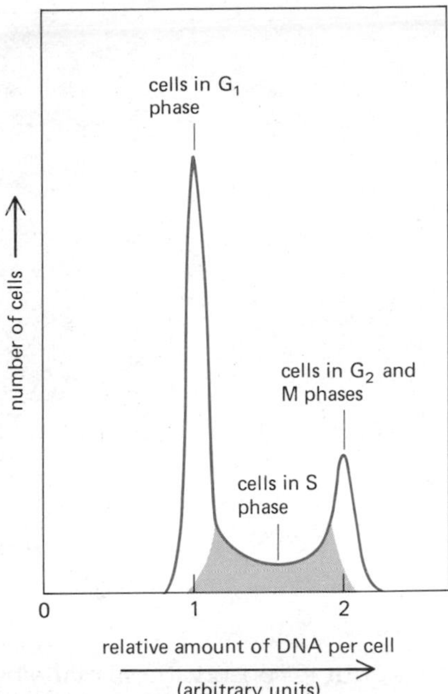

Figure 11–8 Typical results obtained for a growing cell population when the DNA content of its individual cells is determined with a fluorescence analyzer. DNA amounts are given in arbitrary units and are determined by the amount of fluorescent dye bound. The fact that the number of cells in G_1 is much greater than the number of cells in G_2 and M indicates that the G_1 phase is longer than the G_2 phase in this population (see Figure 11–5).

How might we account for the action of the specific growth factors that cause only the appropriate cells in tissues to divide? First, it is important to note that cells arrested at R not only fail to synthesize DNA and divide, but they also stop growing. However, this does not mean that they stop biosynthesis completely. In particular, the normal degradation (or "turnover") of proteins in a mammalian cell (p. 334) is so extensive that a growth-arrested cell must have a protein-synthesis rate that is about one-fifth of the growing-cell rate just to maintain the status quo.

It has been proposed that the growth-control mechanisms in tissues act directly on the overall rate of protein synthesis in a cell. According to this view, in the absence of specific stimulatory factors (and/or in the presence of inhibitory factors) cells will synthesize proteins only at some low "maintenance rate." Although proteins with average turnover rates can thereby be kept at the same concentrations as in growing cells, the concentration of any very unstable protein (including the hypothetical U protein) will be reduced in direct proportion to the reduction in its rate of synthesis. On the other hand, in environmental conditions that stimulate an increase in the general rate of protein synthesis, the amount of U protein will increase above the threshold level, allowing the cells to pass through the restriction point (R) in their cycle and to divide (Figure 11–9).

Cultured cells whose division has been arrested at R can remain viable and healthy for long periods, even if severely starved, whereas cells starved at random points in their cycle usually die. This suggests that the growth-control mechanism involving a specific restriction point may have evolved partly because of the need for a safe resting state (at R) for cells whose growth conditions or interactions with other cells demand that they stop dividing. Cells that have been arrested in this stable resting state are sometimes said to have entered a "G_0 phase" of the cell cycle.

The main alternative to the trigger-protein model of growth control is the so-called "transition probability" model. This type of model was proposed

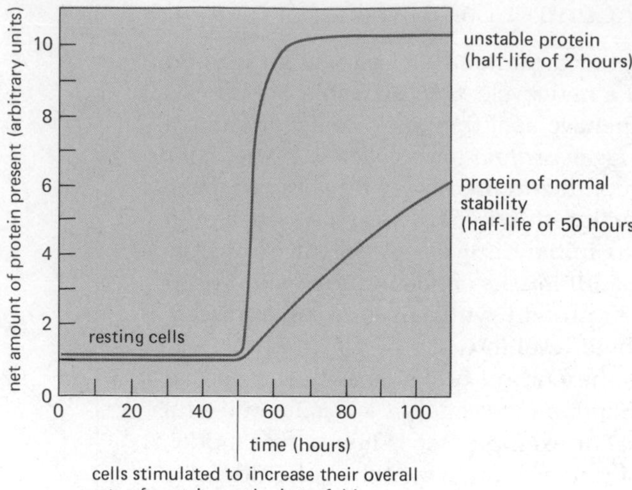

Figure 11–9 Illustration of the principle that the relative amount of highly unstable protein present in a cell will increase compared to the amount of an average protein when the cell is stimulated to synthesize all of its proteins more rapidly (see also p. 750). Because cell growth and division will keep pace with the *average* increase in the amount of cellular protein (*lower curve*), an unstable protein will constitute a much larger fraction of the total cell protein in a growing cell than in a nongrowing cell (and therefore be present at a higher concentration).

to account for observations made by time-lapse cinematography of cloned cell lines growing under uniform conditions in culture. Although such cells are genetically identical, they show a wide range of cell-cycle times. A typical distribution of such times (Figure 11–10) has the appearance that would be expected if cycle times were regulated by a probabilistic or stochastic event. In other words, it appears as though every cell has a constant probability per unit time of passing the restriction point R in its cycle, independently of how much time has elapsed since its last division. The initiation of the S phase of the cycle would thus be a random process, analogous to the radioactive decay of an unstable atom. It is worth noting, however, that the large variation in cell-cycle times in Figure 11–10 can also be accommodated in the more biologically reasonable trigger-protein model, since it has been demonstrated that even genetically identical G_1 cells vary to a surprising degree in the rate at which they synthesize protein.

Cell Division Is Regulated by a Variety of "Feedback Control" Mechanisms

The rate of cell division in tissues is controlled by unknown mechanisms that allow cells to divide if, and only if, new cells are needed. For example, normally quiescent liver cells are stimulated to divide rapidly after part of the liver has been removed, and they stop dividing as soon as the normal liver mass has been restored (see p. 905). The same type of limited cell division is seen in the skin following injury. Without such "feedback control" of cell division, the form and function of a multicellular animal would be quickly destroyed—either by excessive cell division (as occurs in cancer) or by a failure to replace the dead cells in tissues that normally experience continual cell losses (such as epithelial cell sheets). Similar regulative mechanisms are also important for the orderly development of cells and tissues during embryogenesis (see Chapters 15 and 16).

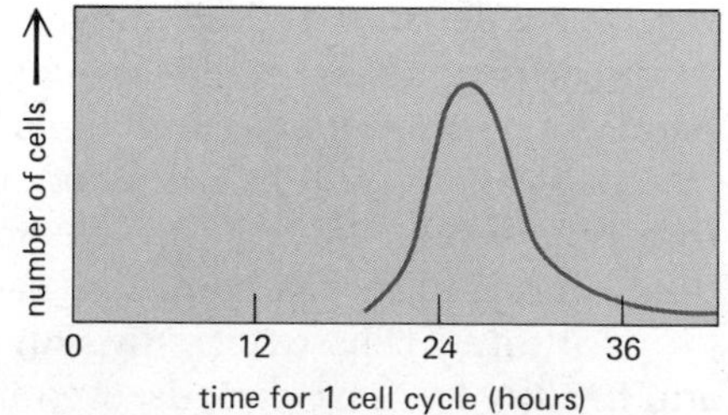

Figure 11–10 A plot illustrating the variation in cell-cycle time typically observed for a homogeneous population of cells growing in tissue culture. Such data are obtained by observing individual cells in the microscope and directly measuring the time between their successive divisions.

Normal Mammalian Cells Grown in Culture Will Stop Dividing When They Run Out of Space[8]

When fibroblasts are plated on dishes containing tissue-culture medium, they adhere to the surface, spread out, and divide. Normally, this process will continue until a confluent monolayer is formed in which neighboring cells touch one another. At this point, with no more free space in which to spread, normal cells stop dividing—a phenomenon known as *contact inhibition of cell division*. If such a monolayer is "wounded" with a needle so as to create a cell-free line on the dish, the cells at the edges of the line spread into the empty space and divide (Figure 11–11).

The term "contact inhibition" may, however, be misleading. Other experiments have been carried out in which the extent of cell spreading has been controlled by altering the adhesiveness of the surfaces on which cells are grown, rather than by cell crowding. The results obtained suggest that the important determinant of cell division in culture is not cell contact but the degree of cell spreading—for even without cell contact, the less spread out a cell is, the longer is its growth cycle.

The rounding up of a cell is associated with a decrease in its total rate of protein synthesis. The experiment diagrammed in Figure 11–11 could, therefore, be interpreted in terms of a trigger-protein mechanism of growth control as follows: when a cell contacts other cells, the limitation thus set on its ability to spread decreases the rate at which it synthesizes protein, including U protein. The concentration of U protein in the cell therefore falls, causing the cell to become arrested at the R point in G_1. However, cells at the boundary of a wound are released from this growth inhibition, inasmuch as their rate of protein synthesis increases as they spread out, causing the concentration of the U protein to rise above threshold.

Why should changes in the shape of a cell change its rate of protein synthesis? Possibly, the large increase in the ratio of cell-surface area to cell volume that can accompany cell flattening influences protein synthesis. Alternatively, increased rates of protein synthesis could be caused by the major changes in the organization of the cytoskeleton that occur whenever a cell flattens.

If these observations on cultured cells also apply to cells in tissues, one factor likely to help control cell division in organisms is the contact between adjacent cells that limits their ability to spread. Conversely, empty spaces created either by wounding or by natural cell loss will stimulate cell division directly by allowing the neighboring cells to spread into the vacated space. However, this cannot fully explain growth control in tissues, as we shall see.

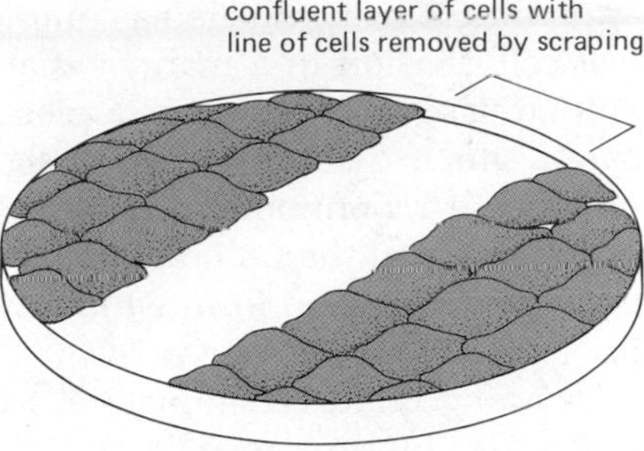

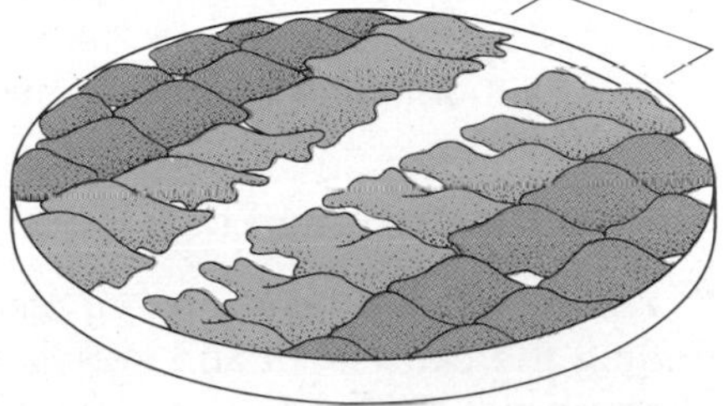

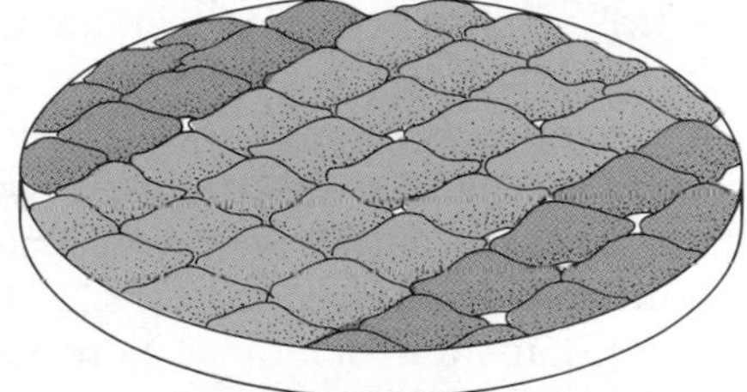

Figure 11–11 One interpretation of a classical experiment in which cultured cells stop dividing when they have formed a confluent monolayer but are stimulated to grow again next to a wound created by scraping.

Positional Signals Also Control Cell Division[9]

In general, cells in tissues divide only when they are in their proper environment, whether or not empty spaces are available. For example, cell division in the multilayered skin epidermis is usually limited to the single layer of basal epidermal cells that sit on the basal lamina overlying the dermis (see p. 912). These dividing cells generate daughter cells that are forced to move outward in the epidermis toward the surface of the skin. As the daughter cells migrate, they stop dividing and start making large amounts of keratin filaments, which accumulate in the cytoplasm, thereby forming a tough protective surface layer. If the basal cells are either injected into the underlying dermis or raised so as to lose contact with the basal lamina, they stop dividing. It seems that these epidermal cells divide only if they are stimulated by some unknown factor available at the border of the dermis and epidermis. This factor could be either a diffusible molecule or part of the basal lamina.

Although they are not understood in detail, such positional requirements for cell division seem to be quite general. For example, in mammals, the developing cells of the embryonic pancreatic epithelium can be shown to require a specific protein factor secreted from the underlying pancreatic mesenchyme for normal division and development. More strikingly, elegant tissue-transplantation experiments carried out with the developing epithelium of insect appendages have revealed a complex pattern of cell-cell interactions that control cell division in a manner that is probably common to many developing animal tissues. Thus, when a piece of epithelium is transplanted to a homologous site, it "heals in" without significant cell division. However, if it is transplanted to a nonhomologous site, both the graft and the adjacent host cells proliferate and then differentiate to generate the cells that would normally lie between the region from which the graft was taken and the region to which it was transplanted (see p. 865).

Many cultured animal cells require minute amounts (as little as 10^{-10} moles per liter) of specific growth factors in order to divide and/or survive, and different cell types require different mixtures of factors. Such growth factors can be proteins or small molecules such as peptides or steroids. While some of these growth factors are hormones that circulate in the blood, others probably act as short-range "local chemical mediators" (see p. 724) of some of the positional effects on cell division that have been observed in tissues.

Cancer Cells Have Lost Their Normal Growth Control[10]

In recent years, a good deal of research in cell biology has been devoted to finding the cause(s) of and cure(s) for cancer. Cancer cells exhibit a number of properties that make them dangerous to the host, often including an ability to invade other tissues and to induce capillary ingrowth (which assures that the proliferating cancer cells have an adequate supply of blood). However, one of the defining features of cancer cells is that they respond abnormally to the control mechanisms that regulate the division of normal cells, and they continue to divide in a relatively uncontrolled fashion until they kill the host. This deadly absence of restraint has provided a great incentive for studying the control of cell division. One result has been a large number of widely advertised "breakthroughs," many of which have been briefly hailed as the answer to understanding the control of cell division and cancer. In the recent past, the fundamental difference between normal and tumor cells has been variously proposed to lie in changes in cellular cyclic nucleotide levels, plasma membrane fluidity, secreted proteins, the cytoskeleton, and ion fluxes, to name just a few. While the actual molecular mechanisms involved remain elusive, it is clear that cancer cells are less subject to most of the feedback mechanisms that control normal cell division, both in tissues and in culture. For example, cancer cells will usually continue to divide in culture beyond the point at which normal cells are stopped by contact inhibition, proliferating and piling up upon one another even when they are no longer able to flatten out on the culture dish (Figure 11–12). Moreover, cancer cells require fewer protein growth factors than do normal cells in order to survive and divide in culture. (In some cases this may be because they produce their own growth factors.)

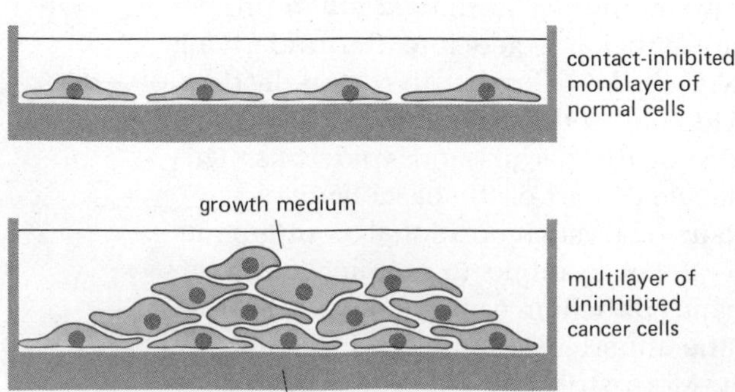

Figure 11–12 Schematic diagram illustrating the pattern of growth shown by many cancer cells in tissue culture. Unlike most normal cells, cells taken from tumors usually continue to grow and pile up on top of each other after they have formed a confluent monolayer.

A second and fundamentally important difference between normal and cancer cells is that cancer cells, as a population, can go on dividing indefinitely. In contrast, nearly all normal cells in mammals seem to die after a limited number of divisions. For example, when normal mammalian fibroblasts are grown in culture, they will divide between 20 and 50 times on average, depending on the animal from which they were taken. As such a culture ages, its individual cells take progressively longer to go through each division cycle, and eventually the entire population stops dividing and dies. In general, cells taken from older animals will divide fewer times in culture than the same cells taken from a young animal, suggesting that the older cells have used up many of their allotted divisions while in the animal.

Such observations have led to the belief that as cells differentiate, they become programmed to die after a certain number of divisions. This *programmed cell death* could conceivably be valuable to the organism as an additional safeguard against the unbridled growth of one particular cell. It means that most cells that escape from the normal controls on cell division should give rise only to a relatively small clone of progeny cells before the whole population dies. In Chapter 16 we shall see, however, that cancer involves something more than just abnormalities of proliferation and that other factors besides controls of division and programmed senescence conspire to make cancer a relatively rare event.

Summary

The integrity of tissues can be maintained only if the growth and division of each individual cell in a multicellular organism are programmed and coordinated with its neighbors. As a result, different cells divide at very different rates, depending on their precise location. Cells that are not actively proliferating have a reduced rate of protein synthesis and are arrested in G_1 phase. Once a cell has become committed to divide by passing a special "restriction point" (R) in its cycle late in G_1, it will make DNA (in the S phase), proceed through the G_2 and M phases, and enter the G_1 phase of the next cycle. Whether or not a mammalian cell will grow and divide is determined by a variety of "feedback control" mechanisms, which include the availability of space in which a cell can flatten and the secretion of specific stimulatory and inhibitory factors by cells in the immediate environment. Cancer cells can be shown to have escaped many of these controls, making them dangerous to their host.

Tumor Viruses as Tools for Studying the Control of the Cell Cycle[11]

Many viruses (Figure 11–13) cause tumors in a variety of vertebrates ranging from reptiles to monkeys. It therefore seems likely that certain specific viruses will be found to be the cause of some human cancers. Thus it is neither surprising nor unreasonable that very large sums of money already have been spent trying to isolate viruses from human tumors. So far, however, the search has proven difficult, and most scientists now doubt that viruses are a major cause of tumors in man. Nevertheless, the study of such *tumor viruses* has been immensely rewarding because it has uncovered clues as to how the growth and division of cells are controlled. A key element in this study has been the development of cell culture systems in which the **cancerous (neoplastic) transformation** caused by tumor viruses can be observed without the almost insuperable difficulties imposed by experiments in animals. At present, there exist a variety of tissue culture cell lines whose infection by

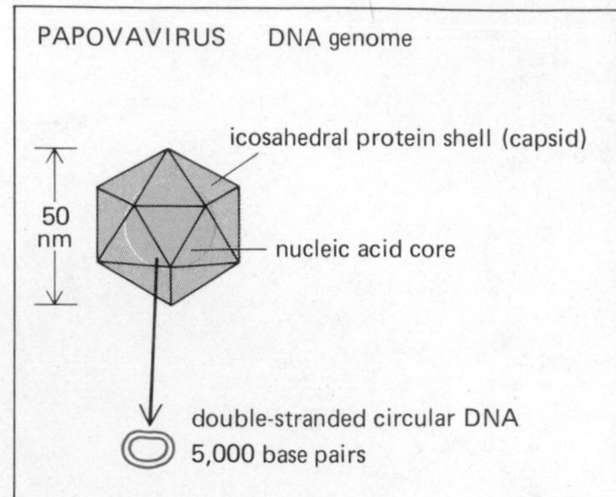

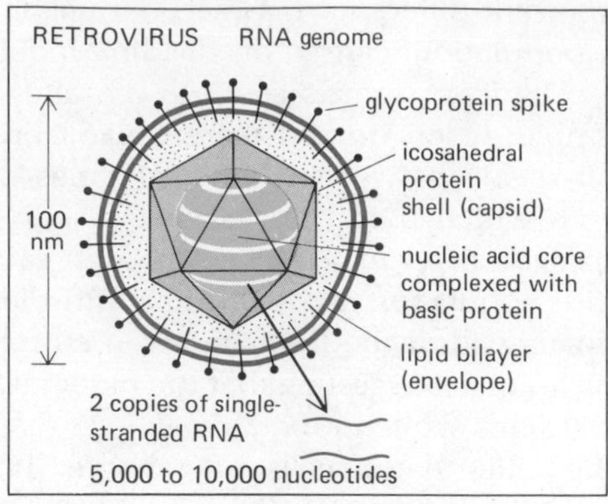

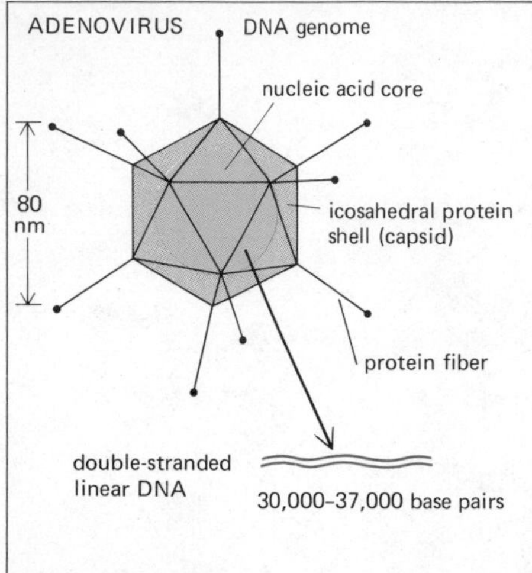

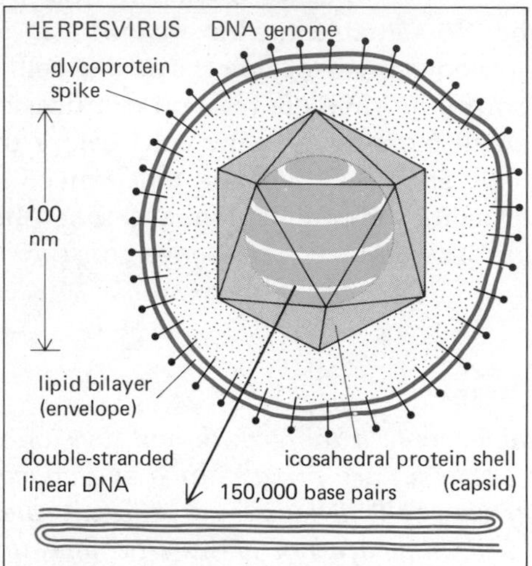

Figure 11–13 Schematic drawings of the four major types of viruses that cause tumors in animals. As indicated, the capsids of the retroviruses and the herpesviruses are surrounded by a lipid bilayer membrane.

given tumor viruses leads to their transformation into their cancerous equivalents. Such "transformed cells" often can be quickly identified by their altered morphology or growth requirements. Some of the major differences between transformed cells and normal cells are listed in Table 11–1.

The first tumor virus to be properly identified was the Rous sarcoma virus (RSV), a virus of chickens whose genetic component is an RNA molecule about 9000 nucleotides long. Through the study of how this virus transforms chicken cells, it was discovered that its RNA genome is transcribed into DNA that subsequently becomes inserted into host chromosomal DNA (see p. 238). Many different **RNA tumor viruses (retroviruses)** are now known, providing to cancer research the means to transform a variety of different normal cell types into their cancerous equivalents. Several classes of **DNA tumor viruses** have also been well-studied. These include several small papovaviruses, in particular the monkey virus SV40 and the mouse virus polyoma, both of which have genomes that contain only about 5000 base pairs. A great deal is also known about the much larger adenoviruses and about the Epstein-Barr virus (EBV), a herpes-like virus that has been implicated in the human cancer Burkitt's lymphoma. Like the oncogenic retroviruses, all of these DNA tumor viruses transform cells through insertion of viral genetic material into host chromosomal DNA. As a consequence, viral genes become part of the genetic dowry of the infected cell, are replicated along with the cell chromosome, and are thus expressed not only in the original cell but in all of its progeny (collectively called a *clone*) (Figure 11–14). This ineradicable genetic endowment can consign an infected cell clone to a cancerous fate.

Table 11–1 Some Changes Commonly Observed When a Normal Tissue-Culture Cell Is Transformed by a Tumor Virus

1. Plasma-Membrane-related Abnormalities
- A. Enhanced transport of metabolites.
- B. High production of plasminogen activator increases amount of extracellular proteolysis.
- C. Excessive blebbing of plasma membrane.

2. Adherence Abnormalities
- A. Diminished adhesion to surfaces; therefore maintains a rounded morphology.
- B. Failure of actin filaments to organize into large bundles.
- C. Low extracellular fibronectin deposition.

3. Growth and Division Abnormalities
- A. Growth to an unusually high cell density.
- B. Lowered requirement for growth factors in serum.
- C. Less "anchorage dependence" (that is, can grow without the need to flatten on a solid surface).
- D. Cells cause tumors when injected into susceptible animals.

How do viruses disturb the control of cell growth and division? Two different general mechanisms are known: (1) The integration of viral DNA into the host genome may in itself cause changes in the structure or level of expression of nearby important host cell genes. This process is known as *insertional mutagenesis* because the changes are inherited like the mutations induced by more conventional means. Several such cases of insertional mutagenesis have recently been documented. Some of these involve the insertion of a highly active RNA polymerase II viral promoter next to a key cellular gene whose overexpression leads to the cancerous phenotype. Because mRNA chains started on the viral promoter incorporate the coding sequence of the adjacent gene, a much larger than normal amount of the cellular control protein is synthesized in the cell. (2) Many tumor viruses carry one or several genetic loci (known as **oncogenes**) that are directly and solely responsible for neoplastic transformation of the host cell; there may be no need in this case to alter a host cell gene. A large number of viral oncogenes are known, and they are active in a wide variety of different species; the ubiquity and properties of these remarkable genes account for much of the utility of tumor viruses as experimental agents.

Oncogenes Are Identified Through the Methods of Molecular Genetics[12]

Various techniques have been used to identify viral oncogenes. A versatile genetic approach exploits mutations in the oncogene that render the gene inactive when the temperature is raised. Such *temperature-sensitive* mutant genes make it possible to suppress and then reinstitute the effect of the gene at will by raising or lowering the temperature. At low temperatures (34°C), susceptible cells containing a viral oncogene will adopt a cancerous phenotype and, for example, grow in the absence of some of the growth factors normally required (Table 11–1). If expression of the oncogene is then prevented by raising the temperature to 39°C, the cell returns promptly (usually within a matter of hours) to the normal phenotype (Figure 11–15). This cycle

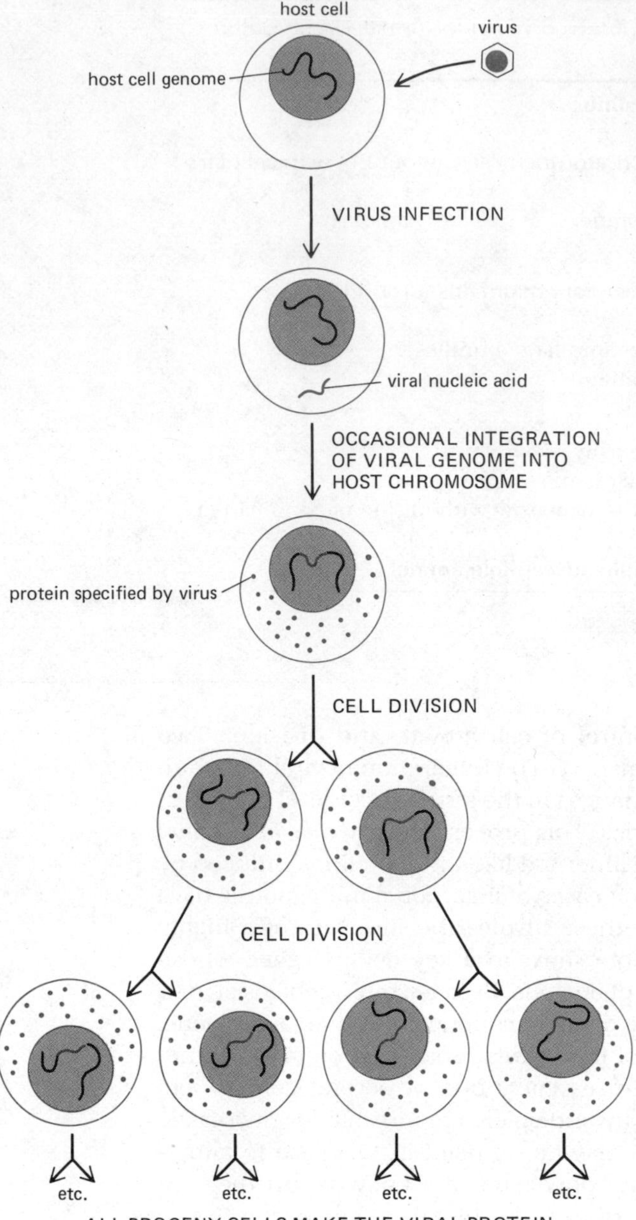

Figure 11–14 Schematic diagram illustrating how the occasional integration of the nucleic acid sequence of a virus genome into the host cell chromosome can lead to the permanent presence of a viral protein in the progeny cells. For retroviruses, a DNA copy of the viral RNA genome is made in the cell by a viral enzyme called *reverse transcriptase,* and it is this DNA copy that is integrated into the host chromosome (see Figure 5–52, p. 239).

of changes in cell phenotype can be repeated at will by lowering and raising the temperature. Such changes demonstrate that the temperature-sensitive viral gene is the gene responsible for the neoplastic transformation of the cell, thereby identifying it as an oncogene.

A method called *DNA-mediated transfection* can also be used to identify a viral oncogene. In this procedure, a fragment of DNA that bears only one of the viral genes is introduced into tissue-culture cells, where it is occasionally integrated into the genome and expressed. Some of these DNA fragments will cause progeny of the recipient cell to become neoplastic, so that they will even form tumors if implanted in a suitable animal host. In this way, an oncogene can be identified in a viral genome and shown to be wholly responsible for neoplastic transformation of infected cells, enabling them to cause tumors in animals.

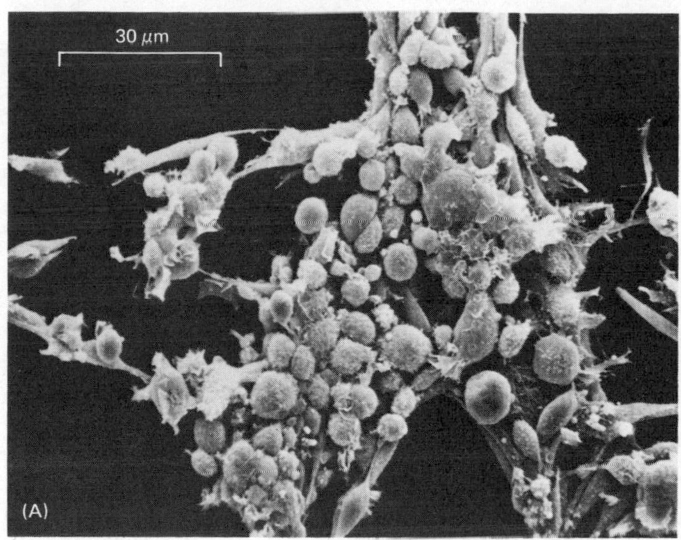

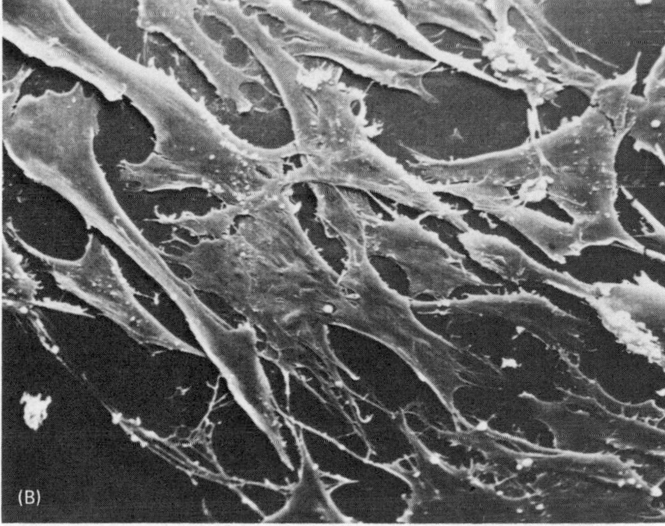

Figure 11–15 Scanning electron micrographs of a tissue-culture cell that has been transformed by a retrovirus carrying a temperature-sensitive mutant oncogene. (A) The rounded morphology of the neoplastically transformed cell observed at low temperature (34°C). (B) The same cell regains a normal morphology when the oncogene product is inactivated by a shift to higher temperature (39°C). (Courtesy of G. Steven Martin.)

The Oncogenes of RNA Tumor Viruses Are the Best Understood[13]

The oncogenic portions of the various DNA tumor genomes are invariably localized in regions that function early in the viral life cycle. The best studied of such genes are those of the SV40 and polyoma viruses. However, the analysis of their function has been complicated by the fact that these oncogenes are part of an overlapping set of genes—one of which (that coding for the large T-antigen protein, see p. 392) is required for viral DNA replication. It is not yet clear exactly how many of the several gene products specified by these overlapping gene sets are involved in neoplastic transformation. However, one of them that is required is the "middle T antigen" of polyoma; some of these protein molecules are found on the inner surface of the plasma membrane, where they seem to be associated with protein kinases that phosphorylate tyrosine (see below). The larger DNA tumor viruses, such as adenovirus, have a much more complex genome; here it seems that several different oncogenic proteins are produced and that combinations of these are required to transform cells fully.

The simplest oncogenes to study are those of retroviruses, in which a single gene can be involved whose coding sequence does not overlap that of other genes of the virus. As a result, the oncogene can often be deleted without adversely affecting the multiplication of the virus. The inherent dispensability of such oncogenic regions was first revealed by the finding of mutant Rous sarcoma virus particles that multiplied normally but did not convert their host cells into cancer cells. Many such mutants were found to have deletions of a large internal segment that coded for a protein of molecular weight 60,000. It is the presence of the product of this so-called ***src* gene** that converts normal chicken cells into their cancerous equivalents.

With the finding that the *src* gene plays no role in viral multiplication, it became clear that retrovirus multiplication requires only three classes of virus-specified polypeptides: one for the outer envelope (*env*) proteins, a second for the inner capsid (gag) proteins, and a third for the reverse transcriptase (*pol*) protein.

In many retroviruses, a large functional oncogenic segment replaces part of the *gag*, *pol*, or *env* sequences of the virus. Such retroviruses are defective, being unable to multiply by themselves and requiring the simultaneous presence of a nononcogenic "helper virus" to provide the missing gene product.

Nondefective oncogenic retroviruses like Rous sarcoma virus are in fact rare compared to the many defective retroviruses (for example, Harvey sarcoma virus and Abelson virus) lacking intact *gag* or *env* segments. The oncogenic proteins produced by such defective viruses are often fusion proteins having *gag* sequences at their amino-terminal end and oncogenic sequences at their carboxyl-terminal end.

When present on a retrovirus, each of the known oncogenic proteins affects many aspects of cell structure and function. How can such diverse effects be caused by the action of a single viral gene? Two alternative explanations have been offered. The protein encoded by the oncogene may alter a single cellular component that in turn affects a wide variety of cellular functions to give rise to the neoplastic phenotype. Or the oncogene product may itself directly alter many different components of the cell. Many investigators believe that the second alternative is more likely, partly because it is possible to mutate the Rous sarcoma virus *src* oncogene so as to suppress only certain of its effects on the cell.

Even though the protein products of many oncogenes seem to affect many different cellular functions, it is possible that only a few of these effects are required for tumorigenesis. If uncontrolled cell division lies at the heart of the neoplastic phenotype, the fundamental change in tumor cells may be their failure to halt at the restriction point (R) in the cell cycle under conditions that arrest the progress of normal cells. Few, if any, neoplastic cells are preferentially arrested at R when grown in tissue culture; instead, they stop growing at various points in the cell cycle when starved or subjected to toxic conditions. How the action of an oncogene (or any other tumorigenic agent) brings about this profound change is unknown.

One way that oncogenes may disturb the normal mechanism of growth regulation is by preventing a cell's normal differentiation into a cell with a limited growth potential. In fact, many tumor cells are less well differentiated than their normal counterparts. This is an especially prominent feature of human leukemias (cancers of blood cells) and has led to the suggestion that these tumors occur because cells in a specific lineage fail to complete their maturation. Since blood cell precursors seem to be capable of incessant division, a continuously expanding population—a tumor—results. This view is further supported by the effects caused by certain viral oncogenes. For example, one type of retrovirus causes cancers of red blood cell precursors. If the viral oncogene is inactivated by a mutation, the previously immature cells begin to differentiate and lose their neoplastic properties.

As mentioned, cancer cells can be immortal: unlike normal cells they do not have a limited life in culture. Many tumor viruses can bestow immortality on their host cells as a consequence of the effects of the viral oncogene. In the case of adenoviruses, it has been possible to identify a portion of the viral genome that converts cells to immortal growth without inducing any other aspect of the neoplastic phenotype. Results such as these indicate the value of viral genes as tools for studying the molecular mechanisms of cell senescence, as well as the mechanisms of neoplastic transformation.

Oncogenes Frequently Code for Protein Kinases[14]

A key breakthrough occurred in 1978 with the discovery that the *src* gene product was a *protein kinase*—an enzyme that catalyzes the transfer of a phosphate residue from a nucleoside triphosphate to an amino acid side chain in selected protein targets (p. 743). Soon afterward, several other oncogenic proteins were found to be protein kinases. Each of the virus-specified kinases so far examined phosphorylates many target proteins; this could explain the multiple effects of their respective oncogenes. Many retroviral kinases are special in that they phosphorylate *tyrosine* side chains rather than the serine and threonine residues phosphorylated by the great majority of protein kinases. Since tyrosine phosphorylation is rare, cells infected with tumor viruses may contain five to ten times more phosphotyrosine on proteins than do normal cells. Establishing the identity and function of these phosphorylated proteins is crucial to any effort to understand how viral oncogenesis occurs. A few candidate proteins have been identified—such as the actin-binding protein vinculin (see pp. 588 and 600)—but it has not yet been possible to prove that any of these are involved in inducing and/or maintaining the transformed phenotype.

Since the effects of viral oncogenes are so numerous, we might expect to find their products at many places in the cell and particularly in the nucleus, where many of the events of cell division are thought to originate. Surprisingly, this has not been true of many of the oncogene products well studied to date. Rather, most of these proteins (representing the products of several very different types of tumor viruses) have been found associated with the plasma membrane (Figure 11–16). In some instances the protein is inserted into the lipid bilayer, in others the precise nature of the membrane association is unclear. These findings support a long-standing hypothesis that states that external controls of cell division act largely at the cell surface, and they raise the possibility that most of the substrates for tumor-virus protein kinases will be found closely associated with the plasma membrane.

Tyrosine phosphorylation also plays a part in the control mechanisms of normal cells. For example, the binding of the protein hormone *epidermal growth factor (EGF)* to its receptor on the surface of a cell induces phosphorylation of tyrosine in the receptor itself and in several other cellular proteins. These phosphorylations are thought to be the signal that triggers the intracellular effects of EGF, causing a cell that was previously resting to divide. Interestingly, several of the proteins whose tyrosines are phosphorylated in response to EGF are the same as those that show enhanced tyrosine phosphorylation in response to certain viral oncogenes. These findings suggest that the same phosphorylations of tyrosines may mediate both a normal (EGF) and an abnormal (oncogene) stimulus to cell division.

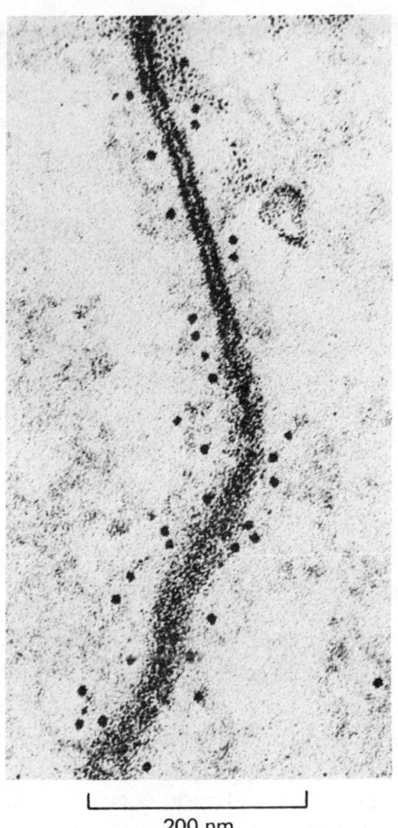

Figure 11–16 Electron micrograph demonstrating that the protein kinase synthesized by the oncogene of Rous sarcoma virus is attached to the inner surface of the plasma membrane in neoplastically transformed cells. The kinase has been localized by reacting it with specific antibodies to which electron-dense ferritin particles are attached. (Courtesy of Ira Pastan; from M. C. Willingham, G. Jay, and I. Pastan, *Cell* 18:125–134, 1979. © M.I.T. Press.)

Increased Levels of a Normal Cellular Protein Can Destroy Normal Cell Growth Regulation[15]

At first it seemed almost perverse that nature had distributed oncogenes so widely among viruses. However, we now realize that the oncogenes of retroviruses are merely normal cellular genes in another guise. By unknown mechanisms, genes from the vertebrate genome appear to have been more or less randomly incorporated at some low frequency into retrovirus genomes. Those few viruses carrying oncogenes were presumably selected from the rest during evolution because they keep the host cell in a proliferative state and thereby maintain the conditions most suitable for virus multiplication (Figure 11–17).

Although the cellular genes incorporated into viruses in this way are oncogenic, neither their structure nor their function seems to have been modified appreciably by the relocation. For example, the oncogene of Rous sarcoma virus and its progenitor gene in normal cells both encode similar tyrosine protein kinases that associate with the plasma membrane. For some oncogenes, the normal progenitor gene has been isolated by DNA cloning techniques and put back into normal cells under circumstances that encourage expression of the cloned gene at a high level; remarkably, such cells become neoplastic!

A new view of viral oncogenesis has therefore emerged. A tumor virus may do no more than overload the host cell with a normal cellular gene product that may be concerned with the control of cell division. This product seems to drive the transformed cell to unrestrained proliferation only because it is present in excessive amounts. It may also be that a chemical carcinogen or a spontaneous mutation need only induce the inappropriate overexpression of such a crucial control gene to set the cell on the path to malignancy (Figure 11–18).

More than fifteen different retrovirus oncogenes have already been identified, and each derives from a different normal gene in the vertebrate genome. To date, about half of these oncogenes have been shown to have a protein kinase activity. Collectively, these genes are likely to constitute a heterogeneous family whose members are related by their involvement in the regulation of normal cell division and development in vertebrates. If correct, this idea represents a substantial windfall for experimental biologists. Retroviruses, by extracting vital regulatory genes from the vertebrate genome, have made these genes readily accessible for study.

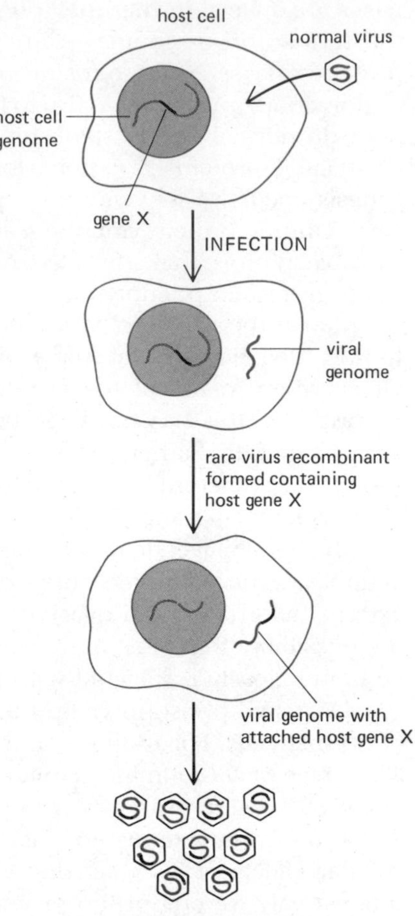

Figure 11–17 Schematic illustration of the rare process by which retroviruses incorporate the nucleotide sequences corresponding to selected host genes into their genomes. Retroviruses that contain oncogenes seem to have originated in this way (see text).

Active Oncogenes Have Been Isolated Directly from Human Tumors[16]

It has recently been possible to isolate active oncogenes from certain tumors of humans and other species without the intervention of viruses. For certain tumors (lung, bladder, colon, neuroblastoma, lymphoma, and leukemia), DNA extracted from the tumor cells is found to transmit the property of cancerous growth when introduced into normal cells in culture. By means of recombinant DNA techniques (p. 185), the DNA sequences corresponding to several of these active oncogenes have been identified and subsequently isolated in highly purified form through cloning in bacterial plasmids. The oncogenes isolated from different types of human tumors are frequently the same (lung and colon), and two of them (from bladder and lung cancers) have been shown to be the cellular homolog of two previously identified retrovirus oncogenes—found respectively in the Harvey sarcoma virus and the Kirsten sarcoma virus. These *ras* genes code for 21,000-dalton proteins that bind GTP and GDP, are present in very small amounts in normal cells, and as yet remain a total mystery at the functional level.

Somehow the cellular genes that act as oncogenes have sustained some genetic damage that has either increased their level of activity or altered the function of their gene products, but the details are not yet known. These recent findings strongly suggest that a relatively small group of genes mediates many forms of cancer.

Summary

DNA and RNA viruses that cause tumors in animals frequently induce neoplastic transformation of cells grown in cell culture. Through our current ability to manipulate tumor virus genomes experimentally, some basic facts about the nature of cancer can be obtained. The oncogenic (cancer-causing) genes of RNA tumor viruses (retroviruses) are normal cellular genes that accidently have been inserted into retroviral genomes, in either their original or a slightly modified form. Several retroviral oncogenes code for membrane-bound protein kinases that phosphorylate tyrosine residues in specific cell proteins. Overproduction of such proteins following retroviral infection must somehow relax the normal control mechanisms that regulate cell growth and division. Very recently, cells from certain human tumors have been found to contain some of the same altered cellular genes that were earlier revealed by analyses of oncogenic retroviruses.

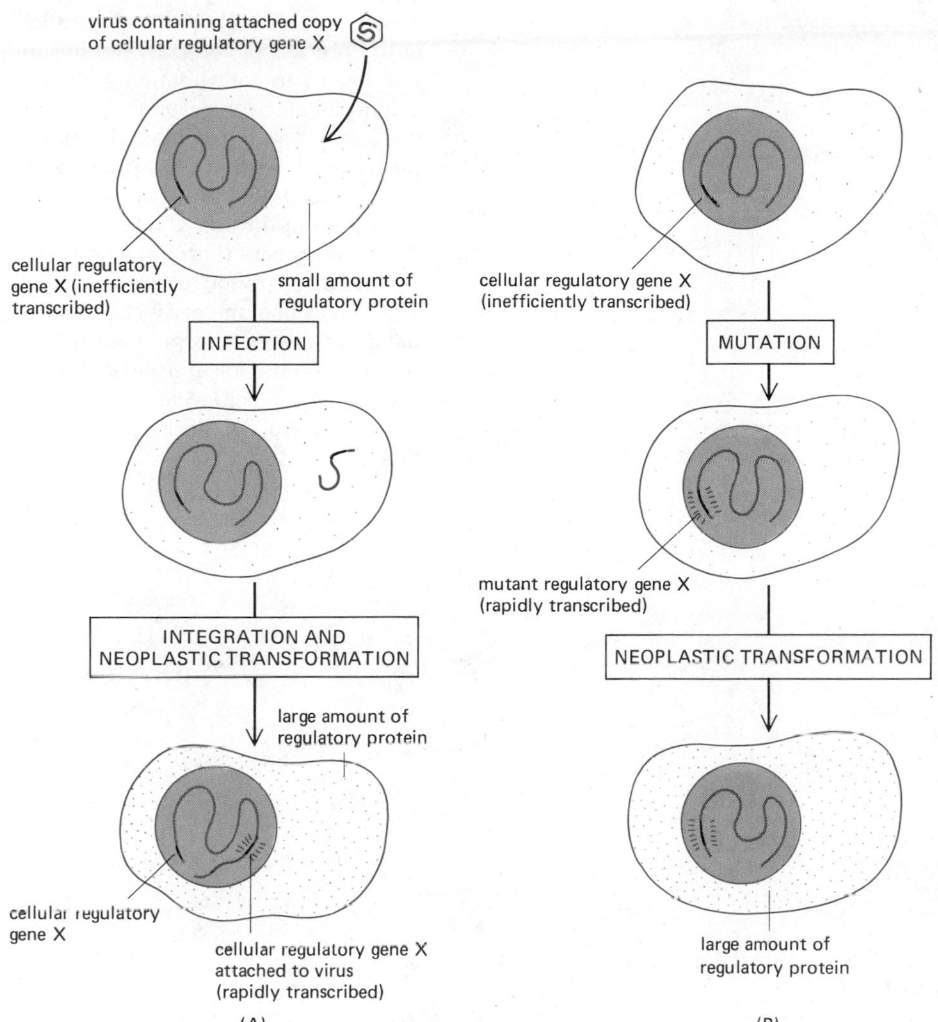

Figure 11–18 Schematic illustration of a possible common mechanism of virus-induced and spontaneous cancer. In both cases shown here, uncontrolled cell growth is attributable to an overproduction of a normal cellular regulatory protein. However, in other cases the cellular regulatory protein is known to be altered by an amino acid change, with no overproduction. (A) A retrovirus can cause an overproduction by incorporating a copy of the normal cellular gene into its genome. The RNA from the cellular gene is transcribed in unusually large amounts from the integrated DNA copy of the viral genome because the level of expression is controlled by the viral promoter. (B) A spontaneous or carcinogen-induced mutation can cause overproduction of the same cellular regulatory protein by directly increasing the rate at which its gene is transcribed. Studies carried out in yeast cells suggest that such major increases in the level of eucaryotic gene expression are more likely to be generated by small deletions or insertions of DNA pieces than by single base-pair mutations.

Events in the S Phase[17,18,19]

Studies with mutant yeast cells have shown that duplication of a structure on the nuclear envelope called the **spindle pole body** is required before DNA synthesis can begin. The analogue of this structure in animal cells is the **centriole,** which acts both as part of a major microtubule organizing center that is closely associated with the nucleus during interphase (the *cell center,* see p. 579) and as a component of each spindle pole during mitosis. The centriole duplicates, by what appears to be a templating process, once in each cell cycle (see Figure 11–19 and p. 580). As with the spindle pole body in yeast, it is possible that the attainment of a particular stage in centriole duplication is a critical event in the initiation of DNA replication. Unfortunately, there is as yet no way to block centriole duplication and thereby test this idea.

In fact, it is not known what actually triggers the initiation of DNA synthesis, nor whether the S phase begins with the sudden replication of DNA at many sites on the genome, or more gradually at only a few sites. It is even possible that synthesis begins with the replication of one special region of the chromosome, which then triggers a cascade of subsequent initiation events elsewhere. Whatever the details, the triggering mechanism is clearly of the "all-or-none" type, since once the S phase has begun, DNA replication will continue until all of the cell's DNA is replicated.

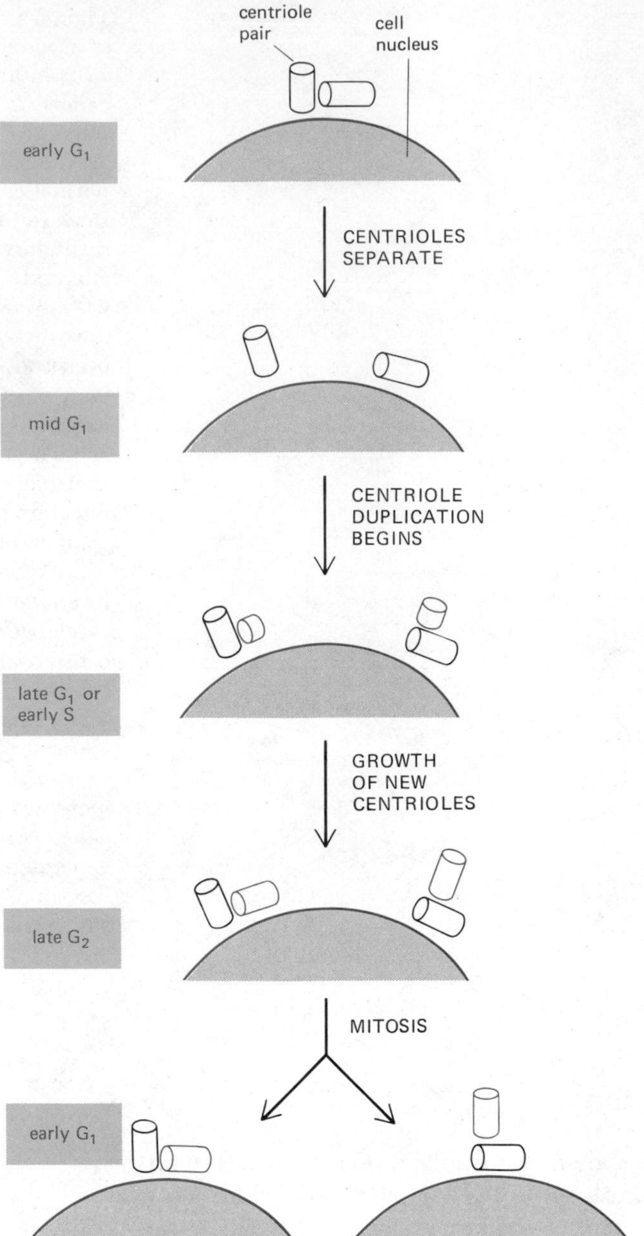

Figure 11–19 Schematic illustration of the process of centriole duplication that occurs in coordination with each cell cycle. Electron microscopy reveals that each daughter centriole undergoes a series of morphological changes as it grows, suggesting that centriole duplication is a much more complex process than indicated here. Centriole separation, as well as the attainment of some early stage of daughter centriole formation, may be required events for entry into the S phase of the cell cycle (see P. Sherline and R. N. Mascardo, *J. Cell Biol.* 93:507–511, 1982).

During the S Phase, Clusters of Replication Forks Become Simultaneously Active on Each Chromosome[19]

As discussed in Chapter 5, DNA is synthesized after a special initiating event at a *replication origin* creates a DNA **replication fork.** This Y-shaped structure is asymmetric because the patterns of DNA synthesis on its leading and lagging strands differ from each other (Figure 11–20). The genome is so small in bacteria that as few as 2 such replication forks can duplicate all of the DNA completely each cell generation. But the DNA in a mammalian chromosome is 50 times longer, so that many replication forks must be moving simultaneously in order to complete replication in a reasonable period. On average, about 100 replication forks are operating on each chromosome throughout the S phase. How are these replication forks positioned along the chromosome, and how is the complete replication of all the DNA in a cell assured during each cycle?

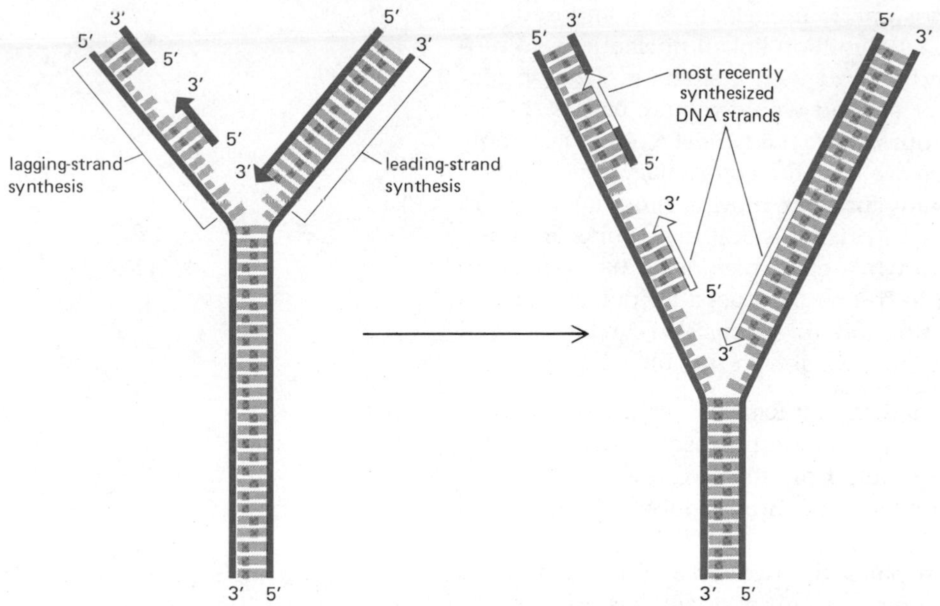

Figure 11–20 The asymmetric structure of a DNA replication fork. As previously discussed in detail (p. 223), the DNA polymerase enzyme can synthesize a DNA chain only in the 5′-to-3′ direction. Because the two strands of the DNA double helix run in opposite directions, one strand of the parental DNA template must be copied by a discontinuous "backstitching" mechanism. Synthesis on this strand (called the *lagging strand*) is therefore quite different from synthesis on the other strand (called the *leading strand*), and the replication fork is asymmetric.

The method first used to determine how eucaryotic chromosomes replicate was developed in the early 1960s. Cells growing in tissue culture are labeled briefly with tritiated thymidine so as to make the DNA synthesized during that period highly radioactive. The cells are then gently lysed and the DNA is streaked onto the surface of a glass slide to obtain long, stretched-out DNA molecules. Finally, the slide is covered with a film of photographic emulsion so that the pattern of labeled DNA can be determined by autoradiography. Since the time allotted for radioactive labeling is chosen to allow each replication fork to move several microns along the DNA, the replicated DNA can be seen as lines of silver grains in the light microscope, even though the DNA itself is invisible (the diameter of the DNA helix is only 2 nm, compared to the 200-nm limit of resolution in a light microscope). The type of result obtained is illustrated in Figure 11–21A: dark tracks of replicated DNA, whose lengths increase with increased labeling time, reveal that the replication forks in eucaryotes travel at a speed of about 50 nucleotides per second. This is only one-tenth the rate at which bacterial replication forks move, possibly reflecting the increased difficulty in replicating DNA packaged in chromatin.

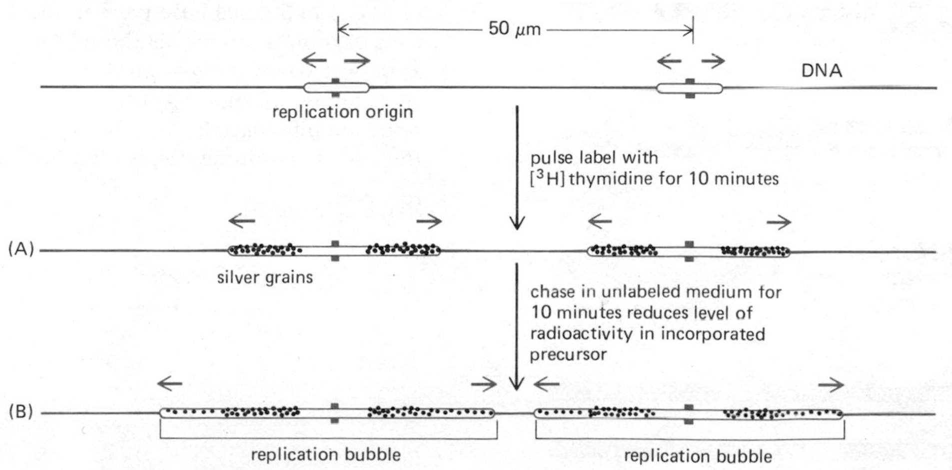

Figure 11–21 Schematic diagram of the experiments that demonstrated the pattern in which replication forks move during the S phase. The new DNA made in human tissue-culture cells was briefly labeled with a pulse of highly radioactive thymidine. In the experiment illustrated in (A), the cells were lysed and the DNA was stretched out on a glass slide that was subsequently covered with a photographic emulsion. After several months the emulsion was developed, revealing a line of silver grains over the radioactive DNA. The experiment in (B) was the same except that a further incubation in unlabeled medium allowed additional DNA, with a lower level of radioactivity, to be replicated. The fact that the dark tracks in (B) were found in pairs, with grains tapering off in opposing directions, first suggested bidirectional fork movement from a central replication origin (see Figure 11–22). Note that the colored DNA in this figure is shown only to help with the interpretation of the autoradiograph; the unlabeled DNA is invisible in such experiments.

Since an average human chromosome is thought to be composed of a single DNA molecule containing about 150 million linked nucleotides, to replicate such a DNA molecule from end to end with a single replication fork moving at a rate of 50 nucleotides per second would require $0.02 \times 150 \times 10^6 = 3.0 \times 10^6$ seconds (about 800 hours)! In fact a typical S phase lasts only 8 to 10 hours. It is not surprising, therefore, that the autoradiographic method described above demonstrates that many forks are moving simultaneously on each eucaryotic chromosome. What is surprising is that many forks are often close together in the same DNA region while other regions of the same chromosome have none. With a variation in the method used for detecting replication-fork rates and spacings, replication-fork directions also can be analyzed (Figure 11–21B). Experiments of this type have shown the following:

1. Replication-fork origins tend to be activated in clusters (called **replication units**) of perhaps 20 to 80 origins; within such units each origin is spaced at intervals of 30,000 to 300,000 base pairs from the next one.
2. New clusters of replication units are activated throughout the S phase until all of the DNA has been replicated.
3. Most replication forks are found in pairs: the two forks of a pair move in opposite directions away from a common point of origin, forming a structure referred to as a *replication bubble.*
4. The forks in a replication unit terminate when they meet an adjacent fork moving in the opposite direction. In this way, all of the DNA in the region is replicated to form two complete daughter DNA helices (Figure 11–22).
5. Replication forks move at comparable rates throughout the S phase.

New Histones Are Assembled into Chromatin as DNA Replicates[20]

The hierarchical structure of eucaryotic chromatin is briefly reviewed in Figure 11–23. As detailed in Chapter 8, two copies of each of the four nucleosomal histones (H2A, H2B, H3, and H4) combine to form a disc-shaped, octameric protein core around which the double-stranded DNA of the chromosome is

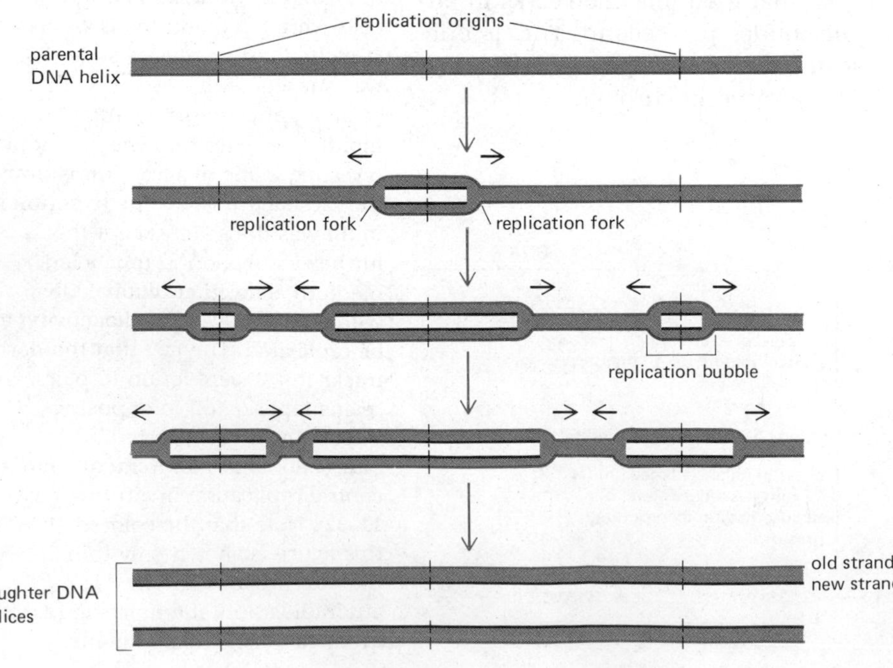

Figure 11–22 The pattern by which DNA is replicated in eucaryotic chromosomes as initially determined by the type of experiments illustrated in Figure 11–21. The replication origins shown are spaced at intervals of 30,000 to 300,000 base pairs in most cells. A replication fork is thought to stop only when it encounters a replication fork moving in the opposite direction; in this way, all of the DNA is eventually replicated.

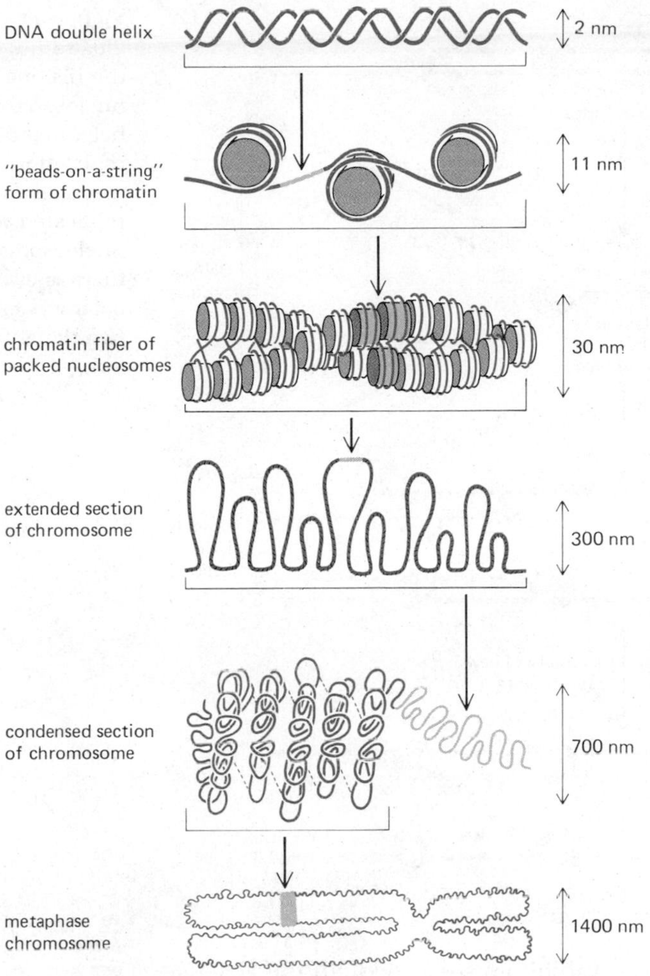

Figure 11–23 A review of chromatin structure. From the top down, the structures shown are the DNA double helix; a string of three nucleosomes; a region of 30-nm-diameter chromatin fiber; a string of ten contiguous looped domains formed by the folding of this 30-nm fiber; a model for a portion of a metaphase chromosome; and the entire metaphase chromosome. The most important entities for a discussion of DNA replication are the nucleosome itself, which appears never to leave the DNA to which it is bound, and the looped domain, which is suspected to serve as a functional unit for gene expression and may contain a single replication origin.

wound at intervals of 200 base pairs, creating a **nucleosome.** Neighboring nucleosomes are packed together to form a chromatin fiber of about 30-nm diameter whose exact structure is not known. This fiber is, in turn, thought to be folded into large **looped domains,** each of which contains a length of chromatin corresponding to tens of thousands of DNA base pairs. By analogy with specialized polytene and lampbrush chromosomes, looped domains are thought to function as units in RNA transcription—the whole domain decondensing and condensing as genes are turned on and off, respectively (p. 450). These looped domains are condensed in unknown ways during interphase, and during mitosis they condense further to form the metaphase chromosome (Figure 11–23).

A large amount of new histone, approximately equal in mass to the newly synthesized DNA, is required to make new chromatin. If each histone were encoded only by a single gene, histone synthesis would presumably lag behind DNA synthesis. Histone synthesis keeps pace because there are multiple copies of the gene for each histone. Vertebrate cells have about 40 repeated sets, each set containing all five histone genes. These gene sets are thought to be clustered together at one place on the chromosome.

Unlike most proteins, which are made continuously throughout interphase (see p. 640), the histones are synthesized mainly during the S phase. It seems that histone synthesis is regulated both at the level of transcription and at the level of mRNA degradation. It can be shown that the histone mRNAs become unusually unstable whenever DNA synthesis stops (as it does at the

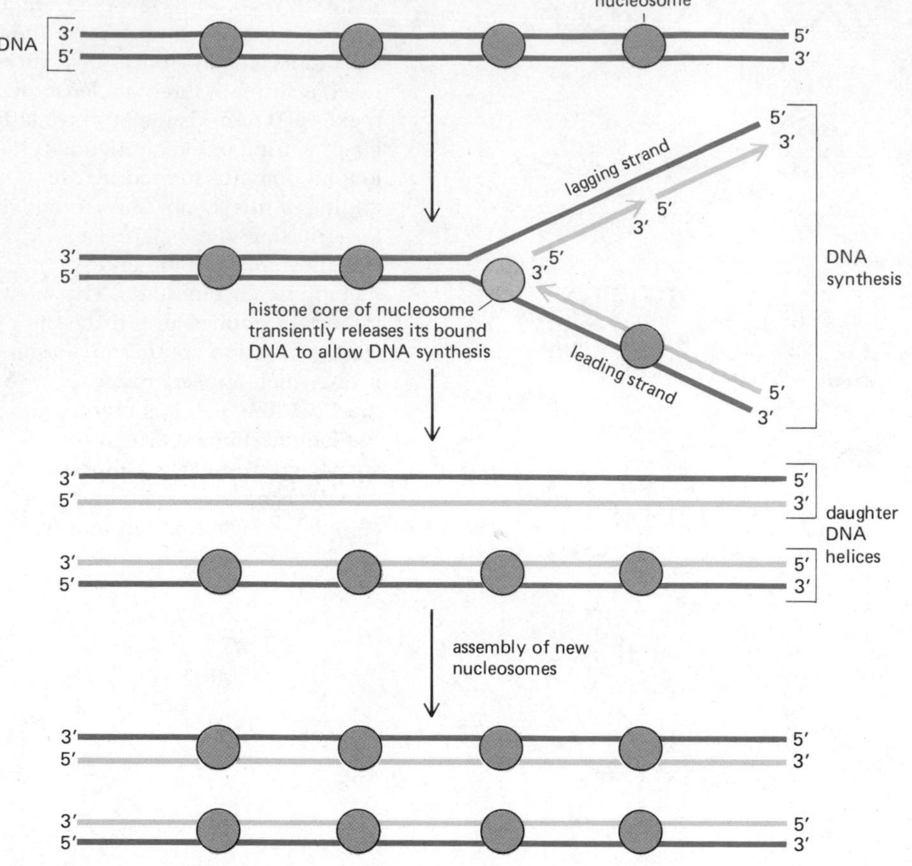

Figure 11–24 Schematic diagram illustrating the proposed transfer of the histone octamers of preexisting nucleosomes to the daughter DNA helix made on the leading strand at a replication fork. Although the mechanism by which DNA is replicated while still bound to the nucleosomal histones is not known, there is evidence that the old histone octamers end up only on one of the two daughter helices made at the fork.

end of S phase or when DNA synthesis inhibitors are added): in the absence of DNA synthesis, any histone mRNA present is degraded within minutes. In contrast, the histones themselves are remarkably stable proteins, which may survive for the entire life of a cell. The linkage between DNA synthesis and histone synthesis seems to reflect a feedback mechanism that monitors the level of free histone and guarantees that the amount of new histone made is appropriate for the amount of new DNA synthesized.

Once they are assembled in nucleosomes, histones appear never to leave the DNA to which they are bound. Therefore, as a replication fork proceeds, it must somehow pass through the parental nucleosomes. How this is done is not known, but one experiment has suggested that the old histone octamers are passed directly to the daughter DNA helix that is synthesized on the leading side of the fork, as illustrated in Figure 11–24. If this is true, then all of the new histone octamers that form the core of new nucleosomes (composed of the new histones H2A, H2B, H3, and H4) would be deposited onto a histone-free daughter DNA helix generated on the lagging side of the fork.

These new histones are normally deposited in the first minutes after the replication fork passes. Thus, when replicating chromatin is spread and examined in the electron microscope, Y-shaped replication forks are observed in which both daughter DNA helices are uniformly packaged into nucleosomes. However, if protein synthesis (including histone synthesis) is blocked with the drug cycloheximide, DNA synthesis continues for 30 minutes or so, and electron micrographs of the chromatin show long lengths of histone-free DNA emanating from one side of a replication fork (Figure 11–25). When the cycloheximide is removed, histone synthesis restarts and all of the accumu-

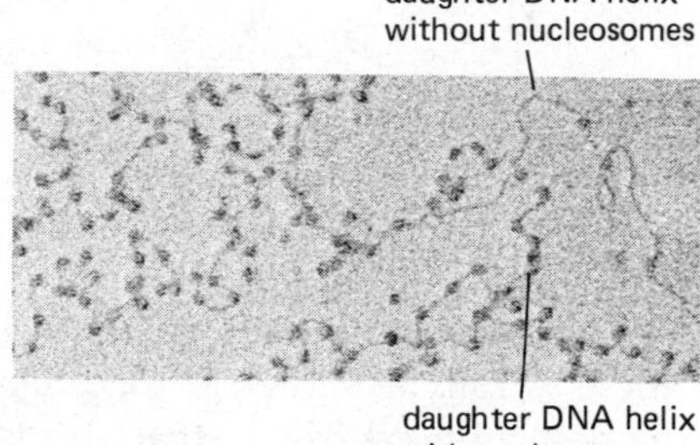

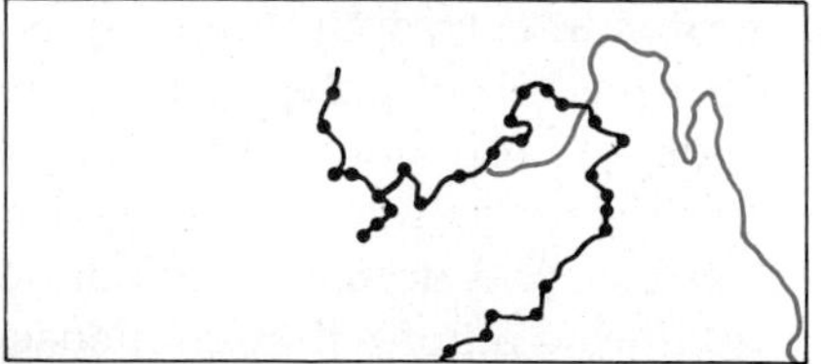

Figure 11–25 Electron micrograph of spread chromatin showing the abnormal accumulation of nucleosome-free DNA that occurs when new histone synthesis is inhibited with cycloheximide. In favorable areas of the spread, DNA that lacks histones can be seen to issue from only one side of a replication fork, in agreement with the schematic representation in Figure 11–24. (Micrograph courtesy of Harold Weintraub.)

lated bare DNA quickly becomes packaged into nucleosomes. The fact that S-phase cells survive brief treatments with cycloheximide demonstrates that newly synthesized DNA need not be immediately packaged in nucleosomes for a cell to remain viable.

The Orientation of Replication Origins Relative to Genes May Have Important Biological Consequences[20,21]

By analogy with the unique replication origins determined for several animal viruses, it seems likely that replication origins are special DNA sequences and, therefore, that the replication bubbles previously diagrammed in Figure 11–22 normally start only at certain sites on the DNA. Moreover, the average spacing between replication origins is at least roughly comparable to the average distance thought to separate adjacent looped domains of chromatin. In general, therefore, there may be only a single replication origin in each looped chromatin domain.

According to our previous discussion, when two separate replication forks move away from a replication origin, a different daughter DNA helix should receive the histone octamers from the parental nucleosomes on each side of the origin (see Figure 11–26 and its legend). In this case, the precise location of an origin relative to a transcription unit, or gene, will determine how the preexisting parental histones are transferred to the two daughter genes. Not all nucleosomes are exactly the same, and different chromatin structures are present in different genetic regions (p. 404). The precise location of a replication origin relative to a gene could thereby have important biological consequences, determining the chromatin structure of that gene in the next cell generation, as outlined in the legend to Figure 11–27.

Different Genetic Regions on the Same Chromosome Replicate at Distinct Times During the S Period[22]

The replication of DNA in the region between one replication origin and the next should normally require less than 30 minutes to complete, given the rate at which a replication fork moves and the typical distances measured between the replication origins in a replication unit. Yet the S phase usually lasts for

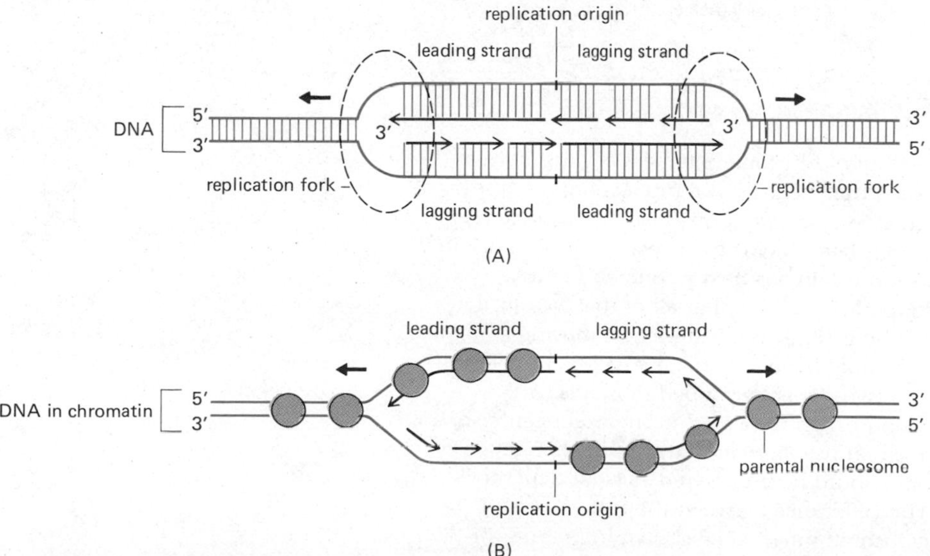

Figure 11–26 Illustration of the expected segregation of old histones to opposite daughter DNA helices on either side of an origin of DNA replication, if old histone octamers behave as proposed in Figure 11–24. In (A), the leading strand of a replication fork is seen to synthesize a different daughter DNA helix, if one compares the two forks that move away from a single replication origin. (Due to the antiparallel alignment of the two strands in the DNA double helix, a different DNA template strand must be copied in the 5′-to-3′ direction at each of these two forks.) In (B), the expected segregation pattern for the old histones at these forks is shown, with the new histones that assemble on the lagging strand omitted for clarity.

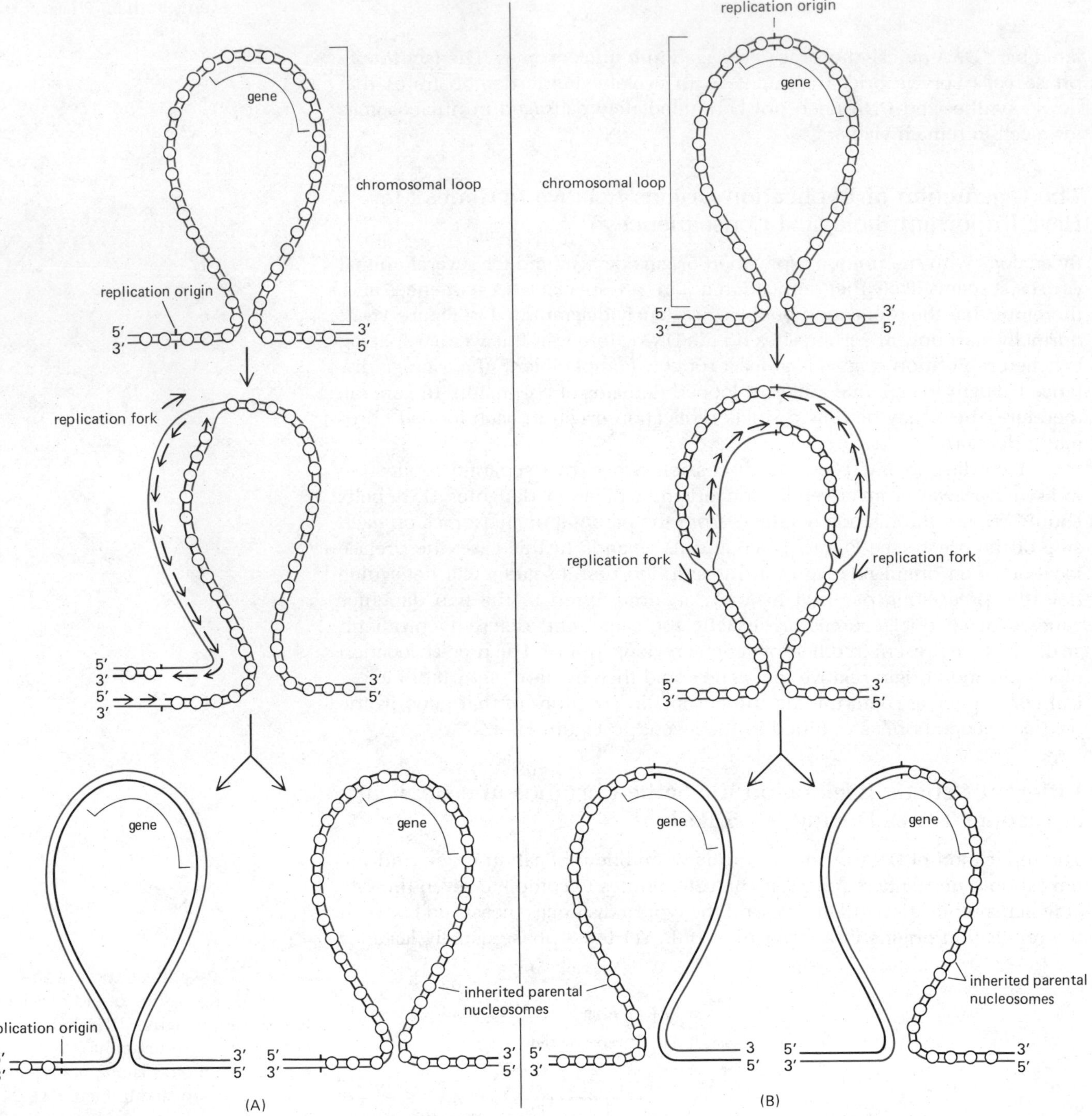

Figure 11–27 If the old histone octamers end up only on one of the two daughter helicies at a replication fork, the pattern of inheritance of parental histones in a genetic region will depend on the position of the nearest replication origin.

(A) If a replication origin lies next to one end of a chromosomal looped domain so that all of the DNA in that loop is replicated by a single replication fork moving along it, all of the loop's original histone octamers will be transferred only to one of the two daughter looped domains.

(B) If a replication origin is located in the exact center of a looped domain, equal numbers of that domain's original histone octamers should be transferred to each daughter chromosome. The difference between the two extreme patterns of nucleosome inheritance shown in (A) and (B) could have important biological consequences (pp. 459–460).

about 8 hours in a mammalian cell. This means that the DNA in each *replication unit* (which contains a cluster of perhaps 20 to 80 replication origins) should be replicating for only a small fraction of the total S-phase interval.

Are replication units activated at random, or is there a definite order in which different regions of the genome are replicated? This question has been most clearly answered by using the thymidine analogue 5-bromodeoxyuridine (BrdU) to label synchronized cell populations for different one-hour periods throughout the S phase. Those regions of mitotic chromosomes that have incorporated BrdU into their DNA can be recognized by their decreased staining with certain dyes. Therefore, examination of the condensed chromosomes that form at the next mitosis reveals the regions that replicated in each period during the preceding S phase. A typical result is illustrated in Figure 11–28 for an arbitrarily selected single chromosome. Detailed analysis of chromosomes treated in this way shows that chromosomal regions are replicated in large units (as one would expect from the clusters of replication forks seen in DNA autoradiographs), and that different regions of each chromosome are replicated in a reproducible order during the S phase. With these techniques, at least three temporal groups of replication units can be distinguished on most chromosomes.

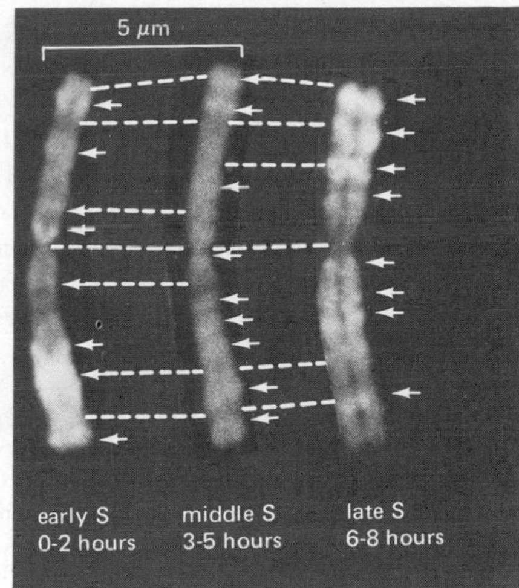

Figure 11–28 Light micrographs of a stained mitotic chromosome in which the replicating DNA has been differentially labeled during different defined intervals of the preceding S phase. In these particular experiments, cultured cells grown in the presence of the synthetic nucleoside 5-bromodeoxyuridine were briefly pulsed with thymidine during early, middle, or late S phase. Because the DNA made during the thymidine pulse is a double helix with T on one strand and BrdU on the other strand, it stains more darkly than the remaining DNA (which has BrdU on both strands) and shows up as a bright band (*arrows*) on these negatives. Dashed lines connect corresponding positions on the three copies of the chromosome shown. (Courtesy of Elton Stubblefield.)

Replication Times During the S Phase Are Correlated with Interphase Chromatin Structure[23]

In most higher eucaryotic cells, some regions of the DNA are significantly more condensed than others. This **heterochromatin** remains in a relatively condensed form even during interphase, when the chromosomes have assumed the more dispersed form in which they are actively synthesizing RNA. Since the DNA in heterochromatin is not transcribed into RNA, it is said to be inactive. Indirect evidence suggests that the unusual condensation of this chromatin prevents the expression of its genes.

An important clue to the significance of the timing of replication lies in the observation that the blocks of condensed heterochromatin, including the regions near the centromere that remain condensed throughout interphase, are replicated very late in the S phase. Late replication thus seems to be related to the packing of the DNA in chromatin. An important example of this occurs in the inactive X chromosome in female mammals. This entire chromosome replicates only late in the S phase, whereas its active homolog replicates throughout the S phase. While these two X chromosomes contain exactly the same DNA sequences, only the inactive X chromosome is condensed into heterochromatin (p. 458). The order in which replication origins are activated must therefore be based, at least in part, on the exact chromatin structure in which the origins reside. The available evidence suggests that those regions of the genome whose chromatin is least condensed during interphase, and therefore most accessible to the replication machinery, are replicated first.

Consistent with this view is the finding that many of the replication units seen on the metaphase chromosome in Figure 11–28 correspond to a distinct chromosome band visualized by the various fixation and staining procedures used for karyotyping. As many as 2000 distinct bands can be detected at mitosis in the haploid set of mammalian chromosomes, and each band may represent the condensation during mitosis of neighboring regions of the DNA with similar chromatin structures (see p. 397). The fact that these neighboring regions seem to be activated for replication *as units* during the S phase implies that the chromatin in a band has a structural unity even during interphase.

Do some of the neighboring regions of chromatin that form the chromosome bands seen during mitosis also serve as functional units for gene expression? If so, one might expect that the DNA in such a band would be packed into a less condensed chromatin structure in a cell type in which it is expressed than in cell type in which it is silent. In fact, there are reports of differences in the replication times of selected chromosome bands when the cells from different tissues are compared. If these reports are corroborated, they imply that the process of cell differentiation in a multicellular animal is accompanied by subtle changes in the chromatin structure of selected chromosome bands. These changes could in turn play a central part in embryogenesis by helping to determine which sets of genes are expressed in each type of cell as it differentiates (see p. 459).

How Is the Timing of DNA Replication Controlled?[19,24]

While the mechanics of DNA replication at a replication fork are well understood at the molecular level, it is still not clear how replication bubbles are formed nor how the timing of replication is regulated. How, for example, are the many replication origins in each replication unit (or chromosome band) activated all at once during the S phase? A replication unit probably contains on average roughly 50 contiguous looped domains (each of which might have a separate replication origin), specially condensed together even in interphase chromatin. Thus replication could be controlled in at least two different ways: (1) some global system might specifically recognize each chromosome band and act to decondense it, thereby making all of its 50 replication origins simultaneously available to the replication proteins that form replication bubbles; or (2) once these replication proteins find a few origins in a set, the ensuing local replication might perturb the chromatin structure in the rest of the replication unit so as to trigger replication at all of its remaining origins. Either of these models could explain an all-or-none replication of the DNA in different chromosome bands during the S phase.

We have yet to account for the regular ordering of band replication. Two different hypotheses have been advanced to explain this ordering. According to one, different replication proteins, each specific for a particular type of chromosomal band, are synthesized at different times during the S phase. The other, which now seems more likely, proposes that the replication proteins merely act on those regions of DNA to which they have greatest access; for example, there may be a continuous decondensation of the chromosomes during interphase, with chromosome bands becoming accessible to the replication proteins one by one as the bands uncoil.

Chromatin-bound Factors Ensure That Each Region of the DNA Is Replicated Only Once During Each S Phase[24]

New origins of DNA replication are activated throughout the S phase. Since the successive origins in each replication unit are spaced at intervals of 30,000 to 300,000 base pairs, the time required to complete the synthesis initiated from any single origin should range from 5 to 50 minutes. Because a typical S phase lasts for 8 hours, an enormous "bookkeeping" problem arises during the middle and late stages of the S phase. Those replication origins already used have been fully duplicated and, at least in respect to their DNA sequences, are presumably identical to other replication origins not yet used. Yet every replication origin must be used once, and only once, in each S phase. How is this accomplished?

While the molecular mechanism is not known, the general nature of the solution of the bookkeeping problem has been revealed by cell fusion experiments. When a cell in the S phase is fused with a cell in G_1, DNA synthesis is induced in the G_1 nucleus, suggesting that the transition from G_1 to the S phase is mediated by a diffusible initiator of DNA synthesis. In contrast, the

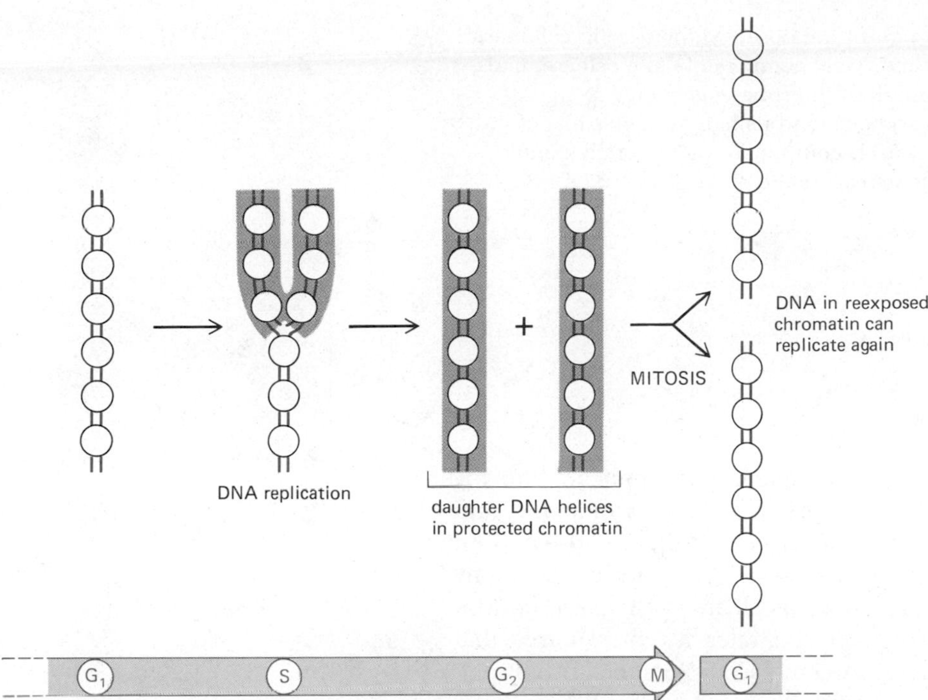

Figure 11–29 A diagram illustrating the protection of DNA against further replication in the same cell cycle in which it has already replicated. This protection is crucial for replication "bookkeeping," but its molecular nature is not known.

fusion of cells in S with cells in G_2 fails to induce any DNA synthesis in the G_2 nucleus, even though the S-phase nucleus continues to synthesize DNA. This indicates that the G_2 chromatin is unable to replicate because some nondiffusible factor is tightly bound to it.

The existence of such a nondiffusible inhibitor nicely solves the bookkeeping problem mentioned; acting perhaps by modifying the chromatin of freshly replicated DNA, it assures that once-replicated DNA is not replicated again in the same S period. Since all of the chromatin in a G_2 nucleus is blocked in this way, its DNA is unable to replicate further even if combined with S-phase cytoplasm. On the other hand, the inhibitor must be removed at or near the time of mitosis, since the DNA in the G_1 nuclei that emerge in the daughter cells will replicate if provided with S-phase cytoplasm (Figure 11–29).

Summary

At an origin of DNA replication, a replication bubble forms as two replication forks move in opposite directions away from each other. The chromatin structure is rapidly reformed after the replication fork passes, and it appears that the old histones remain bound in nucleosomes on one daughter DNA helix while new histones assemble into nucleosomes on the other. The parental histones are thereby thought to be directly inherited.

Clusters of many neighboring replication origins, spaced about one looped domain of chromatin apart, are activated more or less simultaneously during the S phase. Since the replication fork moves at about 50 nucleotides per second, less than an hour should be required to complete the DNA synthesis in each cluster of origins, known as a replication unit. Throughout a typical eight-hour S phase, different replication units are activated in a sequence determined in part by their chromatin structure, the most condensed regions of chromatin being replicated last. The correspondence between replication units and the bands seen on mitotic chromosomes suggests that these are functional as well as structural units of chromatin.

Figure 11–30 A plot of the observed increase in cell mass versus cell-cycle time. Most components in a cell are made continuously throughout interphase, generally at an increasing rate as the cell (and its biosynthetic capacity) enlarges. Note that each component must exactly double during the cycle if a steady state of growth is to be maintained.

The Logic of the Cycle

Most Proteins Are Synthesized Continuously Throughout Interphase

More than just the DNA and its associated proteins must be replicated during the cell cycle. A cell must double each and every one of its components, and thereby its mass, between successive mitoses. In view of this requirement, it is perhaps not surprising that, unlike DNA, the vast majority of the many different protein and RNA molecules present in a cell are synthesized continuously throughout interphase. Figure 11–30 illustrates a typical curve that would describe the increase in cell mass, the increase in total cell protein, or the increase in total cell RNA observed during the cycle. Since the synthetic capacity of the cell increases as it gets larger, each of these three parameters rises at an increasing rate as the cell passes from G_1 to G_2. A two-dimensional gel electrophoretic analysis of the proteins produced at various stages of the cell cycle in a synchronized mammalian cell population indicates that, of the more than 1000 major protein species detected, only a few are made primarily at a specific time in the cycle (Figure 11–31).

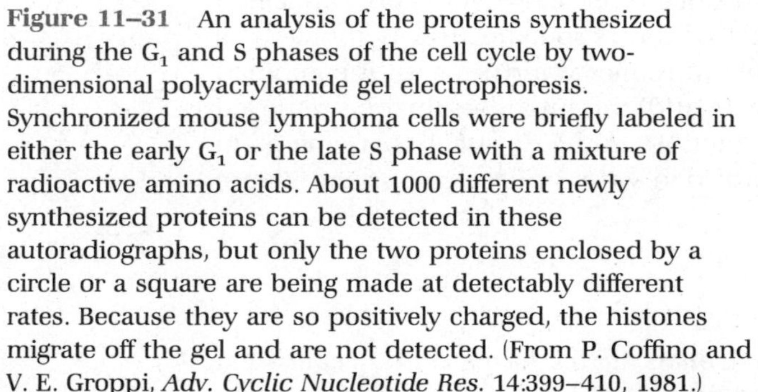

Figure 11–31 An analysis of the proteins synthesized during the G_1 and S phases of the cell cycle by two-dimensional polyacrylamide gel electrophoresis. Synchronized mouse lymphoma cells were briefly labeled in either the early G_1 or the late S phase with a mixture of radioactive amino acids. About 1000 different newly synthesized proteins can be detected in these autoradiographs, but only the two proteins enclosed by a circle or a square are being made at detectably different rates. Because they are so positively charged, the histones migrate off the gel and are not detected. (From P. Coffino and V. E. Groppi, *Adv. Cyclic Nucleotide Res.* 14:399–410, 1981.)

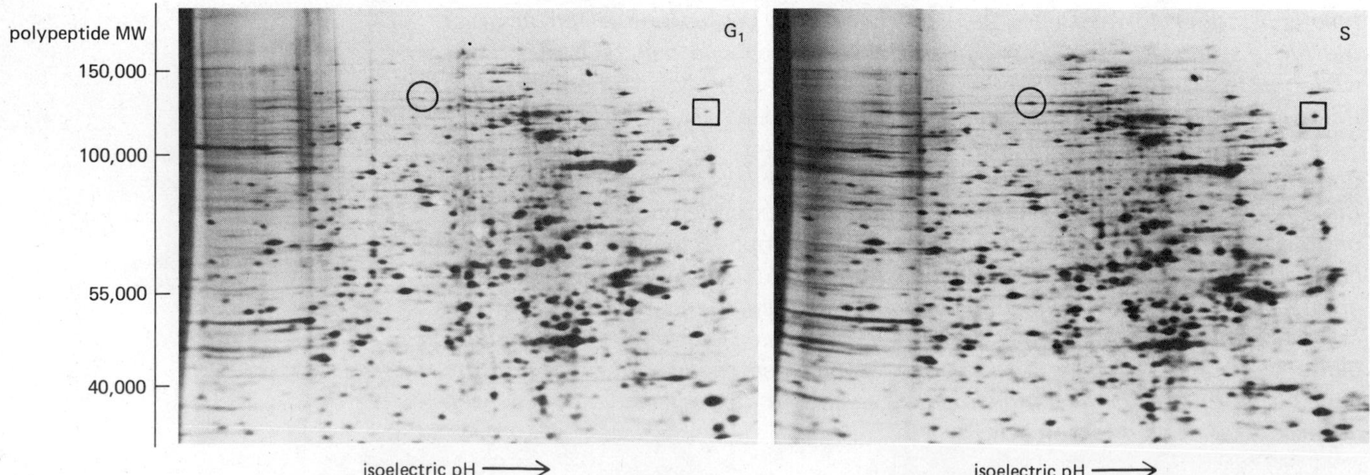

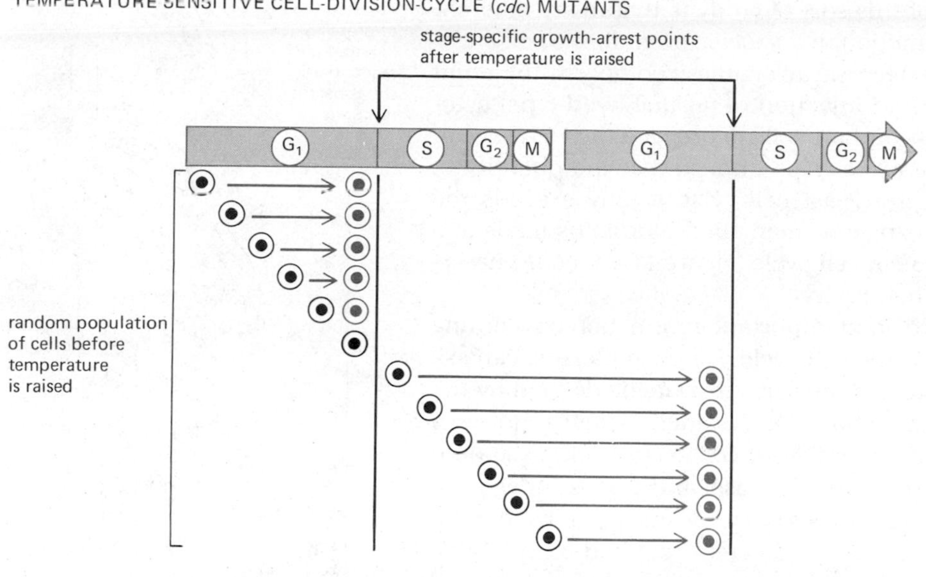

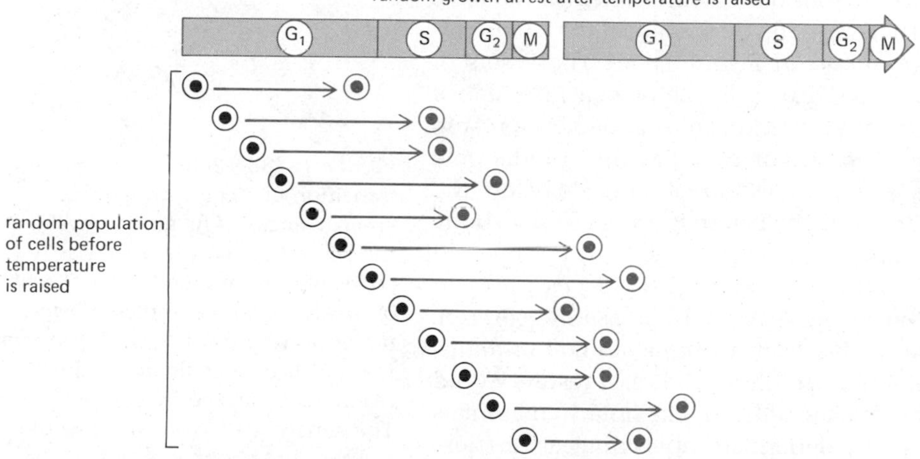

Figure 11–32 Schematic diagram illustrating how a temperature-sensitive cell-division-cycle (*cdc*) mutant is distinguished from other temperature-sensitive mutants. Most of the mutant cells obtained that fail to grow at high temperature will be defective in some process needed for general cell viability at all cell-cycle stages (for example, ATP production). As indicated, these mutant cells tend to stop cycling after a fixed time at high temperature and thus stop at random points in their cell cycle. In contrast, the cdc mutants stop cycling only at a fixed cell-cycle stage that is characteristic for each particular mutant (at the beginning of the S phase in the example shown).

The Cell Cycle Proceeds by an Obligatory Series of Sequential Reactions[18,25]

The nature of the cell cycle has been clarified by studies on mutant strains of cells that grow and divide at low temperatures (34°C for mammalian cells, 23°C for yeast cells) but not at high temperatures (39°C for mammalian cells, 36°C for yeast cells). Such "temperature-sensitive" mutant cells generally contain a single altered protein that functions only at the low temperature. Most such mutants show defects in growth shortly after the temperature has been raised. But there are rarer mutants in which cell division stops only when it reaches a particular stage-specific event, such as the beginning of DNA synthesis, nuclear division, or cytokinesis (Figure 11–32).

Such *cell-cycle mutants* have been most extensively analyzed in *Saccharomyces cerevisiae* (baker's yeast), in which mutants that define more than 35 different **cell-division-cycle (*cdc*) genes** have been isolated. These mutants have been exploited to investigate the general relationship between specific protein functions and the cell cycle.

One of the main conclusions of this sort of study is that many different proteins in yeast have an essential function in a specific event in the cell cycle. Furthermore, these proteins are not necessarily synthesized only at the point in the cycle at which they act, since an injection of normal (wild-type) cytoplasm has been found to enable most mutant cells to proceed through *several* normal cell divisions at high temperature. This implies that the quantity of these proteins present throughout the yeast cell cycle greatly exceeds the amount needed in a single cycle. By analogy, many of the proteins made at a uniform rate throughout the *mammalian* cell cycle (Figure 11–31) could nevertheless act at specific points in the cell cycle.

The yeast mutants have also provided important information concerning the biochemical pathways involved in the cell cycle. For example, one can ask at how many different points in the cycle the various proteins defined by the *cdc* mutants act. Consider DNA replication, for instance, which requires a large number of different proteins; many different *cdc* gene products should be required simultaneously to complete the S phase, and this is indeed observed. Similarly, experiments so far have shown that the cdc proteins in yeast fall into at least seven "sets" that define seven different required biosynthetic events in the cycle. It has been possible to determine the order of the times of action of these sets of proteins and to ask whether the completion of one event in the series is required before any protein in the next set can act.

If we confine our attention to events that occur within the cell nucleus, the picture that emerges is one of a strict, sequential ordering of the cell-cycle-specific reactions, as schematically outlined in Figure 11–33. These sets of reactions in the yeast cycle are organized like a linear puzzle in which a unique "piece" must be put in place before the next can be added. As a result, events always occur in a strict cyclical order. For example, the spindle pole body must duplicate before DNA can replicate; DNA must replicate before the nucleus can divide; and the nucleus must divide before the spindle pole body can duplicate again. In principle, the timing of a cycle thus constituted could be simply controlled by the rate of addition of preformed "pieces"; protein syntheses specific to one portion of the cell cycle are not necessarily required.

From the limited information available from mammalian *cdc* mutants, as well as from the use of specific metabolic inhibitors, the yeast results would seem to be generally applicable to mammalian cells. For example, in mammals as well as in yeast, cytokinesis is normally dependent upon nuclear division, which in turn is dependent upon completion of DNA replication. On the other hand, in both yeast and mammals, the initiation of a second nuclear cycle consisting of DNA replication and nuclear division is not dependent upon cytokinesis occurring in the preceding cycle: if cytokinesis is prevented, a single cell with two nuclei results. This represents an important exception to the general rule of linear order described above.

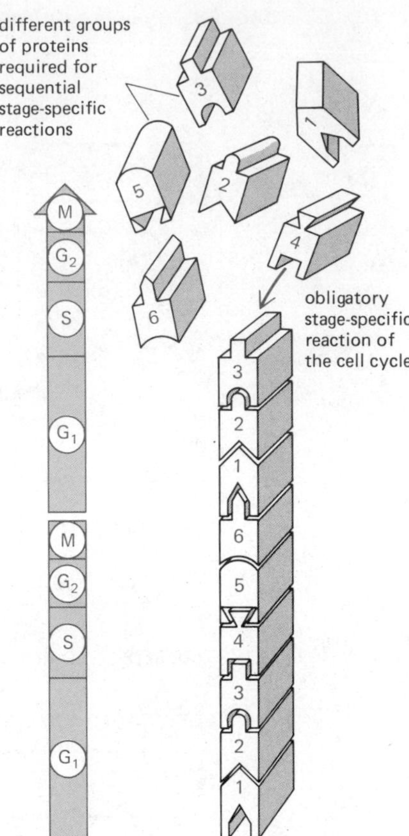

Figure 11–33 Schematic view of the transition of a cell through the various stages of its growth cycle. In this view, the cycle is driven by a series of biosynthetic "events," *each of which must be completed before the next event can begin.* These events are the stage-specific biosynthetic reactions catalyzed by cdc proteins. The genetic data from yeast would suggest that about 10% of the total proteins in a cell are involved in such stage-specific reactions and that the vast majority of these proteins are present at all times, even though they are required only at one point in the cycle.

Controls Operating Early in the Cycle Largely Determine the Cell Mass at Mitosis[7,26]

The cell mass at mitosis has been measured for a variety of cultured cells in order to determine how precisely cell mass is regulated. A rather wide variation in mass at mitosis is observed, even when cells are genetically identical. Nevertheless, when related tetraploid, diploid, and haploid cell lines are compared, the average relative cell masses at mitosis turn out to be 4, 2, and 1, respectively. This suggests that there exists a mechanism for regulating cell mass with respect to nuclear DNA content, but that the regulation is also rather imprecise at the level of the individual cell.

Newly formed daughter cells vary much more in size than they do later when they enter mitosis. The large size differences between daughter cells

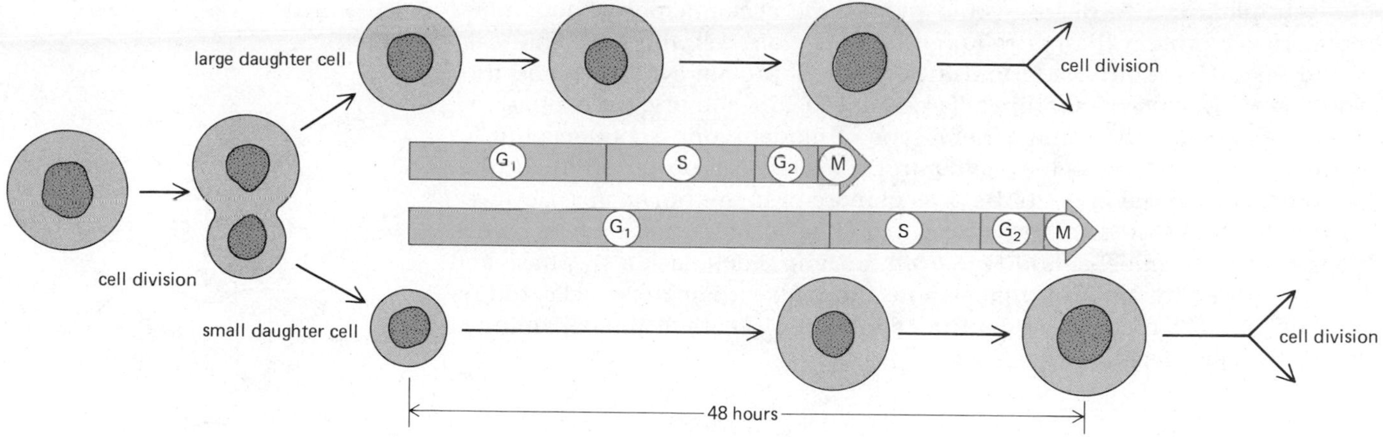

Figure 11–34 Schematic illustration of the lengthening of the G_1 phase observed when cell division accidentally produces a small daughter cell. Since the small cell will synthesize protein at a reduced rate compared to the large cell, this type of cell behavior is in accord with a trigger-protein type of model for growth control.

disappear by the start of the S phase because the smaller daughter cells stay in G_1 until they reach approximately the same size as the larger ones (Figure 11–34). From that point on, the phases of division proceed at the same rate. Similarly, in cells in which the rate of protein synthesis has been reduced by half with low levels of an inhibitor of protein synthesis, the average G_1 period lengthens about threefold, while the lengths of the S and G_2 periods remain about the same. Thus, by the time a mammalian cell has started to synthesize DNA, its final mass at the time of mitosis seems to be determined.

These observations indicate the existence of a special monitoring system that controls the cell cycle at the restriction point in G_1. This monitoring system appears to be designed to delay entry into DNA synthesis until enough components are available to finish each of the sequential biosyntheses required in the S, G_2, and M phases still to come. In terms of the schematic view of the cell cycle presented previously (Figure 11–33), the cells would stop cycling at a restriction point in G_1 if step 3 were a biosynthetic reaction that was very much more sensitive to a reduction in the overall protein synthesis rate than any of the other cell-cycle-specific reactions in the series.

How is the cell size at cell division controlled at the molecular level? It has been postulated that some trigger protein must accumulate to a threshold level to allow progress past a restriction point in each cell cycle (p. 617). How might this threshold level be measured? Some general ways in which all-or-none decisions can be made by cells will be discussed later, in the context of cell signaling (see p. 751). However, the direct proportionality observed between the number of sets of chromosomes and cell size suggests an alternative possibility: perhaps the postulated trigger protein binds to chromosomal sites, with progression into the S phase (and thus to a new cell cycle) requiring an amount of this protein in excess of that needed to bind to the chromosomes. In this case, the chromosomes would serve as an inhibitor of trigger-protein action, thereby providing a yardstick for measuring the total amount of trigger protein present in a cell (Figure 11–35). In addition, because the commitment

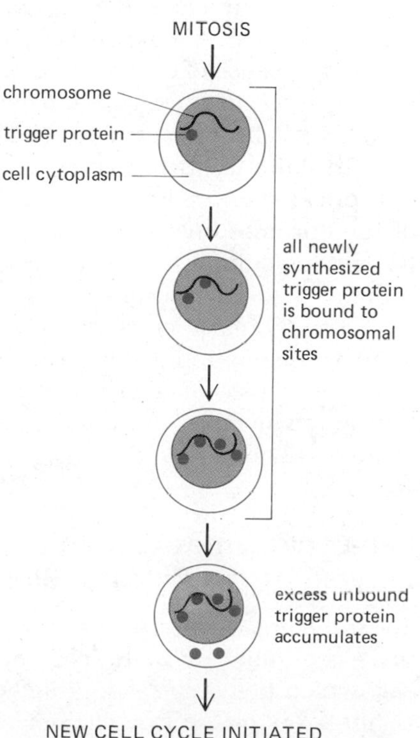

Figure 11–35 A scheme postulated to account for models in which a threshold amount of a trigger protein must accumulate before each new cell cycle can begin. In this hypothetical scheme, only the free trigger protein is active. By binding the trigger protein very tightly, the chromosomes (or some related nuclear structure) serve as inhibitors that prevent subthreshold levels of the trigger protein from acting.

of a tetraploid cell to divide would require the accumulation of four times more trigger protein than is required for a haploid cell, the tetraploid cells would need to acquire a fourfold higher rate of protein synthesis (and thus would grow to a correspondingly larger size) before entering the S phase.

Interestingly, the same general type of model proposed to account for features of mammalian cell-cycle control has also been proposed to account for control of the cell cycle in the cells of more primitive organisms. However, some of the best studied of these cells (including *Schizosaccharomyces pombe* [a fission yeast], amoeba, and *Physarum* [a slime mold]) lack a G_1 phase and thus progress directly from mitosis into the S phase. For these cells, the restriction point that seems to control cell division occurs at the beginning of the M phase rather than in the late G_1 phase.

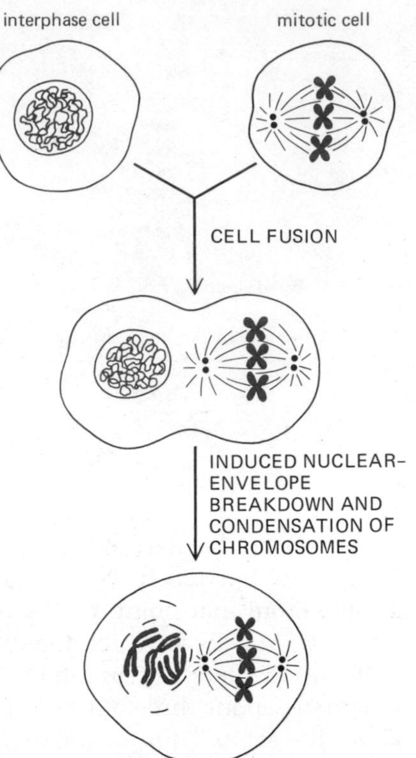

Figure 11–36 Schematic diagram of an experiment in which cells in the M phase are fused with cells in interphase. As indicated, a premature condensation of the chromosomes of the interphase cell results (see Figure 11–37).

The G_2 Phase Serves to Prepare Cells for Mitosis[27]

Preventing protein synthesis in cells even late in G_2 prevents their entry into the mitotic phase, indicating that some proteins synthesized during this period are essential for cell division. A clue to the identity of some of these proteins has been provided by experiments in which cells in M are fused with interphase cells. In such cells, the chromosomes of the interphase nucleus quickly condense and the surrounding nuclear membrane disintegrates (Figures 11–36 and 11–37). This result suggests that soluble factors that are absent from interphase cells appear in late G_2.

It has been suggested that a soluble protein kinase (which catalyzes protein phosphorylation) is activated near the end of G_2 and drives the transition from G_2 to M. For example, this kinase could be responsible for the phosphorylation of the proteins of the nuclear lamina, which may in turn cause the nuclear-envelope breakdown observed during the M phase (see p. 659). In addition, the kinase is thought to be responsible for the extensive phosphorylation of histone H1 molecules found in mitotic chromosomes (up to 6 phosphates per molecule). Since histone H1 is present in about one molecule per nucleosome and is known to be involved in packing nucleosomes together, its extensive phosphorylation just prior to the M phase could be a major cause of chromosome condensation. While still quite tentative, this type of molecular explanation illustrates the level at which one must ultimately explain the entire cell cycle.

In contrast to the above results, new spindles are not induced when interphase cells are fused with cells in the M phase, despite the condensation of the chromosomes. Therefore, crucial parts of the elaborate machinery of the mitotic spindle (to be described below) are prepared late in G_2 and used up once a spindle forms.

The G_2 phase ends as the prophase stage of M begins. *Prophase* is defined as the point in the cycle at which condensed chromosomes first become visible (see p. 648). Inasmuch as the extent of chromosome condensation appears to increase continuously during late G_2, the beginning of the M phase is arbitrarily defined.

Cell-Cycle Times Can Be Drastically Shortened When Not Limited by the Rates of Biosynthesis[28]

For some organisms, most of the macromolecules (except for DNA) needed to make the initial 10^3 to 10^4 cells of the embryo are present in the unfertilized egg, which has a correspondingly large mass at the time of fertilization (see Chapter 14). In the frog egg, for example, after the first division, the next 12 cell divisions (called *cleavages*) occur at 30-minute intervals, producing the

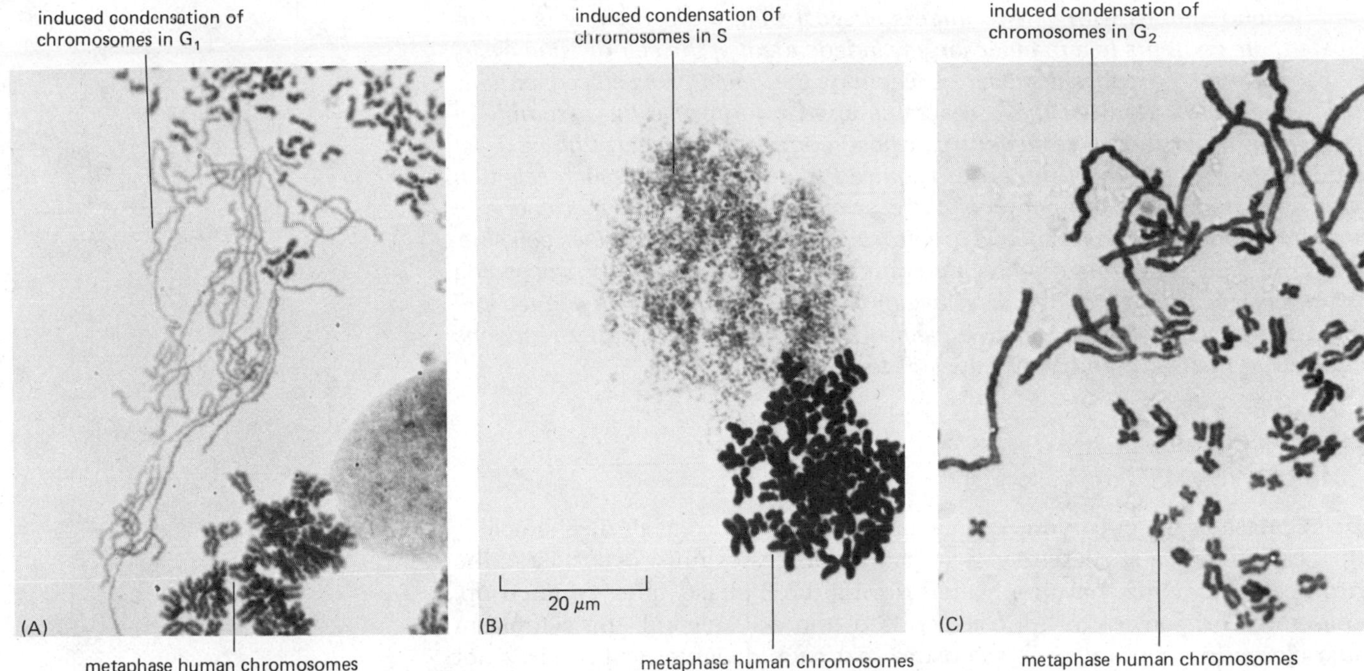

Figure 11–37 Premature condensation of interphase chromosomes following fusion of interphase rat kangaroo (PTK) cells with mitotic HeLa (human) cells. In (A), the PTK cell was in G_1 phase; consequently, its prematurely condensed chromosomes are still single chromatids. In (B), the PTK cell was in S phase, and its chromatin now adopts a "pulverized" appearance. In (C), the PTK cell was in G_2 phase, and now the chromatids, although very long compared to the normal metaphase chromosome, are double. (From K. Sperling and P. Rao, *Humangenetik* 23:235–258, 1974.)

8000 (2^{13}) cells found in the midblastula. With each division the cells become progressively smaller until the normal size of adult cells is attained. These 30-minute cell cycles consist entirely of back-to-back S and M phases.

As illustrated in the electron micrograph of Figure 11–38, the rapid S phase found in several types of early embryos is made possible by a very large number of replication origins spaced at intervals of only a few thousand base pairs (rather than the tens of thousands of base pairs found between the origins of replication later in development). Any DNA injected into a frog egg is replicated to some extent, including small pieces of bacterial DNA that would not be expected to contain a site normally used for initiating DNA replication. It seems that, in contrast to the situation thought to exist in adult cells, many different DNA sequences can serve as replication origins in the cells of early embryos. But the main factor allowing these rapid division rates is the presence in an egg of all of the components needed to make new cells (except for DNA), including spindle components, histones, ribosomes, mitochondria, and enzymes.

An important general conclusion to be drawn from the existence of these rapid cell cycles in embryos is that there is nothing about cell division itself that requires a long time to accomplish. For example, chromosomes can be condensed in preparation for mitosis in minutes, rather than in the hours it takes in a normal G_2 period. Most of the time required between successive cell divisions is needed to complete the biosynthesis of the many different components, other than DNA, that are needed to generate two cells from one.

Summary

The very rapid cell divisions that occur after the fertilization of large eggs indicate that the orderly events of the cell cycle can be completed in 30 minutes or less when all components needed to make a new cell are present in great excess. Normally, however, cells must double their mass before they divide, and this process seems to require 10 to 20 hours or more in mammalian cells, during which time most of the cell's proteins are continuously synthesized.

Genetic studies with yeast mutants suggest that, while growing cells duplicate their contents in a manner largely independent of the cell-division cycle, their growth also triggers a subset of reactions that must proceed sequentially during the cycle. Many of these reactions appear to involve the assembly of preexisting molecules into structures whose completion triggers the next assembly event. The sum of the times required to complete each such reaction determines the total time between successive cell divisions. In general, one of these reactions serves as a special monitoring system that determines cell size. In mammalian cells, this monitoring system prevents a cell from proceeding past the restriction point in G_1 (and beginning DNA synthesis) until enough cell components are available to finish each of the sequential assembly reactions required in the S, G_2, and M phases still to come.

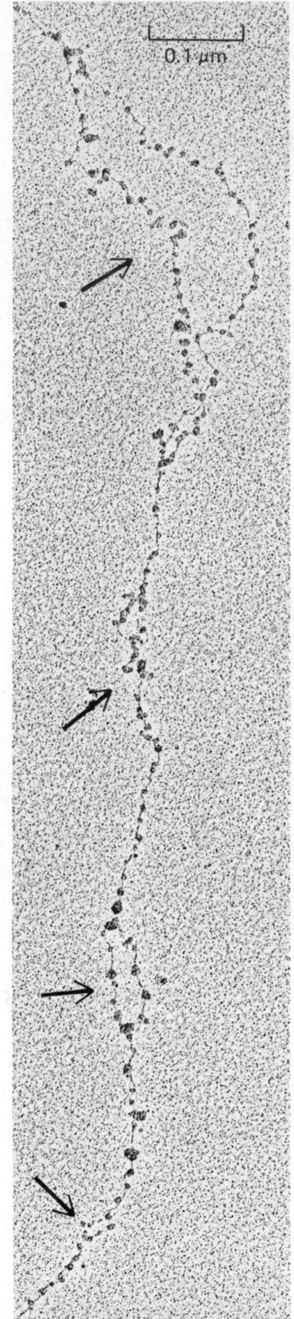

Figure 11–38 Electron micrograph of spread chromatin from an early embryo of *Drosophila melanogaster* showing the closely spaced clusters of replication bubbles observed. Only about 10 minutes elapse between some of the successive nuclear divisions in this embryo. (Courtesy of Victoria Foe.)

Cell Division[29]

The M phase of the eucaryotic cell cycle is a very complex affair mechanically. For a cell to divide successfully, two distinct processes must occur. First, the chromosomes, which have replicated during the S phase, must be lined up, separated, and moved to opposite ends of the cell. Second, the cytoplasm must cleave in such a way as to ensure that each daughter cell receives not only one complete set of chromosomes, but also the necessary cytoplasmic constituents and organelles. Although these two processes, nuclear division (mitosis) and cytoplasmic division (cytokinesis) are experimentally separable, they usually occur in close succession, so that cytokinesis starts toward the end of mitosis (Figure 11–39).

The need for a complex mitotic machinery arose from the evolution of cells having greatly increased amounts of DNA packaged in a number of discrete chromosomes. The function of the machinery is to ensure that the replicated chromosomes are precisely divided between the two daughter cells at division. This machinery is called the **mitotic apparatus.** The accuracy with which it works has been estimated in yeast cells, where an error in chromosome segregation is made only about once every 10^5 divisions.

Although detailed descriptions of the mitotic apparatus and its actions in a wide variety of cell types are now available, present understanding of events at the molecular level is still fragmentary. Nevertheless, we shall attempt an analysis of cell division in terms of the molecular components involved, since only in this way can cellular functions eventually be understood. If we can learn how microtubules align and move chromosomes and how actin filaments divide the cytoplasm in two, we shall also be in a better position to understand how the same structures function in other cellular processes.

Before discussing the mechanisms involved in cell division, it is necessary to outline the various stages of mitosis and cytokinesis as they occur in a typical cell of a higher animal.

Cell Division Is Traditionally Divided into Six Stages[30]

The first five stages of cell division constitute mitosis, while the sixth is cytokinesis. *In vivo* these six stages form a continuous dynamic sequence, the complexity and beauty of which are hard to appreciate from written descriptions or from a set of static pictures. The description of cell division is based on observations from two sources: light microscopy of living cells (often combined with microcinematography) and light and electron microscopy of fixed and stained cells. In Figures 11–40 and 11–41 we present drawings and a brief summary of the various stages of cell division. Light micrographs of cell division in a typical animal and a typical plant cell are shown in Figures 11–42 and 11–43, respectively.

The description of cell division in Figure 11–41 is highly schematic. Innumerable variations on all stages of cell division occur in the animal and plant kingdoms. We shall mention some of these variations when we take a closer look at the mechanisms of cell division since they help us to understand the different parts of the mammalian mitotic apparatus.

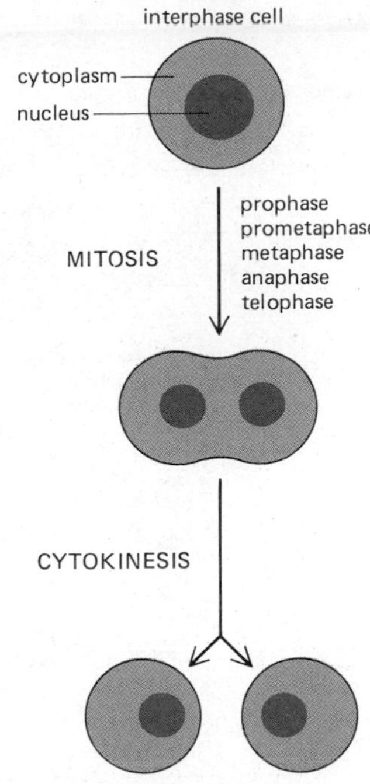

Figure 11–39 The M phase of the cell cycle (the cell-division phase) is divided into five stages of mitosis followed by cytokinesis.

Some Events in Mitosis Depend on the Mitotic Spindle While Others Do Not[31]

The complex sequence of events during nuclear division is visually dominated by the formation and activity of the **mitotic spindle,** which originates in the cytoplasm at prophase (Figure 11–40). And yet not all of these events depend on the spindle. As might be expected, the drug colchicine, which destroys all the microtubules within a cell, prevents the spindle-dependent alignment of chromosomes and the movement of daughter chromatids to opposite poles of the cell. However, in some cells, such as those of the sea urchin, the behavior of the chromatin and nuclear envelope is unaffected by colchicine treatment: in the absence of the mitotic spindle, the condensation and subsequent decondensation of the chromatin, the sudden separation of the two chromatids at their centromere, and the dissolution and re-formation of the nuclear envelope all occur normally. (In contrast, in mammalian cells only the early events of mitosis occur, and each chromosome remains as a pair of condensed sister chromatids during colchicine treatment.)

As previously described, cell-fusion experiments also suggest that the chromatin and nuclear-envelope changes that lead up to metaphase are controlled by soluble factors unrelated to the mitotic spindle, since the fusion of a metaphase cell with one in interphase will cause condensation of the chromosomes and disruption of the nuclear envelope of the interphase nucleus, without formation of a new spindle.

The Spindle Contains Organized Arrays of Microtubules in Dynamic Equilibrium Between Assembly and Disassembly[32]

The mitotic spindle consists of fibers constructed of microtubules and microtubule-associated proteins. The spindle fibers fall into two categories defined by their attachments. The **polar fibers,** which in most spindles are the most numerous, extend from the two poles of the spindle toward the equator; the **kinetochore fibers** are attached to the centromere of each chromatid and extend toward the spindle poles (Figure 11–44). An average spindle contains about 10^8 tubulin molecules assembled into microtubules. However, this figure does not account for all of the tubulin in the cell, and the function of the spindle in mitosis can be shown to depend on a dynamic equilibrium between the spindle microtubules and a pool of soluble tubulin molecules.

This dynamic equilibrium can be demonstrated by observing mitotic cells under conditions that reversibly shift the equilibrium between tubulin polymerization and depolymerization; the behavior of the spindle fibers can be directly observed in living cells by illumination with polarized light since the fibers are birefringent (Figures 11–45 and 11–46). If mitotic cells are placed in heavy water (D_2O) or treated with taxol, either of which inhibits microtubule disassembly, the spindle fibers lengthen (Figure 11–46). Such stabilized spindles cannot move chromosomes, and mitosis is arrested. At the other extreme, mitosis is blocked when the spindle fibers are reversibly disrupted by any one

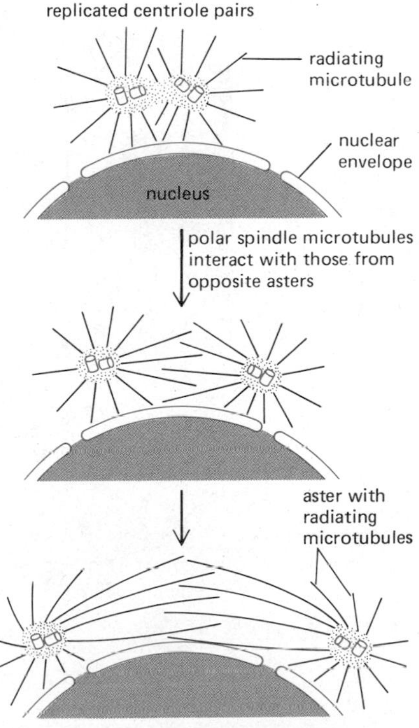

Figure 11–40 The origin of the bipolar spindle in the cytoplasm at prophase from two mitotic centers.

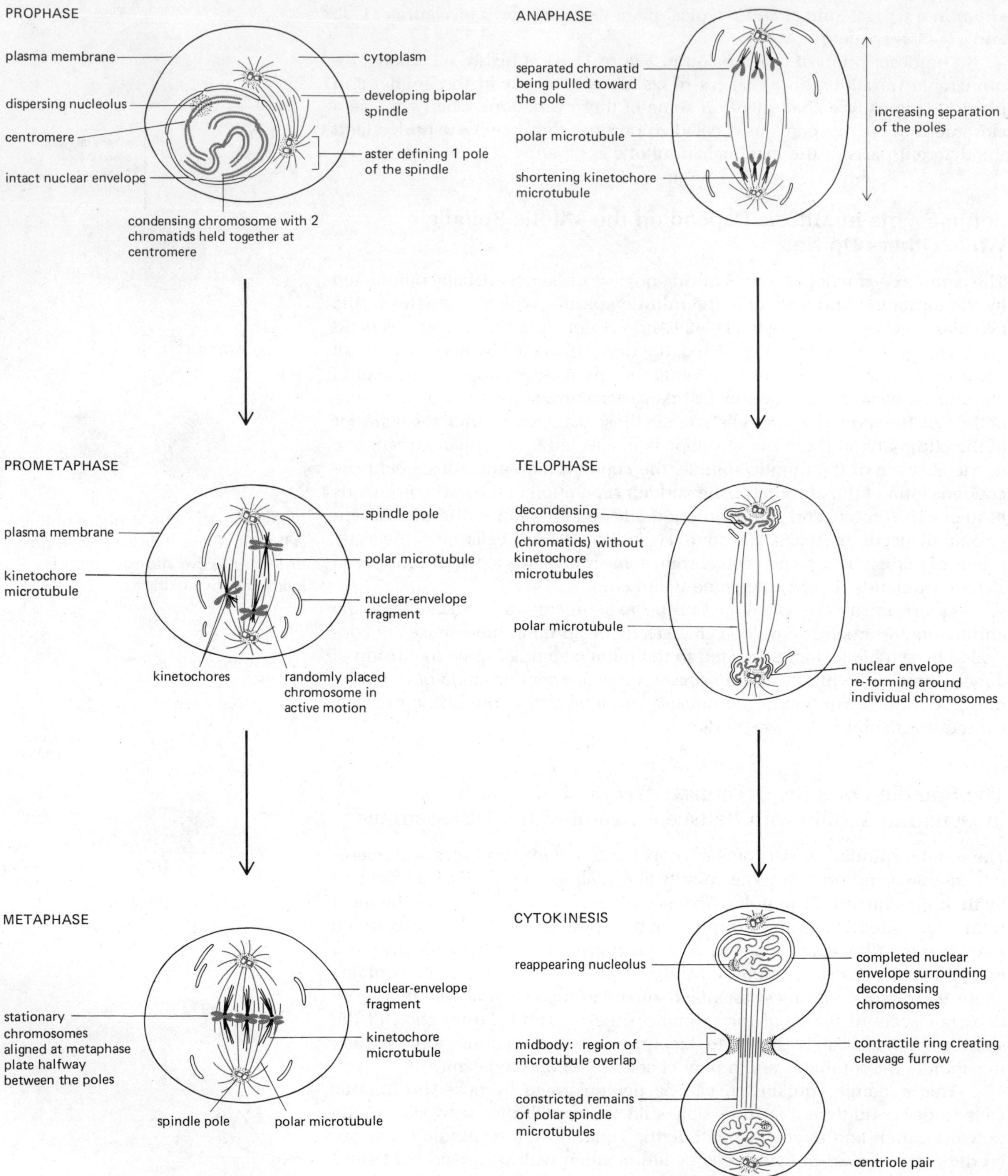

Figure 11–41 The six stages of cell division.

PROPHASE

The transition from the G_2 phase to the M phase of the cell cycle is not a sharply defined event. The chromatin, which is diffuse in interphase, slowly condenses into well-defined chromosomes, the exact number of which is a characteristic of the particular species; each chromosome has duplicated during the preceding S phase and consists of two sister *chromatids* joined at a specific point along their length by a region known as the *centromere.* While the chromosomes are condensing, the nucleolus begins to disassemble and gradually disappears.

At the beginning of prophase, the mass of cytoplasmic microtubules that are part of the cytoskeleton disassemble, forming a large pool of tubulin molecules. These molecules are then presumably reused in the construction of the main component of the mitotic apparatus, the *mitotic spindle.* This is a bipolar fibrous structure, largely composed of microtubules, that assembles initially outside the nucleus. The focus for the formation of the spindle is marked in most animal cells by the centrioles. The cell's original pair of centrioles replicates by a process that begins just prior to the S phase to give rise to two pairs of centrioles. Each centriole pair now becomes part of a *mitotic center* that forms the focus for a radial array of microtubules, the *aster* (aster = star). The two asters initially lie side by side close to the nuclear envelope. By late prophase the bundles of *polar microtubules* that interact between the two asters (seen as *polar fibers* in the light microscope) preferentially elongate and appear to push the two centers apart along the outside of the nucleus. In this way, a bipolar mitotic spindle is formed (see Figure 11–40).

PROMETAPHASE

Prometaphase starts abruptly with the disruption of the nuclear envelope, which breaks up into membrane fragments indistinguishable from bits of endoplasmic reticulum. These fragments remain visible around the spindle during mitosis. The spindle, which has been lying outside the nucleus, can now enter the nuclear area. Specialized structures called *kinetochores* develop on either face of the centromeres and become attached to a special set of microtubules, called *kinetochore fibers* or *kinetochore microtubules.* These fibers radiate in opposite directions from each side of each chromosome and interact with the fibers of the bipolar spindle. The chromosomes are thrown into agitated motion due to the interactions of their kinetochore fibers with other components of the spindle.

METAPHASE

As a result of their prometaphase oscillations, the chromosomes become arranged so that their centromeres all lie in one plane. The kinetochore fibers seem to be responsible for aligning the chromosomes halfway between the spindle poles and for orienting them with their long axes at right angles to the spindle axis. Each chromosome is held in tension at the *metaphase plate* by the paired kinetochores and their associated fibers pointing to opposite poles of the spindle.

ANAPHASE

Often metaphase lasts for a long time. As if triggered by a special signal, anaphase begins abruptly as the paired kinetochores on each chromosome separate, allowing each chromatid to be pulled slowly toward a spindle pole. All chromatids are moved (toward the pole they face) at the same speed, about 1 μm per minute. During these anaphase movements, kinetochore fibers shorten as the chromosomes approach the poles. At about the same time, the spindle fibers elongate and the two poles of the polar spindle move further apart. Anaphase typically lasts only a few minutes.

TELOPHASE

As the separated daughter chromatids arrive at the poles, the kinetochore fibers disappear. The polar fibers elongate still further, and a new nuclear envelope re-forms around each group of daughter chromatids. The condensed chromatin expands once more, nucleoli begin to reappear, and mitosis is at an end.

CYTOKINESIS

The cytoplasm divides by a process known as *cleavage,* which usually starts sometime during late anaphase or telophase. The membrane around the middle of the cell, perpendicular to the spindle axis and between the daughter nuclei, is drawn inward to form a *cleavage furrow,* which gradually deepens until it encounters the narrow remains of the mitotic spindle between the two nuclei. This narrow bridge, or *midbody,* may persist for some time before it narrows and finally breaks at each end, leaving two completed, separated daughter cells.

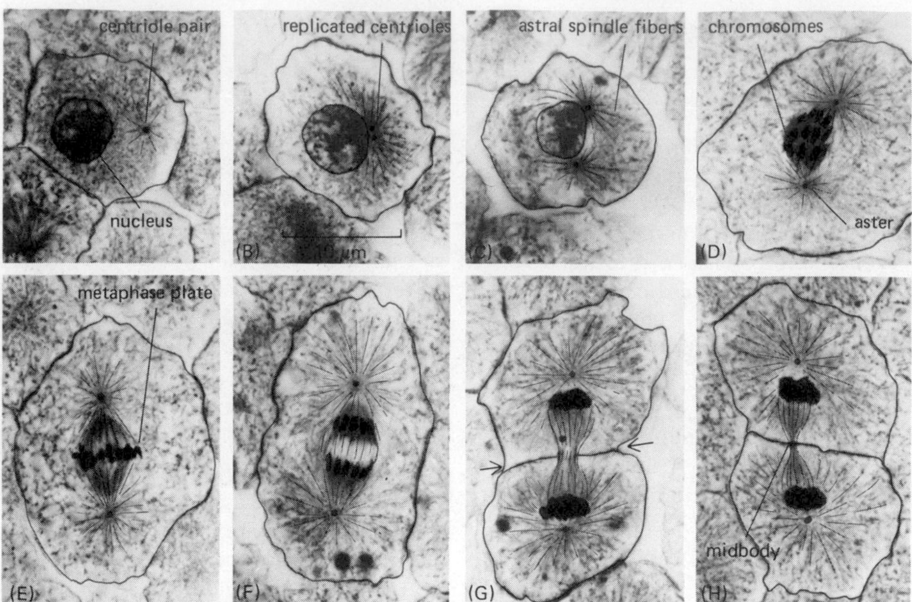

Figure 11–42 The course of mitosis in typical animal cells (whitefish). (A) Interphase: the cell center containing the centriole is often small and somewhat removed from the nucleus. (B) Early prophase: the cell center has doubled and moved near the nucleus as increasing numbers of fibers radiate from it. (C) Midprophase: the two asters are separating around the nucleus. (D) Prometaphase: the nuclear envelope has broken down, allowing the spindle fibers to interact with the chromosomes. (E) Metaphase: the bipolar spindle structure is clear, and all the chromosomes are aligned across the middle of the spindle. (F) Anaphase: the chromatids all separate synchronously to the poles under the influence of the spindle fibers. (G) Early telophase: the chromatids are massed at each pole, while the cleavage furrow (*arrows*) has compressed the spindle fibers remaining between the chromatids. (H) Telophase: the daughter nuclei are re-formed, although still compact; cytokinesis is almost complete; and the midbody persists between daughter nuclei. (Courtesy of Jeremy D. Pickett-Heaps.)

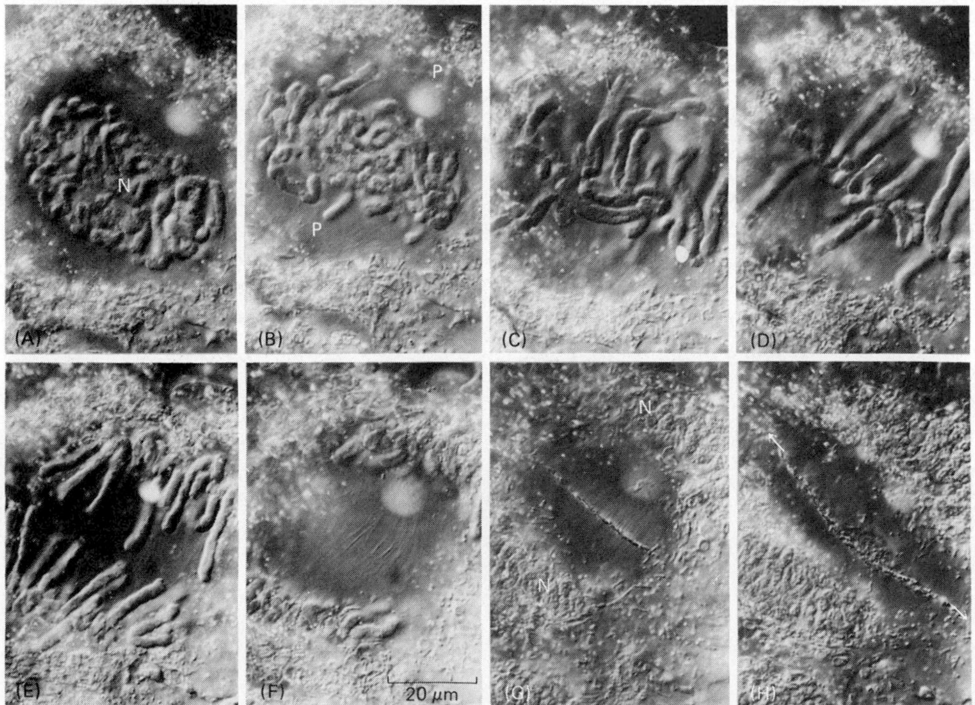

Figure 11–43 The course of mitosis in typical plant cells. These micrographs were taken of a living *Haemanthus* cell by using differential-interference-contrast microscopy. In (A) and (G), N denotes the center of a nucleus, and P in (B) marks each spindle pole. Arrows in (H) indicate the direction of growth of the new cell plate, which is also seen in (G). Time sequence in minutes is as follows: A, 0; B, 15; C, 17; D, 54; E, 83; F, 124; G, 169; H, 199. (Courtesy of Andrew Bajer.)

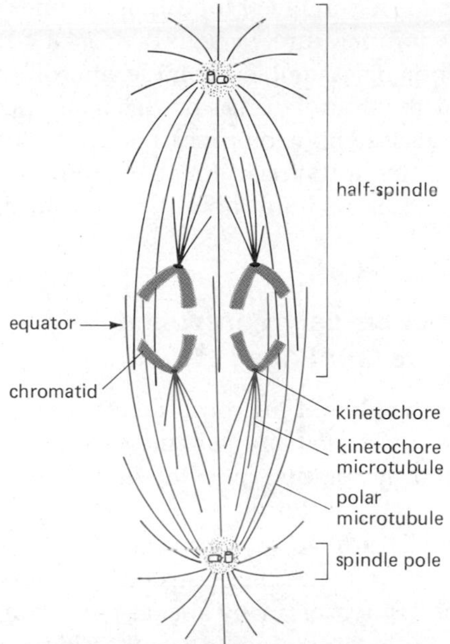

Figure 11–44 The spindle at early anaphase, showing its construction from two half-spindles (*black* and *color*), each composed of kinetochore and polar microtubules. For clarity, the sister chromatids of only two chromosomes are shown and most of the polar microtubules are omitted.

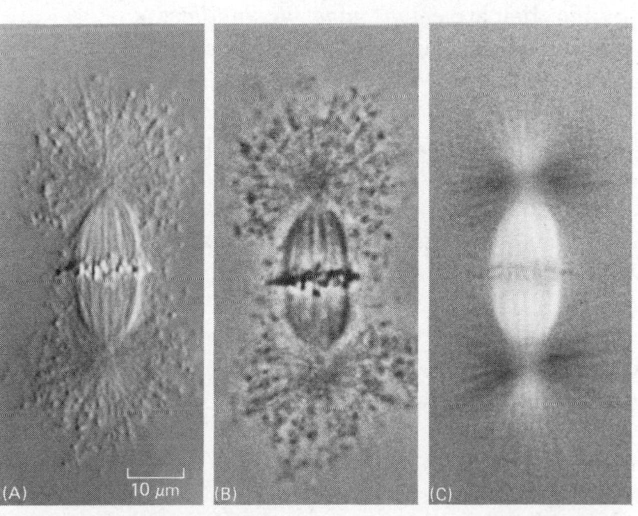

Figure 11–45 An isolated metaphase spindle viewed by three different techniques of light microscopy: (A) differential-interference-contrast microscopy, (B) phase-contrast microscopy, and (C) polarized light microscopy. (Courtesy of E. D. Salmon and R. R. Segall, from *J. Cell Biol.* 86:355–365, 1980. Reproduced by copyright permission of the Rockefeller University Press.)

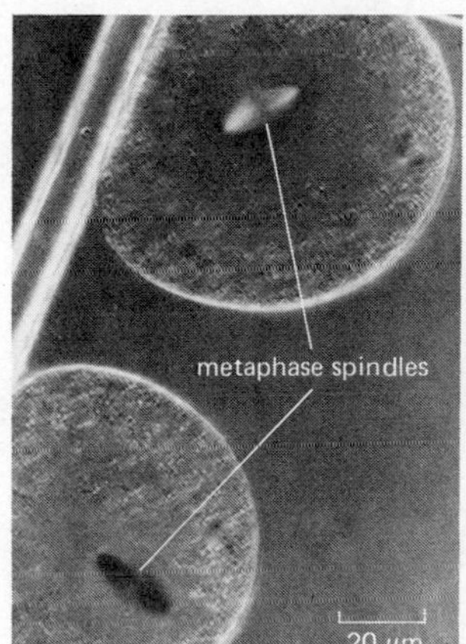

(A) CELLS IN SEA WATER: NORMAL SPINDLE

(B) CELLS PLACED IN 45% HEAVY WATER IN SEA WATER FOR 2 MINUTES: INCREASED NUMBER OF SPINDLE FIBERS

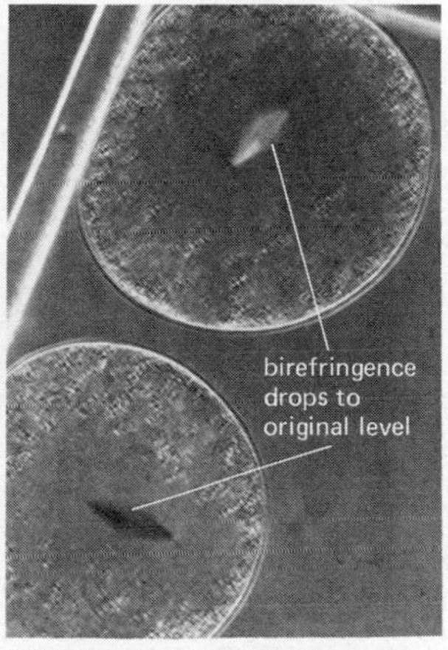

(C) CELLS BACK IN SEA WATER FOR 3 MINUTES: NORMAL SPINDLE

Figure 11–46 Effect of heavy water (D_2O) on the spindle of the oocyte of *Pectinaria*, viewed in the living cell by polarized light microscopy. Whether birefringence is seen as positive or negative contrast depends on the orientation of the structure with respect to the plane of polarized light. Since one spindle is at right angles to the other, the two spindles appear with opposite contrast. Chromosomes are not visible with these optics. (From S. Inoué and H. Sato, *J. Gen. Physiol.* 50:259–292, 1967. Reproduced by copyright permission of the Rockefeller University Press.)

of three treatments that interfere with the assembly of tubulin molecules into microtubules: the drug colchicine, low temperature, or high hydrostatic pressure. The fact that neither stabilized nor disassembled spindle microtubules can move chromosomes suggests that the spindle must be delicately poised at equilibrium between assembly and disassembly in order to perform mitotic movements. Before discussing in more detail the basis for these movements, we must describe how the spindle is organized and how the chromosomes are positioned.

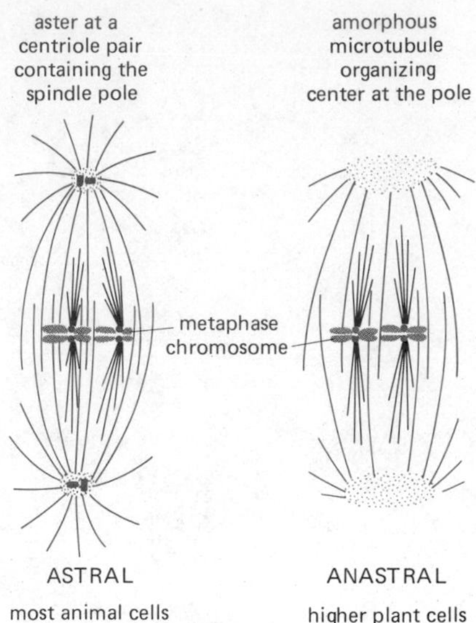

Figure 11–47 Metaphase spindle organization in cells containing astral and anastral spindle types.

Spindle Formation at Prophase Is Largely Controlled by the Mitotic Centers That Define the Poles[33,34]

We have seen in Chapter 10 that the assembly of microtubule arrays is organized by specialized regions known as *microtubule organizing centers.* In the dividing cell, this function is performed by the two **mitotic centers** that ultimately form the poles of the spindle.

The mitotic centers of animal cells are usually associated with **centrioles,** and for a long time it was assumed that these structures were the organizing centers for spindle assembly. However, many species, including all higher plants, form fully functional spindles in the complete absence of centrioles. Such spindles have no asters (hence they are called *anastral spindles*) and are less focused at the poles than astral spindles (Figure 11–47). Moreover, when the centrioles in mitotic animal cells are destroyed with a laser microbeam, the spindle continues to function normally. It therefore seems that centrioles are not essential for spindle microtubule assembly, although they probably act as a focus for it when they are present.

If the centriole itself is not the microtubule organizing center of the spindle, then what is? There is now clear evidence that the true organizing center is an ill-defined cloud of lightly staining material, visible in the electron microscope, at the poles of both astral (centriole-containing) and anastral (noncentriole-containing) spindles. When the mitotic centers from animal cells are isolated and used as nucleating agents *in vitro,* the polar microtubules of the aster grow from the amorphous material surrounding the centrioles and not from the centrioles themselves.

Rather than being required to organize spindles, the position of the duplicated centrioles at the two opposite poles of the spindle may have evolved to ensure that these structures, which are important in animal cell polarity and movement (see p. 601), are properly segregated (one to each daughter cell) during cell division.

Each Mitotic Chromosome Consists of Two Chromatids and a Pair of Kinetochores That Bind to the Ends of Microtubules[35]

The replicated chromosomes are linked to the mitotic spindle by specialized structures known as **kinetochores.** Each replicated chromosome consists of two sister chromatids joined at a relatively rigid region called the **centromere** (Figure 11–48). During late prophase, a mature kinetochore develops on each of two opposite faces of the centromere, so that the two kinetochores (one on each sister chromatid) are oriented in opposite directions. The kinetochores can be seen in the electron microscope as multilayered structures (Figure 11–49), but little else is known about their composition.

At the beginning of prometaphase, when the nuclear envelope breaks down, a separate set of spindle fibers becomes associated with each chromatid. These fibers are made of microtubules that radiate outward in opposite directions from the two kinetochore regions of each chromosome (Figure 11–48). They serve to orient the chromosomes with respect to the spindle at meta-

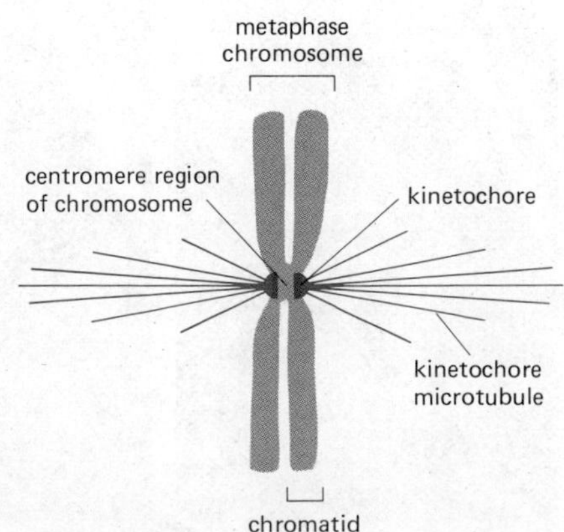

Figure 11–48 Schematic drawing of a metaphase chromosome and its kinetochore microtubules.

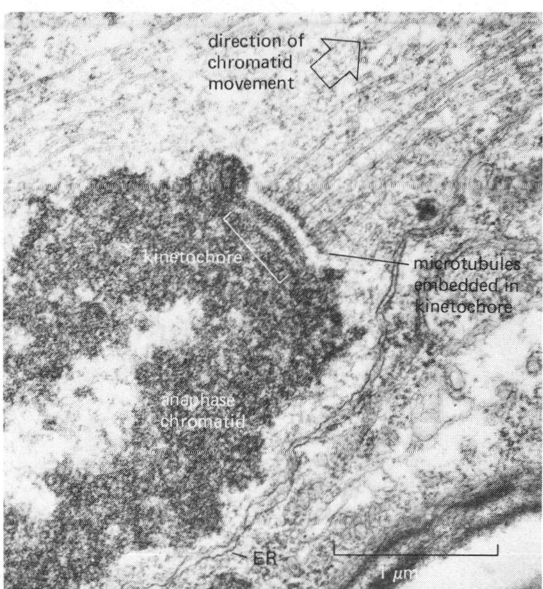

Figure 11–49 Kinetochore with microtubules inserted in an anaphase chromatid of the green alga *Oedogonium*. Most kinetochores have a trilaminar structure. This kinetochore has an unusually complex structure with additional layers. (From J. D. Pickett-Heaps and L. C. Fowke, *Aust. J. Biol. Sci.* 23:71–92, 1970. Reproduced by permission of CSIRO.)

phase and eventually, at anaphase, transmit the pole-directed forces responsible for moving the chromatids to each pole. The number of microtubules associated with each kinetochore varies widely among species: while some fungi have only one, human kinetochores have 20 to 40.

The Chromosome Alignment at Metaphase Is Generated by Interactions Between the Kinetochore and Polar Spindle Fibers[34,36]

The breakdown of the nuclear envelope, which signals the beginning of prometaphase, enables the chromosomes to engage with the spindle machinery. They do this in such a way that their eventual segregation is assured, one chromatid of each chromosome faithfully ending up in each daughter nucleus. It has been said that the chromosomes in mitosis are like the corpse at a funeral; they provide the reason for the proceedings but do not take an active part in them. It is only the kinetochores that link the chromosomes to the mitotic machinery, and all chromosomal movements appear to result from the interactions between the kinetochore fibers and the rest of the spindle. This interaction has two results: (1) it orients each chromosome with respect to the spindle axis so that one kinetochore faces each pole; and (2) it moves each chromosome into a plane (the *metaphase plate*) at the center of, and at right angles to, the bipolar spindle. In mammalian cells this process takes between 10 and 20 minutes, a period known as *prometaphase.*

Prometaphase is characterized by a period of apparently frantic activity while the spindle appears to be trying to contain and align the chromosomes at the metaphase plate. In actuality, the individual chromosomes are violently rotating and oscillating to and fro between the poles because their kinetochore fibers are interacting randomly with the polar spindle fibers. Eventually one or the other set of polar fibers succeeds in "capturing" one kinetochore permanently; the other kinetochore on the chromosome soon associates with the fibers of the other pole.

Information about this important set of events in mitosis has come from delicate micromanipulation experiments in which extremely fine glass needles are used to poke and pull at chromosomes inside a living mitotic cell. By this

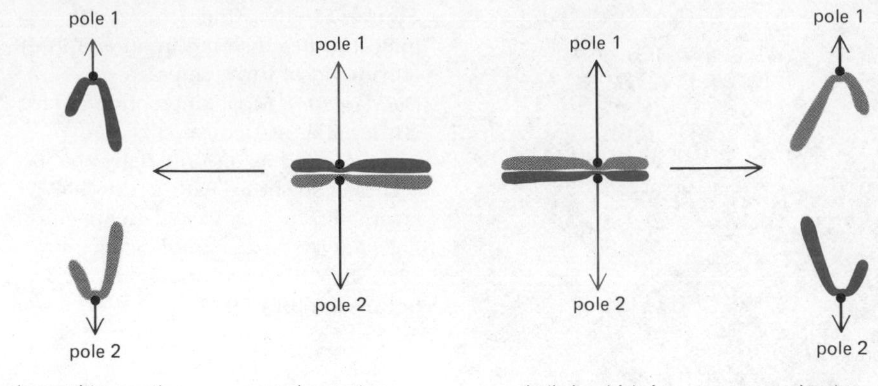

Figure 11–50 The distribution of chromatids to daughter cells is normally random and depends only on the orientation of the chromosome at metaphase. Both this orientation and the subsequent distribution can be altered by micromanipulation.

means it has been demonstrated that kinetochores are not committed to face one particular spindle pole as if polarized like a magnet, since a kinetochore of a manipulated chromosome will reengage to either pole if it is turned around. Moreover, by micromanipulation during prometaphase, it is possible to force the two sets of kinetochore fibers on a single chromosome to engage with the same spindle pole. If the chromosome remains like this, the entire chromosome (with joined sister chromatids) is drawn toward the pole thus engaged. (In general, such an arrangement is unstable, however, since the kinetochore fibers spring back to their preferred opposite orientation.) It is therefore clear that a steady poleward tension is exerted on each kinetochore fiber as soon as it engages with the spindle. Normally, each chromosome will have one kinetochore fiber engaged with each pole, and the opposing tension thereby generated eventually aligns each chromosome on the metaphase plate (see below). But it is the random prometaphase movements, and the final chance orientation of the chromosome that results, that guarantee the random segregation of chromatids to the daughter cells, which is so important during the analogous nuclear division in meiosis (Figure 11–50).

Metaphase Chromosomes Are Held in a Deceptively Static State by Balanced Bipolar Forces[34]

At metaphase the spindle and the chromosomes aligned on the metaphase plate appear to be resting. This apparent quiescence, however, is deceptive, for any cytoplasmic particle that finds itself in either half of the metaphase spindle is carried toward the associated pole. Such particles move slowly and steadily at a rate of about 1 μm per minute, which is about the rate of anaphase chromosome movement itself. This indicates that during metaphase continuous poleward-directed forces are generated within the spindle. The chromosomes remain stationary because the two forces are equal and opposite. It is therefore not surprising that as soon as the two kinetochores on each chromosome separate at anaphase the two chromatids are carried toward opposite poles.

Why do all of the chromosomes line up at an equal distance from the two spindle poles at metaphase, thereby defining the metaphase plate? The alignment is probably a simple consequence of a force-generating system in which the pull on each kinetochore fiber decreases the closer its kinetochore approaches to a pole (Figure 11–51).

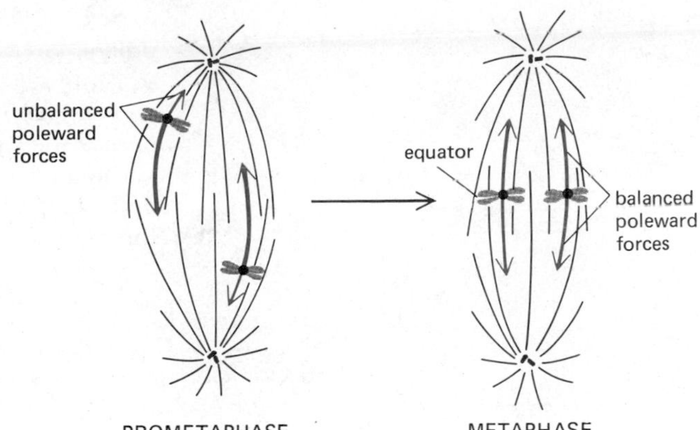

Figure 11–51 Chromosomes randomly enter the spindle during prometaphase and become centered because the force on each kinetochore decreases the closer it gets to a pole. This results in the chromosomes lining up at the equator, where they are held under tension by balanced poleward forces.

The Sister Chromatids Separate Suddenly at Anaphase

Anaphase is initiated by a sudden splitting of each chromosome, caused by the breaking apart of the sister chromatids at their point of union at the centromere. This splitting, which separates the kinetochores, is an independent event that will occur even on a chromosome not attached to the spindle. It allows the bipolar spindle forces acting on the metaphase plate to initiate the slow, stately movement of each chromatid toward one or the other pole.

What type of attachment holds sister chromatids together until anaphase? Although the answer is not known, one idea is that the DNA sequence that specifies the existence of a centromere might encode a special signal that blocks its own DNA replication during the S phase. According to this view, the unreplicated centromeric DNA holds chromatids together, and the triggering of this last trace of DNA synthesis at the centromere provides the final impetus for chromatid separation at anaphase (Figure 11–52).

At Anaphase the Chromosomes Are Pulled Toward the Poles and the Two Poles Are Pushed Apart[37]

The all-important chromatid movements at anaphase are the product of two independent sets of events within the spindle: (1) a poleward movement of the kinetochore fibers that *pulls* the attached chromatids closer to the poles, and (usually starting a little later) (2) the elongation and sliding of the polar fibers in each half-spindle that *pushes* the two poles further apart (Figure 11–53). It is possible to distinguish between these two events by their differential sensitivity to certain drugs. A low concentration of chloral hydrate, for

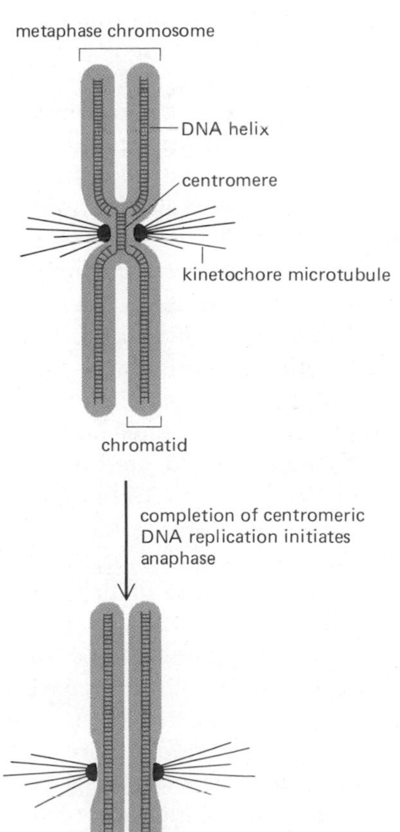

Figure 11–52 One of many possible schemes postulated to account for the tight attachment of sister chromatids to each other at metaphase and for their sudden detachment at anaphase. Whatever the actual mechanism, a centromeric DNA sequence must have two special effects: it must cause each sister chromatid to (1) organize a kinetochore and (2) remain paired to its partner chromatid until anaphase. Recently, DNA molecules corresponding to several different yeast centromeres have been isolated through the use of recombinant DNA technology. Centromeric functions have thereby been shown to be specified by a DNA sequence of less than 1000 base pairs.

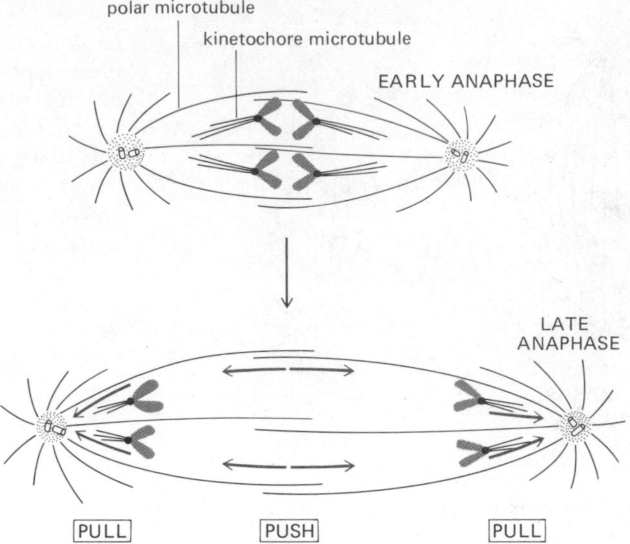

Figure 11–53 Both pushing and pulling forces act at anaphase to separate sister chromatids. Elongation and sliding of the polar microtubules *pushes* the two poles apart, while the chromatids are *pulled* toward opposite poles by forces that act on their kinetochore fibers.

example, prevents the movement and lengthening of the polar fibers but has no effect on the kinetochore fibers or the poleward movement of the chromatids. Moreover, in different organisms the relative contribution of each of the two events to the final separation of the two sets of chromosomes varies considerably. In some cells the polar separation hardly changes at all during anaphase, whereas in others the spindle elongates to such an extent that the poles become separated by 15 times the distance of their original separation in the metaphase spindle!

By painstakingly reconstructing the three-dimensional architecture of complete spindles from hundreds of serial thin sections examined in the electron microscope, it has been shown that the polar microtubules from each half-spindle overlap in a central region near the spindle equator (Figure 11–54). During anaphase these two sets of antiparallel polar microtubules slide away from each other in the region of overlap and, presumably, grow longer by the addition of subunits to their free ends (Figures 11–55 and 11–56). The two poles are apparently pushed apart by forces generated by this combination of microtubule sliding and growth.

We have explained in Chapter 10 that microtubules have a well-defined polarity and grow faster from one end—the (+) end—than from the other—the (−) end. As expected, the polar spindle fibers have the same polarity as the microtubules that are organized by the cell center during interphase: all the polar microtubules have their slow-growing (−) ends at the poles, where they are apparently "capped" and prevented from depolymerizing (see p. 577). This means that the polar microtubules have free (+) ends near their area of overlap at the spindle equator, and it is these fast-growing (+) ends that are thought to elongate by the addition of tubulin subunits during anaphase. The kinetochore microtubules have recently been discovered to have their (+) ends attached to the kinetochore and their (−) ends near the poles. This means that the kinetochore microtubules have the same structural polarity as the polar microtubules in each half-spindle. It has been suggested that each kinetochore microtubule is protected from depolymerization by the capping of *both* of its ends, which could explain why kinetochore microtubules are much more stable than polar microtubules under a variety of experimental conditions.

What generates the forces that pull the kinetochore fibers poleward and enables the polar fibers of each half-spindle to push away from each other in

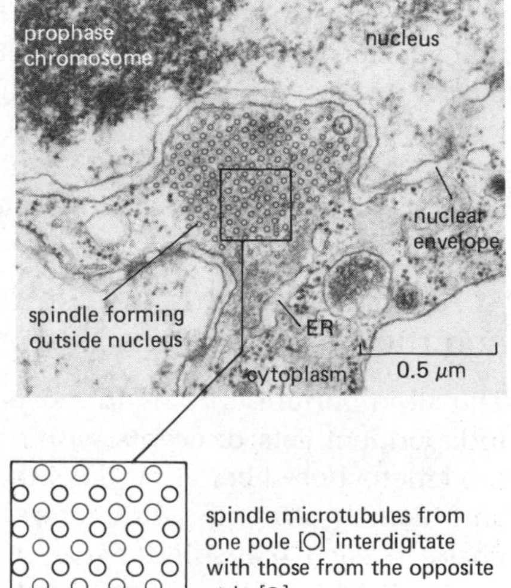

Figure 11–54 A transverse section showing the spindle of the diatom *Melosira* at prophase, prior to nuclear membrane breakdown. This section passes through the central region of microtubule overlap in this spindle, which has an unusually orderly and precise packing of the polar microtubules originating from opposite poles. The microtubule interactions in the spindles of higher eucaryotes are less regular, although the same general principles apply. (From D. H. Tippit, K. L. McDonald, and J. D. Pickett-Heaps, *Cytobiologie*, 12:52–73, 1975.)

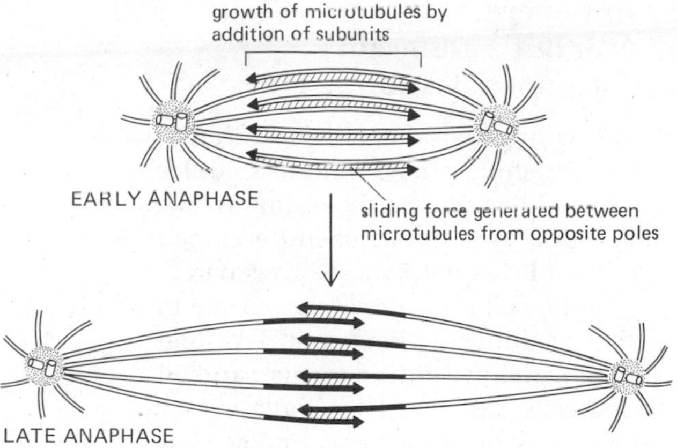

Figure 11–55 Illustration of the prophase components that probably contribute to the increasing separation of the spindle poles during mitosis. Both the chromosomes and the kinetochore microtubules are omitted for clarity.

their central region of overlap? As we have seen, microtubule assembly and disassembly are important for spindle function. However, it seems unlikely that the forces that could in principle be generated by microtubule assembly and disassembly processes constitute the entire "motor" that moves the chromosomes, since there is evidence that the spindle contains other proteins that are capable of generating a force.

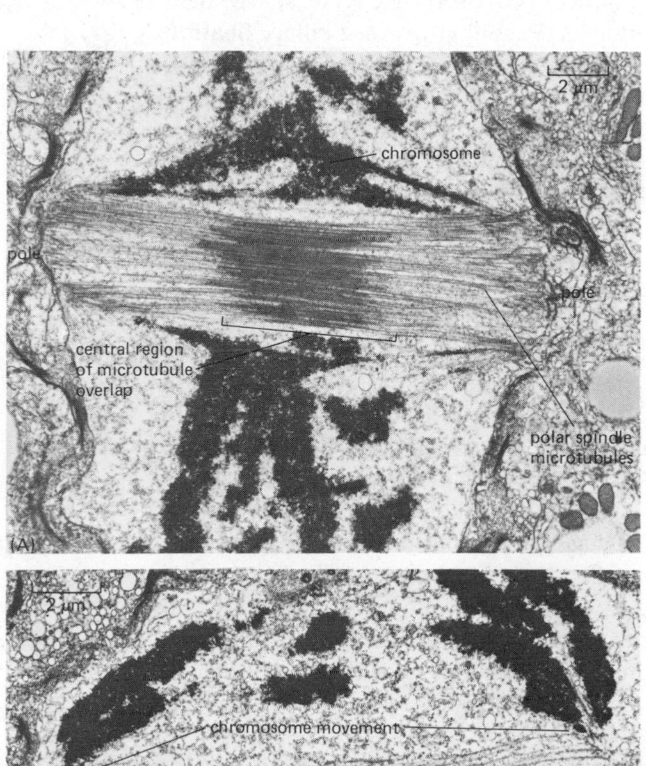

Figure 11–56 Electron micrographs showing both spindle elongation and the reduction in the degree of polar microtubule overlap during mitosis in a diatom. (A) Metaphase. (B) Late anaphase. (Courtesy of Jeremy D. Pickett-Heaps.)

A Dyneinlike ATPase Is Thought to Generate the Force That Pushes the Poles Apart in Anaphase, But a Different Mechanism Seems to Move the Chromosomes to the Poles[38]

In order to study the molecular mechanisms involved in anaphase chromosome movements, it would be ideal to have highly purified mitotic spindles that function perfectly *in vitro.* While this goal has thus far been impossible to achieve, a technically less demanding model system has been developed that is more accessible to study than a normal living mitotic cell. To generate this system, the plasma membrane of a mitotic cell is rendered permeable to macromolecules by treatment with a dilute solution of detergent. Provided that microtubule-stabilizing buffers are used, anaphase movements can still be induced to start and stop in such lysed cells. Because the spindle is now freely accessible to macromolecules, the effects of various macromolecular probes (including specific antibodies) on the movement of the mitotic spindle can be tested.

In this way it has been shown that inhibitors that bind to either actin or myosin (such as anti-myosin antibodies) inhibit cytokinesis but have no effect on anaphase chromosome movements. This result makes it very unlikely that a force-generating system involving actin and myosin like that in muscle is responsible for the movements. Another obvious candidate for the force-generating system is *dynein,* a protein associated with microtubules in cilia and flagella that generates a sliding force by hydrolyzing ATP (see p. 567). There are several inhibitors of the dynein ATPase that prevent ciliary beating; these inhibitors also block the pole-to-pole separation at anaphase in crude spindle isolates. Moreover, "cross-bridges" have been seen between adjacent microtubules in some spindles that are analogous to the dynein bridges seen between the outer doublet microtubules in a cilium (see Figure 10–32). It therefore seems likely that a dyneinlike molecule uses the energy of ATP hydrolysis to cause sliding movements of the two sets of polar microtubules past each other where they overlap near the spindle equator, thereby driving the pole-to-pole separation at anaphase.

The spindle dynein is unlikely to be the same molecule that mediates ciliary beating. First, the interacting ciliary microtubules are aligned in parallel with the same polarity, whereas the polar microtubules that slide past each other in the spindle are arranged with an opposite polarity (Figure 11–55). In addition, antibodies that bind to ciliary dynein fail to stain the mitotic spindle.

The mechanism that generates the force that drives the chromosome-to-pole movement during anaphase has not yet been elucidated. Micromanipulation experiments demonstrate that each chromatid is individually anchored to the spindle, with its kinetochore fibers being most strongly attached to the spindle near the poles (Figure 11–57). The kinetochore fibers get shorter and shorter as the chromosomes move toward the poles, and eventually disappear at telophase. Thus, at the same time that the polar microtubules are lengthening at their free (+) ends near the spindle equator, the (−) ends of the kinetochore microtubules are depolymerizing near a spindle pole. This depolymerization would seem to play at least some role in causing the chromosome-to-pole movement. Not only is this movement inhibited if microtubule depolymerization is blocked by the addition of taxol or D_2O, but it is *accelerated* if the rate of microtubule depolymerization is increased by the addition of very small amounts of colchicine. Moreover, unlike the pole-to-pole separation at anaphase, chromosome-to-pole movement is not strongly inhibited when dynein inhibitors are added to lysed cell preparations. It therefore seems that different force-generating assemblies cause the two different anaphase movements. According to one commonly held view, the "motor"

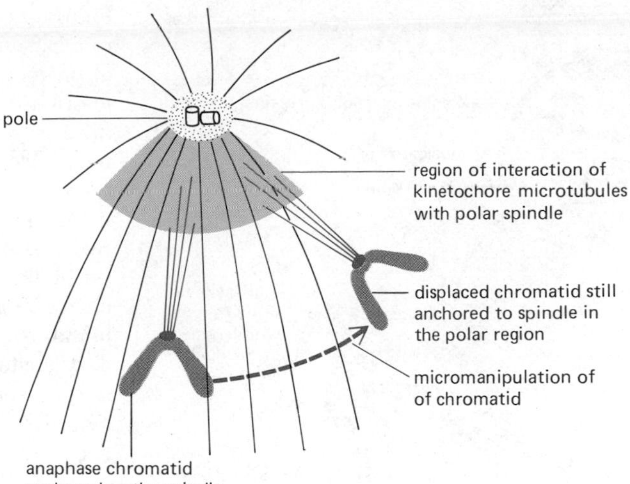

Figure 11–57 Micromanipulation experiments show that chromatids are primarily attached to the spindle by the interactions of the kinetochore microtubules with the spindle near the poles. A half-spindle of a cell in anaphase is shown.

that drives the chromosome-to-pole movement is a complex one, and its rate is limited by the rate of disassembly of the kinetochore microtubules when they reach a pole.

Analysis of the machinery that causes anaphase movements is no doubt made unusually difficult by the fact that the forces required are small. One can calculate that the movement of each chromatid at anaphase should require only the amount of energy available from the hydrolysis of about 20 ATP molecules, assuming that the chromatid is transported by an energy-efficient process.

At Telophase the Nuclear Envelope Initially Re-forms Around Individual Chromosomes[39]

By the end of anaphase, the chromosomes have fully separated into two equal groups, one at each pole of the spindle. In telophase, the final stage of mitosis, a nuclear envelope re-forms around each group of chromosomes to form the two daughter interphase nuclei. At least three parts of the nuclear-envelope complex must be considered during its breakdown and reassembly in mitosis.

1. The *outer and inner nuclear membranes* proper, which are continuous with the endoplasmic reticulum membrane (p. 429).
2. The underlying *nuclear lamina,* a thin structural assembly composed of three major proteins that interacts with the inner nuclear membrane, with chromatin, and with the nuclear pores (p. 430).
3. The *nuclear pores,* which are formed by large complexes of several proteins, all as yet poorly characterized (p. 432).

At prophase, the three lamina proteins are observed to become highly phosphorylated. The phosphorylation is believed to cause them to disassemble, thereby disrupting the nuclear lamina. Perhaps as a consequence, the nuclear envelope proper breaks up into closed membrane fragments (morphologically indistinguishable from bits of endoplasmic reticulum) that remain visible around the spindle throughout mitosis. Although this is not certain, it has been suggested that some of the nuclear-pore complexes that are liberated remain with the chromosomes during mitosis.

At telophase, fragments of the nuclear membrane associate with the surface of individual chromosomes, partially enclosing each of them before fusing to re-form the complete nuclear envelope (Figure 11–58). During this

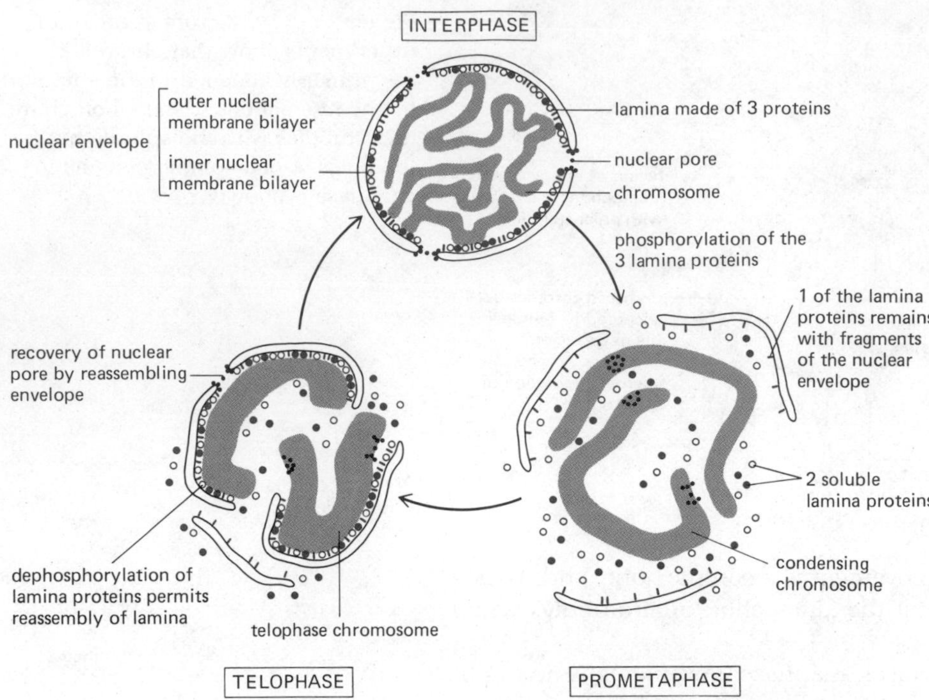

Figure 11–58 Schematic view of the nuclear-envelope cycle—breakdown at prometaphase and reassembly at telophase—that occurs during mitosis. Between prometaphase and telophase the nuclear envelope is disrupted and all of the mitotic movements take place that move the two sets of chromosomes to opposite poles. Because the new nuclear envelope is formed by the juxtaposition of membrane fragments and clustered individual chromosomes, as shown, most of the cytoplasmic components will be excluded from the new nucleus.

process the nuclear pores are retrieved and the dephosphorylated lamina proteins reassociate to form the nuclear lamina. This process of nuclear envelope reassembly may be orchestrated by the repolymerization of the lamina proteins, especially since one of them remains with the nuclear membrane fragments after prophase and may, thereby, mark this membrane for retargeting to the nucleus. The lamina interacts not only with the inner nuclear membrane, but also with the chromatin and with the nuclear pores; therefore, the cycle of phosphorylation and dephosphorylation of the lamina proteins during the M phase could well initiate nuclear-envelope breakdown and reformation, respectively.

After the nuclear envelope reassembles, RNA synthesis resumes, causing the nucleolus to reappear as the condensed chromatin uncoils and assumes its dispersed interphase conformation.

It is interesting that in some more primitive eucaryotes the nuclear envelope does not disassemble during mitosis; these organisms are said to have a "closed," rather than an open, spindle. Thus, nuclear-envelope breakdown is not an absolute requirement for a successful M phase.

The Mitotic Spindle Determines the Site of Cytoplasmic Cleavage During Cytokinesis[40]

Although nuclear and cytoplasmic division are generally linked, they are quite separable events. In fact, nuclear division is not always followed by cytokinesis. Some fungi and algae, for example, may undergo many rounds of nuclear division without cytoplasmic division, forming huge multinucleated cells that may reach several feet in length! Uninucleated cells are generated later by cytoplasmic cleavage around the nuclei.

Although cleavage does not necessarily accompany mitosis, the mitotic spindle plays an important part in determining its positioning and timing. Cytokinesis usually starts in anaphase, continuing through telophase and into the following interphase period. The first visible sign of cleavage in animal cells is a slight puckering of the plasma membrane seen during anaphase

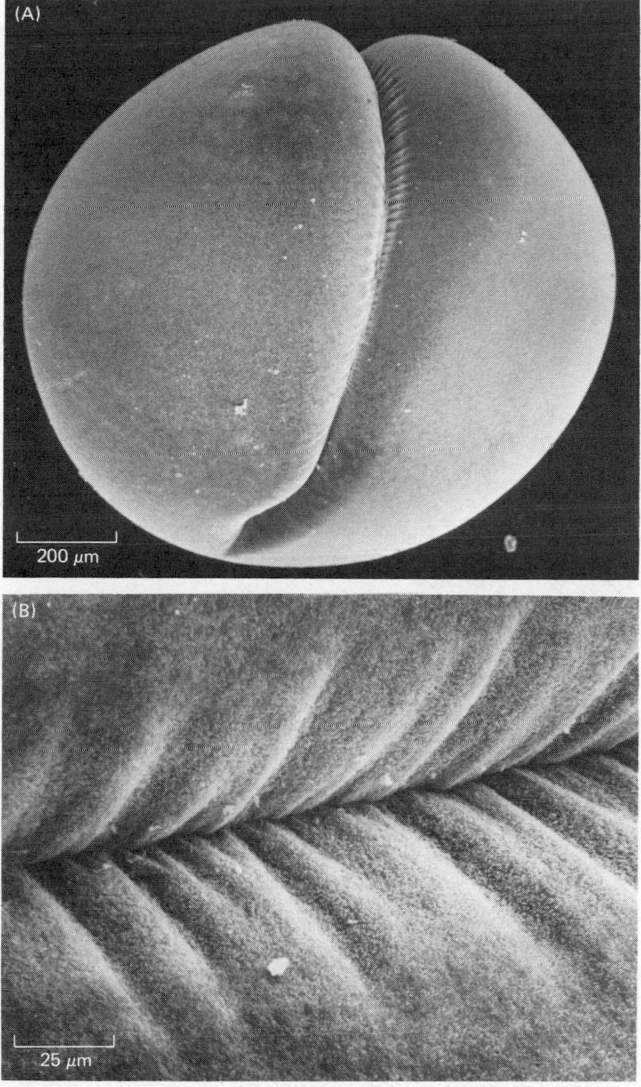

Figure 11–59 Scanning electron micrographs of early cleavage in a frog egg. The furrowing of the cell membrane is caused by the activity of the contractile ring underneath it. (A) Low-magnification view of egg surface; (B) surface of furrow at higher magnification. (From H. W. Beams and R. G. Kessel, *Am. Sci.* 64:279–290, 1976.)

known as **furrowing** (Figure 11–59). Furrowing invariably occurs in the plane of the metaphase plate, at right angles to the long axis of the mitotic spindle. If, by micromanipulation, the spindle is purposely moved early enough, an incipient furrow disappears and a new one develops at the new spindle site. Later, once the furrowing process is well under way, cleavage proceeds even if the spindle is removed by suction or destroyed by colchicine.

During mitosis the microtubules and actin-filament bundles of the cytoskeleton are disassembled to subunits, which appear to be reused to construct the machinery of mitosis and cytokinesis respectively. The remaining filaments of the cell, the *intermediate filaments,* are not used in cell division and are not disassembled. Instead, in most cells, the meshwork of intermediate filaments that surrounds the interphase nucleus elongates during mitosis to enclose the two daughter nuclei and is finally separated by the action of the cleavage furrow (Figure 11–60).

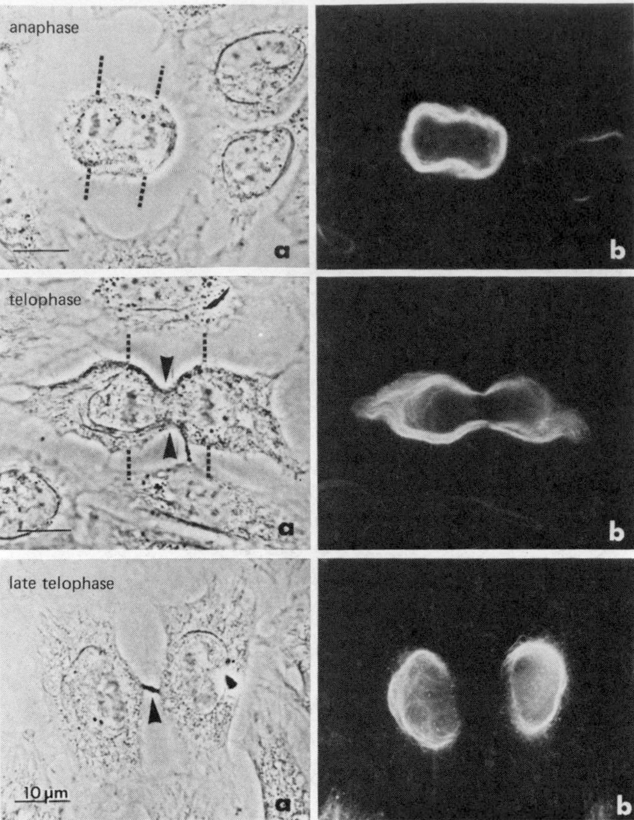

Figure 11–60 A band of intermediate filaments surrounds the nuclear region during mitosis and presumably helps to organize the mitotic processes in the cell. The micrographs shown here were obtained by staining permeabilized cells with fluorescently labeled antibodies that bind to intermediate filaments. (a) Phase-contrast micrographs; dotted lines demark the positions of chromosomes, while the arrowheads indicate the position of the contractile ring (telophase) or the midbody (late telophase). (b) Fluorescence micrographs. (From S. H. Blose, *Proc. Natl. Acad. Sci. USA* 76:3372–3376, 1979.)

Actin and Myosin Generate the Forces for Cleavage[41]

Cleavage is accomplished by the contraction of a ring composed mainly of actin filaments, like a purse string being drawn tight around the center of the cell. This bundle of filaments, known as the **contractile ring,** must be attached in some way to the cytoplasmic face of the plasma membrane (Figure 11–61). The contractile ring spontaneously assembles by an unknown mechanism in early anaphase; once assembled, it can be activated prematurely by electrical stimulation, showing that the contractile ring is ready for action before it is actually used.

The force exerted by the contractile ring during cleavage is large enough to bend a fine glass needle inserted into the cell, and its magnitude can be measured in this way. There can be little doubt that the musclelike sliding of actin and myosin filaments in the contractile ring generates this force. For example, in lysed mitotic cells, addition of an inactivated myosin subfragment blocks sites on actin that normally bind myosin, thereby stopping cleavage. Similarly, in sea urchin eggs, injection of anti-myosin antibodies causes the relaxation of the cleavage furrow without affecting nuclear division.

The contractile ring maintains the same thickness during contraction, which suggests that it continuously reduces its volume by a loss of filaments. The ring is finally dispensed with altogether when cleavage ends, as the plasma membrane of the cleavage furrow narrows to constrict the **midbody,** which remains as a tether between the two daughter cells. The midbody is constructed from the remains of the two sets of polar microtubules tightly packed together with dense matrix material (Figure 11–62). Since it persists for some hours, the midbody may be designed to retain some connection between the

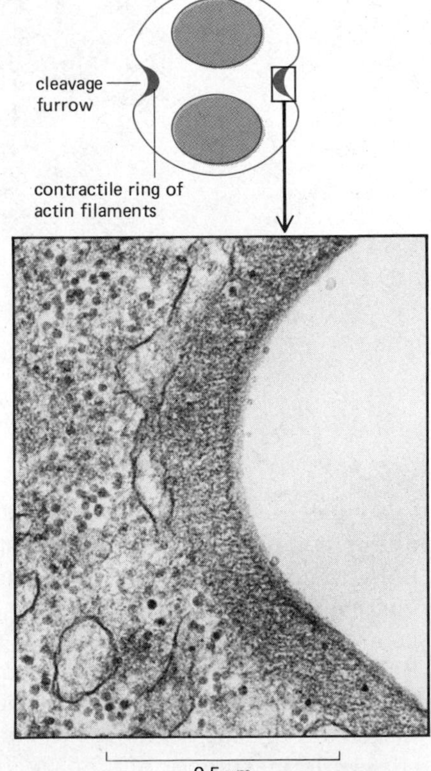

Figure 11–61 Electron micrograph of the ingrowing edge of the cleavage furrow of a dividing animal cell. (From H. W. Beams and R. G. Kessel, *Am. Sci.* 64:279–290, 1976.)

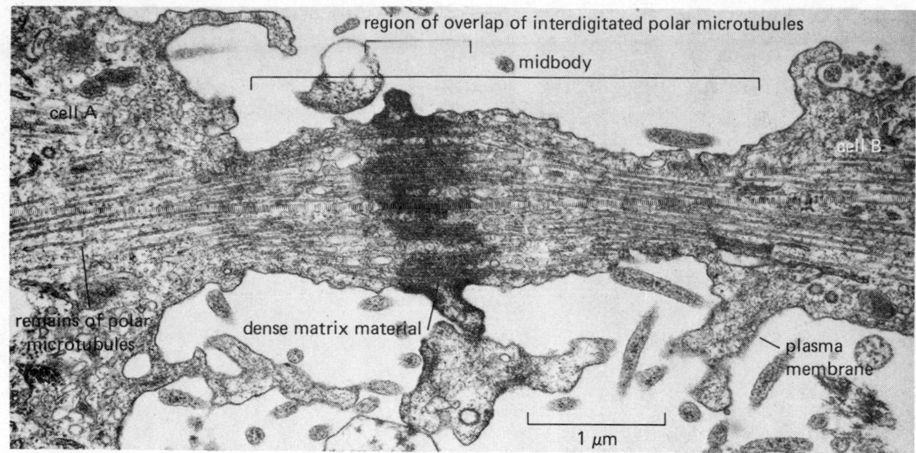

Figure 11–62 Electron micrograph of the midbody of a dividing animal cell. Cleavage is virtually complete, but the daughter cells remain attached by this thin strand of cytoplasm (see also Figure 11–40). (Courtesy of J. M. Mullins.)

two daughter cells prior to their complete separation, and it may also play an active role in the final "pinching-off" process (Figure 11–63).

In most dividing cells the cleavage furrow is more or less symmetrically placed so that the two daughter cells produced are of equal size. In some special cases, however, the contractile ring develops in such a position as to create two cells of different sizes. Such asymmetric cell divisions are of great importance in oogenesis (see Chapter 14) and are commonly seen in the early development of some embryos, in which an inhomogeneous egg cytoplasm is distributed in a precise way to groups of cells that will later give rise to different parts of the embryo. The mechanisms that control the orientation and spatial positioning of such asymmetric contractile rings are not understood.

Cytokinesis greatly increases the total cell-surface area, as two cells form from one. It is therefore clear that the two cells resulting from cytokinesis require more plasma membrane than the original cell. In animal cells net membrane biosynthesis increases just before cell division. This extra membrane seems to be stored as *blebs* on the surface of cells about to divide.

Cytokinesis Occurs by a Completely Different Mechanism in Plant Cells with Cell Walls[42]

Most higher plant cells are enclosed by a rigid *cell wall* and are, therefore, immotile. Their mechanism for cytokinesis is markedly different from that in animal cells. Following mitosis, the cytoplasm is divided from inside the cell by the construction of a new cell wall between the two daughter cells rather

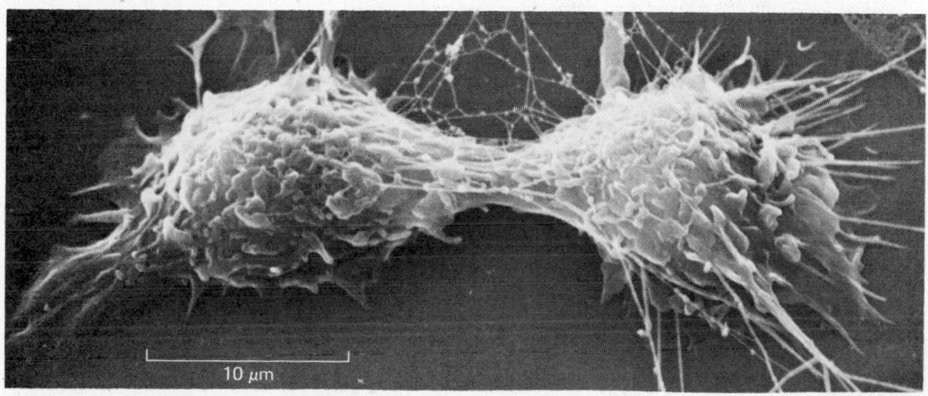

Figure 11–63 Scanning electron micrograph of an animal cell in tissue culture shortly after its division. (Courtesy of Guenter Albrecht-Buehler.)

than by the pinching off of two daughter cells by means of a contractile ring. This partition precisely determines the relative positions that the two daughter cells will subsequently occupy in the plant. An obvious corollary is that the plane of cell division, together with cell enlargement, determines plant form (see Chapter 19).

As in animal cells, cytokinesis can occur separately from the events of mitosis. For example, in the endosperm tissue of seeds, mitoses occur without cytokinesis to give rise to a giant multinucleated cell. Much later, cross-walls are laid down between the separate nuclei to create new cells, with no involvement of the mitotic spindle remnants, which have long since disassembled. The factors that control the accurate construction of a new cell wall in plant cells are largely unknown, but the sequence of events is now well documented.

The new cross-wall, or **cell plate,** usually assembles in association with the residual polar spindle microtubules, which have formed an open, cylindrical structure containing microtubules organized in parallel arrays called the **phragmoplast** (Figure 11–64). As outlined in Figure 11–65, small membrane-bounded vesicles, largely derived from the Golgi apparatus and filled with wall precursors, contact the microtubules of the phragmoplast and are transported inward along the microtubules toward the equatorial region. Here they fuse to form a disclike, membrane-bounded structure, the *early cell plate.* The polysaccharide precursor molecules released from these vesicles assemble within the early cell plate to form pectin, hemicellulose, and other con-

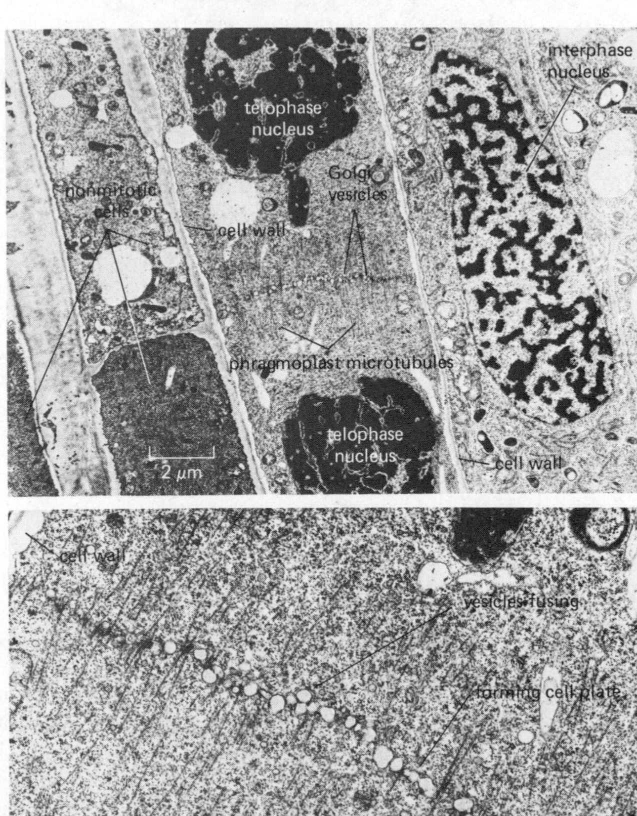

Figure 11–64 Electron micrographs of sectioned plant cells showing the phragmoplast and its associated vesicles, which fuse to form the new cell plate. (From J. D. Pickett-Heaps, *Dev. Biol.* 15:206–236, 1967.)

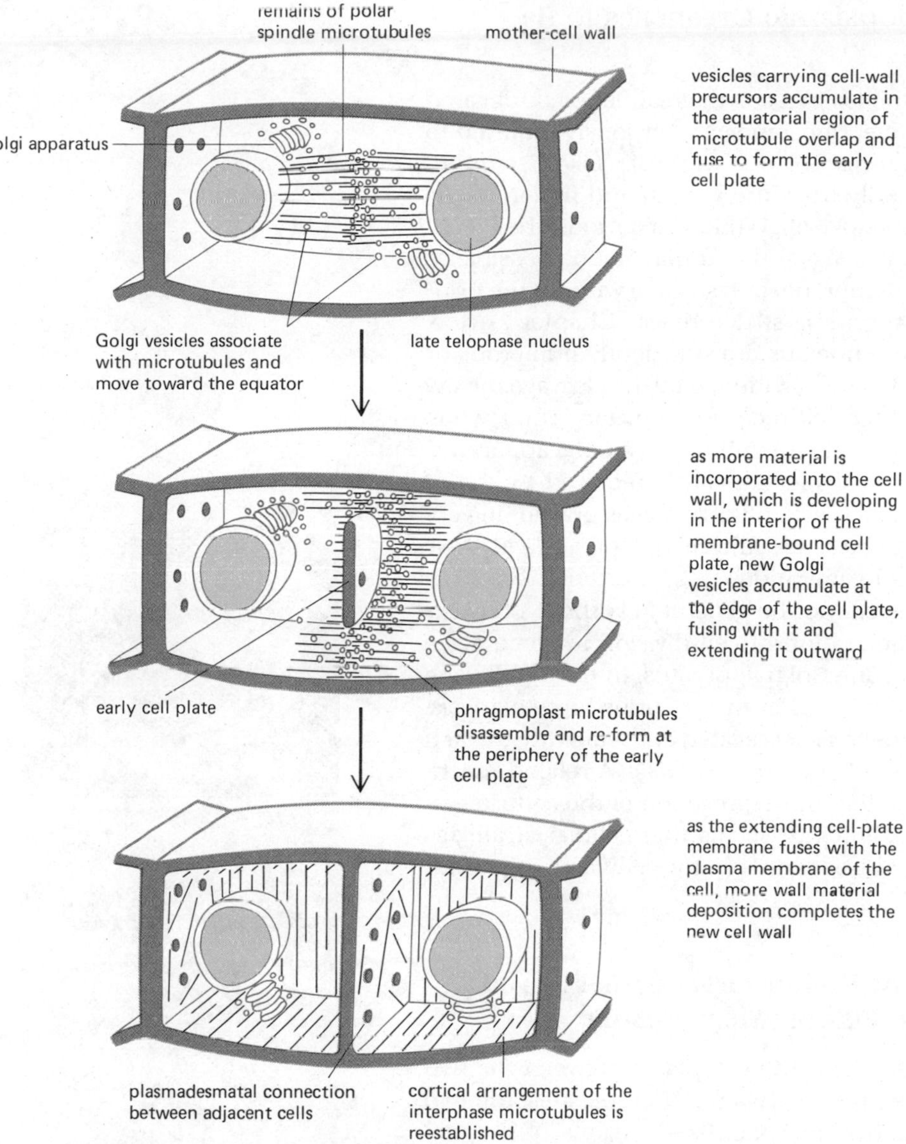

Figure 11–65 An outline of the process of cytokinesis in a higher plant cell with a rigid cell wall.

stituents of the primary cell wall. This disc now has to expand laterally to reach the original cell wall. To make this possible, the microtubules of the early phragmoplast disassemble and reassemble at the periphery of the early cell plate. There they attract a second set of vesicles; these again fuse at the equator, extending the edge of the plate. This process is repeated until the cell plate reaches the plasma membrane and completely separates the two new cells. Finally, cellulose microfibrils are laid down within the cell plate to complete the new cell wall (Figures 11–65 and 11–66).

The assembly of microtubules in the phragmoplast appears to be nucleated by the electron-dense material found in the equatorial region. As a putative microtubule organizing center, this dense material would seem to be responsible for the remarkable reorganizations of the phragmoplast microtubules that accompany cell-plate formation.

Associated with the vesicles of the forming cell plate are elements of the endoplasmic reticulum, which often remain trapped across the plate. These subsequently are transformed into *plasmodesmata,* the complex pores that traverse the mature cell wall, interconnecting the cytoplasm of all the cells in the plant (see p. 1110).

Cytokinesis Must Allow the Cytoplasmic Organelles to Be Faithfully Inherited

The nucleus is only one of many organelles in a cell that cannot be duplicated without a preexisting copy. For example, ribosomes are obviously required to make more ribosomes, since this requires protein synthesis. Likewise, mitochondria and chloroplasts can arise only from the growth and fission of the corresponding preexisting organelle (see p. 529). While less obvious, the mechanisms by which other organelles grow suggest that it may not be possible to make a new Golgi apparatus, plasma membrane, or lysosome without the prior presence of at least part of the corresponding structure (see Chapter 7). Normally, all of these organelles except the nucleus are sufficiently numerous or extensive that each would be expected to be partitioned by the cleavage furrow during cytokinesis so that each daughter cell receives a portion. Thus, while no cell is likely to survive unless it inherits the membrane of a Golgi apparatus, for example, it is possible that no special mechanism is required to assure that each daughter cell ends up with one. Certainly, organelles present in very large numbers, such as ribosomes, will be safely inherited if, on average, their numbers merely double once each cell generation.

Indirect evidence that organelles cannot be generated without preexisting copies can be found from an examination of cell division in some algae that have only one chloroplast or only one Golgi apparatus. In these cells, the organelle that is present in a single copy splits in half prior to cytokinesis. One of the two halves of the old organelle is segregated subsequently to each daughter cell when the cell divides (see Figure 9–63, p. 530). A related mechanism operates for the orderly duplication and segregation of the centriole in all animal cells (see Figure 11–19). It therefore seems that cellular strategies can be designed for the organized segregation of any cellular component whenever necessary.

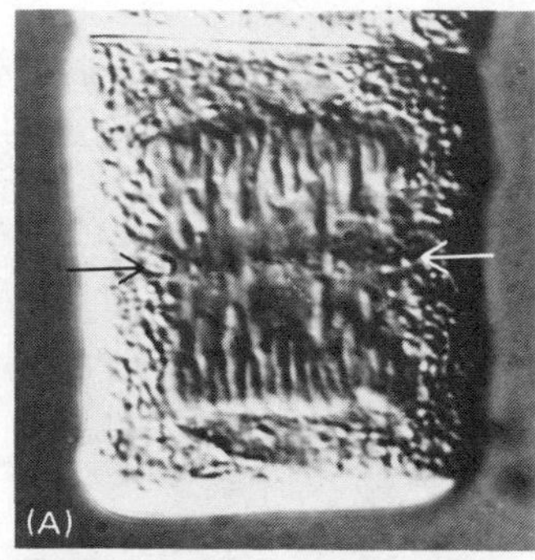

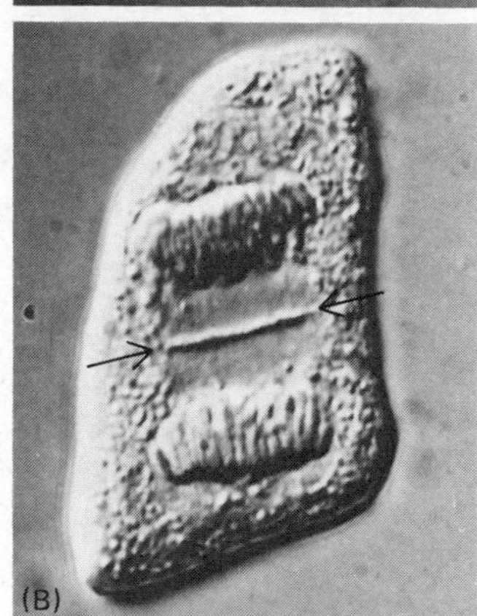

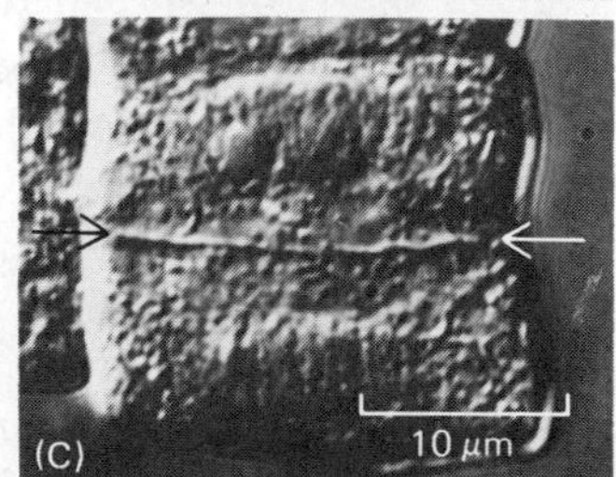

Figure 11–66 Light micrographs of cytokinesis in a plant cell. (A), (B), and (C) show progressively more advanced stages of the development of the cell plate, which is forming in a plane perpendicular to the micrograph; the edge of this plane is demarked by the two arrows. (Courtesy of Jeremy D. Pickett-Heaps.)

The Elaborate Mitotic Process of Higher Organisms Evolved Gradually from the Procaryotic Fission Mechanisms[43]

In procaryotic cells, division of the DNA and the cytoplasm between the two daughter cells is mediated by a single process. When DNA replicates, the two copies of the chromosomes are attached to specialized regions of the cell membrane that are gradually separated by the inward growth of the membrane between them. Fission takes place between the two attachment sites, so that each daughter cell captures one chromosome (Figure 11–67). With the evolution of the eucaryotes, the genetic apparatus advanced in complexity and the chromosomes increased in number and in size. It became necessary to develop a more elaborate mechanism for dividing the chromosomes between daughter cells. Clearly, the mitotic apparatus did not evolve all at once. Mitosis in many primitive eucaryotes is seen to retain a membrane-attachment mechanism, with the nuclear membrane taking over the part played by the plasma membrane in procaryotes. One of the clearest cases of such an intermediate mechanism occurs in the dinoflagellate *Crypthecodinium cohnii,* a large, single-cell alga. The intermediate status of this organism is also reflected in the biochemistry of its chromosomes, which, like those of procaryotes, have relatively little associated protein.

In *C. cohnii,* the nuclear membrane remains intact throughout mitosis, and the spindle microtubules remain entirely outside the nucleus. Where these spindle microtubules press on the outside of the nuclear envelope, the envelope becomes indented in a series of parallel channels (Figure 11–67). The chromosomes become attached to the inner membrane of the nuclear enve-

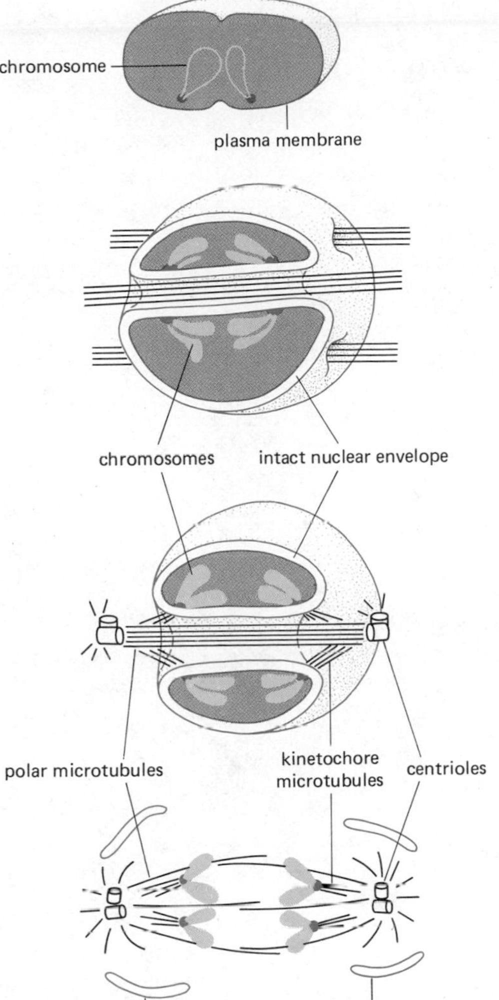

Figure 11–67 Diagram illustrating some stages in the evolution of the mitotic spindle of higher organisms. These stages have been identified through studies of the chromosome-separation mechanisms utilized in more primitive cells.

lope opposite these channels, and the separation of the chromosomes is entirely mediated on the inside of this channeled nuclear membrane. Thus, the extranuclear "spindle" (which forms a rigid rod with no dynamic properties) is merely used to order the nuclear membrane and thereby define the plane of division.

A somewhat more advanced, though still extranuclear, spindle is seen in hypermastigotes. These large protozoa from the gut of insects provide a particularly clear illustration of the independence of spindle elongation and the chromosome movements that separate the chromatids, since the sister kinetochores are first pulled apart by the growth of the nuclear membrane (to which they are attached) prior to becoming attached to the spindle. Only when the kinetochores are near the poles of the spindle do they acquire the kinetochore fibers required for attaching them to the polar spindle fibers. Because the spindle fibers remain separated from the chromosomes by the nuclear envelope, the kinetochore fibers, which are formed outside the nucleus, must somehow attach to the chromosomes through the nuclear membranes. After this attachment has been made, the kinetochores are drawn poleward in a conventional manner (Figure 11–67).

It seems likely that the present role of the kinetochore fibers evolved in parallel with the dissolution of the nuclear envelope at mitosis; at this stage of evolution, the last remnants of the procaryotic mechanism disappeared and more intimate interactions between the mitotic spindle and the chromosomes became possible.

Summary

The process of cell division consists of nuclear division (called mitosis) followed by cytoplasmic division (called cytokinesis). Mitosis is dominated by the formation of the highly ordered polar spindle, composed of microtubules and associated proteins that are organized by the two mitotic centers at the opposite spindle poles. The chromosomes condense during prophase and develop kinetochore fibers that begin to interact with the polar fibers of the spindle after the nuclear envelope breaks down in prometaphase. Under tension due to opposing pole-directed forces pulling on the kinetochore fibers, the chromosomes line up at the spindle equator during metaphase. At anaphase, this tension is suddenly released as sister chromatids detach from each other and are pulled to opposite poles. During the final, telophase stage of mitosis, the nuclear envelope re-forms on the surface of each group of separated chromosomes. Cell division ends as the cytoplasmic contents are divided by the process of cytokinesis, and the chromosomes decondense and resume RNA synthesis.

References

General

Inoué, S. Cell division and the mitotic spindle. *J. Cell Biol.* 91:132s–147s, 1981. (A review.)

John, P.C.L., ed. The Cell Cycle. Cambridge, Eng.: Cambridge University Press, 1981.

Lewin, B. Gene Expression, Vol. 2, Eucaryotic Chromosomes, 2nd ed. New York: Wiley, 1980. (Chapters 4, 7, and 20 on mitosis, the cell division cycle, and DNA replication, respectively.)

Mitchison, J.M. The Biology of the Cell Cycle. Cambridge, Eng.: Cambridge University Press, 1971.

Prescott, D.M. Reproduction of Eucaryotic Cells. New York: Academic Press, 1976.

Zimmerman, A.M.; Forer, A., eds. Mitosis/Cytokinesis. New York: Academic Press, 1981. (A collection of reviews.)

Cited

1. Pardee, A.B.; Dubrow, R.; Hamlin, J.L.; Kletzien, R.F. Animal cell cycle. *Annu. Rev. Biochem.* 47:715–750, 1978.
2. Yanishevsky, R.M.; Stein, G.H. Regulation of the cell cycle in eukaryotic cells. *Int. Rev. Cytol.* 69:223–259, 1981.

   Lloyd, D.; Poole, P.K.; Edwards, S.W. The Cell Division Cycle. New York: Academic Press, 1982.
3. Cheng, H.; LeBlond, C.P. Origin, differentiation and renewal of the four main epithelial cell types in the mouse small intestine. *Am. J. Anat.* 141:461–480, 1974.

   Potten, C.S.; Schofield, R.; Lajtha, L.G. A comparison of cell replacement in bone marrow, testis, and three regions of surface epithelium. *Biochim. Biophys. Acta* 560:281–299, 1979.
4. Pardee, A.B. A restriction point for control of normal animal cell proliferation. *Proc. Natl. Acad. Sci. USA* 71:1286–1290, 1974.

   Smith, J.A.; Martin, L. Do cells cycle? *Proc. Natl. Acad. Sci. USA* 70:1263–1267, 1973.
5. Mitchison, J.M. The Biology of the Cell Cycle. Cambridge, Eng.: Cambridge University Press, 1971.
6. Van Dilla, M.A.; Trujillo, T.T.; Mullaney, P.F.; Coulter, J.R. Cell microfluorometry: a method for rapid fluorescence measurement. *Science* 163:1213–1214, 1969.

7. Rossow, P.W.; Riddle, V.G.H.; Pardee, A.B. Synthesis of labile, serum-dependent protein in early G_1 controls animal growth. *Proc. Natl. Acad. Sci. USA* 76:4446–4450, 1979.

Tyson, J.; Garcia-Herdugo, G.; Sachsenmaier, W. Control of nuclear division in *Physarum polycephalum:* comparison of cycloheximide pulse treatment, UV irradiation and heat shock. *Exp. Cell Res.* 119:87–98, 1979.

8. Folkman, J.; Moscona, A. Role of cell shape in growth control. *Nature* 273:345–349, 1978.

9. Rutter, W.J.; Pictet, R.L.; Morris, P.W. Toward molecular mechanisms of developmental processes. *Annu. Rev. Biochem.* 42:601–646, 1973.

French, V.; Bryant, P.J.; Bryant, S.V. Pattern regulation in epimorphic fields. *Science* 193:969–981, 1976.

Bryant, P.J.; Bryant, S.V.; French, V. Biological regeneration and pattern formation. *Sci. Am.* 237(1):67–81, 1977.

Sato, G., ed. Hormones and Cell Culture. Cold Spring Harbor, N.Y.: Cold Spring Harbor Laboratory, 1979.

10. Pierce, G.B.; Shikes, R.; Fink, L.M. Cancer: A Problem of Developmental Biology. Englewood Cliffs, N.J.: Prentice-Hall, 1978.

Cairns, J. Cancer: Science and Society. San Francisco: Freeman, 1978.

Medrano, E.E.; Pardee, A.B. Prevalent deficiency in tumor cells of cycloheximide-induced cycle arrest. *Proc. Natl. Acad. Sci. USA* 77:4123–4126, 1980.

11. Tooze, J., ed. DNA Tumor Viruses, 2nd. ed. Cold Spring Harbor, N.Y.: Cold Spring Harbor Laboratory, 1980.

Weiss, R.; Teich, N.; Varmus, H.; Coffin, J., eds. RNA Tumor Viruses. Cold Spring Harbor, N.Y.: Cold Spring Harbor Laboratory, 1982.

12. Eckhardt, W. Properties of temperature-sensitive mutants of polyoma virus. *Cold Spring Harbor Symp. Quant. Biol.* 39:37–40, 1975.

Hassell, J.A.; Topp, W.C.; Rifkin, D.B.; Moreau, P.E. Transformation of rat embryo fibroblasts by cloned polyoma virus DNA fragments containing only part of the early region. *Proc. Natl. Acad. Sci. USA* 77:3978–3982, 1980.

Houweling, A.; Van Den Elsen, P.J.; Van Der Eb, A.J. Partial transformation of primary rat cells by the leftmost 4.5% fragment of adenovirus 5 DNA. *Virology* 105:537–550, 1980.

13. Smith, A.E.; Smith, R.; Paucha, E. Characterization of different tumor antigens present in cells transformed by simian virus 40. *Cell* 18:335–346, 1979.

Duesberg, P.H.; Vogt, P.K. Differences between the ribonucleic acids of transforming and nontransforming avian tumor viruses. *Proc. Natl. Acad. Sci. USA* 67:1673–1680, 1970.

Purchio, A.F.; Erikson, E.; Brugge, J.S.; Erikson, R.L. Identification of a polypeptide encoded by the avian sarcoma virus *src* gene. *Proc. Natl. Acad. Sci. USA* 75:1567–1571, 1978.

14. Collett, M.S.; Erikson, R.L. Protein kinase activity associated with the avian sarcoma virus *src* gene product. *Proc. Natl. Acad. Sci. USA* 75:2021–2024, 1978.

Levinson, A.D.; Oppermann, H.; Levintow, L.; Varmus, H.E.; Bishop, M.J. Evidence that the transforming gene of avian sarcoma virus encodes a protein kinase associated with a phosphoprotein. *Cell* 15:561–572, 1978.

Hunter, T.; Sefton, B.M. Transforming gene product of Rous sarcoma virus phosphorylates tyrosine. *Proc. Natl. Acad. Sci. USA* 77:1311–1315, 1980.

Ushiro, H.; Cohen, S. Identification of phosphotyrosine as a product of epidermal growth factor-activated protein kinase in A-431 cell membranes. *J. Biol. Chem.* 255:8363–8365, 1980.

15. Bishop, J.M. Enemies within: the genesis of retrovirus oncogenes. *Cell* 23:5–6, 1981.

Oskarsson, M.; McClements, W.L.; Blair, D.G.; Maizel, J.V.; Vande Woude, G.F. Properties of a normal mouse cell DNA sequence (sarc) homologous to the src sequence of Moloney sarcoma virus. *Science* 207:1222–1224, 1980.

16. Shih, C.; Padhy, L.C.; Murray, M.; Weinberg, R.A. Transforming genes of carcinomas and neuroblastomas introduced into mouse fibroblasts. *Nature* 290:261–264, 1981.

Krontiris, T.C.; Cooper, G.M. Transforming activity of human tumor DNAs. *Proc. Natl. Acad. Sci. USA* 78:1181–1184, 1981.

Perucho, M.; et al. Human-tumor-derived cell lines contain common and different transforming genes. *Cell* 27:467–476, 1981.

Cooper, G.M. Cellular transforming genes. *Science* 217:801–806, 1982.

Parada, L.F.; Tabin, C.J.; Shih, C.; Weinberg, R.A. Human EJ bladder carcinoma oncogene is homologue of Harvey sarcoma virus *ras* gene. *Nature* 297:474–478, 1982.

Chang, E.H.; Furth, M.E.; Scolnick, E.M.; Lowy, D.R. Tumorigenic transformation of mammalian cells induced by a normal human gene homologous to the oncogene of Harvey murine sarcoma virus. *Nature* 297:479–483, 1982.

17. Robbins, E.; Jentzsch, G.; Micali, A. The centriole cycle in synchronized HeLa cells. *J. Cell Biol.* 36:329–339, 1968.

Vorobjev, I.A.; Chentsov, Y.S. Centrioles in the cell cycle: epithelial cells. *J. Cell Biol.* 98:938–949, 1982.

18. Hartwell, L. Cell division from a genetic perspective. *J. Cell Biol.* 77:627–637, 1978.

Byers, B. Cytology of the yeast life cycle. In The Molecular Biology of the Yeast *Saccharomyces* (J.N. Strathern, E.W. Jones, J.R. Broach, eds.), pp. 59–96. Cold Spring Harbor, N.Y.: Cold Spring Harbor Laboratory, 1981.

19. Huberman, J.A.; Riggs, A.D. On the mechanism of DNA replication in mammalian chromosomes. *J. Mol. Biol.* 32:327–341, 1968.

Hand, R. Eucaryotic DNA: organization of the genome for replication. *Cell* 15:317–325, 1978.

20. Riley, D.; Weintraub, H. Conservative segregation of parental histones during replication in the presence of cycloheximide. *Proc. Natl. Acad. Sci. USA* 76:328–332, 1979.

Hereford, L.M.; Osley, M.A.; Ludwig, J.R.; McLaughlin, C.S. Cell-cycle regulation of yeast histone mRNA. *Cell* 24:367–376, 1981.

Gallwitz, D. Kinetics of inactivation of histone mRNA in the cytoplasm after inhibition of DNA replication in synchronised HeLa cells. *Nature* 257:247–248, 1975.

21. Russev, G.; Hancock, R. Assembly of new histones into nucleosomes and their distribution in replicating chromatin. *Proc. Natl. Acad. Sci. USA* 79:3143–3147, 1982.

22. Stubblefield, E. Analysis of the replication pattern of Chinese hamster chromosomes using 5-bromodeoxyuridine suppression of 33258 Hoechst fluorescence. *Chromosoma* 53:209–221, 1975.

23. Lewin, B. Gene Expression, Vol. 2, Eucaryotic Chromosomes, 2nd ed., pp. 428–447. New York: Wiley, 1980.

Brown, S.W. Heterochromatin. *Science* 151:417–425, 1966.

24. Harland, R. Initiation of DNA replication in eukaryotic chromosomes. *Trends Biochem. Sci.* 6:71–74, 1981.

Rao, P.N.; Johnson, R.T. Mammalian cell fusion: studies on the regulation of DNA synthesis and mitosis. *Nature* 225:159–164, 1970.

25. Pringle, J.R.; Hartwell, L.H. The *Saccharomyces cerevisiae* cell cycle. In The Molecular Biology of the Yeast *Saccharomyces* (J.N. Strathern, E.W. Jones, J.R. Broach, eds.), pp. 97–142. Cold Spring Harbor, N.Y.: Cold Spring Harbor Laboratory, 1981.

Simchen, G. Cell cycle mutants. *Annu. Rev. Genet.* 12:161–191, 1978.

26. Prescott, D.M. Reproduction of Eucaryotic Cells. New York: Academic Press, 1976.

Mitchison, J.M. The Biology of the Cell Cycle. Cambridge, Eng.: Cambridge University Press, 1971.

27. Johnson, R.T.; Rao, P.N. Mammalian cell fusion: induction of premature chromosome condensation in interphase nuclei. *Nature* 226:717–722, 1970.

Bradbury, E.M.; Inglis, R.J.; Matthews, H.R. Control of cell division by very lysine rich histone (F1) phosphorylation. *Nature* 247:257–261, 1974.

Isenberg, I. Histones. *Annu. Rev. Biochem.* 48:159–191, 1979.

28. Kreigstein, H.J.; Hogness, D.S. Mechanism of DNA replication in *Drosophila* chromosomes: structure of replication forks and evidence for bidirectionality. *Proc. Natl. Acad. Sci. USA* 71:135–139, 1974.

Callan, H.G. DNA replication in the chromosomes of eukaryotes. *Cold Spring Harbor Symp. Quant. Biol.* 38:195–203, 1974.

29. Wolfe, S.L. Biology of the Cell, 2nd ed., pp. 398–431. Belmont, Ca.: Wadsworth, 1981.

Mazia, D. Mitosis and the physiology of cell division. In The Cell, Vol. 3 (J. Brachet, A.E. Mirsky, eds.), pp. 77–412. London: Academic Press, 1961.

Inoué, S. Cell division and the mitotic spindle. *J. Cell Biol.* 91:131s–147s, 1981.

Dustin, P. Microtubules, pp. 340–397. New York: Springer-Verlag, 1978.

30. Nicklas, R.B. Mitosis. In Advances in Cell Biology, Vol. 2 (D.M. Prescott, L. Goldstein, E.H. McConkey, eds.), pp. 225–298. New York: Appleton-Century-Crofts, 1971.

31. Matsui, S.I.; Yoshida, H.; Weinfeld, H.; Sandberg, A.A. Induction of prophase in interphase nuclei by fusion with metaphase cells. *J. Cell Biol.* 54:120–132, 1972.

32. Inoué, S.; Sato, H. Cell motility by labile association of molecules: the nature of mitotic spindle fibers and their role in chromosome movement. *J. Gen. Physiol.* 50:259–292, 1967.

Bajer, A.S.; Molé-Bajer, J. Spindle dynamics and chromosome movements. *Int. Rev. Cytol. (Suppl.)* 3:1–271, 1972.

Salmon, E.D. Pressure-induced depolymerization of spindle microtubules. I. Changes in birefrigence and spindle length. *J. Cell Biol.* 65:603–614, 1975.

33. Roos, U.-P. Light and electron microscopy of rat kangaroo cells in mitosis. III. Patterns of chromosome behavior during prometaphase. *Chromosoma* 54:363–385, 1976.

34. Nicklas, R.B. Chromosome movement: current models and experiments on living cells. In Molecules and Cell Movement (S. Inoué, R.E. Stephens, eds.), pp. 97–118. New York: Raven Press, 1975.

35. Moens, P.B. Kinetochore microtubule numbers of different-sized chromosomes. *J. Cell Biol.* 83:556–561, 1979.

36. Begg, D.A.; Ellis, G.W. Micromanipulation studies of chromosome movement. *J. Cell Biol.* 82:528–541, 1979.

McIntosh, J.R. Cell division. In Microtubules (K. Roberts, J.S. Hyams, eds.), pp. 381–441. New York: Academic Press, 1979.

37. Ris, H. The anaphase movement of chromosomes in the spermatocytes of grasshoppers. *Biol. Bull. (Woods Hole)* 96:90–106, 1949.

Kuriyama, R.; Borisy, G.G. Microtubule-nucleating activity of centrosomes in Chinese hamster ovary cells is independent of the centriole cycle but coupled to the mitotic cycle. *J. Cell Biol.* 91:822–826, 1981.

Euteneuer, U.; McIntosh, J.R. Structural polarity of kinetochore microtubules in PtK_1 cells. *J. Cell Biol.* 89:338–345, 1981.

Pickett-Heaps, J.D.; Tippit, D.H. The diatom spindle in perspective. *Cell* 14:455–467, 1978.

38. Cande, W.Z.; Wolniak, S.M. Chromosome movement in lysed mitotic cells is inhibited by vanadate. *J. Cell Biol.* 79:573–580, 1978.

Pratt, M.M.; Otter, T.; Salmon, E.D. Dynein-like Mg^{2+}-ATPase in mitotic spindles isolated from sea urchin embryos *(Strongylocentrotus droebachiensis)*. *J. Cell Biol.* 86:738–745, 1980.

Hyams, J. Dynein in the spindle? *Nature* 295:648–649, 1982.

39. Mazia, D. How cells divide. *Sci. Am.* 205(3):101–120, 1961.

Gerace, L.; Blobel, G. The nuclear envelope lamina is reversibly depolymerized during mitosis. *Cell* 19:277–287, 1980.

40. Beams, H.W.; Kessel, R.G. Cytokinesis: a comparative study of cytoplasmic division in animal cells. *Am. Sci.* 64:279–290, 1976.

Blose, S.H. Ten-nanometer filaments and mitosis: maintenance of structural continuity in dividing endothelial cells. *Proc. Natl. Acad. Sci. USA* 76:3372–3376, 1979.

41. Mabuchi, I.; Okuno, M. The effect of myosin antibody on the division of starfish blastomeres. *J. Cell Biol.* 74:251–263, 1977.

Schroeder, T.E. Actin in dividing cells: contractile ring filaments bind heavy meromyosin. *Proc. Natl. Acad. Sci. USA* 70:1688–1692, 1973.

Meeusen, R.L.; Cande, W.Z. *N*-Ethylmaleimide-modified heavy meromyosin. *J. Cell Biol.* 82:57–65, 1979.

Pasternak, C.A. Surface membranes during the cell cycle. *Trends Biochem. Sci.* 1:148–151, 1976.

42. Bajer, A. Fine structure studies on phragmoplast and cell plate formation. *Chromosoma* 24:383–417, 1968.

Pickett-Heaps, J.D.; Northcote, D.H. Organization of microtubules and endoplasmic reticulum during mitosis and cytokinesis in wheat meristems. *J. Cell Sci.* 1:109–120, 1966.

Hepler, P.K.; Newcomb, E.H. Fine structure of cell plate formation in the apical meristem of Phaseolus roots. *J. Ultrastruct. Res.* 19:498–513, 1967.

43. Kubai, D.F. The evolution of the mitotic spindle. *Int. Rev. Cytol.* 43:167–227, 1975.

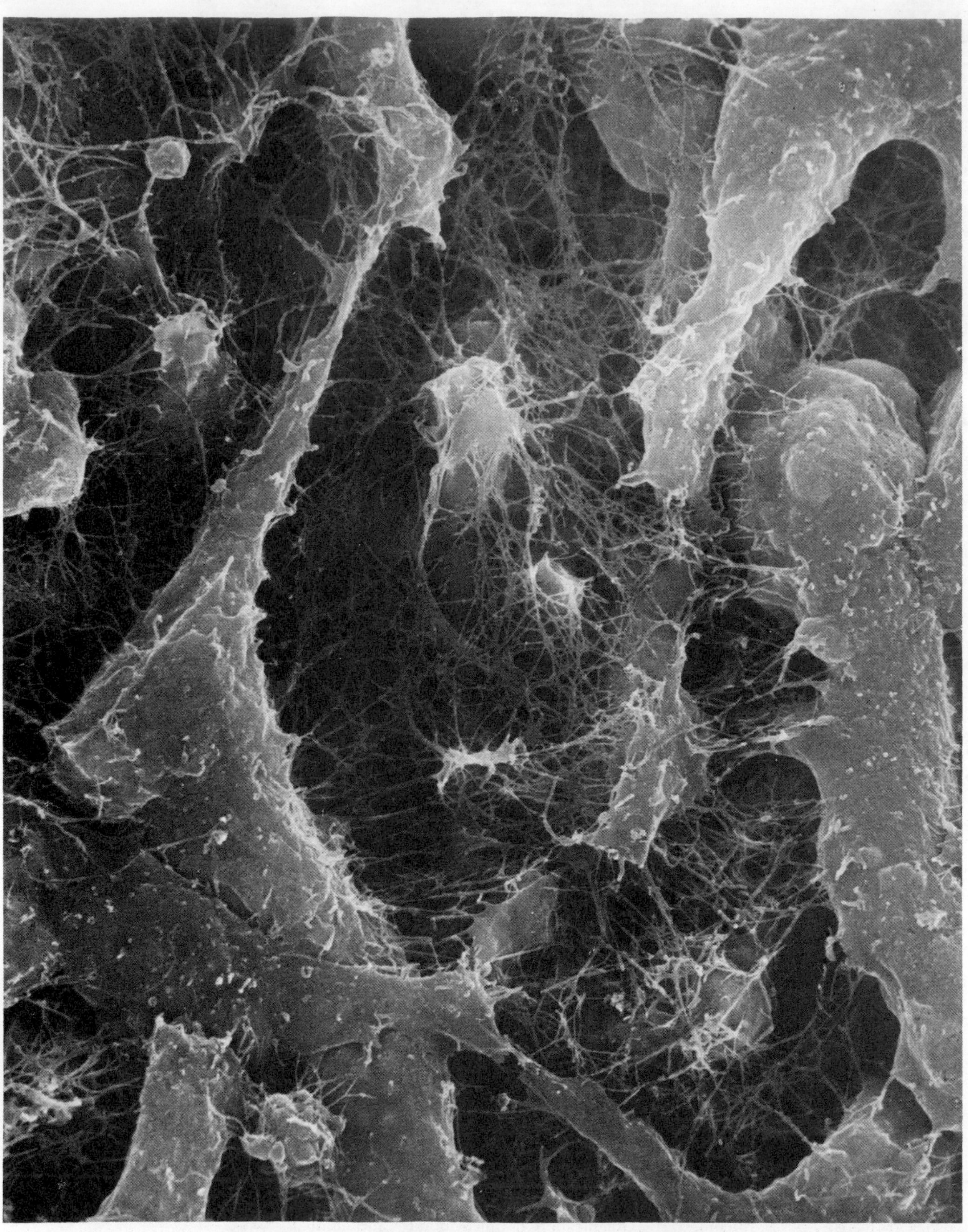

Scanning electron micrograph of neural crest cells migrating through an intricate extracellular matrix composed largely of protein fibers embedded in a hydrated polysaccharide gel. (Courtesy of Jan Löfberg.)

12

Cell-Cell Adhesion and the Extracellular Matrix

One of the great advantages of multicellularity is that it gives cells the freedom to specialize in ways that are incompatible with survival in isolation, but are beneficial to the organism as a whole. Such specialization may even entail the death of the specialized cell, as when epithelial cells in the skin accumulate keratin and die to form a tough, protective layer of cell skeletons on the outside of the organism. The specialized cells of multicellular animals are generally organized on a microscopic scale into cooperative assemblies called *tissues*, and the different varieties of tissues in turn combine to form larger functional units called *organs* (Figure 12–1).

All cells in tissues are in contact with a complex network of extracellular macromolecules referred to as the *extracellular matrix*. Besides helping to hold cells and tissues together, the extracellular matrix provides a highly organized lattice within which cells can migrate and interact with each other. In addition to being held together by the extracellular matrix, cells in direct physical contact with neighboring cells are often linked to them at specialized regions of their plasma membranes referred to as *cell junctions*. Some of these junctions serve mainly to hold cells together, while others allow small mole-

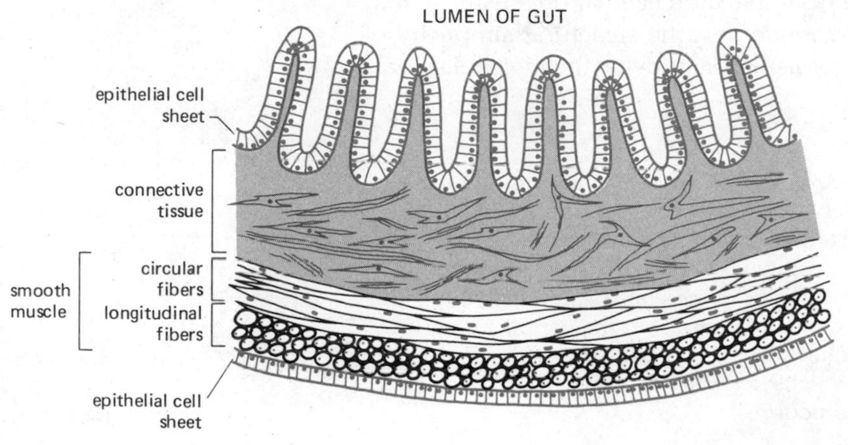

Figure 12–1 Schematic drawing of a cross-section through the intestine showing how this organ is constructed from epithelial, connective, and muscle tissues. Each tissue is an organized assembly of cells held together by cell junctions and/or extracellular matrix.

cules to pass from the inside of one cell to the inside of a neighboring cell, which may help to coordinate the activities of the cells forming the tissue.

In this chapter we first consider what is known about how cells recognize each other in the process of assembling into tissues and organs and then we discuss the structure and organization of intercellular junctions and of the extracellular matrix.

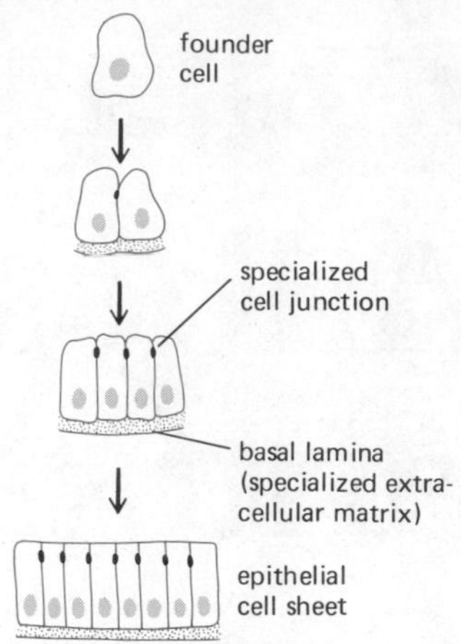

Figure 12–2 The simplest mechanism by which cells assemble to form a tissue. The progeny of the founder cells are retained in the epithelial sheet by the basal lamina and by the formation of intercellular junctions.

Intercellular Recognition and Cell Adhesion

There are two very different ways in which specialized cells become associated with one another to form a tissue. In the simplest of these, the tissue forms from the progeny of one or more "founder cells," which are prevented from wandering away by extracellular-matrix macromolecules and by the formation of specialized cell junctions (Figure 12–2). This is how the cells in epithelial cell sheets remain together, and much of animal development involves the formation, folding, and differentiation of such cell sheets.

The other strategy, which seems far more complex, involves the migration of individual cells over some distance and their subsequent assembly with local or other migrant cells into a tissue. In vertebrate embryos, for example, cells from the *neural crest* migrate to a large number of different regions, where they differentiate and assemble into a variety of tissues, including those of the peripheral nervous system (Figure 12–3). Such a process requires some mechanism for directing the cells to their final destination, such as the secretion of a chemical that attracts migrating cells (by *chemotaxis*) or the laying down of a specific pathway in the extracellular matrix that guides migrating cells (by *contact guidance*).

Given either strategy for tissue formation, groups of cells must recognize each other in order to stay together and remain distinct from the cells of surrounding tissues. Moreover, the tissues that result are not randomly arranged in an organ. How do cells recognize each other, and how do groups of cells become ordered in organs? As yet we know very little about the mechanisms involved in these processes. But there are clues in the relatively primitive multicellular behavior displayed by some unicellular organisms.

Myxobacteria Exploit the Advantages of Social Behavior[1]

Myxobacteria are one of the few social procaryotes. They are rod-shaped bacteria that live in the soil and feed on insoluble macromolecules, which they break down by secreting hydrolytic enzymes. To increase the efficiency of feeding, cells stay together and multiply in loose multicellular aggregates, or "swarms," which enables the cells to pool their secreted digestive enzymes (the "wolf-pack" effect). Myxobacteria resemble multicellular organisms in that they move and feed in groups. The genetic and the structural simplicity of myxobacteria make them attractive organisms for studying the molecular basis of primitive cell-cell interactions.

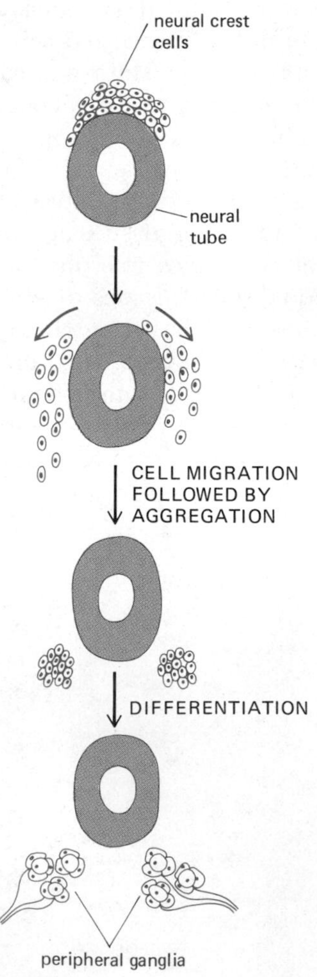

Figure 12–3 An example of a more complex mechanism by which cells assemble to form a tissue. Neural crest cells migrate from the upper surface of the neural tube to form a variety of different cell types and tissues throughout the embryo. Here they are shown assembling and differentiating to form two collections of neurons in the peripheral nervous system. Such a collection of nerve cells is called a *ganglion.* Other neural crest cells become supporting (satellite) cells that surround the neurons.

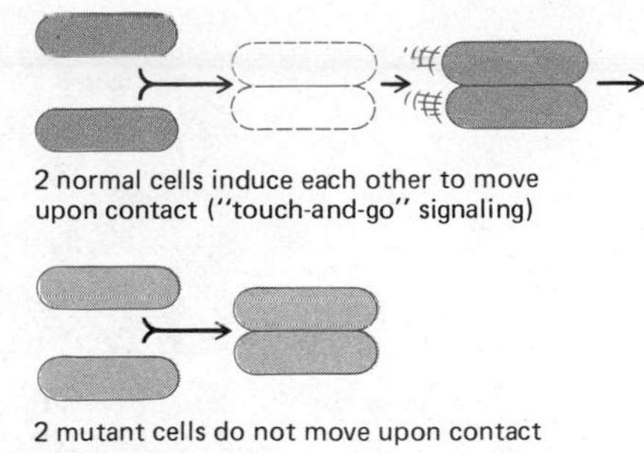

2 normal cells induce each other to move upon contact ("touch-and-go" signaling)

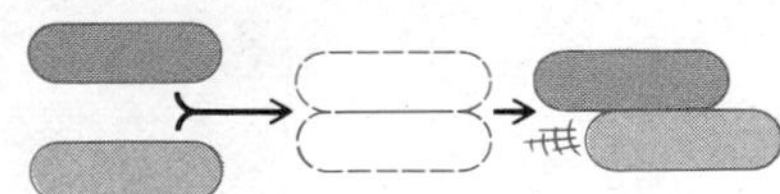

2 mutant cells do not move upon contact

a mutant cell defective in "touch-and-go" signaling moves when in contact with a normal cell

Figure 12–4 Schematic diagram showing how mutant myxobacteria demonstrate the existence of a "touch-and-go" group movement system in which cells move only when stimulated by contact with other cells. In the example shown, the mutant is defective in giving a signal but is able to respond to a signal from a normal cell. Other mutants (not shown) are able to give a signal but are unable to respond to one because they are defective either in a receptor or in the "motor" that drives the movement itself; such mutant cells fail to move when in contact with a normal cell.

Within a swarm, the individual myxobacterial cells are able to move independently with an unusual gliding movement that does not depend on flagella or other visible locomotory structures; yet the swarm, which may contain thousands of cells, remains a coherent unit, migrating only slowly. What communication mechanism keeps each bacterium in the swarm? One is a primitive form of contact guidance. When the bacteria move, they leave a trail of slime (a primitive extracellular matrix) along which other myxobacteria move. Although moving cells may occasionally cross a slime trail, they usually join it. A second mechanism has been demonstrated in experiments with mutant forms; these studies show that motility in myxobacteria is controlled by two distinct movement systems, one for single cells and one for cell groups. In the group system, which requires at least nine different gene products, individual cells move only when stimulated by contact with other cells. Some of the gene products seem to signal neighboring cells to move during cell contact, while others seem to act as receptors for such signals (Figure 12–4). This "touch-and-go" system keeps the cells moving together in a swarm.

Myxobacteria display their most impressive multicellular behavior when they are starved. Under such conditions, the cells in a swarm aggregate tightly together in groups and pile on top of one another to form a multicellular *fruiting body* (Figures 12–5 and 12–6). Within the fruiting body, bacterial cells differentiate into spores that are held together by a covering of extracellular matrix. The spores are able to survive for long periods of time, even in extremely hostile environments. Only when conditions are more favorable do they germinate to produce normal bacteria. The creation of a large fruiting body with a high concentration of spores ensures that a dense swarm will be generated (Figure 12–5) rather than a few isolated cells that cannot feed efficiently on their own.

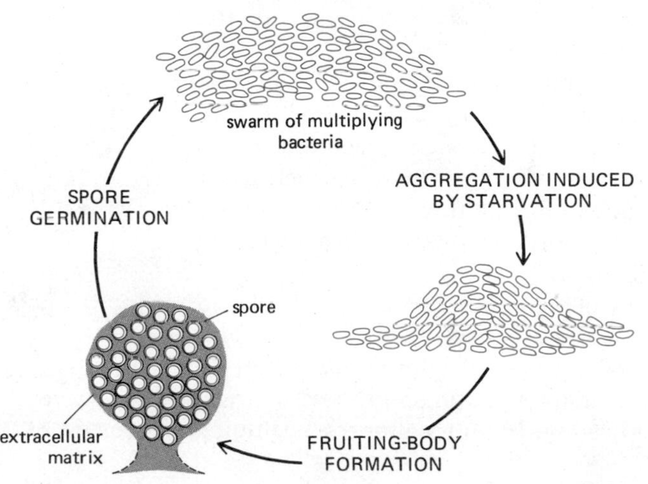

Figure 12–5 The life cycle of myxobacteria. Starvation causes the cells to aggregate and differentiate to form a fruiting body. The spores in the fruiting body germinate when conditions become favorable again.

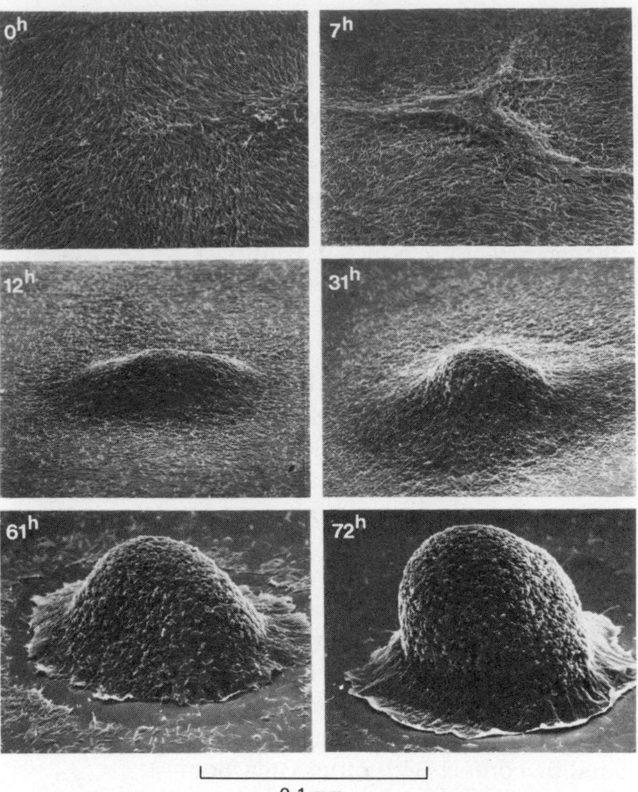

Figure 12–6 Scanning electron micrographs of a swarm of myxobacteria aggregating and forming a fruiting body in response to starvation. The hours elapsed since starvation began are shown on each micrograph. (Courtesy of Jerry Kuner.)

How do cells coordinate their behavior in order to form a fruiting body? The analysis of mutants defective in fruiting reveals that the cells aggregate together in response to at least four different secreted signals produced by the cells themselves. How such signaling leads to fruiting is at present best understood in another organism with a life cycle closely resembling that of myxobacteria—*Dictyostelium discoideum.* This cellular slime mold is a eucaryote whose genome is only four times larger than that of the procaryotic myxobacteria (and 100 times smaller than that of humans).

The Assembly of Slime Mold Amoebae into a Multicellular Slug Involves Chemotaxis and Specific Cell Adhesion[2]

Dictyostelium discoideum amoebae live as independent cells on the forest floor, feeding on bacteria and yeast and dividing every few hours. (In the laboratory they can be maintained in defined liquid media in the absence of other microorganisms.) When their food supply is exhausted, the amoebae stop dividing and gather together in a number of central collecting points, called *aggregation centers,* attracted by a chemotactic signal secreted by the amoebae themselves. In each aggregation center, cells adhere to each other by means of specific cell surface molecules to form a tiny (1–2 mm), vertical, wormlike structure that falls over on its side and crawls about as a glistening slug, leaving a trail of slime behind it (Figure 12–7).

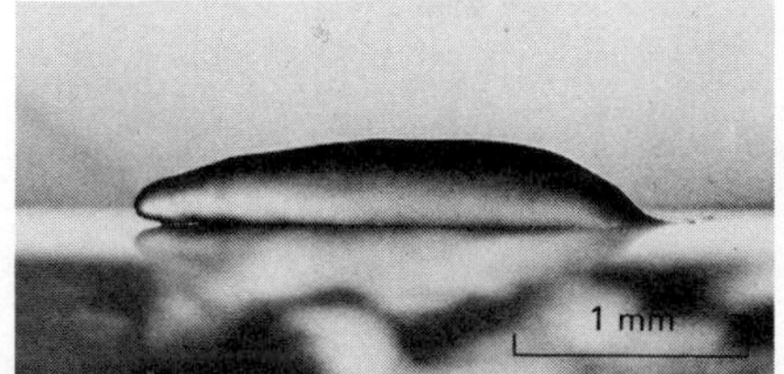

Figure 12–7 Light micrograph of a migrating slug of the cellular slime mold *Dictyostelium discoideum.* (Courtesy of David Francis.)

The multicellular slug shows a variety of behaviors that are not displayed by the free-living amoebae. For example, it is extremely sensitive to light and heat and will migrate toward a light source as feeble as a luminous watch. As the slug migrates, the cells begin to differentiate, initiating a process that will end with the production of a fruiting body some 30 hours after the beginning

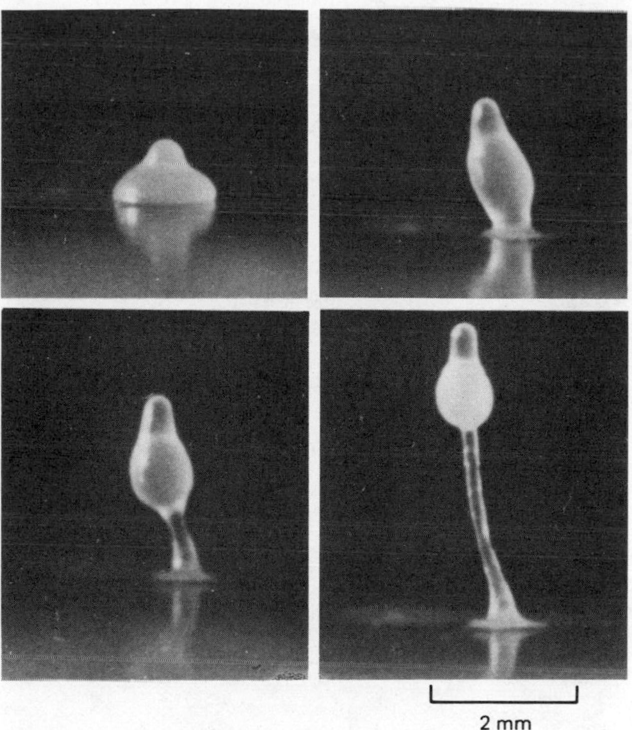

Figure 12–8 Light micrographs of *Dictyostelium discoideum* showing various stages in fruiting-body formation. (Courtesy of John Bonner.)

of aggregation (Figure 12–8). The complex cell migrations that occur in stalk and fruiting body formation are diagrammed in Figure 12–9. The cells in the front of the slug become the stalk region; those behind differentiate into spores, while those at the very rear form the foot plate. Both the stalk cells and the spore cells become covered with extracellular matrix (in the form of cellulose walls) and, in the end, all except the spore cells die. Only when conditions are favorable will the spores germinate to produce the single-cell amoebae that start the cycle again (Figure 12–10).

Chemotactic signaling in *Dictyostelium* is the best-understood example of chemotaxis in eucaryotic cells. In response to starvation, the amoebae start making and secreting cyclic AMP in a pulsatile fashion. For unknown reasons, certain of the cells become aggregation centers; the cyclic AMP they make binds to specific receptors on the surface of other starved amoebae, thereby orienting their normal locomotion in the direction of the source of cyclic AMP. This chemotactic response can be directly demonstrated by applying a tiny amount of cyclic AMP with a micropipette to any point on the surface of a starved amoeba cell. The result is the immediate formation of a pseudopod, which grows toward the micropipette (Figure 12–11). Normally the pseudopod would adhere to the surface on which the cell was placed and would pull the cell along in the same direction.

The area of influence of each aggregation center is enlarged because the initial cyclic AMP signal is released in pulses from the center and is relayed from cell to cell. Each pulse of cyclic AMP induces surrounding cells both to move toward the source of the pulse and to secrete their own pulse of cyclic AMP. In turn, this new pulse, released with a slight delay, orients and induces a pulse of cyclic AMP from cells just beyond and so on. In this way, regular

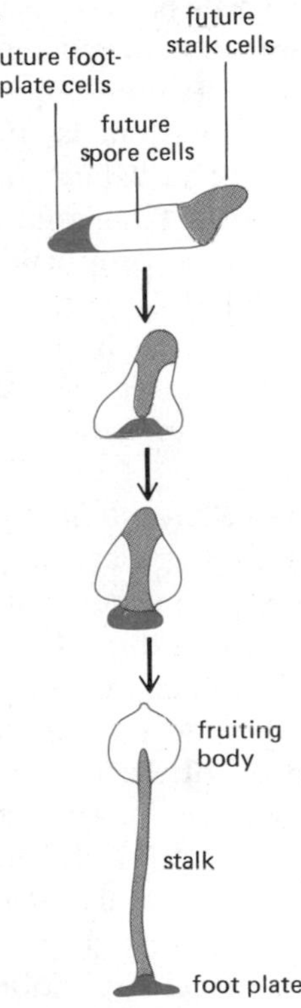

Figure 12–9 Schematic diagram of the cell migrations involved in the formation of a fruiting body in *Dictyostelium discoideum.* Cells in the front of the slug migrate down to become the stalk, while cells in the middle migrate up and differentiate into the collection of spores that form the fruiting body.

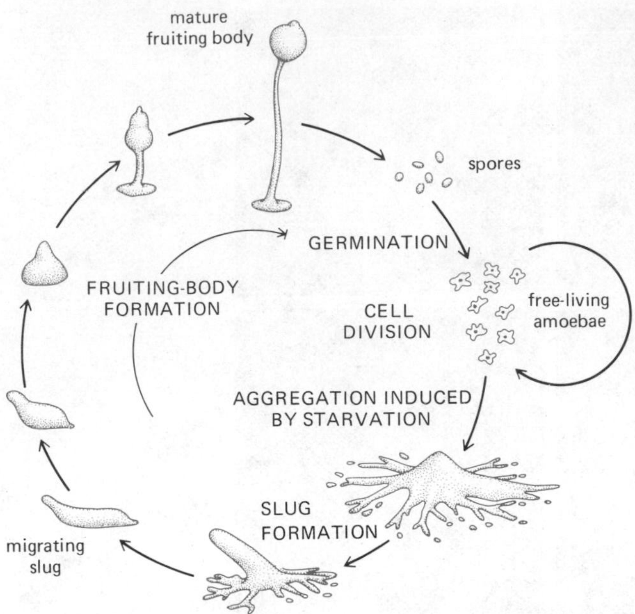

Figure 12–10 The life cycle of *Dictyostelium discoideum.* When starved, slime mold amoebae aggregate and differentiate to form a migrating slug, which in turn forms a fruiting body. The released spores germinate to form amoebae when conditions are favorable again.

pulsatile waves of cyclic AMP flow from each aggregation center, causing more distant amoebae to move inward in surging concentric or spiraling waves that can be seen in time-lapse motion pictures (Figure 12–12). The advantage of such a relay system is that the signal moves outward from the center at a constant speed (distance proportional to time) so that it can influence a large area. By contrast, a signal that merely diffuses progressively, slows down as it moves (distance proportional to the square root of time). The difference can be appreciated by comparing the aggregation process in *Dictyostelium discoideum* with that in a strain that lacks the relay system, *Dictyostelium minutum*. In the latter, the range of the signal released from each aggregation center is greatly reduced, and the slug and fruiting bodies that form are very small.

When *Dictyostelium* amoebae are starved, they not only begin to secrete and respond to pulses of cyclic AMP in the process of chemotactic orientation; thousands of new genes are activated, and they start to synthesize many new molecules, including those utilized in the process of cell-to-cell adhesion. As a result, as the aggregating amoebae form streams moving toward aggregation centers, they adhere to each other tightly, end-to-end. Still later, in the slug stage, they adhere to each other over most of their surfaces. A number of different molecules that are synthesized only by starved amoebae have been implicated in cell adhesion. For example, the initial end-to-end aggregation appears to be mediated by a cell-surface glycoprotein, since antibodies directed against this glycoprotein inhibit the aggregation. However, the molecular basis of cell-cell adhesion is better understood in another simple organism, the sponge.

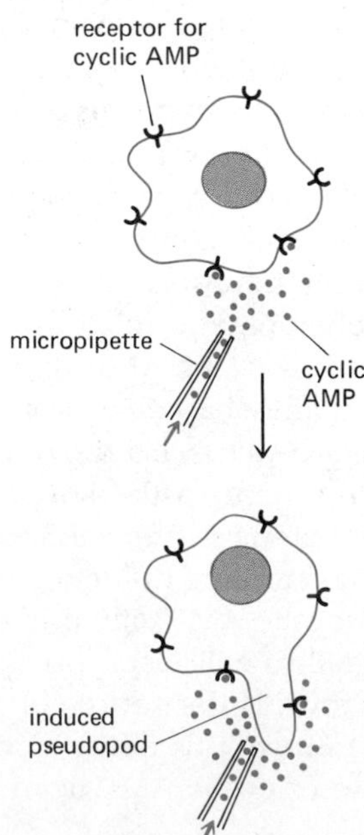

Figure 12–11 The application of a small amount of cyclic AMP to any point on the surface of a starved *Dictyostelium* amoeba (a single cell) induces the immediate formation of a pseudopod at that point. In this way the amoeba is induced to move toward a source of cyclic AMP. The cyclic AMP acts by binding to cell-surface receptors.

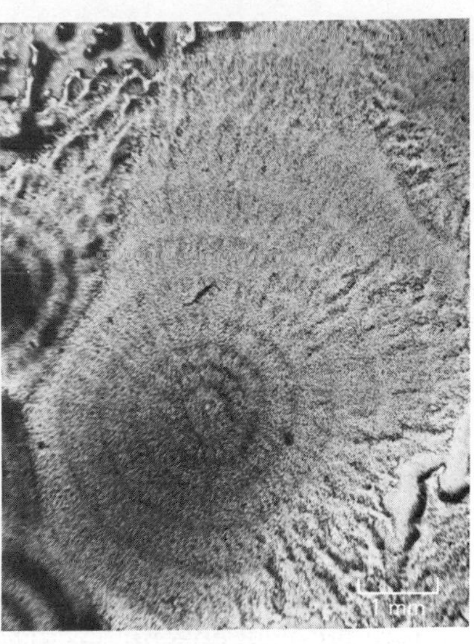

Figure 12–12 Light micrograph of waves of starved *Dictyostelium* amoebae aggregating toward an aggregation center. Individual amoebae cannot be distinguished at this low magnification. (Courtesy of Gunter Gerisch.)

Species-specific Cell Aggregation in Sponges Is Mediated by a Large Extracellular Aggregation Factor[3]

One of the attractions of using social microorganisms such as myxobacteria and *Dictyostelium* for studying cell aggregation is that the phenomenon proceeds quite normally in a culture dish, where it is accessible for investigation. Unfortunately this is rarely the case with cell-to-cell recognition processes occurring in the development of multicellular animals, and it is usually necessary instead to dissociate developed tissues into single cells that can then reassemble *in vitro*. Since such dissociated cells often reassemble into structures resembling the original tissue, it is hoped that studying reassembly will illuminate the processes by which the tissue originally formed in the animal. This type of experiment was first performed in 1907 with sponge cells.

Sponges are the simplest multicellular animals. They consist of only five or six cell types and can be mechanically dissociated into single cells by gently pressing the adult organism through a mesh. When the dissociated cells are mixed together, they rapidly reaggregate and eventually reorder themselves to form a normal sponge. In a classical experiment of this kind, dissociated cells from two species of sponge of different colors were mixed together. The cells adhered to each other in a species-specific manner to form independent aggregates of one color or the other (Figure 12–13). Although this result is obtained with only certain species of sponge, it demonstrates that some adult sponge cells have the capacity to distinguish cells of their own species from those of another.

This species-specific aggregation is mediated by a large extracellular *aggregation factor* thought to be a complex of proteoglycan and/or glycoprotein molecules. Adult sponges spontaneously dissociate into single cells when they are incubated in sea water depleted of Ca^{2+} and Mg^{2+}. During dissociation, the aggregation factor is released into the sea water. If the dissociated cells are washed free of the aggregation factor and put back in normal sea water, they will not reaggregate unless the factor is added back. The factor-induced aggregation is rapid, requires Ca^{2+}, and is species-specific.

The sponge aggregation factor is a huge particle, with a diameter of about 100 nm and a molecular weight of about 20 million, and it polymerizes further in the presence of Ca^{2+}. It binds to specific protein receptors (called *baseplates*) on the surface of sponge cells of the same species but not to sponge cells of other species. The particles seem to cross-link sponge cells by binding first to the cell-surface baseplates and then to each other. It is the binding of the particles to one another that requires the presence of Ca^{2+} (Figure 12–14). Thus, when the aggregation factor is covalently coupled to agar beads, it mediates the aggregation of the beads in the presence of Ca^{2+}. On the other

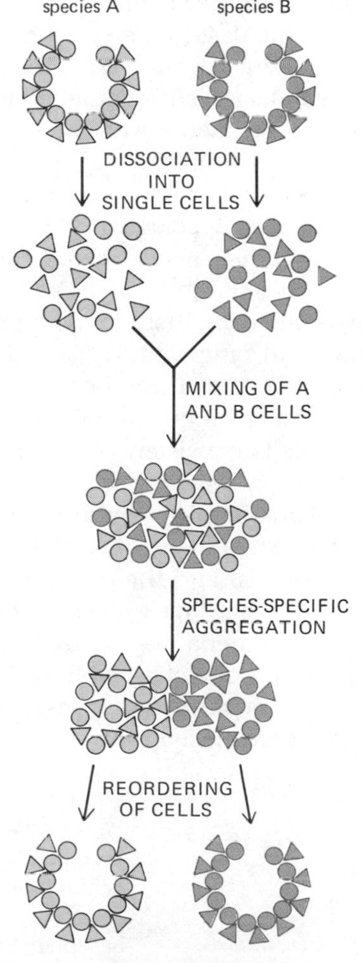

Figure 12–13 The species-specific aggregation of sponge cells. When dissociated cells from two differently colored species of sponge are mixed together, they form species-specific aggregates that eventually re-form individual sponges. A sponge is composed of about five or six different cell types. Only two are diagrammed here, for the sake of simplicity.

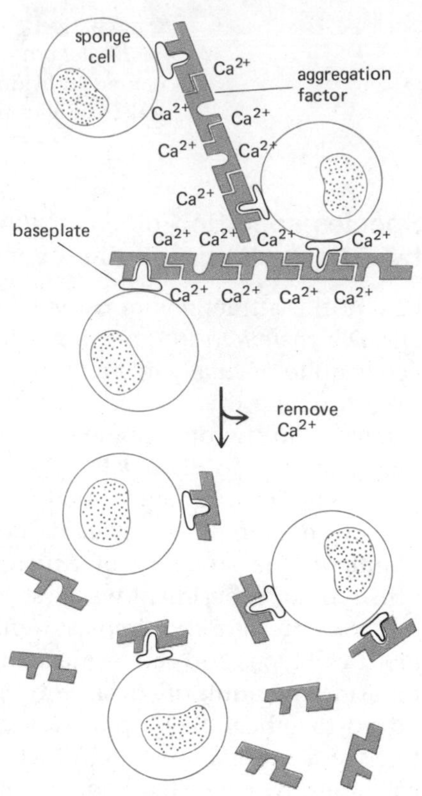

Figure 12–14 Schematic drawing of how a large aggregation factor is thought to cross-link sponge cells by binding first to specific cell-surface receptor proteins (called baseplates) and then to each other in the presence of Ca^{2+}. When cell aggregates are washed in the absence of Ca^{2+}, all of the aggregation factor not attached directly to baseplates is released into the supernatant and removed. The cells now fail to aggregate, even when Ca^{2+} is added back, since the concentration of factor remaining is so low that its association with baseplates cannot be maintained.

hand, if the baseplate protein rather than the aggregation factor is coupled to agar beads, the beads require both the factor and the presence of Ca^{2+} for aggregation (Figure 12–15).

In summary, the sponge cell-recognition mechanism is one in which a large, multivalent linker molecule recognizes the same protein on two different cells and thereby holds them together.

Dissociated Embryonic Vertebrate Cells Preferentially Associate with Cells of the Same Tissue[4]

The approach initially used to study sponge cell aggregation has been widely used to study cell recognition and adhesion in embryonic vertebrate cells. Unlike adult vertebrate tissues, which are extremely difficult to dissociate, embryonic tissues are easily dissociated by low concentrations of the proteolytic enzyme trypsin. When the dissociated cells from two different embryonic tissues are mixed together, they initially form mixed aggregates and subsequently sort themselves into separate regions, grouped by their tissue of origin. When cells from two different species are mixed, they generally ignore the species difference and sort out to form tissue-specific aggregates that contain cells of both species.

A similar result is obtained even when a very different test is used to measure cell adherence. The test involves dissociating cells from one type of tissue, allowing them to assemble into aggregates, and then mixing the aggregates with radioactively labeled cells that have been dissociated from various tissues. The total number of radioactive cells bound to the aggregates is determined at various times after they have been added. In this assay, the radioactive cells consistently bind more rapidly to aggregates of their own tissue than to aggregates of other tissues (Figure 12–16).

Thus, similar results are obtained by two different assays, one determining the extent of cell adherence during prolonged incubation (the cell-sorting experiments) and the other measuring the rate of cell adherence (the radioactive cell-binding experiments), although the mechanisms involved in the two phenomena may be different. Unfortunately, it is still uncertain whether these tissue-specific, cell-to-cell recognition processes involve extracellular aggregation factors, the expression of complementary cell-surface molecules, or some other mechanism.

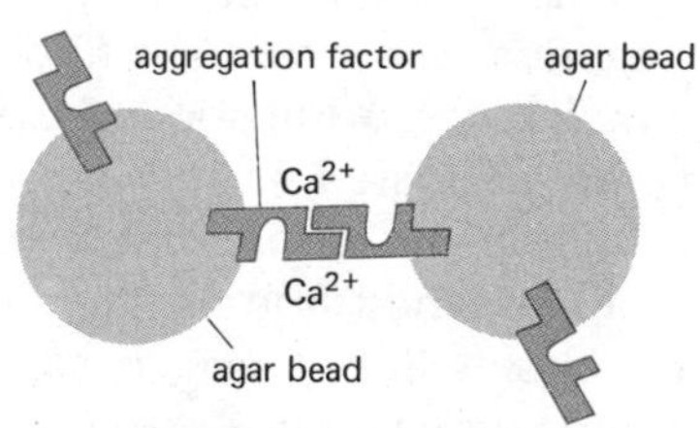

(A) AGGREGATION FACTOR COVALENTLY LINKED TO BEAD

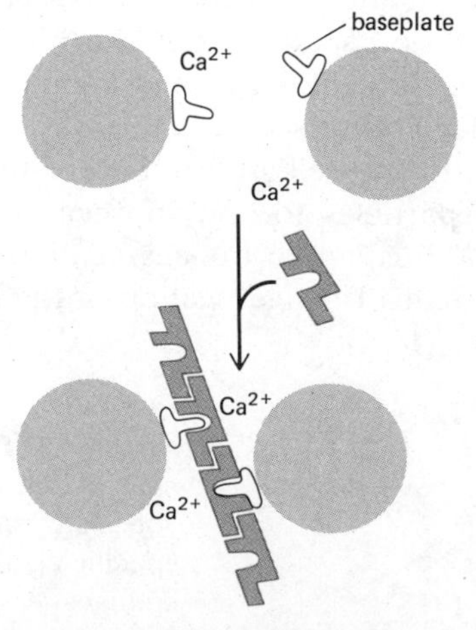

(B) BASEPLATE COVALENTLY LINKED TO BEAD

Figure 12–15 Evidence for the scheme in Figure 12–14 is provided by covalently attaching either the aggregation factor or the baseplate protein to agar beads. Whereas the factor-coupled beads aggregate in the presence of Ca^{2+} (A), the baseplate-coupled beads aggregate only if the aggregation factor is added as well (B).

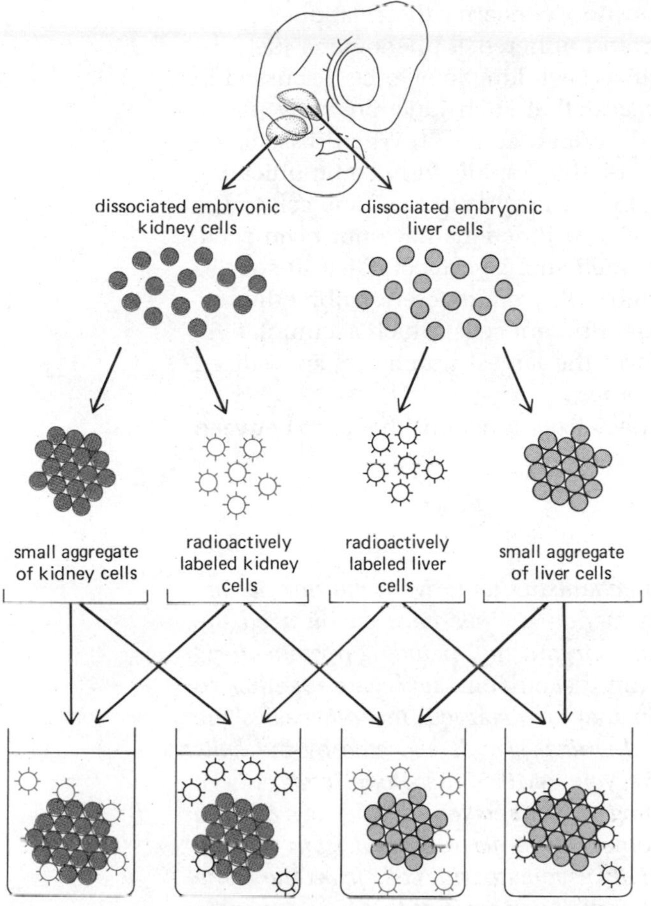

Figure 12–16 Tissue-specific adherence of dissociated vertebrate embryo cells determined by a radioactive cell-binding assay. By determining the number of radioactively labeled cells bound to the cell aggregates after various periods of time, the rate of cell adherence can be measured. The rate of adherence is always greater between cells of the same kind.

Cells from Different Tissues Display a Hierarchy of Adhesiveness[5]

When a mixture of dissociated cells from two different tissues sort out *in vitro,* the cells of one tissue usually end up inside the aggregate, surrounded by cells of the other tissue. According to the *differential adherence hypothesis,* this difference arises because cells from different tissues show a hierarchy of adhesiveness, with the cells on the inside of mixed aggregates adhering so strongly to each other that they eventually exclude less adherent cells from the interior. Therefore, if mixing experiments showed that cells from tissue A are always internal to cells of tissue B at the end of the experiment, while cells of tissue B are always internal to cells of tissue C, the A cells should always be internal to C cells at the end of an experiment in which A and C cells have been mixed. In fact, this is always the result observed (Figure 12–17).

The differential adherence hypothesis provides a satisfying explanation for a remarkable observation made many years ago. When dissociated cells (or pieces of tissue) from the three different germ layers of an amphibian embryo are mixed together, the cells (or tissues) sort out so that the ectoderm ends up on the outside, with the endoderm on the inside, and the mesoderm in between—just as they were in the intact embryo. This observation, which has been made repeatedly with different tissues and different species, suggests that differential cell adhesion may play a part in stabilizing the organization of the germ layers during embryonic development.

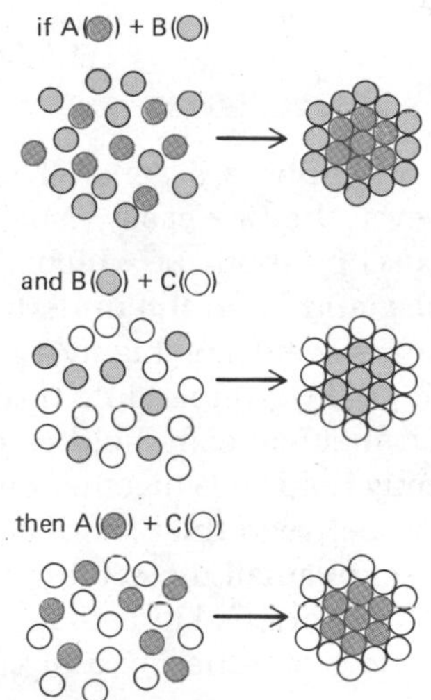

Figure 12–17 The differential adherence hypothesis predicts that if A cells are internal to B cells and B cells are internal to C cells at the end of cell-sorting experiments, then A cells should be internal to C cells when these cells are mixed in the same type of experiment. As shown in the example, this is always what happens.

An important but unresolved question concerns the relationship between the *in vitro* cell-cell recognition and adherence phenomena just discussed and the formation of the specialized cell junctions to be discussed in the next section. Two observations suggest that such junctions may play a part in the cell-adhesion experiments described above: (1) When dissociated sponge cells of the same species aggregate, they rapidly form communicating junctions that allow small ions to pass from the cytoplasm of one cell to the cytoplasm of the other, although typical gap junctions have not been seen. Such ionic coupling does not occur between sponge cells of different species in mixed aggregates. (2) When dissociated cells from different embryonic vertebrate tissues sort out *in vitro*, the most adherent cells, which accumulate in the interior of the aggregate, tend to form the largest number of specialized adhering junctions, called *spot desmosomes*.

We shall next look at the types of specialized junctions that form between cells in tissues.

Summary

Difficulties in studying the molecular mechanisms underlying the normal assembly of cells into complex tissues in higher animals have encouraged the study of simpler systems. Some unicellular organisms, including myxobacteria and the cellular slime mold Dictyostelium discoideum, *aggregate together to form multicellular fruiting bodies when they are starved. In both cases, the starved cells aggregate in response to chemical signals secreted by the cells themselves; in* Dictyostelium *the signal is cyclic AMP. Some sponge cells adhere to cells of their own species by secreting large, species-specific aggregation factors that cross-link similar cells together. Cells disassociated from various tissues of vertebrate embryos preferentially associate with cells from the same tissue when they are mixed together; the molecular basis of this tissue-specific recognition process is unknown.*

Cell Junctions

Cell junctions are generally too small to be resolved by light microscopy. However, they are easily visualized by thin-section and freeze-fracture electron microscopy, both of which demonstrate that the interacting plasma membranes (and often the underlying cytoplasm and the intervening intercellular space as well) are highly specialized in these regions. Such junctions are traditionally grouped into three functional categories: (1) **adhering junctions,** which mechanically hold cells together; (2) **impermeable junctions,** which not only hold cells together but seal them in such a way that molecules cannot leak in between them; and (3) **communicating junctions,** which mediate the passage of small molecules from one interacting cell to the other.

The various kinds of intercellular junctions are listed in Table 12–1. The main type of adhering junction is called a *desmosome. Tight junctions* (and *septate junctions* in invertebrates) are the main impermeable junctions. Communicating junctions are of two types: *gap junctions* and *chemical synapses.* At gap junctions small molecules can pass directly from the interior of one cell to the interior of the other, but at chemical synapses the cells communicate only indirectly, even though they are in physical contact (Figure 12–18). The "sending" cell at a synapse (called the *presynaptic cell*) secretes a chemical signal (called a *neurotransmitter*) that diffuses across the synaptic space and signals the other cell (called the *postsynaptic cell*). Since chemical synapses will be considered in Chapter 18, they will not be discussed here.

Table 12–1 Intercellular Junctions

1. *Adhering junctions*
   - a. belt desmosome
   - b. spot desmosome
   - c. hemidesmosome
2. *Impermeable junctions*
   - a. tight junction
   - b. septate junction (invertebrates only)
3. *Communicating junctions*
   - a. gap junction
   - b. chemical synapse

Chemical synapses should not be confused with the less common *electrical synapse,* at which electrical impulses pass directly from one nerve cell to another via gap junctions.

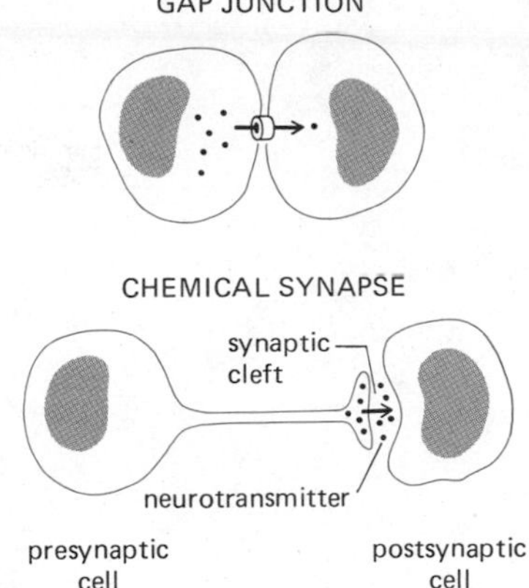

Figure 12–18 Schematic diagram of two types of communicating cell junctions. At a gap junction, small molecules pass directly from the cytoplasm of one cell to the cytoplasm of the other. At a chemical synapse, the presynaptic nerve cell secretes a chemical signal (neurotransmitter) that diffuses across the synaptic cleft and signals the postsynaptic cell at its surface.

Desmosomes Anchor Cells Together[6]

Desmosomes are widely distributed in tissues, where they enable groups of cells to function as structural units. They are most abundant in tissues that are subject to severe mechanical stress, such as cardiac muscle, skin epithelium, and the neck of the uterus, suggesting that they are important for holding cells together. They occur in three different forms: belt desmosomes, spot desmosomes, and hemidesmosomes, all three of which are present in most epithelial cells.

Belt desmosomes form a continuous band around each of the interacting cells in an epithelial sheet, near the cell's apical end (Figure 12–19). The bands in adjacent cells are directly apposed and are separated by a poorly characterized filamentous material (in the intercellular space) that presumably holds the interacting membranes together. Within each cell, contractile bundles of actin filaments run along the belts just under the plasma membrane. These filament bundles probably mediate one of the most fundamental processes in animal morphogenesis—the folding of epithelial cell sheets into tubes (Figure 12–20). For example, the oriented contraction of these bundles of actin filaments is thought to cause an apical narrowing of each of the epithelial cells in the neural plate, resulting in the rolling up of the epithelial plate to form the neural tube (see p. 822).

Spot desmosomes act like rivets to hold epithelial cells together at buttonlike points of contact (Figure 12–21). They also serve as anchoring sites for *keratin filaments* (also called tonofilaments, see p. 595), which extend from one side of the cell to the other across the cell interior, forming a structural framework for the cytoplasm. Since other filaments extend from cell to cell at spot desmosomes (Figure 12–22), the keratin filament networks inside adjacent cells are connected indirectly through these junctions to form a continuous

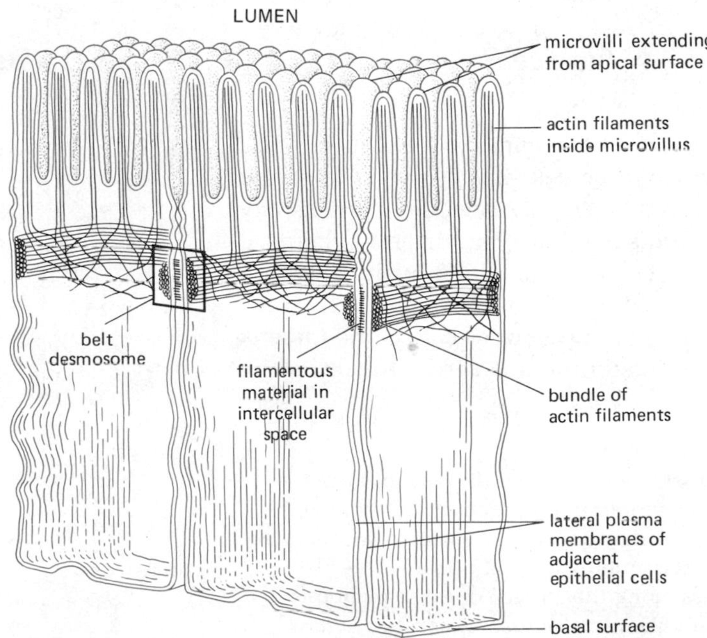

Figure 12–19 Schematic drawing of a belt desmosome between epithelial cells of the small intestine. This junction (also called a *zonula adherens*) encircles each of the interacting cells and is characterized by filamentous material in the intercellular space and a contractile bundle of actin filaments running along the cytoplasmic surface of the junctional plasma membrane.

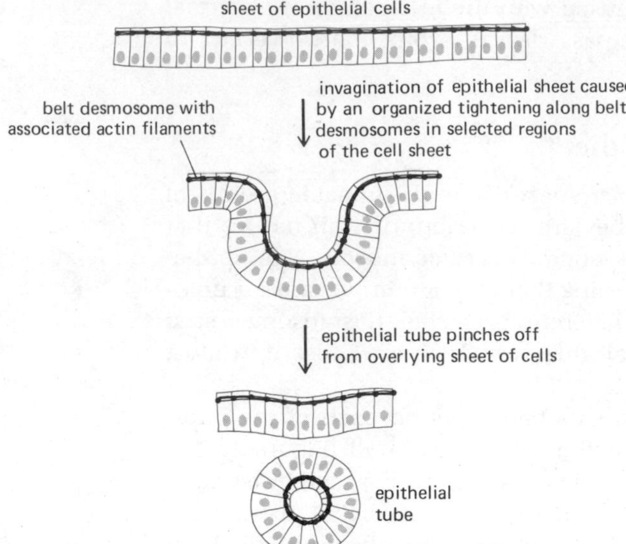

Figure 12–20 Schematic drawing of the folding of an epithelial sheet to form an epithelial tube (as in the formation of the neural tube). It is thought that the oriented contraction of the bundle of actin filaments running along the belt desmosomes causes the apical narrowing of the epithelial cells in selected regions of the cell sheet and that this results in the rolling up of the epithelial sheet into a tube, which then pinches off from the parent epithelial sheet.

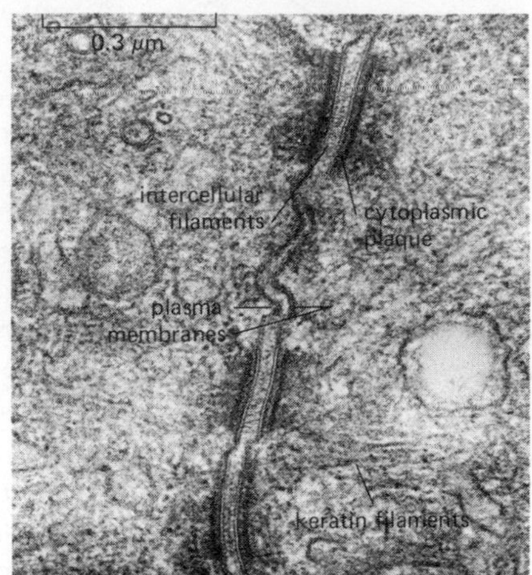

Figure 12–21 An electron micrograph of three spot desmosomes between two epithelial cells in the intestine of a rat. (From N. B. Gilula, in Cell Communication [R. P. Cox, ed.], pp. 1–29. New York: Wiley, 1974.)

network of fibers across the entire epithelial sheet (Figure 12–23). **Hemidesmosomes,** or half-desmosomes, resemble spot desmosomes, but instead of joining adjacent epithelial cell membranes together, they join the basal surface of epithelial cells to the underlying basal lamina (a specialized structure of the extracellular matrix—see below). Together spot desmosomes and hemidesmosomes act as rivets that distribute any shearing forces through the epithelial sheet and its underlying connective tissues as a whole (Figure 12–23).

Tight Junctions Form a Permeability Barrier Across Cell Sheets[6,7]

Epithelial cell sheets cover the surface of the body and line all of its cavities. Despite extensive biochemical differences, these cell sheets have at least one important function in common: they serve as highly selective permeability barriers, separating inside and outside fluids that have very different chemical compositions. **Tight junctions** play a crucial part in maintaining the selective-barrier function of cell sheets. For example, the epithelial cells lining the small intestine must keep most of the gut contents in the inner cavity (the lumen); simultaneously, the cells must pump selected nutrients across the cell sheet

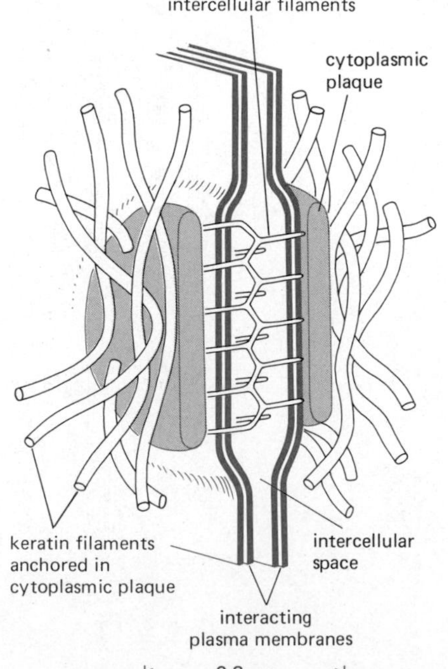

Figure 12–22 A highly schematized drawing of a spot desmosome. On the cytoplasmic surface of each interacting plasma membrane is a dense plaque associated with a thick network of keratin filaments. The keratin filaments either terminate in or pass along the surface of these plaques. Poorly characterized filaments connect adjacent plaques across the intercellular space.

into the extracellular fluid on the other side, from which they are absorbed into the blood. This transport depends on two different sets of specialized membrane transport proteins: one is confined to the apical surface of the epithelial cell (the surface facing the lumen) and pumps selected molecules in; the other, which is confined to the basal and lateral, or so-called basolateral surface, pumps them out again on the other side (Figure 12–24). It is clear that to maintain directional pumping the apical set of pumps must not be allowed to diffuse (in the plasma membrane) to the basolateral surface of the cell, nor must the basolateral set be allowed to diffuse to the apex. Furthermore, the transported molecules must be prevented from leaking back into the lumen. Tight junctions make transport possible in two different ways. First, they act as diffusion barriers within the lipid bilayer of the plasma membrane. They thus prevent the transport proteins in the apical membrane from diffusing into the basolateral membrane, and vice versa. Second, they seal neighboring cells together to create a continuous sheet of cells between which even small molecules are unable to pass.

At a tight junction, the interacting plasma membranes are so closely apposed that there is no intercellular space: if an electron-dense marker is added to one side of a cell sheet, it will not pass beyond the tight junction (Figure 12–25). As these junctions can be disrupted either by treatment with proteolytic enzymes or by agents that chelate Ca^{2+} or Mg^{2+}, both special proteins and divalent cations are required for maintaining their integrity.

In electron micrographs, tight junctions are seen to be formed when specific proteins in the two interacting plasma membranes make direct contact across the intercellular space (Figure 12–26). By freeze-fracture electron microscopy, these proteins are visualized as linear rows of intramembrane particles (Figure 12–27). A beltlike structure composed of many anastomosing rows of such particles completely encircles each cell in the sheet.

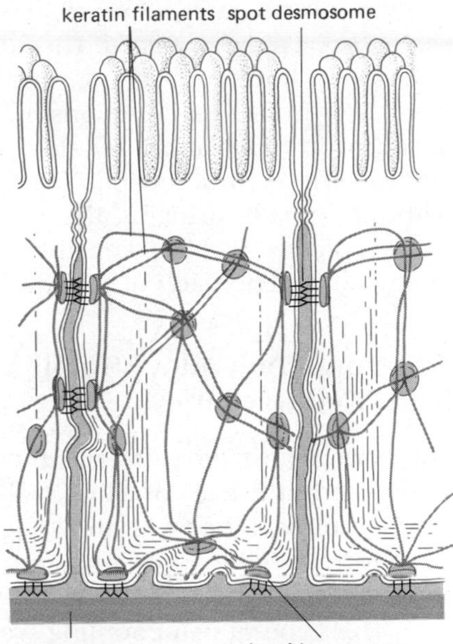

Figure 12–23 A schematic drawing of the distribution of spot desmosomes and hemidesmosomes in epithelial cells of the small intestine. Note how the keratin filament networks of adjacent cells are indirectly connected to each other through spot desmosomes and to the basal lamina through hemidesmosomes.

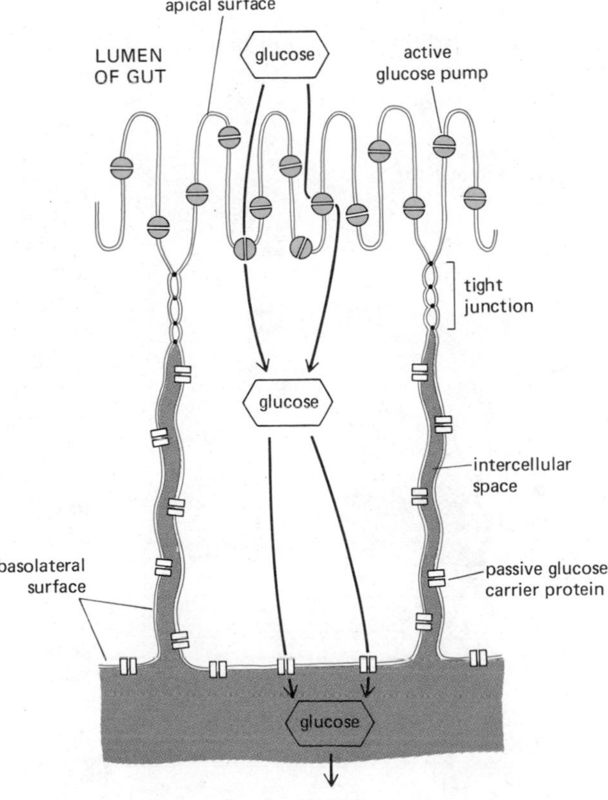

Figure 12–24 Schematic drawing of a small intestinal epithelial cell showing how tight junctions serve to confine different transport proteins to different regions of the plasma membrane. This permits nutrient transfer across the epithelial sheet from the gut lumen to the blood. In the example shown here, glucose is actively transported into the cell by glucose pump proteins at the apical surface and diffuses out of the cell by facilitated diffusion mediated by passive glucose transport proteins in the basolateral membranes.

In invertebrates, **septate junctions,** rather than tight junctions, usually serve to seal sheets of cells. These junctions have a highly characteristic morphology in both thin-section and freeze-fracture electron micrographs and they differ from tight junctions in two ways. First, the junctional proteins are arranged in much more regular, parallel rows that interact across the intercellular space; and second, the junctional proteins themselves form a seal without actually bringing the two apposed plasma membranes into direct contact (Figure 12–28). Like tight junctions, these "septal ribbons" form a complete belt around each of the interacting cells.

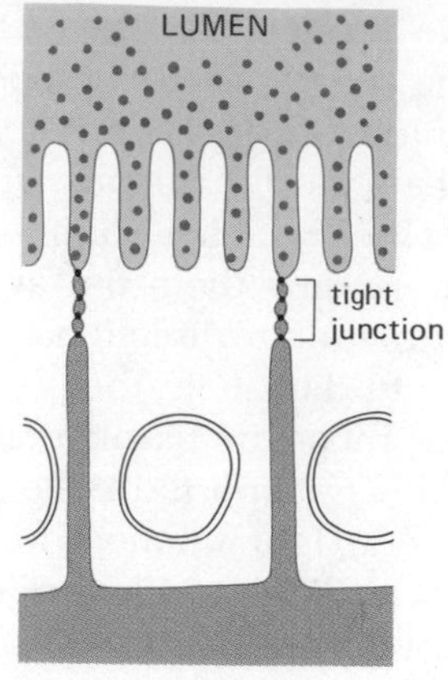

Figure 12–25 Schematic drawing showing that soluble molecules on one side of an epithelial cell sheet cannot pass beyond the tight junctions that seal adjacent cells together.

Gap Junctions Allow Small Molecules to Pass Directly from Cell to Cell[6,8]

The commonest type of cell junction is the **gap junction,** which is widely distributed in tissues of all animals. Such junctions are said to be communicating junctions because they allow small, water-soluble molecules to pass directly from the cytoplasm of one cell to the cytoplasm of the other, thereby coupling the cells both electrically and metabolically.

This type of cell coupling was first demonstrated, in 1958, by inserting microelectrodes into each of two interacting nerve cells in the nerve cord of a crayfish. When a voltage gradient was applied between the two electrodes, current readily passed, indicating that inorganic ions (which carry current in living tissues) could pass freely from one cell interior to the other. In experiments done several years later, small fluorescent molecules were injected into one cell in a tissue and shown to pass readily into adjacent cells without leaking into the extracellular space. When fluorescent molecules of different sizes were injected, it was found that molecules smaller than 1000 to 1500 daltons could pass between cells while larger molecules could not, suggesting a functional pore size for the connecting channels of about 1.5 nm (Figure 12–29). This pore size implies that coupled cells share a variety of small molecules (such as inorganic ions, sugars, amino acids, nucleotides, and vitamins) but do not share their macromolecules (proteins, nucleic acids, and polysaccharides).

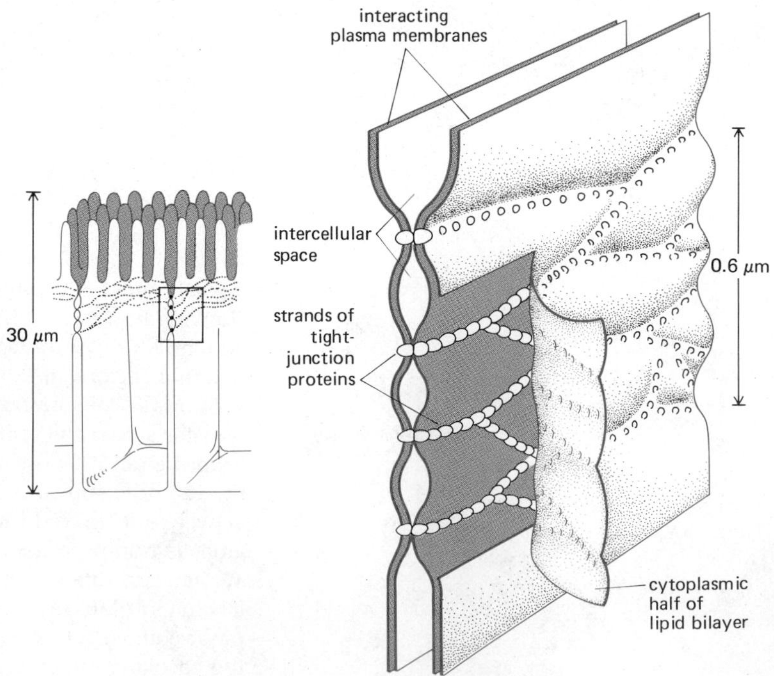

Figure 12–26 Schematic drawing of a tight junction between epithelial cells of the small intestine. The adjacent plasma membranes are held together by continuous strands of junctional proteins that make contact across the intercellular space, creating a complete seal. A beltlike band of anastomosing sealing strands encircles each cell in the sheet. The cytoplasmic half of one membrane has been schematically peeled back to expose the protein strands. In freeze-fracture electron microscopy the tight-junction proteins remain with the cytoplasmic half of the lipid bilayer (see Figure 12–27B) instead of staying in the outer half as shown here.

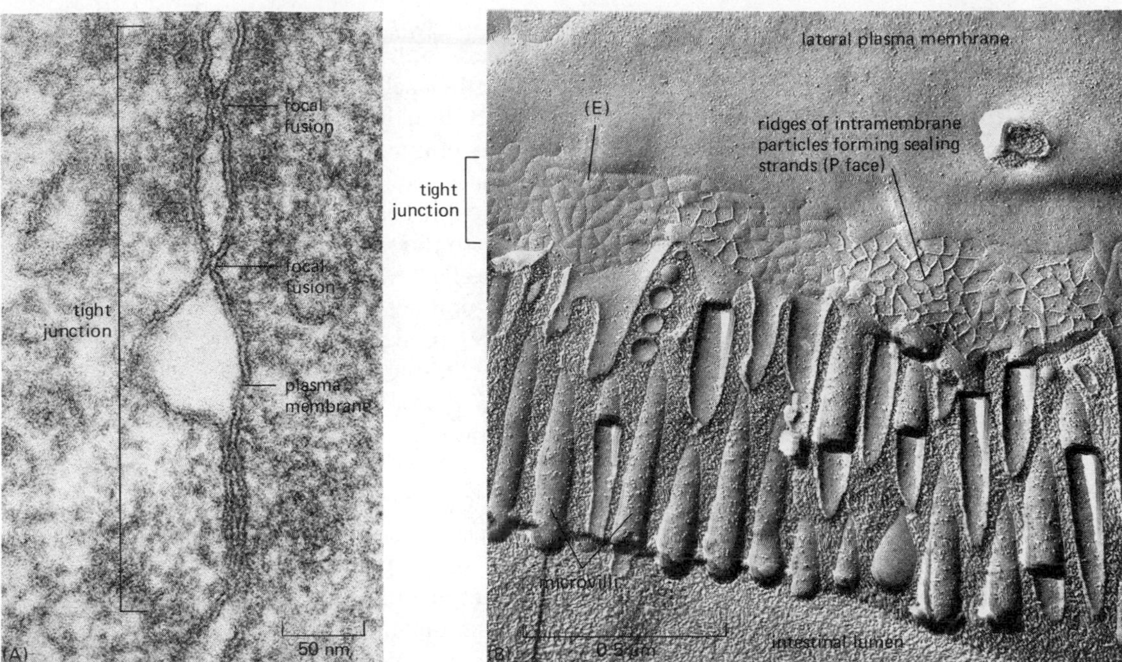

Figure 12–27 Thin-section (A) and freeze-fracture (B) electron micrographs of a tight junction between epithelial cells of the rat small intestine. In (A), the junction is seen in cross-section as a series of focal fusions between the interacting plasma membranes. In (B), as in all freeze-fracture micrographs, the plane of the micrograph is parallel to the plane of the membrane, and the sealing strands are seen as ridges of intramembrane particles on the cytoplasmic (P) fracture face of the membrane or as complementary grooves on the external (E) face of the membrane. Note that the cells in (B) are oriented with their apical ends down. (From N. B. Gilula, in Cell Communication [R. P. Cox, ed.], pp. 1–29. New York: Wiley, 1974.)

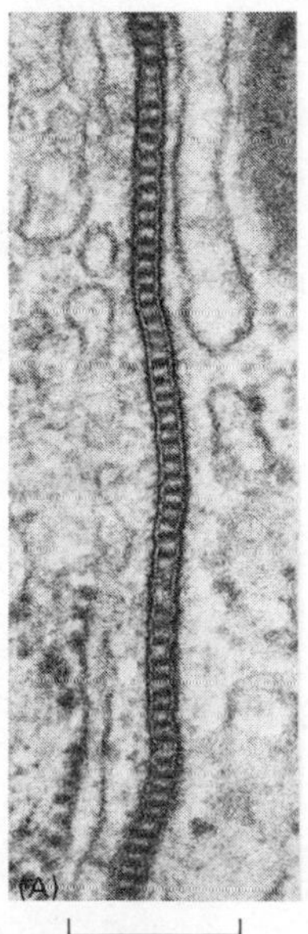

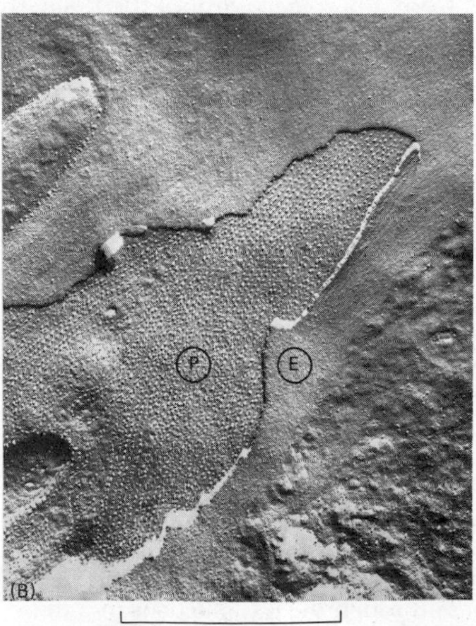

Figure 12–28 Thin-section (A) and freeze-fracture (B) electron micrographs of an invertebrate septate junction between two epithelial cells of a mollusk. The interacting plasma membranes are joined by proteins arranged in parallel rows with a regular periodicity, and they interact across the intercellular space. In the cross-sectional view (A), each row of interacting proteins is seen as a dense bar or septum, while in (B) these proteins are seen as parallel rows of intramembrane particles associated with the cytoplasmic (P) fracture face. (From N. B. Gilula, in Cell Communication [R. P. Cox, ed.], pp. 1–29. New York: Wiley, 1974.)

In principle, this type of coupling should allow *metabolic cooperation* between cells, since small intracellular molecules made only by a subpopulation of cells within a tissue could be shared with other cells in the tissue. Although such metabolic cooperation has not yet been directly demonstrated in organisms, it has been shown to occur in tissue culture. For example, mutant cell lines that lack the enzyme thymidine kinase are unable to incorporate radioactive thymidine into their DNA. But if such cells are cultured together with normal (wild-type) cells (which have thymidine kinase) and are then exposed to radioactive thymidine, autoradiography reveals radioactive labeling in the DNA of those mutant cells in direct contact with wild-type cells (Figure 12–30). This observation implies that a DNA precursor containing the radioactive thymidine is passed directly from the wild-type cells into mutant cells that are in contact with them. When this type of experiment is performed with mutant cells unable to form gap junctions, such metabolic cooperation does not occur.

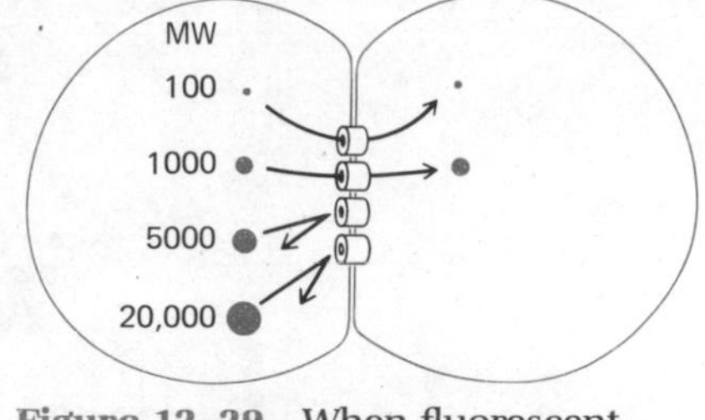

Figure 12–29 When fluorescent molecules of various sizes are injected into one of two cells coupled by gap junctions, molecules smaller than about 1500 daltons can pass into the other cell, but larger molecules cannot. This suggests that the functional diameter of the putative channel connecting the two cells is about 1.5 nm.

Although it is generally assumed that the electrical, dye, and metabolic coupling detected between cells in contact is mediated by gap junctions, the evidence is still circumstantial. Wherever cells are coupled, gap junctions have almost always been found. Conversely, where gap junctions have been absent between vertebrate cells, coupling has not been demonstrable. However, it has not yet been possible to reconstitute functional gap junctions in artificial membranes, as will be necessary for an unambiguous demonstration that molecules pass through them.

Since gap junctions connect most animal cells, it is difficult to exclude the possibility that other types of junctions may also mediate coupling. For example, it has not been possible to find cells joined by tight junctions that do not also have gap junctions, and so the possibility remains that tight junctions may also couple cells.

Why Are So Many Cells Electrically and Metabolically Coupled via Gap Junctions?

In some tissues, cell coupling via gap junctions serves an obvious function. For example, electrical coupling synchronizes the contractions of heart muscle cells and of the smooth muscle cells responsible for the peristaltic movements of the intestine. Similarly, electrical synapses between nerve cells permit action potentials to spread rapidly from cell to cell without the delay that

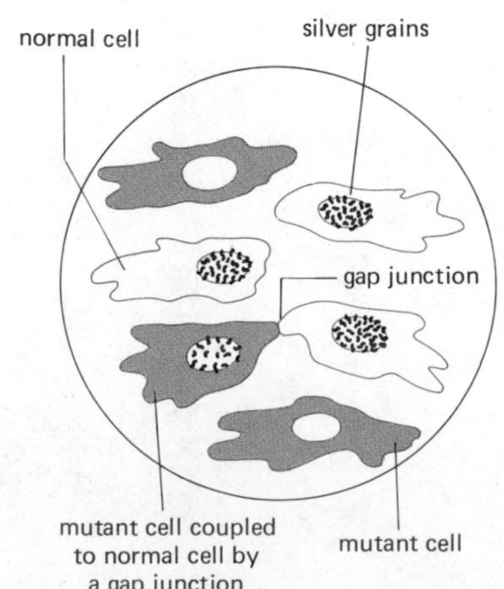

Figure 12–30 Schematic drawing of an autoradiograph demonstrating metabolic cooperation between cells in culture connected by gap junctions. The mutant cells lack the enzyme thymidine kinase and therefore cannot incorporate radioactive thymidine into DNA when thymidine is added to the medium. Normal cells can incorporate the thymidine into their DNA, and their nuclei are therefore stippled with black dots representing developed silver grains in autoradiographs. If in mixed cultures of normal and mutant cells a mutant cell makes contact and forms gap junctions with a normal cell, its nucleus is also labeled in autoradiographs, as shown here. This labeling occurs because the radioactive thymidine is incorporated into small radioactive DNA precursor molecules (the nucleoside triphosphate form of thymidine) in the normal cell; these precursors then pass through the gap junctions into the mutant cell, where they are incorporated into DNA.

occurs at chemical synapses; this is advantageous where speed and reliability are crucial, as in certain escape responses in fish or insects.

The real difficulty lies in understanding why gap junctions occur in tissues that are not electrically active. In principle, the resultant sharing of small metabolites and ions provides a mechanism for coordinating the activities and responses of individual cells in such tissues. For example, the activities of cells in an epithelial cell sheet, such as the beating of cilia or the shape changes associated with rolling up into a tube during embryogenesis (Figure 12–20), might well be coordinated in this way. So far, however, gap junctions have not been directly demonstrated to serve such a function.

Most Cells in Early Embryos Are Ionically Coupled[9]

Cell coupling via gap junctions may well be important in embryogenesis. In early embryos (beginning with the late eight-cell stage in mouse embryos), most cells are electrically coupled to each other. It is thought that metabolic coupling may be an important means of distributing nutrients in these embryos before the blood circulatory system develops. For example, yolk cells of the squid embryo are coupled to all of the other cells in the embryo until blood begins to circulate; after the circulatory system becomes functional, the yolk cells are coupled only to each other.

During differentiation, cells in each developing tissue or organ normally uncouple from surrounding tissue, reflecting the establishment of separate tissue identities. For example, as the amphibian neural tube closes, its cells uncouple from the overlying ectoderm. In some cases cells are known to couple transiently during development: for instance, immature muscle cells (myoblasts) couple just before they fuse to form multinucleated muscle fibers. It is possible that such transient coupling is one mechanism whereby cells recognize each other in order to interact during development. According to this purely hypothetical view, cells could sample each other's intracellular environments by promiscuous coupling, terminating the interactions with cells that do not "taste" right.

It is also possible that the coupling of cells in embryos might provide a pathway for cell signaling over longer distances. Through gap junctions, small molecules could pass from a region of the tissue where their intracellular concentration is kept high to a region where it is kept low, thereby setting up a smooth concentration gradient. The local concentration could provide cells with "positional information" to control their differentiation according to their location in the embryo. In the developing amphibian neural plate, for example, in which cells are electrically coupled, the cells at different positions along the plate have been found to have different membrane potentials. This means that an electric (ionic) current must be flowing through the neural plate between regions of different membrane potential, creating gradients in ion concentration along the plate. Any of these gradients could serve as a reference system for the determination of relative cell position, instructing each cell to develop in the manner appropriate to its location along the plate (Figure 12–31).

Cells May Control the Permeability of Their Gap Junctions[10]

The permeability of gap junctions is rapidly (within seconds) and reversibly decreased by experimental manipulations that decrease the intracellular pH or increase the intracellular concentrations of free Ca^{2+}. While it is still uncertain whether cell coupling is normally regulated by such ionic changes, these observations raise the possibility that gap junctions are dynamic structures whose permeability can be controlled by the cells that form them.

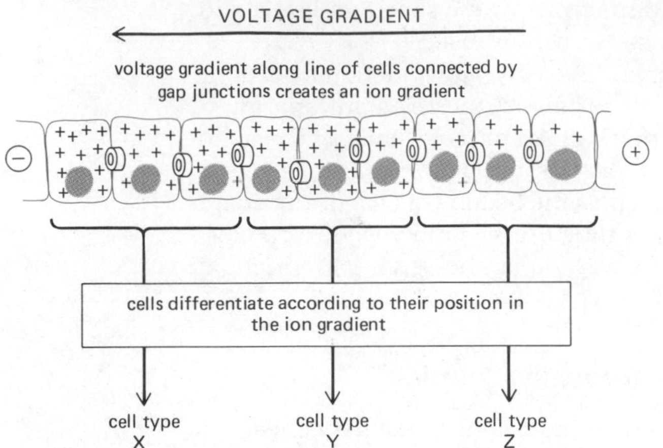

Figure 12–31 A voltage gradient will cause current to flow through gap junctions, thereby creating an ion gradient in coupled cells. In this purely hypothetical scheme, the ion gradient serves as a reference system by which coupled cells in a sheet determine their relative positions.

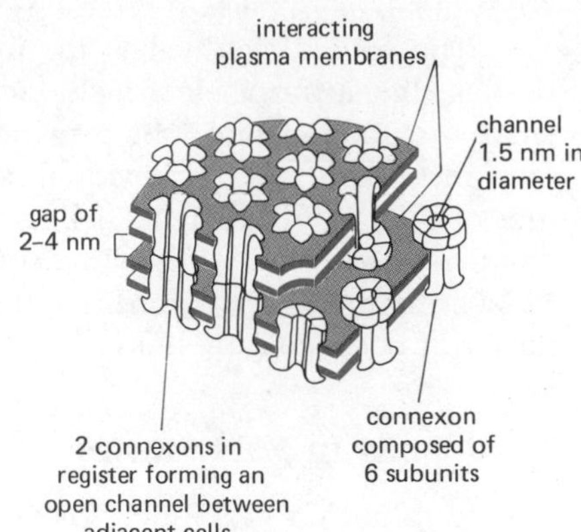

Figure 12–32 Schematic drawing of the currently favored model of a gap junction based on electron microscopic and x-ray diffraction observations. The drawing shows the interacting plasma membranes of two adjacent cells. The apposed lipid bilayers are penetrated by protein assemblies called "connexons," each of which is thought to be composed of six protein subunits. Two connexons join across the intercellular gap to form a channel connecting the two cells.

There is at least one case where ionic control of cell coupling is probably crucial. If a cell dies or is damaged, it is important that it rapidly uncouples from its neighbors. This is thought to be achieved by a large increase in intracellular free Ca^{2+}—either by influx through a damaged plasma membrane or because the cell is no longer able to pump Ca^{2+} out of the cytosol efficiently.

Gap Junctions Are Composed of Channels That Directly Connect the Interiors of the Interacting Cells[6,11]

As schematized in Figure 12–32, gap junctions are constructed from proteins that extend out from the plasma membrane to form structures called *connexons,* which are believed to connect the two cell interiors by a continuous aqueous channel. Presumably each of the two interacting cells contributes enough protein to form one connexon, and this connexon forms half the length of a channel. The connexons join in such a way that, unlike tight junctions, the interacting plasma membranes are separated by a "gap" of 2 to 4 nm (thus, the term "gap junction"), so that even relatively large molecules can easily penetrate between them (Figure 12–33). Each connexon is seen as an intramembrane particle in freeze-fracture electron micrographs, and each gap junction can contain up to several hundred clustered connexons (Figure 12–34).

An unusual resistance to proteolytic enzymes and detergents has made it possible to isolate gap junctions from rodent liver (Figure 12–35). Biochemical analyses of such preparations suggest that the junctions are composed of one major protein of about 27,000 daltons. Gap junctions can form within minutes (even in the absence of new protein synthesis) when dissociated cells are placed in culture. In such cases, these junctions must self-assemble from preformed gap-junction subunits, possibly from subunits diffusing in the plane of the plasma membrane, which are triggered to assemble when they bind to similar subunits in a closely apposed membrane. Cells from one vertebrate will often couple with cells from any other vertebrate in culture, demonstrating that the components of the gap junction have been highly conserved during evolution.

A drawing summarizing all of the various types of junctions formed between cells in an epithelial sheet is shown in Figure 12–36.

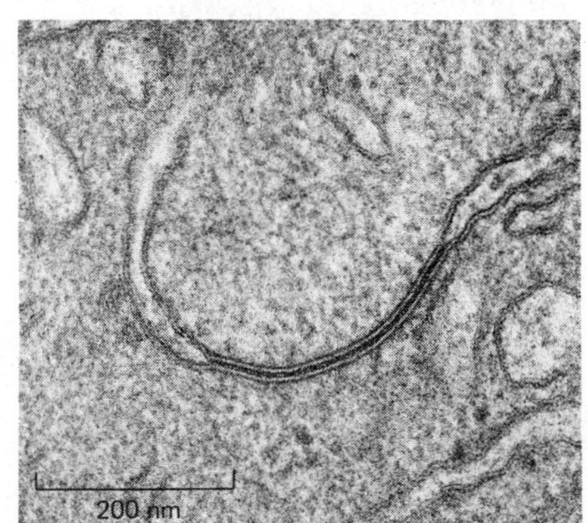

Figure 12–33 A thin-section electron micrograph of a gap junction between two mouse liver cells (hepatocytes). The gap between the interacting membranes is electron-dense because it is filled with an electron-dense dye (procion brown), which has been added to the extracellular fluid. (Courtesy of N. B. Gilula.)

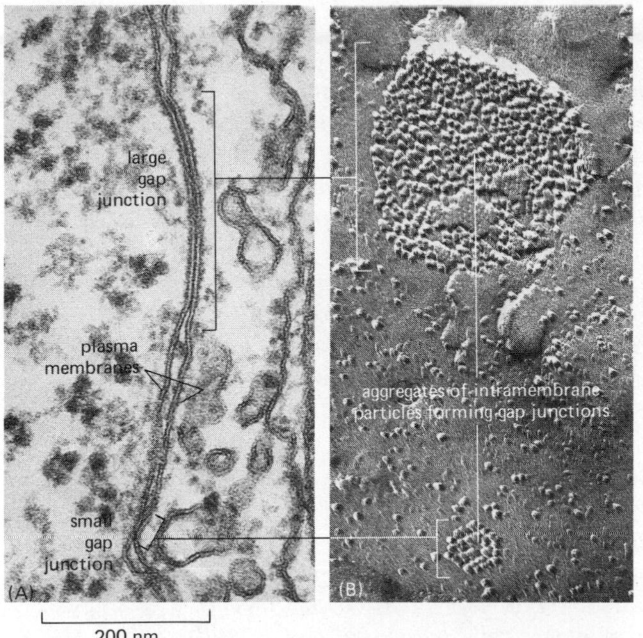

Figure 12–34 Thin-section (A) and freeze-fracture (B) electron micrographs of a large and a small gap junction between fibroblasts in culture. In (B), each gap junction is seen as a cluster of homogeneous intramembrane particles associated exclusively with the cytoplasmic (P) fracture face of the plasma membrane. Each intramembrane particle corresponds to a connexon, illustrated in Figure 12–32. (From N. B. Gilula, in Cell Communication [R. P. Cox, ed.], pp. 1–29. New York: Wiley, 1974.)

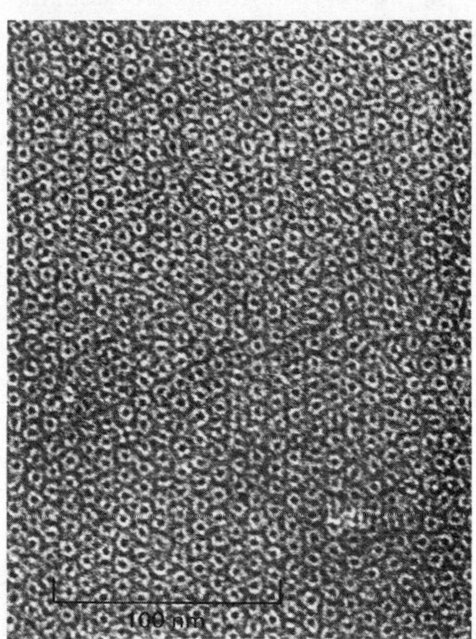

Figure 12–35 Electron micrograph of isolated gap junction from rat liver. The preparation has been negatively stained to show the connexons, which are organized in a hexagonal lattice. The densely stained central hole in each connexon has a diameter of about 2 nm. (From N. B. Gilula, in Intercellular Junctions and Synapses [Receptors and Recognition Series B, Vol. 2; J. Feldman, N. B. Gilula, and J. D. Pitts, eds.], pp. 3–22. London: Chapman and Hall, 1978.)

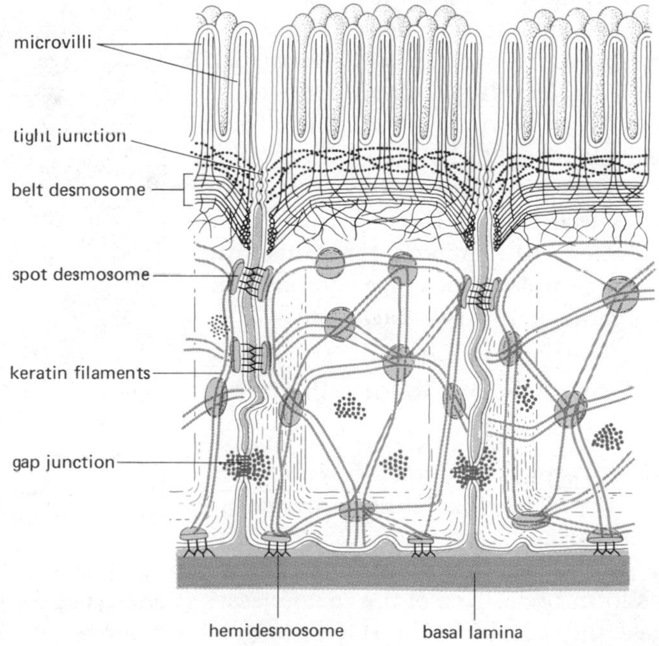

Figure 12–36 Schematic drawing emphasizing the distribution of the various junctions between epithelial cells of the small intestine.

Summary

The plasma membranes of neighboring cells in tissues are linked to each other at specialized contact sites called cell junctions. Three main types of cell junctions occur in most vertebrate tissues: desmosomes, tight junctions, and gap junctions. Desmosomes *mechanically hold cells together—either at buttonlike points of contact (spot desmosomes) or at continuous bands of contact around interacting cells in an epithelial sheet (belt desmosomes). Both types of desmosomes serve in different ways as anchorage sites for components of the cell cytoskeleton.* Tight junctions *play an indirect but critical part in the transport of small hydrophilic molecules across epithelial cell sheets. They do so in two ways: they seal the plasma membranes of adjacent cells together so that even small molecules cannot leak between them, and they serve as diffusion barriers within each lipid bilayer so that specific transport proteins can be restricted to the apical or the basolateral compartments of the epithelial cell plasma membrane.* Gap junctions *are thought to be composed of clusters of protein channels that allow ions and molecules of less than about 1500 daltons to pass directly from the inside of one cell to the inside of the other. Cells connected by gap junctions share many of their small molecules and are said to be metabolically and ionically (electrically) coupled. While gap junctions are clearly important in coordinating the activities of electrically active cells, it is still unclear why so many other types of cells are coupled to each other by gap junctions.*

The Extracellular Matrix[12]

Most cells in multicellular organisms are in contact with an intricate meshwork of interacting, extracellular macromolecules that constitute the **extracellular matrix** (Figure 12–37). These versatile protein and polysaccharide molecules are secreted locally and assemble into an organized meshwork in the extracellular space of most tissues. In addition to serving as a universal biological glue, they also form highly specialized structures such as cartilage, tendons, basal laminae, and (with the secondary deposition of a form of calcium phosphate crystals) bone and teeth. While we shall confine our discussion to the extracellular matrix of vertebrates, unique and interesting related structures are seen in many other organisms, such as the cell walls of bacteria and plants, the cuticles of worms and insects, and the shells of mollusks.

Until recently, the vertebrate extracellular matrix was thought to serve mainly as a relatively inert scaffolding that stabilized the physical structure of tissues. But now it is clear that the matrix plays a far more active and complex role in regulating the behavior of the cells that contact it—influencing their development, migration, proliferation, shape, and metabolic functions. The extracellular matrix has a correspondingly complex molecular composition; unfortunately our understanding of its organization is still fragmentary.

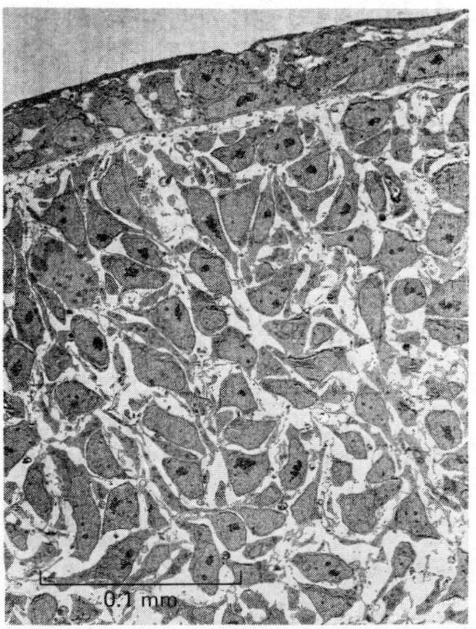

Figure 12–37 Low-power electron micrograph showing cells surrounded by spaces filled with extracellular matrix. The particular cells shown are those in an early chick limb during the time when different cell characters are being determined. (Courtesy of Cheryll Tickle.)

The Extracellular Matrix Consists Primarily of Fibrous Proteins Embedded in a Hydrated Polysaccharide Gel

The macromolecules that constitute the extracellular matrix are secreted by local cells, especially fibroblasts, which are widely distributed in the matrix. In specialized matrix structures, such as cartilage and bone, these macromolecules are secreted locally by more specialized cells: for example, chondroblasts form cartilage, and osteoblasts form bone. Two of the main classes of extracellular macromolecules that make up the matrix are (1) the *collagens*

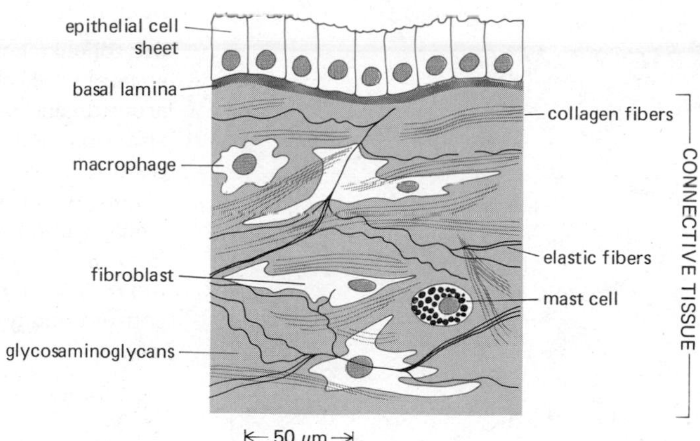

Figure 12–38 Schematic drawing of the connective tissue underlying an epithelial cell sheet.

and (2) the polysaccharide *glycosaminoglycans* (GAGs), which are usually covalently linked to protein to form *proteoglycans.* The glycosaminoglycan and proteoglycan molecules form a highly hydrated, gel-like "ground substance" in which collagen fibers are embedded. While the long collagen fibers strengthen and help to organize the matrix, the aqueous phase of the polysaccharide gel permits the diffusion of nutrients, metabolites, and hormones between the blood and the tissue cells. In many cases, fibers of the rubberlike protein *elastin* are also present and impart resilience to the matrix. In addition, two high molecular weight glycoproteins are among the major components of extracellular matrices: *fibronectin,* which is widely distributed in connective tissues (as well as in the blood), and *laminin,* which has so far been found only in basal laminae. Many other protein components of this type no doubt remain to be discovered.

The term **connective tissue** is often used to describe the extracellular matrix plus the cells found in it, such as fibroblasts, macrophages, and mast cells (Figures 12–38 and 12–39). The amount of connective tissue in organs varies greatly: skin and bone are composed mainly of connective tissue, whereas the brain and spinal cord contain very little. Moreover, the relative amounts of the different types of matrix macromolecules and the way that they are organized within the extracellular matrix vary enormously, giving rise to an amazing diversity of forms, each highly adapted to the functional requirements of the particular tissue. Thus, the matrix can become calcified to form the rock-hard structures of bone or teeth, or it can assume an almost crystalline order to form the transparent matrix of the cornea (the anterior covering of the eye), or it may take on the ropelike organization of the collagen fibers in tendons, which gives them their enormous tensile strength.

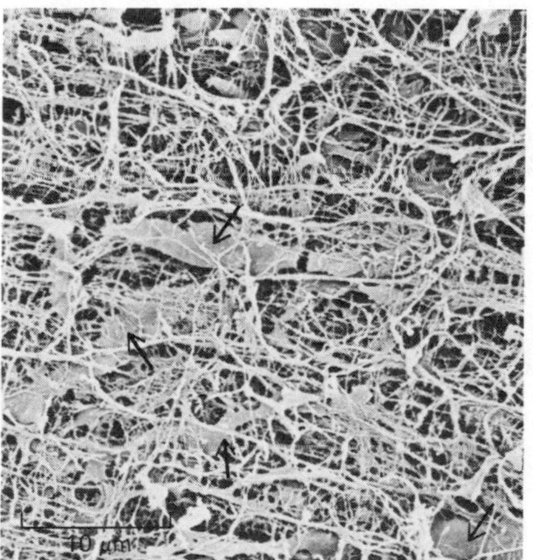

Figure 12–39 Scanning electron micrograph of fibroblasts (*arrows*) in the extracellular matrix of the cornea in a chick embryo. The matrix is largely composed of collagen fibrils (there are no elastic fibers in the cornea). The glycosaminoglycans, which normally form a hydrated gel filling the interstices of the fibrous network, have collapsed onto the surface of the collagen fibers during the dehydration process involved in specimen preparation. (Courtesy of Robert Trelstad.)

Collagen Is the Major Protein of the Extracellular Matrix[12]

The **collagens** are a family of highly characteristic fibrous proteins found in all multicellular animals. They are the most abundant proteins in mammals, constituting 25% of their total protein. The central feature of all collagen molecules is their stiff, triple-stranded helical structure. Three collagen polypeptide chains, called *α-chains,* are wound around each other in a regular helix to generate a ropelike collagen molecule about 300 nm long and 1.5 nm in diameter (Figure 12–40).

So far, seven genetically distinct collagen α-chains, each about 1000 amino acid residues long, have been well defined (Table 12–2). Although in principle more than 100 different types of triple-stranded collagen molecules could be assembled from various combinations of these seven α-chains, fewer than a

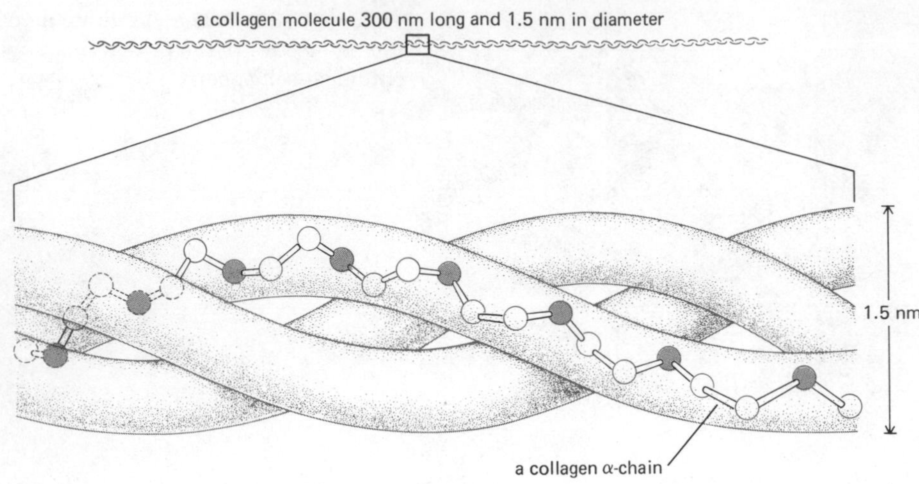

Figure 12–40 Schematic drawing of the ropelike collagen molecule. Three separate helical α-chains are wrapped around each other to form a triple-stranded, helical rod. Every third residue in each α-chain is glycine (*color*), which is the only amino acid small enough to occupy the crowded interior of the triple helix. This 300 nm × 1.5 nm collagen molecule is conventionally referred to as *tropocollagen.*

dozen types of collagen molecules have been described. The major types are referred to as types I, II, III, IV, and V (Table 12–2). Types I, II, and III are the main types of collagen found in connective tissues, and of these, type I is much the most common, constituting 90% of the collagen in the body. After being secreted into the extracellular space, types I, II and III collagen molecules assemble into ordered polymers called **collagen fibrils,** which are long (up to many μm), thin (10 to 300 nm in diameter), cablelike structures clearly visible in electron micrographs (Figures 12–39 and 12–41). Such fibrils are often grouped into larger bundles, which can be seen in the light microscope as *collagen fibers* several μm in diameter. Type IV molecules (the main collagen

Table 12–2 Types of Collagen and Their Properties

Type	Molecular Formula*	Polymerized Form	Distinctive Features	Tissue Distribution
I	$[\alpha1(I)]_2\alpha2(I)$	fibril	low hydroxylysine low carbohydrate broad fibrils	skin, tendon, bone, ligaments, cornea, internal organs (accounts for 90% of body collagen)
II	$[\alpha1(II)]_3$	fibril	high hydroxylysine high carbohydrate usually thinner fibrils than type I	cartilage, intervertebral disc, notochord, vitreous body of eye
III	$[\alpha1(III)]_3$	fibril	high hydroxyproline low hydroxylysine low carbohydrate	skin, blood vessels, internal organs
IV	$[\alpha1(IV)]_3$ (controversial)	basal lamina	very high hydroxylysine high carbohydrate probably retains procollagen extension peptides	basal laminae
V	$[\alpha1(V)]_2\alpha2(V)$	unknown	high hydroxylysine high carbohydrate	widespread (in small amounts)

*The seven different α-chains are designated α1(I) through α1(V), α2(I), and α2(V).

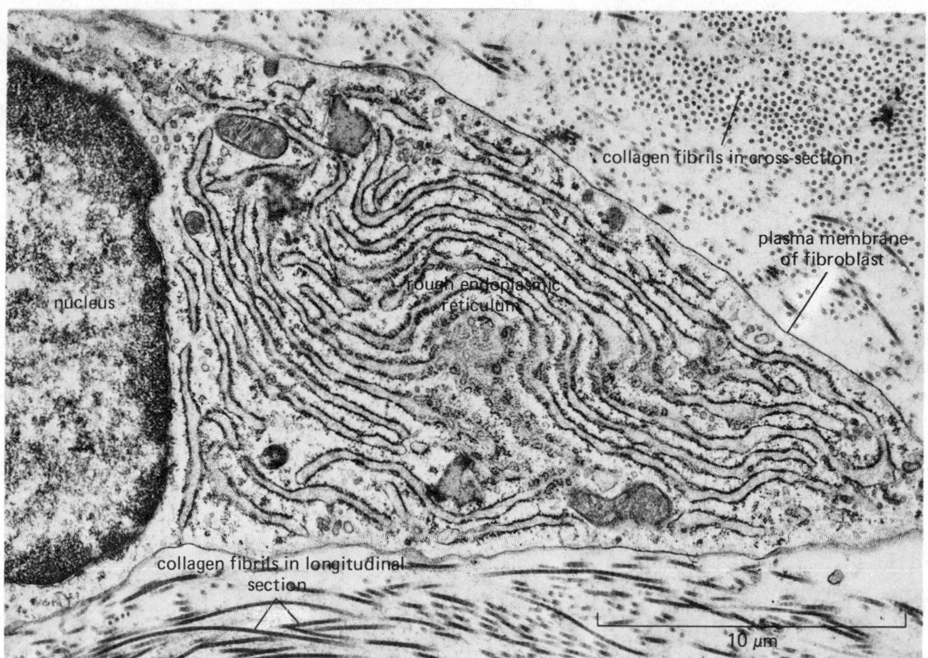

Figure 12–41 Electron micrograph showing part of a fibroblast surrounded by collagen fibrils in connective tissue. The extensive rough endoplasmic reticulum in the fibroblast cytoplasm reflects the cell's active synthesis and secretion of collagen and other extracellular matrix macromolecules. (Courtesy of Russell Ross.)

in basal laminae) and type V (found in small amounts in basal laminae and elsewhere) do not form fibrils, and their arrangement in tissues is uncertain.

The collagen polypeptide chains are synthesized on membrane-bound ribosomes and injected into the lumen of the endoplasmic reticulum (ER) as larger precursors, called *pro-α-chains.* These precursors have not only the "signal peptide" required for threading secreted proteins through the membrane of the ER (see p. 343) but other, extra amino acids, called *extension peptides,* at both their amino- and carboxyl-terminal ends. In the lumen of the ER, each pro-α-chain combines with two others to form a hydrogen-bonded, triple-stranded, helical molecule (Figure 12–42). The extension peptides are probably important in guiding triple-helix formation, since pro-α-chains have

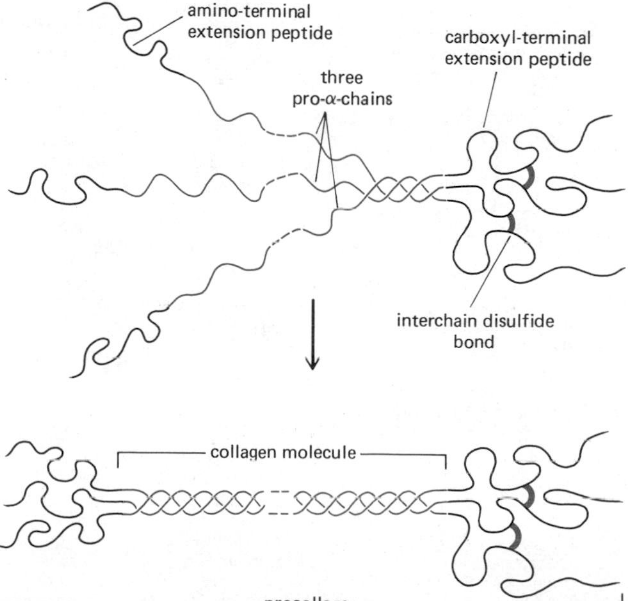

Figure 12–42 Collagen α-chains are initially synthesized in the form of pro-α-chains that contain extra (extension) peptides that will later be removed. One of the functions of these extension peptides is to help guide triple-helix formation during the assembly of the *procollagen* molecule. Note that in the procollagen molecule, the carboxyl-terminal extension peptides are covalently linked together by disulfide bonds.

been reported to assemble spontaneously into triple helices in a test tube under conditions in which α-chains do not. The extension peptides may also be important in the packaging of procollagen molecules together with other matrix macromolecules in the cell prior to secretion.

Collagen Chains Have an Unusual Amino Acid Composition and Sequence[13]

Collagen α-chains are extremely rich in glycine and proline, both of which are important in the formation of a stable triple helix. Glycine is the only amino acid small enough to occupy the crowded interior of the collagen triple helix, and it occurs as every third residue in most regions of the α-chain. This is the most striking example of the fact that collagen has a very much more regular amino acid sequence than a typical globular protein.

Some of the proline (and lysine) residues are hydroxylated in the endoplasmic reticulum before the pro-α-chains associate to form triple-stranded procollagen molecules. Such hydroxyproline (and hydroxylysine) residues are rarely found in other proteins. There is indirect evidence that the hydroxyl groups of hydroxyproline residues form interchain hydrogen bonds that help to stabilize the triple helix. For example, conditions that prevent proline hydroxylation (such as a deficiency of oxygen, of iron, or of ascorbic acid [vitamin C]) inhibit procollagen helix formation. In scurvy, a human disease caused by a dietary deficiency of vitamin C, the nonhydroxylated pro-α-chains are degraded in the cell, and the skin and blood vessels become extremely fragile.

The hydroxylation of lysine residues is crucial to a second posttranslational modification of procollagen. Like most secreted proteins, procollagen molecules are glycosylated in the cell before they are secreted by exocytosis into the extracellular space. However, the glycosylation of procollagen molecules is unusual in that the oligosaccharide side chains are short (only two sugar residues), do not contain sialic acid and are covalently attached to the hydroxyl group of hydroxylysine (Figure 12–43). The amount of carbohydrate added to procollagen varies greatly among different types of collagen (Table

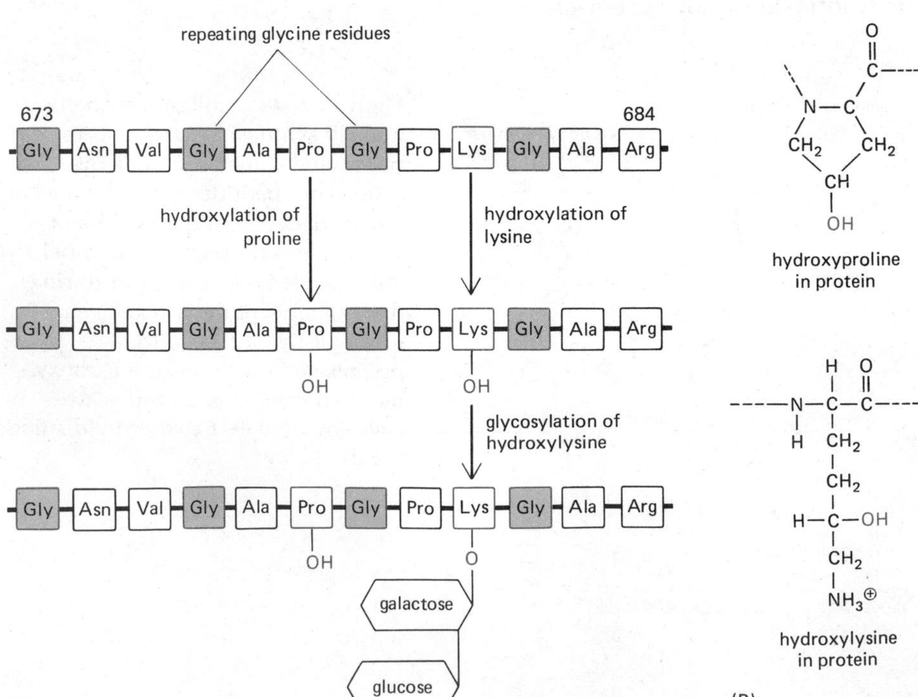

Figure 12–43 A 12-amino-acid segment (from residue 673 to 684) of a pro-α-chain showing the three types of covalent modifications that pro-α-chains undergo before they wrap together to form a triple-stranded procollagen molecule: selected proline and lysine residues are hydroxylated, and some of the resulting hydroxylysine residues are then glycosylated (A). The structures of hydroxyproline and hydroxylysine residues are shown in (B).

12–2), and its function is unknown. As we shall see later, hydroxylysine residues are important not only for glycosylation but also for the extensive cross-linking of collagen molecules that occurs in the extracellular space after secretion.

During Secretion, Procollagen Molecules Are Cleaved to Form Collagen Molecules, Which Self-assemble into Fibrils[13,14]

During the process of secretion, the extension peptides of types I, II, and III procollagen molecules are removed by specific proteolytic enzymes called *procollagen peptidases.* This converts the procollagen molecules to collagen (also called *tropocollagen*) molecules, which then come together in the extracellular space to form the much larger collagen fibrils. The process of fibril formation is driven, in part, by the tendency of the collagen molecules to self-assemble. However, the fibrils form close to the cell surface, and it seems likely that the cell regulates the sites and rates of fibril assembly—somewhat like a spider laying down a web.

The extension peptides have at least two functions: they guide the intracellular formation of the triple-stranded collagen molecules; and they prevent the intracellular formation of large collagen fibrils, which would be catastrophic for the cell. It is equally important, however, that once they have performed their functions, they should be removed. In certain diseases, the proteolysis of procollagen is incomplete; as a result, collagen fibril formation is impaired and the affected individuals have excessively fragile skin and hypermobile joints. Type IV (and possibly type V) procollagen molecules are unusual in that they seem to remain uncleaved after secretion. This is probably why these collagens do not form typical collagen fibrils.

When isolated collagen fibrils are fixed and stained and viewed in an electron microscope, they exhibit cross-striations every 67 nm. This pattern reflects the packing arrangement of the individual collagen molecules in the fibril. Presumably to maximize resistance of the aggregate to tensile (extension) stress, the individual collagen molecules are staggered as shown in Figure 12–44 so that adjacent molecules are displaced longitudinally by almost one-quarter of their length (a distance of 67 nm). Figure 12–45 illustrates how this arrangement gives rise to the striations seen in negatively stained fibrils.

While there is general agreement that collagen molecules are arranged in the two-dimensional "67-nm-stagger" pattern shown in Figure 12–44, there is less agreement about how they are packed in the three dimensions of a cylindrical fibril. One attractive model suggests that groups of collagen molecules first assemble (in their 67-nm-stagger pattern) into long, thin helical subunits (microfibrils) with five collagen molecules making up each turn of the helix. These cylindrical microfibrils become somewhat flattened as they in turn pack together in register to form the much thicker collagen fibril (Figure 12–46). Consistent with such a model is the observation that when

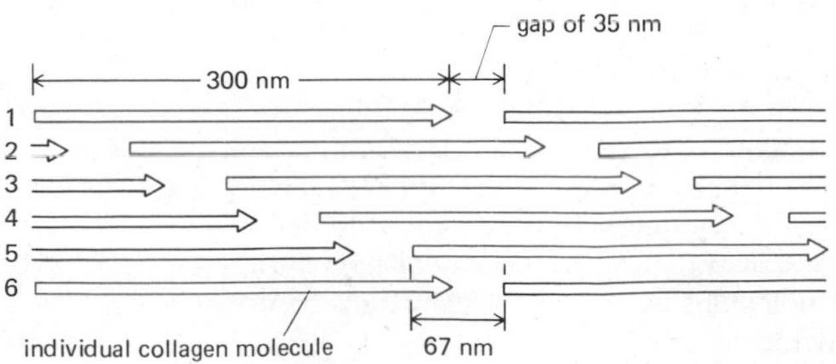

Figure 12–44 Schematic drawing showing the staggered arrangement of collagen molecules in a collagen fibril. Adjacent molecules (shown as arrows) are displaced by 67 nm, with a 35-nm gap between successive molecules in a row. The gap size is such that the pattern repeats after five molecules have been lined up in this staggered fashion; thus the molecules in rows 1 and 6 are in register.

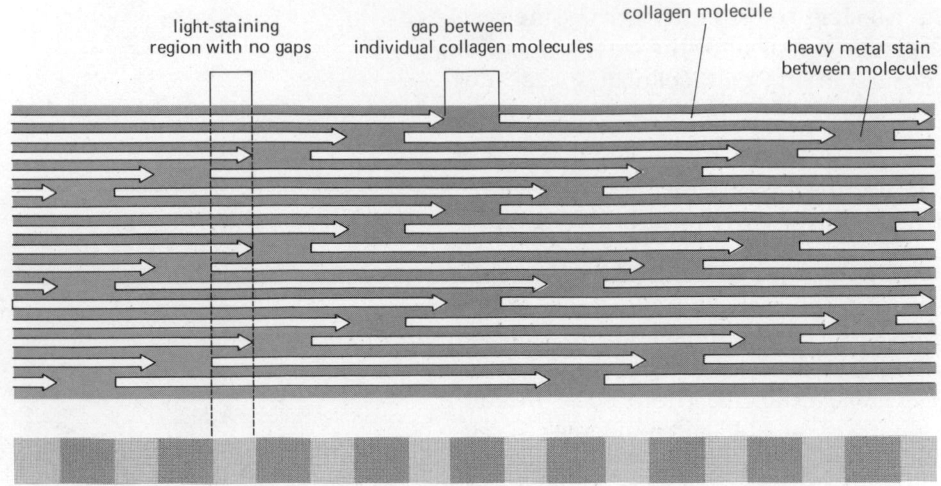

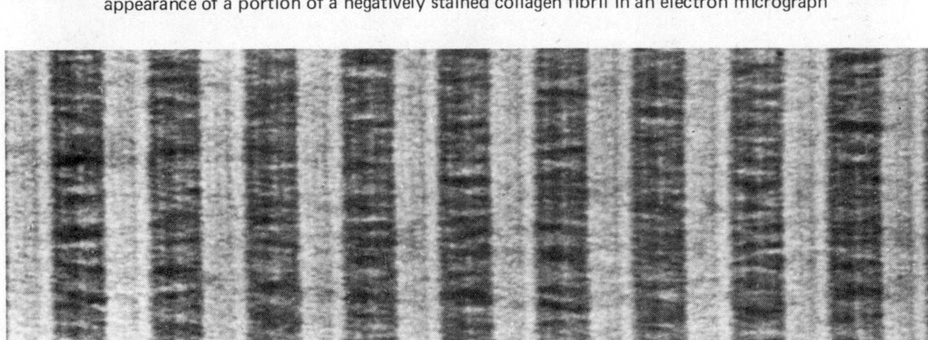

Figure 12–45 Schematic diagram of how the staggered arrangement of collagen molecules gives rise to the striated appearance of a negatively stained fibril. Since the negative stain fills only the space between the molecules, the stain in the gaps between the individual molecules in each row accounts for the dark staining bands. An electron micrograph of a negatively stained fibril is shown at the bottom of the figure. (Electron micrograph courtesy of Robert Horne.)

collagen molecules assemble *in vitro*, thin filaments appear at intermediate stages in the formation of collagen fibrils.

Since collagen molecules can assemble in a test tube to form collagen fibrils in the absence of other proteins, the information for fibril assembly must be contained in the amino acid sequences of the collagen α-chains. While proline, hydroxyproline, and glycine residues are mainly responsible for the formation of the collagen triple helix, the side chains of the remaining amino acids are thought to be largely responsible for fibril formation. In fact, computer analyses of the amino acid sequences of α-chains have shown that (1) charged and uncharged residues are periodically clustered along the chains with a periodicity of about 67 nm (which corresponds to about 234 amino acid residues) and (2) the maximum number of possible interchain electrostatic and hydrophobic bonds can be formed between two α-chains when these chains are displaced by multiples of 67 nm relative to one another (Figure 12–47).

What determines the diameter and arrangement of collagen fibrils in tissues? In a test tube, types I, II, and III collagen molecules each form fibrils

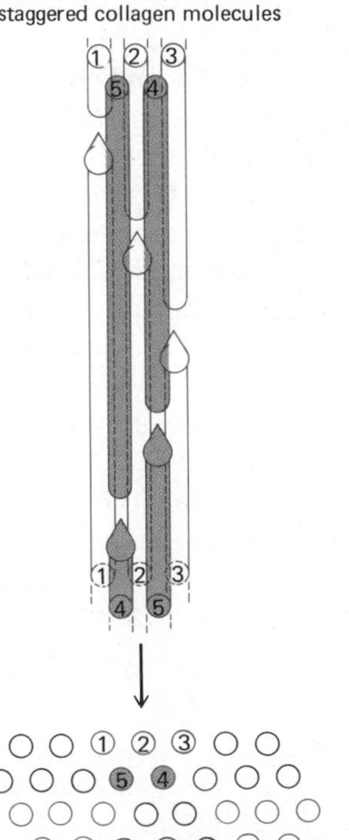

Figure 12–46 A possible model for the three-dimensional packing arrangement of collagen molecules in a fibril. Groups of molecules assemble in a 67-nm-stagger pattern into long microfibrils that are helices with five staggered collagen molecules per turn. These microfibrils become flattened in order to pack together in register to form a collagen fibril. The two lines of molecules closest to the viewer in the upper drawing are shown in color.

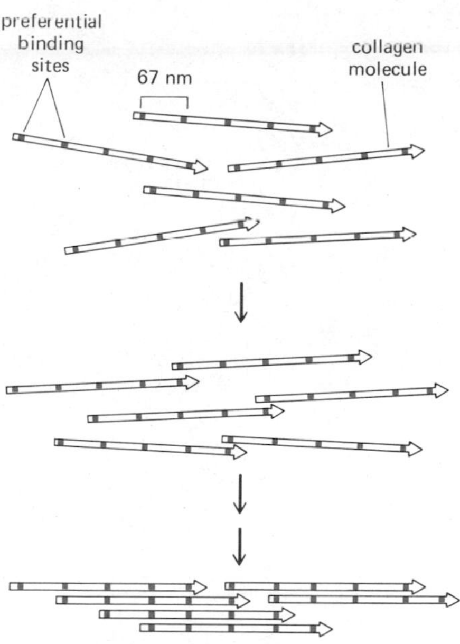

Figure 12–47 Highly schematic diagram showing how the repeating distribution of charged and hydrophobic amino acids along individual collagen molecules helps determine the packing arrangement in a fibril. Since these residues (here referred to as preferential binding sites) are clustered with a periodicity of about 67 nm, the 67-nm-staggered arrangement maximizes the number of interchain electrostatic and hydrophobic bonds that can be formed.

with diameters of about 50 nm. In living organisms, however, a single type of collagen molecule forms fibrils with a uniform diameter of from 10 to 300 nm, depending on the tissue. This suggests that other molecules in the extracellular matrix regulate fiber diameter. While various matrix macromolecules that are secreted together with collagen, such as glycosaminoglycans and fibronectin, have been shown to bind to collagen, it is not clear how they influence fibril formation.

Nor is it clear how the long-range order of collagen fibrils is determined in various tissues. In mammalian skin, collagen fibrils appear to be randomly dispersed in the extracellular matrix. In tendons, however, collagen fibrils are organized into parallel bundles, which for maximum strength are aligned along the major axis of stress that operates on the tendon. And in tadpole skin and the chick cornea, collagen fibrils are arranged in orderly multiple layers, with the fibrils in each layer lying parallel to each other but nearly at right angles to the fibrils in the layers on either side (Figure 12–48). This crystal-like arrangement of fibrils is required in order to provide both strength and transparency, but it is uncertain how it is achieved. As we shall discuss below, the orientation of the extracellular matrix is probably determined, at least in part, by the orientation of the cytoskeleton in the cells that secrete the matrix macromolecules.

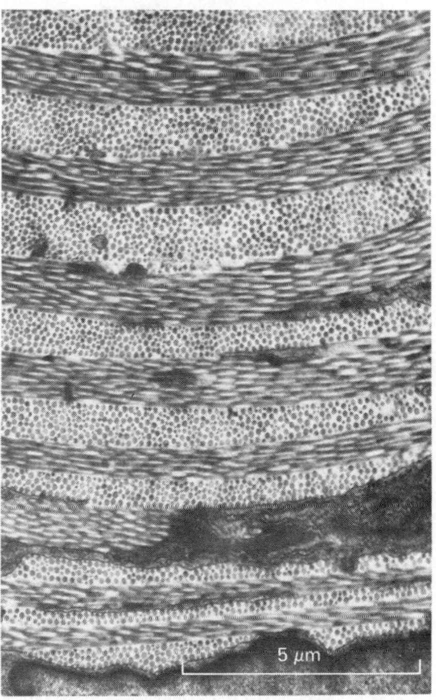

Figure 12–48 Electron micrograph of a cross-section of tadpole skin showing the plywoodlike arrangement of collagen fibrils, in which successive layers of fibrils are laid down at right angles to each other. This arrangement of collagen fibrils, which is also found in the cornea, produces both strength and transparency. (Courtesy of Jerome Gross.)

Once Formed, Collagen Fibrils Are Greatly Strengthened by Covalent Cross-linking[12,14]

After collagen fibrils have formed in the extracellular space, they are greatly strengthened by the formation of covalent cross-links within and between the constituent collagen molecules (Figure 12–49). If cross-linking is inhibited, collagenous tissues become fragile, and structures such as skin, tendons, and blood vessels tend to tear. The types of covalent bonds involved are unique to collagen (and elastin), and they are formed in several steps. First, certain lysine and hydroxylysine residues are deaminated by the extracellular enzyme lysyl oxidase to yield highly reactive aldehyde groups. The aldehydes then form covalent bonds with each other or with other lysine or hydroxylysine residues. Some of these bonds are relatively unstable and are ultimately modified to form a variety of more stable cross-links. The extent and type of cross-linking varies from tissue to tissue. For example, collagen is especially highly cross-linked in the Achilles tendon, where tensile strength is crucial.

We can now see some of the ways in which collagen-secreting cells adapt the collagen component of the extracellular matrix to the needs of the tissue. They can synthesize one or more of the genetically different types of collagen molecule, and each of these can be modified in a regulated manner by post-translational hydroxylation and glycosylation in the endoplasmic reticulum and Golgi apparatus. After secretion into the extracellular space, the collagen molecules assemble into fibrils, which can be cross-linked to a greater or lesser degree depending on the tensile strength required. By secreting varying kinds and amounts of noncollagen matrix macromolecules along with the collagen,

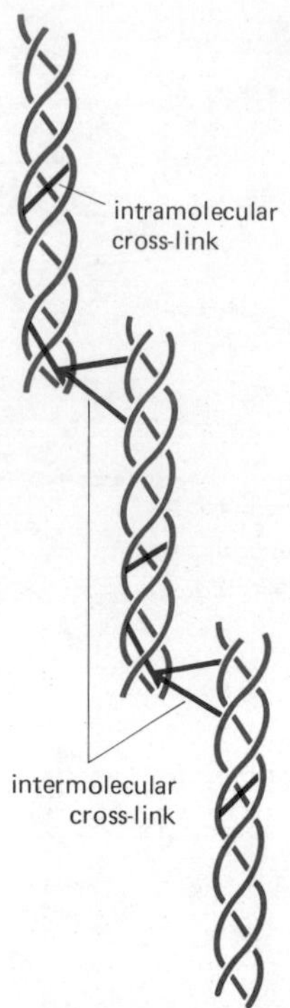

Figure 12–49 Schematic diagram of the covalent intramolecular and intermolecular cross-links formed between modified lysine side chains within a collagen fibril.

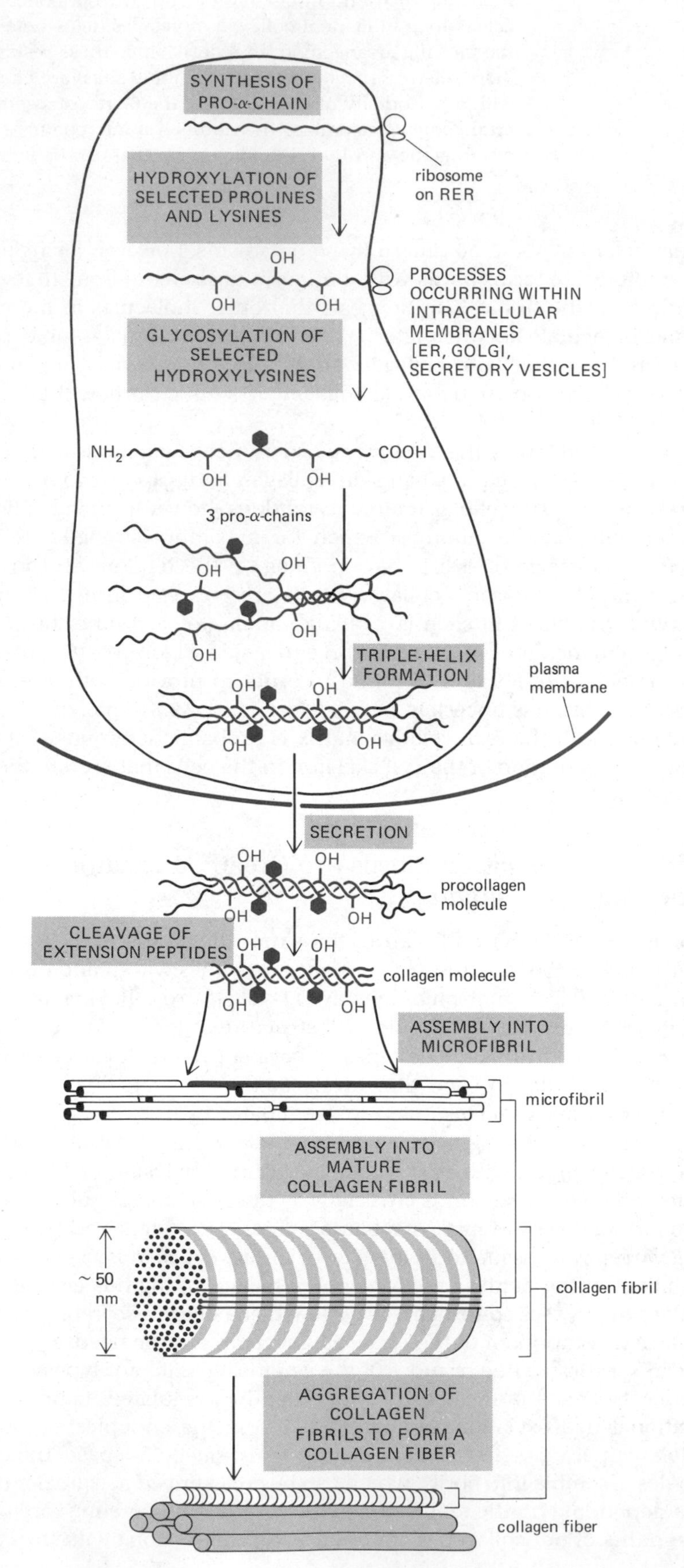

Figure 12–50 Schematic drawing of the various intracellular and extracellular events involved in the formation of a collagen fibril. While extension peptide cleavage and fibril formation are shown occurring after secretion, there is some evidence that cleavage of the amino-terminal peptides and some aggregation of collagen molecules occurs just prior to secretion from the cell. Although this is not shown, the larger extracellular aggregates of collagen molecules are stabilized by covalent cross-links. As an example of how the collagen fibrils can form ordered arrays in the extracellular space, they are shown further assembling into large collagen fibers that are visible in the light microscope.

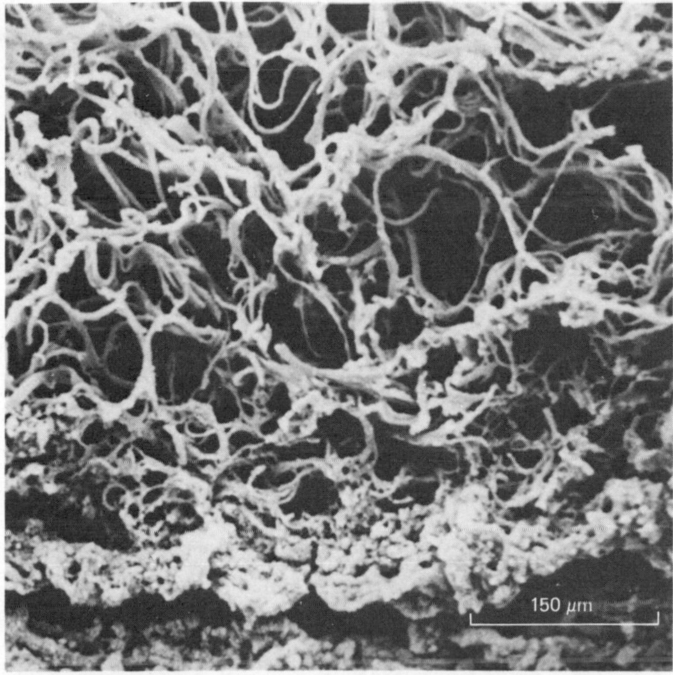

Figure 12–51 Scanning electron micrograph showing the extensive network of elastic fibers in a section of human skin (dermis). The tissue has been heated under pressure to remove the collagen and glycosaminoglycans. (From T. Tsuji, R. M. Lavker, and A. M. Kligman, *J. Microscop.* 115:165–173, 1978.)

cells presumably determine the geometry and properties of the fibrils in their environment. Finally, we shall see that, by initiating collagen fibril assembly at their surfaces, oriented cells can lay down an oriented matrix. A schematic summary of the various steps involved in collagen synthesis and assembly is given in Figure 12–50.

Elastin Is a Cross-linked, Random-Coil Protein That Gives Tissues Their Elasticity[15]

Tissues such as skin, blood vessels, and lungs require elasticity in addition to tensile strength in order to function. An extensive network of **elastic fibers** in the extracellular matrix of these tissues gives them the required ability to recoil after transient stretch (Figure 12–51). The main component of elastic fibers is **elastin,** a 70,000 dalton protein, which, like collagen, is unusually rich in proline and glycine but, unlike collagen, contains little hydroxyproline and no hydroxylysine. The details of elastin synthesis and processing are poorly understood. Elastin molecules are secreted into the extracellular space, where they form filaments and sheets in which the elastin molecules are highly cross-linked to each other to generate an extensive network. Unlike most other proteins, the function of elastin molecules requires that their polypeptide backbone remain unfolded as so-called "random coils" (Figure 12–52). It is the cross-linked, random-coil structure of the elastic fiber network that allows the network to stretch and recoil like a rubber band (Figure 12–53). At the same time, the long, inelastic collagen fibrils interwoven with the elastic fibers limit the extent of stretching and thereby prevent the tissue from tearing.

Figure 12–52 Schematic drawing of an elastin molecule in various "random-coil" conformations. Unlike most proteins, the elastin molecule does not adopt a unique structure but oscillates between a variety of partially extended, random conformations, as illustrated.

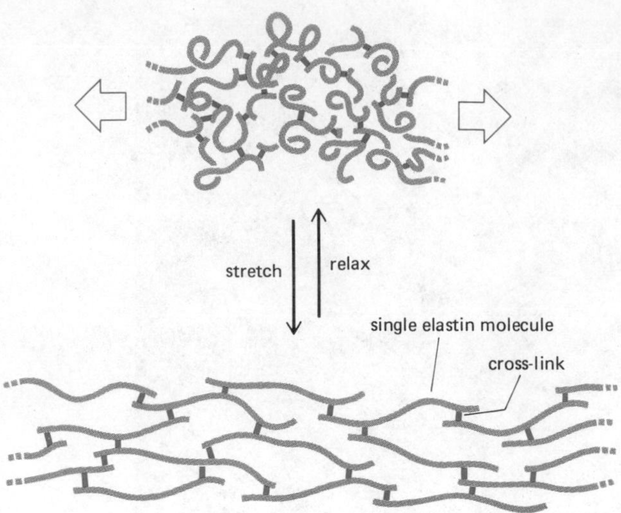

Figure 12–53 Elastin molecules are joined together by covalent bonds to generate an extensive cross-linked network. Because each elastin molecule in the network can expand and contract as a random coil, the entire network can stretch and recoil like a rubber band.

Elastic fibers are not composed solely of elastin, however; they also contain a glycoprotein that is usually distributed as microfibrils on the surface of the elastic fibers. In developing elastic tissues, these glycoprotein microfibrils often appear before elastin does and may serve to organize the secreted elastin molecules into the fibers and sheets which they later form.

Proteoglycans and Hyaluronic Acid Are Major Constituents of the Extracellular Matrix[16]

Glycosaminoglycans (GAGs), formerly known as mucopolysaccharides, are long, unbranched polysaccharide chains composed of repeating disaccharide units. They are now called glycosaminoglycans because one of the two sugar residues in the repeating disaccharide is always an amino sugar (*N*-acetylglucosamine or *N*-acetylgalactosamine). Glycosaminoglycans are highly negatively charged due to the presence of sulfate or carboxyl groups or both on many of the sugar residues (Figure 12–54). Seven groups of glycosaminoglycans have been distinguished by their sugar residues, the type of linkage between these residues, and the number and location of sulfate groups. They are *hyaluronic acid* (the only group in which none of the sugars is sulfated), *chondroitin 4-sulfate, chondroitin 6-sulfate, dermatan sulfate, heparan sulfate, heparin,* and *keratan sulfate* (Table 12–3).

Hyaluronic acid (also called hyaluronate) exists as a single, very long carbohydrate chain of several thousand sugar residues in a regular, repeating

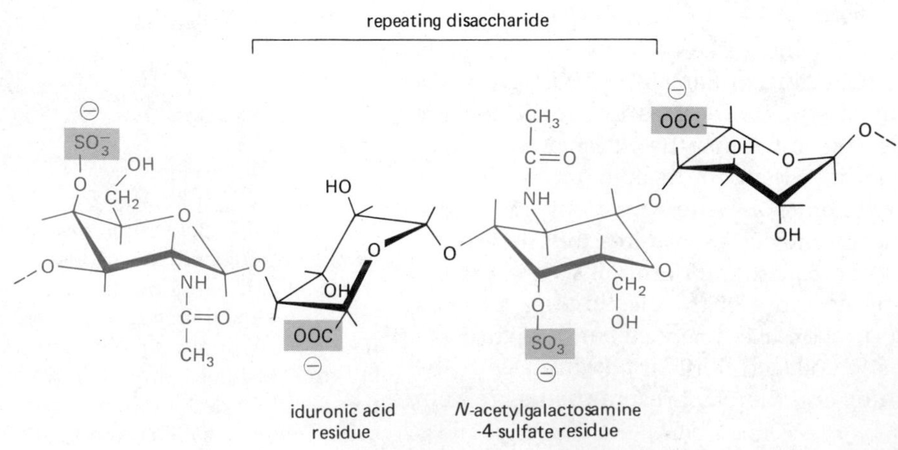

Figure 12–54 A repeating disaccharide sequence of the glycosaminoglycan chain of a dermatan sulfate molecule. Note the high density of negative charges along the chain due to the presence of both carboxyl and sulfate groups.

Table 12–3 The Glycosaminoglycans

Glycosaminoglycan	Molecular Weight	Repeating Disaccharide (A-B)$_n$: Monosaccharide A	Repeating Disaccharide (A-B)$_n$: Monosaccharide B	Sulfates per Disaccharide Unit	Linked to Protein	Other Sugar Components	Tissue Distribution
Hyaluronic acid	4000 to 8×10^6	D-glucuronic acid	*N*-acetyl-D-glucosamine	0	–	0	various connective tissues, skin, vitreous body, cartilage, synovial fluid
Chondroitin 4-sulfate	5000–50,000	D-glucuronic acid	*N*-acetyl-D-galactosamine	0.2–1.0	+	D-galactose D-xylose	cartilage, cornea, bone, skin, arteries
Chondroitin 6-sulfate	5000–50,000	D-glucuronic acid	*N*-acetyl-D-galactosamine	0.2–2.3	+	D-galactose D-xylose	cornea, bone, skin, arteries
Dermatan sulfate	15,000–40,000	D-glucuronic acid or *L-iduronic acid	*N*-acetyl-D-galactosamine	1.0–2.0	+	D-galactose D-xylose	skin, blood vessels, heart, heart valves
Heparan sulfate	5000–12,000	D-glucuronic acid or *L-iduronic acid	*N*-acetyl-D-glucosamine	0.2–3.0	+	D-galactose D-xylose	lung, arteries, cell surfaces
Heparin	6000–25,000	D-glucuronic acid or *L-iduronic acid	*N*-acetyl-D-glucosamine	2.0–3.0	+	D-galactose D-xylose	lung, liver, skin, mast cells
Keratan sulfate	4000–19,000	D-galactose	*N*-acetyl-D-glucosamine	0.9–1.8	+	D-galactosamine D-mannose L-fucose, sialic acid	cartilage, cornea, intervertebral disc

*L-Iduronic acid is produced by the epimerization of D-glucuronic acid at the position where the carboxyl group is located.

sequence of disaccharide units (Figure 12–55). Hyaluronic acid, however, is not typical of the glycosaminoglycans. First, the others tend to contain a number of different disaccharide units arranged in more complex sequences. Second, the others have very much shorter chains, consisting of fewer than 300 sugar residues. Third, all of the other glycosaminoglycans are covalently linked to protein to form **proteoglycan** molecules (formerly called mucoproteins). Like collagen molecules, the glycosaminoglycans are modified before they are secreted: the polysaccharide chains, which are initially formed by stepwise addition of monosaccharide or disaccharide units, are covalently modified in

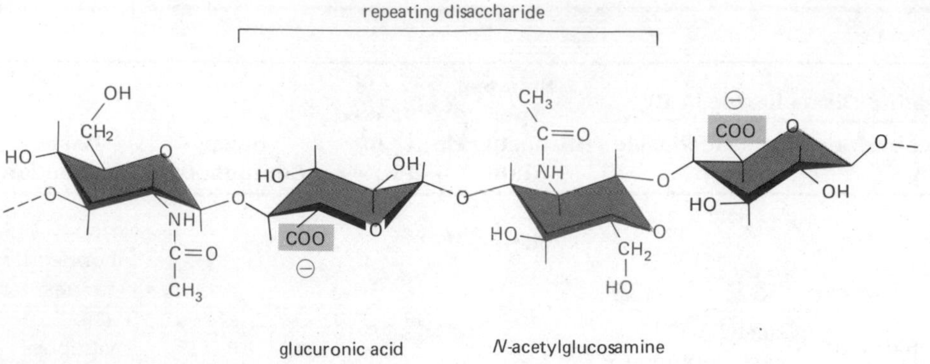

repeating disaccharide

glucuronic acid N-acetylglucosamine

Figure 12–55 The repeating disaccharide sequence in hyaluronic acid, which consists of a very long chain of up to several thousand sugar residues. Note the absence of sulfate groups.

the Golgi apparatus by a sequential and coordinated series of sulfation and epimerization reactions. (Epimerization reactions alter the configuration of the substituents around one of the carbon atoms on a sugar molecule.) Finally, the appropriate glycosaminoglycans are covalently linked to serine residues of a *core protein* to form a proteoglycan molecule before they leave the cell (Figure 12–56).

Proteoglycans are very different from typical glycoproteins. Glycoproteins usually contain from 1% to 60% carbohydrate by weight in the form of numerous, relatively short (generally less than 15 sugar residues), branched oligosaccharide chains of variable composition, which often terminate with sialic acid (p. 285). In contrast, proteoglycans are much larger (up to millions of daltons), and they usually contain 90% to 95% carbohydrate by weight in the form of many long, unbranched glycosaminoglycan chains, usually without sialic acid. For example, a typical proteoglycan molecule in cartilage might consist of about 100 chondroitin sulfate chains and 60 keratan sulfate chains linked to a serine-rich protein core of about 1900 amino acids (approximately one glycosaminoglycan chain for every 12 amino acid residues, Figure 12–57).

In principle, proteoglycans have the potential for almost limitless heterogeneity. They can differ markedly in protein content, molecular size, and the number and types of glycosaminoglycan chains per molecule. Moreover, although there is always an underlying repeating pattern of disaccharides, the length and composition of the glycosaminoglycan chains can vary greatly, as can the spatial arrangement of hydroxyl, sulfate, and carboxyl side groups along the chains. This hopelessly complex picture is likely to be simplified as we learn more about the principles of proteoglycan synthesis and assembly.

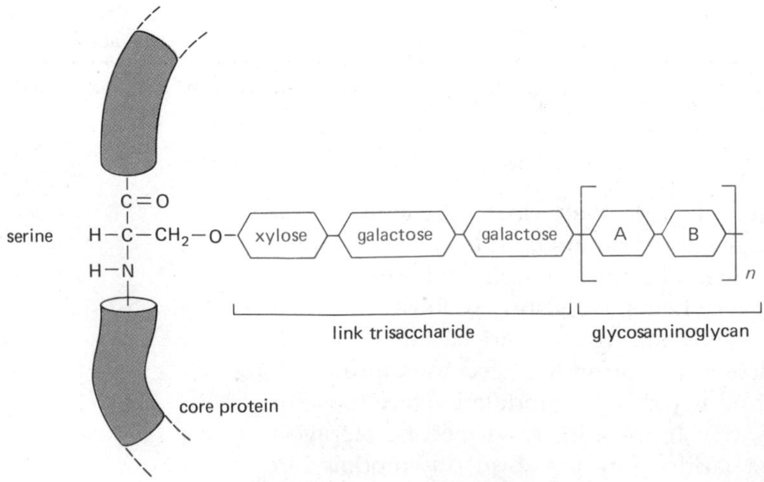

Figure 12–56 Schematic drawing of the linkage between a glycosaminoglycan chain and a serine residue of a core protein in a proteoglycan molecule. A specific "link trisaccharide" at the end of the glycosaminoglycan chain is bonded to the serine. The rest of the glycosaminoglycan chain consists mainly of a repeating disaccharide unit (composed of the two monosaccharides A and B in Table 12–3).

Figure 12–57 Schematic drawing of a typical proteoglycan molecule consisting of many glycosaminoglycan chains covalently linked to a protein core. More than one type of glycosaminoglycan chain is often found in the same proteoglycan molecule. The lower drawing shows a typical glycoprotein molecule (pancreatic ribonuclease B) drawn to scale for comparison.

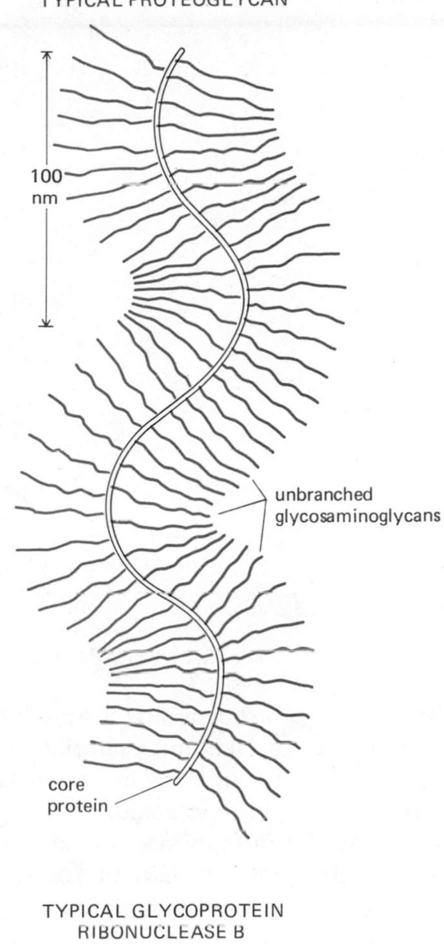

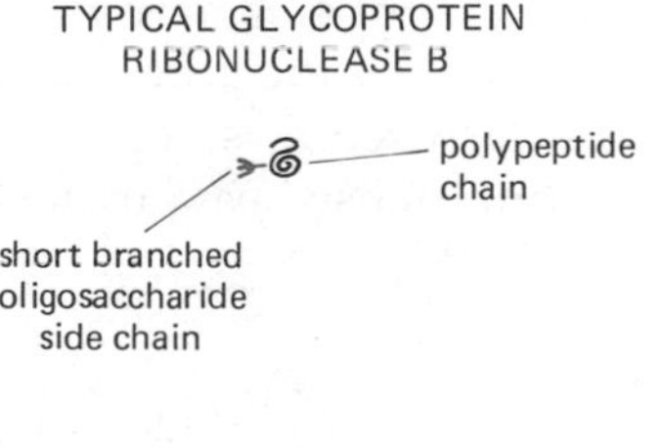

Glycosaminoglycan Chains Occupy Vast Amounts of Space and Form Hydrated Gels[16,17]

Polysaccharide chains are too inflexible to fold back on themselves to form the compact globular structures that many polypeptide chains form. Thus, glycosaminoglycan chains tend to adopt highly extended, random-coil conformations and to occupy a huge volume for their mass (Figure 12–58). Being hydrophilic, they attract large amounts of water, thereby forming hydrated gels even at very low concentrations. This tendency is markedly enhanced by their high density of negative charges, which attract osmotically active cations. This water-attracting property of glycosaminoglycans creates a swelling pressure, or turgor, in the extracellular matrix that resists compressive forces (in contrast to collagen fibrils, which resist stretching forces).

Because of their porous and hydrated organization, the glycosaminoglycan chains allow the rapid diffusion of water-soluble molecules and the migration of cells and cell processes. They effectively fill the extracellular space, even though by weight the amount of glycosaminoglycan in connective tissue is less than 10% of the amount of the fibrous proteins (collagen and elastin).

There is increasing evidence that hyaluronic acid has a special function in tissues through which cells are migrating during development or wound repair. Not only is it produced in large amounts in such tissues, its degradation by the enzyme *hyaluronidase* is associated with the cessation of cellular migration. Such correlations have been demonstrated in a variety of different tissues, suggesting that increased local production of hyaluronic acid, which attracts water and thereby swells the matrix, may be a general strategy used to facilitate cell migration during morphogenesis and repair.

Glycosaminoglycan Chains May Be Highly Organized in the Extracellular Matrix[16]

Given the structural heterogeneity of proteoglycan molecules, it seems highly unlikely that their function is limited to providing hydrated space around and between cells. It may be, for example, that they can form gels of varying pore size and charge density, thus functioning as sieves to regulate the traffic of molecules and cells according to their size and/or charge. There is some evidence that they function in this capacity in the basal lamina of the kidney glomerulus, which filters molecules passing into the urine from the bloodstream (see below). Such functions would require that the glycosaminoglycan chains be highly organized in the matrix.

In fact, the manner in which glycosaminoglycans and proteoglycans are organized in the extracellular matrix is largely unknown. Some of these molecules can bind to each other in specific ways, as well as to other macromolecular components of the matrix, such as collagen, elastin, and fibronectin,

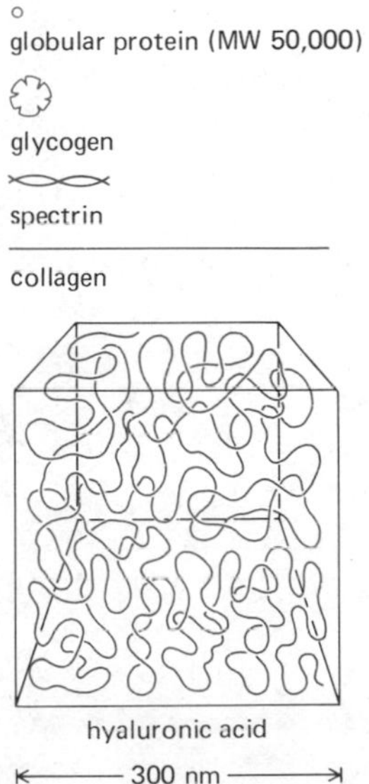

Figure 12–58 The relative volumes occupied by various proteins, a glycogen granule, and a hydrated molecule of hyaluronic acid of about 8×10^6 daltons.

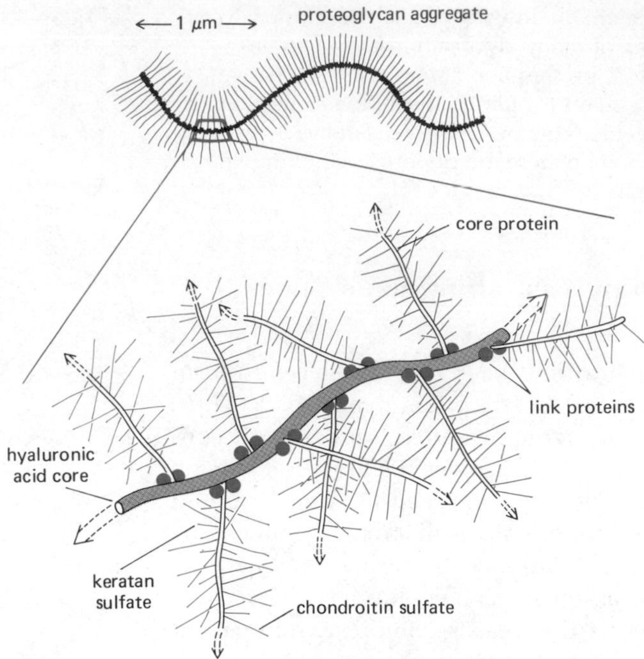

Figure 12–59 Schematic drawing of a giant proteoglycan aggregate in cartilage consisting of about 100 proteoglycan monomers (each like that shown in Figure 12–57) noncovalently bound to a single hyaluronic acid chain through two special link proteins. The molecular weight of such a complex can be 10^8 or more, and it occupies a volume equivalent to that of a bacterium.

and it would be surprising if such interactions were not important in ordering the matrix. To take one example, certain proteoglycans are found to exist in large aggregates that are noncovalently bound through their core proteins to a hyaluronic acid molecule. In some cases, as many as 100 proteoglycan monomers are bound to a single hyaluronic acid chain, producing a giant complex with a molecular weight of 100 million or more and occupying a volume equivalent to that of a bacterium (Figures 12–59 and 12–60).

In several cases the ordering of proteoglycans in tissues can be visualized by electron microscopy, since the glycosaminoglycan chains selectively bind certain electron-dense heavy metals. When the proteoglycans in rat tail tendon

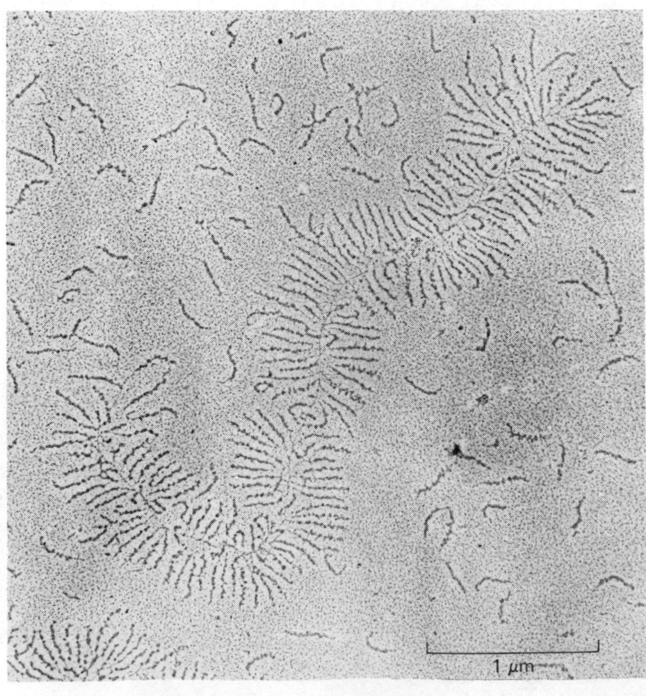

Figure 12–60 Electron micrograph of a proteoglycan aggregate from bovine fetal cartilage. The aggregate is of the type shown schematically in Figure 12–59. It has been shadowed with platinum. (Courtesy of Lawrence Rosenberg.)

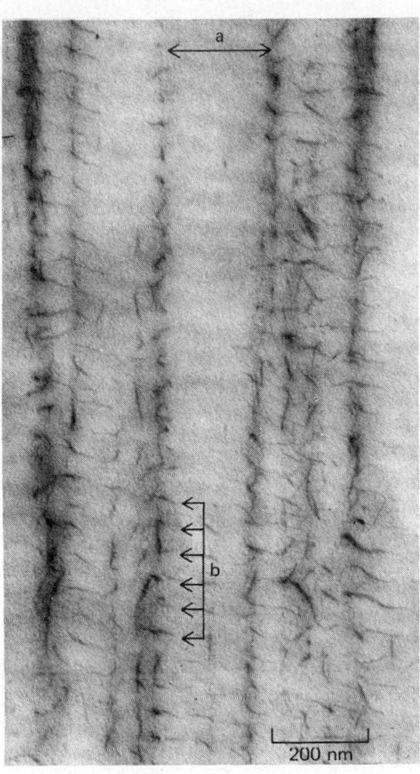

Figure 12–61 Electron micrograph of a longitudinal section of rat tail tendon stained with a copper-containing dye to visualize proteoglycan molecules. The proteoglycan molecules appear to be organized as a meshwork of filaments surrounding the collagen fibrils. Where proteoglycan threads are not seen crossing a collagen fibril (such as the fibril whose width is indicated by the double arrow a), the plane of section presumably cuts through the interior of the fibril. The fact that proteoglycan filaments cross the collagen fibrils at intervals of about 67 nm (for example, those indicated by arrows b), which is the intrinsic collagen-banding repeat distance, suggests a specific interaction between proteoglycan and collagen molecules. (Reproduced with permission from J. E. Scott, *Biochem. J.* 187:887–891, 1980.)

are visualized in this way, they are seen as threadlike structures running between and across collagen fibrils. Remarkably, they cross the collagen fibrils at intervals of 67 nm, which is the same periodicity as that of the intrinsic banding pattern of the collagen fibrils (Figure 12–61). Such ordered patterns of proteoglycans are likely to be widespread in the extracellular matrix. Given the known diversity of both collagen molecules and the proteoglycans, they should be able to form a great variety of different three-dimensional structures.

Although polysaccharide chains do not fold into globular structures in the way that many polypeptide chains do, some spontaneously assemble into highly ordered helical or ribbonlike structures. In higher plants, for example, the cellulose (polyglucose) chains are packed tightly together in ribbonlike crystalline arrays to form the microfibrillar component of the cell wall (Figure 12–62 and see p. 1101). In addition, two *different* polysaccharide chains have been shown to associate specifically with each other, producing defined aggregates with a regular helical structure (Figure 12–63). Glycosaminoglycan chains themselves can adopt a variety of helical conformations when they are in the solid state; and direct polysaccharide-polysaccharide interactions of the type shown in Figure 12–63 could occur in the extracellular matrix as well.

Given the diverse structures and functions of extracellular matrices, it seems probable that proteoglycan molecules can assume structural conformations as diverse as their chemistry. If so, we have hardly begun to understand them.

Fibronectin Is an Extracellular-Matrix Glycoprotein That Promotes Cell Adhesion[18]

The noncollagen glycoproteins of the extracellular matrix have been relatively neglected until recently. A good deal is now known, however, about **fibronectin,** a fiber-forming glycoprotein (about 5% carbohydrate by weight) composed of two disulfide-bonded subunits of 220,000 daltons each (Figure 12–64). Fibronectin exists as large aggregates in the extracellular space. While most of the protein is not directly bound to cells, some of it is bound to the surfaces of fibroblasts and other cells when they are grown in culture. A closely related protein called *cold insoluble globulin* is found in relatively large amounts in the blood and other body fluids.

Fibronectin first attracted attention when it was discovered to be present in greatly reduced amounts on the surface of fibroblasts derived from tumors (neoplastically "transformed" fibroblasts) compared to normal fibroblasts. In general, there is a good correlation between the decrease in cell-surface fi-

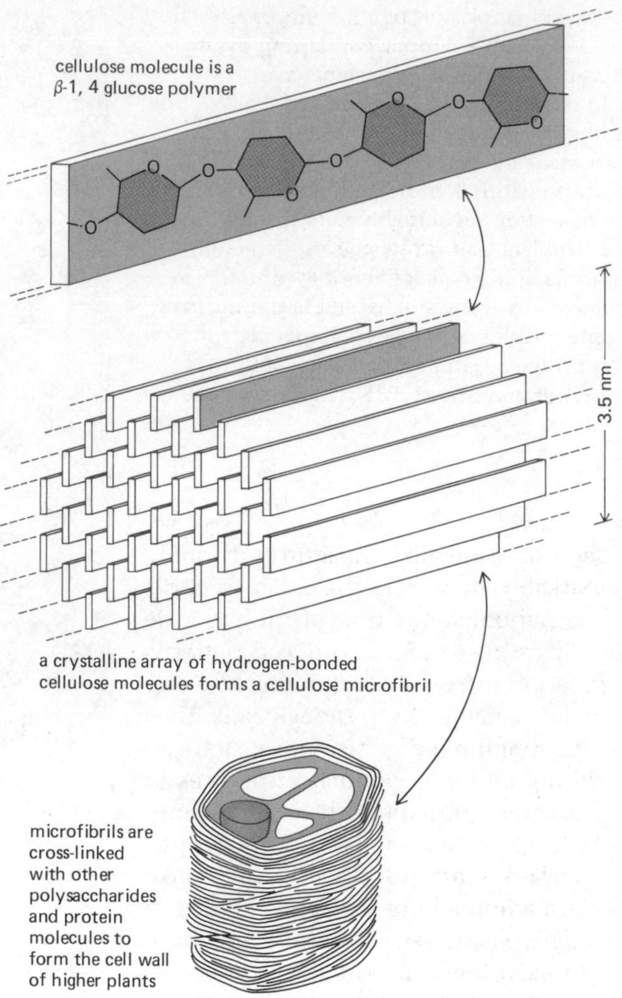

Figure 12–62 Schematic drawing of the organization of cellulose molecules in the cell wall of a higher plant. Although not shown in this figure, the cellulose microfibrils are cross-linked by several other polysaccharides and proteins in the cell wall.

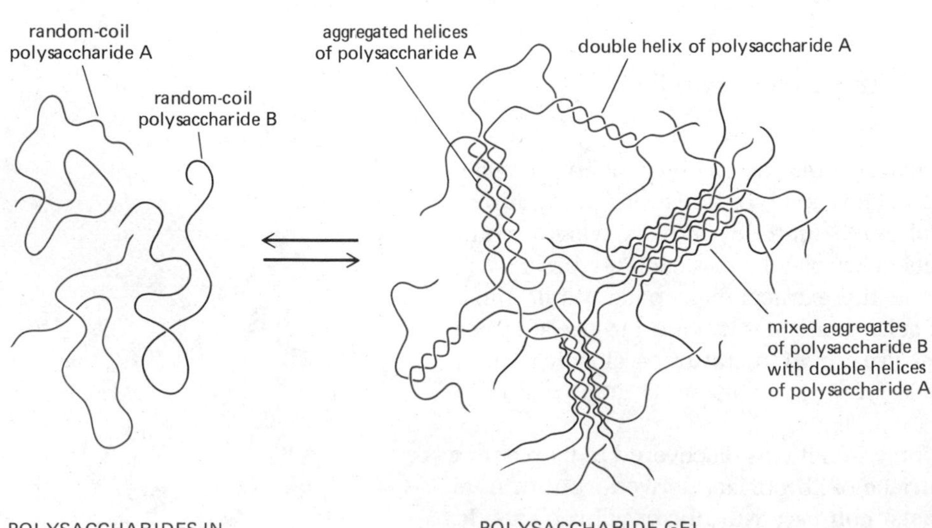

Figure 12–63 Schematic diagram showing some of the ordered conformations that two different polysaccharide chains, A and B, can assume in forming a gel. Since these interactions between molecules are confined to certain regions of the chains (so-called junctional regions) and are not propagated along the entire molecule, each chain can combine with more than one partner and thereby form a gel network. Examples of gel-forming polysaccharides are the agars (of algae) and the pectins (of higher plants).

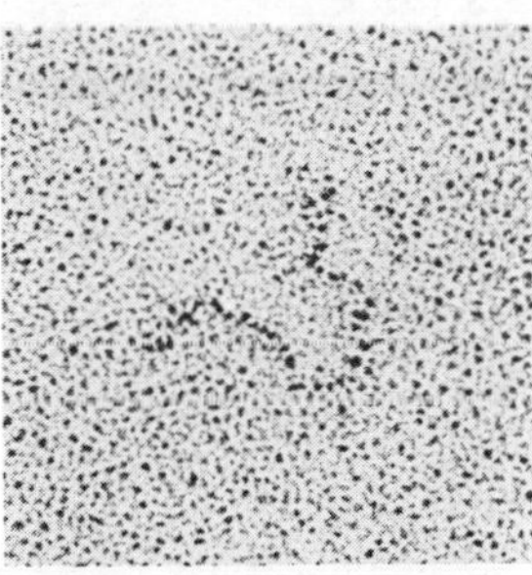

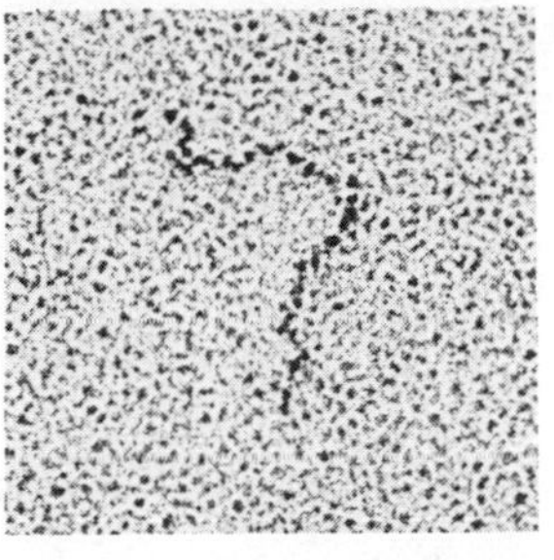

Figure 12–64 Electron micrographs of fibronectin molecules shadowed with platinum. Each molecule consists of two identical fibrous monomers joined at one of their ends. (From J. Engel et al., *J. Mol. Biol.* 150:97–120, 1981. © Academic Press Inc. [London] Ltd.)

bronectin observed and the ability of cultured transformed cells to cause tumors, invade tissues, and metastasize (spread widely) when reinjected into animals.

Transformed cells behave differently from normal cells in culture: they adhere poorly to the substrate and fail to flatten out and develop organized intracellular actin filament bundles called stress fibers (see p. 586); and they grow to a much higher density than do normal cells (see p. 620). If large amounts of fibronectin are added to cultures of transformed cells that themselves make relatively little fibronectin, the cells rapidly adhere, flatten out, and generate well-organized intracellular actin filament bundles (Figure 12–65). While they look like normal fibroblasts, they still grow to abnormally high density. This suggests that fibronectin promotes cell adhesion but does not directly influence the control of cell proliferation. Purified fibronectin has now been shown to promote the adhesion of a variety of cell types to other cells, as well as to collagen and other substrates. Since fibronectin has been found in high concentrations in regions of cell migration during development, this glycoprotein is thought to influence cell migration *in vivo* through its influence on cell adhesion.

The Basal Lamina Is a Specialized Extracellular Matrix That Contains a Unique Type of Collagen[19]

Basal laminae are thin layers of specialized extracellular matrix that underlie all epithelial cell sheets and tubes; they also surround individual muscle cells, fat cells, and Schwann cells (which wrap around peripheral nerve fibers to form myelin). The basal lamina thus separates these cells and cell sheets from the underlying or surrounding connective tissue. In other locations, such as the kidney glomerulus and lung alveolus, a basal lamina lies between two different cell layers, where it functions as a highly selective filter (Figure 12–66). However, there is increasing evidence that basal laminae serve more than simple structural and filtering roles. They seem to be able to induce cell differentiation, influence cell metabolism, organize the proteins in adjacent plasma membranes, and serve as specific "highways" for cell migration.

The basal lamina is synthesized by the cells that rest on it (Figures 12–67 and 12–68). Although the precise composition varies from tissue to tissue, and even from region to region within the same lamina, a major component of all basal laminae is type IV collagen. Type IV pro-α-chains are unusual in having extra-long extension peptides that are probably not cleaved after secretion; for this reason, these procollagen molecules do not form typical collagen fibrils, although they do become covalently cross-linked to each other. In addition to proteoglycans and fibronectin, which are important constituents of basal laminae, the large glycoprotein **laminin** has been shown to be

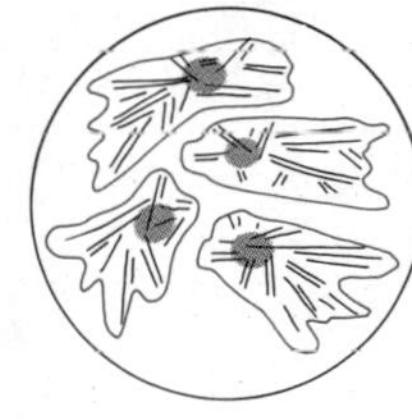

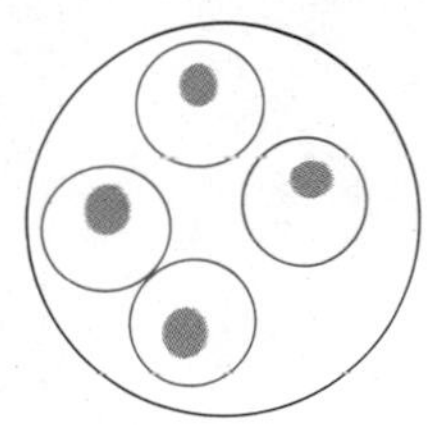

Figure 12–65 Schematic illustration of how the addition of fibronectin to transformed fibroblasts (which are not making fibronectin) induces them to flatten out and assemble well-organized intracellular actin filament bundles. The morphology of these cells now resembles that of normal fibroblasts that make their own fibronectin.

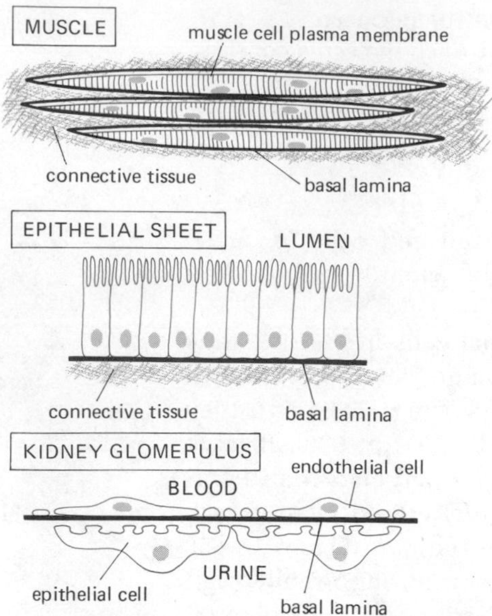

Figure 12–66 Schematic drawing of three different distributions of basal laminae (*black lines*): surrounding cells (such as muscle cells), underlying epithelial cell sheets, and interposed between two cell sheets (as in the kidney glomerulus). Note that in the kidney glomerulus both cell sheets have gaps in them, so that the basal lamina serves as the permeability barrier that determines which molecules will pass into the urine from the blood.

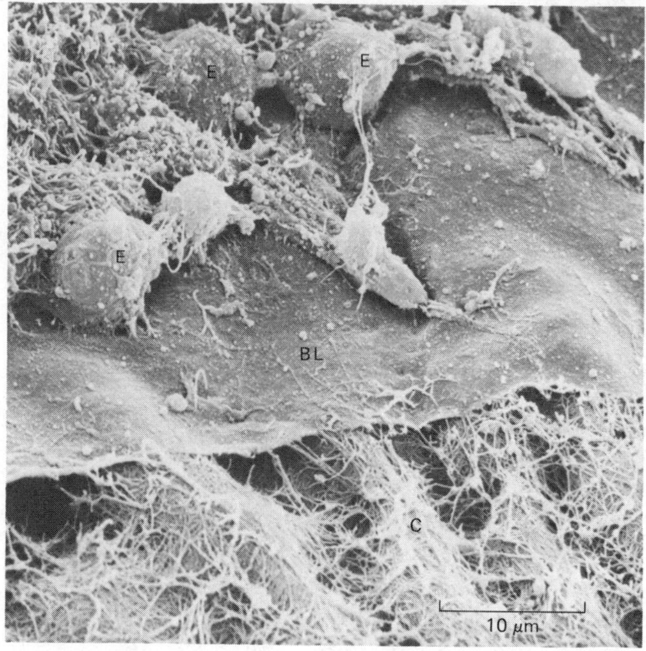

Figure 12–67 Scanning electron micrograph of basal lamina in the cornea of a chick embryo. Some of the epithelial cells (E) have been removed to expose the upper surface of the ruglike basal lamina (BL). Note the network of collagen fibrils (C) interacting with the lower face of the lamina. The macromolecules that comprise the basal lamina are synthesized by epithelial cells that sit on the lamina. (Courtesy of Robert Trelstad.)

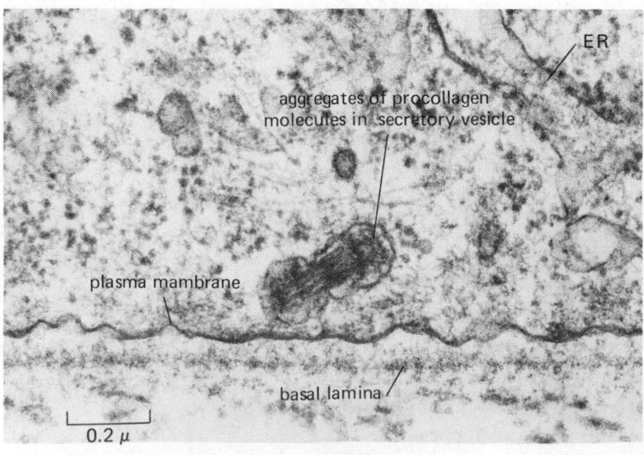

Figure 12–68 Electron micrograph showing the basal lamina in cross-section underlying an epithelial cell in the cornea of a chick embryo. (Courtesy of Robert Trelstad.)

a major component of all basal laminae studied so far. It consists of at least two subunits (220,000 and 440,000 daltons) that are disulfide-bonded to each other (Figure 12–69). Basal laminae undoubtedly contain many other proteins yet to be identified. The detailed molecular organization of basal laminae is unknown, although there is some evidence that laminin and proteoglycan molecules are concentrated along the inner and outer surfaces of the basal lamina, with collagen molecules sandwiched in the middle.

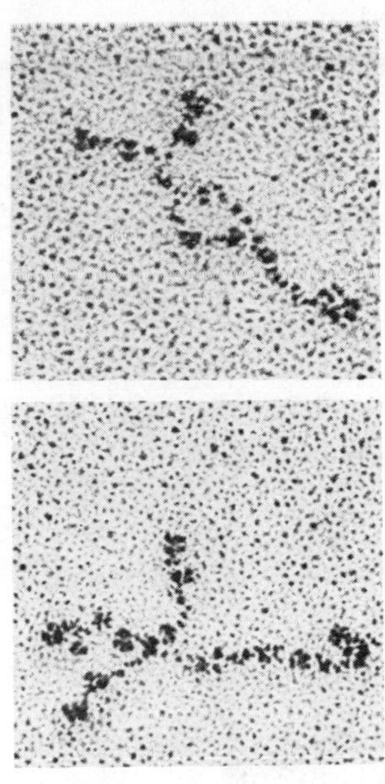

Figure 12–69 Electron micrograph of laminin molecules shadowed with platinum. Each molecule has the form of an asymmetrical cross, with one long and three short arms, consistent with there being at least two different polypeptide chains. (From J. Engel et al., *J. Mol. Biol.* 150:97–120, 1981. © Academic Press Inc. [London] Ltd.)

Basal Laminae Perform Diverse and Complex Functions

Basal laminae have been shown to perform a surprising diversity of functions. In the kidney glomerulus, the basal lamina acts as a semipermeable filter, regulating the passage of macromolecules from the blood into the forming urine. Proteoglycans seem to be important to this function, for when they are removed by specific enzymes, the filtering properties of the lamina are destroyed. The basal lamina may also act as a selective cellular barrier: for example, the lamina beneath epithelial cells prevents fibroblasts in the underlying connective tissue from making contact with the epithelial cells, but it does not stop macrophages, lymphocytes, or nerve processes from passing through it.

It is likely that the basal lamina plays an important part in tissue regeneration after injury. When tissues such as muscle, nerve, and epithelia are damaged, the basal lamina survives and provides a scaffolding along which regenerating cells can migrate. In this way, the original tissue architecture is readily reconstructed. The most dramatic example of the role of the basal lamina in regeneration comes from studies on the *neuromuscular junction,* where a nerve cell makes synaptic contact with the surface of a skeletal muscle cell.

A basal lamina encloses the muscle cell as a whole, but at the site of neuromuscular contact the lamina has a special character. It is, for example, possible to make antibodies that bind to it exclusively in this region. The specialized basal lamina at the neuromuscular junction, ragged and insignificant as it appears in the electron microscope, is the crucial element coordinating the spatial organization of the components on either side of the synapse—where the nerve transmits its stimulus to the muscle. The evidence for this central role of the basal lamina in the construction of a synapse will be discussed in detail in Chapter 18 (see p. 1054). This one well-studied example of specificity in a basal lamina makes it clear that we still have a great deal to learn about these structures. It also suggests that specific (but as yet undefined) components in the extracellular matrix may play a critical part in cell-recognition processes, including those involved in embryonic development.

Intracellular Actin Filament Bundles Direct the Organization of Matrix Macromolecules on Cell Surfaces and Thereby Organize the Extracellular Matrix[20]

Fibronectin, collagen, proteoglycan, and hyaluronic acid molecules, as well as being present in the extracellular matrix, can all be associated with cell surfaces. It is not clear how these molecules are attached to the outside of the plasma membrane, and it is largely a matter of semantics where the plasma-membrane-associated components end and the extracellular matrix begins. The glycocalyx of a cell, for example, includes components of both (see p. 284).

When fluorescent antibodies are used to visualize fibronectin on the surface of cultured fibroblasts, fibronectin is found to be distributed in striking fibrillar arrays that are concentrated between adjacent cells and between cells

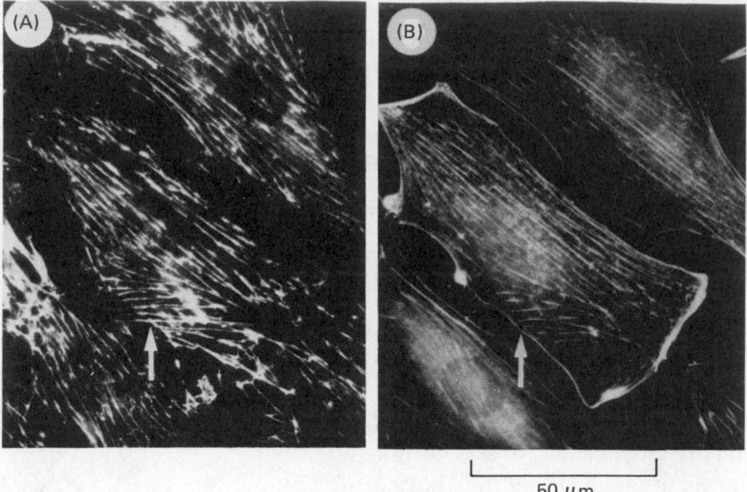

Figure 12–70 Immunofluorescence micrographs of extracellular fibronectin fibers (A) and intracellular actin filament bundles (B) in three cultured rat fibroblasts. The fibronectin is visualized by the binding of fluorescein-coupled anti-fibronectin antibodies and the actin by rhodamine-coupled anti-actin antibodies. (From R. O. Hynes and A. T. Destree, *Cell* 15:875–886, 1978. © M.I.T. Press.)

and the substratum (Figure 12–70A). If cells are treated with the drug cytochalasin, which disrupts the internal actin filament bundles, the fibronectin filaments dissociate from the cell surface (just as they do during mitosis when a cell rounds up). It seems that there exists some indirect connection between extracellular fibronectin and intracellular actin filaments.

It is therefore not surprising that the orientation of the fibronectin strands secreted by a cell coincides with the orientation of the actin filament bundles inside, both sets of fibers being oriented along the long axis of the cell (Figure 12–70B). Similar observations have been made concerning the distribution of newly synthesized collagen molecules on the surface of cultured fibroblasts, which may be bound through their interaction with fibronectin. It is therefore clear that the high degree of order within cells can be transmitted to the extracellular matrix through the orientation of the macromolecules that cells secrete.

In Reciprocal Fashion, an Ordered Extracellular Matrix Influences the Organization and Behavior of the Cells It Contains[20]

The macromolecules of the extracellular matrix have striking effects on the behavior of cells in culture, influencing their shape, movement, metabolism, and differentiation. For example, corneal epithelial cells make very little collagen or proteoglycan when they are cultured on synthetic surfaces; when cultured on basal lamina, collagen, or proteoglycans, however, they accumulate and secrete large amounts of these matrix macromolecules. Moreover, the basal surfaces of epithelial cells cultured on synthetic surfaces are irregular, and the overlying cytoskeletons within the cells are disorganized. When the same cells are cultured on basal lamina or matrix macromolecules, the basal surfaces are smooth and the overlying cytoskeletons are highly organized as they are in the intact tissue.

In a similar way, if cultures of transformed cells are grown on organized fibronectin filaments, the fibronectin induces them to flatten and assemble intracellular actin filament bundles that are aligned with the extracellular fibronectin filaments. Taken together with the evidence that intracellular actin filament bundles can influence the arrangement of secreted fibronectin molecules (see above), these observations indicate that extracellular fibronectin communicates with intracellular actin filaments in both directions across the fibroblast plasma membrane.

Since the cytoskeletons of cells can order the matrix macromolecules they secrete, and the matrix macromolecules can in turn organize the cytoskeletons of cells that contact them, extracellular matrix can in principle propagate order from cell to cell (Figure 12–71). In this way, the matrix could play a central part in generating and maintaining the patterns of cells in tissues and organs during development. The cell-surface-associated matrix macromolecules would serve as both "links" and "adaptors" in this ordering process, mediating the interactions between cells and the matrix around them.

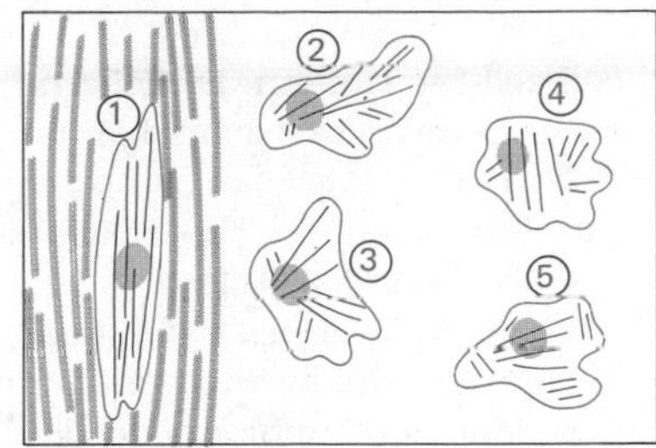

orientation of cytoskeleton in cell ① orients the assembly of secreted extracellular matrix molecules in the vicinity

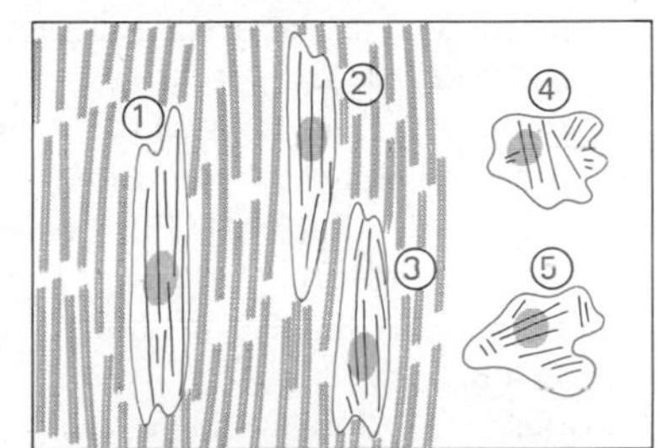

the oriented extracellular matrix reaches cells ② and ③ and orients the cytoskeleton of those cells

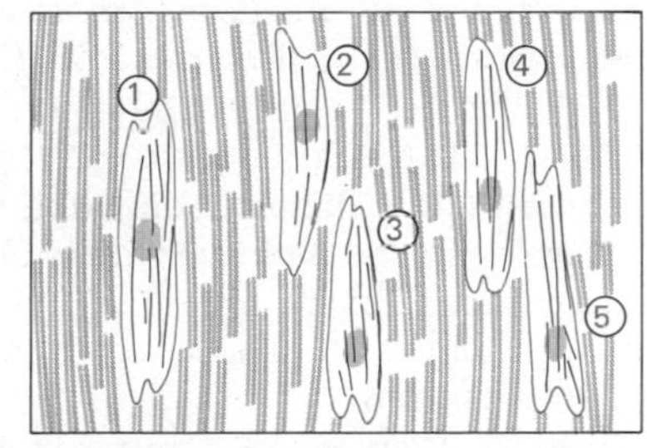

cells ② and ③ now secrete an oriented matrix in their vicinity; in this way the ordering of cytoskeletons is propagated to cells ④ and ⑤

Figure 12–71 A hypothetical scheme showing how the extracellular matrix could propagate order from cell to cell within a tissue. Although, for simplicity, the figure shows one cell influencing the orientation of its neighboring cells, by this scheme cells could mutually affect each others' orientation.

Summary

All cells in tissues are in contact with an intricate extracellular matrix. This matrix not only holds the cells together in tissues, and tissues together in organs, but it also influences the development, polarity, and behavior of the cells it contacts. The matrix is known to contain three major fiber-forming proteins—collagen, elastin, and fibronectin—which are interwoven in a hydrated gel formed by a network of glycosaminoglycan chains. All of these macromolecules are secreted locally by cells in contact with the matrix.

The collagens are ropelike, triple-stranded, helical molecules that aggregate in long cablelike fibrils or sheets in the extracellular space; these in turn can assemble in a variety of highly ordered arrays. Elastin molecules form an extensive cross-linked network of fibers and sheets that can stretch and recoil, imparting elasticity to the matrix. Fibronectin molecules form fibers that promote cell adhesion. The glycosaminoglycans are a heterogeneous group of long, negatively charged polysaccharide chains that (except for hyaluronic acid) are covalently linked to protein to form giant proteoglycan molecules. All of these matrix proteins and polysaccharides are thought to interact and to assemble in a large variety of different three-dimensional structures, ordered in part by the cells secreting the matrix. Since the orientation of the matrix will in turn influence the orientation of the cells it contains, order is likely to be propagated from cell to cell through the matrix.

References

Cited

1. Kaiser, D.; Manoil, C. Myxobacteria: cell interactions, genetics and development. *Annu. Rev. Microbiol.* 33:595–639, 1979.

   White, D. Cell interactions and the control of development in myxobacteria populations. *Int. Rev. Cytol.* 72:203–228, 1981.

2. Loomis, W.F. *Dictyostelium discoideum.* A Developmental System. New York: Academic Press, 1975.

   Gerisch, G. Chemotaxis in *Dictyostelium. Annu. Rev. Physiol.* 44:535–552, 1982.

   Gerisch, G. Univalent antibody fragments as tools for the analysis of cell interactions in *Dictyostelium. Curr. Top. Dev. Biol.* 14:243–270, 1980.

3. Wilson, H.V. On some phenomena of coalescence and regeneration in sponges. *J. Exp. Zool.* 5:245–258, 1907.

   Humphreys, S.; Humphreys, T.; Sano, J. Organization and polysaccharides of sponge aggregation factor. *J. Supramol. Struct.* 7:339–351, 1977.

   Burger, M.M.; Burkart, W.; Weinbaum, G.; Jumblatt, J. Cell-cell recognition: molecular aspects. Recognition and its relation to morphogenetic processes in general. In Cell-Cell Recognition. Society for Experimental Biology Symposium, No. 32 (A.S.G. Curtis, ed.), pp. 1–23. Cambridge, Eng.: Cambridge University Press, 1978.

4. Townes, P.; Holtfreter, J. Directed movements and selective adhesion of embryonic amphibian cells. *J. Exp. Zool.* 128:53–120, 1955.

Moscona, A.A.; Hausman, R.E. Biological and biochemical studies on embryonic cell-cell recognition. In Cell and Tissue Interactions. Society of General Physiologists Series (J.W. Lash, M.M. Burger, eds.), Vol. 32, pp. 173–185. New York: Raven, 1977.

Roth, S.; Weston, J. The measurement of intercellular adhesion. *Proc. Natl. Acad. Sci. USA* 58:974–980, 1967.

5. Steinberg, M.S. Does differential adhesion govern self-assembly processes in histogenesis? Equilibrium configurations and the emergence of a hierarchy among populations of embryonic cells. *J. Exp. Zool.* 173:395–434, 1970.

6. Staehelin, L.A.; Hull, B.E. Junctions between living cells. *Sci. Am.* 238(5):141–152, 1978.

Gilula, N.B. Junctions between cells. In Cell Communication (R.P. Cox, ed.), pp. 1–29. New York: Wiley, 1974.

7. Farquhar, M.G.; Palade, G.E. Junctional complexes in various epithelia. *J. Cell Biol.* 17:375–412, 1963.

Goodenough, D.A.; Revel, J.P. A fine structural analysis of intercellular junctions in the mouse liver. *J. Cell Biol.* 45:272–290, 1970.

8. Furshpan, E.J.; Potter, D.D. Low-resistance junctions between cells in embryos and tissue culture. *Curr. Top. Dev. Biol.* 3:95–127, 1968.

Lowenstein, W.R. Permeable junctions. *Cold Spring Harbor Symp. Quant. Biol.* 40:49–63, 1976.

Hooper, M.L.; Subak-Sharpe, J.H. Metabolic cooperation between cells. *Int. Rev. Cytol.* 69:45–104, 1981.

Gilula, N.B.; Reeves, O.R.; Steinbach, A. Metabolic coupling, ionic coupling and cell contacts. *Nature* 235:262–265, 1972.

9. Furshpan, E.S.; Potter, D.D. Low-resistance junctions between cells in embryos and tissue culture. *Curr. Top. Dev. Biol.* 3:95–127, 1968.

Sheridan, J.D. Cell coupling and cell communication during embryogenesis. In Cell Surface in Animal Embryogenesis and Development. Cell Surface Reviews (G. Poste, G.L. Nicolson, eds.), Vol. 1, pp. 409–448. Amsterdam: Elsevier, 1977.

Warner, A.E. The early development of the nervous system. In British Society of Developmental Biology Symposium 5 (D.R. Garrod, J. Feldman, eds.), pp. 109–127. Cambridge, Eng.: Cambridge University Press, 1981.

10. Lowenstein, W.R. Permeable junctions. *Cold Spring Harbor Symp. Quant. Biol.* 11:49–63, 1976.

Turin, L.; Warner, A.E. Intracellular pH in early Xenopus embryo: its effect on current flow between blastomeres. *J. Physiol. (Lond.)* 300:489–504, 1980.

11. Caspar, D.L.D.; Goodenough, D.; Makowski, L.; Phillips, W.C. Gap junction structures. I. Correlated electron microscopy and x-ray diffraction. *J. Cell Biol.* 74:605–628, 1977.

Hertzberg, E.L.; Lawrence, T.S.; Gilula, N.B. Gap junctional communication. *Annu. Rev. Physiol.* 43:479–491, 1981.

12. Hay, E.D. Extracellular matrix. *J. Cell Biol.* 91:205s–223s, 1981.

Hay, E.D., ed. Cell Biology of Extracellular Matrix. New York: Plenum, 1982.

13. Prockop, D.J.; Kivirikko, K.I.; Tuderman, L.; Guzman, N. The biosynthesis of collagen and its disorders. *N. Engl. J. Med.* 301:13–23, 77–85, 1979.

Eyre, D.R. Collagen: molecular diversity in the body's protein scaffold. *Science* 207:1315–1322, 1980.

Bornstein, P.; Sage, H. Structurally distinct collagen types. *Annu. Rev. Biochem.* 49:957–1003, 1980.

14. Trelstad, R.L.; Hayashi, K. Tendon collagen fibrillogenesis: intracellular subassemblies and cell surface changes associated with fibril growth. *Dev. Biol.* 71:228–242, 1979.

Trus, B.L.; Piez, K.A. Compressed microfibril models of the native collagen fibril. *Nature* 286:300–301, 1980.

15. Sandberg, L.B.; Soskel, N.T.; Leslie, J.G. Elastin structure, biosynthesis and relation to disease states. *N. Engl. J. Med.* 304:566–577, 1981.

Franzblau, C.; Faris, B. Elastin. In Cell Biology of Extracellular Matrix (E.D. Hay, ed.), pp. 65–93. New York: Plenum, 1982.

16. Lindhal, U.; Höök, M. Glycosaminoglycans and their binding to biological macromolecules. *Annu. Rev. Biochem.* 47:385–417, 1978.

Chakrabarti, B.; Park, J.W. Glycosaminoglycans: structure and interaction. *CRC Crit. Rev. Biochem.* 8:225–313, 1980.

Rodén, L. Structure and metabolism of connective tissue proteoglycans. In The Biochemistry of Glycoproteins and Proteoglycans (W.J. Lennarz, ed.), pp. 267–371. New York: Plenum, 1980.

17. Rees, D.A. Polysaccharide Shapes (Outline Studies in Biology), pp. 62–73. London: Chapman and Hall, 1977.

Trelstad, R.L.; Hayashi, K.; Toole, B.P. Epithelial collagens and glycosaminoglycans in the embryonic cornea. Macromolecular order and morphogenesis in the basement membrane. *J. Cell Biol.* 62:815–830, 1974.

Toole, B.P. Morphogenetic role of glycosaminoglycans in brain and other tissues. In Neuronal Recognition (S.H. Barondes, ed.), pp. 275–329. New York: Plenum, 1976.

18. Yamada, K.M.; Olden, K. Fibronectins—adhesive glycoproteins of cell surface and blood. *Nature* 275:179–184, 1978.

Ruoslahti, E.; Engvall, E.; Hayman, E.G. Fibronectin: current concepts of its structure and functions. *Coll. Res.* 1:95–128, 1981.

Hynes, R.O. Fibronectin and its relation to cellular structure and behavior. In Cell Biology of Extracellular Matrix (E.D. Hay, ed.), pp. 295–334. New York: Plenum, 1982.

19. Kefalides, N.A.; Alper, R.; Clark, C.C. Biochemistry and metabolism of basement membranes. *Int. Rev. Cytol.* 61:167–228, 1979.

Timpl, R.; et al. Laminin—a glycoprotein from basement membranes. *J. Biol. Chem.* 254:9933–9937, 1979.

Farquhar, M.G. The glomerular basement membrane: a selective macromolecular filter. In Cell Biology of Extracellular Matrix (E.D. Hay, ed.), pp. 335–378. New York: Plenum, 1982.

20. Bornstein, P.; Duksin, D.; Balian, G.; Davidson, J.M.; Crouch, E. Organization of extracellular proteins on the connective tissue cell surface: relevance to cell-matrix interactions *in vitro* and *in vivo*. *Ann. N.Y. Acad. Sci.* 312:93–105, 1978.

Hynes, R. Structural relationships between fibronectin and cytoplasmic cytoskeletal networks. In Cytoskeletal Elements and Plasma Membrane Organization (G. Poste, G.L. Nicolson, eds.), Vol. 7, pp. 100–137. Amsterdam: Elsevier, 1981.

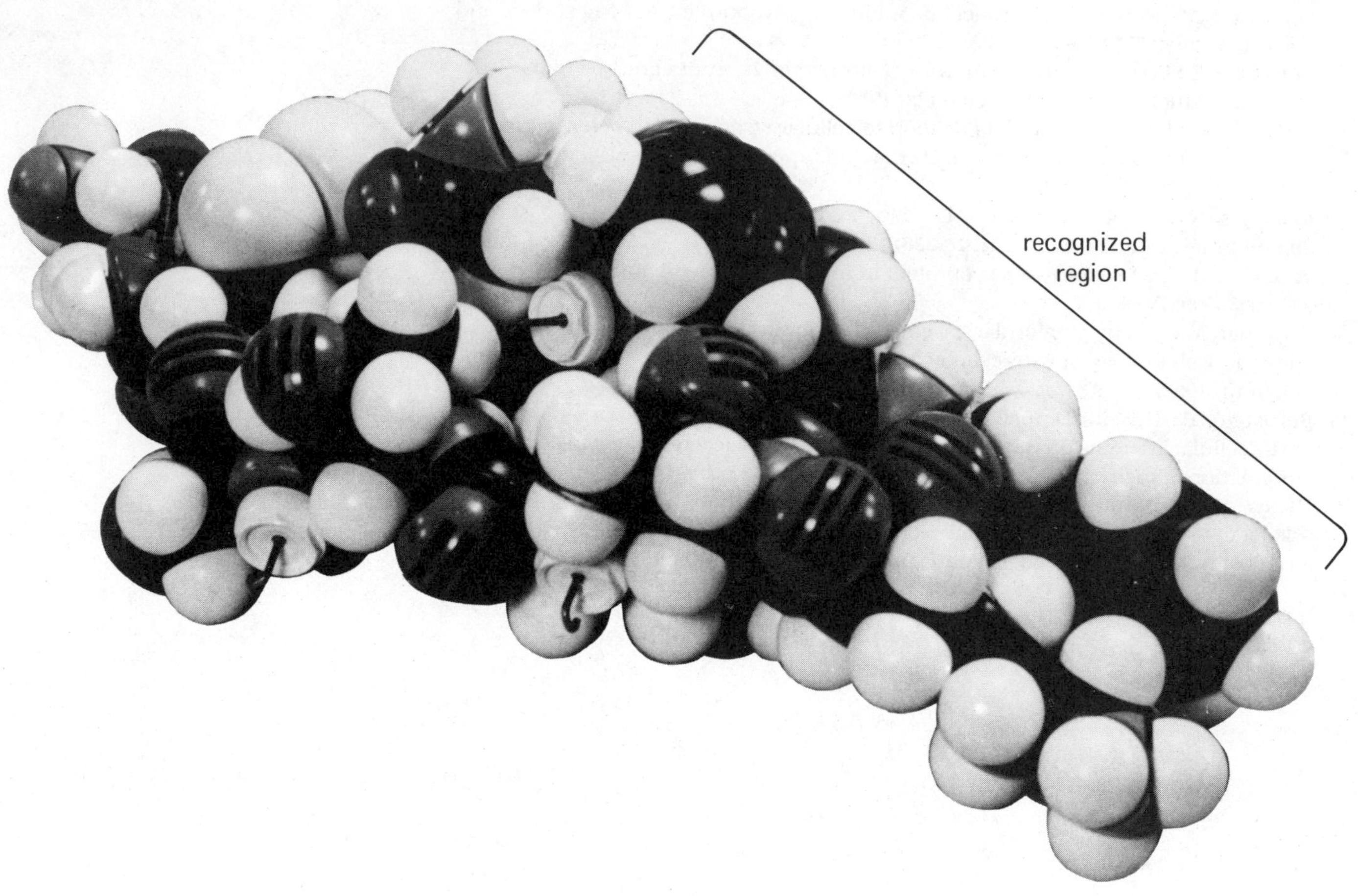

Atomic model of somatostatin, a polypeptide hormone of 14 amino acids. The region recognized by its specific cell-surface receptor is indicated. (Courtesy of Ralph F. Hirschmann.)

13 Chemical Signaling Between Cells

The evolution of multicellular organisms has depended on the ability of cells to communicate with each other. Communication between cells is required to regulate their development and organization into tissues, to control their growth and division, and to coordinate their diverse activities. The importance and complexity of intercellular communication in higher animals suggest that a large proportion of the genes in these organisms are concerned with such processes.

Cells are thought to communicate in three ways: (1) they secrete chemicals that signal cells some distance away; (2) they display plasma-membrane-bound signaling molecules that influence other cells that make direct physical contact; and (3) they form gap junctions that directly join the cytoplasms of the interacting cells (Figure 13–1).

Communication that depends on cell-to-cell contact through gap junctions has been discussed in Chapter 12. Little is known about cell communication via the direct interactions of molecules bound to the outer surface of cells. In principle, it should differ from signaling by secreted chemical signals only in that the signaling molecules are membrane-bound. But membrane-bound molecules are hard to solubilize and purify; consequently, their mechanism of action is difficult to study. Therefore, it is not surprising that unambiguous evidence for this form of communication has yet to be obtained. If we observe that one cell affects an adjacent one, how can we be certain that the signal is relayed via cell-surface molecules rather than through transient gap junctions or by secreted short-range chemical mediators?

In this chapter we shall be concerned primarily with indirect mechanisms of communication—those mediated by secreted chemical signals. Fortunately, secreted molecules are very much easier to study than the membrane-bound variety, and a great deal is known about how they work.

INDIRECT SIGNALING BY SECRETED CHEMICALS

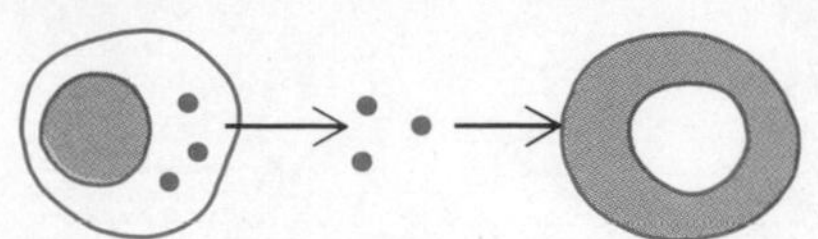

DIRECT SIGNALING BY PLASMA-MEMBRANE-BOUND MOLECULES

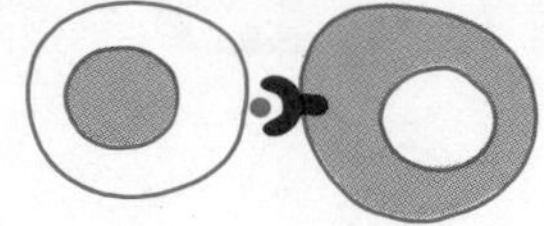

DIRECT SIGNALING VIA GAP JUNCTIONS

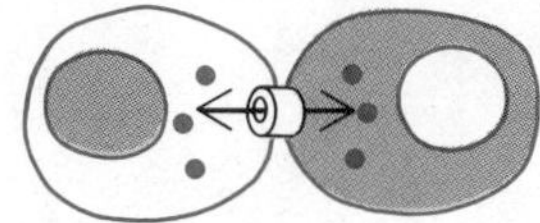

Figure 13–1 Schematic diagram showing three different ways in which cells are thought to communicate with each other.

Three Different Strategies of Chemical Signaling: Local Chemical Mediators, Hormones, and Neurotransmitters

Chemical signaling operates in three different ways: (1) most cells in the body secrete one or more chemical signals that function as **local chemical mediators** because they are so rapidly taken up or destroyed that they act only on cells in the immediate environment; (2) specialized *endocrine cells* secrete **hormones** that travel through the bloodstream to influence target cells widely distributed in the body; and (3) nerve cells form specialized junctions (*chemical synapses*) with the target cells they influence and secrete very short-range chemical mediators called **neurotransmitters,** which act only on the adjoining target cell (Figure 13–2).

Endocrine cells and nerve cells are highly specialized for chemical signaling and work together to coordinate the diverse activities of the billions of cells in a higher animal. Nerve cells transmit information much more rapidly than endocrine cells because they do not depend on diffusion or blood flow to convey information over long distances; instead, electrical impulses carry the signal rapidly along nerve processes. Only at the nerve terminals, when a neurotransmitter is released, are the electrical impulses converted into chemical signals; the neurotransmitter has to diffuse only a microscopic distance to the target cell, a process that takes less than a millisecond (Figure 13–2). While hormones are greatly diluted in the bloodstream and therefore must be able to act at very low concentrations (typically $<10^{-8}$ M), neurotransmitters are diluted much less and can achieve high concentrations in the region of the target cell. For example, the concentration of the neurotransmitter acetylcholine in the synaptic cleft of an active neuromuscular junction is about 5×10^{-4} M. In other respects, however, the mechanisms of chemical signaling by hormones and neurotransmitters are generally similar, and many of the signaling molecules that are used by endocrine cells are also used by nerve cells (neurons).

LOCAL CHEMICAL MEDIATOR

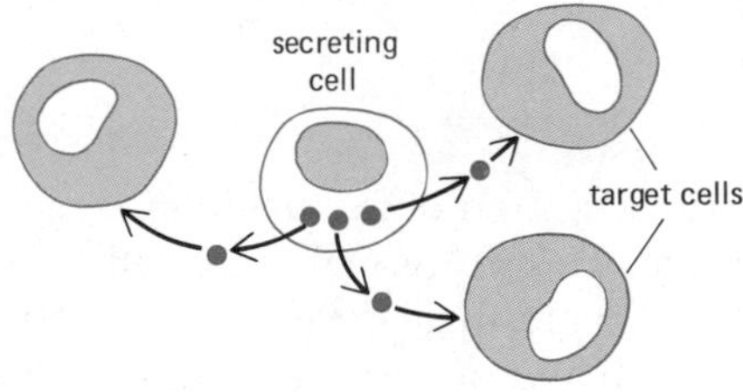

HORMONE

secretory (endocrine) cell

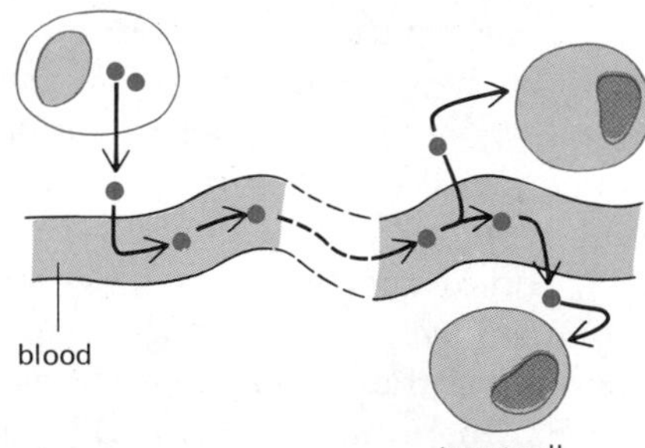

NEUROTRANSMITTER

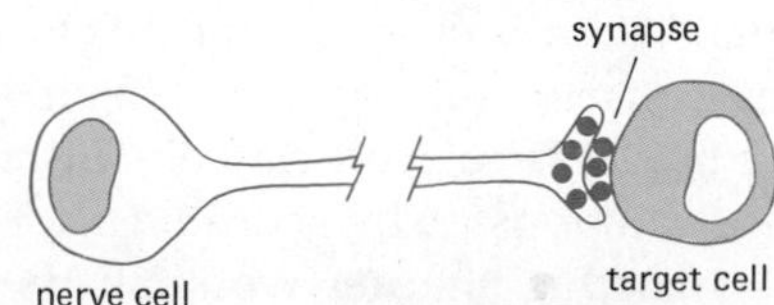

Figure 13–2 Three different classes of extracellular signaling molecules.

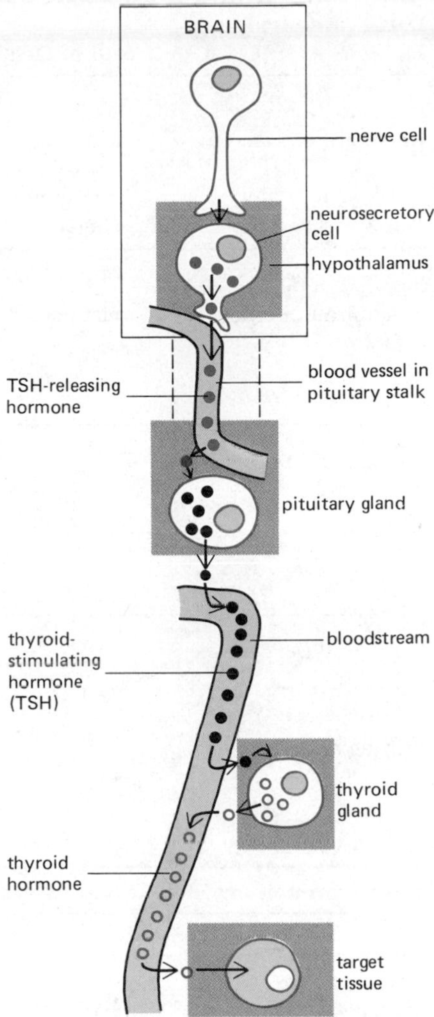

Figure 13–3 Schematic illustration of the indirect manner in which thyroid hormone secretion is regulated by the nervous system. When stimulated by nerve cells in higher centers of the brain, specific neurosecretory cells in the hypothalamus secrete into blood vessels of the pituitary stalk TSH-releasing hormone, which stimulates the release of TSH (thyroid-stimulating hormone) by specific cells in the pituitary gland. TSH in turn stimulates the cells in the thyroid gland to synthesize and secrete thyroid hormone. Thyroid hormone then stimulates a variety of metabolic processes in most cells in the body. Not illustrated in this figure is the fact that the secretion of both TSH-releasing hormone and TSH are suppressed by increased concentrations of thyroid hormone in the blood. This *feedback inhibition* prevents the blood levels of thyroid hormone from rising too high.

The Hypothalamus Is the Main Regulator of the Endocrine System[1]

The endocrine system and nervous system are physically and functionally linked by a specific region of the brain called the **hypothalamus.** The hypothalamus lies immediately above the pituitary gland, to which it is connected by a hypothalamic extension called the *pituitary stalk.* The bridging function of the hypothalamus is mediated by cells that have properties of both nerve cells and endocrine cells: they have nerve processes that carry electrical impulses but release their signaling molecules into the blood; for this reason they are called *neurosecretory cells.* Each of the hypothalamic neurosecretory cells can be stimulated by other nerve cells in higher regions of the brain to secrete a specific peptide hormone into the blood vessels of the pituitary stalk: the hormone then specifically stimulates or suppresses the secretion of a second hormone from the pituitary. Many of the pituitary hormones regulated by the hypothalamus in this way stimulate another endocrine gland to secrete a third hormone into the blood. Consequently, the hypothalamus serves as the main regulator of the endocrine system. Figure 13–3 illustrates how this hierarchy works in the regulation of the secretion of *thyroid hormone.*

Selected examples of local chemical mediators, neurotransmitters, and hormones are given in Table 13–1, together with their sites of origin, structures, and principal actions. For the most part, the examples chosen are discussed elsewhere in this book. It can be seen that these signaling molecules are as varied in structure as they are in function. They include small peptides, larger proteins and glycoproteins, amino acids and related compounds, steroids (molecules derived from cholesterol and closely related in structure), and fatty acid derivatives.

Different Cells Respond in Different Ways to the Same Chemical Signal[2]

The ability of a cell to respond to a particular extracellular signaling molecule depends on its having specific proteins, called **receptors,** that bind the signaling molecule. Many signaling molecules act at very low concentration (typically $\leq 10^{-8}$ M), and their complementary receptors usually bind them with high affinity (affinity constant $K \geq 10^8$ liters per mole; see p. 97). In mature

Table 13–1 Some Examples of Signaling Molecules

Local Chemical Mediators	Site of Origin	Structure	Major Effects
Proteins			
Nerve growth factor	all tissues innervated by sympathetic nerves	2 identical chains of 118 amino acids	survival and growth of sensory and sympathetic neurons
Small Peptides			
Eosinophil chemotactic factor	mast cells	4 amino acids	chemotactic signal for a special type of white blood cell (eosinophilic leukocytes)
Amino Acid Derivatives			
Histamine	mast cells	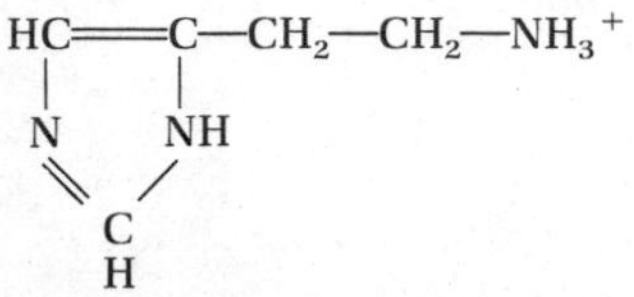 $HC{=}C{-}CH_2{-}CH_2{-}NH_3^+$ (imidazole ring: HC=C–NH–CH=N)	causes blood vessels to dilate and become leaky
Fatty Acid Derivatives			
Prostaglandin E_2	many different cell types	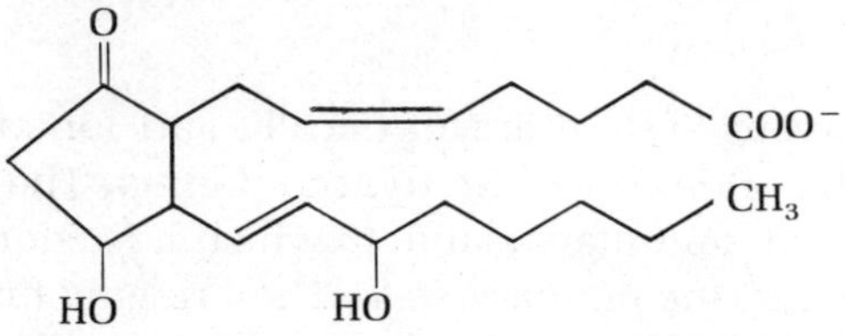 (structure: cyclopentanone ring with O, HO, side chains ending in COO^- and CH_3, HO)	contraction of smooth muscle

Neurotransmitters*	Site of Origin	Structure	Major Effects
Amino Acids and Related Compounds			
Glycine	nerve terminals	$^+H_3N{-}CH_2{-}COO^-$	inhibitory transmitter in central nervous system
Norepinephrine	nerve terminals	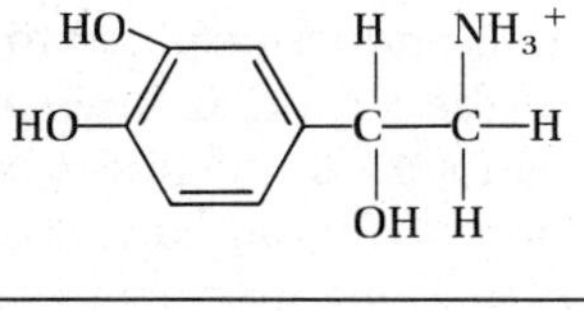 $(HO)_2C_6H_3{-}CH(OH){-}CH_2{-}NH_3^+$	excitatory and inhibitory transmitter in central and peripheral nervous system
γ-Aminobutyric acid (GABA)	nerve terminals	$^+H_3N{-}CH_2{-}CH_2{-}CH_2{-}COO^-$	inhibitory transmitter in central nervous system
Acetylcholine	nerve terminals	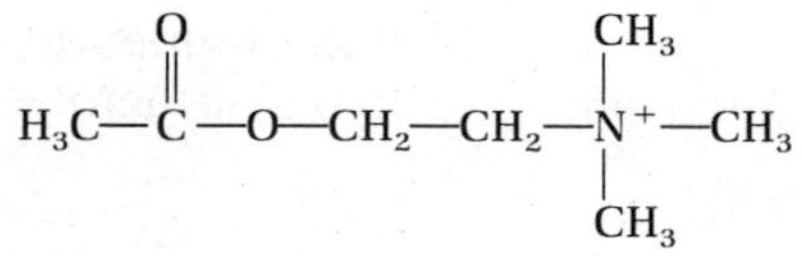 $H_3C{-}C({=}O){-}O{-}CH_2{-}CH_2{-}N^+(CH_3)_3$	excitatory transmitter at neuromuscular junction; excitatory and inhibitory transmitter in central and peripheral nervous system
Small Peptides			
Enkephalin	nerve terminals	5 amino acids	morphine-like action (inhibits pain pathways in central nervous system)

*Some of these molecules may function as local chemical mediators in the nervous system rather than strictly as neurotransmitters (see p. 726). Excitatory neurotransmitters stimulate the postsynaptic cell, while inhibitory neurotransmitters suppress the postsynaptic cell.

Table 13–1 *Continued*

Hormones	Site of Origin	Structure	Major Effects
Proteins and Glycoproteins			
Insulin	beta cells of pancreas	protein α-chain = 21 amino acids β-chain = 30 amino acids	utilization of carbohydrate (including uptake of glucose into cells); stimulation of protein synthesis; stimulation of lipid synthesis in fat cells
Somatotropin (growth hormone)	anterior pituitary	protein 191 amino acids	stimulation of liver to produce somatomedins, which in turn cause growth of muscle and bone
Somatomedins	liver	proteins	growth of bone and muscle; influences metabolism of Ca^{2+}, phosphate, carbohydrate, and lipid
Adrenocorticotropic hormone (ACTH)	anterior pituitary	protein 39 amino acids	stimulation of adrenal cortex to produce cortisol; fatty acid release from fat cells
Parathormone	parathyroid	protein 84 amino acids	increase in bone resorption, thereby increasing blood Ca^{2+} and phosphate, increase in resorption of Ca^{2+} and Mg^{2+} and decrease in resorption of phosphate in kidney tubules
Follicle-stimulating hormone (FSH)	anterior pituitary	glycoprotein α-chain = 92 amino acids β-chain = 118 amino acids	stimulation of ovarian follicles to grow and secrete estradiol; stimulation of spermatogenesis in testis
Luteinizing hormone (LH)	anterior pituitary	glycoprotein α-chain = 92 amino acids β-chain = 115 amino acids	stimulation of oocyte maturation and ovulation and progesterone secretion from ovary; stimulation of testis to produce testerone
Epidermal growth factor	unknown	53 amino acids	stimulation of epidermal and other cells to divide
Thyroid-stimulating hormone (TSH)	anterior pituitary	glycoprotein α-chain = 92 amino acids β-chain = 112 amino acids	stimulation of thyroid to produce thyroxine; fatty acid release from fat cells
Small Peptides			
TSH-releasing factor	hypothalamus	3 amino acids	stimulation of anterior pituitary to secrete thyroid-stimulating hormone (TSH)

(*Continued*)

Table 13–1 *Continued*

Hormones	Site of Origin	Structure	Major Effects
LH-releasing factor	hypothalamus	10 amino acids	stimulation of anterior pituitary to secrete luteinizing hormone (LH)
Vasopressin	posterior pituitary	9 amino acids	elevation of blood pressure by constriction of small blood vessels; increase in water resorption in kidney tubules
Somatostatin	hypothalamus	14 amino acids	inhibition of somatotropin release from anterior pituitary
Amino Acid Derivatives			
Epinephrine	adrenal medulla	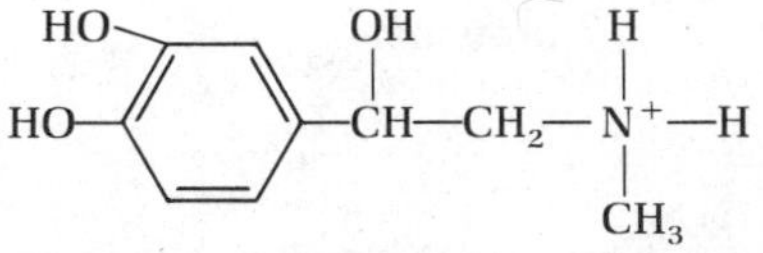	increase in blood pressure and heart rate; increase in glycogenolysis in liver and muscle; fatty acid release from fat cells
Thyroxine	thyroid	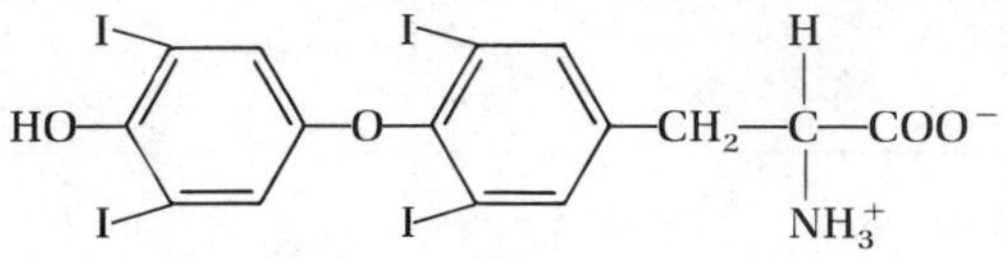	increase in metabolic activity in most cells
Steroids			
Cortisol	adrenal cortex	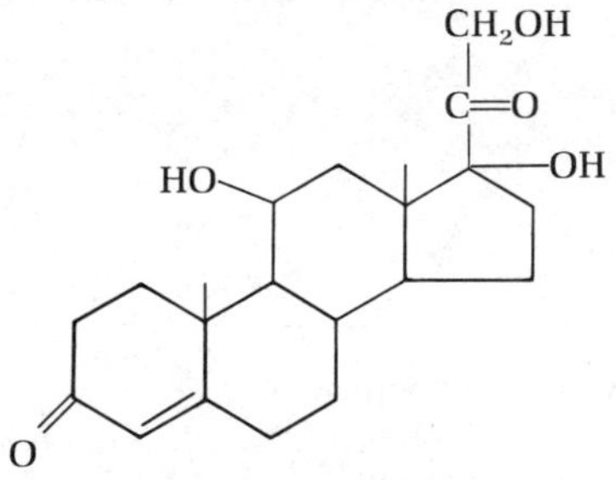	affect on metabolism of proteins, carbohydrates and lipids; suppression of inflammatory reactions
Estradiol	ovary, placenta	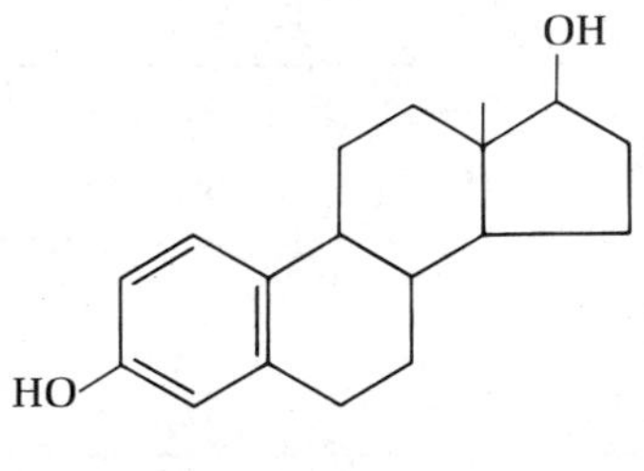	development and maintenance of secondary female sex characteristics; maturation and cyclic function of accessory sex organs; development of duct system in mammary glands
Testosterone	testis	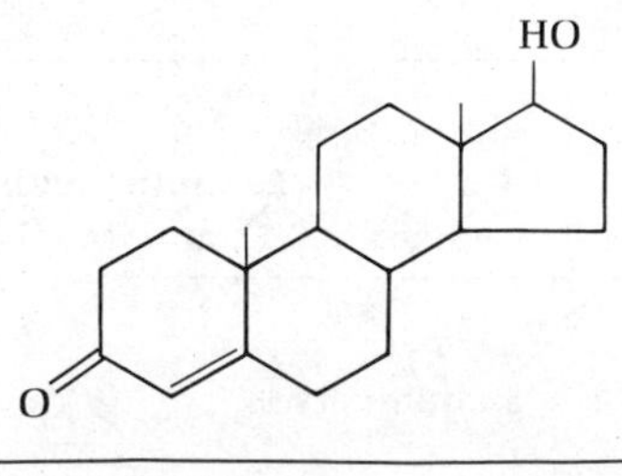	development and maintenance of secondary male sex characteristics; maturation and normal function of accessory sex organs

animals most cells are specialized to perform one primary function, and they contain a characteristic array of receptors that allows them to respond to each of the different chemical signals that initiate or modulate that function.

Most chemical signals ultimately influence target cells either by altering the properties or rates of synthesis of existing proteins or by initiating the synthesis of new ones. In different target cells the same signaling molecule often affects different proteins and therefore has different effects. For example, *acetylcholine* stimulates the contraction of skeletal muscle cells, but it decreases the rate and force of contraction in heart muscle cells. In this particular case the acetylcholine receptor proteins on skeletal muscle cells are different from those on heart muscle cells. But receptor differences are not always the explanation. In many cases the same signaling molecules bind to identical receptor proteins and yet produce very different responses in different target cells. This indicates that target cells are programmed in two ways: (1) they are equipped with a distinctive set of receptors for responding to a complementary set of chemical signals, and (2) they are programmed to respond to each signal in their own characteristic way (Figure 13–4).

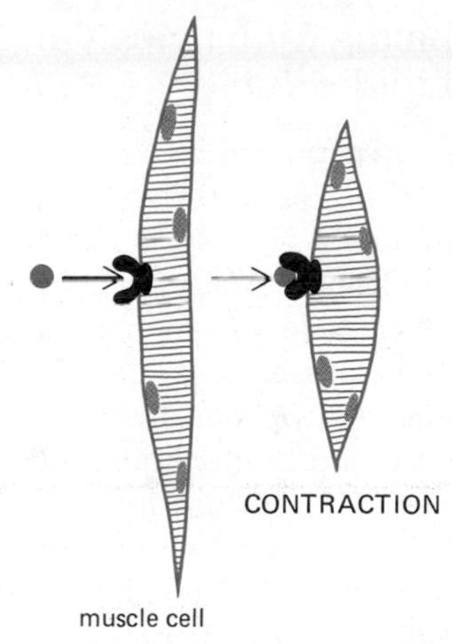

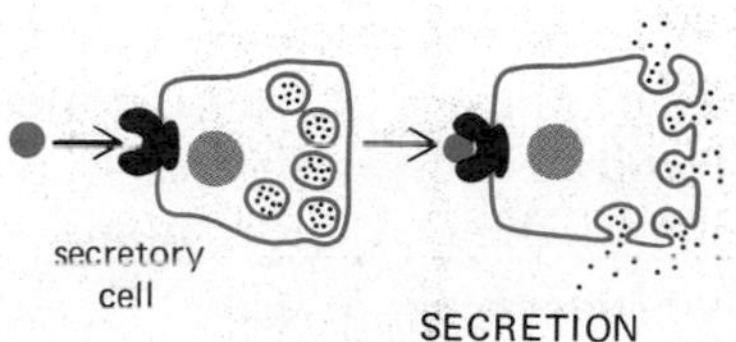

Figure 13–4 Schematic drawing showing how the same signaling molecule binding to identical receptors on two different target cells (muscle and secretory) can induce different responses. Each type of target cell is programmed to respond in a characteristic way to each specific signal.

Some Cellular Responses to Chemical Signals Are Rapid and Transient, While Others Are Slow and Long-lasting

When they coordinate the responses of cells to changes in an animal's environment, chemical signals generally induce rapid and transient responses. For example, an increase in blood glucose levels stimulates endocrine cells in the pancreas to secrete the protein hormone *insulin* into the blood. Within minutes the resulting increase in insulin concentration stimulates liver and muscle cells to take up more glucose, and blood glucose levels fall. Then the rate of insulin secretion and, consequently, the rate of glucose uptake by liver and muscle cells return to their previous levels. In this way a relatively constant blood glucose concentration is maintained. Neurotransmitters elicit even more rapid responses: skeletal muscle cells contract and relax again within milliseconds in response to acetylcholine released from nerve terminals at a neuromuscular junction.

Chemical signals also play an important part in animal development, often influencing when and how certain cells differentiate. These effects are usually slow in onset and long-lasting. For example, the steroid female sex hormone *estradiol* is secreted in large amounts by cells in the ovary around the time of puberty. Estradiol induces changes in a wide variety of cells in different parts of the body, changes that eventually lead to the development of secondary female characteristics, such as breast enlargement. While this effect is slowly reversed if estradiol secretion stops, some of the responses to steroid sex hormones during very early mammalian development are irreversible (see p. 722). Similarly, a tenfold increase in thyroid hormone levels in the blood of a tadpole induces all of the dramatic and irreversible changes that result in its transformation into a frog (Figure 13–5).

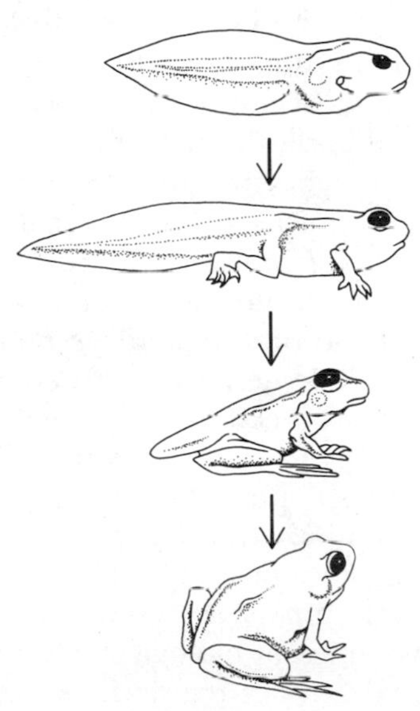

Figure 13–5 Various stages in the metamorphosis of a tadpole into a frog. All of these dramatic changes during metamorphosis are signaled by thyroid hormone. If the presumptive thyroid gland is removed from a developing embryo, the animal fails to undergo metamorphosis and continues to grow as a tadpole. If thyroid hormone is injected into such a giant tadpole, the tadpole transforms into a frog.

Signaling Molecules Can Be Either Water-soluble or Lipid-soluble[1]

All known neurotransmitters, as well as most hormones and local chemical mediators, are water-soluble. The main exceptions are the steroid and thyroid hormones, which are relatively water-insoluble and are made soluble for transport in the bloodstream by binding to specific carrier proteins. This difference in solubility gives rise to a fundamental difference in the mechanism by which the two classes of molecules influence target cells. Water-soluble molecules are too hydrophilic to pass directly through the lipid bilayer of a target-cell plasma membrane; instead they bind to specific receptor proteins on the cell surface. The steroid and thyroid hormones, on the other hand, are hydrophobic, and once released from their carrier proteins, they can pass easily through the plasma membrane of the target cells; these hormones bind to specific receptor proteins *inside* the cell (Figure 13–6).

Another important difference between these two classes of signaling molecules is the length of time that they persist in the bloodstream or tissue fluids. Water-soluble hormones are generally removed and/or broken down within minutes of entering the blood and the local chemical mediators and neurotransmitters are removed even faster, within seconds or milliseconds of entering the extracellular space. By contrast, steroid hormones persist in the blood for hours and thyroid hormone for days. Consequently, water-soluble signaling molecules usually mediate responses of short duration, while the water-insoluble molecules tend to mediate longer lasting responses.

HYDROPHILIC SIGNALS

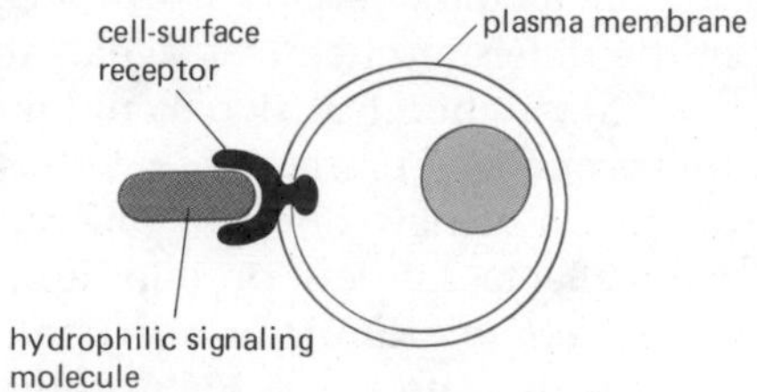

HYDROPHOBIC SIGNALS

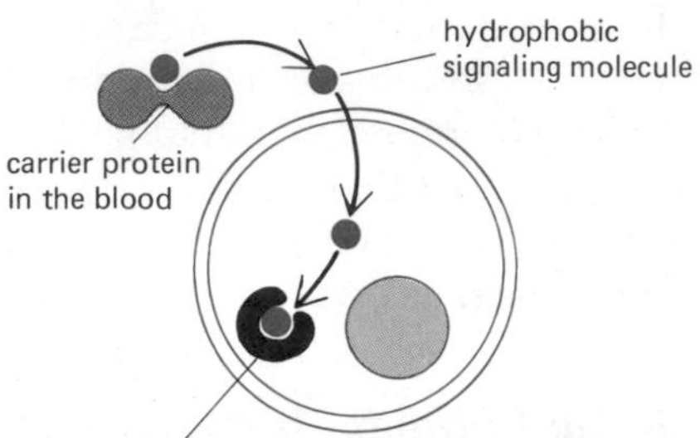

Figure 13–6 The differences in signaling mechanisms between hydrophilic and hydrophobic signaling molecules. Hydrophilic signaling molecules, being unable to cross the plasma membrane directly, bind to receptors on the surface of the target cell. Hydrophobic signaling molecules, being able to diffuse across the plasma membrane, bind to receptors inside the target cell. Because they are insoluble in aqueous solutions, hydrophobic signaling molecules are transported in the bloodstream bound to specific carrier proteins from which they dissociate prior to entering the target cell.

Local Chemical Mediators Are Rapidly Destroyed After They Are Secreted[3]

Many signaling molecules are secreted into the extracellular fluid and act only on cells in their immediate vicinity. These local chemical mediators are distinguished from hormones by the fact that they are so rapidly taken up by cells and/or destroyed that they generally do not enter the blood in significant amounts.

Some local mediators are secreted by cells specialized for that function. For example, *histamine* (a derivative of the amino acid histidine, see Table 13–1) is secreted mainly by mast cells. These cells, which are found in connective tissues throughout the body, store histamine in large secretory vesicles and release it rapidly by exocytosis when stimulated by injury, local infection, or certain immunological reactions. Histamine causes local blood vessels to dilate and become leaky, which facilitates the access of serum proteins (such as antibodies and components of the complement system, see Chapter 17) and phagocytic white blood cells to the sites of injury. Among the other mediators released by mast cells are two tetrapeptides that attract a class of white blood cells called *eosinophils* from the blood to the site of tetrapeptide release; eosinophils contain a variety of enzymes that help inactivate histamine and other chemical mediators released by mast cells.

While some local chemical mediators, like histamine, are secreted by specialized cells, others are of more widespread origin. The **prostaglandins,** a family of 20 carbon fatty acid derivatives, are an important example of such local mediators. Like other local mediators, prostaglandins are rapidly destroyed near the site of their synthesis by specific enzymes. Of the more than 16 different prostaglandins that belong to 9 classes (designated PGA, PGB, PGC, ... PGI), many are known to bind to different cell-surface receptors and to have different biological effects. Unlike most signaling molecules, they are not stored but are continuously released to the cell exterior. Prostaglandins are continuously synthesized in membranes from precursors that have been cleaved

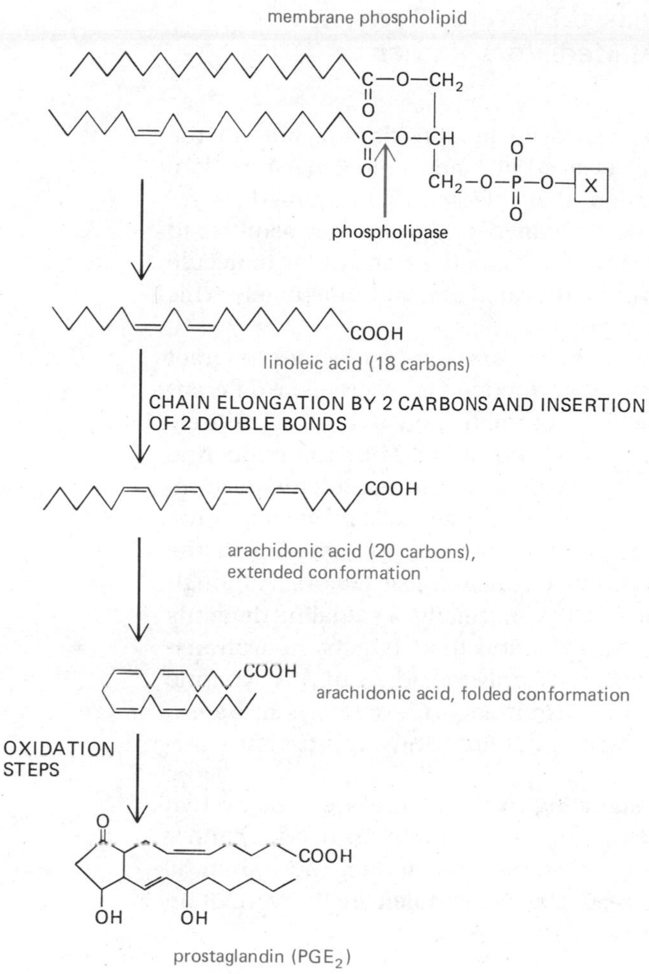

membrane phospholipid

phospholipase

linoleic acid (18 carbons)

CHAIN ELONGATION BY 2 CARBONS AND INSERTION OF 2 DOUBLE BONDS

arachidonic acid (20 carbons), extended conformation

arachidonic acid, folded conformation

OXIDATION STEPS

prostaglandin (PGE_2)

Figure 13–7 Synthesis of the prostaglandin PGE_2. The subscript refers to the two carbon-carbon double bonds outside the ring of PGE_2. Prostaglandins are continuously synthesized by most cells from fatty acid chains cleaved from membrane phospholipids.

from membrane phospholipids by phospholipases (Figure 13–7); they are also continuously degraded. However, when cells are activated by a change in their environment, many of them increase their rates of prostaglandin synthesis. The resulting increase in the local level of prostaglandins influences both the cell that makes the prostaglandins and its immediate neighbors.

A wide variety of biological activities have been ascribed to prostaglandins. They cause contraction of smooth muscle, aggregation of platelets, and inflammation. For example, certain prostaglandins produced in large amounts in the uterus at the time of childbirth seem to be important in stimulating the contraction of the uterine smooth muscle cells; these prostaglandins are now widely used as pharmacological agents to induce abortion. An important recent discovery has been that aspirin and some other anti-inflammatory agents probably work by inhibiting prostaglandin biosynthesis.

Not all local chemical mediators are rapidly destroyed after they have been secreted. Collagen and other macromolecules of the extracellular matrix can be considered as special types of local mediators. They are secreted by local cells and signal other local cells to alter their behavior. These molecules differ from other local chemical mediators in that they are insoluble and therefore do not diffuse from the region where they are synthesized. Consequently, unlike diffusible mediators, they do not have to be destroyed rapidly in order to prevent their effects from spreading.

Some Signaling Molecules Released by Nerve Terminals Probably Act as Local Chemical Mediators Rather Than as Neurotransmitters[4]

Neurotransmitters are either rapidly destroyed by specific enzymes in the synaptic cleft (as is the acetylcholine released at a neuromuscular junction) or rapidly retrieved by the nerve terminal that released them. Immediate removal serves two purposes: it confines the activity of the neurotransmitter to the postsynaptic cell, and it terminates the action of the transmitter molecule so that each signal is very brief and can be repeated almost immediately. This makes the signaling process extremely precise.

Nerve cells and endocrine cells direct their signaling molecules to target cells in very different ways. Each type of endocrine cell secretes a different hormone into the blood, and the specificity of the response depends entirely on which target cells have receptors for each hormone. Thus the endocrine system uses a large number of different hormones (and complementary receptors) to regulate the activities of many different target cells in specific ways. On the other hand, the speed, precision, and intricacy of signaling in the nervous system to a large extent depends on anatomical factors. Although target cells have specific receptors, most of the specificity of signaling depends on synaptic connections between nerve cells and their targets; neurotransmitters are released at synapses and influence only the adjacent postsynaptic cell (Figure 13–8). Some of these neurotransmitters are excitatory and stimulate the postsynaptic cell, while others are inhibitory and suppress the postsynaptic cell.

If all excitatory and inhibitory signaling in the central nervous system were focused on single cells in this way, there should be no need for more than a very small number of signaling molecules. But in fact more than 30 different signaling molecules have already been identified in the vertebrate

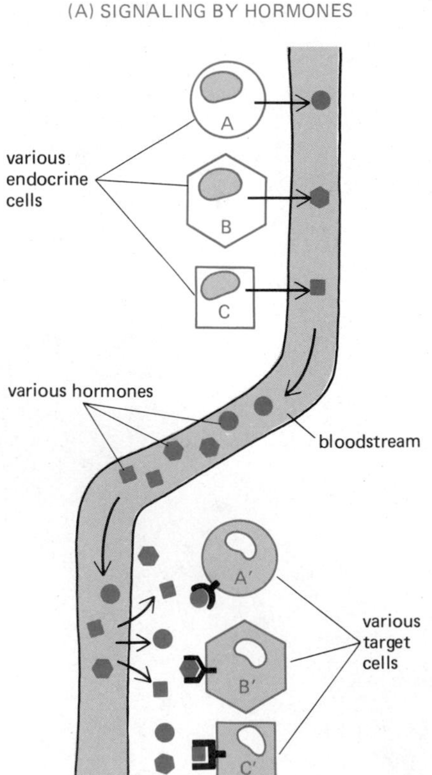

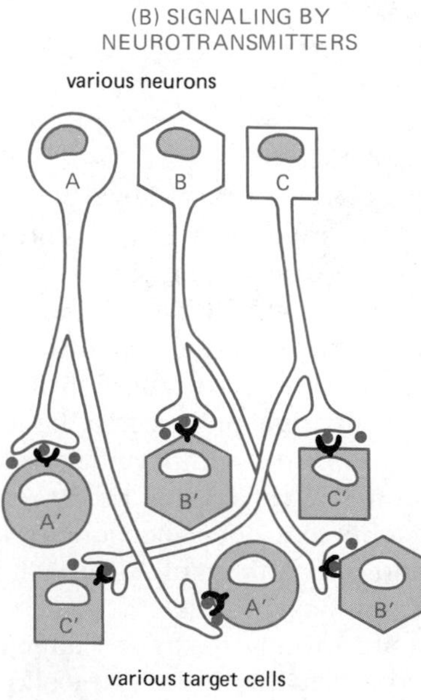

Figure 13–8 The contrast between cell signaling by means of hormones (A) and by means of neurotransmitters (B). Endocrine cells secrete many different hormones into the blood and signal specific target cells, which have receptors for binding specific hormones and thereby "pull" the appropriate hormones out of the extracellular fluid. By contrast, the specificity of signaling by many nerve cells arises from the contacts between their nerve processes and the specific target cells they signal: only a target cell having synaptic contact with a nerve cell is exposed to the neurotransmitter released from the nerve terminal. Whereas different endocrine cells must use different hormones in order to communicate with different target cells, different nerve cells can use the same neurotransmitter and still communicate in a specific manner.

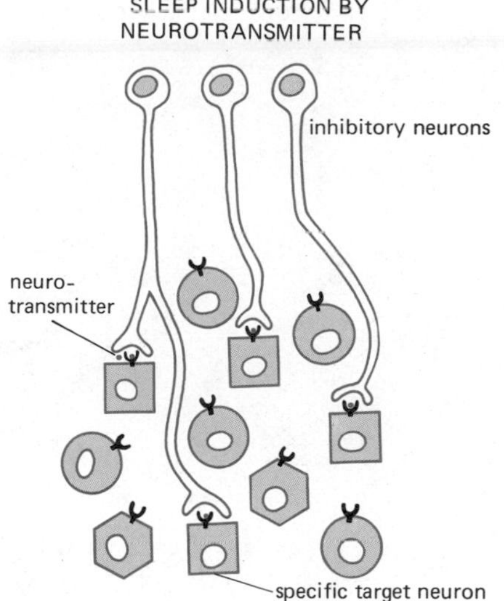

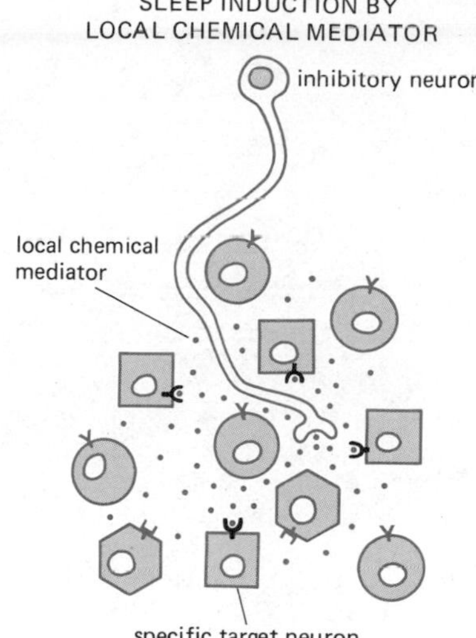

Figure 13–9 Two hypothetical schemes by which sleep might be induced if the prerequisite for sleep were the suppression of certain neurons (shown as squares). In principle, this could be achieved either with an inhibitory neurotransmitter released by neurons that synapse on all of the square cells or, more simply, by a local chemical mediator released from inhibitory nerve terminals in the general region where square cells are found. In the second case, the square cells would be the only ones suppressed because they alone possess receptors for the mediator.

brain, including acetylcholine, various amino acids (glycine, aspartic acid, glutamic acid, and γ-aminobutyric acid, or GABA), amino acid derivatives (norepinephrine, dopamine, serotonin, and histamine), and a large variety of peptides. This may mean that many signaling molecules function not as conventional neurotransmitters but as local chemical mediators (*neuroregulators*) that are released from nerve terminals and then diffuse locally to influence a large number of cells. Such signaling is not synaptic in the strict sense of the word (one terminal—one target cell), and so a large number of signaling molecules (and complementary receptors) are required to assure specificity, as is the case in the endocrine system.

Why might such a local mediator mechanism of neural signaling evolve? Suppose, for example, that sleep requires the suppression of a large number of specific nerve cells in a particular region of the brain. This suppression could be accomplished if specific nerve terminals containing an inhibitory neurotransmitter synapsed on all of these nerve cells. Alternatively, a relatively small number of nerve terminals might release an inhibitory local chemical mediator into the region; as long as all of the relevant nerve cells have receptors for such a "sleep substance," they would be suppressed (Figure 13–9). In fact, alert animals can be put to sleep by the injection (into their brain cavities) of cerebrospinal fluid from the brain cavities of animals that have been kept awake for many days.

Some Hormones and Local Chemical Mediators Act as Specific Growth Factors[5]

The rate of division of certain types of cells is regulated by chemical signals, some of which are conventional hormones. In mammals, estradiol causes breast epithelial cells to divide at puberty; in tadpoles, thyroxine induces particular muscle cells and cartilage cells to proliferate during metamorphosis, while at the same time it induces cells in the tail of the animal to self-destruct. The pituitary hormone *somatotropin* (also called *growth hormone*) indirectly stimulates cell division by inducing liver (and perhaps other) cells to secrete a number of protein hormones that cause certain cells to divide. These latter hormones, collectively called *somatomedins* because they mediate the effects of somatotropin, stimulate the growth and metabolism of muscle and cartilage cells: infants who produce too little somatotropin become dwarfs, while those who produce too much become giants.

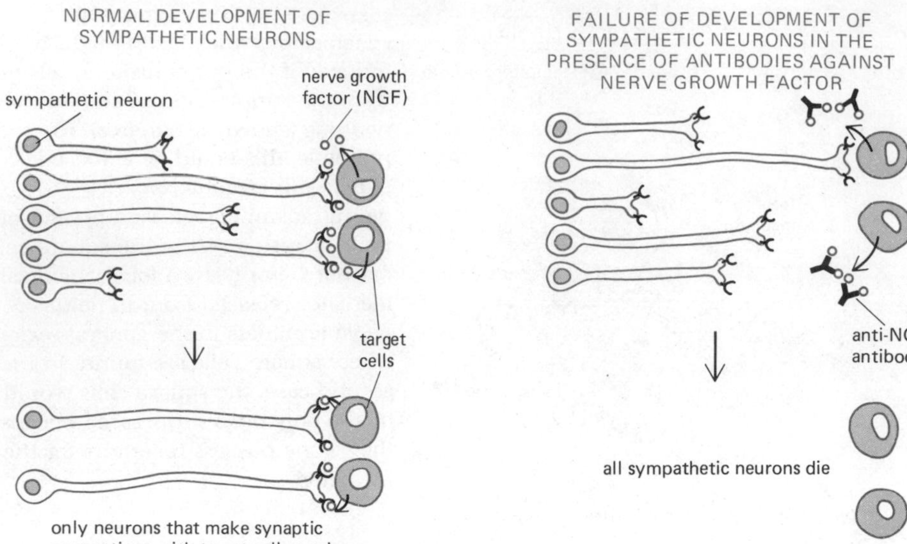

Figure 13–10 Schematic diagram showing that developing sympathetic neurons depend on nerve growth factor for survival. Treating developing animals with antibiodies against nerve growth factor results in the death of sympathetic neurons. It is thought that nerve growth factor is released from target cells and that it binds to receptors present on the nerve terminals of the sympathetic neurons that make synapses on the target cells.

Other growth factors function as local chemical mediators rather than as hormones. The survival and growth of certain classes of nerve cells during development depend on **nerve growth factor (NGF,** a protein dimer composed of two identical polypeptide chains 118 amino acids long), which is thought to be secreted by the target cells of these nerve cells. Three types of observation have demonstrated the importance of NGF for the survival of developing neurons of the sympathetic nervous system: (1) anti-NGF antibodies injected into newborn mice cause the selective death of sympathetic neurons (Figure 13–10); (2) many immature sympathetic neurons survive indefinitely in tissue culture in the absence of other cells if NGF is added to the culture medium; without NGF they die within a few days; (3) developing sympathetic neurons that fail to make synaptic connections with their target cells normally die but can be saved by injections of NGF.

Together these results suggest that many developing sympathetic neurons survive only if they are signaled by small amounts of NGF released by the target cells they innervate (Figure 13–10). Many central and peripheral nerve cells (including sympathetic neurons) produced during normal development are known to die within days of being formed; only those that manage to make synaptic contact with appropriate target cells seem to survive. It is likely that NGF is one of many neuronal survival factors and that different types of neurons require different specific factors produced by the target cells that they innervate in order to survive. The neuronal redundancy that occurs during neurogenesis presumably ensures that all target cells are innervated.

NGF may also play a part in directing sympathetic nerve fibers to their appropriate target cells. When NGF is injected into the brain of a newborn mouse, it attracts sympathetic fibers to grow into the central nervous system,

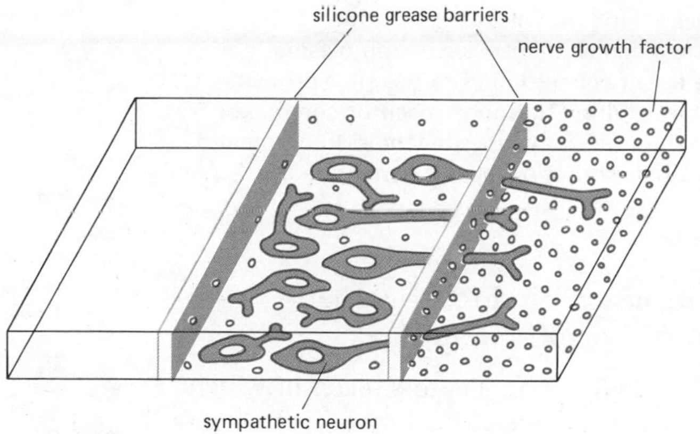

Figure 13–11 An experiment showing that nerve growth factor can influence the direction of nerve process outgrowth. When sympathetic neurons are placed in the central well of a three-chamber culture dish in which the chambers are separated by barriers of silicone grease, the cells are seen to extend nerve processes into the side chamber that contains nerve growth factor but not into the side chamber lacking the factor.

where normally they never go. An analogous phenomenon can be demonstrated *in vitro* if immature sympathetic neurons are placed in the central well of a three-chambered culture dish in which the chambers are separated by barriers of silicone grease, which is impermeable to NGF. Nerve fibers migrate beyond the silicone barriers into an adjacent chamber only if that chamber contains NGF (Figure 13–11).

Summary

Signaling molecules can be subdivided into three general classes according to their mechanism of delivery: (1) local chemical mediators are rapidly taken up or destroyed so that they act only on local cells; (2) hormones are carried in the blood to target cells throughout the body; and (3) neurotransmitters act only on the postsynaptic cell. Each cell type in the body contains a distinctive set of receptor proteins that enables it to bind and respond to a complementary set of signaling molecules in a preprogrammed and characteristic way.

The signaling molecules can also be classified according to their solubility in water. Hydrophobic signaling molecules, such as steroid hormones, pass through the plasma membrane and activate receptor proteins in the cell cytoplasm, while hydrophilic signaling molecules, including all neurotransmitters and the great majority of hormones and local chemical mediators, activate receptor proteins on the surface of the target cell.

Signaling Mediated by Intracellular Receptors: Mechanisms of Steroid Hormone Action

All steroid hormones are synthesized from cholesterol. Being relatively small (molecular weight $\simeq$ 300) hydrophobic molecules, they cross the plasma membrane by simple diffusion. Once inside the target cell, each type of steroid hormone binds tightly but reversibly to a different receptor protein present in the cytoplasm. The binding of the hormone causes the receptor protein to undergo an allosteric change in its conformation that increases its ability to bind to DNA. Since the receptor proteins are able to migrate through the nuclear pores, thier increased binding affinity for DNA causes the hormone-receptor complexes to accumulate in the cell nucleus (Figure 13–12).

The thyroid hormones are also small hydrophobic molecules that bind to intracellular receptors. They are thought to work like steroid hormones except that their receptors are concentrated in the nucleus even before hormone binding.

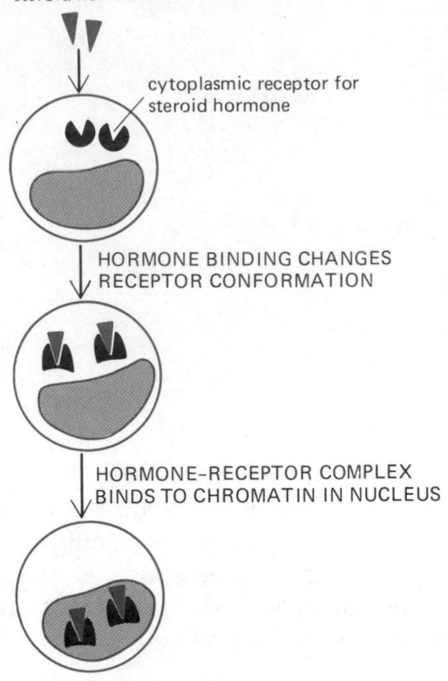

Figure 13–12 Schematic drawing of steroid hormone signaling. Steroid molecules diffuse across the plasma membrane of the target cell and bind to receptor proteins in the cytoplasm. The resulting hormone-receptor complexes then migrate into the nucleus where they bind to chromatin and regulate the transcription of specific genes.

Steroid-Hormone-Receptor Complexes Bind to Chromatin and Regulate the Transcription of Specific Genes[6]

A typical target cell contains about 10,000 steroid receptors, each of which will reversibly bind one molecule of a specific steroid hormone with high affinity (affinity constant $K = 10^8$ to 10^{10} liters per mole). When hormone levels are high, most receptors become complexed to hormone molecules and in this activated form bind to chromatin in the nucleus. As hormone levels fall, the equilibrium shifts so that hormone molecules dissociate from receptors and the freed receptors return to the cytoplasmic pool (Figure 13–13).

Some of the activated hormone receptors bound to chromatin regulate the transcription of specific genes. However, only a small number of genes in any target cell are directly influenced by steroid hormones. For example, 30 minutes after cultured rat liver cells are exposed to *cortisol*, only 6 of the 1000 proteins that can be distinguished by two-dimensional gel electrophoresis are increased in amount, and 1 is decreased. The synthesis rates of these proteins returns to normal once the hormone is removed. Assuming that about 10% of the cell's proteins can be detected on these gels, cortisol probably directly alters the transcription of only about 50 genes. Interestingly, the same hormone receptor is found to influence different sets of genes in different target cells (see following page).

Experiments performed *in vitro* have revealed that the binding of a specific steroid hormone induces an allosteric change in the steroid receptor (*receptor activation*) that greatly increases the affinity with which the receptor protein binds to DNA isolated from any source. Since steroid receptors presumably diffuse continuously in and out of the nucleus, it is thought that this increase in nonspecific binding to DNA is what causes activated receptors to accumulate in the nucleus. Once activated by hormone, the receptor is thought also to bind with a higher affinity to the few specific sites on the chromatin of the target cell that are relevant for regulating gene transcription; however, this cannot be observed directly.

One of the most decisive experiments supporting this view of steroid receptor action involves the use of mutant cell lines selected for an altered response to cortisol. The mutant cells were derived from lymphocyte tumors, called lymphomas. Like some normal lymphocytes, lymphoma cells are killed by the addition of low concentrations of cortisol to their growth medium. The

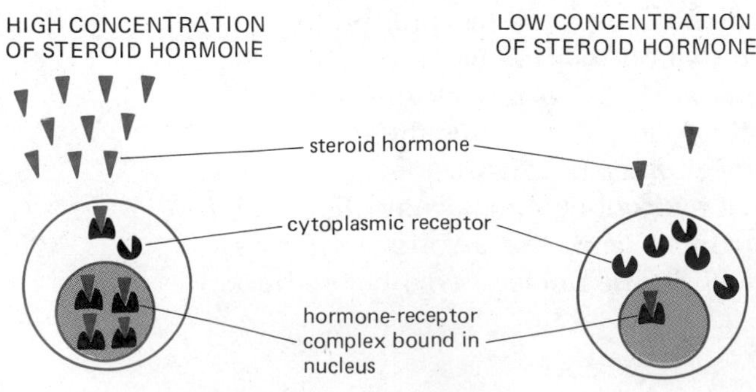

Figure 13–13 Schematic drawing showing that when steroid hormone levels are high, most of the steroid receptor proteins in a target cell are complexed to hormone and bound to chromatin in the nucleus; and when hormone levels are low, most of the receptors are free in the cytoplasm.

mutant cells, however, are resistant to the lethal effects of cortisol, and some of them are found to contain cortisol receptor proteins that fail to migrate from the cell cytoplasm to the nucleus after binding cortisol (Figure 13–14). While these mutant receptors have a normal ability to bind cortisol, they have lost most of their ability to bind to DNA. Experiments with a variety of different mutant receptor proteins show a strong correlation between the ability to bind DNA in the test tube and the extent of nuclear migration upon binding cortisol inside the cell. These experiments therefore suggest that DNA constitutes at least part of the recognition site for the receptor in the cell nucleus.

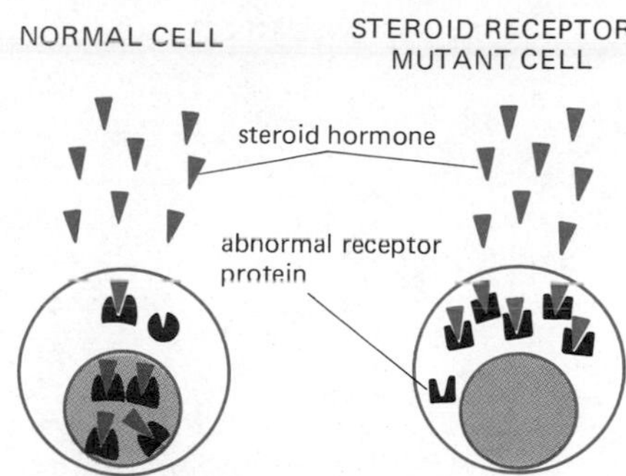

Figure 13–14 One type of steroid receptor mutant cell in which the abnormal receptors bind the hormone but then fail to localize in the nucleus.

While it is generally believed that the binding of steroid-hormone-receptor complexes to specific sites in chromatin regulates specific gene transcription, it is extremely difficult to identify such sites directly. The main difficulty is that the receptors bind nonspecifically to DNA; also, the binding of only a small fraction of the cell's 10,000 hormone-receptor complexes may suffice to regulate 50 genes. Moreover, even with 10,000 receptors per cell, the receptor proteins constitute only about one part in 50,000 by weight of the total cell protein, and so only small amounts of receptor have been obtained in a highly purified form. For these reasons, it has been difficult to determine unambiguously whether steroid hormone receptors derive their specificity from the recognition of special DNA sequences, special chromosomal proteins, or both. Recently, however, recombinant DNA techniques have been used to clone a gene that is regulated by cortisol, making large amounts of specific DNA available. When this DNA was tested, purified cortisol receptors were found to bind specifically to certain DNA sequences within and around the gene. This important experiment suggests that steroid receptors recognize specific DNA sequences when they induce transcription of a gene.

Steroid Hormones Often Induce Both Primary and Secondary Responses[7]

In many cases the response to a steroid hormone takes place in two steps. The direct induction of transcription of a few specific genes is known as the *primary response.* The products of these genes may in turn, however, activate other genes and produce a delayed *secondary response.* The latter may result in a major amplification of the initial hormonal effect.

A striking example is seen in the fruit fly *Drosophila.* Within five to ten minutes of the injection of the steroid insect molting hormone *ecdysone,* six major new sites of RNA synthesis (seen as *puffs*) are induced on the giant polytene chromosomes of the salivary gland (see p. 401). After a delay, some of the proteins produced during this primary response induce an additional 100 or so sites of RNA synthesis, leading to the synthesis of a large group of proteins characteristic of the secondary response. The response is controlled by feedback through one or more of the initial proteins, which shuts off further transcription of all of the primary response genes (Figure 13–15). It is likely that similar mechanisms provide for both amplification and control in many of the responses of mammalian cells to hormones.

Steroid Hormones Regulate Different Genes in Different Target Cells[8]

The responses to steroid hormones, like hormonal responses in general, are determined as much by the nature of the target cell as by the nature of the hormone. In principle, this observation has two possible explanations. Either different types of cells have different receptors for the same hormone, or the receptors are the same but the genes activated by them are different. There is strong evidence that the latter explanation is correct.

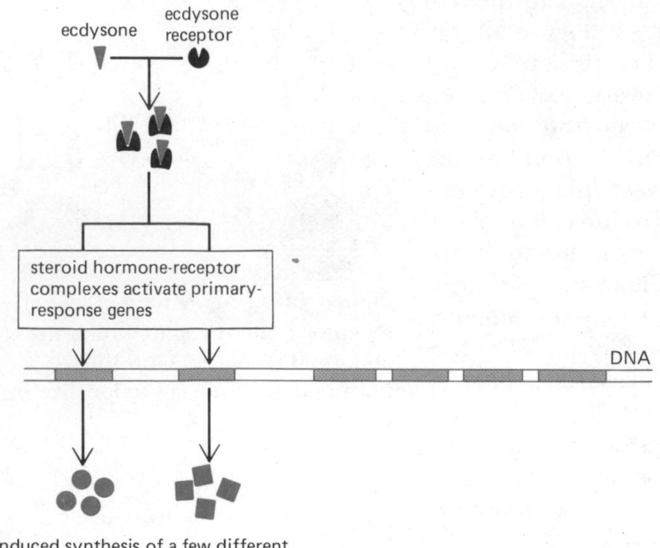

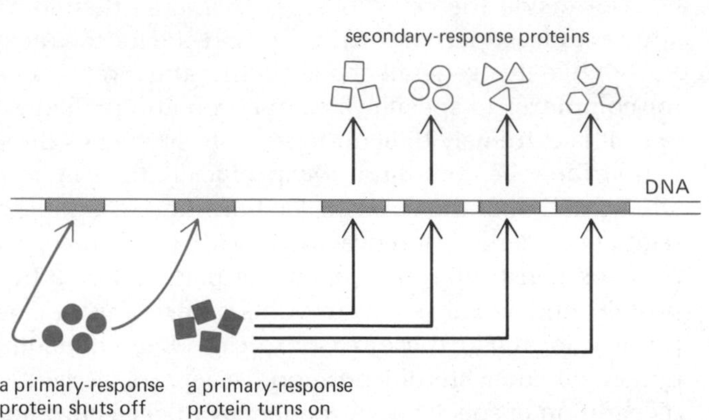

Figure 13–15 Schematic diagram of the early primary response to ecdysone in *Drosophila* cells (A) and the delayed secondary response (B). Some of the primary-response proteins turn on secondary-response genes, while others turn off the primary-response genes. The actual number of primary- and secondary-response genes is greater than shown.

The evidence comes mainly from studies of mammalian mutants in which a particular steroid hormone receptor is abnormal. In particular, a defect in the receptor for the male hormone *testosterone* causes genetically male individuals to appear to be females, since all mammals develop along female lines unless they are exposed to testosterone during embryonic development. Mutant males have normal testosterone-secreting testes, but because their tissues have defective testosterone receptors, they cannot respond to the hormone. They therefore develop all of the secondary sexual characteristics of females, and their testes remain in the abdomen and fail to descend. This **testicular feminization syndrome** occurs in mice, rats, and cattle, as well as in man. Although only the gene coding for testosterone receptors is abnormal, all of the many different cell types in the body that are normally influenced by testosterone are affected (Figure 13–16). It follows that the same testoster-

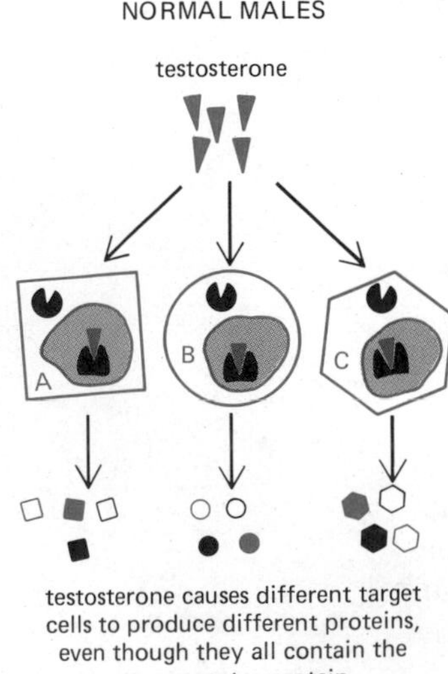

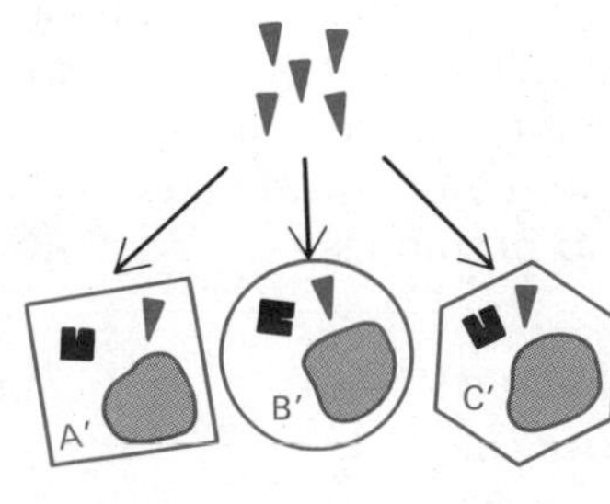

Figure 13–16 Different target cells containing the same receptor protein respond differently to testosterone. Thus a single gene defect in the testicular feminization syndrome, which leads to an abnormal testosterone receptor, results in all target cells failing to respond to testosterone.

one receptor protein must normally be present in all target cells, even though its activation regulates very different sets of genes in each type of cell.

The general principles of steroid hormone signaling are therefore clear. Each of the different steroid hormones induces a characteristic set of responses because (1) only cells designed to respond make receptors for a particular hormone and (2) the chromatin of the target cell is probably organized so that the steroid receptor activates only the appropriate genes (see Chapter 18). Since the genes available for activation are different in each type of target cell, the same hormone binding to the same steroid receptor protein has different effects on different cells.

Summary

Steroid hormones are small, hydrophobic molecules derived from cholesterol that are solubilized by binding reversibly to specific carrier proteins in the blood. Once released from their carrier proteins, they diffuse through the plasma membrane of the target cell and bind reversibly to specific steroid hormone-receptor proteins in the cytosol. By complexing to the hormone, the receptor protein acquires an affinity for DNA that causes it to accumulate in the cell nucleus. There the hormone-receptor complex binds to chromatin and regulates the transcription of a small number of genes. The products of some of these genes may, in turn, activate other genes and produce a delayed secondary response, thereby amplifying the initial effect of the hormone. Each steroid hormone is recognized by a different receptor protein, but the same receptor protein regulates different genes in different target cells. This suggests that the chromatin of each cell type is organized so as to make only the appropriate genes available for regulation by the hormone-receptor complex.

Signaling Mediated by Cell-Surface Receptors: Cyclic AMP and Calcium Ions as Second Messengers

Water-soluble signaling molecules, including all of the known neurotransmitters, protein hormones, and growth factors, bind to specific receptor proteins on the surface of the target cells they influence. These cell-surface receptors bind the signaling molecule (the ligand) with high affinity and convert this extracellular event into an intracellular signal that alters the behavior of the target cell. Since these receptors are insoluble integral membrane proteins and usually constitute less than 1% of the total protein mass of the plasma membrane, they are difficult to isolate and study.

The Use of Labeled Ligands Revolutionized the Study of Cell-Surface Receptors[9]

Attempts to use radiolabeled ligands to demonstrate receptors on the surface of target cells began in the 1950s but were hampered by two problems: (1) the process of coupling the signaling molecule to a radioactive isotope (usually radioactive iodine or hydrogen) greatly reduced its functional activity; (2) most of the binding to the target cell surface was nonspecific, so that only a small fraction of the total amount of bound ligand was attached to specific receptors. Around 1970, solutions to these technical problems made it possible to demonstrate directly the presence of specific receptors on the surface of intact cells and isolated membranes.

By using ligands labeled with radioactive atoms, fluorescent dyes, or electron-dense molecules (such as ferritin), it is now possible to study the

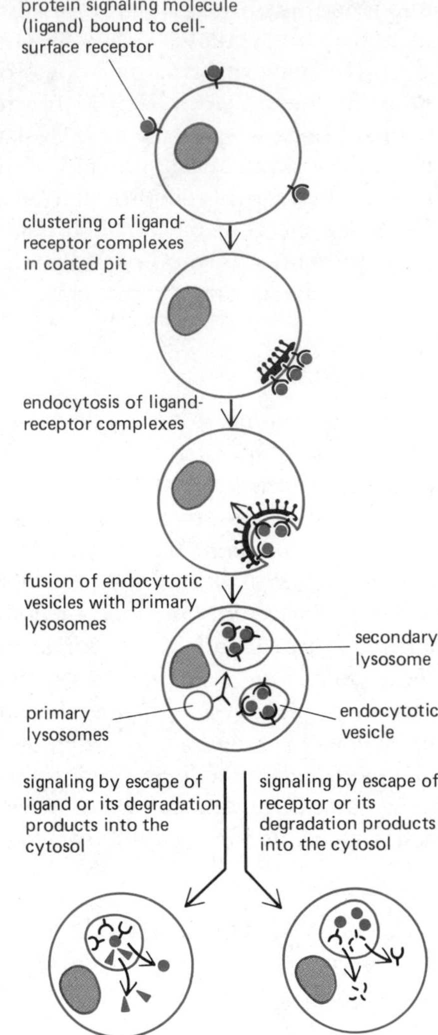

Figure 13–17 Hypothetical scheme showing what would be needed in order for protein ligands or their cell-surface receptors to act as their own intracellular mediators: a special mechanism is required in order for the ligand or receptor (or their degradation products) to escape from an intracellular vesicle compartment into the cytosol.

numbers, distribution, and properties of specific cell-surface receptors as well as to follow the fate of the ligand-receptor complexes after binding. It has been shown that the number of receptors for a specific ligand can vary from 500 to more than 100,000 per cell and that the initial receptor distribution can be either diffuse or localized to specific regions of the plasma membrane.

Protein Hormones and Growth Factors Are Ingested by Receptor-mediated Endocytosis[10]

Experiments with labeled ligands have demonstrated that many protein signaling molecules enter target cells by receptor-mediated endocytosis (see p. 309). For example, insulin binds to receptor proteins diffusely distributed on the surface of fibroblasts; within minutes the insulin-receptor complexes cluster in coated pits and are ingested in endocytotic vesicles. It is therefore conceivable that such protein signaling molecules (or their degradation products) act directly within the cell, much as steroid and thyroid hormones do. However, it must be remembered that receptor-mediated endocytosis usually results in the transfer of extracellular molecules to lysosomes (p. 309). In order to enter the cytosol, these hydrophilic molecules would require some special mechanism for escaping from either the endocytotic vesicle or the lysosome (Figure 13–17).

It is possible to show by a direct experiment that at least some signaling molecules do not have to enter cells in order to influence them. For example, the effects of insulin can be exactly mimicked by specific antibodies that bind to insulin receptors on the surface of target cells. Therefore, although insulin is normally endocytosed by target cells, the ingested hormone cannot itself be the intracellular signal. Similarly, while *thyroid-stimulating hormone* (TSH) normally activates thyroid cells to synthesize and secrete thyroid hormone, antibodies binding to the TSH receptors on the surface of thyroid cells can be just as effective (Figure 13–18). In fact, such antibodies are the usual cause of hyperthyroid disease in man, a condition in which too much thyroid hor-

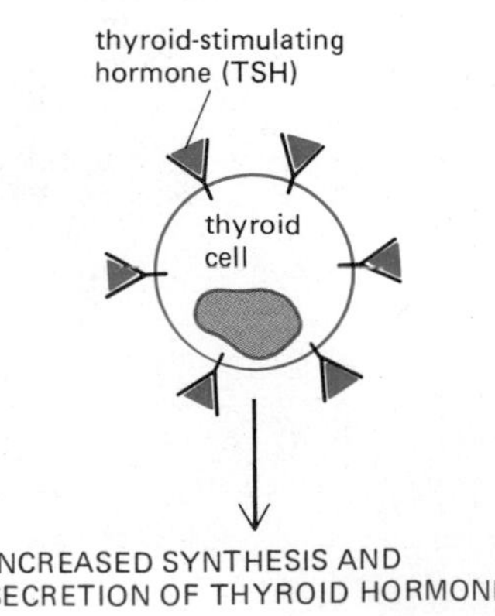

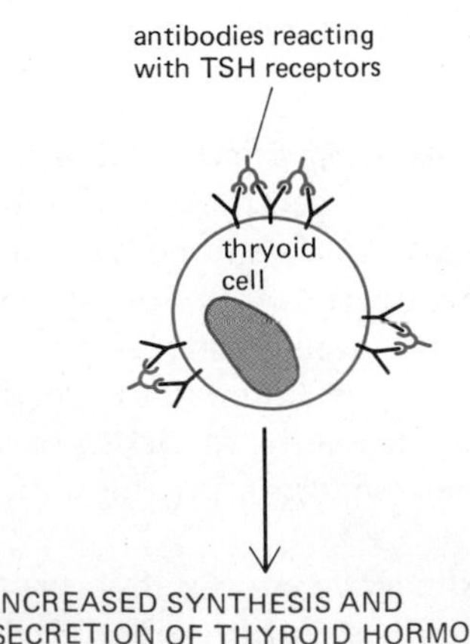

Figure 13–18 Antibodies directed against cell-surface receptors for thyroid-stimulating hormone (TSH) mimic the effects of TSH binding to these receptors. This suggests that TSH is not required for the signaling process and therefore probably does not act as its own intracellular mediator.

mone is produced. While these observations demonstrate that insulin and TSH do not have to enter the cell in order to act, they do not exclude the possibility that the cell-surface receptors for these hormones may act as intracellular signals, since the receptors are normally endocytosed along with either the bound hormone or the antibody (Figure 13–17).

Many hydrophilic signaling molecules induce responses that occur much too rapidly to involve receptor-mediated endocytosis. Mast cells secrete histamine within seconds of ligand binding to surface receptors, while the responses to some neurotransmitters occur in milliseconds. In these cases the receptors must instead act as transmembrane transducers to generate new intracellular signals.

Cell-Surface Receptors Act as Transducers by Regulating Enzymes or Ion Channels in the Plasma Membrane[11]

The great majority of cell-surface receptors that bind hydrophilic signaling molecules are thought to undergo a conformational change when they bind to a ligand at the cell exterior. This change leads to the generation of an intracellular signal that alters the behavior of the target cell. The intracellular signaling molecules are often referred to as **second messengers,** the *first messengers* being the extracellular ligands themselves.

There are two general ways in which cell-surface receptors are known to generate intracellular signals. One is by activating or inactivating a plasma-membrane-bound enzyme. In some cases this enzyme catalyzes the production of a soluble intracellular mediator; the change in the intracellular concentration of the mediator then serves as the signal. An important enzyme that acts in this way is *adenylate cyclase,* which catalyzes the synthesis of *cyclic AMP* (*cAMP*) from ATP on the cytoplasmic side of the plasma membrane (Figure 13–19). In other cases the enzyme activated by an extracellular ligand directly causes the phosphorylation of cellular proteins. For example, *epidermal growth factor* (*EGF*) stimulates epidermal cells and a variety of other cell types to divide by binding to receptor proteins on the cell surface. The receptors are (or are closely associated with) protein kinases that are activated by the binding of EGF to transfer a phosphate from ATP to a tyrosine residue on specific cellular proteins, including the receptor protein itself and other plasma membrane proteins as well as some cytosolic proteins.

Alternatively, cell-surface receptors may open or close gated ion channels in the plasma membrane. This generates a signal in either of two ways: (1) it causes a small and transient flux of ions that briefly changes the voltage across the plasma membrane, or (2) it causes a major influx of ions into the cytosol, which in turn initiates an intracellular response. The first mechanism operates mainly in electrically active cells such as neurons and muscle cells. For example, most neurotransmitters regulate the membrane potential of the postsynaptic target cell by opening or closing ion channels in the target cell plasma membrane: a decrease in the membrane potential below a certain threshold level triggers an explosive depolarization of the membrane (an *action potential*), which rapidly spreads to the rest of the target cell membrane. These changes in membrane potential are not accompanied by an appreciable change in ion concentration in the cytosol, so that the initial signal localized at the postsynaptic plasma membrane is not converted into a truly intracellular signal until the action potential reaches the nerve terminal. Then, because the plasma membrane of the nerve terminal contains voltage-gated Ca^{2+} channels that are transiently opened when the membrane is depolarized by the action potential, Ca^{2+} enters the terminal down its very steep electrochemical gradient and acts as a second messenger to initiate neurotransmitter secretion (see Chapter 18).

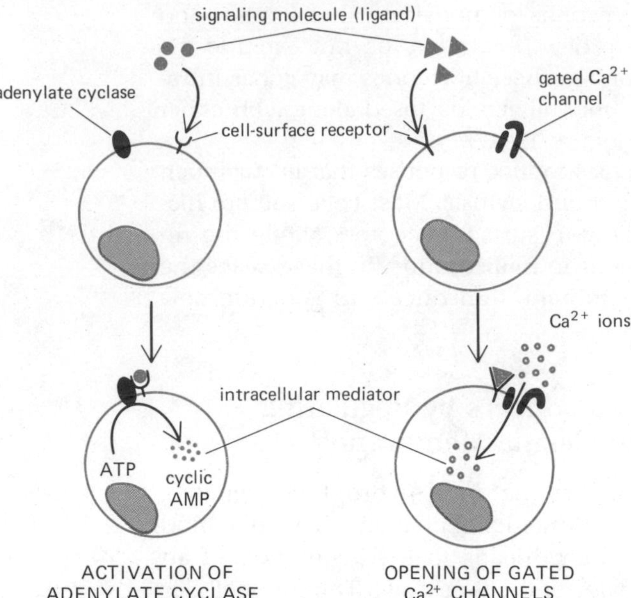

Figure 13–19 Two common mechanisms by which cell-surface receptors generate intracellular signals: (1) the activation of membrane-bound adenylate cyclase molecules increases the intracellular concentration of cyclic AMP; and (2) the opening of membrane-bound, gated Ca^{2+} channels allows Ca^{2+} to enter the cell. For simplicity, the receptors are shown to interact directly with the cyclase molecules or ion channels following ligand binding. In fact, other membrane proteins mediate the coupling of receptors to the cyclase molecules and possibly also to Ca^{2+} channels.

Many animal cells that are not electrically active have cell-surface receptors that are functionally linked to Ca^{2+} channels in the plasma membrane: ligand binding activates these receptors, thereby opening the channels and allowing Ca^{2+} to enter the cytosol, where it then functions as a second messenger (Figure 13–19).

Cyclic AMP Is a Ubiquitous Intracellular Mediator[12]

Cyclic AMP (Figure 13–20) regulates intracellular reactions in all procaryotic and nucleated animal cells that have been studied to date. Although it serves as an important intracellular signaling molecule, it seems not to be required for cell survival or division, since some mutant eucaryotic cell lines that make no detectable cyclic AMP grow normally in culture.

The identification of cyclic AMP as a common intracellular mediator for various hormones was a major advance. The first evidence for such a soluble intracellular mediator came from studies on the effects of the hormone *epinephrine* on glycogen metabolism in liver cells. It was found that epinephrine causes the activation of the enzyme *glycogen phosphorylase,* which catalyzes

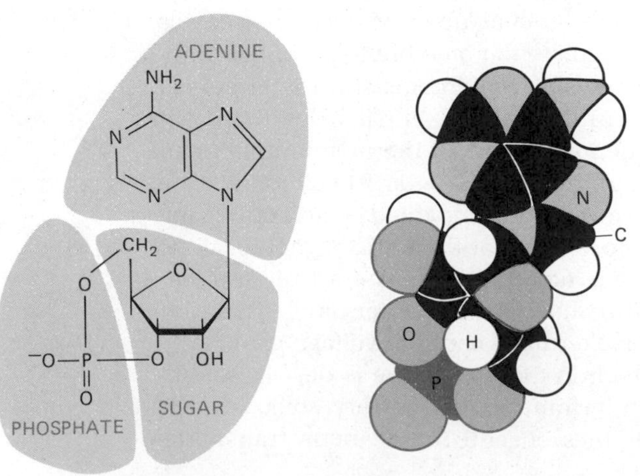

Figure 13–20 Cyclic AMP shown as a formula and as a space-filling model. (C, H, N, O, and P indicate carbon, hydrogen, nitrogen, oxygen, and phosphorus atoms, respectively.)

the breakdown of glycogen. It was then possible to show that treating isolated liver cell membranes with epinephrine (in the presence of ATP) induced the production of a small, heat-labile mediator that could substitute for the hormone and activate the phosphorylase present in a membrane-free extract of liver cells. The mediator was identified in 1959 as cyclic AMP.

For cyclic AMP to function as an intracellular mediator, its intracellular concentration (normally $\leq 10^{-6}$ M) must be tightly controlled and able to change rapidly in response to extracellular signals: upon hormonal stimulation, cyclic AMP levels can change by fivefold in seconds. Cyclic AMP is synthesized from ATP by the plasma-membrane-bound enzyme **adenylate cyclase,** but it is also rapidly destroyed in cells by one or more specific enzymes called **phosphodiesterases,** which hydrolyze cyclic AMP to adenosine 5′-monophosphate (5′-AMP) (Figure 13–21). Cell-surface receptors for which cyclic AMP is the intracellular messenger act by altering (usually stimulating) the activity of adenylate cyclase rather than by altering phosphodiesterase activity. However, the continuous rapid breakdown or removal of any intracellular mediator is also required in order to obtain either a rapid increase or a rapid decrease in its concentration, as explained below (see p. 750).

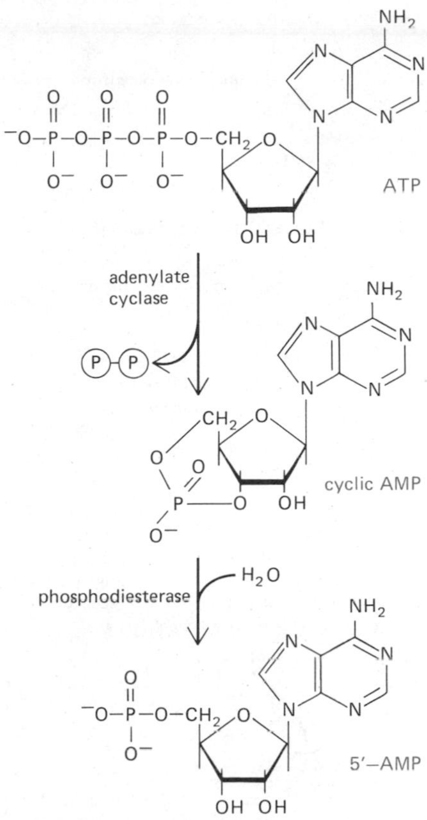

Figure 13–21 The synthesis and degradation of cyclic AMP.

Receptor and Adenylate Cyclase Molecules Are Separate Proteins That Functionally Interact in the Plasma Membrane[13]

Many hormones and local chemcial mediators work by activating adenylate cyclase. Several examples, and the effects they produce, are listed in Table 13–2. Just as the same steroid hormone produces different effects in different target cells, so different target cells respond very differently to changes in their intracellular cyclic AMP levels.

Since each type of animal cell responds to an increase in cyclic AMP in a characteristic way, any ligand that activates adenylate cyclase in a given target cell usually produces the same effect. For example, at least four different hormones activate adenylate cyclase in fat cells, and all of them stimulate the breakdown of triglyceride (the storage form of fat) to fatty acids (Table 13–2). This can be explained in two ways: either each different type of receptor is tightly linked to its own adenylate cyclase molecule in the plasma membrane, or the different receptors share a common pool of adenylate cyclase molecules. That the latter is the case is suggested by receptor "transplantation" experiments. For example, epinephrine receptors isolated from detergent-solubilized plasma membranes can be transplanted to the plasma membrane of

Table 13–2 Some Hormone-induced Effects in Different Target Cells Mediated by Cyclic AMP

Target Tissue	Hormone	Major Response
Thyroid	thyroid-stimulating hormone	thyroxine secretion
Adrenal cortex	adrenocorticotropic hormone (ACTH)	cortisol secretion
Ovary	luteinizing hormone	progesterone secretion
Muscle, Liver	epinephrine	glycogen breakdown
Bone	parathormone	bone resorption
Heart	epinephrine	increase in heart rate
Kidney	vasopressin	water resorption
Fat	epinephrine, ACTH, glucagon, thyroid-stimulating hormone	triglyceride breakdown

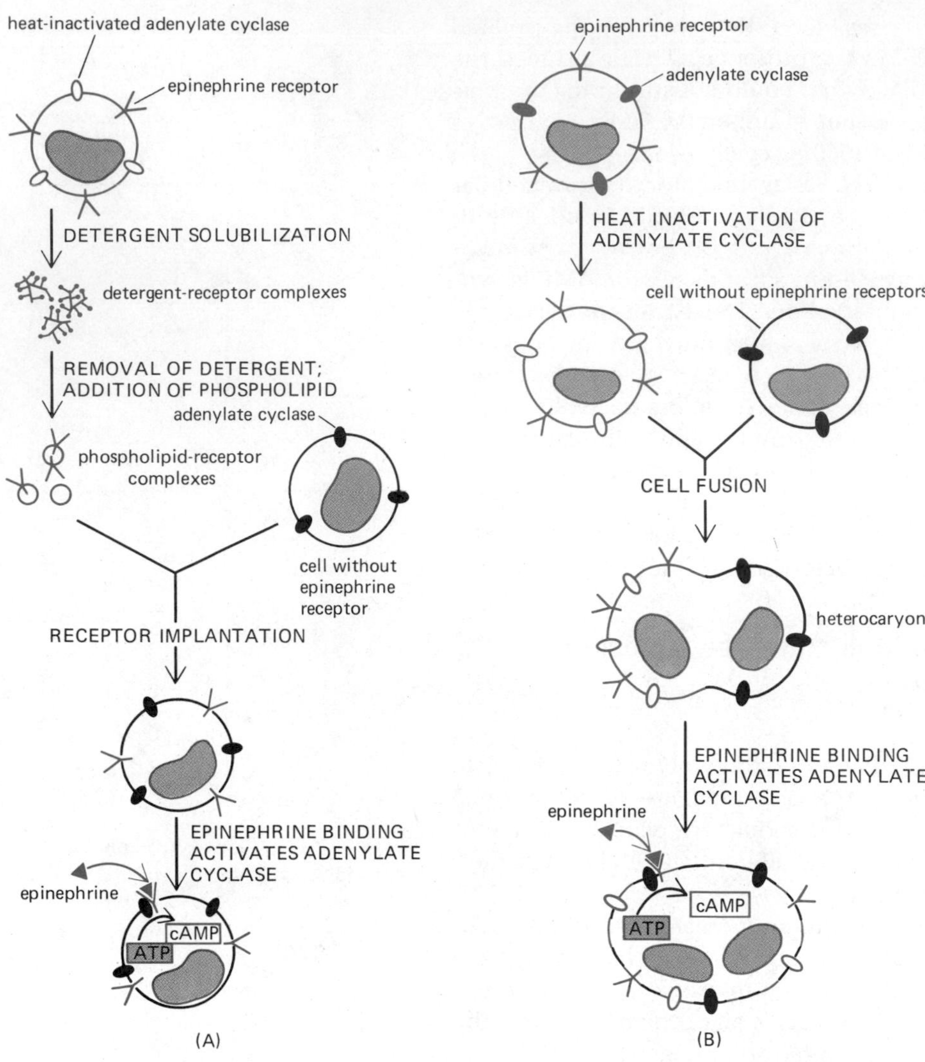

Figure 13–22 Experiments in which receptor proteins and adenylate cyclase molecules from different cells have been demonstrated to interact functionally in the plasma membrane of an intact cell, either by the "transplantation" of solubilized receptors (A) or by cell fusion (B).

other cells that do not have their own epinephrine receptors; such transplanted receptors are able to functionally interact following receptor activation (Figure 13–22A).

Earlier analogous experiments used cell fusion instead of receptor transplantation to unite receptor proteins with cyclase molecules. Cells with epinephrine receptors but whose adenylate cyclase molecules had been inactivated (by heat or chemical treatment) were fused with cells that did not have epinephrine receptors but had intact cyclase molecules. Within minutes after fusion, such heterocaryons showed a marked activation of adenylate cyclase in response to epinephrine, again suggesting that receptors from one cell membrane can functionally interact with cyclase molecules contributed by another cell membrane (Figure 13–22B). Thus receptors and cyclase molecules are thought to be separate proteins that functionally interact following receptor activation.

Receptors Activate Adenylate Cyclase Molecules Indirectly Through a GTP-binding Protein[14]

Activated receptors do not activate adenylate cyclase directly; a third membrane protein, which binds GTP on its cytoplasmic surface, couples the two together. This mechanism was initially suggested by the finding that hormonal activation of adenylate cyclase in disrupted cells requires GTP. Evidence for a GTP-binding protein has since come from the isolation of specific mutant cell lines in which binding of epinephrine fails to activate adenylate cyclase in spite of normal levels of epinephrine receptors and of cyclase. By mixing plasma membrane preparations from such "uncoupled" cells with detergent extracts of plasma membranes from other cells, a hormone-sensitive adenylate cyclase system that requires GTP can be reconstituted. There is good evidence that the detergent extracts in these experiments contain a special membrane protein, the **GTP-binding protein (G protein),** that is missing from the "uncoupled" mutant cells and is required in order to couple receptor activation to adenylate cyclase activation.

Activation of adenylate cyclase by hormone receptors must persist only as long as the ligand is present if cells are to be able to respond rapidly to changes in hormone levels. This type of response is assured because the G protein is a GTPase that self-inactivates by hydrolyzing its bound GTP to GDP and inorganic phosphate. If GTP analogues in which the terminal phosphate cannot be hydrolyzed are used instead of GTP in physically disrupted normal cells, the activation of adenylate cyclase by hormones is increased and greatly prolonged.

The bacterial toxin responsible for the symptoms of cholera inhibits this normal "shut-off" mechanism. **Cholera toxin** is an enzyme that catalyzes the transfer of ADP-ribose from intracellular NAD^+ to the G protein, altering it so that it can no longer hydrolyze its bound GTP. This means that once an adenylate cyclase molecule has been activated by a hormone-receptor complex and a molecule of G protein, it remains irreversibly in the active state. The resulting prolonged elevation in intracellular cyclic AMP levels within intestinal epithelial cells causes a large efflux of Na^+ and water into the gut, which is responsible for the severe diarrhea caused by cholera toxin.

Thus it appears that adenylate cyclase activation requires at least three plasma membrane-bound proteins, which interact in the following sequence: (1) Hormone binding to the receptor protein alters the conformation of the receptor, enabling it to bind to and activate the G protein when the two protein molecules collide in the lipid bilayer. (2) The latter protein then becomes able to bind GTP (in place of GDP) at its cytoplasmic surface; the binding of GTP changes the conformation of the G protein so that it can activate an adenylate cyclase molecule to synthesize cyclic AMP. (3) To complete the cycle, the G protein hydrolyzes the bound GTP to GDP, which returns the cyclase to its original inactive state. One model of how the coupling process may operate is shown in Figure 13–23.

Why have cells evolved such complex, multistep mechanisms for signal transduction? Why interpose the G protein between the receptor and the enzyme it is to activate? One reason is that each activated receptor protein is able to collide with and activate many molecules of G protein, thereby greatly amplifying the initial extracellular signal. Another possibility is that this scheme, in principle, allows a single type of hormone receptor to be functionally coupled to a number of different membrane enzymes and ion channels besides adenylate cyclase. For example, the G protein functions as an adapter, and it may be that different adapters are made in different cell types, allowing the same hormone receptor protein to be coupled to different target proteins in the plasma membrane of different cells.

Ca^{2+} Also Functions as a Ubiquitous Intracellular Mediator[15]

There is increasing evidence that Ca^{2+}, like cyclic AMP, is an important intracellular regulator and that it functions as a second messenger for certain extracellular signaling molecules. The first evidence for this role came from

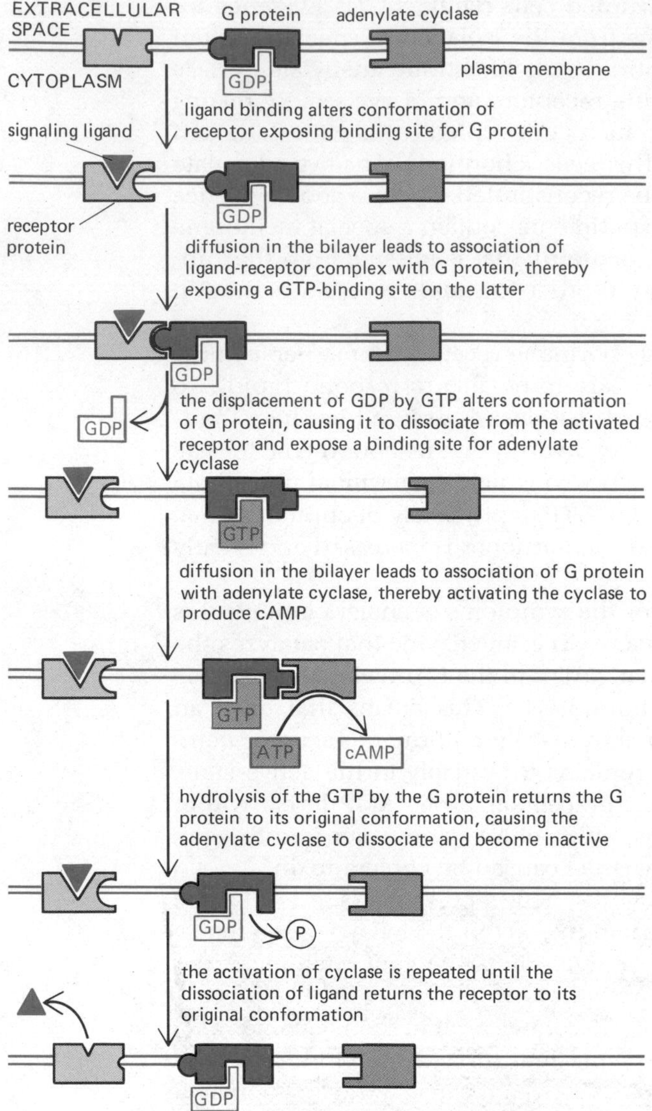

Figure 13–23 The "collision-coupling" model of adenylate cyclase activation. While in this version of the model the receptor proteins, adenylate cyclase molecules, and GTP-binding (G) proteins are all assumed to diffuse independently in the plasma membrane of the target cell and to interact only after ligand binds to the receptor proteins, it is also possible that the G protein and cyclase are permanently associated. Although not shown in the figure, each activated receptor protein activates many molecules of G protein, thereby greatly amplifying the response.

an experiment done in 1947 showing that the intracellular injection of a small amount of Ca^{2+} causes a skeletal muscle cell to contract. Like cyclic AMP, the concentration of free Ca^{2+} in the cytosol is normally very low. While the total concentration of Ca^{2+} in cells is similar to that of extracellular Ca^{2+} ($\geq 10^{-3}$ M), the concentration of free Ca^{2+} in the cytosol ($\leq 10^{-7}$ M) is more than a thousandfold lower because most of the Ca^{2+} in cells is bound to other molecules or sequestered in mitochondria and other intracellular organelles. This means that there is an enormous gradient in the concentration of free Ca^{2+} across the plasma membrane of animal cells that tends to drive Ca^{2+} into the cell.

In order to maintain this gross imbalance, any net influx of Ca^{2+} across the plasma membrane must be matched by a net efflux across this membrane. This is achieved largely by a plasma-membrane-bound Ca^{2+}-ATPase that uses the energy of ATP hydrolysis to pump Ca^{2+} out of the cell. In some cells, Ca^{2+} is also actively expelled by other plasma-membrane-bound pumps that act as Na^+-driven antiports, coupling the efflux of Ca^{2+} to the influx of Na^+ (Figure 13–24A).

However, transport across the plasma membrane is not the only mechanism for removing Ca^{2+} from the cytosol. In fact, the area of the plasma membrane is generally 10 to 100 times less than the combined areas of the various membranes that enclose Ca^{2+}-sequestering organelles in the cell. Membrane-bound Ca^{2+}-ATPases enable the endoplasmic reticulum (especially the sarcoplasmic reticulum of muscle cells) to take up large amounts of Ca^{2+} from the cytosol against a steep concentration gradient, while mitochondria use the electrochemical gradient across the inner mitochondrial membrane, generated during the electron transfer steps of oxidative phosphorylation, to drive the uptake of Ca^{2+} from the cytosol (see p. 497).

In addition to the various membrane transport proteins that actively pump Ca^{2+} out of the cytosol, a variety of Ca^{2+}-binding molecules contribute to the removal of free Ca^{2+}. These include small molecules such as phosphates, as well as macromolecules such as Ca^{2+}-binding proteins (see below). The various mechanisms for maintaining a very low concentration of free Ca^{2+} in the cytosol are summarized in Figure 13–24.

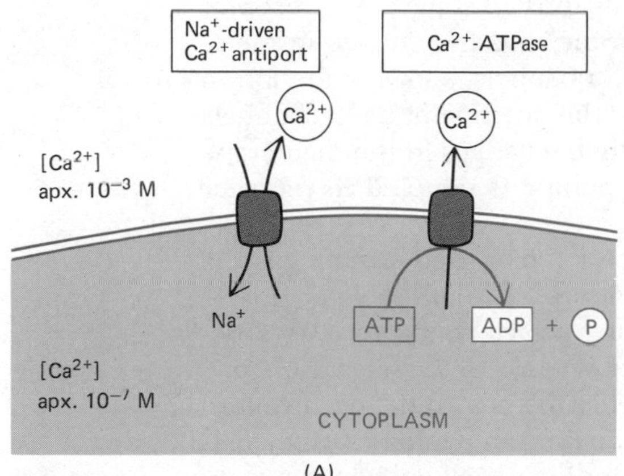

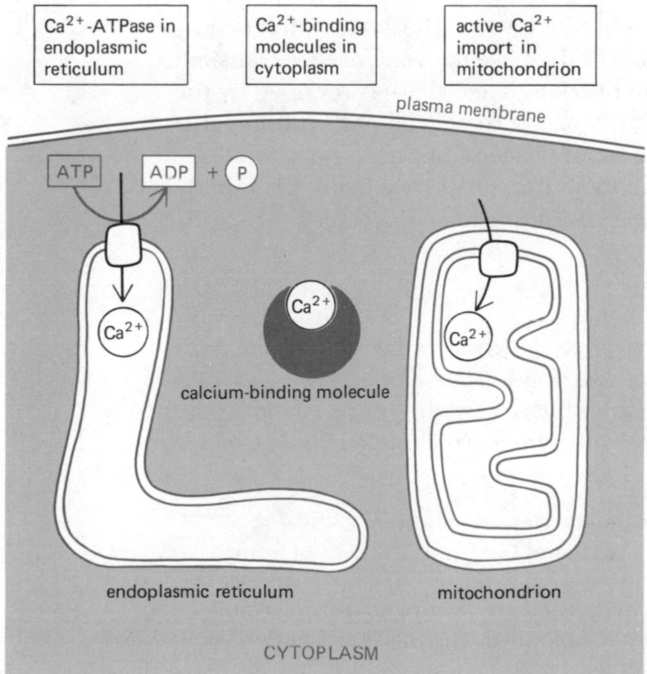

Figure 13–24 The main ways in which cells maintain a very low concentration of free Ca^{2+} in the cytosol in face of high concentrations of Ca^{2+} in the extracellular fluid. Ca^{2+} is actively pumped out of the cytosol to the cell exterior (A), as well as into intracellular, membrane-enclosed organelles such as the endoplasmic reticulum and mitochondria (B). In addition, various molecules in the cell bind free Ca^{2+} tightly.

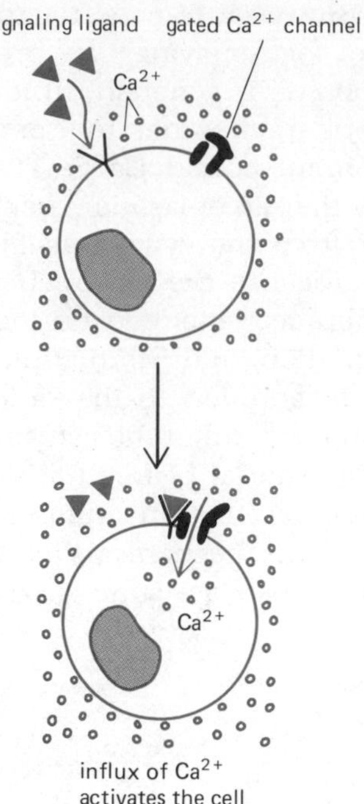

Figure 13–25 The functional coupling of cell-surface receptors to Ca^{2+} channels in the plasma membrane. Ligand binding to such receptors opens the channels, thereby allowing Ca^{2+} to enter the cell down its steep electrochemical gradient. Although for the sake of simplicity ligand-receptor complexes are shown directly binding to the Ca^{2+} channels, the actual mechanism of coupling between the receptors and the channels is unknown.

The Activation of Some Cell-Surface Receptors Opens Membrane-bound Ca^{2+} Channels[16]

Cells utilize the very large Ca^{2+} gradients across their plasma membranes for transducing extracellular signals. Just as some cell-surface receptors are functionally coupled to adenylate cyclase molecules, others are coupled to Ca^{2+} channels in the plasma membrane. The transient opening of such channels following receptor activation allows Ca^{2+} to enter the cytosol, where it acts as a second messenger (Figure 13–25). This mechanism operates in most secretory cells that are activated by extracellular ligands. In some cases, receptor activation first depolarizes the plasma membrane, and the change in membrane potential opens voltage-dependent Ca^{2+} channels, allowing the ion to enter the cell from the outside. In other cases, the opening of Ca^{2+} channels in the plasma membrane occurs independently of changes in the membrane potential, and it is uncertain how channel opening is coupled to receptor activation.

An important alternative mechanism for Ca^{2+}-mediated signaling is the release of Ca^{2+} from intracellular organelles (Figure 13–26). In skeletal muscle cells, for example, the activation of cell-surface acetylcholine receptors depolarizes the plasma membrane, which somehow leads to the release of Ca^{2+} from the sarcoplasmic reticulum of these cells, thereby initiating myofibril contraction. Similarly, in some secretory cells the Ca^{2+} that initiates exocytosis is released from internal stores. How ligand binding to cell-surface receptors results in the opening of Ca^{2+} channels in internal membranes is still a mystery.

The increase in the concentration of free Ca^{2+} in the cytosol that occurs during cell signaling is always transient. In the case of cyclic AMP, a transient response occurs because the activation of adenylate cyclase is reversed by the hydrolysis of GTP by the G protein and because the cyclic AMP produced is rapidly destroyed by phosphodiesterases. In the case of Ca^{2+}, the gated Ca^{2+} channels open only transiently and the Ca^{2+} that enters the cytosol is rapidly pumped out and/or buffered by binding to intracellular molecules.

Summary

Most cell-surface receptor proteins activated by extracellular signaling ligands generate intracellular signals in one of two ways. They alter the activity of a plasma-membrane-bound enzyme, which in turn alters the concentration of an intracellular mediator. Alternatively, they alter the permeability of ion channels

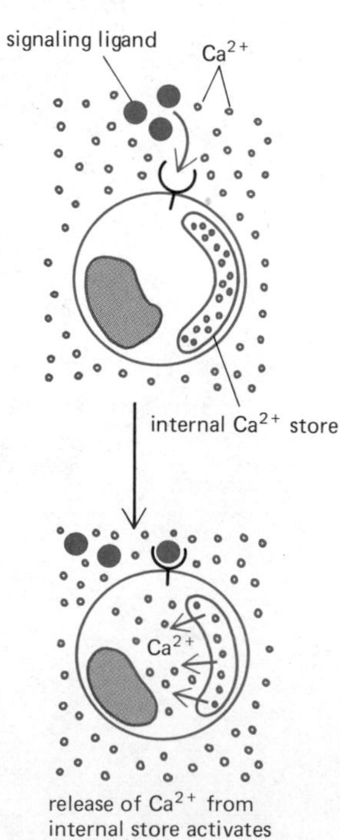

Figure 13–26 Schematic drawing showing cell-surface receptors acting by releasing Ca^{2+} into the cytosol from internal stores. In muscle cells, the internal Ca^{2+} store is known to be the sarcoplasmic reticulum; in other cells the location of the store is unknown. The mechanisms involved in coupling receptor activation on the cell surface to Ca^{2+} release in the cell interior are also unknown.

in the plasma membrane, thereby changing the membrane potential, the intracellular concentration of the transported ion, or both.

Many cell-surface receptors are functionally linked to the membrane-bound enzyme adenylate cyclase. When these receptors are activated by the binding of an extracellular signaling ligand, the ligand-receptor complex initiates a series of protein interactions in the plasma membrane that increases (or in some cases, decreases) the synthesis of cyclic AMP by adenylate cyclase. The resulting increase in intracellular cyclic AMP concentration persists only as long as the ligand is present because the adenylate cyclase activation is short-lived and the cyclic AMP produced is rapidly hydrolyzed by phosphodiesterases.

The activation of other types of cell-surface receptors opens membrane-bound Ca^{2+} channels, resulting in an influx of Ca^{2+} into the cytosol either from the extracellular fluid or from internal Ca^{2+} stores. The rise in intracellular Ca^{2+} is transient because the activated Ca^{2+} channels open only transiently, and the Ca^{2+} that enters the cytosol is rapidly pumped out and/or buffered by binding to Ca^{2+}-binding molecules.

The Mode of Action of Cyclic AMP and Calcium Ions as Second Messengers

Although cyclic AMP and Ca^{2+} are not the only intracellular mediators of extracellular signals, they are so commonly used as second messengers that their modes of action inside the cell merit special consideration. Moreover, they may well turn out to play a part in direct intercellular communication mediated by gap junctions or by plasma-membrane-bound signaling molecules, as well as in communication mediated by secreted signals.

Cyclic AMP Activates Intracellular Protein Kinases[12,17]

Cyclic AMP exerts its effects in animal cells by activating specific cellular enzymes called **cyclic-AMP-dependent protein kinases,** which catalyze the transfer of a phosphate group from ATP to a specific serine or threonine residue of a small group of proteins in the target cell. The covalent phosphorylation in turn regulates the activity of these proteins.

This sequence of events was first revealed by studies of glycogen metabolism in skeletal muscle cells. Glycogen is the major storage form of glucose, and its synthesis and degradation are closely regulated by specific hormones. For example, when an animal is frightened or otherwise stressed, the adrenal gland secretes epinephrine into the blood, "alerting" various tissues in the body. Among other effects, the circulating epinephrine induces muscle cells to break down glycogen to glucose 1-phosphate and at the same time to stop synthesizing new glycogen. Glucose 1-phosphate is converted to glucose 6-phosphate, which is then oxidized by glycolysis to provide ATP for sustained muscle contraction. In this way epinephrine prepares the muscle cells for strenuous activity.

Epinephrine works by activating adenylate cyclase in the muscle cell plasma membrane and thereby increases the level of cyclic AMP in the cytoplasm. The cyclic AMP in turn activates a cyclic-AMP-dependent protein kinase that specifically phosphorylates *glycogen synthase,* the enzyme that performs the final step in glycogen synthesis from glucose. This phosphorylation inactivates glycogen synthase and thereby shuts off glycogen synthesis. The same protein kinase also phosphorylates another enzyme, *phosphorylase kinase,* activating it to phosphorylate another enzyme, *glycogen phosphorylase.* This last phosphorylation activates the phosphorylase, causing it to remove glucose residues from the glycogen molecule.

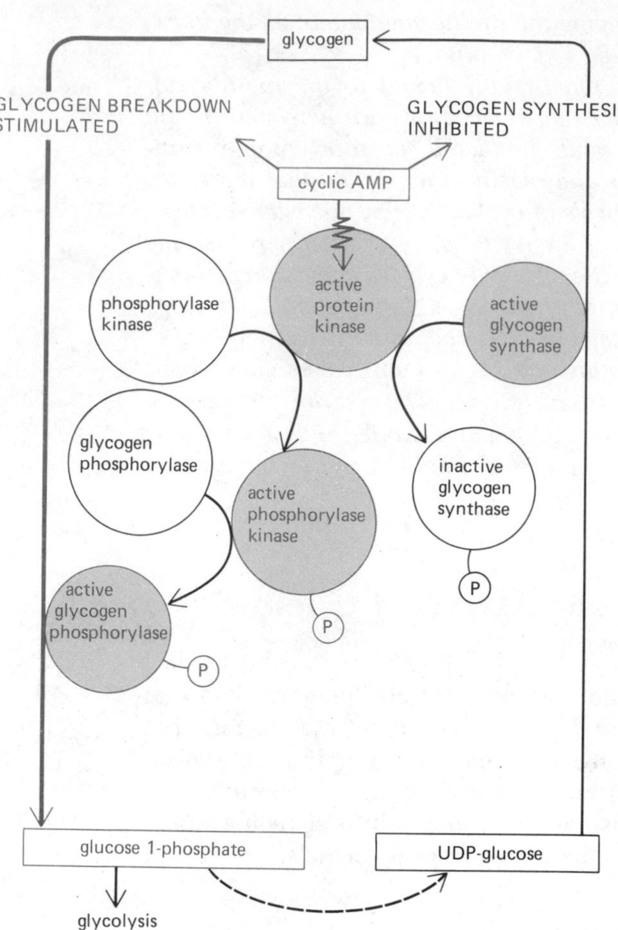

Figure 13–27 How an increase in cyclic AMP in skeletal muscle cells (induced by epinephrine binding to cell-surface receptors) stimulates glycogen breakdown and inhibits glycogen synthesis. The binding of cyclic AMP to a specific protein kinase activates this enzyme to phosphorylate and thereby activate phosphorylase kinase, which in turn phosphorylates and activates glycogen phosphorylase, the enzyme that breaks down glycogen. The same cyclic-AMP-dependent protein kinase phosphorylates and thereby *inactivates* glycogen synthase, the enzyme involved in glycogen synthesis. The active forms of these four enzymes are shown in color.

By means of this cascade of interactions, an increase in cyclic-AMP levels both inhibits glycogen synthesis and stimulates glycogen breakdown, thus maximizing the amount of glucose available to the cell (Figure 13–27). The regulation of glycogen metabolism by phosphorylation illustrates a general rule: enzymes *activated* by phosphorylation are usually concerned with degradative metabolic pathways, while those *inactivated* by phosphorylation are usually concerned with synthetic pathways.

The protein kinase activated by cyclic AMP in muscle cells has been purified and shown to consist of two types of subunit—a regulatory subunit to which cyclic AMP binds and a catalytic subunit that catalyzes the phosphorylation of the substrate protein molecules. In the absence of cyclic AMP, the regulatory and catalytic subunits form an inactive complex. The binding of cyclic AMP alters the conformation of the regulatory subunit so that it dissociates from the complex and releases the catalytic subunit, which is thereby activated (Figure 13–28).

Cyclic-AMP-dependent protein kinases are found in all animal cells, where they probably account for all of the effects of cyclic AMP. While in many cases the substrates for these protein kinases are not known, it is clear that they differ in different cell types, explaining why cyclic AMP effects vary according to the target cell. There are many different protein kinases in cells, only a small fraction of which are regulated by cyclic AMP. While some are regulated by Ca^{2+} or by cyclic GMP (see below), the mechanism of regulation for most of these enzymes has not been identified.

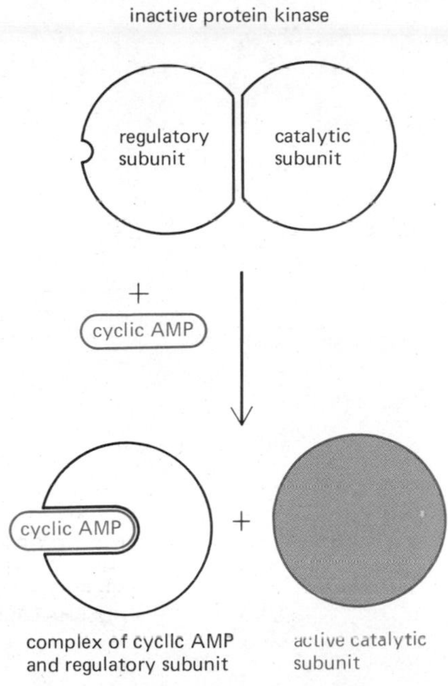

Figure 13–28 The activation of cyclic-AMP-dependent protein kinase. The binding of cyclic AMP to the regulatory subunit induces a conformational change, causing this subunit to dissociate from the complex. The released catalytic subunit is now active and is, therefore, shown in color. Although the protein kinase is shown as a dimer for the sake of clarity, it is actually thought to be a tetramer consisting of two regulatory and two catalytic subunits. Each regulatory subunit has two cyclic-AMP-binding sites, and the release of the catalytic subunits is a cooperative process requiring the binding of more than two cyclic AMP molecules to the tetramer. This greatly sharpens the response of the kinase to changes in cyclic AMP concentration as discussed on page 752.

Cyclic AMP Also Regulates the Dephosphorylation of Cellular Proteins[18]

The effects of cyclic AMP are usually transient: this means that cells must have ways of dephosphorylating the proteins that have been phosphorylated by cyclic-AMP-dependent protein kinases. The dephosphorylation is catalyzed by *phosphoprotein phosphatase,* an enzyme itself regulated by cyclic AMP.

In skeletal muscle cells, a single phosphoprotein phosphatase dephosphorylates each of the three key enzymes in the glycogen pathway regulated by cyclic AMP (phosphorylase kinase, glycogen phosphorylase, and glycogen synthase) (Figure 13–29). These dephosphorylation reactions tend to counter-

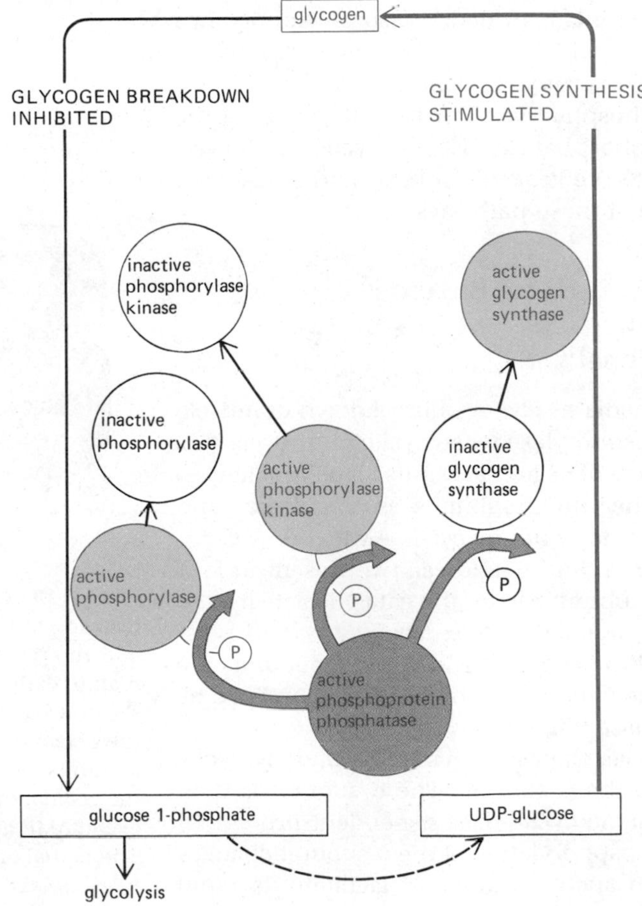

Figure 13–29 When cyclic AMP levels fall, a single phosphoprotein phosphatase dephosphorylates the three key enzymes in the cyclic-AMP-regulated glycogen breakdown cascade. These dephosphorylations reverse the phosphorylation reactions stimulated by cyclic AMP, inactivating the enzymes involved in glycogen breakdown and activating glycogen synthase, the enzyme involved in glycogen synthesis. In contrast, when cyclic AMP levels are high, the protein kinase is active, the reactions reversed, and the phosphoprotein phosphatase inhibited.

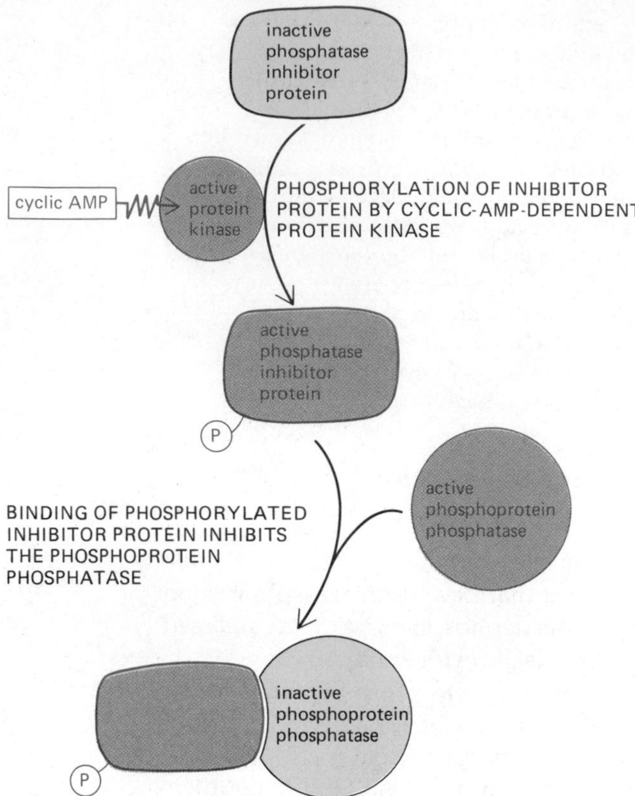

Figure 13–30 Schematic illustration of how cyclic AMP inhibits the phosphoprotein phosphatase that would otherwise immediately reverse the phosphorylation reactions stimulated by cyclic AMP. The cyclic-AMP-dependent protein kinase phosphorylates a phosphatase inhibitor protein, thereby enabling it to bind to and inhibit the phosphoprotein phosphatase.

act the protein phosphorylations stimulated by cyclic AMP. However, when the cyclic-AMP-dependent protein kinase is activated, it also phosphorylates a specific *phosphatase inhibitor protein,* which is thereby activated. This activated inhibitor protein binds to phosphoprotein phosphatase and inactivates it (Figure 13–30). By both activating phosphorylase kinase and inhibiting the opposing action of phosphoprotein phosphatase, rises in cyclic AMP levels have a much larger and sharper effect on glycogen synthesis and breakdown than if cyclic AMP regulated only one of these pathways.

Phosphorylase Kinase Illustrates How an Enzyme Can Be Activated Only Transiently by an Extracellular Chemical Signal[18]

As described, phosphorylase kinase mediates glycogen breakdown in muscle cells by phosphorylating glycogen phosphorylase when cyclic AMP levels rise. It is a multisubunit enzyme, but only one of its four subunits actually catalyzes the phosphorylation reaction: the other three subunits are regulatory and enable the enzyme complex to be activated both by cyclic AMP and by Ca^{2+}. The four subunits are designated α, β, γ, and δ, and each is present in four copies in the enzyme complex. The γ subunit carries the catalytic activity; the δ subunit is the Ca^{2+}-binding polypeptide *calmodulin* (see p. 748) and is responsible for the Ca^{2+}-dependence of the enzyme. The α and β subunits are the targets for cyclic-AMP-mediated regulation, both being phosphorylated by the cyclic-AMP-dependent protein kinase (Figure 13–31).

The activation of the phosphorylase-kinase-enzyme complex by cyclic AMP is caused by phosphorylation of the β subunit. The α subunit is then phosphorylated more slowly by the same cyclic-AMP-dependent protein kinase. It has been proposed that the phosphorylation of the α subunit changes the conformation of the β subunit in such a way as to facilitate its rapid

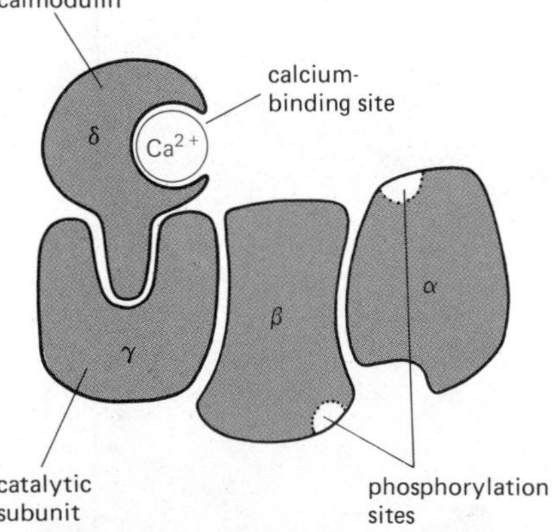

Figure 13–31 Highly schematized drawing of the four subunits of the enzyme phosphorylase kinase from mammalian muscle. While the γ subunit has the catalytic activity of the active enzyme, the α and β subunits and the δ subunit mediate the regulation of the enzyme by cyclic AMP and Ca^{2+}, respectively. The actual enzyme complex contains four copies of each subunit.

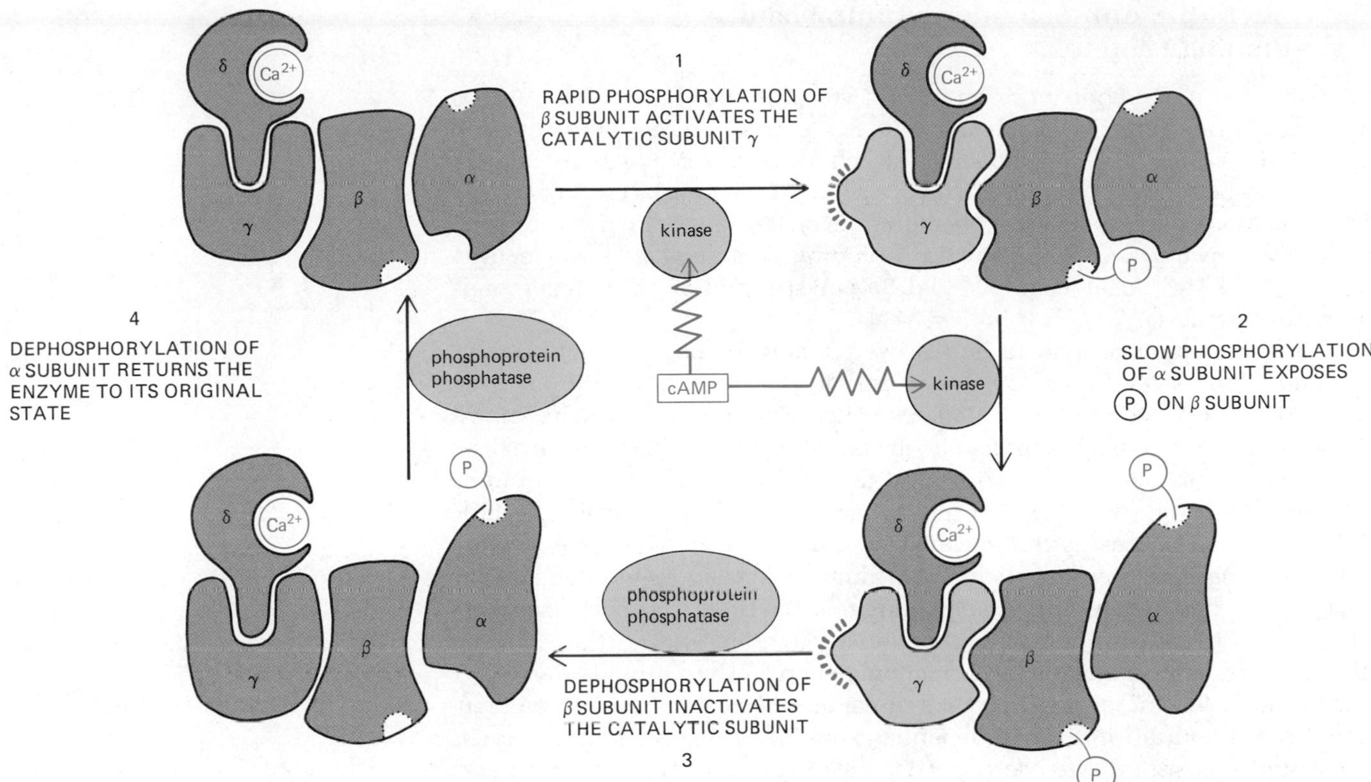

Figure 13–32 One proposed mechanism to explain the switchlike activation and inactivation of phosphorylase kinase following a rise in intracellular cyclic AMP concentration. Ca^{2+} is shown bound to the δ subunit, since the γ subunit can be activated by the protein kinase only when Ca^{2+} is bound.

dephosphorylation by phosphoprotein phosphatase. Dephosphorylation of the β subunit then inactivates the enzyme (Figure 13–32). Thus, a rise in the concentration of cyclic AMP activates phosphorylase kinase (and thereby its substrate glycogen phosphorylase) for a defined period of time (about one or two minutes).

In this way, the relatively simple interaction of only a few proteins creates a biological "switch" that is turned on suddenly and then turned off again after a fixed delay. The cyclic-AMP-dependent protein kinase plays two parts in this proposed cycle: it first activates the phosphorylase kinase by phosphorylating the β subunit and then, by phosphorylating the α subunit, it determines the time at which the enzyme will be inactivated by dephosphorylation.

The Reversible Covalent Modification of Proteins Regulates a Large Number of Cellular Processes[19]

More than 25 different enzymes are now known to be regulated by phosphorylation and dephosphorylation, and the hundreds of different phosphorylated proteins revealed by two-dimensional gel electrophoresis in eucaryotic cells suggest that there must be many more. The elucidation of the phosphorylation-dephosphorylation mechanism has also led to the discovery of other reversible covalent modifications that regulate the activity of proteins: methylation-demethylation, acetylation-deacetylation, uridylation-deuridylation, adenylation-deadenylation, and SS/SH interconversions, among others. Reversible covalent modifications are so abundant that it seems likely that most cellular proteins that catalyze the rate-limiting steps in biological processes are regulated in one or other of these ways. In view of the complexities of the feedback loops known to exist in the glycogen pathways alone, it is clear that the unraveling of the mechanisms and kinetics of these important regulatory processes poses a formidable challenge.

Ca^{2+} Alters the Conformation of Intracellular Ca^{2+}-binding Proteins[20]

Since the free Ca^{2+} concentration in the cytosol is usually less than 10^{-7} M and usually does not rise much above 10^{-5} M even when the cell is activated by an influx of Ca^{2+}, any structure in the cell that is to serve as a direct target for Ca^{2+}-dependent regulation must have a binding affinity for Ca^{2+} of around 10^{-6} M. Moreover, since the concentration of free Mg^{2+} in the cytosol is relatively constant at 10^{-3} M, the Ca^{2+}-binding sites must have a selectivity for Ca^{2+} over Mg^{2+} of at least 1000-fold. Several specific Ca^{2+}-binding proteins fulfill these criteria.

The first such protein to be discovered, and the best characterized, is *troponin C* in skeletal muscle cells; its role in muscle contraction has been discussed in Chapter 10 (see p. 559). A related Ca^{2+}-binding protein, known as **calmodulin,** has been identified in all animal and plant cells that have been examined. It has been highly conserved during evolution and appears to be a ubiquitous intracellular Ca^{2+} receptor, playing a part in the majority of the Ca^{2+}-regulated processes that have been studied in eucaryotic cells. Calmodulin is a single polypeptide chain of 148 amino acid residues whose sequence is related to that of troponin C, suggesting that the latter is a specialized form of calmodulin. Like troponin C, calmodulin has four high-affinity Ca^{2+}-binding sites and undergoes a large conformational change when it binds Ca^{2+}.

Among the increasing numbers of cellular proteins known to be regulated by calmodulin in a Ca^{2+}-dependent manner are some forms of cyclic nucleotide phosphodiesterase and adenylate cyclase, as well as membrane-bound Ca^{2+}-ATPases, phosphorylase kinase, and the myosin light chain kinase of both muscle and nonmuscle cells. A typical animal cell contains more than 10^7 molecules of calmodulin, and it constitutes as much as 1% of the total cell protein. Some of the calmodulin is associated with mitotic spindles, actin filament bundles, and 10-nm intermediate filaments, which suggests that it may regulate their activities as well.

The allosteric activation of calmodulin by Ca^{2+} is analogous to the allosteric activation of protein kinases by cyclic AMP. In the case of phosphorylase kinase, calmodulin has been identified as a permanent regulatory subunit of the enzyme (see Figure 13–31). In most cases, however, the binding of Ca^{2+} allows free calmodulin to become bound to various target proteins in the cell (Figure 13–33). For example, the Ca^{2+}-dependent activation of cyclic nucleotide phosphodiesterase, adenylate cyclase, and some membrane-bound Ca^{2+}-ATPases result from the Ca^{2+}-induced binding of Ca^{2+}-calmodulin complexes to a regulatory subunit of each of these enzymes. Thus the response of a target cell to an increase in free Ca^{2+} concentration in the cytosol depends

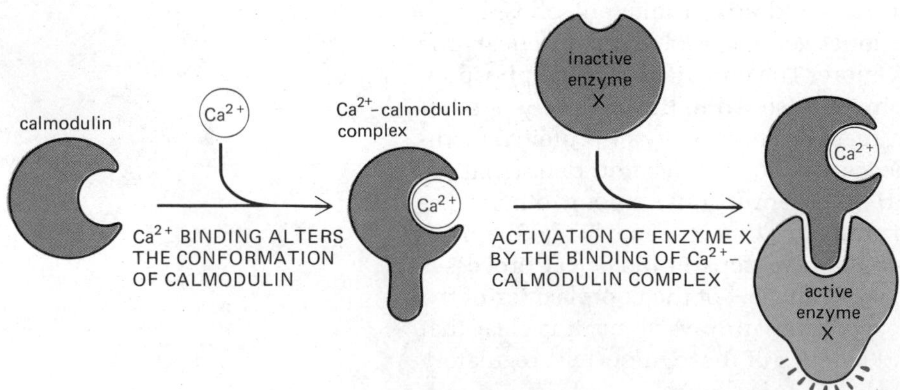

Figure 13–33 Schematic drawing showing how an increase in free Ca^{2+} in the cytosol indirectly activates an enzyme by altering the conformation of calmodulin molecules. Although for simplicity calmodulin is shown with only one binding site for Ca^{2+}, each calmodulin molecule is known to have four Ca^{2+}-binding sites.

on which calmodulin-binding target proteins are present in the cell. Since calmodulin can exist in several different conformations, depending on the number of calcium ions bound per molecule, it is possible that these different conformations interact with different target proteins in a cell. In this way, calmodulin could translate quantitative differences in the concentration of intracellular free Ca^{2+} into different cellular responses.

The cellular activities regulated by cyclic AMP and by Ca^{2+} overlap to a large extent, and the intracellular concentrations of both molecules are often altered by the same extracellular signaling molecules. These two intracellular signaling pathways can interact in at least two ways: (1) Intracellular Ca^{2+} and cyclic AMP levels can influence each other; for example, calmodulin can regulate the enzymes that make and break down cyclic AMP, while cyclic-AMP-dependent kinases may phosphorylate Ca^{2+} channels or pumps. (2) The same protein can be regulated both by Ca^{2+} and cyclic AMP; for example, phosphorylase kinase can be activated by the cyclic-AMP-dependent protein kinase as well as by Ca^{2+} binding to calmodulin (see Figure 13–31).

Does Cyclic GMP Act as an Intracellular Second Messenger?[21]

Another cyclic nucleotide, cyclic guanosine monophosphate (**cyclic GMP**) (Figure 13–34), is present in animal cells, although at a concentration at least ten times lower than that of cyclic AMP. Although cyclic GMP also acts as an intracellular messenger, its role is somewhat different from that of cyclic AMP. Unlike adenylate cyclase, **guanylate cyclase,** which catalyzes the production of cyclic GMP from GTP, is not always membrane-bound. Nor can it be activated by hormones or other extracellular signaling molecules in disrupted cells. Therefore, although cyclic GMP levels increase during the responses of intact cells to a large number of different signaling ligands, it is not thought to be the primary intracellular mediator for any of these responses.

One function of cyclic GMP may be to modulate responses initiated by an increase in the concentration of free intracellular Ca^{2+}. Such increases in cytosolic Ca^{2+} tend to increase cyclic GMP levels either by activating guanylate cyclase or by inhibiting cyclic GMP phosphodiesterase, which degrades cyclic GMP. Like cyclic AMP, cyclic GMP acts primarily by activating specific protein kinases that, in this case, are cyclic-GMP-dependent.

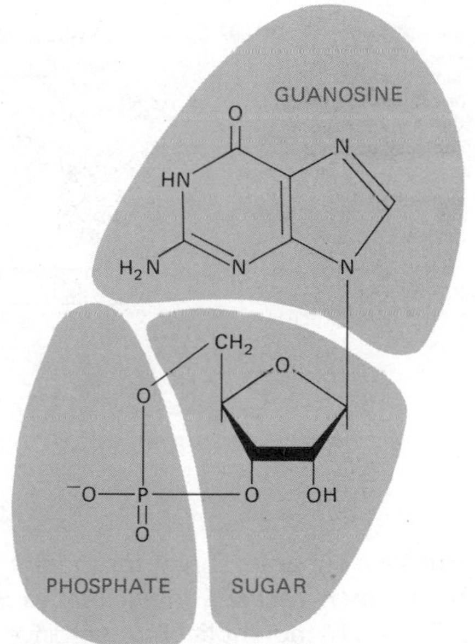

Figure 13–34 Cyclic GMP.

Extracellular Signals Are Amplified Enormously by the Use of Second Messengers and Enzymatic Cascades

The use of second messengers not only allows cell-surface receptors to transduce extracellular signals into intracellular ones, it also provides a mechanism for greatly amplifying the initial signal. We have already seen that when a ligand binds to a receptor and activates adenylate cyclase, each receptor protein activates many molecules of GTP-binding protein and thereby many molecules of adenylate cyclase. Each cyclase molecule, in turn, catalyzes the conversion of a large number of ATP molecules to cyclic AMP molecules. Similarly, when ligand binding to a receptor opens Ca^{2+} channels, a large number of calcium ions enter the cytosol. These second messengers themselves act as allosteric effector molecules to activate specific proteins, such as protein kinases, which in turn convert (by phosphorylation in the case of kinases) a very large number of substrate molecules into third messengers, and so on. Such an enzymatic cascade becomes amplified as it proceeds, so that one extracellular signaling molecule can generate many thousands of effector molecules within the target cell (Figure 13–35).

In addition to amplification, linking several regulatory proteins in sequence provides enormous potential for control. Each protein in the sequence

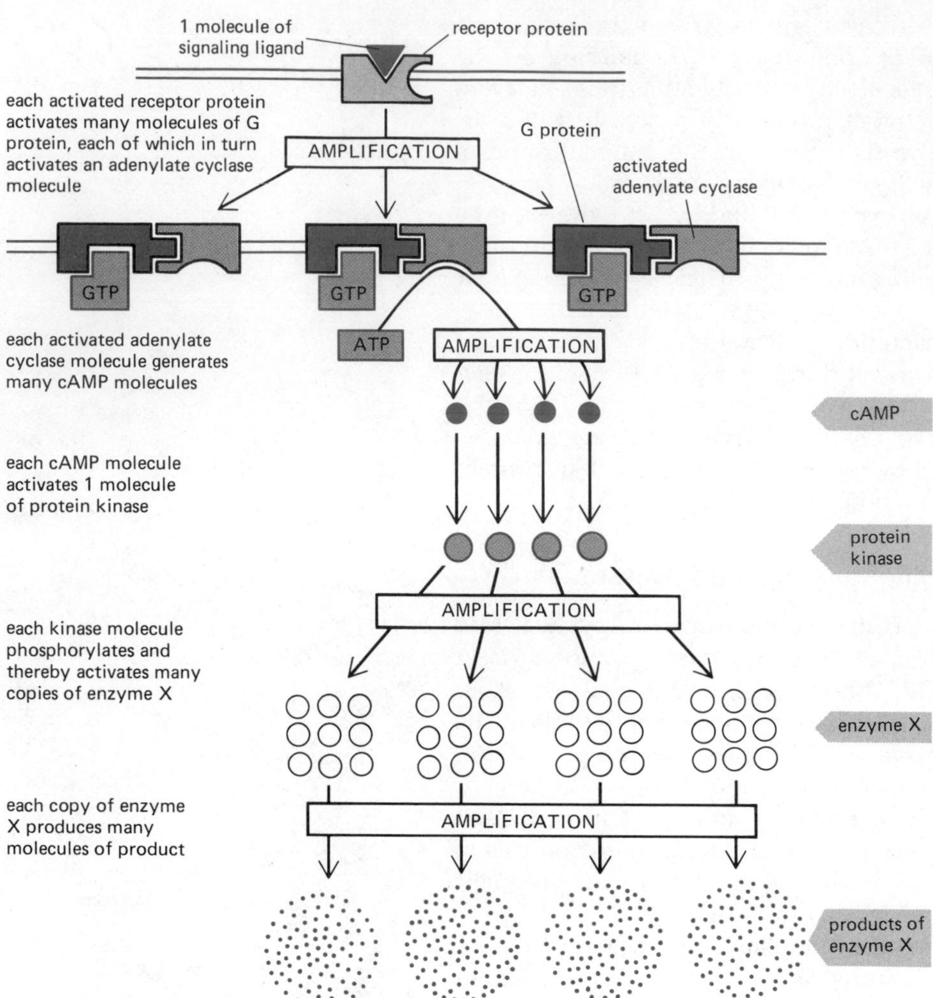

Figure 13–35 Besides transducing an extracellular signal into an intracellular one, the coupling of cell-surface receptors to adenylate cyclase activation greatly amplifies the initial signal.

can be a separate target for metabolic control, as, for example, in the glycogen breakdown cascade in skeletal muscle cells.

Such metabolically explosive cascades require tight regulation. Therefore, it is not surprising that cells have such efficient mechanisms for rapidly degrading cyclic AMP and for buffering and sequestering Ca^{2+}, as well as for inactivating the responding enzymes and transport proteins once they have been activated. However, it is often not appreciated that the rapid removal or inactivation of a signaling molecule is as important for turning on a response rapidly as it is for turning off a response.

The Concentration of a Molecule Can Change Quickly If It Is Continuously Degraded or Removed at a Rapid Rate[22]

If it is important for a cell to alter the concentration of a molecule (such as cyclic AMP or Ca^{2+}) very rapidly, then the molecule must be constantly degraded or otherwise removed at a rapid rate. For example, consider two intracellular molecules X and Y that are both normally maintained at a concentration of 1000 molecules per cell. X has a slow "turnover," being synthesized and degraded at a rate of 10 molecules per second, while Y has a turnover 10 times greater, being synthesized and degraded at a rate of 100 molecules per second. If the rates of synthesis of both X and Y are increased tenfold without any change in their rates of degradation, at the end of one second the concentration of Y will have increased by nearly 900 molecules per cell (10×100

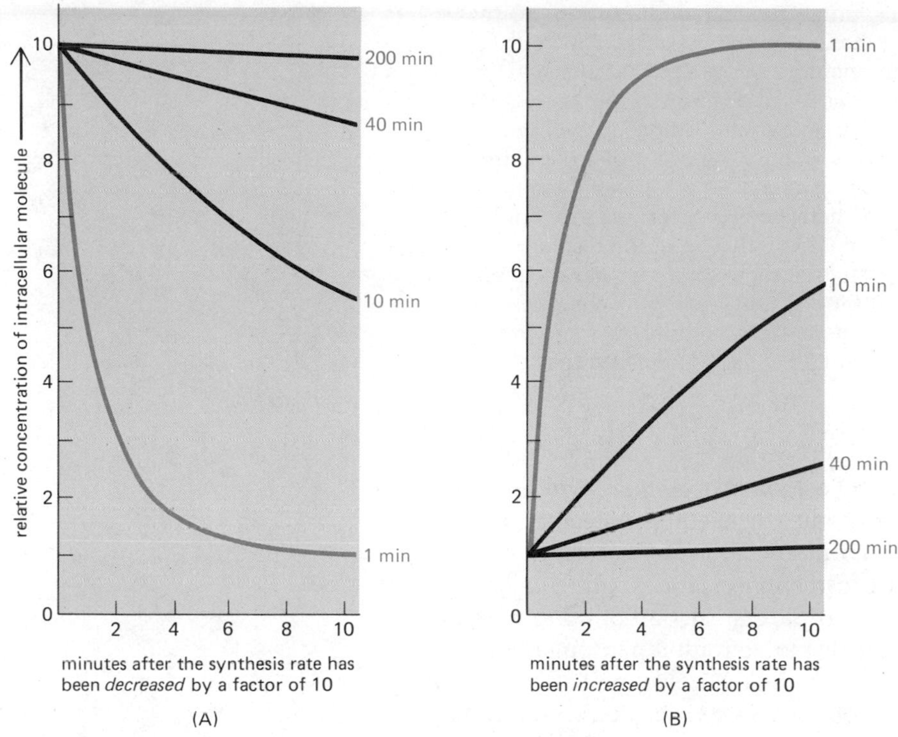

Figure 13–36 Graph showing the predicted relative rates of change in the intracellular concentrations of molecules with differing turnover times when their rates of synthesis are either decreased (A) or increased (B) by a factor of 10. In both cases the concentrations of those molecules that are normally being rapidly degraded in the cell (*colored lines*) change quickly, whereas the concentrations of those that are normally turning over slowly change proportionally more slowly. The numbers in color on the right-hand side are the half-lives assumed for each of the different molecules.

– 100) while the concentration of X will have increased only by 90 molecules per cell. In fact, the time required for a molecule to reach half of its new equilibrium concentration, after its synthesis rate has been either increased or decreased, is equal to its normal half-life, the time required for its concentration to fall by half if all synthesis were stopped (Figure 13–36).

The same principles apply to proteins as well as to small molecules, and to molecules in the extracellular space as well as to those in cells. Many intracellular proteins that are rapidly degraded or covalently modified have half-lives of 10 minutes or less; in most cases these are proteins with key regulatory roles, whose concentrations are rapidly regulated in the cell by changes in their rates of synthesis.

Cells Can Respond Gradually or Suddenly to Signals[23]

Certain cellular responses to signaling ligands are gradual and increase proportionally as the concentration of the ligand increases. The responses to steroid hormones often follow this pattern, presumably because each hormone receptor protein binds a single molecule of hormone and each specific site on the chromatin binds a single hormone-receptor complex. As the concentration of hormone increases, the concentration of hormone-receptor complexes increases proportionally, as does the number of complexes bound to specific sites on chromatin; therefore, the cellular response is a gradual and linear one.

Other responses to signaling ligands, however, begin more abruptly as the concentration of ligand increases, and some may even occur in a nearly "all-or-none" manner. Here no response can be detected below a threshold concentration of ligand, while a maximum response is produced when this concentration is reached. Some growth factors, for example, act as all-or-none signals that tell cells to begin replicating their DNA as a prelude to cell division. What might be the molecular basis for such a switchlike response to a graded signal?

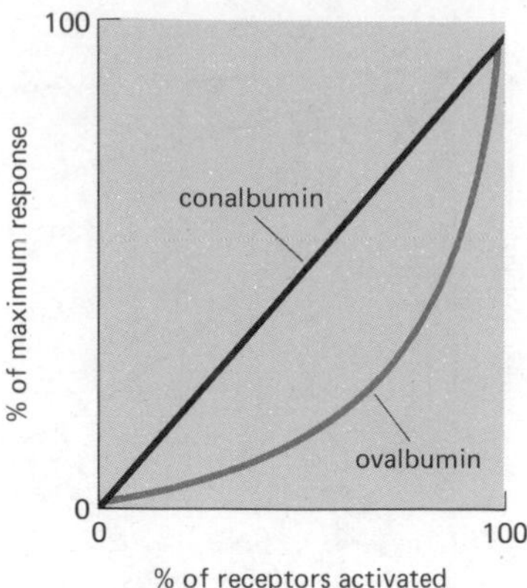

Figure 13–37 Diagram of the observed response of chick oviduct cells to the steroid hormone estradiol. When activated, estradiol receptors migrate to the nucleus and turn on the transcription of several different genes. Dose-response curves for two of these genes, one coding for the egg protein conalbumin and one coding for the egg protein ovalbumin, are shown. The linear response curve for conalbumin indicates that a single activated receptor turns on one conalbumin gene. In contrast, the lag in the response curve for ovalbumin suggests that more than one activated receptor (in this case 2) must bind simultaneously to the ovalbumin gene in order to initiate its transcription. (Adapted from E. R. Mulvihill and R. D. Palmiter, *J. Biol. Chem.* 252:2060–2068, 1977.)

One possible mechanism exploits the principle of *cooperativity,* in which more than one intracellular effector molecule (or effector-molecule-receptor complex) must bind to some target macromolecule in order to induce a response. In some steroid-hormone-induced responses, for example, it appears that more than one hormone-receptor complex must be bound simultaneously to specific sites on chromatin in order to activate a particular gene. As a result, gene activation begins more abruptly as the hormone concentration rises than it would if only one bound complex were required for activation (Figure 13–37). A similar cooperative mechanism operates in the activations mediated by calmodulin: since two or more Ca^{2+} ions must bind before calmodulin adopts its activating conformation, a 50-fold increase in activation occurs when the free intracellular Ca^{2+} concentration increases only tenfold. Such cooperative responses become sharper as the number of cooperating molecules increases, and if the number is large enough, responses of the all-or-none type can be achieved (Figures 13–38 and 13–39).

Another mechanism by which a cell could respond in an all-or-none way to a gradual increase in a signaling ligand is to make a *competitive inhibitor* that binds the intracellular effector molecule so tightly as to make it

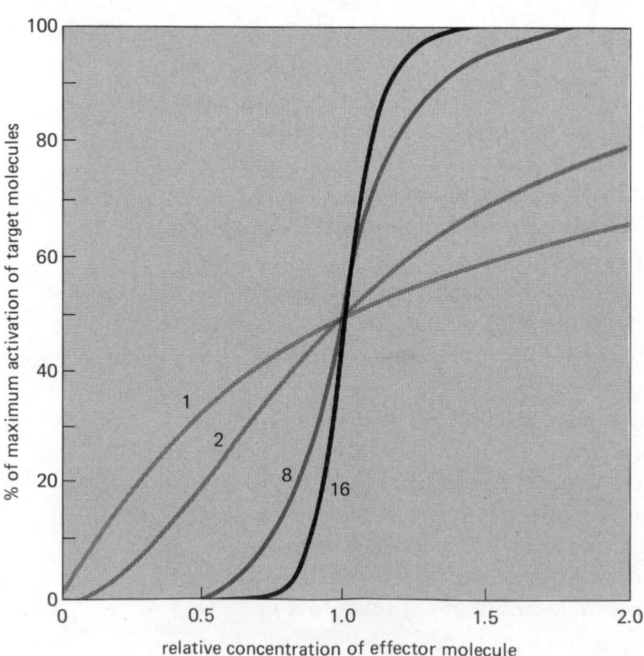

Figure 13–38 Diagram illustrating how sharp "all-or-none" activation curves result if many effector molecules must bind simultaneously to a target macromolecule in order to activate it. The curves shown are those expected if the activation requires the simultaneous binding of 1, 2, 8, and 16 effector molecules, respectively.

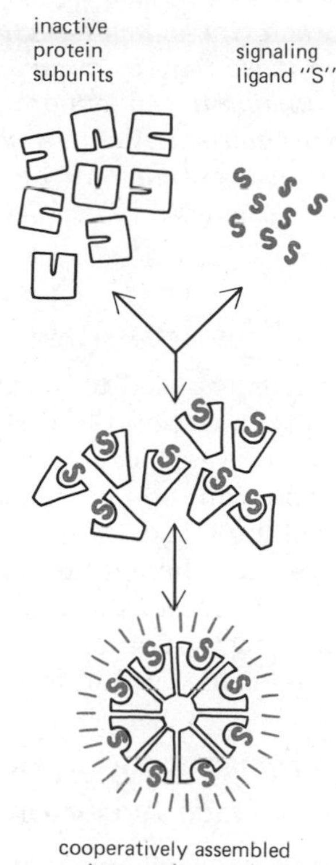

Figure 13–39 Highly schematic illustration of one type of signaling mechanism expected to show a sharp threshold response. Here the simultaneous binding of eight molecules of a signaling ligand is required to form an active protein complex.

unavailable to bind to its target until there is more effector than inhibitor. This type of scheme has been suggested to account for some aspects of the regulation of cell division (see p. 643).

A third possible mechanism involves a principle known as "accelerating positive feedback." By this mechanism, nerve and muscle cells generate all-or-none *action potentials* in response to neurotransmitters. For example, the activation of acetylcholine receptors at a neuromuscular junction opens cation channels in the muscle cell plasma membrane. The result is a net influx of Na^+ that locally depolarizes the membrane. When the depolarization reaches a threshold level, voltage-dependent Na^+ channels open in the same membrane region, producing a further influx of Na^+, which further depolarizes the membrane and thereby opens more Na^+ channels. In this way, a wave of depolarization (an action potential) spreads over the entire muscle membrane.

The same type of accelerating feedback mechanism can involve a receptor protein that is an enzyme rather than an ion channel. Suppose a particular signaling ligand activates an enzyme and that two or more molecules of the product of the enzymatic reaction bind back to the enzyme to activate it further (Figure 13–40). The consequence will be a very slow rise in the concentration of the enzyme product until, at some threshold level of ligand, enough of the product has accumulated to activate the enzyme in an accelerating fashion; the concentration of the enzyme product then suddenly increases to a much higher level. In this way the cell can translate a gradual change in the concentration of a signaling ligand into a switchlike change in the level of a particular enzyme product. This type of mechanism is relatively simple in principle, since it does not require the cooperative interaction of a large number of binding sites to produce an all-or-none response to a signaling ligand.

Responses are also greatly sharpened when a ligand activates one enzyme and at the same time inhibits another that catalyzes the opposite reaction. We have already discussed one example of this common type of regulation in the stimulation of glycogen breakdown in skeletal muscle cells, where a rise in the intracellular cyclic AMP level both activates phosphorylase kinase and inhibits the opposing action of phosphoprotein phosphatase (see p. 745).

Summary

The continuous, rapid removal of both free Ca^{2+} and cyclic AMP from cells makes possible both the rapid increase and decrease in the concentrations of these intracellular mediators when cells respond to signals. Rising cyclic AMP levels affect cells by stimulating cyclic-AMP-dependent protein kinases to phosphorylate specific target proteins. The effects are reversible since phosphorylated proteins are rapidly dephosphorylated when cyclic AMP levels fall. In a similar way, increased free Ca^{2+} levels affect cells by binding to and altering the conformation of calmodulin; the Ca^{2+}-calmodulin complexes in turn activate many different target proteins, including Ca^{2+}-dependent protein kinases.

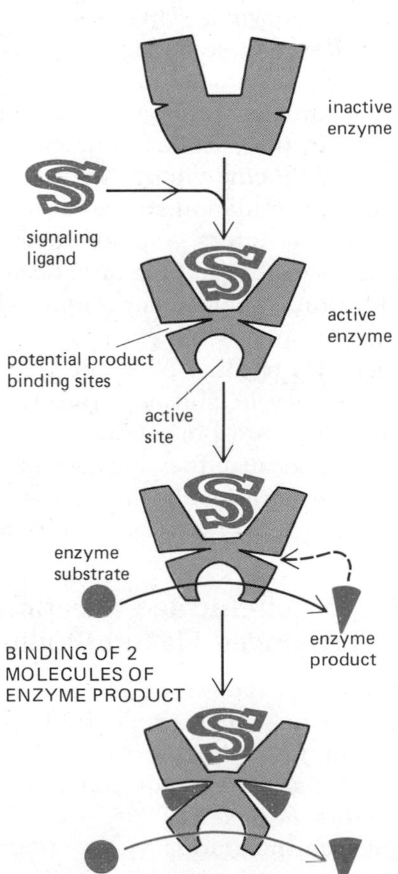

Figure 13–40 Schematic illustration of the "accelerating positive feedback" mechanism described in the text.

Since each type of cell has characteristic sets of target proteins that are regulated in these ways by cyclic-AMP-dependent kinases and/or by calmodulin (or by calmodulin-dependent protein kinases), each cell type responds in a different and characteristic way to a change in intracellular cyclic AMP or Ca^{2+} levels. Thus, by using cyclic AMP or Ca^{2+} as second messengers, extracellular signals are both greatly amplified and made specific for each cell type.

Target Cell Adaptation

Target cells exposed to a signaling ligand for a prolonged period often lose the ability to respond to that ligand. This process of **adaptation** or **desensitization** is reversible, and it makes many cells especially sensitive to *changes* in the concentration of a chemical signal rather than to the absolute concentration of the signal. How is a desensitized target cell different from a normally sensitive one? There is no single answer. In some cases, desensitization results from a decrease in the number of specific cell-surface receptor proteins or from the inactivation of such receptors; in other cases, it is due to changes in the proteins involved in transducing the signal following receptor activation.

Some Cells Become Desensitized by Endocytosing Their Surface Receptors[24]

After protein hormones and growth factors have bound to receptors on the surface of target cells, they are often ingested by receptor-mediated endocytosis. Since endocytotic vesicles generally deliver their contents to lysosomes, the ligand and often the receptor to which it is bound are degraded by hydrolytic enzymes. This process not only represents a major pathway for the breakdown of some signaling ligands, it also plays an important part in regulating the concentration of certain receptor proteins on the surface of target cells. Although receptor degradation and replacement take place continuously, in the absence of ligand a receptor usually has a half-life of a day or so. By inducing endocytosis, some ligands markedly increase the rate of receptor degradation so that at high ligand concentrations the number of cell-surface receptors gradually decreases. The result is a concomitant decrease in the sensitivity of the target cell to the ligand. This type of target cell desensitization is known as **receptor-down regulation.**

An interesting example of receptor-down regulation is seen in individuals who become obese from overeating. They have chronically high blood glucose levels, and as a result, persistently elevated levels of blood insulin. The insulin-responsive cells in such individuals are relatively insensitive to insulin because they have fewer insulin receptors. Such individuals normally secrete more insulin in compensation. But in an obese individual with impaired insulin reserves, insulin deficiency results, causing *diabetes.*

Some Cell-Surface Receptors Are Reversibly Inactivated by Prolonged Ligand Binding[25]

A different type of cell-surface receptor regulation often occurs in response to unusually high concentrations of small signaling ligands such as epinephrine or acetylcholine. Rather than inducing the internalization and digestion of the ligand-receptor complexes, these ligands reversibly inactivate the receptors. For example, frog red blood cells exposed to high concentrations of epinephrine gradually (over hours) lose both their sensitivity to epinephrine and their ability to bind it. This cannot be due to ligand-induced receptor endocytosis and degradation, however, because (1) it also occurs in mem-

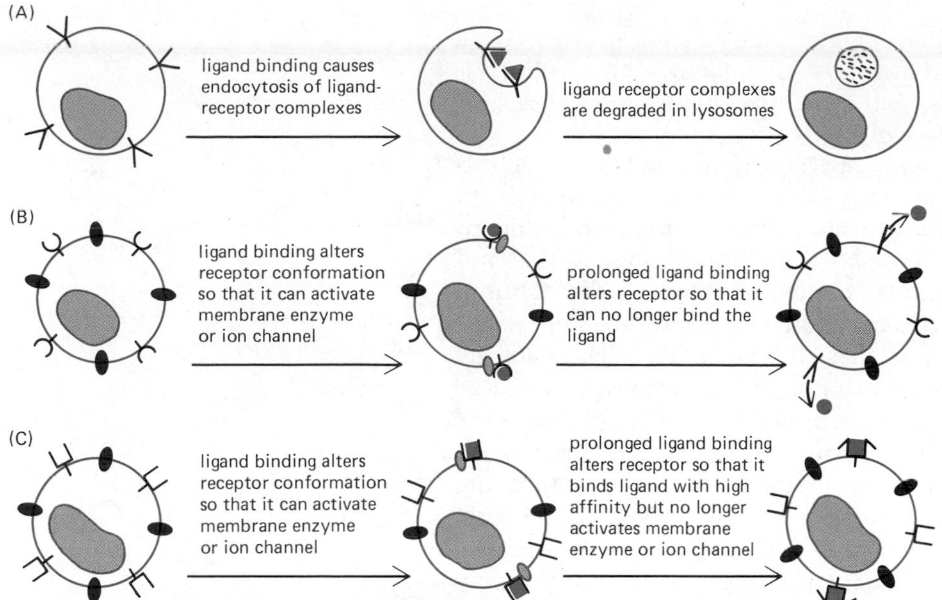

Figure 13–41 Three different ways in which receptors can be inactivated by high concentrations of ligand. Ligand binding induces the endocytosis of ligand-receptor complexes, which are degraded in lysosomes (A). Prolonged ligand binding alters the conformation of the receptor so that it can no longer bind the ligand (B) or so that it binds the ligand without activating a membrane-bound enzyme or ion channel (C).

branes isolated from the red blood cells and (2) epinephrine sensitivity gradually recovers even in the absence of new protein synthesis. Instead, it seems likely that the binding of epinephrine induces a prolonged but reversible change in the conformation of the receptor protein that prevents the protein from binding the ligand again for a long period after the ligand dissociates; whether this involves a reversible covalent modification of the receptor protein is unknown.

Such reversible inactivation of cell-surface receptors does not always involve the loss of the receptor's ability to bind the ligand. Prolonged binding of acetylcholine to skeletal muscle acetylcholine receptors at a neuromuscular junction induces the receptors to adopt an inactive conformation, but such inactivated receptors are still able to bind acetylcholine (with an even higher affinity than normal receptors). However, unlike the normal variety, inactivated receptors with bound acetylcholine are unable to open cation channels in the muscle cell plasma membrane and therefore cannot stimulate muscle contraction. While receptor activation and cation channel opening occur within milliseconds of binding, receptor inactivation occurs more slowly, requiring seconds to minutes.

The three different ways in which ligands inactivate receptors are summarized in Figure 13–41. While they are all thought to be mechanisms whereby target cells decrease their responsiveness to a particular ligand, they may also provide alternative pathways for responding to the ligand. For example, while an inactivated acetylcholine receptor cannot stimulate muscle contraction, it could, in principle, generate an intracellular mediator that regulates some other property of the muscle cell.

Morphine-induced Target Cell Desensitization Is Not the Result of Receptor Inactivation[26]

Not all forms of target cell adaptation are due to ligand-induced receptor degradation or inactivation. In morphine addicts, for example, target cells in the brain become desensitized to morphine (Figure 13–42) and yet have normal levels of functional cell-surface morphine (opiate) receptors. (This explains why addicts must take much higher doses of morphine than normal individuals to achieve the same degree of pain relief or euphoria.) If morphine-in-

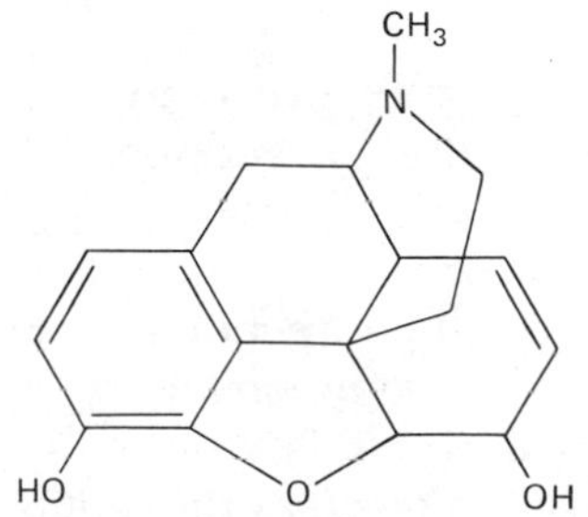

CH_3 N HO O OH

morphine

Figure 13–42 The structure of morphine.

duced desensitization is not related to a change in the receptor proteins, what is the mechanism? Whereas the mechanism is still unknown, the problem is amenable to study using morphine-sensitive neural cell lines in culture.

Morphine receptors on the surface of such cells are functionally coupled to adenylate cyclase molecules by means of GTP-binding proteins, much as epinephrine receptors are coupled to adenylate cyclase in muscle cells. In contrast to epinephrine receptors on muscle cells, however, morphine-induced receptor activation leads to the *inactivation* of adenylate cyclase and thereby causes a decrease in intracellular cyclic AMP levels. If the cultured cells are maintained in the presence of a constant concentration of morphine, they eventually become desensitized, so that both adenylate cyclase activity and intracellular cyclic AMP levels return to normal. This adaptation occurs without a change in the amount of morphine bound to cell-surface receptors, and it does not prevent the cell from responding to a further increase in morphine concentration. If morphine is now removed from the culture medium, there is a marked increase in adenylate cyclase activity; as a result, intracellular cyclic AMP rises to very high levels. The large increase in cyclic AMP concentration in target cells is thought to be responsible for the very unpleasant "cold-turkey" reactions (anxiety, sweating, tremors, etc.) experienced by addicts when morphine is withdrawn.

Why do some of our cells have receptors for a drug like morphine, which comes from poppy seeds? Pharmacologists have long suspected that morphine may mimic some endogenous signaling molecule that regulates pain perception and mood. In 1975, two pentapeptides with morphinelike activity, called **enkephalins,** were isolated from pig brain, and soon thereafter larger polypeptides with similar activity, called **endorphins,** were isolated from the pituitary gland. All of these so-called *endogenous opiates* contain a common four-amino-acid sequence and bind to the same cell-surface receptors that morphine (and related narcotics) bind to. But, unlike morphine, they are rapidly degraded after release and so do not accumulate in large enough quantities to induce the tolerance seen in morphine addicts.

Chemical Signaling in Unicellular Organisms Provides Useful Models for Study[27]

The mechanisms involved in chemical signaling between cells in multicellular animals may have evolved from mechanisms used by unicellular organisms for responding to chemical changes in their environment. In fact, some of the same intracellular mediators are used by both types of organisms.

When certain bacteria, such as *Escherichia coli,* are starved for glucose, they respond by increasing the synthesis of a number of specific enzymes that exploit other sources of energy in their environment. This response is mediated by an increase in intracellular levels of cyclic AMP. The cyclic AMP acts in the way that steroid hormones are thought to act in many animal cells: it binds to specific receptor proteins in the cell, and the cyclic-AMP-receptor complexes in turn bind to specific sites on the bacterial DNA, thereby enhancing the transcription of the specific genes that code for the required enzymes.

When the single-celled paramecium (Figure 13–43) swims into a large obstacle, the direction of beating of its cilia is reversed. This response is mediated by the transient opening of Ca^{2+} channels in the cell's membrane and the resulting influx of Ca^{2+} into the cell. The rise in intracellular Ca^{2+} concentration reverses the motion of the cilia, and the cell swims backward away from the obstacle. Within a few seconds Ca^{2+} is pumped out of the cytosol, so that intracellular Ca^{2+} levels return to normal and the cell resumes forward swimming. Mutants lacking these gated Ca^{2+} channels can move forward but

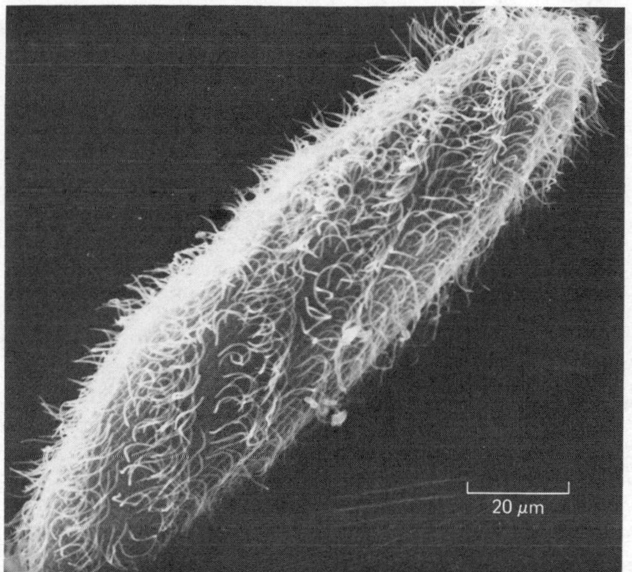

Figure 13–43 Scanning electron micrograph of a paramecium. These organisms swim by synchronously beating their cilia. (Courtesy of Sidney Tamm.)

not backward and are therefore referred to as "pawn" (as in the game of chess) mutants.

Some of the best studied reactions of unicellular organisms to extracellular signals are chemotactic responses, in which cell movement is oriented toward or away from a specific chemical signal. Elsewhere we have briefly discussed chemotaxis in the cellular slime mold *Dictyostelium discoideum* (see p. 676). In bacteria, the availability of a large number of mutants has made it possible to gain important insights into the molecular mechanisms of procaryotic chemotaxis, particularly into the mechanism of adaptation. Moreover, there is increasing evidence that some of the mechanisms of adaptation seen in bacteria also regulate the chemotactic responses of vertebrate white blood cells.

Bacterial Chemotaxis Is a Simple Kind of Intelligent Behavior[28]

Motile bacteria will swim toward higher concentrations of nutrients (attractants), such as sugars and amino acids, and away from higher concentrations of various noxious chemicals (repellents) (Figure 13–44). This simple but intelligent behavior, referred to as **chemotaxis,** has been most intensively studied in *E. coli* and *Salmonella typhimurium.*

Bacteria swim by means of flagella, which are much simpler than the flagella of eucaryotic cells (see p. 561) and consist of a helical tube containing a single type of protein subunit. Each flagellum is attached at its base, by a short flexible hook, to a small protein disc embedded in the bacterial mem-

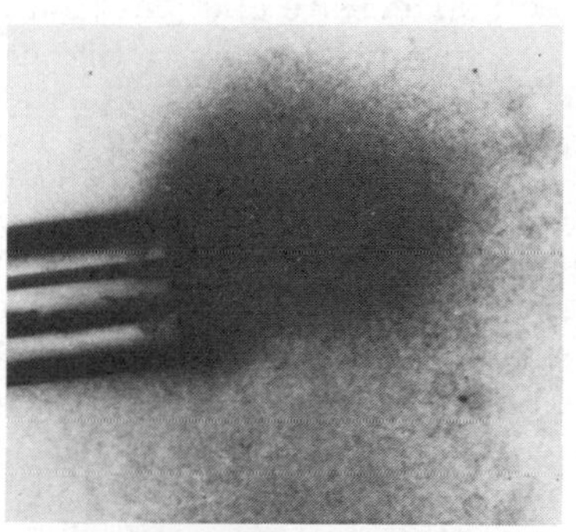
(A)

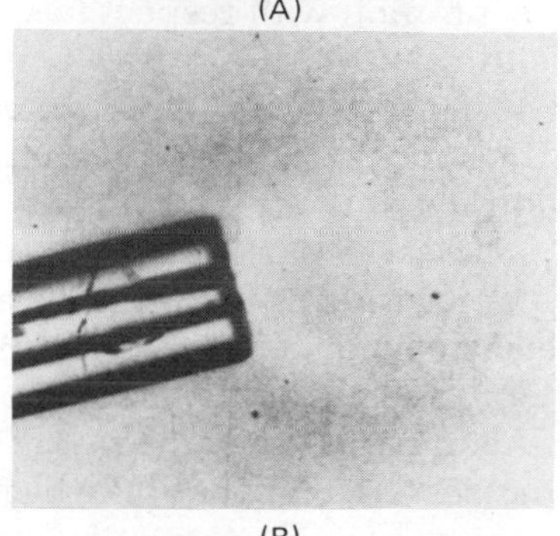
(B)

Figure 13–44 Photographs of *Salmonella typhimurium* bacteria attracted to a small glass capillary tube containing the amino acid serine (A) and repelled from a capillary tube containing phenol (B). The photographs were taken five minutes after the capillary tubes were introduced into the culture dishes containing the bacteria. This capillary tube assay is a simple method for demonstrating bacterial chemotaxis. (From B. A. Rubik and D. E. Koshland, *Proc. Natl. Acad. Sci. USA* 75:2820–2824, 1978.)

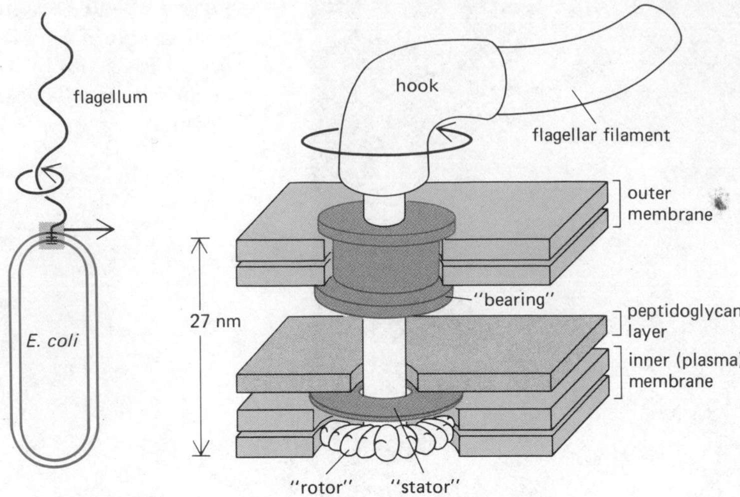

Figure 13–45 Schematic drawing of the flagellar rotatory motor of *E. coli.* The "rotor" is a protein disc integrated into the plasma membrane. Driven by the H^+ gradient across this membrane, it rotates rapidly (~100 revolutions/second) in the lipid bilayer against another protein disc (the "stator") that is somehow anchored to the peptidoglycan layer. A rod links the "rotor" to a hook and flagellum, thereby causing them to rotate. The protein "bearing" serves to seal the outer membrane as the rotating rod passes through it. In this illustration, the parts of the assembly that remain stationary are shaded, while the rotating parts are white.

brane. Incredible though it seems, this disc is part of a tiny "motor" that uses the energy stored in the transmembrane H^+-gradient to rotate rapidly and turn the helical flagellum (Figure 13–45).

Because the flagella on the bacterial surface have a certain "handedness" their direction of rotation makes a difference. Counterclockwise rotation allows all of the flagella to draw together into a coherent bundle so that the bacterium swims uniformly in one direction. Clockwise rotation, however, causes them to fly apart so that the bacterium tumbles chaotically (Figure 13–46). In the absence of any environmental stimulus, the direction of rotation of the disc reverses every few seconds, producing a characteristic pattern of movement in which smooth swimming in a straight line is interrupted by abrupt changes in direction caused by tumbling.

The normal swimming behavior of bacteria is modified by chemotactic attractants or repellents that bind to specific receptor proteins and affect the frequency of tumbling by increasing or decreasing the time that elapses between successive changes in the direction of flagellar rotation. When bacteria are swimming toward a higher concentration of an attractant, they tumble less frequently than when they are swimming in the opposite direction (or when no gradient is present). Since the periods of smooth swimming are longer when a bacterium is traveling up the gradient, it will gradually progress in the direction of the attractant (Figure 13–47). Similarly, upon encountering an increasing concentration of repellent, bacteria tumble more frequently; as a result, they gradually move away from the repellent.

In their natural environment, bacteria respond to spatial gradients of attractants or repellents by monitoring the change in concentration of these chemicals over time as they swim from one place to another. (Monitoring concentration with respect to space would involve comparing concentrations at either end of the cell, which would be extremely difficult given the very small size of a bacterium.) Changes over time can be provided in the laboratory by the sudden addition or removal of a chemical to the culture medium. When an attractant is added in this way, tumbling is rapidly suppressed, as expected. But after some time, even in the continuing presence of the attractant, tumbling frequency returns to normal. This process of desensitization or **adaptation** is specific for the particular type of attractant, since the bacteria still respond to other, unrelated attractants. The bacteria remain in the adapted state as long as there is no gradient of that attractant; removal of the attractant briefly enhances tumbling until the bacteria adapt again to the new level. Adaptation

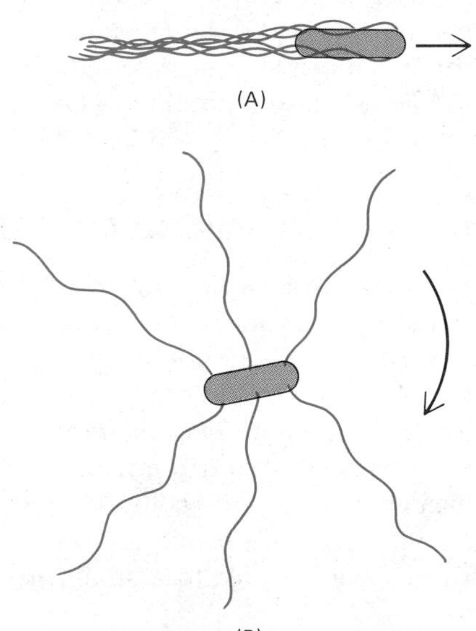

Figure 13–46 Schematic drawing showing the positions of the flagella on *E. coli* during swimming. When the flagella rotate counterclockwise (A), they are drawn together into a single bundle, which acts as a propeller to produce smooth swimming. When the flagella rotate clockwise (B), they fly apart and produce tumbling.

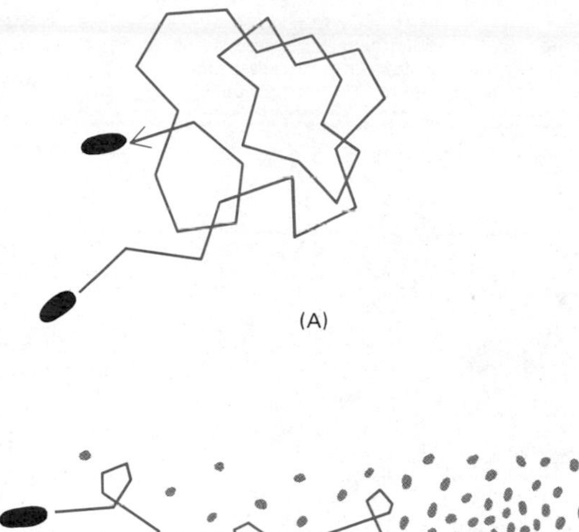

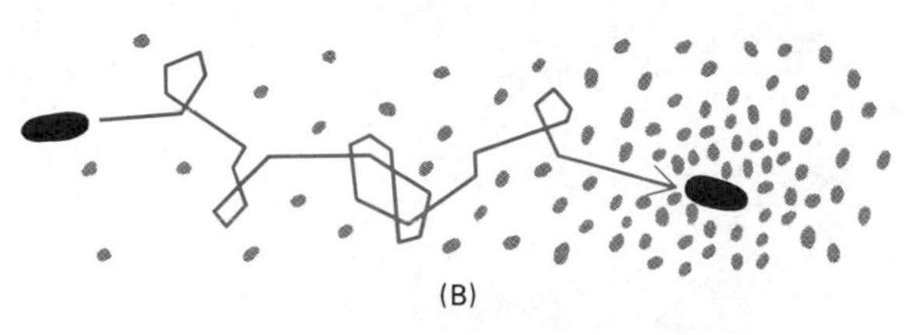

Figure 13–47 Schematic diagram of the tracks of a swimming bacterium. In the absence of a chemotactic signal (A), periods of smooth swimming are interrupted by brief tumbles that randomly change the direction of swimming. In the presence of a chemotactic attractant (B), tumbling is suppressed while the bacterium is swimming toward a higher concentration of the attractant so that it gradually moves in the direction of the attractant.

is an essential part of the chemotactic response, as it enables bacteria to respond to *changes* in concentration rather than to steady-state levels of an attractant and therefore to prolong swimming when moving in a favorable direction.

Chemotaxis-deficient Mutants Have Revealed Four Classes of Proteins Involved in Bacterial Chemotaxis[29]

The molecular mechanisms responsible for bacterial chemotaxis are being unraveled by the isolation and analysis of mutant bacteria defective in different aspects of chemotaxis. The mutants isolated so far fall into four broad classes that reflect the sequential flow of information from the cell-surface receptors to the flagellar motor.

The first class, the *specifically nonchemotactic* mutants, swim normally and respond to most chemotactic stimuli but are unable to respond to one specific chemical or group of closely related chemicals. The lesions in these mutants lie in 1 of the 20 or more genes encoding specific **periplasmic receptor proteins** that bind a specific chemical with high affinity. These receptor proteins are soluble and are found in the periplasmic space (between the outer membrane and the plasma membrane); they are the same proteins that help mediate the transport of specific chemicals across the plasma membrane (see p. 298). Although the transport and chemotaxis systems use a common initial receptor protein, the other parts of their machinery are different, as indicated by mutations that inactivate transport without affecting chemotaxis and vice versa.

The second class of mutants, the *multiply nonchemotactic* mutants, fail to respond to chemicals detected by several different cell-surface receptors but respond normally to chemicals detected by the remaining receptors. The lesions in these mutants involve one of three related transmembrane proteins, which are responsible for transmitting chemotactic signals across the plasma membrane. Because they become methylated during the chemotactic response (see below), they are known as **methyl-accepting chemotaxis proteins (MCPs).** Each MCP is activated by binding its own set of periplasmic receptor proteins: MCP I binds type I receptors; MCP II binds type II receptors; and MCP III binds type III receptors (Figure 13–48). When a bacterium is

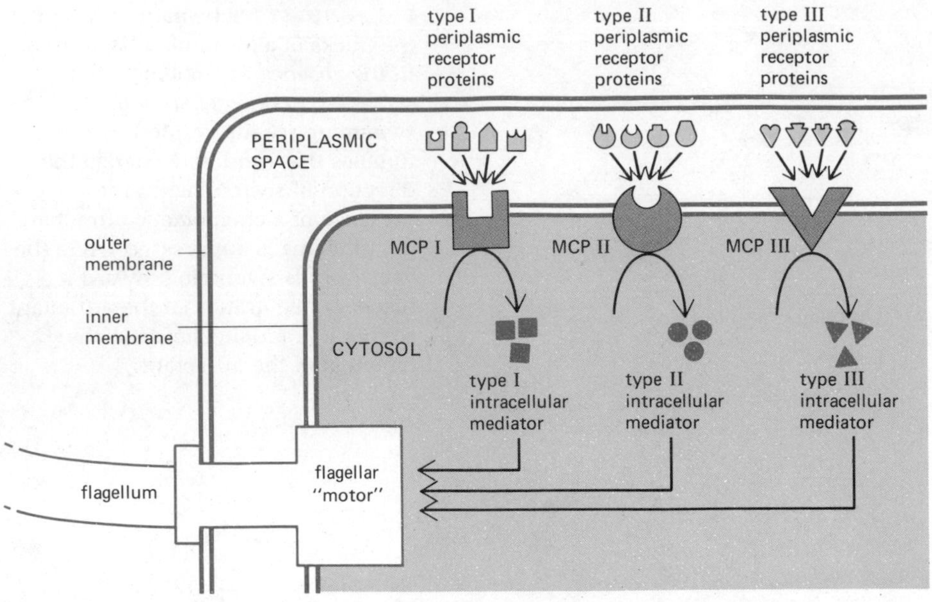

Figure 13–48 The steps in signal transduction during bacterial chemotaxis. Chemical attractants (not shown) bind to specific receptor proteins in the periplasmic space. The receptors then interact with one of three methyl-accepting chemotaxis proteins (MCPs) in the inner (plasma) membrane. The latter interaction activates the MCP to produce an intracellular mediator that causes the flagellar "motor" to continue to rotate counterclockwise, thereby suppressing tumbling and causing continuous smooth swimming. There are three sets of periplasmic receptor proteins (types I, II, and III), each of which interacts both with a specific small molecule and its own MCP. Although not shown in the figure, some chemical attractants directly bind to and activate an MCP; in such cases the MCP acts as both receptor and transducer.

exposed to a chemotactic attractant, the binding of the attractant induces a conformational change in the periplasmic receptor protein, causing the latter to bind to, and thereby activate, the appropriate MCP. The resulting activation of the MCP has two separable effects that correspond to the excitation and adaptation phases of the chemotactic response, respectively: (1) excitation occurs because the activated MCP generates an intracellular signal that causes the flagellar motor to continue to rotate counterclockwise, resulting in the suppression of tumbling and continuous smooth swimming; (2) adaptation occurs because the activated MCP can now be methylated by enzymes in the cytoplasm, reversing the activation of the MCP (see below).

The third class of mutants, the *generally nonchemotactic* mutants, fail to respond to any chemotactic stimuli. They have defects in one of eight different genes (proteins), including defects in the enzymes responsible for methylating (and demethylating) the MCPs and in other proteins required for relaying information between the receptors and the flagellar motor.

The final class of mutants, which also fail to respond to any chemotactic stimuli, are called *nonmotile* mutants. This class includes mutants with defects in 2 genes that control flagellar rotation and defects in 16 genes involved in the synthesis and assembly of flagella.

Protein Methylation Is Responsible for Adaptation[29]

There is compelling evidence that adaptation in bacterial chemotaxis results from the covalent methylation of the MCPs. When methylation is blocked by mutation, adaptation does not occur and exposure of the mutant bacteria to an attractant results in the suppression of tumbling for days instead of for minutes.

The methylation of MCPs is catalyzed by a soluble enzyme (*methyl transferase*) that transfers a methyl group from the common methyl group donor called S-*adenosylmethionine* to a free carboxyl group on a glutamic acid residue of the MCP (Figure 13–49). If an attractant is added that binds to a type I receptor, there is a large increase in the level of methylation of MCP I, while an attractant that binds to a type II or type III receptor induces the methylation of MCP II or MCP III, respectively. The methylation remains at this new level as long as the attractant is present. As many as four methyl groups can

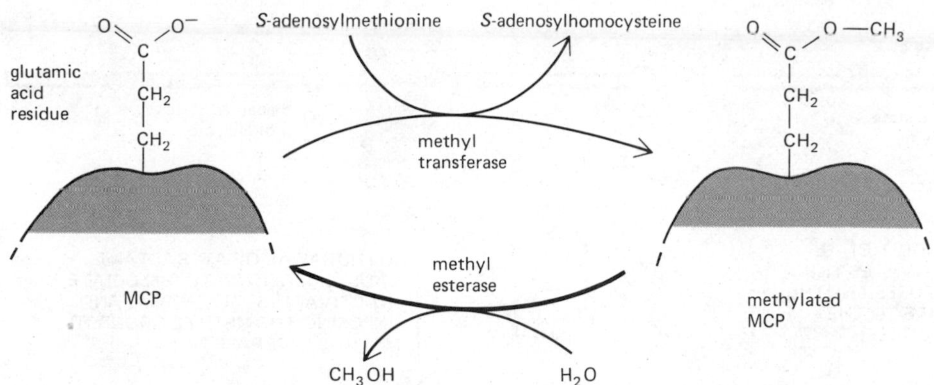

Figure 13–49 Methylation and demethylation reactions involving MCPs. Up to four methyl groups can be added to each MCP by ester linkages.

be transferred to a single MCP, the extent of methylation increasing at higher concentrations of attractant. When the attractant is removed, the MCP is demethylated by a soluble demethylating enzyme (Figure 13–49). Although the level of methylation changes during a chemotactic response, methylation remains constant when a bacterium is adapted because the rates of methylation and demethylation are equal.

At the molecular level this mechanism implies a scheme in which an MCP with one receptor bound is activated until it receives one additional methyl group (Figure 13–50). At this point the MCP can be activated again only if a second receptor molecule binds to it. But this in turn triggers the delayed addition of another methyl group, which returns the MCP to its initial state of low activity. This progression can continue until the MCP has four activated receptors bound and four methyl groups covalently attached. When the attractant is removed, receptors dissociate from the MCP, which is rapidly demethylated to reverse the progression.

Recent evidence suggests that a similar reversible methylation of protein carboxyl groups is also involved in chemotactic responses of mammalian white blood cells, which are attracted by specific signaling molecules released at sites of inflammation. When these cells are exposed to an attractant, certain membrane-bound molecules become methylated, and when methylation is blocked by specific drugs, chemotaxis is inhibited, just as it is in bacteria.

A Model for Bacterial Chemotaxis[29]

The activation of MCPs by activated receptor molecules must lead to the generation of an intracellular signal that affects the direction of rotation of the flagellar motor. The nature of this signal is not known. The process, however, can be illustrated through the use of a hypothetical model for bacterial chemotaxis. The model is based on the assumption that when the intracellular

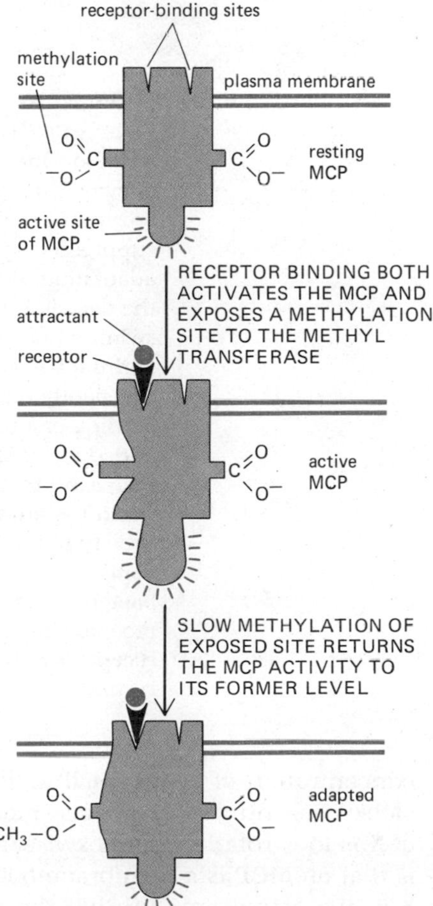

Figure 13–50 Highly schematized drawing of the sequential activation and adaptation (via methylation) of an MCP. Note that the activity of the MCP, and therefore the tumbling frequency of the bacterium, is the same in the resting and adapted state. Although the MCP is shown with two methylation sites (and receptor binding sites) for simplicity, there are thought to be four methylation sites (and probably four receptor binding sites) on each MCP. Also not shown is that the MCPs are methylated to some extent even in the absence of a chemotactic stimulus.

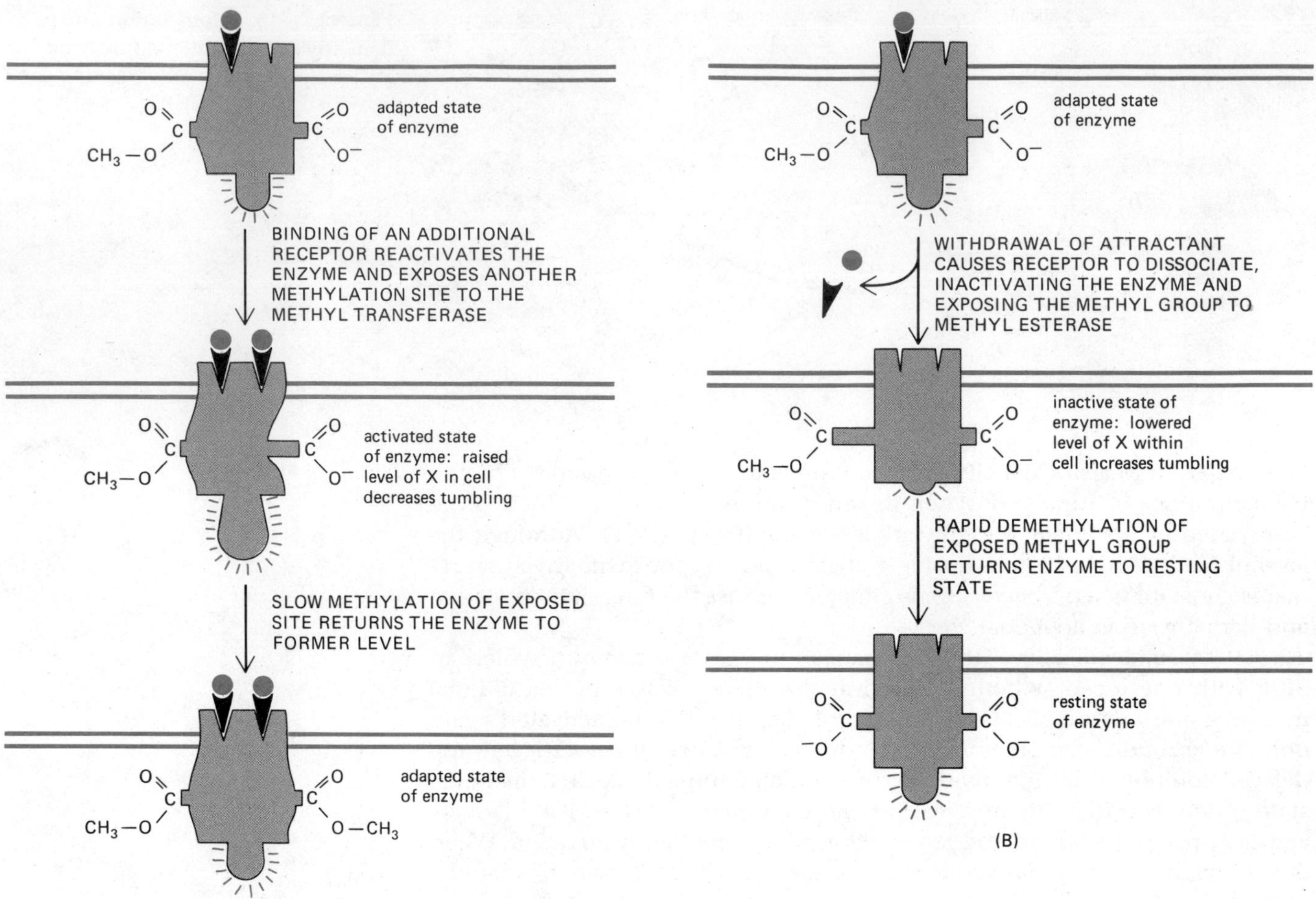

Figure 13–51 A hypothetical model of bacterial chemotaxis based on the assumption that an MCP is a membrane-bound enzyme that synthesizes an intracellular mediator X. The enzyme is rapidly activated by receptor binding and slowly suppressed by subsequent methylation during the process of adaptation. When the intracellular concentration of X is high, the flagella rotate counterclockwise so that the bacterium swims smoothly, while when X is low, rotation is clockwise and the bacterium tumbles. In this figure, an adapted MCP is transiently reactivated (so that tumbling decreases) by the binding of an additional receptor when more attractant is added (A) or is transiently inactivated (so that tumbling increases) by the dissociation of the receptor from the MCP when the attractant is removed (B). Note that the enzymatic activity of the MCP (and therefore the intracellular concentration of X and the tumbling frequency of the bacterium) is the same in the adapted states (with either one receptor bound and one methyl group added or two receptors bound and two methyl groups added) and in the resting state.

concentration of some small molecule X is high, the flagella rotate counterclockwise, and the cell swims smoothly. When the intracellular concentration of X is low, rotation is clockwise, and the cell tumbles. A further assumption is that an MCP is a membrane-bound enzyme that catalyzes the synthesis of X in the cytoplasm and that the process of chemotactic excitation tends to activate this enzyme and thereby increase the intracellular concentration of X, while the process of adaptation (that is, MCP methylation) tends to inactivate the enzyme and thereby lower the concentration of X back to its normal "resting level." At this resting level, X is at a concentration that allows the smooth swimming with occasional tumbling exhibited by cells that are not in a chemotactic environment.

According to this model, when an attractant is added, a periplasmic receptor protein activates an MCP so that the concentration of X increases and tumbling is suppressed. But a delayed increase in the methylation of the MCP returns the enzyme to its normal state so that the concentration of X returns to resting levels. Now the cell is adapted, and the pattern of tumbling is normal. However, the tumbling rate will change again if the attractant level changes—either decreasing so as to release a bound receptor from the MCP (thereby transiently inactivating the enzyme and increasing tumbling) (Figure 13–51B) *or* increasing so as to add a second receptor to the MCP (thereby transiently activating the enzyme and suppressing tumbling) (Figure 13–51A). Only when the concentration of attractant is so high that the receptor-binding and methylation sites on the MCP are saturated will the bacterium be unable to respond to a further increase in attractant concentration.

In discussing the mechanisms of bacterial chemotaxis, we have concentrated on the positive chemotactic responses to attractants and have ignored the negative chemotactic responses to repellents. The genetic studies discussed have shown that bacteria use exactly the same mechanisms to swim away from repellents as they use to swim toward attractants, except that all of the responses are reversed. For example, in the above model a receptor for a repellent would bind to an MCP and inactivate, rather than activate, the MCP, thereby decreasing the intracellular concentration of X. An increased concentration of the repellent would thereby cause the flagellar motor to rotate clockwise, causing the bacterium to tumble.

Summary

By adapting to high concentrations of a signaling ligand in a time-dependent, reversible manner, some cells are able to respond to changes in a ligand's concentration instead of to its absolute level. Adaptation occurs in various ways: ligand binding can inactivate a cell-surface receptor either by inducing its internalization and degradation or by causing the receptor to adopt an inactive conformation. Alternatively, it can result from changes in one of the nonreceptor proteins that are involved in signal transduction following receptor activation. At a molecular level the best understood example of adaptation occurs in bacterial chemotaxis, in which the reversible methylation of a key signal-transducing protein in the plasma membrane enables the cell to swim toward an optimal environment.

References

Cited

1. Williams, R.H., ed. Textbook of Endocrinology, 5th ed. Philadelphia: Saunders, 1974.

   White, A.; et al. Principles of Biochemistry, 6th ed., pp. 1185–1317. New York: McGraw-Hill, 1978.

2. Bradshaw, R.A.; Frazier, W.A. Hormone receptors as regulators of hormone action. *Curr. Top. Cell Regul.* 12:1–37, 1977.

3. Samuelsson, B.; Granström, E.; Green, K.; Hamberg, M.; Hammarström, S. Prostaglandins. *Annu. Rev. Biochem.* 44:669–695, 1975.

Harris, R.H.; Ramwell, P.W.; Gilmer, P.J. Cellular mechanisms of prostaglandin action. *Annu. Rev. Physiol.* 41:653–668, 1979.

4. Pappenheimer, J.R. The sleep factor. *Sci Am.* 235(2):24–29, 1976.
5. Levi-Montalcini, R.; Calissano, P. The nerve-growth factor. *Sci. Am.* 240(6):68–77, 1979.

   Yankner, B.A.; Shooter, E.M. The biology and mechanism of nerve growth factor. *Annu. Rev. Biochem.* 51:845–868, 1982.

   Campenot, R.B. Local control of neurite development by nerve growth factor. *Proc. Natl. Acad. Sci. USA* 74:4516–4519, 1977.
6. Yamamoto, K.R.; Alberts, B.M. Steroid receptors: elements for modulation of eukaryotic transcription. *Annu. Rev. Biochem.* 45:721–746, 1976.

   Ivarie, R.D.; O'Farrell, P.H. The glucocorticoid domain: steroid-mediated changes in the rate of synthesis of rat hepatoma proteins. *Cell* 13:41–55, 1978.

   Payvar, F.; et al. Purified glucocorticoid receptors bind selectively *in vitro* to a cloned DNA fragment whose transcription is regulated by glucocorticoids *in vivo*. *Proc. Natl. Acad. Sci. USA* 78:6628–6632, 1981.
7. Ashburner, M.; Chihara, C.; Meltzer, P.; Richards, G. Temporal control of puffing activity in polytene chromosomes. *Cold Spring Harbor Symp. Quant. Biol.* 38:655–662, 1974.
8. Attardi, B.; Ohno, S. Physical properties of androgen receptors in brain cytosol from normal and testicular feminized (Tfm/y♂) mice. *Endocrinology* 103:760–770, 1978.
9. Lefkowitz, R.J.; Roth, J.; Pricer, W.; Pastan, I. ACTH receptors in the adrenal: specific binding of ACTH-^{125}I and its relation to adenyl cyclase. *Proc. Natl. Acad. Sci. USA* 65:745–752, 1970

   Kahn, C.R. Membrane receptors for hormones and neurotransmitters. *J. Cell Biol.* 70:261–286, 1976.
10. Kahn, C.R. Membrane receptors for hormones and neurotransmitters. *J. Cell Biol.* 70:261–286, 1976.

    Kahn, C.R.; Baird, K.L.; Jarrett, D.B.; Flier, J.S. Direct demonstration that receptor crosslinking or aggregation is important in insulin action. *Proc. Natl. Acad. Sci. USA* 75:4209–4213, 1978.

    Rees Smith, B.; Buckland, P.R. Structure-function relations of the thyrotropin receptor. In Receptors, Antibodies and Disease (Ciba Foundation Symposium 90), pp. 114–132. London: Pitman, 1982.
11. Sutherland, E.W. Studies on the mechanism of hormone action. *Science* 177:401–408, 1972.

    Carpenter, G.; Cohen, S. Epidermal growth factor. *Annu. Rev. Biochem.* 48:193–216, 1979.

    Stevens, C.F. The neuron. *Sci. Am.* 241(3):54–65, 1979.
12. Sutherland, E.W. Studies on the mechanism of hormone action. *Science* 177:401–408, 1972.

    Pastan, I. Cyclic AMP. *Sci. Am.* 227(2):97–105, 1972.
13. Ross, E.M.; Gilman, A.G. Biochemical properties of hormone-sensitive adenylate cyclase. *Annu. Rev. Biochem.* 49:533–564, 1980.

    Schramm, M.; Orly, J.; Eimerl, S.; Korner, M. Coupling of hormone receptors to adenylate cyclase of different cells by cell fusion. *Nature* 268:310–313, 1977.
14. Rodbell, M. The role of hormone receptors and GTP-regulatory proteins in membrane transduction. *Nature* 284:17–22, 1980.

    Helmreich, E.J.M.; Zenner, H.P.; Pfeuffer, T.; Cori, C.F. Signal transfer from hormone receptor to adenylate cyclase. *Curr. Top. Cell Regul.* 10:41–87, 1976.

    Johnson, G.L.; Kaslow, H.R.; Farfel, Z.; Bourne, H.R. Genetic analysis of hormone-sensitive adenylate cyclase. In Advances in Cyclic Nucleotide Research (P. Greengard, G.A. Robison, eds.), Vol. 13, pp. 1–38. New York: Raven, 1980.

    Lai, C.-Y. The chemistry and biology of cholera toxin. *CRC Crit. Rev. Biochem.* 9:171–206, 1980.

    Cassel, D.; Pfeuffer, T. Mechanism of cholera toxin action: covalent modification of the guanyl nucleotide-binding protein of the adenylate cyclase system. *Proc. Natl. Acad. Sci. USA* 75:2669–2673, 1978.
15. Heilbrunn, L.V.; Wiercenski, F.J. The action of various cations on muscle protoplasm. *J. Cell. Comp. Physiol.* 29:15–32, 1947.

Carafoli, E.; Crompton, M. The regulation of intracellular calcium. *Curr. Top. Membr. Transp.* 10:151–216, 1978.

Racker, E. Fluxes of Ca^{2+} and concepts. *Fed. Proc.* 39:2422–2425, 1980.

16. Rubin, R.P. The role of calcium in the release of neurotransmitter substances and hormones. *Pharmacol. Rev.* 22:389–428, 1970.

Rasmussen, H.; Waisman, D. The messenger function of calcium in endocrine systems. In Biochemical Actions of Hormones (G. Litwack, ed.), Vol. 8, pp. 1–115. New York: Academic Press, 1981.

17. Cohen, P. The role of protein phosphorylation in neural and hormonal control of cellular activity. *Nature* 296:613–620, 1982.

Hoppe, J.; Wagner, K.G. Cyclic AMP-dependent protein kinase I, a unique allosteric enzyme. *Trends Biochem. Sci.* 4:282–285, 1979.

Smith, S.B.; White, H.D.; Siegel, J.B.; Krebs, E.G. Cyclic AMP-dependent protein kinase I: cyclic nucleotide binding, structural changes, and release of the catalytic subunits. *Proc. Natl. Acad. Sci. USA* 78:1591–1595, 1981.

18. Cohen, P. The role of cyclic-AMP-dependent protein kinase in the regulation of glycogen metabolism in mammalian skeletal muscle. *Curr. Top. Cell Regul.* 14:117–196, 1978.

19. Krebs, E.G.; Beavo, J.A. Phosphorylation-dephosphorylation of enzymes. *Annu. Rev. Biochem.* 48:923–959, 1979.

Cohen, P. The role of protein phosphorylation in the neural and hormonal control of intermediary metabolism. In Cellular Controls in Differentiation (C.W. Lloyd, D.A. Rees, eds.), pp. 81–105. London: Academic Press, 1981.

20. Means, A.R.; Dedman, J.R. Calmodulin—an intracellular calcium receptor. *Nature* 285:73–77, 1980.

Klee, C.B.; Crouch, T.H.; Richman, P.G. Calmodulin. *Annu. Rev. Biochem.* 49:489–515, 1980.

Cheung, W.Y. Calmodulin. *Sci. Am.* 246(6):48–56, 1982.

Kretsinger, R.H. Mechanisms of selective signaling by calcium. *Neurosci. Res. Program Bull.* 19:213–328, 1981.

21. Goldberg, N.D.; Haddox, M.K. Cyclic GMP metabolism and involvement in biological regulation. *Annu. Rev. Biochem.* 46:823–896, 1977.

22. Schimke, R.T. On the roles of synthesis and degradation in regulation of enzyme levels in mammalian tissues. *Curr. Top. Cell Regul.* 1:77–124, 1969.

23. Mulvihill, E.R.; Palmiter, R.D. Relationship of nuclear estrogen receptor levels to induction of ovalbumin and conalbumin mRNA in chick oviduct. *J. Biol. Chem.* 252:2060–2068, 1977.

Lewis, J.; Slack, J.; Wolpert, L. Thresholds in development. *J. Theor. Biol.* 65:579–590, 1977.

24. Raff, M. Self regulation of membrane receptors. *Nature* 259:255–266, 1976.

Kahn, C.R.; Roth, J. Cell membrane receptors for polypeptide hormones: applications to the study of disease states in mice and men. *Am. J. Clin. Pathol.* 63:656–668, 1975.

25. Lefkowitz, R.J. Regulation of β-adrenergic receptors by β-adrenergic agonists. In Receptors and Hormone Action (L. Birnbaumer, B.W. O'Malley, eds.), Vol. 3, pp. 179–194. New York: Academic Press, 1978.

Katz, B.; Thesleff, S. A study of the "desensitization" produced by acetylcholine at the motor end-plate. *J. Physiol.* 138:63–80, 1957.

26. Klee, W.A.; Sharma, S.K.; Nirenberg, M. Opiate receptors as regulators of adenylate cyclase. *Life Sci.* 16:1869–1874, 1975.

Snyder, S.H. Opiate receptors and internal opiates. *Sci. Am.* 236(3):44–56, 1977.

27. Schiffmann, E.; Gallin, J.I. Biochemistry of phagocyte chemotaxis. *Curr. Top. Cell Regul.* 15:203–261, 1979.

28. Berg, H. How bacteria swim. *Sci. Am.* 233(2):36–44, 1975.

Adler, J. The sensing of chemicals by bacteria. *Sci. Am.* 234(4):40–47, 1976.

Koshland, D.E., Jr. Biochemistry of sensing and adaptation. *Trends Biochem. Sci.* 5:297–301, 1980.

29. Springer, M.S.; Goy, M.F.; Adler, J. Protein methylation in behavioral control mechanisms and in signal transduction. *Nature* 280:279–284, 1979.

Koshland, D.E., Jr. Biochemistry of sensing and adaptation in a simple bacterial system. *Annu. Rev. Biochem.* 50:765–782, 1981.

III
From Cells to Multicellular Organisms

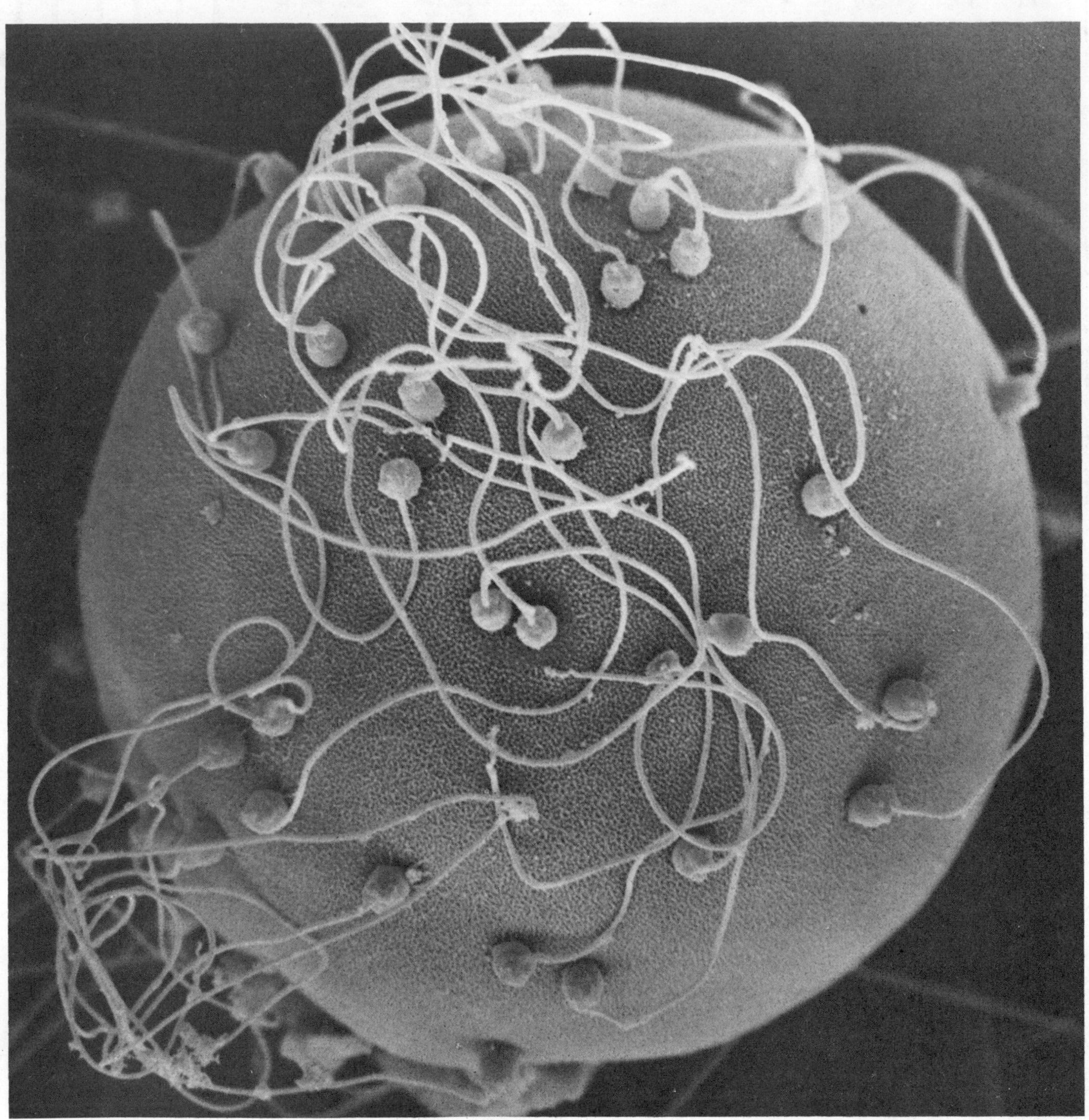

Scanning electron micrograph of a clam egg with many sperm attached to its surface. (Courtesy of David Epel.)

Germ Cells and Fertilization 14

Sex is not necessary for reproduction. Amoebae, for example, reproduce by simple mitotic division; *Hydra* bud off complete offspring from the middle part of their bodies (Figure 14–1); while sea anemones and marine worms split in two, each half-organism regenerating the missing half. But while such **asexual reproduction** is simple and direct, it gives rise to offspring that are genetically identical to the parent organism. **Sexual reproduction,** on the other hand, involves the mixing of genomes from two separate members of the species to produce offspring that usually differ genetically from one another and from both their parents. It seems that sexual reproduction, with the consequent genetic diversification, must have great advantages, for the vast majority of plants and animals have adopted it. Even many procaryotes and single-celled eucaryotes have evolved a capacity to reproduce sexually. This chaper is concerned with the cellular machinery of sexual reproduction. Before discussing in detail how the machinery works, we shall pause to consider the reasons for its existence and the consequences of its operation.

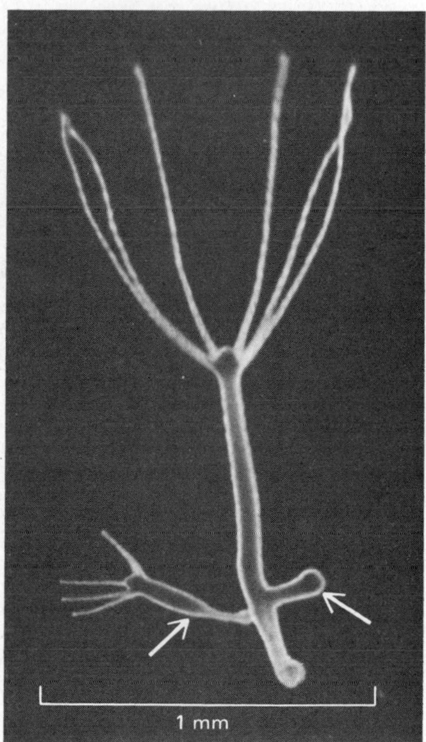

Figure 14–1 Photograph of a *Hydra* from which two new organisms are budding (*arrows*). The offspring, genetically identical to their parent, will eventually detach and live independently. (Courtesy of Amata Hornbruch.)

The Benefits of Sex[1]

The sexual reproductive cycle involves an alternation of *haploid* generations of cells, each carrying a single set of chromosomes, with *diploid* generations of cells, each carrying a double set of chromosomes. The mixing of genomes is achieved by fusion of two haploid cells to form a diploid cell. New haploid cells are created in their turn when a cell of the diploid generation divides by the process of *meiosis,* through which the genes of the double chromosome set are parceled out afresh in single chromosome sets (Figure 14–2). *Genetic recombination* between chromosomes during meiosis provides each cell of the new haploid generation with a novel assortment of genes, originating partly from one ancestral cell of the previous haploid generation and partly from the other. Thus, through cycles of haploidy, fusion, diploidy, and meiosis, old combinations of genes are broken up and new combinations are created.

In Multicellular Animals the Diploid Phase Is Complex and Long, the Haploid Simple and Fleeting

Cells proliferate in the course of the sexual cycle by ordinary mitotic division, most commonly during the diploid phase. Some primitive organisms, such as certain types of yeast, are exceptional in that only the haploid cells proliferate mitotically: the diploid cell, once formed, proceeds directly to meiosis. In lower plants, such as mosses and ferns, both haploid and diploid phases of growth are important. In the flowering plants, however, the haploid phase is very brief and simple, while the diploid phase is extended into a long period of development and proliferation. The same is true, in still greater degree, of almost all multicellular animals, including all vertebrates. Practically the whole of their life cycle is spent in the diploid state; the haploid cells exist only briefly, do not divide at all, and are highly specialized for sexual fusion (Figure 14–3). We shall argue below that the predominantly diploid life cycle has created important opportunities for evolution.

Haploid cells specialized for sexual fusion are called **gametes.** Typically, two types of gametes are formed: one large and nonmotile, referred to as the **egg** (or **ovum**); the other small and motile, referred to as the **sperm** (or **spermatozoon**) (Figure 14–4). During the diploid phase that follows fusion of gametes, the cells proliferate and diversify to form a complex multicellular organism. In most animals, though not in plants, a useful distinction can be drawn between the cells of the **germ line,** from which the next generation of gametes will be derived, and the **somatic** cells, which form the rest of the body and leave no progeny. In a sense, the somatic cells exist only to help the cells of the germ line survive and propagate.

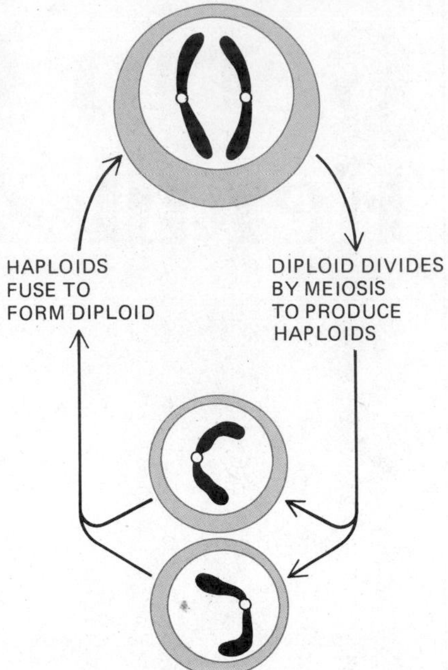

Figure 14–2 The sexual life cycle involves an alternation of haploid and diploid generations of cells.

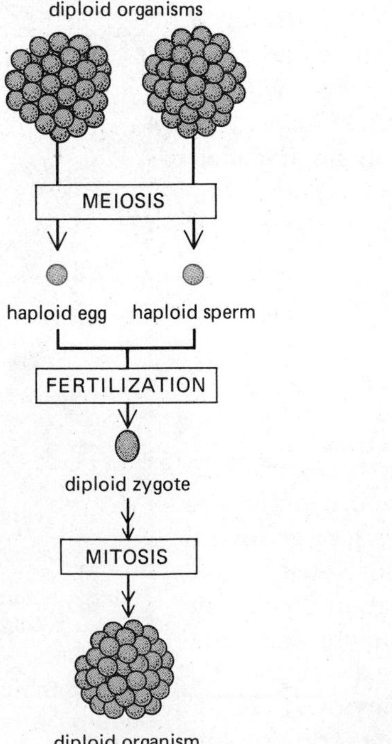

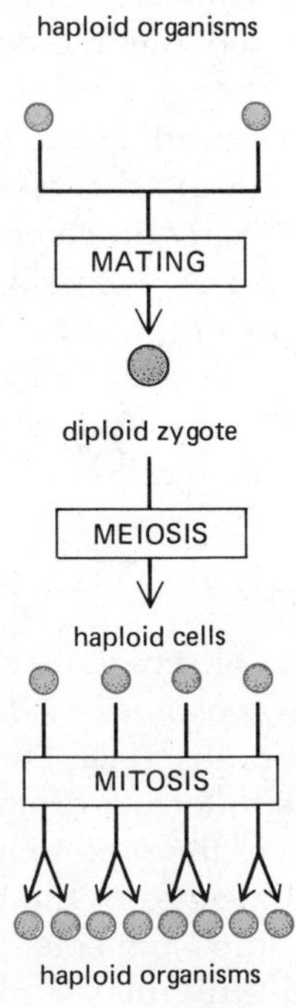

Figure 14–3 Schematic drawing showing how in higher eucaryotic organisms cells proliferate in the diploid phase to form a multicellular organism, and only the gametes are haploid. In some lower eucaryotes, by contrast, it is the haploid cells that proliferate, and the only diploid cell is the *zygote,* which exists transiently following mating. The haploid cells are shown in color.

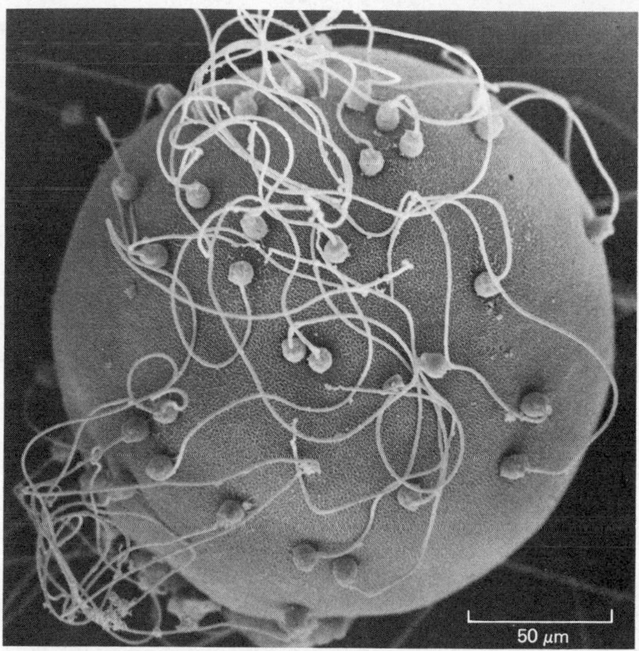

Figure 14–4 Scanning electron micrograph of a clam egg with many spermatozoa bound to its surface. (Courtesy of David Epel.)

Sexual Reproduction Gives a Competitive Advantage to Organisms in an Unpredictably Variable Environment

The machinery of sexual reproduction is elaborate, and the resources spent on it are large. What benefits does it bring and why did it evolve? Sexual individuals beget unpredictably dissimilar offspring, whose haphazard genotypes are at least as likely to represent a change for the worse as a change for the better. Why then should sexual individuals have a competitive advantage over individuals that breed true, by an asexual process? Although this problem continues to perplex population geneticists, the general conclusion seems to be that the reshuffling of genes through sexual reproduction helps a species to survive in an unpredictably variable environment. If a parent produces many offspring with a wide variety of gene combinations there is a better chance that at least one will be well matched to future circumstances, whatever they may be.

But this is not the place for a detailed discussion of the competitive advantages of sexual reproduction. We want to focus instead on a rather different question: given that a species reproduces sexually, what will the consequences be for its evolution in the long term? Why is it that all the most complex organisms have evolved through generations of sexual, rather than asexual, reproduction?

Sexual Reproduction Helps Establish Favorable Alleles in a Large Population

Evolution depends to a large extent on mutations that alter existing genes to create in their place new *alleles*, or variants, of these genes. Suppose that two individuals in a population undergo beneficial mutations affecting in each a different *genetic locus* and therefore a different function. In an asexual species, each of these individuals will give rise to a clone of mutant progeny, and the two clones will compete until one or the other triumphs. One of the two beneficial alleles created by the mutations will thus spread through the population, while the other will eventually be lost. The two mutations cannot both benefit the members of a species unless they both occur, successively, in one and the same cell line; and since beneficial mutations are rare, it will generally be a long while before that happens. In a sexual species, on the other hand, beneficial new alleles created by mutations occurring in different individuals, and affecting different loci, can be brought together into the same genome by mating and recombination. Several such beneficial new alleles can thus spread simultaneously through the population instead of competing against one another (Figure 14–5). Detailed calculations show, however, that the advantage will be significant only for very large populations (greater than about a million).

The evolution of a complex organism, however, requires something more than the introduction of improved forms of existing genes. It requires the creation of new genes to serve new functions.

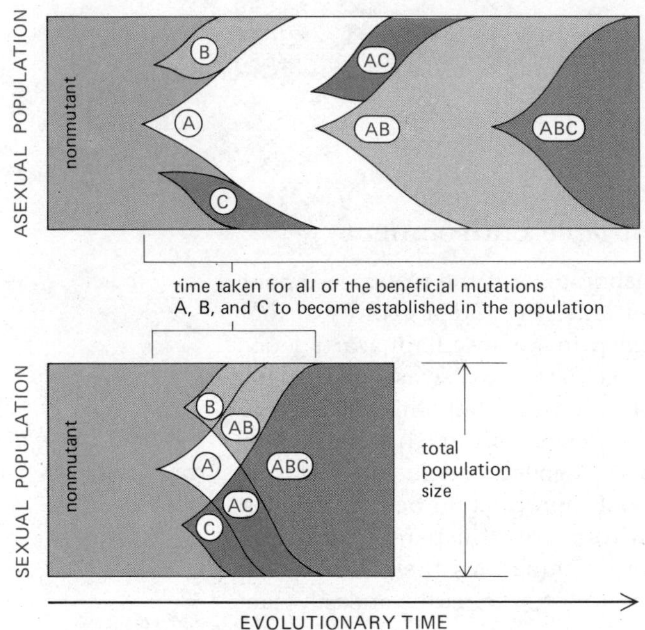

Figure 14–5 Diagram illustrating how sexual reproduction helps the spread of beneficial mutations through a population. A, B, and C represent three beneficial mutations occurring at three different loci; of the three, A confers the greatest fitness, while individuals carrying the three mutations A, B, and C simultaneously have the greatest fitness of all. In the asexual population, mutations A, B, and C arise at first in separate individuals, and these individuals compete with one another as well as with the original nonmutants; A wins and becomes fixed in the population, while B and C become extinct. AB individuals are not produced until the mutation B occurs afresh in an individual descended from the original A mutant, and ABC individuals are not produced until the mutation C occurs in an AB individual. In the sexual population, the mutations A, B, and C arise independently in separate individuals, as before, but now combined AB, AC, and ABC genotypes can be rapidly created by genetic recombination. Thus all three beneficial mutations simultaneously spread through the population, which will rapidly arrive at the ABC genotype.

New Genes Evolve by Duplication and Divergence

Many of the proteins in a multicellular animal can be grouped into families: the collagens, the globins, the actins, the serine proteases, and so on. Proteins in the same family are related both in function and in amino acid sequence. There can be little doubt that each family has evolved from a single ancestral gene by a process of *duplication* and *divergence* (see p. 118). Different members of a protein family are often characteristic of different tissues of the body, where they perform analogous but distinctive tasks. The creation of new genes by diversification and specialization of existing genes has plainly been crucial for the evolution of complex multicellular organisms. We shall see that the detailed sequence of events, however, is very different for diploid and haploid species. Diploid species enjoy an important advantage: they have a spare copy of each gene, and this spare copy can mutate and serve as raw material for innovation. A haploid species does not have this easy means of taking the first step toward evolving a larger and more sophisticated genome. To make the mechanism plain, we must first explain more carefully the relationship between sex and diploidy.

Sexual Reproduction Keeps a Diploid Species Diploid[2]

A diploid organism contains two copies of each gene; yet in most cases a single copy is sufficient for good health and survival. A mutation that disrupts the function of an essential gene will be lethal in a haploid individual but may be harmless in the diploid if only one of the two gene copies is affected. In general, diploid organisms harbor many such *recessive lethal* mutant alleles in their genomes. However, sexual reproduction sets a limit to their frequency. Individuals both of whose parents carry a recessive lethal mutation of the same gene may inherit two mutant copies of the gene and no good copies. Such individuals will die, and their mutant copies of the gene will die with them. The more common the mutant gene in the population, the higher the rate of its elimination by this mechanism. Thus a balance is struck between the rate of elimination of the mutant allele and its rate of production by new mutational events. At equilibrium the recessive lethal mutant allele, though far more common than it would be in a haploid population, is still a rarity: the vast majority of individuals will be truly diploid at the locus in question, with two functional copies of the gene. Essentially the same argument applies to recessive mutations that are merely deleterious rather than lethal.

Consider, by contrast, a population consisting initially of diploid individuals that reproduce asexually. Here there will be no selection against recessive lethal or deleterious mutations affecting one of the two gene copies. Heterozygous individuals will run no risk of producing nonviable homozygous progeny by sexual recombination. Thus in the course of many generations, recessive deleterious mutations will accumulate in the genome until it degenerates from diploidy into a condition in which the total amount of DNA is as large as ever but only a single functional copy of each of the original essential genes remains. The organism in this way becomes functionally haploid. Without sexual reproduction, therefore, the diploid species will not stay diploid: with sexual reproduction it will (Figure 14–6).

A Diploid Species Has a Spare Copy of Each Gene Free to Mutate to Serve a New Function[2]

We have spoken of recessive deleterious mutations, which are indeed the most common type. Let us now return to the consideration of a mutation that modifies an existing gene so that it can perform a valuable new function. As

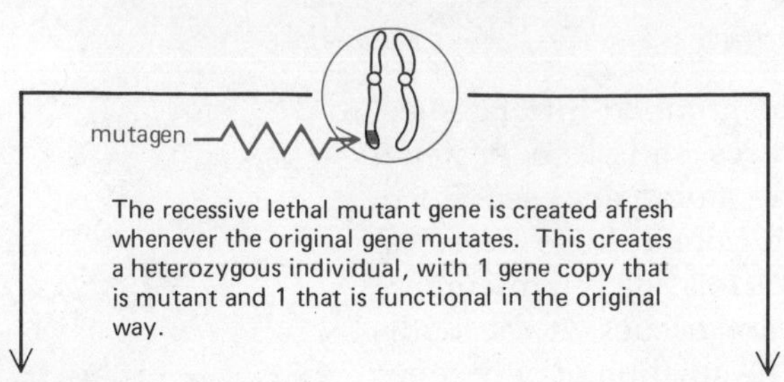

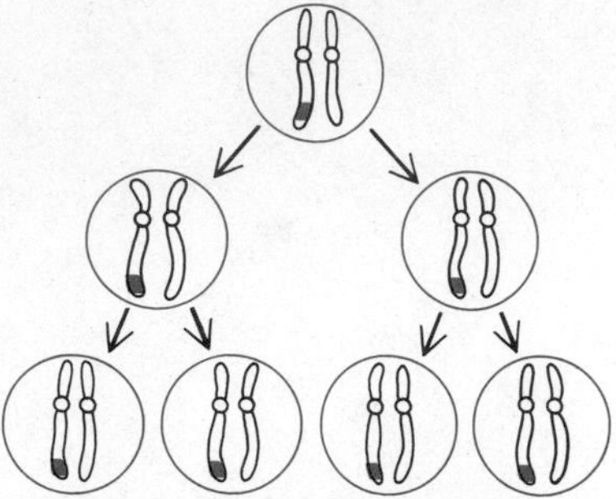

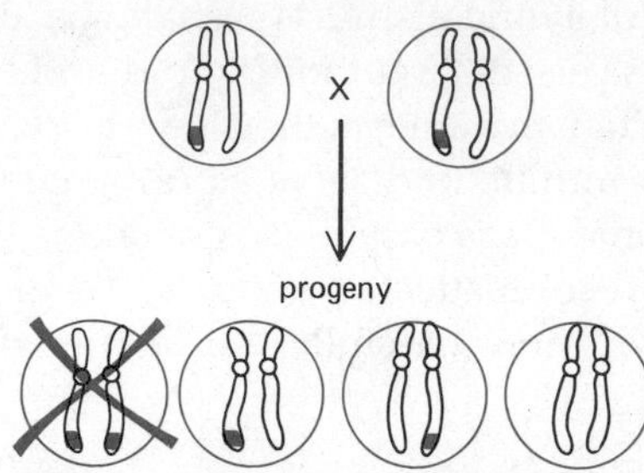

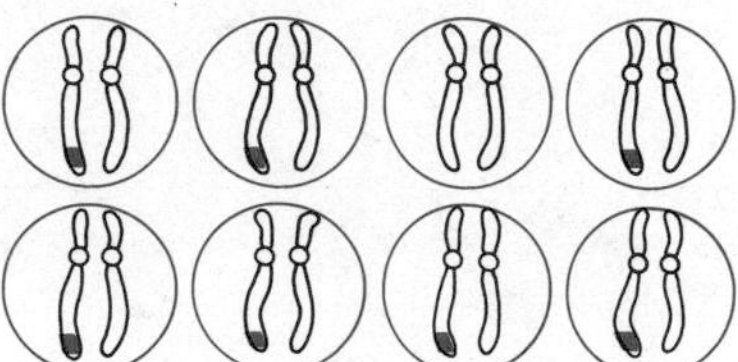

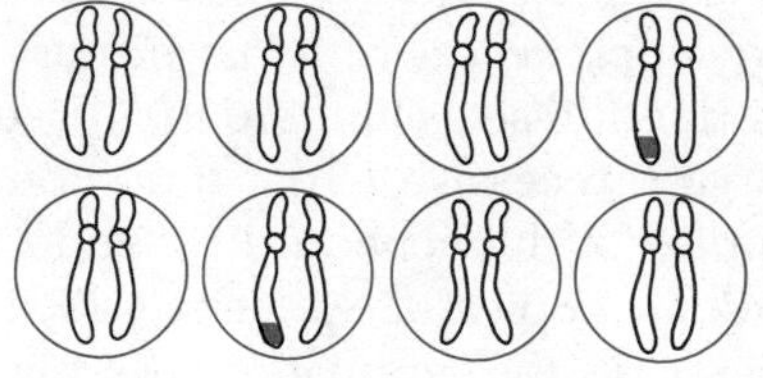

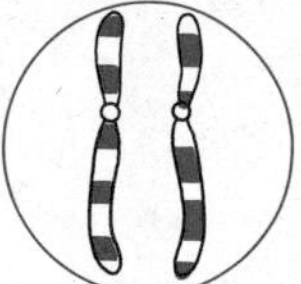

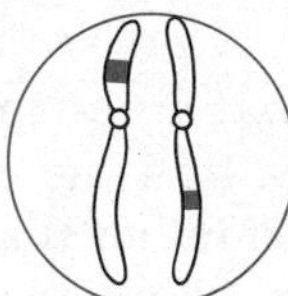

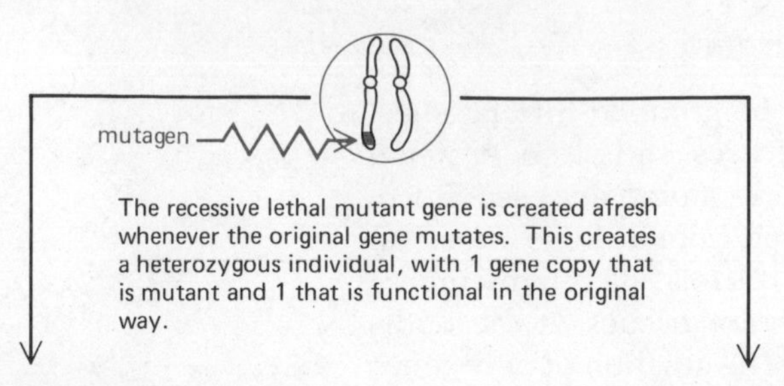

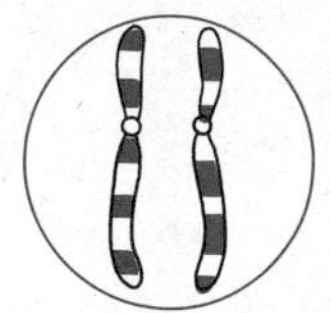

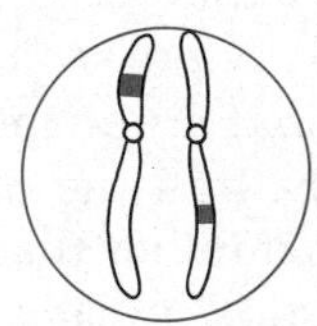

Figure 14–6 Schematic diagram showing how sexual reproduction keeps diploid organisms diploid during the course of evolution. For simplicity, only recessive lethal mutations are considered. The argument for recessive deleterious mutations would be similar.

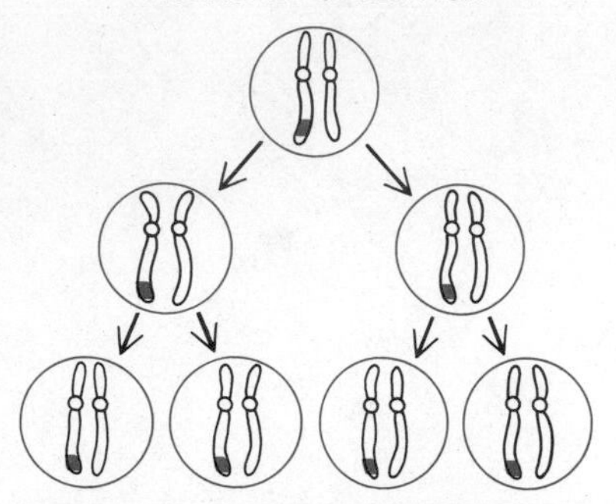

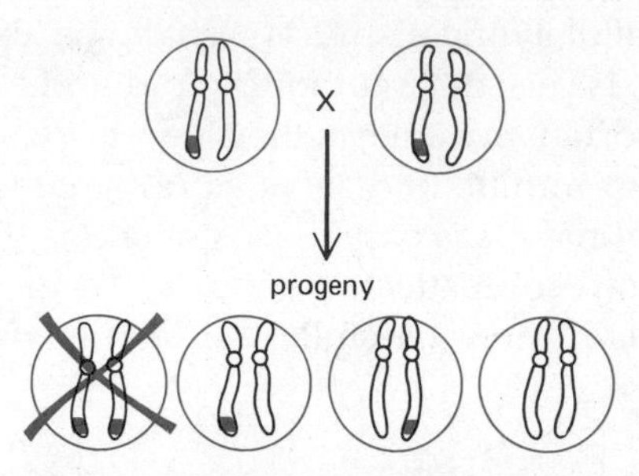

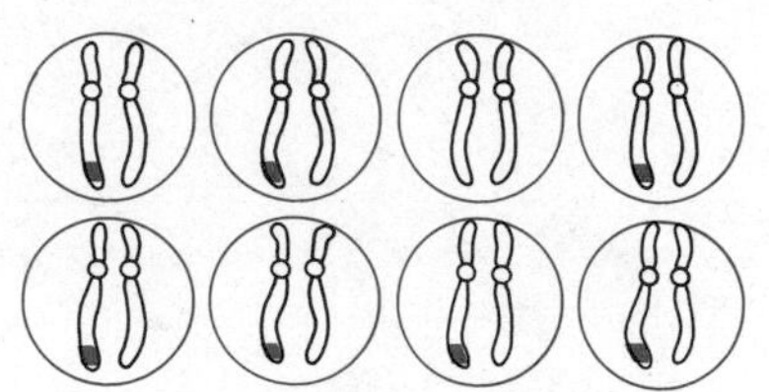

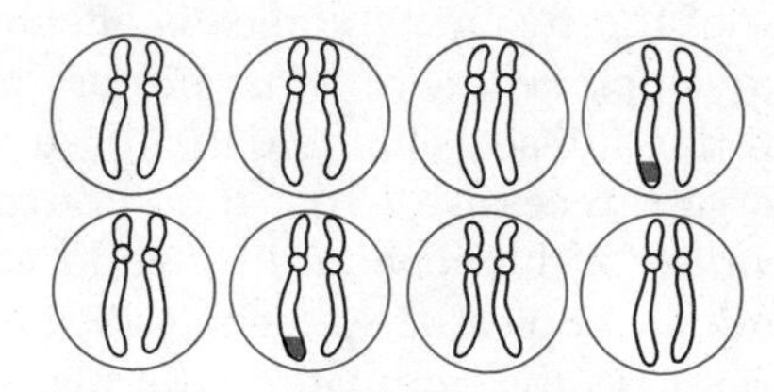

a rule, the mutation will at the same time spoil the gene for its original function. This change spells disaster for a haploid organism if the original function is vital. But in the diploid organism the occurrence of the mutation in one of the two gene copies is not merely tolerable, it is beneficial. The heterozygous individual will have the advantage of both the old gene function and the new. Homozygotes, with two copies of the old allele or two copies of the new, will have inferior fitness. In such cases of *heterozygote advantage,* the mutant gene will spread rapidly through the sexual diploid population until an equilibrium is reached in which both the old and the new alleles are present at high frequency and the number of heterozygous individuals is large. (This phenomenon, known as *balanced polymorphism,* has been well documented experimentally.) There is a penalty to be paid, however; for when two heterozygotes mate, a large proportion of their offspring, according to the usual Mendelian rules, will be homozygotes of inferior fitness. But this state of affairs will not persist forever: there is a way forward.

A Diploid Species Can Rapidly Enrich Its Genome by the Addition of New Genes[3]

From time to time, in all organisms, spontaneous gene duplications occur: a chromosome that contains one copy of a gene G gives rise, through an error of DNA replication, to a chromosome containing two copies of the gene G in tandem. The duplication in itself brings no advantage and will, as a rule, be found only in a very few individuals. Suppose, though, that it occurs at a locus where a beneficial mutant allele G* is maintained by heterozygote advantage at high frequency in the population, in coexistence with the original allele G (Figure 14–7). There is a high probability then that in the diploid cell containing the GG chromosome (carrying the duplication), the homologous companion chromosome will carry the G* allele, giving a genotype GG/G*. By genetic recombination at meiosis (see below), gametes with a GG* chromosome can then be produced. In these, the original gene G and the mutant G* are ar-

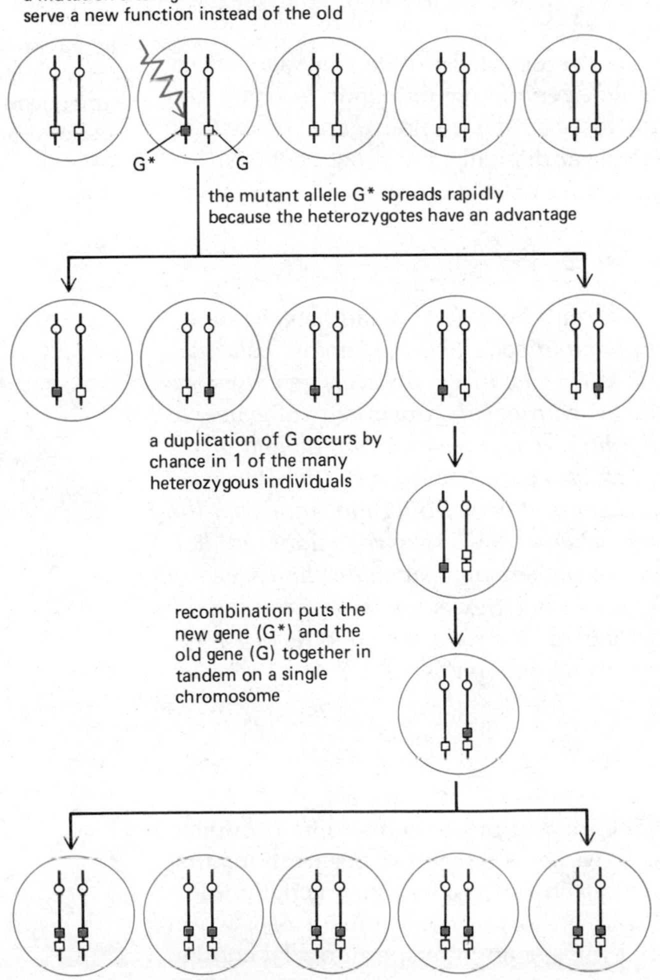

Figure 14–7 The creation of a new gene (G*) in a sexually reproducing diploid organism via a "mutate, spread, and duplicate" pathway.

ranged in tandem, no longer as two alleles competing to occupy the same locus but as two separate genes, each with a locus of its own. This is a winning combination, and it will be rapidly propagated until the whole population consists of GG*/GG* homozygotes (Figure 14–7). With this genotype each individual will not only have the advantage of possessing both the old gene G and the new gene G*, it will also be able to pass on the advantage to all its progeny.

Thus in the sexual diploid species, new genes can be created by mutation of the spare copies of existing genes; the new genes can become common through heterozygote advantage without loss of the original genes; and, finally, they can be inserted as additions to the genome through gene duplication and genetic recombination. This sequence of events is possible only for the diploid species. For a haploid species, innovation is more difficult. If it is not to lose the old gene in the process of acquiring the new, it must wait until the innovative mutation occurs in one of the very few individuals that already carry a duplication at the appropriate locus. Since both the particular mutation and the particular duplication occur only rarely, the haploid species will have to wait a very long time indeed before they occur in conjunction (Figure 14–8). Detailed calculations show that the diploid organism should typically be able to enlarge and enrich its genome with new genes for new purposes at a rate that is hundreds or thousands of times faster than the haploid organism.

In summary, sexual reproduction maintains diploidy, and diploidy, in turn, gives a species special opportunities for the evolution of a larger, more complex, and more versatile genome.

We shall now proceed to examine the detailed cellular mechanisms of sex. First, the events of *meiosis,* in which genetic recombination occurs and diploid cells of the germ line divide to produce haploid gametes, will be considered, then the gametes themselves, and, finally, the process of *fertilization,* in which the gametes fuse to form a new diploid organism.

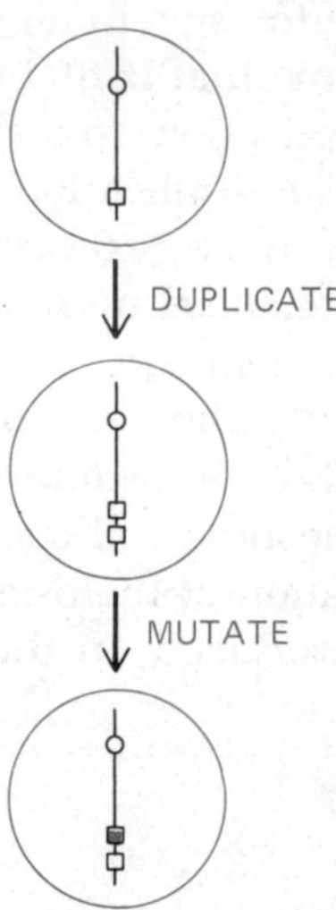

Figure 14–8 The creation of a new gene in a haploid organism. This sequence of events appears much simpler than that shown in Figure 14–7, but it proceeds very much more slowly.

Summary

Sexual reproduction involves a cyclic alternation of diploid and haploid states: diploid cells divide by meiosis to form haploid cells, and the haploid cells fuse in pairs at fertilization to form new diploid cells. In the process, genomes are mixed and recombined to give individuals with novel assortments of genes. In higher plants and animals, most of the life cycle is spent in the diploid phase and the haploid phase is very brief. Sexual reproduction has probably been favored by evolution because random genetic recombination improves the chances of producing at least some offspring that will survive in an unpredictably variable environment. At the same time, sex may facilitate the spread of beneficial mutations through a large population. Sex is necessary also for the maintenance of diploidy and thus has helped to create the conditions for the rapid evolution of new genes in higher plants and animals.

Meiosis[4]

The realization that germ cells are haploid, and must therefore be produced by a special mechanism of cell division, came as a result of observations that were also among the first to suggest that chromosomes carry genetic information. In 1883, during cytological studies on the development of a worm (*Parascaris equorum*), it was found that the egg and the sperm nuclei contain only two chromosomes each, whereas the fertilized egg contains four chromosomes. The chromosome theory of heredity could therefore explain the long-standing paradox that maternal and paternal contributions to the character of the progeny seem often to be equal, despite the enormous difference in the size of the egg and the sperm.

The other important implication of this finding was that germ cells must be formed by a special kind of nuclear division in which the chromosome complement is precisely halved. The behavior of the chromosomes during meiosis, in which this reduction is achieved, turned out to be more complex than at first expected. Consequently, it was not until the early 1930s, as a result of a great deal of painstaking research that combined cytology and genetics, that the important features of the meiotic process became firmly established.

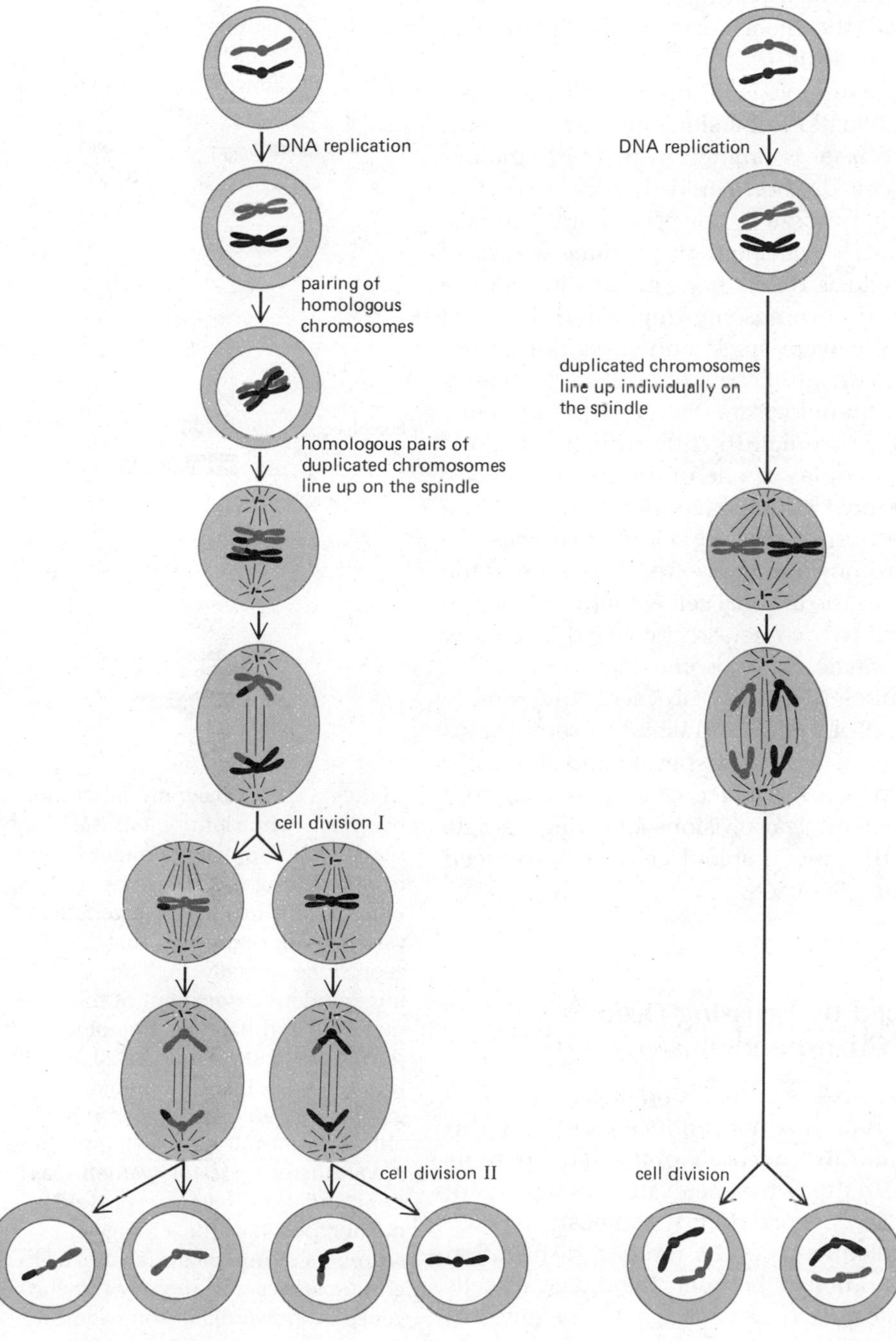

Figure 14–9 Schematic diagram comparing the process of meiosis to a normal mitosis. For clarity, only one set of homologous chromosomes is shown. The pairing of homologous chromosomes (homologs) is unique to meiosis; because each chromosome is duplicated and exists as attached sister chromatids before this pairing occurs, two nuclear divisions are required to produce the haploid gametes. Each diploid cell that enters meiosis therefore produces four haploid cells.

Meiosis Involves Two Nuclear Divisions Rather Than One

With the exception of the sex chromosomes, a diploid nucleus contains two copies of each chromosome, one from the male and one from the female parent. These two copies are called **homologs.** Before an ordinary *mitotic* division, both homologs of each chromosome are duplicated, and the two copies remain together as **sister chromatids.** These sister chromatids line up on the spindle with their kinetochore fibers pointing toward opposite poles. As a result, the sister chromatids are separated from each other at anaphase, and each daughter cell inherits one copy of each homolog (p. 648). However, each haploid gamete produced by the division of a diploid cell during meiosis must contain only one homolog of each pair. This requirement makes an extra demand on the machinery for cell division because it means that homologs must be able to recognize one another and become physically paired before they line up on the spindle. This pairing of the maternal and the paternal copy of each chromosome, the mechanism of which we shall describe in a later section, is unique to meiosis.

Given a mechanism for the pairing of homologous chromosomes, meiosis might in principle take place by a modification of a single mitotic cell cycle in which chromosome duplication (S phase) is omitted, with the homologs pairing before M phase. The ensuing cell division could then produce two haploid cells directly. The actual meiotic process is, however, more complex. Before the homologs pair, they undergo duplication to produce a pair of closely associated sister chromatids. Meiosis differs from mitosis in that the sister chromatids behave as a unit, as if chromosome duplication had not occurred. First, each homolog acts as if it were single and seeks out its homologous partner with which to pair. The resulting pair, or *bivalent,* then lines up on the spindle. At anaphase the two homologs are distributed to opposite poles, each still composed of joined sister chromatids. Thus, when the meiotic cell divides, each daughter inherits two copies of one of the two homologs. The two progeny of this division (**division I of meiosis**) therefore contain a diploid amount of DNA but differ from normal diploid cells in two ways: (1) both of the two DNA copies of each chromosome derive from only one of the two homologous chromosomes present in the original cell (either the paternal or the maternal homolog), and (2) these two copies are inherited as closely associated sister chromatids forming a single chromosome (Figure 14–9).

Formation of the actual gamete nuclei can now proceed quite simply through a second cell division, **division II of meiosis,** in which chromosomes are aligned, without further replication, on a second spindle and the sister chromatids separate, as in normal mitosis, to produce cells with a haploid DNA content. Meiosis thus consists of two nuclear divisions following a single phase of chromosome replication, so that four haploid cells are produced from each cell that enters meiosis (Figure 14–9).

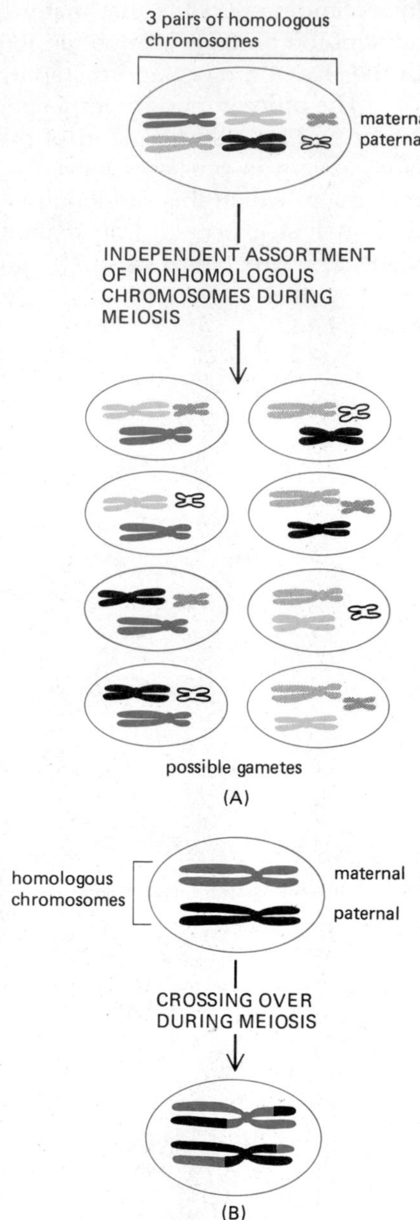

Figure 14–10 Diagram illustrating two major contributions to the reassortment of genetic material that occurs during meiosis. Both mechanisms increase the genetic variability in organisms that reproduce sexually. (A) The independent assortment of the maternal and paternal homologs during the first meiotic division produces 2^n different haploid gametes for an organism with n chromosomes. Here $n = 3$ and there are 8 different possible gametes, as indicated. (B) Crossing over during meiotic prophase I exchanges segments of homologous chromosomes and thereby reassorts genes in individual chromosomes.

Genetic Reassortment Is Enhanced by Crossing Over Between Homologous Nonsister Chromatids[5]

We have already seen that genes can be mixed by the fusion of gametes from two different individuals. But genetic variation is not produced solely by this means. Unless they are identical twins, no two offspring of the same parents are quite alike. This is because long before the two gametes fuse, two different kinds of genetic reassortment have already occurred during meiosis.

One kind of reassortment is a consequence of the random distribution of the different maternal and paternal homologs between the daughter cells at meiotic division I, as a result of which each gamete acquires a different

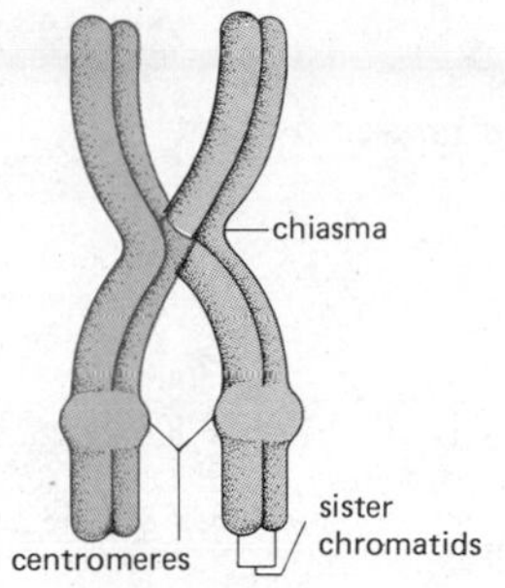

Figure 14–11 Schematic drawing of paired homologous chromosomes during the transition to metaphase of meiotic division I. A single crossover event has occurred earlier in prophase to create one chiasma. Note that the four chromatids present are arranged in two distinct pairs of sister chromatids, each pair being tightly aligned as well as joined at their centromeres. The entire unit shown here is therefore frequently referred to as a *bivalent.*

mixture of maternal and paternal chromosomes (Figure 14–10A). From this fact alone, the cells of any one individual could, in principle, give rise to 2^n genetically different gametes, where n is the haploid number of chromosomes. In humans, for example, each individual can produce at least $2^{23} = 8.4 \times 10^6$ genetically different gametes. But in actuality the number is very much greater than this because of **chromosomal crossing over,** a process that takes place during the long prophase of meiotic division I and in which parts of homologous chromosomes are exchanged. On average, between two and three such crossover events occur on each pair of human chromosomes. This process scrambles the genetic constitution of each of the chromosomes in gametes, as illustrated in Figure 14–10B.

Chromosomal crossing over involves breaking the single maternal and paternal DNA double helices in each of two chromatids and rejoining them to each other in a reciprocal fashion by a process known as **genetic recombination.** What is known about the molecular details of this process is outlined in Chapter 5 (see p. 241). Recombination takes place during prophase of meiotic division I, at a time when the two sister chromatids are so tightly packed together that their individuality cannot be distinguished (see below). Much later in this extended prophase, the two separate chromatids of each chromosome become clearly visible. They are now seen to be connected at their centromeres, and each of the four chromatids in each bivalent can be identified as belonging to one homolog or the other. The two homologs remain attached to each other at points where a crossover between a paternal and a maternal chromatid has occurred. At each such point, called a **chiasma** (plural **chiasmata**), two of the four chromatids are seen to have crossed over between the homologs (Figure 14–11). Chiasmata are thus the morphological consequences of a prior, unobserved crossover event.

At this stage of meiosis, each pair of homologs, or bivalent, is usually held together by at least one chiasma. Many bivalents contain more than one chiasma, reflecting the fact that multiple crossovers can occur between homologs (Figures 14–12 and 14–13).

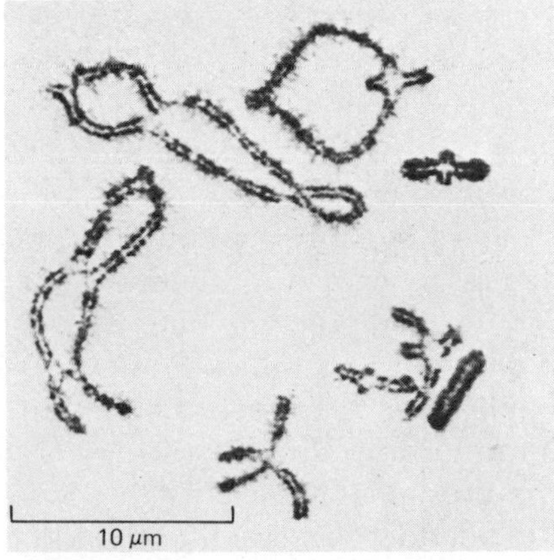

Figure 14–12 Light micrograph of several bivalents containing multiple chiasmata at diplotene. These large grasshopper chromosomes provide particularly favorable material for cytological observations of this kind. (Courtesy of Bernard John.)

A Synaptonemal Complex Mediates Chromosome Pairing[6]

Elaborate morphological changes occur in the chromosomes as they pair (*synapse*) and separate (*desynapse*) during the first meiotic prophase. This prophase is divided into five sequential stages—*leptotene, zygotene, pachytene,*

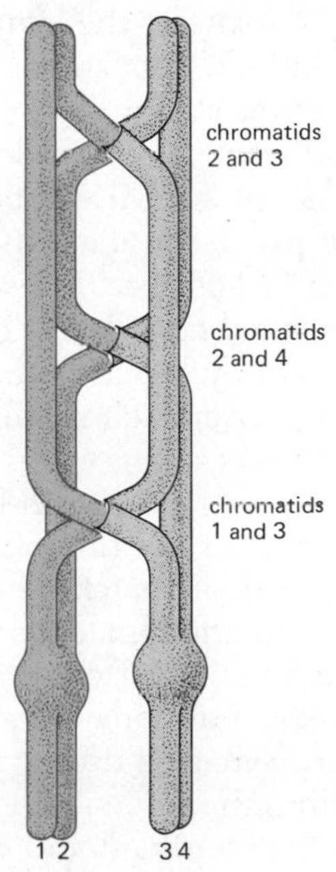

Figure 14–13 Schematic drawing of chromosomes resembling those in Figure 14–11, but containing three different chiasmata resulting from three separate crossover events. Note that each of the two chromatids on each chromosome can cross over with either of the chromatids on the other chromosome in the bivalent. For example, here chromatid 3 has undergone an exchange with both chromatid 1 and chromatid 2.

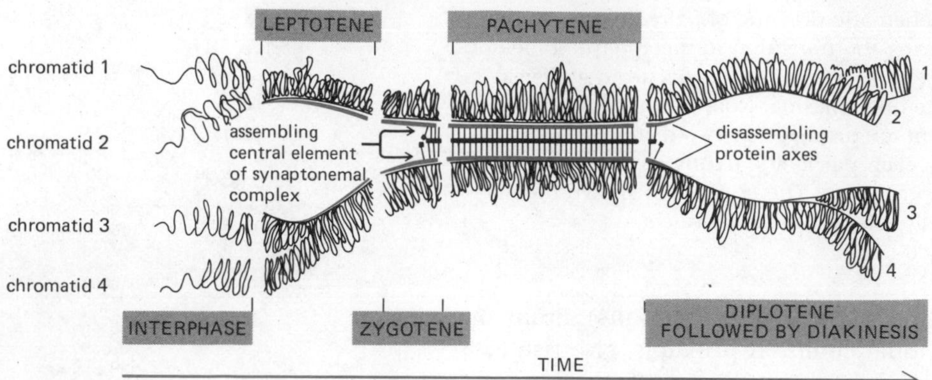

Figure 14–14 Diagrammatic illustration of the time course of chromosome synapsis and desynapsis during meiotic prophase I. Note that a fully formed synaptonemal complex exists throughout the pachytene stage.

diplotene, and *diakinesis*—defined by these morphological changes. The most striking event is the initiation of intimate chromosome synapsis at **zygotene,** when a specialized structure called the **synaptonemal complex** begins to develop between homologous chromosomes. **Pachytene** is said to begin as soon as synapsis is complete, and it generally persists for days, until desynapsis begins the **diplotene** stage, in which the chiasmata are first seen.

Genetic recombination requires a close apposition between the recombining chromosomes. The synaptonemal complex, which forms just before pachytene and dissolves just afterward (Figure 14–14), keeps homologous chromosomes together and closely aligned and is thought to be required for the crossing-over events to occur. The synaptonemal complex consists of a long, ladderlike protein core, on opposite sides of which the two homologs are closely apposed to form a long, linear chromosome pair (bivalent, Figure 14–15). The sister chromatids in each homolog are kept tightly packed together, their DNA extending away from the same side of the protein ladder in a series of loops. Thus, while the homologous chromosomes are closely aligned along their length in the synaptonemal complex, the maternal and paternal chromatids that will recombine with each other are kept separated on either side of the protein ladder by more than 100 nm (Figure 14–16).

From cytological studies, chromosome synapsis is seen to be preceded by the formation of a ropelike proteinaceous axis along each of the homologs. As pairing proceeds, the axes appear to adhere to each other to become the lateral elements of the synaptonemal complex, forming the two sides of the protein ladder. Both the axes and the lateral elements contain a protein with unique silver-staining properties that make these structures visible by both light and electron microscopy (Figure 14–17).

It is not known what causes the homologous parts of chromosomes to become precisely aligned during zygotene. Because the chromatin of one homolog is positioned well apart from the chromatin of its partner in the synaptonemal complex, it has been suggested that the specificity for pairing is mediated by the axes themselves. In one view, the proteins in these axes might first assemble in a conformation dictated by the exact chromatin structure at points along each chromosome. If the axes subsequently pair with each other by a "like-with-like" mechanism, the alignment of the axes might indirectly align the homologous regions of their attached chromosomes according to their matching chromatin structures (Figure 14–18). Some point-by-point matching mechanism is required to explain the observation that the presence of an inverted section of chromosome in one of two pairing homologs usually (but not always) results in a temporary interruption of the normal zipperlike synapsis during zygotene, allowing homologous genes to synapse even within the inversion (Figures 14–19 and 14–20). Each of the various stages of meiosis is outlined and described in detail in Figure 14–21.

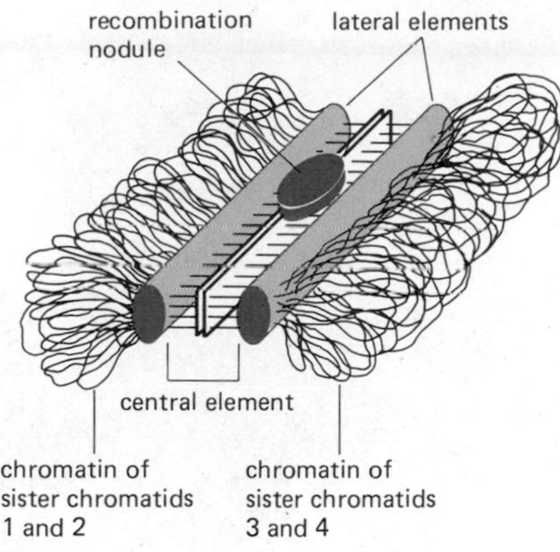

Figure 14–15 Schematic drawing of a typical synaptonemal complex, showing the lateral and central elements of the complex and including a recombination nodule (to be described in the text). Only a short section of the long, ladderlike complex is shown.

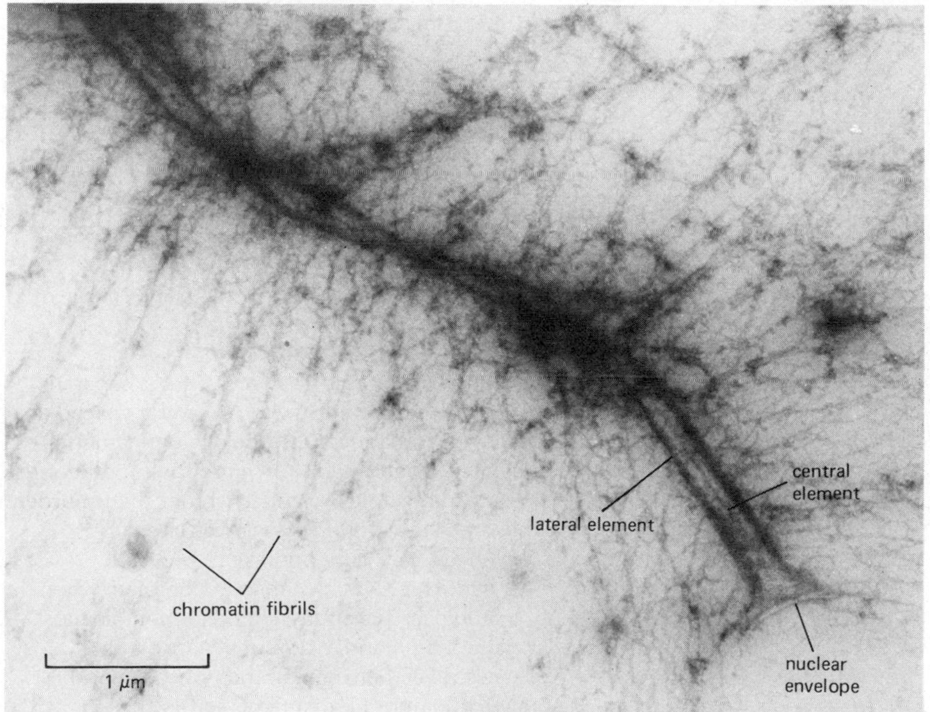

Figure 14–16 Electron micrograph of part of a bivalent from a Syrian hamster spermatocyte at pachytene. A whole-mount preparation was stained and treated to show the synaptonemal complex and associated chromatin spread in two dimensions. Chromatin fibrils radiate as collapsed loops from the two lateral elements of the synaptonemal complex. Each lateral element is the axis of a homolog; the synaptonemal complex is formed by the parallel association of these lateral elements—together with a third, linear, central element—connected by thin transverse filaments. The synaptonemal complex twists along the core of the bivalent, but it is straight and thick near its termination on the nuclear envelope. (From M. J. Moses and A. J. Solari, *J. Ultrastruct. Res.* 54:109, 1976.)

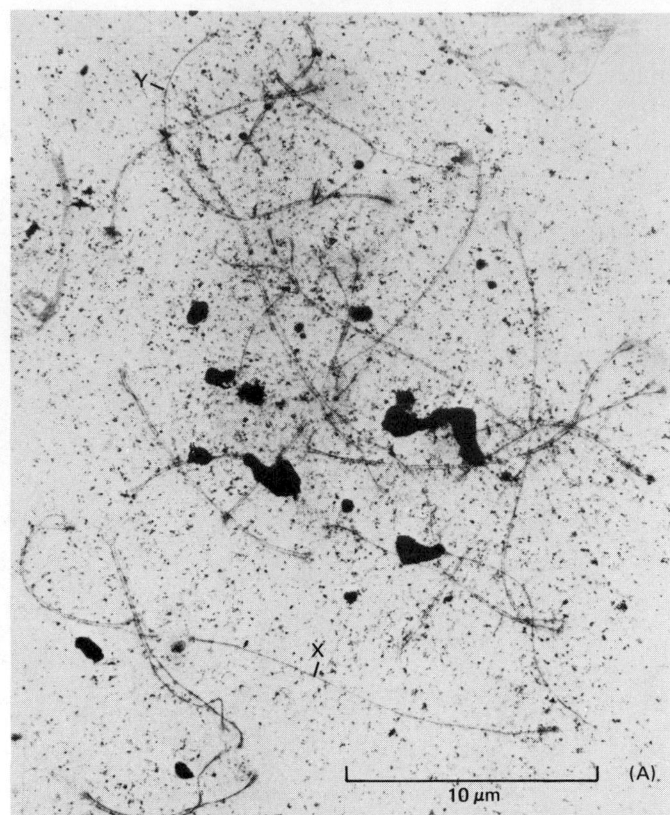

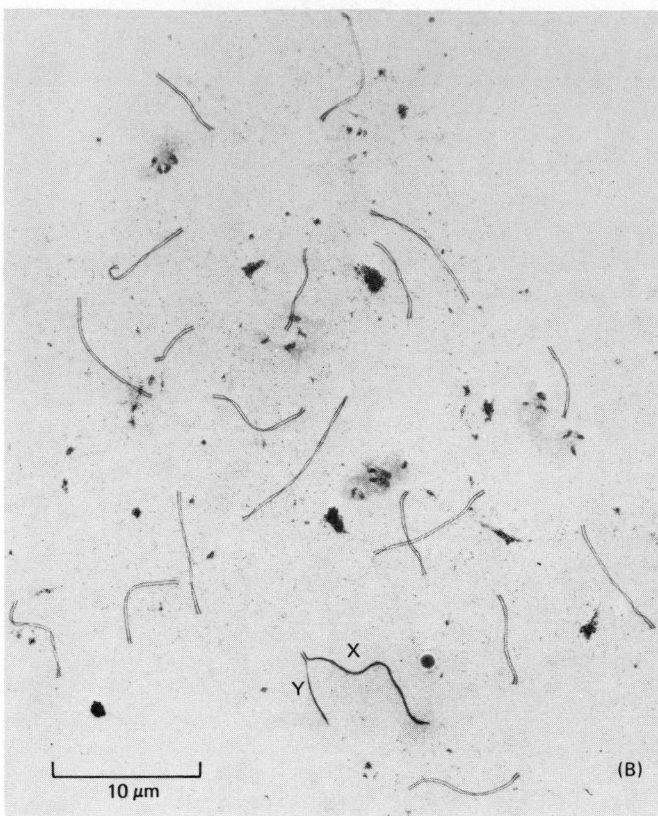

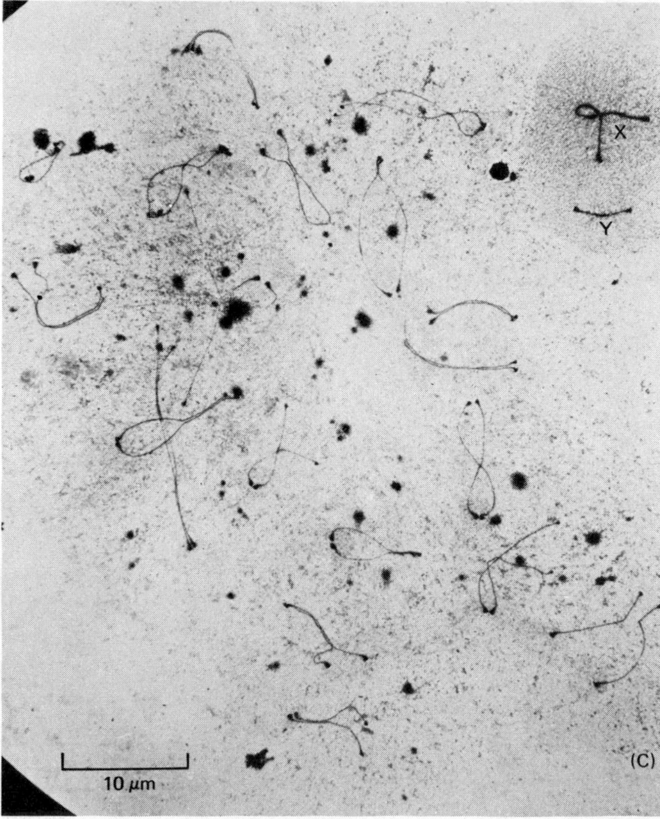

Figure 14–17 Electron micrographs of full complements of synaptonemal complexes from silver-stained, whole-mount spreads of mouse spermatocytes of (A) early (zygotene), (B) middle (pachytene) and (C) later (diplotene) stages of meiotic prophase.

(A) SYNAPSING (ZYGOTENE): As the axes of as yet unsynapsed chromosomes form, they are seen as separate. The axes then move together and, when properly spaced at one or more synaptic initiation sites, the synaptonemal complex forms, often beginning at the end. The separation between the sex chromosomes (X and Y) shows that in order to pair, chromosomes must often travel over large intranuclear distances. The dark bodies are nucleoli.

(B) SYNAPSED (PACHYTENE): Synapsis is complete when synaptonemal complexes have fully joined all homologous autosomes in pairs. The X and Y do not synapse fully. Crossing over takes place between strands of chromatin (chromatids), which cannot be distinguished in these preparations.

(C) DESYNAPSING (DIPLOTENE): Just before the axes disassemble, they separate, marking the end of synapsis. In places they are held together by persistent segments of synaptonemal complex that are thought to represent sites where crossing over has occurred. Later, when the chromatin condenses and chromatids are distinguishable, chiasmata will indicate the crossovers. (Original micrographs courtesy of Montrose J. Moses.)

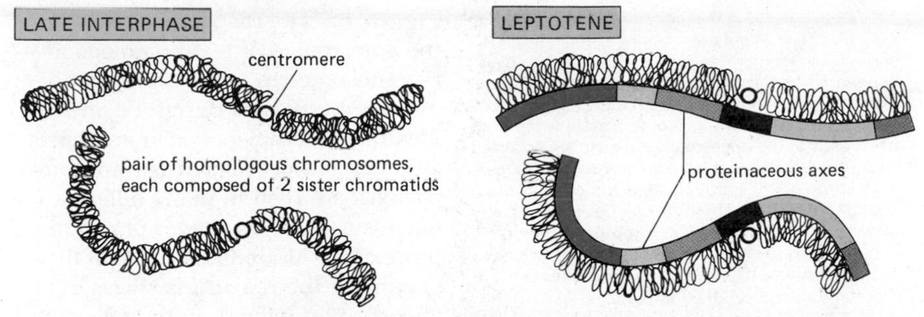

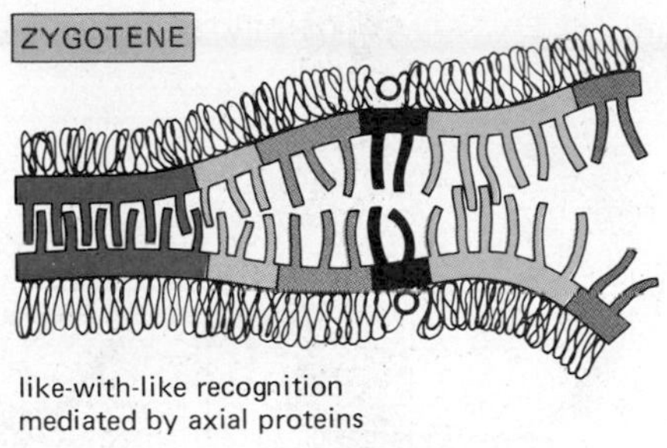

Figure 14–18 Highly schematic illustration of a possible like-with-like matching scheme postulated to explain the homologous chromosome pairings seen at zygotene. In this view, the proteinaceous axes on the leptotene chromosomes acquire a local character that depends on some characteristic feature of chromatin structure along each chromosome.

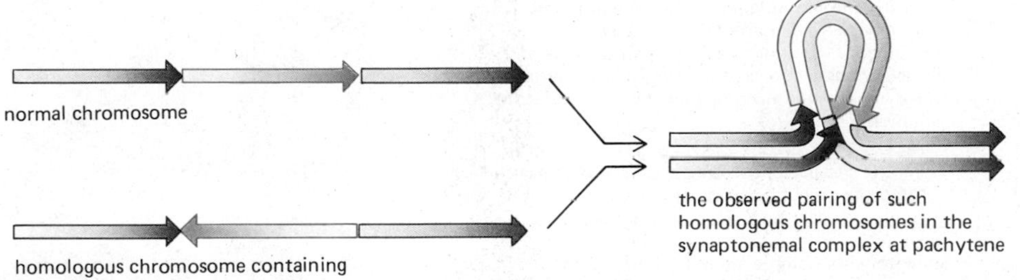

Figure 14–19 Schematic drawing of the formation of the synaptonemal complex between one normal chromosome and its homolog carrying an inverted section. Such structures demonstrate that a local like-with-like pairing scheme brings homologous chromosomes together.

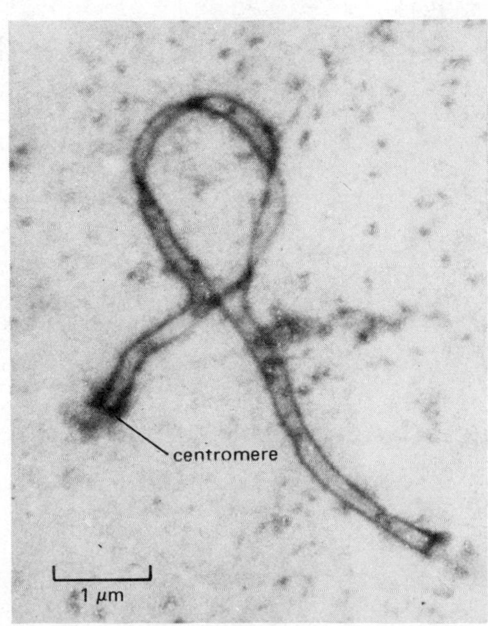

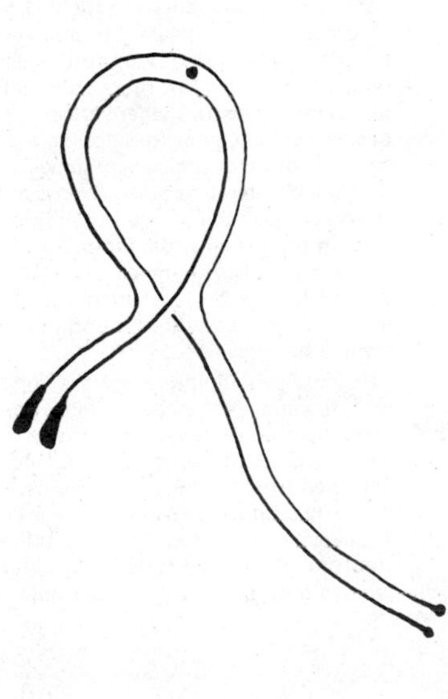

Figure 14–20 Electron micrograph and drawing of two tightly synapsed mouse homologs at pachytene, one of which contains an inversion. A recombination nodule is seen in the loop. (From P. A. Poorman, M. J. Moses, T. H. Roderick, and M. T. Davisson, *Chromosoma* 83:419, 1981.)

(A) THE FIVE STAGES OF MEIOTIC PROPHASE I

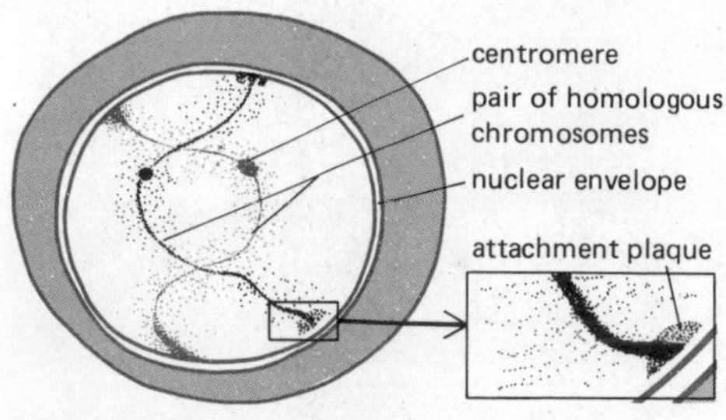

LEPTOTENE Prophase I begins at the leptotene stage, when each chromosome is first seen to have condensed from its interphase conformation to produce a long thin thread with a proteinaceous central *axis*. Each chromosome is attached at both of its ends to the nuclear envelope via a specialized structure called an *attachment plaque*. Although each chromosome has replicated and consists of 2 *sister chromatids*, these chromatids are unusually closely apposed, and each chromosome therefore appears to be single (separate chromatids will not become visible until late in prophase, at either the diplotene stage or in diakinesis).

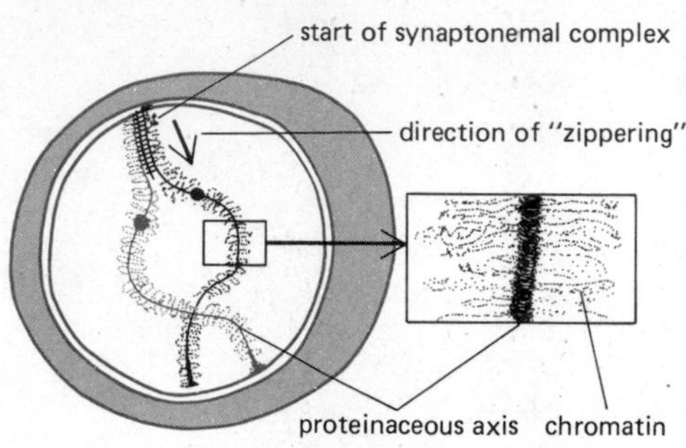

ZYGOTENE Leptotene is considered to end and the zygotene stage of prophase to begin as soon as synapsis, or intimate pairing, between the 2 homologs is initiated. The initial recognition requires that the homologs recognize each other from a distance. Synapsis often starts when the homologous ends of the 2 chromosomes are brought together on the nuclear envelope and continues inward in a zipperlike manner from both ends, aligning the 2 homologous chromosomes side by side. In other cases, synapsis may begin in internal regions of the chromosomes and proceed toward the ends, producing the same type of alignment. Each gene is thus thought to be brought into juxtaposition with its homologous gene on the opposite chromosome. As the homologs pair, their ropelike proteinaceous axes are brought together to form the 2 *lateral elements*, or "sides," of the long ladderlike structure called the *synaptonemal complex*. Each resulting chromosome pair in meiotic prophase I is usually called a *bivalent*, but since each homologous chromosome in the pair consists of 2 closely apposed sister chromatids, it is better to think of each chromosome pair as a *tetrad*, another commonly used term.

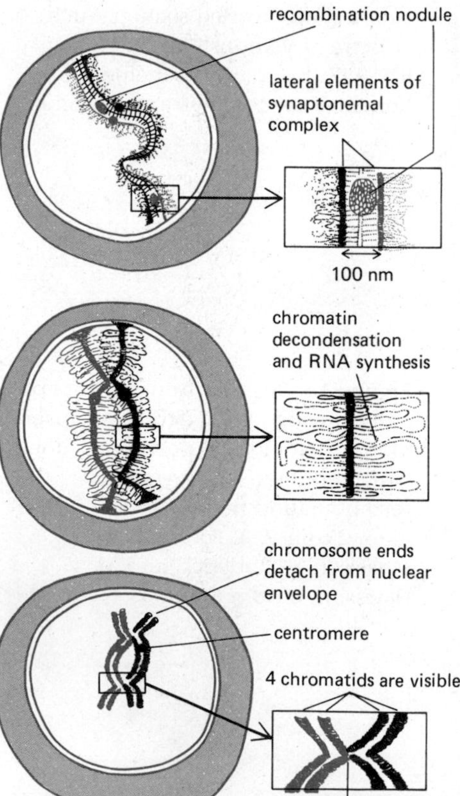

PACHYTENE As soon as synapsis is complete all along the chromosomes, the cells are said to have entered the pachytene stage of prophase, where they may remain for days. At this stage, large *recombination nodules* appear at intervals on the synaptonemal complexes and are thought to mediate chromosomal exchanges. These exchanges result in crossovers between 2 nonsister chromatids, that is, 1 from each of the 2 paired homologous chromosomes. Although invisible at pachytene, each such crossover will appear later as a chiasma.

DIPLOTENE Desynapsis begins the diplotene stage of meiotic prophase I. The synaptonemal complex dissolves, allowing the 2 homologous chromosomes in a bivalent to pull away from each other to some extent. However, each bivalent remains joined by 1 or more chiasmata, representing the sites where crossing over has occurred. In oocytes (developing eggs), diplotene can last for months or years, since it is at this stage that the chromosomes decondense and engage in RNA synthesis to provide storage materials for the egg. In the extreme, the diplotene chromosomes can become highly active in RNA synthesis and expand to an enormous extent, producing the lampbrush chromosomes found in amphibians and some other organisms.

DIAKINESIS Diplotene merges imperceptibly into diakinesis, the stage of transition to metaphase, as RNA synthesis ceases and the chromosomes condense, thicken, and become detached from the nuclear envelope. Each bivalent is clearly seen to contain 4 separate chromatids, with each pair of sister chromatids linked at their centromeres, while nonsister chromatids that have crossed over are linked by chiasmata.

Figure 14–21 Diagrams illustrating the appearance of two homologous chromosomes in a cell undergoing meiosis. Events are described and illustrated as they occur in mammals, although closely related chromosomal changes are seen in many different organisms. The five stages of meiotic prophase I (A) are illustrated on this page, with the remaining stages of meiosis (B) depicted on the following page.

(B) THE REMAINING STAGES OF MEIOSIS

After the long prophase I has ended, 2 successive nuclear divisions without an intervening period of DNA synthesis bring meiosis to an end. These remaining meiotic stages typically occupy only 10% or less of the total time required for meiosis, and they are named in accord with the corresponding stages of mitosis. Thus, the rest of meiotic division I is said to consist of metaphase I, anaphase I, and telophase I. At the end of division I, the chromosomal complement has been reduced from tetraploidy to diploidy, just as in mitosis, and 2 cells have formed from 1. The critical difference is that for each type of chromosome, *2 sister chromatids* joined at their centromeres have been segregated to each cell rather than 2 separate chromatids as at mitosis. Division II quickly follows, consisting of a transient interphase II with no chromosome replication, followed by prophase II, metaphase II, anaphase II, and telophase II. In the end, 4 haploid nuclei are produced from each diploid cell that entered meiosis.

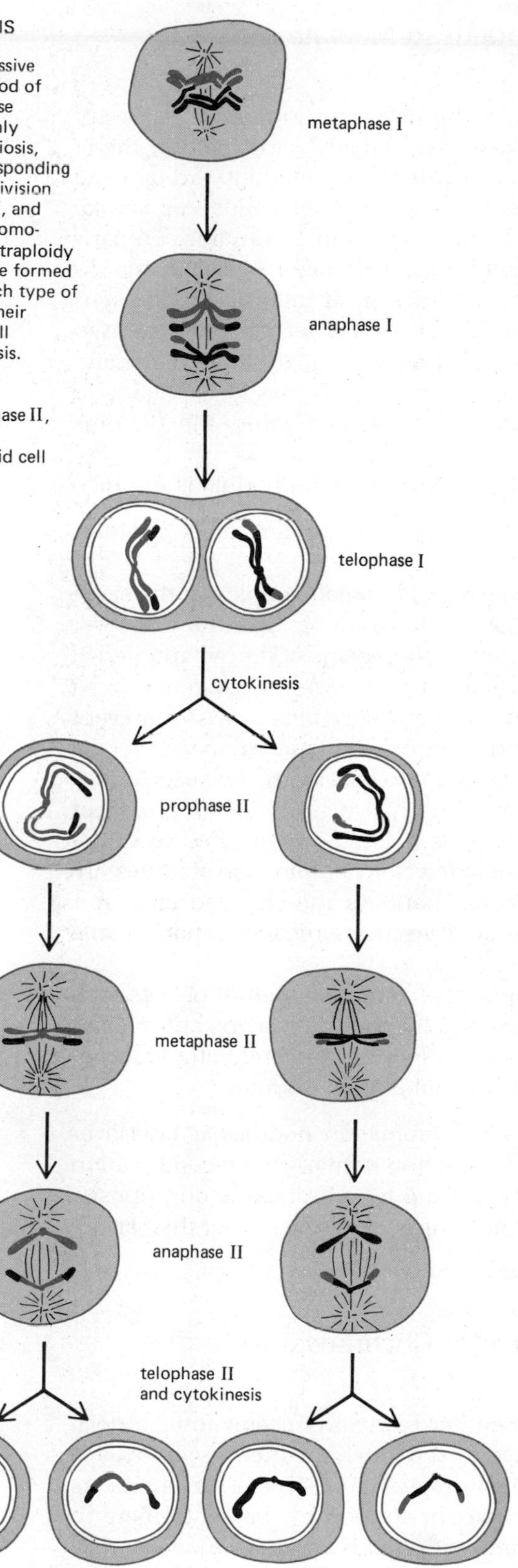

Recombination Nodules Are Thought to Mediate the Chromatid Exchanges[7]

While the synaptonemal complex provides the structural framework necessary for recombination events, it probably does not directly participate in them. The active recombination process is thought to be mediated instead by large **recombination nodules,** which are either spherical, ellipsoidal, or barlike protein-containing assemblies with a diameter of about 90 nm (for comparison, a large globular protein molecule of molecular weight 400,000 has a diameter of about 10 nm). Recombination nodules sit at intervals on the synaptonemal complex, placed like basketballs on a ladder between the two homologous chromatids (see Figure 14–15). They are thought to mark the site of a large multienzyme "recombination machine" that brings local regions of DNA on the maternal and paternal chromatids together across the 100 nm-wide synaptonemal complex.

The evidence for this function for the recombination nodule is indirect:

1. The total number of nodules is about equal to the total number of chiasmata seen later in prophase.
2. The nodules are distributed along the synaptonemal complex in the same way that crossover events are distributed; for example, like the crossover events themselves, they are absent from those regions of the synaptonemal complex that hold heterochromatin together. Moreover, both genetic and cytological measurements reveal that the occurrence of one crossover event will prevent a second crossover event from occurring at any nearby chromosomal site. Similarly, the nodules tend not to occur very near each other.
3. Some *Drosophila* mutations cause an abnormal distribution of crossover events along the chromosomes, as well as a greatly diminished recombination frequency; here correspondingly fewer recombination nodules are found, with a changed distribution that parallels the changed crossover distribution. This correlation strongly suggests that a recombination nodule determines the site of each crossover event.
4. Genetic recombination is thought to involve a limited amount of DNA synthesis at the site of each crossover event. Electron microscope autoradiography shows that radioactive DNA precursors are preferentially incorporated into pachytene DNA at or near recombination nodules.

Because there are about as many recombination nodules as crossover events, their suggested role in crossover initiation implies that recombination nodules must be extremely efficient in causing the chromatids on opposite homologs to recombine. Unfortunately, nothing is yet known about their structure or mechanism of action.

Chiasmata Play an Important Part in Chromosome Segregation in Meiosis

As well as mediating genetic reassortment, crossing over seems to be crucial for the segregation of the two homologs to separate daughter nuclei. This is because it is the chiasmata that hold the maternal and paternal homologs together until anaphase I, playing the part performed by the centromere in an ordinary mitotic division. Thus in mutant organisms that are deficient in meiotic chromosome crossing over, some of the chromosome pairs lack chiasmata at metaphase I and fail to segregate normally. As a result, a high proportion of the resulting gametes contain too many or too few chromosomes.

It is now clear that there are at least two major differences in the way in which chromosomes separate in normal mitosis and in meiotic division I. (1) Whereas during mitosis the kinetochores on each sister chromatid have

Figure 14–22 Comparison of the mechanisms of chromosome alignment (at metaphase) and separation (at anaphase) in meiotic division I and meiotic division II. The mechanisms used in meiotic division II are the same as those in normal mitosis (see Chapter 11).

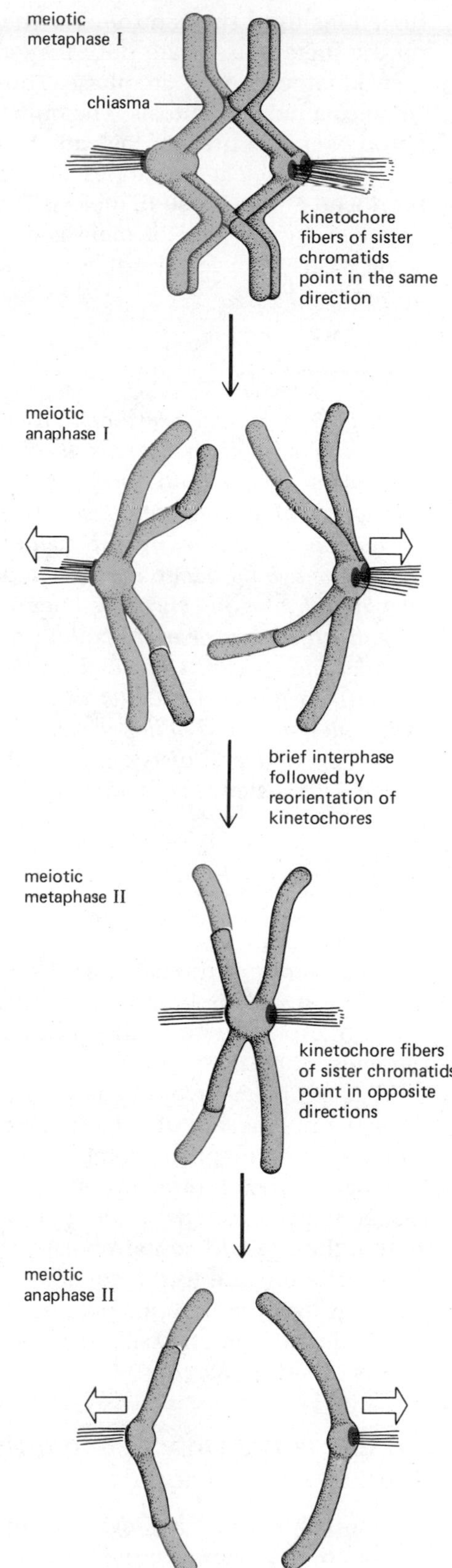

attached kinetochore fibers pointing in opposite directions, at metaphase I of meiosis the kinetochores on both sister chromatids have attached kinetochore fibers pointing in the same direction (Figure 14–22). (2) Whereas the movement of each chromatid to the poles is triggered by the separation of sister kinetochores during mitosis (thus beginning anaphase, see p. 655), meiotic anaphase I movements seem to be initiated by the disruption of the forces keeping the arms of sister chromatids closely apposed, which in turn dissolves the chiasmata that have been holding the homologous maternal and paternal chromosomes together. This presumably explains not only why the chiasmata are necessary for the normal alignment of the chromosomes at metaphase I in many organisms, but also why the chromosomes produced at anaphase I tend to have nonadherent sister chromatid arms, giving them an unusual "splayed-out" appearance compared to normal mitotic chromosomes (Figure 14–22).

Pairing of the Sex Chromosomes Ensures That They Also Segregate[8]

We have explained how homologous chromosomes pair so that they segregate between the daughter cells. But what about the sex chromosomes, which in male mammals are not homologous? Females have two X chromosomes, which pair and segregate like other homologs. But males have one X and one Y chromosome, which must be paired during the first metaphase if the sperm are to contain either one Y or one X chromosome and not both or neither. The necessary pairing is made possible by a small region of homology between the X and Y sex chromosomes, which enables them to pair during the first meiotic prophase. In this way, the even partitioning of X and Y chromosomes on the spindle is assured and only two types of sperm are produced: sperm containing one Y chromosome, which will give rise to male embryos, and sperm containing one X chromosome, which will give rise to female embryos.

Meiotic Division II Resembles a Normal Mitosis

Of the two successive cell divisions that constitute meiosis, division I occupies almost all of the time involved and is by far the more complex (Figure 14–23). It also has a number of unique features. For example, the DNA replication during S phase tends to take much longer than normal. Moreover, cells can remain in the first meiotic prophase for days, months, or even years, depending on the gamete being formed and the species. The nuclear envelope remains intact during this prolonged period and disappears only when the spindle fibers begin to form, as prophase I gives way to metaphase I.

After the end of meiotic division I, nuclear membranes re-form around the two daughter nuclei and a brief interphase begins. During this period, the chromosomes decondense somewhat, but soon the chromosomes recondense and prophase II begins. As there is no DNA synthesis during this interval, in some organisms the chromosomes seem to pass almost directly from one division phase into another. In all organisms, prophase II is brief: the nuclear envelope breaks down as the new spindle forms, after which metaphase II, anaphase II, and telophase II follow in quick succession. As in mitosis, the kinetochore fibers form on sister chromatids and extend in opposite directions from the centromere. The two sister chromatids are kept together on the metaphase plate until they are released by the sudden separation of their kinetochores at anaphase. Thus, unlike division I, division II closely resembles a normal mitosis. The only major difference is that one copy of each chromosome is present instead of two.

After nuclear envelopes have formed around the four haploid nuclei produced at telophase II, meiosis is complete (see Figure 14–21B). As we shall now see, by the end of meiosis a vertebrate egg is fully developed (and in some cases even fertilized), whereas a sperm has only just begun its development.

Summary

In meiosis, two successive cell divisions give rise to four haploid cells from a single diploid cell. In animals, the formation of both eggs and sperm begins in a similar way. In both cases meiosis is dominated by prophase of meiotic division I, which can occupy 90% of the total meiotic period. At this time each chromosome consists of two tightly joined sister chromatids. Chromosomal crossover events occur during the pachytene stage of prophase I, when each pair of homologous chromosomes is held in register by a synaptonemal complex. Each crossover event is thought to be mediated by a large recombination nodule, and it results in the formation of a chiasma, which persists until anaphase I. In the first meiotic cell division, one member of each chromosome pair, still composed of linked sister chromatids, is distributed to each daughter cell. A second cell division, without DNA replication, then rapidly ensues in which each sister chromatid is segregated into a separate haploid cell.

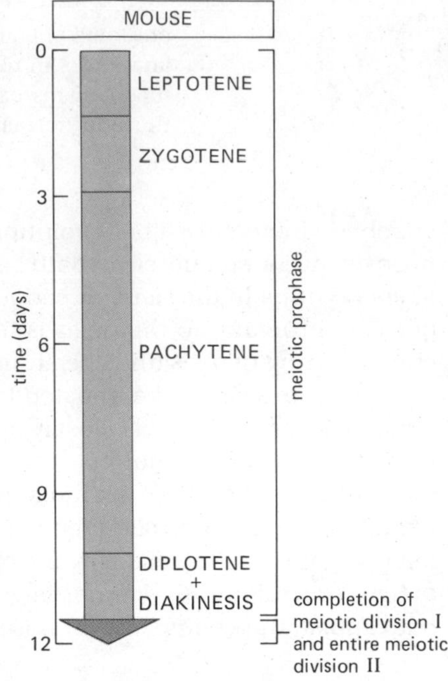

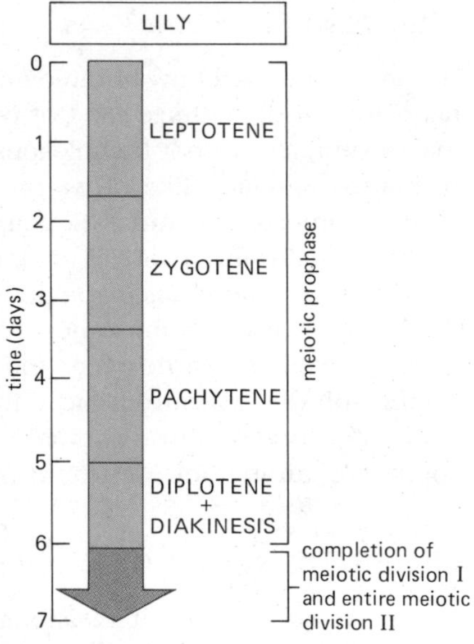

Figure 14–23 Diagram comparing the different amounts of time required for each of the stages of meiosis illustrated previously in Figure 14–21. Approximate time for both a male mammal (mouse) and a plant (lily) are shown. Times differ for male and female gametes of the same species (egg and sperm), as well as for the same gametes of different species. For example, meiosis in a human male lasts for 24 days, compared with 12 days in the mouse. However, in all cases meiotic prophase I is much longer than all other meiotic stages combined.

Gametes

In all vertebrate embryos, certain cells are singled out early in development as progenitors of the gametes. These **primordial germ cells** migrate to the developing gonads (ovaries in females, testes in males), where, after a period of mitotic proliferation, they undergo meiosis and differentiate into mature gametes. The fusion of egg and sperm after mating begins the cycle again.

It is unclear what causes certain cells to become germ cells in mammalian embryos, but in at least one organism the determining factor is known to be a component (or components) of the egg cytoplasm. In *Drosophila,* a specialized region of cytoplasm, the *polar plasm,* at the posterior end of the egg contains small, RNA-rich granules (*polar granules*); cells that form at this end of the egg and contain polar granules become primordial germ cells and eventually migrate to the gonads to form oocytes or sperm. If polar plasm is injected into the anterior pole of an egg, cells that would normally have developed into somatic cells develop into primordial germ cells instead (see Figure 15–41, p. 852).

An Egg Is the Only Cell in a Higher Animal Able to Develop into a New Individual

In one respect at least, eggs are the most remarkable of animal cells: once activated, they can give rise to a complete new individual within a matter of days or weeks. In higher animals this is a property unique to eggs: the fusion with a sperm during fertilization initiates a program of development that progressively unfolds to generate a new individual.

In most nonmammalian eggs, the early part of the developmental program mainly involves rapid cell division, or *cleavage,* in which the total mass of the embryo usually remains constant. The egg is very large to begin with, and the cleaving cells progressively decrease in size with each division until they reach the normal size of an adult somatic cell. Even though a great deal of DNA and protein synthesis is involved in the early cleavage divisions, RNA synthesis (gene transcription) is not required: cleavage proceeds normally in the presence of drugs that inhibit RNA synthesis and can even continue (although in abnormal fashion) after the nucleus of an activated egg has been removed. The explanation is that prior to fertilization these eggs have accumulated large reserves of messenger RNA, ribosomes, tRNA, and all of the precursors needed for macromolecular synthesis. Especially large reserves of nutrients are required by those eggs that undergo a long period of embryonic development outside the body, where they have no external source of nourishment; for this reason, amphibian eggs are much bigger than mammalian eggs, for example. Those freshwater and marine invertebrates, such as sea urchins, which develop from small eggs outside the body generally develop into feeding larvae very rapidly.

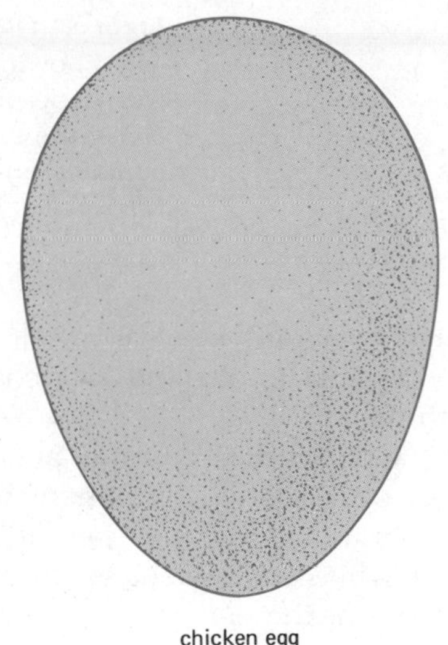

Figure 14–24 The actual size of two different eggs.

Eggs Are Highly Specialized Cells with Unique Features[9]

In terms of development, an egg is the least restricted cell in an animal: it can give rise to every cell type in the organism. However, it is by no means an undifferentiated cell. It is highly specialized for the single function of generating a new individual; as a result, it has many unique characteristics.

The most obvious distinguishing feature of an egg is its large size. An egg is typically spherical or ovoid, and its diameter is 60 μm to 150 μm in man and sea urchins, 1 mm to 2 mm in frogs and fishes, and many centimeters in birds and reptiles (remember that a typical somatic cell has a diameter of only about 20 μm) (Figures 14–24 and 14–25). The size of the nucleus can be equally impressive: for example, in a frog egg with a diameter of 1500 μm, the diameter of the nucleus is about 400 μm.

One reason for the large size of the egg is the need for nutritional reserves mentioned above. This need is supplied largely by **yolk,** which is rich in protein and is usually contained within discrete structures called *yolk granules.* In eggs that develop into large animals outside the body, yolk can account for more than 95% of the volume of the egg, whereas in mammals, whose embryos are largely nourished by their mothers, it constitutes less than 5% of the total egg volume.

Another important egg-specific structure is the outer **egg coat,** a specialized form of extracellular matrix consisting largely of glycoprotein molecules, some secreted by the egg and others by surrounding cells. In all species this structure has an inner layer immediately surrounding the egg plasma membrane, called the **zona pellucida** in mammalian eggs (Figure 14–26) and the **vitelline layer** in other vertebrate and in invertebrate eggs. This layer protects the egg from mechanical damage; in some eggs it also acts as a species-specific barrier to sperm, admitting only sperm of the same or closely related species. Often additional coats that overlie the vitelline layer are secreted by surrounding cells. For example, as frog eggs pass from the ovary through the oviduct (the tube that conveys them to the outside), they acquire several additional layers of gelatinous coating secreted by epithelial cells lining the oviduct. Similarly, the "white" (albumin) and shell of chicken eggs are added (after fertilization) as the eggs pass along the oviduct, while insect eggs are covered by a thick, tough layer called the *chorion,* which is secreted by the specialized cells that surround each developing egg.

Many eggs (including those of mammals) contain specialized secretory vesicles just under the plasma membrane in the outer region, or *cortex,* of the egg cytoplasm (Figure 14–27). When the egg is activated by a sperm, these

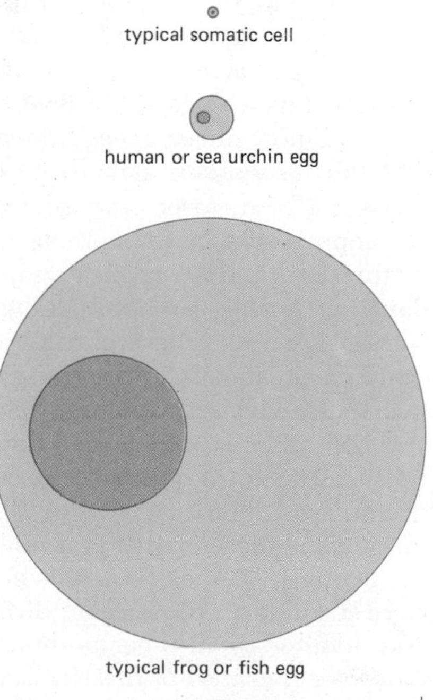

Figure 14–25 The relative sizes of various eggs compared to a typical somatic cell.

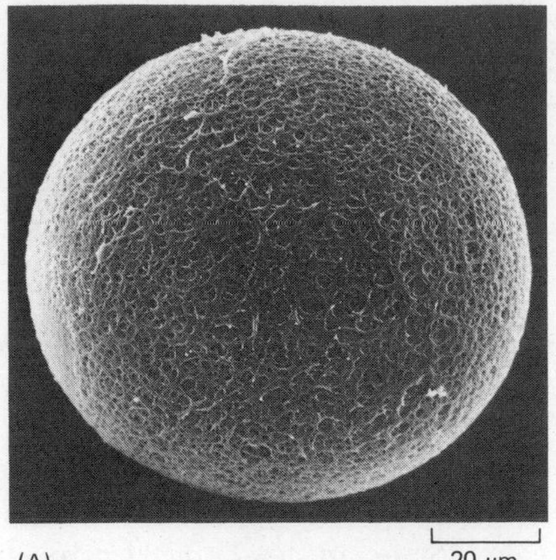

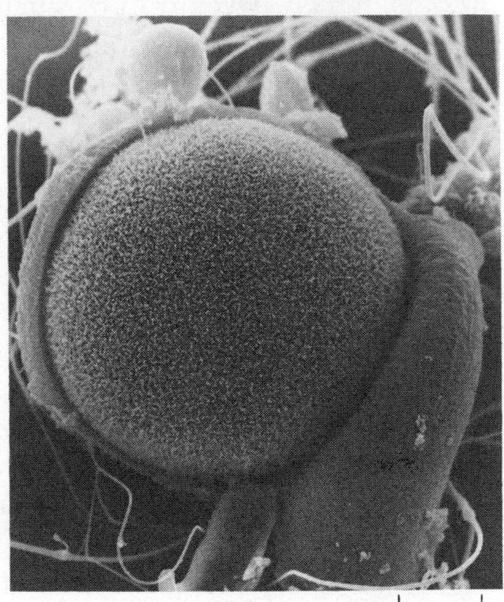

Figure 14–26 Scanning electron micrographs of hamster eggs showing the zona pellucida. In (B) the zona (to which many spermatozoa are attached) has been peeled back to reveal the underlying plasma membrane of the egg, which contains numerous microvilli. (From David M. Phillips, *J. Ultrastruct. Res.* 72:1–12, 1980.)

cortical granules release their contents by exocytosis; the contents of the granules act to alter the egg coat so as to prevent other sperm from fusing with the egg.

While cortical granules are usually distributed throughout the egg cortex, other cytoplasmic components can have a strikingly asymmetrical distribution. In a frog egg, for example, most of the yolk is at one pole (the *vegetal pole*), while the nucleus is closer to the opposite pole (the *animal pole*). The polarity of the egg often determines the polarity of the embryo (see Figure 15–3, p. 815).

Eggs Develop in Stages[9]

While the details of egg development (**oogenesis**) vary in different species, the general stages are similar (Figure 14–28). Primordial germ cells migrate to the forming gonad to become **oogonia;** after a period of mitotic proliferation oogonia differentiate into **primary oocytes,** which begin their first meiotic division. The DNA replicates so that each chromosome consists of two chromatids, the homologous chromosomes pair along their long axes, and crossing over occurs between the chromatids of these paired chromosomes. At this point, prophase is arrested for a time period lasting from a few days to many years, depending on the species. During this phase, the primary oocytes acquire outer coats and cortical granules, accumulate ribosomes, messenger RNA, yolk, glycogen, and lipid, and preparation is made for the unfolding of the developmental program. In many oocytes these activities are reflected in the appearance of the still paired chromosomes, which decondense and form lateral loops, taking on the characteristic "lampbrush" appearance of a chromosome busily engaged in RNA synthesis (see p. 453).

The next phase of egg development, called **egg maturation,** begins only with the onset of sexual maturity. Under the influence of hormones (see below) division I of meiosis proceeds: the chromosomes recondense, the nuclear envelope breaks down (which is generally taken to mark the beginning of maturation), and the replicated homologous chromosomes segregate into two daughter nuclei, each containing half the original number of chromosomes (although each of these chromosomes is unusual in being composed of two sister chromatids). But the cytoplasm divides very asymmetrically to produce two **secondary oocytes** that differ greatly in size: one is a small **polar body,** and the other is a large cell containing all of the developmental potential. Finally, division II of meiosis occurs: the two sister chromatids of each chromosome acquired in division I separate by a process that is analogous to mitotic anaphase except that now there is only half the normal diploid number of chromosomes. After this chromosome separation, the cytoplasm of the large secondary oocyte again divides asymmetrically to give the mature **ovum** and another small polar body, each with a haploid number of single chromosomes. Because of the two asymmetrical divisions of their cytoplasm, oocytes maintain their large size in spite of undergoing the two meiotic divisions. All of the polar bodies are small, and they eventually degenerate. At some point in this developmental sequence, depending on the species, the egg is released from the ovary by the process of **ovulation.**

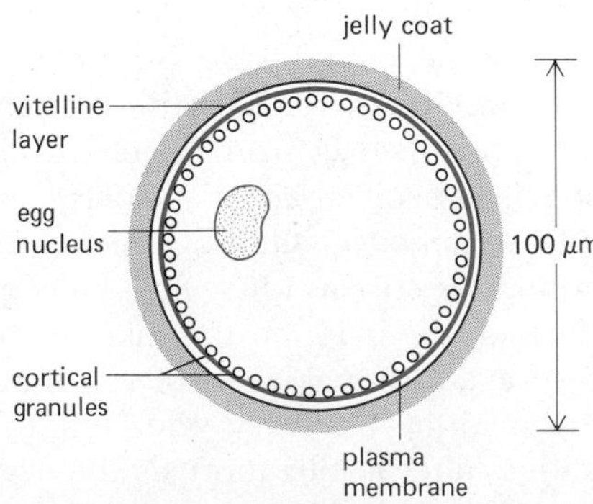

Figure 14–27 Schematic drawing of a sea urchin egg showing the location of cortical granules. Note also that the vitelline layer is covered by a jelly coat.

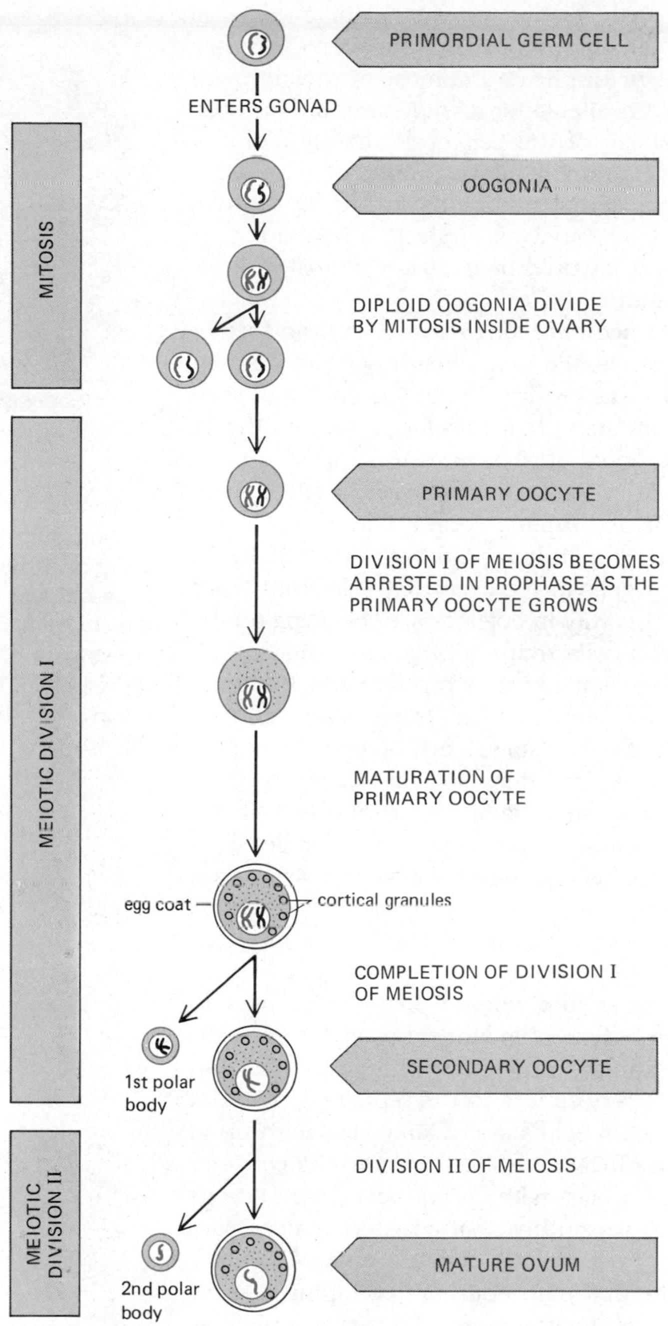

Figure 14–28 The various stages of oogenesis. Oogonia develop from primordial germ cells that migrate into the ovary early in embryogenesis. After a number of mitotic divisions, oogonia begin meiotic division I and are then called primary oocytes. In mammals, primary oocytes are formed very early and remain arrested in prophase of meiotic division I until the female animal becomes sexually mature. At this point, a small number periodically mature under the influence of hormones, completing meiotic division I to become secondary oocytes, which then undergo meiotic division II to become mature eggs. The stage at which the egg is released from the ovary and is fertilized varies from species to species.

Many Eggs Grow to Their Large Size Through Special Mechanisms[9, 10]

A small somatic cell with a diameter of 10 μm typically takes about 24 hours to double its mass in preparation for cell division. With the same mechanisms and rates of macromolecular syntheses, the same cell would take a very long time to reach the thousandfold greater mass of a mammalian egg with a diameter of 100 μm or the millionfold greater mass of an insect egg with a diameter of 1000 μm. Yet some insects live only a few days and manage to produce eggs with diameters even greater than 1000 μm. It is clear that eggs must have special mechanisms for achieving their large size.

One factor that assists such growth is that the eggs of many species delay the completion of meiosis until the end of their maturation, so that they contain the diploid chromosome set in duplicate for most of their growth period. They therefore have more DNA available for RNA synthesis than does an average somatic cell in the G_1 phase of the cell cycle. In addition, by retaining both maternally derived and paternally derived copies of every gene, eggs avoid the risk posed by recessive lethal mutations in one of the two parental chromosome sets; if eggs had to survive for long in a haploid state, with only one copy of each gene, this risk would be large since most individuals carry some recessive lethal mutations.

Some eggs extend the process of accumulating extra DNA even further and produce extra copies of certain genes. We have already seen in Chapter 8 that the somatic cells of most organisms require 100 to 500 copies of ribosomal RNA genes in order to produce enough ribosomes for protein synthesis. Because eggs require even greater numbers of ribosomes to support protein synthesis during early embryogenesis, in some amphibian eggs the rRNA genes are specifically amplified to generate 1 or 2 million copies (Figure 14–29).

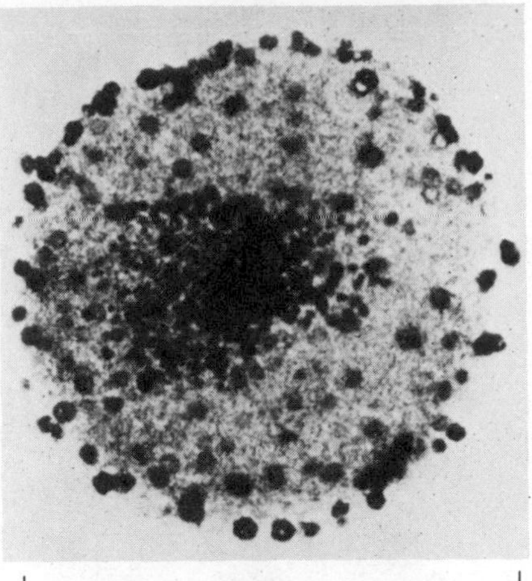

Figure 14–29 Photomicrograph of an isolated nucleus from a frog egg stained with cresyl violet to show the large number of nucleoli, reflecting the enormous amplification of ribosomal RNA genes. (From D. D. Brown and I. Dawid, *Science* 160:273–275, 1968. Copyright 1968 by the American Association for the Advancement of Science.)

The growth of many eggs is partly dependent on the synthetic activities of other cells, particularly of accessory cells in the ovary. Two different types of ovarian accessory cells function in this way in oogenesis, depending on the species. Some invertebrates have **nurse cells** that not only surround the egg but are usually connected to it by cytoplasmic bridges through which macromolecules can pass directly into the egg cytoplasm. The nurse cells manufacture for the invertebrate egg the products—ribosomes, mRNA, protein, and so on—that a vertebrate egg would manufacture for itself. But how are such molecules passed into the egg? One way may be by electrophoresis: it has been possible to demonstrate an electrophoretic movement of molecules from nurse cells into an oocyte, that is driven by a voltage gradient between these cells.

In some species, nurse cells are derived from the same oogonium that gives rise to the oocyte with which they are associated. For example, in *Drosophila* embryos an oogonium undergoes four mitotic divisions to form 16 cells. One of these cells becomes the egg, while the others become nurse cells and remain attached to each other and to the egg by cytoplasmic bridges (Figure 14–30). In the nurse cells, DNA replication occurs repeatedly without cell division, so that eventually each cell reaches a very large size, with up to a thousand times the normal amount of DNA (arranged in *polytene chromosomes*, see p. 401). All 15 of the nurse cells, each with the equivalent of hundreds or thousands of genomes, are enlisted to synthesize the materials necessary for a single egg.

The other kind of accessory cells that help nourish developing oocytes are **follicle cells,** which are found in most vertebrates. They are arranged as an epithelial layer around the egg (Figure 14–31) and are connected to the egg by gap junctions that permit the exchange of small molecules but not macromolecules (see p. 686). While these cells are unable to provide the egg with preformed macromolecules through these communicating junctions, they may help to supply the smaller precursor molecules from which macromolecules are made.

Eggs can also receive nutritive help from cells outside the ovary. For example, a major constituent of large eggs, the yolk, is usually synthesized outside the ovary and imported into the egg. In chickens, amphibians, and insects, yolk proteins are made by liver cells (or their equivalent), which secrete these proteins into the blood. Within the ovaries, oocytes take up the yolk proteins from the extracellular fluid by receptor-mediated endocytosis (see Figure 6–74, p. 308).

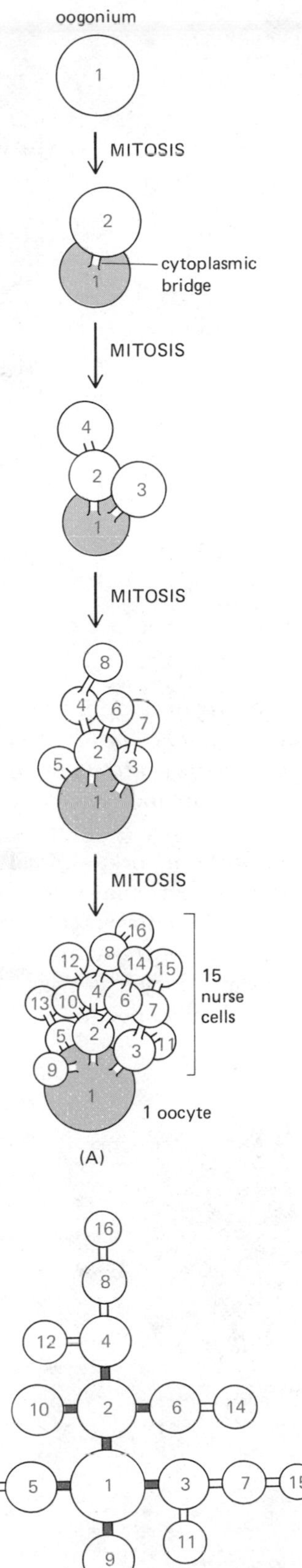

Figure 14–30 Schematic drawing showing how a single *Drosophila* oogonium gives rise to 15 nurse cells and 1 large oocyte, all connected by cytoplasmic bridges (A). At each mitosis, every cell divides once: in the first mitosis, cell 1 divides to produce cells 1 and 2; in the second mitosis, cell 1 divides to produce cells 1 and 3, while cell 2 divides to produce cells 2 and 4, and so on. Only cells 1 or 2 become the egg, perhaps because only these cells are connected by intercellular bridges to four others (B). During egg development, the nurse cells become extremely large and make large amounts of ribosomes and macromolecules and pump them into the oocyte through the cytoplasmic bridges.

Hormones Induce Egg Maturation and Ovulation[9,11]

The contribution of surrounding accessory cells to egg development is not confined to nutrition: in both invertebrates and vertebrates, the accessory cells respond to polypeptide hormones (**gonadotropins**) produced elsewhere in the body so as to control the maturation of the oocyte and eventually (in most species) ovulation.

The hormonal basis of egg maturation and ovulation is especially well understood in starfish and amphibians. In these animals, the gonadotropin hormones stimulate the accessory cells to secrete a secondary mediator that acts on the oocyte to initiate maturation. In starfish the mediator is *1-methyl adenine,* while in amphibians it is the steroid hormone *progesterone.* The secondary mediators bind to cell-surface receptors on the oocyte plasma membrane and induce maturation, probably by increasing the concentration of free Ca^{2+} inside the oocyte through the release of Ca^{2+} from an internal storage site. The evidence for this role of Ca^{2+} in egg maturation comes from the following experiments: (1) Injecting Ca^{2+} into the egg cytosol induces maturation in the absence of hormones, while injecting Ca^{2+} chelators (such as the compound EGTA) prevents maturation even in the presence of hormones. (2) When the Ca^{2+}-binding protein *aequorin* (which emits light when it binds Ca^{2+}) is injected into a starfish or amphibian egg, a transient flash of light accompanies the binding of the appropriate maturation-inducing mediator to receptors on the egg surface.

In humans the processes of oocyte development and of gonadotropin-induced maturation and ovulation are much more complex and less well understood. The primary oocytes of the newborn female are arrested in prophase of meiotic division I and most are surrounded by a single layer of follicle cells; such an oocyte with its surrounding follicle cells constitutes a **primordial follicle** (Figure 14–31).

Beginning sometime before birth, a small proportion of primordial follicles sequentially begin to grow to become **developing follicles:** the follicle cells enlarge and proliferate to form a multilayered envelope around the primary oocyte; the oocyte itself enlarges and develops a zona pellucida and cortical granules (Figure 14–32). The developing follicles grow continuously, and some of them develop a fluid-filled cavity, or *antrum,* to become **antral follicles** (Figure 14–33). Before puberty, all of the primordial follicles that begin to grow degenerate in the ovary at various stages of development without ever releasing their oocytes.

It is not known what causes certain primordial follicles to begin growing, but it is thought not to involve hormonal stimulation. On the other hand, the

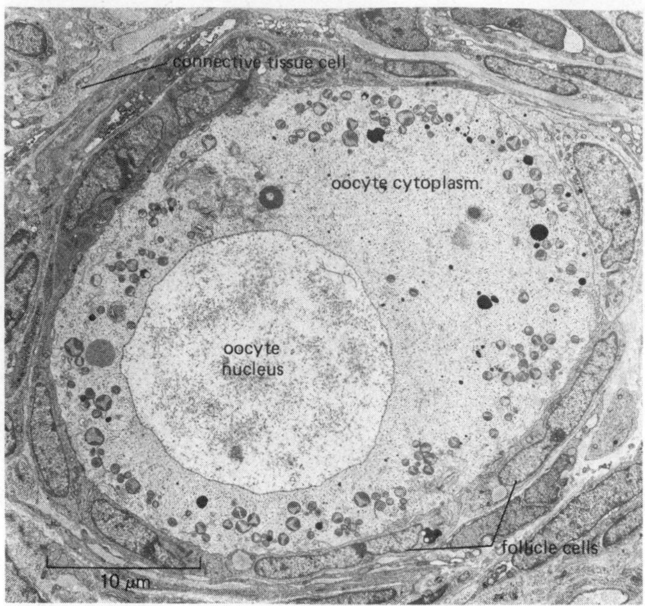

Figure 14–31 Electron micrograph of a rabbit primordial follicle, consisting of a central primary oocyte surrounded by a single layer of flattened follicle cells. The primordial follicle is in turn surrounded by the connective tissue of the ovary. Note that the primary oocyte does not have a zona pellucida or cortical granules at this stage of its development. (From J. Van Blerkom and P. Motta, Cellular Basis of Mammalian Reproduction. Baltimore: Urban & Schwarzenberg, 1979.)

continuing development of such follicles probably depends on gonadotropins (mainly *follicle stimulating hormone (FSH)*) secreted by the pituitary gland and on estrogens secreted by the follicle cells themselves. Starting at puberty, once each month (about halfway through the menstrual cycle), a surge of secretion by the pituitary of another gonadotropin, *luteinizing hormone (LH)*, activates one—and only one—antral follicle to complete its development: the enclosed primary oocyte matures to complete meiotic division I as the stimulated follicle rapidly enlarges and ruptures at the surface of the ovary, releasing the secondary oocyte within (Figure 14–34). As is the case in most mammals, the secondary oocyte is triggered to undergo division II of meiosis only if it is fertilized by a sperm.

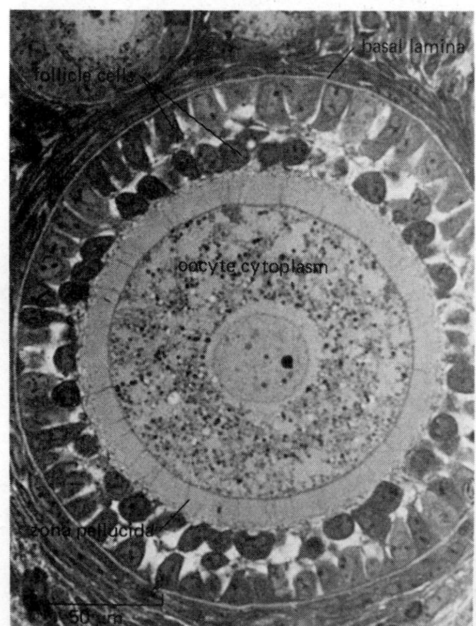

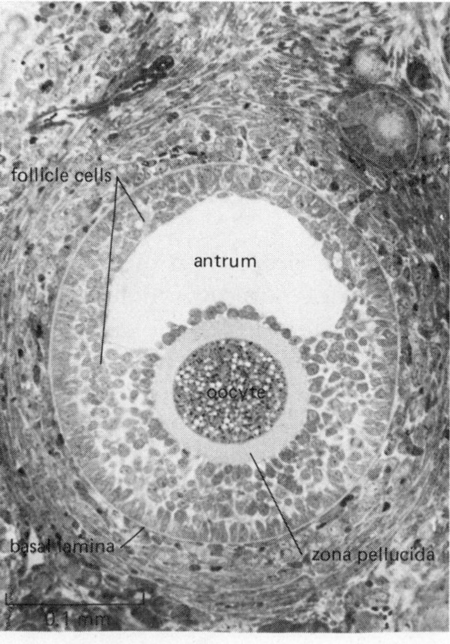

Figure 14–32 Photomicrograph (*left*) of a maturing follicle in the ovary of a rabbit. The primary oocyte has acquired cortical granules (not readily seen in this micrograph) and a thick zona pellucida and is surrounded by a multilayer of follicle cells. The innermost follicle cells extend processes through the zona and form gap junctions with the oocyte. The entire follicle is surrounded by a basal lamina. (From J. Van Blerkom and P. Motta, Cellular Basis of Mammalian Reproduction. Baltimore: Urban & Schwarzenberg, 1979.)

Figure 14–33 Photomicrograph (*right*) of an antral follicle in the ovary of a rabbit. Note the small primordial follicle in the upper right corner of the micrograph. (From J. Van Blerkom and P. Motta, Cellular Basis of Mammalian Reproduction. Baltimore: Urban & Schwarzenberg, 1979.)

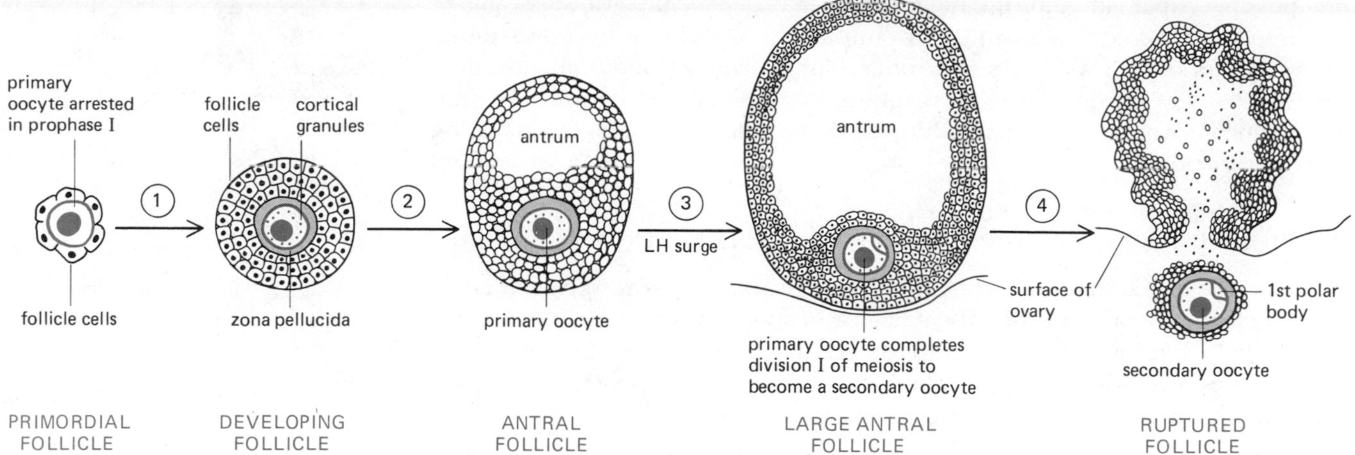

Figure 14–34 Schematic drawing showing the stages in human oocyte development. (1) Starting before birth, a small proportion of primordial follicles sequentially begin to grow and are now called developing follicles. (2) After a period of continuous growth, some developing follicles accumulate fluid to become antral follicles. (3) From the time of puberty, once each month a surge of luteinizing hormone (LH) released by the pituitary activates one antral follicle to mature: the primary oocyte in this follicle completes meiosis I to form a polar body and a secondary oocyte. (4) The secondary oocyte together with the polar body and some surrounding follicle cells are released when the follicle ruptures at the surface of the ovary. The secondary oocyte will only undergo meiotic division II if it is fertilized.

How does the midcycle LH surge initiate oocyte maturation? The follicle cells of antral follicles are coupled to each other and to the oocyte by gap junctions. When such a follicle is activated by the LH surge, one consequence is that the follicle cells uncouple from the oocyte. It has been suggested that follicle cells normally prevent the primary oocyte in a developing antral follicle from maturing into a secondary oocyte by transferring an inhibitory substance through gap junctions into the oocyte. In this view, the LH surge initiates egg maturation by uncoupling the oocyte from its surrounding cells so that the level of the inhibitory substance falls.

One of the enigmatic features of oocyte maturation in primates is that only one of the many antral follicles present in the ovaries at the time of the LH surge each month is stimulated to mature and release its oocyte. The rest are destined to degenerate. Once a follicle has been activated, some feedback mechanism must operate to insure that no other follicles mature during that cycle.

Oogenesis Is Wasteful[12]

In a female human embryo, about 1700 primordial germ cells migrate and invade the developing ovaries during the first months of embryonic development. These oogonia proliferate for several months to produce about 7 million cells, which at this stage stop proliferating and begin their first meiotic prophase to become primary oocytes. However, most oogonia fail to mature into primary oocytes and degenerate in the ovary. In addition, many of the oocytes that do develop also degenerate, so that at birth only about 2 million primary oocytes remain in the ovaries. This degeneration of oocytes continues throughout the long period of meiotic prophase arrest and throughout the reproductive life of a woman: a proportion of the primordial follicles sequentially begin to grow, but more than 99.9% of them fail to complete development and degenerate. At puberty only about 300,000 primary oocytes remain, and by menopause only a few are present.

From the time of puberty, one developing follicle is stimulated each month to mature—to complete development and to ovulate. This means that during the 40 or so years of a woman's reproductive life, only 400 to 500 eggs will have been released. All the rest will have degenerated. It is still a mystery why so many eggs are formed only to die in the ovaries.

Toward the end of a woman's reproductive life, the mature eggs released from the ovary at ovulation will have been arrested in prophase of the first meiotic division for 40 to 50 years. Damage to the eggs during this period is one possible explanation for the high incidence of genetic abnormalities among children born to older women. For example, 1% of children born to women over 40 years old have *Down's syndrome* (Mongolism), a condition caused by an extra copy of chromosome 21, resulting from the failure of the two homologs of chromosome 21 to separate during the nuclear division of the maturing oocyte in meiosis I.

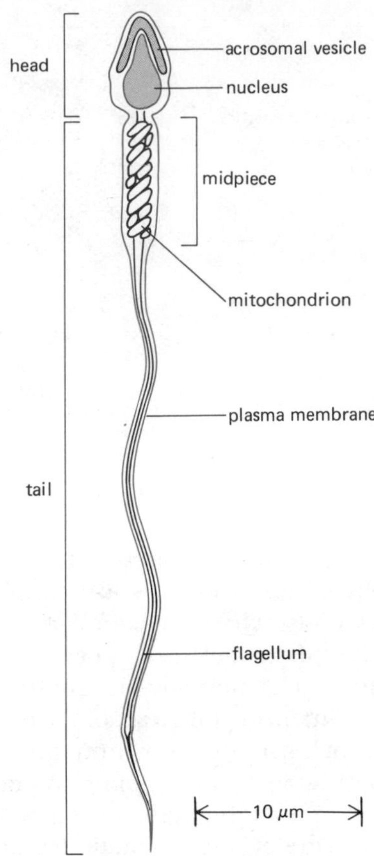

Figure 14–35 Drawing of a human sperm in longitudinal section.

Sperm Are Highly Adapted for Delivering Their DNA to an Egg[13]

Whereas the egg is the largest cell in an organism, the sperm (*spermatozoon,* plural *spermatozoa*) is usually the smallest. A sperm has two main functions: to deliver its haploid set of genes to the egg for sexual recombination and to activate the developmental program of the egg. Highly compact and streamlined, a sperm is equipped with a strong flagellum to propel it through an aqueous medium (Figure 14–35). Even so, the vast majority fail in their mission: of the hundreds of millions of sperm released from the male, only a few ever manage to fertilize an egg.

Most sperm are "stripped-down" cells unencumbered by cytoplasmic organelles such as ribosomes, endoplasmic reticulum, or Golgi apparatus, which are unnecessary for the task of delivering the DNA to the egg. On the other hand, they contain many mitochondria strategically placed where they can most efficiently power the flagellum. Sperm usually consist of two morphologically and functionally distinct regions enclosed by a single plasma membrane: the **head,** which contains an unusually highly condensed haploid nucleus, and the **tail,** which propels the sperm to the egg and helps it burrow through the egg coat. The DNA in the nucleus is inactive and extremely tightly packed, so that its volume is minimized for transport. Indeed, the chromosomes of many sperm have dispensed with the histones of somatic cells and are packed instead with simple, highly positively charged proteins.

In the sperm head, closely apposed to the anterior end of the nuclear envelope, is a specialized secretory vesicle called the **acrosomal vesicle** (Figure 14–35). This vesicle contains hydrolytic enzymes that enable the sperm to penetrate the egg's outer coat. When a sperm contacts an egg, the contents of the vesicle are released by exocytosis in the so-called *acrosomal reaction.* In invertebrate sperm, this reaction also releases specific proteins that bind the sperm tightly to the egg coat.

The motile tail of a sperm is a long flagellum whose central axoneme emanates from a basal body situated just posterior to the nucleus. As described earlier (see p. 562), the axoneme consists of two central singlet microtubules surrounded by nine evenly spaced microtubule doublets. The flagellum of some sperm (including those of mammals) differs from other flagella in that the axoneme is further surrounded by nine outer dense fibers, giving a 9 + 9 + 2 arrangement rather than the more usual 9 + 2 pattern (Figures 14–36 and 14–37). These dense fibers are stiff and noncontractile, and it is not known if they contribute to the active bending of the flagellum, which is

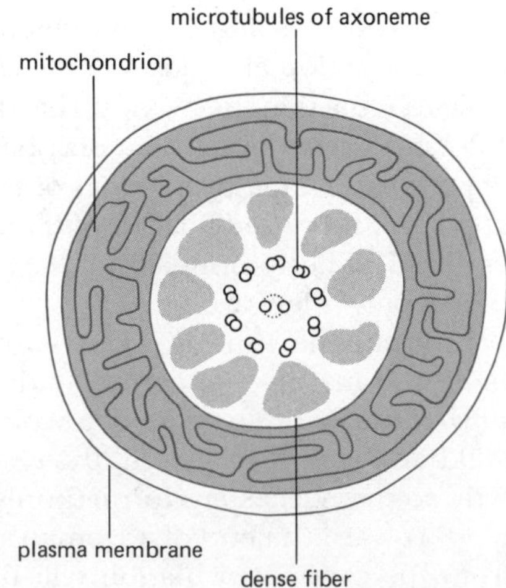

Figure 14–36 Drawing of a mammalian sperm as seen in cross-section through the midpiece in an electron microscope. The flagellum is composed of an axoneme surrounded by nine dense fibers. The axoneme consists of two singlet microtubules surrounded by nine microtubule doublets. Note that the mitochondrion is wrapped around the dense fibers so that it is well placed for providing the ATP required for flagellar movement.

caused by the sliding of the adjacent microtubule doublets past one another (see p. 566). Flagellar movement is powered by the hydrolysis of ATP generated by highly specialized mitochondria in the anterior part of the sperm tail (called the *midpiece*), exactly where it is needed (Figures 14–35 and 14–36).

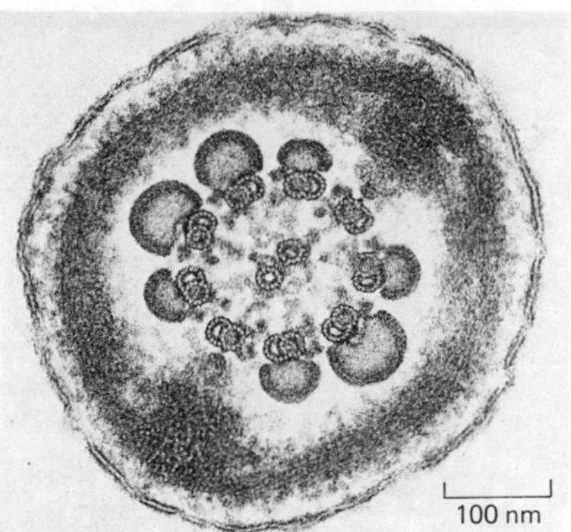

Figure 14–37 Electron micrograph of the flagellum of a guinea pig sperm seen in cross-section. Note that in the region shown two of the dense fibers have terminated and have become continuous with an outer fibrous sheath. (Courtesy of Daniel S. Friend.)

Sperm Are Produced Continuously in Many Mammals[14]

There are important differences between the way eggs are produced (oogenesis) and the way sperm are produced (**spermatogenesis**). In human females, for example, we have described how many primordial germ cells proliferate in the ovary to generate a limited number of oogonia early in embryogenesis but only complete meiosis to produce the final eggs at intervals, one at a time. In human males, on the other hand, spermatogenesis does not begin until puberty and then goes on continuously in the epithelial lining of very long, tightly coiled tubes, called *seminiferous tubules,* in the testes. Immature germ cells, called **spermatogonia,** are located along the outer edge of these tubes next to the basal lamina, where they divide continuously by mitosis. Some of the daughter cells stop proliferating and differentiate into **primary spermatocytes.** These cells enter the first meiotic prophase, in which their paired homologous chromosomes participate in crossing over, and then proceed with division I of meiosis to produce two **secondary spermatocytes,** each containing 22 duplicated autosomal chromosomes and either a duplicated X or a duplicated Y chromosome. Each chromosome still consists of two sister chromatids, and the two secondary spermatocytes undergo division II of meiosis to produce four **spermatids,** each with a haploid number of single chromosomes. These haploid spermatids then undergo morphological differentiation into mature spermatozoa, which escape into the lumen of the seminiferous tubule (Figures 14–38 and 14–39). They subsequently pass into the *epididymis,* a coiled tube overlying the testis, where they are stored and undergo further maturation.

An intriguing and unique feature of spermatogenesis is that the developing male germ cells fail to complete cytoplasmic division (cytokinesis) during mitosis and meiosis, so that all of the daughter cells, except for the least differentiated spermatogonia, remain connected by cytoplasmic bridges (Figure 14–40). These cytoplasmic bridges persist until the very end of sperm differentiation, when individual sperm are released into the tubule lumen. This means that the progeny of a single spermatogonium maintain cytoplasmic continuity throughout their differentiation. (A group of cells joined in this way is known as a *syncytium*). This accounts for the observation that mature sperm arise synchronously in any given area of a seminiferous tubule. But what is the function of the syncytial arrangement?

Sperm Nuclei Are Haploid, But Sperm Cell Differentiation Is Directed by the Diploid Genome[15]

Unlike eggs, sperm undergo most of their differentiation after their nuclei have completed meiosis to become haploid. The cytoplasmic bridges between them provide a device by which each developing haploid sperm, by sharing a common cytoplasm with its neighbors, can be supplied with all the products of a complete diploid genome. This may be necessary for two reasons. First, the diploid genome from which the sperm derives will generally include some defective gene copies, corresponding to recessive lethal mutations; a haploid cell receiving one of these defective gene copies is likely to die unless it is provided with the functional gene products encoded by other nuclei that have the good gene copy. Second, the genetic material is often not evenly divided

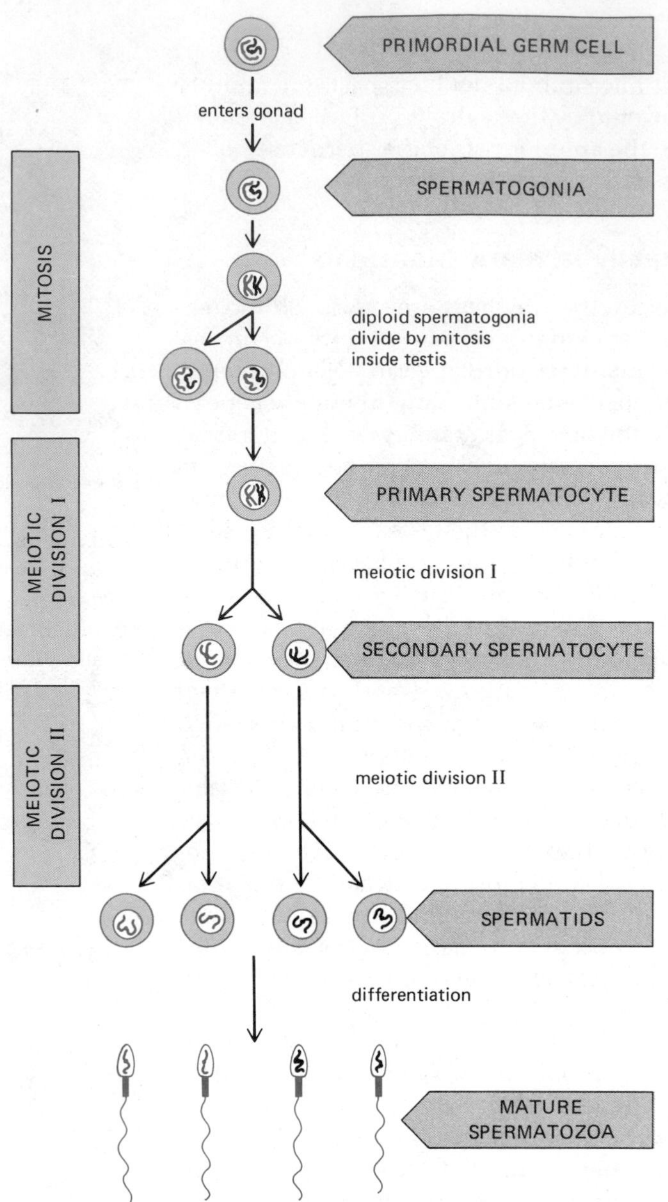

Figure 14–38 The various stages of spermatogenesis. Spermatogonia develop from primordial germ cells that migrate into the testis early in embryogenesis. When the animal becomes sexually mature, the spermatogonia divide by mitosis to renew themselves continually, and some begin meiosis to become primary spermatocytes, which in turn continue through meiotic division I to become secondary spermatocytes. After they complete meiotic division II, these secondary spermatocytes produce haploid spermatids that differentiate into mature sperm. Note that spermatogenesis differs from oogenesis (Figure 14–28) in two ways: (1) new cells enter meiosis continually from the time of puberty, and (2) each cell that begins meiosis gives rise to four mature gametes rather than one.

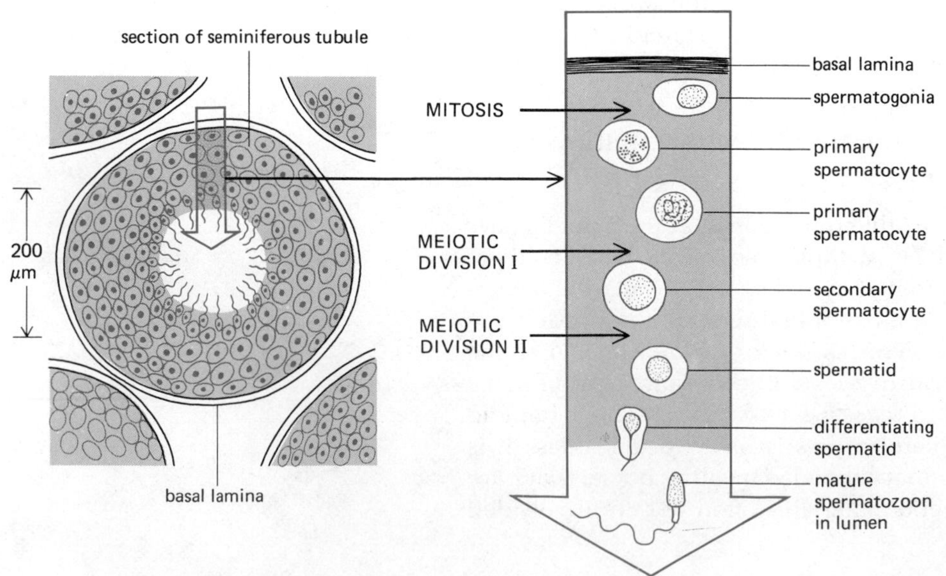

Figure 14–39 Schematic drawing of a cross-section of a seminiferous tubule in a mammalian testis. Dividing spermatogonia are found along the basal lamina. Some of these cells stop dividing and enter meiosis to become primary spermatocytes. Eventually mature sperm are released into the lumen. In man it takes about 24 days for spermatocytes to complete meiosis and about 9 weeks for a spermatogonium to develop into four mature sperm.

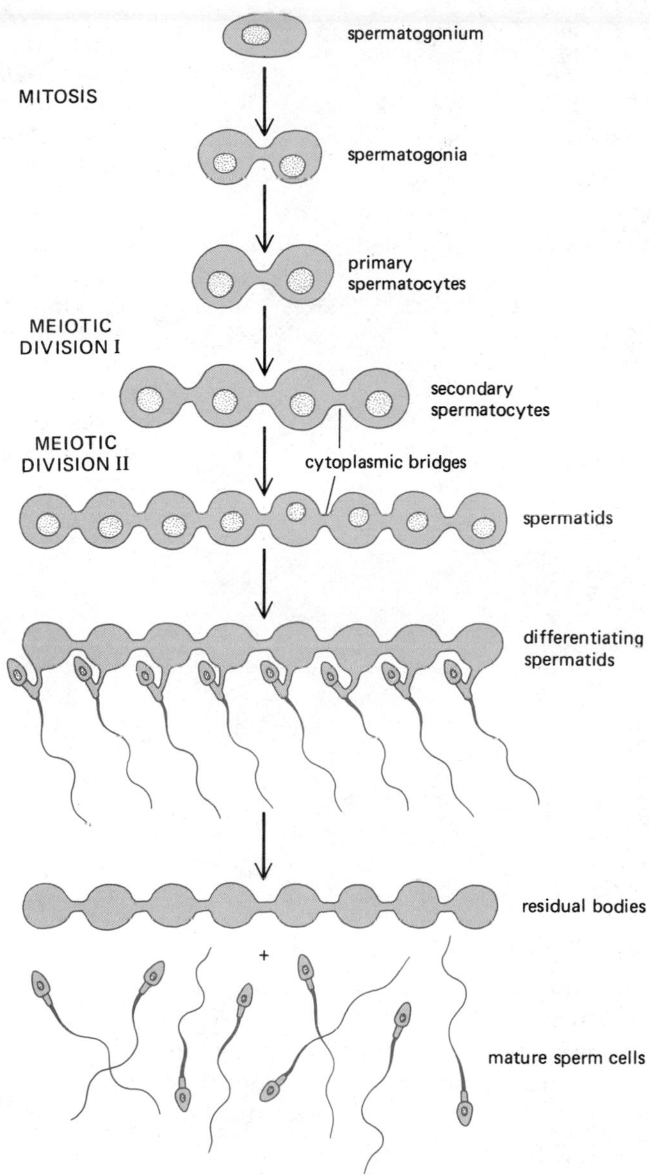

Figure 14–40 Schematic illustration showing how the progeny of a single spermatogonium remain connected to each other by cytoplasmic bridges throughout their differentiation into mature sperm. For the sake of simplicity, only two connected spermatogonia are shown entering meiosis to eventually form eight connected haploid spermatids. In fact, the number of connected cells that go through two meiotic divisions and differentiate together is very much larger than shown here.

among the sperm nuclei produced at meiosis. For example, in man some sperm inherit an X chromosome at meiosis, while others inherit a Y chromosome. Since the X chromosome carries many essential genes that are lacking on the Y chromosome, if it were not for the cytoplasmic bridges between one developing sperm and the next, it seems likely that the Y-bearing sperm would be unable to survive and mature, with the consequence that no males could be produced in the next generation.

In fact, there is direct experimental evidence that sperm differentiation is governed by products of the diploid genome. Some of this comes from studies of *Drosophila* mutants known as *disjunction mutants,* in which chromosomes are divided unequally between daughter cells during meiosis. As a result, some sperm contain too few chromosomes, some too many, and some contain none at all. Remarkably, the major features of sperm differentiation occur in all such cells, even in those without chromosomes (Figure 14–41). This can be explained along the lines suggested above: the products of missing chromosomes could be supplied by diffusion through the cytoplasmic bridges

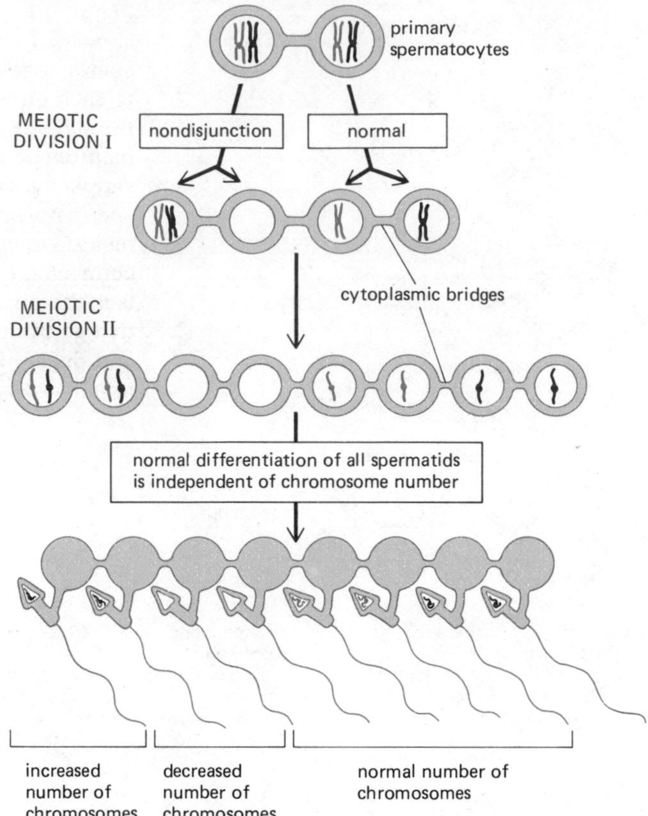

Figure 14–41 Schematic drawing of spermatogenesis in disjunction mutants of *Drosophila,* in which chromosomes are often divided unequally among daughter cells during meiosis so that some cells contain too many chromosomes while others contain too few or none at all. The reason that all of the sperm develop quite normally even when they have no chromosomes may be that the products of missing chromosomes are supplied by neighboring cells via cytoplasmic bridges. For the sake of clarity, only one pair of chromosomes is illustrated.

that connect adjacent germ cells. Alternatively, it is possible that the diploid spermatogonia or primary spermatocytes may produce stable directions for sperm differentiation (presumably in the form of long-lived mRNA) before meiosis, so that there is no need for the haploid sperm genome to function during the period of differentiation. Whichever is the true explanation, it is evident that the differentiation of a sperm, even though it occurs when the sperm nucleus is haploid, uses products of both sets of parental chromosomes.

Summary

An egg is programmed to form a new individual organism when activated by a sperm. Many eggs become enormous by importing macromolecules made elsewhere, such as yolk protein, and by enlisting the help of surrounding accessory cells to nourish the egg. Eggs develop in stages from primordial germ cells that migrate into the ovary very early in development to become oogonia. After mitotic proliferation, oogonia become primary oocytes that begin meiotic division I, becoming arrested at prophase for days or years, depending on the species. During part of this prophase-I arrest period, primary oocytes accumulate ribosomes and macromolecules. Further development (egg maturation) depends on polypeptide hormones (gonadotropins), which act on the surrounding accessory cells, causing them to induce a small proportion of the primary oocytes to mature. These induced primary oocytes complete meiotic division I to form a small polar body and a large secondary oocyte; this secondary oocyte subsequently undergoes meiotic division II to form a second small polar body and a large mature ovum. The stage at which the developing oocyte is released from the ovary and is ready for fertilization differs in different species.

A sperm is highly specialized for the task of delivering its DNA to the egg. It is a small and compact cell, with an unusually condensed nucleus and a long flagellum. Spermatogenesis differs from oogenesis in several important ways: (1) Whereas in many female organisms the total pool of oocytes is produced early in embryogenesis, in males new germ cells enter meiosis continually from the time of sexual maturation. (2) Whereas each primary oocyte produces only one mature egg (while the other three haploid nuclei produced by meiosis degenerate), each primary spermatocyte gives rise to four mature sperm. (3) Since mature spermatogonia and all spermatocytes fail to complete cytokinesis during mitosis and meiosis, respectively, the progeny of a single spermatogonium develop as a syncytium, maintaining cytoplasmic continuity throughout development. As a result, sperm development is capable of being directed by the products of both parental chromosomes, despite the fact that, unlike egg development, it occurs while the nuclei are haploid.

Fertilization[16]

Once released, egg and sperm alike are destined to die within hours unless they find each other and fuse in the process of fertilization. Through fertilization the egg and sperm are saved: the egg is activated to begin its developmental program, and the nuclei of the two gametes fuse to complete the sexual reproductive process. Much of what we know about the mechanism of fertilization has been learned from studies of marine invertebrates—especially sea urchins (Figure 14–42). While fertilization in mammals takes place in the confines of the reproductive tract after copulation, fertilization in sea urchins occurs in sea water, into which both sperm and eggs are released. This makes it much more accessible to study. Moreover, because such *external fertilization* is chancy, these aquatic organisms produce enormous numbers of gametes. A typical female sea urchin contains several million eggs and a typical male several billion sperm, so that sea urchin gametes can be obtained as pure populations in very large numbers, all at the same stage of development. When mixed together, the events of sperm-egg interaction begin synchronously within seconds.

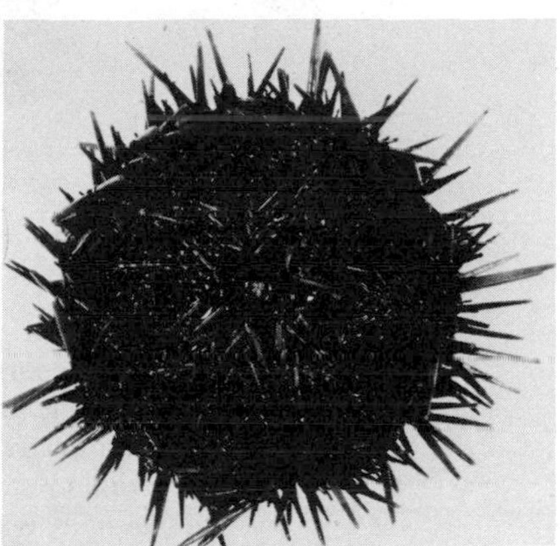

Figure 14–42 Photograph of a sea urchin (shown at approximate actual size). (Courtesy of David Epel.)

Despite the great evolutionary distance between sea urchins and mammals, there is evidence that many of the events of fertilization are similar in both.

A Sperm Must Be Activated Before It Can Fertilize an Egg[17]

Eggs and sperm are specialized for fusion, but it is important that they fuse only with each other and not with other cells of the organism. Both use special mechanisms to ensure specific fusion. The egg can fuse only with sperm because its fusion surface (the plasma membrane) is covered with a coat of specialized extracellular matrix, the vitelline layer or zona pellucida, through which only sperm of the same species can pass. Mammalian sperm are unable to fertilize an egg until they have undergone a process referred to as **capacitation,** induced by secretions in the female genital tract. The mechanism of capacitation is unclear; it seems to involve an alteration in the lipid composition of the sperm plasma membrane. Capacitated sperm bind specifically to a major glycoprotein in the zona pellucida, which is thought to trigger the sperm to undergo the **acrosomal reaction,** releasing the contents of the acrosomal vesicle to the exterior surroundings. Among the molecules released are hydrolytic enzymes that help the sperm penetrate the zona pellucida and thereby gain access to the egg's plasma membrane, with which it can fuse (Figures 14–43 and 14–44).

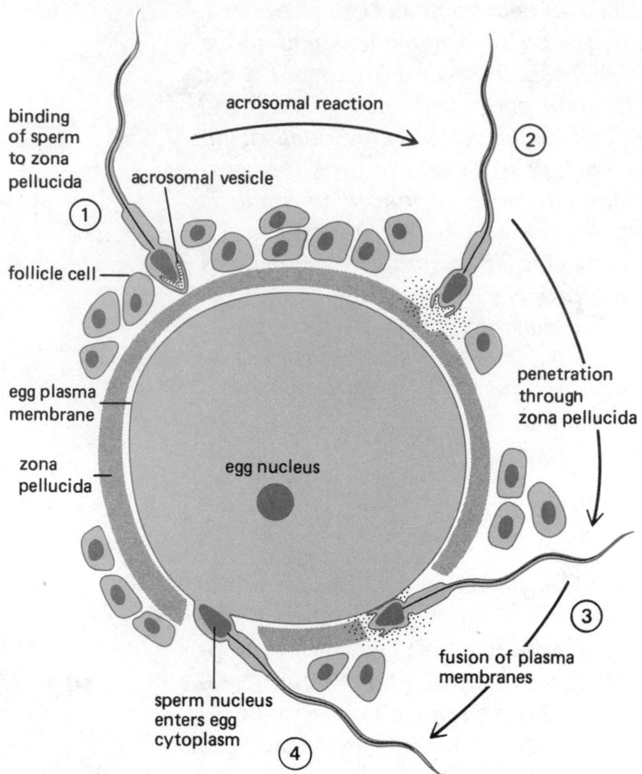

Figure 14–43 Schematic illustration of the acrosomal reaction occurring when a mammalian sperm fertilizes an egg. A single glycoprotein in the zona pellucida is thought to be responsible for both binding the sperm and inducing the acrosomal reaction. Note that a mammalian sperm interacts tangentially with the egg plasma membrane so that fusion occurs at the side rather than at the tip of the sperm head.

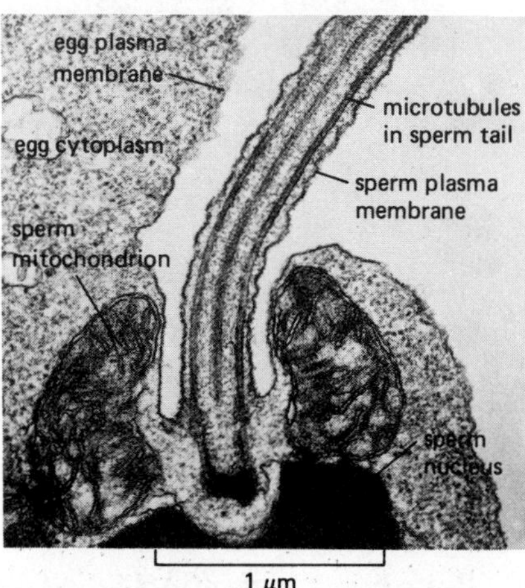

Figure 14–44 Electron micrograph of a sea urchin sperm (seen in cross-section) entering a sea urchin egg after the plasma membranes of the two gametes have fused. (Courtesy of Daniel S. Friend.)

In invertebrates such as sea urchins, the acrosomal reaction also plays an important part in fertilization, and a great deal more is known about the mechanisms involved. While in mammalian sperm it is the equatorial (post-acrosomal) region of the plasma membrane of the sperm head that fuses with the egg (Figure 14–43), in marine invertebrates, it is actually the membrane of the acrosomal vesicle that fuses. This membrane, normally kept sequestered within the sperm, becomes exposed on the sperm surface only when it fuses with the sperm plasma membrane at the time of the acrosomal reaction. This membrane fusion is normally accompanied by the formation of a long, actin-containing **acrosomal process,** which projects from the anterior end of the sperm. As shown in Figure 14–45, the tip of this process becomes covered by the components of the old acrosomal vesicle membrane. The tip is also coated with the secreted contents of the acrosomal vesicle, including both specific binding proteins that mediate its attachment to the vitelline layer (see below) and hydrolytic enzymes that then permit it to bore through this layer to the egg's plasma membrane. At this point the membrane at the tip of the acrosomal process fuses with the egg membrane, allowing the sperm nucleus to enter the egg (Figure 14–46).

The trigger for the acrosomal reaction for sea urchin sperm is a polysaccharide component (a polymer of fucose sulfate) of the egg's jelly coat: when this material is extracted from a sea urchin egg and added to sperm of

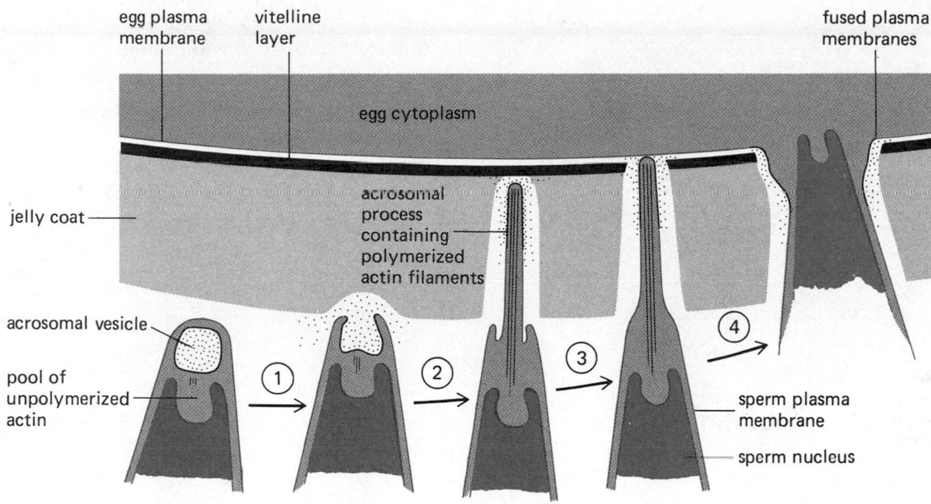

Figure 14–45 Schematic drawing of the details of the acrosomal reaction in sea urchin. When the sperm contacts the jelly coat, exocytosis of the acrosomal vesicle occurs (1), followed by the explosive polymerization of actin to form the long acrosomal process that penetrates the jelly coat (2). Proteins released from the acrosomal vesicle (*black dots*) adhere to the surface of the acrosomal process and serve both to bind the sperm to the vitelline layer and to digest this layer (3). When the old acrosomal vesicle (whose membrane forms the tip of the acrosomal process) contacts the egg plasma membrane (3), the two membranes fuse, the actin filaments disassemble, and the sperm enters the egg (4).

the same species, it induces a normal acrosomal reaction within seconds. The polysaccharide seems to work by inducing an influx of Ca^{2+} into the sperm head, which initiates the exocytotic release of the acrosomal vesicle. At the same time, the Ca^{2+} influx induces an efflux of H^+ in exchange for Na^+; the resulting rise in pH inside the sperm head initiates the formation of the *acrosomal process* by stimulating the explosive polymerization of actin. It has been suggested that the rise in intracellular pH causes unpolymerized actin to dissociate from special actin-binding proteins in the sperm cytoplasm that prevent actin polymerization (see p. 571).

Sperm-Egg Adhesion Is Mediated by Species-specific Proteins[18]

The species-specificity of fertilization is often determined by the specificity of the binding of the sperm to the innermost layer of the egg coat. For example, while sea urchin sperm will usually undergo an acrosomal reaction in response to eggs of different species, they cannot bind to such eggs and therefore cannot fertilize them. Removing the egg coat, however, often removes this

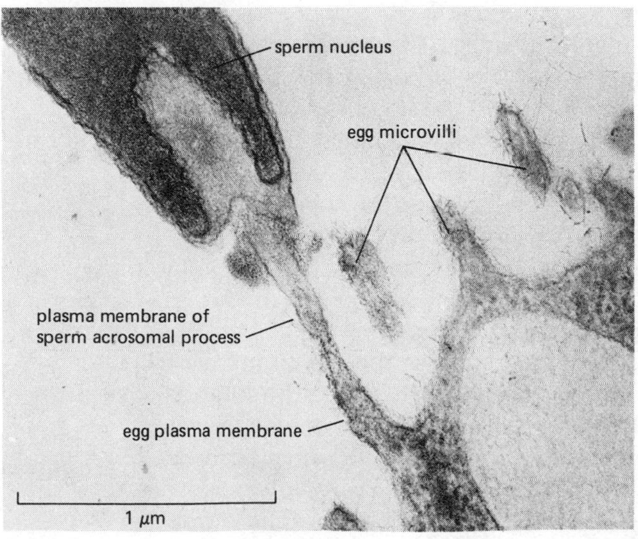

Figure 14–46 Electron micrograph showing a sea urchin sperm in the process of fertilizing an egg. The membrane at the tip of the acrosomal process of the sperm has fused with the egg plasma membrane at the tip of a microvillus on the egg surface. An unfertilized sea urchin egg is covered with more than 100,000 microvilli. (Courtesy of Frank Collins.)

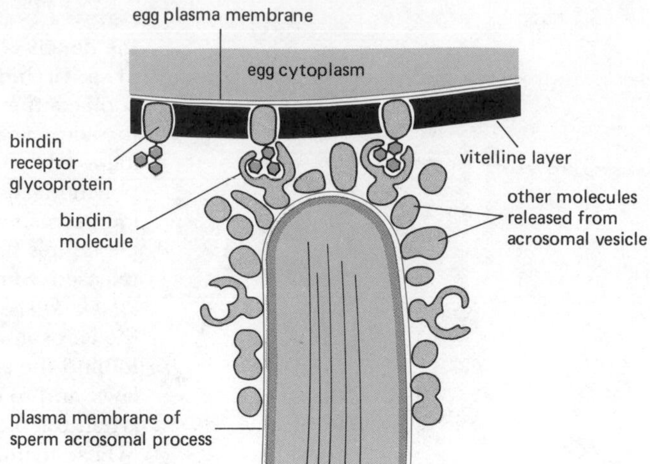

Figure 14–47 A highly schematic diagram of bindin molecules covering the surface of the acrosomal process of a sea urchin sperm. These proteins are thought to bind to a specific oligosaccharide sequence on a receptor glycoprotein associated with the vitelline layer of the egg.

barrier to fertilization between species: for example, hamster eggs from which the zona pellucida has been removed with specific enzymes can be fertilized by human sperm. Not surprisingly, such hybrid "humsters" do not develop.

The molecule in sea urchin sperm that is responsible for the species-specific adherence of sperm to the egg's vitelline layer has been isolated. It is a protein (called *bindin*) of about 30,000 daltons that is normally sequestered in the acrosomal vesicle. After its release in the acrosomal reaction, it coats the surface of the acrosomal process and mediates the attachment of the sperm to the egg. Isolated bindin molecules bind only to the vitelline layer of sea urchin eggs of the same species. The vitelline layer of a sea urchin egg has been found to contain species-specific glycoproteins with which bindin interacts in the adhesion process. There is some evidence that bindin acts as a lectin that recognizes specific carbohydrate determinants on the glycoprotein molecules (Figure 14–47).

Mammalian sperm contain molecules in their plasma membranes that bind directly to a specific glycoprotein in the egg zona pellucida; unlike the situation in sea urchins, the same egg glycoprotein can be shown to activate sperm to undergo the acrosomal reaction.

Egg Activation Is Mediated by Changes in Intracellular Ion Concentrations[16, 19]

Once an activated sea urchin sperm attaches to an egg, the acrosomal process rapidly bores through the vitelline layer. The membrane at the tip of the process fuses with the egg plasma membrane and the sperm nucleus is injected into the egg. Within 30 minutes the sperm and egg nuclei (called *pronuclei*) fuse to recreate a diploid nucleus.

Besides contributing its DNA to this so-called **zygote,** the sperm activates the developmental program of the egg. Before fertilization an egg is metabolically dormant: it does not synthesize DNA, and it synthesizes RNA and protein at very low rates. Once released from the supportive environment of the ovary, an egg will die within hours unless rescued by a sperm.

The binding of a sperm to the egg surface induces the egg to increase its metabolism and to begin DNA synthesis and cleavage. While the mechanism of activation is not known, it is clear that the sperm serves only to trigger a preset program in the egg. The sperm itself is not required. An egg can be activated by a variety of nonspecific chemical or physical treatments; for example, a frog egg can be activated by pricking it with a needle. (The devel-

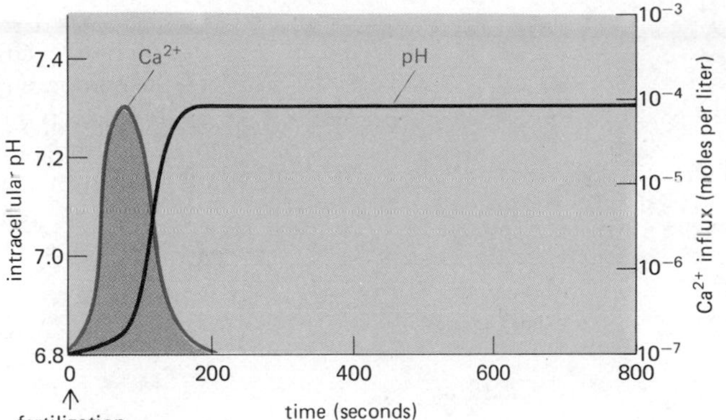

Figure 14–48 Two ionic changes responsible for the activation of a sea urchin egg following fertilization: beginning about 20 seconds after fertilization, Ca^{2+} is released into the cytosol from intracellular stores, increasing the intracellular concentration of free Ca^{2+} for about 2½ minutes; at about 60 seconds, a sustained efflux of H^+ coupled to an influx of Na^+ causes a permanent increase in intracellular pH.

opment of an egg that has been activated in the absence of a sperm is called **parthenogenesis;** some organisms, including a few vertebrates, normally reproduce parthenogenically.) Moreover, the initial stages of egg activation cannot depend on the generation of any new proteins because they occur perfectly normally in the presence of drugs that inhibit protein synthesis.

In sea urchins, all the early steps in egg activation are mediated by changes in ion concentrations within the egg. Three different ionic changes occur within seconds or minutes of the addition of sperm to a suspension of eggs: (1) an increase in the permeability of the plasma membrane to Na^+ causes the membrane to depolarize within a few seconds; (2) a massive release of Ca^{2+} from an unknown intracellular site causes a marked increase in the concentration of Ca^{2+} in the cytosol within 20 to 30 seconds; and (3) an efflux of H^+ coupled to an influx of Na^+ begins within 60 seconds and causes a large increase in intracellular pH (Figure 14–48). As we shall now describe, these three ionic changes have two consequences: first, they cause the egg to become impenetrable to further sperm; and second, they mediate the initial steps in the developmental program of the egg.

The Rapid Depolarization of the Egg Membrane Prevents Further Sperm-Egg Fusions, Thereby Mediating the Fast Block to Polyspermy[20]

Although many sperm can attach to an egg, normally only one fuses with the egg plasma membrane and injects its nucleus into the cell. If more than one sperm fuses (a condition referred to as *polyspermy*), extra mitotic spindles are formed, resulting in the abnormal segregation of chromosomes during cleavage; nondiploid cells are produced and development quickly stops. This means that the egg must provide a block to the entry of extra sperm very soon after fertilization. The mechanism of this **fast block to polyspermy** is not the same in all species. Fish eggs have a small channel, called the *micropyle,* through which sperm must pass in single file. The passage of a single sperm through the channel stimulates the egg so that the cortical granules release their contents, plugging the hole so that no other sperm can enter.

The eggs of most organisms, however, do not have a micropyle and can fuse with sperm anywhere on their surface. In some eggs (such as those of sea urchins and amphibians), the rapid depolarization of the plasma membrane caused by the fusion of the first sperm prevents further sperm from fusing. The membrane potential of a sea urchin egg is about −60 mV. Within a few seconds of adding sperm, the membrane potential falls precipitously and reverses to about +20 mV, where it remains for a minute or so before

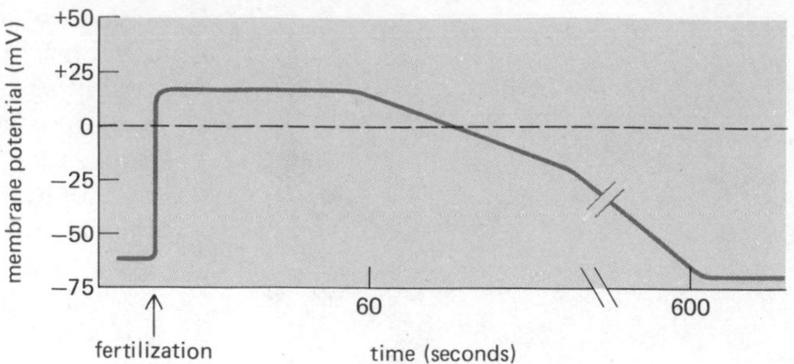

Figure 14–49 Graph showing the changes in the sea urchin egg membrane potential after fertilization. The rapid depolarization somehow prevents further sperm from fusing with the egg plasma membrane, thereby mediating the fast block to polyspermy.

gradually returning to the original prefertilization level (Figure 14–49). If depolarization is prevented by fertilizing eggs in low Na^+, which reduces the sperm-triggered Na^+ influx that is largely responsible for depolarizing the membrane, there is an enhanced incidence of polyspermy. Moreover, if an unfertilized egg is depolarized artificially by a current passed into it through a microelectrode, sperm can attach to the egg but cannot fuse; if the membrane is now repolarized with the microelectrode, the attached sperm fuse with and enter the egg. Although the molecular mechanism is unknown, it seems likely that the membrane depolarization that normally accompanies fertilization alters the conformation of a crucial protein in the egg plasma membrane so that the sperm membrane can no longer fuse with the egg membrane.

The egg membrane potential returns to normal within a few minutes after fertilization; therefore, a second mechanism must provide a longer-term barrier to polyspermy. In most eggs, including human eggs, this barrier is provided by substances released from the cortical granules that are located in the egg cortex.

The Cortical Reaction Is Responsible for the Late Block to Polyspermy[21]

The cortical granules in sea urchin eggs fuse with the plasma membrane and release their contents within 30 seconds of adding sperm. Like the sperm acrosomal reaction, and indeed most triggered secretory processes, this **cortical reaction** is mediated by a large rise in the concentration of free Ca^{2+} in the cytosol. In an activated sea urchin egg, the Ca^{2+} concentration increases by about a hundredfold within 20 or 30 seconds of adding sperm and then after a minute or two drops back toward normal (see Figure 14–48).

The importance of Ca^{2+} in triggering the cortical reaction can be demonstrated directly in plasma membranes isolated from sea urchin eggs with cortical granules still attached to their cytoplasmic surfaces (Figure 14–50). When small amounts of Ca^{2+} are added to such preparations, exocytosis occurs within seconds.

The proteases and other enzymes released in the cortical reaction alter the structure of the egg coat in such a way that additional sperm cannot penetrate. In sea urchin eggs, the cortical reaction has at least two separate effects: (1) proteolytic enzymes released from the cortical granules rapidly destroy the glycoproteins that serve as bindin receptors for sperm attachment, and (2) the released contents of the cortical granules cause the vitelline layer that overlies the plasma membrane to move away from the egg surface and, at the same time, released enzymes cross-link proteins in the vitelline layer, causing it to harden. In this way, a *fertilization membrane* is formed that

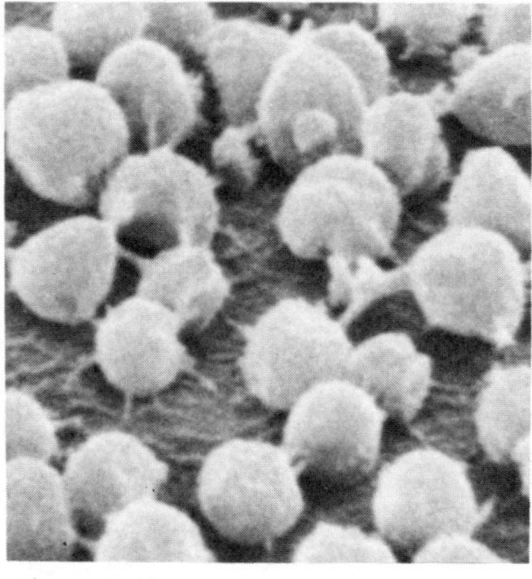

Figure 14–50 Scanning electron micrograph of cortical granules attached to the isolated plasma membrane of an unfertilized sea urchin egg. When Ca^{2+} is added to this preparation, the cortical granules fuse with the plasma membrane and release their contents by exocytosis. Since there are about 18,000 cortical granules in each cell, the cortical reaction causes the surface area of the egg to more than double in less than a minute; the extra membrane is accommodated by a lengthening of each microvillus on the egg surface. (From V. D. Vacquier, *Dev. Biol.* 43:62–74, 1975.)

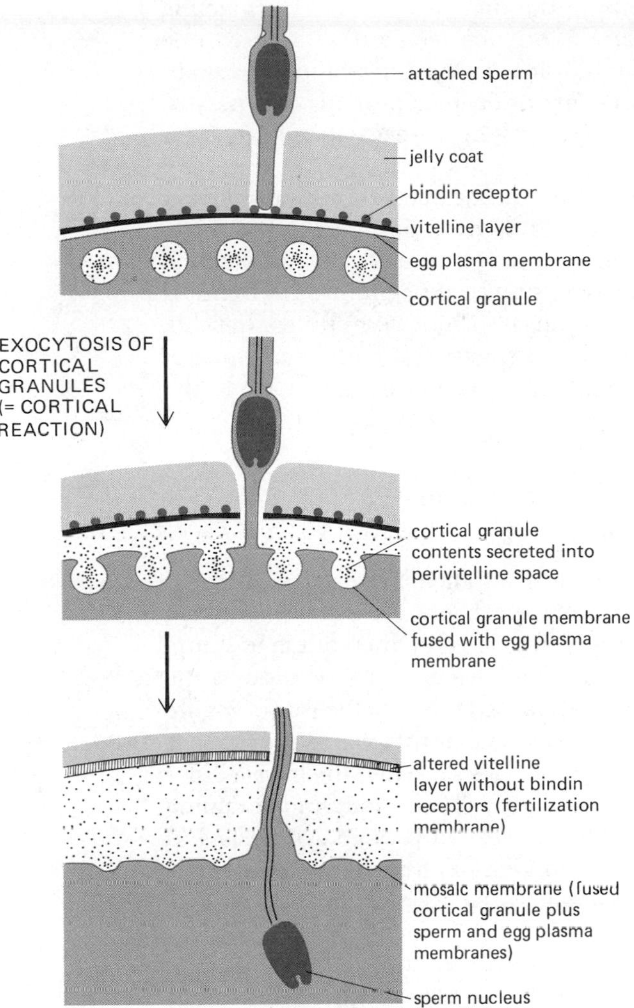

Figure 14–51 Schematic diagram illustrating how the cortical reaction in a sea urchin egg prevents additional sperm from entering the egg. The released contents of the cortical granules raise the vitelline layer and alter it so that it no longer contains bindin receptors and is hardened into a fertilization membrane that sperm cannot penetrate.

sperm cannot penetrate (Figure 14–51). In mammalian eggs, the cortical reaction acts in a similar way to prevent polyspermy—causing the glycoprotein in the zona pellucida (discussed on p. 804) to become altered so that it can no longer bind sperm or activate them to undergo an acrosomal reaction.

An Increase in Intracellular Free Ca^{2+} Initiates Egg Development[16,22]

Although membrane depolarization is the first detectable change following fertilization, it does not activate the egg to begin biosynthesis. Artificially depolarizing the egg membrane does not lead to egg activation; nor does blocking membrane depolarization at the time of fertilization inhibit activation. Thus, membrane depolarization appears to serve only to prevent polyspermy.

There is abundant evidence that it is the transient increase in cytosolic Ca^{2+} concentration (which propagates around the egg from the site of sperm fusion as a ring-shaped wave) that initiates the program of egg development. The cytosolic concentration of Ca^{2+} can be artificially increased either by injecting Ca^{2+} directly into the egg or by the use of Ca^{2+} carrying ionophores, such as A23187 (see p. 301). This activates the eggs of all animals so far tested, including mammals. Moreover, preventing the increase in Ca^{2+} by injecting the Ca^{2+} chelator EGTA inhibits egg activation after fertilization. At least one way in which Ca^{2+} acts in cells is by binding to the Ca^{2+}-binding protein *calmodulin,* which in turn activates a variety of cellular proteins (see p. 748). Calmodulin has been found in large amounts in all eggs that have been studied.

Since the increase in Ca^{2+} concentration in the cytosol is transient, lasting only for 2 to 3 minutes after fertilization, it is clear that it cannot directly mediate the events observed during the later stages of egg activation, which in sea urchins include an increase in protein synthesis beginning at 8 minutes and the initiation of DNA synthesis beginning at about 40 minutes. Instead, the rise in Ca^{2+} concentration serves only to trigger the entire sequence of developmental events; some more permanent change must take place in the egg while the Ca^{2+} level is high.

A Rise in the Intracellular pH in Some Organisms Induces the Late Synthetic Events of Egg Activation[23]

In sea urchins, the transient rise in Ca^{2+} concentration activates specific transport proteins in the egg plasma membrane (probably via calmodulin) that harness the energy stored in the Na^+ gradient across the membrane to pump H^+ out of the cell (see p. 295). The efflux of H^+ leads to an increase in the intracellular pH from 6.6 to 7.2, which is maintained through the rest of zygote development (see Figure 14–48). There is evidence that this rise in pH induces the late synthetic events in activated sea urchin eggs: (1) When the intracellular pH is raised in unfertilized eggs by incubating them in a medium containing ammonia (Figure 14–52), there is a marked increase in protein synthesis and DNA replication, even in the absence of an increase in free intracellular Ca^{2+}. (2) When eggs are placed in Na^+-free sea water just after fertilization, so that there is no Na^+ gradient to drive the H^+ efflux, the intracellular pH does not rise and the late events do not occur. Such eggs can be rescued by adding ammonia to the medium: now the intracellular pH rises and the synthesis of protein and DNA is induced, even in the absence of extracellular Na^+.

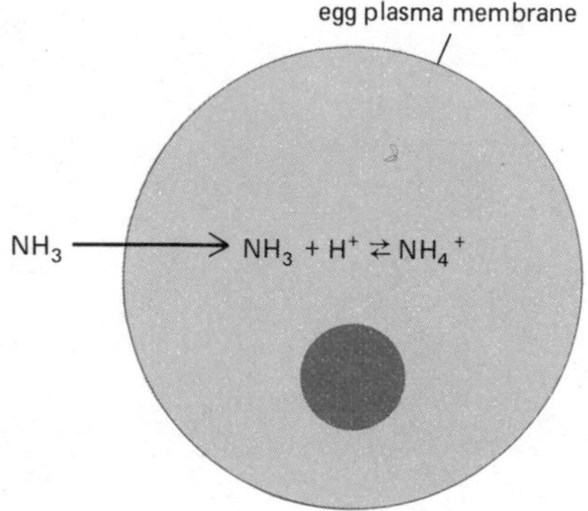

Figure 14–52 Schematic illustration of how incubating cells (such as eggs) in ammonia increases the intracellular pH. The ammonia diffuses through the plasma membrane and combines with H^+ in the cytosol to form NH_4^+, thereby decreasing the intracellular concentration of H^+ and increasing the pH.

The marked increase in protein synthesis in fertilized eggs does not require RNA synthesis, since it is unaffected by the drug actinomycin D, which inhibits RNA synthesis. This increase in protein synthesis normally results from at least two separate changes: (1) preexisting mRNA molecules stored in the egg are unmasked and made available for protein synthesis and (2) the egg ribosomes are activated so that they translate the mRNA molecules more rapidly. By contrast, the increase in protein synthesis in fertilized eggs treated with ammonia results solely from an increased recruitment of preexisting mRNA molecules. This suggests that, while the increase in intracellular pH is responsible for the mRNA recruitment process, some other factor normally increases the rate of ribosome movement along the mRNA strands.

Mammalian Eggs Can Be Fertilized *in Vitro*[24]

Although fertilization is a very special event, occurring only once every generation, it relies on the same types of ion fluxes that are commonly used to regulate intracellular processes in somatic cells (see Chapter 13). The sequence of some of the events in sea urchin egg activation after fertilization is summarized in Table 14–1. The same sequence is not followed in all species. For example, an increase in intracellular pH does not follow activation in some other invertebrate eggs nor in some vertebrate eggs. Nonetheless, the same general principles apply.

Table 14–1 Sequence of Events Following Fertilization of Sea Urchin Eggs

Event	Time After Fertilization	Intracellular Mediator
1. Plasma membrane depolarization	0–3 seconds	sperm-induced increase in plasma membrane permeability to Na^+ (and to some extent to Ca^{2+})
2. Increased concentration of free intracellular Ca^{2+}	20–140 seconds	release of bound Ca^{2+} from intracellular storage sites
3. Cortical granule exocytosis	30–60 seconds	increased intracellular Ca^{2+}
4. Increased protein synthesis	8 minutes	increased intracellular pH
5. Fusion of sperm and egg nuclei	30 minutes	
6. Initiation of DNA replication	40–45 minutes	increased intracellular pH

Compared with sea urchins eggs, mammalian eggs are extremely difficult to study. Whereas sea urchin eggs are readily available by the millions, investigators must be content to work with tens or hundreds of mammalian eggs. Nonetheless, it is now possible to fertilize mammalian eggs *in vitro* and study some of the events that occur when they are activated. Unlike sea urchin and amphibian eggs, where a rapid depolarization of the plasma membrane accompanies fertilization and prevents polyspermy, fertilization of a hamster egg induces a series of transient hyperpolarizations, which seem not to prevent other sperm from entering the egg. On the other hand, the release of Ca^{2+} into the cytosol from intracellular stores seems to initiate mammalian egg activation, just as it does in sea urchins. Mammalian eggs that have been fertilized *in vitro* can develop into normal individuals when transplanted into the uterus; in this way some previously infertile women have been able to produce normal children.

Embryogenesis, in which an apparently simple fertilized egg cell develops into a new individual as complex as a human being, is perhaps the most remarkable phenomenon in all of biology. What is known about this developmental process at the cellular level is the subject of the next chapter.

Summary

Sperm and eggs fuse in the process of fertilization. Contact with the egg coat induces a sperm to undergo the acrosomal reaction; some of the proteins released help the sperm digest its way through the egg coat so that the sperm plasma membrane can fuse with the egg plasma membrane. In many eggs two processes prevent additional sperm from fusing: (1) A rapid depolarization of the egg plasma membrane prevents it from fusing with other sperm, thereby providing a temporary fast block to polyspermy. (2) An influx of Ca^{2+} into the cytosol activates the cortical granules to release their contents; this cortical reaction alters the egg coat so that sperm can no longer bind or penetrate, thus providing a prolonged late block to polyspermy. The influx of Ca^{2+} into the egg cytosol caused by the sperm also activates the egg to begin its developmental program.

References

General

Austin, C.R.; Short, R.V., eds. Germ Cells and Fertilization. Cambridge, Eng.: Cambridge University Press, 1972.

Browder, L. Developmental Biology, Chapters 5, 6, and 8. Philadelphia: Saunders, 1980.

Karp, G.; Berrill, N.J. Development, 2nd ed., Chapters 4 and 5. New York: McGraw-Hill, 1981.

Epel, D. The program of fertilization. *Sci. Am.* 237(11):128–138, 1977.

Cited

1. Williams, G.C. Sex and Evolution. Princeton, N.J.: Princeton University Press, 1975.
   Maynard Smith, J. Evolution of Sex. Cambridge, Eng.: Cambridge University Press, 1978.
2. Ayala, F.; Kiger, J. Modern Genetics. Reading, Ma.: Addison-Wesley, 1980.
3. Lewis, J.; Wolpert, L. Diploidy, evolution and sex. *J. Theor. Biol.* 78:425–438, 1979.
4. Whitehouse, H.L. Towards an Understanding of the Mechanism of Heredity, 3rd ed. London: St. Martins, 1973. (Contains a lucid description of the development of our current understanding of chromosome behavior during meiosis.)
   Wolfe, S.L. Biology of the Cell, 2nd ed., pp. 432–470. Belmont, Ca.: Wadsworth, 1981.
   Lewin, B. Gene Expression, Vol. 2, Eucaryotic Chromosomes, 2nd ed., pp. 102–141. New York: Wiley, 1980.
5. John, B.; Lewis, K.R. The Meiotic Mechanism. Oxford Biology Readers (J.J. Head, ed.). Oxford, Eng.: Oxford University Press, 1976.
   Goodenough, U. Genetics. New York: Holt, Rinehart & Winston, 1978.
6. Moses, M.J. Synaptonemal complex. *Annu. Rev. Genet.* 2:363–412, 1968.
   Moses, M.J. The synaptonemal complex and meiosis. In Molecular Human Cytogenetics ICN-UCLA Symposia on Molecular and Cellular Biology, Vol. 7 (R.S. Sparkes, D.E. Comings, C.F. Fox, eds.), pp. 101–125. New York: Academic Press, 1977.
7. Carpenter, A.T.C. Recombination nodules and synaptonemal complex in recombination-defective females of *Drosophila melanogaster. Chromosoma* 75:259–292, 1979.
8. Solari, A.J. The behavior of the XY pair in mammals. *Int. Rev. Cytol.* 38:273–317, 1974.
9. Karp, G.; Berrill, N.J. Development, 2nd ed., pp. 116–138. New York: McGraw-Hill, 1981.
   Browder, L. Developmental Biology, pp. 173–231. Philadelphia: Saunders, 1980.
   Grant, P. Biology of Developing Systems, pp. 265–282. New York: Holt, Rinehart & Winston, 1978.
   Hart, N.H.; Hopper, A.F. Foundations of Animal Development, pp. 37–58. Oxford, Eng.: Oxford University Press, 1979.
10. Woodruff, R.I.; Telfer, W.H. Electrophoresis of proteins in intercellular bridges. *Nature* 286:84–86, 1980.
11. Masui, Y.; Clarke, H.J. Oocyte maturation. *Int. Rev. Cytol.* 57:185–282, 1979.
    Peters, H.; McNatty, K.P. The Ovary: A Correlation of Structure and Function in Mammals, pp. 11–22, 60–84. Berkeley, Ca.: University of California Press, 1980.
    Richards, J.S. Hormonal control of ovarian follicular development. *Recent Prog. Horm. Res.* 35:343–373, 1979.
12. Peters, H.; McNatty, K.P. The Ovary: A Correlation of Structure and Function in Mammals, pp. 98–106. Berkeley, Ca.: University of California Press, 1980.
13. Bloom, W.; Fawcett, D.W. A Textbook of Histology, 10th ed., pp. 805–815. Philadelphia: Saunders, 1975.
    Fawcett, D.W. The mammalian spermatozoon *Dev. Biol.* 44:394–436, 1975.
14. Bloom, W.; Fawcett, D.W. A Textbook of Histology, 10th ed., pp. 819–830. Philadelphia: Saunders, 1975.
    Browder, L. Developmental Biology, pp. 146–172. Philadelphia: Saunders, 1980.

Clermont, Y. Kinetics of spermatogenesis in mammals: seminiferous epithelium cycle and spermatogonial renewal. *Physiol. Rev.* 52:198–236, 1972.

Karp, G.: Berrill, N.J. Development, 2nd ed., pp. 100–116. New York: McGraw-Hill, 1981.

15. Monesi, V.; Geremia, R.; D'Agostino, A.; Boitani, C. Biochemistry of male germ cell differentiation in mammals: RNA synthesis in meiotic and postmeiotic cells. In Current Topics in Developmental Biology, Vol. 12 (A.A. Moscona, A. Monroy, eds.), pp. 11–36. New York: Academic Press, 1978.

Lindsley, D.L.; Tokuyasu, K.T. Spermatogenesis. In The Genetics and Biology of Drosophila, Vol. 2d (M. Ashburner, T.R.F. Wright, eds.), pp. 225–294. New York: Academic Press, 1980.

16. Epel, D. Fertilization. *Endeavour* (New Series) 4:26–31, 1980.

Epel, D.; Vacquier, V.D. Membrane fusion events during invertebrate fertilization. In Membrane Fusion, Cell Surface Reviews, Vol. 5 (G. Poste, G.L. Nicolson, eds.), pp. 1–63. Amsterdam: Elsevier, 1978.

Shapiro, B.M.; Schackmann, R.W.; Gabel, C.A. Molecular approaches to the study of fertilization. *Annu. Rev. Biochem.* 50:815–843, 1981.

17. Shapiro, B.M.; Eddy E.M. When sperm meets egg: biochemical mechanisms of gamete interaction. *Int. Rev. Cytol.* 66:257–302, 1980.

Bedford, J.M.; Cooper, G.W. Membrane fusion events in the fertilization of vertebrate eggs. In Membrane Fusion, Cell Surface Reviews, Vol. 5 (G. Poste, G.L. Nicolson, eds.), pp. 65–127. Amsterdam: Elsevier, 1978.

18. Metz, C.B. Sperm and egg receptors involved in fertilization. *Curr. Top. Dev. Biol.* 12:107–147, 1978.

Vacquier, V.D. The adhesion of sperm to sea urchin eggs. In The Cell Surface: Mediator of Developmental Processes (S. Subtelny, N.K. Wessells, eds.), pp. 151–168. New York: Academic Press, 1980.

Bleil, J.D.; Wassarman, P.M. Mammalian sperm-egg interaction: identification of a glycoprotein in mouse egg zonae pellucidae possessing receptor activity for sperm. *Cell* 20:873–882, 1980.

19. Epel, D. Mechanisms of activation of sperm and egg during fertilization of sea urchin gametes. In Current Topics in Developmental Biology, Vol. 12 (A.A. Moscona, A. Monroy, eds.), pp. 186–246. New York: Academic Press, 1978.

20. Hagiwara, S.; Jaffe, L.A. Electrical properties of egg cell membranes. *Annu. Rev. Biophys. Bioeng.* 8:385–416, 1979.

21. Schuel, H. Secretory functions of egg cortical granules in fertilization and development: a critical review. *Gamete Research* 1:294–382, 1978.

Vacquier, V.D. Dynamic changes of the egg cortex: a review. *Dev. Biol.* 84:1–26, 1981.

22. Ridgway, E.B.; Gilkey, J.C.; Jaffe, L.F. Free calcium increases explosively in activating medaka eggs. *Proc. Natl. Acad. Sci. USA* 74:623–627, 1977.

23. Johnson, J.D.; Epel, D.; Paul, M. Intracellular pH and activation of sea urchin eggs after fertilization. *Nature* 262:661–664, 1976.

Winkler, M.M.; Steinhardt, R.A.; Grainger, J.L.; Minning, L. Dual ionic controls for the activation of protein synthesis at fertilization. *Nature* 287:558–560, 1980.

24. Gwatkin, R.B.L. Fertilization. In The Cell Surface in Animal Embryogenesis and Development, Cell Surface Reviews, Vol. 1 (G. Poste, G.L. Nicolson, eds.), pp. 1–54. Amsterdam: Elsevier, 1977.

Yanagimachi, R. Sperm-egg association in mammals. In Current Topics in Developmental Biology, Vol. 12 (A.A. Moscona, A. Monroy, eds.), pp. 83–106. New York: Academic Press, 1978.

Grobstein, C. External human fertilization. *Sci. Am.* 240(6):57–67, 1979.

Lopata, A. Successes and failures in human *in vitro* fertilization. *Nature* 288:642–643, 1980.

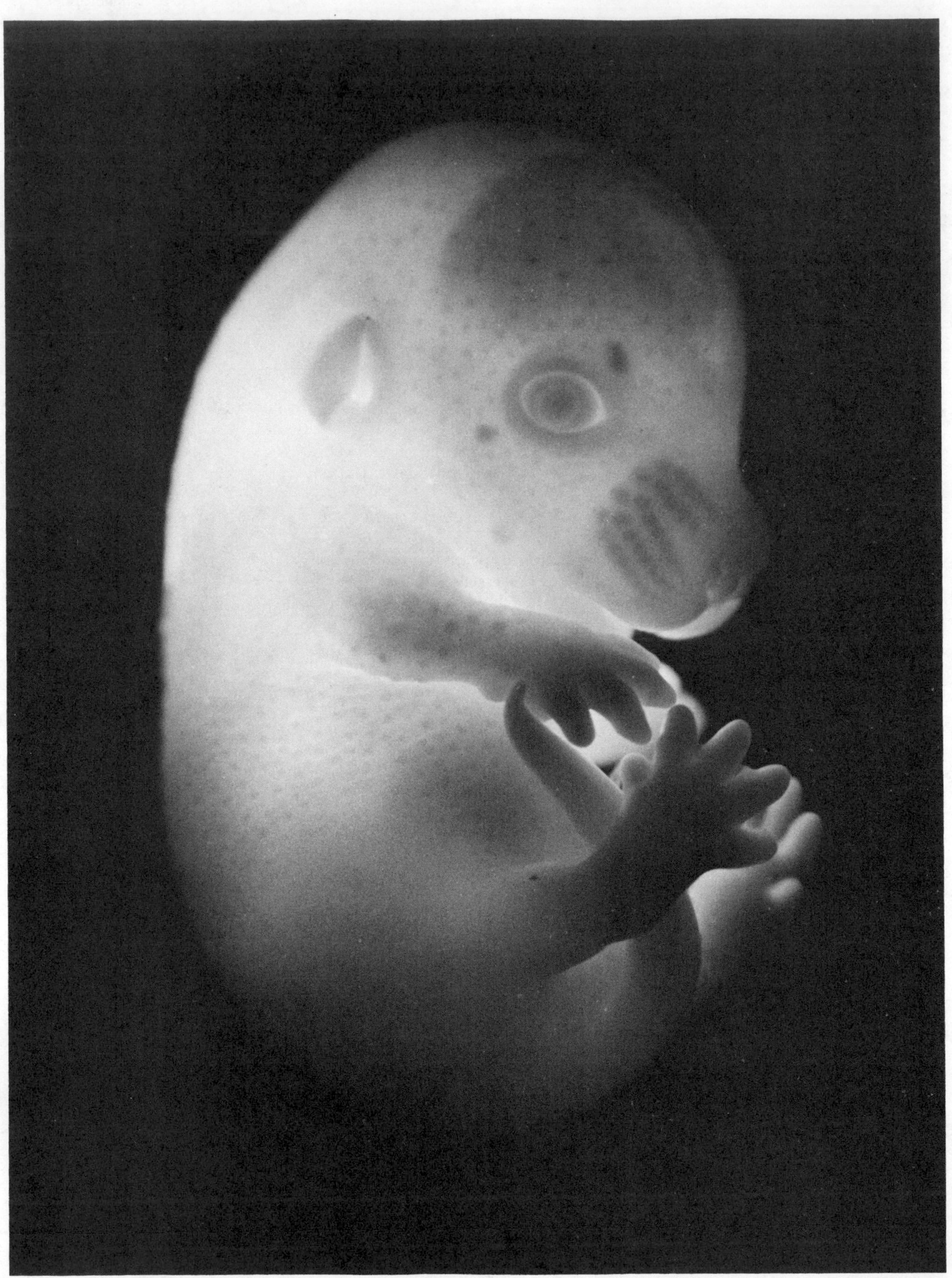

A mouse embryo at 15 days of gestation.

Cellular Mechanisms of Development

15

Almost every multicellular animal is a clone of cells descended from a single original cell, the fertilized egg. Thus the cells of the body, as a rule, are genetically alike. But phenotypically they are different: some are specialized as muscle, others as neurons, others as blood cells, and so on. The different cell types are arranged in a precisely organized pattern, and the whole structure has a well-defined shape. All these features are determined by the DNA sequence of the genome, which is reproduced in every cell. Each cell must act on the same genetic instructions, but it must interpret them with due regard to time and circumstance, so as to play its proper part in the multicellular society.

Multicellular organisms are often very complex, but they are built up through a quite limited repertoire of cellular activities. Cells grow and divide. They die. They form mechanical attachments. They generate forces for cell locomotion and deformation. They fuse with one another. They differentiate, that is, switch on or off the production of certain of the substances for which the genome codes. They secrete or display on their surfaces materials that influence the activities of their neighbors. These forms of cell behavior are the universal basis of animal development. This chapter tries to explain how such activities are brought into play at the right times and places to generate a whole organism.

Rather than follow in detail any one organism from the beginning of its development to the end, we shall consider in turn the different aspects of cell behavior in development, exemplifying the general principles by reference to the animals that display them best. We shall discuss how cell movements and forces shape the embryo, how the genes inside cells and the interactions between cells control the development of the spatial pattern of differentiation, and how some cells migrate along defined paths through the embryo to settle in specific sites. To illustrate these topics, we shall use amphibians, sea urchins, mice, flies, birds, cockroaches, and a nematode worm.

Thus, we begin with the cell movements and forces that control the shaping of the embryo, as exemplified in amphibians and sea urchins. The problem of how cells come to adopt different characters and express different genes according to their positions in the body is first considered in the mouse, then in *Drosophila,* and thirdly with reference to limb development in birds and cockroaches. For comparison, we shall discuss the small nematode worm *Caenorhabditis elegans,* as an example of an animal differing from vertebrates and insects in that all its developmental processes occur with perfect accuracy and predictability, allowing the fate of each individual cell to be precisely specified. Finally, cell migration in vertebrate embryos is briefly considered as a preface to the specialized problems of neural development discussed in Chapter 18.

Cleavage and Blastula Formation

In this section and the next we consider how the geometrical structure of the early embryo is formed and what the physical forces are that mold it. We shall take as our chief example the frog *Xenopus laevis* (Figure 15–1), whose early development has been particularly well studied. Like other amphibian embryos, it is robust and easy to manipulate experimentally.

To put the events that we are about to describe in context, it is helpful to regard the development of a vertebrate—and indeed of many other types of animal—as having three phases. In the first phase, the fertilized egg *cleaves* to form many smaller cells, which become organized into an epithelium and perform a complex series of *gastrulation* movements, whose outcome is the creation of a rudimentary gut cavity. The second phase is that of *organogenesis,* in which the various organs of the body, such as limbs, eyes, heart, and so on, are formed. In the third phase, the pattern of structures that has been generated in this way on a small scale proceeds to grow to its adult size. These phases are not sharply distinct but overlap considerably. We shall follow the mechanics of *Xenopus* development from the fertilized egg to the beginning of organogenesis.

Cleavage Produces Many Cells from One[1]

The amphibian egg is a large cell, about a millimeter in diameter, enclosed in a transparent extracellular capsule or egg coat. Most of the cell's volume is occupied by yolk platelets, which are aggregates chiefly of lipid and protein. The yolk is concentrated toward the lower end of the egg, called the **vegetal pole;** the other end is called the **animal pole.** In the initial cleavages, beginning soon after fertilization, this one large cell subdivides by repeated mitosis into many smaller cells, or **blastomeres,** without any change in total mass. These first cell divisions are extremely rapid, with a cycle time of about 30 minutes, thanks to reserves of RNA, protein, membrane, and other materials that accumulated in the egg while it matured in the mother. The only crucial biosynthesis obviously required is that of DNA, and unusually rapid DNA replication is made possible by an exceptionally large number of replication origins (see p. 645).

Figure 15–1 Synopsis of the development of *Xenopus laevis* from newly fertilized egg to feeding tadpole. The adult male and female are shown mating at the top. The developmental stages are viewed from the side, except for the 10-hour and 19-hour embryos, which are viewed respectively from below and from above. All stages except the adults are drawn to the same scale. (After P. D. Nieuwkoop and J. Faber, Normal Table of *Xenopus laevis* [Daudin]. Amsterdam: North-Holland, 1956.)

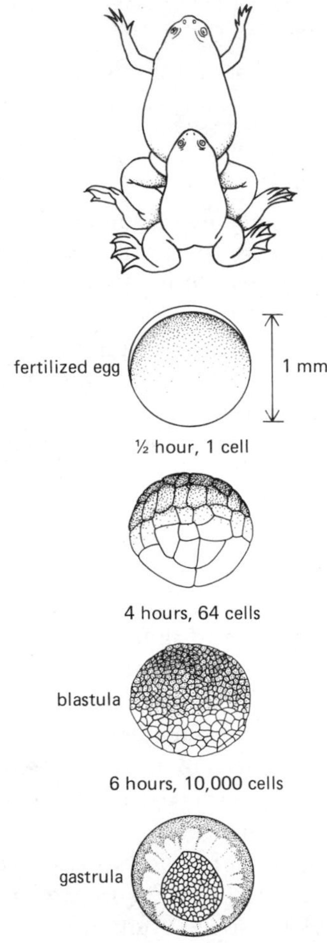

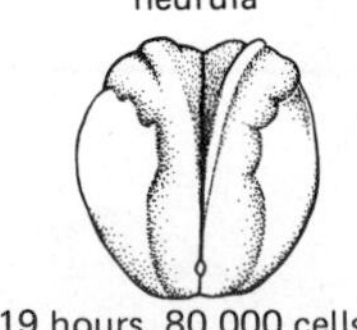

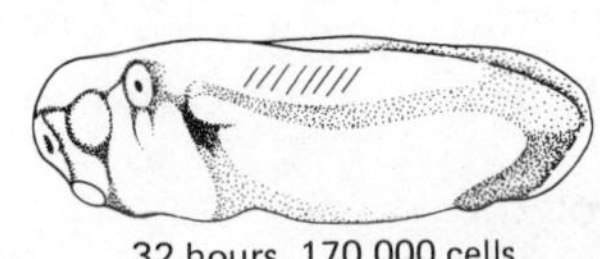

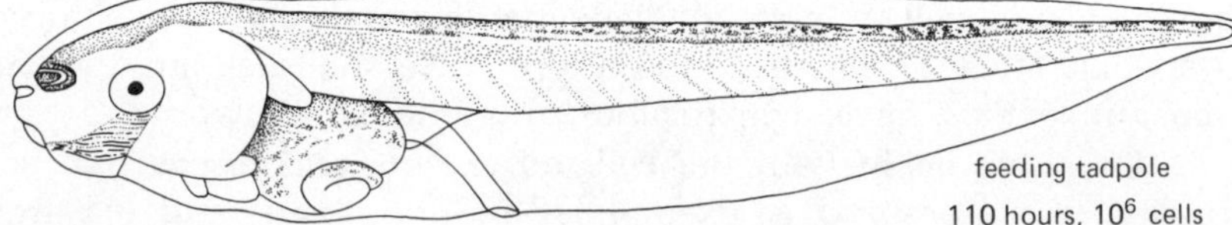

The first cleavage divides the egg vertically, or parallel to the animal-vegetal axis, into two symmetrical halves (Figure 15–2). The next cleavage is again vertical, but perpendicular to the first, giving four cells of similar size. The third cleavage passes horizontally through these four cells, slightly above the midline, to give four small cells stacked on top of four larger, more yolky cells. All the cells divide synchronously for the first 12 cleavages, but the divisions are asymmetric, so that the lower, vegetal cells, encumbered with yolk, are fewer and larger. The synchrony is lost abruptly after about 12 cycles of cleavage.

In other species the relative orientations of the successive cleavage planes may follow different schemes. In very yolky eggs, such as those of birds, cleavage fails to cut through the yolk, and all the nuclei remain clustered at the animal pole; the embryo then develops from a cap of cells on top of the yolk.

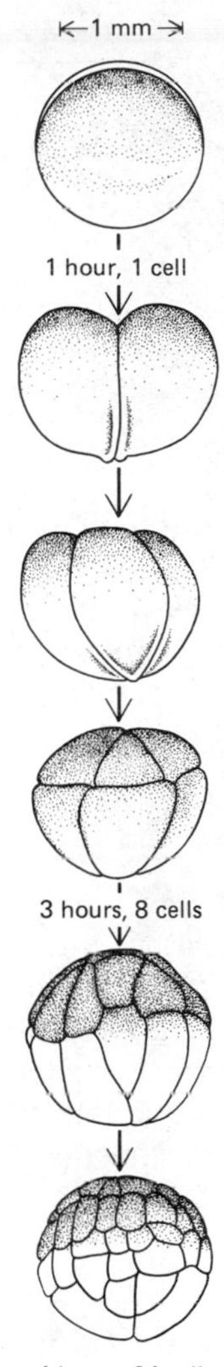

Figure 15–2 The stages of cleavage in *Xenopus*, as seen from the side.

The Polarity of the Embryo Depends on the Polarity of the Egg[1,2]

The animal that develops from the fertilized egg will have a head and a tail, a back and a belly, and a median plane of symmetry dividing the body into a right side and a left side. It is often useful to describe the animal by reference to three axes: **antero-posterior,** from head to tail; **dorso-ventral,** from back to belly; and **medio-lateral,** from the median plane outward to the left or to the right. These polarities are established very early in development. The amphibian egg, though spherically symmetrical in shape, is not spherically symmetrical in its chemical constitution. The cells that form by cleavage at the animal and vegetal poles are different: the yolky vegetal cells will contribute to the gut, while most of the other tissues derive from the cells of the animal pole. In some amphibian eggs another asymmetry becomes visible shortly after fertilization—a band of light pigmentation called the *gray crescent.* This forms opposite the site of sperm entry and reflects a reorganization of the cytoskeleton controlled by the centriole, which the sperm brings into the egg with it (see Chapter 10). Together, these two asymmetries of the egg define the antero-posterior and dorso-ventral axes of the embryo (Figure 15–3). The first cleavage plane normally cuts through the middle of the gray crescent so that the first two blastomeres represent the future right and left sides of the body.

The Blastula Consists of an Epithelium Surrounding a Cavity[3]

From the outset, the cells of the embryo are not only bound together mechanically, but also coupled by gap junctions through which ions and other small molecules can pass (see p. 686). While the significance of the gap junctions is still unclear, other junctions, whose function is more obvious, also form. In the outermost regions of the embryo, tight junctions (see p. 684) between the blastomeres create a seal, isolating the interior of the embryo from the external medium. At about the 16-cell stage, the crevices between the cells deep inside the embryo enlarge to form a single cavity, the **blastocoel,** by a mechanism in which sodium ions are pumped across the cell membranes into the extracellular spaces in the interior of the embryo, and water follows as a result of the consequent osmotic pressure differences. The cells surrounding the blastocoel have become organized into an epithelium, and the embryo is now termed a **blastula** (Figure 15–4). In amphibians, the

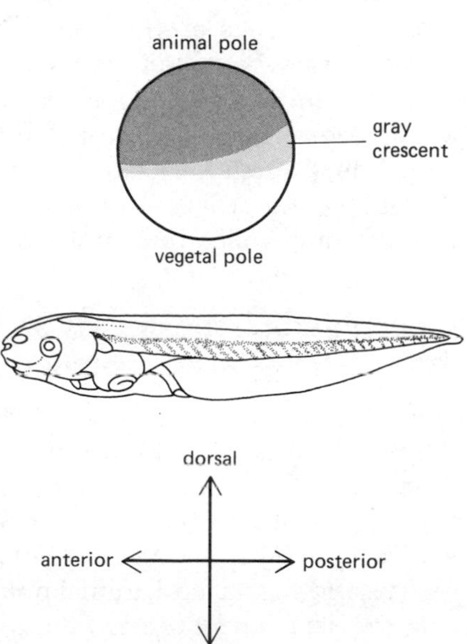

Figure 15–3 The asymmetries of the *Xenopus* egg are shown at the top. These determine the axes of the body of the tadpole, shown below.

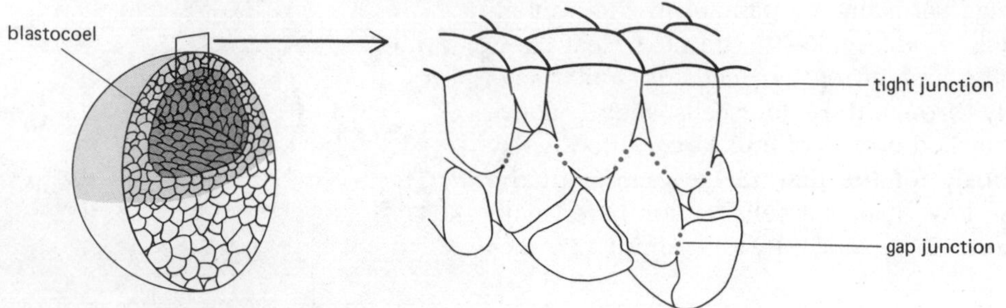

Figure 15–4 The blastula. At this stage the cells are arranged to form an epithelium surrounding a fluid-filled cavity, the blastocoel. The cells are electrically coupled via gap junctions, and tight junctions close to the outer surface create a seal, isolating the interior of the embryo from the external medium.

epithelium of the blastula is several cells thick, though in some other species it may be only one cell thick. As is true of all epithelial sheets, the epithelium of the blastula is polarized, the outer, inner, and lateral faces of the cells differing in composition and function (see p. 685). This organization of the cells of the blastula as an epithelial sheet is crucial in coordinating their subsequent behavior.

Summary

The eggs of most species are large cells, containing stores of nutrients and other cell components specified by the maternal genome. During the cleavage divisions following fertilization the egg subdivides into many smaller cells, but no growth occurs. In the amphibian, tight junctions develop between the cells in the outermost layers of the embryo, sealing off the interior from the external medium. Fluid is drawn into the interior of the embryo to form a cavity, the blastocoel, surrounded by an epithelium. The embryo at this stage is called a blastula.

Gastrulation, Neurulation, and Somite Formation[4]

Once the cells of the blastula have become arranged into an epithelial sheet, the scene is set for the coordinated movements of **gastrulation.** This dramatic process transforms the simple hollow ball of cells into a multilayered structure with a central axis and bilateral symmetry: by a complicated invagination, a large area of cells on the outside of the embryo are brought to lie inside it. Subsequent development will depend on the interactions of the inner, outer, and middle layers thus formed. Gastrulation in one form or another is an almost universal episode of development throughout the animal kingdom. Since the geometry of gastrulation in amphibians is rather contorted, we shall first follow the process in the sea urchin. Gastrulation in this close relative of the vertebrates illustrates the same basic principles.

Gastrulation Transforms a Hollow Ball of Cells into a Three-layered Structure[5]

Because the sea urchin embryo is transparent, both its internal and its external development can be followed in the living animal, and the activities of the individual cells can be analyzed. The starting point for gastrulation is a very simple blastula: a sheet of about 1000 cells, 1 cell thick, surrounding a spherical cavity. The embryo as a whole is enclosed in a thin layer of extracellular matrix, and vegetal and animal poles can be distinguished. Gastrulation begins with the detachment of a few dozen cells—the *primary mesenchyme cells*—from

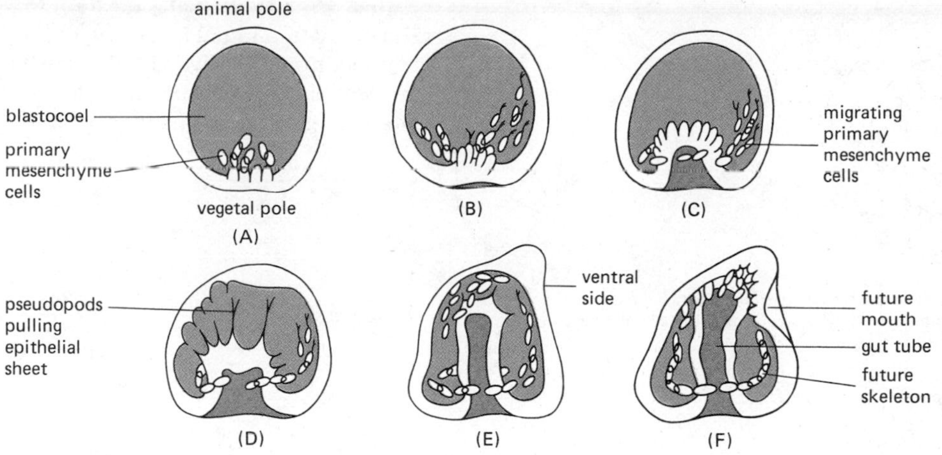

Figure 15–5 Gastrulation in the sea urchin, viewed here as a transparent object, focusing on a vertical plane through the center of the embryo. (A) The primary mesenchyme cells break loose from the epithelium at the vegetal pole of the blastula. (B) These cells then crawl over the inner face of the wall of the blastula. (C) Meanwhile the epithelium at the vegetal pole is beginning to tuck inward. (D and E) Pseudopods extend from the invaginating epithelial sheet, pulling it right into the blastocoel cavity to form the gut tube. (F) The end of the gut tube makes contact with the wall of the blastula; the site of contact coincides with the site of the future mouth opening. (After L. Wolpert and T. Gustafson, *Endeavour* 26:85–90, 1967.)

the epithelium at the vegetal pole (Figure 15–5A). These cells move into the blastocoel cavity and migrate along its lining, pulling themselves along by stretching out long thin processes, or *pseudopods,* with sticky tips (Figure 15–6). When the tip of a pseudopod contacts a surface to which it can firmly attach, it contracts, pulling the cell after it. Old pseudopods seem to be withdrawn and new pseudopods extended almost at random, pulling the cell hither and thither. Yet the cells finally settle in well-defined positions, where they begin to form the skeleton (Figure 15–5F). What controls the pattern of their settling? It may be that the lining of the blastocoel cavity is graded in adhesiveness, so that the cells gather at the sites to which their pseudopods on average stick most strongly. That such a mechanism could work in principle has been shown by studies of fibroblasts moving in tissue culture on a substratum of graded adhesiveness (made by depositing onto cellulose acetate a layer of palladium metal of graded thickness). The fibroblasts, which are migratory, tend to accumulate at the most adhesive end of the gradient.

As the primary mesenchyme cells start to migrate, the epithelium at the vegetal pole begins to invaginate, buckling inward into the blastocoel to form the gut (Figure 15–5C). To initiate the process, the epithelial cells in this region change their shape: the ends facing into the blastocoel become broader than the ends facing the external surface, causing the sheet to bend and bulge up into the blastocoel (Figure 15–7). However, the invagination cannot proceed very far by such means alone. The rest of the movement of gastrulation is accomplished through the activity of certain cells in the rounded tip of the

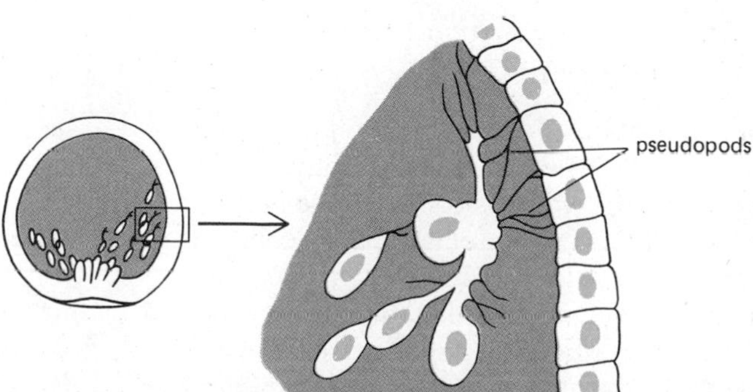

Figure 15–6 The primary mesenchyme cells crawl over the inner face of the wall of the blastula by putting out contractile pseudopods with sticky tips. (After L. Wolpert and T. Gustafson, *Endeavour* 26:85–90, 1967.)

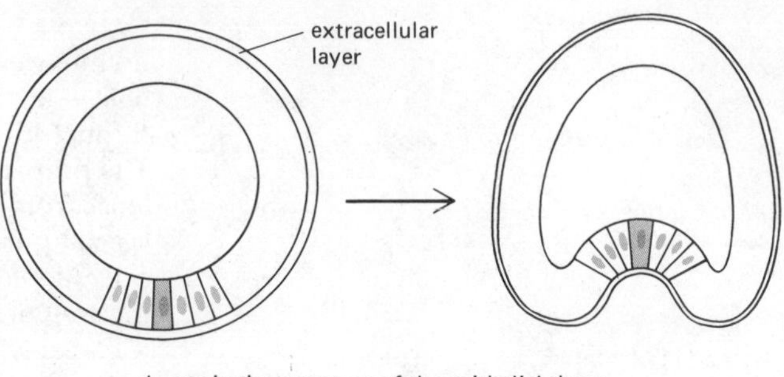

change in the curvature of the epithelial sheet
may be forced by change in the shape of the cells

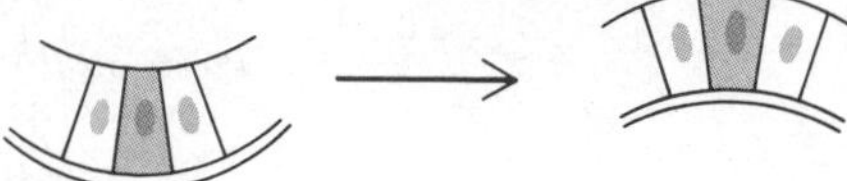

Figure 15–7 Schematic diagram to show how the initial intucking of the epithelium at the vegetal pole may be brought about. For molecular details of a process of this type, see p. 823.

infolding sheet. These, like the primary mesenchyme cells, extend long pseudopods into the blastocoel cavity, but unlike the primary mesenchyme cells they do not become detached from the sheet for the time being. The pseudopods contact the walls of the cavity, adhere there, and contract, thereby pulling the sheet deep into the cavity to form the tube of the gut (Figure 15–5D, E). The movement is complete when the blind end of the gut tube has been brought up against the epithelium near the opposite end of the embryo (Figure 15–5F). Where the two epithelia make contact, the mouth will develop—the epithelia fuse and a hole forms. The cells whose pseudopods dragged the gut into the embryo, once their task is done, break loose from the epithelium and migrate into the cavity as the so-called *secondary mesenchyme*. (This distinction between primary and secondary mesenchyme is a peculiarity of the sea urchin and not a general feature of gastrulation.)

Gastrulation transforms the hollow spherical blastula into a three-layered structure: the innermost layer, the tube of the primitive gut, is the **endoderm;** the outermost layer, the epithelium that has remained external, is the **ectoderm;** and between the two, the looser layer of tissue composed of primary and secondary mesenchyme cells is the **mesoderm.** These are the three primary **germ layers** common to higher animals. The organization of the embryo into the three layers roughly corresponds to the organization of the adult with gut on the inside, epidermis on the outside, and connective tissue in between. Very crudely, these three types of adult tissues may be said to derive, respectively, from the endoderm, the ectoderm, and the mesoderm, although there are exceptions (see p. 882).

The Ability of Cells to Extend, Adhere, and Contract Is the Universal Basis of Morphogenetic Movement[6]

The movements of gastrulation are brought about by just three types of cell activity: extension, adhesion, and contraction. The same three activities in various combinations, together with cell growth and division, underlie almost all morphogenetic movements, both of cells as individuals and of cells in groups. Thus the dragging of the endoderm into the blastocoel illustrates how a few cells in an epithelium can provide the motive power for the entire sheet by putting out adhesive protrusions. A related phenomenon can be demonstrated with some types of cell in tissue culture. If, for example, the leading edge of an epithelium advancing across a petri dish is experimentally detached from the substratum, the whole sheet retracts. This shows that only the cells at the leading edge are stuck to the substratum and that they exert a pull on the rest of the sheet. The initial invagination of the sea urchin endoderm is an example of another sort, showing how epithelia bend and deform autonomously as a consequence of changes in cell shape. Although the geometry is different, extension, adhesion, and contraction of cells again provide the basic mechanism for such deformations, as we shall describe when discussing neurulation.

Gastrulation in Amphibians[7]

Gastrulation is harder to follow in amphibians than in sea urchins for three reasons: the embryo is not transparent; the epithelium of the blastula is more than one cell thick; and the presence of very yolky cells impedes the movement of the invaginating cell sheets, making the geometry more complex. The infolding of the endoderm begins not at the vegetal pole, but to one side of it: the site is marked at first by a short indentation, called the **blastopore,** in the exterior of the blastula. This indentation gradually extends (Figure 15–8), curving round to form a complete circle surrounding a plug of very yolky cells. Sheets of cells meanwhile turn in around the lip of the blastopore and move deep into the interior of the embryo. At the same time, the external epithelium in the region of the animal pole actively spreads to take the place of the cell sheets that have turned inward. Eventually the epithelium of the animal hemisphere extends in this way to cover the whole external surface of the embryo, and, as gastrulation reaches completion, the blastopore circle shrinks almost to a point.

The process of invagination seems to depend on the same basic mechanisms as in the sea urchin. It begins with changes in the shape of the cells at the site of the blastopore. In the amphibian, these are called *bottle cells:* they have narrow necks attaching them to the exterior of the epithelium and broad bodies toward the interior (Figure 15–9). They probably act like wedges to cause the epithelium to curve and so to tuck inward, producing the indentation seen from outside. Once this first tuck has formed, cells can continue to pass into the interior as a sheet to form the gut. The sheet is probably

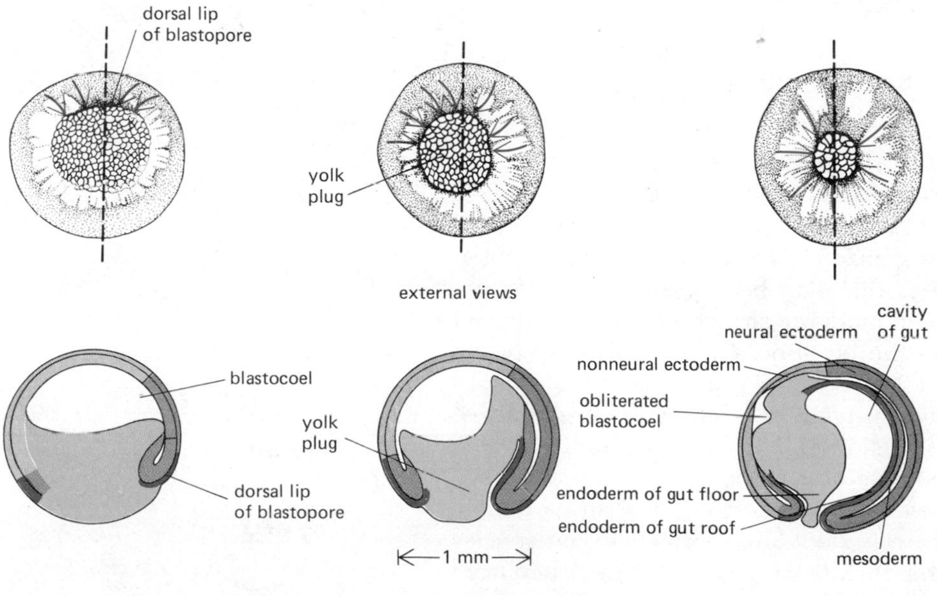

Figure 15–8 Gastrulation in *Xenopus.* The external views show the embryo from the vegetal pole; the accompanying cross-sections are cut in the plane indicated by the broken lines. The directions of cell movement are indicated by arrows. (After R. E. Keller, *J. Exp. Zool.* 216:81–101, 1981.)

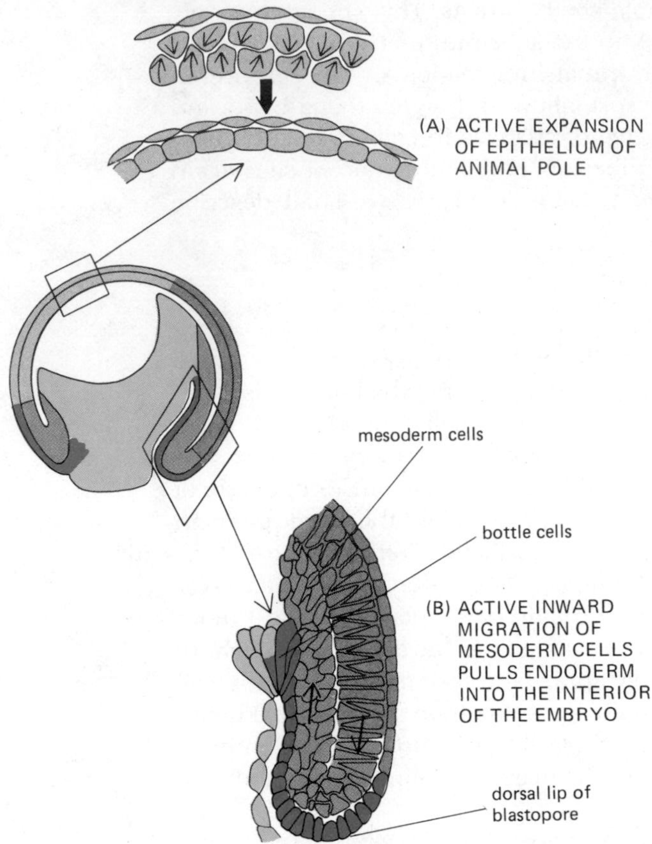

Figure 15–9 Some details of cell behavior during gastrulation in *Xenopus*. (A) The epithelium of the animal pole expands actively, becoming thinner in the process. (B) Bottle cells appear to initiate the invagination at the dorsal lip of the blastopore. Cells move over the lip and crawl deeper into the interior of the embryo. (After R. E. Keller, *J. Exp. Zool.* 216:81–101, 1981.)

drawn in, as in the sea urchin, by the pseudopods of future mesoderm cells crawling over the internal surface of the wall of the blastocoel, pulling the endoderm cells along with them. The outcome is a standard type of three-layered structure: an outermost sheet of ectoderm, an innermost tube of endoderm forming the rudiment of the gut, and between them a layer of mesoderm. Again, the mouth develops as a hole formed at an anterior site where endoderm and ectoderm come into direct contact without intervening mesoderm.

Movements Are Organized About the Blastopore[8]

All these processes, though complex, are orderly, so that it is possible to plot on the surface of the embryo before gastrulation a *fate map* showing which cells of the very early stage will be carried into which parts of the mature animal by the subsequent movements. But how is the whole complex of gastrulation movements set in train and organized? In some amphibians, as noted earlier, one side of the fertilized egg is distinguished by a band of pigment called the gray crescent. This appears to mark the site where invagination will begin, that is, the future dorsal lip of the blastopore. If the dorsal lip of the blastopore is excised from a normal embryo at the beginning of gastrulation and grafted into another embryo but in a different position, the host embryo gastrulates both at the site of its own blastopore and at the site of the grafted blastopore (Figure 15–10). The movements of gastrulation at the second site entail the formation of a second whole set of body structures, and a double embryo (Siamese twins) results. By carrying out such grafts between species with differently pigmented cells, so that host tissue can be distinguished from implanted tissue, it has been shown that the grafted blastopore lip actually

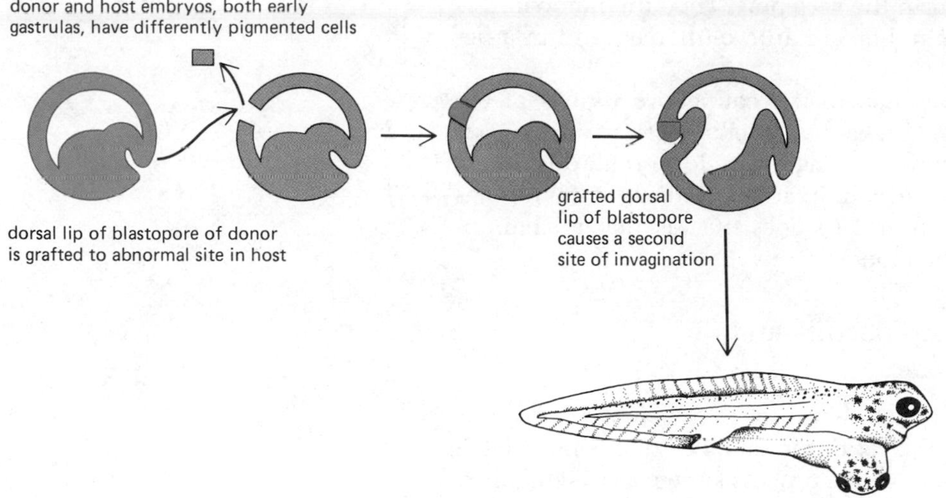

Figure 15–10 Diagram of an experiment showing that the dorsal lip of the blastopore initiates and controls the movements of gastrulation and thereby, if transplanted, organizes the formation of a second set of body structures. The operation is illustrated here for a newt: the results are similar for *Xenopus.* (After L. Saxén and S. Toivonen, Primary Embryonic Induction. London: Logos Press, 1962.)

recruits host epithelium into its own system of invaginating endoderm and mesoderm. It is not known to what extent the recruitment depends on the mechanical attachment of host epithelium to graft blastopore and to what extent on chemical influences of graft on host. Because of its central role in triggering gastrulation and thereby the whole consequent formation of the body, the dorsal lip of the blastopore has been named the *organizer.*

The Endoderm Will Form the Gut and Associated Organs Such as the Lungs and Liver

It is necessary now to outline briefly the later development of the layers of endoderm, mesoderm, and ectoderm that constitute the embryo just after gastrulation. The endoderm forms a tube, the primordium of the digestive tract, from the mouth to the anus. It gives rise not only to the pharynx, esophagus, stomach, and intestines, but also to many associated glands. For example, the salivary glands, the liver, the pancreas, the trachea, and the lungs all develop from extensions of the wall of the originally simple digestive tract and grow to become systems of branching tubes that open into the gut or pharynx. To be more precise, the endoderm forms only the inner, epithelial components of all these structures—the lining of the gut, for example, but not the supporting muscular and fibrous elements of its wall. These latter tissues are among the many that arise from the mesoderm.

The Mesoderm Will Form Connective Tissues, Muscles, and the Vascular and Urogenital Systems[9]

The mesodermal layer is divided from a very early stage into separate parts on the left and right of the body. Defining the central axis of the body, and effecting this separation, is a very early specialization of the mesoderm known as the **notochord.** This is a slender rod of cells, about 80 μm in diameter, with ectoderm above it, endoderm below it, and mesoderm on either side (see Figure 15–13). The cells of the notochord become swollen with vacuoles, so that the rod elongates and stretches out the embryo. In the most primitive chordates, which have no vertebrae, the notochord is retained as a primitive substitute for a vertebral column. In the vertebrates, it serves as a core around which mesodermal cells gather to form the vertebrae. Thus the notochord is the precursor of the vertebral column, both in an evolutionary and in a developmental sense.

In general, the mesoderm gives rise to the connective tissues of the body—at first to the loose, space-filling, three-dimensional mesh of cells known as *mesenchyme,* and ultimately to bone, cartilage, muscle, and fibrous tissue, including the inner, dermal layer of the skin. In addition, most of the tubules of the urogenital system form from it and so does the vascular system, including the heart and the cells of the blood.

The Ectoderm Will Form the Epidermis and the Nervous System

At the end of gastrulation the sheet of ectoderm covers the embryo, and thus eventually forms the outer, epidermal layer of the skin. But that is not all: the entire nervous system also derives from it. In a process known as **neurulation,** a broad central region of the ectoderm thickens, rolls up into a tube, and pinches off from the rest of the sheet. The region of ectoderm that undergoes this transformation seems to be defined by an interaction with the mesoderm that has come to lie beneath it in the course of gastrulation—that is, with the notochord and the mesoderm adjacent to the notochord (see p. 870). The tube thus formed from the ectoderm is called the **neural tube;** it is the rudiment of the brain and the spinal cord. Along the line where the neural tube pinches off from the future epidermis, a number of ectodermal cells break loose from the epithelium and migrate as individuals out through the mesoderm. These are the cells of the **neural crest;** they will form almost all the outlying components of the nervous system (including the sensory and sympathetic ganglia and the Schwann cells that make the myelin sheaths of peripheral nerves) as well as the cells of the adrenal gland that secrete epinephrine and the pigment cells of the skin. In the head, many of the neural crest cells will differentiate into cartilage, bone, and other connective tissues, which elsewhere in the body have a mesodermal origin. This is one of several instances that run counter to the general conception of the three germ layers in terms of three concentric layers of the adult body.

The sense organs, in which lights, sounds, smells, and so forth from the external world impinge on the nervous system, also have ectodermal origins: some derive from the neural tube, some from the neural crest, and some from the exterior layer of ectoderm. The retina, for example, originates as an outgrowth of the brain and so is derived from cells of the neural tube, while the olfactory cells of the nose differentiate directly from the ectodermal epithelium lining the nasal cavity.

The Neural Tube Is Formed Through Coordinated Changes in Cell Shape[10]

The formation of the neural tube (Figure 15–11) is a dramatic event to watch. At first the surface of the gastrula appears more or less uniform. But subtle changes are occurring: the ectoderm close to the midline begins to thicken, forming the *neural plate.* Then the lateral edges of the neural plate start to rear up in folds; these *neural folds* gradually roll together, while the midline of the plate sinks deeper; and eventually the folds meet and fuse to form the hollow neural tube, roofed over by a continuous sheet of ectoderm. As in gastrulation, the whole process depends on extension, adhesion, and contraction on the part of individual cells in an epithelial sheet.

The ectoderm is two cell layers deep in *Xenopus,* only one layer deep in many other species. In either case the neural plate becomes thickened through

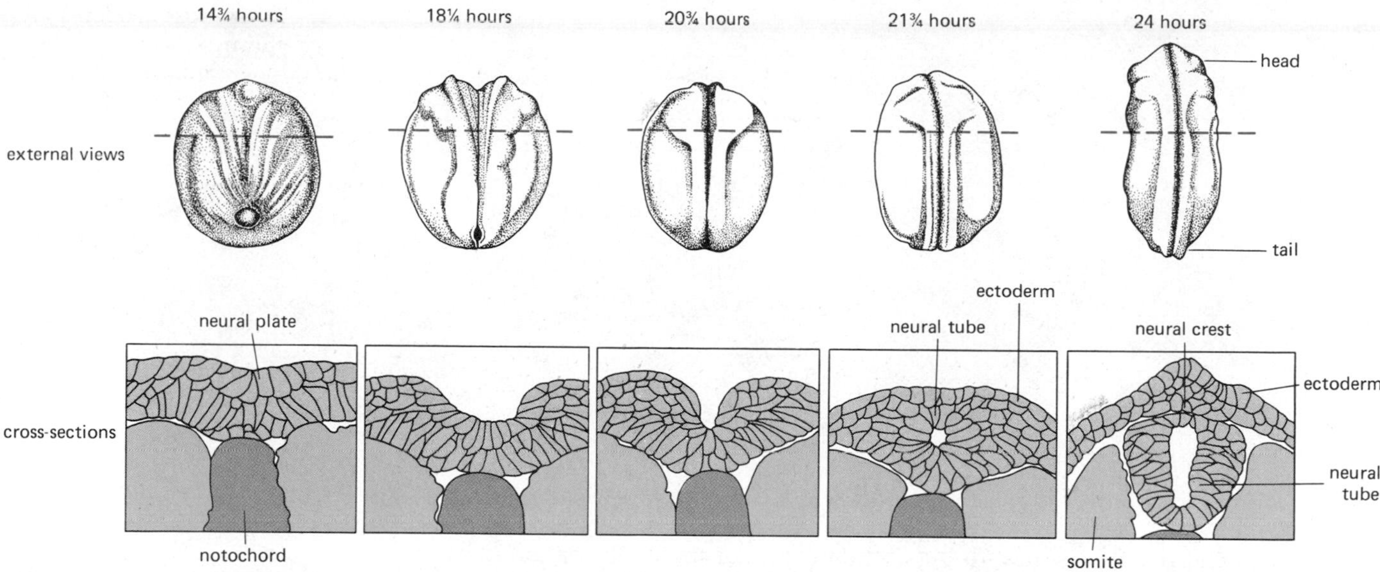

Figure 15–11 Neural tube formation in *Xenopus.* The external views are from the dorsal aspect. The cross-sections are cut in a plane indicated by the broken lines. (After T. E. Schroeder, *J. Embryol. Exp. Morphol.* 23:427–462, 1970.)

changes of cell shape, without change in the number of layers. The cells of the neural plate are bound together by strong lateral attachments. First the cells elongate in a direction perpendicular to the surface of the sheet. This elongation involves the lining up of microtubules and is essential for the continuation of neurulation, since drugs such as colchicine that disassemble microtubules prevent or reverse the elevation of the neural folds. The elongated cells then become wedge-shaped, with the narrower part of the wedges toward the upper, apical surface of the sheet. As the cells are held tightly together at their lateral faces and their width does not change at the base, this results in the rolling up of the cell sheet (Figure 15–12). The narrowing of the cells at their apices is accomplished by the contraction of the actin-filament bundles that run beneath the upper surfaces of the cells (p. 683). As the ends of the cells become narrower, their upper surface membrane becomes puckered (Figure 15–12). This may be a symptom of rapid change in shape due to contraction, occurring in circumstances where strong lateral attachments prevent flow of excess membrane past the junctional complexes and down the sides of the cells.

Blocks of Mesoderm Cells Uncouple to Form Somites on Either Side of the Body Axis[11]

On either side of the newly formed neural tube lies a broad expanse of mesoderm (Figure 15–13). The thicker, more medial part of this mesoderm gives rise to the vertebrae, the ribs, the skeletal musculature, and the dermis. It consists at first of a single continuous slab of tissue on each side of the body. To form the repetitive, segmented structures of the vertebrate body axis, this slab soon breaks up into separate "blocks," or **somites** (Figure 15–14). Each somite corresponds to one unit in the final sequence of articulated elements.

The somites do not form simultaneously but one after another, starting near the head and ending at the tail. Segmentation is accompanied by changes in the connections between the mesoderm cells. The cells in the early unsegmented mesoderm of the amphibian are electrically coupled via gap junctions, but this coupling between the cells disappears at or just before the time when the somites develop. Apparently the cells first alter their mutual attachments and then rotate in groups to form somites.

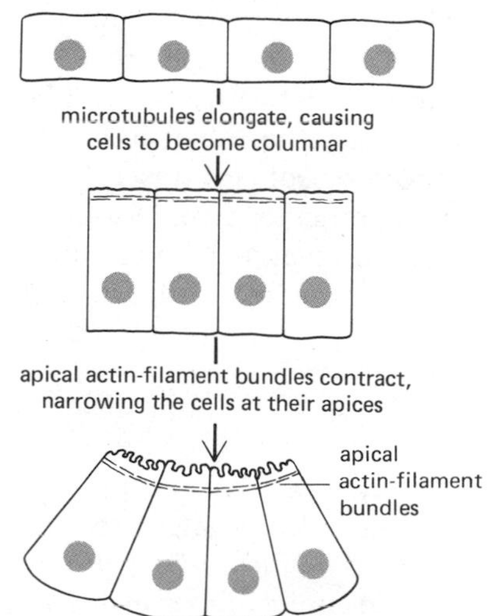

Figure 15–12 The figure shows schematically how microtubules and actin filaments bend an epithelium by changing the shapes of cells.

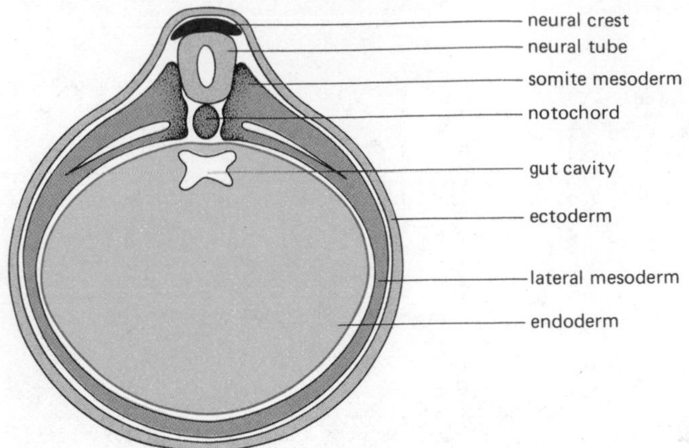

Figure 15–13 A cross-section (schematic) through the trunk of a frog after the neural tube has closed.

Despite appearances, the timing of the process of somite formation does not depend on a stimulus propagated from head to tail along the mesoderm at the time of segmentation. If an embryo is cut in two, so that the unsegmented posterior end is separated from the segmenting anterior end, the posterior mesoderm will nevertheless proceed to split up into somites, following the same schedule that would have been observed if the two parts of the embryo had remained in contact. Each region of the tissue behaves as though it were controlled by an internal clock that determines when the cells in that region shall become mature for somite formation.

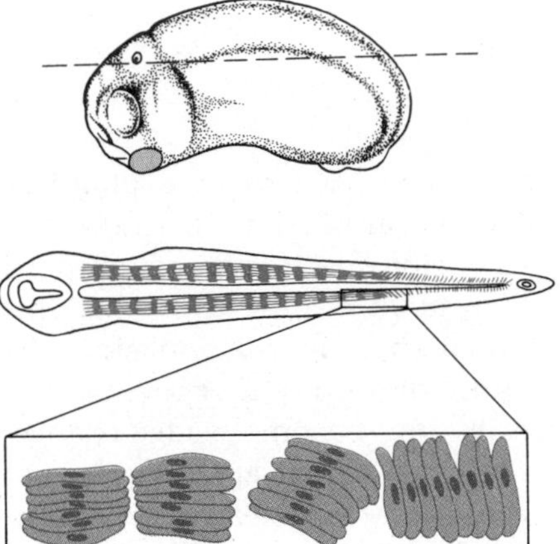

Figure 15–14 Somite formation in *Xenopus*. A side view of the embryo is shown at the top; the broken line indicates the plane of the horizontal section shown below. The bottom drawing is a schematic high-magnification view of the mesoderm cells in the process of regrouping to form somites.

The Vertebrate Body Plan Is First Formed in Miniature and Then Maintained as the Embryo Grows

The embryo at the stage when the somites are forming is typically a few millimeters long and consists of about 10^5 cells. While we have been speaking thus far of *Xenopus*, the scale and general form are much the same for a salamander, a fish, a chick, or a human (Figure 15–15). Later these species of embryo will grow to be very different in size and shape, but for the moment they can all be seen to share the basic vertebrate body plan. The details will be filled in later as the embryo grows. The central nervous system is represented by the neural tube, with an enlargement at one end for the brain; the gut and its derivatives by a tube of endoderm; the segments of the trunk by the somites; the other connective tissues, including the vascular system, by the more peripheral unsegmented mesoderm; and the epidermal layer of the skin by the ectoderm. During subsequent development, all of these components will enlarge, by a factor of as much as a hundred or more in length or a million or more in volume and cell number. But the same basic organization of the body will be preserved.

Summary

In the process of gastrulation, part of the epithelium of the blastula becomes invaginated. The invagination movements appear to be initiated by changes in the shapes of the cells of the blastula, forcing the epithelium to bend inward. The main motive force in the later phases is probably traction applied to the epithelial sheet by the future mesoderm cells. Gastrulation transforms the embryo into a three-layered structure with an internal epithelial tube of endoderm, an external epithelial covering of ectoderm, and a middle layer of mesodermal cells that have broken loose from the original epithelial sheet. The endoderm

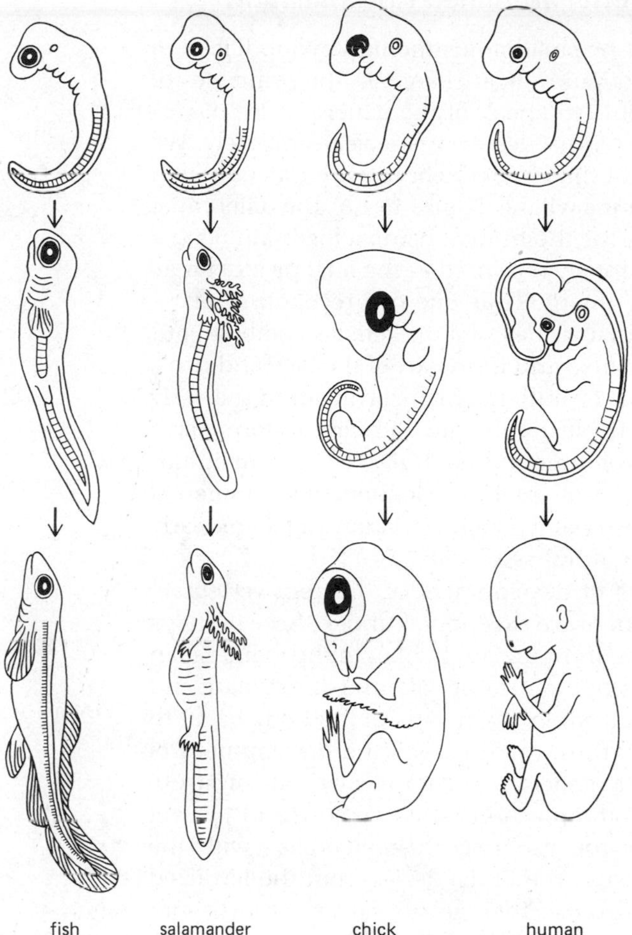

Figure 15–15 Comparison of the embryonic development of a fish, an amphibian, a bird, and a mammal. The early stages (*above*) are closely similar; the later stages (*below*) are more divergent. The earliest stages are drawn roughly to scale; the later stages are not. (After E. Haeckel, The Evolution of Man. London, 1879.)

will form the lining of the gut and its derivatives; the ectoderm will form chiefly the epidermis and the nervous system; and the mesoderm will form most of the muscles, connective tissues, vascular system, and urogenital tract. Since the movements of gastrulation are orderly, it is possible to plot a fate map predicting which parts of the pregastrulation embryo will give rise to the various parts of the body.

Gastrulation brings together originally separate populations of cells, which then interact. For example, the dorsal mesoderm induces the overlying ectoderm to thicken, roll up, and pinch off to form the neural tube and neural crest. The process depends again on changes in the shape of the epithelial cells, involving microtubules and actin filaments. In the middle of the dorsal mesoderm, a rod of specialized cells, called the notochord, constitutes the central axis of the embryo. The long slabs of mesoderm on either side of the notochord become segmented into somites. In somite formation, the cells change their contacts and regroup autonomously according to a predetermined program, starting at the anterior end of the embryo and finishing at the tail.

Early Steps in Pattern Formation: The Mouse[12]

We now turn to the problem of pattern formation—the question of how the behavior of cells is controlled according to their position in the body and how the parts of the embryo interact and differentiate. As an introduction to this topic, we shall briefly consider the early development of the mouse.

Mammalian Development Involves an Added Complication

Evolution proceeds for the most part by small adjustments—by modification of the proportions of the body rather than radical change in the principles of its construction. This makes it possible to speak of the general rules of vertebrate development without having to discuss each species separately. We have seen that the embryonic forms of different vertebrates are indeed much more closely similar than the adult forms will be (Figure 15–15): the differential growth that gives the bird a long beak or the human being a big brain occurs relatively late. Resemblances that are totally obscured in the final product may be plain in the early stages of development. Thus one can recognize the rudiments of a fish's gills in the branchial arches of a mammalian embryo; but in the mammal those rudiments later fuse and form part of the face and neck, instead of the respiratory system. This evolutionary conservatism in the early embryo is not hard to understand. The structures that are first to form serve as the framework on which later development must build; even a small mutation of the initial structures may upset many subsequent developmental processes that depend on them. Presumably, such mutations affecting early development are seldom favored by natural selection.

As a rule, then, the early stages of development of different vertebrate species are surprisingly similar. But there are exceptions, and of these perhaps the most striking is seen when we turn from the amphibian to the mammalian embryo. The mammal *begins* its development rather like the amphibian, and the later stages, following gastrulation, when the organs of the body begin to form, are again similar. But between these two periods the development of the mammal takes a large detour to generate a complicated set of structures—notably the amniotic sac and the placenta—that enclose and protect the embryo proper and provide for the exchange of metabolites with the mother. These structures, like the rest of the body, derive from the fertilized egg but are called *extraembryonic* because they are discarded at birth and form no part of the adult. The presence of the delicate but vital extraembryonic structures and the inaccessibility of the embryo within the mother's uterus mean that surgical manipulation *in situ* is rarely feasible. Experimentation on mammalian embryos has only recently become possible with the development of techniques for maintaining them in culture outside the body of the mother.

The Steps Before Gastrulation

The development of a placenta allows the mammalian embryo to get its nutrition from the mother, so that it is unnecessary for the mammalian egg to contain large stores of raw materials such as yolk. Thus the egg of the mouse has a diameter of only about 80 μm and, therefore, a volume about 2000 times smaller than a typical amphibian egg. It is surrounded initially by a transparent cell coat, the *zona pellucida*. The fertilized egg cleaves within this coat to form a mulberry-shaped cluster of cells, the **morula** (Figure 15–16). Sometime between the 8-cell and 16-cell stages, the surface of the morula becomes smoother and more nearly spherical as the cells change their cohesiveness and become compacted together (Figure 15–17), with tight junctions forming between the outer cells and sealing off the interior of the morula from the external medium. Soon after, the internal intercellular spaces enlarge to create a central fluid-filled cavity—the blastocoel. At this stage, the morula is said to have become a **blastocyst.** The cells of the blastocyst form a spherical shell enclosing the blastocoel, with one pole distinguished by a thicker accumulation of cells. As shown in Figure 15–16, the outer cell layer is the **trophectoderm;** the accumulation of cells inside the trophectoderm at one pole is the **inner cell mass.**

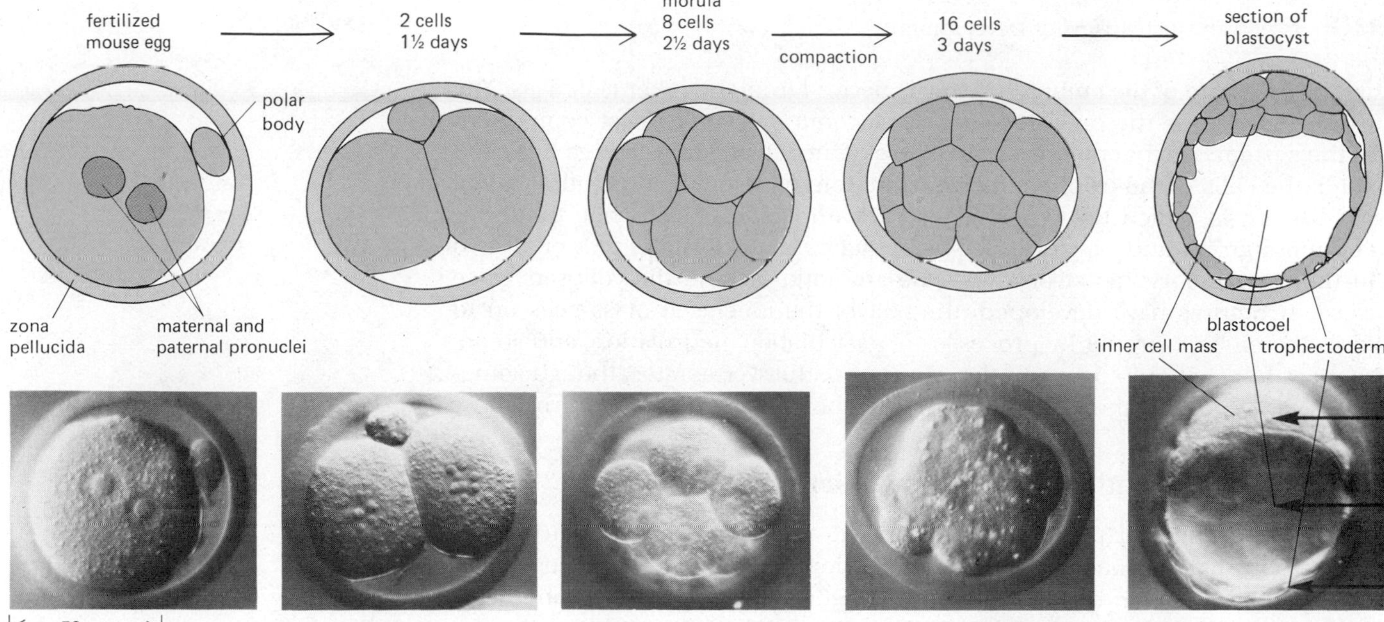

Figure 15–16 The early stages of mouse development. (Photographs courtesy of Patricia Calarco, from G. Martin, *Science* 209:768–776, 1980. Copyright 1980 by the American Association for the Advancement of Science.)

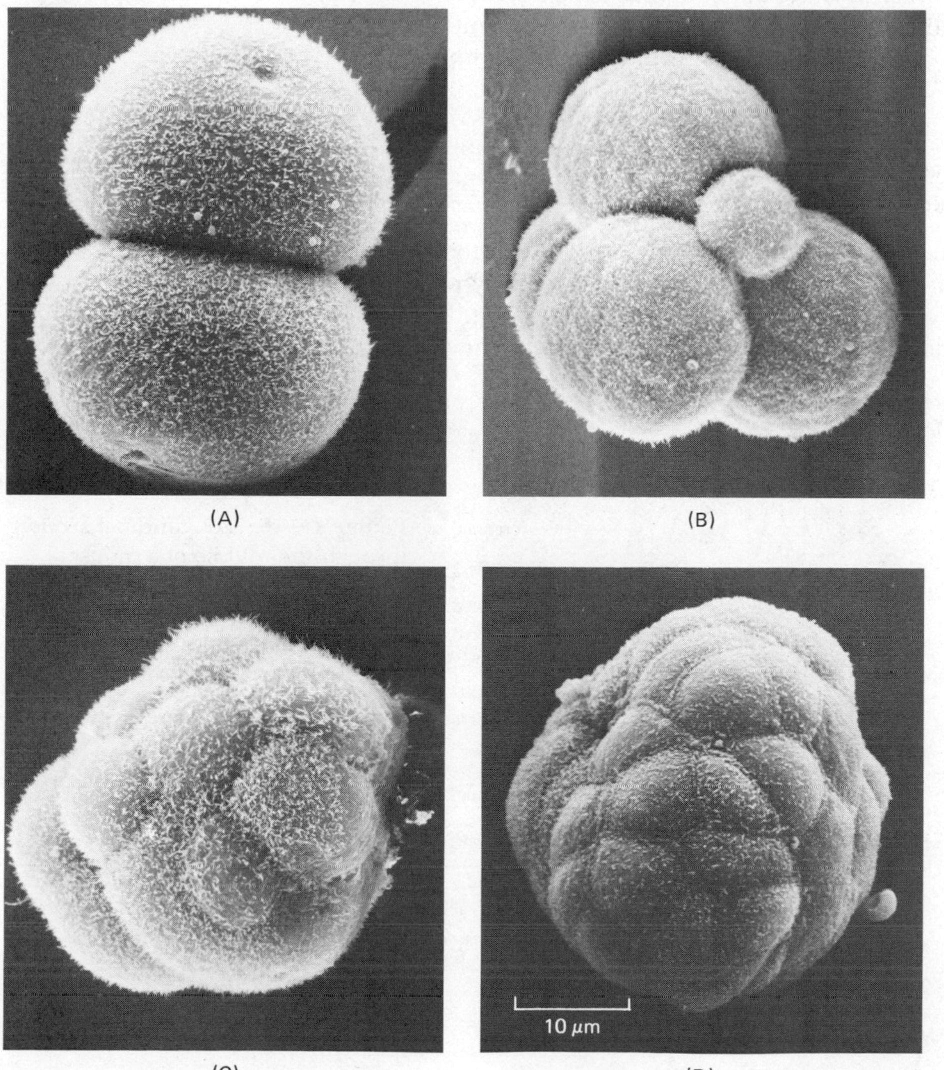

Figure 15–17 Scanning electron micrographs of the early mouse embryo. The zona pellucida has been removed. (A) Two-cell stage. (B) Four-cell stage (a polar body is visible in addition to the four blastomeres). (C) Eight-to-sixteen-cell morula—compaction occurring. (D) Blastocyst. (Courtesy of Patricia Calarco; D, from P. Calarco and C. J. Epstein, *Dev. Biol.* 32:208–213, 1973.)

The whole of the embryo proper is derived from the inner cell mass. The trophectoderm is the precursor of the placenta and the earliest component of the system of extraembryonic structures. Once the zona pellucida has been shed, the cells of the trophectoderm come into close contact with the wall of the uterus, in which the embryo becomes implanted. Meanwhile the inner cell mass grows and part of it begins to differentiate, forming first of all two further extraembryonic structures—the amniotic sac and the yolk sac. Once these structures have developed, the rest of the inner cell mass goes on to form the embryo proper by processes of gastrulation, neurulation, and so on, that are largely homologous to those seen in other vertebrates, though sometimes distortions of the geometry make the homology hard to discern.

Organogenesis and Growth Before Birth[13]

In a mouse, the first phase of development, up to the end of gastrulation, takes about 7 days from the time of fertilization. The period of organogenesis, in which all the main organs are formed (Figure 15–18), takes from about the seventh to about the fourteenth day. The growth phase continues thereafter into adult life. Birth occurs 19 or 20 days after fertilization. In humans, the first and second phases of development take three or four times as long as in the mouse, and the period of growth *in utero* is very much extended.

Figure 15–19 shows the growth in mass of the mouse embryo from conception to birth, while Figure 15–20 illustrates the development of its outward form. Having thus briefly outlined the process of mammalian development, we now return to consider some important experimental results obtained with mouse embryos.

Studies of Chimeras Show That All of the Cells of the Very Early Mammalian Embryo Are Functionally Equivalent[14]

Up to the eight-cell stage, all of the cells of the mouse embryo look alike. After this stage, chemical differences—for example, in respect of the proteins synthesised—can be demonstrated between the cells of the trophectoderm and those of the inner cell mass. How do these differences originate? Might they

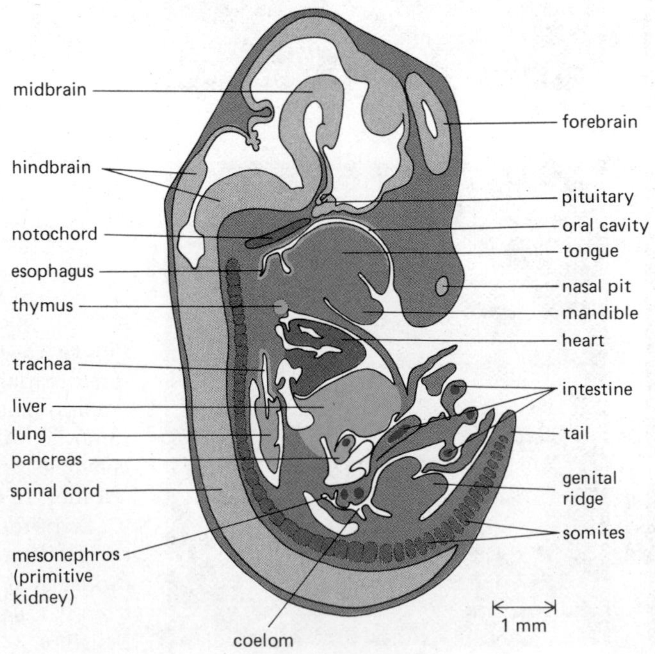

Figure 15–18 Longitudinal section through the middle of a mouse embryo at 13½ days of development. (After R. Rugh, Vertebrate Embryology: The Dynamics of Development. New York: Harcourt Brace World, 1964.)

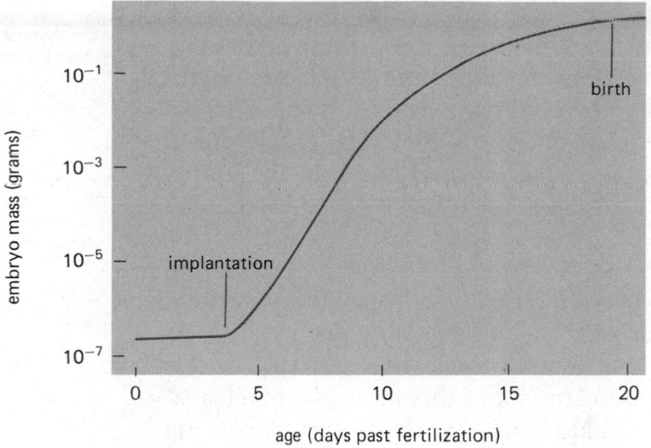

Figure 15–19 Graph of the growth of the mass of the mouse embryo as a function of time elapsed since fertilization. Growth cannot begin until the embryo has implanted in the wall of the uterus. (After R. Rugh, The Mouse: Its Reproduction and Development. Minneapolis: Burgess, 1968.)

be due to chemical differences between the portions of cytoplasm or membrane that the cells inherit from the uncleaved egg? In some lower species there is indeed good evidence for such localized determinants in the egg (see p. 850), but in mammals they seem to play no part. If they were important, subsequent development should be grossly disturbed if the set of cells in the early embryo were rearranged, added to, or diminished artificially. But in fact experiments show that the early mammalian embryo is remarkably adaptable and that each of its cells can form any part of the later embryo or adult.

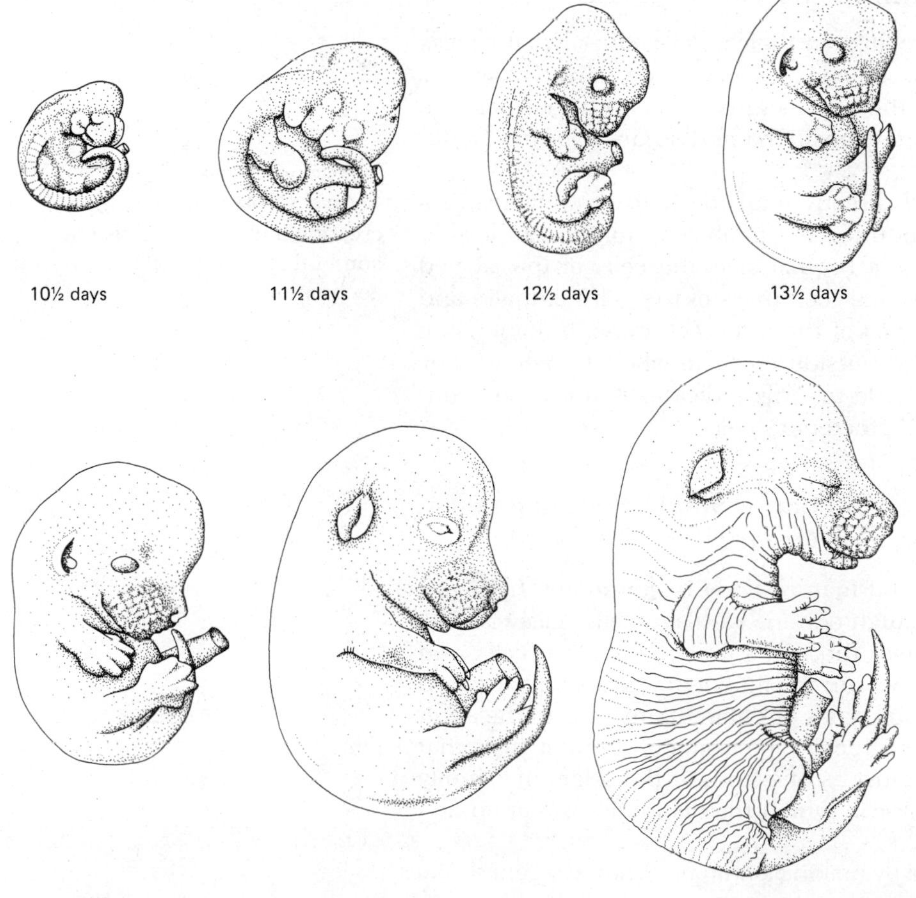

Figure 15–20 External views of the mouse embryo at 10½ through 16½ days of development. (From R. Rugh, Vertebrate Embryology: The Dynamics of Development. New York: Harcourt Brace World, 1964. © 1964 by Harcourt Brace World, Inc. Reproduced by permission of the publisher.)

One example of the versatility of the cells of the early mammalian embryo is seen in the formation of identical twins. This phenomenon shows that two normal individuals can develop from a single fertilized egg—each one originating from only a part of the normal embryo. Experimentally, one can take a mouse embryo at the two-cell stage, destroy one of the cells by pricking it with a needle, and put the resulting "half-embryo" to develop in the uterus of a foster mother. In a fair proportion of cases a perfectly normal mouse will emerge.

A converse experiment is also possible: two eight-cell embryos can be combined to form a single giant morula, which will develop into a mouse of normal size (Figure 15–21). This remarkable animal has four parents, and its parentage can be proved with the help of genetic markers. For example, if one of its pairs of parents is of a white-coated strain and the other pair of a black-coated strain, it will typically have a piebald black and white coat, consisting of patches of cells of the two different constituent genotypes (Figure 15–21). Such creatures, formed from aggregates of genetically different groups of cells, are called **chimeras.** Chimeras can also be made by injecting cells from an early embryo of one genotype into a blastocyst of another genotype. The injected cells become incorporated in the inner cell mass of the host blastocyst, and a chimeric animal develops. It is even possible to make a chimera by injecting a single cell in this way; thus one can assay the developmental capabilities of the single cell. One of the major conclusions derived from these studies is that the cells of the very early mammalian embryo (up to the eight-cell stage) are initially similar and unrestricted in their capabilities: they are all *totipotent.*

8-cell-stage mouse embryo whose parents are white mice

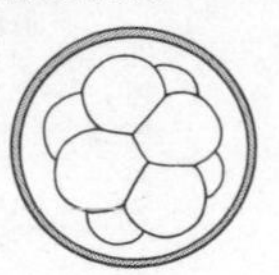

8-cell-stage mouse embryo whose parents are black mice

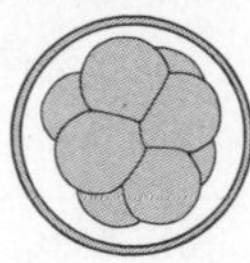

zona pellucida of each egg is removed by treatment with protease

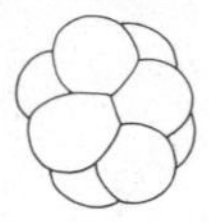

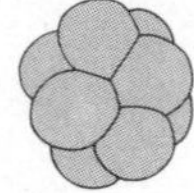

embryos are pushed together and fuse when incubated at 37°C

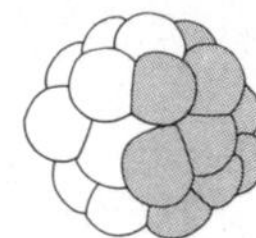

development of fused embryos continues *in vitro* to blastocyst stage

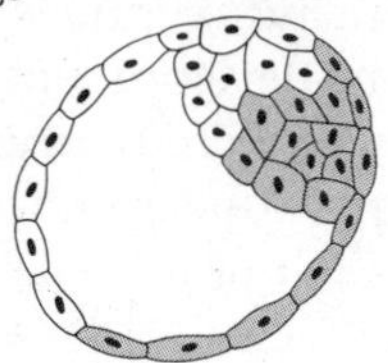

blastocyst transferred to pseudopregnant mouse, which acts as a foster mother

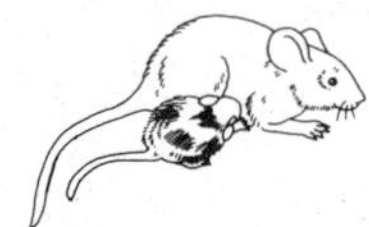

the baby mouse has 4 parents (but its foster mother is not one of them)

Figure 15–21 A procedure for creating a chimeric mouse by combining two morulae of different genotypes.

Position in the Morula Determines the Fate of a Cell[15]

If the cells at the eight-cell stage are alike in their potentialities, what causes them thereafter to go separate ways, some taking on the specialized character of the inner cell mass, others that of the trophectoderm? One obvious possible explanation is that they lie in different positions and so are exposed to different environments.

This hypothesis can be tested directly with the help of chimeras. If a labeled embryo is completely surrounded by unlabeled embryos, to form a giant chimeric morula of 15 times the normal size, the cells of the labeled embryo tend to become part of the inner cell mass of the giant chimera and to acquire the characteristic properties of the inner cell mass. In contrast, if labeled cells are packed around the outside of an unlabeled embryo, they tend to become part of the trophectoderm. This indicates that position controls the assignment of distinctive characters to cells.

A Group of Founder Cells, Rather Than a Single Founder Cell, Gives Rise to a Particular Tissue or Organ[16]

The differentiated state of cells in the mature body is commonly heritable: when a cartilage cell divides, its daughters are cartilage cells. Likewise the daughters of a bone cell are bone cells, those of a liver cell are liver cells, and so on (see Chapter 16). What then if we run the film of development backward and retrace the ancestry of the various types of differentiated cell in the adult body? Shall we find at the origin of each cell type a single "founder" cell, that is, a single distinctive ancestral cell, just as one finds at the origin of a family of mutant organisms a single ancestor in which the distinctive mutation first occurred?

This question can be answered by making a chimera from two genetically distinct embryos and examining the composition of the various differentiated tissues that result (Figure 15–22). If all the cells that are differentiated in a particular way were derived from a single founder cell, they should all be genetically alike; if they are genetically different, they cannot all be derived from a single ancestral cell. In fact, the tissues and organs of chimeric individuals are rarely pure in genotype: that is, they rarely derive from just one of the constituent embryos. Usually a given tissue or organ in the chimeric animal has a mixed composition. The proportions of this mixture vary from one chimera to the next because the cells of the two genotypes mingle and become randomly arranged in the early embryo.

If there are many founder cells for a given part of the body, and they are a random mix of the two genotypes in the chimera, that part of the body will almost always contain some cells of each genotype. If there are only a very few such founder cells, they may often chance to be all of the same genotype so that that part of the body will not be chimeric. A statistical analysis of the data indicates that the numbers of founder cells for practically all the organs and tissues of the body must in fact be considerably greater than two.

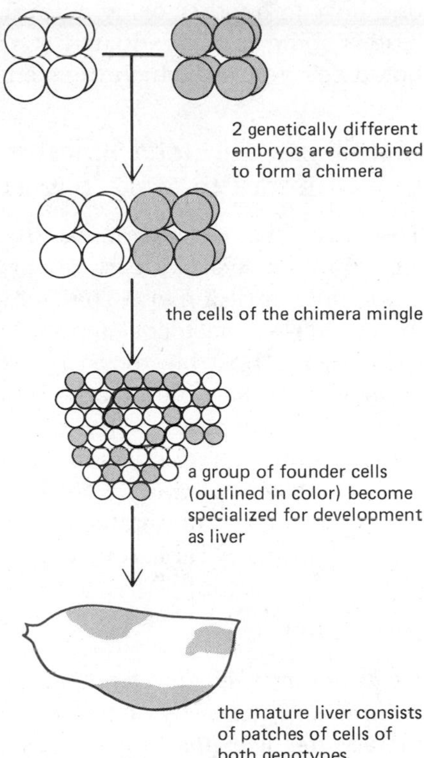

Figure 15–22 Scheme illustrating the type of experiment that will show that it is a group of cells, rather than a single cell, that becomes committed to give rise to a particular organ or tissue.

Teratomas Can Arise from Embryos That Develop in the Wrong Environment[17]

Germ cells and early embryos are delicate systems poised for grand developments. On the one hand, they are vulnerable and easily destroyed; on the other hand, they can go dangerously out of control. In the normal course of events, cleavage and embryogenesis do not begin until the ovum is fertilized by a sperm, and the embryo then develops in the special environment of the uterus. But the normal process can be upset. An unfertilized egg may, for example, be artificially triggered into cleavage without the help of sperm (see Chapter 14). Such a *parthenogenetic* embryo begins development and *in utero* may go as far as the stage of limb-bud formation. For mammals, however, that seems to be the limit—degeneration follows. Despite much interest and experimental effort, there are no convincingly authenticated cases of virgin birth in mammals. The reason for this failure remains unclear: in other vertebrate orders (among lizards, for example), parthenogenesis occurs naturally and yields viable adults.

In mammals, unfertilized eggs also occasionally become activated spontaneously after ovulation. This occurs quite commonly in some animals, such as mice of the LT strain, but the resulting embryo degenerates *in utero*. Occasionally, however, an oocyte is activated to begin development *before* release from the ovary. Such an oocyte gives rise to an almost normal blastocyst within the ovary, but in this abnormal or *ectopic* site, instead of then degenerating, the cells of the parthenogenetic embryo begin to proliferate in a disorganized and uncontrolled way. The result is the bizarre type of growth known as a *teratoma:* a disorganized mass of cells representing many varieties of differentiated tissue—tooth, bone, glandular epithelium, and so on—mixed with undifferentiated stem cells that continue to divide to generate yet more of these differentiated tissues.

Teratomas arise spontaneously in males, too, from the germ cells in the testis; this is common in mice of another strain (called 129). Teratomas can also be provoked artificially by grafting the developing male gonads, which contain primitive germ cells, from a young embryo into an adult. Alternatively, they can be obtained by grafting a perfectly normal early embryo into the kidney or ovary of an adult.

The teratomas originating in all these various ways look similar, and from all of them it is possible to derive transplantable cancers known as *teratocarcinomas.* A teratocarcinoma will grow without limit until it kills its host. It can be maintained indefinitely by grafting samples of the tumor cells serially from one host to another or by culturing the cells *in vitro*. The tumor always includes some undifferentiated stem cells, together with a variety of differentiated cell types to which the stem cells give rise.

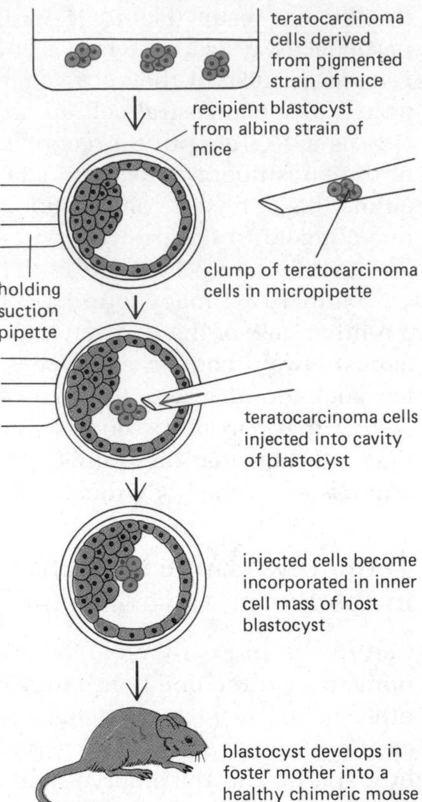

Figure 15–23 Experiment showing that teratocarcinoma cells can combine with the cells of a normal blastocyst to form a healthy chimeric mouse.

Cells from Teratocarcinomas Can Cooperate with Normal Cells in a Chimera to Make a Normal Mouse[17,18]

The stem cells of the teratocarcinoma resemble the cells of early embryos. In fact, they behave like early embryonic cells if they are put into a normal embryonic environment! The technique is to take an ordinary blastocyst and inject into its blastocoel teratocarcinoma cells that are distinguished from those of the host blastocyst by various genetic markers (Figure 15–23). The injected cells become incorporated in the inner cell mass of the blastocyst, and a chimeric mouse often develops. In this mouse, differentiated cells derived from the teratocarcinoma can be found contributing to almost any part of the body, cooperating with the neighboring cells of normal origin to give a healthy animal. The teratocarcinoma stem cells are thus shown to be totipotent, and their cancerous character is shown to be reversible.

Summary

Early mammalian development involves the formation of a set of extraembryonic structures that enclose the embryo proper and allow for exchange of metabolites with the mother. After the fertilized egg has cleaved to form an eight-cell morula, the embryo undergoes compaction and develops a central fluid-filled cavity to become a blastocyst. The outer cells of the blastocyst constitute the trophectoderm, from which the placenta and other extraembryonic structures will develop; the inner cell mass gives rise to the whole of the embryo proper, by a process involving gastrulation, neurulation, and so on, as in the amphibian. The cells of the early mouse embryo are all equivalent in their developmental potential up to the eight-cell stage, and the cells of two early embryos can be combined to form a normally proportioned but chimeric mouse. In a chimera, the cells undergo some random mingling; and in the resulting adult mouse, each of the various specialized tissues and organs is generally itself chimeric. This shows that each type of tissue or organ is derived from a group of founder cells rather than from a single founder cell.

Mammalian eggs activated to begin development in abnormal circumstances may give rise to tumors called teratomas, from which cancerous teratocarcinoma cell lines can be derived. When inserted into a normal blastocyst, a teratocarcinoma cell reverts to normal cell behavior, and its progeny contribute to the formation of a healthy chimeric animal. This is a striking demonstration of the general principle that the development of early embryonic cells is guided and controlled by their environment.

Determination and Differentiation[19]

A fertilized egg may develop into a male or a female, a human being or a mouse. The outcome is governed by the genome. But how? Let us first consider what the final structure is, in terms that we might hope to relate to molecular processes. The body is constructed from a rather limited number of crudely distinguishable types of cell—about 200 in a vertebrate, according to the traditional histological classification (see Chapter 16). These are arranged in a complicated but regular pattern in three dimensions. The cell types are distinct essentially because, in addition to the many "household" proteins that they all require, each makes a different set of specialized or "luxury" proteins: keratin in epidermal cells, hemoglobin in red blood cells, digestive enzymes in gut cells, crystallins in lens cells, and so on. Since the cell types differ in that they contain different sets of gene products, one may ask whether this is simply because they contain different sets of genes. The lens cells, for example, might have lost the genes for hemoglobin, keratin, and so on, while retaining those for crystallins; or they might have selectively amplified the number of copies of the crystallin genes. There are several lines of evidence to show that such is not the case, and that almost every type of cell contains the same complete genome that is to be found in the fertilized egg. Some of the relevant data have been reviewed in Chapter 8. The cells of the body appear to differ not because they contain different genes but because they *express* different genes. Gene activity is subject to control: genes can be switched on and off (see pp. 435–455).

A most powerful piece of evidence that the genome itself remains constant despite the observable changes of differentiation comes from experiments on nuclear transplantation, using amphibian eggs (Figure 15–24). These egg cells are so large that, using a fine glass pipette, one can inject into them nuclei taken from other cells. The nucleus of the egg itself is destroyed beforehand by ultraviolet irradiation. The egg is activated to begin development by the prick of the fine pipette used to inject the transplanted nucleus. Thus one can test whether the nucleus from a differentiated somatic cell contains a complete genome, equivalent to that of a normal fertilized egg and equally serviceable for development. The answer is yes: a normal, fertile adult frog can

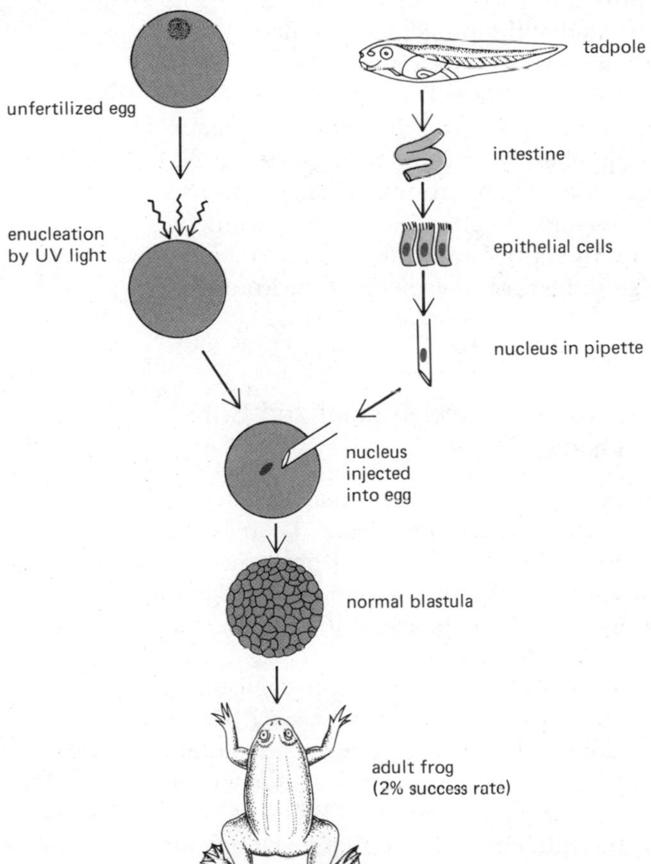

Figure 15–24 Diagram of an experiment showing that the nucleus of a differentiated cell from the gut of a tadpole contains all the genetic material necessary to control the formation of an entire frog. (After J. B. Gurdon, Gene Expression during Cell Differentiation. Oxford, Eng.: Oxford University Press, 1973.)

be produced, for example, from an egg whose own nucleus has been replaced by the nucleus of a differentiated cell from the lining of the gut of a tadpole. There are technical difficulties in these experiments, and one might wish that they had proved successful with nuclei from a wider range of differentiated cell types, and in a wider range of species. But, for all that, the general rule seems to be that, during development, the genome is largely inviolable.

A few exceptions to this rule are known. For example, in certain invertebrates, some of the chromosomes present in cells of the germ line are eliminated early in embryogenesis from the somatic (non-germ-line) cells. In some other species, including *Xenopus laevis,* genes for ribosomal RNA are selectively replicated in the oocyte, and in some insect larvae the chromosomes become nonuniformly polytene, so that certain genes are amplified more than others (see Chapter 8). In mammals, the production of immunoglobulins by lymphocytes involves the splicing together of initially separate segments of DNA in these specialized cells (see Chapter 17). It is possible that further examples of local transpositions and rearrangements of DNA sequences in particular types of cell will come to light with the application of the powerful methods of analysis based on gene cloning.

In Higher Eucaryotes, the Behavior of a Cell Depends on Its History as well as on Its Environment and Its Genome

The experiments on nuclear transplantation show that differences of cell character in development as a rule arise not from changes in the genome but from changes in the molecules associated with the genome. This general principle needs to be stressed because differentiation has features that could otherwise be interpreted in terms of genetic mutation. In particular, as already mentioned, the state of differentiation is heritable: the progeny of a differentiated cell are usually differentiated in the same way as their parent.

The differences of cell type are ultimately due to the different influences that cells have been exposed to in the embryo, but the differences are maintained because the cells somehow remember the effects of those past influences and pass them on to their descendants. The humblest bacterium can rapidly adjust its chemical activities in response to changes in its environment. But the cells of a higher animal are more sophisticated than that. Their behavior is governed not only by their genome and their *present* environment, but also by their history.

Cells Often Become Determined for a Future Specialized Role Long Before They Differentiate Overtly[20]

Cell memory is crucial for both the development and the maintenance of complex patterns of specialization in multicellular organisms. Though the mechanism is largely unknown, it is probable that it depends on some sort of positive feedback loop in the cell's internal system for controlling gene activity. Chapter 8 explains why and presents some possible specific mechanisms in detail.

The most familiar evidence of cell memory is seen in the persistence and stability of the differentiated states of cells in the adult body (see Chapter 16). Through cell memory, nonproliferating cells, such as neurons, maintain their specific characters, and proliferating cells pass on their characters to their progeny. Overt differentiation, however, is generally the last stage in a long process. Thanks to cell memory, the influences that direct a cell toward one mode of differentiation or another may act much earlier. For example, certain cells in the somites become specialized at a very early stage as precursors of muscle cells and migrate from the somites into the regions where the limbs will form (see p. 881 for details). These muscle-cell precursors do not contain the large quantities of specialized contractile proteins found in mature muscle cells; indeed, they look superficially just like the other cells of the limb rudiment, which do not derive from the somites. Only after several days do they become overtly differentiated and begin manufacturing large quantities of specialized muscle proteins. By contrast, the other limb cells with which they mingled differentiate instead as connective tissue. Thus the developmental choice between specialization as muscle and specialization as connective tissue has been made long before the stage at which it is expressed in overt differentiation, and is presumably recorded in the cells as a chemical distinction of a much less obvious sort.

A cell that has made a developmental choice is said to be **determined.** Determination is such an important and subtle concept in embryology, however, that we need a stricter definition. We should say: a cell is determined if it has undergone a self-perpetuating change of internal character that distinguishes it and its progeny from other cells in the embryo and that commits these progeny to a specialized course of development. The terms in this definition are worth considering individually:

1. The change must distinguish the cell and its progeny from other cells: determination involves the establishment of differences that are *heritable* from one cell generation to the next.
2. The change must commit the cell and its progeny to a specialized course of development: a cell is not determined simply because it is a step ahead of others in its maturation. Determination involves the selection of a particular developmental pathway.
3. The change must be a change of internal character, not merely a change of environment. In particular, a cell does not count as determined simply because it has come to occupy a particular position in the body, from which its specialized prospective fate can normally be predicted.
4. The change must be self-perpetuating: the element of memory is essential. A cell is not determined if it loses its distinctive character whenever the external influences that gave rise to the distinction disappear.

Determination is sometimes defined as an *irreversible* change. We prefer to be less absolute and to speak of it as self-perpetuating, since occasionally the cell memory can be upset and the state of determination can be altered, as we shall see later (p. 838).

The term *differentiation* is generally reserved for *overt* cell differentiation, that is, for a specialization of cell character that is grossly apparent. This distinction between differentiation and determination, however, is sometimes blurred.

The Time of Cell Determination Can Be Discovered by Transplantation Experiments[21]

To prove that a cell or group of cells is determined, one must show that it has a distinctive character that is maintained even in circumstances other than those in which it arose. The standard experimental method involves transplantation to a test environment. Thus the time of determination, strictly speaking, is operationally defined with respect to a particular choice of environment for assay of the state of determination, and different tests sometimes give somewhat different results.

A simple example of an experiment to test determination comes from studies on amphibian embryos. As noted earlier, one can plot a fate map for a blastula or an early gastrula, showing which of its parts will normally develop into what. The cells in one region, for example, are clearly fated to become epidermis if development proceeds normally, while those in another region are clearly fated to form brain. But when are these two groups of cells *determined* for their particular modes of differentiation? The answer can be obtained by cutting out a block of cells from one site and exchanging it with a block cut similarly from the other, so that cells from the prospective epidermal region are put in the position of prospective brain and vice versa. If the cells were already determined at the time of transplantation, they should develop autonomously according to their origins: the cells from the prospective epidermal region as misplaced epidermal tissue in the brain, and those from the prospective brain region as misplaced brain tissue in the epidermis. In fact, cells transplanted at the early gastrula stage show no memory of their origins and differentiate in the fashion appropriate to their new locations. They were not then determined with respect to the distinction between epidermis and brain, even though they clearly had, in the normal embryo from which they were taken, distinct presumptive fates. If, however, the same experiment is done at a somewhat later stage, in the late gastrula, the prospective brain cells transplanted to an epidermal site will differentiate as misplaced neural tissue, while the prospective epidermal cells transplanted to a brain site differentiate there as misplaced epidermis. This shows that both groups of cells have undergone some lasting internal change and have thus become determined, sometime between the early and the late gastrula stage.

The Genetic Control of Development Is Best Studied in *Drosophila*[22]

Having explained the basic concept of determination, we shall now discuss how it applies in detail to a particular organism. We shall focus on the fruit fly *Drosophila,* because of the special opportunities it offers for genetic analysis. The study of *Drosophila* will provide some key insights into the fundamental problem of how genes specify the differences between cells in different parts of the body.

Drosophila develops from an egg via a larval stage (Figure 15–25). But the adult, or *imago,* is not just an enlarged and matured form of the larva; it has a radically different structure. The adult fly arises largely from certain groups of cells, called the *imaginal cells,* which are set aside, apparently undifferentiated, in the larval body. The larva may almost be thought of as a walking, feeding counterpart of the extraembryonic structures of a mammal, a vehicle to contain and nourish the imaginal cells from which most of the adult body will develop. The imaginal cells for the head, thorax, and genitalia are organized into structures called **imaginal discs;** the imaginal cells for the abdomen are grouped in clusters called *abdominal histoblast nests.*

Many experimental studies have focused on the imaginal discs (Figure 15–26). There are 19 discs, arranged as 9 pairs on either side of the larva plus 1 disc in the midline. The discs are pouches of epithelium, shaped like crumpled and flattened balloons, that evaginate and differentiate at metamorphosis to form the epidermis of the adult fly and some internal tissues as well. From one pair of discs the eyes and antennae develop; from another, the wings and part of the thorax; from another, the first pair of legs; and so on. Because the cells of one imaginal disc look just like those of another, the question arises whether they are already determined, in the larval stage, for their distinctive modes of development.

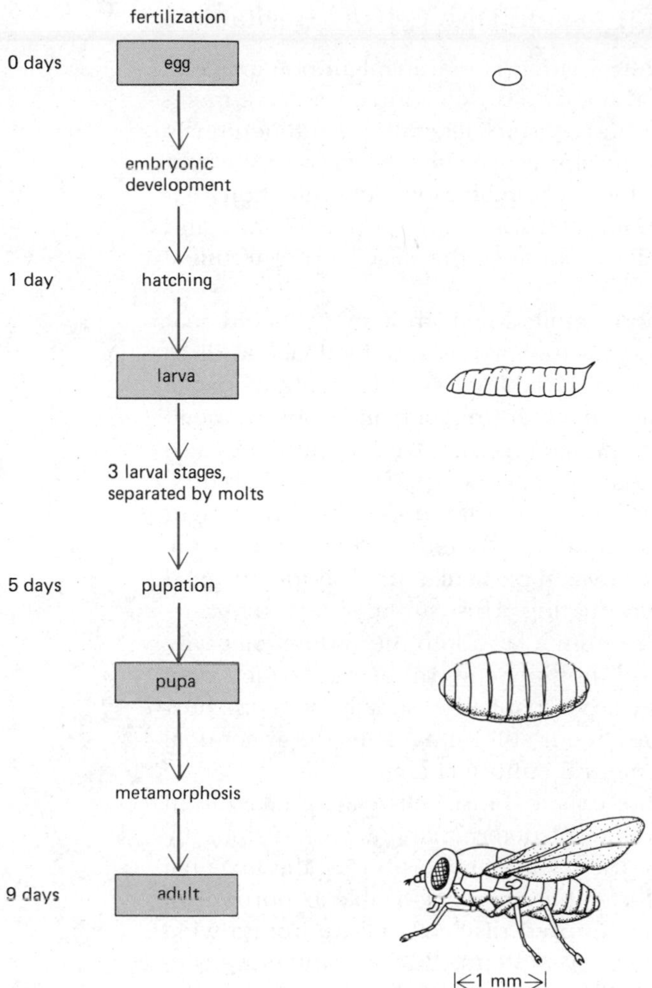

Figure 15–25 Synopsis of *Drosophila* development from egg to adult fly.

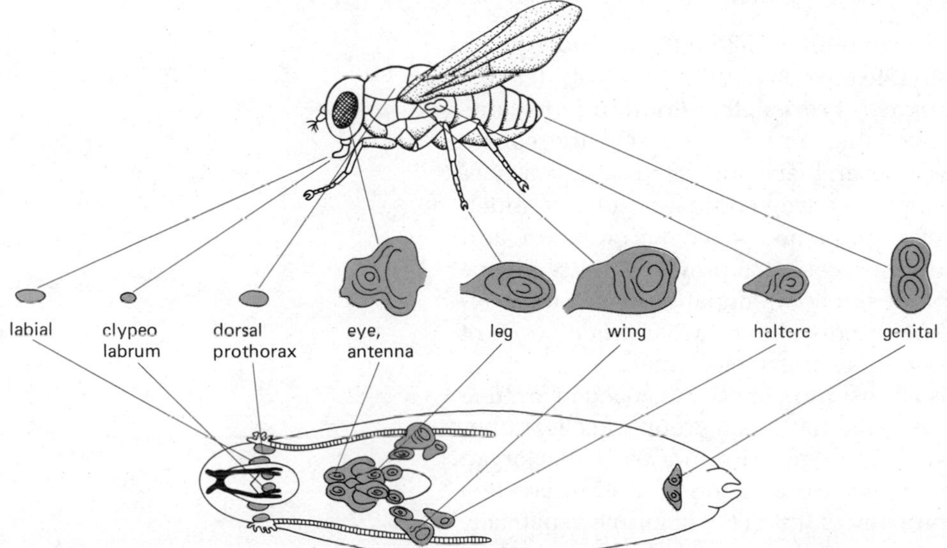

Figure 15–26 The imaginal discs in the *Drosophila* larva (schematic) and the adult structures they give rise to. Only one of the three leg discs is illustrated. (After J. W. Fristom, et al., in Problems in Biology: RNA in Development [E. W. Hanley, ed.], p. 382. Salt Lake City: University of Utah Press, 1969.)

The State of Determination of Imaginal Disc Cells Is Heritable[23]

One approach to the question of determination is to transplant one imaginal disc into the position of another in the larva. In due course the larva metamorphoses, and the grafted disc is then found to differentiate autonomously into the structure appropriate to its origin, irrespective of its new site. But *Drosophila* offers other possibilities for experiments on determination that reveal the heritability of the determined stage in a more dramatic way and show that it can be more than a brief prelude to the dead end of terminal differentiation.

The larva grows by a succession of molts in which it sheds its old coat of cuticle and lays down a larger one. As the larva grows, its imaginal discs grow also: their cells proliferate but remain undifferentiated. Finally, a change in circulating hormones causes the larva to enter pupation and undergo metamorphosis into a fly. The same hormonal change that triggers pupation also triggers the terminal differentiation of the imaginal disc cells. This differentiation, once complete, is irreversible, and the hormonal conditions that brought it about are no longer required. Consequently, hormonal conditions in the adult are similar to those in the young larva. In particular, they permit imaginal disc cells to proliferate without differentiating. This can be shown by transplanting an imaginal disc (or part of it) from a larva into the abdominal cavity of an adult fly. Here the cells grow and multiply and can be maintained even far beyond the life span of a fly, if they are transplanted serially from one host adult when it has grown old to another that is still young. Thus the abdominal cavity of the adult can be used as a natural culture chamber.

At any time, a part of the mass of imaginal disc cells can be taken from the fly abdomen and tested for its state of determination. This is done by grafting the tissue into a larva; when the larva metamorphoses, the imaginal disc cells differentiate to form structures that are recognizable as portions of adult structures (Figure 15–27). If the cultured disc cells derive from a wing disc, they give wing structures; if from an eye/antenna disc, eye/antenna structures; and so on. It is therefore clear that the state of determination of the imaginal disc cells is truly heritable, and heritable through an indefinite number of cell generations, during which it remains cryptic. To this rule there are, however, exceptions, and the exceptions are no less instructive than the rule.

Groups of Cells Occasionally Transdetermine[23,24]

A sample of cultured imaginal disc cells sometimes differentiates into a structure appropriate to a disc other than that from which the culture was derived. Such cells are said to have *transdetermined.* **Transdetermination** represents a switch from one heritable state to another and so resembles the consequence of a genetic mutation. However, several facts indicate that it is a phenomenon of a different sort. For example, transdetermination occurs much more often than one would expect of a spontaneous genetic mutation, and its frequency is not affected by chemical mutagens that provoke DNA sequence changes. But the strongest evidence that transdetermination is not the consequence of mutation comes from experiments showing that it is a *group* of cells, rather than a single cell, that undergoes transdetermination.

The logic of the demonstration is almost the same as was used in arguing from the composition of chimeric mammals that it is a group of cells, rather than a single cell, that become determined to give rise to a particular organ or tissue. In the experiments on *Drosophila,* x-irradiation is used to create a mutant cell in an imaginal disc. The progeny of this cell will form a genetically marked clone, distinguishable from the surrounding cells by their mutant character. When transdetermination subsequently occurs, a sample of the

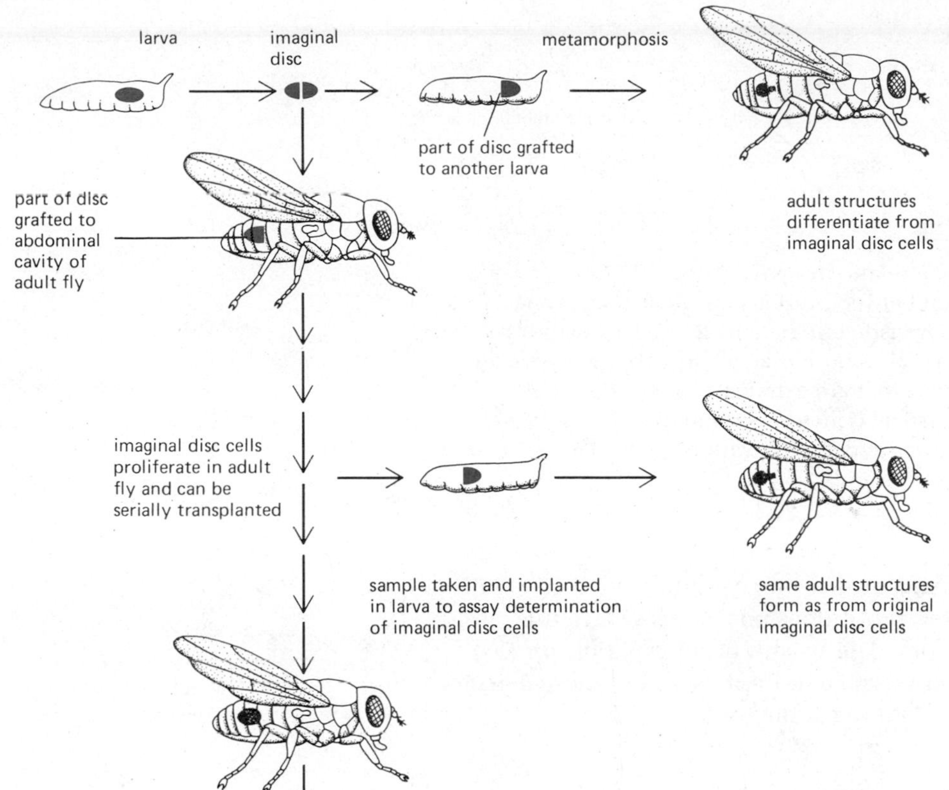

Figure 15–27 Experiments to test the state of determination of imaginal disc cells. The method of assay is to implant the cells in a larva that is about to undergo metamorphosis; the cells then differentiate to form recognizable adult structures, which lie, however, inside the body of the host fly and are not integrated with it. The disc cells can either be assayed immediately or be implanted in the abdomen of adult flies and allowed to proliferate for an indefinite period, without differentiating, before the assay for cell determination is done. In both cases, the cells generally differentiate to form the structures appropriate to the disc from which they derived originally.

transdetermined tissue is sometimes found to include cells of both mutant and normal genotypes (Figure 15–28). One can conclude, therefore, that the original switch in the state of determination must have occurred in more than one cell.

Transdetermined cells may revert to their original state of determination or transdetermine to yet another state. Some transitions occur more frequently than others. For example, antennal cells most often transdetermine to wing; wing cells may transdetermine back to antenna, or they may switch to mesothorax (the body segment to which the wings are attached), or, less often, to leg; and so on. A diagram can be drawn showing which transitions occur and how frequently (Figure 15–29). It thus appears that there are a limited number of standard discrete states of determination in *Drosophila* and that a cell must opt for one of these.

Figure 15–28 Schematic diagram of an experiment showing that transdetermination occurs in a group of cells rather than in a single cell.

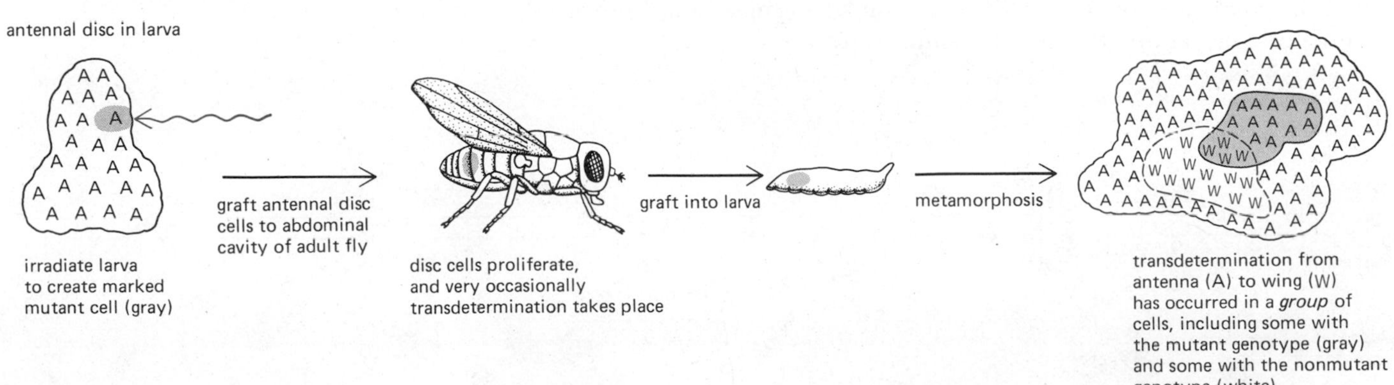

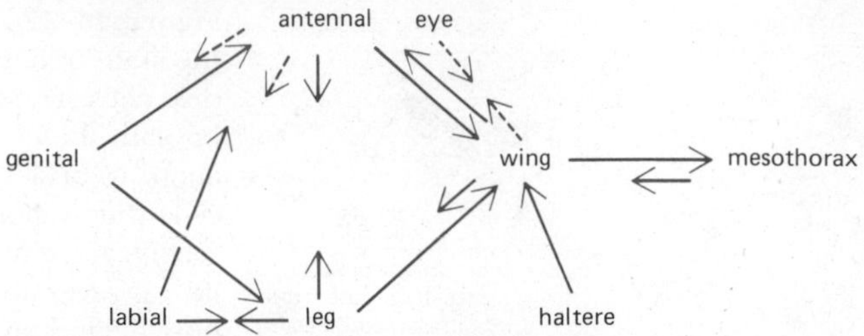

Figure 15–29 The relative frequencies of transdetermination between various imaginal discs. Long arrows symbolize transdeterminations that occur with high frequency; short arrows, those that occur with low frequency; dotted arrows, those that are extremely rare or doubtful. (After E. Hadorn, in The Genetics and Biology of *Drosophila*, Vol. 2C [M. Ashburner and T. F. Wright, eds.], pp. 555–617. London: Academic Press, 1978.)

All this evidence concerning imaginal discs argues for the existence of control genes governing disc character, but it does not tell us where they are on the chromosomes or how they work. The wealth of information on *Drosophila* genetics has, however, made it possible to locate some of these genes and to begin an experimental analysis of their function.

Homoeotic Mutants Reveal Genes Whose Activities Control Cell Determination[25]

Mutant strains of *Drosophila* are occasionally encountered showing bizarre disturbances of the body plan. Wings, for example, may sprout from the head where there should be eyes (the mutation *ophthalmoptera*), or legs may grow in place of antennae (the mutation *Antennapedia*) (Figure 15–30). Such mutations, which transform parts of the body into structures appropriate to other positions, are called **homoeotic.** Since the structures affected derive from imaginal discs, these mutations must convert certain imaginal disc cells to the mode of behavior normally characteristic of imaginal disc cells located elsewhere. The consequent abnormality is as though cells in the affected discs had undergone transdetermination. Homeotic mutations, therefore, appear to be disturbances of control genes of the type involved in determination and transdetermination. In transdetermination the switch from one character to another might occur, for example, when transcription of a control gene is turned off; a homeotic mutation, on the other hand, might alter the DNA sequence of a control gene so that its transcription is permanently defective or so that the product functions inadequately. For several of the conversions of disc character that occur through transdetermination, there are known homoeotic mutations that exert a similar effect.

Figure 15–30 The head of a normal adult *Drosophila* (A) compared with that of a fly carrying the homoeotic mutation *Antennapedia* (B). The fly shown here displays the mutation in an extreme form; usually only parts of the antennae are converted into leg structures (see Figure 15–44). (*Antennapedia* drawing based on a photograph supplied by Peter Lawrence.)

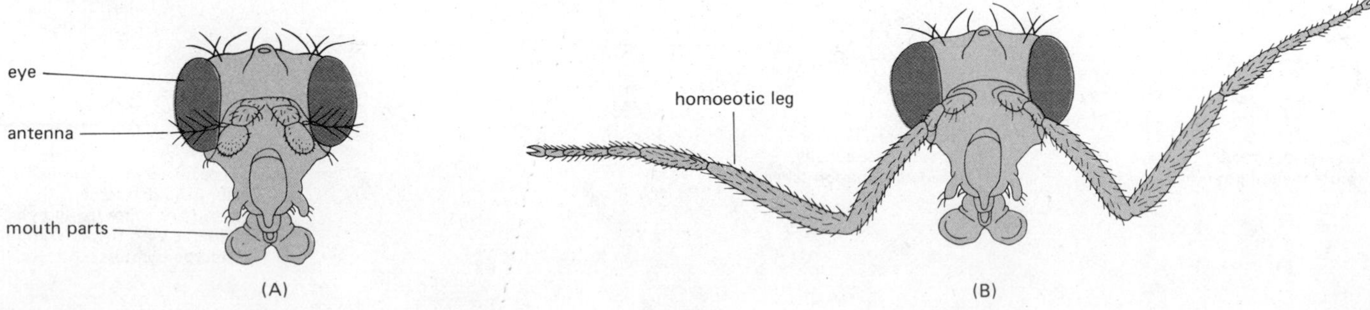

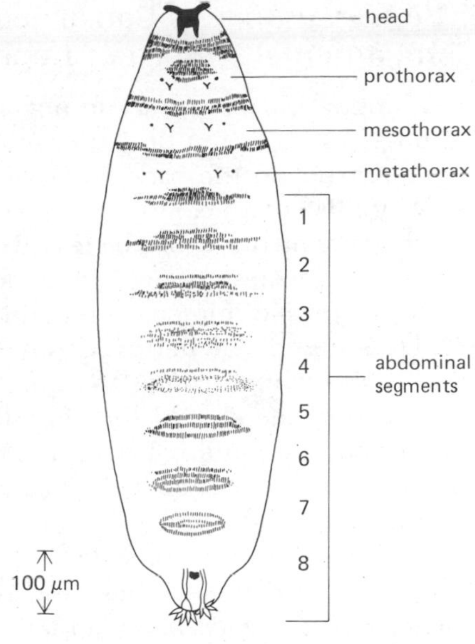

Figure 15–31 Dorsal view of a normal first-stage *Drosophila* larva. (After B. T. Wakimoto and T. C. Kaufman, *Dev. Biol.* 81:51–64, 1981.)

The Bithorax Complex Controls Differences Between Thoracic and Abdominal Segments[26]

Thanks to homoeotic mutants, more than 30 distinct loci of control genes have been identified on the genetic map of *Drosophila.* Some of the homoeotic mutations are found to lie next to each other in the genome in very closely linked clusters. The largest of these clusters is named the **bithorax complex** and comprises at least eight genes that play an essential part in controlling the differences between thoracic and abdominal segments of the body. For example, a mutation of one of these genes, called *bithoraxoid,* yields a fly whose first abdominal segment, which is normally legless, bears a pair of legs and has the appearance of a metathoracic segment (that is, the most posterior segment of the thorax). The role of the bithorax complex is most fully revealed, however, in studies of larval development.

The *Drosophila* larva, like the adult fly, is composed of a series of distinctive segments: first the head, then three thoracic segments (pro-, meso-, and metathorax), then eight abdominal segments (Figure 15–31). The genes of the bithorax complex control the characters of the larval segments as well as those of the imaginal discs. For example, the *bithoraxoid* larva, like the adult, has its first abdominal segment converted to the appearance of a metathorax. Some of the other homoeotic mutations in this group cause still more drastic disorders of structure and are consequently lethal: their effects are never seen in the adult fly because the mutant dies before it can develop that far. Lethal mutations of this sort are propagated only if they are recessive. Heterozygotes, with one mutant copy of the gene and one normal copy, are then viable, and by breeding from a pair of heterozygous parents, one can get homozygous progeny in which both copies of the gene are mutant. These progeny die prematurely as very young larvae but survive long enough to reveal the mutant phenotype. In particular, one can witness in this way the effects of a deletion of the entire bithorax complex.

Larvae deficient in all the genes of the bithorax complex are found to have a simple structure: they consist of a head, a prothorax, a mesothorax, another mesothorax, and another, and so on—to a total of 10 mesothoracic segments. In other words, when all the genes of the bithorax complex have been deleted, all segments posterior to the mesothorax take on one and the same character—the character of mesothorax. Partial deletions of the bithorax complex cause transformations that are less extensive; for example, the metathorax and the segments anterior to it may appear normal, while the segments posterior to metathorax are converted to metathorax (Figure 15–32). These observations justify the initial statement that the bithorax complex plays an essential part in controlling the differences between the thoracic and abdominal segments: when it is missing, the distinctions are not made.

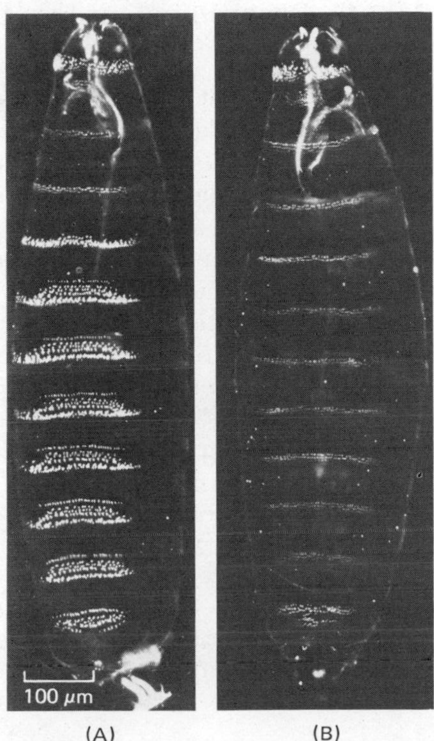

Figure 15–32 A normal *Drosophila* embryo (A) compared with a mutant embryo (B) lacking most of the genes of the birthorax complex. (The precise genotype of the mutant is *Dpbxd*^{100}*DfP9/DfP9.*) In the mutant, the *number* of segments is normal, but those posterior to the metathorax have the appearance of metathorax. (Courtesy of Gary Struhl; A, reprinted by permission from *Nature* 293:36–41. Copyright © 1981 Macmillan Journals Limited.)

The Larval Body Is Constructed by Modulation of a Fundamental Pattern of Repeating Segments[27]

It is important to appreciate not only what the bithorax complex does, but also what it does not do. It does not control the number of segments, and it does not define the basic internal structure of the typical segment. A larva lacking this complex in its genome has the normal number of segments, and each segment has more or less the regular anatomy characteristic of head, prothorax, or mesothorax. There must be an underlying mechanism that generates a specific number of repetitions of a standard basic unit or prototype. The function of the bithorax complex is then to modulate the repetitions so that each segment comes to have its own distinctive character.

Drosophila should therefore have other genes that control, on the one hand, the basic internal plan of the segmental unit and, on the other hand, the total number of such units. A mutation in a gene controlling the basic internal plan of the prototype segment should affect every segment in a similar way. Studies of mutations affecting early developmental stages have revealed such a class of genes, occupying at least six different loci in the genome. In larvae with the *gooseberry* mutation, for example, the posterior half of every segment is similarly altered. A further six mutations are known that appear to cause deletions in every other segment. Finally, there are three mutations that reduce the total number of segments by eliminating a block of contiguous segments. In the mutant *knirps,* for example, the six middle segments of the abdomen are missing, leaving the larva simply with a head, three thoracic segments, a first abdominal segment, and an eighth abdominal segment.

The insect larva, therefore, seems to be built by a modulated repetition of a basic prototype unit, the segment. Many other organisms and parts of organisms appear to be constructed on a similar principle. Examples in the vertebrate body include the rows of somites and of teeth and the segments of the limbs (upper arm, forearm, and so on). But the underlying genetic control mechanisms are still a complete mystery in vertebrates; it is only in *Drosophila* that we are beginning to understand them.

Mitotic Recombination Can Be Exploited to Produce Marked Mutant Clones of Cells[28]

As we have seen, it is useful in the analysis of development to be able to create mutant clones of cells in the body. A powerful technique for doing this in *Drosophila* exploits the phenomenon of mitotic recombination.

To recapitulate from Chapter 14, a normal somatic cell contains two homologous sets of chromosomes, one homolog of each pair being derived from the father and one being derived from the mother. In normal mitotic division there is no exchange of DNA between the paternal and the maternal chromosomes, and each daughter cell receives a full intact set of paternal genes plus a full intact set of maternal genes. Normally, an exchange of genes by crossing over between the paternal and the maternal homologs happens only in germ cells, at meiosis. But occasionally crossing over between homologs occurs during the division cycle of an ordinary somatic cell. This is called **mitotic recombination.** It will be imperceptible if the fragments exchanged between the paternal and the maternal chromosomes are identical, that is, if the cell is homozygous at the loci that are exchanged. Mitotic recombination can, however, have a striking effect if the dividing cell is heterozygous at these loci. For example, recombination may produce daughter cells that have different pigmentation and proliferate to form patches of tissue of a distinctive color. The mechanism is explained in Figure 15–33 and the accompanying

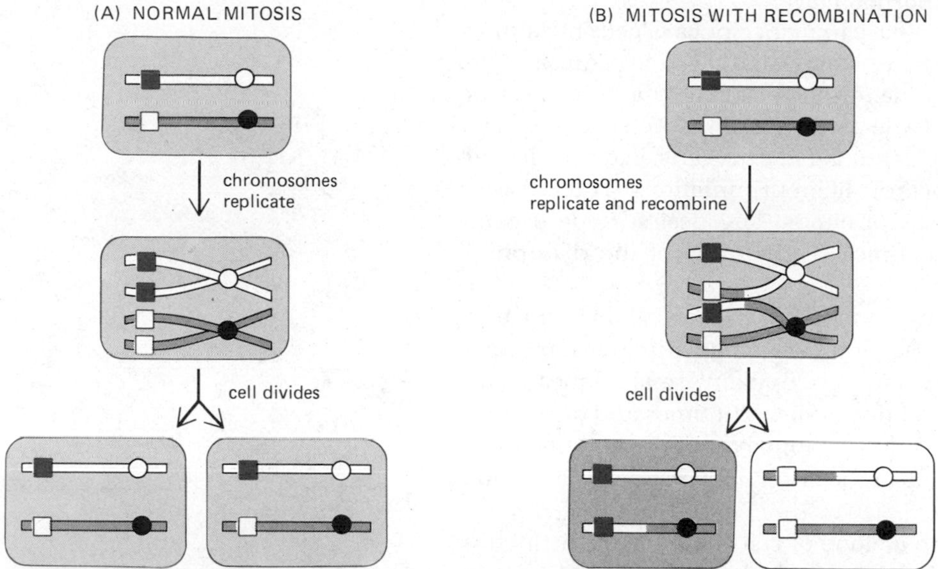

Figure 15–33 Mitotic recombination compared with normal mitosis. The paternal chromosomes are shaded; the maternal, unshaded. In this example, we suppose that the genome contains a locus for a pigmentation gene with two alleles, *R* (*small colored square*) and *r* (*small open square*), such that a homozygous *R/R* cell appears red, a heterozygous *R/r* cell appears pink, and a homozygous *r/r* cell appears white.

(A) In the normal division cycle, the maternal chromosome of the heterozygous cell duplicates to form two chromatids coupled at the maternal centromere, both carrying the *R* allele; similarly, the paternal chromosome duplicates to form two chromatids coupled at the paternal centromere, both carrying the *r* allele. At mitosis, the two chromatids of each pair pull apart, and each daughter cell receives, at random, one or other of the two identical chromatids that had been coupled at the maternal centromere, plus one or other of the two identical chromatids that had been coupled at the paternal centromere; thus each daughter cell inherits the heterozygous *R/r* genotype.

(B) In a division cycle in which mitotic recombination occurs after the chromosomes have replicated, the two chromatids coupled at the maternal centromere are different; while one still carries the allele *R*, the other has undergone an exchange with one of the paternal chromatids and thereby acquired the allele *r*. Similarly, the two paternally coupled chromatids are different, carrying, respectively, the alleles *r* and *R*. Now each daughter cell receives, at random, one or other of the two maternally coupled chromatids, plus one or other of the two paternally coupled chromatids. Thus, by mitotic recombination, one daughter cell can receive both copies of the *R* allele while the other receives both copies of the *r* allele. In this way a heterozygous *R/r* cell (*light color*) gives rise to two daughters of different genotype, one being homozygous *R/R* (*darker color*), the other homozygous *r/r* (*white*.) The two homozygous daughters will then reproduce themselves in the normal fashion, giving rise to a *twin spot*, consisting of a clone of red (*R/R*) cells next to a clone of white (*r/r*) cells, amid the field of pink (*R/r*) cells that have not undergone mitotic recombination.

Mitotic recombination depends on random collisions between the two homologous chromosomes and is therefore rare; in the mitotic cycle there is no regular controlled pairing of the homologs, such as occurs in meiosis, to facilitate recombination.

legend, where it is shown how a single mitotic recombination event can give rise to a *twin spot,* consisting of two genetically different marked clones of neighboring cells in a background of normal cells.

Two features of *Drosophila* make it easy to mark clones of cells by mitotic recombination. First, the process can be provoked artificially by exposing the embryo or larva to x-rays (the recombination is presumably a side effect of damage to the chromosomes). Second, the genotype of the heterozygous animal, and in particular the genes for which it is heterozygous, can be selected from a large and well-catalogued stock of different mutations. Thus readily identifiable clones of homozygous cells of almost any desired type can be introduced into the body at specified times without any of the disruption attendant upon surgery.

The size of a marked clone depends on the number of cell divisions that have elapsed since it was created. In general, large marked clones are produced by x-irradiation early in embryonic development, small clones if the irradiation is done later. Thus by x-irradiating at different times and measuring the size of the resultant clones in the adult fly, one can derive the timetable of growth and the sizes of the rudiments of the adult body parts at the various stages of development.

Where the aim is simply to create a clone of cells that have a distinctive appearance, a variety of genetic markers are available. For example, mitotic recombination in a fly that is heterozygous for the recessive mutation *yellow* (symbolized *y,* with corresponding wild-type allele y^+) will produce a pair of mutant clones with genotypes y/y and y^+/y^+, respectively, amid the heterozygous y/y^+ tissue. Since the *yellow* mutation is recessive, only the y/y clone will have a mutant appearance: it will be visible as a yellow y/y patch on a brown background of y/y^+ and y^+/y^+ cells. Similarly, using the recessive mutation *multiple wing hairs* (*mwh*), one of the recombinant clones will be detectable (if it lies in the wing) as a patch of *mwh/mwh* abnormal cells, each of which gives rise to several hairs on the wing cuticle instead of the usual single hair (Figure 15–34). X-ray-induced mitotic recombination, using markers such as these, provides a powerful method for analyzing the clonal construction of the body.

Sharp Demarcation Lines Separate Polyclonal Compartments[29]

The *Drosophila* wing, being a broad, flat, epithelial structure, is very convenient for clonal analysis. When marked clones are generated by x-irradiation, they are found to be located at random and generally have rather irregular outlines. Comparisons of clones in different animals show that the spatial arrangement of the progeny of a single cell is variable from one animal to another and not strictly predetermined. Clones that lie close to the central axis of the wing, however, behave as though they were constrained by an invisible demarcation line, similarly positioned in every fly, that divides the wing into anterior and posterior *compartments.* Some clones lie on one side of the line and some on the other, but they never straddle it; and where they abut it, the clone outline is unusually sharp and straight (Figure 15–35). Each compartment is a *polyclone*: it consists of several entire clones of cells.

Compartment boundaries of this type are most clearly revealed by a further genetic trick. In a class of *Drosophila* mutants known as *Minutes,* cell proliferation is abnormally slow. *Minute* mutations are dominant, so that heterozygous cells with the M/M^+ genotype proliferate at a lower rate than homozygous wild-type M^+/M^+ cells. It is possible furthermore to breed flies in which the *Minute* locus is closely linked to the locus of a recessive marker mutation such as *multiple wing hairs* (*mwh*). A heterozygous $M\ mwh^+/M^+\ mwh$ fly will have wild-type wing hairs but a slow growth rate. In this fly it is

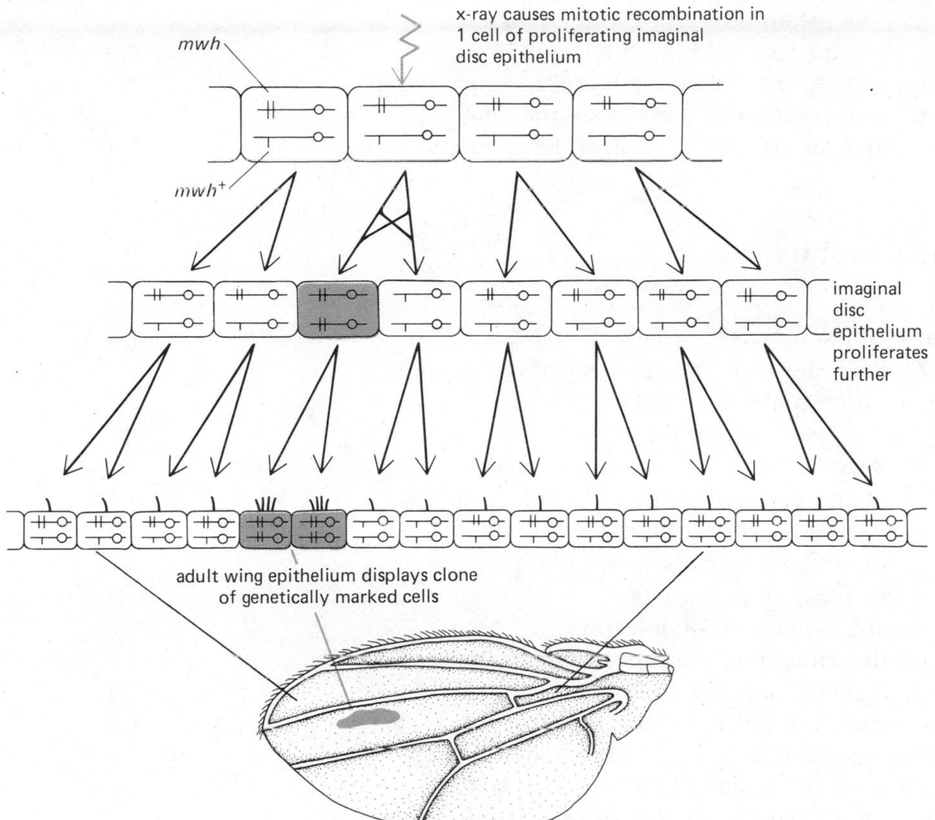

Figure 15–34 Mitotic recombination can be used to produce a clone of mutant cells in the *Drosophila* wing, displaying the multiple wing hair (*mwh*) marker.

possible to produce recombinant clones with the genotype M^+ *mwh*/M^+ *mwh* by x-irradiation. These clones will grow much faster than the surrounding tissue and will be easy to see because of their aberrant wing hairs. Thus one observes enormous clones of M^+ *mwh*/M^+ *mwh* cells occupying a large fraction of the entire wing even if they have been induced relatively late in development (Figure 15–35). Yet the compartment boundary running down the middle of the wing is strictly maintained: every clone is confined entirely to the anterior half of the wing or entirely to the posterior half, and the dividing line is sharp and straight. Can this behavior be explained in terms of the presence of some fold or mechanical barrier in the imaginal disc constraining clone growth? Careful microscopic examination provides no support for this conjecture. The compartment boundary does not, for example, coincide with the central wing vein or with any other obviously visible structure. It seems

Figure 15–35 The shapes of marked clones in the *Drosophila* wing reveal the existence of a compartment boundary. The border of each marked clone is straight where it abuts the boundary. Even when a marked clone has grown more rapidly than the rest of the wing, and so is very large, it respects the boundary in the same way (*last drawing*). Note that the compartment boundary does not coincide with the central wing vein. (After F. H. C. Crick and P. A. Lawrence, *Science* 189:340–347, 1975.)

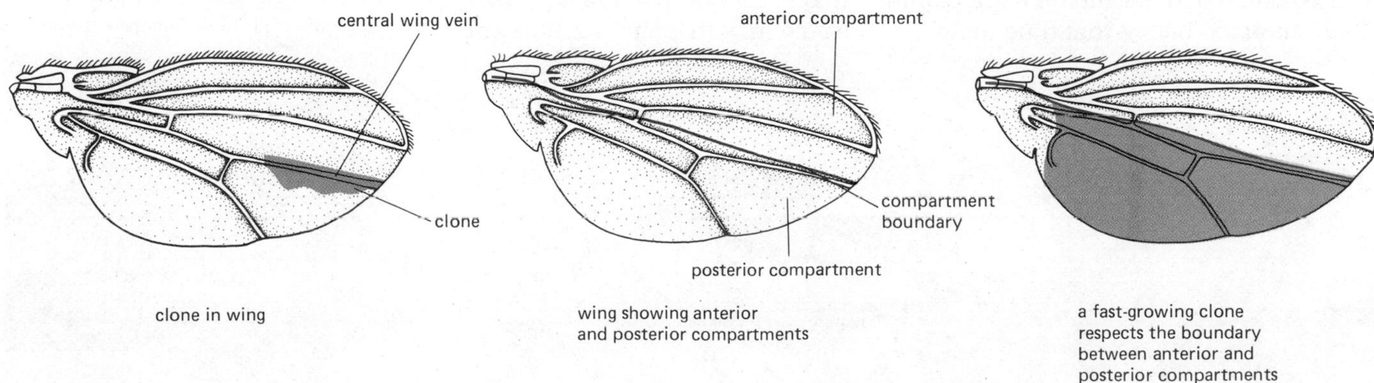

instead that some distinction has been established in the intrinsic character of the imaginal disc cells very early in development, dividing them into two categories, the one set strictly confining itself to the anterior half of the wing, the other to the posterior half. In other words, at some early stage the cells seem to have become determined for either an anterior or a posterior character.

Different Sets of Genes Are Active in the Cells of Different Compartments[29]

How can a difference of cell character give rise to such a sharp demarcation line in the wing? One possibility is that cells in the same state of determination stick more strongly to one another than to cells in a different state of determination. The sharp compartment boundary would then be due to a sort of surface-tension effect, like an interface between oil and water. This suggestion is supported by experiments on sorting out and selective cohesion in mixtures of imaginal disc cells in culture: in artificial aggregates formed from such a mixture, cells from the same compartment are found to adhere preferentially to one another.

Another, and perhaps more powerful, argument for interpreting compartments as due to differences in cell determination comes from genetics. For example, a mutation called *engrailed* affects only the cells that form the posterior half of the disc (both in the wing disc and in several other discs), converting them to an anterior character. As a result, the posterior half of the wing is transformed more or less into a mirror image of the anterior half. Another mutation, *bithorax* (belonging, incidentally, to the bithorax complex discussed earlier), converts just the anterior half of the haltere—a small knob-shaped balancer important in flight and placed one segment behind the wing—into the anterior half of a wing (Figure 15–36). Homoeotic mutations such as these imply that the cells of the anterior and posterior compartments of imaginal discs normally have different sets of genes active.

The State of Determination Is Built Up Combinatorially[29]

These observations have naturally prompted a search for other compartment boundaries, and they have been found. For example, in the wing disc (which gives rise not only to the wing but also to an adjacent region of the thorax called the *notum*), at least three different compartment boundaries have been identified, intersecting to define eight distinct compartments (Figure 15–37). Similar systems of compartments have been found in the eye/antenna region, in the leg, and elsewhere.

The time of determination of a compartment, as well as its existence, can be discovered by clonal analysis. Mutant clones created by x-irradiation after the time of determination are confined to one compartment or the other. Clones created before that time may have progeny in both compartments and

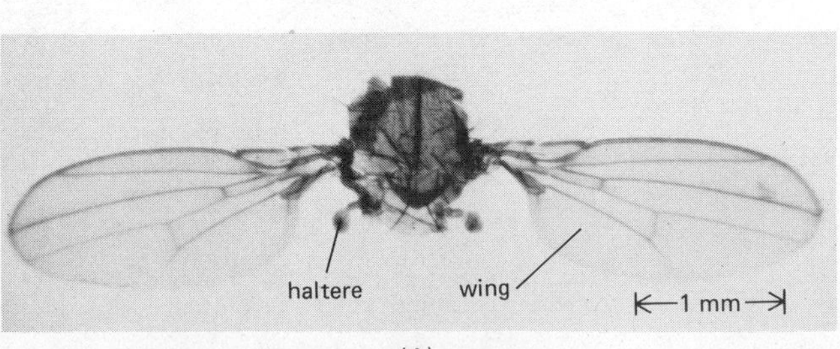

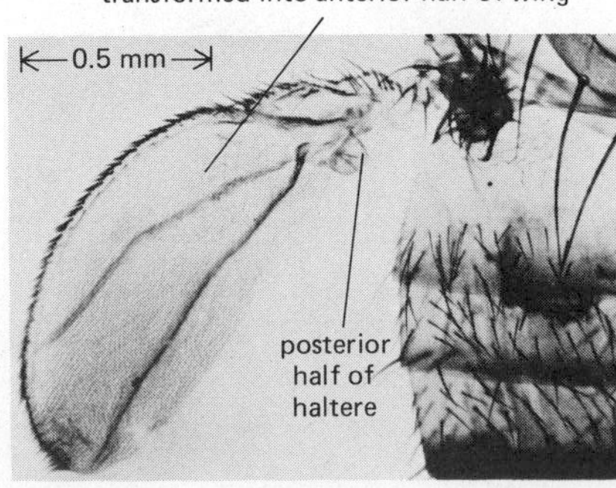

Figure 15–36 The haltere of a fly with the *bithorax* mutation (B) compared with a normal wing and a normal haltere (A). The anterior compartment of the mutant haltere is transformed into the anterior compartment of a wing. (A, courtesy of Peter Lawrence; B, from F. H. C. Crick and P. A. Lawrence, *Science* 189:340–347, 1975. Copyright 1975 by the American Association for the Advancement of Science.)

do not respect the boundary. In this way it has been shown that the anterior/posterior boundary in the wing is established first and that the dorsal/ventral and proximal/distal (that is, wing/notum) subdivisions are established later. It therefore appears that the state of determination of the cells in any part of the fly is built up by a sequence of decisions between alternative pathways, corresponding to the different discs and the different compartments within them. Homoeotic mutations affect the control genes that record and express these decisions.

The cells in a given compartment have acquired in effect an "address" represented by a specific combination of activities of control genes. Alteration of the activity of one of the control genes in the combination can alter the address to that of some other compartment elsewhere in the fly. The combinatorial method of determination makes economical use of control genes: for instance, the same piece of genetic machinery may serve to register the anterior/posterior distinction in each of several different imaginal discs. Thus the mutation *engrailed* converts not only the posterior half of a wing to an anterior wing character, but also the posterior half of a leg to an anterior leg character. We have seen the same principles at work in the larva: one set of genes acts repeatedly in many successive segments to lay out in each of them the prototype segment plan, distinguishing posterior from anterior and so on, while another set of genes controls the differences between one segment and another. Several sets of genes act jointly to define for each cell a detailed address.

In principle, if there were n control genes each of which independently could be either active or inactive, it would be possible to use them in different combinations to specify 2^n different cell states or addresses, just as we can identify any one of 2^4 different compartments by combinations of just four words, drawn from the four pairs of alternatives anterior/posterior, proximal/distal, dorsal/ventral and wing/leg.

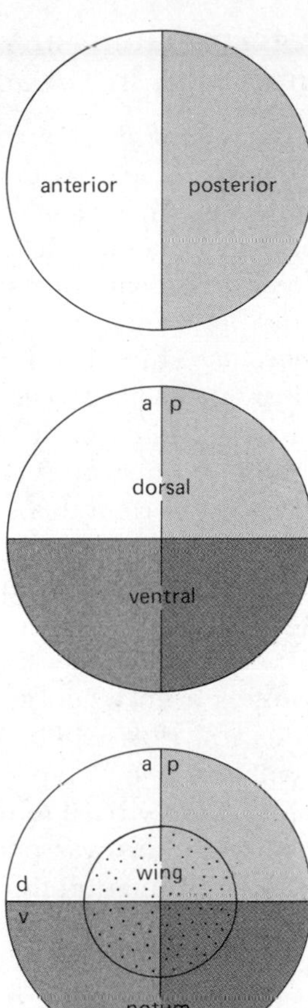

Figure 15–37 A set of three dichotomous distinctions, corresponding to three intersecting compartment boundaries, can jointly define eight distinct compartments.

The Extent of Cell Proliferation in *Drosophila* Is Not Determined by Counting Cell Divisions: Fast-growing Clones May Nearly Fill Their Compartment But Do Not Make It Big[29,30]

We have already described how it is possible, by irradiating heterozygous *Minute* (M/M^+) flies, to create clones of M^+/M^+ cells that divide much faster than the surrounding cells. Such a fast-growing clone becomes abnormally large and may almost fill the compartment in which it lies. Yet the compartment itself remains practically normal in both size and pattern: the left and right wings, for example, have nearly the same size and shape even if the one contains an M^+/M^+ clone and the other does not. To form their unfairly large share of the adult structure, the fast-growing cells on the slow-growing background have gone through more cell divisions than they would have done on a fast-growing background, and the slow-growing cells in the same compartment have gone through fewer divisions than they would have done if they had not been competing for space with the fast-growing cells. For both sets of cells, the number of division cycles has been abnormal; evidently the cells have continued to proliferate until the compartment as a whole has attained normal size, instead of autonomously counting their way through a preset number of division cycles and then stopping. It follows that the size of the compartment in *Drosophila* is determined not by the counting of cell divisions but by some spatial signal that tells a cell when the structure in which it lies has reached full size. This signal presumably arises from interactions between the cells within the compartment and depends on their distances from one another.

Cell Determination in Vertebrates Resembles Cell Determination in *Drosophila*[31]

Studies of *Drosophila* probably offer the best prospect for elucidating the molecular biology of animal development. If they succeed, shall we have understood only this fly, or shall we have discovered principles that apply universally, and in particular to vertebrates? For those who hope for universal principles, it is disappointing to note that practically no clear-cut instances of homoeotic mutations or of compartment boundaries have been observed in vertebrates. This does not necessarily mean that they do not exist. They may simply be easier to see in *Drosophila*.

Cell determination in vertebrates, however, does seem to be essentially the same phenomenon as cell determination in *Drosophila*. For example, determination in vertebrates somewhat analogous to that of imaginal discs in *Drosophila* can be demonstrated by transplantation of early embryonic rudiments. Thus in the early chick embryo, one can graft somites from one position to another long before the time of their differentiation. At this stage thoracic somites and neck somites are practically indistinguishable, yet thoracic somites even when grafted into the neck will give rise to thoracic vertebrae bearing ribs, while neck somites even when grafted into the future thorax will give rise to typical neck vertebrae.

It is generally hard to uncouple the time of differentiation from the time of determination, as was possible for *Drosophila* imaginal discs by culture in the abdomen of an adult fly. One does, however, encounter cells in vertebrates that remain determined but undifferentiated throughout life. In particular, stem cells, such as those in the basal layer of the skin or in the bone marrow, while proliferating continuously, throw off progeny all committed to a particular form of overt differentiation, such as keratinized skin or as blood cells, respectively (see Chapter 16).

Experiments on vertebrates have revealed a great deal about cell interactions in development, but our ignorance of the genetics of vertebrate development is still profound. The insights to be gained from *Drosophila* will surely have implications that extend far beyond the fruit fly.

Summary

The cells of the body appear to contain identical genomes and to become different only because of the differential expression of their genes. These changes in gene expression are often heritable, and in this sense cells have a memory: the present character of a cell depends on its history as well as on its genome and on its present environment. Cells may become heritably determined for a specialized course of development long before they differentiate overtly.

In Drosophila, *the epidermis of the adult fly develops largely from the imaginal discs of the larva. The cells of different imaginal discs are in different states of determination, specific for the formation of particular structures, such as wing and antenna. The cells can remain determined but undifferentiated through many cycles of proliferation, though groups of cells may occasionally transdetermine. Homoeotic mutations affect the genes that control the differences between one disc and another. Studies of homoeotic mutations in the larva suggest that the body is constructed of homologous segments that represent modulated repetitions of a basic prototype segment. The genes identified by homoeotic mutations govern the modulations that distinguish one segment from another; other genes control the basic plan of the prototype segment and the total number of segments.*

The pattern of cell determination in Drosophila *has been analyzed using marked clones of mutant cells created in the body by x-ray-induced mitotic recombination. The shapes of the marked clones reveal that the wing (and other structures besides) is subdivided into a number of compartments consisting of cells in different states of determination. The cells in a given compartment have in effect a distinctive address, represented by a particular combination of activities of control genes. Cells in different compartments do not mix.*

The genetics of vertebrate development are less well understood than those of Drosophila, *but many of the same principles of developmental control are thought to apply.*

Patterns in Space

Cells Are Assigned Different Characters According to Their Positions[32]

We have considered determination from the point of view of the individual cell and its internal state. Thus far we have touched only briefly on the central question of **pattern formation:** how is the spatially organized arrangement of cells in different states generated? What signals, if any, tell a cell where it is, and how are these signals interpreted?

In theory, there are two extreme possibilities. On the one hand, neighboring cells might start out identical and the distinctions between cells might be generated in a spatially ordered fashion, each cell being assigned the character appropriate to its position (Figure 15–38A). On the other hand, an orderly final pattern might develop from an initially chaotic mixture of different types of cell (Figure 15–38B): in this case, either the cells would sort out according to their characters, or those cells that were inappropriately placed would die while those that were appropriately placed would survive and proliferate. In fact, in almost all cases the first possibility is closer to the truth: cells become determined according to their position—there are hardly any instances in normal development in which spatial order is created out of an initially random mixture of cell types.

Before discussing the general mechanisms by which differences of cell character are imposed on a field of initially similar cells, we will consider the phenomenon of asymmetrical cell division, observed in the early cleavage

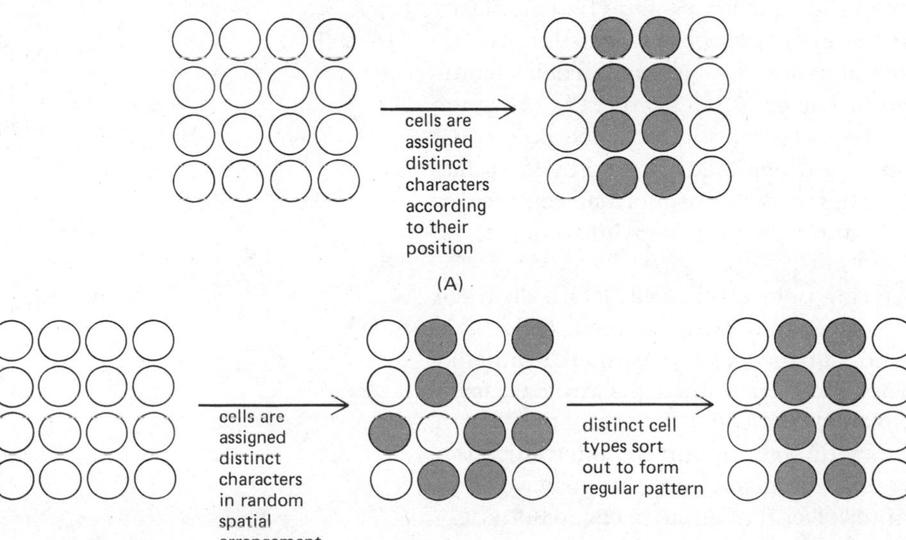

Figure 15–38 Two alternative ways of generating a regular spatial pattern of distinct cell types. (A) The upper pair of diagrams represent what usually happens; (B) the lower three diagrams represent a sequence of events that is rare in normal development.

stages of many species and involving localized determinants of cell character in the egg cytoplasm (Figure 15–39). This phenomenon represents a particularly direct way of specifying differences between cells in a spatially organized manner.

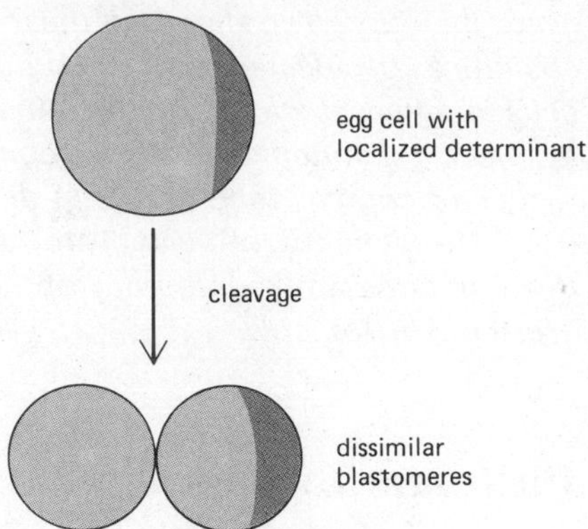

Figure 15–39 Localized determinants in the cytoplasm of an egg, if parceled out asymmetrically during cleavage, produce blastomeres that differ from one another from the outset.

Localized Determinants Are Sometimes Identifiable in the Cytoplasm of the Egg[33]

It was pointed out earlier that the polarity of the amphibian embryo is governed by the polarity of the egg. The egg itself has a spatial pattern, with certain components concentrated in certain regions. As a result of this asymmetry, the cells that form by cleavage are different because they inherit different portions of these localized materials. The importance of localized determinants in the egg varies from species to species. At one extreme are mammals, in which it seems that such determinants are of no importance, and all the cells of the early morula are alike totipotent. At the other extreme are the so-called mosaic eggs of mollusks, ascidians (sea squirts), and some other phyla, which cleave to form blastomeres whose potencies are different from the outset. There is evidence that these very early differences derive from the inheritance of different determinants in the cytoplasm of the egg. For example, if the uncleaved egg of the ascidian *Styela* is centrifuged and its contents thereby visibly rearranged, the subsequent pattern of differentiation of the embryo is disturbed in the way that would be expected if the rearranged regions of cytoplasm contained the controlling factors. The clearest evidence, however, relates to the determination of germ cells, that is, of the precursors of eggs and sperm.

The Determinant of Germ-Cell Character Is Localized at One End of the *Drosophila* Egg[34]

Germ cells have a special status in the body and are often determined very early in development. In *Drosophila,* the determinant of germ-cell character can be shown to be located in a specialized region of cytoplasm at one pole of the egg.

The *Drosophila* egg, like that of most other insects, cleaves in a peculiar way, passing through a syncytial stage, in which several thousand nuclei occupy a single undivided volume of cytoplasm (Figure 15–40). The cytoplasm at the posterior end of the egg, called the *polar plasm,* has a distinctive appearance characterized by the presence of small (0.5–1.0 μm) particles, composed partly of RNA, called *polar granules.* The cells that form at the posterior end of the egg, incorporating a part of the polar plasm, are called *pole cells:* they are distinguished from the other cells by their large size and by the polar granules enclosed in their cytoplasm. They are the primordial germ cells, which will later migrate to the gonads and there develop into oocytes or sperm.

Ultraviolet irradiation of the posterior pole of the egg before cleavage gives rise to animals that are sterile because they have no germ cells. This effect apparently results from inactivation of the polar plasm: an irradiated egg can be rescued from infertility by an injection of polar plasm taken from an unirradiated egg. The function of polar plasm as the determinant of germ cell character has been proved still more directly by further microinjection experiments (Figure 15–41). Polar plasm can be taken from the posterior end of a donor egg and injected into the anterior end of another egg, where only somatic cells would normally form. This causes the formation of cells that look like pole cells. These cells are then removed from the second embryo

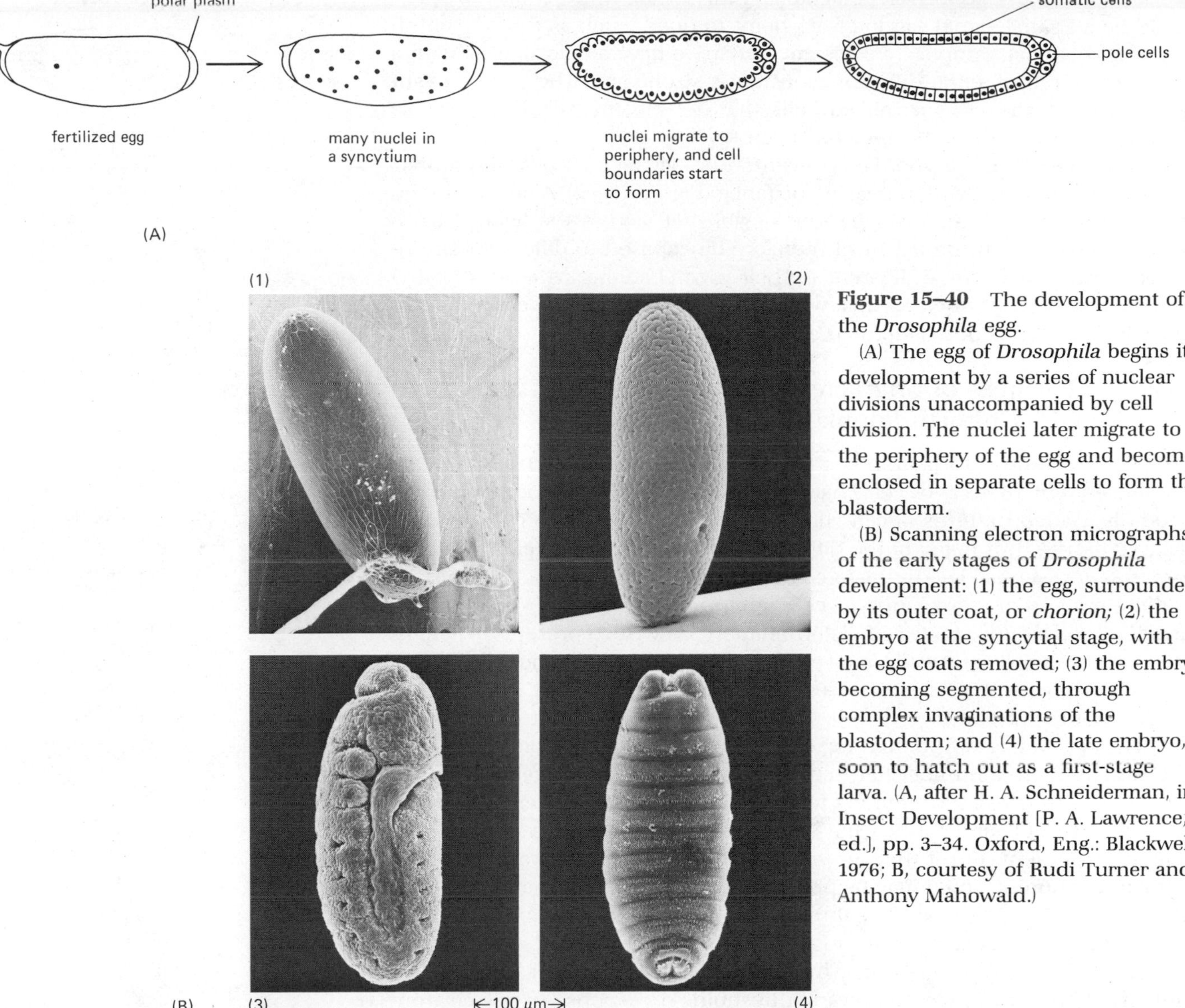

Figure 15–40 The development of the *Drosophila* egg.

(A) The egg of *Drosophila* begins its development by a series of nuclear divisions unaccompanied by cell division. The nuclei later migrate to the periphery of the egg and become enclosed in separate cells to form the blastoderm.

(B) Scanning electron micrographs of the early stages of *Drosophila* development: (1) the egg, surrounded by its outer coat, or *chorion;* (2) the embryo at the syncytial stage, with the egg coats removed; (3) the embryo becoming segmented, through complex invaginations of the blastoderm; and (4) the late embryo, soon to hatch out as a first-stage larva. (A, after H. A. Schneiderman, in Insect Development [P. A. Lawrence, ed.], pp. 3–34. Oxford, Eng.: Blackwell, 1976; B, courtesy of Rudi Turner and Anthony Mahowald.)

and injected into a third embryo of a different genotype. Some of the progeny of the third fly turn out to have the genotype of the injected cells rather than that of the natural germ cells of the parent fly. Thus polar plasm injected into the anterior end of an egg causes nuclei that would otherwise have belonged to somatic cells to develop instead as nuclei of germ cells, which are capable of development into viable oocytes and sperm. There is some evidence that amphibian germ cells may be determined in a similar way.

Cell Character Is Controlled by Spatial Cues[32]

The localization of determinants in the egg cannot be precise enough to specify directly more than a few early distinctions within the embryo, and in some species, as in mammals, it may specify none at all. Sooner or later, as the cells proliferate, regular patterns of cell character difference must be generated by other means, within fields of cells that are at first alike. For that to happen, cells in different locations must experience different influences, providing them with some sort of cue as to their position.

The most straightforward mechanism for setting up a pattern in space would be a spatial signal varying from point to point, such as the concentration of a chemical. Suppose, for instance, that we have a field of cells with a localized source of some diffusible substance at one site. The source might be a group of already specialized cells that secrete the substance. Let us assume that, once the substance has been secreted, it is slowly degraded as it diffuses through the tissue. The concentration will be high near the source, gradually decreasing with increasing distance (Figure 15–42). A concentration gradient of the substance will thus be established across the field. Cells in different positions in the field will therefore be exposed to different concentrations and may become different in their own character as a result. A hypothetical substance such as this, whose concentration is "read" by cells to determine their position, is termed a **morphogen.**

Sharp Differences of Character Emerge Gradually in an Initially Uniform Population[35]

If the concentration gradient in such a case is smooth, one might expect that the consequent pattern of cell characters should also be smoothly graded. Smoothly graded patterns of cell character on a small scale do indeed occur in some tissues. But many of the differentiations of greatest interest in development are discrete. The ultimate cell types are sharply distinct; there is no graded series of mature kinds of cells intermediate between cartilage and muscle, for example. States of determination likewise represent switchlike choices between discrete alternatives. How can sharp distinctions arise in a population of cells that are initially uniform and in response to a spatial signal that is smoothly graded?

The answer is not known for certain, but a possible explanation is as follows. There are a number of discrete alternative self-perpetuating states in which a cell may exist in the long term. An array of cells that are all alike to start with will all tend toward the same ultimate state if all experience the same environment. But if they are exposed to some graded spatial variation in the environment, some may be pushed toward one self-perpetuating state, others to another. There will be a threshold level of any relevant environmental variable such that cells exposed to levels above the threshold are launched on one course of development, while those exposed to lower levels follow another. There may indeed be several thresholds of response to one environmental variable, so that a single variable may control the pattern of several different choices. Once a cell is well launched toward a given self-perpetuating state, it will persist on that course even in the absence of the environmental influence that initially controlled the choice. In this way, transient, position-dependent influences can have effects that are "remembered" as discrete choices of cell state and thereby define the spatial pattern of determination.

Spatial influences may be said to supply the cell with **positional information.** The consequent choice of cell state represents the cell's record of the positional information supplied. This record, registered as an intrinsic feature of the cell itself, may be called its **positional value.**

Positional Information Is Refined by Installments[36]

The concepts of positional information and positional value help to clarify the analysis of pattern formation in many different systems. Several important general principles are neatly illustrated by a simple experiment on the developing limbs of a chick. In the chick embryo, the leg and the wing originate at about the same time in the form of small tongue-shaped buds projecting from the flank. The cells in the two pairs of limb buds appear similar and are undifferentiated at first, showing no hint of the subsequent skeletal pattern.

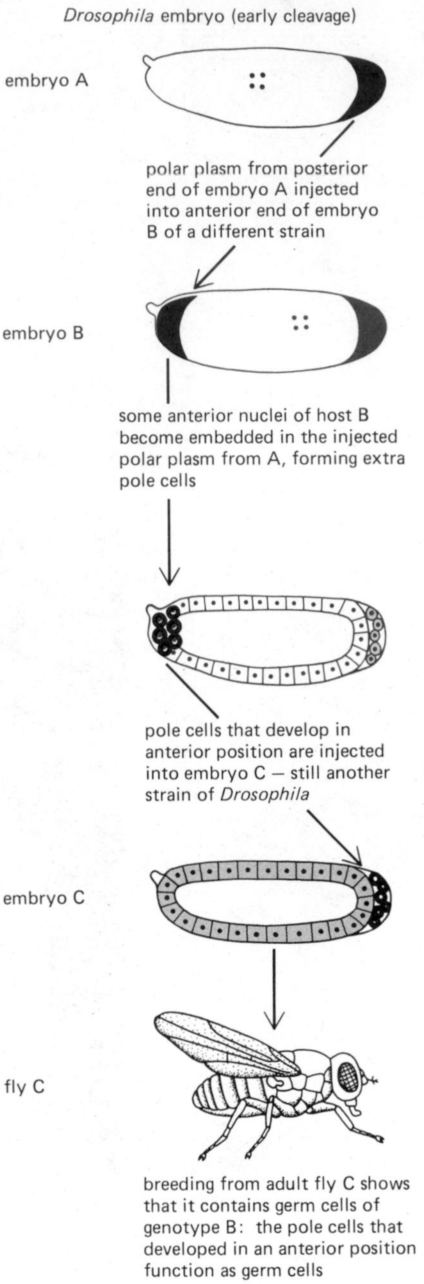

Figure 15–41 Scheme of experiment to show that the polar plasm in the *Drosophila* egg is the determinant of germ-cell character.

A small block of undifferentiated tissue at the base of the leg bud, from the region that would normally give rise to part of the thigh, can be cut out and grafted into the tip of the wing bud. Developing there, the graft forms not the appropriate part of the wing, nor indeed a misplaced piece of thigh tissue, but a toe (Figure 15–43). This experiment shows, first of all, that early leg cells are intrinsically different from wing cells: their positional information determines them as leg even though leg and wing will both consist ultimately of the same few differentiated types of cell. Second, the experiment shows that the grafted cells, though already determined as leg, will still respond to cues indicating their position along the limb axis, so as to form the tip of a leg instead of the base of a leg. We can thus conclude that the full specification of position in vertebrates is not supplied all at once but is built up from a series of items of positional information registered in the cell memory at different times. As in the determination of compartments in insect imaginal discs, the final cell state is arrived at by a sequence of decisions.

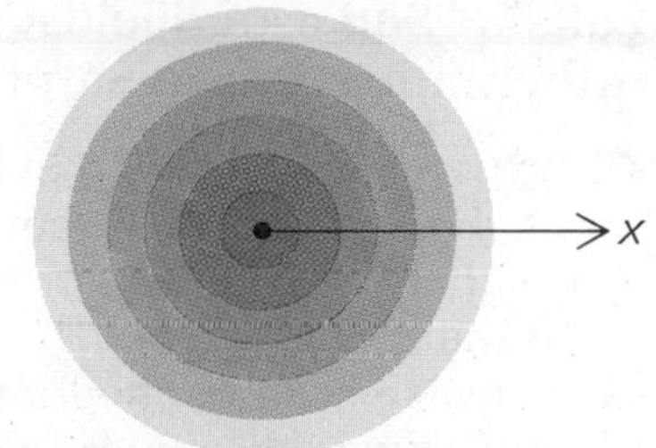

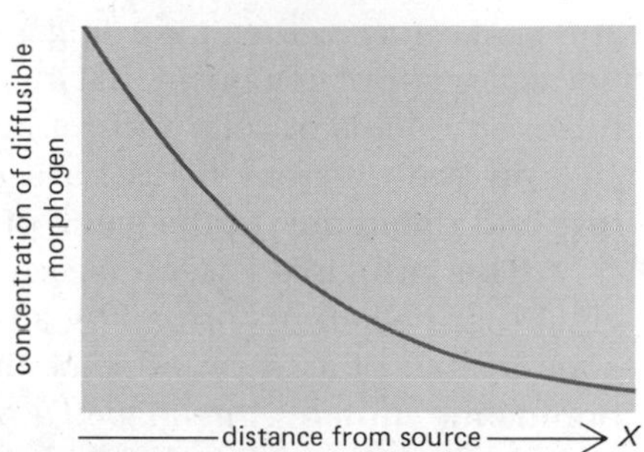

Figure 15–42 If a substance is produced at a point source and is degraded as it diffuses from that point, a concentration gradient results with a maximum at the source. The substance can serve as a morphogen, whose local concentration controls the behavior of cells according to their distance from the source.

Nonequivalence: Cells That Ultimately Differentiate in the Same Way Can Have Different Positional Information[32,37]

The development of the limbs, and of many other organs such as teeth or vertebrae, involves relatively few modes of differentiation. For example, in the case of the limbs, the chief cell types are those of muscle, cartilage, bone, and loose connective tissue. But these few differentiated types are arranged in a complex spatial pattern. The forelimb differs from the hind limb, not because it gives rise to different types of tissue, but because it gives rise to a different spatial arrangement of tissues. As the transplantation experiments show, the intrinsic differences between limbs with respect to the pattern that will be generated are determined long before differentiation begins. The cells of the forelimb bud and the hind-limb bud, though they will all give rise to the same range of differentiated types of cell, are *nonequivalent:* they have different positional values. As we have seen (p. 848), the same is true of somites. The difference between forelimb and hind limb, or between one somite and another, is not created by a signal that simply directs some cells to become cartilage, others bone, others muscle, and so on. Rather, it is that the cells are first supplied with an indication of where they lie on the antero-posterior axis of the body, and the complexity of the pattern they later form arises from the way they interpret this information and use it in conjunction with subsequent positional cues.

If, as we would suggest, cells can retain their positional values even after they have differentiated, it follows that cells of one and the same differentiated type, such as cartilage, can be nonequivalent. This implies a much greater variety among the cells of the body than histologists conventionally recognize. If only we could analyze the molecular composition of cells in sufficient detail and examine the states of activity of all their control genes, we should surely be able to distinguish in chemical terms far more than the 200-odd categories into which the cells of a vertebrate are traditionally classified.

There are several pieces of evidence for the existence of such subtle differences between cells of the same differentiated type in different parts of the body. For example, the cartilage cells that form the rudiments of different bones grow at different rates; they do so even if the rudiments are kept in culture, isolated from one another and from the rest of the embryo. (The experiment has been done with the rudiments of the tibia and the fibula from the lower leg of the chick embryo, which at first are of similar size but develop into bones of very different sizes.) The cells of the skin and of the nervous system provide other important examples of nonequivalence, to be discussed later in this chapter (p. 871) and in Chapter 18.

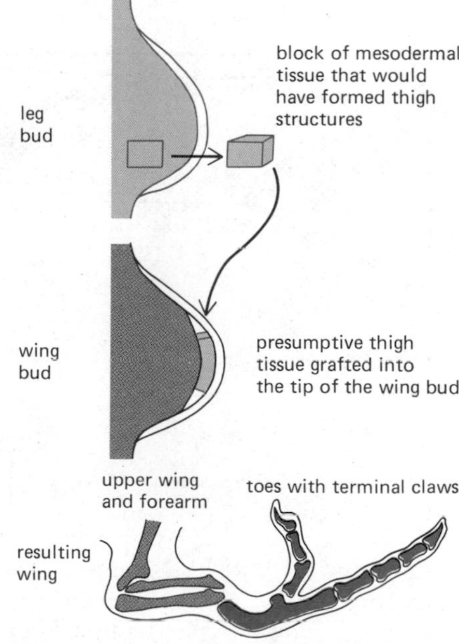

Figure 15–43 Prospective thigh tissue grafted into the tip of a chick wing bud forms toes. (After J. W. Saunders, et al., *Dev. Biol.* 1:281–301, 1959.)

Cells in Separate Fields May Be Supplied with Positional Information in the Same Way But Interpret It Differently[32,38]

The experiment illustrated in Figure 15–43 serves to emphasize a further point about the way in which complex positional specifications are built up. It shows that the cues that provide cells with information about their position along the axis of the vertebrate limb are effectively the same in the leg and the wing. Leg cells grafted into the tip of the wing bud can correctly "read" the indications that their position is distal and that they should therefore make digits. But they interpret that positional information in their own way and make toes rather than fingers. With the help of cell memory, the same device for supplying positional information can be used repeatedly in different regions or fields of cells and yet produce in each field a different pattern: the cells in the different fields have different histories, and they consequently respond differently to the same positional cues.

This principle is seen clearly in *Drosophila.* A group of *Drosophila* imaginal disc cells transformed by a homoeotic mutation shows behavior analogous to that of a group of chick limb cells grafted to a strange position. The homoeotic mutant *Antennapedia* provides an example. In this type of fly, the transformation of antenna into leg is usually only partial, so that it is only a patch of antennal tissue, and not the whole antenna, that is converted to the appearance of leg. The location of the transformed tissue is variable. In cases where it lies at the tip of the antenna, however, it regularly forms structures appropriate to the tip of a leg; if it lies at the base of the antenna, it regularly forms structures appropriate to the base of a leg; and so on, for all the inter-

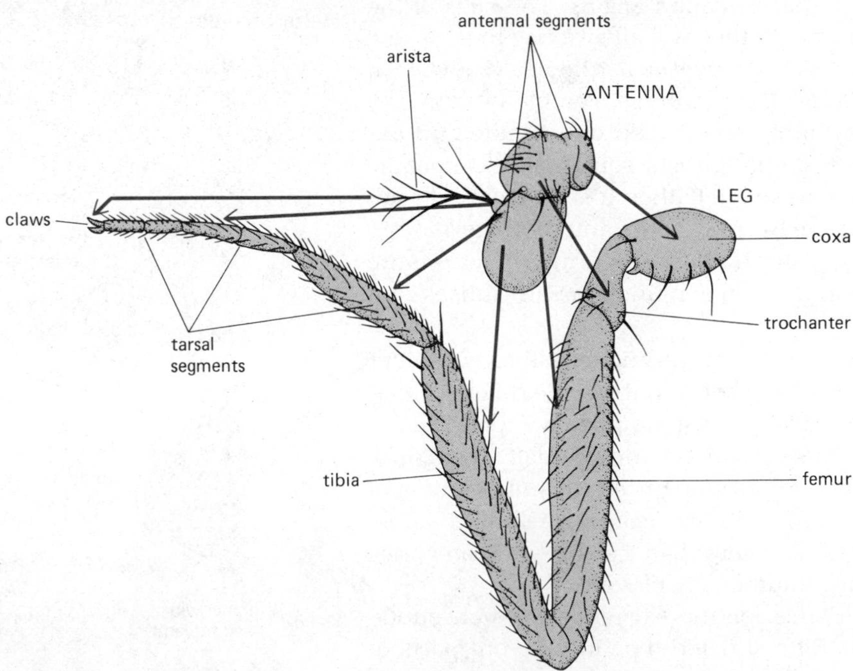

Figure 15–44 In the mutant *Antennapedia,* parts of the antenna are homoeotically transformed into parts of the leg. The transformed cells form the parts of the leg that correspond to their positions in the antenna. The colored arrows show the correspondences between leg and antenna parts. (After J. H. Postlethwait and H. A. Schneiderman, *Dev. Biol.* 25:606–640, 1971.)

mediate locations (Figure 15–44). Thus the cues that indicate position in the insect appendage seem to be equally effective for cells with an antennal character and for cells with a leg character, but each of those two types of cells interprets the cues in its own way.

Since Embryonic Fields Are Small, Gross Features of the Adult Must Be Determined Early[39]

The mechanisms in animal embryos for supplying positional information, whatever they are, appear to have a rather short range: they are generally found to act only over small fields of the order of a millimeter long (or about 100 cell diameters) or less. There is clearly a limit to the amount of detail that can be defined in so small a space. This is the fundamental reason why the final positional specification of a cell has to be built up as a composite of a sequence of items of positional information registered at different times. It also makes determination involving cell memory crucial for the development of large complex animals. The distinction between head and tail has to be established when the rudiments of the head and the tail are no more than about a millimeter apart. The circumstances that gave rise to that distinction are ancient history by the time the animal is a centimeter or a meter long. And if the distinction between head and tail is to be maintained, it must be through cell memory.

Thus the gross plan of the body is specified early, and successive levels of detail are filled in later as the rudiment of each part grows to a size at which mechanisms for supplying additional positional information can conveniently act.

Summary

The eggs of many animals contain localized cytoplasmic determinants, which are inherited by different blastomeres in the early cleavage divisions and govern their different modes of development. However, in later development, differences of cell character are usually produced by the exposure of initially similar cells to different external influences, which vary with the cells' positions in the embryo. In the simplest case, a graded concentration of a diffusible substance may serve to control the character of cells according to their distance from the source of the substance. Discrete differences of cell character would correspond to thresholds in the response to the substance. The full positional specification of a cell may be built up from a combination of items of positional information supplied at different times. For example, early cells in the forelimb and hind-limb regions of a vertebrate embryo acquire different positional values, making forelimb and hind-limb cells nonequivalent in their intrinsic character, even before the detailed pattern of cell differentiation has been determined. This detailed pattern of cell differentiation is subsequently specified by means of information as to position within the organ. Homologous organs, such as the leg and the wing of a chick or the leg and the antenna of a fly, appear to use the same system for supplying positional information. Because of cell memory, the cells in these different fields interpret the same positional information differently according to their different prior histories.

Since the fields over which the spatial signals act are generally small, the large-scale features of the adult organism are generally determined early, when the embryo is small, and the more detailed features are specified later. The different positional values of cells in different regions of the body control their different programs of subsequent growth as well as their specific patterns of differentiation.

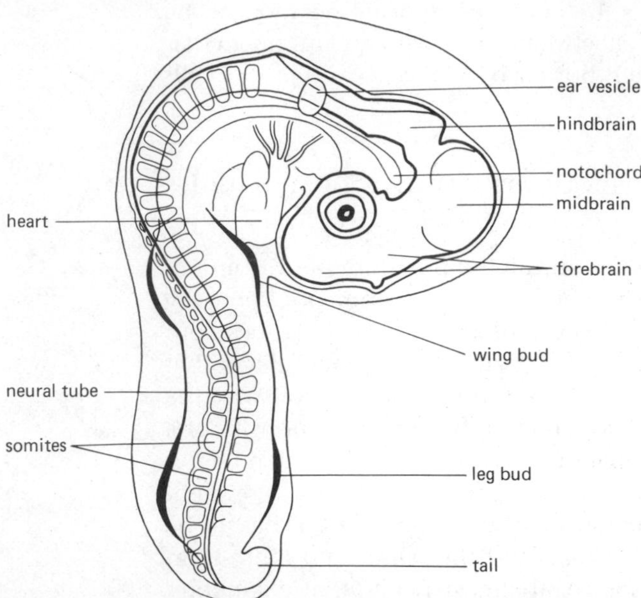

Figure 15–45 A chick embryo after three days of incubation, illustrating the positions of the early limb buds. (After W. H. Freeman and B. Bracegirdle, An Atlas of Embryology. London: Heinemann, 1967.)

Positional Information in Limb Development

We now consider in detail some ways in which positional information may be supplied, taking our examples from limb development—first, the embryonic development of limbs in the chick, and then, for a different perspective, the regeneration of limbs in the cockroach.

The Developing Chick Limb Can Be Analyzed in Terms of a Three-dimensional System of Coordinates[32,40]

Each limb bud of a chick embryo (Figure 15–45) has initially a very simple structure: a thin epithelial jacket of ectoderm is filled with a tissue of apparently uniform and undifferentiated cells of mesodermal origin called **mesenchyme** (Figure 15–46). From this mesenchyme the musculature, skeleton, tendons, dermis, and other connective tissues eventually develop. The chief landmark in the limb bud at early stages is a thickening in the ectoderm, called the **apical ectodermal ridge,** which forms a rim around the tip of the bud (Figure 15–47). Three days later, a hazy outline of the entire skeleton can be seen upon staining for cartilage, and three days after that, the skeleton stands out sharp and precise (Figure 15–48). How is this intricate and highly organized pattern of cells generated?

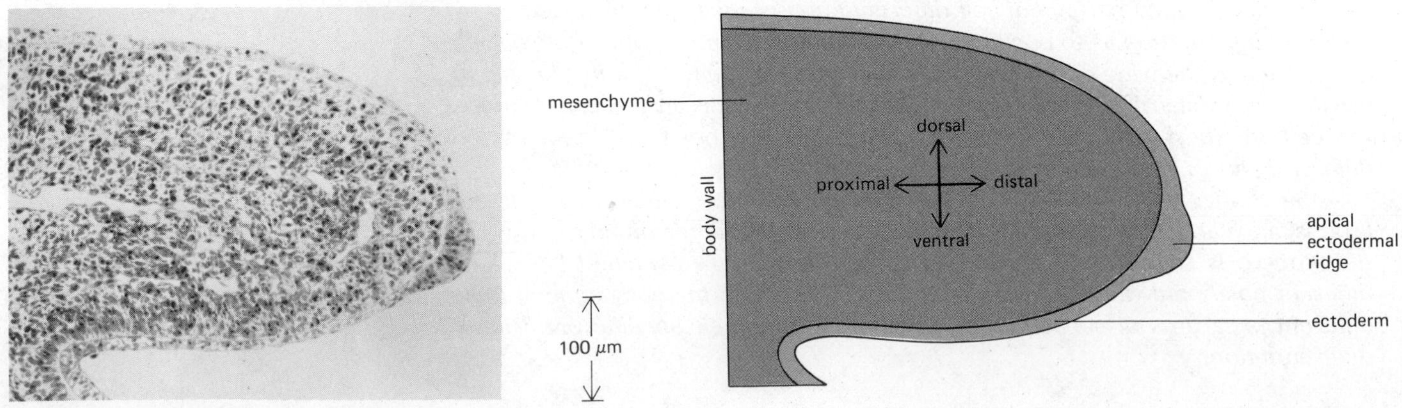

Figure 15-46 A chick wing bud after 3½ of incubation, sectioned in a plane containing the proximo-distal and dorso-ventral axes of the bud (that is, sectioned in a plane at right angles to the trunk of the embryo).

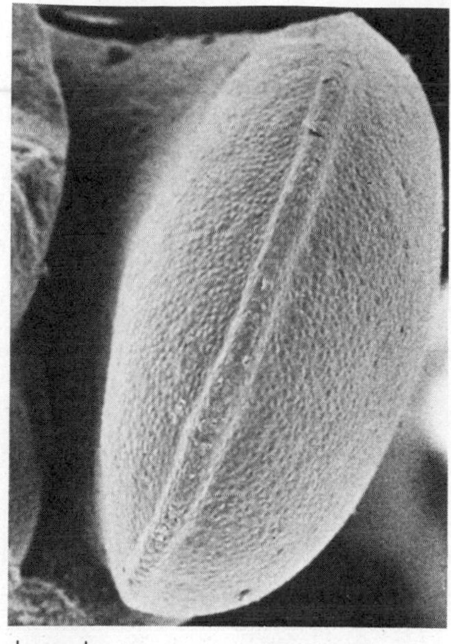

Figure 15–47 Scanning electron micrograph of a chick leg bud after four days of incubation. The bud is viewed end-on, showing the apical ectodermal ridge. The appearance of a wing bud is almost identical at this stage. (Courtesy of J. R. Hinchcliffe and D. S. Dawd, from J. R. Hinchcliffe and D. R. Johnson, The Development of the Vertebrate Limb. Oxford, Eng.: Oxford University Press, 1980.)

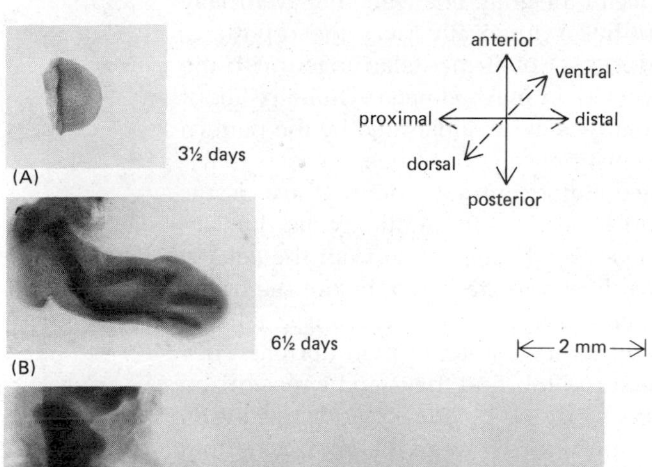

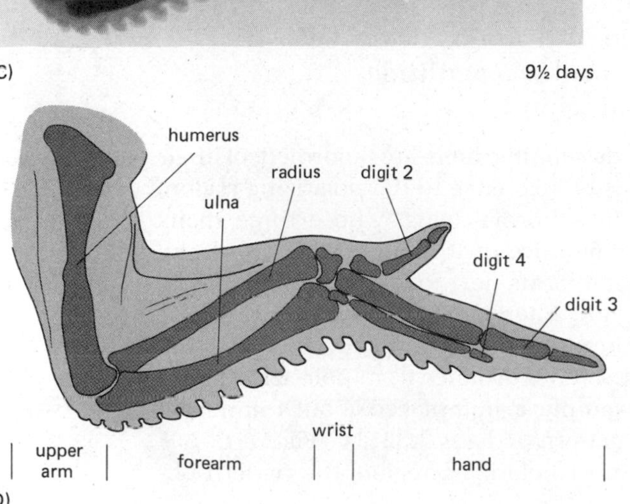

Figure 15–48 The normal developing chick wing, viewed from its dorsal aspect. (A) At 3½ days of incubation. (B) At 6½ days of incubation. (C) At 9½ days of incubation. (D) A drawing of the wing photographed in (C).

It is convenient to describe the limb by reference to three axes: the proximo-distal axis, from base to tip; the antero-posterior axis, from "thumb" to "little finger"; and the dorso-ventral axis, at right angles to these. For each axis there seems to be a distinct mechanism to specify a component of positional information. Cells of the mesenchyme may thus be assigned positional values in a three-dimensional system of coordinates, and these positional values are thought then to govern the choice of differentiated state in each region so as to define the pattern of the skeleton and of the other tissues too. We shall concentrate on the antero-posterior and proximo-distal axes since these are the best understood.

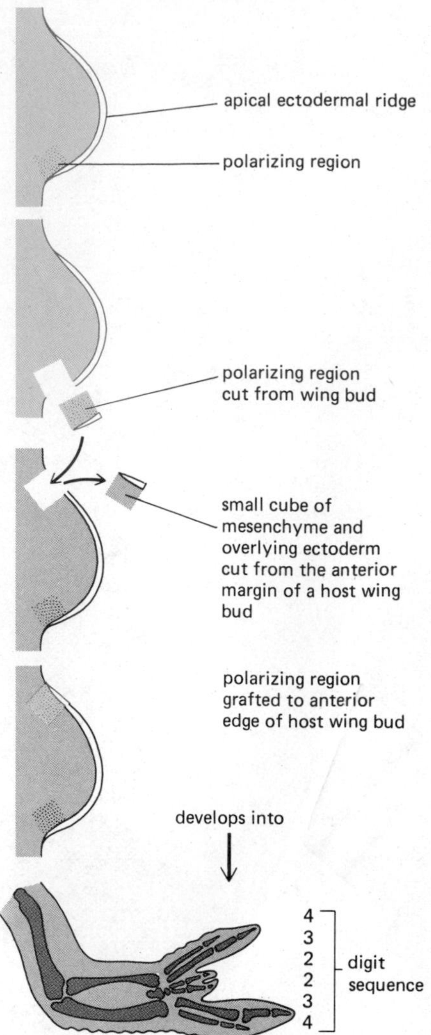

Figure 15–49 A graft of polarizing tissue causes a mirror-image duplication of the pattern of the host wing.

The Cells of the Polarizing Region Control the Antero-Posterior Patterning of Adjacent Tissue[41]

Transplantation experiments have revealed the existence of a small group of mesenchyme cells at the posterior margin of the developing limb that have a particularly important effect on the adjacent limb tissues. The effect is best seen by transplanting the cells from the posterior margin to an anterior position in another wing bud (Figure 15–49). Within a day, the host bud is found to have grown wider under the influence of the graft. The wing that eventually results, however, is not only bigger but has a drastically reorganized pattern: its skeleton is duplicated, having a second set of elements arranged with the sequence "thumb" to "little finger" reversed, in mirror-image symmetry about the midline of the limb. This is particularly well demonstrated by the pattern of digits, since each digit has a clearly distinctive appearance.

The duplicate elements are formed almost entirely from the host tissue, and not from the graft itself. This can be proved by grafting cells that are marked in one way or another and so can be distinguished from the cells of the host. If the donor embryo is given a dose of x-irradiation before the transplant operation, the grafted cells are crippled by radiation damage so that they cannot divide, though they become somewhat larger than normal. They nevertheless cause a duplication of the host limb, and they can be recognized as a small group of abnormally large cells making practically no contribution to the structures whose formation they provoke. From all this it appears that a graft of tissue from the posterior margin of a donor limb bud specifies a new pattern of structures in the host limb, ordered in a definite antero-posterior sequence in relation to itself. The duplication is manifest both in the skeleton and in the pattern of all the other tissues of the host limb. The region from which the graft is taken is called the **zone of polarizing activity** or **polarizing region.**

Positional Information Along the Antero-Posterior Axis May Be Supplied by Gradations in the Magnitude of a Signal from the Polarizing-Region Cells[42]

It seems as though the cells of the developing limb are informed of their position along the antero-posterior axis by reference to the polarizing region, according to their distance from it. In a normal wing (whose three digits correspond to the three middle digits of a five-digit hand and so are designated by the numbers 2, 3, and 4), digit 4 forms next to the polarizing region, digit 3 farther away, and digit 2 farther still. When additional polarizing regions are grafted into different positions around the rim of the bud, the pattern of digits obtained conforms with that expected if distance from polarizing tissue determines the pattern of digits. For example, a graft placed about 1 mm from the host polarizing region produces a pattern of digits "432234" (Figure 15–50). If the graft is now moved nearer the host polarizing region, the pattern ob-

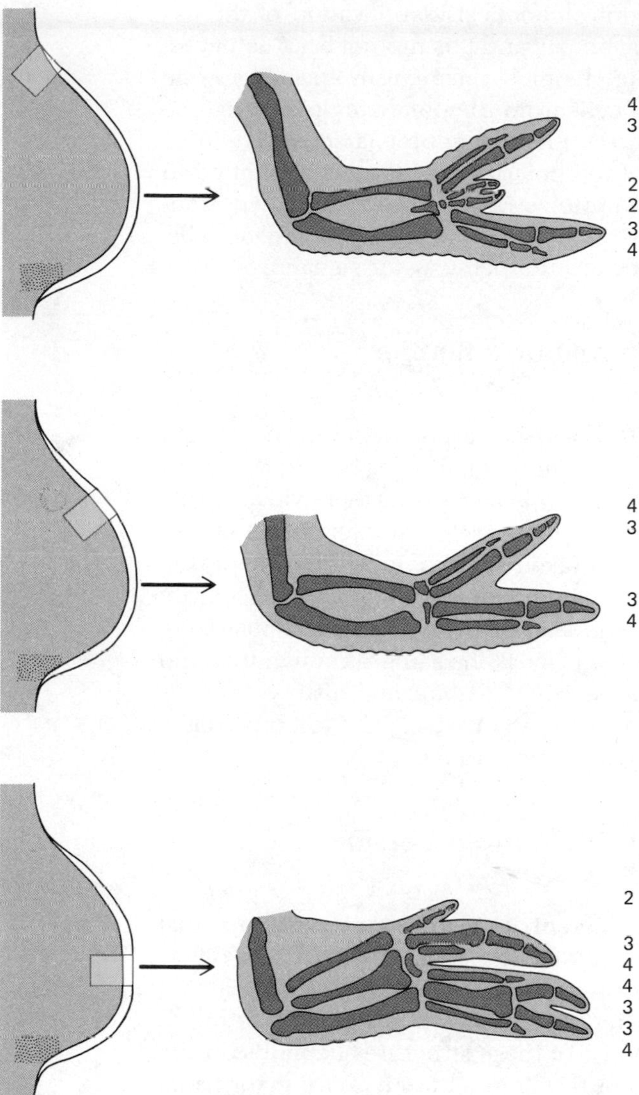

Figure 15–50 Distance from the grafted polarizing region determines the type of structure formed by the cells of the host wing bud.

tained is "43234" or even "4334." If the polarizing region is grafted to the center of the limb bud rim, the pattern of the digits between the graft and host polarizing regions is now "4334" or "434"; in addition, a limb with the normal sequence of digits develops from the part of the bud anterior to the graft. This shows that the polarizing region can signal in both directions across the antero-posterior axis of the limb. An obvious suggestion is that the polarizing region might act by producing a substance that diffuses across the limb, setting up a concentration gradient. As outlined earlier (p. 853), cells might measure their distance from the polarizing region by the concentration of this substance that they experience. Thus digit 4, which is nearest the polarizing region, would be specified by the highest concentration; digit 2, which is farthest away, by a low concentration; and digit 3, which lies between digit 4 and digit 2, by an intermediate concentration. The graded nature of the signal from the cells of the polarizing region is demonstrated directly by grafting these cells in small numbers. On average, a graft of 30 cells will cause formation of an extra digit 2. With about 80 polarizing cells, a digit 3 will be produced, and with about 130, a digit 4.

Proof of this hypothesis will require chemical identification of the hypothetical signaling substance, or morphogen, and this has not been achieved. As yet, there are very few clues to the chemical mechanism apart from one recent finding: the action of a graft of cells from the polarizing region can be mimicked by an implant of an inert carrier material impregnated with retinoic acid. It is not known how the effect of the polarizing region is transmitted to the adjacent limb tissue. Cell-cell communication could occur via gap junctions, but although such junctions have been seen between mesenchyme cells, there is no evidence to prove that they are the pathway for signaling.

The Polarizing Region of a Mammal or a Reptile Is Effective in the Chick Also[43]

The chick leg bud, like the wing bud, possesses a polarizing region, which causes formation of extra toes when grafted into another leg bud, and of extra wing digits when grafted into a wing bud. This effect is further evidence that the two limbs use the same system to supply positional information (see p. 854). Furthermore, small pieces of tissue taken from the posterior margins of pheasant, mouse, human, and even turtle limb buds exhibit polarizing activity in the chick wing bud. These findings show that the signal from the polarizing regions of different limbs and different species is the same, although the limb cells interpret the signal according to their own genome and history. No matter what the source of the implanted polarizing tissue, the additional digits formed from chick wing mesenchyme are chick wing digits.

The Parts of the Limb Are Laid Down in Succession Along the Proximo-Distal Axis[44]

As the limb bud grows and elongates, differentiation starts in its most proximal part, while the cells at its tip remain undifferentiated. The cells that differentiate first form the most proximal structures of the limb, and the remaining parts are then laid down, in succession, by proliferation from the undifferentiated cells at the tip. The laying down of these structures depends on the thickened rim of ectoderm at the tip of the limb bud, the apical ectodermal ridge mentioned earlier (Figure 15–51). If the ridge is removed, the distal structures yet to be laid down are not formed. Thus, excising the ridge from a wing bud at an early stage produces a wing consisting of an upper arm only; the same operation at a slightly later stage produces a wing with upper arm and forearm, but no hand (Figure 15–51); and so on.

The Apical Ectodermal Ridge Delimits the Special Region of Mesenchyme from Which Successive Distal Parts Develop, But It Does Not Instruct the Mesenchyme as to Which Parts It Should Form[45]

Removal of the apical ridge halts the laying down of structures along the proximo-distal axis. The importance of the ridge to limb growth can be further demonstrated by grafting an additional apical ridge to the dorsal surface of a limb bud. A secondary outgrowth develops, with elements arranged in correct proximo-distal sequence. The ridge defines the site of outgrowth; it maintains a **progress zone,** an undifferentiated population of mesenchyme cells beneath it from which successive parts of the limb are progressively laid down. But does the ridge also specify what those parts shall be, perhaps by issuing an instruction "be upper arm" at an early stage and an instruction "be hand" somewhat later?

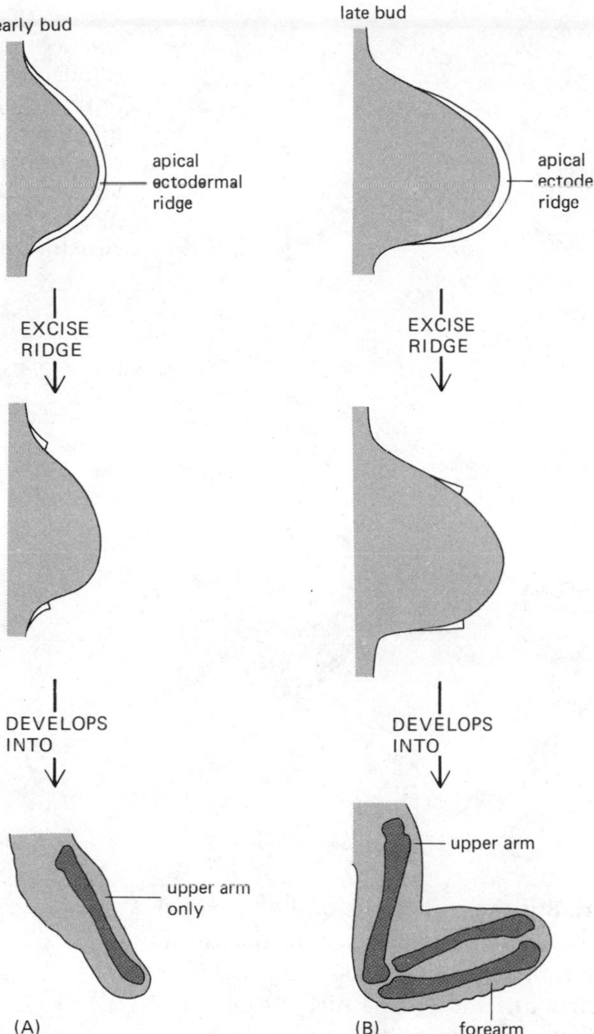

Figure 15–51 If the apical ectodermal ridge is cut off, the distal parts of the limb fail to develop from the underlying mesenchyme. (A) If the operation is done early, the level of truncation of the resulting limb is proximal. (B) If the operation is done later, the truncation is more distal.

This question can be investigated by separating the mesenchyme tissue from the ectodermal casing bordered by the ridge, and then recombining these components from limbs of different ages (Figure 15–52). If the mesenchyme of a late bud is capped with the ectodermal jacket of an early bud, that mesenchyme proceeds to develop its normal sequence of elements. If the mesenchyme of an early bud is stuffed into the ectodermal jacket of a late bud, the wing that develops has again the full normal sequence of parts along the proximo-distal axis. In short, the age of the ectoderm makes no difference to the character of the structures laid down in the mesenchyme. Therefore, the ridge cannot be telling the mesenchyme which parts it should form at any given instant; instead, it is merely marking out the site of the progress zone and telling the mesenchyme to proceed with its own developmental program.

Positional Specification Along the Proximo-Distal Axis Depends on the Amount of Time Spent in the Progress Zone[46]

These experiments show that the character of the limb segment to be laid down next in the proximal-to-distal progression must be determined by some mechanism intrinsic to the mesenchyme. But how? All the regions of the limb are composed of descendants of cells that were once close to the apical ridge,

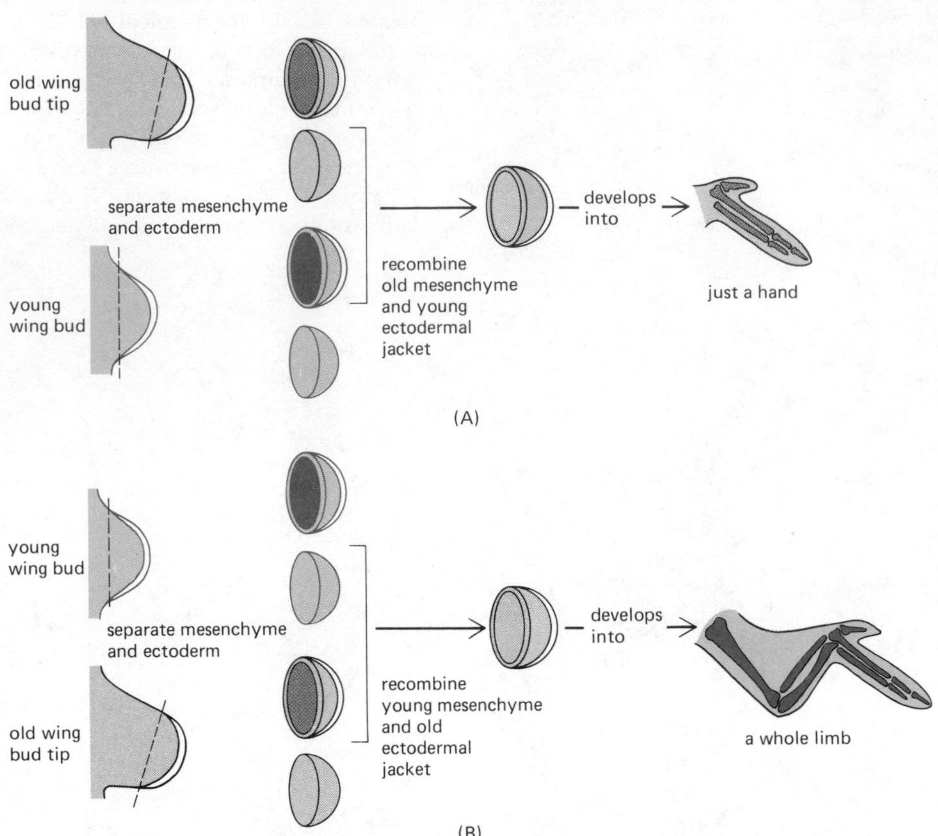

Figure 15–52 By replacing the ectoderm of an old wing bud with ectoderm from a young wing bud (A), or the ectoderm of a young bud with that from an old bud (B), it can be shown that the age of the ectoderm is immaterial. It is the age of the mesenchyme that determines what structures develop.

in the progress zone at the tip of the bud. But the rudiments of the proximal structures emerged from that zone early, and the rudiments of the distal structures emerged late. Time spent in the progress zone could therefore be the factor that determines which limb structures are to be made. Cells may acquire a proximal or distal positional value and so form structures of a proximal or a distal character, depending on whether they (or their ancestors) have spent a short or a long time in the progress zone.

According to this idea, structures already laid down in the stump do not determine which structure will form next, and if the progress zone is transplanted, it should behave independently. Indeed, when a progress zone from an early bud is grafted in place of the progress zone of a late bud, the resulting wing has repeated parts. For example, an upper arm and forearm may develop from the stump of the late bud, followed in sequence by upper arm, forearm, and hand from the progress zone of the early bud grafted onto it (Figure 15–53). Similarly, if the progress zone from a late bud is grafted in place of

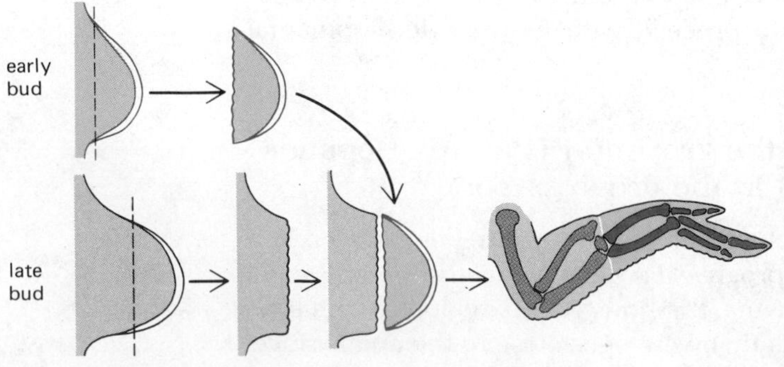

Figure 15–53 When the tip of an early wing bud is grafted in place of the tip of a late wing bud, a composite wing develops with two forearms in series along the proximodistal axis: both the host stump and the grafted tip behave more or less as they would have if they had developed separately in their normal situations.

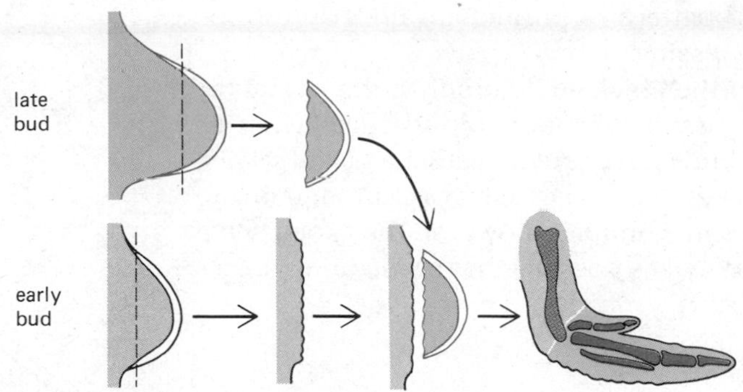

Figure 15–54 The converse of the operation shown in Figure 15–53. When the tip of a late wing bud is grafted in place of the tip of an early wing bud, the resulting limb consists of an upper arm and a hand, with no forearm: each part of the composite wing bud behaves more or less autonomously, as in the previous operation.

that of an early bud, the wing that develops has parts missing: no more than the upper arm, joined directly to a hand, may develop from the stump of the early bud combined with the progress zone of the late bud (Figure 15–54).

The progress zone appears to link pattern specification with limb growth: cells of the progress zone change their positional value as they proliferate. The rates of these two processes could be neatly coordinated if the cells measure the time spent in the progress zone by counting cell-division cycles and thereby fix their positional value. It turns out that the number of cell cycles required to lay down all the parts along the proximo-distal axis of the chick wing is about seven. This equals the number of segments of the wing, if the wrist, with its array of small carpal bones, is counted as the equivalent of two segments. Therefore, each segment might, so to speak, correspond to one tick of a cell-division clock by which positional value is determined. The quasi-repetitive aspect of the limb pattern, with its alternation of bones and joints, might thus reflect the operation of a *cyclical* timing mechanism.

It must be emphasized, however, that the cells in the limb bud do not divide in synchrony; neighboring cells do not clock exactly the same number of ticks while in the progress zone. Only the *average* number of division cycles clocked by the cells in any region is a smoothly graded function of the distance along the limb. Thus a tidy pattern could be defined only if neighboring cells interact with one another over short distances so as to smooth out their local differences before they differentiate.

Neighboring Mesenchyme Cells in Early Chick Limb Buds May Interact so as to Smooth Out Discontinuities in the Pattern of Positional Values[47]

Whether cells actually measure time by counting cell-division cycles is a matter for speculation. But there is evidence that "smoothing" interactions between mesenchyme cells do in fact occur. The evidence has come from experiments much like those illustrated in Figures 15–53 and 15–54, except that the grafts are between limb buds at approximately the same early stage rather than between limb buds at different stages. One such experiment involves amputating two early limb buds (one at a proximal level, close to its base, and the other at a more distal level) and then grafting the small distal portion removed from the second bud onto the short proximal stump of the first. This operation can be used, for example, to create an abnormally short composite bud that lacks the cells from which the forearm would normally develop. It seems that some adjustment occurs to produce a limb more nearly normal than would be expected if each part of the composite bud behaved autonomously. Thus, when the mesenchyme cells from different proximo-distal levels are juxtaposed, there is evidently an interaction between them that tends to smooth out the discontinuities in the pattern of positional values and to interpolate the values that are initially missing.

This sort of *regulation* is seen in the chick embryo only if the operation is done early, well before any of the tissues of the limb bud have begun to differentiate. When fragments of limb buds are grafted together at slightly later stages, practically no interaction is seen: each part behaves autonomously, according to its individual history. In some animals, however, the capacity for regulation persists throughout life and makes possible the regeneration of lost parts, as we shall see.

Some Limbs Can Regenerate[48]

In general, it is not to be expected that the body will generate a replacement for a large portion that has been cut off. The conditions after amputation will usually be quite different from those that led to the formation of the original part. But regeneration does occur in some cases. A much studied example is the regeneration of limbs in amphibians. If the leg of a newt or axolotl, for instance, is amputated at any distance from its base, the missing parts are replaced. A mound of apparently undifferentiated mesenchyme cells, called a *regeneration blastema* and covered by a cap of epidermis, forms on the end of the amputation stump. The blastema grows and differentiates to develop precisely the parts that should lie distal to the cut: a hand, for example, if the hand was cut off, and a whole forearm and hand if the whole forearm and hand were cut off. The blastema derives from the cells that lie close to the cut surface, and it is the intrinsic character of these cells that determines what structures are regenerated. It does not matter whether the blastema develops, as usual, on the end of the proximal stump of the limb or, by experimental manipulation, on the cut face of the *distal* detached portion of the amputated limb (see Figure 15–55). In either case, the cells of the blastema give rise to just those limb parts that normally lie *distal* to the level of the cut, even though in the latter instance this means creating a mirror-image duplicate of what is already present.

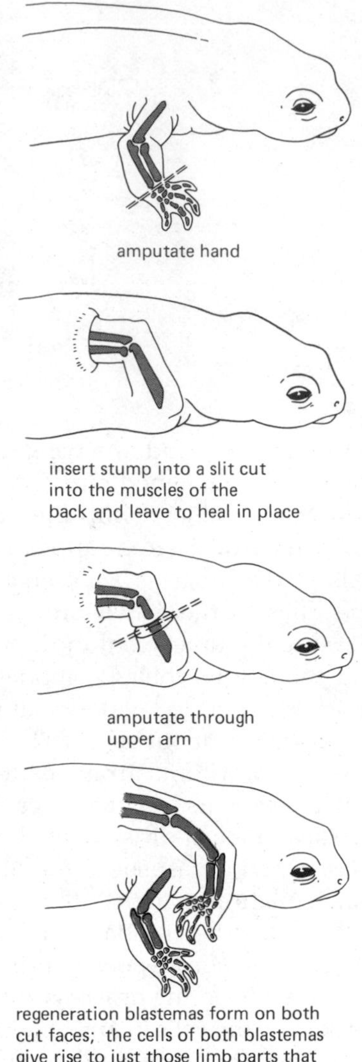

Figure 15–55 Scheme of an experiment on the salamander showing that the limb parts produced by a regeneration blastema depend on the level of the cut and not on the structures present in the limb stump. A forearm and a hand are generated from both the distal and the proximal portion of the original limb.

A remarkable disturbance is seen, however, when the blastema is treated with retinoic acid or related compounds: it will then frequently form *more* than the appropriate set of parts, giving, for example, after amputation through the wrist, a limb comprising original upper arm, original forearm, regenerated upper arm, regenerated forearm, and regenerated hand, in series. It seems that retinoic acid somehow alters the proximo-distal component of the positional value of the cells in the amphibian regeneration blastema, while in the embryonic chick limb bud it alters only the antero-posterior component! These effects represent important but still baffling clues to the biochemistry of pattern formation. They also highlight the uncertainties as to how far limb regeneration in amphibians follows the same rules as initial limb development or uses the same system of positional cues. The blastema of the regenerating amphibian limb does, however, look similar to the embryonic limb bud, and interactions between the mesenchyme and the ectoderm (that is, the epidermis) as before seem to be important.

Cockroach Legs Undergo Intercalary Regeneration[49]

The limbs of insects such as cockroaches also regenerate and are in some ways simpler to study. In these animals, it seems, a system of positional cues capable of controlling the development of the limb pattern is maintained after the limb has been formed. The differentiated cells are able to respond to those cues and to regenerate the pattern if it is disturbed. The workings of the pattern formation system can thus be studied by operations done long after the end of embryonic development. Experiments on regeneration in the leg of the juvenile cockroach have revealed some simple principles that may have a much more general application.

The experiments to be described involve the epidermal sheet of cells and cuticle that covers the cockroach and forms the externally visible parts of the limbs. This outer covering grows by successive molts, in which the juvenile cockroach sheds its old cuticle and lays down a new and larger cuticle in its place. The cuticle is secreted by the epidermal cells, which are arranged underneath in a sheet one cell layer thick. Positional values in the epidermal sheet of cells are displayed in the pattern of the overlying cuticle that they lay down; the effect of experimental manipulation on the patterning of epidermal cells is detected in the cuticle after the animal has molted. Regeneration can only be observed in juveniles, since fully mature adults do not grow or molt.

The cockroach leg consists of several segments called (in sequence from base to tip) coxa, trochanter, femur, tibia, and tarsus, the tarsus itself being a composite of several smaller segments and terminating in a pair of claws (Figure 15–56). If two legs are amputated through the tibia, say, but at different levels, the distal fragment of the one can be grafted onto the proximal stump of the other in such a way that the composite leg heals with the middle part of the tibia missing. Yet the leg that emerges after the animal has molted appears normal: the missing middle part has regenerated (Figure 15–57A). More surprising is the result of a variant of this operation. The tibia of one cockroach leg is cut through near the proximal end and that of another leg near the distal end. The large detached portion of the first leg is then stuck onto the large remaining stump of the second leg to give an excessively long leg with a middle part present in duplicate (Figure 15–57B). The animal is left to molt. The leg that results, far from being more nearly normal, is now even longer because a third middle part of a tibia has developed between the two already present! As shown in Figure 15–57B, the bristles on this freshly formed region point in the direction opposite to that of the bristles on the rest of the tibia.

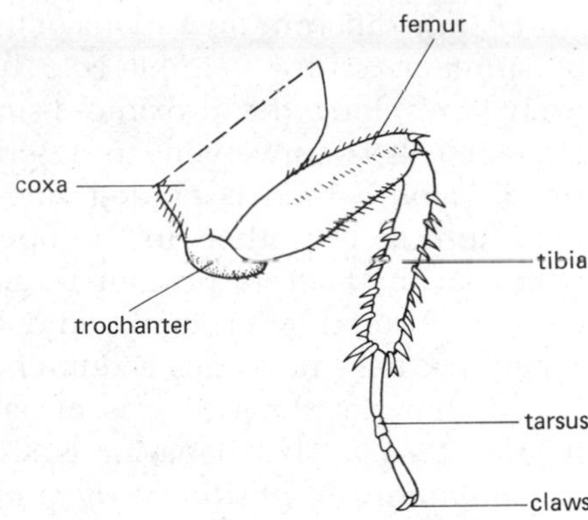

Figure 15–56 The cockroach leg. With each successive molt, the leg grows bigger but does not change its basic structure.

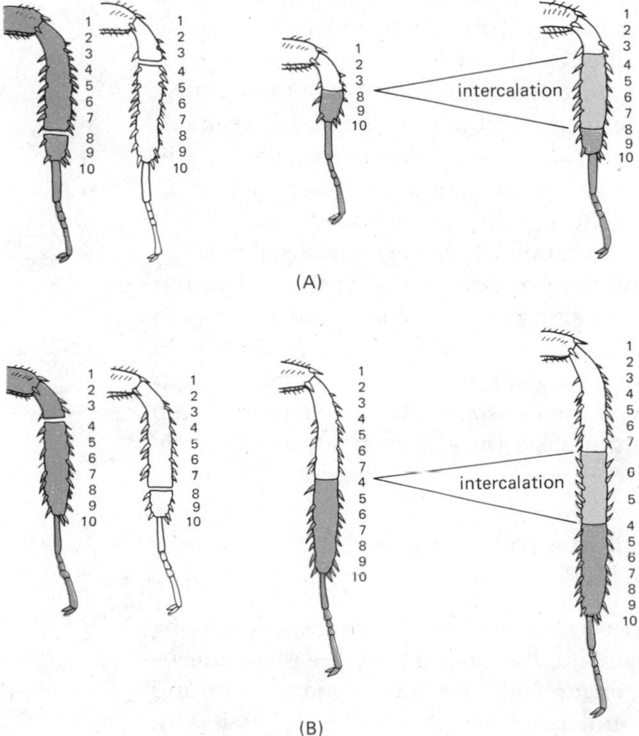

Figure 15–57 When mismatched portions of the cockroach tibia are grafted together, new tissue (*light color*) is intercalated to fill in the gap in the pattern of positional values (numbered from 1 to 10). In case (A), intercalation restores the missing part. In case (B), intercalation generates a third middle part of a tibia between the two middle parts already present. The bristles indicate the polarity of the intercalated tissue. In both cases, continuity is restored in the final pattern of positional values.

Many different operations of this type can be performed. The results can be summarized by a simple rule, based on the supposition that the cells at each level along the proximo-distal axis of the segment have a distinctive character. It is convenient to describe this character by a number, the positional value, which is graded smoothly from a maximum at one end to a minimum at the other. In the operations described above, epidermal cells with sharply different positional values are brought together. As a result, new cells are formed by proliferation of the epidermis in the neighborhood of the junction. These new cells acquire positional values smoothly interpolated between those of the two sets of cells that were brought into confrontation (Figure 15–57). This behavior is summed up in the **rule of intercalation:** *discontinuities of positional value provoke local growth, and the newly grown cells take on intermediate positional values so as to restore continuity in the pattern.* Growth ceases only when cells with all the missing positional values have been intercalated in the initial gap and have become spread out to the normal spatial separation from one another. This process as a whole is called **intercalary regeneration.**

The Same Pattern of Positional Values Is Repeated in Successive Segments of the Cockroach Leg[49]

As an alternative to grafting together parts of one leg segment, as described above, one can graft together cells from different segments. For example, one cockroach leg can be amputated through the tibia and another through the femur, and the detached distal part of the tibia can be placed on the proximal stump of the femur so as to make a hybrid limb segment that is part femur and part tibia. We can denote the successive levels of a normal femur by the symbols F_1, F_2, ... F_{10} (counting from the proximal to the distal end of the segment), and the corresponding levels of the normal tibia by the symbols T_1, T_2, ... T_{10}. If the two legs have been cut through at corresponding levels in the different segments—the host leg, say, between F_5 and F_6 and the donor leg between T_5 and T_6, so that the hybrid segment of the composite limb has the structure $F_1F_2F_3F_4F_5/T_6T_7T_8T_9T_{10}$—no new tissue is intercalated at the junction. If, however, the limbs have been cut through at noncorresponding levels, so that a hybrid segment is put together with the structure $F_1F_2F_3/T_8T_9T_{10}$, for example, intercalation occurs: the missing midsegment levels 4567 are regenerated from femoral or tibial cells, or both, to give a leg segment patterned according to the scheme $F_1F_2F_{34567}T_8T_9T_{10}$. This new segment is still part femur and part tibia, but it has all the positional values from 1 to 10 represented. Thus the rule of intercalation applies as before, except that in effect it pays no regard to the "letter" denoting which segment a cell belongs to, but only to the "number" denoting the positional value of the cell within its segment. We conclude that the same system of positional values serves to organize the epidermal cells in each segment. There is a close analogy here with the chick, where the same system for supplying positional information operates in both leg and wing, and with the *Drosophila* larva, where the same genes are used repeatedly to define the basic plan of each successive body segment.

Circumferential Intercalation Obeys the Same Rule as Proximo-Distal Intercalation[49]

The rule of intercalation correctly describes patterns of regeneration along the proximo-distal axis of the cockroach leg, but does it also describe patterns of regeneration around the circumference? There are several markers around the circumference—bristles, ridges, and coloring—that make it possible to

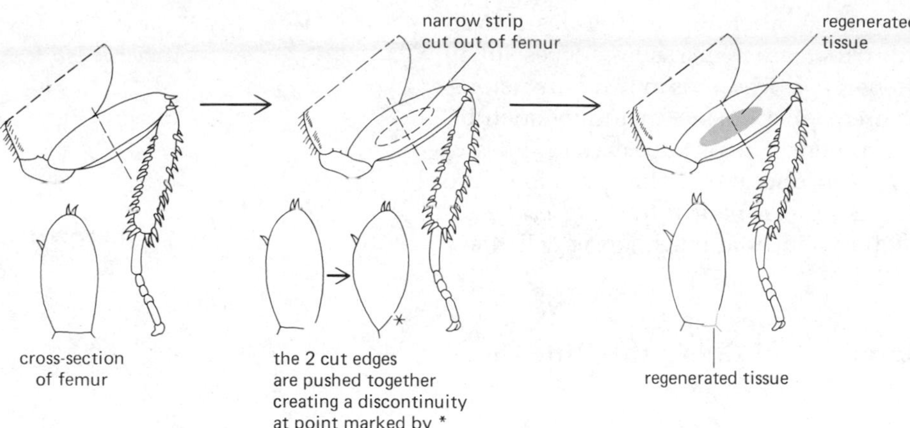

Figure 15–58 A gap in the pattern of circumferential positional values can be created by cutting out a long narrow strip. New tissue (*light color*) is intercalated to repair the defect and restore continuity in the pattern.

distinguish one region from another and help in the interpretation of experiments. If, for example, a long narrow strip of epidermis and cuticle is cut out, creating a small discontinuity in the circumferential pattern of positional values, the leg after molting is found to have regenerated the appropriate tissue to repair the defect (Figure 15–58). If this long narrow strip is cut out and immediately replaced with a strip taken from the same region of another leg, there is no regeneration, only healing. But if the replacement consists of a strip taken from a *different* region of another leg, two discontinuities are created initially in the circumferential pattern (Figure 15–59), and now, after molting, the circumference of the leg is found to have enlarged by generating, at each of the two sites where graft and host tissue meet, all of the tissue that should normally lie between the regions that abutted one another (Figure 15–59). In short, the behavior is apparently the same as that observed along the proximo-distal axis. The cells behave as though there is a distinct positional value for each position around the circumference, and they respond to rearrangements according to the same rule: discontinuities of positional value provoke local growth of the epidermis, the newly grown cells taking on intermediate positional values so as to restore continuity of the pattern.

Intercalation in the Epidermis Is a Two-dimensional Problem[50]

It follows from all this that just as the epidermal cells of a given leg segment form a two-dimensional sheet rolled up into a cylinder, so also their positional values must be represented by a two-dimensional variable. Thus each position in the normal limb epidermis is associated with a unique and distinctive positional value. In fact it is rather misleading to talk of the system in terms of each dimension considered separately—disturbances of one dimension of

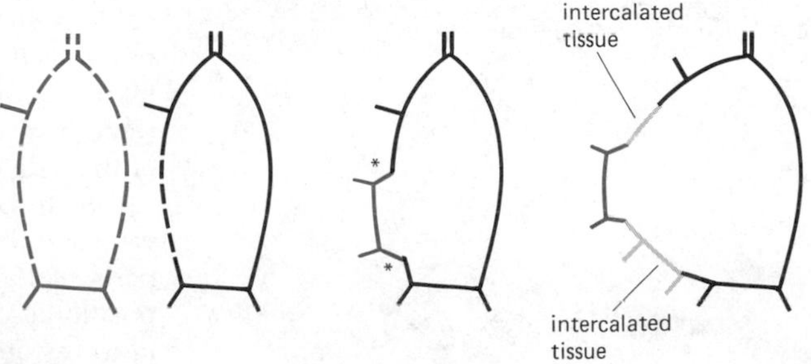

Figure 15–59 Cross-sections of the cockroach femur showing the effect of excising a portion of the circumference and grafting in its place a portion (*dark color*) from another region of the circumference of another leg. This creates two discontinuities (*) where normally separate parts of the circumference confront one another. At each of these, new tissue (*light color*) is intercalated to form the regions of the circumference that would normally intervene.

a pattern generally involve disturbances of the other. However, it has proved possible to devise a simple set of rules that are remarkably successful in describing intercalation and related regenerative processes in two dimensions. These rules correctly predict many different phenomena, including such bizarre effects as the production of supernumerary limbs after certain types of grafting procedure. In two dimensions, as in one, the central principles are contained in the rule of intercalation. The crucial part of the rule is the requirement of continuity, that is, the requirement that neighboring cells have closely similar positional values.

Regeneration of a Two-dimensional Patch Obeys the Rule of Intercalation[49,50]

Suppose that a two-dimensional patch is cut from the side of a cockroach leg. The observed result in such cases is that the excised part of the pattern is regenerated: the epidermal cells from around the perimeter of the wound move in to cover the area of the wound, and there they proliferate and reconstruct the parts of the limb that have been removed. This is just what the rule of intercalation predicts. For when the cells from around the perimeter move in to cover the wound area, they encounter cells from other parts of that perimeter, with different positional values, giving rise to discontinuities in the pattern of positional values. The discontinuities, by the rule of intercalation, stimulate growth, and the newly grown cells take on intermediate positional values that restore continuity in the pattern. In two dimensions, as in one, the requirement of continuity is enough to guarantee that everything that has been cut out will be regenerated.

Precisely the same argument can be used to show that the rule of intercalation accounts also for the regeneration of amputated legs (see Figure 15–60). After amputation, cells grow in from the perimeter of the wound to cover the cut surface, just as they do when a patch is excised from the *side* of a leg. Again, the cells reconstruct the parts of the epidermis that normally lie inside the perimeter of the wound. It simply happens in this case that the region of the wound includes the tip of the leg, and the regenerated epithelium is poked out into an elongated shape. The one essential novelty in the regeneration of the amputated leg is that the cells are no longer restricted to forming parts of the segment from which they originate but can reconstruct more distal segments that have been removed in their entirety. Why cells change their segment character in these circumstances, but not in the graft combinations described earlier (p. 866), is not known.

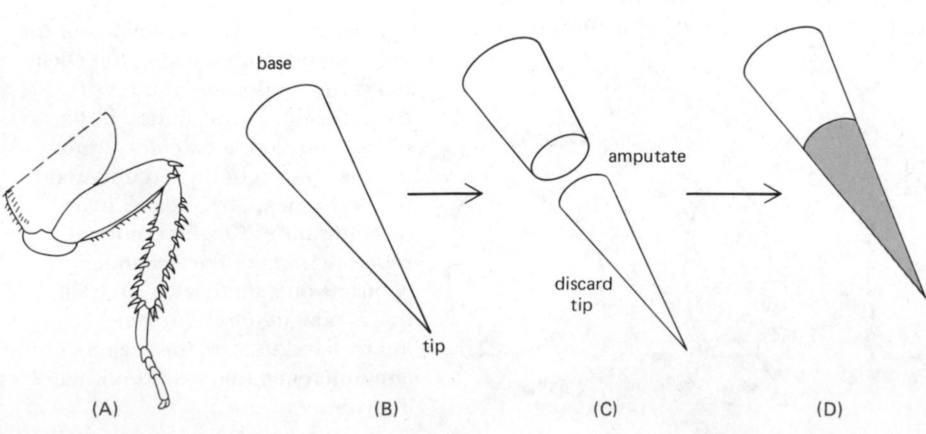

Figure 15–60 The epithelium of the cockroach leg (A) can be represented as the surface of a cone (B): the two structures are topologically equivalent. The epithelium forms a single continuous surface. The tip of the leg is just one point on this surface. Amputation of the leg (C) amounts to removal of the patch of epithelium that includes the tip. New tissue (*light color*) is intercalated to fill in the missing patch (D) in the same way as when a strip is cut out of the side of the leg (see Figure 15–58). In both cases the positional values of the cells forming the perimeter of the cut determine the positional values that must be filled in to restore continuity.

The Rule of Intercalation May Apply to Many Different Systems

The rule of intercalation represents something of a triumph for the concept of positional values. It predicts correctly a surprisingly wide range of regeneration phenomena, not only in cockroaches but also in amphibians, flies, and crustaceans. But what does the rule mean in terms of cellular mechanisms, and how is it related to the process of pattern formation in embryonic development?

The rule of intercalation, as presented here, consists of two essential clauses, which we may call the *growth rule* and the *continuity rule*. The first states that discontinuities in the pattern of positional values provoke local growth. The second states that the newly grown cells take on intermediate positional values that restore continuity in the pattern. One might expect that the continuity rule would apply to almost any system in which the positional values of the cells were smoothly variable and maintained by smoothly graded influences passing between susceptible neighboring cells. On the other hand, this rule would not be expected to apply to systems in which the behavior of each cell is fixed by cell memory according to past experiences and is not variable according to present circumstance. In the chick limb bud, for example, we have already noted that at early stages neighboring cells may interact so as to smooth out discontinuities in the pattern of positional values (p. 863), as though in obedience to the rule of intercalation. At later stages this ability is lost—the cells behave autonomously, and discontinuities created experimentally will persist.

The growth rule, stating that discontinuities in the pattern of positional values provoke local growth, can be viewed as a consequence of a slightly more general rule stating that any close crowding in the pattern of positional values provokes local growth. The latter rule might explain why intercalary regenerates grow to the correct final size, and it may be the basis of a simple mechanism for the control of growth in normal development: a pattern of positional values laid out at first on a small scale should grow according to the rule until a certain mature spacing is attained between the cells with the different positional values. This might, for example, explain how the sizes of compartments in *Drosophila* are controlled (p. 847).

Summary

In the developing chick limb, a combination of different mechanisms supplies the mesenchyme cells with three-dimensional positional information. For the antero-posterior axis, the pattern appears to be controlled by a signal whose magnitude depends on the distance from the polarizing region at the posterior margin of the limb bud: when a polarizing region is grafted into an anterior position, it produces a mirror-image duplication of the antero-posterior pattern. The structures along the proximo-distal axis are laid down in succession under the influence of the apical ectodermal ridge: when the ridge is cut off, the distal parts of the limb fail to develop. The character of the cells at the different levels along this axis, as manifest in the types of skeletal element that they form, appears to depend on the time that the cells have spent in the progress zone close beneath the apical ridge before beginning their overt differentiation. There seems to be also some cell-cell interaction in the mesenchyme at early stages, tending to smooth out any discontinuities in the pattern of positional values.

Limbs can regenerate in amphibians such as the newt and in insects such as the cockroach. The regenerate is formed by proliferation of the cells in the neighborhood of the cut, and its structure depends on their positional values.

When normally separate parts of a limb are grafted together, intercalary regeneration occurs. This is most clearly seen in the cockroach leg. The process is described by the rule of intercalation: discontinuities of positional value provoke local growth, and the newly grown cells take on intermediate positional values that restore continuity in the pattern. Positional value in the cockroach leg epithelium is a two-dimensional variable. The rule of intercalation can be applied in two dimensions and correctly predicts the regeneration of amputated limbs. The rule may apply to many different systems in which the pattern of positional values and the controls on cell growth depend on continuing interactions between cells rather than on cell memory alone.

Inductive Interactions in the Development of Epithelia

Many organs contain two or more distinct populations of cells that originate separately and later come together and interact. The process by which the pattern of differentiation of the cells in one tissue is controlled by the influence of a second tissue in close contact with them is called **induction.** Interactions of this sort are particularly important in the development of epithelia in vertebrates.

Mesoderm Induces Ectoderm to Form the Different Parts of the Neural Tube[51]

Experiments on amphibians done in the early decades of this century show that the formation of neural structures from the ectoderm (p. 822) is due to an inductive influence from the underlying mesoderm. If a piece of mesoderm is taken from the area just beneath the future neural tube of one gastrulating amphibian embryo and directly implanted beneath the ectoderm of another gastrulating embryo in, say, the belly region, the ectoderm in that region will thicken and roll up to form a piece of misplaced neural tube. Moreover, the character of this piece of neural tube depends on the origin of the grafted mesoderm. If the mesoderm has been taken from an anterior site, a piece of forebrain will be formed; if it is taken from a posterior site, a piece of spinal cord forms instead. This suggests that distinctive positional values are imprinted on the ectodermal cells according to the positional values of the underlying mesoderm cells.

The pattern of differentiation of many epithelia at later stages is also largely controlled by the underlying mesenchyme. The skin and the gut, for example, with all their specializations and derivatives, are constructed from epithelial and mesenchymal components in this way. We shall consider first the structures that grow out from the skin.

The Dermis Controls the Nature and Pattern of the Structures That Form from the Epidermis[52]

The skin consists of two layers: the epidermis, which is an epithelium derived from ectoderm, and the dermis, which is a connective tissue derived chiefly from mesoderm. Keratinized appendages, such as hairs, feathers, scales, and claws, form from epidermis, and so do many types of gland. The different types of appendages and their local arrangements are characteristic of particular regions of the body: the skin of a chick, for example, has feathers on the back and wings but scales on the legs. Moreover, the feathers on the back are arranged in orderly hexagonal arrays forming well-defined tracts (Figure 15–61).

Figure 15–61 The feather tracts on the back of a chick embryo at nine days of incubation. Note the regular spacing of the feather rudiments in each tract. (Courtesy of A. Mauger and P. Sengal.)

If embryonic leg epidermis, which normally forms scales, is removed from its normal site and combined with embryonic back dermis, which normally underlies feathers, then feathers, rather than scales, develop from the leg epidermis (Figure 15–62). In general, it is found that the dermis controls not only the character of the appendages formed by the epidermis, but also their exact arrangement. Again we encounter the phenomenon of nonequivalence (p. 853): superficially similar dermal tissues from different regions of the body prove to be different in that they induce different behavior in overlying epidermis.

The signal from dermis to epidermis that results in formation of a feather in the chick seems to be nearly the same as that which results in the formation of a hair in a mouse. Thus if dermis from the whiskered region of the snout of a mouse embryo is combined with embryonic chick epidermis, the chick epidermis will begin to form feather buds, but they will be arranged in the pattern typical of mouse whiskers.

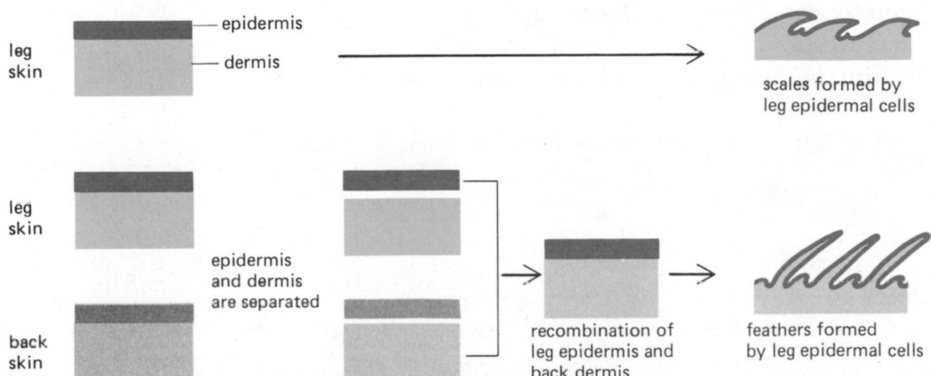

Figure 15–62 Scheme of experiments showing that the dermis controls the type of appendage formed by the epidermis.

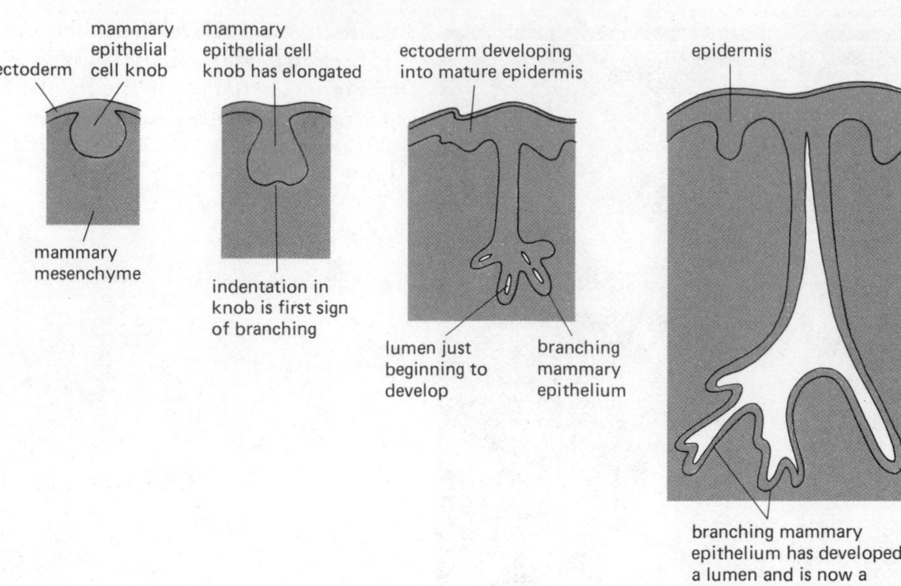

Figure 15–63 The development of the mammary gland in a female mouse embryo.

Epithelium Invades Mesenchyme to Form the Tubules of a Gland[53]

Most glands form as ingrowths of an epithelium into an underlying mesenchyme. Each gland is characterized both by its distinctive geometrical pattern and by the chemistry of its secretions. The mammary gland, for instance, originates as a bud of ectoderm that invades the mesenchyme beneath, branching repeatedly to form a "tree" of tubules (Figure 15–63); the epithelial cells that develop at the ends of the tubules will produce milk, given an appropriate hormonal stimulus, and they are the only cells in the body that will do so.

Inductive cues from the mesenchyme are necessary for gland development, but their interpretation depends also on the nature of the responding epithelium. For example, the prospective epithelial and mesenchymal components of different glands can be separated at an early stage in gland formation and recombined in culture, where the control relationships in the subsequent development can be studied. If mammary gland epithelium is combined with salivary gland mesenchyme, the epithelial branching pattern that results will resemble that of a salivary gland rather than a mammary gland. Nevertheless, if such a gland is exposed to appropriate hormones by grafting it into a pregnant female, the epithelial cells lining the ends of the tubules will produce milk, the secretion corresponding to their origin. In this particular case, though not in all others, the mesenchyme appears to control the geometrical pattern of the gland, whereas the chemistry of the secretion is already determined by the characters of the cells in the early epithelial rudiment.

Summary

Many organs develop through interactions that occur when two or more groups of cells with separate origins are brought into contact. In these cases, the one group of cells may induce specialized patterns of behavior in the other. Thus mesoderm induces overlying ectoderm to form a neural tube and also controls its more detailed regional specialization. In the skin, the dermis governs the character and arrangement of the appendages formed by the epidermis. In the formation of mammary and salivary glands, the pattern of ingrowth and branching of the epithelium is controlled by the mesenchyme it invades, though the chemistry of the secretion may have been determined already in the epithelium.

Multicellular Development Studied Cell by Cell: The Nematode Worm[54]

We have now discussed many of the mechanisms involved in animal development and have seen how cells proliferate, move, interact, and differentiate to create an orderly structure. So far our examples have come mostly from rather complex creatures containing many millions of cells. In such organisms it is not possible to follow the fate of every cell individually. This feat has, however, been achieved in studies of a nematode worm, *Caenorhabditis elegans.*

C. elegans is small, anatomically simple, and easy to manipulate genetically. It has, furthermore, another feature that makes it particularly amenable to detailed cell-by-cell analysis. In most of the other animal species that we have been dealing with, keeping track of individual cells is difficult not only because the organism is large and complex, but also because the cells move in somewhat disorderly ways: they may mingle haphazardly at certain stages of development, and the positions of the progeny of a given cell are variable from animal to animal (see pp. 831 and 844). Thus a pair of adjacent cells in one individual may be sisters, while the pair of cells at the corresponding sites in another individual may be fourth cousins: there is some randomness in the relationship between cell lineage and cell position. By contrast, in nematodes and some other phyla (such as mollusks and annelids), cell divisions and displacements are ordered with extreme precision, and the lineage relationships between the cells forming the various parts of the body are very nearly the same in every individual. It is this feature in particular that has made it possible to follow the development of *C. elegans* cell by cell, all the way from the egg to the adult form.

Caenorhabditis elegans Is Anatomically and Genetically Simple

As an adult, *C. elegans* is about 1 mm long and consists of only some 1000 somatic cells and 2000 germ cells (Figure 15–64). Its anatomy has been reconstructed, cell by cell, by electron microscopy of serial sections. The body plan of this simple worm is fundamentally the same as that of most higher animals in that it has a bilaterally symmetrical, elongate body that develops from three

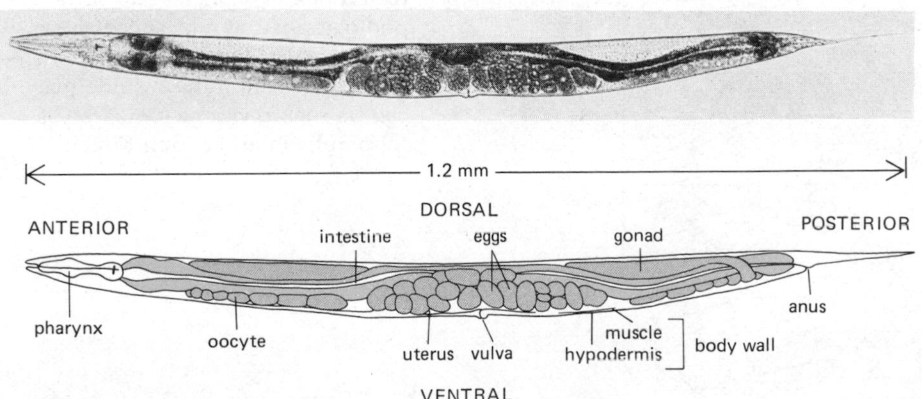

Figure 15–64 *Caenorhabditis elegans:* a side view of the adult hermaphrodite form. (From J. E. Sulston and H. R. Horvitz, *Dev. Biol.* 56:110–156, 1977.)

germ layers. The worm has a pharynx at the anterior end that pumps bacteria into the intestine, and an anus near the posterior end. The outer body wall is composed of two layers: the protective hypodermis, or "skin," and the underlying muscular layer. Inside the body wall, a long simple tube of endodermal cells forms the intestine. A second tube, located between the intestine and the body wall, constitutes the gonad. This gonadal tube is composed of somatic cells and contains the germ-line cells within it. *C. elegans* can exist as either a hermaphrodite or a male. Thus, the animal can either reproduce by self-fertilization of the hermaphrodite or by cross-fertilization between male and hermaphrodite. This greatly facilitates genetic studies of *C. elegans*.

The relative simplicity of the *C. elegans* anatomy is reflected in a similar simplicity of its genome. The animal has an estimated total of 3000 essential genes on its six homologous pairs of chromosomes. The quantity of DNA in its haploid genome is about 20 times more than in *E. coli* and about 35 times less than in man. Currently, about 350 genes have been identified by mutation. These include genes that influence visible features such as the shape or behavior of the worm, genes that code for known proteins such as acetylcholinesterase or myosin, and genes that affect the course of development.

Nematode Development Is Essentially Invariant[55]

C. elegans begins life as a single cell, the fertilized egg. During embryogenesis, this cell gives rise, by repeated cell divisions, to about 550 cells that are organized to form a small worm inside the egg shell. After hatching, further divisions result in the growth and sexual maturation of the worm as it passes through four successive larval stages separated by molts. After its final molt to the adult stage, it begins to produce its own eggs. The entire developmental sequence, from egg to egg, takes only about three days.

Since *C. elegans* is transparent, cells can be watched as they divide, migrate, and differentiate in living animals (Figure 15–65). By this simple technique of direct observation, the behavior and lineage of all of the cells from the single-cell egg to the adult animal have been described. In both the embryonic and postembryonic stages of development, most cells behave identically in all the individuals examined.

These descriptive studies have revealed that the somatic structures of the animal develop by invariant cell lineages. This means that a given precursor cell follows the same pattern of cell divisions in every individual, and the fate of each descendant cell is predictable from its ancestry or position in the lineage tree (Figure 15–66). (The only exceptions are a few cells that

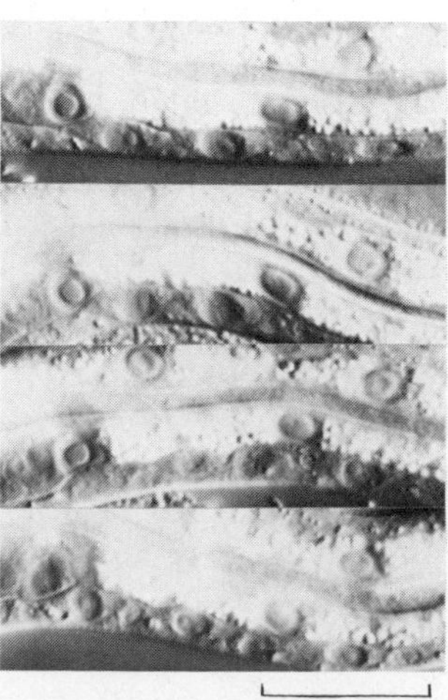

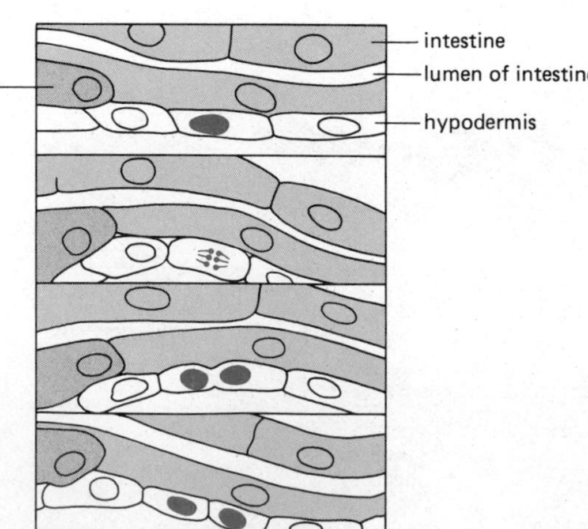

Figure 15–65 Part of the midventral region of a living *C. elegans* larva photographed at four successive times, separated by intervals of a few minutes. A cell in the hypodermis can be seen dividing. (Courtesy of John Sulston and Judith Kimble.)

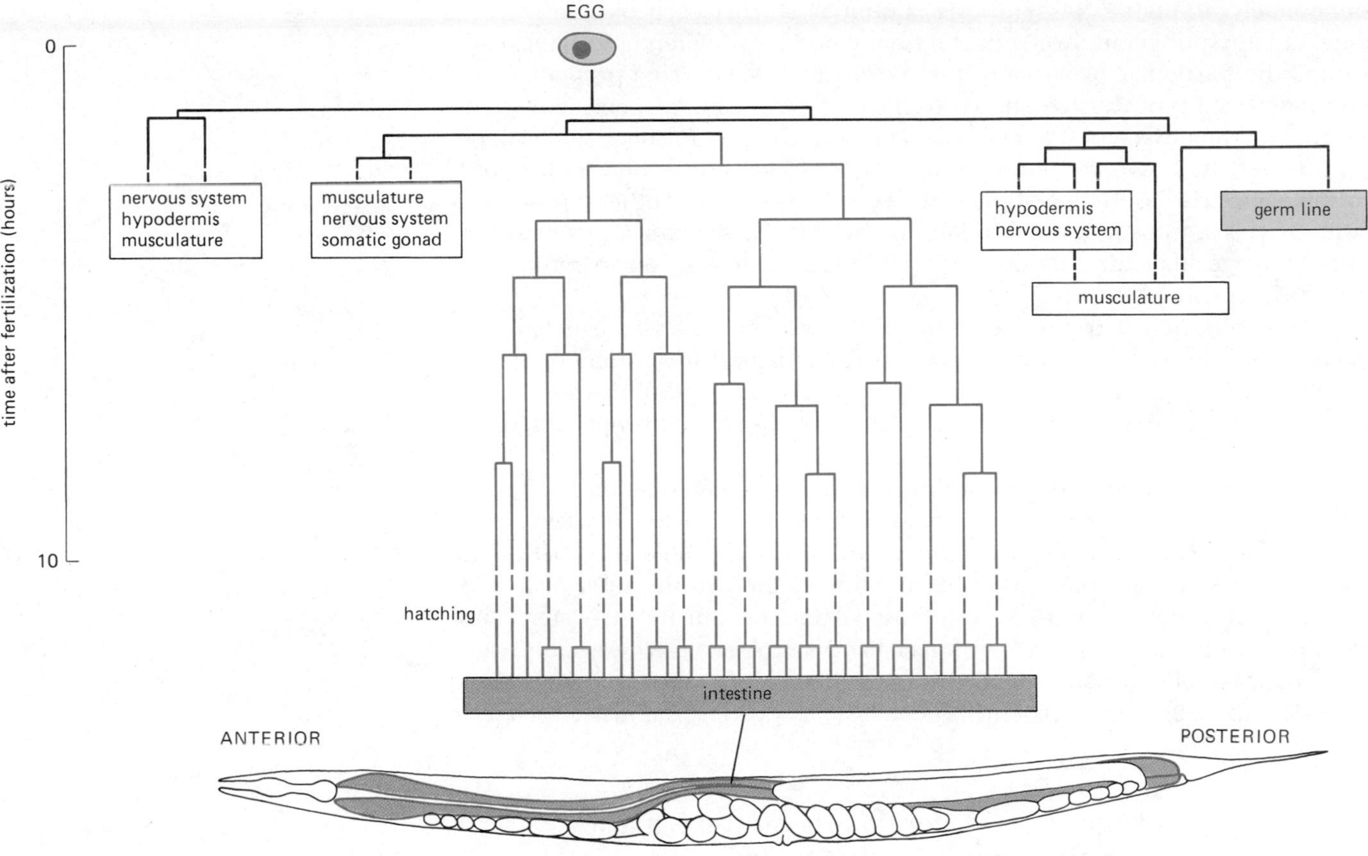

Figure 15–66 The lineage tree for the cells that form the intestine of *C. elegans*. The egg (*top*) is drawn to the same scale as the adult (*bottom*).

have either of two possible fates.) In contrast to the somatic cells, the cells of the germ line do not have a precisely invariant lineage: after hatching, the two germ-line precursor cells undergo a variable sequence of cell divisions, and the fate of their progeny depends on the position that they come to occupy in the gonad.

The cell lineage of the early embryo shows that in *C. elegans,* as in other animals (p. 830), most structures do not arise as clones descended from a single founder cell. Thus the hypodermis, the nervous system, the musculature, and the somatic component of the gonad all have a multicellular origin: each is composed of cells from several separate lineages (Figure 15–66). Moreover, in a few cases, a nerve cell and a muscle cell can even arise as sisters in a final division in the lineage tree. Some parts of *C. elegans* are, however, exceptions to the rule, consisting of a single entire clone of cells. In particular, the intestinal cells and the germ-line cells each constitute a clone derived from just one founder cell (Figure 15–66).

Cytokinesis Is Not Required for Cell Differentiation[56]

As the cells of the nematode embryo proliferate, they begin to differentiate, manufacturing various specialized gene products. The specialized genes can still be expressed on schedule, however, even if cell division—or, to be precise, cytokinesis—is blocked. Furthermore, when cleavage is prevented, genes normally characteristic of distinct types of differentiated cell may be expressed together in a single cell.

This phenomenon can be demonstrated by the use of drugs to block cell division in the early embryo. When either two-cell or four-cell embryos are treated with a mixture of colchicine and cytochalasin B, cytokinesis is arrested immediately, though DNA synthesis continues so that each blastomere becomes highly polyploid. At about the time when cell differentiation normally occurs, the particular blastomere that normally gives rise to a particular differentiated cell type can be shown to begin synthesizing a specialized gene product characteristic of that cell type. For example, if all further cell divisions are arrested in a two-cell embryo, one of these cells—the one that would normally give rise to the intestinal cells, as well as to cells of other types—will exhibit signs of intestinal differentiation. And if cell division is arrested at the four-cell stage instead, only one out of the four cells goes on to synthesize the intestinal product.

As an alternative to drugs, certain genetic mutations can be used to block cell divisions during larval development. Again, some features of differentiation are observed in the arrested precursors in the absence of cell division. Here, the arrested precursor cell exhibits a mixture of specialized properties, each characteristic of one of its normal descendants.

From these experiments, it seems that in the differentiation of the nematode, cytokinesis serves to segregate the different specialized features into separate cells. One attractive hypothesis is that certain molecules critical to cell determination are passed on to only one of the two daughter cells of a division. These determinants would endow that cell, but not its sister, with the ability to express a particular specialized set of genes. The evidence for localized intracellular determinants in some other organisms has already been discussed (p. 850). Their existence in *C. elegans*, however, is a matter of speculation.

The Influence of Local Cell Interactions on the Behavior of Cells During Development Can Be Studied by Laser Microsurgery[54,57]

A cell's development is controlled both by factors inherited from its mother cell and by extrinsic cues picked up from its neighbors. If the invariance observed in *C. elegans* development is a consequence of reproducible external signals received from neighboring cells, it should be possible to disturb a cell's normal environment and alter its highly predictable fate by killing individual neighbor cells. This can be done using a focused laser beam with a diameter of about 0.5 μm (the average nucleus is about 2 μm in diameter in *C. elegans*). When the nucleus of a cell is exposed to repeated pulses of laser light, the cell dies with no apparent damage to other cells in the animal.

If the developmental fate of a particular cell is reproducibly changed by the destruction of a specific neighboring cell, it can be assumed that the missing cell normally helps to determine that cell's fate in the intact animal. Such experiments have revealed that development in the nematode is largely controlled through cell lineage rather than intercellular interactions: usually a cell will persist in its normal developmental pathway even if its neighbors are destroyed. However, there are exceptions, showing that cell signaling is also important. The development of the egg-laying apparatus provides an example.

An "Anchor Cell" Controls Vulva Development[54,57]

Eggs are laid through a ventral orifice, the *vulva*, in the hypodermis (skin) of the hermaphrodite (Figure 15–64). Precursor cells in the hypodermis, just under the tube of somatic cells that form the gonad, give rise to the cells of the vulva by specific lineages. A single nondividing cell in the gonad, called the *anchor cell*, attaches or "anchors" the overlying gonad (the uterus) to the developing vulva to create a passageway through which the eggs will pass

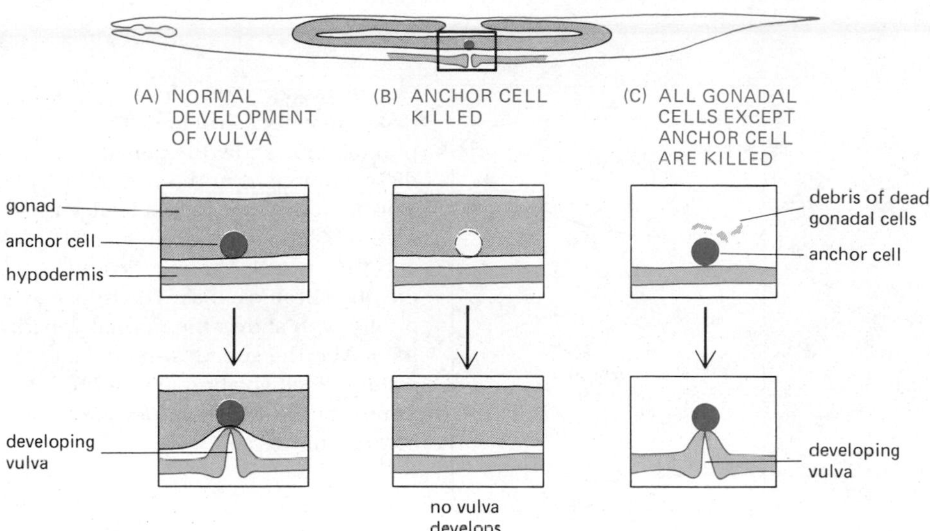

Figure 15–67 Schematic diagram of experiments showing that an inductive influence from the anchor cell is required for the development of the vulva.

from the uterus to the outside world. Laser destruction studies show that this anchor cell is responsible for inducing the nearby hypodermal cells to form a vulva. If the anchor cell is killed, the precursor cells that normally divide to produce vulval cells give rise to ordinary hypodermal cells instead. Conversely, if all of the gonadal cells except this anchor cell are killed, the vulva is still formed normally (Figure 15–67).

These experiments indicate that the anchor cell signals the vulval precursors to form a vulva. Other experiments show that the anchor cell is required for more than an initiating signal. If the anchor cell is killed several hours before the vulval divisions normally begin, no vulval development occurs. If this cell is killed instead somewhat later, just before the time of these divisions, the hypodermal precursor cells generate the normal number of vulval cells, but these cells are unable to assemble into a normally shaped vulva. Thus, the continued presence of the anchor cell is required for the complete development of the vulva. This could mean that the anchor cell provides one type of inductive signal for activation of the divisions that generate the vulval cells and another type for the coordinated assembly of the vulval cells into the vulva. Alternatively, a single "vulva-inducing substance" could be continuously produced by the anchor cell, provoking a succession of different responses in the cells of the vulval lineage as they mature.

A "Distal Tip Cell" Causes Continued Proliferation of Nearby Germ Cells[54,58]

During larval development, the two germ-line precursor cells present at hatching give rise to about 2000 germ cells, which fill the gonad in the adult. The germ cells near the distal tip of the gonad continue to proliferate by mitosis, while the rest of the germ cells enter meiosis (Figure 15–68). The proliferating cells, like spermatogonia in a male mammal (p. 797), serve as germ-line stem cells, replenishing the supply of meiotic germ cells as they mature into gametes and are used for reproduction.

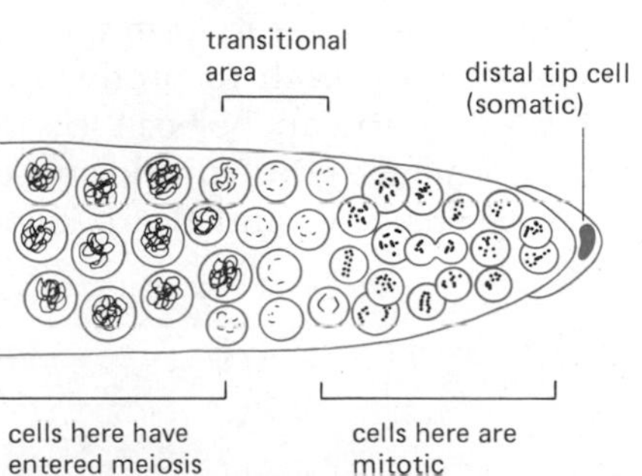

Figure 15–68 The normal organization of germ cells in the gonad of *C. elegans*.

In both larvae and adults, the distal, mitotic end of the gonad is marked by the presence of one or two nondividing somatic cells. (The number depends on the sex.) These are called the *distal tip cells.* If they are killed, all the germ-line stem cells soon enter meiosis (Figure 15–69). Evidently the distal tip cells are necessary to maintain mitotic proliferation in the adjacent part of the germ cell population and to inhibit those adjacent cells from entering

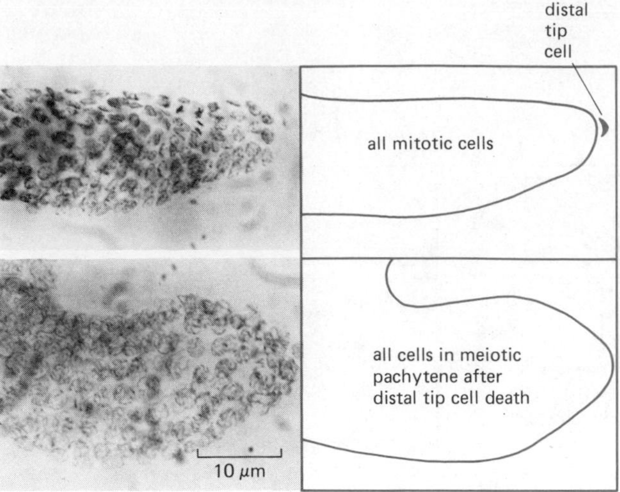

Figure 15–69 The consequences of distal tip cell destruction. The photographs show the gonad dissected free from the rest of the worm and stained by the Feulgen method, which is specific for DNA and thus reveals the mitotic and meiotic chromosomes. The upper photograph shows the normal gonad; the lower, the gonad 24 hours after distal tip cell ablation. (From J. E. Kimble and J. G. White, *Dev. Biol.* 81:208–219, 1981.)

meiosis and maturing into gametes. As an alternative to destroying the distal tip cell, it is possible to displace it from its normal position. If this is done at an early stage in the formation of the gonad, the germ-line stem cell population develops in a correspondingly altered location. Thus the position of the distal tip cell establishes the polar arrangement of mitotic and maturing meiotic germ cells in the normal animal. One can see here an analogy with the role of the apical ectodermal ridge of the chick limb bud, controlling the location and maintenance of the progress zone.

Cell Fate Can Be Controlled by Inhibitory Interactions Among Cells in an Equivalence Group[54,57,59]

In a few cases, a cell will abandon its normal course of development and adopt that of a neighboring cell that has been ablated: this suggests that the two neighboring cells are originally equivalent and that they become different only as a consequence of some interaction between them. This type of cell replacement is seen only within certain discrete groups of neighboring cells called **equivalence groups.** Each such group of cells serves to make a specialized structure in the animal. Two well-documented and one possible equivalence group in the hypodermis are shown diagrammatically in Figure 15–70.

The equivalence of cells in an equivalence group is best shown in experiments carried out on the midventral group of hypodermal cells in the hermaphrodite. This group consists of 6 precursor cells (cells D through I, Figure 15–70) in the region of the future vulva. The 3 cells F, G, and H normally divide several times to form the vulva itself, while the other 3 cells undergo a single division each to produce 6 hypodermal cells. Vulval development is activated by the anchor cell (p. 876); if the anchor cell is killed, all 6 cells in the equivalence group divide only once to produce 12 hypodermal cells. Moreover, if cells F, G, and H are killed, cells D, E, and I abandon their normal

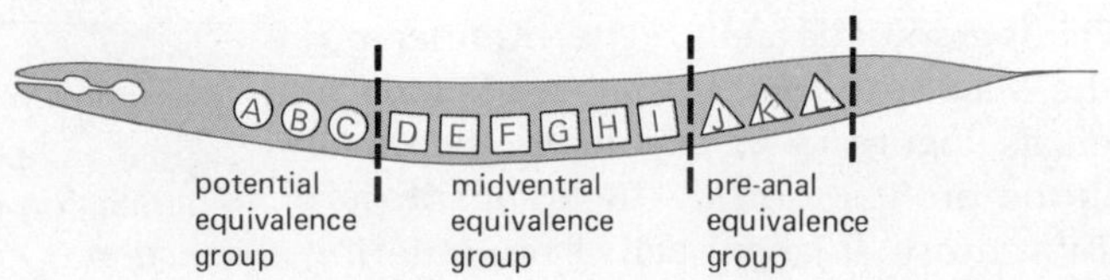

Figure 15–70 Equivalence groups in the ventral hypodermis.

course of development and generate vulval cells instead. Thus, all 6 cells are equivalent in that each can have either a hypodermal or a vulval fate.

The nature of the interactions between the cells in an equivalence group is most clearly demonstrated in the pre-anal equivalence group in the male hypodermis. The precursor cells in this group cooperate to make structures in the male tail that are necessary for mating. Extensive destruction of cells around the group indicates that they are not affected by any cell outside the group (that is, no cell external to this posterior-most group affects it in the way in which the anchor cell affects the midventral group). The equivalence group consists of 3 cells (cells J, K, and L, Figure 15–70) that normally have distinct fates J, K, and L, respectively. However, if cell K is killed, cell J will abandon the course leading to fate J and will follow the course leading to fate K instead. Moreover, if cell L is killed, cell K replaces cell L to undergo fate L, and cell J replaces cell K to undergo fate K, while fate J is abandoned. Reciprocal replacements are never seen; cell L never replaces cell K, and cell K never replaces cell J. Thus, the order in which cells are replaced defines a hierarchy of fates (L before K before J). This hierarchy is maintained when cells K and L are killed, since cell J invariantly assumes fate L in this situation.

From these results, it appears that the presence of cell L must normally influence the fate of cells J and K and that the presence of cell K must similarly influence the fate of cell J (Figure 15–71). The nature of the influence is unknown. One possibility is that cells L and K secrete some inhibitory substance that normally prevents their neighbors from following the same course of development. Another possibility is that fate is governed by position: the mere presence of a cell at a particular site might, by preventing a neighbor from occupying that site, channel that neighbor into a different course of development.

The idea of equivalence groups is supported by genetic data. Mutations exist that alter the lineages of cells in one equivalence group without changing the lineages of cells outside the group. For example, a mutation known as *multivulva* causes all 6 cells of the vulval equivalence group to become vulval precursors. As a result, up to 48 vulval cells are made in this mutant rather than the normal 22. These cells fail to assemble into a single large vulva but instead invaginate locally in small clusters to produce multiple "mini-vulvae." In this mutant, destruction of the anchor cell does not affect the vulval lineages and all 6 cells behave as if they are locked into the anchor-cell-activated state.

An equivalence group consists of a set of neighboring precursor cells that are, in the simplest hypothesis, all initially in the same state of determination. The equivalence group may perhaps be regarded as the fundamental "unit of assembly," analogous to the set of founder cells of a compartment in *Drosophila*. It seems that cell determination in an animal that develops by invariant lineages may depend on many of the same principles that operate in animals where cell lineage, cell character, and cell position are not so predictably related.

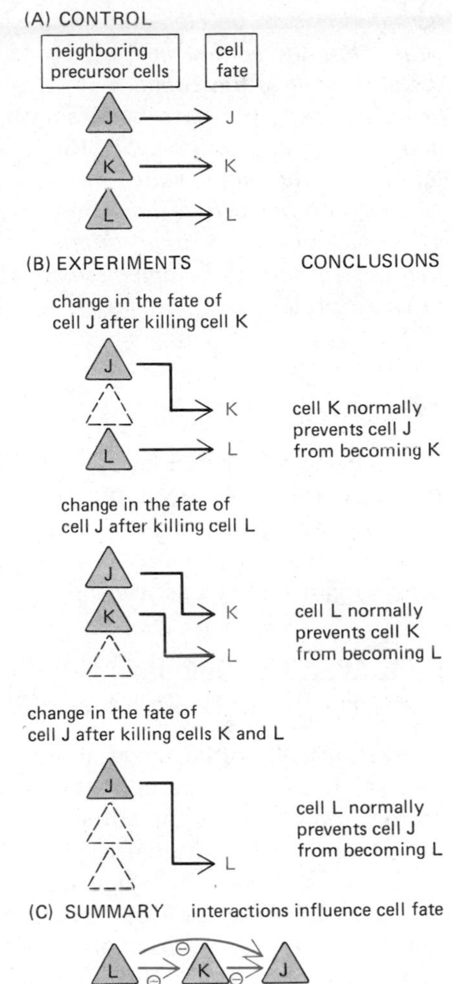

Figure 15–71 Diagram of experiments on the pre-anal equivalence group, indicating that cell interactions influence cell fate.

Summary

Nematodes, like some other invertebrate phyla, but unlike insects and vertebrates, develop by such a precisely predictable pattern of cell divisions that a somatic cell at a particular position in the body has the same lineage in every individual. The normal developmental lineage of all the cells of the nematode Caenorhabditis elegans *has been mapped out in detail, and the consequences of experimental interference have been studied at the level of individual identified cells. If cell division is blocked, differentiation can still occur, and the specialized genes normally characteristic of distinct types of differentiated cells may be expressed together in one and the same cell. While most cells of the nematode worm develop autonomously, some depend on interactions with other cells. Thus the anchor cell induces the development of the vulval cells, and the distal tip cell of the gonad maintains proliferation of the germ-line stem cells. Several structures have been shown to develop from equivalence groups of neighboring precursor cells. The cells of an equivalence group appear to be initially in the same state of determination, to the extent that one cell can replace another within the group; later the cells of the group become different as a consequence of interactions with one another. The equivalence group in the nematode may be analogous to the set of founder cells of a compartment in* Drosophila.

Migratory Cells

It is necessary to consider one last important phenomenon of general importance in embryology: cell migration. Cell migration is a central feature of the development of the nervous system (to be discussed in Chapter 18). Neurons, however, are not the only cells that migrate; cell migration is important for other organ systems as well.

Cells Wander: Selective Cohesion May Stop Them from Straying from Their Proper Place[60]

Individual cells often move about relative to their neighbors. In a chimeric mouse embryo, for example, the cells of the two component morulae mingle so that the tissues of the adult are a chaotic patchwork of the two genotypes. In *Drosophila,* the boundaries of each clone of cells detected after x-irradiation are again somewhat irregular and disorderly. But haphazard displacements of cells after they have been determined will disrupt the organized spatial pattern of cell types. Thus once cells have been assigned a character appropriate to their position, they must stay in their proper territory. In *Drosophila,* a compartment boundary seems to be established through selective cohesion: cells with the same character stick more strongly to one another than to cells with a different character. Experiments *in vitro* provide some evidence that the same principle may keep differentiated cells from straying in vertebrates. For example, embryonic heart cells and liver cells can be disaggregated and mixed together so that the mixture re-forms into a solid ball of cells. The two types of cell often sort out from one another, as though each had a stronger affinity for its own kind than for other kinds of cells (see p. 680). This affinity would naturally tend to restrain cells from wandering from their site of origin.

There are, however, some cells in the embryo that do not merely wander locally but undertake long migrations from their site of origin to colonize distant parts.

Germ Cells Leave the Yolk Sac and Settle in the Genital Ridges[61]

Germ cells are determined very early in development (pp. 789 and 850). In vertebrates they then migrate from their site of origin in the neighborhood of the gut or yolk sac to the epithelium of the genital ridges, which develop much later. In amphibians the germ cells cover the long distance from the gut to the genital ridges by actively moving through the tissues: like the primary mesenchyme cells of the sea urchin, these cells apparently extend processes that shorten to pull the cell body forward. In the chick, on the other hand, a large part of the journey is by passive travel in the bloodstream. But whatever the means of locomotion, the puzzle is how the germ cells come to colonize the genital ridges specifically. One possibility is that the ridges secrete a chemical that attracts germ cells by chemotaxis. In the chick, however, the germ cells seem to settle initially not only in the genital ridges but also in some other, inappropriate sites—even in the head—and the specificity of localization may thus be partly due to the death of cells that settle in the wrong place. In any case, the germ cells must at least recognize the genital ridges as the specific site in which to settle and survive, whether they encounter them by chemotaxis or by chance.

Muscle Cells in the Chick Limb Originate by Migration from the Somites[62]

As mentioned earlier, the precursors of the muscle cells in the chick limb are also determined very early and migrate into the region of the future limb from the somites. The migration has been demonstrated by experiments in which cells are grafted from quail embryos into chick embryos.

Although the quail is similar in most respects to the chick, its cells can be distinguished in histological sections by a large, strongly staining mass of heterochromatin associated with the nucleolus. This nucleolar marker makes it possible to identify grafted cells that have migrated from the site where they were implanted. If quail somite tissue is substituted for the somite tissue of a very young chick embryo whose wing buds have yet to appear (Figure 15–72), it is found that all the muscle cells, and only the muscle cells, in the wings that subsequently develop have a quail origin. Evidently, the future muscle cells migrate into the prospective wing region and remain there inconspicuously until the time comes for them to differentiate. These muscle-precursor cells are determined but undifferentiated and are practically indistinguishable in appearance from the other mesenchymal cells in the early limb bud.

How then do the future muscle cells become localized to form muscles in the appropriate places? Most probably, the surrounding connective tissue cells, which do not arise from the somites, control the final distribution of the migratory muscle cells.

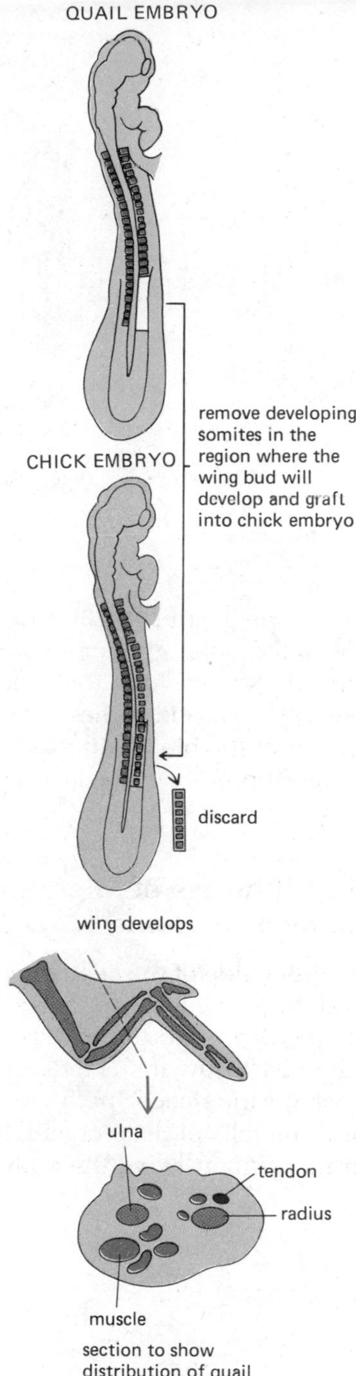

Figure 15–72 If quail somite cells are substituted for the somite cells of a chick embryo at two days of incubation and the wing of the chick is sectioned a week later, it is found that the muscle cells in the chick wing derive from the transplanted quail somites.

Cells Disperse from the Neural Crest and Form Many Different Tissues[63]

The epithelium at the site of closure of the neural tube (see p. 822) gives rise to the important group of migratory cells known as the **neural crest cells.** These break loose from the epithelium and migrate along specific pathways through the embryo to give rise to a great variety of tissues, including peripheral neurons, Schwann cells, pigment cells, and, in the head, various connective tissues. This immediately raises a number of questions. What decides the pathway of migration for a particular neural crest cell? Are the neural crest cells all equivalent before they begin their migrations, or are they already determined as different? What effect does the environment at the site where a neural crest cell finally settles have on its character?

Most of the neural crest derivatives were identified by early experiments in which the crest was simply removed and the resulting defect recorded. In more recent experiments, the destiny of the neural crest cells has been demonstrated in the chick more directly by following cells that have been marked before they begin their migration. The marked cells have been of two kinds—either from a chick embryo labeled with radioactive thymidine or from a quail embryo. Marked neural crest cells of either kind can be transplanted into the appropriate location in a host chick embryo, in place of the host's

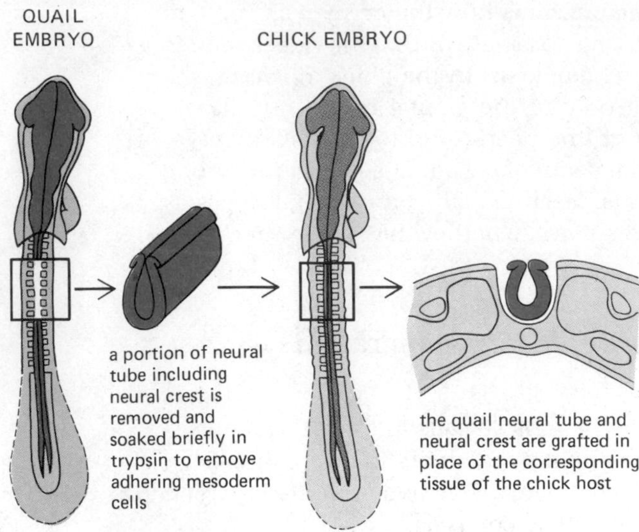

Figure 15–73 Diagram showing how quail neural crest cells can be substituted for those of the chick. The sketch on the right represents a transverse section through the host chick embryo containing the grafted quail tissue. (After N. Le Douarin, *Nature* 286:663–669, 1980.)

own tissue (Figure 15–73), and the migrant cells can then be identified several days later. Such grafting experiments have extended the list of neural crest derivatives to include the cells that secrete the hormone calcitonin and the cells of the carotid bodies (internal sense organs that monitor the oxygenation and pH of the blood). In this way it has also been possible to answer some of the questions about the factors that govern the movements and differentiation of neural crest cells.

The Pathways of Migration Are Defined by the Host Connective Tissue[63,64]

On either side of the central axis or trunk of the embryo, cells leave the neural crest by two main pathways, one just below the ectoderm and the other leading deep into the body via the somites (Figure 15–74). The cells that migrate just below the ectoderm give rise to the pigment cells of the skin, and those on the deeper path give rise to the various nervous tissues and to the pigment cells of deep organs. The site that a crest cell will colonize depends on its position along the body axis. This relationship has been clearly dem-

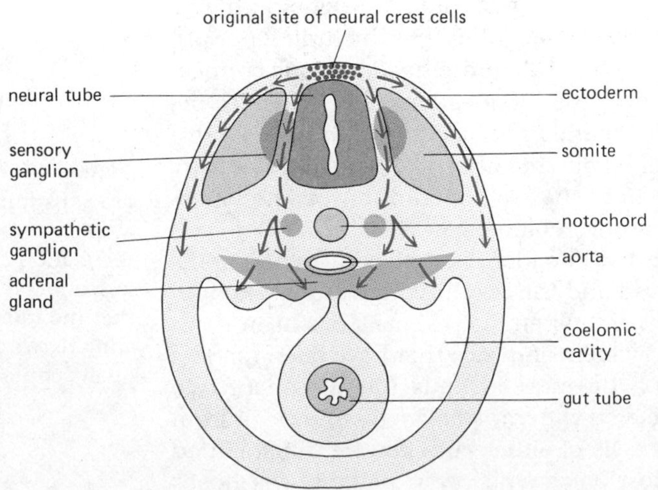

Figure 15–74 Schematic diagram showing the main pathways of neural crest cell migration in a chick embryo, cross-sectioned through the middle part of the trunk. The cells that take the superficial pathway, just beneath the ectoderm, will form pigment cells of the skin; those that take the deep pathway via the somites will form sensory ganglia, sympathetic ganglia, and parts of the adrenal gland. Neural crest cells from this level do not contribute to enteric ganglia.

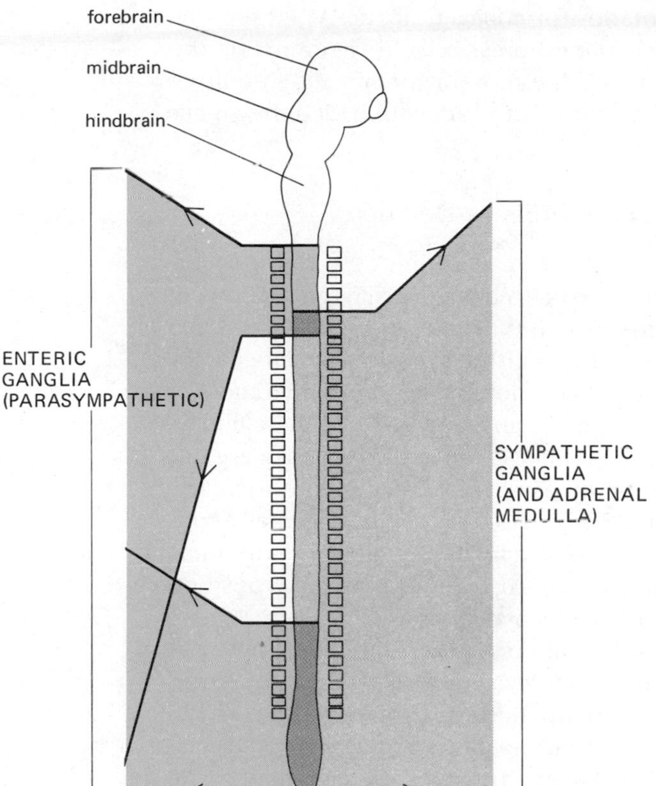

Figure 15–75 The origins of the different groups of autonomic neurons from the neural crest overlying different regions of the future spinal cord. (Neural crest cells and autonomic neurons of the head are not indicated in this diagram.) Neural crest cells from the different levels contribute as follows: first somite to fifth somite—enteric (parasympathetic) ganglia but *not* sympathetic ganglia; sixth and seventh somites—both enteric and sympathetic ganglia; eighth to twenty-eighth somite—sympathetic ganglia (and adrenal medulla) but *not* enteric ganglia; twenty-ninth somite and posterior levels—both enteric and sympathetic ganglia. For clarity, the enteric ganglia are shown only on the left side of the diagram and the sympathetic only on the right. In reality, the two sides of the body are, of course, symmetrical with respect to the fates of the neural crest cells.

onstrated for the crest cells that take the deep pathways and differentiate into the peripheral neurons of the autonomic nervous system. These neurons become clustered in ganglia such as the sensory ganglia, the ciliary ganglion beside the eye, the chain of sympathetic ganglia close beside the vertebral column, and the enteric (parasympathetic) ganglia in the wall of the gut.

Different regions of the neural crest normally give rise to different ganglia, as shown in Figure 15–75. The different destinations of the crest cells from these different regions are governed not by differences in the initial character of the cells, but simply by the differences in their starting positions. If neural crest tissue from an anterior level that normally goes to form enteric ganglia is grafted into a more posterior position, it will go to form sympathetic ganglia instead of enteric; similarly, if the crest from this more posterior position is grafted to the anterior level, it will form enteric ganglia instead of sympathetic. Thus the cells migrate along pathways defined for them by the host connective tissue and settle at whatever sites those pathways may lead to: they show no tendency to seek out the particular pathway appropriate to the site from which they were originally taken.

How are the pathways of neural crest cell migration specified? There are various possibilities. It is unlikely that the crest cells home in on the target site simply under the influence of some diffusible chemical attractant released by the target. The routes of migration seem too long and devious for that. For example, the anterior regions of the crest that do contribute to the enteric ganglia lie further from those ganglia than the thoracic regions of the crest that do not. The pathways are more likely defined by local features of the connective tissue through which the cells migrate. One widely favored hypothesis is that the distribution of fibronectin and glycosaminoglycans in the extracellular matrix marks out the preferred pathways of migration. The cells might also be directed by the orientation of collagen fibers. Whether such guidance systems actually operate on neural crest cells *in vivo* remains uncertain. It has, however, been shown that the movements of cells in culture can be guided by analogous features of the substratum with which they make contact (see p. 602).

The Differentiation of Neural Crest Cells Is Decided by the Local Environment[63,65]

Whatever the mechanism of pathway guidance, cells from different regions of the neural crest settle at different sites, where they mature into different cell types. Most of the cells that become neurons of the sympathetic ganglia, for example, will synthesize the neurotransmitter norepinephrine, while most of the cells that become neurons of the enteric ganglia (which are classified as parasympathetic) will synthesize acetylcholine. The neural crest cells are not intrinsically determined as sympathetic or parasympathetic before their migration. When neural crest tissue from the anterior region that normally contributes to the enteric but not to sympathetic ganglia is grafted to a thoracic level, its cells differentiate in the manner appropriate to their new position and synthesize norepinephrine rather than acetylcholine.

Moreover, the responsiveness of neural crest cells to the local environment is retained even at very late stages in development. When cultured in isolation, individual cells taken from sympathetic ganglia of newborn rats can mature as norepinephrine-synthesizing neurons. But when grown in association with certain nonneuronal cell types, such as muscle, they mature instead as acetylcholine-synthesizing neurons. When the culture conditions are manipulated, the single neurons can be observed to switch from the one phenotype to the other, passing through a phase in which they synthesize both neurotransmitters simultaneously. The influence of the nonneuronal cells on the choice of neurotransmitter does not require cell-to-cell contact. Isolated ganglion cells can be induced to synthesize acetylcholine simply by exposure to medium in which the appropriate nonneuronal cells have been growing. This suggests that the switch is controlled by a soluble chemical released into the medium by the nonneuronal cells.

The Development of the Nervous System Poses Special Problems

Discussion of the neural crest has brought us to a topic so far neglected in this chapter: the development of the nervous system. All the questions considered up to this point can be summarized as follows: how do the different kinds of cells in the body arise and come to be arranged in their proper places? But the nervous system poses an additional problem: how do the nerve cells come to be properly connected? In most other departments of embryology, cells can be considered as pointlike objects, each one having a well-defined position and intrinsic character. But the essence of a neuron is that it is not pointlike but enormously extended, with a long axon and dendrites connecting it to other cells. Its function in controlling and integrating the activities of the body depends on these connections. If the connections are wrong, the nervous system malfunctions. We can explain how the different kinds of neuron arise and how their cell bodies come to lie in a regular arrangement in terms of the same principles that apply to the rest of the body. But the orderly outgrowth of axons and dendrites and the formation of a regular system of synapses are phenomena of another class. The advancing tip of a growing axon or dendrite does indeed crawl along in rather the same manner as a migratory cell: it could be called a migratory organ of a sessile cell. To some extent, the factors controlling its movement are the same as for the migratory cell, involving contact guidance and so on. But its connection to its cell of origin, its involvement with other nerve fibers, and its capacity to form synapses raise new problems requiring special treatment. We therefore leave the construction of the nervous system, the *tour de force* of development, for analysis in Chapter 18.

Summary

A number of types of cell migrate for long distances through other tissues of the embryo to reach their final locations. Germ cells are one example; their final restricted distribution in the body depends in part on the death of those that settle in inappropriate sites. Muscle cells in the vertebrate limb also derive from migratory precursors. Neural crest cells are another important example. These are the ancestors of many different types of cell, including melanocytes, peripheral neurons and glia, and connective tissue cells of the head. Neural crest cells arising from different levels along the body axis migrate along different pathways, probably defined by mechanical contact guidance or by chemical markers in the extracellular matrix or on cell surfaces. Neural crest cells are not fully determined before they begin their migration: those that ordinarily give rise to parasympathetic neurons, for example, can give rise to sympathetic neurons instead if they are transplanted to a different level. It can be shown that the differentiation of these migratory cells is controlled by the environment in which they settle. Migratory behavior is a characteristic general feature of neurons, and it plays a central part in the development of the nervous system.

References

General

Browder, L. Developmental Biology. Philadelphia: Saunders, 1980.

Ham, R.G.; Veomett, M.J. Mechanisms of Development. St. Louis: Mosby, 1979.

Karp, G.; Berrill, N.J. Development, 2nd ed. New York: McGraw-Hill, 1981.

Spemann, H. Embryonic Development and Induction. New Haven: Yale University Press, 1938. (Reprinted, New York: Hafner, 1967.)

Weiss, P.A. Principles of Development. New York: Holt, 1939.

Wessells, N.K. Tissue Interactions and Development. Menlo Park, Ca.: Benjamin-Cummings, 1977.

Cited

1. Browder, L. Developmental Biology, pp. 322–351. Philadelphia: Saunders, 1980.

   Gerhart, J.C. Mechanisms regulating pattern formation in the amphibian egg and early embryo. In Biological Regulation and Development (R.F. Goldberger, ed.), Vol. 2, pp. 133–316. New York: Plenum, 1980.

   Hara, K.; Tydeman, P.; Kirschner, M. A cytoplasmic clock with the same period as the division cycle in *Xenopus* eggs. *Proc. Natl. Acad. Sci. USA* 77:462–466, 1980.

2. Gerhart, J.; Ubbels, G.; Black, S.; Hara, K.; Kirschner, M. A reinvestigation of the role of the grey crescent in axis formation in *Xenopus laevis*. *Nature* 292:511–516, 1981.

   Maller, J.; Poccia, D.; Nishioka, D.; Kidd, P.; Gerhart, J.; Hartman, H. Spindle formation and cleavage in *Xenopus* eggs injected with centriole-containing fractions from sperm. *Exp. Cell Res.* 99:285–294, 1976.

3. Furshpan, E.J.; Potter, D.D. Low-resistance junctions between cells in embryos and tissue culture. *Curr. Top. Dev. Biol.* 3:95–128, 1968.

Slack, C.; Warner, A.E. Intracellular and intercellular potentials in the early amphibian embryo. *J. Physiol.* 232:313–330, 1973.

Kalt, M.R. The relationship between cleavage and blastocoel formation in *Xenopus laevis*. II. Electron microscopic observations. *J. Embryol. Exp. Morphol.* 26:51–66, 1971.

4. Karp, G.; Berrill, N.J. Development, 2nd ed., pp. 322–431. New York: McGraw-Hill, 1981.

Browder, L. Developmental Biology, pp. 452–506. Philadelphia: Saunders, 1980.

5. Gustafson, T.; Wolpert, L. Cellular movement and contact in sea urchin morphogenesis. *Biol. Rev.* 42:442–498, 1967.

6. Trinkaus, J.P. Cells into Organs: The Forces That Shape the Embryo. Englewood Cliffs, N.J.: Prentice-Hall, 1969.

7. Keller, R.E. An experimental analysis of the role of bottle cells and the deep marginal zone in the gastrulation of *Xenopus laevis*. *J. Exp. Zool.* 216:81–101, 1981.

8. Spemann, H. Embryonic Development and Induction. New Haven: Yale University Press, 1938. (Reprinted, New York: Hafner, 1962.)

9. Kitchin, I.C. The effects of notochordectomy in *Amblystoma mexicanum*. *J. Exp. Zool.* 112:393–411, 1949.

10. Burnside, B. Microtubules and microfilaments in amphibian neurulation. *Am. Zool.* 13:989–1006, 1973.

Karfunkel, P. The mechanisms of neural tube formation. *Int. Rev. Cytol.* 38:245–271, 1974.

11. Blackshaw, S.E.; Warner, A.E. Low resistance junctions between mesoderm cells during development of trunk muscles. *J. Physiol.* 255:209–230, 1976.

Pearson, M.; Elsdale, T. Somitogenesis in amphibian embryos. I. Experimental evidence for an interaction between two temporal factors in the specification of somite pattern. *J. Embryol. Exp. Morphol.* 51:27–50, 1979.

12. Austin, C.R.; Short, R.V., eds. Reproduction in Mammals, Book 2, Embryonic and Fetal Development. Cambridge, Eng.: Cambridge University Press, 1972.

Johnson, M.H., ed. Development in Mammals. 3 vols. Amsterdam: Elsevier, 1977.

13. Langman, J. Medical Embryology, 4th ed. Baltimore: Williams & Wilkins, 1981.

Rugh, R. The Mouse: Its Reproduction and Development. Minneapolis: Burgess, 1968.

14. Tarkowski, A.K. Experiments on the development of isolated blastomeres of mouse eggs. *Nature* 184:1286–1287, 1959.

McLaren, A. Mammalian Chimaeras. Cambridge, Eng.: Cambridge University Press, 1976.

Kelly, S.J. Studies of the developmental potential of 4- and 8-cell stage mouse blastomeres. *J. Exp. Zool.* 200:365–376, 1977.

15. Hillman, N.; Sherman, M.I.; Graham, C. The effect of spatial arrangement on cell determination during mouse development. *J. Embryol. Exp. Morphol.* 28:263–278, 1972.

16. McLaren, A. Mammalian Chimaeras. Cambridge, Eng.: Cambridge University Press, 1976.

Gardner, R.L. The relationship between cell lineage and differentiation in the early mouse embryo. In Genetic Mosaics and Cell Differentiation (W.J. Gehring, ed.), pp. 205–241. New York: Springer-Verlag, 1978.

Nesbitt, M.N.; Gartler, S.M. The applications of genetic mosaicism to developmental problems. *Annu. Rev. Genet.* 5:143–162, 1971.

17. Tarkowski, A.K. Induced parthenogenesis in the mouse. In The Developmental Biology of Reproduction (C.L. Markert, J. Papaconstantinou, eds.), Society for Developmental Biology Symposium No. 33, pp. 107–129. New York: Academic Press, 1975.

Kaufman, M.H.; Barton, S.C.; Surani, M.A.H. Normal postimplantation development of mouse parthenogenetic embryos to the forelimb bud stage. *Nature* 265:53–55, 1977.

Illmensee, K.; Stevens, L.C. Teratomas and chimeras. *Sci. Am.* 240(4):120–132, 1979.

Martin, G.R. Teratocarcinomas and mammalian embryogenesis. *Science* 209:768–776, 1980.

18. Mintz, B.; Illmensee, K. Normal genetically mosaic mice produced from malignant teratocarcinoma cells. *Proc. Natl. Acad. Sci. USA* 72:3585–3589, 1975.

Papaioannou, V.E.; Gardner, R.L.; McBurney, M.W.; Babinet, C.; Evans, M.J. Participation of cultured teratocarcinoma cells in mouse embryogenesis. *J. Embryol. Exp. Morphol.* 44:93–104, 1978.

19. Gurdon, J.B. The Control of Gene Expression in Animal Development. Cambridge: Harvard University Press, 1974.

Gurdon, J.B. Transplanted nuclei and cell differentiation. *Sci. Am.* 219(6):24–35, 1968.

Browder, L. Developmental Biology, pp. 34–57. Philadelphia: Saunders, 1980.

20. Weiss, P.A. Principles of Development, pp. 289–437. New York: Holt, 1939.

21. Spemann, H. Über die Determination der ersten Organanlagen des Amphibienembryo I-VI. *Arch. Entw. Mech. Org.* 43:448–555, 1918.

22. Schneiderman, H.A. New ways to probe pattern formation and determination in insects. In Insect Development (P.A. Lawrence, ed.), Royal Entomological Society of London Symposium No. 8, pp. 3–34. Oxford, Eng.: Blackwell, 1976.

23. Hadorn, E. Transdetermination in cells. *Sci. Am.* 219(5):110–123, 1968.

Gehring, W.; Nöthiger, R. The imaginal discs of *Drosophila.* In Developmental Systems: Insects (S. Counce, C.H. Waddington, eds.), Vol. 2, pp. 211–290. New York: Academic Press, 1973.

24. Gehring, W. Clonal analysis of determination dynamics in cultures of imaginal disks in *Drosophila melanogaster. Dev. Biol.* 16:438–457, 1967.

Kauffman, S.A. Control circuits for determination and transdetermination. *Science* 181:310–317, 1973.

25. Gehring, W.; Nöthiger, R. The imaginal discs of *Drosophila.* In Developmental Systems: Insects (S. Counce, C.H. Waddington, eds.), Vol. 2, pp. 211–290. New York: Academic Press, 1973.

Morata, G.; Lawrence, P.A. Homoeotic genes, compartments and cell determination in *Drosophila. Nature* 265:211–216, 1977.

Postlethwait, J.H.; Schneiderman, H.A. Developmental genetics of *Drosophila* imaginal discs. *Annu. Rev. Genet.* 7:381–433, 1973.

26. Lewis, E.B. A gene complex controlling segmentation in *Drosophila. Nature* 276:565–570, 1978.

Struhl, G. A gene product required for correct initiation of segmental determination in *Drosophila. Nature* 293:36–41, 1981.

27. Nüsslein-Volhard, C.; Wieschaus, E. Mutations affecting segment number and polarity in *Drosophila. Nature* 287:795–801, 1980.

28. Stern, C. Genetic Mosaics and Other Essays. Cambridge: Harvard University Press, 1968.

Nöthiger, R. Clonal analysis in imaginal discs. In Insect Development (P.A. Lawrence, ed.), Royal Entomological Society of London Symposium No. 8, pp. 109–117. Oxford, Eng.: Blackwell, 1976.

Gehring, W.J., ed. Genetic Mosaics and Cell Differentiation. New York: Springer-Verlag, 1979.

29. Morata, G.; Lawrence, P.A. Homoeotic genes, compartments and cell determination in *Drosophila. Nature* 265:211–216, 1977.

García-Bellido, A.; Lawrence, P.A.; Morata, G. Compartments in animal development. *Sci. Am.* 241(1):102–111, 1979.

Crick, F.H.C.; Lawrence, P.A. Compartments and polyclones in insect development. *Science* 189:340–347, 1975.

30. Simpson, P.; Morata, G. Differential mitotic rates and patterns of growth in compartments in the *Drosophila* wing. *Dev. Biol.* 85:299–308, 1981.

31. Kieny, M.; Mauger, A.; Sengel, P. Early regionalization of the somitic mesoderm as studied by the development of the axial skeleton of the chick embryo. *Dev. Biol.* 28:142–161, 1972.

32. Wolpert, L. Positional information and pattern formation. *Curr. Top. Dev. Biol.* 6:183–224, 1971.

Wolpert, L. Pattern formation in biological development. *Sci. Am.* 239(4):154–164, 1978.

33. Browder, L. Developmental Biology, pp. 370–405. Philadelphia: Saunders, 1980.

34. Illmensee, K.; Mahowald, A.P. Transplantation of posterior polar plasm in *Drosophila.* Induction of germ cells at the anterior pole of the egg. *Proc. Natl. Acad. Sci. USA* 71:1016–1020, 1974.

35. Lewis, J.; Slack, J.M.W.; Wolpert, L. Thresholds in development. *J. Theor. Biol.* 65:579–590, 1977.
36. Saunders, J.W., Jr.; Gasseling, M.T.; Cairns, J.M. The differentiation of prospective thigh mesoderm grafted beneath the apical ectodermal ridge of the wing bud in the chick embryo. *Dev. Biol.* 1:281–301, 1959.
37. Lewis, J.H.; Wolpert, L. The principle of non-equivalence in development. *J. Theor. Biol.* 62:479–490, 1976.
38. Postlethwait, J.H.; Schneiderman, H.A. Pattern formation and determination in the antenna of the homoeotic mutant *Antennapedia* of *Drosophila melanogaster*. *Dev. Biol.* 25:606–640, 1971.
39. Crick, F. Diffusion in embryogenesis. *Nature* 225:420–422, 1970.
40. Ede, D.A.; Hinchliffe, J.R.; Balls, M., eds. Vertebrate Limb and Somite Morphogenesis. Cambridge, Eng.: Cambridge University Press, 1977.
41. Saunders, J.W., Jr.; Gasseling, M.T. Ectodermal-mesenchymal interactions in the origin of limb symmetry. In Epithelial-Mesenchymal Interactions (R. Fleischmajer, R.E. Billingham, eds.), pp. 78–97. Baltimore: Williams & Wilkins, 1968.
   Smith, J.C. Evidence for a positional memory in the development of the chick wing bud. *J. Embryol. Exp. Morphol.* 52:105–113, 1979.
42. Tickle, C.; Summerbell, D.; Wolpert, L. Positional signalling and specification of digits in chick limb morphogenesis. *Nature* 254:199–203, 1975.
   Tickle, C. The number of polarizing region cells required to specify additional digits in the developing chick wing. *Nature* 289:295–298, 1981.
   Tickle, C.; Alberts, B.; Wolpert, L.; Lee, J. Local application of retinoic acid to the limb bud mimics the action of the polarizing region. *Nature* 296:564–566, 1982.
43. Tickle, C.; Shellswell, G.; Crawley, A.; Wolpert, L. Positional signalling by mouse limb polarizing region in the chick wing bud. *Nature* 259:396–397, 1976.
   Fallon, J.F.; Crosby, G.M. Polarizing zone activity in limb buds of amniotes. In Vertebrate Limb and Somite Morphogenesis (D.A. Ede, J.R. Hinchliffe, M. Balls, eds.), pp 55–69. Cambridge, Eng.: Cambridge University Press, 1977.
44. Saunders, J.W. The proximo-distal sequence of origin of the parts of the chick wing and the role of the ectoderm. *J. Exp. Zool.* 108:363–403, 1948.
45. Rubin, L.; Saunders, J.W., Jr. Ectodermal-mesodermal interactions in the growth of limb buds in the chick embryo: constancy and temporal limits of the ectodermal induction. *Dev. Biol.* 28:94–112, 1972.
46. Summerbell, D.; Lewis, J.H.; Wolpert, L. Positional information in chick limb morphogenesis. *Nature* 244:492–496, 1973.
   Lewis, J.H. Fate maps and the pattern of cell division: a calculation for the chick wing-bud. *J. Embryol. Exp. Morphol.* 33:419–434, 1975.
47. Kieny, M. Proximo-distal pattern formation in avian limb development. In Vertebrate Limb and Somite Morphogenesis (D.A. Ede, J.R. Hinchliffe, M. Balls, eds.), pp. 87–104. Cambridge, Eng.: Cambridge University Press, 1977.
   Summerbell, D. Regulation of deficiencies along the proximal distal axis of the chick wing bud: a quantitative analysis. *J. Embryol. Exp. Morphol.* 41:137–159, 1977.
48. Wallace, H. Vertebrate Limb Regeneration. New York: Wiley, 1981.
   Goss, R.J. Principles of Regeneration. New York: Academic Press, 1968.
   Butler, E.G. Regeneration of the urodele forelimb after reversal of its proximodistal axis. *J. Morphol.* 96:265–282, 1955.
   Maden, M. Vitamin A and pattern formation in the regenerating limb. *Nature* 295:672–675, 1982.
49. Bohn, H. Tissue interactions in the regenerating cockroach leg. In Insect Development (P.A. Lawrence, ed.), Royal Entomological Society of London Symposium No. 8, pp. 170–185. Oxford, Eng.: Oxford University Press, 1976.
   Bryant, P.J.; Bryant, S.V.; French, V. Biological regeneration and pattern formation. *Sci. Am.* 237(1):66–81, 1977.
   Bryant, S.V.; French, V.; Bryant, P.J. Distal regeneration and symmetry. *Science* 212:993–1002, 1981.
   Mittenthal, J.E. Intercalary regeneration in legs of crayfish: distal segments. *Dev. Biol.* 88:1–14, 1981.
50. Lewis, J. Simpler rules for epimorphic regeneration: the polar-coordinate model without polar coordinates. *J. Theor. Biol.* 88:371–392, 1981.

51. Spemann, H. Embryonic Development and Induction, pp. 260–296. New Haven: Yale University Press, 1938. (Reprinted, New York: Hafner, 1962.)

Jacobson, A.G. Inductive processes in embryonic development. *Science* 152:25–34, 1966.

52. Sengel, P. Morphogenesis of Skin. Cambridge, Eng.: Cambridge University Press, 1975.

Sengel, P. Feather pattern development. In Cell Patterning, Ciba Foundation Symposium 29 (new series), pp. 51–70. Amsterdam: Elsevier, 1975.

53. Wessells, N.K. Tissue Interactions and Development. Menlo Park, Ca.: Benjamin-Cummings, 1977.

Sakakura, T.; Nishizuka, Y.; Dawe, C.J. Mesenchyme-dependent morphogenesis and epithelium-specific cytodifferentiation in mouse mammary gland. *Science* 194:1439–1441, 1976.

54. Kimble, J.E. Strategies for control of pattern formation in *Caenorhabditis elegans. Philos. Trans. R. Soc. Lond. (Biol.)* 295:539–551, 1981.

55. Sulston, J.E.; Horvitz, H.R. Post-embryonic cell lineages of the nematode, *Caenorhabditis elegans. Dev. Biol.* 56:110–156, 1977.

56. Laufer, J.S.; von Ehrenstein, G. Nematode development after removal of egg cytoplasm: absence of localized unbound determinants. *Science* 211:402–405, 1981.

Laufer, J.; Bazzicalupo, P.; Wood, W.B. Segregation of developmental potential in early embryos of *Caenorhabditis elegans. Cell* 19:569–577, 1980.

57. Sulston, J.E.; White, J.G. Regulation and cell autonomy during postembryonic development of *Caenorhabditis elegans. Dev. Biol.* 78:577–597, 1980.

Kimble, J. Alterations in cell lineage following laser ablation of cells in the somatic gonad of *Caenorhabditis elegans. Dev. Biol.* 87:286–300, 1981.

58. Kimble, J.E.; White, J.G. On the control of germ cell development in *Caenorhabditis elegans. Dev. Biol.* 81:208–219, 1981.

59. Sulston, J.E.; Horvitz, H.R. Abnormal cell lineages in mutants of the nematode *Caenorhabditis elegans. Dev. Biol.* 82:41–55, 1981.

60. McLaren, A. Mammalian Chimaeras, pp. 104–117. Cambridge, Eng.: Cambridge University Press, 1976.

Townes, P.L.; Holtfreter, J. Directed movements and selective adhesion of embryonic amphibian cells. *J. Exp. Zool.* 128:53–120, 1955.

Steinberg, M.S. Does differential adhesion govern self-assembly processes in histogenesis? Equilibrium configurations and the emergence of a hierarchy among populations of embryonic cells. *J. Exp. Zool.* 173:395–434, 1970.

61. Nieuwkoop, P.D.; Sutasurya, L.A. Primordial Germ Cells in the Chordates, pp. 113–127. Cambridge, Eng.: Cambridge Unviersity Press, 1979.

Ham, R.G.; Veomett, M.J. Mechanisms of Development, pp. 573–579. St. Louis: Mosby, 1979.

Heasman, J.; Hynes, R.O.; Swan, A.P.; Thomas, V.; Wylie, C.C. Primordial germ cells of *Xenopus* embryos: the role of fibronectin in their adhesion during migration. *Cell* 27:437–447, 1981.

Meyer, D.B. The migration of primordial germ cells in the chick embryo. *Dev. Biol.* 10:154–190, 1964.

Saunders, J.W., Jr. Death in embryonic systems. *Science* 154:604–612, 1966.

62. Chevallier, A.; Kieny, M.; Mauger, A. Limb-somite relationship: origin of the limb musculature. *J. Embryol. Exp. Morphol.* 41:245–258, 1977.

Christ, B.; Jacob, H.J.; Jacob, M. Experimental analysis of the origin of the wing musculature in avian embryos. *Anat. Embryol.* 150:171–186, 1977.

63. Le Douarin, N.M. The ontogeny of the neural crest in avian embryo chimaeras. *Nature* 286:663–669, 1980.

64. Noden, D.M. Interactions directing the migration and cytodifferentiation of avian neural crest cells. In The Specificity of Embryological Interactions, Receptors and Recognition, Series B, Vol. 4 (D. Garrod, ed.), pp. 3–49. London: Chapman and Hall, 1978.

Erickson, C.A.; Tosney, K.W.; Weston, J.A. Analysis of migratory behavior of neural crest and fibroblastic cells in the chick. *Dev. Biol.* 77:142–156, 1980.

Tosney, K.W. The segregation and early migration of cranial neural crest cells in the avian embryo. *Dev. Biol.* 89:13–24, 1982.

Greenberg, J.H.; Seppä, S.; Seppä, H.; Hewitt, A.T. Role of collagen and fibronectin in neural crest cell adhesion and migration. *Dev. Biol.* 87:259–266, 1981.

65. Patterson, P.H. Environmental determination of autonomic neurotransmitter functions. *Annu. Rev. Neurosci.* 1:1–17, 1978.

Patterson, P.H.; Potter, D.D.; Furshpan, E.J. The chemical differentiation of nerve cells. *Sci. Am.* 239(1):50–59, 1978.

16 Differentiated Cells and the Maintenance of Tissues

In the space of a few days or weeks, a single fertilized egg gives rise to a complex multicellular organism consisting of differentiated cells arranged in a precise pattern. As a rule, the pattern of the body is set up on a small scale and then grows. During embryonic development, the different cell types become determined, each in its proper place. In the subsequent period of growth, the cells proliferate but, with certain exceptions, their specialized characters remain more or less fixed. The organism may continue to get bigger throughout life, as do most crustaceans and fish, or it may halt at a certain size, as do birds and mammals. In some types of animals with a fixed body size, such as flies and nematodes, proliferation of somatic cells ceases once the adult state has been attained. In many other such animals, however, and in particular in the higher vertebrates, cells continue to proliferate in the adult, replacing cells that die.

As the cells of vertebrate tissues such as skin, blood, and lung become worn out and are destroyed, new cells of the appropriate types take their places. Thus the adult body can be likened to a stable ecosystem in which one generation of individuals succeeds another but the organization of the system as a whole remains unchanged. This chapter will concentrate on the higher vertebrates, and in discussing the problems of tissue maintenance and renewal will try to convey something of the remarkable variety of structure, function, and life history to be found among their specialized cell types.

Maintenance of the Differentiated State[1]

Although the various tissues of the body differ greatly in many ways, they all have certain basic requirements. They all need mechanical strength, which is very often provided by a supporting framework of extracellular matrix. This connective tissue scaffolding is found, for example, in muscles, glands, and bone marrow and beneath epithelia such as the epidermal layer of the skin (Figure 16–1). It is produced largely by *fibroblast cells,* which live in the matrix. In addition, almost all tissues need a blood supply to provide nutrients and

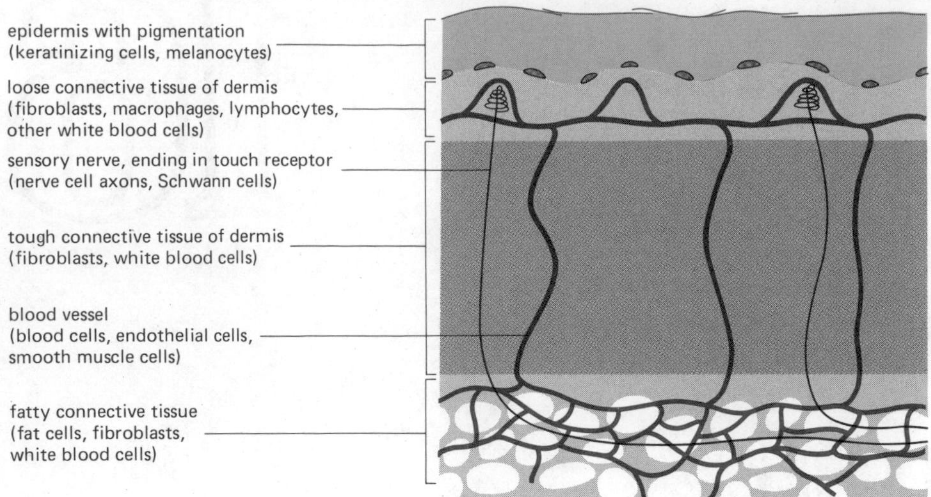

Figure 16–1 Schematic diagram of the skin, showing some of the components of the complex mixture of cell types that it consists of.

remove waste products, and so they are pervaded by blood vessels lined with *endothelial cells.* Likewise, most tissues are innervated, containing axons from *nerve cells,* together with the *Schwann cells* that ensheath them. *Macrophages* are often present to dispose of debris from dying cells and remove unwanted matrix, as are *lymphocytes* and other white blood cells that combat infection. *Melanocytes* may be present to provide pigmentation. Most of these different cells, ancillary to the specialized function of the tissue, originate outside it, invading the tissue in the course of its development (in the case of endothelial cells, nerve cells, Schwann cells, and melanocytes) or continually during life (in the case of macrophages and white blood cells). Amidst all this supporting apparatus lie the principal specialized cells of the tissue: the contractile cells of the muscle, the secretory cells of the gland, or the blood-forming cells of the bone marrow, for example.

Almost every tissue, therefore, is an intricate mixture of many cell types, which must remain different from one another while coexisting in the same environment. This is made possible largely through *cell memory* (p. 834), whereby differentiated cells autonomously maintain their specialized character and pass it on to their progeny. Experiments performed in tissue culture show directly that cells have this fundamentally important capacity: taken out of their usual context, they and their progeny still remain true to their original instructions.

Differentiated Cells Commonly Remember Their Character Even in Isolation: Pigment Epithelium of the Retina[2]

One clear demonstration of the heritability of the differentiated state has come from tissue-culture studies of the epithelial cells that form the pigmented layer of the retina (Figure 16–2). Because these cells display their specialized character by manufacturing dark brown granules of melanin, it is easy to monitor their state of differentiation. The pigmented epithelial cells from the retina of a chick embryo can be isolated and grown in culture, where they proliferate to form clones. Single cells taken from these clones breed true to give subclones of similar pigmented epithelial cells. The differentiated state can be maintained in this way through more than 50 cell generations.

However, the behavior of the cells is not completely independent of their environment. It requires care to devise a culture medium in which pigment cells can survive. In certain media, or in conditions of extreme crowding, they may survive but synthesize little or no pigment. But even while failing to

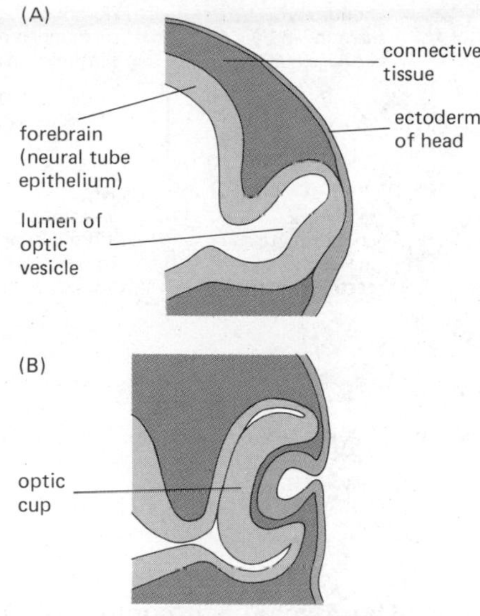

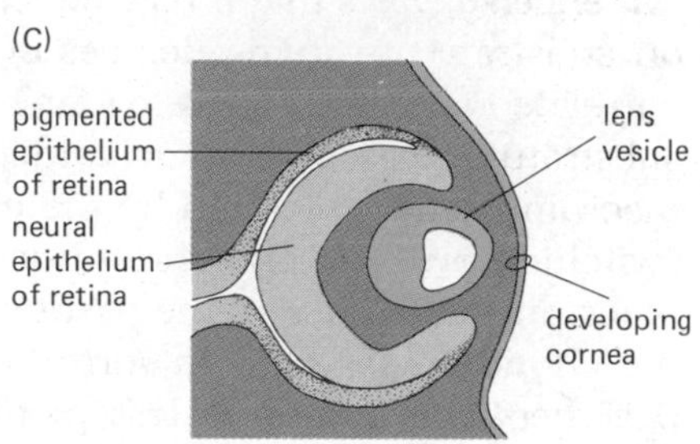

Figure 16–2 The development of the vertebrate eye. (A) The retina develops from the *optic vesicle,* an epithelial outpocketing of the forebrain. (B) This epithelium makes contact with the ectoderm covering the exterior of the head, inducing it to invaginate to form a lens. (C) The part of the optic vesicle that faces outward invaginates at the same time so that its lumen is reduced to an interface between two layers that together form a cuplike structure. The layer of the optic cup closest to the lens differentiates into the *neural retina,* which comprises the light receptors themselves and the neurons that relay the visual stimuli to the brain (see Figure 16–8). The other layer differentiates into the retinal *pigment epithelium.* Its cells are heavily loaded with melanin granules and thus form a dark enclosure for the photoreceptive system (serving to reduce the amount of scattered light, much as a coat of black paint inside a camera does). In addition, the pigment cells are bound together by tight junctions and help to isolate the neural retina from the fluid that pervades the connective tissues around the eyeball.

express their differentiated character, they remain *determined* as pigment cells: when they are returned to more favorable culture conditions, they synthesize pigment once again. No manipulation of the medium or of the culture conditions has been found to cause them to differentiate instead into blood cells, for example, or into liver cells or heart cells.

The Extracellular Matrix That a Cell Secretes Helps Maintain the Cell's Differentiated State[3]

Experiments analogous to those on the pigment epithelium of the retina have been done with several other types of cells, some of which exhibit a richer repertoire of behavior in culture than do pigment cells. Cartilage cells are a particularly interesting case, on account of their interactions with the extracellular matrix that surrounds them.

Early studies revealed a close parallel with retinal pigment epithelium. Differentiated cartilage cells, or **chondrocytes,** will grow in an appropriate medium to give rise to clones of differentiated chondrocytes, which are easily recognized because they synthesize large quantities of highly distinctive cartilage matrix. The cartilage phenotype is retained through repeated subclonings of the cells if the conditions of growth are right. Even cells grown for many generations in an uncongenial medium, one that does not permit them to synthesize cartilage matrix, will revert to synthesizing that type of matrix when they are returned to a "permissive" medium.

However, when the cells are cultured at low density in another slightly different medium for several weeks, a steadily increasing proportion of them undergo a fundamental change: they cease to make type II collagen, which is the type characteristic of cartilage, and instead begin to make type I collagen, which is characteristic of fibroblasts. The two types of collagen (which can be distinguished by means of fluorescently labeled antibodies) are the products of different genes. It seems that some of the chondrocytes have been converted into fibroblasts. By the end of a month, almost all the cells in the low-density culture have switched to making type I collagen. The switch must occur abruptly, since very few cells are ever observed to make both types of collagen simultaneously.

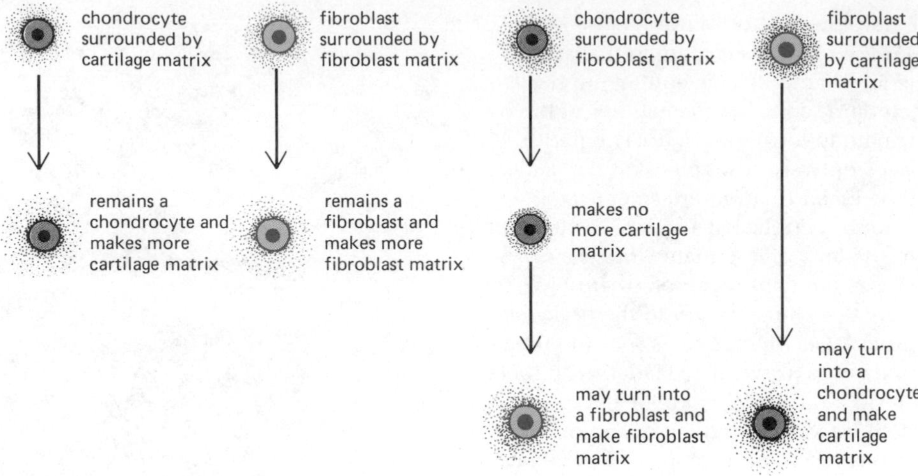

Figure 16–3 The hypothesized effects of the extracellular matrix on the differentiation of fibroblasts and chondrocytes. The evidence that fibroblasts can be converted into chondrocytes comes from observations of the growth of cartilage *in vivo* by recruitment of cells resembling fibroblasts from the perichondrium (see p. 934).

The mechanism controlling the switch is not known for certain. Some evidence suggests that it may depend on cell shape. When the cells are grown on substrata of varying degrees of stickiness, the switching of phenotype is correlated with the extent to which the cells become flattened down onto the substratum. Another line of evidence suggests that the extracellular matrix macromolecules secreted by the chondrocytes and fibroblasts influence the switching process. The chrondrocytes grow as clones, forming separate colonies on the surface of the tissue-culture dish. In the centers of the colonies the chondrocytes tend to surround themselves with cartilage matrix; at the periphery this matrix is less plentiful. The peripheral cells tend to be the earliest to switch to type I collagen synthesis, and the central cells tend to be the last. It has been proposed, therefore, that the matrix secreted by the chondrocytes helps to maintain the chondrocyte phenotype.

Other experiments appear to support this hypothesis. When the characteristic proteoglycans of cartilage matrix are added to a culture of chondrocytes, they stimulate the cells to synthesize yet more matrix of the same type. This suggests that the extracellular material acts as part of a positive feedback loop, making the synthesis of cartilage matrix a self-sustaining process. Hyaluronic acid also affects chondrocyte differentiation, though in the reverse manner. Fibroblasts secrete large quantities of hyaluronic acid, while chondrocytes secrete relatively little. When free hyaluronic acid is added to cultured chondrocytes, it powerfully inhibits the synthesis of cartilage matrix.

Under normal circumstances in an adult animal, chondrocytes and fibroblasts appear to be stably differentiated cell types. The above experiments suggest that the matrices secreted by these cells act on the cells themselves and thus help maintain their distinctive states of differentiation (Figure 16–3).

Cell-Cell Interactions Can Modulate the Differentiated State[4]

If a cell's state of differentiation is affected by extracellular material, it follows that it must be affected by surrounding cells that secrete the material. Hence there is presumably a cooperative interaction between chondrocytes, such that they stimulate one another to make cartilage matrix, and an antagonistic interaction with fibroblasts, whose presence tends to inhibit chondrocyte differentiation. Thus not all aspects of a cell's state of differentiation are maintained autonomously.

Even in the adult, therefore, the character of a cell may alter when its environment changes. These alterations of character, however, are rarely very great. Most of them can be classified as **modulations** of the differentiated state—that is, reversible interconversions between closely related cell phenotypes. The modulations may depend on short-range interactions with neighboring cells, similar to the interactions that control cell character in the embryo, or on other signals, such as hormones secreted into the circulation. For example, liver cells adjust their synthesis of specific enzymes (through changes in specific mRNA levels) according to the ambient concentration of the steroid hormone hydrocortisone. But radical transformations—say from liver cell to nerve cell—in general are prohibited.

The skin provides a good example of modulation of cell character due to the influence of neighboring cells. In the embryo, the development of the *epidermis*, or ectoderm, is governed by the embryonic *dermis* beneath it (p. 871). But is this interaction transient, a purely embryonic affair of which the adult cells have only memories, or is it sustained throughout life? The question can be answered by repeating with adult tissues the types of experiments performed with embryos. For example, one can combine the epidermal layer of the skin from one region, say the ear, with the dermal layer from another region, such as the sole of the foot, where the skin has a markedly different character. It is found that, just as in the embryo, the dermis dictates the behavior of the epidermal cells: thus epidermis from the ear becomes converted to the distinctive thick, ridged appearance of the foot epidermis if it is combined with foot dermis. The regional specializations of the epidermis are therefore governed by regional cues from the dermis that operate continuously in the adult.

It should be emphasized that the state of differentiation is modulated in this way only to a limited extent: the epidermis is not entirely dependent on cues from its environment to maintain its character. Epidermis of the tongue, for instance, keeps the appearance of tongue epidermis even when it is grafted in combination with ear dermis. Furthermore, although epidermal cells may change their regional state of specialization, they remain epidermal cells even in the most alien surroundings: if a suspension of dissociated epidermal cells is prepared from the tail of a rat and injected beneath the capsule of its kidney, the cells grow there to form epidermal cysts, which contain hair follicles and sebaceous glands, like skin on the surface of the body.

Some Structures Are Maintained by a Continuing Interaction Between Their Parts: Taste Buds and Their Nerve Supply[5]

Taste buds provide one of the most extreme examples of a state of differentiation that depends on a continuing cell-cell interaction. These little structures by which we perceive sweetness, sourness, saltiness, and bitterness are formed chiefly in the epithelium on the upper surface of the tongue. Each consists of about 50 cells that are easily distinguished by their shape from the epithelial cells that surround them (Figure 16–4). The elongated cells of the taste bud are arranged like the staves of a barrel, extending through the full thickness of the epithelium and forming a small opening (the taste pore) to the exterior. Through this pore, presumably, must pass the molecules to be tasted. Two types of cells can be distinguished in the taste bud, one appearing pale and the other dark; and these, in a manner still unknown, act as the taste transducers. The sensory signal is conveyed to the brain by nerves whose fibers penetrate the taste bud and end in contact with its cells. If the nerves are severed, the taste buds disappear entirely. When the nerves regenerate, they induce epithelial cells to change their state of differentiation to form new taste buds. Taste buds can be caused to form even in a region of epithelium that normally has none, such as that on the undersurface of the tongue, if it is cultured together with an appropriate sensory ganglion and thereby becomes innervated.

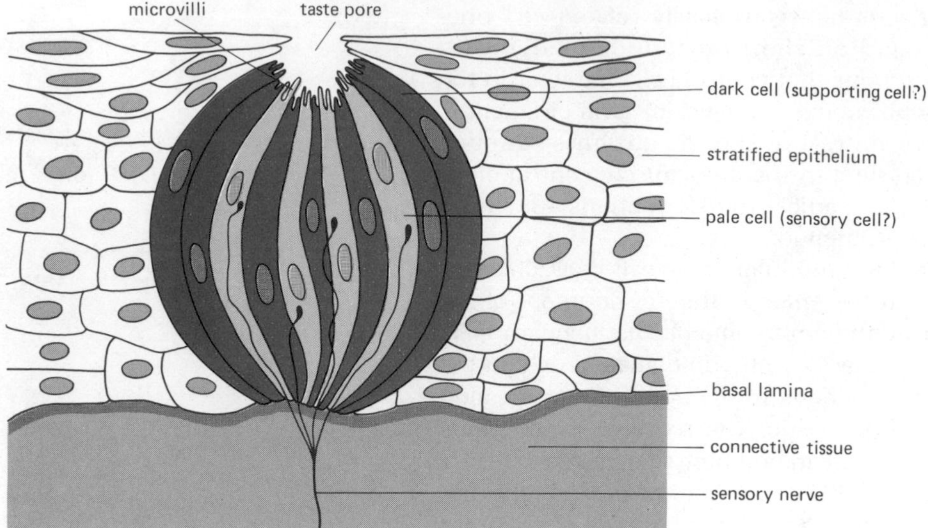

Figure 16–4 Schematic diagram of a taste bud.

An Agent That Causes Changes in DNA Methylation Can Cause Radical Alterations of Differentiated Character[6]

A few other instances of extreme transformations from one differentiated state to another are known. In an adult newt, for example, some of the pigmented epithelial cells of the iris will alter their character to form a lens if the existing lens is removed. This type of change is called *transdifferentiation* or *metaplasia.* Such cases, however, are exceptional. For the most part, transitions from one terminally differentiated state to another that is radically different do not occur in normal cells, though they are occasionally observed in cancer cells.

The striking stability of the differentiated state of most cells would be no mystery if the differentiated cells had actually lost the genes that they do not express. But, as discussed earlier, that is not the case: most differentiated cells still contain a complete genome. We might gain some insight into the mechanisms that ordinarily prohibit metaplasia, if only we could discover some extraordinary treatment that could cause the prohibited transitions to occur. A few agents are known that have such effects on certain cell types.

In particular, some recent experiments on cultured cell lines indicate that metaplasia can be caused by substances that interfere with DNA methylation, and so support the idea that DNA methylation plays a part in keeping some genes stably active or stably repressed, presumably by the mechanism discussed on page 461. A simple test of the hypothesis entails artificially altering the state of methylation of the DNA and looking for any change in the differentiated character of the cells. The method used involves growing the cells for one or more cell cycles in the synthetic nucleotide analogue 5-aza C, which becomes incorporated in DNA in place of some C residues. The 5-aza C is not only incapable of being methylated, but it also powerfully inhibits the activities of the methylating enzyme. This breaks the chain of events by which the pattern of DNA methylation of a gene is passed from one cell generation to the next. When cultured cell lines resembling fibroblasts are treated in this way, they differentiate subsequently into a variety of cell types, including skeletal muscle cells, to which they never give rise under normal conditions. It has also been shown that 5-aza C treatment of cells containing an inactive X chromosome from a human female can sometimes cause reactivation and expression of some of the genes on that chromosome. These and other pieces of evidence suggest that DNA methylation can be important in the maintenance of the differentiated state.

Summary

During embryonic development, the cells of the body become irreversibly differentiated. Thus retinal pigment cells remain specialized as such even through 50 cell generations of growth in vitro. *Though states of differentiation are generally stable and not interconvertible, some cell types undergo certain limited alterations. For example, chondrocytes can apparently turn into fibroblasts. The extracellular matrix that surrounds these cells seems to govern the type of matrix that they synthesize, in such a way that production of cartilage matrix or of fibroblast matrix normally becomes a self-sustaining process. Minor reversible changes, or modulations, of the differentiated state can occur in many other cell types, and they reflect the importance of continuing cell-cell interactions. Taste buds provide an extreme example of a state of differentiation that depends on a continuing interaction, for their specialized cells disappear completely in the absence of nerves and reappear when innervation is restored. In a few cases, radical transformations of the state of cell differentiation have been provoked artificially by certain chemical treatments.*

Tissues with Permanent Cells

Not all the populations of differentiated cells in the body are subject to turnover and renewal. Some cell types, having been generated in appropriate numbers in the embryo, are retained throughout adult life; they are never seen to divide, and, if lost, they cannot be replaced. Almost all the varieties of nerve cells are permanent in this sense. So are a few other types of cells, including, in mammals, the muscle cells of the heart and the lens cells of the eye.

While all these cells have an extremely long life span, and necessarily live in protected environments, they are quite dissimilar in other respects, and it is difficult to give a general reason why they should be permanent when so many other cell populations are subject to renewal. In fact, for heart muscle cells it is difficult to give any reason at all. In the case of nerve cells (which will be discussed in detail in Chapter 18), it seems likely that extensive cell turnover in the adult would as a rule be disadvantageous, since it would be difficult to reestablish in the adult the precise and complex pattern of nerve connections, set up under quite different circumstances during development, on which the function of the nervous system depends. Moreover, any memories recorded in the form of slight modifications of the structure or interconnections of individual nerve cells would presumably be obliterated in the course of cell turnover. In the lens, on the other hand, the permanence of the cells appears to be simply an inevitable consequence of the mode of growth of the tissue.

The Cells at the Center of an Adult Lens Are Remnants of the Embryo[7]

Very little of the adult body consists of the same molecules that were laid down in the embryo. The **lens** of the eye is one of the few structures whose cells are not only preserved but are preserved without turnover of their contents.

The lens is formed from the ectoderm at the site where the developing optic vesicle makes contact with it. The ectoderm here thickens, invaginates, and finally pinches off as a *lens vesicle* (Figure 16–2). The lens thus originates as a spherical shell of cells formed from an epithelium, one cell layer thick, surrounding a central cavity. The rear part of the epithelium—that is, the part facing the retina—soon undergoes a striking transformation. Its cells synthe-

Figure 16–5 Scanning electron micrograph of lens fibers in a fragment of an adult human lens. The lens fibers are the closely stacked objects resembling planks in a timber yard. Each one is a single, lifeless, elongated cell. Individual lens fibers have lengths of up to 12 mm. (From R. G. Kessel and R. H. Kardon, Tissues and Organs: A Text-Atlas of Scanning Electron Microscopy. San Francisco: Freeman, 1979. © 1979 W. H. Freeman and Company.)

size and become filled with *crystallins,* the characteristic proteins of the lens. In the process they elongate enormously, differentiating into *lens fibers.* Eventually their nuclei disintegrate and protein synthesis ceases. In this way, the part of the lens vesicle epithelium facing the retina is expanded into a thick refractile body, consisting of many tall, prism-shaped, lifeless cells packed side by side (Figure 16–5), and the central cavity of the vesicle is obliterated (Figure 16–6). Meanwhile the front part of the epithelium of the lens vesicle—the part facing the external world—remains as a thin sheet of low cuboidal cells (Figure 16–7). Growth of the lens depends on the proliferation of these cells at the front, pushing some of the cells from this region around the rim of the lens,

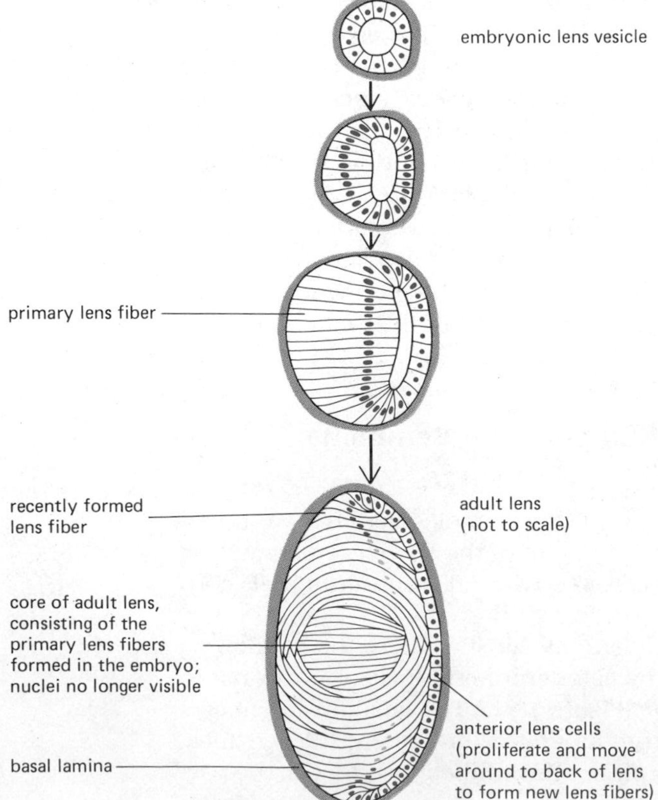

Figure 16–6 The development of the human lens (schematic).

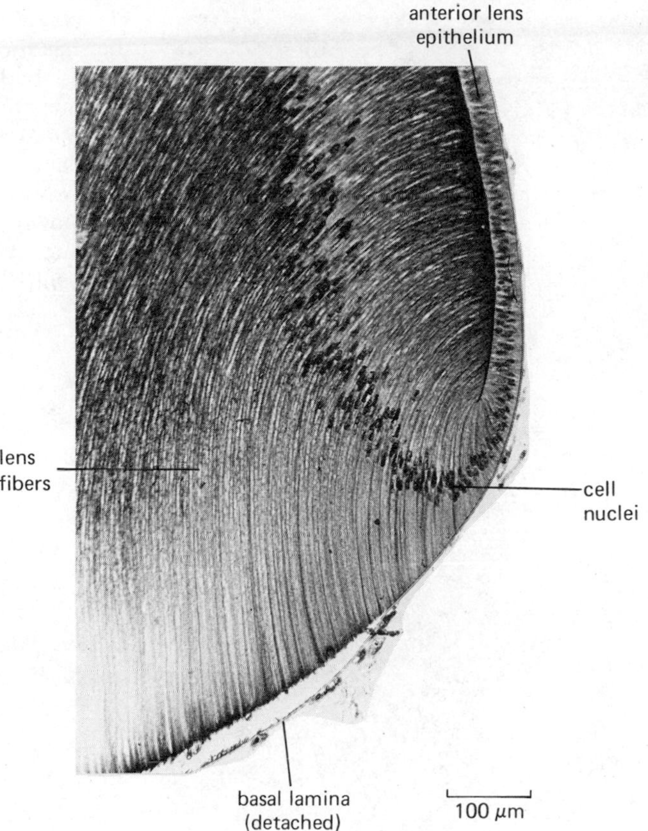

Figure 16–7 Light micrograph of part of the rim of a mature lens, showing the junction between the thin sheet of anterior lens epithelium that covers the front of the lens and the differentiated lens fibers to the rear. Continuing growth of the anterior lens epithelium pushes additional cells back around the rim of the lens to become lens fibers. (Courtesy of Peter Gould.)

toward the back. As cells move to the rear, they stop dividing, begin to synthesize crystallins, and differentiate into lens fibers. Additional lens fibers continue to be recruited in this way throughout life, though at an ever decreasing rate.

The types of crystallins filling the earliest generations of lens fibers are different from those of the later generations, just as the hemoglobins of fetal red blood cells are different from those of adult red blood cells. But whereas red blood cells are discarded, lens fibers are not. Thus at the core of the adult lens lie fibers that were laid down in the embryo and are still packed with the distinctive types of crystallins manufactured in that earlier period. Differences of refractive index between the early embryonic types of crystallins and those that are laid down later help to free the lens of the eye from the optical aberrations that bedevil simple lenses made out of homogeneous media such as glass.

Most Permanent Cells Renew Their Parts: Photoreceptor Cells of the Retina[8]

There are few cells as immutable as the lens fibers. As a rule, even those cells that persist throughout life without dividing undergo renewal of their component parts. Thus, while they do not divide, heart muscle cells and nerve cells are metabolically active and capable not only of synthesizing new RNA and protein, but also of altering their size and structure during adult life. Heart muscle cells, for example, grow bigger if the load on them is increased, while nerve cells can regenerate axons and dendrites that have been cut off (see Chapter 18).

The process of turnover of cell components is illustrated in a particularly striking way in the highly specialized neural cells that form the **photorecep-**

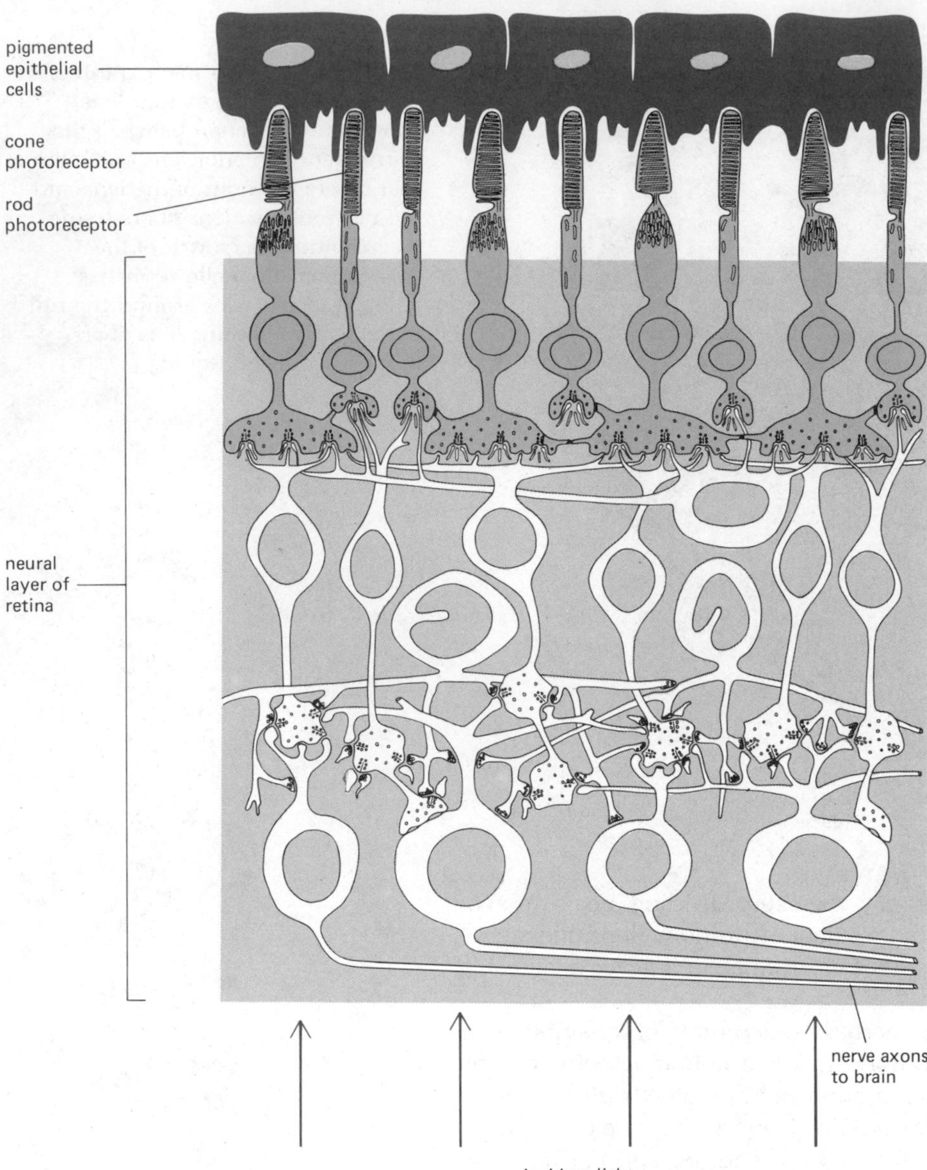

Figure 16–8 Schematic diagram of the structure of the retina. The spaces shaded pale gray between neurons and between photoreceptors in the neural retina are occupied by a population of specialized supporting cells, whose individual outlines are not shown here. (Modified from J. E. Dowling and B. B. Boycott, *Proc. R. Soc. Lond. (Biol.)* 166:80–111, 1966.)

tors of the retina. The neural retina consists of several cell layers, organized in a way that seems perverse. The neurons that transmit visual signals to the brain lie closest to the external world, so that the light, focused by the lens, must pass through them to reach the photoreceptor cells (Figure 16–8). These cells lie with their photoreceptive ends, or *outer segments,* partly buried in the pigment epithelium. The photoreceptors are classified as **rod cells** or **cone cells,** according to their shape. They contain different photosensitive complexes of protein with visual pigment. Rods are especially sensitive at low light levels, while the cones, of three different varieties with different spectral responses, serve for color perception. The outer segment of each type of photoreceptor appears to be a modified cilium with a characteristic ciliumlike arrangement of microtubules in the region where the outer segment is connected with the rest of the cell (Figure 16–9). The rest of the outer segment is almost entirely filled, however, with a dense stack of membranes in which the photosensitive proteins carrying the visual pigment are embedded. Meanwhile, at their opposite, innermost ends, the photoreceptor cells form synapses on a set of retinal interneurons.

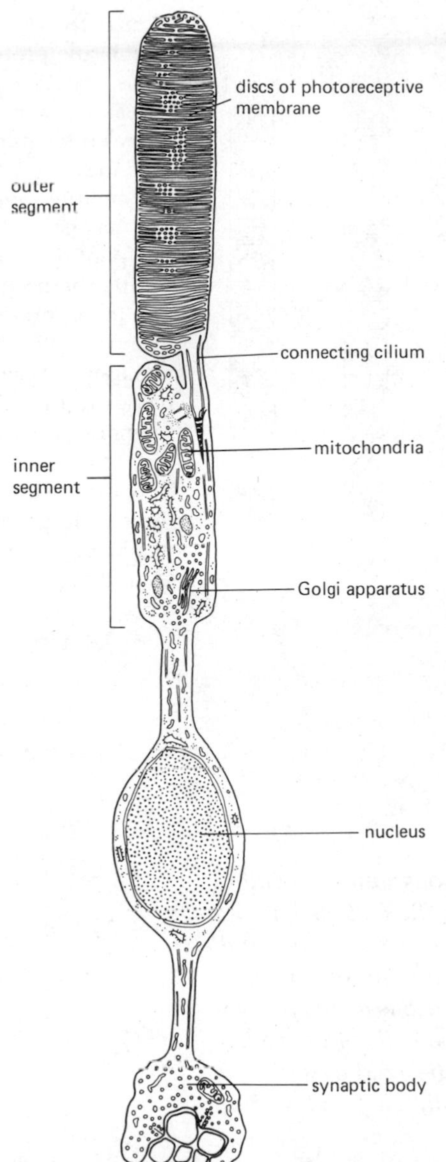

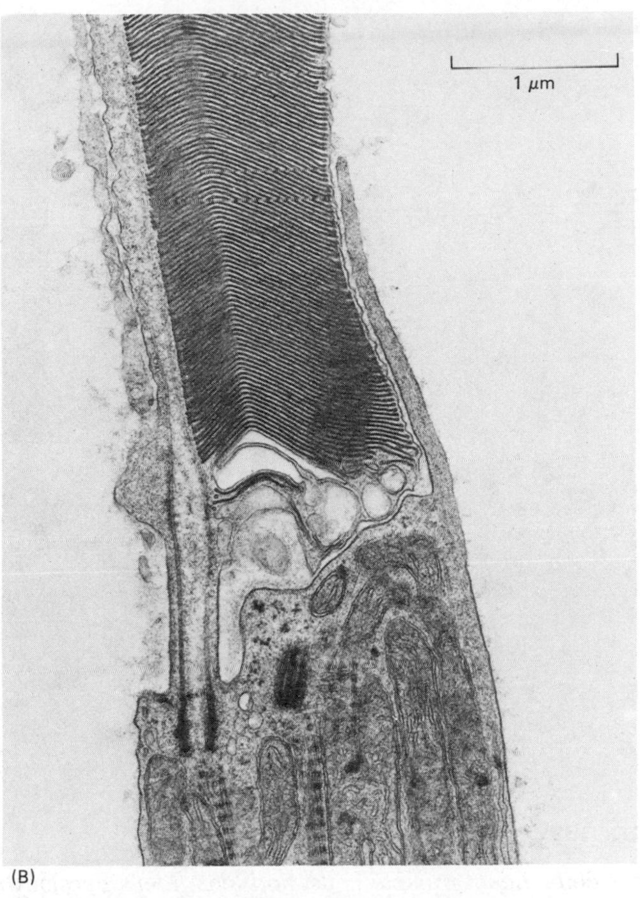

Figure 16–9 (A) Schematic diagram of a rod photoreceptor cell. The actual number of photoreceptive discs in the outer segment is about 1000. (B) Electron micrograph of part of a rod photoreceptor, showing the base of the outer segment and the modified cilium that connects it to the inner segment. (A, from T. L. Lentz, Cell Fine Structure. Philadelphia: Saunders, 1971; B, from M. J. Hogan, J. A. Alvarado, and J. E. Weddell, Histology of the Human Eye: An Atlas and Textbook. Philadelphia: Saunders, 1971.)

The photoreceptors are permanent cells and are unable to divide. But the photosensitive protein molecules are not permanent. There is a steady turnover, which can be demonstrated by the continuing incorporation of radioactive amino acids into these molecules. In the rods (though not, curiously, in the cones), this turnover is organized in an orderly production line, which can be analyzed by following the passage of a cohort of labeled protein molecules through the cell after a short pulse of radioactive amino acid has been given (Figure 16–10). After the usual stages of incorporation into protein and packaging in the Golgi apparatus in the inner segment of the cell, the radioactivity appears first at the base of the stack of membranes in the outer segment. From here it is gradually displaced toward the tip as new material is fed into the base of the stack. Finally (after about 10 days in the rat), on reaching the tip of the outer segment, the labeled proteins and the layers of membrane in which they are embedded are phagocytosed (chewed off and digested) by the cells of the pigment epithelium.

We shall have more to say about photoreceptors and their function in the nervous system in Chapter 18.

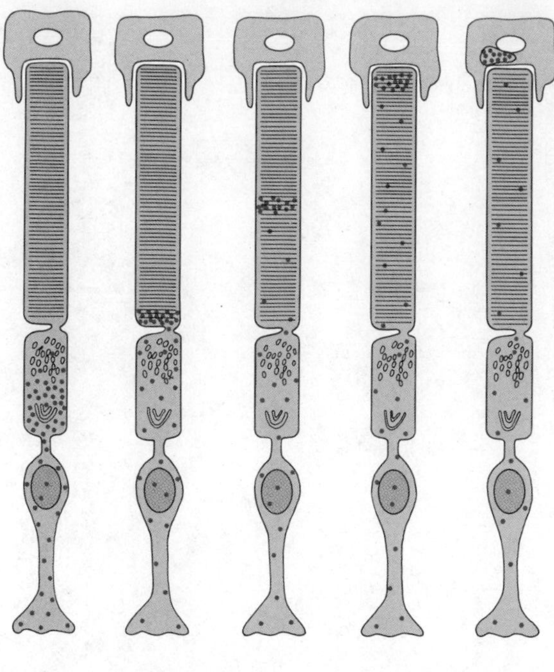

Figure 16–10 Turnover of membrane protein in a rod cell. A pulse of [^{3}H]leucine is supplied and its passage through the cell is followed by autoradiography. Colored dots indicate sites of radioactivity. The method only reveals the leucine that has been incorporated into polypeptides; the rest is washed out during the preparation of the tissue. The incorporated leucine is first seen in the neighborhood of the Golgi apparatus (1), and from there it passes to the base of the outer segment into a newly synthesized disc of photoreceptive membrane (2). New discs are formed at a rate of three or four per hour (in a mammal), displacing the older discs toward the pigment epithelium (3–5).

Summary

Nerve cells, heart muscle cells, and lens fibers persist throughout life without dividing and without being replaced. In mature lens fibers, the cell nuclei have degenerated and protein synthesis has stopped, so that the core of the adult lens consists of lens proteins laid down early in embryonic life. But in most other permanent cells, metabolic activity continues, and there is a steady turnover of cell components. This is clearly displayed in the rod cells of the retina, in which new layers of photoreceptive membrane are synthesized close to the nucleus, are steadily displaced outward, and are eventually engulfed and digested by cells of the pigment epithelium of the retina.

Renewal by Simple Duplication[9]

Most of the differentiated cell populations in a vertebrate are not permanent, but instead are subject to renewal. New differentiated cells can be produced during adult life in either of two ways: (1) they can form by the *simple duplication* of existing differentiated cells, which divide to give pairs of daughter cells of the same type; or (2) they can be generated from undifferentiated *stem cells* by a process that involves a change of cell phenotype, as will be explained later in this chapter.

Rates of renewal vary from one tissue to another. The turnover time may be as short as a week or less, as in the epithelial lining of the small intestine (which is renewed by means of stem cells), or as long as a year or more, as in the pancreas (which is renewed by simple duplication). Many tissues whose normal rates of renewal are very slow can be stimulated to produce new cells at higher rates when the need arises. We shall discuss the liver and the endothelial cells that line blood vessels as two examples of cell populations that are renewed by simple duplication. Both these tissues normally have a slow cell turnover but can regenerate new cells rapidly after damage.

The Liver Is an Interface Between the Digestive Tract and the Blood[10]

Digestion is a complex process. The cells that line the digestive tract secrete into the lumen of the gut a variety of substances, such as hydrochloric acid and digestive enzymes, to break down food molecules into simpler nutrients. The cells absorb these nutrients from the gut lumen, process them, and then release them into the blood for utilization by other cells of the body. All of these activities are adjusted according to the composition of the food consumed and the levels of metabolites in the circulation. The complex set of tasks is performed by a division of labor (Figure 16–11): some of the cells are

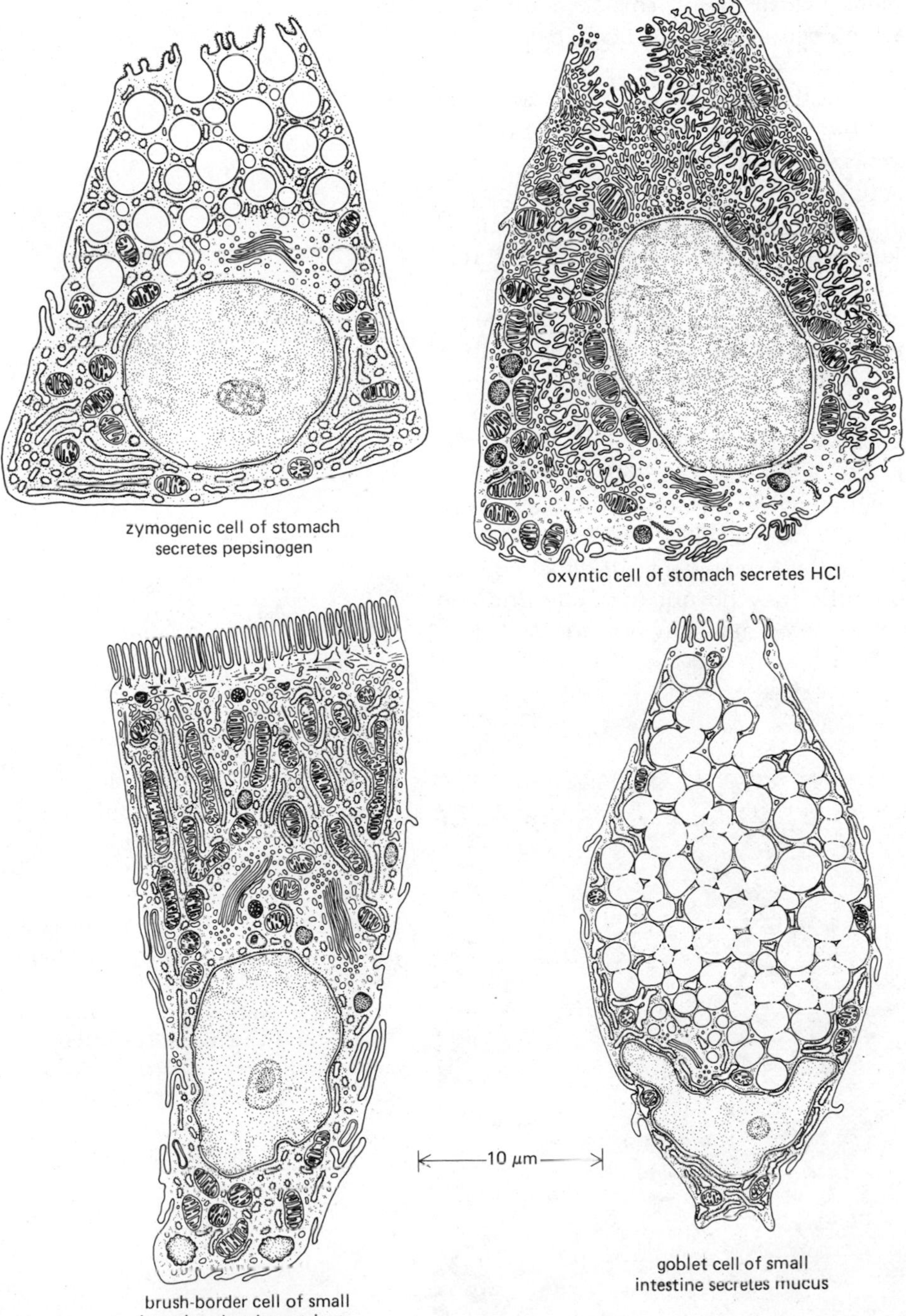

Figure 16–11 Some of the specialized cell types found in the epithelial lining of the gut. Neighboring positions in the epithelial sheet are often occupied by cells of dissimilar types (see Figure 16–20B). (After T. L. Lentz, Cell Fine Structure. Philadelphia: Saunders, 1971.)

specialized for the secretion of HCl, others for the secretion of enzymes, others for absorption of nutrients, others for the production of peptide hormones, such as gastrin, that regulate digestive and metabolic activities, and so on. Some of these different cell types lie closely intermingled in the wall of the gut; others are segregated in large glands that communicate with the gut and originate in the embryo as outgrowths of the gut epithelium.

The liver is the largest of the glands that communicate with the gut. In the embryo it develops at a site where a major vein runs close to the wall of the primitive gut tube, and the adult organ retains a singularly close relationship with the blood. The cells in the liver that derive from the primitive gut epithelium—the **hepatocytes**—are arranged in folded sheets, facing blood-filled spaces called *sinusoids* (Figure 16–12). The blood is separated from the surface of the hepatocytes by a single layer of flattened endothelial cells that covers the sides of each hepatocyte sheet (Figure 16–13). This structure facilitates the chief functions of the liver, which center on the exchange of metabolites between hepatocytes and the blood.

The liver is the main site at which nutrients that have been absorbed from the gut and then transferred to the blood are processed for use by other cells of the body. Hepatocytes are thus responsible for the synthesis, degradation, and storage of a vast number of different substances. At the same time, the hepatocytes remain connected with the lumen of the gut via a system of minute channels (or *canaliculi*) and larger ducts (Figure 16–13), and secrete into the gut by this route an emulsifying agent, *bile*, which helps in the absorption of fats. Within the population of hepatocytes there seems to be (in contrast to the rest of the digestive tract) remarkably little division of labor: each hepatocyte appears to be able to perform the same broad range of metabolic and secretory tasks.

The hepatocytes in the liver have also a different life-style from the cells that line the lumen of the gut itself. The latter are exposed to very harsh conditions; they cannot live for long in contact with the abrasive and corrosive contents of the gut, and they must be rapidly replaced by a continual supply of newborn cells. The hepatocytes, however, are removed from direct contact with the contents of the gut; consequently, they do not normally undergo such rapid turnover. They are normally renewed at a slow but precisely controlled rate.

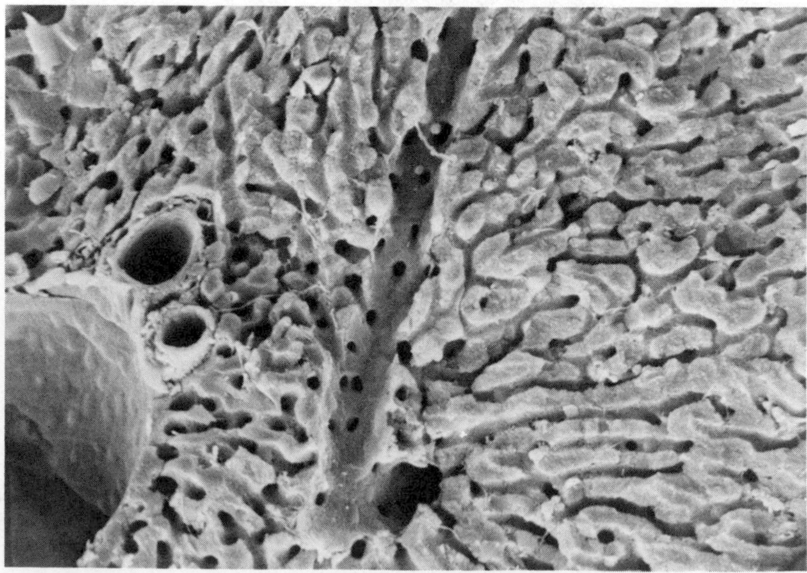

Figure 16–12 Scanning electron micrograph of a portion of the liver, showing the irregular sheets of hepatocytes and the many small channels, or sinusoids, for the flow of blood. The larger channels are vessels that distribute and collect the blood that flows through the sinusoids. (From R. G. Kessel and R. H. Kardon, Tissues and Organs: A Text-Atlas of Scanning Electron Microscopy. San Francisco: Freeman, 1979 © 1979 W. H. Freeman and Company.)

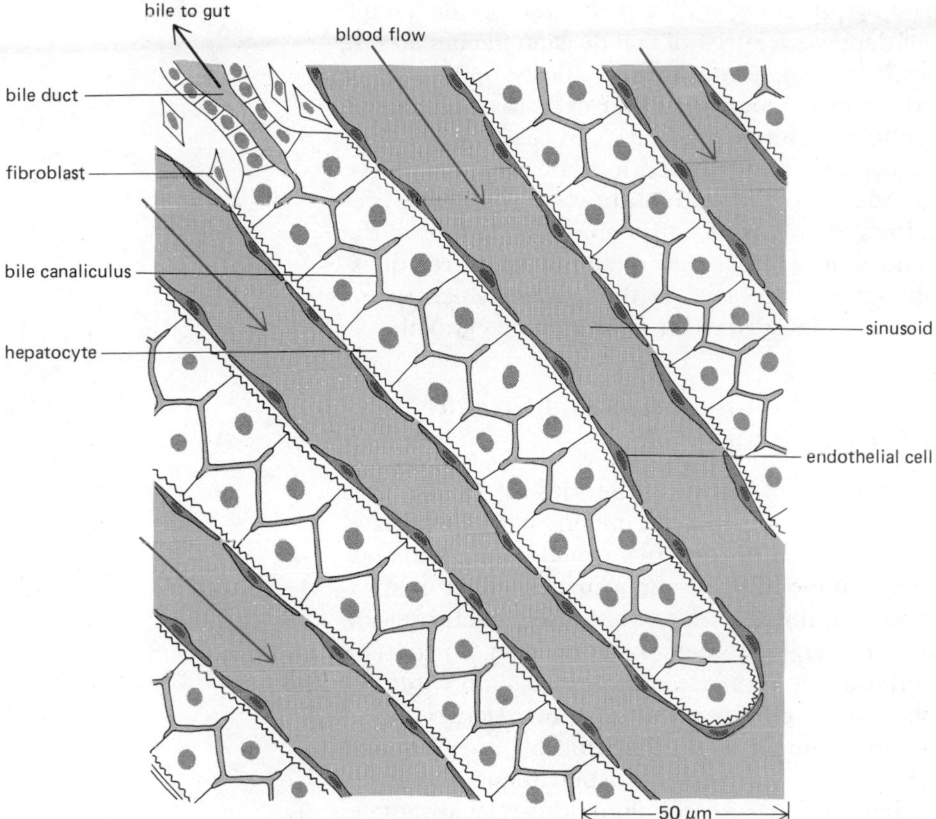

Figure 16–13 Schematic diagram of the fine structure of the liver. The hepatocytes are separated from the bloodstream by a single thin sheet of endothelial cells. Small holes in the endothelial sheet allow exchange of molecules and small particles between the hepatocytes and the bloodstream without exposing the hepatocytes to buffeting by direct contact with the circulating blood cells. Besides exchanging materials with the blood, the hepatocytes form a system of minute bile canaliculi into which they secrete bile, which is ultimately discharged into the gut via bile ducts. The real structure is less regular than this diagram suggests.

Liver Cell Loss Stimulates Liver Cell Proliferation[11]

Even in a slowly renewing tissue, a small but persistent imbalance between the rate of cell production and the rate of cell loss will lead to disaster. If 2% of the liver cells in a human were to divide each week, but only 1% were lost, the liver would grow to exceed the weight of the whole of the rest of the body within eight years. Some homeostatic mechanism must operate to adjust the rate of cell proliferation according to the mass of tissue present. The necessity for such control is all the greater in an organ like the liver, whose cells may from time to time be destroyed by poisons (such as alcohol) in the diet.

Firm evidence for homeostatic control of liver cell proliferation comes from experiments in which large numbers of hepatocytes are removed surgically or are intentionally killed by poisoning with carbon tetrachloride. Within a day or so after either sort of damage, a surge of cell division occurs among the surviving hepatocytes, and the lost tissue is very quickly replaced. For example, if two-thirds of a rat's liver is removed, a liver of nearly normal size can regenerate from the remainder within a week or so. In cases of this kind, a signal for liver regeneration can be demonstrated in the circulation: if two rats have their circulations surgically connected, and two-thirds of the liver of one of them is excised, mitosis is provoked in the unmutilated liver of the other. What the circulating factor is and how it acts remain unanswered questions. Similar regenerative phenomena are seen in the kidney, which seems to have its own separate but analogous system for controlling its growth.

Regeneration May Be Hindered by Uncoordinated Growth of the Components of a Mixed Tissue[12]

Like most tissues, the liver is a mixture of cell types. Besides the hepatocytes and the endothelial cells that line its sinusoids, it contains both specialized macrophages (*Kupffer cells*), which engulf particulate matter in the bloodstream and dispose of worn-out red blood cells, and a small number of fibroblasts, which provide a tenuous supporting framework of connective tissue (see Figure 16–13). All of these cell types are capable of division. For perfect regeneration, their proliferation must be properly coordinated. Embryonic development generates a balanced and well-organized mixture; regeneration in the adult may fail to do so. For example, if the hepatocytes are poisoned repeatedly with carbon tetrachloride or with alcohol at such frequent intervals that they cannot recover fully between attacks, the fibroblasts take advantage of the situation and the liver becomes irreversibly clogged with connective tissue, leaving little space for the hepatocytes to grow even after the toxic agents are withdrawn. This condition, called *cirrhosis*, is common in chronic alcoholics.

In a similar way, glial cells in the brain may proliferate to form a special kind of scar tissue, which blocks the sprouting of new nerve cell processes after damage. And the regeneration of skeletal muscle is often seriously hindered by the too rapid growth of its connective tissue component, so that scar tissue replaces the muscle fibers.

Endothelial Cells Constitute the Fundamental Component of All Blood Vessels[13]

By contrast with the above examples of ill-coordinated behavior of fibroblasts and glial cells, the **endothelial cells** that form the lining of blood vessels have a remarkable capacity to adjust their numbers and arrangement to suit local requirements. Almost all tissues depend on a blood supply, and the blood supply depends on endothelial cells. They create an adaptable life-support system, ramifying into every region of the body. If it were not for endothelial cells extending and remodeling the network of blood vessels, tissue growth and repair would be impossible.

The largest blood vessels are the arteries and the veins, which have a thick tough wall of connective tissue and smooth muscle (Figure 16–14). The wall is lined by an exceedingly thin, single layer of endothelial cells, separated from the surrounding outer layers by a basal lamina. The thickness of the connective tissue component of the vessel wall varies according to the vessel's diameter and function, but the endothelial lining is always present (Figure 16–15). In the finest branches of the vascular tree—the capillaries and sinusoids—the walls consist of nothing but endothelial cells and a basal lamina (Figure 16–16). Thus endothelial cells line the entire vascular system, from the

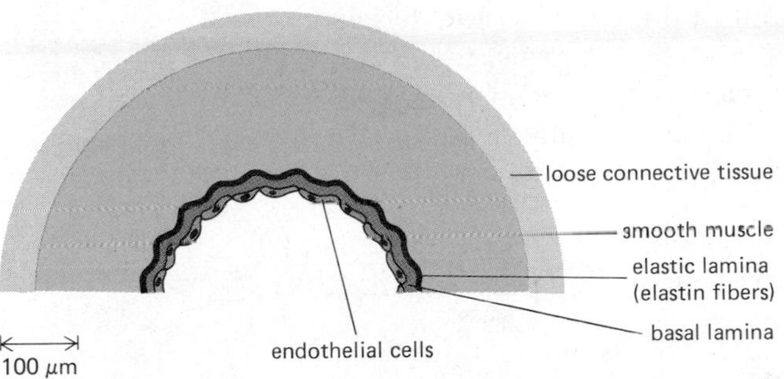

Figure 16–14 Schematic diagram of a part of the wall of a small artery. The endothelial cells, though inconspicuous, are the fundamental component. Compare with the capillary in Figure 16–16.

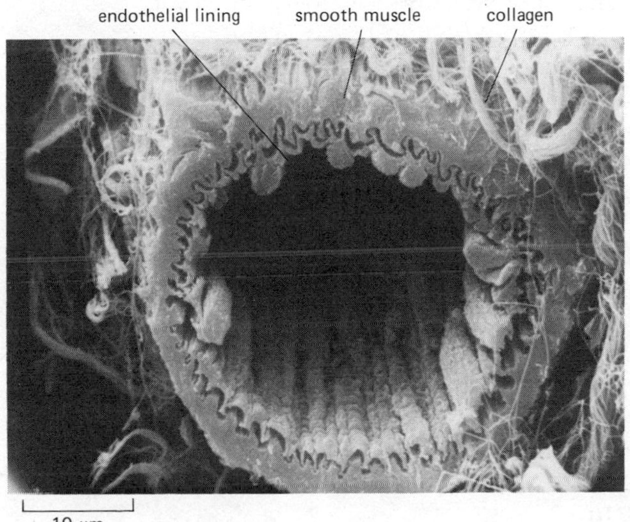

Figure 16–15 Scanning electron micrograph of a cross-section through an arteriole (very small artery), showing the inner lining of endothelial cells and the surrounding layer of smooth muscle and collagenous connective tissue. A slight contraction of the smooth muscle has thrown the endothelial lining of the vessel into folds. In fixation, the endothelial lining has shrunk away from the muscular wall, leaving a small gap. (From R. G. Kessel and R. H. Kardon, Tissues and Organs: A Text-Atlas of Scanning Electron Microscopy. San Francisco: Freeman, 1979. © 1979 W.H. Freeman and Company.)

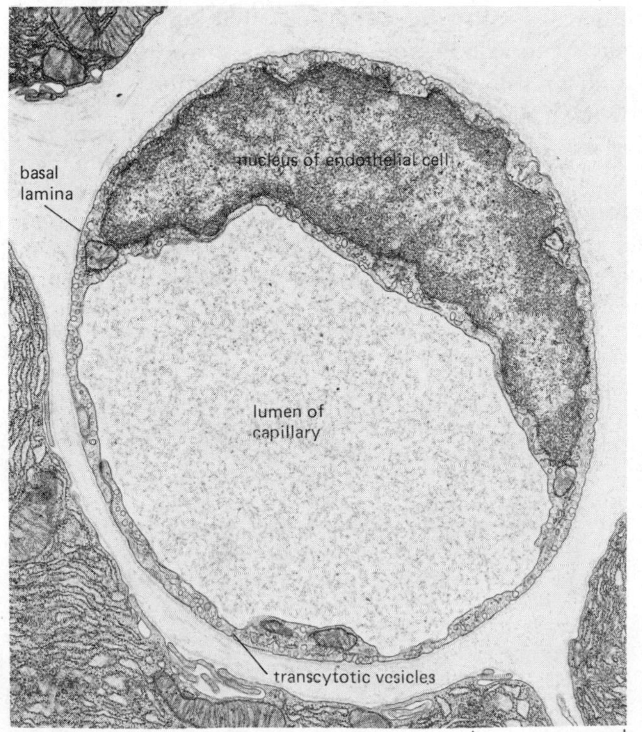

Figure 16–16 Electron micrograph of a small capillary from the pancreas, in cross-section. The wall is formed by a single endothelial cell. Note the small (80-nm) "transcytotic" vesicles, which are believed to provide transport in and out of this type of capillary: soluble materials are taken up into the vesicles by endocytosis at the luminal surface of the cell and discharged by exocytosis at the external surface, or vice versa. (From R. P. Bolender, *J. Cell Biol.* 61:269–287, 1974. Reproduced by permission of the Rockefeller University Press.)

heart to the smallest capillary, and control the passage of materials into and out of the bloodstream. A study of the embryo reveals, moreover, that the arteries and veins themselves have developed from small simple vessels constructed solely of endothelial cells and a basal lamina. The endothelial cells are the pioneers—thick layers of connective tissue and smooth muscle form around them later where required.

New Endothelial Cells Are Generated by Simple Duplication of Existing Endothelial Cells[14]

Throughout the vascular system of the adult, the endothelial cells retain a capacity for cell division and movement. If, for example, a part of the wall of the aorta is damaged and denuded of endothelial cells, new cells are formed in the endothelium surrounding the denuded area and migrate in to cover the exposed surface. The new cells are even capable of covering the inner surface of the plastic tubing used by surgeons to replace parts of damaged blood vessels.

The proliferation of endothelial cells can be demonstrated by using [^{3}H]thymidine to label cells in S phase. In normal vessels the proportion of endothelial cells that become labeled is especially high at branch points in arteries, where turbulence and the resulting wear on the endothelial cells seem to stimulate cell turnover. On the whole, however, endothelial cells turn over rather slowly, with roughly one in a hundred cells being replaced each day.

Endothelial cells not only repair the lining of established blood vessels, they also create new blood vessels. They must do this in the embryo to keep pace with the growth of the body; in normal adult tissues, such as bone or the wall of the uterus, which undergo recurrent cycles of tissue remodeling and reconstruction; and in the repair of damaged tissue.

New Capillaries Form by Sprouting[15]

New vessels originate as capillaries, which sprout from existing small vessels. This process of **angiogenesis** has been demonstrated in rabbits by punching a small hole in the ear and fixing glass cover slips on either side to create a thin viewing chamber with transparent walls into which the cells that surround the wound can grow. Angiogenesis can also be conveniently observed in naturally transparent structures such as the cornea of the eye. Irritants applied to the cornea induce the growth of new blood vessels from the rim of the cornea, which has a rich blood supply, in toward the center, which normally has almost none. Thus the cornea becomes vascularized through an invasion of endothelial cells into the tough collagen-packed corneal tissue.

Observations such as these reveal that the endothelial cells that will form a new capillary grow out from the side of an existing capillary or small venule by stretching out pseudopodia (Figure 16–17). The cells at first form a solid

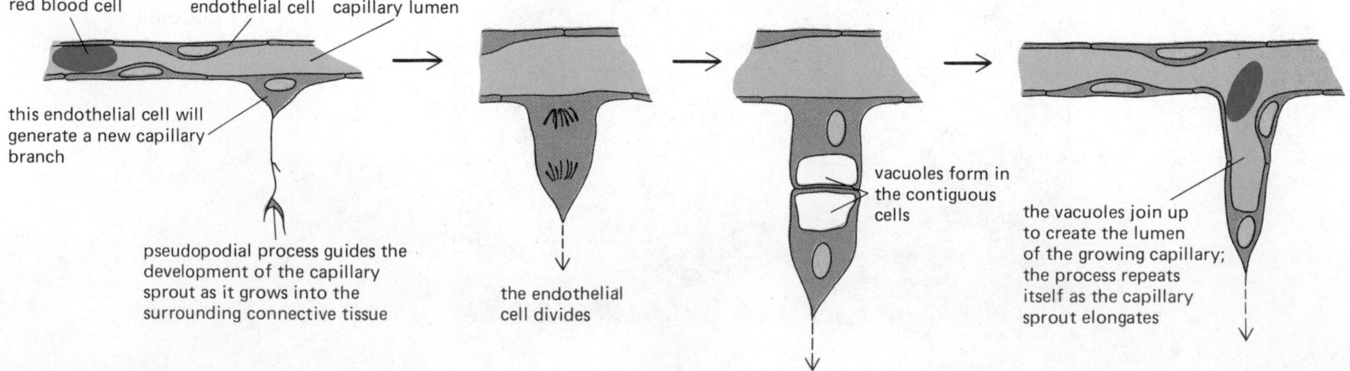

Figure 16–17 A new blood capillary forms by the sprouting of an endothelial cell from the wall of an existing small vessel. This schematic diagram is based on observations of cells in the transparent tail of a living tadpole. (After C. C. Speidel, *Am. J. Anat.* 52:1–79, 1933.)

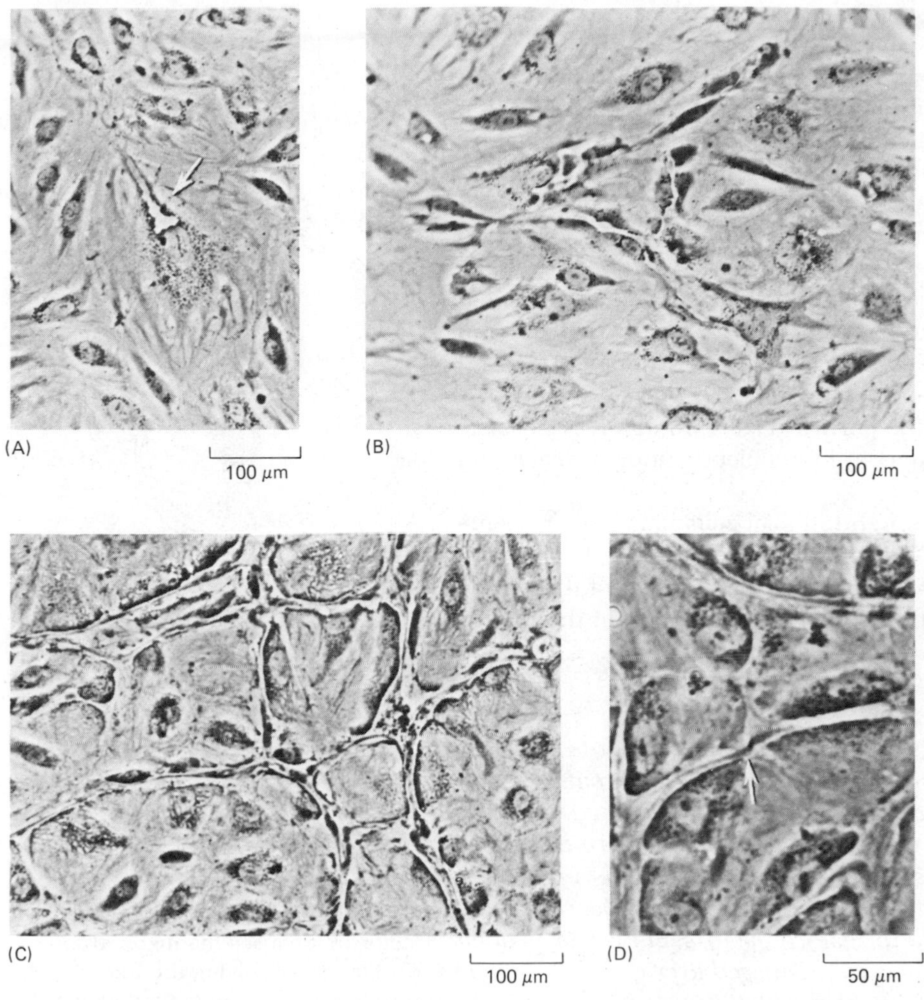

Figure 16–18 Endothelial cells in culture spontaneously develop internal vacuoles that join up, giving rise to a network of capillary tubes. Photographs (A), (B), and (C) show successive stages in the process; the arrow in (A) indicates a vacuole forming initially in a single endothelial cell. (D) shows the junction between two endothelial cells at higher magnification (the arrow marks the site where the capillary channels formed by the two cells connect).

The cultures are set up from small patches of two to four endothelial cells taken from short segments of capillary. These cells will settle on the surface of a collagen-coated culture dish and form a small flattened colony that enlarges gradually as the cells proliferate. The colony spreads across the dish, and eventually, after about 20 days, capillary tubes begin to form in the central regions. Once tube formation has started, branches soon appear, and after 5 to 10 more days, an extensive network of tubes is visible. Time-lapse photography shows that the established capillary tubes undergo remodeling, with new branches appearing while others retract into the parent vessel. (From J. Folkman and C. Haudenschild, *Nature* 288:551–556, 1980. © MacMillan Journals Ltd.)

sprout, which then becomes hollowed out to form a tube. This process continues until the sprout encounters another capillary, with which it connects, allowing blood to circulate. Experiments in tissue culture have shown that endothelial cells will spontaneously form capillary tubes even if they are isolated from all other types of cell (Figure 16–18). The first sign of tube formation in culture is the appearance in a cell of an elongated vacuole that is at first completely encompassed by cytoplasm. Contiguous cells develop similar vacuoles, and eventually the cells arrange their vacuoles end to end so that the vacuoles become continuous from cell to cell, forming a capillary channel. The capillary tubes that develop in a pure culture of endothelial cells do not contain blood, and nothing travels through them. Clearly, blood flow and pressure are not required for the formation of a capillary network.

Growth of the Capillary Network Is Controlled by Factors Released by the Surrounding Tissues[16]

In the living animal, endothelial cells form new capillaries only where there is a need for them. In wound repair, a short burst of capillary growth is stimulated in the neighborhood of the damaged tissue. Local irritants and local infections also cause a proliferation of new capillaries. There is some evidence that macrophages, which gather at sites of damage and infection, secrete a factor that induces endothelial cells to form new capillary sprouts.

Many of the newly formed capillaries will regress and disappear when the process of repair is complete.

Perhaps the most striking demonstration that tissues can produce signals for angiogenesis comes from studies on tumor growth. A tumor that grows as a solid mass remains small unless it is provided with capillaries: without a blood supply that extends into its interior, the tumor must rely on diffusion of nutrients from its exterior and so cannot enlarge beyond a diameter of a few millimeters. But if the tumor cells can induce the formation of a capillary network that invades the tumor mass, there need be no limit to the tumor's growth. There is good evidence that tumors capable of unlimited growth release a substance, called *tumor angiogenesis factor,* that acts on endothelial cells in just this way. A small sample of such tumor tissue implanted in the cornea will cause blood vessels to grow quickly toward the implant from the vascular margin of the cornea (Figure 16–19). It is possible that normal cells deprived of oxygen may attract a blood supply by secreting the same angiogenic factor.

The dependence of tumor cells on endothelial cells illustrates a theme to which we shall return at the end of this chapter. It shows that the problem of cancer must be considered not only in terms of the behavior of the cancer cell itself, but also in terms of its relationships with other cells in the body.

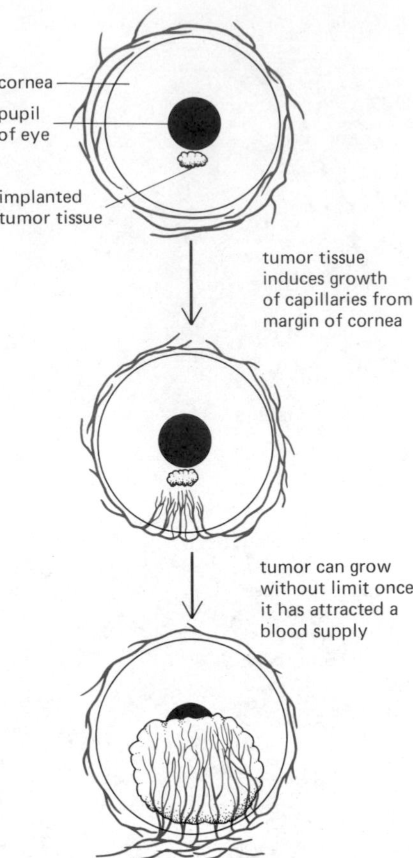

Figure 16–19 Tumor tissue implanted in the cornea releases a factor that causes the ingrowth of capillaries, supplying the tumor with blood-borne nutrients that allow it to grow. The ingrowth of capillaries is called angiogenesis.

Summary

Most populations of differentiated cells in vertebrates are subject to turnover through cell death and renewal. In some cases, the fully differentiated cell simply divides to produce daughter cells of the same differentiated type. Hepatocytes in the liver and endothelial cells lining blood vessels are examples. The rate of proliferation of such cells is controlled to maintain appropriate total cell numbers. Thus if a large part of the liver is destroyed, the remaining hepatocytes increase their division rate to restore the loss. But repair is often imperfect, as when the fibroblasts in a severely damaged liver grow too rapidly in relation to the hepatocytes and replace them with fibrous tissue.

Endothelial cells form a single cell layer that lines all blood vessels and regulates exchanges between the bloodstream and the surrounding tissues. New blood vessels develop from the walls of existing small vessels by the outgrowth of these endothelial cells, which have the capacity to form hollow capillary tubes even when isolated in culture. In the living animal, damaged tissues and some tumors attract a blood supply by secreting factors that stimulate nearby endothelial cells to construct new capillary sprouts. Tumors that fail to attract a blood supply are severely limited in their growth.

Renewal by Stem Cells: Epidermis

From cell populations that are renewed by simple duplication, we turn now to those that are renewed by means of **stem cells.** These populations vary widely, not only in cell character and rate of turnover, but also in the geometry of the process of cell replacement. In the lining of the small intestine, for example, cells are arranged as a single-layered epithelium. This epithelium covers the surfaces of the *villi* that project into the lumen of the gut, and it lines the deep *crypts* that descend into the underlying connective tissue (Figure 16–20). The stem cells lie in a protected position in the depths of the crypts. The differentiated cells generated from them (see p. 614) are carried upward by a sliding movement of the epithelial sheet until they reach the exposed surfaces of the villi, from whose tips they are finally shed. A contrasting example is found in the skin: here the epidermis is a many-layered epi-

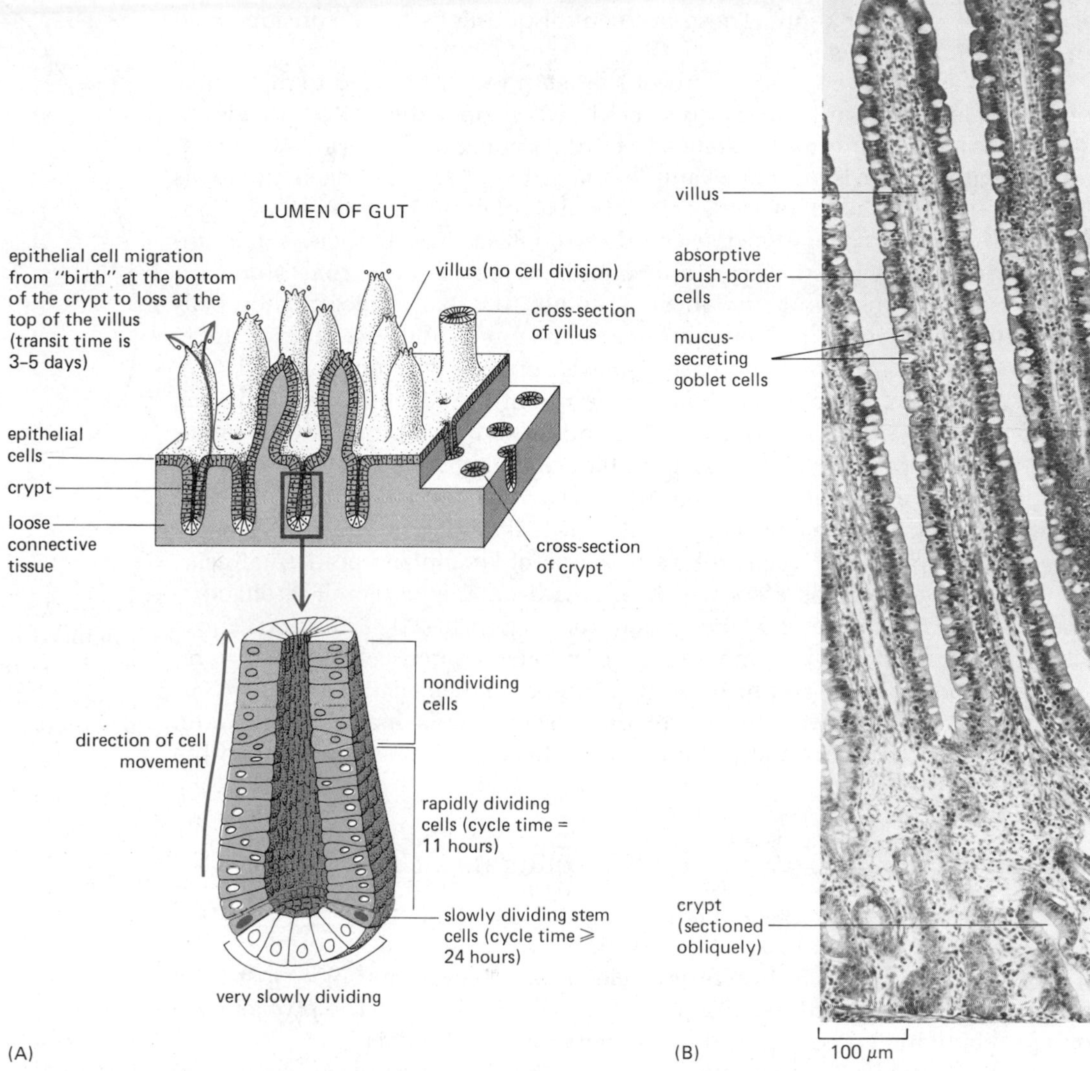

Figure 16–20 (A) Schematic diagram showing the pattern of cell turnover and the proliferation of stem cells in the lining of the small intestine. (B) Photograph of a section of part of the lining of the small intestine, showing the villi and crypts. Note how mucus-secreting goblet cells (visible as pale ovals) are interspersed among the absorptive brush-border cells in the epithelium of the villi. (Courtesy of Peter Gould.)

thelium, and the differentiating cells travel outward from their site of origin in a direction perpendicular to the plane of the cell sheet. In the case of blood cells, the spatial pattern of production appears chaotic. Before going further into such details, however, we must pause to consider what a stem cell is.

Stem Cells Have the Ability to Divide Without Limit and to Give Rise to Differentiated Progeny[17]

The defining properties of a stem cell are as follows:

1. It is not itself terminally differentiated (that is, it is not at the end of a pathway of differentiation).
2. It can divide without limit.
3. When it divides, each daughter has a choice: it can either remain a stem cell like its parent, or it can embark on a course leading irreversibly to terminal differentiation (Figure 16–21).

What factors determine whether stem cells exercise their ability to divide or stay quiescent? What governs the choice that a daughter cell must make between terminal differentiation and life as a stem cell? And what range of possibilities does a daughter cell have when it embarks on a pathway leading to terminal differentiation? These are central questions to be considered in the following sections.

Stem cells are required wherever there is a recurring need to make new differentiated cells and the differentiated cells cannot themselves divide. In several tissues, the terminal state of cell differentiation is obviously incompatible with cell division. For example, the cell nucleus may disintegrate, as in the outermost layers of the skin, or be extruded, as in the case of mammalian red blood cells. Alternatively, the cytoplasm may be heavily encumbered with materials, such as the myofibrils of muscle cells, that would get in the way of mitosis and cytokinesis. In other terminally differentiated cells the chemistry of differentiation may be in some more subtle way incompatible with cell division. In any such case, renewal must depend on stem cells.

The job of the stem cell is not to carry out the differentiated function, but to produce cells that will. Consequently, stem cells often have a rather nondescript appearance, making them hard to identify. But that is not to say that stem cells are all alike. Though not overtly differentiated, they are nevertheless *determined* (see p. 835): the muscle satellite cell, as a source of skeletal muscle; the epidermal basal cell, as a source of keratinized epidermal cells; the spermatogonium, as a source of spermatozoa; the basal cell of olfactory epithelium, as a source of olfactory neurons (Figure 16–22); and so on. Those stem cells that give rise to only one type of differentiated cell are called *unipotent;* those that give rise to more than one type are called *pluripotent.* We begin our discussion with the **epidermis,** for its simple spatial organization makes it relatively easy to study the natural history of its stem cells and the fate of their progeny.

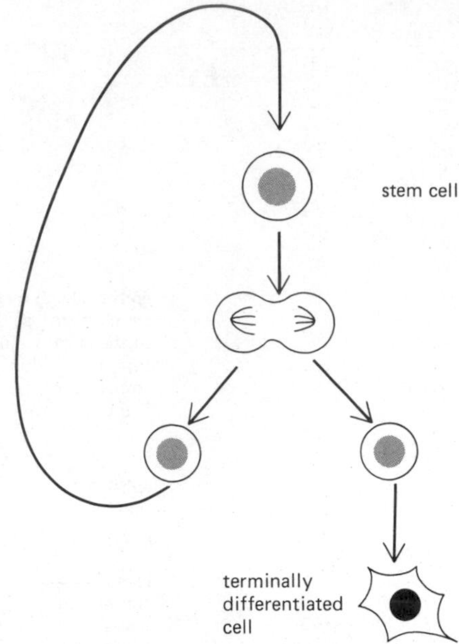

Figure 16–21 Each daughter produced when a stem cell divides can either remain a stem cell itself or go on to become terminally differentiated.

The Epidermis Is Organized into Proliferative Units[18,19]

The epidermal layer of the skin and the epithelial lining of the digestive tract are the two tissues that suffer the most direct and damaging encounters with the external world. In both, mature differentiated cells are rapidly lost from the most exposed positions and just as rapidly replaced by the proliferation of less differentiated cells that occupy more sheltered niches.

The epidermis comprises several layers that differ in appearance (Figure 16–23). The inner layers consist of metabolically active cells, strongly bound together by spot desmosome junctions; the cells of the outer layers are dead relics, packed full with the fibrous protein keratin. The innermost of the inner layers is composed of *basal cells* that sit on the basal lamina that separates the epidermis from the underlying dermis. It is chiefly these cells that undergo mitosis. Above the basal cells are several layers of larger, flatter *prickle cells.*

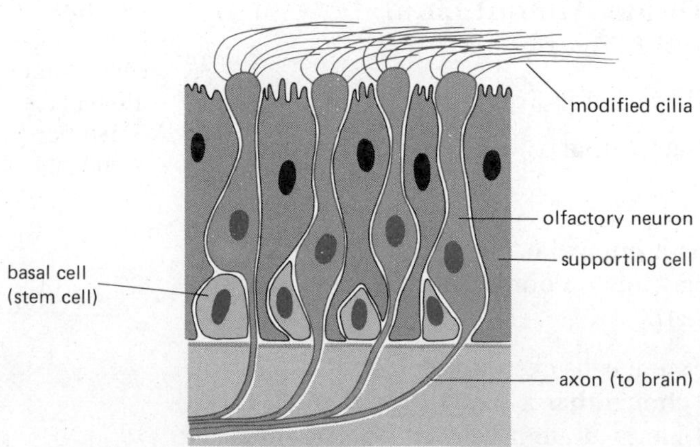

Figure 16–22 Schematic diagram of a section of olfactory epithelium (specialized for sensing smells). Three cell types can be distinguished: supporting cells, basal cells, and olfactory neurons. Autoradiographic experiments show that the basal cells are the stem cells for production of the olfactory neurons, which constitute one of the very few exceptions to the rule that neurons are permanent cells. Each olfactory neuron survives for about a month (in a mammal) before it is replaced. Six to eight modified cilia project from the globular head of the olfactory neuron and are believed to contain the smell receptors. The axon extending from the other end of the neuron conveys the message to the brain. A new axon must grow out and make appropriate connections whenever a basal cell differentiates into an olfactory neuron.

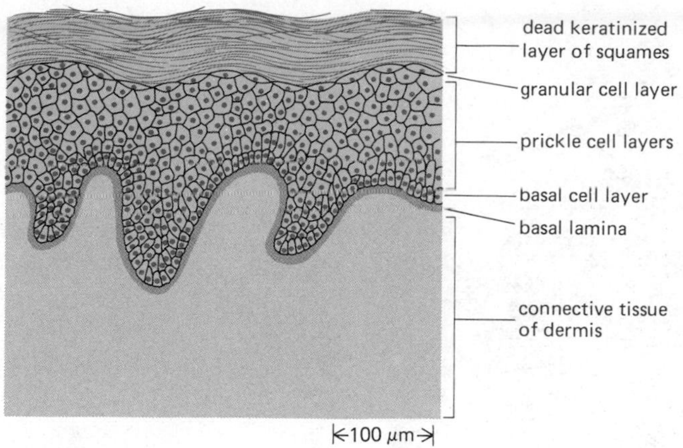

Figure 16–23 Schematic drawing of a section of moderately thick mammalian epidermis. The *granular cells* between the prickle cells and the flattened squames are in the penultimate stages of keratinization and contain darkly staining aggregates of a poorly characterized material called *keratohyalin* (quite distinct from keratin). In addition to the cells destined for keratinization, the deep layers of the epidermis include small numbers of cells of quite different character—specifically, macrophagelike *Langerhans cells,* derived from the bone marrow; *melanocytes,* derived from the neural crest; and *Merkel cells,* which are associated with nerve endings in the epidermis.

They take their name from their appearance in the light microscope: their innumerable spot desmosomes, with thick tufts of keratin filaments (also known as *tonofilaments*) inserted in each, are just visible as tiny prickles around the surfaces of the cells (Figure 16–24).

Still further out lie cells in which all the intracellular organelles have disappeared and which are thus reduced to flattened scales or *squames* containing practically nothing but keratin. The squames are so compressed and thin that their boundaries can hardly be made out in ordinary light microscope preparations. A soaking in sodium hydroxide, however, makes them swell slightly, and with suitable staining a remarkably orderly geometrical

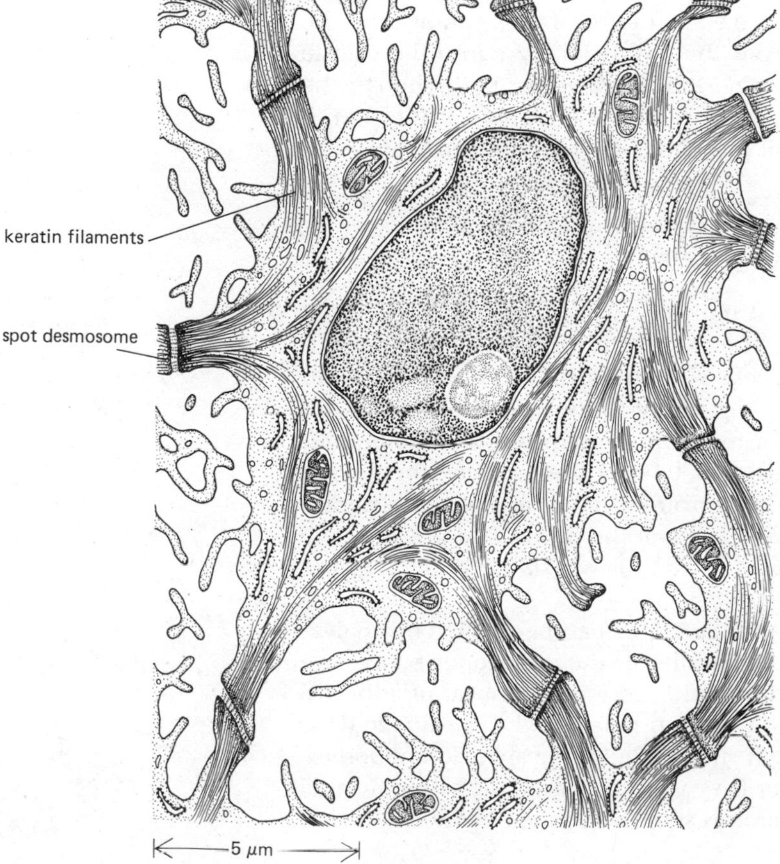

Figure 16–24 Section of a prickle cell from the epidermis, showing the bundles of keratin filaments that traverse the cytoplasm and that are inserted at the spot desmosome junctions that bind the cell to its neighbors. (From R. V. Krstić, Ultrastructure of the Mammalian Cell: An Atlas. Berlin: Springer, 1979.)

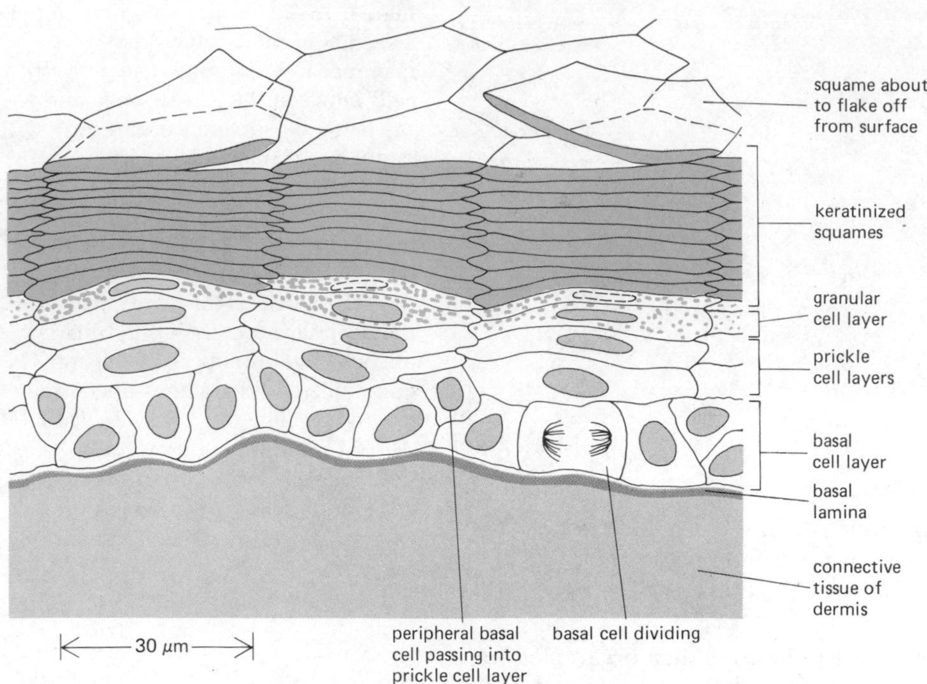

Figure 16–25 Epidermal proliferative units or columns in the thin skin of the mouse ear. The structure is revealed by swelling the keratinized squames in a solution containing NaOH. Studies of guinea pigs show that this type of columnar organization occurs where the epidermis is thinner than about 40 μm.

arrangement can be seen in regions where the skin is thin. The squames are found to be stacked in tidy hexagonal columns that interlock neatly at their edges, as shown in Figure 16–25. The diameter of each column is such that about 10 basal cells form the foundation on which it rests. The basal cells can be classified as central or peripheral according to whether they lie beneath the center or the periphery of their column. The peripheral cells, but not the central (according to the account that we shall follow here), can at times be seen in the act of passing upward from the basal cell layer into the prickle cell layer. Each column is called an *epidermal proliferative unit.* Though the orderly columnar arrangement is found only in some regions of the skin, it will serve well to illustrate the general principles of epidermal cell renewal.

Differentiating Epidermal Cells Synthesize a Sequence of Different Keratins as They Mature[18,20]

Having described the static picture, let us now set it in motion. The central basal cell divides, and some of its daughters, in turn dividing, shift to peripheral basal positions. Peripheral basal cells slip out of the basal cell layer into the prickle cell layer, onto the first step of the outward-moving escalator. Prickle cells flatten and eventually transform into keratinized squamous cells, losing their nuclei as they are carried out toward the surface. Keratinized squamous cells finally flake off and drift through the air as dust. The period from the time a cell is born in the basal layer of the human skin to the time it is shed from the surface varies from two to four weeks, depending on the region of the body.

The accompanying chemical transformations can be studied by analyzing either thin slices of epidermis cut parallel to the surface, or successive layers of cells stripped off by repeated application and removal of adhesive tape. The keratin molecules can be extracted and identified according to their electric charge, their molecular weight, their affinity for specific antibodies, and the pattern of small peptides that they yield on partial digestion. In this way it has been shown that keratins are present in all layers of the epidermis.

But there are many different types of keratins, encoded by a large family of genes that has presumably evolved by duplications and mutations of some ancestral gene; and in different layers of the epidermis, different types of keratins are produced. Thus the keratins in the prickle cells are different from those that fill the dead keratinized squamous cells. As the stem cell at the base of the column is transformed into the squame at the top, it expresses a succession of different selections from its repertoire of homologous keratin genes.

For Each Proliferative Unit, There Is an "Immortal" Stem Cell[19,21]

According to the picture presented above, each epidermal proliferative unit shelters beneath it a central basal cell from which the future cells of the unit are derived. The line of descendants of such a stem cell will not die out in the lifetime of the animal. We may, with a slight twist of language, call this stem cell *immortal* (Figure 16–26). Each time the immortal stem cell divides, one of its daughters inherits the mantle of immortality, while the other, sooner or later, perhaps after a few divisions, passes into the column of differentiating cells and is finally shed from the skin. What then distinguishes an immortal stem cell from the others? From what has been said so far, there is no reason to think that the immortal stem cell is inherently different from neighboring basal cells. They too may be stem cells in character and become mortal only because they are jostled from the central position and swept out into a current that will carry them away. Immortality, in short, is a property defined for stem cells in terms of prospective fate, not intrinsic character. Still, somewhere on its trajectory of development, each mortal daughter of an immortal stem cell must pass a point of no return beyond which its intrinsic character is so changed that it could no longer serve as a stem cell even if it were put back in the central basal position. Where is that point?

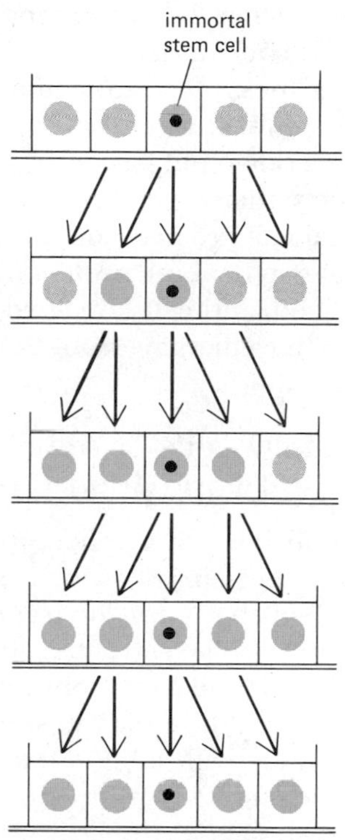

Figure 16–26 Each proliferative unit must contain, in each cell generation, at least one "immortal" stem cell, whose descendants will still be present in the unit in the distant future. The arrows indicate lines of descent. The immortal stem cell is shown here occupying the central basal position in each cell generation.

Stem Cell Potential May Be Maintained by Contact with the Basal Lamina[22]

In principle, the divisions of an immortal stem cell could always be asymmetric, so that one and only one of its daughters inherits the character required for immortality, while the other is somewhat altered already at the time of its birth in a way that forces it to differentiate and ultimately to die. In this case, there could never be any increase in the existing number of immortal stem cells. However, if a patch of epidermis is destroyed, one observes that the damage is repaired by surrounding healthy epidermal cells that migrate and proliferate to cover the denuded area (see p. 619). In this process, new epidermal proliferative units are formed whose central basal cells must have arisen ultimately from divisions that generated two immortal stem cells from one.

Thus the fate of the daughters of a stem cell must be governed by external circumstances, and not simply by their internal characters. What then is the determining factor? One attractive hypothesis is that stem cell character is maintained by contact with the basal lamina and that the changes leading to terminal differentiation begin as soon as a cell loses contact with the lamina. Tissue-culture experiments lend some support to this suggestion: epidermal cells continue to proliferate if they are grown in contact with an appropriate substratum, such as a carpet of fibroblasts, but promptly differentiate if they are kept in suspension. The hypothesis remains controversial and is probably an oversimplification; it does, however, neatly explain how the supply of stem cells might be adjusted to keep the surface of the body covered.

Basal Cell Proliferation Is Regulated According to the Thickness of the Epidermis[23]

While contact with the basal lamina may decide the choice between survival as a stem cell and death through terminal differentiation, other controls must also operate to regulate the rate of production of epidermal cells. A variety of hormones and growth factors (see p. 727) are thought to be involved. For example, if the outer layers of the epidermis are stripped away, the division rate of the basal cells increases. After a transient overshoot, normal thickness is restored, and the division rate in the basal layer declines to normal. It is as though the cells in the proliferative basal layer were released from an inhibitory influence by the removal of the outer, differentiated layers and were brought back under that inhibition as soon as the outer layers regained their full thickness. According to one hypothesis, a factor called an epidermal *chalone* is synthesized in the epidermis and exerts just such a self-inhibitory effect, slowing down mitosis in the basal layers so as to adjust the rate of production of differentiated cells according to need. The consequences of faulty control of basal cell proliferation are seen in *psoriasis*. In this common skin disorder, the rate of basal cell proliferation is greatly increased—the epidermis becomes thickened, and cells are shed from the surface of the skin within as little as a week after emerging from the basal layer, before they have had time to keratinize fully.

Secretory Cells of the Skin Are Secluded in Glands and Have Different Population Kinetics[24]

The skin has other functions besides providing a protective barrier, and in certain specialized regions other types of cell besides the keratinized cells described above develop from the epidermis. In particular, secretions are produced by cells segregated in deep-lying glands, which have patterns of renewal quite different from those of keratinizing regions.

A *sweat gland* is the simplest example of such a structure. It consists of a long tube with a blind end and develops as an ingrowth of the epidermis. Sweat is secreted by the cells in the bottom portion of the tube and is conveyed to the surface of the skin via the excretory duct (Figure 16–27). The cells of the secretory terminal portion form an epithelium one layer thick, embraced by a small number of contractile *myoepithelial cells* (see Figure 16–29). The cells of the excretory duct form an epithelium two layers thick, with no myoepithelial component. Two varieties of sweat glands are recognized, and several other types of glands are probably evolutionary modifications of the same prototype. The glands producing tears, ear wax, saliva, and milk all belong in this category. In all of them, there is a difference between the secretory cells and the cells that line the ducts: in the salivary glands and the mammary glands at least, it is the ducts that contain the stem cells for the renewal of the secretory population.

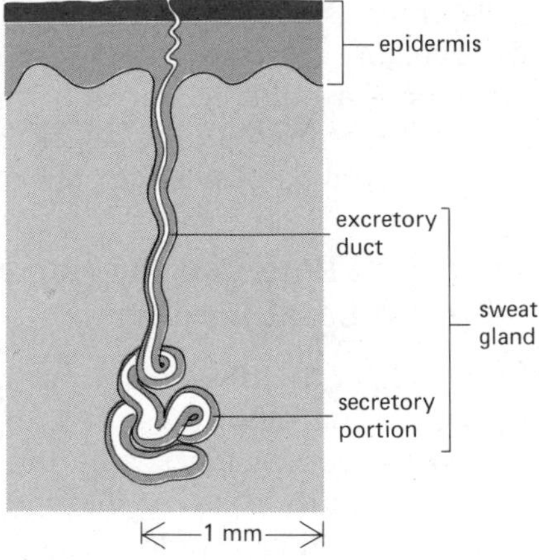

Figure 16–27 Schematic diagram of a sweat gland.

The mammary gland has been extensively studied because of the striking hormonal control of its cell division and differentiation. Milk production must be switched on when a baby is born and switched off when the baby is weaned. In a mammary gland that is neither making milk nor preparing to do so, the glandular tissue consists of branching systems of excretory ducts embedded in connective tissue and lined, in their secretory portions, by a single layer of relatively inactive epithelial cells, including some myoepithelial cells. As a first step toward large-scale milk production, the hormones that circulate during pregnancy cause the duct cells to proliferate and the terminal portions of the ducts to grow and branch, forming little dilated outpocketings, or *alveoli* (Figure 16–28). The cells lining the alveoli (Figure 16–29) are the secretory cells, but they do not start to secrete milk (Figure 16–30) until they

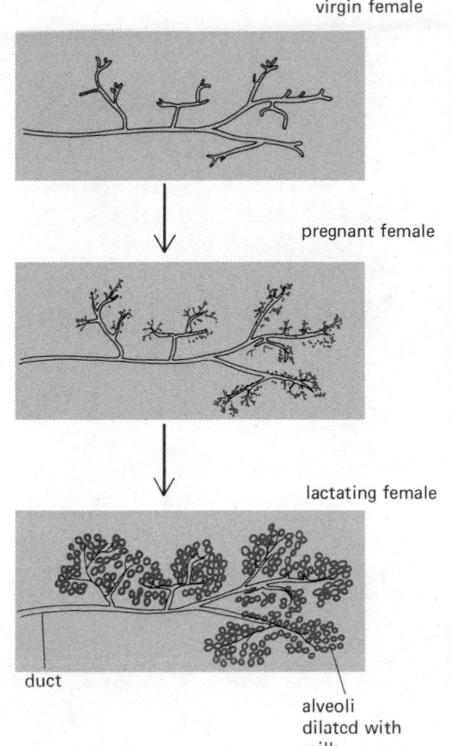

Figure 16–28 Schematic diagram of the growth of alveoli from the ducts of the mammary gland during pregnancy and lactation. Only one small part of the gland is shown. The resting gland contains a small amount of inactive glandular tissue embedded in a large amount of fatty connective tissue (*pale gray*). During pregnancy an enormous proliferation of the glandular tissue takes place at the expense of the fatty connective tissue, with the secretory portions of the gland developing preferentially to create alveoli.

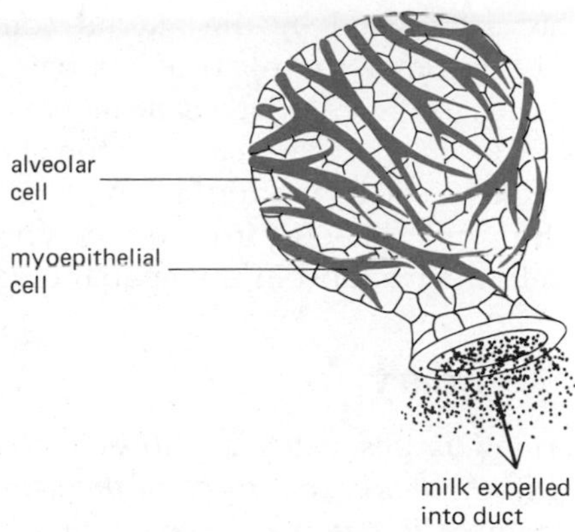

Figure 16–29 One of the milk-secreting alveoli of the mammary gland, with a basket of myoepithelial cells embracing it. The myoepithelial cells contract and expel milk from the alveolus in response to the hormone oxytocin, which is released as a reflex response to the stimulus of suckling.

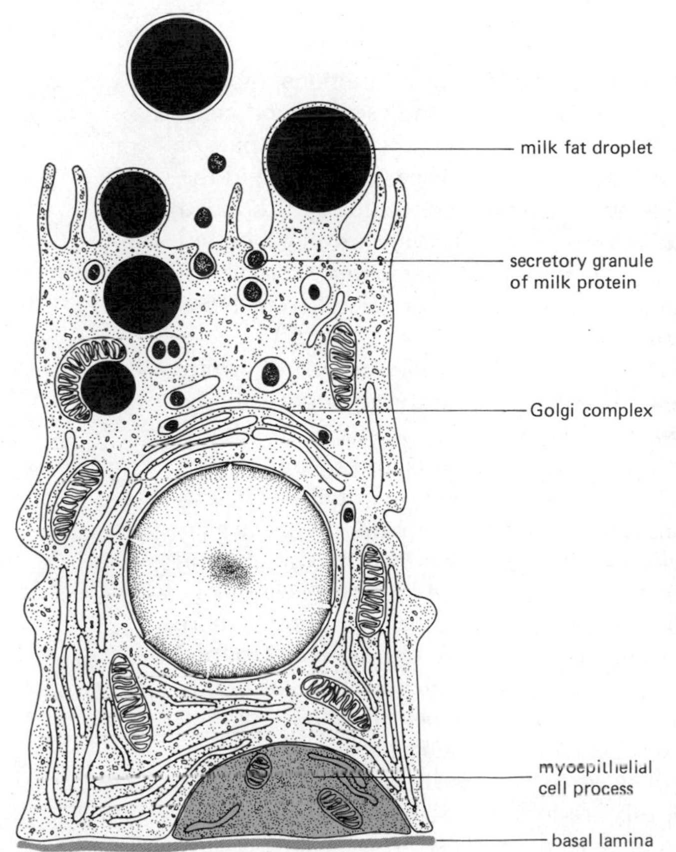

Figure 16–30 Schematic diagram of a milk-secreting cell from an active mammary gland. A single type of cell produces both the milk proteins and the milk fat. The proteins are secreted in the normal way by exocytosis, while the fat is released as droplets surrounded by plasma membrane detached from the cell. (From W. Bloom and D. W. Fawcett, A Textbook of Histology, 10th ed. Philadelphia: Saunders, 1975.)

are stimulated by the altered combination of hormones circulating in the mother after the birth of the baby. When the baby is weaned and suckling stops, the secretory cells degenerate, macrophages clear away the debris, most of the alveoli disappear, and the gland reverts to a resting state until the whole cycle is set in motion again at the next pregnancy. The mammary gland is thus very different from the epidermis both in the control and periodicity of cell renewal and in the spatial organization of the process.

Summary

Many tissues, especially those with a rapid turnover—such as the lining of the gut, the epidermal layer of the skin, and the blood—are renewed by means of stem cells. Stem cells, by definition, have the ability to divide without limit, yielding some progeny that differentiate and others that remain stem cells. In the skin, the stem cells of the epidermis lie in the basal layer, in contact with the basal lamina. The progeny of the stem cells differentiate on leaving this layer and, as they move outward, synthesize a succession of different types of keratin until eventually their nuclei degenerate, producing an outer layer of dead keratinized cells that are finally shed from the surface. In regions where the epidermis is thin, it is neatly organized into proliferative units or columns, with an "immortal" stem cell at the base of each. The fate of the daughters of a stem cell is controlled by extrinsic factors, such as contact with the basal lamina. The rate of stem cell proliferation is regulated homeostatically according to the thickness of the epidermis. In the glands connected to the epidermis, such as sweat glands and mammary glands, stem cells exist, but the pattern of cell renewal is different.

Renewal by Pluripotent Stem Cells: Blood Cell Formation[25]

The blood contains many types of cells with very different functions, ranging from the transport of oxygen to the production of antibodies. All blood cells, however, have certain similarities in their life history. They all spend part of their existence enmeshed in other tissues and a part freely circulating in the bloodstream. They all have limited life spans and are continuously produced throughout the life of the animal. And, most remarkably, they are all generated ultimately from the same type of stem cell. This *hematopoietic* (or blood-forming) *stem cell* is thus pluripotent, giving rise to all of the different types of terminally differentiated blood cells.

The cells of the blood (Figure 16–31) can be classified as red or white. The **red blood cells,** or **erythrocytes,** carry hemoglobin throughout the body. The **white blood cells,** or **leucocytes,** combat infection and engulf and digest debris, making their way across the walls of blood vessels and migrating into the tissues to perform these tasks. In addition, the blood contains *platelets,* which are not entire cells, but small detached cell fragments or "minicells" derived from large cells called *megakaryocytes.* Platelets control the clotting of the blood and help to repair breaches in the walls of blood vessels.

Whereas one red blood cell or one platelet is similar to another, there are many different classes of white blood cell. These are traditionally grouped into five major categories on the basis of their appearance in the light microscope: *neutrophils* (also called *polymorphonuclear leucocytes* because of their irregularly lobed nucleus), *eosinophils, basophils, lymphocytes,* and *monocytes* (see Table 16–1). The fine structure of some of these is shown in Figure 16–32. Neutrophils, eosinophils, and basophils are jointly classified as *granulocytes.* They contain numerous lysosomes and secretory vesicles or granules, and

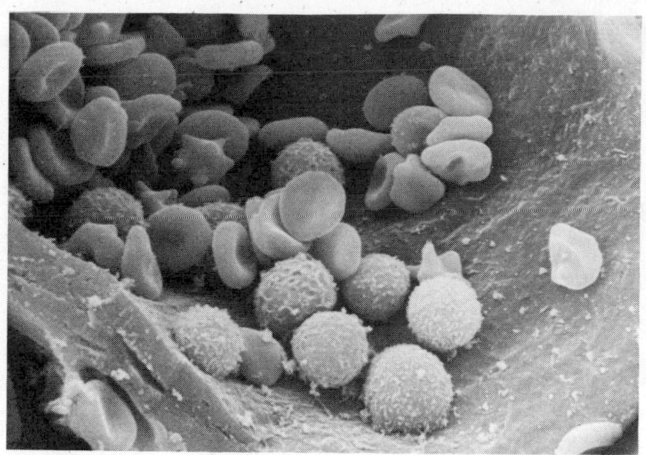

Figure 16–31 Scanning electron micrograph of mammalian blood cells in a small blood vessel. The larger, more spherical cells with a rough surface are white blood cells; the smaller, smoother flattened cells are red blood cells. (From R. G. Kessel and R. H. Kardon, Tissues and Organs: A Text-Atlas of Scanning Electron Microscopy. San Francisco: Freeman, 1979. © 1979 W. H. Freeman and Company.)

they take their individual names from the different staining properties of the granules. The differences of staining reflect major differences of chemistry and function. Neutrophils, the commonest type, engulf, kill, and digest bacteria. Lymphocytes comprise a functionally heterogeneous group of cells all concerned with immune responses; in addition, there are *killer cells* that look like lymphocytes and function as accessory cells in immune responses but are not part of the immune system proper. Monocytes, on leaving the bloodstream, become *macrophages,* which can dispose of invading microorganisms, foreign bodies, and cellular debris by phagocytosis. Neutrophils and macrophages are the main "professional phagocytes" in the body.

Table 16–1 Blood Cells

Type of Cell	Main Functions
Red blood cells (erythrocytes)	transport O_2 and CO_2
White blood cells (leucocytes)	
Granulocytes	destroy invading bacteria
Neutrophils (polymorphonuclear leucocytes)	
Eosinophils	destroy larger parasites and modulate allergic inflammatory reactions
Basophils	release histamine and serotonin in certain immune reactions
Lymphocytes	make immune responses
Killer cells	kill virally infected cells and some tumor cells
Monocytes	become macrophages in the tissues
Megakaryocytes, giving rise to *platelets*	initiate blood clotting

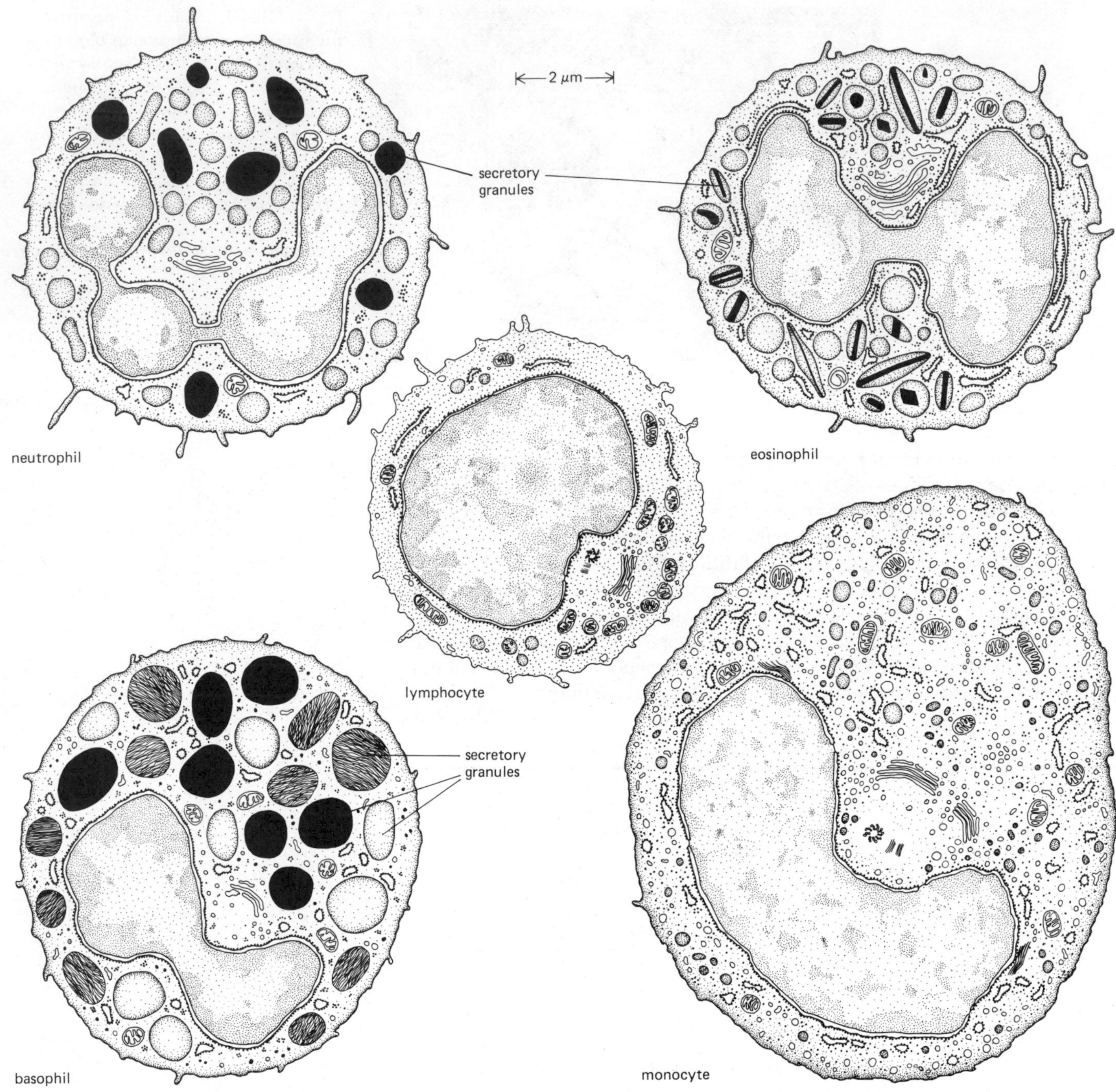

Figure 16–32 Thin sections of the major types of white blood cells (leucocytes) found in the circulation, showing the variety of internal structures observed. All of these cells develop from the same pluripotent stem cell. (Slightly modified from T. L. Lentz, Cell Fine Structure. Philadelphia: Saunders, 1971.)

The different types of blood cells are produced in different numbers, and the production of each must be regulated individually to meet changing needs. Thus hematopoiesis necessarily involves some complex controls, and our understanding of these is still very incomplete. The process is more difficult to analyze than turnover of a tissue such as the epidermal layer of the skin. In epidermis, there is a regular spatial organization that makes it easy to follow the process of renewal and to identify the stem cells; this is not true of the blood. The study of cell turnover in the blood has depended on more sophisticated experimental techniques, involving radioactive tracers, the transfer of cells from one animal to another, and the study of individual cells and their progeny in culture.

New Blood Cells Are Generated in the Bone Marrow[25]

The erythrocyte is the most common type of cell in the blood. When mature, it is packed full of hemoglobin and contains practically none of the usual cell organelles. In an erythrocyte of an adult mammal, even the nucleus, endoplasmic reticulum, mitochondria, and ribosomes are absent, having been extruded from the cell in the course of its development (Figure 16–33). The erythrocyte therefore cannot grow or divide; the only possible way of making more erythrocytes is by means of stem cells. Furthermore, erythrocytes have a limited life span. This can be demonstrated with a brief dose of radioactive iron, which is incorporated into protein in the erythrocytes that are being formed at the time and remains in them until they die. The proportion of circulating erythrocytes that are radioactively labeled stays roughly constant for a period of a few months after administration of the radioisotope and then begins to decline, until finally none are left. By this means it has been established that the average human erythrocyte survives for about 120 days in the circulation. The worn-out erythrocytes are captured and destroyed by macrophages in the liver and spleen.

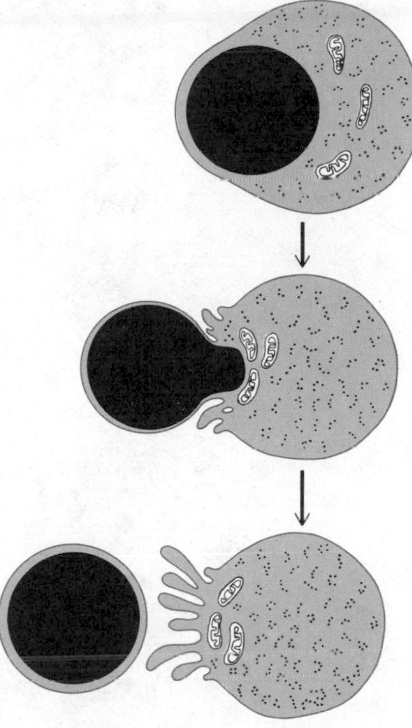

Figure 16–33 Schematic diagram of an immature red blood cell (erythrocyte) extruding its nucleus shortly before leaving the bone marrow and passing into circulation. The discarded nucleus will be engulfed and digested by one of the macrophages residing in the bone marrow. The remaining mitochondria and ribosomes will be lost within a day or two.

Leucocytes (white blood cells) are fewer in number than erythrocytes (in a ratio of about 1 to 1000 in the bloodstream), and the different types have quite different rates of turnover. Most granulocytes circulate in the blood for only a few hours before migrating into the connective tissues, where they reside somewhat longer. For example, neutrophils survive for a few days after leaving the bloodstream and then die. Monocytes, by contrast, can persist as macrophages outside the bloodstream for months or perhaps even years. The life histories of the various types of lymphocytes are more complex and will be discussed in Chapter 17; while some of them can survive for years, moving back and forth between the blood and other tissues, most of them die within days or weeks after forming.

Replacements for all these types of cells must be generated at appropriate rates. To keep erythrocyte numbers steady, new cells must be formed at a furious rate (more than 2 million per second in humans). In mammals, these new erythrocytes are produced chiefly in the bone marrow, where one can identify erythrocyte precursors that contain hemoglobin but still have nuclei. These precursors, which can be ordered according to hemoglobin content, represent successive stages in the development of the mature anucleate erythrocyte. It is fairly easy to recognize also in the bone marrow the immature precursors of the three types of granulocytes and of megakaryocytes. The megakaryocytes remain in the bone marrow when mature and are one of its most striking features (Figure 16–34). They are extraordinarily large (diameter up to 60 μm), with a highly polyploid nucleus and a cytoplasm subdivided by meandering layers of membranes (Figure 16–35). Platelets originate as vesicles that detach in large numbers from the outer regions of the megakaryocyte.

The different types of blood cell precursors in the bone marrow are intermingled with one another, as well as with fat cells and with the fibroblasts that form a delicate supporting meshwork of collagen fibers. In addition, the whole tissue is richly supplied with thin-walled blood vessels into which the newborn blood cells are discharged (Figure 16–36).

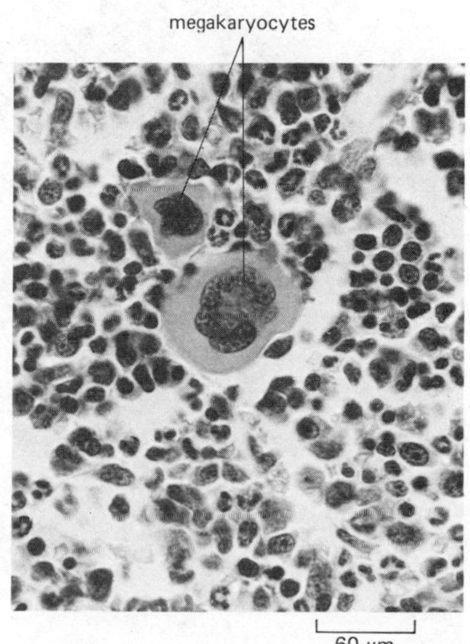

Figure 16–34 Light micrograph showing two megakaryocytes in the bone marrow. The enormous size of megakaryocytes compared to the surrounding bone marrow cells is due to the fact that they have highly polyploid nuclei. (Courtesy of Peter Gould.)

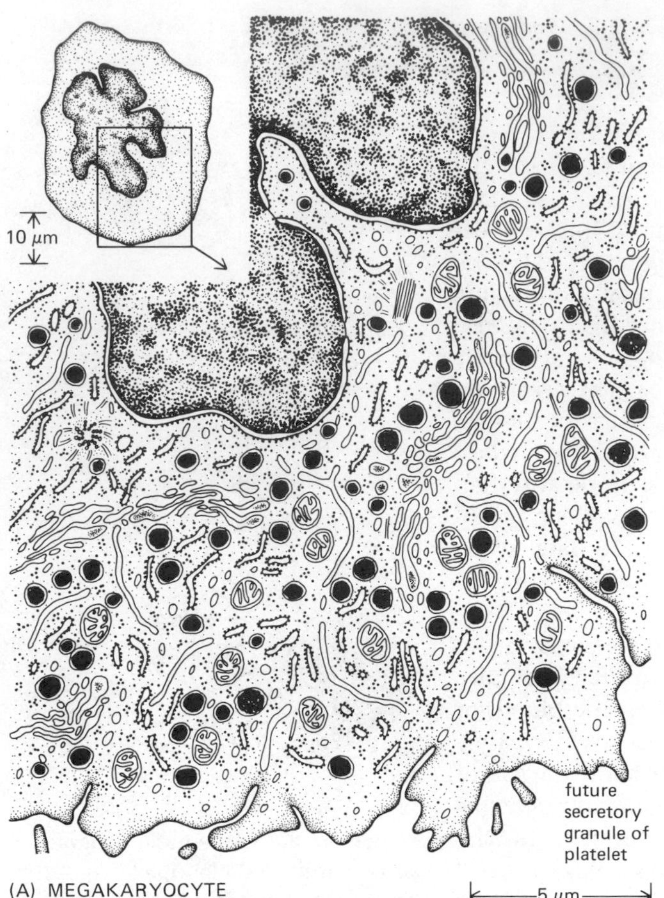

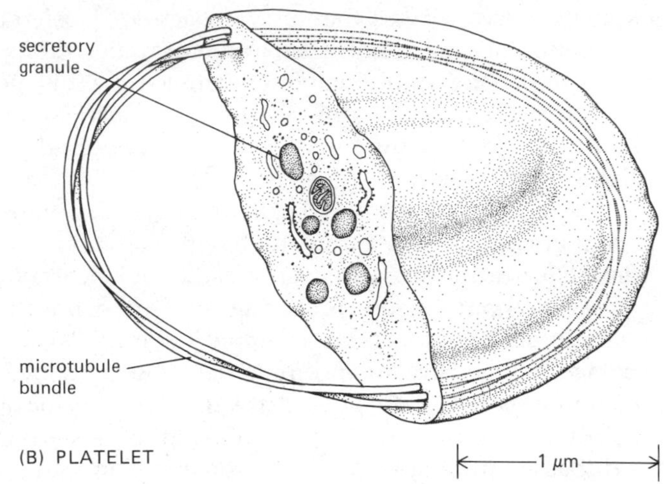

Figure 16–35 (A) A megakaryocyte (illustrated in section) and (B) one of the platelets (depicted in three-dimensional cut-away view) that split off from it. Note the circumferential band of microtubules that give the platelet its shape. (From R. V. Krstić, Ultrastructure of the Mammalian Cell: An Atlas. Berlin: Springer, 1979.)

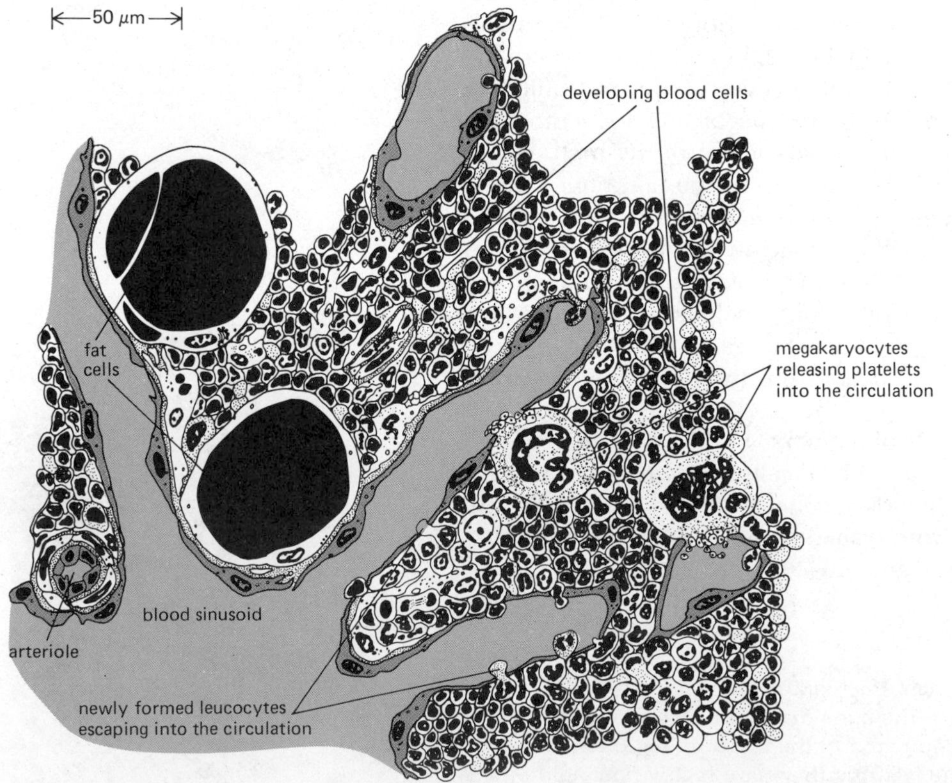

Figure 16–36 Section of a region of bone marrow. This tissue is the source of new blood cells in adult humans. It is not easy to distinguish the different types of blood cell precursor in the marrow, which therefore appears more chaotic than it is in fact. (Slightly modified from L. Weiss and R. O. Greep, Histology, 4th ed. New York: McGraw-Hill, 1977.)

Bone Marrow Contains Pluripotent Stem Cells That Can Establish Hematopoietic Colonies[25,26]

Since the arrangement of cells in the bone marrow is so disorderly, it is difficult to identify any but the immediate precursors of the mature blood cells. The corresponding cells at still earlier stages in their development, before any overt differentiation has begun, are confusingly similar in appearance, and the appearance of the ultimate stem cells is still a matter of conjecture. Indeed, from merely descriptive studies, it is not possible to prove that these stem cells are actually located in the bone marrow or to determine whether there is a distinct type of stem cell for each type of blood cell. These questions can, however, be answered by experiment. Most of the key information has come from studies on mice.

If an animal is exposed to a large dose of x-irradiation, cell division is halted in all the tissues of the body, including those where blood cells are produced. The x-rays exert this effect by causing breaks and other irreparable damage in the chromosomes, and the animal dies within a few days as a result of its inability to manufacture new cells, and in particular new blood cells. An irradiated animal can, however, be saved by a transfusion of cells taken from the bone marrow of a healthy, immunologically compatible donor. Among these cells there are evidently some that can colonize the irradiated host and reequip it with hematopoietic tissue.

One of the tissues where colonies develop is the spleen, which in a normal mouse is an important additional site of hematopoiesis. When the spleen of the irradiated mouse is examined a week or two after the transfusion of cells from the healthy donor, a number of distinct nodules are seen in it, each of which is found to contain a mass of proliferating hematopoietic tissue (Figure 16–37). The discreteness of the nodules suggests that each might be, like a bacterial colony on a culture plate, a clone of cells descended from a single founder cell; and with the help of genetic markers, it can be established that this is indeed the case. The founder of such a colony is called a *colony-forming unit*, or *CFU*. Some, if not all, CFUs must be stem cells, since some of the colonies derived from them renew themselves indefinitely while producing at the same time new terminally differentiated blood cells.

Experiments of this type show that hematopoietic stem cells are present in the circulation as well as in the bone marrow, though in smaller numbers, and that these cells can leave the bloodstream and settle and proliferate in certain tissues such as the spleen. This colonizing ability is important in the embryonic development of hematopoietic tissues. Blood cell formation occurs first in outlying extraembryonic regions of the mesoderm, then in the liver and spleen, and finally in the bone marrow, where it persists throughout adult life. It appears likely that in normal development the spleen, liver, and bone marrow are successively colonized by stem cells originating elsewhere.

To show that at least some CFUs are *pluripotent* stem cells, able to give rise to many different types of blood cells, it is enough to examine the composition of individual well-developed spleen colonies. Some of these are found to contain a mixture of maturing erythrocytes, megakaryocytes, granulocytes, and macrophages. By slightly more complicated techniques, it is possible to prove in addition that lymphocytes, which develop mainly in other regions of the body, can originate from the same clone as the other blood cell types. (There is evidence of a similar origin for *mast cells*, which are not normally encountered in the blood but lodge in connective tissue where they secrete heparin and histamine in inflammatory reactions.) Thus there are CFU stem cells from the bone marrow that are capable of producing all the different classes of blood cells (Figure 16–38). Their progeny consequently have a choice

x-irradiation halts blood cell production; mouse would die if no further treatment was given

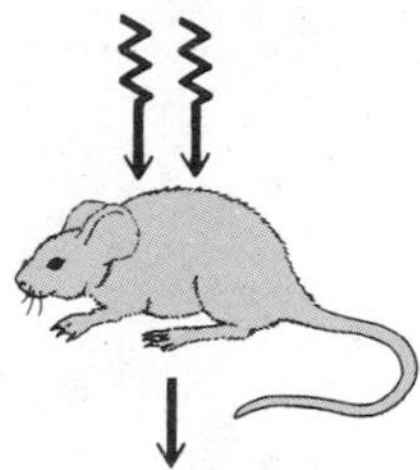

inject bone marrow cells from healthy donor

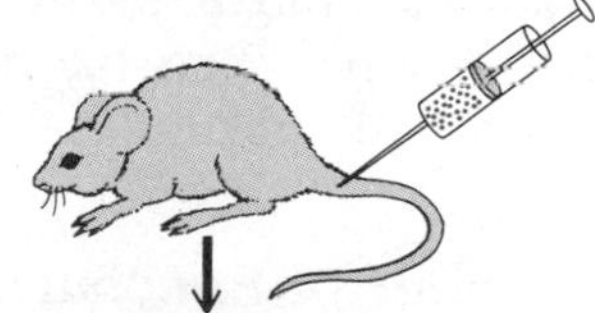

mouse survives: 2 weeks after injection, many healthy blood cells are in circulation

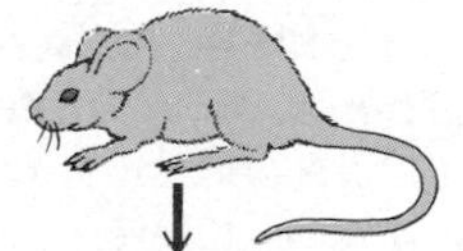

examination of spleen reveals unusual nodules on its surface

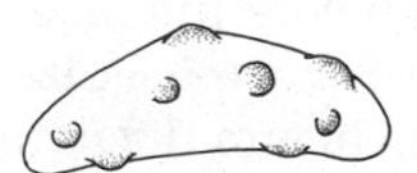

each spleen nodule contains a clone of hematopoietic cells, descended from 1 of the injected bone marrow cells

Figure 16–37 Schematic diagram of an experiment in which the spleen of a heavily irradiated animal is seeded by hematopoietic stem cells transfused from a healthy donor.

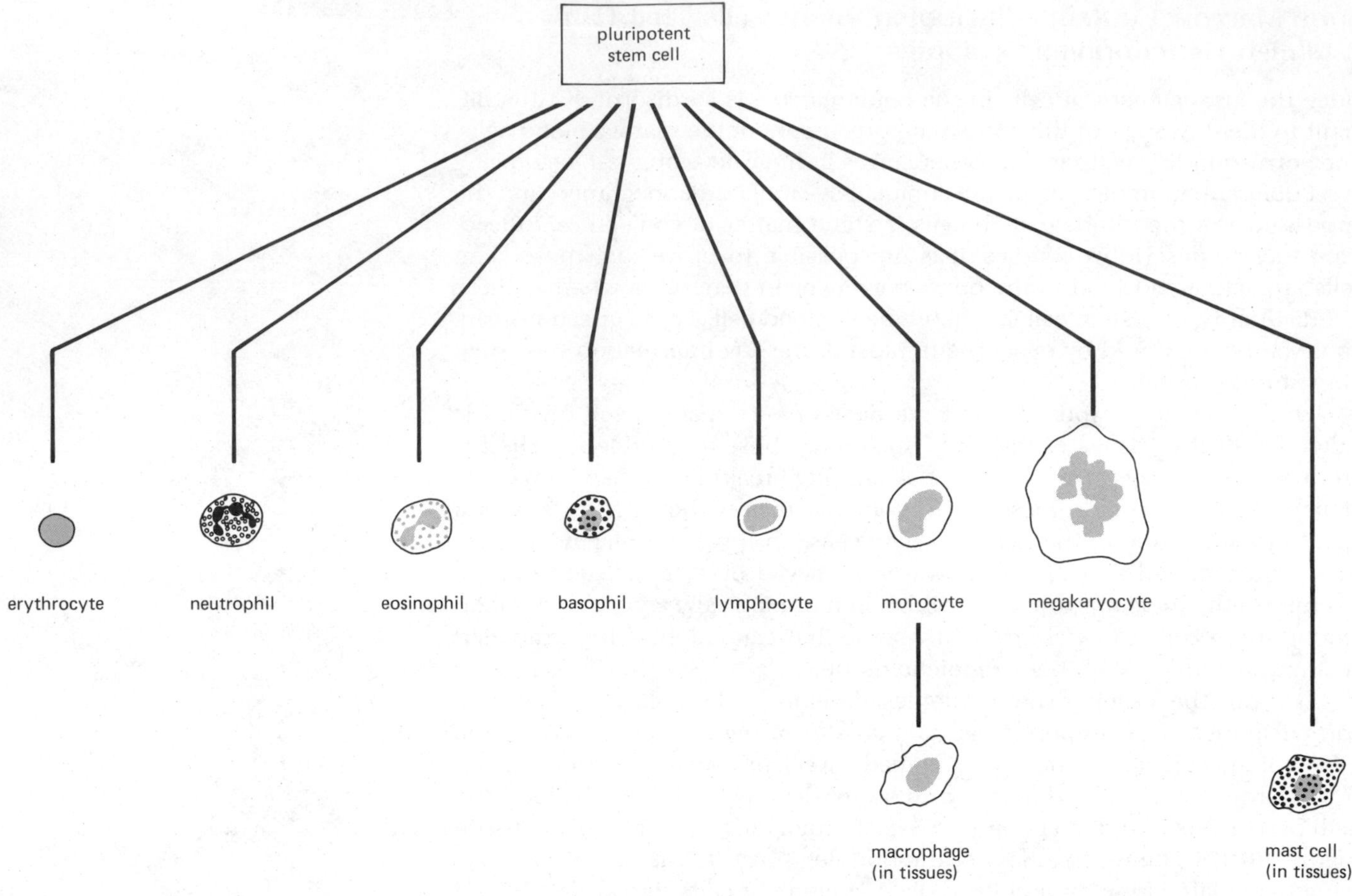

Figure 16–38 The different classes of cells that derive from the pluripotent hematopoietic stem cell.

between several alternative lines of differentiation. This choice might be made at random, or it might, for example, be controlled by the environment of the stem cells. Though there has been much debate, the problem of what governs the choice is still not resolved.

The Number of Specialized Blood Cells Is Amplified by Divisions That Follow Commitment[27]

Once a cell has differentiated as an erythrocyte or a granulocyte or some other type of blood cell, there is no going back: the state of differentiation is not reversible. Therefore, at some stage in their development, the progeny of the pluripotent stem cell must become irreversibly committed or determined for a particular line of differentiation. At what stage does this commitment occur? It is clear from simple microscopic examination of the bone marrow that it happens well before the final division in which the mature differentiated cell is formed: one can recognize specialized precursor cells that already show signs of having begun differentiation but are still proliferating. It thus appears that commitment to a particular line of differentiation is followed by a series of cell divisions that amplify the number of cells of a given specialized type. In this way, a very small number of pluripotent stem cells serve to generate very large numbers of differentiated blood cells. Furthermore, it turns out that the amplifying divisions are subject to important controls that regulate the production of each type of blood cell according to need. Such controls are especially well documented for the cell lineage committed to erythrocyte formation.

Production of Erythrocytes Is Controlled Through Hormonal Regulation of the Cell Divisions That Follow Commitment[25,28]

Erythropoietin is a glycoprotein hormone, with a molecular weight of about 46,000, produced chiefly in the kidney. A shortage of erythrocytes (after hemorrhage, for example) stimulates cells in the kidney to synthesize and secrete erythropoietin into the bloodstream. (Conversely, an excess of erythrocytes depresses output.) Erythropoietin in turn stimulates the production of erythrocytes. Since a change in the rate of release of new erythrocytes into the bloodstream is observed as early as one or two days after an increase in erythropoietin levels in the bloodstream, the hormone must act on cells that are very close precursors of the mature erythrocytes. Cells become sensitive to erythropoietin after they have become committed to the erythrocyte line of differentiation, and this sensitivity can be used to indicate how far they have traveled along that line of differentiation.

When dispersed bone marrow cells are grown in culture, colonies of erythrocytes develop only if erythropoietin is included in the medium. If moderately low concentrations of the hormone are present, relatively small colonies, comprising no more than about 60 erythrocytes, develop within a few days. Each of these colonies is evidently founded by a precursor cell (known as an *erythrocyte colony-forming unit,* or *CFU-E*) that is highly sensitive to erythropoietin and gives rise to mature erythrocytes after about six division cycles or less. Moreover, the number of such precursor cells in a sample of bone marrow depends on its previous exposure to erythropoietin in the intact animal, before the culture was set up. If the animal had an abnormally high level of erythropoietin in its circulation, its bone marrow is found to contain an abnormally large number of the CFU-E cells that give rise to erythroid colonies in culture. It thus seems that the CFU-E cells in the bone marrow must themselves derive from an earlier type of precursor cell whose proliferation is also stimulated by erythropoietin.

This conclusion is confirmed by further studies in culture. If the concentration of erythropoietin in the culture medium is increased tenfold above the optimum for production of the small erythrocytic colonies, a new class of much larger colonies is observed, comprising up to 5000 erythrocytes each (Figure 16–39). These colonies take a week or ten days to develop, as opposed to the two days required for the small erythrocytic colonies. The precursor cell from which they derive is called the *erythrocytic burst-forming unit,* or *BFU-E.* The BFU-E is distinct from the pluripotent stem cell in that it responds to erythropoietin by proliferating to generate erythrocytes. It is also distinct from the CFU-E in that it requires a higher level of hormone for stimulation, and its progeny go through as many as 12 division cycles before they become mature erythrocytes. The cell also differs in size from the CFU-E and can be separated from it by sedimentation. Cells of a character intermediate between the BFU-E and the CFU-E have also been observed. Thus the BFU-E appears to be a cell committed to erythrocytic differentiation and an early ancestor of the CFU-E (Figure 16–40).

Committed erythrocyte precursors evidently go through successive rounds of cell division during which they become progressively more sensitive to erythropoietin. The manufacture of large quantities of mRNA coding for hemoglobin, and of hemoglobin itself, is delayed until after the CFU-E stage. The large number of cell divisions that occur in the erythrocyte lineage under the influence of erythropoietin provides a powerful means of controlling the production of erythrocytes without upsetting the production of other types of blood cells.

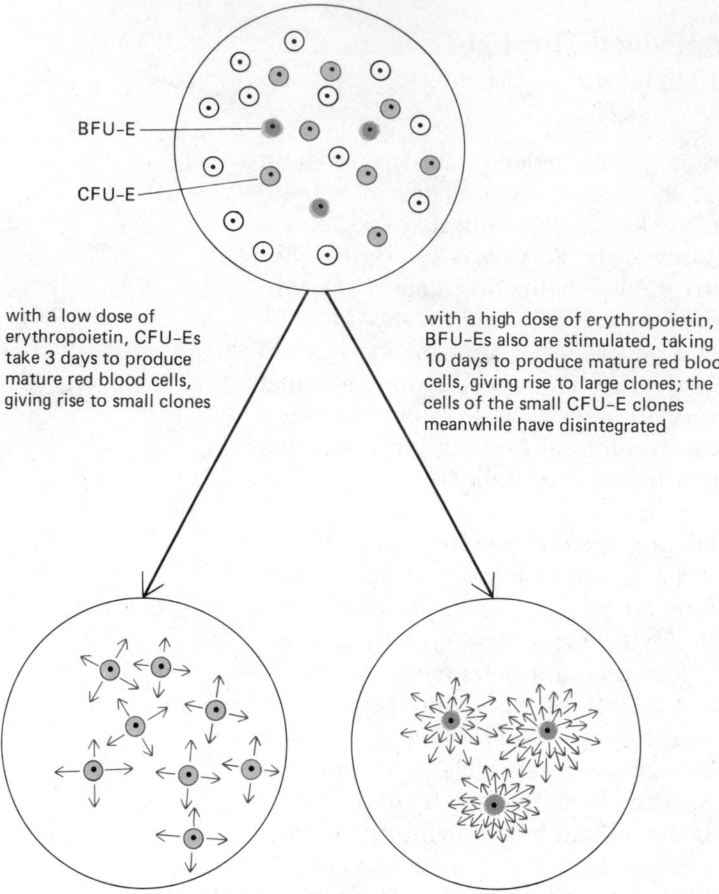

Figure 16–39 Schematic diagram of a tissue-culture experiment showing the effect of erythropoietin on the differentiation of bone marrow cells into red blood cells (erythrocytes). The BFU-Es and the CFU-Es are two types of red blood cell precursor, distinguishable by their response to the hormone.

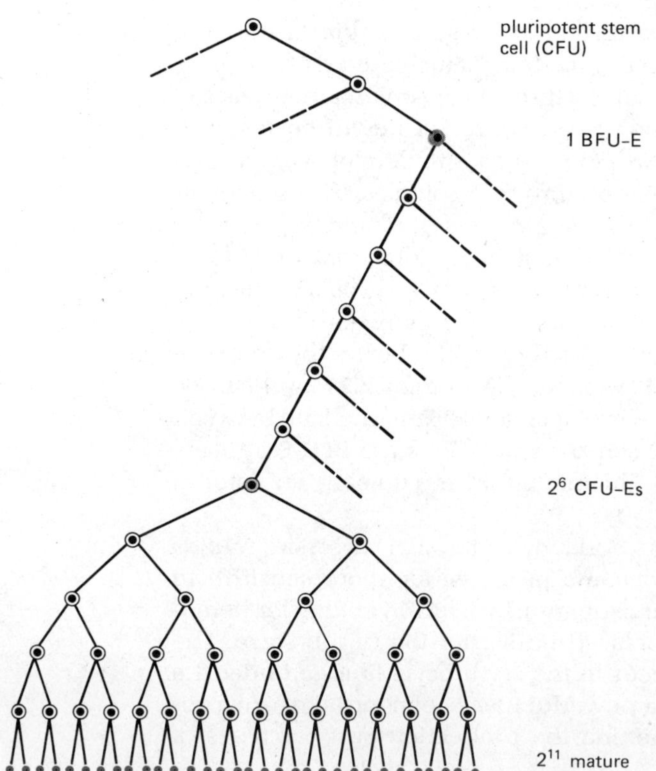

Figure 16–40 Pedigree showing the relationship of the pluripotent stem cell (CFU), the BFU-E, the CFU-E, and the mature red blood cell.

Specific Glycoprotein Hormones Control the Survival and Fate of the Different Classes of Committed Hematopoietic Precursor Cells[29]

There is increasing evidence for the existence of other hormones analogous to erythropoietin, produced in response to needs for the various types of white blood cells. It appears that a specific glycoprotein factor is required for each variety of committed precursor cell. Without the appropriate factor, the cell will die; with the factor, it will proliferate to give differentiated progeny.

The production of neutrophilic granulocytes and macrophages provides an interesting example. Both these cell types derive from the same precursor, which is committed in the sense that it can give rise to no other cell types than these two, and has a limited capacity to divide. The precursor depends for its survival on the *granulocyte/macrophage colony-stimulating factor* (or *GM-CSF*). This is a glycoprotein of molecular weight of about 23,000 that is secreted by many different types of cells in the body and whose concentration in the bloodstream rises sharply in response to infection. When bone marrow cells are cultured in a dish containing GM-CSF, colonies of up to about 10,000 cells develop, consisting of granulocytes and macrophages exclusively. When a single precursor cell is taken (by micromanipulation under a microscope) and placed alone in a tiny culture well with GM-CSF, the clone of cells that develops from it may consist of granulocytes, or macrophages, or a mixture of both. To test whether the GM-CSF itself directs the choice, one can take the two undifferentiated daughters of a single dividing precursor cell and transfer one to a culture well containing a high concentration of GM-CSF and the other to a well with a low concentration. Sometimes the two sister cells both yield colonies of one and the same cell type. But when they differ, they do so in a predictable way: the colony growing in the high concentration of GM-CSF consists of granulocytes, while the colony in the low concentration consists of macrophages. This suggests that there is a step in the life history of the common granulocyte/macrophage precursor at which it becomes committed more narrowly, to one or other of the two fates, and that the choice is determined by the concentration of GM-CSF in its environment. The development of techniques such as these for studying single hematopoietic cells in controlled conditions promises to clarify many of the problems that have proved difficult to solve by experiments on mixed populations of cells.

Summary

The many different types of blood cells all derive from a common pluripotent stem cell and are generated chiefly in the bone marrow in the adult. During embryonic development, pluripotent stem cells circulating in the bloodstream are able to settle in the bone marrow, spleen, or liver and establish new hematopoietic colonies there. The rate of production of mature blood cells of each type is largely regulated through a series of divisions undergone by the different blood precursor cells after they have become committed to a particular line of differentiation, but before they are fully differentiated. In the erythrocyte lineage, the committed cells as they divide become increasingly sensitive to erythropoietin. This hormone is produced by the kidney in response to the need for red blood cells and stimulates the production of red blood cells by causing the committed precursor cells to divide and complete their maturation. Analogous factors control the survival and behavior of the various committed precursors of white blood cells.

Quiescent Stem Cells: Skeletal Muscle

The term "muscle" covers a multitude of cell types, all specialized for contraction, but in other respects dissimilar. As noted in Chapter 10, a contractile apparatus involving actin and myosin and regulated by Ca^{2+} is a basic feature of eucaryotic cells in general, but there are several distinct ways in which specialized cells have developed this apparatus to a high degree. Mammals possess four main categories of cells specialized for contraction: skeletal muscle cells, heart (or cardiac) muscle cells, smooth muscle cells, and myoepithelial cells (Figure 16–41). These differ in function, structure, and development. Although all of them appear to generate contractile forces by means of actin and myosin, the types of actin and myosin employed are somewhat different in amino acid sequence, are differently arranged in space, and are associated with different sets of proteins to control contraction.

Skeletal muscle cells, the most familiar, are responsible for practically all movements that are under voluntary control. These cells can be huge (up to half a meter long and 100 μm in diameter in the adult human) and are often referred to as *muscle fibers* because of their shape. Each one is a syncytium, containing many nuclei within a common cytoplasm. Heart muscle cells, smooth muscle cells, and myoepithelial cells are more conventional, having only a single nucleus. **Heart muscle cells** resemble skeletal muscle cells in having their actin and myosin filaments aligned in orderly arrays, giving the cells a striated appearance. **Smooth muscle cells** are so called because they, in contrast, do not appear striated. The functions of smooth muscle are highly varied, ranging from the propulsion of food along the digestive tract to the erection of hairs as a result of cold or fear. **Myoepithelial cells,** unlike all the other muscular types, lie in epithelia and are derived from the ectoderm; they have no striations. They form the dilator muscle of the iris and also serve to expel saliva, sweat, and milk from the corresponding glands (see Figure 16–29).

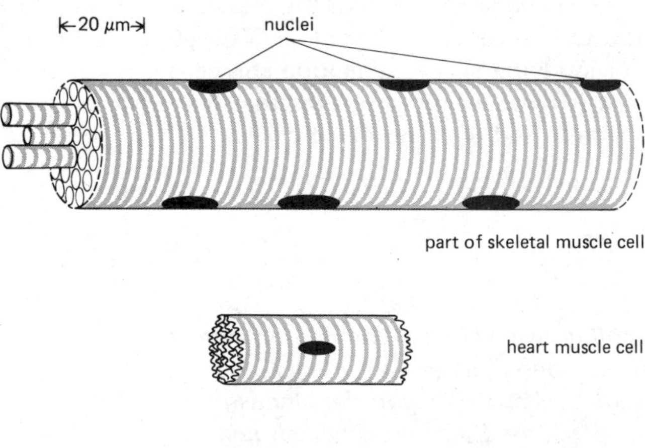

smooth muscle cell

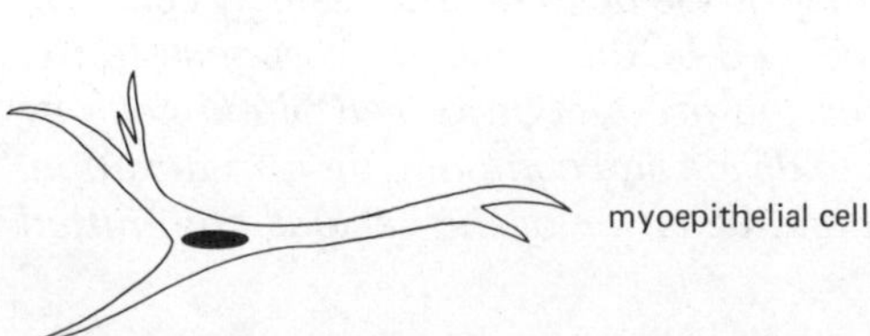

Figure 16–41 The four classes of muscle cells of a mammal.

The four main categories of muscle cell can be further divided into distinctive subtypes, each with its own peculiar features. Rather than continue with comparisons and contrasts, we focus here on the skeletal muscle cell, which has a curious mode of development as well as an unusual strategy for repair.

Skeletal Muscle Cells Do Not Divide[30]

Cell division requires precisely coordinated transformations and movements of nucleus and cytoplasm. The process would be unusually complex and difficult for a skeletal muscle cell, which contains many nuclei instead of one and also has its cytoplasm crammed with highly ordered arrays of actin and myosin. In fact, skeletal muscle cells do not divide. Each nucleus contains the diploid quantity of DNA and is unable to replicate its DNA. Most skeletal muscle cells probably survive for the lifetime of the animal, but some are likely to be destroyed in one way or another. Since the cells do not divide, those that are lost cannot be replaced by simple duplication of survivors. Replacements can be produced only by a reactivation of the process by which skeletal muscle is formed in the embryo.

New Skeletal Muscle Cells Form by Fusion of Myoblasts[31]

The previous chapter described how certain cells, originating from the somites of a vertebrate embryo at a very early stage, become determined as *myoblasts* (that is, as precursors of skeletal muscle cells). Myoblasts can and do proliferate, remaining all the while apparently undifferentiated and scarcely distinguishable from neighboring mesenchyme cells. To form the multinucleate skeletal muscle cells, the myoblasts in due course fuse with one another, and as they fuse, they abruptly begin to manufacture the specialized proteins characteristic of differentiated muscle (Figure 16–42). Fusion involves some form of specific mutual recognition between myoblasts: they do not fuse with the adjacent nonmuscle cells.

In tissue culture, myoblasts have been kept proliferating for as long as two years, retaining for that period the ability to fuse and differentiate into muscle cells in response to a suitable change in the culture conditions. The process of fusion is cooperative: fusing myoblasts alter the composition of the culture medium in such a way as to encourage other myoblasts to fuse. The preparations of individual myoblasts for fusion appear to be coupled also to the events of the cell-division cycle, and fusion occurs only during the G_1 phase.

Muscle Differentiation Requires Coordinated Changes in the Expression of Many Different Genes[32]

Since cultured myoblasts can be triggered to differentiate synchronously by an appropriate change of culture medium, they provide a convenient system for biochemical study of the control of gene expression during differentiation. The mature muscle fiber is distinguished from other cells by a large number of characteristic proteins, including specific types of actin, myosin, tropomyosin, and troponin (all part of the contractile apparatus); creatine phosphokinase (for the specialized metabolism); and acetylcholine receptor (to make the membrane sensitive to neural stimulation). In proliferating myoblasts these proteins are absent, or present only in very low concentrations. The muscle-specific myosin, tropomyosin, and troponin subunits, for example, are undetectable in proliferating myoblasts from a bird. The synthesis of these proteins first becomes detectable when the myoblasts begin to fuse,

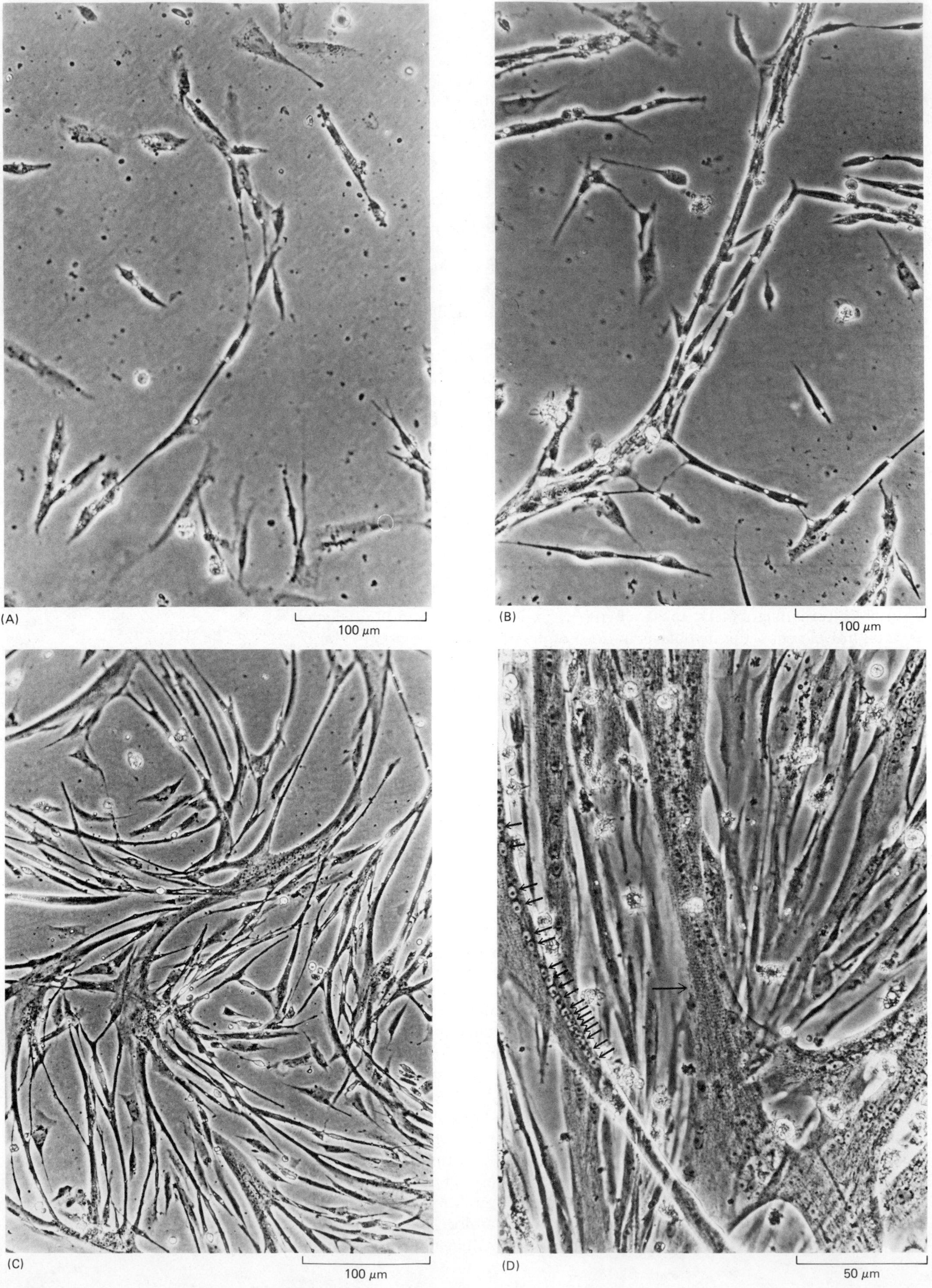
(A)
100 μm
(B)
100 μm
(C)
100 μm
(D)
50 μm

◄ **Figure 16–42** Myoblasts will proliferate in culture, line up, and then fuse to form multinucleate muscle cells. Photographs (A) through (D) are phase-contrast micrographs of living cultures, illustrating successive stages of the process. (D) is at higher magnification, showing the cross-striations that are just beginning to be visible as the contractile apparatus develops (*long arrow*), and the accumulations of many nuclei within a single cell (*short arrows*). (Courtesy of Rosalind Zalin.)

about 12 hours after triggering by switching to a culture medium that reduces the rate of proliferation and promotes differentiation. The production of the muscle-specific proteins closely parallels an increase in the concentrations of the corresponding species of messenger RNA: gene expression here seems to be controlled at the level of transcription. Within 40 hours after the change of medium, the rate of synthesis of many of the muscle-specific proteins has increased by a factor of at least 500, to a maximum of about 40,000 molecules per cell nucleus per minute. This rate is the same, within 10%, for each of seven types of subunit (myosin heavy chain, two myosin light chains, and two subunits each for tropomyosin and troponin). The time course of the increase is also practically identical. The general pattern of events is similar for other muscle-specific proteins, though the quantitative details are somewhat different. Moreover, as the differentiating myoblasts begin to synthesize muscle-specific proteins, the rates of production of many of the other proteins detected by two-dimensional polyacrylamide gel analysis also change: some syntheses are switched off, others rise to a peak and then fall, others shift from one steady level to another, and so on. We seem to be a long way from understanding how this complex pattern of events is controlled and coordinated.

Some Myoblasts Persist as Satellite Cells in the Adult[30,33]

With respect to tissue growth and renewal, the essential point is that once myoblasts have fused, further division is impossible, though the immature syncytial muscle cell can enlarge by recruiting more myoblasts to fuse with it. If all the myoblasts in an embryo were to fuse with one another simultaneously, there would be no myoblasts left thereafter and hence no possibility of increasing the number of skeletal muscle cells as the fetus grows. In fact, the process of fusion is staggered over a long period of development, and the stock of myoblasts is replenished by mitosis so that it is never entirely consumed. Even in the adult, a few myoblasts persist as small, flattened, and inactive cells lying in close contact with the mature muscle fibers (Figure 16–43). If the muscle is damaged, these so-called *satellite cells* are roused into activity—they begin to proliferate and their progeny fuse to form new muscle fibers. Satellite cells are at the same time a self-renewing population and a source of terminally differentiated cells: they are the stem cells of skeletal muscle.

The State of Differentiation of Skeletal Muscle Fibers Can Be Modulated by Electrical Stimulation[34]

Terminally differentiated skeletal muscle fibers are not all alike. But the distinctions between them, unlike the distinctions between blood cell types, are not irreversibly determined in the process of differentiation from the common stem cell.

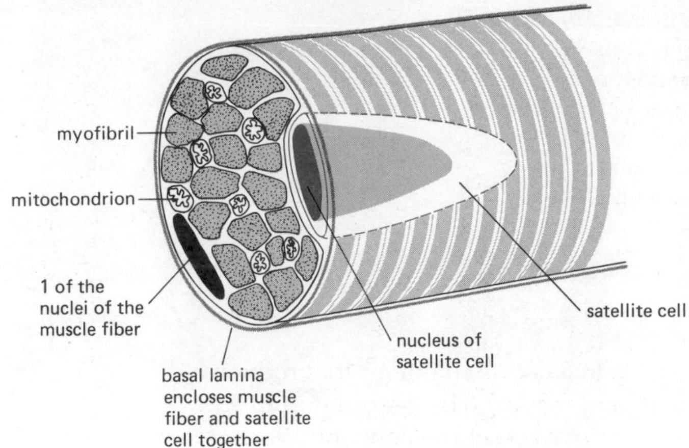

Figure 16–43 The muscle satellite cell (a quiescent myoblast) lies pressed close against the surface of the multinucleate muscle fiber.

Two main types of fibers are easily recognized. *Red muscle fibers,* such as those of the dark meat of a chicken, are rich in the oxygen-binding protein myoglobin. *White muscle fibers,* such as those of the white meat of the chicken, have much less myoglobin. (There are intermediate fibers as well, but we shall concentrate on the red and the white.) The different content of myoglobin—a molecule similar to hemoglobin—reflects a different metabolism with different oxygen requirements: the red fibers are specialized for oxidative phosphorylation, the white for anaerobic glycolysis. The different types of metabolism, in turn, support different types of contractile activity. The red fibers give a slow-twitch response to stimulation, are more resistant to fatigue, and are most efficient for generating sustained forces. The white fibers give a fast twitch, fatigue more easily, and are most efficient for quick, intermittent movements. Red fibers and white fibers contain different forms of the contractile proteins (such as myosin), transcribed from different genes. While most muscles contain a mixture of fiber types, some muscles are predominantly red (or slow) and others are predominantly white (or fast).

Muscle fibers are called into action by nerves, and the specializations in contractile properties just noted would be futile if the type of muscle were not matched to the pattern of neural commands received. How, then, is the observed matching brought about so that axons that transmit commands for sustained contraction innervate red fibers and the axons that transmit commands for quick, intermittent contraction innervate white fibers? An answer is given by experiments on two adjacent muscles in the leg of a rat—one slow, the other fast (Figure 16–44). The nerves to the two muscles are cut and then transposed so that each nerve grows back to reinnervate the muscle of contrary character. The properties of the muscles thereupon change: the fast becomes slow, and the slow becomes fast. The nerves evidently dictate the muscles' choice of differentiated state. Whatever other difference there might be between the two nerves, it is certain at least that they signal different patterns of excitation. The "slow" nerve tends to transmit prolonged bursts of action potentials with a low rate of repetition of action potentials within each burst; the "fast" nerve tends to transmit brief bursts of action potentials with a high rate of repetition in each burst. These patterns of muscle excitation can be mimicked by cutting the nerve and stimulating the muscle directly through implanted metal electrodes. A muscle stimulated artificially in this way for a few weeks with a slow pattern of shocks becomes slow, and one stimulated with a fast pattern of shocks becomes fast. It therefore appears that the pattern of electrical stimulation controls the pattern of gene expres-

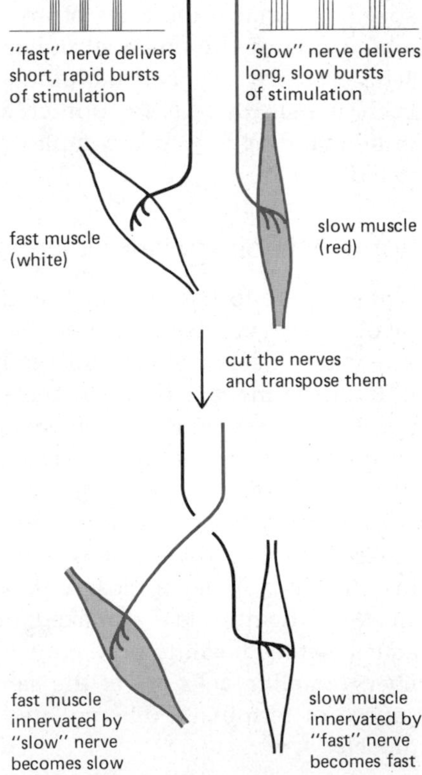

Figure 16–44 The consequences of transposing a "fast" nerve onto a slow muscle, and vice versa. The pattern of electrical stimulation delivered by each nerve controls the differentiated character of the muscle that it innervates.

sion in a muscle cell. This is another example of modulation of the differentiated state: the alterations of gene expression are limited and reversible, and the muscle fiber remains a muscle fiber, though it may change its myosin, its myoglobin content, and its complement of metabolic enzymes.

Summary

Skeletal muscle cells represent one of four main categories of vertebrate cells specialized for contraction. They are responsible for voluntary movement. Each skeletal muscle cell is a syncytium and develops by fusion of uninucleate myoblasts. Myoblasts can divide mitotically, but the multinucleate skeletal muscle cells cannot. The act of myoblast fusion is generally coupled with the beginning of muscle cell differentiation. In the process, many different genes are switched on coordinately. In adult life, some myoblasts persist in a quiescent state as satellite cells. When a muscle is damaged, these serve as stem cells, being reactivated to proliferate and to fuse to replace muscle cells that have been lost. The state of differentiation of mature skeletal muscle cells is modulated according to the type of electrical stimulation they receive from nerve cells.

Soft Cells and Tough Matrix: Growth, Turnover, and Repair in Skeletal Connective Tissue[35]

The jointed framework of rigid struts on which the skeletal muscles act is made of bone. Yet bone, for all its rigidity, is by no means a permanent and immutable tissue. Throughout its hard extracellular matrix are channels and hollows occupied by living cells. These cells are engaged in an unceasing process of remodeling. One class of cells demolishes old bone matrix while another class of cells deposits new bone matrix. This mechanism provides for turnover and replacement of the matrix in the interior of the bone.

The rigidity of the matrix is such that bone can grow only by *apposition,* that is, by the laying down of additional matrix and cells on the free surfaces of the hard tissue. This restriction on the mode of growth of bone helps to preserve the structure of the skeleton in adult life. In the embryo, appositional growth of bone must occur in coordination with the growth of other tissues so that the pattern of the body can be scaled up without its proportions being radically disturbed.

For most of the skeleton, and in particular for the long bones of the limbs and trunk, coordinated growth is achieved by a complex strategy. In the embryo, a set of minute "scale models" of the bones are first formed out of cartilage. Each scale model grows, and as new cartilage is formed, the older cartilage is replaced by bone. Cartilage growth and erosion and bone deposition are so ingeniously coordinated during development that the adult bone, though it may be half a meter long, is almost the same shape as the initial cartilaginous model, which was no more than a few millimeters long. Rather than go into the detailed geometry of this process, we shall concentrate on the forms of cell behavior underlying the growth and turnover of cartilage and bone, which are relevant both to the embryo and to the adult.

Cartilage Can Grow by Swelling[35,36]

The collaboration of bone and cartilage depends on their contrasting properties. Both types of tissues are formed from mesenchyme cells that secrete large quantities of collagenous extracellular matrix (see p. 692). But whereas bone matrix is rigid, cartilage matrix is deformable. Consequently cartilage,

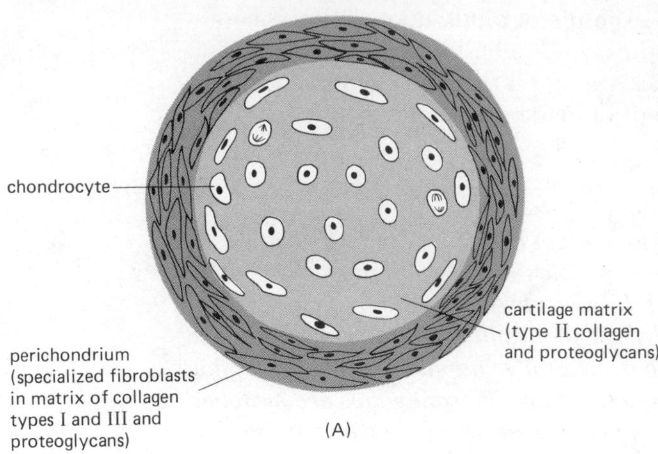

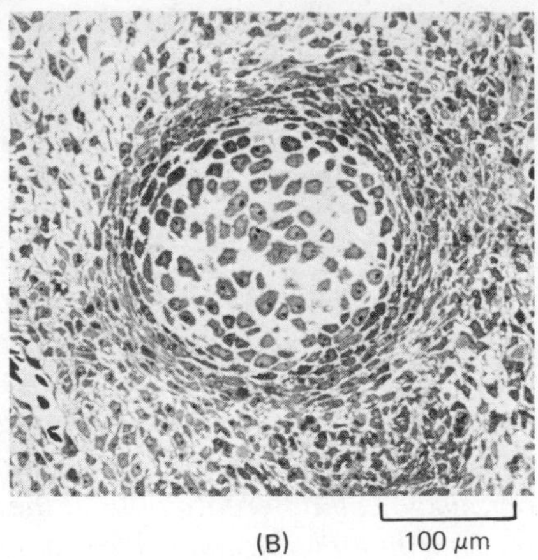

Figure 16–45 (A) Schematic diagram of a section through a rod of cartilage, showing the surrounding fibrous perichondrium. Each chondrocyte fills a lacuna in the cartilage matrix. (B) Cross-section through a rod of cartilage from a chick embryo. This photomicrograph represents an early stage of development. As the tissue grows, the quantity of cartilage matrix per chondrocyte will become much greater and the boundary between cartilage and perichondrium will become more sharply demarcated. (Courtesy of Peter Gould.)

unlike bone, is able to grow by swelling, as cells already embedded in the matrix secrete more matrix around themselves.

Cartilage cells, or chondrocytes, are isolated from one another, each occupying a small cavity, or *lacuna,* in the matrix (Figure 16–45). Cartilage usually contains no blood capillaries, and its cells are sustained by diffusion of nutrients and gases through the matrix to and from blood vessels that lie far away. A compact layer of collagenous connective tissue, the *perichondrium,* surrounds most of the cartilage mass (Figure 16–45). The cartilage expands from within as the chondrocytes secrete new matrix, while the fibrous perichondrium acts like a corset, constraining the consequent changes of shape. New cells are formed in this growth process: a chondrocyte, isolated in its lacuna in the matrix, will divide to give birth to two cells, each of which then proceeds to secrete more matrix so that a layer of matrix is soon formed between them. The resulting pairs of adjacent sister cells are a characteristic feature of the microscopic appearance of cartilage. Each member of the pair may divide again, giving rise to a family group of cells, each one secreting matrix and thereby gradually moving away from its relatives (Figure 16–46).

New cells may also be recruited into the cartilage from the perichondrium. Perichondrial cells resembling fibroblasts divide and undergo a conversion in which they begin to secrete cartilage matrix around themselves, and quickly become full-fledged chondrocytes. This appears to be the converse of the phenomenon described earlier in the chapter (p. 894), in which chondrocytes transform into cells resembling fibroblasts.

Osteoblasts Secrete Bone Matrix While Osteoclasts Erode It[35,37]

Bone is a more complex tissue than cartilage. The bone matrix is secreted by **osteoblasts** that lie at the surface of the existing matrix and deposit fresh layers of bone onto it (Figure 16–47). Some of the osteoblasts remain free at the surface, while others gradually become embedded in their own secretion. This freshly formed material (consisting chiefly of collagen) is called *osteoid.* It is rapidly converted into hard bone matrix by the deposition of calcium phosphate (more precisely, hydroxyapatite) crystals in it. A bone-specific protein, *osteonectin,* which binds strongly both to collagen and to hydroxyapatite, appears to provide sites for the growth of the crystals and to anchor them to the organic matrix. Once imprisoned in hard matrix, the original bone-forming cell, now called an **osteocyte,** has no opportunity to divide or to secrete further matrix in appreciable quantities. The osteocyte, like the chondrocyte, occupies a small cavity or lacuna in the matrix, but unlike the chondrocyte it is not isolated from its fellows. Tiny channels, or *canaliculi,* radiate from each lacuna and contain cell processes from the resident osteocyte, enabling it to form gap junctions with adjacent osteocytes. Though the networks of osteocytes do not themselves secrete or erode matrix, they probably play a major part in controlling the activities of the cells that do.

While bone matrix is deposited by osteoblasts, it is eroded by **osteoclasts** (Figure 16–48). These large multinucleated cells are a type of macrophage. Like other macrophages, they develop from monocytes that originate in the hematopoietic tissue of the bone marrow. These precursor cells are released into the bloodstream and collect at sites of bone resorption, where they fuse to form the multinucleated osteoclasts, which cling to surfaces of the bone matrix and eat it away.

There are many unsolved problems about this process. In particular, what determines whether matrix will be deposited by osteoblasts or eroded by osteoclasts at a given bone surface? Bones have a remarkable ability to remodel their structure in such a way as to adapt to the load imposed on them, and this implies that the deposition and erosion of the matrix are somehow controlled by local mechanical stresses. According to one theory, the stresses may act on the cells by giving rise to local electric fields to which the cells are sensitive: the collagen fibers in the matrix could mediate such an effect, since they are piezoelectric—that is, they become electrically polarized when subject to a mechanical stress. Whatever the mechanism may be, it seems likely that osteocytes are somehow involved: any region of bone matrix whose osteocytes have been killed (for example, by interruption of the blood supply) is promptly eroded.

Osteoclasts are capable of tunneling deep into the substance of compact bone (Figure 16–49), forming cavities that are then invaded by other cells. A blood capillary grows down the center of such a tunnel, and the walls of the tunnel become lined with a layer of osteoblasts. These lay down concentric layers of new bone, which gradually fill the cavity, leaving only a narrow canal surrounding the new blood vessel. Many of the osteoblasts become trapped in the bone matrix and survive as concentric rings of osteocytes. At the same time that some tunnels are filling up with bone, others are being bored by osteoclasts, cutting through older concentric systems. The consequences of this perpetual remodeling are beautifully displayed in the layered patterns of matrix observed in compact bone (Figure 16–50).

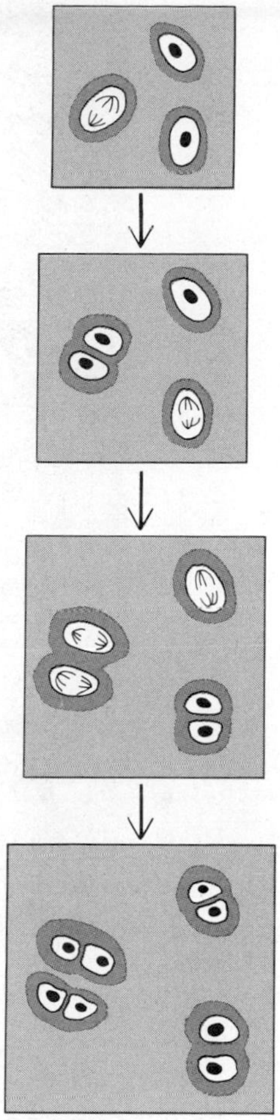

Figure 16–46 The growth of cartilage. The tissue expands as the chondrocytes divide and make more matrix. The freshly synthesized matrix with which each cell surrounds itself is shaded dark. Cartilage also grows by the recruitment of fibroblasts from the perichondrium and their conversion into chondrocytes (see Figure 16–45).

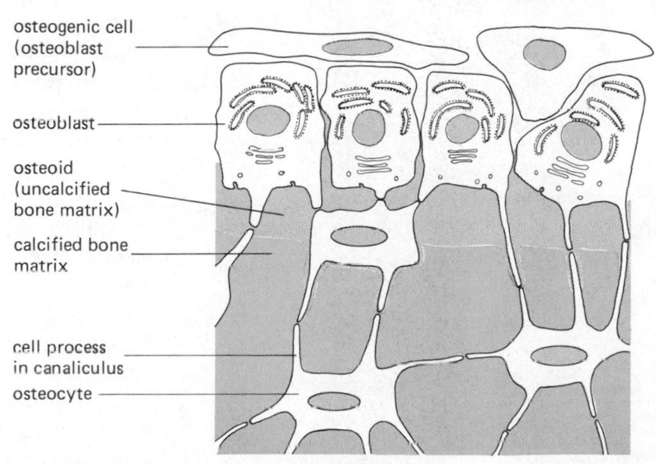

Figure 16–47 Schematic diagram showing how osteoblasts lining the surface of bone secrete the organic matrix of bone (osteoid) and are converted into osteocytes as they become embedded in this matrix. The matrix calcifies soon after it has been deposited.

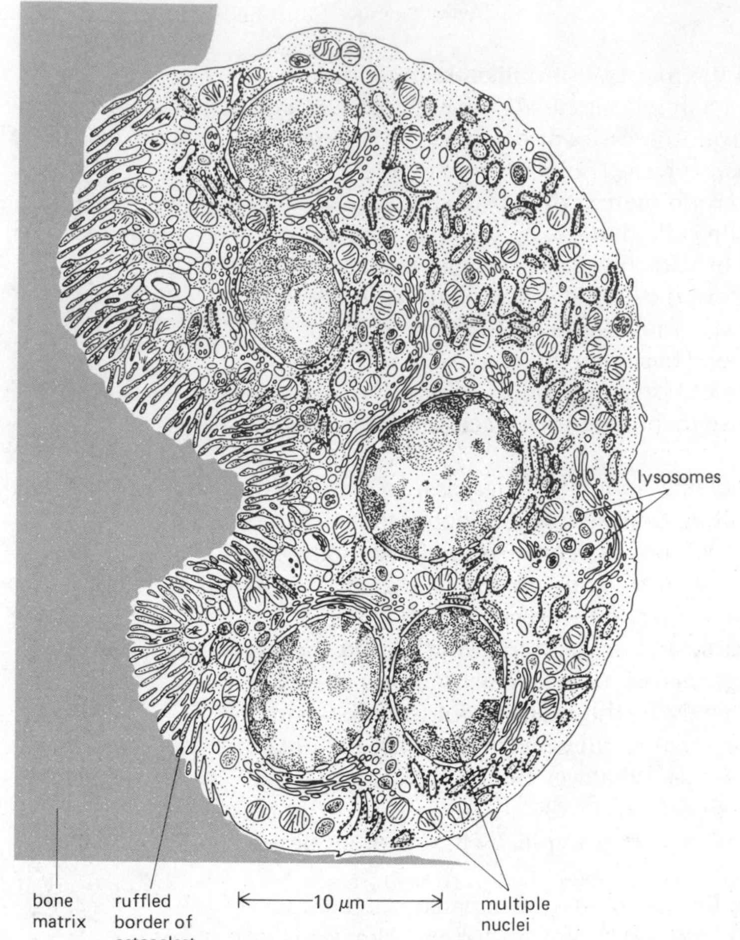

Figure 16–48 An osteoclast is a giant cell that erodes bone matrix. The cell is shown here in cross-section. (From R. V. Krstić: Ultrastructure of the Mammalian Cell: An Atlas. Berlin: Springer, 1979.)

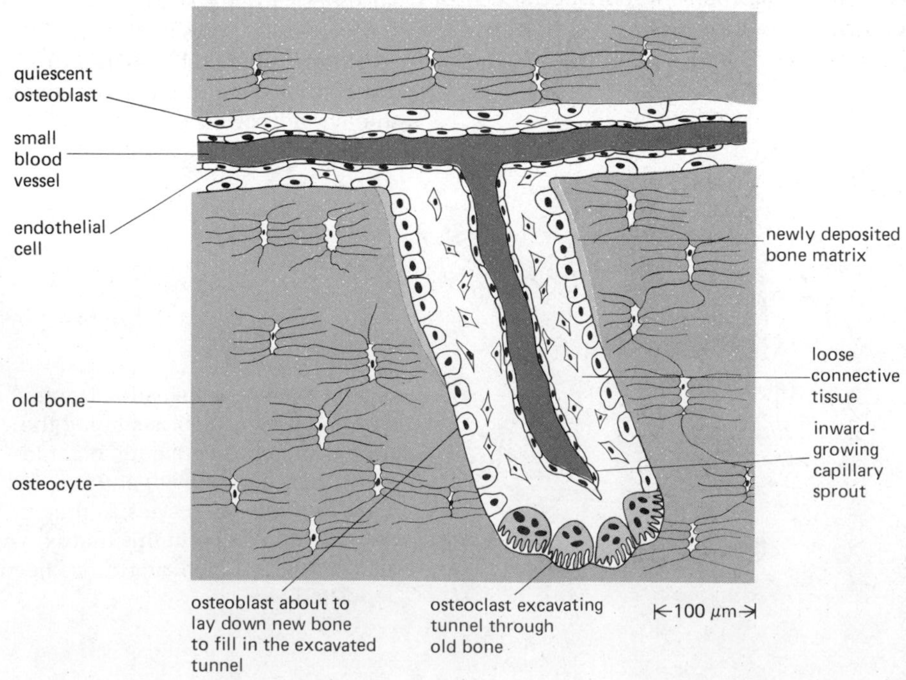

Figure 16–49 Schematic diagram of the remodeling of compact bone. Osteoclasts acting together in a small group excavate a tunnel through the old bone, advancing at a rate of about 50 μm per day. Osteoblasts enter the tunnel behind them, line its walls, and begin to form new bone, depositing matrix at a rate of one or two μm per day. At the same time, a capillary sprouts down the center of the tunnel. The tunnel will eventually become filled with concentric layers of new bone, with only a narrow central canal remaining. Each such canal, besides providing a route of access for osteoclasts and osteoblasts, contains one or more blood vessels bringing the nutrients that the bone cells must have to survive. Typically, about 5% to 10% of the bone in a healthy adult mammal is replaced in this way per year. (After Z. F. G. Jaworski, B. Duck, and G. Sekaly, *J. Anat.* 133:397–405, 1981.)

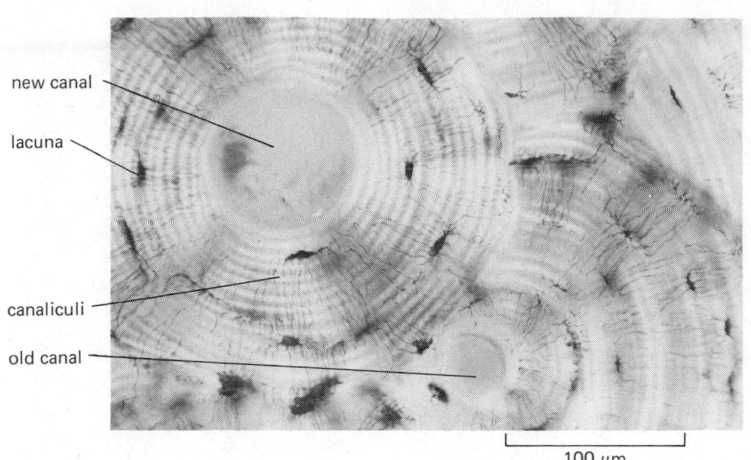

Figure 16–50 Photomicrograph of a transverse section through part of a long bone. The section has been prepared by grinding; the hard matrix has been preserved but not the cells. Lacunae and canaliculi that were occupied by osteocytes are clearly visible, however. The alternating bright and dark concentric rings correspond to the alternating orientation of the collagen fibers in the successive layers of bone matrix laid down by the osteoblasts that lined the wall of the canal during life. (This pattern is revealed here by viewing the specimen between partly crossed polaroid filters.) Note how the older system of concentric layers of bone at lower right, with the narrow central canal, has been partly cut through and replaced by the newer system, whose central canal is presumably still large because it is still in the process of being filled in.

Cartilage Is Eroded by Osteoclasts to Make Way for Bone[35]

The replacement of cartilage by bone in the course of development (Figure 16–51) is also thought to depend on the activities of osteoclasts. As it matures, cartilage begins in certain regions to become mineralized, like bone, by deposition of calcium phosphate crystals in its matrix. At the same time, the chondrocytes in these regions swell and die, leaving large empty cavities. Osteoclasts and blood vessels invade the cavities and erode the mineralized cartilage matrix, while osteoblasts following in their wake begin to deposit bone matrix. The only surviving remnant of cartilage in the adult long bone is a thin layer that forms a smooth covering on the bone surfaces at joints, where one bone articulates with another.

Some cells capable of forming new cartilage persist in the connective tissue that surrounds a bone. If the bone is broken, the cells in the neighborhood of the fracture will carry out a repair by a rough-and-ready recapitulation of the original embryonic process, in which cartilage is first laid down to bridge the gap and is then replaced by bone. A bone, like the body as a whole, is a dynamic system, maintaining its structure through a balance between the opposed activities of a variety of specialized cells.

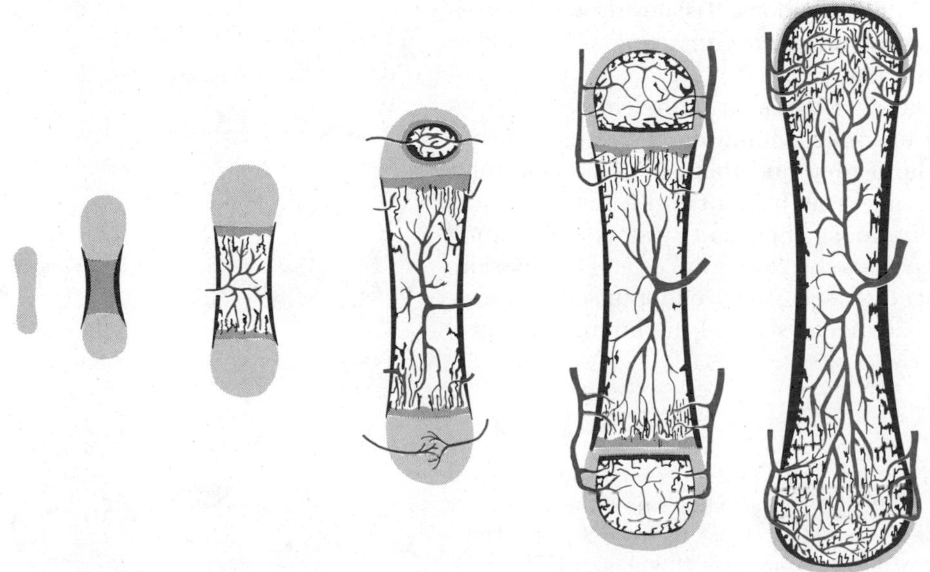

Figure 16–51 Schematic diagram of the development of a bone from a miniature cartilage model. Uncalcified cartilage is shown in pale gray, calcified cartilage in dark gray, bone in black, and blood vessels in color. The cartilage is not converted to bone but is gradually replaced by it through the action of osteoclasts and osteoblasts, which invade the cartilage in association with blood vessels. Osteoclasts erode cartilage and bone matrix, while osteoblasts secrete bone matrix. The process of ossification begins in the embryo and is not completed until the end of puberty. Note that not all bones develop in this way. The *membrane bones* of the skull, for example, are formed directly as bony plates, without any prior cartilage model.

Summary

Cartilage and bone both consist of cells embedded in a solid matrix. Cartilage has a deformable matrix and can grow by swelling, whereas bone is rigid and can grow only by accretion at its surfaces. Bone is, nonetheless, subject to perpetual remodeling through the combined action of osteoclasts (specialized macrophages), which erode matrix, and osteoblasts, which secrete it. Some osteoblasts become trapped in the matrix as osteocytes and play a part in regulating the turnover of bone matrix. Most long bones develop from miniature cartilage "models," which, as they grow, serve as templates for the deposition of bone by the combined action of osteoblasts and osteoclasts. Similarly, in the repair of a bone fracture in the adult, the gap is first bridged by cartilage, which is later replaced by bone.

Territorial Stability in the Adult Body[38]

We have discussed how cells in various types of tissues maintain their differentiated state, how new cells are produced to replace those that are lost, and how the extracellular matrix is remodeled and renewed. But why do the different types of cells not become progressively jumbled and misplaced? Why does the whole structure not sag, warp, or otherwise change its proportions as new parts are substituted for old?

Of course, to some extent, the body does sag and warp with the passage of time: that is a part of aging. But it does so remarkably little. The skeleton, despite constant remodeling, provides a rigid framework whose dimensions scarcely change. That is presumably because the parts of a bone are renewed not all at once but little by little, rather like a building whose bricks are replaced one at a time. If large components were removed en masse, the structure of the remainder would no doubt alter. For example, the shape of the jaw bone will change if the teeth are pulled out (Figure 16–52).

The growth and renewal of many of the soft parts of the body are homeostatically controlled so that each component is adjusted to fit its niche. The epidermis spreads to cover the surface of the body, and the cells halt their migration by *contact inhibition* when that end is achieved (p. 619); connective tissue grows to just the extent necessary to fill the gap created by a wound; and so on. But something more than this is required. The various types of differentiated cells must be maintained not only in the correct relative quantities, but also in the correct relative positions. Tissue turnover necessarily involves cell movements. Somehow those movements must be limited; the cells must be subject to territorial restraints.

These restraints are of various kinds. Glands and other masses of specialized cells are often contained, for example, within tough capsules of connective tissue. Some types of cells die if they find themselves outside their normal environment, deprived of specific growth factors on which their survival depends (see p. 619). Perhaps the most important strategy for keeping the different cells in their places, however, is the strategy of selective adhesion: cells of the same type tend to stick together (see p. 680), either in solid masses, such as smooth muscle, or in epithelial sheets, such as the lining of the gut.

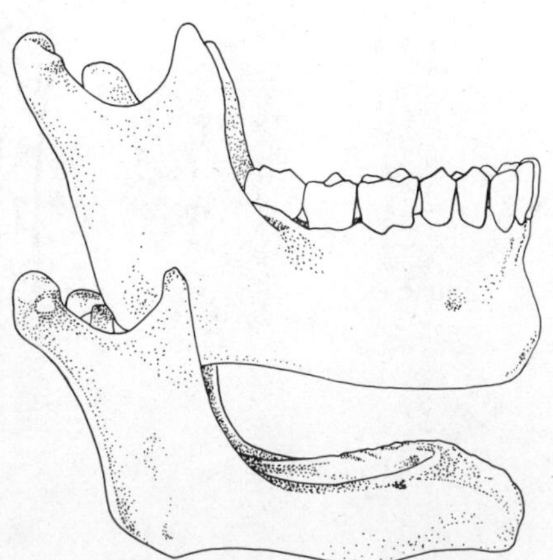

Figure 16–52 The upper part of the drawing shows a normal human jaw; the lower part shows how the bone changes its shape if the teeth are lost. (From R. J. Goss, The Physiology of Growth. New York: Academic Press, 1978.)

Epithelial Organization Helps to Keep Cells in Their Proper Territories

Epithelial cells are held in place by their attachments to each other and to the basal lamina. The basal lamina marks a strictly respected boundary between two compartments—between the epithelium and the tissues beneath it. In normal individuals, only a few specialized types of cells, such as lymphocytes, macrophages, and nerve cell processes, can cross this barrier.

An epithelial sheet, besides being a sort of prison for the cells that constitute it and for their progeny, can form an enclosure to contain other cells. Thus the sheet of endothelial cells that lines blood vessels contains the blood cells within the vascular system. Similarly, the epithelium lining the gut prevents the connective tissue cells of the gut wall from wandering into the lumen. A majority of the cell types in the body are arranged as epithelial sheets, including many that undergo very rapid turnover. Thus the compartmentation of the body created by epithelia plays a crucial part in keeping cells properly segregated and confined to their correct territories.

Normal Somatic Cells Are Destined to Die for the Sake of the Survival of the Germ Cell[39]

At the beginning of this chapter, we suggested that the body can be viewed as a society or ecosystem whose individual members are cells. The central concerns of the chapter have been similar to those of the ecologist: births, deaths, habitats, territorial limitations, the maintenance of population sizes, and the like. But there is one ecological topic conspicuously absent: the topic of natural selection. We have said nothing of competition or mutation among somatic cells. The reason is that a healthy body is in this respect a very peculiar society: it is a society in which absolute altruism is the supreme rule of behavior for every class of individuals except one. Every somatic cell is committed to die, leaving no progeny, in order that the germ cells alone shall have a chance of survival. There is no mystery here, however. Although the somatic cells die, they help to perpetuate the existence of the genes that they carry; for the body is a clone, and the genes of the somatic cells are the same as the genes of the germ cells. The genome that produces dead-end differentiated cells is successful because it also produces germ cells, which survive because of the self-sacrifice of their relatives.

The evolution of altruistic behavior that benefits close relatives has been much discussed by sociobiologists. We see it in the care of parents for their children, and, to a more remarkable degree, in the care of worker bees for their queen-bee sisters. The peculiarly altruistic behavior displayed by social insects, such as ants, bees, and wasps, can be interpreted in terms of the peculiar genetics of their sexual reproduction cycle, which results in the sister insects of these species having more similar genomes than sisters or brothers—or, indeed, parents and children—of most other species. Thus an altruistic act for the benefit of a sister ant, bee, or wasp does more to promote the survival of the genes that dictated it than the same altruistic act would in another species. In the society of cells that form the body of a multicellular plant or animal, the genomes are not merely similar, but identical, and it is therefore not surprising that altruism is here carried to the ultimate degree.

Of course, the individual cells of a clone do not have to follow a policy of specialization and altruism: the thousands of genetically identical *E. coli* bacteria that descend from a single parent bacterium compete with one another rather than cooperate. But once a collaborative strategy for the propagation of the genome has evolved, any mutation that gives rise to nonaltruistic behavior by individual members of the cooperative takes on a peculiarly sinister aspect. Selfish behavior by a mutant cell in the body jeopardizes the future of the whole multicellular enterprise. In other words, mutation and natural selection acting *within* the population of somatic cells amount to a recipe for extinction. How great then is the risk of disaster from such a cause, and what kind of protection has been evolved against it?

Cancer Cells Break the Rules of Altruistic Social Behavior[40]

Something on the order of 10^{16} cell divisions take place in a human body in the course of a lifetime. Mutations will occur spontaneously from time to time: the average rate is hard to estimate accurately for human cells, but it is probably about 10^{-6} mutations per gene per cell division. Thus, in a lifetime, a typical gene is likely to have undergone mutation on about 10^{10} separate occasions in any individual. Among the mutant somatic cells there must surely be many that proliferate in disregard of normal controls or faster than their nonmutant counterparts. From this point of view, the problem of cancer seems to be not why it occurs, but why it occurs so infrequently. Why are we not unfailingly overrun by mutant clones of cells that have a selective advantage over our healthy normal cells?

The answer is complex. To take one example, suppose that a cell in the epithelium of the small intestine undergoes a mutation that increases its division rate. The consequences will depend on the position of the cell when the mutation occurs. If the cell lies in the neck of a crypt, for example, it and its mutant progeny will still be swept out in the usual way onto the exposed surface of a villus, and so be shed into the lumen of the gut and discarded (see Figure 16–20). To establish a clone that persists in the body, the mutation must originate in a stem cell. However, even this by itself presents no immediate danger. Each crypt represents an independent proliferative unit, effectively isolated from each other crypt by an exposed zone where cells become terminally differentiated and are destroyed. Thus a mutant stem cell that simply proliferates faster than normal can populate its own crypt and villi, but its progeny cannot pass from one crypt to another, as would be required if they were to colonize the rest of the gut. To be dangerous, mutant stem cells of the gut epithelium must violate the territorial restraints that keep normal cells in their proper places. They must *invade* or *metastasize.* Somehow, the mutant cells must grow and spread into the underlying connective tissue from which the normal cells are excluded by the basal lamina and the rules of epithelial cell behavior. Or they must send out colonists that can travel through alien territories and continue their proliferation in circumstances where normal cells would not. Cells that proliferate to excess but stay put form *benign tumors,* which can usually be completely removed by local surgery. Cancers, by contrast, are *malignant tumors,* consisting of cells that defy not simply the normal controls on their proliferation, but also the normal controls on their position. The more widely such cells spread in the body, the harder they become to eradicate surgically.

It is generally agreed that human cancers, with the exception perhaps of teratomas (p. 831) and a few that are caused by viruses, develop from mutant clones of cells. But the body is organized in such a way that while many cells mutate, few mutant cells are dangerous. To generate a cancer, a mutation has to arise in a cell that is appropriately placed and to confer on that cell complex new properties. Indeed, there is evidence that a single mutation in general is not enough: most cancers apparently develop only after multiple mutations have occurred in a single cell lineage. Cancer research has the difficult task of analyzing the many different strategies by which mutant cells break the rules of altruistic behavior, invading and colonizing habitats normally reserved for other types of cells.

Summary

The body of an adult vertebrate exists in a state of dynamic equilibrium, equipped with homeostatic controls on cell proliferation and with various mechanisms to keep cells from straying from their proper places. Although huge numbers of somatic cell mutations must occur in the life of an animal, producing cells that disobey some of the normal controls, cancers are relatively rare. This is probably in large part because the genesis of a malignant tumor requires the coincidental occurrence of several specific mutations in a single cell, enabling it to proliferate in disregard of the usual constraints and to invade regions of the body from which it would normally be excluded.

Appendix

Cells of the Adult Human Body: A Catalog

How many distinct cell types are there in an adult human being? A large textbook of histology will mention about 200 that qualify for individual names. These traditional names are not, like the names of colors, labels for parts of a continuum that has been subdivided arbitrarily: they represent, for the most part, discrete and distinctly different categories. Within a given category there is often some variation—the skeletal muscle fibers that move the eyeball are small, while those that move the leg are big; auditory hair cells in different parts of the ear may be tuned to different frequencies of sound; and so on. But there is no continuum of adult cell types intermediate in character between, say, the muscle cell and the auditory hair cell.

The traditional histological classification is based on the shape and structure of the cell as seen in the microscope and on its chemical nature as assessed very crudely from its affinities for various stains. Subtler methods reveal new subdivisions within the traditional classification. Thus modern immunology has shown that the old category of "lymphocyte" includes more than 10 quite distinct cell types (see Chapter 17). Similarly, pharmacological and physiological tests reveal that there are many different varieties of smooth muscle cell—those in the wall of the uterus, for example, are highly sensitive to estrogen, and in the later stages of pregnancy to oxytocin, while those in the wall of the gut are not. In these cases, cell biologists already have an inkling of the chemical basis of the distinctions. Another major type of diversity is revealed by embryological experiments of the sort discussed in Chapter 15. These show that, in many cases, apparently similar cells from different regions of the body are nonequivalent, that is, they are inherently different in their developmental capacities and in their effects on other cells. For example, connective tissue cells from different regions of the dermis must be nonequivalent, since they provoke the overlying epidermal cells to behave differently (see p. 895). Thus, within categories such as "fibroblast" there are probably many distinct cell types, different chemically in ways that we cannot yet perceive directly.

For these reasons, any classification of the cell types in the body must be somewhat arbitrary with respect to the fineness of its subdivisions. In the catalog given here, we list only the adult human cell types that a large modern histology textbook would recognize to be different, grouped into families roughly according to function. In some cases where it is clear that a traditional category needs to be subdivided, but the subtypes are not yet well characterized, we have simply written "(various)" after the name of the cell type. In particular, we have not attempted to subdivide the class of neurons of the central nervous system. Also, where a single cell type such as keratinizing epidermal cell is conventionally given a succession of different names as it matures, we give only two entries—one for the differentiating cell and one for the stem cell. With these serious provisos, the 210 varieties of cells in the catalog represent a more or less exhaustive list of the distinctive ways in which a given mammalian genome can be expressed in the phenotype of a normal cell of the adult body. It is noteworthy that more than 60% of the cell types listed are epithelial.

Keratinizing Epithelial Cells

keratinocyte of epidermis (=differentiating epidermal cell)
basal cell of epidermis (stem cell)
keratinocyte of fingernails and toenails
basal cell of nail bed (stem cell)
hair shaft cells
  medullary
  cortical
  cuticular
hair-root sheath cells
  cuticular
  of Huxley's layer
  of Henle's layer
  external
hair matrix cell (stem cell)

Cells of Wet Stratified Barrier Epithelia

surface epithelial cell of stratified squamous epithelium of tongue, oral cavity, esophagus, anal canal, distal urethra, vagina
basal cell of these epithelia (stem cell)
cell of external corneal epithelium
cell of urinary epithelium (lining bladder and urinary ducts)

Epithelial Cells Specialized for Exocrine Secretion

cells of salivary gland
  mucous cell (secretion rich in polysaccharide)
  serous cell (secretion rich in glycoprotein enzymes)
cell of von Ebner's gland in tongue (secretion to wash over taste buds)
cell of mammary gland, secreting milk
cell of lacrimal gland, secreting tears
cell of ceruminous gland of ear, secreting wax
cell of eccrine sweat gland, secreting glycoproteins (dark cell)
cell of eccrine sweat gland, secreting small molecules (clear cell)
cell of apocrine sweat gland (odoriferous secretion, sex-hormone sensitive)
cell of gland of Moll in eyelid (specialized sweat gland)
cell of sebaceous gland, secreting lipid-rich sebum
cell of Bowman's gland in nose (secretion to wash over olfactory epithelium)
cell of Brunner's gland in duodenum, secreting alkaline solution of mucus and enzymes
cell of seminal vesicle, secreting components of seminal fluid, including fructose (as fuel for swimming sperm)
cell of prostate gland, secreting other components of seminal fluid
cell of bulbourethral gland, secreting mucus
cell of Bartholin's gland, secreting vaginal lubricant
cell of gland of Littré, secreting mucus
cell of endometrium of uterus, secreting mainly carbohydrates
isolated goblet cell of respiratory and digestive tracts, secreting mucus
mucous cell of lining of stomach
zymogenic cell of gastric gland, secreting pepsinogen

oxyntic cell of gastric gland, secreting HCl
acinar cell of pancreas, secreting digestive enzymes and bicarbonate
Paneth cell of small intestine, secreting lysozyme
type II pneumocyte of lung, secreting surfactant
Clara cell of lung (function unknown)

Cells Specialized for Secretion of Hormones

cells of anterior pituitary, secreting
- growth hormone
- follicle-stimulating hormone
- luteinizing hormone
- prolactin
- adrenocorticotropic hormone
- thyroid-stimulating hormone

cell of intermediate pituitary, secreting melanocyte-stimulating hormone
cells of posterior pituitary, secreting
- oxytocin
- vasopressin

cells of gut, secreting
- serotonin
- endorphin
- somatostatin
- gastrin
- secretin
- cholecystokinin
- insulin
- glucagon

cells of thyroid gland, secreting
- thyroid hormone
- calcitonin

cells of parathyroid gland, secreting
- parathyroid hormone
- oxyphil cell (function unknown)

cells of adrenal gland, secreting
- epinephrine
- norepinephrine
- steroid hormones
  - mineralocorticoids
  - glucocorticoids

cells of gonads, secreting
- testosterone (Leydig cell of testis)
- estrogen (theca interna cell of ovarian follicle)
- progesterone (corpus luteum cell of ruptured ovarian follicle)

cells of juxtaglomerular apparatus of kidney
- juxtaglomerular cell (secreting renin)
- macula densa cell, peripolar cell, mesangial cell } (uncertain but probably related in function; possibly involved in secretion of erythropoietin)

Epithelial Absorptive Cells in Gut, Exocrine Glands, and Urogenital Tract

brush border cell of intestine (with microvilli)
striated duct cell of exocrine glands
gall bladder epithelial cell
brush border cell of proximal tubule of kidney
distal tubule cell of kidney
nonciliated cell of ductulus efferens
epididymal principal cell
epididymal basal cell

Cells Specialized for Metabolism and Storage

hepatocyte (liver cell)
fat cells
  white fat
  brown fat
  lipocyte of liver

Epithelial Cells Serving Primarily a Barrier Function, Lining the Lung, Gut, Exocrine Glands, and Urogenital Tract

type I pneumocyte (lining air space of lung)
pancreatic duct cell (centroacinar cell)
nonstriated duct cell of sweat gland, salivary gland, mammary gland, etc. (various)
parietal cell of kidney glomerulus
podocyte of kidney glomerulus
cell of thin segment of loop of Henle (in kidney)
collecting duct cell (in kidney)
duct cell of seminal vesicle, prostate gland, etc. (various)

Epithelial Cells Lining Closed Internal Body Cavities

vascular endothelial cells of blood vessels and lymphatics
  fenestrated
  continuous
  splenic
synovial cell (lining joint cavities, secreting largely hyaluronic acid)
serosal cell (lining peritoneal, pleural, and pericardial cavities)
squamous cell lining perilymphatic space of ear
cells lining endolymphatic space of ear
  squamous cell
  columnar cells of endolymphatic sac
    with microvilli
    without microvilli
  "dark" cell
  vestibular membrane cell (resembling choroid plexus cell)
  stria vascularis basal cell
  stria vascularis marginal cell
  cell of Claudius
  cell of Boettcher
choroid plexus cell (secreting cerebrospinal fluid)
squamous cell of pia-arachnoid
cells of ciliary epithelium of eye
  pigmented
  nonpigmented
corneal "endothelial" cell

Ciliated Cells with Propulsive Function

of respiratory tract
of oviduct and of endometrium of uterus (in female)
of rete testis and ductulus efferens (in male)
of central nervous system (ependymal cell lining brain cavities)

Cells Specialized for Secretion of Extracellular Matrix

epithelial:
  ameloblast (secreting enamel of tooth)
  planum semilunatum cell of vestibular apparatus of ear (secreting proteoglycan)
  interdental cell of organ of Corti (secreting tectorial "membrane" covering hair
    cells of organ of Corti)

nonepithelial (connective tissue)
  fibroblasts (various—of loose connective tissue, of cornea, of tendon, of reticular tissue of bone marrow, etc.)
  pericyte of blood capillary
  nucleus pulposus cell of intervertebral disc
  cementoblast/cementocyte (secreting bonelike cementum of root of tooth)
  odontoblast/odontocyte (secreting dentin of tooth)
  chondrocytes
    of hyaline cartilage
    of fibrocartilage
    of elastic cartilage
  osteoblast/osteocyte
  osteoprogenitor cell (stem cell of osteoblasts)
  hyalocyte of vitreous body of eye
  stellate cell of perilymphatic space of ear

Contractile Cells

skeletal muscle cells
  red (slow)
  white (fast)
  intermediate
  muscle spindle—nuclear bag
  muscle spindle—nuclear chain
  satellite cell (stem cell)
heart muscle cells
  ordinary
  nodal
  Purkinje fiber
smooth muscle cells (various)
myoepithelial cells
  of iris
  of exocrine glands

Cells of Blood and Immune System

red blood cell
megakaryocyte
macrophages
  monocyte
  connective tissue macrophage (various)
  Langerhans cell (in epidermis)
  osteoclast (in bone)
  dendritic cell (in lymphoid tissues)
  microglial cell (in central nervous system)
neutrophil
eosinophil
basophil
mast cell
T lymphocyte
  helper T cell
  suppressor T cell
  killer T cell
B lymphocyte
  IgM
  IgG
  IgA
  IgE
killer cell
stem cells for the blood and immune system (various)

Sensory Transducers

photoreceptors
- rod
- cones
  - blue sensitive
  - green sensitive
  - red sensitive

hearing
- inner hair cell of organ of Corti
- outer hair cell of organ of Corti

acceleration and gravity
- type I hair cell of vestibular apparatus of ear
- type II hair cell of vestibular apparatus of ear

taste
- type II taste bud cell

smell
- olfactory neuron
- basal cell of olfactory epithelium (stem cell for olfactory neurons)

blood pH
- carotid body cell
  - type I
  - type II

touch
- Merkel cell of epidermis
- primary sensory neurons specialized for touch (various)

temperature
- primary sensory neurons specialized for temperature
  - cold sensitive
  - heat sensitive

pain
- primary sensory neurons specialized for pain (various)

configurations and forces in musculoskeletal system
- proprioceptive primary sensory neurons (various)

Autonomic Neurons

cholinergic (various)
adrenergic (various)
peptidergic (various)

Supporting Cells of Sense Organs and of Peripheral Neurons

supporting cells of organ of Corti
- inner pillar cell
- outer pillar cell
- inner phalangeal cell
- outer phalangeal cell
- border cell
- Hensen cell

supporting cell of vestibular apparatus
supporting cell of taste bud (type I taste bud cell)
supporting cell of olfactory epithelium
Schwann cell
satellite cell (encapsulating peripheral nerve cell bodies)
enteric glial cell

Neurons and Glial Cells of Central Nervous System

neurons (huge variety of types—still poorly classified)
glial cells
- astrocyte (various)
- oligodendrocyte

Lens Cells

anterior lens epithelial cell
lens fiber (crystallin-containing cell)

Pigment Cells

melanocyte
retinal pigmented epithelial cell

Germ Cells

oogonium/oocyte
spermatocyte
spermatogonium (stem cell for spermatocyte)

Nurse Cells

ovarian follicle cell
Sertoli cell (in testis)
thymus epithelial cell

References

General

Bloom, W.; Fawcett, D.W. A Textbook of Histology, 10th ed. Philadelphia: Saunders, 1975.
Clark, W.E. Le Gros The Tissues of the Body, 6th ed. Oxford, Eng.: Clarendon Press, 1971.
Goss, R.J. The Physiology of Growth. New York: Academic Press, 1978.
Ham, A.W.; Cormack, D.H. Histology, 8th ed. New York: Harper & Row, 1979.
Weiss, L.; Greep, R.O. Histology, 4th ed. New York: McGraw-Hill, 1977.

Cited

1. Wessells, N.K. Tissue Interactions and Development. Menlo Park, Ca.: Benjamin-Cummings, 1977.
2. Cahn, R.D.; Cahn, M.B. Heritability of cellular differentiation: clonal growth and expression of differentiation in retinal pigment cells *in vitro*. *Proc. Natl. Acad. Sci. USA* 55:106–114, 1966.
3. Coon, H.G. Clonal stability and phenotypic expression of chick cartilage cells *in vitro*. *Proc. Natl. Acad. Sci. USA* 55:66–73, 1966.
   von der Mark, K.; Gauss, V.; von der Mark, H.; Müller, P. Relationship between cell shape and type of collagen synthesized as chondrocytes lose their cartilage phenotype in culture. *Nature* 267:531–532, 1977.
   Archer, C.W.; Rooney, P.; Wolpert, L. Cell shape and cartilage differentiation of early chick limb bud cells in culture. *Cell Differ.* 11:245–251, 1982.
   von der Mark, K. Immunological studies on collagen type transition in chondrogenesis. *Curr. Top. Dev. Biol.* 14:199–225, 1980.
   Lash, J.W.; Vasan, N.S. Somite chondrogenesis *in vitro*. Stimulation by exogenous extracellular matrix components. *Dev. Biol.* 66:151–171, 1978.
4. Billingham, R.E.; Silvers, W.K. Studies on the conservation of epidermal specificities of skin and certain mucosas in adult mammals. *J. Exp. Med.* 125:429–446, 1967.
5. Zalewski, A.A. Neuronal and tissue specifications involved in taste bud formation. *Ann. N.Y. Acad. Sci.* 228:344–349, 1974.
6. Eguchi, G. "Transdifferentiation" of vertebrate cells in cell culture. In Embryogenesis in Mammals, *Ciba Symp.* 40:241–257, 1976.

Razin, A.; Riggs, A.D. DNA methylation and gene function. *Science* 210:604–610, 1980.

Jones, P.A.; Taylor, S.M. Cellular differentiation, cytidine analogs and DNA methylation. *Cell* 20:85–93, 1980.

Felsenfeld, G.; McGhee, J. Methylation and gene control. *Nature* 296:602–603, 1982.

7. Goss, R.J. The Physiology of Growth, pp. 210–225. New York: Academic Press, 1978.

Clayton, R.M. Divergence and convergence in lens cell differentiation: regulation of the formation and specific content of lens fibre cells. In Stem Cells and Tissue Homeostasis (B. Lord, C. Potten, R. Cole, eds.), pp. 115–138. Cambridge, Eng.: Cambridge University Press, 1978.

8. Young, R.W. Visual cells. *Sci. Am.* 223(4):80–91, 1970.

Bloom, W.; Fawcett, D.W. A Textbook of Histology, 10th ed., pp. 917–963. Philadelphia: Saunders, 1975.

9. Cameron, I.L. Cell proliferation and renewal in the mammalian body. In Cellular and Molecular Renewal in the Mammalian Body (I.L. Cameron, J.D. Thrasher, eds.), pp. 45–86. New York: Academic Press, 1971.

10. Moog, F. The lining of the small intestine. *Sci. Am.* 245(5):154–176, 1981.

Bloom, W.; Fawcett, D.W. A Textbook of Histology, 10th ed., pp. 688–716. Philadelphia: Saunders, 1975.

11. Goss, R.J. The Physiology of Growth, pp. 251–266. New York: Academic Press, 1978.

Holder, N. Regeneration and compensatory growth. *Br. Med. Bull.* 37:227–232, 1981.

12. Anderson, J.R., ed. Muir's Textbook of Pathology, 11th ed., pp. 683–692. London: Edward Arnold, 1980.

13. Bloom, W.; Fawcett, D.W. A Textbook of Histology, 10th ed., pp. 386–426. Philadelphia: Saunders, 1975.

Zweifach, B.W. The microcirculation of the blood. *Sci. Am.* 200:54–60, 1959.

14. Goss, R.J. The Physiology of Growth, pp. 120–137. New York: Academic Press, 1978.

15. Folkman, J.; Haudenschild, C. Angiogenesis *in vitro*. *Nature* 288:551–556, 1980.

16. Folkman, J. The vascularization of tumors. *Sci. Am.* 234(5):58–73, 1976.

17. Cheng, H.; Leblond, C.P. Origin, differentiation, and renewal of the four main epithelial cell types in the mouse small intestine. V. Unitarian theory of the origin of the four epithelial cell types. *Am. J. Anat.* 141:537–562, 1974.

Graziadei, P.P.C.; Monti Graziadei, G.A. Continuous nerve cell renewal in the olfactory system. In Handbook of Sensory Physiology, Vol. IX, Development of Sensory Systems (M. Jacobson, ed.), pp. 55–82. New York: Springer-Verlag, 1978.

18. Sengel, P. Morphogenesis of Skin. Cambridge, Eng.: Cambridge University Press, 1975.

19. MacKenzie, J.C. Ordered structure of the stratum corneum of mammalian skin. *Nature* 222:881–882, 1969.

Allen, T.D.; Potten, C.S. Fine-structural identification and organization of the epidermal proliferative unit. *J. Cell Sci.* 15:291–319, 1974.

20. Fuchs, E.; Green, H. Changes in keratin gene expression during terminal differentiation of the keratinocyte. *Cell* 19:1033–1042, 1980.

21. Potten, C.S. The epidermal proliferative unit: the possible role of the central basal cell. *Cell Tissue Kinet.* 7:77–88, 1974.

22. Ham, A.W.; Cormack, D.H. Histology, 8th ed., pp. 635–639. New York: Harper & Row, 1979.

Green, H. Terminal differentiation of cultured human epidermal cells. *Cell* 11:405–415, 1977.

Watt, F.M.; Green, H. Stratification and terminal differentiation of cultured epidermal cells. *Nature* 295:434–436, 1982.

Potten, C.S.; Schofield, R.; Lajtha, L.G. A comparison of cell replacement in bone marrow, testis and three regions of surface epithelium. *Biochim. Biophys. Acta* 560:281–299, 1979.

23. Potten, C.S.; Allen, T.D. The fine structure and cell kinetics of mouse epidermis after wounding. *J. Cell Sci.* 17:413–447, 1975.

24. Bloom, W.; Fawcett, D.W. A Textbook of Histology, 10th ed., pp. 588–592, 907–916. Philadelphia: Saunders, 1975.

Patton, S. Milk. *Sci. Am.* 221(1):58–68, 1969.

Vonderhaar, B.K.; Topper, Y.J. A role of the cell cycle in hormone-dependent differentiation. *J. Cell Biol.* 63:707–712, 1974.

Cowie, A.T.; Forsyth, I.A.; Hart, I.C. Hormonal Control of Lactation. New York: Springer-Verlag, 1980.

Richards, R.C.; Benson, G.K. Ultrastructural changes accompanying involution of the mammary gland in the albino rat. *J. Endocrinol.* 51:127–135, 1971.

25. Wintrobe, M.M. Blood, Pure and Eloquent. New York: McGraw-Hill, 1980.

Ham, A.W.; Cormack, D.H. Histology, 8th ed., pp. 295–322. New York: Harper & Row, 1979.

Clarkson, B.; Marks, P.A.; Till, J.E., eds. Differentiation of Normal and Neoplastic Hematopoietic Cells. Cold Spring Harbor, N.Y.: Cold Spring Harbor Laboratory, 1978.

Weiss, L.; Greep, R.O. Histology, 4th ed., pp. 433–502. New York: McGraw-Hill, 1977.

26. Till, J.E.; McCulloch, E.A. A direct measurement of the radiation sensitivity of normal mouse bone marrow cells. *Radiat. Res.* 14:213–222, 1961.

Wu, A.M.; Till, J.E.; Siminovitch, L.; McCulloch, E.A. A cytological study of the capacity for differentiation of normal hemopoietic colony-forming cells. *J. Cell Physiol.* 69:177–184, 1967.

Lala, P.K.; Johnson, G.R. Monoclonal origin of B lymphocyte colony-forming cells in spleen colonies formed by multipotential hemopoietic stem cells. *J. Exp. Med.* 148:1468–1477, 1978.

Harrison, P.R. Stem cell regulation in erythropoiesis. *Nature* 295:454–455, 1982.

27. Till, J.E.; McCulloch, E.A. Hemopoietic stem cell differentiation. *Biochim. Biophys. Acta* 605:431–459, 1980.

28. Goldwasser, E. Erythropoietin and the differentiation of red blood cells. *Fed. Proc.* 34:2285–2292, 1975.

Adamson, J.W.; Brown, J.E. Aspects of erythroid differentiation and proliferation. In Molecular Control of Proliferation and Differentiation (J. Papaconstantinou, W.J. Rutter, eds.), *Symp. Soc. Dev. Biol.* 35:161–179, 1978.

Heath, D.S.; Axelrad, A.A.; McLeod, D.L.; Shreeve, M.M. Separation of the erythropoietin-responsive progenitors BFU-E and CFU-E in mouse bone marrow by unit gravity sedimentation. *Blood* 47:777–792, 1976.

Gregory, C.J. Erythropoietin sensitivity as a differentiation marker in the hemopoietic system: studies of the three erythropoietic colony responses in culture. *J. Cell Physiol.* 89:289–301, 1976.

29. Metcalf, D. Clonal analysis of proliferation and differentiation of paired daughter cells: action of GM-CSF on granulocyte-macrophage precursors. *Proc. Natl. Acad. Sci. USA* 77:5327–5330, 1980.

Metcalf, D. Hemopoietic colony stimulating factors. In Handbook of Experimental Pharmacology, Vol. 57, Tissue Growth Factors (R. Baserga, ed.), pp. 343–384. New York: Springer-Verlag, 1981.

30. Goldspink, G. Development of muscle. In Differentiation and Growth of Cells in Vertebrate Tissues (G. Goldspink, ed.), pp. 69–99. London: Chapman and Hall, 1974.

Stockdale, F.E.; Holtzer, H. DNA synthesis and myogenesis. *Exp. Cell Res.* 24:508–520, 1961.

31. Yaffe, D. Cellular aspects of muscle differentiation *in vitro*. *Curr. Top. Dev. Biol.* 4:37–77, 1969.

Konigsberg, I.R. Diffusion-mediated control of myoblast fusion. *Dev. Biol.* 26:133–152, 1971.

Zalin, R. The cell cycle, myoblast differentiation, and prostaglandin as a developmental signal. *Dev. Biol.* 71:274–288, 1979.

32. Merlie, J.P.; Buckingham, M.E.; Whalen, R.G. Molecular aspects of myogenesis. *Curr. Top. Dev. Biol.* 11:61–114, 1977.

Devlin, R.B.; Emerson, C.P., Jr. Coordinate regulation of contractile protein synthesis during myoblast differentiation. *Cell* 13:599–611, 1978.

Bowman, L.H.; Emerson, C.P. Formation and stability of cytoplasmic mRNAs during myoblast differentiation: pulse-chase and density labeling analyses. *Dev. Biol.* 80:146–166, 1980.

33. Moss, F.P.; Leblond, C.P. Satellite cells as the source of nuclei in muscles of growing rats. *Anat. Rec.* 170:421–435, 1971.

Carlson, B.M. The regeneration of skeletal muscle. A review. *Am. J. Anat.* 137:119–149, 1973.

Konigsberg, U.R.; Lipton, B.H.; Konigsberg, I.R. The regenerative response of single mature muscle fibers isolated *in vitro*. *Dev. Biol.* 45:260–275, 1975.

34. Buller, A.J.; Eccles, J.C.; Eccles, R.M. Interactions between motoneurones and muscles in respect of the characteristic speeds of their responses. *J. Physiol.* 150:417–439, 1960.

Lømo, T.; Westgaard, R.H.; Dahl, H.A. Contractile properties of muscle: control by pattern of muscle activity in the rat. *Proc. R. Soc. Lond. (Biol.)* 187:99–103, 1974.

Rubinstein, N.A.; Kelly, A.M. Myogenic and neurogenic contributions to the development of fast and slow twitch muscles in rat. *Dev. Biol.* 62:473–485, 1978.

35. Goss, R.J. The Physiology of Growth, pp. 64–89. New York: Academic Press, 1978.

Ham, A.W.; Cormack, D.H. Histology, 8th ed., pp. 367–462. New York: Harper & Row, 1979.

Bloom, W.; Fawcett, D.W. A Textbook of Histology, 10th ed., pp. 233–287. Philadelphia: Saunders, 1975.

36. Skoog, T.; Ohlsén, L.; Sohn, S.A. Perichondrial potential for cartilagenous regeneration. *Scand. J. Plast. Reconstr. Surg.* 6:123–125, 1972.

37. Termine, J.E.; et al. Osteonectin, a bone-specific protein linking mineral to collagen. *Cell* 26:99–105, 1981.

Bassett, C.A.L. Electrical effects in bone. *Sci. Am.* 213(4):18–25, 1965.

Brighton, C.T.; Black, J.; Pollack, S.R., eds. Electrical Properties of Bone and Cartilage: Experimental Effects and Clinical Applications. New York: Grune & Stratton, 1979.

Jotereau, F.V.; Le Douarin, N.M. The developmental relationship between osteocytes and osteoclasts: a study using the quail-chick nuclear marker in endochondral ossification. *Dev. Biol.* 63:253–265, 1978.

38. Sinclair, D. Human Growth after Birth, 3rd ed., pp. 161–233. Oxford, Eng.: Oxford University Press, 1978.

39. Krebs, J.; May, R.M. Social insects and the evolution of altruism. *Nature* 260:9–10, 1976.

40. Cairns, J. Mutation selection and the natural history of cancer. *Nature* 255:197–200, 1975.

Nicholson, G.L. Cancer metastasis. *Sci. Am.* 240(3):66–76, 1979.

Cairns, J. Cancer: Science and Society. San Francisco: Freeman, 1978.

17 The Immune System

Our immune system saves us from certain death by infection. Any child born with a severely defective immune system will soon die unless the most extraordinary measures are taken to isolate it from a host of infectious agents—bacterial, viral, fungal, and parasitic. Indeed, any vertebrate that is immunologically deficient runs the same deadly risk.

All vertebrates have an immune system. The defense systems found among invertebrates are more primitive, often relying chiefly on phagocytic cells. The so-called professional phagocytes—mainly macrophages and polymorphonuclear leucocytes—also play an important role in defending vertebrates against infection, but they are only one part of a much more complex and sophisticated defense strategy.

Immunology, the study of the immune system, grew out of the common observation that people who recover from certain infections are "immune" to the disease thereafter; that is, they rarely develop the same disease again. Immunity is highly specific: an individual who recovers from measles is protected against the measles virus but not against other common viruses, such as mumps or chicken pox. Such specificity is a fundamental characteristic of immune responses.

The various responses of the immune system destroy and eliminate invading organisms and any toxic molecules produced by them. Because immune reactions are destructive, it is essential that they be made in response only to molecules that are foreign to the host and not to those of the host itself. This ability to distinguish *foreign* molecules from *self* molecules is another fundamental feature of the immune system. Occasionally it fails to make this distinction and reacts against the host's own cells; such *autoimmune reactions* can be fatal.

While the immune system evolved to protect vertebrates from infection by microorganisms and larger parasites, most of what we know about immunity has come from studies of the responses of laboratory animals to injections of noninfectious substances, such as foreign proteins and polysaccharides. Almost any macromolecule, as long as it is foreign to the recipient, can induce an immune response; any substance capable of eliciting an immune response is referred to as an **antigen.** Remarkably, the immune system can distinguish between antigens that are very similar to each other—such as between two proteins that differ in only a single amino acid or between two optical isomers.

There are two broad classes of immune responses: (1) **Humoral antibody responses** involve the production of antibodies, which circulate in the bloodstream and bind specifically to the foreign antigen that induced them. The binding of antibody to the antigen makes it easier for phagocytic cells to ingest the antigen and often activates a system of blood proteins, collectively called *complement,* that helps destroy the antigen. (2) **Cell-mediated immune responses** involve the production of specialized cells that react mainly with foreign antigens on the surface of host cells, either killing the host cell if the antigen is an infecting virus or inducing other host cells, such as macrophages, to destroy the antigen.

The main challenges in immunology are to understand (1) how the immune system specifically recognizes and responds to millions of different foreign antigens, (2) how it distinguishes these foreign molecules from self molecules, and (3) how it distinguishes different classes of invading microorganisms so that it can tailor its responses to eliminate the invaders efficiently. We begin our discussion of how the immune system accomplishes these three feats by considering the cellular basis of immunity. We shall then consider in detail the function and structure of antibodies, the complement system, and the special features of cell-mediated immunity.

The Cellular Basis of Immunity

The Immune System Is Composed of Billions of Lymphocytes[1]

The cells responsible for immunity are white blood cells known as **lymphocytes.** They are found in large numbers in the blood and the lymph (the colorless fluid in the lymphatic vessels that connect the lymph nodes in the body) and in specialized **lymphoid tissues,** such as the thymus, lymph nodes, spleen, and appendix (Figure 17–1).

The total number of lymphocytes in the body is very large ($\sim 2 \times 10^{12}$ in man); in cell mass, the immune system is comparable to the liver or brain. Although lymphocytes have long been recognized as a major cellular component of the blood, it was not until the late 1950s that their central role in immunity was demonstrated. The proof came from experiments in which rats were heavily irradiated in order to kill most of their white blood cells, including lymphocytes. Since such rats are unable to make immune responses, it is possible to transfer various types of cells into them and determine which ones reverse the deficiency. Only lymphocytes restored the immune response of irradiated animals (Figure 17–2). Since both antibody and cell-mediated responses were restored, these experiments established that lymphocytes are responsible for both classes of immune response.

B Lymphocytes Make Humoral Antibody Responses; T Lymphocytes Make Cell-mediated Immune Responses[2]

During the 1960s, it was discovered that the two major classes of immune responses are mediated by two different classes of lymphocytes: **T cells,** which develop in the *thymus,* are responsible for cell-mediated immunity; **B cells,** which develop independently of the thymus, produce antibodies. Thus the removal of the thymus from a newborn animal markedly impairs its cell-mediated immune responses but has much less effect on its antibody re-

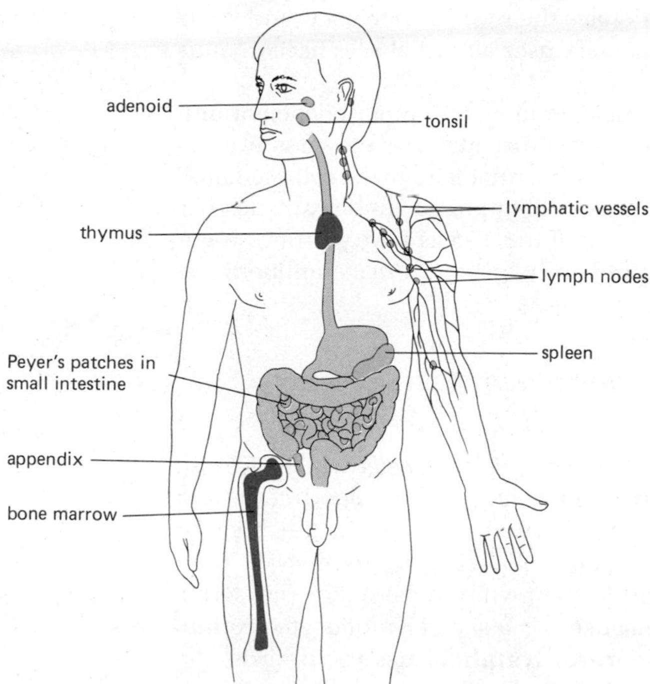

Figure 17–1 Diagram of human lymphoid tissues. Lymphocytes develop in both the thymus and bone marrow *(dark color),* which are therefore referred to as the *central lymphoid tissues.* The newly formed lymphocytes migrate from these central tissues to the *peripheral lymphoid tissues (light color),* where they can react with antigen.

sponses. In birds it is possible to demonstrate the converse effect: because B lymphocytes develop in a gut-associated lymphoid organ unique to birds, the *bursa of Fabricius,* the removal of the bursa of Fabricius at hatching impairs the bird's ability to make antibodies but has little effect on cell-mediated immunity. Studies of children born with impaired immunity reveal the same type of dichotomy: whereas some children cannot make antibodies but have

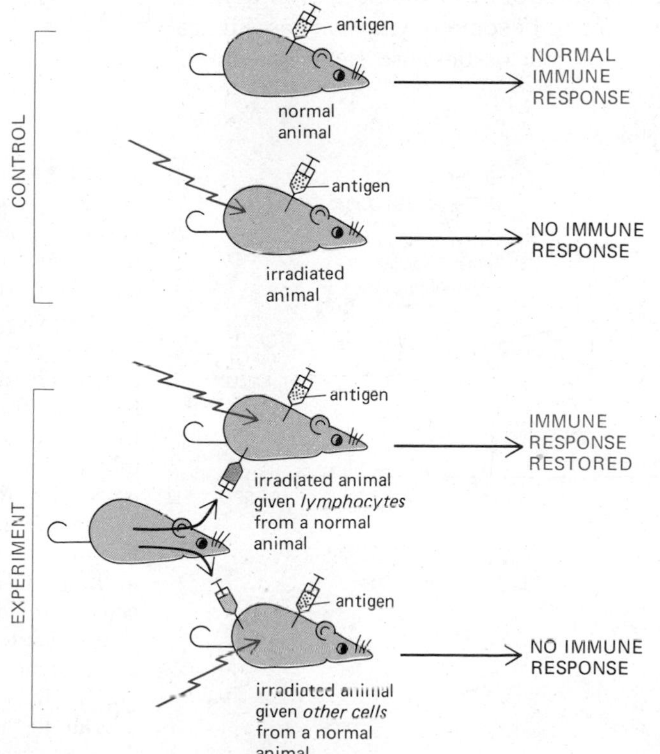

Figure 17–2 The classic experiment showing that lymphocytes are responsible for recognizing and responding to foreign antigens. An important feature of all such cell-transfer experiments is that cells are transferred between animals of the same *inbred strain.* Members of an inbred strain are genetically identical. If lymphocytes are transferred to a genetically different animal that has been irradiated, they react against the "foreign" antigens of the host and can kill the animal (see p. 999).

normal cell-mediated immunity, others have the reverse deficiency; and those with selectively impaired cell-mediated responses almost always have thymus abnormalities.

One of the puzzling features of these studies on immunodeficient animals was that individuals deficient in T cells (because their thymus was removed at birth or was abnormal) were not only unable to make cell-mediated immune responses, but they also had somewhat impaired antibody responses. As we shall see later, this is because some of the T cells have a crucial regulatory role in immunity and are required to help B cells make antibody responses.

All Lymphocytes Develop from Pluripotent Hemopoietic Stem Cells[3]

Lymphocytes develop from *pluripotent hemopoietic stem cells,* which give rise to all of the blood cells, including red blood cells, white blood cells, and platelets (see p. 923). The stem cells are located primarily in the liver (in fetuses) and bone marrow (in adults). Some of their progeny migrate from these *hemopoietic tissues* via the blood to the thymus, where they proliferate and differentiate into lymphocytes. Because it is a site of lymphocyte production, the thymus is referred to as a **central lymphoid tissue.** In birds, the bursa of Fabricius is also a site of lymphocyte production and hence is also a distinct central lymphoid tissue.

Although many lymphocytes die soon after they have developed in a central lymphoid tissue, others migrate via the blood to the **peripheral lymphoid tissues**—the lymph nodes, spleen, and gut-associated lymphoid tissues (Peyer's patches in the small intestine, appendix, tonsils, and adenoids)—where they become *thymus-derived (T) lymphocytes* and (in birds) *bursa-derived (B) lymphocytes.* It is in the peripheral lymphoid tissues that T cells and B cells react with foreign antigens.

Mammals have no bursa of Fabricius, and their hemopoietic stem cells develop into lymphocytes in the hemopoietic tissues themselves and then migrate to the peripheral lymphoid tissues to become B lymphocytes (Figure 17–3). In mammals, therefore, the hemopoietic tissues also serve as central lymphoid tissues.

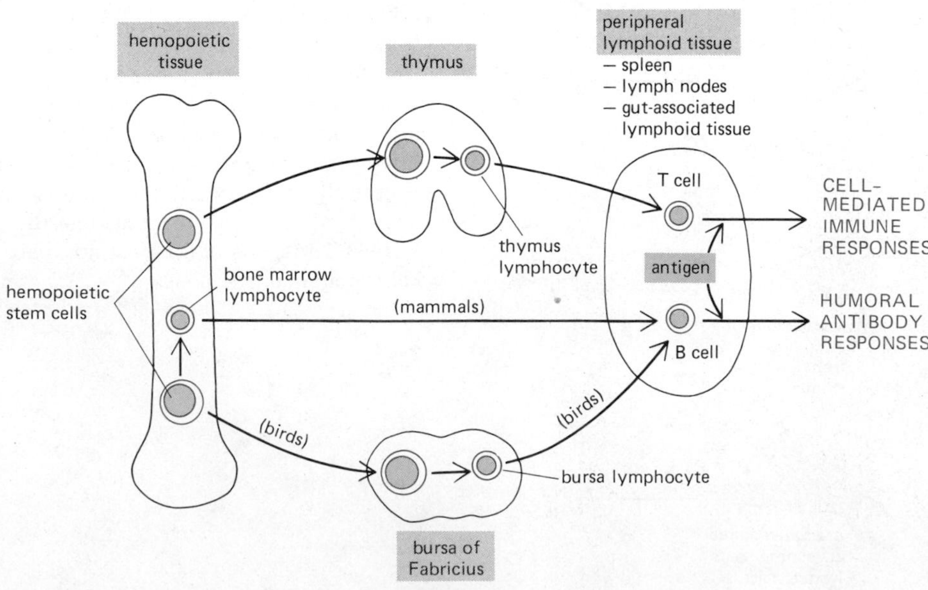

Figure 17–3 The development of T and B lymphocytes. In both mammals and birds, hemopoietic stem cells migrate via the blood to the thymus, where they differentiate into thymus lymphocytes. While some of these lymphocytes die in the thymus, others migrate to the peripheral lymphoid tissues to become T cells. In birds, other hemopoietic stem cells migrate to the bursa of Fabricius, where they differentiate into bursa lymphocytes; some of these lymphocytes die, while others migrate to the peripheral tissues to become B cells. In mammals, the hemopoietic stem cells destined to become B cells differentiate into lymphocytes in the hemopoietic tissue itself and then migrate to the peripheral lymphoid tissues to become B cells.

Because most of the migration of lymphocytes from thymus and bursa (and in mammals, from hemopoietic tissues) occurs early in development, removing either of these organs from *adult* animals has relatively little effect on immune responses. That is why their role in immunity remained undiscovered for so long. Nonetheless, there is a slow, continuous turnover of lymphocytes in mature animals, and new lymphocytes continue to develop from stem cells in the central lymphoid tissues throughout life.

Cell-Surface Markers Make It Possible to Distinguish and Separate T and B Cells[4]

T and B cells become morphologically distinguishable only after they have been stimulated by antigen. Unstimulated ("resting") T and B cells look very similar, even in an electron microscope: both are usually small, only marginally bigger than red blood cells, and are filled by the nucleus (Figure 17–4A). Both are activated by antigen to proliferate and differentiate. Activated B cells develop into antibody-secreting cells, the most mature of which are *plasma cells,* filled with an extensive rough endoplasmic reticulum and having a characteristic morphology (Figure 17–4B). In contrast, activated T cells contain very little endoplasmic reticulum and do not secrete antibody (Figure 17–4C).

Since both T and B lymphocytes occur in all peripheral lymphoid tissues, it has been necessary to find convenient means of distinguishing and separating the two cell types in order to study their individual properties. Fortunately, the many differences in the plasma membrane proteins of T and B cells can serve as distinguishing markers. One of the most widely used markers is the *Thy-1* glycoprotein, which is found on T but not B lymphocytes in mice; anti-Thy-1 antibodies can thus be used to remove or purify T cells from a

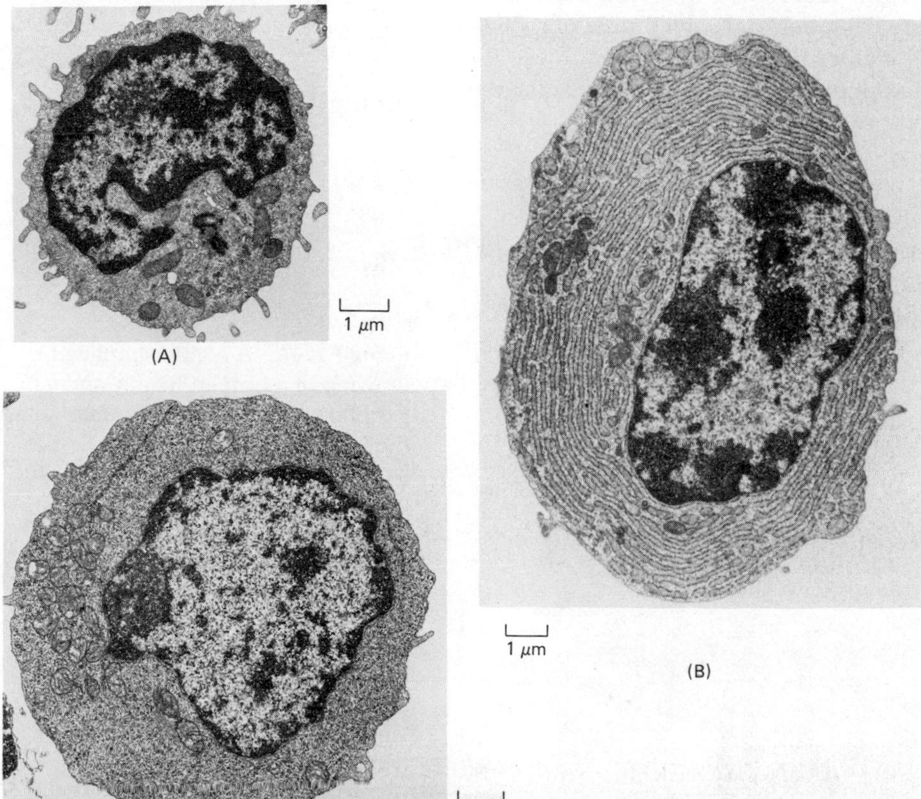

Figure 17–4 Electron micrographs of a small lymphocyte (A), an activated B cell (B), and an activated T cell (C). The small lymphocyte could be a T cell or a B cell, for these cells are difficult to distinguish morphologically until they have been activated. Note, however, that the activated B cell (a plasma cell) is filled with an extensive rough endoplasmic reticulum (ER) distended with antibody molecules, while the activated T cell has relatively little rough ER but is filled with free ribosomes. Note that the three cells are shown at the same magnification. (A, courtesy of Dorothy Zucker-Franklin; B, courtesy of Carlo Grossi; A and B, from D. Zucker-Franklin, et al., Atlas of Blood Cells: Function and Pathology. Philadelphia: Lea and Febiger, 1981; C, courtesy of Stefanello dePetris.)

mixed population of mouse lymphocytes. The use of cell-surface antigenic markers for distinguishing and separating T and B cells has revolutionized cellular immunology and has played an important part in the rapid advances in this field in recent years. New markers defining an ever increasing number of functionally distinct subpopulations of T and B lymphocytes are continually being found, both in experimental animals and in man.

Most Lymphocytes Continuously Recirculate Between the Blood and Lymph[5]

The great majority of T and B lymphocytes continuously recirculate between the blood and lymph. They leave the bloodstream, squeezing out between specialized endothelial cells found in certain small veins, and enter various tissues, including all lymph nodes. After percolating through a tissue, they accumulate in small lymphatic vessels that connect to a series of lymph nodes downstream. Passing into larger and larger vessels, the lymphocytes eventually enter the main lymphatic vessel (the *thoracic duct*), which carries them back into the blood. This continuous recirculation presumably ensures that the appropriate lymphocytes will come into contact with antigen (and with each other, see below) and serves to disperse the resulting activated T and B cells to lymphoid tissues throughout the body.

In the various peripheral lymphoid tissues, T and B cells are largely segregated in separate areas (Figure 17–5) to which they will also specifically migrate if they are injected into another animal. The molecular basis of this area-specific homing is unknown. It is probable, however, that these areas lack rigid boundaries, for, as we shall see, T and B cells must interact with one another in the course of most antibody responses.

In addition to these general differences in migration pathways between T and B cells, there are equally important differences in traffic patterns between different sets of cells within each class. For example, B cells of a certain subclass leave the bloodstream in the wall of the small intestine; these cells constitute, in effect, a gut-specific subsystem of lymphocytes, specialized for responding to antigens that enter the body from the intestine.

The Immune System Works by Clonal Selection[6]

The most remarkable feature of the immune system is that it can respond to millions of different foreign antigens in a highly specific way. Historically, two hypotheses have been proposed to explain how the immune system produces such a diversity of specific antibodies. The *instruction hypothesis*, which dom-

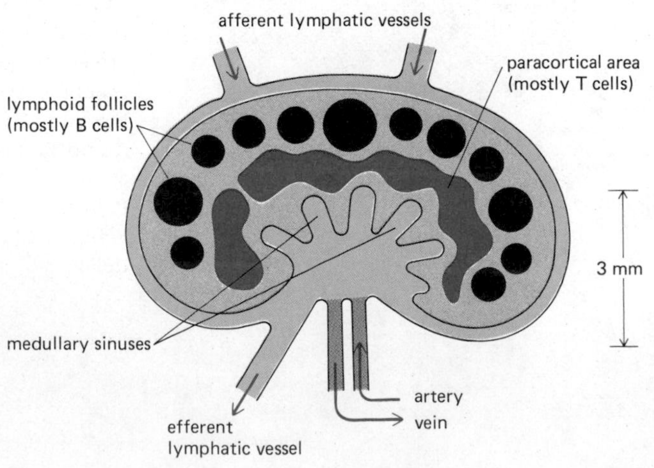

Figure 17–5 A highly simplified drawing of a human lymph node. B lymphocytes are located primarily in the lymphoid follicles, while T lymphocytes are found mainly in the paracortical area. Both enter the lymph node from the blood via specialized small veins in the paracortical area. While T cells remain in this area, the B cells migrate to the lymphoid follicles. Eventually both T cells and B cells migrate to the medullary sinuses and leave the node via the efferent lymphatic vessel. The lymphatic vessels ultimately empty into the bloodstream, allowing the lymphocytes to begin another cycle.

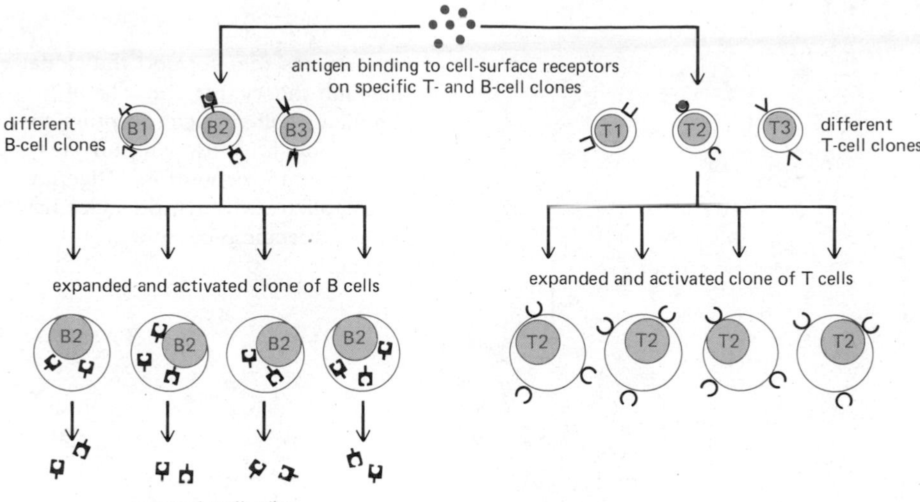

Figure 17–6 Schematic diagram of the clonal selection theory. An antigen activates only those T- and B-cell clones already committed to respond to it. The immune system is thought to consist of millions of different lymphocyte clones, hundreds of which may be activated by a particular antigen.

inated immunological thinking in the 1940s, was that antibodies are made as unfolded polypeptide chains whose final conformation is determined by the antigen around which they become folded. At the time this seemed the simplest explanation for the observation that animals can make specific antibodies to man-made substances that do not exist in nature. However, the instruction hypothesis had to be abandoned when protein chemists discovered that the three-dimensional, folded structure of a protein molecule, such as an antibody, is determined solely by its amino acid sequence. In fact, a denatured (unfolded) antibody molecule can refold to form its orginal antigen-binding site even in the absence of antigen.

The instruction hypothesis was replaced in the 1950s by the **clonal selection theory,** which is now an established part of immunological dogma. This theory is based on the proposition that during development each lymphocyte becomes committed to react with a particular antigen before ever being exposed to it. A cell expresses this commitment in the form of surface receptor proteins that specifically fit the antigen. The binding of antigen to the receptors activates the cell, causing it both to multiply and mature. Thus a foreign antigen selectively stimulates those cells that bear complementary antigen-specific receptors and are thus already committed to respond to it. This is what makes immune responses antigen-specific (Figure 17–6).

The term "clonal" in clonal selection derives from the postulate that the immune system is composed of millions of different families, or *clones,* of cells, each consisting of T or B lymphocytes descended from a common ancestor. Since each ancestral cell is already committed to make one particular antigen-specific receptor protein, all cells in a clone have the same antigen specificity.

Thus, according to the clonal selection theory, the immune system functions on the "ready-made" rather than the "made-to-measure" principle. The question of how an animal makes so many different antibodies becomes, therefore, a problem of genetics rather than of protein chemistry.

There is compelling evidence supporting the clonal selection theory. For example, when lymphocytes from an animal that has not been immunized are incubated in a test tube with any of a number of radioactively labeled antigens—say, A, B, C, and D—only a very small proportion ($<0.01\%$) bind each antigen, suggesting that only a few cells bear specific receptors for A, B, C, or D. This interpretation is confirmed by making antigen A so radioactive that any cell that binds it is lethally irradiated; the remaining population of lymphocytes is then no longer able to produce an immune response to A,

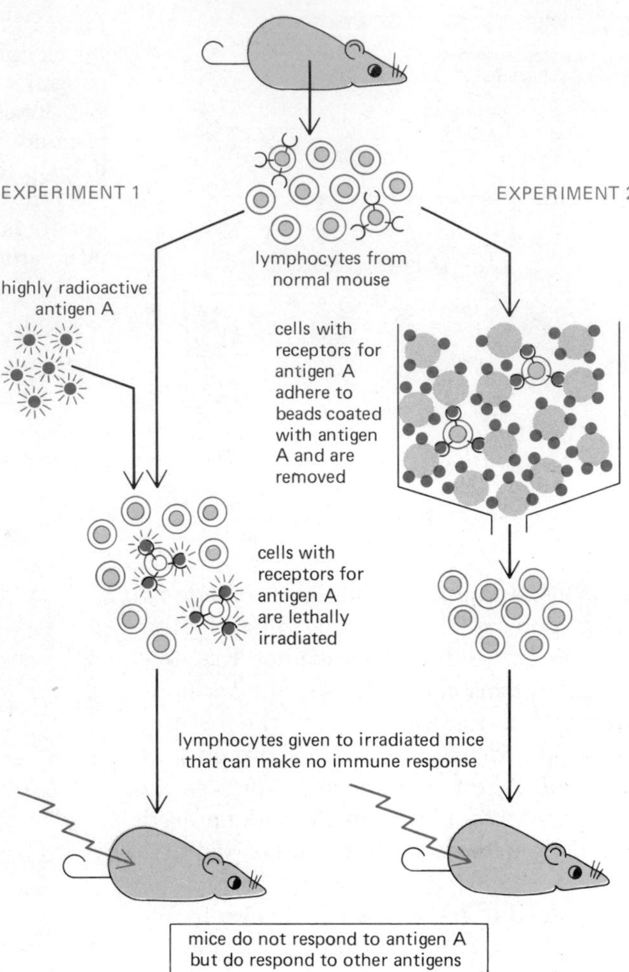

Figure 17–7 Two types of experiments that support the clonal selection theory. For the sake of simplicity, cell-surface receptors are shown only on those lymphocytes committed to respond to antigen A; in fact, all T and B lymphocytes have antigen-specific receptors on their surface.

whereas it can still respond normally to antigen B, C, or D. The same effect can be achieved by coating a column of glass beads with antigen A and then passing the lymphocytes through the column. The cells with receptors for A stick to the beads, while other cells pass through; as a result, the cells that emerge from the column no longer respond to A, whereas they respond normally to other antigens (Figure 17–7).

These two experiments indicate that (1) lymphocytes are committed to respond to a particular antigen before they have been exposed to it, and (2) lymphocytes have surface receptors that specifically bind the antigen. Two major predictions of the clonal selection theory are therefore confirmed. Although most experiments of this kind have involved B cells and antibody responses, similar experiments on T cells suggest that the clonal selection hypothesis holds for T-cell-mediated responses as well.

Most Antigens Stimulate Many Different Lymphocyte Clones[7]

Most macromolecules, including virtually all proteins and most polysaccharides, can serve as antigens. Those parts of an antigen's surface that combine with the antigen-binding site on an antibody molecule or on a lymphocyte receptor are called **antigenic determinants.** Molecules that bind specifically to antibody or to a lymphocyte receptor but cannot induce immune responses are called **haptens.** Haptens can be made antigenic by coupling them to a suitable macromolecule, called a *carrier*. A hapten used commonly in immunological experiments is the *dinitrophenyl (DNP)* group, which is usually coupled to a protein in order to make it antigenic (Figure 17–8).

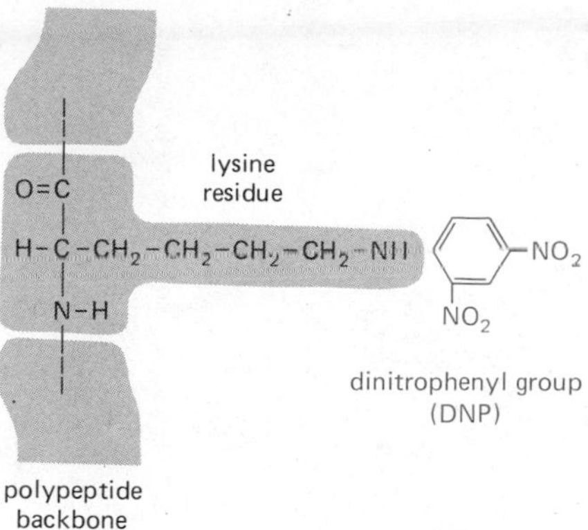

Figure 17–8 The simple hapten DNP shown covalently coupled to a lysine side chain on a protein. Only when coupled to such a macromolecular carrier can haptens induce an immune response.

Most antigens have a variety of different antigenic determinants on their surfaces that stimulate the production of antibodies or T-cell responses. Some determinants are more *immunogenic* (immunity-inducing) than others, so that the reaction to them may dominate the overall response; such determinants are said to be *immunodominant.*

As one might expect of a system that works by clonal selection, even a single antigenic determinant will, in general, activate many different clones, each of which produces an antigen-binding site with a different affinity for the determinant. For example, even the relatively simple structure of the DNP group can be "looked at" in many different ways. Thus when it is coupled to a protein carrier, it usually stimulates the production of hundreds of different species of anti-DNP antibodies, each made by a different B-cell clone. Such responses are said to be *polyclonal.* When only a few clones respond, the response is said to be *oligoclonal;* and when the total response is made by a single B- or T-cell clone, it is said to be *monoclonal.* The responses to most antigens are polyclonal.

Very little is known about the early biochemical events involved in lymphocyte activation. There are three reasons for our ignorance. First, although hundreds of different lymphocyte clones are usually activated by a given antigen, those hundreds still represent only a tiny fraction of the millions of clones that constitute the entire immune system and are therefore very difficult to separate out for study. Second, immune responses do not become detectable until days after antigen has bound to the lymphocyte, by which time the cells are many steps removed from the early events of activation. The third and most important reason, however, is that virtually all lymphocyte responses involve complex interactions among a variety of cell types. Consequently it is extremely difficult to study the early biochemical changes in any one cell type.

Immunological Memory Is Due to Clonal Expansion and Lymphocyte Differentiation[8]

The immune system, like the nervous system, can remember. That is why we develop lifelong immunity to many common viral diseases after our initial exposure to the virus. The same phenomenon can easily be demonstrated in experimental animals. If an animal is injected once with antigen A, its immune response (either humoral or cell-mediated) will appear after a lag period of several days, rise rapidly and exponentially, and then, more gradually, fall again. This is the characteristic course of a **primary immune response,** occurring on an animal's first exposure to an antigen. If some weeks or months or even years are allowed to pass and the animal is reinjected with antigen A, it will produce a **secondary immune response** very different from the primary response: the lag period is shorter, the response is greater, and its duration is longer (Figure 17–9). These differences indicate that the animal has "remembered" its first exposure to antigen A. If, instead of a second injection of antigen A, the animal is given a different antigen (for example, antigen B), the response is typical of a primary, and not a secondary, immune response; therefore, the secondary response reflects antigen-specific memory for antigen A.

The clonal selection theory provides a useful conceptual framework for understanding the cellular basis of immunological memory. In a mature animal, the T and B cells in the peripheral lymphoid tissues are a mixture of cells in at least three discrete stages of differentiation that can be designated *virgin cells, memory cells,* or *effector cells.* When **virgin cells** encounter an-

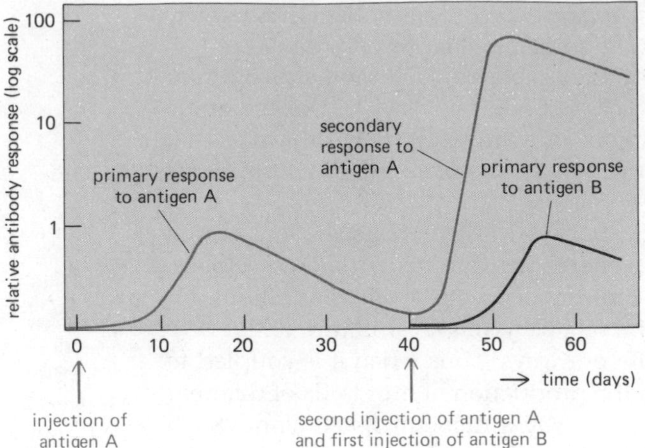

Figure 17–9 Diagram showing a primary and secondary antibody response induced by a first and second exposure, respectively, to antigen A. Note that the secondary response is faster and greater than the primary response and is specific for A, indicating that the immune system has specifically "remembered" encountering antigen A before. Evidence for the same type of immunological memory is obtained if T-cell-mediated responses rather than B-cell antibody responses are measured.

tigen for the first time, some of them are stimulated to multiply and become **effector cells**—that is, cells actively engaged in making a response (T effector cells carry out cell-mediated responses, while B effector cells secrete antibody). Other virgin T and B cells are stimulated to multiply and differentiate instead into **memory cells**—that is, cells that do not themselves make a response but are readily induced to become effector cells by a later encounter with the same antigen (Figure 17–10). There is some evidence that virgin lymphocytes tend to remain in peripheral lymphoid tissues and do not recirculate between blood and lymph. Moreover, they are relatively short-lived, probably dying within days or weeks unless they meet their specific antigen. Memory cells, on the other hand, recirculate and may live for many months or even years without dividing.

According to this scheme, immunological memory is generated during the primary response because (1) the proliferation of each antigen-triggered virgin cell creates many memory cells—a process known as *clonal expansion;* (2) the memory cells have an increased life-span and recirculate between the blood and lymph; and (3) each memory cell is prepared to respond more readily to antigen than does a virgin cell. The changes induced during the primary response thus ensure that most of the cells in the recirculating pool of lymphocytes are appropriate to the antigenic environment of the animal and are already primed and ready for action.

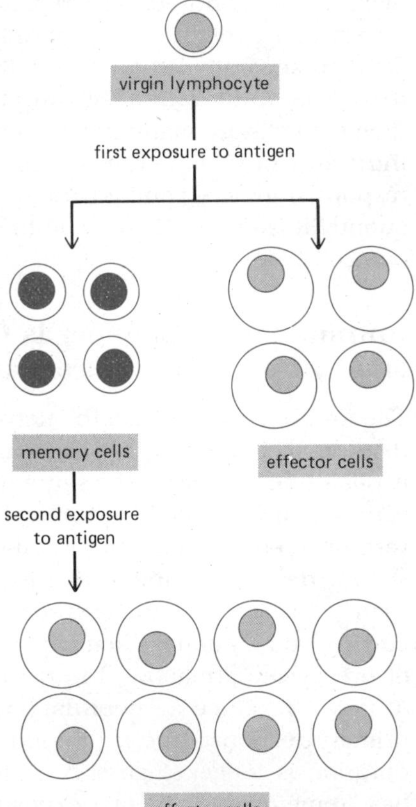

Figure 17–10 When virgin T or B cells are activated by their specific antigen, they usually proliferate and differentiate either into effector cells or into memory cells. During a subsequent exposure to antigen, the memory cells proliferate, and some of them differentiate into effector cells more efficiently than do virgin cells.

The Failure to Respond to Self Antigens Is Due to Acquired Immunological Tolerance[9]

How is the immune system able to distinguish foreign molecules from self molecules? One possibility is that an animal inherits genes that encode receptors for foreign antigens but not self antigens so that its immune system is genetically constituted to respond only to the former. Alternatively, the immune system may be inherently capable of responding to both foreign and self antigens but "learns" not to respond to self early in development. The latter explanation is almost certainly correct. The first evidence for this was an observation made in 1945. Normally, when tissues are transplanted from one individual to another, they are recognized as foreign by the immune system and are destroyed. But dizygotic (that is, nonidentical) cattle twins that had exchanged blood cells *in utero* as a result of the spontaneous fusion of their placentas were found to accept skin grafts from each other (Figure 17–11). These findings were later reproduced experimentally—in chicks, by allowing the blood vessels of two different embryos to fuse; and in mice, by introducing

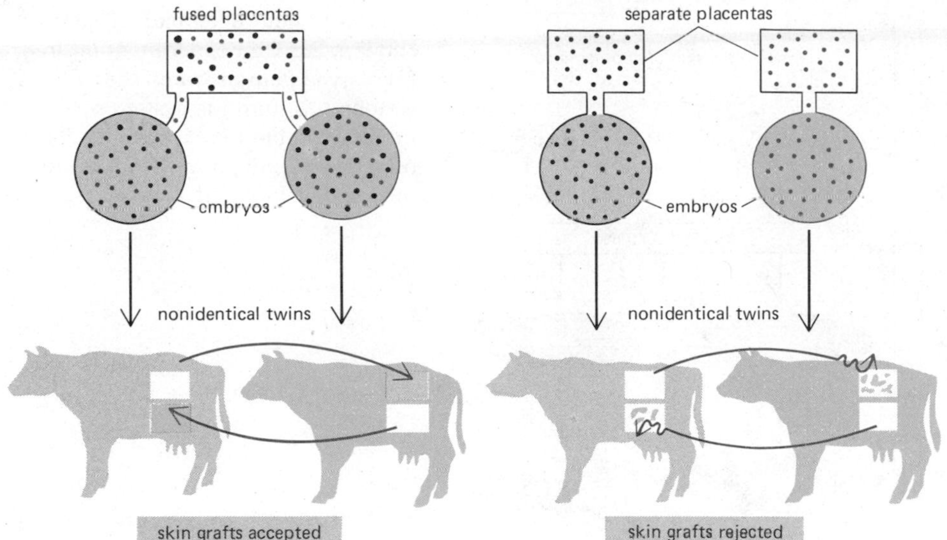

Figure 17–11 The observation that initially suggested that immunological self-tolerance is acquired and not genetic. Adult nonidentical twin cattle that had exchanged blood cells *in utero* through a common placenta do not recognize each other's tissues as foreign and accept skin grafts from one another. Nonidentical twins that had separate placentas *in utero* reject each other's skin grafts.

spleen cells from a different mouse strain into neonatal mice, where they survived for most of the recipient animal's life. In both cases, when the animals matured, grafts from the joined or donor animal were accepted (Figure 17–12), while "third-party" grafts were rejected. Thus, the continuous presence of nonself antigens starting before the immune system has matured leads to a permanent unresponsiveness to the specific nonself antigens. The resulting state of induced antigen-specific immunological unresponsiveness is known as **acquired immunological tolerance.**

An elegant experiment performed in 1962 demonstrated that the failure of an animal's immune system to respond to its own tissues (*natural immunological tolerance*) is acquired in the same way and is not inborn. The experiment depended on the fact that most tissues bear some unique tissue-specific antigens that the developing immune system must learn to recognize as self. The pituitary gland of a tree-frog larva was removed so that the larva contained no pituitary-specific antigens while its immune system matured. The pituitary was kept alive by transplanting it under the skin of another larva whose immune system was immature and therefore unable to reject the foreign gland. The two larvae were allowed to mature, and the pituitary was then returned to its original owner; remarkably, it was rejected as if it were foreign. On the other hand, when only half of the pituitary was removed and stored under the skin of another larva, this pituitary tissue was not rejected when

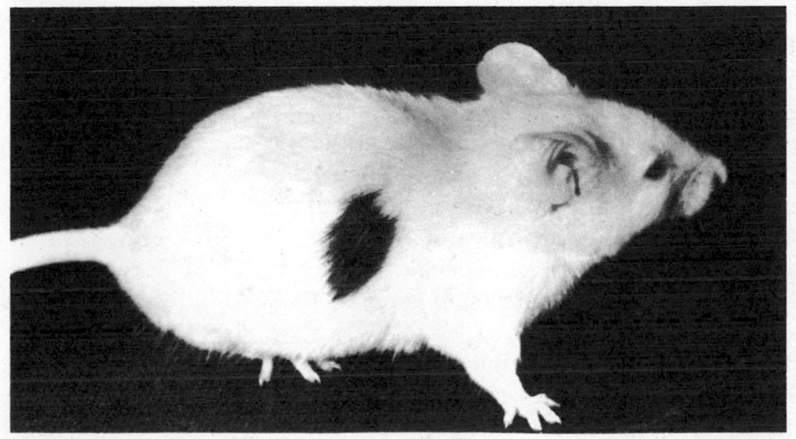

Figure 17–12 The skin graft seen here, transplanted from an adult brown mouse to an adult white mouse, has survived for many weeks only because the latter was made immunologically tolerant by injecting blood cells from the brown mouse into it at the time of birth. (Courtesy of Leslie Brent, from I. Roitt, Essential Immunology. Oxford, Eng.: Blackwell Scientific, 1980.)

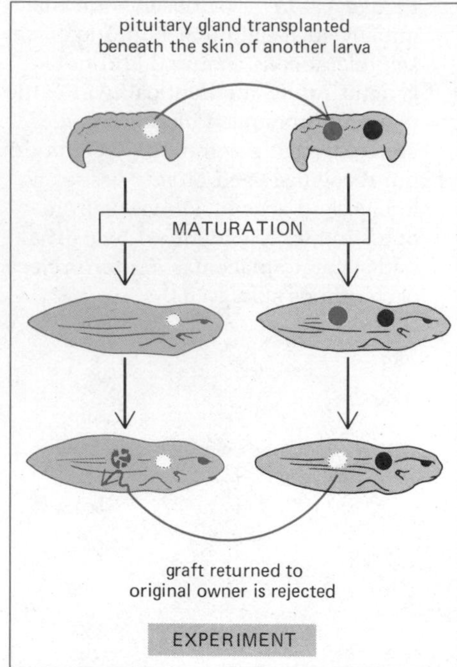

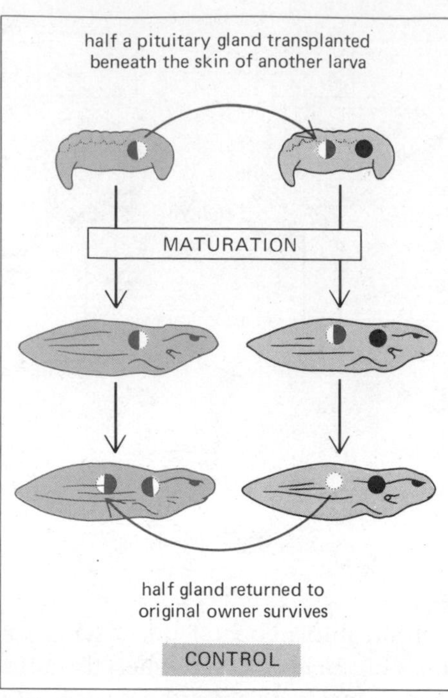

Figure 17–13 Schematic diagram showing how an experiment on tree frog larvae demonstrated that a vertebrate's immune system is inherently able to react against its own tissues but "learns" not to do so during development.

returned to its original owner (Figure 17–13). Thus it is clear that the immune system is genetically capable of responding to self but learns not to do so. Other experiments have shown that the maintenance of self-tolerance requires the constant presence of the self antigens; if a particular antigen is removed, an animal will regain the ability to respond to it within weeks or months.

Tolerance to self antigens sometimes breaks down, causing T or B cells (or both) to react against their own tissue antigens. Such **autoimmune reactions** are responsible for a number of diseases. For example, *myasthenia gravis* is an autoimmune disease in which individuals make antibodies against the acetylcholine receptors on their own skeletal muscle cells; the antibodies interfere with the normal functioning of the receptors so that such patients become weak and can die from being unable to breathe.

Immunological Tolerance to Foreign Antigens Can Also Be Induced in Mature Animals[10]

It is generally much more difficult to induce immunological tolerance to foreign antigens in adult than in immature animals. But with some antigens it can be done experimentally by injecting the antigen (1) in very high doses, (2) in repeated very low doses, (3) together with an immunosuppressive drug, or (4) intravenously after the antigen has been ultracentrifuged to remove all aggregates so that the normal mechanisms of antigen presentation (see below) are bypassed. In these cases, a second exposure to the same antigen (under conditions that would normally induce a response) not only fails to elicit a secondary immune response, it often fails to elicit any response at all. Such tolerant animals still respond normally to other antigens, indicating that immunological tolerance, like immunological memory, is antigen-specific (Figure 17–14). While both B and T cells can be made tolerant in adults, in general, T cells are more susceptible than B cells.

The molecular mechanisms involved in tolerance either to self or to foreign antigens are still unknown. There is evidence that the cellular mechanism varies: in some cases the lymphocyte clones that would normally re-

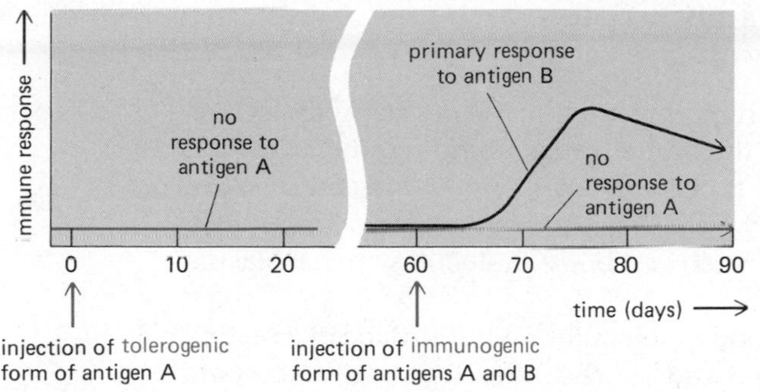

Figure 17–14 The experimental induction of immunological tolerance to a foreign antigen. The injection of a tolerance-inducing (tolerogenic) dose and/or form of antigen A (see text) not only fails to induce an immune response but also renders the animal specifically unresponsive to further injections of antigen A given in a form and dose that would normally induce a response. Note that the response to a different antigen, B, is unaffected.

spond to a particular antigen are eliminated; in other cases they survive, but their responses are specifically suppressed by a subclass of T cells known as *suppressor T cells* (see p. 997).

In summary, the binding of an antigen to its complementary receptors on a T or B lymphocyte can have any one of at least three consequences: (1) the lymphocyte may divide and differentiate to become an effector cell or a memory cell; (2) it may become tolerant; or (3) it may be unaffected by the encounter. The "decision" to *turn on, turn off,* or *ignore* depends largely on the nature and concentration of the antigen and upon complex interactions between different classes of lymphocytes and between lymphocytes and specialized macrophagelike *antigen-presenting cells,* which will be discussed in a later section. The decision also depends on the maturity of the lymphocyte. For example, newly formed B cells are highly susceptible to the induction of tolerance, while mature B cells are relatively resistant; this means that developing B cells with a high affinity for self molecules in their environment will become tolerant and never be activated.

Summary

The immune system evolved to defend vertebrates against infection. It is composed of billions of lymphocytes comprising millions of different clones. The lymphocytes in each clone share a unique cell-surface receptor that enables them to bind a particular "antigenic determinant" consisting of an arrangement of atoms on a part of a molecule. There are two classes of lymphocytes: B cells, which make antibodies, and T cells, which make cell-mediated immune responses.

Beginning early in lymphocyte development, those B and T cells with receptors for antigenic determinants on self molecules are eliminated or suppressed; as a result, the immune system is normally able to respond only to foreign antigens. The binding of a foreign antigen to a lymphocyte initiates a response by the cell that helps to eliminate the antigen. As part of the response, some of the lymphocytes proliferate and differentiate into memory cells, so that the next time that the same antigen is encountered the immune response is faster and much greater.

The Functional Properties of Antibodies[11]

The only known function of B lymphocytes is to make antibodies. A unique feature of antibodies, one that distinguishes them from all other known proteins, is that they can exist in millions of different forms, each with its own unique binding site for antigen. Collectively called **immunoglobulins** (abbreviated as **Ig**), they represent one of the major classes of proteins found in the blood, constituting about 20% of the total plasma protein by weight.

The Antigen-specific Receptors on B Cells Are Antibody Molecules[12]

As predicted by the clonal selection hypothesis, all of the antibody molecules made by an individual B cell have the same antigen-binding site. The first antibodies made by a newly formed B cell are not secreted; instead they are inserted into the plasma membrane, where they serve as receptors for antigen. Each B cell has approximately 10^5 such antibody molecules in its plasma membrane.

When antigen binds to the antibody molecules on the surface of a resting B cell, it usually initiates a complicated and poorly understood series of events culminating in cell proliferation and differentiation to produce antibody-secreting cells. Such cells now make large amounts of soluble (rather than membrane-bound) antibody with the same antigen-binding site as the cell-surface antibody and secrete it into the blood. While activated B cells can begin secreting antibody while they are still small lymphocytes, the end stage of this differentiation pathway is the large plasma cell (see Figure 17–4B), which secretes antibodies at the rate of about 2000 molecules per second. Plasma cells seem to have committed so much of their protein-synthesizing machinery to making antibody that they are incapable of further growth and division and die after several days of antibody secretion.

B Cells Can Be Stimulated to Make Antibodies in a Culture Dish[13]

Two significant advances in the 1960s revolutionized research on B cells. The first was the development of the **hemolytic plaque assay,** which made it possible to identify and count individual B cells secreting antibody against a specific antigen. In the simplest form of this assay, lymphocytes (commonly from the spleen) are taken from animals that have been immunized against sheep red blood cells (SRBC). They are then embedded in agar together with an excess of SRBC so that the dish contains a "lawn" of immobilized SRBC with occasional lymphocytes in it. Under these conditions, the cells are unable to move, but any anti-SRBC antibody secreted by a B cell will diffuse outward and coat all SRBC in the vicinity of the secreting cell. Once the SRBCs are coated with antibody, they can be killed by adding complement (see p. 988). In this way, the presence of each antibody-secreting cell is indicated by the presence of a clear spot, or *plaque,* in the opaque layer of SRBC. The same assay can be used to count cells making antibody to other antigens, such as proteins or polysaccharides, simply by coupling these antigens to the surface of the SRBC.

The second important advance was the demonstration that B lymphocytes can be induced to make antibody by exposing them to antigen in culture, where the cell interactions can be manipulated and the environment controlled. This led to the discovery that both T lymphocytes and specialized *antigen-presenting cells* are required for antibody production by B lymphocytes against most antigens; the cell interactions involved will be described in a later section of this chapter.

Antibodies Have Two Identical Antigen-binding Sites[11]

The simplest antibody molecules are Y-shaped molecules with two identical antigen-binding sites—one at the tip of each arm of the Y (Figure 17–15). Because of their two antigen-binding sites, they are said to be *bivalent.* Such antibody molecules can cross-link antigen molecules into a large lattice, as long as the antigen molecules each have three or more antigenic determinants

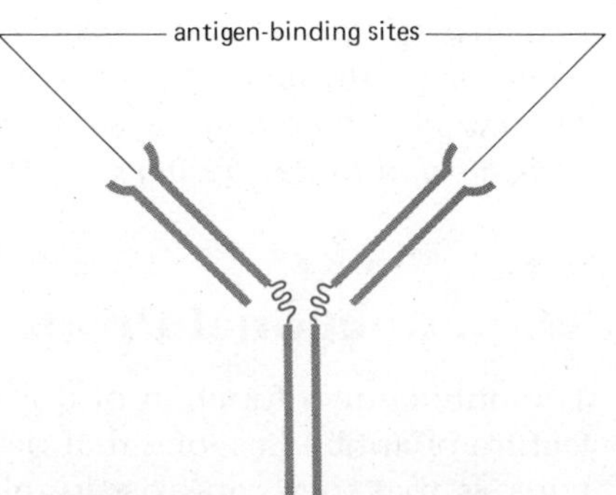

Figure 17–15 A highly schematic diagram of an antibody molecule with two identical antigen-binding sites.

(see Figure 17–30). Once it reaches a certain size, such a lattice precipitates out of solution. This tendency of large immune complexes to precipitate is useful for detecting the presence of antibodies and antigens, as we shall see later. The efficiency of antigen-binding and cross-linking reactions by antibodies is greatly increased by a flexible *hinge region* where the arms of the Y join the tail, allowing the distance between the two antigen-binding sites to vary (Figure 17–16).

The protective effect of antibodies is not due simply to their ability to bind antigen. They engage in a variety of biological activities that are mediated by the tail of the Y. This part of the molecule determines what will happen to the antigen once it is bound. Antibodies with the same antigen-binding sites can have a variety of different tail regions and, therefore, different functional properties.

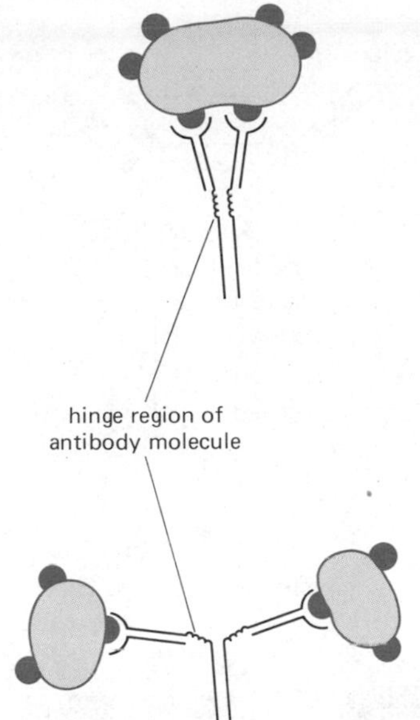

Figure 17–16 The hinge region of an antibody molecule improves the efficiency of antigen binding and cross-linking.

An Antibody Molecule Is Composed of Four Polypeptide Chains—Two Identical Light Chains and Two Identical Heavy Chains[14]

The basic structural unit of an antibody molecule consists of four polypeptide chains, two identical **light (L) chains** (each containing about 220 amino acids), and two identical **heavy (H) chains** (each usually containing about 440 amino acids). The four chains are held together by a combination of noncovalent interactions and covalent bonds (disulfide linkages). The molecule is composed of two identical halves in which both L and H chains contribute almost equally to the two identical antigen-binding sites (Figure 17–17).

The proteolytic enzymes papain and pepsin split antibody molecules into different characteristic fragments: *papain* produces two separate and identical **Fab** (*f*ragment *a*ntigen *b*inding) **fragments,** each with one antigen-binding site, and one **Fc fragment** (so called because it readily crystallizes). *Pepsin,* on the other hand, produces one **F(ab′)$_2$ fragment,** so called because it consists of two covalently linked F(ab′) fragments (each slightly larger than a Fab fragment); the rest of the molecule is broken down into smaller fragments (Figure 17–18). Because F(ab′)$_2$ fragments are bivalent, they can still cross-link antigens and form precipitates, unlike the univalent Fab fragments.

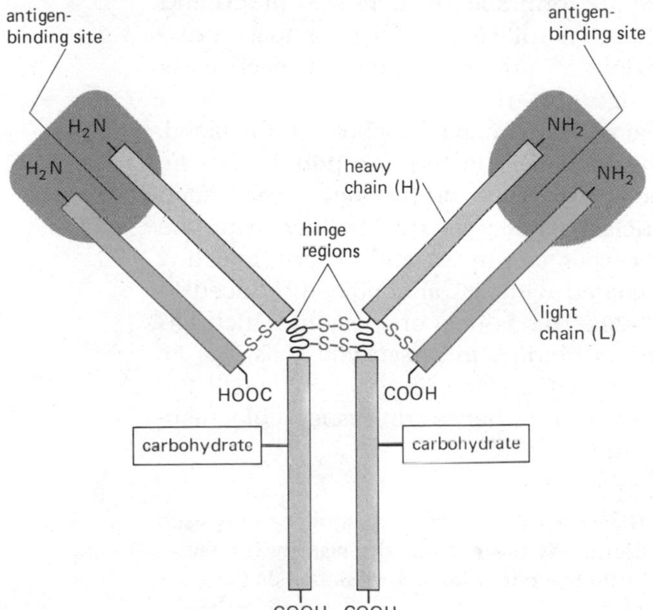

Figure 17–17 Schematic drawing of a typical antibody molecule composed of two identical heavy (H) chains and two identical light (L) chains. Note that the antigen-binding sites are formed by a complex of the amino-terminal regions of both L and H chains, but the tail region is formed by H chains alone. Each H chain contains one or more oligosaccharide chains of unknown function.

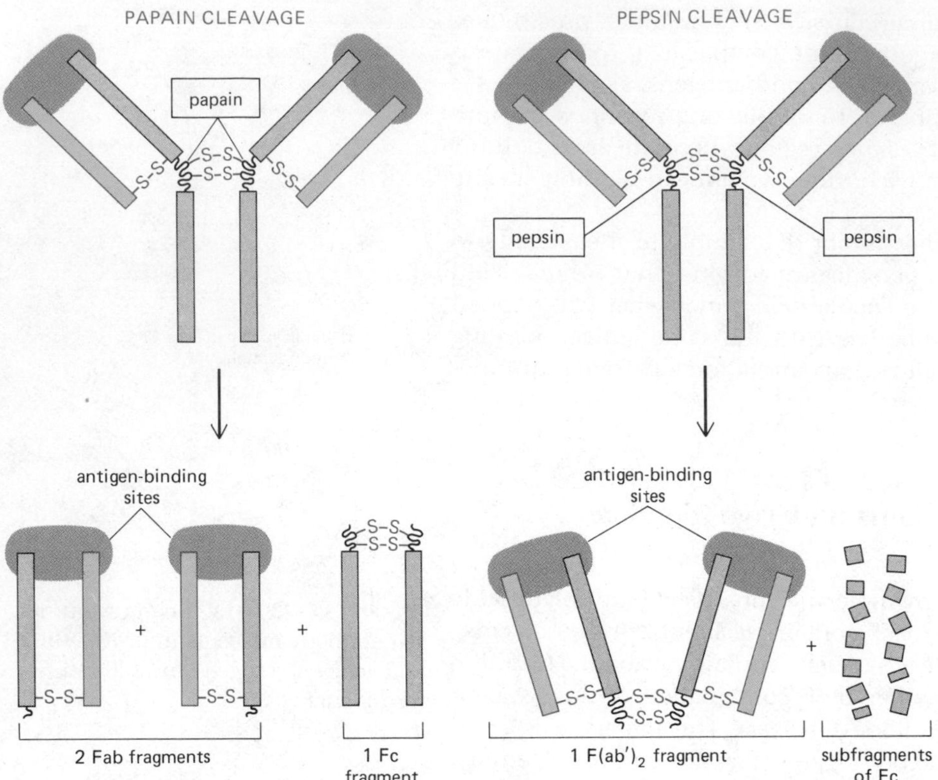

Figure 17–18 The different fragments produced when antibody molecules are cleaved with two different proteolytic enzymes (papain and pepsin) provided important clues for the investigators who determined the four-chain structure of antibodies.

Neither of these fragments has the other biological properties of intact antibody molecules because they lack the tail (Fc) region that mediates these properties.

There Are Five Different Classes of H Chains, Each with Different Biological Properties[11,15]

In higher vertebrates, there are five different *classes* of antibodies, IgA, IgD, IgE, IgG, and IgM, each with its own class of H chain—α, δ, ϵ, γ, and μ, respectively; IgA molecules have α-chains, IgG molecules have γ-chains, and so on (Table 17–1). In addition, there are a number of subclasses of IgG and of some of the other immunoglobulins. The different H chains impart a distinctive conformation to the tail regions of antibodies and give each class characteristic properties of its own (Figure 17–19).

IgG antibodies constitute the major class of immunoglobulin in the blood. They are copiously produced during *secondary* immune responses. The Fc region of IgG molecules binds to specific receptors on phagocytic cells, such as macrophages and polymorphonuclear leucocytes, thereby increasing the efficiency with which the phagocytic cells can ingest and destroy infecting microorganisms that have become coated with IgG antibodies produced in response to the infection (Figure 17–20). This is only one way in which IgG molecules combat infection. As well as binding to phagocytic cells, the Fc region of IgG can bind to and thereby activate the first component of the *complement system,* which under these circumstances unleashes a biochemical attack that kills the microorganism (see p. 988).

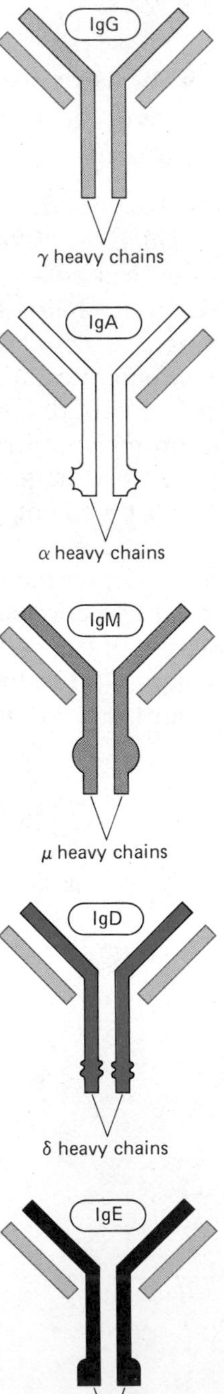

Figure 17–19 Highly schematic diagram showing how each different class of antibody has a distinctive class of H chain that imparts a distinctive conformation to its tail, or Fc region.

Table 17–1 Properties of the Major Classes of Antibody in Man

Properties	IgM	IgD	IgG	IgA	IgE
	Class of Antibody				
Heavy chains	μ	δ	γ	α	ε
Light chains	κ or λ	κ or λ	κ or λ	κ or λ	κ or λ
Number of 4-chain units	5	1	1	1 or 2	1
% of total Ig in blood	5	< 1	80	15	< 1
Activates complement	+ + + +	−	+ +	−	−
Crosses placenta	−	−	+	−	−
Binds to professional phagocytes (macrophages and polymorphs)	−	−	+	−	−
Binds to mast cells and basophils	−	−	−	−	+

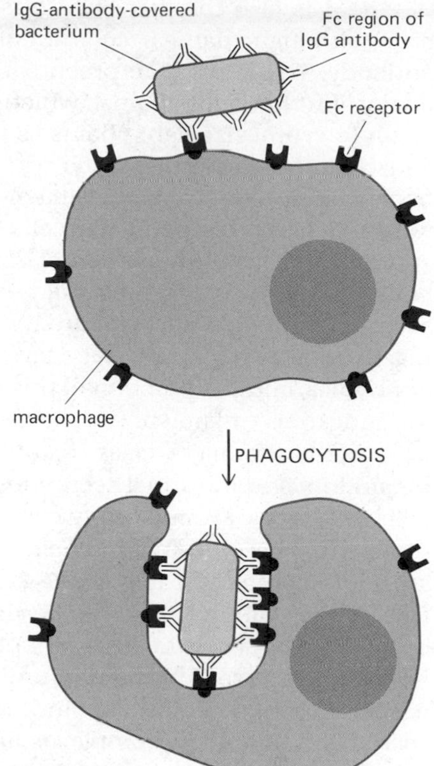

Figure 17–20 Schematic diagram showing how an IgG-antibody-coated bacterium is efficiently phagocytosed by macrophages that have cell-surface receptors able to bind the Fc region of IgG molecules. The binding of the antibody-covered bacterium to these Fc receptors on the macrophage activates the phagocytotic process.

IgG molecules are the only antibodies that can pass from mother to fetus. Cells of the placenta that are in contact with maternal blood have receptors that bind the Fc region of IgG molecules and mediate their passage to the fetus. The antibodies are first ingested by receptor-mediated endocytosis and then transported across the cell and released by exocytosis into the fetal blood. Other classes of antibodies do not bind to these receptors and therefore cannot pass across the placenta.

Although IgG is by far the predominant antibody produced in most secondary antibody responses, **IgM** is the major class of antibody secreted into the blood in the early stages of a *primary* antibody response. In its secreted form, IgM is a pentamer composed of five four-chain units and thus has a total of 10 antigen-binding sites; such pentamers are even more efficient than IgG molecules in activating the complement system when they bind to antigen. Each pentamer contains one copy of another polypeptide chain, called a *J (joining) chain* (~20,000 daltons), which is produced by IgM-secreting cells and is covalently inserted between two adjacent Fc regions, where it presumably initiates the process of oligomerization (Figure 17–21).

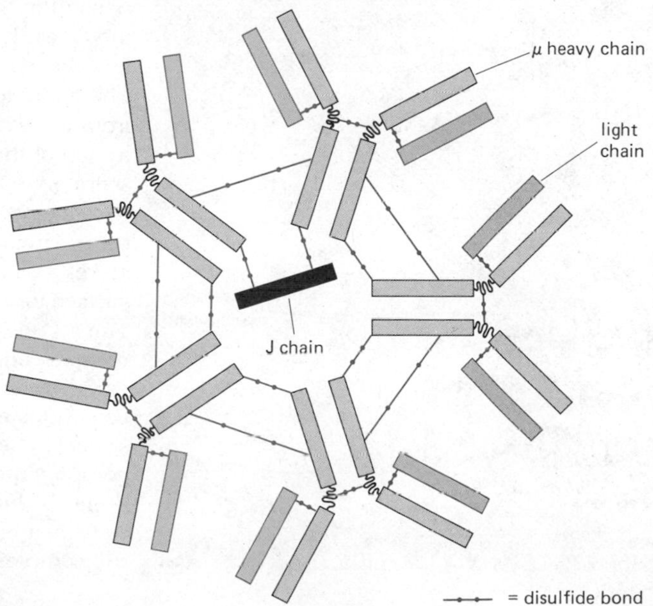

Figure 17–21 A pentameric IgM molecule. The five subunits are held together by disulfide bonds. A single J chain, disulfide-bonded between two μ heavy chains, is thought to initiate the assembly of the pentamer.

IgM is also the first class of antibody to be produced by developing B cells, although many B cells eventually switch to making other classes of antibody. The immediate precursors of B cells, so-called *pre-B cells,* make μ-chains (but not light chains), which accumulate in the cells. When pre-B cells begin to synthesize light chains as well, these combine with μ-chains to form four-chain IgM molecules (each with two μ-chains and two light chains), which become inserted into the plasma membrane, where they function as receptors for antigen. At this point the cells have become B lymphocytes and can respond to antigen. Whereas all classes of antibodies can exist in a membrane-bound form (as antigen-specific, cell-surface receptors) or a water-soluble, secreted form, IgM and **IgD** are the predominant classes found on most resting B cells. Surprisingly, very few B cells are ever activated to secrete IgD antibodies, and the function of this class of antibodies, other than as receptors for antigen, is unknown.

IgA is the major class of antibody in secretions (milk, saliva, tears, and respiratory and intestinal secretions). It exists either as a four-chain monomer (like IgG) or as a dimer of two such units. IgA molecules in secretions are dimers that carry a single J chain and an additional polypeptide chain called *secretory component* (Figure 17–22), which IgA dimers pick up from the surface of the epithelial cells lining the intestine, bronchi, or the milk, salivary, or tear ducts. Secretory component is synthesized by the epithelial cells and is initially exposed on the nonluminal (external) surface of these cells, where it serves as a receptor for binding IgA from the blood. The resulting IgA-secretory-component complexes are ingested by receptor-mediated endocytosis, transferred across the epithelial cell cytoplasm, and secreted into the lumen on the opposite side of the cell (Figure 17–23). In addition to this transport role, secretory component may also protect the IgA molecules from being digested by proteolytic enzymes in the secretions.

The Fc region of **IgE** molecules binds with very high affinity (>10^{10} liters/mole) to specific receptor proteins on the surface of mast cells in tissues and of basophilic leucocytes in the blood. The bound IgE molecules in turn serve as receptors for antigen: antigen binding to them triggers the cells to secrete a variety of biologically active amines (particularly histamine and, in some species, serotonin) (Figure 17–24). These amines cause dilation and increased

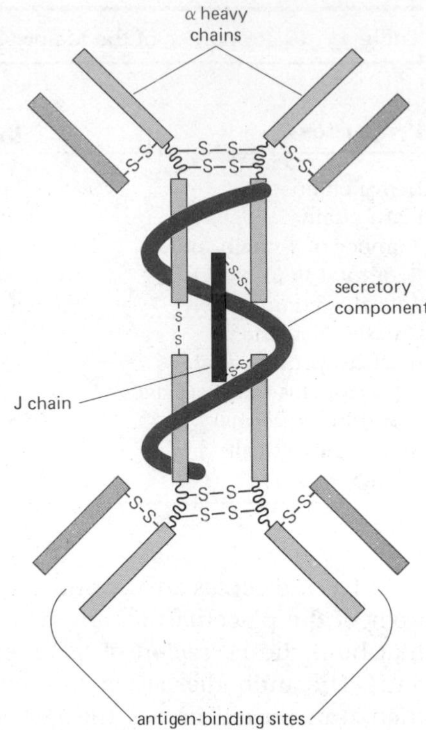

Figure 17–22 A highly schematized diagram of a dimeric IgA molecule found in secretions. In addition to the two IgA monomers that are disulfide-bonded through one of their α heavy chains, there is a single J chain and an additional polypeptide chain of 71,000 daltons called the secretory component.

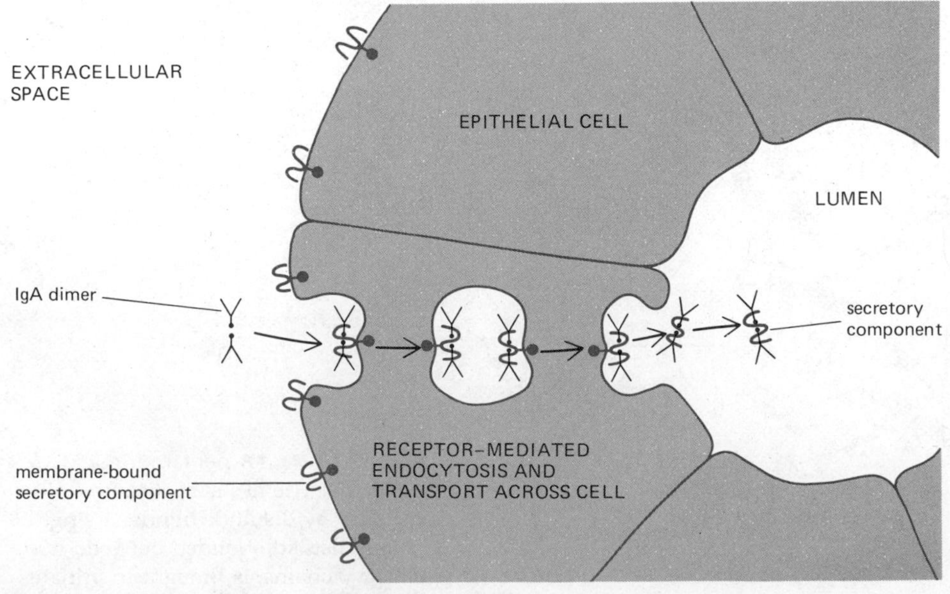

Figure 17–23 The mechanism by which the secretory component mediates the transport of a dimeric IgA molecule across an epithelial cell. The entire complex is transported from the extracellular fluid into the lumen of the epithelial tube. The secretory component is synthesized by the epithelial cell as a transmembrane glycoprotein and serves as a receptor on its basolateral surface for binding the IgA dimer. The secretory-component-IgA complex enters the cell in an endocytotic vesicle, which crosses the cell and is exocytosed at the apical surface. The part of the secretory component that is bound to the IgA dimer is then cleaved from its transmembrane tail, thereby releasing the complex into the lumen.

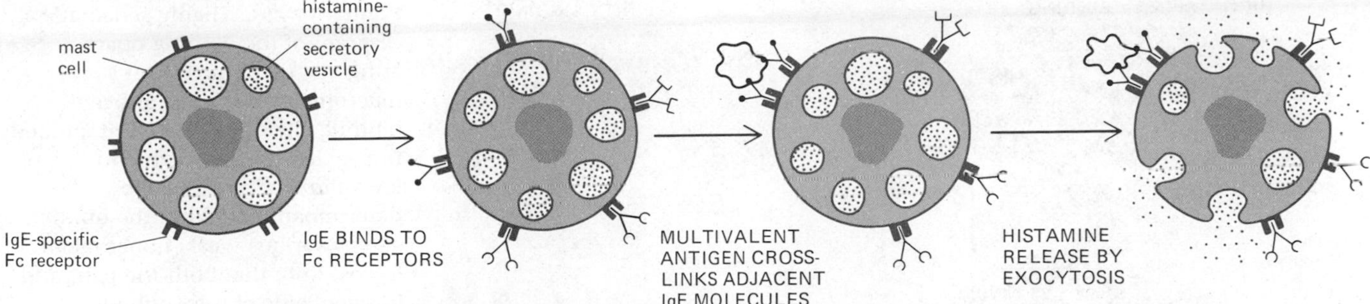

Figure 17–24 Schematic illustration of how mast cells (and basophils) passively acquire cell-surface receptors that bind antigen. IgE antibodies secreted by activated B lymphocytes enter the tissues and bind to receptor proteins on the mast cell surface that specifically recognize the Fc region of these antibodies. Thus, unlike B cells, individual mast cells and basophils have cell-surface antibodies with a variety of different antigen-binding sites. Antigen molecules cross-link those membrane-bound IgE antibodies with complementary antigen-binding sites, thereby activating the mast cell to release its histamine by exocytosis.

permeability of blood vessels and are largely responsible for the clinical manifestations of such *allergic* reactions as hay fever, asthma, and hives. Their protective function is less clear, but it is believed that they may play a part in allowing white blood cells, antibodies, and complement components to enter sites of inflammation.

Antibodies Can Have Either κ or λ Light Chains But Never Both

In addition to the five classes of H chains, higher vertebrates have two types of L chains, κ and λ, either of which can be associated with any of the H chains (Figure 17–25). However, an individual antibody molecule always consists of identical L chains and identical H chains; therefore, its antigen-binding sites are always identical. This symmetry is crucial for the cross-linking function of antibodies. Consequently, an immunoglobulin molecule may have κ or λ L chains, but never both. No difference in the biological function of these two types of L chain has yet been identified.

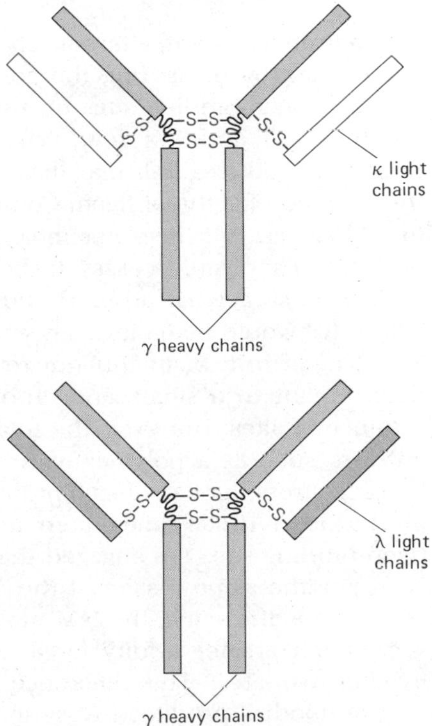

Figure 17–25 An IgG antibody molecule can have either κ or λ light chains, but in any one molecule the two light chains are always identical.

The Strength of an Antibody-Antigen Interaction Depends on Both the Affinity and the Number of Binding Sites[11]

The binding of an antigen to antibody, like the binding of a substrate to an enzyme, is reversible. It is mediated by the sum of many relatively weak noncovalent forces, including hydrophobic and hydrogen bonds, van der Waals forces, and ionic interactions. These weak forces are effective only when the antigen molecule is close enough to allow some of its atoms to fit into complementary recesses on the antibody surface. The complementary regions of a four-chain antibody unit are its two identical antigen-binding sites, while the corresponding region on the antigen is an antigenic determinant (Figure 17–26). Most antigenic macromolecules have many different antigenic determinants; if two or more of them are identical (as in a polymer with a repeating structure), the antigen is said to be *multivalent* (Figure 17–27).

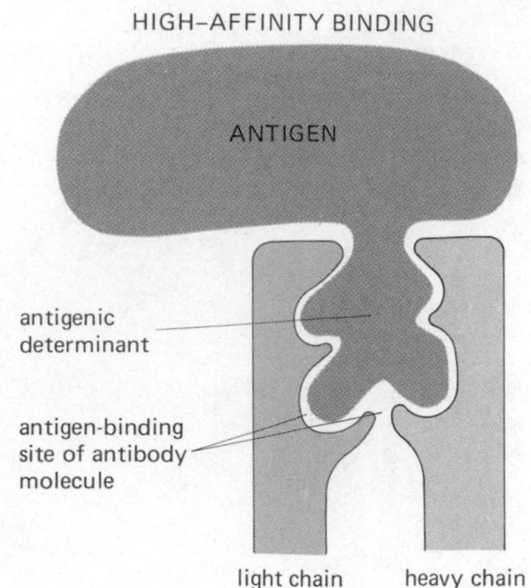

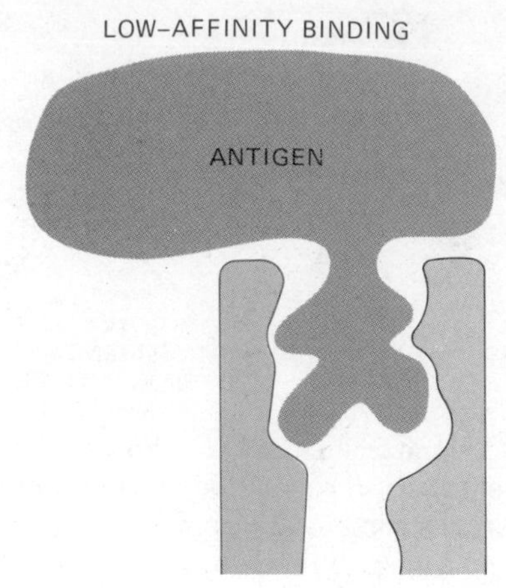

Figure 17–26 Highly schematic diagram of the binding of an antigenic determinant on a macromolecule to the antigen-binding site of two different antibody molecules—one of high and one of low affinity. The antigenic determinant is held in the binding site by various weak, noncovalent forces. Note that both the light and heavy chains of the antibody molecule contribute to the antigen-binding site.

The reversible binding reaction between an antigen with a single antigenic determinant (denoted as Ag) and a single antigen-binding site (denoted Ab) can be expressed as:

$$\mathrm{Ag} + \mathrm{Ab} \rightleftharpoons \mathrm{AgAb}$$

The equilibrium point depends both on the concentrations of Ab and Ag and on the strength of their interaction. Clearly, a larger fraction of Ab will become associated with Ag as the concentration of Ag is increased. The strength of the interaction is generally expressed as the **affinity constant (*K*)** (see Figure 3–5, p. 97), where

$$K = [\mathrm{AgAb}]/[\mathrm{Ag}][\mathrm{Ab}].$$

This affinity constant, alternatively called an association constant (K_a), can be determined by measuring the concentration of free Ag required to fill half of the antigen-binding sites on the antibody. When half the sites are filled, [AgAb] = [Ab] and $K = 1/[\mathrm{Ag}]$. Thus, the reciprocal of this antigen concentration that produces half maximal binding is equal to the affinity constant of the antibody for the antigen. Common values range from as low as 5×10^4 to as high as 10^{12} liters per mole. The affinity constant at which an immunoglobulin molecule ceases to be considered an antibody for a particular antigen is somewhat arbitrary, but it is unlikely that an antibody with a K below 10^4 would be biologically effective.

The **affinity** of an antibody reflects the goodness of the fit of an antigenic determinant to a single antigen-binding site, and it is independent of the number of sites. However, the total **avidity** of an antibody for a multivalent antigen, such as a polymer with repeating subunits, is defined as the total binding strength of all of its binding sites together. A typical IgG molecule will bind at least 10,000 times more strongly to a multivalent antigen if both antigen-binding sites are engaged than if only one site is involved.

For the same reason, if the affinity of the sites in an IgG and an IgM molecule is the same, the IgM molecule (having 10 binding sites) will have a very much greater avidity for a multivalent antigen than an IgG molecule (having two sites). This difference in avidity is important in view of the fact that antibodies produced early in an immune response usually have much lower affinities than those produced later. (The increase in the average affinity of antibodies produced with time after immunization is called *affinity maturation.*) Because of its high total avidity, IgM—the major Ig class produced early in immune responses—can function even when each of its binding sites has only a low affinity.

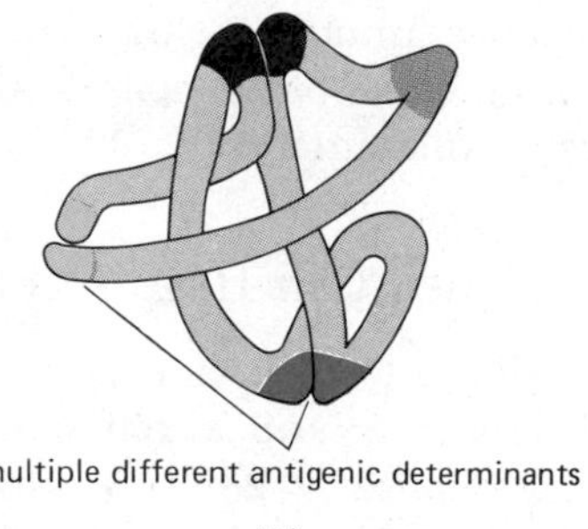

(A)

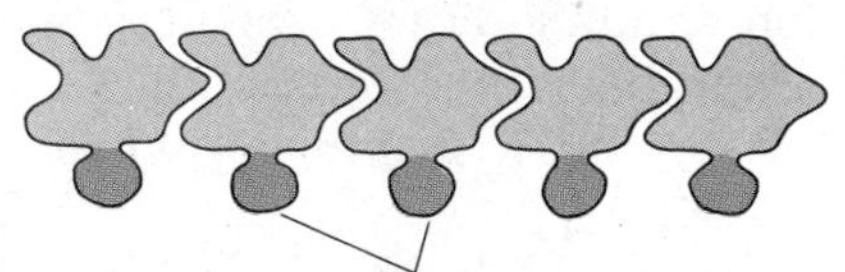

(B)

Figure 17–27 Schematic drawings of molecules with multiple antigenic determinants. (A) A globular protein having a number of *different* antigenic determinants. Note that different regions of a polypeptide chain can come together in the folded structure to form a single antigenic determinant on the surface of the protein. (B) A polymeric structure with many *identical* antigenic determinants; such a molecule is called a *multivalent antigen.*

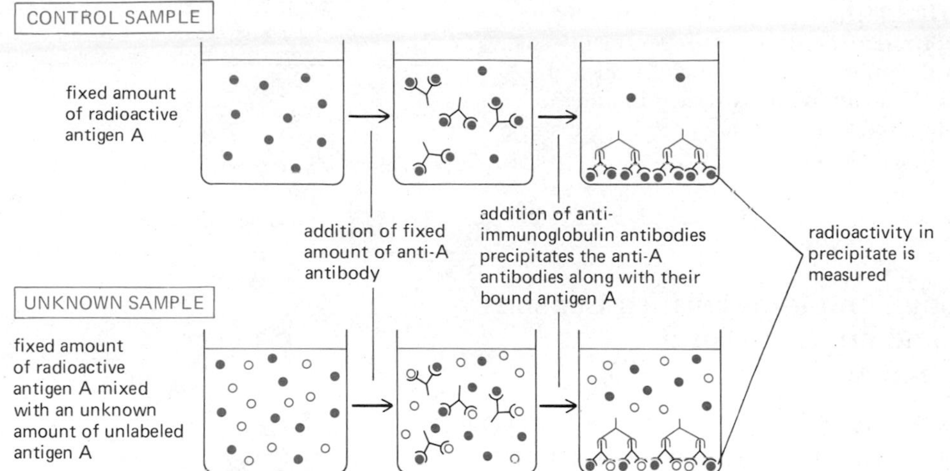

Figure 17–28 The principle of a radioimmunoassay. Unlabeled antigen competes with radioactive antigen for binding to antibody. This reduces the amount of radioactivity in the antibody-antigen precipitate. The amount by which the precipitated radioactivity is reduced compared to the control sample indicates the concentration of antigen in the unknown sample.

Antibody-Antigen Interactions Can Be Measured in Many Ways[11,16]

The precise antigen specificity of antibodies makes them versatile and powerful tools that can be used to detect, quantify, and localize a large variety of biologically interesting molecules. But how does one detect or measure antibody-antigen interactions? The initial binding reaction of antigen and antibody—the so-called *primary reaction*—can be measured in many different ways. In the *radioimmunoassay,* which is an invaluable technique for measuring even minute quantities of a substance, a known amount of radioactive antigen is added, with a fixed amount of antibody, to a sample containing an unknown quantity of the same antigen in nonradioactive form. The unlabeled antigen competes with the labeled antigen for antibody-binding sites so that the greater the amount of antigen in the unknown sample, the smaller the amount of radioactive antigen bound to antibody. The free and bound radioactive antigen can be separated and measured by a variety of methods that depend on the different properties of free and bound antigen; one general approach is to precipitate the antibody-antigen complexes with anti-immunoglobulin antibody (Figure 17–28).

Alternatively, radiolabeled, fluorescent, or enzyme-coupled antibodies can be used to detect and locate specific molecules in cells or tissues. Here the bound antibodies are visualized by autoradiography or by fluorescence microscopy, or by the colored product of the enzyme-substrate reaction, respectively (see Chapter 4).

Many tests, however, depend on the *secondary reactions* that follow as a consequence of the primary interaction of antibody with antigen. These secondary reactions include *precipitation, cell agglutination* (clumping), and *complement fixation.* The last of these can be exploited because components of the complement system bind only to antibody that is complexed with antigen; thus the disappearance of complement components can be used as a measure of the amount of antibody-antigen complex formed. But the most commonly used secondary assay involves the detection of antibody-antigen precipitates that form in fluids or gels. For example, in the *Ouchterlony assay,* antigen and antibody are placed in separate wells cut in an agar gel. They are then left to diffuse outward from the wells until they meet in optimal proportions to form a large precipitate, which becomes visible as an opaque line because of the light it scatters (Figure 17–29).

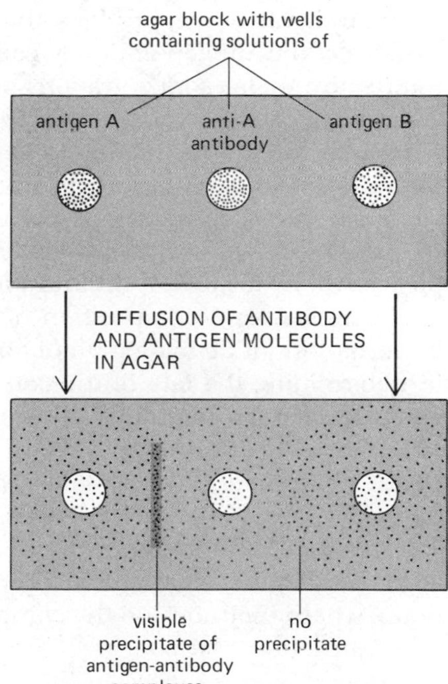

Figure 17–29 Schematic drawing of an Ouchterlony assay for detecting antigen-antibody interactions. Antibodies and antigens are placed in different wells in an agar gel and allowed to diffuse toward each other. When antigen and complementary antibodies meet in optimal proportions, large antibody-antigen complexes precipitate, and these can be detected by the light that they scatter.

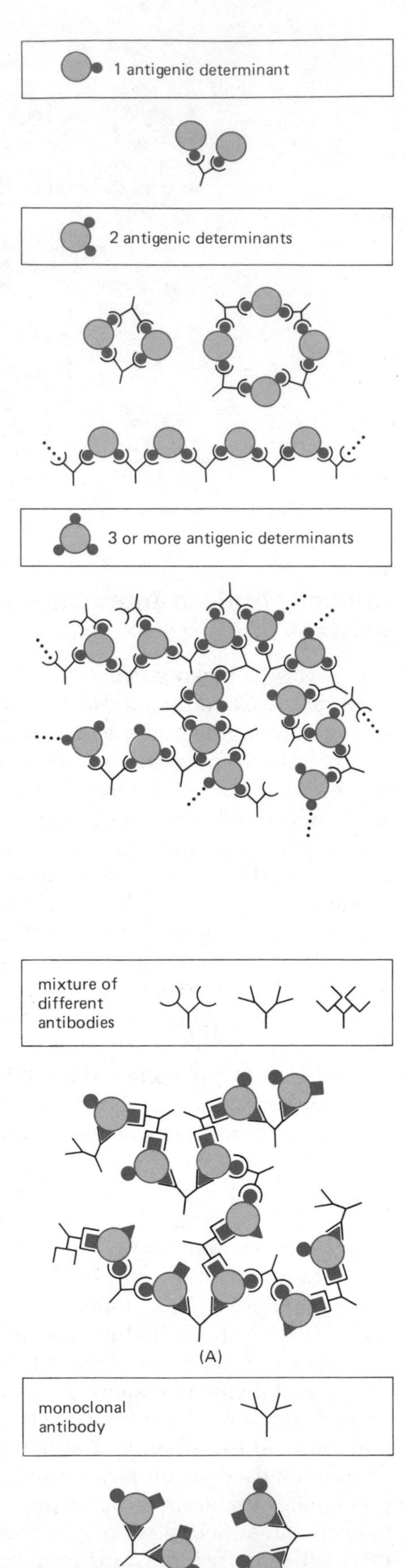

Figure 17–30 The different types of antibody-antigen complexes that form depend on the number of antigenic determinants on the antigen. Here a single species of antibody (a monoclonal antibody) is shown binding to antigens containing one, two, and three copies of a single type of antigenic determinant.

The Size of the Antigen-Antibody Complexes Formed Depends on the Valence of the Antigen and on the Relative Concentrations of the Antigen and Antibody[11,17]

The basis of precipitation reactions is the cross-linking of multivalent antigens by bivalent antibodies. If only one species of antibody is present, molecules with only one antigenic determinant cannot be cross-linked. If an antigen is bivalent, it can form small cyclic complexes or linear chains with antibody, while an antigen with three or more antigenic determinants can form large three-dimensional lattices that readily precipitate (Figure 17–30). In fact, however, the majority of antisera prepared against an antigen contain a variety of different antibodies that react with different determinants on the antigen and can cooperate in cross-linking the antigen. By contrast, homogeneous (monoclonal) antibodies can only precipitate molecules with repeating identical antigenic determinants (a multivalent antigen, Figure 17–31).

Given valence conditions that allow the formation of large aggregates, the size of the antigen-antibody complexes that form will depend critically on the relative molar concentrations of the two reactants. If there is an excess of either antigen or antibody, large complexes are unlikely to form: with *antigen excess,* most complexes will contain only a single antibody molecule with a molecule of antigen bound to each of its antigen-binding sites; with large *antibody excess,* most complexes will consist of single molecules of antigen with antibodies bound to each of its antigenic determinants. The largest complexes will be formed at around molar *equivalence* (Figure 17–32).

The size and composition of antibody-antigen complexes are not only important in influencing precipitation reactions in test tubes, they are crucial in determining the fate of the complexes in the body. Complexes formed at equivalence or in antibody excess have multiple protruding Fc regions (Figure 17–32) and therefore bind strongly to Fc receptors on macrophages, which ingest and degrade them. Small complexes, formed in antigen excess, have only one Fc region per complex (Figure 17–32). Therefore, they bind poorly to Fc receptors on macrophages and are less efficiently destroyed. Instead they are often deposited in small blood vessels in the skin, kidneys, joints, and brain, where they activate the complement system, causing inflammation and the destruction of tissue.

Figure 17–31 (A) A mixture of different antibodies binding to different antigenic determinants on the same antigen molecule can cooperate to form a lattice (precipitate). Such mixtures of antibodies are found in conventional antisera produced against most macromolecules. (B) A monoclonal antibody is less effective because it precipitates only antigens that have multiple identical determinants, as shown in Figure 17–30.

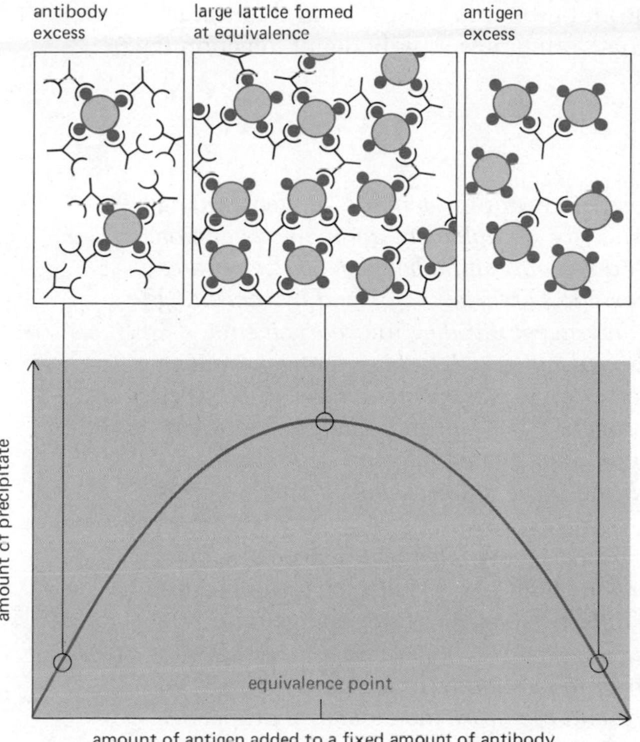

Figure 17–32 Antibody and antigen concentrations influence the size of antigen-antibody complexes formed. The largest complexes form when both molecules are present at about the same molar concentration ("equivalence"), while the smallest complexes form when antigen is present in great excess. Note that the small complexes formed in antigen excess have only one antibody molecule per complex; for this reason they are inefficiently cleared from the extracellular fluids by macrophages.

Antibodies Recruit Complement and Various Cells to Fight Infection[18]

We have already discussed how antibodies initiate the destruction of invading microorganisms by phagocytes and by complement. But these are not the only ways that antibodies defend vertebrates against infection. Antibody-coated cells can be killed without being phagocytosed by various cells with receptors that recognize the Fc region of antibodies. The cells most active in this process are called **K (killer) cells,** which look like lymphocytes but are not T or B cells. The mechanism of killing by K cells is unknown. Thus, although antibodies on their own are unable to kill invading organisms, they become lethal by recruiting complement, phagocytic cells, and K cells (Figure 17–33). Moreover, antibodies can combine with viruses or bacterial toxins (such as tetanus or

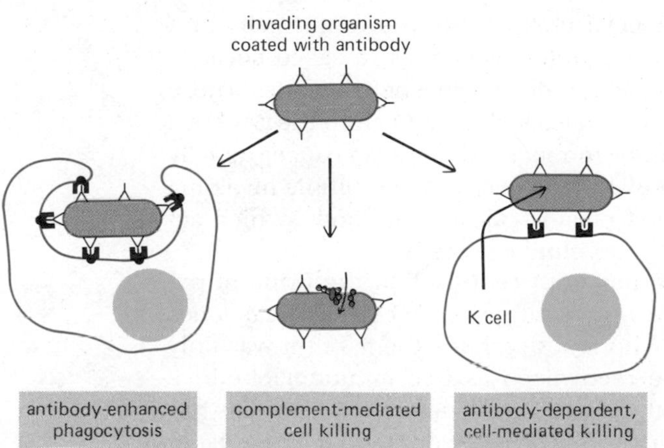

Figure 17–33 Three ways in which antibodies can help eliminate invading organisms. While antibody-enhanced phagocytosis and complement-mediated killing are known to be important in the defense against infection, K-cell killing has so far been demonstrated mainly in a test tube, with antibody-coated vertebrate cells as targets.

botulinum toxin) and prevent them from binding to receptors on their target cells. It is, therefore, not surprising that vertebrates rapidly die of infection if they are unable to make antibodies.

Summary

An antibody molecule is a Y-shaped protein with two identical antigen-binding sites at the tips of the Y and binding sites for complement components and/or various cell-surface receptors on its Fc region. Antibodies defend vertebrates against infection by inactivating viruses and bacterial toxins and by recruiting complement and various cells to kill and ingest invading microorganisms.

Each B-cell clone makes antibody molecules with a unique antigen-binding site. Initially, the molecules are inserted into the plasma membrane, where they serve as cell-surface receptors for antigen. When antigen binds to the membrane-bound antibodies, B cells are activated to multiply and to synthesize a large amount of soluble antibody with the same antigen-binding site, which is secreted into the blood.

Each antibody molecule is composed of two identical heavy (H) chains and two identical light (L) chains. Parts of both the H and L chains form the antigen-binding sites. There are five different classes of antibodies (IgA, IgD, IgE, IgG, and IgM), each with a distinctive H chain (α, δ, ϵ, γ, and μ, respectively). The H chains also form the Fc region of the antibody, which determines what other proteins will bind to the antibody and therefore the biological properties of the class. Either type of L chain (κ or λ) can be associated with any class of H chain.

The Fine Structure of Antibodies

The unique feature of antibodies is that they exist in so many different forms: each class of immunoglobulin contains millions of different antibodies, each with a different antigen-binding site and a different amino acid sequence. Any one of these antibody molecules therefore, constitutes less than one part in a million of the immunoglobulin molecules in the blood. This fact presented immunochemists with a uniquely difficult problem in protein chemistry: how to obtain enough of any one antibody to determine its amino acid sequence and three-dimensional structure.

Myeloma Proteins Are Homogeneous Antibodies Made by Plasma-Cell Tumors[11]

The problem was resolved by the special character of the tumor cells in a cancer known as **multiple myeloma.** Multiple myeloma is so called because multiple tumors develop in the bone marrow, or "myelogenous" tissues. These tumors secrete large amounts of a single species of antibody into the patient's blood. The antibody is homogeneous, or monoclonal, because cancer usually begins with the uncontrolled growth of a single cell, and in multiple myeloma the single cell is an antibody-secreting plasma cell. The antibody, which accumulates in the blood, is known as a **myeloma protein.**

It had been known since the nineteenth century that the urine of patients with this disease often contains unusual proteins, called *Bence Jones proteins* after the English physician who first described them; but it was only in the 1950s that the proteins were recognized as free immunoglobulin L chains. Much of what we know about the detailed structure of antibodies has come from studying myeloma proteins from the urine or blood of patients, or from mice in which similar cancers have been purposely induced.

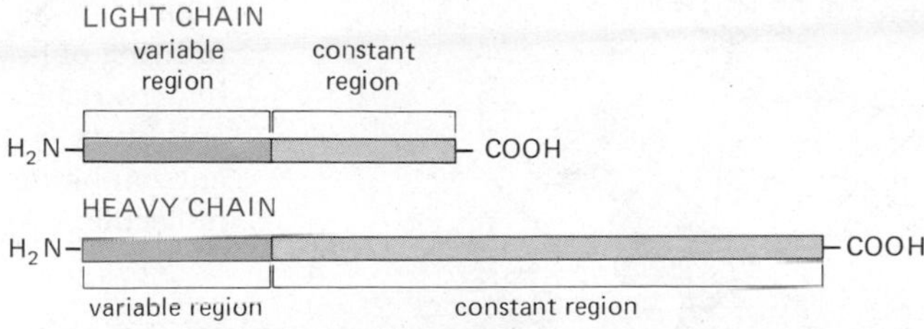

Figure 17–34 Both light and heavy chains have distinct constant and variable regions. For light chains, the carboxyl-terminal halves of chains of the same type (either κ or λ) all have the same sequence (with occasional minor differences), while the amino-terminal halves are all different. For heavy chains, the amino-terminal variable region is of similar size to that in light chains (about 110 amino acid residues), whereas the constant region is three or four times the size of a light-chain constant region (depending on the class).

L and H Chains Consist of Constant and Variable Regions[11,19]

The comparison of the amino acid sequences of many different myeloma proteins revealed a striking feature with important and surprising genetic implications. Both L and H chains have an extremely variable sequence at their amino-terminal ends but a constant sequence at their carboxyl-terminal ends. For example, when the amino acid sequences of many different myeloma κ chains (each about 220 amino acids long) are compared, the carboxyl-terminal halves are the same, or show only minor differences, whereas the amino-terminal halves are all different. Thus, L chains have a **constant region** about 110 amino acids long and a **variable region** of the same size. The amino-terminal variable region of the H chains is also about 110 amino acids long, but the H-chain constant region is about 330 or 440 amino acids long, depending on the class (Figure 17–34).

It is the amino-terminal ends of the L and H chains that come together to form the antigen-binding site, and the variability of their amino acid sequences provides the structural basis for the diversity of antigen-binding sites. The existence of the variable and constant regions in antibody molecules raises important genetic questions that we shall discuss later. But before it became possible to investigate these genetic questions directly, other important features of antibody structure emerged from structural studies on myeloma proteins.

The L and H Chains Each Contain Three Hypervariable Regions That Together Form the Antigen-binding Site[19,20]

Only part of the variable region participates directly in the binding of antigen. This conclusion was first deduced from estimates of the maximum size of an antigen-binding site. The measurements, which involved the use of oligomers of increasing size as "molecular rulers," were initially made on antibodies reactive against dextran, a polymer of D-glucose. When disaccharides, trisaccharides, and higher oligosaccharides of glucose are used to inhibit the binding of dextran to anti-dextran antibodies, inhibition increases with chain length up to about six glucose units; larger oligosaccharides have no greater effect. This suggests that the largest antigen-binding sites can contact at most five or six sugar residues of an antigen. Thus it is most unlikely that all 220 amino acids of the variable regions of both L and H chains contribute directly to the antigen-binding site.

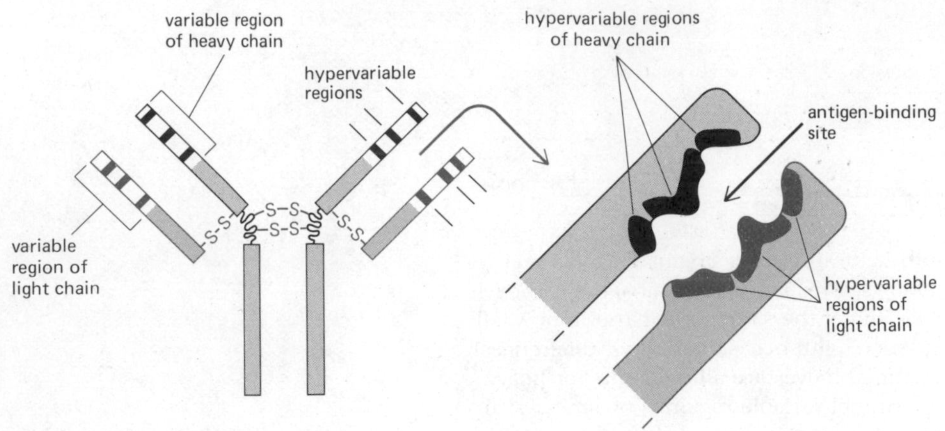

Figure 17–35 Highly schematic drawing of how the three hypervariable regions in each light and heavy chain together form the antigen-binding site of an antibody molecule.

In fact, it is now clear that the binding site of antibodies is formed by only about 20 to 30 of the amino acid residues in the variable region of each chain. This was first suggested by amino acid sequence data, which showed that the variability in the variable regions of both L and H chains is for the most part restricted to three small **hypervariable regions** in each chain. The remaining parts of the variable region, known as *framework regions,* are relatively constant. These findings led to the prediction that only the 5 to 10 amino acids in each hypervariable region form the antigen-binding site (Figure 17–35), a prediction that has since been confirmed by x-ray diffraction studies of antibody molecules (see below).

The L and H Chains Are Folded into Repeating Domains[11,21]

With the sequencing of the first H chain, which was completed in the late 1960s, another important feature of immunoglobulin structure became apparent. The new insight came, in the first place, from the sequence of the constant region, which is about three times as long in most H chains as it is in L chains. It turned out that the H-chain constant region consists of three homologous segments, each about 110 amino acids long and each containing one intrachain disulfide bond. The amino acid sequences of these three segments not only have a significant degree of homology to each other, but also to the constant region of L chains. Similarly, the single variable domains in both the L and H chains are homologous to each other and, to a lesser extent, to the constant domains.

It was correctly predicted from these findings that both L and H chains are made up of repeating segments, or *domains,* each of which folds independently to form a compact functional unit. Accordingly, as shown in Figure 17–36, an L chain consists of one variable (V_L) and one constant (C_L) domain, while most H chains consist of a variable domain (V_H) and three separate constant domains (C_H1, C_H2, and C_H3). (The μ- and ϵ-chains each have one variable and four constant domains.) While the variable domains are responsible for antigen binding, the constant domains of the H chains (excluding C_H1) form the Fc region that determines the biological properties of the antibody.

The homology between their domains suggests that immunoglobulin chains probably arose in evolution by a series of gene duplications beginning with a primordial gene coding for a single 110-amino-acid domain of unknown function. The recent evidence that each domain of the constant region of an H chain is encoded by a DNA sequence separated from the next coding region by an intervening sequence (that is, an *intron*—see p. 414) is consistent with

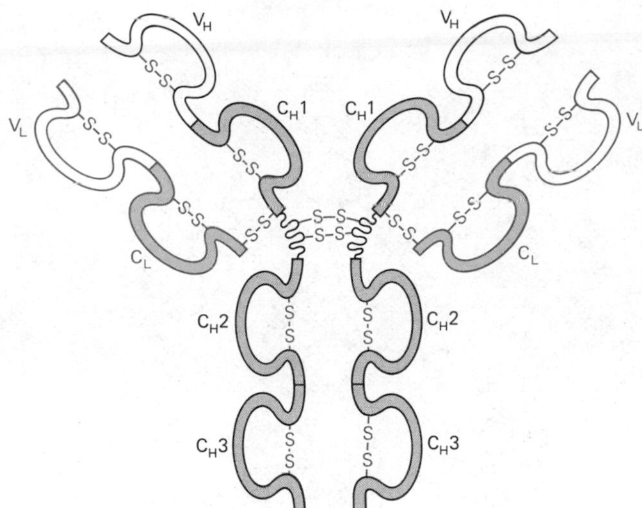

Figure 17–36 Schematic illustration showing that the light and heavy chains are folded into repeating domains that are similar to one another. While the variable domains of the light and heavy chains (V_L and V_H) make up the antigen-binding sites (see Figure 17–35), the constant domains of the heavy chains (mainly C_H2 and C_H3) determine the other biological properties of the molecule. The heavy chains of IgM and IgE antibodies have an extra constant domain (C_H4).

this hypothesis. Although the intron sequences are removed when the primary RNA transcripts are spliced into mRNA molecules (Figure 17–37), the presence of introns in DNA may have facilitated the accidental duplications of DNA segments that gave rise to the antibody genes during evolution (see p. 471).

X-ray Diffraction Studies Have Revealed the Structure of Immunoglobulin Domains and Antigen-binding Sites in Three Dimensions[19,22]

Even when the complete amino acid sequence of a protein is known, it is not possible to deduce its three-dimensional structure. To determine the three-dimensional structure requires x-ray diffraction studies of protein crystals. To date, several myeloma protein fragments and one intact IgG molecule have been crystallized, and x-ray studies of their structures have confirmed the predictions of the immunochemists. More important, these studies have revealed the way in which millions of different antigen-binding sites are constructed on a common structural theme.

As illustrated in Figure 17–38, all Ig domains have very similar three-dimensional structures based on what is now called the *immunoglobulin fold.*

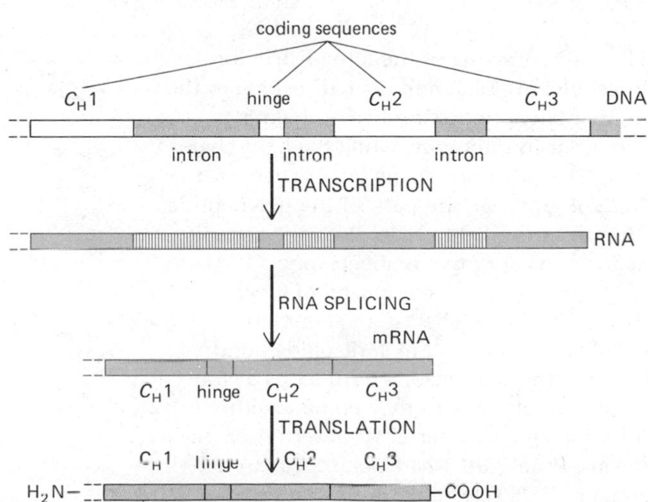

Figure 17–37 The organization of the DNA sequences that encode the constant regions of an immunoglobulin heavy chain. Note that the coding sequences for each domain and for the hinge region are separated by noncoding sequences (introns). The intron sequences are removed by splicing of the primary RNA transcripts to form mRNA. The DNA encoding the variable region of the heavy chain is not shown.

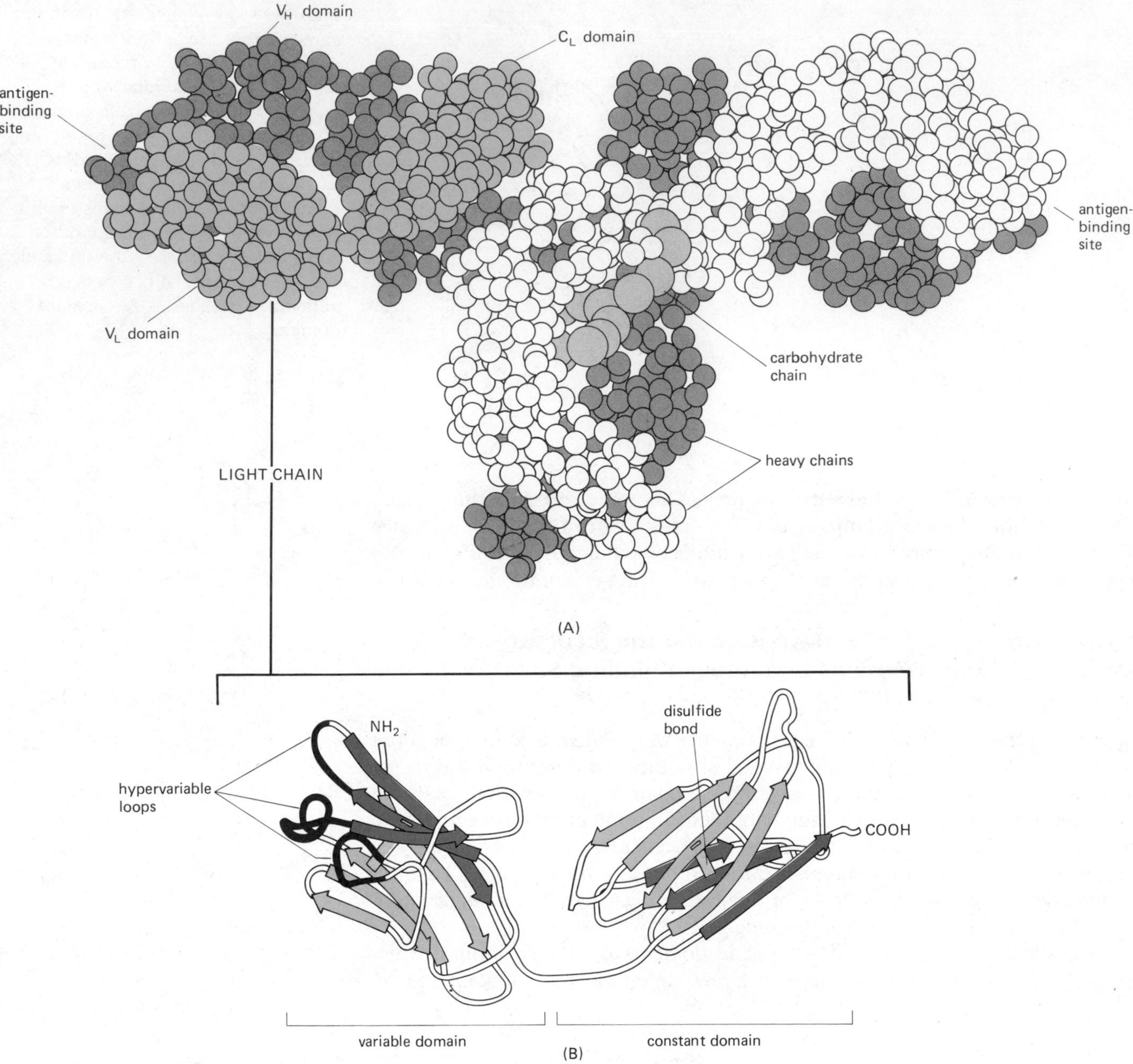

Figure 17–38 Schematic drawing of the folded structure of an IgG antibody molecule. (A) Each amino acid residue in the protein is drawn as a small sphere. One heavy chain is shown in white, the other in dark gray, while the light-chain domains are in color. The oligosaccharide chain attached to a C_H2 domain is in light gray. (B) The path of the polypeptide chain is sketched for an entire light chain. Both the variable and constant domains consist of two β sheets (one composed of three strands and one composed of four strands). The chains in these two sheets are shown in different colors; the sheets are joined by a disulfide bond. Note that all of the hypervariable regions form loops at the far end of the variable domain, where they come together to form the antigen-binding site. (A, after E. W. Silverton, M. A. Navia, and D. R. Davies, *Proc. Natl. Acad. Sci. USA* 74:5140, 1977; B, after M. Schiffer, R. L. Girling, K. R. Ely, and A. B. Edmundson, *Biochemistry* 12:4620, 1973.)

Each domain is roughly a cylinder 4 × 2.5 × 2.5 nm and composed of a "sandwich" of two extended protein layers: one layer contains three strands of polypeptide chain, while the other contains four. In each layer the adjacent strands are antiparallel and form a β-sheet (see p. 113). The two layers are aligned roughly parallel to each other and are connected by a single intrachain disulfide bond.

The variable domains are unique in that each has its particular set of three hypervariable regions, which are arranged in three *hypervariable loops.* The hypervariable loops of both the L and H variable domains are clustered together to form the antigen-binding site, as had been predicted (Figure 17–38). An important principle to emerge from these studies is that the variable region of an antibody molecule consists of a highly conserved, rigid framework, with hypervariable loops attached at one end. Therefore, through changes in only the hypervariable amino acids, an enormous diversity of antigen-binding sites can be generated without disturbing the common overall three-dimensional structure necessary for antibody function.

X-ray analysis of crystals with an antigenic determinant (hapten) bound to the antigen-binding sites has revealed exactly how the hypervariable loops of the L and H variable domains cooperate to form an extensive and continuous antigen-binding surface in particular cases. The dimensions and shape of each different site vary depending on the conformation of the polypeptide chain in the hypervariable loops, which, in turn, is determined by the sequence of the amino acid side chains contained in the loops. Thus, while the general principles of antibody structure are now clear and the detailed structures of several antigen-binding sites have been determined, it is likely that we shall never know the fine structure of most of the millions of antigen-binding sites that exist.

Summary

Each immunoglobulin L and H chain consists of a variable region of about 110 amino acid residues at its amino-terminal end, followed by a constant region, which is the same size in the L chain and three or four times larger in the H chain. Each chain is composed of repeating, similarly folded domains: an L chain has one variable region (V_L) and one constant region (C_L) domain, while an H chain has one variable region (V_H) and three or four constant region (C_H) domains. The amino acid sequence variation in the variable regions of both L and H chains is for the most part confined to several small hypervariable regions, which come together at one end of the molecule to form the antigen-binding site. Each antigen-binding site is only large enough to contact an antigenic determinant the size of five or six sugar residues.

The Generation of Antibody Diversity[23]

It is estimated that a mouse is able to make between 10^6 and 10^9 different antibody molecules, which are collectively referred to as its *antibody repertoire.* This repertoire is apparently large enough to ensure that there will be an antigen-binding site to fit almost any antigenic determinant. Since antibodies are proteins and proteins are encoded by genes, the ability of an animal to make millions of different antibodies poses a unique genetic problem: how to make millions of different proteins without requiring an unreasonably large number of genes. Not surprisingly, the solution to the problem involves some unique genetic mechanisms.

More Than One Gene Segment Codes for Each L and H Chain[24]

The operation of unusual genetic mechanisms in the production of antibodies was apparent long before it became clear how those mechanisms contribute to the diversity of antigen-binding sites. As noted earlier, the amino acid sequence studies of myeloma proteins indicated that each immunoglobulin chain consists of a distinct variable (V) and constant (C) region and raised the question of how such chains were genetically encoded. It was suspected at the time that the V and C regions of each chain might be encoded by two separate genes that were somehow joined together in the DNA before they were expressed.

The first direct evidence that DNA is rearranged during B-cell development was obtained in 1976 by experiments in which DNA from early mouse embryos, which do not make antibodies, was compared with the DNA of mouse myeloma cells, which do. The two kinds of DNA were digested with a restriction nuclease and the resulting fragments hybridized to radioactive DNA sequences prepared by the *in vitro* copying of the *V* sequence or the *C* sequence of the L-chain mRNA molecules isolated from the myeloma cells (see p. 188). The results showed that the specific *V* and *C* coding sequences were on different DNA restriction fragments in the embryos but on the same restriction fragment in myeloma cells (Figure 17–39). Thus, in embryonic DNA, in which immunoglobulin genes are not expressed, the DNA sequences encoding the *V* and *C* regions of an immunoglobulin chain are located in different parts of the genome, while in a myeloma cell, which produces the immunoglobulin chain, these two sequences are brought together.

It is now known that for each type of immunoglobulin chain—κ light chains, λ light chains, and heavy chains—there exists a separate pool of genes from which a single polypeptide chain is eventually synthesized. Each gene pool contains a set of different ***V* genes** located hundreds of thousands of nucleotides upstream (that is, on the 5′ side as measured on the coding DNA strand) from one or more ***C* genes.** During B-cell development any one of the *V* genes can be translocated so that it lies close to a particular *C* gene. Only after such a DNA rearrangement has occurred can an immunoglobulin chain be synthesized.

The gene pools coding for the κ, λ, and H chains are each on different chromosomes. In the mouse, the κ gene pool is on chromosome 6 and contains a single *C* gene (C_κ) plus a large set of *V* genes ($V_\kappa1$, $V_\kappa2$, $V_\kappa3$ . . .). The λ pool is on chromosome 16, and it contains only two *V* genes ($V_\lambda1$ and $V_\lambda2$),

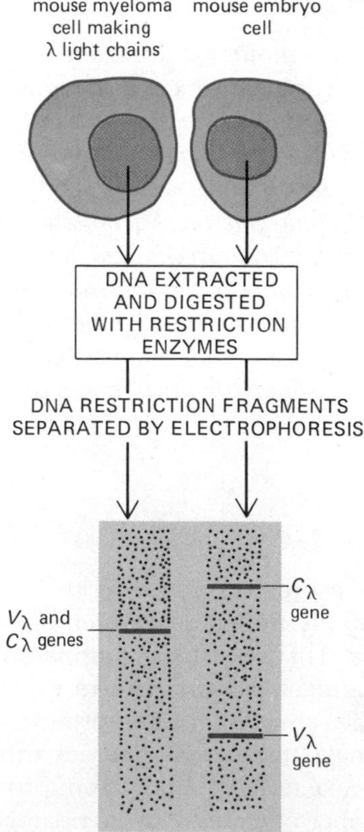

Figure 17–39 The experiment that directly demonstrated that DNA is rearranged during B-cell development. DNA was extracted from a mouse plasma-cell tumor (a myeloma) synthesizing a specific λ light chain and from a 13-day mouse embryo. The two DNA preparations were digested with a restriction nuclease and electrophoresed through an agar gel. Those separated fragments carrying the C_λ coding sequence and those carrying the particular V_λ coding sequence were detected by hybridization to a radioactive DNA sequence prepared *in vitro* by copying the *V*-region sequence or the *C*-region sequence of the mRNA molecules specifying the specific myeloma λ chain. Whereas the V_λ and C_λ sequences were found on the same DNA fragments in the myeloma-cell DNA, they were found on separate fragments in the DNA extracted from the embryo (and in DNA extracted from a different myeloma tumor making a different light chain).

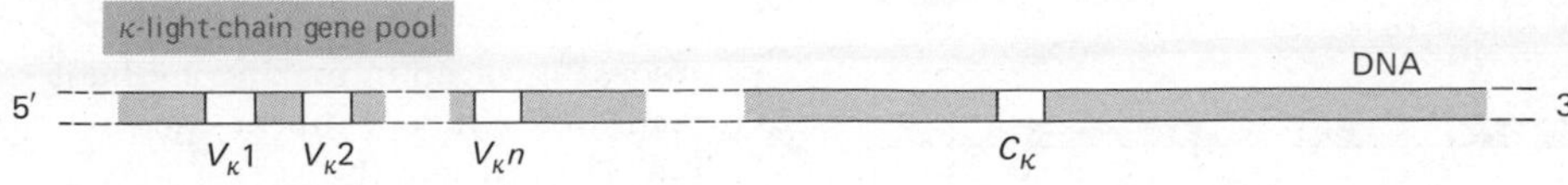

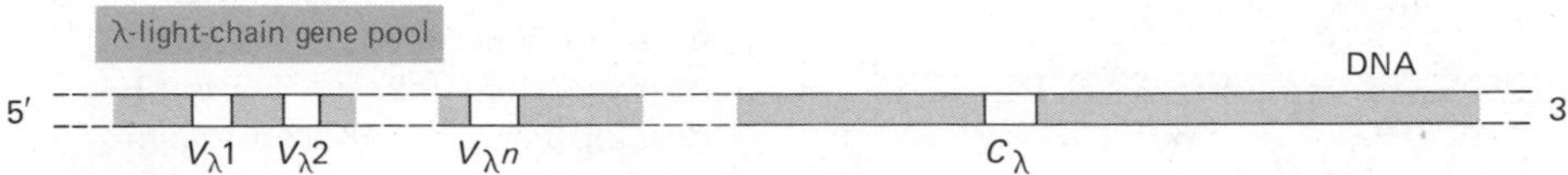

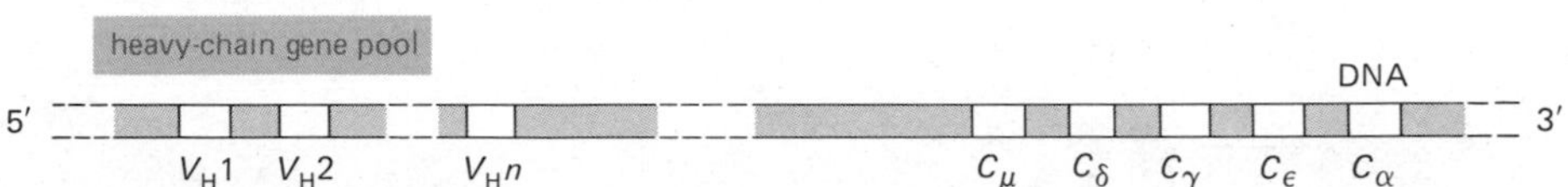

Figure 17–40 The three different immunoglobulin gene pools found in mammals. The figure is not drawn to scale, and many details are omitted.

each associated with one or two different *C* genes (C_λ genes). The heavy-chain pool is on chromosome 12, and it contains a large set of *V* genes (V_H genes) plus an ordered cluster of different *C* genes, each encoding a different class of heavy chain (C_μ, C_δ, C_λ, C_ϵ, and C_α) (Figure 17–40). For the heavy chain, the initial translocation of a V_H gene always brings it into proximity with the C_μ gene, so that an IgM molecule is always the first antibody produced by a developing B cell.

In fact, these gene pools are more complicated than indicated in Figure 17–40. Each V region of the polypeptide chain, instead of being encoded by a single *V* gene as originally believed, is encoded by two or three distinct ***V* gene segments,** which are only united into a functional *V* gene when the DNA rearrangement occurs that brings together the *V* and *C* coding regions. As we shall see, this joining of various *V* gene segments to generate a *V*-region gene makes an important contribution to the diversity of the antigen-binding sites.

Two Gene Segments Code for the V Region of Each L Chain[25]

The fact that V regions are encoded by more than one gene segment was discovered in 1978, when the first nucleotide sequence analysis of an immunoglobulin gene—a V_λ gene isolated from a mouse embryo—was completed. Instead of encoding all 110 of the amino acids of the V_λ region, the V_λ gene coded for only the amino-terminal 97 amino acids. The remaining 13 amino acids of the V_λ region proved to be encoded by a separate, short DNA segment hundreds of thousands of nucleotides downstream, which has become known as a joining, or **J_λ gene segment** (not to be confused with the protein *J chain,* which is encoded elsewhere in the genome—see p. 967). At all times, the J_λ gene segment lies adjacent to the C_λ gene, from which it is separated by an intron. During B-cell development, the V_λ gene (which we shall from now on refer to as the V_λ *segment,* since it encodes only part of the V_λ region of the polypeptide chain) is translocated so that it comes to lie precisely next to the J_λ gene segment, generating a V_λ-J_λ-intron-C_λ DNA sequence. This sequence is transcribed into RNA molecules from which the introns are later removed by RNA splicing to produce mRNA molecules in which the *V*, *J*, and *C* se-

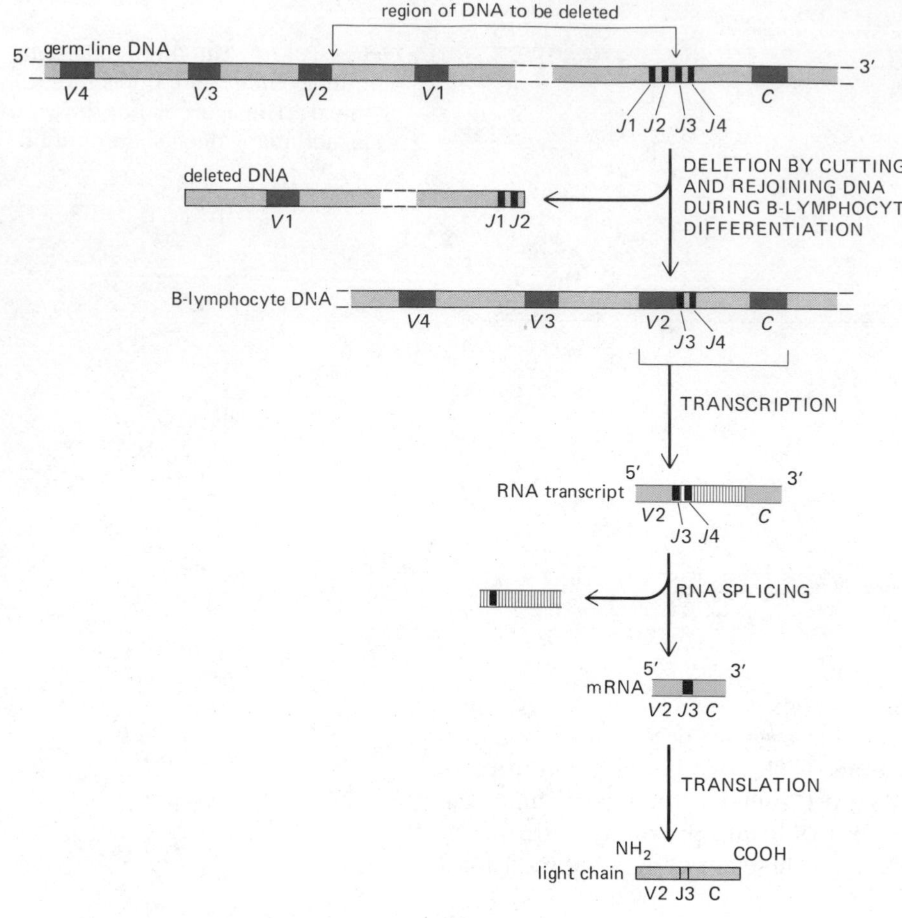

Figure 17–41 The *V-J* joining process involved in making a κ light chain in the mouse. In the "germ-line" DNA (where the immunoglobulin genes are not being expressed and are therefore not rearranged), the four *J* gene segments are separated from each other and from the *C* gene by short introns, and from the *V* gene segments by hundreds of kilobases. During B-cell development, the chosen *V* gene segment (*V2* in this case) is moved to lie precisely next to one of the *J* gene segments (*J3* in this case) by deleting the intervening DNA. The "extra" *J* gene (*J4*) and intron sequences are transcribed and then removed by RNA splicing.

quences are contiguous. It is these mRNA molecules that are translated into light-chain polypeptides.

It was subsequently established that there are several different *J* gene segments in each immunoglobulin gene pool in mice: one associated with each *C* gene in the λ-chain gene pool, and four in both the κ- and H-chain gene pools (each separated from its neighboring *J* gene segments by an intron). In the κ- and H-chain gene pools, any of the *V* gene segments can be joined to any of the *J* gene segments during B-cell development, which increases the number of V regions that these pools can make by a factor of four. Furthermore, there is some variation in the exact site of the *V-J* join, generating still more diversity in the amino acid sequence. Significantly, the *V-J* joining site encodes part of the third hypervariable region of the light chain.

After *V-J* joining, all of the "extra" *J* gene segments downstream from the joined J gene segment are transcribed and their sequences, along with the intron sequences, are then removed by RNA splicing (Figure 17–41).

There is increasing evidence that the joining of specific *V* and *J* gene segments during B-cell development involves the deletion of all of the DNA that lies between them, as shown in Figure 17–41. The actual mechanism involved in the joining of *V* and *J* gene segments, which can be hundreds of thousands of nucleotides apart, remains a mystery. However, there are specific, highly conserved DNA sequences just downstream from *V* gene segments and upstream from *J* gene segments, and it is thought that they may serve as recognition sites for site-specific DNA recombination enzymes (Figure 17–42). Such enzymes have not yet been identified.

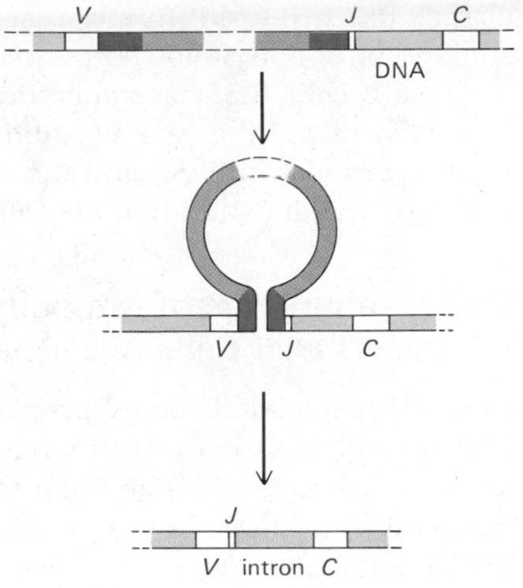

Figure 17–42 Schematic diagram of how specific DNA sequences *(color)* on the downstream side of a *V* gene segment and on the upstream side of a *J* gene segment are thought to mediate *V-J* joining. The interaction between these sequences is probably mediated by a site-specific, genetic recombination system that catalyzes the breaking and rejoining of the DNA double helices at these sequences (see p. 247).

Three Gene Segments Code for the V Region of Each H Chain[26]

The assembly of the gene encoding the variable region of an H chain (V_H) during B-cell development is even more complex than the process involved in assembling a gene for an L-chain variable region (V_L). A *J* gene segment is again involved, but in addition some of the amino acids in the third hypervariable region of the V_H region are coded for by yet another separate gene segment, called a ***D* (diversity) gene segment.** The number of different *D* gene segments is unknown, but there cannot be less than 10 in the mouse. A series of site-specific recombination events joins a *D* gene segment to any V_H and any J_H gene segment to create a functional V_H gene. The existence of separate *D* segments further increases the number of V_H regions that a mouse can make by at least a factor of 10.

Antibody Diversity Is Increased by Somatic Recombination, by the Combinatorial Joining of Light and Heavy Chains, and by Somatic Mutation[23,27]

The immune system has evolved a variety of different mechanisms for diversifying the antigen-binding sites of antibodies, only some of which depend on the above-described somatic rearrangement of DNA during B-cell development. Gene-counting experiments, using the technique of DNA hybridization (see p. 189), suggest that a mouse inherits several hundred V_κ gene segments, a similar number of V_H gene segments, and only two V_λ gene segments. A reasonable estimate would be that, by variously combining the different *V*, *D*, and *J* gene segments that it inherits, a mouse can make at least 10,000 different V_H regions and 1000 different V_L regions.

One simple but important mechanism that greatly increases antibody diversity is the combining of different L and H chains that occurs when an immunoglobulin molecule is assembled. Since the variable regions of both the L and H chains contribute to an antibody's antigen-binding site, an animal with 1000 genes encoding V_L regions and 10,000 genes encoding V_H regions could combine their products in $1000 \times 10{,}000$ different ways to make 10^7 different antigen-binding sites (assuming that any L chain can combine with any H chain to make an antigen-binding site).

Somatic mutations have recently been shown to occur in and around *V*-region genes and probably increase the number of different antibodies by a factor of at least 10 to 100. The mechanism by which mutations are induced specifically in and around *V*-region genes is unknown. However, mutations have been found to occur much more frequently in IgG and IgA antibodies than in IgM antibodies made from the same V_H gene. This is perhaps not surprising, as IgM is produced early in immune responses, while IgG and IgA appear relatively late. Therefore, B cells that have switched from IgM to IgG or IgA (see p. 986) will generally have undergone more cell divisions and might be expected to have accumulated more mutations than B cells that are still making IgM. But there may also be a mechanism for increasing the rate of mutation of the *V* genes after the switch from IgM to other immunoglobulin classes has occurred. In any case, the alteration of antigen-binding sites by somatic mutation, followed by the preferential antigen-induced proliferation of those B cells that have altered sites with high affinity, may account for at least part of the increase in antibody affinity that is observed to occur after immunization (affinity maturation). Thus, in addition to increasing antibody diversity, somatic mutation may also serve to fine-tune the antibody response.

The Mechanisms of Antibody Gene Expression Ensure That B Cells Are Monospecific[28]

The clonal selection theory predicted, and various experiments have shown, that individual B cells are monospecific—that is, they make antibody with only a single type of antigen-binding site. This means that there must be some mechanism for limiting the possibilities open to individual B cells when their immunoglobulin genes are activated during development, so that they make only one type of L chain and one type of H chain. For example, a B cell cannot make κ and λ light chains and maintain monospecificity, since the two chains would almost always have two different variable regions and therefore would form antibody molecules with more than one kind of antigen-binding site. In fact, one or the other of these gene pools is activated in each B cell, but never both.

Moreover, since B cells, like other somatic cells, are diploid, each cell has six gene pools to encode antibodies: an H-chain pool, a λ-chain pool, and a κ-chain pool from each parent. The monospecificity of B cells means that each cell must activate genes in only two of these six pools: one of the four light-chain gene pools and one of the two heavy-chain gene pools (Figure 17–43). Thus, as well as choosing between κ and λ, a B cell must choose between maternal and paternal antibody gene pools. The expression of only the maternal or the paternal allele of an immunoglobulin gene in any given B cell is called **allelic exclusion.** For other proteins encoded by autosomal genes, both maternal and paternal genes in a cell appear to be expressed about equally. The only other known exception in vertebrates results from the inactivation of one of the two X chromosomes in females.

Why is it important that B cells be monospecific? Monospecificity ensures that each antibody molecule is composed of two identical halves and, therefore, that it contains two identical antigen-binding sites. It seems likely that this property was selected for during evolution because it endows antibodies with the ability to form large lattices of cross-linked antigens (see Figures 17–30 and 17–31).

The mechanisms involved in allelic exclusion and in the choice of one L-chain type during B-cell development are unknown. One obvious possibility is that antibody gene segments simply are not rearranged in those gene pools that are not expressed. However, there is now good evidence that this is not the case; often rearrangements do occur on the unexpressed chromosomes, but they are abnormal and, consequently, unable to produce a functional immunoglobulin chain. This observation raises the possibility that the joining of V-region gene segments during B-cell development is a somewhat haphazard process that generates unproductive rearrangements more often than productive ones. In that case, allelic exclusion could occur simply because the probability of a successful rearrangement occurring in more than one gene pool for each chain is very low. This possibility implies that in many cells there would be no successful rearrangement at all and therefore no antibody synthesis. Because such cells could not be stimulated by any antigen, they would soon die. Such a mechanism seems very wasteful, but it may be the price paid for monospecificity.

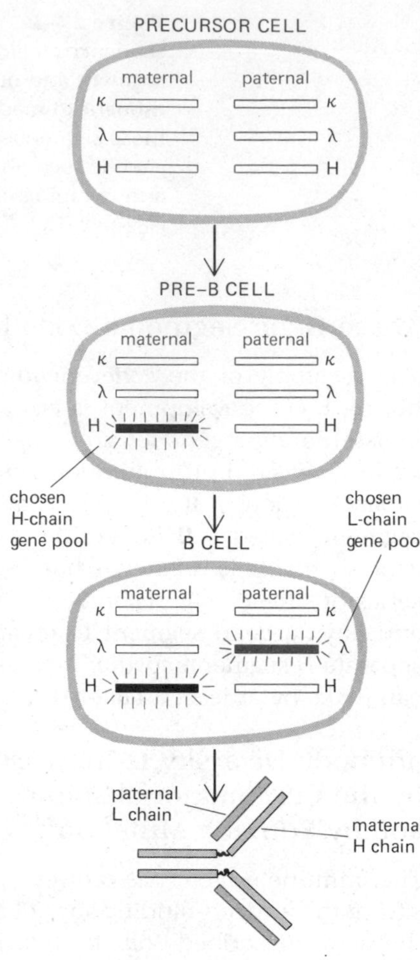

Figure 17–43 The sequential choices in immunoglobulin gene activation that developing B cells must make in order to produce antibodies with only one type of antigen-binding site. Each cell must choose one of four L-chain gene pools and one of two H-chain gene pools. During development, a precursor cell first activates one heavy-chain gene pool to become a *pre-B cell,* making free μ heavy chains. After a period of extensive proliferation, a pre-B cell activates one κ- or λ-light-chain pool to become a B cell that makes a unique IgM molecule.

The Switch from a Membrane-bound to a Secreted Form of the Same Antibody Occurs Through a Change in the H-Chain RNA Transcripts[29]

We now turn from the genetic mechanisms that determine the antigen-binding site of an antibody to those that determine its biological properties—the genetic mechanisms that determine what form of heavy-chain constant region is synthesized. While the choice of particular DNA segments to encode the antigen-binding site is a commitment for the life of a B cell and its progeny, the type of C_H region made can change during B-cell development.

We have already mentioned that all classes of antibody can be made in a membrane-bound form as well as in a soluble, secreted form. The membrane-bound antibodies serve as antigen receptors on the B-cell surface, while after stimulation by antigen the same antibodies are produced in secreted form. In the case of IgM, the sole difference between the two forms resides in the carboxyl terminus of the μ chain: whereas membrane-bound μ chains have a hydrophobic carboxyl terminus that anchors them in the lipid bilayer of the B-cell plasma membrane, the μ chains of secreted IgM molecules have instead a hydrophilic tail that allows them to escape from the cell. Since B cells contain one copy of the C_μ gene per haploid genome, their ability to make μ chains with two different types of constant regions at first seemed paradoxical.

The paradox was resolved with the discovery that the activation of B cells by antigen induces a change in the μ-chain RNA transcripts in the nucleus. These new transcripts are somewhat shorter than those encoding the membrane-bound μ chain. We have already discussed in Chapter 8 (see p. 419) how the RNA sequence encoding the hydrophilic tail of the secreted IgM molecule is removed from the longer transcript that produces the membrane-bound molecule. The process is illustrated in Figure 17–44. It is likely that the switch from a membrane-bound to a secreted form of the other classes of antibodies involves a similar mechanism.

Figure 17–44 A B cell switches from making a plasma-membrane-bound form to a secreted form of the same antibody molecule by altering the H-chain RNA transcripts it produces when it is activated by antigen. It is not known whether this results from a change in transcription or from a change in the manner in which poly A is added to the 3′ end of the primary transcript. The two forms of H chain differ only in their carboxyl terminals: the membrane-bound form has a hydrophobic tail that holds it in the membrane, while the secreted form has a hydrophilic carboxyl tail that enables it to escape from the cell. While the long RNA transcript specifying the membrane-bound form of H chain has a donor and acceptor site permitting the RNA sequence encoding the hydrophilic tail of the secreted form to be removed by RNA splicing, the short RNA transcript specifying the secreted form only has a donor site, and therefore RNA splicing cannot occur.

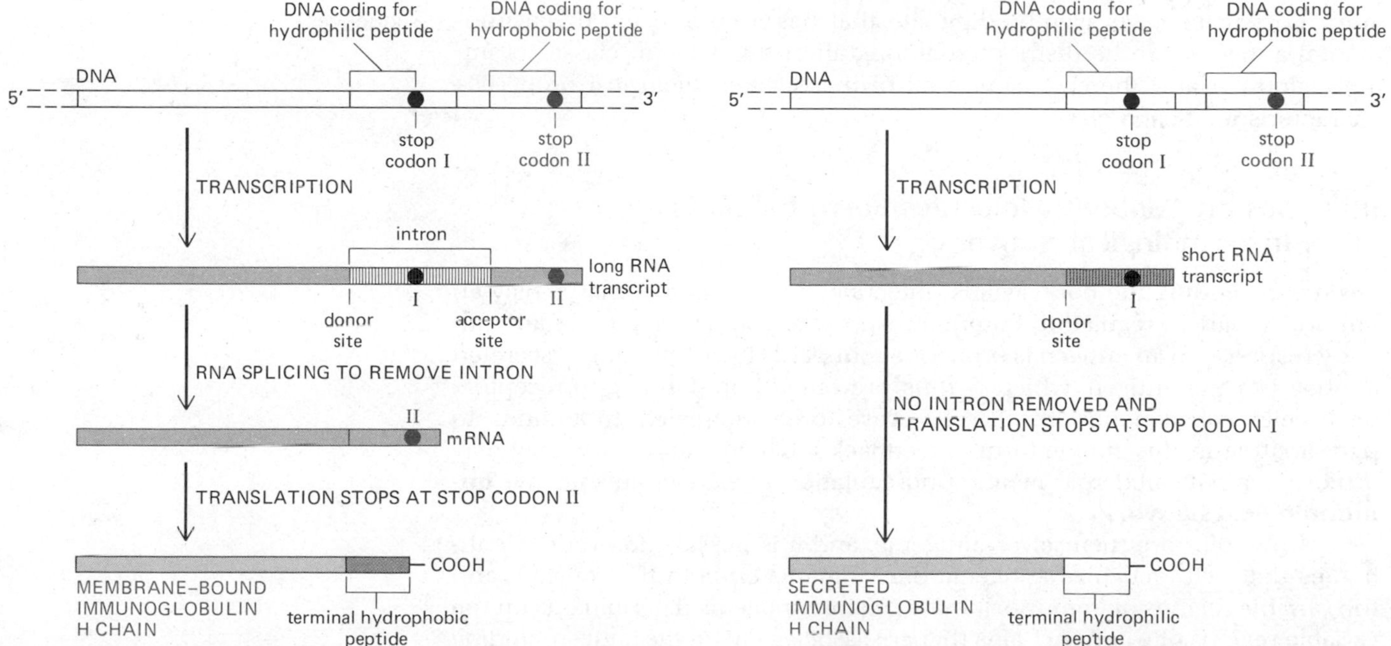

B Cells Can Switch the Class of Antibody They Make[30,31]

The switch from a membrane-bound form to a secreted form of antibody is not the only type of change that can occur in the C region of the heavy chain during B-cell development. All B cells begin their antibody-synthesizing lives by making IgM, but many eventually switch to making other classes of antibody, such as IgG or IgA—a process called **class switching.** The fact that B cells can switch the class of antibody they make without a change in the antigen-binding site implies that the same assembled V_H gene can sequentially associate with different C_H genes.

Class switching seems to occur in two sequential steps. First, a cell that is making membrane-bound IgM can switch to the simultaneous production of membrane-bound IgM and the membrane-bound form of another antibody class, such as IgD. Such a B cell is thought to produce large primary RNA transcripts that contain the assembled V_H-region sequence and both the C_μ and C_δ sequences. These RNA transcripts are then spliced in two different ways to produce two different species of mRNA molecules bearing the same V_H sequence: one with a C_μ sequence and one with a C_δ sequence (Figure 17–45). The same mechanism is thought to operate when a B cell switches to the simultaneous production of membrane-bound forms of IgM and one of the other classes of antibody, such as IgG, IgE, or IgA.

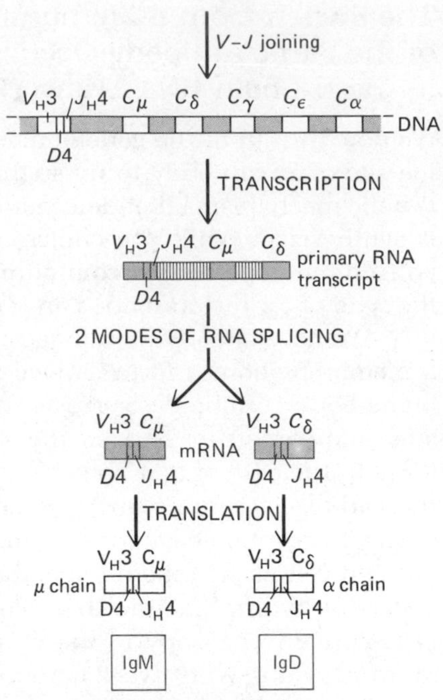

Figure 17–45 B cells that simultaneously make plasma-membrane-bound IgM and IgD molecules having the same antigen-binding sites are thought to produce long RNA transcripts containing both C_μ and C_δ sequences. These transcripts are spliced in two different ways to produce mRNA molecules that have the same V_H sequence joined to either a C_μ or a C_δ sequence. It is possible that the RNA transcripts produced by such cells are even longer than shown and contain all of the different C_H sequences.

The second step in class switching occurs when a B cell that is simultaneously making membrane-bound IgM plus a second class of membrane-bound antibody is stimulated by antigen to start secreting the second class of antibody. This step involves DNA deletion. For example, a cell making both membrane-bound IgM and IgA from a long RNA transcript containing all of the C_H gene sequences as well as an assembled V_H gene sequence with the structure V_H2-$D1$-J_H3 can begin secreting IgA by deleting most of the DNA between J_H3 and C_α, including the C_μ, C_δ, C_γ and C_ϵ genes (Figure 17–46). Evidence that this step in class switching involves DNA deletion comes from experiments on myeloma cells: myeloma cells secreting IgG lack the DNA coding for C_μ and C_δ, and those secreting IgA lack the DNA coding for all of the other classes of heavy-chain constant regions.

The ability of any assembled V_H gene to associate with any of the C_H genes has important functional implications: it means that in an individual animal a particular antigen-binding site that has been selected by environmental antigens can be distributed among all of the different classes of immunoglobulin and thereby acquire all of the different biological properties characteristic of each class.

Idiotypes on Antibody Molecules Form the Basis of an Immunological Network[32]

Besides defending the body against infection, antibodies themselves play an important part in regulating immune responses. The termination of an antibody response to an antigen is brought about partly by the binding of secreted antibody to the antigen, which is thus prevented from binding to receptors on B cells; consequently, the B cells cease to be stimulated. In addition to participating in this simple form of feedback inhibition, antibodies may play a more sophisticated role in immunoregulation as part of an intricate **immunological network.**

Antibodies are themselves antigenic, and it is possible to produce antibodies that will recognize antigenic determinants on both the constant and the variable regions of immunoglobulin chains. Antigenic determinants on the variable regions of L and H chains that are associated with the antigen-binding site of an antibody are called **idiotypes** (Figure 17–47). Each specific antigen-

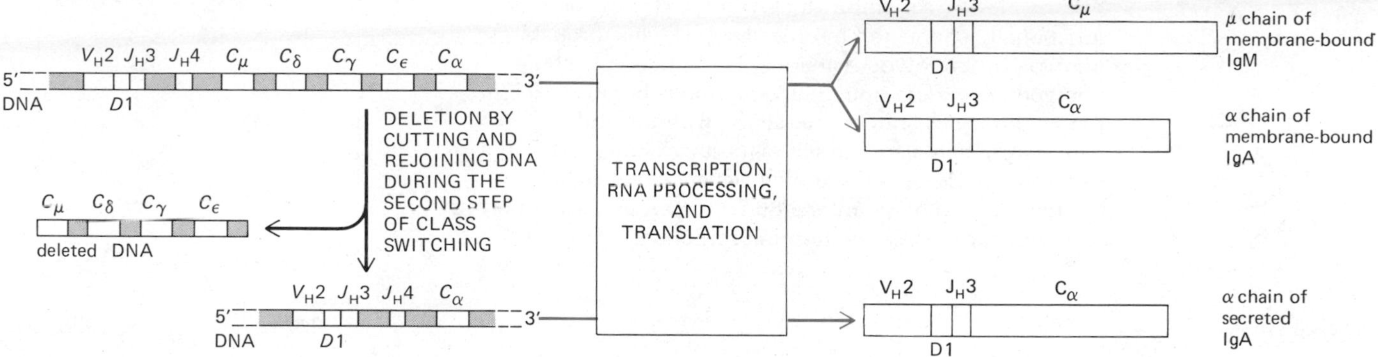

Figure 17–46 The DNA rearrangement involved in the second step of class switching. It is thought that a B cell making both membrane-bound IgM and membrane-bound IgA antibody from a V_H gene with the structure V_H2-$D1$-J_H3 produces long RNA transcripts containing all of the C_H gene sequences; these transcripts are spliced in two different ways to produce mRNA molecules with the same V_H sequence joined to either a C_μ or a C_α sequence as illustrated in Figure 17–45. When such a cell is stimulated by antigen, it begins to secrete IgA antibody by deleting the DNA between J_H3 and C_α. Although not shown, the heavy-chain constant regions (C_α) of the secreted and membrane-bound IgA antibodies are slightly different, as illustrated for IgM in Figure 17–44.

binding site has its own characteristic set of idiotypes; thus an animal with millions of different antigen-binding sites will contain millions of different idiotypes. Since any individual idiotype is present in only minute amounts in the body, an animal is not tolerant to its own idiotypes and will make both T- and B-cell responses against them if immunized appropriately with any one of its own antibodies.

One might expect that an animal immunized with antigen A would first produce large amounts of anti-A antibodies and then produce antibodies against the idiotypes of these anti-A antibodies—and in turn antibodies against the anti-idiotype antibodies, and so on. In fact, this type of *network* reaction has been demonstrated in circumstances in which most of the antibodies produced in the initial response to antigen express the *same* idiotype. In such *restricted* responses, both antibodies and T cells that specifically recognize the dominant idiotype are activated, and these can either inhibit or enhance the response of lymphocytes whose receptors express the idiotype. While such anti-idiotype interactions can be shown to be important in regulating restricted responses, it is still uncertain whether they regulate the more common responses in which antibodies expressing many different idiotypes are produced.

It is intriguing to consider the implications of an animal being able to make antibodies against any of its own idiotypes. Since presumably there are at least as many different idiotypes as different antigen-binding sites, it follows that an average antigen-binding site must recognize at least one idiotype in its own immune system. Thus, all of the antigen-binding sites of an immune system are potentially linked together in a complex network of idiotype-anti-idiotype interactions (Figure 17–48). Since T and B cells appear to share at least some idiotypes (see p. 993), both classes of lymphocyte probably participate in such a network. An immune response might therefore be viewed as a reverberating perturbation of an immunological network rather than as a response of independent, antigen-reactive lymphocytes.

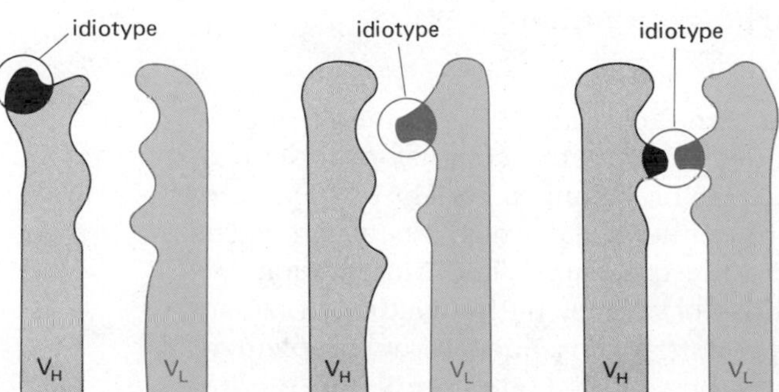

Figure 17–47 An idiotype is an antigenic determinant associated with the antigen-binding site of an antibody molecule. It can be formed by V_H, V_L, or both. Each different antigen-binding site has its own unique set of idiotypes.

Figure 17–48 Any individual lymphocyte may be functionally connected to other lymphocytes through idiotype–anti-idiotype reactions. The extent of such an idiotype network is potentially enormous because each of the lymphocytes shown interacting with the anti-A lymphocyte can also interact with other lymphocytes in a similar way. There is increasing evidence that such idiotype–anti-idiotype interactions play an important part in regulating at least some immune responses.

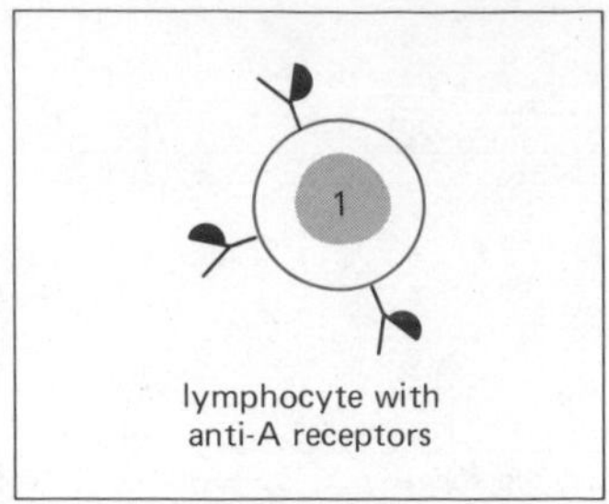

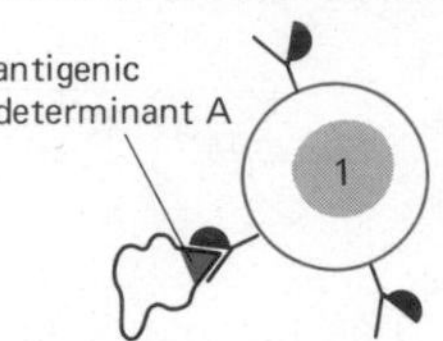

lymphocyte 1 can bind antigenic determinant A

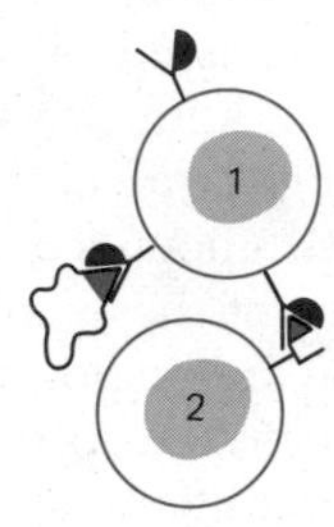

lymphocyte 1 can also interact with receptors on lymphocyte 2 that express an idiotype that resembles antigenic determinant A

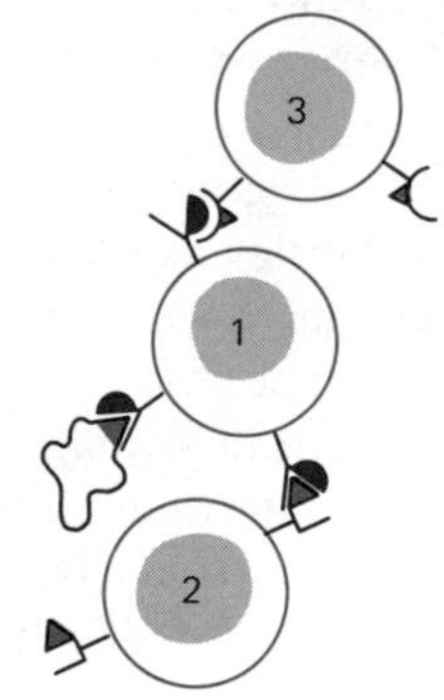

lymphocyte 1 can also interact with lymphocyte 3, which has a receptor that recognizes an idiotype on the anti-A receptor of lymphocyte 1

Summary

Antibodies are produced from three separate gene pools encoding the κ, λ, and H chains, respectively. In each pool, separate gene segments that code for different parts of the variable regions of L and H chains can be brought together by site-specific recombination events during B-cell differentiation. The L-chain gene pools contain one or more constant (C) genes and sets of variable (V) and joining (J) gene segments. The H-chain gene pool contains a set of C genes and sets of V, diversity (D), and J gene segments. To make an antibody molecule, a V gene segment is recombined with a J gene segment, to produce a V gene for the light chain, and a V_H gene segment is recombined with a D and J_H gene segment to produce a V gene for the heavy chain. Each of the assembled gene segments is then co-transcribed with the appropriate C-region sequence to produce an mRNA molecule that codes for the complete polypeptide chain. By variously combining inherited gene segments coding for V_L and V_H regions, vertebrates can make thousands of different L chains and thousands of different H chains that can associate to form millions of different antibody molecules. This number is probably further increased by at least ten to a hundredfold by somatic mutations that occur in the gene segments coding for V regions.

All B cells initially make IgM antibodies. Some later switch to make antibodies of other classes that have the same antigen-binding site as the original IgM antibodies. Such class switching allows the same antigen-binding sites to be distributed among antibodies with many different biological properties.

The Complement System[33]

Complement *complements* the action of antibody in killing cells. In fact, it is the principal means by which antibodies defend vertebrates against most bacterial infections, and individuals with a deficiency in one of the central complement components (C3) are subject to repeated infections. Besides its role in antibody-mediated cell lysis, complement attracts phagocytic cells to sites of infection and enhances the ability of these cells to ingest and destroy microorganisms.

Complement Activation Involves a Sequential Proteolytic Cascade

Complement is not one protein but a complex system of proteins composed of about 20 interacting components, designated C1 (a complex of three proteins), C2, C3, ... up to C9, factor B, factor D, and a variety of regulatory proteins. They are all soluble proteins of molecular weights between 24,000 and 400,000 that circulate in the blood and extracellular fluid. Most are inactive unless triggered by an immune response or directly by an invading organism or by some other means. One of the eventual consequences of complement activation is the sequential assembly of the so-called **late complement com-**

ponents (C5, C6, C7, C8, and C9) into a large protein complex that mediates cell lysis (the *lytic complex*).

The aggregation of late components is triggered by a sequence of proteolytic activation reactions involving the **early complement components** (C1, C2, C3, C4, factor B, and factor D). Most of these early components are proenzymes that are activated sequentially by proteolytic cleavage: as each proenzyme in the sequence is cleaved, it is activated to generate a proteolytic enzyme that cleaves the next proenzyme in the sequence, and so on. Because many of the activated components bind tightly to membranes, most of these events take place on cell surfaces.

The pivotal component of this *proteolytic cascade* is C3, and its activation by cleavage is the central reaction in the entire complement-activation sequence. C3 can be activated by two different pathways, the *classical pathway* and the *alternative pathway;* in both cases, C3 is cleaved by an enzyme complex called a **C3 convertase.** A different C3 convertase is produced by each pathway, but in both cases it is formed by the spontaneous assembly of two of the complement components activated earlier in the cascade. C3 convertase cleaves C3 into two fragments. The larger of these (C3b) binds to the target-cell membrane next to the C3 convertase to form an even larger enzyme complex with an altered specificity—**C5 convertase.** The C5 convertase then cleaves C5 to initiate the spontaneous assembly of the late components—C5 through C9—that creates the lytic complex (Figure 17–49).

Since each activated enzyme cleaves many molecules of the next proenzyme in the chain, the activation of the early components consists of an amplifying cascade: each molecule activated at the beginning of the sequence leads to the production of many lytic complexes.

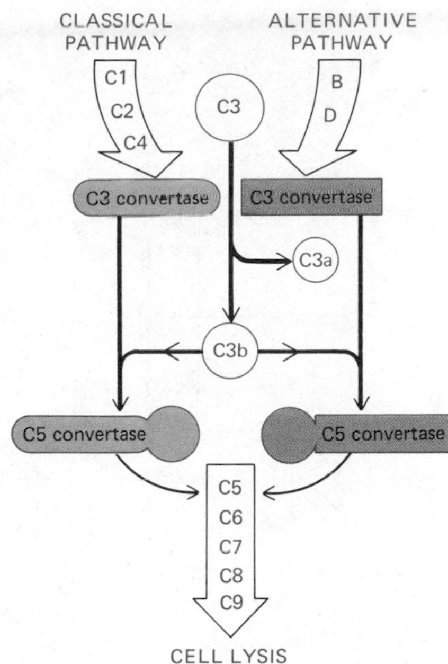

Figure 17–49 A simplified diagram showing the generation of C3 and C5 convertases by the classical and alternative pathways of complement activation.

The Classical Pathway Is Activated by Antibody-Antigen Complexes

The **classical pathway** involves components C1, C2, and C4, and it is usually activated when IgG or IgM antibodies bind to antigens on the surface of microorganisms. The first component in the classical pathway is C1, which consists of three subcomponents—C1q, C1r, and C1s. C1q is a large and unique protein, with a shape that resembles a bunch of six tulips, each tulip having a globular protein head and a collagenlike tail (Figure 17–50). Each globular head will bind to a single constant region of an IgG or an IgM antibody (on the γ or μ chain, respectively), providing that the other end of the antibody has already bound to antigen. This binding to antibody activates C1q to start the early proteolytic cascade of the classical pathway. However, more than one head must be bound in this way before activation occurs; consequently, a *cluster* of foreign antigenic determinants is required to trigger the classical pathway. Such clusters occur frequently on the surface of microorganisms.

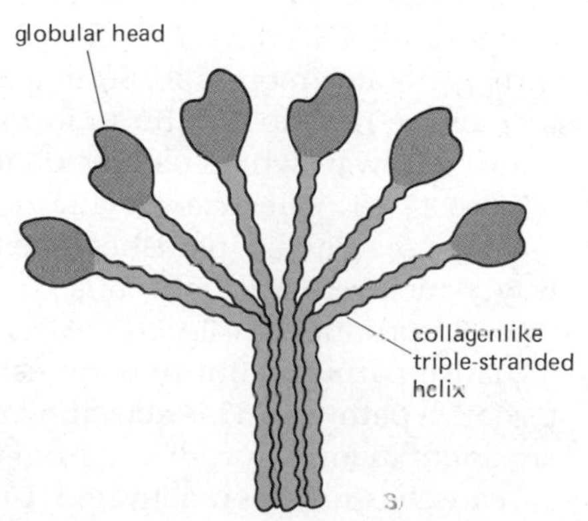

Figure 17–50 Schematic drawing of the unusual structure of C1q. It is a large protein (~400,000 daltons) made up of six identical subunits, each composed of three different polypeptide chains. The carboxyl-terminal halves of each of the three polypeptide chains in a subunit are folded into a globular structure, while the amino-terminal halves have a typical collagen amino acid sequence and wind together to form a collagenlike triple-stranded helix (see p. 691). The six subunits are covalently linked together by disulfide bonds between their triple-helical stems to form a structure resembling a bunch of tulips. It is the heads of the tulips that bind to IgG or IgM antibody; thus, each C1q molecule has six antibody-binding sites.

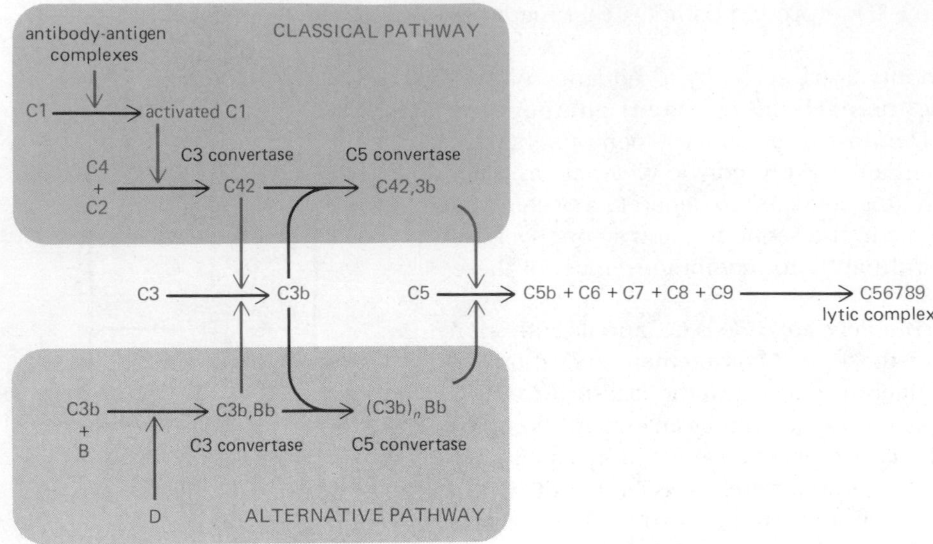

Figure 17–51 A schematic drawing comparing the generation of C3 convertase and C5 convertase by the classical and alternative pathways. While the classical pathway is triggered by antibody-antigen complexes, the alternative pathway is triggered both by the presence of C3b and by cell-wall polysaccharides and other activators. Activated complement components are often designated with a superscript bar—for example, activated C1 is $\bar{C1}$. To simplify the text we have omitted these bars.

The activation of the C1q subcomponent of the C1 complex activates C1r to become proteolytic, which in turn cleaves and thereby activates C1s. Activated C1s then sequentially cleaves C4 and C2; activated C4 immediately binds to a nearby membrane and then binds activated C2. This forms the complex C42, which is the C3 convertase produced by the classical pathway. C42 cleaves C3 to produce two fragments, C3a and C3b; the latter rapidly binds to the target membrane next to C42 to form C42,3b, the C5 convertase for the classical pathway. This C5 convertase cleaves C5 into C5a and C5b, and C5b combines with C6 to initiate the assembly of the late components to form the lytic complex (Figure 17–51).

The Alternative Pathway Can Be Directly Activated by Microorganisms

When the classical pathway is activated, the **alternative pathway** is also called into action to serve as a positive feedback loop that amplifies the initial production of C3b. However, it can also be activated in the absence of antibody by polysaccharides present in the cell envelopes of bacteria, yeast, and protozoa and is therefore thought to provide a first line of defense against infection before an immune response can be mounted.

The main components of the alternative pathway are C3b, factor B, and factor D. The first step in activating the pathway is the binding of factor B to membrane-bound C3b. Factor D, which circulates in the blood in an active form, cleaves the bound factor B to generate the active fragment, Bb and thereby the C3 convertase for the alternative pathway—C3b,Bb—which, in turn, generates more C3b. Several additional C3b molecules bind to the target membrane next to C3b,Bb to form $(C3b)_n$Bb, the C5 convertase for the alternative pathway, which cleaves C5 to initiate the assembly of the lytic complex (Figure 17–51). Because the alternative pathway requires Mg^{2+} but not Ca^{2+}, it can be readily distinguished in experimental studies from the classical pathway, which requires both ions.

By producing C3b, the classical pathway automatically activates the alternative pathway. But how do cell-envelope polysaccharides activate the alternative pathway? This activation depends on the fact that C3b-like molecules are spontaneously produced from C3 at low rates even when the complement cascade has not been activated. Unlike the large clusters of C3b produced by

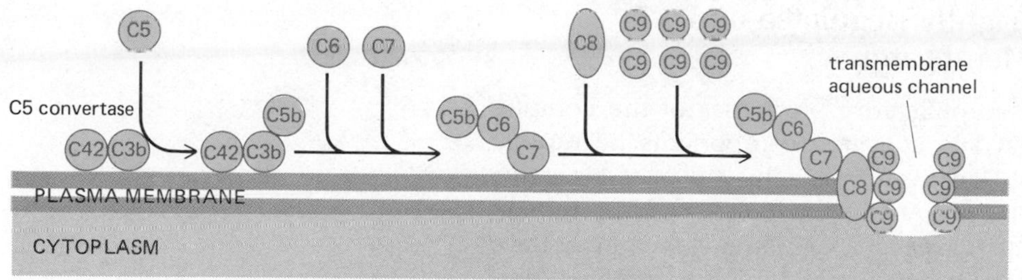

Figure 17–52 Schematic drawing showing the assembly of the late complement components to form a transmembrane channel. The actual channel is formed by 12 molecules of C9 in the lytic complex, which consists of a dimer of two C56789 complexes.

the classical pathway, most of the spontaneously formed C3b-like molecules bind as single molecules to randomly selected membranes and are rapidly destroyed by specific inhibitor proteins (see p. 992). However, the polysaccharides in the cell envelopes of certain types of microorganisms protect these membrane-bound C3b-like molecules from degradation. As a result, some of these molecules survive to activate the alternative pathway.

C3b, produced by either the classical or the alternative pathways, has a number of important properties. As we have seen, it activates the alternative pathway to produce more C3b, and it combines with C3 convertase to form C5 convertase. In addition, it binds to specific receptor proteins on macrophages and polymorphonuclear leucocytes, thereby enhancing the ability of these cells to phagocytose a cell to which C3b has attached. In this way, C3b plays a major role in the defense against microorganisms even in the absence of the lytic complex.

In the course of the complement cascade, several small, biologically active protein fragments are generated by the proteolysis of the various components. These include C3a, C4a, and C5a, all of which stimulate mast cells and basophilic leucocytes to secrete histamine. C5a is also a chemotactic attractant for polymorphonuclear leucocytes. These various protein fragments are the cause of the local inflammatory response usually associated with the activation of complement.

The Assembly of the Late Complement Components Generates a Transmembrane Lytic Complex[33,34]

The assembly of the late components begins with the splitting of C5 (to give C5a and C5b) by the C5 convertase that is produced by either the classical or alternative pathway. C5b remains bound to the C5 convertase and has the transient capacity to bind C6 to form C56 and then C7 to form C567. The C567 complex then binds firmly to a membrane close to the site where the complement activation was initiated. This complex adds one molecule of C8 and six molecules of C9 to form the complex C56789, two of which combine to form the large **lytic complex** having a molecular weight of about 2 million (Figure 17–52). This complex has a characteristic doughnutlike morphology in negatively stained electron micrographs, reflecting the annular arrangement of the complement components in the complex (Figure 17–53). The lytic complexes make the membrane leaky—both because they destabilize the lipid bilayer and because they can form aqueous channels through it. Because small molecules leak into and out of the cell around and through the lytic complexes, while macromolecules remain inside, water is drawn into the cell by osmosis, causing it to swell and burst. The process is so efficient that a very small number of C56789 complexes (perhaps even one) can kill a cell. Even cells that do not have large osmotic pressure gradients across their plasma membranes and are therefore not susceptible to such osmotic lysis are still killed by the lytic complex, presumably because it disorganizes their plasma membranes.

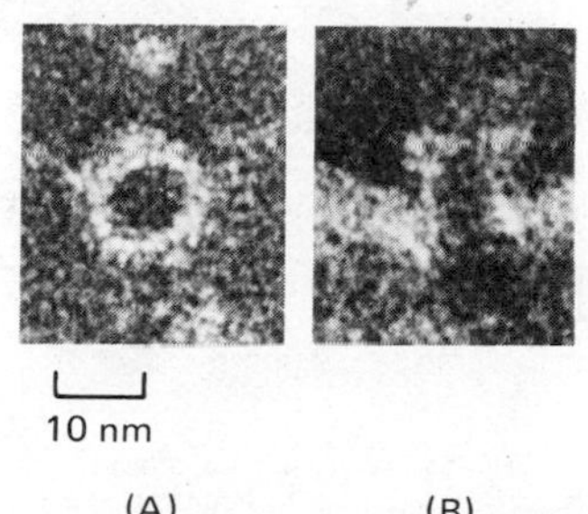

Figure 17–53 Negatively stained electron micrographs of complement lesions in the plasma membrane of a red blood cell. The lesion in (A) is seen *en face*, while that in (B) is seen from the side as an apparent transmembrane channel. The negative stain fills the individual channels, which therefore look black. (From R. Dourmashkin, *Immunology* 35:205–212, 1978.)

The Complement Cascade Is Tightly Regulated and Designed to Attack a Nearby Membrane

The inflammatory, destructive, and amplificatory properties of the complement cascade make it essential that key activated components be rapidly inactivated after they are generated. Deactivation is achieved in at least two ways. First, specific inhibitor proteins present in the blood act to terminate the cascade by either binding or cleaving certain components once they have been activated by proteolytic cleavage. For example, inhibitor proteins bind to the activated components of the C1 complex and block their further action, while other inhibitor proteins in the blood cleave C3b and thereby inactivate it. Without these inhibitors, all of the serum C3 might be depleted by the positive feedback loop created by the alternative pathway.

A second, important mechanism of regulation is based on the instability of some of the activated components in the cascade; unless they bind immediately to an appropriate component in the chain, or to a nearby membrane, they rapidly become inactivated. Especially dramatic is the case of activated C4 and C3b. When either of these components is formed by cleavage, it undergoes a series of rapid conformational changes that creates a short-lived active form. This active form has a hydrophobic site as well as a highly reactive glutamic acid side chain that is produced by the mechanical breakage of an unusual thioester bond in the protein (Figure 17–54). As a result, this glutamic acid forms a covalent bond with a protein or polysaccharide on a nearby membrane. Because of the very short half-life of their active forms (less than 0.1 millisecond), both C4 and C3b normally become bound only to membrane sites very close to the complement components that activate them. The complement attack is thereby confined to the surface membrane of a microorganism and prevented from spreading to the normal host cells in the vicinity.

How did such a complex system of complement components ever evolve? Clearly, it must have developed in gradual steps, with many of the more elaborate components, such as the large lytic complex (components C5 through

Figure 17–54 The proteolytic activation of either C3 or C4 induces a conformational change in the protein, breaking the unusual intramolecular covalent bond shown. The breaking of this thioester bond between protein side chains generates a very reactive carbonyl group that couples covalently to another macromolecule, forming an ester or an amide linkage. However, the ability of the protein to react in this way decays with a half-life of 60 μsec or so, confining the reaction to membranes that are very near the site of the proteolytic activation.

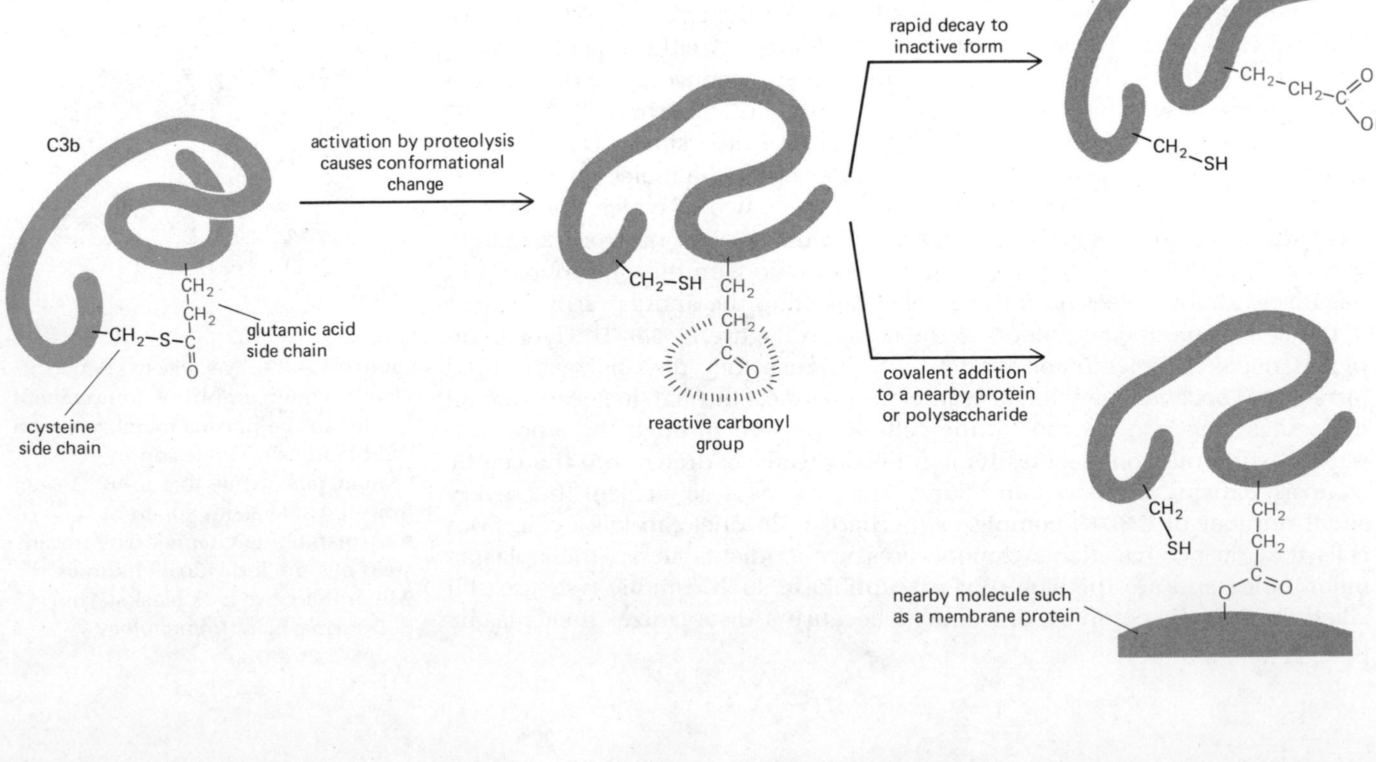

C9), being added rather late in the evolution of the system. It seems likely that the system originally evolved around C3, primarily to create the covalent complex between C3b and foreign cell membranes. This complex alone markedly enhances the ability of professional phagocytes to ingest and destroy microorganisms. Indeed, humans who lack one of the late components, and therefore are unable to assemble the lytic complex, are still protected against infection by all but a few types of bacteria.

Summary

The complement system acts on its own and in cooperation with antibodies in defending vertebrates against infection. It is composed mainly of inactive blood proteins that are sequentially activated in an amplifying series of reactions either by the classical pathway, which is triggered by IgG or IgM antibodies binding to antigen, or by the alternative pathway, which can be triggered directly by the cell envelopes of invading microorganisms. The most important complement component is the C3 protein, which can be activated by proteolytic cleavage and then binds covalently to nearby membranes; microorganisms with activated C3 (C3b) on their surface are readily ingested and destroyed by professional phagocytes. In addition, C3b initiates the assembly of the late complement components, which form a large lytic complex in the membrane that can kill cells. Complement activation also releases a variety of small soluble peptide fragments that attract polymorphonuclear leucocytes and stimulate mast cells to secrete histamine: this results in an inflammatory response at sites of complement activation. The complement proteolytic cascade is focused on the membranes of target cells by the fact that several of its components, including C3b, remain activated for less than one-tenth of a millisecond.

T Lymphocytes and Cell-mediated Immunity

The diverse responses of T cells are collectively called *cell-mediated immune reactions.* Like antibody responses, they are important in defending vertebrates against infection, particularly by certain viruses and fungi. Also like antibody responses, they are exquisitely antigen-specific. But they do not involve the secretion of antibody. In fact, much less is known about T cells and their responses than about B cells, mainly because their receptors and products are still poorly characterized in comparison with antibodies.

The T-Cell-Receptor Enigma[35]

While it is known that T cells have antigen-specific receptors on their surfaces, the biochemical nature of these receptors, and in particular whether they are related to antibodies, has been the subject of a prolonged controversy. On the one hand, there is indirect evidence that at least some T cells use antibodylike receptors to recognize antigen. On the other hand, there is more direct molecular genetic evidence that they do not.

The indirect evidence has come from studies using anti-idiotype antibodies. As we have already discussed, it is possible to produce antibodies that recognize antigenic determinants associated with the antigen-binding site of an antibody; such antigenic determinants are called *idiotypes.* Anti-idiotype antibodies that react with the antigen-binding site of a soluble antibody made against an arbitrary antigen X will bind not only to anti-X antibodies in solution, but also to B cells that have the same antibodies on their surface (as

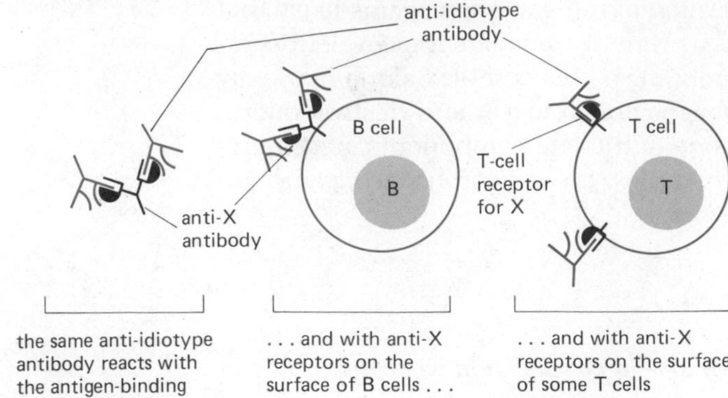

Figure 17–55 Schematic diagram of an experiment suggesting that B cells and T cells that react with the same antigen may sometimes express a similar idiotype (an antigenic determinant in the antigen-binding site of an antibody) on their cell-surface receptors.

receptors for antigen X). Not surprisingly, the binding of the anti-idiotype antibodies to these receptors on the B-cell surface can inhibit the B cell's ability to recognize and respond to antigen X. In some cases, the same anti-idiotype antibodies have been shown to bind to T cells and inhibit their ability to respond to antigen X (Figure 17–55). Genetic studies suggest that the idiotypes shared by B-cell and T-cell receptors may be encoded by gene segments that specify the variable regions of immunoglobulin H chains. Anti-idiotype antibodies have been used to isolate small amounts of receptor from T-cell plasma membranes. Although these receptors consist of polypeptides similar in size to conventional H chains, they do not react with antibodies made against the constant regions of any of the known immunoglobulin H or L chains. These findings suggest that T-cell receptors may be composed of a new class of immunoglobulin H chains encoded by a special set of constant-region genes and perhaps by some of the same gene segments that code for the V_H regions of conventional antibodies. The difficulty with this hypothesis is that experiments using recombinant DNA technology have failed to demonstrate the expected rearrangements in V_H, D, or J_H gene segments in functional T cells. This suggests that T cells may have their own set of genes (or gene segments) encoding their antigen binding sites. At the moment, the T-cell-receptor problem remains unsolved.

Different T-Cell Responses Are Mediated by Different T-Cell Subpopulations[36]

When T cells are stimulated by antigen, they divide and differentiate into activated effector cells that are responsible for various cell-mediated immune reactions. At least three different reactions are carried out by T cells: (1) they specifically kill foreign or virus-infected vertebrate cells; (2) they help specific T or B lymphocytes respond to antigen and can activate some nonlymphocyte cells, such as macrophages; and (3) they suppress the responses of specific T or B lymphocytes. These different functions are carried out by different subpopulations of T cells—called *cytotoxic T cells, helper* (or *inducer*) *T cells,* and *suppressor T cells,* respectively—that can be distinguished from each other by the cell-surface antigens they express. Moreover, there is evidence that the populations of both the helper and suppressor T cells are themselves heterogeneous. For example, the helper T cells that activate B cells appear to be different from those that activate suppressor T cells. Because both helper T cells and suppressor T cells act as regulators of the immune response, these two types of T cells are referred to as **regulatory T cells.**

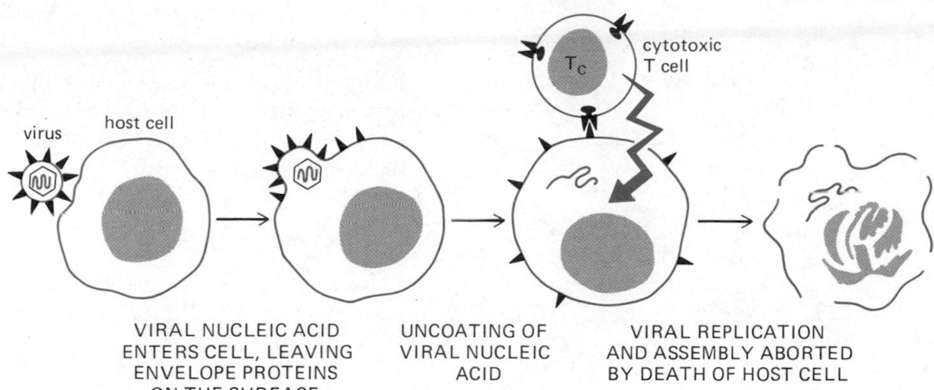

Figure 17–56 Schematic drawing showing how a cytotoxic T cell can recognize viral antigens on the surface of an infected cell and kill them before the virus begins to replicate.

Cytotoxic T Cells Kill Virus-infected Cells[37]

When **cytotoxic T cells** are exposed to foreign or virus-infected vertebrate cells, they become activated over the course of several days to become effector cells that specifically bind to and kill the target cells that activated them. The ability of activated cytotoxic T cells to kill is most conveniently measured by incubating them with ^{51}Cr-containing target cells for several hours and measuring the amount of ^{51}Cr radioactivity released from killed target cells; ^{51}Cr is taken up by living cells and is only released when the cells die.

Cytotoxic T cells defend us against certain viral diseases. By reacting against foreign viral antigens that are expressed on the surface of virus-infected cells before viral replication has begun, they kill the infected cells and thereby prevent the multiplication of the virus (Figure 17–56). Although the molecular events responsible for the killing of the target cell are unknown, it is clear that contact between the cytotoxic T cell and the target cell is required and that this produces an irreversible lesion in the latter within minutes. Single cytotoxic T cells have been observed to kill many individual target cells without being harmed themselves.

Helper T Cells Are Required for Most B Cells and T Cells to Respond to Antigen[38]

Although T cells themselves do not secrete antibody, they are essential for B-cell antibody responses to most antigens. This was first discovered in the mid-1960s through experiments in which either thymus cells or bone marrow cells were injected into irradiated mice together with antigen. Mice that had received only bone marrow or only thymus cells were unable to make antibody; but if a mixture of thymus and bone marrow cells was injected, large amounts of antibody were produced. It was later shown that the thymus provides T cells, while the bone marrow provides B cells in this type of experiment (Figure 17–57). The use of a specific chromosome marker to distinguish between the injected T and B cells showed that all of the antibody-secreting cells are B cells, leading to the conclusion that T cells must help B cells respond to antigen.

The T cells that provide this help are now recognized as a special subclass of **helper T cells.** It has also become clear that helper T cells are also required to enable cytotoxic T cells and suppressor T cells to respond to antigen.

Although the antibody response to most antigens is dependent on helper T cells, there are some antigens that can activate B lymphocytes without T-cell help. Such *T-cell-independent* antigens are usually large polymers with repeating, identical antigenic determinants.

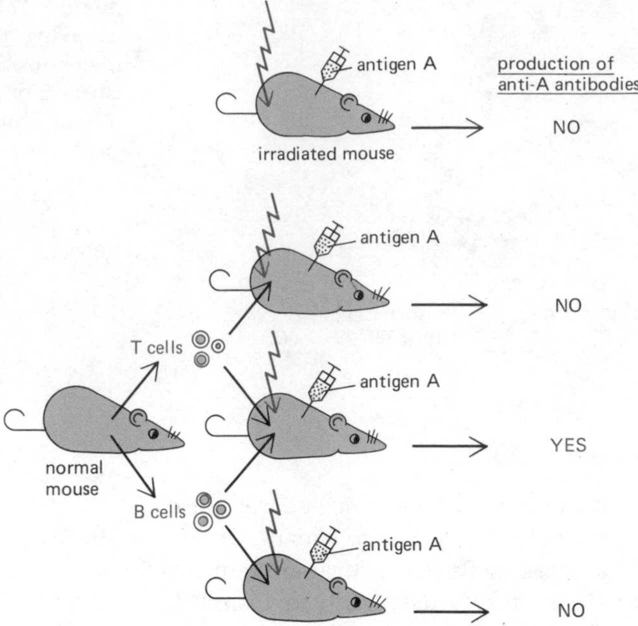

Figure 17–57 The experiment that first suggested that both T cells and B cells are required for an animal to make antibody responses.

Helper T Cells Activate Macrophages by Secreting Lymphokines[39]

Some helper T cells respond to antigen by secreting a variety of substances that activate various white blood cells, including other lymphocytes. These factors are collectively called **lymphokines** or, in some cases, **interleukins** (between leucocytes). Unlike antibodies, they do not specifically recognize or react with the antigen that induces their production.

One of the most important and best studied lymphokine, **macrophage migration inhibition factor (MIF),** activates nearby macrophages to become more efficient at phagocytosing and digesting invading organisms. Another function, as its name suggests, is to inhibit the migration of macrophages. Consequently, they accumulate in regions where T cells have been activated. This ability of T cells to activate macrophages is especially important in defense against infections by organisms that can survive simple phagocytosis by nonactivated macrophages: tuberculosis is one such infection.

The antigen-triggered secretion of lymphokines by helper T cells underlies the familiar tuberculin skin test. If tuberculin (an extract of the bacterium responsible for tuberculosis) is injected into the skin of individuals who have been immunized against tuberculosis, or who have had tuberculosis, a characteristic immune response occurs in the skin. It is initiated at the site of injection by the secretion of lymphokines by memory helper T cells reactive to tuberculin. The lymphokines attract macrophages into the site, causing the characteristic swelling of a positive reaction to tuberculin.

Another important lymphokine secreted by certain helper T cells is **T-cell growth factor** (or **interleukin 2**). It binds to receptors on the surface of *activated* T cells and stimulates them to proliferate. If T cells are activated by an antigen in culture in the continuous presence of interleukin 2, they will proliferate indefinitely. In this way, antigen-specific *T cell lines,* derived from either cytotoxic T cells, helper T cells, or suppressor T cells, can be generated. Some of these cell lines can be cloned to provide homogeneous populations of functional, antigen-specific T cells, which promise to revolutionize the study of cell-mediated immunity and immunoregulation.

Suppressor T Cells Inhibit the Responses of Other Lymphocytes[40]

The discovery that T lymphocytes can *help* B cells make antibody responses was followed several years later by the discovery that they can also *suppress* the response of B cells or other T cells to antigens. Such T-cell suppression was first demonstrated in mice that had been made specifically unresponsive (tolerant) to sheep red blood cell (SRBC) antigens. When T cells from tolerant mice were transferred to normal mice, the latter also became specifically unresponsive to SRBC antigens. This implied that the tolerant state was due to suppression of the response by T cells. Subsequent experiments using surface antigenic markers showed that the so-called **suppressor T cells** responsible are a different population from helper T cells.

Suppressor T cells, like most T and B lymphocytes, function only if they are continually prodded by helper T cells. But the helper T cell that activates a suppressor T cell is itself inhibited by the suppressor cell. This feedback circuit is useful because it means that the activity of both types of cells is self-regulating; this is but one example of many self-regulating circuits in the large network of interacting lymphocytes.

Most of the billions of T lymphocytes of the immune system are thought to be helper or suppressor T cells whose primary function is regulating the activity of both T and B lymphocytes. How do these regulatory T cells recognize the specific T and B cells that they influence?

Helper and Suppressor T Cells Can Recognize Foreign Antigens on the Target Lymphocyte Surface[41]

One way in which regulatory T cells may interact with B cells or with other T cells is by recognizing foreign antigen that has become attached to the surface of the interacting cell. For example, this seems to account for the collaboration that occurs between helper T cells and B lymphocytes in mice immunized with a hapten (such as DNP) covalently linked to a protein carrier (X). In this collaboration the B cells recognize the hapten on the hapten-carrier complex, while the T cells recognize the carrier. This was first demonstrated by an experiment utilizing irradiated mice that cannot respond to antigen unless they are given lymphocytes.

Ordinarily, an irradiated mouse given a mixture of T and B cells from mice immunized with DNP coupled to X (DNP-X) will produce large amounts of antibody against the DNP hapten in response to an injection of DNP-X. However, if the same mouse is challenged with DNP-Y, consisting of the same hapten on a different protein carrier (Y), it will not make anti-DNP antibodies. This "carrier effect" shows that the specific carrier is somehow recognized in the course of an antibody response to the hapten. The ingenious experiment that showed that the carrier is recognized by helper T cells involved giving an irradiated mouse T and B cells from mice immunized with the DNP-X conjugate as before, but transferring in addition T cells from mice immunized with the carrier Y alone. Such a mouse now makes anti-DNP antibodies in response to both DNP-X and DNP-Y (Figure 17–58). This important experiment demonstrates that T cells reacting against one determinant on an antigen (in this example, protein Y) are required to help B cells make antibody against a different determinant on the same antigen (in this case, DNP). It is thought that helper T cells in such cases are guided to the B cells by recognizing antigen on the B-cell surface. The simplest model of how this might happen is illustrated in Figure 17–59. Similar results for suppressor T cells suggest that they may recognize their target cells in the same way.

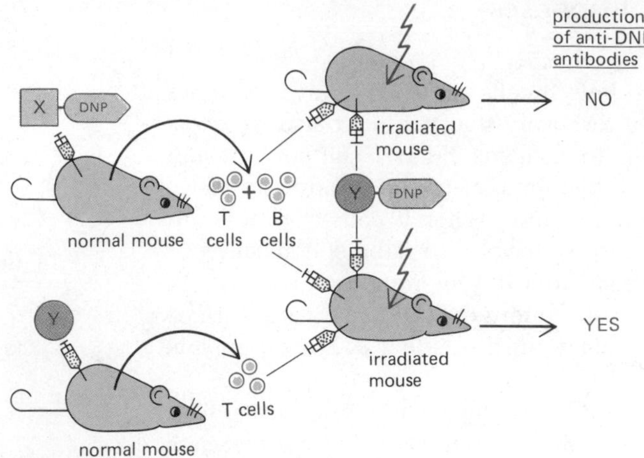

Figure 17–58 The experiment that first showed that T cells reacting against one antigenic determinant on a macromolecule are required to help B cells make antibodies against a different antigenic determinant on the same macromolecule. In this experiment, a mixture of T and B lymphocytes from an animal immunized with DNP coupled to the protein X (DNP-X) does not make anti-DNP antibodies when it is transferred to an irradiated animal and exposed to DNP coupled to a second protein, Y (DNP-Y). However, the same B cells are helped to make anti-DNP antibodies by T cells from an animal previously immunized with protein Y alone. Unlike the experiment shown in Figure 17–57, the conditions of this experiment were such that unimmunized (virgin) lymphocytes did not make detectable antibody responses when transferred to an irradiated host and exposed to antigen for the first time.

Another way in which regulatory T lymphocytes can recognize their target cells is by means of idiotype-anti-idiotype interactions. In those antibody responses in which most of the antibodies produced express the same idiotype, two types of helper T cells have been found: one that recognizes foreign antigen on the B-cell surface and another that recognizes the idiotype on the membrane-bound antibody molecules that function as receptors on the B-cell surface (Figure 17–60). It is uncertain whether such idiotype-reactive helper T cells are also involved in those immune responses in which antibodies with many different idiotypes are produced.

Regulatory T Cells May Communicate with Their Target Lymphocytes by Secreting Soluble Helper or Suppressor Factors[42]

Very little is known about the molecular mechanisms that lymphocytes use to communicate with one another. One possibility is that they interact directly via antigen or idiotype bridging, as illustrated in Figure 17–60, so that very short-range signaling molecules—either membrane-bound or secreted—can act between them. But the interacting lymphocytes responding to a particular antigen make up such a small fraction of the total lymphocyte population that one might wonder whether antigen and/or idiotype recognition would suffice to bring the two relevant cells together. For this reason many immunologists have been attracted to the idea that lymphocytes may communicate by secreting specific signaling molecules that operate over relatively long distances.

In fact, a variety of soluble, protein regulatory factors have been isolated from T lymphocytes—**helper factors** from helper T cells and **suppressor factors** from suppressor T cells. For the most part, these factors have the same antigen-specificity and activity as the T cells that produced them. (In this they differ from lymphokines or interleukins, which are not antigen-specific.) It is still uncertain, however, whether these antigen-specific factors normally function in soluble form; it is possible, for example, that they represent membrane-bound receptors that have been shed from cells. In fact, like some T-cell receptors, some of the isolated factors have been shown to react with anti-idiotype antibodies, suggesting that their antigen-binding sites may be encoded, in part at least, by *V* gene segments in the H-chain gene pool.

Another intriguing feature of many of these factors is that they react not only with anti-idiotype antibodies but also with antibodies against some of the proteins encoded by genes in the *major histocompatibility complex*, a large

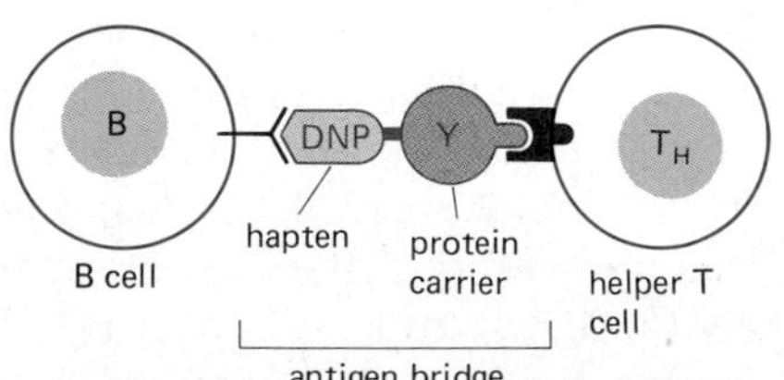

Figure 17–59 One possible way in which Y-reactive helper T cells might interact with DNP-reactive B cells, thereby activating the B cells to make anti-DNP antibodies in the experiment shown in Figure 17–58. The two cells are shown making contact through an *antigen bridge* between the B-cell receptors and T-cell receptors.

chromosomal region intimately concerned with T-cell function. As will now be discussed, this complex of genetic loci was originally discovered when biologists began to investigate why tissues are rejected when they are transplanted between two different individuals.

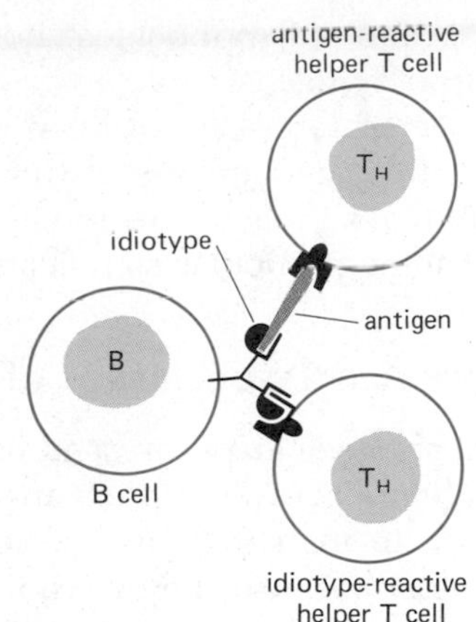

Figure 17–60 Some regulatory T cells interact with their target lymphocytes by recognizing antigen on the lymphocyte surface, while others recognize receptor idiotypes.

Transplantation Reactions Are T-Cell-mediated Immune Responses[43]

When tissue grafts are exchanged between individuals of the same species (*allografts*) or of a different species (*xenografts*), they are usually rejected. In the 1950s, experiments involving skin grafting between different strains of mice demonstrated that **graft rejection** is an immune response to the foreign antigens on the surface of the grafted cells. It is now known that these reactions are mediated mainly by T cells—probably by both cytotoxic and helper cells, although this is controversial. Immunological rejection is the main obstacle to organ transplantation in humans: kidney and heart transplants usually do not survive unless the donor and recipient are genetically identical (identical twins). Alternatively, the recipient's immune system can be suppressed with drugs and the donor and recipient appropriately matched so as to minimize their antigenic differences.

The rejection of transplanted tissue by the recipient is a familiar reaction. Less familiar is the converse response—the rejection of recipient tissue by the graft. This often occurs in patients who receive bone marrow grafts as treatment for their immunodeficiency. A normal individual given such a graft will mount an immune response against the transplanted cells (including the lymphocytes) and destroy them. The immunodeficient patient cannot do this, and frequently the grafted lymphocytes react against the recipient's own antigens in a **graft-versus-host response** that can be fatal. This response, which is thought to be mediated primarily by helper T cells, is the main obstacle to bone marrow transplantation in man.

Both graft rejection and graft-versus-host reactions can be modeled by analogous reactions that occur when cells of one individual are mixed with lymphocytes of another in tissue culture. In such mixed cultures, helper T cells in both populations are mutually stimulated to start dividing—a reaction called the *mixed-lymphocyte response.* In the course of a mixed lymphocyte response, cytotoxic T cells from each individual are activated to kill the lymphocytes from the other individual—a reaction called the *cytotoxic-T-cell response.*

Collectively the two *in vivo* and the two *in vitro* responses are called **transplantation reactions.** All of these reactions are directed against foreign versions of cell-surface antigens called *transplantation* or *histocompatibility antigens.* By far the most important of these are the major histocompatibility antigens, a family of antigens encoded by a complex of genes called the **major histocompatibility complex (MHC).**

T Cells Appear to Be Obsessed with Foreign MHC Antigens[44]

The antigens of the major histocompatibility complex (MHC) are remarkable for at least two reasons. First, they are overwhelmingly preferred as target antigens for the T-cell transplantation reactions, even though a large number of other antigens on cell surfaces are recognized by T cells. Second, an unusually large fraction of T cells is able to recognize foreign MHC antigens: whereas fewer than 0.1% of an individual's T cells respond to any one conventional antigen, approximately 5 to 10% of them respond to the MHC antigens of any other individual. The latter observation constitutes an important challenge to the clonal selection theory, which predicts that only a very small proportion of lymphocytes should be reactive to any one antigen, or even any group of antigens.

Vertebrates do not need to be protected against invasion by the foreign cells of other members of the species. Therefore, the apparent obsession of an animal's T cells with foreign MHC antigens suggests that MHC molecules have some particular significance in ordinary T-cell-mediated immunity.

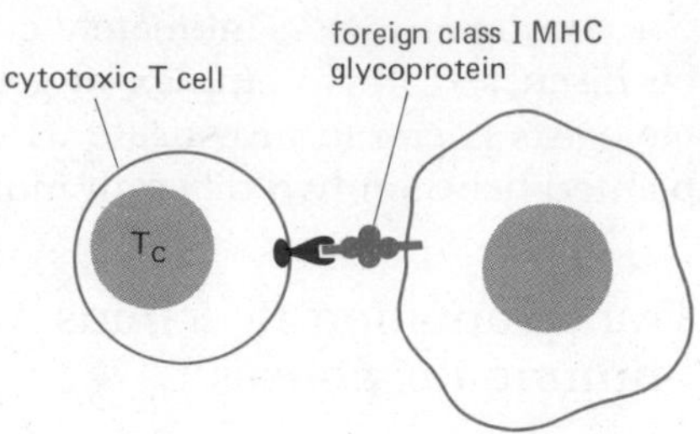

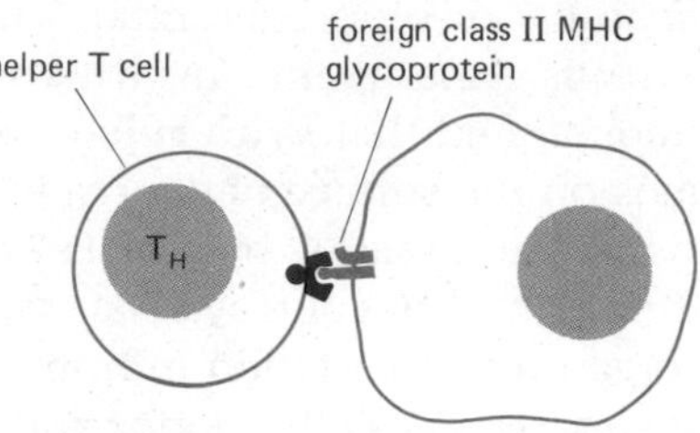

Figure 17–61 Cytotoxic T cells react mainly against foreign class I MHC glycoproteins, while helper T cells react mainly against foreign class II MHC glycoproteins in various transplantation reactions.

There Are Two Classes of MHC Molecules[45]

MHC antigens are expressed on the cells of all higher vertebrates. They were first demonstrated in mice and called **H-2 antigens** (histocompatibility-2 antigens). In man they are called **HLA antigens** (human-leucocyte-associated antigens) because they were first demonstrated on leucocytes. The genes of the *H-2* complex are on mouse chromosome 17; those of the *HLA* complex are on the short arm of human chromosome 6.

The two principal classes of MHC antigens, *class I* and *class II,* each comprise a set of cell-surface glycoproteins. In transplantation reactions, cytotoxic T cells respond mainly against foreign class I glycoproteins, while helper T cells respond mainly against foreign class II glycoproteins (Figure 17–61). The fact that the two classes of MHC antigens stimulate these two different subpopulations of T cells provides an important clue to their function.

Class I MHC Glycoproteins Are Found on Virtually All Nucleated Cells and Are Extremely Polymorphic[45,46]

Class I MHC glycoproteins are encoded by at least three separate genetic loci, called *H-2K, H-2D,* and *H-2L* in the mouse and *HLA-A, HLA-B,* and *HLA-C* in man (Figure 17–62). Each of these loci encodes a single polypeptide chain with a molecular weight of about 45,000 (~345 amino acid residues). Each polypeptide is inserted in the plasma membrane with a short hydrophilic carboxyl-terminal segment inside the cell, followed by a short hydrophobic segment that traverses the lipid bilayer, and then a large amino-terminal segment that is exposed to the cell exterior. The last segment, which represents about 80% of the total mass, is folded into three separate domains, two of which contain single intrachain disulfide bridges (Figure 17–63). Amino acid sequence analysis suggests that the three different class I MHC loci evolved by gene duplication.

Class I MHC glycoproteins have three unusual features: (1) They are found on the surface of almost all nucleated somatic cells, where they can make up as much as 1% of the plasma membrane protein (about 5×10^5 molecules/cell). (2) They are noncovalently associated with a small protein called β_2*-microglobulin,* which is coded for by a gene on a different chromosome. Amino acid sequence analysis has shown this 11,500-dalton protein

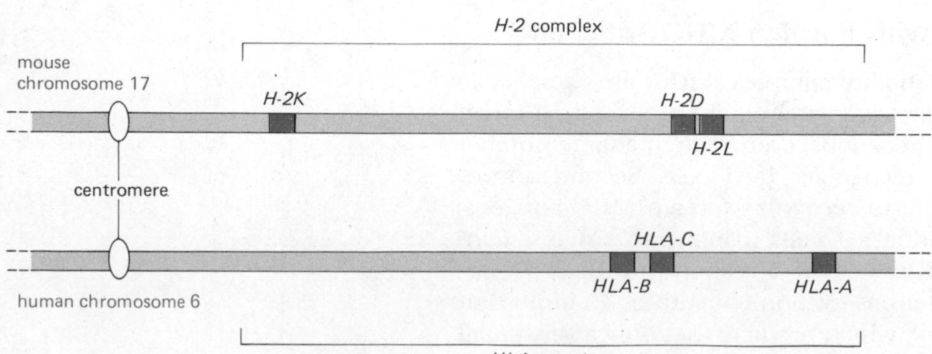

Figure 17–62 Schematic drawing of the major histocompatibility complex (MHC) in mouse and man, showing the location of loci coding for class I MHC glycoproteins.

to be homologous to a single immunoglobulin domain. This suggests an evolutionary link between class I MHC glycoproteins and immunoglobulin, which is further supported by the recent finding of amino acid sequence homology between one of the disulfide-bonded loops in class I glycoproteins and those in immunoglobulins. (3) The loci that code for these glycoproteins are the most *polymorphic* known in higher vertebrates; that is, within a species, there is an extraordinarily large number of different alleles (alternative forms of the same gene) at each locus, each allele being present in a relatively high frequency. For example, more than 50 such alleles are known at both the *H-2K* and the *H-2D* locus, and one allele can differ from another at the same locus by as many as 25% of the proteins' amino acid residues.

The diversity of class I MHC glycoproteins is of a very different nature from the diversity of antibodies. Although hundreds of different class I glycoproteins can be made by a species as a whole, any individual inherits only a single allele at each locus from each parent and therefore can make at most two different forms of each class I glycoprotein. By contrast, an individual can make millions of different antibodies. While the function of antibody diversity is obvious, the enormous polymorphism of the MHC glycoproteins in vertebrate populations has been one of the many mysteries associated with the MHC.

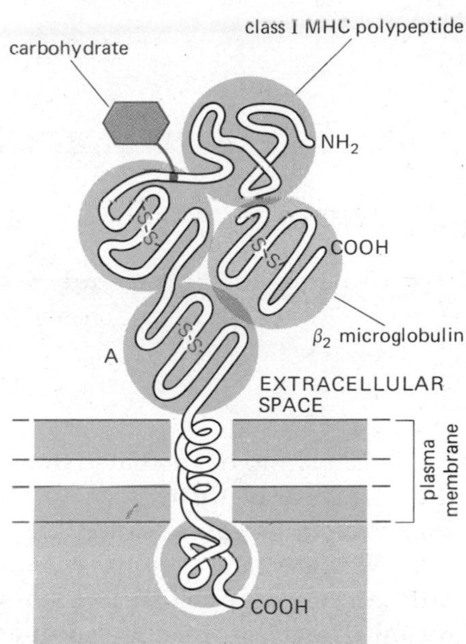

Figure 17–63 Schematic drawing of a class I MHC glycoprotein molecule expressed on the surface of virtually all nucleated somatic cells. It is noncovalently associated with a smaller polypeptide chain, β_2 microglobulin, which is homologous to an immunoglobulin domain and not encoded within the MHC. The domain labeled A is also homologous to an immunoglobulin domain.

The Genes Coding for Class II MHC Glycoproteins Were Originally Discovered as Immune Response (*Ir*) Genes[47]

The discovery of class II MHC glycoproteins began with the observation that certain T-cell dependent immune responses were controlled by specific genes that did not code for antibodies. When different guinea pigs are immunized with the same simple antigen (for example, a synthetic polypeptide consisting of only lysine residues, called polylysine), some animals make vigorous immune responses while others do not respond at all. The mating of nonresponders always produces nonresponders, indicating that the ability to respond to this simple antigen is genetically controlled. When nonresponders (*rr*) are mated with heterozygous responders (*Rr*), approximately 50% of the progeny are nonresponders (*rr*), suggesting that responsiveness to polylysine is controlled by a single dominant gene, which is therefore known as an **immune response (*Ir*) gene.** Responses to different antigens are often controlled by different *Ir* genes.

The demonstration of *Ir* genes depends on using simple antigens, such as synthetic polymers, which have only a small number of different antigenic determinants. Since natural proteins are complex antigens with many different antigenic determinants, the chance that an individual will be unable to respond to any of them, and thus be revealed as a nonresponder, is very small.

Similar experiments with different *inbred strains* of mice (that is, strains in which the individual mice are genetically homogeneous) produce results similar to those of the earlier experiments with guinea pigs: some strains give strong T-cell-dependent immune responses when immunized with a simple synthetic polymer, while other strains do not respond at all. Genetic mapping studies in specially bred strains of mice that differ only in restricted regions of their genome (so-called *congenic strains*) located these *Ir* genes within the *H-2* gene complex in a region between *H-2K* and *H-2D*, subsequently named the ***I* region.** A number of different *Ir* genes, each controlling T-cell-dependent responses to different antigenic determinants, have now been defined in mice and have been mapped to one or other of several different *subregions* within the *I* region (Figure 17–64). While for most of these genes, responsiveness to the antigenic determinant is inherited as a dominant character, in a few, nonresponsiveness is dominant. In these cases, the genetic failure to respond

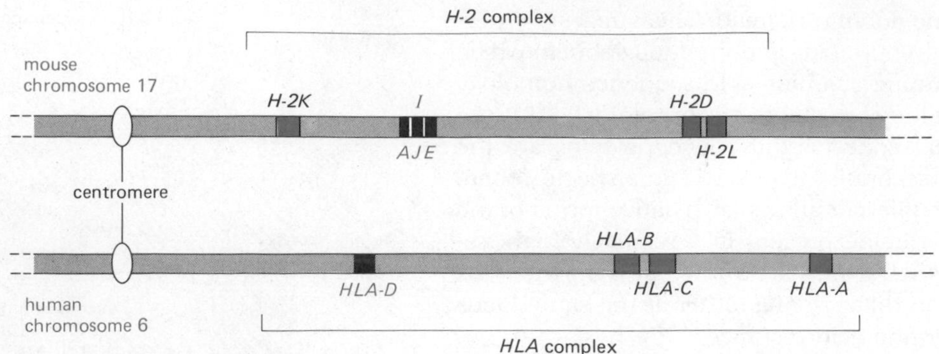

Figure 17–64 Schematic drawing of the *H-2* and *HLA* complexes, showing the location of loci encoding class I *(color)* and class II *(black)* MHC glycoproteins. The *I* region in the mouse is thought to contain three subregions, although in only two of these (*A* and *E*) have the protein products been characterized (see Figure 17–65). The *HLA-D* region in man also contains several loci.

can be shown to be mediated by suppressor T cells, and the genes controlling the response of these cells to a specific determinant are called *immune suppression* (Is) *genes* rather than *Ir* genes.

Several years after the *Ir* genes had been defined and mapped to the MHC, a family of cell-surface antigens was discovered to map to the same region. These so-called ***I*-region-associated (Ia) antigens,** now called **class II MHC antigens,** are highly polymorphic cell-surface glycoproteins. However, they differ from class I MHC antigens in having a much narrower tissue distribution: they are expressed only by certain cell types, including most B cells, some T cells, some macrophages, and macrophagelike cells that present antigen to T cells (so-called *antigen-presenting cells*).

The class II MHC glycoproteins are composed of two noncovalently bonded polypeptide chains—an α-chain having a molecular weight of about 33,000 and a β-chain having a molecular weight of about 28,000. Both chains are encoded by genes in the *I* region, are transmembrane in orientation, and are glycosylated (Figure 17–65). Amino acid sequence studies suggest that these glycoproteins may also contain antibodylike domains. In man, the *HLA-D* region encodes class II glycoproteins (see Figure 17–64).

While initially the relationship between class II glycoproteins and *Ir* genes was uncertain, there is increasing evidence that at least some *Ir* genes code for class II glycoproteins. But how do class II MHC molecules influence the response of T cells to specific antigenic determinants?

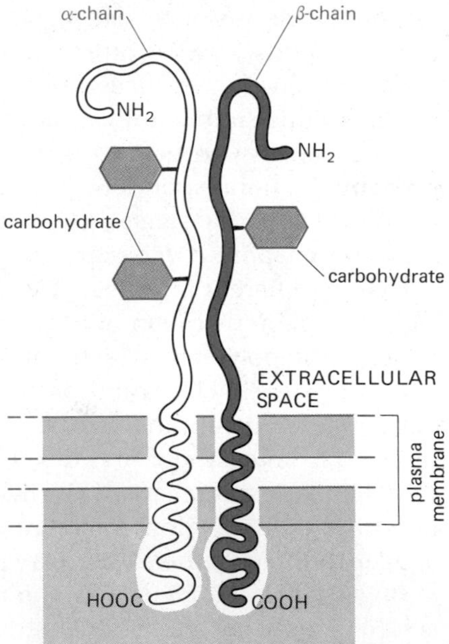

Figure 17–65 Schematic drawing of a class II MHC glycoprotein. The *I-A* subregion in mouse and the *HLA-D* region in man encode both α-chains (33,000 daltons) and β-chains (28,000 daltons), while the *E* subregion in mouse encodes only α-chains that combine with β-chains encoded in the *A* subregion.

T Cells Recognize Foreign Antigens in Association with Self MHC Molecules[48]

For many years the MHC has posed a number of puzzling questions for immunologists: (1) Why are T cells obsessed with foreign MHC antigens in transplantation reactions? (2) Why are the MHC glycoproteins so polymorphic? (3) Why are class I MHC molecules on almost all nucleated somatic cells, while class II MHC molecules are mainly on cells concerned with immune responses? (4) How do class II MHC glycoproteins control the ability of an animal to make T-cell-dependent responses to specific antigenic determinants? Apparently the MHC glycoproteins play an important part in T-cell function—but in what way?

Perhaps the most important single experiment that helped to clarify the MHC puzzle was performed in 1974. Mice of strain X were infected with virus A. Seven days later, the spleens of these mice contained activated cytotoxic T cells that could kill virus-infected, strain-X fibroblasts within several hours in tissue culture. They would kill the fibroblasts only if they were infected with virus A and not if they were infected with virus B; thus the cytotoxic T cells were virus-specific. Unexpectedly, however, the same T cells were unable to kill fibroblasts from strain-Y mice infected with the same virus A (Figure 17–66).

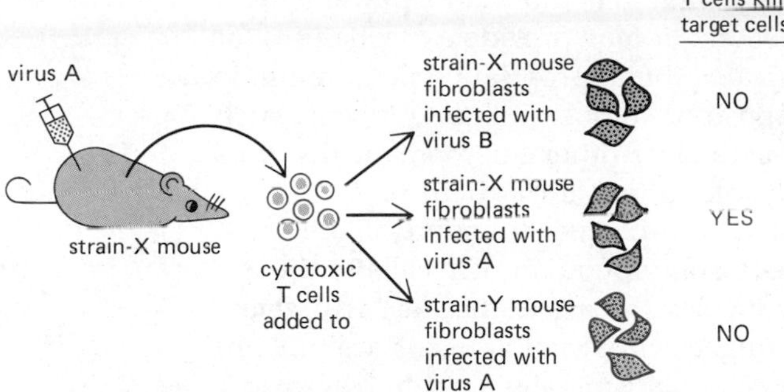

Figure 17–66 The crucial experiment showing that cytotoxic T cells recognize some aspect of the host target-cell surface in addition to the foreign (viral) antigen. By repeating this experiment with target cells that differ from the cells of the infected mouse only in limited regions of their genome, the feature of the target-cell surface that the cytotoxic T cells recognize was shown to be a class I MHC glycoprotein.

The cytotoxic T cells were clearly recognizing more than just the virus in the killing process; they were also recognizing some host component that was present on strain X but not on strain Y fibroblasts. Further experiments showed that the host components in question were the class I MHC glycoproteins: if the target fibroblasts were genetically different from the infected mice everywhere except at any of the class I MHC loci, they were readily killed; but if the target cells differed at these loci but were identical with the infected mice elsewhere in their genome, they could not be killed. Thus, the class I MHC glycoproteins are somehow involved in presenting cell-surface-bound virus antigens to cytotoxic T cells.

An important clue to the function of class II MHC molecules came from experiments on the proliferative response of memory helper T cells in culture. This response depends on the presence in the culture of **antigen-presenting cells:** helper T cells respond to antigen only if it is presented on the surface of an antigen-presenting cell. A crucial observation was that antibodies binding to class II MHC glycoproteins on the surface of antigen-presenting cells prevented these cells from presenting antigen to the helper T cells. Antibodies directed against other cell-surface molecules on antigen-presenting cells, including class I MHC molecules, had no effect.

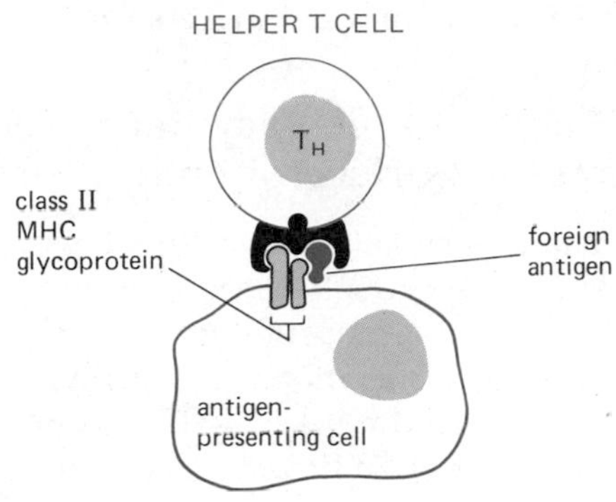

Several important conclusions can be drawn from these and related experiments: (1) Most T cells do not respond to antigen in solution but only to antigen on the surface of cells. For example, helper T cells respond to antigen on the surface of antigen-presenting cells, and cytotoxic T cells respond to antigen on the surface of their target cells. (2) Most T cells recognize antigens on the surface of cells only in association with *self* MHC glycoproteins expressed on the same cell surface; this property of T cells is referred to as **MHC associative recognition.** (3) Different T-cell subpopulations recognize antigen in association with different classes of MHC glycoproteins; some helper T cells respond to antigen in association with class II glycoproteins, while cytotoxic T cells respond to antigen in association with class I glycoproteins (Figure 17–67). Moreover, there is evidence that at least some suppressor T cells respond to antigen in solution as B cells do, without MHC associative recognition, while other suppressor T cells may respond to antigen in association with a special subclass of class II MHC molecules whose chemistry is unknown.

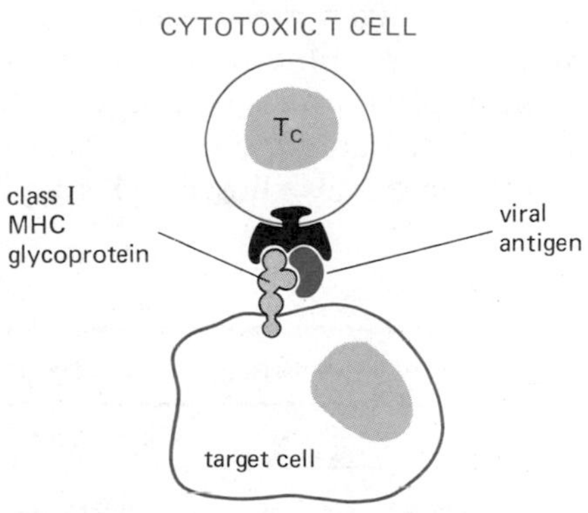

Figure 17–67 Helper T cells recognize foreign antigens on the surface of an antigen-presenting cell in association with class II MHC glycoproteins, while cytotoxic T cells recognize foreign (usually viral) antigens on the surface of any host cell in association with class I MHC glycoproteins. It is not known how antigen is associated with the MHC glycoproteins, nor whether T cells use one receptor (as shown) or two different receptors to recognize the antigen and MHC glycoprotein.

MHC Glycoproteins May Serve as Guides for Activating the Appropriate Subpopulations of T Cells[49]

Given that the immune system has evolved to prevent infection by microorganisms, there would seem to be two obvious advantages in MHC associative recognition. First, it focuses the attention of T cells on cell surfaces; for example, it would be inefficient for cytotoxic T cells to bind free virus (or soluble viral antigens), since that would occupy their receptors and prevent them from destroying virally infected cells. Second, it may provide a mechanism for assuring that each class of antigen activates the appropriate type of immune response: cytotoxic T cells cannot dispose of soluble foreign antigens, such as bacterial toxins, nor can they kill bacteria or other microorganisms; thus there would be no point in their being able to recognize them.

It seems likely that at least part of the normal function of the MHC glycoproteins is to guide the appropriate subpopulation of T cells to the appropriate antigens. The guidance may involve the association of MHC glycoproteins only with specific classes of antigen. Thus viral antigens are thought to associate with class I MHC glycoproteins and thereby activate cytotoxic T cells. (Since any nucleated somatic cell can be infected by viruses, this would explain why virtually all such cells express class I molecules.) Other antigens, such as those produced by bacteria, are thought to associate with class II glycoproteins on antigen-presenting cells and thereby to stimulate helper T cells; the T cells in turn activate B cells and macrophages, leading to both the phagocytosis and the complement-mediated killing of the bacteria.

Some of the properties of class I and class II MHC genes and glycoproteins are compared in Table 17–2.

Helper T Cells May Recognize Fragments of Foreign Antigens on the Surface of Antigen-presenting Cells[50]

The molecular mechanisms involved in MHC associative recognition are unknown. Although there is indirect evidence that foreign antigens associate with MHC glycoproteins to form complexes on the surface of cells, the nature of these associations is uncertain.

Since the majority of immune responses are initiated by helper T cells, the question of how antigens activate these cells is especially critical. We have already seen that helper T cells respond primarily to antigen that has been taken up by specialized antigen-presenting cells. But what is the molecular form of antigen recognized by the helper T cells? There is indirect evidence that at least some of them recognize digested fragments of foreign macromolecules complexed with class II MHC glycoproteins on the surface of the antigen-presenting cell and thereby become activated. A hypothetical outline of this process is shown in Figure 17–68. Activated T cells may next recognize

Table 17–2 Properties of Class I and Class II MHC Glycoproteins

	Class I	Class II
Genetic loci	*H-2K, H-2D, H-2L* in mouse *HLA-A, HLA-B, HLA-C* in man	*I* region in mouse *HLA-D* in man
Chain structure	45,000-dalton glycoprotein + β_2-microglobulin (11,500 daltons)	α-chain (33,000 daltons) + β-chain (28,000 daltons)
Cell distribution	almost all nucleated somatic cells	most B cells, some T cells, some macrophages, antigen-presenting cells, thymus epithelial cells
Involved in presenting antigen to	cytotoxic T cells	helper T cells
Polymorphism	+ + + +	+ +

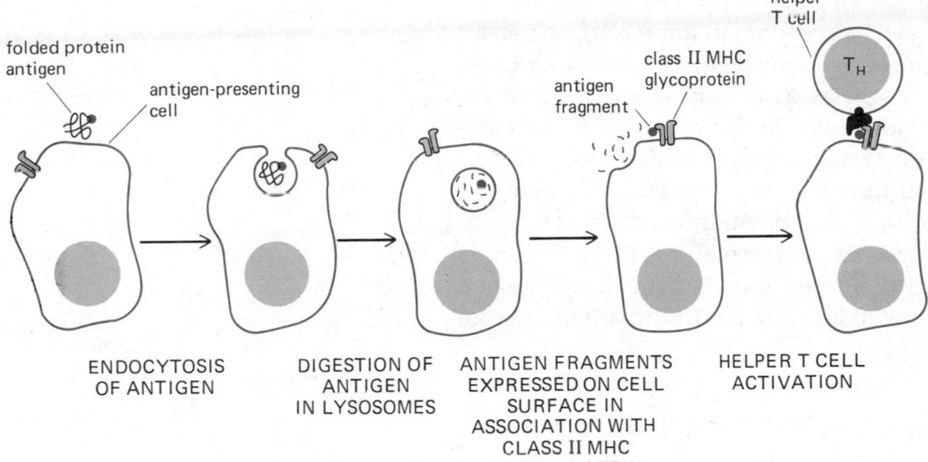

Figure 17–68 A hypothetical scheme of how antigen may be "processed" by an antigen-presenting cell.

a similar complex of antigenic determinant and class II MHC glycoprotein on the surface of B cells in order to activate the B cells.

The question of how T cells simultaneously recognize a foreign antigen and an MHC glycoprotein is still unresolved. The main issue is whether they have two different cell-surface receptors, one to recognize the foreign antigen and another to recognize the MHC glycoprotein (*dual recognition*) or a single receptor that recognizes a complex of the antigen and MHC molecule (*altered-self recognition*) (Figure 17–69). In either case, the obsession of T cells with *foreign* MHC molecules in transplantation reactions probably reflects the fact that they can recognize conventional foreign antigens only in association with their own (self) MHC glycoproteins. In particular, it is possible that many T cells react to foreign MHC glycoproteins because these molecules (either alone or in combination with other molecules on the foreign cell surface) resemble various combinations of self MHC molecules complexed with fragments of conventional foreign antigens.

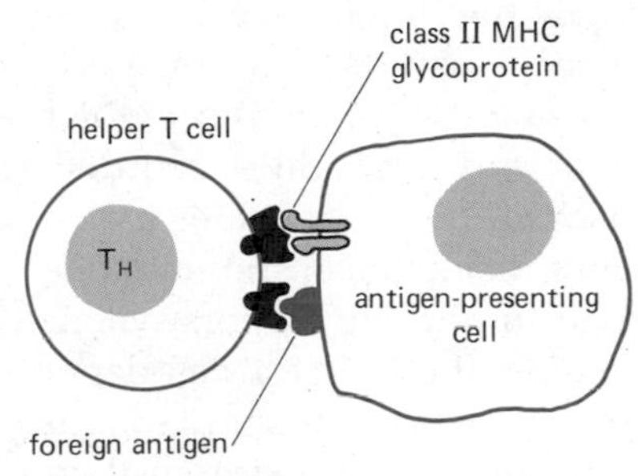

Why Are MHC Glycoproteins So Polymorphic?[51]

Now that we know that MHC glycoproteins are involved in presenting antigen to T cells, the *Ir* genes have lost much of their mystery. We can at least imagine how a particular allelic form of a class II MHC glycoprotein (P) could be ineffective at presenting a specific antigenic determinant (A) to helper T cells but still be able to present other antigenic determinants (B, C, D, etc.). For example, P might not be able to bind to the antigen molecule in such a way as to present determinant A effectively. Alternatively, helper T cells able to recognize the particular combination P + A might be missing from the T-cell repertoire in individuals carrying the P allele. This could occur, for example, if the combination P + A happened to resemble a self MHC glycoprotein (either alone or when this MHC glycoprotein is associated with some other self molecule): in that case, helper T cells reactive to P + A would have been eliminated during lymphocyte development, when tolerance to self develops (we shall discuss self-tolerance briefly below).

By the same arguments, there should be some allelic forms of class I glycoproteins that affect the responses of cytotoxic T cells to specific antigenic determinants in the same way that the common *Ir* genes affect the responses of helper T cells. Similarly, there should be MHC alleles that are particularly effective at presenting specific antigenic determinants to certain suppressor

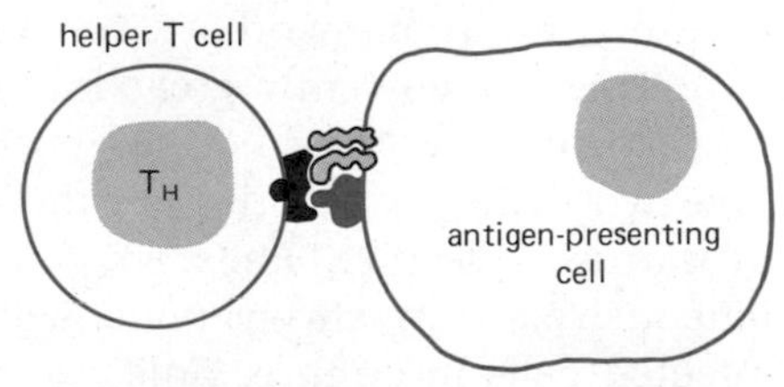

Figure 17–69 Alternative hypotheses of how T cells recognize antigen in association with MHC glycoproteins. The *dual recognition* hypothesis states that T cells recognize foreign antigen and self MHC glycoproteins by means of two different receptors. The *altered-self recognition* hypothesis states that T cells have a single class of receptor that recognizes a complex of foreign antigen and self MHC glycoprotein.

T cells, which should reveal themselves as immune suppression (*Is*) genes. In fact, both of these types of alleles are being discovered in increasing numbers.

The hypothesis that MHC glycoproteins associate with foreign antigens and present these antigens to T cells suggests a reasonable explanation for the extensive polymorphism of MHC molecules. In the evolutionary war between microorganisms and the immune system, microorganisms will tend to change their antigens to avoid associating with MHC molecules. When one succeeds, it will be able to sweep through a population as an epidemic. In such circumstances, the few individuals that produce a new MHC molecule that can associate with an antigen of the altered microorganism will have a large selective advantage. In this way, selection will tend to maintain a large diversity of MHC molecules in the population.

The Immune System Is Ineffective Against Most Tumors[52]

Until recently it was widely held that tumors arise frequently in all of us but are detected and eliminated early in their development by the immune system, particularly by T-cell-mediated responses. This is referred to as the *immune surveillance hypothesis.* That occasional malignant tumors spontaneously regress was taken as evidence to support it. The hypothesis predicts that immunodeficient patients or experimental animals should have an unusually high incidence of cancer. There is now compelling evidence that, for the most part, this is not the case. The most convincing observations have been made in *nude mice,* which are both congenitally hairless (hence their name) and markedly T-cell deficient (because they have very abnormal thymuses.) When kept under infection-free conditions, these mice do not have a higher incidence of spontaneous tumors than normal mice, nor are they more prone to develop tumors when they are treated with chemical carcinogens. It seems, then, that T-cell-dependent immunity normally plays little part in the control of most spontaneous or chemically induced cancer. On the other hand, the immune system does play an important part in defending vertebrates against the great majority of virally induced tumors in experimental animals, and probably against the few tumors in man that are viral in origin. The immunity in virally induced tumors is in no way different from the immunity seen in many other virus infections.

There is currently great interest in the possibility that other, nonimmunological mechanisms may be important in the defense against cancer. The two main possibilities are the killing of tumor cells by macrophages or by **natural killer (NK) cells.** NK cells are small lymphocytelike cells that spontaneously, and relatively nonspecifically, kill a variety of tumor and virus-infected cells in culture. Unlike K cells (see p. 973), they do not require the target cell to be coated with antibodies in order to kill it. How they distinguish abnormal cells from normal ones is unknown.

The Immune System Has Had to Solve Three Major Problems in Antigen Recognition

As we indicated at the start of this chapter, the immune system is faced with three problems in antigen recognition. The first is how to distinguish between the apparently infinite array of foreign antigens and ensure that a specific response can be made against most of them—even when they are present in low concentration. The second is how to ensure that this response is appropriate to the particular class of invading antigen so that the antigen is eliminated. And the third is how to avoid responding to the large number of self antigens.

The immune system solves the first problem by producing millions of different immunoglobulin molecules, each with a different antigen-binding site, and distributing them among millions of different lymphocyte clones, each committed to making one particular antigen-binding site. Although the universe of foreign antigens is still larger than the number of different antigen-binding sites, each site can react with a variety of related but different antigenic determinants; such reactions are called *cross-reactions*. Furthermore, various mechanisms operate to concentrate antigen in the vicinity of responding lymphocytes and thus increase the sensitivity of the response; antigen-presenting cells are thought to play an important part in this process.

The solution to the second problem, that of *appropriateness*, may, in part at least, lie in the MHC glycoproteins. The immune system is composed of a variety of different classes and subclasses of lymphocytes: B cells making different classes of antibodies and T cells making different types of cell-mediated immune responses. It is thought that the MHC glycoproteins help to channel each class of antigen into the appropriate immune response pathway. This hypothesis, if correct, implies that the MHC glycoproteins function in a primitive way as antigen-recognition molecules distinguishing between different classes of antigens; but if and how they do so remains to be shown.

We know even less about the third problem, that of how the immune system discriminates between self and nonself antigens. Some self molecules are sequestered in cells and tissues and never make contact with lymphocytes, while others fail to stimulate lymphocytes, either because they are present in too low a concentration or because they do not associate with MHC glycoproteins. The immune system need not bother with these self molecules. On the other hand, many self molecules are accessible to lymphocytes and able to associate with MHC glycoproteins and activate lymphocytes. During lymphocyte development, the lymphocytes that recognize these self molecules must be inactivated. Inactivation is achieved by killing or suppressing (by means of suppressor T cells) either the appropriate B cells and cytotoxic T cells, or the helper T cells that are required to activate them, or both. Although the process is poorly understood, it is likely that the choice between killing and suppression is determined by the nature, location, and concentration of the self molecule.

These remarkable powers of recognition make the immune system unique among cellular systems—except perhaps for the nervous system. In fact, the immune system and nervous system have a number of properties in common. Most important, they are both composed of very large numbers of phenotypically distinct cells that are organized into intricate networks. Within the network, individual cells can interact either positively or negatively, and the response of one cell reverberates through the system by affecting many other cells. While the neural network is relatively fixed in space, the cells constituting the immunological network are constantly changing their locations and interact with each other only transiently. In the next chapter, we shall consider the cells of the vertebrate nervous system, which is by far the most complex and sophisticated cellular system known.

Summary

There are at least three functionally distinct subclasses of T cells: (1) cytotoxic T cells, which can directly kill foreign or virus-infected cells; (2) helper T cells, which can help B cells make antibody responses, help other T cells make cell-mediated immune responses, and activate macrophages; and (3) suppressor T cells, which can inhibit the responses of B cells and other T cells. Helper and suppressor T cells are the main regulators of immune responses. They interact with their target lymphocytes by recognizing either foreign antigen or receptor idiotypes on the target-cell surface.

Most T cells recognize foreign antigens only when these antigens are associated on cell surfaces with the membrane glycoproteins encoded by genes in the major histocompatibility complex (MHC). There are two main classes of MHC glycoproteins: (1) class I glycoproteins are expressed on almost all nucleated somatic cells and function in presenting viral antigens to cytotoxic T cells; and (2) class II MHC glycoproteins—which are recognized in association with foreign antigens by helper T cells—are expressed on most B cells, on some T cells and macrophages, and on specialized antigen-presenting cells. On all of these various cell surfaces, the MHC glycoproteins are thought to serve as primitive antigen-binding receptors that help each class of foreign antigen to activate the appropriate type of T-cell response. The fact that certain allelic forms of the class I and class II MHC glycoproteins are ineffective in presenting particular antigenic determinants to T cells may explain why these are the most polymorphic vertebrate proteins known.

References

General

Benacerraf, B.; Unanue, E.R. Textbook of Immunology. Baltimore: Williams & Wilkins, 1979.

Cunningham, A.J. Understanding Immunology. New York: Academic Press, 1978.

Eisen, H.N. Immunology, 3rd ed. New York: Harper & Row, 1981.

Golub, E. The Cellular Basis of the Immune Response: An Approach to Immunobiology, 2nd ed. Sunderland, Ma.: Sinauer, 1981.

Hood, L.E.; Weissman, I.L.; Wood, W.B. Immunology, 2nd ed. Menlo Park, Ca.: Benjamin-Cummings, 1982.

McConnell, I.; Munro, A.; Waldmann, H. The Immune System: A Course on the Molecular and Cellular Basis of Immunity, 2nd ed. Oxford, Eng.: Blackwell, 1981.

Roitt, I. Essential Immunology, 4th ed. Oxford, Eng.: Blackwell Scientific, 1980.

Cited

1. Gowans, J.L.; McGregor, D.D. The immunological activities of lymphocytes. *Prog. Allergy* 9:1–78, 1965.
2. Greaves, M.F.; Owen, J.J.T.; Raff, M.C. T and B Lymphocytes: Origins, Properties and Roles in Immune Responses. Amsterdam: Excerpta Medica, 1973.
3. Cooper, M.; Lawton, A. The development of the immune system. *Sci. Am.* 231(5):59–72, 1974.

   Owen, J.J.T. Ontogenesis of lymphocytes. In B and T Cells in Immune Recognition (F. Loor, G.E. Roelants, eds.), pp. 21–34. New York: Wiley, 1977.
4. Raff, M.C. Cell-surface immunology. *Sci. Am.* 234(5):30–39, 1976.

   Reinherz, E.L., Schlossman, S.F. The differentiation and function of human T lymphocytes. *Cell* 19:821–827, 1980.
5. Gowans, J.L.; Knight E.J. The route of re-circulation of lymphocytes in the rat. *Proc. R. Soc. Lond. (Biol.)* 159:257–282, 1964.

   Sprent, J. Migration and lifespan of lymphocytes. In B and T Cells in Immune Recognition (F. Loor, G.E. Roelants, eds.), pp. 59–82. New York: Wiley, 1977.
6. Burnet, F.M. The Clonal Section Theory of Acquired Immunity. Nashville, Tn.: Vanderbilt University Press, 1959.

   Ada, G. Antigen binding cells in tolerance and immunity. *Transplant. Rev.* 5:105–129, 1970.

   Wigzell, H. Specific fractionation of immunocompetent cells. *Transplant. Rev.* 5:76–104, 1970.
7. Pink, J.R.L.; Askonas, B.A. Diversity of antibodies to cross-reacting nitrophenyl haptens in inbred mice. *Eur. J. Immunol.* 4:426–429, 1974.

8. Greaves, M.F.; Owen, J.J.T.; Raff, M.C. T and B Lymphocytes: Origins, Properties and Roles in Immune Responses, pp. 117–186. Amsterdam: Excerpta Medica, 1973.

9. Owen, R.D. Immunogenetic consequence of vascular anastomoses between bovine twins. *Science* 102:400–401, 1945.

Billingham, R.E.; Brent, L.; Medawar, P.B. Quantitative studies on tissue transplantation immunity. III. Activity acquired tolerance. *Philos. Trans. R. Soc. Lond. (Biol.)* 239:357–414, 1956.

Triplett, E.L. On the mechanism of immunologic self-recognition. *J. Immunol.* 89:505–510, 1962.

Lindstrom, J. Autoimmune response to acetylcholine receptors in myasthenia gravis and its animal model. *Adv. Immunol.* 27:1–50, 1979.

10. Howard, J.G.; Mitchison, N.A. Immunological tolerance. *Prog. Allergy* 18:43–96, 1975.

11. Kabat, E.A. Structural Concepts in Immunology and Immunochemistry, 2nd ed. New York: Holt, Rinehart & Winston, 1976.

Nisonoff, A.; Hopper, J.E.; Spring, S.B. The Antibody Molecule. New York: Academic Press, 1975.

12. Warner, N.L. Membrane immunoglobulins and antigen receptors on B and T lymphocytes. *Adv. Immunol.* 19:67–216, 1974.

13. Jerne, N.K.; et al. Plaque forming cells: methodology and theory. *Transplant. Rev.* 18:130–191, 1974.

Dutton, R.W.; Mishell, R.I. Cellular events in the immune response. The *in vitro* response of normal spleen cells to erythrocyte antigens. *Cold Spring Harbor Symp. Quant. Biol.* 32:407–414, 1967.

14. Edelman, G.M. The structure and function of antibodies. *Sci. Am.* 223(2):34–42, 1970.

Porter, R.R. Structural studies of immunoglobulins. *Science* 180:713 716, 1973.

15. Spiegelberg, H.L. Biological activities of immunoglobulins of different classes and subclasses. *Adv. Immunol.* 19:259–294, 1974.

Fisher, M.M.; Nagy, B.; Bazin, H.; Underdown, B.J. Biliary transport of IgA: role of secretory component. *Proc. Natl. Acad. Sci. USA* 76:2008–2012, 1979.

Ishizaka, T.; Ishizaka, K. Biology of Immunoglobin E. *Prog. Allergy* 19:60–121, 1975.

16. Yalow, R.S. Radioimmunoassay. *Annu. Rev. Biophys. Bioeng.* 9:327–345, 1980.

17. Theofilopoulos, A.N.; Dixon, F.J. The biology and detection of immune complexes. *Adv. Immunol.* 28:89–221, 1979.

18. Roitt, I. Essential Immunology, 4th ed., pp. 173–196. Oxford, Eng.: Blackwell, 1980.

19. Capra, J.D.; Edmundson, A.B. The antibody combining site. *Sci. Am.* 236(1):50–59, 1977.

20. Wu, T.T.; Kabat, E.A. An analysis of the sequences of the variable regions of Bence Jones proteins and myeloma light chains and their implications for antibody complementarity. *J. Exp. Med.* 132:211–250, 1970.

21. Edelman, G.M. The structure and function of antibodies. *Sci. Am.* 223(2):34–42, 1970.

Sakano, H.; et al. Domains and the hinge region of an immunoglobulin heavy chain are encoded in separate DNA segments. *Nature* 277:627–633, 1979.

22. Silverton, E.W.; Navia, M.A.; Davies, D.R. Three-dimensional structure of an intact human immunoglobulin. *Proc. Natl. Acad. Sci. USA* 74:5140–5144, 1977.

Amzel, L.M.; Poljak, R.J. Three-dimensional structure of immunoglobulins. *Annu. Rev. Biochem.* 48:961–967, 1979.

23. Adams, J.M. The organization and expression of immunoglobulin genes. *Immunol. Today* 1:10–17, 1980.

Tonegawa, S.; et al. Somatic reorganization of immunoglobulin genes during lymphocyte differentiation. *Cold Spring Harbor Symp. Quant. Biol.* 45:839–858, 1981.

Leder, P. The genetics of antibody diversity. *Sci. Am.* 246(5):72–83, 1982.

24. Dreyer, W.J.; Bennett, J.C. The molecular basis of antibody formation: a paradox. *Proc. Natl. Acad. Sci. USA* 54:864–869, 1965.

Hozumi, N.; Tonegawa, S. Evidence for somatic rearrangement of immunoglobulin genes coding for variable and constant regions. *Proc. Natl. Acad. Sci. USA* 73:3628–3632, 1976.

25. Tonegawa, S.; Maxam, A.M., Tizard, R.; Bernard, O.; Gilbert, W. Sequence of a mouse germ-line gene for a variable region of an immunoglobulin light chain. *Proc. Natl. Acad. Sci. USA* 75:1485–1489, 1978.

26. Davis, M.M.; et al. An immunoglobulin heavy-chain gene is formed by at least two recombinational events. *Nature* 283:733–739, 1980.

27. Leder, P. The genetics of antibody diversity. *Sci. Am.* 246(5):72–83, 1982.

Gearhart, P.J.; Johnson, N.D.; Douglas, R.; Hood, L. IgG antibodies to phosphorylcholine exhibit more diversity than their IgM counterparts. *Nature* 291:29–34, 1981.

28. Perry, R.P.; Coleclough, C.; Weigert, M. Reorganization and expression of immunoglobulin genes: status of allelic elements. *Cold Spring Harbor Symp. Quant. Biol.* 45:925–933, 1981.

Early, P.; Hood, L. Allelic exclusion and nonproductive immunoglobulin gene rearrangements. *Cell* 24:1–3, 1981.

29. Early, P.; et al. Two mRNAs can be produced from a single immunoglobulin μ gene by alternative RNA processing pathways. *Cell* 20:313–319, 1980.

30. Lawton, A.R.; Kincade, P.W.; Cooper, M.D. Sequential expression of germ line genes in development of immunoglobulin class diversity. *Fed. Proc.* 34:33–39, 1975.

31. Yaoita, Y.; Kumagai, Y.; Okumura, K.; Honjo, T. Expression of lymphocyte surface IgE does not require switch recombination. *Nature* 297:697–699, 1982.

Honjo, T.; et al. Rearrangements of immunoglobulin genes during differentiation and evolution. *Immunol. Rev.* 59:33–67, 1981.

32. Jerne, N.K. The immune system. *Sci. Am.* 229(1):52–60, 1973.

Jerne, N.K. Toward a network theory of the immune system. *Ann. Immunol. Inst. Pasteur (Paris)*, 125C:378–389, 1974.

Raff, M. Immunological networks. *Nature* 265:205–207, 1977.

33. Lachmann, P.J. Complement. In Clinical Aspects of Immunology, 4th ed. (P.J. Lachmann, K. Peters, eds.), pp. 18–49. Oxford, Eng.: Blackwell, 1982.

Müller-Eberhard, H.J.; Schreiter, R.D. Molecular biology and chemistry of the alternative pathway of complement. *Adv. Immunol.* 29:1–53, 1980.

Reid, K.B.M.; Porter, R.R. The proteolytic activation systems of complement. *Annu. Rev. Biochem.* 50:433–464, 1981.

34. Mayer, M.M. The complement system. *Sci. Am.* 229(5):54–66, 1973.

35. Binz, H.; Wigzell, H. Antigen-binding, idiotypic receptors from T lymphocytes: an analysis of their biochemistry, genetics, and use as immunogens to produce specific immune tolerance. *Cold Spring Harbor Symp. Quant. Biol.* 41:275–284, 1977.

Rajewsky, I.; Eichmann, K. Antigen receptors of T helper cells. *Contemp. Top. Immunobiol.* 7:69–112, 1977.

Williamson, A.R. Genes coding for T-lymphocyte receptors. *Immunol. Today* 3:68–72, 1982.

36. Cantor, H.; Boyse, E.A. Regulation of cellular and humoral immune responses by T-cell subclasses. *Cold Spring Harbor Symp. Quant. Biol.* 41:23–32, 1977.

Cantor, H.; Gershon, R.K. Immunological circuits: cellular composition. *Fed. Proc.* 38:2058–2064, 1979.

37. Blanden, R.V. T cell response to viral and bacterial infection. *Transplant. Rev.* 19:56–88, 1974.

Zinkernagel, R.M. Major transplantation antigens in host responses to infection. *Hosp. Pract.* 13(7):83–92, 1978.

38. Claman, H.N.; Chaperon, E.A. Immunologic complementation between thymus and marrow cells—a model for the two-cell theory of immunocompetence. *Transplant Rev.* 1:92–113, 1969.

Davies, A.J.S. The thymus and the cellular basis of immunity. *Transplant. Rev.* 1:43–91, 1969.

39. Waksman, B.H. Overview: biology of the lymphokines. In Biology of the Lymphokines (S. Cohen, E. Pick, J. Oppenheim, eds.), pp. 585–616. New York: Academic Press, 1979.

Watson, J.; Mochizuki, D.; Gillis, S. T-cell growth factors: interleukin 2. *Immunol. Today* 1:113–116, 1980.

Schreier, M.H.; Iscove, N.N.; Tees, R.; Aarden, L.; von Boehmer, H. Clones of killer and helper T cells: growth requirements, specificity and retention of function in long-term culture. *Immunol. Rev.* 51:315–336, 1980.

Paul, W.E.; Sredni, B.; Schwartz, R.H. Long-term growth and cloning of non-transformed lymphocytes. *Nature* 294:697–699, 1981.

40. Gershon, R.K. T-cell control of antibody production. *Contemp. Top. Immunobiol.* 3:1–40, 1974.

41. Mitchison, N.A.; Rajewsky, K.; Taylor, R.B. Co-operation of antigenic determinants and of cells in the induction of antibodies. In Developmental Aspects of Antibody Formation and Structure, Vol. 2 (J. Stertzl, I. Ríha, eds.), pp. 547–561. New York: Academic Press, 1970.

Woodland, R.; Cantor, H. Idiotype-specific T helper cells are required to induce idiotype-positive B memory cells to secrete antibody. *Eur. J. Immunol.* 8:600–606, 1978.

42. Tada, T.; Okumura, K. The role of antigen-specific T cell factors in the immune response. *Adv. Immunol.* 28:1–87, 1979.

Germain, R.N.; Benacerraf, B. Helper and suppressor T-cell factors. *Springer Sem. Immunopathol.* 3:93–127, 1980.

43. Billingham, R.; Silvers, W. The Immunobiology of Transplantation. Foundations of Immunology Series. Englewood Cliffs, N.J.: Prentice-Hall, 1971.

44. Simonsen, M. On the nature and measurement of antigenic strength. *Transplant. Rev.* 3:22–35, 1970.

Wilson, D.B.; Howard, J.C.; Nowell, P.C. Some biological aspects of lymphocytes reactive to strong histocompatibility alloantigens. *Transplant. Rev.* 12:3–29, 1972.

45. Klein, J. Biology of the Mouse Histocompatibility-2 Complex: Principles of Immunogenetics Applied to a Single System. New York: Springer-Verlag, 1975.

Shreffler, D.C.; David, C.S. The H-2 major histocompatibility complex and the I immune response region: genetic variation, function and organization. *Adv. Immunol.* 20:125–195, 1975.

van Rood, J.J.; de Vries, R.R.P.; Bradley, B.A. Genetics and biology of the HLA system. In The Role of the Major Histocompatibility Complex in Immunobiology (M.E. Dorf, ed.), pp. 59–113. New York: Garland STPM, 1981.

Klein, J.; Juretic, A.; Constantin, N.B.; Nagy, Z.A. The traditional and a new version of the mouse H-2 complex. *Nature* 291:455–460, 1981.

46. Nathenson, S.G.; Uehara, H.; Ewenstein, B.M.; Kindt, T.J.; Coligan, J.E. Primary structural analysis of the transplantation antigens of the murine H-2 major histocompatibility complex. *Annu. Rev. Biochem.* 50:1025–1052, 1981.

Ploegh, H.L.; Orr, H.T.; Strominger, J.L. Major histocompatibility antigens: the human (HLA-A, -B, -C) and murine (H-2K, H-2D) class 1 molecules. *Cell* 24:287–299, 1981.

47. McDevitt, H.O.; Benacerraf, B. Genetic control of specific immune responses. *Adv. Immunol.* 11:31–74, 1969.

McDevitt, H.O.; Delovitch, T.L.; Press, J.L.; Murphy, D.B. Genetic and functional analysis of the Ia antigens: their possible role in regulating the immune response. *Transplant Rev.* 30:197–235, 1976.

Strominger, J.L.; et al. Biochemical analysis of products of the MHC. In The Role of the Major Histocompatibility Complex in Immunobiology (M.E. Dorf, ed.), pp. 115–172. New York: Garland STPM, 1981.

48. Zinkernagel, R.M.; Doherty, P.C. Restriction of *in vitro* T cell-mediated cytotoxicity in lymphocytic choriomeningitis within a syngeneic or semiallogeneic system. *Nature* 248:701–702, 1974.

Zinkernagel, R.M.; Doherty, P.C. MHC-restricted cytotoxic T cells: studies on the biological role of polymorphic major transplantation antigens determining T-cell restriction-specificity, function and responsiveness. *Adv. Immunol.* 27:51–177, 1979.

Shevach, E.M.; Paul, W.E.; Green, I. Histocompatibility-linked immune response gene function in guinea pigs. Specific inhibition of antigen-induced lymphocyte proliferation by alloantisera. *J. Exp. Med.* 136:1207–1221, 1972.

Schwartz, R.H.; David, C.S.; Sachs, D.H.; Paul, W.E. T lymphocyte-enriched murine peritoneal exudate cells. III. Inhibition of antigen-induced T lymphocyte proliferation with anti-Ia antisera. *J. Immunol.* 117:531–540, 1976.

49. Zinkernagel, R.M. Major transplantation antigens in host responses to infection. *Hosp. Pract.* 13(7):83–92, 1978.

50. Benacerraf, B. A hypothesis to relate the specificity of T lymphocytes and the activity of I region-specific Ir genes in macrophages and B lymphocytes. *J. Immunol.* 120:1809–1812, 1978.

Sprent, J. Role of H-2 gene products in the function of T helper cells from normal and chimeric mice measured *in vivo. Immunol. Rev.* 42:108–137, 1978.

51. Jerne, N.K. The generation of self tolerance and of antibody diversity. *Eur. J. Immunol.* 1:1–9, 1971.

von Boehmer, H.; Haas, W.; Jerne, N.K. Major histocompatibility complex-linked immune-responsiveness is acquired by lymphocytes of low-responder mice differentiating in thymus of high-responder mice. *Proc. Natl. Acad. Sci. USA* 75:2439–2442, 1978.

Schwartz, R.H. A clonal deletion model for Ir gene control of the immune response. *Scand. J. Immunol.* 7:3–10, 1978.

52. Möller, G.; Möller, E. The concept of immunological surveillance against neoplasia. *Transplant. Rev.* 28:3–16, 1976.

Rygaard, J.; Povlsen, C.O. Is immunological surveillance not a cell-mediated immune function? *Transplantation* 17:135–136, 1974.

Lachmann, P.J.; Mitchison, N.A. Immune response to tumors. In Clinical Aspects of Immunology, 4th ed. (P. Lachmann, K. Peters, eds.), pp. 1263–1278. Oxford, Eng.: Blackwell, 1982.

Kiessling, R.; Wigzell, H. An analysis of murine NK cell as to structure, function and biological relevance. *Immunol. Rev.* 44:165–208, 1979.

The Nervous System

18

How can we hope to understand the workings of our own brains? How are we ever to decipher the circuitry of such vast and intricate networks, composed of more than 10^{10} nerve cells, with more than a thousand times that number of interconnections? Even the largest modern computers are less complex and in many ways less powerful, or so it seems. Our understanding of the brain is so rudimentary that we scarcely know whether it makes sense to draw the comparison; we cannot even say as yet how many functionally distinct categories of nerve cells there are.

Nonetheless, important progress has been made, especially by studying the individual nerve cell and the molecules that compose it. Here, at least, some simple unifying principles can be discerned, and they provide the foundation on which any explanation of the multicellular system must be built. Indeed, it is a paradoxical fact that, although the brain as a whole remains the most baffling organ in the body, the properties of the individual nerve cells are understood better than those of almost any other cell type. From these properties we can begin to explain the operation of small parts of the much larger system in the whole organism. This chapter, therefore, will focus on the nerve cell, and the approach will be from molecules up. We shall describe how a relatively small set of membrane proteins, mostly those forming ion channels, enable nerve cells to conduct, transmit, and respond to signals. We shall then see how ion channels enable nerve cells to combine and process the information transmitted to them by these signals. Finally, we shall consider how neurons develop to form the orderly network of connections that constitutes a functioning nervous system.

Cells of the Nervous System: A Preliminary Sketch

Before embarking on a more minute account of the cells of the nervous system, it is helpful to survey very briefly some of their main features.

Nerve Cells Carry Electrical Signals

Nerve cells, or **neurons,** receive, conduct, and transmit signals. The significance of the signals varies according to the part played by the individual cell in the functioning of the nervous system as a whole (Figure 18–1). In a *motor*

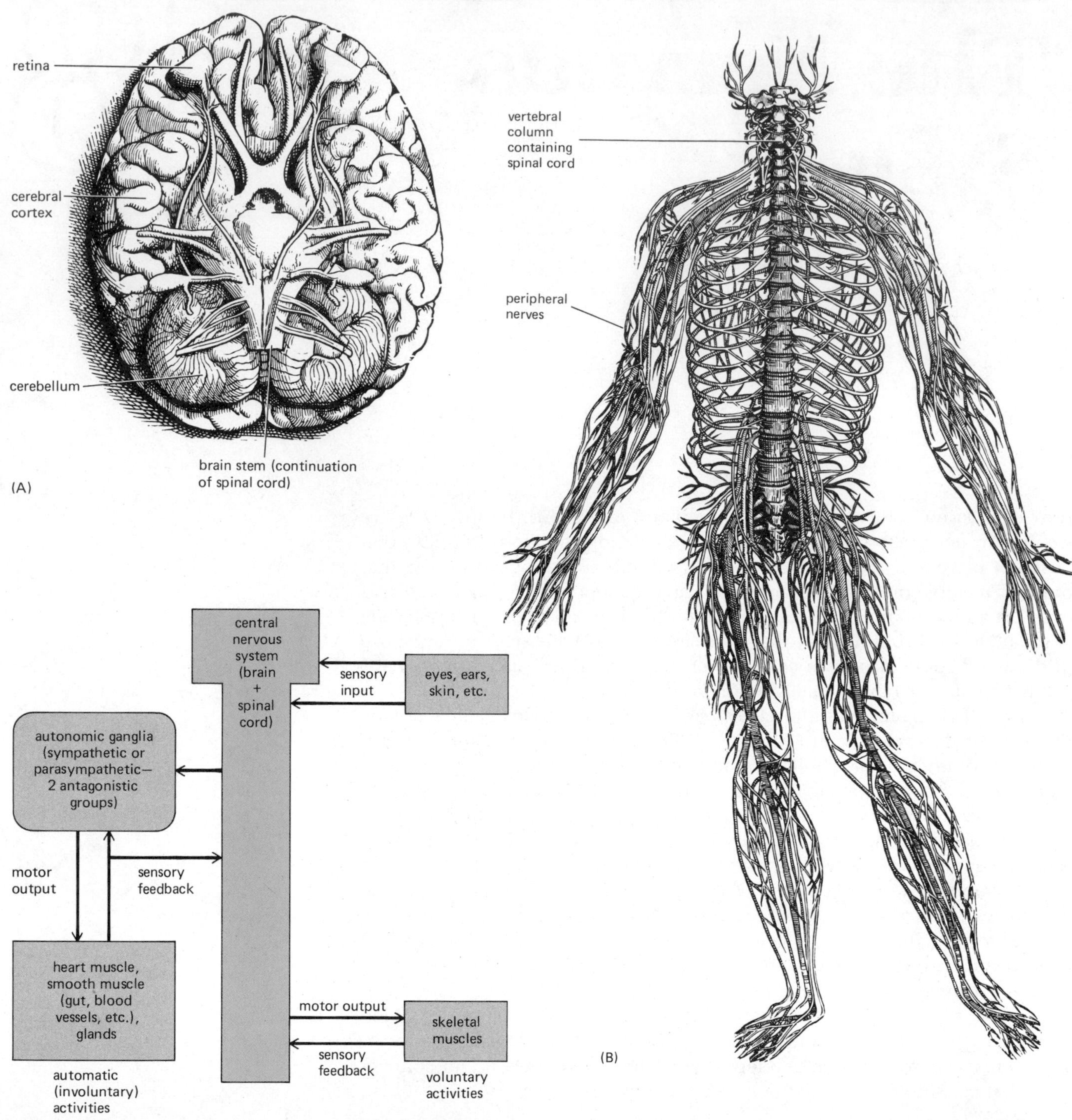

Figure 18–1 The general organization of the nervous system of a vertebrate. (A) The human brain, viewed from below. (B) The peripheral nerves. (C) A block diagram of the vertebrate nervous system as a whole. Each peripheral nerve consists of many enormously long nerve cell processes, some belonging to sensory neurons and carrying information inward to the central nervous system, others belonging to motor neurons and carrying commands outward to the muscles. The nerve cell bodies lie either inside the central nervous system (in the case of the motor neurons that control skeletal muscles) or outside it clustered in *ganglia* (in the case of sensory neurons and autonomic motor neurons). In the central nervous system, the vast majority of the nerve cells are interneurons, which both receive their inputs from and deliver their outputs to other nerve cells. The interneurons form a complex network to process sensory information and control motor output. (A and B, from A. Vesalius, De Humani Corporis Fabrica. Basel, Switz.: Oporinus, 1543.)

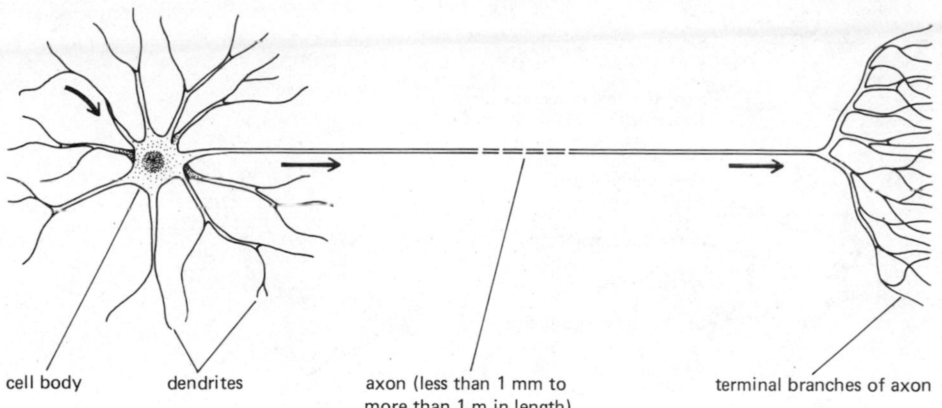

Figure 18–2 Schematic diagram of a typical neuron of a vertebrate. The arrows indicate the direction in which signals are conveyed.

neuron, the signals represent commands for the contraction of a particular muscle. In a *sensory neuron,* they represent the information that a specific type of stimulus, such as a light, a mechanical force, or a chemical substance, is present at a certain site in the body. In an *interneuron,* they represent parts of a computation that combines sensory information from many different sources and generates an appropriate set of motor commands in response. Yet despite the varied significance of the signals, their *form* in all of these cases is the same, consisting of changes of the electrical potential across the neuron's plasma membrane. Communication depends on the fact that an electrical disturbance produced in one part of the cell spreads to other parts. Without active amplification, the disturbance becomes attenuated with increasing distance from its source. Over short distances the attenuation is negligible, and in fact many small neurons conduct their signals passively, without amplification. For long-distance communication, however, such passive spread is inadequate. Thus the larger neurons have evolved an active signaling mechanism that represents one of their most striking and characteristic features. An electrical stimulus that exceeds a certain threshold strength triggers an explosion of electrical activity that is rapidly propagated along the neuron's plasma membrane. This traveling wave of electrical excitation is known as an *action potential,* or *nerve impulse.* It can carry a message without attenuation from one end of a neuron to the other at a speed of up to 100 meters per second, or even faster in some cells.

The function of a nerve cell depends on its shape, for this determines the sites from which signals can be received and the targets to which they can be relayed. Neurons in general are extremely elongated—more so than any other class of cells in the body. A motor neuron of a human being, sending out a process from the spinal cord to a muscle in the foot, may be a meter long. Typically, three major portions of the neuron are distinguishable: the **cell body,** the **dendrites,** and the **axon** (Figure 18–2). The cell body is the biosynthetic center: it contains the nucleus, together with almost all of the ribosomes, the endoplasmic reticulum, and the Golgi apparatus. The dendrites are a set of branching, tubular cell processes that extend like antennae from the cell body and provide an enlarged surface area for the reception of signals from other cells. The axon is a cell process, generally single and longer than the dendrites, that conducts action potentials away from the cell body to distant targets. It commonly divides at its far end into many branches, distributing its signals to many destinations simultaneously.

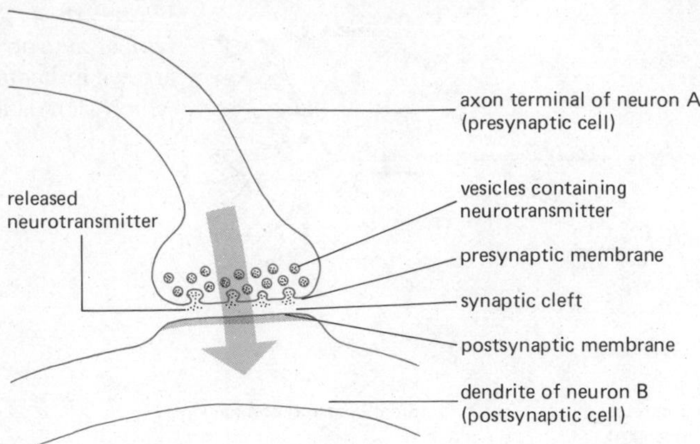

Figure 18–3 Schematic diagram of a typical synapse. An electrical signal arriving at the axon terminal of neuron A triggers the release of a chemical messenger (the neurotransmitter), which crosses the synaptic cleft and causes an electrical change in the membrane of a dendrite of neuron B. A broad arrow indicates the direction of signal transmission. The axon of a single neuron such as that shown in Figure 18–2 may make thousands of synaptic connections with other cells. Conversely, the neuron may receive signals through thousands of synaptic connections on its dendrites and cell body.

Nerve Cells Communicate Chemically at Synapses

Neuronal signals are transmitted from one cell to another at specialized sites of contact known as **synapses** (Figure 18–3). The usual mechanism of transmission appears surprisingly indirect. The cells are electrically isolated from one another, the *presynaptic cell* being separated from the *postsynaptic cell* by a *synaptic cleft.* A change of electrical potential in the presynaptic cell triggers it to release a chemical known as a **neurotransmitter,** which diffuses across the synaptic cleft and provokes an electrical change in the postynaptic cell. Thus the communication involves converting an electrical into a chemical signal and then converting the chemical signal back to an electrical one again.

Neurons differ both in the neurotransmitter they release and in their shape and size (Figure 18–4). The diverse types of neurons are woven together to form circuits of daunting complexity. But there are often simplifying features. In many cases the structures are orderly and to a large extent repetitious. For example, the visual cortex of the brain has been shown to be constructed on a modular principle: it is composed of groups of neurons receiving information from different regions of the visual field, the neurons in each group being similarly interconnected so that each group performs an equivalent computation on its own part of the visual input.

Neural Tissue Consists of Neurons and Glial Cells

Neural tissue does not consist of neurons alone but always includes supporting or **glial cells** (Figure 18–5). In the mammalian brain, the glial cells outnumber the neurons by about ten to one, taking up practically all the space that is not occupied by neurons or by blood vessels. Glial cells in the central nervous system are grouped into four major classes: *astrocytes, oligodendrocytes, ependymal cells,* and *microglial cells.* Astrocytes provide both mechanical and metabolic support for the delicate and complex neuronal circuits; they synthesize and degrade neuronally important compounds and help to

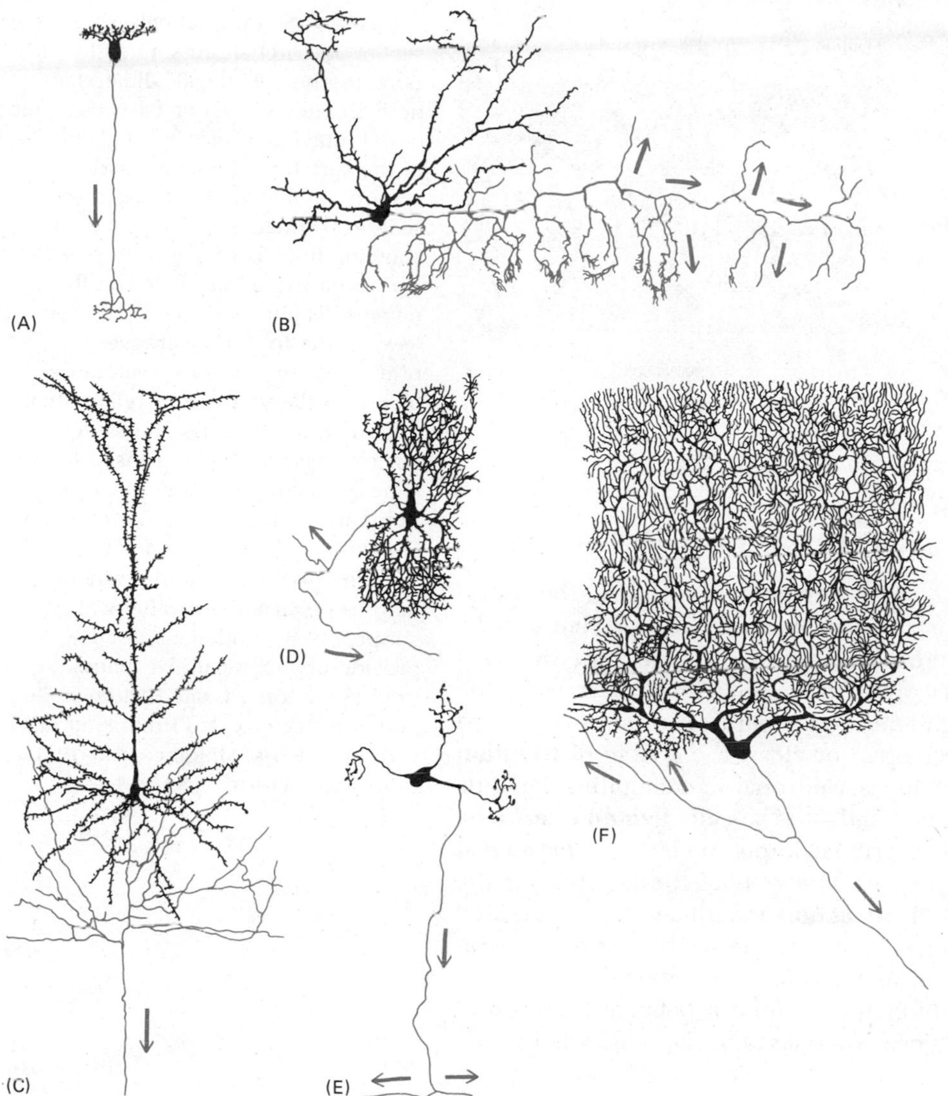

Figure 18–4 A few of the many types of neurons in the vertebrate nervous system as they appear when stained by the Golgi technique. The axon is drawn in color, the cell body and dendrites in black. Cells (A) and (B) have short axons, which are shown in their entirety. Cells (C), (D), (E), and (F) have long axons, of which only the initial portion is shown. (A) is a bipolar cell from the retina of a lizard; (B) is a basket cell from the cerebellum of a mouse; (C) is a pyramidal cell from the cerebral cortex of a rabbit; (D) is a cell from the brainstem of a human being; (E) is a granule cell from the cerebellum of a cat; and (F) is a Purkinje cell from the cerebellum of a human being. The drawings are not to scale: cell (A) is about 100 μm long, while the part of cell (F) shown in the drawing is about 400 μm across (the length of its axon—not shown—is on the order of centimeters). (From S. Ramón y Cajal, Histologie du Système Nerveux de l'Homme et des Vertébrés. Paris: Maloine, 1909–1911; reprinted, Madrid: C.S.I.C., 1972.)

control the ionic composition of the fluids surrounding nerve cells. Oligodendrocytes form insulating sheaths of myelin around neuronal processes in the central nervous system (see Figure 18–22). Ependymal cells line the internal cavities of the central nervous system, and microglial cells are a specialized type of macrophage. In embryonic development, glial cells appear to guide the migrations of neurons and the growth of axons and dendrites. They probably play other roles too, that are less well understood.

This completes our preliminary survey. We now take up the main thread of the chapter, beginning with an account of the molecular basis of electrical signaling in the neuron.

Summary

Nerve cells, or neurons, are highly elongated cells that conduct electrical signals. Typically, the signals are received on the dendrites and cell body and are then conveyed outward along the axon as action potentials, for communication to other cells via synapses, where the electrical signals are relayed by means of a chemical neurotransmitter. In addition to neurons, neural tissue always includes various glial cells, which play supporting roles.

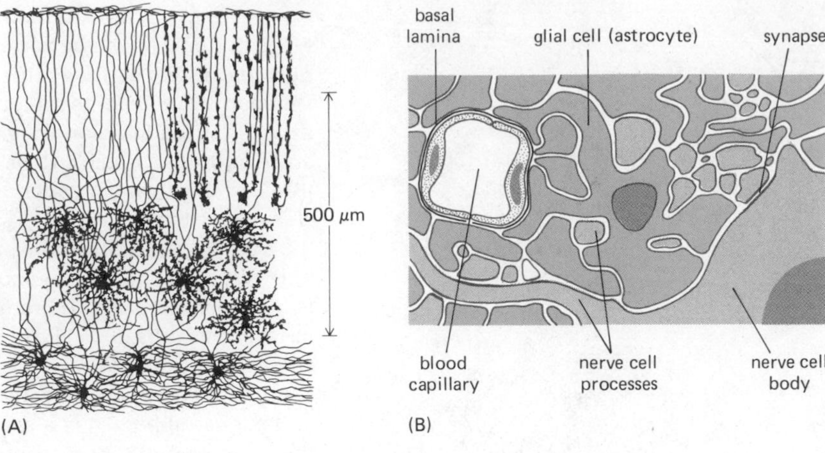

Figure 18–5 (A) Glial cells in a section of cerebellum stained by the Golgi method. The cells illustrated here are all varieties of astrocytes, one of the four major classes of glial cells in the vertebrate central nervous system. (B) Schematic diagram of a section through a part of the brain, showing how astrocyte processes *(color)* fill the spaces between the nerve cells and nerve cell processes *(gray)*. Note that the astrocyte processes surround the wall of the blood capillary; this may reflect their role in controlling the chemical environment of the neurons. Astrocytes contain especially large numbers of specialized intermediate filaments (glial fibrillary acidic protein—see p. 595) and probably help to provide the mechanical support that in other tissues is provided by extracellular matrix. (A, from S. Ramón y Cajal, Histologie du Système Nerveux de l'Homme et des Vertébrés. Paris: Maloine, 1909–1911; reprinted, Madrid: C.S.I.C., 1972.)

Voltage-gated Channels and the Action Potential[1]

The voltage difference across a cell's plasma membrane—that is, the **membrane potential**—depends on the distribution of electric charge (Figure 18–6). Charge is carried back and forth across the nerve cell membrane by small inorganic ions—chiefly Na^+, K^+, Cl^-, and Ca^{2+}—but these can traverse the lipid bilayer only by passing through special protein channels, as discussed in Chapter 6. When the ion channels open or close, the charge distribution shifts and the membrane potential changes. Neuronal signaling thus depends on channels whose permeability is regulated—the so-called **gated channels.** Two classes of gated channels are of crucial importance: (1) *voltage-gated channels*—especially voltage-gated *Na^+ channels*—play the key role in the explosions of electrical activity by which action potentials are propagated along a nerve cell process; and (2) *ligand-gated channels,* which convert extracellular chemical signals into electrical signals, play a central role in the operation of synapses. These two types of channels are not peculiar to neurons: they are also found in other cell types, such as muscle cells, where they perform similar functions.

The Na^+-K^+ Pump Charges the Battery That Powers the Action Potential[2]

The most important ions involved in the propagation of the action potential in most types of neurons are Na^+ and K^+. Neurons, like all other cells, expend a great deal of metabolic energy pumping Na^+ out and K^+ in by means of an Na^+-K^+ ATPase in the plasma membrane (see p. 291). As a result, the concentration of Na^+ is about ten times lower inside the cell than it is outside, while the distribution of potassium is roughly the reverse. These concentration differences represent a store of free energy available to drive fluxes of ions across the plasma membrane. The store is large in the sense that it is depleted only slightly by the transient ion fluxes associated with the passage of a single action potential. Even if the Na^+-K^+ pump is inactivated by an inhibitor such as ouabain, a typical nerve cell, like a well-charged storage battery, can still propagate many thousands of action potentials before it runs down. It can do so because a very small flow of ions into the cell carries sufficient charge to cause a large change in the membrane potential (Figure 18–7). One can therefore take the intracellular and extracellular concentrations of Na^+ and K^+ to be practically constant even when a cell is electrically active: the ion fluxes responsible for the action potential are so small that they cause only tiny concentration changes.

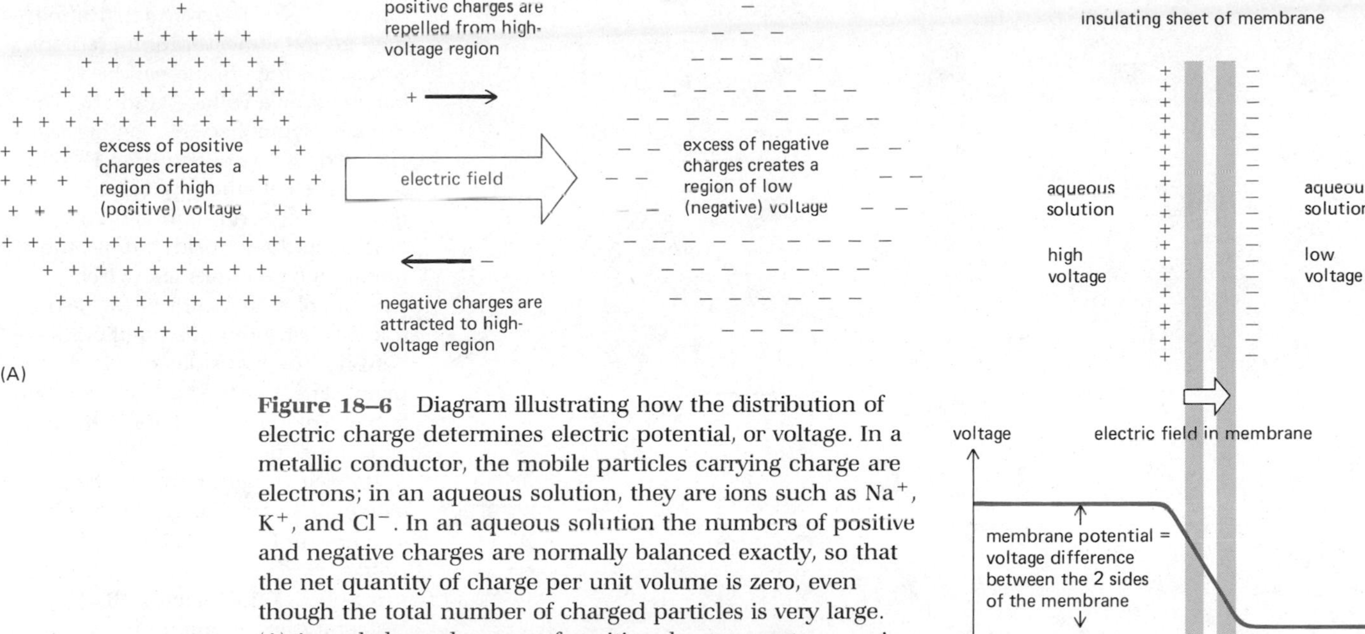

Figure 18–6 Diagram illustrating how the distribution of electric charge determines electric potential, or voltage. In a metallic conductor, the mobile particles carrying charge are electrons; in an aqueous solution, they are ions such as Na^+, K^+, and Cl^-. In an aqueous solution the numbers of positive and negative charges are normally balanced exactly, so that the net quantity of charge per unit volume is zero, even though the total number of charged particles is very large. (A) An unbalanced excess of positive charges creates a region of high electrical potential (high voltage), repelling other positive charges and attracting negative charges. An excess of negative charges has the opposite effect. (B) When an accumulation of positive charges on one surface of a membrane is balanced by an equal and opposite accumulation of negative charges on the other surface, a difference of electrical potential is set up between the two sides of the membrane.

The Membrane Potential Depends on Selective Membrane Permeability[3]

The dependence of membrane potential on membrane permeability is fundamental to all the electrical activities of the nerve cell. This dependence has already been discussed briefly in Chapter 6; here we shall analyze it in greater detail. It can best be explained by considering the nerve cell at rest and with intracellular concentrations of Na^+ and K^+ set by a balance between the fluxes driven by the Na^+-K^+ ATPase and the rates of leakage back through Na^+ and

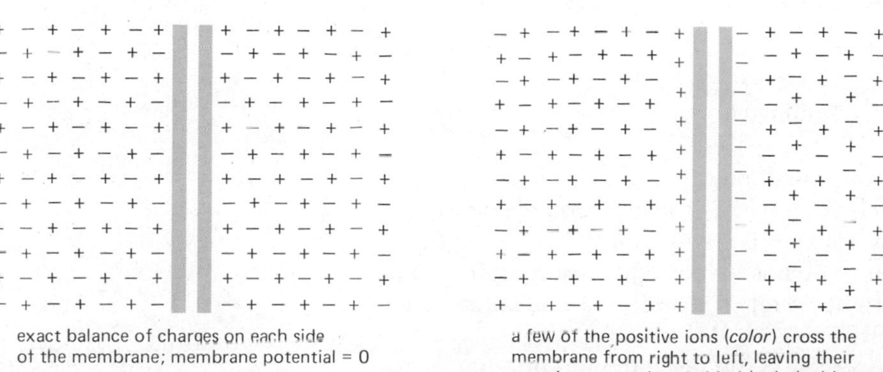

Figure 18–7 A small flow of ions carries sufficient charge to cause a large change in the membrane potential. The ions that give rise to the membrane potential lie in a surface layer close to the membrane, held there by their electrical attraction to their counterparts (counterions) on the other side of the membrane. For a typical neuron, 1 microcoulomb of charge (6×10^{12} ions) per cm^2 of membrane, transferred from one side of the membrane to the other, would change the membrane potential by roughly 1 volt; that is, the membrane has a *capacitance* of about 1 microfarad per cm^2. This means that, for example, in an axon of diameter 1 μm, the quantity of K^+ ions that have to flow out to alter the membrane potential by 100 millivolts is only about 1/10,000 of the total quantity of K^+ in the cytoplasm.

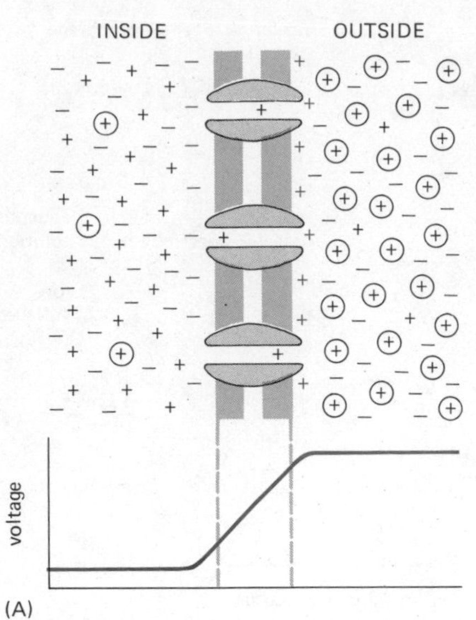

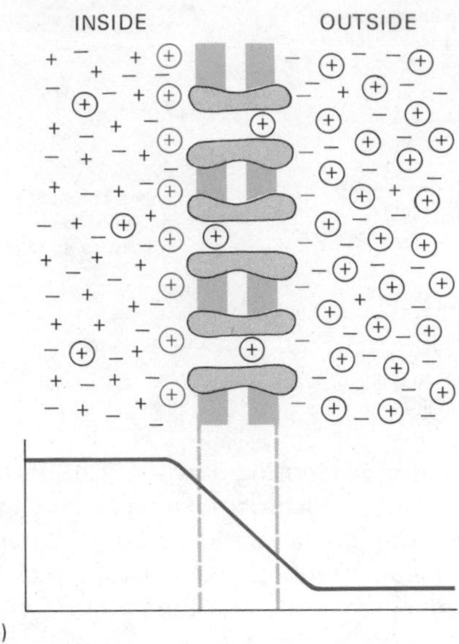

Figure 18–8 Diagram illustrating how an ion concentration gradient across the membrane can be balanced by a voltage gradient. The + signs symbolize K^+, the ⊕ signs Na^+. The K^+ concentration is high inside the cell and low outside, while the Na^+ concentration is high outside and low inside. In (A), the membrane channels allow free passage of K^+ but not of Na^+, so that the K^+ distribution is equilibrated between the two sides of the membrane, but the Na^+ distribution is not. The passage of a few K^+ ions across the membrane to the outside of the cell creates a charged layer *(color)* on each side of the membrane. The flow of K^+ continues until a voltage difference is created that just opposes the effect of the K^+ concentration gradient. At equilibrium, the electrochemical gradient for K^+ is zero, and the net flux of K^+ is zero. In (B), conversely, the membrane channels allow free passage to Na^+ but not to K^+. This results in a voltage difference being set up that is opposite in direction to that shown in (A) and just balances the Na^+ concentration difference between the two sides.

K^+ **leak channels** (see p. 290). The resting condition is defined in electrical terms: the steady-state or **resting potential** is the membrane potential at which the *net* flow of current across the plasma membrane is zero. In other words, at the resting potential, the flows of Na^+, K^+, Cl^-, and other ions, though they need not be individually zero, are exactly balanced with respect to the total charge carried across the membrane.

The flow of any particular species of ion through a membrane channel is driven by its *electrochemical gradient.* This gradient represents the combination of two influences: the voltage gradient across the membrane, and the concentration gradient of the ion. When these two influences just balance each other, the electrochemical gradient is zero and there is no net flow of the ion in question through the channel (Figure 18–8). A simple formula, the *Nernst equation,* states the equilibrium condition quantitatively. If the inside of the membrane is at a voltage V relative to the outside, and the internal and external concentrations of the ion are respectively c_i and c_o, there will be no net flow of the ion across the membrane when

$$V = \frac{RT}{zF}\ln\frac{c_o}{c_i} = \frac{RT}{zF} \times 2.3 \log_{10}\frac{c_o}{c_i}$$

where

R = the gas constant
T = the absolute temperature
F = Faraday's constant
z = the valence of the ion

For a univalent ion at room temperature

$$\frac{RT}{zF} \times 2.3 \approx 58 \text{ millivolts}$$

Thus there is no net flow of Na^+ if the membrane potential has a value of $58 \log_{10} ([Na^+]_o/[Na^+]_i)$ millivolts, which is known as the *Na^+ equilibrium potential,* V_{Na}. Similarly, the net flow of K^+ is zero if the membrane potential has the value $58 \log_{10} ([K^+]_o/[K^+]_i)$ millivolts, which is the K^+ equilibrium potential, V_K. For a typical cell, V_{Na} is somewhere between +50 and +65 millivolts, and V_K is between −70 and −100 millivolts.

For any particular membrane potential V, the net force tending to drive ions out of the cell is proportional to the difference $V - V_{Na}$ for Na^+, and

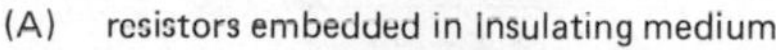

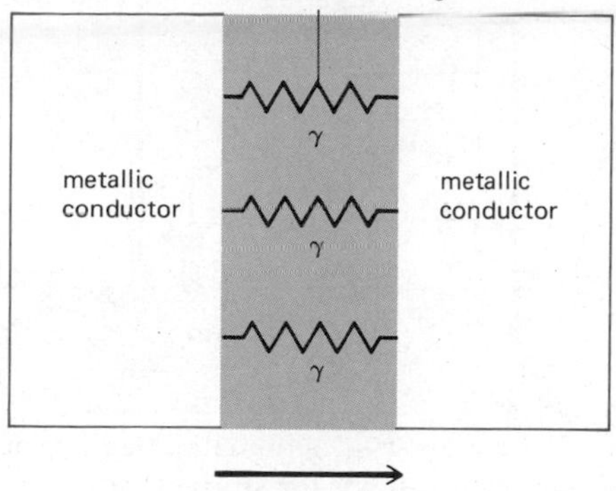

current = I
voltage drop = V

Ohm's law: $I = \frac{1}{r} \times V$ where r = resistance

or equivalently $I = g \times V$ where g = conductance $= \frac{1}{r}$

Through one resistor:

$$\text{current} = I^{(1)} = \gamma \times V$$

where γ = conductance of single resistor

Through many resistors in parallel:

$$\text{total current} = I = (\gamma + \gamma + \ldots + \gamma)V$$
$$= gV$$

where g = total conductance

(B) channels embedded in plasma membrane

CYTOPLASM (INSIDE)
EXTRACELLULAR SPACE (OUTSIDE)
solution of ions (low [Na^+])
solution of ions (high [Na^+])
γ_{Na}
γ_{Na}
γ_{Na}

Na^+ current = I_{Na}
voltage drop = V

reverse driving force for Na^+ due to concentration difference = V_{Na}
net driving force for Na^+ = $V - V_{Na}$

Through one channel:

$$Na^+ \text{ current} = I^{(1)}_{Na} = \gamma_{Na}\,(V - V_{Na})$$

Through 1 cm^2 area of membrane:

$$\text{total } Na^+ \text{ current} = I_{Na} = g_{Na}(V - V_{Na})$$

where $g_{Na} = \gamma_{Na} \times$ number of channels per cm^2

Figure 18–9 The rules relating current, voltage, resistance, and conductance (A) for a conventional electric curcuit and (B) for ion flow through a cell membrane. (B) shows the equations for the part of the electric current that is carried by Na^+; the equations for other ions are exactly analogous, and the total current is the sum of the contributions made by all ion species. Note that the conductance of a membrane channel, unlike that of an ordinary resistor, is not strictly a constant but itself depends on V and on the ion concentrations; for example, if the Na^+ concentration is very low, the Na^+ conductance will be very small. Note also that the symbol g_{Na} (or g_K) appearing in the main text refers to the combined conductance of the whole set of Na^+ (or K^+) channels in an area of membrane; the conductance of a single channel is symbolized γ_{Na} (or γ_K).

$V - V_K$ for K^+. The actual current of each ion depends not only on this driving force, but also on the ease with which that ion passes through its membrane channels (see Figure 18–9). If the *conductances* of the sets of channels for Na^+ and K^+ are respectively g_{Na} and g_K, the Na^+ and K^+ currents are respectively $g_{Na}\,(V - V_{Na})$ and $g_K\,(V - V_K)$. (Conductance is the reciprocal of resistance, and the unit of measurement is the reciprocal ohm, or *siemens*, S).

At the resting potential, by definition, the flows of the individual ion species balance so that the net electric current is zero. Consequently, if V is the resting potential, it must satisfy the equation:

$$g_{Na}\,(V - V_{Na}) + g_K\,(V - V_K) + \text{current driven by the } Na^+\text{-}K^+ \text{ ATPase pump} + \text{currents carried by other ions} = 0$$

Since the leak channels are much more permeable to K^+ than to Na^+, g_K is relatively large. Moreover, the current driven by the Na^+-K^+ pump and the currents carried by other ions are relatively small. Thus the K^+ current will be too big to be balanced by the other terms in this equation unless $V - V_K$ is close to zero. The resting potential V therefore must be close to the K^+ equilibrium potential V_K, which is between -70 and -100 millivolts. If the membrane potential differs from this resting value V, a net current will flow, tending to return the membrane potential to the resting value.

By the same argument, if the Na^+ conductance g_{Na} were to become large, the membrane potential would shift toward a new level close to V_{Na} (Figure 18–8B). In fact, this is just what happens for a brief moment during the passage of the action potential as a consequence of the opening of the voltage-gated Na^+ channels. Unlike the leak channels, these Na^+ channels are mostly closed while the neuron is at rest; they are forced open only by a change in the membrane potential. We shall now consider the properties of the voltage-gated Na^+ channels in detail.

Ion Channels Are Characterized by Their Selectivity, Their Gating, and Their Sensitivity to Specific Toxins[4]

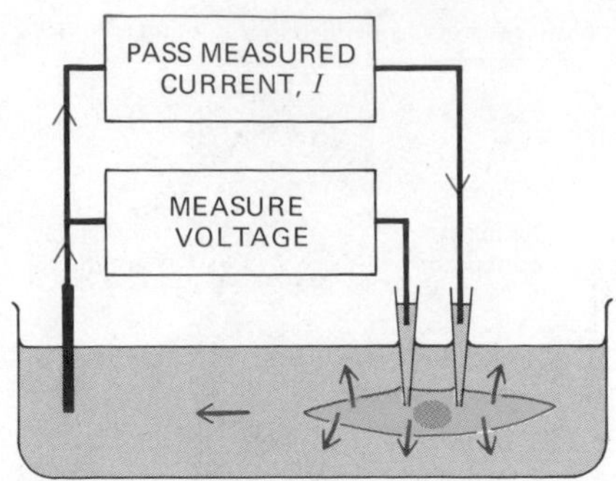

Figure 18–10 Schematic diagram of an arrangement for studying the relationship between the voltage across a cell membrane and the current passing through it. The arrows represent flow of current. The most commonly used intracellular electrodes are made of fine glass tubing, pulled to a tip diameter of a fraction of a micrometer and filled with a conducting solution of an electrolyte such as KCl. When a cell is impaled with such an electrode, the cell membrane adheres to the glass, forming a tight seal so that the interior of the electrode communicates with the interior of the cell but not with the medium in the bath. A problem with this method is that, for a highly elongated cell, the voltage measured at the tip of the recording electrode may not be the same as the voltage in distant parts of the cell. With some very large cells, such as the giant neuron of the squid, this problem can be avoided by using intracellular electrodes in the form of fine metal wires that extend along the length of the axon.

Methods for studying ion channels largely depend on the fact that a flow of ions amounts to a flow of electric current, which can be measured almost instantaneously with great precision and sensitivity. A common procedure involves inserting two microelectrodes into a cell whose membrane contains the channels of interest (Figure 18–10). One of the two intracellular electrodes—the voltage electrode—is used to measure the voltage across the membrane, relative to a third electrode that is placed in the medium in which the cell is immersed. The other intracellular electrode—the current electrode—is used to pass a measured current. If the current is directed *into* the cell, so as to add a positive contribution to the charge in the interior, the membrane potential becomes less negative than it is in the normal resting state. A shift of membrane potential in this direction is a *depolarization.* Conversely, if the electrode current flows in the opposite direction, the membrane potential becomes more negative; such a shift is a *hyperpolarization.* In either case, the altered membrane potential causes current to flow through the membrane channels, balancing the current injected through the electrode. A steady membrane potential is maintained when, and only when, there is no net loss or accumulation of charge in the cell—that is, when, and only when, the current flowing through the membrane channels is exactly equal and opposite to the current injected. Therefore, if the membrane potential is kept steady, the current through the membrane channels can be deduced from the current through the current electrode. In this way, the current electrode serves at the same time both to control the membrane potential and to measure the current through the membrane channels. As an added refinement, it is possible with suitable electronic circuitry to adjust the injected current automatically according to the signal from the voltage electrode, so as to hold the membrane potential steady at any chosen voltage, V. This arrangement is known as a **voltage clamp,** and the designated value of V is called the *command voltage.* By setting different command voltages and measuring the injected current that is required to maintain them, one can systematically investigate the membrane conductance as a function of the membrane potential.

The ions that carry the transmembrane current can be identified by monitoring the effect of changing the concentration of specific ions in the bath. For example, any current carried through the membrane channels by Na^+ will be dependent on the extracellular Na^+ concentration and will disappear if the extracellular Na^+ concentration is adjusted so that the Na^+ equilibrium potential, V_{Na}, equals the membrane potential, V. Currents through both gated and leak channels will be similarly affected by such changes in concentration gradients across the membrane and therefore can be analyzed in the same way. A voltage-gated channel will make its presence felt by an abrupt change in membrane *conductance* for a specific ion when V is abruptly changed.

Such methods make it possible to distinguish the contributions of different ions to the total current and to identify flows that are gated. But if the current is carried partly by Na^+ and partly by K^+, are these ions traveling through the same or different channels? A simple test has been made possible by the discovery of channel-specific toxins. Thus, if *tetrodotoxin* (abbreviated *TTX*)—a poison extracted from the puffer fish—is added to the external medium, the voltage-gated component of the Na^+ current is blocked, while the K^+ current is unaffected. *Tetraethylammonium* (*TEA*) ions, on the other hand, block a voltage-gated component of the K^+ current that many neurons display, leaving the Na^+ current unaffected. These observations, together with other evidence, indicate that there are at least two distinct types of voltage-gated channels: one selective for Na^+ and blockable by tetrodotoxin, the other selective for K^+ and blockable by tetraethylammonium ions.

Depolarization Causes Na^+ Channels First to Open and Then to Become Inactivated[5]

By definition, a voltage-gated channel is one that opens and closes in response to changes in the voltage across the membrane. This description suggests a simple on/off "switch." But the voltage-gated Na^+ channel responsible for the action potential is slightly more complex, and time-delays play a crucial part in its operation. The behavior of the channel can be explored using the voltage-clamp technique described above. If the membrane potential is held at the normal resting value of about −70 mV (millivolts) by setting the command voltage at that level, practically no Na^+ current flows, indicating that the Na^+ channels are largely closed. If the command voltage is now shifted abruptly to a value less negative than the normal resting potential—say to 0 mV—and the cell is held in this depolarized state, the voltage-gated Na^+ channels open, and Na^+ flows into the cell, driven by the Na^+ concentration gradient across the membrane. This Na^+ current reaches a maximum about half a millisecond after the shift to the new voltage. However, instead of remaining large, it then falls off, returning to nearly zero within a few milliseconds, even though the membrane remains depolarized (Figure 18–11). The channels have opened for a moment, only to slam shut again. Having shut, they stay shut, in an *inactivated* state that is plainly different from their initial closed state, when they were capable of opening in response to membrane depolarization. The channels do not become responsive again until the membrane voltage has been brought back toward its original negative value and a recovery period of a few milliseconds has elapsed.

Fluctuations in the Transmembrane Current Suggest That Individual Channels Are Opening and Closing Randomly[6]

The nerve cell membrane contains many thousands of voltage-gated Na^+ channels, and the Na^+ current crossing the membrane is the sum of the currents flowing through all of these. How then do the individual channels behave? Since they are all presumably identical in structure, one possibility

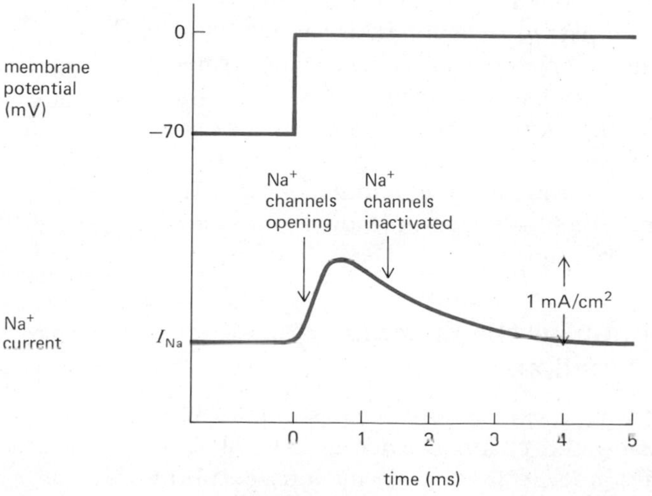

Figure 18–11 When the membrane of an axon is abruptly depolarized from the resting potential to a new fixed value, a transient Na^+ current flows, as voltage-gated Na^+ channels first open (rapidly) and then become inactivated (slowly). Data are shown here for a squid giant axon at 6°C.

is that each might pass a current that rises and falls with the same smooth and predictable time course as the total current, so acting in strict synchrony with the other channels of the same type. Alternatively, the individual channels might open and close in an abrupt all-or-none fashion, but asynchronously, with somewhat haphazard timing, after the membrane is depolarized. Such individual variations could average out in the aggregate to give the smooth record observed. If so, a closer examination of the record should reveal that it is not quite perfectly smooth or predictable: superimposed on the idealized curve there should be small, jerky, random fluctuations, reflecting the opening and closing of individual channels. Such fluctuations can in fact be observed. Evidently, the individual channels are not opening smoothly or in perfect unison.

In the absence of more direct information, individual channels can thus be studied by analyzing the current fluctuations—a procedure called **fluctuation analysis** or **noise analysis** (see Figure 18–12). On the assumption that each channel has only two possible conductance states, one fully open and the other fully closed, and that it switches instantaneously between them, it has been estimated from noise analysis that a single open Na^+ channel (in a frog nerve) has a conductance of about 10^{-11} S (siemens). This means that an electrochemical potential difference of 100 mV will drive a current of 10^{-12} ampere through the open channel, corresponding to the passage of about 6000 Na^+ ions in the course of one millisecond. This estimate, and the assumptions on which it is based, have been confirmed by the more direct method of *patch recording* discussed in the following section.

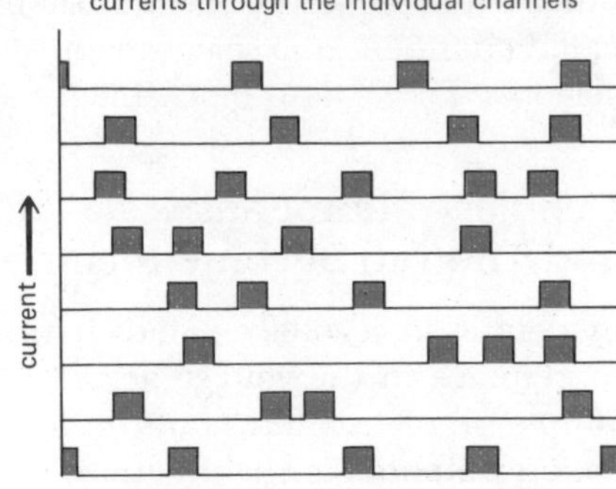

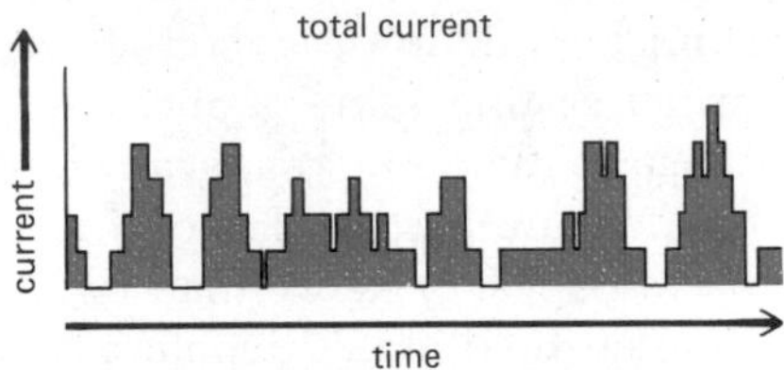

Figure 18–12 The principles of fluctuation (noise) analysis. The total current across a membrane is the sum of currents through many individual channels. If the individual channels randomly open and close (*upper set of graphs*), the total current (*lower graph*) will show random fluctuations. The number of individual channels contributing to the current and the current flowing through a single open channel can be calculated from the magnitude of the fluctuations and of the average total current. If the current through a single open channel is I and the average number of channels open at any time is n, then the average total current is nI and (from simple statistical calculations) the magnitude of the fluctuations (that is, the standard deviation of the total current about its mean) is approximately $\sqrt{n}\,I$. The value of n and of I, therefore, can be derived from measurements of the total current and its fluctuations. For the Na^+ channel immediately after a switch to a depolarized voltage, the probability of the open state changes with time (that is, n is not a constant), and thus the calculation procedure is somewhat more complex than in the steady-state situation, although the principles are the same.

Gated Channels Open and Close in an All-or-None Fashion[7]

Patch recording provides a rare—indeed, an almost unparalleled—opportunity to observe the kinetic behavior of a single protein molecule. The idea is simple, though the execution is tricky. A glass micropipette filled with saline is pressed onto the surface of a cell, and a gentle suction applied, causing the membrane to balloon into the micropipette (Figure 18–13); if the glass is clean and the cell membrane is not covered with extracellular materials, a tight electrical seal is formed.

Current then can enter the pipette only by passing through the protein channels in the patch of membrane that covers the mouth of the pipette. Given a cell with a low density of channels in its membrane, and using a pipette tip diameter of less than 1 μm, there will be only a handful of channels, and sometimes none or only one, in the patch. With modern electronic equipment the single-channel currents, of only about 10^{-12} ampere, can be recorded and measured as the voltage across the patch of membrane is varied. Figure 18–14 shows some typical observations of the response of single voltage-gated Na^+ channels from a rat muscle cell. It is plain that each channel opens in an all-or-none fashion. When open, each has the same conductivity, but the times of opening and closing are random. Therefore the aggregate current crossing the membrane of an entire cell through a large population of channels does not indicate the *degree* to which a typical individual channel is open, but rather the *probability* that it is open.

The Membrane Electric Field Controls the Energies of the Different Channel Conformations[8]

The properties of the voltage-gated Na^+ channels are worth summarizing since they are typical of other voltage-gated channels, such as the voltage-gated K^+ channels. First, the channels are selective for specific ions. Second, they do not open smoothly but make abrupt transitions between a number

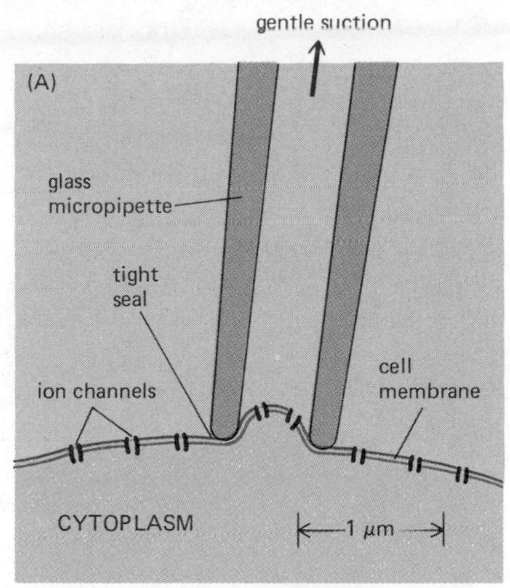

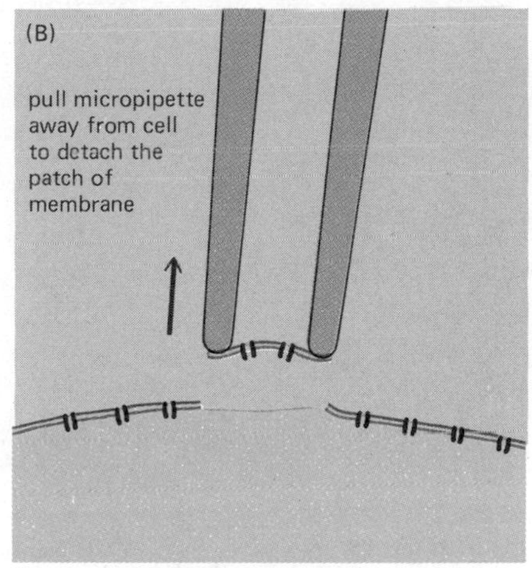

Figure 18–13 The method of patch recording. Current can enter or leave the micropipette only by passing through the channel or channels in the patch of membrane covering its tip. Recordings of the current through these channels can be made with the patch still attached to the rest of the cell, as in (A), or detached, as in (B). The advantage of the detached patch is that it is easy to alter the composition of the solution on either side of the membrane and observe the effect on channel behavior.

of discrete conformations. They can be either open or closed, but not partially open. There are at least two distinct closed conformations—at least one in which the channel is capable of opening in response to a change of membrane voltage and at least one in which it is inactivated. Transitions from one conformation to another occur probabilistically: a given conformation under given conditions has a certain probability per unit time of making a transition to another state, just as a radioactive isotope has a certain probability per unit time of undergoing radioactive decay. The rate of transition can be characterized by a *relaxation time,* analogous to the half-life of the radioisotope. In this terminology, one can say that the Na^+ channel opens and then becomes inactivated in response to membrane depolarization because, while both transitions are favored by the depolarization, the transition from the closed but responsive state to the open state occurs with a short relaxation time, while that from either of these states to the inactivated state occurs with a long relaxation time.

The phenomena of voltage-gating can be understood in terms of simple physical principles. The interior of the resting neuron is at a voltage about 50 to 100 mV more negative than the external medium. This potential difference may seem small, but since it exists across a cell membrane only 5 nm thick, the resulting voltage gradient is of the order of 100,000 V/cm. Proteins in the

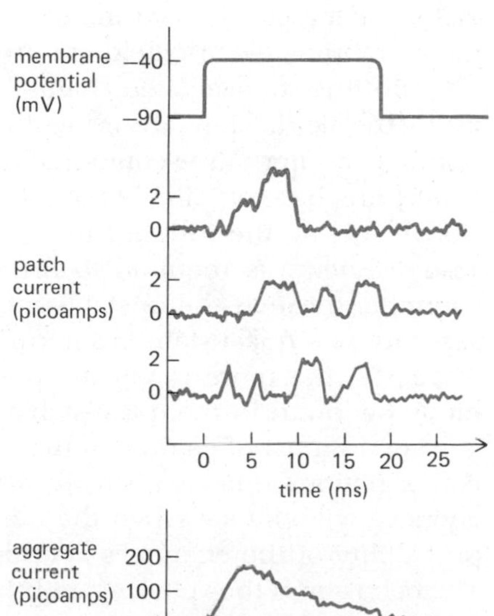

Figure 18–14 Recordings of the current through individual voltage-gated Na^+ channels in a patch of membrane detached from an embryonic rat muscle cell (see Figure 18–13B). The patch probably contained just two channels. The membrane was abruptly depolarized, as indicated in the uppermost trace, causing the Na^+ channels to open. The three current records shown are from three separate repetitions of the experiment on the same patch of membrane. Each major episode of current flow represents the opening of a single channel, except in the upper record, where two channels in the patch evidently opened together, giving for a short period a current of double amplitude (4 picoamps instead of 2 picoamps). The minor fluctuations in the current arise from electrical noise in the recording apparatus. The bottom graph shows the sum of the currents measured in 144 repetitions of the experiment; this aggregate current is equivalent to the usual Na^+ current that would be observed flowing through a relatively large region of the cell membrane, containing 288 channels in parallel. The time course of the aggregate current reflects the changing probability that any individual channel will be in the open state. (Based on data from R. Horn, J. Patlak, and C. F. Stevens, *Nature* 291:426–427, 1981.)

(A) MEMBRANE POTENTIAL = 100 MILLIVOLTS (INSIDE NEGATIVE)

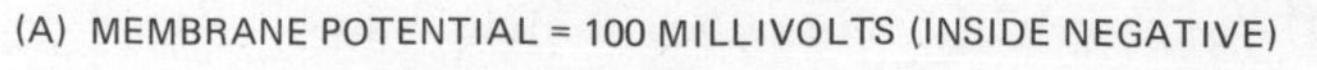

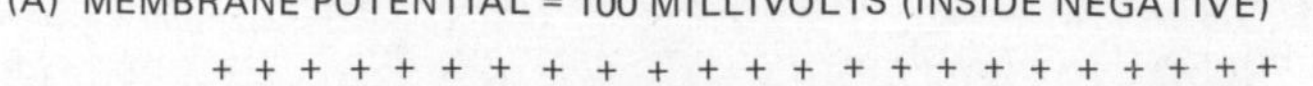

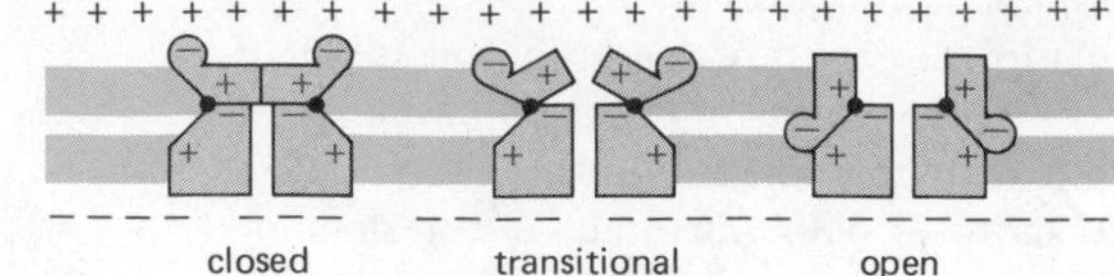

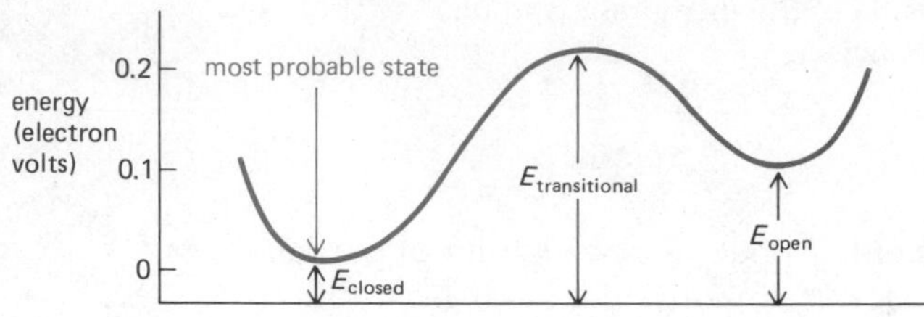

(B) MEMBRANE POTENTIAL = 0

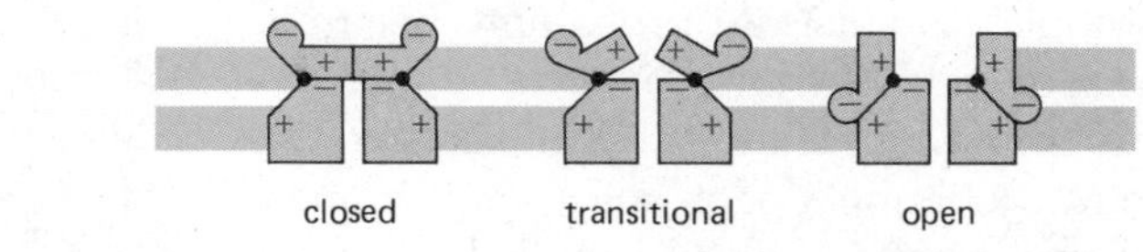

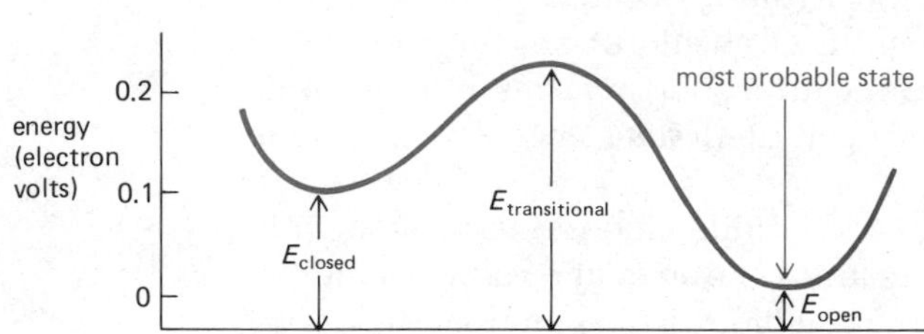

Figure 18–15 A simple model illustrating the relationship between the energy of a voltage-gated channel and its conformation during transitions between a closed and an open state. Internal forces, here represented by attractions between charges on different parts of the channel, stabilize the open and the closed conformations against small jolts, but large jolts can cause the channel to snap from one state to the other. The open and closed conformations correspond to energy minima; the transitional conformations have higher energy and are unstable. The total energy E of a given conformation depends on the membrane potential. When the membrane is strongly polarized (A), the closed conformation has the lowest energy; when the membrane is depolarized (B), the open conformation has lowest energy. The minute charge movements involved in the change of channel conformation can be detected experimentally; they are known as *gating currents*.

membrane are thus subjected to a very large electric field. Membrane proteins, like all others, have a number of charged groups on their surfaces and polarized bonds (giving rise to bond dipole moments) between their various atoms. The electric field therefore exerts forces on the molecular structure. On the other hand, the internal forces between the parts of a protein molecule are relatively strong and act to make a particular conformation stable against such distorting forces. Thus for many membrane proteins, the effects of changes in the membrane electric field are probably insignificant.

But the voltage-gated channels have evolved a delicately poised sensitivity to the field. They are evidently composed of proteins that can adopt a number of alternative conformations, each of which is stable against small distortions but can "flip" to another conformation if it is given a sufficiently violent jolt by the random thermal movements of its surroundings (Figure 18–15). Energy is required to drive the protein (or the complex of protein subunits) through the unstable intermediate transitional conformations that separate one quasi-stable conformation from another. The larger this energy barrier is, the more rarely the transition will occur. The channel will only rarely be found in the quasi-stable conformations that have a high energy; it will spend most of its time in the low energy conformations. If the alternative conformations differ with respect to their charge distribution, their relative energies will change when the membrane electric field changes, and so the probability of the channel's adopting a particular conformation will change. The behavior of the voltage-gated Na^+ channel can be easily interpreted along these lines (Figure 18–16).

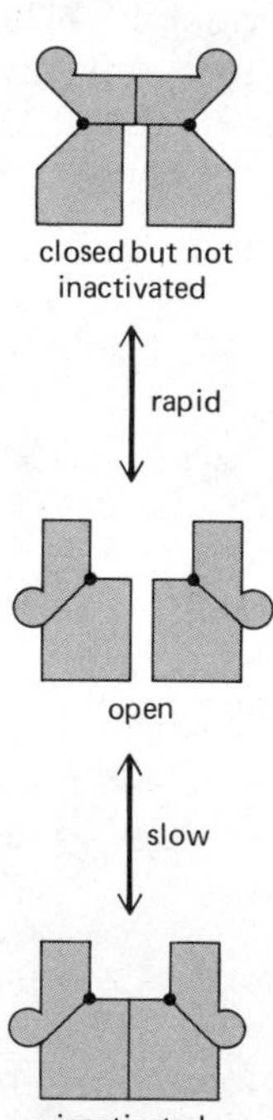

Figure 18–16 The voltage-gated Na^+ channel has at least three different quasi-stable conformations that it can adopt. In fact, detailed measurements suggest that there are probably more than three possible conformations.

Voltage-gated Na^+ Channels Are Responsible for the Action Potential[9]

Voltage-gated Na^+ channels make nerve cells electrically excitable and enable them to conduct action potentials. To explain how, we shall first examine the sequence of events in an excitation in which all parts of the cell membrane are activated at the same time. We shall then show how an excitation initiated at one end of a cell is propagated along its length.

When the membrane of a cell with many Na^+ channels is partially depolarized by a momentary stimulus, some of the channels promptly open, allowing Na^+ ions to enter the cell. The influx of positive charge depolarizes the membrane further, thereby opening more channels, which admit more Na^+ ions, causing still further depolarization. This process continues in a self-amplifying fashion until the membrane potential has shifted from its resting value of about −70 mV all the way to the Na^+ equilibrium potential of about +50 mV. At that point, where the net electrochemical driving force for the flow of Na^+ is zero, the cell would come to a new resting state with all its Na^+ channels permanently open, if the open channel conformation were stable. The cell is saved from such a permanent electrical spasm by the automatic inactivation of the Na^+ channels, which now gradually close and stay closed until the membrane potential has returned to its initial negative resting value. The whole cycle, from initial stimulus to return to the original resting state, takes a few milliseconds or less (Figure 18–17).

In many types of neurons, though not all (mammalian myelinated axons being a major exception), the recovery is hastened by the presence of voltage-gated K^+ channels in the plasma membrane. Like the Na^+ channels, these channels open in response to membrane depolarization, but they do so relatively slowly. By increasing the permeability of the membrane to K^+ just as the Na^+ channels are closing through inactivation, the K^+ channels help to bring the membrane rapidly back toward the K^+ equilibrium potential, so returning it to the resting state (Figure 18–18). The repolarization of the membrane causes the K^+ channels to close again and allows the Na^+ channels to recover from their inactivation. In this way the cell membrane can be made ready in less than a millisecond to respond to a second depolarizing stimulus.

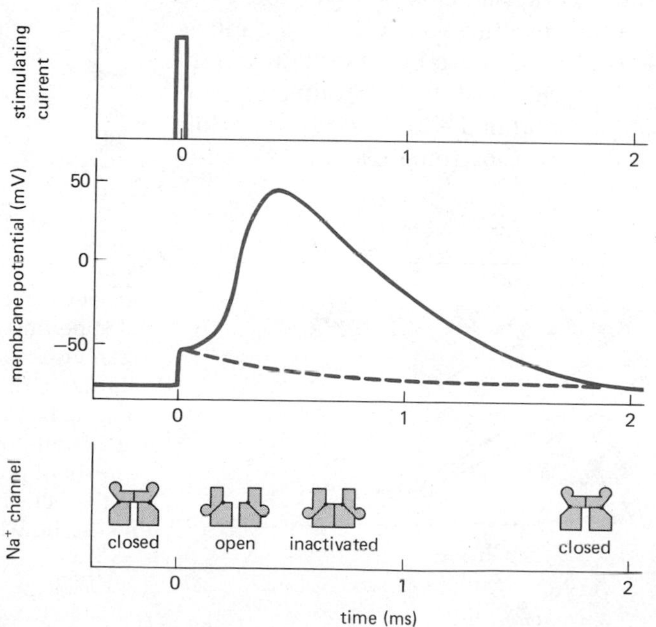

Figure 18–17 The triggering of an action potential by a brief pulse of current that partially depolarizes the membrane. In the middle graph, the solid curve shows the course of the action potential due to the opening and subsequent inactivation of the voltage-gated Na^+ channels; the dashed curve shows how the membrane potential would have simply relaxed back to the resting value after the initial depolarizing stimulus if there had been no voltage-gated channels in the membrane. Note that the membrane cannot fire a second action potential until the Na^+ channels have recovered from the state of inactivation that terminated the previous action potential. Until then the membrane is *refractory* to stimulation.

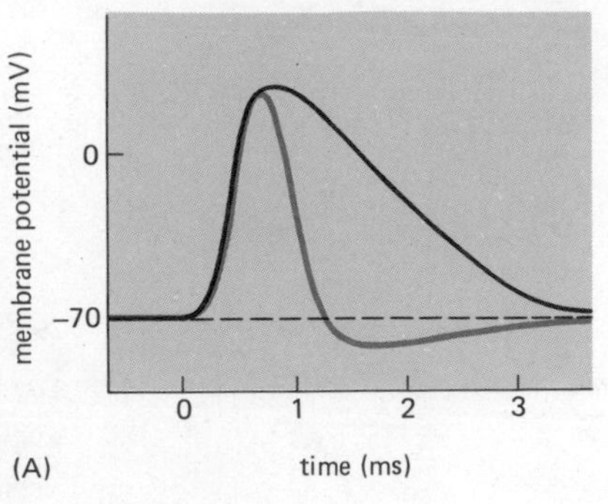

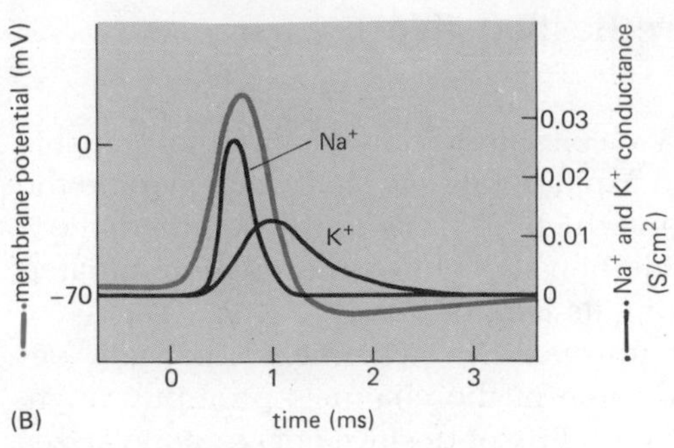

Figure 18–18 The time course of an action potential. (A) The black line shows the form of the action potential due to voltage-gated Na^+ channels alone; the colored line shows the form of the action potential when voltage-gated K^+ channels are also present, helping to bring the membrane potential back to a negative value more rapidly. Note that when voltage-gated K^+ channels are present, there is a slight hyperpolarization in the aftermath of the action potential, due to the increased permeability of the membrane to K^+. (B) The colored line shows the same action potential shown in color in (A), while the black lines show how the Na^+ and K^+ conductances of the membrane change in the course of this action potential. (Adapted from A. L. Hodgkin and A. F. Huxley, *J. Physiol.* 117:500–544, 1952.)

Action Potentials Are All or None

The sequence of events just described, in which an explosive depolarization is set off by a small decrease of the membrane potential, is equivalent to the firing of an action potential. To open enough Na^+ channels to trigger the action potential, the initial decrease in membrane potential must be sufficient to depolarize the membrane to a certain *threshold* voltage. Provided that this threshold is reached, an increase in the strength of the depolarizing stimulus makes no difference to the peak voltage change achieved by the membrane: the system, once it is triggered, drives itself to saturation irrespective of the size of the trigger stimulus (Figure 18–19). This *all-or-none* character of action potentials will be seen later to contrast with the graded nature of the voltage changes due to opening of ligand-gated channels at synapses. It is their all-or-none property that enables action potentials to convey signals over long distances without attenuation or distortion.

To see why this is so and how the mechanism works, it is helpful to consider first how electrical disturbances spread along a nerve cell in the absence of action potentials.

Voltage Changes Can Spread Passively Within a Neuron[10]

It is possible to depolarize an axon locally by injecting current through a microelectrode inserted into it (Figure 18–20). If the current is small, the depolarization will be subthreshold, practically no Na^+ channels will open, and no action potential will be triggered. In the resulting steady state, the inflow of current through the microelectrode will be balanced by an outflow of current across the membrane: the total inflow will equal the total outflow. Some of the current will flow out in the neighborhood of the microelectrode, while some will travel down the interior of the axon for some distance in either

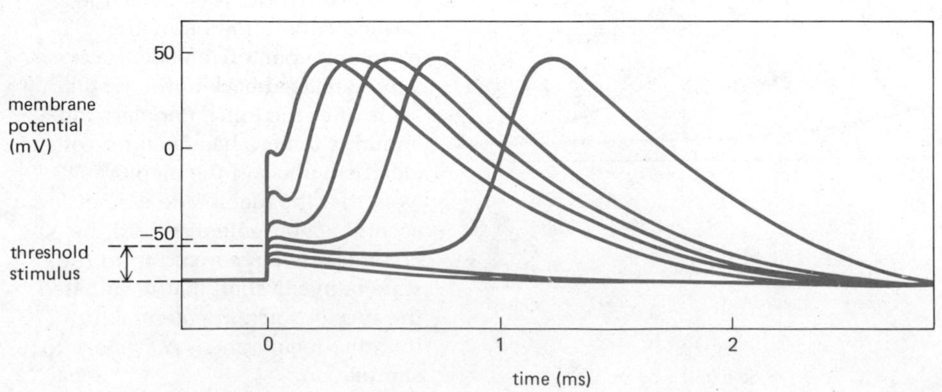

Figure 18–19 The all-or-none property of action potentials. The set of superimposed curves shows the responses to a set of triggering stimuli of different magnitudes. Stimuli below the threshold produce no action potential; stimuli above threshold produce an action potential of the full standard amplitude no matter how large the stimulus.

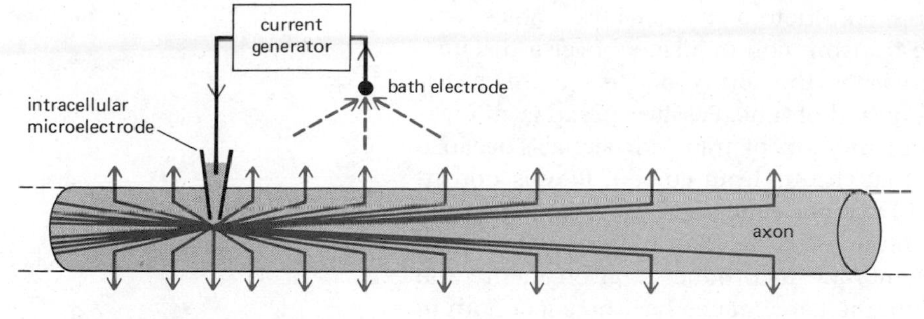

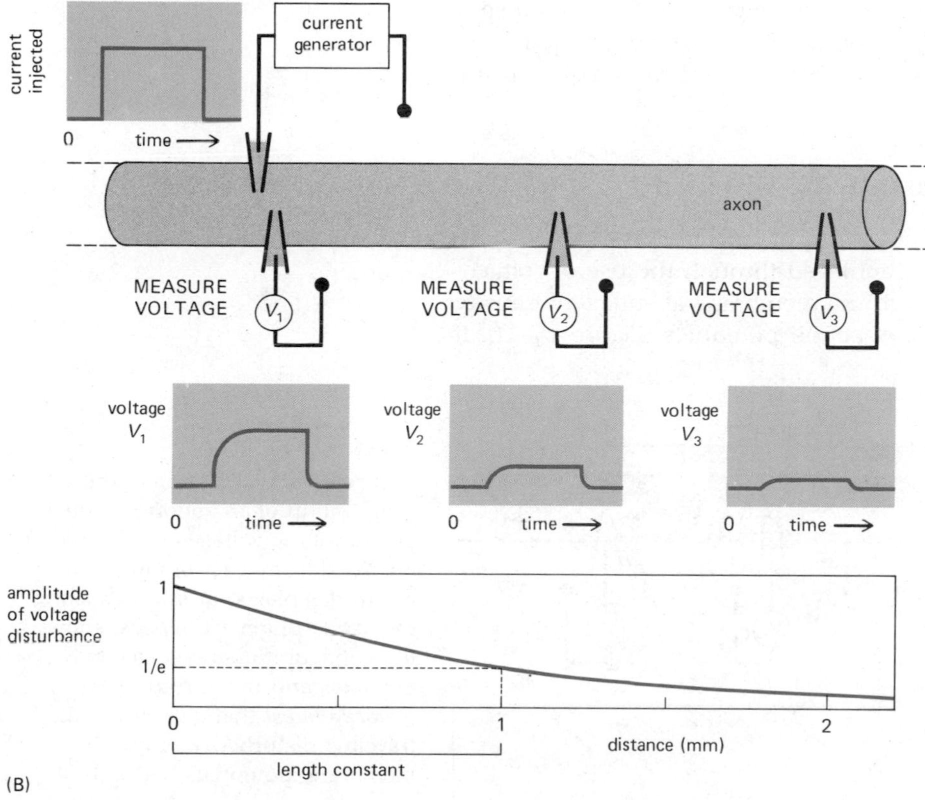

Figure 18–20 Current injected into an axon through a microelectrode flows out again across the plasma membrane; the magnitude of the outflowing current falls off exponentially with distance from the microelectrode. The current flow is assumed to cause a subthreshold depolarization of the membrane. (A) is a schematic diagram of the pattern of current flow. (B) shows how the disturbance of membrane potential produced by injection of a pulse of current falls off with distance from the source of the disturbance. The length constant is the distance over which the amplitude of the disturbance of the membrane potential falls off by a factor of $1/e$. The length constant ranges from about 0.1 mm (for a very small axon with a relatively leaky membrane) to about 5 mm (for a very large axon with a relatively nonleaky membrane). Here it is 1 mm.

direction before escaping. The amount of current flowing along each pathway will depend on the resistance of that pathway. Because the cytoplasm in the axon offers some resistance, the largest current outflow will occur in the neighborhood of the microelectrode, with progressively smaller outflows in more remote regions. The consequence of this pattern of current flow is that the magnitude of the disturbance of the membrane potential falls off exponentially with the distance from the source of the disturbance. This **passive spread** of an electrical signal along a nerve cell process—that is, spread of a signal without any amplification through the opening of voltage-gated channels—is analogous to the spread of a signal along an undersea telegraph cable; as the current flows down the central conductor (the cytoplasm), some leaks out through the sheath of insulation (the membrane) into the external medium so that the signal fed in at one end becomes progressively attenuated. For this reason, the electrical characteristics involved in passive spread are often referred to as **cable properties** of the axon.

Axons, though, are much worse conductors than electric cables, and passive spread is inadequate for the transmission of a signal over a distance of more than a few millimeters, even when the source of the signal is a disturbance that is sustained for a long period of time. Passive spread is still less satisfactory for the long-distance transmission of transient signals, because the change in membrane potential that results from current flow is not instantaneous but takes a while to build up. The time required depends on the membrane *capacitance,* that is, on the quantity of charge that has to be accumulated on either side of the membrane to produce a given membrane potential (see Figure 18–7). The membrane capacitance has the effect both of slowing down the passive transmission of signals along the axon, and of distorting them, so that a sharp pulselike stimulus delivered at one point can be detected a few millimeters away only as a slow, gradual rise and fall of the potential, with greatly diminished amplitude (see Figure 18–20). To transmit faithfully over more than a few millimeters, therefore, an axon requires, in addition to its passive cable properties, an active mechanism to maintain the strength of the signal as it travels.

Action Potentials Provide for Rapid Long-Distance Communication[11]

Rapid long-distance neural signaling is achieved through the use of voltage-gated Na^+ channels, which are present in the membrane at sufficient density all along the axon to permit the firing of action potentials (Figure 18–21). If

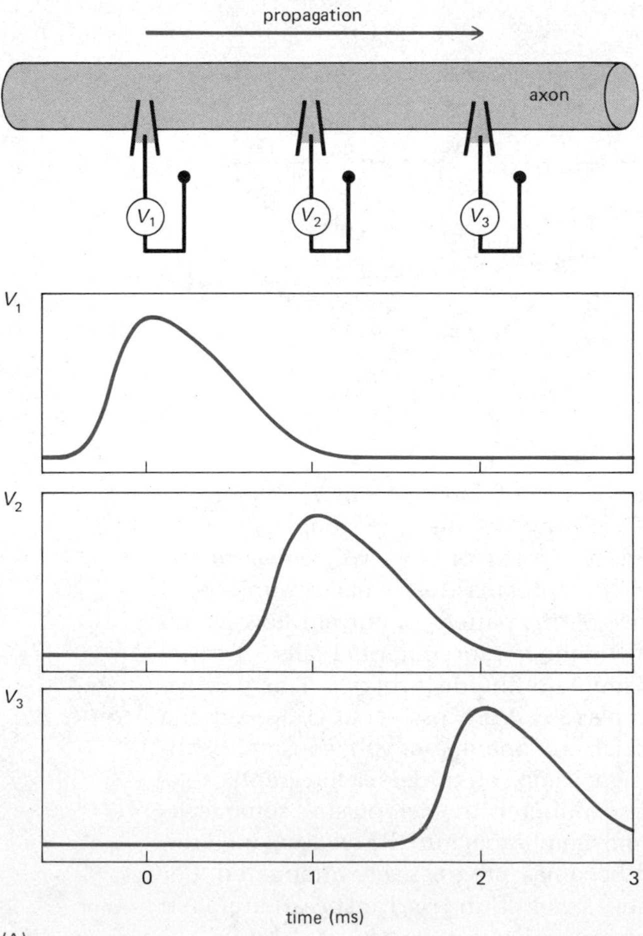

Figure 18–21 Diagram of the propagation of an action potential. (A) shows the voltages that would be recorded from a set of intracellular electrodes placed at intervals along the axon. (B) (*opposite page*) shows the configurational changes in Na^+ channels and the current flows (*colored lines*) that give rise to the traveling disturbance of the membrane potential.

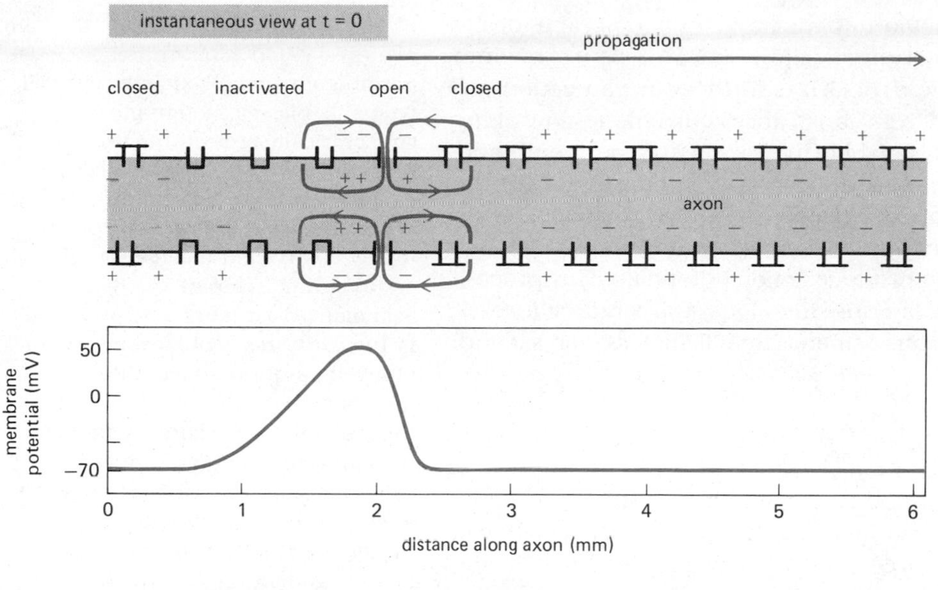
instantaneous view at t = 0
propagation
closed
inactivated
open
closed
axon
membrane potential (mV)
50
0
−70
0
1
2
3
4
5
6
distance along axon (mm)

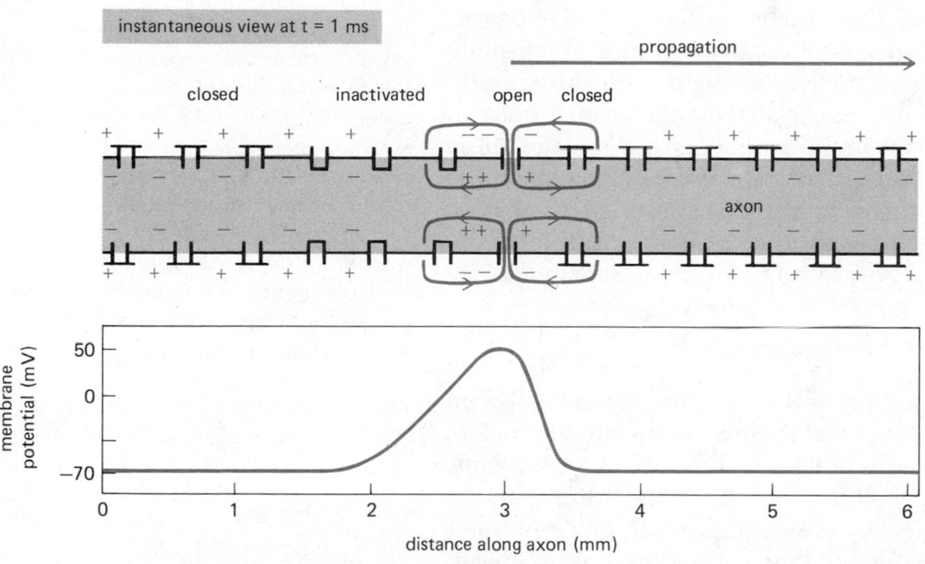
instantaneous view at t = 1 ms
propagation
closed
inactivated
open
closed
axon
membrane potential (mV)
50
0
−70
0
1
2
3
4
5
6
distance along axon (mm)

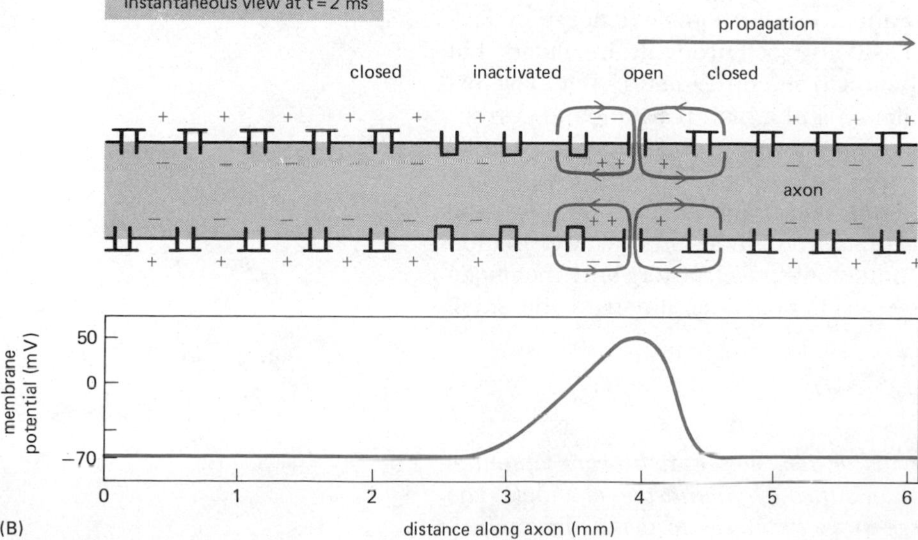
instantaneous view at t = 2 ms
propagation
closed
inactivated
open
closed
axon
membrane potential (mV)
50
0
−70
0
1
2
3
4
5
6
distance along axon (mm)

(B)

one injects enough current into an axon to depolarize the membrane locally beyond its threshold, the Na^+ channels open, allowing Na^+ ions to flow into the cell; thus the patch of membrane depolarizes further, and an action potential is observed. The large influx of Na^+ ions causes currents to flow along the length of the axon, depolarizing neighboring regions of the membrane, just as the injection of current through the microelectrode did under the conditions of passive spread. But now the depolarization of the neighboring regions of membrane through current spread is sufficient to excite them also to threshold, so that they in their turn produce action potentials. This process continues down the axon, much as fire spreads along a firecracker fuse, at speeds that in a vertebrate range from 1 meter to 100 meters per second, depending on the type of axon.

Figure 18–22 (A) Schematic diagram of a myelinated axon from a peripheral nerve. Each Schwann cell wraps its plasma membrane concentrically around the axon to form a segment of myelin sheath about 1 mm long. For clarity, the layers of myelin are not shown so tightly compacted together as they are in reality (see part D). (B) Schematic diagram of a Schwann cell in the early stages of forming a spiral of myelin around an axon during development. Note that it is the inner tongue of the Schwann cell (marked with an arrow) that continues to extend around the axon, thereby adding turns of membrane to the myelin sheath. (C) Schematic diagram of an oligodendrocyte, which forms myelin sheaths in the central nervous system. A single oligodendrocyte myelinates several separate axons. (D) Electron micrograph of a section from a nerve in the leg of a rat. Two Schwann cells can be seen: one is just beginning to myelinate its axon, the other has formed an almost mature myelin sheath. (E) Electron micrograph of an oligodendrocyte in the spinal cord of a kitten, extending processes to myelinate at least two axons. (D and E, from C. Raine, in Myelin [P. Morell, ed]. New York: Plenum, 1976.)

Myelination Speeds Conduction[12]

Anything that increases the speed and efficiency of passive spread of depolarization along the membrane will also increase the speed and efficiency of propagation of action potentials. A large axon diameter is one such factor. Thus invertebrates such as the squid have evolved giant axons, with a diameter as large as 1 mm, for rapid signaling. The vertebrates have a better device, however. They achieve equally high speeds of conduction in a much more compact manner, by insulating many of their axons with a **myelin sheath.** The myelin sheath is formed by specialized glial cells—**Schwann cells** in peripheral nerves and **oligodendrocytes** in the central nervous system. These cells wrap layer upon layer of their own plasma membrane in a tight spiral around the axon (Figure 18–22). Each myelinating Schwann cell devotes itself to a single axon, forming a segment of sheath that is about 1 mm long; oligodendrocytes form similar segments of sheath but for many separate axons simultaneously.

The myelin sheath is so tight and thick—consisting of up to 100 concentric layers of plasma membrane in some cases—that it prevents almost all current leakage across the covered portion of the axon membrane. Between one segment of sheath and the next, small regions of axon membrane remain bare (Figure 18–23). These so-called *nodes of Ranvier,* only about 0.5 μm long, are foci of electrical activity. Practically all the Na^+ channels of the axon are concentrated at the nodes, giving a density of several thousand channels per μm^2 there, while the regions of axonal membrane covered by myelin sheath are almost entirely devoid of these channels. When an action potential is triggered at a node, it depolarizes the neighboring regions as usual. The ensheathed portions of the axon membrane are not excitable, since they lack the necessary channels and, in any case, are so thoroughly insulated that hardly any current can flow across them. On the other hand, they have excellent cable properties—a low capacitance and a high resistance to current leakage. Consequently, the currents associated with the nodal action potential are promptly and efficiently funneled by passive spread to the next node, where another action potential is triggered. Thus conduction is *saltatory:* the signal propagates along the axon by leaping from node to node. Myelination brings two main advantages: action potentials travel faster, and metabolic energy is conserved because the active excitation is confined to the small nodal regions.

Summary

Electrical signaling in nerve cells depends on changes of membrane potential due to movements of small numbers of ions through membrane channels. The Na^+*-*K^+ *ATPase pump builds up a large store of energy to drive these move-*

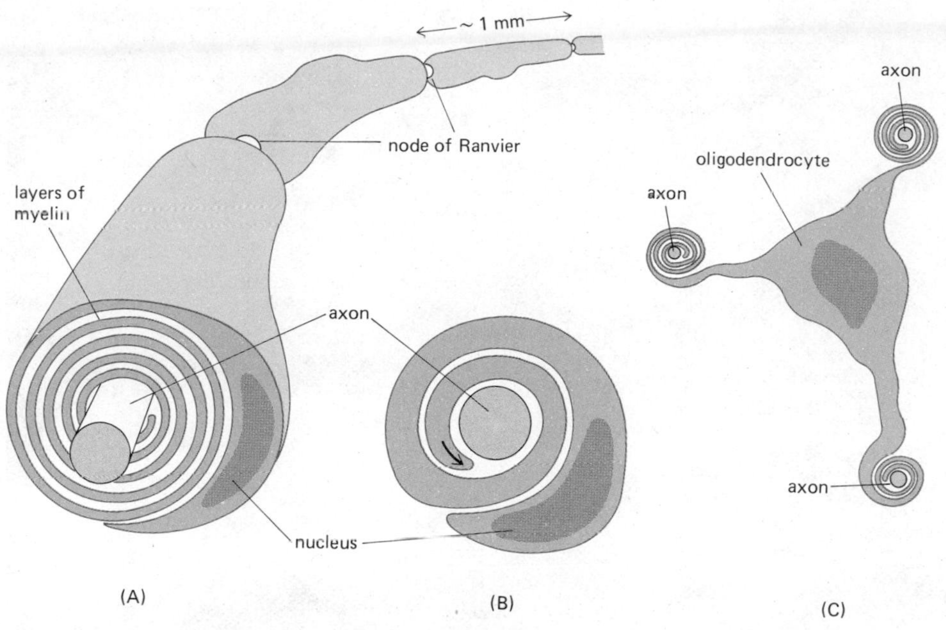
~ 1 mm
node of Ranvier
layers of myelin
axon
nucleus
(A)
(B)
axon
oligodendrocyte
axon
axon
(C)

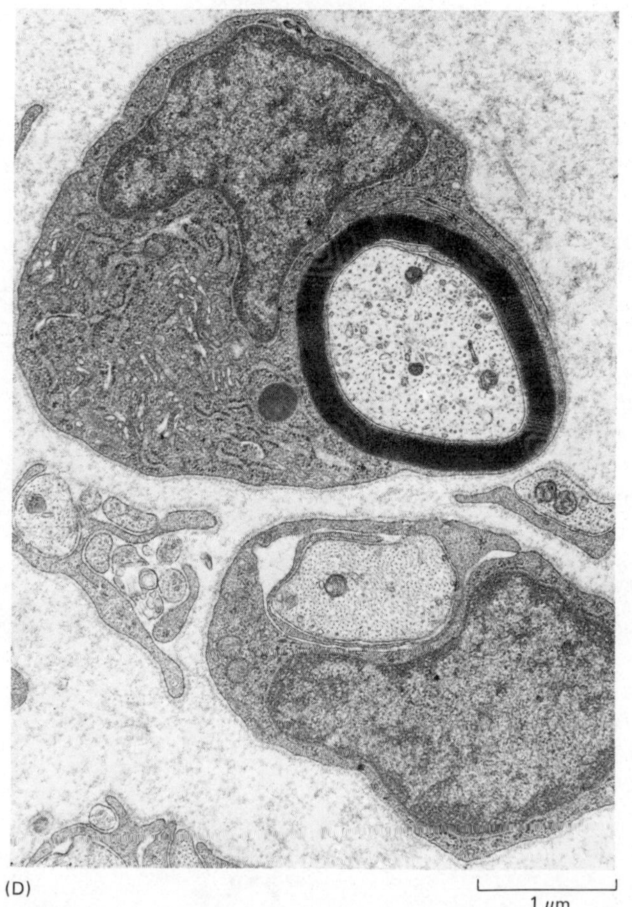
(D)
1 μm

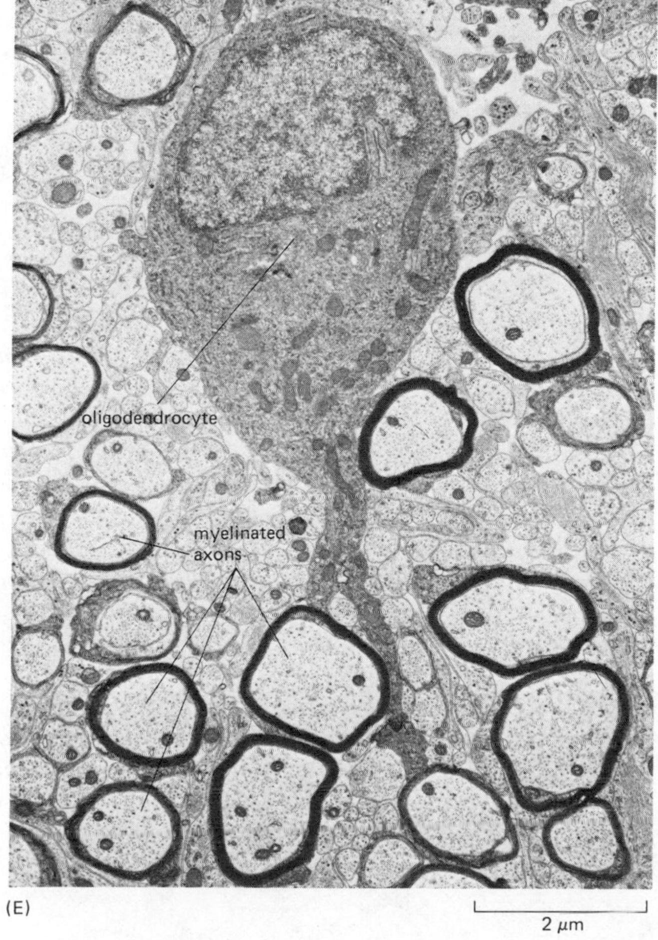
oligodendrocyte
myelinated axons
(E)
2 μm

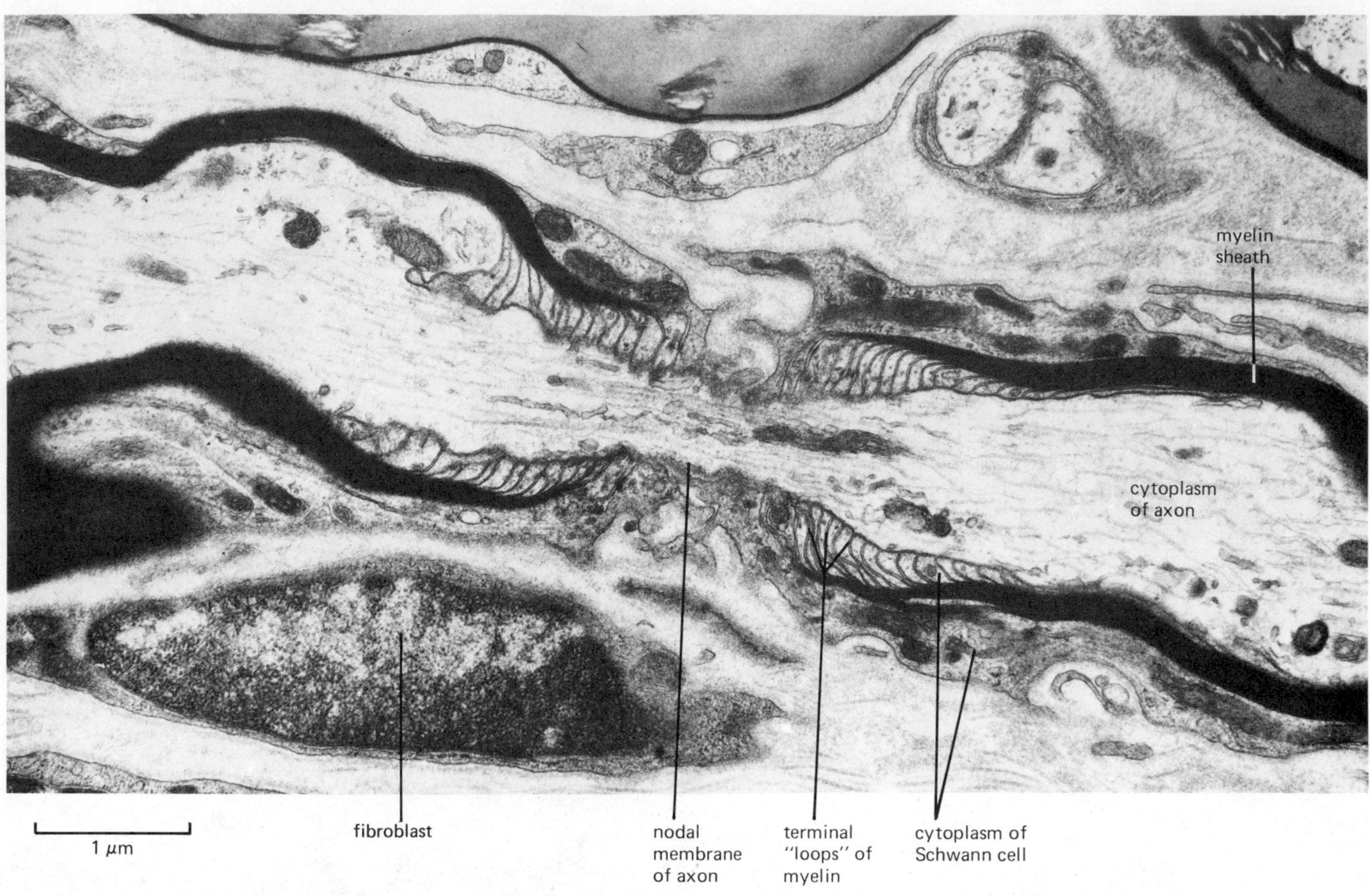

Figure 18–23 Electron micrograph of a longitudinal section of an axon from a peripheral nerve, showing a node of Ranvier, where a small portion of the axon's plasma membrane is left exposed between the ends of two adjacent segments of myelin sheath. (Courtesy of A. R. Lieberman.)

ments by creating inside the cell a concentration of Na^+ that is much lower than outside the cell and a concentration of K^+ that is much higher. In the resting neuron, the K^+-selective leak channels in the membrane make it more permeable to K^+ than to other ions, and the membrane potential is consequently close to the K^+ equilibrium potential of about −70 mV. Sudden depolarization of the membrane alters its permeability by causing voltage-gated Na^+ channels to open; but if the depolarization is maintained, these channels become inactivated. The individual channels make abrupt transitions between their alternative conformational states under the influence of the membrane electric field. An action potential is triggered when a brief depolarizing stimulus causes some of the voltage-gated Na^+ channels to open, making the membrane more permeable to Na^+ and further shifting the membrane potential toward the Na^+ equilibrium potential. This positive feedback causes still more Na^+ channels to open, so as to give an all-or-none action potential. The action potential is rapidly terminated by the inactivation of the Na^+ channels and, in many neurons, by the opening of voltage-gated K^+ channels. The propagation of an action potential along a nerve fiber depends on the fiber's passive cable properties. When the membrane is locally depolarized and fires an action potential, the current entering through open Na^+ channels at that site depolarizes neighboring regions of the membrane, where action potentials are triggered in turn. In many vertebrate axons, the speed and efficiency of propagation of action potentials are increased by insulating sheaths of myelin that leave only small regions of excitable membrane exposed.

Synaptic Transmission[13]

The simplest way for one neuron to pass its signal to another is by direct electrical coupling through gap junctions. Such **electrical synapses** between neurons occur at a number of sites in the nervous systems of many different species, including vertebrates. Electrical synapses have the virtue that transmission occurs without delay. On the other hand, they are not adaptable to such a range of different functions, or so rich in possibilities for adjustment and control, as are the **chemical synapses** that provide the majority of nerve cell connections. Electrical communication through gap junctions was considered in Chapter 12 (pp. 686–692). Here we shall confine our discussion to chemical synapses.

The principles of chemical communication at a synapse are the same as those of chemical communication by means of water-soluble hormones, as discussed in Chapter 13. In both cases, a cell releases a chemical messenger by exocytosis into the extracellular medium, and this messenger then acts on another cell, or set of cells, by binding to membrane receptor proteins. At the synapse, the messenger is the neurotransmitter, which travels only a fraction of a micrometer, by diffusion, from its source to its target; the hormone, by contrast, travels in the bloodstream over long distances. This, however, is not a radical difference. Some chemicals do double duty, acting both as circulating hormones released from endocrine cells and as neurotransmitters released from nerve endings. Moreover, there are nerve cells, typically neuronal in shape and able to conduct action potentials, whose terminals discharge their contents as hormones into the circulation. The neurosecretory cells of the hypothalamus belong in this category (see p. 719).

Synaptic transmission differs from most hormonal signaling systems in that it requires specialized mechanisms to transform electrical signals into chemical ones, and vice versa. The release of the neurotransmitter must be coupled to the arrival of an action potential in the presynaptic terminal, and the binding of the neurotransmitter to cell-surface receptors must be coupled to the generation of a voltage change in the postsynaptic cell. According to the type of receptor to which they bind, neurotransmitters produce effects that are rapid in onset and brief in duration or slow in onset and more prolonged. The former effects depend on receptors that behave as gated ion channels, so that the binding of the transmitter almost instantaneously causes a flow of current across the membrane of the postsynaptic cell; this is the most typical synaptic response and the best understood. But some neurotransmitters act more like hormones or local chemical mediators (see p. 726); these bind to receptors that are coupled to enzymes, such as adenylate cyclase, and produce more prolonged changes in the postsynaptic cell by altering the concentration of intracellular second messengers such as cyclic AMP (see Chapter 13). Synapses, therefore, can operate on different time scales. On the one hand, they provide for rapid transmission of signals; on the other, they can serve as sites where electrical signals produce longer lasting neuronal changes and even the very long-term changes that are believed to be the cellular basis of learning and memory.

In this section we shall discuss first the molecular mechanisms of the typical form of rapid synaptic transmission based on ligand-gated ion channels. We shall then consider how synapses are used for neuronal computations, that is, for processes in which signals from many different sources converge on a single postsynaptic neuron and jointly control its firing. Following this, in the next section we shall consider the mechanisms of long-term synaptic change. In all these processes, as in the propagation of the action potential, membrane channels play a central part.

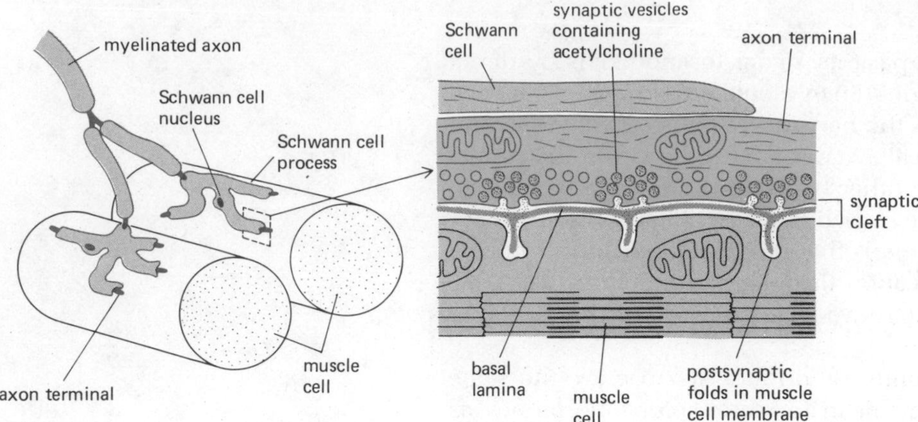

Figure 18–24 Schematic diagram of the neuromuscular junction. The neuromuscular junction is often referred to as the "end plate" because of the appearance of the axon terminal in some species.

The Neuromuscular Junction Is the Best Understood Synapse[14]

The brain is so densely packed with neurons that it is extremely difficult to perform experiments on single brain synapses. A detailed understanding of synaptic function has come instead chiefly from work on the junctions between nerve and skeletal muscle in the frog and, to a lesser extent, on synapses between giant neurons in mollusks.

Skeletal muscle cells in vertebrates, like nerve cells, are electrically excitable, and the **neuromuscular junction** (Figure 18–24) has proved to be a good model for chemical synapses in general. Figure 18–25 compares its fine structure with that of a typical synapse between two neurons in the brain. A motor nerve and its muscle can be dissected free from the surrounding tissue and maintained in a bath of controlled composition. The nerve can be stimulated with extracellular electrodes, and the response of a single muscle cell can be monitored with an intracellular microelectrode (Figure 18–26). The microelectrode is relatively easy to insert because skeletal muscle cells are very large (on the order of 100 μm in diameter).

This simple arrangement has been the basis for a long and fruitful series of investigations that began in the 1950s. The background to the early experiments was a discovery made in the 1930s that stimulation of a motor nerve causes the release of acetylcholine and that acetylcholine in its turn stimulates skeletal muscle to contract. Thus acetylcholine was identified as the neurotransmitter at the neuromuscular junction. But how is the release of acetylcholine brought about, and how does it exert its effect on the muscle?

Voltage-gated Ca^{2+} Channels Couple Action Potentials to Exocytosis[15]

The action potential is propagated along the axon by the opening and closing of Na^+ channels until it reaches the axon terminal, the site of contact with the muscle cell. Here the action potential opens **voltage-gated Ca^{2+} channels,** allowing Ca^{2+} to enter the axon terminal and elicit the exocytotic release of acetylcholine.

Two simple observations provide evidence that a gated influx of Ca^{2+} into the axon terminal is essential for synaptic transmission. First, if there is no Ca^{2+} in the extracellular medium bathing the axon terminal, no transmitter is released and transmission fails. Second, if Ca^{2+} is injected artificially into the cytoplasm at the axon terminal through a micropipette, transmitter is released even without electrical stimulation of the axon. (This microinjection

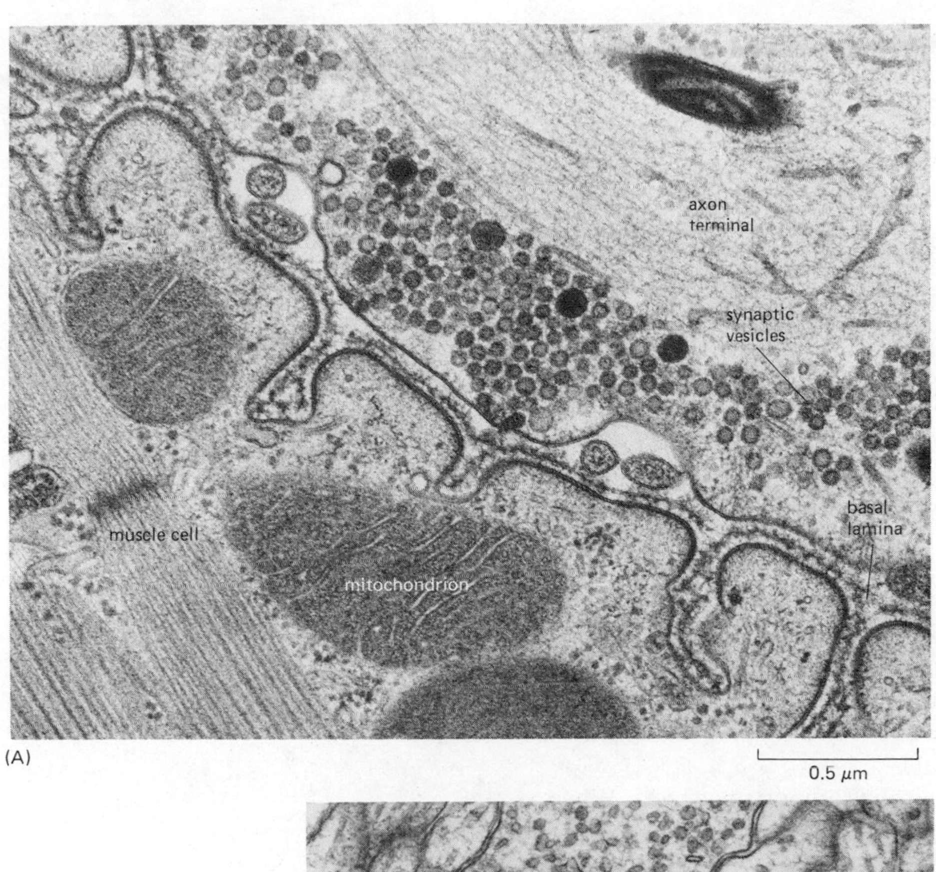

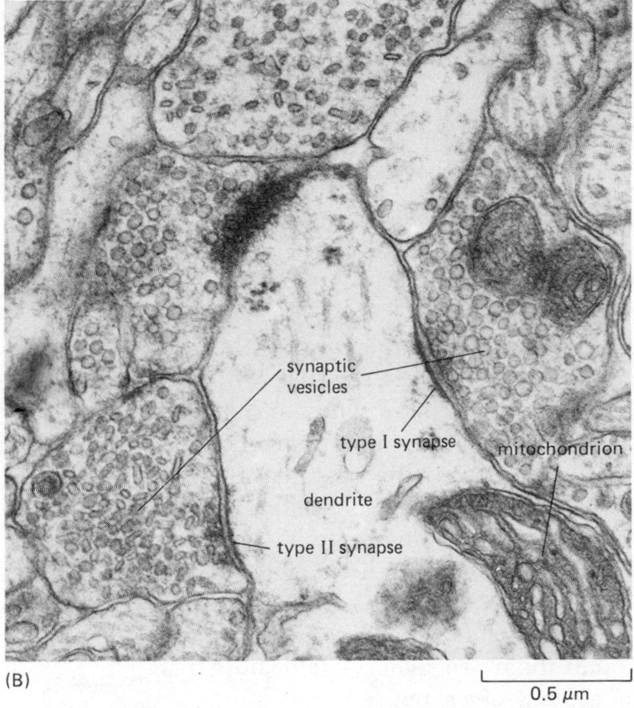

Figure 18–25 (A) Electron micrograph of part of a neuromuscular junction. (B) Electron micrograph of a small region from the brain of a rat (part of the olivary pretectal nucleus). Two synapses are clearly visible in (B), each showing pre- and postsynaptic membranes, a synaptic cleft between them, and synaptic vesicles on the presynaptic side, as in (A). The two synapses labeled in (B) differ from one another in the size and shape of their vesicles: the vesicles at the *type I* synapse are round, while those at the *type II* synapse are flattened and are believed to contain a different neurotransmitter. Note that there is no basal lamina interposed between the pre- and postsynaptic membranes at synapses in the brain, though some extracellular material, reminiscent of that seen at a desmosome junction, is faintly apparent in the cleft. The absence of a basal lamina represents the chief structural difference between a synapse in the central nervous system and a neuromuscular junction. Note also the characteristic "thickened" appearance of the postsynaptic membrane and, to a lesser extent, of the presynaptic membrane in both (A) and (B). (A, courtesy of John Heuser; B, courtesy of G. Campbell and A. R. Lieberman.)

experiment is difficult to do at the neuromuscular junction because the axon terminal is so small, but it has been done at a synapse between giant neurons in the squid.) Such observations have made it possible to build up the following picture of events in the axon terminal.

The Ca^{2+} channels open in response to membrane depolarization much as Na^+ channels do, but they differ from Na^+ channels in two important respects: first, they are selectively permeable to Ca^{2+} rather than Na^+ or other

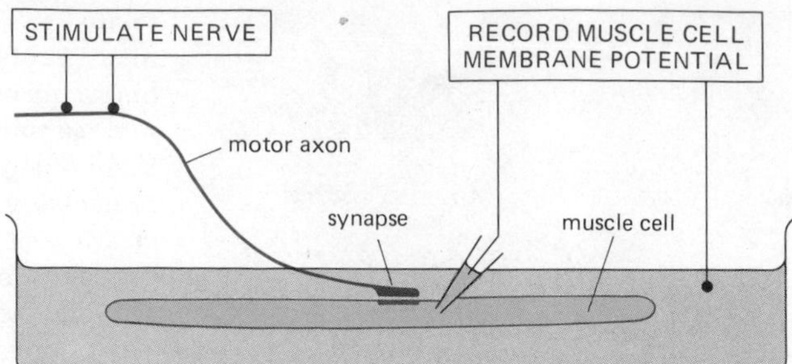

Figure 18–26 Schematic diagram of an experimental arrangement used to study synaptic transmission at the neuromuscular junction.

ions; second, they are not so rapidly inactivated and generally remain open for the duration of the depolarization. The Ca^{2+} channels in most axons are confined to the presynaptic membrane, and even there they are present only in relatively small numbers. For this reason, and because the extracellular and intracellular concentrations of Ca^{2+} are relatively low, the current flowing through the Ca^{2+} channels is usually fairly small compared with the currents flowing through the abundant voltage-gated Na^+ and K^+ channels in the axon terminal. The voltage changes in the terminal are then governed chiefly by the properties of the Na^+ and K^+ channels. (There are exceptions to this rule, however: in some cases the voltage-gated Ca^{2+} channels are more plentiful, boosting transmitter release and acting as a dominant factor in the electrical excitability of the presynaptic terminal.)

The influx of Ca^{2+}, though small, has important effects. Whereas the extracellular Ca^{2+} concentration is generally more than 10^{-3} M, the concentration of free Ca^{2+} in the cytosol is in the region of 10^{-7} M or less. The very steep Ca^{2+} concentration gradient across the plasma membrane, together with the voltage difference, drives enough Ca^{2+} through the open channels to raise the free Ca^{2+} concentration inside the axon terminal by a factor on the order of ten or a hundred, and this triggers the release of neurotransmitter. The increase of free Ca^{2+} concentration is short-lived because Ca^{2+}-binding proteins, Ca^{2+}-sequestering vesicles, and mitochondria rapidly take up the Ca^{2+} that has entered the axon terminal.

The neurotransmitter is stored in small secretory vesicles called **synaptic vesicles** inside the axon terminal, near the presynaptic membrane. The rise in Ca^{2+} concentration in the terminal causes the vesicles to fuse with the presynaptic membrane, discharging their contents into the synaptic cleft (see Figure 7–55, p. 364). The number of vesicles discharging per unit time increases very steeply with increasing Ca^{2+} concentration inside the axon terminal. At the neuromuscular junction—an extreme case—the increase roughly follows a fourth power law, so that a rise of only 20% in the internal free Ca^{2+} concentration is enough to double the rate of transmitter release. This makes transmission acutely sensitive to factors, such as the duration of the action potential, that affect the length of time that the Ca^{2+} channels remain open and the magnitude of the ion flux through them. We shall see later how this sensitivity is exploited to modulate the efficacy of a synapse.

Neurotransmitter Release Is Quantal and Random[16]

The synaptic vesicles range from about 40 nm in diameter, for those that contain acetylcholine, up to 200 nm, for other neurotransmitters, and they are easily recognized in electron micrographs of synapses (see Figure 18–25). A typical axon terminal at a neuromuscular junction contains many thousands of these vesicles, of which only a few hundred are released in response to a single action potential. Recently developed techniques have made it possible

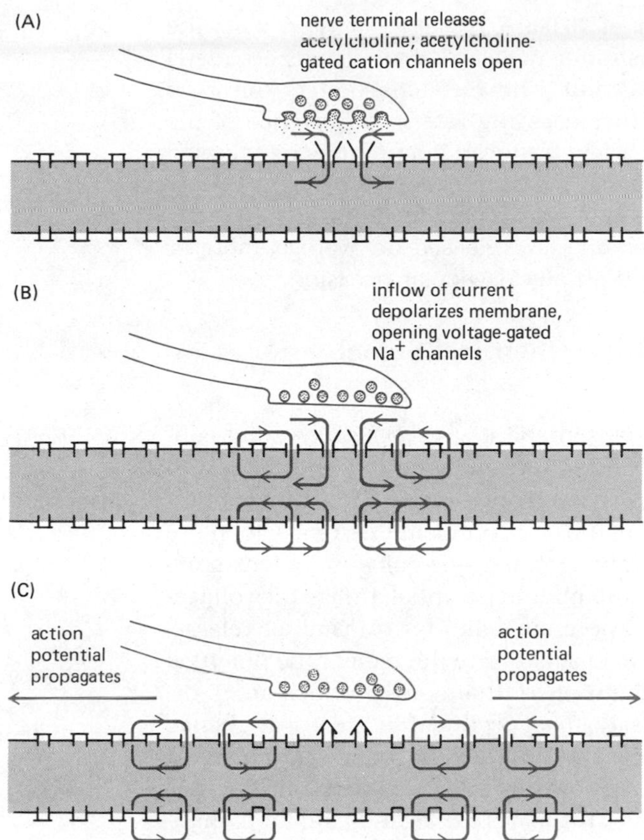

Figure 18–27 Diagram showing how the opening of ion channels gated by acetylcholine at the neuromuscular junction initiates an action potential that propagates along the muscle cell membrane, causing contraction of the muscle cell.

to catch the vesicles in the act of fusing with the presynaptic membrane; this is achieved by rapidly freezing the tissue within milliseconds after stimulating the nerve (see p. 363). However, our present knowledge of the mode of transmitter release has come first and foremost from electrophysiological experiments.

Each vesicle, by spewing its contents into the synaptic cleft, produces a voltage change in the postsynaptic cell that can be recorded with an intracellular electrode. Stimulation of the nerve generally causes the sudden release of so many vesicles that the muscle cell membrane is depolarized beyond its threshold and fires an action potential. This excitation sweeps over the cell (Figure 18–27), causing a contraction, as described on page 558. Even when the axon terminal is electrically quiet, occasional brief depolarizations of the muscle membrane are observed in the neighborhood of the synapse. These **miniature synaptic potentials** (Figure 18–28) have a rather uniform amplitude of only about 1 mV—far below threshold—and they occur at random, with a certain low probability per unit time—typically about once per second. Each miniature potential results from a single synaptic vesicle fusing with the presynaptic membrane so as to discharge its contents. The amplitude is uniform because each vesicle contains practically the same number of mol-

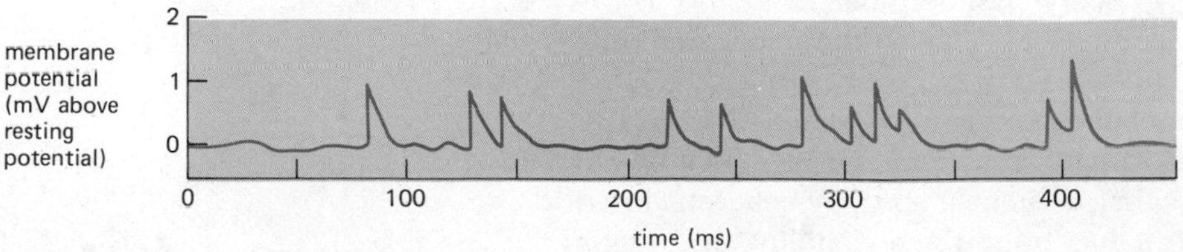

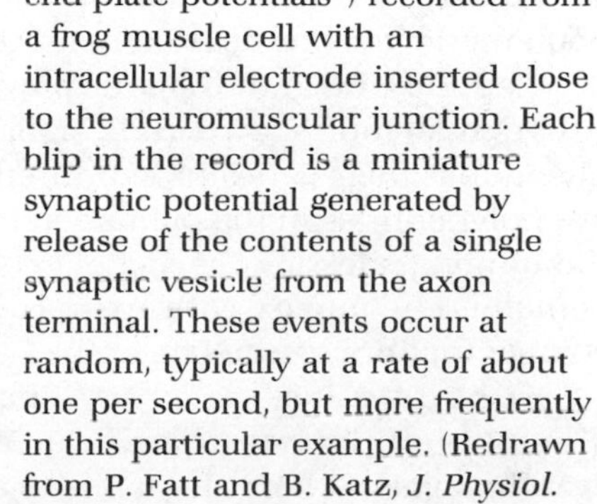

Figure 18–28 Miniature synaptic potentials (often called "miniature end-plate potentials") recorded from a frog muscle cell with an intracellular electrode inserted close to the neuromuscular junction. Each blip in the record is a miniature synaptic potential generated by release of the contents of a single synaptic vesicle from the axon terminal. These events occur at random, typically at a rate of about one per second, but more frequently in this particular example. (Redrawn from P. Fatt and B. Katz, *J. Physiol.* 117:109–128, 1952.)

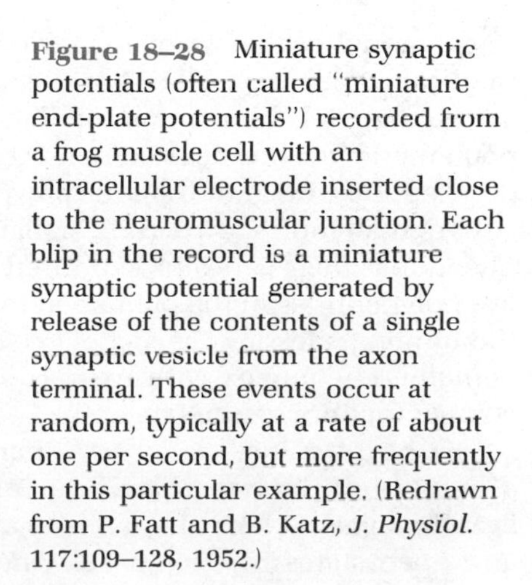

ecules of acetylcholine, on the order of 10,000. This number represents the minimum packet or *quantum* of transmitter release. Larger signals are made up of integral multiples of this basic unit. The Ca^{2+} that enters the axon terminal during an action potential increases the rate of occurrence of the exocytotic events to the point where a few hundred quanta may be released within a fraction of a millisecond. Nonetheless, the process remains probabilistic and identical stimulations of the nerve do not always produce exactly the same postsynaptic effect: if 300 quanta are released on average, more or less than this number may be released on any particular occasion.

Ligand-gated Channels Convert the Chemical Signal Back into Electrical Form

The muscle cell membrane at the synapse behaves as a *transducer* that converts a chemical signal in the form of a neurotransmitter concentration into an electrical signal. The conversion is achieved by ligand-gated ion channels in the postsynaptic membrane: when the neurotransmitter binds to these channels externally, they change their conformation—opening to let ions cross the membrane—and thereby alter the membrane potential. Unlike the voltage-gated channels responsible for action potentials and for transmitter release, the ligand-gated channels are relatively insensitive to the membrane potential (see Figure 18–29). They cannot by themselves, therefore, produce an all-or-none self-amplifying excitation. Instead, they produce an electrical change that is graded according to the intensity and duration of the external chemical signal—that is, according to how much transmitter is released into the synaptic cleft and how long it stays there. This feature of ligand-gated ion channels is important in information processing at synapses, as will be discussed later.

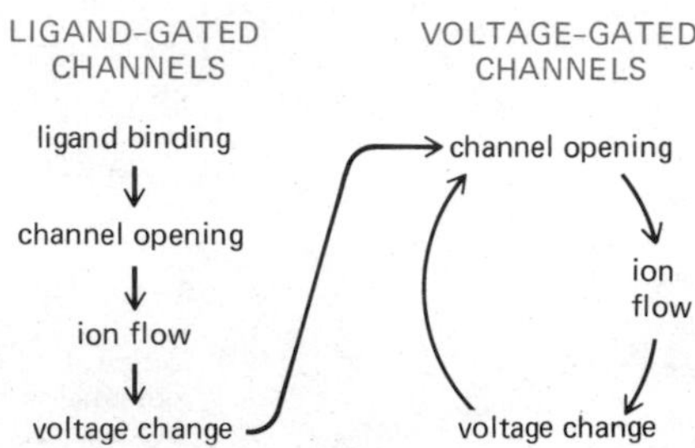

Figure 18–29 Diagram summarizing the function of ligand-gated and voltage-gated channels. The arrows indicate causal connections.

Postsynaptic ligand-gated channels have two other important properties. First, the receptors associated with them have an enzymelike specificity for particular ligands so that they respond only to one neurotransmitter—the one released from the presynaptic terminal; other transmitters are virtually without effect. Second, different types of channels are characterized by different ion selectivities: some may be selectively permeable to Na^+, others to K^+, others to Cl^-, and so on, while others may, for example, be relatively unselective among the cations but exclude anions. However, ion selectivity is fixed for a particular postsynaptic membrane: all of the channels at a synapse usually have the same unchanging selectivity.

The Acetylcholine Receptor Is a Ligand-gated Cation Channel[17]

The channel in the skeletal muscle cell membrane gated by acetylcholine is the best understood of all ligand-gated ion channels. This molecule, known as the **acetylcholine receptor,** has a molecular weight of 250,000 and is a pentameric glycoprotein composed of four distinct types of transmembrane polypeptides. Each of the four polypeptides is coded for by a separate gene, although all four show strong similarities in their amino acid sequences, implying that their genes have evolved from a single ancestral gene. Two of the five polypeptides in the pentamer, the alpha chains, are the same and provide the binding sites for acetylcholine. Two acetylcholine molecules bind to the pentameric complex with weak cooperativity and cause the conformational change that opens the channel.

Like the voltage-gated Na^+ channel, the acetylcholine-gated channel has a number of discrete alternative conformations and in the presence of the ligand jumps randomly from one to another, switching abruptly between closed and open states (Figure 18–30). Once it has bound acetylcholine and made

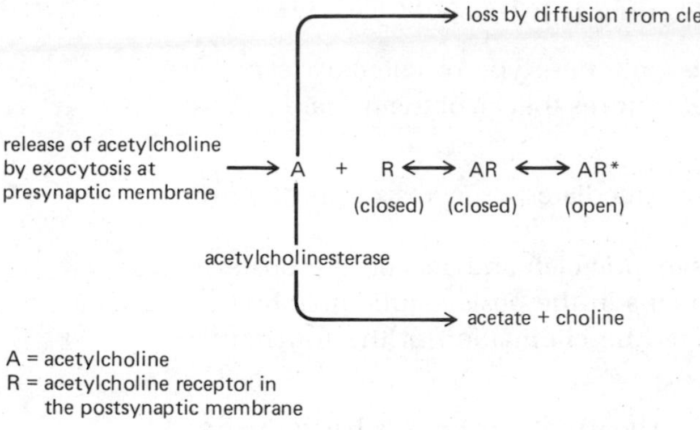

Figure 18–30 Summary of processes involving acetylcholine and the acetylcholine receptor at the neuromuscular junction.

the transition to the open state, it remains open for a randomly variable length of time, averaging about 1 ms. Greatly prolonged exposure to acetylcholine (which rarely occurs in normal circumstances) causes the channel to enter a *desensitized state,* analogous to the inactivated state of the Na^+ channel. In the open conformation, the channel has a lumen at its extracellular end about 2.5 nm in diameter, narrowing toward the cell interior to a small pore about 0.65 nm in diameter. The charge distribution in the channel wall is such that negative ions are excluded, while any positive ion of diameter less than 0.65 nm can pass through. The normal traffic consists chiefly of Na^+ and K^+, together with some Ca^{2+}. Since there is little selectivity among cations, their relative contributions to the current through the channel depend chiefly on their concentrations and on the electrochemical driving forces. If the muscle cell membrane is at its resting potential, the net driving force for K^+ is near zero, since the voltage gradient nearly balances the K^+ concentration gradient across the membrane. For Na^+, on the other hand, the voltage gradient and the concentration gradient both act in the same direction to drive ions into the cell. (The same is true for Ca^{2+}, though the extracellular concentration of Ca^{2+} is so much lower than that of Na^+ that Ca^{2+} makes only a small contribution to the total inward current.) Opening of the acetylcholine receptor channel therefore leads to a large net influx of positive ions, causing membrane depolarization.

Acetylcholine Is Removed from the Synaptic Cleft by Diffusion and by Hydrolysis[18]

If the postsynaptic cell is to be accurately controlled by the pattern of signals sent from the presynaptic cell, the postsynaptic excitation must be switched off promptly when the presynaptic cell falls quiet. At the neuromuscular junction, this is achieved by the rapid removal of acetylcholine from the synaptic cleft, through two mechanisms (Figure 18–30). First, the acetylcholine disperses by diffusion, a rapid process because the dimensions involved are small. Second, the acetylcholine is hydrolyzed to acetate and choline by *acetylcholinesterase.* This enzyme is secreted by the muscle cell and becomes anchored by a short collagenlike "tail" to the basal lamina that lies between the nerve terminal and the muscle cell membrane. Each acetylcholinesterase molecule can hydrolyze up to 10 molecules of acetylcholine per millisecond, so that all of the transmitter is eliminated from the synaptic cleft within a few hundred microseconds after its release from the nerve terminal. Thus acetylcholine is available only for a fleeting moment to bind to its receptors and drive them into the open conformation that produces the conductance change in the postsynaptic membrane (Figure 18–31). The sharply defined timing of presynaptic signals is thus preserved in sharply timed postsynaptic responses.

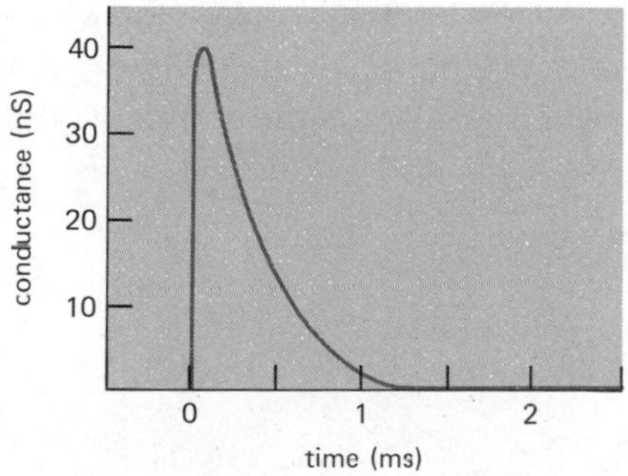

Figure 18–31 The conductance change produced in the postsynaptic membrane by a single quantum (one vesicle) of acetylcholine at the frog neuromuscular junction. About 1600 channels are open at the time of peak conductance, and each channel remains open for an average of 400 microseconds.

Some Synapses Are Excitatory, Others Inhibitory[19]

Though the neuromuscular junction is only one type of chemical synapse among many, it exemplifies the essential features that all of them share (Figure 18–32):

1. A neurotransmitter is released by Ca^{2+}-mediated exocytosis from the presynaptic nerve terminal.
2. The transmitter diffuses across the synaptic cleft and acts on the postsynaptic cell by binding to receptor proteins in the postsynaptic membrane.
3. Transmission is rapidly terminated by the elimination of the transmitter from the cleft.

We now have to consider briefly some of the variations on this basic theme.

There are many neurotransmitters besides acetylcholine. They can be divided into two major groups: on one hand, the small molecules, such as acetylcholine itself and certain *monoamines* and *amino acids;* on the other hand, the **neuropeptides** (Figure 18–33). The traditional view has been that each adult neuron secretes only one species of transmitter—the same at every synapse that it makes. But this view is now the subject of active research and dispute, and it seems that at many synapses a neuropeptide is secreted together with a neurotransmitter of the other type.

The significance of the different neurotransmitters lies not so much in their own chemistry as in the different behavior of the postsynaptic receptor proteins to which they bind. We have seen that the acetylcholine receptor in the skeletal muscle cell membrane is a cation channel, so that acetylcholine, by opening it, depolarizes the cell toward the threshold for firing an action potential. This transmitter receptor therefore mediates an *excitatory* effect. Some other receptors, such as those for the amino acid transmitter *γ-aminobutyrate* (*gamma-aminobutyric acid,* or *GABA*), mediate *inhibitory* effects by stabilizing the membrane against electrical excitation (Figure 18–34). The GABA receptor, like the acetylcholine receptor, is a gated ion channel, but it has a different ion selectivity: it admits small negative ions—chiefly Cl^-—and is impermeable to positive ions. The concentration of Cl^- is much higher outside the cell than inside, corresponding to an equilibrium potential for Cl^- that is close to the normal resting potential or even more negative. The opening of these Cl^- channels, therefore, tends to hold the membrane potential at a very negative or even a hyperpolarized value, making it difficult to depolarize the membrane and hence difficult to excite the cell.

Not only are there different types of receptors for different neurotransmitters; there can also be several different types of receptors for one and the same neurotransmitter. Acetylcholine, for example, acts in opposite ways on

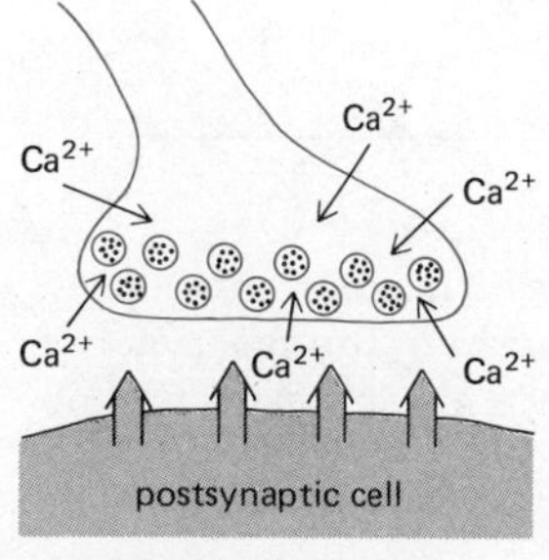

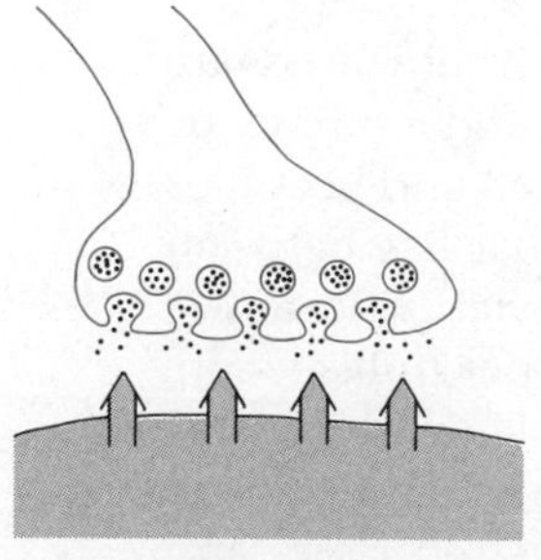

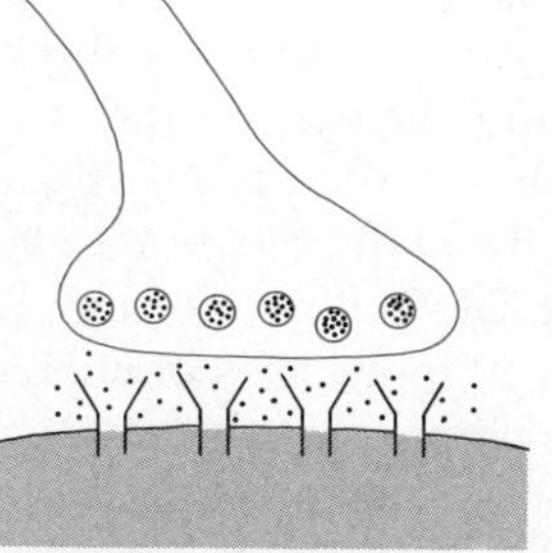

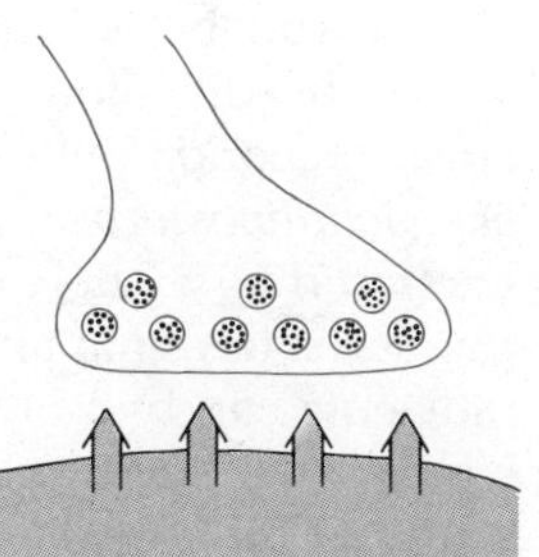

Figure 18–32 Summary of the essential events at a chemical synapse. The receptor proteins in the postsynaptic membrane can be either ion channels, as illustrated, or enzymes.

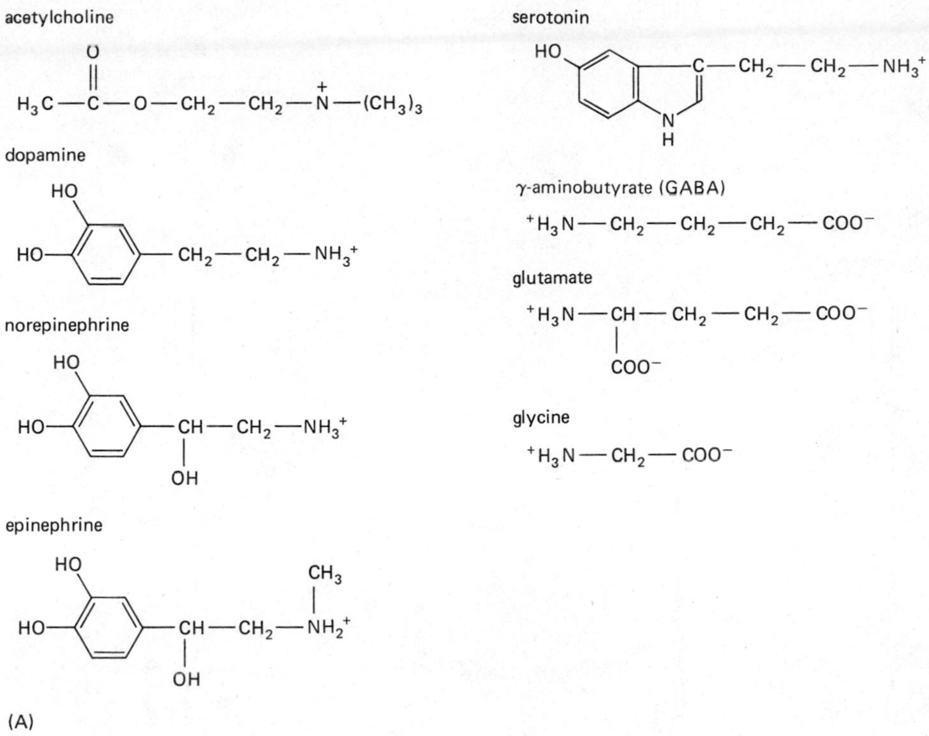

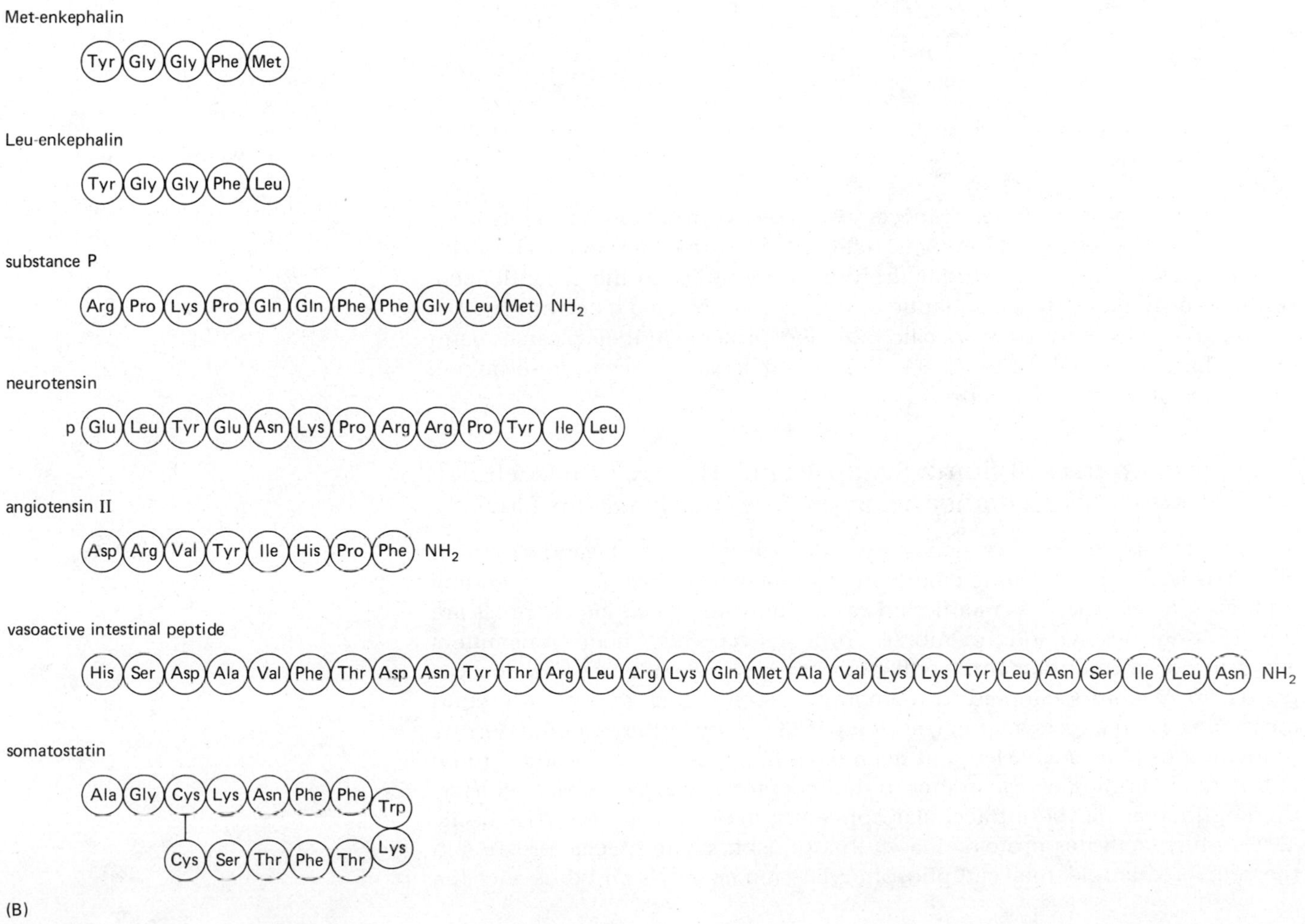

acetylcholine

$H_3C-\overset{O}{\overset{\|}{C}}-O-CH_2-CH_2-\overset{+}{N}-(CH_3)_3$

dopamine

HO-(ring, HO)-$CH_2-CH_2-NH_3^+$

norepinephrine

HO-(ring, HO)-$CH(OH)-CH_2-NH_3^+$

epinephrine

HO-(ring, HO)-$CH(OH)-CH_2-NH_2^+(CH_3)$

(A)

serotonin

HO-(indole ring, N H)-$C-CH_2-CH_2-NH_3^+$

γ-aminobutyrate (GABA)

$^+H_3N-CH_2-CH_2-CH_2-COO^-$

glutamate

$^+H_3N-CH(COO^-)-CH_2-CH_2-COO^-$

glycine

$^+H_3N-CH_2-COO^-$

Met-enkephalin

Tyr Gly Gly Phe Met

Leu-enkephalin

Tyr Gly Gly Phe Leu

substance P

Arg Pro Lys Pro Gln Gln Phe Phe Gly Leu Met NH_2

neurotensin

p Glu Leu Tyr Glu Asn Lys Pro Arg Arg Pro Tyr Ile Leu

angiotensin II

Asp Arg Val Tyr Ile His Pro Phe NH_2

vasoactive intestinal peptide

His Ser Asp Ala Val Phe Thr Asp Asn Tyr Thr Arg Leu Arg Lys Gln Met Ala Val Lys Lys Tyr Leu Asn Ser Ile Leu Asn NH_2

somatostatin

Ala Gly Cys Lys Asn Phe Phe Trp Lys Thr Phe Thr Ser Cys

(B)

Figure 18–33 (A) Some of the small molecules that have been identified as neurotransmitters. Acetylcholine, dopamine, norepinephrine, epinephrine, and serotonin (also known as 5-hydroxytryptamine, or 5-HT) are monoamines; GABA, glutamate, and glycine are amino acids. (B) A small selection of the many neuropeptides that are thought to act as neurotransmitters. (The suffix NH_2 marks an amino acid that is modified to the amide form; the prefix p marks one modified to the pyro form.)

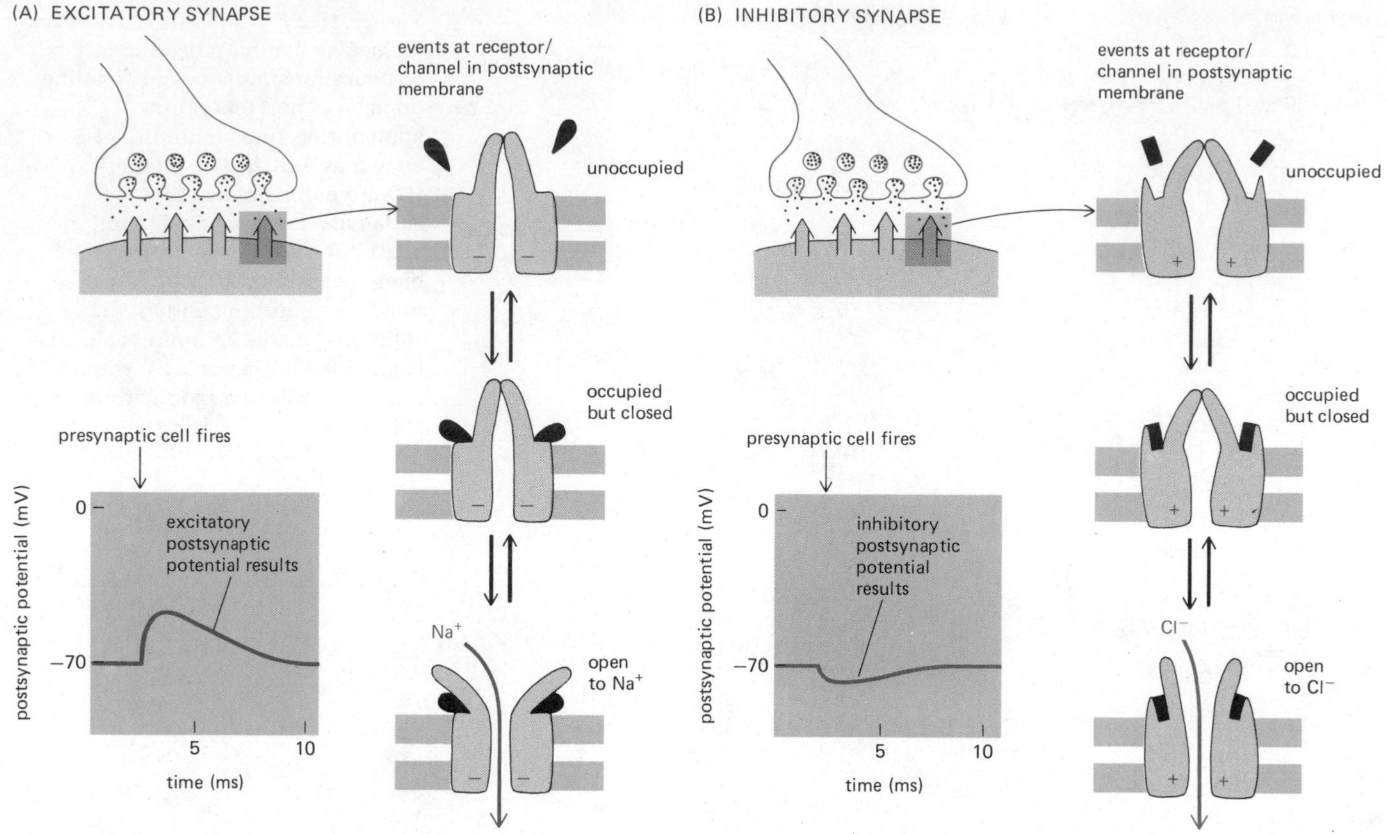

Figure 18–34 Excitatory (A) and inhibitory (B) synapses compared. The effect of the neurotransmitter depends on the ion selectivity of the gated channel associated with the receptor.

sketetal muscle cells and on heart muscle cells, exciting the former and inhibiting the latter, because the acetylcholine receptors are different in the two cases.

Synapses vary in other respects also. For example, the mechanism for terminating the action of GABA is different from that described above for acetylcholine. Instead of being hydrolyzed while it lies in the synaptic cleft, GABA is retrieved by the presynaptic terminals that secreted it or by the neighboring glial cells. Both have specific transport proteins in their plasma membranes for actively taking up GABA. A similar mechanism serves to terminate the action of many other neurotransmitters.

Neurotransmitters at Some Synapses Act Through Intracellular Second Messengers Rather Than by Directly Gated Ion Flows[20]

Synapses, then, can be classified as excitatory or inhibitory. Ligand-gated ion channels in the postsynaptic membrane can mediate either effect, depending on their ion selectivity. As mentioned earlier, however, gated ion channels are not the only postsynaptic membrane proteins on which neurotransmitters act. Another and fundamentally different mode of synaptic transmission depends on receptors coupled to membrane proteins that generate a second messenger in the postsynaptic cell (see p. 735). Many of the receptors for the monoamines *norepinephrine* and *dopamine*, for example, are believed to be of this type. Binding of transmitter to the receptor activates *adenylate cyclase*, thereby increasing the intracellular concentration of cyclic AMP. The cyclic AMP in turn activates protein kinases that phosphorylate specific proteins in the cell: for example, they can phosphorylate ion channels and thus alter the cell's electrical behavior. The ultimate effects can be either excitatory or inhibitory. Indeed, cyclic AMP can, in principle, trigger changes at almost any level in the cell's control machinery, even to the extent of altering the pattern of gene expression.

Through the variety of different receptor proteins, neurotransmitters produce many different effects on the postsynaptic cell. As a general rule, ligand-gated ion channels are responsible for actions on a time scale of milliseconds to seconds, while second-messenger systems operate on a time scale of seconds, minutes, and even longer. We shall now examine how the many types of synapse—excitatory, inhibitory, fast, and slow—participate in the processing of neural information.

Many Synaptic Inputs Combine to Drive a Single Neuron

A vertebrate skeletal muscle cell typically receives only one synaptic connection, from a single motor neuron whose cell body is located in the spinal cord. By contrast, several thousand nerve terminals from hundreds or thousands of different neurons make synapses on the motor neuron itself; its cell body and dendrites are almost completely covered with them (Figure 18–35). Some of these synapses transmit signals from the brain, others bring sensory information from muscles or from the skin, and others supply the results of computations made by interneurons in the spinal cord. The motor neuron must combine the information received from these many different sources and react by firing signals along its own axon or by remaining quiet.

The motor neuron provides a typical example of the way in which neurons individually play their part in the fundamental task of computing an output from a complex set of inputs. Of the many synapses on the motor neuron, some will tend to excite it, others to inhibit it. Although the motor neuron makes only one type of neurotransmitter, it makes many different

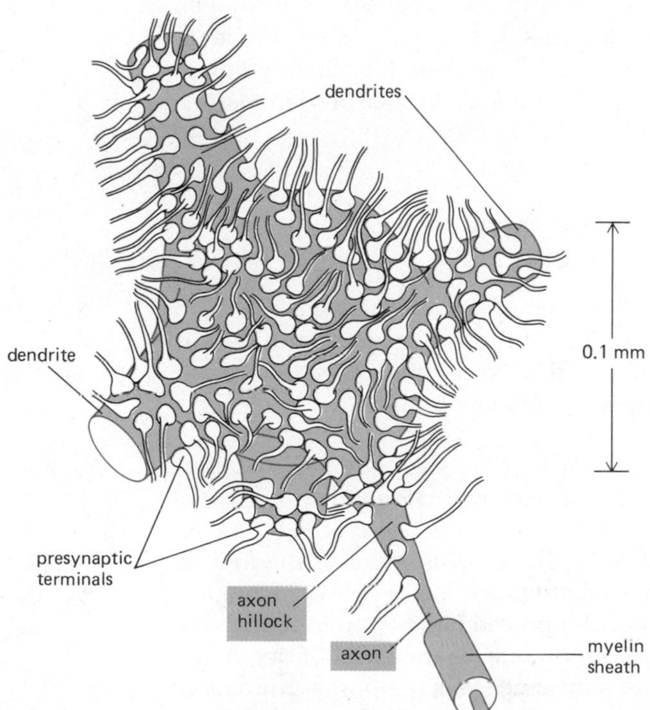

Figure 18–35 Schematic diagram of a motor neuron cell body in the spinal cord, showing some of the many thousands of nerve terminals that synapse on the cell and deliver signals from other parts of the organism to control its firing. The regions of the motor neuron plasma membrane that are not covered with synaptic endings are covered by glial cells (not shown).

types of receptor proteins, concentrating them at different postsynaptic sites on its surface. Each synapse produces a characteristic voltage change or **postsynaptic potential (PSP)** when the associated presynaptic cell fires. A depolarization corresponds to an excitatory PSP; a hyperpolarization corresponds to an inhibitory PSP. Whereas action potentials are fairly uniform from one neuron to another, the PSPs generated at different synapses on a single neuron are enormously variable in size as well as duration. At one synapse on the motor neuron an incoming nerve impulse might produce a depolarization of 0.1 mV, while at another there might be a depolarization of 20 mV. Other things being equal, the effect is stronger the larger the area of synaptic contact, but the nature of the system is such that even small PSPs can combine to produce a large effect.

The Membrane Potential in the Cell Body Represents a Spatial Summation of Postsynaptic Potentials[21]

The membrane of the dendrites and cell body, though rich in receptor proteins, contains few voltage-gated Na^+ channels and so is relatively inexcitable. Individual PSPs generally do not cause it to fire action potentials. Thus each incoming signal is faithfully reflected in a PSP of graded magnitude, which falls off with distance from the site of the synapse. If signals arrive simultaneously at several synapses in the same region of the dendritic tree, the total PSP in that neighborhood will be, roughly speaking, the sum of the individual PSPs, with inhibitory PSPs making a negative contribution to the total. Moreover, the net electrical disturbance produced in one postsynaptic region will spread to other regions, through the passive cable properties of the dendritic membrane. The cell body, at the center of this mass of inputs, is relatively small (generally less than 100 μm in diameter) compared with the dendritic tree (whose branches may extend for millimeters). The membrane potential in the cell body and its immediate neighborhood will therefore be roughly uniform and will be a composite of the effects of all the signals impinging on the cell, weighted according to the distances of the synapses from the cell body. The **grand postsynaptic potential** of the cell body is thus said to represent a **spatial summation** of all the stimuli received. If excitatory inputs predominate, it will be a depolarization; if inhibitory inputs predominate, a hyperpolarization.

Temporal Summation Translates the Frequency of Presynaptic Signals into the Size of a PSP

Spatial summation is one of two key features of the mechanism by which neurons process the information converging on them; **temporal summation** is the other. Spatial summation combines the effects of signals received at different sites on the membrane; temporal summation combines the effects of signals received at different times.

When an action potential arrives at a synapse, it evokes a PSP that rises rapidly to a peak and then declines to the baseline with a roughly exponential time course. If a second action potential arrives before the first PSP has decayed completely, the second PSP adds to the remaining tail of the first. If, after a period of inactivity, a long train of action potentials is delivered in quick succession, each PSP adds to the tail of the preceding PSP, building up to a large sustained average PSP whose magnitude reflects the rate of firing of the presynaptic neuron (Figure 18–36). This is the essence of temporal summation: it translates the frequency of incoming signals into the magnitude of a net PSP.

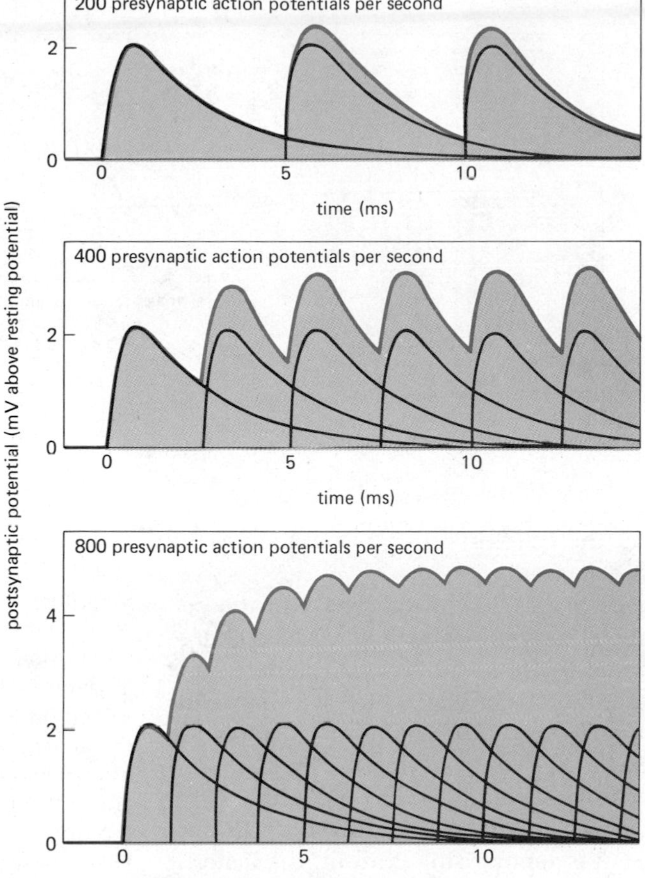

Figure 18–36 Temporal summation. The overlapping curves within the shaded region of each graph represent the individual contributions to the total postsynaptic potential evoked by the arrival of the successive presynaptic action potentials.

The Grand PSP Is Translated into Nerve Impulse Frequency for Long-Distance Transmission[22]

Temporal and spatial summation together provide the means by which the rates of firing of many neurons jointly control the membrane potential in the body of a single postsynaptic cell. The final step in the neuronal computation is the generation of an output by the postsynaptic cell, usually in the form of action potentials, for signaling to other cells that are often far away. The output signal must represent the magnitude of the grand PSP in the cell body. While the grand PSP is a continously graded variable, action potentials are all or none and uniform in size. The only free variable in signaling by action potentials is the time interval between one action potential and the next. The magnitude of the grand PSP therefore has to be translated or *encoded* in a new form for long-distance transmission. Specifically, it is encoded in terms of the *frequency* of firing of action potentials (Figure 18–37). This encoding is achieved by a special set of voltage-gated ion channels at the base of the axon, adjacent to the cell body, in a region known as the **axon hillock** (see Figure 18–35).

Before we explain how these channels operate, a word of qualification is necessary. We have presented the neuronal computation as a two-step process in which the synaptic inputs produce a grand PSP in the cell body and the grand PSP then triggers action potentials in the axon hillock. Although it is convenient to picture the mechanism in this way, it is an oversimplification. The firing of an action potential itself causes drastic changes of the membrane potential of the cell body, which therefore no longer directly re-

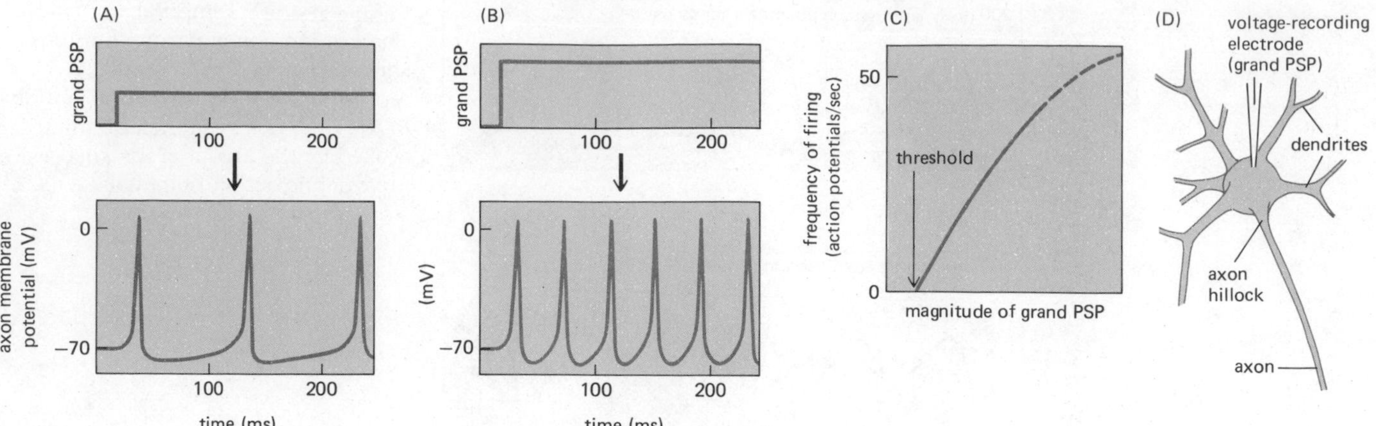

Figure 18–37 The encoding of the grand PSP in the form of the frequency of firing of action potentials by an axon. A comparison of (A) and (B) shows how the firing frequency of an axon increases with an increase in the grand PSP, while (C) summarizes the general relationship. In (D), the experimental setup for measuring the grand PSP is shown. In (A) and (B), the upper graphs (marked "grand PSP") show the net intensity of synaptic stimulation as received by the cell body, while the lower graphs show the resulting trains of action potentials that are transmitted along the axon. The upper graphs can be thought of as representations of the grand PSP that would be observed if the firing of action potentials were somehow blocked.

flects the net synaptic stimulation that the cell is receiving. How then is the strength of that stimulation to be characterized?

One way to think of it is in terms of the flow of current into the cell body: synaptic stimulation causes depolarization by allowing positive charge to enter the cell, and the net effect can be equated with that of a depolarizing current injected into the cell through a microelectrode, which likewise leads to repetitive firing of action potentials. Through the inflow of current, the membrane of the axon hillock is gradually driven up to the threshold level of depolarization, then rapidly discharges by firing an action potential and returns to the baseline (as will be described below); then it is driven up toward threshold again; and so on. The larger the depolarizing current, the faster it drives the hillock membrane back to the firing threshold, and the more rapid the succession of action potentials. It is a complex problem to give a rigorous analysis of the encoding mechanism. In the simple qualitative account that follows, we shall loosely refer to "the strength of synaptic stimulation" or to "the grand PSP," meaning the grand PSP that would be observed if action potentials were somehow prevented from firing.

Encoding Requires a Combination of Different Voltage-sensitive Channels[23]

The propagation of action potentials depends chiefly, and in many vertebrate axons almost entirely, on voltage-gated Na^+ channels. The membrane of the axon hillock is where action potentials are initiated, and Na^+ channels are plentiful there. But to perform its special function of encoding, the axon hillock membrane typically contains in addition at least four other classes of gated channels—three selective for K^+ and one selective for Ca^{2+}. The three varieties of gated K^+ channels have quite different properties; we shall refer to them as the *delayed*, the *early*, and the *Ca^{2+}-activated K^+ channels.* The workings of this system of channels for encoding have been most thoroughly studied in giant neurons of mollusks, but the principles appear to be similar for vertebrate nerve cells.

To understand the necessity for multiple types of channel, consider first the behavior that would be observed if the only voltage-sensitive channels present in the nerve cell were the Na^+ channels. Below a certain threshold level of synaptic stimulation, the depolarization of the axon hillock membrane would be insufficient to trigger an action potential. With gradually increasing stimulation, the threshold would be crossed: the Na^+ channels would open, and an action potential would fire. The action potential would be terminated in the usual way by inactivation of the Na^+ channels. Before another action potential could fire, these channels would have to recover from their inactivation. But that requires a return of the membrane voltage to a very negative value, which would not occur as long as the strong depolarizing stimulus was maintained. An additional channel type is needed, therefore, to repolarize the membrane after each action potential so as to make it ready to fire another. This task is performed by the **delayed K^+ channels,** which we have already encountered (p. 1027) in the discussion of the propagation of the action potential. They respond to membrane depolarization in much the same way as the Na^+ channels, but with a longer time delay. By opening during the falling phase of the action potential, they permit an efflux of K^+, which short-circuits the effect of the depolarizing stimulus and drives the membrane back toward the K^+ equilibrium potential. This potential is so far negative that the Na^+ channels recover from their inactivated state. In addition, the K^+ conductance turns itself off: repolarization of the membrane causes the delayed K^+ channels themselves to close again (before they have had time to enter an inactivated state). Once repolarization has occurred, the depolarizing stimulus from synaptic inputs becomes capable of raising the membrane voltage to threshold so as to cause another action potential to fire. In this way, sustained stimulation of the dendrites and cell body leads to repetitive firing of the axon.

Early K^+ Channels Help to Make the Firing Rate Proportional to the Stimulus[24]

However, repetitive firing in itself is not enough: the frequency of the firing has to reflect the intensity of the stimulation. Detailed calculations show that a simple system of Na^+ and delayed K^+ channels is inadequate for the purpose. Below a certain threshold level of steady stimulation the cell will not fire at all; above that threshold it will abruptly begin to fire at a relatively rapid rate. The **early K^+ channels** solve the problem. When open, they oppose the effect of a depolarizing stimulus and hinder the triggering of action potentials, and their opening is controlled so that they act to reduce the rate of firing at levels of stimulation that are only just above the threshold. In this way they help to remove the discontinuity in the relationship between the firing rate and the intensity of stimulation.

Specifically, the early K^+ channels are gated in much the same way as the voltage-gated Na^+ channels: they rapidly open when the membrane is depolarized and then become completely inactivated. But they differ from the Na^+ channels in that recovery from inactivation is slower and requires a return to an even more negative membrane potential. The consequences for the behavior of the membrane are difficult to deduce by simple qualitative arguments, but a full mathematical analysis of one or two types of neurons has shown that the outcome of the whole complex sequence of voltage-dependent channel openings, closings, and inactivations is a firing rate proportional to the strength of the depolarizing stimulus over a very broad range (Figure 18–37). It is likely that the many types of neurons that show a similarly graded rate of firing achieve it in a similar way.

Adaptation Lessens the Response to an Unchanging Stimulus[25]

One further commonly encountered refinement of the process of encoding is mediated by the two other types of channels mentioned at the outset—the voltage-gated Ca^{2+} channels and the **Ca^{2+}-activated K^+ channels.** The former are the same as (or very similar to) the Ca^{2+} channels that mediate release of transmitter at presynaptic nerve terminals: those present in the neighborhood of the axon hillock open when an action potential fires, allowing Ca^{2+}

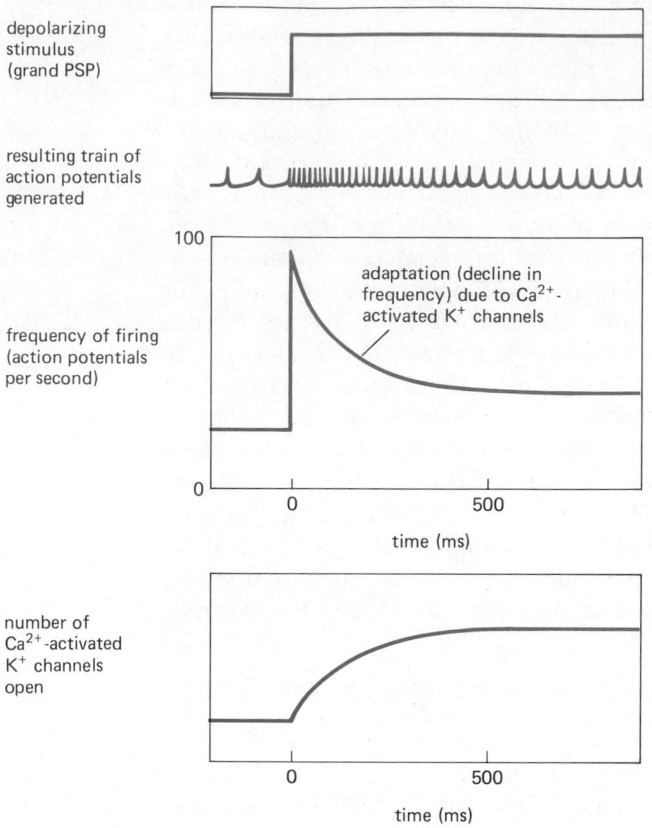

Figure 18–38 Adaptation. When steady stimulation is prolonged, the stimulated cell gradually reduces the strength of its response, as expressed in the rate of firing of action potentials.

into the axon. The Ca^{2+}-activated K^+ channel is different from any of the channel types described earlier. It opens in response to a raised concentration of Ca^{2+} at the *internal* face of the membrane.

Suppose that a strong depolarizing stimulus is applied for a long time, triggering a long train of action potentials (Figure 18–38). Each action potential permits a brief influx of Ca^{2+} through the voltage-gated Ca^{2+} channels, so that the internal Ca^{2+} concentration gradually builds up to a high level. This opens the Ca^{2+}-activated K^+ channels, and the increased permeability of the membrane to K^+ makes the membrane harder to depolarize and increases the delay between one action potential and the next. In this way, a neuron that is stimulated continuously for a prolonged period becomes gradually less responsive to the constant stimulus—a phenomenon known as **adaptation.** Adaptation allows a neuron, and indeed the nervous system generally, to react sensitively to *change,* even against a high background level of steady stimulation. It is one of the main mechanisms that enable us, for example, to ignore the constant pressure of clothing on the body and yet be alert to a touch on the shoulder, or to hear a sudden noise against the roar of traffic.

Not All Signals Are Delivered via the Axon[26]

In the typical neuron that we have been describing, the dendrites and cell body *receive* signals by synaptic transmission, while the axon *sends* signals to other cells; electrical signals propagate along the dendrites by passive spread, and they are converted to action potentials at the axon hillock. All these features are subject to variation: many types of neurons do not conform to this model, although the molecular principles of their operation are the same. For example, in most invertebrates the majority of neurons have a *unipolar* organization (Figure 18–39): the cell body is connected by a single stalk to a

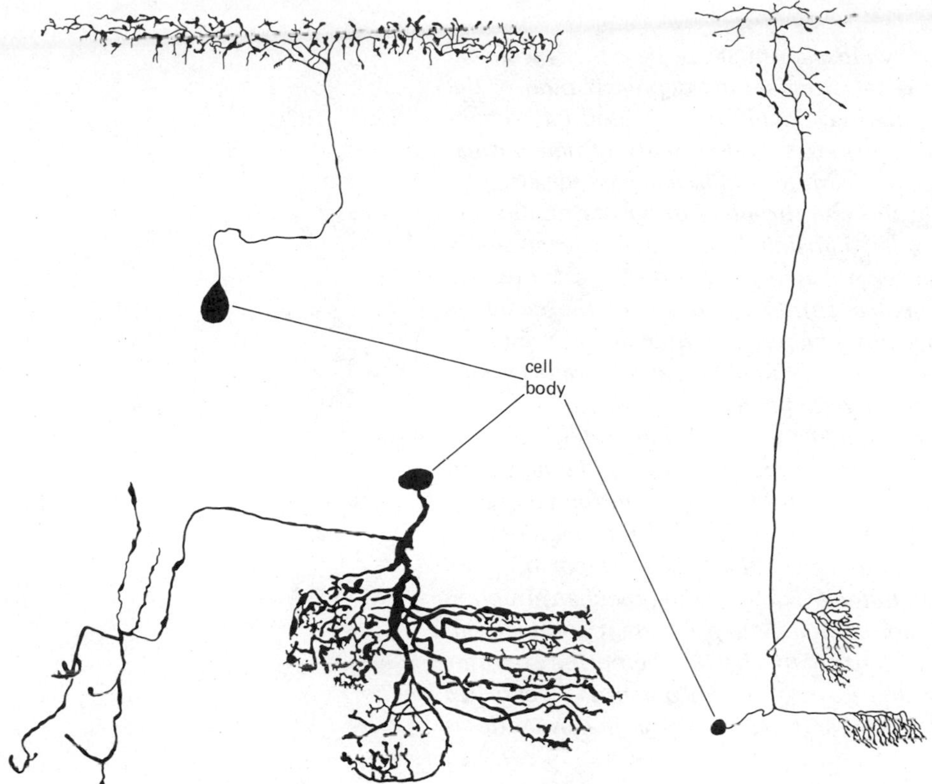

Figure 18–39 Neurons from a fly, showing the structure typical of most neurons in invertebrates, in which the nerve cell body is connected by a stalk to the system of nerve cell processes and does not have dendrites projecting from it directly. The sensory neurons in the spinal ganglia of vertebrates have a similar organization. (From N. Strausfeld, Atlas of an Insect Brain. New York: Springer, 1976.)

branching system of cell processes, among which it is not always easy to distinguish dendrite from axon. In vertebrates and invertebrates alike it is commonplace for processes that appear to be dendrites to form presynaptic as well as postsynaptic structures and to deliver signals to other cells as well as receive them. Conversely, synaptic inputs are sometimes received at strategic sites along the axon—for example, close to the axon terminal, where they can enhance or inhibit the release of neurotransmitter from that terminal (Figure 18–40).

Synapses at which a dendrite delivers a stimulus to another cell play a large part in communication between neurons that lie close together, within a few millimeters or less. Over such distances, electrical signals can be propagated passively, from postsynaptic sites on the dendritic membrane where they are received, to presynaptic sites where they control transmitter release. Indeed, there are neurons that possess no axon, do not conduct action potentials, and perform all of their signaling via dendrites. If the dendritic tree is large, separate parts of it can behave as more or less independent pathways for communication and for information processing. In some neurons, the range of possibilities is still further complicated by the presence of voltage-gated channels in the dendritic membrane, making the dendrites electrically excitable.

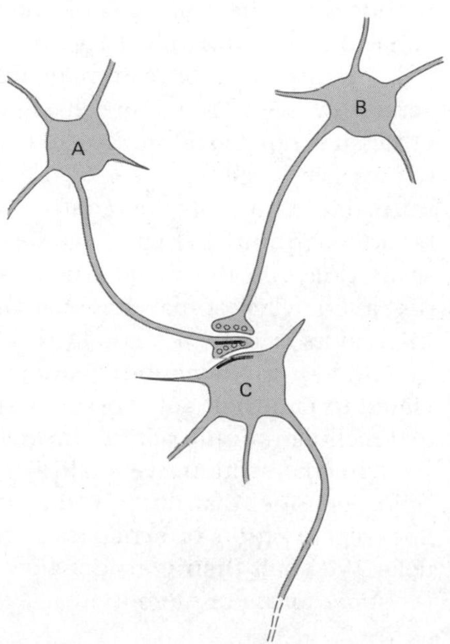

Figure 18–40 An axo-axonic synapse. The neurotransmitter relased from the axon terminal of cell B acts on channels in the axon terminal of cell A, thereby altering the number of quanta of neurotransmitter released onto C when A fires. If firing of B causes a reduction in the stimulus delivered by A to C, B is said to exert a *presynaptic inhibition;* the contrary effect is called *presynaptic facilitation.*

Summary

Neural signals pass from cell to cell at synapses, which can be either electrical (gap junctions) or chemical. At a chemical synapse, the depolarization of the presynaptic membrane by an action potential opens voltage-gated Ca^{2+} channels, allowing an influx of Ca^{2+} that triggers exocytotic release of neurotransmitter from synaptic vesicles. The neurotransmitter diffuses across the synaptic cleft and binds to receptor proteins in the membrane of the postsynaptic cell; it is eventually eliminated from the cleft by diffusion, by enzymatic degradation, or by reuptake. Receptor proteins that form ligand-gated ion channels mediate rapid postsynaptic effects of the neurotransmitter; opening of the channels produces an excitatory or an inhibitory postsynaptic potential according to the channel ion selectivity. Receptors coupled to enzymes such as adenylate cyclase generally mediate slower and more prolonged effects.

A typical neuron receives on its dendrites and cell body many different excitatory and inhibitory synaptic inputs, which combine, by spatial and temporal summation, to produce a grand postsynaptic potential in the cell body. The magnitude of the grand postsynaptic potential is encoded, for long-distance transmission, in the rate of firing of action potentials, by a system of gated channels in the membrane of the axon hillock. The encoding mechanism often shows adaptation, so that the cell responds weakly to a constant stimulus but strongly to a change. There are many variants of this basic scheme: for example, not all neurons produce an output in the form of action potentials, dendrites can be presynaptic as well as postsynaptic, and axons can be postsynaptic as well as presynaptic.

Channel Regulation and Memory

The few types of ion channels discussed in this chapter appear sufficient to account for all the basic forms of neural signaling. Through distribution of these channels in different combinations in different regions of the nerve cell membrane, an enormous variety of functions are performed. Each type of neuron, by virtue of its particular assortment and distribution of channels, has its own style of computation and signal transmission. The factors that control the spatial arrangement of ion channels and regulate changes in their numbers are therefore among the most important determinants of nerve cell behavior. Unfortunately, they are very poorly understood.

Regulation occurs at many different levels, beginning with the control of gene expression. In some cells a particular channel protein is synthesized, in others it is not. Once synthesized, it must be directed to the appropriate region of the plasma membrane. Like other membrane proteins, it will then be subject to turnover through degradation and replacement. The rate of turnover can be adjusted and is highly variable. Typically, a channel protein survives for several days in the membrane before being internalized by endocytosis and degraded by lysosomal enzymes. On a more rapid time scale, of minutes rather than hours or days, the numbers of channels that are functional can be changed by covalent modification: channel proteins can be phosphorylated or methylated to jam their gates open or shut. Such modifications can be initiated by extracellular stimuli acting through second messengers such as cyclic AMP.

In this section, we shall discuss first some insights into the regulation of ion-channel distribution that have come from studies of the degeneration and regeneration of synapses between motor neurons and skeletal muscle cells. We shall then consider how the regulation of channels might account for some forms of memory.

The Distribution of Ion Channels in a Muscle Cell Changes in Response to Denervation[27]

Throughout life, synapses are liable to be eliminated, and new synapses are capable of forming. Such changes are most clearly observable in the vertebrate neuromuscular system. If the nerve that innervates a muscle in an adult is cut, the detached terminal portions of the axons degenerate and the muscle is deprived of its synaptic inputs. However, the cell bodies of the motor neurons (which lie in the spinal cord) generally survive, and their truncated axons grow back again toward the denervated muscle. The mechanism of axon growth will be the subject of a later section; the essential point in the present context is that, on reaching the muscle, the regenerating axons halt their growth and form synapses. Though these axons show a strong preference for the sites where synapses existed before, they can also make junctions at entirely new locations on the muscle cell. Thus it is possible to study the changing patterns of membrane specialization associated with the destruction and creation of synapses.

In an adult mammal, each skeletal muscle cell normally has just one synapse on it, and acetylcholine receptors are strictly localized in the membrane beneath the axon terminal, where their concentration is more than a thousand times greater than in regions remote from the axon terminal. Fluorescence bleaching experiments (see p. 281) show that the receptors at the synapse are somehow tethered in place and not free to diffuse in the plane of the membrane. They have also a low rate of turnover, continuing in service for five days or more before they are degraded and replaced.

When a muscle cell is denervated by cutting the nerve controlling it, the junctional receptors remain in place, but radical changes occur in the extrajunctional regions. Within a few days, large quantities of new acetylcholine receptors are synthesized and inserted in the membrane over the whole surface of the cell, making it *supersensitive* to the transmitter (Figure 18–41). These *extrajunctional receptors* have a relatively rapid turnover time, on the order of one day. They are present at a density intermediate between the normal density of receptors at a synapse and the normal density in extrajunctional regions. The electrical excitability of the membrane is altered also by the insertion of voltage-gated Ca^{2+} channels, which permit the propagation of Ca^{2+}-dependent action potentials even when the Na^+ channels are blocked with tetrodotoxin. At the same time, the muscle cell membrane changes its

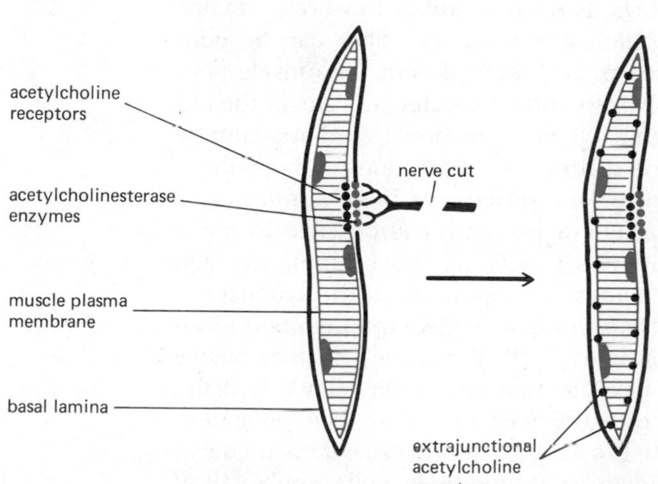

Figure 18–41 Schematic diagram showing how the distribution of acetylcholine receptors in the muscle cell membrane changes as a consequence of denervation. Note that a high concentration of receptors remains at the old neuromuscular junction, but, in addition, new extrajunctional receptors are distributed over the entire surface.

receptivity to new innervation: while a denervated muscle cell will permit regenerating axons to form new synapses on it, a normally innervated muscle cell will not. Presumably this whole complex of changes is centered on the need for the denervated muscle cell to become innervated again.

The Receptivity of a Muscle Cell Can Be Controlled by Electrical Stimulation[28]

Denervation appears to cause the changes just described largely by depriving the muscle cell of electrical stimulation. The effects of denervation can be mimicked by anaesthetizing the nerve, thereby blocking the passage of action potentials. If the denervated muscle is then stimulated artificially through implanted metal electrodes, the extrajunctional sensitivity to acetylcholine is suppressed and new synapses are prevented from forming. This illustrates a profoundly important principle, of which we shall say more later: electrical activity can regulate the development of synaptic connections.

Once a denervated muscle has become reinnervated, the diffuse distribution of acetylcholine receptors disappears, although a high concentration of them remains at the sites of the newly formed neuromuscular junctions. More surprisingly, a high concentration of the receptors also remains at the sites of the old neuromuscular junctions, even if there are no longer any axon terminals synapsing there. These old sites, moreover, continue to be exceptional in that they remain receptive to the formation of synapses by axons that arrive subsequently, while the surrounding membrane, where acetylcholine sensitivity has been suppressed, is not. This indicates that there is some durable structure associated with the muscle cell membrane that preserves a high concentration of acetylcholine receptors at a synapse and marks it out as a privileged site where axon terminals can synapse even while the muscle is electrically exercised. But what is this something that holds the parts of the synapse in place?

The Site of a Synaptic Contact Is Marked by a Persistent Specialization of the Basal Lamina[29]

Each muscle fiber is enveloped in a basal lamina. At the neuromuscular junction, this lamina separates the muscle plasma membrane from the axon terminal so that acetylcholine molecules released from the terminal have to pass through the lamina to reach the postsynaptic receptors (see Figure 18–25A). If the muscle fiber is badly damaged, it degenerates and dies, and macrophages move in to clear away the debris. The basal lamina, however, remains and provides a scaffolding within which a new muscle fiber can be constructed from surviving stem cells (see p. 931). Even if both the muscle fiber and the axon terminal have been destroyed and eliminated, the site of the old neuromuscular junction is still recognizable from the corrugated appearance of the basal lamina there—a relic of the distinctive corrugations of the defunct muscle cell membrane at the synapse. This **junctional basal lamina,** furthermore, has a specialized chemical character, and it is possible to make antibodies that selectively bind to it. The junctional basal lamina, in fact, actually *controls* the localization of the other components of the synapse.

This control has been demonstrated in a series of experiments on amphibians (Figure 18–42). First, by destroying both the nerve and the muscle cells, leaving only an empty shell of basal lamina, it is easily shown that the junctional basal lamina specifically holds the acetylcholinesterase molecules, which, at a normal synapse, hydrolyze the acetylcholine released by the nerve terminal. The acetylcholinesterase remains tethered by its collagenlike tail to the junctional basal lamina, even after the muscle cell is dead and gone.

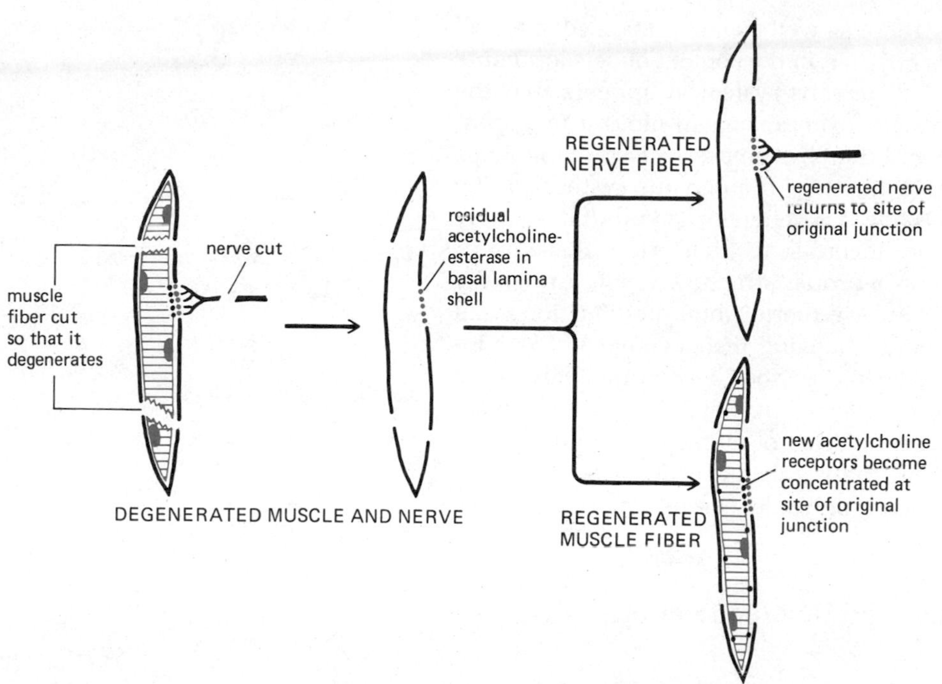

Figure 18–42 Diagram of experiments showing that the specialized character of the basal lamina at the neuromuscular junction controls the localization of the other components of the synapse.

Second, it can be shown that the junctional basal lamina holds the axon terminal in place. Thus, if the muscle cell but not the nerve is destroyed, the axon terminal remains attached to the basal lamina for many days. On the other hand, removing the basal lamina with collagenase causes the axon terminal to detach even if the muscle cell is still present. More remarkable is the fact that the basal lamina by itself can mark out for a regenerating axon the site of the old neuromuscular junction and induce the membrane of the incoming axon to differentiate there to form a mature synaptic terminal. This has been demonstrated by destroying both the muscle and the nerve and then allowing the nerve to regenerate while the basal lamina remains empty: the regenerating axon regularly seeks out the original synaptic site and differentiates there into a synaptic ending, as though the original muscle cell were present.

Lastly, the junctional basal lamina has been shown to control the postsynaptic specialization of the muscle cell membrane: it is responsible for localizing the acetylcholine receptors at the junctional region of the muscle plasma membrane. This is shown by the converse of the experiment just described. The muscle and the nerve are both destroyed, leaving an empty shell of basal lamina, but now the muscle is allowed to regenerate while the nerve is prevented from doing so. The acetylcholine receptors synthesized by the regenerated muscle localize predominantly in the region of the old junction, even though the nerve is absent.

It is likely that most of the junctional basal lamina is secreted by the muscle cell, although the axon terminal may also make a contribution. The early interaction of the two cells apparently creates a structure that stabilizes the synaptic connection between them.

Synaptic Plasticity Provides a Mechanism for Memory[30]

The formation or elimination of a synapse is an event whose consequences may last a lifetime. The example of the neuromuscular junction shows, moreover, that synapse formation can be regulated by electrical activity. Though studies of the central nervous system are much more difficult, there is reason to think that similar principles apply there as well (some of the evidence will be discussed in the last section of this chapter, in the context of development). In both the peripheral and the central nervous system, it appears that the pattern of synaptic connections is *plastic:* experience can mold it by stimulating or repressing electrical activity and thereby can exert a lasting influence on subsequent patterns of behavior. In this way the nervous system can be endowed with a long-term memory. While it is generally agreed that memory depends on synaptic changes, it is still uncertain to what extent these occur at a gross level, through alterations of neuronal structure visible under the microscope, and to what extent they involve more subtle modifications that change the efficacy of synapses without changing their geometry. Clear instances of plasticity at both levels have been described. Structural changes are certainly important for some long-term effects, but they take too long to account for short-term memory, on the time scale of minutes or hours. Short-term effects are believed to depend on the regulation of ion channels. The molecular details have been worked out in only a few cases, one of which we shall now describe.

A Short-Term Memory Is Registered by Modification of Channel Proteins[31]

The sea snail *Aplysia,* a type of mollusk, withdraws its gill if its siphon is touched (Figure 18–43). If the siphon is touched repeatedly, the animal becomes **habituated:** it learns to ignore the stimulus, withdrawing its gill only slightly, if at all, in response. Habituation is similar in function to adaptation, though it operates on a longer time scale and, as we shall see, at a different point in the neural pathway. A nasty experience, such as a hard bang or an electric shock, removes the habituation and **sensitizes** the animal so that it responds again very readily to being touched. The sensitization persists for many minutes, hours, or sometimes even days after the brief noxious stimulus that caused it, and represents a simple form of short-term memory. The modification of behavior can be traced to a change occurring in a particular class of synapses in the neural circuit that controls the gill-withdrawal reflexes.

Touching of the siphon stimulates a set of *sensory neurons* to fire. These neurons have excitatory synapses on another set of neurons that drive muscles for gill withdrawal. The responses of the *gill-withdrawal neurons* to firing of the sensory neurons can be recorded with an intracellular electrode: during habituation, the postsynaptic potential is observed to decrease with repeated firing of the sensory cell. Sensitization has the reverse effect, increasing the postsynaptic potential. In both cases, the changes are due to alterations in the amount of neurotransmitter released from the presynaptic terminals of the sensory neurons when they fire. Transmitter release is controlled by the amount of Ca^{2+} that enters the terminals during the action potential. In habituation, repeated firing of the sensory cells leads to a modification of channel proteins in the terminals such that Ca^{2+} entry is reduced; in sensitization, by contrast, Ca^{2+} entry is increased. The detailed molecular changes are best understood in the case of sensitization.

In sensitization, the alteration in transmitter release from the sensory neurons is triggered by the firing of another set of neurons responsive to the noxious stimulus. These *facilitator neurons* synapse on the presynaptic terminals of the sensory neurons (Figure 18–44). The facilitator neurons release serotonin, which acts on the membrane of the sensory neuron terminals by binding to receptors that are coupled to an adenylate cyclase. The activation of the adenylate cyclase increases the concentration of cyclic AMP in the terminal, and the cyclic AMP activates a protein kinase. The protein kinase is thought to phosphorylate K^+ channels in the membrane of the sensory neu-

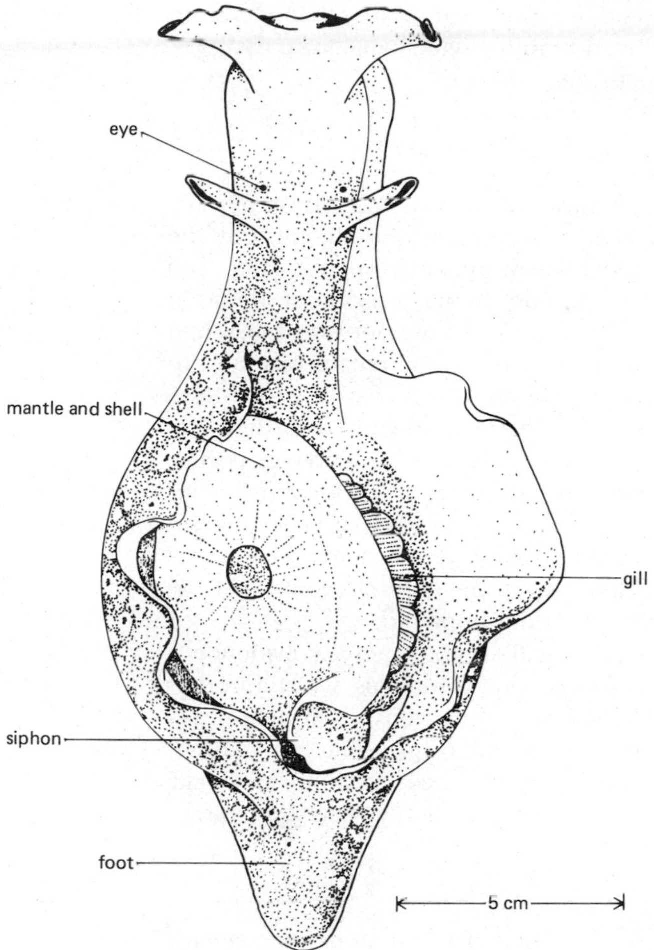

Figure 18–43 The sea snail *Aplysia punctata* viewed from above. An overlying flap of tissue has been drawn aside to reveal the gill under the protective mantle and shell. (After J. Guiart, *Mem. Soc. Zool. France* 14.219, 1901.)

ron terminal, locking them shut. The blocking of these K^+ channels means that action potentials invading the terminal decay more slowly than usual. The prolonged action potentials hold the voltage-gated Ca^{2+} channels open for a longer time, permitting a greater influx of Ca^{2+}, which in turn triggers the release of a larger number of synaptic vesicles, producing a larger postsynaptic potential and thereby a more vigorous withdrawal of the gill.

It has been possible to analyze this train of events in such detail because the neurons involved are large (diameter on the order of 100 μm) and easily identifiable, permitting both intracellular recording and intracellular injections. For example, the role of phosphorylation by the protein kinase has been

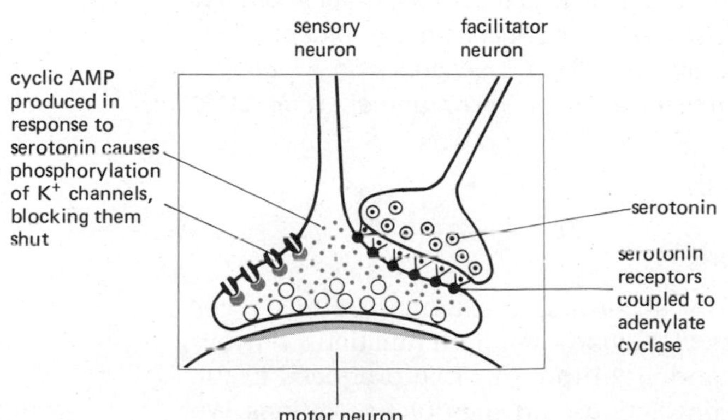

Figure 18–44 Diagram of the synaptic mechanism underlying facilitation of the gill-withdrawal reflex in *Aplysia*.

demonstrated by injecting the protein kinase directly into the sensory neuron and showing that it produces a sensitization indistinguishable from that produced by stimulating the facilitator neurons.

Memory Remains Mysterious[30,32]

In the mammalian brain also, synapses have been identified whose efficacy can be altered promptly and lastingly by appropriate stimulation, but the technical problems in analyzing the mechanism are very much greater. The most striking examples are found in the **hippocampus,** a region of the brain known from other evidence to be somehow involved in memory storage. When the hippocampus is destroyed, the capacity to recall the recent past and to lay down memories of new experiences is lost, though preexisting long-term memories are retained. Presumably the long-term memories are recorded elsewhere, perhaps as structural alterations of synapses in the cerebral cortex. Other observations also suggest that different mechanisms are responsible for short-term and long-term memory. For example, a person who has been knocked unconscious for a short period shows, on recovery, a loss of memory for events that occurred shortly before the accident, while memory of events that occurred more than, say, half an hour before is not impaired.

Yet despite various scraps of physiological and biochemical evidence, a vast mass of psychological data, and a few general principles, we still understand almost nothing about the cellular basis of memory in vertebrates—neither the detailed anatomy of the neural circuits responsible nor the molecular biology of the changes that experience produces in them. We shall, however, present a few more clues when we discuss neural development.

Summary

Neural signaling depends on a precisely regulated distribution of ion channels in the plasma membrane. The distribution changes when synapses are created or destroyed. A normal, innervated skeletal muscle cell has its acetylcholine receptors concentrated at the neuromuscular junction, conducts its action potentials by means of voltage-gated Na^+ channels, and will not permit new synapses to form on its surface. If the muscle cell is denervated, acetylcholine receptors and voltage-gated Ca^{2+} channels appear throughout the plasma membrane and the whole surface becomes receptive to the formation of new synapses. These changes are largely controlled by the amount of electrical stimulation that the cell receives. The site of an established neuromuscular junction is marked out by a specialization of the junctional basal lamina, which appears to govern both the localization of acetylcholine receptors in the muscle cell membrane and the positioning of the presynaptic axon terminal.

Synapses, besides being created and eliminated, can have their efficacy altered. Synaptic modifications due to neural activity provide a basis for memory. A simple form of learning in the mollusk Aplysia *has been shown to depend on phosphorylation of ion channels at a synapse, by a mechanism in which the release of a neurotransmitter stimulates intracellular formation of cyclic AMP, which activates a protein kinase.*

Sensory Input

The nervous system regulates behavior according to external circumstances and coordinates the internal activities of the body. For both functions, sensory information must enter the system, and an output must be produced in the form of signals that control muscle contractions and glandular secretions. We have already briefly discussed the output mechanism, at least as far as skeletal muscle is concerned. We have now to consider the mechanisms of sensory input.

Any signal that is to be fed into the nervous system must first be converted to an electrical form. The significance of the electrical message will depend on the device that has effected the conversion—that is, on the **transducer.** Each transducer is responsive to a specific aspect of the environment or a particular type of event—such as light, temperature, a specific chemical, or a mechanical force or displacement. In some cases, the transducer is part of a neuron that propagates action potentials. In other cases, it is part of a sensory cell specialized for transduction but not for long-distance communication; such a cell then passes its signal to an adjacent neuron via a synapse.

A vast flood of sensory information is supplied to the nervous system by the transducers. The brain must process this information to extract the significant features: it must pick out words from the hubbub of sound, recognize a face in the pattern of light and dark, and so on. This represents a second or neural stage of sensory processing, more subtle and complex by far than that which occurs at the level of transduction.

In general, both the molecular mechanisms of transduction and the subsequent processing of sensory data by the brain are poorly understood. The best insights at both levels come from studies of the vertebrate visual system. Before discussing vision, however, we shall consider two other sense receptors—the stretch receptors in muscle and the sound receptors in the ear—that illustrate several important general principles of sensory transduction.

Stimulus Magnitude Is Reflected in the Receptor Potential[33]

In a sense, practically every neuron behaves as a chemosensory transducer: it receives chemical stimulation at synapses and generates an electrical signal in response. As we have seen, the chemical stimulus acts on ion channels in the membrane, altering their permeability and thereby producing a shift in the membrane voltage (the postsynaptic potential). This electrical effect is graded according to the intensity of the stimulus. For long-distance transmission, the graded electrical signal is encoded in the frequency of firing of action potentials. Commonly, the cell shows adaptation, attenuating its output when a stimulus is held constant but generating a strong signal in response to change.

Almost exactly the same principles apply to transduction in sense organs. The neurons that serve as muscle **stretch receptors** illustrate them well for the case where the initial stimulus producing the change in membrane permeability is mechanical rather than chemical. The stretch receptors provide the nervous system with information about muscle length and its rate of change. This sensory feedback (together with signals from the brain and from other parts of the spinal cord) helps to control the firing of motor neurons, as explained in the caption to Figure 18–45. Each muscle contains sets of modified muscle fibers grouped together in *muscle spindles,* and the individual spindle fibers have the terminals of sensory neurons wrapped around them (Figure 18–45). When the spindle fibers are stretched, the sensory neurons fire action potentials, which are conveyed to the spinal cord. The electrical behavior of a single sensory neuron can be studied with an intracellular electrode inserted close to its site of attachment to a spindle fiber. The rate of firing of action potentials is graded according to the extent or rate of stretch. Most of the sensory neurons show marked adaptation, responding strongly to rapid change but weakly to steadily maintained stretch (Figure 18–46). When action potentials are prevented by blocking the Na^+ channels with tetrodotoxin, it becomes plain that the primary effect of stretching is to produce a

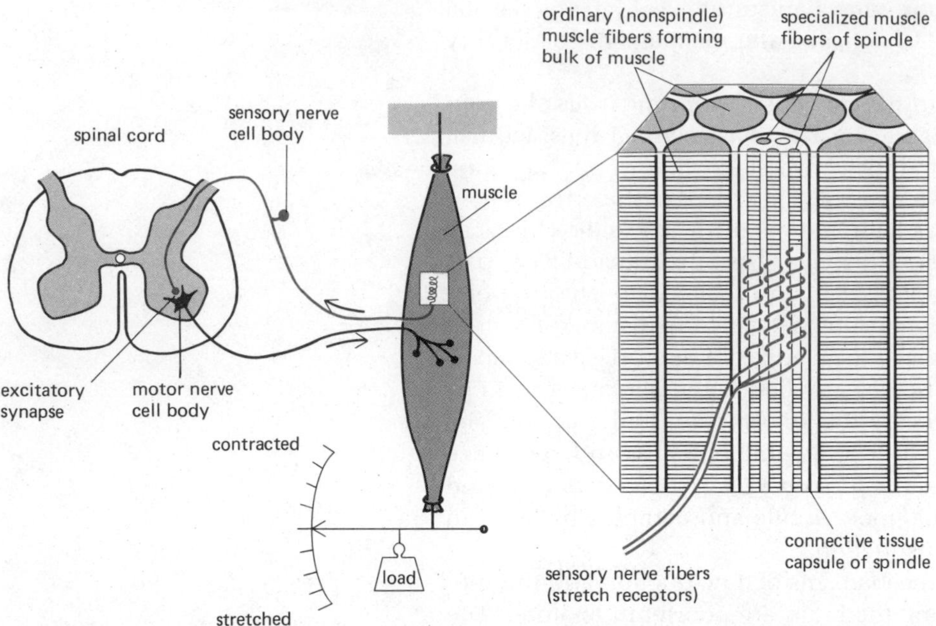

Figure 18–45 Stretch receptors in a skeletal muscle provide sensory feedback, via an excitatory synapse in the spinal cord, to control the firing of motor neurons. The basic circuitry is shown on the left, and a simplifed schematic diagram of a muscle spindle, where the sensory transduction occurs, on the right. If the muscle is stretched, a signal from the stretch receptors stimulates the motor neurons to fire, producing a command for muscle contraction, which resists further stretch. This *monosynaptic reflex arc* is particularly important in the maintenance of posture. It is also the basis for the "knee jerk," in which a sharp blow just under the kneecap pulls on a tendon, stretching an extensor muscle and stimulating its spindle fibers to produce rapid contraction of the muscle and extension of the leg. The specialized muscle fibers of the spindle lie in parallel with the other muscle fibers and are stretched when the muscle as a whole is stretched. The stretchable middle portions of the spindle fibers, around which the sensory nerve terminals are wrapped, contain cell nuclei but practically no contractile apparatus. The spindle fibers are shorter and smaller than the ordinary muscle fibers (15 μm to 30 μm in diameter, as against 50 μm to 100 μm). They have their own motor innervation (not shown in this figure), which has a similar function to that of the zero adjustment on a meter—it sets the spindle "zero" point at the level of muscle stretch required at any given time, so that a posture specified by the motor control areas of the brain is automatically maintained. There are several varieties of spindle fiber, differing somewhat in their structure and innervation.

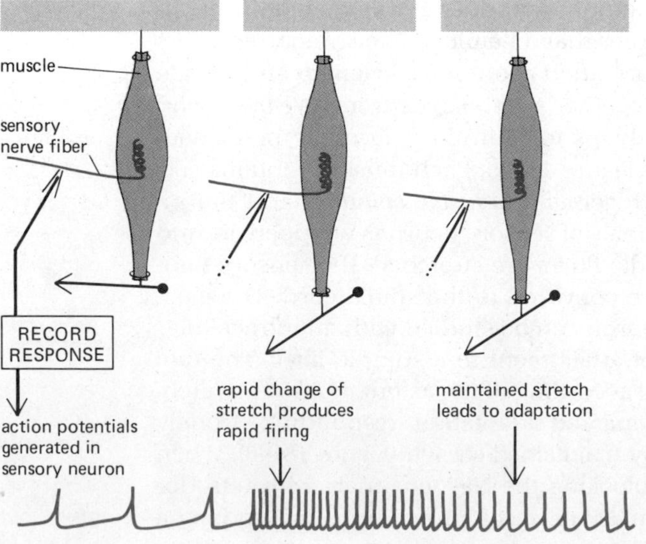

Figure 18–46 Adaptation in a muscle stretch receptor.

graded local depolarization of the sensory nerve terminal. This voltage change is called the **receptor potential.** It is analogous to a postsynaptic potential, and it controls the rate of firing of action potentials in a similar way. Receptor potentials are a universal feature of the sensory transducers associated with the nervous system.

Sense Receptors Are Tuned to Detect Specific Stimuli[34]

The more narrowly defined the sensitivity of a transducer, the more precise the information it supplies to the nervous system. Even among muscle stretch receptors there is specialization: for example, some respond to maintained stretch, while others respond only to a change. Perhaps the most delicately selective of all receptors, however, are the auditory **hair cells** that detect sound in the ears of vertebrates. Like most other receptors, these owe their selectivity partly to the filtering action of the structures in which they are embedded and partly to their own intrinsic properties.

The auditory hair cells take their name from the *stereocilia* (Figure 18–47) that project from their upper surfaces, and whose structure is discussed in Chapter 10 (p. 584). The cells are arrayed in rows along the **basilar membrane,** a thin resilient sheet of tissue forming a dividing partition between two parallel fluid-filled channels in the cochlea. The tips of the stereocilia of the majority of the cells are partly embedded in an overhanging sheet of extracellular material known as the *tectorial membrane* (Figure 18–48). Sound waves propagating along the basilar membrane cause a minute movement, shifting the auditory hair cells relative to the tectorial membrane and so tilting their stereocilia. A receptor potential is produced in the auditory hair cells in response to this deformation, and the signal is passed on across a chemical synapse to nerve endings that make contact with them.

The mechanics of the cochlea are such that sound waves change their amplitude as they propagate along it. High-frequency waves attain their maximum amplitude at one end, low-frequency waves at the other. Thus the cochlea behaves like a spectroscope, distributing sounds of different frequencies to auditory hair cells in different positions. Receptor potentials localized in a particular group of these cells therefore signal the presence of sound of a particular frequency.

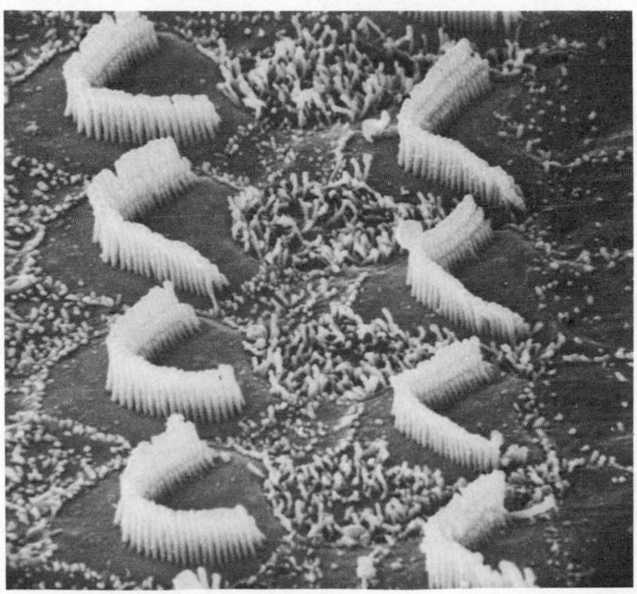

Figure 18–47 Scanning electron micrograph of a part of the cochlea, showing some of the outer hair cells, with stereocilia arranged in a V-shaped pattern on the top of each. (From R. G. Kessel and R. H. Kardon, Tissues and Organs: A Text-Atlas of Scanning Electron Microscopy. San Francisco: Freeman, 1979. © 1979 W. H. Freeman and Company.)

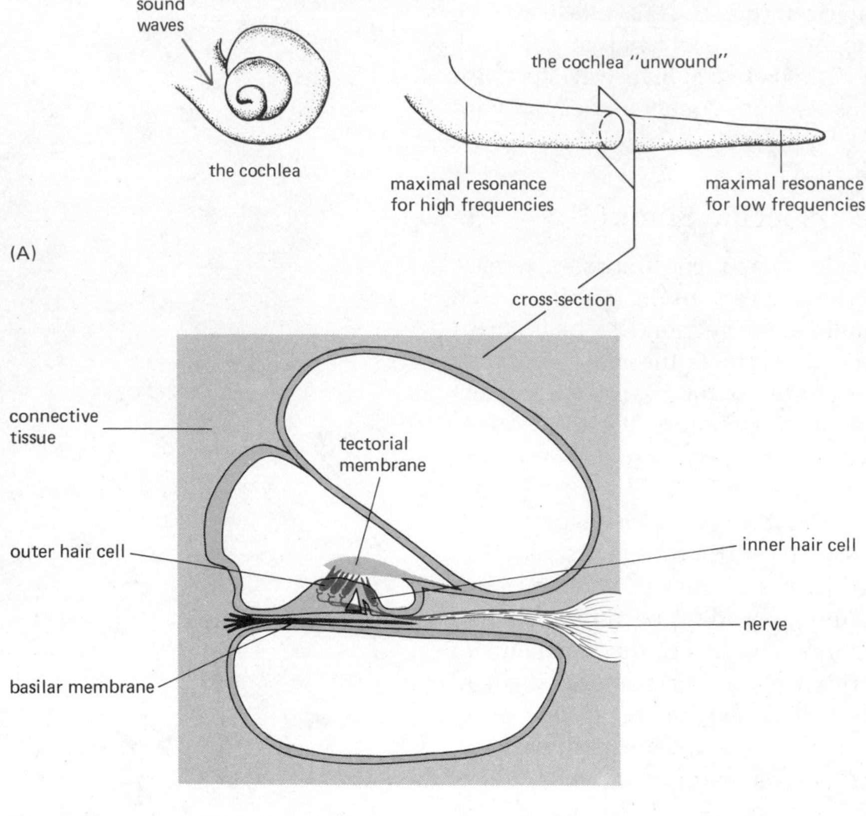

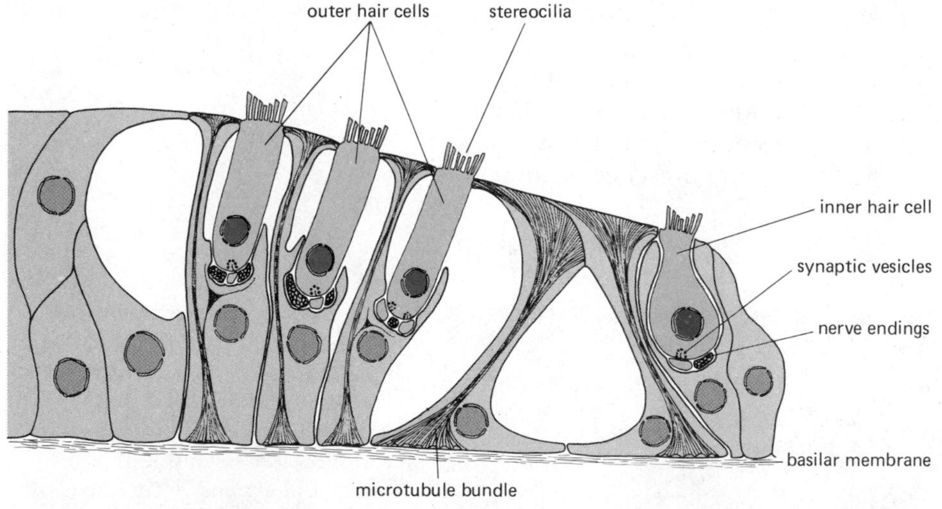

Figure 18–48 Schematic diagram of part of the mammalian auditory system, showing the complex auxiliary apparatus through which sound vibrations are transmitted to the auditory hair cells. (A) Low-magnification external view of the cochlea. (B) View of a section through the cochlea at intermediate magnification, showing how the auditory hair cells are situated on the basilar membrane, with their stereocilia partly embedded in the overlying tectorial membrane. (C) High-magnification view of the auditory hair cells *(color)* and of the highly specialized supporting cells *(gray)* that surround them. The hair cells have synaptic connections through which they deliver the receptor potential to neurons; there are also synapses delivering signals in the opposite direction, from the central nervous system to hair cells, perhaps modifying their sensitivity. The auditory input is received chiefly through the inner hair cells; although there are only about 4000 of them, as compared with about 20,000 outer hair cells, over 90% of the fibers in the auditory nerve are connected to them. (C, from A. W. Ham and D. H. Cormack, Histology, 8th ed. Philadelphia: Lippincott, 1979.)

This mechanism, however, accounts for only part of the very precise selectivity of the receptors. The response of a hair cell is not simply proportional to the amplitude of vibration of the part of the basilar membrane on which it sits, but is still more narrowly tuned to sound of a specific frequency. This suggests that there is a second stage of frequency-selective filtering associated with the auditory hair cells. The nature and location of this second filter are not known in mammals, in spite of much experimental effort.

In reptiles such as turtles, the anatomy of the cochlea is somewhat different, and the basilar membrane plays a minor part in frequency discrimination. The whole burden of selectivity falls on the auditory hair cells, and it appears that each one of these is individually tuned to respond to a certain frequency of vibration. Evidence of such tuning has come from recordings of the electrical behavior of auditory hair cells in the ear of the turtle. An intracellular electrode can be used both to monitor the potential inside a hair cell and to inject current into it. When a small steady current is abruptly switched on or abruptly switched off, the membrane potential is set oscillating (or "ringing") like a gong that has been banged (Figure 18–49). The cell behaves, in other words, as a damped electrical resonator. The characteristic frequency of the electrical oscillation in each hair cell exactly matches the frequency of sound to which it is most receptive. The mechanism of the oscillation is unknown, though it is possible in principle for certain combinations of voltage-gated ion channels to give rise to such behavior.

The precise nature of the coupling between the electrical response and the mechanical deformation in the cochlea remains a mystery and a source of astonishment. This transducing system enables the mammalian ear to operate over a range of sound intensities spanning more than seven orders of magnitude and allows us to hear sounds so faint that they make the basilar membrane vibrate with an amplitude of no more than a fraction of the diameter of a hydrogen atom.

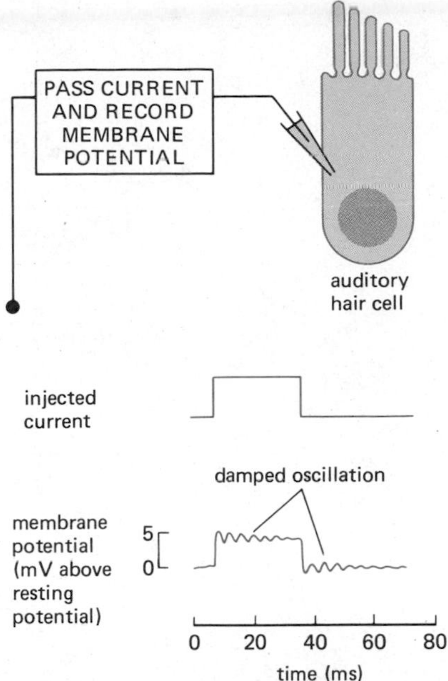

Figure 18–49 Tuned electrical resonance in an auditory hair cell in the ear of a turtle. Injection of a pulse of current provokes a damped electrical oscillation whose frequency is the same as the frequency of sound to which the cell is most sensitive. (After A. C. Crawford and R. Fettiplace, *J. Physiol.* 312:377–412, 1981.)

Rod Cells Can Detect a Single Photon[35]

The transducers by which we perceive light are also phenomenally sensitive, and their functioning is much better understood. As described in Chapter 16, the photoreceptors in the vertebrate eye are of two classes: the **cone cells,** which serve for color vision and require fairly bright light; and the **rod cells,** which provide for monochromatic vision in dim light. A rod cell can produce a measurable electrical signal in response to a single photon—a human being can perceive five photons as a flash of light. Rods and cones appear to operate on similar principles, but rods have been more intensively studied.

The rod cell (see p. 901) consists of an *outer segment,* containing the photoreceptive apparatus, an *inner segment,* containing many mitochondria, a *nuclear region,* and, at the base, a *synaptic body* that makes contact with nerve cells of the retina. In the dark, paradoxically, the cell is quite strongly depolarized; the depolarization holds voltage-gated Ca^{2+} channels open in the synaptic body, and the resulting influx of Ca^{2+} produces a steady release of neurotransmitter. The depolarization is due to open Na^+ channels in the plasma membrane of the outer segment. Illumination causes these channels to close, so that the receptor potential takes the form of a *hyperpolarization,* which leads to a decrease in the rate of transmitter release (Figure 18–50). The transmitter has an inhibitory action on many of the postsynaptic neurons, which therefore are freed from an inhibition—and thus, in effect, excited—by the illumination. The rate of release of transmitter from the photoreceptors is graded according to the intensity of the light: the brighter the light, the greater the hyperpolarization and the greater the decrease in transmitter release. When the cell is in its most sensitive state, the absorption of a single photon reduces the Na^+ influx by a million ions or more, generating a hyperpolarization of about 1 mV. How then does light cause the Na^+ channels to close?

The outer segment, where the crucial events occur, is a cylindrical structure containing within a sheath of plasma membrane a stack of about a thousand *discs.* Each disc is a closed, flattened sac of membrane in which photosensitive molecules of **rhodopsin** are embedded, packed closely together

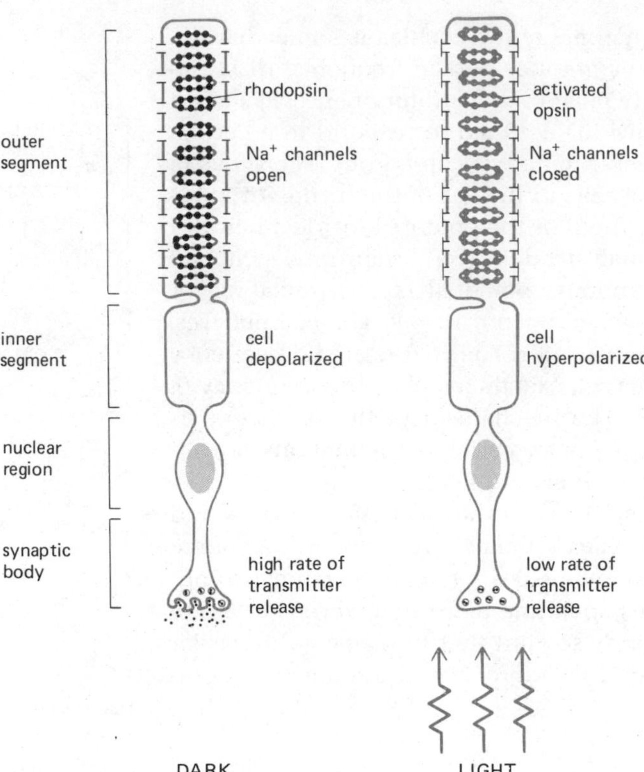

Figure 18–50 Schematic diagram of the response of a rod photoreceptor cell to illumination.

at a density of about 10^5 per μm^2. Each rhodopsin molecule consists of a transmembrane polypeptide, *opsin,* with a prosthetic group, 11-*cis*-retinal, which absorbs light. Absorption of a photon causes the 11-*cis*-retinal to isomerize to all-*trans*-retinal, which then dissociates from the opsin. As a consequence, the opsin undergoes a conformational change, and this somehow results in the closing of Na^+ channels in the plasma membrane. It seems, therefore, that there must be a second messenger in the cytoplasm of the outer segment to couple these two spatially separate events. Though the nature of the second messenger is not known for certain, there is evidence strongly suggesting that Ca^{2+} plays this role. Light apparently causes a release of Ca^{2+} from the discs into the cytoplasm, and the raised cytoplasmic Ca^{2+} causes the Na^+ channels to close. At the same time, the conformational change in rhodopsin initiates a cascade of enzymatic reactions that causes a fall in the concentration of cyclic GMP in the cytosol through an increase in the activity of the phosphodiesterase enzyme that breaks down cyclic GMP. The function of the cyclic GMP change is still unclear.

The Visual World Is Mapped onto a Sequential Hierarchy of Arrays of Neurons[36]

The state of the external world is represented in the nervous system by voltages in ordered arrays of cells—separate arrays for different aspects of the world, monitored by different types of transducers. For a given modality, such as vision, the first representation is in the receptor cells themselves. The voltage in each photoreceptor represents the brightness of the particular part of the world that it "sees." The information from the photoreceptors is relayed through successive arrays of neurons and processed at each step until eventually, in the highest centers in the brain, it is combined with input delivered through other sensory channels; here, at length, the transmuted sensory input serves to control the production of the output signals that govern behavior.

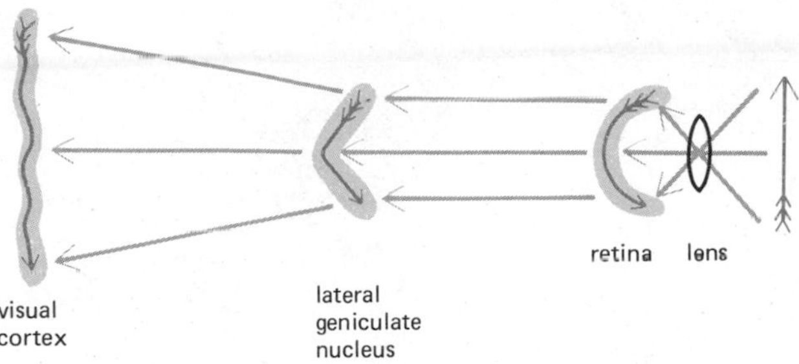

Figure 18–51 Schematized plan of the major visual pathway in a mammal. Signals from adjacent regions of the visual world are focused onto adjacent groups of photoreceptors and passed along on parallel neural paths to successively higher levels in the brain. At each level, the visual world (symbolized by a vertical arrow) is mapped in an orderly fashion onto sheetlike arrays of neurons. The diagram is highly simplified and makes no attempt to show how the inputs from right and left eyes are brought together to provide stereoscopic vision.

Signals from adjacent regions of the external world are focused on adjacent regions of the retina, detected by adjacent photoreceptors, and carried along parallel pathways to adjacent neurons in each level of the hierarchy (Figure 18–51). Thus the visual world is *mapped* onto successive sheets of nerve cells, starting in the retina and proceeding to the visual cortex of the brain. Before considering how the visual information is sifted and transformed from one map to the next, we must examine the anatomical connections between successive maps.

The photoreceptors pass on their information via synapses to the system of neurons in the middle layer of the retina. This layer consists of *bipolar, horizontal,* and *amacrine cells* (Figure 18–52). All these three classes of neurons are small enough to conduct their signals by passive spread: they do not fire action potentials. The horizontal and amacrine cells (Figure 18–53) extend their processes laterally, parallel to the plane of the retina, while the bipolars are oriented perpendicularly and provide a direct link to the next array in the hierarchy—the *retinal ganglion cells.* The ganglion cells send their axons to the brain, encoding the visual information as action potentials for transmis-

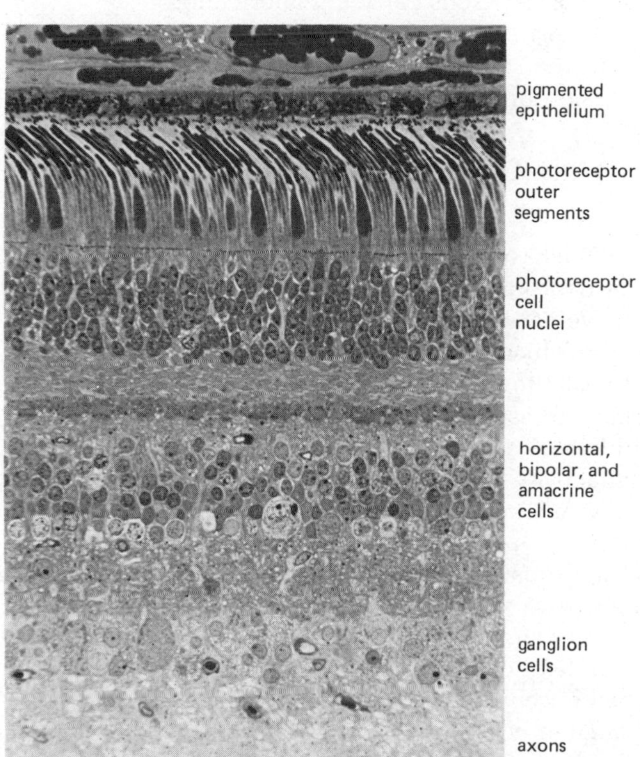

Figure 18–52 Photomicrograph of a section through the retina, showing how the cells are arranged in a series of well-defined layers. The direction of the incident light is from bottom to top through these layers. Thus, the light must pass through the neurons involved in transmitting visual signals to the brain before it reaches the photoreceptor cells where it is detected. The signals generated by these photoreceptors then pass through an intermediate layer of neurons to the ganglion cells, which relay the information to the brain in the form of action potentials. For a schematic diagram of the structure of the retina, see Figure 16–8 on page 900. (Courtesy of John Marshall.)

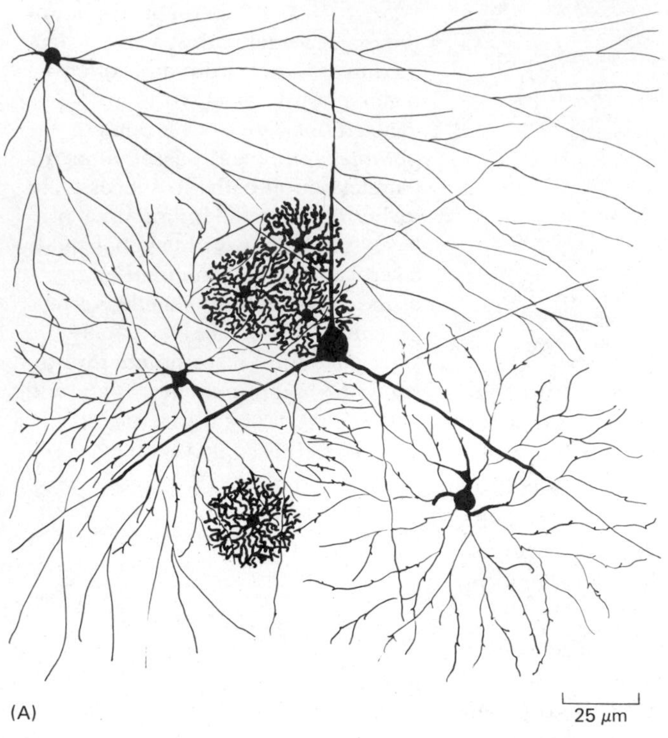

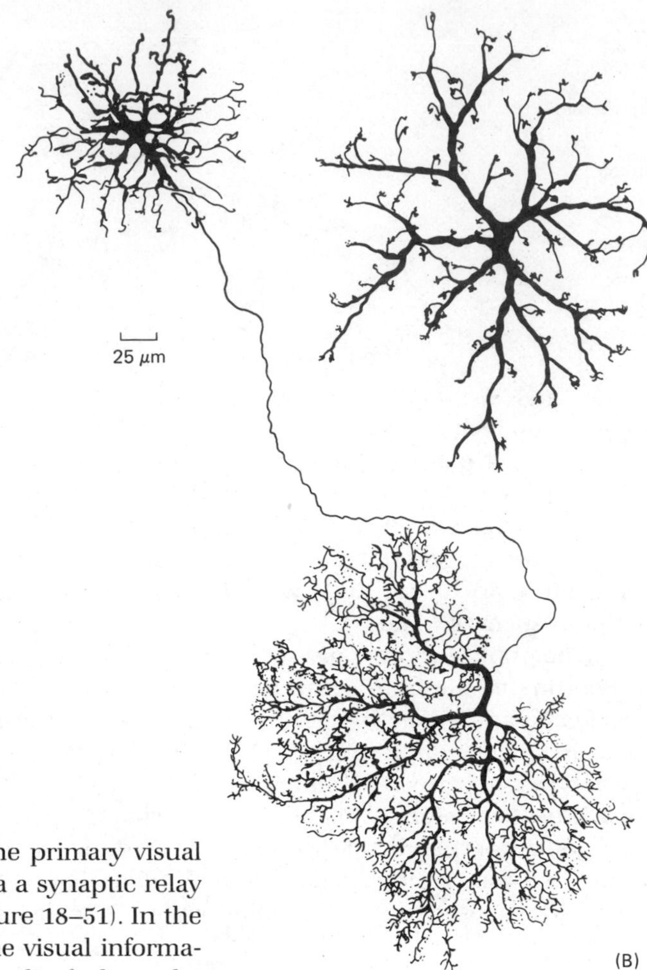

Figure 18–53 Golgi-stained amacrine cells (A) and horizontal cells (B) from the mammalian retina, as seen face-on in a flat preparation. (The amacrine cells are from a monkey, the horizontals from a cat.) Both classes of cells have short axons or none. The eight amacrine cells shown in (A) represent at least four distinct types, all without axons, while the two horizontal cells shown in (B) differ in that one has an axon (with a profuse mass of terminal branches) and the other does not. Amacrine cells with different shapes are also chemically different from one another: at least six different types can be distinguished by the different neuropeptides they contain. (A, from B. B. Boycott and J. E. Dowling, *Philos. Trans R. Soc. Lond. (Biol.)* 255:109–184, 1969; B, from B. B. Boycott, in Essays on the Nervous System [R. Bellairs and E. G. Gray, eds.]. Oxford, Eng.: Clarendon Press, 1974.)

sion. In mammals, the input from the eye goes chiefly to the primary visual region of the **cerebral cortex,** (or *visual cortex,* for short) via a synaptic relay station in the brain called the *lateral geniculate nucleus* (Figure 18–51). In the visual cortex, which consists of several layers of neurons, the visual information is again passed from layer to layer, more or less perpendicularly to the cortical surface. From the primary visual cortex, axons carry the information to yet other cortical areas. (The picture is further complicated by the arrival of inputs from both left and right eyes in the same region of cortex, but we shall defer discussion of that aspect of vision until the end of the chapter.)

Neurons at Higher Levels Detect More Complex Features of the Visual World[37]

From a purely anatomical examination of the tissues involved, it is not possible to deduce how the successive arrays of neurons process visual information. To understand the neural computation, we need to know the meaning of the electrical signal from each cell. The problem sounds vague, but it can be restated in precise operational terms: to ask the meaning of the signal from a given cell in the visual system is, in effect, to ask what specific type of event or circumstance in the external world excites that cell. The answer can be discovered by monitoring the electrical activity of the cell with a microelectrode while presenting various types of visual stimuli at various locations in the animal's field of view.

The region within which a stimulus must fall in order to influence a particular cell is known as the cell's **receptive field** (Figure 18–54). For each photoreceptor, the effective stimulus is very simple: the visual field must be illuminated in the appropriate region with light of an appropriate wavelength. But the effective stimulus becomes progressively more complex at higher levels in the visual system. The retinal ganglion cells provide an important example. The receptive fields of these cells are generally larger than those of the

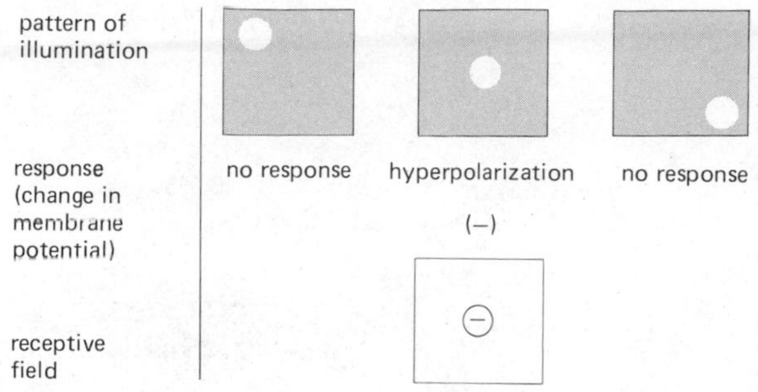

Figure 18–54 The receptive field of a rod photoreceptor cell. The receptive field is discovered by presenting different patterns of illumination in the visual field and recording the responses of the cell. The receptive field of the photoreceptor is very simple: the cell responds (by hyperpolarizing) if and only if a particular spot in the visual field is illuminated (irrespective of the illumination of other parts of the visual field). The receptive field is therefore mapped by marking a ⊖ (representing hyperpolarization) at the sensitive spot in the visual field.

photoreceptors and they partly overlap. A typical ganglion cell responds very weakly to uniform illumination. Moreover, a small spot of light that occupies only part of the cell's receptive field has contrary effects according to whether it lies in the center or the periphery of the receptive field: for example, it may excite firing of the ganglion cell if it lies at the center but inhibit firing if it lies in the periphery. For such a ganglion cell, the most effective stimulus is a bright circular patch with a dark circular surround (Figure 18–55). Other ganglion cells respond most strongly to the converse pattern—a dark patch with a bright surround. The ganglion cells, in short, are detectors of spots of *contrast* in the visual scene.

Cells in the cortex have still more specialized requirements. One large class of cortical neurons only responds well to a *bar* of light or dark, on a contrasting background; for a given neuron, the bar must be in a particular region of the visual field and have a particular orientation relative to the horizontal (Figure 18–56). To take another example, certain other cortical neurons require, in order to respond, not only that the stimulus have a particular orientation but also that it be moving in a particular direction and be of a specific length and breadth.

Information Is Processed Through Local Computations Involving Convergence, Divergence, and Lateral Inhibition[38]

The behavior of the high-level neurons, which serve as detectors of complex features of the visual world, is determined by the pattern of neural connections along the visual pathway. As an example, consider a retinal ganglion cell responsive to a spot of light with a dark surround. How is it endowed with this specific sensitivity?

The anatomy of the retina suggests the sort of circuitry that is involved. As shown in Figure 18–52, individual bipolar cells gather their input from several adjacent photoreceptors and distribute their output to several adjacent

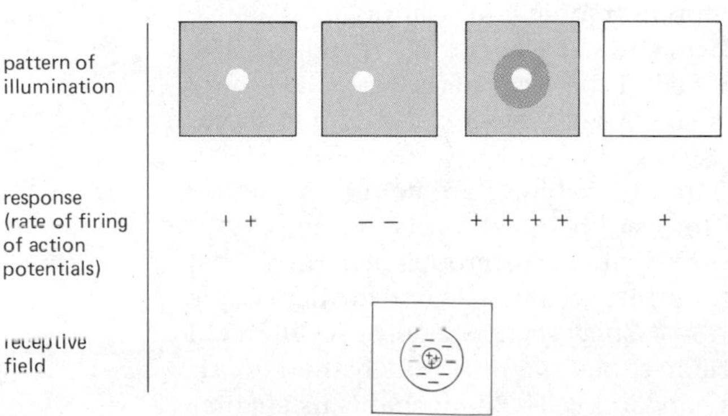

Figure 18–55 The more complex receptive field of a retinal ganglion cell, mapped out by the method described in Figure 18–54. The ganglion cell shown here is excited by switching on a small spot of light that falls on the center of its receptive field and is inhibited by a spot of light falling on the periphery of its receptive field. Excitation and inhibition take the form of increase (+) and decrease (−) in the rate of firing of action potentials. The cell fires action potentials spontaneously, at a low rate, even in the dark. The strongest excitation is produced by a bright spot of light with a dark surround. Other ganglion cells exhibit reverse behavior, being excited most strongly by a dark spot with a bright surround. Both classes of ganglion cells respond only very weakly to uniform illumination: they are detectors of contrast.

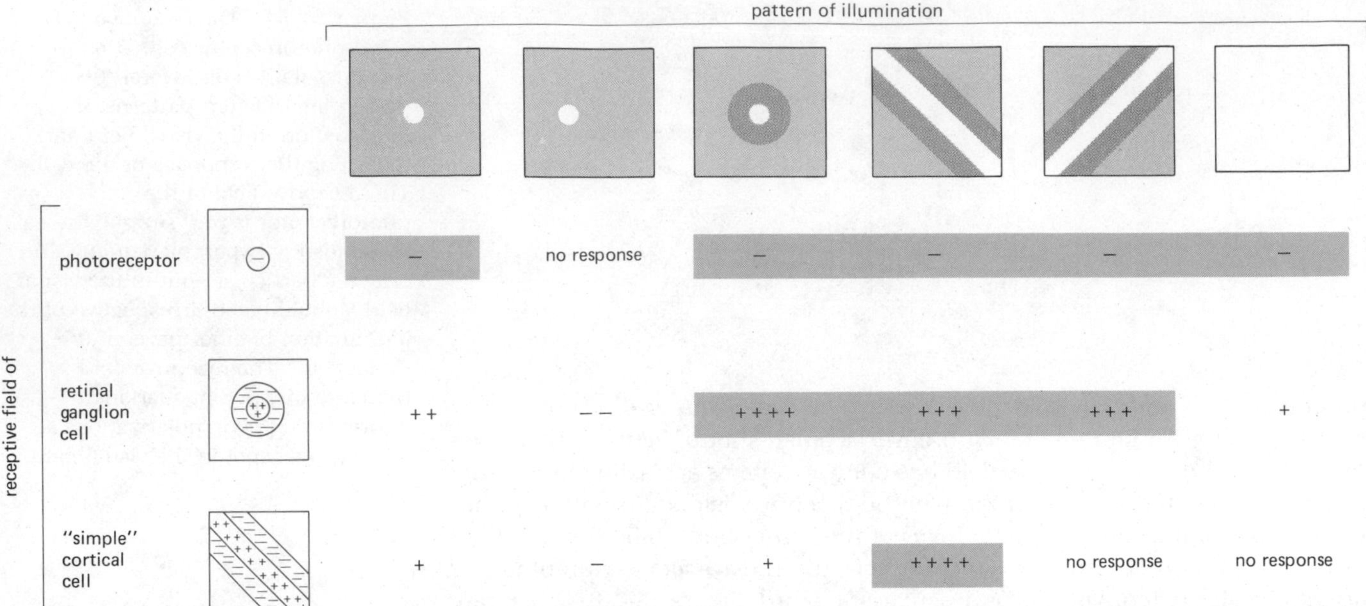

Figure 18–56 Summary of receptive fields and responses to different patterns of light at successive levels in the visual pathway of a mammal. The square represents a certain small part of the visual field. The left-hand column shows the receptive fields of three cells receiving input from that region: a photoreceptor, a retinal ganglion cell, and a "simple" cell in the visual cortex. The response of each type of cell to each of six different patterns of illumination is tabulated. Note that a strong response from a photoreceptor only signals that a certain precisely defined spot in the visual field is illuminated, irrespective of the illumination of neighboring points. A strong response from the cortical cell, on the other hand, represents much more information as to the local structure of the pattern of illumination (though somewhat less precise information is conveyed about the location of the stimulus in the visual field). The diagram shows only one of the types of retinal ganglion cell and only one of the many types of cortical cell.

ganglion cells: both *convergence* and *divergence* are found in the pattern of connections. Moreover, besides the bipolars passing signals perpendicularly through the depth of the retina, there are horizontal and amacrine cells mediating lateral interactions between neighboring bipolar and ganglion cells. While the electrophysiological details are complex, the basic principles are simple. With an appropriate distribution of inhibitory and excitatory synapses, the perpendicular and lateral connections can deliver opposite types of signals to the ganglion cells. Thus light falling on a given photoreceptor can excite the ganglion cell directly below it while causing *lateral inhibition* of the surrounding ganglion cells (Figure 18–57). Light falling on a neighboring photoreceptor will have an analogous effect, exciting a different ganglion cell and inhibiting a set that includes the cell previously excited. If all the photoreceptors are illuminated uniformly, excitation and inhibition will roughly balance each other for each ganglion cell, giving practically no net response. The optimal stimulus for a given ganglion cell of the class described is a spot of light, providing excitation, with a dark surround, providing freedom from lateral inhibition.

Convergence, divergence, and lateral inhibition are common themes throughout the nervous system, and they are believed to play an important part in the local computations performed by the many groups of neurons that lie close together and deal with closely correlated input. As one further simple (though speculative) example, Figure 18–58 suggests how a cortical cell might be made selectively responsive to oriented bar stimuli through the convergence of inputs from a row of retinal ganglion cells. By mechanisms such as

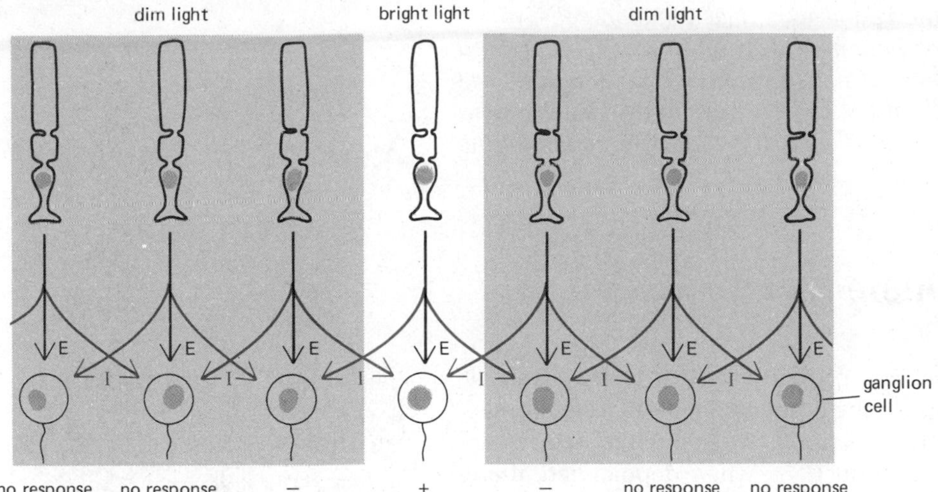

Figure 18–57 Schematic diagram of lateral inhibition. In the idealized system shown here, the arrangement of inhibitory (I) and excitatory (E) synapses is such that light falling on a photoreceptor excites the ganglion cell directly beneath, while inhibiting the ganglion cells on either side. Ganglion cells in regions of uniform illumination show practically no response because inhibition and excitation balance each other. Ganglion cells show a response only if they lie in a neighborhood where there is a *contrast* of bright with dim illumination, as is the case for the ganglion cells in the middle of the diagram.

these, a typical neuron at a higher level in the visual system, driven by a combination of inputs from a cluster of neurons at the level below, is enabled to detect more abstract and complex features of the local pattern of visual stimulation. Thus the information content of the electrical signal carried by each neuron becomes progressively richer as the visual input ascends the visual pathway.

Summary

Sensory stimuli are translated into neural signals by specialized transducers. In a muscle stretch receptor, for example, the sensory nerve terminal is depolarized by stretching, and the magnitude of the depolarization—the receptor potential—is encoded for transmission in the rate of firing of action potentials. Auditory hair cells, which respond selectively to sound of a particular frequency, do not themselves fire action potentials but signal the magnitude of their receptor potentials to adjacent neurons via chemical synapses. The same is true of photoreceptors in the eye. In a photoreceptor, light causes a conformational change in molecules of rhodopsin, and this, through an intracellular second messenger, causes Na^+ channels in the plasma membrane to close, thereby hyperpolarizing the cell and decreasing its output of neurotransmitter. The

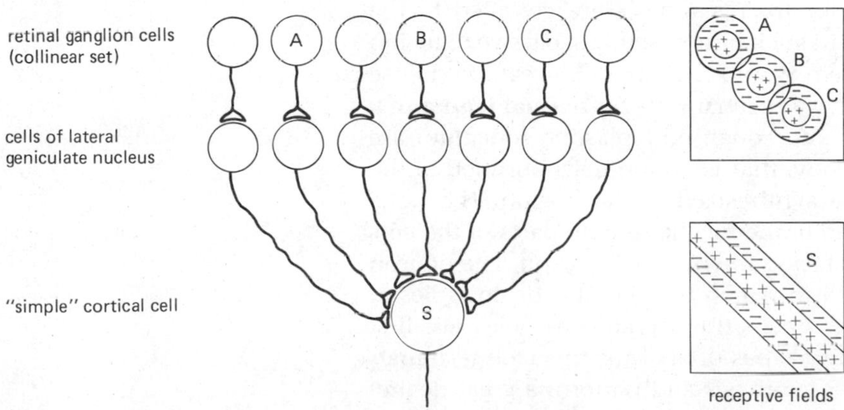

Figure 18–58 Hypothetical diagram to show how a set of cells lying along a straight line, each having a circularly symmetric receptive field, could combine to drive a neuron at a higher level (such as the "simple" cortical cell illustrated in Figure 18–56) in such a way that the high-level cell becomes a detector of bar-shaped stimuli with a specific orientation. The real patterns of connection are believed to be more complex than this, and inhibitory synapses play a major part in determining the specific responsiveness of cortical cells.

signal is passed via interneurons in the retina to retinal ganglion cells, which relay it to the brain in the form of action potentials. The information is processed as it passes through the neural network of convergent, divergent, and lateral inhibitory connections, so that the cells at the higher levels in the visual system act as detectors of more complex features of the pattern of illumination of the visual field.

Maintenance and Development of Nerve Cell Structure[39]

Neurons are not unique in their electrical properties. Most types of muscle cells also conduct action potentials, as do the epidermal cells of frog tadpoles, and even some plant cells, such as the green alga *Nitella.* Some oocytes are electrically excitable, and furthermore admit Ca^{2+} when depolarized, thereupon releasing their cortical granules in a manner that closely parallels the release of transmitter from a nerve terminal (see p. 806).

On the other hand, neurons are truly exceptional in their geometrical structure (see Figure 18–4). No other type of cell can be 10 meters long, as a motor neuron of a whale is, or can put out processes to connect with 100,000 other cells, as a Purkinje neuron in the cerebellum does (Figure 18–59). This section considers the specialized apparatus that allows mature neurons to maintain such shapes and embryonic neurons to develop them. In doing so, it will serve as the starting point for a general discussion of the cellular mechanisms of neural development.

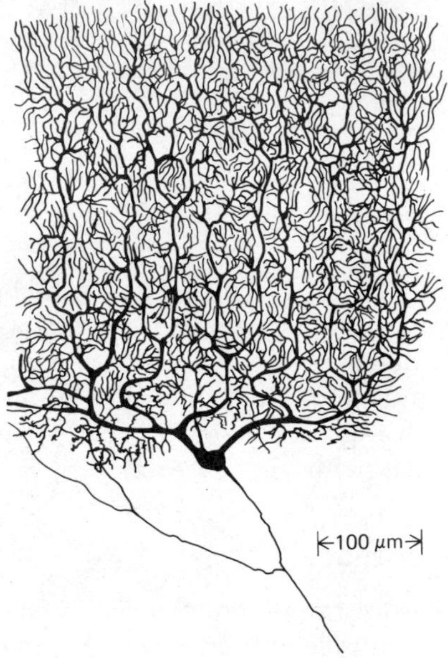

Figure 18–59 A Purkinje neuron from the human cerebellum. The whole of the "tree" of the dendrites and the initial part of the axon are shown. This cell receives synaptic inputs from about 100,000 other neurons. It forms part of the brain's machinery for controlling complex movements. (From S. Ramón y Cajal, Histologie du Système Nerveux de l'Homme et des Vertébrés. Paris: Maloine, 1909–1911; reprinted, Madrid: C.S.I.C., 1972.)

Axonal Transport Maintains Chemical Communication Between the Cell Body and the Outlying Parts of a Neuron[40,41]

A long thin process such as an axon, whose diameter may be only a millionth of its length, runs a serious risk of breakage. It requires some sort of internal strengthening, which appears to be provided by the plentiful *neurofilaments* and *microtubules* that course through the cytoplasm of practically all nerve cell processes (Figure 18–60). In fact, these two classes of cytoskeletal proteins, together with actin, probably account for about half of the total protein in the brain.

The neuronal cytoskeleton is doubly important: besides providing mechanical support, it plays a crucial part in the chemical transactions of the cell, whose extremely elongated shape creates severe problems of internal communication. It is one thing to transmit action potentials, quite another to transport molecules. Proteins, for example, are needed throughout the axon and dendrites, but the instructions for their synthesis are confined to the nucleus, which may lie a meter away. In such circumstances, passive diffusion is totally inadequate: an average protein would take about 50 years to diffuse a meter. Neurons therefore require active mechanisms for **axonal transport.**

In a typical neuron, ribosomes and rough endoplasmic reticulum are confined almost entirely to the cell body, that is, to the neighborhood of the nucleus, and so all proteins must be synthesized in that region. The Golgi apparatus, which is the source of membrane for the remote parts of the cell, also lies close to the nucleus, usually facing the base of the axon. The neuron can thus be viewed as a secretory cell in which the site of exocytosis lies at an enormous distance from the site where the secretory vesicles are first formed (Figure 18–61). Although in most types of neurons the axon terminals contain enzymes for neurotransmitter synthesis and perform a great deal of local recycling of synaptic vesicle membrane, there is a continuing need for

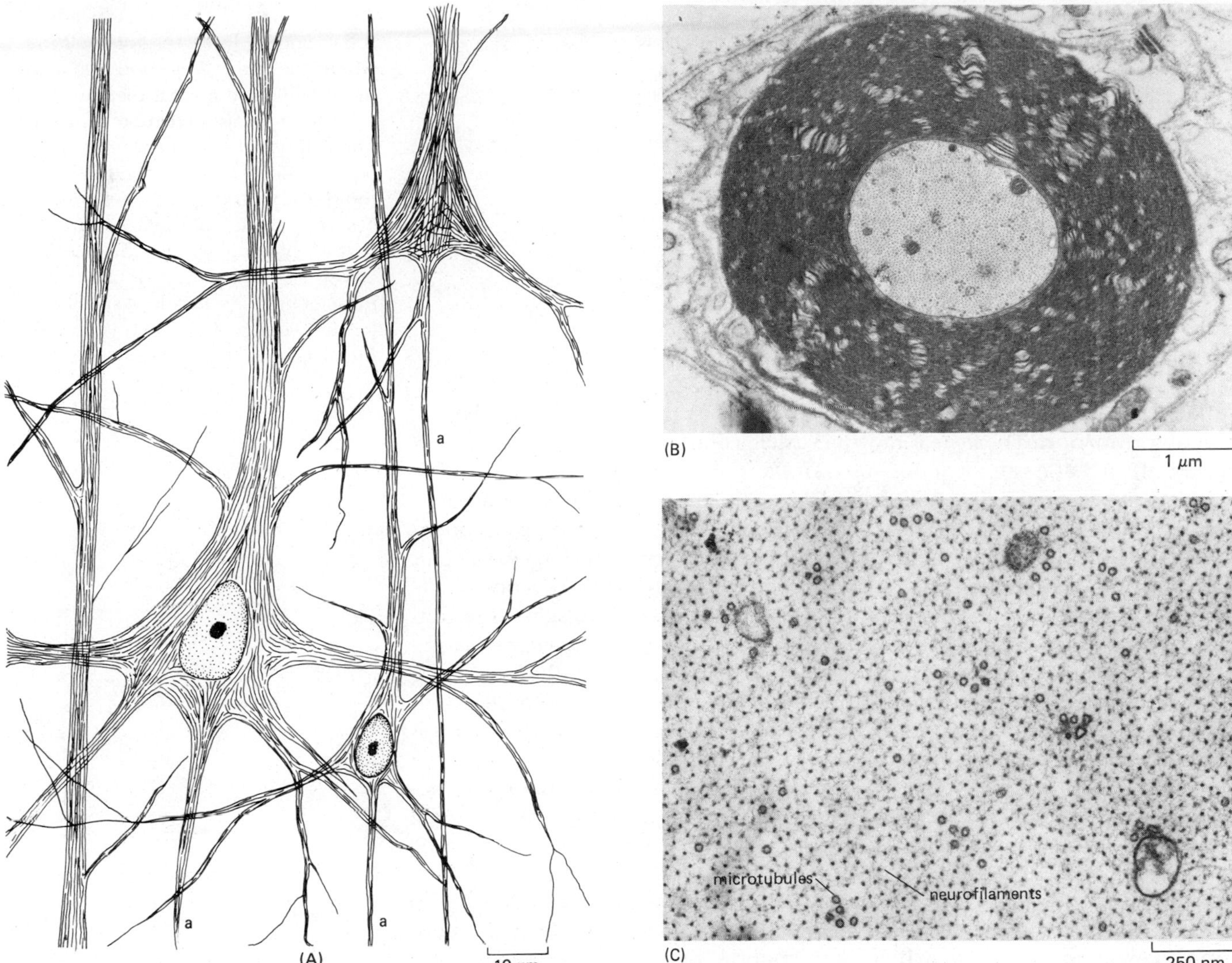

Figure 18–60 The neuronal cytoskeleton. (A) Pyramidal cells from the visual cortex, viewed under the light microscope after silver staining. Many fine fibrils can be seen coursing along the dendrites and axons (a) and through the cell bodies. (B) Electron micrograph of a cross-section through a heavily myelinated sensory axon of a cat. The cytoplasm of the axon contains many neurofilaments and microtubules, as well as some mitochondria and small membranous vesicles. The myelin sheath has been distorted in fixation, revealing its lamellar structure. (C) High-magnification view of a portion of the cytoplasm of an axon similar to that shown in (B). Note that the membranous vesicles are closely associated with small groups of microtubules. (A, from S. Ramón y Cajal, Histologie du Système Nerveux de l'Homme et des Vertébrés. Paris: Maloine, 1909–1911; reprinted, Madrid: C.S.I.C., 1972; B and C, courtesy of John Hopkins.)

supplies of freshly synthesized membrane and enzymes from the cell body. Membranous vesicles of various shapes and sizes can be detected in transit along the axon, and if an axon is artifically constricted, accumulations of vesicles pile up rapidly at the site of the constriction. Moreover, they pile up on both sides, implying that there is both an outward or *anterograde transport* from the cell body and a return or *retrograde transport* from the far end of the axon. The returning vesicles, which are generally larger than the outgoing vesicles, contain debris destined for degradation in lysosomes. They also often contain molecules taken up by endocytosis from the extracellular medium at the axon terminal.

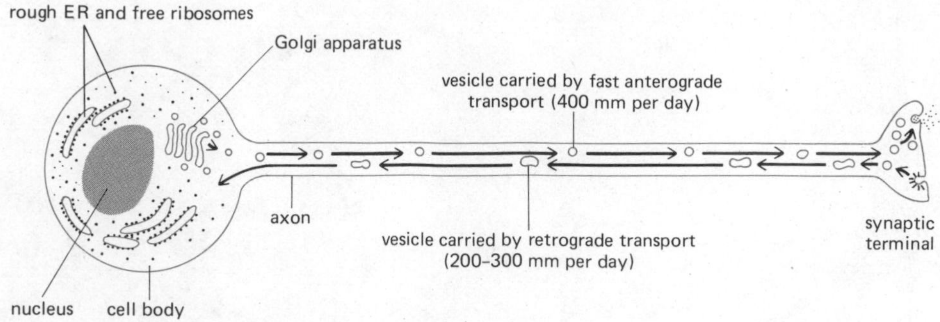

Figure 18–61 A neuron viewed schematically as a secretory cell in which the site of secretion (the axon terminal) lies at a great distance from the site of macromolecular synthesis (the cell body). This mode of organization creates a need for a rapid axonal transport mechanism. The diagram is not meant to imply that all synaptic vesicles have to be transported from the cell body; in most neurons, synaptic vesicles are formed largely by local recycling of membrane in the axon terminal.

Axonal Transport Has Both Fast and Slow Components[41]

By labeling the molecules involved, it is possible to measure the speed of vesicular transport. The fastest-moving vesicles of the anterograde component travel outward at a rate of about 400 mm per day (in warm-blooded animals). This rate is practically the same regardless of species, size of axon, or amount of electrical activity. The fastest-moving vesicles of the retrograde component travel at about a half or two-thirds of the anterograde rate. The larger vesicular bodies can be watched moving in living cells in culture. They advance by fits and starts, in saltatory fashion, as though each one were engaging intermittently with some driving mechanism. It is not known in detail how the driving force is generated, although it has been shown to depend on the hydrolysis of ATP. The axon contains actin and myosin as well as microtubules and neurofilaments, and vesicular transport is halted both by agents that disrupt actin filaments and by agents that disrupt microtubules.

The vesicular traffic along the axon, conveying lipids, membrane glycoproteins, and materials for secretion, constitutes the *fast component of axonal transport.* In addition, there are *slow components of axonal transport,* whereby the proteins of the cytoskeleton are themselves steadily exported from the cell body, together with enzymes of the cytosol, as discussed on page 599. Tubulin and neurofilament proteins move out together at the slowest rate of all, about 1 mm per day or even less, which corresponds roughly to the rate at which microtubules lengthen by addition of tubulin monomers in other structures such as the mitotic spindle. Similar fast and slow transport mechanisms operate in dendrites. Thus the neuron is able to renew, maintain, and repair its far-flung processes and to keep up chemical communication between the nucleus and the most distant parts of the cell.

We turn now from the problem of the maintenance of neuronal structure to the problem of its development.

Developing Axons and Dendrites Terminate in a Growth Cone[42]

Given that a nerve cell can conduct action potentials and transmit and receive signals across synapses, its specific role is determined by its pattern of contacts with other cells. To understand how the cell comes to perform its function, one must therefore consider how it manages to send out long processes to appropriate destinations and to establish a precisely organized set of synaptic connections with other cells. Attention focuses in particular on two structures: the *growth cone,* by means of which the developing nerve cell process (axon or dendrite) advances toward its destination, and the *synapse,* which the process forms when it arrives there. The growth cone plays the central part in the formation of nerve connections, and an account of its behavior as studied in isolation will provide the basis for our subsequent discussion of the development of systems of nerve cells.

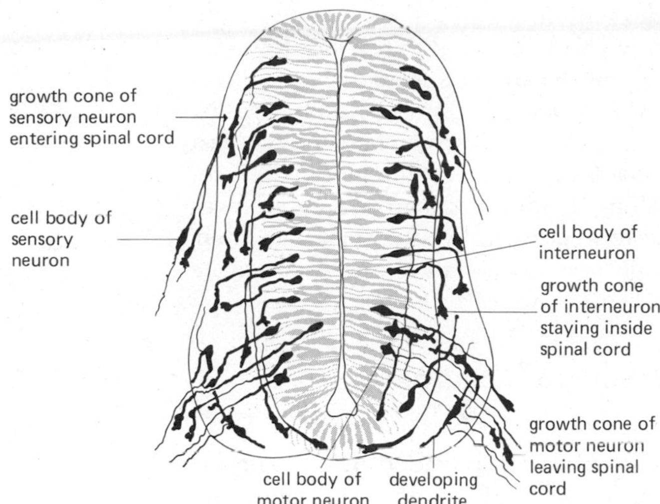

Figure 18–62 Growth cones in the developing spinal cord of a three-day chick embryo, as seen in a Golgi-stained cross-section. Most of the neurons, apparently, have as yet only one elongated process—the future axon. The growth cones of the interneurons remain inside the spinal cord, those of the motor neurons emerge from it (to make their way toward muscles), and those of the sensory neurons grow into it from outside (where their cell bodies lie). Many of the cells in the more central regions of the embryonic spinal cord are still proliferating and have not yet begun to differentiate as neurons or glial cells. (From S. Ramón y Cajal, Histologie du Système Nerveux de l'Homme et des Vertébrés. Paris: Maloine, 1909–1911; reprinted, Madrid: C.S.I.C., 1972.)

The mature neuron develops from a relatively small precursor cell, which stops dividing before it extends processes. As a rule, the axon begins to grow out from the nerve cell body first, the dendrites forming slightly later. The sequence of events was originally observed in intact embryonic tissue by Golgi staining (Figure 18–62). This technique reveals a peculiar enlargement, with an irregular, spiky shape, at the tip of the developing nerve cell process. The structure appears to be forcing its way forward through the surrounding tissue. It is the **growth cone.**

Most of our present information about the properties of growth cones has come from studies in tissue culture. Embryonic nerve cells in tissue culture send out processes that are hard to identify as axon or dendrite and are therefore given the noncommittal name of *neurite.* The growth cone at the end of each neurite in the culture dish can be likened to a hand on the end of an arm: it consists of a broad flat expansion of the neurite, like the palm of a hand, with many long microspikes extending from it like fingers (Figure 18–63). The microspikes are typically 0.1 μm to 0.2 μm in diameter and can be up to 50 μm in length; the broad flattened region of the growth cone is typically about 5 μm wide and 5 μm long, though its precise shape changes from minute to minute. The "webs" or "veils" of the growth cone between the microspikes have a ruffling membrane, like the leading edge of a moving fibroblast (see p. 604). The microspikes themselves are also continually active.

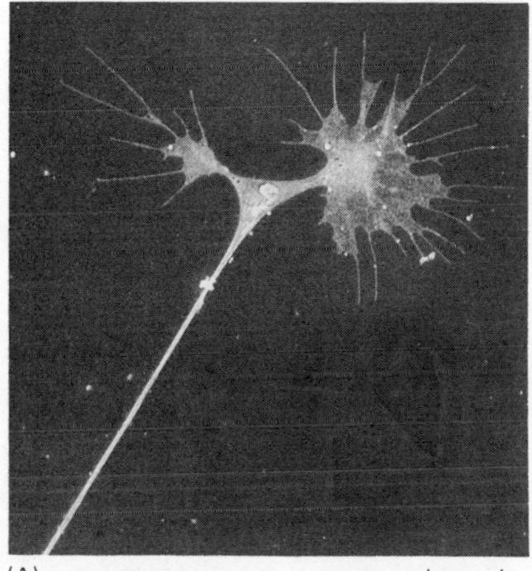

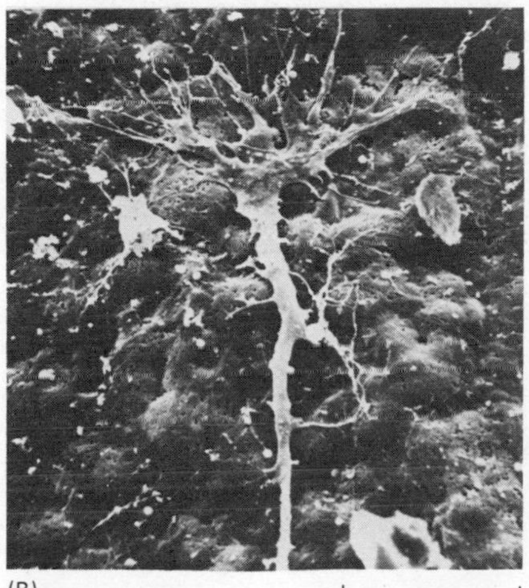

Figure 18–63 (A) Scanning electron micrograph of growth cones at the end of a neurite put out by a chick sympathetic neuron in culture. A previously single growth cone has here recently divided in two. Note the many microspikes and the taut appearance of the neurite. (B) Scanning electron micrograph of the growth cone of a sensory neuron *in vivo,* crawling over the inner surface of the epidermis of a *Xenopus* tadpole. (A, from D. Bray, in Cell Behaviour [R. Bellairs, A. Curtis, and G. Dunn, eds.]. Cambridge, Eng.: Cambridge University Press, 1982; B, from A. Roberts, *Brain Res.* 118:526–530, 1976.)

As some are retracting back into the growth cone, others are extending out from it, waving about, and touching down and adhering to the substratum.

These observations, and others to be discussed subsequently, suggest that the growth cone is crawling forward in much the same way as the leading edge of a cell such as a fibroblast (see Chapter 10). There is, however, at least one essential difference: whereas the leading edge of a fibroblast is a locomotive organ that drags its cell body along behind it, the growth cone leaves its cell body in place, causing the neurite to grow in length. What part, then, does the growth cone play in the assembly of the materials required for neurite growth?

Cytoskeletal Proteins and Membrane for Outgrowth Are Inserted at Different Sites[42]

Some clues to the mode of growth of neurites are provided by the internal structure of the developing neuron. As in an adult neuron, ribosomes are largely confined to the cell body, which must therefore be the site of protein synthesis. The neurite contains microtubules and neurofilaments, together with occasional membranous vesicles and mitochondria. The broad "palm" of the growth cone, by contrast, is full of small, irregular anastomosing membranous vesicles, rather like smooth endoplasmic reticulum (Figure 18–64). Immediately beneath the ruffling margins of the growth cone, and filling the microspikes, is a dense irregular meshwork of actin filaments. The growth cone also contains mitochondria. Finally, the microtubules and neurofilaments of the neurite come to an end in this region.

In principle, the neurite could grow by insertion of new material at its base (close to the cell body), at its tip, or along its length. In fact, it appears that different components of the neurite are inserted in different places (Figure

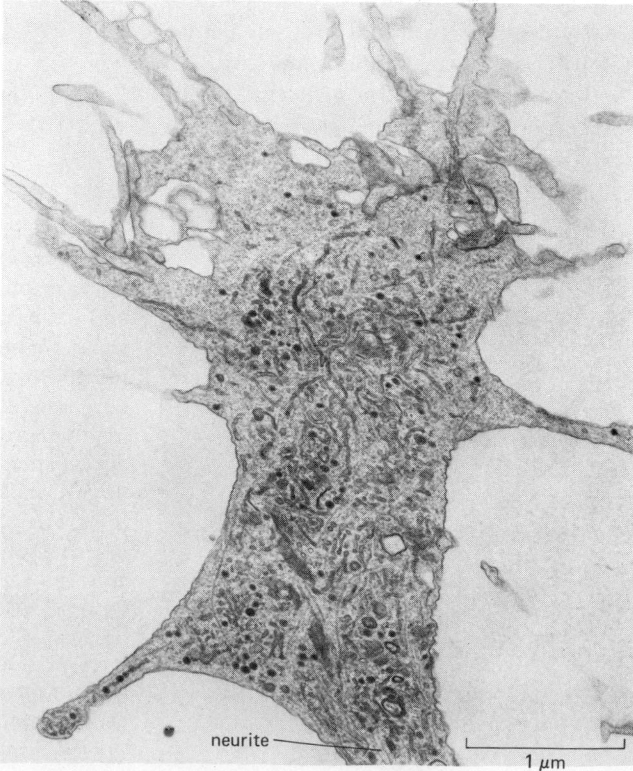

Figure 18–64 Electron micrograph of a section through a growth cone, showing the irregular masses of vesicles and smooth endoplasmic reticulum that it contains. (Courtesy of Gerald Shaw.)

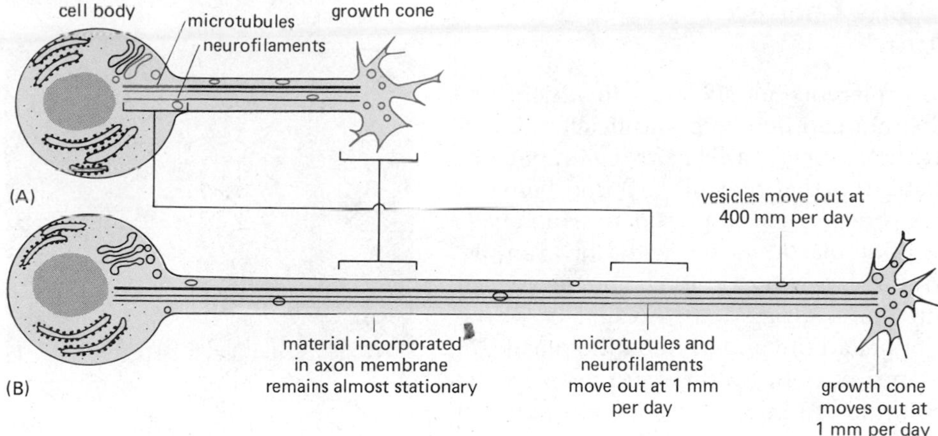

Figure 18–65 Schematic diagram suggesting how the parts of a developing neurite are assembled as it grows. Newly synthesized materials are shown in color in (A). (B) shows in color the positions of the same materials some time later. Microtubules and neurofilaments are slowly extruded from the cell body by the addition of new subunits there, while vesicles of membrane are transported rapidly towards the growth cone, where they become incorporated in the plasma membrane. Unlike the microtubules and neurofilaments, the plasma membrane of the neurite, once formed, appears to remain practically stationary in relation to the cell body: specks of extracellular debris adhering to it do not move.

18–65). The microtubules and neurofilaments are probably spun out from the cell body largely by addition of newly synthesized subunits at the proximal end of the neurite. They move outward at a rate of about 1 mm per day, which is the same as the rate of the slowest component of axonal transport in a mature neuron. Since this is also roughly the average rate of advance of the growth cone, it is possible that, while the neurite is growing, neither disassembly nor assembly of microtubules and neurofilaments occurs at the distal end of the neurite.

New membrane, on the other hand, is probably added at the tip. The growth cone is a site of rapid exocytosis and endocytosis, as might be guessed from the large quantities of membranous vesicles it contains; and small vesicles of membrane are carried down the neurite from the cell body to the growth cone in the fast component of axonal transport. It appears from these and other observations that membrane is synthesized in the cell body, transported to the growth cone as vesicles, and there inserted by exocytosis to allow elongation of the neurite (Figure 18–65).

Without Microtubules, the Developing Neurite Retracts; Without Actin Filaments, It Cannot Advance[43]

The growth cone is primarily an organ of locomotion and therefore must adhere to the substratum over which it advances. In fact, in culture it is often the only firm site of attachment of the neurite to the substratum. Thus it is possible to pull the middle part of the neurite slightly to one side, introducing a kink, while the cell body and the growth cone remain fixed. When the neurite is released, it "twangs" back into a straight line. If the growth cone is cut off, the rest of the neurite retracts rapidly toward the cell body. Evidently the neurite is under tension: the growth cone is pulling it forward.

The advance of the growth cone depends on actin filaments. If cytochalasin B is added to the culture medium, to prevent the polymerization of actin into filaments, the growth cone halts its microspike activity and locomotion. Nevertheless, it continues to adhere to the substratum, and the neurite maintains its length. By contrast, if colchicine, which disrupts microtubules, is added to the culture medium, the neurite retracts toward the cell body. At the same time, new microspikes and even new growth cones develop from the proximal regions of the neurite, which were previously smooth and straight. It has been suggested, therefore, that microtubules serve to stabilize the elongating neurite and to restrict the sites where growth cone activity can occur, just as in a fibroblast they appear to control the location of regions of membrane ruffling (see p. 601).

Growth Cones Can Be Guided Along Chemically Defined Tracks in the Substratum[44]

Like fibroblasts, growth cones show a preference for substrata to which they can adhere strongly. A choice of substrata can be offered artificially by culturing nerve cells in a dish whose surface has been partly coated with patches of palladium (Figure 18–66). The substratum that remains exposed between the patches of palladium can be either ordinary naked plastic, to which cells adhere less strongly than to palladium, or plastic coated with, for example, the positively charged polymer polyornithine, to which cell membranes (which are negatively charged) adhere much more strongly. In the first case, the growth cones remain on the patches of palladium and keep off the plastic; in the second, they advance along the lanes of polyornithine and keep off the palladium. Materials such as polyornithine are in fact so adhesive that not only the growth cone but also the rest of the neurite remains stuck to the substratum when it is coated with them. The track taken by the growth cone is then recorded in the shape of the neurite behind it. On sticky substrata, furthermore, growth cones branch more frequently as they advance.

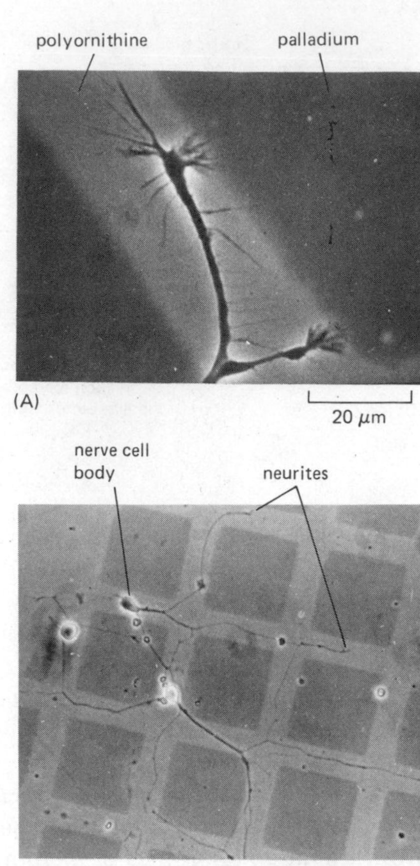

Figure 18–66 Growth cones on a dish whose surface has first been coated with polyornithine and then has had patches of palladium deposited on top of that. Because cell surfaces are negatively charged they strongly adhere to polyornithine, which is positively charged. The growth cones advance along the lanes of polyornithine and stay off the palladium. (A) is a phase-contrast photograph at high magnification, showing growth cones at the boundary between the two substrata. (B) is a phase-contrast photograph at lower magnification; the routes taken by the growth cones are recorded in the disposition of the neurites laid out behind them, which have remained adherent to the polyornithine. (From P. Letourneau, *Dev. Biol.* 44:92–101, 1975.)

The substrata over which growth cones travel in the living animal are not well characterized, and the factors that guide growth cones in normal development and control their branching are for the most part poorly understood. There is, however, one form of **contact guidance** of growth cones that is clearly established as important, both *in vivo* and *in vitro:* guidance by other nerve fibers. Growth cones will cling to existing neurites and advance along them. Thus the first neurites to grow out serve as pioneers, creating a track that later fibers can follow. Since there is a strong cohesion between neurite and neurite as well as between neurite and growth cone, the consequence is that nerve fibers in a mature animal are usually found grouped together in tight parallel bundles, or *fascicles.* The macroscopically visible *peripheral nerves,* containing axons that go from the central nervous system to outlying parts of the body, originate in this way (though subsequently the axons become individually enveloped and insulated from one another by Schwann cells). A neuronal cell-surface glycoprotein (known as the cell-adhesion molecule, or CAM) that helps to mediate these cohesive interactions between developing neurites has been identified. Anti-CAM antibodies have been shown to inhibit the tendency of developing neurites to become bundled together in fascicles.

But this still leaves us with the problem of discovering what guides the pioneer growth cones. One attractive possibility, for which there is some evidence, is that the microspikes act as "feelers" by which a growth cone can explore its surroundings and recognize markers on the surfaces of other cells that serve as signposts along its path (see Figure 18–67). The longer the microspikes, the greater the distance at which the presence of a signpost cell or target cell can be detected by the growth cone and cause a change in its direction of movement.

The Growth Cones of Some Types of Neurons Are Also Guided by Chemotactic Molecules Such as Nerve Growth Factor[45]

Besides being guided by contact interactions, a growth cone is susceptible to the effects of molecules dissolved in the extracellular fluid. This has been clearly demonstrated in the case of **nerve growth factor (NGF).** As explained in Chapter 13 (p. 728), certain classes of neurons, in particular peripheral sensory neurons and some of the peripheral motor neurons that innervate smooth muscle and glands—the so-called *sympathetic neurons*—require NGF for their survival. They require it also, and in slightly higher doses, in order to send out neurites (Figure 18–68). In the latter case NGF must act directly

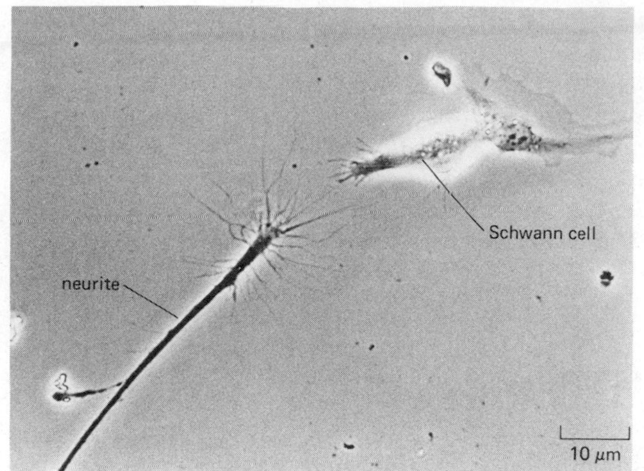

Figure 18–67 Phase-contrast photomicrograph showing the growth cone of a sensory neuron in culture advancing toward a Schwann cell. One of the microspikes sent out by the growth cone has just made contact with the Schwann cell. (Courtesy of Jennifer Pinder.)

on the growth cone itself: if NGF-sensitive neurons are placed in a chambered culture dish, they will extend neurites from one chamber into another only if the second chamber contains NGF (p. 729). The growth cones of such neurons can actually be guided in their direction of movement by the local concentration gradient of NGF, as another *in vitro* experiment shows: if a micropipette containing NGF is placed close to the growth cone but out of its direct line of advance, the growth cone will turn toward this source of NGF. We shall see later the importance of such effects in the living animal.

The Growth Cone Must Recognize Its Target and Finally Halt

A growth cone must eventually undergo a transformation. On reaching its target, it must recognize that it has arrived: synapses must be formed, and outgrowth must cease. Since, in the mature neuron, new subunits continue to be added to the microtubules and neurofilaments in the cell body and transported outward, the creation of a synaptic terminal requires a switch in the operation of the molecular machinery at the end of the neurite so that the microtubules and neurofilaments terminating there are disassembled or degraded as fast as they arrive (see Chapter 10). This switch in the behavior of the cytoskeleton at the nerve terminal must be accompanied by a change in membrane turnover also. In the case of a developing dendrite that forms a postsynaptic specialization, exocytosis and endocytosis must largely cease, while in a developing axon that forms a presynaptic terminal, the perpetual rapid exocytosis and endocytosis of the growth cone must give way to the Ca^{2+}-triggered exocytosis and subsequent endocytotic membrane retrieval that underlie synaptic transmission.

It remains to be discovered how the growth cone recognizes its proper target, and how the transformation from growth cone to synaptic terminal is brought about. These unanswered questions lie at the heart of the next problem to be considered—the problem of how neurons become connected to form a functional nervous system.

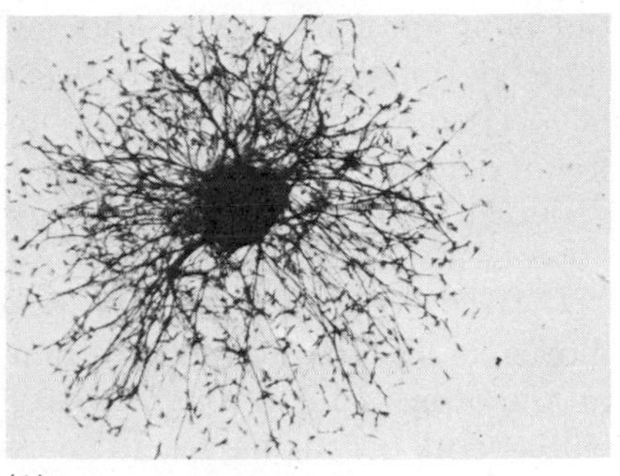
(A)

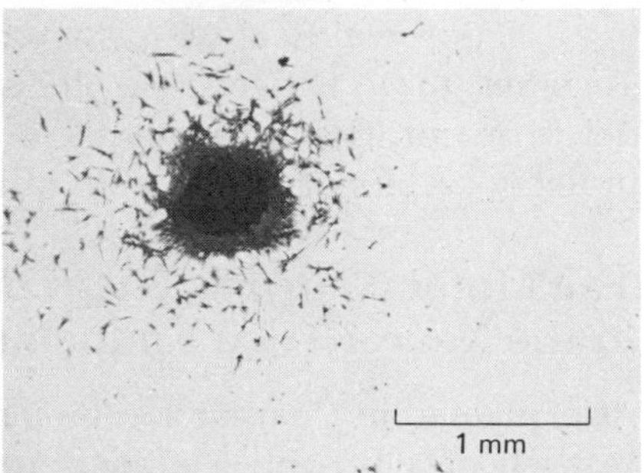

(B)

Figure 18–68 Photomicrographs of a sensory ganglion cultured for 36 hours (A) with nerve growth factor and (B) without. Neurites grow out from the sensory neurons only if nerve growth factor is present in the medium. Each culture also contains Schwann cells that have migrated out of the ganglion; these are unaffected by nerve growth factor. (Courtesy of Clive Thomas.)

Summary

The neuronal cytoskeleton consists chiefly of neurofilaments, microtubules, and actin. It maintains the elongated structure of the neuron and provides for transport of materials to and from the cell body, where proteins and lipids are synthesized for use elsewhere in the cell. Axonal transport has fast anterograde and retrograde components, consisting of vesicles moving at speeds of up to 400 mm per day, and slow anterograde components, carrying proteins of the cytoskeleton and cytosol at speeds of a few mm per day. In a developing neuron, the cytoskeleton is essential for the movements by which the growth cone crawls forward, pulling out an elongating axon or dendrite behind it. The locomotion of a growth cone resembles that of a fibroblast and appears to be guided both by contact interactions with the substratum and by the chemotactic influence of substances such as nerve growth factor that are dissolved in the extracellular fluid.

The Development of Neuromuscular Connections[46]

The system of nerve connections in a higher animal is vast and complex, and the central problem in neural development is to understand how certain specific connections develop and not others. Several different mechanisms are involved. To a large extent neurons are programmed to connect with appropriate partners, but they do not do so with complete accuracy. The system is first roughed out, with an excess of neurons and synapses, and the initial set of connections is then whittled down, revised, and corrected to give the final precisely ordered pattern. This process of adjustment of synaptic connections is affected in many instances by electrical activity: the firing of the neurons can decide whether a synapse is consolidated or eliminated. Thus, by causing neurons to fire, external stimuli influence the development of the pattern of nerve connections. As a result, the nervous system retains traces of its past experiences, expressed in the fine details of a structure whose general plan is genetically determined. The part played by experience in some aspects of neural organization seems insignificant, in other aspects crucial. The ordering of connections by which a chick flaps its wings for flight, or moves its legs in alternation when walking, is largely inborn and not learned; on the other hand, a mammalian eye deprived of visual experience in infancy will lose its connections with nerve cells in the brain and become blind.

This section focuses on the vertebrate neuromuscular system and in particular on the motor neurons that innervate the muscles of the limbs. The final section will deal with the construction of the vertebrate visual system. Together, these two systems illustrate most of the basic principles of neural development. The fundamental forms of nerve cell behavior, as far as they are understood, appear to be much the same in invertebrates as in vertebrates.

The Motor Neurons That Control the Limbs Are Generated in the Neural Tube Epithelium

As explained in Chapter 15, the nervous system of a vertebrate develops from two sets of cells—those of the **neural tube** (see p. 822) and those of the **neural crest** (see p. 881)—both originating from the ectoderm. The neural tube grows to form the central nervous system (that is, the brain and spinal cord), while the neural crest is the source of the neurons whose cell bodies lie outside the central nervous system and of Schwann cells, which form the myelin sheaths of the peripheral nerves. The neural tube, with which we shall be mainly concerned here, consists initially of a single-layered epithelium whose cells proliferate to give rise to both the neurons and the glial cells of the central nervous system. In the process, the simple epithelium is transformed into a thicker and more complex structure, with many layers of cells of various types. Among these are the motor neurons that send out axons to connect with the muscles of the limbs.

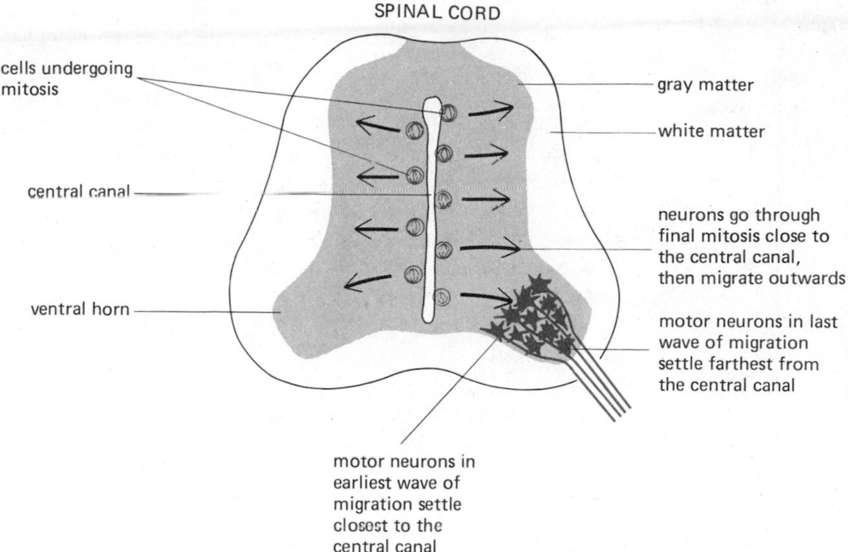

Figure 18–69 Schematic diagram of a cross-section of the developing spinal cord, showing how neuronal precursor cells go through their final mitosis close to the central canal and then migrate radially outward. The region of spinal cord shaded gray contains many cell bodies and corresponds to the *gray matter* of the adult nervous system. The unshaded region, corresponding to the *white matter* of the adult system, consists chiefly of bundles of axons traveling along the length of the spinal cord and connecting one region of gray matter to another. (These regions appear white in the adult because of the large amount of myelin.)

Radial Glial Cells Form a Temporary Scaffold to Guide the Migrations of the Immature Neurons[47]

In principle, it does not matter where the body of a nerve cell is located, as long as it makes the right connections. However, in order to make the right connections, the cell body must be in approximately the right place. Thus, the growth of axons and dendrites is commonly preceded by a phase of cell migration in which the immature neurons move from their birthplace to settle in some other location.

It is possible to trace these neuronal migrations by labeling the dividing neuronal precursor cells with tritiated thymidine. The motor neurons that will innervate the limbs, after passing through their last mitosis close to the lumen of the neural tube, move outward radially to settle in the *ventral horn* of the future spinal cord (Figure 18–69). There is a regular relationship between the "birth date" of a motor neuron (that is, the time of its final division) and the site where it comes to rest: surprisingly, the lastborn cells migrate out past the firstborn and settle in the most peripheral positions.

The nerve cell bodies are guided in their migrations by a specialized class of cells in the neural tube—the **radial glial cells** (see Figure 18–70). These can be considered as cells of the original columnar epithelium of the neural tube, which become extraordinarily stretched as the wall of the tube grows and thickens. Each of the radial glial cells extends all the way from the inner to the outer surface of the tube, a distance which in some regions of the developing brain of a primate may be as much as 20 mm. Three-dimensional reconstructions from serial electron microscope sections reveal that the immature migrating neurons cling closely to the radial glial cells and evidently crawl along them (Figure 18–71).

The radial glial cells remain for many days—in some species for months—as a nondividing population, clearly distinct from the neurons and their precursors. Eventually, toward the end of development, they disappear from most regions of the brain and spinal cord: it has been suggested that many of them transform into *astrocytes* (see Figure 18–5). Thus the radial glial cells can be viewed as developmental apparatus, necessary—like scaffolding—for the complex process of construction, but not retained in most parts of the final structure, the mature nervous system.

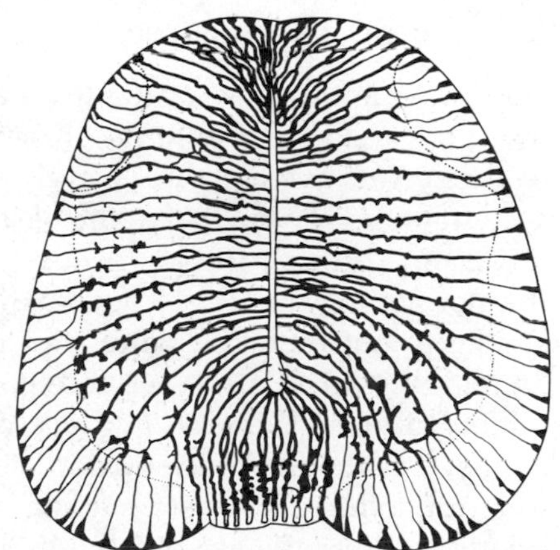

Figure 18–70 Golgi-stained section of the spinal cord of a five-day chick embryo, showing the radial glial cells extending from the lumen to the outer surface. Cells similar to these help to guide the migrations of the immature nerve cells throughout the central nervous system. (From S. Ramón y Cajal, Histologie du Système Nerveux de l'Homme et des Vertébrés. Paris: Maloine, 1909–1911; reprinted, Madrid: C.S.I.C., 1972.)

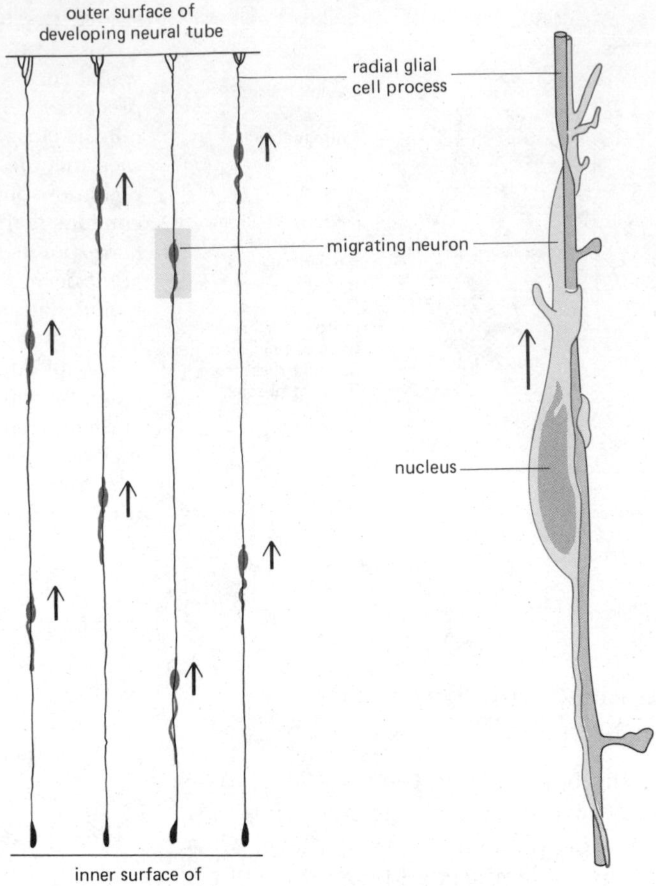

Figure 18–71 Schematic diagram showing immature nerve cells migrating along radial glial cell processes. The diagram is based on reconstructions from serial electron microscope sections of the developing cerebral cortex of a monkey. (After P. Rakic, *J. Comp. Neurol.* 145:61–84, 1972.)

Axons Grow Out Along Precisely Defined Pathways to Specific Target Areas[48]

Once a neuron has migrated to its appropriate position, it sends out an axon that must find its way to its proper target. Thus, as the motor neurons that will innervate the limb arrive at the end of their migration, they begin to form growth cones. These pierce the basal lamina surrounding the neural tube and travel out through the connective tissues of the embryo toward the sites of the developing muscles. They follow well-defined routes: witness the precise similarity between the pattern of nerves on the right and left sides of the animal (Figure 18–72). Even foreign axons that have been experimentally induced to enter the limb in place of the normal innervation are confined to almost exactly the same standard set of paths, which can therefore be likened to a public highway system along which growth cones are free to travel. Evidently these paths are defined by the limb's intrinsic structure, but the molecular basis of the guidance remains a mystery. Similar predetermined paths appear to guide axon outgrowth in the central nervous system, where they are probably defined by the local properties of the embryonic glial cells.

The paths branch along their course, with different branches leading to different targets. Thus individual growth cones face a series of choices at successive branch points. The choices seem to be made according to precise rules, with the consequence that a highly ordered system of connections is set up between neurons and their target cells. This is clearly demonstrable for motor neurons and limb muscles. The innervation of each muscle can be mapped by a method that exploits the phenomenon of retrograde axonal transport. The procedure is to inject into the muscle a tracer substance, some

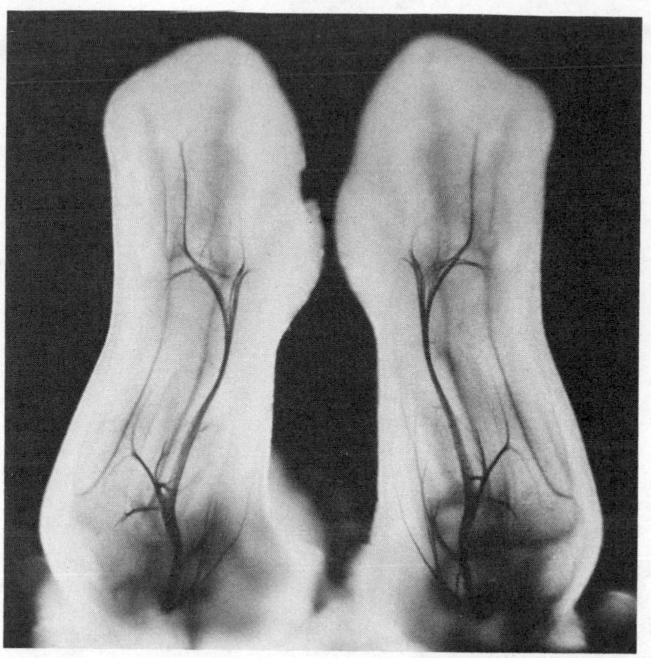

Figure 18–72 Light micrograph of the wings of an eight-day chick embryo, silver-stained to show the patterns of nerves. Compare the right wing with the left: the routes followed by the nerves are almost exactly symmetrical on the two sides of the body, implying the existence of a precise guidance system for nerve outgrowth.

of which is taken up by endocytosis into the nerve terminals in that muscle. A useful tracer for this purpose is the enzyme *horseradish peroxidase* (*HRP*) since it can be detected in very small quantities by the colored products of the reaction that it catalyzes. The tracer is carried back along the axons to the spinal cord, where it reveals the location of the nerve cell bodies that innervate the muscle in question (Figure 18–73). These are found to lie in a cluster that occupies the same position in every animal but is different for different muscles.

The Pattern of Nerve Connections Is Governed by the Nonequivalent Characters of Cells, Not Merely by Their Positions[49]

The importance of the precise pattern of connections between spinal cord and musculature can be demonstrated by disturbing it experimentally. It is possible, for example, to transpose nerves in the leg of an adult rat so that the neurons normally innervating a main extensor muscle connect instead

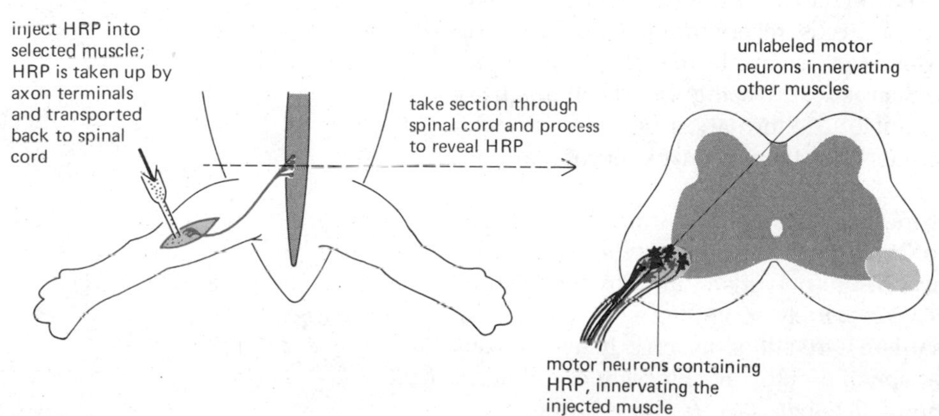

Figure 18–73 Schematic diagram showing how retrograde transport of horseradish peroxidase (HRP) can be used to identify which motor neurons in the spinal cord innervate a particular muscle. Note that each muscle is supplied by a nerve that contains the processes of many individual nerve cells.

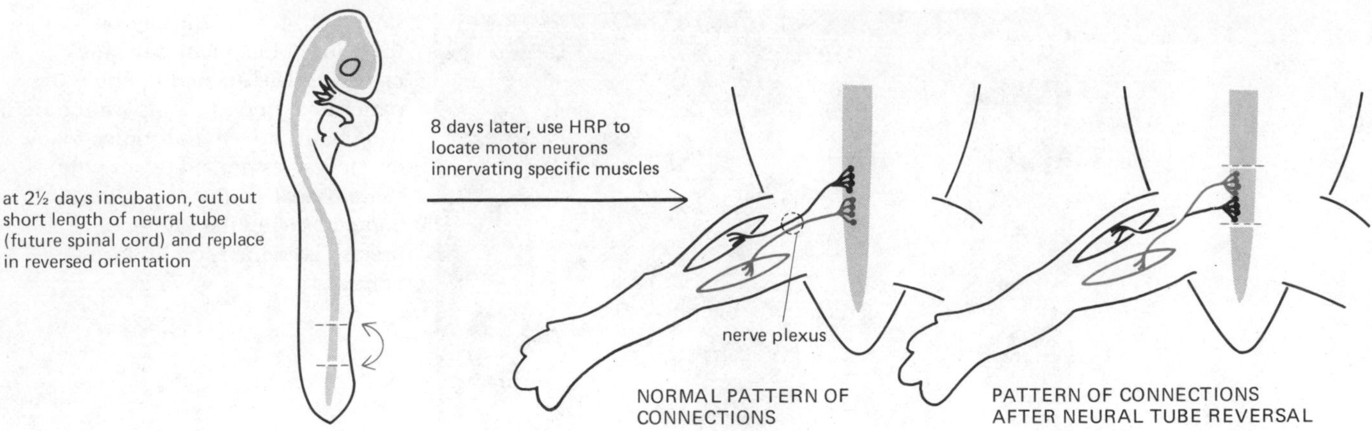

Figure 18–74 Schematic diagram of an experiment on a chick embryo, demonstrating that motor neurons, even when misplaced, nevertheless send their axons to the muscles appropriate to their original positions in the embryonic spinal cord. Note that the axons from motor neurons at different levels along the spinal cord are funneled together into a *plexus* at the base of the limb and then separate again to innervate their separate targets. A growth cone passing through the region of the plexus has a large choice of targets open to it.

with a flexor muscle, and vice versa. The animal then shows permanently reversed reflexes: instead of withdrawing its leg from a painful stimulus, it extends it, causing the pain and damage to become still worse. Since wrong connections lead to persistent malfunction, it is crucial that the right connections should be established during development.

How then is the correct pattern of connections set up? Are growth cones simply channeled to their different destinations as a direct consequence of their different starting positions? This hypothesis can be tested by studying the connections that develop when the starting positions are altered. For example, a short portion of the neural tube of a chick embryo can be cut out at an early stage, before axon outgrowth, and replaced with its anteroposterior axis reversed (Figure 18–74). Thus the precursors of the neurons originally destined to innervate muscle A are put in the place of the neurons originally destined to innervate muscle B, and vice versa. In such circumstances, provided that the shift of position is not too extreme, the growth cones of the misplaced neurons will usually travel out by altered routes to connect with the muscle appropriate to their *original* position in the neural tube. This implies that the neurons destined to innervate different muscles are nonequivalent (see p. 853): they are distinguished from one another not simply by their positions, but by their intrinsic chemical characters, and these chemical differences control the specific choice of target. The pattern of connections in such a case is said to be governed by *neuronal specificity*.

While neuronal specificity apparently causes a certain set of connections to be preferred, it does not absolutely dictate that those and only those connections must form. There is a limit to how wide a detour axons will make to synapse with their preferred targets: if the motor neurons are moved very far from their original positions, they may synapse with foreign targets. Moreover, in adult mammals, as opposed to embryos, regenerating axons will form synapses indiscriminately with any denervated muscle that they may be artificially caused to encounter. This is scarcely surprising: the mechanisms of neuronal specificity have evolved to control the formation of connections in normal development and not in animals that have been experimentally rearranged.

There is good evidence for neuronal specificity in the visual system also. The same device, analogous to the color coding of wires in a telephone cable, appears to be important in many parts of the nervous system in establishing a regular pattern of connections between widely separated sets of cells. Although the detailed molecular mechanisms are still a mystery, it seems likely that growth cones are directed along specific pathways by contact guidance involving a system of distinctive chemical labels, are perhaps attracted to specific target areas by chemotactic signals, and ultimately find their particular target cells by recognizing specific molecules on the target cell surface. Neuronal specificity probably has a part to play in controlling the choice of path, the responses to signals, and the recognition of the target.

Cells That Fail to Make Connections Die[50]

No biological system operates with perfect accuracy, and in the nervous system, where accuracy is especially necessary, it seems especially difficult to achieve. For example, it is inevitable that some individuals will carry genes for, say, muscles that are larger than average, while others will have genes for big brains or short arms. Since sexual recombination will shuffle these genes, there is no guarantee that the set of variant genes affecting the size and pattern of one part of the body will be accompanied by genes bringing about precisely correlated changes elsewhere. Necessarily, there will often be a mismatch between one part of the nervous system and another or between the nervous system and the other parts of the body with which it must connect. The more complex the organism and its nervous system, the greater the opportunities for disorder as a result of such independent variability in the component parts.

The mechanisms of development in the nervous system have evolved in such a way as to allow for adjustments. Many types of target cells are programmed to die if they do not become innervated within a certain time. Cell death also disposes of developing embryonic neurons that fail to connect to a target cell. This eliminates loose ends in the system of connections. The phenomenon is easily demonstrated experimentally. If the neural tube of the early chick embryo is destroyed, the muscles that should have received innervation from it go through the initial stages of differentiation in its absence, but they then atrophy and disappear. Conversely, if a limb bud is cut off at an early stage, before it has begun to be innervated, the corresponding motor neurons in the developing spinal cord send out their axons toward the site where the limb should be, creating for a while a tangled mass of nerve fibers at the base of the amputated limb bud; but the deprived motor neurons then all die, shortly after the time at which they would normally have made connections with muscles.

The mechanism that causes the death of a neuron that fails to make a connection is still the subject of speculation. One widely favored suggestion is that death may result from the absence of a "survival factor" provided by the normal target cell. According to this theory, the survival factor—presumably some molecule—would be picked up only by axon terminals that succeed in making contact with the target and would then be carried by retrograde axonal transport back to the cell body, which would thereby be saved from dying. Nerve growth factor apparently acts as a survival factor of this sort for some classes of neurons (see p. 728).

Normal Motor Neuron Death Is Prevented by Toxins That Block Neuromuscular Transmission[51]

From what has just been said, some neurons should be expected to die during normal development. But in fact they die in many parts of the nervous system in surprisingly large numbers, and for reasons that are by no means clear. For example, vertebrate embryos produce roughly twice as many motor neurons as they will finally require, thinning out this surplus by neuronal death shortly after neuromuscular connections have been established. There is evidence that most of the nerve cells that die have made contact with the muscles appropriate to their positions in the spinal cord. Motor neurons, however, besides synapsing on muscle cells, must themselves receive synapses from other neurons in the spinal cord, and it is possible that the motor neurons that die are those that have not received the right connections.

Whatever the function of the adjustment of motor neuron numbers in the embryo, its mechanism shows an interesting dependence on muscle activity. Vertebrate embryos start to writhe and squirm and wave their limbs about haphazardly almost as soon as neuromuscular connections begin to form. These movements result from spontaneous firing of action potentials in the central nervous system and occur even in embryos whose sensory neurons have been destroyed. If an embryo is treated with a toxin, such as curare, that blocks synaptic transmission at neuromuscular junctions, the movements cease. One might guess that this treatment should either have no effect on motor neuron death or perhaps increase it. In fact, it has a dramatic effect in just the opposite direction: as long as the block is maintained, practically all the motor neurons survive. Whatever the mechanism, it is clear that muscle activity is important in the normal development of the motor system, just as sensory input is important in the normal development of the sensory system (see below): in both cases, electrical signaling affects the maintenance of nerve connections.

Surplus Synapses Are Eliminated by a Competitive Process[46,52]

Even after the surplus motor neurons have died, the developing muscles are left with a large excess of synapses on them. The excess is due to the extensive branching of motor axons once they enter muscle tissue. Each motor axon with its many branches makes synapses on many muscle fibers, apparently at random, and most of the muscle fibers become innervated by more than one axon (Figure 18–75). While this guarantees that every muscle fiber becomes innervated, it also entails making many more synapses than are ultimately required; for in the adult, each skeletal muscle fiber has only one synapse on it, formed by one of the branches of a single motor axon. The process of elimination of the surplus synapses during development has been well studied in the rat. In the soleus muscle of the rat leg, for example, about five axons on average innervate each muscle fiber at birth, and this number is whittled down to precisely one in the next two or three weeks. Clearly, if the surplus synapses were eliminated at random, some muscle fibers would be left with no synapse at all while others would remain innervated by several axons. The fact that each muscle fiber retains one, and only one, synapse implies that the process of synapse elimination is competitive: one axon wins the permanent place, while all the others, however numerous they may be, lose their hold on the muscle fiber. Essentially the same phenomenon of

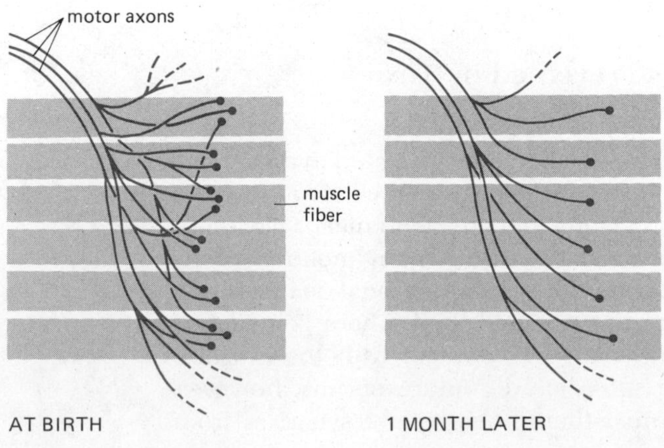

Figure 18–75 Elimination of surplus synapses in a mammalian skeletal muscle in the period after birth. In this schematic diagram, for the sake of clarity, the number of terminal branches of each motor axon is underrepresented; in reality, a single motor axon in a mature muscle typically branches to innervate several hundred muscle fibers.

overproduction of synapses followed by competitive elimination is observed in many other parts of the nervous system: it seems to be a standard way of setting up a precise distribution of connections in spite of some randomness in the behavior of individual cells. Again, the molecular mechanism is unknown, though we shall see later (p. 1090) that electrical activity can have an important influence on the outcome of the competition.

Denervated Muscle Cells Release a Factor That Stimulates Nerve Cells to Sprout[53]

Developmental processes do not end with the attainment of maturity. As discussed earlier (p. 1053), neuromuscular connections in the adult can regenerate after a nerve has been cut. The ends of the severed axons undergo a transformation into growth cones and make their way back to the denervated muscle. There they form synapses preferentially at the sites where synapses existed before, which are marked out by the special character of the junctional basal lamina. While the junctional basal lamina evidently causes the formation of a stable synaptic terminal from a growing axon, other factors can bring about a converse transformation and cause growth cones to sprout from a mature terminal. Such factors play an important part in the regeneration of neuromuscular connections after some but not all of the nerve cells that innervate a muscle have been destroyed. In these circumstances the denervated muscle fibers secrete a diffusible factor that stimulates the sprouting of new growth cones from the surviving nerve terminals on the neighboring innervated muscle fibers. The sprouts grow out to reinnervate the denervated muscle fibers. In this instance, the nerve supply is adjusted to match the size of the target not by the elimination of a surplus, but by an expansion of the supply.

In the case of skeletal muscle, the sprouting factor has not yet been identified. For smooth muscle, where analogous phenomena are observed, it has been shown that nerve growth factor regulates the supply of innervation according to the requirements of the target. Denervation induces release of NGF from the smooth muscle, and the NGF stimulates growth of axons toward the muscle so as to restore its innervation.

The chemical identification of NGF, as the first to be discovered of many substances that must govern neural development, has made possible the beginnings of an analysis of its action at a molecular level. Whereas we are thus on the way to understanding the molecular basis of synapse formation in some parts of the peripheral nervous system, we are still far from that goal in the central nervous system. Even here, however, a number of general rules of synapse formation are beginning to emerge, as described in the following section.

Summary

The neurons and glial cells of the central nervous system of a vertebrate are generated from the cells of the neural tube epithelium. Having completed their final cell division, the young neurons generally migrate in an orderly fashion along radial glial cell processes to settle in new positions, from which they send out axons and dendrites along precisely defined routes to set up a regular system of connections. The formation of neuromuscular connections seems to be guided by neuronal specificity: the motor neurons destined to innervate a particular muscle behave as though they have a distinctive character that causes them to innervate that target preferentially even if they are artificially misplaced. Neurons that fail to make connections usually die, as do large numbers of even the motor neurons that do make connections. The death of these cells appears to depend on electrical activity: it is prevented by toxins that block synaptic transmission at the neuromuscular junction. Surviving motor neurons at first make an excess of synapses, such that each muscle cell receives axon branches from several different motor neurons. The surplus synapses are subsequently eliminated by a competitive process, leaving each muscle cell with one and only one synapse on it. If a muscle cell is completely denervated, it releases a factor that stimulates axons in the vicinity to sprout so as to provide it again with innervation.

Neural Maps and the Development of the Visual System[54]

The vast numbers of neurons in the brain belong to a relatively small number of functional classes that can be crudely distinguished according to their shape, their connections, and the neurotransmitters they secrete. Neurons in the same class generally have their cell bodies grouped together in the same region, make similar connections, and develop in the same way. Such a group of neurons is known as a *nucleus* if they lie in a cluster or as a *lamina* if they form a sheetlike array. A typical nucleus in the brain might comprise half a dozen distinct types of neurons, forming a large functional unit connected to other nuclei and laminae in the brain by large bundles of axons forming *fiber tracts.* The first level of anatomical description of the brain therefore consists in listing the nuclei and laminae and specifying which is connected to which.

At the next level of detail, one can compare neurons that belong to the same class but occupy different positions within a given nucleus or lamina: what precise sites do they project to in the next such group of neurons? As a rule, the projection is orderly and obeys a simple but powerful organizing principle: neighboring cells in one group project to neighboring sites in the next. As a consequence, the pattern of connections establishes a **continuous map** of one array of neurons onto another (see Figure 18–51).

As discussed earlier, this arrangement is clearly important for sensory processing in the visual system, where the two-dimensional image of the world as focused on the retina is mapped out through a succession of relays onto the sheetlike visual cortex of the brain. The same principles apply to other sensory modalities: the brain contains maps of the body surface, as perceived through receptors for touch, maps of the spectrum of possible sounds, laid out according to pitch, and so on. In all of these cases, the many neurons in each large array act in parallel to process many bits of information that are of the same general character but arise from different regions of the world of experience. The continuity of the mapping ensures that neurons dealing with closely related items of information lie close together and so can interact for purposes of information processing. Moreover, the orderliness of the map at each stage guarantees that each item of information as it is processed is kept in context, identifiably associated with a particular region of the world of experience. Continuous neural maps, therefore, are of fundamental importance in the organization of the vertebrate brain. How then do they develop? This question will be the starting point for our discussion of the formation of nerve connections in the visual system.

Neuronal Specificity Appears to Control the Mapping of the Visual World from the Retina onto the Tectum[55]

The neural map best studied from the point of view of development is that projected from the retina onto the *optic tectum* in amphibians and fish. (The optic tectum is a laminar region of the brain that receives the main visual

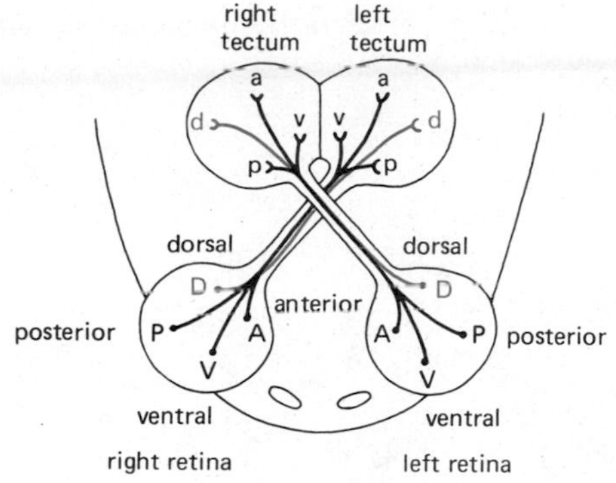

the ganglion cells in each retina send their axons to the opposite tectum, setting up an orderly map (A→a, V→v, etc.)

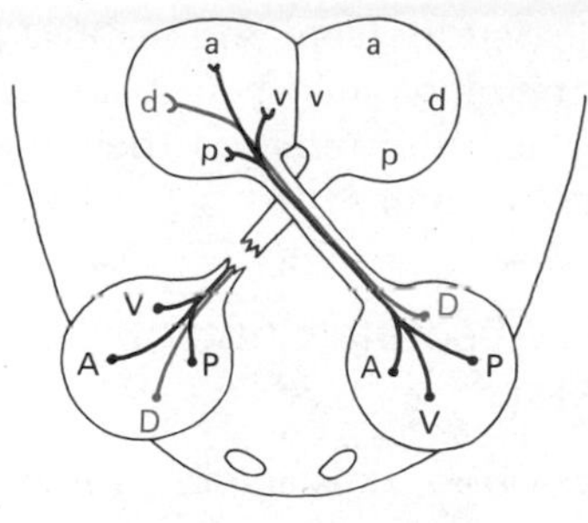

cut the right optic nerve and rotate the right eye; the severed ends of the axons from the retina degenerate

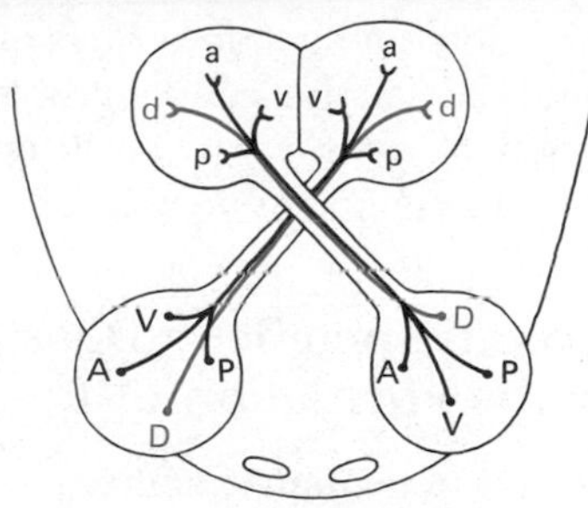

the ganglion cells in each part of the retina, even though now misplaced, regenerate connections with the same part of the tectum that they were connected with before

Figure 18–76 The regeneration of connections between eye and brain in an amphibian after one eye has been rotated. The axons from each part of the rotated retina regenerate so as to reconnect with the part of the tectum appropriate to the *original* positions of the retinal cell bodies. Thus, for example, light falling on the ventral part of the rotated retina is perceived as though it were falling on the dorsal part.

input in lower vertebrates.) In many of these species the optic nerve, in which the retinal axons travel from eye to brain, will regenerate if it is cut. Cutting the nerve blinds the animal; regeneration restores vision after a few weeks or months. The regenerated axons, like those in the normal animal, set up a continuous map from the retina onto the tectum. The pattern of connections can be plotted electrophysiologically by inserting a recording electrode at a selection of sites in the tectum and waving a small object about in front of the animal to discover for each tectal site the position in the visual field from which that site receives its stimulation.

If the optic nerve is cut and the eye is at the same time turned upside down in its socket, the inverted retina again regenerates connections with the tectum, and the animal recovers its sight. Now, however, it sees the world upside down: if food is dangled above it, it makes a lunge downward instead of upward, and so on. Electrophysiological mapping shows that this is because each part of the retina has reconnected with the part of the tectum that was appropriate to its *original* position (Figure 18–76). There is a close parallel here with the phenomena seen when a portion of the neural tube is reversed (see p. 1082). Again, an interpretation can be given in terms of **neuronal specificity.** It is as though each cell in the retina has been assigned, during development, a specific positional label that causes it, no matter where it is put thereafter, to connect with a certain region of the tectum. According to the basic hypothesis of neuronal specificity, which was originally proposed to explain these observations, the axon of the retinal cell is able to recognize its proper target site in the tectum because the tectal cells themselves carry a corresponding (or complementary) set of chemical labels according to their position. By this theory, the rule for forming a connection is that the label of the presynaptic (retinal) cell must match that of the postsynaptic (tectal) cell.

In principle, this neuronal specificity mechanism could be of universal importance, serving throughout the nervous system to determine which cell connects with which. In practice, although there is good evidence for neuronal specificity in many places, it has been very hard to define precisely how large a part it plays in neural organization. An important step forward toward discovering the molecular basis of specificity in the visual system has recently been made, however. By means of monoclonal antibodies, it has been possible to identify, on retinal cells in the chick embryo, a cell-surface glycoprotein that, like the hypothetical label of neuronal specificity, distinguishes the retinal cells according to their position. The concentration of the marker is smoothly graded across the retina, being 35 times higher at one pole than at the other, and it is present on most, if not all, of the retinal cell types. The gradient is already detectable at four days of embryonic development and is subsequently maintained as the retina grows. It may therefore reflect a positional label assigned to cells at an early embryonic stage and retained thereafter as a guiding factor in the formation of neural connections.

Neuronal Specificity Does Not Exert Absolute Control over the Formation of the Neural Map[56]

Although neuronal specificity seems to provide a satisfactory explanation of the development of the retinotectal map, more detailed studies show that the rules in fact are not quite so simple or so clear-cut. It is possible, for example, to destroy half of the retina and then, at different times during the succeeding weeks and months, to record the pattern of tectal connections made by the remaining half retina. The axon terminals from the surviving retinal cells are found gradually to shift and spread out, still in an orderly array, so as to cover the entire tectum. Conversely, if one half of the tectum is destroyed, the projection from the entire retina gradually becomes compressed in an orderly way onto the remaining half of the tectum.

Such observations can be explained by supposing that several different mechanisms combine to control the pattern of connections, as follows: (1) incoming axons compete to form synapses; (2) uninnervated or denervated target cells in the tectum, like denervated muscle cells, produce sprouting stimuli that serve to adjust the distribution of presynaptic nerve terminals according to the quantity of target tissue; and (3) some similarity between axons from neighboring retinal cells tends to make them synapse at neighboring sites in the tectum, keeping the system of connections knitted together as a continuous map even when the projection as a whole shifts (we shall discuss a possible mechanism for this on p. 1092). Neuronal specificity, acting in addition to these controls, would serve to define the orientation of the map. In this view of specificity, neurons from a particular region of the retina, by virtue of some distinctive chemical character, have a preference to connect with a corresponding region of the tectum, but not a strict compulsion.

A set of organizing principles such as these, while acting together in normal development to give a regular map, may have conflicting effects in the artificial circumstances of an experiment. The interpretation of many of the experimental phenomena is therefore difficult, and the account given above is partly speculative. Nevertheless, the evidence from the retinotectal system leads to the same general conclusion as the evidence from the neuromuscular system: neuronal specificity exists but it does not reign absolute; the formation of connections is regulated by other influences as well.

Visual Connections in Young Mammals Are Adjustable and Sensitive to Visual Experience[57]

In mammals, the early developmental processes by which visual maps are set up before birth are not easily accessible to experimental study, and visual connections do not regenerate if the optic nerve is cut. But the visual system is not mature at birth, and it has been possible to show that in the first few months or years of life it undergoes subtle but important adjustments that depend on visual experience. When visual experience is deficient during a certain *sensitive period,* these adjustments can go badly awry. A common example is the "lazy eye." Children with a squint often fall into the habit of using one eye only, while the other eye is perpetually misdirected, and rarely receives a sharply focused image on its retina. If the squint is corrected, and the child is taught to use both eyes, it will grow up with both eyes functioning well. But if the squint goes uncorrected throughout childhood, the unused eye loses permanently and almost completely the power of vision, in a way that no lens can correct. The eyeball itself remains normal in appearance and structure. The defect lies in the brain.

Detailed studies of the functions of individual neurons in the mammalian visual system, using the electrophysiological techniques described earlier (p. 1066), have made it possible to pinpoint the cellular basis of such lesions. The findings have a significance that goes beyond the visual system, for they show very clearly how experience can affect the structure of the brain, and thereby its subsequent function. They highlight the importance of sensory stimulation for the development of the child, and they suggest a cellular mechanism by which experiences in adult life may also leave their mark in the brain.

In the Visual Cortex of the Brain, the Projections from the Two Eyes Are Mapped in Alternating Stripes[58]

In discussing the effects of experience on the visual system, we shall concentrate on the development of the synaptic connections by which the inputs from the two eyes are combined to provide binocular vision. To explain the developmental phenomena, it is necessary to describe the anatomy of the system in the adult. Each eye in a mammal such as a monkey or a cat sees almost the same visual field and sends its axons back along paths such that the two channels of information relating to the same region of the external world are brought together in the same region of the brain (Figure 18–77). Thus on the left visual cortex there are two orderly maps of the right half of the visual field, one received from the left eye, the other from the right. In the adult brain these two maps are almost, but not exactly, in register. The inputs from the two eyes are segregated in a pattern of narrow (0.4 mm) alternating stripes known as **ocular dominance columns.** These can be demonstrated by injecting radioactive amino acids into one eye. The labeled molecules are taken up by the retinal neurons and carried by axonal transport to the cerebral cortex, somehow crossing the synapses in a relay station (the lateral geniculate nucleus) on the way. Autoradiographs of sections of the cortex clearly show labeled bands, receiving their input from the labeled eye, alternating with unlabeled bands, receiving their input from the unlabeled eye (Figure 18–78).

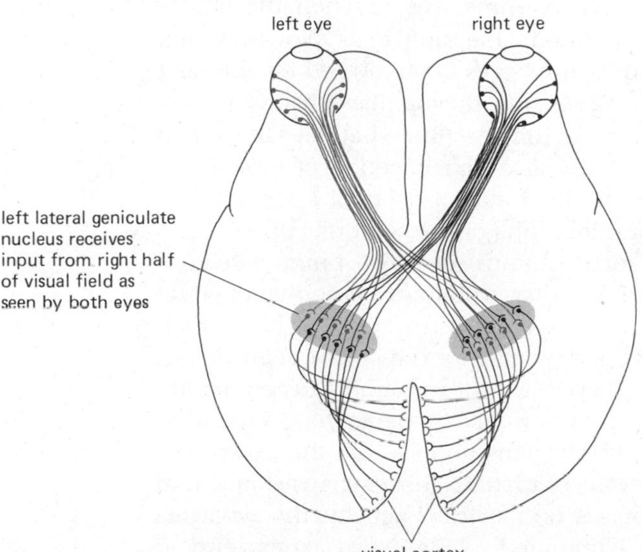

Figure 18–77 Schematic diagram of the major human visual pathway, showing how the inputs from the right and left eyes are distributed so that related streams of information are brought together in the same region of the brain. Note that all the information obtained by the left side of each eye (relating to the right side of the visual field) is relayed to the left side of the brain, and vice versa.

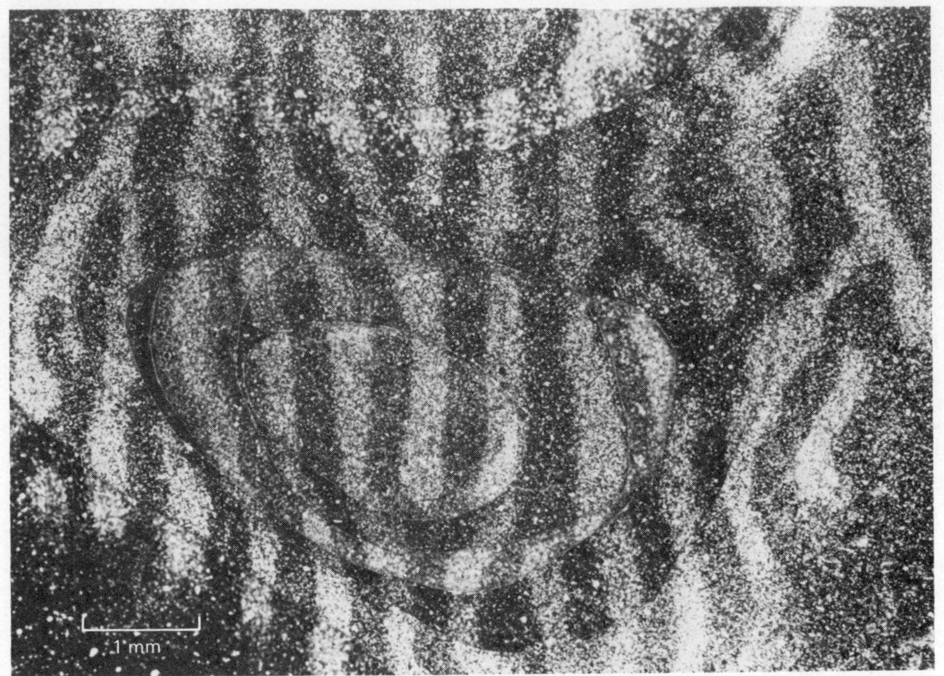

Figure 18–78 Ocular dominance columns in the visual cortex of a normal monkey. Radioactive proline is injected into one eye and the animal is then allowed to survive for 10 days, in which time the radioactive label is transported to the parts of the cortex that receive their input from that eye. Sections of the cortex are cut tangentially to its surface, and autoradiographs are prepared. With dark-field illumination, the silver grains covering the radioactive regions appear bright against a dark background. The picture is a montage composed of photographs of several successive sections cut at slightly different depths through the thickness of the cortex. The ocular dominance columns connected to the labeled eye *(bright bands)* are of the same width as those connected to the unlabeled eye *(dark bands)*. (From D. H. Hubel, T. N. Wiesel, and S. Le Vay, *Philos. Trans. R. Soc. (Biol.)* 278:377–409, 1977.)

Active Synapses Tend to Displace Inactive Synapses[58,59]

The same autoradiographic method has been used, chiefly in cats and monkeys, to study how these ocular dominance columns develop. At first no ocular dominance columns can be seen: the projections from the two eyes overlap so that each region in the visual cortex receives inputs from both eyes. Only later (typically during the first few weeks after birth) do the projections sort themselves into alternating stripes, with neurons in adjacent stripes receiving input predominantly from opposite eyes. The process provides yet another example of initial overproduction of synapses, followed by elimination of the surplus as the overlapping sets of axon terminals retract into separate territories. There is some disagreement as to how far this step in development depends on visual experience. But it is clearly established that even after the ocular dominance columns have formed, there is a period when visual deprivation can disturb them, as the following experiment shows.

Two monkeys are taken at the age of three weeks, when the ocular dominance columns are already well defined. One monkey is allowed to develop normally, while the other has one of its eyes kept covered for the next few weeks. The consequence of covering one eye during this sensitive period is permanent blindness or semiblindness in that eye. But what does this mean in cellular terms? Each of the monkeys is given an injection of radioactive amino acid in one of its eyes and then killed after a delay of a few days to allow for transport of the label. Autoradiographs of the cerebral cortex of the normal monkey reveal ocular dominance columns roughly 0.4 mm wide connected to the right eye, alternating with ocular dominance columns of equal width connected to the left eye (Figure 18–78). But in the monkey in which one eye has been deprived of visual experience, there is a severe disturbance: the columns connected to the deprived eye have shrunk almost to zero width, while those connected to the eye that had normal experience have expanded to a width of almost 0.8 mm (Figure 18–79). This implies that the axons carrying the projections from the two eyes compete to form synapses and that the outcome of the competition depends on electrical activity: the synapses made by inactive axons tend to be eliminated, while active axons tend to

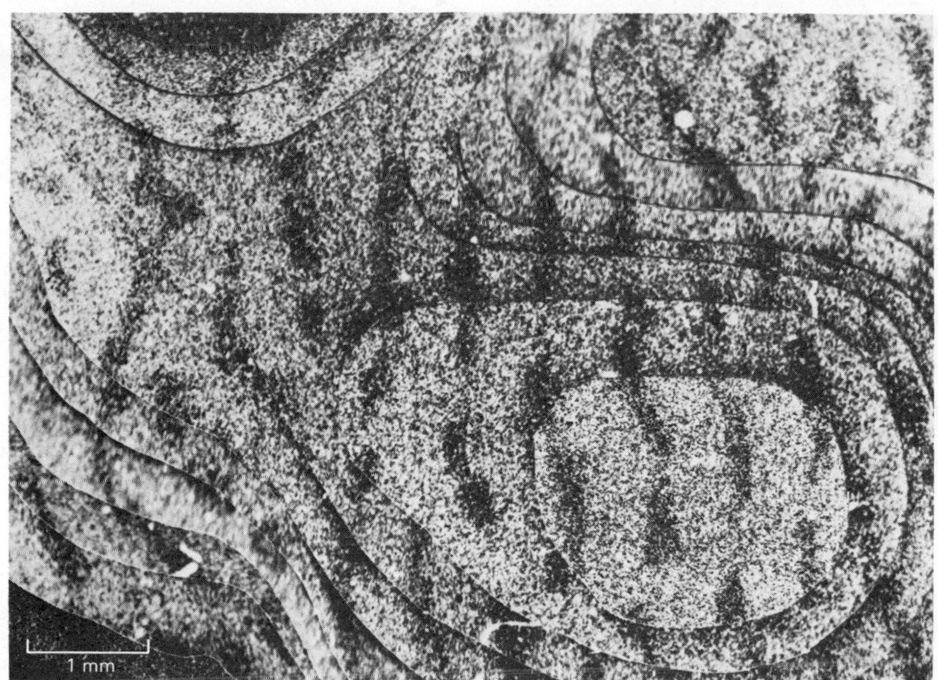

Figure 18–79 Ocular dominance columns in the visual cortex of a monkey that has had one eye covered during the sensitive period of development. The other eye has received an injection of radioactive proline, and autoradiographs have been prepared as described in the caption to Figure 18–78. The ocular dominance columns connected to the eye deprived of visual experience *(dark bands)* are abnormally narrow, while those connected to the other eye are abnormally wide. If the deprived eye is labeled, a converse picture is seen, with narrow bright bands alternating with broad dark bands. (From D. H. Hubel, T. N. Wiesel, and S. Le Vay, *Philos. Trans. R. Soc. (Biol.)* 278:377–409, 1977.)

maintain their synapses and make more. Because visual experience exerts its effect by biasing a competition, the consequences of depriving both eyes similarly, for example by raising the animal in the dark, are much less marked.

Synchronously Active Synapses on the Same Cell Reinforce One Another; Asynchronously Active Synapses Compete for Survival[57,60]

A more detailed understanding of the development of the neuronal connections underlying binocular vision has come from studies using the electrophysiological techniques described earlier (p. 1066) to assess the function of individual cells in the visual cortex. The axons bringing visual information to the cortex synapse in one of its middle layers, in which the individual neurons (in monkeys, at least) are mostly *monocularly driven;* that is, a typical neuron in this layer fires in response to a stimulus seen through one eye (either left or right) but is unresponsive to stimuli seen with the other. It is in this layer that the pattern of ocular dominance columns is most sharply defined. Above and below it in the cortex are layers of cells that are involved in subsequent steps of visual processing. These cells are mostly *binocularly driven:* they receive inputs from both eyes and fire at a maximal rate only when the appropriate stimulus is presented to both eyes simultaneously.

What are the consequences of early visual deprivation for the cortical cells? First of all, as would be expected from the previous evidence, if one eye is kept covered until the time of testing, there is a deficit of cortical cells driven by that eye: in all layers, the majority of neurons are driven by the active eye only. But what if one deprives the young animal specifically of binocular stimulation, by allowing each eye individually to have visual experience but never the two together? One method is to cover different eyes on alternate days. If this is done, each eye when tested separately is found to drive cortical cells in each layer in the usual way; when the two eyes are tested together, however, it turns out that practically no cortical cells are binocularly driven. Evidently binocular visual experience is required in the young animal for the establishment or maintenance of binocular inputs to cortical cells. This means that, to form lasting binocular connections, separate interneurons relaying the monocular input from each eye must not only synapse together on the same postsynaptic cell, but also must be synchronously stimulated during early life.

How precise must the synchrony be? This question can be answered by rearing a kitten so that its two eyes regularly see the same scenes but at slightly different times. If the delay interval between viewings with the right eye and viewings with the left is a few seconds or less, cortical connections develop normally; if it is 10 seconds or more, the kitten is left with a marked deficit of binocularly driven cells.

Such studies of the intact brain provide evidence for the following general rule of cell behavior: neurons that fire synchronously, or within a few seconds of one another, tend to maintain and reinforce one another's synapses on a shared postsynaptic cell; those that do not fire with this degree of synchrony tend to compete, until the postsynaptic cell is controlled by one of the neurons alone. In short, *synchronous firing establishes convergent connections.* We shall call this the *rule of associative synaptogenesis.* As a consequence of it, the fact that the two eyes of the animal regularly see the same scene becomes represented in the wiring diagram of the visual cortex. In this respect, the structure of the brain reflects the individual animal's experience.

The Pattern of Connections That Develops in the Brain Reflects the Regular Associations Between Events in the External World[61]

There is good evidence that the rule of associative synaptogenesis operates in other developing neural systems, where it can serve other important purposes. It has been suggested, for example, that it could be the means of ensuring continuity in neural maps from one array of nerve cells to another. Axons usually divide into many terminal branches to innervate neighborhoods that comprise many target cells. Two axons whose cell bodies lie close together in the same array will often be triggered to fire at the same time and thus will tend to form synapses on the same target cells in a second array. Consequently, adjacent neurons in the first array will be constrained to innervate adjacent, partly overlapping neighborhoods in the second.

The rule of associative synaptogenesis would seem to represent a powerful organizing device for sensory systems. In principle, it provides a general mechanism by which the connectivity of the brain can be adjusted to create cells that respond specifically to sets of sensory phenomena, of any sort, that are regularly associated. In this way, perhaps, the brain is able to learn to recognize regular combinations of features of the outside world and can come to represent in its own structure and function the existence of connections between one external occurrence and another. Such a capacity is the essence of associative memory.

It is not certain whether the acquisition of memories in adult life depends on the creation and elimination of synapses or on more subtle modifications. Nonetheless, the study of the development of binocular visual connections has clearly shown the power of a simple rule of cell behavior in explaining the complexities of the construction of the nervous system. Practically nothing is known about the molecular mechanisms involved in associative synaptogenesis, so that the rule presents us with an unsolved problem in the molecular biology of the cell. But it also holds out the hope that by solving this problem we shall gain new insights into the functional organization of the brain.

Summary

In amphibians and fish, the cells of the retina send their axons chiefly to the optic tectum, onto which they project an orderly map of the visual world. If the retinal axons are cut, they regenerate to form an orderly map once more. If the retina is rotated, the regenerated axons link the misplaced retinal cells with the tectal cells appropriate to their original positions, as though guided by neuronal specificity: it seems that retinal cells are distinguished by labels characteristic of their original positions in the retina and connect preferentially with tectal cells carrying positional labels that correspond. At the same time, there appear to be competitive and cooperative interactions between the retinal axons, causing them to spread out in an orderly array to innervate the whole available surface of the tectum.

In mammals, the development of the visual system is highly sensitive to visual experience during a critical period after birth. If one eye is deprived of experience, its territory in the visual cortex is taken over by the other eye. If the two eyes are never used together, the cortical connections required for normal binocular vision do not become established. These phenomena suggest a general rule of nerve cell behavior: different neurons contacting the same target cell tend to maintain their synapses on the shared target if, and only if, they commonly fire at the same time. This rule of associative synaptogenesis represents a mechanism by which the structure of the brain may be adjusted to reflect the connections between events in the external world.

References

General

The Brain. *Sci. Am.* 241(3), 1979. (A whole issue devoted to neurobiology.)

Cooke, I.; Lipkin, M., eds. Cellular Neurophysiology: A Source Book. New York: Holt, Rinehart and Winston, 1972. (An anthology of the most important original papers, 1921–1967.)

Jacobson, M. Developmental Neurobiology, 2nd ed. New York: Plenum, 1978.

Kandel, E.R.; Schwartz, J.H. Principles of Neural Science. New York: Elsevier, 1981.

Katz, B. Nerve, Muscle, and Synapse. New York: McGraw-Hill, 1966.

Kuffler, S.W.; Nicholls, J.G. From Neuron to Brain. Sunderland, Ma.: Sinauer, 1976.

Patterson, P.H.; Purves, D. Readings in Developmental Neurobiology. Cold Spring Harbor, N.Y.: Cold Spring Harbor Laboratory, 1982. (An anthology, mainly of recent papers.)

Ramón y Cajal, S. Histologie du Système Nerveux de l'Homme et des Vertébrés. Paris: Maloine, 1909–1911. Reprinted, Madrid: Consejo Superior de Investigaciones Científicas, Instituto Ramón y Cajal, 1972.

Schmidt, R.F., ed. Fundamentals of Neurophysiology, 2nd ed. New York: Springer-Verlag, 1978.

Cited

1. Stevens, C.F. The neuron. *Sci. Am.* 241(3):55–65, 1979.
   Keynes, R.D. Ion channels in the nerve-cell membrane. *Sci. Am.* 240(3):126–135, 1979.
   Kuffler, S.W.; Nicholls, J.G. From Neuron to Brain, pp. 88–144. Sunderland, Ma.: Sinauer, 1976.
   Hodgkin, A.L. The Conduction of the Nervous Impulse. Liverpool, Eng.: Liverpool University Press, 1964.
2. Hodgkin, A.L.; Keynes, R.D. Active transport of cations in giant axons from *Sepia* and *Loligo*. *J. Physiol.* 128:28–60, 1955.
3. Baker, P.F.; Hodgkin, A.L.; Shaw, T.I. The effects of changes in internal ionic concentrations on the electrical properties of perfused giant axons. *J. Physiol.* 164:355–374, 1962.

Hodgkin, A.L.; Katz, B. The effect of sodium ions on the electrical activity of the giant axon of the squid. *J. Physiol.* 108:37–77, 1949.

4. Hodgkin, A.L.; Huxley, A.F.; Katz, B. Measurement of current-voltage relations in the membrane of the giant axon of *Loligo. J. Physiol.* 116:424–448, 1952.

Hodgkin, A.L.; Huxley, A.F. Currents carried by sodium and potassium ions through the membrane of the giant axon of *Loligo. J. Physiol.* 116:449–472, 1952.

Stevens, C.F. Ionic channels in neuromembranes: methods for studying their properties. In Molluscan Nerve Cells: From Biophysics to Behavior (J. Koester, J.H. Byrne, eds.), pp. 11–31. Cold Spring Harbor, N.Y.: Cold Spring Harbor Laboratory, 1980.

5. Hodgkin, A.L.; Huxley, A.F. The dual effect of membrane potential on sodium conductance in the giant axon of *Loligo. J. Physiol.* 116:497–506, 1952.

6. Stevens, C.F. Study of membrane permeability changes by fluctuation analysis. *Nature* 270:391–396, 1977.

7. Sigworth, F.J.; Neher, E. Single Na^+ channel currents observed in cultured rat muscle cells. *Nature* 287:447–449, 1980.

Horn, R.; Patlak, J.; Stevens, C.F. Sodium channels need not open before they inactivate. *Nature* 291:426–427, 1981.

8. Ritchie, J.M. A pharmacological approach to the structure of sodium channels in myelinated axons. *Annu. Rev. Neurosci.* 2:341–362, 1979.

Armstrong, C.M. Sodium channels and gating currents. *Physiol. Rev.* 61:644–683, 1981.

Stevens, C.F. Interactions between intrinsic membrane protein and electric field: an approach to studying nerve excitability. *Biophys. J.* 22:295–306, 1978.

Hille, B. Ionic channels in excitable membranes: current problems and biophysical approaches. *Biophys. J.* 22:283–294, 1978.

9. Hodgkin, A.L.; Huxley, A.F. A quantitative description of membrane current and its application to conduction and excitation in nerve. *J. Physiol.* 117:500–544, 1952.

Chiu, S.Y.; Ritchie, J.M.; Rogart, R.B.; Stagg, D. A quantitative description of membrane currents in rabbit myelinated nerve. *J. Physiol.* 292:149–166, 1979.

10. Hodgkin, A.L.; Rushton, W.A.H. The electrical constants of a crustacean nerve fibre. *Proc. R. Soc. Lond. (Biol.)* 133:444–479, 1946.

Jack, J. An introduction to linear cable theory. In The Neurosciences; Fourth Study Program (F.O. Schmitt, F.G. Worden, eds.), pp. 423–437. Cambridge, Ma.: MIT Press, 1979.

11. Rogart, R. Sodium channels in nerve and muscle membrane. *Annu. Rev. Physiol.* 43:711–725, 1981.

12. Morell, P.; Norton, W.T. Myelin. *Sci. Am.* 242(5): 88–118, 1980.

Bray, G.M.; Rasminsky, M.; Aguayo, A.J. Interactions between axons and their sheath cells. *Annu. Rev. Neurosci.* 4:127–162, 1981.

13. The Synapse. *Cold Spring Harbor Symp. Quant. Biol.* 40, 1976.

Katz, B. Nerve, Muscle and Synapse, pp. 97–158. New York: McGraw-Hill, 1966.

Kuffler, S.W.; Nicholls, J.G. From Neuron to Brain, pp. 145–236. Sunderland, Ma.: Sinauer, 1976.

Kandel, E.R.; Schwartz, J.H. Principles of Neural Science, pp. 63–120. New York: Elsevier, 1981.

14. Peters, A.; Palay, S.L.; Webster, H. de F. The Fine Structure of the Nervous System. Philadelphia: Saunders, 1976. (Contains an excellent collection of electron micrographs.)

Heuser, J.E.; Reese, T. Structure of the synapse. In Handbook of Physiology. The Nervous System, Vol. 1, Cellular Biology of Neurons (E.R. Kandel, ed.), pp. 261–294. Baltimore: Williams & Wilkins, 1977.

Dale, H.H.; Feldberg, W.; Vogt, M. Release of acetylcholine at voluntary motor nerve endings. *J. Physiol.* 86:353–380, 1936.

Fatt, P.; Katz, B. An analysis of the end-plate potential recorded with an intracellular electrode. *J. Physiol.* 115:320–370, 1951.

15. Katz, B. The Release of Neural Transmitter Substances. Liverpool, Eng.: Liverpool University Press, 1969.

Katz, B.; Miledi, R. The timing of calcium action during neuromuscular transmission. *J. Physiol.* 189:535–544, 1967.

Miledi, R. Transmitter release induced by injection of calcium ions into nerve terminals. *Proc. R. Soc. Lond. (Biol.)* 183:421–425, 1973.

Kelly, R.B.; Deutsch, J.W.; Carlson, S.S.; Wagner, J.A. Biochemistry of neurotransmitter release. *Annu. Rev. Neurosci.* 2:399–446, 1979.

16. Heuser, J.E.; et al. Synaptic vesicle exocytosis captured by quick freezing and correlated with quantal transmitter release. *J. Cell Biol.* 81:275–300, 1979.

Fatt, P.; Katz, B. Spontaneous subthreshold activity at motor nerve endings. *J. Physiol.* 117:109–128, 1952.

del Castillo, J.; Katz, B. Quantal components of the end plate potential. *J. Physiol.* 124:560–573, 1954.

17. Lester, H.A. The response to acetylcholine. *Sci. Am.* 236(2):106–118, 1977.

Raftery, M.A.; Hunkapiller, M.W.; Strader, C.D.; Hood, L.E. Acetylcholine receptor: complex of homologous subunits. *Science* 208:1454–1457, 1980.

Klymkowsky, M.W.; Stroud, R.M. Immunospecific identification and three-dimensional structure of a membrane-bound acetylcholine receptor from *Torpedo californica. J. Mol. Biol.* 128:319–334, 1979.

Sakmann, B.; Patlak, J.; Neher, E. Single acetylcholine-activated channels show burst-kinetics in presence of desensitizing concentrations of agonist. *Nature* 286:71–73, 1980.

18. Massoulié, J.; Bon, S. The molecular forms of cholinesterase and acetylcholinesterase in vertebrates. *Annu. Rev. Neurosci.* 5:57–106, 1982.

19. Iversen, L.L. The chemistry of the brain. *Sci. Am.* 241(3):134–149, 1979.

Bloom, F.E. Neuropeptides. *Sci. Am.* 245(4):148–168, 1981.

Cooper, J.R.; Bloom, F.E.; Roth, R.H. The Biochemical Basis of Neuropharmacology, 3rd ed. New York: Oxford University Press, 1978.

20. Hartzell, H.C. Mechanisms of slow postsynaptic potentials. *Nature* 291:539–544, 1981.

Kehoe, J.; Marty, A. Certain slow synaptic responses: their properties and possible underlying mechanisms. *Annu. Rev. Biophys. Bioeng.* 9:437–465, 1980.

Greengard, P. Cyclic nucleotides, phosphorylated proteins, and the nervous system. *Fed. Proc.* 38:2208–2217, 1979.

21. Kuffler, S.W.; Nicholls, J.G. From Neuron to Brain, pp. 333–353. Sunderland, Ma.: Sinauer, 1976.

Eccles, J.C. The Physiology of Synapses. New York: Springer, 1964.

Barrett, J.N. Motoneuron dendrites: role in synaptic integration. *Fed. Proc.* 34:1398–1407, 1975.

22. Coombs, J.S.: Curtis, D.R.; Eccles, J.C. The generation of impulses in motoneurones. *J. Physiol.* 139:232–249, 1957.

Fuortes, M.G.F.; Frank, K.; Becker, M.C. Steps in the production of motoneuron spikes. *J. Gen. Physiol.* 40:735–752, 1957.

23. Koester, J.; Byrne, J.H., eds. Molluscan Nerve Cells: From Biophysics to Behavior, pp. 125–180. Cold Spring Harbor, N.Y.: Cold Spring Harbor Laboratory, 1980. (A collection of short papers on the role of membrane channels in neuronal firing behavior.)

24. Connor, J.A.; Stevens, C.F. Prediction of repetitive firing behaviour from voltage clamp data on an isolated neurone soma. *J. Physiol.* 213:31–53, 1971.

25. Meech, R.W. Calcium-dependent potassium activation in nervous tissues. *Annu. Rev. Biophys. Bioeng.* 7:1–18, 1978.

26. Bullock, T.H.; Horridge, G.A. Structure and Function in the Nervous Systems of Invertebrates, pp. 38–124. San Francisco: Freeman, 1965.

Shepherd, G.M. Microcircuits in the nervous system. *Sci. Am.* 238(2):92–103, 1978.

27. Cotman, C.W.; Nieto-Sampedro, M.; Harris, E.W. Synapse replacement in the nervous system of adult vertebrates. *Physiol. Rev.* 61:684–784, 1981.

Fambrough, D.M. Control of acetylcholine receptors in skeletal muscle. *Physiol. Rev.* 59:165–227, 1979.

Pumplin, D.W.; Fambrough, D.M. Turnover of acetylcholine receptors in skeletal muscle. *Annu. Rev. Physiol.* 44:319–335, 1982.

28. Lømo, T.; Rosenthal, J. Control of ACh sensitivity by muscle activity in the rat. *J. Physiol.* 221:493–513, 1972.

Frank, E.; Jansen, J.K.S.; Lømo, T.; Westgaard, R.H. The interaction between foreign and original motor nerves innervating the soleus muscle of rats. *J. Physiol.* 247:725–743, 1975.

Lømo, T.; Jansen, J.K.S. Requirements for the formation and maintenance of neuromuscular connections. *Curr. Top. Dev. Biol.* 16:253–281, 1980.

Lømo, T.; Slater, C.R. Acetylcholine sensitivity of developing ectopic nerve-muscle junctions in adult rat soleus muscles. *J. Physiol.* 303:173–190, 1980.

29. Sanes, J.R.; Marshall, L.M.; McMahan, U.J. Reinnervation of muscle fiber basal lamina after removal of myofibers. *J. Cell Biol.* 78:176–198, 1978.

Burden, S.J.; Sargent, P.B.; McMahan, U.J. Acetylcholine receptors in regenerating muscle accumulate at original synaptic sites in the absence of the nerve. *J. Cell Biol.* 82:412–425, 1979.

Sanes, J.R.; Hall, Z.W. Antibodies that bind specifically to synaptic sites on muscle fiber basal lamina. *J. Cell Biol.* 83:357–370, 1979.

30. Tsukahara, N. Synaptic plasticity in the mammalian central nervous system. *Annu. Rev. Neurosci.* 4:351–379, 1981.

31. Kandel, E.R. Small systems of neurons. *Sci. Am.* 241(3):66–76, 1979.

Kandel, E.R. Calcium and the control of synaptic strength by learning. *Nature* 293:697–700, 1981.

32. Bliss, T.V.P.; Lømo, T. Long-lasting potentiation of synaptic transmission in the dentate area of the anaesthetized rabbit following stimulation of the perforant path. *J. Physiol.* 232:331–356, 1973.

Chung, S.-H. Synaptic memory in the hippocampus. *Nature* 266:677–678, 1977.

Squire, L.R. The neuropsychology of human memory. *Annu. Rev. Neurosci.* 5:241–273, 1982.

33. Schmidt, R.F., ed. Fundamentals of Sensory Physiology. New York: Springer-Verlag, 1978.

Kuffler, S.W.; Nicholls, J.G. From Neuron to Brain, pp. 307–331. Sunderland, Ma.: Sinauer, 1976.

Katz, B. Depolarization of sensory terminals and the initiation of impulses in the muscle spindle. *J. Physiol.* 111:261–282, 1950.

34. Johnstone, J.R. Basic problems of cochlear physiology. *Trends Neurosci.* 4:106–109, 1981.

Dallos, P. Cochlear physiology. *Annu. Rev. Psychol.* 32:153–190, 1981.

Crawford, A.C.; Fettiplace, R. An electrical tuning mechanism in turtle cochlear hair cells. *J. Physiol.* 312:377–412, 1981.

35. Stryer, L. Biochemistry, 2nd ed., pp. 896–905. San Francisco: Freeman, 1981.

Bownds, M.D. Molecular mechanisms of visual transduction. *Trends Neurosci.* 4:214–217, 1981.

Baylor, D.A.; Lamb, T.D.; Yau, K.-W. Responses of retinal rods to single photons. *J. Physiol.* 288:613–634, 1979.

36. Kuffler, S.W.; Nicholls, J.G. From Neuron to Brain, pp. 16–73. Sunderland, Ma.: Sinauer, 1976.

37. Hubel, D.H.; Wiesel, T.N. Brain mechanisms of vision. *Sci. Am.* 241(3):150–162, 1979.

Kuffler, S.W. Discharge patterns and functional organization of mammalian retina. *J. Neurophysiol.* 16:37–68, 1953.

Hubel, D.H.; Wiesel, T.N. Ferrier Lecture: functional architecture of macaque monkey visual cortex. *Proc. R. Soc. Lond. (Biol.)* 198:1–59, 1977.

Hubel, D.H. Exploration of the primary visual cortex, 1955–78. *Nature* 299:515–524, 1982.

38. Dowling, J.E. Information processing by local circuits: the vertebrate retina as a model system. In The Neurosciences; Fourth Study Program (F.O. Schmitt, F.G. Worden, eds.), pp. 163–181. Cambridge Ma.: MIT Press, 1979.

Barlow, H.B. David Hubel and Torsten Wiesel: their contributions towards understanding the primary visual cortex. *Trends Neurosci.* 5:145–152, 1982.

39. Scott, B.I.H. Electricity in plants. *Sci. Am.* 207(4):107–117, 1962.

Hagiwara, S.; Jaffe, L.A. Electrical properties of egg cell membranes. *Annu. Rev. Biophys. Bioeng.* 8:385–416, 1979.

Roberts, A. Conducted impulses in the skin of young tadpoles. *Nature* 222:1265–1266, 1969.

40. Bray, D.; Gilbert, D. Cytoskeletal elements in neurons. *Annu. Rev. Neurosci.* 4:505–523, 1981.

41. Schwartz, J.H. The transport of substances in nerve cells. *Sci. Am.* 242(4):152–171, 1980.

Grafstein, B.; Forman, D.S. Intracellular transport in neurons. *Physiol. Rev.* 60:1167–1283, 1980.

Tytell, M.; Black M.M.; Garner, J.A.; Lasek, R.J. Axonal transport: each major rate component reflects the movement of distinct macromolecular complexes. *Science* 214:179–181, 1981.

Adams, R.J. Organelle movement in axons depends on ATP. *Nature* 297:327–329, 1982.

42. Johnston, R.N.; Wessells, N.K. Regulation of the elongating nerve fiber. *Curr. Top. Dev. Biol.* 16:165–206, 1980.

Ramón y Cajal, S. Recollections of My Life (E.H. Craigie, trans.). In Memoirs of the Americal Philosophical Society, Vol. 8. Philadelphia, 1937. Reprinted, Cambridge Ma.: MIT Press, 1960. (The discovery of the growth cone is described in Part 2, Chapter 7.)

Harrison, R.G. On the origin and development of the nervous system studied by the methods of experimental embryology. *Proc. R. Soc. Lond. (Biol.)* 118:155–196, 1935.

Bray, D.; Bunge, M.B. The growth cone in neurite extension. In Locomotion of Tissue Cells, *Ciba Symp.* 14:195–209, 1973.

43. Yamada, K.M.; Spooner, B.S.; Wessells, N.K. Ultrastructure and function of growth cones and axons of cultured nerve cells. *J. Cell Biol.* 49:614–635, 1971.

Bray, D.; Thomas, C.; Shaw, G. Growth cone formation in cultures of sensory neurons. *Proc. Natl. Acad. Sci. USA* 75:5226–5229, 1978.

44. Letourneau, P.C. Cell-to-substratum adhesion and guidance of axonal elongation. *Dev. Biol.* 44:92–101, 1975.

Wessells, N.K.; et al. Responses to cell contacts between growth cones, neurites and ganglionic non-neuronal cells. *J. Neurocytol.* 9:647–664, 1980.

Rutishauser, U.; Gall, W.E.; Edelman, G.M. Adhesion among neural cells of the chick embryo. IV. Role of the cell surface molecule CAM in the formation of neurite bundles in cultures of spinal ganglia. *J. Cell Biol.* 79:382–393, 1978.

Ho, R.K.; Goodman, C.S. Peripheral pathways are pioneered by an array of central and peripheral neurones in grasshopper embryos. *Nature* 297:404–406, 1982.

45. Levi-Montalcini, R.; Calissano, P. The nerve-growth factor. *Sci. Am.* 240(6):68–77, 1979.

Campenot, R.B. Local control of neurite development by nerve growth factor. *Proc. Natl. Acad. Sci. USA* 74:4516–4519, 1977.

Gundersen, R.W.; Barrett, J.N. Characterization of the turning response of dorsal root neurites toward nerve growth factor. *J. Cell Biol.* 87:546–554, 1980.

Greene, L.A.; Shooter, E.M. The nerve growth factor: biochemistry, synthesis, and mechanism of action. *Annu. Rev. Neurosci.* 3:353–402, 1980.

46. Cowan, W.M. The development of the brain. *Sci. Am.* 241(3):112–133, 1979.

Purves, D.; Lichtman, J.W. Elimination of synapses in the developing nervous system. *Science* 210:153–157, 1980. (A very good brief review.)

Anderson, H.; Edwards, J.S.; Palka, J. Developmental neurobiology of invertebrates. *Annu. Rev. Neurosci.* 3:97–139, 1980.

47. Hollyday, M. Motoneuron histogenesis and the development of limb innervation. *Curr. Top. Dev. Biol.* 15:181–216, 1980.

Rakic, P. Neuronal-glial interaction during brain development. *Trends Neurosci.* 4:184–187, 1981.

Caviness, V.S.; Rakic, P. Mechanisms of cortical development: a view from mutations in mice. *Annu. Rev. Neurosci.* 1:297–326, 1978.

48. Katz, M.J.; Lasek, R.J.; Nauta, H.J.W. Ontogeny of substrate pathways and the origin of the neural circuit pattern. *Neuroscience* 5:821–833, 1980.

Constantine-Paton, M. Axonal navigation. *Bioscience* 29:526–532, 1979.

Landmesser, L. The development of motor projection patterns in the chick hind limb. *J. Physiol.* 284:391–414, 1978.

Lance-Jones, C.; Landmesser, L. Pathway selection by chick lumbosacral motoneurons during normal development. *Proc. R. Soc. Lond. (Biol.)* 214:1–18, 1981.

49. Sperry, R.W. The growth of nerve circuits. *Sci. Am.* 201(5):68–75, 1959.

Landmesser, L.T. The generation of neuromuscular specificity. *Annu. Rev. Neurosci.* 3:279–302, 1980.

Lance Jones, C.; Landmesser, L. Motoneurone projection patterns in embryonic chick limbs following partial deletions of the spinal cord. *J. Physiol.* 302:559–580, 1980.

Sperry, R.W. Chemoaffinity in the orderly growth of nerve fiber patterns and connections. *Proc. Natl. Acad. Sci. USA* 50:703–710, 1963.

50. Hamburger, V. Regression versus peripheral control of differentiation in motor hypoplasia. *Am. J. Anat.* 102:365–410, 1958.

Katz, M.J.; Lasek, R.J. Evolution of the nervous system: role of ontogenetic mechanisms in the evolution of matching populations. *Proc. Natl. Acad. Sci. USA* 75:1349–1352, 1978.

51. Oppenheim, R.W. Neuronal cell death and some related regressive phenomena during neurogenesis: a selective historical review and progress report. In Studies in Developmental Neurobiology: Essays in Honor of Viktor Hamburger (W.M. Cowan, ed.), pp. 74–133. New York: Oxford University Press, 1981.

Jacobson, M. Developmental Neurobiology, 2nd ed., pp. 279–307. New York: Plenum, 1978.

Hamburger, V. Some aspects of the embryology of behavior. *Q. Rev. Biol.* 38:342–365, 1963.

Pittman, R.H.; Oppenheim, R.W. Neuromuscular blockade increases motoneurone survival during normal cell death in the chick embryo. *Nature* 271:364–366, 1978.

52. Brown, M.C.; Jansen, J.K.S.; Van Essen, D. Polyneuronal innervation of skeletal muscle in new-born rats and its elimination during maturation. *J. Physiol.* 261:387–422, 1976.

Dennis, M.J. Development of the neuromuscular junction: inductive interactions between cells. *Annu. Rev. Neurosci.* 4:43–68, 1981.

53. Brown, M.C.; Holland, R.L.; Hopkins, W.G. Motor nerve sprouting. *Annu. Rev. Neurosci.* 4:17–42, 1981.

Ebendal, T.; Olson, L.; Seiger, A.; Hedlund, K.-O. Nerve growth factors in the rat iris. *Nature* 286:25–28, 1980.

54. Kandel, E.R.; Schwartz, J.H. Principles of Neural Science, pp. 170–198. New York: Elsevier, 1981.

55. Gaze, R.M. The Formation of Nerve Connections. New York: Academic Press, 1970.

Gottlieb, D.I.; Glaser, L. Cellular recognition during neural development. *Annu. Rev. Neurosci.* 3:303–318, 1980.

Trisler, G.D.; Schneider, M.D.; Nirenberg, M. A topographic gradient of molecules in retina can be used to identify neuron position. *Proc. Natl. Acad. Sci. USA* 78:2145–2149, 1981.

56. Gaze, R.M. The problem of specificity in the formation of nerve connections. In Receptors and Recognition, Vol. B4, Specificity of Embryological Interactions (D.R. Garrod, ed.), pp. 51–93. London: Chapman and Hall, 1978.

Yoon, M.G. Progress of topographic regulation of the visual projection in the halved optic tectum of adult goldfish. *J. Physiol.* 257:621–643, 1976.

57. Barlow, H.B. Visual experience and cortical development. *Nature* 258:199–204, 1975.

Wiesel, T.N. Postnatal development of the visual cortex and the influence of environment. *Nature* 299:583–591, 1982.

58. Hubel, D.H.; Wiesel, T.N.; Le Vay, S. Plasticity of ocular dominance columns in monkey striate cortex. *Philos. Trans. R. Soc. (Biol.)* 278:377–409, 1977.

59. Rakic, P. Prenatal genesis of connections subserving ocular dominance in the rhesus monkey. *Nature* 261:467–471, 1976.

Swindale, N.V. Rules for pattern formation in mammalian visual cortex. *Trends Neurosci.* 4:102–104, 1981.

60. Le Vay, S.; Wiesel, T.N.; Hubel, D.H. The development of ocular dominance columns in normal and visually deprived monkeys. *J. Comp. Neurol.* 191:1–51, 1980.

Hubel, D.H.; Wiesel, T.N. Binocular interaction in striate cortex of kittens reared with artificial squint. *J. Neurophysiol.* 28:1041–1059, 1965.

Blasdel, G.G.; Pettigrew, J.D. Degree of interocular synchrony required for maintenance of binocularity in kitten's visual cortex. *J. Neurophysiol.* 42:1692–1710, 1979.

Keating, M.J. The role of visual function in the patterning of binocular visual connexions. *Br. Med. Bull.* 30:145–151, 1974.

61. Purves, D.; Lichtman, J.W. Elimination of synapses in the developing nervous system. *Science* 210:153–157, 1980.

Chung, S.H. In search of the rules for nerve connections. *Cell* 3:201–205, 1974.

Willshaw, D.J.; von der Malsburg, C. How patterned neural connections can be set up by self-organization. *Proc. R. Soc. Lond. (Biol.)* 194:431–445, 1976.

Special Features of Plant Cells

19

Anyone can distinguish a flowering plant from a mammal. Even deciding whether an individual cell is of plant or animal origin is usually a simple matter, although there are problematic cases. But as the level of analysis reaches down into the cell—to cytoplasm, organelle, and molecule—the similarities between the two kingdoms begin to outweigh the differences. It requires sophisticated procedures to distinguish plant mitochondria, nuclei, or ribosomes from their animal counterparts; many intracellular components, such as microtubules, are essentially indistinguishable. The disparities between plants and animals tend *not* to be in fundamental molecular features like DNA replication, protein synthesis, mitochondrial ATP production, or the basic molecular design of cell membranes. Rather they relate to higher order functions of cells and tissues. Most of the disparities between the two kingdoms developed through evolutionary diversifications that can be traced back to two fundamental events: the ability to fix carbon dioxide by photosynthesis (discussed in Chapter 9), and the production of a rigid *cell wall* by the progenitors of plants, the consequences of which are discussed in this chapter.

The Central Significance of the Cell Wall

The plant cell wall is a specialized form of extracellular matrix that is closely applied to the external surface of the plant cell plasma membrane. While most animal cells also have extracellular matrix components on their surface as part of their cell coat or glycocalyx (see p. 692), the plant cell wall is generally much thicker, stronger, more organized, and—most important—more rigid. In evolving relatively rigid cell walls, which vary from 0.1 μm to many micrometers in thickness, plants forfeited the ability to move about and, therefore, did not develop muscles, bones, or nervous systems. In fact, most of the differences between plants and animals—in nutrition, digestion, osmoregulation, growth, reproduction, intercellular communication, defense mechanisms, as well as in morphology—can be traced to the plant cell wall.

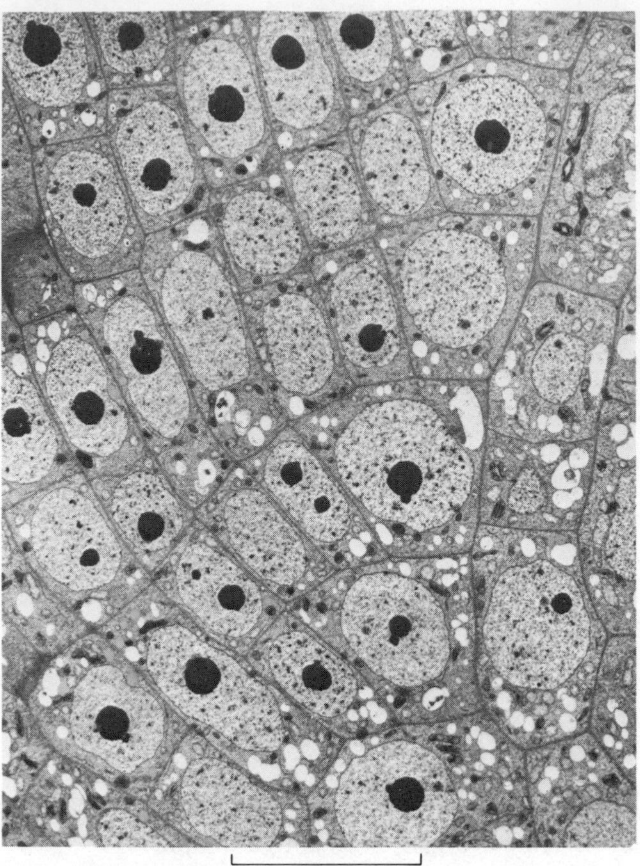

Figure 19–1 Electron micrograph of the root tip of a rush, showing the organized pattern of cells that results from an ordered sequence of cell divisions in cells with rigid cell walls. (Courtesy of B. Gunning.)

The cell wall provides a home for the plant cell proper. Each cell wall interacts with that of its neighbors, binding the cells together to form the intact plant (Figure 19–1). The cell walls also form channels for the circulation of fluids within the plant and for intercellular communication. The walls are consequently responsible for functions that in animals are provided by the skeleton, the skin, and the circulatory system. It is not surprising that to perform all these functions cell walls with highly varied compositions and structures have evolved. Indeed, the different cell types in a plant are recognized and classified by the shape and nature of their cell walls. It was, in fact, the thick cell walls of cork, visible under a simple microscope, that in 1663 enabled Robert Hooke to distinguish cells clearly and to name them as such.

In this section we shall examine the ways in which plant cells have evolved to exploit their walled environment. The starting point is a description of the nature of the wall itself.

The Cell Wall Is Composed of Cellulose Fibers Embedded in a Matrix of Polysaccharide and Protein[1]

In a multicellular plant, the newly formed cells are small in relation to their final size. To accommodate their enlargement, the cell walls of young growing plant cells (Figure 19–2) are thinner and are only semirigid, unlike those of cells that have stopped growing. The cell walls of growing cells are called **primary cell walls.** The fully grown cell may either retain its primary cell wall, sometimes thickening it considerably, or, in certain cases, deposit new tough wall layers of a different composition, called **secondary cell walls.**

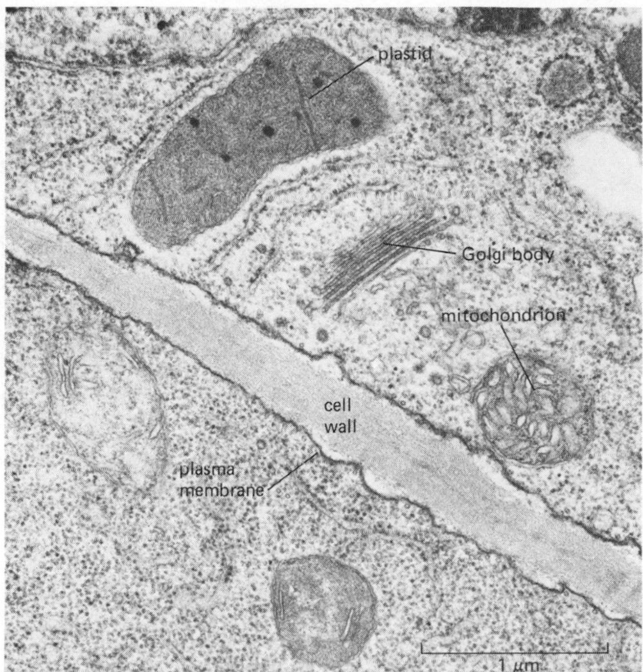

Figure 19–2 Electron micrograph showing the primary cell wall separating two cells in the root tip of a cress plant. (From B. Gunning and M. Steer, Ultrastructure and the Biology of Plant Cells. London: Arnold, 1975.)

Although the primary cell walls of higher plants vary greatly in their detailed organization, like all extracellular matrices they are constructed according to a common principle: they achieve their strength from long, tough fibers that are held together by a matrix of protein and polysaccharide. The underlying architectural principle involved here, that of strong fibers embedded in an amorphous matrix (Figure 19–3), is also used in such common building materials as fiber glass or reinforced concrete. In the higher plant cell wall the fibers are generally made from the polysaccharide *cellulose,* the most abundant organic macromolecule on earth. The matrix, however, is composed predominantly of two other sorts of polysaccharide—*hemicellulose* and *pectin* (Figure 19–4). The exact composition varies considerably both within and between species, but the basic construction remains the same: the fibers and matrix molecules are cross-linked by covalent bonds and weak interactions into a very complex structure. Because of this immense complexity, we still do not have a complete description of all the molecules within any one cell wall; nor do we know exactly how they are cross-linked.

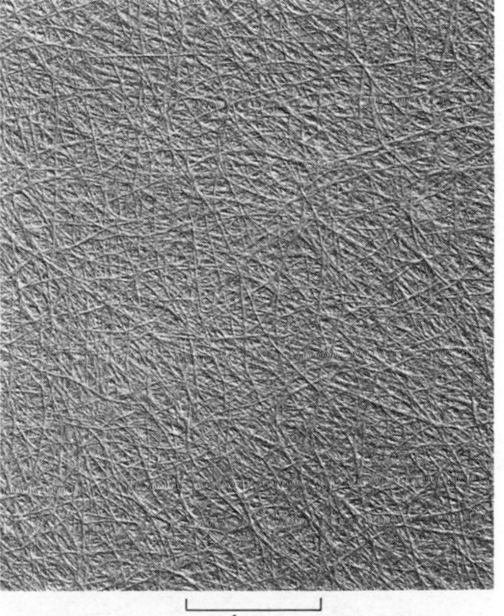

Figure 19–3 The cellulose microfibrils of the primary cell wall. In this metal-shadowed preparation the matrix molecules of the wall have been dissolved away, revealing an apparently random network of cellulose microfibrils. In some primary cell walls the microfibrils are much more ordered. (Courtesy of A. B. Wardrop.)

Fiber and Matrix Polysaccharides Form a Complex Cross-linked Structure[1,2]

A **cellulose** molecule consists of a linear chain of several thousand glucose units, each covalently linked by a characteristic $\beta 1 \rightarrow 4$ glycosidic bond. This linkage gives each molecule a flat ribbonlike structure that is stabilized by internal hydrogen bonds (Figure 19–5). Other hydrogen bonds between adjacent cellulose molecules cause these ribbonlike molecules to adhere strongly to one another in parallel arrays of 60 to 70 cellulose chains all having the same polarity, thereby forming very long, highly ordered, crystalline aggregates called **microfibrils.** These microfibrils (see Figure 12–62) are surrounded by a larger number of less precisely packed chains of cellulose as well as by certain hemicellulose molecules.

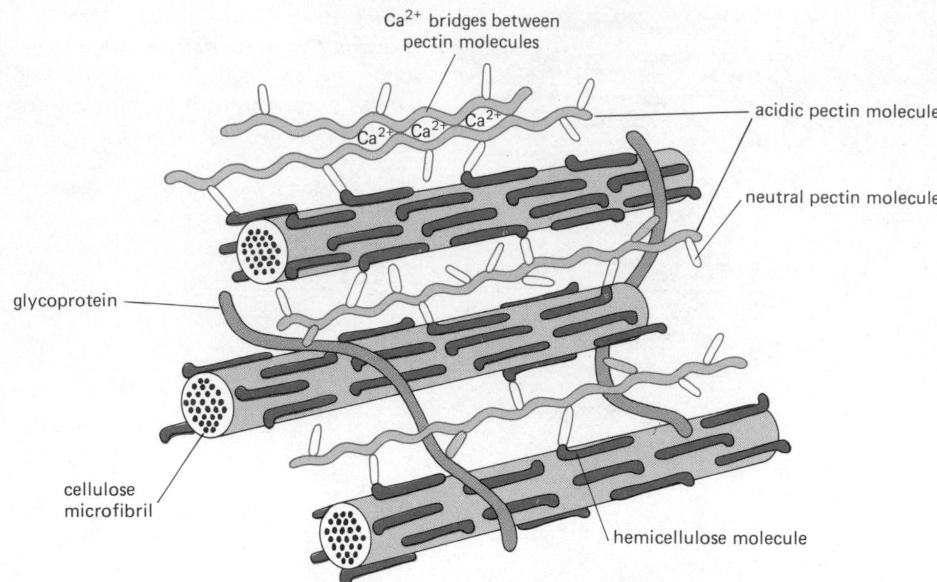

Figure 19–4 Schematic diagram showing how the two major components of the primary cell wall, the fibers and matrix, might be interconnected. Hemicellulose molecules (for example, xyloglucans) are linked by hydrogen bonds to the surface of the cellulose microfibrils. Some of these hemicellulose molecules are cross-linked in turn to acidic pectin molecules (for example, rhamnogalacturonans) by short neutral pectin molecules (for example, arabinogalactans). Cell-wall glycoproteins are probably attached to the pectin molecules. Details of the structure of cellulose microfibrils are shown in Figures 12–62 and 19–5.

Hemicellulose is the name given to a heterogeneous group of branched matrix polysaccharides that bind tightly but noncovalently to the surface of the cellulose microfibrils and to each other, thereby coating the microfibrils and cross-linking them via hydrogen bonds into a complex network. There are many different classes of hemicelluloses, but they all have a long $\beta 1 \rightarrow 4$ linked linear backbone of one sugar from which short side chains of other sugars protrude (Figure 19–6). The particular sugars vary with the type of hemicellulose. It is the backbone sugars of the hemicellulose molecules that form hydrogen bonds along the outside of the cellulose microfibrils. The particular hemicelluloses made by a cell vary widely, depending both on the plant species and its stage of development.

The third major type of cell-wall polysaccharide, the **pectins,** are heterogeneous, branched, and highly hydrated polysaccharides that contain many negatively charged galacturonic acid residues (Figure 19–7). Because of their negative charge, pectins avidly bind cations; and when Ca^{2+} is added to a solution of pectin molecules, it cross-links them to produce a semirigid gel. There is evidence that similar Ca^{2+} cross-links play a role in holding the cell-wall components together, but the exact nature of the bonding that gives the wall its integrity is unknown. Pectin is particularly abundant in the *middle*

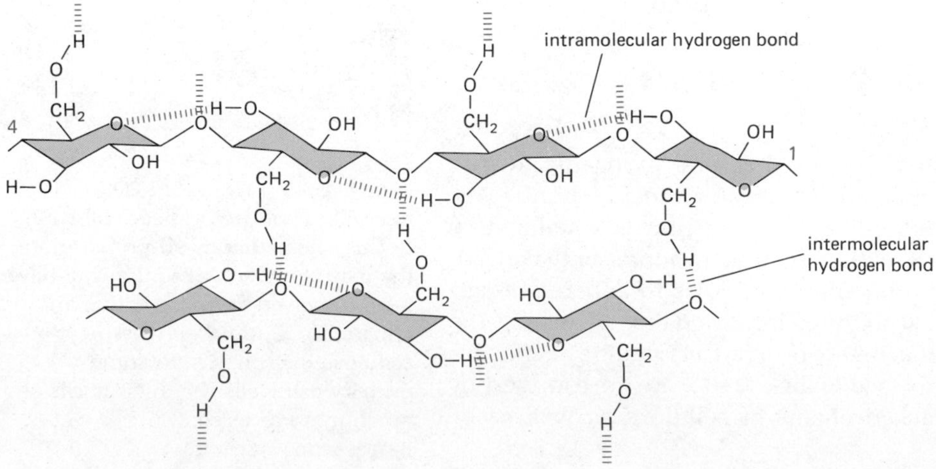

intramolecular hydrogen bond
intermolecular hydrogen bond

Figure 19–5 Two $\beta 1 \rightarrow 4$ linked glucose chains of cellulose. Intramolecular hydrogen bonds stabilize each chain; intermolecular hydrogen bonds tightly cross-link adjacent chains within a microfibril.

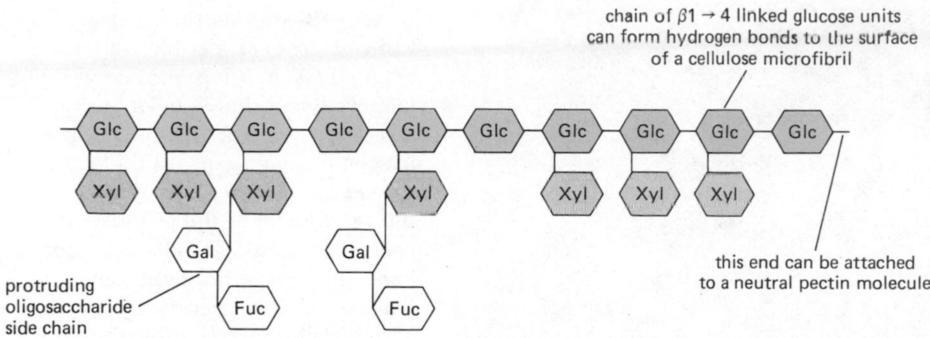

Figure 19–6 An example of a hemicellulose molecule from the cell wall of a typical flowering plant. The backbone consists of a celluloselike chain of glucose residues, which is hydrogen-bonded to the surface of cellulose microfibrils. In this example, a xyloglucan is shown, which has xylose attached to several of the glucose units. Other sugars, such as galactose and fucose, may also be present.

lamella, the region that serves to cement together the cell walls of adjacent cells. It is this layer that ruptures in some places, forming the intercellular air spaces found in many tissues (Figure 19–8).

In addition to the three classes of polysaccharides, the primary cell wall also contains small amounts of protein. The major protein contains a large number of residues of the unusual amino acid hydroxyproline. We have encountered this amino acid in collagen, the extracellular matrix protein in animals (see p. 696). In the case of plants, however, the hydroxyproline as well as many serine residues are linked to short oligosaccharide side chains, thus forming glycoproteins. Since it is difficult to extract the glycoprotein without destroying the structure of the cell wall, it seems that these molecules—together with cellulose, hemicellulose, and the pectins—are tightly integrated in the complex polysaccharide matrix of the wall.

In order for a plant cell to grow or change its shape, the cell wall has to stretch or deform. Because cellulose microfibrils are highly inelastic, such changes must involve the movement of microfibrils past one another. The possible types of microfibril movements allowed will depend on the orientation of the microfibrils within the primary wall, as well as on the bonding interactions between matrix macromolecules and between these macromolecules and the cellulose microfibrils. We shall return to this important topic in later sections.

The Limited Porosity of Cell Walls Limits Molecular Exchange Between Plant Cells and Their Environment[3]

All cells take in nutrients and expel waste products across their plasma membranes. They also respond to chemical signals in their environment. In the case of plant cells, such molecules and signals must, in addition, penetrate the cell wall. Since the matrix of the wall is a highly hydrated polysaccharide gel (the primary cell wall being 60% water by weight), water, gases, and small water-soluble molecules penetrate rapidly. Compared to the plasma mem-

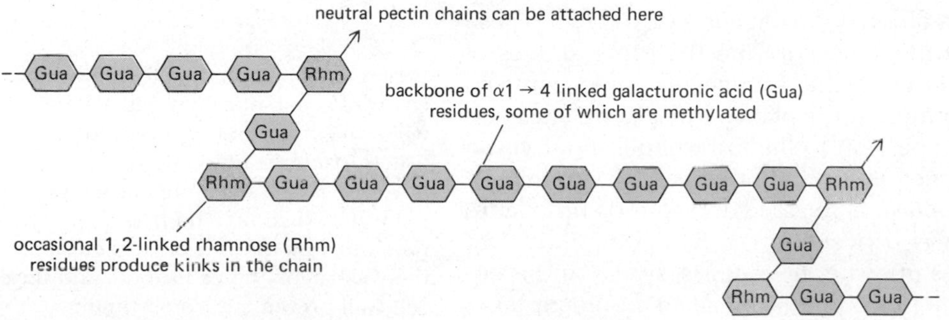

Figure 19–7 An example of an acidic pectin molecule (a rhamnogalacturonan) from the cell wall of a higher plant. Kinks in the straight backbone chain of negatively charged galacturonic acid residues are introduced by the occasional rhamnose residues, which also serve as attachment sites for the neutral pectins that cross-link them to hemicellulose (see Figure 19–4).

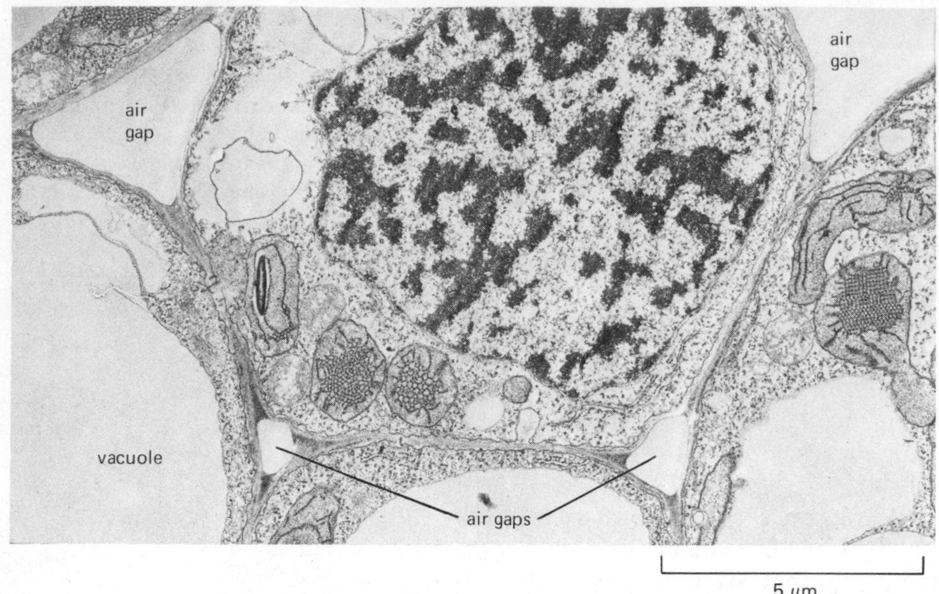

Figure 19–8 Electron micrograph of leaf cells from an oat plant that has been grown in the dark. The middle lamella, which holds neighboring cells together, has ruptured in places, leaving large air gaps. In the leaf these gaps allow carbon dioxide to diffuse directly to the photosynthetic cells. Because these cells have not been exposed to the light, their plastids have not yet developed into mature chloroplasts. (Courtesy of B. Gunning, from T. W. Goodwin, ed., Biochemistry of Chloroplasts. London: Academic Press, 1967. © 1967 Academic Press Inc. [London] Ltd.)

brane, the cross-linked structure of the cell wall only slightly impedes the diffusion of small molecules like water, sucrose, or K^+. (Even in the case of a cell with a wall 15 μm thick, only 10% of the resistance to water flow between the cytoplasm and the external medium is contributed by the wall; the remaining 90% is due to the plasma membrane.) Macromolecules, on the other hand, penetrate most plant cell walls very slowly. Measurements of the pores in the cell wall reveal diameters in the range of 3.5 nm to 5.2 nm. This is sufficiently small that the movement across the wall of molecules with a molecular weight much above 15,000 to 20,000 will be extremely slow. In some cases, the deposition of a waxy cuticle (see Figure 19–13C) over the wall further reduces the diffusion of molecules. In other cases, the walls are more porous and permit the outward passage of secretory products such as mucilages, which may have molecular weights in excess of 100,000. However, the general case is clear: plants must subsist on molecules of low molecular weight, and any intercellular signaling molecules that have to pass through the cell wall must also be small. In fact, most of the known plant signaling molecules, such as the growth-regulating substances—auxins, cytokinins, and gibberellins—have molecular weights of less than 500 (see p. 1139).

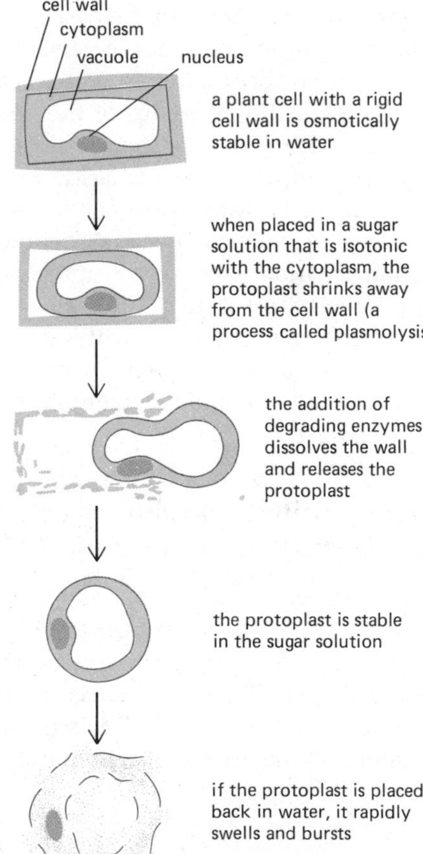

Figure 19–9 A plant cell without its cell wall is osmotically unstable and will swell and burst if placed in water. Inside its rigid cell wall, however, it can swell only as far as the wall will allow. The pressure developed by the cell pressing against the wall keeps the cell turgid, and the cell wall prevents it from bursting.

The Tensile Strength of Cell Walls Allows Plant Cells to Generate an Internal Hydrostatic Pressure Called Turgor[4]

Their cell walls enable plant cells to function in the hypotonic environment of the plant. The extracellular fluid inside higher plants is confined to the space occupied by the aqueous phase of all of the cell walls, and the long tubes formed by the empty cell walls of dead *xylem cells* (see p. 1109). These xylem cells carry water (the *transpiration stream*) from the roots to sites of evaporation, mainly in the leaves. Although this extracellular fluid contains more solutes than does the dilute solution in the plant's external milieu, such as the soil, it is still hypotonic in comparison to the intracellular fluid. If the wall of a plant cell is digested using cellulases and other wall-degrading enzymes, the cell proper, called a *protoplast,* is released and rounds up (Figure 19–9). If such a spherical protoplast is exposed to the hypotonic fluid that normally bathes the plant cell, it takes up water by osmosis, swells, and eventually bursts. In contrast, a walled cell placed in the same environment takes

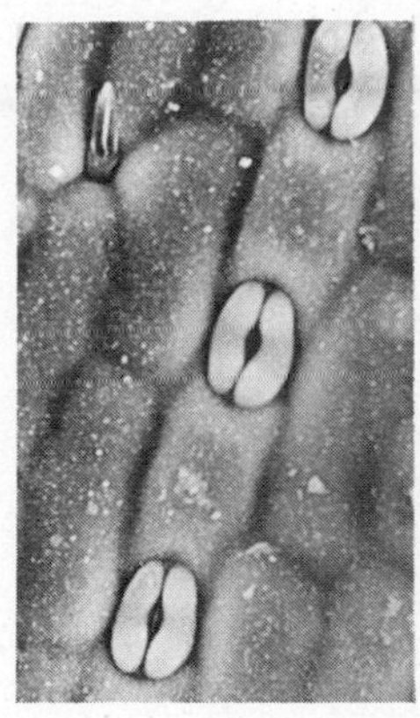

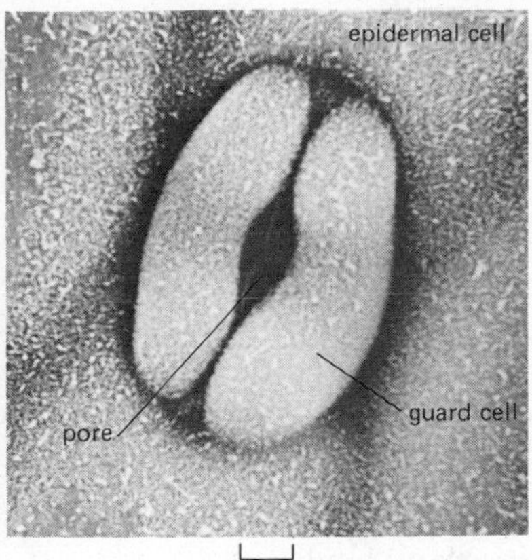

Figure 19–10 Scanning electron micrographs of stomata in the leaf epidermis of a tropical grass, shown at two different magnifications. The turgor-regulated movements of the two guard cells control the size of the enclosed pore, thus regulating the gas exchange between the leaf and the environment. (Courtesy of H. W. Woolhouse and G. J. Hills.)

up water but can only swell to a limited extent. The cell develops an internal hydrostatic pressure pushing outward on the rigid cell wall, thus preventing any further net influx of water. Therefore, most plant cells, in contrast to those of animals, do not need to match the osmolarity of their intracellular and extracellular tissue fluids.

The outward **turgor pressure** (or **turgor**) in a plant cell, caused by the osmotic imbalance between its intracellular and extracellular fluids, is vital to plants. Turgor is the main driving force for cell expansion during growth and is the cause of much of the mechanical rigidity of plant tissues (compare a wilted leaf of a dehydrated plant with a turgid leaf of a well watered one). Turgor also generates the limited movements seen in plants, such as the essential movements of stomatal guard cells that regulate the exchange of gases between leaves and the surrounding air (Figure 19–10), and the movement of plant parts such as traps in carnivorous plants and leaves of "sensitive" plants that close when touched.

Turgor Is Regulated by Feedback Mechanisms That Control the Concentration of Intracellular Solutes[5]

In view of the importance of turgor pressure to the plant, it is not surprising that plant cells have evolved sensitive mechanisms for regulating its magnitude. The turgor pressure varies greatly from plant to plant and from cell to cell over a range equivalent to half an atmosphere (in some large-celled algae) to nearly 50 atmospheres (in some stomatal guard cells). Cells can increase their turgor pressure by increasing the concentration of osmotically active solutes in the cytosol—either by pumping them in from the extracellular fluid across the plasma membrane or by generating them from osmotically inactive polymeric stores. In both cases, feedback loops monitor the level of turgor and regulate it.

How do such feedback control systems work? Experiments on the rapid responses of plant cells to turgor pressure changes suggest that a "turgor-pressure detector" is located in the plasma membrane. Thus, a sudden fall in turgor pressure induces inward solute transport, most commonly of K^+, while a sudden increase in turgor brings about a solute efflux. These responses are very rapid and presumably reflect changes in specific transport proteins in the plasma membrane. In contrast, alterations in the rate at which osmotically active solutes are generated in the cytoplasm from polymeric stores occur more slowly.

Turgor-sensing systems are especially important in plant cells in environments with extreme or fluctuating osmotic properties. Plants subjected to drought or to soil containing very high salt concentrations have evolved a variety of special adaptations for turgor regulation. For example, plants living in a high-salt habitat must accumulate very high internal solute concentrations to maintain turgor. Since the accumulation of ions such as K^+ to such high levels would probably alter the activities of vital enzymes, plant cells in these environments accumulate specially compatible organic solutes: polyhydroxylic compounds such as glycerol or mannitol, amino acids such as proline, or *N*-methylated derivatives of amino acids such as glycinebetaine. These solutes can reach very high concentrations in the cytosol. Thus, the turgor-pressure detector can regulate organic biosyntheses in the cytoplasm as well as ion transport across the plasma membrane.

The Cell Wall Is Modified During the Creation of Specialized Cell Types[6]

There are relatively few basic cell types within the body of a flowering plant, and they are all easily distinguished by the shape and structure of their cell wall (Figure 19–11). All these cell types arise from cells with a *primary cell wall,* by a process of cell growth that is usually followed by a period of cell-

Figure 19–11 (*Below and on opposite page*) Some of the cells and tissues of a higher plant. There are several other specialized cell types that are not illustrated here. These include stomatal guard cells and transfer cells, both of which are discussed elsewhere in the chapter. But the important point is that all plants are constructed from a relatively small number of basic cell types.

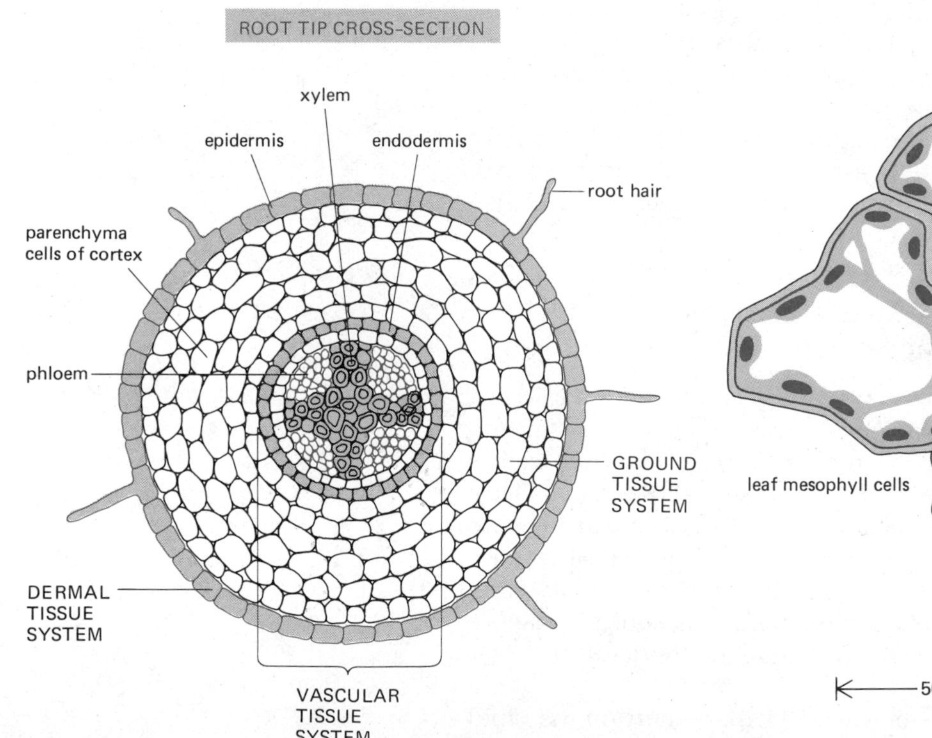

The various organs of a higher plant (e.g., leaf, stem, and root) are each composed of 3 easily recognizable *tissue systems*—vascular, ground, and dermal. In this schematic drawing of a cross-section of a root tip, the *vascular tissue system* is embedded in the *ground tissue system,* which is, in turn, enclosed by the *dermal tissue system.* The same three tissue systems, in different arrangements, make up all the parts of a higher plant. Each is composed of a relatively small number of common cell types, five of which are illustrated on this page and the next.

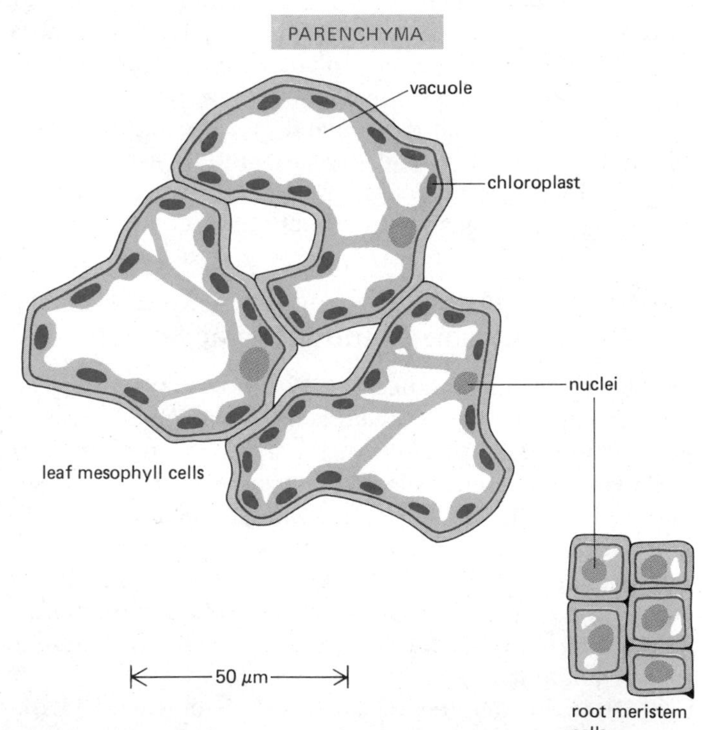

Parenchyma cells are found in all tissue systems. They are living cells, generally capable of further division, and are the least obviously specialized type of cell, having a thin primary cell wall. They include the apical meristematic cells of roots and shoots and the green photosynthetic cells of a leaf.

PHLOEM

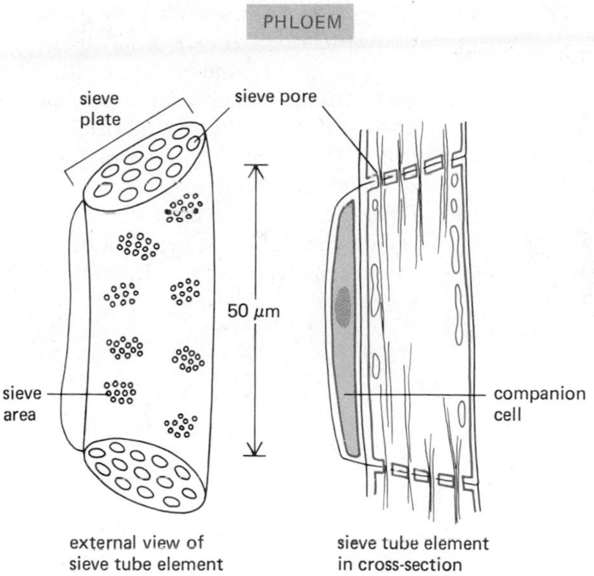

Phloem is a complex set of cells in the vascular tissue involved in the transport of organic solutes in the plant. The main conducting cells (elements) are aligned to form tubes called *sieve tubes.* The sieve tube elements at maturity are living cells interconnected by perforations in their end walls (sieve plates). Since these cells have lost their nuclei and much of their cytoplasm, they rely on associated *companion cells* for their maintenance. These companion cells have the additional function of transporting soluble food molecules in and out of the sieve tube elements through porous sieve areas in the wall.

XYLEM

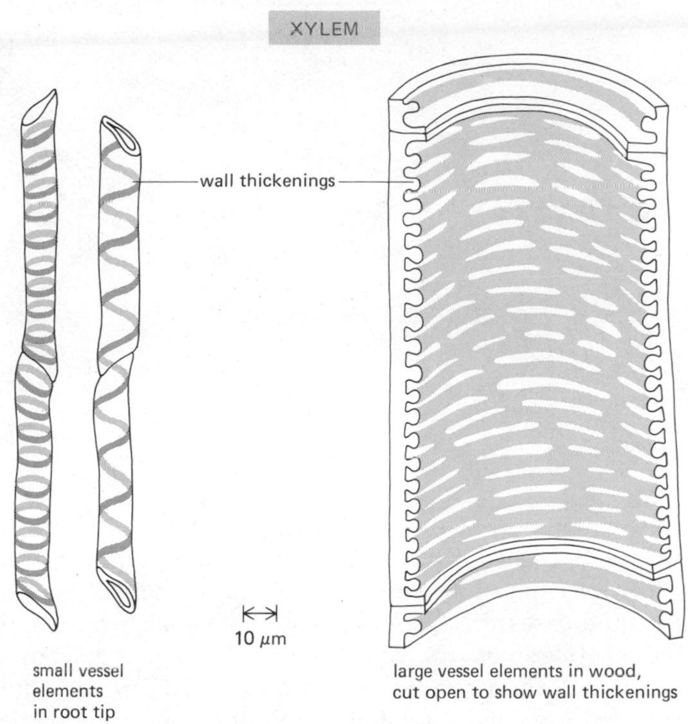

Xylem is a second complex set of cells in the vascular tissue system. The main conducting cells are the vessel elements shown here, which carry water and dissolved ions in the plant. The vessel element is a dead cell at maturity, with a cell wall that has been secondarily thickened and heavily lignified. Its end wall is largely removed, enabling very long continuous tubes to be formed.

SCLERENCHYMA

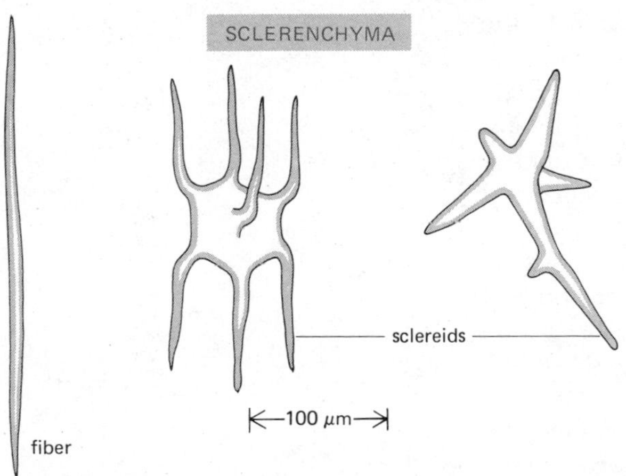

Sclerenchyma are usually dead cells with thick, lignified secondary walls, which have strengthening and supporting functions. Two common types are *fibers* (see Figure 19–12), which are often found in bundles, and *sclereids,* which are shorter branched cells found in seed coats and fruit.

COLLENCHYMA

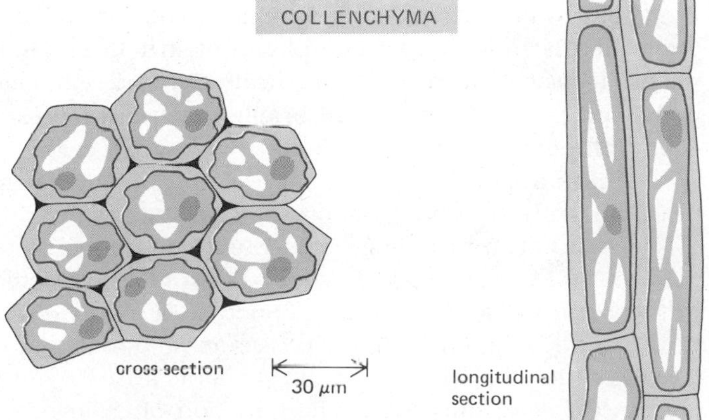

Collenchyma are living cells similar to parenchyma cells, except that they are usually elongated and have unevenly thickened cell walls. They function as support cells in the ground tissue system of the plant.

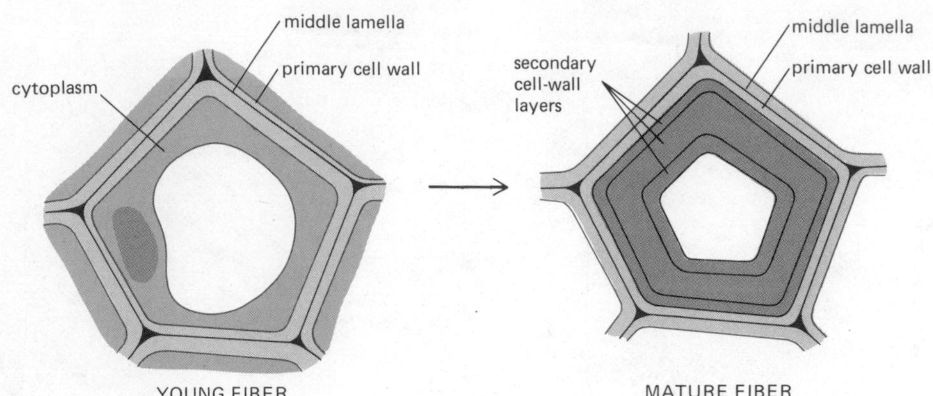

Figure 19–12 Secondary cell-wall deposition shown schematically in a cross-section of a fiber cell. In this case, three new cell-wall layers have been laid down within the primary cell wall. Because the net orientation of the cellulose microfibrils is different in each layer, a strong plywoodlike effect is created. In many mature fibers, as shown here, the cell inside the wall dies.

wall elaboration. Once cell growth stops, there is a relaxation of the constraints that were placed on the composition of the primary wall by the necessity to allow for expansion. This then allows the cell to deposit new wall materials. In some cases this deposition simply consists of new primary cell-wall material, while in others whole new wall layers of different composition are laid down to form a *secondary cell wall.* The form and composition of the final cell wall are closely related to the function of the particular specialized cell type, a fact that makes each cell type readily distinguishable by its morphology. The secondary cell wall is usually deposited between the plasma membrane and the primary cell wall, sometimes in successive layers with different orientations (Figure 19–12). In some cases, however, special macromolecules are deposited within the existing wall (lignin in xylem cells) or on its outer surface (cutin and waxes in epidermal cells). New polymers deposited in the secondary cell wall replace the highly hydrated pectin components characteristic of the primary cell wall, with the result that the secondary wall is considerably more dense and less hydrated than the primary wall. The secondary cell walls provide most of the plant's mechanical support and form the basis for such useful products as wood and paper.

The Wall of the Mature Cell Is Adapted to Its Function[6,7]

The wall of the mature or differentiated plant cell is uniquely suited to the function that the cell performs. The required tailoring is achieved by depositing new wall material and partially removing old material in a precisely controlled fashion so that complex wall structures can be created in the appropriate places. In order to illustrate the variety of cell walls that a mature plant cell can make, we shall briefly discuss three examples of specialized cells—phloem cells, xylem cells, and epidermal cells.

Phloem is a complex tissue responsible for transporting the products of photosynthesis, usually sucrose, from the photosynthetic cells to the rest of the plant. The principal conducting component is the *sieve tube,* a long column of living cylindrical cells that are interconnected by the perforations in their end walls (Figure 19–11). Sucrose is pumped into the cells at the top of the sieve tube, and it then passes in solution from cell to cell down the length of the tube. The sieve tube elements arise from thin-walled *procambial* or *cambial cells,* and their differentiation often involves the deposition of a relatively thick primary cell wall whose main components are simply a greatly increased amount of cellulose and hemicellulose. Numerous pores lined by plasma membrane are formed in the end walls by the localized removal of wall material (Figure 19–13).

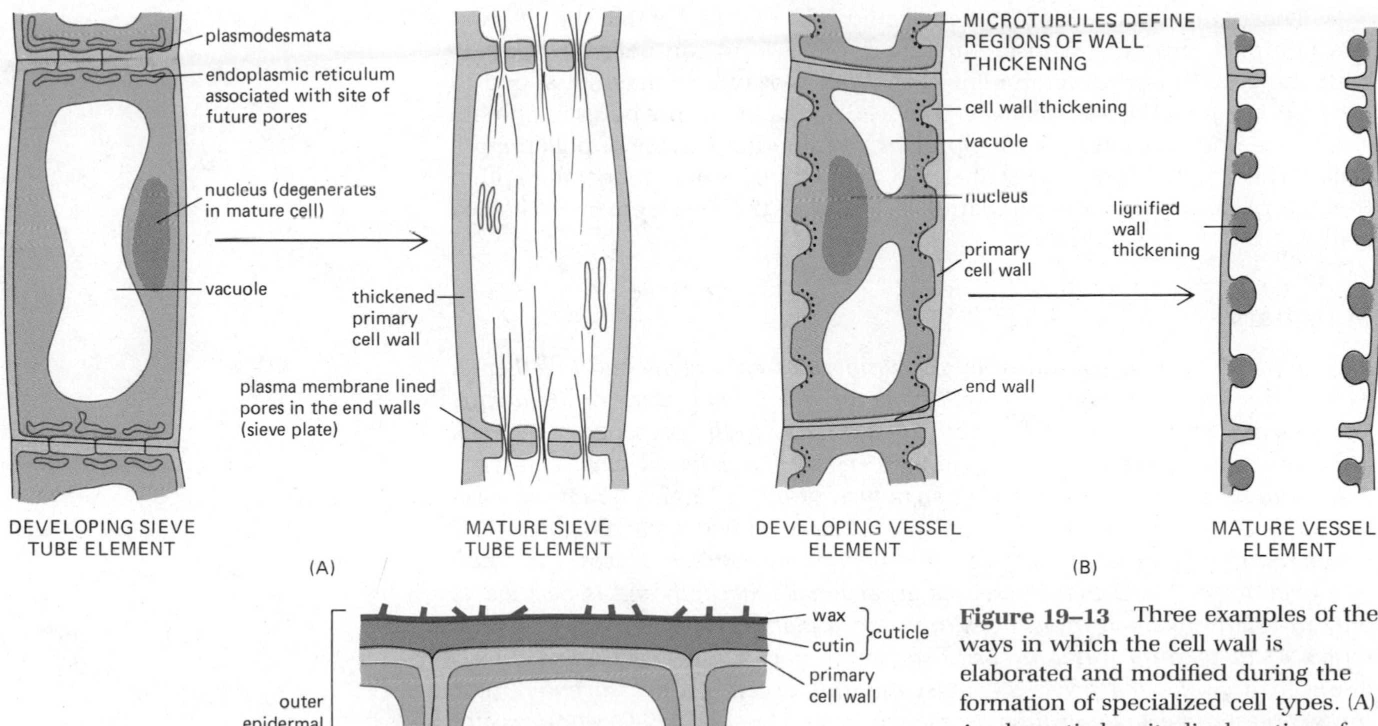

Figure 19–13 Three examples of the ways in which the cell wall is elaborated and modified during the formation of specialized cell types. (A) A schematic longitudinal section of the development of a sieve tube element in phloem. The primary cell wall becomes thickened and the end walls become perforated to form the *sieve plates* that connect adjacent elements of the sieve tube. The mature cell retains its plasma membrane, but the nucleus and much of the cytoplasm are lost. (B) A schematic longitudinal section of a small developing vessel element of the xylem. This cell forms annular wall thickenings, but many other patterns are found. Ultimately the protoplast and end walls disappear to create an open-ended tube. The mature element is dead, having lost its protoplast. (C) Drawing of a section of a typical mature epidermal cell of a leaf. On the outer face of the thick primary cell wall a waterproof cutin layer has been deposited together with a layer of wax to produce a cuticle. The wax layer often gives the cuticle an elaborate sculptured shape.

Xylem is another complex tissue that arises from the division of thin-walled cambial cells and that is composed of tubes. It is responsible for conducting water and dissolved inorganic ions from the roots to the rest of the plant body. The main conducting cells of the xylem are the *vessels* and the *tracheids.* These tubular cells have unusually thick secondary cell walls that are strengthened by high local concentrations of *lignin,* which account for 20% to 30% of the weight of the wall (Figure 19–13). Unlike phloem cells, the cells inside these walls die, leaving only the cell wall behind. During the initial differentiation of the xylem cells in young growing tissues, cellulose thickenings are laid down in patterns defined by groups of microtubules that appear along the plasma membrane. Often, sheets of endoplasmic reticulum membrane are found between the successive bands of microtubules, and these define the areas of the wall that will not be thickened. It is the regions of thickened wall that will later be strengthened by the deposition of lignin, a highly insoluble polymer of aromatic phenolic units that forms an immense cross-linked network within the wall and—in bulk—produces the familiar material wood.

Epidermal cells cover the external surface of the plant. They usually have thick primary cell walls. In addition, during their differentiation they deposit a thick, tough *cuticle* on their outer face that protects the plant from infection, damage, and water loss (Figure 19–13). The cuticle is made primarily of *cutin* (or suberin in bark), which is a polymer of long-chain fatty acids that forms an extensive cross-linked network on the plant surface. The cutin layer is frequently impregnated and overlaid by a complex mixture of *waxes.* Thus the plant cell cuticle is chemically different from insect and crustacean cell cuticles, which are composed of proteins and polysaccharides.

These three examples of differentiated cells emphasize that the cell wall is a complex structure whose composition and form can change markedly with the growth and development of the cell. Not only is material added to the wall, but material is specifically removed (for example, the pores in phloem sieve tube end walls and the sculpturing of the outer surface of pollen grain walls). These reflect a precise spatial and temporal control that operates within the cytoplasm of each differentiating cell during the development of its cell wall.

Summary

Higher plants are composed of large numbers of cells cemented together in fixed positions by the rigid cell walls that surround them. Many of the unique features of plants are related directly or indirectly to the presence of this cell wall, whose composition and appearance reflect the different plant cell types and their function. The underlying structure of all cell walls, however, is remarkably consistent. Tough fibers of cellulose are embedded in a highly cross-linked matrix of polysaccharides, such as pectins and hemicelluloses. The result is a primary cell wall that possesses great tensile strength and is permeable only to relatively small molecules. In water, a plant cell without a wall (a protoplast) will take up water by osmosis, swell, and burst. Inside its cell wall, however, it swells and presses against the wall, creating a pressure known as turgor, rather like an inner tube pressing against a bicycle tire. Turgor pressure is closely regulated and is vital for both cell expansion and the mechanical rigidity of the young plant.

During the formation of specialized cell types—for example, cells of the two conducting tissues, xylem and phloem—the cell wall is modified. Greatly strengthened regions of the wall may be formed, which can include the addition of one or more new layers (the secondary cell wall), and regions may be selectively removed, as in the formation of a conducting tube from a long row of cylindrical cells.

The Interactions and Communication Between Cells

The previous section has shown that rigid cell walls pose very special problems for the general growth and development of plant cells. These same cell walls also severely limit the ways in which plant cells, immobilized within the plant, can interact and communicate, not only with each other but also with their environment. As one example, a plant does not have a nervous system to enable one part to communicate rapidly with another. In this section we examine the special problems of intercellular communication within plants and look at some of the ways that plants have evolved to solve them.

Plant Cells Are Connected to Their Neighbors by Special Cytoplasmic Channels Called Plasmodesmata[8]

Except for a very few specialized cell types, every living cell in a higher plant is connected to its living neighbors by fine cytoplasmic channels, each of which is called a **plasmodesma** (plural, **plasmodesmata**), which pass through the intervening cell walls. As shown in Figure 19–14, the plasma membrane of one cell is continuous with that of its neighbor at each plasmodesma. A plasmodesma is a roughly cylindrical, membrane-lined channel with a diameter of 20 nm to 40 nm. Running from cell to cell through the center of

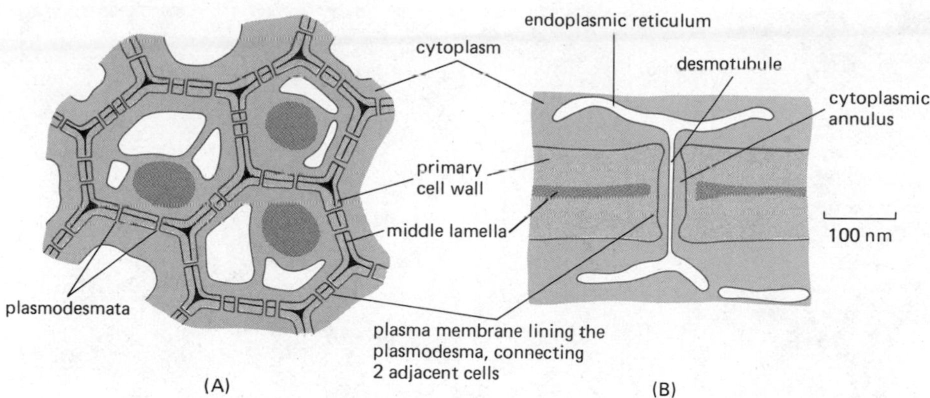

Figure 19–14 (A) Cytoplasmic channels called plasmodesmata pierce the cell wall and connect all cells in the plant together. (B) The plasmodesma is lined with plasma membrane common to two connected cells, and usually contains a fine tubular structure, the desmotubule, derived from endoplasmic reticulum.

most plasmodesmata is a narrower cylindrical structure, the *desmotubule,* which electron micrographs show to be continuous with elements of the endoplasmic reticulum membrane of each of the connected cells (Figure 19–15). Between the outside of the desmotubule and the inner face of the cylindrical plasma membrane is an annulus of cytosol (Figure 19–14). The annulus often appears to be constricted at each end of the plasmodesma. These constrictions may be of great significance, for they are located at sites where each cell could, in principle, regulate the flux of molecules through the annulus that joins the two cytosols. Collars of specialized wall material have been detected around these constrictions, giving rise to speculations about valvelike controls analogous to sphincter muscles.

The great majority of plasmodesmata are formed at the time of cell division, when the new cell wall that will bisect the parental cell is being constructed (see Figure 11–65). Although some alterations may be made later, the number and distribution of plasmodesmata across the cell wall are largely determined at this time.

Plasmodesmata Allow Molecules to Pass Directly from Cell to Cell[8,9]

Apart from their suggestive structure, what evidence is there that plasmodesmata function in intercellular communication? Some circumstantial evidence is that plasmodesmata are especially frequent in the walls of columns of cells that lead toward sites of intense secretion, such as in nectar-secreting glands. In such cells, there may be 15 or more plasmodesmata per square micrometer of wall surface, whereas there is often less than 1 per square micrometer in other sites. Other evidence includes the fact that the measured rates of solute movement from plant cell to plant cell are faster than can be accounted for by plasma membrane permeability alone.

The most direct evidence for intercellular transport via plasmodesmata comes from experiments involving intracellular injections of dyes or pulses of electric current. For example, procion dyes do not easily cross the plasma membrane; yet if introduced through a fine capillary into one cell of an *Elodea* leaf, they readily pass into neighboring cells. Similarly, if pulses of electrical current are applied to the interior of one cell, receiver electrodes in adjacent cells detect the same pulses, albeit in attenuated form. The degree of attenuation is found to vary with the density of plasmodesmata and the number of cells between the injection and receiver electrodes (Figure 19–16). Moreover, since a receiver electrode placed outside the plasma membrane of the injected cell does not detect these pulses, they must have followed a pathway that bypasses the high electrical resistance of this membrane.

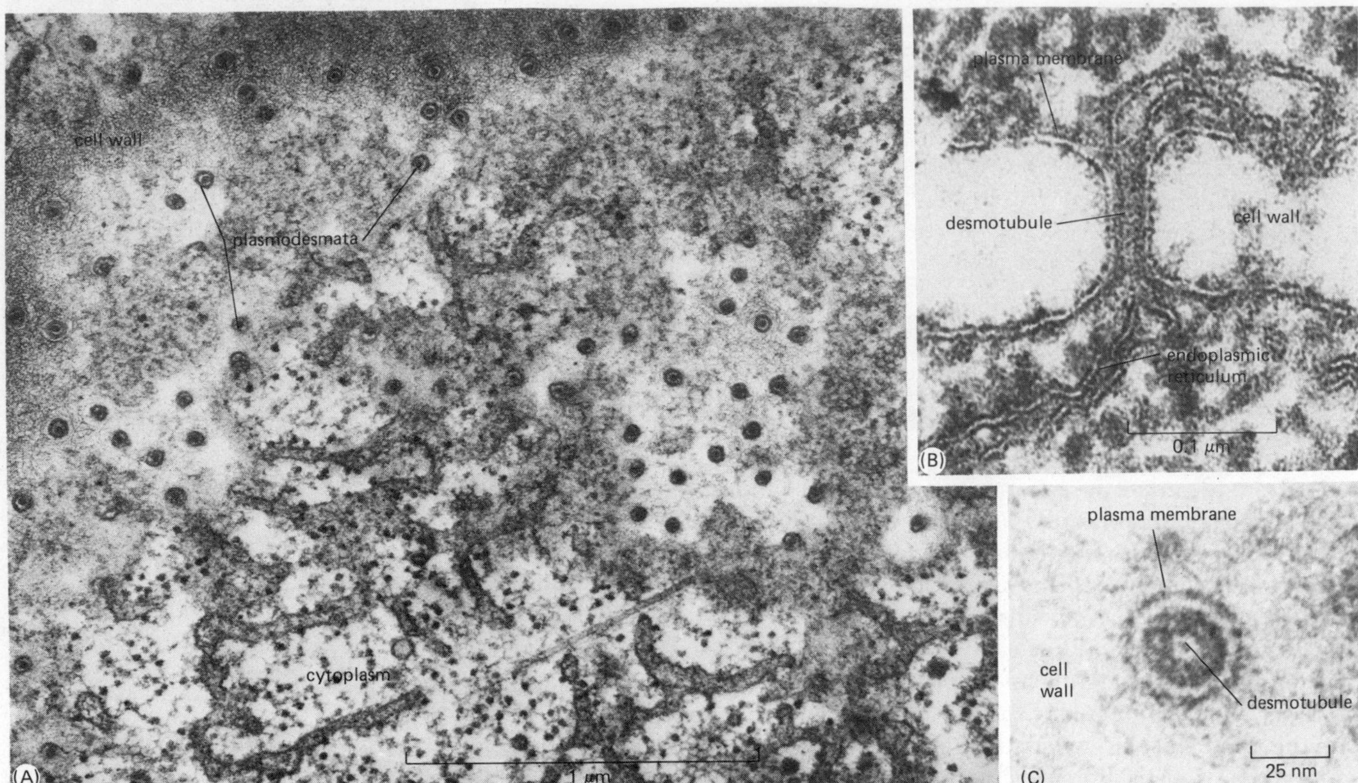

Figure 19–15 Plasmodesmata as seen in the electron microscope. (A) A glancing section of a cell wall from an *Abutilon* flower. The wall is pierced by numerous plasmodesmata, which function in this case in the transport of nectar between cells. (B) Vertical section of a plasmodesma from a water fern. The plasma membrane lines the pore and is continuous from one cell to the next. Endoplasmic reticulum and its association with the central desmotubule can be seen. (C) A similar plasmodesma in cross-section. (A, courtesy of B. Gunning and J. E. Hughes, from *Aust. J. Plant Physiol.* 3:619–637, 1976; B and C, courtesy of R. Overall.)

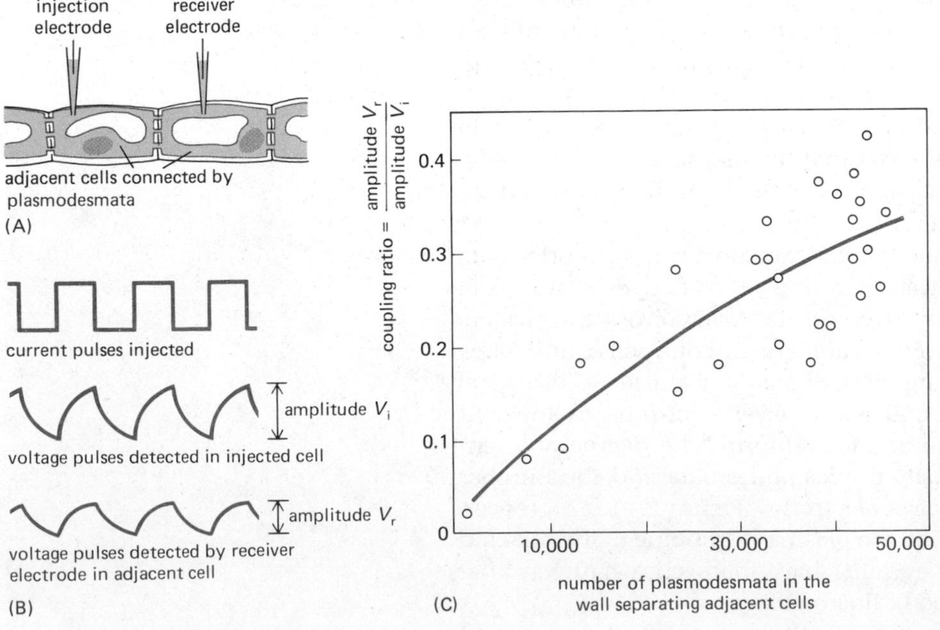

Figure 19–16 An experiment that demonstrates plasmodesmatal conductivity. Electrodes are placed in two adjacent cells of a water fern (A) and current pulses are injected into one. The amplitude of the voltage changes detected in the injected cell and the adjacent cell are then measured (B), and the attenuation of the pulse, or *coupling ratio,* is plotted against the total number of plasmodesmata that connect the two cells (C). Electrical coupling between cells varies with the number of plasmodesmata. (Data from R. L. Overall and B. Gunning, *Protoplasma* 111:151–160, 1982.)

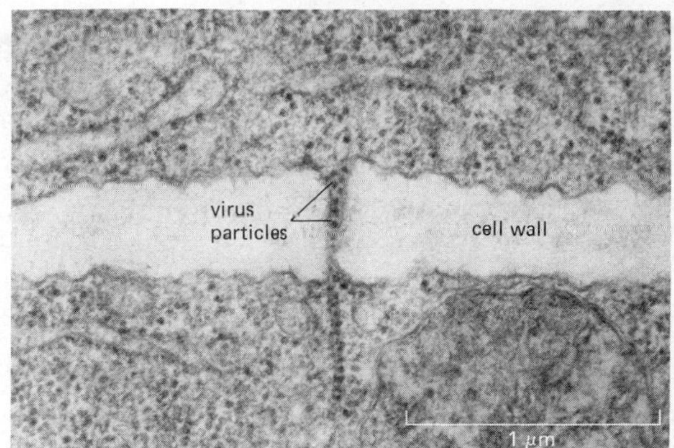

Figure 19–17 An electron micrograph showing a small spherical plant virus passing from one cell to another through a plasmodesma. (Courtesy of K. Plaskitt.)

The weight of the evidence therefore strongly suggests that plasmodesmata mediate transport between adjacent plant cells, much as gap junctions appear to mediate transport between adjacent animal cells (see p. 686). Although the lumen of a plasmodesma is at least ten times wider than that of a gap junction, it is unlikely to allow a free exchange of macromolecules between adjacent cells: in many cases, despite being connected by plasmodesmata, neighboring cells can differentiate in very different ways and can have very different internal solute concentrations. Plasmodesmatal transport between plant cells is thus thought to be selectively controlled. The experimental evidence suggests that molecules with a molecular weight larger than 800 cannot pass freely through a plasmodesma. Certain plant viruses, however, appear to be able to overcome these controls and enlarge plasmodesmata in order to use this route to pass from cell to cell (Figure 19–17).

The Fluids in a Plant Are Segregated into One Large Intracellular Compartment and One Large Extracellular Compartment

Plasmodesmata convert a plant from being just a collection of individual cells into being a large interconnected commune of living protoplasts. It follows that the whole plant body can be viewed as consisting of two compartments: (1) an intracellular compartment, known as the **symplast,** that is made up of the total protoplast commune—including the phloem tubes—bounded by the combined plasma membranes of all these living cells; and (2) an extracellular compartment, or **apoplast,** comprising all of the cell walls, the empty dead cells of the xylem tubes, and the water contained in both (Figure 19–18). Both of these compartments have their own internal transport processes, can be locally sealed off, and are subject to local modifications that regulate the flow of fluids and materials between them.

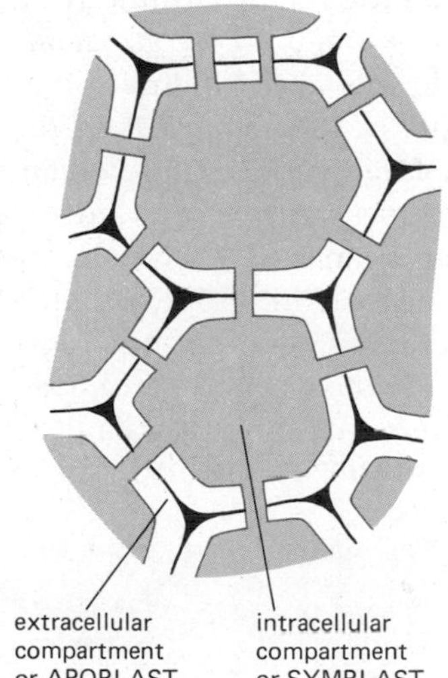

Figure 19–18 Highly schematic diagram of a group of plant cells interconnected by plasmodesmata that are lined by plasma membrane. This membrane divides the plant into an extracellular space (apoplast) and an intracellular space (symplast). For clarity, no organelles are shown.

Fluids Are Transported Throughout the Plant by Xylem and by Phloem[10]

The spectrum of complexity in the plant kingdom extends from unicellular organisms to large flowering plants containing up to 10^{13} cells—about as many as in an adult human. As in animals, multicellularity permits a division of labor, with different cell types supporting one another by virtue of the specialized functions that emerge during their differentiation.

Two of the major functions of plants are carried out in **photosynthetic cells,** which contain chloroplasts and are sources of assimilated carbon-containing compounds, and **absorptive cells,** which take up mineral nutrients

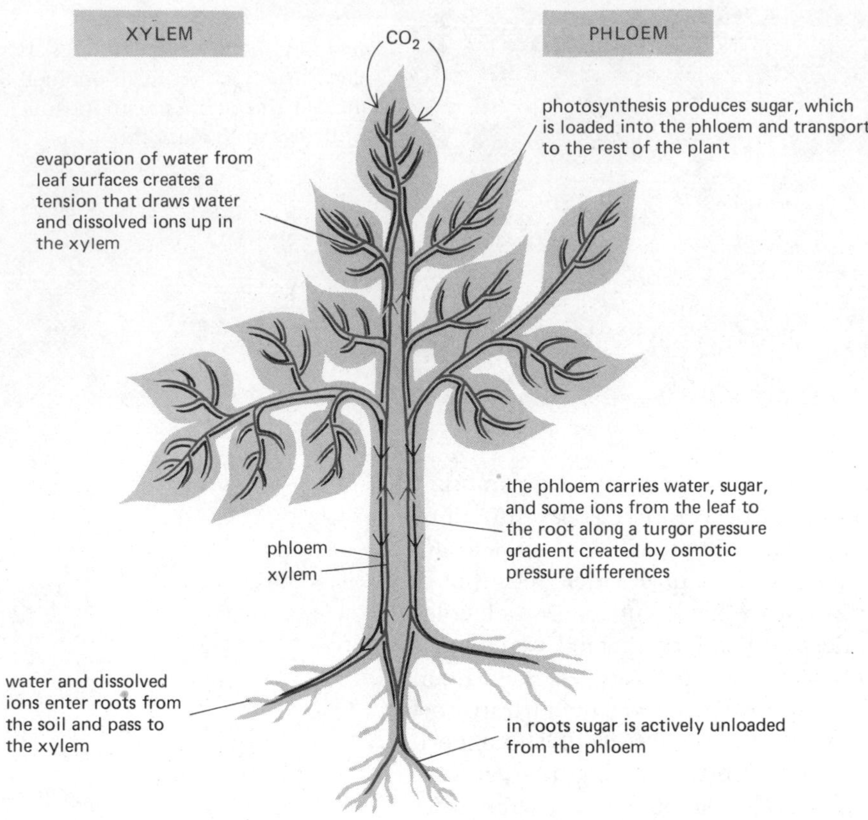

Figure 19–19 Diagram showing the two major routes, the xylem and the phloem, by which water and solutes are transported throughout the plant. These routes have been highly simplified; for example, an extensive lateral water exchange that occurs between the xylem and the phloem is not shown.

from the environment. In the majority of higher plants, these two functions cannot be carried out by the same cells because one requires light and the other occurs below the soil surface in darkness. Each process also has other special requirements. Photosynthesis requires a special microenvironment in which the relative humidity and carbon dioxide supply are closely controlled. This is achieved by *stomatal pores,* which are located in a cuticle-covered epidermis and which are opened and closed by the turgor-regulated movements of their guard cells (Figure 19–10). In contrast, absorption requires a very large surface area, which is provided by the roots, as well as membrane transport systems that are often augmented by symbiotic associations with microorganisms. The photosynthetic and absorptive cells cross-feed each other, in addition to supplying other regions of the plant with both the minerals and organic compounds required for biosynthesis (Figure 19–19).

Photosynthetic and absorptive tissues are linked by the xylem and the phloem that form vascular networks for long-distance transport (Figures 19–19 and 19–20). Transport in the xylem is unidirectional, toward surfaces of evap-

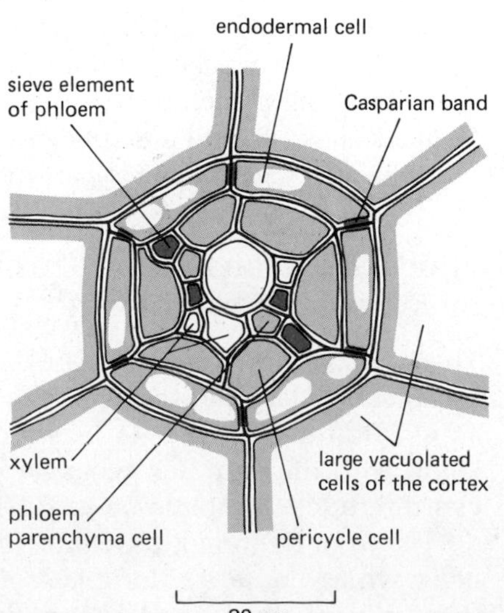

Figure 19–20 The vascular network of a simple plant—the water fern *Azolla*—depicted in cross-section through the center of the root. While this example shows an extremely simple, but defined, arrangement of conducting cells, with four xylem and four phloem sieve tube elements, most higher plants have much more complex arrangements of cells in their vascular tissue. Note that the xylem and phloem are surrounded by endodermal cells, and that the strategically placed Casparian bands in the endodermal cell walls prevent water leakage outward through the apoplast. Details of this modified cell wall are shown in Figure 19–21.

oration. The solutes carried in the xylem fluid, mainly mineral salts and nitrogen-containing compounds, are probably secreted into it by specialized parenchyma cells in the roots. Back leakage through the apoplast near the site of xylem loading is inhibited by apoplastic seals (*Casparian bands*), which are analogous in function to tight junctions between animal epithelial cells (Figure 19–21). At the other end of the system, other types of parenchyma cells equipped with specific membrane transport proteins pump solutes from the xylem fluid into receiver cells in the photosynthetic tissues. Most of the solvent (water) percolates on through the xylem and eventually evaporates, primarily from the surface of the photosynthetic tissue in the leaf. (For a description of some of the various types of cells involved, see Figure 19–11.)

Transport in the phloem is more complex and is not limited to one direction: it takes solutes, mainly sucrose, from various sources of production to sites of consumption and storage, irrespective of location. Again, there are different specialized cells in the system for loading and unloading the solutes.

In regions of extensive transport, animal cells greatly increase the total area of their plasma membrane by extending outwardly directed protuberances called microvilli (p. 582). In plants the cell wall precludes such a mechanism, and plant cells are forced to use an alternative strategy. Instead, specialized *transfer cells* increase their surface area by infoldings of wall material, which are lined by the plasma membrane (Figure 19–22). These transfer cells occur in many sites in the plant body where the rates of transport across the plasma membrane are especially high, for example in those regions—such as leaf veins—where sucrose is pumped into the phloem (Figure 19–23) and in regions where solutes are pumped from the xylem into the tissues.

It is worth noting that the transport of fluid from one part of a plant to another differs in two important respects from fluid transport in animals. First, animals have only one transport system—the blood system—while plants have two distinct systems—the phloem and the xylem. Second, in plants, fluid does not circulate the way that blood does in an animal; instead, there is a continuous net movement of water from the roots to the leaves.

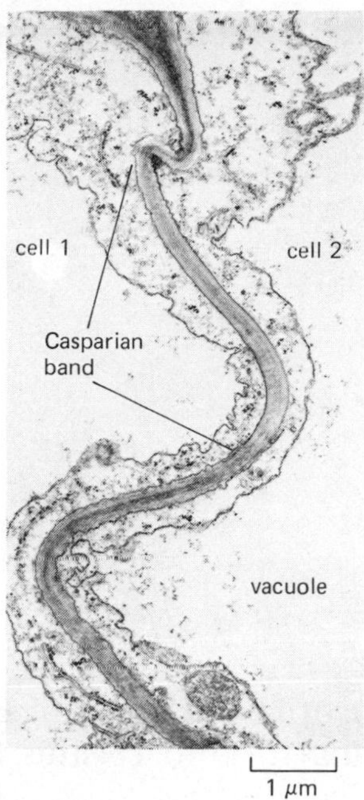

Figure 19–21 The Casparian band from a root endodermal cell similar to that shown in Figure 19–20. Compared with a normal primary cell wall, the band has a smooth texture, the result of heavy impregnation with water-resistant cutin that makes this specialized wall impermeable to water. (From B. Gunning and M. Steer, Ultrastructure and the Biology of Plant Cells. London: Arnold, 1975.)

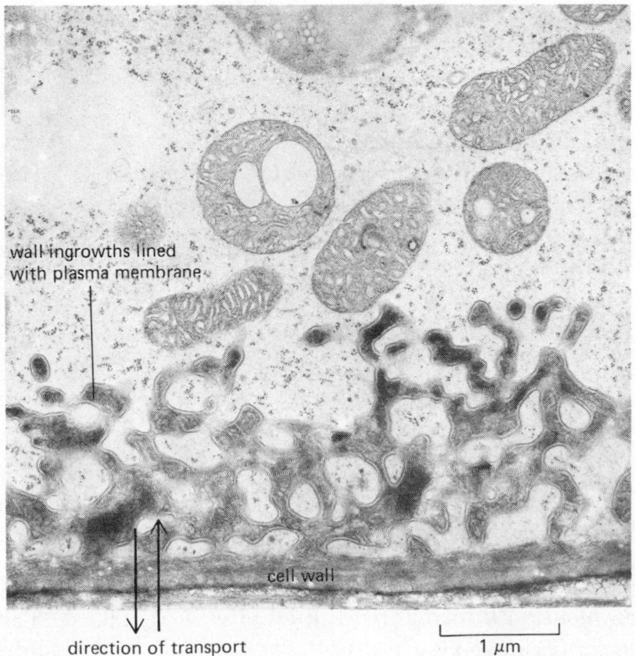

Figure 19–22 An electron micrograph of a transfer cell involved in the transport of material from the leafy part of the moss *Funaria* to the sporophyte. The surface area of the plasma membrane is vastly increased by the elaborate ingrowths of the cell wall. (Courtesy of A. J. Browning, from A. J. Browning and B. E. S. Gunning, *Protoplasma* 93:7–26, 1977.)

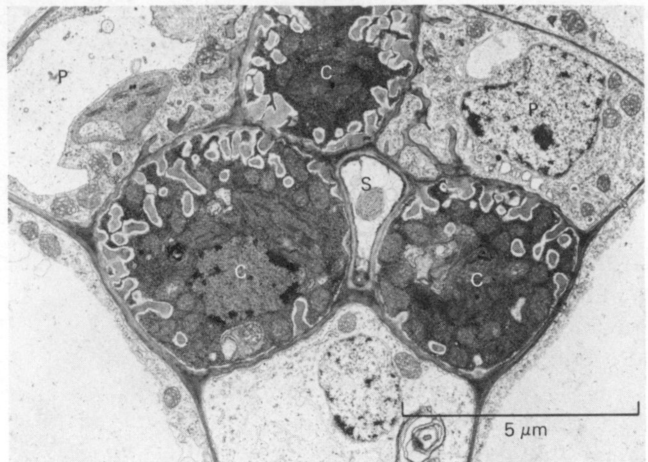

Figure 19–23 Transfer cells in a leaf vein. A phloem sieve tube element (S) is shown surrounded by phloem companion cells (C) and phloem parenchyma cells (P). Note the extensive wall ingrowths characteristic of transfer cells. (Courtesy of B. Gunning.)

Chloroplasts Use Light Energy to Fix CO_2, Enabling the Photosynthetic Tissues to Export Sucrose

The green tissues of a plant derive their color from high concentrations of the photosynthetic pigment **chlorophyll,** which is located in chloroplasts. Light absorption by chlorophyll molecules begins an electron-transport process in the chloroplast that drives the pumping of protons across its thylakoid membranes and leads to the production of biologically useful energy (as ATP) and reducing power (as NADPH). The ATP and NADPH produced are in turn used by the chloroplast to convert atmospheric CO_2 into sugars (see Chapter 9). It is the synthetic activity of their chloroplasts that enables the photosynthetic tissues to export large amounts of organic carbon to the cells in the rest of the plant. Most of this carbon is exported as the disaccharide sucrose; as a result, the fluid in the phloem tubes (the phloem sap) typically contains 10% to 25% sucrose by weight.

Symbiotic Bacteria Allow Some Plants to Use Atmospheric Nitrogen[11]

With one exception, all the mineral nutrients taken up by the roots and transported through the plant by the xylem are present in the soil, derived mainly from weathered rocks. The one exception is nitrogen. All of the nitrogen in living organisms ultimately derives from atmospheric nitrogen that has been incorporated (fixed) into organic compounds by processes that require a good deal of energy (hence the high cost of artificially produced nitrogen fertilizers). The only organisms that can fix atmospheric nitrogen are procaryotes (certain of the eubacteria, and cyanobacteria). While some of these are free-living soil organisms, others—for example, the bacterium *Rhizobium*—form symbiotic associations with the roots of certain plants, such as the *legumes*—peas, beans, and clover.

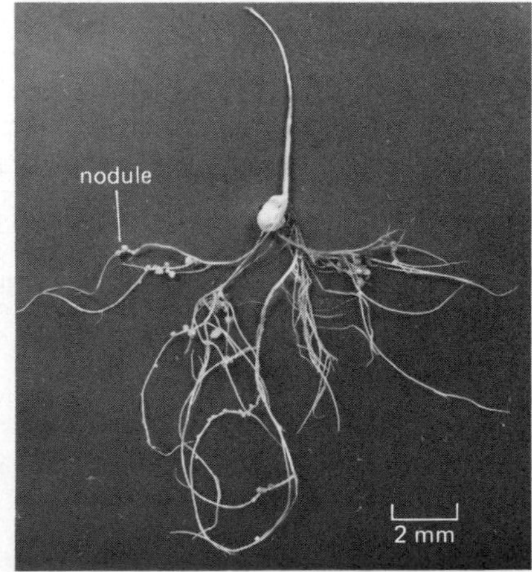

Figure 19–24 A young pea plant in symbiotic association with the nitrogen-fixing bacterium *Rhizobium.* The root nodules containing the bacteria are clearly visible. (Courtesy of A. Johnston.)

For such a symbiotic association to occur, the bacteria and the root hairs of the host plant must first recognize each other. There is some evidence that this recognition event (along with many other such host-pathogen interactions—see below) may involve a host *lectin,* a protein that recognizes a species-specific carbohydrate component on the surface of the bacterium (see p. 1118). Whatever the mechanisms of recognition, however, the specific binding of bacteria to the root triggers a complex chain of events, as a result of which the bacteria enter the root cells via an *infection thread* and stimulate the host cortical cells to divide and form a large *root nodule* (Figure 19–24). Each nodule

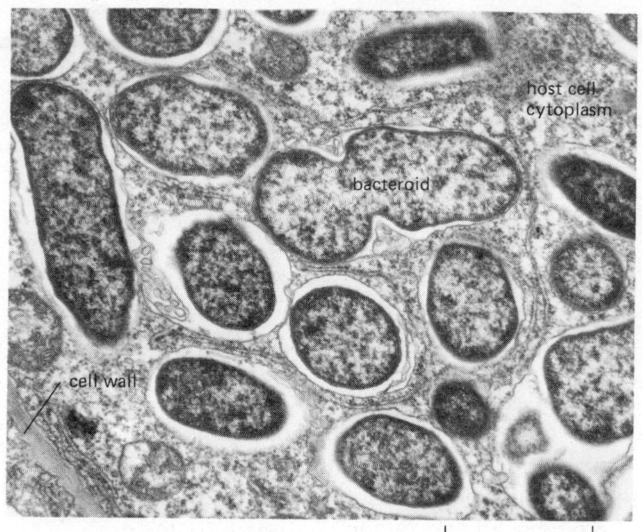

Figure 19–25 Electron micrograph of a thin section through a pea root nodule like those shown in Figure 19–24. The nitrogen-fixing *Rhizobium* bacteroids, surrounded by host-cell-derived membrane, fill the host cell cytoplasm. (Courtesy of B. Huang and Q. S. Ma.)

contains about half its weight in intracellular bacteria, which have lost most of their own cell wall. The plasma membrane of each of these bacteria remains and is surrounded in turn by a host-cell-derived membrane (Figure 19–25). It is these altered bacteria, called *bacteroids,* that fix the nitrogen eventually used by the plant.

The bacterial enzyme catalyzing the fixation of nitrogen is a complex protein molecule called *nitrogenase.* In symbiotic *Rhizobia,* this enzyme catalyzes the conversion of atmospheric nitrogen to ammonia, which is released into the cytoplasm of the host cell, where it is converted to glutamine, glutamate, and eventually all the other amino acids. Genetic analyses reveal that the successful symbiotic fixation of nitrogen requires the coordinated expression of a large number of separate genes in the bacterium and many genes in the host plant. Most of the *Rhizobium* genes involved in nitrogen fixation, called *nif genes,* are clustered together on a plasmid carried by this bacterium.

The bacterial nitrogenase is inactivated by free oxygen, and the enzyme therefore has an extremely short half-life in air. This means that elaborate mechanisms have had to evolve to ensure that the bacteroids in the root are kept in an anaerobic environment while the root itself maintains an adequate oxygen supply. As part of this mechanism, the *Rhizobium* induces the host cells to produce *leghemoglobin,* an oxygen-binding molecule analogous to mammalian myoglobin, which surrounds the bacteria and helps to restrict their oxygen supply.

Nitrogen fixation consumes a great deal of energy, supplied ultimately by the sun, through photosynthesis. It is estimated that the fixation of 1 molecule of nitrogen (N_2) by *Rhizobium* requires between 25 and 35 ATP molecules.

Cell-Cell Recognition in Plants Involves Specific Sequences of Sugar Residues[12]

The recognition of *Rhizobium* by legume roots is just one example of a specific interaction between a plant cell and another cell. In some cases the other cell may be from a different species, for example, a fungal pathogen, while in other cases it may be from the same species, for example, a pollen grain landing on the stigma of a flower. There is now a persuasive body of evidence to suggest that such recognition events depend for their specificity on particular sequences of sugars contained in cell-surface polysaccharides, glycoproteins,

Table 19–1 Commonly Used, Commercially Available Lectins and the Specific Sugar Residues That They Recognize

Lectin	**Sugar Specificity**
Concanavalin A (from Jack beans)	α-D-glucose, and α-D-mannose
Soybean lectin	D-galactose and *N*-acetyl-D-galactosamine
Wheat germ lectin	*N*-acetylglucosamine
Lotus seed lectin	fucose
Potato lectin	*N*-acetylglucosamine

and glycolipid molecules. In some cases the sugar sequence itself can substitute for one of the two cells in eliciting a characteristic sequence of reactions in the other cell. For example, cell wall oligosaccharides from some plant pathogens can by themselves elicit the typical defense reactions of the host plant. Although there are no instances where all the molecules involved in any one plant recognition system have been isolated and characterized, it is generally assumed that many of the molecules responsible for recognizing the sugar sequences at the cell surface are lectins.

Lectins are proteins or glycoproteins with two or more binding sites that recognize a specific sequence of sugar residues. They were originally isolated from plants, where they are found in large quantities in many seeds, but they have subsequently been found in all types of organisms. Many plant-seed lectins are highly toxic storage proteins that serve to deter animals from eating the seeds, while others seem to be involved in cell-cell recognition. Since lectins bind to cell-surface glycoproteins or glycolipids, they are widely used as biochemical tools in cell biology to localize and isolate these sugar-containing plasma membrane molecules. Some commonly used plant lectins and their sugar specificities are listed in Table 19–1.

The specific interaction of a pollen grain with the appropriate stigma is a well-studied example of lectin function. This interaction triggers the stigma cell to release water, thereby hydrating the pollen grain and inducing it to extend the long pollen tube required for fertilization (Figure 19–26). About half of all known flowering plants have a genetically specified mechanism to prevent self-pollination, thereby assuring outbreeding. In the cabbage family, for example, the molecular components of the recognition system are encoded by the *S* gene complex: the components include a large glycoprotein on the sticky surface of the stigma and a lectin that recognizes it on the surface of the pollen grain. A pollen grain will germinate on a stigma, grow, and fertilize an egg cell of a plant only if the particular plant expresses different alleles in its *S* gene complex. If a pollen grain is pretreated with the purified glycoprotein from a stigma expressing the same set of alleles, it is prevented from germinating on a normally compatible stigma. Presumably, therefore, the interaction of the pollen grain lectin with its self-glycoprotein triggers a response in the pollen grain that actively prevents germination, thus ensuring the self-incompatibility. Specific lectins and their endogenous receptors are presently being isolated and studied in a variety of plant cell recognition systems.

Summary

The presence of the tough, relatively impermeable cell wall profoundly influences the ways in which plant cells can interact and communicate, not only with each other but also with their environment. All of the living cells in a plant are connected to each other by plasmodesmata—small, regulated cytoplasmic

Figure 19–26 Pollen grains of *Brassica* on the surface of the stigma. This scanning electron micrograph shows the distinctively sculptured surface of the pollen grains together with the growing pollen tubes of those grains that have germinated. (Courtesy of H. G. Dickinson.)

channels, lined by plasma membrane, that cross the cell wall and through which many solutes are transported. All of the living protoplasts of the plant body are thereby linked to form an interconnected system called the symplast. The remaining wall space, through which most water transport takes place, is called the apoplast. The photosynthetic cells of the plants produce sugars that are transported to the rest of the plant through the living cells of the phloem, part of the symplast. The cells of the root absorb water and dissolved minerals from the soil and transport them to the leaves through the dead cells of the xylem, part of the apoplast. Nearly all of the nitrogen found in the molecules of living organisms derives ultimately from atmospheric nitrogen fixed by procaryotes, many of which can form complex symbiotic associations with plant roots. Plant cell recognition phenomena—including symbiotic plant-bacteria interactions, plant-pathogen interactions, and the specific pollination of flowers—are thought to involve the recognition of molecules containing specific sugar sequences. Lectins, which are naturally occurring proteins that recognize sugar residues, are thought to be involved in the recognition event.

The Internal Organization of Plant Cells

The cells of higher plants contain all the intracellular compartments previously described for animal cells, including the cytosol, Golgi apparatus, endoplasmic reticulum, nucleus, mitochondria, peroxisomes, and lysosomes. Similarly, plant cells have cytoskeletons composed of actin filaments and microtubules that are indistinguishable from those found in animal cells. However, plant cells can be distinguished from those of animals by the presence of two types of membrane-bounded compartment—the *vacuole* and the *plastids.* Both organelles are related to the immobile life-style of plant cells. These and other special features of the interior of plant cells are the subject of this section.

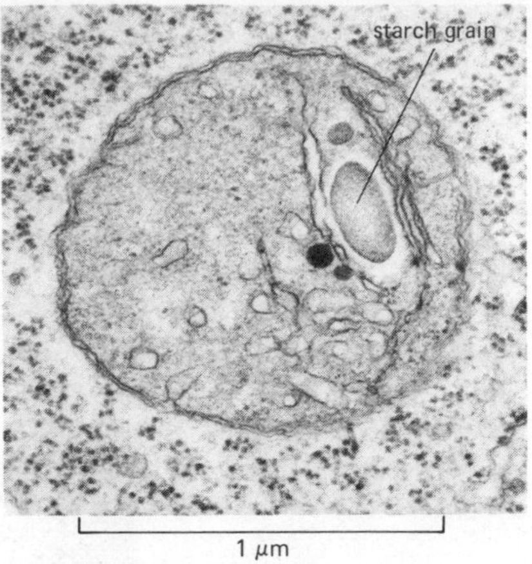

Figure 19–27 Electron micrograph of a typical proplastid from a root-tip cell of a bean plant. It is surrounded by the plastid envelope, which consists of two membranes; the inner membrane gives rise to the sparse internal membrane system. (From B. Gunning and M. Steer, Ultrastructure and the Biology of Plant Cells. London: Arnold, 1975.)

The Chloroplast Is One Member of a Family of Organelles, the Plastids, Unique to Plants[13]

Because their cell walls prevent plants from feeding in the same way that animals do, plant cells are dependent upon their own organic nutrients produced by photosynthesis. In green plants photosynthesis takes place in **chloroplasts,** which serve as a permanent internal food source (p. 510). The products of photosynthesis can be used directly by the cell for biosyntheses, stored as an osmotically inert polysaccharide (usually starch), or converted to a low molecular weight sugar (usually sucrose) that is exported to meet the metabolic needs of other tissues in the plant, such as the roots.

Chloroplasts are but one member of a family of closely related organelles, the **plastids.** All plastids share certain features—most notably, they have their own small genome and are enclosed by an envelope composed of a double membrane. Since the structure and function of the chloroplast was discussed in detail in Chapter 9, the focus here will be on other members of the family.

All plastids, including chloroplasts, develop from **proplastids,** which are relatively small organelles present in *meristematic* cells (Figure 19–27). Proplastids develop according to the requirements of each differentiated cell. If the leaf is grown in darkness, they develop into **etioplasts.** These have a semicrystalline array of internal membranes that contain *protochlorophyll* (a yellow chlorophyll precursor) instead of chlorophyll (Figure 19–28). If exposed to light, the etioplasts develop into chloroplasts by converting protochlorophyll

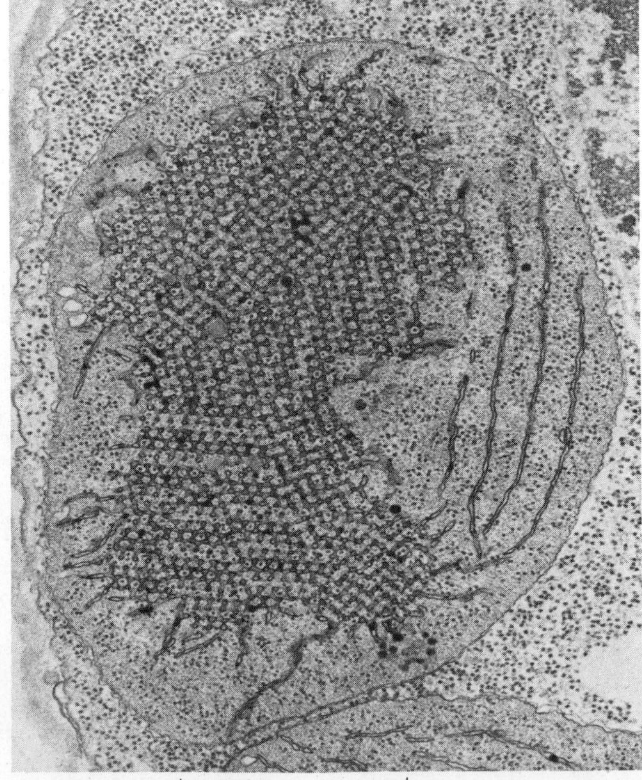

Figure 19–28 Electron micrograph of an etioplast from an oat seedling grown in the dark. The semicrystalline array of internal membranes contains protochlorophyll. (Courtesy of B. Gunning.)

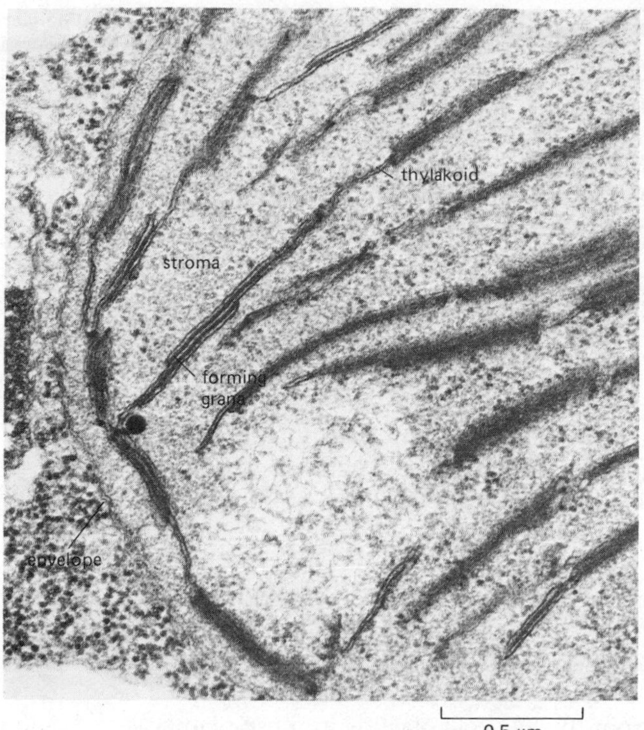

Figure 19–29 An electron micrograph of a portion of a young greening chloroplast from an oat seedling. The rearrangement of the etioplast membranes is underway: chlorophyll is now present, and small grana are beginning to form. (Courtesy of B. Gunning.)

to chlorophyll and synthesizing new membrane, pigments, photosynthetic enzymes, and components of the electron-transport chain (Figure 19–29).

Plastids of one form or another are present in all living plant cells. Other forms of plastid are **chromoplasts** (Figure 19–30), which accumulate carotenoid pigments and are responsible for the yellow-orange-red coloration of petals and fruits in many species, and **leucoplasts,** which are little more than enlarged proplastids and occur in many epidermal and internal tissues that do not become green and photosynthetic. A common form of leucoplast is the *amyloplast* (Figure 19–31), which stores starch in storage tissues and, in certain cells of the stems, leaves, and roots, functions in a plant's responses to gravity.

It is unclear precisely how the choice of developmental pathway from the proplastid to the different forms of plastid is regulated, but it is clear that

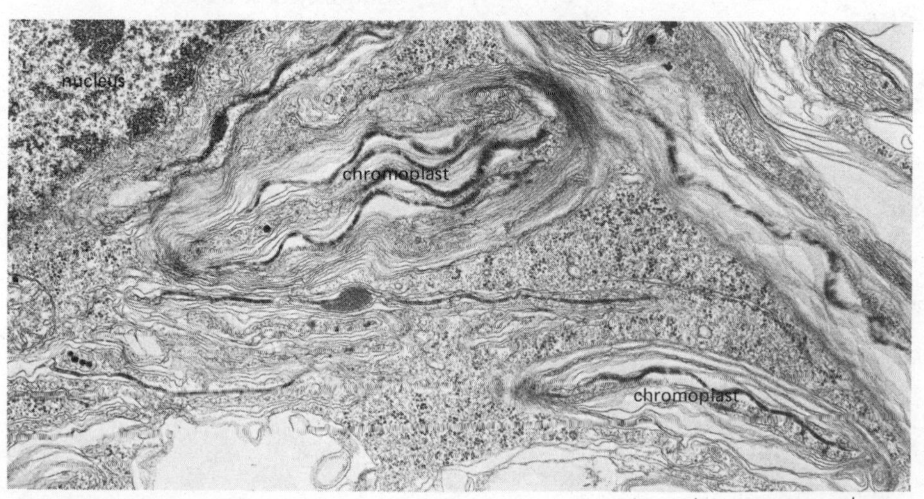

Figure 19–30 Pigment-containing chromoplasts from a cell in the yellow-orange petals of a daffodil. The outlines of these plastids are convoluted, and dissolved in their chaotic internal membranes is the pigment β-carotene, which gives the petals their color. (Courtesy of B. Gunning and M. Steer, Ultrastructure and the Biology of Plant Cells. London: Arnold, 1975.)

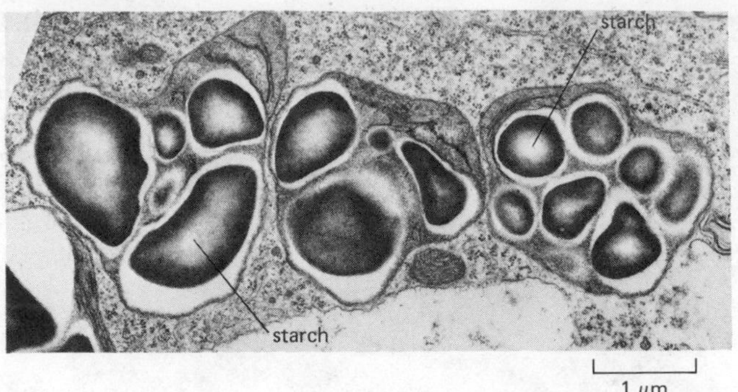

Figure 19–31 Three amyloplasts, or starch-storing plastids, in a root-tip cell of soybean. (Courtesy of B. Gunning.)

the nuclear genome exerts a large measure of control. Nuclear mutations can switch development between chromoplast and chloroplast pathways, or they can block development, giving rise either to the various forms of leucoplast or to the immature chloroplasts with abnormal pigmentation that are found in many ornamental plants.

All plastids contain multiple copies of the plastid genome, and most, if not all, are capable of division within the cell (see Chapter 9, p. 530). The only type of cell in higher plants that loses its population of plastids is the male sperm cell in certain species. As a consequence, such plants (for example, maize) acquire their plastids solely from the egg cell: plastids in these plants, like mitochondria in animals, are thus maternally inherited.

The Plant Cell Vacuole Is a Remarkably Versatile Organelle[14]

The most conspicuous compartment in most plant cells consists of one or more very large vesicles, collectively called **vacuoles** (Figure 19–32), which are separated from the cytoplasm by a single membrane called the **tonoplast.** A vacuole generally occupies more than 50% of the cell volume, but the range of variation may be from 5% to 95%, depending on the cell type. Plant cells

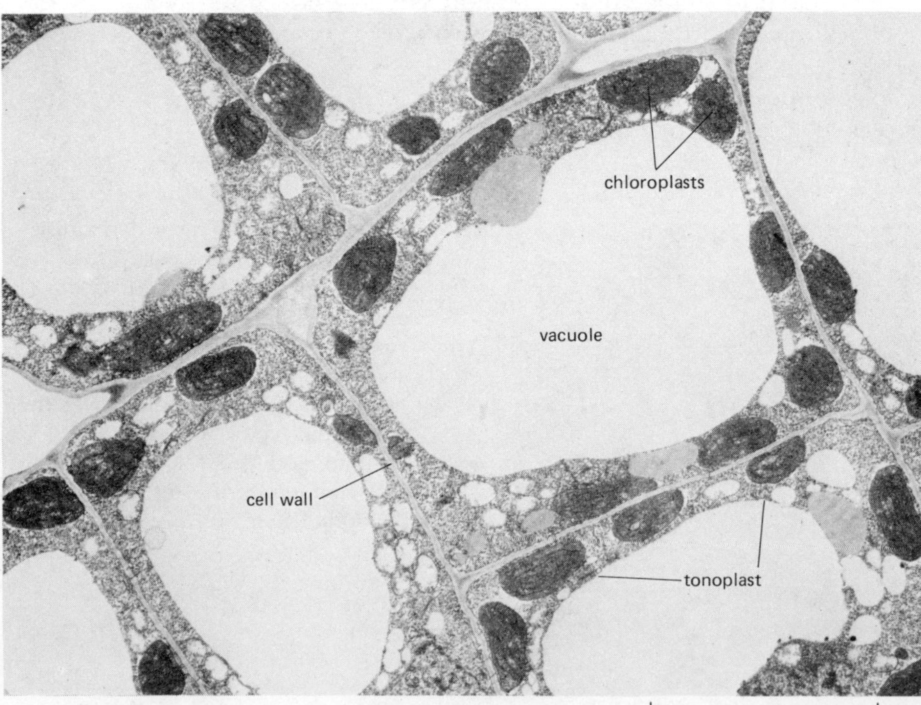

Figure 19–32 Electron micrograph of cells in a young tobacco leaf. The cytoplasm in these highly vacuolated cells is confined to a thin layer, containing numerous chloroplasts, pressed against the cell wall. (Courtesy of J. Burgess.)

use their vacuoles for transporting and storing nutrients, metabolites, and waste products. In a sense, the vacuole can be regarded as equivalent to the extracellular space of animals.

Vacuoles arise initially in young dividing cells, probably by the progressive fusion of vesicles derived from both the endoplasmic reticulum and the Golgi apparatus. They are structurally and functionally related to the lysosomes of animal cells, and they contain a wide range of hydrolytic enzymes. But plant cells use their vacuoles in a marvelous variety of ways. We shall consider only two of these—space filling and storage.

As Plant Cells Grow, They Accumulate Water in Their Vacuoles[14]

The simple *space-filling* function of the vacuole is of great importance to plants, which have to grow to capture energy from the sun rather than move to capture food. The mechanical stability provided by the combination of a cell wall and turgor pressure allows plant cells to grow to a relatively large size, so that they generally occupy a much larger volume than animal cells. However, to produce large cells by filling them with cytoplasm would be costly, both in terms of maintenance and initial synthesis. Although some plant cells, such as those at sites of intense cell division, make more cytoplasm as they grow, the majority accumulate water as they become larger through turgor-driven cell-wall expansion, usually in small vacuoles that then coalesce to form a large vacuole (Figure 19–33). The large vacuole in such cells may be crisscrossed by numerous strands of cytoplasm. The presence of vacuoles means that most mature plant cells have a large ratio of surface to cytoplasmic volume, with the cytoplasm forming a thin layer pressed against the cell wall. In photosynthetic cells, for example, chloroplasts are arranged in a thin layer of cytoplasm at the cell periphery, facilitating exchanges (Figure 19–32). In some cells the nucleus and the bulk of the cytoplasm lie at one end and most of the vacuolar compartment at the other. This asymmetry often develops prior to developmentally important asymmetrical cell divisions. The vacuole, though, does not push the cytoplasm to one end of the cell. Because the contents of the cell are at hydrostatic pressure equilibrium, vacuoles are unable to act like spatially directed pistons that push the cytoplasm; rather it is the plant cell cytoskeleton that organizes the cytoplasm, as will become apparent later in this chapter.

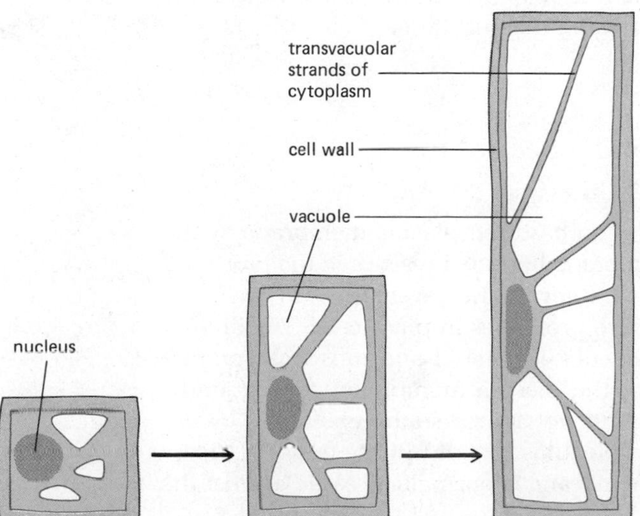

Figure 19–33 Schematic diagram showing how a large increase in cell volume can be obtained with no increase in cytoplasmic volume. Turgor-driven cell expansion accompanies the uptake of water into an expanding vacuole. The cytoplasm is eventually confined to a thin peripheral layer interconnected by transvacuolar strands of cytoplasm that emanate from the region of the nucleus.

Vacuoles Can Function as Storage Organelles[14]

Vacuoles can store many types of molecules, in particular essential substances that are potentially harmful if present in bulk in the cytoplasm. For example, in the appropriate plant, the vacuoles of certain specialized cells contain such interesting products as rubber and opium. Even ubiquitous molecules like Na^+ are stored in these organelles, where their osmotic activity contributes to turgor pressure. Analysis of *Nitella* cells indicates that Na^+ pumps located in the tonoplast maintain low concentrations of Na^+ in the cytosol and four- to fivefold higher concentrations in the vacuole; since the vacuole occupies a much greater volume than the cytoplasm in *Nitella*, the great bulk of cellular Na^+ is in the vacuole.

The different permeability properties of the plasma membrane and tonoplast govern the different solute compositions of the cytoplasm and the vacuole. The permeability of these two membranes is regulated by turgor pressure and is determined by the various membrane transport proteins that transfer specific sugars, amino acids, and other metabolites across the lipid bilayer (see Chapter 6). The substances in the vacuole differ qualitatively and quantitatively from those in the cytoplasm. However, because the tonoplast is a lipid bilayer with little mechanical strength, the hydrostatic pressure must remain roughly equal in cytoplasm and vacuole. The two compartments must act together in osmotic balance to maintain turgor.

Among the products stored by vacuoles are those with a metabolic function. For example, succulent plants take up carbon dioxide at night and store it as malate in vacuoles until light energy can be used to convert it to sugar the next day. Vacuoles can also store organic molecules for much longer periods: for example, proteins are stored in the vacuoles of storage cells of many seeds (Figure 19–34). Different vacuoles with distinct functions, for example, lysosomal and storage, are often present in the same cell.

Other molecules stored in vacuoles are involved in the interactions of the plant with animals or with other plants. For example, the anthocyanin pigments color flower petals so that they attract pollinating insects. Other molecules participate in defense mechanisms. Plants cannot move to avoid being eaten by herbivores; instead they synthesize an enormous variety of noxious metabolites that are released from vacuoles when the cells are eaten or otherwise damaged. These range from poisonous alkaloids to unpalatable inhibitors of digestion. During evolution, both plants and animals have used such weapons in a subtle chemical warfare, the balance shifting when a new potent deterrent to herbivores develops or, conversely, when an insect evolves a means of breaking down a toxic plant metabolite so that it can feed upon, and even be attracted to, plants possessing the substance.

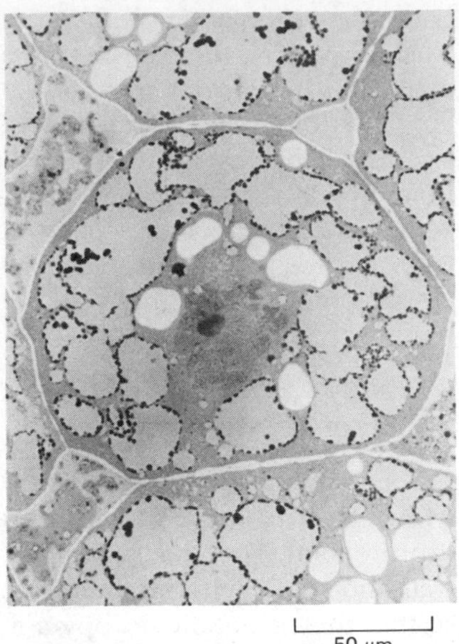

Figure 19–34 A light micrograph of a cell from a pea seed cotyledon. The early stages of protein deposition can be seen at the edges of the extensive system of vacuoles. Legumes such as peas and beans store large amounts of protein in such vacuoles in order to sustain embryo growth when the seed germinates. (Courtesy of S. Craig.)

Plant Cells Exocytose But Generally Seem Not to Endocytose Macromolecules[15]

Animal cells transport macromolecules across their plasma membrane by the sequential formation and fusion of membrane-bounded vesicles in the processes of endocytosis and exocytosis (see Chapter 6). However, the cell wall and turgor pressure both severely limit these processes in plant cells. The limited porosity of the cell wall prevents particulate material and most macromolecules from reaching the outer face of the plasma membrane; consequently, with very few exceptions, plant cells do not ingest such substances by endocytosis. Even the uptake of small molecules by fluid-phase endocytosis (p. 306) is difficult because the plasma membrane is normally pressed against the

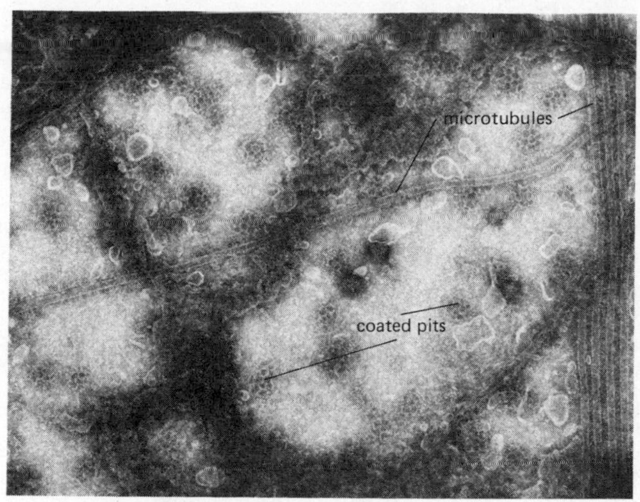

Figure 19–35 The plasma membrane of a tobacco protoplast. The protoplast, attached to the electron microscope grid, has been burst open, washed, and negatively stained to reveal the cytoplasmic face of the plasma membrane. Cortical microtubules and numerous coated pits are clearly visible. (Courtesy of L. C. Fowke.)

cell wall by turgor pressure. Nonetheless, the plasma membrane of a plant cell has numerous *coated pits,* which are thought to pinch off to form coated endocytotic vesicles (Figure 19–35; see also p. 309).

While endocytosis is limited in plant cells, exocytosis is not. In particular, most plant cells secrete macromolecules packaged by the Golgi apparatus, as we shall now discuss.

Golgi Vesicles Deliver Cell-Wall Material to Specific Regions of Plasma Membrane[16]

Most of the cell-wall matrix components are transported via vesicles derived from the Golgi apparatus to the plasma membrane, where they are secreted by exocytosis (Figure 19–36). Because the wall varies in composition and morphology at different locations around the cell, the Golgi vesicles must be directed to specific regions of the plasma membrane. This is accomplished, at least in part, by the cytoskeleton. One example is provided by the formation of the new primary cell wall after mitosis (as described in detail in Chapter 11, p. 664). At the end of telophase, a set of microtubules remains between the two daughter nuclei, parallel to the spindle axis. This set consists of the two groups of polar spindle microtubules, of opposite polarity, whose ends interdigitate in a disclike region, called the **phragmoplast,** in the plane of the former spindle equator. Golgi-derived vesicles containing wall precursors, especially pectin, are guided inward along these oriented microtubules until they reach the central disc, where they fuse with each other to form the **cell plate** (Figure 19–37). The cell plate is extended at its edge by the addition of further Golgi vesicles, which are guided there along new sets of microtubules that assemble around the periphery of the disc as the more central microtubules disassemble. Finally, the growing cell plate fuses with the mother cell wall to create two separate daughter cells (Figure 19–38).

The later delivery of the contents of numerous Golgi vesicles to the plasma membrane during secondary wall formation is probably guided in a similar way. However, a problem arises with the large amounts of new membrane that are added to the existing plasma membrane by the vesicle fusions. In some actively secreting cells, the number of Golgi vesicles involved in exocytosis would lead to a doubling of the plasma membrane area every 20 minutes. Clearly a system for membrane retrieval must operate, and it is probable that

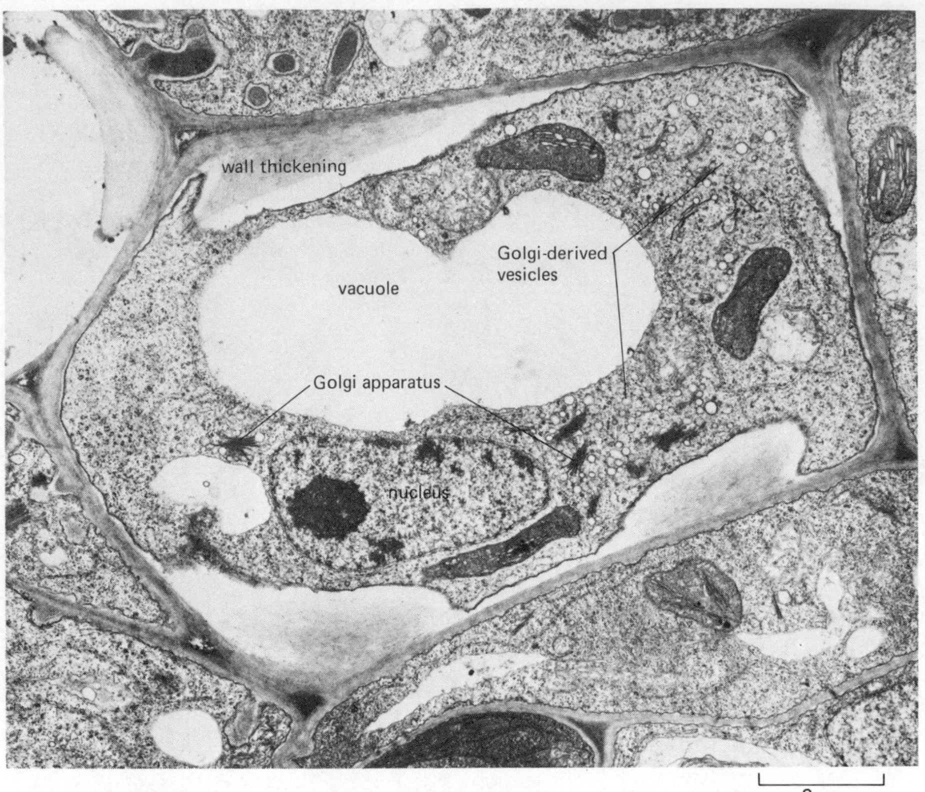

Figure 19–36 Electron micrograph of a young developing xylem element from goose grass. Several cell-wall thickenings are seen, together with the numerous Golgi apparatuses and their derived vesicles that are contributing material to the new wall. (From B. Gunning and M. Steer, Ultrastructure and the Biology of Plant Cells. London: Arnold, 1975.)

the numerous coated pits seen in plant plasma membranes function in membrane recycling, just as they do in animal cells (see p. 309).

The Golgi apparatus in plant cells consists of stacks of cisternae, which are engaged mainly in the production of a very wide range of extracellular polysaccharides. Most of these polysaccharides are insoluble cell-wall components, but others are more soluble, such as those that form the slime secreted by root tips to lubricate their passage through the soil. Each Golgi stack is separate, rather than connected to other stacks as in animal cells (Figures 19–39 and 19–40). This may allow individual Golgi stacks to be specialized for different secretions.

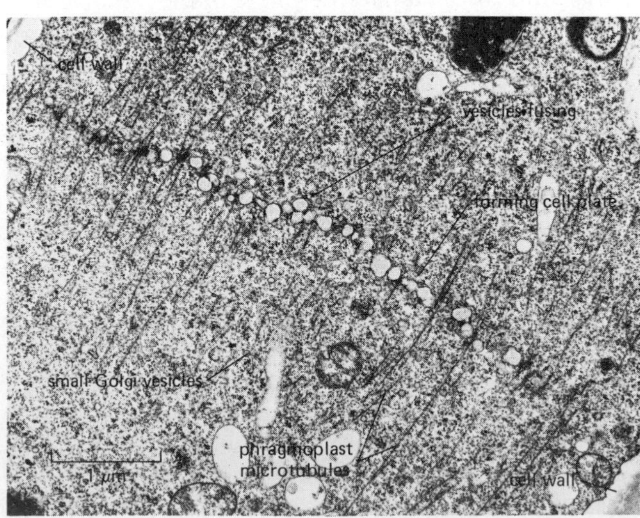

Figure 19–37 Electron micrograph of the phragmoplast in a dividing plant cell. The microtubules guide vesicles containing cell-wall precursors toward the growing cell plate. For details, see Figure 11–65, page 665. (Courtesy of J. D. Pickett-Heaps.)

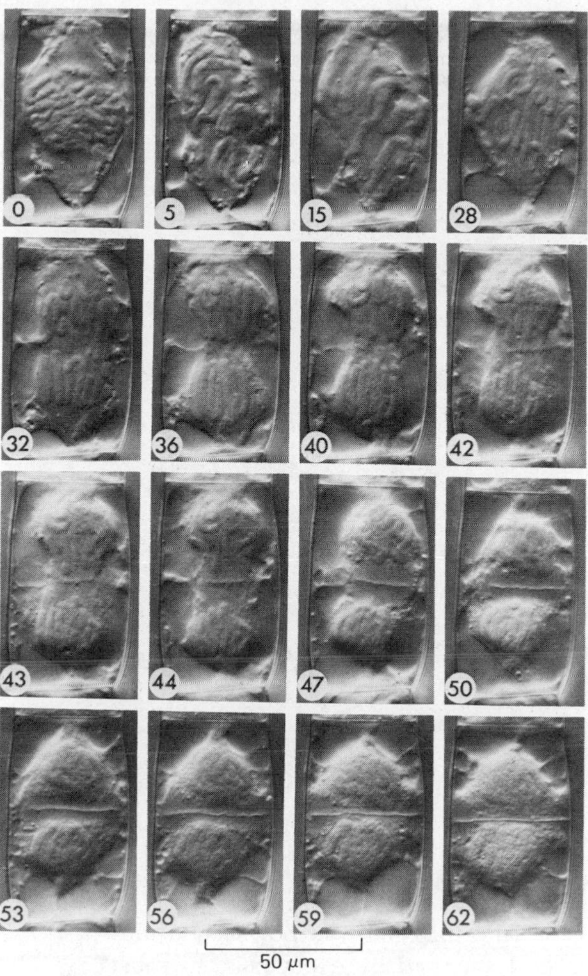

Figure 19–38 Sequential light micrographs of a dividing stamen hair cell. The elapsed time in minutes is shown at the bottom left corner of each photograph. The vesicles that align to form the cell plate can be seen after 42 minutes. The plate then extends sideways until it reaches and fuses with the mother cell wall. (Courtesy of P. K. Hepler.)

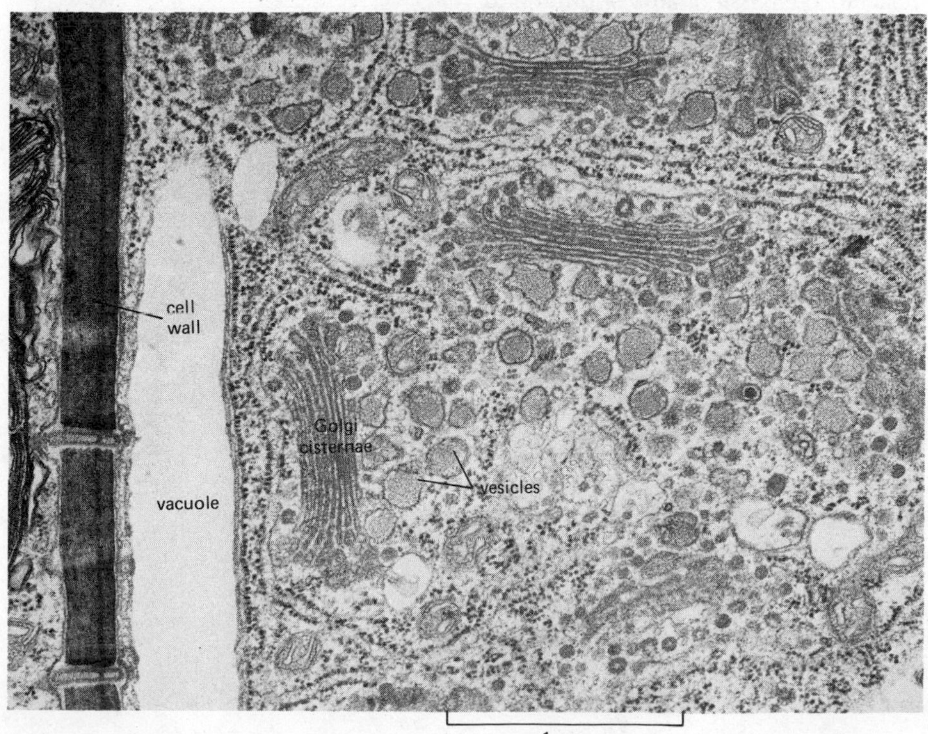

Figure 19–39 Electron micrograph of a hair cell from the green alga *Bulbochaete*. Numerous discrete stacks of Golgi cisternae can be seen, together with associated vesicles containing amorphous material. Unlike those of most animal cells, the Golgi cisternae in different stacks are not interconnected. (Courtesy of T. W. Frazer, from B. Gunning and M. Steer, Ultrastructure and the Biology of Plant Cells. London: Arnold, 1975.)

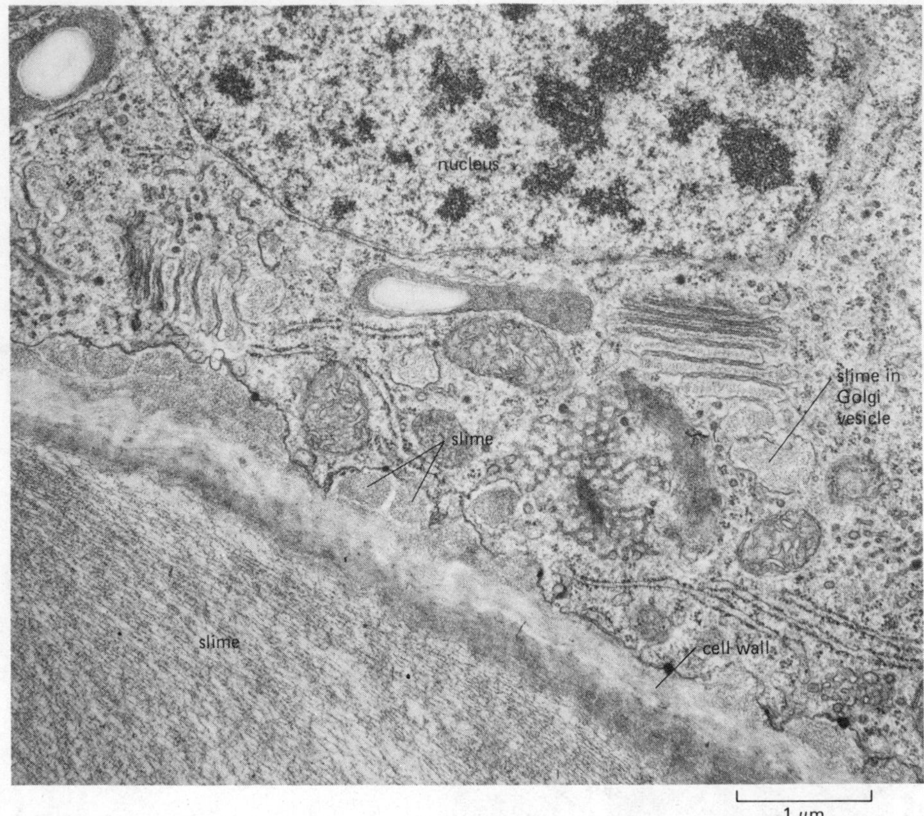

Figure 19–40 Electron micrograph of a cell from the root cap of Timothy grass. In such cells the Golgi apparatus functions mainly in manufacturing, packaging, and delivering mucilage (slime) to the surface of the root tip to lubricate its passage through the soil. Material identical to that seen in the Golgi cisternae can also be seen in vesicles and just outside the plasma membrane. (From B. Gunning and M. Steer, Ultrastructure and the Biology of Plant Cells. London: Arnold, 1975.)

Cellulose Synthesis Occurs at the Surface of Plant Cells[17]

There is one important exception to the rule that cell-wall polysaccharides are produced in the Golgi apparatus and secreted by exocytosis. In most plants, cellulose is synthesized on the exterior surface of the cell by a plasma-membrane-bound enzyme complex that uses a sugar nucleotide precursor, probably UDP-glucose. The resulting nascent cellulose chains spontaneously self-assemble into microfibrils, which become integrated into the general fabric of the cell wall.

Certain primitive algae are covered by elaborate, cellulose-containing, cell-wall *scales.* These scales are produced in Golgi vesicles and are then secreted intact by exocytosis. In such organisms the cellulose synthetic machinery is associated with the Golgi vesicle membrane, and it becomes part of the plasma membrane following vesicle fusion (Figure 19–41). By analogy, it has been suggested that the enzyme complexes responsible for cellulose synthesis in higher plants are also delivered via Golgi vesicles to the plasma membrane, but that in this case the enzyme complexes become active only after they have been exposed to the extracellular environment.

Cortical Microtubules Orient the Extracellular Deposition of Cellulose Microfibrils[18]

Cellulose microfibrils are often specifically oriented in the cell wall. But what controls their orientation? This is an important question because, as we shall see later, the orientation of cellulose microfibrils is crucial in determining the shape of plant cells. An important clue was provided by the discovery that in a plant cell the majority of cytoplasmic microtubules are arranged in the cortex of the cell with the same orientation as the cellulose microfibrils that are currently being deposited in that region.

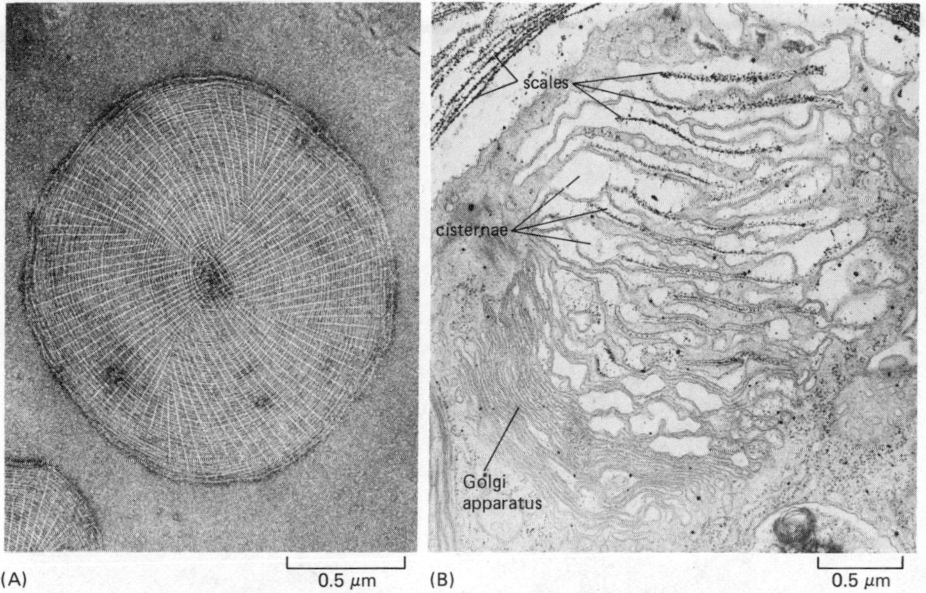

Figure 19–41 The cell wall of the small alga *Pleurochrysis* is made up of numerous discrete scales. (A) Each scale, seen here negatively stained, is a disc composed of a spiral arrangement of cellulose microfibrils. (B) The Golgi apparatus can be seen producing cisternae, each of which contains one scale that is eventually added to the wall by exocytosis. The scales here are stained specifically for carbohydrate. (From R. M. Brown and D. K. Romanovicz, *Applied Polymer Symposium* 28:537–585, 1976. Copyright © 1976 John Wiley and Sons, Inc. Reprinted by permission of John Wiley and Sons, Inc.)

Lying close to the inner face of the plasma membrane (Figure 19–42), the cortical microtubules are oriented mainly perpendicular to the axis of elongation of the cell (Figure 19–43). This microtubule array, which encircles the cell, is composed of short overlapping lengths of microtubules. In certain cases bridges between the plasma membrane and the underlying cortical microtubules can be seen.

The congruent orientation of microtubules (lying just inside the plasma membrane) and cellulose microfibrils (lying just outside the plasma membrane) is seen in many different types and shapes of cell (Figure 19–44). It is present during both primary and secondary cell-wall deposition, and the positions of cortical microtubules correlate with regions of local wall deposition—as when xylem cell-wall thickenings are formed at specific sites on the cell surface (see Figure 19–13).

What happens, then, if the entire system of cortical microtubules is depolymerized by treating a plant tissue with the drug colchicine (p. 570)? The consequences for subsequent cellulose deposition are not as straightforward as might have been expected. The drug treatment has no effect on the production of new cellulose microfibrils, and in some cases cells can continue to deposit new microfibrils in a preexisting orientation. However, any developmental change in the microfibril pattern is invariably blocked by the drug. For example, a cell that would normally start to develop regular wall thickenings to form a xylem vessel instead deposits a disorganized smear of wall material in the presence of colchicine. The conclusion is that a preexisting orientation of microfibrils can survive, even in the absence of microtubules, but that any stage of cell development that involves a change in the deposition of cellulose microfibrils requires that intact microtubules be present to determine the new orientation (Figure 19–45).

The ability of cortical microtubules to influence the orientation of newly formed cellulose microfibrils suggests that these microtubules are directly or indirectly connected to extracellular components across the plasma membrane. However, the molecular basis of this connection is unknown. It is worth recalling that the cytoskeleton of an animal cell can also communicate with components of the extracellular matrix to influence their mutual orientation (see p. 712).

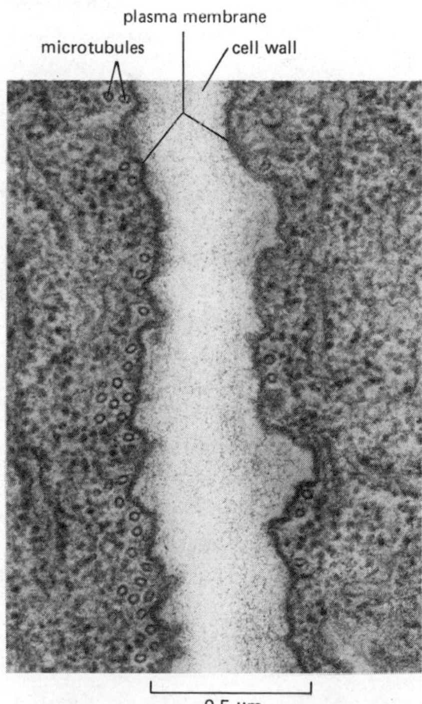

Figure 19–42 Electron micrograph of two adjacent wheat root-tip cells, showing the numerous cortical microtubules typical of interphase cells.

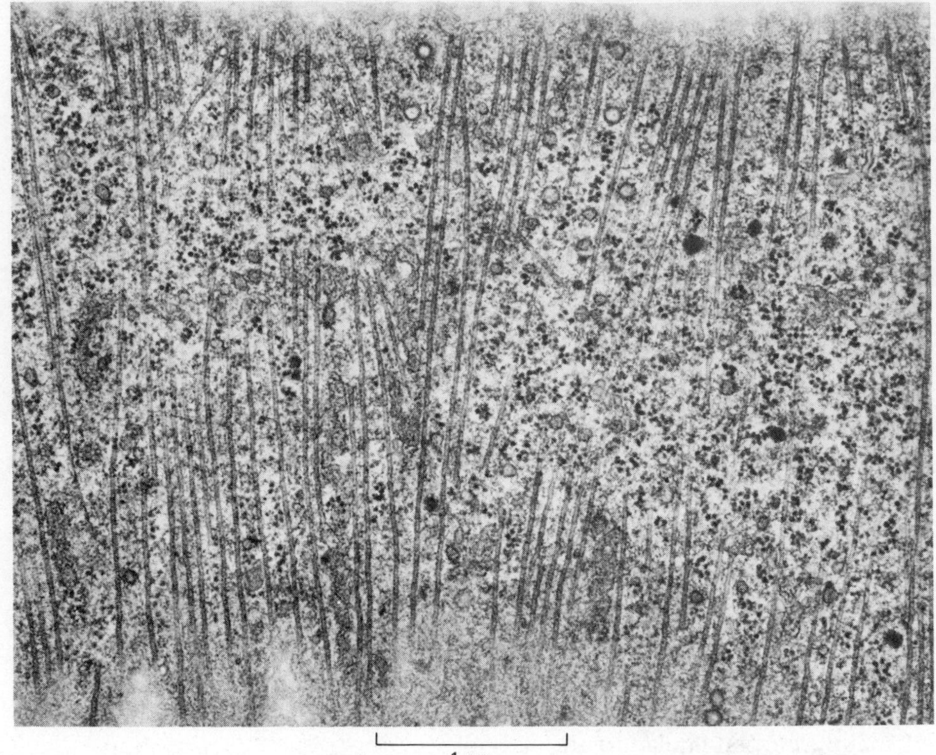

Figure 19–43 A grazing section of a root-tip cell from Timothy grass, showing cortical microtubules lying just below the plasma membrane. These microtubules are oriented perpendicular to the long axis of the cell. (Courtesy of B. Gunning.)

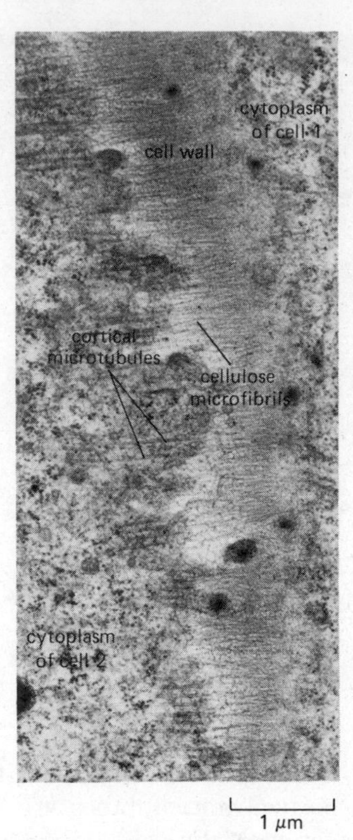

Figure 19–44 A grazing section of a root-tip cell from Timothy grass, showing that the orientation of the cortical microtubules is parallel to the orientation of the cellulose microfibrils in the primary cell wall. (Courtesy of B. Gunning.)

The Movement of Materials in Large Plant Cells Is Driven by Cytoplasmic Streaming[19]

Cell metabolism demands that substrates, intermediates, cofactors, messengers, and enzymes must be able to move from one part of the cytoplasm to another. In small cells, such as bacteria or even most animal cells, diffusion enables small solutes to move over distances comparable to the size of the cell in fractions of a second. However, plant cells, because of their cell walls, vacuoles, and turgor, are able to become very large. They are commonly more than 100 μm long, while some are a few millimeters or even centimeters long. Diffusion is relatively ineffective over such distances, as the time taken for a molecule to reach its destination by diffusion alone varies with the square of the distance involved. It is not surprising, therefore, that large plant cells display an extensive **cytoplasmic streaming** that stirs their cytoplasm and moves material around.

Examination of living plant cells reveals that the larger the cell, the more extensive are the movements of its cytoplasm. Small cells exhibit agitated

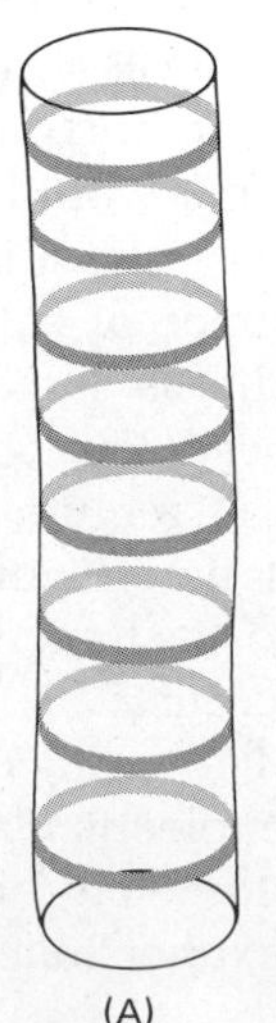

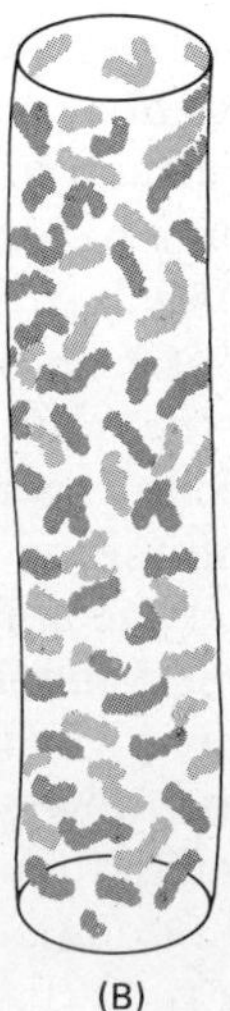

Figure 19–45 The organized pattern of wall thickenings formed during normal xylem cell differentiation (A) depends on the presence of organized arrays of cortical microtubules. In the presence of colchicine the cortical microtubules are depolymerized and a disorganized pattern of wall thickenings results (B).

movements of their organelles known as *saltations* (from *saltare*, Latin for to dance or to jump). As in animal cells, particles stop and go, being suddenly propelled at a high speed in a particular direction. In larger plant cells, there is more directionality to the cytoplasmic movements; and in cells with a thin rim of cytoplasm surrounding a huge central vacuole, it is common for the cytoplasm to rotate almost continuously, at a rate of several micrometers per second. Such cytoplasmic movements promote not only intracellular traffic but also intercellular transport by delivering solutes to the openings of plasmodesmata connecting adjacent cells.

The thin cells that form the hairs on the surface of plants are transparent; therefore, their cytoplasmic movements can be readily observed in the living cell. These cells contain large vacuoles through which run thin strands of cytoplasm about 1 μm in diameter (Figure 19–46). Individual particles, such as mitochondria, can be seen moving rapidly through the cytoplasmic strands. The strands seem to emanate from a region near the cell nucleus but continuously change their shape and position, merging, branching, collapsing, and forming anew.

The cytoplasmic streaming responsible for such organelle movements in higher plant cells probably involves actin. However, a better understood example of streaming is that found in certain giant algal cells.

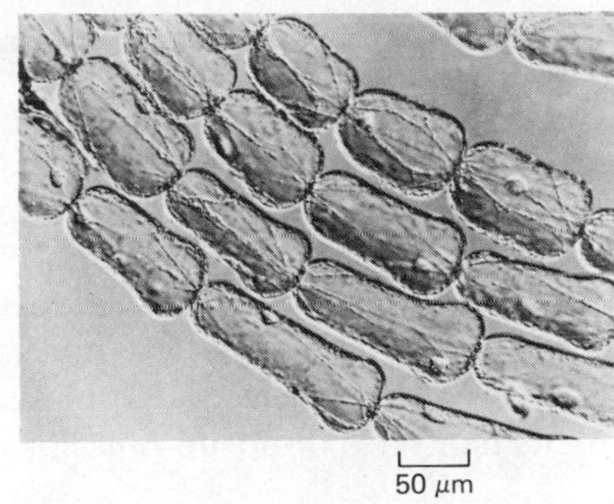

Figure 19–46 Stamens from the flower of *Tradescantia* are covered in long thin hairs, each made from a row of large single cells. Several hairs are shown here by Nomarski differential-interference-contrast microscopy. The fine cytoplasmic strands along which streaming occurs can be clearly seen crossing the very large vacuoles. (From H. Stebbings and J. S. Hyams, Cell Motility. London and New York: Longman, 1979. © 1979 Longman Group, Ltd.)

The Interaction of Actin and Myosin Drives Cytoplasmic Streaming in Giant Algal Cells [19,20]

The cylindrical cells of the green algae *Chara* and *Nitella* are enormous, being 2 cm to 5 cm in length. These giant, multinucleated cells provide the most dramatic examples of cytoplasmic streaming. A continuous ribbon of cytoplasm streams along a gentle helical path down one side of each cell and back across the other side in an endless belt. The streaming cytoplasm moves in only one direction, at speeds of up to 75 μm per second, sweeping internal membranes, mitochondria, nuclei, and cytosol around and around the cell (Figure 19–47).

Not all of the cytoplasm flows in these giant cells. The cell cortex is static; it consists of the cytoplasm just beneath the plasma membrane, which contains a monolayer of chloroplasts that are aligned in rows parallel to the direction of streaming. The motility-generating system is known to lie between this static, chloroplast-containing cortex and the moving layer of cytoplasm. In the light microscope, thin fibrils 0.2 μm in diameter can be seen just under the aligned rows of chloroplasts. The electron microscope reveals that each of these fibrils consists of a bundle of actin filaments, all aligned with the same polarity. The polarity of the actin filaments is such that the movement of myosin filaments along them could produce the observed cytoplasmic stream-

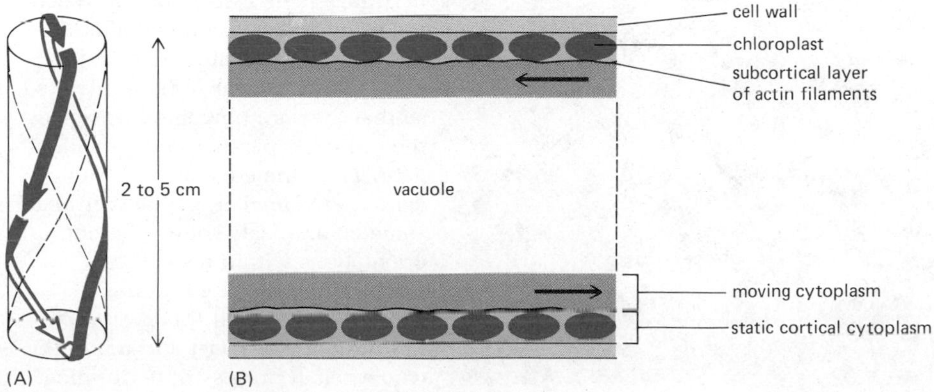

Figure 19–47 Schematic diagram showing the pattern of streaming cytoplasm in the giant algal cell *Nitella.* (A) The path followed by the moving cytoplasm in the cylindrical cell. For clarity the cell's diameter is somewhat exaggerated relative to its strength. (B) A longitudinal section through part of a cell, showing the arrangement of the static and moving layers of cytoplasm. The static cortical cytoplasm contains the chloroplasts, which are attached to underlying bundles of actin filaments. Internal to the actin filaments is the moving layer of cytoplasm containing the nuclei, mitochondria, and other organelles. In the living cell the relative size of the vacuole is much larger than is shown here.

ing (see p. 591). It seems possible, therefore, that organelles in the moving cytoplasm are indirectly attached to the actin filaments by myosin molecules, which use the energy of ATP hydrolysis to slide along the actin filaments, pulling the organelles with them.

Our knowledge of the structure of plant myosin is still meager, but actin filament bundles have been found in a variety of plant cells, including various hair cells. The fact that the cytoplasmic streaming in most higher plant cells can be multidirectional, in contrast to the unidirectional streaming seen in the giant algae, suggests that in such cases neighboring bundles of actin filaments can have opposite polarities.

Regions of the Plant Cell Cytoskeleton Can Be Reorganized in Response to Local Stimuli[21]

Many plant cells can respond to changes in the intensity and direction of light by altering the position of their chloroplasts. At low light intensities, the chloroplasts tend to become aligned in a monolayer perpendicular to the incident light, thereby maximizing their exposure to the light. High light intensities induce a protective response in which the chloroplasts become aligned against the cell walls that are parallel to the incident light, thereby minimizing their exposure (Figure 19–48). These movements, which almost certainly involve actin filaments, have been most thoroughly studied in two species of algae.

In *Vaucheria,* a large filamentous algal cell that contains many chloroplasts, the illumination of a small region of an otherwise darkened cell induces the chloroplasts to migrate into the illuminated area. The migration begins within 10 minutes and is accompanied by the localized formation of a network of actin filaments in the illuminated region.

Mougeotia is an alga in which each cylindrical cell contains a single platelike chloroplast. In this organism, the response to light involves a rotation that adjusts the orientation of the chloroplast until it is either edge on or perpendicular to the incident light, depending on the light intensity. The photoreceptor molecule involved in the response seems to be a *phytochrome,* a type of pigment known to mediate many different light responses in plants. It is located on or very near the plasma membrane, and its illumination with a microbeam causes a local influx of Ca^{2+} that activates the molecular ma-

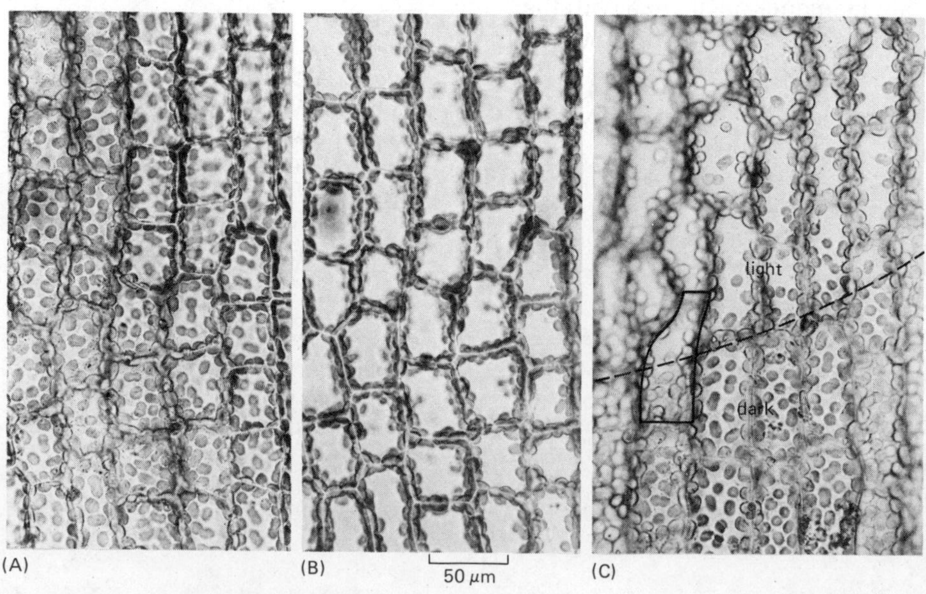

Figure 19–48 Light micrographs of leaf cells from a moss, showing how the chloroplasts move in response to light. The direction of illumination is perpendicular to the plane of the micrograph. (A) In low light, the disclike chloroplasts orient to maximize light absorption. (B) When the same area is examined after 30 minutes of bright illumination, the chloroplasts appear to have migrated, so that they are now lined up against the cell walls parallel to the incident light. (C) At the edge of the beam of light (*dotted line*), it can be seen that some chloroplasts show different orientations within a single cell (*outlined*), suggesting that the response in (B) is at the level of the individual chloroplast and not of the whole cell. (Courtesy of B. Gunning.)

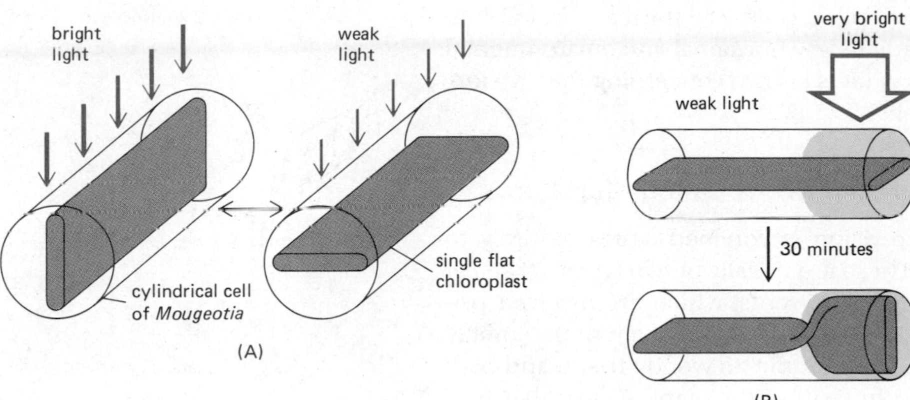

Figure 19–49 Schematic drawing of chloroplast movements in the green alga *Mougeotia*. This cylindrical cell contains a single, large, flat chloroplast, which orients itself so as to regulate the amount of the light absorbed (A). A portion of the chloroplast can be induced to orient independently of the rest if exposed to a very bright light, showing that the response is locally mediated (B).

chinery responsible for rotating the chloroplast. If a localized microbeam is used to illuminate only a small part of the cell, just the illuminated part of the chloroplast bends (Figure 19–49).

These examples of light-induced organelle movements, together with others, indicate that the plant cell cytoskeleton can respond in a specific manner to external stimuli. Moreover, different regions of the cytoskeleton of a single cell can respond independently. Such a responsive cytoskeleton may be especially important in plant cells, which are confined to one place by their walls.

Summary

Two conspicuous organelles are unique to plant cells, the plastids and vacuoles. The plastids are a varied group of organelles, the most familiar of which is the photosynthetic chloroplast found in all green tissue. The vacuole is a large, water-filled cytoplasmic space surrounded by a membrane called the tonoplast. Vacuoles are used in a variety of ways by plant cells—for example, as an economical means of space filling during cell enlargement and for the storage of food reserves or toxic waste products. Although plant cells themselves cannot move, their cytoplasm, particularly in cells with very large vacuoles, is kept stirred by means of active cytoplasmic streaming. In some cells this streaming has been shown to be based on cytoplasmic actin filaments.

The internal organization of the cell and its cytoskeleton is important in the production of the cell wall. The matrix components of the cell wall are made and exported by the Golgi apparatus, but the cellulose microfibrils are synthesized in situ at the cell surface. Both the site of deposition of the wall components and the specific orientation of the cellulose fibers are controlled by arrays of cortical microtubules. The components of the plant cytoskeleton are also responsive to the environment, as is illustrated by the light-mediated movements of chloroplasts.

Cell Growth and Division

The various parts of a plant arise by a complex process in which a genetically determined pattern of cell division is followed by selected cell growth and, finally, by cell differentiation. Because plant cells have rigid cell walls and cannot move about, two questions in plant morphogenesis are especially important: (1) what determines the precise sequence of planes of cell division, and (2) what controls the magnitude and orientation of cell elongation? We shall see that the answers to both questions lie, at least in part, in special arrays of microtubules that are unique to plant cells. The third aspect of plant development, cell differentiation, is regulated by hormonal and environmental factors. In this section we shall outline what is known concerning the division, growth, and differentiation of plant cells.

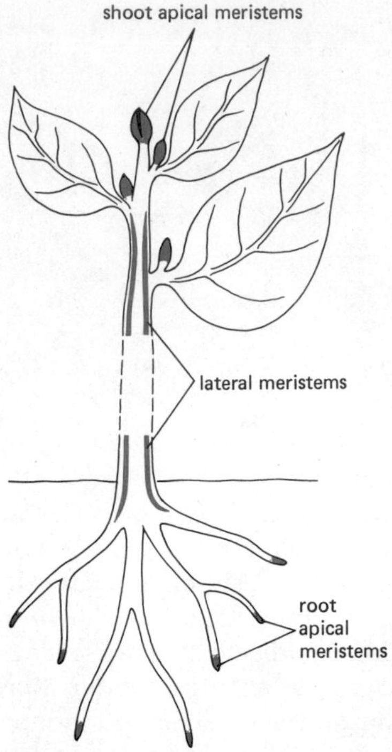

Figure 19–50 Highly schematic diagram of a higher plant, showing the locations of the main meristematic regions, the areas of most rapid cell division. The activity of the shoot and root apical meristems is responsible for increasing the length, while the activity of the lateral meristems is responsible for increasing the girth of the various parts of the plant.

Most New Plant Cells Arise in Special Areas Called Meristems[22]

During the growth of the plant, cell division is confined almost entirely to specialized regions—the **meristems.** These are usually of two types: (1) *apical meristems,* at the tips of growing shoots and roots, which are involved primarily in growth by extension, and (2) *lateral meristems,* circumferentially arranged groups of cells that give rise, for example, to woody tissue and bark and are involved primarily in increases in girth of the plant (Figure 19–50).

Both types of meristem consist of cells analogous to animal stem cells, such as those of the blood or epidermis (p. 910). The stem cell (or *initial,* as it is often called in plants) has a thin wall, and divides to give two progeny cells: one that remains a stem cell capable of unlimited further divisions and another that will usually undergo a limited number of further divisions before finally differentiating into a specialized cell. In some cases—for example, the root tip of *Azolla* (a water fern)—exact cell lineages have been described (Figure 19–51) in much the same way as they have been described for a nematode worm (p. 875).

Most cells produced in the root-tip meristem go through three distinct phases of development: (1) division, (2) growth or elongation, and (3) differentiation. These three steps, which are spatially as well as temporally separated, give rise to the characteristic architecture of a root tip (Figure 19–52). Despite a certain amount of overlap, it is comparatively easy to distinguish in a root tip a zone of cell division, a zone of cell elongation (which accounts for the growth in length of the root), and a zone of cell differentiation. The process of cell differentiation, however, often begins while a cell is still enlarging. When the differentiation process is complete, some of the cell types created remain alive (for example, phloem cells), while others die (for example, xylem vessel members and tracheids).

Cell division is not absolutely confined to the cells in a meristem. Some very large, highly vacuolated, and, in some cases, fully differentiated cells can divide, either naturally or as a consequence of an artificial stimulus such as a nearby wound. The retention of the ability to divide in mature cells is a general feature of plants. While many mature cells in animals can divide, mature plant cells are unusual in the degree to which they are able to "dedifferentiate" and give rise to pluripotent cells whose progeny can produce completely different mature cell types. In some experimental cases, a switch

Figure 19–51 Drawing of a longitudinal section through a root tip of the water fern *Azolla.* The arrows indicate cell lineage relationships, with each branching pair of arrows indicating a cell division. The large apical stem cell divides sequentially at three faces. (In this section only two are visible.) The derivatives undergo a programmed sequence of divisions in predetermined planes, such that the lineage of all 9000 or so cells in each root can be mapped. A drawing of a cross-section of a similar root can be seen in Figure 19–20. Most roots of higher plants are very much more complicated and contain more cells, but the same principles govern their final form. For simplicity, the cells of the root cap have been omitted.

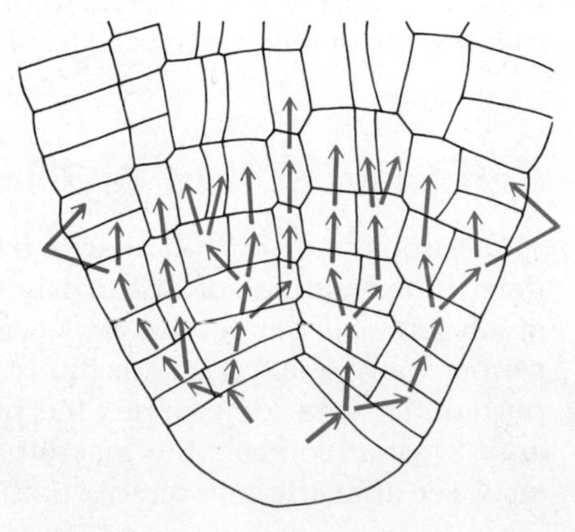

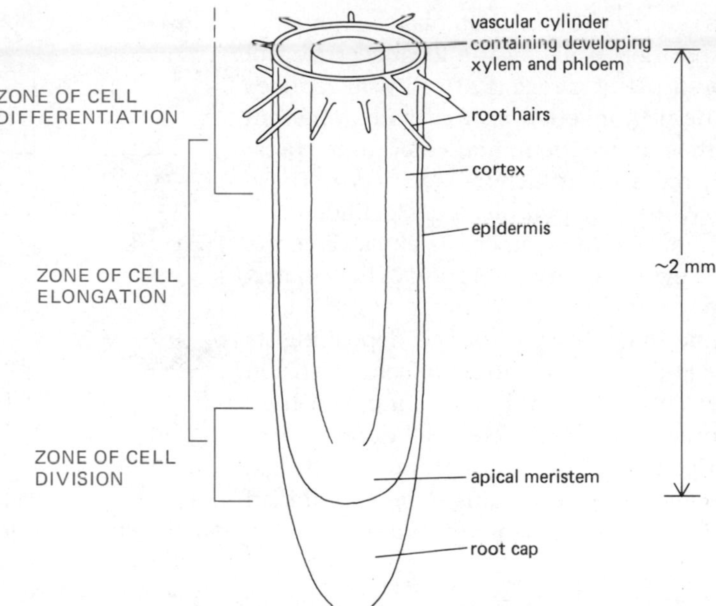

Figure 19–52 The organization of the final 2 mm of a growing root tip. The approximate zones in which cells can be found dividing, elongating, and differentiating are indicated.

in cell type can even take place without an intervening cell division (Figure 19–53). This feature may be a particular adaptation of plants, which, being immobile and unable to avoid injury, have been under greater selective pressure to evolve efficient tissue and cell repair mechanisms.

The Shape of a Growing Plant Cell Is Determined by the Organization of Cellulose Microfibrils[23]

The shape of a growing plant cell is determined by controlled and directed cell expansion, which plays a major part in determining the final form of a plant. The driving force for cell growth or expansion is turgor, which exerts a uniform outward pressure. How is this nondirectional pressure harnessed to produce cells with asymmetrical shapes? Again the answer lies in the cell wall.

We can illustrate the process by examining a cylindrical cell, one of the most common cell shapes in plants. Normally, as a cell grows, the cylinder elongates much more than it increases in girth, as was illustrated previously in Figure 19–33 (p. 1123). Since the cellulose microfibrils in the cell wall are unstretchable, they must slide past each other to allow the cell to grow. The microfibrils lie in the plane of the primary cell wall, often aligned in specific orientations. In spherical cells the arrangement tends to be random, but in elongating cylindrical cells the most recently laid down microfibrils in the side walls commonly lie perpendicular to the axis of elongation, surrounding the cylinder with numerous hoops of cellulose. As successive layers of wall are deposited (the most recently formed being closest to the plasma membrane), a gradation develops in the degree of orientation of the microfibrils: the microfibrils in the outermost layers have often been so reoriented by wall stretching that they have lost all trace of their initial organization.

The oriented hoops of cellulose microfibrils in a newly formed cell wall prevent major increases in the width of a growing cell while permitting the turgor pressure to cause a gradual increase in cell length (Figure 19–54). However, cell growth can occur only if the turgor pressure in the cell exceeds the local tensile strength of the wall. In principle, then, the plant cell could use two different strategies to grow: it could increase its turgor, or it could weaken

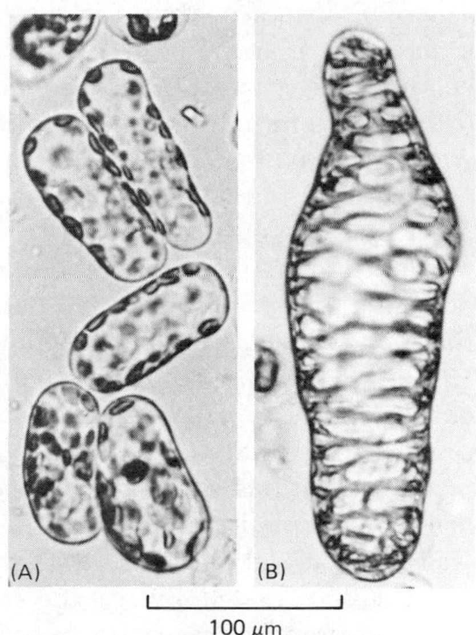

Figure 19–53 An example of a switch in mature cell type *in vitro*. (A) Mechanically isolated cells from a *Zinnia* leaf. These are fully mature photosynthetic cells with large vacuoles and numerous chloroplasts. (B) After incubation in growth medium for eight days, the cells, which have not divided, grow, lose their chloroplasts, and form the elaborate cell-wall thickenings that are characteristic of xylem vessel cells. (Courtesy of J. Burgess.)

the cell wall in local areas. These is good evidence that plants adopt the second strategy. Plant cells can weaken their walls by the local secretion of protons (H^+) into the wall. The lowered pH of a region of cell wall reduces the number of weak bonds holding the wall together so that its component macromolecules can slip past each other under the influence of turgor pressure. To facilitate wall growth, other, more complex changes occur. These changes include the activation of enzymes that hydrolyze glycosidic bonds and other covalent linkages. At the same time that the cell is expanding, the cell wall is also being augmented by the synthesis and secretion of new matrix and microfibrillar components.

There is often a considerable time delay between a cell depositing its wall so as to influence its future shape and the actual phase of cell expansion that brings about that shape change. Thus plant cells can anticipate their future morphology. Exactly how the precise pattern of cell-wall deposition is determined is not clear but, as discussed earlier, it is apparently controlled by the pattern of microtubules in the cell cortex. It remains to be determined how the plant cell controls the organization of these microtubules.

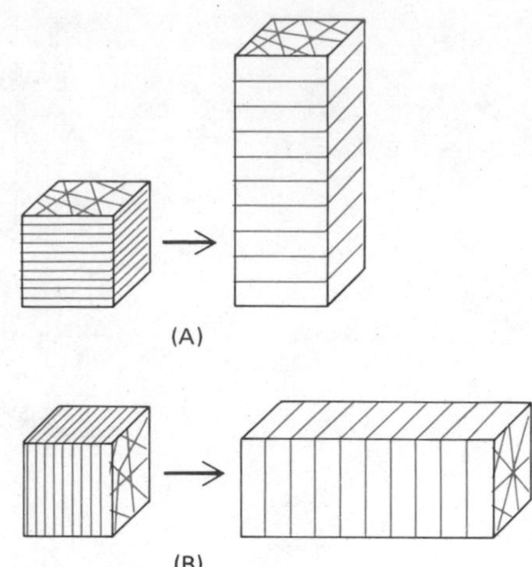

Figure 19–54 Schematic illustration of how the net orientation of cellulose microfibrils within the cell wall influences the direction in which the cell will elongate. (A) and (B) start as identically shaped cells (shown here as cubes for simplicity) with different net cellulose microfibril orientations within their walls. Turgor pressure causes each cell to elongate in a direction perpendicular to its orientated microfibrils. In turn, the final shape of an organ, such as a shoot, is profoundly influenced by the direction in which its constituent cells expand.

A Preprophase Band of Microtubules Marks the Future Plane of Cell Division[24]

The organized growth of a plant requires that cells in selected sites divide in a particular plane, and at a particular time, so as to set up correctly oriented cell lineages. Microtubules play a crucial part not only in the mitotic spindle, in cell-plate formation, and in orienting the cellulose microfibrils in the cell wall (Figures 19–55 and 19–56), but also in influencing the site and plane of cell division.

An early sign that a higher plant cell has become committed to divide in a particular plane is the congregation, within the cell cortex, of microtubules (often over a hundred in number) in a narrow band 1 μm to 3 μm wide that passes right around the cell just beneath the plasma membrane (Figures 19–56A and 19–57). This dense band of microtubules replaces the usual sparse cortical array and appears just before prophase. It is therefore called the **preprophase band.** By the time mitosis finally begins, the band has disappeared. Yet it has somehow left its mark in the cell cortex, for when the new cell plate forms in late telophase, it grows outward to fuse with the parental wall precisely at the zone that was formerly occupied by the preprophase band (Figure 19–58). Even if the cell contents are displaced by centrifugation after the preprophase band has disappeared, the growing cell plate will find its way back to the plane defined by the former preprophase band.

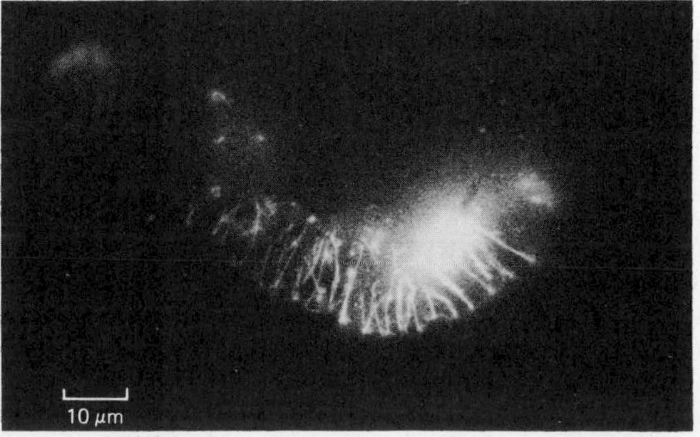

Figure 19–55 Immunofluorescence micrograph of a single carrot cell growing *in vitro*. This highly elongated cell has been stained with anti-tubulin antibody, revealing the hoops of microtubules around the cylindrical cell, perpendicular to its long axis. The orientation of microtubules in such cells mirrors the orientation of cellulose microfibrils deposited in the cell wall. (Courtesy of C. W. Lloyd.)

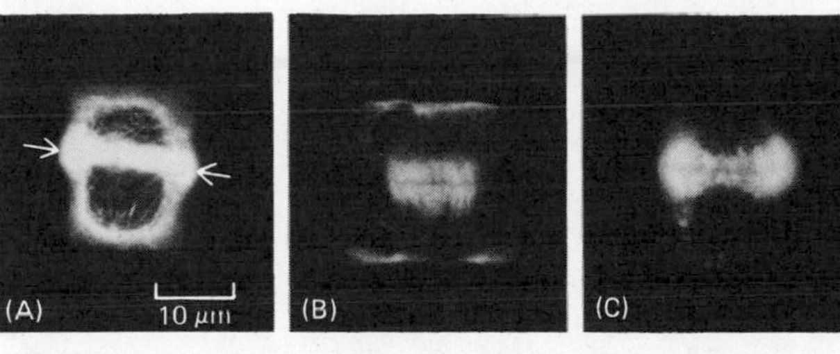

Figure 19–56 Immunofluorescence micrographs of isolated root-tip cells stained with anti-tubulin antibody to show the location of microtubules at different stages of plant cell division. (A) A preprophase band of microtubules (*arrows*) encircles the cell just prior to mitosis, while other microtubules can be seen surrounding the nucleus. (B) After the cell has entered mitosis and the chromosomes have separated, microtubules in the young phragmoplast initiate the formation of the cell plate as cytokinesis begins. No trace of the preprophase band remains. (C) At a later stage of cytokinesis, microtubules are concentrated where the cell plate is extending at its edges. Compare with Figure 11–65. (Courtesy of S. Wick.)

Irrespective of whether the division is symmetrical or asymmetrical, whether it is transverse, longitudinal, or tangential, the plant cell specifies where it is going to divide *before* it even enters mitosis (Figure 19–59). Although the microtubules of the preprophase band themselves disappear, they must leave some molecular memory in the place they once occupied. The importance of such controls is particularly clear when one considers that very asymmetric divisions commonly create two daughter cells with different developmental fates: for example, stomatal cells, root hair cells, and generative cells of pollen grains all develop from the smaller daughter cells produced by asymmetrical division (Figure 19–60). It may be that this behavior, peculiar to plant cells, reflects the lack of cell mobility in plants. Being static, plant cells have to ensure that they divide in specified sites and planes, because spatial reorganization after division is precluded by the cell walls.

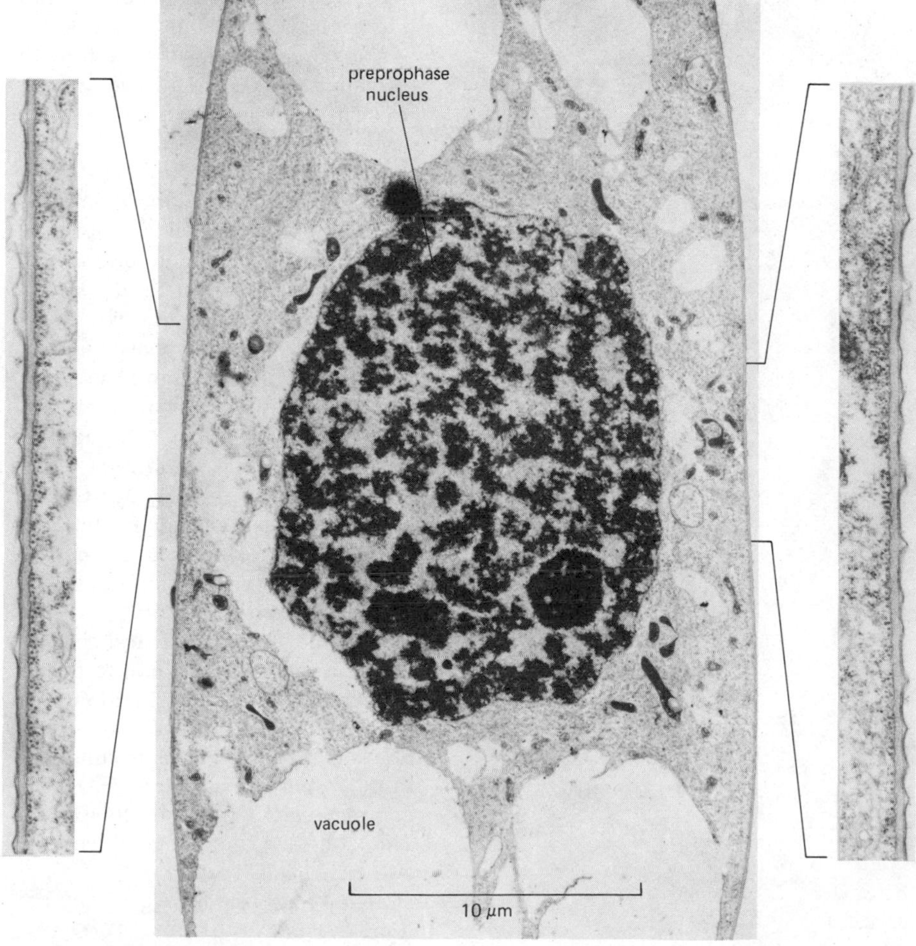

Figure 19–57 An electron micrograph of a cell from a stamen hair that is about to divide. On either side of the cell, a preprophase band of microtubules can be seen encircling the cell. The position of this band predicts where the new cell wall will meet the old wall during cytokinesis. Enlargements of the indicated areas show the preprophase band of microtubules in greater detail. (Courtesy of C. Busby, from C. Busby and B. Gunning, *Eur. J. Cell. Biol.* 21:214–223, 1980.)

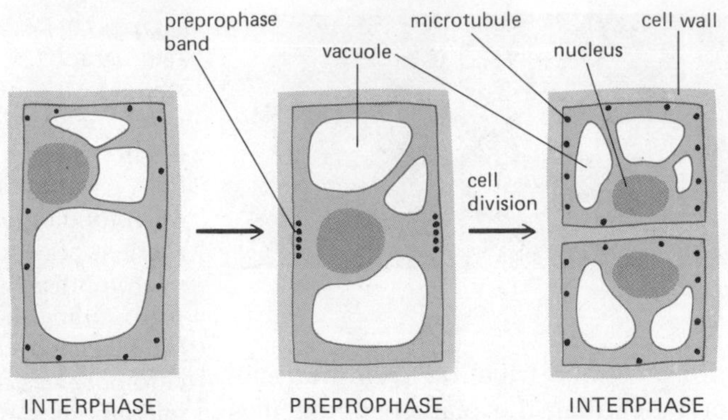

Figure 19–58 Schematic diagram showing the arrangement of the cortical array of microtubules during plant cell division. In interphase, cortical microtubules are distributed along the length of the cell wall. During preprophase, however, they congregate in a discrete band circling the cell. This preprophase band of microtubules accurately predicts where the new cell wall will join the old one when the cell later divides.

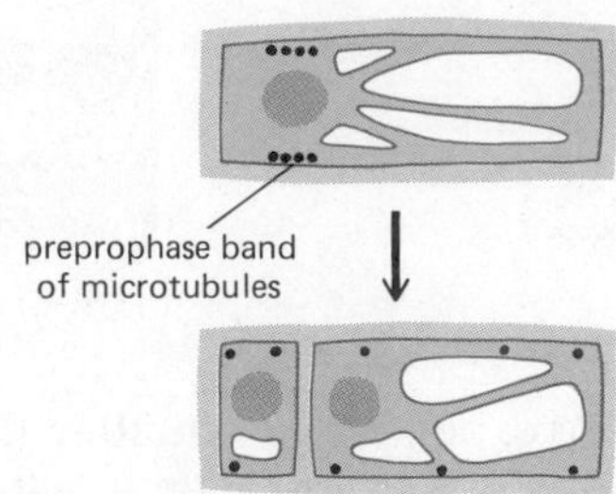

Figure 19–59 Schematic drawing of an elongated epidermal cell from a root, showing how the plane of asymmetric cell division is predicted by the preprophase band of microtubules. This epidermal cell divides asymmetrically to form a large daughter cell that will continue as an epidermal cell, and a small daughter cell that will become a root hair cell.

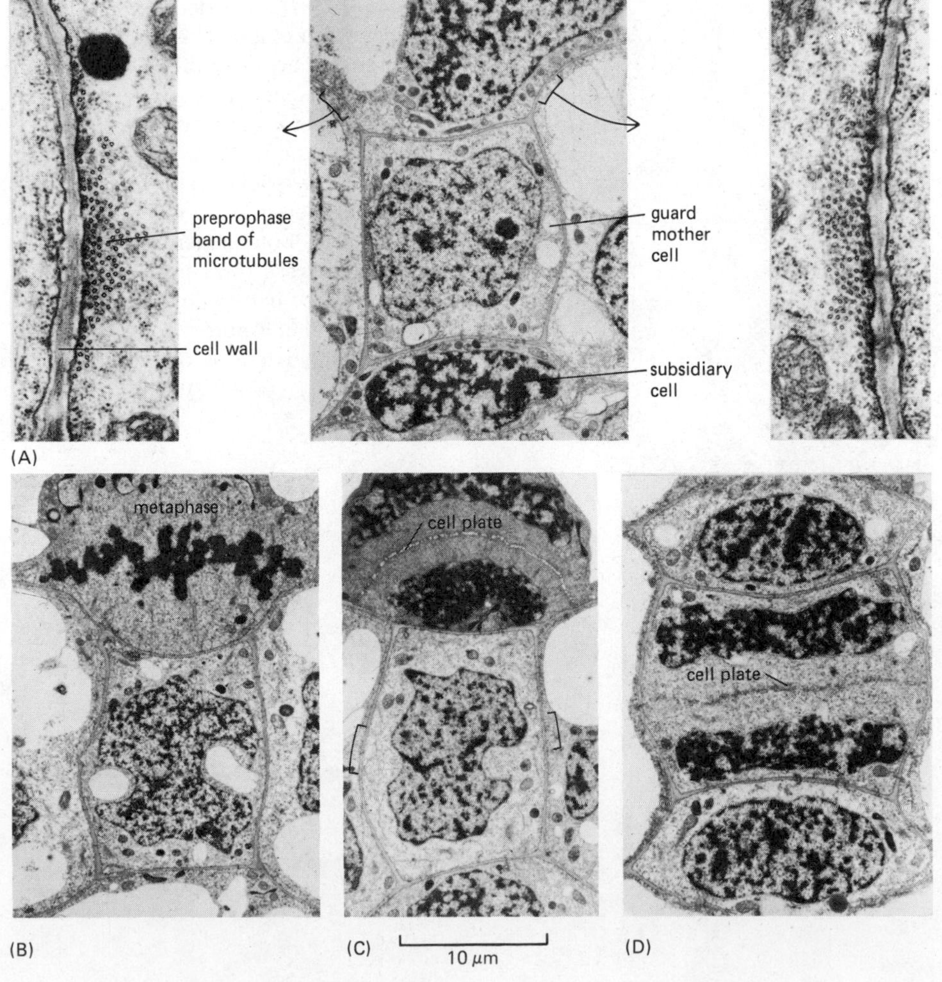

Figure 19–60 Sequence of electron micrographs of cells in the epidermis of a sugar cane leaf, showing how the formation of stomata involves a precise sequence of symmetric and asymmetric cell divisions. The plane of each division is accurately predicted by a preprophase band. (A) The uppermost cell is about to divide asymmetrically to form a subsidiary cell (the one at the bottom has already been formed). The position of the preprophase band is indicated by brackets, and the band is shown in greater detail on either side. (B) Somewhat later, the upper cell is in metaphase, and the preprophase band has gone. (C) The cell plate is now forming in the upper cell during cytokinesis and is curving down toward the position of the former preprophase band. Meanwhile, the large central cell, the guard mother cell, is about to divide symmetrically, and the position of its preprophase band is indicated by brackets. (D) Cytokinesis is complete in the upper cell and almost complete in the guard mother cell. The latter division will produce the two guard cells that flank the stomatal pore (see Figure 19–10). (Courtesy of C. Busby.)

Hormones Help Control the Growth and Shape of Plants[25]

The coordinated process of cell division, expansion, and differentiation that underlies plant development is controlled at two levels: external and internal. *External controls* operate via such environmental factors as gravity, temperature, and light intensity and duration; these controls are extremely complex and will not be discussed further. The *internal controls* that regulate plant growth and development are mediated by a variety of plant growth regulatory molecules, commonly called **plant hormones.**

Only five classes of plant hormones are known: **auxins, gibberellins, cytokinins, abscisic acid,** and the gas **ethylene.** As shown in Figure 19–61, all of these are small molecules that readily penetrate cell walls. They are produced by plant cells and are usually transported along specific paths to influence their target cells. For example, the net transport of auxin in shoots is from the tip to the base at a rate of about 1 cm per hour. Despite the small number of hormones, plants have evolved ways to maximize their use for regulating cell responses: cells respond primarily to combinations of two or more hormones, and some of the hormones act at concentrations that can differ by a factor of 10^4 or 10^5, depending on the target cell. For example, auxin and gibberellin both stimulate cell elongation in stems, while auxin acts in conjunction with cytokinin to control the growth of buds behind the stem apex. Some of the hormones can also collaborate with small metabolites to influence target cells: the balance between auxin and sucrose, for example, is thought to be crucial in determining the developmental pathways of phloem and xylem tissues.

All of the hormones have both rapid and longer term effects. In each case, the rapid effects are thought to be mediated by changes in plasma membrane permeability. As an example, auxin acts within a few minutes to activate a H^+ pump in the plasma membranes of cells; the efflux of H^+ acidifies the cell wall, thereby weakening bonds within it, and a turgor-driven cell expansion results (Figure 19–62). Auxin also has longer term effects on such expanding cells, involving alterations in gene transcription and the synthesis and secretion of new, consolidating, cell-wall material. These effects appear not to be mediated by the activation of the H^+ pump, and the mechanisms involved are still unknown.

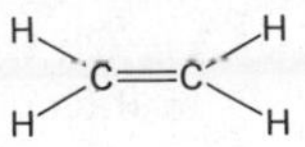

ethylene

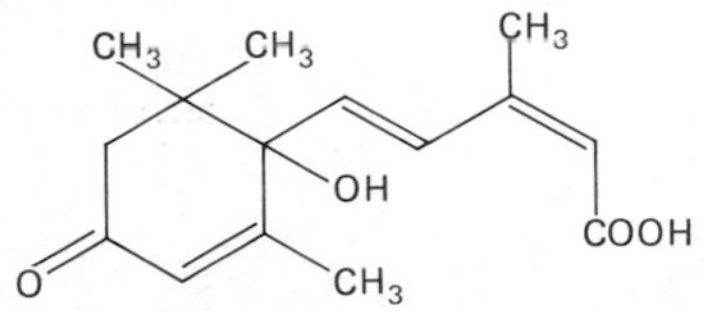

abscisic acid (ABA)

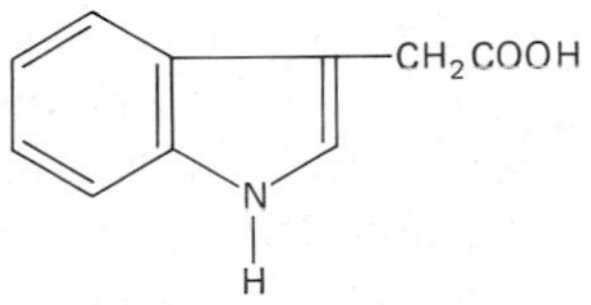

indole-3-acetic acid (IAA) [an auxin]

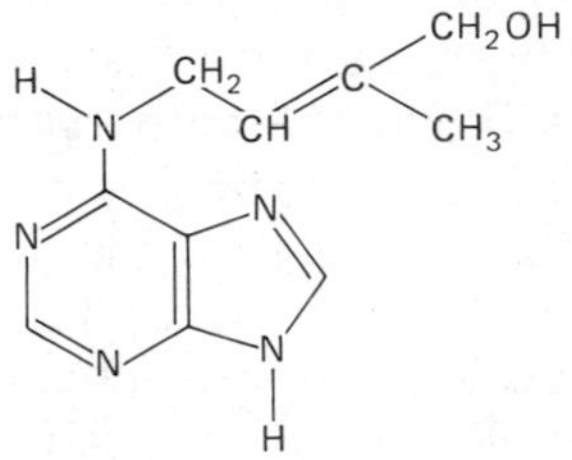

zeatin [a cytokinin]

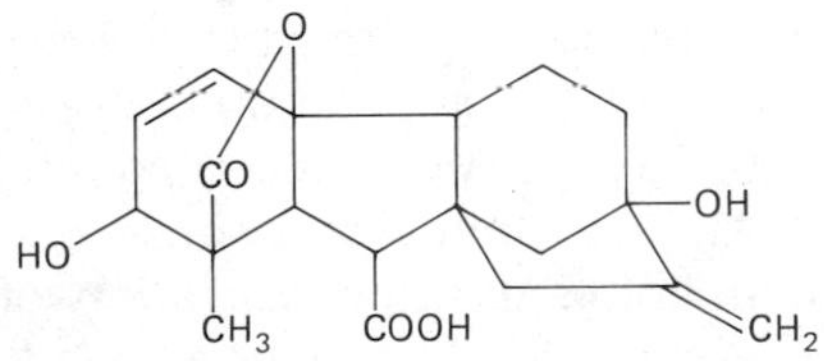

gibberellic acid (GA3) [a gibberellin]

Figure 19–61 The formulas of one representative molecule from each of the plant hormone classes discussed.

Tissue Culture Facilitates Studies of Mechanisms of Cell Determination in Plants[26]

In plant cells, differentiation is not necessarily a terminal process, as it is in most animal cells. Under certain conditions, it is possible for many plant cells either to resume cell division or to switch their developmental pathway. This

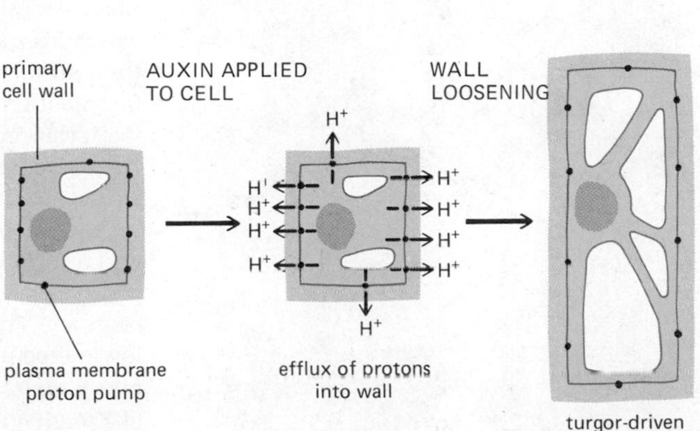

Figure 19–62 Schematic illustration of how auxin activates a plasma membrane proton pump, producing an efflux of protons into the cell-wall space. Bonds within the primary wall are loosened, and turgor-driven cell expansion results. The direction of expansion is determined by the net orientation of the cell-wall microfibrils. Other, longer-term effects follow this response.

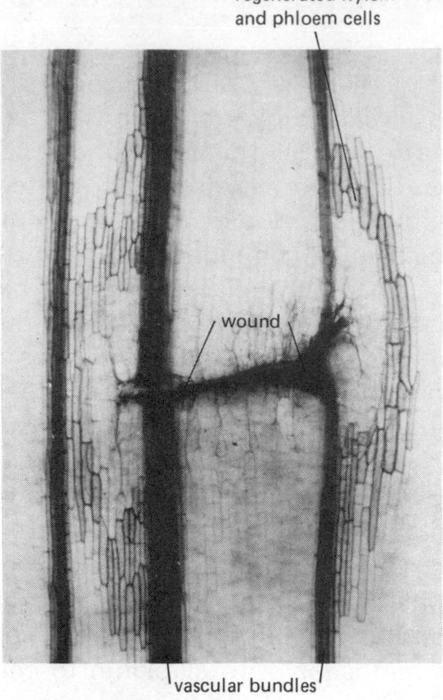

Figure 19–63 Light micrograph showing a wound response in the common houseplant *Coleus*, seen in a longitudinal section. The vascular bundles have been severed at the wound. Seven days later, after the stimulation of division and redifferentiation in the nearby cortical cells, a series of new xylem and phloem cells have been regenerated that restore vascular continuity around the wound. (Courtesy of N. P. Thomson.)

property of plant cells is clearly seen in wounded tissue, such as a cut stem. Some of the cells in the stem are stimulated to divide and repair the wound, while other cells may redifferentiate. Cortical cells, for example, may become xylem cells to restore vascular continuity around the wound (Figure 19–63). Such processes of cell redifferentiation do not depend on the cells or tissues being part of an intact plant: when plant tissue is cultured in a suitable medium containing nutrients and hormones, many cells are stimulated to divide. Such cells will proliferate indefinitely, producing a mass of relatively undifferentiated cells called a *callus.*

Callus cultures from a wide variety of plants, such as tobacco, sycamore, and carrot, have proved extremely useful for studying the cellular basis of morphogenesis and have enabled the complex influences of plant hormones on this process to be described. For example, by altering the auxin-to-cytokinin ratio, either root or shoot formation can be induced in such cultured cells (Figure 19–64). In addition, small plant embryos can be induced in callus cultures, each of which can regenerate a complete plant (Figure 19–65).

A callus can be mechanically dissociated into a suspension of single cells and small clumps that can be maintained in a *cell suspension culture.* The cells in such cultures all look very much alike, having thin primary walls and large vacuoles crossed by thin cytoplasmic strands (Figure 19–66). In several cases, single isolated plant cells from such cultures have been grown into clumps of cells from which whole plants have been regenerated. This requires the careful manipulation of hormone balance, and it has been accomplished with cells of such species as potato, tobacco, petunia, and carrot (see Figure 8–75). The ability of a single mature cell to give rise to a whole new plant containing all the usual variety of differentiated cell types demonstrates that genes are not lost or permanently inactivated during the normal process of cell differentiation in these plants.

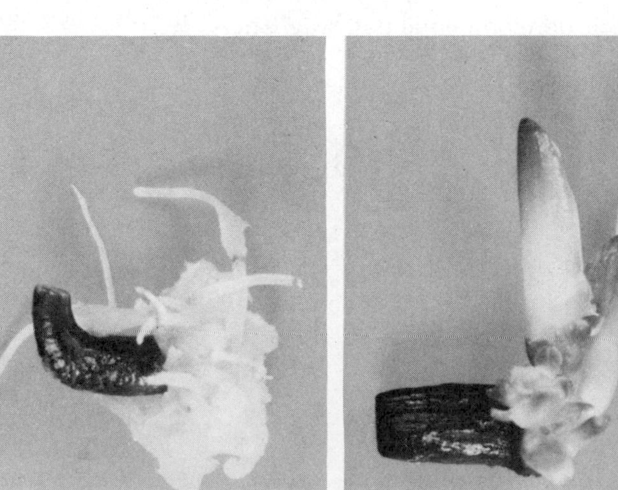

Figure 19–64 The response to hormones in plant callus cultures. Small pieces of stem from a *Freesia* plant were induced to produce a callus from one end, and these were then cultured with an auxin (indolebutyric acid) and a cytokinin (benzylaminopurine). The balance of the two hormones determines their effect. (A) The callus was exposed to a high concentration of auxin (2 mg/l) and a low concentration of cytokinin (0.25 mg/l) and, as a result, produced roots. (B) The callus was exposed to a low concentration of auxin (0.25 mg/l) and a high concentration of cytokinin (0.5 mg/l) and has produced shoots. (Courtesy of G. Hussey.)

Figure 19–65 The production of plants from callus cultures. The *Freesia* callus culture in (A) was induced to produce shoots (B) and then roots (C) by altering the balance of hormones so that a self-supporting plant was generated (C). (Courtesy of G. Hussey.)

An especially important example of the dedifferentiation of single plant cells in culture is found in the case of immature pollen (the microspore). These highly specialized cells, which have only the haploid number of chromosomes, can be manipulated in culture so as to proliferate and regenerate into whole plants (Figure 19–67). Haploid plants generated in this way can be of great practical use for plant breeding and genetics. Rice and tobacco are two important examples where this approach has been beneficial.

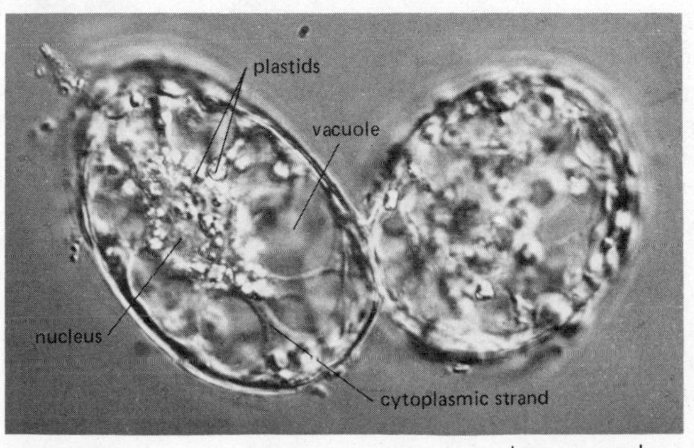

Figure 19–66 Callus cells can be grown as free-living, single cells suspended in liquid media. Two such cells derived from a sycamore callus are shown here. They are highly vacuolated cells with cytoplasmic strands radiating from the region of the nucleus.

Figure 19–67 Plants with only the haploid number of chromosomes can be regenerated from pollen grains. On the left is a normal diploid tobacco plant; on the right is a haploid version of the same plant derived from a single pollen grain. Haploid plants, which sometimes occur in nature, are generally smaller than the diploid. (Courtesy of J. Dunwell.)

Plant Cells Without Their Walls Can Be Manipulated Much Like Animal Cells[27]

One of the difficulties of working with plants or even with isolated plant cells is that the presence of the cell wall precludes direct access to the plasma membrane. However, specific enzymes are available that hydrolyze the glycosidic bonds holding the cell-wall polysaccharides together; plant cells treated with these enzymes can survive the complete removal of their cell walls if kept in a medium whose osmotic strength matches that of their cytoplasm (see Figure 19–9). The resulting **protoplasts** are naked spherical plant cells that will remain metabolically active for long periods in culture, where they can be manipulated in much the same way as animal cells (Figure 19–68). For example, studies of plant virus infection have been greatly facilitated by the ability to inoculate protoplasts synchronously with virus particles, and the preparation of such intact organelles as chloroplasts and vacuoles from protoplasts is now routine. A great deal of current interest centers on the observation that protoplasts can be made to fuse with other protoplasts of both the same and, in some cases, different species to produce stable fusion products (heterocaryons). These fused protoplasts (as well as normal protoplasts) can regenerate a cell wall (Figure 19–69) and begin cell division in culture. In many

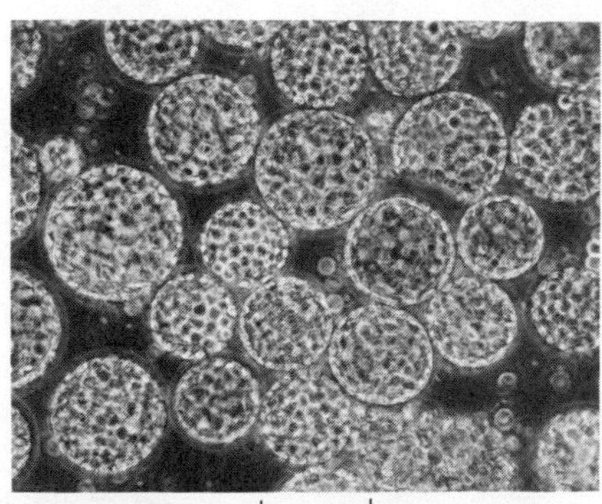

Figure 19–68 Light micrograph of protoplasts prepared from the green leaf cells of a tobacco plant. Without their walls the cells round up and need to be stabilized in a sugar solution that matches the osmotic pressure of their cytoplasm. Numerous chloroplasts can be seen arranged around the periphery of each protoplast. (Courtesy of J. Burgess.)

cases, if nuclear as well as cytoplasmic fusion occurs, the resulting callus can be used to regenerate complete new plants, forming **somatic hybrids** from the fusion product of two species. By fusing protoplasts from species that do not normally mate, new plants can be generated. In most cases, however, poorly understood factors such as those governing chromosome elimination or compatibility prevent the formation of a genetically stable hybrid plant.

Summary

Plant cells are immobilized by their walls. Consequently, during plant cell growth and development the form of the adult plant arises, not as a result of the complex movements and migrations of cells, as in animals, but as a result of precisely oriented planes of cell division coupled with controlled cell expansion and cell differentiation. Cell division takes place largely in specialized areas of the plant called meristems. The plane of division is predicted by a group of cortical microtubules called the preprophase band, a feature particularly clearly seen in several important asymmetric cell divisions. The final shape of the cell is determined by the specific orientation of cellulose microfibrils in the wall, which is determined by microtubule orientation during wall deposition. The coordinated processes of cell division and differentiation are controlled both by environmental factors and by plant growth substances, or hormones. The latter are small molecules—for example, auxins and cytokinins—that usually act in concert to produce their effects.

Many aspects of plant cell growth and development have been studied using plant cells grown in vitro, *either in a callus or as single cells. A dramatic demonstration of the totipotency of many plant cells is the regeneration of a complete plant from a single somatic cell. Protoplasts, or plant cells without their cell walls, can be manipulated in culture in much the same way that animal cells can, and they have the added property that whole plants can be regenerated from them.*

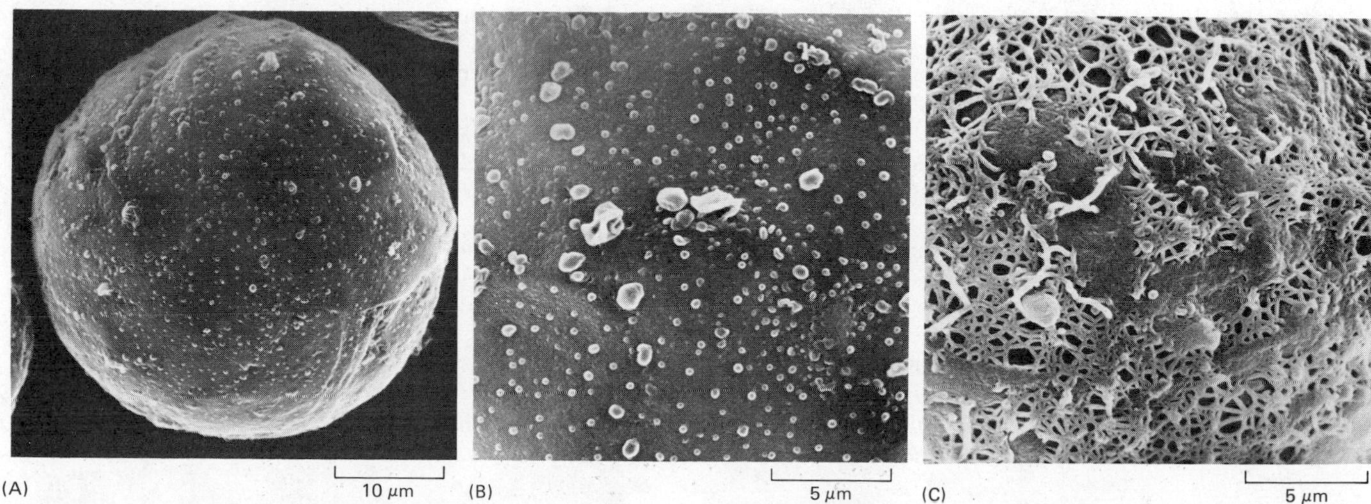

Figure 19–69 Scanning electron micrographs of tobacco protoplasts. The freshly isolated spherical protoplast (A) is similar to those shown in Figure 19–68. The cell wall has been completely removed, and the naked plasma membrane, seen in more detail in (B), is revealed, interrupted only by a few membrane protrusions. After culturing the protoplasts for some time, they begin to regenerate a new cell wall (C). At the early stage shown here the new meshwork of cellulose microfibrils can be seen on the cell surface, covering the plasma membrane. (Courtesy of J. Burgess.)

References

General

Cutter, E.G. Plant Anatomy, 2nd ed., Part 1, Cells and Tissues; Part 2, Organs. London: Arnold, 1978.

Esau, K. Anatomy of Seed Plants, 2nd ed. New York: Wiley, 1977.

Gunning, B.E.S.; Steer, M.W. Ultrastructure and the Biology of Plant Cells. London: Arnold, 1975.

Raven, P.H.; Evert, R.F.; Curtis, H. Biology of Plants, 3rd ed. New York: Worth, 1981.

Stumpf, P.K.; Conn, E.E., eds. The Biochemistry of Plants—A Comprehensive Treatise. 8 vols. New York: Academic Press, 1980. (Volume 1, The Plant Cell [N.E. Tolbert, ed.], is of particular relevance.)

Wareing, P.F.; Phillips, I.D.J. Growth and Differentiation in Plants, 3rd ed. London: Pergamon, 1981.

Cited

1. Rees, D.A. Polysaccharide Shapes. London: Chapman and Hall, 1977.
   Bauer, W.D. Plant cell walls. In The Molecular Biology of Plant Cells (H. Smith, ed.), pp. 6–23. Oxford, Eng.: Blackwell, 1977.
   Roland, J.C.; Vian, B. The wall of the growing plant cell: its three-dimensional organization. *Int. Rev. Cytol.* 61:129–166, 1979.
2. Aspinall, G.O. Chemistry of cell wall polysaccharides. In The Biochemistry of Plants—A Comprehensive Treatise, Vol. 3 (J. Preiss, ed.), pp. 473–500. New York: Academic Press, 1980.
   Darvill, A.; McNeil, M.; Albersheim, P.; Delmer, D. The primary cell walls of flowering plants. In The Biochemistry of Plants—A Comprehensive Treatise, Vol. 1 (N.E. Tolbert, ed.), pp. 91–162. New York: Academic Press, 1980.
   Tanner, W.; Loewus, F.A., eds. Encyclopedia of Plant Physiology, New Series. Vol. 13B, Plant Carbohydrates II, Extracellular Carbohydrates. Heidelberg: Springer-Verlag, 1982.
3. Milburn, J.A. Water Flow in Plants. London: Longman, 1979.
   Carpita, N.; Sabularse, D.; Montezinos, D.; Delmer, D.P. Determination of the pore size of cell walls of living plant cells. *Science* 205:1144–1147, 1979.
4. Raven, P.H.; Evert, R.F.; Curtis, H. Biology of Plants, 3rd ed. New York: Worth, 1981. (Chapters 3 and 28 are particularly relevant.)
5. Baker, D.A.; Hall, J.L. Ion Transport in Plant Cells and Tissues. Amsterdam: North-Holland, 1975.
   Zimmerman, U. Cell turgor pressure regulation and turgor pressure-mediated transport processes. *Symp. Soc. Exp. Biol.* 31:117–154, 1977.
6. Esau, K. Anatomy of Seed Plants, 2nd ed. New York: Wiley, 1977.
   Tanner, W.; Loewus, F.A., eds. Encyclopedia of Plant Physiology, New Series. Vol. 13B, Plant Carbohydrates II, Extracellular Carbohydrates. Heidelberg: Springer-Verlag, 1982.
7. Cronshaw, J. Phloem structure and function. *Annu. Rev. Plant Physiol.* 32:465–484, 1981.
   Cutter, E.G. Plant Anatomy, 2nd ed., Part 1, Cells and Tissues. London: Arnold, 1978.
8. Gunning, B.E.S.; Robards, A.W., eds. Intercellular Communication in Plants: Studies on Plasmodesmata. New York: Springer-Verlag, 1976.
9. Gunning, B.E.S.; Hughes, J.E. Quantitative assessment of symplastic transport of pre-nectar into the trichomes of *Abutilon* nectaries. *Aust. J. Plant Physiol.* 3:619–637, 1976.
10. Gunning, B.E.S. Transfer cells and their roles in transport of solutes in plants. *Sci. Prog. (Oxford)* 64:539–568, 1977.
    Milburn, J.A. Water Flow in Plants. London: Longman, 1979.
    Baker, D.A. Transport Phenomena in Plants. London: Chapman and Hall, 1978.
11. Bauer, W.D. Infection of legumes by Rhizobia. *Annu. Rev. Plant Physiol.* 32:407–484, 1981.

Brill, W.J. Agricultural microbiology. *Sci. Am.* 245:146–156, 1981.

12. Heslop-Harrison, J. Cellular Recognition Systems in Plants, London: Arnold, 1978.
Loewus, F.A.; Ryan, C.A., eds. The Phytochemistry of Cell Recognition and Cell Surface Interactions. Recent Advances in Phytochemistry, Vol. 15. New York: Plenum, 1980.
Barondes, S.H. Lectins: their multiple endogenous cellular functions. *Annu. Rev. Biochem.* 50:207–231, 1981.
Ferrari, T.E.; Bruns, D.; Wallace, D.H. Isolation of a plant glycoprotein involved with control of intercellular recognition. *Plant Physiol.* 67:270–277, 1981.
Ayers, A.R.; Ebel, J.; Finelli, F.; Berger, N.; Albersheim, P. Host pathogen interactions. *Plant Physiol.* 57:751–759, 1976. (This should be read in conjunction with the three related papers that follow it.)

13. Thomson, W.W. Development of nongreen plastids. *Annu. Rev. Plant Physiol.* 31:375–399, 1980.

14. MacRobbie, E.A.C. Accumulation of ions in plant cell vacuoles. In Perspectives in Experimental Biology, Vol. 2 (N. Sunderland, ed.), pp. 369–380. New York: Pergamon, 1976.
Matile, P. Biochemistry and function of vacuoles. *Annu. Rev. Plant Physiol.* 29:193–213, 1978.
Boller, T.; Kende, H. Hydrolytic enzymes in the central vacuole of plant cells. *Plant Physiol.* 63:1123–1132, 1979.

15. Cram, W.J. Pinocytosis in plants. *New Phytol.* 84:1–17, 1980.

16. Mollenhauer, H.H.; Morré, D.J. The Golgi apparatus. In The Biochemistry of Plants—A Comprehensive Treatise, Vol. 1 (N.E. Tolbert, ed.), pp. 438–489. New York: Academic Press, 1980.
Northcote, D.H. Macromolecular aspects of cell wall differentiation. In Encyclopedia of Plant Physiology, New Series. Vol. 14A, Nucleic Acids and Proteins in Plants I (D. Boulter, B. Parthier, eds.), pp. 637–655. Heidelberg: Springer-Verlag, 1982.
Robinson, D.G. Plant cell wall synthesis. *Adv. Bot. Res.* 5:89–151, 1977.

17. Ross Colvin, J. The biosynthesis of cellulose. In The Biochemistry of Plants—A Comprehensive Treatise, Vol. 3 (J. Preiss, ed.), pp. 544–570. New York: Academic Press, 1980.

18. Lloyd, C.W., ed. The Cytoskeleton in Plant Growth and Development. London: Academic Press, 1982.

19. Kamiya, N. Physical and chemical basis of cytoplasmic streaming. *Annu. Rev. Plant Physiol.* 32:205–236, 1981.
Williamson, R.E. Actin in motile and other processes in plant cells. *Can. J. Bot.* 58:766–772, 1980.
Palevitz, B.A. Actin cables and cytoplasmic streaming in green plants. In Cell Motility (R. Goldman, T. Pollard, J. Rosenbaum, eds.), pp. 601–611. Cold Spring Harbor, N.Y.: Cold Spring Harbor Laboratory, 1976.

20. Allen, N.S. Cytoplasmic streaming and transport in the characean alga *Nitella*. *Can. J. Bot.* 58:786–796, 1980.
Kersey, Y.M.; Hepler, P.K.; Palevitz, B.A.; Wessels, N.K. Polarity of actin filaments in Characean algae. *Proc. Natl. Acad. Sci. USA* 73:165–167, 1976.
Williamson, R.E.; Ashley, C.C. Free Ca^{2+} and cytoplasmic streaming in the alga *Chara*. *Nature* 296:647–651, 1982.

21. Virgin, H.I. Light and chloroplast movements. *Symp. Soc. Exp. Biol.* 22:329–352, 1968.
Haupt, W. Light-mediated movement of chloroplasts. *Annu. Rev. Plant Physiol.* 33:205–233, 1982.
Wagner, G.; Klein, K. Mechanism of chloroplast movement in *Mougeotia*. *Protoplasma* 109:169–185, 1981.

22. Gunning, B.E.S. Microtubules and cytomorphogenesis in a developing organ: the root primordium of *Azolla pinnata*. In Cytomorphogenesis in Plants (O. Kiermayer, ed.), pp. 301–325. New York: Springer, 1981.
Cutter, E.G. Plant Anatomy, 2nd ed., Part 2, Organs. London: Arnold, 1978.

23. Green, P.B. Organogenesis—a biophysical view. *Annu. Rev. Plant Physiol.* 31:51–82, 1980.

Gunning, B.E.S.; Hardham, A.R. Microtubules. *Annu. Rev. Plant Physiol.* 33:651–698, 1982.

24. Pickett-Heaps, J.D.; Northcote, D.H. Organization of microtubules and endoplasmic reticulum during mitosis and cytokinesis in wheat meristems. *J. Cell Sci.* 1:109–120, 1966.

Wick, S.M.; Seagull, R.W.; Osborn, M.; Weber, K.; Gunning, B.E.S. Immunofluorescence microscopy of organized microtubule arrays in structurally stabilized meristematic plant cells. *J. Cell Biol.* 89:685–690, 1981.

Gunning, B.E.S.; Hardham, A.R.; Hughes, J.E. Pre-prophase bands of microtubules in all categories of formative and proliferative cell division in *Azolla* roots. *Planta* 143:145–160, 1978.

Gunning, B.E.S.; Hardham, A.R. Microtubules. *Annu. Rev. Plant Physiol.* 33:651–698, 1982.

Lloyd, C.W., ed. The Cytoskeleton in Plant Growth and Development. London: Academic Press, 1982.

25. Wareing, P.F.; Phillips, I.D.J. Growth and Differentiation in Plants, 3rd ed. New York: Pergamon, 1981. (Chapters 3, 4, and 5 cover hormones and their action.)

Smith, H.; Grierson, D., eds. Molecular Biology of Plant Development. Berkeley, Ca.: University of California Press, 1982.

Trewavas, A.J. Growth substance sensitivity: the limiting factor in plant development. *Plant Physiol.* 55:60–72, 1982.

26. Vasil, I.K., ed. Perspectives in plant cell and tissue culture. *Int. Rev. Cytol.*, Suppl. 11A and B, 1980.

Vasil, I.K.; Ahuja, M.R.; Vasil, V. Plant tissue cultures in genetics and plant breeding. *Adv. Genet.* 20:127–215, 1979.

Tran Thanh Van, K.M. Control of morphogenesis in *in vitro* cultures. *Annu. Rev. Plant Physiol.* 32:291–311, 1981.

Guha, S.; Maheshwari, S.C. *In vitro* production of embryos from anthers of *Datura*. *Nature* 204:497, 1964.

27. Fowke, L.C.; Gamborg, O.L. Applications of protoplasts to the study of plant cells. *Int. Rev. Cytol.* 68:9–51, 1980.

Cocking, E.C.; Davey, M.R.; Pental, D.; Power, J.B. Aspects of plant genetic manipulation. *Nature* 293:265–270, 1981.

Shepard, J.F. The regeneration of potato plants from leaf cell protoplasts. *Sci. Am.* 246:112–121, 1982.

Index

Page numbers in **bold face** refer to a major text discussion of the entry; page numbers in *italics* refer to figures and tables; an italicized page number followed by "ff" indicates that two or more figures follow consecutively after the indicated page reference; "cf." means "compare."

Index to Illustrative Review Panels and Other Special Features